Food Microbiology

Fundamentals and Frontiers

Food Microbiology
Fundamentals and Frontiers

EDITED BY

Michael P. Doyle
Center for Food Safety and Quality Enhancement, Department of Food Science and Technology, University of Georgia, Georgia Station, Griffin, Georgia 30223-1797; Department of Food Science, University of Georgia, Athens, Georgia 30602

Larry R. Beuchat
Center for Food Safety and Quality Enhancement, Department of Food Science and Technology, University of Georgia, Georgia Station, Griffin, Georgia 30223-1797

Thomas J. Montville
Department of Food Science, Rutgers, The State University of New Jersey, New Brunswick, New Jersey 08903-0231

ASM Press • Washington D.C.

Cover image: Actin-based motility of *Listeria monocytogenes.* Confocal laser scanning photograph of Vero cells infected with *L. monocytogenes.* The bacteria (red) move by inducing the formation of filamentous actin tails (green) in the cytoplasm of the cell. See chapter 18. (Courtesy of Iñigo Lasa and Pascale Cossart, Institut Pasteur, Paris.)

1325 Massachusetts Ave., NW
Washington, DC 20005

Library of Congress Cataloging-in-Publication Data

Food microbiology : fundamentals and frontiers / edited by Michael P. Doyle, Larry R. Beuchat, Thomas J. Montville.
p. cm.
Includes bibliographical references and index.
ISBN 1-55581-117-5
1. Food—Microbiology. I. Doyle, Michael P. II. Beuchat, Larry R. III. Montville, Thomas J.
QR115.F654 1997
576'.163—DC20 96-36638
CIP

Printed in the United States of America

Contents

III Foodborne Pathogenic Bacteria 127

IV Mycotoxigenic Molds 391

V *Viruses* 435

VI *Foodborne and Waterborne Parasites* 447

VII *Preservatives and Preservation Methods* 495

VIII *Food Fermentations* 579

IX Advanced Techniques in Food Microbiology 695

Contributors

GARY R. ACUFF
Department of Animal Science, Texas A&M University, College Station, Texas 77843-2471

JOHN W. AUSTIN
Microbiology Research Division, Bureau of Microbial Hazards, Food Directorate, Health Protection Branch, Health Canada, Banting Research Centre, Ottawa, Ontario PL 2204A1, K1A 0L2, Canada

W. MARK BARBOUR
DuPont Central Research and Development, Experimental Station, P.O. Box 80228, Wilmington, Delaware 19880

DANE T. BERNARD
National Food Processors Association, 1401 New York Avenue, N.W., Washington, DC 20005

LARRY R. BEUCHAT
Center for Food Safety and Quality Enhancement, Department of Food Science and Technology, University of Georgia, Georgia Station, Griffin, Georgia 30223-1797

RIJKELT BEUMER
Department of Food Science, Wageningen Agricultural University, P.O. Box 8129, 6700 EV Wageningen, The Netherlands

GREGORY A. BOHACH
Department of Microbiology, Molecular Biology and Biochemistry, University of Idaho, Moscow, Idaho 83843

ROBERT E. BRACKETT
Center for Food Safety and Quality Enhancement, Department of Food Science and Technology, University of Georgia, Griffin, Georgia 30223-1797

ROBERT L. BUCHANAN
Food Safety Research Unit, Agricultural Research Service, U.S. Department of Agriculture, 600 E. Mermaid Lane, Wyndmoor, Pennsylvania 19038

HERBERT J. BUCKENHÜSKES
Gewürzmüller GmbH, Klagenfurter Strasse 1-3, D-70469 Stuttgart, Germany

LLOYD B. BULLERMAN
Department of Food Science and Technology, 143 H. C. Filley Hall, East Campus, P.O. Box 830919, University of Nebraska, Lincoln, Nebraska 68583-0919

IAIN CAMPBELL
Department of Biological Sciences, Heriot-Watt University, Riccarton, Edinburgh EH14 4AS, Scotland

DEAN O. CLIVER
Department of Population Health and Reproduction, School of Veterinary Medicine, University of California, Davis, Davis, California 95616-8743

P. COSSART
Unité des Interactions Bactéries-Cellules, Centre National de Recherche Scientifique URA 1300, Institut Pasteur, 28 rue du Dr. Roux, 75724 Paris Cedex 15, France

JEAN-YVES D'AOUST
Food Directorate, Health Protection Branch, Health Canada, Sir F. G. Banting Research Centre, Postal Locator 22.04.A2, Tunney's Pasture, Ottawa, Ontario K1A 0L2, Canada

P. MICHAEL DAVIDSON
Department of Food Science and Toxicology, Food Research Center, University of Idaho, Moscow, Idaho 83844-1053

JAMES S. DICKSON
Department of Microbiology, Immunology, and Preventive Medicine, Iowa State University, Ames, Iowa 50011-3211

KAREN L. DODDS
Evaluation Division, Bureau of Microbial Hazards, Food Directorate, Health Protection Branch, Health Canada, Banting Research Centre, Ottawa, Ontario PL 2204A1, K1A 0L2, Canada

MICHAEL P. DOYLE
Center for Food Safety and Quality Enhancement, Department of Food Science and Technology, University of Georgia, Georgia Station, Griffin, Georgia 30223-1797; Department of Food Science, University of Georgia, Athens, Georgia 30602

JÓZSEF FARKAS
Department of Refrigeration and Livestock Products Technology, University of Horticulture and Food, 1052 Budapest, pf. 53, Hungary

GRAHAM H. FLEET
Department of Food Science and Technology, University of New South Wales, Sydney, New South Wales 2052, Australia

JOSEPH F. FRANK
Center for Food Safety and Quality Enhancement, Department of Food Science and Technology, University of Georgia, Athens, Georgia 30602

SILVIA GONZALEZ AYALA
Faculty of Medicine, National University, La Plata, Argentina

PER EINAR GRANUM
Department of Pharmacology, Microbiology and Food Hygiene, Norwegian College of Veterinary Medicine, P.O. Box 8146 Dep., N-0033 Oslo, Norway

PAUL A. HARTMAN
Department of Microbiology, Immunology and Preventive Medicine, 205 Science I, Iowa State University, Ames, Iowa 50014

EUGENE G. HAYUNGA
Office of Research on Women's Health, National Institutes of Health, Building 1, Room 207,
Bethesda, Maryland 20892

AILSA D. HOCKING
CSIRO Division of Food Science and Technology, P.O. Box 52, North Ryde, New South Wales 2113, Australia

LYNN M. JABLONSKI
Department of Biological Sciences, Northern Illinois University, DeKalb, Illinois 60115

TIMOTHY C. JACKSON
Nestle Research and Development Center, 201 Housatonic Avenue, New Milford, Connecticut 06776

MARK E. JOHNSON
Center for Dairy Research, Department of Food Science, University of Wisconsin-Madison, 1605 Linden Drive,
Madison, Wisconsin 53706-1565

ERIC A. JOHNSON
Food Research Institute, University of Wisconsin-Madison, Madison, Wisconsin 53706

JAMES B. KAPER
Center for Vaccine Development, Division of Geographic Medicine, Department of Medicine,
University of Maryland School of Medicine, 685 West Baltimore Street, Baltimore, Maryland 21201

JIMMY T. KEETON
Department of Animal Science, Texas A&M University, College Station, Texas 77843-2471

CHARLES W. KIM
Division of Infectious Diseases, T-15, 080, Health Science Center, State University of New York at Stony Brook,
Stony Brook, New York 11794-8153

SYLVIA M. KIROV
Division of Pathology, University of Tasmania, Clinical School, 43 Collins Street, Hobart, Tasmania 7000,
Australia

KEITH A. LAMPEL
Center for Food Safety and Applied Nutrition, Food and Drug Administration, 200 C Street, S.W.,
Washington, DC 20204

ALEX LOPEZ
c/o Malaysian Cocoa Board, Jen. Tunka Abd. Rahman, Beg Berkunci 211, Kota Kimbalu 88999, Sabah, Malaysia

ANTHONY T. MAURELLI
Department of Microbiology and Immunology, F. Edward Hébert School of Medicine, Uniformed Services University
of the Health Sciences, 4301 Jones Bridge Road, Bethesda, Maryland 20814-4799

BRUCE A. MCCLANE
Department of Molecular Genetics and Biochemistry, University of Pittsburgh School of Medicine, E1240
Biomedical Science Tower, Pittsburgh, Pennsylvania 15261-2072

JIANGHONG MENG
Center for Food Safety and Quality Enhancement, University of Georgia, Georgia Station,
Griffin, Georgia 30223-1797

KENNETH B. MILLER
Microbiology Research, Technical Center, Hershey Foods Corporation, 1025 Reese Avenue,
Hershey, Pennsylvania 17033-0805

THOMAS J. MONTVILLE
Department of Food Science, Rutgers, The State University of New Jersey, New Brunswick, New Jersey 08903-0231

IRVING NACHAMKIN
Department of Pathology and Laboratory Medicine (4115), University of Pennsylvania Medical Center, 3400 Spruce Street, Philadelphia, Pennsylvania 19104-4283

SERVÉ NOTERMANS
National Institute of Public Health and The Environment, P.O. Box 1, 3720 BA Bilthoven, The Netherlands

JAMES D. OLIVER
Department of Biology, The University of North Carolina at Charlotte, 9201 University City Boulevard, Charlotte, North Carolina 28223-0001

MERLE D. PIERSON
Department of Food Science and Technology, Virginia Polytechnic Institute and State University, Blacksburg, Virginia 24061

JOHN I. PITT
CSIRO Division of Food Science and Technology, P.O. Box 52, North Ryde, New South Wales 2113, Australia

MORRIS E. POTTER
Division of Bacterial and Mycotic Diseases, National Center for Infectious Diseases, Centers for Disease Control and Prevention, Atlanta, Georgia 30333

STEVEN C. RICKE
Department of Poultry Science, Texas A&M University, College Station, Texas 77843-2472

ROY M. ROBINS-BROWNE
Department of Microbiology and Infectious Diseases, Royal Children's Hospital, and Department of Microbiology, University of Melbourne, Parkville, Victoria 3052, Australia

J. ROCOURT
Centre National de Reference des *Listeria* and WHO Collaborating Center for Foodborne Listeriosis, Institut Pasteur, 28 rue du Dr. Roux, 75724 Paris Cedex 15, France

FRANK ROMBOUTS
Department of Food Science, Wageningen Agricultural University, P.O. Box 8129, 6700 EV Wageningen, The Netherlands

PETER SETLOW
Department of Biochemistry, University of Connecticut Health Center, Farmington, Connecticut 06030

NARUMOL SILARUG
Division of Epidemiology, Permanent Secretary Office, Building 6-F, Ministry of Public Health, Muang District, Nonthaburi, Thailand, 11000

L. MICHELE SMOOT
Department of Food Science and Technology, Virginia Polytechnic Institute and State University, Blacksburg, Virginia 24061

C. A. SPEER
Veterinary Molecular Biology Laboratory, Montana State University, Bozeman, Montana 59717-0360

JAMES L. STEELE
Department of Food Science, University of Wisconsin-Madison, 1605 Linden Drive, Madison, Wisconsin 53706-1565

STERLING S. THOMPSON
Microbiology Research, Technical Center, Hershey Foods Corporation, 1025 Reese Avenue, Hershey, Pennsylvania 17033-0805

GEORGE TICE
DuPont Central Research and Development, Experimental Station, P.O. Box 80228, Wilmington, Delaware 19880

RICHARD C. WHITING
Microbial Food Safety Research Unit, Agricultural Research Service, U.S. Department of Agriculture,
600 E. Mermaid Lane, Wyndmoor, Pennsylvania 19038

KAREN WINKOWSKI
Department of Food Science, Rutgers, The State University of New Jersey, New Brunswick, New Jersey 08903-0231

SHAOHUA ZHAO
Center for Food Safety and Quality Enhancement, University of Georgia, Georgia Station,
Griffin, Georgia 30223-1797

TONG ZHAO
Center for Food Safety and Quality Enhancement, University of Georgia, Georgia Station,
Griffin, Georgia 30223-1797

Reviewers

Walter H. Andrews
Charles W. Bacon
James N. Bacus
J. Stanley Bailey
Susan F. Barefoot
Timothy J. Barrett
Robert B. Beelman
Merlin S. Bergdoll
Thomas E. Besser
Jacques S. Bille
J. Russell Bishop
Edward J. Bottone
Robert R. Brubaker
Robert L. Buchanan
Donald H. Burr
Marisa Caipo
John A. Carpenter
Yuhuan Chen
David Collins-Thompson
Allison D. Crandall
Sandra L. Curtis
Mark A. Daeschel
Bibhuti R. Dasgupta
David W. Dreeson
Herbert L. DuPont
Charles G. Edwards
Melvin W. Eklund
Jeffery M. Farber
John J. Farmer, III
Henry P. Fleming
Charles P. Gerba
David A. Golden
David E. Gombas
Mansel W. Griffiths
John H. Hanlin
Linda J. Harris
Mark A. Harrison
Eugene G. Hayunga
John J. Iandolo
George J. Jackson
Lee Ann Jaykus
Michael G. Johnson
Eric A. Johnson
Sam W. Joseph
Charles A. Kaysner
Charles W. Kim
Ronald G. Labbe
Douglas W. Lehrian
James A. Lindsay
John E. Linz
Hermy C. Lior
Joseph M. Madden
Robert T. Marshall
Alejandro Mazzotta
Richard J. Meinersmann
J. David Miller
Marguerite A. Neill
Norman F. Olson
James J. Pestka
Johnny Peterson
N. Rukma Reddy
Lee J. Romanczyk
Noel G. Rudie
Philippe J. Sansonetti
Gerald M. Sapers
Donald W. Schaffner
Patrick M. Schlievert
Gregory R. Siragusa
James L. Smith
John N. Sofos
Myron Solberg
Clarence A. Speer
Don F. Splittstoesser
Keith H. Steinkraus
Michael E. Stiles
Susan C. Straley
Bala Swaminathan
Mark L. Tamplin
Phillip I. Tarr
Robert V. Tauxe
Fred C. Tenover
David M. Wilson
Amy C. L. Wong

Preface

THE FIELD OF FOOD MICROBIOLOGY IS AMONG THE MOST DIVERSE of the areas of study within the discipline of microbiology. Its scope encompasses a wide variety of microorganisms including spoilage, probiotic, fermentative, and pathogenic bacteria, molds, yeasts, viruses, and parasites; a diverse composition of foods; a broad spectrum of environmental factors that influence microbial survival and growth; and a multitude of research approaches that range from very applied studies of survival and growth of foodborne microorganisms to basic studies of the mechanisms of pathogenicity of harmful foodborne microorganisms. Several excellent books address many different aspects of food microbiology. The purpose of *Food Microbiology: Fundamentals and Frontiers* is to complement these books by providing new, state-of-the-art information that emphasizes the molecular and mechanistic aspects of food microbiology, and not to dwell on other aspects well covered in food microbiology texts.

This advanced text fulfills the need of research microbiologists, graduate students, and professors of food microbiology courses for an in-depth treatment of food microbiology. It provides current, definitive factual material written by experts on each subject. The book is written at a level which presupposes a general background in microbiology and biochemistry needed to understand the "how and why" of food microbiology at a basic scientific level.

The book is composed of nine major sections that address each of the major areas of the field. "Factors of Special Significance to Food Microbiology" provides a brief history of food microbiology, a perspective on and description of the basic principles that affect the growth, survival, and death of microbes, coverage of spores, and the use of indicator microorganisms and microbiological criteria. "Microbial Spoilage of Foods" covers the principles of spoilage, dominant microorganisms, and spoilage patterns for each of three major food categories. The 13 chapters in the "Foodborne Pathogenic Bacteria" section provide a current molecular understanding of foodborne bacterial pathogens in the context of their pathogenic mechanisms, tolerance to preservation methods, and underlying epidemiology as well as basic information about each microorganism's metabolic characteristics, symptoms of illness, and common food reservoirs. Similar perspectives are given by chapters in the sections "Mycotoxigenic Molds," "Viruses," and "Foodborne and Waterborne Parasites."

"Preservatives and Preservation Methods" presents information on mechanisms, models, and kinetics in three chapters which elucidate physical, chemical, and biological methods of food preservation. The "Food Fermentations" section emphasizes the genetics and physiology of microorganisms involved in fermentation of foods and beverages. The influence of fermentation on product characteristics is also examined.

Since rapid, genetic, and immunological methods for detecting foodborne microorganisms, predictive modeling, and hazard analysis and critical control points are key issues to the future of food microbiology, it is appropriate that these topics are covered in the closing section, "Advanced Techniques in Food Microbiology." An appendix to chapter 40 introduces the reader to a valuable computer tool, the United States Department of Agriculture (USDA) Pathogen Modeling Program, which is available from the USDA Eastern Regional Research Center via the Internet.

We are grateful to all of our coauthors for their dedication to producing a book that is at the cutting edge of food microbiology, and to our reviewers whose critical evaluations enabled us to fine tune each chapter.

Michael P. Doyle
Larry R. Beuchat
Thomas J. Montville

Factors of Special Significance to Food Microbiology

I

Paul A. Hartman

1 The Evolution of Food Microbiology

For centuries, humans have taken advantage of, as well as been plagued by, microorganisms and their activities in foods. This book describes the current state of food microbiology, whose ultimate goal is to provide an adequate, organoleptically satisfying, wholesome, and safe food supply. By necessity, this book represents the "state of the science" at a certain point in time. The current practice of food microbiology is, however, the culmination of decades of research and experience. Thus, food microbiology must be understood as an evolving science connected to both a history and a future. The evolution of food microbiology is an exciting story which is outlined in Fig. 1.1 and presented below.

EARLY DEVELOPMENTS IN FOOD PRODUCTION AND PRESERVATION

Food spoilage and food poisoning caused by microorganisms were problems that must have continually preoccupied early humans. Imagine the predicament of a Neanderthal (Europe, 50,000 B.C.) or Cro-Magnon (North America, 40,000 B.C.) who had just slaughtered a woolly mammoth and needed to preserve a ton of meat (20, 22). During the retreat of the last ice age, 10,000 to 20,000 years ago, what were formerly nomadic populations of hunters and gatherers domesticated both food crops and animals and turned to production agriculture in four major river civilizations—Egypt on the Nile, Sumer on the Tigris and Euphrates in Mesopotamia (now Iraq), India on the Indus, and China on the Yellow (21).

Studies continue even today to determine more precisely when food production originated. A recent study (24) of archaeological sites dated between 18,300 and 17,000 years ago revealed that barley flourished in the Nile Valley, near Aswan in Egypt. The practice of animal husbandry originated about 8,000 to 10,000 years ago. Evidence based on nucleic acid sequences indicates that there were two independent domestications of cattle (15) and a single domestication of chickens (10). The spread of agricultural practices from one society to another made it more possible to assure a stable food supply and encouraged community development. By 3,000 B.C., the people of Sumer (now Iraq) had developed a sophisticated agricultural economy. They constructed irrigation canals and grew a wide variety of crops and livestock (13, 20). The herders of Sumeria did not face food preservation problems as pressing as did the hunters that preceded them. Most livestock could be moved from place to place and slaughtered when needed. Livestock also had monetary value and could be considered as an early form of currency, as was salt. The ability to preserve food and store it from times of plenty to times of need was prerequisite for the shift from hunter-gatherers to an agricultural society.

The production of bread, alcoholic beverages (chapters 36 and 37), and a variety of acid-fermented foods

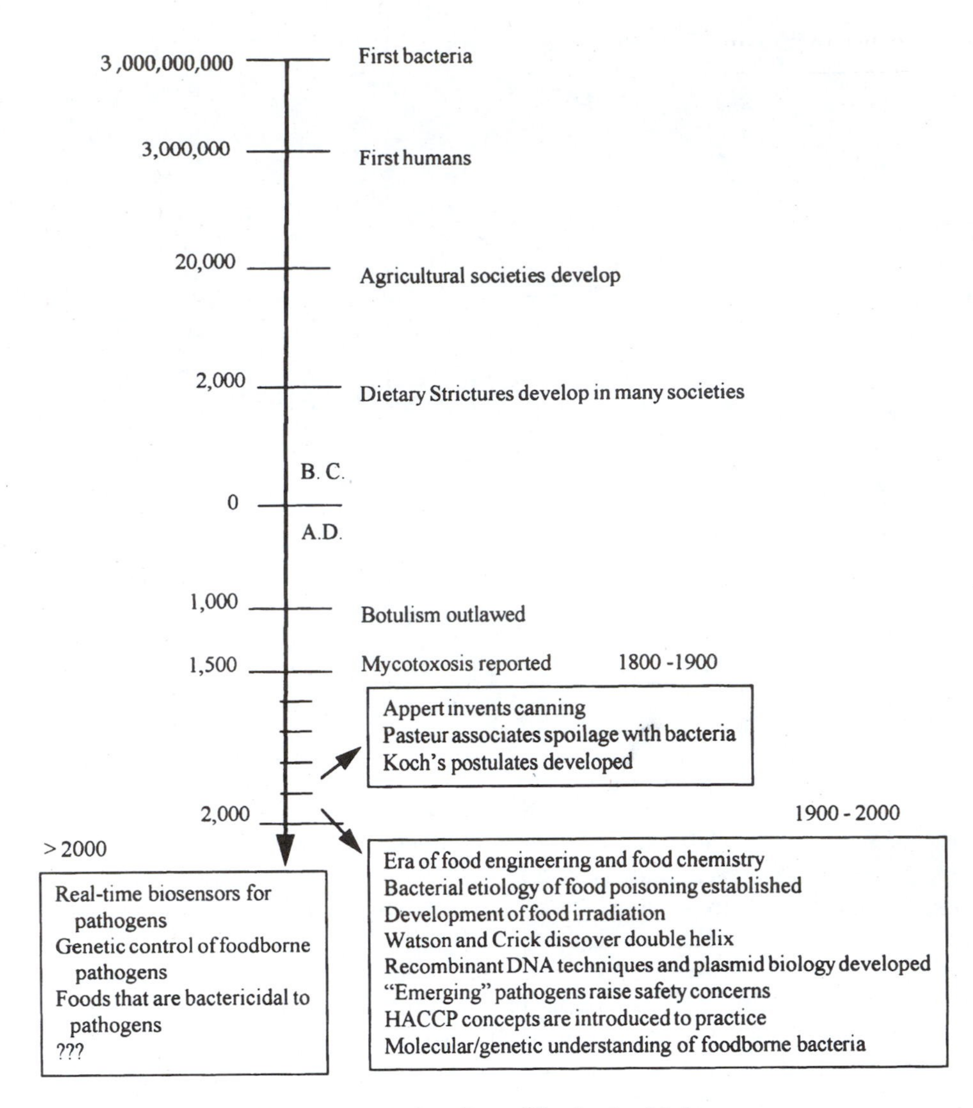

Figure 1.1 A time line of food microbiology.

(chapters 31 through 34), the preservation of meat and fish products by drying or adding salt (chapters 28 and 29), and the production of other indigenous foods (chapters 34 and 35) were critical to the development of stable societies. These microbial processes had multiple origins in different parts of the world (19). People also began to understand (although they did not yet know why) that foods should be kept away from contact with air, light and moisture. Some foods were preserved in early times by coating them with honey or clay and later with olive oil (22). Salt became an especially valuable commodity because it was essential to human health and useful for food preservation. However, salt was not available in quantity in many locales. According to some historians (22), the availability of salt, which was recognized as an important nutrient and food preservative, influenced the course of history. The salt of the Dead Sea was one reason for the Romans' interest in Palestine. If they had not colonized that country, the story of Christ and Christianity would have been quite different.

For thousands of years, people recognized that diseases could be spread by foods. Prohibitions on eating pork—in the Jewish and Muslim religions, for example—had their origins in medical doctrine. Certainly, pork is a dangerous meat in a hot climate, and this must have been realized when dietary regulations were being formulated (20). Nevertheless, although the peoples of the Near East knew all about pork and its dangers, no taboos were placed on it until about 1,800 B.C. By the middle of the first millennium B.C., religious laws in

India had also begun to list many "impure" items of diet. "Unclean food" included meat which had been cut with a sword, dog meat, human meat, and the meat of carnivorous animals, locusts, camels, and hairless or excessively hairy animals. Rice which had turned sour through being left to stand overnight, ready-made food from the market, and dishes which had been sullied by insects or mice or sniffed by a dog, cat, or human were all regarded as unfit for eating. Although most of these edicts carried the weight of religious sanction, they were fundamentally laws of simple hygiene and reflected similar practices derived in other countries (20).

AWARENESS OF BOTULISM AND OTHER FATAL FOODBORNE DISEASES

It was not until the 10th century A.D. that microbiological food poisoning was recognized in civil law. In 900 A.D., Emperor Leo VI of Byzantium issued an edict that forbade eating blood sausage prepared by stuffing blood into a pig stomach and preserving it by smoking. Those caught preparing blood sausage lost all their property and were exiled. Also, to keep civic authorities vigilant, the edict stated that the chief magistrate of the city would be fined 10 pounds of gold (about $35,000 in 1995 United States dollars). Emperor Leo's edict was proclaimed because of an association between botulism and blood sausage (20). The exact cause was unknown. At about that time, a common belief was that miasmas (infectious vapors) transmitted disease.

Poisoning by spoiled grains was recognized by the ancient Greeks and Romans, and many epidemics of major proportions occurred through the Middle Ages in Europe, Russia, and elsewhere. Ergotism, caused by growth of a mold, *Claviceps purpurea*, in grain, was recognized in 1582 and reported again around 1600. In the mid-16th century, epidemics were associated with scabrous rye and other grains infested with *C. purpurea*, and precautions were taken to avoid the contaminated grains. With few exceptions, these precautions were effective; the last major outbreak of ergot poisoning in the United States was in 1825.

Because its underlying causes were unknown, microbiological food poisoning was recurrent. Botulism reappeared many times. For example, in 1793 (9 centuries after Emperor Leo's edict), 13 were affected and 6 died in Germany after eating blood sausage. It was believed that the illness was caused by a fatty acid. The disease was made reportable and the product again was regulated (18).

THE ADVENT OF THERMAL PROCESSING

Because the difficulties with transportation and storage of foods were aggravated by war, in 1795 the French government offered a substantial award for a new preservation method. Nicholas Appert, a Paris confectioner, accepted the challenge and developed a method whereby wide-mouth glass bottles were filled with food, corked, and heated in boiling water. Appert won the prize in 1805 and, at the insistence of Napoleon I, published a book describing his process in 1810. That same year P. Durand of England patented the use of tin cans for thermally processed foods. Neither Appert nor Durand understood *why* thermally processed food did not spoil. Appert only recognized that the container must be sealed and heated to eliminate what he called "agents of putrefaction" and "fermentable principles." In an improvement on Appert's original canning process, I. Solomon, a Baltimore canner, developed in 1860 a simple process that enabled his cannery to increase its output from 2,500 to 20,000 cans per day. In this process, calcium chloride was added to the water in which cans were heated to increase the water temperature from boiling temperature (100°C) to 116°C. This modest increase in temperature reduced the cooking time from 6 h to only 25 to 40 min. R. Chevallier-Appert was issued a patent in 1853 for sterilization of food at even higher temperatures by using steam under pressure in an autoclave, and by 1874 commercial retorts had been introduced.

During this period, the causes of other types of food poisoning became known. For example, the cause of trichinosis in animals (chapter 25) was discovered in 1835 by J. Paget, and the first human case was described in 1859 by F. A. von Zenker. Little did anyone realize that Pasteur and other scientists would soon lay the foundation leading to the discovery of many microbes that caused food poisoning.

FOOD MICROBIOLOGY BECOMES A SCIENCE

Between 1854 and 1864, Louis Pasteur placed heat preservation methods on a scientific basis. He made many discoveries, including experimental proof that certain bacteria were associated with food spoilage and caused specific diseases. Thus, this food scientist became the father of food microbiology and the infant science was born. The first use of what we now know as "pasteurization," the heating of wine to destroy undesirable organisms, was introduced commercially in 1867–1868. Meanwhile, in the 1800s, methods to cultivate microorganisms in pure culture and to associate specific bacteria as the causative agents of specific diseases were developed by R. Koch (who, in 1884, first isolated the cholera vibrio during a worldwide pandemic), J. Lister, and others (3, 13). The isolation and study of pure bacterial monocultures in the laboratory remained at the center of food microbiology for the next hundred years.

Although the cause of botulism, *Clostridium botulinum*, was discovered by E. Van Ermengem in 1896, botulism remained a major problem in thermally processed foods until the 1920s (18). The United States Public Health Service appointed a Botulism Commission, composed of a combination of industrial, academic, and federal experts, to recommend appropriate thermal processes for the prevention of botulism. Their report, published in 1922, emphasized the importance of heating canned foods sufficiently to destroy the most thermal-resistant spores. Consumer education for safely preparing home-canned foods also reduced the incidence of botulism. Nonetheless, to this day, the majority of botulism cases are caused by home canning. The importance of consumer education to the microbial safety of foods cannot be overemphasized. The Food and Drug Administration' general consumer education program through fiscal year 1998 targets the prevention of foodborne illness as the highest priority for its Center for Food Safety and Applied Nutrition.

In 1963, botulism again won notoriety when several outbreaks were attributed to smoked fish from the Great Lakes. Extensive studies by E. M. Foster and colleagues at the University of Wisconsin detailed the ecology of fish-borne *C. botulinum*. One result of the 1963 outbreaks was the realization that foods encased in plastic wraps that exclude oxygen to prevent the growth of aerobic spoilage bacteria and extend shelf life provide ideal conditions for the growth of anaerobic *C. botulinum* (chapter 15).

OTHER PIONEERING DEVELOPMENTS IN FOOD PRESERVATION

Many other pioneering developments on other food preservation processes occurred in the 1800s (13). In 1842, an English patent was issued to H. Benjamin for a process in which an ice-salt brine mixture was used to depress the freezing point to freeze foods more rapidly. An 1861 United States patent on freezing fish was issued to E. Piper of Maine. These patents were not used extensively because refrigeration was in its infancy and there were problems in keeping the foods frozen. Clarence Birdseye, a Massachusetts inventor, when on an expedition to Labrador in 1915 noticed that fish and meat left out at −40 to −50°F (−40 to −46°C) froze solid almost immediately and tasted fresh when thawed months later. Birdseye sold his quick-freeze process to General Foods Corp. in 1929, but it was not until the 1950s that rapid-frozen foods gained popular acceptance. Powdered milk was produced in England in 1855. Pasteurized milk was sold in Germany by 1880 and in the United States by 1890. Commercially dried fruits and vegetables appeared in 1886.

Studies on the use of ionizing energy to preserve foods were initiated in 1925 by F. Ludwig and H. Hopf in Germany, but it was not until after World War II that intensive research commenced at the Massachusetts Institute of Technology and three industry locations. This consortium disbanded within a few years, however, when it became apparent that the project was becoming too expensive. The United States Army started, in 1948, an extensive study in collaboration with industry on the wholesomeness of irradiated foods. The Atomic Energy Commission joined the program in 1953. This National Food Irradiation Program made much progress, but it received a setback upon passage in 1958 of the Delaney Clause (described later), which defined ionizing energy as a food additive rather than a process. Some petitions for food-irradiation processes have been approved by the Food and Drug Administration, and several are presently in use (chapter 28). However, 70 years after its inception, food irradiation remains an underutilized technology.

If the period from the 1890s to the 1940s is described as the era of food preservation, then the era extending from the 1950s through the 1980s can be characterized as the era of "food science," based on chemistry and engineering. As the importance of water activity became clear, intermediate-moisture foods were introduced. Spray-dried and freeze-dried foods soon appeared in many pantries.

RECOGNITION OF OTHER BIOLOGICAL AGENTS OF FOOD POISONING (5)

A. A. Gärtner, in 1888, isolated from meat incriminated in a large food-poisoning outbreak a bacterium subsequently named *Salmonella enteritidis*. The genus *Salmonella* was named in 1900 after a U.S. Department of Agriculture (USDA) bacteriologist, Dr. Salmon, who first described a member of the group, *Salmonella choleraesuis*, which he thought caused hog cholera. (It was later discovered that a virus caused hog cholera and Salmon's bacterium was an incidental isolation. Nevertheless, Salmon's name remains a part of the history of microbiology!) Salmonellosis remains a major problem (chapter 8). For example, one of the worst food poisoning incidents in the history of the United States occurred in 1985 when 16,284 cases and 7 deaths were documented when pasteurized milk somehow became recontaminated with a potent strain of *Salmonella typhimurium*. In 1994, this was exceeded by a national outbreak of *Salmonella enteritidis* affecting 225,000 people who consumed contaminated ice cream products (12). K. Shiga discovered in 1898 an enteric pathogen closely related to the salmonellae. Shiga's bacterium, which causes bacillary dysentery (chapter 12), was later named *Shigella dysenteriae.*

In 1906, an aerobic, spore-forming bacillus (chapter 17) was recognized as a cause of food poisoning. The true significance of this discovery did not become apparent until the taxonomy of the genus *Bacillus* was clarified by N. R. Smith and R. E. Gordon in 1946 and S. Hauge identified from among the numerous *Bacillus* species that the pathogen was always *Bacillus cereus.*

Staphylococci (chapter 19) were first recognized by Pasteur, who found them in pus. They were not associated with food poisoning until studies by T. Denys in 1894 and M. A. Barber in 1914. Barber used himself as a test subject to show that milk that he obtained on a farm in the Philippines contained staphylococci that induced vomiting. The existence of an exotoxin was confirmed by G. M. Dack and coworkers, who studied cream-filled Christmas cakes in 1930.

In 1939, J. Schleifstein and M. B. Coleman described gastroenteritis caused by a bacterium that, in 1965, was named *Yersinia enterocolitica* by R. Sakazaki (chapter 11). However, it was not until 1969 that B. Nilehn documented a foodborne outbreak and 1971 when T. Dadisman and coworkers documented the first United States outbreak, which occurred in New York. Major attention was focused on yersiniosis in 1982, when pasteurized milk from a plant in Tennessee was implicated in a large interstate outbreak; 172 culture-positive infections were identified. Most patients required hospitalization, and 17 underwent appendectomies when their symptoms were mistakenly diagnosed as appendicitis.

Clostridium perfringens (formerly *C. welchii*), the causative agent of human gas gangrene infections, was first implicated as a cause of foodborne illness by E. Klein in 1885. This type of food poisoning (chapter 16) went almost unrecognized until R. Knox and E. K. Macdonald in England in 1943 and L. S. McClung in the United States in 1945 alerted the scientific community about perfringens food poisoning. In 1953, B. Hobbs and coworkers in England reported that perfringens food poisoning was common but had been overlooked by most laboratories because anaerobic techniques were not used in food microbiology unless botulism was suspected. C. L. Duncan and D. W. Strong demonstrated, in 1969, that perfringens food poisoning was caused by an enterotoxin.

Milk had been implicated in the transmission of several diseases, so it was not surprising that G. Jubb, in 1915, reported milk to be the vehicle responsible for a small outbreak of poliomyelitis in England. Raw milk also was implicated in the transmission of infectious hepatitis (hepatitis A) virus by M. D. Campbell in 1943 in England, and an outbreak caused by contaminated shellfish in Sweden was described by B. Roos in 1956. In 1961, several large outbreaks of hepatitis A caused by clams and oysters occurred in Mississippi, Alabama, New Jersey and Connecticut, focusing attention on foodborne viruses and especially the shellfish problem (see chapter 24).

In 1951, T. Fujino showed that *Vibrio parahaemolyticus* (chapter 13) caused food poisoning associated with seafood consumption in Japan. Within only a few years it became apparent that *V. parahaemolyticus* was responsible for over 70% of Japan's investigated cases of foodborne gastroenteritis. Twenty years after Fujino's work, in 1971, the first United States outbreak was documented in Maryland.

Food poisoning caused by molds again attracted attention when, in 1960–61, a massive outbreak of lethal hepatic "turkey X disease" occurred in England, involving over 100,000 turkey poults at 500 locations. *Aspergillus flavus* was isolated from Brazilian ground nut (peanut) meal and was shown by K. Sargeant and coworkers in 1961 to produce a toxin that caused acute hepatitis at high levels. Later, it was shown that the toxin was carcinogenic at lower levels. Within less than 10 years many different mycotoxigenic molds (chapters 21 to 23) were identified.

Various nematodes and cestodes (chapter 27) had been known for many decades to be transmitted by foods. Antonie van Leeuwenhoek, in 1681, examined his own stools during a bout with diarrhea and observed a protozoan, *Giardia lamblia,* in large numbers. He subsequently observed similar microbes in the guts of rodents and frogs, but did not associate animal reservoirs with disease transmission. This association was not fully appreciated until 1965, when a major waterborne outbreak of giardiasis occurred in Aspen, Colorado, and a series of other waterborne outbreaks occurred in the 1970s. Subsequently, other parasites have been confirmed as causes of major outbreaks of water- and foodborne infections and intoxications.

EMERGING FOODBORNE AND WATERBORNE PATHOGENS

The discovery (or sometimes rediscovery) of foodborne pathogens continues to this day. One of these is *Cryptosporidium.* This parasite was first described in 1907 and the first case was diagnosed in a human in 1976. Several small outbreaks occurred during the next 17 years, but the true danger of cryptosporidia was revealed when, in 1993, an estimated 403,000 persons in Milwaukee, Wisconsin, had prolonged diarrhea and approximately 4,400 required hospitalization during a waterborne outbreak.

Four relatively unknown bacteria became associated with foodborne illness in the 1980s (5, 6). One of these, *Vibrio fetus,* was identified as a cause of abortion in cattle and sheep in 1913 by J. McFadyean and S. Stockman in England. Although sporadic outbreaks of enteritis from milk had been described as early as 1938, *V. fetus* (now named *Campylobacter jejuni*) was considered mostly a veterinary pathogen. This status changed in 1977 following the publication of a report in the *British Medical Journal* by M. B. Skirrow, who described appropriate methods to isolate the bacterium. Once workers knew what to look for and how to conduct the search, campylobacteriosis was recognized as a leading cause of acute bacterial diarrhea worldwide (chapter 9). Another bacterium, enterohemorrhagic *Escherichia coli* O157:H7 (chapter 10), was first recognized as a pathogen in 1982. It causes bloody diarrhea and kidney failure. By 1987, *E. coli* O157:H7 was recognized as more common than *Shigella* spp. in the United States. In 1993 a large multistate outbreak from eating fast-food hamburgers occurred on the West Coast. The outbreak received wide media coverage, which greatly influenced public attitudes about meat sanitation and preparation. The reverberations of these outbreaks are still being felt. The public no longer accept the presence of pathogens in raw meat but are slow to accept their own role in assuring the microbial safety of food "from farm to fork."

Beginning in the 1970s, workers began to associate additional marine vibrios with severe tissue infections and death. *Vibrio vulnificus* (chapter 13), named in 1979, was identified as a cause of seafood-related illness with a high mortality rate in people who are immunologically compromised or have a preexisting hepatic disease. Foodborne *Listeria monocytogenes* (chapter 18) also affects the immunologically compromised. The first recognized outbreak of listeriosis occurred in Nova Scotia in 1981. The vehicle was coleslaw made from cabbage grown in a field in Maine that had been fertilized with sheep manure. After several other outbreaks, a large outbreak that occurred in California in 1985 focused attention on the unique epidemiology of listeriosis. Of 86 confirmed cases of listeriosis among people who had eaten Mexican-style soft cheese, over half were mother-infant pairs. The adoption of a zero tolerance in ready-to-eat foods and renewed attention to plant sanitation appear to have ended multicase listeriosis outbreaks, although sporadic cases still occur.

There have been no large confirmed outbreaks of food poisoning caused by *Aeromonas hydrophila* (named by M. Y. Popoff and coworkers in 1976) or *Plesiomonas shigelloides* (first described by W. W. Ferguson and N. D. Henderson in 1947 and named by H. Habs and R. H. W. Schubert in 1962). Nevertheless, much indirect evidence exists (chapter 14) that at least some strains cause gastroenteritis ranging from mild to severe, especially in the very young, aged, or otherwise immunologically compromised. An aging population and increasing prevalence of immunodeficiency diseases suggest that additional opportunistic pathogens will be discovered and have led to renewed interest in foodborne infections and intoxications. The struggle between humans and microbes is a continuing feature of the evolution of food microbiology. On the one hand, we become more knowledgeable about bacteria and develop better tools with which to identify and combat them. On the other hand, the current age of microbiology is barely a moment in the evolutionary history of bacteria, which extends 3 billion years into prehistory. Recent developments such as the spread of multiple drug resistance among clinically important pathogens warn against complacence.

MODERN LEGISLATION AND THE DEVELOPMENT OF PASTEURIZATION

Although the first British Food and Drugs Act was passed in 1860 and was revised and strengthened in 1872, it was not until 1890 that the United States passed its first food law, a National Meat Inspection Law. This law, which required the inspection of meats for export, was strengthened in 1895. Then, in 1906, a comprehensive Federal Food and Drug Act was approved by Congress; in 1939 a revised version became the Food, Drug and Cosmetics Act. Probably the most rigorous of the early United States food ordinances was the first U.S. Pasteurized Milk Ordinance, published in 1924 (25). It was a culmination of a century of experimentation and development; exactly 100 years earlier (in 1824), William Dewees recommended heating milk just to the boiling point to increase shelf life. In the United States, G. Borden, a Texas inventor, obtained a patent in 1853 for a process wherein milk was condensed under vacuum and heated to 50 to 60°C and sugar was added. The first commercial milk pasteurizer was made in Germany in 1882, and a variety of equipment and pasteurization times and temperatures ensued. In 1900, Russell and Hastings had defined the thermal death point of tubercle bacilli under commercial conditions, but their recommendations often were not followed or were unattainable because of faulty or poorly designed equipment. This oversight was corrected by the 1924 U.S. Pasteurized Milk Ordinance, which specified not only time and temperature, but most importantly, that the process must be conducted in approved equipment. Also during the 1920s, three associations representing sanitarians, manufacturers, and the milk industry worked with regulatory agents to formulate what are known as the "3-A" (for three associations) standards for the performance (such as cleanability) of dairy equipment (2). Included in a revised 1933 U.S. Public Health Service Milk Ordinance and Code was provision for high-temperature short-time (HTST) treatment of milk (chapter 28). Subsequently, in the 1950s, ultra-high-temperature (UHT) processing with aseptic packaging (chapter 28) became available.

During the past 70 years or so, milk pasteurization, together with a cattle-testing program, was so successful in reducing the incidence of tuberculosis in developed countries that it has been called "man's greatest victory over tuberculosis." This is attested to by the fact that tuberculosis is not included as a major food pathogen in this book (14). North America has also achieved freedom from brucellosis (undulant fever), another disease that once was transmitted by meat and milk. These are prime examples of what has been accomplished in food protection by cooperation and collaboration among farmers, veterinarians, food microbiologists, food scientists, food technologists, engineers, industry, and regulatory officials.

A U.S. Compulsory Poultry and Poultry Products Law was passed in 1957, a comprehensive Food Additives Amendment was added to the Food, Drug and Cosmetics Act (often referred to as the Delaney Clause) in 1958, a Wholesome Meat Act was enacted in 1967, and a Poultry Inspection Bill was passed in 1968.

THE "INDICATOR" CONCEPT AND THE DEVELOPMENT OF ANALYTICAL METHODS

When microbiological standards and regulations were being formulated, it was realized that tests could not be conducted for each and every enteric pathogen that might be present in a sample. Instead, a surrogate must be selected. A. von Fritsch of Germany suggested, in 1880, that certain klebsiellae were characteristic of human contamination of water supplies. Five years later, T. Escherich described a fecal bacterium, *Bacillus coli* (now named *Escherichia coli*), and in 1892, F. Schardinger suggested that *E. coli* would be useful as an indicator of fecal pollution (see chapter 4 on indicators and microbial criteria). At that time, methods to detect *E. coli* among a large group of related bacteria termed "coliforms" were not readily available. C. Eijkman, in 1904, determined that incubating test samples at 46°C would differentiate "fecal coliforms" that grow at high temperatures from coliforms that arose from other environments and do not grow at 46°C. Nevertheless, the U.S. Public Health Service, in 1914, changed the indicator standard from *E. coli* to the coliform group. Elaborate and time-consuming tests were conducted on isolates from positive coliform tests to determine if *E. coli*, the coliform bacterium whose presence correlated most closely with fecal pollution, was present. It was not until the 1980s, however, that relatively rapid, simple, and reliable methods for the simultaneous detection of both total coliforms and *E. coli* became available. Similarly, rapid methods were developed to detect and identify other indicator organisms such as enterococci, another bacterial group primarily of intestinal origin first described by M. E. Thiercelin in 1899. The need for indicator organisms was born at a time when the isolation and culture of specific pathogens was difficult, if not impossible. With the plethora of rapid and automated methods now available to test for specific pathogens and toxins it is difficult to justify continued reliance on indicators.

STANDARDIZATION OF METHODS

The promulgation of laws governing water and foods required uniform and efficient analytical methods, such as those described in chapters 38 and 39. At the turn of the century, workers became acutely aware that analytical results were almost useless for comparative purposes because of large laboratory-to-laboratory variations in methods. In 1895, the American Public Health Association recognized this need and appointed a committee to draft uniform procedures for the determination of several chemical attributes of water. In 1899, another committee was charged with extending standard procedures to all methods, including bacteriological, involved in water analysis. This committee's report, published in 1905, constituted the first edition of a manual, *Standard Methods of Water Analysis*. The book, now in its 18th edition, is entitled *Standard Methods for the Examination of Water and Wastewater* (11). Also in 1905, S. C. Prescott of the Massachusetts Institute of Technology proposed that differences in composition of culture media, temperature and time of incubation, and other variables involved in the bacteriological examination of milk needed to be reconciled. A committee was formed and a report on *Standard Methods of Bacterial Milk Analysis* was published; this report evolved into another manual, *Standard Methods for the Examination of Dairy Products*, which is in its 16th edition (16). Another manual, *Recommended Methods for the Microbiological Examination of Foods*, was published in 1958. This manual eventually became the *Compendium of Methods for the Microbiological Examination of Foods*, which is now in its 3rd edition (23). Other organizations also publish compilations of microbiological procedures applicable to water and foods. One of these is the Association of Official Analytical Chemists International (AOAC), which since 1916 has validated methods used for regulatory purposes by subjecting the methods to collaborative studies to substantiate their accuracy and reproducibility. Approved methods are published in AOAC's Official Methods of Analysis (1, 4). In addition, the U.S. Food and Drug Administration publishes a *Bacteriological Analytical Manual*, which is now in its 8th edition (8). Also, throughout the years, as methodology has evolved, governmental agencies have published in the *Federal Register* methods acceptable for regulatory purposes (7, 9).

FROM DETECTION METHODS TO IN-PROCESS PREVENTION

In 1959, H. E. Baumann and others at the Pillsbury Co., in cooperation with the National Aeronautics and Space Agency, the U.S. Army Natick Laboratories, and the U.S. Air Force Space Laboratory Group, set out to produce almost completely safe foods for the space program. However, the sample-consumptive nature of food analysis was fundamentally different from quality assurance in electronics, where every item could be tested and then used. The amount of sample required to assure the desired safety level using statistically based microbial testing of finished products would require so much product that little would be left for use on the space flights. An alternative, preventive system of quality control, now known as the Hazard Analysis—Critical Control Point (HACCP) system (chapter 41), was developed. HACCP involves complete control over raw materials, process, environment, personnel, distribution, and storage (17). All of these had previously received attention, but not as an integrated system. Although the original emphasis was placed on raw materials as they arrived at the manufacturing plant on through to the consumer level, more recently the HACCP concept has included production agriculture, starting with the plant seed or individual animal production unit.

The HACCP system was first made public in 1971, but it was not seriously considered by the food industry until it was recommended in 1985 by the National Academy of Sciences Subcommittee on Microbiological Criteria for Foods and Food Ingredients. In 1989, the Subcommittee published a pamphlet (revised in 1992) titled "HACCP Principles for Food Production." HACCP has become widely accepted; for example, in 1995 the USDA Food Safety Inspection Service published in the *Federal Register* proposed regulations under which all slaughter and processing plants will be required to develop and implement an HACCP program within 3 years. Previously, in 1993, HACCP became international in scope when it was approved by the Committee of Food Hygiene of the Codex Alimentarius Commission. The Codex Alimentarius Commission is an international body responsible for the execution of the Joint FAO/WHO Food Standards Program, which was created in 1962 and is aimed at protecting the health of consumers and facilitating international trade in foods. The Codex consists of a large collection of food standards presented in a uniform manner. A well-documented HACCP plan also is required for certification of a food company under the recently formed European International Standards Organization (ISO) 9000 standards. The ISO 9000 is a voluntary process in which a third-party assessment certifies that a high-level quality assurance system is in place. Because of increased international trade, United States companies are discovering that ISO 9000 certification is necessary.

The International Commission on Microbiological Specifications for Foods (ICMSF), a standing committee of the International Association of Microbiological Societies, was formed in 1962. The ICMSF establishes internationally acceptable microbiological criteria and attempts to reach agreement on the essential supporting methods from among the plethora of methods in the literature.

THE ERA OF MOLECULAR BIOLOGY AND GENETICS

Until the 1960s, the practice of food microbiology was relatively unchanged since the time of Pasteur. It was a descriptive, qualitative science that focused on *what* happens with relatively little emphasis on, or understanding of, the *why* or underlying mechanisms. This observation is not meant to denigrate those pioneers of microbiology on whose foundations we build. One must know "what" before one can ask "why." The knowledge of molecular biology and modern experimental tools did not exist prior to the 1960s. Molecular biology was born in the 1940s and 1950s, when scientists from the physical sciences entered the field of biology with the express intent of applying the methods of physical science to biological phenomena. They brought with them a more quantitative and mechanistic approach to science which now permeates all areas of biology, including food microbiology. *Food Microbiology: Fundamentals and Frontiers* has been written from this perspective. Wherever possible, the detailed mechanisms responsible for the topic at hand are stressed. It is clear, however, that in many cases, we have just begun to understand "what" happens, let alone "why."

Only within the last 30 years have fundamental concepts of biology, such as the genetic code, the structure-function relationship of proteins, the chemiosmotic coupling of energy-generating and -requiring reactions, and the transfer of genetic information been developed. Some of these are reviewed in chapter 2. In most cases, the general principles have been developed using relatively simple, well-studied bacteria such as *E. coli*. Frequently, foodborne microbes turn out to be quite different. Lactose catabolism is a case in point. The genes for lactose catabolism by *E. coli* were among the first to be studied in detail. The resultant *lac* operon is frequently used to teach the concepts of induction, derepression, carbon catabolite repression, and the role of protein kinases in the synthesis of the β-galactosidase which is ultimately excreted and cleaves lactose to glucose and galactose. However, from a practical standpoint, the most important application of lactose catabolism is in the dairy industry, where the lactose in milk is fermented by lactic acid bacteria such as *Lactococcus lactis*. Lactose catabolism by *L. lactis* is by a completely different mechanism from the *lac* operon model. *L. lactis* phosphorylates lactose as it is translocated across the cell membrane by the phosphoenolpyruvate:phosphotransferase system. The intracellular lactose phosphate is then hydrolyzed by an intracellular phospho-β-galactosidase.

The evolution of the dairy industry from a farm-based "art" to a highly technological industry provides other excellent examples of how basic science affects food microbiology. A fundamental understanding of the plasmid biology of fermentative organisms has reduced the incidence of "stuck" fermentations that have lost the ability to metabolize lactose. An understanding of the complex process by which bacteriophage attack and kill starter culture bacteria has generated many strategies for development of phage-resistant fermentations. Through the use of recombinant DNA technology, *E. coli* is rapidly replacing the fourth stomach of a milk-fed calf as the source of the rennet (chymotrypsin) used to make many cheeses.

Advances in molecular biology and genetics have revolutionized analytical food microbiology. Pasteur would be lost in a modern food microbiology laboratory. "Plate and count" microbiology is rapidly giving way to thermocyclers, gel boxes, microtiter plates, and ELISA readers which allow direct quantification of pathogens and their toxins (see chapters 38 and 39). "Rapid" salmonella tests have reduced analysis time from 5 days to less than 48 h. Methods being developed by companies from all over the world will consecutively break the 24-, 12-, 8-, and probably 4-h barriers. The ideal of a real-time biosensor for microbial contamination may one day be achieved.

CONCLUSION

The history of food microbiology is rich and exciting. It has taken us from the slow realization that certain diseases are caused by microorganisms that grow in foods to the empirical control of these microbes using physical, chemical, and biological manipulation. A mechanistic understanding of microbial physiology and metabolism has provided new approaches to food preservation and laid the foundation for genetic control of foodborne pathogens. We may one day be able to control foodborne pathogens by direct regulation of their genes. Control of insect pests in plants has progressed from the use of extrinsic chemical insecticides, to biocontrol using *Bacillus thuringiensis*, to the cloning of *B. thuringiensis* genes directly into plants to make them intrinsically insecticidal. Perhaps one day there will be salmonella-resistant chicken or listeria-resistant milk. Real-time biosensors may ultimately replace postprocess sampling.

Food microbiology stands on a scientific footing not more than 100 years old. Its current practice is being transformed by knowledge and tools generated by molecular biology and genetics. Its future, though hard to predict, will certainly be bright.

References

1. **Andrews, W. H.** 1994. Update on validation of microbiological methods by AOAC International. *J. AOAC Int.* **77:** 925–931.

2. **Atherton, H. V.** 1986. The 3-A story. *Dairy Food Sanit.* 6:96–98.
3. **Chung, K.-T., S. F. Stevens, and D. H. Ferris.** 1995. A chronology of events and pioneers of microbiology. *SIM News* **45**:3–13.
4. **Cunniff, P. A. (ed.).** 1995. *Official Methods of Analysis of AOAC International,* 16th ed., vol. I and II. AOAC International, Arlington, Va.
5. **Doyle, M. P. (ed.).** 1989. *Foodborne Bacterial Pathogens.* Marcel Dekker, Inc., New York.
6. **Doyle, M. P.** 1994. The emergence of new agents of foodborne disease in the 1980s. *Food Res. Int.* **27**:219–226.
7. **Environmental Protection Agency.** 1986–1994. National primary drinking water regulations. *Fed. Regist.* **51**:37608–37612; **52**:42224–42245; **53**:16348–16358; **54**:27544–27568, 29998–30002; **55**:22752–22756; **56**:636–643, 49153–49154; **57**:1850–1852, 24744–24747; **58**:65622–65632; **59**:6332–6444, 62456–62471.
8. **Food and Drug Administration.** 1992. *FDA Bacteriological Analytical Manual* (BAM), 7th ed. AOAC International, Arlington, Va.
9. **Food and Drug Administration.** 1993. Quality standards for foods with no identity standards: bottled water. *Fed. Regist.* **58**:52042–52050.
10. **Fumihito, A., T. Miyake, S.-I. Sumi, M. Takada, S. Ohno, and N. Kondo.** 1994. One subspecies of the red junglefowl (*Gallus gallus gallus*) suffices as the matriarchic ancestor of all domestic breeds. *Proc. Natl. Acad. Sci. USA* **91**: 12505–12509.
11. **Greenberg, A. E., L .S. Clesceri, and A. D. Eaton (ed.).** 1992. *Standard Methods for the Examination of Water and Wastewater,* 18th ed. American Public Health Association, Washington, D.C.
12. **Hennessy, T. W., C. W. Hedberg, L. Slutsker, K. E. White, J. M. Besser-Wiek, M. E. Moen, J. Feldman, W. W. Coleman, L. M. Edmonson, K. L. MacDonald, M. T. Osterholm, and the Investigation Team.** 1996. A national outbreak of *Salmonella enteritidis* infections from ice cream. *N. Engl.* J. Med. **334**:1281–1286.
13. **Jay, J. M.** 1992. *Modern Food Microbiology,* 4th ed. Van Nostrand Reinhold, New York.
14. **Kapur, V., T. S. Whittam, and J. M. Musser.** 1994. Is *Mycobacterium tuberculosis* 15,000 years old? *J. Infect. Dis.* **170**:1348–1349.
15. **Loftus, R. T., D. E. MacHugh, D. G. Bradley, P. M. Sharp, and P. Cunningham.** 1994. Evidence for two independent domestications of cattle. *Proc. Natl. Acad. Sci. USA* **91**: 2757–2761.
16. **Marshall, R. T. (ed.).** 1993. *Standard Methods for the Examination of Dairy Products,* 16th ed. American Public Health Association, Washington, D.C.
17. **Pierson, M. D., and D. A. Corlett, Jr.** 1992. *HACCP Principles and Applications.* Chapman & Hall, New York.
18. **Smith, L. D. S.** 1977. *Botulism: the Organism, Its Toxins, the Disease.* Charles C Thomas Publishers, Springfield, Ill.
19. **Steinkraus, K. H. (ed.).** 1983. *Handbook of Indigenous Fermented Foods.* Marcel Dekker, Inc., New York.
20. **Tannahill, R.** 1973. *Food in History.* Stein and Day Publishers, New York.
21. **Thomas, H.** 1979. *A History of the World.* Harper & Row, Publishers, Inc., New York.
22. **Toussaint-Samat, M. (translated by A. Bell).** 1992. *History of Food.* Blackwell Publishers, Cambridge, Mass.
23. **Vanderzant, C., and D. F. Splittstoesser (ed.).** 1992. *Compendium of Methods for the Microbiological Examination of Foods,* 3rd ed. American Public Health Association, Washington, D.C.
24. **Wendorf, F., R. Schild, N. El Hadidi, A. E. Close, M. Kobusiewicz, H. Wieckowska, B. Issawi, and H. Haas.** 1979. Use of barley in the Egyptian late paleolithic. *Science* **205**:1341–1348.
25. **Westhoff, D. C.** 1978. Heating milk for microbial destruction: a historical outline and update. *J. Food Prot.* **41**:122–130.

Thomas J. Montville

2

Principles Which Influence Microbial Growth, Survival, and Death in Foods

A former president of the American Society for Microbiology defined "microbiology" as "an artificial subdiscipline of the field of biology" (based on size). If this is true, then food microbiology is certainly indistinguishable from the larger field of microbiology, except for its context. The purpose of this chapter is to provide some of that context and to review microbiological concepts prerequisite to the understanding of foodborne microbes.

Food microbiologists must understand the basic biophysical principles of microbiology, must have a firm knowledge of food systems, and must be able to integrate both areas to solve the microbiological problems that occur in extremely complex food ecosytems. The first part of this chapter examines foods as ecosystems and discusses those intrinsic and extrinsic environmental factors that control bacterial growth in foods. Ecology teaches that the *interaction* of environmental factors determines which organisms can or cannot grow in that environment. The manipulation of multiple environmental factors (i.e., pH, salt concentration, temperature, etc.) to inhibit microbial growth is the essence of multiple "hurdle" technology (40), which is increasingly being adopted as a technology for food preservation. Faced with changing environments, cells must maintain homeostasis in a variety of vital functions such as internal pH and membrane fluidity. The ability of a cell to maintain homeostasis may ultimately determine its fate in foods (29).

Because many microbial processes are governed by first-order kinetics, fundamental kinetic concepts are explained in the second part of this chapter. The log phase of microbial growth, the dependence of enzyme velocity on substrate concentration, and many types of lethality follow first-order or pseudo-first-order kinetics. The doubling time, D values, and z values used by food microbiologists are kinetic constants based on first-order kinetics. Since food science is interdisciplinary and encompasses aspects of chemistry, engineering, and biology, food microbiologists should understand how "their" constants relate to the kinetic constants used by chemists and engineers.

The third part of this chapter focuses on physiology and metabolism of foodborne microbes. The pathways for carbohydrate catabolism are especially important. The catabolic pathways used by the lactic acid bacteria dictate the characteristics of many fermented foods. In nonfermented foods, spoilage is usually associated with the appearance of a specific catabolic product(s). Lactic acid makes milk sour; hydrogen causes cans to swell. Foodborne microbes have a variety of catabolic pathways for processes as simple as glucose catabolism. The pathway used is determined by genetic and environmental parameters. The ability of bacteria to use different biochemical pathways which generate different amounts of ATP influences their ability to grow under adverse conditions in foods. For example, the pump that removes protons from the cytoplasm to maintain pH homeostasis is fueled by ATP. Because facultative anaerobes generate more ATP by aerobic respiration than by anaerobic fermentation (see below), organisms such as *Staphylococcus aureus* can grow at a lower pH and at lower water activities under aerobic conditions than under anaerobic conditions. The generation and utilization of energy, "bioenergetics," are critically important to the cell.

The limitations of classical microbiology are reviewed in the last section of this chapter. Just as our inability to see colors having wavelengths longer than violet or shorter than red limits our perception of the physical world, the techniques of classical microbiology can limit our perception of the microbial world. The inability of injured cells and cells that are "viable but not culturable" to form colonies on petri dishes does not mean that these cells do not exist. Bacteria living in biofilms on processing equipment or attached to surfaces are also physiologically different from the free-living cells studied in the laboratory.

FOOD ECOSYSTEMS, HOMEOSTASIS, AND HURDLE TECHNOLOGY

Foods as Ecosystems

Foods are complex ecosystems. This is so widely accepted that the International Commission on Microbial Specifications for Foods chose *The Microbial Ecology of Foods* as the title of their books subtitled *Factors Affecting Life and Death of Microorganisms* and *Food Commodities* (33, 34). Ecosytems are composed of the environment and the organisms that live in it. The food environment is composed of intrinsic factors inherent to the food (e.g., pH, water activity, and nutrients) and extrinsic factors external to it (e.g., temperature, gaseous environment, the presence of other bacteria). Intrinsic and extrinsic factors can be manipulated to preserve food, and food preservation can be viewed as "the ecology of zero growth" (7).

When applied to microbiology, ecology can be defined as "the study of the *interactions* between the chemical, physical, and structural aspects of a niche and the composition of its specific microbial population" (53). "Interactions" is emphasized to highlight the multivariable nature of ecosystems. The complex relationship between multiple environmental parameters in foods almost dictates a computer modeling approach (see chapter 40). A very complete set of reviews about food ecosystems, from which this chapter draws heavily, has been published by the Society for Applied Bacteriology (6).

Foods can be heterogeneous on a micrometer scale. Heterogeneity and its associated gradients of pH, oxygen, nutrients, etc., are key ecological factors in foods (7). Foods may contain several distinct microenvironments. This is well illustrated by the food poisoning outbreaks in "aerobic" foods caused by the "obligate anaerobe" *Clostridium botulinum*. Growth of *C. botulinum* in potatoes, sautéed onions, and cole slaw exposed to air has caused botulism outbreaks (45). The oxygen in these foods is driven out during cooking and diffuses back in so slowly that the bulk of the product remains anaerobic. The growth of certain fungi and bacilli at the surface of tomato products can produce oxygen and pH gradients such that *C. botulinum* growth and toxin production can occur in a very limited area of the product (32, 52). This is also a good example of how ecosystems are nested, one inside the next, inside the next. Microbes exist in microenvironments that are in food macroenvironments contained in human environments (where a person forgets to refrigerate a product) in a global environment (where the weather may provide a temperature conducive to microbial growth).

Intrinsic Factors That Influence Microbial Growth

Those factors inherent to the food itself are considered "intrinsic" factors. These include naturally occurring compounds that may stimulate or retard microbial growth, compounds added as preservatives, the oxidation-reduction potential, water activity, and pH. Most of these factors are covered separately in the chapters on physical and chemical methods of food preservation. The influence of pH on gene expression is a relatively new area and is covered briefly below.

The expression of genes governing proton transport, amino acid degradation, adaptation to acidic or basic conditions, and even virulence can be regulated by the external pH (pH_o) (60). Cells sense changes in pH_o through several different mechanisms. pH-induced protonation and deprotonation of amino acids can change the protein's secondary or tertiary protein structure, altering the protein's function which signals the change. The cell may respond to only one form of small signal molecules. For example, organic acids cross the cytoplasmic membrane only in the protonated form. An increased intracellular concentration would indicate increased environmental acidity. The transmembrane proton gradient (i.e., the ΔpH) itself can serve as a sensor and up- or down-regulate energy-dependent processes.

Intracellular pH (pH_i) must be maintained above some critical pH_i at which intracellular proteins become irreversibly denatured. In *Salmonella typhimurium*, where

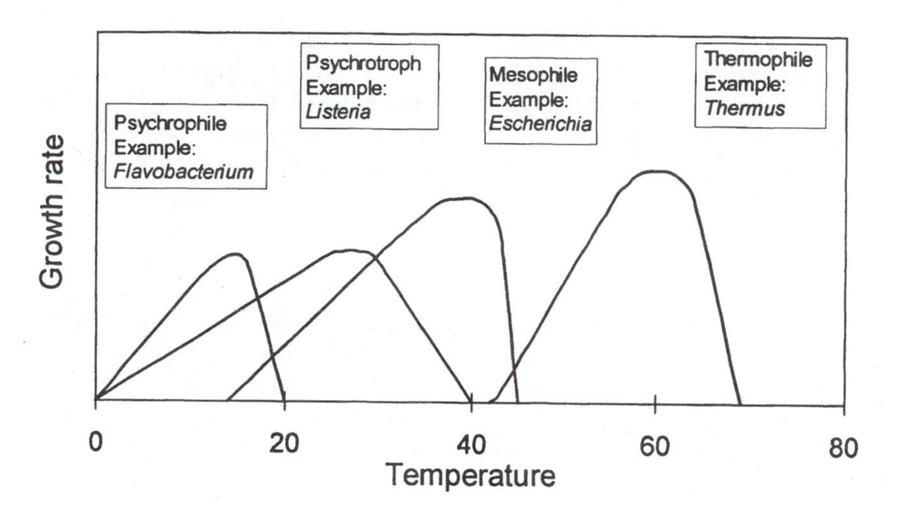

Figure 2.1 Relative growth rates of bacteria at different temperatures.

it has been studied most extensively, there are three progressively more stringent mechanisms to maintain a pH_i consistent with viability (27, 28). These three mechanisms are the homeostatic response, the acid tolerance response, and the synthesis of acid shock proteins.

At $pH_o > 6.0$, salmonella cells adjust their pH_i through the homeostatic response. The homeostatic response maintains pH_i by allosterically modulating the activity of proton pumps, antiports, and symports to increase the rate at which protons are expelled from the cytoplasm. The homeostatic mechanism is constitutive and functions in the presence of protein synthesis inhibitors.

The acid tolerance response (ATR) is triggered by pH_o of 5.5 to 6.0 (27). This mechanism is sensitive to protein synthesis inhibitors; at least 18 ATR-induced proteins have been identified. ATR appears to involve the membrane-bound ATPase proton pump and maintains $pH_i >$ 5.0 at pH_o values as low as 4.0. The loss of ATPase activity caused by gene disruption mutations or metabolic inhibitors abolishes the ATR, but not the pH homeostatic mechanism described above. The ATR may confer cross-protection to other environmental stressors. The exposure of *S. typhimurium* cells to pH 5.8 for a few doublings induces 12 proteins, represses 6 proteins, and renders the cells less sensitive to salt and heat (41). Acid adaptation also occurs in *Escherichia coli* O157:H7 (42).

The synthesis of acid shock proteins is the third way that cells regulate pH_i. The synthesis of these proteins is triggered by pH_o from 3.0 to 5.0. They constitute a set of *trans*-acting regulatory proteins distinct from the ATR proteins. They may be similar to cold shock proteins which help confer acid resistance in *Listeria monocytogenes* (28).

Other phenotypes can also be regulated by pH. The expression of the *Yersinia enterocolitica inv* gene in laboratory media at 23°C but not at 37°C seems paradoxical since its expression is required for infection of warm-blooded animals. However, at the pH of the small intestine (5.5), the *inv* gene is expressed at 37°C (62). The *yst* gene, which codes for a heat-stable enterotoxin in *Y. enterocolitica*, is regulated similarly (50).

Extrinsic Factors That Influence Microbial Growth

Temperature and gas composition are the primary extrinsic factors influencing microbial growth. Controlled and modified atmospheres are covered at length in chapter 28. The influence of temperature on microbial growth and physiology cannot be overemphasized. While the influence of temperature on growth kinetics is obvious and covered here in some detail, the influence of temperature on gene expression is equally important. Cells grown at refrigerated temperature are not just "slower" than those grown at ambient temperature; they express different genes and are physiologically different. Later chapters provide organism-specific detail about how temperature regulates phenotypes ranging from motility to virulence.

A rule of thumb in chemistry suggests that reaction rates double with every 10°C increase in temperature. This simplifying assumption is valid for bacterial growth rates only over a limited range of organism-dependent temperatures (Fig. 2.1). Above the optimal growth temperature, the growth rates decrease precipitously. Below the optimum, growth rates also decrease, but more gradually. Bacteria can be classified as psychrophiles, psychrotrophs, mesophiles, and thermophiles according to how temperature influences their growth.

Both psychrophiles and psychrotrophs grow, albeit slowly, at 0°C. True psychrophiles have optimum growth rates at 15°C and cannot grow above 25°C. Psychrotrophs, such as *L. monocytogenes* and *C. botulinum* type E, have an optimum of about 25°C and cannot grow above 40°C. Because these foodborne pathogens, and even some mesophilic *Staphylococcus aureus* strains, can grow at <10°C, conventional refrigeration is not sufficient to assure the safety of a food (61). Additional barriers to microbial growth should be incorporated into refrigerated foods containing no other inhibitors (55).

Several metabolic capabilities are important for growth in the cold. Homeoviscous adaptation enables cells to maintain membrane fluidity at low temperatures. As temperatures decrease, the cell synthesizes increasing amounts of mono- and di-unsaturated fatty acids (15, 67). The "kinks" caused by the double bonds prevent tight packing of the fatty acids into a more crystalline array. The accumulation of compatible solutes at low temperatures (37) is analogous to their accumulation under conditions of low water activity, as discussed in chapter 28. The production of cold shock proteins also contributes to an organism's ability to grow at low temperatures.

Temperature regulates the expression of virulence genes in several pathogens. The expression of 16 proteins on seven distinct operons on the *Y. enterocolitica* virulence plasmid is high at 37°C, weak at 22°C, and undetectable at 4°C (68). Similarly, the genes required for virulence of *Shigella* spp. are expressed at 37°C, but not at 30°C. The expression of genes required for *L. monocytogenes* virulence is also regulated by temperature (39). Cells grown at 4, 25, and 37°C all synthesize internalin, a protein required for penetration of the host cell. Cells grown at 37°C produce hemolytic activity, whereas this gene product is not produced at 4°C or 25°C. However, the hemolytic activity is restored during the infection process (14).

Growth temperature can also influence a cell's thermal sensitivity. *L. monocytogenes* cells preheated at 48°C have increased thermal resistance (24). Holding listeria cells at 48°C for 2 h in sausages increases their *D* values at 64°C 2.4-fold. This thermotolerance is maintained for 24 h at 4°C (22). The role of heat shock proteins in increased thermal resistance is discussed further in chapter 28. Shock proteins synthesized in response to one stressor may provide cross-protection against other stressors (41).

Homeostasis and Hurdle Technology

Consumer trends towards blander foods have decreased the use of intrinsic factors such as acidity and salt as the sole means of providing food safety (29). Many food products use multiple hurdle technology to inhibit microbial growth. Instead of setting one environmental parameter to the extreme limit for growth, hurdle technology "deoptimizes" a variety of factors (40). For example, a limiting water activity of 0.85 or a limiting pH of 4.6 prevents the growth of foodborne pathogens. Hurdle technology might obtain similar inhibition at pH 5.2 and a water activity of 0.92. Mechanistically, hurdle technology assaults multiple distinct homeostatic processes (30). Intracellular pH must be maintained within relatively narrow limits. This is done by using energy to pump out protons, as described below. In low-water-activity environments, cells must use energy to accumulate compatible solutes, as described in chapter 28. Membrane fluidity must be maintained through homeoviscous adaptation (67), another energy-requiring process. The expenditure of energy to maintain homeostasis is fundamental to microbial life. When cells channel the energy needed for biosynthesis into maintenance of homeostasis, their growth is inhibited. When the energy demands of homeostasis exceed the cell's energy-producing capacity, the cell dies.

THE IMPORTANCE OF FIRST-ORDER KINETICS

Growth Kinetics

Food microbiology is concerned with all four phases of microbial growth. Growth curves showing the lag, exponential, stationary, and death phases of a culture are normally plotted as the number of cells on a log scale or $\log_{10}$ cell number versus time. These plots represent the state of microbial populations rather than individual microbes. Thus, both the lag phase and stationary phase of growth represent periods when the growth rate equals the death rate to produce no net change in cell numbers.

During the lag phase, cells adjust to their new environment by inducing or repressing enzyme synthesis and activity, initiating chromosome and plasmid replication, and, in the case of spores, differentiating into vegetative cells (see chapter 3). The length of the lag phase depends on the current temperature, the inoculum size (larger inocula usually have shorter lag phases), and the physiological history of the organism. If actively growing cells are inoculated into an identical fresh medium at the same temperature, the lag phase may vanish. Conversely, these factors can be manipulated to extend the lag phase beyond the time where some other food quality attribute (such as proteolysis or browning) becomes unacceptable. Foods are generally considered microbially safe if obvious spoilage precedes microbial

Table 2.1 First-order kinetics equations to describe exponential growth and inactivation[a]

Growth	Thermal inactivation	Irradiation
1a. $N = N_0e^{\mu t}$	1b. $N = N_0e^{-kt}$	1c. $N = N_0e^{-D/D_0}$
2a. $2.3\log(N/N_0) = \mu\Delta t$	2b. $2.3\log(N/N_0) = -(k\Delta t)$	
3a. $\Delta t = [2.3\log(N/N_0)]/\mu$	3b. $\Delta t = -[2.3\log(N/N_0)]/k$	
4a. $t_d = 0.693/\mu$	4b. $D = 2.3/k$	
	5b. $E_a = \frac{2.3RT_1T_2}{z} \times \frac{9}{5}$	

[a]Abbreviations: N, cell number (CFU per gram); N_0, initial cell number (CFU per gram); t, time (hours); μ, specific growth rate (hours^{-1}); t_d, doubling time (hours); k, rate constant (hours^{-1}); D, decimal reduction time (hours); E_a, activation energy (kilocalories per mole); T_1T_2, reference temperature and test temperature (degrees Kelvin); D_0, rate constant (hours^{-1}); D, dose (grays).

growth. However, "spoiled" is a subjective and culturally biased concept. It is safer to manipulate factors to produce conditions where the cell cannot grow regardless of time (such as reduction of pH to <4.6 to inhibit botulinal growth).

During the log, or exponential, phase of growth, bacteria reproduce by binary fission. One cell divides into two cells, which divide into four cells, which divide into eight cells, etc. Thus, during exponential growth, first-order reaction kinetics can be used to describe the change in cell numbers. Food microbiologists often use doubling times as the kinetic constant to describe the rate of logarithmic growth. Doubling times (t_d), which are also referred to as "generation" times (t_{gen}), are related to classical kinetic constants as shown in Table 2.1.

The influence of different parameters on a food's final microbial load can be illustrated by manipulating the equations in Table 2.1. Equation 1a states that the number of organisms (N) at any time is directly proportional to the initial number of organisms (N_0). Thus, decreasing the initial microbial load 10-fold will reduce the cell number at any time by 10-fold, although at extended times the population from the lower inoculum may reach the same final number. Because the instantaneous specific growth rate (μ) and time are in the power function of the equation, they have more marked effects on N. Consider a food where $N_0 = 10^4$ CFU/g and $\mu = 0.2\ h^{-1}$ at 37°C. After 24 h, the cell number would be 1.2×10^6 CFU/g. Reducing the initial number by 10-fold will reduce the number after 24 h 10-fold to 1.2×10^5 CFU/g. However, reducing the temperature from 37 to 7°C has a much more profound effect. If one makes the simplifying assumption that the growth rate decreases twofold with every 10°C decrease in temperature, then μ will be decreased eightfold to $0.025\ h^{-1}$ at 7°C. When equation 1a is solved using these values (i.e., $N = 10^4e^{0.025\times 24}$), then N at 24 h is 1.8×10^4 CFU/g. Both time and temperature have much greater influence over the final cell number than does the initial microbial load.

Equation 3a can be used to determine how long it will take a microbial population to reach a certain level. Consider the case of ground meat manufactured with an N_0 of 10^4 CFU/g. How long can it be held at 7°C before reaching a level of 10^8 CFU/g? According to equation 3a, $t = [2.3(\log 10^8/10^4)]/0.025$, or 368 h.

Food microbiologists frequently use doubling times (t_d) to describe growth rates of foodborne microbes. The relationship between t_d and μ is more obvious if equation 2a is written using natural logs (i.e., $\ln[N/N_0] = \mu\Delta t$) and solved for the condition where $t = t_d$ and $N = 2N_0$. Since the natural log of 2 is 0.693, the solution for equation 2a is $0.693/\mu = t_d$ (equation 4a). The average rate constant k, defined as the number of generations per unit time (i.e., $1/t_{gen}$), is also used by applied microbiologists. The instantaneous growth rate constant μ is related to k by the equation $\mu = 0.693k$. Both rate constants characterize populations in the exponential phase of growth. A more detailed explanation of the differences between these rate constants is given by Brock and Madigan (9). Some typical specific growth rates and doubling times are given in Table 2.2.

Table 2.2 Representative specific growth rates and doubling times of microorganisms

Organism	μ (h^{-1})	t_d (h)
Bacteria		
Optimal conditions	2.3	0.3
Limited nutrients	0.20	3.46
Psychrotroph (5°C)	0.023	30
Molds, optimal	0.1 to 0.3	6.9 to 20

Death Kinetics

The killing of microbes by energy input (equations 1b and 1c, Table 2.1), acid, bacteriocins, and other lethal agents is also governed by first-order kinetics. If one knows the initial microbial number, the first-order rate constant, and the time of exposure, one can predict the number of viable cells remaining. In food microbiology, the *D* value (amount of time required to reduce N_0 by 90%) is the most frequently used kinetic constant. The use of *D* values in thermobacteriology is more fully covered in chapter 28. *D* values are inversely proportional to the rate constant *k*, as shown in equation 4b. Both *D* and *k* values are defined for a given temperature. The relationship between *k* and *T* is related to the activation energy E_a, as determined by the Arrhenius equation $k = s^{-E_a/RT}$, where *s* is the frequency constant, *R* is the ideal gas constant, and *T* is degrees Kelvin. In thermobacteriology, the relationship between *D* and *T* is given by the *z* value. The *z* value is defined as the number of degrees Fahrenheit required to change the *D* value by a factor of 10. The *z* value is related to the E_a by the equation $z = 2.3RT_1T_2/E_a \times (9/5)$ where T_1 and T_2 are actual and reference temperatures. A *z* value of 18°F equals an E_a of about 40 kcal/mol.

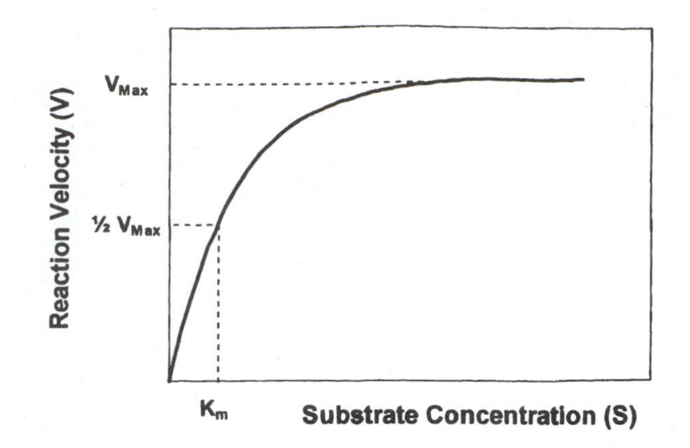

Figure 2.2 Plot of enzyme activity rate (v) versus substrate concentration (s) showing the maximum fate (V_{max}) and the affinity of an enzyme for its substrate (K_m).

Enzyme Activity

Enzymatic reactions can also follow first-order reaction kinetics. These biological catalysts decrease the E_a of reactions and thereby accelerate reaction rates by factors up to 10^{20}. The classic equation used to describe enzyme action is:

$$E + S \Leftrightarrow ES \Leftrightarrow EP \Rightarrow E + P$$

where *E* is enzyme, *S* is substrate, *P* is product, and *ES* and *EP* are the enzyme substrate and enzyme product intermediates, respectively. As written, the reaction is second order and the rate is determined by the enzyme concentration and the substrate concentration. If, however, the substrate concentration is much higher than the enzyme concentration, the reaction becomes pseudo-first order and is proportional to the amount of enzyme present. The rate is also pseudo-first order if substrate is limiting and the amount of enzyme is in excess. In this case, the rate is determined by the amount of substrate present. The key constants in enzymology are K_m, an affinity constant of the enzyme for its substrate, and V_{max}, the maximum velocity when an enzyme is saturated with substrate (Fig. 2.2). These concepts, while not developed further here, are important to food microbiology because (i) specific substrates are often the rate-limiting factor in microbial growth, (ii) beneficial foodborne microbes are often used as sources of crude enzyme preparation, and (iii) many enzymatic reactions such as thermostable nucleases, coagulase, and β-galactosidase are characteristic of specific microorganisms.

MICROBIAL PHYSIOLOGY AND METABOLISM

The Second Law of Thermodynamics dictates that all things progress to the state of maximum randomness in the absence of energy input. Since life is a fundamentally ordered process, all living things must generate energy to maintain their ordered state. Foodborne bacteria do this through the oxidation of reduced compounds. Oxidation occurs in a chemical couple only where the oxidation of one compound is linked to the reduction of another. In the case of aerobic bacteria, the initial carbon source, glucose, is oxidized to carbon dioxide, oxygen is reduced to water, and 38 mol of ATP are generated per mol of glucose catabolized. Most of the ATP is generated through oxidative phosphorylation in the electron transport chain. In oxidative phosphorylation, the energy of the electrochemical gradient generated when oxygen is used as the terminal electron acceptor drives the formation of a high-energy bond between P_i and an adenine nucleotide. Anaerobic bacteria, which lack functional electron transport chains, must reduce an internal compound through the process of fermentation and generate only 1 or 2 mol of ATP per mol of hexose catabolized. In this case, ATP is formed by substrate-level phosphorylation and the phosphate group is transferred from an organic compound to the adenine nucleotide.

Glycolytic Pathways: Carbon Flow and Substrate Level Phosphorylation

EMP Pathway

The most commonly used pathway for glucose catabolism (glycolysis) is the Embden-Meyerhof-Parnas (EMP) pathway (Fig. 2.3). In many organisms, the pathway is bidirectional (i.e., amphibolic) and can work in the direction of glucose, glycogen, and starch synthesis. The overall rate of glycolysis in this pathway is regulated by the activity of phosphofructokinase. This enzyme converts fructose 6-phosphate to fructose 1,6-bisphosphate. Phosphofructokinase activity is subject to allosteric regulation, where the binding of AMP or ATP at one site inhibits or stimulates (respectively) the phosphorylation of fructose 6-phosphate at the enzyme's active site. Fructose 1,6-bisphosphate activates lactate dehydrogenase

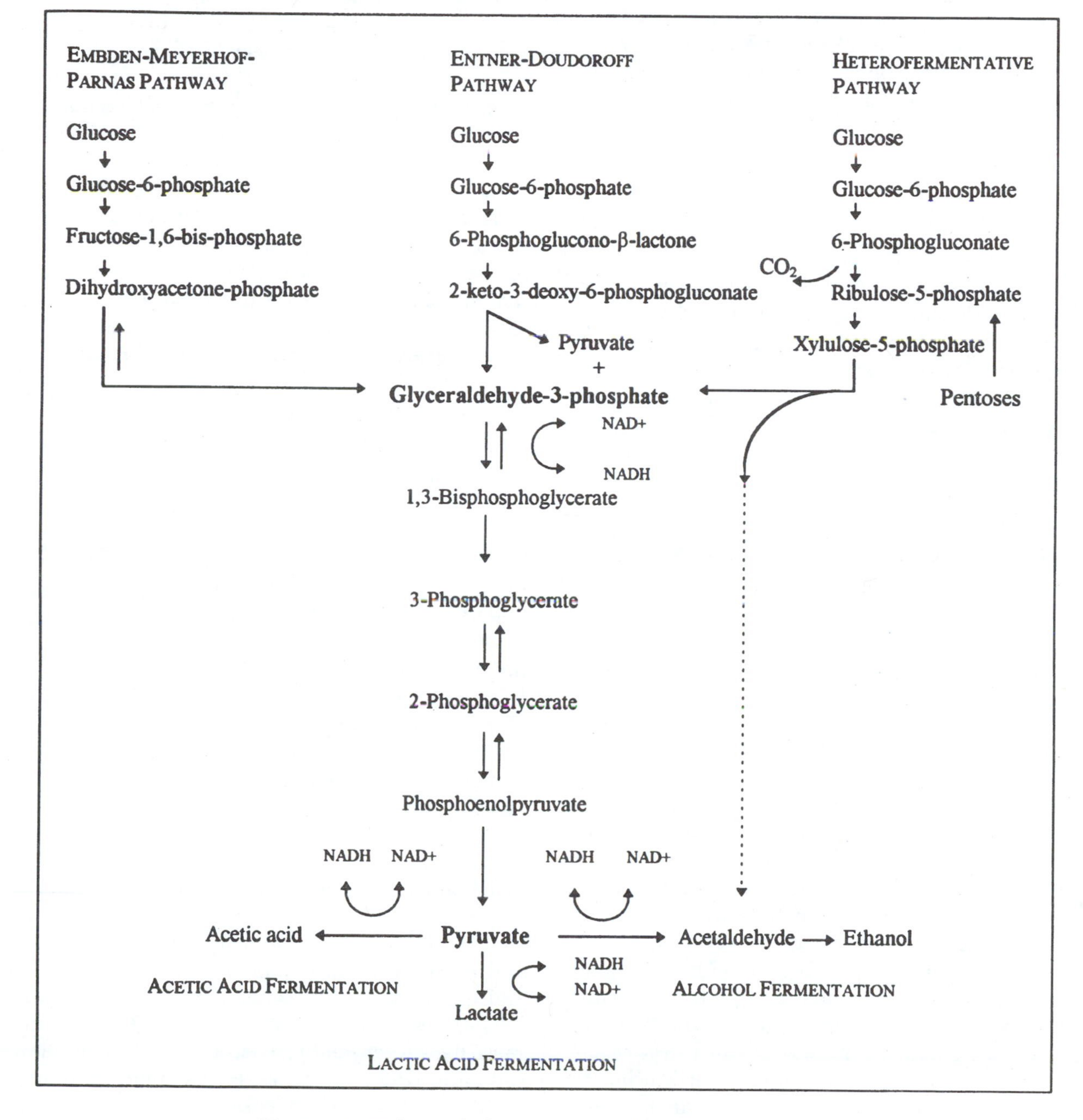

Figure 2.3 Major catabolic pathways used by foodborne bacteria.

(see below) so that the flow of carbon to pyruvate is tightly linked to the regeneration of NAD when pyruvate is reduced to lactic acid.

Another key enzyme of the EMP pathway is aldolase. The ultimate fermentation end products generated by the catabolism of pentoses and hexoses are partially determined by which enzyme converts the sugars to smaller units. Aldolase cleaves one molecule of fructose 1,6-bisphosphate to two three-carbon units, dihydroxyacetone phosphate and glyceraldehyde 3-phosphate. Other glycolytic pathways use ketodeoxyphosphogluconate aldolase to make two three-carbon units or phosphoketolase to produce one two-carbon compound and one three-carbon units. Substrate-level phosphorylation generates a net gain of two ATP when 1,3-diphosphoglycerate and phosphoenolpyruvate donate phosphoryl groups to ADP.

Entner-Doudoroff Pathway

The Entner-Doudoroff pathway is an alternate glycolytic pathway that yields one ATP per molecule of glucose and diverts one three-carbon unit to biosynthetic pathways. In aerobes that use this pathway, such as *Pseudomonas* species, the difference between forming one ATP by this pathway versus the two ATP formed by the EMP pathway is inconsequential compared to the 34 ATP formed from oxidative phosphorylation. In the Entner-Doudoroff pathway, glucose is converted to 2-keto-3-deoxy-6-phosphogluconate. The enzyme ketodeoxyphosphogluconate aldolase cleaves this to one molecule of pyruvate (directly, without the generation of an ATP) and one molecule of 3-phosphoglyceraldehyde. The 3-phosphoglyceraldehyde is then catabolized by the same enzymes used in the EMP pathway with the generation of one ATP by substrate-level phosphorylation using phosphoenolpyruvate as the phosphoryl group donor.

Heterofermentative Catabolism

Heterofermentative bacteria, such as leuconostocs and some lactobacilli, have neither aldolases nor ketodeoxyphosphogluconate aldolase. The heterofermentative pathway is based on pentose catabolism. The pentose can be obtained by transport into the cell or by intracellular decarboxylation of hexoses. In either case, the pentose is converted to xylulose 5-phosphate with ribulose 5-phosphate as an intermediate. The xylulose 5-phosphate is cleaved by phosphoketolase to a glyceraldehyde 3-phosphate and a two-carbon unit which can be converted to acetaldehyde, acetate, or ethanol. Although this pathway yields only one ATP, it offers cells a competitive advantage by allowing them to utilize pentoses which homolactic organisms cannot catabolize.

Homofermentative Catabolism

Homofermentative bacteria in the genera *Lactococcus* and *Pediococcus* and some *Lactobacillus* species produce lactic acid as the sole fermentation product. The EMP pathway is used to produce pyruvate, which is then reduced by lactate dehydogenase, forming lactic acid and regenerating NAD. Lactic acid bacteria form either D-(−), L-(+), or DL-lactic acid, depending on the stereospecificity of their lactate dehydrogenase. Some species have a racemase which converts the D stereoisomer to the L form until the stereoiosomers reach equilibrium.

Some lactobacillus species, such as *Lactobacillus plantarum* (73), are characterized as facultatively heterofermentative. Hexoses are their preferred carbon source and are metabolized by the homofermentative pathway. If only pentoses are available, the cell shifts to a heterofermentative mode. When grown at low hexose concentrations, these bacteria do not make enough fructose 1,6-bisphosphate to activate their lactate dehydrogenase. This also causes them to shift to heterofermentative catabolism.

The TCA Cycle

The tricarboxylic acid (TCA) cycle links gylcolytic pathways to respiration. It generates $NADH_2$ and FADH as substrates for oxidative phosphorylation while providing additional ATP through substrate-level phosphorylation. With each turn of the TCA cycle, 2 pyruvate + 2 ADP + 2 FAD + 8 NAD $\Rightarrow$ 6 CO_2 + 2 ATP + 2 $FADH_2$ + 8 NADH. Succinic acid, oxaloacetate, and α-ketoglutarate link the TCA cycle to amino acid biosynthesis. The TCA cycle is used by all aerobes, but some anaerobes lack all of the enzymes required to have a functional TCA cycle.

The TCA cycle is also the basis for two industrial fermentations important to the food industry. The industrial fermentations for the acidulant citric acid and the flavor enhancer glutamic acid have a similar biochemical basis (18). Both fermentations take advantage of impaired TCA cycles. The production of citric acid uses mutants of *Aspergillus niger* and *Aspergillus wentii* which do not express α-ketoglutarate dehydrogenase and have decreased activities of isocitrate dehydrogenase and aconitase during the stationary phase of growth. This causes citric acid to accumulate. Other reactions must then regenerate oxaloacetate so that the flow of carbon to citrate continues. For example, pyruvate carboxylase mediates a condensation reaction between pyruvate and CO_2 to provide oxaloacetate.

The production of glutamic acid is similar in principle. Glutamic acid producers such as *Corynebacterium glutamicum* are also deficient in α-ketoglutarate dehydrogenase. They have increased levels of glutamate dehydrogenase so that glutamic acid accumulates while other shunts are used to continue the flow of carbon into the TCA cycle. Because membranes are impermeable to charged compounds, glutamic acid accumulates and inhibits its own synthesis unless steps to permeabilize the membrane are taken. The membrane can be permeabilized by using biotin or oleic acid auxotrophs impaired in membrane synthesis. When fed biotin or oleic acid in limiting amounts, these cells have leaky membranes. Addition of saturated fatty acids or penicillin also permeabilizes membranes.

Aerobes, Anaerobes, Regeneration of NAD, and Respiration

The flow of carbon to pyruvate always consumes NAD. Regardless of the pathway used to generate pyruvate, NAD must be regenerated for continued catabolism. When $NADH_2$ is oxidized to NAD, another compound must be reduced, i.e., serve as an electron acceptor. Aerobes having functional electron transport chains use molecular oxygen as the terminal electron acceptor during the oxidative phosphorylation that is the hallmark of respiration. As electrons travel down the electron transport chain, protons are pumped out, forming a proton gradient across the membrane. This proton gradient can be converted to ATP by the action of the BF_0F_1 ATPase. Oxidation of $NAD(P)H_2$ yields three ATP. Oxidation of $FADH_2$ yields two ATP. ATP and NADH are thus, in a sense, interconvertible. Sulfur and nitrite can also serve as terminal electron acceptors in "anaerobic respiration."

Anaerobes, in contrast, have a fermentative metabolism. Fermentations oxidize carbohydrates in the absence of an external electron acceptor. The final electron acceptor is an organic compound produced from the degradation of the carbohydrate. In the most obvious case, pyruvic acid is the terminal electron acceptor when it accepts an electron from NADH and is reduced to lactic acid. Paradoxically, some anaerobes are aerotolerant and can generate more energy in the presence of low levels of oxygen than in its absence. For example, some lactic acid bacteria have inducible NADH oxidases that regenerate NAD by reducing molecular oxygen to H_2O_2 (73). This spares the use of pyruvate as an electron acceptor and allows it to be converted to acetic acid with the concomitant generation of an additional ATP. Catalase-negative cells are inhibited by H_2O_2, but some lactic acid bacteria have NADH peroxidase which detoxifies the H_2O_2. Obligate anaerobes such as *C. botulinum* have no means to detoxify H_2O_2 and die rapidly when exposed to air.

Bioenergetics

All catabolic pathways generate energy with which the bacteria can perform useful work. Energy generation and utilization play a critical role in the life of bacteria. Several excellent reviews (36, 46) and books (31, 56) on bioenergetics provide additional depth and clarity on this topic. The preceding section on microbial biochemistry stressed the role of ATP in the cell's energy economy, but transmembrane gradients of other compounds play an equally important role. Transmembrane gradients release energy when one compound moves from high concentration to low concentration (i.e., "with the gradient"). This energy can be coupled to the transport of a second compound from a low concentration to a high concentration (i.e., "against the gradient").

While the flow of electrons during respiration was recognized as a cellular energy source in the early 1960s, the nature of the link between the energy generated by the electron transport chain and its conservation in the form of ATP was elusive. Many thought that the link was an unstable high-energy intermediate. However, Peter Mitchell won the Nobel Prize for his formulation of a radically different mechanism, the chemiosmotic hypothesis. According to this theory, the link between the energy generated by respiration and ATP was not a chemical intermediate but the transmembrane gradient of protons. This energy is now known as the proton motive force (PMF). ATP and the PMF are the fundamental currencies of cellular energetics. They are interconvertible by a membrane-bound BF_0F_1 ATPase that can use ATP to generate a proton gradient or use the proton gradient to make ATP.

According to Mitchell's chemiosmotic theory (51), the PMF has two components. An electrical component, the membrane potential ($\Delta\Psi$), represents the charge potential across the membrane. The pH gradient (ΔpH) across the membrane is the second component. Together, these make up the PMF as stated by the equation: $PMF = \Delta\Psi - z\Delta pH$. In this equation, $z = 2.3\ RT/F$, where R is the gas constant, T is the absolute temperature, and F is the Faraday constant. The factor z converts the pH gradient into millivolts and has a value of 59 mV at 25°C. The PMF is defined as being interior negative and alkaline, resulting in a negative value. (In the equation above, $z\Delta pH$ is not being subtracted from $\Delta\Psi$, but makes this negative term more negative.) There also is some interconversion of the $\Delta\Psi$ and the ΔpH components of

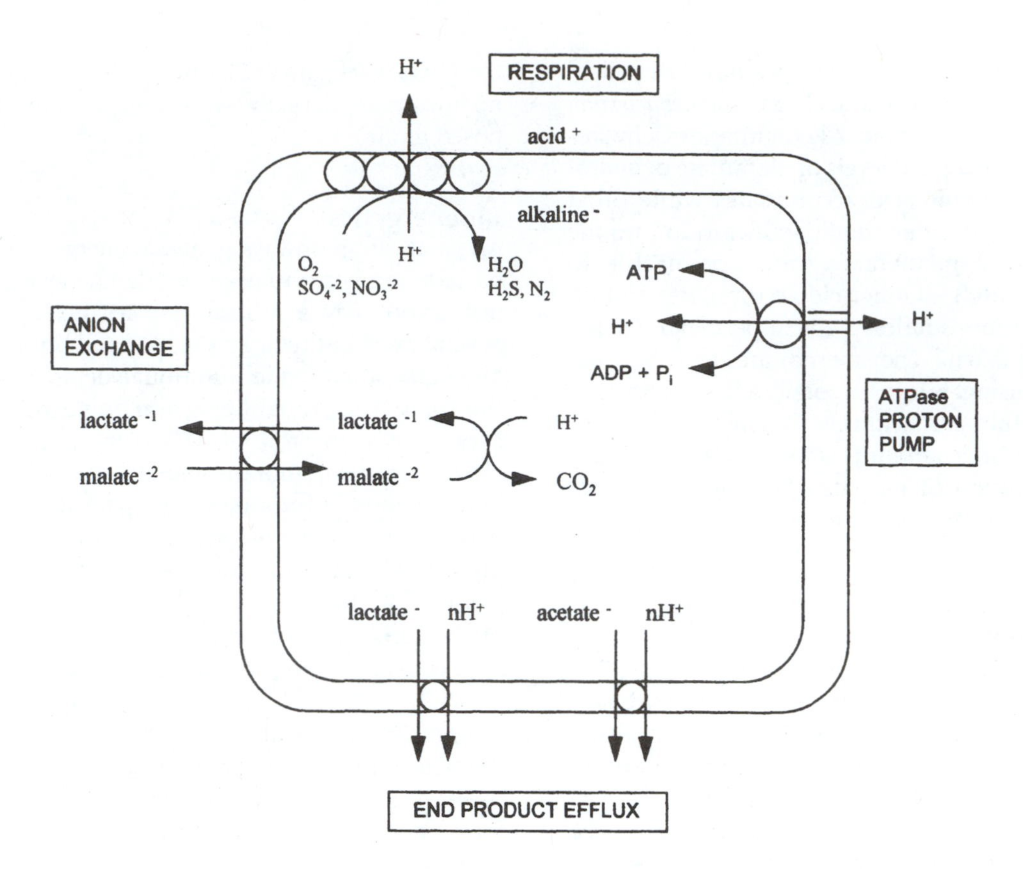

Figure 2.4 PMF can be generated by respiration, ATP hydrolysis, end-product efflux, or anion-exchange mechanisms. (Modified from reference 73.)

PMF. If, for example, the ΔpH component decreases when an organism is transferred to a more neutral environment, the cell compensates by increasing ΔΨ so that the total PMF remains relatively constant. PMF values can be as high as −200 mV for aerobes, or in the range of −100 mV to −150 mV for anaerobes. Protein phosphorylation, flagellar synthesis and rotation, reversed electron transfer, and protein transport use PMF as an energy source (31).

PMF is generated by several mechanisms (Fig. 2.4). The translocation of protons down the electrochemical gradient during respiration generates a proton gradient when oxygen is used as the terminal electron acceptor. The oxidation of NADH is accompanied by the export of enough protons to make three ATP. Proton gradients are also established during anaerobic respiration when nitrate or sulfate serves as the electron acceptor. The proton gradient is converted to ATP by the BF_0F_1 ATPase when it is driven in the direction of ATP synthesis (46). The bacterial BF_0F_1 ATPase is nearly identical to chloroplast and mitochondrial BF_0F_1 ATPases.

The BF_0F_1 ATPase is reversible. Aerobes use it to convert PMF to ATP. In anaerobes, it converts ATP to PMF. Maintaining internal pH homeostasis may be the principal role of the BF_0F_1 ATPase in anaerobes (36). Internal pH not only influences the activity of cytoplasmic enzymes, but also regulates the expression of genes responsible for functions ranging from amino acid degradation to virulence (60). Anaerobes deacidify their cytoplasm by using the BF_0F_1 ATPase to pump protons out. The proton pumping is driven by ATP hydrolysis. Some of the energy lost from ATP hydrolysis can be recovered if the resultant proton gradient is used to perform useful work, such as transport (see below). Most bacteria maintain their internal pH (pH_i) near neutrality, but lactic acid bacteria can tolerate lower pH_i values and expend less ATP on pH homeostasis. Acid-induced death is the direct result of an excessively low pH_i (27).

Given their limited capacity for ATP generation, it is not unexpected that some lactic acid bacteria can also generate ΔpH by ATP-independent mechanisms. The electropositive excretion of protons with acidic end products (49) has been demonstrated for lactate and acetate. For example, under some conditions, *Lb. plantarum* excretes three protons per molecule of acetate, thus sparing one ATP (73). The antiport (see below)

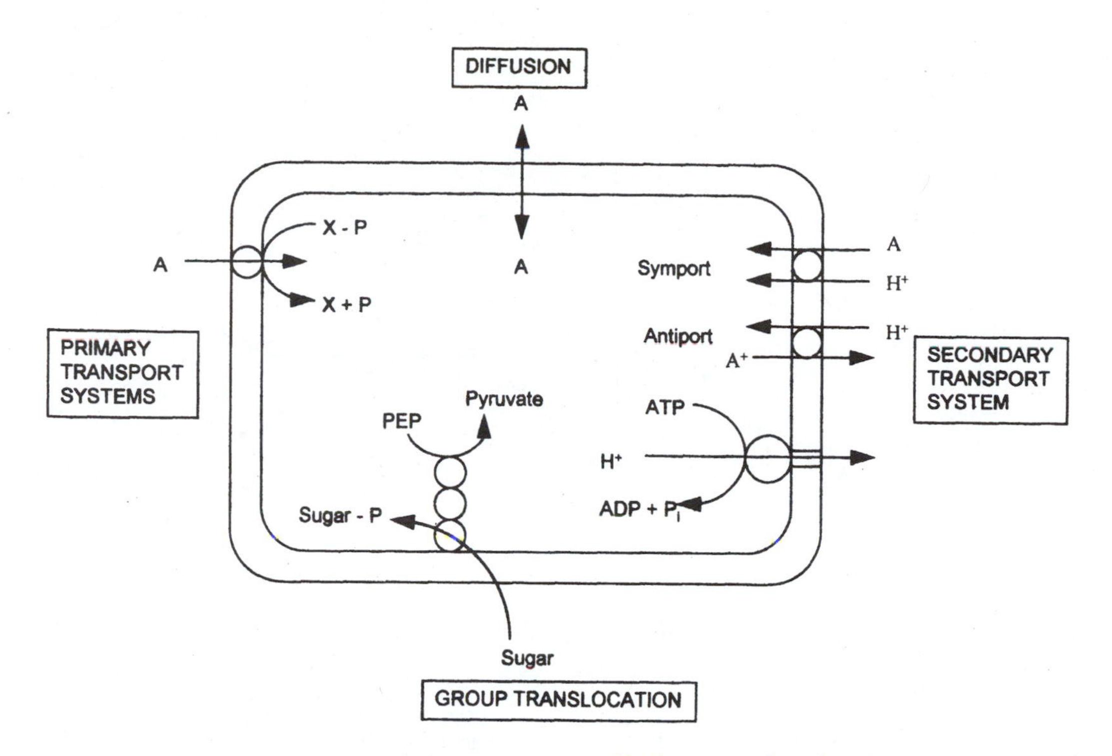

Figure 2.5 Transport can be at the direct expense of high-energy phosphate bonds or can be linked to the proton gradient of the PMF.

exchange of precursor and product in anion-degrading systems, such as the malate^{2-}:lactate^{1-} exchange of the malolactic fermentation, might contribute to the generation of $\Delta\Psi$ (46).

Bacteria have evolved several mechanisms to achieve similar ends. The accumulation of compounds against a gradient (i.e., transport) is work and requires energy. In the case of primary transport systems and group translocation, this work is done by phosphoryl group transfer (Fig. 2.5). Secondary transport systems are fueled by the energy stored in the gradients which form the PMF.

LIMITATIONS OF CLASSICAL MICROBIOLOGY

Limitations of Plate Counts

All methods based on the plate count and pure culture microbiology have the same limitations. The plate count is based on the assumptions that every cell forms one colony and that every colony originates from one cell. Accepting for the moment the underlying assumption that the free monocultured cell is an appropriate unit of study, the ability of a given cell to form a colony depends on a myriad of factors including the physiological state of the cell, the medium used for enumeration, the incubation temperature, etc. Table 2.3 (44) illustrates these points by providing *D* values at 55°C for *L. monocytogenes* with different thermal histories (heat shocked for 10 min at 80°C or not heat shocked) on selective (McBride's) or nonselective (TSAY) media under aerobic or anaerobic atmospheres. The *D* values for *E. coli* O157:H7 are affected in a similar fashion (54). Injured cells and cells that are viable but nonculturable (VNC) pose special problems to food microbiologists, as discussed below.

Injury

Microorganisms may be injured rather than killed by *sublethal* levels of stressors such as heat, radiation, acid, or sanitizers. This injury is characterized by decreased resistance to selective agents or by increased nutritional

Table 2.3 Influence of thermal history and enumeration protocols on experimentally determined *D* values at 55°C for *L. monocytogenes*[a]

	D_{55} value (min)			
	TSAY medium		McBride's medium	
Atmosphere	+ Heat shock	– Heat shock	+ Heat shock	– Heat shock
Aerobic	18.7	8.8	9.5	6.6
Anaerobic	26.4	12.0	NG	NG

[a]Data from reference 44. Details given in text. NG, no growth.

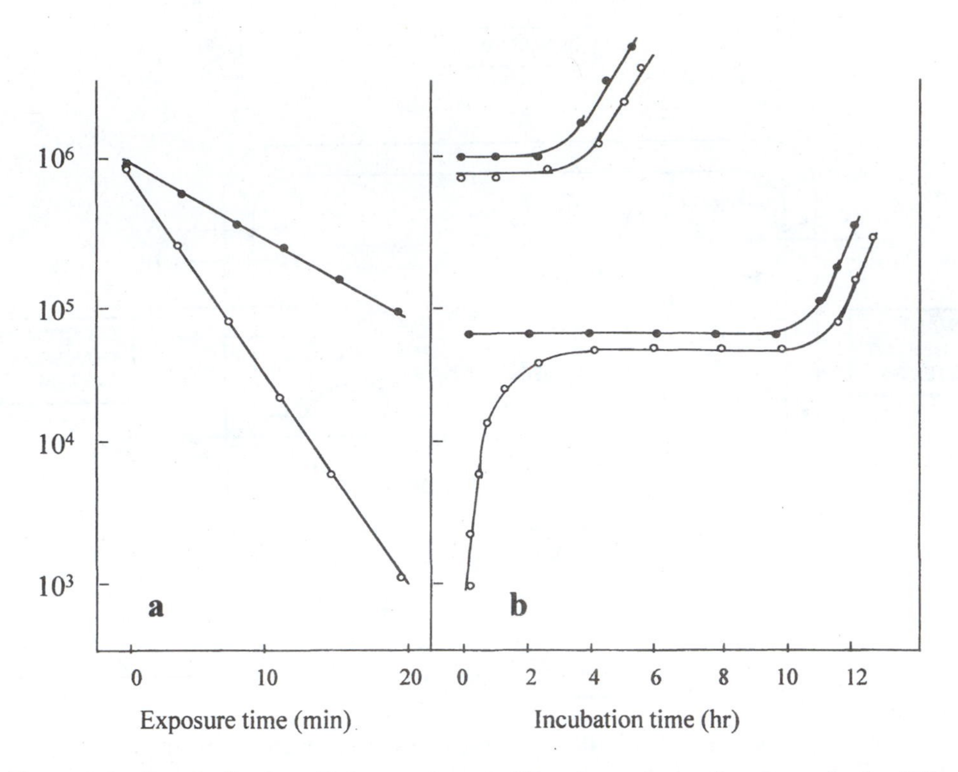

Figure 2.6 Data indicative of injury and repair. When bacteria are plated on selective (○) or nonselective (●) media during exposure to some stressor (e.g., heat, panel a), the decrease in CFU on a nonselective medium represents the true lethality, while the difference between the values obtained on each medium is defined as "injury." During "repair" (panel b), resistance to selective agents is regained and the value obtained on the selective medium approaches that of the nonselective medium. Unstressed controls are shown at the top of panel b. (Modified and redrawn from reference 10.)

requirements (35). Food microbiologists became aware of the importance of microbial injury in the 1970s. There are excellent reviews from this pre-computer-database era dealing with injury in general (10) and injury in bacterial spores (1, 26), yeasts and molds (5, 72), and gram-negative bacteria (4).

Injury is a complex process influenced by time, temperature, concentration of injurious agent, strain of target pathogen, experimental methodology, etc. For example, while a standard sanitizer test indicates that several sanitizers kill listeria, viable cells can be recovered using listeria repair broth (66). The degree of injury decreases and the extent of lethality increases as the time and sanitizer concentration increase. For example, *L. monocytogenes* cells grown at <28°C undergo a 3- to 4-log kill when exposed to 52°C. However, if grown at 37 or 42°C, there is little death but 2 to 3 logs of injury when the cells are then heated to 52°C (69).

Data suggestive of injury are illustrated in Fig. 2.6. Cells subjected to a mild stress are plated on a rich nonselective medium and a medium containing 6% salt. The difference between the populations of cells able to form colonies on each medium represents injured cells. (If 10^7 CFU/ml of a population is found per ml on the nonselective medium and 10^4 CFU/ml can grow on the selective medium, then 10^3 CFU/ml are injured). A shoulder on a lethality curve may represent injury that is easily repaired. As the time of heating increases, the extent of injury increases, but during the post-stress period the injured population undergoes repair, regaining its ability to form colonies on the selective medium, and the number of injured cells decreases.

Injury is important to food safety for several reasons. (i) If injured cells are mistakenly classified as dead during the determination of thermal resistance, the organism's thermal sensitivity will be overestimated and the resultant *D* values will be errantly low. (ii) Injured cells that escape detection at the time of post-processing sampling may repair prior to consumption and present a safety or spoilage problem. (iii) The "selective agent" may be common food ingredients such as salt, organic acids, or humeticants or can even be suboptimal temperature. For example, *Staphylococcus aureus* cells injured by acid during sausage fermentation can grow on tryptic soy agar, but not on the same medium containing 7.5% salt. These injured cells can repair in sausage if held at 35°C (but not if held at 5°C) and then grow and produce enterotoxin (70).

Injury in spores is complex. The many biochemical steps of sporulation, germination, and outgrowth explained in chapter 3 provide a plethora of targets which can be damaged. Thermal injury is the best-studied form of injury and can occur during extrusion as well as during conventional thermal processing (43). Spores can also be injured by chemicals and irradiation. Extensive reviews of spore injury (1, 23, 26) indicate that DNA, RNA, enzymes, and membranes may be damaged during injury. Spore injury can be manifest as increased sensitivity to salt, surfactants, acidity, incubation temperature, and oxygen toxicity (25). Irradiation-induced injury of spores is primarily caused by single-strand breaks in DNA and is also manifest as increased sensitivity to pH, salt, and heat (23). *rec* systems can repair injury caused by single-strand DNA breaks. Radiation-induced heat sensitivity is caused by damage to the spore cortex peptidoglycan and can last for several weeks to several months (23).

Injured spores can be revived using several different methods. The addition of soluble starch or charcoal to medium binds surfactants which would otherwise enter the spore through the damaged cytoplasmic membrane. Increasing the osmolarity of the recovery medium stabilizes injured spores by minimizing leakage of spore constituents. Lysozyme enhances the recovery of spores whose germination systems have been damaged.

Vegetative cells injured by heat, freezing, and detergents usually leak intracellular constituents from damaged membranes (71). The reestablishment of membrane integrity is an important event during repair. Osmoprotectants can prevent or minimize freeze injury in *L. monocytogenes* (20, 21). Oxygen toxicity also causes injury. Recovery of injured cells is often enhanced by adding peroxide detoxifying agents such as catalase or pyruvate to the recovery medium or by excluding oxygen through the use of anaerobic incubation conditions or adding Oxyrase (which enzymatically reduces oxygen) to the recovery medium.

Repair is the process by which cells recover from injury. Repair requires de novo synthesis of RNA and protein (9, 63) and often is manifest as an extension of the lag phase of growth. The extent and rate of repair are influenced by a variety of environmental factors. *L. monocytogenes* cells injured at 55°C for 20 min start to repair immediately at 37°C and are completely recovered by 9 h (48). Heat-injured *L. monocytogenes* do not replicate in milk at 4°C. Repair at 4°C is delayed for 8 to 10 days, and full recovery requires 16 to 19 days (17). Studies in cured luncheon meat (3) confirm that its microbial stability is due to the extended lag period required for the repair of spoilage organisms. The magnitude of the extension is determined by both the severity of the thermal process and the product's salt concentration.

VNC Cells

Salmonella, Campylobacter, Escherichia, and *Vibrio* species, and perhaps species from other genera, can exist in a state where they are viable, but cannot be cultured by normal microbiological methods. This differentiation of vegetative cells into a dormant viable but nonculturable (VNC) state is a survival strategy for many nonsporulating species. The VNC state is morphologically different from the "normal" vegetative cell. During the transition to the VNC state, rod-shaped cells shrink and become small spherical bodies which are totally different from bacillus and clostridial spores (38, 57). It takes from 2 days to several weeks for an entire population of vegetative cells to become VNC (57, 65).

Although VNC cells cannot be cultured, their viability can be determined through direct microscopic observation in the presence of appropriate dyes. The differential interaction of acridine orange with DNA and RNA distinguishes between live and dead cells. RNA is more abundant than DNA in live cells and causes acridine orange to fluoresce red. Acridine orange fluoresces green when it reacts with DNA, which is more abundant in dead cells because dead cells have no transcriptional activity and RNA has a relatively short half-life. Thus, under direct microscopic observation, populations which appear to have "died off," on the basis of the criteria of plate counts, fluoresce red if they are VNC, but green if they are truly dead (47). Iodonitrotetrazolium violet can also stain VNC cells. Respiring cells reduce iodonitrotetrazolium violet to form an insoluble compound detectable by microscopic observation (57). Unculturable (<10 CFU/ml) *Salmonella enteritidis* populations starved at 7°C have been quantified as 10^4 viable cells per ml using these methods (13). Experimental data (57) in Fig. 2.7 illustrate a *Vibrio vulnificus* population that appears to have died off (i.e., gone through a 6-log reduction in CFU) at a time when almost all the cells (>10^5/ml) are quantified as viable.

VNC cells can also be identified by their substrate-responsive metabolism. When VNC cells are incubated with yeast extract (as a nutrient) and nalidixic acid (an

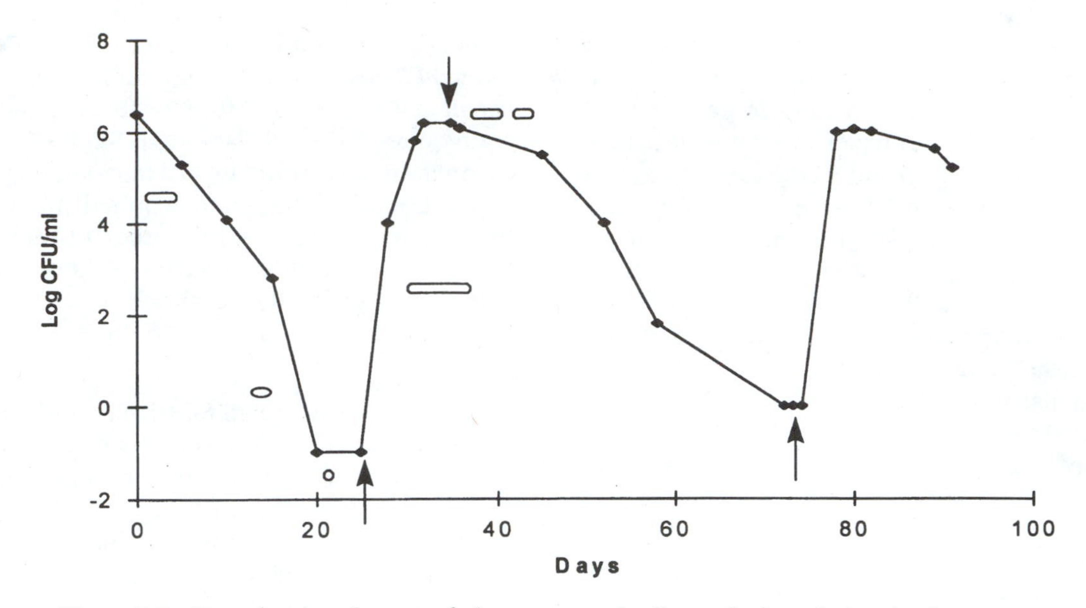

Figure 2.7 Data showing changes of plate count and cell morphology during development of the VNC state induced by temperature downshifts (at time 0 and ↓) and resuscitation of VNC cells by temperature upshifts (↑). (Reprinted from reference 57, with permission.)

inhibitor of cell division), their elongation can be quantified microscopically. Because this widely used method does not work with gram-positive bacteria (which are insensitive to nalidixic acid), one might infer from the literature that the VNC state is limited to gram-negative species. However, other methods have demonstrated the VNC state for *Streptococcus faecalis, Micrococcus flavus,* and *Bacillus subtilis* (11).

Because the VNC state is most often induced by nutrient limitation in aquatic environments, it might appear irrelevant to the nutrient-rich milieu of food. However, the VNC state can also be induced by changes in salt concentration, exposure to hypochlorite, and shifts in temperature (47, 59). *V. vulnificus* populations shifted to refrigeration temperatures are still lethal to mice when the entire population of 10^5 viable cells becomes nonculturable (<0.04 CFU/ml)(58). The bacteria resuscitate in the mice and can be isolated postmortem using culture methods. Increases or decreases in temperature can induce the VNC state in different organisms. When starved at 4 or 30°C for more than a month, *Vibrio harveyi* becomes VNC at 4°C, but remains culturable at 30°C. In contrast, *E. coli* enter the VNC state at 30°C, but die at 4°C (19). Foodborne pathogens in nutritionally rich media can become VNC when shifted to refrigerated temperature (47, 57, 58). This has chilling implications for the safety of refrigerated foods.

Resuscitation of VNC cells is demonstrated by an increase in culturability that is not accompanied by an increase in the total number of cells. The return to culturability can be induced by temperature shifts or gradual return of nutrients. It can take several days for the population to fully recover its culturability. The same population of bacteria can go through multiple cycles of the VNC and culturable states in the absence of growth (57). Inhibitors of protein synthesis or of peptidoglycan synthesis prevent VNC cells from resuscitating.

Increased awareness of the VNC state should lead to a reexamination of our concept of viability, our dependence on enrichment culture to isolate pathogens, and our reliance on established cultural methods to monitor microbes in the environment (2, 47, 64). The mechanisms of VNC formation, the mechanisms which make VNC cells resistant to environmental stress, and the nature of the event that signals resuscitation are largely unknown. Clearly, food microbiologists need to do more research in the area of VNC foodborne pathogens.

Biofilms

To suggest that more research is needed about biofilms would be a gross understatement. Although planktonic (i.e., free, single) cells are easy to study and pure culture is the foundation of microbiology as we know it, "in all natural habitats studied to date bacteria prefer to reproduce on any available surface rather than in the liquid phase" (12). Furthermore, most definitions of biofilms reveal that they exist as *communities* of microbial species embedded in a biopolymer matrix on some substratum. Biofilms are also heterogeneous in time and space, frequently appearing as collections of mushroom-shaped microcolonies with moving water channels between them (12, 74).

Biofilm formation is a multistep process in which the substratum first undergoes a conditioning process that allows cells to be adsorbed by weak reversible electrostatic forces. Adhesion or anchoring of the cells via some biopolymer follows rapidly. The synthesis of the matrix polymer may be up-regulated by adsorption of the cell. In *Pseudomonas aeruginosa,* transcription of alginate biosynthetic genes is activated by response regulators which increase synthesis of a sigma-like factor which regulates transcription of the *algD* promoter (8). (The sensing part of this two-component signaling system has not yet been identified.) The *algD* promoter regulates virtually all of the alginate biosynthetic genes, which are in a single operon. This system also contains an alginate lyase which allows for cell dispersion when the environment threatens communal life. The microcolonies have defined boundaries which allow fluid channels to run through the biomatrix. This requires higher-level differentiation, quorum sensing, or some kind of cell-to-cell communication to prevent undifferentiated growth from filling in these channels which bring nutrients and remove wastes. Costerton (16) paints a vivid picture of this system, concluding "that the highly structured biofilm mode of growth provides bacteria with a measure of homeostasis, a primitive circulatory system, a framework for the development of cooperative and specialized cell functions, and a large measure of protection from antimicrobial agents." Two special issues (volume 15, numbers 3 and 4) of the *Journal of Industrial Microbiology* provide an up-to-date understanding of biofilms in medical, dental, agricultural, and environmental settings.

Cells in biofilms are more resistant to heat, chemicals, and sanitizers. This has been attributed to the diffusional barrier created by the biomatrix. The confocal scanning laser microscopy images of circulatory channels in hydrated biofilms shatter the barrier hypothesis. The increased resistance is now attributed to the very slow growth rates of cells in biofilms (16). Indeed, cells in the nutrient-depleted interior of the microcolony may be in the VNC state. From a pragmatic standpoint, reviews on biofilms in the food industry (12, 74) emphasize the importance of cleaning prior to sanitation of process equipment. While there are no materials inherently resistant to biofouling, true biofilms take days to weeks to reach equilibrium. Proper cleaning ensures that the cells in the nascent biofilm can be reached by sanitizers. The design of equipment with smooth, highly polished surfaces also impedes biofilm formation by making the initial adsorption step more difficult.

CONCLUSION

Microbial growth in foods is a complex process governed by genetic, biochemical, and environmental factors. Much of what we "know" about foodborne microbes must be viewed with the detached objectivity required of an unproven hypothesis. Developments in molecular biology and microbial ecology will change or deepen our perspective about the growth of microbes in foods. Some of these developments are detailed in this book. Others developments will unfold over the coming decades, perhaps by readers whose journey begins now.

ACKNOWLEDGMENTS

This is manuscript F-10974-1-96 of the New Jersey Agricultural Experiment Station. Research in the author's laboratory and preparation of this manuscript were supported by state appropriations and U.S. Hatch Act Funds.

References

1. **Adams, D. M.** 1978. Heat injury of bacterial spores. *Adv. Appl. Microbiol.* **23**:245–261.
2. **Barer, M. R., L. T. Gribbon, C. R. Harwood, and C. E. Nwaguh.** 1993. The viable but non-culturable hypothesis and medical bacteriology. *Rev. Med. Microbiol.* **4**:183–191.
3. **Bell, R. G., and K. M. De Lacy.** 1984. Heat injury and recovery of *Streptococcus faecium* associated with the souring of chub-packed luncheon meat. *J. Appl. Bacteriol.* **57**:229–236.
4. **Beuchat, L. R.** 1978. Injury and repair of Gram-negative bacteria with special consideration of the involvement of the cytoplasmic membrane. *Adv. Appl. Microbiol.* **23**:219–243.
5. **Beuchat, L. R.** 1984. Injury and repair of yeasts and moulds. *J. Appl. Bacteriol.* **12**:293–308.
6. **Board, R. G., D. Jones, R. G. Kroll, and G. L. Pettipher (ed.).** 1992. Ecosystems: Microbes: Food. *J. Appl. Bacteriol.* 73:1S–178S.
7. **Boddy, L., and J. W. T. Wimpenny.** 1992. Ecological concepts in food microbiology. *J. Appl. Bacteriol.* **73**:23S–38S.
8. **Boyd, A., and A. M. Chakrabarty.** 1995. *Pseudomonas aeruginosa* biofilms: role of the alginate exopolysaccharide. *J. Ind. Microbiol.* 15:162–168.
9. **Brock, T. D., and M. T. Madigan.** 1988. *Biology of Microorganisms,* p. 793–795. Prentice Hall, Englewood Cliffs, N.J.
10. **Busta, F. F.** 1978. Introduction to injury and repair of microbial cells. *Adv. Appl. Microbiol.* **23**:195–201.
11. **Byrd, J. J., H.-S. Xu, and R. R. Colwell.** 1991. Viable but nonculturable bacteria in drinking water. *Appl. Environ. Microbiol.* **57**:875–878.

12. **Carpentier, B., and O. Cerf.** 1993. Biofilms and their consequences, with particular reference to hygiene in the food industry. *J. Appl. Bacteriol.* **75:**499–511.
13. **Chmielewski, R., and J. F. Frank.** 1995. Formation of viable but nonculturable Salmonella during starvation in chemically defined solutions. *Lett. Appl. Microbiol.* **20:** 380–384.
14. **Conte, M. P., C. Longhi, G. Petrone, M. Polidoro, P. Valenti, and L. Seganti.** 1994. *Listeria monocytogenes* infection of Caco-2 cells: role of growth temperature. *Res. Microbiol.* 145:677–682.
15. **Cossins, A. R., and M. Sinensky.** 1984. Adaptation of membranes to temperature, pressure and exogenous lipids, p. 1–20. *In* M. Shinitzky (ed.), *Physiology of Membrane Fluidity.* CRC Press, Boca Raton, Fla.
16. **Costerton, J. W.** 1995. Overview of microbial biofilms. *J. Ind. Microbiol.* 15:137–140.
17. **Crawford, R. W., C. M. Belizeau, T. J. Poeler, C. W. Donnelly, and U. K. Bunning.** 1989. Comparative recovery of uninjured and heat-injured *Listeria monocytogenes* cells from bovine milk. *Appl. Environ. Microbiol.* **55:**1490–1494.
18. **Crueger, W., and A. Crueger.** 1984. *Biotechnology: a Textbook of Industrial Microbiology.* Sinauer Associates, Inc., Sunderland, Mass.
19. **Duncan, S., L. A. Glover, K. Killham, and J. I. Prosser.** 1994. Luminescence-based detection of activity of starved and viable but nonculturable bacteria. *Appl. Environ. Microbiol.* **60:**1308–1316.
20. **El-Kest, S. E., and E. H. Marth.** 1991. Injury and death of frozen *Listeria monocytogenes* as affected by glycerol and milk components. *J. Dairy Sci.* **74:**1201–1208.
21. **El-Kest, S. E., and E. H. Marth.** 1991. Strains and suspending menstrua as factors affecting death and injury of *Listeria monocytogenes* during freezing and frozen storage. *J. Dairy Sci.* **74:**1209–1213.
22. **Farber, J. M., and B. E. Brown.** 1990. Effect of prior heat shock on heat resistance of *Listeria monocytogenes* in meat. *Appl. Environ. Microbiol.* **56:**1584–1587.
23. **Farkas, J.** 1994. Tolerance of spores to ionizing radiation: mechanisms of inactivation, injury, and repair. *J. Appl. Bacteriol.* 76:81S–90S.
24. **Fedio, W. M., and H. Jackson.** 1989. Effect of tempering on the heat resistance of *Listeria monocytogenes. Lett. Appl. Microbiol.* **9:**157–160.
25. **Feeherry, F. E., D. T. Munsey, and D. B. Rowley.** 1987. Thermal inactivation and injury of *Bacillus stearothermophilus* spores. *Appl. Environ. Microbiol.* **53:**365–370.
26. **Foegeding, P. M., and F. F. Busta.** 1981. Bacterial spore injury—an update. *J. Food Prot.* **44:**776–786.
27. **Foster, J. W., and H. K. Hall.** 1991. Inducible pH homeostasis and the acid tolerance response of *Salmonella typhimurium. J. Bacteriol.* **173:**5129–5135.
28. **Foster, J. W., Y. K. Park, L. S. Bang, K. Karem, H. Betts, H. K. Hall, and E. Shaw.** 1994. Regulatory circuits involved with pH-regulated gene expression in *Salmonella typhimurium. Microbiology* **140:**341–352.
29. **Gould, G. W.** 1992. Ecosystems approaches to food microbiology. *J. Appl. Bacteriol.* **73:**58S–68S.
30. **Gould, G. W.** 1995. Homeostatic mechanisms during food preservation by combined methods, p. 397–410. *In* G. V. Barbosa-Canovas and J. Welti-Chanes (ed.), *Food Preservation by Moisture Control.* Technomic Publishing Co., Inc., Lancaster, Pa.
31. **Harold, F. M.** 1981. *The Vital Force: a Study of Bioenergetics.* W. H. Freeman and Co., New York.
32. **Huhtanen, C. N., J. Naghski, C. S. Custer, and R. W. Russel.** 1976. Growth and toxin production by *Clostridium botulinum* in moldy tomato juice. *Appl. Environ. Microbiol.* **32:**711–715.
33. **International Commission on Microbiological Specifications for Foods.** 1980. *Microbial Ecology of Foods,* vol. 1. *Factors Affecting Life and Death of Microorganisms.* Academic Press, Inc., New York.
34. **International Commission on Microbiological Specifications for Foods.** 1980. *Microbial Ecology of Foods,* vol. 2. *Food Commodities.* Academic Press, Inc., New York.
35. **International Commission on Microbiological Specifications for Foods.** 1980. Injury and its effect on recovery, p. 205–214. *In Microbial Ecology of Foods,* vol. 1. *Factors Affecting Life and Death of Microorganisms.* Academic Press, Inc., New York.
36. **Kashket, E. R.** 1987. Bioenergetics of lactic acid bacteria: cytoplasmic pH and osmotolerances. *FEMS Microbiol. Rev.* 46:233–244.
37. **Ko, R., L. T. Smith, and G. M. Smith.** 1994. Glycine betaine confers enhanced osmotolerance and cryotolerance in *Listeria monocytogenes. J. Bacteriol.* **176:**426–431.
38. **Kondo, K., A. Takade, and K. Amako.** 1994. Morphology of the viable but nonculturable *Vibrio cholerae* as determined by the freeze fixation technique. *FEMS Microbiol. Lett.* **123:**179–184.
39. **Leimeister-Wachter, M., E. Donnan, and T. Chakraborty.** 1992. The expression of virulence genes in *Listeria monocytogenes* is thermal regulated. *J. Bacteriol.* **174:**947–952.
40. **Leistner, L.** 1994. Principles and applications of hurdle technology, p. 1–21. *In* G. W. Gould (ed.), *New Methods of Food Preservation.* Blackie Academic and Professional, Glasgow.
41. **Leyer, G. J., and E. A. Johnson.** 1993. Acid adaptation induces cross-protection against environmental stresses in *Salmonella typhimurium. Appl. Environ. Microbiol.* **59:**1842–1847.
42. **Leyer, G. J., L.-L. Wang, and E. A. Johnson.** 1995. Acid adaptation of *Escherichia coli* O157:H7 increases survival in acid foods. *Appl. Environ. Microbiol.* 61:3752–3755.
43. **Likimani, T. A., and J. N. Sofos.** 1990. Bacterial spore injury during extrusion cooking of corn/soybean mixtures. *Int. J. Food Microbiol.* **11:**243–249.
44. **Linton, R. H., J. B. Webster, M. D. Pierson, J. R. Bishop, and C. R. Hackney.** 1992. The effect of sublethal heat shock and growth atmosphere on the heat resistance of *Listeria monocytogenes* Scott A. *J. Food. Prot.* **55:**84–87.
45. **Lund, B. M.** 1992. Ecosystems in vegetable foods. *J. Appl. Bacteriol.* **73:**115S–126S.
46. **Maloney, P. C.** 1990. Microbes and membrane biology. *FEMS Microbiol. Rev.* **87:**91–102.

47. **McKay, A. M.** 1992. Viable but non-culturable forms of potentially pathogenic bacteria in water. *Lett. Appl. Microbiol.* **14:**129–135.
48. **Meyer, D. H., and C. W. Donnelly.** 1992. Effect of incubation temperature on repair of heat-injured *Listeria* in milk. *J. Food Prot.* **55:**579–582.
49. **Michels, P. A. M., J. P. J. Michels, J. Boonstra, and W. L. Konings.** 1979. Generation of an electrochemical proton gradient in bacteria by the excretion of metabolic end products. *FEMS Microbiol. Lett.* **5:**357–364.
50. **Mikulskis, A. V., I. Delor, V. H. Thi, and G. R. Cornelis.** 1994. Regulation of *Yersinia enterocolitica* enterotoxin *Yst* gene. Influence of growth phase, temperature, osmolarity, pH and bacterial host factors. *Mol. Microbiol.* **14:**905–915.
51. **Mitchell, P.** 1966. Chemiosmotic coupling in oxidative and photosynthetic phosphorylation. *Biol. Rev. Cambridge Philos. Soc.* **41:**445–502.
52. **Montville, T. J.** 1982. Metabiotic effect of *Bacillus licheniformis* on *Clostridium botulinum:* implications for home-canned tomatoes. *Appl. Environ. Microbiol.* **44:**334–338.
53. **Mossel, D. A. A., and C. B. Struijk.** 1992. The contribution of microbial ecology to management and monitoring of the safety, quality and acceptability (SQA) of foods. *J. Appl. Bacteriol.* **73:**1S–22S.
54. **Murano, E. A., and M. O. Pierson.** 1993. Effect of heat shock and incubation atmosphere on injury and recovery of *Escherichia coli* **O157:**H7. *J. Food Prot.* 56:568–572.
55. **National Food Processors Association.** 1988. Factors to be considered in establishing good manufacturing practices for the production of refrigerated food. *Dairy Food Sanit.* **8:**288–291.
56. **Nicholls, D. G., and S. J. Ferguson.** 1992. Bioenergetics 2. Academic Press, Inc., San Diego.
57. **Nilsson, L., J. D. Oliver, and S. Kjelleberg.** 1991. Resuscitation of *Vibrio vulnificus* from the viable but nonculturable state. *J. Bacteriol.* **173:**5054–5059.
58. **Oliver, J. D., and R. Bocklan.** 1995. In vivo resuscitation, and virulence towards mice, of viable but nonculturable cells of *Vibrio vulnificus. Appl. Environ. Microbiol.* **61:**2620–2623.
59. **Oliver, J. D., F. Hite, D. McDougald, N. L. Andon, and L. M. Simpson.** 1995. Entry into, and resuscitation from, the viable but nonculturable state by *Vibrio vulnificus* in an estuarine environment. *Appl. Environ. Microbiol.* **61:**2624–2630.
60. **Olson, E. R.** 1993. Influence of pH on bacterial gene expression. *Mol. Microbiol.* **8:**5–14.
61. **Palumbo, S. A.** 1986. Is refrigeration enough to restrain foodborne pathogens? *J. Food Prot.* **49:**1003–1009.
62. **Pepe, J. C., J. L. Badger, and V. L. Miller.** 1994. Growth phase and low pH affect the thermal regulation of the *Yersinia enterocolitia inv* gene. *Mol. Microbiol.* **11:**123–135.
63. **Pierson, M. D., R. F. Gomez, and S. E. Martin.** 1978. The involvement of nucleic acids in bacterial injury. *Adv. Appl. Microbiol.* **23:**263–285.
64. **Rollins, D. M., and R. R. Colwell.** 1986. Viable but nonculturable stage of *Campylobacter jejuni* and its role in survival in the natural aquatic environment. *Appl. Environ. Microbiol.* **52:**531–538.
65. **Roszak, D. B., D. J. Grimes, and R. R. Colwell.** 1984. Viable but nonrecoverable stage of *Salmonella enteritidis* in aquatic systems. *Can. J. Microbiol.* **30:**334–338.
66. **Sallam, S. S., and C. W. Donnelly.** 1992. Destruction, injury and repair of *Listeria* species exposed to sanitizing compounds. *J. Food Prot.* **59:**771–776.
67. **Sinensky, M.** 1974. Homeoviscous adaptation—a homeostatic process that regulates the viscosity of membrane lipids in *Escherichia coli. Proc. Natl. Acad. Sci. USA* **71:**522–525.
68. **Skurnik, M.** 1985. Expression of antigens encoded by the virulence plasmid of *Yersinia enterocolitica* under different growth conditions. *Infect. Immun.* **47:**183–190.
69. **Smith, J. L., B. S. Marmer, and R. C. Benedict.** 1991. Influence of growth temperature on injury and death of *Listeria monocytogenes* Scott A during a mild heat treatment. *J. Food Prot.* **54:**166–169.
70. **Smith, J. L., and S. A. Palumbo.** 1978. Injury to *Staphylococcus aureus* during sausage fermentation. *Appl. Environ. Microbiol.* **36:**857–860.
71. **Smith, J. L., and S. A. Palumbo.** 1982. Microbial injury reviewed for the sanitarian. *Dairy Food Sanit.* **2:**57–63.
72. **Stevenson, K. E., and T. R. Graumlich.** 1978. Injury and recovery of yeasts and molds. *Adv. Appl. Microbiol.* **23:**203–217.
73. **Tseng, C.-P., and T. J. Montville.** 1993. Metabolic regulation of end product distribution in lactobacilli: causes and consequences. *Biotechnol. Prog.* 9:113–121.
74. **Zottola, E. A., and K. C. Sasahara.** 1994. Microbial biofilms in the food processing industry—should they be a concern? *Int. J. Food Microbiol.* **23:**125–148.

Peter Setlow
Eric A. Johnson

3 Spores and Their Significance

Members of the gram-positive *Bacillus* and *Clostridium* spp. and some closely related genera respond to slowed growth or starvation by initiating the process of sporulation. The molecular biology of sporulation (95, 136, 199, 216) and spore resistance (13, 63, 69, 145, 198) has been extensively and elegantly studied in the genus *Bacillus* (78, 83, 199, 201, 202). Spores formed by the genera *Bacillus, Clostridium, Desulfotomaculum*, and *Sporolactobacillus* present practical problems in food microbiology. This chapter describes the fundamental basis of sporulation and problems that spores present to the food industry.

The first notable morphological event in sporulation is an unequal cell division. This creates the smaller nascent spore or forespore compartment and the larger mother cell compartment. As sporulation proceeds, the mother cell engulfs the forespore, resulting in a cell (the forespore) within a cell (the mother cell), each with a complete genome (65, 199). Since the spore is formed within the mother cell, it is termed an endospore.

Throughout sporulation there is a defined pattern of gene expression which is ordered not only temporally but also spatially, as some genes are expressed only in the mother cell or forespore (57, 216). The gene expression pattern is controlled by the ordered synthesis and activation of new sigma (σ; specificity) factors for RNA polymerase. At least four of these are synthesized specifically for modulation of gene expression during sporulation. A number of DNA binding proteins, both repressors and activators, also regulate gene expression during sporulation (57, 214, 216). As sporulation proceeds there are striking morphological and biochemical changes in the developing spore. It becomes encased in two novel layers, a peptidoglycan layer (the spore cortex) and a number of layers of spore coats which contain the proteins unique to the spore (22, 200). The spore also accumulates a huge depot ($\geq$10% of dry weight) of pyridine-2,6-dicarboxylic acid (dipicolinic acid [DPA]; Fig. 3.1), found only in spores, as well as a large amount of divalent cations (152). The developing forespore also synthesizes a large amount of small acid-soluble proteins (SASP), some of which coat the spore chromosome and protect the DNA from damage (200–202). In addition, the spore becomes both metabolically dormant and extremely resistant to a variety of harsh treatments including heat, radiation, and chemicals (69, 74, 128). As a consequence of these physiological changes, spores can survive extremely long periods in the absence of exogenous nutrients (28, 29, 103).

Despite the spore's extreme dormancy, if given the appropriate stimulus (often a nutrient such as a sugar or amino acid) the spore can rapidly return to life via spore germination (73, 193). Within minutes of exposure to a germinant, spores lose most of their unique characteristics, including loss of DPA by excretion and loss of the cortex and SASP by degradation. The spore's resistance is

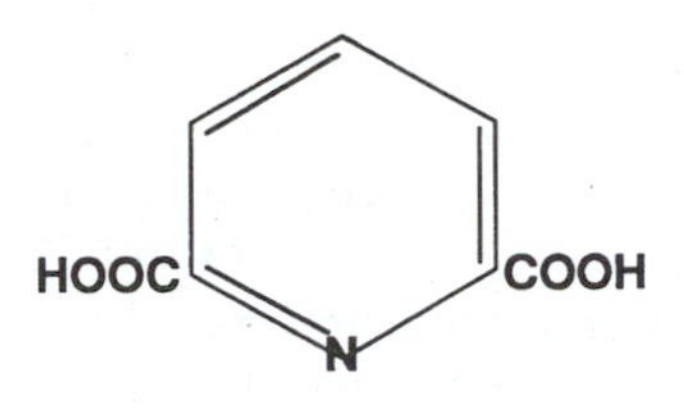

Figure 3.1 Structure of dipicolinic acid. Note that at physiological pH both carboxyl groups will be ionized.

also lost in the first minutes of germination, when active metabolism of both endogenous and exogenous compounds begins and macromolecular synthesis is initiated. Eventually the germinated spore is converted back to a growing vegetative cell through the process of outgrowth.

Detailed study of sporulation, spores, and their "return to life" via spore germination and outgrowth has been motivated by a number of factors, one of which is the intrinsic attraction of this model developmental system. This attraction has led to an increasingly detailed understanding of sporulation mechanisms responsible for the elaborate networks that modulate gene expression during this process, as well as factors involved in spore resistance. An additional motivating factor behind work on this system is the major role played by sporeformers in food spoilage and foodborne diseases. While a tremendous amount of knowledge has been gained in studies motivated by these two disparate factors, the amount of cross-talk between the workers in these two fields has never been optimal, to the detriment of both fields. Thus one purpose of this chapter is to highlight the present state of knowledge concerning the molecular mechanisms behind sporulation, spore resistance and dormancy, and spore germination and outgrowth. Hopefully, it will provide a counterpoint to more applied aspects of this system. This review will focus on molecular mechanisms, most of which have been examined in *Bacillus subtilis.* This organism is neither an important pathogen nor an important agent of food spoilage. However, its natural transformability, as well as an abundance of molecular biological and genetic information, has made *B. subtilis* the organism of choice for mechanistic studies on sporulation, spore germination, and spore resistance. Although much less mechanistic work has been carried out with other sporeformers, those studies indicate that the fundamental mechanisms regulating gene expression during sporulation, bringing about spore resistance and dormancy, and involved in spore germination are probably similar to those in *B. subtilis.*

SPORULATION

Distribution of Sporeformers

The sporulating organisms discussed in this chapter form heat-resistant endospores which contain DPA and are refractile or phase bright under phase-contrast microscopy. Most studies on sporulation, spores, and spore germination have been carried out with species of either the aerobic bacilli or the anaerobic clostridia. However, other genera form similar spores; among these are the genera *Sporosarcina, Sporolactobacillus,* and *Ther-moactinomyces* (206). Studies using rRNA sequence analysis to determine evolutionary relatedness have shown that the genera *Bacillus, Sporosarcina, Sporolactobacillus,* and *Thermoactinomyces* (206, 212) are quite closely related. *Bacillus* and *Clostridium* are more distantly related (239) but clearly derived from a common ancestor, most likely a sporeformer. A number of other genera, including *Staphylococcus,* are also derived from the same common ancestor (239) yet cannot sporulate. Sporulation-specific genes whose sequences are highly conserved among sporeformers have disappeared from these latter organisms (20, 134). Some of these nonsporing species, for example *Planococcus citreus,* are more closely related to present-day sporeformers than are other sporeformers (212). Unfortunately, the genetic changes causing the loss of sporulation ability in these organisms have not been well characterized.

Induction of Sporulation

The commonest mechanism for inducing sporulation in the laboratory is limitation for one or more nutrients, including carbon or nitrogen. This is achieved either by exhausting one or more nutrients during cell growth, by shifting cells from a rich to a poor medium, or by adding an inhibitor (decoyinine) of guanine nucleotide biosynthesis. Although these laboratory methods cause the great majority of cells in a culture to sporulate, this is undoubtedly not the way sporulation is induced in nature. Indeed, even cells in a growing culture produce a finite number of spores, with the percentage of spores increasing as the culture's growth rate decreases (47). The time from initiation of sporulation to its completion may take as little as 8 h.

Sporulation is sensitive to repression by good carbon sources. This suggests that catabolite repression may regulate sporulation. Unfortunately, the precise mechanism for modulation of carbon catabolite repression of sporulation is still not clear, despite significant insight in recent years (32, 49). The precise intracellular signal that triggers initiation of sporulation is unknown. Molecules such as cyclic AMP (cAMP) and cGMP have been ruled out and a possible role for guanine nucleotides has been suggested. Indeed, despite the impressive elucidation of the pathways regulating gene expression during sporulation (see below), the physiological signals which initiate it remain generally obscure. However, induction of synthesis of enzymes of the tricarboxylic acid cycle is required (92). Sporulation may be modulated in some cell-density-dependent fashion by small molecules secreted into the medium (78, 232), but the mechanism and significance of such possible regulation are unclear.

Sporulation-Associated Phenomena

In addition to sporulation, the cessation or slowing of growth of sporeformers can be associated with (i) synthesis of degradative enzymes such as amylases and proteases (116); (ii) synthesis of antibiotics such as gramicidin, bacitracin, or surfactin (242); (iii) in some species, synthesis of protein toxins active against insects (7), animals, or humans; (iv) development of motility (160); and (v) in a few species (i.e., *B. subtilis*), development of genetic competence (54) (Fig. 3.2). Although these phenomena are not necessary for sporulation, in most cases they are regulated at least partially by the mechanisms that modulate gene expression during sporulation (54, 116, 160, 242). In some cases, for example protease synthesis and competence development in *B. subtilis*, the expression of the genes involved is under multiple controls by both positive and negative effectors (54, 116). There is very good evidence that competence is regulated in a cell-density-dependent manner via the secretion of small peptides termed competence pheromones (143, 209).

There are drastic changes in the metabolism of the sporulating cell. Some of these changes include catabo-

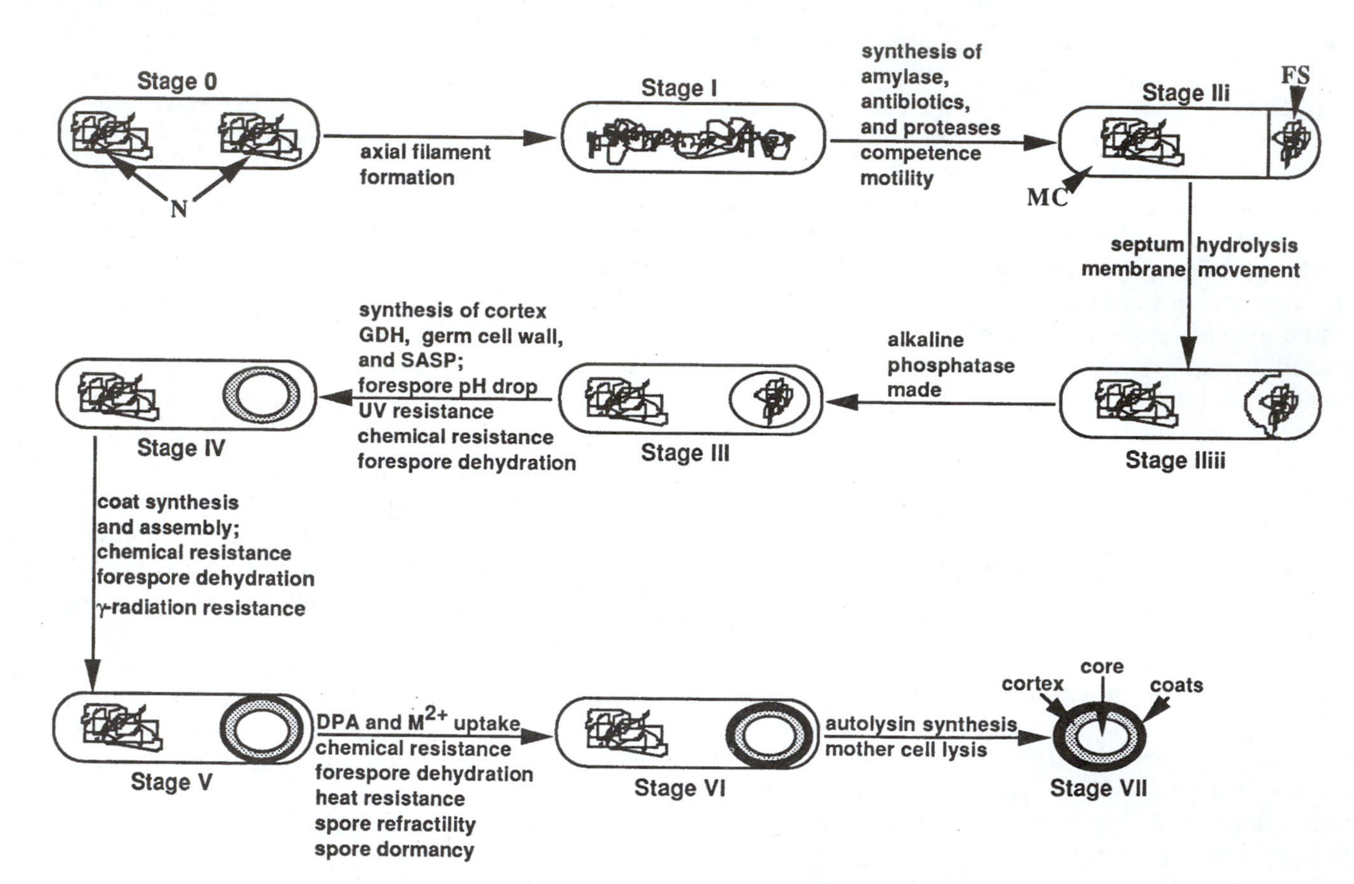

Figure 3.2 Morphological, biochemical, and physiological changes during sporulation of a rod-shaped *Bacillus*. In stage 0, a cell with two nucleoids (N) is shown; in stage IIi the mother cell and forespore are designated MC and FS, respectively. Note that the forespore nucleoid is more condensed than that in the mother cell. Stage IIii is not shown in this scheme, and the forespore nucleoid is not shown after stage III for clarity. The time of some biochemical and physiological events, such as forespore dehydration and acquisition of types of resistance to different chemicals (all lumped together as "chemical resistance"), stretches over a number of stages. The data for this figure are taken from references 57, 65, 153, 166, and 201.

lism of previously formed polymers, such as poly-β-hydroxybutyrate, and initiation of oxidative metabolism due to synthesis of tricarboxylic acid cycle enzymes (210). Many of these latter enzymes are not present in the developing forespore (193). Consequently, the mother cell and forespore have different metabolic capabilities. However, the precise source(s) of energy in the forespore is not clear, although the mother cell may provide much of the forespore's energy, either directly or indirectly.

Morphological, Biochemical, and Physiological Changes during Sporulation

Sporulation is generally divided into seven stages based primarily on the morphological characteristics of cells throughout the developmental process (Fig. 3.2). Growing cells are in stage 0, although sporulation is initiated only following completion of chromosome replication, i.e., only in binucleate cells (91). Older literature had suggested that the first morphological feature unique to sporulation was the presence of the two nucleoids in an axial filament which could be observed in electron micrographs; this was defined as stage I. Because of concerns that this axial filament was an artifact of preparation and because no single mutant has ever been found which is blocked in the transition between stage 0 and stage I, stage I has often been ignored. However, there is recent evidence that stage I is a discrete stage in sporulation (166), although its significance remains obscure.

The first morphological feature of sporulation is the formation of an asymmetric septum. It divides the sporulating cell (now at stage IIi [57, 65, 87]) into the larger mother cell and smaller forespore compartments. Despite the isolation of many genes involved in both septum formation (11, 12, 126) and the siting of its placement (mid-cell for normal cell division or asymmetric for sporulation) (125, 127, 230), the mechanism for asymmetric placement of the sporulation septum is not clear. Similarly, although the sporulation septum differs in some respects from the normal cell division septum, the nature of the differences is not known. There is evidence that a chromosome is not present in the forespore as the septum is formed, but is somehow injected into the forespore at the end of or following completion of septation (240). Both mother cell and forespore compartments then contain complete and apparently identical chromosomes. Because the forespore is much smaller than the mother cell, the forespore chromosome is more condensed than the mother cell chromosome (184, 199). However, the significance of this condensation is presently unclear. A biochemical marker for late stage II is the synthesis of high levels of alkaline phosphatase. Following stage IIi, genes in the two compartments of the sporulating cell may be expressed differentially.

Following septum formation, the mother cell membrane grows around and eventually engulfs the forespore. Thus, the forespore is surrounded by two complete membranes termed the inner and outer forespore membranes (see below). These two membranes have opposite polarities (56, 65, 236). Progression to this stage of sporulation, termed stage III, has been finely characterized by electron microscopy. There are a number of specific changes which occur in the transition from stage II to stage III, leading to the subdivision of stage II into three substages (87), only two of which are shown in Fig. 3.2.

In the transition from stage III to stage IV, a large peptidoglycan structure termed the cortex is laid down between the inner and outer forespore membranes. The cortex has a structure similar to that of cell wall peptidoglycan, but with a number of differences (discussed later). The spore's nascent germ cell wall is also made at about the same time as the cortex, but appears to have the same structure as cell wall peptidoglycan and is formed between the cortex and the inner forespore membrane, probably by forespore enzymes (152). During the stage III–IV transition, the forespore also synthesizes two biochemical markers, glucose dehydrogenase and SASP. Although the function of glucose dehydrogenase in spores is not known, the SASP are involved in spore resistance. The developing forespore acquires full UV resistance and some chemical resistance at this time (201). Late in stage III the forespore pH falls by 1 to 1.3 units and forespore dehydration begins (141, 153).

In the stage IV-to-V transition, a series of proteinaceous coat layers are laid down outside the outer forespore membrane (200). Although coat synthesis is normally ascribed only to the stage IV-to-V transition, this is too narrow a distinction as some coat proteins are clearly made in the stage III–IV transition and even after stage V (57). The amount and variety of the spore coat proteins appear to vary considerably between species. *B. subtilis* has a complex coat structure, and other species have somewhat simpler coats (200). The reason for this is not clear, but many *B. subtilis* coat proteins can be lost with no apparent phenotypic effect. Forespore γ-radiation resistance begins to be acquired during this period, as is further chemical resistance. Forespore dehydration also continues during this time (153, 201). During the stage V-to-VI transition, the spore core's depot of DPA is accumulated following DPA synthesis in the mother cell (57). DPA uptake is paralleled by uptake of an enormous amount of divalent cations, predominantly Ca^{2+}, but with much Mg^{2+} and Mn^{2+} as well (152). The great majority of these cations are also in the spore core, presumably associated with DPA (152). The precise state of these compounds in the spore core is not known, although the amount of DPA accumulated exceeds its solubility. During this period the spore's central region or core also undergoes the final process of dehydration (153). As a consequence of the high ratio of solids to water in the spore core at this stage, the spore appears phase bright or white under the phase-contrast microscope. In addition, because of permeability changes in the spore membranes, the spore stains poorly if at all with common bacteriological stains. The spore also becomes metabolically dormant during this period and acquires further γ-radiation and chemical resistance (201). Finally, in the transition to stage VII, one or more autolysins are produced in the mother cell, resulting in its lysis and release of the spore.

The seven stages are useful in characterizing various asporogenous or *spo* mutants which have no defect in growth but are blocked in a particular stage in sporulation. These *spo* mutants are given an added designation denoting the stage in which they are blocked. Thus *spo0* and *spoII* mutants are blocked in stages 0 and II, respectively. The various stages have also allowed facile correlation of biochemical changes with various morphological changes (Fig. 3.2). However, the sporulation scheme outlined above is an oversimplification for several reasons. First, since the various stages are intermediates in a continuous developmental process, it is probably misleading to think of the stages as discrete entities. Second, the scheme given is only for rod-shaped organisms which sporulate without terminal swelling of the forespore compartment. There are many sporeformers in which the forespore compartment swells considerably and the mother cell elongates (65). Similarly, some sporeformers grow as cocci (206). These differences undoubtedly modify the sporulation process. However, detailed knowledge of sporulation in these of organisms is limited.

Regulation of Gene Expression during Sporulation

Much of our knowledge about the regulation of gene expression during sporulation is derived from *spo* mutants. Asporogeny can be caused by mutations in any one of more than 75 distinct genetic loci (57, 214). The identification of biochemical or physiological markers for various stages of sporulation provides another major aid in understanding gene regulation during sporulation (Fig. 3.2). Analysis of these markers in *spo* mutants provides strong evidence that sporulation is primarily a linear process of sequential events. Thus, *spo0* mutants generally exhibit no sporulation-specific marker events, *spoII* mutants generally demonstrate only stage 0-specific marker events, and so on. Analysis of *spo* mutants has furnished a broad outline of regulation of gene expression during sporulation. However, the advent of molecular cloning technology was necessary to provide detailed understanding of this process. The great majority of genes in which mutations block sporulation have been cloned and sequenced. In addition, many sporulation-specific genes in which mutation has little or at most a subtle effect on sporulation have been isolated and analyzed. For most of these genes, their precise functions, the location of their expression (mother cell or forespore), and the factors involved in regulation of their expression are known (57, 214, 216).

It was hypothesized that sporulation requires transcription of many new genes and that this change in transcriptional pattern was modulated by changes in the specificity of the cell's RNA polymerase due to synthesis of new σ (specificity) factors (135). This hypothesis has been expanded over the past 15 years because changes in gene expression during sporulation require modulation of transcription by many mechanisms including (i) synthesis of new RNA polymerase σ factors with altered promoter specificity at various times and in the different compartments of the sporulating cell; (ii) activation or inactivation of new σ factors by both covalent and noncovalent modification; (iii) intricate regulatory communication between mother cell and forespore to ensure coordinate development in both compartments; (iv) synthesis of new DNA binding proteins to activate or repress transcription; (v) degradation of repressors; and (vi) modulation of transcription factor activity by phosphorylation. The information explosion from this work has now provided an extremely detailed picture of the control of gene expression during sporulation. This picture is striking not only in its complexity, but also in the redundancy of its control mechanisms. The following discussion of these control mechanisms has been simplified and concentrates on major regulatory gene products.

Initiation of sporulation requires expression of *spo0* gene products at a significant level in vegetative cells (57, 83). The most important of these *spo0* gene products is Spo0A. This protein is the response-regulator half of a two-component (signal transduction) regulatory system. These signal transduction systems transmit signals

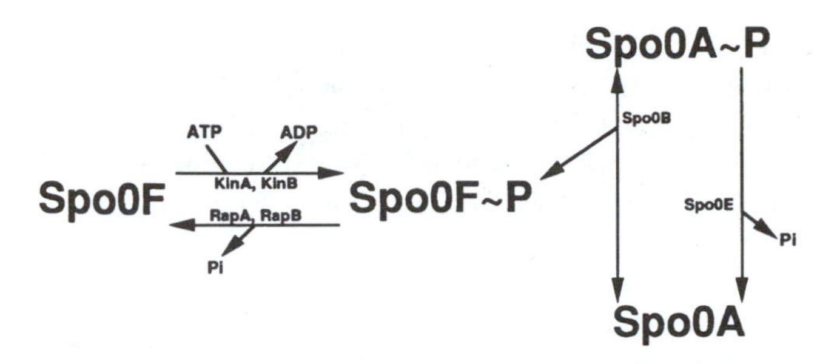

Figure 3.3 Gene products and reactions which affect levels of Spo0A~P. Spo0E is a phosphatase specific for Spo0A~P; RapA and RapB are phosphatases specific for Spo0F~P and are different names for Spo0L and Spo0P, respectively (83, 158, 162).

(often by binding to DNA and affecting transcription) when an aspartyl residue in the response regulator is phosphorylated by a sensor protein which has kinase activity (25). In growing cells, Spo0A is primarily in the dephosphorylated state. However, under conditions which initiate sporulation, the degree of Spo0A phosphorylation rises through action of multiple sensor kinases. In *B. subtilis* the majority of phosphate on Spo0A appears to be derived via a phosphorelay initiated by phosphorylation of Spo0F (Fig. 3.3 and 3.4). The phosphate is then transferred from Spo0F~P to Spo0A by the Spo0B protein. There are at least two major kinases (KinA and KinB) which can initiate the phosphorelay. These enzymes can also phosphorylate Spo0A directly but with low efficiency (83). Mutations in either *kinA* or *kinB* generally have little to no effect on sporulation, but a *kinA kinB* double mutant causes asporogeny. A third kinase (KinC) acts preferentially on Spo0A, but its precise function in wild-type cells is unclear (110, 123, 124). In addition to the many kinases feeding phosphate into Spo0A, there are at least three phosphatases that can dephosphorylate Spo0A~P. It can be dephosphorylated either directly or through the dephosphorylation of Spo0F~P (158, 162) (Fig. 3.3). This multiplicity of kinases and phosphatases allows a variety of environmental signals to be integrated so that the precise intracellular level of Spo0A~P can be set in response to the environment in toto. A specific threshold concentration of Spo0A~P is probably needed to initiate sporulation (35). While the level of Spo0A~P is clearly

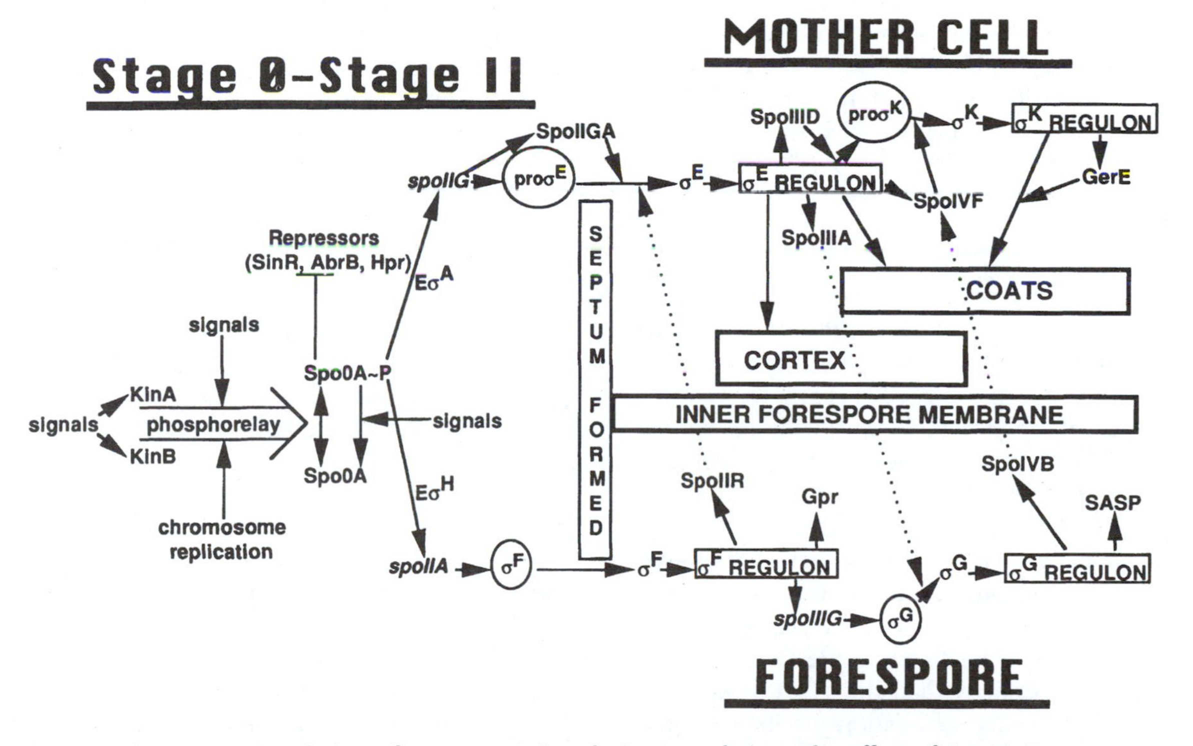

Figure 3.4 Regulation of gene expression during sporulation. The effect of Spo0A~P on repressors is negative; other effects of regulatory molecules on reactions are generally positive, although the effect of signals may be positive or negative. The enclosure of the pro-σ factors and σ factors denotes that at this time these factors are inactive. This figure is adapted from that of Piggot et al. (166) and includes data from references 9, 57, 83, 101, 116, 132, 166, 214, and 215.

crucial in determining whether a cell sporulates or not, it is unclear how the kinases and phosphatases that modulate Spo0A~P levels are themselves regulated (83, 217). This is crucial information that is needed for understanding the physiological regulation of sporulation.

The phosphorylation of Spo0A greatly increases its affinity for binding to sites upstream of several key genes, although different sites have much different affinities for Spo0A~P (83). The *abrB* gene has a very high affinity for Spo0A~P, and its binding of Spo0A~P greatly decreases *abrB* transcription. Since AbrB is a labile protein, the decrease in *abrB* transcription rapidly decreases the intracellular AbrB concentration. AbrB is a repressor of several genes normally expressed in the stage 0–II transition, including the *spo0H* gene which codes for a minor σ factor for RNA polymerase termed σ^H. AbrB is also an activator for synthesis of a second repressor, termed Hpr, which represses additional stage 0-expressed genes, especially the genes for several proteases. In addition to AbrB and Hpr, there is a third repressor termed SinR. SinR binds to and represses other genes expressed in the stage 0–II transition, including those of the *spoIIG* and *spoIIA* operons (9, 83). SinR action is blocked by synthesis of a protein termed SinI, which binds to SinR and blocks its action. Synthesis of SinI is stimulated by Spo0A~P and probably repressed by AbrB and Hpr. Thus, early in the stage 0–II transition, many genes normally repressed during vegetative growth are derepressed through action of Spo0A~P (Fig. 3.4).

Important among the stage 0–II genes is *spo0H*, which encodes a minor sigma factor, σ^H. There is some transcription of *spo0H* during vegetative growth by RNA polymerase containing the cell's main sigma factor ($E\sigma^A$). However, levels of active σ^H during vegetative growth are maintained at a low level, probably by a posttranscriptional mechanism (83). Early in sporulation, rising levels of functional σ^H promote transcription by $E\sigma^H$. As Spo0A~P levels increase even further, concentrations become high enough to bind to the promoter regions of the *spoIIG* and *spoIIA* operons which have been derepressed by removal of SinR as a SinI-SinR complex (57, 83). The binding of Spo0A~P stimulates transcription of these genes by RNA polymerase containing σ^A or σ^H, respectively (10, 16, 57, 83). $E\sigma^H$ has a different promoter specificity than does $E\sigma^A$, as do RNA polymerases with all other sporulation-specific σ factors (79). This allows these different RNA polymerases to recognize and transcribe different sets of genes. The *spoIIA* operon transcribed by $E\sigma^H$ encodes three proteins, with the third cistron (*spoIIAC*) encoding another sigma factor termed σ^F. The *spoIIAB* gene encodes an inhibitor of σ^F function, while *spoIIAA* encodes an antagonist of SpoIIAB (50, 55, 79, 177). Prior to septation, σ^F is inactive in the sporulating cell due to its interaction with SpoIIAB. Interaction of SpoIIAB with σ^F rather than with SpoIIAA may be promoted by the high ATP/ADP ratio in the preseptation sporulating cell (3). Phosphorylation of SpoIIAA by SpoIIAB also influences the interaction between these two proteins (148). The *spoIIG* operon transcribed by $E\sigma^A$ encodes two proteins, with the second cistron, *spoIIGB*, also encoding an alternative σ factor for RNA polymerase, termed σ^E (57, 79). However, unlike σ^F, σ^E is made as a precursor termed pro-σ^E, which is inactive in directing transcription. The first gene of the *spoIIG* operon (*spoIIGA*) is thought to encode the protease responsible for processing pro-σ^E to σ^E (113, 215). However, action of SpoIIGA on pro-σ^E requires gene expression under σ^F control in the forespore compartment (57, 203) (Fig. 3.4).

Although the *spoIIG* operon is transcribed before septum formation, σ^E is not generated until after septation, and then possibly only in the mother cell (87, 113, 129, 217). Similarly, σ^F is also produced before septum formation and is maintained in an inactive state by interaction with inhibitory SpoIIAB. However, following septation, σ^F becomes active in the forespore. In vitro studies have indicated that a decrease in the ATP/ADP ratio causes SpoIIAB to switch partners from σ^F to SpoIIAA, relieving the inhibition of σ^F by SpoIIAB (3). There may be a decrease, perhaps only transient, in the forespore's ATP/ADP ratio, although this remains to be demonstrated. As noted above, phosphorylation of SpoIIAA by SpoIIAB also influences these proteins' interactions (148). In any event, $E\sigma^F$ now initiates transcription of several genes in the forespore including *spoIIR* (or *csfX*), *gpr*, and *spoIIIG* (101, 132, 196, 222) (Fig. 3.4). The *spoIIR* gene product appears to act by promoting pro-σ^E processing in a vectorial fashion such that pro-σ^E is processed to σ^E, presumably by SpoIIGA, only in the mother cell. The precise mechanism whereby SpoIIR influences SpoIIGA action is not yet clear. Indeed, it is possible that pro-σ^E is processed to σ^E in both the mother cell and the forespore, but that σ^E action in the forespore is somehow inhibited (203).

It is clear that during sporulation pro-σ^E processing requires synthesis of SpoIIR in the forespore (101, 132). This requirement for an event in the forespore being prerequisite to an event in the mother cell is the first of three examples of regulatory cross-talk between the two compartments of the sporulating cell which ensures coordinate development of the mother cell and forespore (136). Of the other two $E\sigma^F$-transcribed genes, *gpr* encodes a protease that acts on SASP in the first minutes of spore germination and *spoIIIG* encodes another σ factor termed σ^G, which is responsible for the bulk of forespore-specific transcription (196). However, σ^G is not active immediately following its synthesis in stage II. Instead, some mother cell-specific event(s), including transcription of the *spoIIIA* operon by $E\sigma^E$, is required for generation of active $E\sigma^G$, which now acts in stage III (57, 88, 171) (Fig. 3.3 and 3.4). This is the second example of regulatory cross-talk between mother cell and forespore. The precise mechanism for regulation of σ^G activity is not yet clear but may involve regulation by SpoIIAB in the same manner as for σ^F (171). The generation of active σ^E only in the mother cell, and the synthesis and activation of σ^G only in the forespore, firmly establish the basis for compartment-specific transcription during sporulation.

As noted above, $E\sigma^G$ transcribes several genes expressed only in the forespore (57, 196). Some of these are genes essential for sporulation, but their function is unknown. As the third example of regulatory cross-talk, at least one such gene (*spoIVB*) is also responsible for communicating with the mother cell and regulating gene expression in this compartment (42) (Fig. 3.4; see below). The *ssp* genes are a large set of genes transcribed by $E\sigma^G$. These genes encode the SASP that are major protein components of the spore core. The majority of the SASP are of the α/β type and are DNA binding proteins which confer upon spore DNA resistance to a variety of treatments (201, 202) (see below). In contrast to $E\sigma^G$, $E\sigma^E$ transcribes genes only in the mother cell, including genes needed specifically for spore cortex biosynthesis and some genes (termed *cot* genes) needed for spore coat formation and assembly (57). $E\sigma^E$ also transcribes the *spoIIID* gene, which encodes a DNA binding protein that modulates $E\sigma^E$ action, resulting in different classes of $E\sigma^E$-transcribed genes. SpoIIID is needed for transcription of the *sigK* gene by $E\sigma^E$ (57, 114). The *sigK* gene encodes a final new σ factor termed σ^K (79, 214). In *B. subtilis,* this gene has a large intervening sequence which is removed only from the mother cell genome by a recombinase. Expression of the recombinase is regulated such that generation of an intact *sigK* gene just precedes *sigK* transcription. However, this intervening sequence is absent in the *sigK* genes of other *Bacillus* species (1). Consequently, removal of the intervening sequence does not appear to be a general mechanism for regulating σ^K function. As is the case with σ^E, σ^K is also synthesized as an inactive pro-σ^K and is processed proteolytically to σ^K about 1 h after its synthesis (113). Although the details of this proteolytic processing are not yet clear, conversion of pro-σ^K to σ^K in the mother cell requires expression of one σ^G-controlled gene (*spoIVB*) in the forespore (42, 130) (the third example of regulatory cross-talk) and participation of at least one additional σ^E-controlled operon (*spoIVF*) in the mother cell (43) (Fig. 3.4), one cistron of which may encode a protease. The products of the *spoIVB* gene and *spoIVF* operon may somehow interact between or across the inner and outer forespore membranes to cause pro-σ^K cleavage to σ^K (113). $E\sigma^K$ itself also transcribes *sigK* (in conjunction with SpoIIID) (46), the gene for DPA synthase (45), other *cot* genes (214), and one gene termed *gerE* which encodes a DNA binding protein that modulates $E\sigma^K$ activity, causing a large change in the pattern of *cot* gene expression (57) (Fig. 3.4). Essentially all of our detailed knowledge of gene expression during sporulation ends at this point. There is little definitive information about genes which may be required for mother cell lysis. However, synthesis of one autolytic enzyme (CwlC) is known to be transcribed by $E\sigma^K$ (117).

The preceding picture of regulation of gene expression during sporulation is derived predominantly from studies with *B. subtilis.* Is the picture similar in organisms from other species or genera including *Clostridium*? Although there is no definitive answer to this question, a number of key regulatory genes including *spo0A, spoIIGA, spoIIGB, spoIIIG,* and *sigK* have been analyzed in other sporeformers, including *Clostridium* spp. (1, 20, 233). The striking conservation of both the sequences of the encoded proteins and the organization of these genes strongly suggests that regulation of sporulation across species is by similar if not identical mechanisms.

THE SPORE

Spore Structure

The spore as released from the mother cell at the end of sporulation is biochemically and physiologically different from the vegetative cell. The spore structure is also extremely different from that of a cell (Fig. 3.5). Many spore structures, including the exosporium and coats, have no counterparts in the vegetative cell. The outermost spore layer, the exosporium, varies significantly in size between species and its various components are not well characterized (152, 225). Underlying the exosporium are the spore coats. The complexity of this structure varies significantly between species. In *B. subtilis* spores, a number of distinct coat layers can be easily seen in electron micrographs (200). Where individual coat proteins have been analyzed, they are unique to the spore stage of bacterial existence. Some coat proteins have an

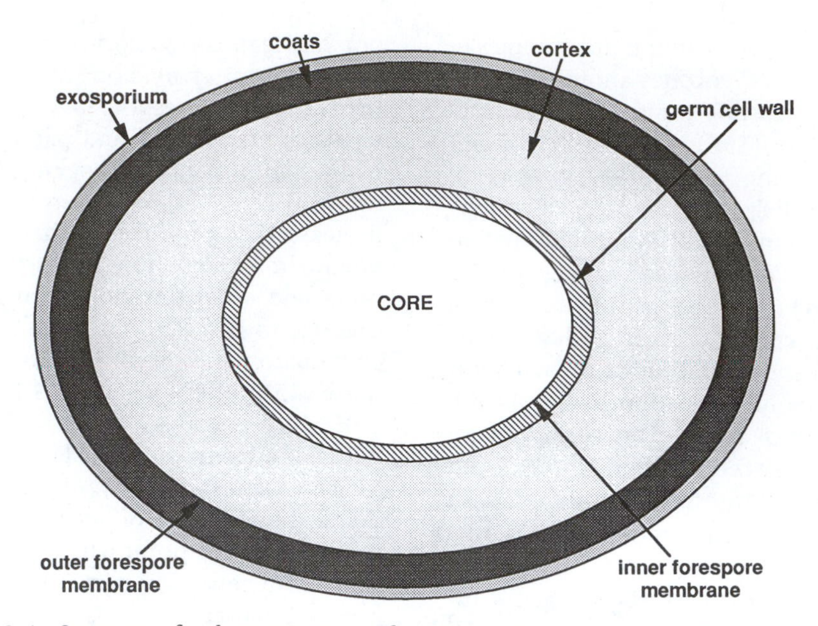

Figure 3.5 Structure of a dormant spore. The various structures are not drawn precisely to scale, especially the exosporium, whose size varies tremendously between spores of different species. The relative size of the germ cell wall is also generally smaller than shown. The positions of the inner and outer forespore membranes, between the core and germ cell wall and between the cortex and coats, respectively, are also noted.

extremely unusual amino acid composition (200). Several of these coat proteins appear active primarily in assembly of the final coat structure. Posttranslational modification of some coat proteins, in particular formation of dityrosine cross-links, may be important in spore coat structure. The spore coats protect the spore cortex from attack by lytic enzymes. The coats also may provide an initial barrier to chemicals such as oxidizing agents (69, 200, 201). However, the spore coats play no significant role in maintenance of spore resistance to heat or radiation (69, 111, 198, 201).

Underlying the spore coats is the outer forespore membrane. This is a complete functional membrane in the developing forespore and may also be so in the spore (56). If it is a complete functional membrane, it may play a large role in the extreme impermeability of the spore to small molecules. The protein composition of this membrane is different from that of the inner forespore membrane (56).

Underlying the outer forespore membrane is the spore's cortex. This large peptidoglycan layer is structurally similar to cell wall peptidoglycan, but with several differences (22). The cortical peptidoglycan always contains diaminopimelic acid, even if the vegetative cell wall peptidoglycan contains lysine. A significant percentage (~65%) of the muramic acid residues in cortical peptidoglycan also lack any peptide residues. Although a few muramic acid residues contain a single D-alanine, the majority are present as muramic acid lactam which is not present in cell wall peptidoglycan. A few genes for cortex synthesis have been identified and characterized (21, 23, 44). Although the precise mechanism of cortex synthesis is not known, the required gene products are made only in the mother cell. Spore cortical peptidoglycan exhibits a similar degree of peptide cross-linking to vegetative cell peptidoglycan (169). However, there is no knowledge of the timing of cross-link formation in the cortex, nor of the distribution of cross-links throughout the cortical volume, as the three-dimensional structure of the spore cortex is not known. This is a major deficiency, as the spore cortex is thought to be the structure responsible for the dehydration of the spore core and thus much of spore resistance (see below).

Between the cortex and the inner forespore membrane is the germ cell wall. Its structure may be identical to that in vegetative cells. The next structure, the inner forespore membrane, is a complete membrane and is an extremely strong permeability barrier to hydrophilic molecules and to most molecules of ≥150 molecular weight (70). This membrane's phospholipid content is similar to that of growing cells (56). However, the spore core volume surrounded by the inner forespore membrane appears smaller than is predicted based on its phospholipid content. Since the core's volume can expand significantly upon spore germination in the absence of membrane synthesis, the "excess" membrane in the inner forespore membrane may allow for the expansion.

Finally, the central region, or core, contains the spore's DNA, ribosomes, and most enzymes, as well as the depots of DPA and divalent cations. There are also many unique gene products in the dormant spore, including the large SASP pool (10 to 20% of spore protein), much of which is bound to spore DNA (196, 202). A notable feature of the spore core is its low water content (69). While vegetative cells have ~4 g of water per g of dry weight, the spore core has only 0.4 to 1 g of water per g of dry weight. The core's low water content is thought to play a major role in spore dormancy and in spore resistance to a variety of agents (69, 201). In contrast to the low water content in the spore core, other regions of the spore have a more normal water content (69). Although the spore cortex is clearly essential for generating and maintaining spore core dehydration, the precise mechanism(s) involved is not clear.

Spore Macromolecules

Spores are biochemically different from vegetative cells. Some proteins found in the spore coats and core are not present in the vegetative cell. The SASP are particularly noteworthy, as some of these proteins play a major role in spore resistance (196, 201, 202). SASP are synthesized in the forespore during stage III of sporulation, when their coding genes are transcribed by $E\sigma^G$. There are two kinds of SASP in *Bacillus* spp., termed γ type and α/β type. The γ-type SASP are 75- to 105-residue proteins which do not bind to any other spore macromolecule. Their only known function is to be degraded during germination, thus providing amino acids for the germinating spore. Their degradation is initiated by the sequence-specific protease GPR (germination protease; the product of the *gpr* gene). The γ-type SASP are encoded by a single gene in bacilli. However, neither this gene nor γ-type SASP have been found in clostridial spores. The sequences of γ-type SASP are not homologous to other proteins in currently available databases, and the proteins' sequence has diverged significantly during evolution. The α/β-type SASP are named for the two major proteins of this type in *B. subtilis* spores. In contrast to γ-type SASP, α/β-type SASP are coded for by at least seven genes. All of these are expressed, although in most species two proteins (major α/β-type SASP) are expressed at high levels and the remainder at much lower levels (minor α/β-type SASP). However, both minor and major α/β-type SASP have similar properties in vitro and in vivo (see below).

The α/β-type SASP are also found in clostridial spores (26, 27, 84). The amino acid sequences of these small (60 to 75-residue) proteins are highly conserved both within and across species, although they too have no homology to other proteins in current databases (26, 27, 195, 202). The α/β-type SASP, like the γ type, are also degraded to amino acids in the first minutes of spore germination. Their degradation is initiated by GPR. However, the α/β-type SASP are also DNA binding proteins, both in vivo and in vitro (191, 202). In vivo these proteins saturate the spore chromosome and provide much of the spore DNA's resistance to various treatments. Binding of α/β-type SASP to DNA is not sequence specific but exhibits a preference for G+C-rich regions, although A+T-rich regions are bound. These proteins bind to the outside of the DNA helix, interacting primarily with the sugar phosphate backbone (77). This binding alters the DNA structure from a B-like helix to an A-like helix, although this is not the classical A-like helix of a poorly hydrated DNA fiber (197). The precise structure of the α/β-type SASP-DNA complex is unknown. The binding of these proteins to DNA has striking effects on DNA properties, including providing great resistance to chemical and enzymatic cleavage of the DNA backbone, altering the DNA's UV photochemistry, and greatly slowing DNA depurination (see below).

Not only do spores contain unique proteins, but a number of proteins present in vegetative cells are absent from spores. These include amino acid and nucleotide biosynthetic enzymes, which are degraded during sporulation and then synthesized at defined times during spore outgrowth (193). Other enzymes, in particular the enzymes of amino acid and carbohydrate catabolism, are present at similar levels in spores and cells (193). Spores also contain all of the enzymes needed for RNA and protein synthesis and many DNA repair enzymes (193). However, at least one enzyme needed for initiation of DNA replication may be absent from spores (193). Enzyme activities common to both cell and spore enzymes are invariably the same gene product. In general, spore and cell rRNAs and tRNAs are similar if not identical, although much tRNA in spores lacks the 3′-terminal A residue and little if any spore tRNA is aminoacylated (193). However, spores lack most if not all functional mRNA. Spore DNA appears identical to cell DNA but has different proteins associated with it.

Spore Small Molecules

Spores are quite different from cells in their content of small molecules (Table 3.1), which are located in the core. The low amount of spore core water and the huge

Table 3.1 Small molecules in cells and spores of *Bacillus* species

Molecule	Content (mmol/g dry wt) in:	
	Cells[a]	Spores[b]
ATP	3.6	≤0.005
ADP	1	0.2
AMP	1	1.2–1.3
Deoxynucleotides	0.59[c]	<0.025[d]
NADH	0.35	<0.002[e]
NAD	1.95	0.11[e]
NADPH	0.52	<0.001[e]
NADP	0.44	0.018[e]
Acyl-CoA	0.6	<0.01[e]
CoASH[f]	0.7	0.26[e]
CoASSX[g]	<0.1	0.54[e]
3PGA	<0.2	5–18
Glutamic acid	38	24–30
DPA	<0.1	410–470
Ca^{2+}		380–916
Mg^{2+}		86–120
Mn^{2+}		27–56
H^+	7.6–8.1[h]	6.3–6.9[h]

[a]Values for *B. megaterium* in mid-log phase are from references 48, 141, 185, 186, 192, 193, 201, and 205.
[b]Values are the range from spores of *B. cereus*, *B. subtilis*, and *B. megaterium* and are from references 141, 185, 186, 192, 193, and 201.
[c]Value is the total of all four deoxynucleoside triphosphates.
[d]Value is the sum of all four deoxynucleotides.
[e]Values are for *B. megaterium* only.
[f]CoASH, free CoA.
[g]CoASSX, CoA in disulfide linkage to CoA or a protein.
[h]Values are expressed as pH and are the range with *B. cereus*, *B. megaterium*, and *B. subtilis*.

depot of DPA and divalent cations have already been noted (Table 3.1). Interestingly, physical methods have been used to show that the ions in the spore core are quite immobile (30). This is consistent with a dearth of free water in spores. The pH in the spore core is 1 to 1.5 units lower than that in a growing cell (141, 186). The pH of the forespore changes relative to that of the mother cell during the stage III–IV transition (Fig. 3.2). Spores of most species have a large depot of 3-phosphoglyceric acid (3PGA) (Table 3.1), which is accumulated shortly after (and probably in response to) the forespore pH decrease (115, 142, 201). In contrast to growing cells, spores have little if any of the common "high-energy" compounds found in cells, including deoxynucleoside triphosphates, ribonucleoside triphosphates, reduced pyridine nucleotides, and acyl coenzyme A (acyl-CoA) (Table 3.1). However spores do contain significant amounts of ribonucleoside mono- and diphosphates (but not deoxynucleotides), oxidized pyridine nucleotides, and CoA (Table 3.1). The high-energy forms of these latter compounds are lost from the forespore late in sporulation, as the spore becomes dormant at about the time of DPA uptake (Fig. 3.2). Interestingly, much of the CoA in spores is in disulfide linkage, some as a CoA disulfide and some linked to protein (Table 3.1). The function of these disulfides is not known, but they are reduced in the first minutes of spore germination (185). In addition to lacking many amino acid biosynthetic enzymes, spores have very low levels of most free amino acids (133), but often have high levels of glutamic acid (Table 3.1).

Spore Dormancy

Spores are metabolically dormant, catalyzing no detectable metabolism of endogenous or exogenous compounds and having no high-energy compounds (128). The major cause of the spore's metabolic dormancy is undoubtedly the low water content of the core, which may preclude enzyme action (201). In a further reflection of this metabolic dormancy, the spore core contains at least two enzyme-substrate pairs which are stable in the dormant spore for months to years but which interact in the first 15 to 30 min of spore germination, resulting in complete substrate degradation. These two enzyme-substrate pairs are 3PGA-phosphoglycerate mutase (PGM) and SASP-GPR (201). Although special regulatory mechanisms other than dehydration stabilize 3PGA and SASP in the developing forespore despite the presence of PGM and GPR (201), these mechanisms do not explain the extreme stability of 3PGA and SASP in dormant spores. The reason for the lack of PGM and GPR action on their substrates in spores may again be enzyme inactivity caused by spore core dehydration. Although this explanation seems reasonable, in the absence of precise data on the amount of free water in the spore core, it is difficult to test this idea definitively.

Spore Resistance

The spore's metabolic dormancy is undoubtedly one factor in its ability to survive extremely long periods in the absence of nutrients. A second factor in long-term spore survival is the spore's extreme resistance to potentially lethal treatments including heat, radiation, chemicals, and desiccation (17, 63, 69, 198, 201). Spores are much more resistant than vegetative cells to a variety of killing treatments. Table 3.2 presents representative data for cells and spores of *B. subtilis*. Note, however, that some organisms form much more resistant spores than does *B. subtilis*. Spore resistance is due to a variety of factors, with factors such as spore core dehydration and α/β-type SASP involved in many types of resistance, while factors such as spore impermeability may be involved in only one. Since different and sometimes multiple factors contribute to different types of spore resistance, it is not surprising that spore resistance to different treatments is acquired at different times in sporulation (Fig. 3.2). Because resistance of spores to one treatment is often caused by multiple factors, elucidation of detailed mechanisms of spore resistance is difficult. However, in recent years the mechanisms of spore heat, UV, and H_2O_2 resistance have been elucidated. The following detailed discussion of spore resistance concentrates on *B. subtilis* because of the copious detailed mechanistic data available for this organism. However, studies with spores of other organisms, in particular on the mechanism of heat resistance, have indicated that factors involved in resistance of *B. subtilis* spores are also involved in resistance of spores from other species and genera. However, the relative importance of particular factors in spore resistance may vary significantly between species.

Spore Freezing and Desiccation Resistance

Growing bacteria generally incur some lethality during freezing and greater lethality during desiccation, unless special precautions are taken to prevent killing. The precise mechanism(s) of killing is not clear, but one cause may be DNA damage; freeze-drying cells can cause significant mutagenesis (8). However, neither the precise type of DNA damage nor the reason for DNA damage caused by freeze-drying is clear. In contrast to the sensitivity of cells to freeze-drying, spores are resistant to multiple cycles of freeze-drying (74) (Table 3.2). A complete explanation for spore desiccation resistance is not yet available. However, the α/β-type SASP provide one component of spore desiccation resistance by preventing DNA damage caused by freeze-drying (60). Spores lacking these proteins (termed $\alpha^-\beta^-$ spores) are much more sensitive to killing by freeze-drying than are wild-type spores (Table 3.2), with the killing of $\alpha^-\beta^-$ spores due in large part to DNA damage. However, $\alpha^-\beta^-$ spores are still significantly more resistant to freeze-drying than are vegetative cells, indicating that spore resistance to freeze-drying has causes in addition to α/β-type SASP. Accumulation of various carbohydrates such as the disaccharide trehalose is often important in the resistance of yeast or other fungal spores to freezing or desiccation (40). However, bacterial spores do not accumulate such sugars (193).

Spore Pressure Resistance

Spores are much more resistant to high pressures (≥12,000 atm) than are cells (208). Although spores are

Table 3.2 Killing and mutagenesis of spores and cells of *B. subtilis* by various treatments[a]

A. Freeze-drying

No. of freeze-drying cycles	Survival (%)		
	Cells[b]	Wild-type spores	$\alpha^-\beta^-$ spores[c]
1	2		
3		100 (<0.5)	7 (14)

B. 10% Hydrogen peroxide[a]

Time of treatment (min)	Survival (%)		
	Cells[b]	Wild-type spores	$\alpha^-\beta^-$ spores[c]
2.5	0.3	92	
5		88	50
10			10 (14)
20		50	0.1
60		6 (≤0.5)	

C. UV[d]

	Dose to kill 90% of the population (J/m²)		
	Cells[b]	Wild-type spores	$\alpha^-\beta^-$ spores[c]
	40	315	25

D. Moist heat

Treatment temp (°C)	*D* value		
	Cells[b]	Wild-type spores	$\alpha^-\beta^-$ spores[c]
95		14 min (≤0.5)	
85		360 min (≤0.5)	15 min (13)
65	< 15 s	105 h	10 hr
22		2.5 yr (≤0.5)	2.8 months (18)

E. Dry heat

Treatment temp (°C)	*D* value		
	Cells[b]	Wild-type spores	$\alpha^-\beta^-$ spores[c]
120		33 min (12)	
90	5 min		5.5 min (12)

[a]Data taken from references 59, 60, 146, 187, 190, and 196. Values in parentheses are the percentage of survivors with asporogenous or auxotrophic mutations when spores undergo 30–99% killing.
[b]Cells in the log phase of growth. Similar results have been obtained with wild-type or $\alpha^-\beta^-$ cells.
[c]These spores lack the two major α/β-type SASP and thus ~70% of the α/β-type SASP pool.
[d]UV irradiation with light predominantly at 254 nm.

more resistant than cells to lower pressures, spores are killed more rapidly at lower pressures than at higher pressures (36, 75). This apparent anomaly is caused by the promotion of spore germination at lower pressures; the germinated spores are then rapidly killed by the pressure treatment. In contrast, spore germination is not promoted by very high pressures. There is no good understanding of the factors involved in spore resistance to high pressure or the mechanism whereby lower pressures promote spore germination.

Spore γ-Radiation Resistance

Spores are generally more resistant to γ-radiation than are vegetative cells (63). In the few organisms where it has been studied, γ-radiation resistance is acquired 1 to 2 h before acquisition of heat resistance. The precise factors involved in spore γ-radiation resistance are not known, although SASP do not appear to be involved (201). The low water content in the spore core would be expected to provide protection against γ-radiation. However, no analysis has correlated the degree of spore core dehydration with spore γ-radiation resistance. One significant impediment to the understanding of spore γ-radiation resistance is the lack of knowledge about the damage caused by γ-radiation that results in spore death. Presumably this damage is to spore DNA. However, the precise nature of this damage and whether it is similar to that generated in vegetative cells is not known. Given the very different environments inside the vegetative cell and dormant spore, it is certainly possible that the γ-radiation damage to DNA differs significantly in these two states.

Spore UV Radiation Resistance

Spores of many species are 7 to 50 times more resistant than are vegetative cells to UV radiation at 254 nm (194, 198, 201) (Table 3.2), the wavelength giving maximal killing. Spores are also more resistant than cells at both longer and shorter UV wavelengths. UV resistance is acquired by the developing forespore 2 h before acquisition of heat resistance (Fig. 3.2), in parallel with synthesis of α/β-type SASP. These latter proteins are essential, and may also be sufficient, for spore UV resistance. Coats, cortex, and core dehydration are not necessary for spore UV resistance.

The major reason for spore UV resistance is the different UV photochemistry of DNA in spores and in cells. The major photoproducts formed upon UV irradiation of cells or purified DNA are cyclobutane-type dimers between adjacent pyrimidines (64). The most abundant of these are between adjacent thymine residues (TT) (Fig. 3.6A), with smaller amounts between adjacent cytosine and thymine residues (CT) and adjacent cytosine residues (CC). In addition, UV irradiation of cells or purified DNA generates the various 6-4 photoproducts which are also formed between adjacent pyrimidines (234). All of these photoproducts can be lethal as well as mutagenic, although 6-4 photoproducts may be the most mutagenic (71). In contrast to the UV photochemistry of purified DNA or DNA in cells, UV irradiation of spores generates few cyclobutane-type pyrimidine dimers, and 6-4 photoproduct formation, if it takes place, is with a much lower yield as a function of UV fluence (52, 194, 198). However, UV irradiation of spores does generate large amounts of a thyminyl-thymine adduct initially termed "spore photoproduct" (SP) (52) (Fig. 3.6B). The yield of SP as a function of UV fluence in spores is similar to the yield of TT as a function of UV fluence in cells. SP is a potentially lethal photoproduct (194, 198, 202). Thus the difference in UV photochemistry between DNA in cells and spores is, by itself, not sufficient to explain spore UV resistance, as there must be a difference in the capacity of cells and spores to efficiently repair TT and SP, respectively. Indeed, spores have at least two mechanisms for SP repair, both of which operate in the first minutes of spore germination. One mechanism is via the same excision repair system that repairs TT and other lesions in growing cells (151). Spores lacking this repair system are two- to threefold more sensitive to UV than are wild-type spores (62, 194). The second repair

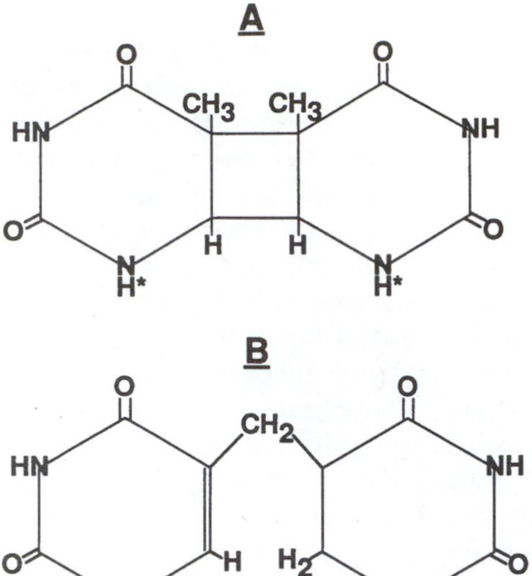

Figure 3.6 Structures of (A) cyclobutane-type TT dimer and (B) 5-thyminyl-5,6-dihydrothymine adduct (spore photoproduct). The positions of the hydrogens noted by the asterisks are the locations of the glycosylic bond in DNA.

system, unique to both spores and SP, monomerizes SP to two thymines without excision of the lesion and is more error free than is excision repair (235). Spores lacking this SP-specific repair system are 5- to 10-fold more UV sensitive than are wild-type spores. Spores lacking both repair systems are 20- to 40-fold more UV sensitive than are wild-type spores (62, 194). The gene in which mutation abolishes SP-specific repair has been cloned and sequenced (62). The product of this gene, termed spore photoproduct lyase (Spl), shows some sequence homology to DNA-photoreactivation enzymes which cleave TT using light energy. However, SP monomerization does not require light, and Spl lacks at least one residue highly conserved in photoreactivating enzymes that is thought to be involved in transfer of light energy. In contrast to enzymes of excision repair which are present in both growing cells and spores, Spl is synthesized only in the developing forespore with transcription of the *spl* gene directed by $E\sigma^G$ (161).

The major factor causing the altered UV photochemistry of spore DNA is the saturation of spore DNA with α/β-type SASP (196, 201, 202). Spores lacking ~80% of these proteins ($\alpha^-\beta^-$ spores) are more UV sensitive than are vegetative cells (Table 3.2), and UV irradiation of $\alpha^-\beta^-$ spores generates significant amounts of TT and reduced amounts of SP. Generation of TT is the reason for the UV sensitivity of $\alpha^-\beta^-$ spores. The interchangeable role of α/β-type SASP in spore UV resistance has been shown by the restoration of UV resistance to $\alpha^-\beta^-$ spores by synthesis of sufficient levels of either major or minor α/β-type SASP from the same or a different species. Strikingly, UV irradiation of a complex between DNA and any of a number of purified α/β-type SASP generates SP and no TT. The yield of SP as a function of UV fluence in these complexes is ~10-fold lower than the yield of SP in vivo as a function of UV fluence (188). This difference is due to the large DPA depot in spores which acts as a photosensitizer. Binding of α/β-type SASP to DNA also blocks formation of CT and CC, as well as a variety of 6-4 photoproducts (61). The change in the UV photochemistry of DNA upon binding of α/β-type SASP strongly indicates that the DNA in this complex is in an altered structure. Indeed, DNA complexed with α/β-type SASP appears to be in an A-like helix, as has been suggested to be the case in spores (197).

During spore germination, α/β-type SASP are degraded to free amino acids which can then support protein synthesis. SASP degradation is initiated by the sequence-specific protease termed GPR (176). However, SASP degradation takes place more slowly than DPA release. Because of the photosensitizing action of DPA, spores early in germination are actually more UV resistant than are dormant spores due to release of DPA prior to significant SASP degradation (201). However, as SASP degradation proceeds, this elevated UV resistance falls to that of the vegetative cell. In a GPR mutant in which α/β-type SASP degradation is greatly slowed, the elevated UV resistance during spore germination lasts much longer than in wild-type spores (176).

Spore Chemical Resistance

Spores are much more resistant than cells to a variety of chemical compounds, including cross-linking agents such as glutaraldehyde, oxidizing agents (Table 3.2), phenols, chloroform, octanol, ethylene oxide, iodine, and detergents, as well as to pH extremes and lytic enzymes such as lysozyme (17, 173). Resistance of spores to these various agents is acquired at different times in sporulation (Fig. 3.2). For some compounds, spore coats play a role in chemical resistance, possibly by providing an initial barrier against attack. This is clearly true for lytic enzymes, as spores with coats removed by chemical treatment or with coats altered due to mutation can be sensitive to lysozyme degradation of their cortex. The very low permeation of most molecules into the spore core also plays an important role in spore resistance to many chemicals. However, for most chemicals, good data are not available for the temporal correlation between acquisition of spore resistance and other biochemical changes in the developing spore, nor for the correlation of spore chemical resistance with parameters such as core dehydration and cortex size, nor on the mechanism(s) whereby various chemicals cause spore death.

Some detailed information is available about these points for oxidizing agents, in particular hydrogen peroxide. Data on hypochlorite are more limited. Hydrogen peroxide kills cells by several mechanisms, but a major one is the generation of hydroxyl radicals which can cause mutagenic or lethal DNA damage (89). However, spore killing by hydrogen peroxide or hypochlorite is not accompanied by significant DNA damage or mutagenesis (Table 3.2). Consequently, DNA in spores must be extremely well protected against damage by these agents. One reason for the high resistance of spores to these agents may be that spore coats form a protective barrier (17, 145, 204, 231). The spore core may also be relatively impermeable to these hydrophilic compounds (70); the decreased spore core water content also plays a role in spore hydrogen peroxide resistance (168). However, these mechanisms do not specifically protect spore DNA against these agents, as this is accomplished through the saturation of the spore chromosome with α/β-type SASP (187). In contrast to wild-type spores, which are extremely resistant to hydrogen peroxide and hypochlorite, $\alpha^-\beta^-$ spores are much more sensitive to these agents (175, 187) (Table 3.2). Furthermore, while there is no increase in mutation frequency among survivors of wild-type spores killed 90 to 99% by hydrogen peroxide or hypochlorite, $\alpha^-\beta^-$ spores treated similarly have a 5 to 15% frequency of obvious mutations among the survivors (175, 187) (Table 3.2). DNA from $\alpha^-\beta^-$ spores killed by hydrogen peroxide has a high frequency of single-strand breaks, whereas DNA from wild-type spores killed similarly exhibits no such damage (187). These data suggest that, in wild-type spores, the saturation of DNA with α/β-type SASP provides such good protection against the DNA damage caused by oxidizing agents that spore killing by these agents is by other mechanisms. However, in $\alpha^-\beta^-$ spores where much of the spore DNA is no longer covered with α/β-type SASP, the rates of DNA damage caused by such agents are greatly increased, and DNA damage is a significant cause of spore death. In support of this simple model, it has been shown that one component of spore hydrogen peroxide resistance is acquired during sporulation in parallel with accumulation of α/β-type SASP. This component of resistance is not acquired during sporulation of an $\alpha^-\beta^-$ strain (187). In addition, saturation of DNA with α/β-type SASP blocks hydrogen peroxide cleavage of the DNA backbone in vitro (187). However, α/β-type SASP binding to DNA is not the only factor involved in spore resistance to hydrogen peroxide. Indeed, one additional component of resistance is acquired during sporulation at about the time of final spore core dehydration (187). This may reflect a role for spore core dehydration in hydrogen peroxide resistance or, possibly, a mechanism in which the loss of reduced pyridine nucleotides from the developing spore (which also occurs at this time) blocks extensive production of hydroxyl radicals through the Fenton reaction (89, 187). The mechanism(s) of inducible resistance of *B. subtilis* cells to hydrogen peroxide has been studied, and roles for catalases in this process have been demonstrated (18). However, the role of these enzymes in spore hydrogen peroxide resistance is unclear. In *Escherichia coli* a DNA binding protein, Dps, is involved in resistance to hydrogen peroxide (2). Dps exhibits no homology to α/β-type SASP, and the role of such a protein in hydrogen peroxide resistance in *B. subtilis* is unclear.

Spore Heat Resistance

Heat resistance, probably the spore resistance most familiar to food microbiologists, has the most implications for the food industry and is probably the best studied form of resistance in spores. Spore heat resistance is truly remarkable, as spores of many species can withstand 100°C for several minutes. Heat resistance is often quantified as a D_t value, which is the time in minutes at temperature (t) needed to kill 90% of a cell or spore population. Generally, D values for spores at a temperature of t+40°C are approximately equal to those for their vegetative cell counterparts at temperature t. An often overlooked feature of spore heat resistance is that the extended survival of spores at elevated temperatures is paralleled by even longer survival times at lower temperatures. Spore D values increase 4- to 10-fold for each 10°C fall in temperature (69). Consequently, a spore with a D value at 90°C ($D_{90^\circ C}$) of 30 min may have a $D_{20^\circ C}$ value of many years. Indeed, there are several reports of spores surviving over periods of 70 to 100 years and others suggesting survival after millions of years (28, 29, 103). The mechanisms causing spore survival at elevated temperatures are presumed to be the same as those extending spore survival at lower temperatures. However, this assumption has not been rigorously tested.

One deficiency in our understanding of the spore heat resistance mechanism is the identity of the target(s) whose damage results in heat killing of spores. There are good data suggesting that this target is not spore DNA, as spore heat killing is associated with neither DNA damage nor mutagenesis (59) (Table 3.2). There are also data consistent with a protein or proteins being the target of spore heat killing (15). However, the identity of the protein(s) is not clear. It has not been proven that the changes in proteins associated with spore heat killing are the cause rather than the effect of heat killing. Sublethal heat treatment can also damage spores in some way, with this damage being repairable during spore germination and outgrowth (85). Again, the nature of this damage, including what macromolecule is damaged, remains obscure. In contrast to our lack of knowledge about the mechanism(s) of spore heat killing, there is much more information on factors which modulate spore heat resistance. Several of these factors are discussed below.

Sporulation Temperature

Elevated sporulation temperatures increase spore heat resistance (13, 69). Indeed, spores of thermophiles generally have much higher heat resistance than spores of mesophiles. Since spore macromolecules are generally identical to cell macromolecules, spore macromolecules are not intrinsically heat resistant. Presumably, the total macromolecular content of spores from thermophiles is

Table 3.3 Heat resistance of *B. subtilis* spores prepared at different temperatures with different ions and with or without α/α-type SASP[a]

Spore	Prepn temp (°C)	Mineralization	H_2O (g/g of spore core wet wt)	$D_{100°C}$
B. subtilis 4673	50	Native	0.335	45
	37	Ca^{2+}	0.425	37
	37	Native	0.50	8.9
	37	H^+	0.571	2.7
	20	Native	0.55	4.9
B. subtilis 168 wild-type	37	Native	0.37	360[b]
B. subtilis 168 $\alpha^-\beta^-$	37	Native	0.37	15[b]

[a]Data from references 13 and 59.
[b]$D_{85°C}$.

more heat stable than that from mesophiles, accounting for the higher heat resistance of spores from thermophiles. However, spores of the same strain prepared at various temperatures are most heat resistant when prepared at the highest temperature (Table 3.3). This is probably not due to temperature-related changes in total macromolecular composition and may be due to reduced core water content in spores prepared at higher temperatures (13) (Table 3.3). However, the mechanism whereby sporulation temperature affects spore core water content is not known.

In growing bacteria, adaptation to heat stress involves the proteins of the heat shock response. The levels of these proteins would be expected to increase with increasing sporulation temperature. Heat shock proteins could play some role in spore heat resistance (201), but this topic has not been well studied.

α/β-Type SASP

One striking finding about the killing of spores by heat in water is that neither general mutagenesis (Table 3.2) (except possibly in the *gly* region of the *B. subtilis* chromosome [100]) nor DNA damage (201, 202) occurs. This is despite the fact that the elevated temperatures would be expected to cause significant DNA depurination. Therefore spore DNA must be remarkably well protected against heat damage, and the thermal inactivation of spores must be due to mechanisms other than DNA damage. The major cause of spore DNA protection against heat damage appears to be the saturation of spore DNA by α/β-type SASP. Consequently, $\alpha^-\beta^-$ spores of *B. subtilis* have *D* values 5 to 10% of those for wild-type spores (Tables 3.2 and 3.3). In addition, while heat killing of wild-type spores generates <1% obvious mutations in the survivors, heat killing of $\alpha^-\beta^-$ spores produces ~20% obvious mutations among survivors, including auxotrophic, asporogenous, and colony morphology mutations (Table 3.2). Killing of $\alpha^-\beta^-$ spores by heat is also accompanied by a large amount of DNA damage. This DNA damage includes generation of abasic sites which may be the primary heat-induced lesion (189), as well as single-strand breaks which may be generated by secondary cleavage reactions at abasic sites (59). As would be predicted from these latter findings, α/β-type SASP slow DNA depurination in vitro at least 20-fold (59).

Spores treated with dry heat are somewhat different from the spores heated in water discussed above. First, wild-type spores are much more resistant to dry heat than to aqueous heat, as *D* values are 2 to 3 orders of magnitude higher in dry versus hydrated spores (Table 3.2) (34, 241). Second, wild-type spores exhibit a rather high level of mutagenesis (~12% of survivors) upon killing by dry heat (34, 241) (Table 3.2), and this mutagenesis is associated with generation of damage in spore DNA (190). $\alpha^-\beta^-$ spores are much more sensitive to dry heat than are wild-type spores (with survivors of dry heat killing of $\alpha^-\beta^-$ spores also exhibiting a high percentage of mutations) and exhibit heat resistance similar to that of dry vegetative cells (190) (Table 3.2). These latter findings suggest that (i) α/β-type SASP are a major factor increasing the dry heat resistance of wild-type spores over that of vegetative cells; (ii) α/β-type SASP provide significant DNA protection against dry heat damage; but (iii) dry spores are so well protected against mechanisms of heat killing other than DNA damage that at elevated temperatures DNA damage does eventually kill dry spores. In support of these findings, α/β-type SASP slow DNA depurination caused by dry heat in vitro. However, in contrast to the ≥20-fold decrease in DNA depurination caused by α/β-type SASP in solution, these proteins only slow dry heat-induced depurination ~3-fold (190).

Spore Mineralization

Spores accumulate large amounts of divalent cations late in sporulation, approximately in parallel with DPA. DPA itself was thought initially to be involved in spore heat resistance. However, the isolation of DPA^- spores which were heat resistant, and the removal of the majority of a spore's DPA with retention of heat resistance, have refuted this idea (14, 69, 201). DPA may be required for the stability of the dormant state; DPA^- spores are unstable and germinate readily. In contrast to the lack of a role for DPA in spore heat resistance, spore mineralization is implicated in heat resistance. Both the amount and type of mineral ions accumulated affect spore heat resistance (13, 69). The best data come from analyses of spores of several species from which mineral ions have been removed by titration with acid and the spores are then back-titrated with a mineral hydroxide. Analyses of such spores give the order of spore heat resistance with different cations as: $H^+ < Na^+ < K^+ < Mg^{2+} < Mn^{2+} < Ca^{2+}$ < untreated. Despite the clear role for spore mineral ions in spore heat resistance, the complete mechanism of this effect remains unclear. Alteration of spore mineralization can alter spore core water content (Table 3.3), which presumably causes a significant effect on heat resistance. However, mineralization may also affect spore heat resistance independently of effects on core water content (Table 3.3).

Spore Core Water Content

Low core water content is a major factor causing spore heat resistance. The dehydration of the spore begins in the stage III–IV transition and continues throughout stages IV and V, with final dehydration taking place approximately in parallel with acquisition of spore heat resistance (153). Synthesis of the spore cortex is essential both for affecting this dehydration and for maintaining the dehydrated state of the spore core. This is undoubtedly due to the ability of peptidoglycan to change its volume markedly upon changes in ionic strength and/or pH. If an expansion in cortex volume is restricted to one direction, i.e., towards the spore core, core water would be extruded via mechanical action. Although the precise mechanism of this process is unclear, the important role of the cortex in heat resistance is shown by the inverse correlation between spore heat resistance and the volume occupied by the spore cortex relative to that of the core (111) (Fig. 3.7). This correlates not only across species, but also in a single species in which spore cortex biosynthesis has been altered by mutation (167). Presumably, the volume of the spore cortex influences the degree of core dehydration. It is also possible that the amount of cortex and thus its mechanical strength are also crucial to the spore's ability to maintain core dehydration during heat treatment.

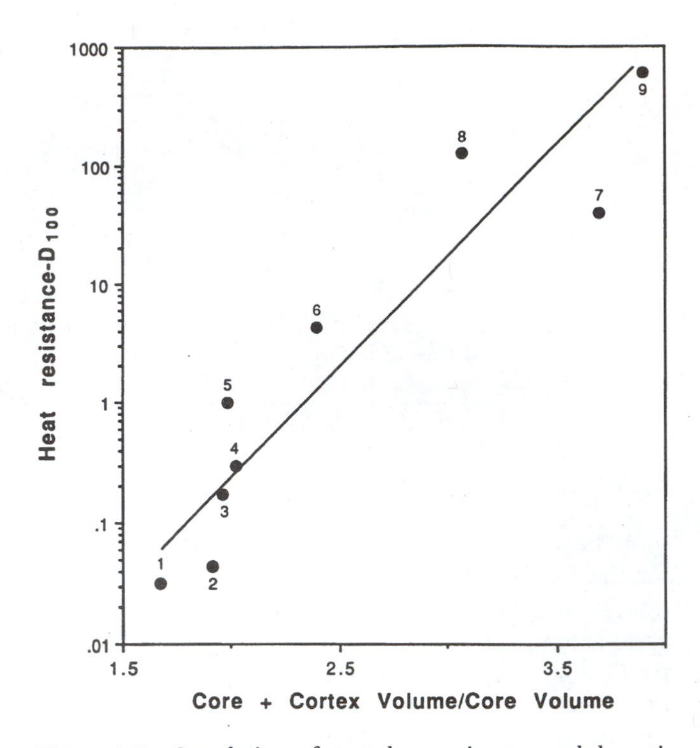

Figure 3.7 Correlation of spore heat resistance and the ratio of the core + cortex volume/core volume. Data have been replotted from the work of Koshikawa et al. (111). The numbers correspond to the following spores: (1) *B. megaterium* exosporiumless variant, coat stripped; (2) *B. megaterium* exosporiumless variant; (3) *B. megaterium*, coat stripped; (4) *B. megaterium;* (5) *B. cereus* T, low calcium; (6) *B. cereus* T, high calcium; (7) *B. subtilis* niger; (8) *B. stearothermophilus* smooth; and (9) *B. stearothermophilus* rough.

Studies of spores from a large number of species have revealed a good correlation between spore core water content and heat resistance over a 20-fold range of *D* values (13, 69) (Fig. 3.8). However, at the extremes of core water contents, *D* values vary widely, presumably reflecting the importance of other factors such as sporulation temperature, cortex structure, etc., in modulating spore heat resistance (69). One value that is unfortunately missing from analyses of spore water content is the amount of core water that is free water. Studies of ion movement in spores have indicated that core ions are immobile, consistent with the presence of little if any free water in the core. However, no good values are available. Presumably, low water content in the spore core causes heat resistance as well as long-term spore survival by slowing water-driven chemical reactions such as DNA depurination, protein deamidation, etc. Low

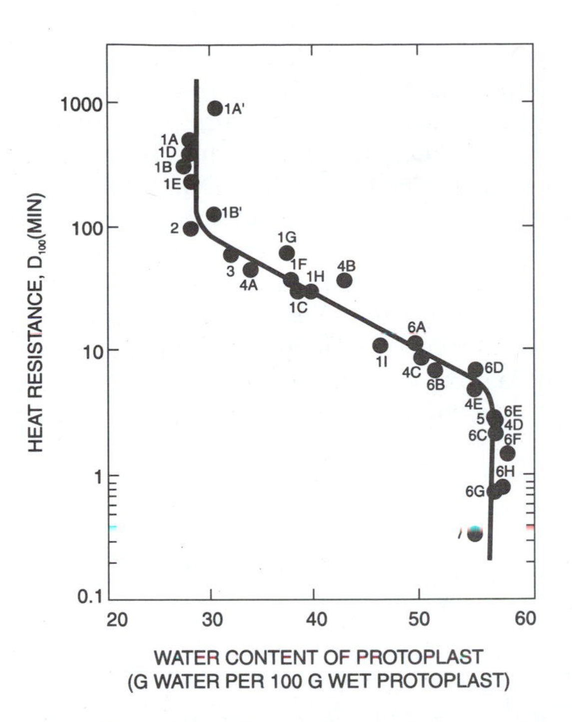

Figure 3.8 Correlation of spore heat resistance and protoplast (core) water content of lysozyme-sensitive sensitive spore types from seven *Bacillus* species which vary in thermal adaptation and mineralization. The figure is from the work of Beaman and Gerhardt (13) with permission. The numbers refer to spores of various species: (1) *B. stearothermophilus;* (2) *"B. caldolyticus";* (3) *B. coagulans;* (4) *B. subtilis;* (5) *B. thuringiensis;* (6) *B. cereus;* and (7) *B. macquariensis.* The letters denote the sporulation temperature or the mineralization of the spores of various species as described in the original publication.

water content also stabilizes macromolecules such as proteins against denaturation by restricting their molecular motion. It would be most informative to know the precise amount of water associated with spore core macromolecules in order to calculate the degree of their stabilization by this process.

SPORE ACTIVATION, GERMINATION, AND OUTGROWTH

Activation

Although spores are metabolically dormant and can remain in this state for many years, if given the proper stimulus they can return to active metabolism within minutes through the process of spore germination (Fig. 3.9). A spore population will often initiate germination more rapidly and completely if activated prior to addition of a germinant (104). However, the requirement for activation varies widely among spores of different species. A number of agents cause spore activation, including low pH and many chemicals, although the most widely used agent is sublethal heat. The precise changes induced by spore activation are not clear, although, in some species, the activation process is reversible. In most species, heat activation releases a small amount of the spore's DPA. However, in *Bacillus stearothermophilus* spores, heat activation releases essentially all of the DPA (14). There is no clear picture of the mechanism of this spore activation.

Germination

The precise period encompassed by spore germination, as distinguished from spore outgrowth, has been given a number of different definitions. For the purposes of this review, we consider spore germination to occur during the first 20 to 30 min following mixing of spores and germinant. During this period, a resistant dormant spore with a cortex and a large pool of DPA, minerals, and SASP is transformed to a sensitive, actively metabolizing germinated spore in which the cortex and SASP have been degraded and DPA and most minerals have been excreted (73, 193). These changes, along with initiation of RNA and protein synthesis, can take place in the complete absence of exogenous nutrients. However, further con-

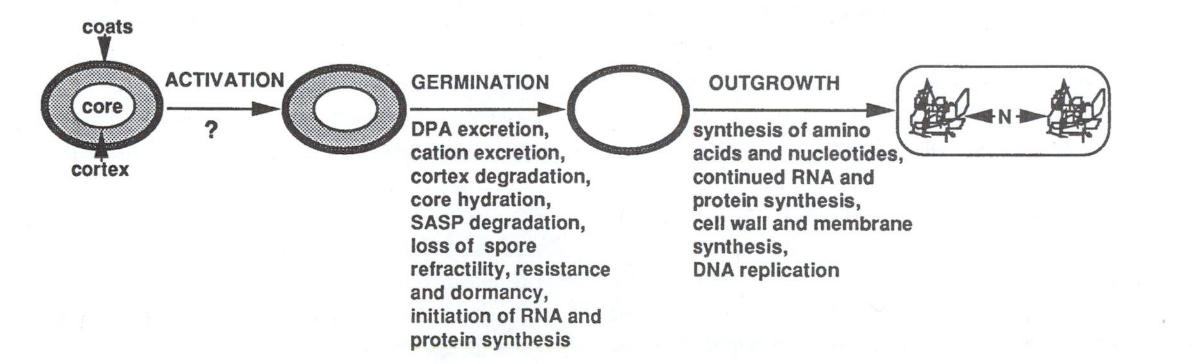

Figure 3.9 Spore activation, germination, and outgrowth. The events in activation are not known, hence the ?. The loss of the spore cortex and the hydration and swelling of the core are shown in the germinated spore. Data are taken from references 73 and 193.

version of this germinated spore into a growing cell via the process of outgrowth requires exogenous nutrients.

The initiation of spore germination in different species can be triggered by a wide variety of compounds, including nucleosides, amino acids, sugars, salts, DPA, and long-chain alkyamines (73), although within a species the requirements are more specific. The precise mechanism whereby these compounds trigger spore germination is not clear. Metabolism of the germinant is not required. Indeed, spores of some species are germinated by inorganic salts (73). Detailed analyses of spores of several *Bacillus megaterium* strains provide strong evidence that triggering of spore germination does not require metabolism of endogenous compounds (51, 179, 180). However, the stereospecificity exhibited by germinants (e.g., L-alanine is a germinant, whereas D-alanine often inhibits germination) strongly suggests that at least some germinants interact directly with a specific protein (73). The precise identity of such proteins is not clear. However, analysis of *B. subtilis* mutants whose spores are defective in their response to specific germinants has provided some suggestive information. These mutations define a class of genes called *ger*, which are defective in initiating spore germination in response to one or more, but not all, germinants (149). The *ger* genes are expressed in the developing forespore during sporulation. Many have been cloned and sequenced (196). Notably, the *gerA* locus, in which mutations prevent spore germination with alanine, encodes three proteins, each of which is probably a membrane protein (149, 243). The precise location in the spore of the *gerA* gene products has yet to be determined. They may form a complex in the inner forespore membrane which acts as a germinant receptor; upon the binding of a germinant this receptor somehow initiates a cascade of reactions leading to initiation of germination (149, 243). Unfortunately, this is only a model. Two other *ger* loci (*gerB* and *gerK*), in which mutations alter the response to other germinants, encode proteins rather similar to those of the *gerA* operon (38, 93). One interesting problem faced by a putative germinant receptor in the inner forespore membrane is that at least part of the protein would have to be outside the membrane. How is this protein protected against heat, given its location outside the inner forespore membrane where the water content is similar to that of growing cells?

In the first minutes of spore germination, the earliest events are release of protons and some divalent cations, in particular Zn^{2+} (99, 223). Release of DPA, loss of spore refractility, and cortex degradation follow (73, 193). In this process, the spore excretes up to 30% of its dry weight and the core increases its water content to that of the vegetative cell (73). These events clearly require large changes in the permeability of the inner forespore membrane. However, neither the nature nor the mechanism of these changes is understood. The time for the changes accompanying the initiation of spore germination may be extremely fast; an individual spore can lose refractility in as little as 30 s to 2 min. However, for a spore population the time can be much longer, as individual spores initiate germination after widely different lag times. Indeed, some spores in a population appear superdormant and may require special conditions or treatment to induce their germination (73). One concern about precise determination of the timing of various events during spore germination is that some of the methods used may overestimate the time for a particular event. In particular, degradation of spore cortex has often been measured by appearance of cortical fragments in the germination supernatant fluid, a measurement that may greatly underestimate the true rate of cortex degradation. Whatever the precise order of these events, initiation of cortex degradation is a crucial event. It relieves the pressure on the spore core, allowing influx of water and a rapid two- to threefold increase in spore core volume. This increase in water content restores the activity of enzymes in the spore core.

The importance of cortex lysis in spore germination has focused attention on a lytic (lysozyme-like) enzyme as a key player in spore germination. Several lytic enzymes have been isolated from spores. One enzyme purified by Johnstone and coworkers from *B. megaterium* spores appears to be a prime candidate for a germination lytic enzyme (66–68, 98). A similar enzyme has also been purified from *Bacillus cereus* spores (144). In the dormant spore, this enzyme appears to be located in the spore cortex in an inactive or zymogen form. Upon germinant addition, the zymogen is proteolytically activated and can degrade the spore cortex. The inhibition of spore germination initiation by protease inhibitors suggests the involvement of a protease in the early events of spore germination (19). While the cortex-lytic enzyme remains an attractive candidate for action in the spore germination cascade, it is somewhat surprising that no protease or lytic enzyme has been identified by analysis of *ger* genes. While there are a number of possible explanations for this negative finding, there may be multiple independent pathways for initiating cortex lysis during spore germination, so that no single mutation will have a large effect. It is to be hoped that molecular genetic analyses of the gene encoding cortex lytic enzymes will provide a definitive answer.

Following the initial events in spore germination, a number of enzymatic reactions begin in the spore core (193). These include utilization of the spore's depot of 3PGA to generate ATP and NADH, degradation of SASP initiated by GPR action and completed by peptidases, catabolism of much of the amino acids produced by SASP degradation for production of high-energy phosphate and reduced pyridine nucleotides, and initiation of catabolism of exogenous compounds. The spore lacks a number of enzymes of the citric acid cycle, so most metabolism in germination and early outgrowth uses glycolysis and/or the hexose monophophate shunt. In *B. megaterium*, endogenous energy reserves completely support ATP production during the first 10 to 15 min of spore germination, but exogenous catabolites are needed subsequently.

RNA synthesis is initiated in the first few minutes of germination, using nucleotides stored in the spore or generated by breakdown of preexisting spore RNA. The identity of the first RNAs made during germination is not clear, but they are probably mRNA. The precise structure of RNA polymerase acting during germination and early outgrowth is also not clear. There are fewer detailed analyses of RNA polymerase from germinated spores than from sporulating cells. Protein synthesis begins shortly after RNA synthesis. Again, the identity of the first proteins made is not clear. All the components of the protein-synthesizing machinery stored in the dormant spore appear functional, with the exception of some tRNA lacking the 3′-terminal adenosine residue. However, this tRNA is rapidly repaired by tRNA nucleotidyltransferase in the first minutes of spore germination, when aminoacyl-tRNA is also generated. As was the case with RNA synthesis early in germination, endogenous amino acids derived largely from α/β-type SASP breakdown support most protein synthesis in the first 20 to 30 min of germination.

Outgrowth

Although the transition from spore germination to spore outgrowth is not absolutely distinct, we will consider this as the time from ~25 min after the initiation of spore germination until the first cell division. Note that most processes during germination are supported from endogenous reserves, while exogenous sources are required to support outgrowth. Thus, outgrowth requires exogenous nutrients and can take as little as 90 min in a rich medium (193). Exogenous reserves of carbon, nitrogen, etc., are required to support spore outgrowth. During outgrowth, the spore regains the ability to synthesize amino acids, nucleotides, and other small molecules due to synthesis of biosynthetic enzymes at defined times. However, the factors regulating gene expression during spore outgrowth are not well characterized.

DNA replication is not generally initiated until at least 60 min after the start of germination. However, the mechanisms controlling initiation of chromosomal replication during outgrowth have not been studied in detail. DNA repair can occur well before DNA replicative synthesis, and even in the first minutes of germination, spores contain deoxynucleoside triphosphates. However, the importance of DNA repair in the first minutes of germination, other than of UV damage, has not been well studied. During spore outgrowth, the volume of the outgrowing spore continues to increase, requiring the synthesis of membrane and cell wall components.

One question that remains intriguing is whether there are genes which are needed for outgrowth but no other stage of growth. A number of mutations affecting outgrowth (termed *out* mutants) have been isolated in *B. subtilis* (155, 182). Some of the genes targeted by these mutations have been cloned, sequenced, and analyzed. While the functions of some of these genes have been established, to date no outgrowth-specific gene has been definitely identified.

PRACTICAL PROBLEMS OF SPORES IN THE FOOD INDUSTRY

Spore-forming bacteria pose specific problems for the food industry. Three species of sporeformers, *Clostridium botulinum*, *Clostridium perfringens*, and *B. cereus*, are well known to produce toxins that can cause illness in humans and animals (41), and many species of sporeformers cause spoilage of foods (90). Sporeformers causing foodborne illness and spoilage are particularly important in low-acid foods packaged in cans, bottles, pouches, or other hermetically sealed containers ("canned" foods), which are processed by heat. Psychrophilic sporeformers have also been increasingly recognized to cause spoilage of refrigerated foods.

Food spoilage by sporeforming bacteria was discovered by Pasteur during investigations of butyric acid fermentation in wines (24, 238). Pasteur was able to isolate an organism he termed *Vibrion butyrique*, which is now referred to as *Clostridium butyricum*, the type species of the genus *Clostridium*. Endospores were discovered independently by Ferdinand Cohn and Robert Koch in 1876, soon after Pasteur had made microbiology famous (105). Investigations by Pasteur and Koch led to the association of microbial activity with the safety and quality of foods. During investigation of the anthrax bacillus, the famous Koch's postulates were born; these proved that a disease is caused by a specific microorganism and also led to the development of pure culture techniques. Diseases and spoilage problems caused by sporeformers have traditionally been associated with foods that are thermally processed, since heat selects for survival and subsequent growth of spore-forming organisms.

The process of appertizing, or preserving food sterilized by heat in a hermetically sealed container, was created in the late 1700s by Appert (5), who believed that the elimination of air was responsible for the long shelf lives of thermally processed foods. The empirical use of thermal processing gradually developed into modern-day thermal processing industries. In the late 1800s and early 1900s several scientists in the United States were instrumental in developing scientific principles to assure the safety and to prevent spoilage of thermally processed foods (72). As a result of these studies, thermal processing of foods in hermetically sealed containers became an important industry in Europe and the United States (170, 174). Prescott and Underwood at the Massachusetts Institute of Technology and Russell at the University of Wisconsin found that endospore-forming bacilli caused the spoilage of thermally processed clams, lobsters, and corn (170, 174). The classic study of Esty and Meyer (58) in California provided definitive values of the heat resistance of *C. botulinum* type A and B spores and helped rescue the United States canning industy from its near demise from botulism in commercially canned olives and other foods. Quantitative thermal processes, understanding of spore heat resistance, aseptic processing, and implementation of the HACCP (Hazard Analysis-Critical Control Point) concept also resulted from developments in the canned food industry.

Low-Acid Canned Foods

The U.S. Food and Drug Administration (FDA) and the U.S. Department of Agriculture (USDA) Food Safety Inspection Service define a low-acid canned food as one with a finished equilibrium pH greater than 4.6 and a water activity (a_w) greater than 0.85. The regulations regarding thermal processing of canned foods are described in the U.S. *Code of Federal Regulations.* USDA has regulatory oversight of products that contain at least 3% raw red meat or 2% cooked poultry, and the FDA supervises other products. A process must be filed with the proper governmental agency prior to commercial processing.

Low-acid foods are packaged in hermetically sealed containers, often cans or glass jars but also plastic pouches and other types of containers. These "cans" (as collectively defined in this chapter) containing low-acid food products must be processed by heat to achieve commercial sterility, which is a condition achieved by application of heat that inactivates microorganisms of public health significance, as well as any microorganisms of non-health significance capable of reproducing in the food under normal nonrefrigerated conditions of storage and distribution. Commercial sterility is an empirical term to indicate a low level of microbial survival and provision of shelf stability (164). Acidification or preservation procedures such as using lower water activity can be used to attain commercial sterility. These procedures are often combined with a mild heat treatment.

In canned low-acid foods, the primary goal is to inactivate spores of *C. botulinum* since this organism has the highest heat resistance of microbial pathogens. The degree of heat treatment applied varies considerably according to the class of food, its spore content, pH, storage conditions, and and other factors. For example, canned low-acid vegetables and uncured meats receive a 12*D* process (see below) or "botulinum cook." Lesser heat treatments are applied to shelf-stable canned cured meats as well as to foods with reduced water activity or other antimicrobial factors which inhibit growth of spore-forming bacteria. In practice, the spore content of ingredients and the cleanliness of the cannery environment are of major importance for successful heat treatment of low-acid canned foods. Certain foods and food ingredients such as mushrooms, potatoes, spices, sugars, and starches may contain high levels of *C. botulinum* and spores of other microbial species.

Pflug (163–165) refined the semilogarithmic microbial destruction model and designed a strategy to achieve the required heat process F_T value. He also introduced the concept of the probability of a nonsterile unit (PNSU) on a one-container basis. This reasoning logically explains the 12*D* term, which designates the time required in a thermal process for a 12-log reduction of *C. botulinum* spores. In thermal processing two values in particular describe the thermal inactivation of an organism: the *D* value is the time required for a 1-log reduction of a microorganism, and the *z* value is the temperature change required to change the *D* value by a factor of 10. While the *D* value represents the resistance of an organism to a specific temperature, *z* represents the relative resistance of an organism to inactivation at different temperatures (138). The thermal process for the 10^{-11} to 10^{-13} level of probability of a botulism incident

Table 3.4 Heat resistance of sporeformers of importance in foods[a]

Organism	Approx. *D* value[b] at temp:						
	80°C	85°C	90°C	95°C	100°C	110°C	120°C
Spores of public health significance							
Group I *Clostridium botulinum* types A and B			50		7–30	1–3	0.1–0.2
Group II *C. botulinum* type B		1–30	0.1–3	0.03–2			
C. botulinum type E	0.3–3				0.01		
Bacillus cereus					3–200	0.03–2.4	
Clostridium perfringens			3–145		0.3–18	2.3–5.2	
Mesophilic aerobes							
Bacillus subtilis					7–70	6.9	0.5
Bacillus licheniformis					13.5	0.5	
Bacillus megaterium					1		
Bacillus polymyxa			4–5		0.1–0.5		
Bacillus thermoacidurans			11–30		2–3		
Thermophilic aerobes							
Bacillus stearothermophilus					100–1,600		1–6
Bacillus coagulans					20–300		2–3
Mesophilic anaerobes							
Clostridium butyricum	4–5	0.4–0.8					
Clostridium sporogenes					80–100	21	0.1–1.5
Clostridium tyrobutyricum	13						
Thermophilic anaerobes							
Desulfotomaculum nigrificans					<480		2–3
Clostridium thermosaccharolyticum					400		3–4

[a]Sources: references 154 and 173.
[b]At pH ca. 7 and $a_w > 0.95$.

occurring will depend on the initial number of *C. botulinum* spores present in a container. This value can be quite high, for example 10^4 for a container of mushrooms, or very low, such as 10^{-1} for a container of meat product (76). Pflug (163–165) emphasized that the thermal processing industry should prioritize process design to protect against (i) public health hazard (botulism) from *C. botulinum* spores, (ii) spoilage from mesophilic spore-forming organisms, and (iii) spoilage from thermophilic organisms in containers stored in warm climates or environments. Generally, low-acid foods in hermetically sealed containers are heated to achieve 3 to 6 min at a temperature of 121°C (250°F) or equivalent at the center or most heat-impermeable region of a food. This assures the inactivation of the most heat-resistant *C. botulinum* spores with a $D_{121°C(250°F)}$ = 0.21 min and a *z* of 10°C (18°F). Economic spoilage is also avoided by achieving ~5*D* killing of mesophilic spores that typically have a $D_{121°C}$ of ~1 min (46). Foods that are distributed in warm climates of tropical or desert areas require a particularly severe thermal treatment of ~20 min at 121°C to achieve a 5*D* killing of *Clostridium thermosaccharolyticum*, *B. stearothermophilus*, and *Desulfotomaculum nigrificans* since these organisms have a $D_{121°C}$ of ~3 to 4 min (Table 3.4). Such severe treatment can have a detrimental impact on nutrient content and organoleptic qualities, but it assures a shelf-stable food.

Heat treatments are also applied to aseptically processed low-acid canned foods in which commercially sterilized cooled product is filled into presterilized containers followed by aseptic hermetic sealing with a closure in a sterile environment. This technology was initially used for commercial sterilization of milk and creams in the 1950s and then encompassed other food products such as soups, eggnog, cheese spreads, sour cream dips, puddings, and high-acid products such as fruit and vegetable drinks (46). Aseptic processing and packaging systems have the potential to reduce energy and packaging and distribution costs. Developments in aseptic packaging have also renewed interest in thermal processing and in the mechanisms of spore, cell, and enzyme inactivation (46).

Table 3.5 Growth requirements of sporeformers of public health significance

Organism	Minimum pH	Inhibitory NaCl concn (%)	Minimum a_w	Temp range for growth (°C)
Group I *C. botulinum*	4.6	10	0.94	10–50
Group II *C. botulinum*	5.0	5	0.97	3.3–45
B. cereus	4.35–4.9	~10	0.91–0.95	5–50
C. perfringens	5.0	~7	0.95–0.97	15–50

Bacteriology of Sporeformers of Public Health Significance

Three species of sporeformers, *C. botulinum, C. perfringens,* and *B. cereus,* are well known to cause foodborne illness (Table 3.5). Certain other species of *Bacillus* such as *B. licheniformis, B. subtilis,* and *B. pumilus* have also been reported to sporadically cause foodborne disease (112). Devastating incidences of intestinal anthrax caused by ingestion of contaminated raw or poorly cooked meat have been reported (213).

The principal microbial hazard in heat-processed foods and in minimally processed refrigerated foods is *C. botulinum.* The genus *Clostridium* consists of gram-positive, anaerobic, spore-forming bacilli that obtain energy by fermentation (31, 82). The species *C. botulinum* is a heterogeneous collection of strains that differ widely in genetic relatedness and phenotypic properties but which all have the property of producing a characteristic neurotoxin of extraordinary potency (220). *C. botulinum* and other pathogenic bacteria produce spores that swell the mother sporangium, giving a "tennis racket" or club-shaped appearance (Fig. 3.10). Spores of *C. botulinum* types B and E frequently possess an exosporium, and type E characteristically produces appendages (Fig. 3.11). *C. botulinum* is commonly divided into four physiological groups (I through IV) on the basis of phenotypic properties (207). Group I (strongly proteolytic strains producing neurotoxin types A, B, and F) and group II (nonproteolytic strains producing neurotoxin serotypes B, E, and F) are the two groups of concern in food safety. Strains in groups I and II have certain properties that affect their ability to grow in foods (140, 207). The spores of group II have considerably less heat resistance than group I spores, but they can grow and produce toxin at refrigerator temperatures. The ability of *C. botulinum* to grow at low temperatures has generated considerable concern that refrigerated food products could lead to botulism outbreaks. Additional inhibitory conditions would be valuable in assuring the safety of refrigerated products from toxin production by *C. botulinum.*

C. perfringens is widespread in soils and is a normal resident of the intestinal tracts of humans and certain animals. *C. perfringens* can grow extremely rapidly in cooked foods and produce an enterotoxin that causes diarrheal disease (118). It also produces a variety of other extracellular toxins and degradative enzymes, but

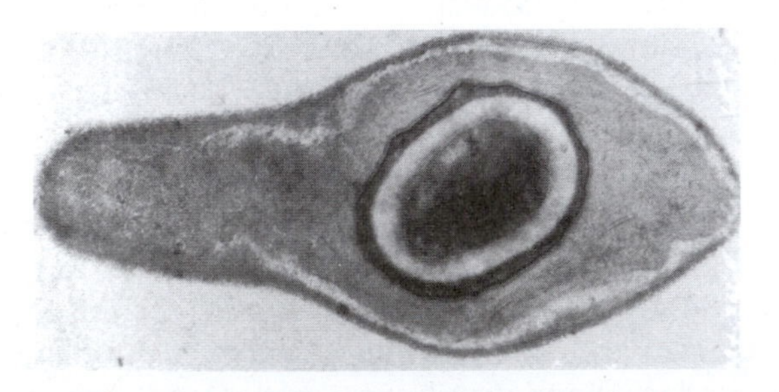

Figure 3.10 Transmission electron micrograph (×50,000) of a longitudinal section through a spore and sporangium of *C. botulinum* type A, showing the characteristic club-shaped morphology. (From reference 212a.)

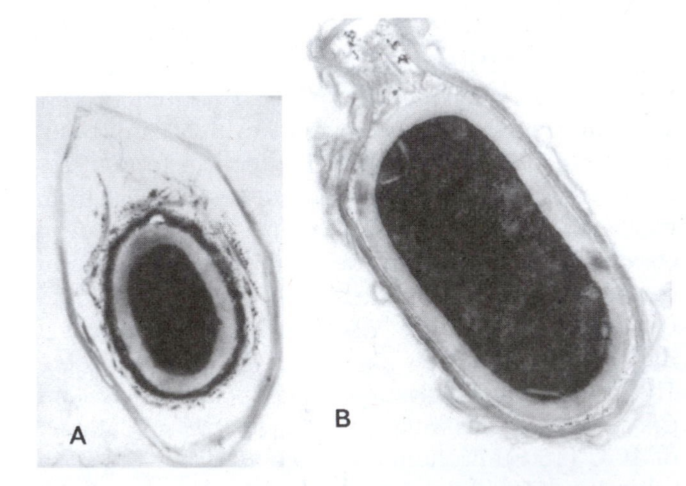

Figure 3.11 Electron micrographs of *C. botulinum* type B (A) and E (B) showing characteristic exosporium in types B and E and appendages in type E. Micrographs courtesy of Philipp Gerhardt from spores produced in E.A.J.'s laboratory.

these are mainly of significance in gas gangrene and diseases in animals (211). *C. perfringens* differs from many other clostridia in being nonmotile, reducing nitrate to nitrite, and carrying out a stormy fermentation of lactose in milk. Contributing to the ability of *C. perfringens* to cause foodborne illness is its ubiquitous distribution in foods and food environments, formation of resistant endospores that survive cooking of foods, an extremely rapid growth rate in warm foods (6 to 9 min at 43 to 45°C), and synthesis of enterotoxins during sporulation in the human intestine.

C. perfringens is the cause of a relatively common type of food poisoning in the United States (39) and in several other countries where surveillance has been conducted. Many if not most foods contain spores of *C. perfringens*. In addition to vegetables and fruits that acquire spores from soil, foods of animal origin are contaminated during slaughtering with spores in the environment or spores residing in the intestinal tract. Dried foods such as spices are a common source of *C. perfringens* and other sporeformers (147). Sporulation of vegetative cells is often difficult to obtain in many laboratory media. When spores do occur they are large, oval, and centrally or subterminally located and swell the cells. The optimum temperature for growth of vegetative cells is about 43 to 45°C (109 to 113°F), and in rich media or in certain foods at the optimum temperature for growth, doubling times as short as 6 to 9 min have been observed (119). Germination of spores and growth of cells will take place at up to 50°C (ca. 122°F). *C. perfringens* does not generally grow below 20°C (68°F), and true psychrophilic strains of the organism have not been isolated. *C. perfringens* will grow over the pH range 5.5 to 8.0 or 8.5, and the optimum is ca. 6.5. At temperatures below 45°C (113°F) certain strains will grow at pH 5. Spores of *C. perfringens* were reported to germinate and outgrow in 5 to 8% NaCl (118), but most strains are inhibited by 5 to 6.5% salt (Table 3.5).

The heat resistance of *C. perfringens* spores varies considerably among strains. In general, two classes of heat sensitivity are common. Heat-resistant spores have $D_{90°C}$ ($D_{194°F}$) values of 15 to 145 min and z values of 9 to 16°C (16 to 29°F), compared to heat-sensitive spores which have $D_{90°C}$ of 3 to 5 min and z of 6 to 8°C (11 to 14°F). The spores of the heat-resistant class generally require a heat shock of 75 to 100°C (167 to 212°F) for 5 to 20 min in order to germinate. The basis of the wide variation in heat resistance is not currently understood. The spores of both classes may survive cooking of foods and may be stimulated by heat shock for germination during the heating procedures.

Food poisoning by *C. perfringens* nearly always involves temperature abuse of a cooked food, and the great majority of food poisonings caused by *C. perfringens* could be avoided if cooked foods were eaten immediately after cooking or rapidly chilled and reheated before consumption to inactivate vegetative cells. The objective in prevention of food poisonings is to limit the multiplication of vegetative cells in the food. Since the spores are widespread and are resistant to heat they will often survive the cooking procedure and subsequently grow if permissive temperatures are allowed. It is not generally practical to eliminate spore contamination of the food, although monitoring spore levels is a good practice in the quality control of raw food ingredients. Hot foods should be held at 60°C (140°F) or higher. Foods that must be stored should be cooled to <7°C (45°F) as rapidly as possible and reheated to 71 to 100°C (160 to 212°F) to kill vegetative cells before consumption.

The genus *Bacillus* comprises rod-shaped, gram-positive bacteria that aerobically form refractile endospores. The spores of the bacilli are more resistant than vegetative cells to heat, drying, food preservatives, and other environmental challenges. Strains of *Bacillus* produce catalase, which in addition to the aerobic production of spores distinguishes *Bacillus* from *Clostridium*. Of the 34 species currently named in the genus, only two species, *B. anthracis* and *B. cereus*, are recognized as common pathogens. *B. cereus* can produce a heat-labile enterotoxin causing diarrheal illness and a heat-stable toxin giving an emetic response in humans (112). Generally the organism must grow to very high numbers (>10^6/g of food) to cause human illness. *B. cereus* is closely related to *B. megaterium*, *Bacillus thuringiensis*, and *B. anthracis*, but *B. cereus* can be distinguished from these species by biochemical tests and the absence of toxin crystals.

B. cereus spores occur widely in foods and are commonly found in milk, cereals, starches, herbs, spices, and other dried foodstuffs. They are also frequently found on the surfaces of meats and poultry, probably because of soil or dust contamination. Investigators in Sweden reported isolation of *B. cereus* from of 47.8% of 3,888 different food samples. In the U.K., *B. cereus* was isolated from 98/108 (91%) of rice samples. The organism causes spoilage of raw and pasteurized milk (33), and foods containing dried milks, such as infant formulas, may possess fairly high levels of spores or cells.

The organism grows over the temperature range of approximately 10 to 48°C (50 to 118.4°F), with an optimum of 28 to 35°C (82.4 to 95°F) (Table 3.5). Psychrophilic strains that produce enterotoxin in milk have been isolated (33, 229). The doubling time at the optimum temperature in a rich medium is 18 to 27 min. Several strains can grow slowly in sodium chloride concentrations of 7.5%. The minimum water activity for growth is 0.95. The organism grows over a pH range of approximately 4.9 to 9.3, but these environmental limits for growth are dependent on water activity, temperature, and other interrelated parameters.

Spores of *B. cereus* are ellipsoidal and central to subterminal and do not distend the sporangium. Spore germination can occur over the temperature range of 8 to 30°C (46.4 to 86°F). Spores from strains associated with food poisoning had a heat resistance of $D_{95°C}$ (203°F) of ~24 min. Other strains were shown to have a wider range of heat resistances. It has been suggested that the strains involved in food poisoning have higher heat resistances and therefore will be more apt to survive cooking. Spores are hydrophobic and attach to food contact surfaces (86).

Since *B. cereus* is widespread in nature and survives extended storage in dried food products, it is not practical to eliminate low numbers of spores from foods. Control against food poisoning should be directed at preventing germination of spores and preventing multiplication of large populations of the bacterium. Cooked foods should rapidly and efficiently be cooled to less than 7°C (45°F), or maintained above 60°C (140°F), and should be thoroughly reheated before serving.

Heat Resistance of *C. botulinum* Spores

Group I *C. botulinum* type A and B strains can produce spores of remarkable heat resistance and are the most important sporeformers in public health safety of thermally processed foods. The classic investigation on their heat resistance was carried out by Esty and Meyer (58) in California as a result of commercial outbreaks of botulism in canned olives and certain other canned vegetables. They examined 109 type A and B strains at five heating temperatures over the range 100 to 120°C (212 to 248°F). They found that the inactivation rate was logarithmic between 100 and 120°C, and the inactivation rate depended on the spore concentration, the pH, and the heating menstruum. Esty and Meyer demonstrated that 0.15 M phosphate buffer (Sorensen's buffer), pH 7.0, gave the most consistent heat-inactivation results, and their use of a standardized system enables comparisons of heat resistance by researchers today. The use of a reproducible system is valuable in periodically determining the heat resistance of new spore crops (94). The data of Esty and Meyer can be extrapolated to give a maximum value of $D_{121.1°C} = 0.21$ min for *C. botulinum* type A and B spores in phosphate buffer (94, 226). The thermal processing industries have used $D_{121°C}$ as a standard in calculating process requirements. Proteolytic type F *C. botulinum* spores had a heat resistance of $D_{98.9°C} = 12.2–23.2$ min and $D_{110°C} = 1.45–1.82$ min (140), which is much lower than that of type A spores. Spores of nonproteolytic type B and E *C. botulinum* have much lower heat resistance than proteolytic A and B strains. Ito et al. (94) reported that type E spores have a $D_{70°C(158°F)}$ varying from 29 to 33 min and a $D_{80°C(176°F)}$ from 0.3 to 2 min depending on the strains. The *z* value ranged from 13 to 15°F. These values are comparable to those of Ohye and Scott (159), who obtained $D_{80°C}$ values of 3.3 and 0.4 min for two type E strains. Spores of nonproteolytic *C. botulinum* type B can have heat resistance considerably higher than that of type E. Scott and Bernard (181) showed that the $D_{82.2°C}$ of nonproteolytic type B strains ranged from 1.5 to 32.3 min as compared with $D_{82.2°C} = 0.33$ min for a type E strain. Media containing lysozyme can significantly enhance recovery of group II *C. botulinum* spores since lysozyme substitutes for spore lytic enzymes that are inactivated by heat (137). *D* values at 85 and 95°C were 100 and 4.4 min for strain 17B and 45.6 and 2.8 min for strain Beluga E on medium plus lysozyme. The thermal resistance of *C. botulinum* spores is strongly dependent on environmental and recovery conditions. Heat resistance is markedly affected by acidity (58). Esty and Meyer found that spores had maximum resistance at pH 6.3 and 6.9, and resistance decreased markedly at pH values below 5 or above 9. The level of sodium chloride (58) or sucrose (218) and decreased a_w increased the heat resistance of *C. botulinum* spores. Sugiyama (219) found that spores grown in media containing fatty acids increased their heat resistance. *C. botulinum* spores coated in oil were more resistant to heat (121). In common with *Bacillus* spp., sporulation of *C. botulinum* at higher temperatures results in spore crops with greater heat resistance, possibly by acquired thermotolerance through the formation of heat shock proteins (227).

Little is known of the compositional factors contributing to heat resistance of *C. botulinum* spores. The metal composition of purified group I *C. botulinum* spores is different from spores of *Bacillus* (107). The minerals required for sporulation and mechanisms of heat resistance of *C. botulinum* and *Bacillus* spp. are probably not the same (106, 107). Unlike *Bacillus* spp., which require manganese for sporulation, type B *C. botulinum* sporulation was enhanced by zinc and inhibited by copper (106). During sporulation, *C. botulinum* accumulated relatively high concentrations of transition metals, particularly zinc (~1% of cell dry weight) and iron and copper (0.05 to 1%). Spores containing increased contents of iron or copper were more rapidly inactivated by heat than were native spores or spores containing increased manganese or zinc (107). In the anaerobic growth environment of clostridia, transition metals would be expected to have important roles in the sporulation and resistance properties of spores. Metals that undergo redox changes, including copper, iron, and manganese, tend to precipitate as the hydroxides or oxides in aerobic environments, but they are more soluble and biologically available at low redox potentials. The mechanisms by which iron and copper accelerate heat inactivation and zinc and manganese protect *C. botulinum* spores against thermal energy have not been elucidated. Iron and copper are redox-active transition metals and may catalyze hydrolytic reactions (237) and may also spontaneously react with oxygen, generating toxic oxygen species that inflict mutations in DNA (131). Manganese and zinc can associate with nucleic acids, and Mn has been demonstrated to provide protection against heat denaturation (228) and also to act as an effective scavenger of free radicals in biological systems (6). Studies have also indicated that heat inactivation of *C. botulinum* spores is accelerated in modified gas atmospheres (108).

C. botulinum spores of groups I and II are highly resistant to irradiation compared with vegetative cells of most microorganisms, and it is not practical to inactivate them in foods by irradiation. *C. botulinum* spores have a $D = 0.1$ to 0.45 Mrad (2.0 to 4.5 kGy). Irradiation resistance depends on *C. botulinum* type; proteolytic types of A, B, and F appear to be most resistant (109). *C. botulinum* spores are also highly resistant to ethylene oxide, but are inactivated by halogen sanitizers and by hydrogen peroxide (109). Hydrogen peroxide is commonly used for sanitizing surfaces in aseptic packaging, and halogen sanitizers are used in cannery cooling waters.

Incidence of Foodborne Illness Caused by *C. botulinum*

The epidemiology of botulism has been thoroughly reviewed (81, 97, 172) and only aspects pertaining to spore survival and outgrowth are presented in this chapter. Fortunately, the incidence of botulism in commercial foods is very low. It has been estimated that about 30 billion cans, bottles, and pouches of low-acid foods are consumed in the United States each year. Since 1940, when heat-processing principles were firmly established, through 1975, less than 10 botulism outbreaks and fewer than four deaths were caused by inadequately commercially canned foods in the the United States (139, 156). From 1971 through 1982, however, botulinum toxin was detected in several commercial canned foods such as mushrooms, salmon, soups, peppers, tuna fish, beef stew, and tomatoes. Survival of spores and toxin production during this period were caused mainly by underprocessing or by container leakage following processing. The detection of botulinal toxin in canned mushrooms and in canned salmon in the 1970s and 1980s prompted United States regulatory agencies to recommend that chlorine or sanitizers be used in cooling water. Recently botulism has been transmitted in several commercial low-acid foods including chopped garlic in oil, cheese sauce, bean dip, and clam chowder (80, 97, 172). These incidences resulted from temperature abuse of products labeled "keep refrigerated" and the absence of inhibitory conditions other than temperature. Following an outbreak of botulism in the United States transmitted in garlic in oil, the FDA mandated that such products be acidified (150).

Botulism in commercial foods can have enormous medical and economic consequences. Outbreaks of botulism in commercial foods have been estimated to cost ca. $30 million per human case, which is a much higher cost estimate than for other foodborne illnesses such as *Salmonella* (ca. $10,000) and *Listeria monocytogenes* (ca. $12,000) (39). Outbreaks of botulism also generate considerable media publicity that negatively affects the food industry. Although commercial botulism has been quite rare due to excellent control during thermal processing, botulism in home-canned and improperly fermented products occurs relatively frequently throughout the world (96). The heat resistance of *C. botulinum* spores is often not appreciated by home canners, and 20 to 30 cases occur each year in the United States with a current case fatality rate of about 10%. Botulism is more common in certain countries such as Poland and China, where improper home canning of meats and poor fermentation of soybean curd occur relatively frequently (96). In recent years, the United States and certain other countries have seen a resurgence of botulism in restaurant-prepared foods, most often caused by poor temperature control of the prepared foods (97, 172).

Inadvertent temperature abuse of foods has resulted in botulism outbreaks. Botulism has occurred from potatoes that were wrapped in foil, baked, and then held at room temperature until they were used for preparing salads (4, 183). During cooking, vegetative organisms are killed but the spores of *C. botulinum* survive and grow in the anaerobic environment created by wrapping the potatoes in foil (221). In April 1994, an outbreak of botulism in Texas affected 23 individuals, 17 of whom were hospitalized (4). This was the largest botulism outbreak in the United States since 1983. The food vehicle was a potato-based dip (skordalia), which was prepared using foil-wrapped baked potatoes that were left at room temperature after baking. These examples illustrate ways in which changes in food processing, elimination of antimicrobials, and relying solely on refrigeration can result in incidences of botulism. Changes in formulation, processing, or packaging can lead to botulism. To assure the botulinogenic safety of a food with potential of supporting *C. botulinum* growth, it is recomended that laboratory challenge tests be perfomed. Guidelines have been recommended for such challenge studies (53, 102, 157).

HACCP and Prevention of Foodborne Disease by Sporeformers

The safety of thermally processed low-acid foods is enhanced by application of a Hazard Analysis and Critical Control Point (HACCP) program. HACCP entails a systematic and quantitative risk assessment program to assure the safety of foods. The HACCP system received attention in the late 1950s when the Pillsbury Company was consulted by the United States government to produce foods that could be eaten by personnel under zero-gravity conditions in space capsules. The most challenging aspect of this project was to come as close as possible to 100% assurance that the foods would not be contaminated with microbial pathogens, toxins, chemicals, or physical hazards that could cause illness or injury and an aborted or crashed vessel mission due to astronauts ill with diarrhea or biological or chemical intoxication. The system was designed to have strict control over all aspects of the safety of food production including raw materials, processing methods, the food plant environment, personnel, storage, and distribution. In practice, the identification of potential hazards for a given process and meticulous control of critical control points is required. Methods for HACCP and quality assurance programs for thermally processed foods have been outlined (46).

Spoilage of Acid and Low-Acid Canned Foods

Thermally processed low-acid foods receive a heat treatment adequate to kill spores of *C. botulinum* but not sufficient to kill more heat-resistant spores of mesophiles and thermophiles. Acid and acidified foods with an equilibrium pH of ≤4.6 are not processed sufficiently to inactivate all spores, since most species of sporeformers do not grow under acid conditions and inactivation of all spores would be detrimental to food quality and nutritional composition. Certain foods such as cured meats and hams do not receive a thermal process sufficient to inactivate sporeformers and thus must be kept under refrigerated conditions for microbial stability. These classes of foods present opportunities for the growth of sporeformers that do not present a public health hazard but which can cause economic spoilage (154).

In practice, the inherent spore contamination of foods and food ingredients and of the cannery environment contributes to spoilage problems. Dry ingredients such as sugar, starches, flours, and spices often contain high levels of sporeformers. Spore populations can also accumulate in a food plant, such as thermophilic spores on heated equipment and saccharolytic clostridia in plants processing sugar-rich foods such as fruits.

The principal spoilage organisms and spoilage manifestations are presented in Table 3.6. The principal classes of sporeformers causing spoilage are thermophilic flat-sour organisms, thermophilic anaerobes not producing hydrogen sulfide, thermophilic anaerobes forming hydrogen sulfide, putrefactive anaerobes, facultative *Bacillus* mesophiles, butyric clostridia, lactobacilli, and heat-resistant molds and yeasts (154). Practical control of these organisms includes monitoring of raw foods entering the cannery, particularly sugars, starches, spices, onions, mushrooms, and dried foods, to limit the initial spore load in a food product, adequate thermal processing depending on subsequent storage and distribution conditions, rapid cooling of products, chlorination of cooling water, and implementing and maintaining good manufacturing practices within the food plant.

Certain species of sporeforming psychrophiles have the ability to spoil refrigerated foods and have caused spoilage of meats and dairy products in recent years. Psychrophilic strains of *Bacillus* have been isolated from spoiled dairy products (33). Psychrophilic clostridia have also been associated with the spoilage of meats (37, 122).

Modeling Growth of Sporeformers in Foods

Abundant information exists on the behavior of microorganisms in foods, and it is often useful to generate statistical models to quantify safety risks in foods. Two general types of models have been used in food microbiology: (i) those analyzing experimental growth and survival data using simple and higher-order polynomials and (ii) theoretic models derived from basic scientific principles and computer analysis, used to predict micro-

Table 3.6 Spoilage of canned foods by sporeformers[a]

Type of spoilage	pH	Major sporeformers responsible	Spoilage defects
Flat-sour	≥5.3	*B. coagulans, B. stearothermophilus*	No gas, pH lowered. May have abnormal odor and cloudy liquor.
Thermophilic anaerobe	≥4.8	*C. thermosaccharolyticum*	Can swells, may burst. Anaerobic anaerobe end products give sour, fermented, or butyric odor. Typical foods are spinach, corn.
Sulfide spoilage	≥5.3	*D. nigrificans, C. bifermentans*	Hydrogen sulfide produced, giving rotten egg odor. Iron sulfide precipitate gives blackened appearance. Typical foods are corn, peas.
Putrefactive anaerobe	≥4.8	*C. sporogenes*	Plentiful gas. Disgusting putrid odor. pH often increased. Typical foods are corn, asparagus.
Aerobic sporeformers	≥4.8	*Bacillus* spp.	Gas usually absent except for cured meats, milk is coagulated. Typical foods are milk, meats, beets.
Butyric spoilage	≥4.0	*C. butyricum, C. tertium*	Gas, acetic and butyric odor. Typical foods are tomatoes, peas, olives, cucumbers.
Acid spoilage	≥1.2	*B. thermoacidurans*	Flat (*Bacillus*) or gas (butyric anaerobes). Off odors depend on organism. Common foods are tomatoes, tomato products, other fruits.

[a]Sources: references 120, 154, and 178.

bial survival. Statistical models can be particularly useful to define important variables and predict microbial behavior in advance of practical testing (157). Certain companies and institutions are beginning to use models involving neural nets. These are valuable since they are adaptive and improve in precision and predictive capability over time. An example of a model used extensively in the dairy industry is that of Tanaka et al. (224) for preventing growth of *C. botulinum* in process cheese. Although models can provide guidelines for food safety, assurance of sterility generally needs to be ascertained by a laboratory challenge study.

CONCLUSION

The scientific investigation of sporeformers has greatly contributed to the development of microbiology for enhancement of food safety and quality. The fundamental understanding of sporulation in *B. subtilis* provides an elegant model of cellular differentiation. Advances in the understanding of the mechanisms of spore heat resistance have contributed to a greater knowledge of dormancy and the ecological success of sporeformers. The remarkable resistance properties of spores and their impact on human disease, particularly botulism, tetanus, and anthrax, have led to the development of microbiology and its importance in medicine and industry.

Acknowledgments

Work in the laboratory of P.S. has received generous support from both the Army Research Office and the National Institutes of Health (GM-19698). Research in the laboratory of E.J. has been supported by the industrial sponsors of the Food Research Institute.

References

1. **Adams, L. F., K. L. Brown, and H. R. Whiteley.** 1991. Molecular cloning and characterization of two genes encoding sigma factors that direct transcription from a *Bacillus thuringiensis* crystal protein gene promoter. *J. Bacteriol.* **173:**3846–3854.
2. **Almiron, M., A. J. Link, D. Furlong, and R. Kolter.** 1992. A novel DNA-binding protein with regulatory and protective roles in starved *Escherichia coli*. *Genes Dev.* **6:**2646–2654.
3. **Alper, S., L. Duncan, and R. Losick.** 1994. An adenosine nucleotide switch controlling the activity of a cell type-specific transcription factor in *B. subtilis*. *Cell* **77:**195–205.
4. **Angulo, F.** 1994. Personal communication.
5. **Appert, N.** 1810. L'Art de conserver pendant plusieurs anées toutes les substances animales et végétables. (Translated by K. G. Bitting, Chicago, Ill., 1920.) *In* S. A. Goldblith, M. A. Joslyn, and J. T. R. Nickerson (ed.), *Introduction to the Thermal Processing of foods*. 1961. AVI Publishing Co., Westport, Conn.
6. **Archibald, F. S, and I. Fridovich.** 1981. Manganese and defenses against oxygen toxicity in *Lactobacillus plantarum*. *J. Bacteriol.* **145:**442–451.
7. **Aronson, A. I.** 1993. Insecticidal toxins, p. 953–964. *In* A. L. Sonenshein, J. A. Hoch, and R. Losick (ed.), *Bacillus subtilis and Other Gram-Positive Bacteria: Biochemistry, Physiology, and Molecular Genetics*. American Society for Microbiology, Washington, D.C.
8. **Ashwood-Smith, M. J., and E. Grant.** 1976. Mutation induction in bacteria by freeze-drying. *Cryobiology* **13:**206–213.

9. **Bai, U., I. Mandic-Mulec, and I. Smith.** 1993. SinI modulates the activity of SinR, a developmental switch protein of *Bacillus subtilis,* by protein-protein interaction. *Genes Dev.* **7:**139–148.
10. **Baldus, J. M., B. D. Green, P. Youngman, and C. P. Moran, Jr.** 1994. Phosphorylation of *Bacillus subtilis* transcription factor Spo0A stimulates transcription from the *spoIIG* promoter by enhancing binding to weak 0A boxes. *J. Bacteriol.* **176:**296–306.
11. **Beall, B., and J. Lutkenhaus.** 1991. FtsZ in *Bacillus subtilis* is required for vegetative septation and for asymmetric septation during sporulation. *Genes Dev.* **5:**447–455.
12. **Beall, B., and J. Lutkenhaus.** 1992. Impaired cell division and sporulation of a *Bacillus subtilis* strain with the *ftsA* gene deleted. *J. Bacteriol.* **174:**2398–2403.
13. **Beaman, T. C., and P. Gerhardt.** 1986. Heat resistance of bacterial spores correlated with protoplast dehydration, mineralization, and thermal adaptation. *Appl. Environ. Microbiol.* **52:**1242–1246.
14. **Beaman, T. C., H. S. Pankratz, and P. Gerhardt.** Heat shock affects permeability and resistance of *Bacillus stearothermophilus* spores. *Appl. Environ. Microbiol.* **54:** 2515–2520.
15. **Belliveau, B. H., T. C. Beaman, H. S. Pankratz, and P. Gerhardt.** 1992. Heat killing of bacterial spores analyzed by differential scanning calorimetry. *J. Bacteriol.* **174:** 4463–4474.
16. **Bird, T. H., J. K. Grimsley, J. A. Hoch, and G. B. Spiegelman.** 1993. Phosphorylation of Spo0A activates its stimulaton of in vitro transcription from the *Bacillus subtilis spoIIG* operon. *Mol. Microbiol.* **9:**741–749.
17. **Bloomfield, S. F., and M. Arthur.** 1994. Mechanisms of inactivation and resistance of spores to chemical biocides. *J. Appl. Bacteriol.* **76:**91S–104S.
18. **Bol, D. K., and R. Yasbin.** 1994. Analysis of the dual regulatory mechanisms controlling expression of the vegetative catalase gene of *Bacillus subtilis. J. Bacteriol.* **176:** 6744–6748.
19. **Boschwitz, H., H. O. Halvorson, A. Keynan, and Y. Milner.** 1985. Trypsin-like enzymes from dormant and germinated spores of *Bacillus cereus* T and their possible involvement in germination. *J. Bacteriol.* **164:**302–309.
20. **Brown, D. P., L. Ganova-Raeva, B. D. Green, S. R. Wilkinson, M. Young, and P. Youngman.** 1994. Characterization of *spo0A* homologues in diverse *Bacillus* and *Clostridium* species identifies a probable DNA-binding domain. *Mol. Microbiol.* **14:**411–426.
21. **Buchanan, C. E., and A. Gustafson.** 1992. Mutagenesis and mapping of the gene for a sporulation-specific penicillin-binding protein in *Bacillus subtilis. J. Bacteriol.* **174:** 5430–5435.
22. **Buchanan, C. E., A. O. Henriques, and P. J. Piggot.** 1994. Cell wall changes during bacterial endospore formation, p. 167–186. *In* J.-M. A. R. Hakenbeck (ed.), *Bacterial Cell Wall.* Elsevier Science Publishers, New York.
23. **Buchanan, C. E., and M.-L. Ling.** 1972. Isolation and sequence analysis of *dacB,* which encodes a sporulation-specific penicillin-binding protein in *Bacillus subtilis. J. Bacteriol.* **174:**1717–1725.
24. **Bulloch, W.** 1938. *The History of Bacteriology.* Oxford University Press, Oxford.
25. **Burbulys, D., K. A. Trach, and J. A. Hoch.** 1991. Initiation of sporulation in *B. subtilis* is controlled by a multicomponent phosphorelay. *Cell* **64:**545–552.
26. **Cabrera-Martinez, R., J. M. Mason, B. Setlow, W. M. Waites, and P. Setlow.** 1989. Purification and amino acid sequence of two small, acid-soluble proteins from *Clostridium bifermentans* spores. *FEMS Microbiol. Lett.* **61:**139–144.
27. **Cabrera-Martinez, R. M., and P. Setlow.** 1991. Cloning and nucleotide sequence of three genes coding for small, acid-soluble proteins of *Clostridium perfringens* spores. *FEMS Microbiol. Lett.* **77:**127–132.
28. **Cano, R. J.** 1994. *Bacillus* DNA in amber: a window to ancient symbiotic relationships. *ASM News* **60:**129–134.
29. **Cano, R. J., and M. K. Borucki.** 1995. Revival and identification of bacterial spores in 25–40 million year old Dominican amber. *Science* **268:**1060–1064.
30. **Carstensen, E. L., and R. E. Marquis.** 1975. Dielectric and electrochemical properties of bacterial spores, p. 563–571. *In* P. Gerhardt, R. N. Costilow, and H. L. Sadoff (ed.), Spores VI. American Society for Microbiology, Washington, D.C.
31. **Cato, E. P., W. L. George, and S. M. Finegold.** 1986. The genus *Clostridium,* p. 1141–1200. *In* H. A. Sneath, N. S. Mair, and M. E. Sharpe (ed.), *Bergey's Manual of Systematic Bacteriology,* vol. 2. The Williams & Wilkins Co., Baltimore.
32. **Chambliss, G. H.** 1993. Carbon source-mediated catabolite repression, p. 213–219. *In* A. L. Sonenshein, R. Losick, and J. A. Hoch (ed.), *Bacillus subtilis and Other Gram-Positive Bacteria: Biochemistry, Physiology, and Molecular Genetics.* American Society for Microbiology, Washington, D.C.
33. **Champagne, C. P., R. R. Laing, D. Roy, A. A. Mafu, and M. W. Griffiths.** 1994. Psychrotrophs in dairy products: their effects and their control. *Crit. Rev. Food Sci. Nutr.* **34:**1–30.
34. **Chiasson, L. P., and S. Zamenhof.** 1966. Studies on induction of mutations by heat in spores of *Bacillus subtilis.* Can. J. Microbiol. **12:**43–46.
35. **Chung, J. D., G. Stephanopoulos, K. Ireton, and A. D. Grossman.** 1994. Gene expression in single cells of *Bacillus subtilis:* evidence that a threshold mechanism controls the initiation of sporulation. *J. Bacteriol.* **176:**1977–1984.
36. **Clouston, J. G., and P. A. Wills.** 1969. Initiation of germination and inactivation of *Bacillus pumilus* spores by hydrostatic pressure. *J. Bacteriol.* **97:**684–690.
37. **Collins, M. D., U. M. Rodrigues, R. A. Edwards, and T. A. Roberts.** 1992. Taxonomic studies on a psychrophilic *Clostridium* from vacuum-packed beef: description of *Clostridium estertheticum* sp. nov. *FEMS Microbiol. Lett.* **96:**235–240.
38. **Corfe, B. M., R. L. Sammons, D. A. Smith, and C. Mauel.** 1994. The *gerB* region of the *Bacillus subtilis* 168 chromosome encodes a homologue of the *gerA* spore germination operon. *Microbiology* **140:**471–478.
39. **Council for Agricultural Science and Technology.** 1994. *Foodborne Pathogens: Risks and Consequences.* Task

Force Report no. 122. Council for Agricultural Science and Technology.

40. **Crowe, J. H., E. A. Hoekstra, and L. M. Crowe.** 1992. Anhydrobiosis. *Annu. Rev. Physiol.* **54:**579–599.

41. **Crowther, J. S., and A. C. Baird-Parker.** 1984. The pathogenic and toxigenic spore-forming bacteria, p. 275–311. *In* A. Hurst and G. W. Gould (ed.), *The Bacterial Spore,* vol. 2. Academic Press Ltd., London.

42. **Cutting, S., A. Driks, R. Schmidt, B. Kunkel, and R. Losick.** 1991. Forespore-specific transcription of a gene in the signal transduction pathway that governs pro-σ^K processing in *Bacillus subtilis. Genes Dev.* **5:**456–466.

43. **Cutting, S., S. Roels, and R. Losick.** 1991. Sporulation operon *spoIVF* and the characterization of mutations that uncouple mother-cell from forespore gene expression in *Bacillus subtilis. J. Mol. Biol.* **221:**1237–1256.

44. **Daniel, R. A., S. Drake, C. E. Buchanan, R. Scholle, and J. Errington.** 1994. The *Bacillus subtilis spoVD* gene encodes a mother-cell-specific penicillin-binding protein required for spore morphogenesis. *J. Mol. Biol.* **235:**209–220.

45. **Daniel, R. A., and J. Errington.** 1993. Cloning, DNA sequence, functional analysis and transcriptional regulation of the genes encoding dipicolinic acid synthetase required for sporulation in *Bacillus subtilis. J. Mol. Biol.* **232:**468–483.

46. **David, J. R. D., R. H. Graves, and V. R. Carlson.** 1996. *Aseptic Processing and Packaging of Foods. A Food Industry Perspective.* CRC Press, Boca Raton, Fla.

47. **Dawes, I. W., and J. Mandelstam.** 1970. Sporulation of *Bacillus subtilis* in continuous culture. *J. Bacteriol.* **103:**529–535.

48. **Decker, S. J., and D. R. Lang.** 1978. Membrane bioenergetic parameters in uncoupler-resistant mutants of *Bacillus megaterium. J. Biol. Chem.* **253:**6738–6743.

49. **Deutscher, J., E. Kuster, U. Bergstedt, V. Charrier, and W. Hillen.** 1995. Protein kinase-dependent Hpr/CcpA interaction links glycolytic activity to carbon catabolite repression in Gram-positive bacteria. *Mol. Microbiol.* **15:**1049–1053.

50. **Diederich, B., J. F. Wilkinson, T. Magnin, S. M. A. Najafi, J. Errington, and M. D. Yudkin.** 1994. Role of interactions between SpoIIAA and SpoIIAB in regulating cell-specific transcription factor σ^F of *Bacillus subtilis. Genes Dev.* **8:**2653–2663.

51. **Dills, S. S., and J. C. Vary.** 1978. An evaluation of respiratory chain associated functions during initiation of germination of *Bacillus megaterium* spores. *Biochim. Biophys. Acta* **541:**301–311.

52. **Donnellan, J. E., Jr., and R. B. Setlow.** 1965. Thymine photoproducts but not thymine dimers are found in ultraviolet irradiated bacterial spores. *Science* **149:**308–310.

53. **Doyle, M. P.** 1991. Evaluating the potential risk from extended shelf-life refrigerated foods by *Clostridium botulinum* inoculation studies. *Food Technol.* **April:**154–156.

54. **Dubnau, D.** 1991. The regulation of genetic competence in *Bacillus subtilis. Mol. Microbiol.* **5:**11–18.

55. **Duncan, L., and R. Losick.** 1993. SpoIIAB is an anti-σ factor that binds to and inhibits transcription by the regulatory protein σ^F from *Bacillus subtilis. Proc. Natl. Acad. Sci. USA* **90:**2325–2329.

56. **Ellar, D. J.** 1978. Spore specific structures and their functions, p. 295–325. *In* R. Y. Stanier, H. J. Rogers, and J. B. Ward (ed.), *Relations between Structure and Function in the Prokaryotic Cell.* Cambridge University Press, London.

57. **Errington, J.** 1993. *Bacillus subtilis* sporulation: regulation of gene expression and control of morphogenesis. *Microbiol. Rev.* **57:**1–33.

58. **Esty, J. R., and K. F. Meyer.** 1922. The heat resistance of the spores of *Bacillus botulinus* and allied anaerobes. XI. *J. Infect. Dis.* **31:**650–663.

59. **Fairhead, H., B. Setlow, and P. Setlow.** 1993. Prevention of DNA damage in spores and in vitro by small, acid-soluble proteins from *Bacillus* species. *J. Bacteriol.* **175:**1367–1374.

60. **Fairhead, H., B. Setlow, W. M. Waites, and P. Setlow.** 1994. Small, acid-soluble proteins bound to DNA protect *Bacillus subtilis* spores from killing by freeze-drying. *Appl. Environ. Microbiol.* **60:**2647–2649.

61. **Fairhead, H., and P. Setlow.** 1992. Binding to DNA of α/β-type small, acid-soluble proteins from spores of *Bacillus* or *Clostridium* species prevents formation of cytosine dimers, cytosine-thymine dimers and bipyrimidine photoadducts upon ultraviolet irradiation. *J. Bacteriol.* **174:**2874–2880.

62. **Fajardo-Cavazos, P., C. Salazar, and W. L. Nicholson.** 1993. Molecular cloning and characterization of the *Bacillus subtilis* spore photoproduct lyase (*spl*) gene, which is involved in the repair of UV radiation-induced DNA damage during spore germination. *J. Bacteriol.* **175:**1735–1744.

63. **Farkas, J.** 1994. Tolerance of spores to ionizing radiation: mechanisms of inactivation, injury and repair. *J. Appl. Bacteriol.* **76:**81S–90S.

64. **Fisher, G. J., and H. E. Johns.** 1976. Pyrimidine photodimers, p. 225–294. *In* S. Y. Wang (ed.), *Photochemistry and Photobiology of Nucleic Acids,* vol. I. Academic Press, Inc., New York.

65. **Fitz-James, P., and E. Young.** 1969. Morphology of sporulation, p. 39–72. *In* G. W. Gould and A. Hurst (ed.), *The Bacterial Spore.* Academic Press, Ltd.,, London.

66. **Foster, S. J., and K. Johnstone.** 1987. Purification and properties of a germination-specific cortex-lytic enzyme from spores of *Bacillus megaterium. Biochem. J.* **242:**573–579.

67. **Foster, S. J., and K. Johnstone.** 1989. Germination-specific cortex-lytic enzyme is activated during triggering of *Bacillus megaterium* KM spore germination. *Mol. Microbiol.* **2:**727–733.

68. **Foster, S. J., and K. Johnstone.** 1989. The trigger mechanism of bacterial spore germination, p. 89–108. *In* I. Smith, R. Slepecky, and P. Setlow (ed.), *Regulation of Procaryotic Development.* American Society for Microbiology, Washington, D.C.

69. **Gerhardt, P., and R. E. Marquis.** 1989. Spore thermoresistance mechanisms, p. 17–63. *In* I. Smith, R. Slepecky, and P. Setlow (ed.), *Regulation of Procaryotic*

Development. American Society for Microbiology, Washington, D.C.

70. **Gerhardt, P., R. Scherrer, and S. H. Black.** 1972. Molecular sieving by dormant spore structures, p. 68–74. *In* H. O. Halvorson, R. Hanson, and L. L. Campbell (ed.), *Spores V.* American Society for Microbiology, Washington, D.C.
71. **Glickman, B. W., R. M. Schaaper, W. A. Haseltine, R. L. Dunn, and D. E. Brash.** 1986. The C-C (6–4) UV photoproduct is mutagenic in *Escherichia coli. Proc. Natl. Acad. Sci. USA* **83:**6945–6949.
72. **Goldblith, S. A., M. A. Joslyn, and J. T. R. Nickerson.** 1961. *An Anthology of Food Science,* vol. 1. *Introduction to the Thermal Processing of Foods.* The AVI Publishing Co., Westport, Conn.
73. **Gould, G. W.** 1969. Germination, p. 397–444. *In* G. W. Gould and A. Hurst (ed.), *The Bacterial Spore.* Academic Press, Ltd., London.
74. **Gould, G. W.** 1983. Mechanisms of resistance and dormancy, p. 173–210. *In* A. Hurst and G. W. Gould (ed.), *The Bacterial Spore,* vol. 2. Academic Press, Ltd., London.
75. **Gould, G. W., and A. J. H. Sole.** 1970. Initiation of germination of bacterial spores by hydrostatic pressure. *J. Gen. Microbiol.* **60:**335–346.
76. **Greenberg, R. A., R. B. Tompkin, B. O. Blade, R. S. Kittaka, and A. Anelis.** 1966. Incidence of mesophilic spores in raw pork, beef, and chicken in processing plants in the United States and Canada. *Appl. Microbiol.* **14:**789–793.
77. **Griffith, J., A. Makhov, L. Santiago-Lara, and P. Setlow.** 1994. Electron microscopic studies of the interaction between a *Bacillus* α/β-type small, acid-soluble spore protein with DNA: protein binding is cooperative, stiffens the DNA and induces negative supercoiling. *Proc. Natl. Acad. Sci. USA* **91:**8224–8228.
78. **Grossman, A. D., and R. Losick.** 1988. Extracellular control of spore formation in *Bacillus subtilis. Proc. Natl. Acad. Sci. USA* **85:**4369–4373.
79. **Haldenwang, W. G.** 1995. The sigma factors of *Bacillus subtilis. Microbiol. Rev.* **59:**1–30.
80. **Hatheway, C. L.** 1994–96. Personal communications.
81. **Hatheway, C. L.** 1995. Botulism: the present status of the disease, p. 55–75. *In* C. Montecucco (ed.), *Clostridial Neurotoxins.* Springer Verlag, Berlin.
82. **Hippe, H., J. R. Andreesen, and G. Gottschalk.** 1992. The genus *Clostridium*—nonmedical, p. 1800–1866. *In* A. Balows, H. G. Truper, M. Dworkin, W. Harder, and K. H. Schleifer (ed.), *The Prokaryotes,* 2nd ed., vol. II. Springer Verlag, New York.
83. **Hoch, J. A.** 1994. The phosphorelay signal transduction pathway in the initiation of sporulation, p. 41–60. *In* P. J. Piggot, C. Moran, Jr., and P. Youngman (ed.), *Regulation of Bacterial Differentiation.* American Society for Microbiology, Washington, D.C.
84. **Holck, A., H. Blom, and P. E. Granum.** 1990. Cloning and sequencing of the genes encoding acid-soluble spore proteins from *Clostridium perfringens. Gene* **91:**107–111.
85. **Hurst, A.** 1983. Injury, p. 255–274. *In* A. Hurst and G. W. Gould (ed.), *The Bacterial Spore,* vol. 2. Academic Press Ltd., London.
86. **Husmark, U., and U. Ronner.** 1992. The influence of hydrophobic, electrostatic and morphologic properties on the adhesion of *Bacillus* spores. *Biofouling* **5:**335–344.
87. **Illing, N., and J. Errington.** 1991. Genetic regulation of morphogenesis in *Bacillus subtilis:* roles of σ^E and σ^F in prespore engulfment. *J. Bacteriol.* **173:**3159–3169.
88. **Illing, N., and J. Errington.** 1991. The *spoIIIA* operon of *Bacillus subtilis* defines a new temporal class of mother-cell-specific sporulation genes under the control of the σ^E form of RNA polymerase. *Mol. Microbiol.* **5:**1927–1940.
89. **Imlay, J. A.., and S. Linn.** 1988. DNA damage and oxygen radical toxicity. *Science* **240:**1302–1309.
90. **Ingram, M.** 1969. Sporeformers as foods spoilage organisms, p. 549–610. *In* G. W. Gould and A. Hurst (ed.), *The Bacterial Spore.* Academic Press Ltd., London.
91. **Ireton, K., and A. D. Grossman.** 1994. A developmental checkpoint couples the initiation of sporulation to DNA replication in *Bacillus subtilis. EMBO J.* **13:**1566–1573.
92. **Ireton, K., S. Jin, A. D. Grossman, and A. L. Sonenshein.** 1995. Krebs cycle function is required for activation of the Spo0A transcription factor in *Bacillus subtilis. Proc. Natl. Acad. Sci. USA* **92:**2845–2849.
93. **Irie, R., Y. Fujita, and T. Okamoto.** 1993. Cloning and sequencing of the *gerK* spore germination gene of *Bacillus subtilis* 168. *J. Gen. Appl. Microbiol.* **39:**453–465.
94. **Ito, K. A., M. L. Seeger, C. W. Bohrer, C. B. Denny, and M. K. Bruch.** 1970. The thermal and germicidal resistance of *Clostridium botulinum* types A, B, and E spores, p. 410–415. *In* M. Herzberg (ed.), *Proceedings of the First U.S.-Japan Conference on Toxic Micro-Organisms.* UJNR Joint Panels on Toxic Micro-Organisms and the U.S. Department of the Interior, Washington, D.C.
95. **Jenal, U., and C. Stephens.** 1996. Bacterial differentiation: sizing up sporulation. *Curr. Biol.* **6:**111–114.
96. **Johnson, E. A.** 1991. Microbiological safety of fermented foods, p. 135–169. *In* J. G. Zeikus and E. A. Johnson (ed.), *Mixed Cultures in Biotechnology.* McGraw Hill, Inc., New York.
97. **Johnson, E. A., and M. C. Goodnough.** 1996. Botulism. *In* W. J. Hausler and M. Sussman (ed.), *Topley and Wilson's Microbiology and Microbial Infections,* 9th ed., vol. 3. Edward Arnold, London (in press).
98. **Johnstone, K.** 1994. The trigger mechanism of spore germination: current concepts. *J. Appl. Bacteriol.* **76:** 175S–245S.
99. **Johnstone, K., G. S. A. B. Stewart, I. R. Scott, and D. J. Ellar.** 1982. Zinc release and the sequence of biochemical events during triggering of *Bacillus megaterium* KM spore germination. *Biochem. J.* **208:**407–411.
100. **Kadota, H., A. Uchida, Y. Sako, and K. Harada.** 1978. Heat-induced DNA injury in spores and vegetative cells of *Bacillus subtilis,* p. 27–30. *In* G. Chambliss and J. C. Vary (ed.), *Spores VII.* American Society for Microbiology, Washington, D.C.
101. **Karow, M. L., P. Glaser, and P. J. Piggot.** 1995. Identification of a gene, *spoIIR,* which links the activation of σ^E to the transcriptional activity of σ^F during sporulation in *Bacillus subtilis. Proc. Natl. Acad. Sci. USA* **92:**2012–2016.
102. **Kautter, D. A., R. K. Lynt, Jr., and H. M. Solomon.** 1981. Evaluation of the botulism hazard from nitrogen-packed sandwiches. *J. Food Prot.* **44:**59–61.

103. **Kennedy, M. J., S. L. Reader, and L. M. Swierczynski.** 1994. Preservation records of micro-organisms: evidence of the tenacity of life. *Microbiology* **140:**2513–2529.

104. **Keynan, A., and Z. Evenchik.** 1969. Activation, p. 359–396. *In* G. W. Gould and A. Hurst (ed.), *The Bacterial Spore.* Academic Press Ltd., London.

105. **Keynan, A., and N. Sandler.** 1984. Spore research in historical perspective, p. 1–48. *In* A. Hurst and G. W. Gould (ed.), *The Bacterial Spore,* vol. 2. Academic Press Ltd., London.

106. **Kihm, D. J., M. T. Hutton, J. H. Hanlin, and E. A. Johnson.** 1988. Zinc stimulates sporulation in *Clostridium botulinum* 113B. *Curr. Microbiol.* **17:**193–198.

107. **Kihm, D. J., M. T. Hutton, J. H. Hanlin, and E. A. Johnson.** 1990. Influence of transition metals added during sporulation on heat resistance of *Clostridium botulinum* 113B spores. *Appl. Environ. Microbiol.* **56:**681–685.

108. **Kihm, D. J., and E. A. Johnson.** 1990. Hydrogen gas accelerates thermal inactivation of *Clostridium botulinum* spores. *Appl. Microbiol. Biotechnol.* **33:**705–708.

109. **Kim, J., and P. M. Foegeding.** 1993. Principles of control, p. 121–176. *In* A. H. W. Hauschild and K. L. Dodds (ed.), *Clostridium botulinum. Ecology and Control in Foods.* Marcel Dekker, Inc., New York.

110. **Kobayashi, K., K. Shoji, T. Shimizu, K. Nakano, T. Sato, and Y. Kobayashi.** 1995. Analysis of a suppressor mutation *ssb* (*kinC*) of *sur0B20* (*spo0A*) mutation in *Bacillus subtilis* reveals that *kinC* encodes a histidine protein kinase. *J. Bacteriol.* **177:**176–182.

111. **Koshikawa, T., T. C. Beaman, H. S. Pankratz, S. Nakashio, T. R. Corner, and P. Gerhardt.** 1984. Resistance, germination, and permeability correlates of *Bacillus megaterium* spores successively divested of integument layers. *J. Bacteriol.* **159:**624–632.

112. **Kramer, J. H., and R. J. Gilbert.** 1989. *Bacillus cereus* and other *Bacillus* species, p. 21–70. *In* M. P. Doyle (ed.), *Foodborne Bacterial Pathogens.* Marcel Dekker, Inc., New York.

113. **Kroos, L., and S. Cutting.** 1994. Intercellular and intercompartmental communication during *Bacillus subtilis* sporulation, p. 155–180. *In* P. J. Piggot, C. P. Moran, Jr., and P. Youngman (ed.), *Regulation of Bacterial Differentiation.* American Society for Microbiology, Washington, D.C.

114. **Kroos, L., B. Kunkel, and R. Losick.** 1989. Switch protein alters specificity of RNA polymerase containing a compartment-specific sigma factor. *Science* **243:**526–529.

115. **Kuhn, N. J., B. Setlow, and P. Setlow.** 1993. Manganese (II) activation of 3-phosphoglycerate mutase of *Bacillus megaterium:* pH-sensitive interconversion of active and inactive forms. *Arch. Biochem. Biophys.* **306:**342–349.

116. **Kunst, F., T. Msadek, and G. Rapoport.** 1994. Signal transduction network controlling degradative enzyme synthesis and competence in *Bacillus subtilis,* p. 1–20. *In* P. J. Piggot, C. P. Moran, Jr., and P. Youngman (ed.), *Regulation of Bacterial Differentiation.* American Society for Microbiology, Washington, D.C.

117. **Kuroda, A., Y. Asami, and J. Sekiguchi.** 1993. Molecular cloning of a sporulation-specific cell wall hydrolase gene of *Bacillus subtilis. J. Bacteriol.* **175:**6260–6268.

118. **Labbe, R. G.** 1989. *Clostridium perfringens,* p. 191–234. *In* M. P. Doyle (ed.), *Foodborne Bacterial Pathogens.* Marcel Dekker, Inc., New York.

119. **Labbe, R. G., and T. H. Huang.** 1995. Generation times and modeling of enterotoxin-positive and enterotoxin-negative strains of *Clostridum perfringens* in laboratory media and ground beef. *J. Food Prot.* **58:**1303–1306.

120. **Landry, W. L., A. H. Schwab, and G. A. Lancette.** 1995. Examination of canned foods, p. 21.01–21.29. *In* Food and Drug Administration, *Bacteriological Analytical Manual,* 8th ed. AOAC International, Gaithersburg, Md.

121. **Lang, O. W.** 1935. Thermal processes for canned marine products. *Univ. Calif. Publ. Public Health* **2:**1–182.

122. **Lawson, P., R. H. Dainty, N. Kristiansen, J. Berg, and M. D. Collins.** 1994. Characterization of a psychrotrophic *Clostridium* causing spoilage in vacuum-packed cooked pork: description of *Clostridium algidicarnis* sp. nov. *Lett. Appl. Microbiol.* **19:**153–157.

123. **LeDeaux, J. R., and A. D. Grossman.** 1995. Isolation and characterization of *kinC,* a gene that encodes a sensor kinase homologous to the sporulation sensor kinases KinA and KinB in *Bacillus subtilis. J. Bacteriol.* **177:**166–175.

124. **LeDeaux, J. R., N. Yu, and A. D. Grossman.** 1995. Different roles for KinA, KinB, and KinC in the initiation of sporulation in *Bacillus subtilis. J. Bacteriol.* **177:**861–863.

125. **Lee, S., and C. W. Price.** 1993. The *minCD* locus of *Bacillus subtilis* lacks the *minE* determinant that provides topological specificity to cell division. *Mol. Microbiol.* **7:** 601–610.

126. **Levin, P. A., and R. Losick.** 1994. Characterization of a cell division gene from *Bacillus subtilis* that is required for vegetative and sporulation septum formation. *J. Bacteriol.* **176:**1451–1459.

127. **Levin, P. A., P. S. Margolis, P. Setlow, R. Losick, and D. Sun.** 1992. Identification of *Bacillus subtilis* genes for septum placement and shape determination. *J. Bacteriol.* **174:**6717–6728.

128. **Lewis, J. C.** 1969. Dormancy, p. 301–358. *In* G. W. Gould and A. Hurst (ed.), *The Bacterial Spore.* Academic Press Ltd., London.

129. **Lewis, P. J., S. R. Partridge, and J. Errington.** 1994. σ factors, asymmetry, and the determination of cell fate in *Bacillus subtilis. Proc. Natl. Acad. Sci. USA* **91:**3849–3853.

130. **Liu, S., S. Cutting, and L. Kroos.** 1995. Sporulation protein SpoIVFB from *Bacillus subtilis* enhances processing of the sigma factor precursor pro-σ^K in the absence of other sporulation gene products. *J. Bacteriol.* **177:**1082–1085.

131. **Loeb, L. A., E. A. James, A. M. Waltersdorph, and S. J. Klebanoff.** 1988. Mutagenesis by the autoxidation of iron with isolated DNA. *Proc. Natl. Acad. Sci. USA* **85:** 3918–3922.

132. **Londono-Vallejo, J.-A., and P. Stragier.** 1995. Cell-cell signalling pathway activating a developmental transcription factor in *Bacillus subtilis. Genes Dev.* **9:**503–508.

133. **Loshon, C. A., and P. Setlow.** 1993. Levels of small molecules in dormant spores of *Sporosarcina* species and comparison with levels in spores of *Bacillus* and *Clostridium* species. *Can. J. Microbiol.* **39:**259–262.

134. **Loshon, C. A., and P. Setlow.** 1994. Unpublished results.

135. **Losick, R., and J. Pero.** 1981. Cascades of sigma factors. *Cell* **25**:582–584.

136. **Losick, R., and P. Stragier.** 1992. Crisscross regulation of cell-type specific gene expression during development in *Bacillus subtilis. Nature* (London) **355**:601–604.

137. **Lund, B. M., and M. W. Peck.** 1994. Heat resistance and recovery of spores of nonproteolytic *Clostridium botulinum* in relation to refrigerated, processed foods with extended shelf-life. *J. Appl. Bacteriol. Symp.* **76**:115S–128S.

138. **Lund, D.** 1975. Thermal processing, p. 31–92. *In* M. Karel, O. R. Fennema, and D. B. Lund (ed.), *Principles of Food Science.* Part II. *Physical Principles of Food Preservation.* Marcel Dekker, Inc., New York.

139. **Lynt, R. K., D. A. Kautter, and R. B. Read, Jr.** 1975. Botulism in commercially canned foods. *J. Milk Food Technol.* **38**:546–550.

140. **Lynt, R. K., D. A. Kautter, and H. M. Solomon.** 1982. Differences and similarities among proteolytic strains of *Clostridium botulinum* types A, B, E and F: a review. *J. Food Prot.* **45**:466–474.

141. **Magill, N. G., A. E. Cowan, D. E. Koppel, and P. Setlow.** 1994. The internal pH of the forespore compartment of *Bacillus megaterium* decreases by about 1 pH unit during sporulation. *J. Bacteriol.* **176**:2252–2258.

142. **Magill, N. G., A. E. Cowan, M. A. Leyva-Vazquez, M. Brown, D. E. Koppel, and P. Setlow.** 1996. Analysis of the relationship between the decrease in pH and accumulation of 3-phosphoglyceric acid in developing forespores of *Bacillus* species. *J. Bacteriol.* **178**:2204–2210.

143. **Magnuson, R., J. Solomon, and A. D. Grossman.** 1994. Biochemical and genetic characterization of a competence pheromone from *B. subtilis. Cell* **77**:207–216.

144. **Makino, S., N. Ito, T. Inoue, S. Miyata, and R. Moriyama.** 1994. A spore lytic enzyme released from *Bacillus cereus* spores during germination. *Microbiology* **140**: 1403–1410.

145. **Marquis, R. E., J. Sim, and S. Y. Shin.** 1994. Molecular mechanisms of resistance to heat and oxidative damage. *J. Appl. Bacteriol.* **70**:40S–48S.

146. **Mason, J. M., and P. Setlow.** 1986. Essential role of small, acid-soluble spore proteins in resistance of *Bacillus subtilis* spores to UV light. *J. Bacteriol.* **167**:174–178.

147. **McKee, L. H.** 1995. Microbial contamination of spices and herbs: a review. *Lebensm.-Wiss. Technol.* **28**:1–11.

148. **Min, K.-T., C. M. Hilditch, B. Diederich, J. Errington, and M. D. Yudkin.** 1993. σ^F, the first compartment specific transcription factor of *Bacillus subtilis,* is regulated by an anti-sigma factor which is also a protein kinase. *Cell* **74**:735–742.

149. **Moir, A., and D. A. Smith.** 1990. The genetics of bacterial spore germination. *Annu. Rev. Microbiol.* **44**:531–533.

150. **Morse, D. L., L. K. Leonard, J. J. Guzewich, B. D. Devine, and M. Shaygani.** 1990. Garlic-in-oil associated botulism: episode leads to product modification. *Am. J. Publ. Health* **80**:1372–1373.

151. **Munakata, N., and C. S. Rupert.** 1974. Dark repair of DNA containing "spore photoproduct" in *Bacillus subtilis. Mol. Gen. Genet.* **130**:239–250.

152. **Murrell, W. G.** 1969. Chemical composition of spores and spore structures, p. 215–273. *In* G. W. Gould and A. Hurst (ed.), *The Bacterial Spore.* Academic Press, London.

153. **Murrell, W. G.** 1981. Biophysical studies on the molecular mechanisms of spore heat resistance and dormancy, p. 64–77. *In* H. S. Levinson, A. L. Sonenshein, and D. J. Tipper (ed.),). *Sporulation and Germination.* American Society for Microbiology, Washington, D.C.

154. **Murrell, W. G.** 1987. Microbiology of canned foods. *Food Res. Q.* **45**:73–89.

155. **Nessi, C., A. M. Albertini, M. L. Speranza, and A. Galizzi.** 1995. The *outB* gene of *Bacillus subtilis* codes for NAD synthetase. *J. Biol. Chem.* **270**:6181–6185.

156. **NFPA/CMI Container Integrity Task Force, Microbiological Assessment Group Report.** 1984. Botulism risk from post-processing contamination of commercially canned foods in metal containers. *J. Food Prot.* **47**:801–816.

157. **Notermans, S., P. in't Veld, T. Wijtzes, and G. C. Mead.** 1993. A user's guide to microbial challenge testing for ensuring the safety and stability of food products. *Food Microbiol.* **10**:145–157.

158. **Ohlsen, R. L., J. K. Grimsley, and J. A. Hoch.** 1994. Deactivation of the sporulation transcription factor Spo0A by the SpoeE protein phosphatase. *Proc. Natl. Acad. Sci. USA* **91**:1756–1760.

159. **Ohye, D. F., and W. J. Scott.** 1957. Studies in the physiology of *Clostridium botulinum* type *E. Aust. J. Biol. Sci.* **10**:85–94.

160. **Ordal, G. W., L. Marquez-Magana, and M. J. Chamberlin.** 1993. Motility and chemotaxis, p. 765–784. *In* A. L. Sonenshein, J. A. Hoch, and R. Losick (ed.), *Bacillus subtilis and Other Gram-Positive Bacteria: Biochemistry, Physiology, and Molecular Genetics.* American Society for Microbiology, Washington, D.C.

161. **Pedraza-Reyes, M., and F. Gutierrez-Corona, and W. L. Nicholson.** 1994. Temporal regulation and forespore-specific expression of the spore photoproduct lyase gene by sigma-G RNA polymerase during *Bacillus subtilis* sporulation. *J. Bacteriol.* **176**:3983–3991.

162. **Perego, M., C. Hanstein, K. M. Welsh, T. Djavakhishvili, P. Glaser, and J. A. Hoch.** 1994. Multiple protein-aspartate phosphatases provide a mechanism for the integration of diverse signals in the control of development in *Bacillus subtilis. Cell* **79**:1047–1055.

163. **Pflug, I. J.** 1987. Endpoint of a preservation process. *J. Food Prot.* **50**:347–351.

164. **Pflug, I. J.** 1987. Factors important in determining the heat process value, F_T, for low acid canned foods. *J. Food Prot.* **50**:528–533.

165. **Pflug, I. J.** 1987. Calculating F_T-values for heat preservation of shelf-stable, low acid canned foods using the straight-line semilogarithmic model. *J. Food Prot.* **50**: 608–615.

166. **Piggot, P. J., J. E. Bylund, and M. L. Higgins.** 1994. Morphogenesis and gene expression during sporulation, p. 113–138. *In* P. J. Piggot, C. P. Moran, Jr., and P. Youngman (ed.), *Regulation of Bacterial Differentiation.* American Society for Microbiology, Washington, D.C.

167. **Popham, D. L., B. Illades-Aguiar, and P. Setlow.** 1995. The *Bacillus subtilis dacB* gene, encoding penicillin-binding protein 5*, is part of a three-gene operon required for proper spore cortex synthesis and spore core dehydration. *J. Bacteriol.* **177:**4721–4729.

168. **Popham, D. L., S. Sengupta, and P. Setlow.** 1995. Heat, hydrogen peroxide, and UV resistance of *Bacillus subtilis* spores with increased core water content with or without major DNA binding proteins. *Appl. Environ. Microbiol.* **61:**3633–3638.

169. **Popham, D. L., and P. Setlow.** 1993. The cortical peptidoglycan from spores of *Bacillus megaterium* and *Bacillus subtilis* is not highly cross-linked. *J. Bacteriol.* **175:**2767–2769.

170. **Prescott, S. C., and W. L. Underwood.** 1897. Micro-organisms and sterilizing processes in the canning industries. *Technol. Q.* **10:**183–199.

171. **Rather, P. N., R. Coppolecchia, H. DeGrazia, and C. P. Moran, Jr.** 1990. Negative regulator of σ^G controlled gene expression in stationary-phase *Bacillus subtilis. J. Bacteriol.* **172:**709–715.

172. **Rhodehamel, E. J., N. R. Reddy, and M. D. Pierson.** 1992. Botulism: the causative agent and its control in foods. *Food Control* **3:**125–143.

173. **Russell, A. D.** 1982. *The Destruction of Bacterial Spores,* p. 169–231. Academic Press Ltd., London.

174. **Russell, H. L.** 1896. Gaseous fermentations in the canning industry, p. 227–231. *In Twelth Annual Report of the Agricultural Experiment Station of the University of Wisconsin.* University of Wisconsin, Madison.

175. **Sabli, M. Z. H., P. Setlow, and W. M. Waites.** 1996. The effect of hypochlorite on spores of *Bacillus subtilis* lacking small acid-soluble proteins. *Lett. Appl. Microbiol.* **22:** 405–407.

176. **Sanchez-Salas, J.-L., M. L. Santiago-Lara, B. Setlow, M. D. Sussman, and P. Setlow.** 1992. Properties of mutants of *Bacillus megaterium* and *Bacillus subtilis* which lack the protease that degrades small, acid-soluble proteins during spore germination. *J. Bacteriol.* **174:**807–814.

177. **Schmidt, R., P. Margolis, L. Duncan, R. Coppolecchia, C. P. Moran, Jr., and R. Losick.** 1990. Control of developmental transcription factor σ^F by sporulation regulatory proteins SpoIIAA and SpoIIAB in *Bacillus subtilis. Proc. Natl. Acad. Sci. USA* **87:**9221–9225.

178. **Schmitt, H. P.** 1966. Commercial sterility in canned foods, its meaning and determination. *Assoc. Food Drug Off. U. S. Q. Bull.* **30:**141–151.

179. **Scott, I. R., and D. J. Ellar.** 1978. Metabolism and the triggering of germination of *Bacillus megaterium,* use of L-[^{3}H] alanine and tritiated water to detect metabolism. *Biochem. J.* **174:**635–640.

180. **Scott, I. R., G. S. A. B. Stewart, M. A. Koncewicz, D. J. Ellar, and A. Crafts-Lighty.** 1978. Sequence of biochemical events during germination of *Bacillus megaterium* spores, p. 95–103. *In* G. Chambliss and J. C. Vary (ed.), *Spores VII.* American Society for Microbiology, Washington, D.C.

181. **Scott, V. N., and D. T. Bernard.** 1982. Heat resistance of spores of non-proteolytic type B *Clostridium botulinum. J. Food Prot.* **45:**909–912.

182. **Scotti, C., M. Piatti, A. Cuzzoni, P. Perani, A. Tognoni, G. Grandi, A. Galizzi, and A. M. Albertini.** 1993. A *Bacillus subtilis* large ORF coding for a polypeptide highly similar to polyketide synthases. *Gene* **130:**65–71.

183. **Seals, J. E., J. E. Snijder, T. A. Edell, C. L. Hatheway, C. J. Johnson, R. C. Swanson, and J. M. Hughes.** 1981. Restaurant associated type A botulism: transmission by potato salad. *Am. J. Epidemiol.* **113:**436–444.

184. **Setlow, B., N. Magill, P. Febbroriello, L. Nakhimovsky, D. E. Koppel, and P. Setlow.** 1991. Condensation of the forespore nucleoid early in sporulation of *Bacillus* species. *J. Bacteriol.* **173:**6270–6278.

185. **Setlow, B., and P. Setlow.** 1977. Levels of acetyl coenzyme A, reduced and oxidized coenzyme A, and coenzyme A in disulfide linkage to protein in dormant and germinated spores and growing and sporulating cells of *Bacillus megaterium. J. Bacteriol.* **132:**444–452.

186. **Setlow, B., and P. Setlow.** 1980. Measurements of the pH within dormant and germinated bacterial spores. *Proc. Natl. Acad. Sci. USA* **77:**2474–2476.

187. **Setlow, B., and P. Setlow.** 1993. Binding of small, acid-soluble spore proteins to DNA plays a significant role in the resistance of *Bacillus subtilis* spores to hydrogen peroxide. *Appl. Environ. Microbiol.* **59:**3418–3423.

188. **Setlow, B., and P. Setlow.** 1993. Dipicolinic acid greatly enhances the production of spore photoproduct in bacterial spores upon ultraviolet irradiation. *Appl. Environ. Microbiol.* **59:**640–643.

189. **Setlow, B., and P. Setlow.** 1994. Heat inactivation of *Bacillus subtilis* spores lacking small, acid-soluble spore proteins is accompanied by generation of abasic sites in spore DNA. *J. Bacteriol.* **176:**2111–2113.

190. **Setlow, B., and P. Setlow.** 1995. Small, acid-soluble proteins bound to DNA protect *Bacillus subtilis* spores from killing by dry heat. *Appl. Environ. Microbiol.* **61:**2787–2790.

191. **Setlow, B., D. Sun, and P. Setlow.** 1992. Studies of the interaction between DNA and α/β-type small, acid-soluble spore proteins: a new class of DNA binding protein. *J. Bacteriol.* **174:**2312–2322.

192. **Setlow, P.** 1973. Deoxyribonucleic acid synthesis and deoxynucleotide metabolism during bacterial spore germination. *J. Bacteriol.* **114:**1099.

193. **Setlow, P.** 1983. Germination and outgrowth, p. 211–254. *In* A. Hurst and G. W. Gould (ed.), *The Bacterial Spore,* vol. 2. Academic Press Ltd., London.

194. **Setlow, P.** 1988. Resistance of bacterial spores to ultraviolet light. *Commun. Mol. Cell. Biophys.* **5:**253–264.

195. **Setlow, P.** 1988. Small, acid-soluble spore proteins of *Bacillus* species: structure, synthesis, genetics, function and degradation. *Annu. Rev. Microbiol.* **42:**319–338.

196. **Setlow, P.** 1989. Forespore-specific genes of *Bacillus subtilis:* function and regulation of expression, p. 211–221. *In* I. Smith, R. Slepecky, and P. Setlow (ed.), *Regulation of Procaryotic Development.* American Society for Microbiology, Washington, D.C.

197. **Setlow, P.** 1992. DNA in dormant spores of *Bacillus* species is in an A-like conformation. *Mol. Microbiol.* **6:**563–567.

198. **Setlow, P.** 1992. I will survive: protecting and repairing spore DNA. *J. Bacteriol.* **174:**2737–2741.

199. **Setlow, P.** 1993. DNA structure, spore formation and spore properties, p. 181–194. *In* P. J. Piggot, P. Youngman, and C. P. Moran, Jr. (ed.), *Regulation of Bacterial Differentiation.* American Society for Microbiology, Washington, D.C.

200. **Setlow, P.** 1993. Spore structural proteins, p. 801–809. *In* J. A. Hoch, R. Losick, and A. L. Sonenshein (ed.), *Bacillus subtilis and Other Gram-Positive Bacteria: Biochemistry, Physiology, and Molecular Genetics.* American Society for Microbiology, Wasington, D.C.

201. **Setlow, P.** 1994. Mechanisms which contribute to the long-term survival of spores of *Bacillus* species. *J. Appl. Bacteriol.* **176:**49S–60S.

202. **Setlow, P.** 1995. Mechanisms for the prevention of damage to the DNA in spores of *Bacillus* species. *Annu. Rev. Microbiol.* **49:**29–54.

203. **Shazand, K., N. Frandsen, and P. Stragier.** 1995. Cell-type specificity during development in *Bacillus subtilis:* the molecular and morphological requirements for σ^E activation. *EMBO J.* **14:**1439–1445.

204. **Shin, S.-Y., E. G. Calvisi, T. C. Beaman, H. S. Pankratz, P. Gerhardt, and R. E. Marquis.** 1994. Microscopic and thermal characterization of hydrogen peroxide killing and lysis of spores and protection by transition metal ions, chelators, and antioxidants. *Appl. Environ. Microbiol.* **60:**3192–3197.

205. **Shioi, J.-I., S. Matsuura, and Y. Imae.** 1980. Quantitative measurements of proton motive force and motility in *Bacillus subtilis. J. Bacteriol.* **144:**891–897.

206. **Slepecky, R. A., and E. R. Leadbetter.** 1994. Ecology and relationships of endospore forming bacteria: changing perspectives, p. 195–206. *In* P. J. Piggot, C. P. Moran, Jr., and P. Youngman (ed.), *Regulation of Bacterial Differentiation.* American Society for Microbiology, Washington, D.C.

207. **Smith, L. D. S., and H. Sugiyama.** 1988. *Botulism. The Organism, Its Toxins, the Disease,* 2nd ed. Charles C Thomas Publisher, Springfield, Ill.

208. **Sole, A. J. H., G. W. Gould, and W. A. Hamilton.** 1970. Inactivation of bacterial spores by hydrostatic pressure. *J. Gen. Microbiol.* **60:**323–334.

209. **Solomon, J. M., R. Magnuson, A. Srivastava, and A. D. Grossman.** 1995. Convergent sensing pathways mediate response to two extracellular competence factors in *Bacillus subtilis. Genes Dev.* **9:**547–558.

210. **Sonenshein, A. L.** 1989. Metabolic regulation of sporulation and other stationary-phase phenomena, p. 109–130. *In* I. Smith, R. A. Slepecky, and P. Setlow (ed.), *Regulation of Procaryotic Development.* American Society for Microbiology, Washington, D.C.

211. **Songer, J. G.** 1996. Clostridial enteric diseases of domestic animals. *Clin. Microbiol. Rev.* **9:**216–234.

212. **Stackebrandt, E., W. Ludwig, M. Weizenegger, S. Dorn, T. J. McGill, G. E. Fox, C. R. Woese, W. Schubert, and K.-H. Schleifer.** 1987. Comparative 16S rRNA oligonucleotide analyses and murein types of round-spore-forming *Bacilli* and non-spore-forming relatives. *J. Gen. Microbiol.* **133:**2523–2529.

212a. **Stevenson, K. E., and R. H. Vaughn.** 1972. Exosporium formation in sporulating cells of *Clostridium botulinum* 78A. *J. Bacteriol.* **112:**618–621.

213. **Stiles, M. E.** 1989. Less recognized or presumptive pathogenic bacteria, p. 673–733. *In* M. P. Doyle (ed.), *Foodborne Bacterial Pathogens,* Marcel Dekker, Inc., New York.

214. **Stragier, P.** 1994. A few good genes, p. 207–246. *In* P. J. Piggot, C. P. Moran, Jr., and P. Youngman (ed.), *Regulation of Bacterial Differentiation.* American Society for Microbiology, Washington, D.C.

215. **Stragier, P., C. Bonamy, and C. Karmazyn-Campelli.** 1988. Processing of a sporulation sigma factor in *Bacillus subtilis:* how morphological structure could control gene expression. *Cell* **52:**697–704.

216. **Stragier, P., P. Margolis, and R. Losick.** 1994. Establishment of compartment-specific gene expression during sporulation in *Bacillus subtilis,* p. 139–154. *In* P. J. Piggot, C. P. Moran, Jr., and P. Youngman (ed.), *Regulation of Bacterial Differentiation.* American Society for Microbiology, Washington, D.C.

217. **Strauch, M. A., D. de Mendoza, and J. A. Hoch.** 1992. *cis*-Unsaturated fatty acids specifically inhibit a signal-transcribing protein kinase required for initiation of sporulation in *Bacillus subtilis. Mol. Microbiol.* **6:**2909–2917.

218. **Sugiyama, H.** 1951. Studies on factors affecting the heat resistance of spores of *Clostridium botulinum. J. Bacteriol.* **62:**81–96.

219. **Sugiyama, H.** 1952. Effect of fatty acids on the heat resistance of *Clostridium botulinum* spores. *Bacteriol. Rev.* **16:**125–126.

220. **Sugiyama, H.** 1980. *Clostridium botulinum* neurotoxin. *Microbiol. Rev.* **44:**419–448.

221. **Sugiyama, H., M. Woodburn, K. H. Yang, and C. Movroydis.** 1983. Production of botulinum toxin in inoculated pack studies of foil-wrapped potatoes. *J. Food Prot.* **44:**896–898.

222. **Sun, D., R. M. Cabrera-Martinez, and P. Setlow.** 1991. Control of transcription of the *Bacillus subtilis spoIIIG* gene which codes for the forespore-specific transcription factor σ^G. *J. Bacteriol.* **173:**2977–2984.

223. **Swerdlow, B. M., B. Setlow, and P. Setlow.** 1981. Levels of H^+ and other monovalent cations in dormant and germinated spores of *Bacillus megaterium. J. Bacteriol.* **148:**20–29.

224. **Tanaka, N., E. Traisman, P. Plantinga, L. Finn, W. Flom, L. Meske, and J. Guffisberg.** 1986. Evaluation of factors involved in antibotulinal properties of pasteurized process cheese spreads. *J. Food Prot.* **49:**526–531.

225. **Tipper, D. J., and J. J. Gauthier.** 1972. Structure of the bacterial endospore, p. 3–12. *In* H. O. Halvorson, R. Hanson, and L. L. Campbell (ed.), *Spores V.* American Society for Microbiology, Washington, D.C.

226. **Townsend, C. T., J. R. Esty, and F. C. Baselt.** 1938. Heat-resistance studies on spores of putrefactive anaerobes in relation to the determination of safe processes for canned foods. *Food Res.* **3:**323–346.

227. **Trent, J. D., M. Gabrielson, B. Jensen, J. Neuhard, and J. Olsen.** 1994. Acquired thermotolerance and heat shock proteins in thermophiles from the three phylogenetic domains. *J. Bacteriol.* **176:**6148–6152.

228. **Vamvakopoulos, N. C., J. N. Vournakis, and S. J. Marcus.** 1977. The effect of magnesium and manganous ions on the structure and template activity for reverse transcriptase of polyribocytidine and its 2′-O-methyl derivative. *Nucleic Acids Res.* **4**:3589–3597.

229. **Van Netten, P., A. Van de Moosdijk, P. Van de Hoensel, D. A. A. Mossel, and I. Perales.** 1990. Psychrotrophic strains of *Bacillus cereus* producing enterotoxin. *J. Appl. Bacteriol.* **69**:73–79.

230. **Varley, A. W., and G. C. Stewart.** 1992. The *divIVB* region of the *Bacillus subtilis* chromosome encodes homologs of *Escherichia coli* septum placement (MinCD) and cell shape (MreBCD) determinants. *J. Bacteriol.* **174**:6729–6742.

231. **Waites, W. M., and C. E. Bayliss.** 1979. The effect of changes in spore coat on the destruction of *Bacillus cereus* spores by heat and chemical treatments. *J. Appl. Biochem.* **1**:71–76.

232. **Waldburger, C., D. Gonzalez, and G. H. Chambliss.** 1993. Characterization of a new sporulation factor in *Bacillus subtilis. J. Bacteriol.* **175**:6321–6327.

233. **Wang, J., C. Sass, and G. N. Bennett.** 1995. Sequence and arrangement of genes encoding sigma factors in *Clostridium acetobutylicum. Gene* **153**:89–92.

234. **Wang, S. Y.** 1976. Pyrimidine bimolecular photoproducts, p. 295–396. *In* S. Y. Wang (ed.), *Photochemistry and Photobiology of Nucleic Acids*, vol. I. Academic Press, Inc., New York.

235. **Wang, T.-C., and C. S. Rupert.** 1977. Evidence for the monomerization of spore photoproduct to two thymines by the light-independent "spore repair" process in *Bacillus subtilis. Photochem. Photobiol.* **25**:123–127.

236. **Wilkinson, B. J., J. A. Deans, and D. J. Ellar.** 1975. Biochemical evidence for the reversed polarity of the outer membrane of the bacterial forespore. *Biochem. J.* **152**:561–569.

237. **Williams, R. J. P.** 1981. The Bakerian lecture. Natural selection of the chemical elements. *Proc. R. Soc. Lond. Ser. B Biol. Sci.* **213**:361–397.

238. **Willis, A. T.** 1969. *Clostridia of Wound Infection.* Butterworths, London.

239. **Woese, C.** 1987. Bacterial evolution. *Microbiol. Rev.* **51**:221–271.

240. **Wu, L. J., and J. Errington.** 1994. *Bacillus subtilis* SpoIIIE protein required for DNA segregation during asymmetric cell division. *Science* **264**:572–575.

241. **Zamenhoff, S.** 1960. Effects of heating dry bacteria and spores on their phenotype and genotype. *Proc. Natl. Acad. Sci. USA* **46**:101–105.

242. **Zuber, P., M. M. Nakano, and M. A. Marahiel.** 1993. Peptide antibiotics, p. 897–916. *In* A. L. Sonenshein, J. A. Hoch, and R. Losick (ed.), *Bacillus subtilis and Other Gram-Positive Bacteria: Biochemistry, Physiology, and Molecular Genetics.* American Society for Microbiology, Washington, D.C.

243. **Zuberi, A. R., I. M. Feavers, and A. Moir.** 1987. The nucleotide sequence and gene organization of the *gerA* spore germination operon of *Bacillus subtilis* 168. *Gene* **162**:756–762.

L. Michele Smoot
Merle D. Pierson

4 Indicator Microorganisms and Microbiological Criteria

INTRODUCTION

Purpose of Microbiological Criteria

The numbers and types of microorganisms present in or on a food product may be used to judge the microbiological safety and quality of that product. Safety is determined by the presence or absence of pathogenic microorganisms or their toxins, the number of pathogens, and the expected control or destruction of these agents. Tests for indicator organisms may be used to assess either microbiological quality or safety when a relationship between the occurrence of the indicator organism and the likely presence of a pathogen or toxin has been established. The level of spoilage microorganisms reflects the microbiological quality, or wholesomeness, of a food product as well as the effectiveness of measures used to control or destroy such microorganisms. Currently, microbiological criteria are used to assess (i) the safety of food, (ii) adherence to Good Manufacturing Practices (GMPs), (iii) the keeping quality (shelf life) of certain perishable foods, and (iv) the utility (suitability) of a food or ingredient for a particular purpose (28). When appropriately applied, microbiological criteria can be a useful means for ensuring the safety and quality of foods, which in turn elevates consumer confidence. Microbiological criteria provide the food industry and regulatory agencies with guidelines for control of food processing systems. In addition, internationally accepted criteria can advance free trade through standardization of food safety and quality requirements.

Need To Establish Microbiological Criteria

A microbiological criterion should be established and implemented only when there is a need and when it can be shown to be both effective and practical. There are many considerations to be taken into account when establishing meaningful microbiological criteria. Listed below are factors considered important when assessing whether or not there is a need to establish a microbiological criterion (28):

- evidence of a hazard to health based on epidemiological data or a hazard analysis
- the nature of the natural and commonly acquired microflora of the food and the ability of the food to support microbial growth
- the effect of processing on the microflora of the food
- the potential for microbial contamination and/or growth during processing, handling, storage, and distribution
- the category of consumers at risk
- the state in which the food is distributed
- the potential for abuse at the consumer level
- spoilage potential, utility, and GMPs

- the manner in which the food is prepared for ultimate consumption
- reliability of methods available to detect and/or quantify the microorganism(s) and toxins of concern
- the costs and benefits associated with the application of the criterion

When establishing microbiological criteria for those foods intended for international trade, materials provided by the International Commission on Microbiological Specifications for Foods (ICMSF) (16) and the *Codex Alimentarius* (9) should be consulted for more detailed discussions.

Who Establishes Microbiological Criteria?

Different scientific organizations have been involved in developing general principles for application of microbiological criteria by regulatory agencies and the food industry. The scientific organizations which have influenced the United States food industry the most include the Joint Food and Agricultural Organization and World Health Organization Codex Alimentarius International Food Standards Program, the ICMSF, the U.S. National Academy of Sciences, and the U.S. National Advisory Committee on Microbiological Criteria for Foods.

The Codex Alimentarius Program first formulated General Principles for the Establishment and Application of Microbiological Criteria in 1981 (9). These principles have been well accepted internationally. In 1984, the National Research Council Subcommittee on Microbiological Criteria for Foods and Food Ingredients formulated general principles for the application of microbiological criteria to food and food ingredients as requested by four United States regulatory agencies (28). The basic principles are well established; however, these principles employ traditional testing, which is based on lot acceptance criteria. Improvements in the development and execution of microbiological criteria will continue to be made as the Hazard Analysis and Critical Control Point (HACCP) system, a science-based preventative system for food control, is implemented on a large scale.

Definitions

The National Research Council of the National Academy of Sciences addressed the issue of microbiological criteria in a 1985 report entitled *An Evaluation of the Role of Microbiological Criteria for Foods and Food Ingredients* (28). In that text, it was established that a microbiological criterion will stipulate that a type of microorganism, group of microorganisms, or toxin produced by a microorganism must either not be present at all, be present in only a limited number of samples, or be present at no more than a specified number or amount in a given quantity of a food or food ingredient. In addition, a microbiological criterion should include the following information (28):

- statement describing the identity of the food or food ingredient
- statement identifying the contaminant of concern
- analytical method to be used for the detection, enumeration, or quantification of each contaminant of concern
- sampling plan
- microbiological limits considered appropriate to the food and commensurate with the sampling plan

Criteria may be either mandatory or advisory. A mandatory criterion is a criterion that may not be exceeded and the food that does not meet the specified limit is required to be subjected to some action, including rejection, destruction, reprocessing, or diversion. An advisory criterion permits acceptability judgments to be made, and it should serve as an alert to deficiencies in processing, distribution, storage, or marketing. For application purposes, the three categories of criteria that are employed include standards, guidelines, and specifications. The following definitions were recommended in the National Research Council's Subcommittee on Microbiological Criteria for Foods and Food Ingredients (28). *Codex Alimentarius* (9) does not give specific definitions; however, they are implied in the Application of Microbiological Criteria section of their criteria principles document, as follows.

Standard: A microbiological criterion that is part of a law, ordinance, or administrative regulation. A standard is a mandatory criterion. Failure to comply constitutes a violation of the law, ordinance, or regulation and will be subject to the enforcement policy of the regulatory agency having jurisdiction (28). . . . [A] criterion contained in a Codex Alimentarius standard. Wherever possible it should contain limits only for pathogenic microorganisms of public health significance in the food concerned. Limits for nonpathogenic microorganisms may be necessary when the methods of detection for the pathogens of concern are cumbersome or unreliable. Standards based on fixed numbers of nonpathogenic microorganisms may result in the recall or downgrading of otherwise wholesome food. To minimize the effect of this approach penalty provisions could be applied when a lot is rejected. Such penalties would result in suspension of the privilege to process food only after repeated violations occur over a specified time period (9).

Guideline: A microbiological criterion often used by the food industry or regulatory agency to monitor a manufacturing process. Guidelines function as alert mechanisms to signal whether microbiological conditions prevailing at critical control points or in the finished product are within the normal range. Hence, they are used to assess processing efficiency at critical control points and conformity with Good Manufacturing Practices. A microbiological guideline is advisory (28). . . . is intended to increase assurance that the provisions of hygienic significance in the Code have been met. It may include microorganisms which are not of direct public health significance (9).

Specification: A microbiological criterion that is used as a purchase requirement whereby conformance becomes a condition of purchase between buyer and vendor of a food ingredient. A microbiological specification may be advisory or mandatory (28). . . . is applied at the establishment at a specified point during or after processing to monitor hygiene. It is intended to guide the manufacturer and is not intended for official control purposes (9).
The Codex use of "specification" only refers to end products and does not include raw materials, ingredients, or foods in contractual agreement between two parties.

It should be noted that there are some differences in the terminology used by Codex Alimentarius and the National Research Council. Currently, the Codex Committee is revising their document "General Principles for the Establishment and Application of Microbiological Criteria for Foods."

SAMPLING PLANS

A sampling plan includes both the sampling procedure and the decision criteria. To examine a food for the presence of microorganisms, either the entire lot or a representative sample is examined by defined methods. A lot is that quantity of product produced, handled, and stored within a limited time period under uniform conditions. Since it is impractical to examine the entire lot, statistical concepts of population probability and sampling must be used to determine the number and size of sample units from the lot and to provide conclusions drawn from the analytical results. The sampling plan is designed so that inferior lots within a set level of confidence are rejected. Detailed information regarding statistical concepts of population probabilities and sampling, choice of sampling procedures, decision criteria, and practical aspects of application as applied to microorganisms in food can be found in a publication by the ICMSF (16).

A simplified example of a sampling plan described in the aforementioned publication is given below. Suppose 10 samples were taken and analyzed for the presence of a particular microorganism. Based on the decision criterion, only a certain number of the sample units could be positive for the presence of that microorganism and the lot still be considered acceptable. If in the criterion the maximum allowable positive units had been set at 2 ($c = 2$), then a positive result for more than 2 of the 10 sample units ($n = 10$) would result in rejection of the lot. Ideally, the decision criterion is set to accept lots that are of the desired quality and reject lots that are not. However, since the entire lot is not examined, there is always the risk that an acceptable lot may be rejected or an unacceptable lot is accepted. The more samples examined, or the larger n, the lower the risk of making an incorrect decision about the lot quality. However, as n increases, sampling becomes more time-consuming and costly. Generally, a compromise is made between the size of n and the level of risk that is acceptable.

The level of risk that is associated with a particular sampling plan can be determined by an operating characteristic curve as seen in Fig. 4.1. The two vertical scales in the graph show (i) the probability of acceptance, P_a, or the ratio of the number of times that the results will indicate a lot should be accepted to the number of times a lot of the given quality is sampled for a decision, and (ii) the probability of rejection, P_r, or the ratio of the number of times that the result will indicate a lot should be rejected to the number of times a lot of that given quality is sampled for a decision. The horizontal axis shows the percent of defective sample units (p) that are in the lot. This probability is usually expressed as percentages of "defectives." The operating characteristic curve in Fig. 4.1, in which the $n = 10$ and $c = 2$, shows that as p increases, P_a will decrease. In other words, if the probability that a test unit will yield a positive result is great, then the probability that the lot will be accepted becomes small. The risks to be considered include those for the consumers or buyers and those for the vendors or producers. The vendors' risk is that probability that a lot of acceptable quality will be rejected, while the probability that a lot of defective quality will be accepted is referred to as the consumers'

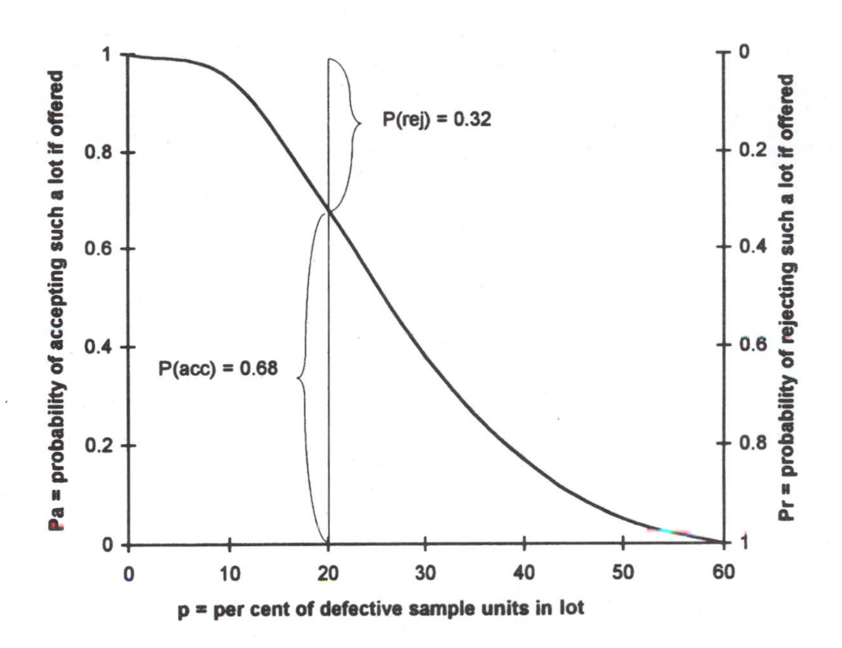

Figure 4.1 Operating characteristic curve for $n = 10$, $c = 2$. (From reference 16.)

risk. The level of risk that a vendor is willing to accept relative to himself and the consumer is then used to design the sampling plan.

Types of Sampling Plans

Microbiological sampling may be used (i) to determine the suitability of a food or food ingredient for its intended purpose and (ii) to monitor performance relative to accepted GMPs (28). Sampling plans are divided into two main categories: variables and attributes. Variables plans depend on the frequency distribution of organisms in the food. For correct application of a variables plan the organisms must be distributed log-normally (i.e., counts transformed to logarithms are normally distributed). When the food is from a common source and it is known to be produced or processed under uniform conditions, log-normal distribution of the organisms present is reasonably assumed (19, 20). However, when there is little or no knowledge of how the food was processed, or when no past performance record is available, an attributes plan is the preferred sampling plan. For this reason, attributes plans are widely used to determine acceptance or rejection of product at ports or other points of entry. Attributes sampling plans may also be used to monitor performance relative to accepted GMPs. Attributes sampling, however, is not appropriate when there is no defined lot or when random sampling is not possible, as might occur when monitoring cleaning practices.

Variables Sampling Plans

Assuming that there is a normal distribution of microorganisms within the lot, characteristics of this distribution can be used to develop acceptance sampling plans. A lot is rejected if $\bar{x} + k_1 s > V$, where $\bar{x}$ is the sample mean computed from the log concentration measurements, s is the standard deviation of these measurements, and V is the log concentration related to safety or quality limits. The value k_1 determines the stringency of the plan for a given number of sample units n and is obtained from the appropriate reference tables. Examples of k_1 values for sample unit numbers between 3 and 10 are given in Table 4.1. First, a decision must be made on the maximum proportion (p_d) of the units in the lot which may exceed the concentration V. The probability of rejecting a lot containing at least a proportion p_d above V is then selected. Based on the chosen P and p_d values, the table gives the value of k_1 over the sample unit numbers 3 to 10.

An example of a variables sampling plan taken from an ICMSF publication (16) follows. Suppose five sample units are to be analyzed per lot. If a lot in which 10% of the sample units exceed V was to be rejected with a probability of 0.95, then the k_1 value of 3.4 is chosen. The larger n is, the smaller the chance of rejecting a lot of acceptable quality and safety.

When evaluating a Good Manufacturing Practice (GMP) limit, the criterion used to determine lot acceptance is $\bar{x} + k_2 s < V$. Separate reference tables for k_2 values are available (Table 4.2). The limit values and V are often

Table 4.1 k_1 values calculated using the noncentral *t* distribution: safety and quality specification (reject if $\bar{x} + k_1s > V$)[a]

Probability (P) of rejection	Proportion (p_d) exceeding V	k_1 value for no. of sample units:							
		3	4	5	6	7	8	9	10
0.95	0.05	7.7	5.1	4.2	3.7	3.4	3.2	3.0	2.9
	0.1	6.2	4.2	3.4	3.0	2.8	2.6	2.4	2.4
	0.3	3.3	2.3	1.9	1.6	1.5	1.4	1.3	1.3
0.90	0.1	4.3	3.2	2.7	2.5	2.3	2.2	2.1	2.1
	0.25	2.6	2.0	1.7	1.5	1.4	1.4	1.3	1.3

[a]Source, reference 16.

numerically similar to the three-class attribute values *M* and *m*, respectively. For more detailed discussion on variables plans the reader is referred to Kilsby (19), Kilsby et al. (20), and Malcolm (22).

Attributes Sampling Plans

Two-Class Plans

The two-class attributes sampling plan assigns the concentration of microorganisms of the sample units tested to a particular attribute class depending on whether the microbiological counts are above or below some preset concentration represented by the letter *m*. The decision criterion is based on (i) the number of sample units tested, *n*, and (ii) the maximum allowable number of sample units yielding unsatisfactory tests results, *c*. For example, when $n = 5$ and $c = 2$ in a two-class sampling plan designed to make a presence/absence decision on the lot (i.e., $m = 0$), the lot is rejected if more than 2 of the 5 sample units tested are positive. As *n* increases for the set number *c*, the stringency of the sampling plan also increases. Conversely, for a set sample size *n*, as *c* increases, the stringency of the sampling plan decreases, allowing for a higher probability of accepting (P_a) food lots of a given quality.

Three-Class Plans

Three-class sampling plans use the concentration of microorganisms in the sample units to determine levels of quality and/or safety. Counts above a preset concentration *M* for any of the *n* sample units tested are considered unacceptable, and the lot is rejected. The level of the test organism which is acceptable in the food is denoted by *m*. This concentration in a three-class attribute plan separates acceptable lots (i.e., counts less than *m*) from marginally acceptable lots (i.e., counts greater than *m* but not exceeding *M*). Counts above *m* and up to and including *M* are not desirable, but the lot can be accepted provided the number of *n* samples that exceed *m*

Table 4.2 k_2 values calculated using the noncentral *t* distribution: GMP limit (accept if $\bar{x} + k_2s < V$)[a]

Probability (*P*) of rejection	Proportion (*p*) exceeding V	k_2 value[b] for no. of sample units:							
		3	4	5	6	7	8	9	10
0.90	0.05	0.84	0.92	0.98	1.03	1.07	1.10	1.12	1.15
	0.10	0.53	0.62	0.68	0.72	0.75	0.78	0.81	0.83
	0.20	0.11	0.21	0.27	0.32	0.35	0.38	0.41	0.43
	0.30	0.26*	0.13*	0.05*	0.01	0.04	0.07	0.10	0.12
	0.40	0.65*	0.46*	0.36*	0.30*	0.25*	0.21*	0.17*	0.16*
	0.50	1.09*	0.82*	0.69*	0.60*	0.54*	0.50*	0.47*	0.44*
0.75	0.01	1.87	1.90	1.92	1.94	1.96	1.98	2.00	2.01
	0.05	1.25	1.28	1.31	1.33	1.34	1.36	1.37	1.38
	0.10	0.91	0.94	0.97	0.99	1.01	1.02	1.03	1.04
	0.25	0.31	0.35	0.38	0.41	0.42	0.44	0.45	0.46
	0.50	0.47*	0.38*	0.33*	0.30*	0.27*	0.25*	0.24*	0.22*

[a]Source, reference 16.
[b]*, negative value.

is no greater than the preset number, c. Thus, in a three-class sampling plan, the food lot will be rejected if any one of the sample units exceeds M or if the number of sample units with contamination levels above m exceeds c. Similar to the two-class sampling plan, the stringency of the three-class sampling plan is also dependent on the two numbers designated by n and c. The larger the value of n for a given value of c, the better the food quality must be to have the same chance of passing, and vice versa. From n and c it is then possible to find the probability of acceptance, P_a, for a food lot of a given microbiological quality.

INDICATORS OF MICROBIOLOGICAL QUALITY

The shelf life of a perishable product is often determined by the number of microorganisms initially present. As a general rule, a food containing a large population of spoilage organisms will have a shorter shelf life than the same food containing fewer numbers of the same spoilage organisms. However, the relationship between total counts and shelf life is not without exceptions. Some types of microorganisms have a greater impact on the organoleptic characteristics of a food than others due to the presence of different enzymes acting upon the food constituents. In addition to the effect of certain levels and/or types of spoilage microorganisms, changes in perceptible quality characteristics will also vary depending on the food and the conditions of storage such as temperature and gaseous atmosphere. Use of microbiological criteria to determine shelf life requires an understanding of the processing conditions and expected microflora of the product.

Microbiological criteria may also be used to monitor adherence to GMPs. Foods produced and stored under GMPs may be expected to have a different microbiological profile than those foods produced and stored under poor conditions. The use of poor-quality materials, improper handling, or unsanitary conditions may result in higher bacterial counts in the finished product. However, low counts in the finished product do not necessarily mean that GMPs were adhered to. Processing steps such as heat treatments, fermentation, freezing, or frozen storage can reduce the counts of bacteria that have resulted from noncompliance with GMPs. Other products, such as ground beef, may normally contain high microbial counts even under the best conditions of manufacture due to the growth of psychrotrophic bacteria during refrigeration. Therefore, a working knowledge of the types and levels of microorganisms present at the different processing steps is needed to establish a relationship between the microbiology of the food and adherence to GMPs.

Table 4.3 Some organisms that are highly correlated with product quality[a]

Organism	Product(s)
Acetobacter spp.	Fresh cider
Bacillus spp.	Bread dough
Byssochlamys spp.	Canned fruits
Clostridium spp.	Hard cheeses
Flat-sour spores	Canned vegetables
Lactic acid bacteria	Beer, wine
Lactococcus lactis	Raw milk (refrigerated)
Leuconostoc mesenteroides	Sugar (during refinery)
Pectinatus cerevisiiphilus	Beer
"Pseudomonas putrefaciens"	Butter
Yeasts	Fruit juice concentrates
Zygosaccharomyces bailii	Mayonnaise, salad dressing

[a]Source, reference 18.

Specifications are often used in microbiological criteria to determine the usefulness of a food or food ingredient for a particular purpose. For example, specifications are set for thermophilic spores in sugar and spices that are to be used in the canning industry. Lots of sugar failing to meet specifications may not be suitable for use in low-acid canning but could be diverted for other uses.

Indicator Microorganisms

Indicator microorganisms can be used in microbiological criteria. These criteria might be used to address existing product quality or to predict the shelf life of the food. It has been suggested that the ideal indicators of product quality or shelf life should meet the following requirements (18):

- they should be present and detectable in all foods whose quality is to be assessed
- their growth and numbers should have a direct negative correlation with product quality
- they should be easily detected and enumerated and be clearly distinguishable from other organisms
- they should be enumerable in a short period of time, ideally within a working day
- their growth should not be affected adversely by other components of the food flora

Some examples of indicators and the products in which they are used are shown in Table 4.3.

Those microorganisms listed in Table 4.3 are the primary spoilage organisms corresponding to each specific product. Loss of quality in other products may not be due to one organism only but to a variety of organisms, owing to the unrestricted environment of the food. In those types of products it is often more practical to determine the counts of groups of microorganisms most likely to cause spoilage in that particular food.

The aerobic plate count (APC) or the standard plate count (SPC) may be a component of microbiological criteria assessing product quality when those criteria are used to (i) monitor foods for compliance with standards or guidelines set by various regulatory agencies, (ii) monitor foods for compliance with purchase specifications, and (iii) monitor adherence to GMPs (28). By modifying the environment of incubation or the medium used, the APC can be used to preferentially screen for groups of microorganisms such as thermodurics, mesophiles, psychrotrophiles, thermophiles, and proteolytic or lipolytic microorganisms. The APC of refrigerated perishable foods such as milk, meat, poultry, and fish may be used to indicate the condition of equipment and utensils used, as well as the time-temperature profile of storage and distribution of the food.

When evaluating results of APC for a particular food, it is important to remember that (i) APCs only measure live cells, and therefore it would not be of value, for example, to determine the quality of raw materials used for a heat-processed food; (ii) APCs are of little value in assessing organoleptic quality, since high microbial counts are generally required prior to organoleptic quality loss, and (iii) since different bacteria vary in their biochemical activities, quality loss may also occur at low total counts depending on the predominant organisms present. With any food, specific causes of unexpected high counts can be identified by examination of samples at control points and by plant inspection. Reliable interpretation of the APC of a food requires knowledge of the expected microbial population at the point in the process or distribution at which the sample is collected. If counts are higher than expected, this will point to the need to determine why there has been a violation of the criterion. A review by Silliker (32) provides a detailed discussion of the use of total counts as an index of sanitary quality, organoleptic quality, and safety.

The direct microscopic count (DMC) is used to give an estimate of both viable and nonviable cells in samples containing a large number of microorganisms (i.e., $>10^5$ CFU/ml). Considering that the DMC does not differentiate between live and dead cells (unless a fluorescent dye such as acridine orange is employed) and it requires that the total cell count exceed 10^5, the use of the DMC is limited as part of microbiological criterion for quality issues. However, the DMC is used as part of microbiological criteria for foods or ingredients such as raw, non-grade A milk; dried milks; liquid and frozen eggs; and dried eggs (28).

Other methods commonly used to indicate quality of different food products include the Howard mold count, yeast and mold count, heat-resistant mold count, and thermophilic spore count. The Howard mold count is used to detect the inclusion of moldy material in canned fruit and tomato products (13) as well as to evaluate the sanitary condition of processing machinery in vegetable canneries (12). Yeasts and molds frequently become predominant on foods when conditions for bacterial growth are less favorable. Therefore, they can potentially be a problem in fermented dairy products, fruits, fruit beverages, and soft drinks. Yeast and mold counts are used as part of microbiological criteria of various dairy products such as cottage cheese and frozen cream (34) and sugar (29). Heat-resistant molds, such as *Byssochlamys fulva* and *Aspergillus fisheri,* that may survive the thermal processes applied to fruit and fruit products may need limits in purchase specifications for ingredients such as fruit concentrates. Concern for thermophilic spores in ingredients used in the canning industry is related to their ability to cause defects in foods held at elevated temperatures because of inadequate cooling and/or storage at too high temperatures. Purchase specifications and verification criteria are often used for thermophilic spore counts in ingredients intended for use in low-acid, heat-processed canned foods.

Metabolic Products

In certain cases, bacterial populations in a food product can also be estimated by testing for metabolic products produced by the microorganisms present in the food. When a correlation is established between the presence of a metabolic product and product quality loss, tests for the metabolite may be a part of a microbiological criterion.

An example of the use of metabolic products as part of a microbiological criterion is the organoleptic evaluation of imported shrimp. Trained personnel are able to classify the degree of decomposition (i.e., quality loss) into one of three classes through organoleptic examination. The shrimp are placed into one of the following quality classes: class 1, passable; class 2, decomposed (slight but definite); class 3, decomposed (advanced). Limits of acceptability of a lot are based on the number of shrimp in a sample that are placed into each of the three classes (4). Other commodities in which organoleptic examination is used to determine quality deterioration include raw milk, meat, poultry, and fish and other seafoods. The food industry also uses these examinations to classify certain foods into quality grades. For

Table 4.4 Some microbial metabolic products that correlate with food quality[a]

Metabolite(s)	Applicable food product(s)
Cadaverine and putrescine	Vacuum-packaged beef
Diacetyl	Frozen juice concentrate
Ethanol	Apple juice, fishery products
Histamine	Canned tuna
Lactic acid	Canned vegetables
Trimethylamine (TMA)	Fish
Total volatile bases (TVB), total volatile nitrogen (TVN)	Seafoods
Volatile fatty acids	Butter, cream

[a]Source, reference 16.

additional information about organoleptic examination of foods the reader is referred to Amerine et al. (3) and Larmond (21).

Other examples of metabolic products used to assess product quality are listed in Table 4.4.

INDICATORS OF FOODBORNE PATHOGENS AND TOXINS

Microbiological criteria as they apply to product safety should only be developed when the application of the criterion can reduce or eliminate a potential foodborne hazard. Each food type should be carefully evaluated through risk assessment to determine the potential hazards and their significance to consumers.

When a food is repeatedly implicated as a vehicle in foodborne disease outbreaks, application of microbiological criteria may be useful. Public health officials and the dairy industry responded to widespread outbreaks of milk-borne disease occurring around the turn of the century in the United States. By imposing controls on milk production, developing safe and effective pasteurization procedures, and setting microbiological criteria, the safety of commercial milk supplies was greatly improved. Epidemiological evidence alone, however, does not necessitate imposing microbiological criteria. The criteria should only be applied when their use results in a safer food (1).

Food products frequently subject to contamination by harmful microorganisms may benefit from the application of microbiological criteria. The National Shellfish Sanitation Program utilizes microbiological criteria in this manner to prevent use of shellfish from polluted waters which may contain various intestinal pathogens (2). Depending on the type and level of contamination anticipated, imposition of microbiological criteria may or may not be justified. Food contaminated with pathogens that do not have the opportunity to grow to levels that would potentially result in a health hazard does not warrant microbiological criteria. Though fresh vegetables are often contaminated with small numbers of *Clostridium botulinum, Clostridium perfringens,* and *Bacillus cereus,* epidemiological evidence indicates that this contamination presents no health hazard. Thus imposing microbiological criteria would not be beneficial for protection against these microorganisms. However, criteria may be appropriate for enteric pathogens on produce since there have been several outbreaks of foodborne illness resulting from fresh produce contaminated with enteric pathogens (6).

Often food processors alter the intrinsic or extrinsic parameters of a food (nutrients, pH, water activity, inhibitory chemicals, gaseous atmosphere, temperature of storage, and the presence of competing organisms) to prevent growth of undesirable microorganisms. If control over one or more of these parameters is lost, then there may be a risk of a health hazard. For example, in the manufacture of cheese or fermented sausage, a lactic acid starter culture is relied upon to produce acid quickly enough to inhibit the growth of *Staphylococcus aureus* to levels that are potentially harmful. Critical control points in a process such as rate of acid formation are implemented to assure prevention of growth of, or contamination of the food by, harmful microorganisms. Associated with each critical control point are one or more critical limits, such as a certain level of acid production within a given amount of time, to assure that control of the critical control point is maintained. These limits are set in order to assure that a microbiological criterion such as preventing a certain level of growth of *S. aureus* is met (30).

Depending on the pathogen, low levels of the microorganism in the food product may or may not be of concern. Some microorganisms have such a low infective dose that their mere presence in a food presents a significant public health risk. For such organisms the concern is not about the ability of the pathogen to grow in the food, but that the microorganism could survive for any length of time in the food. Foods in which the environment is hostile enough to prevent survival of the pathogens or toxigenic microorganisms of concern may not be a candidate for microbiological criteria related to safety. For example, the acidity of certain foods such as fermented meat products might be assumed sufficient for pathogen control. In fact, the growth and toxin production of *S. aureus* might be prevented, but enteric pathogens such as *Escherichia coli* O157:H7 could survive, and thus the product could be unsafe for consumption.

Table 4.5 Suggested sampling plans for combinations of degrees of health and conditions of use[a]

Degree of concern relative to utility and health hazard	Conditions in which food is expected to be handled and consumed after sampling, in the usual course of events[b]		
	Conditions reduce degree of concern	Conditions cause no change in concern	Conditions may increase concern
No direct health hazard Utility, e.g., shelf life and spoilage	Increase shelf-life Case 1 3-class $n = 5$, $c = 3$	No change Case 2 3-class $n = 5$, $c = 2$	Reduce shelf-life Case 3 3-class $n = 5$, $c = 1$
Health hazard Low, indirect (indicator)	Reduce hazard Case 4 3-class $n = 5$, $c = 3$	No change Case 5 3-class $n = 5$, $c = 2$	Increase hazard Case 6 3-class $n = 5$, $c = 1$
Moderate, direct, limited spread	Case 7 3-class $n = 5$, $c = 2$	Case 8 3-class $n = 5$, $c = 1$	Case 9 3-class $n = 10$, $c = 1$
Moderate, direct, potentially extensive spread	Case 10 2-class $n = 5$, $c = 0$	Case 11 2-class $n = 10$, $c = 0$	Case 12 2-class $n = 20$, $c = 0$
Severe, direct	Case 13 2-class $n = 15$, $c = 0$	Case 14 2-class $n = 30$, $c = 0$	Case 15 2-class $n = 60$, $c = 0$

[a]Source, reference 16.
[b]More stringent sampling plans would generally be used for sensitive foods destined for susceptible populations. *n*, number of sample units drawn from a lot; *c*, maximum allowable number of positive results.

An important and sometimes overlooked consideration when evaluating the potential microbiological risks associated with a food is the consumer. More rigid microbiological requirements may be needed if the food is intended for use by infants or elderly or immunocompromised people since they are more susceptible to infectious agents than are healthy adults.

The sampling plan specified in a microbiological criterion should be appropriate to the hazard expected to be associated with the food, the consumer of the food, and the severity of the illness. The hazard associated with a food is determined by (i) the type of organism expected to be encountered and (ii) the expected conditions of handling and consumption after sampling. A more stringent sampling plan is desired for products expected to contain higher degrees of hazards. ICMSF (16) proposed a system for classification of foods according to risk into 15 hazard categories, called cases, with suggested appropriate sampling plans as shown in Table 4.5.

The stringency of sampling plans for foods is based either on the hazard to the consumer from pathogenic microorganisms and their toxins or toxic metabolites or on the potential for quality deterioration to an unacceptable state, and it should take account of the types of microorganisms present and their numbers (16). Foodborne pathogens are grouped into one of three categories based on the severity of the potential hazard (i.e., severe hazards, moderate hazards with potentially extensive spread, and moderate hazards with limited spread), as shown in Table 4.6. Pathogens with potential for extensive spread are often initially associated with specific foods; however, secondary spread to other foods commonly occurs from environmental contamination and cross-contamination within processing plants and food preparation areas, including homes. An example is fresh beef that is contaminated with *E. coli* O157:H7. One or a few contaminated pieces of meat can lead to widespread contamination of product during processing such as grinding to produce ground beef. There can also be cross-contamination if the fresh beef is improperly stored with ready-to-eat foods. Microbial pathogens in the lowest-risk group (moderate hazards, limited spread) are found in many foods, usually in small numbers. Generally, illness is caused only when ingested foods contain large numbers of the pathogen, e.g., *C. perfringens*, or have at some time contained large enough numbers to produce sufficient toxin to cause illness, e.g., *S. aureus*. Outbreaks are usually restricted to consumers of a particular meal or a particular kind of food (25, 28). Summaries of those vehicles associated with outbreaks of foodborne diseases that occurred in the United States, and the microorganisms involved, have been prepared by Bryan (7) for the period 1977–1984 and by Bean et al. (5) for the period 1983–1987.

Table 4.6 Hazardous microorganisms and parasites grouped on the basis of risk severity[a]

Severe hazards
Clostridium botulinum types A, B, E, and F
Shigella dysenteriae
Salmonella typhi serotypes paratyphi A and B
Enterohemorrhagic *Escherichia coli*
Hepatitis A and E virus
Brucella abortus, Brucella suis
Vibrio cholerae O1
Vibrio vulnificus
Taenia solium
Moderate hazards: potentially extensive spread[b]
Listeria monocytogenes
Salmonella spp.
Shigella spp.
Other enterovirulent *Escherichia coli*
Streptococcus pyogenes
Rotavirus
Norwalk virus group
Entamoeba histolytica
Diphyllobothrium latum
Ascaris lumbricoides
Cryptosporidium parvum
Moderate hazards: limited spread
Bacillus cereus
Campylobacter jejuni
Clostridium perfringens
Staphylococcus aureus
Vibrio cholerae non-O1
Vibrio parahaemolyticus
Yersinia enterocolitica
Giardia lamblia
Taenia saginata

[a]Source, reference 31.
[b]Although these organisms are classified as moderate hazards, complications and sequelae may be severe in certain susceptible populations.

Indicator Organisms

Microbiological criteria for food safety may use tests for indicator organisms which suggest the possibility of a microbial hazard. *E. coli* in drinking water, for example, indicates possible fecal contamination and, therefore, the potential presence of enteric pathogens. Direct tests for pathogenic microorganisms and their toxins, excepting *Salmonella* spp. and *S. aureus*, are not routinely applied to foods for quality control purposes. Jay (18) suggested that indicators used to assess food safety should ideally meet the following criteria:

- be easily and rapidly detectable
- be easily distinguishable from other members of the food flora
- have a history of constant association with the pathogen whose presence it is to indicate
- always be present when the pathogen of concern is present
- be an organism whose number ideally should correlate with those of the pathogen of concern
- possess growth requirements and a growth rate equaling those of the pathogen
- have a die-off rate that at least parallels that of the pathogen of concern and ideally persists slightly longer
- be absent from foods that are free of the pathogen except perhaps at certain minimum numbers

Buttiaux and Mossel (8) suggested additional criteria for fecal indicators used in food safety. They include the following:

- ideally the bacteria selected should demonstrate specificity, occurring only in intestinal environments
- they should occur in very high numbers in feces so as to be encountered in high dilutions
- they should possess a high resistance to the external environment, the pollution of which is to be assessed
- they should permit relatively easy and fully reliable detection even when present in low numbers

With the exception of *Salmonella* spp. and *S. aureus*, most tests for assuring safety use indicator organisms rather than direct tests for the specific hazard. An overview of some of the more common indicator microorganisms used for assuring food safety is given below.

Fecal Coliforms and *E. coli*

Fecal coliforms, including *E. coli*, are easily destroyed by heat and may die during freezing and frozen storage of foods. Microbiological criteria involving *E. coli* are useful in those cases where it is desirable to determine if fecal contamination may have occurred. Contamination of a food with *E. coli* implies a risk that other enteric pathogens may be present in the food. Fecal coliform bacteria are used as a component of microbiological standards to monitor the wholesomeness of shellfish and the quality of shellfish growing waters (15, 38). The purpose of this is to reduce the risk of harvesting shellfish from waters polluted with fecal material. The fecal coliforms have a higher probability of containing organisms of fecal origin than do coliform groups which comprise organisms of both fecal and nonfecal origins. Fecal coliforms can become established on equipment and utensils in the food processing environments and contaminate processed foods. At present, *E. coli* is the most widely used indicator of fecal contamination. The failure to detect *E. coli* in a food, however, does not guarantee the absence of enteric pathogens (15, 33). In many raw foods of animal origin, small numbers of *E. coli* can be expected because of the close association of these foods with the animal environment and the likelihood of contamination of carcasses from fecal material, hides, or feathers during slaughter and dressing procedures. The presence of *E. coli* in a heat-processed food means either process failure or, more commonly, postprocessing contamination from equipment or employees or from contact with contaminated raw foods. Rapid direct plating methods for *E. coli* now exist, and in some foods it may be advantageous to use *E. coli* rather than fecal coliforms as a component of microbiological criteria for foods.

Enterococci

Sources of enterococci include fecal material from both warm-blooded and cold-blooded animals and plants (24). Enterococci differ from coliforms in that they are salt tolerant (grow in the presence of 6.5% NaCl) and relatively resistant to freezing (18). Certain enterococci (*Enterococcus faecalis* and *Enterococcus faecium*) are also relatively heat resistant and may survive usual milk pasteurization temperatures. Enterococci can establish and persist in food processing environments for long periods. Because many foods contain small numbers of enterococci, a thorough understanding of the role and significance of enterococci in a food is required before any meaning can be attached to their presence and population numbers. Enterococcal counts have few useful applications in microbiological criteria for food safety. This indicator may be applicable in specific cases to identify poor manufacturing practices.

Metabolic Products

Certain microbiological criteria related to safety rely on tests for metabolites to indicate a potential hazard rather than on direct tests for pathogenic or indicator microorganisms. Examples of the use of metabolites as a component of microbiological criteria include (i) tests for thermonuclease or thermostable DNase in foods containing or suspected of containing $\geq 10^6$ cells of *S. aureus* per ml or per g; (ii) illuminating grains under UV light to detect the presence of aflatoxin produced by *Aspergillus* spp.; and (iii) assaying for the enzyme alkaline phosphatase, a natural constituent of milk that is inactivated during pasteurization, to detect postpasteurization contamination or contamination of pasteurized milk with raw milk (18, 23, 28).

APPLICATION AND SPECIFIC PROPOSALS FOR MICROBIOLOGICAL CRITERIA FOR FOOD AND FOOD INGREDIENTS

It has been recommended that the application of useful microbiological criteria should address the following issues: (i) the sensitivity of the food product(s) relative to safety and quality; (ii) the need for a microbiological standard(s) and/or guideline(s); (iii) assessment of information necessary for establishment for a criterion if one seems to be indicated; and (iv) where the criterion should be applied (28). In this section, examples of the application of microbiological criteria to various foods and food ingredients are presented. No general criteria suitable for all food product groups exist. The relevant background literature that relates to the quality and safety of a specific food product should be consulted prior to implementing microbiological criteria.

The utilization of microbiological criteria, as stated previously, may be either mandatory (standards) or advisory (guidelines, specifications). One obvious example of a mandatory criterion (standard) as it relates to food safety is the "zero tolerance" set for *Salmonella* spp. in all ready-to-eat foods. The U.S. Department of Agriculture Food Safety Inspection Service (USDA/FSIS) has mandated a zero tolerance for *E. coli* O157:H7 in fresh ground beef. Since the emergence of *Listeria monocytogenes* as a foodborne pathogen, the U.S. Food and Drug Administration and USDA have also applied a zero tolerance for this organism in all ready-to-eat foods.

Table 4.7 Summary of cases and sampling plans for *L. monocytogenes*[a]

		Conditions in which food is expected to be handled and consumed after sampling in the usual course of events		
Intended consumer	Health hazard	Reduce degree of hazard	Cause no change in hazard	May increase hazard
Normal individuals	Moderate, direct, potentially extensive spread	**Case 10** $n = 5$ (≤100/g)[b]	**Case 11** $n = 10$ (≤100/g)	**Case 12** $n = 20$ (≤100/g)
Highly susceptible individuals	Severe, direct	**Case 13** $n = 15$ (<1/375 g)	**Case 14** $n = 30$ (<1/750 g)	**Case 15** $n = 60$ (< 1/1,500 g)

[a]Source, reference 17.

[b]Parentheses indicate maximum permissible organism count.

Table 4.8 Examples of food products for which advisory microbiological criteria have been established

Product category	Test parameters	Case	Plan class	*n*	c	Organism limit per g		Reference
						m	*M*	
"Roast" beef	*Salmonella*	12	2	20	0	0		16
Pâté	*Salmonella*	12	2	20	0	0		
Raw chicken	APC	1	3	5	3	5×10^5	10^7	16
Cooked poultry, frozen, ready to eat	*S. aureus*	8	3	5	1	10^3	10^4	16
Cooked poultry, frozen, to be reheated	*S. aureus*	8	3	5	1	10^3	10^4	16
	Salmonella	10	2	5	0	0		
Chocolate/ confectionery	*Salmonella*	11	2	10[a]	0	0		16
Dried milk	APC	2	3	5	2	3×10^4	3×10^5	16
	Coliforms	5	3	5	1	10	10^2	
	Salmonella[b] (normal routine)	10	2	5	0	0		
		11	2	10	0	0		
		12	2	20	0	0		
	Salmonella (high-risk populations)	10	2	15	0	0		
		11	2	30	0	0		
		12	2	60	0	0		
Fresh cheese[c]	*S. aureus*			5	2	10^2	10^3	10
	Coliforms			5	2	10^2	10^3	
Soft cheese[c]	*S. aureus*			5	2	10^2	10^3	10
	Coliforms			5	2	10^2	10^3	
Pasteurized liquid, frozen and dried egg products	APC	2	3	5	2	5×10^4	10^6	16
	Coliforms	5	3	5	2	10^3	10^3	
	Salmonella[a] (normal routine)	10	2	5	0	0		
		11	2	10	0	0		
		12	2	20	0	0		
	Salmonella[a] (high-risk populations)	10	2	15	0	0		
		11	2	30	0	0		
		12	2	60	0	0		
Fresh and frozen fish, to be cooked before eating	APC	1	3	5	3	5×10^3	10^7	16
	E. coli		4		3	11	500	
	Salmonella[d]	10	2	5	0	0		
	V. parahaemolyticus[d]	7	3	5	2	10^2	10^3	
	S. aureus[d]	7	3	5	2	10^3	10^4	
Coconut	*Salmonella*	1	3	5	3	5×10^5	10^7	16
	Growth not expected	11	2	10	0	0		
	Growth expected	12	2	20	0	0		

[a]The 25-g analytical unit may be composited.
[b]The case is to be chosen based on whether the hazard is expected to be reduced, unchanged, or increased.
[c]Requirements only for fresh and soft cheese made from pasteurized milk.
[d]For fish known to derive from inshore or inland waters of doubtful bacteriological quality, or where fish are to be eaten raw, additional tests may be desirable.

Table 4.9 Examples of food products for which mandatory microbiological criteria have been established

Product category	Test parameters	Comments	Reference
United States			
Dairy products			
Raw milk	Aerobic bacteria	Recommendations of U.S. Public Health Service	39
Grade A pasteurized milk	Aerobic bacteria; coliforms	Recommendations of U.S. Public Health Service	
Grade A pasteurized (cultured) milk	Aerobic bacteria; coliforms	Recommendations of U.S. Public Health Service	39
Dry milk (whole)	Standard plate count; coliforms	Recommendations of U.S. Public Health Service	39
Dry milk (non-fat)	Standard plate count; coliforms	Standards of Agriculture Marketing Service (USDA)	34
Frozen desserts	Standard plate count; coliforms	Recommendations of U.S. Public Health Service	39
Starch and sugars	Total thermophilic spore count; flat-sour spores; thermophilic anaerobic spores; sulfide spoilage spores	National Canners Assoc. (NFPA)	26
Breaded shrimp	Aerobic plate counts; *E. coli*; *S. aureus*	FDA Compliance Policy Guide	14
International			
Caseins and caseinates	Total bacterial count; thermophilic organisms; coliforms	Europe	28
Natural mineral waters	Aerobic mesophilic count; coliforms; *E. coli*; fecal streptococci; sporulating sulfite-reducing anaerobes; *Pseudomonas aeruginosa*, parasites, and pathogenic organisms	*Codex*	28
Tomato juice	Mold count	Canada	28
Fish protein	Total plate count; *E. coli*	Canada	28
Gelatin	Total plate count; coliforms; *Salmonella*	Canada	28

However, there is considerable debate as to whether a zero tolerance is warranted for *L. monocytogenes* (11). The ICMSF (17) has recommended that the international community allow for a tolerance of *L. monocytogenes* as outlined in Table 4.7. These recommendations are an example of advisory criteria. Other advisory criteria which have been established to address the safety of foods are shown in Table 4.8.

Mandatory criteria in the form of standards have also been employed for quality issues. Presented in Table 4.9 are examples of foods and food ingredients for which federal, state, and city as well as international microbiological standards have been developed. Some of the advisory criteria that have been developed and applied to foods for quality monitoring are given in Table 4.8.

CURRENT STATUS AND LEGISLATIVE BASIS

The ICMSF is currently recommending that one criterion, a specification, be used. Such a specification, however, could be both mandatory and advisory depending upon its use by either regulatory authorities or food processors. The concept of one criterion with both mandatory and advisory functions is simpler. It has been suggested that a term other than "specification" be used to avoid confusion with the current meaning of the word in the United States (28).

The USDA/FSIS has recently proposed regulations intended to reduce the level of pathogenic microorganisms in meat and poultry products (35). As part of the proposed rule, microbiological testing of carcasses, ground meat, and poultry for the presence of *Salmonella* spp. would be required. The results of daily *Salmonella* testing, presence or absence, would be evaluated over a specific time period using a statistical procedure known as the "moving window sum" to determine if the process is in control. The presence of *Salmonella* spp. is compared to a target frequency by examining a group of consecutive days (the window) of a specific duration in relation to a prespecified acceptable limit. The proposed targets are summarized in Table 4.10. If the number of positive samples is less than or equal to the acceptable limit, the process is in control. As each new sampling day is added, the window moves 1 day. The stringency

Table 4.10 Moving sum rules for various commodities[a]

Commodity	Target (% positive *Salmonella*)	Window size (n) in days	Acceptable limit[b]
Steers/heifers	1	82	1
Cows/bulls	1	82	1
Raw ground beef	4	38	2
Fresh sausage	12	19	3
Turkeys	18	15	3
Hogs	18	17	4
Broilers	25	16	5

[a]Source, reference 36.

[b]There is an approximately 80% probability of meeting the acceptable limit when the process percent positive equals the target.

and sensitivity of the evaluation can be altered by changing the size of the window, the acceptable limit, or the number of samples to be taken each day. The current acceptable limits are based on the criterion that there is an 80% probability that the plant is actually exceeding the target value if it exceeds the acceptable limit. As in any statistical approach, one needs to develop a program that (i) has a low probability of exceeding the limit when the producer is meeting the target and (ii) has a high probability of exceeding the limit when the producer is not meeting the target. The 80% level was deemed as being a reasonable balance between the two, thus providing reasonable decision criteria. This verification procedure will give meat and poultry producers an opportunity to measure their process performance in relation to baseline data and national targets for pathogen reduction. Establishments whose production process exceeds the national targets will have to reevaluate their processing controls and, with FSIS oversight, initiate corrective actions (37).

References

1. **Acuff, G. R.** 1993. Microbiological criteria, p. A6.01–A6.07. *In Proceedings of the World Congress on Meat and Poultry Inspection,* 10–14 October 1993. U.S. Department of Agriculture, Food Safety Inspection Service, Washington, D.C.
2. **Ahmed, F. E. (ed.).** 1991. *Seafood Safety.* National Academy Press, Washington, D.C.
3. **Amerine, M. A., R. M. Pangborn, and E. B. Roessler.** 1965. *Principles of Sensory Evaluation of Food.* Academic Press, Inc., New York.
4. **Anonymous.** 1979. Shrimp decomposition workshop. National Shrimp Breaders and Processors Association, National Fisheries Institute, and Food and Drug Administration, Tampa, Fla.
5. **Bean, N. H., P. M. Griffin, J. S. Goulding, and C. B. Ivey.** 1990. Food borne disease outbreaks, 5-year summary, 1983–1987. *J. Food Prot.* **53:**711–728.
6. **Beuchat, L. R.** 1996. Pathogenic microorganisms associated with fresh produce. *J. Food Prot.* **59:**204–216.
7. **Bryan, F. L.** 1988. Risks associated with vehicles of food borne pathogens and toxins. *J. Food Prot.* **51:**498–508.
8. **Buttiaux, R., and D. A. A. Mossel.** 1961. The significance of various organisms of faecal origin in foods and drinking water. *J. Appl. Bacteriol.* **24:**353–364.
9. **Codex Alimentarius Commission, 14th Session.** 1981. *Report of the 17th Session of the Codex Committee on Food Hygiene.* Alinorm 81/13. Food and Agriculture Organization, Rome.
10. **Codex Alimentarius Commission, 20th Session.** 1993. *Report of the 20th Session of the Codex Commission on Food Hygiene.* Alinorm 93/13A. Food and Agriculture Organization, Rome.
11. **Doyle, M. P.** 1991. Should regulatory agencies reconsider the policy of zero-tolerance of *Listeria monocytogenes* in all ready-to-eat foods? *Food Safety Noteb.* **2:**98.
12. **Eisenberg, W. V., and S. M. Cichowicz.** 1977. Machinery mold—indicator organism in food. *Food Technol.* **31(2):** 52–56.
13. **Food and Drug Administration.** 1978. *The Food Defect Action Levels.* HFF-342. Food and Drug Administration, Washington, D.C.
14. **Food and Drug Administration.** 1989. Raw breaded shrimp—microbiological criteria for evaluating compliance with current good manufacturing practice regulations, chapter 8. *In Compliance Policy Guides,* 7108.25. Food and Drug Administration, Washington, D.C.
15. **Hackney, C. R., and M. D. Pierson (ed.).** 1994. *Environmental Indicators and Shellfish Safety.* Chapman and Hall, New York.
16. **International Commission on Microbiological Specifications for Food.** 1986. *Microorganisms in Foods 2. Sampling for Microbiological Analysis: Principles and Applications,* 2nd ed. University of Toronto Press, Toronto.
17. **International Commission on Microbiological Specifications for Food.** 1993. *Choice of Sampling Plan and Criteria for Listeria monocytogenes.* ICMSF, Papaendal, The Netherlands.
18. **Jay, J. M.** 1992. *Modern Food Microbiology,* 4th ed. Chapman and Hall, New York.
19. **Kilsby, D.** 1982. Sampling schemes and limits, p. 387–421. *In* M. H. Brown (ed.), *Meat Microbiology.* Applied Science Publishers, London.
20. **Kilsby, D., L. J. Aspinall, and A. C. Baird-Parker.** 1979. A system for setting numerical microbiological specifications for foods. *J. Appl. Bacteriol.* **46:**591–599.
21. **Larmond, E.** 1977. *Laboratory Methods for Sensory Evaluation of Food.* Publication no. 1937. Research Branch, Canada Department of Agriculture, Ottawa.
22. **Malcolm, S.** 1984. A note on the use of the non-central t-distribution in setting numerical specifications for foods. *J. Appl. Bacteriol.* **57:**175–177.
23. **Marshall, R. T. (ed.).** 1992. *Standard Methods for the Examination of Dairy Products,* 16th ed. American Public Health Association, Washington, D.C.
24. **Mundt, J. O.** 1970. Lactic acid bacteria associated with raw plant food materials. *J. Milk Food Technol.* **33:**550–553.

25. **National Advisory Committee on Microbiological Criteria for Foods.** 1993. Generic HACCP for raw beef. *Food Microbiol.* **10**:449–488.
26. **National Canners Association.** 1968. *Laboratory Manual for Food Canners and Processors,* vol. 1. AVI Publishing, Westport, Conn.
27. **National Marine Fisheries.** 1995. *Proposed International Commission on Microbiological Specifications for Food Changes to the Current Codex Alimentarius General Principles for the Establishment and Application of Microbiological Criteria for Foods.* National Seafood Inspection Laboratory, U.S. Department of Commerce, Pascagoula, Miss.
28. **National Research Council.** 1985. *An Evaluation of The Role of Microbiological Criteria for Foods and Food Ingredients.* National Academic Press, Washington, D.C.
29. **National Soft Drink Association.** 1975. *Quality Specifications and Test Procedures for "Bottler's Granulated and Liquid Sugar."* National Soft Drink Association, Washington, D.C.
30. **Pierson, M. D.** 1996. Critical limits: significance and determination of critical limits, p. 72–78. *In* D. Peters and R. Peters (ed.), *Proceedings of the 2nd Australian HACCP Conference.* Sydney, Australia.
31. **Pierson, M. D., and D. A. Corlett, Jr. (ed.).** 1992. *HACCP: Principles and Applications.* Van Nostrand Reinhold, New York.
32. **Silliker, J. H.** 1963. Total counts as indexes of food quality, p. 102–112. *In* L. W. Slanetz, C. O. Chichester, A. R. Gaufin, and Z. J. Ordal (ed.), *Microbiological Quality of Foods.* Academic Press, Inc., New York.
33. **Silliker, J. H., and D. A. Gabis.** 1976. ICMSF method studies. VII. Indicator tests as substitutes for direct testing of dried foods and feeds for *Salmonella. Can. J. Microbiol.* **22**:971–974.
34. **U.S. Department of Agriculture.** 1975. General specifications for approved dairy plants and standards for grades of dairy products. *Fed. Regist.* **40(198)**:47910–47940.
35. **U.S. Department of Agriculture.** 1995. Pathogen reduction; hazard analysis critical control point (HACCP) systems; proposed rule. *Fed. Regist.* **60(23)**:6774–6889.
36. **U.S. Department of Agriculture.** 1995. *Moving Sum Procedures for Microbial Testing in Meat and Poultry Establishments.* FSIS, Science & Technology Program, Washington, D.C.
37. **U.S. Department of Agriculture–Food Safety and Inspection Service.** 1995. Pathogen reduction; Hazard Analysis and Critical Control Point (HACCP) systems; proposed rule. *Fed. Regist.* **60**:6774–6889.
38. **U.S. Department of Health, Education and Welfare.** 1965. *National Shellfish Sanitation Program, Manual of Operations.* Part 1. *Sanitation of Shellfish Growing Areas.* U.S. Government Printing Office, Washington, D.C.
39. **U.S. Public Health Service–Food and Drug Administration.** 1978. *Grade A Pasteurized Milk Ordinance. 1978 Recommendations.* PHS/FDA Publication no. 229. U.S. Government Printing Office, Washington, D.C.

Microbial Spoilage of Foods

II

Timothy C. Jackson
Gary R. Acuff
James S. Dickson

5 Meat, Poultry, and Seafood

INTRODUCTION

Meat, poultry, and seafood (muscle foods) are described as spoiled if organoleptic changes make them unacceptable to the consumer. Factors associated with spoilage may include color defects or changes in texture, the development of off flavors, off odors, and slime, or any other characteristic making the food undesirable for consumption. While enzymatic activity within muscle tissues contributes to changes during storage, organoleptically detectable spoilage is generally a result of decomposition and the formation of metabolites resulting from the growth of microorganisms.

The definition of the onset of spoilage is to some extent subjective, being influenced by cultural and economic factors as well as by the sensory acuity of the consumer and the intensity of the defect. However, despite variations in expectations, most consumers would agree that gross discoloration, strong off odors, and the development of slime constitute spoilage (49). The types of spoilage defects produced in muscle foods vary with the type of microflora, muscle species, product composition, and storage environment.

ECOLOGY OF THE SPOILAGE MICROFLORA OF MUSCLE FOODS

Origin of Microflora in Red Meats

Grau (57) stated that meat animals may be regarded as a source of edible tissue sandwiched between two regions that are heavily contaminated with microorganisms. While there has been some debate as to the sterility of muscle tissue, the relative ease with which sterile muscle tissue can be obtained (48) suggests that, if present, populations of bacteria in muscle tissues of healthy live animals are extremely low (62). High numbers of bacteria are present on the hide, hair, and hooves of red meat animals, as well as in the gastrointestinal tract (96). Microorganisms on the hide include *Staphylococcus, Micrococcus,* and *Pseudomonas* species, yeasts and molds, which are normally associated with skin microflora, and species contributed by fecal material and soil (62). The population and composition of this microflora are influenced by environmental conditions. Wet or muddy hides may contain high populations of bacteria indigenous to soil, and contamination of the hide with fecal material may increase the proportion of microorganisms of fecal origin (62).

It is generally agreed that the majority of bacteria on a dressed red meat carcass originate from the hide (57, 88). Initially, the tissue surface beneath the hide is essentially free of bacteria; however, once exposed, this tissue may be inoculated with bacteria from processing activities and the environment. During hide removal, bacteria are carried from the hide onto the underlying tissue with the initial incision. Further transfer of bacteria can occur from the aerosols and dust generated from the hide during removal, from contact with workers' hands, or from contact of the hide or fleece with the exposed tissue surface (57).

Unlike the case for cattle and sheep, the skin of hogs usually is not removed but rather is scalded and left on the carcass (57). While scalding reduces the numbers of microorganisms on the skin, recontamination can occur during dehairing because of the presence of debris in dehairing machines (57, 67). The singeing process, used to burn hair remaining on the carcass after the dehairing process, kills a portion of microorganisms that are naturally present; however, some areas of the carcass surface may receive inadequate heat treatment (57). Microorganisms which may be present in deeper layers of the surface tissue are also protected from the heat applied in the singeing process.

Microorganisms may also be introduced to the carcass surface during the evisceration process. Contamination can occur if the intestinal tract is pierced or if fecal material is introduced from the rectum during the removal of abdominal contents, and handling can result in cross-contamination of other carcasses. Careful evisceration will reduce the potential for contamination (57). In addition to being present in the hide and viscera, bacteria may originate from sources in the processing environment, such as floors, walls, contact surfaces, knives, and workers' hands. Rapid chilling of carcasses at low temperatures and relative humidities and at high air velocities may result in a reduction of bacterial populations, whereas milder chilling conditions may allow the growth of psychrotrophs, increasing their population compared with that of mesophiles. Mesophiles may grow if carcasses are chilled at temperatures of ≥15°C (62).

During the fabrication of meat into subprimal and retail cuts, bacteria present on the tissue surface are transferred by knives and workers' hands onto newly exposed meat surfaces. This is especially true for comminuted meats, in which case considerable new surface areas are created, and microorganisms are distributed throughout the product from the grinding equipment and the use of trimmings. Microorganisms in processed meats originate not only from the meat itself but also from ingredients such as spices, salt, and dried milk powder. Of concern in the processing of injected meats, such as hams, is the microbiological quality of the brine solutions used.

Origin of Microflora in Poultry

As with other meat animals, the internal tissues of healthy poultry are essentially free from bacteria. The skin, feathers, and feet of the bird harbor microorganisms resident to the skin as well as those from litter and feces. Although present on the skin, psychrotrophic microorganisms consisting primarily of *Acinetobacter* and *Moraxella* species are primarily associated with the feathers (57). Contamination and cross-contamination with fecal material may occur during transportation of birds from growing houses to slaughter facilities. Crowding and poor weather conditions during transport may stress the birds, leading to more frequent excretion of fecal material and cecal contents (83). Frequently, birds are withheld from food for several hours before slaughter to minimize intestinal contents and reduce the extent of fecal contamination during slaughter (83, 120).

Additional cross-contamination may occur as birds are hung and bled; the flapping of wings may generate dust and aerosols transferred onto nearby birds or carcasses (14). To facilitate the removal of feathers, carcasses are scalded, usually by immersion in a continuous-flow water tank at 60 to 63°C. During this operation, bacteria from the carcasses are washed into the scald water. Many bacteria are removed from the carcasses, and some are killed by the high temperature of the scald water. Psychrotrophs are readily inactivated in the scald water (57). The defeathering process may spread microorganisms between carcasses or from the defeathering equipment, and consequently there may be an increase in the numbers of psychrotrophs and aerobic mesophiles on the carcasses (57, 115). As with other meat animals, the evisceration process provides an opportunity for cross-contamination from knives, equipment, and workers' hands (14). Spray washing of carcasses after picking and evisceration, but before chilling, will remove organic material and will also remove some of the bacteria potentially introduced during evisceration (14), although bacteria firmly attached to the carcass surface will remain (57, 83).

Freshly eviscerated carcasses are chilled rapidly to limit the multiplication of spoilage microorganisms and restrict the growth of pathogens. Slush ice chilling, continuous-immersion chilling, spray chilling, air chilling, and carbon dioxide chilling systems have been utilized or proposed for this purpose (14). The characteristics of each may influence the microbial population.

Origin of Microflora in Finfish

The population and composition of microflora on finfish are influenced by the environment from which they are taken, the season, and conditions of harvesting, handling, and processing. Water temperature has a significant influence on the initial number and types of bacteria on the surfaces of fish. Higher numbers of bacteria are generally present on fish from warm subtropical or tropical waters than on fish from colder waters (105).

Fish taken from temperate waters harbor predominantly psychrotrophic bacteria, while mesophilic bacteria predominate on fish taken from tropical areas (76). In addition, populations of bacteria may be influenced by the water quality, with higher numbers occurring on the surfaces of fish taken from polluted waters. The activity of the fish will influence populations in the intestine, higher numbers being present in feeding fish than in nonfeeding fish (75). Bacteria from the genera *Acinetobacter, Cytophaga, Flavobacterium, Moraxella, Pseudomonas, Shewanella* (formerly *Alteromonas*), and *Vibrio* dominate on fish and shellfish taken from temperate waters, while *Bacillus* species, coryneforms, and *Micrococcus* species frequently predominate on fish taken from subtropical and tropical waters (76, 105). The composition of the microflora of fish from freshwater environments is also influenced by temperature and will vary from that in marine environments, reflective of the influence of bacteria present in the surrounding terrestrial environment (76).

The initial microflora on fish is influenced by the method of harvesting. Trawled fish generally have larger microbial populations than those that are line caught. In trawling, the dragging of fish and debris along the ocean bottom stirs up mud, thus contaminating the fish. In addition, the compaction of fish in trawling nets may cause expression of intestinal contents, with subsequent contamination of the fish surface (105). Handling and storage of fish aboard the fishing vessel will also affect bacterial populations. Fish are often stored in ice or chilled brine or are frozen while awaiting transportation to the processing plant (76). The microbiological quality of brine water and ice is a concern, as is the variation in temperature of the fish (105). A delay in icing fish also enhances the possibility of rapid microbial growth.

Nets and rough handling may result in penetration or bruising of the fish, providing an avenue for penetration of bacteria into the muscle tissue. Large catches require more time to stow and, while awaiting storage, may be subject to physical abuse and high temperatures on deck. Liston (76) noted that unlike the control over time of death possible in meat and poultry processing, methods of harvesting and shipboard working conditions allow little control over those variables in fish and shellfish. Additionally, contamination may occur from equipment and handling during processing.

Origin of Microflora in Shellfish

Unlike other crustacean shellfish which are alive until heat processed, shrimp die soon after harvesting. Decomposition begins soon after death and involves bacteria on the shrimp surface which originate from the marine environment or from contamination during handling and washing (41). Removal of heads before storage reduces the overall bacterial population (41); however, in the process bacteria may be transferred to the tail meat (84). Crabs are cooked before meat is removed, but bacteria are reintroduced from the environment or from raw crabs during picking and processing. Molluscan shellfish are sessile and filter feeders, and thus their microflora varies, reflecting the quality of water in which they reside (65), the quality of wash water, and other factors.

Bacterial Attachment

Microbial spoilage of meat, poultry, and seafood generally occurs as a result of the growth of bacteria that have colonized muscle surfaces. The first stage in colonization and growth involves the attachment of microbial cells to the surface (12, 33, 43). Bacterial attachment to muscle surfaces involves two stages (43). The first is a loose, reversible sorption which may be related to van der Waals forces or other physicochemical factors (78). One of the factors that influences attachment at this point is the population of bacteria in the water film (23, 43). The second stage consists of an irreversible attachment to surfaces involving the production of an extracellular polysaccharide layer known as a glycocalyx (25).

In addition to cell density, factors such as type of surface, growth phase, temperature, and motility may also influence bacterial attachment to muscle surfaces (15, 23, 32, 40, 74, 90). Bacteria already present on surfaces may influence the ability of other bacteria to attach to surfaces (80). However, Farber and Idziak (40) reported that minimal competition occurs between meat spoilage bacteria during attachment to beef longissimus dorsi muscle. Significant competition does not occur in the attachment of spoilage and pathogenic bacteria to fat and lean surfaces of beef (23). Differences have been reported in the rates of attachment of certain bacterial strains. Compared with several other spoilage bacteria, *Pseudomonas* species have been observed to attach more rapidly to meat surfaces (15, 44).

MICROBIAL PROGRESSION DURING STORAGE

The initial microflora of muscle foods is highly variable, arising from the resident microorganisms in and on the live animal and from environmental sources such as vegetation, water and soil, ingredients used in meat products, workers' hands, and contact surfaces in processing facilities. Despite this variability and differences between muscle tissues of different species, characteristics of microbial spoilage are remarkably similar. A large portion of marketed perishable meat and poultry products are stored at refrigeration temperatures to prolong their shelf life. Refrigeration will restrict the growth of mesophiles, generally a major component of the initial microflora, and allow psychrotrophic microorganisms to grow and eventually dominate the microflora. As microbial growth occurs during storage, the composition of the microflora is altered so that it is dominated by a few, or often a single, microbial species, usually of the genus *Pseudomonas, Lactobacillus, Moraxella*, or *Acinetobacter*, or *Brochothrix thermosphacta* (50). Although these bacteria often make up only a small proportion of the initial microflora, the types of bacteria that ultimately predominate during storage are reflective of these genera as well as characteristics of the muscle tissue and skin and characteristics of the storage environment.

Gill (50) noted that the final composition of the spoilage microflora may be influenced by the proportion of specific spoilage microorganisms in the initial population. High initial numbers of a slowly growing species may compete successfully with lower numbers of a species with a faster growth rate. If the number of spoilage microorganisms in the initial population is high, a slower growth rate may not be an important factor, since less growth may be necessary before spoilage occurs (50).

Pseudomonas species are typically able to compete successfully on aerobically stored, refrigerated muscle foods for several reasons (70). The genus *Pseudomonas* is characterized by a competitive growth rate, even at refrigeration temperatures (49). In addition, pseudomonads are able to grow within the usual pH range of muscle foods (5.5 to 7.0), while many bacteria, e.g., *Moraxella* and *Acinetobacter* species, are less capable of competing under refrigeration temperatures at the lower pH in this range. However, muscle foods of higher pH may allow for bacteria such as *Moraxella* and *Acinetobacter* species to constitute a higher proportion of the microflora (49, 54). Because pseudomonads are highly oxidative, they are able to utilize low-molecular-weight nitrogen compounds as sources of energy. This is a clear competitive advantage for pseudomonads, since meats contain relatively low levels of simple sugars, and more complex energy sources such as protein and fat do not serve as significant substrates for growth until later in spoilage when high bacterial populations are attained.

Dominance of certain types of aerobic spoilage bacteria during spoilage is likely a result of their ability to utilize meat constituents under certain storage conditions rather than due to direct interactions between the competing species themselves. Gill and Newton (51) concluded that the interaction occurs only when high population densities of pseudomonads are able to restrict the growth of competitors, perhaps by sequestering available oxygen.

During storage under vacuum and modified atmospheres, the growth of the aerobic spoilage microflora is suppressed (Table 5.1). Under these conditions, lactic acid bacteria and *B. thermosphacta* predominate. Lactic acid bacteria are favored by their growth rate, their fermentative metabolism, and also their ability to grow at the pH range of meat (37). Some strains produce substances (bacteriocins) which inhibit the growth of competitors (85, 99). At higher pH, *B. thermosphacta* may grow and contribute to spoilage; however, it cannot grow anaerobically when the pH is below 5.8 (17, 56). The frequent occurrence of high populations of *B. thermosphacta* in vacuum-packaged pork and lamb has been attributed to the higher pH of these products. A higher fat content in muscle tissue and more frequent occurrence of high-pH meat have been cited as contributing factors (37, 104). A higher pH may also allow *Shewanella putrefaciens*, which does not grow below pH 6.0, to contribute to the spoilage microflora (86).

There is a recent report of an unusual spoilage of vacuum-packaged refrigerated fresh and cooked beef caused by a new *Clostridium* species, named *C. laramie* (66), which has the ability to grow at 0°C or below and can sporulate and germinate at 2°C. Occurring in normal-pH product during storage at 2°C or below, spoilage is characterized by an initial color of pinkish red, changing to green and production of large quantities of hydrogen sulfide and purge with considerable proteolysis of the meat.

The water activity (a_w) of some types of processed meats is lowered by dehydration or the addition of solutes such as salt or sugar. As a_w is lowered, the growth of some spoilage microorganisms is restricted and the characteristic spoilage microflora is altered. Below a_w 0.98, growth of gram-negative spoilage bacteria is restricted (20). Lactic acid bacteria may grow at a_w as low as 0.93 to 0.94, and micrococci may grow at even lower a_w (117). When low a_w restricts the growth of bacteria, growth of fungi may occur. Molds and yeasts may be involved in the spoilage of products in the a_w range of 0.85 to 0.93, and below a_w 0.85, xerophilic molds or osmophilic yeasts may be involved (20). Microbial growth does not occur on products with a_w below 0.60 (20).

MUSCLE TISSUE AS A GROWTH MEDIUM

Fresh muscle tissue is a highly favorable environment for microbial growth and, as a result, is subject to rapid spoilage unless modified or stored in an environment

Table 5.1 Percentage distribution of microflora of beef knuckles vacuum packaged and stored for 21 days at 0 to 2°C in packages with oxygen transmission rates ranging from 1 to 400 $cm^2/m^2/24\ h$[a]

Microbial type	% Distribution[b] of microflora of beef knuckles packaged in films with oxygen transmission rates ($cm^2/m^2/24$ h) of:					
	1	10	12	13	30	400[c]
Micrococcus					2.6	
Lactobacillus coryneformis				6.1		
Lactobacillus plantarum	11.9		5.6	15.1	14.0	
Lactobacillus cellobiosus	45.7	57.0	65.7	5.5	6.6	3.9
All *Lactobacillus* spp.	57.6	57.0	71.3	26.7	20.6	3.9
Leuconostoc mesenteroides	1.1	38.3	15.0	39.8	51.3	12.4
Leuconostoc paramesenteroides	35.9		12.5	28.5	6.7	
All *Leuconostoc* spp.	37.0	38.3	27.5	68.3	58.0	12.4
Streptococcus		2.3				
Brochothrix thermosphacta			1.0	0.1	4.6	3.8
Coryneform bacteria	1.1			1.5		
Staphylococcus				1.8		
Moraxella-Acinetobacter		0.3	0.2	0.1		
Flavobacterium	2.2					
Pseudomonas		1.6		1.5	14.2	79.9
Erwinia herbicola	1.1					
Aeromonas	1.0	0.3				
All gram-negative rods	4.3	2.2	0.2	1.6	14.2	79.9

[a]From reference 102.
[b]Figures are averages of three knuckles.
[c]Figures are for microflora at 14 days; storage of knuckles packaged and stored in this film was not extended beyond 14 days.

designed to retard microbial activity and reproduction (119). The usable water content of fresh muscle tissue is high, and readily available glycogen, peptides, and amino acids, as well as metal ions and soluble phosphorus, contribute to the suitability of muscle tissue as a substrate to support microbial growth. Detailed reviews of the composition and spoilage processes of postmortem muscle tissues have been published (48, 49, 62, 65, 70, 72, 91).

Because the onset of spoilage of meats, poultry, and seafood is a subjective judgment, there has been some disagreement as to the number of bacteria present at the point when spoilage is detected. It is generally agreed, however, that spoilage defects in meat become evident when the number of bacteria at the surface reaches 10^7 CFU/cm^2 (62). During aerobic spoilage, off odors are first detected when populations reach 10^7 CFU/cm^2. When numbers reach 10^8 CFU/cm^2, the muscle tissue surface will begin to feel tacky, representing the first stage in slime formation (61). Slime formation is attributable to the growth of bacteria and synthesis of polysaccharide, which gradually form a confluent, sticky layer on the surface of the tissue (70). Since spoilage characteristics do not become evident until amino acids are degraded, the concentration of glucose present in the tissue is a primary factor governing the time necessary for onset of aerobic spoilage (48, 70).

Composition and Spoilage of Red Meats

The a_w of lean muscle tissue of red meats is 0.99, with a corresponding water content of 74 to 80%. The protein content may vary from 15 to 22% on a wet weight basis. The lipid content of intact red meats varies from 2.5 to 37%; carbohydrate composition ranges from 0 to 1.2% (70). While the bulk composition of muscle tissue does not change as a result of muscle rigor, significant changes in the concentration of some low-molecular-weight compounds do occur (54). The cessation of muscle cell respiration results in an end to ATP synthesis. Glycolysis leads to the accumulation of lactic acid, and as a result, the pH of the muscle tissue decreases. The final pH and residual glycogen content of the tissue are influenced by the initial glycogen content in the muscle. In tissue with a high initial glycogen store, the pH may decrease to 5.5 before enzymatic activity associated with glycolysis ends as a result of an inability to maintain ATP concentration. Where low amounts of glycogen are initially present, there is a direct correlation between glycogen content and final pH down to 5.8; however, tissue with a lower ultimate pH always contains some residual glycogen when glycolytic activity ceases, since ATP concentration is the limiting factor. In addition to glycogen, glycolytic intermediates such as glucose 6-phosphate and glucose are reduced to low levels following rigor (54).

The decrease in muscle pH and accumulation of various metabolites following rigor facilitate denaturation of some proteins. The release of proteolytic enzymes, such as cathepsins, from lysosomes results in a small amount of protein breakdown (54, 72). Additionally, the reticuloendothelial system in muscle cells ceases to scavenge and therefore ceases to restrict the growth of microorganisms (72). Soluble low-molecular-weight compounds constitute 1.2 to 3.5% of muscle tissue.

Spoilage characteristics of meat products will be influenced by the type of microflora involved. The composition of such microflora is related to intrinsic characteristics of the product as well as to extrinsic factors such as temperature and the composition of the atmosphere in the storage environment. It is well established that the shelf life of perishable meat and other muscle tissues is extended by storage at refrigeration temperatures; consequently, much of the research on meat spoilage has been done on products stored at these temperatures. Storage temperature will influence the type and rate of growth of microorganisms that develop and, consequently, their patterns of substrate utilization and production of metabolites.

The spoilage of meats stored at ambient temperature results from the growth of mesophiles, predominantly *Clostridium perfringens* and members of the family *Enterobacteriaceae* (61, 65). Spoilage deep within muscle tissues, known as sours or bone taint, has been attributed to a slow cooling of carcasses, resulting in the growth of anaerobic mesophiles said to be already present in muscle tissues (16, 91). The natural presence of bacteria within meat tissues has been debated. Some have considered that low populations of bacteria are universally present in muscle tissue freshly removed from the carcass and that this is why bone taint develops. Ingram and Dainty (61) suggested that at ambient temperatures, spoilage occurs as a result of the growth of bacteria within muscle tissue which is so rapid that it precedes spoilage at the meat surface. Because these bacteria are seldom psychrotrophic, spoilage of meats stored at refrigeration temperatures is caused by bacteria on the meat surface. Others have argued that populations of bacteria isolated from tissues involved in bone taint are frequently too low to be directly responsible for such spoilage, and other processes may be involved (103). Gill (48) asserted that the spoilage of meats at ambient temperatures is a surface phenomenon, since the ease with which sterile muscle tissue can be obtained suggets that the interior of intact meat is normally sterile. While possible, spoilage of meat by bacteria present in deep muscle tissues is rare under hygienic practices. A condition similar to bone taint in pork has been attributed to the introduction of spoilage bacteria into tissues in the curing brine, since the condition is most often observed in cured meats (91).

Storage at reduced temperatures will restrict the growth of mesophiles, allowing psychrotrophs to grow and dominate the spoilage microflora. If the surface of whole carcasses or fresh meat cuts becomes dry, bacterial growth may be restricted and fungal spoilage may occur. The growth of molds in the genera *Thamnidium*, *Mucor*, and *Rhizopus* may result in the production of a whiskery, airy, or cottony gray to black growth on beef due to the presence of mycelia. Other fungal genera may produce discolored areas on superficial layers of connective tissue or fat layers covering the muscle tissue (7). Black spot has been attributed to the growth of *Cladosporidium* species, white spot has been attributed to the growth of *Sporotrichum* and *Chrysosporium* species, and green patches have been attributed to the growth of *Penicillium* species (77). Molds will not grow on beef held at temperatures below −5°C (77).

Spoilage of meat surfaces with high moisture content when held under conditions of high relative humidity generally results from bacterial activity. Under aerobic storage conditions, the spoilage microflora is predominated by species of *Pseudomonas*, *Alcaligenes*, *Acinetobacter*, *Moraxella*, and *Aeromonas*.

Most spoilage bacteria, including *Pseudomonas* species, preferentially utilize glucose as a carbon source. As glucose is utilized at the surface of meat, it must diffuse from the bulk of the meat to the surface to replenish the supply for microbial growth (48). When glucose diffusion from deeper tissues declines, lactate and amino acids are utilized. Degradation of amino acids by the spoilage microflora results in the production of ammonia, hydrogen sulfide, indole, skatole, amines, and other compounds, resulting in undesirable odors, flavors, and colors when their concentration becomes detectable to human senses (62, 64, 119). *Acinetobacter* and *Moraxella* species are thought to contribute little to spoilage defects since, unlike *Pseudomonas* species, they apparently lack the ability to readily produce off-odor compounds from the breakdown of amino acids (50).

Composition and Spoilage of Poultry Muscle

The nutritive composition of poultry muscle is similar to that of red meats, and thus the mechanism of microbial spoilage is also similar. The composition of raw poultry muscle tissue varies with age, sex, anatomy, and species (14). Fat is not distributed throughout the muscle tissue as in red meat but is present in the abdominal cavity and beneath the skin. Spoilage of poultry meat has been associated with the growth of pseudomonads, *S. putrefaciens, Acinetobacter* species, and *Moraxella* species (9, 81, 82). Poultry skin and muscle provide excellent growth substrates for spoilage microorganisms. Spoilage is generally restricted to the outer surfaces of the skin and cuts and has been characterized by off odors, sliminess, and various types of discolorations. Skin may provide a barrier to the introduction of spoilage microorganisms to underlying muscle tissue.

The pH of poultry muscle varies with muscle type and age of the bird. Breast muscle typically has a pH of 5.7 to 5.9 (14), and its spoilage patterns have been equated to those of fresh red meats. Under aerobic storage, pseudomonads make up the dominant spoilage microflora (9, 81, 86). The spoilage microflora of leg muscle, typically having a pH of 6.4 to 6.7 (14), has been compared with that of dark, firm, dry (DFD) meat (86). Pseudomonads are also the predominant spoilage microorganisms in leg muscle; however, the high pH also facilitates the growth of aeromonads and *S. putrefaciens* to populations at which they can contribute to spoilage (9, 82).

Clark (24) reported that in contrast to differences in spoilage patterns between DFD meat and normal red meat, there was no difference in the rate of spoilage of breast and leg muscle, which led Newton and Gill (86) to conclude that despite a high pH, chicken leg muscle must contain sufficient available glucose for bacterial growth. For poultry carcasses packaged in oxygen-impermeable films, spoilage may be caused by *S. putrefaciens, B. thermosphacta,* and atypical lactobacilli. The potential for the production of sulfide compounds such as hydrogen sulfide, dimethyl sulfide, and methyl mercaptan makes *S. putrefaciens* an important component of the spoilage microflora (45, 86).

New York dressed poultry is stored uneviscerated before cooking. Under refrigeration, such carcasses have a longer shelf life than those that have been eviscerated. If the carcass skin is not broken and remains dry, the growth of psychrotrophic spoilage microorganisms on the surface is restricted. Instead, spoilage results from the production of hydrogen sulfide by bacteria in the intestines, which diffuses into the muscle tissue. A reaction occurs between the hydrogen sulfide and blood and muscle pigments upon exposure to air to form sulphmyoglobin, a green pigment (83).

Composition and Spoilage of Finfish

The composition of fish muscle is highly variable between species and may fluctuate widely depending on size, season, and fishing grounds (105). The average composition of nonfatty fish, such as cod, has been characterized as 80% water, 18% protein, <1% lipid, and 1% carbohydrate (70). In contrast, the fat and water content of fatty fish, such as herring, varies widely. Lipid content may range from 1 to 30%, with water content varying so that fat and water constitute approximately 80% of the muscle tissue (70). According to Shewan (105), nonprotein soluble components constitute about 1.5% of fish muscle; their composition and concentration vary with species, and within species may vary with size, season, and fishing ground (105). Such components consist of sugars, minerals, vitamins, and nonprotein nitrogen compounds such as free amino acids, ammonia, trimethylamine oxide, creatine, taurine, anserine, uric acid, betaine, carnosine, and histamine (13, 65, 105). Elasmobranchs (sharks, rays) contain about twice as much soluble components as other fish.

As with other muscle foods, the spoilage microflora of fresh ice-stored fish consist largely of *Pseudomonas* species. *S. putrefaciens* may also contribute to the spoilage of seafood (18, 59, 76, 86), and *Acinetobacter* and *Moraxella* species may compose a portion of the spoilage population (2, 59). Shewan (106) outlined the development of spoilage characteristics of fresh fish, such as cod, stored in ice. During the first phase (0 to 6 days), no marked spoilage occurs. The second phase (7 to 10 days) is marked by a lack of odor in the muscle tissue. During the third phase (11 to 14 days), some sourness or slightly sweet to fruity odors become evident, while during the fourth phase (>14 days), sulfur odors, such as of hydrogen sulfide, are apparent. Fecal or potent ammonia odors may also develop. The low a_w of dried and salted fish will suppress the growth of bacteria; therefore, any spoilage that occurs with these products will likely be the result of mold growth (65).

The early stages of spoilage involve utilization of nonprotein nitrogen, resulting in the formation and accumulation of fatty acids, ammonia, and volatile amines (76, 106). An important example is trimethylamine, which is produced when trimethylamine oxide in the muscle tissue is reduced by bacterial activity. Trimethylamine oxide has widespread occurrence in marine animals but is not commonly found in freshwater fish (13).

The compound can be converted by microbial activity into trimethylamine, a major contributor to the fishy odor characteristic of some seafood spoilage (13). As proteolysis proceeds, spoilage occurs. Hydrogen sulfide and other sulfur compounds such as mercaptans and dimethyl sulfide, produced by *S. putrefaciens* and some pseudomonads, may contribute to spoilage (70, 105). While autolytic activities of enzymes present in seafood muscle tissue may contribute to spoilage, such contributions are difficult to estimate since the activities of these enzymes and those of spoilage bacteria are not easily distinguishable (59, 105).

The internal muscle tissue of a healthy, live fish is generally considered to be sterile. Bacteria are present on the outer slime layer of the skin, gill surfaces, and, in feeding fish, the intestines (79, 105). During the spoilage of intact fish, bacterial activity is greater in the gill region than elsewhere on the carcass (105). Disagreement exists regarding the role of intestinal bacteria of uneviscerated fish in spoilage of muscle tissue. Jay (65) stated that bacteria from the intestinal tracts of uneviscerated fish will penetrate the intestinal walls and enter the tissues surrounding the intestinal cavity. However, Liston (76) concluded that there is little evidence that adjacent tissues or blood vessels are invaded by intestinal bacteria.

Composition and Spoilage of Shellfish

Crustacean and molluscan shellfish generally contain larger amounts of free amino acids than do finfish (41, 108). Trimethylamine oxide is present in crustacean shellfish but, with the exception of cephalopods, scallops, and cockles, is absent in molluscan tissue (108). Cathepsin-like enzymes which rapidly break down proteins have been reported to be present in shrimp (41, 65). Crustacean shellfish spoilage occurs similarly to that of fish flesh; however, the higher quantity of free amino acids and other soluble nitrogenous compounds facilitates rapid bacterial spoilage, accompanied by a production of large quantities of volatile base nitrogen (41, 65).

Molluscan shellfish contain a lower total nitrogen concentration in their flesh than do fish or crustacean shellfish. Additionally, while crustacean shellfish contain approximately 0.5% carbohydrate, molluscan shellfish contain considerably larger quantities of carbohydrate, e.g., 3.4% in clam meat and scallops and 5.6% in oysters, mostly in the form of glycogen. As a result, the spoilage pattern of molluscan shellfish differs from that of other seafoods and is generally fermentative (65), the pH of tissues declining as spoilage progresses (70).

FACTORS INFLUENCING SPOILAGE

Proteolytic and Lipolytic Activity

Although pseudomonads and other aerobic spoilage bacteria are able to produce proteolytic enzymes, the production of such enzymes is delayed until the late logarithmic phase of growth. There is general agreement that proteolysis occurs only at populations greater than 10^8 CFU/cm^2, when spoilage is well advanced and bacteria are approaching their maximum cell density (28, 48).

Oxidative rancidity of fat occurs when unsaturated fatty acids react with oxygen from the storage environment. Stable compounds such as aldehydes, ketones, and short-chain fatty acids are produced, resulting in the eventual development of rancid flavors and odors (48). Autoxidation, independent of microbial activity, occurs in muscle foods stored under aerobic environments, the rate being influenced by the proportion of unsaturated fatty acids in the fat (48). Autoxidation of fat is of particular importance in the deterioration of fatty fish and pork, which contain highly unsaturated lipids (70). The phospholipid component of muscle tissue membranes is also rich in unsaturated fatty acids which are susceptible to oxidation. Many spoilage microorganisms are able to produce lipases, which catalyze the hydrolysis of triacylglycerol into glycerol and free fatty acids which may, along with their oxidation products, contribute to rancidity. The production of lipases will be limited or inhibited by the presence of carbohydrates, lipids, and proteins in the medium (3, 4, 70). With a restriction of lipase production while carbohydrate substrates in muscle tissue are being utilized, it is unlikely that microbial lipolytic activity would occur until glucose at the muscle surface is depleted. At this point, amino acids would also be degraded, and the resulting spoilage characteristics would likely mask the effects of rancidity (48).

Spoilage of Adipose Tissue

Adipose tissue consists predominantly of insoluble fat, which cannot be effectively utilized as a substrate for microbial growth until it is broken down and emulsified. As previously noted, lipolytic enzymes are not produced until carbohydrates are exhausted. Spoilage of adipose tissue, therefore, is not dependent on the ability of the microflora to produce lipases but is a process similar to the spoilage of muscle tissue (92), in which glucose and glycolytic intermediates are first utilized, followed by degradation of amino acids, which results in spoilage. However, the amount of soluble components in adipose tissue is lower than in muscle tissue. Adipose tissue lacks significant levels of glycolytic intermediates, and soluble nutrients from the underlying tissue are not readily replenished by diffusion (48, 92). While the spoilage process and rate of growth of spoilage bacteria are similar on adipose and muscle tissues, the low level of carbohydrates in adipose tissue means that spoilage odors will be detected when lower numbers of bacteria are present; most, if not all, available glucose is depleted when populations exceed 10^6 CFU/cm^2. In addition to containing limited amounts of carbohydrates, adipose tissue has a lower lactic acid content than does muscle tissue and therefore the surface pH is higher, approaching 7.0. As a result, the spoilage of adipose tissue has been compared with that of DFD meat (92), and spoilage bacteria such as *S. putrefaciens* may potentially grow. The growth rates of some psychrotrophic microorganisms, e.g., *Hafnia alvei*, *Serratia liquefaciens*, and *Lactobacillus plantarum*, have been reported to be higher on fat than on lean beef and pork tissues, but there is little difference in growth rates of other microorganisms (118). In practice, however, spoilage of fat before lean tissue is unlikely, given the presence of muscle tissue fluids in vacuum-packaged cuts and the restriction of bacterial growth as a result of the drying of carcass surfaces (92).

Spoilage under Anaerobic Conditions

The spoilage microflora of muscle foods is dominated by lactic acid bacteria when oxygen is excluded from the storage environment. If the pH of the muscle tissue is high or residual amounts of oxygen are present, other microorganisms such as *B. thermosphacta* and *S. putrefaciens* may make substantial contributions to product spoilage. The growth rate of bacteria under anaerobic conditions is considerably reduced compared with growth under aerobic conditions. In addition, the maximum cell density achieved under anaerobic conditions (about 10^8 CFU/cm^2) is considerably less than that achieved under aerobic conditions (>10^9 CFU/cm^2) (48). Gill (47) reported that maximum populations of spoilage microorganisms under anaerobic conditions are determined by the rate of diffusion of fermentable substrates, such as glucose and arginine, from underlying tissues. It is not until the maximum density of microflora is reached that obvious spoilage slowly develops (49, 87).

The sour, acid, cheesy odors or the cheesy and dairy flavors that develop in muscle tissue under anaerobic conditions can be attributed, at least in part, to the accumulation of short-chain fatty acids and amines (111). *B. thermosphacta* present in significant numbers will cause more rapid spoilage than lactic acid bacteria. As with lactic acid bacteria, the spoilage characteristics result primarily from the production of organic acids (27, 49).

DFD Meats

Animals subjected to excessive stress or exercise before slaughter can have depleted levels of muscle glycogen. This results in the production of lower amounts of lactic acid and a higher ultimate pH. Muscles with a pH >6.0 will appear darker than normal red meat. This is due to a higher respiration rate, which reduces the depth of oxygen penetration and therefore reduces the level of visible oxymyoglobin (58). The resulting condition, referred to as DFD muscle, occurs most often in beef but can also occur in pigs and other meat animals. In addition to having a higher ultimate pH resulting from lower lactic acid content, such tissue is also deficient in glucose and glycolytic intermediates (86).

Spoilage of DFD meat occurs more quickly than spoilage of meat with normal pH. Rapid spoilage is a function of the absence of glucose in the tissues. With glucose unavailable, there is little delay in degradation of amino acids by pseudomonads. Spoilage will occur at lower bacterial cell densities than in normal meat (>10^6 CFU/cm^2) when metabolites are produced in a sufficient quantity to result in noticeable defects (86).

DFD meat stored vacuum packaged or in modified atmospheres will also spoil rapidly, typically resulting in the development of a green discoloration (52). The high pH of the meat and the absence of glucose and glucose 6-phosphate have been reported to allow *Serratia liquefaciens* and *S. putrefaciens* to compete successfully with the normally dominant lactic acid bacteria to constitute a significant portion of the spoilage microflora (52, 95). The green discoloration observed in DFD meat results from the production of hydrogen sulfide from cysteine or glutathione by *S. putrefaciens* (52), reacting with myoglobin in the muscle tissue to form sulphmyoglobin, a green pigment (89). *S. liquefaciens* does not contribute to greening even though it produces small quantities of hydrogen sulfide; however, its presence is significant, since low numbers produce spoilage odors on DFD meat (52, 86, 95).

PSE Meats

Pale, soft, exudative (PSE) muscle tissue is a condition that occurs in pork and turkey, and to a lesser extent in beef, in which accelerated postmortem glycolysis decreases the muscle pH to its ultimate level while the muscle temperature is still high (48, 58). The condition, occurring in 5 to 20% of pig carcasses, is characterized by the development of a pale color, soft texture, and exudation of fluids from the muscle (48, 58). The occurrence of PSE meat is directly related to porcine stress syndrome, in which animals may die as a result of mild stress, or to malignant hyperthermia, in which death may be caused by exposure to certain anesthetics (58). There is debate over whether PSE meats spoil more slowly than meat with normal pH. Rey et al. (98) reported that the rate of bacterial growth was higher for DFD beef and slower for PSE pork, suggesting that pH affects the growth of spoilage microorganisms under various conditions. Gill (48) concluded that even though an ultimate pH of 5.1 or below has been reported for PSE pork, there is little difference in the ultimate pH and chemical composition of PSE and normal meat. Since the concentrations of low-molecular-weight soluble compounds are likely similar in PSE and normal meats, the development and characteristics of spoilage would likely be similar (48).

Comminuted Products

The limited shelf life of comminuted muscle foods has been attributed both to a higher initial number of bacteria, due to use of a poorer-quality product for grinding and contamination during processing, and to the effects of the comminution of the muscle tissue. Comminution ruptures tissue cells, releasing fluids and nutrients which provide a ready source of nutrients for bacteria. Additionally, bacteria that have been restricted to the surface of meats are distributed throughout the mass by the process (62, 112). Although initial bacterial populations are higher, the type of microflora and spoilage resembles that of noncomminuted intact tissue. On the surface of aerobically stored comminuted products, *Pseudomonas*, *Acinetobacter*, and *Moraxella* species comprise the dominant microflora, while in the interior, because of the limited availability of oxygen, lactic acid bacteria are dominant (62). Occasional contaminants such as members of the *Enterobacteriaceae* or *Aeromonas* occur more often on comminuted products than on intact tissue.

Like other comminuted meats, fresh sausage undergoes rapid spoilage. The addition of salt and spices is not sufficient to delay spoilage, which results from the activity of many of the same microorganisms involved in spoilage of ground beef. The spoilage microflora of refrigerated pork sausage often consists of *B. thermosphacta* (112).

Cooked Products

Cooking muscle foods results in destruction of vegetative cells of resident bacteria, although endospores may survive. For perishable, cooked, uncured meats, spoilage is dependent on the microflora surviving heat processing or contribution of contaminants after cooking which are able to grow under storage conditions. Microorganisms responsible for spoilage of these products may include psychrotrophic micrococci, streptococci, lactobacilli, and *B. thermosphacta* (62). Recontamination is a concern if handling occurs following cooking. Where recontamination is minimal, spoilage is usually caused by nonproteolytic bacteria and involves the development of a sour odor. Recontamination with proteolytic microorganisms results in amino acid breakdown and a putrid odor (62). Retorted muscle food products are subject to spoilage resulting from defects produced before they are heat processed, inadequate processing which enables the survival of heat-resistant mesophilic sporeformers, and reintroduction of microorganisms from postprocessing leakage (62).

Processed Products

As with fresh muscle foods, microbial spoilage of processed products depends on the nature of the product and the ingredients used. Growth of microorganisms in processed products is also affected by the conditions of processing and storage. Spoilage of these products is influenced by whether or not they are cured, heat processed, or fermented and by the a_w and pH.

Jay (65) characterized the spoilage of processed meats such as frankfurters, bologna, sausage, and luncheon meats as belonging to three types: slimy spoilage, souring, and greening. Slimy spoilage, which is usually confined to the outside surfaces of product casings, may develop as discrete colonies which then expand to form a uniform gray slime layer. Slime formation generally requires the presence of moisture. Microorganisms implicated in slime formation include yeasts, two genera of lactic acid bacteria, *Lactobacillus* and *Enterococcus*, and *B. thermosphacta* (65). Souring occurs, generally beneath casings, when bacteria such as lactobacilli, enterococci, and *B. thermosphacta* utilize lactose and other sugars to produce acids. While greening of fresh meats may be a result of hydrogen sulfide production by certain bacteria, greening of cured meat products may also develop in the presence of hydrogen peroxide, which may form on the surface of vacuum- or modified atmosphere-packaged meats upon exposure to air. Hydrogen peroxide may then react with nitrosohemochrome to produce choleglobin, which has a greenish color (55). Greening occurs in cured meats as a result of growth of bacteria. Catalase-negative bacteria are involved, and muscle catalase may be inactivated by the cooking process or the presence of nitrite in the cured meat. As a result, and because of the low oxidation-reduction potential in the meat interior, hydrogen peroxide may accumulate. As with surface greening, reaction of the hydrogen peroxide with nitrosohemochrome may result in the production of a green pigment. *Lactobacillus viridescens* is the most common cause of this type of greening, but species of *Streptococcus* and *Leuconostoc* may also produce this defect. The presence of these bacteria often is a result of poor sanitation before or following processing (55, 65).

Detailed reviews of the microbiology of cured meats are provided by Tompkin (116), Ingram and Simonsen (62), Kraft (70), and Gardner (46). Putrefaction of cured meats usually does not occur, mainly because of reduced a_w, which, at refrigeration temperatures, inhibits spoilage by gram-negative psychrotrophic bacteria (62). Cured meats are spoiled by microorganisms that tolerate a low a_w, such as lactobacilli or micrococci (117). According to Ingram and Dainty (61), aerobic micrococci will predominate on cured meats stored aerobically, while lactobacilli will predominate in the spoilage microflora of vacuum- or modified atmosphere-packaged products. If sucrose is added to cured meats, a slimy dextran layer may form as a result of the activity of *Leuconostoc* species or other bacteria such as *L. viridescens. B. thermosphacta* may also be involved in the spoilage of these products (116). While some differences may exist in the rates of spoilage of cooked and cured products, vacuum-packaged bacon has a rate of spoilage similar to that of many cooked cured products, involving many of the same microorganisms. Ingram and Dainty (61) noted that different spoilage characteristics may develop on fat and lean tissue of aerobically stored bacon; a smoked fish flavor may develop on the lean tissue, and a rancid, cheesy flavor may develop on fat. Dry-cured products owe their stability to low a_w, the presence of nitrite, and smoke added before drying. As a result, bacterial growth is suppressed and spoilage occurs mainly as a result of growth of yeasts and molds if products are stored in a humid environment (116). Spoilage of dry-cured meats may occur if spoilage microorganisms are allowed to grow in the product before salt penetration restricts their growth. Tompkin (116) reported that this is particularly true for hams that are not sufficiently cold at the time they are packed in salt and cure.

If dried meats are properly prepared and stored, their low a_w renders them microbiologically stable. To restrict the growth of some species of bacteria, yeasts, and molds that can grow at low a_w, the water content of dried meat products must be reduced to about 20%. To inhibit the growth of xerotolerant molds, the water content must be further reduced to about 15% (62). Microbial spoilage of these products should not occur unless exposure to high relative humidity or other high moisture conditions allows the uptake of moisture.

CONTROL OF SPOILAGE OF MUSCLE FOODS

The growth of spoilage microorganisms on muscle foods can be controlled by modifying intrinsic characteristics of products or extrinsic characteristics of the storage environment. The shelf life of processed meats can be extended by the use of processing procedures and ingredients that either prevent the growth of spoilage microorganisms or select for a specific spoilage microflora which produces a less objectionable or slower development of spoilage. For fresh meats, the extension of shelf life has become especially important as the food industry moves toward centralization of processing activities and products are being distributed to more distant domestic and international markets. Intrinsic and extrinsic factors influencing growth are discussed in chapter 2, and various methods of food preservation are presented in section VII. Some specific methods used to prevent spoilage and prolong the shelf life of muscle foods are noted here.

In his review of the control and evaluation of spoilage, Dainty (26) categorized methods used to control the spoilage of muscle foods into three strategies which could be used in combination. Strategies were focused on prevention of initial microbial contamination, inactivation of microorganisms which are present on the product, or the use of storage conditions to prevent or slow the growth of microorganisms present in or on the product.

Initial Populations

The presence of higher numbers of bacteria on products before storage results in shorter product shelf life. If the initial population of spoilage microorganisms on a product is high, less time will be required before high numbers are attained and spoilage defects become evident. High populations of microorganisms on carcass surfaces may also allow for attachment of larger numbers of bacteria (23). Attached bacteria are less likely to be removed by washing or other decontamination procedures and may be more resistant to processing conditions. If microbial populations are sufficiently high, spoilage microorganisms, and also pathogens, may survive processing treatments and, if storage conditions permit, proliferate and produce defects.

Many of the factors that influence populations of microorganisms during processing of muscle foods were noted earlier in the discussion of sources of microorganisms. Obviously, effective good manufacturing practices will result in lower populations on products entering storage and distribution.

Water Rinsing

Several methods have been evaluated for their effectiveness in reducing initial microbial populations on muscle foods. Rinsing with water, immersion, and spray systems have been used to remove physical and microbial contaminants from carcasses (33). While the use of both hot and cold water has been shown to reduce microbial populations, greater reductions have been demonstrated when the temperature of water is elevated (5). At lower temperatures, reductions may be primarily a result of removal of cells, while at higher temperatures, inactivation of cells by heat is also involved. High-pressure water sprays will also promote the removal of bacteria; however, there is some concern that higher pressures will drive bacteria into carcass tissues (31).

Spray washing is used in poultry processing after defeathering and evisceration to remove organic material and reduce bacterial populations before carcasses enter immersion chillers (14, 57). High-pressure washes have been demonstrated to be effective in reducing bacterial populations on the surface of whole fish by removing the slime layer (79).

Antimicrobial Treatments

In addition to washing, treatments with antimicrobial compounds such as chlorine and organic acids have been used to sanitize muscle foods. Various concentrations and degrees of effectiveness of chlorine compounds have been reported. In poultry processing, further reductions have been demonstrated during spray washing by the addition of 40 to 60 μg of chlorine per ml to the spray water (101). Chlorine (5 to 20 μg/ml) may also be added to poultry chill water in slush ice systems to inhibit multiplication of psychrotrophs (8). For finfish, the effectiveness of chlorine in dips and sprays has been evaluated; however, Mayer and Ward (79) concluded that the limited reduction of spoilage microflora on the surface of finfish treated with chlorine makes this chemical only marginally effective in increasing shelf life. Similar observations have been made with respect to the use of chlorination to extend the shelf life of meats (50, 68, 110). Kelly et al. (68) reported an additive antimicrobial effect of chlorine and elevated water temperature (>80°C).

Short-chain fatty acids have been used in sprays to reduce microbial populations on carcasses and extend the shelf life. The use of lactic, acetic, propionic, and citric acids has been investigated, although lactic and acetic acids are most often used to reduce microbial populations on carcass surfaces. The effectiveness of organic acids is influenced by such factors as type of acid, concentration, temperature, and point of application in processing (33). The point of application of organic acids in beef carcass processing is important. Organic acids cause the greatest reductions in populations when applied to hot carcass surfaces following dehiding and evisceration (97) and are relatively ineffective when applied to individual cuts following fabrication (1, 36).

Although some washing and decontamination treatments have been demonstrated to be effective in reducing bacterial populations on animal carcasses, the use of such treatments cannot replace effective sanitation. The use of strict sanitary procedures, including organic acid rinses and control of handling, vacuum packaging, and storage temperature, has been shown to increase the shelf life of carcasses and subprimal cuts over those produced from animals slaughtered and processed by conventional procedures (19, 35, 109). Samuels et al. (100) suggested that the shelf life and quality of finfish products could be improved by using an effective sanitation program in addition to an initial decontamination step before processing.

Refrigeration

Temperature is the most important environmental parameter influencing the growth of microorganisms in muscle foods. As the temperature is decreased below the optimum for growth of microorganisms, generation times and lag times are extended, and growth is therefore slowed (Fig. 5.1). As the temperature is reduced to the minimum for growth, extension of the lag time continues to increase until multiplication ceases (70). Changes in growth rate as the temperature is reduced, as well as the minimum temperature for growth, differ for each species of microorganism (50, 70).

The temperature at which muscle foods are stored ultimately influences the type of microorganisms that will grow and eventually cause spoilage, as well as the rate at which growth will occur. Under aerobic storage conditions, the comparatively rapid growth rate of *Pseudomonas* species allows successful competition with mesophiles and other psychrotrophs at temperatures below 20°C. Likewise, under anaerobic conditions at temperatures below 20°C, the rapid growth rate of psychrotrophic lactobacilli enables successful competition with

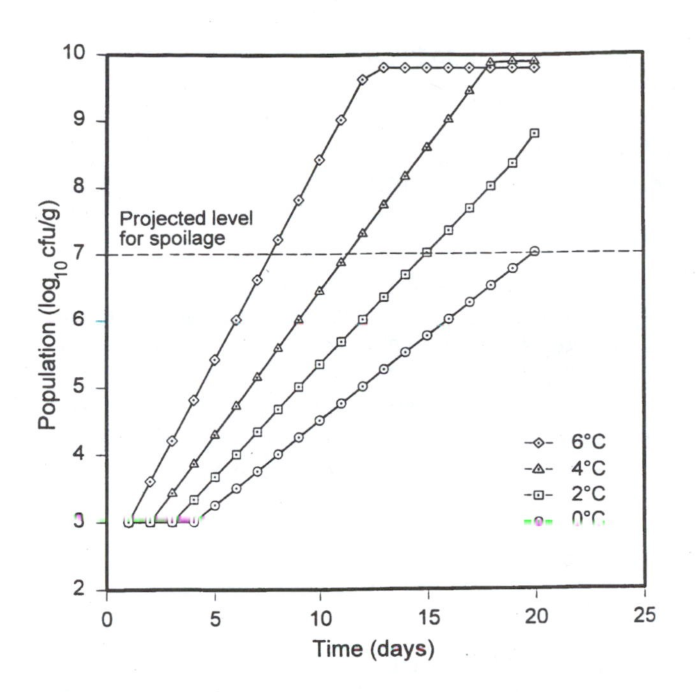

Figure 5.1 Predicted populations of total aerobic bacteria in ground beef as affected by storage temperature (32a).

other psychrotrophic spoilage microorganisms (8, 50, 53).

Storage of foods at temperatures at which mesophiles will grow allows these microorganisms to contribute to spoilage. Higher storage temperatures may also allow the growth of psychrotrophic spoilage microorganisms which would otherwise be restricted by the pH or lactic acid concentration of meat tissue at reduced temperatures (50, 53). Likewise, the minimum temperature for growth of psychrotrophic spoilage microorganisms may be increased by other factors influencing growth, e.g., pH, a_w, or oxidation-reduction potential, or by inhibitory agents such as food additives or increased carbon dioxide (70).

Modified Atmosphere Storage and Vacuum Packaging

The shelf life of muscle foods can be extended by storage under vacuum or modified atmospheres. Modified atmosphere packaging involves the storage of products under high-oxygen barrier film with a headspace of different gaseous composition than air. Typically, this headspace contains elevated amounts of carbon dioxide and may also contain nitrogen and oxygen in various proportions. The use of carbon monoxide, nitrous oxide, and sulfur dioxide in the gaseous atmosphere of packaged muscle foods has also been suggested as a way to control microbial growth. Much of the extended shelf life of such products results from a modification of the spoilage microflora from an aerobic psychrotrophic population consisting of bacteria such as *Pseudomonas, Moraxella,* and *Acinetobacter* species to one consisting predominantly of lactic acid bacteria and *B. thermosphacta* (Fig. 5.2).

In vacuum-packaged products, exclusion of air from the package and slow diffusion of atmospheric oxygen through the high-oxygen barrier film leave only residual oxygen remaining. In fresh muscle foods, respiration of the tissue utilizes the remaining oxygen in the package and generates carbon dioxide (11, 50). While the aerobic or facultative component of the bacterial microflora also utilizes residual oxygen in the package, because of the relatively low populations initially present, their impact on oxygen content is insignificant compared with activity in the surrounding muscle tissue (50). Residual oxygen remaining within vacuum-packaged products is a function of the rate of oxygen diffusion through the

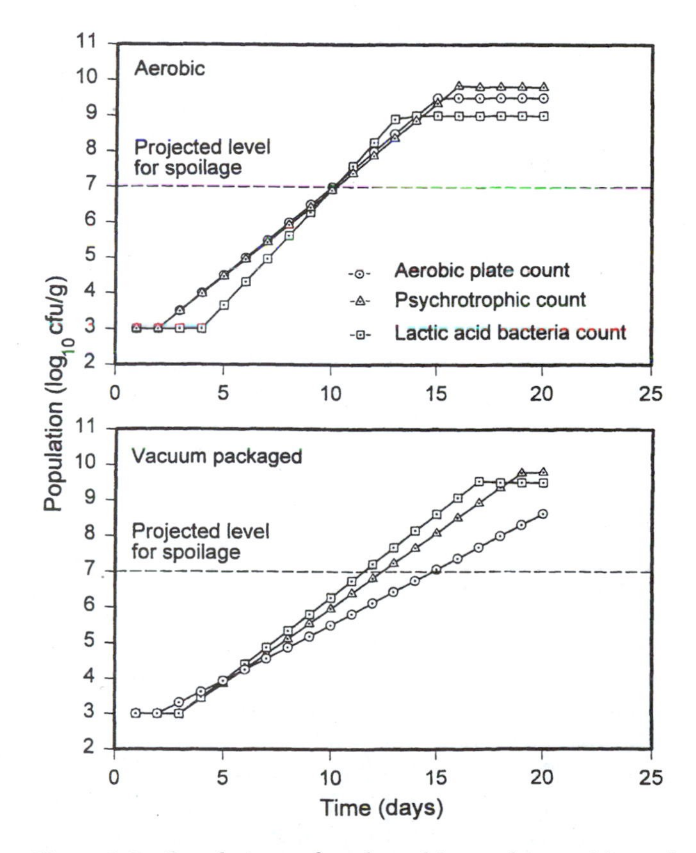

Figure 5.2 Populations of total aerobic, psychrotrophic, and lactic acid bacteria in aerobically packaged and vacuum-packaged ground pork stored at 0°C. The growth curves are based on the derived values for lag phase duration and generation time (119a).

packaging film as well as the rate at which oxygen is depleted by the muscle tissue (50).

Carbon dioxide is commonly used in modified atmosphere environments because of its bacteriostatic activity. The presence of carbon dioxide will result in an increase in the lag phase and generation times of many microorganisms (29). While the specific mechanism of bacteriostatic activity of carbon dioxide is unknown, several mechanisms have been suggested, including exclusion of oxygen from the atmosphere, thus restricting the growth of aerobes, altering cell membrane properties, altering enzyme systems, and acidifying intracellular pH (29, 34).

The inhibitory activity of carbon dioxide is influenced by many factors (39, 121). The temperature at which packaged products are stored will significantly influence the bacteriostatic effect of carbon dioxide, and solubility decreases as the storage temperature is increased. Therefore, the storage temperature should be maintained as low as possible because elevated temperatures limit the effectiveness of carbon dioxide (29, 38, 39, 42, 93).

Oxygen is occasionally included in modified atmospheres used to package fresh red meats to maintain the oxymyoglobin responsible for fresh red or "bloomed" color desired by the consumer (113, 122). While the inclusion of oxygen in modified atmospheres may improve the appearance of products, it may also reduce the shelf life. If significant amounts of oxygen are present, aerobic microorganisms and *B. thermosphacta* may grow, although the concentration of oxygen necessary for growth varies among microorganisms (21, 22, 63, 87, 107).

While nitrogen has no antimicrobial activity, its inclusion displaces oxygen in the package headspace, resulting in a delay in oxidative rancidity and an inhibition of the growth of aerobic microorganisms (39).

Cook-In Bag and Postpasteurization

Meat products can be cooked in flexible plastic packaging materials and distributed in the same packaging. Such a packaging approach avoids the reintroduction of spoilage microorganisms following cooking, and as a result, the shelf life of the product is extended (30). A shortened shelf life may occur if cooked meats are removed from packaging and further processed or portioned for distribution (30). In some cases, surface pasteurization of muscle products following repackaging may be used to inactivate reintroduced spoilage bacteria and thereby increase shelf life. Such a process typically involves exposure of the product surface to temperatures of 82 to 96°C for periods of 30 s to 6 min (30).

Irradiation

Irradiation of meats enhances microbiological safety and quality by significantly reducing populations of pathogenic and spoilage bacteria. The effect of irradiation is dose dependent, with higher doses resulting in a greater reduction of bacterial populations. At present, most interest is in relatively low doses of irradiation, i.e., less than 5 kGy.

The doses of radiation required to inactivate 90% of the vegetative cells of pathogenic bacteria vary somewhat with processing conditions (temperature, presence or absence of air) but typically range from a low of <0.2 kGy for *Aeromonas* (94) and *Campylobacter* (71) species to as high as 0.77 kGy for *Listeria* (60) and *Salmonella* (114) species. Although there is some variation among species, doses required to inactivate 90% of the vegetative cells of spoilage bacteria typically fall in the range of 0.2 to 0.8 kGy. In contrast, the dose required for a 90% inactivation of *Clostridium* endospores is as high as 3.5 kGy (6).

Low-dose (<5-kGy) irradiation has been demonstrated to be effective in extending the shelf life of fresh meats. Several studies have reported shelf life extensions of 14 to 21 days for irradiated meats, as determined by microbiological and organoleptic assays (10, 69, 73). Because of significant reductions in both pathogenic and spoilage bacteria, low-dose irradiation has been suggested as a pasteurization process for fresh meats.

References

1. **Acuff, G. R., C. Vanderzant, J. W. Savell, D. K. Jones, D. B. Griffin, and J. G. Ehlers.** 1987. Effect of acid decontamination of beef subprimal cuts on the microbiological and sensory characteristics of steaks. *Meat Sci.* **19:**217–226.
2. **Adams, R., L. Farber, and P. Lerke.** 1964. Bacteriology of spoilage of fish muscle. II. Incidence of spoilers during storage. *Appl. Microbiol.* **12:**277–279.
3. **Alford, J. A., and L. E. Elliott.** 1960. Lipolytic activity of microorganisms at low and intermediate temperatures. I. Action of *Pseudomonas fluorescens* on lard. *Food Res.* **25:** 296–303.
4. **Alford, J. A., J. L. Smith, and H. D. Lilly.** 1971. Relationship of microbial activity to changes in lipids in foods. *J. Appl. Bacteriol.* **34:**133–146.
5. **Anderson, M. E., R. T. Marshall, W. C. Stringer, and H. D. Naumann.** 1979. Microbial growth on plate beef during extended storage after washing and sanitizing. *J. Food Prot.* **42:**389–392.
6. **Anellis, A., E. Shattuck, M. Morin, B. Srisara, S. Qvale, D. B. Rowley, and E. W. Ross.** 1977. Cryogenic gamma

irradiation of prototype pork and chicken and antagonistic effect between *Clostridium botulinum* types A and B. *Appl. Microbiol.* **34**:823–831.

7. **Ayres, J. C., J. O. Mundt, and W. E. Sandine.** 1980. *Microbiology of Foods.* W. H. Freeman & Company. San Francisco.
8. **Barnes, E. M.** 1976. Microbiological problems of poultry at refrigerator temperatures—a review. *J. Sci. Food Agric.* **27**:777–782.
9. **Barnes, E. M., and C. S. Impey.** 1968. Psychrotrophic spoilage bacteria of poultry. *J. Appl. Bacteriol.* **31**:97–107.
10. **Basker, D., I. Klinger, M. Lapidot, and E. Eisenberg.** 1986. Effect of chilled storage of radiation-pasteurized chicken carcasses on the eating quality of the resultant cooked meat. *J. Food Technol.* **21**:437–441.
11. **Bendall, J. R., and A. A. Taylor.** 1972. Consumption of oxygen by the muscles of beef animals and related species. *J. Sci. Food Agric.* **23**:707–719.
12. **Benedict, R. C.** 1988. Microbial attachment to meat surfaces. *Reciprocal Meat Conf. Proc.* **41**:1–6.
13. **Brown, W. D.** 1986. Fish muscle as food, p. 405–451. *In* P. J. Bechtel (ed.), *Muscle as Food.* Academic Press, Orlando, Fla.
14. **Bryan, F. L.** 1980. Poultry and poultry meat products, p. 410–469. *In* J. H. Silliker, R. P. Elliot, A. C. Baird-Parker, F. L. Bryan, J. H. B. Christian, D. S. Clark, J. C. Olsen, Jr., and T. A. Roberts (ed.), *Microbial Ecology of Foods,* vol. 2. *Food Commodities.* Academic Press, New York.
15. **Butler, J. L., J. C. Stewart, C. Vanderzant, Z. L. Carpenter, and G. C. Smith.** 1979. Attachment of microorganisms to pork skin and surfaces of beef and lamb carcasses. *J. Food Prot.* **42**:401–406.
16. **Callow, E. H., and M. Ingram.** 1955. Bone-taint. *Food* **24**:52–55.
17. **Campbell, R. J., A. F. Egan, F. H. Grau, and B. J. Shay.** 1979. The growth of *Microbacterium thermosphactum* on beef. *J. Appl. Bacteriol.* **47**:505–509.
18. **Chai, T., C. Chen, A. Rosen, and R. E. Levin.** 1968. Detection and incidence of specific species of spoilage bacteria on fish. II. Relative incidence of *Pseudomonas putrefaciens* and fluorescent pseudomonads on haddock fillets. *Appl. Microbiol.* **16**:1738–1741.
19. **Chandran, S. K., J. W. Savell, D. B. Griffin, and C. Vanderzant.** 1986. Effect of slaughter-dressing, fabrication and storage conditions on the microbiological and sensory characteristics of vacuum-packaged beef steaks. *J. Food Sci.* **51**:37–53.
20. **Christian, J. H. B.** 1980. Reduced water activity, p. 70–111. *In* J. H. Silliker, R. P. Elliot, A. C. Baird-Parker, F. L. Bryan, J. H. B. Christian, D. S. Clark, J. C. Olsen, Jr., and T. A. Roberts (ed.), *Microbial Ecology of Foods,* vol. 1. *Factors Affecting the Life and Death of Microorganisms.* Academic Press, New York.
21. **Christopher, F. M., S. C. Seideman, Z. L. Carpenter, G. C. Smith, and C. Vanderzant.** 1979. Microbiology of beef packaged in various gas atmospheres. *J. Food Prot.* **42**:240–244.
22. **Christopher, F. M., C. Vanderzant, Z. L. Carpenter, and G. C. Smith.** 1979. Microbiology of pork packaged in various gas atmospheres. *J. Food Prot.* **42**:323–327.
23. **Chung, K.-T., J. S. Dickson, and J. D. Crouse.** 1989. Attachment and proliferation of bacteria on meat. *J. Food Prot.* **52**:173–177.
24. **Clark, D. S.** 1970. Growth of psychrotolerant pseudomonads and achromobacteria on various chicken tissues. *Poultry Sci.* **49**:1315–1318.
25. **Costerson, J. W., R. T. Irvin, and K.-J. Cheng.** 1981. The bacterial glycocalyx in nature and disease. *Annu. Rev. Microbiol.* **35**:299–324.
26. **Dainty, R. H.** 1971. The control and evaluation of spoilage. *J. Food Technol.* **6**:209–224.
27. **Dainty, R. H., and C. M. Hibbard.** 1980. Aerobic metabolism of *Brochothrix thermosphacta* growing on meat surfaces and in laboratory media. *J. Appl. Bacteriol.* **48**: 387–396.
28. **Dainty, R. H., B. G. Shaw, K. A. De Boer, and E. S. J. Scheps.** 1975. Protein changes caused by bacterial growth on beef. *J. Appl. Bacteriol.* **39**:73–81.
29. **Daniels, J. A., R. Krishnamurthi, and S. S. H. Rizvi.** 1985. A review of effects of carbon dioxide on microbial growth and food quality. *J. Food Prot.* **48**:532–537.
30. **DeMasi, T. W., and K. R. Deily.** 1990. Cooked meats packaging technology. *Reciprocal Meat Conf. Proc.* **43**: 117–121.
31. **De Zuniga, A. G., M. E. Anderson, R. T. Marshall, and E. L. Iannotti.** 1991. A model system for studying the penetration of microorganisms into meat. *J. Food Prot.* **54**:256–258.
32. **Dickson, J. S.** 1991. Attachment of *Salmonella typhimurium* and *Listeria monocytogenes* to beef tissue: effects of inoculum level, growth temperature and bacterial culture age. *Food Microbiol.* **8**:143–151.

32a. **Dickson, J. S.** Unpublished data.

33. **Dickson, J. S., and M. E. Anderson.** 1992. Microbiological decontamination of food animal carcasses by washing and sanitizing systems: a review. *J. Food Prot.* **55**: 133–140.
34. **Dixon, N. M., and D. B. Kell.** 1989. The inhibition by CO_2 of the growth and metabolism of micro-organisms. *J. Appl. Bacteriol.* **67**:109–136.
35. **Dixon, Z. R., G. R. Acuff, L. M. Lucia, C. Vanderzant, J. B. Morgan, S. G. May, and J. W. Savell.** 1991. Effect of degree of sanitation from slaughter through fabrication on the microbiological and sensory characteristics of beef. *J. Food Prot.* **54**:200–207.
36. **Dixon, Z. R., C. Vanderzant, G. R. Acuff, J. W. Savell, and D. K. Jones.** 1987. Effect of acid treatment of beef strip loin steaks on microbiological and sensory characteristics. *Int. J. Food Microbiol.* **5**:181–186.
37. **Egan, A. F.** 1983. Lactic acid bacteria of meat and meat products. *Antonie Leeuwenhoek* **49**:327–336.
38. **Enfors, S. O., and G. Molin.** 1980. Effect of high concentrations of carbon dioxide on growth rate of *Pseudomonas fragi, Bacillus cereus,* and *Streptococcus cremoris. J. Appl. Bacteriol.* **48**:409–416.
39. **Farber, J. M.** 1991. Microbiological aspects of modified-atmosphere packaging technology—a review. *J. Food Prot.* **54**:58–70.

40. **Farber, J. M., and E. S. Idziak.** 1984. Attachment of psychrotrophic meat spoilage bacteria to muscle surfaces. *J. Food Prot.* **47:**92–95.

41. **Feiger, E. A., and A. F. Novak.** 1961. Microbiology of shellfish deterioration, p. 561–611. *In* G. Borgstrom (ed.), *Fish as Food,* vol. 1. *Production, Biochemistry, and Microbiology.* Academic Press, New York.

42. **Finne, G.** 1982. Modified- and controlled-atmosphere storage of muscle foods. Food Technol. **36(2):**128–133.

43. **Firstenberg-Eden, R.** 1981. Attachment of bacteria to meat surfaces: a review. *J. Food Prot.* **44:**602–607.

44. **Firstenberg-Eden, R., S. Notermans, and M. Van Schothorst.** 1978. Attachment of certain bacterial strains to chicken and beef meat. *J. Food Safety* **1:**217–228.

45. **Freeman, L. R., G. J. Silverman, P. Angelini, C. Merritt, Jr., and W. B. Esselen.** 1976. Volatiles produced by microorganisms isolated from refrigerated chicken at spoilage. *Appl. Environ. Microbiol.* **32:**222–231.

46. **Gardner, G. A.** 1982. Microbiology of processing: bacon and ham, p. 129–178. *In* M. H. Brown (ed.), *Meat Microbiology.* Applied Science Publishers Ltd., New York.

47. **Gill, C. O.** 1976. Substrate limitation of bacterial growth at meat surfaces. *J. Appl. Bacteriol.* **41:**401–410.

48. **Gill, C. O.** 1982. Microbial interaction with meats, p. 225–264. *In* M. H. Brown (ed.), *Meat Microbiology.* Applied Science Publishers Ltd., Inc. New York.

49. **Gill, C. O.** 1983. Meat spoilage and evaluation of the potential storage life of fresh meat. *J. Food Prot.* **46:**444–452.

50. **Gill, C. O.** 1986. The control of microbial spoilage in fresh meats, p. 49–88. *In* A. M. Pearson and T. R. Dutson (ed.), *Advances in Meat Research,* vol. 2. *Meat and Poultry Microbiology.* AVI Publishing Co., Inc., Westport, Conn.

51. **Gill, C. O., and K. G. Newton.** 1977. The development of aerobic spoilage flora on meat stored at chill temperatures. *J. Appl. Microbiol.* **43:**189–195.

52. **Gill, C. O., and K. G. Newton.** 1979. Spoilage of vacuum-packaged dark, firm, dry meat at chill temperatures. *Appl. Environ. Microbiol.* **37:**362–364.

53. **Gill, C. O., and K. G. Newton.** 1980. Growth of bacteria on meat at room temperatures. *J. Appl. Bacteriol.* **49:**315–323.

54. **Gill, C. O., and K. G. Newton.** 1982. Effect of lactic acid concentration of growth on meat of gram-negative psychrotrophs from a meatworks. *Appl. Environ. Microbiol.* **43:**284–288.

55. **Grant, G. F., A. R. McCurdy, and A. D. Osborne.** 1988. Bacterial greening in cured meats: a review. *Can. Inst. Food Sci. Technol. J.* **21:**50–56.

56. **Grau, F. H.** 1980. Inhibition of the anaerobic growth of *Brochothrix thermosphacta* by lactic acid. *Appl. Environ. Microbiol.* **40:**433–436.

57. **Grau, F. H.** 1986. Microbial ecology of meat and poultry, p. 1–47. *In* A. M. Pearson and T. R. Dutson (ed.), *Advances in Meat Research,* vol. 2. *Meat and Poultry Microbiology.* AVI Publishing Co., Inc., Westport, Conn.

58. **Greaser, M. L.** 1986. Conversion of muscle to meat, p. 37–102. *In* P. J. Bechtel (ed.), *Muscle as Food.* Academic Press, New York.

59. **Herbert, R. A., M. S. Hendrie, D. M. Gibson, and J. M. Shewan.** 1971. Bacteria active in the spoilage of certain sea foods. *J. Appl. Bacteriol.* 34:41–50.

60. **Huhtanen, C. N., R. K. Jenkins, and D. W. Thayer.** 1989. Gamma radiation sensitivity of *Listeria monocytogenes. J. Food Prot.* **52:**610–613.

61. **Ingram, M., and R. H. Dainty.** 1971. Changes caused by microbes in spoilage of meats. *J. Appl. Bacteriol.* **34:**21–39.

62. **Ingram, M., and B. Simonsen.** 1980. Meats and meat products, p. 333–409. *In* J. H. Silliker, R. P. Elliot, A. C. Baird-Parker, F. L. Bryan, J. H. B. Christian, D. S. Clark, J. C. Olsen, Jr., and T. A. Roberts (ed.), *Microbial Ecology of Foods,* vol. 2. *Food Commodities.* Academic Press, New York.

63. **Jackson, T. C., G. R. Acuff, C. Vanderzant, T. R. Sharp, and J. W. Savell.** 1992. Identification and evaluation of volatile compounds of vacuum and modified atmosphere packaged beef strip loins. *Meat Sci.* **31:**175–190.

64. **Jay, J. M.** 1971. Mechanism and detection of microbial spoilage in meats at low temperatures: a status report. *J. Milk Food Technol.* **35:**467–471.

65. **Jay, J. M.** 1992. *Modern Food Microbiology,* 4th ed. Van Nostrand Reinhold, New York.

66. **Kalchayanand, N., B. Ray, and R. A. Field.** 1993. Characteristics of psychrotrophic *Clostridium laramie* causing spoilage of vacuum-packaged refrigerated fresh and roasted beef. *J. Food Prot.* **56:**13–17.

67. **Kampelmacher, E. H., P. A. M. Guinée, K. Hofstra, and A. van Keulen.** 1961. Studies on Salmonella in slaughterhouses. *Zentralbl. Veterinaermed.* **8:**1025–1042.

68. **Kelly, C. A., J. F. Dempster, and A. J. McLoughlin.** 1981. The effect of temperature, pressure and chlorine concentration of spray washing water on numbers of bacteria on lamb carcasses. *J. Appl. Bacteriol.* **51:**415–424.

69. **Klinger, I., V. Fuchs, D. Basker, B. J. Juven, M. Lapidot, and E. Eisenberg.** 1986. Irradiation of broiler chicken meat. *Isr. J. Vet. Med.* **42:**181–192.

70. **Kraft, A. A.** 1992. *Psychrotrophic Bacteria in Foods: Disease and Spoilage.* CRC Press, Inc., Boca Raton, Fla.

71. **Lambert, J. D. and R. B. Maxcy.** 1984. Effect of gamma radiation on *Campylobacter jejuni. J. Food. Sci.* **49:**665–667.

72. **Lawrie, R. A.** 1985. *Meat Science,* 4th ed. Pergamon Press, New York.

73. **Lescano, G., P. Narvaiz, E. Kairiyama, and N. Kaupert.** 1991. Effect of chicken breast irradiation on microbiological, chemical and organoleptic quality. *Lebensm. Wiss. Univ. Technol.* **24:**130–134.

74. **Lillard, H. S.** 1986. Role of fimbriae and flagella in the attachment of *Salmonella typhimurium* to poultry skin. *J. Food Sci.* **51:**54–56, 65.

75. **Liston, J.** 1956. Quantitative variations in the bacterial flora of flatfish. *J. Gen. Microbiol.* **15:**305–314.

76. **Liston, J.** 1980. Fish and shellfish and their products, p. 567–605. *In* J. H. Silliker, R. P. Elliot, A. C. Baird-Parker, F. L. Bryan, J. H. B. Christian, D. S. Clark, J. C. Olsen, Jr., and T. A. Roberts (ed.), *Microbial Ecology of Foods,* vol. 2. *Food Commodities.* Academic Press, New York.

77. **Lowry, P. D., and C. O. Gill.** 1984. Temperature and water activity minima for growth of spoilage moulds from meat. *J. Appl. Bacteriol.* **56:**193–199.

78. **Marshall, K. C., R. Stout, and R. Mitchell.** 1971. Mechanism of the initial events in the sorption of marine bacteria to surfaces. *J. Gen. Microbiol.* **68:**337–348.

79. **Mayer, B. K., and D. R. Ward.** 1991. Microbiology of finfish and finfish processing, p. 3–17. *In* D. R. Ward and C. R. Hackney (ed.), *Microbiology of Marine Food Products.* Van Nostrand Reinhold, New York.

80. **McEldowney, S., and M. Fletcher.** 1987. Adhesion of bacteria from mixed cell suspension to solid surfaces. *Arch. Microbiol.* **148:**57–62.

81. **McMeekin, T. A.** 1975. Spoilage association of chicken breast muscle. *Appl. Microbiol.* **29:**44–47.

82. **McMeekin, T. A.** 1977. Spoilage association of chicken leg muscle. *Appl. Microbiol.* **33:**1244–1246.

83. **Mead, G. C.** 1982. Microbiology of poultry and game birds, p. 67–101. *In* M. H. Brown (ed.), *Meat Microbiology.* Applied Science Publishers Ltd., New York.

84. **Miget, R. J.** 1991. Microbiology of crustacean processing: shrimp, crawfish, and prawns, p. 65–87. *In* D. R. Ward and C. R. Hackney (ed.), *Microbiology of Marine Products.* Van Nostrand Reinhold, New York.

85. **Newton, K. G., and C. O. Gill.** 1978. The development of the anaerobic spoilage flora of meat stored at chill temperatures. *J. Appl. Bacteriol.* **44:**91–95.

86. **Newton, K. G., and C. O. Gill.** 1980. The microbiology of DFD fresh meats: a review. *Meat Sci.* **5:**223–232.

87. **Newton, K. G., J. C. L. Harrison, and K. M. Smith.** 1977. The effect of storage in various gaseous atmospheres on the microflora of lamb chops held at –1°C. *J. Appl. Bacteriol.* **43:**53–59.

88. **Newton, K. G., J. C. L. Harrison, and A. M. Wauters.** 1978. Sources of psychrotrophic bacteria on meat at the abattoir. *J. Appl. Bacteriol.* **45:**75–82.

89. **Nichol, D. J., M. K. Shaw, and D. A. Ledward.** 1970. Hydrogen sulfide production by bacteria and sulfmyoglobin formation in prepacked chilled beef. *Appl. Microbiol.* **19:**937–939.

90. **Notermans, S., and E. H. Kampelmacher.** 1974. Attachment of some bacterial strains to the skin of broiler chickens. *Br. Poultry Sci.* **15:**573–585.

91. **Nottingham, P. M.** 1982. Microbiology of carcass meats, p. 13–65. *In* M. H. Brown (ed.), *Meat Microbiology.* Applied Science Publishers Ltd., New York.

92. **Nottingham, P. M., C. O. Gill, and K. G. Newton.** 1981. Spoilage at fat surfaces of meat, p. 183–190. *In* T. A. Roberts, G. Hobbs, J. H. B. Christian, and N. Skovgaard (ed.), *Psychrotrophic Microorganisms in Spoilage and Pathogenicity.* Academic Press, New York.

93. **Ogrydziak, D. M., and W. D. Brown.** 1982. Temperature effects in modified-atmosphere storage of seafoods. *Food Technol.* **36(5):**86–96.

94. **Palumbo, S. A., R. K. Jenkins, R. L. Buchanan, and D. W. Thayer.** 1986. Determination of irradiation D-values for *Aeromonas hydrophila. J. Food Prot.* **49:**189–191.

95. **Patterson, J. T., and P. A. Gibbs.** 1977. Incidence and spoilage potential of isolates from vacuum-packaged meat of high pH value. *J. Appl. Bacteriol.* **43:**25–38.

96. **Patterson, J. T., and P. A. Gibbs.** 1978. Sources and properties of some organisms isolated in two abattoirs. *Meat Sci.* **2:**263–273.

97. **Prasai, R. K., G. R. Acuff, L. M. Lucia, D. S. Hale, J. W. Savell, and J. B. Morgan.** 1991. Microbiological effects of acid decontamination of beef carcasses at various locations in processing. *J. Food Prot.* **54:**868–872.

98. **Rey, C. R., A. A. Kraft, D. G. Topel, F. C. Parrish, Jr., and D. K. Hotchkiss.** 1976. Microbiology of pale, dark and normal pork. *J. Food Sci.* **41:**111–116.

99. **Roth, L. A., and D. S. Clark.** 1975. Effect of lactobacilli and carbon dioxide on the growth of *Microbacterium thermosphactum* on fresh beef. *Can. J. Microbiol.* 21:629–632.

100. **Samuels, R. D., A. DeFeo, G. J. Flick, D. R. Ward, T. Rippen, J. K. Riggens, C. Coale, and C. Smith.** 1984. *Demonstration of a Quality Maintenance Program for Fresh Fish Products.* A report submitted to Mid-Atlantic Fisheries Development Foundation. Sea Grant, VPI-SG-84-04R. Virginia Polytechnic Institute and State University, Blacksburg.

101. **Sanders, D. H., and C. D. Blackshear.** 1971. Effect of chlorination in the final washer on bacterial counts of broiler chicken carcasses. *Poultry Sci.* **50:**215–219.

102. **Savell, J. W., D. B. Griffin, C. W. Dill, G. R. Acuff, and C. Vanderzant.** 1986. Effect of film oxygen transmission rate on lean color and microbiological characteristics of vacuum-packaged beef knuckles. *J. Food Prot.* **49:**917–919.

103. **Shank, J. L., J. H. Silliker, and P. A. Goeser.** 1962. The development of a nonmicrobial off-condition in fresh meat. *Appl. Microbiol.* **10:**240–246.

104. **Shaw, B. G., C. D. Harding, and A. A. Taylor.** 1980. The microbiology and storage stability of vacuum packed lamb. *J. Food Technol.* **15:**397–405.

105. **Shewan, J. M.** 1961. The microbiology of sea-water fish, p. 487–560. *In* G. Borgstrom (ed.), *Fish as Food,* vol. 1. *Production, Biochemistry, and Microbiology.* Academic Press, New York.

106. **Shewan, J. M.** 1971. The microbiology of fish and fishery products—a progress report. *J. Appl. Bacteriol.* **34:**299–315.

107. **Silliker, J. H., R. E. Woodruff, J. R. Lugg, S. K. Wolfe, and W. D. Brown.** 1977. Preservation of refrigerated meats with controlled atmospheres: treatment and post-treatment effects of carbon dioxide on pork and beef. *Meat Sci.* **1:**195–204.

108. **Simidu, W.** 1961. Nonprotein nitrogenous compounds, p. 353–384. *In* G. Borgstrom (ed.), *Fish as Food,* vol. 1. *Production, Biochemistry, and Microbiology.* Academic Press, New York.

109. **Smulders, F. J. M., and C. H. J. Woolthuis.** 1983. Influence of two levels of hygiene on the microbiological condition of veal as a product of two slaughtering/processing sequences. *J. Food Prot.* **46:**1032–1035.

110. **Stevenson, K. E., R. A. Merkel, and H. C. Lee.** 1978. Effects of chilling rate, carcass fatness and chlorine spray on microbiological quality and case-life of beef. *J. Food Sci.* **43:**849–852.

111. **Sutherland, J. P., P. A. Gibbs, J. T. Patterson, and J. G. Murray.** 1976. Biochemical changes in vacuum packaged beef occurring during storage at 0–2°C. *J. Food Technol.* **11:**171–1180.

112. **Sutherland, J. P., and A. Varnam.** 1982. Fresh meat processing, p. 103–128. *In* M. H. Brown (ed.), *Meat Microbiology.* Applied Science Publishers Ltd., New York.

113. **Taylor, A. A., N. F. Down, and B. G. Shaw.** 1990. A comparison of modified atmosphere and vacuum skin packing for the storage of red meats. *Int. J. Food Sci. Technol.* **25:**98–109.

114. **Thayer, D. W., G. Boyd, W. S. Muller, C. A. Lipson, W. C. Hayne, and S. H. Bayer.** 1990. Radiation resistance of *Salmonella. J. Ind. Microbiol.* **5:**383–390.

115. **Thomas, C. J., and T. A. McMeekin.** 1980. Contamination of broiler carcass skin during commercial processing procedures: an electron microscopic study. *Appl. Environ. Microbiol.* **40:**133–144.

116. **Tompkin, R. B.** 1986. Microbiology of ready-to-eat meat and poultry products, p. 89–121. *In* A. M. Pearson and T. R. Dutson (ed.), *Advances in Meat Research,* vol. 2. *Meat and Poultry Microbiology.* AVI Publishing Co., Inc., Westport, Conn.

117. **Troller, J. A.** 1979. Food spoilage by microorganisms tolerating low-a_w environments. *Food Technol.* **33(1):**72–75.

118. **Vanderzant, C., J. W. Savell, M. O. Hanna, and V. Potluri.** 1986. A comparison of growth of individual meat bacteria on the lean and fatty tissue of beef, pork and lamb. *J. Food Sci.* **51:**5–8, 11.

119. **van Laak, R. L. J. M.** 1994. Spoilage and preservation of muscle foods, p. 378–405. *In* D. M. Kinsman, A. W. Kotula, and B. C. Breidenstien (ed.), *Muscle Foods: Meat, Poultry and Seafood Technology.* Chapman & Hall, New York.

119a. **Venugopal, R. J., and J. S. Dickson.** Unpublished data.

120. **Wabek, C. J.** 1972. Feed and water withdrawal time relationship to processing yield and potential fecal contamination of broilers. *Poultry Sci.* **51:**1119–1121.

121. **Wodzinski, R. J., and W. C. Frazier.** 1961. Moisture requirements of bacteria. IV. Influence of temperature and increased partial pressure of carbon dioxide on requirements of three species of bacteria. *J. Bacteriol.* **81:**401–418.

122. **Young, L. L., R. D. Reviere, and A. B. Cole.** 1988. Fresh red meats: a place to apply modified atmospheres. *Food Technol.* **42(9):**65–69.

Joseph F. Frank

6 Milk and Dairy Products

Being both highly perishable and nutritious, milk has since prehistoric times been subject to a variety of preservation treatments. Modern dairy processing utilizes pasteurization, heat sterilization, fermentation, dehydration, refrigeration, and freezing as preservation treatments. The result is an assortment of dairy foods having vastly different tastes and textures and a complex variety of spoilage microflora. Spoilage of dairy foods is manifest as off flavors and odors and as changes in texture and appearance. Some defects of milk and cheese caused by microorganisms are listed in Tables 6.1 and 6.2. In this chapter, the discussion of spoilage is organized by types of microorganisms associated with various defects: gram-negative psychrotrophic microorganisms, coliform and lactic acid bacteria, spore-forming bacteria, and yeasts and molds. The major objective of this chapter is to describe the interactions of these microorganisms in dairy foods which lead to commonly encountered product defects.

MILK AND DAIRY PRODUCTS AS GROWTH MEDIA

Milk

Milk is a good growth medium for many microorganisms because of its high water content, near neutral pH, and variety of available nutrients. Milk, however, is not an ideal growth medium since, for example, the addition of yeast extract or protein hydrolysates often increases growth rates. Table 6.3 lists the major nutritional components of milk and their normal concentrations. These components consist of lactose, fat, protein, minerals, and various nonprotein nitrogenous compounds. Many microorganisms cannot utilize lactose and therefore must rely on proteolysis or lipolysis to obtain carbon and energy. In addition, freshly collected raw milk contains various growth inhibitors which decrease in effectiveness with storage.

Carbon and Nitrogen Availability

Carbon sources in milk include lactose, protein, and fat. The citrate in milk can be utilized by many microorganisms but is not present in sufficient amount to support significant growth. A sufficient amount of glucose is present in milk to allow initiation of growth by some microorganisms, but for fermentative microorganisms to continue growth, they must have the appropriate sugar transport system and hydrolytic enzymes for lactose utilization. These are described in chapter 31. Other spoilage microorganisms may oxidize lactose to lactobionic acid. The amount of lactose in milk is in great excess of that needed to support extensive microbial growth.

Although milk has a high fat content, few spoilage microorganisms utilize it as a carbon or energy source. This is because the fat is in the form of globules surrounded by a protective membrane composed of glyco-

Table 6.1 Some defects of fluid milk which result from microbial growth

Defect	Associated microorganisms	Metabolic product	Reference(s)
Bitter	Psychrotrophic bacteria, *Bacillus cereus*	Bitter peptides	75, 80
Rancid	Psychrotrophic bacteria	Free fatty acids	100
Fruity	Psychrotrophic bacteria	Ethyl esters	87
Coagulation	*Bacillus* spp.	Casein destabilization	75
Sour	Lactic acid bacteria	Lactic, acetic acids	38, 99
Malty	Lactic acid bacteria	3-Methyl butanal	77
Ropy	Lactic acid bacteria	Exopolysaccharides	14

Table 6.2 Some defects of cheese which result from microbial growth

Defect	Associated microorganisms	Metabolic product	Reference
Open texture, fissures	Heterofermentative lactobacilli	Carbon dioxide	60
Early gas	Coliforms, yeasts	Carbon dioxide, hydrogen	69
Late gas	*Clostridia* spp.	Carbon dioxide, hydrogen	25
Rancidity	Psychrotrophic bacteria	Free fatty acids	63
Fruity	Lactic acid bacteria	Ethyl esters	9
White crystalline surface deposits	*Lactobacillus* spp.	Excessive D-lactate	89
Pink discoloration	*Lactobacillus delbrueckii* subsp. *bulgaricus*	High redox potential	97

Table 6.3 Approximate concentrations of some nutritional components of milk[a]

Component	Amt
Water	87.3 g/100 g
Lactose	4.6 g/100 g
Fat	3.9 g/100 g
Casein	2.6 g/100 g
Whey protein	0.6 g/100 g
Salt cations	
Sodium	58 mg/100 g
Potassium	140 mg/100 g
Calcium	118 mg/100 g
Magnesium	12 mg/100 g
Salt anions	
Citrate	176 mg/100 g
Chloride	104 mg/100 g
Phosphorus	74 mg/100 g
Nonprotein N	
Total nonprotein N	296 mg/liter
Urea N	142 mg/liter
Peptide N	32 mg/liter
Amino acid N	4 mg/liter
Creatine N	25 mg/liyer

[a]Adapted from reference 55.

proteins, lipoproteins, and phospholipids. Milk fat is available for microbial metabolism only if the globule membrane is physically damaged or enzymatically degraded (2). Generally, milk will be spoiled by other mechanisms before this occurs.

There are primarily two types of proteins in milk, caseins and whey proteins. Caseins are present in the form of highly hydrated micelles and are readily susceptible to proteolysis. Whey proteins (β-lactoglobulin, α-lactalbumin, serum albumin, and immunoglobulins) remain soluble in the milk after precipitation of casein. They are less susceptible to microbial proteolysis. Milk contains nonprotein nitrogenous compounds such as urea, peptides, and amino acids that are readily available for microbial utilization (Table 6.3). These compounds are present in insufficient quantity to support the extensive growth required for spoilage.

Minerals and Micronutrients

Milk is a good source of B vitamins and mineral nutrients. The major salt cations and anions present in milk are listed in Table 6.3. Although milk contains many mineral nutrients such as iron, cobalt, copper, and molybdenum, some of these, such as iron, may not be

1a. $2SCN^- + H_2O_2 + 2H^+ \xrightarrow{\text{(lactoperoxidase)}} (SCN)_2 + H_2O$

1b. $(SCN)_2 + H_2O \longrightarrow HOSCN + SCN^- + H^+$

1c. $HOSCN \longleftrightarrow H^+ + OSCN^-$

2. $SCN^- + H_2O_2 \xrightarrow{\text{(lactoperoxidase)}} OSCN^- + H_2O$

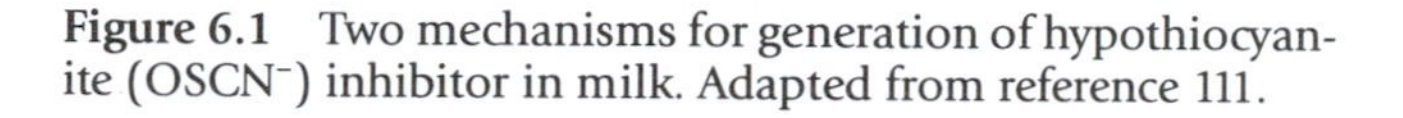

Figure 6.1 Two mechanisms for generation of hypothiocyanite ($OSCN^-$) inhibitor in milk. Adapted from reference 111.

present in a readily usable form. Supplementation of milk with trace elements may be necessary to achieve maximum microbial growth rates. Milk also contains growth stimulants, of which orotic acid (a metabolic precursor for pyrimidines) is the most significant.

Natural Inhibitors

The major microbial inhibitors in raw milk are lactoferrin and the lactoperoxidase system. Natural inhibitors of lesser importance include lysozyme, specific immunoglobulins, and folate and vitamin B_{12} binding systems. Lactoferrin, a glycoprotein, acts as an antimicrobial agent by binding iron. Human milk contains over 2 mg of lactoferrin per ml, but it is of lesser importance in cow milk, which contains only 20 to 200 µg/ml (70). Psychrotrophic aerobes which commonly spoil refrigerated milk are inhibited by lactoferrin, but the presence of citrate in cow milk limits its effectiveness, as the citrate competes with lactoferrin for binding the iron (8).

The most effective natural microbial inhibitor in cow milk is the lactoperoxidase system. Lactoperoxidase catalyzes the oxidation of thiocyanate and simultaneous reduction of hydrogen peroxide, resulting in the accumulation of hypothiocyanite. Two mechanisms for this reaction are illustrated in Fig. 6.1. Hypothiocyanite oxidizes sulfhydryl groups of proteins, resulting in enzyme inactivation and structural damage to the cytoplasmic membrane (111). Lactoperoxidase and thiocyanate are present in milk during synthesis, whereas hydrogen peroxide is formed in milk when oxygen is metabolized by lactic acid bacteria. Since hydrogen peroxide is the limiting substrate of the reaction, the effective use of this inhibitor system for preserving milk relies on adding a hydrogen peroxide-generating system or, less effectively, hydrogen peroxide to the milk (27). Lactic acid bacteria, coliforms, and various pathogens are inhibited by this system (111).

Effect of Heat Treatments

The minimum required heat treatment for milk to be sold for fluid consumption in the United States is 72°C for 15 s, though most processors use slightly higher temperatures and longer holding times. Pasteurization affects the growth rate of the spoilage microflora by destroying inhibitor systems. More severe heat treatments affect microbial growth by increasing available nitrogen through protein hydrolysis and by the liberation of inhibitory sulfhydryl compounds. Lactoperoxidase is only partially inactivated by normal pasteurization treatments (111).

Dairy Products

Dairy products provide substantially different growth environments than fluid milk because these products have nutrients removed or concentrated or have lower pH or water activity (a_w). The composition, pH, and a_w of selected dairy products are presented in Table 6.4. Yogurt is essentially acidified milk and therefore provides a nutrient-rich low-pH environment. Cheeses are less acidified than yogurt but have less water and added salt, resulting in lower a_w. In addition, the solid nature of cheeses limits mobility of spoilage microorganisms. Liquid milk concentrates such as evaporated skim milk do not have sufficiently low a_w to inhibit spoilage and must

Table 6.4 Approximate composition, pH, and a_w of selected dairy products[a]

Product	Component (g/100 g)				a_w	pH
	Water	Fat	Protein	Carbohydrate		
Butter	15.9	81.1	3.6	0.06		6.3
Cheddar cheese	36.7	33.1	24.9	1.3	0.90–0.95	5.2
Swiss cheese	37.2	27.4	28.4	3.4		5.6
Nonfat dried milk	3.2	0.8	36.2	52.0	0.2	
Evaporated skim milk	79.4	0.2	7.5	11.3	0.93–0.98	
Yogurt	89	1.7	3.5	5.1		4.3

[a]Compiled from references 6, 7, 20, and 57.

be canned or refrigerated for preservation. Milk-derived powders have sufficiently low a_w to completely inhibit microbial growth. Butter is a water-in-oil emulsion, and so microorganisms are trapped within water droplets. If butter is salted, the mean salt content of the water droplets will be 6 to 8%, sufficient to inhibit gram-negative spoilage organisms that could grow during refrigeration. However, individual droplets will have significantly higher or lower salt content if the salt is not uniformly distributed during manufacture. This can result in psychrotrophic bacteria growing in the low-salt droplets. Unsalted butter is usually made from acidified cream and relies on low pH and refrigeration for preservation.

PSYCHROTROPHIC SPOILAGE

Preservation of fluid milk relies on effective sanitation, timely marketing, pasteurization, and refrigeration. Raw milk is rapidly cooled after collection and is kept cold until pasteurized, after which it is kept cold until consumption. There is often sufficient time between milk collection and consumption for psychrotrophic bacteria to grow. Many of the flavor defects detected by milk consumers result from this growth. Pasteurized milk is expected to have a shelf life of 14 to 20 days, and so contamination of the contents of a container with even one rapidly growing psychrotrophic microorganism can lead to spoilage. Growth of psychrotrophic bacteria in raw milk can lead to defects in products made from that milk because of the residual activity of degradative enzymes.

Psychrotrophic Bacteria in Milk

Psychrotrophic bacteria which spoil raw and pasteurized milk are primarily aerobic gram-negative rods in the family *Pseudomonadaceae,* with occasional representatives from the family *Neisseriaceae* and the genera *Flavobacterium* and *Alcaligenes.* It is typical that 65 to 70% of raw milk psychrotrophic isolates are in the genus *Pseudomonas* (41). Although representatives of other genera, including *Bacillus, Micrococcus, Aerococcus, Staphylococcus,* and the family *Enterobacteriaceae* may be present in raw milk and may be psychrotrophic, they are usually outgrown by the gram-negative obligate aerobes (especially *Pseudomonas* spp.) when milk is held at its typical storage temperature of 3 to 7°C. The psychrotrophic spoilage microflora of milk is generally proteolytic, with many isolates able to produce extracellular lipases, phospholipase, and other hydrolytic enzymes but unable to utilize lactose. The bacterium most often associated with flavor defects in refrigerated milk is *Pseudomonas fluorescens* (30), with *P. fragi, P. putida,* and *P. aeruginosa* also commonly encountered. Jaspe et al. (53) observed that incubating raw milk for 3 days at 7°C selected for a population of *Pseudomonas* spp. which had a 10-fold-higher growth rate at 7°C and a lower growth rate at 21°C than the initial *Pseudomonas* population. These strains also exhibited 1,000-fold-greater proteolytic activity and 280-fold-greater lipolytic activity than the initial *Pseudomonas* population. Psychrotrophic bacteria commonly found in raw milk are inactivated by pasteurization.

Sources of Psychrotrophic Bacteria in Milk

Soil, water, animals, and plant material constitute the natural habitat of psychrotrophic bacteria found in milk (21). Plant materials, such as grass and hay used for animal feed, may contain over 10^8 psychrotrophs per g (103). Water used on the dairy farm has been found to contain psychrotrophic bacteria, even when it has been chlorinated (104). Although farm water usually contains only low populations of psychrotrophic microorganisms, its use to clean and rinse milking equipment provides a direct means for their entry into milk. Psychrotrophic bacteria isolated from water are often very active producers of extracellular enzymes and grow rapidly at refrigeration temperatures (21). Consequently, water is an important source of milk spoilage bacteria. The teat and udder area of the cow can harbor high levels of psychrotrophic bacteria, even after washing and sanitizing (78). These psychrotrophs probably originate from soil.

Milking equipment, utensils, and storage tanks are the major sources of psychrotrophic contamination of raw milk (21). Proper cleaning and sanitizing procedures can effectively reduce contamination from these sources. Milk residues on unclean equipment provide a growth niche for psychrotrophic bacteria which enter milking machines, pipelines, and holding tanks with water rinses or milk. Milking equipment is generally fabricated from stainless steel, which is readily cleaned and sanitized; however, for some parts, rubber or other nonmetal materials must be used. These materials are difficult to sanitize, since only moderate use results in the formation of microscopic pores or cracks. Bacteria attached to these parts are difficult to inactivate by chemical sanitization (79).

Pasteurized milk products become contaminated with psychrotrophic bacteria by exposure to contaminated equipment or air. Schröder (96) determined that filling equipment was most often the source of psychrotrophs in packaged milk. Although the levels of psychrotrophic bacteria in air are generally quite low, only one viable cell per container is required to spoil the product. Product is often exposed to contaminated air during the packaging process. Pasteurized milk packaged by use of aseptic systems has a much longer shelf life than conventionally packaged milk.

Growth Characteristics and Defects

Generation times in milk of the most rapidly growing psychrotrophic *Pseudomonas* spp. isolated from raw milk are 8 to 12 h at 3°C and 5.5 to 10.5 h at 3 to 5°C (102). These growth rates are sufficient to cause spoilage within 5 days if the milk initially contains only 1 CFU/ml. However, most psychrotrophic pseudomonads present in raw milk grow much more slowly, causing refrigerated milk to spoil in 10 to 20 days.

Defects of fluid milk associated with the growth of psychrotrophic bacteria are related to the production of extracellular enzymes. Sufficient enzyme to cause defects is usually present when the population of psychrotrophs reaches 10^6 to 10^7 CFU/ml (31), but this depends on the specific product; for example, the shelf life of UHT milk (milk treated at ultrahigh temperature [UHT]) is limited by the presence of less enzyme than is the shelf life of pasteurized milk. Bitter and putrid flavors and coagulation result from proteolysis. Rancid and fruity flavors result from lipolysis. The production of extracellular enzymes by psychrotrophic bacteria in raw milk also has implications for the quality of products produced from that milk.

Proteases

Factors Affecting Protease Production

P. fluorescens and other psychrotrophs that may be present in raw milk generally produce protease during the late exponential and stationary phases of growth (42, 93, 109), although there are reports of protease production throughout the growth phase (66). This discrepancy may be due to strain and species differences. Increases in protease activity associated with stationary and decline phases of growth may result from the release of preformed enzyme from the cells (42).

The effect of temperature on protease production does not parallel its effect on growth. The temperature for the optimum production of protease by psychrotrophic *Pseudomonas* spp. is lower than the temperature for the optimum rate of growth (72). Relatively high amounts of protease are produced at temperatures as low as 5°C (71). Rowe (93) and Griffiths (42) indicate that protease production by *Pseudomonas* spp. is inhibited in milk held at 2°C.

Since *Pseudomonas* spp. are obligate aerobes, it is expected that oxygen is required for protease synthesis. Myhara and Skura (82) reported optimum protease production for *P. fragi* in a medium containing 7.4 μg of dissolved oxygen per ml, slightly less than the 8.4 μg/ml contained in saturated water.

The effect of calcium and iron ions on protease production by *Pseudomonas* spp. is relevant to dairy spoilage. Ionic calcium is required for protease synthesis. McKellar and Cholette (73) reported that in the absence of ionic calcium, an inactive precursor to proteinase was produced by *P. fluorescens*. This precursor could not be activated, indicating that calcium is required to stabilize the enzyme. Iron, which may be at a growth-limiting concentration in milk, will repress protease production by *Pseudomonas* spp. when added to milk (32). Maximum protease production will occur only if iron is growth limiting, although this may be an indirect effect of reduced cytochrome synthesis and decreased energy levels (72). The presence of pyoverdine, an iron-chelating pigment produced by *P. fluorescens*, stimulates protease production (74).

Most evidence indicates that *P. fluorescens* regulates protease production as a means to provide carbon to the cell rather than amino acids for protein synthesis (31). Protease production is induced by various low-molecular-weight protein degradation products and subject to end product and catabolite repression. Asparagine is the most effective amino acid inducer, and citric acid is an effective inhibitor of synthesis (72). Protease production by *P. fluorescens* in raw milk is preceded by the depletion of glucose, galactose, lactate, glutamine, and glutamic acid. Supplementation of milk with these compounds delays or inhibits protease production (54).

Properties of Proteases from Psychrotrophic Pseudomonads

Fox et al. (35) summarized properties of proteinases produced by *P. fluorescens*. Properties most relevant to dairy product spoilage include temperature optima from 30 to 45°C, with significant activity at 4°C, and pH optima near neutrality or in the alkaline range; all are metalloenzymes containing either Zn^{2+} or Ca^{2+}. Perhaps the most important characteristic of these enzymes in regard to dairy product spoilage is their extreme heat stability. Decimal reduction times at 140°C range from 50 to 200 s, sufficient to retain significant activity after UHT milk processing (58). This degree of heat stability is surprising, considering that the enzymes are produced and active at refrigeration temperatures. Another unexpected property of these enzymes is their susceptibility to autodegradation at 55 to 60°C, apparently as a result of an unfolding of the protein chain into a more sensitive conformation (101). This low-temperature inactivation phenomenon is highly dependent on pH (21) and varies with species biotype. Milk with high populations of psychrotrophs which has been subjected to a dual heat treatment (low-temperature treatment for enzyme inactivation followed by a UHT treatment) has been reported to have a longer shelf life than conventional UHT milk (11, 46), but this effect is not consistent.

Protease-Induced Product Defects

Proteases of psychrotrophic bacteria cause product defects either at the time they are produced in the product or as a result of enzyme surviving a heat process. Most investigators have observed that these proteases preferentially hydrolyze κ-casein, although some show preference for β- or α_{s-1}-casein (22). No specific cleavage sites have been identified, since specificity depends on reaction conditions and the specific enzyme tested. Degradation of casein in milk by enzymes produced by psychrotrophs results in the liberation of bitter peptides. Bitterness is a common off flavor in pasteurized milk that has been subject to postpasteurization contamination with psychrotrophic bacteria. Continued proteolysis results in putrid off flavors associated with lower-molecular-weight degradation products such as ammonia, amines, and sulfides. Bitterness in UHT (commercially sterile) milk develops when sufficient psychrotrophic bacterial growth occurs in raw milk (estimated at 10^5 to 10^7 CFU/ml) to leave residual enzyme after heat treatment (80). Low-level protease activity in UHT milk can also result in coagulation or sediment formation. UHT milk appears to be more sensitive to protease-induced defects than raw milk, probably a result of either heat-induced changes in casein micelle structure or heat inactivation of protease inhibitors (88).

The effect of proteases of psychrotrophic origin on quality of cheese and other cultured dairy products is minimal, because the combination of low pH and low storage temperature inhibits their activity (61). In addition, proteases are removed with the whey fraction during cheese manufacture. However, growth of proteolytic bacteria in raw milk lowers cheese yield because proteolytic products of casein degradation are lost to the whey rather than becoming part of the cheese (112).

Lipases

Psychrotrophic *P. fluorescens* isolated from milk often produces extracellular lipase in addition to protease. Other commonly found lipase-producing psychrotrophs include *P. fragi* and *P. aeruginosa*.

Factors Affecting Lipase Production

Lipases of psychrotrophic pseudomonads, like proteases, are produced in the late log or stationary phase of growth (1, 72). As with protease, optimal synthesis of lipase generally occurs below the optimum temperature for growth. For example, Andersson (3) reported optimum production of lipase at 8°C by a *P. fluorescens* isolate which exhibited optimum growth at 20°C. Milk is an excellent medium for lipase production by pseudomonads. Although *Pseudomonas* spp. can utilize inorganic nitrogen, lipase production requires an organic nitrogen source. However, some amino acids, especially those which serve as nitrogen but not carbon sources, repress lipase production. This phenomenon has been related to accumulation of metabolic intermediates (33).

Ionic calcium is required for lipase activity, as activity inhibited by EDTA is reversed by addition of Ca^{2+} (1). *P. fluorescens* can produce lipase in calcium-free media, although the amount produced is less than in the presence of calcium (73). Lipase production by *P. fluorescens*, to a greater degree than protease production, is stimulated by limiting iron availability (74). Supplementation of milk with iron delays the onset of lipase production by *P. fluorescens* in raw milk (32).

Although there is conflicting evidence, most reports indicate that lipase production is subject to catabolite repression (72, 100). Lipases are produced by *P. fluorescens* and *P. fragi* in the absence of triglycerides. The presence of triglycerides can either inhibit or stimulate production of lipase, depending on the specific strain, triglyceride concentration, and growth conditions. Surface-active agents such as polysaccharides and lecithin often stimulate the release of lipase from the cell surface (100).

Properties of Lipase from Pyschrotrophic *Pseudomonas* spp.

Properties of lipases produced by *Pseudomonas* spp. have been summarized by Stead (100) and Fox et al. (35). Temperatures for optimal activity of these lipases range from 22 to 70°C, with most between 30 and 40°C. Optimal pH of activity is from 7.0 to 9.0, with most having optima between 7.5 and 8.5. *P. fluorescens* lipase in milk is active at refrigeration temperatures, and significant activity remains at subfreezing temperatures and at low a_w (4).

Heat stability of lipases from psychrotrophic pseudomonads is similar to that of their proteases. Decimal reduction times at 140°C range from 48 to 437 s (58), sufficient to provide residual activity after UHT treatment. Most of these lipases are also subject to accelerated irreversible inactivation at temperatures below 100°C. The optimal conditions for low-temperature inactivation vary considerably with strain, with optimal inactivation between 52 and 80°C being reported. Low-temperature inactivation is markedly affected by milk components, with milk salts increasing susceptibility and casein increasing stability (59). Histidine and $MgCl_2$ were required for the low-temperature inactivation of a heat-stable lipase isolated from *P. fluorescens* (19). The mechanism for low-temperature inactivation of lipase in milk is unknown, but the phenomenon may involve proteolysis of unfolding lipase molecules (58).

Lipase-Induced Product Defects

The triglycerides in raw milk are present in globules that are protected from enzymatic degradation by a membrane. Milk becomes susceptible to lipolysis if this membrane is disrupted by excessive shear force (from pumping, agitation, freezing, etc.). Raw milk contains a mammalian lipase (milk lipase) which will rapidly act on the fat if the globule membrane is disrupted. Most cases of rancidity in raw and pasteurized milk result from this process rather than from the growth of lipase-producing microorganisms. Phospholipase C and protease produced by psychrotrophic bacteria can degrade the fat globule membrane, resulting in the enhancement of milk lipase acitivity (2, 18). Since milk lipase is heat labile, most milk products will not have residual activity. The exceptions are some hard cheeses made from unheated milk or milk given a subpasteurization heat treatment. These cheeses depend on milk lipase activity to develop a characteristic flavor.

Sufficient bacterial lipase can be produced in raw milk to cause defects in products manufactured from that milk. Since residual activities are usually low and the reaction environment is less than optimum, usually only products with long storage times or high storage temperatures are affected. Such products include UHT milk, some cheeses, butter, and whole milk powder. Lipase-producing bacteria can also recontaminate pasteurized milk and cream and grow in these products during refrigerated storage. The rancid flavor and odor resulting from lipase action are usually from the liberation of C_4 to C_8 fatty acids. Fatty acids of higher molecular weight produce a flavor described as soapy. Low levels of unsaturated fatty acids liberated by enzymatic activity may be oxidized to ketones and aldehydes to produce oxidized or "cardboardy" off flavor (26). *P. fragi* produces a fruity off flavor in milk by esterifying free fatty acids with ethanol (87). Ethyl butyrate and ethyl hexanoate are the esters produced at highest amounts, with low levels of ethyl esters of acetate, propionate, and isovalerate also produced. Low levels of ethanol in milk, present as a result of microbial activity, stimulate ester production.

Residual activity from heat-stable microbial lipases can cause off flavors in UHT milk, but lipase-induced defects are not as common as those resulting from microbial protease (80). Lipolytic off flavors in UHT products generally take several weeks to months to develop because of the low amounts of lipase present, or they appear in products made from raw milk with excessively high populations of psychrotrophs (>10^6 CFU/ml) (100).

Rancid defect in butter may result from growth of lipolytic microorganisms during storage, residual heat-stable microbial lipase originating from the growth of psychrotrophic bacteria in the milk or cream, or milk lipase activity in the raw milk or cream. When butter is manufactured from rancid cream, low-molecular-weight free fatty acids are removed with the watery portion of the cream (buttermilk); consequently, the resulting butter will have not a typical rancid flavor but rather a less pronounced soapy off flavor associated with C_{10} to C_{12} free fatty acids. However, the typical odor of rancid butter is associated with lower-molecular-weight fatty acids (C_4 to C_8). Microbial lipases present in butter will exhibit activity even if the product is stored at −10°C (83). Growth of psychrotrophic bacteria in butter occurs only if the product is made from sweet rather than ripened (sour) cream. Sweet cream butter is preserved by salt and refrigeration. Since butter is a water-in-fat emulsion, moisture and salt will not equilibrate during storage. Consequently, if salt and moisture are not evenly distributed in the product during manufacture, lipolytic psychrotrophs will have pockets of high a_w in which to grow (26).

Cheese is more susceptible to defects caused by bacterial lipases than those caused by proteases because lipases, unlike most proteases, are concentrated along with the fat in the curd. The acidic environment of most cheeses limits, but may not eliminate, lipase activity. Some cheeses, such as Camembert and Brie, increase in pH to near neutrality during ripening. Camembert is, in fact, susceptible to defects associated with microbial lipase (29). More acidic cheeses, e.g., Cheddar, are susceptible if they are cured for several months or if high amounts of lipase are present (62). Law et al. (63) reported that Cheddar cheese made from milk containing over 10^6 CFU of *P. fluorescens* per ml before pasteurization developed rancid flavor in 6 to 8 months. Psychrotrophic bacteria do not grow in cured cheeses because of the low pH and salt content but may grow in high-moisture fresh products such as cottage cheese.

Whole milk powder containing bacterial lipase will develop rancidity, indicating a potential problem if milk supports growth of lipolytic psychrotrophs before processing. Low-fat powders such as nonfat dried milk, whey, and whey protein concentrate may contain residual lipase which becomes active when these products are used in fat-containing food formulations (100).

Control of Product Defects Associated with Psychrotrophic Bacteria

Raw Milk

Preventing product defects which result from growth of psychrotrophic bacteria in raw milk involves limiting contamination levels, rapid cooling immediately after milking, and maintenance of cold storage temperatures. Limiting populations of bacteria primarily involves cleaning, sanitizing, and drying cow teats and udders before milking and cleaning and sanitizing equipment. Removal of residual milk solids from milk contact surfaces is critical to psychrotroph control, since these residues protect cells from the action of chemical sanitizers and provide nutrients for growth. Subsequent growth over a period of days results in a biofilm which, in addition to containing high numbers of bacteria, is highly resistant to chemical sanitizers (36). Rapid cooling of milk after collection is important because contamination of the product with psychrotrophic bacteria is unavoidable. Milk, fresh from the cow, enters the farm storage tank at 30 to 37°C. Sanitary standards in the United States require raw milk to be cooled to 7°C within 2 h after milking (5), but most farm systems achieve rapid cooling to less than 4°C. Since milk is often picked up from the farm every 48 h, three additional milkings will be added to the previously collected milk. The second milking should not warm the previously collected milk to over 10°C. As previously indicated, psychrotrophic activity in milk is inhibited at 2°C, but freezing of milk causes disruption of the fat globule membrane, making it highly susceptible to lipolysis. Therefore, the challenge of farm storage systems is to rapidly cool milk to as low a temperature as possible while avoiding ice formation.

Pasteurized Products

Preventing contamination of pasteurized dairy products with psychrotrophic bacteria is primarily a matter of equipment cleaning and sanitation, although airborne psychrotrophs may also limit product shelf life. Even when filling equipment is effectively cleaned and sanitized, it can still become a source of psychrotrophic microorganisms which accumulate during the hours of continuous use which normally occurs (96). These microorganisms probably enter the filler through the vacuum system or from containers. Tanks used for holding pasteurized products before packaging can also be a source of psychrotrophic microorganisms. Tank walls may have microscopic fissures or pits which protect microorganisms from cleaning and sanitizing procedures. Complete elimination of psychrotrophic microorganisms from products can be achieved only by using aseptic packaging technologies whereby equipment and air which contact the pasteurized product are maintained free of vegetative cells.

SPOILAGE BY FERMENTATIVE NONSPOREFORMERS

Spoilage of milk and dairy products resulting from growth of acid-producing fermentative bacteria occurs when storage temperatures are sufficiently high for these microorganisms to outgrow psychrotrophic bacteria or when product composition is inhibitory to gram-negative aerobic organisms. For example, the presence of lactic acid in fluid milk is a good indication that the product was exposed to an unacceptably high storage temperature which allowed growth of lactic acid bacteria. Fermented dairy foods, though manufactured by using lactic acid bacteria, can be spoiled by the growth of "wild" lactic acid bacteria which produce unwanted gas, off flavors, or appearance defects. Fluid milk, cheese, and cultured milks are the major dairy products susceptible to spoilage by non-spore-forming fermentative bacteria.

Non-spore-forming bacteria responsible for fermentative spoilage of dairy products are mostly in either the lactic acid-producing or coliform group. Lactic acid bacteria involved in dairy fermentations can spoil fluid milk, but the strains involved are often environmental types which produce defect-inducing metabolites in addition to lactic acid. Genera of lactic acid bacteria involved in spoilage of milk and fermented products include *Lactococcus, Lactobacillus, Leuconostoc, Enterococcus, Pediococcus,* and *Streptococcus* (16). Coliforms can spoil milk, but this is seldom a problem since they are usually outgrown by either the lactic acid or the psychrotrophic bacteria. Coliform spoilage is more common with fermented products, especially certain cheese varieties. Members of the genera *Enterobacter* and *Klebsiella* are most often associated with coliform spoilage, while

Escherichia spp. only occasionally exhibit sufficient growth to produce a defect.

Sources of Fermentative Spoilage Bacteria

Lactic acid bacteria are normal inhabitants of the skin and streak canal of the cow teat. Consequently, all raw milk contains at least low numbers of these microorganisms. Lactic acid bacteria are also associated with silage and other animal feeds and feces. Coliform bacteria are present on udder skin as a result of fecal contamination; therefore, ineffective cleaning of this area before milking will contribute to high coliform populations in milk. Coliform bacteria in raw milk are also associated with milk residue buildup on inadequately cleaned milking equipment (10).

Defects of Fluid Milk Products

The most common fermentative defect in fluid milk products is souring caused by the growth of lactic acid bacteria. Lactic acid by itself has a clean, pleasant acid flavor and no odor. The unpleasant "sour" odor and taste of spoiled milk is a result of small amounts of acetic and propionic acids (99). Sour odor can be detected before a noticeable acid flavor develops. For discussion of lactic acid production in milk see chapter 31. Other defects may occur in combination with acid production, and these may be more noticeable to the consumer. A malty flavor results from growth of *Lactococcus lactis* subsp. *lactis* var. maltigenes. This strain is unique among lactococci in its ability to produce 2-methylpropanal, 3-methylbutanal, and the corresponding alcohols (77). The aldehydes are produced by decarboxylation of α-ketoisocaproic and α-ketoisovaleric acids. These keto acids are also concurrently used to synthesize leucine and valine by transamination with glutamic acid. Alcohols corresponding to the aldehydes are formed by the action of alcohol dehydrogenase in the presence of $NADH_2$. Malty flavor is primarily from 3-methylbutanal.

Another defect associated with growth of lactic acid bacteria in milk is "ropy" texture. Most dairy-associated species of lactic acid bacteria have strains that produce exocellular polymers which increase the viscosity of milk, causing the ropy defect (14). Some of these strains are used to produce high-viscosity fermented products such as yogurt and Scandinavian ropy milk (vilia, skyr). The defect in noncultured fluid milk products is usually caused by growth of specific strains of lactococci. The polymer produced by these organisms is a polysaccharide containing glucose and galactose with small amounts of mannose, rhamnose, and pentose (15). The polysaccharide is possibly associated with protein (65).

Defects in Cheese

Lactic Acid Bacteria

Some strains of lactic acid bacteria produce flavor and appearance defects in cheese. Lactobacilli are a normal part of the dominant microflora of aged Cheddar cheese. If heterofermentative lactobacilli predominate, the cheese is prone to develop an "open" texture or fissures, a result of gas production during aging (60). Off flavors are also associated with the growth of these organisms (108). Gassy defects in aged Cheddar cheese are more often associated with growth of lactobacilli than with growth of coliforms, yeasts, or sporeformers. The use of elevated ripening temperatures for Cheddar cheese, e.g., 15 rather than 8°C, encourages growth of heterofermentative lactobacilli but not that of non-lactic acid bacteria (24). This phenomenon limits the use of high-temperature storage to accelerate ripening. *Lactobacillus brevis* and *L. casei* subsp. *pseudoplantarum* have been associated with gas production in retail Mozzarella cheese (52). *L. casei* subsp. *casei* produces a soft body defect in Mozzarella cheese as a result of its proteolytic ability. The softened cheese cannot be readily sliced or grated and does not melt properly.

Some cheese varieties occasionally exhibit a pink discoloration. Pink spots in Swiss-type varieties result from the growth of pigmented strains of propionibacteria. In Italian cheese varieties, a pink discoloration may occur either in a band near the surface or throughout the whole cheese. This defect is associated with certain strains of *Lactobacillus delbrueckii* subsp. *bulgaricus* that fail to lower the redox potential of the cheese (97). Another common defect of aged Cheddar cheese is the appearance of white crystalline deposits on the surface. Although they do not affect flavor, these deposits reduce consumer acceptability. Rengipat and Johnson (89) observed an atypical strain of a facultatively heterofermentative *Lactobacillus* species associated with the deposits. This strain produces an unusually high amount of D-lactic acid during cheese aging, resulting in the formation of insoluble calcium lactate crystals, the primary component of the white deposits. *L. casei* subsp. *alactosus* and *L. casei* subsp. *rhamnosus* have been associated with the development of a phenolic flavor in Cheddar cheese, described as being similar to horse urine (52). The flavor develops after 2 to 6 months of aging.

Fruity off flavor in Cheddar cheese is usually not caused by growth of psychrotrophic bacteria, as it is in milk, but rather is a result of growth of lactic acid bacteria (usually *Lactococcus* spp.) which produce esterase. Fruity-flavored cheeses contain high levels of ethanol, a substrate for esterification (9). The major esters contributing to fruity flavor in cheese are ethyl hexanoate and ethyl butyrate.

Coliform Bacteria

As early as 1885, coliform bacteria were recognized as causing gassy defect in Cheddar and related cheese varieties (94). Growth usually occurs during the cheese manufacture process or during the first few days of storage and therefore is referred to as early gas (or early blowing) defect. In hard cheeses, such as Cheddar, this defect occurs when a slow lactic acid fermentation fails to rapidly lower pH or when highly contaminated raw milk is used. Cheese varieties in which acid production is purposely delayed by washing the curds are highly susceptible to coliform growth (40). Soft, mold-ripened cheeses, such a Camembert, increase in pH during ripening, with a resulting susceptibility to coliform growth (39, 95). Frank and Marth (37) reported that 17% of commercial soft and semisoft cheeses tested contained over 10^4 coliforms per g. In a more recent survey, 13 and 17% of Brie and Camembert samples, respectively, contained more than 10^7 CFU of *Enterobacteriaceae* per g on their mold rind (84). Gas formation in retail Mozzarella cheese has been associated with growth of *Klebsiella pneumoniae* (69). Coliform growth in retail cheese is often manifest as a swelling of the plastic package. Approximately 10^7 CFU of coliforms per g is needed to produce a gassy defect.

Defects in Fermented Milk Products

Fermented milk products such as cultured buttermilk, sour cream, and cottage cheese rely on diacetyl produced during fermentation for their typical "buttery" flavor and aroma. These products lose consumer appeal when this flavor is lost as a result of reduction of diacetyl to acetoin and 2,3-butanediol (38). Lactococci, capable of growing at 7°C, may produce sufficient diacetyl reductase to destroy diacetyl in cultured milks (50). Other psychrotrophic contaminants in cultured milks, including yeasts and coliforms, may also be involved in diacetyl reduction (110).

Control of Defects Caused by Lactic Acid and Coliform Bacteria

Defects in fluid milk caused by coliforms and lactic acid bacteria are controlled by good sanitation practices during milking, maintaining raw milk at temperatures below 7°C, pasteurization, and refrigeration of pasteurized products. These microorganisms seldom grow in refrigerated pasteurized milk because of their slow growth rates compared with those of psychrotrophic bacteria. Control of coliform growth in cheese is achieved by using pasteurized milk, encouraging a rapid fermentation, and good sanitation during manufacture. Controlling defects produced by undesirable lactic acid bacteria in cheese and fermented milks is more difficult, since growth of lactic acid bacteria must be encouraged during manufacture, and the final products often provide suitable growth environments. Undesirable strains of lactic acid bacteria are readily isolated from the manufacturing environment, so their control requires attention to plant cleanliness and protecting the product during manufacture.

SPORE-FORMING BACTERIA

Spoilage by spore-forming bacteria can occur in low-acid fluid milk products that are preserved by heat with little chance for recontamination with vegetative cells. Products in this category include aseptically packaged milk and cream as well as sweetened and unsweetened concentrated canned milks. Nonaseptic packaged refrigerated fluid milk may spoil as a result of growth of psychrotrophic *Bacillus cereus* in the absence of more rapidly growing gram-negative psychrotrophs (85). Hard cheeses, especially those with low interior salt concentrations, are also susceptible to spoilage by spore-forming bacteria.

Spore-forming bacteria that spoil dairy products usually originate in the raw milk. Populations present in raw milk are generally quite low (<5,000 CFU/ml), and the occurrence of a sporeformer-induced defect does not always correlate with initial numbers of sporeformers in the raw product (76). This is because products prone to support sporeformer growth are stored for sufficiently long periods of time so that outgrowth of small numbers of cells can eventually cause a defect. Spore-forming bacteria in raw milk are predominantly *Bacillus* spp., with *Bacillus licheniformis, B. cereus, B. subtilis,* and *B. megaterium* most commonly isolated (68, 98). *Clostridium* spp. are present in raw milk at such low levels that enrichment and most-probable-number techniques must be used for quantification (91). Populations of spore-forming bacteria in raw milk vary seasonally. In temperate climates, *Bacillus* and *Clostridium* spp. are at higher levels in raw milk collected in the winter than that collected in the summer, because in the winter cows lie on spore-contaminated bedding materials and are more likely to consume spore-laden silage (10). The occurrence of psychrotrophic *Bacillus* spp. in raw milk does not follow this seasonal pattern (75), but there may be a seasonal occurrence of germination factors (86).

Defects in Fluid Milk Products

Pasteurized milk packaged under conditions which limit recontamination can spoil as a result of the growth of psychrotrophic *B. cereus*. This topic has been reviewed by Meer et al. (75). Psychrotrophic *B. cereus* organisms are present in over 80% of raw milk samples. There is also evidence that psychrotrophic *Bacillus* spp. are introduced into the milk at the processing plant as postpasteurization contaminants (43). Psychrotrophic *B. cereus* can reach populations exceeding 10^6 CFU/ml in milk held for 14 days at 7°C, although slower growth is more common (76). Germination of spores in raw milk occurs soon after pasteurization, indicating heat activation. The defect produced by subsequent growth is described as sweet curdling, since it first appears as coagulation without significant acid or off flavor being formed. Coagulation is caused by a chymosin-like protease (17). Eventually, the enzyme degrades casein sufficiently to produce a bitter-flavored product. Growth may become visible as "buttons" at the bottom of the carton; these are actually bacterial colonies. Psychrotrophic *B. cereus* also produces phospholipase C (lecithinase) which degrades the fat globule membrane, resulting in the aggregation of the fat in cream (75). The result is described as "bitty" cream defect.

Psychrotrophic *Bacillus* spp. other than *B. cereus* are also capable of spoiling heat-treated milk. Cromie et al. (23) observed that psychrotrophic *Bacillus circulans* was the predominant spoilage organism in aseptically packaged heat-treated milk. This microorganism produces acid from lactose, giving the milk a sour flavor. *Bacillus mycoides* is another frequently isolated psychrotrophic sporeformer in milk (86).

Most bacterial spores present in raw milk are moderately heat labile and destroyed by UHT treatments. The major heat-resistant species in milk is *Bacillus stearothermophilus* (81). Other, less heat-resistant *Bacillus* spp., especially *B. subtilis* and *B. megaterium* (47), have been isolated from UHT milk.

Defects in Canned Condensed Milk

Canned condensed milk may be either sweetened with sucrose to lower the a_w or left unsweetened. The unsweetened product must be sterilized by heat treatment. Defects associated with growth of surviving spore-forming organisms in this product have been described by Carić (13). "Sweet coagulation" is caused by growth of *Bacillus coagulans, B. stearothermophilus,* and *B. cereus*. This defect is similar to the sweet curdling defect caused by psychrotrophic *B. cereus* in pasteurized milk. Protein destruction, in addition to curdling, can also occur and is usually caused by growth of *B. subtilis* or *B. licheniformis*. Swelling or bursting of cans can be caused by growth of *Clostridium sporogenes*. "Flat sour" defect (acidification without gas production) can result from growth of *B. stearothermophilus, B. licheniformis, B. coagulans, B. macerans,* and *B. subtilis* (56). Sweetened condensed milk should have sufficiently low a_w to inhibit bacterial spore germination. However, if a_w is not well controlled, *Bacillus* spp. may produce acid or acid-proteolytic spoilage.

Control of Sporeformer-Associated Defects in Fluid Products

Methods for controlling growth of sporeformers in fluid products mainly involve the use of appropriate heat treatments. UHT treatments produce products microbiologically stable at room temperature. However, when sub-UHT heat treatments are more severe than that required for pasteurization, the shelf life of cream and milk can actually decrease, a phenomenon attributed to spore activation (81). Use of a double heat treatment, the first to activate spores and the second to inactivate them, does not result in the expected increase in shelf life (45). Adding hippuric acid, a naturally occurring germinant, to raw milk does not germinate sufficient spores before heat treatment to provide a consistent increase in shelf life (44), nor does adding lysozyme to milk inhibit outgrowth of *Bacillus* spores sufficiently to be of practical significance. A practical means to prevent sporeformers from spoiling nonfermented liquid dairy products given sub-UHT heat treatments has not been developed.

Defects in Cheese

The major defect in cheese caused by spore-forming bacteria is gas formation usually resulting from growth of *Clostridium tyrobutyricum* and occasionally resulting from growth of *C. sporogenes* and *C. butyricum*. This defect is often called late blowing or late gas because it occurs after the cheese has aged for several weeks. Emmental, Swiss, Gouda, and Edam cheeses are most often affected because of their relatively high pH and moisture content and low interior salt content. The defect can also occur in Cheddar and Italian cheeses. Processed cheeses are susceptible to late blowing because spores are not inactivated during heat processing (57). Late gas defect results from the fermentation of lactate to butyric acid, acetic acid, carbon dioxide, and hydrogen gas. Flavor as well as appearance are affected. Populations of *C. tyrobutyricum* spores of less than one per milliliter of milk can produce the defect, because the spores are concentrated in the cheese curd during manufacture (92). The number of spores in raw milk needed to cause a defect varies with size and shape of the cheese in addition to pH and moisture, because cheese size and shape limit salt penetration after brining. The number of spores required to cause late blowing in 9-kg wheels of rinded Swiss cheese was estimated at >100 per liter of raw milk (25). The presence of *C. tyrobutyricum* spores in milk has been traced to the consumption of contaminated silage, which increases levels in cow feces (25). Contaminated silage generally has a high pH and is of low quality.

Control of Sporeformer-Associated Defects in Cheese

Ideally, control of late blowing defect would occur at the farm by instituting feeding and management practices which would reduce the number of spores entering the milk supply (48). In practice, this approach has not achieved the required results, and so cheese manufacturers have tried to control the defect by removing spores from the milk at the plant or inhibiting their growth in the cheese. Numbers of bacterial spores can be reduced in milk by a centrifugation process known as bactofugation (57). Germination of spores in the cheese can be inhibited by addition of nitrate and/or lysozyme (25). Nitrate as a cheese additive is prohibited in many countries, and lysozyme by itself does not provide complete protection. Bacteriocins produced by lactic acid bacteria may provide a highly specific means of inhibiting anaerobic spore germination (105).

YEASTS AND MOLDS

Growth of yeasts and molds is a common cause of spoilage of fermented dairy products because these microorganisms are able to grow well at low pH. Yeast spoilage is manifest as fruity or yeasty odor and/or gas formation. Hard (or cured) cheeses, when properly made, have very low amounts of lactose, and thus the potential for yeast growth is limited. Cultured milks, such as yogurt and buttermilk, and fresh cheeses, such as cottage cheese, normally contain fermentable levels of lactose and therefore are prone to yeast spoilage. A "fermented/ yeasty" flavor observed in Cheddar cheese spoiled by growth of a *Candida* sp. was associated with elevated ethanol, ethyl acetate, and ethyl butyrate (51). The affected cheese had a high moisture content (associated with low starter activity and therefore high residual lactose) and low salt content, which contributed to allowing yeast growth. Yeast spoilage can also occur in dairy foods with low a_w, such as sweetened condensed milk and butter. The most common yeasts present in dairy products are *Kluyveromyces marxianus* and *Debaryomyces hansenii* and their asporogenous counterparts, *Candida famata* and *Candida kefyr* (34). Also prevalent are *Rhodotorula mucilaginosa, Yarrowia lipolytica,* and *Candida parapsilosis* (90). Dairy products provide a highly specialized ecological niche for yeasts, selecting for those that can utilize lactose or lactic acid and that tolerate high salt concentrations (34). Yeasts able to produce proteolytic or lipolytic enzymes may also have a selective advantage for growth in dairy products.

Mold Spoilage of Cheese

Growth of spoilage molds on cheese is a problem which still has significance, though it dates back to prehistory. The most common molds found on cheese are *Penicillium* spp. (12, 106); others, including *Aspergillus, Alternaria, Mucor, Fusarium, Cladosporium, Geotrichum,* and *Hormodendrum* species, are occasionally found. Mold species commonly isolated from processed cheese include *Penicillium roqueforti, P. cyclopium, P. viridicatum,* and *P. crustosum* (107). Vacuum-packaged Cheddar cheese supports the growth of *Cladosporium cladosporiodes, P. commune, Cladosporium herbarum, P. glabrum,* and *Phoma* species (49). *Candida* yeasts have also been isolated from vacuum-packaged cheese.

Controlling Mold Spoilage

Yeasts and molds that spoil dairy products can usually be isolated in the processing plant on packaging equipment, in the air, in salt brine, on manufacturing equipment, and in the general environment (floors, walls, ventilation ducts, etc.). Successful control efforts must start with limiting exposure of pasteurized products to these sources. Mold spores do not survive pasteurization (28). If the initial contamination level is limited, strategies to inhibit growth are more likely to succeed. These strategies include packaging to reduce oxygen, cold storage, and the use of antimycotic chemicals such as sorbate, propionate, and natamycin (pimaracin). None of these control measures is completely effective. Vacuum-packaged cheese is susceptible to thread mold defect, whereby the fungi grow in the wrinkles of the plastic film (49). Some molds are resistant to antimycotic additives. Sorbate-resistant molds are commonly isolated from sorbate-treated cheese but not from untreated cheese (64). Some *Penicillium* spp. not only are resistant to sorbate but will degrade it by decarboxylation, producing 1,3-pentadiene (67). This imparts a kerosene-like odor to the cheese. Some *Mucor* spp. degrade sorbate to 4-hexenol, and some *Geotrichum* spp. degrade it to 4-hexenoic acid. Sorbate can also be used as a carbon source or be oxidized to carbon dioxide and water (64). The ability of some molds to degrade sorbate explains why cheeses with high levels of mold contamination are not effectively preserved by this additive.

References

1. **Abad, P., A. Villafafila, J. D. Frias, and C. Rodriguez-Fernandez.** 1993. Extracellular lipolytic activity from *Pseudomonas fluorescens* biovar I (*Pseudomonas fluorescens* NC1). *Milchwissenschaft* **48:**680–683.
2. **Alkanhal, H. A., J. F. Frank, and G. L. Christen.** 1985. Microbial protease and phospholipase C stimulate lipolysis of washed cream. *J. Dairy Sci.* **68:**3162–3170.
3. **Andersson, R. E.** 1980. Lipase production, lipolysis and formation of volatile compounds by *Pseudomonas fluorescens* in fat containing media. *J. Food Sci.* **45:**1694–1701.
4. **Andersson, R. E.** 1980. Microbial lipolysis at low temperatures. *Appl. Environ. Microbiol.* **39:**36–40.
5. **Anonymous.** 1993. *Grade A Pasteurized Milk Ordinance.* Food and Drug Administration, Washington, D.C.
6. **Banwart, G. J.** 1981. *Basic Food Microbiology.* AVI Publishing Co., Westport, Conn.
7. **Bassette, R., and J. S. Acosta.** 1988. Composition of milk products, p. 39–79. *In* N. P. Wong (ed.), *Fundamentals of Dairy Chemistry.* Van Nostrand Reinhold Co., New York.
8. **Batish, V. K., H. Chander, K. C. Zumdegni, K. L. Bhatia, and R. S. Singh.** 1988. Antibacterial activity of lactoferrin against some common food-borne pathogenic organisms. *Aust. J. Dairy Technol.* **43:**16–18.
9. **Bills, D. D., M. E. Morgan, L. M. Reddy, and E. A. Day.** 1965. Identification of compounds responsible for fruit flavor defect of experimental Cheddar cheeses. *J. Dairy Sci.* **48:**1168–1170.
10. **Bramley, A. J., and C. H. McKinnon.** 1990. The microbiology of raw milk, p. 163–208. *In* R. K. Robinson (ed.), *Dairy Microbiology,* vol. 1. Elsevier Applied Science, New York.
11. **Bucky, A. R., P. R. Hayes, and D. S. Robinson.** 1988. Enhanced inactivation of bacterial lipases and proteinases in whole milk by a modified ultra high temperature treatment. *J. Dairy Res.* **55:**373–380.
12. **Bullerman, L. B., and F. J. Olivigni.** 1974. Mycotoxin producing potential of molds isolated from Cheddar cheese. *J. Food Sci.* **39:**1166–1168.
13. **Carić, M.** 1994. *Concentrated and Dried Dairy Products.* VCH Publishers, Inc., New York.
14. **Cerning, J.** 1990. Exocellular polysaccharides produced by lactic acid bacteria. *FEMS Microbiol. Rev.* **87:**113–130.
15. **Cerning, J., C. Bouillanne, M. Landon, and M. Desmazeaud.** 1992. Isolation and characterization of exopolysaccharides from slime-forming mesophilic lactic acid bacteria. *J. Dairy Sci.* **75:**692–699.
16. **Chapman, H. R., and M. E. Sharpe.** 1990. Microbiology of cheese, p. 203–289. *In* R. K. Robinson (ed.), *Dairy Microbiology,* vol. 2. Elsevier Applied Science, New York.
17. **Choudhery, A. K., and E. M. Mikolajcik.** 1971. Activity of *Bacillus cereus* proteinases in milk. *J. Dairy Sci.* **53:**363–366.
18. **Chrisope, G. L., and R. T. Marshall.** 1976. Combined action of lipase and microbial phospholipase C on a model fat emulsion and raw milk. *J. Dairy Sci.* **59:**2024–2030.
19. **Christen, G. L., and R. T. Marshall.** 1985. Effect of histidine on thermostability of lipase and protease of *Pseudomonas fluorescens* 27. *J. Dairy Sci.* **68:**594–604.
20. **Christian, J. H. B.** 1980. Reduced water activity, p. 70–91. *In Microbial Ecology of Foods,* vol. 1. Academic Press, New York.
21. **Cousin, M. A.** 1982. Presence and activity of psychrotrophic microorganisms in milk and dairy products: a review. *J. Food Prot.* **45:**172–207.
22. **Cousin, M. A.** 1989. Physical and biochemical effects of milk components, p. 205–225. *In* R. C. McKellar (ed.), *Enzymes of Psychrotrophs in Raw Food.* CRC Press, Inc., Boca Raton, Fla.
23. **Cromie, S. J., T. W. Dommett, and D. Schmidt.** 1989. Changes in the microflora of milk with different pasteurization and storage conditions and aseptic packaging. *Aust. J. Dairy Technol.* **44:**74–77.
24. **Cromie, S. J., J. E. Giles, and J. R. Dulley.** 1987. Effect of elevated ripening temperatures on the microflora of Cheddar cheese. *J. Dairy Res.* **54:**69–76.
25. **Dasgupta, A. R., and R. R. Hull.** 1989. Late blowing of Swiss cheese: incidence of *Clostridium tyrobutyricum* in manufacturing milk. *Aust. J. Dairy Technol.* **44:**82–87.
26. **Deeth, H. C., and C. H. Fitz-Gerald.** 1983. Lipolytic enzymes and hydrolytic rancidity in milk and milk products, p. 195–239. *In* P. F. Fox (ed.), *Developments in Dairy Chemistry,* part II. Applied Science, London.
27. **Dionysius, D. A., P. A. Grieve, and A. C. Vos.** 1992. Studies on the lactoperoxidase system: reaction kinetics and antibacterial activity using two methods for hydrogen peroxide generation. *J. Appl. Bacteriol.* **72:**146–153.
28. **Doyle, M. P., and E. H. Marth.** 1975. Thermal inactivation of conidia from *Aspergillus flavus* and *Aspergillus parasiticus.* I. Effects of moist heat, age of conidia, and sporulation medium. *J. Milk Food Technol.* **38:**678–682.
29. **Dumont, J. P., G. Delespaul, B. Miquot, and J. Adda.** 1977. Influence des bactéries psychrotrophs sur les qualités organoleptiques de fromages à pâte molle. *Lait* **57:**619–630.
30. **Ewings, K. N., R. E. O'Conner, and G. E. Mitchell.** 1984. Proteolytic microflora of refrigerated raw milk in South East Queensland. *Aust. J. Dairy Technol.* **39:**65–68.
31. **Fairbairn, D. J., and B. A. Law.** 1987. The effect of nitrogen and carbon sources on proteinase production by *Pseudomonas fluorescens.* *J. Appl. Bacteriol.* **62:**105–113.

32. **Fernandez, L., J. A. Alvarez, P. Palacios, and C. San Jose.** 1992. Proteolytic and lipolytic activities of *Pseudomonas fluorescens* grown in raw milk with variable iron content. *Milchwissenschaft* **47**:160–163.
33. **Fernandez, L., C. San Jose, and R. C. McKellar.** 1990. Repression of *Pseudomonas fluorescens* extracellular lipase secretion by arginine. *J. Dairy Res.* **57**:69–78.
34. **Fleet, G. H.** 1990. Yeasts in dairy products. *J. Appl. Bacteriol.* **68**:199–211.
35. **Fox, P. F., P. Power, and T. M. Cogan.** 1989. Isolation and molecular characteristics, p. 57–120. *In* R. C. McKellar (ed.), *Enzymes of Psychrotrophs in Raw Food.* CRC Press, Inc., Boca Raton, Fla.
36. **Frank, J. F., and R. A. Koffi.** 1990. Surface-adherent growth of *Listeria monocytogenes* is associated with increased resistance to surfactant sanitizers and heat. *J. Food Prot.* **53**:550–554.
37. **Frank, J. F., and E. H. Marth.** 1978. Survey of soft and semisoft cheese for presence of fecal coliforms and serotypes of enteropathogenic *Escherichia coli. J. Food Prot.* **41**:198–200.
38. **Frank, J. F., and E. H. Marth.** 1988. Fermentations, p. 655–738. *In* N. P. Wong (ed.), *Fundamentals of Dairy Chemistry,* 3rd ed. Van Nostrand Reinhold Co., New York.
39. **Frank, J. F., E. H. Marth, and N. F. Olson.** 1977. Survival of enteropathogenic and non-pathogenic *Escherichia coli* during the manufacture of Camembert cheese. *J. Food Prot.* **40**:835–842.
40. **Frank, J. F., E. H. Marth, and N. F. Olson.** 1978. Behavior of enteropathogenic *Escherichia coli* during manufacture and ripening of brick cheese. *J. Food Prot.* **41**:111–115.
41. **Garcia, M. L., B. Sanz, P. Garcia-Collia, and J. A. Ordonez.** 1989. Activity and thermostability of the extracellular lipases and proteinases from pseudomonads isolated from raw milk. *Milchwissenschaft* **44**:547–550.
42. **Griffiths, M. W.** 1989. Effect of temperature and milk fat on extracellular enzyme synthesis by psychrotrophic bacteria during growth in milk. *Milchwissenschaft* **44**:539–543.
43. **Griffiths, M. W., and J. D. Phillips.** 1990. Incidence, source and some properties of psychrotrophic *Bacillus* spp. found in raw and pasteurized milk. *J. Soc. Dairy Technol.* **43**:62–70.
44. **Griffiths, M. W., and J. D. Phillips.** 1990. Strategies to control the outgrowth of spores of psychrotrophic *Bacillus* spp. in dairy products. I. Use of naturally occurring materials. *Milchwissenschaft* **45**:621–625.
45. **Griffiths, M. W., and J. D. Phillips.** 1990. Strategies to control the outgrowth of spores of psychrotrophic *Bacillus* spp. in dairy products. II. Use of heat treatments. *Milchwissenschaft* **45**:719–721.
46. **Guamis, H., T. Huerta, and E. Garay.** 1987. Heat-inactivation of bacterial proteases in milk before UHT-treatment. *Milchwissenschaft* **42**:651–653.
47. **Hassan, A. N., A. S. Zahran, N. H. Metwalli, and S. I. Shalabi.** 1993. Aerobic sporeforming bacteria isolated from UHT milk produced in Egypt. *Egypt. J. Dairy Sci.* **21**:109–121.
48. **Herlin, A. H., and A. Christansson.** 1993. Cheese-blowing anaerobic spores in bulk milk from loose-housed and tied dairy cows. *Milchwissenschaft* **48**:686–689.
49. **Hocking, A. D., and M. Faedo.** 1992. Fungi causing thread mould spoilage of vacuum packaged Cheddar cheese during maturation. *Int. J. Food Microbiol.* **16**:123–130.
50. **Hogarty, S. L., and J. F. Frank.** 1982. Low-temperature activity of lactic streptococci isolated from cultured buttermilk. *J. Food Prot.* **43**:1208–1211.
51. **Horwood, J. F., W. Stark, and R. R. Hull.** 1987. A "fermented, yeasty" flavour defect in Cheddar cheese. *Aust. J. Dairy Technol.* **42**:25:26.
52. **Hull, R., S. Toyne, I. Haynes, and F. Lehman.** 1992. Thermoduric bacteria: a re-emerging problem in cheesemaking. *Aust. J. Dairy Technol.* **47**:91–94.
53. **Jaspe, A., P. Oviedo, L. Fernandez, P. Palacios, and C. Sanjose.** 1995. Cooling raw milk: change in the spoilage potential of contaminating *Pseudomonas. J. Food Prot.* **58**:915–921.
54. **Jaspe, A., P. Palacios, P. Matias, L. Fernandez, and C. Sanjose.** 1994. Proteinase activity of *Pseudomonas fluorescens* grown in cold milk supplemented with nitrogen and carbon sources. *J. Dairy Sci.* **77**:923–929.
55. **Jenness, R.** 1988. Composition of milk, p. 1–38. *In* N. P. Wong (ed.), *Fundamentals of Dairy Chemistry.* Van Nostrand Reinhold Co., New York.
56. **Kalogridou-Vassiliadou, D.** 1992. Biochemical acitivities of *Bacillus* species isolated from flat sour evaporated milk. *J. Dairy Sci.* **75**:2681–2686.
57. **Kosikowski, F. V.** 1982. *Cheese and Fermented Milk Foods,* 2nd ed. F. V. Kosikowski and Associates, Brooktondale, N.Y.
58. **Kroll, S.** 1989. Thermal stability, p. 121–152. *In* R. C. McKellar (ed.), *Enzymes of Psychrotrophs in Raw Food.* CRC Press, Inc., Boca Raton, Fla.
59. **Kumura, H., K. Mikawa, and Z. Saito.** 1993. Influence of milk proteins on the thermostability of the lipase from *Pseudomonas fluorescens* 33. *J. Dairy Sci.* **76**:2164–2167.
60. **Lalaye, L. C., R. E. Simard, B.-H. Lee, R. A. Holley, and R. N. Giroux.** 1987. Involvement of heterofermentative lactobacilli in development of open texture in cheeses. *J. Food Prot.* **50**:1009–1012.
61. **Law, B. A.** 1979. Reviews of the progress of dairy science: enzymes of psychrotrophic bacteria and their effects on milk and milk products. *J. Dairy Res.* **46**:573–588.
62. **Law, B. A., C. M. Cousins, M. E. Sharpe, and F. L. Davies.** 1979. Psychrotrophs and their effects on milk and dairy products, p. 137–152. *In* A. D. Russell and R. Fuller (ed.), *Cold Tolerant Microbes in Spoilage and the Environment.* Academic Press, New York.
63. **Law, B. A., M. E. Sharpe, and H. R. Chapman.** 1976. Effect of lipolytic Gram negative psychrotrophs in stored milk on the development of rancidity in Cheddar cheese. *J. Dairy Res.* **43**:459–468.
64. **Liewen, M. B., and E. H. Marth.** 1985. Growth and inhibition of microorganisms in the presence of sorbic acid: a review. *J. Food Prot.* **48**:364–375.
65. **Macura, D., and P. M. Townsley.** 1984. Scandinavian ropy milk—identification and characterization of endogenous ropy lactic streptococci and their extracellular excretion. *J. Dairy Sci.* **67**:734–744.
66. **Malik, R. K., R. Prasad, and D. K. Mathur.** 1985. Effect of some nutritional and environmental factors on extracellular protease production by *Pseudomonas* sp. B-25. *Lait* **65**:169–183.

67. **Marth, E. H., C. M. Capp, L. Hasenzahl, H. W. Jackson, and R. V. Hussong.** 1966. Degradation of potassium sorbate by *Penicillium* species. *J. Dairy Sci.* **49:**1197–1205.
68. **Martin, J. H., D. P. Stahly, W. J. Harper, and I. A. Gould.** 1962. Sporeforming microorganisms in selected milk supplies. *Proc. XVI Int. Dairy Congr.* **C:**295–304.
69. **Massa, S., F. Gardini, M. Sinigaglia, and M. E. Guerzoni.** 1992. *Klebsiella pneumoniae* as a spoilage organism in Mozzarella cheese. *J. Dairy Sci.* **75:**1411–1414.
70. **Masson, P. L., and J. F. Heremans.** 1971. Lactoferrin in milk from different species. *Comp. Biochem. Physiol.* **39B:**119–129.
71. **McKellar, R. C.** 1982. Factors influencing the production of extracellular proteinase by *Pseudomonas fluorescens*. *J. Appl. Bacteriol.* **53:**305–316.
72. **McKellar, R. C.** 1989. Regulation and control of synthesis, p. 153–172. *In* R. C. McKellar (ed.), *Enzymes of Psychrotrophs in Raw Foods*. CRC Press, Inc., Boca Raton, Fla.
73. **McKellar, R. C., and H. Cholette.** 1986. Possible role of calcium in the formation of active extracellular proteinase by *Pseudomonas fluorescens*. *J. Appl. Bacteriol.* **60:**37–44.
74. **McKellar, R. C., K. Shamsuzzaman, C. San Jose, and H. Cholette.** 1987. Influence of iron (III) and pyoverdine, a siderophore produced by *Pseudomonas fluorescens* B52, on its extracellular proteinase and lipase production. *Arch. Microbiol.* **147:**225–230.
75. **Meer, R. R., J. Baker, F. W. Bodyfelt, and M. W. Griffiths.** 1991. Psychrotrophic *Bacillus* spp. in fluid milk products: a review. *J. Food Prot.* **54:**969–979.
76. **Mikolojcik, E. M., and N. T. Simon.** 1978. Heat resistant psychrotrophic bacteria in raw milk and their growth at 7°C. *J. Food Prot.* **41:**93–95.
77. **Morgan, M. E.** 1976. The chemistry of some microbially induced flavor defects in milk and dairy foods. *Biotechnol. Bioeng.* **18:**953–965.
78. **Morse, P. M., H. Jackson, C. H. McNaughton, A. G. Leggatt, G. B. Landerkin, and C. K. Johns.** 1968. Investigation of factors contributing to the bacteria count of bulk tank milk. II. Bacteria in milk from individual cows. *J. Dairy Sci.* **51:**1188–1191.
79. **Mosteller, T. M., and J. R. Bishop.** 1993. Sanitizer efficacy against attached bacteria in milk biofilm. *J. Food Prot.* **56:**34–41.
80. **Mottar, J. F.** 1989. Effect on the quality of dairy products, p. 227–243. *In* R. C. McKellar (ed.), *Enzymes of Psychrotrophs in Raw Food*. CRC Press, Inc., Boca Raton, Fla.
81. **Muir, D. D.** 1989. The microbiology of heat treated fluid milk products, p. 209–270. *In* R. K. Robinson (ed.), *Dairy Microbiology*, vol. 1. Elsevier Applied Science, New York.
82. **Myhara, R. M., and B. Skura.** 1990. Centroid search optimization of cultural conditions affecting the production of extracellular proteinase by *Pseudomonas fragi* ATCC 4973. *J. Appl. Bacteriol.* **69:**530–538.
83. **Nashif, S. A., and F. E. Nelson.** 1953. The lipase of *Pseudomonas fragi*. III. Enzyme action in cream and butter. *J. Dairy Sci.* **36:**481–488.
84. **Nooitgedagt, A. J., and B. J. Hartog.** 1988. A survey of the microbiological quality of Brie and Camembert cheese. *Neth. Milk Dairy J.* **42:**57–72.
85. **Overcast, W. W., and K. Atmaran.** 1974. The role of *Bacillus cereus* in sweet curdling of fluid milk. *J. Milk Food Technol.* **37:**233–236.
86. **Phillips, J. D., and M. W. Griffiths.** 1986. Factors contributing to the seasonal variation of *Bacillus* species in pasteurized products. *J. Appl. Bacteriol.* **61:**275–285.
87. **Reddy, M. C., D. D. Bills, R. C. Lindsay, and L. M. Libbey.** 1968. Ester production by *Pseudomonas fragi*. I. Identification and quantification of some esters produced in milk cultures. *J. Dairy Sci.* **51:**656–659.
88. **Reimerdes, E. H.** 1982. Changes in the proteins of raw milk during storage, p. 271. *In* P. F. Fox (ed.), *Developments in Dairy Chemistry*, part I. Applied Science, London.
89. **Rengipat, S., and E. A. Johnson.** 1989. Characterization of a *Lactobacillus* strain producing white crystals on Cheddar cheese. *Appl. Environ. Microbiol.* **55:**1579–2582.
90. **Rohm, H., F. Eliskases-Lechner, and M. Bräuer.** 1992. Diversity of yeasts in selected dairy products. *J. Appl. Bacteriol.* **72:**370–376.
91. **Rosen, B., U. Merin, and I. Rosenthal.** 1989. Evaluation of clostridia in raw milk. *Milchwissenschaft* **44:**355–357.
92. **Rosen, B., G. Popel, and I. Rosenthal.** 1990. The affinity of *Clostridium tyrobutyricum* to casein in raw milk. *Milchwissenschaft* **45:**152–154.
93. **Rowe, M. T.** 1990. Growth and extracellular enzyme production by psychrotrophic bacteria in raw milk stored at low temperature. *Milchwissenschaft* **45:**495–499.
94. **Russell, H. L.** 1885. Gas producing bacteria and the relation of the same to cheese, p. 139–150. *In Wisconsin Agricultural Experiment Station 12th Annual Report*. University of Wisconsin, Madison.
95. **Rutzinski, J. L., E. H. Marth, and N. F. Olson.** 1979. Behavior of *Enterobacter aerogenes* and *Hania* species during the manufacture and ripening of Camembert cheese. *J. Food Prot.* **42:**790–793.
96. **Schröder, M. J. A.** 1984. Origins and levels of post pasteurization contamination of milk in the dairy and their effects on keeping quality. *J. Dairy Res.* **51:**59–67.
97. **Shannon, E. L., N. F. Olson, and J. H. von Elbe.** 1969. Effect of lactic starter culture on pink discoloration and oxidation-reduction potential in Italian cheese. *J. Dairy Sci.* **52:**1557–1561.
98. **Shehata, A. E., M. N. I. Magdoub, N. E. Sultan, and Y. A. El-Samragy.** 1983. Aerobic mesophilic and psychrotrophic sporeforming bacteria in buffalo milk. *J. Dairy Sci.* **66:**1228–1231.
99. **Shipe, W. F., R. Bassette, D. D. Deane, W. L. Dinkley, E. G. Hammond, W. J. Harper, D. H. Klein, M. E. Morgan, J. H. Nelson, and R. A. Scanlan.** 1978. Off flavors in milk: nomenclature, standards, and bibliography. *J. Dairy Sci.* **61:**855–869.
100. **Stead, D.** 1986. Microbial lipases: their characteristics, role in food spoilage and industrial uses. *J. Dairy Res.* **53:**481–505.
101. **Stepaniak, L., E. Zakrzewski, and T. Sorhaug.** 1991. Inactivation of heat-stable proteinase from *Pseudomonas fluorescens* P1 at pH 4.5 and 55°C. *Milchwissenshaft* **46:** 139–142.
102. **Suhren, G.** 1989. Producer microorganisms, p. 3–34. *In* R. C. McKellar (ed.), *Enzymes of Psychrotrophs in Raw Foods*. CRC Press, Inc., Boca Raton, Fla.

103. **Thomas, S. B.** 1966. Sources, incidence, and significance of psychrotrophic bacteria in milk. *Milchwissenschaft* **21**:270–275.

104. **Thomas, S. B., and B. F. Thomas.** 1973. Psychrotrophic bacteria in refrigerated bulk-collected raw milk. Part I. *Dairy Ind.* **38**:11–15.

105. **Thualt, D., E. Beliard, J. Le Guern, and C.-M. Bourgeois.** 1991. Inhibition of *Clostridium tyrobutyricum* by bacteriocin-like substances produced by lactic acid bacteria. *J. Dairy Sci.* **74**:1145–1150.

106. **Torrey, G. S., and E. H. Marth.** 1977. Isolation and toxicity of molds from foods stored in homes. *J. Food Prot.* **40**:187–190.

107. **Tsai, W.-Y. J., M. B. Liewen, and L. Bullerman.** 1988. Toxicity and sorbate sensitivity of molds isolated from surplus commodity cheese. *J. Food Prot.* **51**:457–462.

108. **Turner, K. W., and T. D. Thomas.** 1980. Lactose fermentation in Cheddar cheese and the effect of salt. *N. Z. J. Dairy Sci. Technol.* **15**:265–276.

109. **Vilafafila, A., J. D. Frias, P. Abad, and C. Rodriguez-Fernandez.** 1993. Extracellular proteinase activity from psychrotrophic *Pseudomonas fluorescens* biovar 1 (*Ps. fluorescens* NC1). *Milchwissenschaft* **48**:435–438.

110. **Wang, J. J., and J. F. Frank.** 1981. Characterization of psychrotrophic bacterial contamination of commercial buttermilk. *J. Dairy Sci.* **64**:2154–2160.

111. **Wolfson, L. M., and S. S. Sumner.** 1993. Antibacterial activity of the lactoperoxidase system: a review. *J. Food Prot.* **56**:887–892.

112. **Yan, L., B. E. Langlois, J. O'Leary, and C. Hicks.** 1983. Effect of storage conditions of grade A raw milk on proteolysis and cheese yield. *Milchwissenschaft* **38**:715–719.

Robert E. Brackett

7 Fruits, Vegetables, and Grains

INTRODUCTION

Foods of plant origin are diverse in composition. Consequently, patterns of microbiological spoilage differ substantially, sometimes dramatically. Despite their differences, plant products share some fundamental characteristics. They are all horticultural or agronomic products whose structural integrity depends on cellulose and pectin. This chapter discusses the basic microflora and mechanisms involved in the spoilage of plant products and provides information on specific commodities.

Scientific literature describing the spoilage of fruits, vegetables, and grains is often confusing to readers not familiar with the topic. This is because spoilage of foods of plant origin often crosses the interface between traditional plant pathology and food microbiology. Both subdisciplines have distinct viewpoints and philosophies and use unique jargon. These differences often confuse readers as to the role of microflora in the spoilage of plant products and appropriate means of prevention.

For the purpose of this chapter, plant pathology as it relates to degradation of plant materials will be restricted to spoilage problems that arise before harvest, whereas food microbiology will deal with spoilage after harvest. However, the reader should understand that scientific literature makes no such distinction, and differences between plant pathology and food microbiology are usually subtle. Consequently, some definitions of terms often found in the literature are worth discussing. Fruits are defined as the seed-bearing organs of plants and include not only well-known commodities such as apples, citrus fruits, and berries but also items sometimes thought of as vegetables, such as tomatoes, bell peppers, and cucumbers. In contrast, vegetables are defined as all other edible portions of plants, including leaves, roots, and seeds.

Pathogens

Although most microbiologists recognize that a pathogen is a microorganism capable of causing disease, the host upon which the disease is inflicted is often assumed. Most clinical, veterinary, and food microbiologists understand pathogens to mean microorganisms that cause illness in humans or animals. In contrast, plant pathologists and other agricultural scientists who work with horticultural and agronomic commodities often think of pathogens as microorganisms that cause disease or decay of plants. Such assumptions are evident in the scientific literature and can confuse readers not familiar with this specialty area. However, confusion can easily be avoided by specifying the type of pathogen, i.e., plant pathogen or human pathogen, being discussed.

Plant pathogens can also be subdivided into those that are true plant pathogens and those that are opportunistic plant pathogens. Although exceptions exist, true plant pathogens are usually understood to refer to microorganisms which can actively infect plant tissues. The ability to infect is derived from the ability of microorganisms to produce one of several degradative enzymes which enable them to penetrate protective external layers of cells. In contrast, opportunistic plant pathogens infect tissues only when normal defenses of the plant have been compromised in some way, e.g., by mechanical damage or insects.

Field diseases, storage diseases, and market diseases are descriptive terms for spoilage used in older or nonscientific literature. Field diseases include defects related to microbiological spoilage which occur either before or soon after harvest. In contrast, storage or market diseases refer to spoilage which is manifested some time after harvest, particularly during storage or marketing (27). Both terms are inadequate for scientific discussion because they describe the time at which spoilage becomes apparent but do not address specific etiologies or microorganisms involved. Moreover, some microorganisms that can cause field diseases often do not manifest themselves until the fruit or vegetable is harvested. One example is the mold *Phytophthora infestans,* which causes late blight of potatoes. This microorganism can spread to potatoes either via airborne spores or by infecting neighboring potato plants. In most cases, infection causes death of the plant before potato tubers are harvested and thus can be defined as a storage disease. However, *P. infestans* can remain dormant in the tubers, only to rot the tubers during storage or reinfect plants of a new crop (9).

Spoilage

The term "spoilage" connotes different meanings to different people. In its broadest sense, spoilage refers to any change that occurs in a food whereby the food is made unacceptable for human consumption. This definition can include safety- as well as quality-related defects. Although microbiological safety is an important concern, this aspect of spoilage will be discussed only briefly here, since it will be dealt with in subsequent chapters. This chapter will focus on changes in color, flavor, texture, or aroma brought about by the growth of microorganisms on fruits, vegetables, and grains.

Types of Spoilage

In general, plant products exhibit three broad types of spoilage. The first type is active spoilage, caused by plant-pathogenic microorganisms actually initiating infection of otherwise healthy and uncompromised products, resulting in reduction in sensory quality. A second type of spoilage is passive or wound-induced spoilage, in which opportunistic microorganisms gain access to internal tissues via damaged epidermal tissue, i.e., peels or skins. This type of spoilage often occurs soon after the product has been damaged by harvesting or processing equipment or by insects. For example, the common fruit fly (*Drosophila melanogaster*) can inoculate vegetables with *Rhizopus* species as it deposits eggs in wounds or other breaks in the epidermis, i.e., the outer protective tissue system of plant parts (9). Similarly, and third, passive spoilage can occur when opportunistic spoilage microorganisms gain entry to internal tissues via lesions caused by plant pathogens or via natural openings such as lenticles, stomata, or hydathodes (25).

Spoilage of plant products can be manifested in a variety of ways, depending on the specific product, the environment, and microorganisms involved. Traditionally, spoilage has been described by symptoms most often associated with a particular product. Listed in Table 7.1 are examples of some common types of fungal spoilage of fruits and vegetables and the molds normally associated with them. For some fruits or vegetables, this system is sufficient to adequately describe a spoilage problem. In other cases, however, the system is inadequate because more than one type of microorganism may produce identical or similar symptoms. For example, the best-known type of spoilage in vegetables is soft rot, which is usually evidenced by obvious softening and deliquescence of the plant tissue, especially around an initial point of infection. However, the term "soft rot" often does not fully describe the disease or spoilage problem because softening can be caused by various plant pathogenic bacteria, most notably *Erwinia carotovora* and several species of *Pseudomonas* (25). However, bacteria such as *Bacillus* and *Clostridium* species, as well as yeasts and molds, are also occasionally implicated in soft rot spoilage (28).

Types of spoilage are also known by the names of the microorganisms which cause them. For example, *Fusarium* rot and *Rhizopus* soft rot describe not only the symptoms of the spoilage but also the mold that causes them. This system of describing spoilage is preferable to naming the spoilage on the basis of symptoms alone. However, to date no organized system of applying names to specific types of spoilage has been adopted.

Mechanism of Spoilage

Intact healthy plant cells possess a variety of defense mechanisms to resist microbial invasion. Thus, before microbial spoilage can occur, these defense mechanisms must be overcome. Agrios (1) provides a good general overview of the physical and chemical mechanisms by which microorganisms are able to invade plant tissues.

Fruits, vegetables, grains, and legumes have an epidermal layer of cells, i.e., skin, peel, or testa, that provides a

Table 7.1 Types of fungal spoilage of fruits and vegetables

Product	Type of spoilage	Genus or species of mold responsible	Product(s) involved
Fruits[a]	*Alternaria* rot	*Alternaria*	Citrus fruits
	Antrancnose (bitter rot)	*Colletotrichum musae*	Bananas
	Black rot	*Aspergillus niger, Ceratocystis fimriata*	Onions, sweet potatoes
	Brown rot	*Monilinia fructicola*	Peaches
	Crown rot	*Collectotrichum musae, Fusarium roseum, Verticillium theobromae, Ceratocystis paradoxa*	Bananas
	Gray mold rot	*Botrytis cinerea*	Grapes
	Pineapple black rot	*Ceratocystis paradoxa*	Pineapples
	Sour rot	*Geotrichum candidum*	Tomatoes, citrus fruits
	Lenticel rot	*Cryptosporiopsis malicorticus, Phlyctaena vagabunda*	Apples, pears
	Green mold rot	*Penicillium digitatum*	Citrus fruits
	Blue rot	*Penicillium*	Oranges
	Cladosporium rot	*Cladosporium herbarum*	Peaches, cherries
Vegetables	Black mold rot	*Aspergillus*	Onions
	Black rot	*Alternaria*	Carrots, cauliflower
	Downy mildew	*Bremia, Phytophthora*	Lettuce, spinach
	Fusarium rot	*Fusarium*	Asparagus
	Gray mold rot	*Botrytis*	Cabbage
	Rhizopus soft rot	*Rhizopus*	Green beans
	Smudge (anthracnose)	*Colletotrichum*	Onions
	Tuber rot	*Fusarium*	Potatoes
	Watery soft rot	*Sclerotinia*	Celery
	Wilt	*Pythium*	Green beans
	Blue rot	*Penicillium*	Oranges
	Blight	*Phomopsis*	Eggplant
	Finger rot	*Pestalozzia, Fusarium, Gleosporium*	Bananas
	Pink rot	*Trichothecium*	Peaches

[a]Adapted from reference 22.

protective barrier against infection of internal tissues. The composition of the epidermal tissue varies, but walls of cells in this tissue usually consist of cellulose and pectic materials, and the outermost cells are covered by a layer of waxes (cutin). This tissue forms a barrier which is resistant to penetration by most microorganisms. If this barrier is compromised in some way, however, the possibility exists for easy entry of microorganisms to internal tissues.

External barriers can be compromised in a number of ways. One obvious way is if the epidermal tissue is damaged by external sources. Damage due to relatively uncontrollable factors such as insect infestation, wind-blown sand, or rubbing against neighboring surfaces can occur before harvest. After harvest, the product is most often damaged by harvesting or processing equipment, particularly devices that are poorly designed or maintained. In addition, some microorganisms, especially plant-pathogenic molds, possess mechanisms to penetrate external tissues of plants. Once external barriers are penetrated, not only do these microorganisms quickly invade internal tissues, but other opportunistic microorganisms often take advantage of the availability of nutrients present in damaged tissue and cause additional spoilage.

Once microorganisms penetrate the outer tissues of fruits, vegetables, or grains, they must still gain access to internal areas and individual cells to extract nutrients. Plant tissues are held together by the middle lamella, which is composed primarily of pectic substances. These substances, composed of pectic acid, pectinic acid, and pectin, are polymers of α-1,4-linked D-galactopyranosyluronic acid units interspersed with 1,2-linked rhamnopyranose units (13). Any degradation of the middle lamella results in detachment of cells from one another. The integrity of individual plant cells is due to the cell wall, which consists of two main layers: the primary cell wall layer, composed of cellulose and pectates, and the secondary cell wall layer, which consists almost entirely of cellulose.

Changes in the physiological state of plants can also result in openings in external tissues. For example, dehydration of vegetables can result in separation of cells, thereby allowing cracks to develop and facilitating the entry of microorganisms to internal tissues. In such cases, microorganisms that normally would not cause spoilage problems can utilize fermentable carbohydrates and produce metabolites which cause undesirable changes in flavor, aroma, or color.

Degradative Enzymes

Degradative enzymes play an important role in postharvest spoilage of plant products. Five classes of microbial enzymes are primarily responsible for degradation of plant materials (25): pectinases, cellulases, proteases, phosphatidases, and dehydrogenases. Because pectin and cellulose constitute the main structural components of plant cells, pectinases and cellulases are the most important degradative enzymes involved in spoilage.

Pectinases are enzymes that cause depolymerization of the pectin chain. Although pectinases are often discussed as a single enzyme, three main types are recognized for their role in plant spoilage (13, 15). Differences exist primarily in the type and site of reaction on the pectin polymer. Pectin methyl esterase (PME; EC 3.1.1.11) (21) hydrolyzes ester groups from pectin chains, with subsequent production of methyl alcohol (1, 14). Although PME does not directly act to decrease chain length, it does affect pectin solubility and the rate at which other types of pectinases react. PME is produced by several plant pathogens, including *Botrytis cinerea, Monilinia fructicola, Penicillium citrimum,* and *E. carotovora* (13). However, PME is also produced by some microorganisms which are not considered plant pathogens (15).

In contrast to PME, polygalacturonases (PG; EC 3.2.1.15) and pectin lyase (PL; EC 4.2.2.2) (21) are chain-splitting pectinases which reduce the overall length of the pectin chain. Their mechanisms of action differ primarily in the way in which pectin chains are cleaved. PG cleaves the pectin chain by hydrolyzing the linkage between two galacturonan molecules, whereas PL depolymerizes by β-elimination of the linkage. Both PG and PL can exist as endopectinases, which act on middle portions of the pectin chain, or exopectinases, which act on terminal ends of the chains. The ultimate degradation of the pectin chain results in liquefaction of the pectin and complete maceration of plant tissues. The production of pectic enzymes by microorganisms does not necessarily mean that spoilage will result. For example, *Flavobacterium* species produce pectic enzymes without causing marked softening of plant tissues (27). In such cases, the enzymes themselves may not be active in the target tissues. In addition, some plant tissues contain pectic enzyme inhibitors (15). The regulation and genetics of pectinase production by bacteria are described in detail elsewhere (15, 24, 32).

Cellulases constitute the second major class of degradative enzymes which can lead to spoilage. Cellulases function by degrading cellulose, which is a glucose polymer, to glucose. As with pectinases, several types of cellulases exist, some of which attack native cellulose by cleaving cross-linkages between chains, while others act by breaking the cellulose into shorter chains (1). Cellulase activity is important in plant product spoilage not only by contributing to tissue softening and maceration but also by producing glucose, which can be used by microorganisms devoid of degradative enzymes. Although cellulases are important in enabling plant pathogens to initiate infection, they are less important than pectinases in postharvest spoilage of plant products (25).

Influence of Physiological State

The physiological state of plant products, especially those consisting of fruits or vegetables, can have a dramatic effect on susceptibility to microbiological spoilage (20). Fruits, vegetables, and grains usually possess some sort of defense mechanism to resist infection by microorganisms. Usually these mechanisms are most effective when the plant is at peak physiologic health. Once plant tissues begin to age or are in a suboptimal physiologic state, resistance to infection diminishes (31).

Fruits and vegetables differ in the way in which they change physiologically after detachment from the plant. Nonclimacterate fruits and vegetables, such as strawberries, beans, and lettuce, cease to ripen once they have been harvested. In contrast, climacterate fruits and vegetables, such as bananas and tomatoes, continue to mature and ripen after harvest. Most consumers note development of color or softening of texture as the most obvious indications of ripening. However, ripening can continue to the point where fruits and vegetables are overripe, i.e., when normal cell integrity begins to diminish and tissues deteriorate (31). This process, known as senescence, arises from the accumulation of degradative enzymes produced by the fruit or vegetable and is unrelated to microbial decay. However, loss of cellular integrity brought about by senescence makes fruits and vegetables even more susceptible to microbial infection and spoilage. Consequently, climacterate fruits and vegetables are usually among the most perishable of plant products.

MICROBIOLOGICAL SPOILAGE OF VEGETABLES

Although vegetables, fruits, grains, and legumes are all plant-derived products, they possess inherent differences which influence both the type of microorganisms that form the natural microflora and the type of spoilage encountered. Important intrinsic factors which influence the microflora that develop on plant products include pH and water activity (a_w). The a_w of fresh fruits and vegetables is high enough to support growth of most bacteria and fungi and is therefore not considered a limiting factor. Similarly, with the exception of tomatoes, the pH of most vegetables is between 5.0 and 6.0 (Table 7.2), which does not inhibit the growth of most microorganisms.

Natural Microflora

In general, the natural microflora of vegetables includes bacteria, yeasts, and molds representing many genera. However, microflora can vary considerably depending on the type of vegetable, environmental considerations, seasonality, and whether the vegetables were grown close to the soil. Bacteria normally present on vegetables at the time of harvest include both gram-positive and gram-negative forms. However, the manner in which vegetables are stored will often influence subsequent development of particular groups of microorganisms. For example, refrigeration tends to select for psychrotrophic bacteria, such as *Pseudomonas* species (10, 11). Many indigenous microorganisms are considered to be not plant pathogens but rather opportunistic spoilage microorganisms which are frequently associated with spoilage.

Table 7.2 Typical pH values for various vegetables and fruits[a]

Vegetable	pH range	Fruit	pH range
Asparagus	5.4–5.8	Apple	3.1–3.9
Bean	5.0–6.0	Apricot	3.3–4.4
Broccoli	5.2–6.0	Blackberry	3.0–4.2
Cabbage	5.2–5.4	Blueberry	3.2–3.4
Carrot	5.2–5.8	Cantaloupe	6.2–6.5
Pea	5.8–6.5	Cherry	3.2–4.0
Potato	5.4–6.0	Cranberry	2.5–2.7
Pumpkin	4.8–5.5	Grape	3.0–4.0
Rhubarb	2.9–3.3	Grapefruit	2.9–3.4
Spinach	4.8–5.8	Lemon	2.2–2.6
Squash	5.0–5.4	Lime	2.3–2.4
Sweet potato	5.3–5.6	Orange	3.3–4.0
Tomato	4.0–4.4	Peach	3.3–4.2
Turnip	5.2–5.6	Strawberry	3.0–3.9

[a]Adapted from references 9, 28, and 36.

Spoilage Microflora

A variety of microorganisms can cause spoilage of vegetables. Table 7.1 lists some molds that cause spoilage and the type of spoilage associated with each of them. Yeasts, molds, and bacteria cause spoilage of vegetables; however, bacteria are more frequently isolated from initial spoilage defects (10, 11, 25–27). The reason for this is that bacteria are able to grow faster than yeasts or molds in most vegetables and therefore have a competitive advantage, particularly at refrigeration temperatures.

Extensive postharvest spoilage of raw vegetables occurs in the absence of other treatments. Spoilage can be influenced by the history of the land on which vegetables are grown (10). For example, repeated planting of one type of vegetable on the same land over several seasons can lead to the accumulation of plant pathogens in the soil and increased potential for spoilage (27). Likewise, soil contaminated by floodwater or poor-quality irrigation water may expose vegetables to spoilage microorganisms or human pathogens.

Virtually all vegetables receive at least some type of processing or handling before they are consumed. It is important to realize that many common processing steps increase the chances for spoilage (29). Vegetables normally undergo washing and rinsing treatments to remove surface debris. Once washing is complete, vegetables are typically allowed to air dry or are partially dried in centrifugal spin dryers. In addition to removing surface soil and debris, the rinsing step also often reduces the number of microorganisms on the surfaces of vegetables (12).

In most cases, processors of vegetables destined to be sold as fresh produce add 5 to 250 µl of chlorine per liter to wash water. Chlorine is an effective antimicrobial agent. However, the chlorine wash has only a limited antimicrobial effect on the microflora of the produce. Senter et al. (34) observed that 90 to 280 µg of chlorine per liter had a minimal effect on the microflora of tomatoes. In contrast, Beuchat and Brackett (5) reported that dipping tomatoes in a 200 to 250-µg/liter chlorine solution significantly reduced populations of aerobic mesophilic microorganisms but not psychrotrophs or fungi. However, differences in microflora on treated and untreated tomatoes were not evident after 4 days of storage at 10°C. It has been also observed that carrots washed in water containing 200 to 260 µg of chlorine per liter contained about 10-fold less total aerobic microorganisms than carrots washed in water containing no chlorine (4). In this case, however, chlorine-treated carrots developed significantly higher populations of mesophiles, psychrotrophs, and fungi than did untreated carrots.

Although many processors consider washing fresh produce with chlorinated water as a disinfection step, this is not the real purpose for the inclusion of chlorine. Rather, chlorine is more effective in killing microorganisms in the water and minimizing contamination of the vegetables by the rinse water. Washing can actually contribute to spoilage problems if the rinse water contains a large amount of organic matter or if chlorine concentrations are not closely monitored and maintained (10).

Cutting, slicing, chopping, and mixing are other important processing steps to which fresh vegetables are commonly exposed. These operations are becoming even more important as the demand for ready-to-eat products has increased. These processes can sometimes have a dramatic influence on populations of microorganisms and their growth on fresh vegetables. For example, Splittstoesser (35) showed that cutting corn, slicing green beans, and chopping spinach increased populations of microorganisms by about 1 $\log_{10}$ CFU/g. Priepke et al. (30) similarly observed that populations of microorganisms increased faster on cut than on intact salad vegetables.

Processing operations that damage vegetable tissues can lead to increased microbial populations in several ways. First, poorly sanitized equipment can harbor contaminants that are transferred to the vegetables during the cutting operation. For example, *Geotrichum candidum*, sometimes called machinery mold or dairy mold, can accumulate on processing equipment and contaminate vegetables upon contact (10). Second, cutting and slicing allow vegetable tissue fluids to be expressed onto outer surfaces of the vegetable as well as the processing equipment. These fluids can then serve as substrates for growth of microorganisms, allowing them to accumulate to higher populations (10).

Storage, packaging, particularly modified-atmosphere packaging (MAP), and transportation are other important processing steps that can influence the development of microbiological spoilage of vegetables. Most modified-atmosphere techniques involve reducing the concentration of oxygen while increasing the concentration of carbon dioxide. MAP functions by reducing respiration and the senescence process, thereby delaying undesirable changes in sensory quality. The specific composition of gas used for MAP varies according to the product. Produce usually requires carbon dioxide concentrations of at least 5%; however, concentrations in excess of 20% can be detrimental to quality (10).

Microorganisms respond to MAP differently, depending on their tolerance to oxygen and carbon dioxide. MAP can affect obligate aerobic microorganisms by displacing needed oxygen. In addition, carbon dioxide can directly affect microorganisms by causing a reduction in pH of the cytoplasm and interfering with normal cellular metabolism (17). In general, aerobic gram-negative bacteria are most sensitive to carbon dioxide, whereas obligately and facultatively anaerobic microorganisms are more resistant. Likewise, molds are more sensitive than fermentative yeasts to carbon dioxide (17).

Despite the ease with which one can often demonstrate the influence of carbon dioxide on microbial activity in the laboratory, the same is not always true in MAP produce. For example, storage of lettuce (3), carrots (2), and tomatoes (5) resulted in extension of shelf life but did not have an appreciable effect on populations of microorganisms. Similarly, Priepke et al. (30) observed that 10.5% carbon dioxide had only minimal effects on populations of aerobic microorganisms on salad vegetables stored in MAP. In contrast, Berrang et al. (2) reported that a modified atmosphere (10% carbon dioxide, 11% oxygen) significantly inhibited the growth of total aerobic microorganisms in fresh broccoli.

More extensive processing techniques such as thermal processing (canning) and freezing have been commonly used for many years to preserve vegetables. Microbiological spoilage of frozen vegetables is essentially caused by improper storage temperatures. The microflora in frozen vegetables is essentially the same as that in raw products. Therefore, spoilage patterns can be expected to be similar to those of raw products if the products are not properly maintained in a frozen state. However, some molds have been reported to grow at temperatures as low as −5°C (9). Thus, it is at least theoretically possible for frozen vegetable products to succumb to microbiological spoilage.

Preservation of vegetables by canning is accomplished by placing vegetables in hermetically sealed containers and then heating them sufficiently to destroy microorganisms. For most vegetables, processing temperatures frequently exceed 120°C, which eliminates all but the most heat-resistant bacterial endospores. Consequently, spoilage of canned vegetables is usually caused by thermophilic spore-forming bacteria unless container integrity has been compromised in some way (19). Acidification without gas production is one of the most common types of spoilage observed in canned vegetables. This defect, called a flat sour, is caused by *Bacillus stearothermophilus* or *B. coagulans.* The defect ordinarily occurs when cans receive a marginally adequate thermal treatment or if cans are stored at temperatures above 40°C. Another type of spoilage observed with canned vegetables is evidenced by gas production and consequent swelling of cans. Swelling is caused when thermophilic spore-forming anaerobes, such as *Clostridium thermosaccharolyticum,* grow and produce large amounts of hydrogen and carbon dioxide. The production of hydrogen sulfide in some canned vegetables can lead to a spoilage problem known as sulfide stinker. This type of spoilage usually occurs in the absence of can swelling (19).

MICROBIOLOGICAL SPOILAGE OF FRUITS

Fresh fruits are similar to vegetables in that they usually have a high enough a_w to support growth of all but the most xerophilic or osmophilic fungi. However, most fruits differ from vegetables in that they have a more acidic pH (<4.4) (Table 7.2), the exception being melons, as well as a higher sugar content. In addition, fruits usually possess more effective defense mechanisms such as thicker epidermal tissues and higher concentrations of antimicrobial organic acids.

Normal and Spoilage Microflora

As with vegetables, the normal microflora of fruits is varied and includes both bacteria and fungi (19). Sources of microorganisms include all those mentioned for vegetables, including air, soil, and insects. Unlike vegetables, however, spoilage of fruits is most often due to yeasts or molds, except for occasional *Erwinia* rots of pears (22), rather than bacteria. Although molds are often identified as causes of spoilage before and after harvest of fruits (Table 7.1), yeasts are also important (16).

Many of the handling and processing techniques involved in the processing and handling of fresh vegetables also apply to fresh fruits. Fruits are usually washed or rinsed immediately after harvest and then usually packed into shipping cartons. The washing step can play an important role in reducing microbial contamination. Splittsoesser (36) reported that rinsing and scrubbing can reduce populations of microorganisms on apples by more than 99.9%.

Fresh cut or minimally processed fruit products have become increasingly popular in recent years and will likely become more popular because of their convenience in preparation. Despite their popularity, however, cut fruits offer some additional challenges that are less often associated with whole fruits. For example, cutting and slicing will eliminate the protection normally offered by peels and skins. Moreover, high concentrations of sugars in fluids expressed from internal tissues of cut fruits enhance the growth of microorganisms that can tolerate acidic environments. Although both molds and yeasts have this capability, the latter are more often associated with spoilage of cut fruits because they can grow faster than molds (22).

Several fruits are dried to yield intermediate-moisture products which normally rely on low a_w as the main mechanism for preservation. Raisins, prunes, dates, and figs are most commonly consumed as dried fruit products. Dried apples, apricots, and peaches are also popular with consumers. Depending on the means by which fruits are dehydrated, moisture levels can range from less than 5% to 35% (36). In general, most dried fruits have a sufficiently low a_w to inhibit the growth of most bacteria. Spoilage is usually limited to osmophilic yeasts or xerotolerant molds (36). Populations of yeasts and molds on dried fruit usually average under 10^3 CFU/g. Yeasts normally associated with spoilage of dried fruits include *Zygosaccharomyces rouxii* and *Hanseniaspora, Candida, Debaryomyces,* and *Pichia* species. Molds capable of growth below a_w 0.85 include several *Penicillium* and *Aspergillus* species (especially the *A. restrictus* series), *Eurotium* species (the *A. glucus* series), and *Wallemia sebi.*

Fruit concentrates, jellies, jams, preserves, and syrups owe their resistance to spoilage to low a_w. Unlike the case for dried fruits, however, the reduced a_w of these products is achieved by adding sufficient sugar to achieve a_w values of 0.82 to 0.94. In addition to having reduced a_w, these products are usually heated to temperatures of 60 to 82°C, which kills most xerotolerant fungi. Consequently, spoilage of these products usually occurs when containers have been improperly sealed or after they are opened by consumers.

Various heat treatments are used to preserve fruit products. Fruits generally require less severe treatment than vegetables because of the enhanced lethality brought about by their acidic pH. Many canned fruits, whether halved, sliced, or diced, are processed by heating the products to a can center temperature of 85 to 90°C (37). Some processed fruit juices and nectars are rapidly heated to 93 to 110°C and then aseptically filled into containers. In either case, processes are sufficient to kill most vegetative bacteria, yeasts, and molds. However, several genera of molds produce heat-resistant ascospores or sclerotia. Molds typically associated with spoilage of thermally processed fruit products include *Byssochlamys fulva* (anamorph: *Paecilomyces fulvus*), *Byssochlamys nivea* (anamorph: *Paecilomyces niveus*), *Neosartorya fischeri* (anamorph: *A. fischeri*), and *Talaromyces flavus* (anamorphs: *Penicillium vermiculatum* and *P. dangeardii*) (6, 37). Symptoms typically observed with spoiled heat-processed fruit products include visible mold growth, off odors, breakdown of fruit texture, or solubilization of starch or pectin in the suspending medium (7).

MICROBIOLOGICAL SPOILAGE OF GRAINS AND GRAIN PRODUCTS

Although grains are similar to fruits and vegetables in that they are of plant origin, they differ dramatically in many ways. Unlike fruits and vegetables, grains are typically thought of as primarily agronomic or field crops rather than horticultural crops. Grains are planted and cultivated on a larger scale than fruits and vegetables and are harvested only after they have reached full maturity and often have dried to a desired moisture level. After harvest, grains are stored in bins or silos that facilitate additional drying and protection from the weather. Much of the grain produced in developed countries is used as animal feed. Grains destined for human food are ground into flour or meal for bakery or pasta products or further processed into snacks or breakfast cereals. Final products, except doughs, often have a_w values (<0.65) below which most microorganisms will not grow.

Natural Microflora

Grains and grain products normally contain several genera of bacteria, molds, and yeasts, the specific species being dependent on conditions encountered during production, harvesting, storage, and processing. Although the low a_w of grains and grain products might lead one to believe that fungi, especially molds, are the predominant natural microflora, this is not always the case. Seiler (33), for example, surveyed microbial populations in wheat over a 2-year period and observed that bacterial populations were generally about 10-fold higher than mold populations. Of the many different types of microorganisms present on grains, only a few invade the kernel itself. Molds such as *Alternaria*, *Fusarium*, *Helminthosporium*, and *Cladosporium* species are primarily responsible for invading wheat in the field. However, infection by these molds is minimal and does little damage unless the grains are allowed to become too moist. For additional discussions of molds naturally occurring on grains, see chapters 21, 22, and 23.

Effects of Processing

Grains intended for human consumption are rarely used in their native state but rather undergo various processing treatments. For example, grains destined for milling are washed, tempered, screened, and aspirated before being milled. The pretreated grains are then subjected to milling and sifting to separate the hull (bran), germ, and endosperm. The endosperm is then crushed to produce flour. Each step in the pretreatment and milling operations reduces populations of microorganisms such that the flours usually contain lower populations than do grains from which they are made (18). Aside from fewer microorganisms being present in flour, the profile of microflora in milled grains is similar to that of whole grains. The final processing step to which most grain products, specifically flours, are subjected is the addition of liquids (e.g., water or milk) and other ingredients to produce doughs. The type and amount of these ingredients differ depending on the desired dough and final product.

Types of Spoilage

Properly dried and stored grains and grain products are inherently resistant to spoilage due to their low a_w. However, despite attempts to protect grains from uptake of water during storage, the a_w can increase to a level that enables xerotolerant molds to grow. Moreover, it should be kept in mind that reduced a_w often will only slow the growth of fungi and that spoilage will ultimately occur, given enough time. For example, a moisture content of 11 to 14% may be low enough to prevent spoilage of grains stored for less than 1 year, but a maximum of 10 to 12% may be required for long-term storage. However, temperature differentials of 0.5 to 1°C in large bulk quantities of grains can cause some areas of the bulk to reach moisture levels sufficient to allow mold growth. This happens when moisture evaporates from warm areas of bulk-stored grain and then condenses in cooler areas. In such cases, molds such as *Aspergillus* and *Eurotium* species and *W. sebi* may grow and cause spoilage.

Molds are the primary spoilage microorganisms in grains because they can grow at reduced a_w. The primary concern with mold spoilage is the potential production of mycotoxins. However, molds can also affect the sensory quality of grain products. One of the first and most obvious indications of mold growth on grains is a change in appearance. Molded grains may have a powdery appearance. In addition, many molds produce colored spores that can discolor grain-based foods such as breads, crackers, or cakes.

Molds can also cause adverse changes in flavor and aroma of grain products as a result of the production of aromatic volatile compounds. Examples of some flavor compounds typically produced by molds in grains include 3-methyl-butanol, 3-octanone, 3-octanol, 1-octen-3-ol, 1-octanol, and 2-octen-1-ol (23). Although the specific compounds produced are somewhat affected by the substrate, production of volatile metabolites depends more on fungal species than on grain type (8).

Another important adverse affect of mold growth on grains is an increase in the free fatty acid value (FAV). Increases in FAV can result from fungal lipase activity. Because some of these lipases are heat stable, FAV is sometimes used as an indication of fungal spoilage of grains.

Spoilage of grain-based products often differs significantly from that of whole grains. For the most part, properly processed refrigerated doughs have few spoilage problems. However, improper storage conditions may result in growth of yeasts or lactic acid bacteria and cause splitting of bread loaves or the development of slimy texture or undesirable aroma or flavor. Elliott (18) found that the microflora of spoiled doughs consisted of 53% *Leuconostoc dextranicum* and 35% *Leuconostoc mesenteroides*. *Lactobacillus*, *Streptococcus*, *Micrococcus*, and *Bacillus* species and gram-negative rods made up 1.1 to 4.5% of the microflora.

Most grain-based foods receive some type of heat treatment before they are considered edible. These treatments are usually sufficient to inactivate all but the most heat-resistant microorganisms. For example, during baking, the internal temperature of most breads and cakes reaches or slightly exceeds 100°C. Consequently, spoilage microorganisms usually consist of airborne molds which contaminate the product after baking. A notable exception to this is the development of a spoilage condition known as ropiness. This defect is caused by polysaccharide-producing strains of *Bacillus subtilis* or *B. licheniformis*. Spores of these bacteria survive the baking process, germinate, and produce a stringy, brown mass within the bread loaf. This defect usually only occurs when heavily contaminated loaves are improperly cooled. However, modern bread-making methods make ropy bread a rare occurrence. A defect referred to as bloody bread results from the growth of the red-pigmented bacterium *Serratia marcescens*. This defect likewise occurs only rarely in commercially produced breads.

References

1. **Agrios, G. N.** 1988. How pathogens attack plants, p. 63–86. *In Plant Pathology*, 3rd ed. Academic Press, Inc., San Diego, Calif.
2. **Berrang, M. E., R. E. Brackett, and L. R. Beuchat.** 1990. Microbial, color and textural qualities of fresh asparagus, broccoli, and cauliflower stored under controlled atmosphere. *J. Food Prot.* **53:**391–395.
3. **Beuchat, L. R., and R. E. Brackett.** 1990. Growth of *Listeria monocytogenes* on lettuce as influenced by shredding, chlorine treatment, modified atmosphere packaging, temperature and time. *J. Food Sci.* **55:**755–758, 870.
4. **Beuchat, L. R., and R. E. Brackett.** 1990. Inhibitory effects of raw carrots on *Listeria monocytogenes*. *Appl. Environ. Microbiol.* **56:**1734–1742.
5. **Beuchat, L. R., and R. E. Brackett.** 1991. Behavior of *Listeria monocytogenes* inoculated into raw tomatoes and processed tomato products. *Appl. Environ. Microbiol.* **57:**1367–1371.
6. **Beuchat, L. R., and J. I. Pitt.** 1992. Detection and enumeration of heat-resistant molds, p. 251–263. *In* C. Vanderzant and D. F. Splittoesser (ed.), *Compendium of Methods for the Microbiological Examination of Foods*. American Public Health Association, Washington, D.C.
7. **Beuchat, L. R., and S. L. Rice.** 1979. *Byssochlamys* spp. and their importance in processed fruits. *Adv. Food Res.* **25:**237–288.
8. **Börjesson, T., U. Stöllman, and J. Schnürer.** 1992. Volatile metabolites produced by six fungal species compared with other indicators of fungal growth on cereal grains. *Appl. Environ. Microbiol.* **58:**2599–2605.
9. **Brackett, R. E.** 1987. Vegetables and related products, p. 129–154. *In* L. R. Beuchat (ed.), *Food and Beverage Mycology*. Van Nostrand Reinhold, New York.
10. **Brackett, R. E.** 1993. Microbial quality, p. 125–148. *In* R. L. Shewfelt and S. E. Prussia (ed.), *Postharvest Handling: a Systems Approach*. Academic Press, New York.
11. **Brackett, R. E.** 1994. Microbiological spoilage and pathogens in minimally processed fruits and vegetables, p. 269–312. *In* R. C. Wiley (ed.), *Minimally Processed Refrigerated (MPR) Fruits and Vegetables*. Van Nostrand Reinhold, New York.
12. **Brackett, R. E., and D. L. Splittstoesser.** 1992. Fruits and vegetables, p. 287–293. *In* C. Vanderzant and D. L. Splittstoesser (ed.), *Compendium of Methods for the Microbiological Examination of Food*. American Public Health Association, Washington, D.C.
13. **Cheeson, A.** 1980. Maceration in relation to the post-harvest handling and processing of plant material. *J. Appl. Bacteriol.* **48:**1–45.
14. **Codner, R. C.** 1971. Pectinolytic and cellulytic enzymes in the microbial modification of plant tissues. *J. Appl. Bacteriol.* **34:**147–160.

15. **Collmer, A., and N. T. Keen.** 1986. The role of pectic enzymes in plant pathogenesis. *Annu. Rev. Phytopathol.* **24:**383–409.

16. **Deak, T., and L. R. Beuchat.** 1996. Yeasts in specific types of foods, p. 61–96. *In Handbook of Food Spoilage Yeasts.* CRC Press, Boca Raton, Fla.

17. **Daniels, J., A. R. Krishnamurthi, and S. S. H. Rizvi.** 1985. A review of effects of carbon dioxide on microbial growth and food quality. *J. Food Prot.* **48:**532–537.

18. **Elliott, R. P.** 1980. Cereals and cereal products, p. 669–730. *In* J. H. Silliker, R. P. Elliot, A. C. Baird-Parker, F. L. Bryan, J. H. B. Christian, D. S. Clark, J. C. Olson, Jr., and T. A. Roberts (ed.), *Microbial Ecology of Foods,* vol. II. Academic Press, New York.

19. **Goepfert, J. M.** 1980. Vegetables, fruits, nuts and their products, p. 606–642. *In* J. H. Silliker, R. P. Elliot, A. C. Baird-Parker, F. L. Bryan, J. H. B. Christian, D. S. Clark, J. C. Olson, Jr., and T. A. Roberts (ed.), *Microbial Ecology of Foods,* vol. II. Academic Press, New York.

20. **Hao, Y. Y., and R. E. Brackett.** 1994. Pectinase activity of vegetable spoilage bacteria in modified atmosphere. *J. Food Sci.* **59:**175–178.

21. **Hanklin, L., and G. H. Lacy.** 1992. Pectinolytic microorganisms, p. 176–183. *In* C. Vanderzant and D. F. Splittstoesser (ed.), *Compendium of Methods for the Microbiological Examination of Foods.* American Public Health Association, Washington, D.C.

22. **Jay, J. M.** 1992. *Modern Food Microbiology,* 4th ed., p. 187–198. Van Nostrand Reinhold, New York.

23. **Kinderlerer, J. L.** 1989. Volatile metabolites of filamentous fungi and their role in food flavor. *J. Appl. Bacteriol Symp. Suppl.* **67:**133S–144S.

24. **Kotoujansky, A.** 1987. Molecular genetics of pathogenesis by soft-rot erwinias. *Annu. Rev. Phytopathol.* **25:**405–430.

25. **Lund, B. M.** 1971. Bacterial spoilage of vegetables and certain fruits. *J. Appl. Bacteriol.* **34:**9–20.

26. **Lund, B. M.** 1982. The effect of bacteria on post-harvest quality of vegetables and fruits, with particular reference to spoilage. *Soc. Appl. Bacteriol. Symp. Ser.* **10:**133–153.

27. **Lund, B. M.** 1983. Bacterial spoilage, p. 219–257. *In* C. Dennis (ed.), *Post-Harvest Pathology of Fruits and Vegetables.* Academic Press, London.

28. **Lund, B. M.** 1992. Ecosystems in vegetable foods. *J. Appl. Bacteriol. Symp. Suppl.* **73:**115S–126S.

29. **Nguyen-the, C., and F. Carlin.** 1994. The microbiology of minimally processed fresh fruits and vegetables. *Crit. Rev. Food Sci. Nutr.* **34:**371–401.

30. **Priepke, P. E., L. Wei, and A. I. Nelson.** 1976. Refrigerated storage of prepackaged salad vegetables. *J. Food Sci.* **41:**379–382.

31. **Rolle, R. S., and G. W. Chism III.** 1987. Physiological consequences of minimally processed fruits and vegetables. *J. Food Qual.* **10:**157–177.

32. **Rombouts, F. M., and W. Pilnik.** 1980. Pectin enzymes, p. 227–282. *In* A. H. Rose (ed.), *Microbial Enzymes and Bioconversions.* Academic Press, London.

33. **Seiler, D. A. L.** 1986. The microbial content of wheat and flour, p. 241–255. *In* B. Flannigan (ed.), *Spoilage and Mycotoxins of Cereals and Other Stored Products.* Crown, New York.

34. **Senter, S. D., N. A. Cox, J. S. Bailey, and W. R. Forbus, Jr.** 1985. Microbiological changes in fresh market tomatoes during packing operations. *J. Food Sci.* **50:**254–255.

35. **Splittstoesser, D. F.** 1973. The microbiology of frozen vegetables. *Food Technol.* **27(1):**54–60.

36. **Splittstoesser, D. F.** 1987. Fruits and vegetable products, p. 101–128. *In* L. R. Beuchat (ed.), *Fruit and Beverage Mycology,* 2nd ed. Van Nostrand Reinhold, New York.

37. **Splittstoesser, D. F.** 1991. Fungi of importance in processed fruits, p. 201–219. *In* D. K. Arora, K. G. Mukerji, and E. H. Marth (ed.), *Handbook of Applied Mycology.* Marcel Dekker, Inc., New York.

Foodborne Pathogenic Bacteria

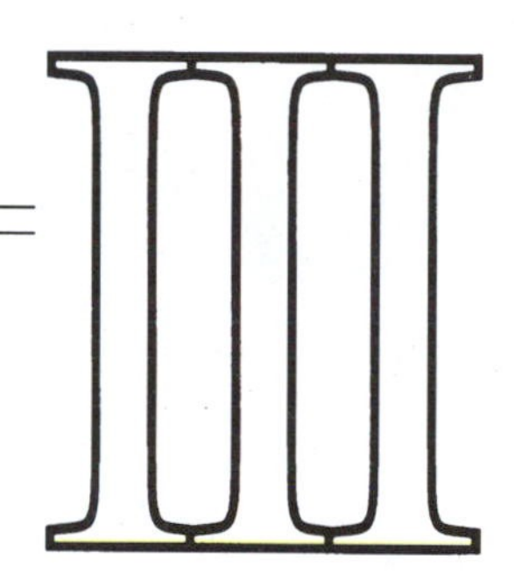

Jean-Yves D'Aoust

Salmonella Species

8

CHARACTERISTICS OF SALMONELLAE

Historical Considerations

In the early 19th century, clinical pathologists in France first documented the association of human intestinal ulceration with a contagious agent which was later identified as typhoid fever. Further investigations by European workers led to the isolation and characterization of the typhoid bacillus and to the development of a serodiagnostic test for the detection of this serious human disease agent (45, 125). Differential clinical and serological traits were subsequently used to identify the closely related paratyphoid organisms. In the United States, contemporary work by Salmon and Smith (1885) had led to the isolation of *Bacillus cholerae-suis* from swine suffering from hog cholera (125). The first quarter of the 20th century witnessed great advances in the serological detection of somatic (O) and flagellar (H) antigens within the *Salmonella* group, a generic term coined by Lignières in 1900 (125). An antigenic scheme for the classification of salmonellae was first proposed by White (1926) and subsequently expanded by Kauffmann (1941) into the Kauffmann-White scheme, which, in 1994, recognized more than 2,300 serovars (161).

Taxonomy

Salmonella spp. are facultatively anaerobic gram-negative rods belonging to the family *Enterobacteriaceae.* Although members of this genus are motile by peritrichous flagella, nonflagellated variants, such as *Salmonella pullorum* and *S. gallinarum*, and nonmotile strains resulting from dysfunctional flagella do occur. Salmonellae are chemoorganotrophic, with an ability to metabolize nutrients by the respiratory and fermentative pathways. The organisms grow optimally at 37°C and catabolize D-glucose and other carbohydrates, with the production of acid and gas. Salmonellae are oxidase negative and catalase positive, grow on citrate as the sole carbon source, generally produce hydrogen sulfide, decarboxylate lysine and ornithine, and do not hydrolyze urea. Many of these traits have traditionally formed the basis for the presumptive biochemical identification of *Salmonella* isolates. According to a contemporary definition, a typical *Salmonella* isolate would produce acid and gas from glucose in triple sugar iron (TSI) agar medium and would not utilize lactose or sucrose in TSI or in differential plating media such as brilliant green, xylose lysine deoxycholate, and Hektoen enteric agars. Additionally, typical salmonellae would readily produce an alkaline reaction from the decarboxylation of lysine to cadaverine in lysine iron agar, generate hydrogen sulfide gas in TSI and lysine iron media, and fail to hydrolyze urea (5, 50). The dynamics of genetic variability arising from bacterial mutations and conjugative intra- and intergeneric exchange of plasmids encoding determinant biochemical traits continue to reduce the proportion of typical *Salmonella* biotypes. From the early studies of Le

Table 8.1 Taxonomical schemes for *Salmonella* spp.

Diagnostic basis	Salient features	Serovar designation	Reference
Biochemical	Five subgenera (I–V) Serovar = species status *S. arizonae*	*S. typhimurium*	114
Biochemical	Three species (*S. typhi, S. choleraesuis, S. enteritidis*) *Arizona* = separate genus	*S. enteritidis* serovar Typhimurium	58
Phenetic analysis-DNA homology	Single species (*S. choleraesuis*) Seven subspecies (*choleraesuis, salamae, arizonae, diarizonae, houtenae, bongori, indica*) Type strain = *S. choleraesuis*	*S. choleraesuis* subsp. *choleraesuis* serovar Typhimurium	129
Phenetic analysis-DNA homology	Single species (*S. enterica*) Seven subspecies (see above) Type strain = *S. typhimurium* LT2	*S. enterica* subsp. *enterica* serovar Typhimurium	128
Multilocus enzyme electrophoresis	Two species (*S. enterica* [six subspecies], *S. bongori*)		167

Minor and coworkers confirming that *Salmonella* utilization of lactose and sucrose was plasmid mediated (126, 127), numerous reports have since underlined the prevalence of Lac^+ and/or Suc^+ biotypes in clinical specimens and food materials. This situation is of public health concern because such atypical microorganisms could easily escape detection on disaccharide-dependent plating media which are commonly used in hospital and food industry laboratories. Bismuth sulfite agar remains a medium of choice because, in addition to its high level of selectivity, it responds solely and most effectively to the production of extremely low levels of hydrogen sulfide gas (45). The diagnostic hurdles afforded by the changing patterns of disaccharide utilization by salmonellae are being further confounded by the increasing occurrence of biotypes that cannot decarboxylate lysine, that possess urease activity, that produce indole, and that readily grow in the presence of potassium cyanide. Clearly, the prevalence of *Salmonella* species as a biochemically homogeneous group of microorganisms is rapidly diminishing. The situation will likely lead to a reassessment of the diagnostic value of these and other biochemical traits and to their likely replacement with molecular technologies targeted at the identification of stable genetic loci and/or their products that are unique to the genus *Salmonella*.

Nomenclature of the *Salmonella* group has progressed through a succession of taxonomical schemes based on biochemical and serological characteristics and on principles of numerical taxonomy and DNA homology (Table 8.1). In the early development of taxonomic schemes, determinant biochemical reactions were used to separate salmonellae into subgroups. The Kauffmann-White scheme stands prominently as the first attempt to systematically classify salmonellae by using these scientific parameters. This major undertaking culminated in the identification of five biochemically defined subgenera (I to V) wherein individual serovars were afforded species status (114). Subgenus III included members of the *Arizona* group (*S. arizonae*). Subsequently, a three-species nomenclatural system using 16 discriminating tests to identify *S. typhi* (single serovar), *S. choleraesuis* (single serovar), and *S. enteritidis* (all other *Salmonella* serovars) was proposed. The latter scheme recognized members of the *Arizona* group as a distinct genus (58). Another system based on numerical taxonomy and DNA relatedness proposed a single species (*S. choleraesuis*) consisting of seven subspecies (129). In numerical taxonomy, a statistical comparison of morphological and biochemical attributes of strains (phenetic analysis) measures the taxonomical proximity of test strains and allows for their separation into distinct taxons. In DNA homology, a high degree of hybridization of *Salmonella* reference DNA with extracts of test strains confirms the genetic relatedness of nucleic acid reactants and supports the inclusion of the test microorganisms into the genus *Salmonella*. Subsequent modification of the former scheme encompassed a change from *S. choleraesuis* to *S. enterica* while retaining the name of the seven recognized subspecies. The type strain was also changed from *S. choleraesuis* to *S. enterica* subsp. *enterica* serovar Typhimurium LT2 (128). The most recent proposal petitions for the elevation of members of *S. enterica* subsp. *bongori* to a new species based on multilocus enzyme electrophoretic patterns (167). The new species would be designated *S. bongori*. According to the WHO Collaborating Centre for Reference and Research on *Salmonella* (Institut Pasteur, Paris), *S. en-*

Table 8.2 Species of the genus *Salmonella*[a]

Species	No. of serovars
Salmonella enterica	
subsp. *enterica*	1,405
subsp. *salamae*	471
subsp. *arizonae*	94
subsp. *diarizonae*	311
subsp. *houtenae*	65
subsp. *indica*	10
Salmonella bongori	19
Total	2,375

[a]From reference 161.

terica and *S. bongori* currently include 2,356 and 19 serovars, respectively (Table 8.2).

The biochemical identification of foodborne and clinical *Salmonella* spp. isolates is generally coupled to serological confirmation, a complex and labor-intensive technique involving the agglutination of bacterial surface antigens with *Salmonella*-specific antibodies. These include O lipopolysaccharides (LPS) on the external surface of the bacterial outer membrane, H antigens associated with the peritrichous flagella, and the capsular (Vi) antigen, which occurs only in *S. typhi*, *S. paratyphi* C, and *S. dublin* (125). The heat-stable O antigens are classified as major or minor antigens. The former category consists of antigens such as the somatic factors O:4 and O:3, which are specific determinants for the somatic groups B and E, respectively. In contrast, minor somatic antigenic components such as O:12 are nondiscriminatory, as evidenced by their presence in different somatic groups. Smooth variants are strains with well-developed serotypic LPS antigens that readily agglutinate with specific antibodies, whereas rough variants exhibit incomplete LPS antigens resulting in weak or no agglutination with *Salmonella* somatic antibodies. H antigens are heat-labile proteins; individual *Salmonella* strains may produce one (monophasic) or two (diphasic) H antigens. Although serovars such as *S. dublin* produce a single H antigen, most serovars can alternatively elaborate two antigens, phase 1 and phase 2 antigens. These homologous surface antigens are chromosomally encoded by the H_1 (phase 1) and H_2 (phase 2) genes and transcribed under the control of the vh_2 locus (125). Capsular antigens commonly encountered in members of the family *Enterobacteriaceae* are limited to the Vi antigen in the genus *Salmonella*. Thermal solubilization of the Vi antigen is necessary for the immunological identification of underlying serotypic LPS.

The aim of serological testing procedures is to determine the complete antigenic formula of a *Salmonella* isolate. We shall use *S. infantis* (6,7:r:1,5) as an example. Commercially available polyvalent somatic antisera each consist of a cocktail of antibodies specific for a limited number of major antigens; e.g., polyvalent B (poly B) antiserum (Difco Laboratories, Detroit, Mich.) recognizes O groups C_1, C_2, F, G, and H. Following a positive agglutination with the poly B antiserum, single-grouping antisera representing the five somatic groupings included in the poly B reagent would be used to define the serogroup of the isolate. The test isolate would react with the C_1-grouping antiserum, indicating that antigens 6,7 are present. H antigens would then be determined by broth agglutination reactions using poly H antisera or the Spicer-Edwards series of antisera. In the former assay, a positive agglutination reaction with one of the five polyvalent antisera (poly A to E; Difco) would lead to testing with single-factor antisera to specifically identify the phase 1 and/or phase 2 flagellar antigens present. Agglutination in poly C flagellar antiserum and reaction with single-grouping H antiserum would confirm the presence of the r antigen (phase 1). The empirical antigenic formula of the isolate would then be (6,7:r). Phase reversal in semisolid agar supplemented with r antiserum would immobilize phase 1 salmonellae, thereby facilitating the recovery of cells in the phase 2 configuration. Serological testing of the phase 2 isolate with poly E and 1-complex antisera would confirm the presence of the flagellar 1 factor. Confirmation of the flagellar 5 antigen with single-factor antiserum would yield the final antigenic formula 6,7:r:1,5, which corresponds to *S. infantis*. A similar analytical approach would be used with the Spicer-Edwards poly H antisera; in this case, the identification of H antigens would arise from the pattern of agglutination reactions among the four Spicer-Edwards antisera and with three additional polyvalent antisera including the L, 1, and e,n complexes.

Physiology

Growth

The genus *Salmonella* consists of resilient microorganisms that readily adapt to extreme environmental conditions. Salmonellae actively grow within a wide temperature range (≤54°C) and also exhibit psychrotrophic properties, as reflected in the ability to grow in foods stored at 2 to 4°C (47) (Table 8.3). Moreover, preconditioning of cells to low temperatures can markedly increase the growth and survival of salmonellae in refrigerated food products (1). Such growth characteristics raise concerns about the efficacy of chill temperatures in ensuring food

Table 8.3 Physiological limits for the growth of *Salmonella* spp. in foods and bacteriological media

Parameter	Limits: Minimum	Limits: Maximum	Product	Serovar	Reference
Temperature	2°C (24 h)		Minced beef[a]	*S. typhimurium*	30
	2°C (2 days)		Minced chicken[b]	*S. typhimurium*	16
	4°C (≤10 days)		Shell eggs[b]	*S. enteritidis*	116
		54.0	Agar medium	*S. typhimurium*	55
pH	3.99[c]		Tomatoes	*S. infantis*	14
	4.05[d]		Liquid medium	*S. anatum*	35
				S. tennessee	
				S. senftenberg	
		9.5	Egg wash water[b]	*S. typhimurium*	100
a_w	0.93[e]		Rehydrated dried soup[b]	*S. oranienburg*	194

[a]Naturally contaminated.
[b]Artificially contaminated.
[c]Growth within 24 h at 22°C.
[d]Acidified with HCl or citric acid; growth within 24 h at 30°C.
[e]Growth within 3 days at 30°C.

safety through bacteriostasis. These concerns are further heightened by the widespread refrigerated storage of foods packaged under vacuum or modified atmosphere to prolong shelf life. Gaseous mixtures consisting of 60 to 80% (vol/vol) CO_2 with various proportions of N_2 and/or O_2 have been found to inhibit the growth of aerobic spoilage microorganisms such as *Pseudomonas* spp. without promoting the growth of *Salmonella* spp. (47). However, the proliferation of salmonellae in inoculated raw minced beef and cooked crab meat stored at 8 to 11°C under modified atmospheres containing low levels of CO_2 (20 to 50% [vol/vol]) warrants caution in the general application of this novel processing technology (22, 106). Studies on the maximum temperature for growth of *Salmonella* spp. in foods are generally lacking. Notwithstanding an early report of growth of salmonellae in inoculated custard and chicken à la king at 45.6°C (6), more recent evidence indicates that prolonged exposure of mesophilic strains to thermal stress conditions results in mutants of *S. typhimurium* capable of growth at 54°C (55). Although the mechanism of this phenomenon has yet to be elucidated, preliminary findings indicate that two separate mutations enable *S. typhimurium* to grow actively at 48°C (*ttl*) and 54°C (*mth*).

The physiological adaptability of *Salmonella* spp. is further demonstrated by their ability to proliferate at pH values ranging from 4.5 to 9.5, with an optimum pH for growth of 6.5 to 7.5 (Table 8.3). It is well established that the bacteriostatic or antibacterial effects of acidic conditions are acidulant dependent (45). Of the many organic acids formed by starter cultures in meat, dairy, and other fermented foods, and of the various organic and inorganic acids used in product acidification, propionic and acetic acids are more bactericidal than the common food-associated lactic and citric acids. Interestingly, the antibacterial action of organic acids decreases with increasing length of the fatty acid chain (45). Early work on the propensity for the growth of salmonellae in acidic environments showed that wild-type strains preconditioned on pH gradient plates could grow in liquid and solid media at considerably lower pH values than the parent strains (104). These findings raise concerns about the safety of fermented foods such as cured sausages and fermented raw milk products, the progressive starter culture-dependent acidification of which could provide a favorable environment for the elevation of endogenous salmonellae to a state of increased acid tolerance. The growth and/or enhanced survival of this human bacterial pathogen during the fermentative process would result in a contaminated ready-to-eat product. A recent report on the increased survival of acid-adapted *Salmonella* spp. in fermented milk and during storage (5°C) of derived Cheddar, Swiss, and Mozzarella cheese is noteworthy (131). Moreover, the presence of acid-tolerant salmonellae in such foods further heightens the level of public health hazard because this acquired physiological trait could minimize the antimicrobial action of gastric acidity (pH 2.5) and promote the survival of salmonellae within the acidic cytoplasm of mononuclear and polynuclear phagocytes of the human host (46).

High salt concentrations have long been recognized for their ability to extend the shelf life of foods by inhibiting the growth of endogenous microflora (157). This bacteriostatic effect, which can also engender cell death, results from a dramatic decrease in water activity

(a_w) and from bacterial plasmolysis commensurate with the hypertonicity of the suspending medium. Studies have shown that foods with a_w of ≤0.93 do not support the growth of salmonellae (45) (Table 8.3). Although salmonellae are generally inhibited in the presence of 3 to 4% NaCl, bacterial salt tolerance increases with increasing temperature in the range of 10 to 30°C. However, the latter phenomenon is associated with a protracted log phase and a decreased rate of growth. Evidence further suggests that the magnitude of this adaptive response is food and serovar specific (45). A recent report on anaerobiosis and its potentiation of greater salt tolerance in salmonellae raises concerns on the safety of modified-atmosphere-packaged and vacuum-packaged foods that contain high levels of salt (8).

The pH, salt concentration, and temperature of the microenvironment can exert profound effects on the growth kinetics of *Salmonella* spp. Several studies have underlined the increased ability of the microorganisms to grow under acidic (pH ≤5.0) conditions or in environments of high salinity (≥2% NaCl) with increasing temperature (45, 62, 189). Similar work on the interrelationship between pH and NaCl at temperatures near optimal for growth (20 to 30°C) has underlined the dominance of medium pH on the growth of salmonellae (189). Interestingly, the presence of salt in acidified foods can reduce the antibacterial action of organic acids: low concentrations of NaCl or KCl stimulate the growth of *S. enteritidis* in broth medium acidified to pH 5.19 with acetic acid (164). Such findings and other reports on the enhanced salt-dependent survival of salmonellae in rennet whey (pH 4.8 to 5.6) and mayonnaise (189) indicate that low levels of salt can undermine the preservative action of organic acids and potentially compromise the safety of fermented and acidified foods. Although the mechanisms for these salt-related phenomena remain elusive, the role of salinity in restoring cellular homeostasis may be linked to the Na^+/K^+ proton antiport systems (68, 69, 152).

The study of interactive forces generated by temperature, pH, and salt on the growth and survival of salmonellae has led to the development and application of mathematical models to predict the fate of salmonellae in foods. In this approach, the growth, survival, or inactivation of a target microorganism under various conditions of pH, NaCl, temperature, or other environmental factors of interest is laboriously characterized in laboratory media. The generated data are then used to derive mathematical models that depict the response of the microorganism under different combinations of environmental factors (80). Predictive models are not without limitations. For example, the use of a model to predict the growth of salmonellae in a food whose salt content lies beyond the range of values originally studied for the derivation of the mathematical model would likely lead to erroneous conclusions. Moreover, models are generally based on the behavior of a few *Salmonella* strains under selected environmental conditions. The physiological diversity among the large number of serovars (Table 8.2) may unduly challenge the reliability of models in predicting bacterial growth responses. Additionally, the lot-to-lot variations in the composition of a given food and in the types and numbers of background microflora are variables that could seriously undermine the predictive capability of mathematical models. A special issue of the *International Journal of Food Microbiology* (volume 23, 1994) on predictive modeling provides invaluable information on the merits and limitations of the modeling approach for *Salmonella* species and other foodborne bacterial pathogens and should be consulted for an in-depth review of the subject.

Survival

Although the potential growth of foodborne salmonellae is of primary importance in safety assessments, the propensity for these pathogens to persist in hostile environments further heightens public health concerns. The survival of salmonellae for prolonged periods of time in foods stored at freezer and ambient temperatures is well documented (45). It is noteworthy that the composition of the freezing menstruum, the kinetics of the freezing process, the physiological state of foodborne salmonellae, and the serovar-specific responses to extreme temperatures determine the fate of salmonellae during freezer storage of foods (38). The viability of salmonellae in dry foods stored at ≥25°C decreases with increasing storage temperature and with increasing moisture content (39, 45).

Heat is widely used in food manufacturing processes to control the bacterial quality and safety of end products. Factors that potentiate the greater heat resistance of salmonellae and other foodborne bacterial pathogens in food ingredients and finished products have been studied extensively (39). Although the heat resistance of *Salmonella* spp. increases as the a_w of the heating menstruum decreases, detailed studies have demonstrated that the solutes used to alter the a_w of the heating menstruum play a determinant role in the level of acquired heat resistance (45). For example, the heating of *S. typhimurium* in menstrua adjusted to a_w 0.90 with sucrose and glycerol conferred different levels of heat resistance, as evidenced by $D_{57.2}$ values of 40 to 55 and 1.8 to 8.3

min, respectively (86). The *D* value represents the amount of time required to effect a 90% kill (1.0 $\log_{10}$ reduction) in the number of viable cells upon heating at 57.2°C. Other important features associated with this adaptive response include the greater heat resistance of salmonellae grown in nutritionally rich media than of salmonellae grown in minimal media, of cells derived from stationary- rather than logarithmic-phase cultures, and of salmonellae previously stored in a dry environment (86, 117, 148). The ability of salmonellae to acquire greater heat resistance following exposure to sublethal temperatures is equally notable. The phenomenon stems from a rapid adaptation of the organism to rising temperatures in the microenvironment to a level of enhanced thermotolerance quite distinct from that described in conventional time-temperature curves of thermal lethality. This adaptive response has potentially serious implications with respect to the safety of thermal processes that expose or maintain food products at marginally lethal temperatures. Exposure of salmonellae to sublethal temperatures (≤50°C) for 15 to 30 min enhances their heat resistance through a rapid chloramphenicol-sensitive synthesis of heat shock proteins (105, 134, 135). Changes in the fatty acid composition of cell membranes in heat-stressed salmonellae to provide a greater proportion of saturated membrane phospholipids reduce the fluidity of the bacterial cell membrane, with an attendant increase in membrane resistance to heat damage (105). The likelihood that other protective cellular functions are triggered by heat shock stimuli cannot be discounted.

The complexities of the foregoing considerations can best be illustrated by the following scenario involving the *Salmonella* contamination of a chocolate confectionery product. Experience has shown that survival of salmonellae in dry-roasted cocoa beans can lead to contamination of in-line and finished products. Thermal inactivation of salmonellae in molten chocolate is difficult because the time-temperature conditions that would be required to eliminate the pathogen effectively in this sucrose-containing product of low a_w would likely result in an organoleptically unacceptable product. The problem is further compounded by the ability of salmonellae to survive for many years in the finished product when stored at ambient temperature (44). Clearly, effective decontamination of raw cocoa beans and stringent in-plant control measures to prevent cross-contamination of in-line products are of prime importance to this food industry.

Recent evidence indicates that brief exposure of *S. typhimurium* to mild acid environments of pH 5.5 to 6.0 (preshock) followed by exposure of the adapted cells to pH ≤4.5 (acid shock) triggers a complex acid tolerance response (ATR) that potentiates the survival of the microorganism under extreme acid environments (pH 3.0 to 4.0). The response translates into an induced synthesis of 43 acid shock proteins and outer membrane proteins (OMPs), reduced growth rate, and pH homeostasis, as demonstrated by the bacterial maintenance of internal pH values of 7.0 to 7.1 and 5.0 to 5.5 upon sequential exposure of cells to external pHs of 5.0 and 3.3, respectively (69, 99). In the ATR, bacterial Mg^{2+}-dependent proton-translocating ATPase encoded by the *atp* operon plays an important role in maintaining cellular pH at ≥5.0 through an energy-dependent transport of intracellular protons to the cell exterior (69). Other transmembrane mechanisms that putatively operate in pH homeostasis include the H^+-coupled ion transport systems (antiport) for the intra- and extracellular transfer of K^+, Na^+, and H^+ ions, the electron transport chain-dependent efflux of H^+, and transport systems committed to the symport of H^+ and solutes (152). The Fe^{2+}-binding regulatory protein encoded by the *fur* (ferric uptake regulator) gene also affects bacterial acid tolerance, as evidenced by the inability of *fur* mutant strains to survive under highly acidic conditions (76). Acid-induced activation of amino acid decarboxylases in salmonellae provides an additional protective mechanism whereby cadaverine and putrescine from the enzymic breakdown of lysine and ornithine, respectively, potentiate acid neutralization and enhanced bacterial survival (152). Further characterization of the *Salmonella* response to acid stress has led to the identification of two additional protective mechanisms that operate in salmonellae in the stationary phase of growth and which are distinct from the previously discussed ATR, which prevails in log-phase cells (69, 122). One of these pH-dependent responses, designated stationary-phase ATR, provides greater acid resistance than the log-phase ATR; it is induced at pH <5.5 and functions maximally at pH 4.3. This stationary-phase ATR induces the synthesis of only 15 shock proteins and is not affected by mutations in the *atp* and *fur* genes (122). The induction of the remaining acid-protective mechanism associated with salmonellae in the stationary phase is independent of external pH and dependent on the alternative sigma factor (σ^S) encoded by the *rpoS* locus. The mechanism seemingly reinforces the ability of stationary-phase cells to survive in hostile environmental conditions. We have thus seen that three possibly overlapping cellular systems confer acid tolerance in *Salmonella* spp.: (i) the pH-dependent, *rpoS*-independent log-phase ATR; (ii) the pH-dependent stationary-phase ATR; and (iii) the pH-independent, *rpoS*-dependent stationary-phase ATR. These systems likely operate in the acidic environments that prevail in fermented and in acidified foods and in phagocytic cells of the infected host.

Acid stress can also trigger enhanced bacterial resistance to other adverse environmental conditions. The growth of *S. typhimurium* at pH 5.8 engendered an increased thermal resistance at 50°C, an enhanced tolerance to high osmotic stress (2.5 M NaCl) ascribed to the induced synthesis of the OmpC proteins, a greater surface hydrophobicity, and an increased resistance to the antibacterial lactoperoxidase system and surface-active agents such as crystal violet and polymyxin B (132).

RESERVOIRS

The ubiquity of salmonellae in the natural environment, coupled with the intensive husbandry practices used in the meat, fish, and shellfish industries and the recycling of offal and inedible raw materials into animal feeds, has favored the continued prominence of this human bacterial pathogen in the global food chain (45, 48). Of the many sectors within the meat industry, poultry products remain the principal reservoirs of salmonellae in many countries, dominating other meat products such as pork, beef, and mutton as potential vehicles of infection (45, 93, 94, 193, 198). The absence of major improvements in the *Salmonella* status of raw poultry products, notably chicken, turkey, and waterfowl, in the North American and European continents stems, in part, from the difficulty in establishing and maintaining *Salmonella*-free supply flocks and distributing noncontaminated feeds to the various sectors of this vertically integrated industry. The housing of birds on premises whose construction may not effectively block entry of feral *Salmonella* carriers, including rodents, insects, and avian species, exacerbates the magnitude of the problem. Moreover, the feeding habits and close proximity of multiplier breeder, layer, and broiler birds in rearing facilities encourage the rapid and widespread dissemination of vertically and horizontally introduced salmonellae in poultry houses.

The continuing pandemic of human *S. enteritidis* phage type 4 (Europe) and 8 (North America) infections associated with the consumption of raw or lightly cooked shell eggs and egg-containing products further bears witness to the importance of poultry products as vehicles of human salmonellosis and to the need for sustained and stringent bacteriological control of poultry husbandry practices. This egg-related pandemic is of particular concern because the problem arises from the transovarian transmission of the infective agent into the interior of the egg prior to shell deposition. The viability of these internalized *S. enteritidis* organisms thus remains unaffected by the egg surface-sanitizing practices currently applied in egg-grading stations. The significant public health and societal costs associated with egg-borne outbreaks of *S. enteritidis* together with the economic losses sustained by the agricultural sectors as a result of depopulation of infected layer flocks and mandatory pasteurization of shell eggs from infected commercial layer flocks will undoubtedly result in dreadful losses. Although several affected countries are currently reporting some abatement in the egg-borne *S. enteritidis* problem (136), many continue to suffer the socio-economic impact of this complex and well-entrenched bacterial epidemic (28, 60, 171).

The persistence of salmonellae in the porcine, bovine, and ovine meat industries originates from the exposure of livestock to environmental sources of contamination and contaminated feeds and from parental transmission of infection. It is abundantly clear that future improvement in the *Salmonella* status of meat animals hinges on coordinated and sustained efforts by all sectors of the meat industry to implement stringent control measures not only at the farm level but also within the processing, distribution, and retailing sectors of the industry (198).

Rapid depletion of feral stocks of fish and shellfish in recent years has greatly increased the importance of the international aquaculture industry as an alternate source for these popular food items. The high-density farming conditions required to maximize biological yields and to satisfy growing market demands open gateways to the widespread infection of species reared in earthen ponds and other unprotected facilities that are continuously exposed to environmental contamination. It is noteworthy that much of the currently available aquacultural products originate from the Asiatic, African, and South American continents. The feeding to reared species of raw meat scraps and offal, of night soil potentially contaminated with typhoid and paratyphoid organisms, and of animal feeds that may harbor a variety of *Salmonella* serovars is not uncommon in these geographic areas. Such practices represent a situation that could lead to the perpetual bacterial contamination of rearing facilities and to severe economic losses through suboptimal growth or death of aquacultural populations (48). Accordingly, aquaculture farmers are relying heavily on antibiotics applied at subtherapeutic levels to safeguard the vigor of farmed fish and shellfish. A serious public health concern arises from the use of antimicrobial agents such as ampicillin, chloramphenicol, sulfa drugs, and quinolones, which currently provide the mainstay for the clinical management of systemic salmonellosis in humans (10, 48, 175). This antibiotic-dependent husbandry practice is shortsighted and irresponsible upon consideration that the misuse of medically important therapeutic agents in the livestock industry continues to select for antibiotic-resistant salmonellae (10, 45, 48, 51). The health consequences of consuming contaminated aquacultural fish in a sushi dish or lightly cooked fish or shellfish that contain viable salmonellae are predictable.

Fruits and vegetables have gained notoriety in recent years as vehicles of human salmonellosis. The situation has developed from the increased global export of fresh and dehydrated fruits and vegetables from countries that enjoy tropical and subtropical climates. The prevailing hygienic conditions during the production, harvesting, and distribution of products in these countries do not always meet minimum standards, and they facilitate product contamination. More specifically, the fertilization of crops with untreated sludge or sewage effluents potentially contaminated with antibiotic-resistant *Salmonella* spp., the irrigation of garden plots and fields and the washing of fruits and vegetables with contaminated waters, the repeated handling of product by local workers, and the propensity for environmental contamination of spices and other condiments during drying in unprotected facilities represent weak links that undermine product safety. Operational changes in favor of field irrigation with treated effluents, washing of fruits and vegetables with disinfected waters, education of local workers on the hygienic handling of fresh produce, and greater protection of products from environmental contamination during all phases of production and marketing would markedly enhance the bacterial quality and safety of these frequently ready-to-eat products.

FOODBORNE OUTBREAKS

National epidemiological registries continue to highlight the importance of *Salmonella* spp. as the leading cause of foodborne bacterial diseases in humans, in whom reported incidents of foodborne salmonellosis tend to dwarf those associated with other foodborne pathogens (Table 8.4). It is noteworthy that the problem of human salmonellosis from the consumption of contaminated foods generally remains on the increase worldwide (Table 8.5). Notwithstanding recent improvements in procedures for the epidemiological investigation of foodborne incidents in many countries (193), the global increases in foodborne salmonellosis

Table 8.4 Epidemiology of foodborne bacterial pathogens

Country	Period	Mean no. of reported incidents[a]							Reference
		Salmonella spp.	*Staphylococcus aureus*	*Clostridium perfringens*	*Campylobacter* spp.	*Bacillus cereus*	*Escherichia coli*	*Listeria* spp.	
Canada	1986–87	48.0	15.5	19.0	11.0	16.0	6.0[b]	0.0	94
Denmark	1985–89	5.2	2.4	5.8	0.4	2.0	0.2[c]	0.4	199
England and	1986–89	438.3	10.0	53.0	53.3	19.3	0.0	0.3	199
Wales	1989–91	922.0	8.3	50.7	5.0	24.7	1.0	0.0	180
Finland	1985–89	7.8	5.6	8.8	0.2	3.8	0.0	0.0	199
France	1993	200.0	37.0	32.0	0.0	3.0	0.0	0.0	130
Germany (former	1985–89	18.0	3.2	1.4	0.6	0.4	0.8[c,d]	0.4	199
Federal Republic	1991–92	71.0	1.0	1.0	0.0	0.5	0.0	0.5	202
of Germany)									
Hungary	1985–89	131.2	16.0	5.0	1.2	5.2	0.0	0.0	199
Japan[e]	1987–90	110.8	120.3	19.0	31.8	10.3	20.0	0.0	188
	1991	159.0	95.0	21.0	24.0	9.0	30.0[c]	0.0	144
Netherlands	1985–89	8.0	1.6	2.4	2.0	4.8	0.0	0.0	199
Romania	1985–89	20.6	10.2	0.0	0.0	0.2	0.4[c]	0.0	199
Scotland	1985–89	154.0	2.4	3.3	23.3	3.0	0.0	0.0	199
Spain	1985–89	467.6	34.6	5.0	0.4	0.6	3.4[c]	0.0	199
Sweden	1985–89	7.0	2.8	4.6	0.8	0.6	0.0	0.0	199
United States	1985–87	64.0	7.3	3.7	5.3	4.3	0.7	0.0	17

[a]Mean annual number of reported incidents where applicable.
[b]Enterohemorrhagic *E. coli* O157:H7.
[c]Generic reporting only.
[d]Human disease caused by *E. coli* and another bacterial pathogen.

Table 8.5 Trends in foodborne salmonellosis

Country	No. of reported incidents				No. of cases[a]	Reference(s)
	1985	1987	1989	1991		
Austria	124	151	440	—[b]	19.2–61.5	199
Canada	59	53	—	—	1.9–4.4	93, 94
Czechoslovakia	94	120	258	—	94.2–258.0	199
England and Wales	372	421	935	936	22.0–45.0	180, 199
France	7	178	462	—	NA[c]	199
Hungary	116	122	131	—	86.7–127.6	199
Japan	—	90	146	159	2.9–8.2	188
Poland	380	690	709	—	44.1–79.0	199
Scotland	133	180	151	—	35.0–46.0	199
United States	79	52	—	—	0.76–8.2	17

[a]Range of cases per 100,000 population within the review period.
[b], data not published.
[c]NA, Not available.

are considered real, not a result of enhanced surveillance programs and/or greater resourcefulness of medical and public health officials. Events such as the ongoing pandemic of egg-borne *S. enteritidis* clearly have an impact on current diseases statistics. Major outbreaks of foodborne salmonellosis in the last few decades are of interest because they underline the multiplicity of foods and *Salmonella* serovars that have been implicated in human disease (Table 8.6). In 1974, temperature abuse of egg-containing potato salad served at an outdoor barbecue led to an estimated 3,400 human cases of *S. newport* infection; cross-contamination of the salad by an infected food handler was suspected (103). The large Swedish outbreak of *S. enteritidis* PT4 in 1977 was attributed to the consumption of a mayonnaise dressing in a school cafeteria (98). In 1984, Canada experienced its largest outbreak of foodborne salmonellosis, which was attributed to the consumption of Cheddar cheese manufactured from heat-treated and pasteurized milk; the episode resulted in no fewer than 2,700 confirmed cases of *S. typhimurium* PT10 infection. Manual override of the flow diversion valve reportedly led to the entry of raw milk into vats of thermized and pasteurized cheese-milk (52). The following year witnessed the largest outbreak of foodborne salmonellosis in the United States, involving 16,284 confirmed cases of illness (121, 173). Although the cause of this outbreak was never ascertained, a cross-connection between raw and pasteurized milk lines seemingly was at fault. A large outbreak of salmonellosis affecting more than 10,000 Japanese consumers was attributed to a cooked egg dish (144). In 1993, paprika imported from South America was incriminated as the contaminated ingredient in the manufacture of potato chips distributed in Germany (124). The latest major outbreak of foodborne salmonellosis occurred in the United States and involved ice cream contaminated with *S. enteritidis.* The transportation of ice cream mix in an unsanitized truck that had previously carried raw eggs was identified as the source of contamination (9).

CHARACTERISTICS OF DISEASE

Symptoms and Treatment

Human *Salmonella* infections can lead to several clinical conditions, including enteric (typhoid) fever, uncomplicated enterocolitis, and systemic infections by nontyphoid microorganisms. Enteric fever is a serious human disease associated with the typhoid and paratyphoid strains which are particularly well adapted for invasion and survival within host tissues. Clinical manifestations of enteric fever appear after a period of incubation ranging from 7 to 28 days and may include diarrhea, prolonged and spiking fever, abdominal pain, headache, and prostration (46). Diagnosis of the disease relies on the isolation of the infective agent from blood or urine samples in the early stages of the disease or from stools after the onset of clinical symptoms (45). An asymptomatic chronic carrier state commonly follows the acute phase of enteric fever. The treatment of enteric fever is based on supportive therapy and/or the use of chloramphenicol, ampicillin, or trimethoprim-sulfamethoxazole to eliminate the systemic infection. Marked global increases in the resistance of typhoid and paratyphoid organisms to these antibacterial drugs in the last decade have greatly undermined their efficacy in human therapy. The problem is particularly serious in underdevel-

Table 8.6 Major foodborne outbreaks of human salmonellosis

Year	Country(ies)	Vehicle	Serovar	No.		Reference(s)
				Cases[a]	Deaths	
1973	Canada, United States	Chocolate	*S. eastbourne*	217	0	40, 49
1973	Trinidad	Milk powder	*S. derby*	3,000[b]	NS[c]	196
1974	United States	Potato salad	*S. newport*	3,400[b]	0	103
1976	Spain	Egg salad	*S. typhimurium*	702	6	7
1976	Australia	Raw milk	*S. typhimurium* PT9	>500	NS	177
1977	Sweden	Mustard dressing	*S. enteritidis* PT4	2,865	0	98
1981	The Netherlands	Salad base	*S. indiana*	600[b]	0	18
1981	Scotland	Raw milk	*S. typhimurium* PT204	654	2	36
1984	Canada	Cheddar cheese	*S. typhimurium* PT10	2,700	0	52
1984	France, England	Liver pâté	*S. goldcoast*	756	0	24, 154
1984	International	Aspic glaze	*S. enteritidis* PT4	766	2	27
1985	United States	Pasteurized milk	*S. typhimurium*	16,284	7	121
1987	Republic of China	Egg drink	*S. typhimurium*	1,113	NS	200
1987	Norway	Chocolate	*S. typhimurium*	361	0	113
1988	Japan	Cuttlefish	*S. champaign*	330	0	151
1988	Japan	Cooked eggs	*Salmonella* spp.	10,476	NS	144
1991	United States, Canada	Cantaloupes	*S. poona*	>400	NS	70
1991	Germany	Fruit soup	*S. enteritidis*	600	NS	79
1993	France	Mayonnaise	*S. enteritidis*	751	0	169
1993	Germany	Paprika chips	*S. saintpaul* *S. javiana* *S. rubislaw*	>670	0	124
1994	United States	Ice cream	*S. enteritidis*	>645	0	9
1994	Finland, Sweden	Alfalfa sprouts	*S. bovismorbificans*	492	0	160

[a]Confirmed cases unless stated otherwise.
[b]Estimated number of cases.
[c]NS, not specified.

oped countries, where multiple antibiotic-resistant salmonellae are frequently implicated in outbreaks of enteric fever and are the cause of unusually high fatality rates (29, 166).

Human infections with nontyphoid salmonellae commonly result in enterocolitis which appears 8 to 72 h after contact with the invasive pathogen. The clinical condition is generally self-limiting, and remission of the characteristic nonbloody diarrheal stools and abdominal pain usually occurs within 5 days of onset of symptoms. The successful treatment of uncomplicated cases of enterocolitis may require only supportive therapy such as fluid and electrolyte replacement. The use of antibiotics in such episodes is contraindicated because it tends to prolong the carrier state and the intermittent excretion of salmonellae (45). This asymptomatic persistence of salmonellae in the gut likely results from a marked antibiotic-dependent repression of native gut microflora which normally competes with salmonellae for nutrients and intestinal binding sites. Human infections with nontyphoid strains can also degenerate into systemic infections and precipitate various chronic conditions. In addition to *S. dublin* and *S. choleraesuis*, which exhibit a predilection toward septicemia, similarly high levels of virulence have been observed with other nontyphoid strains. Preexisting physiological, anatomical, and immunological disorders in human hosts can also favor severe and protracted illness through the inability of host defense mechanisms to respond effectively to the presence of invasive salmonellae (45). More frequent reports, in recent years, on the chronic and debilitating sequelae of nontyphoid systemic infections are of concern because the emerging pattern may be linked to increased levels of virulence among nontyphoid salmonellae, increased susceptibility of human populations to chronic bacterial diseases, or synergy between the two factors (34).

Salmonella-induced chronic conditions such as aseptic reactive arthritis, Reiter's syndrome, and ankylosing spondylitis are noteworthy. Bacterial prerequisites for the onset of these chronic diseases include the ability of the bacterial strain to infect mucosal surfaces, presence of outer membrane LPS, and a propensity to invade host cells (26, 179). Recent evidence indicates that development of these arthropathies is linked to a genetic predisposition in individuals that carry the class I HLA-B27 histocompatibility antigen. Other antigens encoded by the HLA-B locus, such as the B7, B22, B40, and B60 antigens, which serologically cross-react with the B27 antigen have also been associated with reactive arthritis (26, 137, 190). It is disconcerting that therapeutic eradication of a human *Salmonella* infection in the intestinal tract does not preclude the onset of chronic rheumatoid diseases in a distal, noninfected limb (190). Although the underlying mechanisms for these chronic conditions have yet to be fully elucidated, one of several theories suggests that phagocytosis of salmonellae results in the chemical alteration and slow phagocyte release of bacterial LPS. The altered LPS moieties are then reinserted into the cell membranes of synoviocytes and monocytes associated with articulations and synovial fluids, thereby rendering these surface-altered host cells susceptible to lysis by antibody and complement. This autoimmune response would result in rheumatoid tissue damage caused by a localized inflammation of host tissues and release of cytokines and enzymes from antigen-specific activated phagocytes (26, 179). Another suggested mechanism for the bacterial induction of arthropathies involves bacterial heat shock (stress) proteins. These molecular entities are synthesized in response to adverse environmental conditions and to the release of antibacterial substances such as the reactive oxygen metabolites produced during the oxidative burst of activated phagocytes (46). Limited data indicate that heat shock proteins from invasive pathogens trigger a proliferative response in synovial T lymphocytes of the host which, in turn, engenders rheumatoid disease (179).

Prophylactics

The major global impact of typhoid and paratyphoid salmonellae on human health led to the early development of parenteral vaccines consisting of heat-, alcohol-, or acetone-killed cells (54). The phenol heat-killed typhoid vaccine which continues to be widely used in protective immunity can precipitate adverse reactions including fever, headache, pain, and swelling at the site of injection. Although the need for immunogenic preparations against nontyphoid salmonellae could be argued, the development of prophylactics against the multiplicity of serovars and bacterial surface antigens in the global ecosystem, the rapid succession of serovars in human populations and in reservoirs of infection, and the unpredictable pathogenicity of infective strains have dampened research interest and hampered significant advances in this field of clinical sciences. Nevertheless, the pandemic of poultry- and egg-borne *S. enteritidis* infections that continues to afflict consumers in many countries underlines the potential benefit of specific vaccines for human and veterinary applications (45, 136, 171).

Live attenuated vaccines continue to generate great research interest because such preparations induce strong and durable humoral and cell-mediated responses in vaccinees (31, 43). The identification of bacterial genes wherein mutations lead to the attenuation of the carrier strain has greatly accelerated the development of live oral vaccines (31, 54). The avirulent mutant strain of *S. typhi* (Ty21a) is a product of chemical mutagenesis whereby the underlying molecular basis for attenuation is poorly understood. The absence of Vi antigens in the Ty21a strain and mutation in the *galE* locus which significantly reduces the number of biosynthetic LPS enzymes have been proposed as determinants for attenuation (31, 54). More recent evidence suggest that a mutation in the *rpoS* gene also contributes to the safety of this strain in humans (170). The *rpoS* gene, a component of the *Salmonella* virulence arsenal, directs the synthesis of numerous protective proteins in response to environmental stress conditions. The *rpoS* mutation in Ty21a may be linked to a single nucleotide change in genomic DNA and may result in a frameshift and gene transcription into an unstable RNA polymerase designated the σ^S factor (170). The Ty21a vaccine in the form of enteric coated capsules is manufactured by the Swiss Serum and Vaccine Institute (Berne) under the trade name Vivotif Berna. The preparation, which has successfully met the challenge of intense field testing in Chile and Egypt (60 and 96% efficacy, respectively), currently enjoys wide usage in European and North American medical communities. Immunization results from the administration of three capsules on alternate days and provides up to 3 years of protection.

An injectable vaccine prepared from the Vi capsular polysaccharide antigen of *S. typhi* was recently released by Pasteur Mérieux Sérums et Vaccins (Lyon, France) under the trade name Typhim Vi. A single dose of phenol-preserved antigen provides for a rapid rise in serum levels of Vi antibodies and confers protection for up to 3 years. In the presence of an active typhoid infection, the Vi antibodies facilitate the bactericidal action of serum complement (classical pathway). Two large-scale clinical studies in Nepal and eastern Transvaal (South Africa) confirmed vaccine efficacy, which ranged from 65 to 75% with low levels of adverse reactions (115). In contrast to the Ty21a vaccine, whose heterogeneous molecular composition predisposes recipients to stronger and varied secondary reactions, the use of a single, chemically defined antigen in Typhim Vi minimizes the risks of undesirable side effects. Developmental research may soon culminate in the marketing of new typhoid vaccines exhibiting greater levels of immunogenicity, biological safety, and storage stability (54). For example, a temperature-sensitive strain of *S. typhi* that grows optimally at 29°C but whose proliferation ceases at 37°C shows great potential as the immunogen in a novel live oral typhoid vaccine (21). Research at the National Institutes of Health (Bethesda, Md.) has also yielded promising results on the development of a conjugate vaccine consisting of O-acetylated pectin linked to a tetanus toxoid that raises antityphoid antibodies in mice (184a). The preparation is of particular interest as a human prophylactic because it is strongly immunogenic, is divalent, and carries no endotoxic moiety. Another rapidly evolving and productive field of research involves the use of attenuated *Salmonella* strains as vectors for the delivery of heterologous antigens to mammalian immune systems (31, 54). For example, the administration of a live attenuated *Salmonella* strain harboring the gene encoding fragment C of the tetanus toxin would provoke a potent immunological response against the carrier *Salmonella* strain and the tetanus toxin (31). Clearly, the field of vaccine development has witnessed great advances in recent years, and such success will undoubtedly continue as new knowledge on the molecular complexities of bacterial attenuation facilitates the delivery of defined, effective, and stable prophylactic preparations.

Antibiotic Resistance

Current global trends depicting sustained increases in the number of antibiotic-resistant *Salmonella* strains in humans and farm animals are most disquieting (123, 146, 191). The liberal administration of antimicrobial agents in hospitals and other treatment centers has led to the emergence and persistence of resistant strains. The widespread use and abuse of poorly controlled antibiotics in the meat animal and aquaculture industries is unacceptable (107, 150, 198). In the United States, estimates based on available data indicate that almost half of the 44 million lb (20 million kg) of antibiotics produced in 1986 were used in animal husbandry. A large proportion (≤90%) of these antibiotics were administered at subtherapeutic levels (107). Although the wisdom and merits of subtherapeutic (prophylactic) and growth-promoting antibiotic regimens in animal husbandry remain a highly controversial issue, recent evidence suggests that such treatments facilitate the emergence and persistence of drug-resistant *Salmonella* strains and other foodborne bacterial pathogens in farm animals and derived-meat products (45, 153). Moreover, the magnitude of the problem is amplified by the current inability of slaughtering plants to effect significant pathogen reductions on meat carcasses and to eliminate potential sources of product cross-contamination within the plant environment (198). The importance of these sobering realities is reflected in several epidemiological investigations in which human salmonellosis attributed to antibiotic-resistant strains was linked to the use of antimicrobial agents in farm animals (101, 181). Inappropriate antibiotic treatment of a septicemic patient infected with a resistant *Salmonella* strain could lead to a fatal outcome if the therapeutic drug administered upon hospitalization subsequently proved to be identical to the resistance phenotype (186). Such an occurrence would likely stem from the antibiotic-dependent inactivation of resident microflora in the intestinal tract and from the rapid colonization, growth, and dissemination into deeper tissues of antibiotic-resistant salmonellae to a point beyond therapeutic management (48, 51).

The genetic determinants of antibiotic resistance are encoded in the bacterial chromosome or in cytoplasmic plasmids (46, 150). Resistance (R) plasmids are autonomous and self-replicating fragments of DNA whose transcription products confer resistance to the carrier cell. The propensity for the conjugative intra- and intergeneric transfer of R plasmids carrying single or multiple resistance codons between compatible cells sharing the same ecological niche is a serious public health issue. Such exchanges of genetic materials can favor the emergence and spread of foodborne *Salmonella* spp. that are resistant to one or more medically important therapeutic agents such as chloramphenicol and trimethoprim-sulfamethoxazole. Current practices in animal husbandry further amplify the resistance problem by effectively selecting for resistant strains through the inclusion of subtherapeutic doses of antibiotics in feed rations. For example, feeding a tetracycline-supplemented formulation to a *Salmonella*-infected meat animal could select for strains that carry R plasmids that encode resistance not only to tetracycline but also to several other antibacterial agents. The consumer product would then bear the markings of this unheralded experiment in gene selection. It is widely held that the major problems of tetracycline, streptomycin, and sulfa drug resistance currently encountered in meat-borne salmonellae and other bacterial pathogens are the legacy of sectors of the agricultural industry that continue to practice animal prophylaxis using subtherapeutic doses of antimicrobial agents (45, 51, 107, 150, 198).

Several antibiotics such as bacitracin, virginiamycin, and apramycin are considered bona fide veterinary drugs because they are reportedly devoid of any therapeutic value in human medicine. However, recent evidence indicates that the agricultural use of apramycin may lead to the emergence of gentamicin-resistant *Salmonella* strains in treated livestock (158, 191). The phenomenon, which stems from the linkage of the apramycin and gentamicin resistance codons on a single R plasmid, leads to a new realization that on-farm use of any antibiotics may, in good time, engender major public health problems. The consideration that genetic determinants for *Salmonella* virulence and antibiotic resistance can occur on the same plasmid dramatically increases the disease potential of this emerging bacterial elite, precursors to the "super bug" era (174, 192). In the next few decades (or sooner), we may well witness the international scientific community feverishly engaged in the development of new drugs to replace those currently in use, and whose spectra of activity would then have slipped into obsolescence in the face of a rising tide of highly virulent and multiple antibiotic-resistant *Salmonella* strains. Most of the foregoing considerations also apply to the aquaculture industry, whose prominence in the global fish and shellfish market has soared as a result of the rapid depletion of feral stocks and increasing consumer demand for fish and shellfish products (48, 183). The aqueous environment and warm climates generally associated with aquacultural operations promote the proliferation of *Salmonella* spp. in rearing ponds and basins and provide favorable conditions for the exchange of R plasmids and cross-contamination of reared species. Moreover, the addition of subtherapeutic doses of antibiotics of human significance into holding facilities together with the sedimentation of detritus and medicated feeds provide a uniquely favorable habitat for the selection and growth of antibiotic-resistant salmonellae.

Recent increases in the rates of human isolation of chloramphenicol-, ampicillin-, and trimethoprim-sulfamethoxazole-resistant *Salmonella* spp. have necessitated a major shift in the antibiotic treatment of human systemic salmonelloses and chronic carriers. Although expanded-spectrum cephalosporins currently retain some level of prominence in the treatment of systemic infections, medical interest is rapidly turning toward fluoroquinolones for the timely and successful management of deeply seated *Salmonella* infections (13, 25). The efficacy of this class of drugs derives from the inhibition of the active (A) subunits of the bacterial gyrase enzyme, which is essential for the replication and supercoiling of genomic DNA (13). The future of these novel therapeutic agents is uncertain, as they are being overshadowed by increased isolations of fluoroquinolone-resistant *Salmonella* strains from human and animal sources (88, 109). The bacterial resistance to fluoroquinolones can arise from a chromosomal mutation that alters the N terminus of the gyrase A subunit or from compositional changes in the OMP porins which reduce the influx of the antibiotic into the bacterial cytoplasm (197). Current husbandry practices in the international meat and aquaculture industries are sustaining the momentum of the emerging problem of quinolone resistance. The consequences of the wide usage of fluoroquinolones in veterinary settings are already visible, as evidenced by the increasing prevalence of resistant *Salmonella* and *Campylobacter* spp. in livestock (48, 57, 88, 109). The predictable epilog to the aquacultural use of these invaluable human drugs for prophylactic and therapeutic purposes has yet to be written (48). Recent reports on the cross-resistance to nalidixic acid and fluoroquinolones in *Salmonella* and *Campylobacter* spp. intimate that difficult times lie ahead (88, 109).

The use of medically important drugs at subtherapeutic levels in the meat and aquacultural industries should be reassessed in terms of their purported value and of their potentially negative impact on public health and on the safety of the global food supply (51, 107, 201).

INFECTIOUS DOSE

It is well established that newborns, infants, the elderly, and immunocompromised individuals are more susceptible to *Salmonella* infections than healthy adults (45). The incompletely developed immune system in newborns and in infants, the frequently weak and/or delayed immunological responses in the elderly and debilitated persons, and the generally low gastric acid production in infants and seniors facilitate the intestinal colonization and systemic spread of salmonellae in these segments of the population (23, 46). Antibiotic treatment of subjects prior to their encounter with salmonellae enhances bacterial virulence through an antibiotic-mediated clearance of native gut microflora which reduces the level of bacterial competition for nutrients and attachment sites in the intestinal tract of the host (101, 181).

Detailed investigations of foodborne outbreaks have indicated that the ingestion of just a few *Salmonella* cells can be infectious (Table 8.7). From early reports that large numbers of salmonellae inoculated into eggnog

Table 8.7 Human infectious doses of *Salmonella* spp.[a]

Food	Serovar	Infectious dose	Reference
Eggnog	*S. meleagridis*	10^6–10^7	138
	S. anatum	10^5–10^7	
Goat cheese	*S. zanzibar*	10^5–10^{11}	168
Carmine dye	*S. cubana*	10^4	120
Imitation ice cream	*S. typhimurium*	10^4	11
Chocolate	*S. eastbourne*	10^2	49
Hamburger	*S. newport*	10^1–10^2	66
Cheddar cheese	*S. heidelberg*	10^2	67
Chocolate	*S. napoli*	10^1–10^2	87
Cheddar cheese	*S. typhimurium*	10^0–10^1	52
Chocolate	*S. typhimurium*	$\leq 10^1$	113
Paprika potato chips	*S. saintpaul*	$\leq 4.5 \times 10^1$	124
	S. javiana		
	S. rubislaw		

[a]Adapted from reference 48.

and fed to human volunteers produced overt disease (138), more recent evidence suggests that 1 to 10 cells can constitute a human infectious dose (52, 113). Determinant factors in salmonellosis are not limited to the immunological heterogeneity within human populations and to the virulence of infecting strains but may also include the chemical composition of incriminated food vehicles. A common denominator of the foods associated with low infectious doses (Table 8.7) is the high fat content in chocolate (cocoa butter), cheese (milk fat), and meat (animal fat). Entrapment of salmonellae within hydrophobic lipid micelles may possibly afford protection against the bactericidal action of gastric acidity. Following a bile-mediated dispersion of the lipid moieties in the duodenum, the viable salmonellae would resume their infectious course in search of suitable points of attachment in the lower portion of the small intestine (colonization). The rapid emptying of gastric contents could also provide an alternate mechanism for the successful infection of susceptible hosts. The swift passage of a liquid bolus through an empty stomach would minimize bacterial exposure to gastric acidity and sustain the migration of viable salmonellae in the intestinal tract (23).

Recent publications on the dynamics of human *Salmonella* infections are of singular interest (84, 85). An in-depth epidemiological study of a large outbreak of *S. typhimurium* involving chicken served to delegates at a medical conference showed that the clinical course in patients was directly related to the number of ingested salmonellae (85). The incubation period for the onset of symptoms was inversely related to the infectious dose. Patients with short (≤22-h) periods of incubation suffered more frequent diarrheal bowel movements, higher maximum body temperatures, greater persistence of clinical symptoms, and greater frequency of hospitalization. Interestingly, no association between the age of infected individuals and the length of the incubation period was noted. Similar findings were reported in retrospective dose-response studies of foodborne salmonellosis (84, 143).

The compelling evidence that ingestion of only a few *Salmonella* cells can develop into a variety of clinical conditions and deteriorate into septicemia and even death highlights the unpredictable pathogenicity of this large and heterogeneous group of human bacterial pathogens. Food producers, processors, and distributors need to be reminded that low levels of salmonellae in a finished food product can lead to serious public health consequences and undermine the reputation and economic viability of the incriminated food manufacturer (45).

PATHOGENICITY AND VIRULENCE FACTORS

Specific and Nonspecific Human Responses

The presence of viable salmonellae in the human intestinal tract confirms the successful evasion of ingested organisms from nonspecific host defenses. Antibacterial lactoperoxidase in saliva, gastric acidity, mucoid secretions from intestinal goblet cells, intestinal peristalsis, and sloughing of luminal epithelial cells synergistically oppose bacterial colonization of the intestinal mucosa. In addition to these constitutive hurdles to bacterial infection, the antibacterial actions of nonspecific phagocytic cells (neutrophils, macrophages, monocytes) coupled with the immune responses associated with specific T and B lymphocytes, the epitheliolymphoid tissues (Peyer's patches), and the classical or alternative pathways for complement inactivation of invasive pathogens mount a formidable defense against the systemic spread of salmonellae. The interplay between these immunological defense factors was recently reviewed (46).

The human diarrheagenic response to foodborne salmonellosis results from the migration of the pathogen in the oral cavity to intestinal tissues and mesenteric lymph follicles (enterocolitis). The event coincides with bacterial enterotoxin production, extensive leukocyte influx into the infected tissues, increased mucus secretion by goblet cells, and mucosal inflammation triggered by the leukocytic release of prostaglandins. The latter occurrence also activates the adenyl cyclase in intestinal epithelial cells, resulting in increased fluid secretion into the intestinal lumen (65, 159). The failure of host defense systems to hold the invasive salmonellae in check can degenerate into septicemia and other chronic clinical conditions.

Attachment and Invasion

The establishment of a human *Salmonella* infection rests on the ability of the organism to attach to (colonization) and enter (invasion) intestinal columnar epithelial cells (enterocytes) and specialized M cells overlying Peyer's patches. Salmonellae must successfully compete with indigenous gut microflora for suitable attachment sites on the luminal surface of the intestinal wall and evade capture by secretory immunoglobulin A that may also be present on the surface of epithelial cells. Colonization of enterocytes arises, in part, from the interaction of bacterial type 1 (mannose-sensitive) or type 3 (mannose-resistant) fimbriae, surface adhesins, nonfimbriate (mannose-resistant) hemagglutinins, or enterocyte-induced polypeptides with host glycoprotein receptors located on the microvilli or glycocalyx of the intestinal surface (46, 159). Although the role of *Salmonella* motility in the adherence and invasion processes is uncertain, motility may not be essential for bacterial internalization and may simply increase the frequency of productive contacts between the pathogen and its targeted epithelial cell (73). An authoritative review on the morphology of interactions between salmonellae and the enterocytes, M cells, and immune apparatus in the mucosal and submucosal layers of the host intestine is recommended as additional reading (159). Fine-structure studies have also shown that proteinaceous appendages develop on the surface of salmonellae upon contact with epithelial cells (73, 83). The ATPase and translocase enzymes encoded by the *invC* and *invG* loci, respectively, may contribute to the formation of these structures (37, 73) (Fig. 8.1). The inability of mutants defective in the assembly of these appendages to enter cultured epithelial cells suggests further that these bacterial organelles are essential for invasion (83). The appendages are 0.3 to 1.0 nm in length and ca. 60 nm in diameter, considerably thicker than flagella (ca. 20 nm) and type 1 fimbriae (ca. 7 nm). The assembly of *Salmonella* appendages, which may be linked to the *invI* and *invJ* loci, is energy dependent but independent of de novo bacterial protein synthesis (37, 83). The appendages are short-lived and shed concomitantly with the appearance of membrane ruffles on colonized epithelial cells (83). Following bacterial attachment, signal transduction between the pathogen and host cell culminates in an energy-dependent *Salmonella* invasion of enterocytes and M cells (75, 83). Although bacterial protein synthesis is not a prerequisite for the onset of invasion, its prolonged inhibition compromises the ability of salmonellae to enter cultured epithelial cells (133). The role of the *invE* locus in *Salmonella* spp. is of singular importance in the invasive mechanism because it triggers two profound changes in enterocytes and M cells: a Ca^{2+} influx and cytoskeleton rearrangement in the targeted host cells (Fig. 8.1). The cytoskeleton of eukaryotic cells is a highly organized network of contractile and supportive filaments consisting of actin or other protein elements that define the apical-basal polarity of intestinal epithelial cells and control the movement of intracellular organelles (65, 159). Uncharacterized gene products from the *invE* of an adherent *Salmonella* cell promote the influx of luminal Ca^{2+} into the mammalian cell (Table 8.8). This ionic translocation engenders a polymerization of host cell actin into microfilaments in the vicinity of the invading pathogen (81) (Fig. 8.1). Histologically, the latter event is seen as an evagination of the apical cytoplasm of epithelial cells (i.e., membrane ruffle) around the adherent salmonellae which subsequently mediates the pinocytotic uptake of the bacterial pathogen (111). Several experimental findings lend support to this proposed mechanism of *Salmonella* invasion. Growth of *S. typhimurium* in Ca^{2+}-supplemented medium increases the invasive capacity of the strain for HeLa cells (155). Mutants in the *invE* locus cannot trigger the influx of luminal Ca^{2+} into epithelial cells, a prerequisite for the polymerization of actin and cytoskeleton rearrangement (68, 75). Other work has shown that disruption of actin microfilaments with cytochalasins B and D preempts the *Salmonella* invasion of host cells (73). A similar end result was observed with intraepithelial chelators of Ca^{2+}

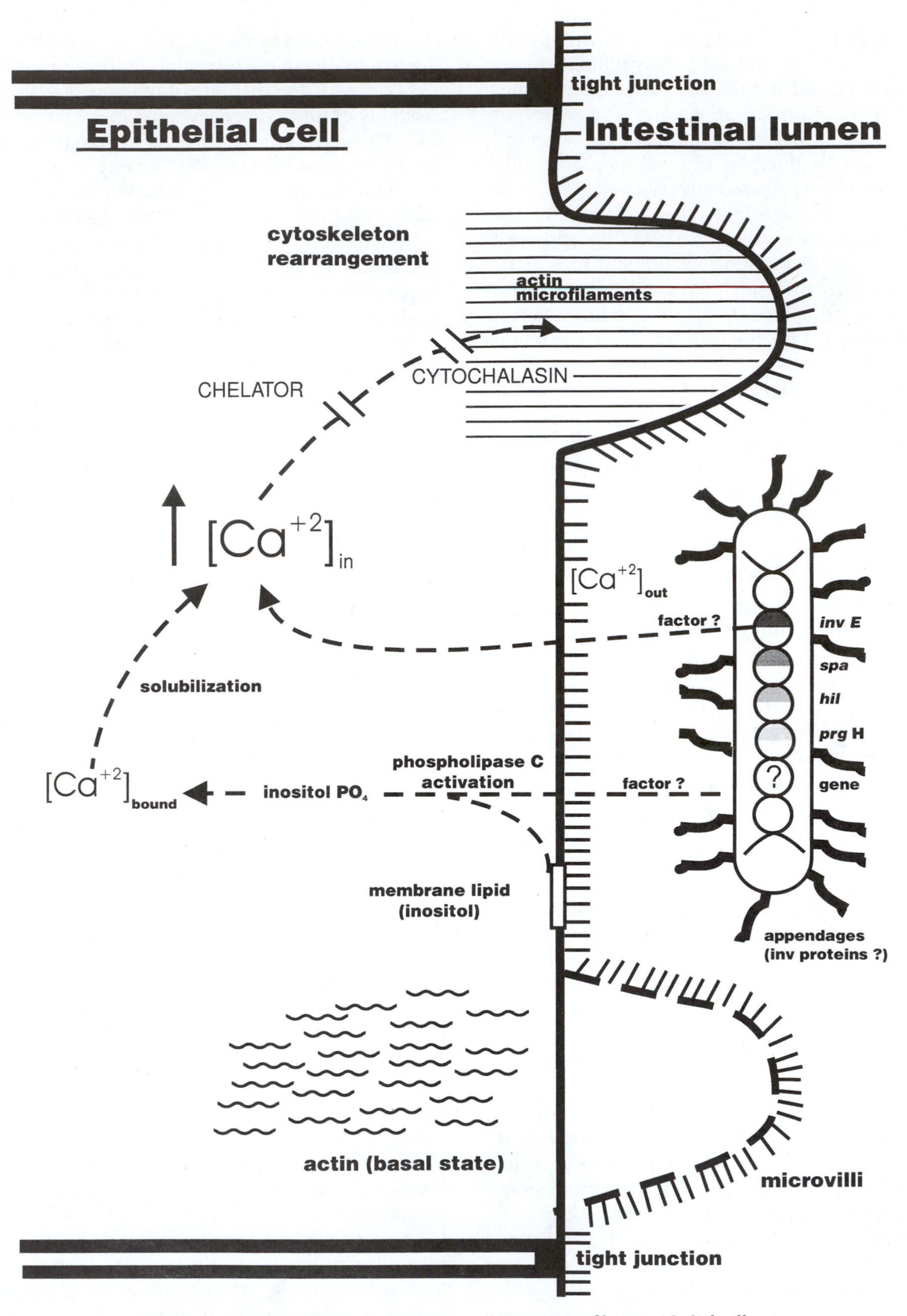

Figure 8.1 Mechanism for the *Salmonella* invasion of host epithelial cells.

Table 8.8 Characteristics of *Salmonella* invasion genes

Locus	Mutant phenotype[a]		Gene product			Reference(s)
	Adherence	Invasion	Size (kDa)	Identity[b]	Putative function	
invA	+	–	75.9	Translocase	Translocates InvE proteins involved in membrane ruffle formation	74
invB	+	+	14.8	Unknown	Unknown	56
invC	+	–	47.4	ATPase[c]	Energizes *invG*-dependent translocation of InvJ proteins for assembly of surface appendage	37, 56
invD	NA[d]	NA	30.0	Glycine decarboxylase	Unknown	72, 142
invE	+	–	42.8	Unknown	Regulates Ca^{2+} influx into host cells required for membrane ruffle formation	81
invF	+	–	24.4	Unknown	Transcriptional regulator	112
invG	+	–	62.3	Translocase	Translocation of InvJ required for surface appendage formation	37, 112
invH	(–)[e]	(–)[e]	16.5	Adhesin	Encodes membrane protein for adhesion and invasion	4, 112
invI	+	–	18.1	Synthetase	Assembly of surface appendages	37
invJ	+	–	36.4	Unknown	Assembly of surface appendages	37

[a]Expression of invasion and adherence phenotypes retained (+) or reduced or lost (-) in mutant strains.
[b]Putative gene product.
[c]Experimentally confirmed gene product.
[d]NA, not available.
[e]Serovar dependent.

ions (65, 75). Salmonellae can also activate a second host-pathogen system to increase the intracellular Ca^{2+} ion concentration in epithelial cells and enhance the polymerization of actin filaments. Through gene products that remain unknown, adherent salmonellae stimulate phospholipase C activity in the host cells. This enzyme activation results in the breakdown of lipids in the plasma membrane of host cells and in the intracytoplasmic release of inositol triphosphate (Fig. 8.1) which, in turn, triggers the solubilization of Ca^{2+} bound in the cytoplasmic matrix (172). Therefore, two distinct systems, the *invE*-dependent translocation of Ca^{2+} from the external medium into the epithelial cell and the inositol phosphate pathway, synergistically contribute to the high cytoplasmic levels of Ca^{2+} required for actin polymerization and subsequent formation of membrane ruffles that are essential for the internalization of salmonellae into epithelial cells.

We have thus far seen that salmonellae attach to the microvilli of the intestinal epithelium, an intercellular contact that activates salmonellae into the production of transient, proteinaceous appendages deemed essential for *Salmonella* invasion. These events are closely followed by the *invE*- and inositol phosphate-dependent concentration of Ca^{2+} within the cytoplasm of host cells, which results in actin polymerization and formation of membrane ruffles. The appearance of membrane ruffles coincides with the disappearance of the bacterial appendages and with the pathogen-directed aggregation of host plasma membrane proteins, predominantly class I major histocompatibility complex heavy-chain proteins. The latter protein aggregates are incorporated into the membrane of infected pinocytotic vacuoles to camouflage the intravacuolar presence of a foreign bacterial antigen, thereby preempting host cell-mediated immune responses (77). Clearly, such a deceptive but seemingly effective molecular ploy enhances the virulence of the invasive *Salmonella* spp. The completed internalization of salmonellae into the epithelial cell precipitates a reversion of actin microfilaments and membrane ruffles to their basal states.

The invasion (*inv*), the surface presentation of antigens (*spa*), the hyperinvasive (*hil*), and the *phoP*-repressed (*prgH*) regions in the *Salmonella* chromosome are contiguous and functionally related loci located within a 40- to 50-kb segment of chromosomal DNA (i.e., pathogenicity island) that encode determinant factors for the facilitated entry of salmonellae into host cells (89, 142). The *inv* genome is a multigenic locus consisting of no fewer than 15 genes (*invA* to -*O*). Many of these loci code for the synthesis of virulence determinants or enzymes responsible for the translocation of virulence determinants to the surface of host cells (Table 8.8). It should be noted that several *inv* loci represent new designations for former *spa* genes, including *spaM* (*invI*), *spaN* (*invJ*), and *spaP* (*invL*) (37, 82). Although the *inv* genes were originally thought to be unique to *Salmonella* spp., more recent evidence has shown that the nucleotide sequences of the *invA* and *invE* loci of *Salmonella* spp. can also be found in homologous genes of *Yersinia* spp. (89). Lesions in the *inv* locus markedly reduce the ability of salmonellae to invade cultured epithelial cells but have no effect on bacterial adherence to host cells (73).

The *invA*, *-B*, and *-C* genes are contiguous on the *Salmonella* chromosome and lie within the same transcriptional unit, whereas *invD* is located upstream from this unit (71, 142). Evidence also suggests that these four genomic traits are commonly encountered in the genus *Salmonella* (72) (Table 8.1). Environmental factors such as high osmolarity and low pO_2 enhance bacterial invasiveness by altering the superhelicity of chromosomal DNA, which consequently affects the level of transcription of invasion-related genes such as *invA* (59, 75). Lesions in the *invA* and *invC* loci but not in *invB* and *invD* severely impede the invasiveness of carrier strains (56, 71). Moreover, mutations in *invA* to *-D* did not significantly alter the ability of salmonellae to attach to host cells (37, 56, 71, 73) (Table 8.8). The invasion phenotype is restored in *invA* mutants upon coinfection of epithelial cells with the mutant and wild-type strains (74). Similar results were obtained upon coinfection with *Shigella flexneri* containing the *mxiA* gene, a structural and functional homolog of the *Salmonella invA* locus (73, 82). Recent work has also shown that mutations in the *invA* gene translate into an inability of the mutant strain to increase the Ca^{2+} levels in mammalian cells and to induce cytoskeletal rearrangement and actin polymerization, which are essential for the internalization of salmonellae (74) (Fig. 8.1). The foregoing considerations suggest that *invA* triggers signal transduction between the invasive pathogen and the targeted epithelial cell. Evidence that the phenotypes of *invA* and *invE* mutants are very similar, the adjacent positions of these two loci on the *Salmonella* chromosome (i.e., *invGEABC*), and the finding that the activity of the *invA* locus could not be complemented by coinfection of host cells with *invA* and *invE* mutants suggest that the two genes are associated with the same entry pathway. It is further speculated that *invA* encodes a factor for the translocation of the *invE* gene product which actively participates in the import of external Ca^{2+} into epithelial cells (Table 8.8 and Fig. 8.1).

The roles of the *invC*, *-G*, and *-J* loci are equally important in *Salmonella* invasiveness, as demonstrated by a functional linkage between these genes in providing for the synthesis and translocation of the InvJ protein (37). Perhaps the *invC* locus encodes a functional ATPase that provides the necessary energy for the *invG*-mediated translocation of the InvJ protein from the bacterial cytoplasm to the external environment (37, 56) (Table 8.8). Mutations in *invC* or *invG* preclude the translocation of InvJ, whereas mutations in *invJ* produce a noninvasive phenotype. Although the effector function of InvJ remains elusive, it is tempting to speculate that this secreted protein constitutes a structural component of the previously described surface appendages that form upon contact of the pathogen with epithelial cells, or that InvJ actively participates in the assembly of such appendages (37, 73). Limited data on the *invI* locus suggest that the gene product may be a synthetase that also participates in the formation of *Salmonella* surface appendages (37) (Table 8.8).

The impact of the highly conserved *invH* gene on *Salmonella* adherence and invasion of host tissues appears to be serovar dependent, given that mutations in this locus markedly reduced the invasive capacity of host-adapted serovars such as *S. typhi*, *S. choleraesuis*, and *S. gallinarum* (4). Levels of adhesion varied widely among mutant strains. Interestingly, the InvH protein appeared to be restricted to the bacterial membrane fraction and could not be detected in the periplasm or the culture supernatant. These findings suggest that the *invH* gene may translate into a surface adhesin necessary for *Salmonella* invasion of epithelial cells. In contrast to other *inv* loci, the *invH* gene seemingly is unique to *Salmonella* species and could not be identified in related enteric bacteria (4).

The foregoing discussion indicates that the *inv* locus plays a determinant role in the attachment to and invasion of mammalian epithelial cells by salmonellae. The *inv* region maps at 59 min on the *Salmonella* chromosome and consists of no fewer than 15 genomes located within a 40- to 50-kb segment that also carries *hil* and *spa* genetic determinants for invasion. Recent characterization of several *inv* loci has produced the following gene order: *invD.....HFGEABCIJ* (4, 37, 142). The transcription of *invF* to *-J* progresses from left to right, whereas *invH* is transcribed in the opposite direction (4, 112). The precise location and direction of transcription of the *invD* locus have yet to be ascertained (72). The functional linkages between *invA* and *-E* in the formation of membrane ruffles and Ca^{2+}-dependent internalization of *Salmonella* spp. and linkage of *invC*, *-G*, and *-J* in the assembly of bacterial surface appendages are notable. *Salmonella* species are particularly well endowed with stable chromosomal determinants to effectively adhere to targeted cells, a prerequisite for the invasion of human and animal hosts and the onset of acute or chronic disease. The importance of surface pili (fimbriae) in the enhancement of *Salmonella* adhesion cannot be minimized (45, 46).

Growth and Survival within Host Cells

In contrast to several bacterial pathogens such as *Yersinia*, *Shigella*, and enteroinvasive *Escherichia coli* strains, which replicate within the cytoplasm of host cells, salmonellae are confined to endocytotic vacuoles wherein bacterial replication begins within hours following internalization (75). The infected vacuoles translocate from the apical to the basal pole of the host cell, where salmonellae are released into the lamina propria (108, 159). During their proliferation within epithelial vacuoles, salmonellae also induce the formation of stable filamentous structures within the epithelial cytosol (78). These organelles, which contain lysosomal membrane glycoproteins and acid phosphatase, are seemingly connected to the infected vacuoles. The formation of these filaments requires viable salmonellae within the membranous vacuoles and is blocked by inhibitors of vacuolar acidification. Although the role of these induced filamentous structures is poorly understood, preliminary findings point to their involvement in the intravacuolar replication of *Salmonella* spp. Recent work has tentatively identified a bacterial surface mechanism that would facilitate the migration of salmonellae into deeper layers of tissues upon its release into the lamina propria. The demonstrated ability of thin aggregrate fimbriae on the outer surface of *Salmonella* cells to bind host plasminogen and the tissue-type plasminogen activator could markedly increase the invasiveness of infecting strains. The zymogen would be converted into its proteolytic (plasmin) form on the bacterial surface, thereby providing salmonellae with an effective means to breach host tissue barriers and facilitate transcytosis into deeper tissues (178).

The systemic migration of salmonellae exposes the organism to phagocytosis by macrophages, monocytes, and polymorphonuclear leukocytes and to the antibacterial conditions that prevail in the cytoplasm of these host defense cells (46). For example, the membrane oxidase-dependent formation of toxic oxygen products such as singlet oxygen, superoxide anion, hydrogen peroxide, and hydroxyl radicals during the oxidative metabolic burst of phagocytes is countered by the protective activity of several bacterial enzymes, including superoxide dismutase, peroxidase, and catalase. Recent findings have also shown that 30 bacterial proteins are synthesized in response to the toxic oxygen products of macrophages. Of these, nine are dependent on the synthesis of a nucleic acid activator encoded by the chromosomal *oxyR* locus, which induces the transcription of genetic loci responsible for the synthesis of protective proteins (162).

The *phoP*/*phoQ* regulon is a two-component transcriptional regulator system that enables survival of salmonellae within the hostile environment of phagocytes, notably the high acidity within phagolysosomes (host cell construct of an infected phagosome with incorporated lysosomal granules), and release of antibacterial defensins by phagocytic cells (139, 162). Defensins, which also occur in epithelial cells, are small, nonspecific cationic peptides that inactivate salmonellae by inserting into the outer bacterial membrane, thereby creating transmembrane channels that increase bacterial permeability to ions and precipitate cell death (63). The antidefensin gene product(s) encoded by the *Salmonella phoP*/*phoQ* system remains elusive (90). Nonetheless, the importance of this gene product is reflected in the inability of *phoP* or *phoQ* mutants to resist the antibacterial action of defensins. The product of the *phoQ* gene is a kinase that reportedly senses the hostile phagolysosomal environment by means of a short chain of 20 of its 487 amino acid residues that extends into the periplasmic space of the *Salmonella* cell envelope. Interestingly, half of the amino acids in this periplasmic chain are acidic and form an anionic box that putatively functions as a receptor for the cationic defensins or as a sensor of highly acidic conditions in the phagocyte. For example, an interaction between defensins and the PhoQ kinase anionic box could trigger the autophosphorylation of the kinase sensor at a histidine residue, followed by a kinase-dependent transfer of phosphate to an aspartate residue in the amino terminus of the PhoP transcriptional activator protein (139, 140). The phosphorylated PhoP protein would then activate the transcription of several *phoP*-activated genes, including *pagA*, *pagB*, *pagC*, *psiD*, and the *phoN* locus, which encodes the periplasmic nonspecific acid phosphatase. A recent study on the effect of defensins on *pag* gene activation failed to support the putative role of the PhoQ anionic box in the acquired *Salmonella* resistance to defensins (141). Current evidence also indicates that the unlinked *pagA*, *pagB*, and *psiD* loci are not involved in *Salmonella* virulence, whereas the *pagC* gene encodes an OMP that promotes survival within macrophages. Although the amino acid sequence of PagC is similar to that of the *ail* gene product in *Yersinia enterocolitica*, this structural homology between the two gene products does not extend to functional homology, as evidenced by the loss of invasiveness in *ail* but not *pagC* mutants (162). The transcription of *pagC* within the macrophage is maximal at pH 4.9, and its protein product reportedly contributes to serum resistance but affords no protection against defensins or highly acidic environments (139, 195). Interestingly, transcription of the *pag* genes is induced within acidified *Salmonella*-infected phagocytes but not in infected epithelial vacuoles (3). In addition to the five previously mentioned *phoP*-activated genes, 13 new positively regulated loci (*pagD* to *-P*) were recently identified (20). Genetic determinations using transposon insertions have suggested that these newly identified *pag* genes do not contribute to the *phoP/phoQ*-dependent resistance to defensins, whereas the *pagD, pagJ, pagK,* and *pagM* loci participate in mouse virulence and in bacterial survival in macrophages.

In contrast to the *pag* genes that are expressed under adverse environmental conditions such as low pH, nutrient deficiency, and presence of defensins and during the stationary phase of growth, the *phoP/phoQ* system also regulates the expression of *prg* (*phoP*-repressed) genes which are induced under nonstress conditions (19). Stated differently, conditions that activate *pag* genes generally repress *prg* expression. Recent work identified the unlinked *prgA, -B, -C, -E,* and *-H* genes that encode 20 protective polypeptides and whose repression likely occurs at the level of transcription. Preliminary findings indicate that the *prgH* locus plays a determinant role in the invasion of host epithelial cells by salmonellae (19) (Fig. 8.1). Other proteins arising from the transcription of *prg* genes are required for the diffuse membrane ruffling of macrophages, increased pinocytosis, and formation within macrophages of spacious phagosomes wherein slow acidification favors greater bacterial survival and attendant virulence (2). It is noteworthy that only a limited number of *phoP*-activated genes contribute to *Salmonella* virulence (195) and that mutations in this regulon (*pagC, -D, -J, -K,* and *-M*) result in phenotypes exhibiting decreased survival within macrophages, increased susceptibility to acidic pH, serum complement (*pagC*), and defensins, and reduced invasiveness of epithelial cells (*prgH*).

Other Virulence Factors

The virulence of *Salmonella* spp. as reflected in the ability to cause acute and chronic diseases in humans and in a variety of animal hosts stems from structural and physiological attributes that act synergistically or independently in promoting bacterial adhesion and invasiveness (45, 46). The previously discussed ATR is a good example of *Salmonella* physiological responsiveness to adverse environmental conditions which, concurrently, potentiates greater bacterial virulence in hosts and greater acid tolerance in fermented foods. In addition to the *inv* locus and *phoP/phoQ* system, which, respectively, encode determinants for *Salmonella* invasiveness and resistance to the antibacterial conditions within phagocytes, several other virulence factors affect *Salmonella* pathogenicity.

Virulence Plasmids

Virulence plasmids are large cytoplasmic DNA structures that replicate independently from the bacterial chromosome. These autonomous organelles contain many virulence loci ranging from 30 to 60 MDa in size, and they occur with a frequency of one to two copies per chromosome (91). The presence of virulence plasmids within the genus *Salmonella* is limited and has been confirmed in *S. typhimurium, S. dublin, S. gallinarum-pullorum, S. enteritidis, S. choleraesuis,* and *S. abortusovis* (91). The absence of a virulence plasmid in the host-adapted and highly infectious *S. typhi* is noteworthy (46, 91). Gene products from the transcription of this plasmid potentiate systemic spread and infection of extraintestinal tissues but have no effect on *Salmonella* adhesion and invasion of epithelial and M cells (45). More specifically, it is suggested that virulence plasmids enable carrier strains to rapidly multiply within host cells and overwhelm host defense mechanisms (119, 187). The *Salmonella* virulence plasmids contain highly conserved nucleotide sequences and exhibit functional homology in that plasmid transfer from a wild-type serovar to another plasmid-cured serovar restores the virulence of the recipient strain (46, 91).

The *Salmonella* plasmid virulence (*spv*) region consists of a gene cluster that encodes products for the prolific growth of salmonellae in host reticuloendothelial tissues. This genetic entity was formerly identified as *mka* (mouse killing agent), *mkf* (mouse killing factor), or *vir* (virulence) (162). Signals that trigger *spv* transcription include the hostile environment within host phagocytes, iron limitation, elevated temperatures, low pH, and nutrient deprivation associated with the stationary phase of growth (97, 187). The *spv* regulon is approximately 8.0 kb in size and contains at least five genes (*spvR, -A, -B, -C,* and *-D*) and two principal promoters, one for the *spvR* locus and another for the *spvABCD* transcriptional unit (92). These loci are transcribed in a single direction under the combined regulatory action of the *spvR* gene product and a chromosomal *rpoS* (formerly *katF*)-encoded sigma (σ^S) factor (97). The proteinaceous sigma factor, a subunit within the RNA polymerase dedicated to *spvR* transcription, recognizes the *spvR* promoter region and activates transcription of the *spv* regulon (119). The SpvR protein positively regulates the expression of the downstream *spvABCD* transcriptional unit through the *spvA* promoter sequence (91, 184). The role of the gene products encoded by the *spvABCD* genes in *Salmonella* virulence remains elusive (91). The nucleotide sequences of the *spv* loci are highly conserved within the genus *Salmonella* (184, 187). Mutations in this region strongly attenuate or inactivate the ability of salmonellae to establish deeply seated infections. Recent evidence suggests that minor differences in the nucleotide sequences of the *spvR* loci in *S. dublin* and *S. typhimurium* markedly alter the capacity of the *spvR* gene to induce the *spvA* promoter and transcription of the *spvABCD* loci. These findings provide some insight on the molecular basis for the comparatively greater virulence of *S. dublin* in humans (187).

Siderophores

Siderophores are yet another facet of the *Salmonella* virulence armamentarium. These elements retrieve essential iron from host tissues to drive key cellular functions such as the electron transport chain and enzymes associated with iron cofactors (41). To this end, salmonellae must compete with host transferrin, lactoferrin, and ferritin ligands for available iron (41, 46). For example, transferrin scavenges tissue fluids for Fe^{3+} ions to form Fe^{3+}-transferrin complexes that bind to surface host cell receptors. Upon internalization, the complexes dissociate and the released Fe^{3+} is complexed with ferritin for intracellular storage (Fig. 8.2). In response to a limited availability of Fe^{3+} in host tissue, salmonellae sequester Fe^{3+} ions by means of a high-affinity phenolate enterochelin (also designated enterobactin), consisting of a cyclic trimer of dihydroxybenzoic acid and L-serine, and/or a low-affinity hydroxamate aerobactin chelator, an anabolic product derived from one citrate and two lysine (one hydroxylated, one acetylated) residues (147, 162). The synthesis of these bacterial siderophores is

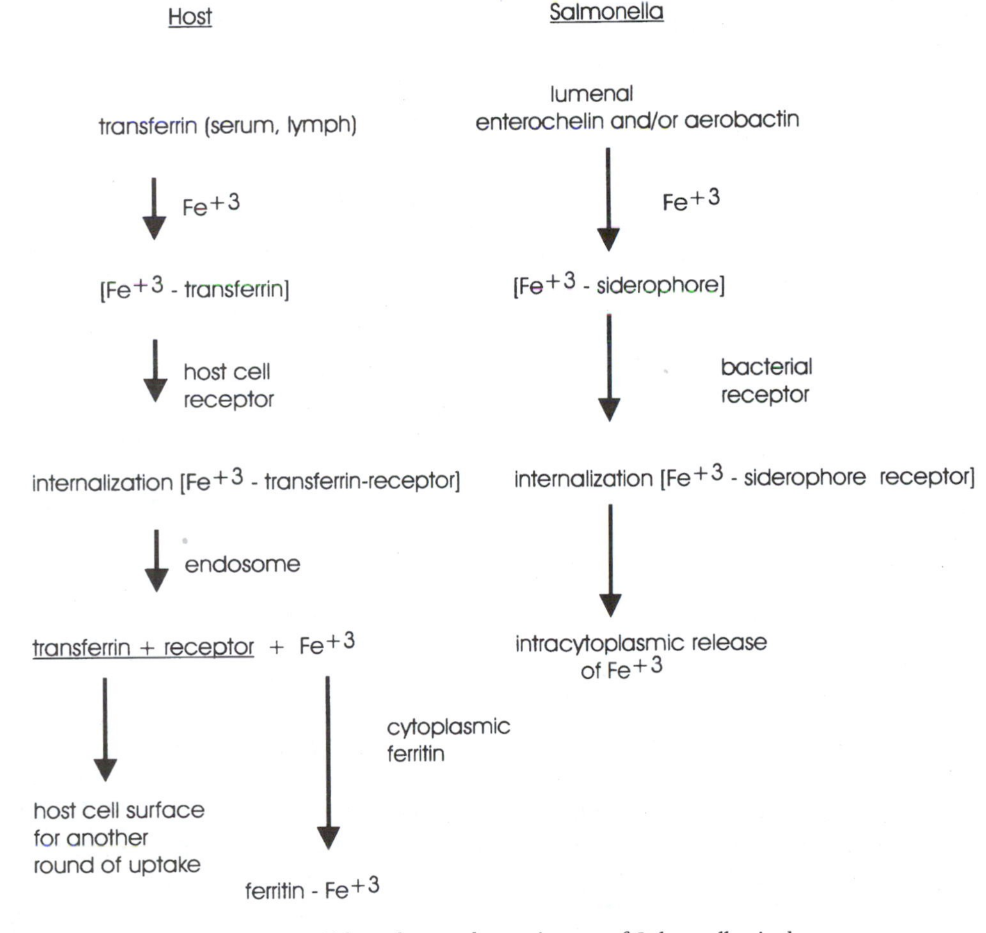

Figure 8.2 Siderophores, determinants of *Salmonella* virulence.

chromosomally regulated by the *fur* gene (41). The *Salmonella* binding of trivalent iron proceeds with the interaction of a ferri-siderophore complex with an OMP receptor that was induced in response to limiting concentrations of intracellular iron. The complex is then transposed into the bacterial cytoplasm, where the ferric moiety is reduced to the ferrous state. The low affinity of the siderophore for Fe^{2+} results in the release of the divalent ion into the bacterial cytoplasm for subsequent use in key metabolic functions (41) (Fig. 8.2). It is notable that the degree of *Salmonella* virulence is directly related to the enterochelin content of the infecting strain. Experimental insight into this relationship follows from the reduced virulence of auxotrophic strains deficient in the aromatic pathway which is responsible for the formation of dihydrobenzoic acid, the precursor of enterochelin (45).

Toxins

Diarrheagenic enterotoxin figures prominently as a *Salmonella* virulence factor because of its ability to produce overt clinical symptoms in human cases of salmonellosis. The release of toxin into the cytoplasm of infected host cells precipitates an activation of adenyl cyclase localized in the epithelial cell membrane and a marked increase in the cytoplasmic concentration of cyclic AMP in host cells. The concurrent fluid exsorption into the intestinal lumen results from a net secretion of Cl^- ions in the crypt regions of the intestinal mucosa and depressed Na^+ absorption at the level of the intestinal villi (46, 182). Enterotoxigenicity is a virulence phenotype that prevails in *Salmonella* serovars, including *S. typhi*, and is expressed within hours following bacterial contact with the targeted host cells (45, 61).

Salmonella enterotoxin is a thermolabile protein with a molecular mass of 90 to 110 kDa (46). The narrow pH range (6.0 to 8.0) for active enterotoxin suggests that the delivery of a functional toxin may be impaired or inhibited in acidic phagolysosomes (pH 4.5 to 5.0) and in epithelial endosomes (pH 5.0 to 6.5). The ability of enterotoxin to bind to GM_1 ganglioside receptors on host cell surfaces remains a controversial issue (45, 46, 61, 165). The *Salmonella* enterotoxin is encoded by a 6.3-kb chromosomal gene (*stx*) which regulates the synthesis of three proteins with molecular masses of 45, 26, and 12 kDa (32). Detailed studies on the quaternary structure of cholera toxin have confirmed an A (active) and B (binding) subunit structure in which A activates the host cell adenylate cyclase and B binds to the epithelial GM_1 ganglioside (102). The reactivity of the three *Salmonella* enterotoxin-related proteins with monospecific antisera raised against cholera A and B subunits and the identical electrophoretic gel mobility patterns of enterotoxin and cholera A and B subunits suggest that the *Salmonella* enterotoxin presents a structural configuration similar to that in *Vibrio* spp. (32, 64). The lack of cross-reactivity between cholera antitoxin and *Salmonella* enterotoxin (165), contrary to findings of earlier studies (32, 45, 46), together with the recent demonstration that the chromosomal enterotoxin (*stx*) and cholera toxin (*ctx*) genes exhibit weak DNA sequence homology (32, 33), brings into question the close antigenic and genetic relatedness of the two toxins. Genetic coding of enterotoxin seemingly is not limited to the bacterial chromosome, as evidenced by a plasmid-mediated transfer of enterotoxigenicity (110). Additional studies with *Salmonella* mutants harboring defective toxin genes would further elucidate the role of enterotoxin in intestinal infection.

In addition to expressing enterotoxin, *Salmonella* strains generally produce a thermolabile cytotoxic protein of 56 to 78 kDa which is localized in the bacterial outer membrane (12, 46). The cytotoxin is not inactivated with antisera raised against Shiga toxin or *E. coli* Shiga toxins 1 and 2, as determined in green monkey kidney (Vero) and HeLa cell bioassays (12, 53). These findings are at variance with an early report on the inactivation of *Salmonella* cytotoxin with Shiga antitoxin (46). Moreover, DNA hybridization studies also support the nonidentity of *Salmonella* cytotoxin and Shiga toxins (12). Maximum production of cytotoxin in laboratory media occurs at pH 7.0 and 37°C during the early stationary phase of growth. Hostile environments such as acidic pH and elevated (42°C) temperature precipitate an extracelluar release of toxin, possibly as a result of induced bacterial lysis (53). The virulence attribute of cytotoxin stems from its inhibition of protein synthesis and lysis of host cells, thereby promoting the dissemination of viable salmonellae into host tissues (46, 118). The ability of added Ca^{2+} ions to block the cytotoxin-dependent disruption of cell monolayers suggests that cytotoxin may function as a chelator of divalent cations that normally contribute to the structural integrity of monolayers (155).

Vi Antigen, LPS, and Porins

Three virulence determinants located within or on the external surface of the *Salmonella* outer membrane will be discussed briefly. The capsular polysaccharide Vi antigen occurs in most strains of *S. typhi*, in a few strains of *S. paratyphi* C, and rarely in *S. dublin* (162). The Vi antigen significantly increases the virulence of carrier strains by inhibiting the opsonization of the C3b host complement factor to surface LPS, a critical event in the induction of macrophage phagocytosis of invasive salmonellae (46). Two genes *(viaA* and *-B)* are associated with the formation of the Vi antigen. However, the *viaB* locus, which encodes no fewer than six proteins, appears to be the key genetic element in the control of Vi antigen synthesis, as evidenced by the presence of *viaA* in members of the *Enterobacteriaceae* and in *Salmonella* strains that do not express the Vi antigen (162). Recent data further indicate that the *ompR* locus also affects Vi capsule formation in *S. typhi* (156).

The length of serotypic LPS that protrudes from the bacterial outer membrane not only defines the rough (short LPS) and smooth (long LPS) phenotypes but also plays an important role in repelling the potentially lytic attack of the host complement system (46). In effect, the LPSs of smooth variants sterically hinder the stable insertion of the C5b-9 complement factor into the inner cytoplasmic membrane that would precipitate bacteriolysis. The general inability of the short LPS in rough variants to protect against the C5b-9 lytic insertion renders such variants more susceptible to lysis and consequently less virulent (46). The primary carbohydrate composition of LPS can also affect the level of serum complement activation. For example, an isogenic pair of *Salmonella* spp., one containing the somatic B (4,12) and one containing C (6,7) LPS antigens, activated serum complement at slow and rapid rates, respectively (176). Correspondingly, serogroup B exhibits greater virulence than serogroup C.

Porins are OMPs that function as transmembrane (outer membrane) channels in regulating the influx of nutrients, antibiotics, and other small-molecule species (42, 149). To date, four porins, encoded by the *ompF, ompC, ompD,* and *phoE* genes, have been identified in *S. typhimurium* (149). Gene transcription of the *ompF* and *ompC* loci is induced by changes in the microenvironment. Low osmolarity, low nutrient availability, and low temperature can trigger the transcription of the *ompF* gene, with a concomitant repression of the *ompC* locus. Conversely, favorable environmental conditions induce the transcription of the *ompC* gene and repression of the *ompF* locus (42, 145, 149). This is analogous to the previously described *phoP*/*phoQ*-dependent activation of *pag* genes under adverse conditions and concomitant repression of *prg* genes (19). The transmembrane channels formed by OmpF and OmpC are 1.1 to 1.3 nm in diameter and consist of trimers formed from monomeric subunits. The *envZ* gene product activates OmpR through a phosphorylation-dependent mechanism, whereas the activated *ompR* locus regulates the transcription of the *ompF* and *-C* loci (42, 145). This mechanism is reminiscent of the previously discussed *phoP*/*phoQ* regulon, where phosphorylation of *phoP* by the *phoQ*-encoded kinase activates the transcription of *pag* genes. Mutations in the *ompR*/*envZ* (designated *ompB*) regulon dramatically attenuate the virulence of carrier strains; in contrast, mutations in the *ompC, -F,* or *-D* locus have little effect on *Salmonella* virulence (156). The reliability of attenuated *ompB* mutants as vaccine candidates has yet to be established. Interestingly, reports on the ability of *Salmonella* porins to elicit immunological host defense responses suggest that porins are of limited importance in *Salmonella* pathogenicity (46, 185). Recent evidence indicates that the *rck* (resistance to complement killing) locus, which is located on the *Salmonella* virulence plasmid, encodes gene products that protect both smooth and rough variants against the lytic C5b-9 complement factor (95, 96). Nucleotide sequencing of *rck* and amino acid analysis of the Rck protein showed homology with the *phoP*/*phoQ*-dependent *pagC* locus and gene product (96).

CONCLUSION

Salmonella spp. continue as the leading cause of foodborne bacterial diseases. The situation has endured because of the ubiquity of salmonellae in the natural environment and their prevalence in many sectors of the global food chain (45). Intense animal husbandry practices and major difficulties in controlling the spread of salmonellae in the vertically integrated meat and poultry production and processing industries continue to render raw poultry and meats principal vehicles of human foodborne salmonellosis. The widespread administration of prophylactic doses of medically important antibiotics to reared animal species may promote on-farm selection of antibiotic-resistant strains and markedly increase the human health risks associated with handling and consumption of contaminated meat products. A similar scenario can also be drawn for the expanding aquacultural industry, where intensive rearing of fish and shellfish in generally unprotected facilities, together with a liberal use of subtherapeutic regimens of antibiotic treatment, is common practice. The growing worldwide popularity of fluoroquinolones as prophylactic drugs in the agricultural and aquacultural industries is of serious concern because such a husbandry approach may undermine the clinical efficacy of these novel and invaluable drugs (45, 46). Reports on the acquired resistance of foodborne *Salmonella* and *Campylobacter* spp. to fluoroquinolones are already published (45). Moreover, the propensity for bacterial cross-resistance to fluoroquinolones adds yet another dimension to this problem (88, 109). The importance of *Salmonella* vehicles other than raw poultry, meats, and derived products cannot be minimized. The incidence of human salmonellosis associated with the consumption of fresh fruits and vegetables, spices, chocolate, and milk products (Table 8.6) highlights the importance of sanitary practices during the harvesting, processing, and distribution of raw foods and food ingredients (44, 48).

The arsenal of virulence factors that enable *Salmonella* species to evade the various antibacterial host defense mechanisms is remarkable and disconcerting. The physiological adaptability of salmonellae to hostile conditions in the natural environment safeguards their survival and infectious potential (Table 8.3). In addition, the microorganism benefits from chromosome- and plasmid-encoded virulence determinants that facilitate its attachment to and invasion of host cells (*inv, prgH*) (Fig. 8.1), resistance to intraphagocyte acidity (*phoP*/*phoQ*), complement lysis (Vi antigen, LPS, *phoP*/*phoQ* and *rck*), and antibacterial substances (*ompF, ompC, ompD, phoE*), widespread invasion of deep host tissues (*spv*), competition for available Fe^{3+} (siderophores), and toxin production (enterotoxin, cytotoxin).

It is clear that the occurrence of salmonellae in the global food chain and its current and projected repercussions on human health are cause for concern. Unless significant changes in agricultural and aquacultural practices are implemented, human foodborne salmonellosis will prevail in the next century.

References

1. **Airoldi, A. A., and E. A. Zottola.** 1988. Growth and survival of *Salmonella typhimurium* at low temperature in nutrient deficient media. *J. Food Sci.* **53**:1511–1513.
2. **Alpuche-Aranda, C. M., E. L. Racoussin, J. A. Swanson, and S. I. Miller.** 1994. *Salmonella* stimulate macrophage macropinocytosis and persist within spacious phagosomes. *J. Exp. Med.* **179**:601–608.
3. **Alpuche-Aranda, C. M., J. A. Swanson, W. P. Loomis, and S. I. Miller.** 1992. *Salmonella typhimurium* activates virulence gene transcription within acidified macrophage phagosomes. *Proc. Natl. Acad. Sci. USA* **89**:10079–10083.
4. **Altmeyer, R. M., J. K. McNern, J. C. Bossio, I. Rosenshine, B. B. Finlay, and J. E. Galan.** 1993. Cloning and molecular characterization of a gene involved in *Salmonella* adherence and invasion of cultured epithelial cells. *Mol. Microbiol.* **7**:89–98.
5. **Andrews, W. H., V. R. Bruce, G. June, F. Satchell, and P. Sherrod.** 1992. *Salmonella*, p. 51–69. *In* L. A. Tomlinson (ed.), *Bacteriological Analytical Manual*, 7th ed. AOAC International, Arlington, Va.
6. **Angelotti, R., M. J. Foter, and K. H. Lewis.** 1961. Time-temperature effects on salmonellae and staphylococci in foods. *Am. J. Public Health* **51**:76–88.
7. **Anonymous.** 1977. *Aviation Catering. Report of Working Group.* World Health Organization Regional Office for Europe, Copenhagen.
8. **Anonymous.** 1986. Microbiological safety of vacuum-packed, high salt foods questioned. *Food Chem. News* **28**: 30–31.
9. **Anonymous.** 1995. Ice-cream firm reaches tentative *Salmonella* case agreement. *Food Chem. News* **36**:53.
10. **Anonymous.** 1995. Aquaculture cited by ASM as problem for antibiotic resistance. *Food Chem. News* **37**: 20–21.
11. **Armstrong, R. W., T. Fodor, G. T. Curlin, A. B. Cohen, G. K. Morris, W. T. Martin, and J. Feldman.** 1970. Epidemic *Salmonella* gastroenteritis due to contaminated imitation ice cream. *Am. J. Epidemiol.* **91**:300–307.
12. **Ashkenazi, S., T. G. Cleary, B. E. Murray, A. Wanger, and L. K. Pickering.** 1988. Quantitative analysis and partial characterization of cytotoxin production by *Salmonella* strains. *Infect. Immun.* **56**:3089–3094.
13. **Asperilla, M. A., R. A. Smego, Jr., and L. K. Scott.** 1990. Quinolone antibiotics in the treatment of *Salmonella* infections. *Rev. Infect. Dis.* **12**:873–889.
14. **Asplund, K., and E. Nurmi.** 1991. The growth of salmonellae in tomatoes. *Int. J. Food Microbiol.* **13**:177–182.
15. **Aznar, R., C. Amaro, E. Alcaide, and M. L. Lemos.** 1989. Siderophore production by environmental strains of *Salmonella* species. *FEMS Microbiol. Lett.* **57**:7–12.
16. **Baker, R. C., R. A. Qureshi, and J. H. Hotchkins.** 1986. Effect of an elevated level of carbon dioxide containing atmosphere on the growth of spoilage and pathogenic bacteria. *Poult. Sci.* **65**:729–737.
17. **Bean, N. H., P. M. Griffin, J. S. Goulding, and C. B. Ivey.** 1990. Foodborne disease outbreak, 5-year summary, 1983–1987. *J. Food Prot.* **53**:711–728.
18. **Beckers, H. J., M. S. M. Daniels-Bosman, A. Ament, J. Daenen, A. W. J. Hanekamp, P. Knipschild, A. H. H. Schuurmann, and H. Bijkerk.** 1985. Two outbreaks of salmonellosis caused by *Salmonella indiana*. A survey of the European Summit outbreak and its consequences. *Int. J. Food Microbiol.* **2**:185–195.
19. **Behlau, I., and S. I. Miller.** 1993. A PhoP-repressed gene promotes *Salmonella typhimurium* invasion of epithelial cells. *J. Bacteriol.* **175**:4475–4484.
20. **Belden, W. J., and S. I. Miller.** 1994. Further characterization of the PhoP regulon: identification of new PhoP-activated virulence loci. *Infect. Immun.* **62**:5095–5101.
21. **Bellanti, J. A., B. J. Zeligs, S. Vetro, Y. H. Pung, S. Luccioli, M. J. Malavasic, A. M. Hooke, T. R. Ubertini, R. Vanni, and L. Nencioni.** 1993. Studies of safety, infectivity and immunogenicity of a new temperature-sensitive (ts) 51-1 strain of *Salmonella typhi* as a new live oral typhoid fever vaccine candidate. *Vaccine* **11**:587–590.
22. **Bergis, H., G. Poumeyrol, and A. Beaufort.** 1994. Etude de développement de la flore saprophyte et de *Salmonella* dans les viandes hachées conditionnées sous atmosphère modifiée. *Sci. Aliments* **14**:217–228.

23. **Blaser, M. J., and L. S. Newman.** 1982. A review of human salmonellosis. 1. Infective dose. *Rev. Infect. Dis.* **4**:1096–1106.
24. **Bouvet, E., C. Jestin, and R. Ancelle.** 1986. Importance of exported cases of salmonellosis in the revelation of an epidemic, p. 303. *In Proceedings of the Second World Congress on Foodborne Infections and Intoxications.* Institute of Veterinary Medicine, Berlin.
25. **Bryan, J. P., H. Rocha, and W. M. Scheld.** 1986. Problems in salmonellosis: rationale for clinical trials with newer β-lactam agents and quinolones. *Rev. Infect. Dis.* 8:189–203.
26. **Bunning, V. K., R. B. Raybourne, and D. L. Archer.** 1988. Foodborne enterobacterial pathogens and rheumatoid disease. *J. Appl. Bacteriol. Symp. Suppl.* **17**:87S–107S.
27. **Burslem, C. D., M. J. Kelly, and F. S. Preston.** 1990. Food poisoning—a major threat to airline operations. *J. Soc. Occup. Med.* **40**:97–100.
28. **Caffer, M. I., and T. Eiguer.** 1994. *Salmonella enteritidis* in Argentina. *Int. J. Food Microbiol.* **21**:15–19.
29. **Carmen Palomino, W., R. Lucia Aguad, L. Manuel Rodriguez, G. Graciela Cofre, and J. Villanueva.** 1986. Clinical course of infections by *Salmonella typhi*, paratyphus A and paratyphus B in relation to the sensitivity of the etiological agent to chloramphenicol. *Rev. Med. Chile* **114**:919–927.
30. **Catsaras, M., and D. Grebot.** 1984. Multiplication des *Salmonella* dans la viande hachée. *Bull. Acad. Vet. France* **57**:501–512.
31. **Chatfield, S., M. Roberts, P. Londono, I. Cropley, G. Douce, and G. Dougan.** 1993. The development of oral vaccines based on live attenuated *Salmonella* strains. *FEMS Immunol. Med. Microbiol.* **7**:1–8.
32. **Chopra, A. K., C. W. Houston, J. W. Peterson, R. Prasad, and J. J. Mekalanos.** 1987. Cloning and expression of the *Salmonella* enterotoxin gene. *J. Bacteriol.* **169**: 5095–5100.
33. **Chopra, A. K., and J. W. Peterson.** 1994. Molecular characterization of *Salmonella* enterotoxin. Presented at the 7th International Congress of Bacteriology and Applied Microbiology Division, Prague, Czechoslovak Republic.
34. **Christmann, D., T. Staub, and Y. Hansmann.** 1992. Manifestations extra-digestives des salmonelloses. *Med. Mal. Infect.* **22**:289–298.
35. **Chung, K. C., and J. M. Goepfert.** 1970. Growth of *Salmonella* at low pH. *J. Food Sci.* **35**:326–328.
36. **Cohen, D. R., I. A. Porter, T. M. S. Reid, J. C. M. Sharp, G. I. Forbes, and G. M. Paterson.** 1983. A cost benefit study of milk-borne salmonellosis. *J. Hyg.* **91**:17–23.
37. **Collazo, C. M., M. K. Zierler, and J. E. Galan.** 1995. Functional analysis of the *Salmonella typhimurium* invasion genes *inv* I and *inv* J and identification of a target of the protein secretion apparatus encoded in the *inv* locus. *Mol. Microbiol.* **15**:25–38.
38. **Corry, J. E. L.** 1971. The water relations and heat resistance of microorganisms, p. 1–42. *In Scientific and Technical Survey no. 73.* The British Food Manufacturing Industries Research Association, Leatherhead, Surrey, U.K.
39. **Corry, J. E. L.** 1976. The safety of intermediate moisture foods with respect to *Salmonella*, p. 215–238. *In* R. Davies, G. G. Birch, and K. J. Parker (ed.), *Intermediate Moisture Foods.* Applied Science Publishers Ltd., London.
40. **Craven, P. C., D. C. Mackel, W. B. Baine, W. H. Barker, E. J. Gangarosa, M. Goldfield, H. Rosenfeld, R. Altman, G. Lachapelle, J. W. Davies, and R. C. Swanson.** 1975. International outbreak of *Salmonella eastbourne* infection traced to contaminated chocolate. *Lancet* **i**:788–793.
41. **Crosa, J. H.** 1989. Genetics and molecular biology of siderophore-mediated iron transport in bacteria. *Microbiol. Rev.* **53**:517–530.
42. **Csonka, L. N.** 1989. Physiological and genetic responses of bacteria to osmotic stress. *Microbiol. Rev.* **53**:121–147.
43. **Curtiss, R., III, S. M. Kelly, and J. O. Hassan.** 1993. Live oral avirulent *Salmonella* vaccines. *Vet. Microbiol.* **37**:397–405.
44. **D'Aoust, J.-Y.** 1977. *Salmonella* and the chocolate industry. A review. *J. Food Prot.* **40**:718–727.
45. **D'Aoust, J.-Y.** 1989. *Salmonella*, p. 327–445. *In* M. P. Doyle (ed.), *Foodborne Bacterial Pathogens.* Marcel Dekker, Inc., New York.
46. **D'Aoust, J.-Y.** 1991. Pathogenicity of foodborne *Salmonella*. *Int. J. Food Microbiol.* **12**:17–40.
47. **D'Aoust, J.-Y.** 1991. Psychrotrophy and foodborne *Salmonella*. *Int. J. Food Microbiol.* **13**:207–216.
48. **D'Aoust, J.-Y.** 1994. *Salmonella* and the international food trade. *Int. J. Food Microbiol.* **24**:11–31.
49. **D'Aoust, J.-Y., B. J. Aris, P. Thisdele, A. Durante, N. Brisson, D. Dragon, G. Lachapelle, M. Johnston, and R. Laidley.** 1975. *Salmonella eastbourne* outbreak associated with chocolate. *Can. Inst. Food Sci. Technol. J.* **8**:181–184.
50. **D'Aoust, J.-Y., and U. Purvis.** 1995. *Isolation and Identification of Salmonella from Foods.* MFHPB-20. Health Protection Branch, Health Canada, Ottawa.
51. **D'Aoust, J.-Y., A. M. Sewell, E. Daley, and P. Greco.** 1992. Antibiotic resistance of agricultural and foodborne *Salmonella* isolates in Canada: 1986–1989. *J. Food Prot.* **55**:428–434.
52. **D'Aoust, J.-Y., D. W. Warburton, and A. M. Sewell.** 1985. *Salmonella typhimurium* phage-type 10 from cheddar cheese implicated in a major Canadian foodborne outbreak. *J. Food Prot.* **48**:1062–1066.
53. **Dewanti, R., and M. P. Doyle.** 1992. Influence of cultural conditions on cytotoxin production by *Salmonella enteritidis*. *J. Food Prot.* **55**:28–33.
54. **Dougan, G.** 1994. The molecular basis for the virulence of bacterial pathogens: implications for oral vaccine development. *Microbiology* **140**:215–224.
55. **Droffner, M. L., and N. Yamamoto.** 1992. Procedure for isolation of *Escherichia, Salmonella*, and *Pseudomonas* mutants capable of growth at the refractory temperature of 54°C. *J. Microbiol. Methods* **14**:201–206.
56. **Eichelberg, K., C. C. Ginocchio, and J. E. Galan.** 1994. Molecular and functional characterization of the *Salmonella typhimurium* invasion genes *invB* and *invC:* homology of *invC* to the F_0F_1 ATPase family of proteins. *J. Bacteriol.* **176**:4501–4510.

57. **Endtz, H. P., G. J. Ruijs, B. van Klingeren, W. H. Jansen, T. van der Reyden, and R. P. Mouton.** 1991. Quinolone resistance in campylobacter isolated from man and poultry following the introduction of fluoroquinolones in veterinary medicine. *J. Antimicrob. Chemother.* **27:**199–208.
58. **Ewing, W. H.** 1972. The nomenclature of *Salmonella*, its usage, and definitions for the three species. *Can. J. Microbiol.* **18:**1629–1637.
59. **Falkow, S., R. R. Isberg, and D. A. Portnoy.** 1992. The interaction of bacteria with mammalian cells. *Annu. Rev. Cell Biol.* **8:**333–363.
60. **Fantasia, M., and E. Filetici.** 1994. *Salmonella enteritidis* in Italy. *Int. J. Food Microbiol.* **21:**7–13.
61. **Fernandez, M., J. Sierra-Madero, H. de la Vega, M. Vazquez, Y. Lopez-Vidal, G. M. Ruiz-Palacios, and E. Calva.** 1988. Molecular cloning of a *Salmonella typhi* LT-like enterotoxin gene. *Mol. Microbiol.* **2:**821–825.
62. **Ferreira, M. A. S. S., and B. M. Lund.** 1987. The influence of pH and temperature on initiation of growth of *Salmonella* spp. *Lett. Appl. Microbiol.* **5:**67–70.
63. **Fields, P. I., E. A. Groisman, and F. Herron.** 1989. A *Salmonella* locus that controls resistance to microbicidal proteins from phagocytic cells. *Science* **243:**1059–1060.
64. **Finkelstein, R. A., B. A. Marchlewicz, R. J. McDonald, and M. Boesman-Finkelstein.** 1983. Isolation and characterization of a cholera-related enterotoxin from *Salmonella typhimurium. FEMS Microbiol. Lett.* **17:**239–241.
65. **Finlay, B. B.** 1994. Molecular and cellular mechanisms of *Salmonella* pathogenesis. *Curr. Top. Microbiol. Immunol.* **192:**163–185.
66. **Fontaine, R. E., S. Arnon, W. T. Martin, T. M. Vernon, E. J. Gangarosa, J. J. Farmer, A. B. Moran, J. H. Silliker, and D. L. Decker.** 1978. Raw hamburger: an interstate common source of human salmonellosis. *Am. J. Epidemiol.* **107:**36–45.
67. **Fontaine, R. E., M. L. Cohen, W. T. Martin, and T. M. Vernon.** 1980. Epidemic salmonellosis from cheddar cheese: surveillance and prevention. *Am. J. Epidemiol.* **111:** 247–253.
68. **Foster, J. W., and B. Bearson.** 1994. Acid-sensitive mutants of *Salmonella typhimurium* identified through a dinitrophenol lethal screening strategy. *J. Bacteriol.* **176:** 2596–2602.
69. **Foster, J. W., and H. K. Hall.** 1991. Inducible pH homeostasis and the acid tolerance response of *Salmonella typhimurium. J. Bacteriol.* **173:**5129–5135.
70. **Francis, B. J., J. V. Altamirano, M. G. Stobierski, W. Hall, B. Robinson, S. Dietrich, R. Martin, F. Downes, K. R. Wilcox, C. Hedberg, R. Wood, M. Osterholm, G. Genese, M. J. Hung, S. Paul, K. C. Spitalny, C. Whalen, and J. Spika.** 1991. Multistate outbreak of *Salmonella poona* infections—United States and Canada, 1991. *Morbid. Mortal. Weekly Rep.* **40:**549–552.
71. **Galan, J. E., and R. Curtiss III.** 1989. Cloning and molecular characterization of genes whose products allow *Salmonella typhimurium* to penetrate tissue culture cells. *Proc. Natl. Acad. Sci. USA* **86:**6383–6387.
72. **Galan, J. E., and R. Curtiss III.** 1991. Distribution of the *invA, -B, -C,* and *-D* genes of *Salmonella typhimurium* among other *Salmonella* serovars: *invA* mutants of *Salmonella typhi* are deficient for entry into mammalian cells. *Infect. Immun.* **59:**2901–2908.
73. **Galan, J. E., and C. Ginocchio.** 1994. The molecular genetic bases of *Salmonella* entry into mammalian cells. *Biochem. Soc. Trans.* **22:**301–306.
74. **Galan, J. E., C. Ginocchio, and P. Costeas.** 1992. Molecular and functional characterization of the *Salmonella* invasion gene *invA:* homology of InvA to members of a new protein family. *J. Bacteriol.* **174:**4338–4349.
75. **Garcia-del Portillo, F., and B. B. Finlay.** 1994. Invasion and intracellular proliferation of *Salmonella* within nonphagocytic cells. *Microbiologia SEM* **10:**229–238.
76. **Garcia-del Portillo, F., J. W. Foster, and B. B. Finlay.** 1993. Role of acid tolerance response genes in *Salmonella typhimurium* virulence. *Infect. Immun.* 61:4489–4492.
77. **Garcia-del Portillo, F., M. G. Pucciarelli, W. A. Jeffries, and B. B. Finlay.** 1994. *Salmonella typhimurium* induces selective aggregation and internalization of host cell surface proteins during invasion of epithelial cells. *J. Cell Sci.* **107:**2005–2020.
78. **Garcia-del Portillo, F., M. B. Zwick, K. Y. Leung, and B. B. Finlay.** 1994. Intracellular replication of *Salmonella* within epithelial cells is associated with filamentous structures containing lysosomal membrane glycoproteins. *Infect. Agents Dis.* **2:**227–231.
79. **Geiss, H. K., I. Ehrhard, A. Rösen-Wolff, H. G. Sonntag, J. Pratsch, A. Wirth, D. Krüger, I. Knollmann-Schanbacher, H. Kühn, and C. Treiber-Klötzer.** 1993. Foodborne outbreak of a *Salmonella enteritidis* epidemic in a major pharmaceutical company. *Gesundh.-Wes.* **55:**127–132.
80. **Gibson, A. M., N. Bratchell, and T. A. Roberts.** 1988. Predicting microbial growth: growth responses of salmonellae in laboratory medium as affected by pH, sodium chloride and storage temperature. *Int. J. Food Microbiol.* **6:**155–178.
81. **Ginocchio, C., J. Pace, and J. E. Galan.** 1992. Identification and molecular characterization of a *Salmonella typhimurium* gene involved in triggering the internalization of salmonellae into cultured epithelial cells. *Proc. Natl. Acad. Sci. USA* **89:**5976–5980.
82. **Ginocchio, C. C., and J. E. Galan.** 1995. Functional conservation among members of the *Salmonella typhimurium* InvA family of proteins. *Infect. Immun.* 63:729–732.
83. **Ginocchio, C. C., S. B. Olmsted, C. L. Wells, and J. E. Galan.** 1994. Contact with epithelial cells induces the formation of surface appendages on *Salmonella typhimurium. Cell* **76:**717–724.
84. **Glynn, J. R., and D. J. Bradley.** 1992. The relationship of infecting dose and severity of disease in reported outbreaks of salmonella infections. *Epidemiol. Infect.* **109:** 371–388.
85. **Glynn, J. R., and S. R. Palmer.** 1992. Incubation period, severity of disease, and infecting dose: evidence from a *Salmonella* outbreak. *Am. J. Epidemiol.* **136:**1369–1377.
86. **Goepfert, J. M., I. K. Iskander, and C. H. Amundson.** 1970. Relation of the heat resistance of salmonellae to the water activity of the environment. *Appl. Microbiol.* **19:**429–433.

87. **Greenwood, M. H., and W. L. Hooper.** 1983. Chocolate bars contaminated with *Salmonella napoli:* an infectivity study. *Br. Med. J.* **286:**1394.

88. **Griggs, D. J., M. C. Hall, Y. F. Jin, and L. J. V. Piddock.** 1994. Quinolone resistance in veterinary isolates of *Salmonella. J. Antimicrob. Chemother.* **33:**1173–1189.

89. **Groisman, E. A., and H. Ochman.** 1993. Cognate gene clusters govern invasion of host epithelial cells by *Salmonella typhimurium* and *Shigella flexneri. EMBO J.* **12:**3779–3787.

90. **Groisman, E. A., and M. H. Saier, Jr.** 1990. *Salmonella* virulence: new clues to intramacrophage survival. *Trends Biochem. Sci.* **15:**30–33.

91. **Guiney, D. G., F. C. Fang, M. Krause, and S. Libby.** 1994. Plasmid-mediated virulence genes in non-typhoid *Salmonella* serovars. *FEMS Microbiol. Lett.* **124:**1–10.

92. **Gulig, P. A., H. Danbar, D. G. Guiney, A. J. Lax, F. Norel, and M. Rhen.** 1993. Molecular analysis of *spv* virulence genes of the *Salmonella* virulence plasmids. *Mol. Microbiol.* **7:**825–830.

93. **Health Canada.** 1991. *Foodborne and Waterborne Disease in Canada. Annual Summary 1985–86.* Polyscience Publications Inc., Morin Heights, Québec.

94. **Health Canada.** 1994. *Foodborne and Waterborne Disease in Canada. Annual Summary 1987.* Polyscience Publications Inc., Morin Heights, Québec.

95. **Hefferman, E. J., J. Harwood, J. Fierer, and D. Guiney.** 1992. The *Salmonella typhimurium* virulence plasmid complement resistance gene *rck* is homologous to a family of virulence-related outer membrane protein genes, including *pagC* and *ail. J. Bacteriol.* **174:**84–91.

96. **Hefferman, E. J., L. Wu, J. Louie, S. Okamoto, J. Fierer, and D. G. Guiney.** 1994. Specificity of the complement resistance and cell association phenotypes encoded by the outer membrane protein genes *rck* from *Salmonella typhimurium* and *ail* from *Yersinia enterocolitica. Infect. Immun.* **62:**5183–5186.

97. **Heiskanen, P., S. Taira, and M. Rhen.** 1994. Role of *rpo* S in the regulation of *Salmonella* plasmid virulence (*spv*) genes. *FEMS Microbiol. Lett.* **123:**125–130.

98. **Hellström, L.** 1980. Food-transmitted *S. enteritidis* epidemic in 28 schools, p. 397–400. *In Proceedings of the World Congress on Foodborne Infections and Intoxications.* Institute of Veterinary Medicine, Berlin.

99. **Hickey, E. W., and I. N. Hirshfield.** 1990. Low pH-induced effects of patterns of protein synthesis and on internal pH in *Escherichia coli* and *Salmonella typhimurium. Appl. Environ. Microbiol.* **56:**1038–1045.

100. **Holley, R. A., and M. Proulx.** 1986. Use of egg washwater pH to prevent survival of *Salmonella* at moderate temperatures. *Poult. Sci.* **65:**922–928.

101. **Holmberg, S. D., M. T. Osterholm, K. A. Senger, and M. L. Cohen.** 1984. Drug-resistant *Salmonella* from animal fed antimicrobials. *N. Engl. J. Med.* **311:**617–622.

102. **Holmgren, J., and I. Lönnroth.** 1980. Structure and function of enterotoxins and their receptors, p. 88–103. *In* V. Ouchterlony and J. Holmgren (ed.), *Cholera and Related Diarrheas.* Nobel Symposium (Stockholm, 1978). Karger, Basel.

103. **Horwitz, M. A., R. A. Pollard, M. H. Merson, and S. M. Martin.** 1977. A large outbreak of foodborne salmonellosis on the Navajo nation Indian reservation, epidemiology and secondary transmission. *Am. J. Public Health* **67:**1071–1076.

104. **Huhtanen, C. N.** 1975. Use of pH gradient plates for increasing the acid tolerance of salmonellae. *Appl. Microbiol.* **29:**309–312.

105. **Humphrey, T. J., N. P. Richardson, K. M. Statton, and R. J. Rowbury.** 1993. Effects of temperature shift on acid and heat tolerance in *Salmonella enteritidis* phage type 4. *Appl. Environ. Microbiol.* **59:**3120–3122.

106. **Ingham, S. C., R. A. Alford, and A. P. McCown.** 1990. Comparative growth rates of *Salmonella typhimurium* and *Pseudomonas fragi* on cooked crab meat stored under air and modified atmosphere. *J. Food Prot.* **53:**566–567.

107. **Institute of Medicine.** 1988. *Report of a Study. Human Health Risks with the Subtherapeutic Use of Penicillin and Tetracyclines in Animal Feed.* National Academy Press, Washington, D.C.

108. **Isberg, R. R., and G. T. V. Nhieu.** 1994. Two mammalian cell internalization strategies used by pathogenic bacteria. *Annu. Rev. Genet.* **27:**395–422.

109. **Jacobs-Reitsma, W. F., P. M. F. J. Koenraad, N. M. Bolder, and R. W. A. W. Mulder.** 1994. In vitro susceptibility of *Campylobacter* and *Salmonella* isolates from broilers to quinolones, ampicillin, tetracycline, and erythromycin. *Vet. Q.* **16:**206–208.

110. **Jayasheela, M., U. K. Sharma, A. K. Mondal, and S. N. Saxena.** 1987. Plasmid mediated enterotoxigenicity in *Salmonella* strains isolated from patients with gastroenteritis. *Indian J. Med. Res.* **85:**496–499.

111. **Jones, B. D., H. F. Paterson, A. Hall, and S. Falkow.** 1993. *Salmonella typhimurium* induces membrane ruffling by a growth factor-receptor-independent mechanism. *Proc. Natl. Acad. Sci. USA* **90:**10390–10394.

112. **Kaniga, K., J. C. Bossio, and J. E. Galan.** 1994. The *Salmonella typhimurium* invasion genes *inv* F and *inv* G encode homologues of the AraC and PulD family of proteins. *Mol. Microbiol.* **13:**555–568.

113. **Kapperud, G., S. Gustavsen, I. Hellesnes, A. H. Hansen, J. Lassen, J. Hirn, M. Jahkola, M. A. Montenegro, and R. Helmuth.** 1990. Outbreak of *Salmonella typhimurium* infection traced to contaminated chocolate and caused by a strain lacking the 60-megadalton virulence plasmid. *J. Clin. Microbiol.* **28:**2597–2601.

114. **Kauffmann, F.** 1966. *The Bacteriology of Enterobacteriaceae.* Munksgaard, Copenhagen.

115. **Keitel, W. A., N. L. Bond, J. M. Zahradnik, T. A. Cramton, and J. B. Robbins.** 1994. Clinical and serological responses following primary and booster immunization with *Salmonella typhi* Vi capsular polysaccharide vaccines. *Vaccine* 12:195–199.

116. **Kim, C. J., D. A. Emery, H. Rinke, K. V. Nagaraja, and D. A. Halvorson.** 1989. Effect of time and temperature on growth of *Salmonella enteritidis* in experimentally inoculated eggs. *Avian Dis.* **33:**735–742.

117. **Kirby, R. M., and R. Davies.** 1990. Survival of dehydrated cells of *Salmonella typhimurium* LT 2 at high temperatures. *J. Appl. Microbiol.* **68:**241–246.

118. **Koo, F. C. W., J. W. Peterson, C. W. Houston, and N. C. Molina.** 1984. Pathogenesis of experimental salmonello-

sis: inhibition of protein synthesis by cytotoxin. *Infect. Immun.* **43**:93–100.

119. **Kowarz, L., C. Coynault, V. Robbe-Saule, and F. Norel.** 1994. The *Salmonella typhimurium katF* (*rpoS*) gene: cloning, nucleotide sequence, and regulation of *spvR* and *spvABCD* virulence plasmid genes. *J. Bacteriol.* **176**:6852–6860.
120. **Lang, D. J., L. J. Kunz, A. R. Martin, S. A. Schroeder, and L. A. Thomson.** 1967. Carmine as a source of nosocomial salmonellosis. *N. Engl. J. Med.* **276**:829–832.
121. **Lecos, C.** 1986. Of microbes and milk: probing America's worst *Salmonella* outbreak. *Dairy Food Sanit.* **6**:136–140.
122. **Lee, I. S., J. L. Slonczewski, and J. W. Foster.** 1994. A low-pH-inducible, stationary-phase acid tolerance response in *Salmonella typhimurium. J. Bacteriol.* **176**:1422–1426.
123. **Lee, L. A., N. D. Puhr, E. K. Maloney, N. H. Bean, and R. V. Tauxe.** 1994. Increase in antimicrobial-resistant *Salmonella* infections in the United States, 1989–1990. *J. Infect. Dis.* **170**:128–134.
124. **Lehmacher, A., J. Bockemühl, and S. Aleksic.** 1995. A nationwide outbreak of human salmonellosis in Germany due to contaminated paprika and paprika-powdered potato chips. *J. Infect. Dis.* **115**:501–511.
125. **Le Minor, L.** 1981. The genus *Salmonella,* p. 1148–1159. *In* M. P. Starr, H. Stolp, H. G. Truper, A. Balows, and H. G. Schlegel (ed.), *The Prokaryotes.* Springer-Verlag, New York.
126. **Le Minor, L., C. Coynault, and G. Pessoa.** 1974. Déterminisme plasmidique du caractère atypique "lactose positif" de souches de *S. typhimurium* et de *S. oranienburg* isolées au Brésil lors d'épidémies de 1971 à 1973. *Ann. Microbiol.* **125A**:261–285.
127. **Le Minor, L., C. Coynault, R. Rhode, B. Rowe, and S. Aleksic.** 1973. Localisation plasmidique de déterminant génétique du caractère atypique "saccharose +" des *Salmonella. Ann. Microbiol.* **124B**:295–306.
128. **Le Minor, L., and M. Y. Popoff.** 1987. Request for an opinion. Designation of *Salmonella enterica* sp. nov., nom. rev., as the type and only species of the genus *Salmonella. Int. J. Syst. Bacteriol.* **37**:465–468.
129. **Le Minor, L., M. Y. Popoff, B. Laurent, and D. Hermant.** 1986. Individualisation d'une septième sous-espèce de *Salmonella: S. choleraesuis* subsp. indica subsp. nov. *Ann. Inst. Pasteur/Microbiol.* **137B**:211–217.
130. **Lepoutre, A., J. Salomon, C. Charley, and F. LeQuerrec.** 1994. Les toxi-infections alimentaires collectives en 1993. *Bull. Epidemiol. Hebd.* **52**:245–247.
131. **Leyer, G. J., and E. A. Johnson.** 1992. Acid adaptation promotes survival of *Salmonella* spp. in cheese. *Appl. Environ. Microbiol.* **58**:2075–2080.
132. **Leyer, G. J., and E. A. Johnson.** 1993. Acid adaptation induces cross-protection against environmental stresses in *Salmonella typhimurium. Appl. Environ. Microbiol.* **59**:1842–1847.
133. **MacBeth, K. J., and C. A. Lee.** 1993. Prolonged inhibition of bacterial protein systhesis abolishes *Salmonella* invasion. *Infect. Immun.* **61**:1544–1546.
134. **Mackey, B. M., and C. M. Derrick.** 1986. Elevation of the heat resistance of *Salmonella typhimurium* by sublethal heat shock. *J. Appl. Bacteriol.* **61**:389–393.
135. **Mackey, B. M., and C. Derrick.** 1990. Heat shock protein synthesis and thermotolerance in *Salmonella typhimurium. J. Appl. Bacteriol.* **69**:373–383.
136. **Mason, J.** 1994. *Salmonella enteritidis* control programs in the United States. *Int. J. Food Microbiol.* **21**:155–169.
137. **Mattila, L., M. Leirisalo-Repo, S. Koskimies, K. Granfors, and A. Sütonen.** 1994. Reactive arthritis following an outbreak of *Salmonella* infection in Finland. *Br. J. Rheum.* **33**:1136–1141.
138. **McCullough, N. B., and C. W. Eisele.** 1951. Experimental human salmonellosis. 1. Pathogenicity of strains of *Salmonella meleagridis* and *Salmonella anatum* obtained from spray-dried whole egg. *J. Infect. Dis.* **88**:278–289.
139. **Miller, S. I.** 1991. Pho P/Pho Q: macrophage-specific modulators of *Salmonella* virulence. *Mol. Microbiol.* **5**:2073–2078.
140. **Miller, S. I., A. M. Kukral, and J. J. Mekalanos.** 1989. A two-component regulatory system (phoP/phoQ) controls *Salmonella typhimurium* virulence. *Proc. Natl. Acad. Sci. USA* **86**:5054–5058.
141. **Miller, S. I., W. S. Pulkkinen, M. E. Selsted, and J. J. Mekalanos.** 1990. Characterization of defensin resistance phenotypes associated with mutations in the *phoP* virulence regulon of *Salmonella typhimurium. Infect. Immun.* **58**:3706–3710.
142. **Mills, D. M., V. Bajaj, and C. A. Lee.** 1995. A 40 kb chromosomal fragment encoding *Salmonella typhimurium* invasion genes is absent from the corresponding region of the *Escherichia coli* K-12 chromosome. *Mol. Microbiol.* **15**:749–759.
143. **Mintz, E. D., M. L. Cartter, J. L. Hadler, J. T. Wassell, J. A. Zingeser, and R. V. Tauxe.** 1994. Dose-response effects in an outbreak of *Salmonella enteritidis. Epidemiol. Infect.* **112**:13–23.
144. **Miyagawa, S., and A. Miki.** 1992. The epidemiological data of food poisoning in 1991. *Food Sanit. Res.* **42**:78–104.
145. **Mizuno, T., and S. Mizushima.** 1990. Signal transduction and gene regulation through the phosphorylation of two regulatory components: the molecular basis for the osmotic regulation of the porin genes. *Mol. Microbiol.* **4**:1077–1082.
146. **Murray, B. E.** 1986. Resistance of *Shigella, Salmonella,* and other selected enteric pathogens to antimicrobial agents. *Rev. Infect. Dis.* **8**(Suppl. **2**):S172–S181.
147. **Nassif, X., and P. Sansonetti.** 1987. Les systèmes bactériens de captation du fer: leur rôle dans la virulence. *Bull. Inst. Pasteur* **85**:307–327.
148. **Ng, H., H. G. Bayne, and J. A. Garibaldi.** 1969. Heat resistance of *Salmonella:* the uniqueness of *Salmonella senftenberg* 775W. *Appl. Microbiol.* **17**:78–82.
149. **Nguyen Van, J. C., and L. Gutman.** 1994. Résistance aux antibiotiques par diminution de la perméabilité chez les bactéries à Gram négatif. *Presse Med.* **23**:522–531.
150. **Novick, R. P.** 1981. The development and spread of antibiotic-resistant bacteria as a consequence of feeding antibiotics to livestock. *Ann. N. Y. Acad. Sci.* **368**:23–59.
151. **Ogawa, H., H. Tokunou, M. Sasaki, T. Kishimoto, and K. Tamura.** 1991. An outbreak of bacterial food poisoning caused by roast cuttlefish "yaki-ika" contaminated

with *Salmonella* spp. (1) Champaign. *Jpn J. Food Microbiol.* 7:151–157.

152. **Olson, E. R.** 1993. MicroReview. Influence of pH on bacterial gene expression. *Mol. Microbiol.* **8**:5–14.

153. **Pacer, R. E., J. S. Spika, M. C. Thurmond, N. Hargrett-Bean, and M. E. Potter.** 1989. Prevalence of *Salmonella* and multiple antimicrobial-resistant *Salmonella* in California dairies. *J. Am. Vet. Med. Assoc.* **195**:59–63.

154. **Palmer, S. R., and B. Rowe.** 1986. Trends in *Salmonella* infections. *Public Health Lab. Serv. Microbiol. Dig.* **3**:18–21.

155. **Peterson, J. W., and D. W. Niesel.** 1988. Enhancement by calcium of the invasiveness of *Salmonella* for HeLa cell monolayers. *Rev. Infect. Dis.* **10**:S319–S322.

156. **Pickard, D., J. Li, M. Roberts, D. Maskell, D. Hone, M. Levine, G. Dougan, and S. Chatfield.** 1994. Characterization of defined *ompR* mutants of *Salmonella typhi*: *ompR* is involved in the regulation of Vi polysaccharide expression. *Infect. Immun.* **62**:3981–3993.

157. **Pivnick, H.** 1980. Curing salts and related materials, p. 136–159. *In* J. H. Silliker, R. P. Elliott, A. C. Baird-Parker, F. L. Bryan, J. H. B. Christian, D. S. Clark, J. C. Olson, Jr., and T. A. Roberts (ed.), *Microbial Ecology of Foods,* vol. 1. *Factors Affecting Life and Death of Microorganisms.* Academic Press, New York.

158. **Pohl, P., Y. Glupczynskj, M. Marin, G. van Robaeys, P. Lintermans, and M. Couturier.** 1993. Replicon typing characterization of plasmids encoding resistance to gentamicin and apramycin in *Escherichia coli* and *Salmonella typhimurium* isolated from human and animal sources in Belgium. *Epidemiol. Infect.* **111**:229–238.

159. **Polotsky, Y., E. Dragunsky, and T. Khavkin.** 1994. Morphologic evaluation of the pathogenesis of bacterial enteric infections. *Crit. Rev. Microbiol.* **20**:161–208.

160. **Pönkä, A., Y. Andersson, A. Sütonen, B. de Jong, M. Jahkola, O. Haikala, A. Kuhmonen, and P. Pakkala.** 1995. *Salmonella* in alfalfa sprouts. *Lancet* **345**:462–463.

161. **Popoff, M. Y., J. Bockemuhl, and A. McWhorter-Murlin.** 1994. Supplement 1993 (no. 37) to the Kauffmann-White scheme. *Res. Microbiol.* **145**:711–716.

162. **Popoff, M. Y., and F. Norel.** 1992. Bases moléculaires de la pathogénicité des *Salmonella*. *Med. Mal. Infect.* **22**:310–324.

163. **Rabsch, W., P. Paul, and R. Reissbrodt.** 1987. A new hydroxamate siderophore for iron supply of *Salmonella*. *Acta Microbiol. Hung.* **34**:85–92.

164. **Radford, S. A., and R. G. Board.** 1995. The influence of sodium chloride and pH on the growth of *Salmonella enteritidis* PT 4. *Lett. Appl. Microbiol.* **20**:11–13.

165. **Rahman, H., V. B. Singh, and V. D. Sharma.** 1994. Purification and characterization of enterotoxic moiety present in cell-free culture supernatant of *Salmonella typhimurium*. *Vet. Microbiol.* **39**:245–254.

166. **Rajajee, S., T. B. Anandi, S. Subha, and B. R. Vatsala.** 1995. Patterns of resistant *Salmonella typhi* infection in infants. *J. Trop. Pediatr.* **41**:52–54.

167. **Reeves, M. W., G. M. Evins, A. A. Heiba, B. D. Plikaytis, and J. J. Farmer III.** 1989. Clonal nature of *Salmonella typhi* and its genetic relatedness to other salmonellae as shown by multilocus enzyme electrophoresis, and proposal of *Salmonella bongori* comb. nov. *J. Clin. Microbiol.* **27**:313–320.

168. **Reitler, R., D. Yarom, and R. Seligmann.** 1960. The enhancing effect of staphylococcal enterotoxin on *Salmonella* infection. *Med. Off.* **104**:181.

169. **Richard, F., E. Pons, B. Lelore, V. Bleuze, B. Grandbastien, C. Collinet, R. Mathis, J.-F. Diependale, and P. Legrand.** 1994. Toxi-infection alimentaire collective du 8 juin 1993 à Douai. *Bull. Epidemiol. Hebd.* **3**:9–11.

170. **Robbe-Saule, V., C. Coynault, and F. Norel.** 1995. The live oral typhoid vaccine Ty21a is a rpoS mutant and is susceptible to various environmental stresses. *FEMS Microbiol. Lett.* **126**:171–176.

171. **Roberts, J. A., and P. N. Sockett.** 1994. The socioeconomic impact of human *Salmonella enteritidis* infection. *Int. J. Food Microbiol.* **21**:117–129.

172. **Ruschkowski, S., I. Rosenshine, and B. B. Finlay.** 1992. *Salmonella typhimurium* induces an inositol phosphate flux in infected epithelial cells. *FEMS Microbiol. Lett.* **95**:121–126.

173. **Ryan, C. A., M. K. Nickels, N. T. Hargrett-Bean, M. E. Potter, T. Endo, L. Mayer, C. W. Langkop, C. Gibson, R. C. McDonald, R. T. Kenney, N. D. Puhr, P. J. McDonnell, R. J. Martin, M. L. Cohen, and P. A. Blake.** 1987. Massive outbreak of antimicrobial-resistant salmonellosis traced to pasteurized milk. *J. Am. Med. Assoc.* **258**:3269–3274.

174. **Sameshima, T., H. Ito, I. Uchida, H. Danbara, and N. Terakado.** 1993. A conjugative plasmid pTE195 coding for drug resistance and virulence phenotypes from *Salmonella naestved* strain of calf origin. *Vet. Microbiol.* **36**:197–203.

175. **Sandaa, R. A., V. L. Torsvik, and J. Goksoyr.** 1992. Transferable drug resistance in bacteria from fish-farm sediments. *Can. J. Microbiol.* **38**:1061–1065.

176. **Saxen, H., I. Reima, and P. H. Mäkelä.** 1987. Alternative complement pathway activation by *Salmonella* O polysaccharide as a virulence determinant in the mouse. *Microb. Pathog.* **2**:15–28.

177. **Seglenieks, Z., and S. Dixon.** 1977. Outbreak of milkborne *Salmonella* gastroenteritis—South Australia. *Morbid. Mortal. Weekly Rep.* **26**:127.

178. **Sjöbring, U., G. Pohl, and A. Olsen.** 1994. Plasminogen, adsorbed by *Escherichia coli* expressing curli or by *Salmonella enteritidis* expressing thin aggregative fimbriae, can be activated by simultaneously captured tissue-type plasminogen activator (t-PA). *Mol. Microbiol.* **14**:443–452.

179. **Smith, J. L.** 1994. Arthritis and foodborne bacteria. *J. Food Prot.* **57**:935–941.

180. **Sockett, P. N., J. M. Cowden, S. LeBaigne, D. Ross, G. K. Adak, and H. Evans.** 1993. Foodborne disease surveillance in England and Wales: 1989–1991. *Commun. Dis. Rep.* **3**:R159–R173.

181. **Spika, J. S., S. H. Waterman, G. W. Soo Hoo, M. E. St. Louis, R. E. Pacer, S. M. James, M. L. Bissett, L. W. Mayer, J. Y. Chiu, B. Hall, K. Greene, M. E. Potter, M. L. Cohen, and P. A. Blake.** 1987. Chloramphenicol-resistant *Salmonella newport* traced through hamburger to dairy farms. *N. Engl. J. Med.* **316**:565–570.

182. **Stephen, J., T. S. Wallis, W. G. Starkey, D. C. A. Candy, M. P. Osborne, and S. Haddon.** 1985. Salmonellosis: in retrospect and prospect. *Ciba Found. Symp.* **112**:175–192.

183. **Stickney, R. R.** 1990. A global overview of aquaculture production. *Food Rev. Int.* **6**:299–315.

184. **Suzuki, S., K. Komase, H. Matsui, A. Abe, K. Kawahara, Y. Tamura, M. Kijima, H. Danbara, M. Nakamura, and S. Sato.** 1994. Virulence region of plasmid pN2001 of *Salmonella enteritidis. Microbiology* **140**:1307–1318.

184a. **Szu, S. C.** 1994. Presented at the Asia-Pacific Symposium on Typhoid Fever, Bangkok, Thailand.

185. **Tabaraie, B., B. K. Sharma, P. R. nee Sharma, R. Sehgal, and N. K. Ganguly.** 1994. Stimulation of macrophage oxygen free radical production and lymphocyte blastogenic response by immunization with porins. *Microbiol. Immunol.* **38**:561–565.

186. **Tacket, C. O., L. B. Dominguez, H. J. Fisher, and M. L. Cohen.** 1985. An outbreak of multiple-drug-resistant *Salmonella* enteritis from raw milk. *JAMA* **253**:2058–2060.

187. **Taira, S., P. Heiskanen, R. Hurme, H. Heikkilä, P. Riikonen, and M. Rhen.** 1995. Evidence for functional polymorphism of the *spv* R gene regulating virulence gene expression in *Salmonella. Mol. Gen. Genet.* **246**:437–444.

188. **Tanaka, N.** 1993. Food hygiene in Japan—Japanese food hygiene regulations and food poisoning incidents. *Dairy Food Environ. Sanit.* **13**:152–156.

189. **Thomas, L. V., J. W. T. Wimpenny, and A. C. Peters.** 1992. Testing multiple variables on the growth of a mixed inoculum of *Salmonella* strains using gradient plates. *Int. J. Food Microbiol.* **15**:165–175.

190. **Thomson, G. T. D., D. A. DeRubeis, M. A. Hodge, C. Rajanayagam, and R. D. Inman.** 1995. Post-*Salmonella* reactive arthritis: late clinical sequelae in point source cohort. *Am. J. Med.* **98**:13–21.

191. **Threlfall, E. J.** 1992. Antibiotics and the selection of food-borne pathogens. *J. Appl. Bacteriol. Symp. Suppl.* **73**: 96S–102S.

192. **Threlfall, E. J., M. D. Hampton, H. Chart, and B. Rowe.** 1994. Identification of a conjugative plasmid carrying antibiotic resistance and salmonella plasmid virulence *(spv)* genes in epidemic strains of *Salmonella typhimurium* phagetype 193. *Lett. Appl. Microbiol.* **18**:82–85.

193. **Todd, E. C. D.** 1994. Surveillance of foodborne disease, p. 461–536. *In* Y. H. Hui, J. R. Gorham, K. D. Murrell, and D. O. Cliver (ed.), *Foodborne Disease Handbook.*, vol. 1. *Diseases Caused by Bacteria.* Marcel Dekker, Inc., New York.

194. **Troller, J. A.** 1986. Water relations of foodborne bacterial pathogens—an updated review. *J. Food Prot.* **49**:656–670.

195. **Vescovi, E. G., F. C. Soncini, and E. A. Groisman.** 1994. The role of the PhoP/PhoQ regulon in *Salmonella* virulence. *Res. Microbiol.* **145**:473–480.

196. **Weissman, J. B., R. M. A. D. Deen, M. Williams, N. Swanton, and S. Ali.** 1977. An island-wide epidemic of salmonellosis in Trinidad traced to contaminated powdered milk. *West Indies Med. J.* **26**:135–143.

197. **Wiedemann, B., and P. Heisig.** 1994. Mechanisms of quinolone resistance. *Infection* **22**(Suppl. 2):S73–S79.

198. **World Health Organization.** 1988. *Salmonellosis Control: the Role of Animal and Product Hygiene.* Technical Report Series 774. World Health Organization, Geneva.

199. **World Health Organization.** 1992. *WHO Surveillance Programme for Control of Foodborne Infections and Intoxications in Europe. Fifth Report 1985–1989.* Institute of Veterinary Medicine—Robert von Ostertag Institute, Berlin.

200. **Ye, X. L., C. C. Yan, H. H. Xie, X. P. Tan, Y. Z. Wang, and L. M. Ye.** 1990. An outbreak of food poisoning due to *Salmonella typhimurium* in the People's Republic of China. *J. Diarrheal Dis. Res.* **8**:97–98.

201. **Young, H. K.** 1994. Do nonclinical uses of antibiotics make a difference? *Infect. Control Hosp. Epidemiol.* **15**: 484–487.

202. **Zastrow, K. D., and I. Schöneberg.** 1993. Outbreaks of food-borne infections and intoxications in the Federal Republic of Germany 1991. *Gesundh.-Wes.* **55**:250–253.

Irving Nachamkin

9

Campylobacter jejuni

CHARACTERISTICS OF THE ORGANISM

Campylobacter jejuni subsp. *jejuni* (hereafter referred to as *C. jejuni*) is one of many species and subspecies within the genus *Campylobacter*, family *Campylobacteraceae*, and has recently been reviewed (58). During the 1980s there was an explosion of information published on these bacteria, as evident from the 1984 edition of *Bergey's Manual of Systematic Bacteriology* (43a), published at a time when there were only eight species and subspecies within the genus *Campylobacter*. Since then, many investigators have become interested in studying the taxonomy and clinical importance of campylobacters and have greatly expanded the number of genera and species associated with this group of bacteria. The family *Campylobacteraceae* includes 20 species and subspecies within the genus *Campylobacter* (Table 9.1) and four species in the genus *Arcobacter*. Now included in the genus *Campylobacter* are organisms formerly classified in the genus *Wolinella* (*W. curva, W. recta*). *C. hyoilei* was recently added to the family and is associated with proliferative enteritis in pigs (1). The genus *Arcobacter* now contains several former *Campylobacter* species as well as some new species and includes *A. cryaerophilus, A. nitrofigilis, A. butzleri*, and *A. skirrowii*. Former species in the genus *Campylobacter, C. cinaedi* and *C. fennelliae*, are now classified in the genus *Helicobacter*. *Bacteroides gracilis* was recently reclassified as *C. gracilis* (1), and *[Bacteroides] ureolyticus* is thought to be closely related to the genus *Campylobacter*; however, this classification is still uncertain.

Campylobacter is the type genus within the family *Campylobacteraceae*. Organisms are curved, S-shaped, or spiral rods that are 0.2 to 0.9 μm wide and 0.5 to 5 μm long. They are gram-negative, non-spore-forming rods that may form spherical or coccoid bodies in old cultures or cultures exposed to air for prolonged periods. Organisms are motile by means of a single polar unsheathed flagellum at one or both ends. The various species are microaerophilic with a respiratory type of metabolism. Some strains grow aerobically or anaerobically. An atmosphere containing increased hydrogen may be required by some species for microaerobic growth. *C. jejuni* and *C. coli* have genomes approximately 1.7 Mb in size, as determined by pulsed-field gel electrophoresis, which is about one-third the size of the *Escherichia coli* genome (92).

Arcobacters are gram-negative curved, S-shaped, or helical non-spore-forming rods that are 0.2 to 0.9 μm wide and 1 to 3 μm long. Organisms are motile with a single polar unsheathed flagellum. Arcobacters grow at 15, 25, and 30°C but have variable growth at 37 and 42°C. Organisms are microaerophilic and do not require hydrogen for growth. Arcobacters may grow aerobically at 30°C and anaerobically at 35 to 37°C. Most strains are nonhemolytic. *A. skirrowii* may be alpha-hemolytic.

Table 9.1 Reservoirs and disease-associated species in the family *Campylobacteraceae*[a]

Organism	Reservoirs	Disease, sequelae, or comments: Human	Disease, sequelae, or comments: Animal
C. jejuni subsp. *jejuni*	Humans, other mammals, birds	Diarrhea, systemic illness, GBS	Diarrhea in primates
C. jejuni subsp. *doylei*	Unknown	Diarrhea	
C. fetus subsp. *fetus*	Cattle, sheep	Systemic illness, diarrhea	Abortion
C. fetus subsp. *venerealis*	Cattle		Infertility
C. coli	Pigs, birds	Diarrhea	
C. lari	Birds, dogs	Diarrhea	
C. upsaliensis	Domestic pets	Diarrhea	Diarrhea
C. hyointestinalis	Cattle, pigs	Rare, proctitis, diarrhea	Proliferative enteritis
C. mucosalis	Pigs	Rare, diarrhea	Proliferative enteritis
C. hyoilei	Pigs		Proliferative enteritis
C. sputorum biovar sputorum	Humans	Isolated from oral cavity and abscesses	
C. sputorum biovar bubulus	Cattle	Isolated from oral cavity and abscesses	Isolated from genital tract
C. sputorum biovar faecalis	Cattle, sheep		Enteritis
C. concisus	Humans	Periodontal disease	
C. curvus	Humans	Periodontal disease	
C. rectus	Humans	Periodontal disease, pulmonary infections	
C. showae	Humans	Periodontal disease	
C. helveticus	Domestic pets		Diarrhea
C. hyoilei	Pigs		Proliferative enteritis
C. gracilis	Humans	Infections of head, neck, and other sites	
A. butzleri	Cattle, pigs	Diarrhea, other illness	Diarrhea, abortion
A. cryaerophilus	Cattle, sheep, pigs	Diarrhea, bacteremia	Abortion
A. skirrowii	Cattle, sheep, pigs		Abortion, diarrhea; isolated from genital tract of bulls
A. nitrofigilis	Isolated from plants		

[a]Adapted from references 1, 86, and 96.

Most strains are susceptible to nalidixic acid but variable in susceptibility to cephalothin.

ENVIRONMENTAL SUSCEPTIBILITY

C. jejuni is susceptible to a variety of environmental conditions that make it unlikely to survive for long periods of time outside the host. The organism does not grow at temperatures below 30°C, is microaerophilic, and is sensitive to drying, high-oxygen conditions, and low pH (22). The decimal reduction time for campylobacters at 55°C is about 1 min, and the *z* value is about 5°C (68). Thus, the organism should not survive in food products brought to adequate cooking temperatures. Organisms are susceptible to gamma irradiation (1 kGy), but the rate of killing is dependent on the type of product being processed. Irradiation is less effective for frozen materials than for refrigerated or room temperature meats. Early-log-phase cells are more susceptible than organisms grown to log or stationary phase (77). Recent studies suggest that *Campylobacter* spp. are more radiation sensitive than other foodborne pathogens such as salmonellae and *Listeria monocytogenes* and that irradiation treatment that is effective against the latter organisms should be sufficient to kill *Campylobacter* spp. as well (69).

Disinfectants such as sodium hypochlorite (Clorox), *o*-phenylphenol (Amphyl), iodine-polyvinylpyrrolidine (Betadine), alkylbenzyl dimethylammonium chloride (Zephiran), glutaraldehyde (Cidex), formaldehyde, and ethanol have antibacterial activity at commonly used concentrations (99).

Campylobacter spp. are susceptible to low pH and are killed readily at pH 2.3 (13). Organisms have been shown to remain viable and multiply in bile at 37°C and to survive better in feces, milk, water, and urine held at

4°C than in material held at 25°C. The maximum periods of viability of *Campylobacter* spp. at 4°C were 3 weeks in feces, 4 weeks in water, and 5 weeks in urine (13).

RESERVOIRS AND FOODBORNE OUTBREAKS

C. jejuni is zoonotic, with many animals serving as reservoirs for human disease. Reservoirs for infection include rabbits, rodents, wild birds, sheep, horses, cows, pigs, poultry, and domestic pets (3, 22, 88). Contaminated vegetables and shellfish may also be sources of infection (3, 22). Campylobacters are frequently isolated from water, and water supplies have been sources of infection in some reported outbreaks. Organisms may remain dormant in water in a state that has been termed "viable but nonculturable" (81); that is, under unfavorable conditions, the organisms essentially remain dormant and cannot be easily recovered on artificial media. Under favorable conditions, the organisms are able to multiply (52). The role of these forms as a source of infection for humans is not clear; Medema et al. (52) found that viable but nonculturable *C. jejuni* were not able to colonize chicks.

From 1980 to 1982, 23 foodborne outbreaks occurring in 14 states were reported to the Centers for Disease Control and Prevention. Most outbreaks occurred from May to October and peaked during the spring and fall months, with only one outbreak occurring during July and August. Attack rates varied from 14 to 79% (mean, 45%; median, 41%). Raw milk was suspected in over half of the outbreaks. Other vehicles for infection included cake icing, poultry, eggs, and beef (27). Up to 1987, a total of 57 outbreaks affecting almost 6,300 individuals were reported. Most of the foodborne outbreaks were attributed to ingestion of raw milk. Contaminated community water supplies accounted for the majority of reported waterborne outbreaks (91). Unpasteurized raw milk continues to be a source of outbreaks of *Campylobacter* infection in the United States (76, 102). In a 10-year review of outbreaks from 1981 to 1990, 20 outbreaks occurred that affected 1,013 individuals who drank raw milk. The attack rate was 45%. At least one outbreak occurred each year, and most of the outbreaks occurred in children who had gone on field trips to dairy farms (102).

In contrast to the relatively low occurrence of outbreaks caused by campylobacters, many studies have established campylobacters as among the most common causes of sporadic bacterial enteritis in the United States (16, 28, 90). The Centers for Disease Control and Prevention estimates that in the United States, the overall infection rate is 1,000 per 100,000 population, accounting for over 2 million cases annually. This is similar to the incidence of sporadic infection in the United Kingdom and other developed nations (90). Sporadic cases occur most often in the summer months and usually follow ingestion of improperly handled or cooked food, primarily poultry products (88).

CHARACTERISTICS OF DISEASE

C. jejuni and *C. coli*

Most *Campylobacter* species are associated with lower gastrointestinal tract infection; however, extraintestinal infections are common with some species, such as *C. fetus* subsp. *fetus* (hereafter referred to as *C. fetus*), and sequelae of *Campylobacter* infection are being more frequently recognized. *C. jejuni* and *C. coli* have been recognized since the 1970s as agents of gastrointestinal tract infection. As noted above, campylobacters have been established as common causes of sporadic bacterial enteritis in the United States (90).

C. jejuni and *C. coli* are the most common *Campylobacter* species associated with diarrheal illness and are clinically indistinguishable. The relative ratio of *C. jejuni* to *C. coli* is not known because most laboratories do not routinely distinguish these organisms. In the United States, an estimate of approximately 5 to 10% of cases reported as due to *C. jejuni* are due to *C. coli;* this figure may be higher in other parts of the world (93).

A spectrum of illness is seen during *C. jejuni* or *C. coli* infection, and patients may be asymptomatic to severely ill. Symptoms and signs usually include fever, abdominal cramping and diarrhea (with or without blood or fecal leukocytes) that lasts several days to more than 1 week. Symptomatic infections are usually self-limited, but relapses may occur in 5 to 10% of untreated patients. *Campylobacter* infection may mimic acute appendicitis and result in unnecessary surgery. Extraintestinal infections and sequelae do occur and include bacteremia, bursitis, urinary tract infection, meningitis, endocarditis, peritonitis, erythema nodosum, pancreatitis, abortion and neonatal sepsis, reactive arthritis, and Guillain-Barré syndrome (GBS). Deaths attributable to *C. jejuni* infection have been reported but rarely occur (12).

C. jejuni and *C. coli* are susceptible to a variety of antimicrobial agents, including macrolides, fluoroquinolones, aminoglycosides, cholamphenicol, and tetracycline. Erythromycin has been the drug of choice for treating *C. jejuni* gastrointestinal tract infections, but ciprofloxacin is a good alternative drug. Early therapy of *Campylobacter* infection with erythromycin or ciprofloxacin is effective in eliminating the organism from stool and may also reduce the duration of symptoms associated with infection (12).

C. jejuni is generally susceptible to erythromycin, with resistance rates of less than 5%. Rates of erythromycin resistance in *C. coli* vary considerably, with up to 80% of strains showing resistance in some studies. Although ciprofloxacin has been effective in treating *Campylobacter* infections, emergence of fluoroquinolone resistance during therapy has been reported. Several in vitro studies suggest that there is increasing resistance to fluoroquinolones (24, 78, 79) and that in Europe, resistance may be associated with use of antibiotics in poultry; thus, the effectiveness of these drugs may be diminished in the future (12, 97).

Other *Campylobacter* and *Arcobacter* Species

In contrast to *C. jejuni*, *C. fetus* is primarily associated with bacteremia and extraintestinal infections in patients with underlying diseases and may be associated with a poor outcome in some patients. *C. fetus* is also associated with septic abortions, septic arthritis, abscesses, meningitis, endocarditis, mycotic aneurysm, thrombophlebitis, peritonitis, and salpingitis. Although gastroenteritis does occur with this species, the incidence is probably underestimated because the organism does not grow well at 42°C and is usually susceptible to cephalothin, an antimicrobial agent used in some common selective media for stool culture. *C. fetus* subsp. *venerealis* is not associated with human infection (58).

C. upsaliensis is a recently described thermotolerant species that is an important cause of diarrhea and bacteremia. *C. lari* (formerly *C. laridis*) is a thermophilic species isolated first from gulls (of the genus *Larus*) and subsequently from other avian species, dogs, cats, and chickens. *C. lari* has been infrequently found in humans with bacteremia and gastrointestinal and urinary tract infections (58). A waterborne outbreak of infection affecting over 100 individuals occurred in 1985 (16). Other *Campylobacter* species have been isolated from clinical specimens of patients with a variety of diseases, but their pathogenic role has not been determined (58). *C. jejuni* subsp. *doylei* is a nitrate-negative subspecies of *C. jejuni* rarely recovered from patients with upper gastrointestinal tract infections and gastroenteritis. *C. hyointestinalis* has been occasionally associated with proctitis and diarrhea in human infection. *C. concisus* is associated primarily with periodontal disease but has also been isolated from patients with bacteremia, foot ulcer, and upper and lower gastrointestinal tract infections. *C. sputorum* biovar sputorum and biovar bubulus have been associated with lung, axillary, scrotal, and groin abscesses. *C. mucosalis* was isolated from two children with enteritis. *C. helveticus* is a newly described species recovered from domestic cats and dogs but not from humans. The species appears to be phenotypically most closely related to *C. upsaliensis*. *C. rectus* is primarily isolated from patients with active periodontal infections but was also isolated from a patient with pulmonary infection. *C. showae* is a recently described species isolated from human gingival crevice that is somewhat distinctive in its morphologic characteristics from other *Campylobacter* species, appearing as straight rods and multiple flagella. *C. hyoilei* is associated with proliferative enteritis in pigs (1). *C. gracilis*, formerly considered an anaerobic bacterium, has been isolated from gingival crevices of humans and is involved in visceral, head, and neck infections (96).

Arcobacters are aerotolerant, *Campylobacter*-like organisms frequently isolated from bovine and porcine abortion and enteritis. Two of the four *Arcobacter* species have been associated with human infection. *A. butzleri* has been isolated from patients with bacteremia, endocarditis, peritonitis, and diarrhea. In addition, *A. butzleri* has been associated with diarrheal disease in monkeys (*Macaca mulatta*). *A. cryaerophilus* group 1B has been isolated from patients with bacteremia and diarrhea (58).

EPIDEMIOLOGIC TYPING SYSTEMS USEFUL FOR INVESTIGATING FOODBORNE DISEASES

Many typing systems have been devised to study the epidemiology of *Campylobacter* infections; such systems vary in complexity and ability to discriminate between strains. These methods include biotyping, serotyping, bacteriocin sensitivity, detection of preformed enzymes, auxotyping, lectin binding, phage typing, multilocus enzyme electrophoresis, and molecular biology-based methods such as chromosomal restriction endonuclease analysis, ribotyping, and PCR (70).

The most frequently used systems are biotyping and serotyping (70). Several biotyping schemes that have been published can, on the basis of only a few biochemical tests, group *C. jejuni*, *C. coli*, and *C. lari* into major categories. While the discrimination of strains is low with such schemes, biotyping is useful as a first step for epidemiologic investigation. The two major serotyping schemes used worldwide detect heat-labile (49) and O (72) antigens. The heat-labile serotyping scheme originally described by Lior et al. (49) can detect over 100 serotypes of *C. jejuni*, *C. coli*, and *C. lari*. Uncharacterized bacterial surface antigens and, in some serotypes, flagella are the serodeterminants for this serotyping system (2). The Penner O serotyping scheme (72) detects over 60 types of *C. jejuni* and *C. coli* (70) and is based on detection of lipopolysaccharide (LPS) antigens. Both serotyping systems, while simple to perform, have good ability to discriminate between strains. These systems, however, are available in only a few reference laboratories because of the time and expense needed to maintain quality serotyping antisera. Several attempts have been made by commercial enterprises to market limited serotyping antisera for *Campylobacter* species; however, either too few antisera were included or the quality was poor (63). In the future, a combination of phenotypic and genotypic tests such as flagellin gene typing (59) will be used for routine epidemiologic investigation of *Campylobacter* infections.

INFECTIVE DOSE AND SUSCEPTIBLE POPULATIONS

C. jejuni is susceptible to low pH, and thus the gastric environment is sufficient to kill most organisms (13). The infective dose of *C. jejuni*, however, does not appear to be high, with <1,000 organisms being capable of causing illness (11). The only controlled study to examine the infective dose of *C. jejuni* was conducted by Black and colleagues (11). In a study using two different challenge strains of *C. jejuni*, only 18% of human volunteers became ill when infected with 10^8 CFU of strain A3249; however, 46% of the volunteers became ill when infected with another strain, 81-176. In an interesting experiment, Robinson ingested 500 CFU of *C. jejuni* in milk, which resulted in abdominal cramps and nonbloody diarrhea occurring 4 days after ingestion and lasting 3 days (80). Although not studied directly, there appeared to be a dose-related effect on both rate of infection and severity of illness in individuals involved in an outbreak of *Campylobacter* infection after ingesting raw milk (14).

Young children and young adults, 20 to 40 years old, show the highest incidence of sporadic infections in the United States (90). The incidence of *Campylobacter* infection in developing countries such as Mexico and Thailand may be orders of magnitude higher than in the United States (90, 93). In contrast to developed countries, campylobacters are frequently isolated from individuals who may or may not have diarrheal disease. Most symptomatic infections occur in infancy and early childhood, and the incidence decreases with age (18, 93, 94). Age-related increases in humoral immune responses to *Campylobacter* antigens are associated with a decrease in symptomatic illness (50, 94). Travelers to developing countries may acquire *Campylobacter* infection, with isolation rates from 0 to 39% reported in different studies (93).

Bacteremia has been reported to occur at a rate of 1.5 per 1,000 intestinal infections, with the highest rate in the elderly (87). Persistent diarrheal illness and bacteremia may occur in immunocompromised hosts, such as in patients with human immunodeficiency virus infection or hypogammaglobulinemia, and are difficult to treat (12).

VIRULENCE FACTORS AND MECHANISMS OF PATHOGENICITY

Little is known about the mechanism by which *C. jejuni* causes human disease. *C. jejuni* can cause an enterotoxigenic-like illness with loose or watery diarrhea or an inflammatory colitis with fever and the presence of fecal blood and leukocytes and occasionally bacteremia that suggests an invasive mechanism of disease. A major problem in elucidating the pathogenesis of *Campylobacter* infection has been the lack of suitable animal models (30).

Cell Association and Invasion

C. jejuni has been shown to invade in vitro cell lines (21, 40) and/or translocate polarized intestinal epithelial cell lines such as Caco-2 (25, 43). *C. jejuni* may also interact with M cells in the Peyer's patches, which may be an additional pathway for entry of the organism into the intestinal submucosa (98). Ketley identified four distinct phenotypes among clinical strains: noninvasive, invasive, invasive with transcytosis, and transcytosis without invasion (36). Strain 81-176, which has been used by numerous investigators, invades an intestine-derived cell line, Caco-2, and appears within membrane-bound vacuoles (84). Invasion could be inhibited by sugars such as D-glucose, D-mannose, and D-fucose (85). The mechanism of internalization is not known, but uptake of strain 81-176 into another intestinal cell line, INT407, was blocked by microtubule depolymerization and inhibitors of coated-pit formation but not by microfilament depolymerization (65). Inhibitors of endosome acidification had no impact on intracellular survival of the organism. Uptake by cells appears to be an active process and is dependent on bacterial protein synthesis (65). In this regard, when cultured in association with eukaryotic cells, *C. jejuni* appears to produce proteins that are not apparent upon growth in culture medium alone and that may be important in internalization of the organism (42, 67). After uptake, there is some evidence for intracellular survival of the organism (37, 39). Pesci et al. (74) recently cloned a superoxide dismutase (*sodB*) gene containing iron cofactors from *C. jejuni*. A *sodB* mutant had a 12-fold decrease in ability to survive in INT407 cells, which suggests that the enzyme may have some role in intracellular survival (74).

Other proteins that may be important in the pathogenesis of infection include other outer membrane proteins. PEB1, a conserved surface protein in *C. jejuni* and *C. coli* that was recently cloned from *C. jejuni* 81-176, has homology with *Enterobacteriaceae* glutamine-binding protein (GlnH), lysine/arginine/ornithine-binding protein (LAO), and histidine-binding protein (HisJ) and may be involved in adherence (71).

Flagella and Motility

Campylobacter species are motile and have a single polar, unsheathed flagellum at one or both ends. Motility and flagella appear to be important determinants for the invasion-translocation process (31, 100). *Campylobacter* colonization and/or infection in a variety of animal models appear to be dependent on intact motility and full-length flagella (62); however, other colonization factors may be involved (54). Two genes, *flaA* and *flaB*, are involved in the expression of the flagellar filament and are arranged in tandem in both *C. jejuni* and *C. coli* (29, 32, 64). Motility and *flaA* appear to be essential for colonization (31, 62, 101). Components of intestinal mucin, particularly L-fucose, are chemotactic for *C. jejuni*, and motility toward these components may be important in the pathogenesis of infection (34). Miller et al. (55) cloned a gene, *flbA*, that codes for a protein that is involved in the synthesis of *Campylobacter* flagella and shows homology with virulence-related proteins of *Yersinia pestis* LcrD and *Salmonella typhimurium* InvA.

Toxins

C. jejuni has been reported to produce a cholera-like enterotoxin (20, 48, 83); however, other studies have refuted the significance of these findings (41, 66, 73). Daikoku et al. (20) more recently reported the isolation of a toxin with cholera toxin-like properties and found the material to contain three subunits with molecular masses of 68, 54, and 43 kDa. The material enhanced adenylate cyclase activity in HeLa cells, and the 68-kDa protein had immunologic cross-reactivity with cholera toxin. Further, it was reported that both the 68- and 54-kDa proteins had putative ganglioside binding activity (20). Small amounts of this toxin were detected in strains studied by using a rabbit ileal loop model and produced mucosal hemorrhage, inflammation with a polymorphonuclear leukocyte infiltrate, tissue edema, cell damage, and submucosal bleeding. Cholera toxin activity was not found in the tissues of infected animals (26).

The lack of genetic evidence for the presence of this putative toxin, shown by hybridization with probes directed against the A and B subunits of cholera toxin and *E. coli* heat-labile toxin genes (7, 66) or by low-stringency hybridization (17), raises serious doubts about the existence of a classical enterotoxin. *C. jejuni* infection of rabbit ileal loops caused an elevation of cyclic AMP (cAMP), prostaglandin E_2, and leukotriene B_4 levels in tissue and fluids. This fluid caused elevated cellular cAMP in Caco-2 cells and was inhibited by antiserum against prostaglandin E_2 (24). This finding suggests that inflammatory mediators elicited by *C. jejuni* might be involved in the pathogenesis of infection.

Other toxins such as Shiga toxins (57), cytolethal distending toxin (35), and hepatoxins (38) have been described. Cover et al. (19) studied the presence of cytotoxic activity in fecal filtrates of patients with *Campylobacter* enteritis. Cytotoxic activity was found in both patients with disease and healthy asymptomatic subjects, raising doubts about the clinical relevance of the toxin and its role in pathogenicity (19). The role of cytolethal distending toxin (35) in pathogenesis is unknown, but the toxin was reported to be active in CHO, Vero, HeLa, and HEp-2 cells but not in Y-1 cells. Forty-one percent of over 700 strains produced the toxin, and it produced a hemorrhagic response in rat ligated ileal loops. The production of a hepatotoxin described by Kita et al. (38) is interesting in that a strain of *C. jejuni* (GIFU 8734) produced hepatitis in mice, and specific toxin activity could be isolated from the organism.

Other Factors

Several investigators have recently examined the ability of *C. jejuni* to acquire iron from exogenous sources. *C. jejuni* does not appear to produce its own siderophores but can utilize exogenous siderophores from other bacteria as iron carriers (6). *C. jejuni* can obtain iron from hemin and hemoglobin. Hemolytic activity found in *C. jejuni* does not appear to be iron regulated (75). One of the iron-responsive regulatory genes, *fur* (ferric uptake regulator), has recently been cloned from *C. jejuni* and may have some role in the regulation of yet to be determined virulence factors (36, 103).

Autoimmune Sequelae

During the past few years, evidence showing an association of *Campylobacter* infection with acute inflammatory demyelinating polyneuropathy, or GBS, has accumulated (56). Recent studies have also shown an association of *Campylobacter* infection with an illness clinically similar to GBS, known as acute motor axonal neuropathy, that occurs primarily in northern China and may occur in other parts of the world (51).

Mishu and Blaser (56) estimate that the annual incidence of GBS preceded by *C. jejuni* infection ranges from 0.17 to 0.51 cases per 100,000 population and accounts for 425 to 1,272 cases per year in the United States. The pathogenesis of GBS induced by *C. jejuni* is not clear, but a working hypothesis is at least apparent for further investigation.

Kuroki et al. (44) determined that a certain serotype of *C. jejuni*, O:19, accounted for 83% of 12 strains isolated from patients with GBS, compared with 1.7% of over 1,000 sporadic strains. Of the 10 O:19 strains from GBS patients, all belonged to a specific lectin type, type 8, compared with only 1 of the sporadic O:19 strains. This study strongly suggested that O:19 strains were particularly unique and that some specific virulence factor(s) that could induce GBS was associated with these strains. Aspinall et al. (4, 5) and Yuki et al. (105, 106) have found that strains belonging to O:19 as well as other serotypes (O:4, O:1) of *Campylobacter* have core oligosaccharide LPS structures that mimic ganglioside structures such as GM1 and GD1a, a component of motor neurons. Host genetic factors such as HLA type may also play an important role in the development of disease (104).

GENETICS OF *CAMPYLOBACTER* SPECIES

Studying *Campylobacter* species at the genetic level has been difficult because of the lack of cloning and expression systems for identifying virulence determinants. Difficulties in genetic analysis, combined with the lack of suitable in vitro and in vivo models of infection, make the task of studying pathogenesis even more formidable. To date, the only virulence factors that have been studied in any great detail are the flagellar genes, primarily in *C. jejuni* 81116 and *C. coli* VC167, although other strains have now been studied (92). Other genes, including housekeeping genes and antibiotic resistance genes, have been cloned with success (92).

Taylor (92) suggested that difficulties in cloning and expressing genes from *C. jejuni* might be caused by the presence of unusual *Campylobacter* promoter sequences not utilized effectively in *E. coli*, that *E. coli* may lack accessory genes that are needed to process *Campylobacter* gene products, or that DNA may be unstable because of differences in methylation. Direct transposon mutagenesis of *Campylobacter* genes has also been unsuccessful (45). To facilitate cloning and expression of *Campylobacter* genes, several plasmid vectors have been developed (92). Shuttle and suicide vectors suitable for cloning *Campylobacter* genes were first developed by Labigne-Roussel and colleagues (46, 47), and additional cloning vectors have now been developed (45, 92). However, these vectors have not solved problems encountered with expression of *Campylobacter* proteins.

The identification of useful in vivo models as well as advances in studying *Campylobacter* virulence factors at the molecular level will be important for our future understanding of the pathogenesis of infection (95).

IMMUNITY

Protective immunity appears after infection with *Campylobacter* species and seems to be antibody mediated. Black et al. (11) showed that rechallenge of homologous strains 28 days after the initial volunteer challenge resulted in protection against illness but not necessarily against colonization with the organism. Blaser et al. (14) found that 76% of acutely exposed individuals who had not been previously exposed to raw milk became acutely ill, compared with none of 10 individuals who were regular milk drinkers and who drank the implicated milk. Humoral immunity appears to be an important component of protective immunity, as suggested by studies on persistent infection in immunocompromised hosts with human immunodeficiency virus infection or hypogammaglobulinemia (12).

In developing countries where *Campylobacter* infections are endemic, immunity to *Campylobacter* infection appears to be age dependent. In a cohort of Mexican children studied by Calva et al. (18), the ratio of symptomatic to asymptomatic infection decreased from infancy to 6 years of age. An inverse relationship between serum antiflagellin antibodies and diarrheal illness in another cohort of children also supports the role of antibodies in protective immunity (50). Production of cytokines during intestinal infection may play some role in immunity and immune responses. Using a mouse colonization model, Baqar et al. (10) found that oral administration of interleukin-5 and interleukin-6 reduced the level of gut colonization with *C. jejuni*.

Flagellin in an important immunogen during *Campylobacter* infection (61), and antibodies against this protein correlate to some degree with protective immunity. Breast feeding has been shown to provide protection against *Campylobacter* infection in studies conducted in developing countries (50, 53, 60, 82). Antibodies against flagellin present in breast milk appear to be associated with protection of infants against infection (60).

Several groups are working toward developing a vaccine for *Campylobacter* infection. Baqar et al. (9) tested an oral whole-cell killed vaccine coadministered with *E. coli* heat-labile enterotoxin as an immunoadjuvant to rhesus monkeys. The vaccine elicited both humoral and cellular responses; however, the ability to protect against infection was not studied. Killed whole-cell vaccine was shown to provide colonization protection in a mouse model (8). Guerry et al. (33) described the production of a *recA* mutant of *C. jejuni* 81-176 that was able to colonize rabbits and induce colonization protection. Whether this strain was attenuated in any way is not known, but *recA* mutants of other bacteria such as *S. typhimurium* are avirulent (33). Other strategies to prevent the transmission of infection to humans include improved hygiene practices during broiler production, such as decontamination of water supplies, use of competitive exclusion flora which may prevent colonization of young chicks, and immunological approaches through the use of animal vaccines (88, 89).

CONCLUSIONS

Campylobacter species are among the most important human bacterial enteric pathogens, yet little is known about how these intriguing organisms cause disease. Difficult challenges remain in identifying suitable in vitro and in vivo models of infection that will allow investigators to study *Campylobacter* species at both the biological and genetic levels. Elucidation of the genetic basis for pathogenesis of *Campylobacter* infection is still in the embryonic stage; however, several groups in the United States and in the United Kingdom and other European countries are making progress in developing new genetic systems for studying *Campylobacter* species. Finally, it is to be hoped that increasing awareness on the part of government and industry of the importance of *Campylobacter* infection will lead to infusion of resources for research in the future.

References

1. **Alderton, M. R., V. Korolik, P. J. Coloe, F. E. Dewhirst, and B. J. Paster.** 1995. *Campylobacter hyoilei* sp. nov., associated with porcine proliferative enteritis. *Int. J. Syst. Bacteriol.* **45:**61–66.
2. **Alm, R. A., P. Guerry, M. E. Power, H. Lior, and T. J. Trust.** 1991. Analysis of the role of flagella in the heat-labile Lior serotyping scheme of thermophilic campylobacters by mutant allele exchange. *J. Clin. Microbiol.* **29:**2438–2445.
3. **Altekruse, S. F., J. M. Hunt, L. K. Tollefson, and J. M. Madden.** 1994. Food and animal sources of human *Campylobacter jejuni* infection. *J. Am. Vet. Med. Assoc.* **204:**57–61.
4. **Aspinall, G. O., A. G. McDonald, H. Pang, L. A. Kurjanczyk, and J. L. Penner.** 1994. Lipopolysaccharides of *Campylobacter jejuni* serotype O:19: structures of core oligosaccharide regions from the serostrain and two bacterial isolates from patients with the Guillain-Barré syndrome. *Biochemistry* **33:**241–249.
5. **Aspinall, G. O., A. G. McDonald, T. S. Raju, H. Pang, A. P. Molan, and J. L. Penner.** 1993. Chemical structures of the core regions of *Campylobacter jejuni* serotypes O:1, O:4, O:23, and O:36 lipopolysaccharides. *Eur. J. Biochem.* **213:**1017–1027.
6. **Baig, B. H., I. K. Wachsmuth, and G. K. Morris.** 1986. Utilization of exogenous siderophores by *Campylobacter* species. *J. Clin. Microbiol.* **23:**431–433.
7. **Baig, B. H., I. K. Wachsmuth, G. K. Morris, and W. E. Hill.** 1988. Probing of *Campylobacter jejuni* DNA coding for *Escherichia coli* heat-labile enterotoxin. *J. Infect. Dis.* **154:**542. (Letter.)
8. **Baqar, S., L. A. Applebee, and A. L. Bourgeois.** 1995. Immunogenicity and protective efficacy of a prototype *Campylobacter* killed whole-cell vaccine in mice. *Infect. Immun.* **63:**3731–3735.
9. **Baqar, S., A. L. Bourgeois, P. J. Schultheiss, R. I. Walker, D. M. Rollins, R. L. Haberberger, and O. R. Pavlovskis.** 1995. Safety and immunogenicity of a prototype oral whole-cell killed *Campylobacter* vaccine administered with a mucosal adjuvant in non-human primates. *Vaccine* **13:**22–28.
10. **Baqar, S., N. D. Pacheco, and F. M. Rollwagen.** 1993. Modulation of mucosal immunity against *Campylobacter jejuni* by orally administered cytokines. *Antimicrob. Agents Chemother.* **37:**2688–2692.
11. **Black, R. E., M. M. Levine, M. L. Clements, T. P. Hughs, and M. J. Blaser.** 1988. Experimental *Campylobacter jejuni* infections in humans. *J. Infect. Dis.* **157:**472–480.
12. **Blaser, M. J.** 1995. *Campylobacter* and related species, p. 1948–1956. *In* G. L. Mandell, J. E. Bennett, and R. Dolin (ed.), *Principles and Practice of Infectious Diseases.* Churchill Livingstone, New York.
13. **Blaser, M. J., H. L. Hardesty, B. Powers, and W. L. Wang.** 1980. Survival of *Campylobacter fetus* subsp. *jejuni* in biological milieus. *J. Clin. Microbiol.* **11:**309–313.
14. **Blaser, M. J., E. Sazie, and P. Williams.** 1987. The influence of immunity on raw milk-associated *Campylobacter* infection. *JAMA* **257:**43–46.
15. **Blaser, M. J., J. G. Wells, R. A. Feldman, M. A. Pollard, and J. R. Allen.** 1983. Campylobacter enteritis in the United States: a multicenter study. *Ann. Intern. Med.* **98:**360–365.
16. **Borczyk, A., S. Thompson, D. Smith, and H. Lior.** 1987. Water-borne outbreak of *Campylobacter laridis*-associated gastroenteritis. *Lancet* **i:**164–165.
17. **Calva, E., J. Torres, M. Vázquez, V. Angeles, H. de la Vega, and G. M. Ruíz-Palacios.** 1989. *Campylobacter*

jejuni chromosomal sequences that hybridize to *Vibrio cholerae* and *Escherichia coli* LT enterotoxin genes. *Gene* **75**:243–251.

18. **Calva, J. J., G. M. Ruiz-Palacios, A. B. Lopez-Vidal, A. Ramos, and R. Bojalil.** 1988. Cohort study of intestinal infection with *Campylobacter* in Mexican children. *Lancet* **i**:503–505.
19. **Cover, T. L., G. I. Perez-Perez, and M. J. Blaser.** 1990. Evaluation of cytotoxic activity in fecal filtrates from patients with *Campylobacter jejuni* or *Campylobacter coli* enteritis. *FEMS Microbiol. Lett.* **58**:301–304.
20. **Daikoku, T., M. Kawaguchi, K. Takama, and S. Susuki.** 1990. Partial purification and characterization of the enterotoxin produced by *Campylobacter jejuni*. *Infect. Immun.* **58**:2414–2419.
21. **de Melo, M. A., G. Gabbiani, and J. Pechere.** 1989. Cellular events and intracellular survival of *Campylobacter jejuni* during infection of HEp-2 cells. *Infect. Immun.* **57**:2214–2222.
22. **Doyle, M. P., and D. M. Jones.** 1992. Food-borne transmission and antibiotic resistance of *Campylobacter jejuni*, p. 45–48. *In* I. Nachamkin, M. J. Blaser, and L. S. Tompkins (ed.), *Campylobacter jejuni: Current Status and Future Trends*. American Society for Microbiology, Washington, D.C.
23. **Endtz, H. P., G. J. Ruijs, B. van Klingeren, W. H. Jansen, T. van der Reyden, and R. P. Mouton.** 1991. Quinolone resistance in campylobacter isolated from man and poultry following the introduction of fluoroquinolones in veterinary medicine. *J. Antimicrob. Chemother.* **27**:199–208.
24. **Everest, P. H., A. T. Cole, C. J. Hawkey, S. Knutton, H. Goossens, J. P. Butzler, J. M. Ketley, and P. H. Williams.** 1993. Role of leukotriene B_4, prostaglandin E_2, and cyclic AMP in *Campylobacter jejuni*-induced intestinal fluid secretion. *Infect. Immun.* **61**:4885–4887.
25. **Everest, P. H., H. Goossens, J. P. Butzler, D. Lloyd, S. Knutton, J. M. Ketley, and P. H. Williams.** 1992. Differentiated Caco-2 cells as a model for enteric invasion by *Campylobacter jejuni* and *C. coli*. *J. Med. Microbiol.* **37**:319–325.
26. **Everest, P. H., H. Goossens, P. Sibbons, D. R. Lloyd, S. Knutton, R. Leece, J. M. Ketley, and P. H. Williams.** 1993. Pathological changes in the rabbit ileal loop model caused by *Campylobacter jejuni* from human colitis. *J. Med. Microbiol.* **38**:316–321.
27. **Finch, M. J., and P. A. Blake.** 1985. Foodborne outbreaks of campylobacteriosis: the United States experience, 1980–1982. *Am. J. Epidemiol.* **122**:262–268.
28. **Finch, M. J., and L. W. Riley.** 1984. Campylobacter infections in the United States: results of an 11-state surveillance. *Arch. Intern. Med.* **144**:1610–1612.
29. **Fischer, S. H., and I. Nachamkin.** 1991. Common and variable domains of the flagellin gene, flaA, in *Campylobacter jejuni*. *Mol. Microbiol.* **5**:1151–1158.
30. **Fox, J. G.** 1992. In vivo models of enteric campylobacteriosis: natural and experimental infections, p. 131–138. *In* I. Nachamkin, M. J. Blaser, and L. S. Tompkins (ed.), *Campylobacter jejuni: Current Status and Future Trends*. American Society for Microbiology, Washington, D.C.
31. **Grant, C. C. R., M. E. Konkel, W. Cieplak, and L. S. Tompkins.** 1993. Role of flagella in adherence, internalization, and translocation of *Campylobacter jejuni* in nonpolarized and polarized epithelial cells. *Infect. Immun.* **61**:1764–1771.
32. **Guerry, P., R. A. Alm, M. E. Power, and T. J. Trust.** 1992. Molecular and structural analysis of *Campylobacter* flagellin, p. 267–281. *In* I. Nachamkin, M. J. Blaser, and L. S. Tompkins (ed.), *Campylobacter jejuni: Current Status and Future Trends*. American Society for Microbiology, Washington, D.C.
33. **Guerry, P., P. M. Pope, D. H. Burr, J. Leifer, S. W. Joseph, and A. L. Bourgeois.** 1994. Development and characterization of *recA* mutants of *Campylobacter jejuni* for inclusion in vaccines. *Infect. Immun.* **62**:426–432.
34. **Hugdahl, M. B., J. T. Beery, and M. P. Doyle.** 1988. Chemotactic behavior of *Campylobacter jejuni*. *Infect. Immun.* **56**:1560–1566.
35. **Johnson, W. M., and H. Lior.** 1988. A new heat-labile cytolethal distending toxin (CLDT) produced by *Campylobacter* spp. *Microb. Pathog.* **4**:115–126.
36. **Ketley, J. M.** 1995. Virulence of *Campylobacter* species: a molecular genetic approach. *J. Med. Microbiol.* **42**:312–327.
37. **Kiehlbauch, J. A., R. A. Albach, L. L. Baum, and K. P. Chang.** 1985. Phagocytosis of *Campylobacter jejuni* and its intracellular survival in mononuclear phagocytes. *Infect. Immun.* **48**:446–451.
38. **Kita, E., D. Oku, A. Hamuro, F. Nishikawa, M. Emoto, Y. Yagyu, N. Katsui, and S. Kashiba.** 1990. Hepatotoxic activity of *Campylobacter jejuni*. *J. Med. Microbiol.* **33**:171–182.
39. **Konkel, M. E., S. F. Hayes, L. A. Joens, and W. Cieplak.** 1992. Characteristics of the internalization and intracellular survival of *Campylobacter jejuni* in human epithelial cell cultures. *Microb. Pathogen.* **13**:357–370.
40. **Konkel, M. E., and L. A. Joens.** 1989. Adhesion to and invasion of HEp-2 cells by *Campylobacter* spp. *Infect. Immun.* **57**:2984–2990.
41. **Konkel, M. E., Y. Lobet, and W. Cieplak.** 1992. Examination of multiple isolates of *Campylobacter jejuni* for evidence of cholera toxin-like activity, p. 193–198. *In* I. Nachamkin, M. J. Blaser, and L. S. Tompkins (ed.), *Campylobacter jejuni: Current Status and Future Trends*. American Society for Microbiology, Washington, D.C.
42. **Konkel, M. E., D. J. Mead, and W. Cieplak.** 1993. Kinetic and antigenic characterization of altered protein synthesis by *Campylobacter jejuni* during cultivation with human epithelial cells. *J. Infect. Dis.* **168**:948–954.
43. **Konkel, M. E., D. J. Mead, S. F. Hayes, and W. Cieplak, Jr.** 1992. Translocation of *Campylobacter jejuni* across human polarized epithelial cell monolayer cultures. *J. Infect. Dis.* **166**:308–315.

43a. **Krieg, N. R., and J. G. Holt.** 1984. *Bergey's Manual of Systematic Bacteriology*, vol. 1, p. 111. Williams & Wilkins, Baltimore.

44. **Kuroki, S., T. Saida, M. Nukina, T. Haruta, M. Yoshioka, Y. Kobayashi, and H. Nakanishi.** 1993. *Campylobacter jejuni* strains from patients with Guillain-Barré syndrome belong mostly to Penner serogroup 19 and contain β-N-acetylglucosamine residues. *Ann. Neurol.* **33**:243–247.

45. **Labigne, A., A. Courcoux, and L. Tompkins.** 1992. Cloning of *Campylobacter jejuni* genes required for leucine biosynthesis, and construction of leu-negative mutant of *C. jejuni* by shuttle transposon mutagenesis. *Res. Microbiol.* **153:**15–26.

46. **Labigne-Roussel, A., P. Courcoux, and L. Tompkins.** 1988. Gene disruption and replacement as a feasible approach for mutagenesis of *Campylobacter jejuni. J. Bacteriol.* **170:**1704–1708.

47. **Labigne-Roussel, A., J. Harel, and L. Tompkins.** 1987. Gene transfer from *Escherichia coli* to *Campylobacter* species: development of shuttle vectors for genetic analysis of *Campylobacter jejuni. J. Bacteriol.* **169:**5320–5323.

48. **Lindblom, G. B., M. Johny, K. Khalil, K. Mazhar, G. M. Ruiz-Palacios, and B. Kaijser.** 1990. Enterotoxigenicity and frequency of *Campylobacter jejuni, C. coli* and *C. laridis* in human and animal stool isolates from different countries. *FEMS Microbiol. Lett.* **54:**163–167.

49. **Lior, H., D. L. Woodward, J. A. Edgar, L. J. Laroche, and P. Gill.** 1982. Serotyping of *Campylobacter jejuni* by slide agglutination based on heat-labile antigenic factors. *J. Clin. Microbiol.* **15:**761–768.

50. **Martin, P. M. V., J. Mathiot, J. Ipero, M. Kirimat, A. J. Georges, and M. C. Georges-Courbot.** 1989. Immune response to *Campylobacter jejuni* and *Campylobacter coli* in a cohort of children from birth to 2 years of age. *Infect. Immun.* **57:**2542–2546.

51. **McKhann, G. M., D. R. Cornblath, J. W. Griffin, T. W. Ho, C. Y. Li, Z. Jiang, H. S. Wu, G. Zhaori, Y. Liu, L. P. Jou, T. C. Liu, C. Y. Gao, J. Y. Mao, M. J. Blaser, B. Mishu, and A. K. Asbury.** 1993. Acute motor axonal neuropathy: a frequent cause of acute flaccid paralysis in China. *Ann. Neurol.* **33:**333–342.

52. **Medema, G. J., F. M. Schets, A. W. van de Giessen, and A. H. Gavelaar.** 1992. Lack of colonization of 1 day old chicks by viable, non-culturable *Campylobacter jejuni. J. Appl. Bacteriol.* **72:**512–516.

53. **Megraud, F., G. Boudraa, K. Bessaoud, S. Bensid, F. Dabis, R. Soltana, and M. Touhami.** 1990. Incidence of campylobacter infection in infants in western Algeria and the possible protective role of breast feeding. *Epidemiol. Infect.* **105:**73–78.

54. **Meinersmann, R. J., W. E. Rigsby, N. J. Stern, L. C. Kelley, J. E. Hill, and M. P. Doyle.** 1991. Comparative study of colonizing and noncolonizing *Campylobacter jejuni. Am. J. Vet. Res.* **52:**1518–1522.

55. **Miller, S., E. C. Pesci, and C. L. Pickett.** 1993. A *Campylobacter jejuni* homolog of the LcrD/FlbF family of proteins is necessary for flagellar biogenesis. *Infect. Immun.* **61:**2930–2936.

56. **Mishu, B., and M. J. Blaser.** 1993. Role of infection due to *Campylobacter jejuni* in the initiation of Guillain-Barré syndrome. *Clin. Infect. Dis.* **17:**104–108.

57. **Moore, M. A., M. J. Blaser, G. I. Perez-Perez, and A. D. O'Brien.** 1988. Production of shiga-like cytotoxin by *Campylobacter. Microb. Pathog.* **4:**455–462.

58. **Nachamkin, I.** 1995. *Campylobacter* and *Arcobacter,* p. 483–491. *In* P. R. Murray, E. J. Baron, M. A. Pfaller, F. C. Tenover, and R. H. Yolken (ed.), *Manual of Clinical Microbiology,* 6th ed. ASM Press, Washington, D.C.

59. **Nachamkin, I., K. Bohachick, and C. M. Patton.** 1993. Flagellin gene typing of *Campylobacter jejuni* by restriction fragment length polymorphism analysis. *J. Clin. Microbiol.* **31:**1531–1536.

60. **Nachamkin, I., S. H. Fischer, X. H. Yang, O. Benitez, and A. Cravioto.** 1994. Immunoglobulin A antibodies directed against *Campylobacter jejuni* flagellin present in breast-milk. *Epidemiol. Infect.* **112:**359–365.

61. **Nachamkin, I., and X. H. Yang.** 1992. Immune response to *Campylobacter* flagellin, p. 216–222. *In* I. Nachamkin, M. J. Blaser, and L. S. Tompkins (ed.), *Campylobacter jejuni: Current Status and Future Trends.* American Society for Microbiology, Washington, D.C.

62. **Nachamkin, I., X. H. Yang, and N. J. Stern.** 1993. Role of *Campylobacter jejuni* flagella as colonization factors for three-day-old chicks: analysis with flagellar mutants. *Appl. Environ. Microbiol.* **59:**1269–1273.

63. **Nicholson, M. A., and C. M. Patton.** 1993. Evaluation of commercial antisera for serotyping heat-labile antigens of *Campylobacter jejuni* and *Campylobacter coli. J. Clin. Microbiol.* **31:**900–903.

64. **Nuijten, P. J., F. J. van Asten, W. Gaastra, and B. A. van der Zeijst.** 1990. Structural and functional analysis of two *Campylobacter jejuni* flagellin genes. *J. Biol. Chem.* **265:**17798–17804.

65. **Oelschlaeger, T. A., P. Guerry, and D. J. Kopecko.** 1993. Unusual microtubule-dependent endocytosis mechanisms triggered by *Campylobacter jejuni* and *Citrobacter freundii. Proc. Natl. Acad. Sci. USA* **90:**6884–6888.

66. **Olsvik, O., I. K. Wachsmuth, G. Morris, and J. C. Feeley.** 1984. Genetic probing of *Campylobacter jejuni* for cholera toxin and *Escherichia coli* heat-labile enterotoxin. *Lancet* **i:**449.

67. **Panigrahi, P., G. Losonky, L. J. DeTolla, and J. G. Morris.** 1992. Human immune response to *Campylobacter jejuni* proteins expressed in vitro. *Infect. Immun.* **60:**4938–4944.

68. **Park, R. W., P. L. Griffiths, and G. S. Moreno.** 1991. Sources and survival of campylobacters: relevance to enteritis and the food industry. *J. Appl. Bacteriol.* **20**(Suppl.): 97S–106S.

69. **Patterson, M. F.** 1995. Sensitivity of *Campylobacter* spp. to irradiation in poultry meat. *Lett. Appl. Microbiol.* **20:**338–340.

70. **Patton, C. M., and I. K. Wachsmuth.** 1992. Typing schemes: are current methods useful?, p. 110–128. *In* I. Nachamkin, M. J. Blaser, and L. S. Tompkins (ed.), *Campylobacter jejuni: Current Status and Future Trends.* American Society for Microbiology, Washington, D.C.

71. **Pei, Z., and M. J. Blaser.** 1992. PEB1, the major cell-binding factor of *Campylobacter jejuni,* is a homolog of the binding component in gram-negative nutrient transport systems. *J. Biol. Chem.* **268:**18717–18725.

72. **Penner, J. L., and J. N. Hennessy.** 1980. Passive hemagglutination technique for serotyping *Campylobacter fetus* subsp. *jejuni* on the basis of soluble heat-stable antigens. *J. Clin. Microbiol.* **12:**732–737.

73. **Perez-Perez, G. I., D. N. Taylor, P. D. Echeverria, and M. J. Blaser.** 1992. Lack of evidence of enterotoxin involvement in pathogenesis of *Campylobacter* diarrhea, p. 184–192. *In* I. Nachamkin, M. J. Blaser, and L. S. Tompkins (ed.), *Campylobacter jejuni: Current Status and Future*

Trends. American Society for Microbiology, Washington, D.C.

74. **Pesci, E. C., D. L. Cottle, and C. L. Pickett.** 1994. Genetic, enzymatic, and pathogenic studies of the iron superoxide dismutase of *Campylobacter jejuni*. *Infect. Immun.* **62:**2687–2694.
75. **Pickett, C. L., T. Auffenberg, E. C. Pesci, V. L. Sheen, and S. S. D. Jusuf.** 1992. Iron acquisition and hemolysis production by *Campylobacter jejuni*. *Infect. Immun.* **60:**3872–3877.
76. **Potter, M. E., M. J. Blaser, R. K. Sikes, A. F. Kaufmann, and J. G. Wells.** 1983. Human *Campylobacter* infection associated with certified raw milk. *Am. J. Epidemiol.* **117:**475–483.
77. **Radomyski, T., E. A. Murano, D. G. Olson, and P. S. Murano.** 1994. Eliminaton of pathogens of significance in food by low-dose irradiation: a review. *J. Food Prot.* **57:**73–86.
78. **Rautelin, H., O. V. Renkonen, and T. U. Kosunen.** 1991. Emergence of fluoroquinolone resistance in *Campylobacter jejuni* and *Campylobacter coli* in subjects from Finland. *Antimicrob. Agents Chemother.* **35:**2065–2069.
79. **Reina, J., N. Borrell, and A. Serra.** 1992. Emergence of resistance to erythromycin and fluoroquinolones in thermotolerant *Campylobacter* strains isolated from feces 1987–1991. *Eur. J. Clin. Microbiol. Infect. Dis.* **11:**1163–1166.
80. **Robinson, D. A.** 1981. Infective dose of *Campylobacter jejuni* in milk. *Br. Med. J.* **282:**1584.
81. **Rollins, D. M., and R. R. Colwell.** 1986. Viable but nonculturable stage of *Campylobacter jejuni* and its role in the survival in the natural aquatic environment. *Appl. Environ. Microbiol.* **52:**531–538.
82. **Ruiz-Palacios, G. M., J. J. Calva, L. K. Pickering, Y. Lopez-Vidal, P. Volkow, H. Pezzarossi, and M. S. West.** 1990. Protection of breast-fed infants against *Campylobacter* diarrhea by antibodies in human milk. *J. Pediatr.* **116:**707–713.
83. **Ruiz-Palacios, G. M., N. I. Torres, B. R. Ruiz-Palacios, J. Torres, E. Escamilla, and J. Tamayo.** 1983. Cholera-like enterotoxin produced by *Campylobacter jejuni*. *Lancet* **ii:**250–253.
84. **Russell, R. G., and D. C. Blake.** 1994. Cell association and invasion of Caco-2 cells by *Campylobacter jejuni*. *Infect. Immun.* **62:**3773–3779.
85. **Russell, R. G., M. O'Donnoghue, D. C. Blake, J. Zultry, and L. J. DeTolla.** 1993. Early colonic damage and invasion of *Campylobacter jejuni* in experimentally challenged infant *M. mulatta* monkeys. *J. Infect. Dis.* **168:**210–215.
86. **Skirrow, M. B.** 1994. Diseases due to *Campylobacter*, *Helicobacter* and related bacteria. *J. Comp. Pathol.* **111:** 113–149.
87. **Skirrow, M. B., D. M. Jones, E. Sutcliffe, and J. Benjamin.** 1993. *Campylobacter* bacteremia in England and Wales, 1981–1991. *Epidemiol. Infect.* **110:**567–573.
88. **Stern, N. J.** 1992. Reservoirs for *Campylobacter jejuni* and approaches for intervention in poultry, p. 49–60. *In* I. Nachamkin, M. J. Blaser, and L. S. Tompkins (ed.), *Campylobacter jejuni: Current Status and Future Trends*. American Society for Microbiology, Washington, D.C.
89. **Stern, N.** 1994. Mucosal competitive exclusion to diminish colonization of chickens by *Campylobacter jejuni*. *Poult. Sci.* **73:**402–407.
90. **Tauxe, R. V.** 1992. Epidemiology of *Campylobacter jejuni* infections in the United States and other industrialized nations, p. 9–19. *In* I. Nachamkin, M. J. Blaser, and L. S. Tompkins (ed.), *Campylobacter jejuni: Current Status and Future Trends*. American Society for Microbiology, Washington, D.C.
91. **Tauxe, R. V., N. Hargrett-Bean, C. M. Patton, and I. K. Wachsmuth.** 1988. *Campylobacter* isolates in the United States, 1982–1986. *Morbid. Mortal. Weekly Rep.* **37:**1–13.
92. **Taylor, D. E.** 1992. Genetics of *Campylobacter* and *Helicobacter*. *Annu. Rev. Microbiol.* **46:**35–64.
93. **Taylor, D. N.** 1992. *Campylobacter* infections in developing countries, p. 20–30. *In* I. Nachamkin, M. J. Blaser, and L. S. Tompkins (ed.), *Campylobacter jejuni: Current Status and Future Trends*. American Society for Microbiology, Washington, D.C.
94. **Taylor, D. N., P. Echeverria, C. Pitarangsi, J. Seriwatana, L. Bodhidatta, and M. J. Blaser.** 1988. Influence of strain characteristics and immunity on the epidemiology of *Campylobacter* infections in Thailand. *J. Clin. Microbiol.* **26:**863–868.
95. **Tompkins, L. S.** 1992. Genetic and molecular approaches to *Campylobacter* pathogenesis, p. 2441–254. *In* I. Nachamkin, M. J. Blaser, and L. S. Tompkins (ed.), *Campylobacter jejuni: Current Status and Future Trends*. American Society for Microbiology, Washington, D.C.
96. **Vandamme, P., M. I. Daneshvar, F. E. Dewhirst, B. J. Paster, K. Kersters, H. Goossens, and C. W. Moss.** 1995. Chemotaxonomic analyses of *Bacteroides gracilis* and *Bacteroides ureolyticus* and reclassification of *B. gracilis* as *Campylobacter gracilis* comb. nov. *Int. J. Syst. Bacteriol.* **45:**145–152.
97. **Velazquez, J. B., A. Jimenez, B. Chomon, and T. Villa.** 1995. Incidence and transmission of antibiotic resistance in *Campylobacter jejuni* and *Campylobacter coli*. *J. Antimicrob. Chemother.* **35:**173–178.
98. **Walker, R. I., E. A. Schmauder-Chock, J. L. Parker, and D. Burr.** 1988. Selective association and transport of *Campylobacter jejuni* through M cells of rabbit Peyer's patches. *Can. J. Microbiol.* **34:**1142–1147.
99. **Wang, W. L., B. W. Powers, N. W. Luechtefeld, and M. J. Blaser.** 1980. Effects of disinfectants on *Campylobacter jejuni*. *Appl. Environ. Microbiol.* **45:**1202–1205.
100. **Wassenaar, T. M., N. M. Bleumink-Pluym, and B. A. van der Zeijst.** 1991. Inactivation of *Campylobacter jejuni* flagellin genes by homologous recombination demonstrates that flaA but not flaB is required for invasion. *EMBO J.* **10:**2055–2061.
101. **Wassenaar, T. M., B. A. M. Van Der Zeijst, R. Ayling, and D. G. Newell.** 1993. Colonization of chicks by motility mutants of *Campylobacter jejuni* demonstrates the importance of flagellin A expression. *J. Gen. Microbiol.* **139:** 1171–1175.
102. **Wood, R. C., K. L. MacDonald, and M. T. Osterholm.** 1992. Campylobacter enteritis outbreaks associated with drinking raw milk during youth activities. A 10-year review of outbreaks in the United States. *JAMA* **268:**3228–3230.

103. **Wooldridge, K. G., P. H. Williams, and J. M. Ketley.** 1994. Iron-responsive genetic regulation in *Campylobacter jejuni:* cloning and characterization of a *fur* homolog. *J. Bacteriol.* **176**:5852–5856.

104. **Yuki, N., S. Sato, T. Itoh, and T. Miyatake.** 1991. HLA-B35 and acute axonal polyneuropathy following Campylobacter infection. *Neurology* **41**:1561–1563.

105. **Yuki, N., T. Taki, F. Inagaki, T. Kasama, M. Takahashi, K. Saito, S. Handa, and T. Miyatake.** 1993. A bacterium lipopolysaccharide that elicits Guillain-Barré syndrome has a GM1 ganglioside structure. *J. Exp. Med.* **178**:1771–1775.

106. **Yuki, N., T. Taki, M. Takahashi, K. Saito, T. Tai, T. Miyatake, and S. Handa.** 1994. Penner's serogroup 4 of *Campylobacter jejuni* has a lipopolysaccharide that bears a GM1 ganglioside epitope as well as one that bears a GD1a epitope. *Infect. Immun.* **62**:2101–2103.

Michael P. Doyle
Tong Zhao
Jianghong Meng
Shaohua Zhao

10

Escherichia coli O157:H7

INTRODUCTION

Escherichia coli strains are a common part of the normal facultative anaerobic microflora of the intestinal tracts of humans and warm-blooded animals. Isolates are serologically differentiated on the basis of three major surface antigens, which enable serotyping: the O (somatic), H (flagella), and K (capsule) antigens. At present, a total of 174 O antigens, 56 H antigens, and 80 K antigens have been identified. Most *E. coli* strains are harmless commensals; however, some strains are pathogenic and cause diarrheal disease.

E. coli strains that cause diarrheal illness are categorized into specific groups based on virulence properties, mechanisms of pathogenicity, clinical syndromes, and distinct O:H serogroups. These categories include: enteropathogenic *E. coli* strains (EPEC), enterotoxigenic *E. coli* strains (ETEC), enteroinvasive *E. coli* strains (EIEC), diffuse-adhering *E. coli* strains (DAEC), enteroaggregative *E. coli* strains (EAggEC), and enterohemorrhagic *E. coli* strains (EHEC).

EPEC

EPEC can cause severe diarrhea. The major O serogroups associated with illness include O55, O86, O111ab, O119, O125ac, O126, O127, O128ab, and O142. Humans are an important reservoir. The original definition of EPEC is "diarrheagenic *E. coli* belonging to serogroups epidemiologically incriminated as pathogens but whose pathogenic mechanisms have not been proven to be related to either heat-labile enterotoxin, heat-stable enterotoxin, or *Shigella*-like invasiveness." However, EPEC have been determined to induce the attaching and effacing (AE) lesions in cells to which they adhere and can invade epithelial cells (27). Some EPEC can produce one or more toxins.

ETEC

ETEC are a major cause of infantile diarrhea in developing countries. They are also the agents most frequently responsible for traveler's diarrhea. ETEC colonize the proximal small intestine by fimbrial colonization factors (e.g., CFA/I and CFA/II) and produce a heat-labile or heat-stable enterotoxin that elicits fluid accumulation and a diarrheal response. The most frequent ETEC serogroups include O6, O8, O15, O20, O25, O27, O63, O78, O85, O115, O128ac, O148, O159, and O167. Humans are the principal reservoir of ETEC that cause human illness.

EIEC

EIEC cause nonbloody diarrhea and dysentery similar to that caused by *Shigella* spp. by invading and multiplying within colonic epithelial cells. As for *Shigella* spp., the invasive capacity of EIEC is associated with the presence of a large plasmid (ca. 140 MDa) which encodes several outer membrane proteins (OMPs) involved in invasiveness. The antigenicity of these OMPs and that of the O antigens of EIEC are closely related. The principal site of bacterial localization is the colon, where EIEC invade and proliferate in epithelial cells, causing cell death. Humans are a major reservoir, and the serogroups most frequently associated with illness include O28ac, O29, O112, O124, O136, O143, O144, O152, O164, and O167. Among these serogroups, O124 is commonly encountered.

DAEC

DAEC have been associated with diarrhea in children in Mexico. These strains can produce mild diarrhea without blood or fecal leukocytes and are identified by a characteristic diffuse-adherent pattern of adherence to HEp-2 or HeLa cell lines. These bacteria cover the cell surface uniformly. DAEC generally do not elaborate heat-labile or heat-stable toxins or elevated levels of Shiga toxins (Stxs), nor do they possess EPEC adherence factor plasmids or invade epithelial cells.

EAggEC

EAggEC recently have been associated with persistent diarrhea in infants and children in several countries worldwide (119). These *E. coli* strains are uniquely different from the other types of pathogenic *E. coli* because of their ability to produce a characteristic pattern of aggregative adherence on HEp-2 cells. EAggEC adhere in an appearance likened to stacked bricks to the surface of HEp-2 cells. A gene probe derived from a plasmid associated with EAggEC strains has been developed to identify *E. coli* of this type (25); however, considerably more epidemiologic information is needed to elucidate its significance as an agent of diarrheal disease.

EHEC

EHEC were first identified as human pathogens in 1982, when *E. coli* of serotype O157:H7 was associated with two outbreaks of hemorrhagic colitis. Since then, certain strains and serogroups of *E. coli*, including O26:H11, O103, O104, O111, and sorbitol-fermenting O157:H^{-}, have been associated with cases of bloody diarrhea and so also have been identified as EHEC (15, 16, 117). However, serotype O157:H7 is the predominant cause of EHEC-associated disease in the United States and many other countries. All EHEC produce factors cytotoxic to African green monkey kidney (Vero) cells, which have been described as verotoxins (VTs) or Shiga-like toxins (SLTs) because VT1 is similar to the Shiga toxin (Stx) produced by *Shigella dysenteriae* type 1. The ability of *E. coli* O157:H7 to produce VTs was first reported by Johnson et al. (54). Many serotypes of *E. coli* have subsequently been shown to produce VTs; hence, they have been named VT-producing *E. coli* or SLT-producing *E. coli* (SLTEC). A new nomenclature for the SLT (VT) family has been proposed recently in which SLT (VT) have been renamed after the prototype of the family, Stx (12). However, only those strains that cause hemorrhagic colitis are considered to be EHEC. Since *E. coli* O157:H7 is the most common serotype of the EHEC and because more is known about this serotype than other serotypes of EHEC, this chapter will focus specifically on *E. coli* O157:H7.

CHARACTERISTICS OF *E. COLI* O157:H7

The first confirmed isolation of *E. coli* O157:H7 in the United States was in 1975 from a California woman with bloody diarrhea. The bacterium was first identified as a human pathogen in 1982, when it was associated with two foodborne outbreaks of hemorrhagic colitis (60, 96). Most strains of *E. coli* O157:H7 possess several characteristics uncommon to most other *E. coli:* inability to grow well, if at all, at temperatures of ≥44.5°C, inability to ferment sorbitol within 24 h, inability to produce β-glucuronidase (i.e., inability to hydrolyze 4-methylumbelliferyl-D-glucuronide), possession of an attaching and effacing (*eae*) gene, carriage of a 60-MDa plasmid, and expression of an uncommon 5,000 to 8,000-molecular-weight OMP (85).

Acid Tolerance

Unlike most foodborne pathogens, *E. coli* O157:H7 is uniquely tolerant to acidic environments (8, 21, 42, 87, 127, 128). Inoculation studies have revealed that *E. coli* O157:H7 can survive fermentation, drying, and storage of fermented sausage (pH 4.5) for up to 2 months at 4°C, with only a 100-fold reduction in cell populations (42). Studies using acetic, citric, or lactic acid at concentrations of up to 1.5% as organic acid sprays on beef revealed that *E. coli* O157:H7 populations were not appreciably affected by any of the treatments (8). *E. coli* O157:H7, when inoculated at high levels, survived in mayonnaise (pH 3.6 to 3.9) for 5 to 7 weeks at 5°C and for 1 to 3 weeks at 20°C (127) and survived in apple cider (pH 3.6 to 4.0) for 10 to 31 or 2 to 3 days at 8 or 25°C (128), respectively. Outbreaks of *E. coli* O157:H7 infection have been directly associated with consumption of contaminated dry salami and apple cider. The mechanism of acid tolerance has not been fully elucidated but appears to be associated with a protein(s) that can be induced by preexposing the bacteria to acid conditions.

Table 10.1 Comparison of *D* values for *E. coli* O157:H7 and *Salmonella* spp. in ground beef

Temp (°C)	D value (min)		
	E. coli O157:H7		*Salmonella* spp.[a]
	30.5% fat[b]	17–20% fat[c]	
51.7	115.5	ND[d]	54.3
57.2	5.3	4.5	5.43
62.8	0.47	0.40	0.54

[a]From reference 43.
[b]From reference 68.
[c]From reference 32.
[d]ND, not determined.

Antibiotic Resistance

Early surveys of antibiotic resistance revealed that *E. coli* O157:H7 isolates were sensitive to most antibiotics (88). However, recent studies have revealed a trend toward increased resistance to antibiotics. For example, of 56 *E. coli* O157:H7 isolates collected between 1984 and 1987, all were susceptible to the antibiotics tested; however, 13 of 176 isolates (7.4%) isolated between 1989 and 1991 were resistant to streptomycin, sulfisoxazole, and tetracycline (64).

Thermal Inactivation

Studies on the thermal sensitivity of *E. coli* O157:H7 in ground beef have revealed that the pathogen has no unusual resistance to heat, with *D* values at 57.2, 60, 62.8, and 64.3°C of 270, 45, 24, and 9.6 s, respectively (32). Heating ground beef sufficiently to kill typical strains of *Salmonella* will also kill *E. coli* O157:H7 (Table 10.1). The presence of fat increases the thermal tolerance of *E. coli* O157:H7 in ground beef, with *D* values for lean (2.0% fat) and fatty (30.5% fat) ground beef of 4.1 and 5.3 min, respectively, at 57.2°C and 0.3 and 0.5 min, respectively, at 62.8°C (68). Pasteurization of milk (72°C, 16.2 s) has also been determined to be an effective treatment that will kill more than 10^4 *E. coli* O157:H7 cells per ml (23). Proper heating of foods of animal origin, i.e., heating foods to an internal temperature of at least 68.3°C, is an important critical control point to ensure inactivation of *E. coli* O157:H7.

RESERVOIRS OF *E. COLI* O157:H7

Cattle

Most confirmed human *E. coli* O157:H7 outbreaks have been associated with the consumption of undercooked ground beef and, less frequently, unpasteurized milk; hence, cattle have been the focus of many studies to determine their involvement in transmitting the pathogen (7, 71, 122, 124).

Detection of *E. coli* O157:H7 on Farms

The first reported isolation of *E. coli* O157:H7 from cattle was from a <3-week-old calf with colibacillosis in Argentina in 1977 (84). A study of cattle in herds associated with two cases of human *E. coli* O157:H7 infection revealed that 2.3% of calves and 3.0% of heifers, but only 0.15% of adult cows, shed *E. coli* O157:H7 in feces (123). Hence, young animals tend to carry *E. coli* O157:H7 more frequently than adult cattle.

The first national survey of cattle for carriage of *E. coli* O157:H7 isolated the pathogen from only 0.36% of 6,894 preweaned calves in 19 (1.8%) of 1,068 herds sampled in 28 states (41). However, a follow-up study of many of these herds by using more sensitive detection and isolation procedures revealed that *E. coli* O157:H7 prevalence in calves was much higher than previously reported (129). Results revealed that among control herds (from which *E. coli* O157:H7 was not previously isolated), the pathogen was isolated from 6 of 399 calves (1.5%) that were 24 h of age to weaning and from 13 of 263 calves (4.9%) that were weaned to 4 months of age. Eleven of 50 control herds (22%) were positive. Among case herds (from which *E. coli* O157:H7 was previously isolated), *E. coli* O157:H7 was isolated from 5 of 171 calves (2.9%) that were 24 h of age to weaning and from 7 of 132 calves (5.3%) that were weaned to 4 months of age. Seven of 14 case herds (50%) were positive.

The prevalence of *E. coli* O157:H7 in herds in Washington State was determined to range from 8.3% for 60 dairy herds (0.28% positive of 3,570 fecal samples) to 16% for 25 cow/calf operations (0.21% positive of 1,412 fecal samples) (47). A recent U.S. Department of Agriculture survey of cattle from 100 feedlots in 13 states revealed that *E. coli* O157:H7 was isolated from 1.61% and *E. coli* O157:NM was isolated from 0.4% of 11,881 rectal samples (24).

Methods used to isolate *E. coli* O157:H7 from feces were not equally sensitive; hence, results of some studies may underrepresent the actual prevalence of *E. coli* O157:H7 carriage by cattle. Populations of *E. coli* O157:H7 detected in calf feces ranged from <10^2 CFU/g in 15 of 31 calves to more than 10^5 but less than 10^6 CFU/g in 3 of 31 calves (129). In addition, fecal shedding of *E. coli* O157:H7 is frequently intermittent and of relatively short duration (several weeks). Studies of calves perorally administered *E. coli* O157:H7 revealed that withholding feed from cattle for 2 days results in some animals in large increases in populations of *E. coli* O157:H7 shed in feces (9).

Direct transmission of *E. coli* O157:H7 from calves to children has been reported (92). *E. coli* O157:H7 phage type 23 was isolated from two children 13 months and 5 years of age living on a farm in Ontario, Canada, as well as from two of seven healthy calves on the farm (92).

A small cluster of cases of *E. coli* O157:H7 infection associated with drinking unpasteurized milk from a dairy farm in England prompted an investigation of the farm. *E. coli* O157 was isolated from 10 (9.5%) of the cattle fecal samples, from 1 (10%) of 10 raw milk samples, and from farmyard manure slurry (17). All *E. coli* O157 strains isolated from human patients, farmyard slurry, bovine rectal swabs, and milk samples harbored a single 92-kb plasmid, produced Stx2 but not Stx1, and were phage type 2 (17). Not only were cattle and their milk determined to be sources of *E. coli* O157, but manure in the environment of the farmyard also was identified as a source.

Sources of *E. coli* O157:H7 for cattle have not been clearly identified. Possible sources include contaminated feedstuffs or water, colonized animals in herds, infected wildlife and humans, or contaminated facilities and equipment surfaces from contact with feces. *E. coli* O157:H7 is isolated predominantly from young animals, with the highest rate of isolation from postweaned calves (41, 47, 129).

Studies in England revealed that 84 of 2,103 (4%) of cattle at slaughter were *E. coli* O157:H7 positive (18). Testing cattle before slaughter and carcasses from the same animals after processing revealed that 7 (30%) of 23 carcasses of rectal swab-positive cattle and 2 (8%) of 25 carcasses of rectal swab-negative cattle were *E. coli* O157:H7 positive (18). These results suggest that cross-contamination of carcasses occurs during slaughter and processing or that *E. coli* O157:H7-positive cattle are not always detected by fecal testing.

Factors Associated with Carriage of *E. coli* O157:H7

Several tentative associations between fecal shedding of *E. coli* O157:H7 and feeding have been made from epidemiologic studies of dairy herds. For example, some calf starter feed regimens or environmental factors, and feed components such as whole cottonseed, were associated with reduced prevalence of *E. coli* O157:H7. In contrast, grouping calves before weaning, sharing feeding utensils among calves without sanitation, and early feeding of grain were associated with increased carriage of *E. coli* O157:H7 (41). Manure handling and the use of computerized feeders also were associated with increased carriage of *E. coli* O157:H7 on dairy farms (47). Fasting calves for 48 h increases the populations of *E. coli* O157:H7 in the gastrointestinal tracts of some animals (9). Additional studies are needed to verify those factors influencing the carriage of *E. coli* O157:H7 by cattle.

Cattle Model for Infection of *E. coli* O157:H7

Studies of calves perorally administered a large dose (10^{10} CFU) of *E. coli* O157:H7 revealed that the bacterium does not cause diarrhea or apparent illness in animals. Calves appear normal, and fecal shedding of *E. coli* O157:H7 varies among animals of the same age group but tends to persist longer in calves than in adults (22). Studies by Brown et al. (9) revealed that the initial sites of localization of *E. coli* O157:H7 in cattle are the forestomachs (rumen, omasum, and reticulum).

Domestic Animals

Experimentally, chicks have been readily colonized following peroral administration of small populations (25 cells) of *E. coli* O157:H7 (100). In addition, chicks administered larger populations of *E. coli* O157:H7 continued to excrete the organism in feces of about half of the birds for at least 10 to 11 months postinoculation (100). Considering that chicks can be readily colonized by small populations of *E. coli* O157:H7 and continue to be long-term shedders, it is possible that chickens and hen eggs can serve as vehicles of *E. coli* O157:H7 (100, 106). However, surveys of 50 poultry farms did not detect *E. coli* O157:H7 in any of 500 individual birds. Hence, it appears that at present poultry is not a primary source of *E. coli* O157:H7. Sheep recently have been identified as carriers of *E. coli* O157:H7 (67). Carriage was transient and was most prevalent during the summer months. Although *E. coli* O157:H7 has been isolated from retail poultry, lamb, and pork (33), the source of contamination is unknown. The pathogen could have been introduced by cross-contamination with beef during cutting at retail.

Humans

Several outbreaks of *E. coli* O157:H7 infection have occurred in day care settings in which the pathogen was spread by person-to-person contact (5, 104). Secondary transmission by humans also has occurred and can further amplify a foodborne outbreak (4). This occurred in an outbreak at a nursing home at which staff became a second wave of transmission by person-to-person spread from the patients. The fecal shedding of *E. coli* O157:H7

Table 10.2 Reported outbreaks and clusters of *E. coli* O157:H7 infection in the United States, 1982 to 1994[a]

Yr	No. of outbreaks	No. of cases
1982	2	47
1984	2	70
1986	2	52
1987	1	51
1988	3	153
1989	2	246
1990	2	75
1991	4	54
1992	4	75
1993	17	1,006
1994	30	511
Total	68	2,334

[a]From reference 14.

by patients with hemorrhagic colitis or hemolytic-uremic syndrome (HUS) usually lasts for no more than 13 to 21 days following onset of symptoms (58). However, in some instances, fecal excretion of the pathogen can last for weeks (83). A child infected during a day care center-associated outbreak continued to excrete the pathogen for 62 days after the onset of diarrhea. Studies of persons living on dairy farms to determine carriage of *E. coli* O157:H7 by farm families revealed elevated antibody titers against the surface antigens of *E. coli* O157; however, the pathogen was not isolated from feces. An asymptomatic long-term carrier state has not been identified.

DISEASE OUTBREAKS

Since its identification as a pathogen in 1982, *E. coli* O157:H7 has been the cause of a series of outbreaks in the United States. Through 1994, 68 outbreaks or clusters of *E. coli* O157:H7 infection have been documented (14). They have increased from an average of 2 per year between 1982 and 1992 to 17 in 1993 and 30 in 1994. This dramatic rise may be due in part to improved recognition of *E. coli* O157:H7 infection following the publicity of the large multistate outbreak in the western United States in January 1993. Since January 1, 1994, individual cases of *E. coli* O157:H7 infection became reportable to the National Notifiable Diseases Surveillance System. Numbers of reported outbreaks and clusters and numbers of cases of infection in the United States between 1982 and 1994 are presented in Table 10.2. Details of individual outbreaks occurring between 1990 and 1994 are provided in Table 10.3. However, the precise incidence of *E. coli* O157:H7 foodborne illness in the United States is not known because infected persons presenting mild or no symptoms and persons with nonbloody diarrhea are less likely to seek medical attention than patients with bloody diarrhea. Investigators at the Centers for Disease Control and Prevention estimate that *E. coli* O157:H7 infection accounts for approximately 20,000 cases of illness and 250 deaths annually (13). However, long-term surveillance is necessary before firm conclusions can be made regarding the incidence of disease.

Geographic Distribution

Geographically, the focus of attention for addressing *E. coli* O157:H7 infections has been largely on the North American continent. Most published studies with an appreciable frequency of detection of this pathogen in sporadic diarrheal diseases have been from the United States and Canada. However, recent reports reveal a considerable contribution of *E. coli* O157:H7 or *E. coli* O157:NM to human disease in other areas of the world as well. A large epidemic involving several thousand patients was reported from Swaziland and South Africa, following consumption of surface water contaminated with *E. coli* O157:NM (48). At least 16 countries on six continents have reported human cases of *E. coli* O157:H7 infection or isolation of the pathogen from bovines, indicating the widespread distribution of the pathogen. However, most occurrences of *E. coli* O157:H7 infection outside North America have involved sporadic cases, although some outbreaks have been reported in Europe and Japan (45). Data from the United Kingdom indicate that reported isolations of *E. coli* O157:H7 from patients increased from 178 in 1989 to 532 in 1991 (112). The most common toxin profile of these strains was production of only Stx2, which is quite different from the most common toxin profile, i.e., coproduction of Stx2 and Stx1, reported for isolates in the United States (112).

Seasonality of *E. coli* O157:H7

Outbreaks and clusters of *E. coli* O157:H7 peak during the warmest months of the year (45). Of outbreaks and clusters reported in the United States, 83.4% (57 of 68) occurred from May to October. The reasons for this seasonal pattern are unknown, but they could hypothetically include (i) an increased prevalence of the pathogen in cattle or other livestock or vehicles of transmission during the summer, (ii) greater human exposure to ground beef or other *E. coli* O157:H7-contaminated foods during the "cook-out" months, and/or (iii) greater improper handling (temperature abuse) or incomplete cooking of products such as ground beef during warm months than other months.

Table 10.3 Reported outbreaks of *E. coli* O157:H7 in the United States for the period 1990 to 1994[a]

Yr	Mo	No.	State	Setting	No. of cases	Source
1990	7	1	ND	Community	65	Roast beef
	11	2	MT	School	10	School lunch
1991	7	1	OR	Community	21	Swimming water
	8	2	WA	Picnic	2	Ground beef
	9	3	MN	Fair	8	Ground beef
	11	4	MA	Community	23	Apple cider
1992	5	1	NY	Unknown	5	Unknown
	6	2	NV	Day care	57	Person to person
	9	3	ME	Home	4	Vegetables; person to person
	12	4	OR	Community	9	Raw milk
1993	1	1[b]	ID	Community	13	Ground beef
	1	—[b]	NV	Community	58	Ground beef
	1	—[b]	CA	Community	32	Ground beef
	1	—[b]	WA	Community	629	Ground beef
	3	2	OR	Restaurant	47	Mayonnaise
	6	3	ME	Unknown	4	Unknown
	6	4	OR	Home	6	Raw milk
	7	5	NC	Day care	27	Person to person
	7	6	IL	Community	8	Unknown
	7	7	NM	Party	4	Unknown
	7	8	MA	Community	10	Ground beef
	7	9	WA	Church picnic	16	Pea salad
	7	10	CA	Home	10	Ground beef
	8	11	OR	Restaurant	27	Cantaloupe
	8	12	PA	Community	3	Ground beef
	8	13	WA	Restaurant	53	Salad bar
	9	14	CT	Club barbeque	23	Ground beef
	9	15	MT	Community	8	Ground beef
	10	16	WA	Restaurant	9	Unknown
	10	17	TX	Unknown	10	Unknown
1994	1	1[b]	WA	Home	11	Ground beef
	1	—[b]	OR	Home	10	Ground beef
	2	2	MN	Unknown	8	Ground beef
	4	3	NE	Home, camp	24	Ground beef
	5	4	ND	Restaurant	33	Ground beef
	5	5	CA	Home	9	Ground beef
	5	6	OH	Community	10	Coney dog sauce
	6	7	NY	Home	17	Ground beef
	6	8	CT	Home	21	Retail meats
	6	9	CT	Community	2	Ground beef
	6	10	OH	Day care	8	Person to person
	6	11	PA	Home	4	Ground beef
	7	12	VA	Community	7	Unknown
	7	13	VA	Camp	20	Unknown
	7	14	OH	Community	5	Unknown
	7	15	WI	Day care	43	Person to person
	7	16	OK	Restaurant	4	Unknown
	7	17	HI	Unknown	17	Unknown
	7	18	NY	Day camp	5	Unknown

(continued)

Table 10.3 Reported outbreaks of *E. coli* O157:H7 in the United States for the period 1990 to 1994[a] *(continued)*

Yr	Mo	No.	State	Setting	No. of cases	Source
1994	7	19	MI	Day care	13	Person to person
	7	20	NJ	Homes	89	Unknown
	7	21	NY	Community	12	Swimming water
	8	22	TX	Cafeteria	26	Salad bar
	8	23	KY	Market	5	Unknown
	8	24	FL	Unknown	9	Unknown
	8	25	OH	Day care	6	Person to person
	9	26	MN	College	11	Unknown
	9	27	NY	Oktoberfest	36	Unknown
	10	28	WA	Home	7	Apple cider
	11	29[b]	WA	Home	15	Salami
	11	—[b]	CA	Home	4	Salami
	11	30	NM	School	20	Unknown

[a]From reference 14.
[b]Multistate outbreak.

Age of Patients

The very young and the elderly are at greatest risk of *E. coli* O157:H7 infection. The age-specific prevalence is highest in children under 5 years of age, gradually decreases with age, and then increases again in persons 65 years of age or older. The very young and very old are most susceptible to *E. coli* O157:H7 sequelae (i.e., HUS) and death from complications (45).

Locations of Outbreaks

Outbreaks have occurred in schools, custodial and chronic care institutions, day care centers, homes, and the community at large; the community outbreaks are usually associated with a restaurant exposure.

Transmission of *E. coli* O157:H7 Infection

Although a variety of foods have been implicated in *E. coli* O157:H7-associated illness, most outbreaks have been associated with consumption of raw or undercooked foods of bovine origin. Among the 68 outbreaks in the United States with an identified food vehicle, 22 (32.4%) were traced to ground beef, 2 (2.9%) were traced to roast beef, and 2 (2.9%) were traced to raw milk (Table 10.4). *E. coli* O157:H7 infections also were associated with eating other foods, including vegetables, apple cider, cantaloupe, mayonnaise-containing salad dressing, and salami. Contact of foods with *E. coli* O157:H7-containing meat or feces (human or bovine) is a likely source of cross-contamination. Person-to-person transmission (13.2%) and waterborne (4.4%) outbreaks also have been documented (Table 10.4).

Examples of Outbreaks

The Original Outbreak

The first documented outbreak of *E. coli* O157:H7 infection occurred in Oregon in 1982, with 26 cases and 19 persons hospitalized (96). All patients had bloody diarrhea and severe abdominal pain. The median age was 28 years, with a range of 8 to 76 years. The duration of illness ranged from 2 to 9 days, with a median of 4 days. This outbreak was associated with eating undercooked hamburgers from fast-food restaurants of a specific chain. *E. coli* O157:H7 was recovered from stools of patients. A second outbreak followed 3 months later and was associated with the same fast-food restaurant chain in Michigan, with 21 cases and 14 persons hospitalized. The median age was 17 years, with a range of 4 to 58

Table 10.4 Leading vehicle foods or mode of spread known for *E. coli* O157:H7 outbreaks in the United States, 1982 to 1994[a]

Rank	Vehicle food(s) or source	No. of outbreaks (%)
1	Ground beef	22 (32.4)
2	Person to person	9 (13.2)
3	Vegetables, salad bars	4 (5.9)
4	Water, swimming water	3 (4.4)
5	Roast beef	2 (2.9)
5	Raw milk	2 (2.9)
5	Apple cider	2 (2.9)
	Unknown	19 (27.9)

[a]From reference 14.

years. Contaminated hamburgers again were implicated as the vehicle, and *E. coli* O157:H7 was isolated both from patients and from a frozen ground beef patty.

Drinking Water Outbreak

A large outbreak of *E. coli* O157:H7 infection associated with contamination of municipal water occurred in Cabool, Mo., between December 16, 1989, and January 15, 1990 (107). Among the 243 affected people identified, 86 had bloody diarrhea, 32 were hospitalized, 2 had HUS, and 4 died. The mean age was 38 years, and all four deaths were women 79 years of age or older. The city had an unchlorinated water supply. Shortly before the peak of the outbreak, 45 in-ground water meters were replaced and two large water mains broke as a result of extreme cold weather. The number of new cases declined rapidly after residents were instructed to boil water and after chlorination of the water supply. A case-control study revealed that no food was associated with illness and that those who were ill drank more municipal water than controls ($P = 0.04$).

Swimming Water Outbreak

In the summer of 1991, an outbreak of *E. coli* O157:H7 was traced to a lakeside park near Portland, Ore. (62). Twenty-one cases of park-associated *E. coli* O157:H7 infection were identified from 19 households. All of the patients were children (median age, 6 years; range, 1 to 16 years); seven, including three with HUS, were hospitalized. The vehicle was fecally contaminated lake water ingested by bathers, among whom were many toddlers not yet toilet trained. All 21 patients reported swimming. Among the bathers, a history of swallowing lake water was associated with illness, and the case patients tended to spend more time in the water. Seven additional cases, presumably the result of secondary transmission, also were identified from four of these households.

Apple Cider Outbreak

Among the unusual foodborne *E. coli* O157:H7 outbreaks in the United States was an occurrence in southeastern Massachusetts in the fall of 1991 that was traced to apple cider (6). Twenty-three cases of *E. coli* O157:H7 infection involving 13 families were identified, with one to five cases per family. Four children had a diagnosis of HUS, and six patients were hospitalized.

Although *E. coli* O157:H7 was not isolated from apple cider made by the implicated processor, a case-control study identified the apple cider as the vehicle of transmission. The implicated cider processor may have pressed apples that were contaminated by soil containing *E. coli* O157:H7, because more than 90% of the apples used in the cider were collected from the ground, or contamination may have occurred during processing. Apples were pressed without washing, the cider was not pasteurized, and no preservatives were added. Prior to the outbreak, apple cider was considered safe from pathogens because of its high acidity and low pH (<4.0). Inoculation studies revealed that *E. coli* O157:H7 is acid tolerant and able to survive in apple cider for 20 days at 8°C (128). Since the infectious dose of *E. coli* O157:H7 is low, ingestion of even a small number of surviving *E. coli* O157:H7 could cause an illness.

The Largest Multistate Outbreak

The largest documented outbreak of *E. coli* O157:H7 infection in the United States occurred in Washington, Idaho, California, and Nevada in early 1993 (4, 14, 44). Approximately 90% of primary cases were associated with eating at a single fast-food restaurant chain (chain A), from which *E. coli* O157:H7 was isolated from hamburger patties. Transmission was amplified by secondary spread (48 patients in Washington alone) via person-to-person transmission. In total, 731 cases were identified: 629 in Washington, 13 in Idaho, 57 in Las Vegas, Nevada, and 34 in southern California. The median age of patients was 11 years, with a range of 4 months to 88 years. One hundred seventy-eight were hospitalized, 56 developed HUS, and 4 children died. Because neither specific laboratory testing nor surveillance for *E. coli* O157:H7 was carried out for earlier cases in Nevada, Idaho, and California, the outbreak went unrecognized until a sharp increase in cases of HUS was identified and investigated in Washington.

The outbreak resulted because of insufficient cooking of hamburgers by chain A restaurants. Epidemiologic investigation revealed that 10 of 16 hamburgers cooked according to chain A's cooking procedures in Washington had internal temperatures of below 60°C, which is substantially less than the minimum internal temperature of 68.3°C required by the State of Washington. An internal temperature of 68.3°C should be sufficient to kill this pathogen in ground beef.

Salami Outbreak

Another unusual outbreak was associated with consumption of dry fermented salami in 1994 in Washington and California and involved 19 cases (13). Among the 15 laboratory-confirmed cases in Washington, the median age of the patients was 6 years (range, 23 months to 77 years). Three patients, including a 2-year-old who developed HUS, required hospitalization. Among the four patients in California, two children developed HUS. Dry-cured salami is not cooked but is usually fermented and dried. The implicated product contained 85% pork, 15% beef, and spices and was fermented with a lactic acid bacterial culture to a final pH of 4.9. Inoculation studies have revealed that *E. coli* O157:H7 can survive the fermentation, drying, and storage of fermented sausage (42).

CHARACTERISTICS OF DISEASE

The spectrum of human illness of *E. coli* O157:H7 infection includes nonbloody diarrhea, hemorrhagic colitis, HUS (20), and thrombotic thrombocytopenic purpura (TTP) (31). Some persons are infected but asymptomatic.

Symptoms of hemorrhagic colitis include a prodromal phase consisting of crampy abdominal pain followed within 1 to 2 days by a nonbloody diarrhea which progresses within 1 or 2 days to bloody diarrhea that lasts for 4 to 10 days. Outbreak investigations revealed that more than 90% of cases of diarrhea caused by *E. coli* O157:H7 were bloody (4). Symptoms generally persist for several days to a few weeks.

HUS largely affects children and is the leading cause of acute renal failure in children. The syndrome is characterized by a triad of features: acute renal insufficiency, microangiopathic hemolytic anemia, and thrombocytopenia. Thrombocytopenia is a typical feature but is transient and is therefore occasionally missed. Several different pathogens are suspected agents of HUS, but *E. coli* O157:H7 is considered the primary cause. Approximately 10% of children under 10 years of age develop overt HUS after symptomatic infection with *E. coli* O157:H7 (108). TTP largely affects adults and resembles HUS histologically. It is accompanied by distinct neurological abnormalities resulting from blood clots in the brain. Fortunately, TTP is a rare syndrome of *E. coli* O157:H7 infection.

Pathological Changes

In cases of HUS, significant pathological changes include swelling of endothelial cells, widened subendothelial regions, and hypertrophied mesangial cells between glomerular capillaries. These changes combine to narrow the lumina of the glomerular capillaries and afferent arterioles and result in thrombosis of the arteriolar and glomerular microcirculation. Histopathologic examination reveals renal endothelial cell perturbation resulting from direct platelet aggregation without subendothelial exposure and from platelet subendothelial adhesion, but no consistent clotting abnormalities. Complete obstruction of renal microvessels can produce glomerular and tubular necrosis, with an increased probability of subsequent hypertension or renal failure (46, 74).

In cases of TTP, because of direct, potentially reversible platelet aggregation in the microcirculation of the brain, early arteriolar and capillary microthrombi are produced without perivascular inflammation. The formation of multimers of von Willebrand factor likely promotes the direct aggregation of platelets in this instance by multimeric bridges through the activation of disulfide bonds (66, 74).

Outcome

About one-third of patients infected with *E. coli* O157:H7 require hospitalization (4). Approximately half of the patients with overt symptoms of HUS require blood dialysis, and three-quarters require transfusions of erythrocytes and/or platelets. The death rate of cases of HUS is less than 5% in North America. The use of antibiotics and anticoagulants in treating HUS is controversial.

TOXIC INFECTIVE DOSE AND SUSCEPTIBLE POPULATIONS

Infective Dose

Retrospective analysis of foods associated with outbreaks of *E. coli* O157:H7 infection revealed that the infective dose is low. For example, between 0.3 and 15 *E. coli* O157:H7 cells per g were found in lots of frozen ground beef patties associated with a major outbreak in the western United States (44, 130). Similarly, 0.3 to 0.4 *E. coli* O157:H7 cells per g were detected in several intact packages of salami that were associated with a foodborne outbreak (13). These data suggest that the infectious dose of *E. coli* O157:H7 is quite low, i.e., less than a few hundred cells. Additional evidence for a low infectious dose is the capability for person-to-person transmission of *E. coli* O157:H7 infection. The ability of the pathogen to tolerate acidic conditions likely enables *E. coli* O157:H7 to survive the acidic environment of the stomach.

Susceptible Populations

All age groups can be affected by *E. coli* O157:H7, but infants and young children most frequently experience severe illness. HUS usually occurs in children, whereas

TTP infrequently but principally occurs in adults (66). Population-based studies have suggested that the highest age-specific incidence of *E. coli* O157:H7 infection is in the group 2 to 10 years of age (3, 120). The high rate of infection in this group likely is the result of increased exposure to contaminated foods, contaminated environments, and infected animals, more opportunities for person-to-person spread between infected children with relatively undeveloped hygiene skills, and the lack of adequate protective antibodies to Stxs (90). The fairly high rate of infection among young adults, especially females, suggests that sociological factors are involved; e.g., women are more likely to prepare food and be the primary care providers for infected young children.

Estimates from clinical data indicate that approximately 10% of children under the age of 10 infected with *E. coli* O157:H7 receive medical attention for overt symptoms of HUS (108). Isolation of *E. coli* O157:H7 from feces is much greater from specimens collected during the first 7 days after onset of illness than later. Generally, patients clear *E. coli* O157:H7 from their gastrointestinal tracts rapidly. By the time HUS develops, two-thirds of patients affected no longer have *E. coli* O157:H7 in their stools (58, 109). Patients whose infections with *E. coli* O157:H7 do not progress to HUS appear to have a somewhat longer duration of detectable excretion (5). An infected patient's serological response against surface epitopes of *E. coli* O157:H7 can last from weeks to months. It can be useful for epidemiologic studies to determine the serological responses of patients suspected of *E. coli* O157:H7 infection when stool cultures are negative for recovery of *E. coli* O157:H7 (19). However, such studies are limited by the fact that not all patients who were *E. coli* O157:H7 culture positive have a demonstrable antibody response.

MECHANISMS OF PATHOGENICITY

The precise mechanism of pathogenicity of *E. coli* O157:H7 has not been fully elucidated. Significant virulence factors, however, have been identified on the basis of histopathologic examination of tissues of HUS and hemorrhagic colitis patients, studies with tissue culture and animal models, and studies using genetic approaches. A general body of knowledge of the pathogenicity of *E. coli* O157:H7 has been developed and indicates that the organism causes disease by its ability to adhere to the host cell membrane (possibly invading host cells) and then producing Stx1 and/or Stx2. Adherence factors, Stx1, and Stx2 are critical factors in the pathogenesis of the *E. coli* O157:H7 infection.

Adherence

Although the adherence factors of *E. coli* O157:H7 have not been completely elucidated, substantial progress in characterizing adherence factors and their genes has been made recently (56, 111). Studies in animal models revealed that the pathogen can colonize the ceca and colons of orally infected gnotobiotic piglets, chickens, infant rabbits, and mice by an AE mechanism (Fig. 10.1). The AE lesion is characterized by intimate attachment of the bacteria to intestinal cells, with effacement of the underlying microvilli and accumulation of filamentous actin in the subjacent cytoplasm.

Formation of the AE Lesion

The molecular genetics of the AE lesion was studied first in EPEC and subsequently in *E. coli* O157:H7. A three-stage model of adhesion and AE lesion formation by EPEC has been proposed by Donnenberg and Kaper (28). The three stages are (i) localized adherence, (ii) signal transduction, and (iii) intimate adherence. Localized adherence, which involves the initial attachment of the bacterium to epithelial cells, is mediated by bundle-forming pili and probably other fimbriae. Signal transduction results in an increased level of intracellular calcium, release of inositol phosphates, and tyrosine phosphorylation of a 90-kDa epithelial cell protein that leads to the effacement of microvilli. Signal transduction is proposed to be mediated by a product of the *eaeB* gene (see below) and other EPEC gene products that require products of the *sep* (secretion of EPEC proteins) locus for surface presentation. Intimate adherence is mediated by intimin, a 94-kDa OMP encoded by the *eaeA* locus. The intimate adherence of the bacterium amplifies the accumulation of filamentous actin and other cytoskeletal proteins within the epithelial cell.

Although *E. coli* O157:H7 produces an AE lesion in the large intestine similar to that induced by EPEC, which in contrast occurs predominantly in the small intestine, the identities of the bacterial factors and the genes responsible for each stage of AE lesion formation have not been fully elucidated. It is believed that similar mechanisms, but factors and genes different from those of EPEC, regulate the processes of AE lesion formation by *E. coli* O157:H7. Karch et al. (57) demonstrated that the adherence of *E. coli* O157:H7 to INT407 tissue culture cells is mediated by nonhemagglutinating, mannose-resistant fimbriae encoded by a 60-MDa plasmid. Carriage of the 60-MDa plasmid, however, does not appear to influence attachment of the bacteria to the intestinal mucosa of gnotobiotic piglets, and some nonfimbriate *E. coli* O157:H7 strains can adhere to HEp-2

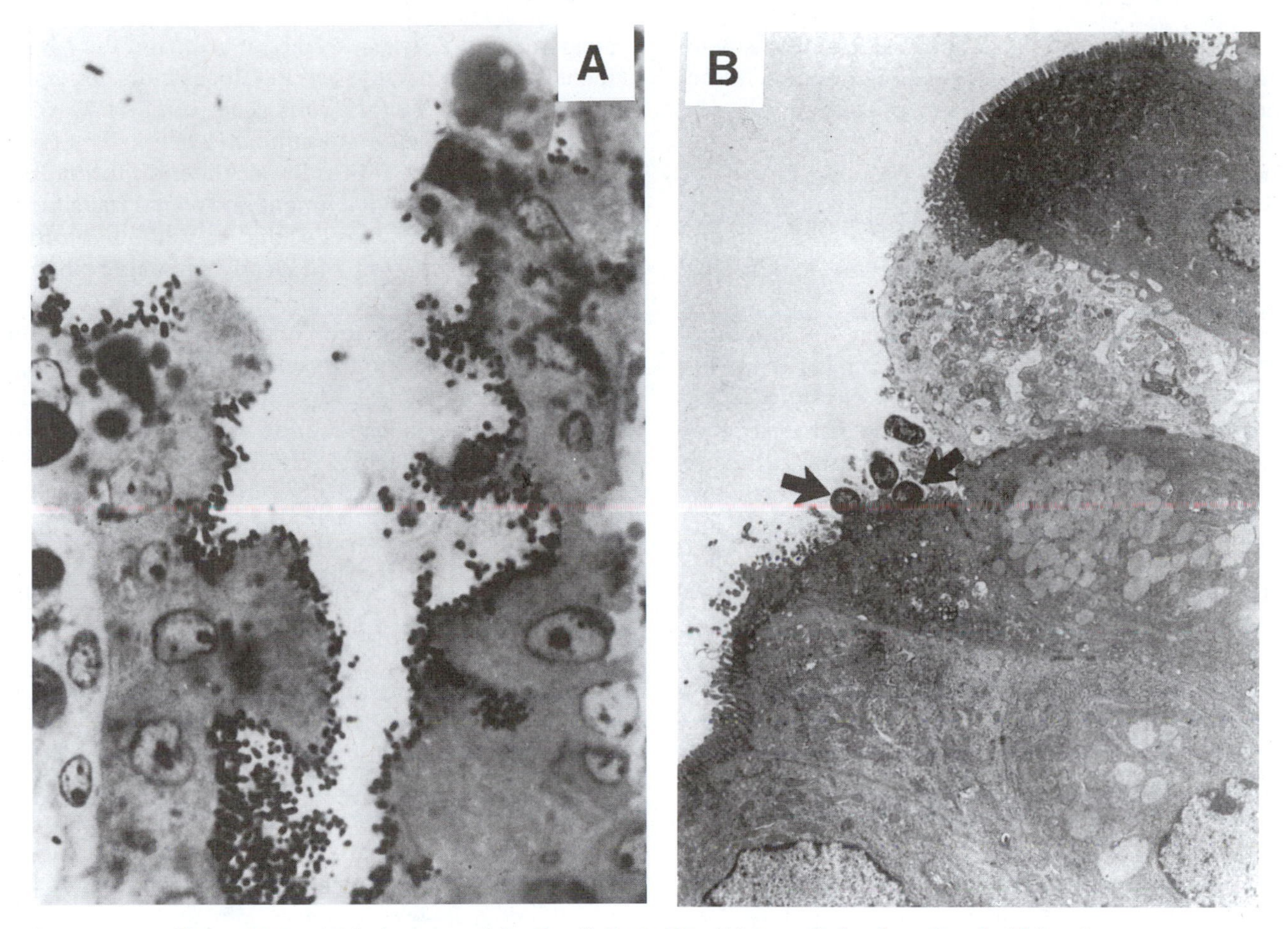

Figure 10.1 AE lesion caused by *E. coli* O157:H7. (A) Irregularly shaped, cuboidal to low columnar crypt neck cells with bacteria adherent to luminal plasma membrane of a gnotobiotic pig (toluidine blue stain). (Reproduced with permission from reference 36.) (B) Transmission electron photomicrograph of rabbit ileum with adherent bacteria (arrow). (Reproduced with permission from reference 103.)

cells (113). Studies by Sherman and Soni (102) revealed that the constituents of outer membranes, but not lipopolysaccharide and H7 flagella, can mediate attachment of *E. coli* O157:H7 to epithelial cells. The ability of antisera raised against the OMPs to inhibit adherence was specific for *E. coli* O157:H7 and had no effect on EPEC adhesion, suggesting that *E. coli* O157:H7 OMPs may serve as adhesins and are antigenically distinct from OMPs expressed by EPEC (101). The bundle-forming pili associated with adherence of EPEC are not produced by *E. coli* O157:H7.

eae Gene and Function

Studies on the mechanism of EPEC adherence revealed that the presence of a 60-MDa pMAR2 plasmid is associated with localized adherence, but transfer of this plasmid into a non-AE *E. coli* strain does not confer upon the recipient strain the ability to attach and efface. EPEC cured of this plasmid can still produce AE lesions on epithelial cells but at much lower frequency than the wild-type strain (56). Other bacterial factors in addition to those encoded by the plasmid appear to be involved with AE lesion formations.

Development of a fluorescent-actin staining test, based on the formation of filamentous actin at sites of bacterial attachment, provided an assay to study the genetics of the AE phenotype. Jerse and Kaper (53) used Tn*phoA* mutagenesis to study this phenotype and isolated mutants that were deficient of actin staining in the fluorescent-actin staining test. The locus into which Tn*phoA* had inserted was an open reading frame encoding a 102-kDa protein. The gene was designated *eae*, for *E. coli* attaching and effacing. The EPEC *eae* gene and its product, which is necessary for production of AE lesions, alone do not confer lesions. Additional loci downstream of the *eae* gene have a role in formation of the AE lesion. Hence, identification of this cluster of genes involved in AE lesion formation has necessitated a change in nomenclature, whereby the *eae* gene is now termed the *eaeA* gene, with subsequent genes termed *eaeB*, etc. The *eaeB* gene in EPEC encodes a 39-kDa secreted protein which is necessary for signal transduction in epithelial cells (30).

By using the *eaeA* gene of EPEC as a probe for homolog hybridization, the *eaeA* gene of *E. coli* O157:H7 has been cloned and sequenced (2, 125). Homologies between the *eaeA* genes of EPEC and *E. coli* O157:H7 were 86 and 83% at the nucleotide and amino acid levels, respectively. Their predicted amino acid sequences have significant similarities with sequences of the invasion proteins of *Yersinia pseudotuberculosis* and *Y. enterocolitica*. Interestingly, the greatest divergence between the predicted protein products of the *E. coli* O157:H7 and EPEC *eaeA* genes is at their C-terminal ends. Starting at the N terminus, the first 704 amino acid residues have 94% identity, whereas the remaining residues have only 49% identity (125). Beebakhee et al. (2) determined that the *eaeA* gene sequence of *E. coli* O157:H7 is 97% homologous to the EPEC *eaeA* gene sequence for the first 2,200 bp and 59% homologous for the last 800 bp.

Differences in the putative receptor-binding regions of EPEC and *E. coli* O157:H7 likely occur because EPEC localize primarily in the small intestine, whereas *E. coli* O157:H7 localizes primarily in the colon and distal ileum. Perhaps the divergence of carboxyl termini of the *eaeA* gene products results in different antigenic and receptor specificities for these putative adhesins. Similar differences in sequence homology also occur in the *eaeA*-like genes of *Hafnia alvei* and *Citrobacter freundii*. *H. alvei* resembles EPEC in its ability to efface rabbit small intestine and is associated with diarrhea in children in Bangladesh, whereas *C. freundii* appears to efface only the colonic mucosae of mice. As in EPEC, an *eaeB* gene is present in *E. coli* O157:H7, but its function has not been elucidated (56).

The function of the *eaeA* gene of *E. coli* O157:H7 and its product (intimin) has been studied by several groups (29, 34, 37, 69). An *eaeA*-deficient mutant of *E. coli* O157:H7 that does not adhere intimately to colonic epithelial cells of newborn piglets and is fluorescent-actin staining test negative on HEp-2 cells was constructed (29). Intimate attachment in vivo was restored when the *E. coli* O157:H7 *eaeA* gene or the EPEC *eaeA* gene was reintroduced into the mutant. These results indicate that the *eaeA* gene is necessary for intimate attachment of *E. coli* O157:H7 in vivo, and the complementation obtained by the EPEC locus indicates that the proteins encoded by the *eaeA* genes of *E. coli* O157:H7 and EPEC are functionally similar (29).

The precise role of intimin in AE lesion formation is uncertain. Intimate attachment and signal transduction events in epithelial cells that result in profound cytoskeletal disruption are clearly separable phenomena. While EPEC *eaeA* deletion mutants are incapable of intimate attachment, they retain the ability to induce host cell tyrosine kinases and actin accumulation. In contrast, other Tn*phoA* mutants of EPEC that are deficient in cytoskeletal disruption remain capable of intimate attachment. Hence, intimin appears to be needed only for the intimate attachment of *E. coli* O157:H7 in forming the AE lesion and is an adhesin that binds to a host cell receptor.

Expressing the *eaeA* gene of *E. coli* O157:H7 in a non-AE *E. coli* strain does not enable intimate attachment either in vitro or in vivo, indicating that the *eaeA* gene by itself is not sufficient for AE lesion formation, and additional bacterial factors and genes likely participate in or regulate this process (69). Dytoc et al. (34) described two Tn*phoA* mutants deficient in bacterial factors that are necessary for *E. coli* O157:H7 attachment and effacement and are likely distinct from the *eaeA* gene product. These factors, however, have not been identified. A low-molecular-weight OMP described by Zhao et al. (126) was determined to be associated with adherence of *E. coli* O157:H7 strain HA1 to human intestinal cells (INT407). A Tn*phoA* mutant deficient in this OMP was substantially less adherent to INT407 cells than the parental strain. It appears that multiple determinants, both regulatory and structural in nature, contribute to the AE phenotype of *E. coli* O157:H7 strains.

Recent studies revealed that the *eaeA*, *eaeB*, and *sep* genes are all located within a ca. 35-kb region that is present in both EPEC and *E. coli* O157:H7 (56, 72). This region, termed LEE, for locus of enterocyte effacement, is not present in normal flora *E. coli*, *E. coli* K-12, or ETEC but is present in other AE bacteria, including *C. freundii* biotype 4280, *H. alvei*, and *E. coli* RDEC-1, which causes diarrhea in rabbits. The site of insertion of the LEE region of EPEC and *E. coli* O157:H7 in the *E. coli* K-12 chromosome is at ca. 82 min, immediately adjacent to the *selC* locus encoding tRNA for selenocysteine. This is also the site of insertion for the retronphage ϕR73 and a large (ca. 70-kb) portion of uropathogenic *E. coli* virulence genes. This large insert for uropathogenic *E. coli* has been termed a pathogenicity island, and insertion of the EPEC and *E. coli* O157:H7 LEE at the same site suggests that this region of the *E. coli* chromosome is a hot spot for insertion of virulence factor genes.

Role of a 60-MDa Plasmid (pO157)

E. coli O157:H7 isolates associated with human illness harbor a plasmid (pO157) of approximately 60 MDa (unrelated to the 60-MDa plasmid found in EPEC) that contains DNA sequences common to plasmids present in other serotypes of SLTEC isolated from patients with hemorrhagic colitis. Because of the unique association of the 60-MDa plasmid with SLTEC, it was hypothesized that the plasmid might encode a virulence factor(s), perhaps involved in the organism's adherence to host epithelial cells. On the basis of this assumption, studies on the association of pO157 with adherence were done with wild-type strains, plasmid-cured strains, and strains of nonpathogenic *E. coli* containing pO157 to determine their ability to infect intestinal epithelial tissue culture cells (e.g., Henle 407 and HEp-2 cells), rabbit intestinal tissues, and gnotobiotic piglets. Studies by Junkins and Doyle (55) revealed that adherence to Henle 407 cells by *E. coli* O157:H7 strain 932 was not dependent on the 60-MDa plasmid. Tzipori et al. (115) determined that plasmid-cured strains produced in gnotobiotic piglets the same mucosal lesions typical of bacterial attachment and effacement as those strains that carried the plasmid. The plasmid also has been associated with expression of fimbriae by some strains of *E. coli* O157:H7 and adhesion to epithelial cells (57, 116); however, others determined that the plasmid-encoded pili are not involved in adherence of the pathogen (55, 113, 115, 116). The plasmid does encode a hemolysin, and the gene encoding the hemolysin has been described by Schmidt et al. (98). Because of the strong conservation of the pO157 plasmid among *E. coli* O157:H7 and other EHEC, and the positive reaction of sera of HUS patients to the hemolysin encoded by pO157, this plasmid is believed to play a role in the pathogenesis of disease, but its function is unclear.

Invasion

E. coli O157:H7 has been considered for many years a noninvasive bacterium. This determination was based on the fact that infection with *E. coli* O157:H7 typically does not result in febrile illness, there are no histopathologic changes of the intestine characteristic of infection with bacteria such as EIEC or *Shigella* spp., fecal leukocytes are not usually present, and the Sereny keratoconjunctivitis assay is negative (95). Also, it has been reported that *E. coli* O157:H7 is not able to invade INT407 and HEp-2 intestinal cell lines (27, 95, 102). However, recently, Oelschlaeger et al. (80) observed that *E. coli* O157:H7 isolates from HUS patients are internalized in T-24 bladder and HCT-8 ileocecal tissue culture cells but not in INT407 or HEp-2 cells. Unlike *Shigella* and *Salmonella* strains, which rely on microfilaments to invade human cells, *E. coli* O157:H7 strains rely on microtubules for invasion. Furthermore, different strains employed different uptake pathways. When wild-type, plasmid-containing *E. coli* O157:H7 strains were compared with their plasmid-cured isogenic derivatives for the ability to invade, wild-type and plasmid-cured strains invaded equally well, thereby indicating that the ability to invade is chromosomally encoded (80). The involvement of an invasive mechanism in *E. coli* O157:H7 infection needs further elucidation.

Shiga Family Toxins (Stxs)

E. coli O157:H7 produces one or two cytotoxins that are cytotoxic to Vero cells, an African green monkey kidney cell line, and thus were originally named VT1 and VT2 (65). VT1 is immunologically and genetically related to Stx, which is produced by *S. dysenteriae* type 1. Hence, these toxins alternately have been named SLTs (51, 52, 77, 78). The terms VT and SLT are used interchangeably. As noted above, it has been proposed that SLTs (VTs) be renamed after the prototype of the family, Stx (12). Other types of Stxs produced by serotypes of *E. coli* other than O157:H7 have been identified (11, 40, 76). The nomenclature of Stxs and their important biological characteristics are listed in Table 10.5.

Molecular studies on Stx1 from different *E. coli* strains revealed that Stx1 is either completely identical to Stx or differs by only one amino acid (76, 78). Unlike Stx1, toxins of the Stx2 group are not neutralized by serum raised against Stx and do not cross-hybridize with Stx1-specific DNA probes (11). There is sequence and antigenic variation within the Stx2 family of toxins produced by *E. coli* O157:H7 and other *E. coli* serotypes. Hence, there are subgroups of Stx2. The Stx2c subgroup is about 97% related to the amino acid sequence of the B subunit of Stx2, whereas the A subunit of Stx2c has 98 to 100% amino acid sequence homology with Stx2. The Stx2-related toxins have only partial serological reactivity with anti-Stx2 serum (11). Another subtype of the Stx2 family is Stx2e (VT2e), which is associated with the edema disease principally in piglets (121). Stx2e has 93.0 and 84.0% amino acid sequence homology with the A and B subunits, respectively, of Stx2. Additional toxins closely related to Stx2 have been found recently in some *E. coli* O157:H7 strains and other *E. coli* serotypes isolated from patients with intestinal infections, suggest-

Table 10.5 Nomenclature and biological characteristics of Stxs and Stx

Nomenclature[a]			Biological characteristics			
SLT	VT	New proposed[b]	Genetic locus	Cross-neutralized by antiserum to:	Receptor	Disease
Stx	Stx	Stx	Chromosome	Stx1	Gb_3	Human diarrhea, HC,[c] HUS
SLT I	VT1	Stx1	Phage	Stx	Gb_3	Human diarrhea, HC, HUS
SLT II	VT2	Stx2	Phage	Stx2c, Stx2e	Gb_3	Human diarrhea, HC, HUS
SLT IIc (SLT IIvha, SLT IIvhb)	VT2c (VT2)	Stx2c	Phage	Stx2, Stx2e	Gb_3	Human diarrhea, HC, HUS
SLT IIe (SLT IIv, SLT IIvp)	VT2e (VT2v, VTe)	Stx2e	Chromosome	Stx2, Stx2c	Gb_4	Pig edema disease

[a]Previous designations are given in parentheses.
[b]From reference 12.
[c]HC, hemorrhagic colitis.

ing that there is great heterogeneity among the Stx2-related toxins (76, 77). Considerable progress has been made within the past few years in our understanding of the biochemistry and molecular biology of Stxs as well as their involvement in the pathogenesis of disease (11, 76, 77).

Structure of Stxs

Stx and Stxs are holotoxins composed of a single enzymatic A subunit of approximately 32 kDa in association with a pentamer of receptor-binding B subunits of 7.7 kDa (11, 76, 77, 82). The Stx family A subunit can be split by trypsin into an enzymatic A1 fragment (approximately 27 kDa) and a carboxyl-terminal A2 fragment (approximately 4 kDa) that links A1 to the B subunits. The A1 and A2 subunits remain linked by a single disulfide bond until the enzymatic fragment is released and enters the cytosol of a susceptible mammalian cell. Each B subunit is composed of six antiparallel β strands forming a closed β barrel capped by a single α helix between strands 3 and 4 (38, 75, 105). The A subunit lies on the side of the B-subunit pentamer, nearest the C-terminal end of the B-subunit α helices. The A subunit interacts with the B-subunit pentamer through a hydrophobic α helix which extends half of the 2.0-nm length of the pore in the B pentamer. This pore is lined by the hydrophobic side chains of the B-subunit α helices (38, 105). The A subunit also interacts with the B subunit via a four-stranded mixed β sheet composed of residues of both the A2 and A1 fragments (38).

The sequences among all members of the Stx family are highly conserved in the regions of active sites. The amino acid residues corresponding to Glu at position 167 (Glu-167), Arg-170, Tyr-114, and Tyr-77 are invariably present (11). Substitution of Asp for Glu-167 of Stx produces a 1,000-fold reduction in toxic activity (49). Interestingly, the 28S rRNA *N*-glycosidase activity of the Stx family is identical to the mode of action of the plant toxin ricin (35, 89). The A chain of ricin and the A subunits of all members of the Stx family have two regions of amino acid sequence homology. Studies on the ricin A chain revealed that individual substitution of the residues described above results in a dramatic reduction in activity, suggesting that the conserved amino acids are important in maintaining enzymatic activity (49, 89). Collectively, these findings indicate that although there is antigenic diversity within the Stx family, sequences which encode active sites necessary for protein synthesis inhibition or receptor binding are highly conserved.

Genetics of Stxs

While all *stx1* operons examined thus far are essentially identical and located on the genomes of lysogenic lambdoid bacteriophages, there is considerable heterogeneity in the *stx2* gene family (76). Unlike the genes for other stx2 variants that are located on bacteriophage that integrate into the chromosome, Stx2e and Stx are encoded on chromosomal genes (70, 76, 99). Subsequently, the genes for additional variants of Stx2 have been isolated from SLTEC (40, 50, 90, 93). A sequence comparison of the growing *stx2* family suggests that genetic recombination among the B-subunit genes, rather than base substitutions, has given rise to the variants of Stx2 found in human and animal strains of *E. coli* (40, 50). However, the operons for every member of the Stx family subgroups are organized identically: the A- and B-subunit genes are arranged in tandem and separated by a 12- to 15-nucleotide gap (76). The operons are transcribed from a promoter which is located 5′ to the A-subunit gene, and each gene is preceded by a putative ribosome-binding site. The existence of an independent promoter for the B-subunit genes has been suggested. The holotoxin stoichiometry suggests that expression of the A- and B-subunit genes is differentially regulated, permitting overproduction of the B polypeptides (76). Finally, Stx and Stx1 production is negatively regulated at the transcriptional level by an iron-Fur protein corepressor complex which binds at the *stx1* promoter but is unaffected by temperature, whereas Stx2 production is neither iron nor temperature regulated (76).

Receptors

All members of the Stx family bind to globoseries glycolipids on the eukaryotic cell surface; Stx, Stx1, Stx2, and Stx2c bind to glycolipid globotriaosylceramide (Gb_3), whereas Stx2e binds primarily to glycolipid globotetraosylceramide (Gb_4) (10, 26, 59, 76). The alteration of binding specificity between Stx2e and the rest of the Stx family is related to the carbohydrate specificity of receptors. The amino acid compositions of B subunits of Stx2 and Stx2e differ at only 11 positions, yet Stx2e binds primarily to Gb_4 whereas Stx2 binds only to Gb_3 (26, 114). High-affinity binding also depends on multivalent presentation of the carbohydrate, as would be provided by glycolipids in a membrane. Studies by Kiarash et al. (63) revealed that the affinity of Stx1 for Gb_3 isoforms is influenced by fatty acyl chain length and by its level of saturation. Stx1 was found to bind preferentially to Gb_3 containing $C_{20:1}$ fatty acid, whereas Stx2c preferred Gb_3 containing $C_{18:1}$ fatty acid (86). The basis for these findings may be related to the ability of different Gb_3 isoforms to present multivalent sugar-binding sites in the optimal orientation and position at the membrane surface. It is also possible that different fatty acyl groups affect the conformation of individual receptor epitopes on the sugar (76).

Mode of Action of the Stxs

Stx and Stxs act by inhibiting protein synthesis. Each of the B subunits is capable of binding with high affinity to an unusual disaccharide linkage (galactose α1–4 galactose) in the terminal trisaccharide sequence of Gb_3 (or Gb_4) (110). Following binding to the glycolipid receptor, the toxin is endocytosed from clathrin-coated pits and transferred first to the *trans* Golgi network and subsequently to the endoplasmic reticulum and nuclear envelope (76, 110). While it appears that transfer of the toxin to the Golgi apparatus is essential for intoxication, the mechanism of entry of the A subunit from the endosome to the cytosol, and particularly the role of the B subunit in the process, remains unclear. In the cytosol, the A subunit undergoes partial proteolysis and splits into a 27-kDa active intracellular enzyme (A1) and a 4-kDa fragment (A2) bridged by a disulfide bond (76, 80, 81). Although the whole toxin is necessary for its toxic effect on whole cells, the A1 subunit is capable of cleaving the *N*-glycoside bond in one adenosine position of the 28S rRNA that makes up 60S ribosomal subunits. This elimination of a single adenine nucleotide inhibits the elongation factor-dependent binding to ribosomes of aminoacyl-bound tRNA molecules. Peptide chain elongation is truncated and overall protein synthesis is suppressed, resulting in cell death (11, 76, 91, 110).

Role of the Toxins in Disease

The precise role of the toxins in mediating colonic disease, HUS, and neurological disorders remains less clear. There is no satisfactory animal model for hemorrhagic colitis or HUS, and the severity of the disease precludes study of experimental infections in humans. Therefore, understanding the role of toxins comes from a combination of studies employing a variety of approaches, including histopathologic examination of diseased human tissues, use of animal models, and use of endothelial tissue culture cells (59, 78). Recent studies support the concept that Stxs contribute to pathogenesis by directly damaging vascular endothelial cells in certain organs, thereby disrupting the homeostatic properties of these cells (110).

The mechanism of *E. coli* O157:H7 pathogenesis is not fully documented, but a likely scenario occurs as follows: after *E. coli* O157:H7 cells are ingested, the bacteria colonize the large intestine and adhere to and possibly invade colonic mucosal epithelial cells, replicate and destroy colonic cells, and damage the underlying tissue and vasculature, possibly by both exotoxin-related and endotoxin-related mechanisms, thereby producing bloody diarrhea (76). Endothelial cells are critical in maintaining transcapillary permeability and the nonthrombogenic state necessary for normal blood flow. When Stxs enter the circulation and attach to Gb_3 molecules on endothelial cells, cell damage results and may not only lead to a procoagulant state by inhibiting the production of prostacyclin but also result in the elicitation of vasoactive substances and cytokines (11, 76). These endogenous mediators may, in turn, augment cell damage, alter hemostatic control, act as autocrine growth factors, and/or cause both endothelial cells and inflammatory cells to express receptors necessary for monocyte- and leukocyte-endothelial cell adhesion

(76). Tissue culture and animal models have revealed that Stxs (i) are directly cytotoxic for certain cell lines, (ii) are enterotoxic and mediate fluid accumulation in ligated ileal loops, and (iii) are paralytic and lethal when injected intravenously into mice and rabbits (110).

Interestingly, *E. coli* O157:H7 strains isolated from patients with hemorrhagic colitis usually produce both Stx1 and Stx2 or Stx2 only; isolates producing only Stx1 are uncommon (76, 77). Patients infected with *E. coli* O157:H7 producing only Stx2 alone or in combination with Stx1 were more likely to develop serious renal or circulatory complications than were patients infected with EHEC producing Stx1 only (97). The relative contributions of Stx1 and Stx2 to the development of HUS and the apparent lack of an amnestic immune response in humans to Stx2 remain to be elucidated. Studies with a rabbit model of hemorrhagic colitis revealed that continuous intraperitoneal infusion of purified Stx2 produces diarrhea and cecal lesions comparable to colonic disease in humans. Rabbits receiving Stx2 also develop acute focal tubular necrosis in the kidneys (1). A study using streptomycin-treated mice to examine the effects of Stx1 and Stx2 by perorally administering an *E. coli* K-12 strain containing the Stx1 or Stx2 gene cloned in a high-copy-number expression vector revealed that death of the infected mice was due solely to Stx2 (118). Extensive histologic examination of organs of the mice revealed abnormalities only in the kidneys, where there was acute bilateral cortical tubular necrosis. Streptomycin-treated mice also died after peroral administration of an *E. coli* O157:H7 derivative that readily colonizes the mouse bowel and produces both Stx1 and Stx2. Passive administration of monoclonal antibodies directed against either the A or B subunit of Stx2 protected mice, whereas passive transfer of anti-Stx1 antibodies did not (118). The animal models used to address the role of Stxs in disease suggest that Stx2 is principally responsible for colonic disease and acute renal failure (76, 111).

Numerous epidemiologic studies have shown a correlation between infection with *E. coli* O157:H7 strains producing Stx(s) and development of HUS in humans (61, 73, 74). Histopathologic examination of kidney tissue from HUS patients revealed profound structural alterations in the glomeruli, the basic filtration unit of the kidney (110). Glomerular endothelial cells were swollen and were often detached from the glomerular basement membrane. Hence, subendothelial matrix components may be exposed and serve as sites for platelet adherence and activation. Microthrombi were deposited in capillary lumina, and there was an influx of inflammatory cells (93). Recently, Stxs were determined to be directly cytotoxic to human endothelial cells derived from large veins (79, 110). Results demonstrated that there is a direct correlation between membrane Gb_3 content and susceptibility to Stx family cytotoxicity, and cells within the human kidney, possibly glomerular endothelial cells, may be enriched with toxin receptors. The direct Stxs-mediated cytotoxic effect on glomerular endothelial cells may represent the initial injury, which leads to a cascade of procoagulant and proinflammatory events characteristic of HUS (110). Damage to glomeruli is the pathophysiologic hallmark of HUS. Glomerular endothelial cells appear swollen and detached from the glomerular basement membrane, whereas adjacent epithelial cell-lined proximal collecting tubules appear normal (110). The damage by Stxs is often not limited to the glomeruli. Arteriolar damage involving internal cell proliferation, fibrin thrombus deposition, and perivascular inflammation occurs. Cortical necrosis also occurs in a small number of HUS cases. In addition, human glomerular endothelial cells are sensitive to the direct cytotoxic action of bacterial endotoxin. Endotoxin in the presence of Stxs also can activate macrophage and polymorphonuclear neutrophils to synthesize and release cytokines, superoxide radicals, or proteinases and thus amplify endothelial cell damage (76, 110).

Neurological symptoms in patients and experimental animals infected with *E. coli* O157:H7 also have been described and are thought to be caused by secondary neuron disturbances that result from endothelial cell damage by Stxs (110). A recent study by Fujii et al. (39) of mice perorally administered an *E. coli* O157:H$^-$ strain revealed that Stx2v impaired the blood-brain barrier and damaged neuron fibers, resulting in death. Presence of the toxin in neurons was verified by immunoelectron microscopy.

CONCLUDING REMARKS

The serious nature of the symptoms of hemorrhagic colitis and HUS caused by *E. coli* O157:H7 place this foodborne pathogen in a category apart from those that typically cause only mild symptoms. Its severe symptoms combined with its apparent low infectious dose (<100 cells) qualify *E. coli* O157:H7 to be among the most serious of known foodborne pathogens. Cattle are a major reservoir of *E. coli* O157:H7, with undercooked ground beef being the single most frequently implicated vehicle of transmission. An important feature of this pathogen is its unique acid tolerance. Outbreaks recently have been associated with consumption of contaminated high-acid foods, including apple cider and fermented dry salami. The mechanisms of pathogenicity of *E. coli* O157:H7 have not been fully elucidated; however, AE adherence and production of one or more Stxs are important virulence factors.

EHEC of serogroups other than O157:H7 have been increasingly associated with cases of HUS. Over 100 non-O157 SLTEC serotypes have been isolated from humans, but not all of these serotypes have been shown to cause illness. *E. coli* O157:H7 is still by far the most important serotype of SLTEC in North America. Isolation of non-O157:H7 SLTEC requires techniques not generally available in clinical laboratories; hence, these bacteria are rarely sought or detected in routine practice. Recent recognition of non-O157 SLTEC necessitates identification of other serotypes of SLTEC in persons with bloody diarrhea and/or HUS and in implicated food. The increased availability in clinical laboratories of techniques such as testing for Stxs or their genes and identification of other virulence markers unique for EHEC may enhance the detection of disease attributable to non-O157 SLTEC.

References

1. **Barrett, T., M. Potter, and I. Wachsmuth.** 1989. Continuous peritoneal infusion of Shiga-like toxin II (SLT II) as a model of SLT II-induced diseases. *J. Infect. Dis.* **159:** 774–777.
2. **Beebakhee, G., M. Louie, J. De Azavedo, and J. Brunton.** 1992. Cloning and nucleotide sequence of the *eae* gene homologue from enterohemorrhagic *Escherichia coli*. *FEMS Microbiol. Lett.* **91:**63–68.
3. **Begue, R. E., M. A. Neill, E. F. Papa, and P. H. Dennehy.** 1994. A prospective study of Shiga-like toxin-associated diarrhea in a pediatric population. *J. Pediatr. Gastroenterol. Nutr.* **19:**164–169.
4. **Bell, B. P., M. Goldoft, P. M. Griffin, M. A. Davis, D. C. Gordon, P. I. Tarr, C. A. Bartleson, J. H. Lewis, T. J. Barrett, J. G. Wells, R. Baron, and J. Kobayashi.** 1994. A multistate outbreak of *Escherichia coli* O157:H7-associated bloody diarrhea and hemolytic uremic syndrome from hamburgers. *JAMA* **272:**1349–1353.
5. **Belongia, E. A., M. T. Osterholm, J. T. Soler, D. A. Ammend, J. E. Braun, and K. L. MacDonald.** 1993. Transmission of *Escherichia coli* O157:H7 in Minnesota child day-care facilities. *JAMA* **269:**883–888.
6. **Besser, R. E., S. M. Lett, J. T. Weber, M. P. Doyle, T. J. Barrett, J. G. Wells, and P. M. Griffin.** 1993. An outbreak of diarrhea and hemolytic uremic syndrome from *Escherichia coli* O157:H7 in fresh-pressed apple cider. *JAMA* **269:**2217–2220.
7. **Borczyk, A. A., M. A. Karmali, H. Lior, and L. M. C. Duncan.** 1987. Bovine reservoir for verotoxin-producing *Escherichia coli* O157:H7. *Lancet* **i:**98.
8. **Brackett, R. E., Y.-Y. Hao, and M. P. Doyle.** 1994. Ineffectiveness of hot acid sprays to decontaminate *Escherichia coli* O157:H7 on beef. *J. Food Prot.* **57:**198–203.
9. **Brown, C., B. Harmon, T. Zhao, and M. P. Doyle.** 1995. Experimental *E. coli* O157:H7 infection in calves. Presented at the 1995 Annual Meeting of ACVP/ASVCP, Nov. 14–17, Atlanta, Ga.
10. **Brunton, J.** 1990. The Shiga toxin family: molecular nature and possible role in disease, p. 377–398. *In* B. Iglewski (ed.), *The Bacteria*, vol. XI. Academic Press, New York.
11. **Brunton, J.** 1994. Molecular biology and role in disease of the verotoxins (Shiga-like toxins) of *Escherichia coli*, p. 391–404. *In* V. L. Miller, J. B. Kaper, D. A. Portnoy, and R. R. Isberg (ed.), *Molecular Genetics of Bacterial Pathogenesis*. American Society for Microbiology, Washington, D.C.
12. **Calderwood, S. B., D. W. K. Acheson, G. T. Keusch, T. J. Barrett, P. M. Griffin, N. A. Strockbine, B. Swaminathan, J. B. Kaper, M. M. Levine, B. S. Kaplan, H. Karch, A. D. O'Brien, T. G. Obrig, Y. Takedo, P. I. Tarr, and I. K. Wachsmuth.** 1996. Proposed new nomenclature for SLT (VT) family. *ASM News* **62:**118–119.
13. **Centers for Disease Control and Prevention.** 1995. *Escherichia coli* O157:H7 outbreak linked to commercially distributed dry-cured salami—Washington and California, 1994. *Morbid. Mortal. Weekly Rep.* **44:**157–160.
14. **Centers for Disease Control and Prevention.** 1995. Surveillance for outbreaks of *Escherichia coli* O157:H7 infection—preliminary summary of 1994 data. Personal communication.
15. **Centers for Disease Control and Prevention.** 1995. Outbreak of acute gastroenteritis attributable to *Escherichia coli* O104:H21—Helena, Montana, 1994. *Morbid. Mortal. Weekly Rep.* **44:**501–503.
16. **Centers for Disease Control and Prevention.** 1995. Community outbreak of hemolytic uremic syndrome attributable to *Escherichia coli* O111:NM—south Australia, 1995. *Morbid. Mortal. Weekly Rep.* **44:**550–558.
17. **Chapman, P. A.** 1993. Untreated milk as a source of verotoxigenic *E. coli* O157. *Vet. Rec.* **133:**171–172.
18. **Chapman, P. A., D. J. Wright, P. Norman, J. Fox, and E. Crick.** 1993. Cattle as a possible source of verotoxin-producing *Escherichia coli* O157:H7 infections in man. *Epidemiol. Infect.* **111:**439–447.
19. **Chart, H.** 1993. Serodiagnosis of infections caused by *Escherichia coli* O157:H7 and other VTEC. *Serodiagn. Immunother. Infect. Dis.* **1:**8–12.
20. **Cleary, T. G.** 1988. Cytotoxin-producing *Escherichia coli* and the hemolytic uremic syndrome. *New Top. Pediatr. Infect. Dis.* **35:**485–501.
21. **Conner, D. E., and J. S. Kotrola.** 1995. Growth and survival of *Escherichia coli* O157:H7 under acidic conditions. *Appl. Environ. Microbiol.* **61:**382–385.
22. **Cray, W. C., Jr., and H. W. Moon.** 1995. Experimental infection of calves and adult cattle with *Escherichia coli* O157:H7. *Appl. Environ. Microbiol.* **61:**1586–1590.
23. **D'Aoust, J. Y., C. E. Park, R. A. Szabo, E. C. D. Todd, D. B. Emmons, and R. C. McKellar.** 1988. Thermal inactivation of *Campylobacter* species, *Yersinia enterocolitica*, and hemorrhagic *Escherichia coli* O157:H7 in fluid milk. *J. Dairy Sci.* **71:**3230–3236.
24. **Dargatz, D. (Centers for Epidemiology and Animal Health, Fort Collins, Colo.).** 1995. *Escherichia coli*

O157:H7 shedding by feedlot cattle. Personal communication.

25. **Debroy, C., B. D. Bright, R. A. Wilson, J. Yealy, R. Kumar, and M. K. Bhan.** 1994. Plasmid-coded DNA fragment developed as a specific gene probe for the identification of enteroaggregative *Escherichia coli*. *Diagn. Microbiol.* **41**:393–398.
26. **DeGrandis, S., H. Law, J. Brunton, C. Gyles, and C. Lingwood.** 1989. Globotetraosyl ceramide is recognized by the pig edema disease toxin. *J. Biol. Chem.* **264**: 12520–12525.
27. **Donnenberg, M. S., A. Donohue-Rolfe, and G. T. Keusch.** 1989. Epithelial cell invasion: an overlooked property of enteropathogenic *Escherichia coli* (EPEC) associated with EPEC adherence factors. *J. Infect. Dis.* **160**:452–459.
28. **Donnenberg, M. S., and J. B. Kaper.** 1992. Enteropathogenic *Escherichia coli*. *Infect. Immun.* **60**:3953–3961.
29. **Donnenberg, M. S., S. Tzipori, M. L. McKee, A. D. O'Brien, J. Alroy, and J. B. Kaper.** 1993. The role of the *eae* gene of enterohemorrhagic *Escherichia coli* in intimate attachment in vitro and in a porcine model. *J. Clin. Invest.* **92**:1418–1224.
30. **Donnenberg, M. S., J. Yu, and J. B. Kaper.** 1993. A second chromosomal gene necessary for intimate attachment of enteropathogenic *Escherichia coli* to epithelial cells. *J. Bacteriol.* **175**:4670–4680.
31. **Doyle, M. P., and V. V. Padhye.** 1989. *Escherichia coli*, p. 235–281. *In* M. P. Doyle (ed.), *Foodborne Bacterial Pathogens.* Marcel Dekker, New York.
32. **Doyle, M. P., and J. L. Schoeni.** 1984. Survival and growth characteristics of *Escherichia coli* associated with hemorrhagic colitis. *Appl. Environ. Microbiol.* **48**:855–856.
33. **Doyle, M. P., and J. L. Schoeni. 1987.** Isolation of *Escherichia coli* O157:H7 from retail fresh meats and poultry. *Appl. Environ. Microbiol.* **53**:2394–2396.
34. **Dytoc, M., R. Soni, F. Cockerill III, J. C. S. De Azavedo, M. Louie, J. Brunton, and P. Sherman.** 1993. Multiple determinants of verotoxin-producing *Escherichia coli* O157:H7 attachment-effacement. *Infect. Immun.* **61**: 3382–3391.
35. **Endo, Y., and K. Tsurugi.** 1988. The RNA N-glycosidase activity of ricin A chain. The characteristics of the enzymatic activity of ricin A chain with ribosomes and with rRNA. *J. Biol. Chem.* **263**:8735–8739.
36. **Francis, D. H., J. E. Collins, and J. R. Duimstra.** 1986. Infection of gnotobiotic pigs with an *Escherichia coli* O157:H7 strain associated with an outbreak of hemorrhagic colitis. *Infect. Immun.* **51**:953–956.
37. **Frankel, G., D. C. A. Candy, P. Everest, and G. Dougan.** 1994. Characterization of the C-terminal domains of intimin-like proteins of enteropathogenic and enterohemorrhagic *Escherichia coli, Citrobacter freundii*, and *Hafnia alvei*. *Infect. Immun.* **62**:1835–1842.
38. **Frazer, M., M. Chernai, Y. Kozlov, and M. James.** 1994. Crystal structure of the holotoxin from *Shigella dysenteriae* at 2.5 A resolution. *Nat. Struct. Biol.* **1**:59–64.
39. **Fujii, J. T. Kita, S. Yoshida, T. Takeda, H. Kobayashi, N. Tanaka, K. Ohsato, and Y. Mizuguchi.** 1994. Direct evidence of neuron impairment by oral infection with verotoxin-producing *Escherichia coli* O157:H- in mitomycin-treated mice. *Infect. Immun.* **62**:3447–3453.
40. **Gannon, V.P.J., C. Teerling, S.A. Masrei, and C.L. Gyles.** 1990. Molecular cloning and nucleotide sequence of another variant of the *Escherichia coli* Shiga-like toxin II family. *J. Gen. Microbiol.* **136**:1125–1135.
41. **Garber, L. P., S. J. Wells, D. D. Hancock, M. P. Doyle, J. Tuttle, J. A. Shere, and T. Zhao.** 1995. Risk factors for fecal shedding of *Escherichia coli* O157:H7 in dairy calves. *J. Am. Vet. Med. Assoc.* **207**:46–49.
42. **Glass, K. A., J. M. Loeffelholz, J. P. Ford, and M. P. Doyle.** 1992. Fate of *Escherichia coli* O157:H7 as affected by pH or sodium chloride and in fermented, dry sausage. *Appl. Environ. Microbiol.* **58**:2513–2516.
43. **Goodfellow, S. J., and W. L. Brown.** 1987. Fate of *Salmonella* inoculated into beef for cooking. *J. Food Prot.* **41**:598–605.
44. **Griffin, P. M., B. P. Bell, P. R. Cieslak, J. Tuttle, T. J. Barrett, M. P. Doyle, A. M. McNamara, A. M. Shefer, and J. G. Wells.** 1994. Large outbreak of *Escherichia coli* O157:H7 in western United States: the big picture, p. 7–12. *In* M. A. Karmali and A. G. Goglio (ed.), *Recent Advances in Verocytotoxin-Producing Escherichia coli Infections.* Proceedings of the 2nd International Symposium and Workshop on Verotoxin (Shiga-Like Toxin)-Producing *Escherichia coli* Infections. Elsevier, Amsterdam.
45. **Griffin, P. M., and R. V. Tauxe.** 1991. The epidemiology of infections caused by *Escherichia coli* O157:H7, other enterohemorrhagic *E. coli*, and the associated hemolytic uremic syndrome. *Epidemiol. Rev.* **13**:60–98.
46. **Habib, R.** 1992. Pathology of the hemolytic uremic syndrome, p. 315–353. In B. S. Kaplan, R. S. Trompeter, and J. L. Moake (ed.), *Hemolytic Uremic Syndrome and Thrombotic Thrombocytopenic Purpura.* Marcel Dekker, New York.
47. **Hancock, D. D., T. E. Besser, M. L. Kinsel, and P. I. Tarr.** 1994. The prevalence of *Escherichia coli* O157:H7 in dairy and beef cattle in Washington State. *Epidemiol. Infect.* **113**:199–207.
48. **Isaacson, M., P. H. Canter, P. Effler, L. Arntzen, P. Bomans, and R. Heenon.** 1993. Haemorrhagic colitis epidemic in Africa. *Lancet* **341**:961.
49. **Hovde, C. J., S. Calderwood, J. Mekalanos, and R. J. Collier.** 1988. Evidence that glutamic acid 167 is an active site residue of Shiga-like toxin I. *Proc. Natl. Acad. Sci. USA* **85**:2568–2572.
50. **Ito, H., T. Yutsudo, T. Hirayama, and Y. Takeda.** 1988. Isolation and some properties of A and B subunits of verotoxin 2 and in vitro formation of hybrid toxins between subunits of verotoxin 1 and verotoxin 2 from *Escherichia coli* O157:H7. *Microbiol. Pathol.* **5**:189–195.
51. **Jackson, M. P., R. J. Neill, A. D. O'Brien, R. K. Holmes, and J. W. Newland.** 1987. Nucleotide sequence analysis and comparison of the structural genes for Shiga-like toxin I and Shiga-like toxin II encoded by bacteriophages from *Escherichia coli* 933. *FEMS Lett.* **44**:109–114.
52. **Jackson, M. P., J. W. Newland, R. K. Holmes, and A. D. O'Brien.** 1987. Nucleotide sequence analysis of the structural genes for Shiga-like toxin I encoded by bacteriophage 933J from *Escherichia coli*. *Microb. Pathog.* **2**: 147–153.

53. **Jerse, A. E., and J. B. Kaper.** 1991. The *eae* gene of enteropathogenic *Escherichia coli* encodes a 94-kilodalton membrane protein, the expression of which is influenced by EAF plasmid. *Infect. Immun.* 59:4302–4309.
54. **Johnson, W. M., H. Lior, and G. S. Bezanson.** 1983. Cytotoxic *Escherichia coli* O157:H7 associated with haemorrhagic colitis in Canada. *Lancet* **i**:76.
55. **Junkins, A. D., and M. P. Doyle.** 1989. Comparison of adherence properties of *Escherichia coli* O157:H7 and a 60-megadalton plasmid-cured derivative. *Curr. Microbiol.* **19**:21–27.
56. **Kaper, J. B.** 1994. Molecular pathogenesis of enteropathogenic *Escherichia coli,* p. 173–195. *In* V. L. Miller, J. B. Kaper, D. A. Portnoy, and R. R. Isberg (ed.), *Molecular Genetics of Bacterial Pathogenesis.* American Society for Microbiology, Washington, D.C.
57. **Karch, H., J. Heesemann, R. Laufs, A. D. O'Brien, C. O. Tacket, and M. M. Levine.** 1987. A plasmid of enterohemorrhagic *Escherichia coli* O157:H7 is required for expression of a new fimbrial antigen and for adhesion to epithelial cells. *Infect. Immun.* **55**:455–461.
58. **Karch, H., H. Russmann, H. Schmidt, A. Schwarzkopf, and J. Heesemann.** 1995. Long-term shedding and clonal turnover of enterohemorrhagic *Escherichia coli* O157 in diarrheal diseases. *J. Clin. Microbiol.* **33**:1602–1605.
59. **Karmali, M.** 1992. The association of verocytotoxins and the classical hemolytic uremic syndrome, p. 199–212. *In* K. Kaplan, R. Tromppeter, and J. Moake (ed.), *Hemolytic Uremic Syndrome and Thrombotic Thrombocytopenic Purpura.* Marcel Dekker, New York.
60. **Karmali, M. A., B. T. Steele, M. Petric, and C. Lim.** 1983. Sporadic cases of hemolytic uremic syndrome associated with fecal cytotoxin and cytotoxin-producing *Escherichia coli. Lancet* **i**:619–620.
61. **Kavi, J., and R. Wise.** 1989. Causes of the hemolytic uraemic syndrome. *Br. Med. J.* **298**:65–66.
62. **Keene, W. E., J. M. McAnulty, F. C. Hoesly, L. P. Williams, K. Hedberg, G. L. Oxman, T. J. Barrett, M. A. Pfaller, and D. W. Fleming.** 1994. A swimming-associated outbreak of hemorrhagic colitis caused by *Escherichia coli* O157:H7 and *Shigella sonnei. N. Engl. J. Med.* **331**:579–584.
63. **Kiarash, A., B. Boyd, and S. Scotland.** 1994. Glycosphingolipid receptor function is modified by fatty acid content: verotoxin 1 and verotoxin 2c preferentially recognize different globotriaosyl ceramide fatty acid homologues. *J. Biol. Chem.* **269**:11138–11146.
64. **Kim, H. H., M. Samadpour, L. Grimm, C. R. Clausen, T. E. Besser, M. Baylor, J. M. Kobayashi, M. A. Neill, F. D. Schoenknecht, and P. I. Tarr.** 1994. Characteristics of antibiotic-resistant *Escherichia coli* O157:H7 in Washington State, 1984–1991. *J. Infect. Dis.* **170**:1606–1609.
65. **Konowalchuk, J., J. Speirs, and S. Stavric.** 1977. Vero response to a cytotoxin of *Escherichia coli. Infect. Immun.* **18**:775–779.
66. **Kovacs, M. J., J. Roddy, S. Gregoire, W. Cameron, L. Eidus, and J. Drouin.** 1990. Thrombotic thrombocytopenic purpura following hemorrhagic colitis due to *Escherichia coli* O157:H7. *Am. J. Med.* **88**:177–179.
67. **Kudva, I. T., P. G. Hatfield, and C. J. Hovde.** 1996. *Escherichia coli* O157:H7 in microbial flora of sheep. *J. Clin. Microbiol.* **34**:431–433.
68. **Line, J. E., A. R. Fain, Jr., A. B. Moran, L. M. Martin, R. V. Lechowich, J. M. Carosella, and W. L. Brown.** 1991. Lethality of heat to *Escherichia coli* O157:H7: D-value and z-value determinations in ground beef. *J. Food Prot.* **54**:762–766.
69. **Louie, M., J. C. S. De Azavedo, M. Y. C. Handelsman, C. G. Clark, B. Ally, M. Dytoc, P. Sherman, and J. Brunton.** 1993. Expression and characterization of the *eaeA* gene product of *Escherichia coli* serotype O157:H7. *Infect. Immun.* **61**:4085–4092.
70. **Marques, L.R.M., J.S.M. Peiris, S.J. Cryz, and A.D. O'Brien.** 1987. *Escherichia coli* strains isolated from pigs with edema disease produce a variant of Shiga-like toxin II. *FEMS Lett.* **44**:33–38.
71. **Martin, M. L., L. D. Shipman, M. E. Potter, I. K. Wachsmuth, J. G. Wells, K. Hedberg, R. V. Tauxe, P. Davis, J. Arnoldi, and J. Tilleli.** 1986. Isolation of *Escherichia coli* O157:H7 from dairy cattle associated with two cases of hemolytic uremic syndrome. *Lancet* **ii**:1043.
72. **McDaniel, T. K., K. G. Jarvis, M. S. Donnenberg, and J. B. Kaper.** 1995. A genetic locus of enterocyte effacement conserved among diverse enterobacterial pathogens. *Proc. Natl. Acad. Sci. USA* **92**:1664–1668.
73. **Milford, D. V., C. M. Taylor, B. Guttridge, S. Hall, B. Rowe, and H. Kleanthous.** 1990. Haemolytic uraemic syndromes in the British isles, 1985–8: association with verocytotoxin producing *Escherichia coli.* Part 1. Clinical and epidemiologic aspects. *Arch. Dis. Child.* **65**:716–721.
74. **Moake, J. L.** 1994. Haemolytic-uraemic syndrome: basic science. *Lancet* **343**:393–397.
75. **Muzzin, A. G.** 1993. OB (oligonucleotide/oligosaccharide binding) fold: common structural and functional solution for non-homologous sequences. *EMBO J.* **12**:861–867.
76. **O'Brien, A., V. Tesh, A. Donohue-Rolfe, M. Jackson, S. Olsnes, K. Sandvig, A. Lindberg, and G. T. Keusch.** 1992. Shiga toxin: biochemistry, genetics mode of action and role in pathogenesis. *Curr. Top. Microbiol. Immunol.* **180**:65–94.
77. **O'Brien, A. D., and R. K. Holmes.** 1987. Shiga and Shiga-like toxins. *Microbiol. Rev.* **51**:206–220.
78. **O'Brien, A. D., G. D. LaVeck, M. R. Thompson, and S. B. Formal.** 1982. Production of *Shigella dysenteriae* type 1-like cytotoxin by *Escherichia coli. J. Infect. Dis.* **146**:763–769.
79. **Obrig, T., P. del Vecchio, J. Brown, T. Moran, B. Rowland, K. Judge, and W. Rothman.** 1988. Direct cytotoxic action of Shiga toxin on human vascular endothelial cells. *Infect. Immun.* **56**:2373–2378.
80. **Oelschlaeger, T. A., T. J. Barrett, and D. J. Kopecko.** 1994. Some structures and processes of human epithelial cells involved in uptake of enterohemorrhagic *Escherichia coli* O157:H7 strains. *Infect. Immun.* **62**:5142–5150.
81. **Olsnes, S., and K. Eiklid.** 1980. Isolation and characterization of *Shigella shigae* cytotoxin. *J. Biol. Chem.* **225**:284–289.

82. **Olsnes, S., R. Reisbig, and K. Eiklid.** 1981. Subunit structure of *Shigella* toxin. *J. Biol. Chem.* **256**:8732–8738.

83. **Orr, P., D. Milley, D. Coiby, and M. Fast.** 1994. Prolonged fecal excretion of verotoxin-producing *Escherichia coli* following diarrheal illness. *Clin. Infect. Dis.* **19**:796–797.

84. **Ørskov, F., I. Ørskov, and J. A. Villar.** 1987. Cattle as reservoir of verotoxin-producing *Escherichia coli* O157:H7. *Lancet* **ii**:276.

85. **Padhye, N. V., and M. P. Doyle.** 1991. Production and characterization of a monoclonal antibody specific for enterohemorrhagic *Escherichia coli* of serotypes O157:H7 and O26:H11. *J. Clin. Microbiol.* **29**:99–103.

86. **Pellizzari, A., H. Pang, and C. Lingwood.** 1992. Binding of verocytotoxin 1 to its receptor is influenced by differences in receptor fatty acid content. *Biochemistry* **31**: 1363–1370.

87. **Rasmussen, M. A., W. C. Cray, T. A. Casey, and S. C. Whipp.** 1993. Rumen contents as a reservoir of enterohemorrhagic *Escherichia coli. FEMS Microbiol. Lett.* **114**:79–84.

88. **Ratnam, S., S. B. March, R. Ahmed, G. S. Bezanson, and S. Kasatiya.** 1988. Characterization of *Escherichia coli* serotype O157:H7. *J. Clin. Microbiol.* **26**:2006–2012.

89. **Ready, M., P. Kim, and J. D. Robertus.** 1991. Site directed mutagenesis of ricin A chain and implications for the mechanism of action. *Proteins* **10**:270–278.

90. **Reida, P., M. Wolff, H. W. Pohls, W. Kuhlmann, A. Lehmacher, S. Aleksic, H. Karch, and J. Bockemuhl.** 1994. An outbreak due to enterohemorrhagic *Escherichia coli* O157:H7 in a children day care center characterized by person-to-person transmission and environmental contamination. *Zentralbl. Bakteriol.* **281**:534–543.

91. **Reisbig, R., S. Olsnes, and K. Eiklid.** 1981. The cytotoxic activity of *Shigella* toxin. Evidence for catalytic inactivation of the 60 S ribosomal subunit. *J. Biol. Chem.* **256**:8739–8744.

92. **Renwick, S. A., J. B. Wilson, R. C. Clarke, H. Lior, A. A. Borczyk, J. Spika, K. Rahn, K. McFadden, A. Brouwer, A. Copps, N. G. Anderson, D. Alvens, and M. A. Karmali.** 1993. Evidence of direct transmission of *Escherichia coli* O157:H7 infection between calves and a human. *J. Infect. Dis.* **168**:792–793.

93. **Richardson, S., T. Rotman, V. Jay, C. Smith, L. Becker, M. Petric, N. Olivieri, and M. Karmali.** 1992. Experimental verocytotoxemia in rabbits. *Infect. Immun.* **60**:4154–4167.

94. **Rietra, P., G. Willshaw, H. Smith, A. Field, S. Scotland, and B. Rowe.** 1989. Comparison of verocytotoxin-encoding phages from *Escherichia coli* of human and bovine origin. *J. Gen. Microbiol.* **135**:2307–2318.

95. **Riley, L. W.** 1987. The epidemiologic, clinical, and microbiological features of hemorrhagic colitis. *Annu. Rev. Microbiol.* **41**:383–407.

96. **Riley, L. W., R. S. Remis, S. D. Helgerson, H. B. McGee, J. G. Wells, B. R. Davis, R. J. Hebert, E. S. Olcott, L. M. Johnson, N. T. Hargrett, P. A. Blake, and M. L. Cohen.** 1983. Hemorrhagic colitis associated with a rare *Escherichia coli* serotype. *N. Engl. J. Med.* **308**:681–685.

97. **Russmann, H., H. Schmidt, J. Heesemann, A. Caprioli, and H. Karch.** 1994. Variants of Shiga-like toxin II constitute a major toxin component in *Escherichia coli* O157 strains from patients with haemolytic uraemic syndrome. *J. Med. Microbiol.* **40**:338–343.

98. **Schmidt, H., L. Beutin, and H. Karch.** 1995. Molecular analysis of the plasmid-encoded hemolysin of *Escherichia coli* O157:H7 strain EDL933. *Infect. Immun.* **63**:1055–1061.

99. **Schmitt, C. K., M. L. McKee, and A. D. O'Brien.** 1991. Two copies of Shiga-like toxin II-related genes common in enterohemorrhagic *Escherichia coli* strains are responsible for the antigenic heterogeneity of the O157:H7 strain E32511. *Infect. Immun.* **59**:1065–1073.

100. **Schoeni, J. L., and M. P. Doyle.** 1994. Variable colonization of chickens perorally inoculated with *Escherichia coli* O157:H7 and subsequent contamination of eggs. *Appl. Environ. Microbiol.* **60**:2958–2962.

101. **Sherman, P., F. Cockerill III, R. Soni, and J. Brunton.** 1991. Outer membranes are competitive inhibitors of *Escherichia coli* O157:H7 adherence to epithelial cells. *Infect. Immun.* **59**:890–899.

102. **Sherman, P., and R. Soni.** 1988. Adherence of verocytotoxin-producing *Escherichia coli* of serotype O157:H7 to human epithelial cells in tissue culture: role of outer membranes as bacterial adhesins. *J. Med. Microbiol.* **26**:11–17.

103. **Sherman, P., R. Soni, and M. Karmali.** 1988. Attaching and effacing adherence of verocytotoxin-producing *Escherichia coli* to rabbit intestinal epithelial in vivo. *Infect. Immun.* **56**:756–761.

104. **Spika, J. S., J. E. Parsons, D. Nordenberg, J. G. Wells, R. A. Gunn, and P. A. Blake.** 1986. Hemolytic uremic syndrome and diarrhea associated with *Escherichia coli* O157:H7 in a day care center. *J. Pediatr.* **109**:287–291.

105. **Stein, P., A. Boodhoo, G. Tyrrell, J. Brunton, and R. J. Read.** 1992. Crystal structure of the cell-binding B oligomer of verotoxin-1 from *E. coli. Nature* (London) **355**: 748–750.

106. **Sueyoshi, M., and M. Nakazawa.** 1994. Experimental infection of young chicks with attaching and effacing *Escherichia coli. Infect. Immun.* **62**:4066–4071.

107. **Swerdlow, D. L., B. A. Woodruff, and R. C. Brady.** 1992. A waterborne outbreak in Missouri of *Escherichia coli* O157:H7 associated with bloody diarrhea and death. *Ann. Intern. Med.* **117**:812–819.

108. **Tarr, P. I.** 1995. *Escherichia coli* O157:H7: clinical, diagnostic, and epidemiological aspects of human infection. *Clin. Infect. Dis.* **20**:1–10.

109. **Tarr, P. I., M. A. Neill, C. R. Clausen, S. L. Watkins, D. L. Christie, and R. O. Hickman.** 1990. *Escherichia coli* O157:H7 and the hemolytic uremic syndrome: importance of early cultures in establishing the etiology. *J. Infect. Dis.* **162**:553–556.

110. **Tesh, V. L., and A. D. O'Brien.** 1991. The pathogenic mechanisms of Shiga toxin and the Shiga-like toxins. *Mol. Microbiol.* **5**:1817–1822.

111. **Tesh, V. L., and A. D. O'Brien.** 1992. Adherence and colonization mechanisms of enteropathogenic and enterohemorrhagic *Escherichia coli. Microb. Pathog.* **12**:245–254.

112. **Thomas, A., H. Chart, T. Cheasty, J. E. Smith, J. A. Frost, and B. Rowe.** 1993. Verocytotoxin-producing *Escherichia coli*, particularly serogroup O157, associated with human infection in the United Kingdom: 1989–1991. *Epidemiol. Infect.* **110:**591–600.

113. **Toth, I., M. L. Cohen, H. S. Rumschlag, L. W. Riley, E. H. White, J. H. Carr, W. W. Bond, and I. K. Wachsmuth.** 1990. Influence of the 60-megadalton plasmid on adherence of *Escherichia coli* O157:H7 and genetic derivatives. *Infect. Immun.* **58:**1223–1231.

114. **Tyrrell, G., K. Ramotar, B. Toye, B. Boyd, C. Lingwood, and J. Brunton.** 1992. Alteration of the carbohydrate binding specificity of verotoxin from gal α 1–4 gal to gal NAC β 1–3 gal α 1–4 gal and vice versa by site-directed mutagenesis of the binding subunit. *Proc. Natl. Acad. Sci. USA* **89:**524–528.

115. **Tzipori, S., R. Gibson, and J. Montanaro.** 1989. Nature and distribution of mucosal lesions associated with enteropathogenic and enterohemorrhagic *Escherichia coli* in piglets and the role of plasmid-mediated factor. *Infect. Immun.* **57:**142–1150.

116. **Tzipori, S., H. Karch, I. K. Wachsmuth, R. M. Robins-Browne, A. D. O'Brien, H. Lior, M. L. Cohen, J. Smithers, and M. M. Levine.** 1987. Role of a 60-megadalton plasmid and Shiga-like toxins in the pathogenesis of infection caused by enterohemorrhagic *Escherichia coli* O157:H7 in gnotobiotic piglets. *Infect. Immun.* **55:**3117–3125.

117. **Wachsmuth, K.** 1994. Summary: public health; epidemiology; food safety; laboratory diagnosis, p. 3–5. *In* M. A. Karmali and A. G. Goglio (ed.) *Recent Advances in Verocytotoxin-Producing Escherichia coli Infections.* Proceedings of the 2nd International Symposium and Workshop on Verocytotoxin (Shiga-Like Toxin)-Producing *Escherichia coli* Infections. Elsevier, Amsterdam.

118. **Wadolkowski, E. A., J. A. Burris, and A. D. O'Brien.** 1990. Mouse model for colonization and disease caused by enterohemorrhagic *Escherichia coli* O157:H7. *Infect. Immun.* **58:**2438–2445.

119. **Wanke, C. A., J. B. Schorling, L. J. Barrett, M. A. Desouza, and R. L. Guerrant.** 1991. Potential role of adherence traits of *Escherichia coli* in persistent diarrhea in an urban Brazilian slum. *Pediatr. Infect. Dis. J.* **10:**746–751.

120. **Waters, J. R., J. C. M. Sharp, and V. J. Dev.** 1994. Infection caused by *Escherichia coli* O157:H7 in Alberta, Canada, and in Scotland: a five-year review, 1987–1991. *Clin. Infect. Dis.* **19:**834–843.

121. **Weinstein, D., M. Jackson, L. Perera, R. Holmes, and A. O'Brien.** 1988. Cloning and sequencing of a Shiga-like toxin II variant from an *Escherichia coli* strain responsible for edema disease of swine. *J. Bacteriol.* **170:**4223–4230.

122. **Wells, J. G., B. R. Davis, I. K. Wachsmuth, L. W. Riley, R. S. Remis, R. Sokolow, and G. K. Morris.** 1983. Laboratory investigation of hemorrhagic colitis outbreaks associated with a rare *Escherichia coli* serotype. *J. Clin. Microbiol.* **18:**512–520.

123. **Wells, J. G., L. D. Shipman, K. D. Greene, E. G. Sowers, J. H. Green, D. N. Cameron, F. P. Downes, M. L. Martin, P. M. Griffin, S. M. Ostroff, M. E. Potter, R. V. Tauxe, and I. K. Wachsmuth.** 1991. Isolation of *Escherichia coli* serotype O157:H7 and other Shiga-like toxin-producing *E. coli* from dairy cattle. *J. Clin. Microbiol.* **29:**985–989.

124. **Whipp, S. C., M. A. Rasmussen, and W. C. Cray.** 1994. Animals as a source of *Escherichia coli* pathogenic for human beings. *J. Am. Vet. Med. Assoc.* **204:**1168–1175.

125. **Yu, J., and J. B. Kaper.** 1992. Cloning and characterization of the *eae* gene of enterohemorrhagic *Escherichia coli* O157:H7. *Mol. Microbiol.* **6:**411–417.

126. **Zhao, S., J. Meng, and M.P. Doyle.** 1995. A unique outer membrane protein associated with colonization of *Escherichia coli* O157:H7 on human intestinal epithelial cells, abstr. B9, p. 167. *In Abstracts of the 95th General Meeting of the American Society for Microbiology 1995.* American Society for Microbiology, Washington, D.C.

127. **Zhao, T., and M. P. Doyle.** 1994. Fate of enterohemorrhagic *Escherichia coli* O157:H7 in commercial mayonnaise. *J. Food Prot.* **57:**780–783.

128. **Zhao, T., M. P. Doyle, and R. E. Besser.** 1993. Fate of enterohemorrhagic *Escherichia coli* O157:H7 in apple cider with and without preservatives. *Appl. Environ. Microbiol.* **59:**2526–2530.

129. **Zhao, T., M. P. Doyle, J. Shere, and L. Garber.** 1995. Prevalence of enterohemorrhagic *Escherichia coli* O157:H7 in a survey of dairy herds. *Appl. Environ. Microbiol.* **61:** 1290–1293.

130. **Zhao, T., M. P. Doyle, and G. Wang.** 1994. Emerging pathogens and rapid detection methods: *Escherichia coli, Listeria, Salmonella, Vibrio cholerae,* p. 386–403. *In Proceedings of the 2nd Asian Conference on Food Safety.* International Life Sciences Institute, Washington, D.C.

Roy M. Robins-Browne

11 Yersinia enterocolitica

CHARACTERISTICS OF THE ORGANISM

Introduction

The genus *Yersinia* comprises 11 species, including one ("*Yersinia ruckeri*") of uncertain status (Table 11.1) (6, 147). As members of the family *Enterobacteriaceae*, yersiniae are oxidase-negative, gram-negative rod-shaped facultative anaerobes which ferment glucose. The genus includes three primary pathogens of humans and several other species which may cause opportunistic infections. The three pathogenic species are *Y. pestis*, the causative agent of bubonic and pneumonic plague (the black death), *Y. pseudotuberculosis*, an intestinal pathogen of rodents which occasionally infects humans, and *Y. enterocolitica*, a common intestinal pathogen of humans. Despite their differences in preferred epidemiological niche and modes of transmission, *Y. pestis*, *Y. pseudotuberculosis*, and *Y. enterocolitica* share a number of essential virulence determinants which enable them to overcome nonspecific immune defenses of their hosts. Analogs of these virulence determinants occur in several other bacterial pathogens, including *Salmonella* and *Shigella* species, providing evidence for horizontal transfer of the genetic information for virulence determinants between enteric pathogens. In addition, yersiniae may have acquired a number of genetic elements from eukaryotic cells. These factors may enhance bacterial virulence by enabling them to undermine essential aspects of the physiological response to infection.

Classification

Y. enterocolitica first emerged as a human pathogen during the 1930s (15). After several unsuccessful attempts to allocate it to a suitable taxonomic position, *Y. enterocolitica* was finally assigned to the family *Enterobacteriaceae*. *Y. enterocolitica* exhibits between 10 and 30% DNA homology with other genera in the *Enterobacteriaceae* and is approximately 50% related to *Y. pestis* and *Y. pseudotuberculosis*. The last two species share greater than 90% DNA homology overall, suggesting that they are different pathotypes of the same species.

Y. enterocolitica is highly heterogeneous, being divisible into a large number of subgroups, chiefly according to biochemical activity and lipopolysaccharide (LPS) O antigens (Tables 11.2 and 11.3). Biotyping is based on the ability of *Y. enterocolitica* to metabolize selected organic substrates and provides a convenient means to subdivide this species into subtypes of variable clinical and epidemiological significance (Tables 11.2 and 11.3) (148). Most primary pathogenic strains of humans and domestic animals occur within biovars 1B, 2, 3, 4, and 5. By contrast, *Y. enterocolitica* strains of biovar 1A are generally obtained from terrestrial and freshwater ecosystems and are often referred to as environmental strains. Not all isolates of *Y. enterocolitica* obtained from soil, water, or unprocessed foods can be assigned to a biovar. These strains invariably lack the characteristic virulence determinants of the primary pathogenic yersiniae (see below) and may represent

Table 11.1 Some biochemical tests used to differentiate *Yersinia* species[a]

Test	Reaction[b]												
	Y. aldovae	*Y. bercovieri*	*Y. enterocolitica* Biovars 1–4	*Y. enterocolitica* Biovar 5	*Y. frederiksenii*	*Y. intermedia*	*Y. kristensenii*	*Y. mollareti*	*Y. pestis*	*Y. pseudo-tuberculosis*	*Y. rohdei*	"*Y. ruckeri*"
Indole	–	–	d	–	+	+	d	–	–	–	–	–
Voges-Proskauer	+	–	+	+[c]	d	+	–	–	–	–	–	–
Citrate (Simmons)	d	–	–	–	d	+	–	–	–	–	+	–
L-Ornithine	+	+	+	–	+	+	+	+	–	–	+	+
Mucate, acid	d	+	–	–	d	d	–	+	–	–	–	–
Pyrazinamidase	+	+	d	–	+	+	+[c]	+	–	–[c]	+	ND
Sucrose	–	+	+	d	+	+	–	+	–	–	+	–
Cellobiose	–	+	+	+	+	+	+	–	–	+	–	+
L-Rhamnose	+	–	–	–	+	+	–	–	–	+	–	–
Melibiose	–	–	–	–	–	+	–	–	d	+	d	–
L-Sorbose	–	–	d	d	+	+	+	+	–	–	ND	ND
L-Fucose	d	+	d	–	+	d	d	–	ND	–	ND	ND

[a]Adapted from reference 147.
[b]+, positive; –, negative; d, different reactions; ND, not determined.
[c]Some reactions may be delayed or weakly positive.

novel nonpathogenic subtypes or even new *Yersinia* species.

The most frequent *Y. enterocolitica* biovar obtained from human clinical material worldwide is biovar 4. Biovar 1B bacteria are usually isolated from patients in the United States and are referred to as American strains, although they have also been identified in a number of countries in Europe, Africa, Asia, and Australasia. Although not common anywhere, biovar 1B yersiniae appear to be inherently more virulent than strains in the other pathogenic categories and have been identified as the cause of several foodborne outbreaks of yersiniosis in the United States.

Serotyping of *Y. enterocolitica*, based on LPS surface O antigens, coincides to some extent with biovar (Table 11.3) and provides a useful additional tool to subdivide this species in a way that relates to pathologic significance (146). Serovar O:3 is the variety most frequently isolated from humans. Almost all of these isolates are biovar 4. Other serovars commonly obtained from humans include O:9 and O:5,27, particularly in northern Europe. Despite the usefulness of serotyping in

Table 11.2 Biotyping scheme of *Y. enterocolitica*[a]

Test	Reaction of biovar[b]:					
	1A	1B	2	3	4	5
Lipase (Tween hydrolysis)	+	+	–	–	–	–
Esculin hydrolysis	d	–	–	–	–	–
Indole production	+	+	(+)	–	–	–
D-Xylose fermentation	+	+	+	+	–	d
Voges-Proskauer reaction	+	+	+	+	+	(+)
Trehalose fermentation	+	+	+	+	+	–
Nitrate reduction	+	+	+	+	+	–
Pyrazinamidase	+	–	–	–	–	–
β-D-Glucosidase	+	–	–	–	–	–
Proline peptidase	d	–	–	–	–	–

[a]Adapted from reference 148.
[b]+, positive; (+), delayed positive; –, negative; d, different reactions.

Table 11.3 Relationship between O serovar and pathogenicity of *Y. enterocolitica* and related species

Species	Serovar(s)[a]
Y. enterocolitica	
Biovar 1A	O:4; O:5; O:6,30; O6,31; O:7,8; O:7,13; O:10; O:14; O:16; O:21; O:22; O:25; O:37; O:41,42; O:46; O:47; O:57; NT
Biovar 1B	**O:4,32; O:8; O:13a,13b;** O:16; **O:18; O:20; O:21;** O:25; O:41,42; NT
Biovar 2	**O:5,27; O:9;** O:27
Biovar 3	**O:1,2,3; O:3; O:5,27**
Biovar 4	**O:3**
Biovar 5	**O:2,3**
Y. bercovieri	O:8; O:10; O:58,16; NT
Y. frederiksenii	O:3; O:16; O:35; O:38; O:44; NT
Y. intermedia	O:17; O:21,46; O:35; O:37; O:40; O:48; O:52; O:55; NT
Y. kristensenii	O:11; O:12,25; O:12,26; O:16, O:16,29; O:28,50; O:46; O:52; O:59; O:61; NT
Y. mollaretti	O:3, O:6,30; O:7,13; O:59; O:62,22; NT

[a]NT, not typeable. Serogroups which include strains considered to be primary pathogens are in boldface.

predicting potential virulence, the scheme is imperfect because of the presence of antigens in other bacteria which cross-react with *Y. enterocolitica* and because some bacteria originally allocated to serogroups of *Y. enterocolitica* were later reclassified as separate species (Table 11.3).

At least 18 flagellar (H) antigens of *Y. enterocolitica*, designated by lowercase letters (a,b; b,c; b,c,e,f,k; m, etc.), have been identified, but complete antigenic characterization of isolates by O and H serotyping is seldom attempted.

Other schemes used to subtype *Yersinia* species include bacteriophage typing, isoenzyme electropherotyping, and the demonstration of restriction fragment length polymorphism of chromosomal and plasmid DNA. These techniques may be helpful in delineating individual isolates or in tracing epidemics but are available in only a few specialized centers and are used infrequently.

Susceptibility and Tolerance

Y. enterocolitica is unusual among pathogenic enterobacteria in being psychrotrophic, as evidenced by its ability to replicate at temperatures between 0 and 44°C. The doubling time at the optimum growth temperature (approximately 28 to 30°C) is around 34 min (Fig. 11.1), which increases to 1 h at 22°C, 5 h at 7°C, and approximately 40 h at 1°C (120). Yersiniae readily withstand freezing and can survive in frozen foods for extended periods even after repeated freezing and thawing (142). *Y. enterocolitica* is susceptible to heat, however, and is destroyed by pasteurization at 71.8°C for 18 s (30, 142). Exposure of surface-contaminated meat to hot water (80°C) for 10 to 20 s reduces bacterial viability by at least 99.9% (130).

Y. enterocolitica is able to grow over a pH range from approximately pH 4 to 10, with an optimum pH of around 7.6 (117). Tolerance of *Y. enterocolitica* to acid depends on the acidulent agent used, the environmental temperature, the composition of the medium, and the growth phase of the bacteria (2). Acid tolerance depends on the activity of urease, which catabolizes urea to release ammonia, which elevates the cytoplasmic pH (34).

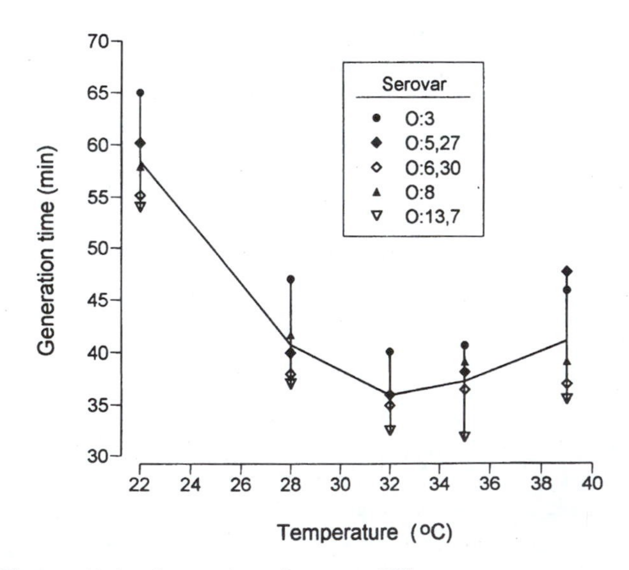

Figure 11.1 Generation times at different temperatures of *Y. enterocolitica* strains of five different serovars, grown in 4% tryptone–1% mannitol salt broth, pH 7.6. (Adapted from reference 120.)

Y. enterocolitica is readily inactivated by ionizing and UV irradiation (20, 38) and by sodium nitrate and nitrite added to food (32). It displays relative resistance to these salts in solution, however, and can tolerate NaCl at concentrations of up to 5% (32, 41). *Y. enterocolitica* is also somewhat resistant to chlorine, particularly when grown under conditions that approximate natural aquatic environments or when cocultivated with predatory aquatic protozoa (55, 75).

Studies of the ability of *Y. enterocolitica* to survive and grow in artificially contaminated foods under various conditions of storage have shown that this species generally survives better at room temperature and refrigeration temperatures than at intermediate temperatures. *Y. enterocolitica* persists longer in cooked foods than in raw foods, probably because of increased availability of nutrients in cooked foods and because the presence of other psychrotrophic bacteria, including environmental strains of *Y. enterocolitica*, in unprocessed food may restrict bacterial growth (120, 131). The number of viable *Y. enterocolitica* may increase more than 10^6-fold on cooked beef or pork within 24 h at 25°C or within 10 days at 7°C (53). Growth is slower on raw beef and pork. *Y. enterocolitica* is able to grow at refrigeration temperature in vacuum-packed meat, boiled eggs, boiled fish, pasteurized liquid eggs, pasteurized whole milk, and tofu (soybean curd) (41, 120). Proliferation also occurs in refrigerated seafoods, such as oysters, raw shrimp, and cooked crabmeat, but more slowly than in pork or beef (100). Bacteria may also persist for extended periods in refrigerated vegetables and cottage cheese, particularly in the presence of chicken meat (123). Although *Y. enterocolitica* tolerates 5% NaCl in culture media (131), the addition of 5% NaCl to foods slows its rate of growth (40).

CHARACTERISTICS OF INFECTION

Infections with *Y. enterocolitica* are remarkable for their wide range of clinical presentations and outcomes. Although most symptomatic infections result in nonspecific, self-limiting diarrhea, yersiniosis may also give rise to a variety of suppurative and autoimmune complications. The risk of these complications is determined largely by host factors, in particular age and underlying immune status.

Acute Infection

Y. enterocolitica enters the gastrointestinal tract after ingestion in contaminated food or water. The infective dose for humans is not known but is likely to exceed 10^4 CFU. Gastric acid appears to be a significant barrier to infection, and in individuals with gastric hypoacidity, the infectious dose may be lower (43).

Most symptomatic infections with *Y. enterocolitica* occur in children, especially in those less than 5 years of age (105). In these patients, yersiniosis presents as diarrhea, often accompanied by low-grade fever and abdominal pain (63, 84). The character of the diarrhea varies from watery to mucoid. A small proportion of children (generally less than 10%) have frankly bloody stools. The illness typically lasts from a few days to 3 weeks, although some patients develop chronic enterocolitis which may persist for several months (29). Rarely, acute enteritis may progress to intestinal ulceration and perforation or to ileocolic intussusception, toxic megacolon, or mesenteric vein thrombosis (29). Children with *Yersinia*-induced diarrhea often complain of abdominal pain and headache. Sore throat is a frequent accompaniment and may dominate the clinical picture in older patients (136).

In children older than 5 years and adolescents, acute yersiniosis often presents as a pseudoappendicular syndrome. The usual features of this syndrome are abdominal pain and tenderness localized to the right lower quadrant. These symptoms are usually accompanied by fever, with little or no diarrhea. The importance of this form of the disease lies in its close resemblance to appendicitis (29). Of those patients with this syndrome who undergo laparotomy, approximately 80% have terminal ileitis, with or without mesenteric adenitis, and a normal or only slightly inflamed appendix (29, 97). *Y. enterocolitica* may be cultured from the distal ileum and the mesenteric lymph nodes (29). The pseudoappendicular syndrome appears to be more frequent in patients infected with more virulent strains of *Y. enterocolitica*, notably American strains of biovar 1B. *Y. enterocolitica* rarely causes true appendicitis.

Although *Y. enterocolitica* is seldom isolated from extraintestinal sites, there appears to be no tissue in which it will not grow. In adults, pharyngitis, sometimes with cervical lymphadenitis, may dominate the clinical presentation (136). Focal disease, in the absence of obvious bacteremia, may present as cellulitis, subcutaneous abscess, pyomyositis, suppurative lymphadenitis, septic arthritis, osteomyelitis, urinary tract infection, renal abscess, sinusitis, pneumonia, lung abscess, or empyema (29).

Bacteremia is a rare complication of infection, except in patients who are immunocompromised or in an iron-overloaded state (29, 43, 74). Factors which predispose to the development of *Yersinia* bacteremia include immunosuppression, blood dyscrasias, malnutrition, chronic renal failure, cirrhosis, alcoholism, diabetes mellitus, and acute and chronic iron overload states, particularly when managed by chelation therapy with desferrioxamine B (29). The more virulent American serovars appear more likely to cause bacteremia than others.

Bacteremia may also result from direct inoculation of *Y. enterocolitica* into the circulation during blood transfusion (116). The bacteria responsible for this form of infection belong to the same bioserovars as those which cause enteric infections. The probable sources of transfusion-acquired infections are blood donors with low-grade, subclinical bacteremia. A small number of bacteria in donated blood will increase during storage at refrigeration temperatures without manifestly altering the appearance of the blood (47). Patients infused with contaminated blood may develop symptoms of a severe transfusion reaction minutes to hours after exposure, depending on the number of bacteria and the amount of endotoxin administered with the blood.

Bacteremic dissemination of *Y. enterocolitica* may lead to various manifestations, including splenic, hepatic, and lung abscesses, catheter-associated infections, osteomyelitis, panophthalmitis, endocarditis, mycotic aneurysm, and meningitis (29). *Yersinia* bacteremia is reported to have a case fatality rate of between 30 and 60%, but the absence of symptoms in some blood donors who presumably had bacteremia at the time of blood donation suggests that some individuals experience low-grade or subclinical bacteremia without serious consequences.

Autoimmune Complications

Although most episodes of yersiniosis remit spontaneously without long-term sequelae, infections with *Y. enterocolitica* are noteworthy for the large variety of immunological complications, including reactive arthritis, erythema nodosum, iridocyclitis, glomerulonephritis, carditis, and thyroiditis, which have been reported to follow acute infection (29). Of these, reactive arthritis is the most widely recognized (3, 14, 80). This manifestation of infection is infrequent before the age of 10 years and occurs most often in Scandinavian countries, where serovar O:3 strains and the human leukocyte antigen HLA-B27 are especially prevalent. Men and women are affected equally. Arthritis typically follows the onset of diarrhea or the pseudoappendicular syndrome by 1 to 2 weeks, with a range of 1 to 38 days. The joints most commonly involved are the knees, ankles, toes, tarsal joints, fingers, wrists, and elbows. The arthritis is usually migratory, the joints being affected one after another. In approximately one in eight patients with arthritis, only one joint is involved. The synovial fluid contains large numbers of inflammatory cells, principally polymorphonuclear leukocytes, and is invariably sterile, although it usually contains bacterial antigens (48). The duration of arthritis is generally less than 3 months, and the long-term prognosis in terms of joint destruction is excellent, although some patients may have symptoms that persist for several years after the acute episode (61, 80). Many patients with arthritis also have extra-articular symptoms, including urethritis, eye inflammation, and erythema nodosum (14).

Y. enterocolitica-induced erythema nodosum occurs predominantly in women and is not associated with HLA-B27, although it is commonly accompanied by arthritis. Other autoimmune complications of yersiniosis, including Reiter's syndrome, iridocyclitis, acute proliferative glomerulonephritis, and rheumatic-like carditis, have been reported almost exclusively from Scandinavian countries (78). Yersiniosis has also been linked to various thyroid disorders, including Graves' disease hyperthyroidism, nontoxic goiter, and Hashimoto's thyroiditis, although the causative role of yersiniae in these conditions is not proven (140).

RESERVOIRS

Y. enterocolitica occupies a broad range of environments and has been isolated from the intestinal tracts of many different mammalian species, as well as from birds, frogs, fish, flies, fleas, crabs, and oysters (29). Foods that may harbor *Y. enterocolitica* include pork, beef, lamb, poultry, and dairy products, notably milk, cream, and ice cream (41, 70, 120). *Y. enterocolitica* is also commonly found in a variety of terrestrial and freshwater ecosystems, including soil, vegetation, lakes, rivers, wells, and streams (70). Most environmental isolates lack markers of bacterial virulence and are of doubtful significance for human or animal health (35, 91). Some domesticated animals, notably sheep, cattle, and deer, may suffer symptoms as a result of infection with *Y. enterocolitica* (129), but in most cases the biovars and serovars of these bacteria differ from those responsible for human infection, indicating a lack of transmission of these strains between animals and humans.

Pigs are the only animal species from which *Y. enterocolitica* of biovar 4 serovar O:3 (the variety most commonly associated with human disease) has been isolated with any degree of frequency (70). Pigs may also carry *Y. enterocolitica* of serovars O:9 and O:5,27, particularly in regions where human infections with these varieties are comparatively common. Individual isolates of *Y. en-*

Table 11.4 Selected foodborne outbreaks of infections with *Y. enterocolitica*

Location	Yr	Mo	No. of cases	Serovar	Source	Reference
Canada	1976	Apr.	138	O:5,27	Raw milk (?)[a]	72
New York	1976	Sept.	38	O:8	Flavored milk	9
Japan	1980	Apr.	1,051	O:3	Milk	85
New York	1981	July	159	O:8	Powdered milk, chow mein	121
Washington	1981	Dec.	50	O:8	Tofu, spring water	135
Pennsylvania	1982	Feb.	16	O:8	Bean sprouts, well water	29
Southern United States	1982	June	172	O:13a,13b	Milk (?)	137
Hungary	1983	Dec.	8	O:3	Pork cheese (sausage)	83
Georgia, United States	1989	Nov.	15	O:3	Pork chitterlings	79

[a](?), the bacteria were not isolated from the incriminated source.

terocolitica from pigs and humans are indistinguishable from each other in terms of serovar, biovar, restriction fragment length polymorphism of chromosomal and plasmid DNA, and carriage of virulence determinants (70).

In countries with a high incidence of human yersiniosis, *Y. enterocolitica* is commonly isolated from pigs at slaughterhouses and butcher shops (24). Tissues which are most frequently culture positive at slaughter include tongue, tonsils, cecum, rectum, feces, and gut-associated lymphoid tissue. *Y. enterocolitica* is seldom isolated from meat offered for retail sale, with the exception of pork tongue (70, 120), although standard methods of bacterial isolation and detection may underestimate the true incidence of contamination (95).

Further evidence of the significance of pigs as a reservoir of human infections is provided by epidemiological studies pointing to the ingestion of raw or undercooked pork as a major risk factor for the acquisition of yersiniosis (98, 138). Susceptible individuals have also become infected after exposure to chitterlings (74, 132). Although *Y. enterocolitica* strains of biovar 4, serovar O:3 have been isolated from domestic dogs and cats (42), which therefore may also serve as a possible source of human infection, transmission from these animals to humans has not been proven.

Several outbreaks of yersiniosis have been linked to the consumption of contaminated cow milk (Table 11.4), but cattle do not appear to be an important reservoir of these bacteria. In these outbreaks, milk may have been contaminated with pig or human feces during or after processing (39, 118). Contamination of pasteurized milk appears to pose a greater threat of infection than raw milk, probably because *Y. enterocolitica* outgrows other fecal microorganisms during storage at refrigeration temperatures most readily when competing psychrotrophic microflora have been eliminated (118).

Food animals are not commonly infected with biovar 1B strains of *Y. enterocolitica*, the reservoir of which reamins unknown (120). The relatively low incidence of human yersiniosis caused by biovar 1B strains, notwithstanding their comparatively high virulence, points to a lack of significant contact between their reservoir and humans. As yersiniae of this biovar are primary pathogens of rodents, it is conceivable that rats or mice are the natural reservoir of these strains (57).

Y. enterocolitica, including pathogenic strains, can persist for extended periods in soil, vegetation, streams, lakes, wells, and spring water, particularly at low environmental temperatures (21). The relatively low optimum growth temperature of *Y. enterocolitica* may account in part for the higher incidence of yersiniosis in temperate regions than in the tropics (in contrast to most other bacterial agents of diarrhea) and the tendency for infections to peak during late autumn and winter in countries where yersiniosis is endemic.

FOODBORNE OUTBREAKS

Considering the widespread occurrence of *Y. enterocolitica* in nature and its ability to colonize food animals, to persist within animals and the environment, and to proliferate at refrigeration temperatures, outbreaks of yersiniosis are surprisingly uncommon. Most foodborne outbreaks in which a source was identified have been traced to milk (Table 11.4). As *Y. enterocolitica* is rapidly destroyed by pasteurization, infection results from the consumption of raw milk or milk that is contaminated after pasteurization (39, 118). During the mid-1970s, two outbreaks of yersiniosis caused by *Y. enterocolitica* O:5,27 occurred among 138 Canadian schoolchildren who had consumed raw milk, but the organism was not recovered from the suspected source (72). In 1976, serogroup O:8 *Y. enterocolitica* was responsible for an outbreak in New York State which affected 217 people, 38 of whom were culture positive (9). The source of infection was chocolate-flavored milk, which evidently became contaminated with *Y. enterocolitica* after pasteurization.

In 1981, an outbreak of infection with *Y. enterocolitica* O:8 affected 35% of 455 individuals at a diet camp in New York State (92). Seven patients were hospitalized as result of infection, five of whom underwent appendectomy. The source of the infection was reconstituted powdered milk and/or chow mein which probably became contaminated during preparation by an infected food handler. During 1982, 172 cases of infection with *Y. enterocolitica* O:13a, 13b occurred in an area which included parts of Tennessee, Arkansas, and Mississippi (137). The suspected source was pasteurized milk which may have become contaminated with pig manure during transport (39).

More recently, an outbreak of infection with *Y. enterocolitica* O:3 affected 15 infants and children in metropolitan Atlanta (79). In this instance, bacteria were transmitted from raw chitterlings (pig intestine) to affected children on the hands of food handlers. Other foods which have been responsible for outbreaks of yersiniosis include pork cheese (a type of sausage prepared from pork chitterlings), bean sprouts, and tofu (29, 83, 135). In the outbreaks associated with bean sprouts and tofu, contaminated well or spring water was the probable source of the bacteria. Water was also the putative source of infection in a case of sporadic *Y. enterocolitica* bacteremia in a 75-year-old man in New York State (73) and a small family outbreak in Ontario, Canada (139). Several outbreaks of presumed foodborne infection with *Y. enterocolitica* O:3 have also been reported from the United Kingdom and Japan, but in most cases the source of these outbreaks was not identified.

MECHANISMS OF PATHOGENICITY

Pathological Changes

Examination of surgical specimens from patients with yersiniosis reveals that *Y. enterocolitica* is an invasive pathogen which induces an inflammatory response in infected tissues (26). The distal ileum, in particular the gut-associated lymphoid tissue, bears the brunt of the infection, although adjacent regions of the intestine and the mesenteric lymph nodes are also frequently involved. In this way, *Y. enterocolitica* displays a tropism for lymphoid tissue similar to that of *Y. pestis* and *Y. pseudotuberculosis* (17).

The paucity of reports of the pathology of yersiniosis in humans stands in contrast to the large number of descriptions of the behavior of *Y. enterocolitica* in in vitro models of virulence. As investigations in volunteers are precluded by the risk of autoimmune sequelae, most information has been obtained from animal models, in particular mice and rabbits (58). Although these animals are not the natural hosts of the serovars of *Y. enterocolitica* that usually infect humans, they have permitted the elucidation of much of the pathogenesis of human disease. Nevertheless, some data derived from animal studies should be interpreted with caution, particularly when death has been used as the endpoint of infection, as this is not the usual outcome of infection in humans (101).

After oral inoculation of mice with a virulent strain of serovar O:8, most bacteria remain within the intestinal lumen, while a minority adhere to the mucosal epithelium, showing no particular preference for any cell type (51). By contrast, invasion takes place almost exclusively through M cells, which are specialized epithelial cells overlying lymphoid follicles (Peyer's patches) in the intestine (Fig. 11.2) (51). After penetrating the epithelium, yersiniae traverse the basement membrane of the relatively porous dome epithelium of the Peyer's patches. Virulent strains then proliferate in the gut-associated lymphoid tissue and the lamina propria, where they cause localized tissue destruction leading to the formation of microabscesses (111, 143). They may also spread through the lamina propria to adjacent villi and via the lymph to more remote sites.

The hallmark lesion of yersiniosis comprises microcolonies of bacteria surrounded by granulocytic and mononuclear inflammatory cells (Fig. 11.3). These lesions occur chiefly within intestinal crypts but may extend as far as the crypt-villus junction. Villus height in the ileum is significantly reduced but is usually normal in other regions of the small intestine (19, 111, 143). Electron microscopic examination of the intestine of experimentally infected rabbits and piglets shows a disorganized or diminished brush border, particularly in the vicinity of microabscesses (Fig. 11.4). These changes are accompanied by reduced mucosal enzyme activity but spontaneous recovery occurs within 14 days of the onset of infection (19).

Y. enterocolitica may spread via the lymph to the draining mesenteric lymph nodes, where the organisms also produce microabscesses. If the bacteria circumvent the lymph nodes to enter the bloodstream, they may disseminate to any organ but continue to show a tropism for lymphoid tissue by preferentially localizing in the reticuloendothelial tissues of the liver and spleen. Although *Y. enterocolitica* strains are classified as facultative intracellular pathogens, because they are innately resistant to killing by macrophages and polymorphonuclear leukocytes, most of the bacteria observed in histological

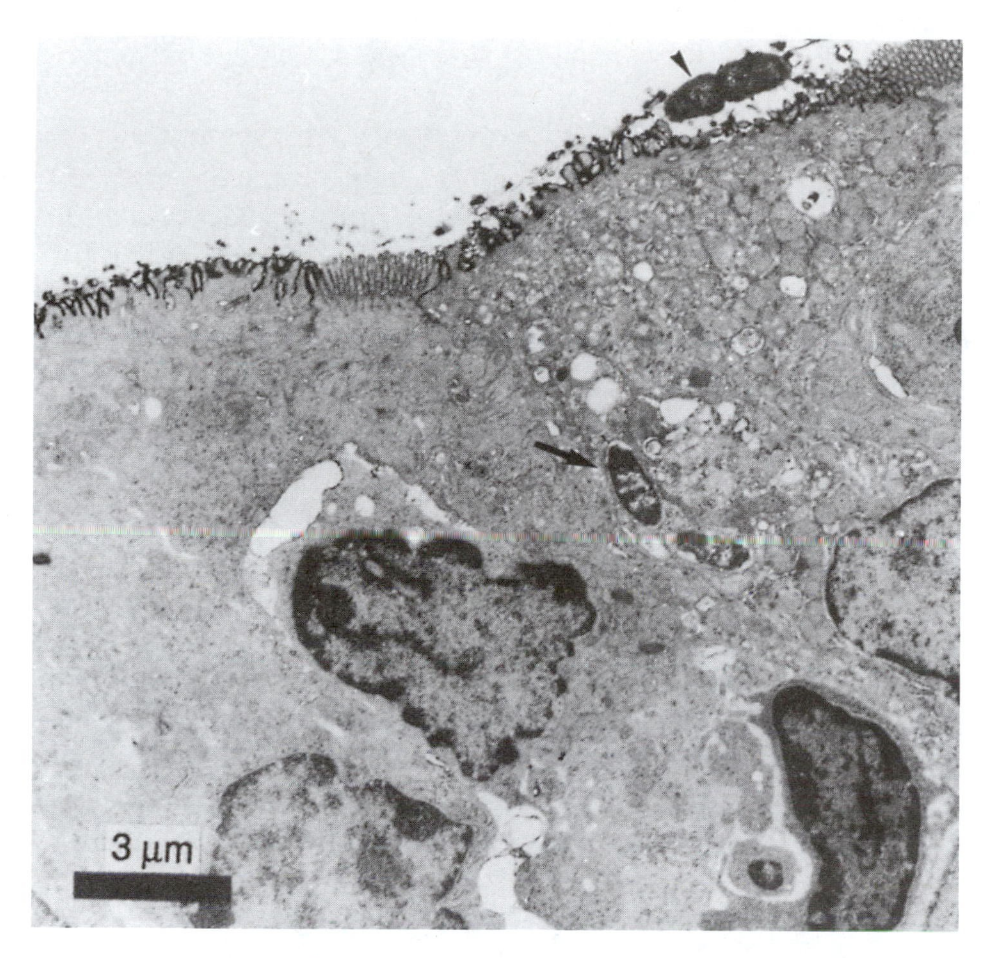

Figure 11.2 Transmission electron micrograph showing the initial interaction (arrowhead) and transport (arrow) of *Y. enterocolitica* through an intestinal M cell, 60 min after inoculation into mouse ileum. Bar = 3 μm. (Reprinted with permission from reference 51.)

sections are located extracellularly (54). The question of whether the bacteria need to pass through host cells to attain virulence has not been resolved.

Virulence Determinants

Y. enterocolitica and *Y. pseudotuberculosis* are paradigms of invasive enteric pathogens whose pathogenesis has been the subject of intensive investigation (13, 17, 26–28). These bacteria possess a panoply of virulence determinants, some with apparently opposing mechanisms of action (13). As expression of virulence is tightly regulated at a genetic level, however, the contradictory actions of these determinants probably reflect the ability of yersiniae to respond to the changing environments they encounter in soil, water, food, and host tissues. Moreover, the differential expression of particular virulence determinants by subpopulations of bacteria within a specific location may offer the bacterial population as a whole alternative forms of parasitism which enhance its overall chances of survival. The virulence determinants of *Y. enterocolitica* are classified into those which are chromosomally encoded and those specified by a 70- to 75-kb virulence plasmid.

Chromosomal Determinants of Virulence

Invasin

All *Y. enterocolitica* strains which belong to virulent serovars are able to produce a 91-kDa surface-expressed outer membrane protein termed invasin. This protein was first identified in *Y. pseudotuberculosis* as a 102-kDa protein product of the chromosomal *inv* gene (68, 103). Despite the difference in size of invasins from *Y. enterocolitica* and *Y. pseudotuberculosis*, the two proteins are functionally highly conserved. When introduced into a noninvasive laboratory strain of *Escherichia coli*, *inv* imbues the recipient with the ability to penetrate cultured

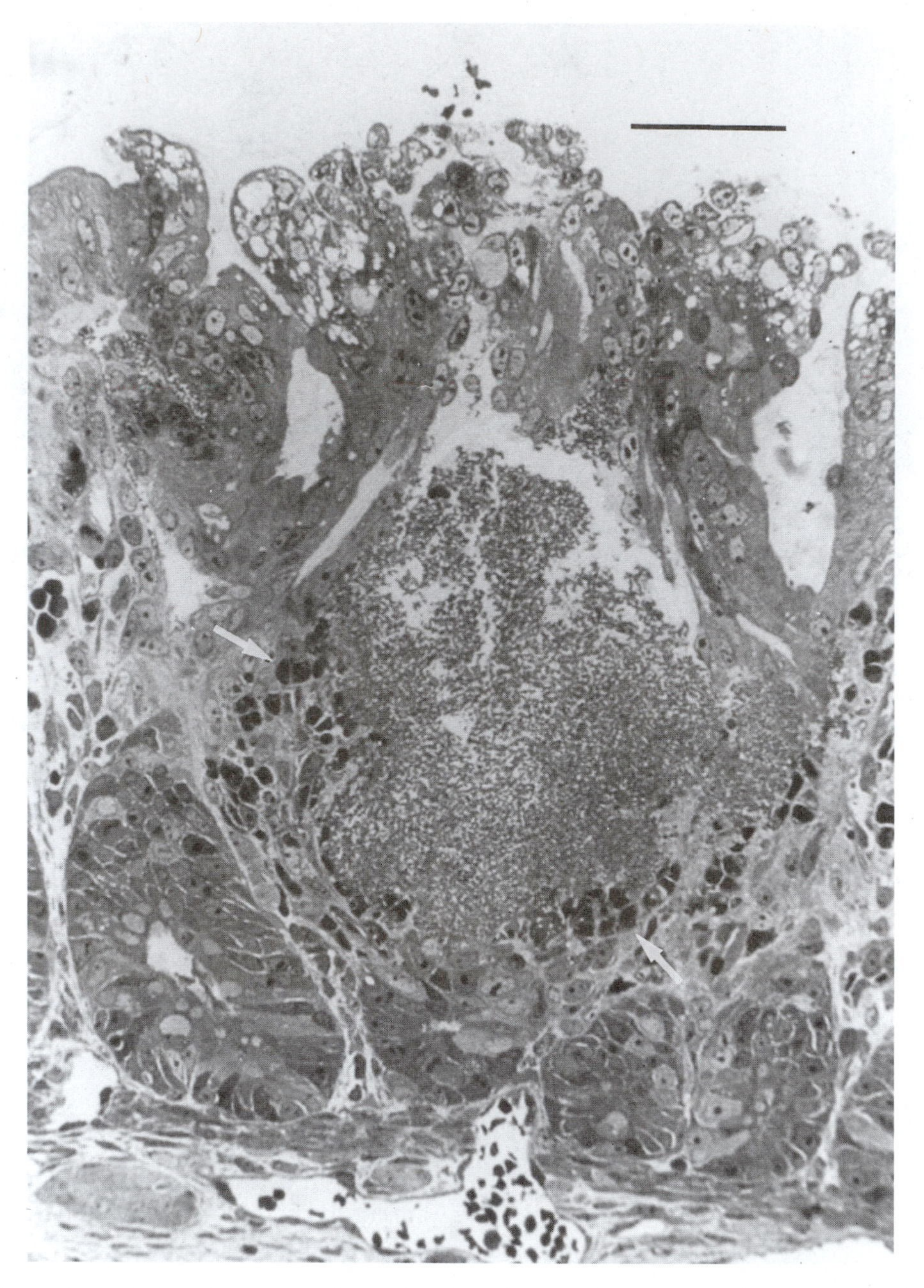

Figure 11.3 Light micrograph of a section though the colon of a gnotobiotic piglet 3 days after inoculation with a virulent strain of *Y. enterocolitica* O:3. Note the microabscess, comprising mostly bacteria, the surrounding inflammatory cells (arrows), and the disrupted epithelium with vacuolated and necrotic cells. Epoxy section; methylene blue stain; bar = 50 μm. (Reprinted with permission from reference 143.)

epithelial cells, such as HeLa, HEp-2, and Henle 407 cells. Analysis of invasin has shown that the amino terminus is inserted in the bacterial outer membrane, and the carboxyl terminus is exposed on the surface. The carboxyl portion specifies binding of invasin to specific ligands, known as β_1 integrins, on host cells (68). Integrins are heterodimeric transmembrane proteins that communicate extracellular signals to the cytoskeleton. They comprise α and β subunits, which form the basis of their classification into families. The β_1 class, which includes the principal receptors for invasin, occurs on many cell types, including epithelial cells, macrophages,

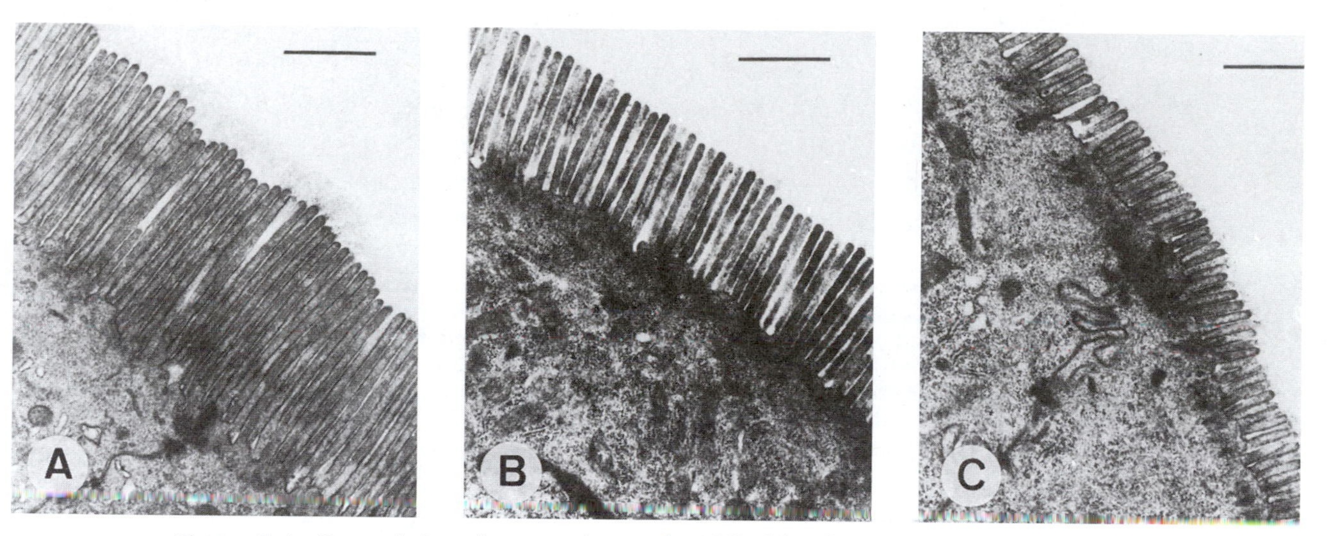

Figure 11.4 Transmission electron micrographs of ileal brush border in the midvillus of control (A), pair-fed (B), and *Y. enterocolitica*-infected (C) rabbits at the same magnification. (Pair-fed animals were fed a diet identical to that ingested by infected animals.) The microvillous brush border is significantly reduced in the infected epithelium compared with the control and pair-fed animals. Bar = 1 μm. (Reprinted with permission from reference 19.)

and T lymphocytes. When invasin binds to its receptors (α_3,β_1, $\alpha_4\beta_1$, $\alpha_5\beta_1$, and $\alpha_6\beta_1$ integrins) on epithelial cells, a sequence of events which results in cytoskeletal alteration and internalization of the bacteria is initiated (68). The internalization process mediated by invasin is propagated entirely by the host cell, because nonviable bacteria and even latex particles coated with invasin are internalized in the same way as living bacteria. Thus, invasion of cells by *Y. enterocolitica* and *Y. pseudotuberculosis* is a prime model of parasite-mediated endocytosis. The uptake of yersiniae by nonphagocytic cells appears to depend on the strong affinity of invasin for integrins, because fibronectin-coated bacteria which also bind to β_1 integrins, but with a lower affinity, are not internalized to any great extent (103).

Although DNA sequences homologous to *inv* occur in all *Yersinia* species, this gene is functional only in *Y. pseudotuberculosis* and the virulent serovars of *Y. enterocolitica,* suggesting that invasin plays a key role in virulence (89). The absence of functional invasin from *Y. pestis* is readily explained by the fact that these bacteria generally enter their hosts by direct inoculation into subcutaneous tissues via a flea bite, obviating the need to penetrate mucous membranes. Although *inv* mutants of *Y. enterocolitica* and *Y. pseudotuberculosis* show a significantly reduced ability to invade epithelial cells in vitro, some doubt has been cast on the contribution of invasin to bacterial virulence. These doubts stem from the observations that (i) expression of invasin by yersiniae is reduced at temperatures above 30°C; (ii) yersiniae do not penetrate the mucosal aspect of polarized epithelia in vitro or the absorptive intestinal epithelium in vivo, preferring to access the tissues via M cells, and (iii) *inv* mutants of *Y. enterocolitica* and *Y. pseudotuberculosis* show only a minor reduction in virulence for orally inoculated mice (103). On the other hand, the findings that (i) invasin is produced in vitro at 37°C when the bacteria are grown at low pH, (ii) invasin is detectable on yersiniae in Peyer's patches during the early stages of infection, and (iii) far fewer *inv* mutants than wild-type bacteria enter Peyer's patches suggest that invasin may play a key role in the initial penetration of the intestinal epithelium by *Y. enterocolitica* (101). The fact that *inv* mutants show only marginally reduced lethality for mice suggests that M cells, which normally function as antigen-sampling cells, are able to internalize the bacteria to some extent in the absence of a specific stimulus or that yersiniae express additional internalins which may compensate in part for the lack of invasin.

Ail

Virulent strains of *Y. enterocolitica* produce an outer membrane protein, unrelated to invasin, which also confers invasive ability on *E. coli.* This 17-kDa peptide with eight putative membrane-spanning domains is specified by a chromosomal *ail* (attachment-invasion) locus, so called because it mediates bacterial attachment to some cultured epithelial cell lines and invasion of others (89, 103). Ail may also allow yersiniae to persist in serum by protecting them from nonspecific destruc-

Y. enterocolitica ST	Q A C D P P S P P A E V S S D W D C C D V C C N P A C A G C
E. coli STa (human type)	N S S N Y C C E L C C N P A C T G C Y
E. coli STa (porcine type)	N T F Y C C E L C C N P A C A G C Y
V. cholerae non-O1 ST	I D C C E I C C N P A C F G C L N
Guanylin	P N T C E I C A Y A A C T G C

Figure 11.5 Alignment of the complete amino acid sequences of mature heat-stable enterotoxins (ST) of *Y. enterocolitica*, enterotoxigenic *E. coli* of human and porcine origin, and *V. cholerae* non-O1 and of the intestinal hormone guanylin. The amino acid residues in boxes are common to all five peptides. Single-letter abbreviations of amino acids: A, alanine; C, cysteine; D, aspartic acid; E, glutamic acid; F, phenylalanine; G, glycine; H, histidine; I, isoleucine; K, lysine; L, leucine; M, methionine; N, asparagine; P, proline; Q, glutamine; R, arginine; S, serine; T, threonine; V, valine; W, tryptophan; Y, tyrosine.

tion by complement (12). The *ail* gene is found only in virulent serovars of *Y. enterocolitica* and, unlike *inv*, is expressed in vitro at host temperatures (37°C). Further circumstantial evidence supporting a role for Ail in virulence stems from the finding that *ail* mutants fail to adhere to or invade cultured cells when the *inv* gene is not expressed. Surprisingly, however, an *ail* mutant of *Y. enterocolitica* showed no reduction in virulence for perorally inoculated mice, indicating that Ail is not required to establish an infection or even to cause systemic infection in these animals (145).

Outer membrane proteins which share significant amino acid homology with Ail have been identified in other *Enterobacteriaceae*. One of these proteins is PagC, a virulence-associated surface expressed protein of *Salmonella typhimurium*, which is required for bacterial survival within macrophages. Despite these structural similarities, Ail and PagC are evidently functionally distinct because PagC does not contribute to the invasive ability of salmonellae, and Ail is not required by yersiniae to resist killing by macrophages (90).

Heat-Stable Enterotoxins

When first isolated from clinical material, most strains of *Y. enterocolitica* secrete a heat-stable enterotoxin, known as Yst, that is reactive in infant mice (99). Yst is made up of 30 amino acids (Fig. 11.5). Its carboxyl terminus is homologous to those of heat-stable enterotoxins from enterotoxigenic *E. coli* and *Vibrio cholerae* non-O1 and to that of guanylin, an intestinal paracrine hormone (Fig. 11.5). These polypeptides also share a common mechanism of action which involves binding to and activation of cell-associated guanylate cyclase and elevation of intracellular concentrations of cyclic GMP. This in turn causes perturbation of fluid and electrolyte transport pathways in intestinal absorptive cells, which results in diarrhea. Yst is encoded by the chromosomal *yst* gene and synthesized as a 71-amino-acid polypeptide, the carboxyl terminus of which becomes the mature toxin (28, 37). The 19 amino acids at the amino terminus contain a signal sequence, which is cleaved and removed during secretion, as is the 23-amino-acid central region. This maturation process via pre and pro precursors resembles that of *E. coli* STa, although these toxins show no significant homology at the amino terminus.

Although the *yst* gene is largely restricted to pathogenic serovars of *Y. enterocolitica* (37), the contribution of Yst to the genesis of diarrhea by this organism is not settled. Doubts regarding the role of Yst in virulence stem from the observations that (i) the toxin is generally not detectable in bacterial cultures incubated at temperatures above 30°C, (ii) production of Yst has not been demonstrated in vivo, and (iii) strains of *Y. enterocolitica* that have spontaneously lost the ability to produce Yst retain full virulence for experimental animals (111). On the other hand, there is a report of a *yst* mutant of *Y. enterocolitica* that causes milder diarrhea in infant rabbits than the wild type (36). In addition, the fact that *yst* (and *inv*) are not normally expressed at 37°C in vitro may indicate that the conditions used to study expression of these genes in vitro do not reflect those to which the bacteria are exposed in vivo. In this regard, the observation that *Y. enterocolitica* produces Yst at 37°C if the bacteria are grown in media with a high osmolarity and pH resembling that found in the intestinal lumen is of interest, although the precise stimulus to Yst synthesis in vivo is not known (87). After repeated passage or prolonged storage, Yst-secreting strains of *Y. enterocolitica* frequently become toxin negative. This phenomenon is not caused by mutation of the *yst* gene but is due to silencing of this gene by YmoA (*Yersinia* modulator). The latter is an 8-kDa protein which appears to down-regulate gene expression in yersiniae by altering DNA topology (86).

Toxins which resemble Yst in terms of heat stability and reactivity in infant mice but differ with respect to molecular weight and/or mechanism of action have been detected in various *Yersinia* species, including non-pathogenic serovars of *Y. enterocolitica* and avirulent *Yersinia* species, such as *Y. bercovieri* and *Y. mollareti* (110, 151). As these bacteria are occasionally obtained from patients with diarrhea, the contribution of these toxins to bacterial virulence cannot be discounted. *Y. enterocolitica* may elaborate these enterotoxins in vitro over a wide range of temperatures from 4 to 37°C (69, 110). Since the toxins are relatively acid stable, they could resist inactivation by stomach acid and thus conceivably cause food poisoning if they were ingested preformed in food. In artificially inoculated foods, however, toxins are synthesized mainly at 25°C during the stationary phase of bacterial growth (119). Thus, the storage conditions required for their production in food would generally result in severe spoilage, making the possibility of Yst ingestion unlikely.

Myf Fibrillae

Many enteric pathogens carry distinctive colonization factors on their surface which mediate their adherence to specific sites on the intestinal epithelium. These adhesins are essential virulence determinants of noninvasive, enterotoxin-secreting bacteria, such as *V. cholerae* and enterotoxigenic *E. coli*, because they allow the bacteria to deliver the enterotoxins close to their target on epithelial cells while resisting removal by peristalsis. Although *Y. enterocolitica* of pathogenic serovars carry a number of afimbrial putative colonization factors, including invasin and YadA (see below), they also produce fibrillar adhesins. These were originally named the C-antigen but renamed Myf (for mucoid *Yersinia* fibrillae) because they bestow a mucoid appearance on the bacterial colonies which express them (67). Myf consists of narrow flexible fimbriae which resemble CS3, an essential colonization factor of many human strains of enterotoxigenic *E. coli*. The structural subunit of Myf is 44% identical at the DNA level to the so-called pH 6 antigen of *Y. pestis*, which also has a fibrillar structure (81). The latter is synthesized within macrophages and may play a role in the interaction between bacteria and phagocytic cells, although it apparently does not contribute to bacterial survival in these cells.

Synthesis of Myf is specified by an operon analogous to that which encodes pyelonephritis-associated (Pap) pili of uropathogenic strains of *E. coli* and includes genes for the production of MyfA (the 21-kDa structural subunit), MyfB, a putative chaperone, and MyfC, a membrane usher (64). Expression of Myf is regulated by temperature (37°C) and acidic pH. In this way Myf resembles invasin, which is not synthesized at 37°C unless the bacteria are grown at low pH. As with Yst, production of Myf is largely restricted to pathogenic serovars of *Y. enterocolitica*. Both Myf and Yst are produced predominantly during the stationary phase of growth, but their production appears not to be coregulated (65). Although Myf may function as a colonization factor of *Y. enterocolitica*, direct proof of this is lacking. The failure to visualize adherent *Y. enterocolitica* on the intestinal epithelium of experimentally infected animals does not discount a role for Myf in pathogenesis, because many colonization factors are host specific and may not mediate bacterial adhesion to the intestinal epithelium of their nonpreferred hosts.

LPS

Like other gram-negative bacteria, *Y. enterocolitica* may be classified as smooth or rough depending on the amount of O-side-chain polysaccharide attached to the inner core region of the cell wall LPS. Synthesis of the O-side chain by *Y. enterocolitica* is specified by the chromosomal *rfb* locus and is regulated by temperature, such that colonies are smooth when grown at temperatures below 30°C but rough when grown at 37°C. *Y. enterocolitica* carrying a mutation in the *rfb* locus displays reduced virulence for mice, indicating that smooth LPS is required for the full expression of virulence (127). Smooth LPS may enhance virulence by increasing bacterial hydrophilicity, thus facilitating passage through the mucous secretions which blanket the intestinal epithelium. On the other hand, elongated O-side chains could conceal surface virulence-associated proteins, such as Ail and YadA, which may be required during a later phase of infection. Hence, repression of O-side chain synthesis at 37°C may increase the chances of bacteria surviving in tissues by allowing them to display essential virulence determinants on their surface at the appropriate stage of infection.

Flagella

Although the assembly of flagella by *Y. enterocolitica* is also regulated by temperature, motility does not appear to contribute to the virulence of these bacteria (66).

Iron Acquisition

Iron is an essential micronutrient of almost all bacteria. Despite the nutrient-rich environment provided to bacteria by mammalian tissues, the availability of iron in some sites may be limited (149). This is because most iron in tissues is bound to high-affinity transport proteins such as transferrin and lactoferrin or is incorporated into organic molecules such as hemoglobin. Several species of pathogenic bacteria produce low-molecular-weight, high-affinity iron chelators known as siderophores (5, 94). These compounds are secreted by the bacteria into the surrounding medium, where they couple with iron. The resultant ferrisiderophore complexes then bind to specific receptors on the bacterial surface and are taken up into the cell. The observation that patients suffering from iron overload show increased susceptibility to severe infections with *Y. enterocolitica* suggested that the availability of iron in tissues may determine the outcome of yersiniosis (109). Investigation of the relationship of yersiniae to iron has revealed that these bacteria express a complex array of processes to acquire iron from inorganic and organic sources (102). Although early studies indicated that virulent *Yersinia* species do not produce siderophores, more recent research has suggested that the more virulent American serovars of *Y. enterocolitica* produce a novel catechol-containing siderophore termed yersiniabactin (5). This compound forms a ferrisiderophore complex with iron and then enters the bacteria after binding to a 65-kDa outer membrane protein receptor named FyuA. The latter also serves as a receptor for pesticin, a bacteriocin produced by *Y. pestis.* American serovars of *Y. enterocolitica* also produce high-molecular-weight, iron-repressible proteins which may be involved in iron acquisition. Production of these proteins requires the *irp2* gene, which is found only in highly virulent yersiniae, including some strains of *Y. pestis* and *Y. pseudotuberculosis,* as well as the American serovars of *Y. enterocolitica* (31).

Transport of ferrisiderophore complexes across the cell wall of *Y. enterocolitica* resembles the analogous pathway in *E. coli* in that it is an energy-dependent process which requires the TonB protein. The latter couples energy provided by inner membrane metabolism to outer membrane protein receptors, such as FyuA. TonB and FyuA mutants of *Y. enterocolitica* show reduced virulence for mice, presumably because of their limited capacity to acquire sufficient iron to grow in tissues (5). Although European serovars of *Y. enterocolitica* do not produce yersiniabactin, they are able to acquire iron from a number of sources, including as ferrisiderophore complexes, in which the siderophore was synthesized by an unrelated microorganism (5, 108). This may have important clinical implications, because siderophores, such as desferrioxamine B, are used therapeutically to reduce iron overload in patients with hemosiderosis and other forms of iron intoxication. When administered to patients, desferrioxamine forms a ferrisiderophore complex which *Y. enterocolitica* can utilize as a growth factor (107). Accordingly, if patients undergoing iron chelation therapy with desferrioxamine B become infected with *Y. enterocolitica,* the bacteria may be able to proliferate in tissues where, under normal circumstances, poor availability of iron would limit their growth.

Urease

All enteric pathogens must negotiate the acid barrier of the stomach to cause disease. In *Y. enterocolitica,* acid tolerance relies on the production of urease which catalyzes the release of ammonia from urea and allows the bacteria to resist pH as low as 2.5 (34). Urease is produced chiefly during the stationary phase of growth at temperatures below 37°C.

The Virulence Plasmid

All fully virulent strains of *Y. pestis, Y. pseudotuberculosis,* and *Y. enterocolitica* carry a 70- to 75-kb plasmid, termed pYV (plasmid for *Yersinia* virulence), which is highly conserved among the three species (26–28). Yersiniae which carry this plasmid exhibit a distinctive phenotype known as the low-calcium response because it manifests only when pYV-bearing bacteria are grown in media containing low concentrations of Ca^{2+}. The principal features of the low-calcium response are the cessation of bacterial growth after one or two generations (calcium dependency) and the appearance of at least 11 new proteins on the bacterial surface or in the culture medium. The secreted proteins are referred to as Yops, because they were once thought to be outer membrane proteins. They are now known to be secreted by the bacteria and then absorbed back onto the bacterial surface, possibly in the form of multimeric structures. The 11 Yops of *Y. enterocolitica* are named YopB, -D, -E, -H, -M, -N, -O, -P, -Q, and -R and LcrV. Although Yops are highly conserved among *Yersinia* species, there is little homology between the Yops of a single species (28).

The genes which specify calcium dependency and the synthesis of Yops are scattered around pYV (Fig. 11.6). The genes for YopO and YopP compose a single operon, as do those for LcrV, YopB, and YopD (the *lcrGVH yopBD* operon). In addition, pYV carries genes for the production of YadA, a true outer membrane protein, and for YlpA, a 29-kDa lipoprotein related to the TraT proteins of the *E. coli* sex plasmid and the virulence plasmid of *S. typhimurium.* Although the contribution of each of these various substances to virulence is not known, evi-

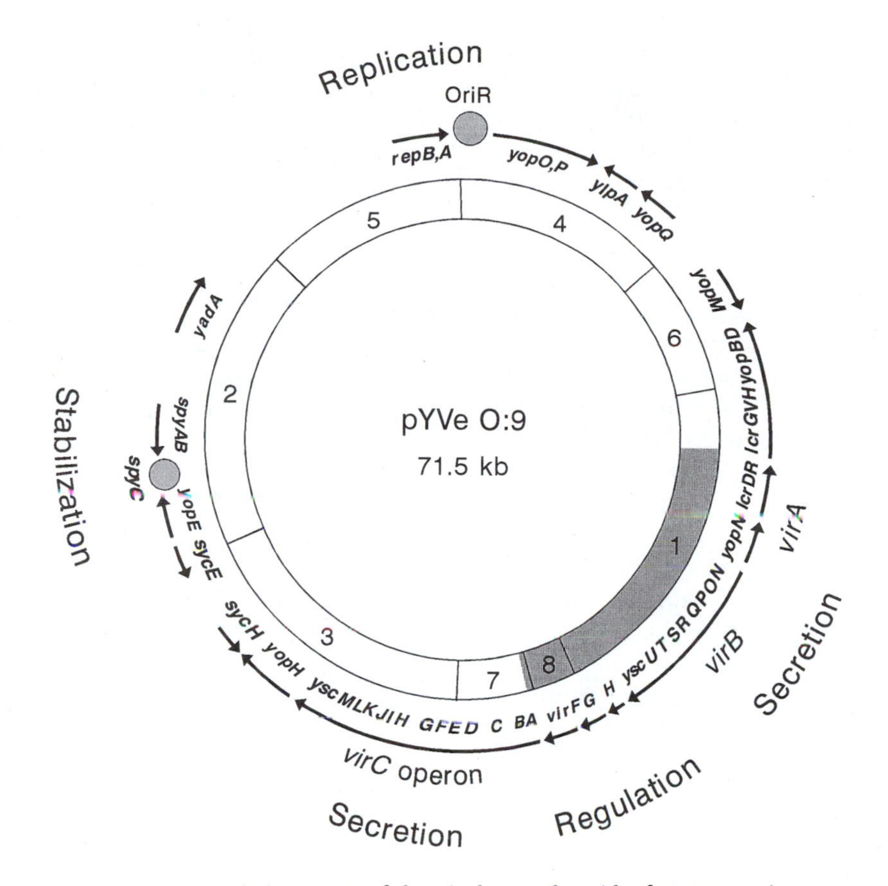

Figure 11.6 *Eco*RI restriction map of the virulence plasmid of *Y. enterocolitica* serovar O:9 showing (i) the region encoding calcium dependency (shaded) and (ii) the locations and directions of transcription (arrows) of the genes encoding YadA; YlpA; YopB, -D, -E, -H, -M, -N, -O, -P, and -Q and LcrV; secretion elements (*virA*, *-B*, and *-C*); regulatory elements (*virF*, *-G*, and *-H*); replicative functions (*repA*, *repB*, and *oriR*); and plasmid stability (*spyA*, *-B*, and *-C*). (Adapted from reference 28.)

dence is accumulating that they act collectively to overcome nonspecific immune defenses of the host, thus allowing the bacteria to proliferate in tissues (17, 28, 113).

YadA

YadA, formerly known as YopA, Yop1, or P1, is a 45-kDa outer membrane protein which polymerizes to form fibril-like structures on the bacterial surface. The amino terminus of this protein is inserted in the outer membrane and contains a typical 25-amino-acid signal sequence, indicating that it is transported across the bacterial membrane via the classical Sec export pathway. YadA acts as an adhesin by mediating binding to intestinal mucus and to certain extracellular matrix proteins, including collagen and cellular fibronectin (126). These proteins in turn may bind to β_1 integrins ($\alpha_1\beta_1$, $\alpha_2\beta_1$, $\alpha_3\beta_1$, and $\alpha_5\beta_1$ integrins) on epithelial cells and stimulate bacterial internalization by parasite-mediated endocytosis in a way similar to that initiated by invasin. Thus, YadA may contribute to bacterial invasion, even though binding to integrins via matrix proteins does not stimulate bacterial uptake to the same extent as that mediated by invasin (13). YadA from *Y. pseudotuberculosis* also evidently promotes bacterial invasion by binding to integrins directly (11, 150). As the contributions of YadA to the virulence of *Y. enterocolitica* and *Y. pseudotuberculosis* appear to differ, however, care should be taken in extrapolating experimental data from one *Yersinia* species to another (71, 114). Evidence has been provided that YadA also conveys resistance to complement-mediated opsonization by binding factor H and reducing deposition of C3b on the bacterial surface (23). In this way, YadA is associated with the ability of yersiniae to resist killing by complement, to resist phagocytosis, and to inhibit the respiratory burst of polymorphonuclear leukocytes, all of which require the bacteria to be preopsonized (22, 23). Given the pluripotential capacity of YadA to increase the likelihood of bacterial survival in host tissues, it is not surprising that YadA mutants of *Y. enterocolitica* show reduced virulence for mice (28, 71). This is in contrast to YadA mutants of *Y. pseudotuberculosis,* which show no attenuation, or *Y. pestis,* which is highly virulent for mice and is naturally defective in YadA production (114, 128).

Synthesis of YadA is regulated at the transcriptional level by temperature but not by the concentration of Ca^{2+}. In contrast, *ylpA,* the *yop* genes, and the *virC* operon (see below) are regulated by both temperature and Ca^{2+}. Expression of *yadA* is controlled by another plasmid-encoded protein, known as VirF, a DNA-binding protein of the AraC family (25). Other members of this family include VirF from *Shigella flexneri* and Rns from enterotoxigenic *E. coli,* both of which are involved with the regulation of virulence in these bacteria. VirF plays a central role in the virulence of *Yersinia* species by governing the transcriptional activation of *yadA, ylpA,* the *yop* genes, and the *virC* operon. Because these genes are coregulated, they have been named the Yop regulon (25). Transcription of *virF* itself is regulated by temperature (but not by Ca^{2+}), probably as a result of combined effects of YmoA and the influence of elevated temperature on DNA supercoiling (112).

Yops

Y. enterocolitica strains which carry mutations in *yop* genes generally display reduced virulence for animals. Although the mechanism of action of each Yop is not known, a number of them act directly on host cells to subvert immune defenses. Some Yops contain amino acid sequences that are homologous to those of eukaryotic proteins, raising the possibility that they originated in eukaryotic cells.

YopM is a 41-kDa protein which shows homology to an external domain of the α chain of membrane glycoprotein 1b (GP1bα), a human platelet-specific receptor for thrombin and the von Willebrand factor (133). Under normal circumstances, binding of these ligands to platelets causes platelet aggregation and the release of several inflammatory mediators. As YopM has sufficient thrombin-binding activity to inhibit thrombin-induced platelet aggregation, it may interfere with the ability of the host to mount a normal inflammatory response to infection and thus may facilitate bacterial survival in tissues.

YopO is an 82-kDa protein, also known as YpkA (*Yersinia* protein kinase A), which catalyzes autophosphorylation on serine (46). This action suggests a regulatory role in yersiniae, but YopO could also act on host cells to undermine signal transduction and protein phosphorylation pathways that are part of the physiological response to infection. Although YopO is an essential virulence determinant of *Y. pseudotuberculosis* (46), YopO mutants of *Y. enterocolitica* show no reduction in their pathogenicity for mice (88), suggesting that the roles of YopO in the pathogenesis of infections caused by these two species may differ.

YopE (25 kDa) and YopH (51 kDa) act synergistically, but independently of each other, to disrupt cellular structure and function. They cause cultured epithelial cells and macrophages to become rounded and to detach from the substratum, a phenomenon known as cytotoxicity (Fig. 11.7). Although no enzymatic activity has been ascribed to YopE, it appears to act by destabilizing actin filaments, thus disrupting the cytoskeleton (Fig. 11.7) (10, 44, 133).

YopH is a potent protein tyrosine phosphatase with a broad substrate range. It exerts profound effects on signal transduction pathways and other cellular responses to bacterial infection by desphosphorylating various phosphotyrosine residues (10, 49, 133). Although most cell types are susceptible to the action of YopE and YopH, their effects on macrophages and polymorphonuclear leukocytes have important implications for resistance to infection.

YopE and YopH act on intracellular targets and are ineffective when added to epithelial cells in culture. Transport of these proteins across the host cytoplasmic membrane requires the presence of at least two more pYV-encoded proteins, YopB (44 kDa) and YopD (37 kDa). These proteins appear to act in concert to transport YopE and YopH into cells, although the way in which they achieve this is not known. YopB and YopD have putative transmembrane domains that may allow them to induce the formation of pores in the cytoplasmic membrane that could permit YopE and YopH to gain access to their intracellular targets (52). YopB may also act as a virulence determinant in its own right by suppressing the production of tumor necrosis factor alpha during infection (8).

LcrV (also known as the V antigen) is a 38-kDa Yop encoded by the same operon as YopB and YopD and has been recognized as a virulence determinant of *Y. pestis* since the 1950s (18). At one time, opinion was divided as to whether LcrV is a bacterial virulence determinant per se or simply a regulator of the low-calcium response (27), but recent evidence obtained by examining nonpolar *lcrV* mutants of *Y. pestis* suggests that LcrV fulfills both of these functions (124). LcrV appears to act by suppressing host inflammatory response mediated by tumor necrosis factor alpha and gamma interferon (93),

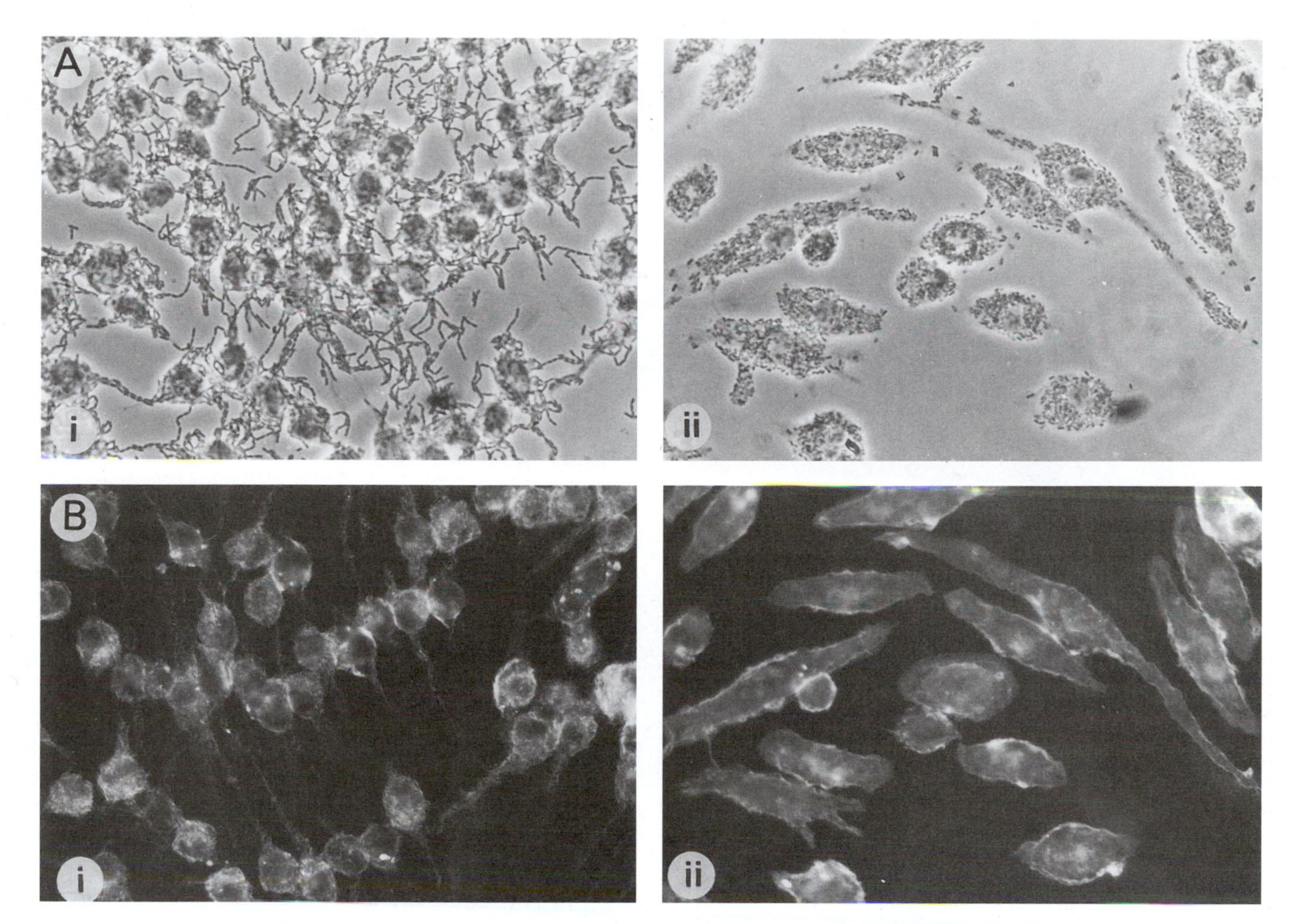

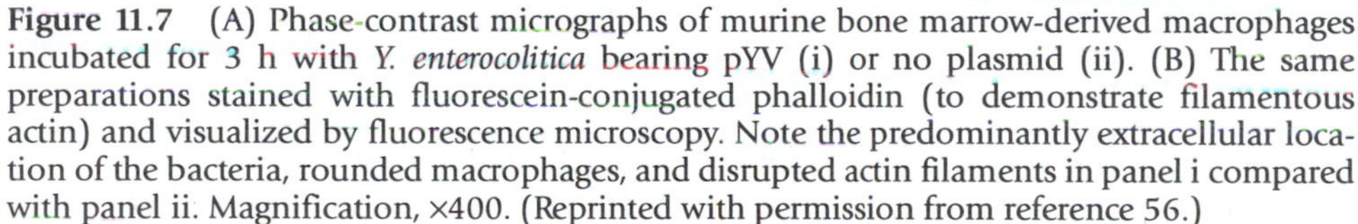
Figure 11.7 (A) Phase-contrast micrographs of murine bone marrow-derived macrophages incubated for 3 h with *Y. enterocolitica* bearing pYV (i) or no plasmid (ii). (B) The same preparations stained with fluorescein-conjugated phalloidin (to demonstrate filamentous actin) and visualized by fluorescence microscopy. Note the predominantly extracellular location of the bacteria, rounded macrophages, and disrupted actin filaments in panel i compared with panel ii. Magnification, ×400. (Reprinted with permission from reference 56.)

but the means by which it achieves this is not known. Mice with circulating antibodies to LcrV, acquired by passive or active immunization, show increased resistance to yersiniosis.

Regulation of Yop Production and Secretion

All Yops and YlpA are produced in vitro at 37°C (but not at temperatures below 30°C) when the concentration of Ca^{2+} is sufficiently low to induce bacteriostasis. The mechanism of regulation by temperature involves VirF (which itself is regulated by temperature) and probably DNA supercoiling (112). The mechanism of regulation by Ca^{2+} is not known, but it may involve feedback inhibition resulting from reduced Yop secretion in the presence of elevated concentrations of Ca^{2+} (4).

There is no doubt that Yops are produced in vivo because animals and humans infected with virulent *Yersinia* species develop antibodies to these proteins during the course of infection (Fig. 11.8). Although host temperature (37°C) is likely to be a key stimulus for the production of Yops in vivo, the second signal equating to low Ca^{2+} is obscure. Originally, when yersiniae were thought to localize within phagocytes, the intraphagocytic environment, which is relatively low in Ca^{2+}, was thought to provide this stimulus (104). Histopathological examination of tissues from experimentally infected animals, however, which showed that *Y. enterocolitica* is located extracellularly (54), was at odds with this suggestion, and an alternative explanation was sought. This was provided to some extent by the finding that the

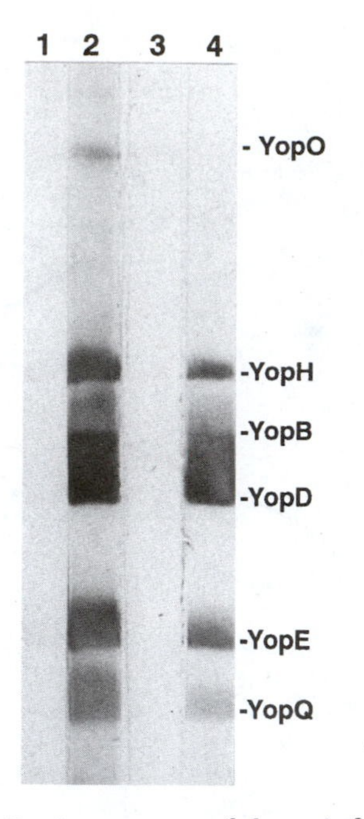

Figure 11.8 Antibody response of sheep infected with *Y. enterocolitica* or *Y. pseudotuberculosis* to Yops. Yops were prepared from *Y. enterocolitica* O:3, separated by polyacrylamide gel elctrophoresis, transferred to a nitrocellulose membrane, and reacted with pre-immune (lanes 1 and 3) or immune (lanes 2 and 4) sera from lambs with naturally acquired infection with pYV-bearing *Y. enterocolitica* (lanes 1 and 2) or *Y. pseudotuberculosis* (lanes 3 and 4). Note (i) the reactive bands only in the lanes incubated with immune sera, and (ii) that infection with *Y. pseudotuberculosis* leads to the formation of antibodies which recognize Yops from *Y. enterocolitica*. (Reprinted with permission from reference 106.)

microabscesses in which yersiniae replicate in vivo are made up of cellular debris that may be low in Ca^{2+} (45). Attempts to mimic the composition of this fluid in vitro have revealed that yersiniae can replicate at 37°C in such media while retaining the ability to elaborate Yops.

In addition, Forsberg et al. (44) have shown that extracellular yersiniae are cytotoxic for HeLa cells even in the presence of millimolar concentrations of Ca^{2+}, provided that the bacteria can attach to the target cells. This observation suggests that a specific, Ca^{2+}-independent interaction between the bacteria and a cell surface ligand stimulates the production and subsequent secretion of Yops. The nature of the stimulus which host cells provide to the bacteria is not known, but it may involve YopN (also known as LcrE), since *yopN* mutants produce Yops without binding to target cells (44).

Yop Secretion

The inner and outer membranes of gram-negative bacteria act as major barriers to protein export. Until recently, only two mechanisms of protein secretion by gram-negative bacteria were known. One of these pathways, termed type I, requires the secreted protein to carry a specific recognition sequence close to the carboxyl terminus and is exemplified by the HlyA hemolysin of *E. coli*. By contrast, type II secretion (also termed the general secretory pathway) involves Sec-dependent recognition of a conserved amino-terminal signal sequence which is removed during secretion. The export of Yops is also governed by an amino-terminal sequence, but the sequence shows no resemblance to classical signal sequences, nor is there any homology between the amino termini of different Yops (28, 144). Furthermore, Yop export is Sec independent and does not involve cleavage of the amino terminus. Hence, Yop secretion represents a novel system of protein export by gram-negative bacteria and has been named type III secretion (115, 144).

The secretion of Yops is thought to occur in two stages, the first of which involves coupling of the Yop to a specific chaperone termed Syc (specific Yop chaperone) that guides it to the export machinery (28). In this way, Yop chaperones are analogous to SecB in the general secretory pathway. They differ from SecB, however, in that the latter binds many different proteins, whereas Yop chaperones are highly specific and are named for the individual Yops with which they associate. To date, specific chaperones have been identified for YopE (SycE), YopH (SycH), and YopB and -D (SycD). The genes encoding these chaperones are located on pYV close to the corresponding *yop* gene (Fig. 11.6). Although they have no significant homology with each other, all Yop chaperones identified to date are acidic, low-molecular-mass (15- to 18-kDa) proteins which probably occur in the bacterial cytoplasm as dimers (28).

The second feature of Yop secretion is the secretory machinery itself. This involves a multicomponent structure which is envisaged to span the inner and outer membranes. It comprises products of the *virC* and *virA* operons, the individual genes of which are named *ysc*, for Yop secretion (Fig. 11.6). The *virC* operon (but not *virA* or *virB* [see below]) is regulated by VirF and forms part of the *yop* regulon (28). The *virC* operon comprises 13 genes, *yscA* to *yscM*, most of which are required for secretion. The *virB* locus includes eight genes, *yscN* to *yscU*, whereas *virA* contains *lcrD* (28). Homologs of *ysc* genes have been identified in several other bacterial pathogens, including shigellae, salmonellae, and the plant pathogens *Pseudomonas solanacearum* and *Xanthomonas campestris* (Table 11.5) (7, 144). YscN is a 48-kDa protein with a putative ATPase domain that could provide the energy required for Yop secretion. YscN has

Table 11.5 Some homologs in the type III secretion pathway

Yersinia spp.	*Shigella* spp.	*Salmonella* spp.	*P. solanacearum*	*X. campestris*
LcrD	MxiA	InvA	HrpO	HrpC2
YpoN (= LrcE)	MxiC	InvE		
YscC	MxiD	InvG	HrpA	HrpA1
YscF	MxiH			
YscJ (= YlpB)	MxiJ		HrpI	HrpB3
YscL			HrpF	
YscN	Spa47	SpaL	HrpE	HrpB6
YscO	Spa13	SpaM		
YscP	Spa32	SpaN		
YscQ	Spa33	SpaO		
YscR	Spa24	SpaP		
YscS	Spa9	SpaQ		
YscT	Spa29	SpaR		
YscU	Spa40	SpaS	HrpN	

significant homology with ATP-binding proteins involved in flagellum-specific export pathways of *E. coli*, *S. typhimurium*, and *Bacillus subtilis*, suggesting that the pathway for the export of virulence proteins may have evolved from that involved in the flagellar assembly.

Pathogenesis of *Yersinia*-Induced Autoimmunity

Arthritis

Following an acute infection with *Y. enterocolitica* or *Y. pseudotuberculosis*, a small proportion of patients develop postinfective reactive arthritis. A similar syndrome also occurs after infections with *Campylobacter*, *Salmonella*, *Shigella*, or *Chlamydia* species (76). The synovial fluid from affected joints of patients with *Yersinia*-induced arthritis is culture negative but contains bacterial antigens, chiefly within inflammatory cells (29, 48). The pathogenesis of this condition is poorly understood, and many explanations have been advanced to account for arthritis and the other autoimmune complications of yersiniosis (59, 140). Although patients with reactive arthritis display higher levels of serum immunoglobulin A antibodies to *Yersinia* antigens than individuals without arthritis, these antibodies are unlikely to contribute to the development of arthritis. Instead, they probably reflect enhanced stimulation of the mucosal immune system as a result of antigen persistence in the intestine or other tissues (33).

Y. enterocolitica and *Y. pseudotuberculosis* may induce polyclonal T-cell stimulation by virtue of their ability to secrete toxins which resemble superantigens, although the precise nature of these toxins is uncertain (1, 134). Yersiniae may also provoke nonspecific immune stimulation when invasin binds to β_1 integrins on T lymphocytes, thus providing a costimulatory signal to these cells (16). Other bacterial antigens which are claimed to contribute to autoimmunity via a more specific interaction with the immune system are YadA and the β subunit of urease (50, 125). The latter is a cationic protein which (i) is recognized by $CD4^+$ T cells from patients with *Yersinia*-induced reactive arthritis and (ii) produces arthritis when injected into the joints of rats (125).

Heat shock proteins are remarkably well conserved among bacteria and mammals and share a number of antigenic determinants. This finding raises the possibility that an immune response to selected epitopes on bacterial heat shock proteins may lead to an autoimmune response at sites where bacterial antigens accumulate. In keeping with this suggestion is the observation that synovial fluid from patients with reactive arthritis contains $CD4^+$ major histocompatibility complex class II-restricted T lymphocytes which recognize epitopes that are shared by a 60-kDa *Yersinia* heat shock protein and its human counterpart (59).

Most individuals with postinfective reactive arthritis are positive for the human leukocyte antigen HLA-B27 (76). In addition, HLA-B27-positive individuals have more severe arthritic symptoms and a more prolonged course than patients who are HLA-B27 negative. Although several explanations have been put forward to account for this relationship, the underlying mechanism remains obscure. Early suggestions that autoimmunity was due to molecular mimicry between YadA and the peptide-binding groove of the HLA-B27 molecule have been discounted (77). On the other hand, evidence for a consensus between the urease β subunit and HLA-B27 is currently under investigation. Hermann et al. (60) have provided a link between yersiniosis, HLA-B27, and tissue damage by showing that T-cell clones derived from the

Table 11.6 Overview of the virulence determinants of *Y. enterocolitica*

Location	Environmental stimulus(i)	Growth phase	Virulence determinants produced
Food, water, the environment	Temperature of ~25°C	Stationary	Urease, smooth LPS, invasin, heat-stable enterotoxin, Myf fibrillae
Host tissues			
Lumen of small intestine	Temperature of 37°C, increasing pH, high osmolarity	Lag, early exponential	Ail, YadA, heat-stable enterotoxin
Wall of small intestine, gut-associated lymphoid tissue, reticuloendothelial tissue	Temperature of 37°C, cell contact, low iron, low Ca^{2+} (?)	Exponential	Yops, yersiniabactin[a]

[a]Highly virulent (American) serovars only.

synovial fluid of patients with *Yersinia*-triggered reactive arthritis are selectively cytotoxic for HLA-B27 cells infected with *Y. enterocolitica.* This finding suggests that $CD8^+$ class I-restricted cytotoxic T cells may recognize specific *Yersinia*-derived peptides when they are presented together with the HLA-B27 molecule.

Thyroid Diseases

Y. enterocolitica has been implicated in the etiology of various thyroid disorders, including autoimmune thyroiditis and Graves' disease hyperthyroidism (140, 141). The latter is an immunological disorder mediated by autoantibodies to the thyrotropin (TSH) receptor. The chief link between *Y. enterocolitica* and thyroid diseases is that patients with these disorders frequently have elevated titers of serum agglutinins to *Y. enterocolitica* O:3 (122). As there is no clear relationship between the incidence or geographic distribution of yersiniosis and that of autoimmune thyroid diseases, the presence of circulating antibodies to *Y. enterocolitica* in such patients is likely to reflect a fortuitous cross-reaction between *Y. enterocolitica* and thyroid antigens rather than a causal relationship (140). In addition, follow-up of patients many years after infection with *Y. enterocolitica* or *Y. pseudotuberculosis* has shown no increased frequency of thyroid disease, nor is there any evidence that hyperthyroidism is exacerbated by infection with *Y. enterocolitica* (140). On the other hand, the observations that the outer membrane of *Y. enterocolitica* carries binding sites for TSH which are recognized by immunoglobulins from patients with Graves' disease (62) and that mice immunized with a component of the human TSH receptor develop antibodies which recognize bacterial surface proteins (82) suggest that infection with *Y. enterocolitica* may trigger an autoimmune response that leads to the development of Graves' disease in susceptible individuals (140).

Although few studies of the pathogenesis of *Yersinia*-induced erythema nodosum or glomerulonephritis have been conducted, experience with other infective agents suggests that these manifestations are caused by the deposition of immune complexes in affected organs.

SUMMARY AND CONCLUSIONS

Y. enterocolitica is a versatile foodborne pathogen with a remarkable ability to adapt to a wide range of environments within and outside its host (Table 11.6). The bacteria typically access their hosts via food or water in which they will have grown to stationary phase at ambient temperature. Under these circumstances, they express factors such as urease and smooth LPS which facilitate their passage through the stomach and the mucus layer of the small intestine. Bacteria in this state may also carry Myf fibrillae and invasin that may promote adherence to and penetration of the dome epithelium overlying the Peyer's patches. The higher infectivity of *Y. enterocolitica* when grown at ambient temperature than at 37°C may account for the small number of reports of human-to-human transmission of yersiniosis (96).

Once *Y. enterocolitica* begins to replicate in the intestine at 37°C, LPS becomes rough, exposing Ail and YadA on the bacterial surface. These factors may promote further invasion while protecting the bacteria from complement-mediated opsonization. When the bacteria make contact with host cells in lymphoid tissue, they are stimulated to synthesize and release Yops, notably YopE and YopH, which further frustrate the efforts of phagocytes to ingest and remove them. Further bacterial replication may lead to tissue damage and the formation of microabscesses. If the highly invasive American serovars of *Y. enterocolitica* become established at sites where iron supplies are growth limiting, the bacteria may produce yersiniabactin so that replication can proceed. Eventually, the cycle is completed when the bacteria rupture microabscesses in intestinal crypts to reenter the intestine and regain access to the environment. This well-defined life cycle of *Y. enterocolitica* with its distinctive temperature-induced phases is reminiscent of the flea-rat-flea cycle of *Y. pestis.*

Although much remains to be learned about *Y. enterocolitica,* investigations into the pathogenesis of yersiniosis have provided fascinating new insights into bacterial pathogenesis as a whole and its genetic regulation. *Y. enterocolitica* and *Y. pseudotuberculosis* were the first invasive human pathogens in which plasmid-mediated virulence was documented, from which internalins (invasin, Ail, and YadA) were cloned and characterized, in which the relationship between iron limitation and ferrisiderophore uptake assumed clinical significance, and in which type III protein secretion was identified. Future research in this area will no doubt lead to new and unexpected discoveries of bacterial strategies to evade host immunity that will further advance our understanding of the interface between microbes and the animal world.

References

1. **Abe, J., T. Takeda, Y. Watanabe, H. Nakao, N. Kobayashi, D. Y. Leung, and T. Kohsaka.** 1993. Evidence for superantigen production by *Yersinia pseudotuberculosis. J. Immunol.* **151:**4183–4188.
2. **Adams, M. R., C. L. Little, and M. C. Easter.** 1991. Modeling the effect of pH, acidulant and temperature on the growth rate of *Yersinia enterocolitica.* J. Appl. Bacteriol. **71:**65–71.
3. **Ahvonen, P., K. Sievers, and K. Aho.** 1969. Arthritis associated with *Yersinia enterocolitica. Acta Rheumatol. Scand.* **15:**232–255.
4. **Allaoui, A., R. Scheen, C. Lambert de Rouvroit, and G. R. Cornelis.** 1995. VirG, a *Yersinia enterocolitica* lipoprotein involved in Ca^{2+} dependency, is related to ExsB of *Pseudomonas aeruginosa. J. Bacteriol.* **177:**4230–4237.
5. **Baumler, A., R. Koebnik, I. Stojiljkovic, J. Heesemann, V. Braun, and K. Hantke.** 1993. Survey on newly characterized iron uptake systems of *Yersinia enterocolitica. Int. J. Med. Microbiol. Virol. Parasitol. Infect. Dis.* **278:**416–424.
6. **Bercovier, H., and H. H. Mollaret.** 1984. Genus XIV. *Yersinia* Van Loghem 1944, 15^{AL}, p. 498–506. *In* N. R. Krieg and J. G. Holt (ed.), *Bergey's Manual of Systematic Bacteriology,* vol. 1. The Williams & Wilkins Co., Baltimore.
7. **Bergman, T., K. Erickson, E. Galyov, C. Persson, and H. Wolf-Watz.** 1994. The *lcrB* (*yscN/U*) gene cluster of *Yersinia pseudotuberculosis* is involved in Yop secretion and shows high homology to the *spa* gene clusters of *Shigella flexneri* and *Salmonella typhimurium. J. Bacteriol.* **176:** 2619–2626.
8. **Beuscher, H. U., F. Rödel, Å. Forsberg, and M. Röllinghoff.** 1995. Bacterial evasion of host immune response: *Yersinia enterocolitica* encodes a suppressor of tumor necrosis alpha expression. *Infect. Immun.* **63:** 1270–1277.
9. **Black, R. E., R. J. Jackson, T. Tsai, M. Medvesky, M. Shayegani, J. C. Feeley, K. I. E. MacLeod, and A. M. Wakelee.** 1978. Epidemic *Yersinia enterocolitica* infection due to contaminated chocolate milk. *N. Engl. J. Med.* **298:**76–79.
10. **Bliska, J. B.** 1994. Yops of the pathogenic *Yersinia* spp., p. 365–381. *In* V. L. Miller, J. B. Kaper, D. A. Portnoy, and R. R. Isberg (ed.), *Molecular Genetics of Bacterial Pathogenesis.* American Society for Microbiology, Washington, D.C.
11. **Bliska, J. B., M. C. Copass, and S. Falkow.** 1993. The *Yersinia pseudotuberculosis* adhesin YadA mediates intimate bacterial attachment to and entry into HEp-2 cells. *Infect. Immun.* **61:**3914–3921.
12. **Bliska, J. B., and S. Falkow.** 1992. Bacterial resistance to complement killing mediated by the Ail protein of *Yersinia enterocolitica. Proc. Natl. Acad. Sci. USA* **89:**3561–3565.
13. **Bliska, J. B., and S. Falkow.** 1994. Interplay between determinants of cellular entry and cellular disruption in the enteropathogenic *Yersinia. Curr. Opin. Infect. Dis.* 7:323–328.
14. **Borg, A. A., J. Gray, and P. T. Dawes.** 1992. *Yersinia*-related arthritis in the United Kingdom. A report of 12 cases and review of the literature. *Q. J. Med.* **84:**575–582.
15. **Bottone, E. J.** 1977. *Yersinia enterocolitica:* a panoramic view of a charismatic microorganism. *Crit. Rev. Microbiol.* **5:**211–241.
16. **Brett, S. J., A. V. Mazurov, I. G. Charles, and J. P. Tite.** 1993. The invasin protein of *Yersinia* spp. provides co-stimulatory activity to human T cells through interaction with beta 1 integrins. *Eur. J. Immunol.* **23:**1608–1614.
17. **Brubaker, R. R.** 1991. Factors promoting acute and chronic diseases caused by yersiniae. *Clin. Microbiol. Rev.* **4:**309–324.
18. **Brubaker, R. R.** 1991. The V antigen of yersiniae: an overview. *Contrib. Microbiol. Immunol.* **12:**127–133.
19. **Buret, A., E. V. O'Loughlin, G. H. Curtis, and D. G. Gall.** 1990. Effect of acute *Yersinia enterocolitica* infection on small intestinal ultrastructure. *Gastroenterology* **98:** 1401–1407.
20. **Butler, R. C., V. Lund, and D. A. Carlson.** 1987. Susceptibility of *Campylobacter jejuni* and *Yersinia enterocolitica* to UV radiation. *Appl. Environ. Microbiol.* **53:**375–378.
21. **Chao, W. L., R. J. Ding, and R. S. Chen.** 1988. Survival of *Yersinia enterocolitica* in the environment. *Can. J. Microbiol.* **34:**753–756.
22. **China, B., B. T. N'Guyen, M. De Bruyere, and G. R. Cornelis.** 1994. Role of YadA in resistance of *Yersinia enterocolitica* to phagocytosis by human polymorphonuclear leukocytes. *Infect. Immun.* **62:**1275–1281.
23. **China, B., M. P. Sory, B. T. N'Guyen, M. De Bruyere, and G. R. Cornelis.** 1993. Role of the YadA protein in prevention of opsonization of *Yersinia enterocolitica* by C3b molecules. *Infect. Immun.* **61:**3129–3136.
24. **Christensen, S. G.** 1987. The *Yersinia enterocolitica* situation in Denmark. *Contrib. Microbiol. Immunol.* **9:**93–97.
25. **Cornelis, G., T. Biot, C. Lambert de Rouvroit, T. Michiels, B. Mulder, C. Sluiters, M.-P. Sory, M. Van Bouchaute, and J.-C. Vanooteghem.** 1989. The *Yersinia yop* regulon. *Mol. Microbiol.* **3:**1455–1459.

26. **Cornelis, G., Y. Laroche, G. Balligand, M.-P. Sory, and G. Wauters.** 1987. *Yersinia enterocolitica*, a primary model for bacterial invasiveness. *Rev. Infect. Dis.* **9**:64–87.
27. **Cornelis, G. R.** 1992. Yersiniae, finely tuned pathogens, p. 231–265. *In* C. Hormaeche, C. W. Penn, and C. J. Smyth (ed.), *Molecular Biology of Bacterial Infection: Current Status and Future Perspectives.* Cambridge University Press, Cambridge.
28. **Cornelis, G. R.** 1994. *Yersinia* pathogenicity factors. *Curr. Top. Microbiol. Immunol.* **192**:243–263.
29. **Cover, T. L., and R. C. Aber.** 1989. *Yersinia enterocolitica. N. Engl. J. Med.* **321**:16–24.
30. **D'Aoust, J. Y., C. E. Park, R. A. Szabo, E. C. Todd, D. B. Emmons, and R. C. McKellar.** 1988. Thermal inactivation of *Campylobacter* species, *Yersinia enterocolitica*, and hemorrhagic *Escherichia coli* O157:H7 in fluid milk. *J. Dairy. Sci.* **71**:3230–3236.
31. **De Almeida, A. M., A. Guiyoule, I. Guilvout, I. Iteman, G. Baranton, and E. Carniel.** 1993. Chromosomal *irp2* gene in *Yersinia:* distribution, expression, deletion and impact on virulence. *Microb. Pathog.* **14**:9–21.
32. **de Giusti, M., and E. de Vito.** 1992. Inactivation of *Yersinia enterocolitica* by nitrite and nitrate in food. *Food Addit. Contam.* **9**:405–408.
33. **de Koning, J., J. Heesemann, J. A. A. Hoogkamp-Korstanje, J. J. M. Festen, P. M. Houtman, and P. L. M. van Oijen.** 1989. *Yersinia* in intestinal biopsy specimens from patients with seronegative spondyloarthropathy: correlation with specific serum IgA antibodies. *J. Infect. Dis.* **159**:109–112.
34. **de Koning-Ward, T. F., and R. M. Robins-Browne.** 1995. Contribution of urease to acid tolerance in *Yersinia enterocolitica. Infect. Immun.* **63**:3790–3795.
35. **Delmas, C. L., and D. J.-M. Vidon.** 1985. Isolation of *Yersinia enterocolitica* and related species from foods in France. *Appl. Environ. Microbiol.* **50**:767–771.
36. **Delor, I., and G. R. Cornelis.** 1992. Role of *Yersinia enterocolitica* YST toxin in experimental infection of young rabbits. *Infect. Immun.* **60**:4269–4277.
37. **Delor, I., A. Kaeckenbeeck, G. Wauters, and G. R. Cornelis.** 1990. Nucleotide sequence of *yst*, the *Yersinia enterocolitica* gene encoding the heat-stable enterotoxin, and prevalence of the gene among pathogenic and non-pathogenic yersiniae. *Infect. Immun.* **58**:2983–2988.
38. **Dion, P., R. Charbonneau, and C. Thibault.** 1994. Effect of ionizing dose rate on the radioresistance of some food pathogenic bacteria. *Can. J. Microbiol.* **40**:369–374.
39. **Doyle, M. P.** 1990. Pathogenic *Escherichia coli*, *Yersinia enterocolitica* and *Vibrio parahaemolyticus. Lancet* **336**:1111–1115.
40. **Erickson, J. P., and P. Jenkins.** 1992. Behavior of psychrotrophic pathogens *Listeria monocytogenes*, *Yersinia enterocolitica*, and *Aeromonas hydrophila* in commercially pasteurized eggs held at 2°, 6.7° and 12.8°C. *J. Food Prot.* **55**:8–14.
41. **Feng, P., and S. D. Weagant.** 1993. *Yersinia*, p. 427–460. *In* Y. H. Hui, J. R. Gorham, K. D. Murrell, and D. O. Cliver (ed.), *Foodborne Disease Handbook*, vol. 1. Marcel Dekker, New York.
42. **Fenwick, S. G., P. Madie, and C. R. Wilks.** 1994. Duration of carriage and transmission of *Yersinia enterocolitica* biotype 4, serotype 0:3 in dogs. *Epidemiol. Infect.* **113**:471–477.
43. **Foberg, U., A. Fryden, E. Kihlstrom, K. Persson, and O. Weiland.** 1986. *Yersinia enterocolitica* septicemia: clinical and microbiological aspects. *Scand. J. Infect. Dis.* **18**:269–279.
44. **Forsberg, A., R. Rosqvist, and H. Wolf-Watz.** 1994. Regulation and polarized transfer of the *Yersinia* outer proteins (Yops) involved in antiphagocytosis. *Trends Microbiol.* **2**:14–19.
45. **Fowler, J. M., and R. R. Brubaker.** 1994. Physiological basis of the low calcium response in *Yersinia pestis. Infect. Immun.* **62**:5234–5241.
46. **Galyov, E. E., S. Håkansson, Å. Forsberg, and H. Wolf-Watz.** 1993. A secreted protein kinase of *Yersinia pseudotuberculosis* is an indispensable virulence determinant. *Nature* (London) **361**:730–732.
47. **Gibb, A. P., K. M. Martin, G. A. Davidson, B. Walker, and W. G. Murphy.** 1994. Modeling the growth of *Yersinia enterocolitica* in donated blood. *Transfusion* **34**:304–310.
48. **Granfors, K., S. Jalkanen, R. von Essen, R. Lahesmaa-Rantala, O. Isomäki, K. Pekkola-Heino, R. Merilahti-Palo, R. Saario, H. Isomäki, and A. Toivanen.** 1989. Yersinia antigens in synovial-fluid cells from patients with reactive arthritis. *N. Engl. J. Med.* **320**:216–221.
49. **Green, S. P., E. L. Hartland, R. M. Robins-Browne, and W. A. Phillips.** 1995. Role of YopH in the suppression of tyrosine phosphorylation and respiratory burst activity in murine macrophages infected with *Yersinia enterocolitica. J. Leukocyte Biol.* **57**:972–977.
50. **Gripenberg-Lerche, C., M. Skurnik, and P. Toivanen.** 1995. Role of YadA-mediated collagen binding in arthritogenicity of *Yersinia enterocolitica* serotype O:8: experimental studies with rats. *Infect. Immun.* **63**:3222–3226.
51. **Grützkau, A., C. Hanski, H. Hahn, and E. O. Riecken.** 1990. Involvement of M cells in the bacterial invasion of Peyer's patches: a common mechanism shared by *Yersinia enterocolitica* and other enteroinvasive bacteria. *Gut* **31**:1011–1015.
52. **Håkansson, S., T. Bergman, J.-C. Vanooteghem, G. Cornelis, and H. Wolf-Watz.** 1993. YopB and YopD constitute a novel class of *Yersinia* Yop proteins. *Infect. Immun.* **61**:71–80.
53. **Hanna, M. O., J. C. Stewart, Z. L. Carpenter, and C. V. Vanderzant.** 1977. Effect of heating, freezing and pH on *Yersinia enterocolitica*-like organisms from meat. *J. Food Prot.* **40**:689–692.
54. **Hanski, C., U. Kutschka, H. P. Schmoranzer, M. Naumann, A. Stallmach, H. Hahn, H. Menge, and E. O. Riecken.** 1989. Immunohistochemical and electron microscopic study of interaction of *Yersinia enterocolitica* serotype O8 with intestinal mucosa during experimental enteritis. *Infect. Immun.* **57**:673–678.
55. **Harakeh, M. S., J. D. Berg, J. C. Hoff, and A. Matin.** 1985. Susceptibility of chemostat-grown *Yersinia enterocolitica* and *Klebsiella pneumoniae* to chlorine dioxide. *Appl. Environ. Microbiol.* **49**:69–72.

56. **Hartland, E. L., S. P. Green, W. A. Phillips, and R. M. Robins-Browne.** 1994. Essential role of YopD in inhibition of the respiratory burst of macrophages by *Yersinia enterocolitica. Infect. Immun.* **62:**4445–4453.
57. **Hayashidani, H., Y. Ohtomo, Y. Toyokawa, M. Saito, K. Kaneko, J. Kosuge, M. Kato, M. Ogawa, and G. Kapperud.** 1995. Potential sources of sporadic human infection with *Yersinia enterocolitica* serovar O:8 in Aomori Prefecture, Japan. *J. Clin. Microbiol.* **33:**1253–1257.
58. **Heesemann, J., K. Gaede, and I. B. Autenrieth.** 1993. Experimental *Yersinia enterocolitica* infection in rodents: a model for human yersiniosis. *APMIS* **101:**417–429.
59. **Hermann, E.** 1993. T cells in reactive arthritis. *APMIS* **101:**177–186.
60. **Hermann, E., D. T. Y. Yu, K.-H. Meyer zum Büschenfelde, and B. Fleischer.** 1993. HLA-B27-restricted CD8 T cells derived from synovial fluids of patients with reactive arthritis and ankylosing spondylitis. *Lancet* **342:**646–650.
61. **Herrlinger, J. D., and J. U. Asmussen.** 1992. Long term prognosis in *Yersinia* arthritis: clinical and serological findings. *Ann. Rheum. Dis.* **51:**1332–1334.
62. **Heyma, P., L. C. Harrison, and R. Robins-Browne.** 1986. Thyrotropin (TSH) binding sites on *Yersinia enterocolitica recognized* by immunoglobulins from humans with Graves' disease. *Clin. Exp. Immunol.* **64:**249–254.
63. **Hoogkamp-Korstanje, J. A. A., and V. M. M. Stolk-Engelaar.** 1995. *Yersinia enterocolitica* infection in children. *Pediatr. Infect. Dis. J.* **14:**771–775.
64. **Iriarte, M., and G. R. Cornelis.** 1995. MyfF, an element of the network regulating the synthesis of fibrillae in *Yersinia enterocolitica. J. Bacteriol.* **177:**738–744.
65. **Iriarte, M., I. Stainier, and G. Cornelis.** 1995. The *rpoS* gene from *Yersinia enterocolitica* and its influence on expression of virulence factors. *Infect. Immun.* **63:**1840–1847.
66. **Iriarte, M., I. Stainier, A. V. Mikulskis, and G. R. Cornelis.** 1995. The *fliA* gene encoding σ^{28} in *Yersinia enterocolitica. J. Bacteriol.* **177:**2299–2304.
67. **Iriarte, M., J. C. Vanooteghem, I. Delor, R. Diaz, S. Knutton, and G. R. Cornelis.** 1993. The Myf fibrillae of *Yersinia enterocolitica. Mol. Microbiol.* **9:**507–520.
68. **Isberg, R. R., and G. T. Van Nhieu.** 1994. Two mammalian cell internalization strategies used by pathogenic bacteria. *Annu. Rev. Genet.* **28:**395–422.
69. **Kapperud, G.** 1982. Enterotoxin production at 4°, 22°, and 37° among *Yersinia enterocolitica* and *Y. enterocolitica*-like bacteria. *Acta Pathol. Microbiol. Immunol. Scand.* **90B:** 185–189.
70. **Kapperud, G.** 1991. *Yersinia enterocolitica* in food hygiene. *Int. J. Food. Microbiol.* **12:**53–65.
71. **Kapperud, G., E. Namork, M. Skurnik, and T. Nesbakken.** 1987. Plasmid-mediated surface fibrillae of *Yersinia pseudotuberculosis* and *Yersinia enterocolitica:* relationship to the outer membrane protein YOP1 and possible importance for pathogenesis. *Infect. Immun.* **55:** 2247–2254.
72. **Kasatiya, S. S.** 1976. *Yersinia enterocolitica* gastroenteritis outbreak—Montreal. *Can. Dis. Weekly Rep.* **2:**73–74.
73. **Keet, E. E.** 1974. *Yersinia enterocolitica* septicemia: source of infection and incubation period identified. *N. Y. State J. Med.* **74:**2226–2229.
74. **Kellogg, C. M., E. A. Tarakji, M. Smith, and P. D. Brown.** 1995. Bacteremia and suppurative lymphadenitis due to *Yersinia enterocolitica* in a neutropenic patient who prepared chitterlings. *Clin. Infect. Dis.* **21:**236–237.
75. **King, C. H., E. B. Shotts, Jr., R. E. Wooley, and K. G. Porter.** 1988. Survival of coliforms and bacterial pathogens within protozoa during chlorination. *Appl. Environ. Microbiol.* **54:**3023–3033.
76. **Kingsley, G. H.** 1993. Reactive arthritis: a paradigm for inflammatory arthritis. *Clin. Exp. Rheumatol.* **11**(Suppl. 8):S29–S36.
77. **Lahesmaa, R., M. Skurnik, K. Granfors, T. Mottonen, R. Saario, A. Toivanen, and P. Toivanen.** 1992. Molecular mimicry in the pathogenesis of spondyloarthropathies. A critical appraisal of cross-reactivity between microbial antigens and HLA-B27. *Br. J. Rheumatol.* **31:**221–229.
78. **Larsen, J. H.** 1980. *Yersinia enterocolitica* infection and rheumatic diseases. *Scand. J. Rheumatol.* **9:**129–137.
79. **Lee, L. A., A. R. Gerber, D. R. Lonsway, J. D. Smith, G. P. Carter, N. D. Puhr, C. M. Parrish, R. K. Sikes, R. J. Finton, and R. V. Tauxe.** 1990. *Yersinia enterocolitica* O:3 infections in infants and children, associated with the household preparation of chitterlings. *N. Engl. J. Med.* **322:**984–987.
80. **Leirisalo-Repo, M.** 1987. *Yersinia* arthritis. Acute clinical picture and long-term prognosis. *Contrib. Microbiol. Immunol.* **9:**145–154.
81. **Lindler, L. E., and B. D. Tall.** 1993. *Yersinia pestis* pH 6 antigen forms fimbriae and is induced by intracellular association with macrophages. *Mol. Microbiol.* **8:**311–324.
82. **Luo, G., G. S. Seetharamaiah, D. W. Niesel, H. Zhang, J. W. Peterson, B. S. Prabhakar, and G. R. Klimpel.** 1993. Immunization of mice with *Yersinia enterocolitica* leads to the induction of antithyrotropin receptor antibodies. *J. Immunol.* **151:**922–928.
83. **Marjai, E., M. Kalman, I. Kajary, A. Belteky, and M. Rodler.** 1987. Isolation from food and characterization by virulence tests of *Yersinia enterocolitica* associated with an outbreak. *Acta Microbiol. Hung.* **34:**97–109.
84. **Marks, M. I., C. H. Pai, L. Lafleur, L. Lackman, and O. Hammerberg.** 1980. *Yersinia enterocolitica* gastroenteritis: a prospective study of clinical, bacteriologic, and epidemiologic features. *J. Pediatr.* **96:**26–31.
85. **Maruyama, T.** 1987. *Yersinia enterocolitica* infection in humans and isolation of the organism from pigs in Japan. *Contrib. Microbiol. Immunol.* **9:**48–55.
86. **Mikulskis, A. V., and G. R. Cornelis.** 1994. A new class of proteins regulating gene expression in enterobacteria. *Mol. Microbiol.* **11:**77–86.
87. **Mikulskis, A. V., I. Delor, V. H. Thi, and G. R. Cornelis.** 1994. Regulation of the *Yersinia enterocolitica* enterotoxin Yst gene. Influence of growth phase, temperature, osmolarity, pH and bacterial host factors. *Mol. Microbiol.* **14:**905–915.
88. **Miliotis, M. D., J. G. Morris, S. Cianciosi, A. Wright, and R. M. Robins-Browne.** 1990. Identification of a con-

junctivitis-associated gene locus from the virulence plasmid of *Yersinia enterocolitica*. *Infect. Immun.* **58**:2470–2477.

89. **Miller, V. L.** 1992. *Yersinia* invasion genes and their products. *ASM News* **58**:26–33.
90. **Miller, V. L., K. B. Beer, W. P. Loomis, J. A. Olson, and S. I. Miller.** 1992. An unusual *pagC*::Tn*phoA* mutation leads to an invasion- and virulence-defective phenotype in salmonellae. *Infect. Immun.* **60**:3763–3770.
91. **Mollaret, H. H., H. Bercovier, and J. M. Alonso.** 1979. Summary of the data received at the WHO Reference Centre for *Yersinia enterocolitica*. *Contrib. Microbiol. Immunol.* **5**:174–184.
92. **Morse, D. L., M. Shayegani, and R. J. Gallo.** 1984. Epidemiologic investigation of a *Yersinia* camp outbreak linked to a food handler. *Am. J. Public Health* **74**:589–592.
93. **Nakajima, R., V. L. Motin, and R. R. Brubaker.** 1995. Suppression of cytokines in mice by protein A-V antigen fusion peptide and restoration of synthesis by active immunization. *Infect. Immun.* **63**:3021–3029.
94. **Neilands, J. B.** 1981. Microbial iron compounds. *Annu. Rev. Biochem.* **50**:715–731.
95. **Nesbakken, T., G. Kapperud, K. Dommarsnes, M. Skurnik, and E. Hornes.** 1991. Comparative study of a DNA hybridization method and two isolation procedures for detection of *Yersinia enterocolitica* O:3 in naturally contaminated pork products. *Appl. Environ. Microbiol.* **57**:389–394.
96. **Nilehn, B.** 1969. Studies on *Yersinia enterocolitica* with special reference to bacterial diagnosis and occurrence in human acute enteric disease. *Acta Pathol. Microbiol. Scand. Suppl.* **206**:1–48.
97. **Nilehn, B., and B. Sjostrom.** 1967. Studies on *Yersinia enterocolitica:* occurrence in various groups of acute abdominal disease. *Acta Pathol. Microbiol. Scand.* **71**:612–628.
98. **Ostroff, S. M., G. Kapperud, L. C. Hutwagner, T. Nesbakken, N. H. Bean, J. Lassen, and R. V. Tauxe.** 1994. Sources of sporadic *Yersinia enterocolitica* infections in Norway: a prospective case-control study. *Epidemiol. Infect.* **112**:133–141.
99. **Pai, C. H., V. Mors, and S. Toma.** 1978. Prevalence of enterotoxigenicity in human and nonhuman isolates of *Yersinia enterocolitica*. *Infect. Immun.* **22**:334–338.
100. **Peixotto, S. S., G. Finne, M. O. Hanna, and C. Vanderzant.** 1979. Presence, growth and survival of *Yersinia enterocolitica* in oyster, shrimp and crab. *J. Food Prot.* **42**:974–981.
101. **Pepe, J. C., and V. L. Miller.** 1993. *Yersinia enterocolitica* invasin: a primary role in the initiation of infection. *Proc. Natl. Acad. Sci. USA* **90**:6473–6477.
102. **Perry, R. D.** 1993. Acquisition and storage of inorganic iron and hemin by the yersiniae. *Trends Microbiol.* **1**:142–147.
103. **Pierson, D. E.** 1994. Mechanisms of *Yersinia* entry into mammalian cells, p. 235–247. *In* V. L. Miller, J. B. Kaper, D. A. Portnoy, and R. R. Isberg (ed.), *Molecular Genetics of Bacterial Pathogenesis.* American Society for Microbiology, Washington, D.C.
104. **Pollack, C., S. C. Straley, and M. S. Klempner.** 1986. Probing the phagolysosomal environment of human macrophages with a Ca^{2+}-responsive operon fusion in *Yersinia pestis*. *Nature* (London) **322**:834–836.
105. **Robins-Browne, R. M.** 1989. *Yersinia enterocolitica*, p. 337–349. *In* M. J. G. Farthing and G. Keusch (ed.), *Enteric Infection. Mechanisms, Manifestations and Management.* Chapman and Hall, London.
106. **Robins-Browne, R. M., A.-M. Bordun, and K. J. Slee.** 1993. Serological response of sheep to plasmid-encoded proteins of *Yersinia* species following natural infection with *Y. enterocolitica* and *Y. pseudotuberculosis*. *J. Med. Microbiol.* **39**:268–272.
107. **Robins-Browne, R. M., and J. K. Prpic.** 1985. Effects of iron and desferrioxamine on infections with *Yersinia enterocolitica*. *Infect. Immun.* **47**:774–779.
108. **Robins-Browne, R. M., J. K. Prpic, and S. J. Stuart.** 1987. Yersiniae and iron. A study in host-parasite relationships. *Contrib. Microbiol. Immunol.* **9**:254–258.
109. **Robins-Browne, R. M., A. R. Rabson, and H. J. Koornhof.** 1979. Generalised infection with *Yersinia enterocolitica* and the role of iron. *Contrib. Microbiol. Immunol.* **5**:277–282.
110. **Robins-Browne, R. M., T. Takeda, A. Fasano, A.-M. Bordun, S. Dohi, H. Kasuga, G. Fang, V. Prado, R. L. Guerrant, and J. G. Morris, Jr.** 1993. Assessment of enterotoxin production by *Yersinia enterocolitica* and identification of a novel heat-stable enterotoxin produced by a noninvasive *Y. enterocolitica* strain isolated from clinical material. *Infect. Immun.* **61**:764–767.
111. **Robins-Browne, R. M., S. Tzipori, G. Gonis, J. Hayes, M. Withers, and J. K. Prpic.** 1985. The pathogenesis of *Yersinia enterocolitica* infection in gnotobiotic piglets. *J. Med. Microbiol.* **19**:297–308.
112. **Rohde, J. R., J. M. Fox, and S. A. Minnich.** 1994. Thermoregulation in *Yersinia enterocolitica* is coincident with changes in DNA supercoiling. *Mol. Microbiol.* **12**:187–199.
113. **Rosqvist, R., K.-E. Magnusson, and H. Wolf-Watz.** 1994. Target cell contact triggers expression and polarized transfer of *Yersinia* YopE cytotoxin into mammalian cells. *EMBO J.* **13**:964–972.
114. **Rosqvist, R., M. Skurnik, and H. Wolf-Watz.** 1988. Increased virulence of *Yersinia pseudotuberculosis* by two independent mutations. *Nature* (London) **334**:522–525.
115. **Salmond, G. P. C., and P. J. Reeves.** 1993. Membrane traffic wardens and protein secretion in Gram-negative bacteria. *Trends Biochem. Sci.* **18**:7–12.
116. **Sazama, K.** 1994. Bacteria in blood for transfusion. A review. *Arch. Pathol. Lab. Med.* **118**:350–365.
117. **Schiemann, D. A.** 1980. *Yersinia enterocolitica:* observations on some growth characteristics and response to selective agents. *Can. J. Microbiol.* **26**:1232–1240.
118. **Schiemann, D. A.** 1987. *Yersinia enterocolitica* in milk and dairy products. *J. Dairy Sci.* **70**:383–391.
119. **Schiemann, D. A.** 1988. Examination of enterotoxin production at low temperatures by *Yersinia* spp. in culture media and foods. *J. Food Prot.* **51**:571–573.
120. **Schiemann, D. A.** 1989. *Yersinia enterocolitica* and *Yersinia pseudotuberculosis*, p. 601–672. *In* M. P. Doyle (ed.), *Foodborne Bacterial Pathogens.* Marcel Dekker, New York.

121. **Shayegani, M., D. Morse, I. DeForge, T. Root, L. M. Parsons, and P. S. Maupin.** 1983. Microbiology of a major foodborne outbreak of gastroenteritis caused by *Yersinia enterocolitica* serogroup O:8. *J. Clin. Microbiol.* **17:**35–40.
122. **Shenkman, L., and E. J. Bottone.** 1976. Antibodies to *Yersinia enterocolitica* in thyroid disease. *Ann. Intern. Med.* **85:**735–739.
123. **Sims, G. R., D. A. Glenister, T. F. Brocklehurst, and B. M. Lund.** 1989. Survival and growth of food poisoning bacteria following inoculation into cottage cheese varieties. *Int. J. Food Microbiol.* **9:**173–195.
124. **Skrzypek, E., and S. C. Straley.** 1995. Differential effects of deletions in *lcrV* on secretion of V antigen, regulation of the low-Ca^{2+} response, and virulence of *Yersinia pestis.* *J. Bacteriol.* **177:**2530–2542.
125. **Skurnik, M., S. Batsford, A. Mertz, E. Schiltz, and P. Toivanen.** 1993. The putative cationic 19-kilodalton antigen of *Yersinia enterocolitica* is a urease beta-subunit. *Infect. Immun.* **61:**2498–2504.
126. **Skurnik, M., Y. el Tahir, M. Saarinen, S. Jalkanen, and P. Toivanen.** 1994. YadA mediates specific binding of enteropathogenic *Yersinia enterocolitica* to human intestinal submucosa. *Infect. Immun.* **62:**1252–1261.
127. **Skurnik, M., and P. Toivanen.** 1993. *Yersinia enterocolitica* lipopolysaccharide: genetics and virulence. *Trends Microbiol.* **1:**148–152.
128. **Skurnik, M., and H. Wolf-Watz.** 1989. Analysis of the *yopA* gene encoding the Yop1 virulence determinants of *Yersinia* spp. *Mol. Microbiol.* **3:**571–529.
129. **Slee, K. J., and N. W. Skilbeck.** 1992. Epidemiology of *Yersinia pseudotuberculosis* and *Y. enterocolitica* infections in sheep in Australia. *J. Clin. Microbiol.* **30:**712–715.
130. **Smith, M. G.** 1992. Destruction of bacteria on fresh meat by hot water. *Epidemiol. Infect.* **109:**491–496.
131. **Stern, N. J., M. D. Pierson, and A. W. Kotula.** 1980. Effects of pH and sodium chloride on *Yersinia enterocolitica* growth at room and refrigeration temperatures. *J. Food Sci.* **45:**64–67.
132. **Stoddard, J. J., D. S. Wechsler, J. P. Nataro, and J. F. Casella.** 1994. *Yersinia enterocolitica* infection in a patient with sickle cell disease after exposure to chitterlings. *Am. J. Pediatr. Hematol. Oncol.* **16:**153–155.
133. **Straley, S. C., E. Skrzypek, G. V. Plano, and J. B. Bliska.** 1993. Yops of *Yersinia* spp. pathogenic for humans. *Infect. Immun.* **61:**3105–3110.
134. **Stuart, P. M., and J. G. Woodward.** 1992. *Yersinia enterocolitica* produces superantigenic activity. *J. Immunol.* **148:**225–233.
135. **Tacket, C. O., J. Ballard, N. Harris, J. Allard, C. Nolan, T. Quan, and M. L. Cohen.** 1985. An outbreak of *Yersinia enterocolitica* infections caused by contaminated tofu (soybean curd). *Am. J. Epidemiol.* **121:**705–711.
136. **Tacket, C. O., B. R. Davis, G. P. Carter, J. F. Randolph, and M. L. Cohen.** 1983. *Yersinia enterocolitica* pharyngitis. *Ann. Intern. Med.* **99:**40–42.
137. **Tacket, C. O., J. P. Narain, R. Sattin, J. P. Lofgren, C. Konigsberg, Jr., R. C. Rendtorff, A. Rausa, B. R. Davis, and M. L. Cohen.** 1984. A multistate outbreak of infections caused by *Yersinia enterocolitica* transmitted by pasteurized milk. *JAMA* **251:**483–486.
138. **Tauxe, R. V., J. Vandepitte, G. Wauters, S. M. Martin, X. Goossens, P. DeMol, R. van Noyen, and G. Thiers.** 1987. *Yersinia enterocolitica* infections and pork: the missing link. *Lancet* **i:**1129–1132.
139. **Thompson, J. S., and M. J. Gravel.** 1986. Family outbreak of gastroenteritis due to *Yersinia enterocolitica* serotype O:3 from well water. *Can. J. Microbiol.* **32:**700–701.
140. **Toivanen, P., and A. Toivanen.** 1994. Does *Yersinia* induce autoimmunity? *Int. Arch. Allergy Immunol.* **104:**107–111.
141. **Tomer, Y., and T. F. Davies.** 1993. Infection, thyroid disease, and autoimmunity. *Endocrine Rev.* **14:**107–120.
142. **Toora, S., E. Budu-Amoako, R. F. Ablett, and J. Smith.** 1992. Effect of high-temperature short-time pasteurization, freezing and thawing and constant freezing, on the survival of *Yersinia enterocolitica* in milk. *J. Food Prot.* **55:**803–805.
143. **Tzipori, S., R. Robins-Browne, and J. K. Prpic.** 1987. Studies on the role of virulence determinants of *Yersinia enterocolitica* in gnotobiotic piglets. *Contrib. Microbiol. Immunol.* **9:**233–238.
144. **Van Gijsegem, F., S. Genin, and C. Boucher.** 1993. Conservation of secretion pathways for pathogenicity determinants of plant and animal bacteria. *Trends Microbiol.* **1:**175–180.
145. **Wachtel, M. R., and V. L. Miller.** 1995. In vitro and in vivo characterization of an *ail* mutant of *Yersinia enterocolitica. Infect. Immun.* **63:**2541–2548.
146. **Wauters, G., S. Aleksic, J. Charlier, and G. Schulze.** 1991. Somatic and flagellar antigens of *Yersinia enterocolitica* and related species. *Contrib. Microbiol. Immunol.* **12:**239–243.
147. **Wauters, G., M. Janssens, A. G. Steigerwalt, and D. J. Brenner.** 1988. *Yersinia mollaretii* sp. nov. and *Yersinia bercovieri* sp. nov., formerly called *Yersinia enterocolitica* biogroups 3A and 3B. *Int. J. Syst. Bacteriol.* **38:**424–429.
148. **Wauters, G., K. Kandolo, and M. Janssens.** 1987. Revised biogrouping scheme of *Yersinia enterocolitica. Contrib. Microbiol. Immunol.* **9:**14–21.
149. **Weinberg, E. D.** 1984. Iron withholding: a defense against infection and neoplasia. *Physiol. Rev.* **64:**65–102.
150. **Yang, Y., and R. R. Isberg.** 1993. Cellular internalization in the absence of invasin expression is promoted by the *Yersinia pseudotuberculosis yadA* product. *Infect. Immun.* **61:**3907–3913.
151. **Yoshino, K., T. Takao, X. Huang, H. Murata, H. Nakao, T. Takeda, and Y. Shimonishi.** 1995. Characterization of a highly toxic, large molecular size heat-stable enterotoxin produced by a clinical isolate of *Yersinia enterocolitica. FEBS Lett.* **362:**319–322.

Anthony T. Maurelli
Keith A. Lampel

12 *Shigella* Species

INTRODUCTION

Bacillary dysentery or shigellosis is caused by members of the *Shigella* species. Dysentery was the term used by Hippocrates to describe an illness characterized by frequent passage of stools containing blood and mucus accompanied by painful abdominal cramps. Perhaps one of the greatest historical impacts of this disease has been its powerful influence in military operations. Long, protracted military campaigns and sieges almost always spawned epidemics of dysentery causing large numbers of military and civilian casualties. With a low infectious dose required to cause disease coupled with oral transmission via fecally contaminated food and water, it is not surprising that dysentery caused by *Shigella* spp. follows in the wake of many natural (earthquakes, floods, famine) and man-made (war) disasters. Apart from these special circumstances, shigellosis remains an important disease in developed countries as well as in underdeveloped countries.

Foodborne shigellosis is a neglected area of study. This chapter will highlight modes of transmission and examples of recent foodborne outbreaks of shigellosis, as well as address the most current understanding of the genetics of *Shigella* pathogenesis. Since no single review can be completely comprehensive, the reader is encouraged to refer to several excellent recent reviews for additional information (26, 54, 63, 69).

CHARACTERISTICS OF THE ORGANISMS

Classification and Biochemical Characteristics

There are four species of the genus *Shigella* serologically grouped (39 serotypes) based on their somatic O antigens: *Shigella dysenteriae* (group A), *S. flexneri* (group B), *S. boydii* (group C), and *S. sonnei* (group D). As members of the family *Enterobacteriaceae*, they are nearly genetically identical to *Escherichia coli* and closely related to *Salmonella* and *Citrobacter* species (62). *Shigella* species are nonmotile, oxidase-negative, gram-negative rods. Some important biochemical characteristics that distinguish these bacteria from other enteric bacteria are their inability to ferment lactose or utilize citric acid as a sole carbon source; they do not produce H_2S and, except for *S. flexneri* 6, do not produce gas from glucose. Shigellae are inhibited by potassium cyanide and do not synthesize lysine decarboxylase (21). Enteroinvasive *Escherichia coli* (EIEC) strains have pathogenic properties similar to those of shigellae and are discussed in chapter 10. EIEC also share certain biochemical properties with shigellae, such as being nonmotile and unable to synthesize lysine decarboxylase. In addition, some serogroups of EIEC have O antigens identical to those of certain *Shigella* serotypes (71). However, while their lactose fermentation phenotype is variable, EIEC express many of the same biochemical characteristics as nonenteroinvasive

E. coli. This biochemical similarity can pose problems in distinguishing these pathogens from *E. coli* found in normal flora.

Shigella spp. are not particularly fastidious in their growth requirements, and in most cases, the organisms are routinely cultivated in the laboratory on artificial medium. Shigellae are easily isolated and grown from analytical samples, including water and clinical samples. In the latter case, shigellae are present in fecal specimens in large populations (10^7/g of stool) at the onset of symptoms, and therefore, identification is easily accomplished by using culture media, biochemical analysis, and serological typing. Shigellae can be detected in samples from convalescent patients for weeks or longer after the initial infection.

However, identification of shigellae in foods is not as facile as in other sources. Foods have many different physical attributes that may affect the successful recovery of shigellae. These factors include (i) composition such as fat content of the food, (ii) physical parameters such as pH and salt, and (iii) natural microbial flora of the food in which other microbes in a sample may overgrow shigellae in broth media. The physiological state of shigellae present in the food is a contributing factor in the successful recovery of these pathogens. The amount of time from the clinical report of a suspected outbreak to the analysis of the food samples can be considerable and thus lessen the chances of identifying the causative agent. Also, shigellae may be present in low populations or in a poor physiological state in the suspected food samples. Under these conditions, special enrichment procedures are required for successful detection of shigellae (6).

Preservation Methods

It is estimated that the number of foodborne illnesses in the United States is 6.5 million to 81 million annually (8) and that the actual number of foodborne outbreaks may be significantly underreported (78). The adverse impact on the public health caused by foodborne pathogens is reflected by their notable morbidity and mortality (66). Foodborne outbreaks of shigellosis continue to be a major public health concern and are increasing worldwide.

Shigella spp. are not associated with any specific foods. From 1975 to 1981, some common foods that were known to be associated with outbreaks caused by shigellae were potato salad, chicken, tossed salad, and shellfish. Establishments that served these foods ranged from the home to restaurants, camps, picnics, schools, airlines, sorority houses, and military mess halls (77). In many cases, the source (food) was not determined. From 1983 to 1987, 2,397 foodborne outbreaks representing 54,453 cases were reported to the Centers for Disease Control and Prevention; in only 38% of the cases was the source of the etiologic agent identified (7). Whereas epidemiological methods may strongly imply a common food source, *Shigella* spp. are not often recovered from foods and identified by standard bacteriological methods. Also, since shigellae are not commonly associated with any particular food, routine inspections of foods to identify these pathogens are not usually performed.

The traditional approach to address the problem of microbially contaminated foods in the processing plant is to inspect the final product. There are several drawbacks to this approach (33); current bacteriological methods are often time-consuming and laborious. An alternative to end product testing is the Hazard Analysis and Critical Control Point (HACCP) system, which would identify certain points of the processing system that may be most vulnerable to microbial contamination and chemical and physical hazards. This approach would be instituted as a preventive program with less reliance on end product testing. A monitoring system such as HACCP may be well suited for pathogenic bacteria, such as *E. coli* O157:H7 and *Salmonella* spp., that are known to be associated with specific foods, e.g., meats and egg products, respectively.

In contrast, establishing specific critical control points for preventing *Shigella* contamination of foods is not always suitable for the HACCP concept. This pathogen is usually introduced into the food supply by an infected person such as a food handler with poor personal hygiene. In some cases, this may occur at the manufacturing site, but more likely it happens at a point between the processing plant and the consumer. Another factor is that foods such as vegetables (lettuce is a good example) can be contaminated at the site of collection and shipped directly to market. Although HACCP is a method for controlling food safety and preventing foodborne outbreaks, pathogens such as shigellae that are not indigenous to but rather are introduced into foods are most likely to be undetected.

FOODBORNE OUTBREAKS

Transmission and Susceptible Populations

The primary means of human-to-human transmission of shigellae is by the fecal-oral route. Most cases of shigellosis are caused by the ingestion of fecally contaminated food or water, and in the case of foods, the major factor for contamination is the poor personal hygiene of food handlers. From carriers, this pathogen can spread by several routes, including food, fingers, feces, and flies. The latter usually transmit the bacteria from fecal matter to foods. The highest number of incidences of shigellosis occur during the warmer months of the year. Improper storage of contaminated foods is the second most common factor that accounted for foodborne outbreaks due to shigellae (77). Other contributing factors are inadequate cooking, contaminated equipment, and food obtained from unsafe sources (7). To reduce the spread of shigellosis, infected patients are monitored until stool samples are negative for shigellae.

Shigellae are frank pathogens capable of causing disease in otherwise healthy individuals. Certain populations, however, may be more predisposed to infection and disease because of the nature of transmission of the organisms. The greatest frequency of illness occurs among children less than 6 years of age. In the United States, outbreaks of shigellosis and other diarrheal diseases in day care centers are increasing as more single-parent families and working women turn to these facilities to care for their children (46, 64). Typical toddler behavior such as oral exploration of the environment and inadequate personal hygiene habits creates conditions ideally suited to bacterial, protozoal, and viral pathogens which are spread by fecal contamination. Transmission of shigellae in this population is very efficient, and the low infectious dose for causing disease increases the risk for shigellosis. Increased risk also extends to family contacts of day care attenders (84).

Shigellosis can be endemic in other institutional settings, such as prisons, mental hospitals, and nursing homes, where crowding and/or insufficient hygienic conditions create an environment for direct fecal-oral contamination. Crowded conditions and poor sanitation contribute to shigellosis being endemic in developing countries as well.

When natural or man-made disasters destroy the sanitary waste treatment and water purification infrastructure, developed countries assume the conditions of developing countries. These conditions place a population at risk for diarrheal diseases such as cholera and dysentery. Recent examples include the war in Bosnia-Herzegovina and famine and political upheaval in Somalia (46). Massive population displacement (e.g., refugees fleeing from Rwanda into Zaire in 1994) can also lead to explosive epidemics of diarrheal disease caused by *Vibrio cholerae* and *S. dysenteriae* 1 (30).

Reservoirs

Humans are the natural reservoir of *Shigella* infections, although several cases of diarrheal disease in monkeys have been reported (41). In one instance, three animal caretakers at a monkey house complained of having diarrhea. Further investigation revealed that *S. flexneri* 1b was present in stool samples from these employees and that the identical serotype was also being shed by four monkeys. The disease was spread by direct contact of the caretakers with excrement from the infected monkeys.

Asymptomatic *Shigella* carriers may exacerbate the spread and maintenance of this pathogen in developing countries. Two studies, one in Bangladesh (37) and the other in Mexico (32), revealed that shigellae were isolated from stool samples collected from asymptomatic children under the age of 5 years. Shigellae were rarely found in infants under the age of 6 months.

Examples of Foodborne Outbreaks

In the cases illustrated below, it becomes obvious that a wide range of foods can be contaminated with shigellae. Disease is caused by the ingestion of these contaminated foods and in some instances subsequently leads to rapid dissemination.

1987—Rainbow Family Gathering

At an annual gathering of the Rainbow family, as many as half of the 12,700 people in attendance may have had shigellosis (85). *S. sonnei* was isolated from stool cultures of tested attendees. Spread of the organism most likely occurred by the fecal-oral route in a crowded environment by contamination of the food or water or both. Outbreaks reported in three other states were due to attendees returning home and secondarily infecting other individuals.

1989 and 1994—Shigellosis aboard Cruise Ships

In October 1989, 14% of the passengers and 3% of the crew members aboard a cruise ship reported having gastrointestinal symptoms (47). A multiple-antibiotic-resistant strain of *S. flexneri* 2a was isolated from several ill passengers and crew. The source of this outbreak was identified as German potato salad. Contamination was introduced by infected food handlers, first in the country where the food was originally prepared and second by a member of the galley crew on the cruise ship. Another outbreak of shigellosis occurred in August 1994 on the cruise ship SS *Viking Serenade* (17). Thirty-seven per cent (586) of the passengers and 4% (24) of the crew reported having diarrhea, and one person died. In this outbreak, *S. flexneri* 2a was isolated from patients. The suspected source of contamination was spring onions.

Table 12.1 Examples of foodborne outbreaks caused by *Shigella* spp.

Yr	Location; source of contamination[a]	Isolate	Reference
1986	Texas; shredded lettuce	*S. sonnei*	18
1987	Rainbow family gathering; food handlers	*S. sonnei*	85
1988–1989	Monroe, N.Y.; multiple sources	*S. sonnei*	86
1988	Outdoor music festival, Michigan; food handlers	*S. sonnei*	45
1988	Commercial airline; cold sandwiches	*S. sonnei*	35
1989	Cruise ship; potato salad	*S. flexneri*	47
1990	Operation Desert Shield (U.S. troops); fresh produce	*Shigella* spp.	38
1991	Alaska; moose soup	*S. sonnei*	28
1992–1993	Operation Restore Hope, Somalia (U.S. troops)	*Shigella* spp.	75
1994	Europe; shredded lettuce from Spain	*S. sonnei*	39
1994	Midwest United States; green onions	*S. flexneri*	16
1994	Cruise ship	*S. flexneri*	17

[a]The source of contamination is listed if known.

1990—Operation Desert Shield

Diarrheal diseases during a military operation can obviously reduce the effectiveness of troops. In Operation Desert Shield, enteric pathogens were isolated from 214 U.S. soldiers, and of those, 113 were diagnosed as having *Shigella* infections; *S. sonnei* was the most prevalent of the shigellae isolated (38). Shigellosis accounted for more time lost from military duties and was responsible for more severe morbidity than enterotoxigenic *E. coli*, the most common enteric pathogen isolated from U.S. troops in Saudia Arabia (38). The suspected source was contaminated fresh vegetables, notably lettuce. Twelve heads of lettuce were examined, and enteric pathogens were isolated from all.

1991—The Alaska Moose Soup Shigellosis

In September 1991, the Alaska Division of Public Health was contacted about a possible gastroenteritis outbreak (28). In Galena, Alaska, 25 people who participated in a gathering of local residents at a community event to which homemade foods were brought contracted shigellosis. The implicated food was homemade moose soup. One of the five women who made the soup reported that she had gastroenteritis before or at the time of preparing the soup. *S. sonnei* was isolated from one hospitalized patient.

1994—Contaminated Lettuce in Norway and the United Kingdom

One hundred ten culture-confirmed cases of shigellosis caused by *S. sonnei* were reported in an outbreak in Norway in 1994 (39). Iceberg lettuce from Spain, served in a salad bar, was suspected as the source of the outbreak in Norway and possibly responsible for increases in shigellosis in other European countries, such as the United Kingdom (25) and Sweden. *S. sonnei* was isolated from patients from several northwest European countries but was not isolated from any foods. Strong epidemiologic evidence indicated that the source of these outbreaks was imported lettuce.

1994—Green Onions

In the summer of 1994 in the Midwest of the United States, an outbreak of *S. flexneri* serotype 6 (mannitol negative) was reported to the Centers for Disease Control and Prevention (16). Although not confirmed, the suspected contaminated food appears to have been green onions (scallions, spring onions) from Mexico. In June at a church potluck meal in Indiana, 17 people developed diarrhea, with 3 having confirmed cases of shigellosis. Twenty-nine people contracted shigellosis at a anniversary reception in Indiana in July. In Illinois, 26 culture confirmed mannitol-negative *S. flexneri* or *Shigella* spp. cases of shigellosis were reported to the state Department of Health. Isolates of *S. flexneri* serotype 6 were also reported from Missouri, Minnesota, Wisconsin, Michigan, and Kentucky. Ingestion of green onions was strongly implicated as the cause of shigellosis, and the food most likely was contaminated by an infected worker at the time of harvest or packing.

One of the striking features of foodborne outbreaks caused by shigellae is that contamination of foods usually is not at the processing plant, but rather, the source can be traced to a food handler. As evident from the examples above and in Table 12.1, these incidents can occur by improper food handling from individuals to small town gatherings and picnics and larger-scale outbreaks such as those on cruise ships and in institutions.

CHARACTERISTICS OF DISEASE

Clinical Presentation

Disease caused by *Shigella* spp. is distinguished from disease caused by most of the other foodborne pathogens described in this volume in at least two important aspects: the production of bloody diarrhea or dysentery and the low infectious dose that can cause clinical symptoms. Bloody diarrhea refers to diarrhea in which the stools contain visible red blood. Dysentery has the same meaning, but the passage of bloody mucoid stools is accompanied by severe abdominal and rectal pain, cramps, and fever. While abdominal pain and diarrhea are experienced by nearly all patients with shigellosis, fever occurs in about one-third and gross blood in the stools occurs in about 40% of the cases (19).

The clinical picture of shigellosis ranges from a mild watery diarrhea to severe dysentery. The dysentery stage of the disease caused by *Shigella* spp. may be preceded by watery diarrhea. This stage reflects the transient multiplication of bacteria as they pass through the small bowel. Jejunal secretions probably are not effectively reabsorbed in the colon as a result of transport abnormalities caused by bacterial invasion and destruction of the colonic mucosa (12). The dysentery stage of disease correlates with extensive bacterial colonization of the colonic mucosa. The bacteria invade the epithelial cells of the colon, spread from cell to cell, but penetrate only as far as the lamina propria. Foci of individually infected cells produce microabscesses which coalesce, forming large abscesses and mucosal ulcerations. As the infection progresses, dead cells of the mucosal surface slough off, thus leading to the presence of blood, pus, and mucus in the stools.

The incubation period for shigellosis is 1 to 7 days, but the illness usually begins within 3 days. Strains of *S. dysenteriae* 1 cause the most severe disease, while *S. sonnei* produces the mildest. *S. flexneri* and *S. boydii* infections can be either mild or severe. Despite the severity of the disease, shigellosis is self-limiting. If left untreated, clinical illness usually persists for 1 to 2 weeks (although it may be as long as a month), and the patient recovers.

Infectious Dose

As mentioned earlier, an important aspect of *Shigella* pathogenesis is the extremely low ID_{50}, i.e., the experimentally determined oral dose required to cause disease in 50% of volunteers challenged with a virulent strain of the organism. The ID_{50} for *S. flexneri*, *S. sonnei*, and *S. dysenteriae* is approximately 5,000 organisms. Volunteers become ill when doses as low as 200 organisms are given (20). The low ID_{50} of *Shigella* spp. underlies the high communicability of bacillary dysentery and gives the disease great explosive potential for person-to-person spread as well as foodborne and waterborne outbreaks of diarrhea.

Complications

Shigellosis can be a very painful and incapacitating disease and is more likely to require hospitalization than other bacterial diarrheas. It is not usually life threatening, and mortality is rare except in malnourished children, immunocompromised individuals, and the elderly (10). However, complications arising from the disease, including severe dehydration, intestinal perforation, toxic megacolon, septicemia, seizures, Reiter's syndrome, and hemolytic uremic syndrome (HUS) have been reported (9). The latter two syndromes are receiving increased research attention. Reiter's syndrome, a form of reactive arthritis, is a postinfection sequela to shigellosis which is strongly associated with individuals of the HLA-B27 histocompatibility group (76). The syndrome consists of three symptoms, urethritis, conjunctivitis, and arthritis, with the last being the most dominant. Infections caused by several other gram-negative enteric pathogens also can lead to this type of sterile inflammatory polyarthropathy (14). HUS is a rare but potentially fatal complication associated with infection by *S. dysenteriae* 1 (65). The syndrome is characterized by hemolytic anemia, thrombocytopenia, and acute renal failure. Epidemiologic studies suggest that Shiga toxin produced by *S. dysenteriae* 1 may be the cause of HUS (48). This hypothesis is supported by the fact that HUS is also caused by strains of *E. coli* O157:H7, which produce high levels of Shiga toxins (40). It has been suggested that Shiga toxin (and the Shiga family of toxins) may cause HUS by entering the bloodstream and damaging vascular endothelial cells such as those in the kidney (40, 48, 79).

Treatment and Prevention

Shigellosis is a self-limiting disease in normally healthy patients, and full recovery can occur even without the use of antibiotics. Although stool fluid losses are not as massive as with other bacterial diarrheas, the diarrhea associated with shigellosis, combined with water loss due to fever and decreased water intake due to anorexia, may result in severe dehydration (10). Oral intake can generally replace fluid losses, although intravenous rehydration may be required in very young and elderly patients.

The antibiotic of choice for treatment of shigellosis is trimethoprim-sulfamethoxazole (19). Antibiotic treatment limits the duration of disease and shortens the period of fecal excretion of bacteria (34). Since an infected person or asymptomatic carrier can be an index case for person-to-person and food- and waterborne spread, antibiotic treatment of these individuals can be a significant public health tool to contain the spread of shigellosis. A clinical problem of growing significance is the rising incidence of multiple drug resistance among *Shigella* isolates. Clinical isolates resistant to sulfonamides, ampicillin, trimethoprim-sulfamethoxazole, tetracycline, chloramphenicol, and streptomycin have been reported (11).

While improvements in sanitary and hygienic conditions can help contain secondary spread of shigellosis, the single most effective means of preventing secondary transmission is hand washing. Despite many years of intensive effort, an effective vaccine against shigellosis still has not been developed.

VIRULENCE FACTORS

Hallmarks of Virulence

Shigella spp. and EIEC are the principal agents of bacillary dysentery and as such belong to the group of enteric pathogens that cause disease by overt invasion of epithelial cells in the large intestine. The clinical symptoms of shigellosis can be directly attributed to the hallmarks of *Shigella* virulence: the ability to invade epithelial cells of the intestine, multiply intracellularly, and spread from cell to cell.

The first hallmark was established by a landmark study that used both in vitro tissue culture assays for invasion and animal models (43). Spontaneous colonial variants of *S. flexneri* 2a that were unable to invade epithelial cells in tissue culture did not cause disease in monkeys. Gene transfer studies using *E. coli* K-12 donors and *S. flexneri* 2a recipients established the second hallmark of *Shigella* virulence. An *S. flexneri* 2a recipient that inherited the *xyl-rha* region of the *E. coli* K-12 chromosome retained the ability to invade epithelial cells but had reduced ability to multiply within these cells (22). This hybrid strain failed to cause a fatal infection in the opiated guinea pig model and was unable to cause disease when fed to rhesus monkeys (24).

It is necessary but not sufficient for shigellae to be able to multiply within the host epithelial cell after invasion. The bacteria must also be able to spread through the epithelial layer of the colon by cell-to-cell spread, which does not require the bacteria to leave the intracellular environment and be reexposed to the intestinal lumen. *Shigella* mutants that are competent for invasion and multiplication but unable to spread between cells have been isolated. These mutants demonstrate intracellular spread as the third hallmark of *Shigella* virulence and will be discussed further below.

Along with the ability to colonize and cause disease, an intrinsic part of a bacterium's pathogenicity is its mechanism for regulating expression of the genes involved in virulence. The characteristic pattern by which *Shigella* spp. regulate expression of their virulence genes is the fourth hallmark of *Shigella* virulence. Virulence in *Shigella* spp. is regulated by growth temperature. After growth at 37°C, virulent strains of *Shigella* are able to invade mammalian cells, but when cultivated at 30°C, they are phenotypically noninvasive. This noninvasive phenotype is reversible by shifting the growth temperature to 37°C. This enables the bacteria to reexpress their virulence properties (52). Temperature regulation of virulence gene expression is a characteristic that shigellae have in common with other human pathogens, such as *E. coli, Salmonella typhimurium, Bordetella pertussis, Yersinia* spp., and *Listeria monocytogenes* (see reference 50 for a review). Regulation of gene expression in response to environmental temperature is a useful bacterial strategy. By sensing the ambient temperature of the mammalian host (e.g., 37°C for humans) to trigger gene expression, this strategy permits shigellae to economize energy that would be expended on the synthesis of virulence products when the bacteria are outside the host. The system also permits the bacteria to coordinately regulate expression of multiple unlinked genes that are required for the full virulence phenotype. Temperature regulation in *S. flexneri* 2a operates at the level of gene transcription (53) and is mediated by both positive and negative transcription factors. A chromosomal gene, *virR (hns)*, encodes a repressor of virulence gene expression (55), whereas two other genes, *virF* and *virB*, encode positive activators (1, 67). These genes will be discussed in a later section. A more thorough treatment of virulence gene regulation in shigellae can be found in several recent review articles (53, 63).

Genetics

Virulence-Associated Plasmid Genes

Shigella virulence is multigenic, involving both chromosomal and plasmid-encoded genes (Table 12.2). Another landmark contribution to the understanding of the pathogenicity of shigellae was the demonstration of the indispensable role of a large plasmid in invasion. A 180-kb plasmid in *S. sonnei* and a 220-kb plasmid in

Table 12.2 Virulence-associated loci of *Shigella* spp.

Locus(i)	Product	Role in virulence
Chromosomal		
rfa, rfb	Enzymes for core and O-antigen biosynthesis	Correct polar localization of IcsA
stx	Shiga toxin	Destruction of vascular tissue
virR (*hns*)	Histone-like protein	Repressor of virulence gene expression
iuc	Synthesis of aerobactin and receptor	Acquisition of iron in the host
sodB	Superoxide dismutase	Inactivation of superoxide radicals; defense against oxygen-dependent killing in host
Plasmid		
icsB	57-kDa protein	Lysis of double membrane for intercellular spread
ipgC	17-kDa protein	Chaperon for IpaB and IpaC
ipaB	62-kDa protein	Invasion; lysis of vacuole; induction of apoptosis
ipaC	43-kDa protein	Invasion
ipaD	38-kDa protein	Invasion
mxi/spa	20 proteins	Secretion of Ipa and other virulence proteins
icsA (*virG*)	120-kDa cell-bound and secreted protein	Actin polymerization for intracellular motility and intercellular spread
virB	Transcriptional activator	Temperature regulation of virulence genes
virF	Transcriptional activator	Temperature regulation of virulence genes

S. flexneri are essential for invasion (72, 73). Other *Shigella* spp. as well as EIEC contain homologous plasmids that are functionally interchangeable and share significant degrees of DNA homology (71). Hence, it is probable that the plasmids of *Shigella* spp. and EIEC are derived from a common ancestor (70).

A 37-kb region of the invasion plasmid of *S. flexneri* 2a contains all of the genes necessary to enable shigellae to penetrate into tissue culture cells. This DNA segment is the minimal region of virulence plasmid needed for a plasmid-cured derivative of *S. flexneri* (and *E. coli* K-12) to invade tissue culture cells (51). The nucleotide sequence of this part of the virulence plasmid from *S. flexneri* and *S. sonnei* is known (GenBank accession number D50601 for *S. sonnei;* see reference 26 for a summary). The region encodes about 33 genes contained in two groups of genes transcribed in opposite orientation (Fig. 12.1). Although a precise transcription map of these genes has not been defined, available evidence and the DNA sequence of the region suggest a multiple-operon organization.

The genes comprising the *ipaBCDA* (invasion plasmid antigens) cluster encode the immunodominant antigens detected with sera from convalescent patients and experimentally challenged monkeys (60). With the exception of *ipaA,* these genes have been experimentally demonstrated to be absolutely required for invasion of mammalian cells (56). The Ipa products are associated with the outer membrane of shigellae. IpaB and IpaC form a complex on the bacterial cell surface and probably are responsible for transducing the signal that leads to entry of shigellae into the host cells via bacterium-directed phagocytosis (58). *ipgC* (invasion plasmid gene) is required for invasion and acts as a chaperon that prevents IpaB and IpaC from forming complexes in the bacterial cytoplasm (58). Although the Ipa proteins have no typical signal sequence, these proteins are secreted into the extracellular medium. Contact of the bacterium with epithelial cells causes increased secretion of the cytoplasmic pool of Ipa products (4, 57). The surface-expressed and cell-free Ipas probably play complementary roles in generating the signals required for uptake by the host cell.

The product of *ipaB* has also been postulated to be the contact hemolysin that is responsible for lysis of the phagocytic vacuole minutes after entry of the bacterium into the host cell (36). The ability of *S. flexneri* to induce apoptosis in infected macrophages is an additional property assigned to IpaB (87).

The products of the *ipa* genes are actively secreted into the extracellular medium even though they contain no signal sequence for recognition by the usual gram-negative bacterial transport system. This secretion requires a dedicated apparatus composed of gene products from the *mxi/spa* complex (Fig. 12.1). The *mxi* (membrane expression of invasion plasmid antigens) genes comprise an operon that encodes several lipoproteins (MxiJ and MxiM), a transmembrane protein (MxiA), and proteins containing signal sequences (MxiD, MxiJ, and MxiM) (2, 3, 5). MxiH, MxiJ, MxiD, and MxiA have homology with proteins involved in secretion of virulence proteins (Yops) in *Yersinia* spp. (2, 3, 5). The *spa*

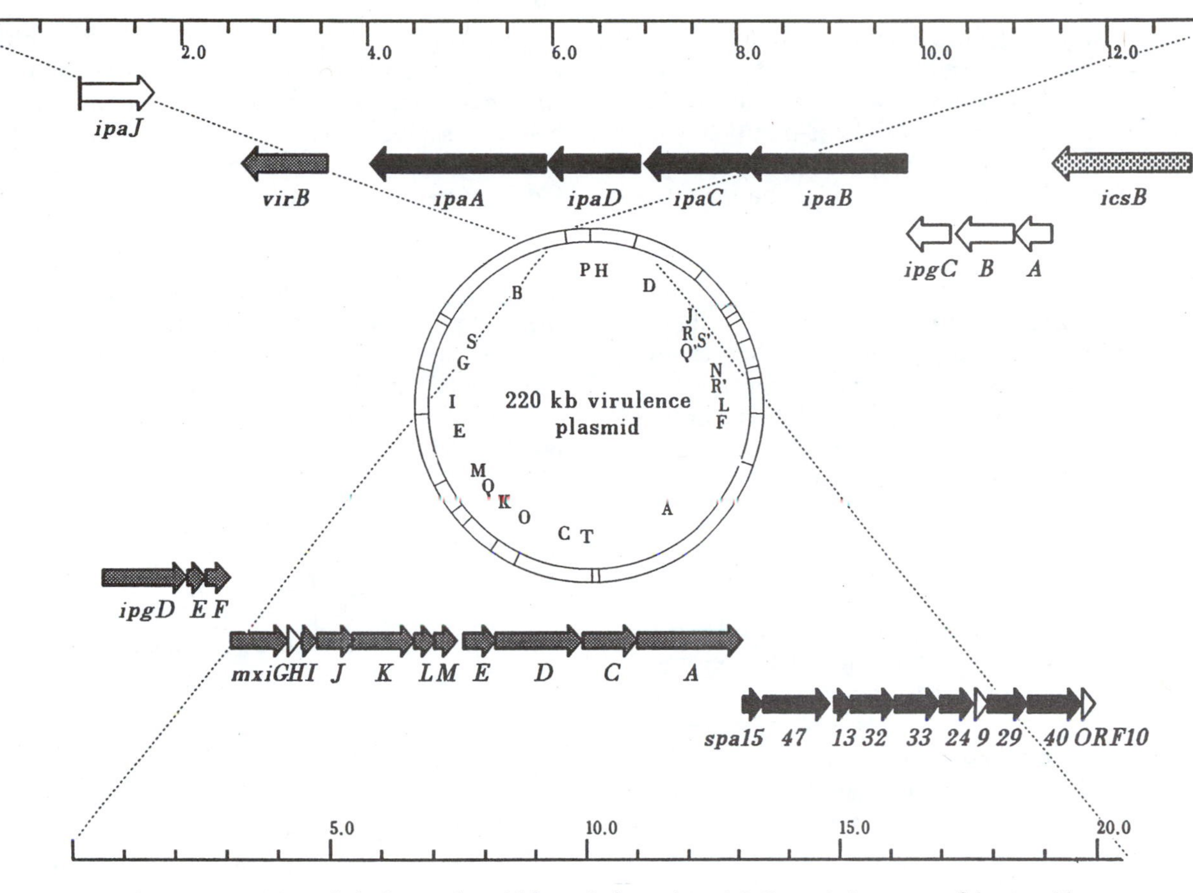

Figure 12.1 Map of virulence plasmid from *S. flexneri* 2a. A *Sal*I restriction map of the 220-kb plasmid is shown in the center. Sections of *Sal*I fragments B and P (upper map) and fragments P, H, and D (lower map) are expanded to illustrate the virulence loci encoded in these regions. Arrows mark the direction of transcription of each open reading frame. The expanded regions are contiguous and include 32 kb. The open reading frame for *icsB* is separated from that of *ipgD* by 314 bp.

(surface presentation of Ipa antigens) genes encode proteins that have significant homologies with proteins involved in flagellar synthesis in *E. coli, Salmonella typhimurium, Bacillus subtilis,* and *Caulobacter crescentus* (74, 83). Included among these genes is *spa47,* which encodes a protein that probably functions as the energy-generating component of the secretion apparatus since it has sequence similarities with ATPases of the flagellar assembly machinery of other bacteria (83).

Apart from the resemblance to genes involved in flagellar synthesis and secretion of Yops, there is an even more striking similarity in both gene organization and predicted protein sequence between the *mxi/spa* region of the *Shigella* virulence plasmid and a virulence-associated chromosomal region from *Salmonella typhimurium* (26, 31). The *Salmonella spa* region encodes homologs of the *Shigella spa* genes in the same gene order. Sequence identities between the protein homologs range as high as 86% (Spa9 versus SpaQ). The relatedness of the *spa* regions strongly suggests that these two human pathogens evolved similar mechanisms for secretion of the virulence proteins required for signal transduction with the mammalian host. Plant pathogens such as *Erwinia carotovora, Xanthomonas campestris,* and *Pseudomonas solanacearum* also contain genes which encode homologs of the *mxi/spa*-encoded proteins (82). This type of virulence protein-dedicated secretion machinery (known as a type III system) is now recognized as a critical element of plant and animal bacterial pathogenesis.

A plasmid-encoded virulence gene that is unlinked to the 37-kb region shown in Fig. 12.1 is not required for invasion but is crucial to the third hallmark of *Shigella* pathogenesis, intra- and intercellular motility. This gene, known as *virG* or *icsA* (intracellular spread), encodes a protein that catalyzes the polymerization of actin in the cytoplasm of the infected cell (12, 49). The IcsA protein is unusual in that it is expressed asymmetrically on the bacterial surface, being present only at one pole (29). The polymerization of actin monomers by IcsA forms a tail leading from the pole and provides the force that propels the bacterium through the cytoplasm. Hence, unipolar expression of IcsA imparts directionality of movement to the bacterium. Although the mechanism of unipolar localization of IcsA is unknown, it is dependent on synthesis of a complete lipopolysaccharide (LPS) (68). LPS mutants of *S. flexneri* 2a express surface IcsA in a circumferential fashion, and while the protein is still capable of polymerizing actin, movement is restricted as the bacterium becomes encased in a shell of actin.

Chromosomal Virulence Loci

In contrast to the genes of the virulence plasmid that are responsible for invasion of mammalian tissues, most of the chromosomal loci associated with *Shigella* virulence are involved in regulation or survival within the host. Synthesis of a complete LPS, which is crucial for correct unipolar localization of IcsA (see above), requires chromosomal loci such as *rfa* and *rfb*. In the case of *S. sonnei* and *S. dysenteriae* 1, plasmid-encoded genes are also necessary for synthesis of the LPS O-side chain (13).

Aerobactin is a hydroxamate siderophore which *S. flexneri* uses to scavenge iron. When the *iuc* locus, which contains the genes for aerobactin synthesis and transport, is inactivated, the aerobactin mutants retain their capacity to invade host cells but are altered in virulence as measured in animal models. These results suggest that aerobactin synthesis is important for bacterial growth within the mammalian host (44, 59).

The *stx* locus in *S. dysenteriae* 1 encodes Shiga toxin (for a review, see reference 61). A mutation in this locus does not alter the ability of the organism to invade epithelial cells or cause keratoconjunctivitis in the Sereny test. However, when tested in macaque monkeys, the mutant strain caused less vascular damage in colonic tissue than the toxin-producing parent (23). Hence, production of Shiga toxin may account for the generally more severe infections caused by *S. dysenteriae* 1 than by the other species of *Shigella*.

Virulence Gene Regulation

The fourth hallmark of *Shigella* pathogenesis is the ability of the organism to modulate expression of its virulence genes in response to growth temperature. The *Shigella* virulence regulon is controlled by both activators and a repressor. The product of the chromosomal *virR* (*hns*) locus is a histone-like protein, H-NS, which behaves as a repressor of *Shigella* virulence gene expression (55). Mutations in *virR* (*hns*) cause deregulation of temperature control such that genes in the virulence regulon are expressed even at the nonpermissive temperature of 30°C. The *virR* (*hns*) locus is allelic with regulatory loci in other enteric bacteria, and like *virR* (*hns*), these alleles act as repressors of their respective regulons (references cited in reference 63). Several different models to explain how H-NS acts as a transcriptional repressor have been proposed and are analyzed in a recent review (63). However, because H-NS is involved in gene regulation in response to diverse environmental stimuli such as osmolarity, pH, and temperature, a comprehensive model to explain its activity has been elusive.

One of the targets regulated by *virR* (*hns*) is the plasmid-encoded transcriptional activator *virB* (81). Expression of genes in the *ipa* and *mxi/spa* clusters is dependent on VirB, and mutations in *virB* abolish the bacterium's ability to invade tissue culture cells (1, 15). Transcription of *virB* is dependent on growth temperature and VirF (80). VirB is probably a DNA-binding protein, as suggested by its homology with the plasmid-partitioning proteins ParB of bacteriophage P1 and SopB of plasmid F. However, targets for VirB binding in the promoter regions of the *ipa* and *mxi/spa* genes have not yet been identified, and the mechanism by which VirB activates virulence gene expression is unknown.

The product of the *virF* locus is a key element in temperature regulation of the *Shigella* virulence regulon. A helix-turn-helix motif in the carboxyl-terminal portion of VirF is characteristic of members of the AraC family of transcriptional activators (27). Consistent with its predicted role as a DNA-binding protein, VirF binds to sequences upstream of *virB* (81). The binding of VirF may act as an antagonist to binding by H-NS and thereby provides a mechanism for responding to temperature. However, expression of *virF* itself is apparently not subject to temperature regulation; hence, an unknown regulator may be involved in activating VirF. Alternatively, temperature may induce conformational changes in VirF that influence its binding affinity and thus its ability to activate promoters.

CONCLUSIONS

While foodborne infections due to *Shigella* spp. may not be as frequent as those caused by other foodborne pathogens, they have the potential for explosive spread because of the extremely low infectious dose needed to cause overt clinical disease. In addition, cases of bacillary dysentery frequently require medical attention and cause lost time from work, as the severity and duration of symptoms can be incapacitating. There is no effective vaccine against dysentery caused by shigellae. These features, coupled with the wide geographic distribution of the strains and sensitivity of the human population, make *Shigella* infection a formidable public health threat.

Acknowledgments

Research on the genetics of *Shigella* virulence in the laboratory of A.T.M. is supported by USUHS protocol RO-7385 and Public Health Service grant AI24656 from the National Institute of Allergy and Infectious Diseases.

References

1. **Adler, B., C. Sasakawa, T. Tobe, S. Makino, K. Komatsu, and M. Yoshikawa.** 1989. A dual transcriptional activation system for the 230 kb plasmid genes coding for virulence-associated antigens of *Shigella flexneri*. *Mol. Microbiol.* **3**:627–635.
2. **Allaoui, A., P. J. Sansonetti, and C. Parsot.** 1992. MxiJ, a lipoprotein involved in secretion of *Shigella* Ipa invasins, is homologous to YscJ, a secretion factor of the *Yersinia* Yop proteins. *J. Bacteriol.* **174**:7661–7669.
3. **Allaoui, A., P. J. Sansonetti, and C. Parsot.** 1993. MxiD, an outer membrane protein necessary for the secretion of the *Shigella flexneri* Ipa invasins. *Mol. Microbiol.* **7**:59–68.
4. **Andrews, G. P., A. E. Hromockyj, C. Coker, and A. T. Maurelli.** 1991. Two novel virulence loci, *mxiA* and *mxiB*, in *Shigella flexneri* 2a facilitate excretion of invasion plasmid antigen. *Infect. Immun.* **59**:1997–2005.
5. **Andrews, G. P., and A. T. Maurelli.** 1992. *mxiA* of *Shigella flexneri* 2a, which facilitates export of invasion plasmid antigens, encodes a homolog of the low-calcium response protein LcrD, of *Yersinia pestis*. *Infect. Immun.* **60**:3287–3295.
6. **Andrews, W. H.** 1989. Methods for recovering injured "classical" enteric pathogenic bacteria (*Salmonella, Shigella* and enteropathogenic *Escherichia coli*) from foods, p. 55–113. *In* B. Ray (ed.), *Injured Index and Pathogenic Bacteria: Occurrence and Detection in Foods, Water and Feeds.* CRC Press, Boca Raton, Fla.
7. **Bean, N. H., and P. M. Griffin.** 1990. Foodborne disease outbreaks in the United States, 1973–1987: pathogens, vehicles, and trends. *J. Food Prot.* **53**:804–817.
8. **Bennett, J. V., S. D. Holmberg, M. F. Rogers, and S. L. Solomon.** 1987. Infectious and parasitic diseases, p. 102–114. *In* R. W. Amlet and H. B. Dull (ed.), *Closing the Gap: The Burden of Unnecessary Illness.* Oxford Press, New York.
9. **Bennish, M. L.** 1991. Potentially lethal complications of shigellosis. *Rev. Infect. Dis.* **13**(Suppl. 4):S319–S324.
10. **Bennish, M. L., J. R. Harris, B. J. Wojtynaik, and M. Struelens.** 1990. Death in shigellosis: incidence and risk factors in hospitalized patients. *J. Infect. Dis.* **161**:500–506.
11. **Bennish, M. L., and M. A. Salam.** 1992. Rethinking options for the treatment of shigellosis. *J. Antimicrob. Chemother.* **30**:243–247.
12. **Bernardini, M. L., J. Mounier, H. d'Hauteville, M. Coquis-Rondon, and P. J. Sansonetti.** 1989. Identification of *icsA*, a plasmid locus in *Shigella flexneri* that governs bacterial intra- and intercellular spread through interaction with F-actin. *Proc. Natl. Acad. Sci. USA* **86**:3867–3871.
13. **Brahmbhatt, H. N., A. A. Lindberg, and K. N. Timmis.** 1992. *Shigella* lipopolysaccharide: structure, genetics, and vaccine development. *Curr. Top. Microbiol. Immunol.* **180**: 45–64.
14. **Bunning, V. K., R. B. Raybourne, and D. L. Archer.** 1988. Foodborne enterobacterial pathogens and rheumatoid disease. *J. Appl. Bacteriol. Symp. Suppl.* **65**:87S–107S.
15. **Buysse, J. M., M. M. Venkatesan, J. Mills, and E. V. Oaks.** 1990. Molecular characterization of a transacting, positive effector (*ipaR*) of invasion plasmid antigen synthesis in *Shigella flexneri* serotype 5. *Microb. Pathog.* **8**:197–211.
16. **Centers for Disease Control and Prevention.** 1994. Personnel communication.
17. **Centers for Disease Control and Prevention.** 1994. Outbreak of *Shigella flexneri* 2a infections on a cruise ship. *Morbid. Mortal. Weekly Rep.* **43**:657.
18. **Davis, H., J. P. Taylor, J. N. Perdue, G. N. Stelma, Jr., J. M. Humphreys, Jr., R. Rowntree III, and K. D. Greene.** 1988. A shigellosis outbreak traced to commercially distributed lettuce. *Am. J. Epidemiol.* **128**:1312–1321.
19. **DuPont, H. L.** 1995. *Shigella* species (bacillary dysentery), p. 2033–2039. *In* G. L. Mandell, J. E. Bennett, and R. Dolin (ed.), *Principles and Practice of Infectious Diseases.* Churchill Livingstone Inc., New York.
20. **DuPont, H. L., M. M. Levine, R. B. Hornick, and S. B. Formal.** 1989. Inoculum size in shigellosis and implications for expected mode of transmission. *J. Infect. Dis.* **159**:1126–1128.
21. **Edwards, P. R., and W. H. Ewing.** 1972. Identification of Enterobacteriaceae. Burgess Publishing Co., Minneapolis.
22. **Falkow, S., H. Schneider, L. Baron, and S. B. Formal.** 1963. Virulence of *Escherichia-Shigella* genetic hybrids for the guinea pig. *J. Bacteriol.* **86**:1251–1258.
23. **Fontaine, A., J. Arondel, and P. J. Sansonetti.** 1988. Role of the Shiga toxin in the pathogenesis of bacillary dysentery studied by using a Tox^- mutant of *Shigella dysenteriae* 1. *Infect. Immun.* **56**:3099–3109.
24. **Formal, S. B., E. H. LaBrec, T. H. Kent, and S. Falkow.** 1965. Abortive intestinal infection with an *Escherichia coli-Shigella flexneri* hybrid strain. *J. Bacteriol.* **89**:1374–1382.
25. **Frost, J. A., M. B. McEvoy, C. A. Bentley, Y. Andersson, and B. Rowe.** 1995. An outbreak of *Shigella* sonnei infection associated with consumption of iceberg lettuce. *Emerging Infect. Dis.* **1**:26–29.
26. **Galán, J. E., and P. J. Sansonetti.** 1996. Molecular and cellular bases of *Salmonella* and *Shigella* interactions with host cells, p. 2757–2773. *In* F. C. Neidhardt, R. Curtiss III, J. L. Ingraham, E. C. C. Lin, K. B. Low, B. Magasanik, W. S. Reznikoff, M. Riley, M. Schaechter, and H. E. Umbarger (ed.), *Escherichia coli and Salmonella typhimurium: Cellular and Molecular Biology,* 2nd ed. American Society for Microbiology, Washington, D.C.

27. **Gallegos, M. T., C. Michan, and J. L. Ramos.** 1993. The XylS/AraC family of regulators. *Nucleic Acids Res.* **21:**807–810.

28. **Gessner, B. D., and M. Beller.** 1994. Moose soup shigellosis in Alaska. *West. J. Med.* **160:**430–433.

29. **Goldberg, M. B., O. Barzu, C. Parsot, and P. J. Sansonetti.** 1993. Unipolar localization and ATPase activity of IcsA, a *Shigella flexneri* protein involved in intracellular movement. *J. Bacteriol.* **175:**2189–2196.

30. **Goma Epidemiology Group.** 1995. Public health impact of Rwandan refugee crisis: what happened in Goma, Zaire, in July, 1994? *Lancet* **345:**339–344.

31. **Groisman, E. A., and H. Ochman.** 1993. Cognate gene clusters govern invasion of host epithelial cells by *Salmonella typhimurium* and *Shigella flexneri. EMBO J.* **12:**3779–3787.

32. **Guerrero, L., J. J. Calva, A. L. Morrow, F. R. Velazquez, F. Tuz-Dzib, Y. Lopez-Vidal, H. Ortega, H. Arroyo, T. G. Cleary, L. K. Pickering, and G. M. Ruiz-Palacios.** 1994. Asymptomatic *Shigella* infections in a cohort of Mexican children younger than two years of age. *Pediatr. Infect. Dis. J.* **13:**597–602.

33. **Hall, P. A.** 1994. Scope for rapid microbiological methods in modern food production, p. 255–267. *In* P. Patel (ed.), *Rapid Analysis Techniques in Food Microbiology.* Blackie Academic & Professional, New York.

34. **Haltalin, K., J. Nelson, and R. Ring.** 1967. Double-blind treatment study of shigellosis comparing ampicillin, sulfadiazone and placebo. *J. Pediatr.* **70:**970–981.

35. **Hedberg, C. W., W. C. Levine, K. E. White, R. H. Carlson, D. K. Winsor, D. N. Cameron, K. L. MacDonald, and M. T. Osterholm.** 1992. An international foodborne outbreak of shigellosis associated with a commercial airline. *JAMA* **268:**3208–3212.

36. **High, N., J. Mounier, M. C. Prevost, and P. J. Sansonetti.** 1992. IpaB of *Shigella flexneri* causes entry into epithelial cells and escape from the phagocytic vacuole. *EMBO J.* **11:**1991–1999.

37. **Hossain, M. A., K. Z. Hasan, and M. J. Albert.** 1994. *Shigella* carriers among non-diarrhoeal children in an endemic area of shigellosis in Bangladesh. *Trop. Geogr. Med.* **46:**40–42.

38. **Hyams, K. C., A. L. Bourgeois, B. R. Merrell, P. Rozmajzl, J. Escamilla, S. A. Thornton, G. M. Wasserman, A. Burke, P. Echeverria, K. Y. Green, A. Z. Kapikian, and J. N. Woody.** 1991. Diarrheal disease during Operation Desert Shield. *N. Engl. J. Med.* **325:**1423–1428.

39. **Kapperud, G., L. M. Rorvik, V. Hasseltvedt, E. A. Hoiby, B. G. Iversen, K. Staveland, G. Johnsen, J. Leitao, H. Herikstad, Y. Andersson, G. Langeland, B. Gondrosen, and J. Lassen.** 1995. Outbreak of *Shigella sonnei* infection traced to imported iceberg lettuce. *J. Clin. Microbiol.* **33:**609–614.

40. **Karmali, M. A., M. Petric, C. Lim, P. C. Fleming, G. S. Arbus, and H. Lior.** 1985. The association between idiopathic hemolytic syndrome and infection by Verotoxin-producing *Escherichia coli. J. Infect. Dis.* **151:**775–782.

41. **Kennedy, F. M., J. Astbury, J. R. Needham, and T. Cheasty.** 1993. Shigellosis due to occupational contact with non-human primates. *Epidemiol. Infect.* **110:**247–257.

42. **Kinsey, M. D., S. B. Formal, G. J. Dammin, and R. A. Giannella.** 1976. Fluid and electrolyte transport in rhesus monkeys challenged intracecally with *Shigella flexneri* 2a. *Infect. Immun.* **14:**368–371.

43. **LaBrec, E. H., H. Schneider, T. J. Magnani, and S. B. Formal.** 1964. Epithelial cell penetration as an essential step in the pathogenesis of bacillary dysentery. *J. Bacteriol.* **88:**1503–1518.

44. **Lawlor, K. M., P. A. Daskeleros, R. E. Robinson, and S. M. Payne.** 1987. Virulence of iron-transport mutants of *Shigella flexneri* and utilization of host iron compounds. *Infect. Immun.* **55:**594–599.

45. **Lee, L. A., S. M. Ostroff, H. B. McGee, D. R. Johnson, F. P. Downes, D. N. Cameron, N. H. Bean, and P. M. Griffin.** 1991. An outbreak of shigellosis at an outdoor music festival. *Am. J. Epidemiol.* **133:**608–615.

46. **Levine, M. M., and O. S. Levine.** 1994. Changes in human ecology and behavior in relation to the emergence of diarrheal diseases, including cholera. *Proc. Natl. Acad. Sci. USA* **91:**2390–2394.

47. **Lew, J. F., D. L. Swerdlow, M. E. Dance, P. M. Griffin, C. A. Bopp, M. J. Gillenwater, T. Mercatante, and R. I. Glass.** 1991. An outbreak of shigellosis aboard a cruise ship caused by a multiple-antibiotic-resistant strain of *Shigella flexneri. Am. J. Epidemiol.* **134:**413–420.

48. **Lopez, E. L., M. Diaz, S. Grinstein, S. Dovoto, F. Mendila-Harzu, B. E. Murray, S. Ashkenazi, E. Rubeglio, M. Woloj, M. Vasquez, M. Turco, L. K. Pickering, and T. G. Cleary.** 1989. Hemolytic uremic syndrome and diarrhea in Argentine children: the role of Shiga-like toxins. *J. Infect. Dis.* **160:**469–475.

49. **Makino, S., C. Sasakawa, K. Kamata, T. Kurata, and M. Yoshikawa.** 1986. A genetic determinant required for continuous reinfection of adjacent cells on large plasmid in *Shigella flexneri* 2a. *Cell* **46:**551–555.

50. **Maurelli, A. T.** 1989. Temperature regulation of virulence genes in pathogenic bacteria: a general strategy for human pathogens? *Microb. Pathog.* **7:**1–10.

51. **Maurelli, A. T., B. Baudry, H. d'Hauteville, T. L. Hale, and P. J. Sansonetti.** 1985. Cloning of virulence plasmid DNA sequences involved in invasion of HeLa cells by *Shigella flexneri. Infect. Immun.* **49:**164–171.

52. **Maurelli, A. T., B. Blackmon, and R. Curtiss III.** 1984. Temperature-dependent expression of virulence genes in *Shigella* species. *Infect. Immun.* **43:**195–201.

53. **Maurelli, A. T., A. E. Hromockyj, and M. L. Bernardini.** 1992. Environmental regulation of *Shigella* virulence. *Curr. Top. Microbiol. Immunol.* **180:**95–116.

54. **Maurelli, A. T., and K. A. Lampel.** 1994. *Shigella,* p. 319–343. *In* J. R. Gorman and K. D. Murrell (ed.), *Foodborne Disease Handbook,* vol. I. Marcel Dekker Publishers, Inc., New York.

55. **Maurelli, A. T., and P. J. Sansonetti.** 1988. Identification of a chromosomal gene controlling temperature regulated expression of *Shigella* virulence. *Proc. Natl. Acad. Sci. USA* **85:**2820–2824.

56. **Ménard, R., P. J. Sansonetti, and C. Parsot.** 1993. Nonpolar mutagenesis of the *ipa* genes defines IpaB, IpaC, and IpaD as effectors of *Shigella flexneri* entry into epithelial cells. *J. Bacteriol.* **175:**5899–5906.

57. **Ménard, R., P. J. Sansonetti, and C. Parsot.** 1994. The secretion of the *Shigella flexneri* Ipa invasins is induced by the epithelial cell and controlled by IpaB and IpaD. *EMBO J.* **13:**5293–5302.
58. **Ménard, R., P. J. Sansonetti, C. Parsot, and T. Vasselon.** 1994. The IpaB and IpaC invasins of *Shigella flexneri* associate in the extracellular medium and are partitioned in the cytoplasm by a specific chaperon. *Cell* **76:**829–839.
59. **Nassif, X., M. C. Mazert, J. Mounier, and P. J. Sansonetti.** 1987. Evaluation with an *iuc*::Tn*10* mutant of the role of aerobactin production in the virulence of *Shigella flexneri*. *Infect. Immun.* **55:**1963–1969.
60. **Oaks, E. V., T. L. Hale, and S. B. Formal.** 1986. Serum immune response to *Shigella* protein antigens in rhesus monkeys and humans infected with *Shigella* spp. *Infect. Immun.* **53:**57–63.
61. **O'Brien, A. D., and R. K. Holmes.** 1996. Protein toxins of *Escherichia coli* and *Salmonella*, p. 2788–2802. *In* F. C. Neidhardt, R. Curtiss III, J. L. Ingraham, E. C. C. Lin, K. B. Low, B. Magasanik, W. S. Reznikoff, M. Riley, M. Schaechter, and H. E. Umbarger (ed.), *Escherichia coli and Salmonella typhimurium: Cellular and Molecular Biology,* 2nd ed. American Society for Microbiology, Washington, D.C.
62. **Ochman, H., T. S. Whittam, D. A. Caugant, and R. K. Selander.** 1983. Enzyme polymorphism and genetic population structure in *Escherichia coli* and *Shigella*. *J. Gen. Microbiol.* **129:**2715–2726.
63. **O'Connell, C. M. C., R. C. Sandlin, and A. T. Maurelli.** 1995. Signal transduction and virulence gene regulation in *Shigella* spp.: temperature and (maybe) a whole lot more, p. 111–127. *In* R. Rappuoli (ed.), *Signal Transduction and Bacterial Virulence.* R. G. Landes Company, Austin, Tex.
64. **Pickering, L. K., A. V. Bartlett, and W. E. Woodward.** 1986. Acute infectious diarrhea among children in day-care: epidemiology and control. *Rev. Infect. Dis.* **8:**539–547.
65. **Raghupathy, P., A. Date, J. C. M. Shastry, A. Sudarsanam, and M. Jadhav.** 1978. Haemolytic-uraemic syndrome complicating shigella dysentery in south Indian children. *Br. Med. J.* **1:**1518–1521.
66. **Roberts, T.** 1989. Human illness costs of foodborne bacteria. *Am. J. Agric. Econ.* **71:**468–474.
67. **Sakai, T., C. Sasakawa, and M. Yoshikawa.** 1988. Expression of four virulence antigens of *Shigella flexneri* is positively regulated at the transcriptional level by the 30 kilodalton *virF* protein. *Mol. Microbiol.* **2:**589–597.
68. **Sandlin, R. C., K. A. Lampel, S. P. Keasler, M. B. Goldberg, A. L. Stolzer, and A. T. Maurelli.** 1995. Avirulence of rough mutants of *Shigella flexneri:* requirement of O antigen for correct unipolar localization of IcsA in bacterial outer membrane. *Infect. Immun.* **63:**229–237.
69. **Sansonetti, P. J.** 1992. Molecular and cellular biology of *Shigella flexneri* invasiveness: from cell assay systems to shigellosis. *Curr. Top. Microbiol. Immun.* **180:**1–19.
70. **Sansonetti, P. J., H. d'Hauteville, C. Ecobichon, and C. Pourcel.** 1983. Molecular comparison of virulence plasmids in *Shigella* and enteroinvasive *Escherichia coli*. *Ann. Microbiol. (Inst. Pasteur)* **134A:**295–318.
71. **Sansonetti, P. J., T. L. Hale, and E. V. Oaks.** 1985. Genetics of virulence in enteroinvasive *Escherichia coli,* p. 74–77. *In* D. Schlessinger (ed.), *Microbiology—1985.* American Society for Microbiology, Washington, D.C.
72. **Sansonetti, P. J., D. J. Kopecko, and S. B. Formal.** 1981. *Shigella sonnei* plasmids: evidence that a large plasmid is necessary for virulence. *Infect. Immun.* **34:**75–83.
73. **Sansonetti, P. J., D. J. Kopecko, and S. B. Formal.** 1982. Involvement of a plasmid in the invasive ability of *Shigella flexneri*. *Infect. Immun.* **35:**852–860.
74. **Sasakawa, C., K. Komatsu, T. Tobe, T. Suzuki, and M. Yoshikawa.** 1993. Eight genes in region 5 that form an operon are essential for invasion of epithelial cells by *Shigella flexneri* 2a. *J. Bacteriol.* **175:**2334–2346.
75. **Sharp, T. W., S. A. Thornton, M. R. Wallace, R. F. Defraites, J. L. Sanchez, R. A. Batchelor, P. J. Rozmajzl, R. K. Hanson, P. Echeverria, A. Z. Kapikian, X. J. Xiang, M. K. Estes, and J. P. Burans.** 1995. Diarrheal disease among military personnel during Operation Restore Hope, Somalia, 1992–1993. *Am. J. Trop. Med. Hyg.* **52:**188–193.
76. **Simon, D. G., R. A. Kaslow, J. Rosenbaum, R. L. Kaye, and A. Calin.** 1981. Reiter's syndrome following epidemic shigellosis. *J. Rheumatol.* **8:**969–973.
77. **Smith, J. L.** 1987. *Shigella* as a foodborne pathogen. *J. Food Prot.* **50:**788–801.
78. **Snyder, O. P.** 1992. HACCP—an industry food safety self-control program. Part IV. *Dairy Food Environ. Sanit.* **12:**230–232.
79. **Tesh, V. L., and A. D. O'Brien.** 1991. The pathogenic mechanisms of Shiga toxin and the Shiga-like toxins. *Mol. Microbiol.* **5:**1817–1822.
80. **Tobe, T., S. Nagai, N. Okada, B. Adler, M. Yoshikawa, and C. Sasakawa.** 1991. Temperature-regulated expression of invasion genes in *Shigella flexneri* is controlled through the transcriptional activation of the *virB* gene on the large plasmid. *Mol. Microbiol.* **5:**887–893.
81. **Tobe, T., M. Yoshikawa, T. Mizuno, and C. Sasakawa.** 1993. Transcriptional control of the invasion regulatory gene *virB* of *Shigella flexneri:* activation by VirF and repression by H-NS. *J. Bacteriol.* **175:**6142–6149.
82. **Van Gijsegem, F., S. Genin, and C. Boucher.** 1993. Conservation of secretion pathways for pathogenicity determinants of plant and animal bacteria. *Trends Microbiol.* **1:**175–180.
83. **Venkatesan, M. M., J. M. Buysse, and E. V. Oaks.** 1992. Surface presentation of *Shigella flexneri* invasion plasmid antigens requires the products of the *spa* locus. *J. Bacteriol.* **174:**1990–2001.
84. **Weissman, J. B., A. Schmerler, P. Weiler, G. Filice, N. Godby, and I. Hansen.** 1974. The role of preschool children and day-care centers in the spread of shigellosis in urban communities. *J. Pediatr.* **84:**797–802.
85. **Wharton, M., R. A. Spiegel, J. M. Horan, R. V. Tauxe, J. G. Wells, N. Barg, J. Herndon, R. A. Meriwether, J. N. MacCormack, and R. H. Levine.** 1990. A large outbreak of antibiotic-resistant shigellosis at a mass gathering. *J. Infect. Dis.* **162:**1324–1328.
86. **Yagupsky, P., M. Loeffelholz, K. Bell, and M. A. Menegus.** 1991. Use of multiple markers for investigation of an epidemic of *Shigella sonnei* infections in Monroe County, New York. *J. Clin. Microbiol.* **29:**2850–2855.
87. **Zychlinsky, A., B. Kenny,** R. Ménard, M. C. Prevost, I. B. Holland, and P. J. Sansonetti. 1994. IpaB mediates macrophage apoptosis induced by *Shigella flexneri*. *Mol. Microbiol.* **11:**619–627.

James D. Oliver
James B. Kaper

13

Vibrio Species

Whereas the 8th edition of *Bergey's Manual* listed five *Vibrio* species, with two recognized as human pathogens, over 20 species have now been described, including at least 12 capable of causing infection in humans. A number of excellent reviews have appeared on the pathogenic vibrios (8, 47, 51, 54, 73, 81, 89, 124, 127, 131), although with the exception of *V. cholerae* and *V. parahaemolyticus,* little is known of the virulence mechanisms they employ. Of the 12 pathogens, 8 have been shown to be directly food-associated, and these are the subject of this review. (*V. carchariae* and *V. damsela* infections appear to result solely from wound infections, and the route of infection for *V. metschnikovii* and *V. cincinnatiensis* is unclear.)

INCIDENCE OF VIBRIOS IN SEAFOOD

One of the most consistent aspects of vibrio infections is a recent history of seafood consumption. Vibrios, which are generally the predominant bacterial genus in estuarine waters, are found associated with a great variety of seafoods. To cite a few recent studies, Prasad and Rao (100) found that 30% of the 129 samples of fresh, frozen, and iced prawn they examined carried vibrios, including *V. parahaemolyticus, V. vulnificus, V. metschnikovii, V. cholerae* non-O1, and *V. fluvialis.* Wong et al. (132, 134) found that 13 to 36% of the shrimp and fish samples they tested harbored pathogenic vibrios, and they also reported high numbers in freshwater clams. Lowry et al. (71) found 100% of the raw oysters they examined to harbor *V. parahaemolyticus,* and 67% contained *V. vulnificus.* Interestingly, 50% of the cooked oysters tested also contained *V. parahaemolyticus* cells, and 25% contained *V. vulnificus.* A recent survey of frozen raw shrimp imported from Mexico, China, and Ecuador found over 63% to harbor *Vibrio* species, including *V. vulnificus* and *V. parahaemolyticus* (5). In a study of the distribution of vibrios in oysters (*Crassostrea virginica*) originating from the coast of Brazil, Matté et al. (75) found, in order of incidence, *V. alginolyticus* (81%), *V. parahaemolyticus* (77%), *V. cholerae* non-O1 (31%), *V. fluvialis* (27%), *V. furnissii* (19%), *V. mimicus* (12%), and *V. vulnificus* (12%).

ISOLATION

Various enrichment broths, frequently coupled with thiosulfate-citrate-bile salts-sucrose (TCBS) agar or other plating media, have been described for the isolation of vibrios. By employing sucrose as a differentiating trait, the 12 human pathogenic vibrios can be separated on TCBS into seven which are generally sucrose positive (*V. cholerae, V. metschnikovii, V. cincinnatiensis, V. fluvialis, V. furnissii, V. alginolyticus,* and *V. carchariae*) and five which are generally sucrose negative (*V. mimicus, V. hollisae, V. damsela, V. parahaemolyticus,* and *V. vulnificus*). Most of the vibrios grow well on TCBS agar, although *V. hollisae* exhibits very poor to no growth on this medium, which is also the case for *V. damsela* when incubated at 37°C (30). Further, the determination of oxidase activity, a crucial test for distinguishing vibrios from the *Enterobacteriaceae*, may give erroneous results when colonies are taken from TCBS agar. Therefore, sucrose-positive colonies on TCBS should be subcultured by heavy inoculation onto a nonselective medium, such as blood agar, and allowed to grow for 5 to 8 h before being tested for oxidase.

IDENTIFICATION

The taxonomy of the vibrios has been described by many authors, and we include here tables which provide the salient differentiating features of the 12 vibrios currently recognized as human pathogens (Table 13.1) and those tests which are of special value in differentiating these pathogens (Table 13.2). Comments about the taxonomy of individual species are noted as each is described. However, while determination of phenotypic traits for the species identification of vibrios remains routine, identifications based on these classic methods have always been problematic. Some researchers have developed immunological assays in an attempt to overcome this problem, while molecular techniques to identify the presence of vibrios in foods are becoming more common and are proving to be a very powerful adjunct to more traditional taxonomic methods.

EPIDEMIOLOGY

The levels of most of the vibrios in both surface waters and shellfish show definite seasonal correlation, generally being greater during the warm weather months between April and October. Similarly, vibrios are more commonly isolated from the warmer waters of the Gulf and East coasts than those of the West and Pacific Northwest. Seasonality is most notable for *V. vulnificus* and *V. parahaemolyticus* infections, whereas those of some vibrios, such as *V. fluvialis*, occur throughout the year.

While there exists considerable variation in the severity of the various vibrio diseases, and indeed even with infections caused by the same *Vibrio* species, the most severely ill patients generally suffer preexisting underlying illnesses, with chronic liver disease being one of the most common. An exception to this generalization is *V. cholerae* O1/O139, which can readily cause disease in noncompromised individuals. In almost all cases, a recent history of seafood consumption, especially raw oysters, is noted (25). *V. cholerae* O1/O139 is also exceptional in having a broader range of vehicles of infection, although seafood is important in the transmission of many cases of cholera. In a report by Bonner et al. (9), covering a 10-year period of vibrio isolates recovered in Alabama, 87% of the patients indicated a recent history of seawater-associated activity. Lowry et al. (71) found that all 51 persons with diarrhea examined who had a *Vibrio* species present in a stool specimen had eaten raw or cooked seafood. Unfortunately, because vibrios are part of the normal estuarine microflora and not a result of fecal contamination, vibrio infections will not likely be controlled through shellfish sanitation programs (8, 131). It is thus essential that raw seafood be adequately refrigerated or iced to prevent significant bacterial growth.

INCIDENCE OF VIBRIO INFECTIONS

The routine use of TCBS agar for screening stool samples for vibrios has generally indicated a low incidence of these infections. Hoge et al. (42), for example, employing this medium during a 15-year survey of all stool specimens obtained in a hospital adjacent to the Chesapeake Bay of Maryland, obtained only 40 *Vibrio* isolates from 32 patients during this time, for an overall incidence of 1.6 per 100,000 patients per year. Similarly, Magalhaes et al. (74) examined 3,250 diarrheal stools received between 1989 and 1991 at a clinical laboratory in Brazil and, despite using enrichment in alkaline peptone water supplemented with 2% NaCl prior to culture on TCBS, isolated vibrios from only 55 (1.7%) of the samples. Bonner et al. (9), who concluded that vibrios were not a major cause of infectious diarrhea, found only one case of vibrio gastroenteritis during the four years TCBS was used in a Gulf Coast survey. However, the routine use of TCBS for screening of stool and wound samples may be warranted in some instances. Desenclos et al. (25) reported an overall annual incidence rate for any vibrio illness to be 11.3 per million for raw-oyster-eating populations in Florida, but only 2.2 for persons who did not eat raw oysters. These rates were greatly affected by several host factors. For example, the annual incidence rate of vibrio illness for raw-oyster eaters with liver disease was 95.4, compared to a rate of 9.2 for raw-oyster eaters without liver disease. Similarly, Levine et al. (68) suggested that, while the reported incidence rate of infections yielding *Vibrio* species is low (0.7/100,000 population) compared to other enteric pathogens (e.g., 7.7/100,000 population for *Shigella* species), the high proportion (45%) of patients with vibrio gastroenteritis who are hospitalized suggests that many milder infections may occur which are not reported.

Table 13.1 Biochemical and other characteristics of 12 *Vibrio* species found in human clinical specimens[a]

Test	% Positive[b]											
	V. cholerae	*V. minicus*	*V. metschnikovii*	*V. cincinnatiensis*	*V. hollisae*	*V. damsela*	*V. fluvialis*	*V. furnissii*	*V. alginolyticus*	*V. parahaemolyticus*	*V. vulnificus*	*V. carchariae*
Indole production (HIB, 1% NaCl)	99	98	20	8	97	0	13	11	85	98	97	100
Methyl red (1% NaCl)	99	99	96	93	0	100	96	100	75	80	80	100
Voges-Proskauer[c] (1% NaCl, Barritt)	75	9	96	0	0	95	0	0	95	0	0	50
Citrate, Simmons	97	99	75	21	0	0	93	100	1	3	75	0
H_2S on TSI	0	0	0	0	0	0	0	0	0	0	0	0
Urea hydrolysis	0	1	0	0	0	0	0	0	0	15	1	0
Phenylalanine deaminase	0	0	0	0	0	0	0	0	1	1	35	NG
Arginine, Moeller[c] (1% NaCl)	0	0	60	0	0	95	93	100	0	0	0	0
Lysine, Moeller[c] (1% NaCl)	99	100	35	57	0	50	0	0	99	100	99	100
Ornithine, Moeller[c] (1% NaCl)	99	99	0	0	0	0	0	0	50	95	55	0
Motility (36°C)	99	98	74	86	0	25	70	89	99	99	99	0
Gelatin hydrolysis (1% NaCl, 22°C)	90	65	65	0	0	6	85	86	90	95	75	0
KCN test (% that grow)	10	2	0	0	0	5	65	89	15	20	1	0
Malonate utilization	1	0	0	0	0	0	0	11	0	0	0	0
D-Glucose, acid production[c]	100	100	100	100	100	100	100	100	100	100	100	50
D-Glucose, gas production[c]	0	0	0	0	0	10	0	100	0	0	0	0
Acid production from:												
D-Adonitol	0	0	0	0	0	0	0	0	1	0	0	0
L-Arabinose[c]	0	1	0	100	97	0	93	100	1	80	0	0
D-Arabitol[c]	0	0	0	0	0	0	65	89	0	0	0	0
Cellobiose[c]	8	0	9	100	0	0	30	11	3	5	99	50
Dulcitol	0	0	0	0	0	0	0	0	0	3	0	0
Erythritol	0	0	0	0	0	0	0	0	0	0	0	0
D-Galactose	90	82	45	100	100	90	96	100	20	92	96	0
Glycerol	30	13	100	100	0	0	7	55	80	50	1	0
myo-Inositol	0	0	40	100	0	0	0	0	0	0	0	0
Lactose[c]	7	21	50	0	0	0	3	0	0	1	85	0
Maltose[c]	99	99	100	100	0	100	100	100	100	99	100	100
D-Mannitol[c]	99	99	96	100	0	0	97	100	100	100	45	50

D-Mannose	78	99	100	100	100	100	100	100	99	100	98	50
Melibiose	1	0	0	7	0	0	3	11	1	1	40	0
α-Methyl-D-glucoside	0	0	25	57	0	5	0	0	1	0	0	0
Raffinose	0	0	0	0	0	0	0	11	0	0	0	0
D-Rhamnose	0	0	0	0	0	0	0	45	0	1	0	0
Salicin[c]	1	0	9	100	0	0	0	0	4	1	95	0
D-Sorbitol	1	0	45	0	0	0	3	0	1	1	0	0
Sucrose[d]	100	0	100	100	0	5	100	100	99	1	15	50
Trehalose	99	94	100	100	0	86	100	100	100	99	100	50
D-Xylose	0	0	0	43	0	0	0	0	0	0	0	0
Mucate, acid production	1	0	0	0	0	0	0	0	0	0	0	0
Tartrate, Jordan	75	12	35	0	65	0	35	22	95	93	84	50
Esculin hydrolysis	0	0	60	0	0	0	8	0	3	1	40	0
Acetate utilization	92	78	25	14	0	0	70	65	0	1	7	0
Nitrate→nitrate[c]	99	100	0	100	100	100	100	100	100	100	100	100
Oxidase[c]	100	100	0	100	100	95	100	100	100	100	100	100
DNase, 25°C	93	55	50	79	0	75	100	100	95	92	50	100
Lipase (corn oil)[c]	92	17	100	36	0	0	90	89	85	90	92	0
ONPG test[c]	94	90	50	86	0	0	40	35	0	5	75	0
Yellow pigment at 25°C	0	0	0	0	0	0	0	0	0	0	0	0
Tyrosine clearing	13	30	5	0	3	0	65	45	70	77	75	0
Growth in nutrient broth with:												
0% NaCl[d]	100	100	0	0	0	0	0	0	0	0	0	0
1% NaCl[d]	100	100	100	100	99	100	99	99	99	100	99	100
6% NaCl[d]	53	49	78	100	83	95	96	100	100	99	65	100
8% NaCl[d]	1	0	44	62	0	0	71	78	94	80	0	0
10% NaCl[d]	0	0	4	0	0	0	4	0	69	2	0	0
12% NaCl[d]	0	0	0	0	0	0	0	0	17	1	0	0
Swarming (marine agar, 25°C)	–	–	–	+	–	–	–	–	+	+	–	100
String test	100	100	100	80	100	80	100	100	91	64	100	100
O/129, zone of inhibition[c]	99	95	90	25	40	90	31	0	19	20	98	100
Polymyxin B, zone of inhibition	22	88	100	92	100	85	100	89	63	54	3	100

[a]HIB, heart infusion broth; 1% NaCl, 1% NaCl added to the standard medium to enhance growth; TSI, triple sugar iron agar. From reference 75a.

[b]After 48 h of incubation at 36°C (unless other conditions are indicated). Most positive reactions occur during the first 24 h. NG, no growth (probably because NaCl concentration is too low); +, most strains (generally about 90 to 100% positive); –, most strains negative (generally about 0 to 10% positive).

[c]Test is recommended as part of the routine set for *Vibrio* identification.

[d]Disk potency, 150 μg.

Table 13.2 Eight key differential tests to divide the 12 clinically significant *Vibrio* species into six groups[a]

Test	Reactions of the species in:											
	Group 1					Group 5			Group 6			
	V. cholerae	*V. mimicus*	Group 2: *V. metschnikovii*	Group 3: *V. cincinnatiensis*	Group 4: *V. hollisae*	*V. damsela*	*V. fluvialis*	*V. furnissii*	*V. alginolyticus*	*V. parahaemolyticus*	*V. vulnificus*	*V. carchariae*
Growth in nutrient broth												
With no NaCl added	+	+	–	–	–	–	–	–	–	–	–	–
With 1% NaCl added	+	+	+	+	+	+	+	+	+	+	+	+
Oxidase			–	+	+	+	+	+	+	+	+	+
Nitrate → nitrite		–	–	+	+	+	+	+	+	+	+	+
myo-Inositol fermentation	–	–	–	+	–	–	–	–	–	–	–	–
Arginine dihydrolase					–	+	+	+	–	–	–	–
Lysine decarboxylase					–				+	+	+	+
Ornithine decarboxylase					–							

[a]All data are for reactions within 2 days at 35 to 37°C unless otherwise specified. Symbols: +, most strains (generally about 90 to 100%) positive; –, most strains negative (generally about 0 to 10% positive). Key test results are boxed. From reference 75a.

These authors also noted, as an additional indication of the severity of these diseases, that 86% of patients with *V. fluvialis* and 35% of those with *V. parahaemolyticus* had bloody stools. In a prospective study, vibrios were isolated from 22% of the 67 stool specimens submitted by attendees who had diarrheal illness during a conference held in New Orleans (71).

SUSCEPTIBILITY TO PHYSICAL AND CHEMICAL TREATMENTS

With the exception of *V. cholerae* and *V. parahaemolyticus*, relatively little is known of the susceptibility of any of the vibrios to various food preservation methods. Summarized here are some of the studies relevant to this point; when detailed studies exist for the various vibrios, they are noted as each is described.

Cold

While reports generally have indicated the vibrios to be sensitive to cold, seafoods have also been reported to be protective for vibrios at refrigeration temperatures. Wong et al. (132) isolated several psychrotrophic strains of *V. mimicus*, *V. fluvialis*, and *V. parahaemolyticus* from frozen seafoods and found these to survive well at 10, 4, and −30°C. *V. parahaemolyticus* has been reported to undergo an initial rapid drop in survival (ca. 99%) when incubated on whole shrimp at 3, 7, 10, or −18°C, although survivors remained at the end of the 8-day study. *V. parahaemolyticus* has also been observed to survive storage in shellstock oysters for at least 3 weeks at 4°C and subsequently to multiply after incubation at 35°C for 2 to 3 days. Similarly, cells of *V. parahaemolyticus* were reduced in cooked fish mince and surimi at 5°C for 48 h, but growth occurred when the product was held at 25°C. At commonly employed refrigeration temperatures of 4 and 8°C, Hood et al. (44, 45) found *V. vulnificus* to increase in numbers, while at a nonrefrigeration temperature (20°C) both *V. parahaemolyticus* and *V. vulnificus* showed large increases. Similar observations have been reported for *V. vulnificus* in oysters stored at 18°C or higher. Such studies clearly suggest that the naturally occurring vibrios are able to multiply in unchilled shellstock oysters. Similarly, studies involving temperature abuse of octopus, cooked shrimp, and crabmeat all have documented growth of *V. parahaemolyticus* to very large numbers when held for even short periods of time under improper refrigeration.

The persistence of *V. vulnificus* in oysters following freezing and storage at −20°C, with or without vacuum packaging, was found to be dependent on the length of frozen storage time for cells packaged without vacuum, with a decrease from ca. 10^5 to ca. 10^1 CFU/g. Vacuum-packaged samples showed significantly lower concentrations of *V. vulnificus* over a 70-day study period as compared to normal-packaged samples (98).

Heat

All of the vibrios are sensitive to heat, although a wide range of thermal inactivation rates have been reported. Inactivation times of 15 to 30 min at 60°C and 5 min at 100°C seem typical, although heat inactivation has been reported to be affected by NaCl levels. Heating of *V. parahaemolyticus* cells at 60, 80, or 100°C for 1 min is said to be lethal to small (5×10^2) populations, although some cells survived heating at 60°C and even 80°C for 15 min when populations of 2×10^5 were used. Cells did not survive boiling for 1 min, however (130). Cook and Ruple (22) reported that decimal reduction times at 47°C averaged 78 s for the 52 strains of *V. vulnificus* examined. Thorough heating of shellfish to provide an internal temperature of at least 60°C for several minutes appears to be sufficient to eliminate the pathogenic vibrios (131). Cook and Ruple (22) found that heating oysters for 10 min in water at 50°C was sufficient to reduce the *V. vulnificus* populations to nondetectable levels.

Irradiation

Doses of 3 kGy of gamma irradiation have been reported to be required for the elimination of vibrios from frozen shrimp (102). High levels (>50 krad) of ^{60}Co have been used to eliminate *V. cholerae* from both fresh and frozen frog legs (109). The use of ionizing radiation for reducing levels of *V. vulnificus* in shellstock oysters has also been studied, with a dose of 1 kGy resulting in a decrease in over 5 logs, with no mortality in the oysters. Higher doses, e.g., 1.5 kGy, resulted in a total loss of *V. vulnificus* but increased oyster mortalities to up to 16% (27).

Polyphosphates

Heated or unheated polyphosphates have been shown to be lethal or highly inhibitory to a variety of foodborne pathogens. However, Wong et al. (132, 133) have reported polyphosphate inhibition of vibrios to vary depending on the species and phosphate type. They found *V. parahaemolyticus* to be significantly protected by heated metaphosphate at 4°C, while pyrophosphate was inhibitory at −30°C. On the other hand, the survival of a psychrotrophic strain of *V. cholerae* was enhanced by heated pyro- and metaphosphate at both 4 and −30°C.

We have observed (unpublished data) that 1% tripolyphosphate has no lethal effect on *V. vulnificus.*

Miscellaneous Inhibitors

The bactericidal effects against *V. parahaemolyticus* of a large variety of dried spices, the oils of several herbs, tomato sauce, and several organic acids have been reported, and many of these substances have been found to be highly toxic at low levels (105). Similarly, *V. vulnificus* has been reported to be totally inactivated when exposed to several fruit or vegetable juices and to a variety of spice extracts. *V. parahaemolyticus* is highly sensitive to as little as 50 ppm butylated hydroxyanisole (BHA) and is inhibited by 0.1% sorbic acid. Of 10 "generally recognized as safe" (GRAS) compounds tested against *V. vulnificus,* only lactic acid and diacetyl were found to be inhibitory, with only diacetyl having an effect on these cells when present in shellstock oysters (117).

Studies on the effect of pH on viability of vibrios indicate that they are highly acid sensitive, although growth in media as low as pH 4.8 has been reported for *V. parahaemolyticus.*

Depuration

Depuration, wherein filter-feeding bivalves are allowed to purify themselves through the pumping of bacteria-free water through their tissues, is of considerable value in removing such contaminating bacteria as *Salmonella* spp. and *Escherichia coli* (12, 103). However, the many studies that have now accumulated on the depuration of oysters and clams clearly show that this method, by itself, offers little hope of significantly reducing the naturally occurring *Vibrio* microflora present in these animals (50, 120). In contrast, oysters artificially infected with vibrios in the laboratory are able to effectively eliminate these added bacteria through depuration (29, 36, 106). This suggests that the normal gut microflora of molluscan shellfish, which is likely obtained at a very early (larval?) stage, is firmly attached to the gut wall and resistant to depuration.

VIBRIO CHOLERAE

V. cholerae O1 is the causative agent of cholera, one of the few foodborne diseases with epidemic and pandemic potential. *V. cholerae* is well defined on the basis of biochemical tests and DNA homology studies but, as recently reviewed (54), this species is not homogeneous with regard to pathogenic potential. Specifically, important distinctions within the species are made on the basis of production of cholera enterotoxin (cholera toxin, or CT), serogroup, and potential for epidemic spread. Until recently, the public health distinction was simple; that is, *V. cholerae* strains of the O1 serogroup which produced CT were associated with epidemic cholera, and all other members of the species either were nonpathogenic or were only occasional pathogens. However, with the recent epidemic of cholera due to strains of the O139 serogroup (see below), such previous distinctions are no longer valid. Two serogroups, O1 and O139, have been associated with epidemic disease, but there are also strains of these serogroups which do not produce CT, do not cause cholera, and are not involved in human disease. Conversely, there are occasional strains of serogroups other than O1 or O139 that are clearly pathogenic, either by the production of CT or by other virulence factors (see below); however, none of these other serogroups has caused large epidemics or pandemics. Therefore, in assessing the public health significance of an isolate of *V. cholerae,* there are two critical properties to be determined beyond the biochemical identification of the species *V. cholerae.* The first of these properties is production of CT, which is the toxin that is responsible for severe, cholera-like disease in epidemic and sporadic forms. The second property is possession of the O1 or O139 antigen, which, since the actual determinant of epidemic or pandemic potential is not known, is at least a marker of such potential. The subject of cholera has recently been reviewed by one of us (J.B.K.), and much of this section is drawn from that review (54). Due to space constraints, most primary references for *V. cholerae* O1/O139 will not be repeated here but can be found in the recent review. Furthermore, unless otherwise stated, all information under this species will refer to *V. cholerae* O1 or O139 strains capable of causing cholera.

Classification

Serogroups

V. cholerae O1

V. cholerae strains of the O1 serogroup that produce CT have long been associated with epidemic and pandemic cholera. Strains isolated from environmental samples in nonepidemic areas are usually CT negative and are considered to be nonpathogenic, based on volunteer studies (see Reservoirs, below). However, CT-negative *V. cholerae* O1 strains have been isolated from occasional cases of diarrhea or extraintestinal infections. This serogroup can be further subdivided into serotypes of the O1 serogroup called Ogawa and Inaba. *V. cholerae* O1 can also be divided into two biotypes, classical and El Tor, which differ in several characteristics. The El Tor biotype is currently the most important biotype; for the past 20 years, occasional strains of the classical biotype have been isolated only in Bangladesh.

V. cholerae Non-O1/Non-O139

In recent years, until the emergence of the O139 serogroup, all isolates that were identified as *V. cholerae* on the basis of biochemical tests but that were negative for the O1 serogroup were referred to as "non-O1 *V. cholerae.*" In earlier years, the non-O1 *V. cholerae* were referred to as NCV (non-cholera vibrios) or NAG (non-agglutinable) vibrios. In view of the emergence of epidemic O139 disease, one might now refer to strains of the O2 through O138 serogroups as nonepidemic *V. cholerae* (NEVC). The basis for serotyping in *V. cholerae* is the lipopolysaccharide (LPS) somatic antigen; H-antigens are not useful in serotyping. The great majority of these strains do not produce CT and are not associated with epidemic diarrhea (reviewed by Morris [80]). These strains are occasionally isolated from cases of diarrhea (usually associated with consumption of shellfish) and have been isolated from a variety of extraintestinal infections. They are regularly found in estuarine environments, and infections due to these strains are commonly of environmental origin. While the great majority of these strains do not produce CT, some strains may produce other toxins (see below); however, for many strains of *V. cholerae* non-O1/non-O139 isolated from cases of gastroenteritis, the pathogenic mechanisms are unknown.

V. cholerae O139 Bengal

The simple distinction between *V. cholerae* O1 and *V. cholerae* non-O1 was rendered obsolete in early 1993 when the first reports appeared of a new epidemic of severe cholera-like disease emerging from eastern India and Bangladesh (reviewed by Albert [2]). Further investigations revealed that this organism did not belong to the O serogroups previously described for *V. cholerae* but to a new serogroup, which was given the designation O139 and a synonym, "Bengal," in recognition of the origin of this strain. This organism appears to be a hybrid of the O1 strains and the non-O1 strains. In important virulence characteristics, specifically, CT and toxin-coregulated pilus (TCP), *V. cholerae* O139 is indistinguishable from typical El Tor *V. cholerae* O1 strains (2). However, this organism does not produce the O1 LPS and lacks several kilobases of the genetic material necessary for production of the O1 antigen (20). Furthermore, like many strains of non-O1 *V. cholerae* and unlike *V. cholerae* O1, this organism produces a polysaccharide capsule (see below).

Biochemical and Serological Identification

Suspected *V. cholerae* isolates can be transferred from primary isolation plates to a standard series of biochemical media used for identification of *Enterobacteriaceae* and *Vibrionaceae.* Both conventional tube tests and commercially available enteric identification systems are suitable for identifying this species. Several key characteristics for distinguishing *V. cholerae* from other species are given in Tables 13.1 and 13.2.

The key confirmation for identification of *V. cholerae* O1 is agglutination in polyvalent antisera raised against the O1 antigen. Polyvalent antiserum for *V. cholerae* O1 is commercially available and can be used in slide agglutination or coagglutination tests. Antiserum against O139 is now available in many reference laboratories and should soon be commercially available. A monoclonal antibody-based coagglutination test suitable for testing isolated colonies or diarrheal stool samples has recently become available commercially (Cholera SMART; New Horizons Diagnostics Corp., Columbia, Md.). This test was recently used to detect *V. cholerae* O1 in stool samples from airline passengers involved in a foodborne outbreak of cholera on a flight from South America to Los Angeles. Oxidase-positive organisms that agglutinate in O1 or O139 antisera can be reported presumptively as *V. cholerae* O1 or O139 and then forwarded to a public health reference laboratory for confirmation. Antisera against the other 137 serogroups of *V. cholerae* are not commercially available.

DNA Probes and PCR Techniques

Nucleic acid probes are not routinely employed for the identification of *V. cholerae* due to the ease of identifying this species by conventional methods. Where DNA probes have been extremely useful is in distinguishing those strains of *V. cholerae* that contain genes encoding CT (*ctx*) from those that do not contain these genes. This distinction is particularly important in examining environmental isolates of *V. cholerae* since the great majority of these strains lack *ctx* sequences.

A number of DNA fragment probes and synthetic oligonucleotide probes have been developed to detect *ctx* sequences (reviewed by Popovic et al. [99]). An oligonucleotide probe labeled with alkaline phosphatase was reported to identify *ctx*-positive strains in just 3 h from picking the colonies from TCBS agar to final hybridization results. The PCR (polymerase chain reaction) technique has also been used to detect *ctx* sequences, and PCR has been used to detect toxigenic *V. cholerae* O1 in food samples. A study of fruit, vegetables, and shellfish which had been seeded with *V. cholerae* O1 found that amplification of the desired fragment was readily observed with fruit and vegetable samples but that seeded shellfish homogenates often inhibited the PCR assay. By diluting the shellfish homogenate to 1%, the desired fragment was readily amplified (61). PCR was used to investigate a small outbreak of cholera in New York due to crabs imported from Ecuador. Although the crabs implicated in the outbreak did not yield *V. cholerae* O1 after culture, the *ctx* PCR technique detected *ctx* sequences in one of four crab samples examined. Restriction fragment length polymorphism (RFLP) analysis of variations in *ctx* genes and in flanking DNA sequences has yielded significant insights into the molecular epidemiology of *V. cholerae*. For example, RFLP analysis demonstrated that toxigenic O1 strains isolated from various states bordering the United States Gulf Coast are clonal. RFLP analysis also showed that a strain isolated from a cholera patient in Maryland was identical to isolates from Louisiana and Texas, which concurred with epidemiological investigations showing that the crabs eaten by the Maryland patient were harvested along the Texas coast. Multilocus enzyme electrophoresis (MEE) analysis (or zymovar analysis) can also distinguish among strains of *V. cholerae* strains. MEE is not always useful in distinguishing strains within a single outbreak, but it appears to be quite useful in investigating the origin of new outbreaks of disease. If maximum divergence among *V. cholerae* isolates is sought, then RFLP analysis of genes encoding rRNA, or ribotyping, shows greater diversity among *V. cholerae* El Tor isolates than does the MEE technique.

Reservoirs

Environment

V. cholerae are part of the normal, free-living (autochthonous) bacterial flora in estuarine areas. Non-O1/non-O139 strains are much more commonly isolated from the environment than are O1 strains, even in epidemic settings in which fecal contamination of the environment might be expected. Outside of epidemic areas (and away from areas that may have been contaminated by cholera patients), O1 environmental isolates are almost always CT negative (78). However, it is clear that CT-producing *V. cholerae* O1 can persist in the environment in the absence of known human disease. An environmental reservoir is the most likely explanation for the persistence of a single strain in the United States Gulf Coast for over 20 years. Periodic introduction of such environmental isolates into the human population through ingestion of uncooked or undercooked shellfish appears to be responsible for isolated foci of endemic disease along the Gulf Coast and in Australia (7).

V. cholerae O1 strains are capable of colonizing the surfaces of zooplankton such as copepods, with between 10^4 and 10^5 *V. cholerae* attached to a single copepod. Other aquatic biota, such as water hyacinths from Bangladeshi waters, have also been shown to be colonized by *V. cholerae* and to promote its growth. *V. cholerae* produces a chitinase and is able to bind to chitin, which is the principal component of crustacean shells; the organism can grow in media with chitin as the sole carbon source.

Persistence of *V. cholerae* within the environment may be facilitated by its ability to assume survival forms, including a viable but nonculturable state and a rugose survival form. The viable but nonculturable state is a dormant state in which *V. cholerae* is still viable but not culturable in conventional laboratory media (for reviews see references 18, 90, and 91). In this dormant state, the cells are reduced in size and become ovoid. The continued viability of the nonculturable *V. cholerae* can be assessed by a direct viable count procedure in which cells are incubated in the presence of yeast extract and nalidixic acid and examined microscopically for cell elongation. The viable but nonculturable state can be induced in the laboratory by incubating a culture of *V. cholerae* in phosphate-buffered saline at 4°C for several days. Although these cells are not culturable with nonselective enrichment broth or plates, nonculturable *V. cholerae* O1 strains injected into ligated rabbit ileal loops or ingested by volunteers have yielded culturable *V. cholerae* O1 in intestinal contents or stool specimens (19).

V. cholerae can also assume a rugose or wrinkled colonial morphology when passaged on a nonselective medium such as Luria agar (reviewed by Kaper et al. [54]). Upon laboratory passage, one in ca. 1,000 smooth colonies will shift to a rugose form; when a rugose strain is passaged, approximately 1 in 1,000 of the resultant colonies will shift back to a smooth morphology. The rate of shift to rugose forms can be increased by stress such as passage in alkaline peptone water. When examined by light and electron microscopy, cells in a rugose culture are small and spherical and are embedded in an amorphous matrix material composed primarily of carbohydrate. The matrix material, or exopolymer, appears to aid aggregation of bacteria in clusters of up to 100 bacteria. In this state, the cells are protected against adverse environmental conditions. Notably, rugose variants survive in the presence of chlorine and other disinfectants and are still capable of causing diarrhea in volunteers.

Humans and Animals

Long-term carriage of *V. cholerae* in humans is extremely rare and is not considered to be significant in transmission of disease. However, short-term carriage of *V. cholerae* by humans is quite important in transmission of disease. Persons with acute cholera excrete 10^7 to 10^8 *V. cholerae* per g of stool; for patients who have 5 to 10 liters of diarrheal stool, total output of *V. cholerae* can be in the range of 10^{11} to 10^{13} CFU. Even after cessation of symptoms, patients who have not been treated with antibiotics may continue to excrete vibrios for 1 to 2 weeks. Furthermore, a high percentage of persons infected with *V. cholerae* in endemic areas have inapparent illness and can still excrete the organism, although excretion generally lasts for less than a week. Asymptomatic carriers are most commonly identified among household members of persons with acute illness: in various studies, the rate of asymptomatic carriage in this group has ranged from 4% to almost 22%. There have also been studies indicating that *V. cholerae* O1 can be sporadically carried by household animals, including cows, dogs, and chickens, but no animal species consistently carries the organism.

In the 1970s it was widely accepted that asymptomatic and convalescent human (and, possibly, animal) carriers were the primary reservoir for cholera. With the recognition that *V. cholerae* can live and multiply in the environment, much greater attention has been given to identification and characterization of environmental reservoirs (reviewed by Colwell and Huq [18]). Nonetheless, CT-producing *V. cholerae* O1 (i.e., disease-causing strains) continue to be isolated almost exclusively from areas that have been contaminated by human feces or sewage from persons or groups of persons known to have had cholera. Similarly, in endemic areas such as Lima, Peru, rates of isolation from the environment correlate primarily with degree of sewage contamination. A dynamic relationship between human and environmental sources of the organism is apparent, with carriage and amplification by human populations playing a critical role in epidemic spread of CT-producing *V. cholerae* (54).

Foodborne Outbreaks

The critical role of water in transmission of cholera has been recognized for more than a century, ever since the London physician and epidemiologist John Snow showed in 1854 that illness was associated with consumption of water from a water system that drew its supply from the Thames at a point below major sewage inflows. In developing countries, ingestion of contaminated water and food is probably the major vehicle for transmission of cholera, while in developed countries, foodborne transmission is more important (35). Such distinctions are often difficult to make since contaminated water is frequently used in food preparation. For example, rice prepared with water contaminated with *V. cholerae* O1 has been implicated in outbreaks from Bangladesh as well as from the United States Gulf Coast. Fruit juices diluted with contaminated water and vegetables irrigated with untreated sewage have been associated with disease in South America. Seafood may acquire the organism from environmental sources (see above) and may serve as a vehicle in both endemic and epidemic disease, particularly if it is uncooked or only partially cooked.

The role of food in transmitting *V. cholerae* O1 has recently been reviewed (57, 79), and the reader is referred to these reviews for more detail and primary references. The spectrum of food items implicated in transmission of cholera includes crabs, shrimp, raw fish, mussels, cockles, squid, oysters, clams, rice, raw pork, millet gruel, cooked rice, street vendor food, frozen coconut milk, and raw vegetables and fruit (79). One shared characteristic of the implicated foods is their neutral or nearly neutral pH. Thus, in investigating a suspected foodborne outbreak with many possible vehicles, one can eliminate the foods with an acid pH and concentrate on neutral or alkaline foods (57). This predilection for neutral foods was demonstrated in an epidemiologic study in West Africa, where boiled rice is commonly prepared in the morning, held unrefrigerated, and eaten with sauce at the midday and evening meals. In a case-control study of illness, tomato sauces with a pH of 4.5 to 5.0 were protective against illness, whereas less acidic sauces (pH 6.0 to 7.0) prepared from ground peanuts were associated with illness due to *V. cholerae* O1 (79). Alkaline conditions are employed in enrichment cultures of food and water samples, wherein the use of alkaline peptone water (pH 8.5) followed by plating on TCBS agar is a common isolation method (57). Survival and growth of *V. cholerae* O1 in food are also enhanced by low temperatures, high organic content, high moisture, and absence of competing flora (79). Survival is increased when foods are cooked before contamination; cooking eliminates competing organisms and has also been suggested to destroy some heat-labile growth inhibitors and produce denatured proteins that the organism uses for growth (79).

As noted below, food buffers *V. cholerae* O1 against killing by gastric acid. While many different food items can provide this buffering capacity, the protection provided by chitin is noteworthy since crustaceans are a frequent vehicle of disease. In dilute hydrochloric acid solutions of approximately the same pH as human gastric acid, survival of *V. cholerae* absorbed to chitin was enhanced compared to *V. cholerae* in the absence of chitin (79).

In the United States, both domestically acquired and imported cases of cholera occur. For domestic cases, crabs, shrimp, and oysters have been the most frequently implicated vehicles, although the largest single outbreak (16 cases) was due to ingestion of contaminated rice. In this outbreak, which occurred on a Gulf Coast oil rig in 1981, cooked rice was moistened with water contaminated by human feces and then held for 8 h after cooking (79). Although the majority of domestic cases cases occur in states bordering the Gulf Coast, seafood shipped from this area has caused disease in both Maryland and Colorado (79). The risk of imported cholera has greatly increased since the establishment of endemic cholera in South America in 1991. The largest such outbreak (75 cases) involved crab salad served on an airplane flying from Peru to California. A smaller outbreak of eight cases occurred in New Jersey due to crabs purchased in Ecuador and carried to the United States in an individual's luggage. Importation from Asia can also occur, even in commercially imported food. A small outbreak of four cases in Maryland was attributed to frozen coconut milk imported from Thailand which was subsequently used in a topping for a rice pudding.

Nearly all cases of gastroenteritis caused by *V. cholerae* of serogroups other than O1 or O139 have been linked to the consumption of raw oysters (reviewed by Morris [80]). Both disease incidence and isolation rate of non-O1/O139 serogroups from oysters are highest in the summer. Such strains were isolated from up to 14% of freshly harvested oysters in one study conducted by the U.S. Food and Drug Administration (FDA). Outside of the United States, outbreaks have also been linked to consumption of contaminated potatoes, chopped eggs, preprepared gelatin, vegetables, and meat samples (80). As with *V. cholerae* O1, survival of non-O1/non-O139 *V. cholerae* is enhanced in foods of alkaline pH.

Characteristics of Disease

The explosive, potentially fatal dehydrating diarrhea characteristic of cholera is actually seen in only a minority of persons infected with CT-producing *V. cholerae* O1/O139. The majority of infections with *V. cholerae* O1 are mild or even asymptomatic. It has been estimated that 11% of patients with classical infections develop severe disease, compared with 2% of those with El Tor infections. An additional 5% of El Tor infections and 15% of classical infections result in moderate illness (reviewed by Kaper et al. [54]). Symptoms of persons infected with *V. cholerae* O139 Bengal appear to be virtually identical to those of persons infected with O1 strains.

The incubation period of cholera can range from several hours to 5 days and depends in part on inoculum size. Onset of illness may be sudden, with profuse, watery diarrhea, or there can be premonitory symptoms such as anorexia, abdominal discomfort, and simple diarrhea. Initially the stool is brown with fecal matter, but soon the diarrhea assumes a pale gray color with an inoffensive, slightly fishy odor. Mucus in the stool imparts the characteristic "rice water" appearance. Vomiting is often present, occurring a few hours after the onset of diarrhea.

In most severe form, termed cholera gravis, the rate of diarrhea may quickly reach 500 to 1,000 ml/h, leading rapidly to tachycardia, hypotension, and vascular collapse due to dehydration. Peripheral pulses may be absent, and blood pressure may be unobtainable. Skin turgor is poor, giving the skin a doughy consistency; the eyes are sunken; and hands and feet become wrinkled, as after long immersion ("washerwoman's hands"). Such severe dehydration can lead to death within hours of the onset of symptoms unless fluids and electrolytes are rapidly replaced. While cholera gravis is a striking clinical entity, milder illnesses are not readily differentiated from other causes of gastroenteritis in cholera-endemic areas.

Gastroenteritis associated with *V. cholerae* non-O1/non-O139 is generally of mild to moderate severity, although severe cholera-like illness has also been seen occasionally. Besides nonbloody and occasionally bloody diarrhea, symptoms can also include abdominal cramps and fever, with nausea and vomiting occurring in a minority of patients (80). Non-O1/O139 *V. cholerae* is also frequently isolated from extraintestinal infections such as septicemia, wound infections, and ear infections; these infections usually involve exposure to fresh or brackish water (80).

Infectious Dose and Susceptible Population

In healthy North American volunteers, doses of 10^{11} CFU of *V. cholerae* were required to consistently cause diarrhea when the inoculum was given in buffered saline (pH 7.2). When stomach acidity was neutralized with 2 g of sodium bicarbonate immediately prior to administration of the inoculum, attack rates of 90% were seen with an inoculum of 10^6. Food has a buffering capacity comparable to that seen with sodium bicarbonate. Ingestion of 10^6 vibrios with food such as fish and rice resulted in the same high attack rate (100%) as when this inoculum was administered with buffer (66). Further studies showed that most volunteers who received as few as 10^3 to 10^4 organisms with buffer developed diarrhea, although lower inocula correlated with a longer incubation period and diminished severity. The incubation time in volunteers between ingestion of vibrios and onset of diarrhea ranged from 8 to 96 h.

The inoculum size in naturally occurring infections is not known with certainty. In one study in Bangladesh, Spira et al. (116) found that it was almost always necessary to use enrichment techniques to isolate *V. cholerae* from household water samples, suggesting that counts were <500 CFU/ml. Attack rates in the households surveyed were only 11%, and only half of infected persons had clinical illness, consistent with the hypothesis that inoculum sizes were relatively low. While conditions will obviously vary widely from one community to another, studies such as these have led to the suggestion that the inoculum in nature is in the range of 10^2 to 10^3 (35).

The volunteer data on the effect of buffer on infectious dose are consistent with epidemiologic data showing that people who are achlorhydric because of surgery, medication (e.g., antacids), or other reasons are at increased risk for cholera. Individuals of blood group O are at increased risk of more severe cholera, which has been shown for natural infection as well as with experimental infection; the mechanism of this increased susceptibility is unknown. In addition to these factors, additional host factors, as yet poorly defined, play a role in disease susceptibility to cholera. The effect of other host factors is illustrated by a study in which the identical inoculum that caused 44 liters of diarrhea in one volunteer caused little or no illness in other individuals. The various factors affecting host susceptibility have recently been reviewed (104).

Strains of non-O1/non-O139 *V. cholerae* have also been studied in volunteers. Of three strains which did not produce CT, only one strain caused diarrhea when fed to volunteers. This strain produced a heat-stable (ST)-like toxin (see below) and caused diarrhea in six of eight volunteers at doses of 10^6 to 10^9 (after neutralization of stomach acid with sodium bicarbonate) (80). The severity of disease was generally mild, but in one volunteer, diarrheal stool volume exceeded 5 liters.

Virulence Mechanisms

Infection due to *V. cholerae* O1/O139 begins with the ingestion of food or water contaminated with the organism. After passage through the acid barrier of the stomach, vibrios colonize the epithelium of the small intestine by means of one or more adherence factors. Invasion into epithelial cells or the lamina propria does not occur. Production of CT (and possibly other toxins) disrupts ion transport by intestinal epithelial cells. The subsequent loss of water and electrolytes leads to the severe diarrhea characteristic of cholera.

Cholera Enterotoxin

Volunteer studies demonstrate that the massive, dehydrating diarrhea characteristic of cholera is induced by CT (cholera enterotoxin, also referred to as cholera toxin or choleragen). CT is among the most extensively studied of bacterial toxins and several recent reviews have been published, including a review on the structure and function of CT (114) and a general review on CT and other toxins produced by *V. cholerae* (53).

Structure

The structure of CT is typical of the A-B subunit group of toxins in which each of the subunits has a specific function. The B subunit serves to bind the holotoxin to the eukaryotic cell receptor, and the A subunit possesses a specific enzymatic function that acts intracellularly. CT consists of five identical B subunits and a single A subunit, and neither of the subunits individually has significant secretogenic activity in animal or intact cell systems. The mature B subunit contains 103 amino acids with a subunit mass of 11.6 kDa. The mature A subunit has a mass of 27.2 kDa and is proteolytically cleaved to yield two polypeptide chains, a 195-residue A_1 peptide of 21.8 kDa and a 45-residue A_2 peptide of 5.4 kDa. After proteolytic cleavage, the A_1 and A_2 peptides are still linked by a disulfide bond before internalization.

Receptor Binding

The receptor for CT is the ganglioside GM_1. Binding of CT to epithelial cells is enhanced by a neuraminidase (NANase) produced by *V. cholerae*. This 83-kDa enzyme enhances the effect of CT by catalyzing the conversion of higher-order gangliosides to GM_1, thereby enhancing the binding of CT and leading to greater fluid secretion. When culture supernatants from a CT-positive, NANase-positive *V. cholerae* strain were compared with supernatants from an isogenic CT-positive, NANase-negative strain, binding of CT to mouse fibroblasts was increased five- to eightfold in the presence of NANase (34). Furthermore, the short-circuit current measured in Ussing chambers (a measure of secretory activity) increased 65% with NANase-positive filtrates compared with

NANase-negative filtrates. These results indicated that NANase plays a subtle but significant role in binding and uptake of CT although this enzyme is not a primary virulence factor of *V. cholerae*.

Enzymatic Activity

The intracellular target of CT is adenylate cyclase, which mediates the transformation of ATP to cyclic AMP (cAMP), a crucial intracellular messenger for a variety of cellular pathways. Regulation of adenylate cyclase occurs via G proteins, which serve to link many cell surface receptors to effector proteins at the plasma membrane. The specific G protein involved is the $G_{s\alpha}$ protein, activation of which leads to increased adenylate cyclase activity. CT (specifically, the A_1 peptide) catalyzes the transfer of the ADP-ribose moiety of NAD to a specific arginine residue in the $G_{s\alpha}$ protein, resulting in the activation of adenylate cyclase and subsequent increases in intracellular levels of cAMP. cAMP activates a cAMP-dependent protein kinase, leading to protein phosphorylation, alteration of ion transport, and ultimately to diarrhea (Fig. 13.1). The alpha subunit of G_s contains a GTP binding site and an intrinsic GTPase activity. Binding of GTP to the α subunit leads to dissociation of the α and the β-γ subunits and subsequent increased affinity of α for adenylate cyclase. The resulting activation of adenylate cyclase continues until the intrinsic GTPase activity hydrolyzes GTP to GDP, thereby inactivating the G protein and adenylate cyclase. ADP-ribosylation of the α subunit by the A_1 peptide of CT inhibits the hydrolysis of GTP to GDP, thus leaving adenylate cyclase constitutively activated, probably for the life of the cell. The ADP-ribosylation activity of A_1 is stimulated in vitro by a family of proteins termed ARFs, for "ADP-ribosylation factors" (reviewed in reference 84). The ARF proteins are ca. 20-kDa GTP binding proteins and constitute a distinct family within the larger group of ca. 20-kDa guanine nucleotide binding proteins which include the *ras* oncogene protein. At least in vitro, ARFs serve as allosteric activators of A_1 and increase the ADP-ribosyltransferase activity of the proteins.

Cellular Response

The increased intracellular cAMP concentrations resulting from the activation of adenylate cyclase by CT lead to increased Cl^- secretion by intestinal crypt cells and decreased NaCl-coupled absorption by villus cells (Fig. 13.1C). The net movement of electrolytes into the lumen results in a transepithelial osmotic gradient which causes water flow into the lumen. The massive volume of water overwhelms the absorptive capacity of the intestine, resulting in diarrhea. The steps between increased levels of cAMP and secretory diarrhea are not known in their entirety, but one crucial step resulting from increased cAMP levels is activation of protein kinase A, which subsequently phosphorylates numerous substrates in the cell.

Regulation of chloride channels by cAMP-dependent protein kinases is well known, and the crucial chloride channel involved in diarrhea due to cholera is the cystic fibrosis gene product CFTR, a Cl^- channel with multiple potential substrate sequences for kinase A. In the cystic fibrosis mouse model, mice that expressed no CFTR protein did not secrete fluid in response to CT, whereas heterozygotes expressed 50% of the normal amount of CFTR and secreted 50% of the normal fluid and chloride ion in response to CT (33).

Alternate Mechanisms of Action

The activation of adenylate cyclase leading to increased cAMP and subsequent altered ion transport is the "classic" mode of action of CT. However, it has recently been appreciated that the increased levels of cAMP and subsequent protein kinase A activation may not explain all of the secretory effects of CT. There is compelling evidence that prostaglandins and the enteric nervous system are involved in the response to CT in addition to the mechanism outlined above. The majority of the studies presenting these alternative mechanisms do not conclude that the above scenario is wrong; rather, they suggest that this scenario simply does not explain all of the secretion due to CT.

Prostaglandins. Several reports have implicated prostaglandins in the pathogenesis of cholera (reviewed in reference 53). Although the exact mechanisms are unclear, the role of prostaglandins, leukotrienes, and other metabolites of arachidonic acid in causing intestinal secretion has been well documented. Cholera patients in the active secretory disease stage have elevated jejunal concentrations of prostaglandin E_2 (PGE_2) compared to patients in the convalescent stage. In one animal study, addition of cAMP induced only a small, transient fluid accumulation in rabbit intestinal loops whereas addition of PGE_2 caused a much stronger fluid accumulation in rabbit loops. Addition of CT led to increases in both cAMP and PGE in rabbit loops and in CHO cells, resulting in the release of arachidonic acid from membrane phospholipids. A model has been suggested in which cAMP levels increased by CT serve not only to activate protein kinase A but to also regulate transcription of a phospholipase or a phospholipase-activating protein. The activated phospholipase could act on membrane

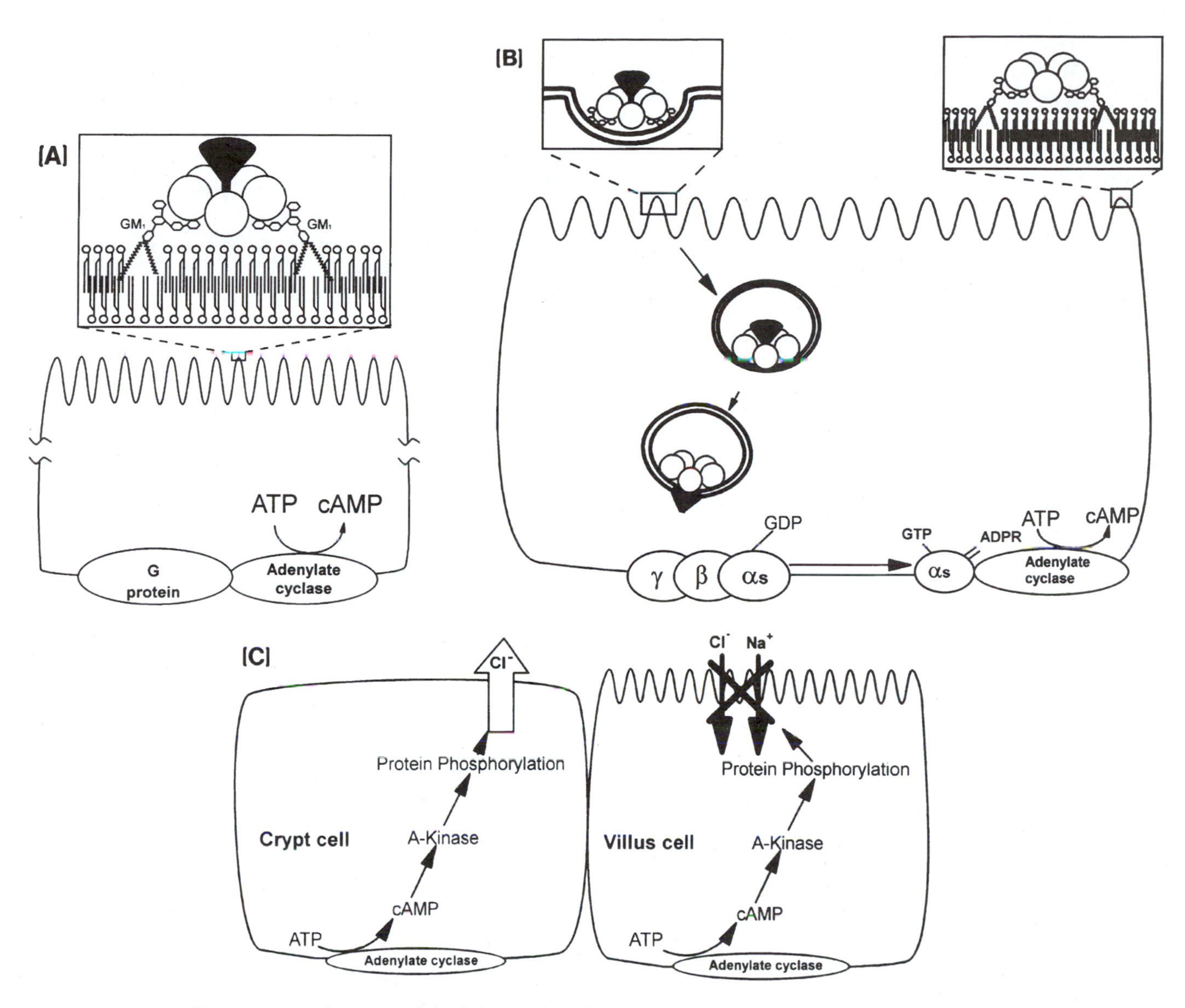

Figure 13.1 Classic model of CT mode of action involving cAMP. More recent evidence indicates that prostaglandins and the enteric nervous system are also involved in the response to CT (see text for details). (A) Adenylate cyclase, located in the basolateral membrane of intestinal epithelial cells, is regulated by G proteins. CT binds via the B subunit pentamer (shown as open circles with the A subunit as the inverted solid triangle) to the GM_1 ganglioside receptor inserted into the lipid bilayer. (B) The toxin enters the cell via endosomes, and the A_1 peptide ADP-ribosylates $G_{s\alpha}$ located in the basolateral membrane. (C) Increased cAMP activates protein kinase A, leading to protein phosphorylation. In crypt cells, the protein phosphorylation leads to increased Cl^- secretion; in villus cells, it leads to decreased NaCl absorption. Adapted from Kaper et al. (53).

phospholipids to produce arachidonic acid, a precursor of prostaglandins and leukotrienes. Consistent with this model, it has been reported that 40 to 60% of the short current response to CT is inhibited by relatively low concentrations of the phospholipase A_2 inhibitor mepacrine. The recent implication of platelet activating factor, a stimulator of phospholipase A_2, in the response to CT (37) further supports a key role for prostaglandin synthesis in the mode of action of CT.

Enteric nervous system. The enteric nervous system plays an important role in intestinal secretion and absorption. The intestine also contains a variety of cells that can produce hormones and neuropeptides such as vasoactive intestinal peptide (VIP) and serotonin (5-hydroxytryptamine) that can affect secretion. It has been estimated that ca. 60% of the effect of CT on intestinal fluid transport could be attributed to nervous mechanisms (53). One proposed mechanism is that CT binds to "receptor cells," namely enterochromaffin cells, which release a substance such as serotonin which activates dendrite-like structures located beneath the intestinal epithelium. This leads to the release of VIP, resulting in electrolyte and fluid secretion. This model is supported by a variety of studies using receptor antagonists or ganglionic or neurotransmitter blockers, as well as direct measurements of increased levels of serotonin (5-hydroxytryptamine) and VIP. Perhaps the most convincing evidence comes from a study in which fluid communication between proximal and distal regions of rabbit intestine was prevented by ligation and fluid secretion was measured in each isolated area (88). Addition of CT to the proximal loop only resulted in changes in fluid secretion in both the proximal and distal loop. However, when the enteric nervous system connecting the two regions was disrupted, addition of CT to the proximal loop had no effect on fluid secretion in the distal region.

Other Toxins Produced by *V. cholerae*

When the first recombinant *V. cholerae* vaccine strains specifically deleted of genes encoding CT were tested in volunteers, it was somewhat surprising that mild to moderate diarrhea was still seen in ca. 50% of the volunteers (67). The volume of diarrhea was not the severe, dehydrating diarrhea seen with wild-type strains, which can exceed 40 liters in volume, but a much milder diarrhea which ranged from 0.3 to 2.1 liters in volume. In addition, some volunteers also experienced abdominal cramps, anorexia, and low-grade fever when fed Δ*ctx V. cholerae* strains. These results prompted a search for additional toxins produced by *V. cholerae*, and it is now known that *V. cholerae* produces a variety of extracellular products that have deleterious effects on eukaryotic cells. Genes encoding two of these toxins, Zot and Ace, are located immediately upstream of the *ctx* genes in a region encoding a filamentous phage called CTXΦ (130a).

Zonula Occludens Toxin (Zot)

The zonula occludens toxin increases the permeability of the small intestinal mucosa by affecting the structure of the intercellular tight junction, or zonula occludens. This activity was discovered by testing culture supernatants of *V. cholerae*, both wild type and Δ*ctx*, in Ussing chambers, a classic technique for measuring trans-epithelial transport of electrolytes across intestinal tissue. In this system, Zot causes a decreased tissue resistance which reflects modification of tissue permeability through the intercellular space, i.e., the paracellular pathway (32). The *zot* gene, encoding a predicted 44.8-kDa polypeptide, is located immediately upstream of the *ctx* locus, and strains that contain *ctx* sequences almost always contain *zot* sequences and vice versa. By increasing intestinal permeability, Zot might cause diarrhea by leakage of water and electrolytes into the lumen under the force of hydrostatic pressure.

Accessory Cholera Enterotoxin (Ace)

The gene encoding accessory cholera enterotoxin is located immediately upstream of *zot* (126). The potential 11.3-kDa product of the *ace* gene causes fluid accumulation in rabbit ligated ileal loops, and like CT, and in contrast to Zot, this toxin increases potential difference rather than tissue conductivity in Ussing chambers. The predicted amino acid sequence of Ace shows a striking similarity to a family of eukaryotic ion-transporting ATPases including the chloride ion channel CFTR. The predicted structure of the Ace protein suggests a model in which multimers of Ace insert into the eukaryotic membrane with hydrophobic surfaces facing the lipid bilayer and hydrophilic sides facing the interior of a transmembrane pore. It is suggested that diarrhea results from Ace monomers aggregating and inserting into the eukaryotic membrane to form an ion channel.

Hemolysin/Cytolysin

Hemolysis of sheep erythrocytes was traditionally used to distinguish between the El Tor and classical biotypes of *V. cholerae*, although more recent El Tor isolates are only poorly hemolytic on sheep erythrocytes. The hemolysin/cytolysin is initially made as an 82-kDa protein and processed in two steps to a 65-kDa active cytolysin. The purified hemolysin is cytolytic for a variety of erythrocytes and cultured mammalian cells and rapidly lethal for mice and causes bloody fluid accumulation in ligated rabbit ileal loops.

Miscellaneous Toxins

In addition to CT, Zot, Ace, and hemolysin/cytolysin, which are widely distributed in *V. cholerae* O1 and O139 and for which genes have been cloned and sequenced, a number of other toxins have been reported for this species (53, 54). A sodium channel inhibitor, a so-called "new cholera toxin," and low levels of a Shiga toxin have been reported for this species, but the responsible proteins and genes have not yet been purified or cloned. A single strain of *V. cholerae* O1 has been reported to produce an ST-like toxin, but this toxin is more common among non-O1 strains (see below).

Role of Additional Toxins in Disease

The role of toxins other than CT in the pathogenesis of disease due to *V. cholerae* is unknown. These toxins clearly cannot cause cholera gravis because the diarrhea seen with Δ*ctx* strains presumably still producing these toxins is not the severe purging seen with wild-type *V. cholerae* strains. When a *V. cholerae* strain specifically deleted of sequences encoding Zot, Ace, and hemolysin/cytolysin as well as the CT A subunit was tested in volunteers, it still caused mild to moderate diarrhea in 7 of 10 volunteers as well as fever and abdominal cramps. While these volunteer studies do not exclude a role for Ace, Zot, and hemolysin/cytolysin in the pathogenesis of cholera, they clearly indicate that there are additional features of *V. cholerae* that result in diarrhea. Toxins other than CT may contribute in part to the diarrhea and other symptoms seen with wild-type and CT-negative *V. cholerae* strains. Such toxins may serve as a secondary secretogenic mechanism when conditions for producing CT are not optimal.

Toxins of *V. cholerae* Non-O1 and Non-O139 Strains

No toxins unique to *V. cholerae* non- O1/non-O139 serogroups have been reported, but some strains of these serogroups may produce one or more toxins that have been previously characterized in *V. cholerae* O1 or other species (43). Most *V. cholerae* non-O1/non-O139 strains produce a cytotoxic protein apparently identical to the El Tor hemolysin (hemolysin/cytolysin) of *V. cholerae* O1. The majority of non-O1/non-O139 strains do not contain genes encoding CT, Zot, or Ace, but a few strains have been shown to possess genes for these toxins.

Some strains of *V. cholerae* non-O1/non-O139 produce a 17-amino-acid heat-stable enterotoxin (designated NAG-ST for non-agglutinable vibrio ST) that shares 50% sequence homology to the STa of enterotoxigenic *E. coli*. In a volunteer study, one subject who ingested a CT-negative *V. cholerae* non-O1 strain producing NAG-ST purged over 5 liters of diarrheal stool (80). This toxin is found only in a minority of *V. cholerae* strains. In one study, 6.8% of *V. cholerae* non-O1 strains from Thailand and none of the strains from the United States or Mexico produced this toxin. In another study, 2.3% of all non-O1 *V. cholerae* isolated from Calcutta, India, contained genes for this toxin. The *tdh* gene encoding the thermostable direct hemolysin (TDH) of *V. parahaemolyticus* (see below) has also been found in *V. cholerae* non-O1/non-O139. In one strain, the *tdh* gene was found on a plasmid, suggesting that this toxin gene could be readily transferred to other species.

Colonization Factors

TCP

The toxin-coregulated pilus (TCP) is the best-characterized intestinal colonization factor of *V. cholerae*. TCPs are long filaments 7 nm in diameter which are laterally associated in bundles. The name of the pilus results from the fact that expression of the pilus is correlated with expression of CT (122). The predicted amino acid sequence of the 20.5-kDa TcpA subunit shows significant homology to the type IV pili of *Pseudomonas aeruginosa*, *Neisseria gonorrhoeae*, *Moraxella bovis*, and *Bacteroides nodosus*. Epitope differences are seen in pili produced by classical and El Tor strains, and the predicted protein sequences share 82 to 83% homology between the biotypes. El Tor strains produce less TCP than classical strains, and culture conditions for optimal TCP expression differ between biotypes. Synthesis of TCP is complex and incompletely understood. Up to 15 open reading frames are found in the *tcp* gene cluster. The sequence and regulation of TCP in the O139 serogroup appear to be identical those seen in O1 El Tor strains.

TCP is the only colonization factor of *V. cholerae* whose importance in human disease has been proven. Volunteers ingesting a classical Ogawa strain specifically mutated in the *tcpA* gene did not experience diarrhea, and no vibrios were recovered from their stools. There is some controversy about the role of TCP in colonization of the El Tor biotype. Some investigators feel that it is unimportant for colonization by this biotype, while others report that El Tor strains specifically mutated in *tcpA* do not colonize and that antibodies directed against the El Tor pilus can protect mice from disease.

ACF

Mutations in a ToxR-regulated locus called *acf*, for accessory colonization factor (ACF), also diminish colonization in mice but not to the same extent as mutations in *tcp* (i.e., ca. 10- to 50-fold decrease for *acf* compared to 1,000-fold for *tcp*). The exact nature of ACF has not been reported, but one of the four open reading frames in this locus (*acfB*) shares similarity to chemotaxis signal-transducing proteins (28) and mutations in another open reading frame (*acfD*) show reduced motility. Thus, *acf* apparently does not encode a fimbrial colonization factor but may be involved in intestinal colonization via motility and/or chemotaxis functions. The *acf* locus is located immediately adjacent to the *tcp* and *toxT* loci (see below).

MFRHA

Mutations in a gene encoding a cell-associated mannose-fucose-resistant hemagglutinin (MFRHA) also lead to decreased colonization in mice (500 to 1,300-fold relative to the parent strain). The exact nature of the MFRHA, expressed by both biotypes, is not known, but it was suggested that the MFRHA is a 27-kDa cationic outer membrane protein which is held on the cell surface primarily by charge interactions with the LPS.

MSHA

The mannose-sensitive hemagglutinin (MSHA) of *V. cholerae* is expressed by strains of the El Tor biotype but is only rarely expressed by strains of the classical biotype. The MSHA appears to be a thin, flexible pilus composed of subunits with a molecular mass of ca. 17 kDa. A monoclonal antibody against MSHA protects against experimental cholera caused by El Tor vibrios in the infant mouse and rabbit intestinal loop models but did not protect against challenge by *V. cholerae* O1 of the classical biotype, thus suggesting that the MSHA is an El Tor-specific protective antigen.

Core-Encoded Pilus

The *cep* (core-encoded pilus) gene is located upstream of the toxin genes *ctx*, *zot*, and *ace* (98). The Cep protein was originally thought to comprise a pilus, but recent evidence indicates that it is the virion capsid protein of the CTXΦ phage (130a). Deletion of the *cep* locus reduced colonization in infant mice 13- to 21-fold. The available evidence suggests that *cep* contributes little, if any, to intestinal colonization of humans since a strain deleted of *cep* as well as *ace*, *zot*, and *ctxA* genes colonized volunteers as well as the parent strain containing *cep*.

OmpU

The 38-kDa outer membrane protein OmpU was originally described as being regulated by ToxR. Recent evidence also indicates that this protein can serve as an adherence factor since antibodies raised against OmpU completely inhibit adherence of *V. cholerae* O1 to cultured epithelial cells and protect mice against challenge with either El Tor or classical strains (115).

Motility and Flagella

V. cholerae cells are motile by means of a single polar, sheathed flagellum. Motility has been shown in a number of studies to be an important virulence property, with nonmotile, fully enterotoxinogenic mutants being diminished in virulence. In several animal and in vitro models, it has been shown that motile *V. cholerae* rapidly enter the mucus gel overlying the intestinal epithelium and can be found in intervillous spaces within minutes to a few hours. In addition to its role in motility, it has also been suggested that the flagellum serves as an adhesin, although it appears that the actual property of motility is more important than the mere presence of the flagellar structure.

LPS

The LPS of *V. cholerae* O1 is the major protective antigen of this pathogen, and its importance in protective immunity greatly outweighs that of CT. The importance of this antigen was seen in India and Bangladesh when the O139 serogroup caused widespread disease in individuals who were presumably immune to the O1 serogroup. There is also evidence to suggest that LPS is involved in adherence of *V. cholerae* O1 to the intestinal mucosa. In one study, purified Inaba LPS significantly inhibited attachment of *V. cholerae* Inaba to rabbit mucosa, and in other in vitro and in vivo studies antibodies against Ogawa or Inaba LPS were shown to prevent adhesion of *V. cholerae* to intestinal mucosa.

Polysaccharide Capsule

Although *V. cholerae* O1 is unencapsulated, strains of *V. cholerae* O139 produce a polysaccharide capsule (48) which has also been termed an O-antigen capsule. The genetic change involved in the conversion of the O1 serogroup to O139 was recently shown to involve the deletion of ca. 22 kb of the *rfb* genes encoding the O1 LPS and the substitution of a ca. 35-kb insert encoding the O139 LPS and polysaccharide capsule (20). In one recent study, there was a suggestion that the polysaccharide capsule of an O139 strain could mediate adherence to culture epithelial cells (115).

Adherence Factors of Non-O1/Non-O139 V. cholerae

Intestinal colonization factors of non-O1/non-O139 *V. cholerae* strains are poorly characterized. Sequences encoding TCP were not found in a collection of *V. cholerae* non-O1 strains nor in environmental isolates of *V. cholerae* O1 which do not produce CT. A variety of fimbria hemagglutinins have been described for these strains, but their role in intestinal adherence is unclear. There is evidence that the capsular polysaccharide may also mediate adherence to intestinal epithelial cells for non-O1 strains. Approximately 70% of such strains produce a polysaccharide capsule. In additional to a potential role in adherence, such a capsule could facilitate septicemia often seen with infections due to non-O1/non-O139 strains.

Regulation

Multiple systems are involved in regulation of virulence in *V. cholerae.* The ToxR regulon controls expression of several critical virulence factors and has been the most extensively characterized. Regulation in response to iron concentration is a distinct regulatory system that controls additional putative virulence factors. Other putative virulence factors such as neuraminidase and various hemagglutinins are apparently not controlled by either regulatory system. There is also a set of poorly characterized genes that are expressed only in vivo and do not belong to the ToxR or iron regulatory systems. These different regulatory systems no doubt allow *V. cholerae* to vary expression of its genes to optimize survival in a variety of environments, from the human intestine to the estuarine environment.

The ToxR Regulon

Expression of several virulence genes in *V. cholerae* O1 and O139 is coordinately regulated so that multiple genes respond in a similar fashion to environmental conditions (reviewed in reference 26). The "master switch" for control of these factors is ToxR, a 32-kDa transmembrane protein that shares homology with other sensory transduction proteins seen in various bacterial pathogens. It has been proposed that the ToxR protein senses environmental conditions and transmits this information to other genes in the ToxR regulon by signal transduction. At least 17 distinct genes are regulated by ToxR including those encoding CT, TCP, ACF, OmpU, OmpT, and three lipoproteins. The effect of ToxR on expression of most of these factors is to increase expression, but expression of OmpT is decreased in the presence of ToxR. These genes make up the ToxR regulon. The importance of ToxR in human disease was demonstrated in volunteer studies wherein a ToxR mutant *V. cholerae* O1 strain was fed to volunteers who subsequently suffered no diarrheal symptoms and did not shed the strain in their stools.

At least two other proteins interact with ToxR to control gene expression in *V. cholerae.* ToxS is a 19-kDa transmembrane protein that helps to assemble or stabilize ToxR monomers into a dimeric form. Expression of at least some of the genes of the ToxR regulon is controlled by another regulatory factor, ToxT, a 32-kDa protein that shares significant sequence homology to the AraC family of transcriptional activators. ToxR controls transcription of the *toxT* gene, and the resulting increased expression of the ToxT protein then leads to activation of other genes in the ToxR regulon. While ToxR has been shown to be capable of binding directly to a tandemly repeated 7-bp DNA sequence found upstream of the *ctxAB* structural genes and thus to increase expression of CT, it does not appear to bind to other sequences near genes encoding other virulence factors. A regulatory cascade has been suggested for control of important virulence factor expression in *V. cholerae* in which ToxR is at the top of the hierarchy, ToxT is at the next level, and a number of virulence genes controlled by ToxT are at the lowest level.

Iron Regulation

Growth of *V. cholerae* under low-iron conditions induces the expression of several new outer membrane proteins that are not seen with cells grown in iron-rich media. Many of these proteins are similar to proteins induced by in vivo growth of *V. cholerae,* indicating that the intestinal site of *V. cholerae* is a low-iron environment. In addition, expression of some outer membrane proteins decreases under iron-limiting conditions. Several proteins whose expression is increased under low-iron conditions have been identified, including hemolysin/cytolysin, vibriobactin, and IrgA.

V. cholerae has at least two high-affinity systems for acquiring iron. The first system involves a phenolate-like siderophore, vibriobactin, which is produced under low-iron conditions. Vibriobactin, which is not essential for virulence, binds iron extracellularly and transports it into the cell through a specific receptor. A second system for acquiring iron utilizes heme and hemoglobin. Genes have been cloned for a 26-kDa inner membrane protein and a 77-kDa outer membrane protein which allow transport of heme into the cell.

Iron-regulation gene expression in *V. cholerae* involves a protein called Fur which binds to a 21-bp operator sequence found in the promoter of iron-regulated genes, thereby repressing transcription (13). In *V. cholerae,* Fur acts as a repressor for the vibriobactin gene (*viuA*) and *irgA,* which encodes an outer membrane protein whose mutation results in a ca. 10-fold decrease in mouse colonization. Regulation of *irgA* also requires a second protein, IrgB, which acts as a positive transcriptional activator; transcription of *irgB* itself is repressed by Fur in the presence of iron.

VIBRIO MIMICUS

Prior to 1981, *V. mimicus* was known as sucrose-negative *V. cholerae* non-O1. This species was shown to be a distinctly different species on the basis of biochemical reactions and DNA hybridization studies, and the name *mimicus* was given because of its similarity to *V. cholerae* (23). This organism is chiefly isolated from cases of gastroenteritis, but occasional strains have been isolated from ear infections as well.

Reservoirs

As with other *Vibrio* species, the reservoir of *V. mimicus* is the aquatic environment. One very interesting study compared the ecology of *V. mimicus* in the tropic, polluted environment of Bangladesh with that in the cleaner, more temperate environment of Okayama, Japan (15). This species was isolated from Bangladesh waters throughout the year, whereas it was not isolated in Okayama when the water temperature fell below 10°C. In Japan, *V. mimicus* was found both in freshwater and in brackish waters with a salinity optimum of 4 ppt; the organism was not recovered from waters with salinity of >10 ppt. Besides being found free in the water column, *V. mimicus* was also isolated from the roots of aquatic plants, from sediments, and from plankton at levels up to 6×10^4 CFU per 100 g of plankton. *V. mimicus* has also been isolated from fish (62) and freshwater prawns (17).

Foodborne Outbreaks

Gastroenteritis due to *V. mimicus* has been linked only to consumption of seafood. In the initial patient series in the United States (110), consumption of raw oysters was significantly associated with disease due to this species; one patient reported eating only shrimp and crab, not oysters, in the week before onset of disease. Cases are usually sporadic, rather than linked with common-source outbreaks, but three patients in Lousiana became ill with *V. mimicus* after attending a company banquet where the foods served included crawfish (110). Disease in Japan is associated with consumption of raw fish, and at least two outbreaks of seafood-borne disease have involved *V. mimicus* of the O41 serogroup (reviewed in reference 62).

Characteristics of Disease

Disease due to *V. mimicus* is characterized by diarrhea, nausea, vomiting, and abdominal cramps in the majority of patients. In a minority of patients, fever, headache, and bloody diarrhea can also be seen. In one series of patients, diarrhea lasted a median of 6 days, with a range of 1.5 to 10 days (110). In this report, disease was associated with consumption of seafood, particularly raw oysters, and the median interval from the time of consumption to onset of illness was 24 h (range, 3 to 72 h). In the 1989 survey of *Vibrio* infections on the Gulf Coast, *V. mimicus* was isolated from 4 cases of gastroenteritis, compared to 26 cases for *V. parahaemolyticus* and 18 cases for non-O1 *V. cholerae* (68).

Infectious Dose and Susceptible Population

There are no volunteer or epidemiological data that would suggest an infectious dose for *V. mimicus.* There does not seem to be a particularly susceptible population, other than people who eat raw oysters. Most patients who developed gastroenteritis with *V. mimicus* were in good health before onset of illness (110).

Virulence Mechanisms

V. mimicus appears to produce no unique enterotoxins, but many strains produce toxins that were first described in other *Vibrio* species, including CT, TDH, Zot, and a heat-stable enterotoxin apparently identical to the NAG-ST produced by *V. cholerae* non-O1/non-O139 strains. One study from Bangladesh (15) reported that 10% of clinical isolates, compared to less than 1% of environmental strains, produced a CT-like toxin. Another study in Japan reported that 94% of clinical isolates produced TDH while none of the environmental strains studied produced this toxin. The *tdh* gene from *V. mimicus* shares 97% homology with the prototypic *tdh* gene from *V. parahaemolyticus* (86). The *zot* gene of *V. cholerae* was found in three of five strains of *V. mimicus* studied (16), only one of which contained *ctx,* suggesting perhaps that unlike *V. cholerae* O1, the *zot* gene may occur in *V. mimicus* in the absence of *ctx.*

There is little information about potential intestinal colonization factors of *V. mimicus.* Like *V. cholerae* non-O1/non-O139, *V. mimicus* does not appear to express TCP or possess *tcp* genes. One study reported that the levels of expression of a cell-associated hemagglutinin (HA) correlated with adherence to human intestinal epithelial cells (128). Pili were also observed on the surface of *V. mimicus* cells, but expression of these pili did not correlate with adherence. The level of adherence to epithelial cells was much lower than that seen with *V. cholerae* O1, suggesting that reduced adherence is responsible for the reduced virulence of *V. mimicus* relative to *V. cholerae* O1. Two other hemagglutinins, one with protease activity and the other without, have been found in culture supernatants, but no relationship to intestinal adherence has been suggested.

VIBRIO PARAHAEMOLYTICUS

Along with *V. cholerae*, *V. parahaemolyticus* is the best described of the pathogenic vibrios, with numerous studies having been reported since the first description of its involvement in a major outbreak of foodborne illness in 1950. Between 1973 and 1987, over 20 outbreaks of *V. parahaemolyticus* in the United States were reported to the Centers for Disease Control and Prevention (CDC), involving over 1,600 persons.

Classification

The taxonomy of *V. parahaemolyticus* has been described by numerous authors; the reader is referred to the excellent reviews of Joseph et al. (51) and Twedt (127), which include the minimal characteristics for identification of *V. parahaemolyticus* as well as tests which serve to distinguish this species from other pathogenic vibrios. As is the case with many of the vibrios, however, strain variation is common, and phenotypic testing is often insufficient for identification to species. A significant percentage of isolates of *V. parahaemolyticus* which are sucrose positive, for example, have been described. In addition, some traits, such as H_2S production, are dependent on the medium or assay method employed, and caution must be exercised in their determination.

V. parahaemolyticus is serotyped according to both its somatic O and capsular polysaccharide K antigens, based on a scheme developed by Sakazaki et al. (107) after a study of 2,720 strains. There are presently 12 O (LPS) antigens and 59 K (acidic polysaccharide) antigens recognized. Although many environmental and some clinical isolates are untypeable by the K antigen, the majority of clinical strains can be classified to their O type. However, there appears to be no correlation between serotype and virulence.

A special consideration in the taxonomy of *V. parahaemolyticus* is the ability of certain strains to produce a hemolysin termed TDH (thermostable direct hemolysin) or Kanagawa hemolysin, which is correlated with virulence in this species (see Virulence Mechanisms, below). The production of this hemolysin is determined on Wagatsuma agar, which contains yeast extract (5 g/liter), peptone (10 g/liter), mannitol (5 g/liter), K_2HPO_4 (0.5 g/liter), NaCl (70 g/liter), agar (15 g/liter), and crystal violet (1 ml of a 0.1% solution). Freshly drawn and washed human blood cells are added to the cooled medium. The use of Wagatsuma agar for determining the production of beta-hemolysis by Kanagawa phenomenon-positive (KP^+) strains, however, has been replaced, to a great extent, by in vitro DNA amplification and gene probe hybridization methods to differentiate KP^+ and KP^- cells (4).

Although a correlation between urea hydrolysis and the ability to produce the Kanagawa hemolysin has been reported (55), Osawa et al. (96) subsequently examined 132 strains of *V. parahaemolyticus* and found urea hydrolysis not to be a reliable marker for the production of TDH. However, they did find that urea hydrolysis may be a marker for the TDH-related hemolysin, an additional virulence factor described under Virulence Mechanisms below.

Reservoirs

V. parahaemolyticus occurs in estuarine waters throughout the world and is easily isolated from coastal waters of the entire United States, as well as from sediment, suspended particles, plankton, and a variety of fish and shellfish. The latter include at least 30 different species, among them clams, oysters, lobster, scallops, shrimp, and crab. In a study carried out by the FDA, 86% of the 635 seafood samples examined were found to be positive for this species. Counts of *V. parahaemolyticus* have been reported as high as 1,300/g of oyster tissue and 1,000/g of crabmeat, although levels of 10/g are more typical for seafood products. Hackney et al. (38), in a 3-year survey of 716 seafood samples taken in North Carolina, found 46% to be positive for *V. parahaemolyticus*. Notable were unshucked oysters (79% positive), unshucked clams (83% positive), unpeeled shrimp (60% positive), and live crabs (100% positive). Another study isolated *V. parahaemolyticus* from ca. 69 to 100% of the commercially obtained or cultured oysters, clams, and shrimps tested, but only 42% of the crabs.

Hackney et al. (38), as well as others, have observed no correlation with fecal coliforms or other indicators, but *V. parahaemolyticus* numbers showed a definite seasonal variation, with samples analyzed in January and February often free of *V. parahaemolyticus*. Others have also reported ecology of *V. parahaemolyticus* to be heavily influenced by water temperature, salinity, and association with certain plankton, with highest numbers occurring in warmer months and in waters of intermediate salinity. The relationships between each of these parameters have been nicely demonstrated by Kaneko and Colwell (52) in their study of the occurrence of this species in Chesapeake Bay waters. *V. parahaemolyticus* has been occasionally isolated from freshwater sites, but only at extremely low levels (<5 CFU/liter) and only during the warmest periods of the year.

It must be emphasized here that KP^+ strains of *V. parahaemolyticus* are of prime importance in human disease, although occasional KP^- strains have also been isolated from diarrheal stools. KP^+ strains constitute a very small percentage (typically <1%) of the *V. parahaemolyticus* strains found in aquatic environments and seafoods. Thus the simple isolation of this species from water or foodstuffs does not, in itself, indicate a health hazard.

The isolation of *V. parahaemolyticus* generally involves a preenrichment step, and many such enrichment broths have been reviewed by Joseph et al. (51) and Twedt (127). Alkaline peptone water has generally been shown to provide superior recovery of *V. parahaemolyticus* from a variety of fish and shellfish, even when the samples have been chilled or frozen. A large number of plating media have similarly been suggested for this species. Indeed, no less than 18 such media are described and reviewed by Joseph et al. (51) and Twedt (127), of which TCBS remains the most commonly employed. Currently, the 8th edition of the *Bacteriological Analytical Manual* of the U.S. FDA (129) prescribes a 16-h enrichment at 35 to 37°C in alkaline peptone water, from which a loopful is streaked onto TCBS agar to obtain isolated colonies of *V. parahaemolyticus*.

Foodborne Outbreaks

Gastroenteritis with *V. parahaemolyticus* is almost exclusively associated with seafood which is consumed raw, inadequately cooked, or cooked but recontaminated. In Japan, *V. parahaemolyticus* is the major cause of foodborne illness, with as great as 70% of all bacterial foodborne illnesses in that country in the 1960s attributable to this species (127). Outbreaks of *V. parahaemolyticus* gastroenteritis in the United States occurring between 1973 and 1987 have been summarized by Bean and Griffin (3). Whereas most Japanese outbreaks involve fish, the United States outbreaks involved primarily crab, shrimp, lobster, and oysters. Shellfish were implicated in 33 of the 42 outbreaks described by Beuchat (6) and Bean and Griffin (3). The first major outbreak (320 persons ill) in the United States occurred in Maryland in 1971, a result of improperly steamed crabs. Subsequent outbreaks have occurred at all United States coasts and Hawaii.

The largest United States outbreak occurred during the summer of 1978 and affected 1,133 of 1,700 persons attending a dinner in Port Allen, La. All stool isolates were KP^+. The food implicated was boiled shrimp, which yielded positive cultures of *V. parahaemolyticus*. The raw shrimp had been purchased and shipped in standard wooden seafood boxes. They were boiled the morning of the dinner, but returned back to the same boxes in which they had been shipped. The warm shrimp were then transported 40 miles in an unrefrigerated truck to the site of the dinner and held an additional 7 to 8 h until served that night.

In their survey of four Gulf Coast states, Levine et al. (68) found *V. parahaemolyticus* to be the most common cause of gastroenteritis (37% of 71 cases) in that area. Similarly, Desenclos et al. (25) found *V. parahaemolyticus* to be the second leading cause (over 26%) of gastroenteritis cases in those persons who had consumed raw oysters in Florida. In a 15-year survey of vibrio infections reported by a hospital adjacent to the Chesapeake Bay, Hoge et al. (42) found 9 (>69%) of 13 vibrio-positive stool specimens to contain *V. parahaemolyticus* as the sole pathogen.

V. parahaemolyticus was the most commonly isolated vibrio (35 cases) from stool samples submitted by 51 persons attending a conference in New Orleans and reporting diarrheal disease (71). Interestingly, among the 30 stool isolates probed for the TDH gene, 7 (23%) were negative, although diarrhea occurred at a similar frequency in these persons as in those whose stools contained TDH^+ cells.

Characteristics of Disease

V. parahaemolyticus has a remarkable ability for rapid growth, and generation times as short as 8 to 9 min at 37°C have been reported. Even in seafoods, generation times of 12 to 18 min have been demonstrated. As a result, *V. parahaemolyticus* has the ability to increase rapidly in number, both in vitro and in vivo, and this is evidenced in the characteristics of the disease it produces. Symptoms reported in the 1971 Maryland outbreak cited above began 4 to over 30 h after food consumption, with a mean time of 23.6 h. The primary symptoms were diarrhea (100%) and abdominal cramps (86%), along with nausea and vomiting (26%) and fever (23%). Symptoms subsided in 3 to 5 days in most individuals, although they lasted 5 to 7 days in 30% and for more than

7 days in another 20%. In most severe cases, diarrhea was watery with mucus, blood, and tenesmus.

A mean incubation period of 16.7 h, with a range of 3 to 76 h, was reported in the Port Allen outbreak. The duration was from less than 1 day to over 8 days, with a mean of 4.6 days. Hospitalization was required for over 7% of the victims. Symptoms included diarrhea (95.1%), cramps (91.5%), weakness (90.2%), nausea (71.9%), chills (54.9%), headache (47.7%), fever (47.5%), and vomiting (12.2%). Ages ranged from 13 to 78 years.

Infectious Dose and Susceptible Population

V. cholerae and *V. parahaemolyticus* are the only vibrios for which experimental evidence exists regarding the dosages required for initiation of gastroenteritis. Studies using human volunteers have shown that ingestion of 2 $\times 10^5$ to 3×10^7 CFU of KP^+ cells can lead to the rapid development of gastrointestinal illness. Conversely, volunteers receiving as many as 1.6×10^{10} CFU of KP^- cells exhibited no signs of diarrheal illness. No particular serotype has been shown to predominate in human illness or environments. Based on typical numbers of *V. parahaemolyticus* present in fish and shellfish and the low incidence of KP^+ cells in these natural samples (see below), it would appear that a meal of raw shellfish would likely contain no greater than 10^4 KP^+ cells. Thus, for disease to result from consumption of contaminated food, it would appear that mishandling at temperatures allowing growth of the cells would be necessary (127).

Virulence Mechanisms

Although the epidemiologic linkage between human virulence and the ability of *V. parahaemolyticus* isolates to produce the Kanagawa hemolysin has long been established, the molecular mechanisms by which this factor can cause diarrhea have only been elucidated recently. Sakazaki et al. (108) observed that 96.5% (2,655 of 2,720) of strains isolated from human patients were KP^+. Surprisingly, such hemolysin-producing strains are rarely found in the environment. Sakazaki et al. (108) reported only 1% (7 of 650) of environmental isolates to be KP^+. Others have reported even lower frequencies (e.g., 4 of 2,218, or 0.18%, were KP^+ strains in Galveston Bay, Tex.). It is now believed that such observations can be explained through a natural selection of KP^+ strains in the intestinal tract and better survival of KP^- strains in the estuarine environment, and some experimental evidence exists for this explanation (51).

At least four hemolytic components actually exist in *V. parahaemolyticus:* a TDH, a thermolabile direct hemolysin, phospholipase A, and lysophospholipase. At present, only the direct hemolysins (i.e., those whose lysis of erythrocytes occurs without additional substituents) have been extensively studied. The Kanagawa hemolysin (TDH) is a protein, yet is only partially inactivated at 100°C for 30 min at pH 6.0. TDH produces edema, erythema, and induration in skin and has capillary permeability activity. TDH is lethal for mice, with a minimum lethal dose reported at ca. 0.6 μg of protein nitrogen. Intravenous injections of 5 μg of TDH kill mice rapidly, whereas intraperitoneal injections may produce only signs of cramping. TDH lyses erythrocytes from a large variety of animals, but not from horses. This hemolytic activity is temperature dependent, is not enhanced by the addition of lecithin, but is inhibited by various gangliosides.

Although the hemolytic activity of TDH is a convenient assay of activity, it does not explain how this toxin might cause diarrheal illness. When administered orally to suckling mice at low levels, TDH causes diarrhea, and at higher levels (e.g., 50 μg) it produces diarrhea and death. When TDH at doses of 100 μg was administered to ligated rabbit ileal loops, fluid accumulation was not observed, in contrast to CT, with which this response was observed following inoculation of only 0.2 μg. At doses of 200 μg, TDH gives a positive response in rabbit ileal loops but with erosive lesions and desquamation of necrotic intestinal mucosa. Such histological changes were not observed when whole bacteria expressing more physiologically relevant TDH levels were tested in ileal loops. Using a KP^+ strain and an isogenic TDH mutant, it was demonstrated that only the KP^+ parent strain was capable of inducing fluid accumulation in the rabbit ileal loop assay (85). When the complete *tdh* gene was returned to the isogenic mutant, restoration of activity was observed. Similar results were seen when culture supernatants of these strains were tested on rabbit ileal tissue mounted in Ussing chambers, a more sensitive measure of secretory activity. In this assay, the ability of TDH to alter ion transport in the intestinal tract was demonstrated at nanogram levels, with no histological changes. Raimondi et al. (101) further investigated the effect of purified TDH in Ussing chambers and demonstrated that TDH induces intestinal chloride ion secretion and that the trisialoganglioside GT_{1b} appears to be the cellular receptor. Furthermore, rather than cAMP or cGMP, TDH uses Ca^{2+} as an intracellular second messenger, thereby being the first bacterial enterotoxin for which the linkage between changes in intracellular calcium and secretory activity has been established.

TDH is encoded by one or more nonidentical copies of the *tdh* gene in *V. parahaemolyticus.* Initial studies found that KP^+ isolates usually contained two copies of the *tdh* gene but that many KP^- or weakly positive isolates contained only one gene copy. The two gene copies (*tdh1* and *tdh2*) of KP^+ strains are not identical and the predicted protein products differ in seven amino acid residues, although the proteins are immunologically indistinguishable. While both gene products contribute to the KP phenotype, >90% of the TDH protein can be attributed to high-level expression of the *tdh2* gene (86). Recent studies have demonstrated that the level of TDH production may be under the control of a regulator similar to the ToxR of *V. cholerae.* A homolog of the *toxRS* genes has been reported in *V. parahaemolyticus,* which promotes the expression of the *tdh2* gene but not of the *tdh1* gene. As occurs in *V. cholerae,* the extent of the *V. parahaemolyticus* ToxR-stimulated increase in *tdh* expression is culture medium dependent. A recent review by Nishibuchi and Kaper (86) describes the regulation of *tdh* gene expression and its involvement in the pathogenesis of *V. parahaemolyticus.*

During one outbreak of gastroenteritis, KP^- isolates of *V. parahaemolyticus* were found to produce a TDH-related hemolysin, termed TRH, but not TDH. The *trh* gene shares 69% identity to the *tdh2* gene, and its biological, immunological, and physicochemical characteristics are similar, but not identical, to those of TDH (87). A survey of 285 strains of *V. parahaemolyticus* revealed that not only were *tdh*-positive strains strongly associated with gastroenteritis, but *trh*-positive strains were also associated. Using a gene probe, *trh* was detected in over 35% of all clinical strains, including 24% of those lacking the *tdh* gene. This suggests that TRH may be an important virulence factor and possibly the cause of diarrhea in those patients from whom only KP^- strains of *V. parahaemolyticus* are isolated from stools. With few exceptions, it appears that strains of *V. parahaemolyticus* produce only one of these two hemolysins.

A novel siderophore, vibrioferrin, which is able to sequester iron from 30% iron-saturated human transferrin, has been described for *V. parahaemolyticus.* Its importance in pathogenesis is unknown, although the ability to utilize such a source of host iron could enhance survival and proliferation in vivo.

While much is understood of the toxins of *V. parahaemolyticus,* little is known about the adherence process, which is an essential step in the pathogenesis of most enteropathogens. Several adhesive factors have been proposed, including outer membrane proteins, lateral flagella, pili, and a mannose-resistant, cell-associated hemagglutinin, but the importance of any of these factors in human disease is unknown.

VIBRIO VULNIFICUS

Of all of the pathogenic vibrios, *V. vulnificus* is the most serious in the United States, alone responsible for 95% of all seafood-related deaths in this country. This bacterium is the leading cause of reported deaths from foodborne illness in Florida (41). Among that portion of the population which is at risk of infection by this bacterium, primary septicemia cases resulting from raw oyster consumption typically carry fatality rates of 60%. This is the highest death rate of any foodborne disease agent in the United States (125). According to CDC and FDA estimates there are 50 foodborne cases per year of *V. vulnificus* serious enough to be recognized by hospital personnel, although estimates of 17,500 to over 41,000 total cases have been calculated (125). The bacterium is unusual in being able to produce wound infections in addition to gastroenteritis and primary septicemias. Wound infections have a 20 to 25% fatality rate, are also seawater and/or shellfish associated, and generally require surgical debridement of the affected tissue or amputation. The biology of *V. vulnificus,* as well as the clinical manifestations of both the primary septicemic and wound forms, has been reviewed (89). Discussion here is limited to the primary septicemic (foodborne) form of infection caused by this bacterium.

Classification

The first detailed taxonomic study of this species was by researchers at the CDC, who in 1976 described 38 strains submitted by clinicians throughout the United States. Originally termed the "lactose-positive" vibrio, its current name was suggested after a series of phenotypic and genetic studies by several laboratories (89). In 1982, a second biotype of *V. vulnificus* was described which can easily be differentiated from biotype 1 by its negative indole reaction. Biotype 2 strains are a major source of fatalities in eels and, although not generally considered to be pathogenic for humans, have been reported to lead to human infection in isolated instances. The discussions here will focus on the originally described biotype 1, which is the major human pathogen.

The phenotypic traits of this species have been fully described in a number of studies (30, 89). The isolation of *V. vulnificus* from blood samples is straightforward, as the bacterium grows readily on TCBS, MacConkey, and blood agar. Isolation from the environment has been much more problematic. Vibrios tend to make up 50%

or more of estuarine bacterial populations, and most of these have not been characterized. It is well recognized that considerable variation exists in the phenotypic traits ascribed to *V. vulnificus*, including lactose and sucrose fermentation, which are considered among the most important in identifying this species. This problem is best exemplified by the isolation from a clinical sample of a bioluminescent strain of this species.

Attempts to characterize the distribution of this species have routinely employed TCBS for its isolation, although the value of this medium for isolating environmental vibrios has been frequently questioned. Various other media have been proposed for isolating *V. vulnificus*, but the best appears to be colistin-polymyxin B-cellobiose agar. This medium has been evaluated for the isolation of *V. vulnificus* from oysters and clams (92), and Sun and Oliver (119) determined that >80% of 1,000 colonies of appropriate morphology that appeared on this medium following plating of oyster homogenates could be identified as *V. vulnificus*. Similar results were obtained by Sloan et al. (113) in their study of five selective enrichment broths and two selective agar media for isolating *V. vulnificus* from oysters. Currently, the 8th edition of *Bacteriological Analytical Manual* of the U.S. FDA (129) prescribes a 16-h enrichment at 35 to 37°C in alkaline peptone water, from which a loopful is streaked onto modified CPC agar to obtain isolated colonies of *V. vulnificus*. This species is best identified by employing a probe against its hemolysin gene (83).

Susceptibility to Environmental Factors

Cook and Ruple (22) determined that *V. vulnificus* cells naturally occurring in oysters undergo a time-dependent inactivation when either shellstock oysters or shucked oyster meats are held at 4, 0, or –1.9°C. Cook (21) found that, after harvest, *V. vulnificus* did not grow within 30 h in shellstock oysters stored at 13°C or below, whereas the pathogen grew when held at 18°C or higher for 12 or 30 h.

In response to a report that *V. vulnificus* could be killed by applying cocktail or Tabasco sauces to raw oysters, Sun and Oliver (118) determined the sensitivity of this species to a commercial horseradish-based sauce and Tabasco sauce. Results revealed that, while Tabasco sauce (but not cocktail sauce) was highly effective in reducing the number of *V. vulnificus* cells present on the oyster meat surface, there was little reduction of vibrios within the oysters by either sauce. The in vitro bactericidal activity of a variety of "generally recognized as safe" compounds against *V. vulnificus* has also been reported. Although several can reportedly kill *V. vulnificus* at low levels, only diacetyl had any significant in vivo effects against this species when naturally present within oysters (117).

Reservoirs

V. vulnificus is a widespread inhabitant of estuarine environments, having been isolated from the Gulf, East, and Pacific Coasts of the United States and from around the world (56, 58, 89, 95, 121). One of the most comprehensive studies on the distribution and ecology of *V. vulnificus* was a three-summer survey of the entire East Coast of the United States (94). More than 6,000 sucrose-negative isolates were examined and, on the basis of phenotypic and DNA-DNA hybridization studies, approximately 1% of the culturable vibrios were identified as *V. vulnificus*. Isolates were obtained from Cape Cod, Mass., to Miami, Fla., from seawater, plankton, fish, and shellfish. Only 7 CFU of *V. vulnificus* were detected per ml of seawater, but an average of 6×10^4 CFU/g was found in oysters. Similar results were obtained by Tamplin et al. (121), who reported from 0.3 to 7,000 CFU of *V. vulnificus* per 100 ml of seawater taken from two estuarine bays in Florida. *V. vulnificus* has also been isolated from crabs, clams, ark shells, tarbos, and seawater surface samples (89). *V. vulnificus* has also been reported in fish; DePaola et al. (24) have recently described its isolation from the intestinal tracts of a variety of bottom-feeding coastal fish and have suggested that such fish may represent a major reservoir of this species.

Oliver et al. (94) found no correlation between the presence of *V. vulnificus* and the presence of fecal coliforms, and this has been repeatedly confirmed by other investigators. A strong correlation has been observed between water temperature and presence of *V. vulnificus* (Fig. 13.2) (58, 89, 123), agreeing with epidemiologic studies on infections caused by *V. vulnificus*. Because of this seasonality and the inability to isolate *V. vulnificus* from water or oysters when water temperatures are low, there has been considerable investigation of the apparent die-off of this species during cold weather months. This is now attributed to a cold-induced viable but nonculturable state, wherein the cells remain viable but are no longer culturable on routine media normally employed for their isolation. This phenomenon, which has now been demonstrated in at least 16 genera, has been the subject of several recent reviews (90, 91). It is also possible that *V. vulnificus* overwinters in certain environments. In this context DePaola et al. (24) isolated *V. vulnificus* during the winter from sheepshead fish at higher densities than in sediment or seawater.

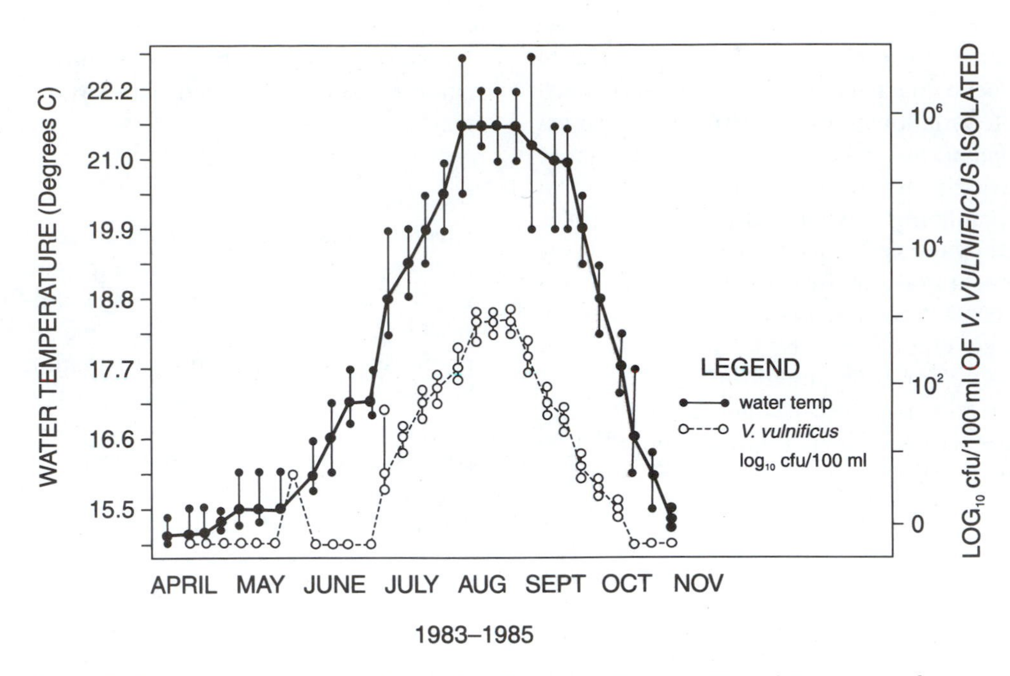

Figure 13.2 Numbers of *V. vulnificus* isolated from northeastern United States coastal waters as a function of time of year and water temperature. From reference 123 with permission of the publisher.

Foodborne Outbreaks

There are no reported cases of more than one person developing *V. vulnificus* infection following consumption of the same lot of oysters. Occasional reports document the consumption of raw oysters and subsequent disease in one family member, while others exhibited no symptoms. Further, because raw oysters are usually eaten whole, there rarely exist any remains of the implicated oyster to sample. Additional raw oysters from the same lot, or even the same serving, may remain, but studies have revealed that two oysters obtained from the same estuarine location often have vastly different numbers of *V. vulnificus* cells. Hence it is difficult to track the source of the *V. vulnificus* strains isolated from the patient. It is possible that newly developed fingerprinting methods can enable molecular epidemiology to be used in tracking infections. Buchrieser et al. (11), employing clamped homogeneous electric field gel electrophoresis to examine 118 strains of *V. vulnificus* isolated from three oysters, reported that no two isolates had the same profile. Hence, it remains to be determined whether all strains are capable of causing infection, or whether only certain strains are pathogenic.

V. vulnificus infections are highly correlated with water temperature, with most cases occurring during the months of April through October. Nearly all *V. vulnificus* infections result from consumption of raw oysters, and most result in primary septicemias.

In the only prospective study on the incidence of vibrios in symptomatic or asymptomatic infections among persons eating raw shellfish, Lowry et al. (71) found 3 *V. vulnificus* stool-positive persons of 479 tested, and none was ill. Coupled with their finding that two-thirds of the raw oysters tested were *V. vulnificus* culture positive, results indicate that exposure to this species can be relatively high and underscores the importance of at-risk persons avoiding raw seafood. Inland clinical laboratories rarely encounter *Vibrio* species, with most isolates coming from coastal areas. Even there, however, the frequency of *V. vulnificus* is not high. In a collaborative 10-year study conducted by five hospitals in Mobile, Ala., only 12 of the vibrios isolated from humans were *V. vulnificus*. No isolates were obtained from stool samples in that study.

Characteristics of Disease

Many studies of infections caused by *V. vulnificus* have been reported. In a review of 57 cases of primary septicemia in which 85% resulted from consumption of raw oysters, Oliver (89) determined that the incubation period ranged from 7 h to several days, with a median of 26 h. The most significant symptoms included fever (94%), chills (86%), nausea (60%), and hypotension (systolic pressure of <85 mm; 43%). While frequently present, symptoms typical of gastroenteritis were not as common: abdominal pain (44%), vomiting (35%), and diarrhea (30%). An unusual symptom that generally (69%) occurs is the development of secondary lesions. These usually occur on extremities, frequently develop into necrotizing fasciitis or vasculitis, and often necessitate surgical debridement or limb amputation.

Gastrointestinal illness with associated diarrhea is relatively infrequent. Johnston et al. (49) were the first to provide evidence of such a syndrome, reporting on three males who presented with abdominal cramps; *V. vulnificus* was isolated from their diarrheal stools. All three had a history of alcohol abuse, routinely took antacids or cimetidine, and had eaten raw oysters during the week prior to their illness. While symptoms in all three subsided without antibiotic treatment, diarrhea continued for a month in one case. Desenclos et al. (25), in a survey of vibrio infections in raw-oyster eaters in Florida, reported eight cases of *V. vulnificus*-induced gastroenteritis, six of which involved consumption of raw oysters. No details of these infections were provided. Levine et al. (68), in a survey of vibrio infections occurring during 1 year in four Gulf Coast states, reported an additional three cases of gastroenteritis from *V. vulnificus*. However, no individual descriptions of these cases were provided.

The time of death in fatal cases of *V. vulnificus* varies considerably, ranging from 2 h post-hospital admission to as long as 6 weeks (89). Most deaths occur within a few days. Hospital stays of several weeks are usual for those surviving primary septicemic disease. Survival largely depends directly on prompt antibiotic administration.

Infectious Dose and Susceptible Population

In almost all *V. vulnificus* infections which follow ingestion of raw oysters, the patient has an underlying chronic disease (89). The most common (80%) of these is a liver- or blood-related disorder, with liver cirrhosis secondary to alcoholism or alcohol abuse being the most typical (25, 89). These diseases typically result in elevated serum iron levels, and laboratory studies of experimental *V. vulnificus* infections have shown that elevated serum iron plays a major role in this disease (135). Other risk factors include hematopoietic disorders, chronic renal disease, gastric disease, use of immunosuppressive agents, and diabetes. Although cases in children have been reported, infections most frequently occur in males (82% of those reviewed by Oliver [89]) whose average age exceeds 50 years.

The infectious dose of *V. vulnificus* is not known. However, some insight to this question may be available using a mouse model. Wright et al. (135) observed that, when mice were injected with 16 μg of iron to produce serum iron overload, the 50% lethal dose (LD_{50}) decreased from ca. 10^6 to a single cell. In a variation of these studies, the administration of small amounts of CCl_4 to produce short-term liver necrosis was found to increase serum iron levels, with an inverse correlation between serum iron levels and LD_{50} values observed. Such data agree with epidemiologic studies indicating that liver damage and sometimes immunocompromising diseases are major underlying factors in the development of *V. vulnificus* infections, and they suggest that extremely low numbers of this pathogen may be sufficient to initiate potentially fatal infections.

V. vulnificus is susceptible to most antibiotics, and while a large variety of antibiotics have been used clinically, laboratory studies suggest tetracycline may be especially effective against this species.

Virulence Mechanisms

Capsule

The polysaccharide capsule produced by some strains of *V. vulnificus* is essential to its ability to initiate infection. Simpson et al. (112) studied 38 strains of *V. vulnificus*, including both clinical and environmental isolates and virulent and avirulent strains, and found that all virulent strains were of the "opaque" (encapsulated) colony type, whereas isogenic cells taken from "translucent" (acapsular) colonies were avirulent (Fig. 13.3). It was further observed that encapsulated cells produce nonencapsulated cells, and this loss of capsule correlates with loss of virulence in

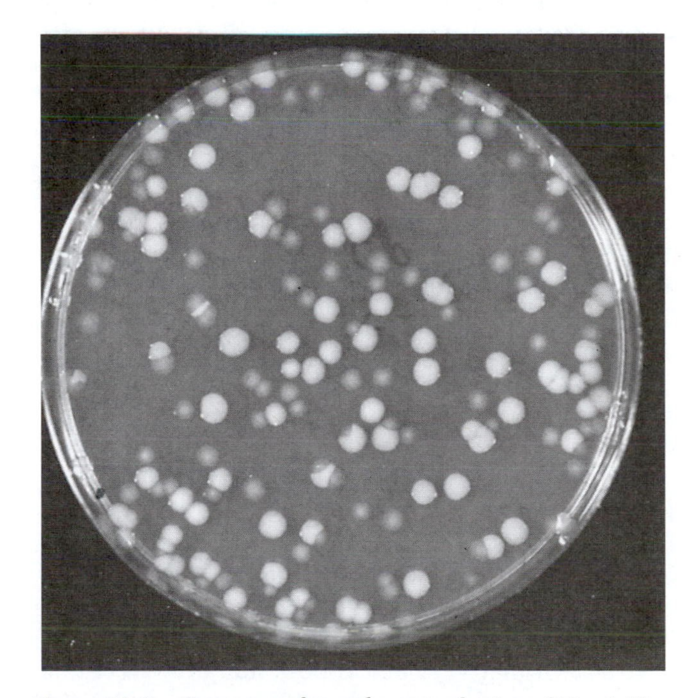

Figure 13.3 Opaque and translucent colonies of *V. vulnificus* C7184. From reference 112.

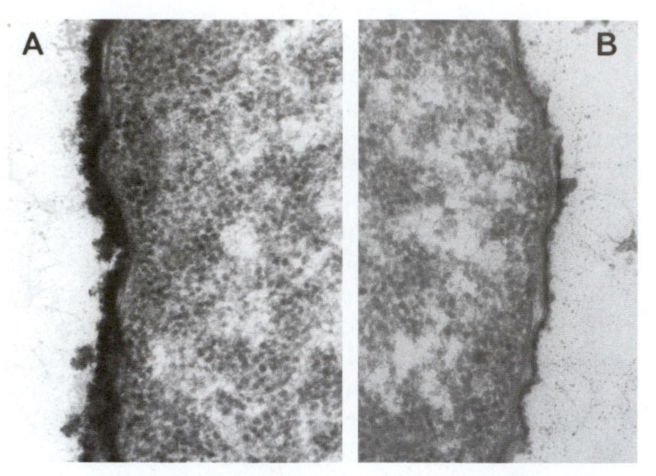

Figure 13.4 Electron micrographs of cells from opaque and translucent colonies of *V. vulnificus* C7184. Cells were stained for acidic polysaccharide with ruthenium red. (A) Opaque cell type, showing thick polysaccharide layer outside the cell envelope. (B) Translucent cell type, demonstrating thin layer of ruthenium red-staining polysaccharide. From reference 112.

otherwise isogenic strains. These authors found that only the encapsulated cells were able to utilize transferrin-bound iron, and only these cells were iron responsive, i.e., were virulent at an inoculum of 10^3 in iron-overloaded mice. The primary reason for the avirulence of translucent cells likely resides in the observation that the capsule (Fig. 13.4) allows these cells to resist phagocytosis.

Simonson et al. (111) used polyclonal antibodies to study the capsule of *V. vulnificus* and found at least 10 distinct serotypes. Of those which were associated with human infection, all 10 serotypes were represented, with 34.6% being of types 2 and 4. In contrast, 84.6% of the typeable environmental strains were of serotype 3 or 5, but only 7.7% were of type 2 or 4. Similarly, Hayat et al. (39) reported that 19% of 21 clinical strains, but none of the 67 environmental isolates examined, agglutinated with antiserum prepared against a clinical isolate. Whether such capsules play a role in the epidemiology of *V. vulnificus* infections remains to be determined.

Iron

Elevated levels of serum iron appear to be essential for *V. vulnificus* to multiply in the human host (89). The effect of elevated serum iron on reduction of LD_{50} values in mice has been described above. Wright et al. (135) have shown that *V. vulnificus* does not grow in normal human serum, suggesting that this bacterium can only produce septicemia in humans with elevated serum iron levels. Although *V. vulnificus* simultaneously produces both hydroxymate and phenolate siderophores, it is unable to compete with serum transferrin for iron, and this is likely a major factor in its inability to initiate infections in healthy individuals (135). A similar result has been reported for other iron-binding proteins such as lactoferrin and ferritin. *V. vulnificus* is able to overcome the binding of haptoglobin to hemoglobin, however, and this may represent another aspect of the importance of iron in the pathogenesis of these infections.

LPS

The symptoms which occur during *V. vulnificus* septicemia, including fever, tissue edema, hemorrhage, and especially hypotension, are those classically associated with gram-negative endotoxic shock. Thus, another product of *V. vulnificus* which may be critical to its virulence is the endotoxic LPS present in cells. McPherson et al. (76) found that intravenous injections of extracted and partially purified LPS from *V. vulnificus* in mice and rats caused decreased arterial blood pressure within 10 min, which further declined, leading to a decrease in heart rate and death within 30 to 60 min. Subsequent studies revealed that an inhibitor of nitric oxide synthase (the LPS-induced enzyme responsible for release of nitric oxide and subsequent host tissue damage) administered 10 min after LPS injection reversed this lethal effect. These results indicate that the classic symptoms of endotoxic shock observed following *V. vulnificus* infection are likely due to the stimulation by LPS of nitric oxide synthase and that inhibition of this enzyme is a possible treatment for the endotoxic shock produced by this organism. In this regard, Simonson et al. (111) identified five LPS serological varieties in *V. vulnificus* by employing monoclonal antibodies. They found that 25% of the clinical isolates expressed LPS antigens 1 and/or 5, whereas only 0.3% of environmental isolates were of these serotypes. The chemical composition of the LPS extracted from one strain of *V. vulnificus* has been reported; however, little is known regarding the relative virulence of the different serotypes.

Other Toxins

In addition to the role of capsule, iron, and endotoxin in the pathogenesis of *V. vulnificus* infections, this bacterium produces a large number of extracellular compounds, including hemolysin, protease, elastase, collagenase, DNase, lipase, phospholipase, mucinase, chondroitin sulfatase, hyaluronidase, and fibrinolysin (89). To date, however, none of these putative virulence factors has definitively been shown to be involved in pathogenesis.

The most studied of these is a potent heat-stable hemolysin/cytotoxin that possesses cytolytic activity against a variety of mammalian erythrocytes, cytotoxic activity against CHO cells, vascular permeability factor activity against guinea pig skin, and lethal activity for mice (64). The toxin has been purified and has a molecular weight of ca. 56,000 and an LD_{50} value for mice of ca. 3 μg/kg following intravenous injection. Mutants having this hemolysin show no reduction in LD_{50} for mice following intraperitoneal injection.

An elastolytic protease has been described that lacks hemolytic activity, but which degrades albumin, immunoglobulin G, complement factors C3 and C4, and elastin. Minimum lethal doses in mice, regardless of route of injection, were reported to be about 25 μg, with extensive hemorrhagic necrosis, edema, and muscle tissue destruction occurring.

While none of these many putative virulence factors is known to be essential to virulence of *V. vulnificus*, they may play a role in pathogenesis or might be essential for the wound infections also produced by this species.

VIBRIO FLUVIALIS

Classification

V. fluvialis was originally described by Lee et al. (65) and referred to as "group F." Subsequently, it was determined that this group was identical to that referred to as "group EF-6" by the CDC, and as a result of further taxonomic study, it was concluded that these isolates were a new species, *V. fluvialis*. Two "biogroups" of group F vibrios were originally described, of which biogroup I was anaerogenic and isolated from aquatic environments and diarrheal cases, whereas biogroup II was aerogenic and not disease associated. Subsequent studies indicated that the aerogenic strains were a unique species, and these were subsequently reclassified as *V. furnissii* (10).

The possibility of confusion with *Aeromonas* strains (especially *Aeromonas hydrophila*) has been recognized, as both are arginine dihydrolase positive. The simplest differentiation is the inability of *V. fluvialis*, being halophilic, to grow in media lacking NaCl. The lack of production of indole by *V. fluvialis* also differentiates this species from *Aeromonas* spp. (51).

Reservoirs

V. fluvialis has been frequently isolated from brackish and marine waters and sediments in the United States (51) as well as other countries. It has been reported in fish and shellfish from the Pacific Northwest and Gulf Coast. Wong et al. (134) reported that about 65 to 79% of the oysters, hard clams, and freshwater clams they examined harbored *V. fluvialis*, but only 25% of the crabs and 6% of the shrimp. This bacterium has only rarely been found in freshwaters.

Foodborne Outbreaks

V. fluvialis was isolated from over 500 patients with diarrheal stools at the Cholera Research Laboratory of Bangladesh during a 9-month period between 1976 and 1977 (46). Approximately half of the patients were less than 5 years old. Since that outbreak, however, *V. fluvialis* has only occasionally been reported to be an enteric pathogen.

In the United States, Levine et al. (68) found *V. fluvialis* to account for 10% of the clinical cases in their survey of vibrio infections along the Gulf Coast. All seven of these cases resulted in gastroenteritis, with three requiring hospitalization. Raw oysters were the implicated vehicle in at least three of the seven, and shrimp in one other.

In the largest clinical series of *V. fluvialis* cases described in the United States, Klontz and Desenclos (60) described 12 persons in Florida from whom this species was recovered between 1982 and 1988. This species was cultured from the stools of 10 of the 12, who presented with gastroenteritis. Eight of the 10 who reported eating seafood during the week before became ill, with raw oysters implicated in five cases, shrimp in two, and cooked fish in one. In a subsequent survey of vibrioses resulting from raw-oyster consumption in Florida during an 8-year period, Desenclos et al. (25) reported that 5.6% of the 125 cases of gastroenteritis were caused by *V. fluvialis*.

Characteristics of Disease

Gastroenteritis with *V. fluvialis* is similar to that of cholera, with diarrhea and vomiting (97%), moderate to severe dehydration (67%), abdominal pain (75%), and fever (35%) being common symptoms (46). Passage of 10 to 12 stools per day has been reported, with individual stool outputs of 0.5 to 7 liters. Diarrhea typically lasts from 16 h to over 3 days. Stools collected during the original Bangladesh outbreak revealed an average of 10^6 *V. fluvialis* cells per ml. A notable difference from cholera is the frequent occurrence of bloody stools in infections due to *V. fluvialis*. Huq et al. (46) reported that 75% of the stool samples they examined contained erythrocytes or leukocytes, and frank bloody stools were observed for several of the cases. Bloody stools were also reported in 86% of the seven cases reported by Levine et al. (68).

All 10 of the gastroenteritis patients reviewed by Klontz and Desenclos (60) had diarrhea (generally watery) with 2 to 20 stools per day (median of 7). Five patients reported at least one episode of bloody stools. Other symptoms included nausea, vomiting, and abdominal cramps. None had hypotension. The medium age of the patients was 37 (range of 1 month to 67 years), with eight being male. The median incubation period for six of the patients who had eaten seafood was 39 h (range of 16 to 60 h). The median duration of illness was 6 days (range of 1 to 60 days); five patients were hospitalized for a median of 4 days (range of 3 to 11 days).

Infectious Dose and Susceptible Population

The infectious dose for this species is not known. Persons with gastroenteritis ranged from 1 month old to >80 years old. It appears that underlying disease plays a major role in infections caused by *V. fluvialis.* Klontz and Desenclos (60) reported that only 4 of 10 patients with *V. fluvialis* gastroenteritis had underlying medical conditions, including diabetes, alcohol abuse, and ulcerative colitis. One patient had a history of cardiopulmonary disease and peptic ulcers and was taking antacids at the time of hospitalization. The fatality described by Klontz et al. (59) did not have a history of alcoholism or liver disease, but had extensive coronary artery disease.

While successful resolution of infection has been reported without antibiotic therapy, such treatment, often along with intravenous fluids, is generally required. Antibiotic therapy was not successful in at least one case.

Virulence Mechanisms

Most of the early studies on virulence of *V. fluvialis* isolates failed to detect evidence of an enterotoxin in this species. This might have been due to the growth conditions employed, as it was subsequently found that production of a heat-labile enterotoxin that induces fluid accumulation in the intestines of suckling mice is dependent on the content of the culture medium. It has been shown that, unlike clinical isolates, most environmental isolates do not induce fluid accumulation.

Lockwood et al. (69) have described three potential enterotoxins for *V. fluvialis,* a CHO cell cytotoxin, a CHO cell-rounding toxin, and a CHO cell elongation factor. Crude preparations of all three heat-labile toxins cause fluid accumulation in infant mice, but only the CHO cell-rounding toxin stimulates fluid secretion in ligated ileal loops.

Chikahira and Hamada (14) studied the effect of cell filtrates and extracts of 39 environmental and clinical isolates of *V. fluvialis* in a variety of assays. A lethal effect on CHO cells was observed to be caused by 84% and 75% of the environmental and clinical isolates, respectively, and 45% and 25%, respectively, caused cell elongation. The cell elongation factor was appreciably neutralized by anti-CT serum, but the cell killing factor was not. Live cultures of all the human and half of the environmental isolates also caused fluid accumulation in rabbit ileal loops. This activity was not neutralized by anti-CT serum. In addition, enzyme-linked immunosorbent assay studies indicated that 64% of the environmental and 67% of the human isolates produced CT. None of the culture filtrates caused fluid accumulation in suckling mice. Culture filtrates of 15 of the 16 strains tested (one human isolate was not lethal) were lethal to mice, with death occurring within 20 min. Less than a third of the isolates were hemolytic for rabbit erythrocytes.

In addition to the above factors, Oliver et al. (93) reported elastase, mucinase, protease, lipase, lecithinase, chondroitin sulfatase, hyaluronidase, DNase, and fibrinolysin production by a strain of *V. fluvialis. V. fluvialis* was comparable to *V. vulnificus* and *V. mimicus* in producing such exoproducts.

VIBRIO FURNISSII

Classification

The taxonomy of *V. furnissii* has been described by Brenner et al. (10) and Farmer et al. (31). Their studies included biochemical reactions and DNA-DNA hybridization relatedness to *V. fluvialis* as well as to *Aeromonas* and *Alteromonas* species and to other vibrios. *V. furnissii* is similar to *Aeromonas hydrophila,* from which it can easily be distinguished by its ability to grow in 6% NaCl. *V. furnissii* can be differentiated from *V. fluvialis* by, among other traits, its production of gas from glucose.

Reservoirs

V. furnissii has been isolated from river and estuarine water, marine molluscs, and crustacea from throughout the world. Wong et al. (134) found a relatively small percentage (ca. 7 to 12%) of the oysters, clams, shrimps, and crabs they examined to harbor this species.

Foodborne Outbreaks

The largest documented outbreaks of *V. furnissii* were in 1969, when this species was isolated during an investigation of two outbreaks of acute gastroenteritis in American tourists returning from the Orient. In the first outbreak, 23 of 42 elderly passengers returning from Tokyo developed gastroenteritis; one woman died and two other persons required hospitalization. Food histories implicated shrimp and crab salad and/or the cocktail sauce served with the salads. *V. furnissii* was recovered from seven stool specimens, two of which also contained *V. parahaemolyticus*. The second outbreak affected 24 of 59 persons returning from Hong Kong. Nine persons were hospitalized. A food vehicle was not identified, but *V. furnissii* was isolated from at least five fecal specimens. However, because several other potentially enteropathogenic bacteria were also isolated from these stool samples, an absolute causal role of *V. furnissii* could not be documented.

Characteristics of Disease

Symptoms described by Brenner et al. (10) for the gastroenteritis outbreaks described above included diarrhea (91 to 100%), abdominal cramps (79 to 100%), nausea (65 to 89%), and vomiting (39 to 78%). There were no reports of fever. Onset of symptoms occurred between 5 and 20 h, with the patients recovering within 24 h.

Infectious Dose and Susceptible Population

Neither the infectious dose of *V. furnissii* nor the susceptible population is known.

Virulence Mechanisms

Chikahira and Hamada (14) provided an extensive description of the toxic products produced by nine environmental strains of *V. furnissii*. Culture filtrates or cell extracts of only 1 (11%) strain caused cell elongation of CHO cells, whereas all of the strains caused cell death, an activity that Lockwood et al. (70) determined to be distinct from elongation activity. Chikahira and Hamada observed that 86% of the 38 *V. furnissii* strains tested produced a factor cross-reacting with CT, whereas 50% of 4 tested produced fluid accumulation in rabbit ileal loops and produced a hemolysin active against rabbit erythrocytes. The culture supernatants of all *V. furnissii* strains were mouse lethal. Strains are not known to be piliated.

VIBRIO HOLLISAE

Classification

V. hollisae was described as a new species in 1982 (40). Of the original 16 strains obtained and characterized by the CDC, 15 were from stool samples or intestinal contents, and many of these patients had diarrhea. The phenotypic traits of this species are described in reference 40.

V. hollisae is unusual among the vibrios in its inability to grow on TCBS agar or MacConkey agar. These two media are routinely employed for the examination of stool samples for vibrios, and hence it is possible that many infections caused by this species are missed in clinical labs. *V. hollisae* grows well on blood agar, and xylose-lysine-desoxycholate agar recovers this species. The API 20E system properly identifies most strains of *V. hollisae,* although the system failed to identify one of the isolates described by Abbott and Janda (1).

Foodborne Outbreaks

Thirty cases of *V. hollisae* have been reported in the literature since the original description of this species by Abbott and Janda (1). Most (87%) have been cases of gastroenteritis in adults, with 13% of the cases being extraintestinal infections. Three cases of septicemia associated with *V. hollisae* have been reported.

There is a strong correlation between *V. hollisae* infections and consumption of raw seafood. Cases also have been associated with consumption of fried catfish and of dried and salted fish. A case of *V. hollisae*-associated diarrhea has been reported involving a 61-year-old male who denied recent travel or seafood consumption (1).

A study of 121 vibrio infections which occurred in 1989 in four Gulf Coast states revealed that 9 were due to *V. hollisae* (68). Eight of these were cases of gastroenteritis, with one being a wound infection. Two of the patients required hospitalization. The foods most commonly consumed by the patients with gastroenteritis were raw oysters, raw clams, crabs, and shrimp. A study of 333 adult cases of vibrio illnesses in Florida over an 8-year period revealed that 32 were due to *V. hollisae* (25). Of these, 20 (62.5%) were associated with ingestion of raw oysters. Seventeen of these 20 cases were gastroenteritis and 3 involved septicemia. Morris et al. (82) described nine cases of diarrhea that were culture positive for *V. hollisae* with no other enteric pathogen identified. All had diarrhea and abdominal pain, and all but one were hospitalized. Six cases were associated with eating raw oysters or clams, and another had eaten raw shrimp. A seventh patient had eaten seafood, but denied eating it raw.

Although *V. hollisae* had previously been isolated from a septicemia case (82), Lowry et al. (72) were the first to describe a case of septicemia from which *V. hollisae* alone was isolated from blood cultures. A 65-year-old male was admitted to hospital with a 12-h history of night fever, vomiting, and abdominal pain. He had passed two loose stools. On the day before hospitalization, the patient ate fried catfish for lunch and again at dinner. His illness began at midnight, and on admission *V. hollisae* was isolated from a blood culture. The patient was given antibiotics and was discharged 8 days after admission. This case is unusual not only because of the septicemia, but also because the infection occurred following consumption of a freshwater fish that had been fried. There was no history of exposure to saltwater or consumption of other seafood by the patient. Catfish are capable of adapting to low salinities, and the Mississippi River in the area where the fish was obtained often has salt concentrations up to 0.5%. Perhaps incomplete cooking or recontamination of the fish occurred. An unusual case of foodborne septicemia also was reported in which the vehicle appeared to be dried and salted fish obtained from a Southeast Asian food store and eaten uncooked.

Characteristics of Disease

Symptoms of gastroenteritis caused by *V. hollisae* are similar to those caused by non-O1 strains of *V. cholerae* (82) and typically include severe abdominal cramping, vomiting, fever, and watery diarrhea (1). In the cases reported by Morris et al. (82), the median duration of diarrhea was 1 day (range of 4 h to 13 days), with occasional reports of bloody diarrhea. Eight of these patients were hospitalized (median duration of 5 days, range of 2 to 9 days), and all recovered.

Infectious Dose and Susceptible Population

Six of the nine cases of gastroenteritis described by Morris et al. (82) were male, with an average age of 35 years (range, 31 to 59). Other reports do not suggest a predilection for males or females or for age differences (1). No seasonality in the infections is apparent, with cases appearing in the winter as well as summer. Most cases of gastroenteritis were in otherwise healthy individuals (1, 82). Abnormal liver function, secondary to alcohol abuse, was present in one case reported by Morris et al. (82). In the septicemia case described by Lowry et al. (72), the patient had consumed a fifth of wine daily for 20 years and had previously undergone pancreatectomy and splenectomy as well as other surgical procedures. Unlike cases of gastroenteritis, underlying disease is associated with septicemia cases (1, 82).

The organism is highly susceptible to most antimicrobial agents (1, 82), and tobramycin and cefamandole have been used successfully in a septicemic case (72).

Virulence Mechanisms

Gene sequences in *V. hollisae* which are homologous to the *tdh* gene of *V. parahaemolyticus* have been reported, although strain-to-strain variation exists (86). The hemolysin has been purified and partially characterized and is related to the *V. parahaemolyticus* TDH.

Intragastric administration of *V. hollisae* into infant mice elicits intestinal fluid accumulation. An enterotoxin which elongates CHO cells and causes fluid accumulation in mice has been purified from this species (63). It has also been detected in extracts of infected mice and in culture fluids from various growth media.

A hydroxamate siderophore, identified as aerobactin, is produced by *V. hollisae* in response to iron limitation. This iron-binding protein may play a role in the ability of this species to cause infection. One bacteremic patient orally took ferrous sulfate as a supplement for chronic anemia, which may have exacerbated illness.

Miliotis et al. (77) have recently described the ability of *V. hollisae* to adhere to and invade cultured epithelial cells, with internalization involving both eukaryotic and prokaryotic factors. The ability to invade epithelial cells is consistent with the invasive disease produced by *V. hollisae* in some patients (72).

VIBRIO ALGINOLYTICUS

Classification

V. alginolyticus was originally classified as a biotype of *V. parahaemolyticus.* The two are genetically quite similar. However, the two can easily be differentiated phenotypically, most readily by the fermentation of sucrose by *V. alginolyticus.* Some differences in taxonomic traits between clinical and environmental isolates of *V. alginolyticus* have been suggested, but whether these are significant has yet to be determined.

Reservoirs

V. alginolyticus inhabits, often at high numbers, seawater and seafood from throughout the world (51). It is easily isolated from fish, clams, crabs, oysters, mussels, and shrimp, as well as water. Results of many surveys have revealed that this species of *Vibrio* is one of the most commonly isolated vibrios. Several studies have revealed a temperature correlation being associated with its isolation, with greatest rates of isolation occurring during the warm water months.

Foodborne Outbreaks

Most *V. alginolyticus* infections are associated with the marine environment and are usually superficial. Farmer et al. (30) determined that only 5.4% of 74 isolates studied at the *Vibrio* Reference Laboratory of the CDC were of fecal or intestinal origin. Only rarely has *V. alginolyticus* been implicated as a foodborne pathogen.

This species has been isolated from 0.5% of healthy people in Japan, with no clinically associated intestinal disease evident. *V. alginolyticus* has been isolated from rice water diarrheal stools of a female patient with acute enterocolitis and also from the trout roe she had consumed. *V. alginolyticus* has also been isolated from blood of a leukemic, who was a 49-year-old female who had consumed raw oysters 1 week earlier. A case of *V. alginolyticus* gastroenteritis in Florida was reported by Desenclos et al. (25). This case was only 1 of the 333 cases of bacteriologically confirmed vibrio illnesses reported in that study.

Characteristics of Disease

There are a few published reports describing the symptoms of gastroenteric disease caused by *V. alginolyticus*. A leukemic patient with *V. alginolyticus* infection was admitted to the hospital with recent confusion, shock, and anemia. She had a temperature of 40°C, with systolic blood pressure of 80 mm Hg. Despite antibiotic administration and treatment for shock, the patient died 12 days after admission. The role of *V. alginolyticus* in this case is unknown, with *Pseudomonas aeruginosa* also being isolated from the patient's blood. The presence of *P. aeruginosa* in the blood was likely a major factor in the fatal outcome.

Infectious Dose and Susceptible Population

Extraintestinal infections of *V. alginolyticus* are typically self-limiting and relatively mild; however, systemic infections are usually severe. Most such cases of *V. alginolyticus* infections occur in patients who are immunocompromised due to severe burns or cancer. A history of alcohol abuse may also be an important factor in *V. alginolyticus* infections.

Virulence Mechanisms

Little is known about the pathogenic mechanisms of *V. alginolyticus*. Production of lipase, lecithinase, chondroitin sulfatase, DNase, and hemolysin has been reported for one strain of *V. alginolyticus;* however, there was no evidence of elastase, mucinase, protease, or hyaluronidase production, nor were culture filtrates cytotoxic for CHO cells (93). The role that these factors may have in infection is unknown.

CONCLUSIONS

The presence of vibrios, especially *V. cholerae, V. vulnificus,* and *V. parahaemolyticus,* in foods represents a serious and growing public health hazard. These species, along with seven other recognized human pathogenic vibrios present in foods, are found in estuarine waters and occur frequently in a variety of fish and shellfish. Studies routinely report 30 to 100% of fresh, frozen, or iced fish and shellfish to harbor vibrios, with *V. parahaemolyticus, V. cholerae* non-O1, *V. vulnificus, V. alginolyticus,* and *V. fluvialis* predominating. Most vibrios demonstrate a seasonality in their ability to be isolated from the environment and foodstuffs and in the infections they cause, both being generally greater during the warmer months. While the overall incidence of infection with vibrios has been estimated to be relatively low, many milder infections may occur which are not reported. Symptoms of infections with the various vibrios range from only mild gastrointestinal upset to death; fatalities occur primarily with two species, *V. cholerae* and *V. vulnificus.* With some exceptions, e.g., *V. cholerae* O1/O139, most severely ill patients suffer preexisting underlying illnesses, with chronic liver disease being the most common.

Cholera produced by *V. cholerae* is one of the few foodborne diseases with epidemic and endemic potential. The seventh pandemic of cholera, which began in 1961, has involved more than 100 countries, affecting more than 3 million persons and killing many thousands. As more has been learned about the pathogenesis of this bacterium in recent years, it has become clear that its pathogenic potential is quite heterogeneous. While a number of toxins are elaborated by *V. cholerae,* it is evident that production of CT is essential to its disease production and that possession of either the O1 or O139 antigen usually correlates with this potential. Disease due to *V. cholerae* often involves seafood, but a wide variety of nonacidic foods have been involved in disease transmission.

Along with *V. cholerae, V. parahaemolyticus* is the best described of the pathogenic vibrios. Occurring in estuarine waters throughout the world, it is generally found in a wide variety of seafoods. In Japan, as many as 70% of all bacterial foodborne illnesses are attributable to this vibrio; outbreaks affecting over 1,100 people have occurred in the United States, and in all cases, seafood (especially shrimp) has been implicated. Disease production is generally limited to strains producing the Kanagawa hemolysin (TDH), which interestingly is not produced by the great majority of environmental isolates.

Of all the pathogenic vibrios, *V. vulnificus* is the most serious in the United States, being responsible for 95% of all seafood-borne deaths in this country. Like *V. cholerae* and *V. parahaemolyticus, V. vulnificus* is found in estuarine waters and the shellfish that inhabit those environments. Development of human infection typically follows consumption of raw oysters, but is generally restricted to persons having certain underlying diseases such as liver cirrhosis. Fatality rates are approximately 60%.

Human infections with the remaining vibrios (*V. mimicus, V. fluvialis, V. furnissii, V. hollisae,* and *V. alginolyticus*) are less common and usually less severe, although deaths have been reported. Seafood as a source of infection is the norm.

Little is known of the susceptibility of the vibrios to food preservation methods. Cold is often considered an effective defense against proliferation of vibrios, although seafoods have been reported to be protective for some of the vibrios at refrigeration temperatures. Others appear able to increase in numbers at refrigeration temperatures. Thorough heating of shellfish appears to be adequate to kill all *Vibrio* spp. and is probably the only effective protective measure currently available.

References

1. **Abbott, S. L., and J. M. Janda.** 1994. Severe gastroenteritis associated with *Vibrio hollisae* infection: report of two cases and review. *Clin. Infect. Dis.* **18:**310–312.
2. **Albert, M. J.** 1994. *Vibrio cholerae* O139 Bengal. *J. Clin. Microbiol.* **32:**2345–2349.
3. **Bean, N. H., and P. M. Griffin.** 1990. Foodborne disease outbreaks in the United States, 1973–1987: pathogens, vehicles, and trends. *J. Food Prot.* **53:**804–817.
4. **Beasley, L., D. D. Jones, and A. K. Bej.** 1994. A rapid method for detection and differentiation of KP^+ and KP^- *Vibrio parahaemolyticus* in artifically contaminated shellfish by in vitro DNA amplification and gene probe hybridization methods, abstr. P-100, p. 386. *In Abstracts of the 94th General Meeting of the American Society for Microbiology.* American Society for Microbiology, Washington, D.C.
5. **Berry, T. M., D. L. Park, and D. V. Lightner.** 1994. Comparison of the microbial quality of raw shrimp from China, Ecuador, or Mexico at both wholesale and retail levels. *J. Food Prot.* **57:**150–153.
6. **Beuchat, L. R.** 1982. *Vibrio parahaemolyticus:* public health significance. *Food Technol.* **March:**80–83, 92.
7. **Blake, P. A.** 1994. Endemic cholera in Australia and the United States, p. 309–319. *In* I. K. Wachsmuth, P. A. Blake, and Ø. Olsvik (ed.), *Vibrio cholerae and Cholera: Molecular to Global Perspectives.* ASM Press, Washington, D.C.
8. **Blake, P. A., R. E. Weaver, and D. G. Hollis.** 1980. Diseases of humans (other than cholera) caused by vibrios. *Annu. Rev. Microbiol.* **34:**341–367.
9. **Bonner, J. R., A. S. Coker, C. R. Berryman, and H. M. Pollock.** 1983. Spectrum of *Vibrio* infections in a Gulf Coast community. *Ann. Intern. Med.* **99:**464–469.
10. **Brenner, D. J., F. W. Hickman-Brenner, J. V. Lee, A. G. Steigerwalt, G. R. Fanning, D. G. Hollis, J. J. Farmer III, R. E. Weaver, S. W. Joseph, and R. J. Seidler.** 1983. *Vibrio furnissii* (formerly aerogenic biogroup of *Vibrio fluvialis*), a new species isolated from human feces and the environment. *J. Clin. Microbiol.* **18:**816–824.
11. **Buchrieser, C., V. V. Gangar, R. L. Murphree, M. L. Tamplin, and C. W. Kaspar.** 1995. Multiple *Vibrio vulnificus* strains in oysters as demonstrated by clamped homogeneous electric field gel electrophoresis. *Appl. Environ. Microbiol.* **61:**1163–1168.
12. **Burkhardt, W., III, S. R. Rippey, and W. D. Watkins.** 1992. Depuration rates of Northern quahogs, *Mercenaria mercenaria* (Linnaeus, 1758), and Eastern oysters, *Crassostrea virginica* (Gmelin, 1791), in ozone- and ultraviolet light-disinfected seawater systems. *J. Shellfish Res.* **11:**105–109.
13. **Butterton, J. R., J. A. Stoebner, S. M. Payne, and S. B. Calderwood.** 1992. Cloning, sequencing, and transcriptional regulation of *viuA,* the gene coding the ferric vibriobactin receptor of *Vibrio cholerae. J. Bacteriol.* **174:**3729–3738.
14. **Chikahira, M., and K. Hamada.** 1988. Enterotoxigenic substances and other toxins produced by *Vibrio fluvialis* and *Vibrio furnissii. Jpn. J. Vet. Sci.* **50:**865–873.
15. **Chowdhury, M. A. R., K. M. S. Aziz, B. A. Kay, and Z. Rahim.** 1987. Toxin production by *Vibrio mimicus* strains isolated from human and environmental sources in Bangladesh. *J. Clin. Microbiol.* **25:**2200–2203.
16. **Chowdhury, M. A. R., R. T. Hill, and R. R. Colwell.** 1994. A gene for the enterotoxin zonula occludens toxin is present in *Vibrio mimicus* and *Vibrio cholerae* O139. *FEMS Microbiol. Lett.* **119:**377–380.
17. **Chowdhury, M. A. R., H. Yamanaka, S. Miyoshi, K. M. S. Aziz, and S. Shinoda.** 1989. Ecology of *Vibrio mimicus* in aquatic environments. *Appl. Environ. Microbiol.* **55:** 2073–2078.
18. **Colwell, R. R., and A. Huq.** 1994. Vibrios in the environment: viable but nonculturable *Vibrio cholerae,* p. 117–133. *In* K. Wachsmuth, P. A. Blake, and Ø. Olsvik (ed.), *Vibrio cholerae and Cholera: Molecular to Global Perspectives.* ASM Press, Washington, D.C.
19. **Colwell, R. R., M. L. Tamplin, P. R. Brayton, A. L. Gauzens, B. D. Tall, D. Herrington, M. M. Levine, S. Hall, A. Huq, and D. A. Sack.** 1990. Environmental aspects of *Vibrio cholerae* in transmission of cholera, p. 327–343. *In* R. B. Sack and Y. Zinnaka (ed.), *Advances in Research on Cholera and Related Diarrheas,* vol. 7. KTK Scientific Publishers, Tokyo.
20. **Comstock, L. E., J. A. Johnson, J. M. Michalski, J. G. Morris, Jr., and J. B. Kaper.** Cloning and sequence of a region encoding surface polysaccharide of *Vibrio cholerae* O139 and characterization of the insertion site in the chromosome of *Vibrio cholerae* O1. *Mol. Microbiol.* **19:** 815–826.
21. **Cook, D. W.** 1994. Effect of time and temperature on multiplication of *Vibrio vulnificus* in postharvest Gulf Coast shellstock oysters. *Appl. Environ. Microbiol.* **60:** 3482–3484.
22. **Cook, D. W., and A. D. Ruple.** 1992. Cold storage and mild heat treatment as processing aids to reduce the numbers of *Vibrio vulnificus* in raw oysters. *J. Food Prot.* **55:**985–989.

23. **Davis, B. R., G. R. Fanning, J. M. Madden, A. G. Steigerwalt, H. B. Bradford, Jr., H. L. Smith, Jr., and D. J. Brenner.** 1981. Characterization of biochemically atypical *Vibrio cholerae* strains and designation of a new pathogenic species, *Vibrio mimicus. J. Clin. Microbiol.* **14:**631–639.
24. **DePaola, A., G. M. Capers, and D. Alexander.** 1994. Densities of *Vibrio vulnificus* in the intestines of fish from the U.S. Gulf Coast. *Appl. Environ. Microbiol.* **60:**984–988.
25. **Desenclos, J.-C. A., K. C. Klontz, L. E. Wolfe, and S. Hoecherl.** 1991. The risk of *Vibrio* illness in the Florida raw oyster eating population, 1981–1988. *Am. J. Epidemiol.* **134:**290–297.
26. **DiRita, V. J.** 1992. Co-ordinate expression of virulence genes by ToxR in *Vibrio cholerae. Mol. Microbiol.* **6:**451–458.
27. **Dixon, W. D.** 1992. The effects of gamma radiation (^{60}Co) upon shellstock oysters in terms of shelf life and bacterial reduction, including *Vibrio vulnificus* levels. M.S. thesis. University of Florida, Gainesville.
28. **Everiss, K. D., K. J. Hughes, M. E. Kovach, and K. M. Peterson.** 1994. The *Vibrio cholerae acfB* colonization determinant encodes an inner membrane protein that is related to a family of signal-transducing proteins. *Infect. Immun.* **62:**3289–3298.
29. **Eyles, M. J., and G. R. Davey.** 1984. Microbiology of commercial depuration of the Sydney rock oyster, *Crassostrea commercialis. J. Food Prot.* **47:**703–706.
30. **Farmer, J. J., III, F. W. Hickman-Brenner, and M. T. Kelly.** 1985. *Vibrio*, p. 282–301. *In* E. H. Lennette, A. Balows, W. J. Hausler, Jr., and H. J. Shadomy (ed.), *Manual of Clinical Microbiology*, 4th ed. American Society for Microbiology, Washington, D.C.
31. **Farmer, J. J., III, R. E. Weaver, S. W. Joseph, and R. J. Seidler.** 1983. *Vibrio furnissii* (formerly aerogenic biogroup of *Vibrio fluvialis*), a new species isolated from human feces and the environment. *J. Clin. Microbiol.* **18:**816–824.
32. **Fasano, A., B. Baudry, D. W. Pumplin, S. S. Wasserman, B. D. Tall, J. M. Ketley, and J. B. Kaper.** 1991. *Vibrio cholerae* produces a second enterotoxin, which affects intestinal tight junctions. *Proc. Natl. Acad. Sci. USA* **88:**5242–5246.
33. **Gabriel, S. E., K. N. Brigman, B. H. Koller, R. C. Boucher, and M. J. Stutts.** 1994. Cystic fibrosis heterozygote resistance to cholera toxin in the cystic fibrosis mouse model. *Science* **266:**107–109.
34. **Galen, J. E., J. M. Ketley, A. Fasano, S. H. Richardson, S. S. Wasserman, and J. B. Kaper.** 1992. Role of *Vibrio cholerae* neuraminidase in the function of cholera toxin. *Infect. Immun.* **60:**406–415.
35. **Glass, R. I., and R. E. Black.** 1992. The epidemiology of cholera, p. 129–154. *In* D. Barua and W. B. Greenough, III (ed.), *Cholera.* Plenum Medical Book Co., New York.
36. **Groubert, T. N., and J. D. Oliver.** 1994. Interaction of *Vibrio vulnificus* and the Eastern oyster, *Crassostrea virginica. J. Food Prot.* **57:**224–228.
37. **Guerrant, R. L., G. D. Fang, N. M. Thielman, and M. C. Fonteles.** 1994. Role of platelet activating factor (PAF) in the intestinal epithelial secretory and Chinese hamster ovary (CHO) cell cytoskeletal responses to cholera toxin. *Proc. Natl. Acad. Sci. USA* **91:**9655–9658.
38. **Hackney, C. R., B. Ray, and M. L. Speck.** 1980. Incidence of *Vibrio parahaemolyticus* in and the microbiological quality of seafood in North Carolina. *J. Food Prot.* **43:**769–773.
39. **Hayat, U., G. P. Reddy, C. A. Bush, J. A. Johnson, A. C. Wright, and J. G. Morris, Jr.** 1993. Capsular types of *Vibrio vulnificus:* an analysis of strains from clinical and environmental sources. *J. Infect. Dis.* **168:**758–762.
40. **Hickman, F. W., J. J. Farmer, III, D. G. Hollis, G. R. Fanning, A. G. Steigerwalt, R. E. Weaver, and D. J. Brenner.** 1982. Identification of *Vibrio hollisae* sp. nov. from patients with diarrhea. *J. Clin. Microbiol.* **15:**395–401.
41. **Hlady, W. G., R. C. Mullen, and R. S. Hopkin.** 1993. *Vibrio vulnificus* from raw oysters. Leading cause of reported deaths from foodborne illness in Florida. *J. Fla. Med. Assoc.* **80:**536–538.
42. **Hoge, C. W., D. Watsky, R. N. Peeler, J. P. Libonati, E. Israel, and J. G. Morris, Jr.** 1989. Epidemiology and spectrum in *Vibrio* infections in a Chesapeake Bay USA community. *J. Infect. Dis.* **160:**985–993.
43. **Honda, T., and T. Miwatani.** 1988. Multi-toxigenicity of *Vibrio cholerae* non-O1, p. 23–32. *In* N. Ohtomo and R. B. Sack (ed.), *Advances in Research on Cholera and Related Diarrheas*, vol. 6. KTK Scientific Publishers, Tokyo.
44. **Hood, M. A., R. M. Baker, and F. L. Singleton.** 1984. Effect of processing and storing oyster meats on concentration of indicator bacteria, vibrios, and *Aeromonas hydrophila. J. Food Prot.* **47:**598–601.
45. **Hood, M. A., G. E. Ness, G. E. Rodrick, and N. J. Blake.** 1983. Effects of storage on microbial loads of two commercially important shellfish species, *Crassostrea virginica* and *Mercenaria campechiensis. Appl. Environ. Microbiol.* **45:**1221–1228.
46. **Huq, M. I., A. K. M. J. Alam, D. J. Brenner, and G. K. Morris.** 1980. Isolation of *Vibrio*-like group, EF-6, from patients with diarrhea. *J. Clin. Microbiol.* **11:**621–624.
47. **Janda, J. M., C. Powers, R. G. Bryant, and S. L. Abbott.** 1988. Current perspectives on the epidemiology and pathogenesis of clinically significant *Vibrio* spp. *Clin. Microbiol. Rev.* **1:**245–267.
48. **Johnson, J. A., C. A. Salles, P. Panigrahi, M. J. Albert, A. C. Wright, R. J. Johnson, and J. G. Morris, Jr.** 1994. *Vibrio cholerae* O139 synonym Bengal is closely related to *Vibrio cholerae* El Tor but has important differences. *Infect. Immun.* **62:**2108–2110.
49. **Johnston, J. M., S. F. Becker, and L. M. McFarland.** 1986. Gastroenteritis in patients with stool isolations of *Vibrio vulnificus. Am. J. Med.* **80:**336–338.
50. **Jones, S. H., T. L. Howell, and K. R. O'Neill.** 1991. Differential elimination of indicator bacteria and pathogenic *Vibrio* sp. from Eastern oysters (*Crassostrea virginica* Gmelin, 1791) in a commercial purification facility in Maine. *J. Shellfish Res.* **10:**105–112.
51. **Joseph, S. W., R. R. Colwell, and J. B. Kaper.** 1982. *Vibrio parahaemolyticus* and related halophilic vibrios. *Crit. Rev. Microbiol.* **10:**77–124.

52. **Kaneko, T., and R. R. Colwell.** 1978. The annual cycle of *Vibrio parahaemolyticus* in Chesapeake Bay. *Microb. Ecol.* **4:**135–155.
53. **Kaper, J. B., A. Fasano, and M. Trucksis.** 1994. Toxins of *Vibrio cholerae,* p. 145–176. *In* I. K. Wachsmuth, P. Blake, and Ø. Olsvik (ed.), *Vibrio cholerae and Cholera: Molecular to Global Perspectives.* ASM Press, Washington, D.C.
54. **Kaper, J. B., J. G. Morris, Jr., and M. M. Levine.** 1995. Cholera. *Clin. Microbiol. Rev.* **8:**48–86.
55. **Kaysner, C. A., C. Abeyta, Jr., P. A. Trost, J. H. Wetherington, K. C. Jinneman, W. E. Hill, and M. M. Wekell.** 1994. Urea hydrolysis can predict the potential pathogenicity of *Vibrio parahaemolyticus* strains isolated in the Pacific Northwest. *Appl. Environ. Microbiol.* **60:**3020–3022.
56. **Kaysner, C. A., C. Abeyta, Jr., M. M. Wekell, A. DePaola, Jr., R. F. Stott, and J. M. Leitch.** 1987. Virulent strains of *Vibrio vulnificus* from estuaries of the United States West Coast. *Appl. Environ. Microbiol.* **53:**1349–1351.
57. **Kaysner, C. A., and W. E. Hill.** 1994. Toxigenic *Vibrio cholerae* O1 in food and water, p. 27–39. *In* I. K. Wachsmuth, P. A. Blake, and Ø. Olsvik (ed.), *Vibrio cholerae and Cholera: Molecular to Global Perspectives.* ASM Press, Washington, D.C.
58. **Kelly, M. T.** 1982. Effect of temperature and salinity on *Vibrio (Beneckea) vulnificus* occurrence in a Gulf Coast environment. *Appl. Environ. Microbiol.* **44:**820–824.
59. **Klontz, K. C., D. E. Cover, F. N. Hyman, and R. C. Mullen.** 1994. Fatal gastroenteritis due to *Vibrio fluvialis* and nonfatal bacteremia due to *Vibrio mimicus;* unusual vibrio infections in two patients. *Clin. Infect. Dis.* **19:**541–542.
60. **Klontz, K. C., and J.-C. A. Desenclos.** 1990. Clinical and epidemiological features of sporadic infections with *Vibrio fluvialis* in Florida USA. *J. Diarrh. Dis. Res.* **8:**1–2.
61. **Koch, W. H., W. L. Payne, B. A. Wentz, and T. A. Cebula.** 1993. Rapid polymerase chain reaction method for detection of *Vibrio cholerae* in foods. *Appl. Environ. Microbiol.* **59:**556–560.
62. **Kodama, H., Y. Gyobu, N. Tokuman, H. Uetake, T. Shimada, and R. Sakazaki.** 1988. Ecology of non-O1 *Vibrio cholerae* and *Vibrio mimicus* in Toyama prefecture, p. 79–88. *In* N. Ohtomo and R. B. Sack (ed.), *Advances in Research on Cholera and Related Diarrheas,* vol. 6. KTK Scientific Publishers, Tokyo.
63. **Kothary, M. H., E. F. Claverie, M. D. Miliotis, J. M. Madden, and S. H. Richardson.** 1995. Purification and characterization of a Chinese hamster ovary cell elongation factor of *Vibrio hollisae. Infect. Immun.* **63:**2418–2423.
64. **Kreger, A., and D. Lockwood.** 1981. Detection of extracellular toxin(s) produced by *Vibrio vulnificus. Infect. Immun.* 33:583–590.
65. **Lee, J. V., P. Shread, and A. L. Furniss.** 1978. The taxonomy of group F organisms: relationships to *Vibrio* and *Aeromonas. J. Appl. Bacteriol.* **45:**ix.
66. **Levine, M. M., R. E. Black, M. L. Clements, D. R. Nalin, L. Cisneros, and R. A. Finkelstein.** 1981. Volunteer studies in development of vaccines against cholera and enterotoxigenic *Escherichia coli:* a review, p. 443–459. *In* T. Holme, J. Holmgren, M. H. Merson, and R. Mollby (ed.), *Acute Enteric Infections in Children. New Prospects for Treatment and Prevention.* Elsevier/North-Holland Biomedical Press, Amsterdam.
67. **Levine, M. M., J. B. Kaper, D. Herrington, G. Losonsky, J. G. Morris, M. L. Clements, R. E. Black, B. Tall, and R. Hall.** 1988. Volunteer studies of deletion mutants of *Vibrio chlorae* O1 prepared by recombinant techniques. *Infect. Immun.* **56:**161–167.
68. **Levine, W. C., P. M. Griffin, and the Gulf Coast *Vibrio* Working Group.** 1993. *Vibrio* infections on the Gulf Coast: result of first year of regional surveillance. *J. Infect. Dis.* **167:**479–483.
69. **Lockwood, D. E., A. S. Kreger, and S. H. Richardson.** 1982. Detection of toxins produced by *Vibrio fluvialis. Infect. Immun.* **35:**702–708.
70. **Lockwood, D. E., S. H. Richardson, A. S. Kreger, M. Aiken, and B. McCreedy.** 1983. *In vitro* and *in vivo* biologic activities of *Vibrio fluvialis* and its toxic products, p. 87–99. *In* S. Kuwahara and N. F. Pierce (ed.), *Advances in Research on Cholera and Related Diarrheas,* vol. 1. KTK Scientific Publishers, Tokyo.
71. **Lowry, P. W., L. M. McFarland, B. H. Peltier, N. C. Roberts, H. B. Bradford, J. L. Herndon, D. F. Stroup, J. B. Mathison, P. A. Blake, and R. A. Gunn.** 1989. *Vibrio* gastroenteritis in Louisiana: a prospective study among attendees of a scientific congress in New Orleans. *J. Infect. Dis.* **160:**978–984.
72. **Lowry, P. W., L. M. McFarland, and H. K. Threefoot.** 1986. *Vibrio hollisae* septicemia after consumption of catfish. *J. Infect. Dis.* **154:**730–731.
73. **Madden, J. M., B. A. McCardell, and J. G. Morris, Jr.** 1989. *Vibrio cholerae,* p. 525–542. *In* M. P. Doyle (ed.), *Foodborne Bacterial Pathogens.* Marcel Dekker, New York.
74. **Magalhaes, V., A. Castello Filho, M. Magalhaes, and T. T. Gomes.** 1993. Laboratory evaluation on pathogenic potentialities of *Vibrio furnissii. Mem. Inst. Oswaldo Cruz Rio de Janeiro* **88:**593–597.
75. **Matté, G. R., M. H. Matté, I. G. Rivera, and M. T. Martins.** 1994. Distribution of pathogenic vibrios in oysters from a tropical region. *J. Food Prot.* **57:**870–873.

75a. **McLaughlin, J. C.** 1995. *Vibrio,* p. 465–476. *In* P. R. Murray, E. J. Baron, M. A. Pfaller, F. C. Tenover, and R. N. Yolken (ed.), *Manual of Clinical Microbiology,* 6th ed. ASM Press, Washington, D.C.

76. **McPherson, V. L., J. A. Watts, L. M. Simpson, and J. D. Oliver.** 1991. Physiological effects of the lipopolysaccharide of *Vibrio vulnificus* on mice and rats. *Microbios* **67:**141–149.
77. **Miliotis, M. D., B. D. Tall, and R. T. Gray.** 1995. Adherence to and invasion of tissue culture cells by *Vibrio hollisae. Infect. Immun.* **63:**4959–4963.
78. **Minami, A., S. Hashimoto, H. Abe, M. Arita, T. Taniguchi, T. Honda, T. Miwatani, and M. Nishibuchi.** 1991. Cholera enterotoxin production in *Vibrio cholerae* O1 strains isolated from the environment and from humans in Japan. *Appl. Environ. Microbiol.* **57:**2152–2157.
79. **Mintz, E. D., T. Popovic, and P. A. Blake.** 1994. Transmission of *Vibrio cholerae* O1, p. 345–356. *In* I. K. Wachsmuth, P. A. Blake, and Ø. Olsvik (ed.), *Vibrio cholerae and Cholera: Molecular to Global Perspectives.* ASM Press, Washington, D.C.

80. **Morris, J. G., Jr.** 1990. Non-O group 1 *Vibrio cholerae:* a look at the epidemiology of an occasional pathogen. *Epidemiol. Rev.* **12**:179–191.
81. **Morris, J. G., Jr.** 1995. "Noncholera" *Vibrio* species, p. 671–685. *In* M. J. Blaser, P. D. Smith, J. I. Ravdin, H. B. Greenberg, and R. L. Guerrant, (ed.), *Infections of the Gastrointestinal Tract.* Raven Press, Ltd., New York.
82. **Morris, J. G., Jr., R. Wilson, D. G. Hollis, R. E. Weaver, H. G. Miller, C. O. Tacket, F. W. Hickman, and P. A. Blake.** 1982. Illness caused by *Vibrio damsela* and *Vibrio hollisae. Lancet* **i**:1294–1296.
83. **Morris, J. G., Jr., A. C. Wright, D. M. Roberts, P. K. Wood, L. M. Simpson, and J. D. Oliver.** 1987. Identification of environmental *Vibrio vulnificus* isolates with a DNA probe for the cytotoxin-hemolysin gene. *Appl. Environ. Microbiol.* **53**:193–195.
84. **Moss, J., and M. Vaughn.** 1991. Activation of cholera toxin and *Escherichia coli* heat-labile enterotoxins by ADP-ribosylation factors, a family of 20-kDa guanine nucleotide-binding proteins. *Mol. Microbiol.* **5**:2621–2627.
85. **Nishibuchi, M., A. Fasano, R. G. Russell, and J. B. Kaper.** 1992. Enterotoxigenicity of *Vibrio parahaemolyticus* with and without genes encoding thermostable direct hemolysin. *Infect. Immun.* **60**:3539–3545.
86. **Nishibuchi, M., and J. B. Kaper.** 1995. Thermostable direct hemolysin gene of *V. parahaemolyticus:* a virulence gene acquired by a marine bacterium. *Infect. Immun.* **63**:2093–2099.
87. **Nishibuchi, M., T. Taniguchi, T. Misawa, V. Khaeomanee-iam, T. Honda, and T. Miwatani.** 1989. Cloning and nucleotide sequence of the gene (*trh*) encoding the hemolysin related to the thermostable direct hemolysin of *Vibrio parahaemolyticus. Infect. Immun.* **57**:2691–2697.
88. **Nocerino, A., M. Iafusco, and S. Guandalini.** 1995. Cholera toxin-induced small intestinal secretion has a secretory effect on the colon of the rat. *Gastroenterology* **108**:34–39.
89. **Oliver, J. D.** 1989. *Vibrio vulnificus,* p. 569–600. *In* M. P. Doyle (ed.), *Foodborne Bacterial Pathogens.* Marcel Dekker, New York.
90. **Oliver, J. D.** 1993. Formation of viable but nonculturable cells, p. 239–272. *In* S. Kjelleberg (ed.), *Starvation in Bacteria.* Plenum Press, New York.
91. **Oliver, J. D.** Public health significance of viable but nonculturable bacteria. *In* R. R. Colwell and D. J. Grimes (ed.), *Noncultural Microorganisms in the Environment.* Chapman and Hall, London, in press.
92. **Oliver, J. D., K. Guthrie, J. Preyer, A. Wright, L. M. Simpson, R. Siebeling, and J. G. Morris, Jr.** 1992. Use of colistin-polymyxin B-cellobiose agar in the isolation of *Vibrio vulnificus* from the environment. *Appl. Environ. Microbiol.* **58**:737–739.
93. **Oliver, J. D., M. B. Thomas, and J. Wear.** 1986. Production of extracellular enzymes and cytotoxicity by *Vibrio vulnificus. Diagn. Microbiol. Infect. Dis.* **5**:99–111.
94. **Oliver, J. D., R. A. Warner, and D. R. Cleland.** 1983. Distribution of *Vibrio vulnificus* and other lactose-fermenting vibrios in the marine environment. *Appl. Environ. Microbiol.* **45**:985–998.
95. **O'Neill, K. R., S. H. Jones, and D. J. Grimes.** 1992. Seasonal incidence of *Vibrio vulnificus* in the Great Bay estuary of New Hampshire and Maine. *Appl. Environ. Microbiol.* **58**:3257–3262.
96. **Osawa, R., T. Okitsu, H. Morozumi, and S. Yamai.** 1996. Occurrence of urease-positive *Vibrio parahaemolyticus* in Kanagawa, Japan, with specific reference to presence of thermostable direct hemolysin (TDH) and the TDH-related-hemolysin genes. *Appl. Environ. Microbiol.* **62**:725–727.
97. **Parker, R. W., E. M. Maurer, A. B. Childers, and D. H. Lewis.** 1994. Effect of frozen storage and vacuum-packaging on survival of *Vibrio vulnificus* in Gulf Coast oysters (*Crassostrea virginica*). *J. Food Prot.* **57**:604–606.
98. **Pearson, G. D. N., A. Woods, S. L. Chiang, and J. J. Mekalanos.** 1993. CTX genetic element encodes a site-specific recombination system and an intestinal colonization factor. *Proc. Natl. Acad. Sci. USA* **90**:3750–3754.
99. **Popovic, T., P. I. Fields, and Ø. Olsvik.** 1994. Detection of cholera toxin genes, p. 41–52. *In* I. K. Wachsmuth, P. A. Blake, and Ø. Olsvik (ed.), *Vibrio cholerae and Cholera: Molecular to Global Perspectives.* ASM Press, Washington, D.C.
100. **Prasad, M. M., and C. C. P. Rao.** 1994. Pathogenic vibrios associated with seafoods in and around Kakinada, India. *Fish. Technol.* **31**:185–187.
101. **Raimondi, D., J. P. Y. Kao, J. B. Kaper, S. Guandalini, and A. Fasano.** 1995. Calcium-dependent intestinal chloride secretion by *Vibrio parahaemolyticus* thermostable direct hemolysin in a rabbit model. *Gastroenterology* **109**:381–386.
102. **Rashid, H. O., H. Ito, and I. Ishigaki.** 1992. Distribution of pathogenic vibrios and other bacteria in imported frozen shrimps and their decontamination by gamma-irradiation. *World J. Microbiol. Biotechnol.* **8**:494–499.
103. **Richards, G. P.** 1988. Microbial purification of shellfish: a review of depuration and relaying. *J. Food Prot.* **51**:218–251.
104. **Richardson, S. H.** 1994. Host susceptibility, p. 273–289. *In* I. K. Wachsmuth, P. A. Blake, and Ø. Olsvik (ed.), *Vibrio cholerae and Cholera: Molecular to Global Perspectives.* ASM Press, Washington, D.C.
105. **Robach, M. C., and C. S. Hickey.** 1978. Inhibition of *Vibrio parahaemolyticus* by sorbic acid in crab meat and flounder homogenates. *J. Food Prot.* **41**:699–702.
106. **Rodrick, G. E., K. R. Schneider, F. A. Steslow, N. J. Blake, and W. S. Otwell.** 1988. Uptake, fate and ultra-violet depuration of vibrios in *Mercenaria campechiensis. Mar. Technol. Soc. J.* **23**:21–26.
107. **Sakazaki, R., S. Iwanami, and K. Tamura.** 1968. Studies on the enteropathogenic, facultatively halophilic bacteria, *Vibrio parahaemolyticus.* II. Serological characteristics. *Jpn. J. Med. Sci. Biol.* **21**:313–324.
108. **Sakazaki, R., K. Tamura, T. Kato, Y. Obara, S. Yamai, and K. Hobo.** 1968. Studies on the enteropathogenic, facultatively halophilic bacteria, *Vibrio parahaemolyticus.* III. Enteropathogenicity. *Jpn. J. Med. Sci. Biol.* **21**:325–331.
109. **Sang, F. C., M. E. Hugh-Jones, and H. V. Hagstad.** 1987. Viability of *Vibrio cholerae* O1 on frog legs under frozen and refrigerated conditions and low dose radiation treatment. *J. Food Prot.* **50**:662–664.

110. **Shandera, W. X., J. M. Johnston, B. R. Davis, and P. A. Blake.** 1983. Disease from infection with *Vibrio mimicus,* a newly recognized *Vibrio* species. *Ann. Intern. Med.* 99:169–171.

111. **Simonson, J. G., P. Danieu, A. B. Zuppardo, R. J. Siebeling, R. L. Murphree, and M. L. Tamplin.** 1995. Distribution of capsular and lipopolysaccharide antigens among clinical and environmental *Vibrio vulnificus* isolates, abstr. B-286, p. 215. *In Abstracts of the 95th General Meeting of the American Society for Microbiology.* American Society for Microbiology, Washington, D.C.

112. **Simpson, L. M., V. K. White, S. F. Zane, and J. D. Oliver.** 1987. Correlation between virulence and colony morphology in *Vibrio vulnificus. Infect. Immun.* **55**:269–272.

113. **Sloan, E. M., C. J. Hagen, G. A. Lancette, J. T. Peeler, and J. N. Sofos.** 1992. Comparison of five selective enrichment broths and two selective agars for recovery of *Vibrio vulnificus* from oysters. *J. Food Prot.* **55**:356–359.

114. **Spangler, B. D.** 1992. Structure and function of cholera toxin and the related *Escherichia coli* heat-labile enterotoxin. *Microbiol. Rev.* **56**:622–647.

115. **Sperandio, V., J. A. Girón, W. D. Silveira, and J. B. Kaper.** 1995. The OmpU outer membrane protein, a potential adherence factor of *Vibrio cholerae. Infect. Immun.* **63**: 4433–4438.

116. **Spira, W. M., M. U. Khan, Y. A. Saeed, and M. A. Sattar.** 1980. Microbiological surveillance of intra-neighbourhood El Tor cholera transmission in rural Bangladesh. *Bull. WHO* **58**:731–740.

117. **Sun, Y., and J. D. Oliver.** 1994. Effects of GRAS compounds on natural *Vibrio vulnificus* populations in oysters. *J. Food Prot.* **57**:921–923.

118. **Sun, Y., and J. D. Oliver.** 1995. Hot sauce: no elimination of *Vibrio vulnificus* in oysters. *J. Food Prot.* **58**:441–442.

119. **Sun, Y., and J. D. Oliver.** 1995. The value of CPC agar for the isolation of *Vibrio vulnificus* from oysters. *J. Food Prot.* **58**:439–440.

120. **Tamplin, M. L., and G. M. Capers.** 1992. Persistence of *Vibrio vulnificus* in tissues of Gulf Coast oysters, *Crassostrea virginica,* exposed to seawater disinfected with UV light. *Appl. Environ. Microbiol.* **58**:1506–1510.

121. **Tamplin, M., G. E. Rodrick, N. J. Blake, and T. Cuba.** 1982. Isolation and characterization of *Vibrio vulnificus* from two Florida estuaries. *Appl. Environ. Microbiol.* **44**:1466–1470.

122. **Taylor, R. K., V. L. Miller, D. B. Furlong, and J. J. Mekalanos.** 1987. Use of *phoA* gene fusions to identify a pilus colonization factor coordinately regulated with cholera toxin. *Proc. Natl. Acad. Sci. USA* **84**:2833–2837.

123. **Tilton, R. C., and R. W. Ryan.** 1987. Clinical and ecological characteristics of *Vibrio vulnificus* in the Northeastern United States. *Diagn. Microbiol. Infect. Dis.* **6**:109–117.

124. **Tison, D. L., and M. T. Kelly.** 1984. *Vibrio* species of medical importance. *Diagn. Microbiol. Infect. Dis.* **2**:263–276.

125. **Todd, E. C. D.** 1989. Preliminary estimates of costs of foodborne disease in the United States. *J. Food Prot.* **52**:595–601.

126. **Trucksis, M., J. E. Galen, J. Michalski, A. Fasano, and J. B. Kaper.** 1993. Accessory cholera enterotoxin (*Ace*), the third toxin of a *Vibrio cholerae* virulence cassette. *Proc. Natl. Acad. Sci. USA* **90**:5267–5271.

127. **Twedt, R. M.** 1989. *Vibrio parahaemolyticus,* p. 543–568. *In* M. P. Doyle (ed.), *Foodborne Bacterial Pathogens.* Marcel Dekker, New York.

128. **Uchimura, M., and T. Yamamoto.** 1992. Production of hemagglutinins and pili by *Vibrio mimicus* and its adherence to human and rabbit small intestines *in vitro. FEMS Microbiol. Lett.* **91**:73–78.

129. **U.S. Food and Drug Administration.** 1995. *Bacteriological Analytical Manual,* 8th ed. AOAC International, Arlington, Va.

130. **Vanderzant, C., and R. Nickelson.** 1972. Survival of *Vibrio parahaemolyticus* in shrimp tissue under various environmental conditions. *Appl. Microbiol.* **23**:34–37.

130a. **Waldor, M. K., and J. J. Mekalanos.** 1996. Lysogenic conversion by a filamentous phage encoding cholera toxin. *Science* **272**:1910–1914.

131. **West, P. A.** 1989. The human pathogenic vibrios—a public health update with environmental perspectives. *Epidemiol. Infect.* **103**:1–34.

132. **Wong, H.-C., L.-L. Chen, and C.-M. Yu.** 1994. Survival of psychrotrophic *Vibrio mimicus, Vibrio fluvialis,* and *Vibrio parahaemolyticus* in culture broth at low temperatures. *J. Food Prot.* **57**:607–610.

133. **Wong, H.-C., L.-L. Chen, and C.-M. Yu.** 1995. Occurrence of vibrios in frozen seafoods and survival of psychrotrophic *Vibrio cholerae* in broth and shrimp homogenates at low temperatures. *J. Food Prot.* **58**:263–267.

134. **Wong, H.-C., S.-H. Ting, and W.-R. Shieh.** 1992. Incidence of toxigenic vibrios in foods available in Taiwan. *J. Appl. Bacteriol.* **73**:197–202.

135. **Wright, A. C., L. M. Simpson, and J. D. Oliver.** 1981. Role of iron in the pathogenesis of *Vibrio vulnificus* infections. *Infect. Immun.* **34**:503–507.

Sylvia M. Kirov

14

Aeromonas and *Plesiomonas* Species

CHARACTERISTICS OF ORGANISMS

The genera *Aeromonas* and *Plesiomonas* comprise gram-negative, facultatively anaerobic, oxidase-positive, glucose-fermenting, rod-shaped bacteria which are generally motile by polar flagella. (The species *Aeromonas salmonicida* and *A. media* are reported to be nonmotile.) *Aeromonas* spp. (aeromonads) usually have a single polar flagellum; *Plesiomonas shigelloides* has two to seven polar flagella and may also produce lateral flagella with a shorter wave length. Both genera are currently classified in the family *Vibrionaceae.* They are commonly found in aquatic environments and have been epidemiologically incriminated as enteropathogens. Thus, they have traditionally been considered together. However, molecular genetic evidence (including 16S rRNA catalog, 5S rRNA sequence, and rRNA-DNA hybridization) suggests that they are not closely related to each other or to other *Vibrio* species. A separate family, *Aeromonadaceae* fam. nov., has been proposed for *Aeromonas,* while *Plesiomonas* may be more closely related to organisms within the family *Enterobacteriaceae* (87, 112).

Although it is accepted that members of both genera can cause serious extraintestinal infections (some of which may be acquired via the oral route), particularly in immunocompromised hosts, *Aeromonas* and *Plesiomonas* species are still regarded as controversial gastrointestinal pathogens. Definite proof (conclusive human volunteer trials and animal models) of enteropathogenicity is lacking. Nevertheless, substantial clinical and microbiological evidence now supports the epidemiologic evidence that at least some strains of *Aeromonas* spp. can cause gastroenteritis in some individuals. The situation with *Aeromonas* spp. may be analogous to that of the various pathotypes of *Escherichia coli* and *Yersinia enterocolitica.* At present, however, as *Aeromonas* virulence mechanisms are not well understood, there is no way to distinguish the strains which pose a threat to human health from the diversity of strains present in foods and water, or even to determine the significance of a given isolate in the clinical laboratory. This chapter reviews what is known about putative *Aeromonas* virulence factors and outlines possible future research directions that may eventually be able to resolve this dilemma. Even less is known about *P. shigelloides* and its possible virulence determinants than is known for *Aeromonas* spp. The role of *P. shigelloides* in gastroenteric disease is therefore still uncertain.

Classification and Identification

Aeromonas species

The taxonomy of *Aeromonas* is complex and has undergone constant change. This has added to confusion in the identification of potentially significant strains (43, 81, 87). *Aeromonas* species can be differentiated from the halophilic vibrios because they are unable to grow in 6% NaCl and from the *Vibrio cholerae* group by their resistance to O/129 vibriostatic agent (150 μg). Other biochemical tests and antibiotic susceptibility testing may also be required for correct genus assignment because of emerging worldwide resistance of *V. cholerae* to O/129 and susceptibility of some aeromonads to this concentration of O/129 (25, 43, 83).

The genus *Aeromonas* was originally divided into a single mesophilic species, *A. hydrophila*, and a single psychrophilic species, *A. salmonicida*. Thus, in the older literature (and unfortunately in some more recent studies), the name *A. hydrophila* is used to include all aeromonads except the latter species, a fish pathogen. DNA-DNA hybridization experiments performed by Popoff in the early 1980s showed that there were at least nine distinct hybridization groups (HGs) among the *A. hydrophila* strains and that these fell into three main phenotypic groups which could be identified on the basis of the reactions of isolates in 8 to 18 biochemical tests. These "phenospecies" were called *A. hydrophila*, *A. sobria*, and *A. caviae* (140). Hence, the name *A. hydrophila* began to be limited to HG1, HG2, and HG3. Most clinical laboratories adopted this simplified phenotypic classification. The genetic complexity of the group, however, meant that not only the *A. hydrophila* phenospecies but also the other two phenospecies each encompassed several HGs (*A. sobria* HG7, HG8, and HG9 and *A. caviae* HG4, HG5, and HG6). The number of DNA hybridization groups ("genospecies") has since expanded to 14. As DNA hybridization technology is not routinely carried out in diagnostic laboratories, more comprehensive biotyping schemes (19 to 24 biochemical tests) which will identify strains to the genospecies level have been proposed in recent years (1, 27). Key tests have been identified so that biochemical typing schemes, such as Aerokey II, now permit preliminary grouping of most clinical isolates (7, 27, 83, 87). Ten phenotypic species have been named, with HG8 and HG10 (now combined as HG8/10) having two biovars (Table 14.1). The name *A. hydrophila* is thus being further limited in some of the recent literature to mean just HG1, which includes the type strain for this species (Table 14.1). Salient features in the identification of *Aeromonas* phenospecies and genospecies in the clinical laboratory are summarized in Table 14.2.

In the discussion to follow, where the name "*A. hydrophila*" has been used in the literature in its broad historical sense to refer to the entire group of motile mesophilic aeromonads, such organisms will generally be referred to as *Aeromonas* spp. or aeromonads or as an unqualified *A. hydrophila* (Tables 14.1, 14.2, and 14.4). The name *A. hydrophila* (HG1, HG2, and HG3) will be used to refer to the more limited group of strains defined by phenotype (criteria of Popoff [140]) or DNA-DNA hybridization. Similarly, "*A. caviae*" refers to strains belonging to DNA HG4, HG5, and HG6, and "*A. sobria*" refers to strains belonging to HG7, HG8, and HG9. Occasionally, names such as *A. hydrophila* (HG1), *A. caviae* (HG5), and *A. jandaei* (HG9) are used; these names refer to the currently recognized phenospecies (Table 14.1) which were previously subgroups of the original three phenospecies described by Popoff (140).

Table 14.1 Currently recognized genospecies and phenospecies of the genus *Aeromonas*[a]

DNA group	Genospecies	Phenospecies
1	*A. hydrophila*	*A. hydrophila*
2	*A. hydrophila*	Unnamed
3	*A. salmonicida*[b]	{ *A. hydrophila* *A. salmonicida*
4	*A. caviae*	*A. caviae*[c]
5A	*A. caviae*	Unnamed
5B	*A. caviae*	*A. media*
6	*A. eucrenophila*	*A. eucrenophila*[d]
7	*A. sobria*	*A. sobria*[d]
8	*A. veronii*	*A. veronii* biovar sobria[e]
9	*A. jandaei*	*A. jandaei*
10	*A. veronii*	*A. veronii* biovar veronii[e]
11	*A. veronii*	Unnamed
12	*A. schubertii*	*A. schubertii*
13	*A. schubertii*	"Group 501"
14	*A. trota*	*A. trota*

[a]For comprehensive coverage, see reference 87.

[b]Includes psychrophilic strains originating from fish. These strains are biochemically distinct (nonmotile and indole negative) from HG3 isolates recovered from clinical material (1, 83).

[c]Some feel that HG4 should be named *A. punctata*, since the type strain for both species is the same and *A. punctata* is the older name in the literature (81).

[d]HG6 and HG7 have not yet been recovered from clinical material (1, 81).

[e]HG8 and HG10 are genetically identical, but the type strains are biochemically distinguishable; hence, they are now called biovars of the first-described species, *A. veronii* (81, 87).

The taxonomy of the genus *Aeromonas* is still evolving. The present nomenclature has become cumbersome and, as seen, is often confusing in the literature. It is important that strains be accurately identified since antibiotic sensitivities and pathogenicity (see below) may vary among strains. Comprehensive biochemical typing schemes which will accurately identify all *Aeromonas* isolates, including environmental strains, are not yet available. The fact that some biochemical characteristics are temperature dependent is a further complication for such biochemical typing schemes (87).

Possible new approaches for the identification of aeromonads based on genetic methods include PCR assays

Table 14.2 Relevant biochemical properties of clinical *Aeromonas* phenospecies or genospecies of *P. shigelloides*[a]

Test	Characteristic[b]							
	A. hydrophila[c]	*A. caviae*[d]	*A. veronii* biovar sobria (HG8/10)	*A. veronii* biovar veronii (HG8/10)	*A. jandaei* (HG9)	*A. schubertii* (HG12)	*A. trota* (HG14)	*P. shigelloides*
O/129, 10 μg/150 μg[e]	R/R	R/R	R/R	R/R	R/R	R/R	R/R	S/S
Indole	+	+	+	+	+	–	+	+
Glucose (gas)	+	–	+	+	+	+	+	–
Voges-Proskauer	+	–	+	+	+	–	–	–
Ornithine decarboxylase	–	–	–	+	–	–	–	+
Acid from:								
L-Arabinose	+	+	–	–	–	–	–	–
Sucrose	+	+	+	+	–	–	–	–
m-Inositol	–	–	–	–	–	–	–	+
D-Mannitol	+	+	+	+	+	–	+	–
Salicin	+	+	–	+	–	–	–	V
Esculin hydrolysis[f]	+	+	–	+	–	–	–	–
H_2S[g]	+	–	+	+	+	–	+	
Cephalothin, 30 μg[h]	R	R	S	S	R	S	R	
Beta-hemolysis (sheep blood)	+	V	+	+	+	+	V	–

[a]Adapted from references 83 and 87.

[b]R, resistant; S, susceptible; +, positive for more than 80% of strains; –, negative for more than 80% of strains; V, variable (20 to 80% positive). Biochemical tests are read daily for 3 to 4 days (87).

[c]Refers to phenospecies; additional biochemical tests are required to separate *A. hydrophila* (HG2, unnamed) and *A. salmonicida* (HG3) originating from clinical material (7, 97).

[d]Refers to both *A. caviae* (HG4) and *A. media* (HG5), which have proven difficult to separate biochemically (1, 81, 87).

[e]99% of strains resistant to 10 μg; >80% resistant to 150 μg.

[f]Agar formulation only (27).

[g]Gelatin-cysteine thiosulfate medium.

[h]Bauer-Kirby disk diffusion. Susceptibility for this method requires a zone size of ≥18 mm.

with 16S rRNA gene-targeted species-specific oligonucleotide primers, determination of rRNA gene restriction patterns, or a recently developed genomic fingerprinting technique (AFLP; European Patent Office patent no. 534858A1) which detects DNA polymorphisms by selective amplifications of restriction fragments (41, 77, 112). Such genetic techniques and other typing techniques, such as multilocus enzyme electrophoresis, electrophoretic typing of esterases, phage typing, gas-liquid chromatography of cell wall fatty acid methyl esters, and polyacrylamide gel electrophoresis (PAGE) of total soluble proteins (5, 136), have already proved useful in the research setting for identifying *Aeromonas* strains and tracing possible sources of infection (6, 138). A serogrouping scheme (97 serogroups) for aeromonads has been developed in the United Kingdom (155). Thus far, however, it has been minimally used in epidemiologic investigations.

Media used for isolation of aeromonads usually exploit their resistance to ampicillin. Blood ampicillin agar (10 to 30 μg of ampicillin per ml) is often used for clinical specimens, and starch ampicillin agar is often used for environmental specimens. (However, *A. trota* is ampicillin sensitive.) Further discussions of these and other selective media for aeromonads are provided elsewhere (43, 84, 87).

P. shigelloides

In contrast to the genus *Aeromonas*, *Plesiomonas* consists of a homogeneous species with a single DNA hybridization group (*P. shigelloides*). Like many *Vibrio* spp., *P. shigelloides* is susceptible to O/129. Like *Aeromonas* spp., it is unable to grow in 6% NaCl. Separation of *P. shigelloides* from aeromonads is based on simple phenotypic tests, exoenzyme production, and a range of carbohydrate fermentations (19, 43, 115). Useful biochemical reactions for differentiating *P. shigelloides* from *Aeromonas* spp. (Table 14.2) include myoinositol fermentation and decarboxylation of ornithine. The L-histidine decarboxylase test (conventional Moeller's tube test) may also be a potentially useful diagnostic feature for differentiating *P. shigelloides* from other members of the family *Vibrionaceae* (127). More than 100 serovars of *P. shigelloides* have been described. Several react with *Shigella* antisera. At present, 76 O and 41 H antigens are included in a proposed international typing scheme (4, 19). Ampicillin-containing selective *Aeromonas* media are not suitable for the isolation of *P. shigelloides*. Suitable isolation media include xylose-sodium deoxycholate-citrate and inositol-brilliant green bile salts and *Plesiomonas* agars (19, 43, 73, 84, 115).

Tolerance or Susceptibility to Preservation Methods

Temperature

Motile aeromonads are said to have an optimal growth temperature of 28°C. However, as they constitute a very heterogeneous group, strains with a very wide temperature growth range (<5 to 45°C) have been described. Thus, the optimal temperature is not 28°C for all strains. The source of strains in foods can influence the rate at which they grow at low temperatures (101). While most strains in foods are mesophiles, many can grow at refrigeration temperatures. Populations of *Aeromonas* species naturally occurring in foods of all types have been shown to increase 10- to 1,000-fold during 7 to 10 days storage at 5°C (15, 26, 131). Psychrotrophic strains which grow very rapidly have also been described. One of these, isolated from goat milk, was found to have a theoretical minimum temperature (T_{min}) for growth of −5.3°C. The T_{min} is determined from growth patterns in broth cultures placed in a temperature gradient incubator. Growth rates are determined as the time required for an increase in optical density of 0.3. The T_{min} is then estimated by plotting the square root of the growth rate versus temperature (92, 144). Most strains isolated from clinical specimens can grow at refrigeration temperatures (42, 93, 131). Temperature alone, therefore, cannot be relied on to control the growth of *Aeromonas* strains in foods (16, 90).

Most strains of *P. shigelloides* do not grow below 8°C, but at least one strain has been reported to grow at 0°C (115, 116). Their optimal temperature for growth, however, is 38 to 39°C, with a maximum around 45°C. About 25% of strains grow at 45°C (43, 115). Temperatures of 42 to 44°C have been recommended for the isolation of *P. shigelloides* from environmental specimens in which the presence of aeromonads poses a problem (73).

Strains of *Aeromonas* and *Plesiomonas* can be recovered from foods stored at −20°C for considerable periods (years) and are successfully stored in culture collections at −70°C.

pH and Salt Concentration

Aeromonas spp. are fairly sensitive to low pH (<5.5). Tolerance to variables such as pH and salt concentration varies with the temperature of growth. Studies have shown that *Aeromonas* spp. are unlikely to present problems in foods with more than 3 to 3.5% (wt/wt) NaCl and pH values below 6.0 when foods are stored at low temperature (90). Acetic acid, lactic acid, tartaric acid, citric acid, H_2SO_4, and HCl, in that order, are effective at restricting growth (2). Polyphosphates could also be useful in controlling the numbers of aeromonads in certain foods (130).

P. shigelloides grows at pH 5 to 8 but may be particularly susceptible to low pH. Strains are reported to be killed rapidly at a pH of ≤4 (19, 115). Tolerance to salt seems similar to that of *Aeromonas* spp. (116).

Atmosphere

Overall, aeromonads grow as well anaerobically as they do aerobically. However, they are more sensitive to $NaNO_2$ under anaerobic conditions. *Aeromonas* growth under modified atmospheres depends on the nature and number of competing microflora. Several reports have cautioned against the use of modified atmospheres to extend the shelf life of packaged meats and fresh vegetables, as this may lead to the consumption of foods in which aeromonads have been able to grow to high levels (15, 16). *Aeromonas* spp. contribute to the spoilage of many foods, but in others, such as milk, they can reach high numbers (up to 10^8) without detectable organoleptic changes (98).

P. shigelloides has been found to be extremely susceptible to modified atmosphere storage (80% CO_2, no O_2) showing no growth in cooked crayfish tails held at 11 and 14°C (78).

Destructive Methods: Heat, Irradiation, Disinfectants, and Chlorine

Although the number of studies is few, *Aeromonas* spp. appear to be readily killed by either heat or irradiation (2, 126, 132, 143). Palumbo et al. concluded that the thermal resistance of aeromonads was similar to that of other gram-negative organisms (132). *D* values at 48°C for different clinical and food isolates heated in raw milk ranged from 3.2 to 6.2 min. A recent study, however, has reported that aeromonads are more sensitive to heat than other food-poisoning bacteria, such as *E. coli* O157:H7, *Staphylococcus aureus*, and *Salmonella typhimurium*. Aeromonads in peptone water were killed within 2 min at 55°C, compared with 15 min for these other pathogens. They were also less resistant to heat in hamburger steaks than the other pathogens (126). These findings require confirmation, but the available information suggests that aeromonads should not present a problem in cooked foods, provided that the food subsequently receives correct handling so as to prevent recontamination. Some *Aeromonas* toxins, such as the "cholera-toxin cross-reactive, cytotonic enterotoxins" produced by some strains of aeromonads, are, however, reportedly heat stable (56°C, 20 min to 100°C, 30 min) (33, 40, 90).

Aeromonas strains are as susceptible to disinfectants, including chlorine, as other gram-negative bacteria found in foods, yet recovery of aeromonads from chlorinated water supplies is commonly reported. This could be the result of posttreatment recontamination, inactivation of chlorine by organic matter, the presence of very high initial numbers, or possibly the persistence of the organism in a "viable but nonculturable" state following treatment. Combinations of a number of variables may be required to restrict *Aeromonas* growth in food and water (2, 90).

More research is needed to determine the effects of factors discussed above on plesiomonads and the ability of the organisms to survive and grow under storage conditions commonly used for many foods, especially those of aquatic origin. Studies to date indicate that adequately cooked foods will not contain viable *P. shigelloides* and that proper refrigeration of foods should control its growth in stored foods (116).

RESERVOIRS

Members of both genera are primarily aquatic organisms. *A. salmonicida* is found associated with freshwater-dwelling fishes (particularly salmonids), which may be either asymptomatic carriers of the organism or diseased (i.e., exhibiting furunculosis). It is regarded as a strict parasite under natural conditions but may exist in a viable but nonculturable state in water. The motile, mesophilic aeromonads are found in fresh, stagnant, estuarine, or brackish water worldwide. As they are also commonly present in drinking water, they are found in sinks, drain pipes, and household effluents. A significant correlation between organic matter content and total numbers of aeromonads in waters has been reported (10). Temperature-dependent seasonal variations in numbers have been observed, with numbers highest in summer (22, 59, 88, 106). The different phenospecies (Popoff criteria) of aeromonads may predominate in different water sources. *A. hydrophila* (HG1, HG2, and HG3) is prevalent in spring water, *A. caviae* is prevalent in marine samples, and *A. sobria* predominates in recreational lakes and river water. All three phenospecies are found in sewage-contaminated waters (11, 43). The distribution of species and the proportion of strains producing toxins can vary in different geographic regions. Clinical strains appear to reflect differences in the distribution of environmental strains expressing virulence-associated properties (95, 97).

Motile aeromonads may associate with or colonize many water-dwelling plants and animals (for example, healthy fish, leeches, and frogs). They have long been recognized as the causal agent of red-leg disease in amphibians and are responsible for diseases in reptiles, fish, shellfish, and snails. In many fish species, they cause a hemorrhagic septicemia (87). They can be found as minor components of fecal flora of a wide variety of animals, including domestic animals used for food (pigs, cows, sheep, and poultry). They have been implicated in outbreaks of bovine abortion and diarrhea in piglets, as well as other infections in a variety of animals (53, 140).

Aeromonads are not generally considered to be normal inhabitants of the gastrointestinal tract of humans. Fecal carriage rates can, however, approach 3% in asymptomatic persons in temperate climates (89, 117). In the tropics or developing regions, carriage rates may reach 30% or more (89, 139). Aeromonads are widespread in foods and are readily found in meat, raw milk, poultry, fish, shellfish, and vegetables. In prepared foods such as pasteurized milk, their incidence is reduced, but they are still found (45, 90, 98). As for water, there are geographic differences in the species isolated and the proportion of isolates possessing putative virulence properties. *A. sobria* is reportedly found most often in poultry, raw meat and offal, and meat products (48, 56, 92). In Japan, the most common (~60%) phenospecies found in sea fish, vegetables, and their products was *A. caviae* (125), while *A. hydrophila* (HG1, HG2, and HG3) and *A. caviae* were predominant in vegetable produce in the United States, comprising 48 and 26%, respectively, of *Aeromonas* isolates (26). A biochemical fingerprinting typing scheme, based on an analysis of the kinetics of reactions in microplates, has been used to analyze *Aeromonas* isolates from drinking water and foods (105). Food isolates were found to be more homogeneous than water isolates, and identical isolates were sometimes found in food of different origins. Kühn et al. proposed that certain *Aeromonas* strains may be especially suited to survive and multiply in particular foods (105).

P. shigelloides is found in fresh and estuarine waters, as well as seawater in warm weather (4a, 43, 113). Its temperature growth restriction means that it is more common in tropical and subtropical climates and also accounts for the seasonal variation (summer incidence) in isolations from river water in temperate climates. *P. shigelloides* has also been isolated from warm- and cold-blooded animals, including dogs, cats, cattle, pigs, snakes, shellfish, and tropical fish. It has been found in healthy humans at very low rates (0.0078%; 3 carriers among 38,454 food handlers and schoolchildren in Japan) (9). In developing countries, a much higher isolation rate (24%, 12 of 51 adults in Thailand) has been reported (19). Few systematic studies of its incidence in foods have been carried out. So far, *P. shigelloides* has been isolated predominantly from fish and seafood (43, 116).

FOODBORNE OUTBREAKS

The main source of *Aeromonas* infection is thought to be water. Increased levels of aeromonads in drinking water have been reported to coincide with increased incidence of *Aeromonas*-associated gastroenteritis (22). Drinking untreated water has also been identified as a significant risk factor for *Aeromonas*-associated gastroenteric disease (119). Similarly, not only gastroenteric but also septicemic and surgical infections in immunocompromised patients hospitalized for more than 2 days appeared to have their origin in the hospital water supply (137, 138). Infection rates increased in the summer and were correlated with the increased number of aeromonads found in the water storage tanks (137). Moreover, control measures, such as the use of mineral water for drinking and sterile water for medical preparations used for such things as gastric lavage and radio-opaque solutions, reduced the number of septicemic cases (138). Increased isolation of aeromonads from human stools has also been correlated with high levels of aeromonads in foods (particularly minced beef, pork, and chicken) in Japan (125). Typing methods have shown, however, that there is little similarity between the majority of water-related *Aeromonas* strains and diarrhea-associated isolates (58, 138). Strains able to colonize or infect the human gastrointestinal tract are probably only a small fraction of environmental strains (58, 129).

There are relatively few published cases in which *Aeromonas* species have been associated with foodborne gastroenteritis. Suspect foods (oysters and other seafoods, edible land snails, egg salad) had been prefrozen and presumably inadequately cooked before consumption or were foods that were consumed after minimal cooking or handling procedures which had allowed rapid growth of aeromonads (18, 90). In only one case (38-year-old male) was the isolate from food (shrimp cocktail) and feces typed (rRNA gene restriction patterns) in a manner to establish definitively the source of the strain (6).

Two outbreaks attributed to *Aeromonas* species have been reported in Japan (102), and recently *Aeromonas* spp. have been implicated in a food-poisoning outbreak in Sweden in which 22 of 27 persons became ill 20 to 30 h after consumption of a smorgasbord containing shrimps, smoked sausage, liver pâté, and boiled ham, all of which contained a high number of aeromonads (log 10^6 to >log 10^7/g of food sample) in virtually pure culture (104). Unfortunately, in this latter outbreak, fecal samples were not submitted for microbiological investigation, and although the foods were tested for other bacterial enteropathogens, they were not tested for the presence of viruses. Wider testing for aeromonads in suspected food-associated infections will be needed to determine the importance of food as a vehicle of *Aeromonas* transmission.

The ability of many strains to grow and produce toxins (some of which are heat stable and can induce fluid accumulation in animal intestines) at low temperatures has led to speculation that disease from *Aeromonas*-contaminated foods may be caused by intoxication (caused by preformed toxins in foods) (42, 109). Although growth and toxin production in bacteriological media under refrigeration conditions are readily detected, toxin production in foods has generally been found to be lower in comparison. Psychrotrophic strains which can produce high levels of toxins in many foods do not appear to be common. Moreover, toxins are inactivated by some foods, such as milk (93, 94, 98). Thus, disease resulting from intoxication seems an unlikely risk in comparison with the possible risk posed by strains able to colonize the human intestine and express virulence properties in vivo.

P. shigelloides has been implicated in two waterborne epidemics (involving ~1,000 people) (158). In one, a single predominant serotype (O17:H2) was identified in patients, and this serotype was also found in tap water from different parts of the area where the outbreak occurred (158). Outbreaks of *Plesiomonas*-associated gastroenteritis have been attributed to *Plesiomonas*-contaminated oysters, chicken, fish (salt mackerel, cuttlefish salad), and shrimp (19, 115, 146). Oysters have been the major food incriminated in the United States (43). Thirty-one patients with gastroenteric symptoms from whom *P. shigelloides* had been isolated were studied by Holmberg et al. (68). Consumption of raw seafood and foreign travel (particularly to Mexico) were identified as major risk factors for these patients with *Plesiomonas*-positive stools (68).

CHARACTERISTICS OF DISEASE

Extraintestinal

Although extraintestinal infections with *Aeromonas* spp. and *P. shigelloides* are uncommon, they tend to be severe and often fatal, particularly in patients with sepsis and meningitis. *Aeromonas* extraintestinal disease is more common and varied (peritonitis, endocarditis, pneumonia, conjunctivitis, and urinary tract infections) than that caused by *P. shigelloides* (cellulitis, arthritis, endophthalmitis, and cholecystitis) (19, 44, 89, 134). The gastrointestinal tract may be the source of some disseminating infections, although some may originate from infected wounds and trauma. *Aeromonas* spp. are well documented as causative agents of wound infections, usually linked to water-associated injuries or aquatic recreational activities (50).

Gastroenteritis

Aeromonas- and *Plesiomonas*-associated cases of gastroenteritis are distinctly seasonal (sharp summer peak). The most common (approximately three quarters of reported cases) clinical presentation with which aeromonads have been associated is a watery diarrhea accompanied by a mild fever. Vomiting may also occur in children under 2 years of age. In more than one-third of patients, diarrhea may last for over 2 weeks. The remaining quarter of incidents are dysentery-like cases with blood and mucus present in the stools. In some cases, the clinical features are suggestive of ulcerative colitis (51). Isolated cases of a cholera-like illness (rice water stools) are also reported in the literature (35).

Patients with stools positive for *P. shigelloides* also have either a watery diarrhea or diarrhea with blood and mucus. The secretory form is usually reported to last from 1 to 7 days but can be prolonged (~3 weeks), with numerous (up to 30) bowel movements per day at the peak of the disease (19, 115). In the study by Holmberg et al., most of the patients (adults) from whom *P. shigelloides* was isolated had symptoms of invasive disease (bloody, mucus-containing stools with polymorphonuclear leukocytes). Patients often had severe abdominal cramps, vomiting, and some degree of dehydration (68). Three culture-positive patients had another known concurrent infection (caused by *Entamoeba histolytica* or *Campylobacter jejuni*) and were excluded from the analysis. It should be noted, however, that this study did not particularly address the etiology of the patients' symptoms.

The lack of suitable animal models which reproduce the features of *Aeromonas*- and *Plesiomonas*-associated diarrhea has made it difficult to establish definitively the enteropathogenicity of these organisms. The intestinal secretory immunoglobulin A (sIgA) response among adults in Mexico with diarrhea was studied as an indicator of enteropathogenicity. Of 12 subjects shedding the phenospecies *A. sobria* or *A. hydrophila* (HG1, HG2, and HG3), 11 had a fourfold or greater sIgA titer rise against the infecting strain. However, no sIgA titer rises were detected in 14 patients shedding *P. shigelloides*. Thus, although this study provided further evidence for the significance of *Aeromonas* species as pathogens in acute diarrhea, it raised questions about the role of *P. shigelloides*, at least in adults with traveler's diarrhea (85).

For *Aeromonas* species, there appears to be a species-associated disease spectrum. *A. veronii* biovar sobria (HG8/10) is more frequently associated with bacteremia than the other species. *A. schubertii* (HG12) is associated with aquatic wound infections but as yet has not been documented in association with gastroenteritis (27, 29). *A. hydrophila* (HG1 and HG3) also features in wound infections and following the use of medicinal leeches (153). *A. veronii* biovar sobria (HG8/10), *A. hydrophila* (HG1 and HG3), *A. caviae* (HG4), and, to a lesser extent, *A. veronii* biovar veronii (HG8/10), *A. trota* (HG14), and *A. jandaei* (HG9) are the species which have been associated with gastroenteritis (27, 28, 81, 87). *A. veronii* biovar sobria has been associated with the dysenteric presentation more frequently than the other species, although studies of invasive *Aeromonas* gastroenteritis are few (107, 161). *A. caviae* (HG4) is most common in pediatric diarrhea (121). *A. media* (HG5) is also occasionally isolated from human feces, but its significance is not known (27, 81).

INFECTIOUS DOSE AND SUSCEPTIBLE POPULATIONS

Extraintestinal Infections

Aeromonas opportunistic extraintestinal infections occur particularly in children and adolescents, who account for approximately one-quarter of all patients presenting with bacteremia. The majority of patients have chronic underlying disorders such as leukemia, solid tumors, aplastic anemia, hemoglobinopathies, cirrhosis, or renal failure (44, 89, 134). *Plesiomonas* meningitis has been reported in neonates, particularly those whose deliveries have been complicated by various medical conditions, including prolonged rupture of the mother's membranes. *Plesiomonas* septicemia has been reported in splenectomized adults (19, 43, 44).

Gastroenteritis

Aeromonas-associated gastroenteritis occurs most commonly in children, the elderly, and the immunocompromised (12, 147). Predisposing risk factors in adults include hospitalization, antimicrobial therapy, neutralization of gastric acid or inhibition of acid secretion, hepatic diseases, and underlying enteric conditions, such as gastric and colonic surgery, gastrointestinal tract bleeding, and idiopathic inflammatory bowel disease (47). Cases of traveler's diarrhea have also been attributed to *Aeromonas* spp. (52, 57). Formula-fed infants, or those with altered gastrointestinal tract flora as a consequence of disease or antibiotic administration, have been reported to favor the survival and colonization of *A. caviae* (121).

The infectious dose of aeromonads is not known. The one human trial, conducted in 1985, was largely unsuccessful. Diarrhea was demonstrated in only 2 of 57 healthy adult volunteers with doses ranging from 10^4 to 10^{10} CFU. One person experienced mild diarrhea with 10^9 CFU (enterotoxin-positive strain, isolated from healthy stool); a second person developed moderate diarrhea with 10^7 CFU of another strain (enterotoxin-positive strain, isolated from diarrheal stool) (118).

The essentially negative results of the human trial should not be taken as evidence negating the enteropathogenicity of *Aeromonas* spp., however. First, the trial was conducted in healthy, human adults who may well have acquired considerable immunity to *Aeromonas* spp. More important, a number of other considerations regarding the temperature-dependent expression of possible virulence factors, such as intestinal colonization factors, have since emerged. Virtually nothing was known in 1985 about these virulence determinants (96, 129) or the "suicide phenomenon" (an acetic acid self-killing phenomenon among mesophilic aeromonads) (120) which could account for the complete lack of recovery of aeromonads from the stools of most challenged volunteers (118). A laboratory accident in the early 1990s in which a stock culture broth of *A. trota* (HG14) (~10^9 CFU) was accidentally ingested by a 28-year-old laboratory worker with no preexisting health problems did correlate with clinical illness. He developed severe gastroenteric symptoms (rice water stools for 2 days) within 24 h, and an *Aeromonas* strain identical to the culture stock was isolated from his stools on two occasions (28).

P. shigelloides has been isolated from the diarrheal feces of both adults and children. It may pose an occupational hazard for veterinarians, zookeepers, fish handlers, and athletes in water sports. Symptomatic individuals from whom *P. shigelloides* has been isolated often have a history of freshwater contact, exposure to amphibia or reptiles, or travel to developing countries. Thus, ~70% of persons who present with plesiomonad-associated diarrhea have either an underlying illness (e.g., cancer or cirrhosis) or an identifiable risk factor (foreign travel, consumption of raw seafood) (19, 68). If *P. shigelloides* is responsible for the gastroenteritis in these individuals, its infectious dose is unknown. In a human volunteer study, none of 33 volunteers (given ampicillin to reduce normal intestinal flora) developed diarrhea, even though some received over 10^9 organisms of a strain originally isolated from a 4-year-old patient with diarrhea whose stool had blood, mucus, and leukocytes. A similar dose in piglets, however, caused moderate to mild diarrhea in three of three infected animals (43, 68, 115).

VIRULENCE FACTORS AND MECHANISMS OF PATHOGENICITY OF *AEROMONAS* SPECIES

Most work on the possible virulence factors of aeromonads has been done with strains isolated from patients with gastroenteric disease. Less is known of the factors which may be significant in the pathogenesis of extraintestinal infections. Some virulence determinants, such as the hemolysin called aerolysin may be common to both forms of infections, while others, such as S-layers, may be more important in extraintestinal infections (see below). However, as the roles of the different putative virulence factors described have not been established for any human *Aeromonas* infections, the proposed virulence determinants will be considered together.

The varied clinical picture of *Aeromonas* infections, and gastroenteric illness in particular, suggests that complex pathogenic mechanisms occur in aeromonads. For the different gastroenteric disease manifestations, it may be that strains possess multiple virulence factors in different combinations which may interact with different levels of the intestine, possibly analogous to the situation that has emerged for the different pathotypes of *E. coli* that cause intestinal illness. Putative *Aeromonas* virulence factors, inferred from comparisons with such better-studied enteropathogenic bacteria, are listed in Table 14.3. As expected, these virulence factors tend to be found more frequently, or are better expressed, in clinical strains than in environmental ones. Yet, as men-

Table 14.3 Potential virulence factors of *Aeromonas* spp.[a]

Factor
Extracellular enzymes (e.g., proteases, lipases, elastase)
Siderophores (enterobactin, amonabactin)
Exotoxins
Cytotoxic (cytolytic enterotoxin, aerolysin, Asao toxin, β-hemolysins)[b]
Cytotonic
Heat stable (56°C, 10–20 min), CT cross-reacting and non-cross-reacting
Heat labile, non-CT cross-reacting
Endotoxins (lipopolysaccharide)
Invasins (invasion of HEp-2, and Caco-2 cells)
Adhesins
Type IV pili
Outer membrane proteins (lectin-like, possibly porins)
S-layers

[a]See references in the text for possible roles of these factors in *Aeromonas* extraintestinal and gastroenteritis-associated infections.
[b]Synonyms.

tioned in the introduction, despite significant advances in this area, there is still no set of virulence factors that can be used to demonstrate that a given clinical or environmental isolate has the capacity to cause human illness. Enterotoxins have been the most studied. Knowledge of invasins and intestinal adhesins is still very scant. The S-layer of *A. salmonicida* is the only *Aeromonas* virulence factor linked to overt pathogenicity for salmonids.

The lack of a suitable animal model of *Aeromonas* diarrhea has slowed progress in the identification of critical *Aeromonas* virulence determinants involved in gastroenteric disease. The removable intestinal tie adult rabbit diarrheal model is currently the most promising available, but it involves moderately complex surgery (135). In this model, histological changes consistent with colonization and infection of the ileum are seen. Rats pretreated with clindamycin and fed aeromonads have also been reported to develop diarrhea and histopathologic evidence of enteritis. An adult streptomycin mouse model (fecal shedding of aeromonads monitored) and 50% lethal dose experiments using infant mice are not diarrheal models but may nevertheless measure the ability of strains to bind to the intestinal mucosa (100). Further work to establish appropriate animal models for studying virulence determinants is of crucial importance if we are to understand *Aeromonas* pathogenicity (81, 100).

S-Layer

S-layers are macromolecular arrays of protein subunits (49 to 58 kDa) found on the bacterial cell surface. Strains of the psychrophilic aeromonad *A. salmonicida* produce an S-layer commonly known as A-layer. A-layer has been shown to have numerous biological activities. It contributes to protection against the bactericidal activities of both immune and nonimmune serum, influences the outcome of interaction of the organism with macrophages, protects against the action of proteases, and binds immunoglobulins. A-layer is also capable of binding a variety of extracellular matrix proteins (for example, basement membrane components collagen IV and laminin) and may therefore play a role in colonization. Transposon mutants with altered ability to produce A-layer have been shown to lose the ability to kill fish (156).

The biological functions of the S-layers of the mesophilic aeromonads are not yet known. S-layer-positive strains of the mesophilic aeromonads *A. hydrophila* (HG1) and *A. veronii* biovar sobria (HG8/10) have been identified. These strains have been principally those from extraintestinal infections, such as wounds and peritonitis. S-layer-producing strains have been particularly associated with a single lipopolysaccharide serogroup, O:11, which contains some highly virulent strains, although there are possibly other groups with S-layers. Like *A. salmonicida*, these S-layer-positive, mesophilic strains are more refractory to the bactericidal activity of pooled human sera than strains which do not possess S-layers, although the effect is probably not directly due to the presence of S-layer. Serogroup O:11 strains, as well as serogroup O:34 strains, which reportedly do not possess S-layers but which also feature in septicemic disease, have recently been found to produce capsular polysaccharide. This may be involved in the virulence of these strains (111). The S-layer may possess antiphagocytic activity, and this could facilitate systemic dissemination after invasion through the gastrointestinal mucosa. The virulence of S-layer-negative strains for mice following intraperitoneal injection increased when the bacteria were mixed with purified S-layer (103). An enzyme-linked immunosorbent assay using specific polyclonal antibodies against S-layer has been developed for the detection of O:11 aeromonads in foods (114).

Extracellular Enzymes

Most aeromonads elaborate a variety of extracellular enzymes: proteases (at least four or five, including a thermolabile serine protease and a thermostable metalloprotease), DNase, RNase, elastase, lecithinase, amylase, lipases, gelatinase, and chitinase. Several of these enzymes have now been cloned, as have two *exe* operons required for extracellular secretion of proteins by *Aeromonas* species (8, 36, 49, 80, 145). The roles of the enzymes in virulence have yet to be determined. Proteases may contribute to pathogenicity by causing direct tissue damage or enhanced invasiveness. In addition, there is proteolytic activation of hemolysin (discussed below). Lipases may be important for bacterial nutrition. They may also constitute virulence factors by interacting with human leukocytes or by affecting several immune system functions by free fatty acids generated by lipolytic activity.

Siderophores

Efficient mechanisms for iron acquisition from the host during an infection are considered essential for virulence. Siderophores are low-molecular-weight compounds with high affinities for various forms of iron. Mesophilic aeromonads produce either of two siderophores, enterobactin or amonabactin. Enterobactin is found in various gram-negative bacteria; amonabactin has so far been found only in *Aeromonas* spp. Two biologically active forms of this latter siderophore exist. The type of siderophore produced by aeromonads varies with hybridization group. Amonabactin is the predominant siderophore in *A. hydrophila* (HG1, HG2, and HG3), *A. caviae* (HG4), *A. media* (HG5), *A. schubertii* (HG12), and *A. trota* (HG14), while *A. veronii* biovar sobria (HG8/10) and *A. jandaei* (HG9) make enterobactin. Amonabactin (*amo*A) and enterobactin (*aeb*C) biosynthetic genes have been cloned. This will facilitate determination of the role(s) of these siderophores in *Aeromonas* virulence (14, 23, 24).

Exotoxins

The toxins have been the best characterized of the putative virulence determinants of *Aeromonas* species. Nevertheless, the toxin story has been confused and controversial. Reports of "toxins" are usually based on the measurement of biological activities in vitro and sometimes in vivo (see below). In recent years, molecular genetic analysis has confirmed that some of the effects described are indeed due to molecules analogous to the toxins of other, better-studied enteropathogenic bacteria. However, for *Aeromonas* species, the role of these molecules as virulence determinants is still unknown, and there is no evidence that production of a particular toxin, for example, correlates with the ability of a given strain to cause disease. Nevertheless, an adhesive-enterotoxic mechanism has been widely postulated as the one most likely to result in the commonest gastroenteric presentation (watery diarrhea) attributed to *Aeromonas* spp.

Factors which have added to the confusion include the difficulties with the taxonomy of the species in this genus and the variety of toxins which may be produced by different strains of aeromonads. Not all strains produce all of the toxins described to date. Even if strains do possess particular toxin genes, these genes may be expressed only under certain growth conditions. Also, at least one toxin possesses multiple activities. Traditionally, the major measurable effects of exotoxins have been their lethality for mice and their enterotoxigenic, hemolytic, and cytotoxic activities. In the laboratory, relatively simple tests have been devised to measure such activities. The most commonly used are illustrated in Fig. 14.1. Bacterium-free broth culture supernatants are tested for enterotoxic activity (fluid accumulation) in a suckling mouse or rabbit ileal loop assay; cytotoxin activity is measured by observing, or quantifying with dye, disruption of a Vero cell monolayer following addition of the broth culture supernatant; hemolysin production is quantified by titration of broth supernatants against a suspension of rabbit erythrocytes (93, 98). A significant correlation in the production of these three exotoxin activities has often been observed. However, this has not always been found, particularly for environmental strains, and the different activities have been shown to peak at different time periods during bacterial growth. Attempts to purify toxins led to reports of both heat-stable (56°C, 10 to 20 min) and heat-labile cytotonic and cytotoxic enterotoxins, some associated and some not associated with hemolytic activity and some related and some unrelated to cholera toxin (CT).

The different types of toxins can cause a problem with the interpretation of toxic activity, especially when studies are done with tissue culture cell lines. If a toxin produces a "cytotoxic" activity, the "cytotonic" toxins from the same strain will be difficult to detect since cytotoxicity is the dominant effect. Proteolytic activities expressed by individual strains may also complicate assays, as these activities can lead to rounding up of cells. In addition, some of the toxins may require enzymatic activation (see below) (70). Hence, the so-called heat lability of a particular toxin may be the result of inactivation of a protease required for its activation. These factors may help to explain some of the conflicting reports in the literature.

Cytotoxic/Cytolytic Enterotoxin

Cytotoxic/cytolytic enterotoxin (usually associated with hemolytic activity) is often found in diarrheal isolates. The observed close association between expression of hemolysin and enterotoxigenic activity in suckling mice

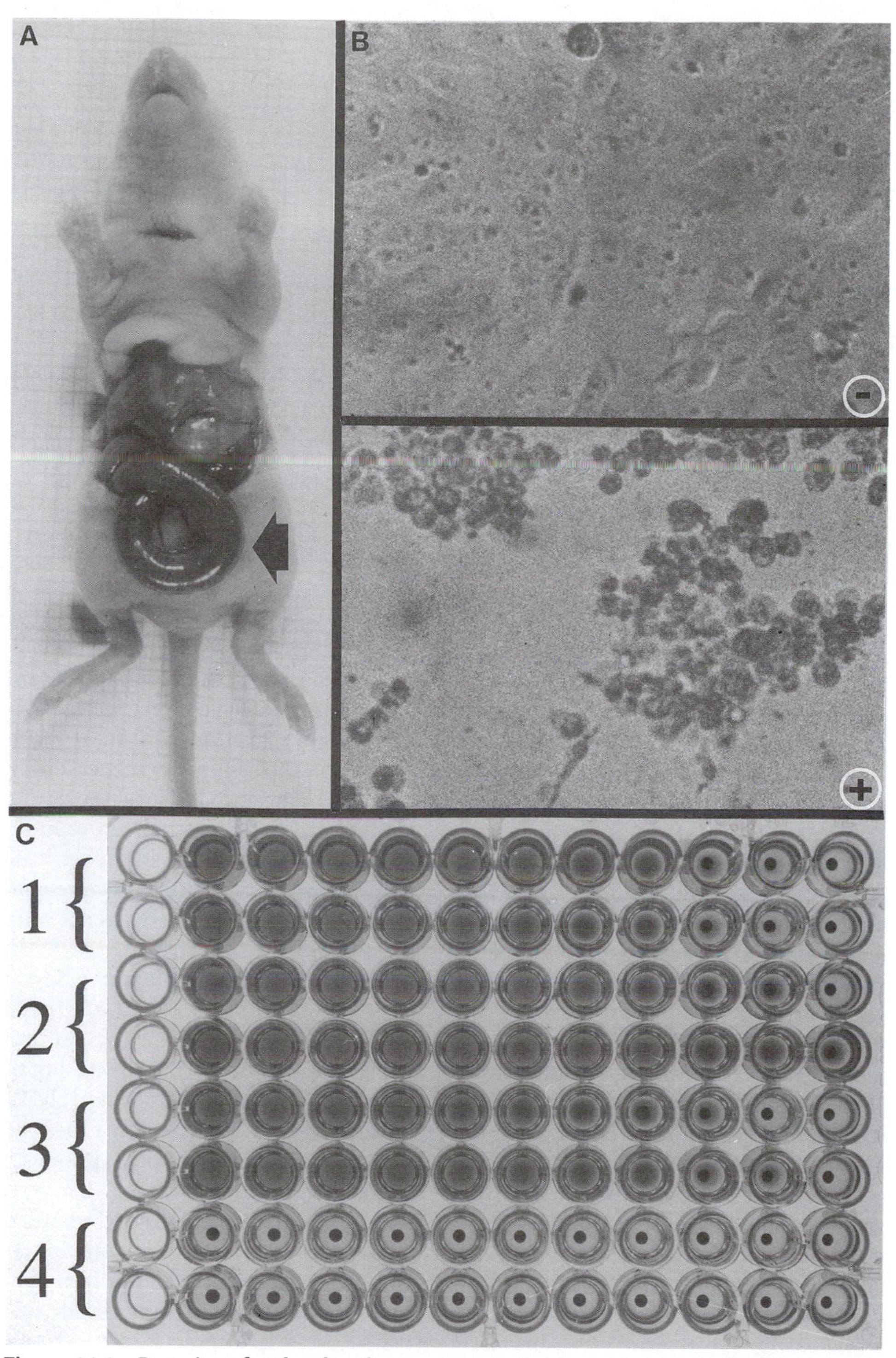

Figure 14.1 Bacterium-free broth culture supernatants are used to detect the ability of an *Aeromonas* isolate to elaborate toxins. (A) Suckling mouse assay to detect enterotoxic activity. Note fluid accumulation in the intestine (arrow). (B) Cytotoxin assay (Vero cells). Cytotoxin causes healthy cells (top) to round up and lift from the monolayer (bottom). (C) Hemolysin titration against rabbit erythrocytes. Strains 1 to 3 have hemolytic titers of >128; strain 4 has not produced detectable levels of hemolysin.

Table 14.4 Molecular characterization of the enterotoxins of *Aeromonas* spp.

Factor	Molecular mass (kDa)	*Aeromonas* spp. from which gene(s) was cloned[a]	Source of strain	Reference
Cytolytic enterotoxin (aerolysin, cytolysin, Asao toxin, β-hemolysin)	49–54	*A. hydrophila* = *A. trota*[b]	Human diarrheal feces	31
		A. hydrophila	Not specified	71
		A. sobria = *A. trota*[b]	Human diarrheal feces	75
		A. hydrophila (HG1, HG2, and HG3)	Eel	62
		A. hydrophila (HG1, HG2, and HG3)	Human	62
		A. sobria	Human	62
		A. hydrophila	Human diarrheal feces	39
Hemolysin	~60–64	*A. hydrophila* (HG1)	ATCC 7966	60
Cytotonic enterotoxin		*A. hydrophila* = *A. trota*[b]	Human diarrheal feces	33
Heat labile	35–40	*A. hydrophila*	Human diarrheal feces	40
Heat stable		Possibly similar to gene from *A. trota* above (33)		

[a]In most cases, the data in the papers reviewed were unclear on the method of classification to the species level. It is assumed that when details were not given, *A. hydrophila* was used in its broad sense to refer to a strain of the motile mesophilic aeromonads. That is its meaning unless noted otherwise in the table. *A. sobria* refers to the phenospecies defined by Popoff (140).

[b]In some studies, strains designated *A. hydrophila* or *A. sobria* were subsequently shown to be *A. trota* (HG14).

prompted studies to detect enterotoxigenic activity with purified hemolysin preparations. Such preparations were shown to be enterotoxigenic in the rabbit ileal loop and suckling mouse assays. Moreover, antibodies raised against purified hemolysin completely neutralized both cytotoxic and enterotoxic activities. Several groups have now cloned and sequenced aerolysin/cytolytic toxin genes (Table 14.4) (31, 39, 62, 71, 75). One of these studies demonstrated conclusively that the product of a cytolytic enterotoxin gene had all of the above cited biological activities (lethality for mice, hemolysis, cytotoxicity, and enterotoxicity) (39). The cytotoxin/aerolysin is a channel-forming protein (34, 160, 162, 163). At very low concentrations, the purified protein forms pores in artificial lipid bilayers. These pores are ~1.5 nm and exhibit weak anion selectivity. This could explain the enterotoxic properties of aerolysin at low concentrations (34, 160).

At the amino acid level, there was no marked homology between the cytolytic enterotoxin cloned in the study discussed above and other known toxin molecules. Some small regions showed homology with the alpha-toxin of *S. aureus*, the cytotoxin of *Pseudomonas aeruginosa*, and a plasmid-encoded hemolysin (HlyA) of *E. coli* (37). A tryptophan-rich region (positions 393 to 403) exhibited some homology with a similar region detected in sulfhydryl-activated hemolysins, such as listeriolysin O, streptolysin O, and pneumolysin produced by *Listeria monocytogenes*, *Streptococcus pyogenes*, and *Streptococcus pneumoniae*, respectively. Likewise, residues 363 to 392 exhibited significant homology with the C-terminal region of *Clostridium perfringens* type A enterotoxin (positions 290 to 319) which has been implicated in the binding of this enterotoxin to its receptor (37, 39). The primary structure of *Clostridium septicum* alpha-toxin has recently been reported to exhibit 27% identity and 72% similarity over a 387-residue region with the primary structure of *Aeromonas* aerolysin (13). There may, therefore, have been a common ancestor for these genes.

In the standard exotoxin assays described above, *A. veronii* biovar sobria (HG8/10) is the species most often found to contain the highest proportion of toxigenic strains and the strains producing the highest titers of toxins. Strains of *A. hydrophila* (HG1, HG2, and HG3) are also frequently positive for these activities. However, strains of *A. caviae* (HG4) tend not to produce exotoxins under the same conditions. Yet, genetic analysis has shown that ~50% of *A. caviae* isolates carry the aerolysin (β-hemolysin/cytolytic enterotoxin) gene (74). Husslein et al. suggested that the gene in these strains is silenced in some way (by mutation, insertion, or deletion) (76). In glucose-free, double-strength Trypticase broth, and in the late stationary phase of growth particularly (that is, under culture conditions different from those used in standard assay), some clinical strains of *A. caviae* have, however, been shown to produce cytotoxic and cytotonic enterotoxin activities (121). Singh and Sanyal have shown that passage of strains of all of the three aforementioned phenospecies through rabbit ileal loops increased the proportion of strains able to produce detectable enterotoxin (152). The conclusion from the studies to date, therefore, is that most strains of *A. veronii* biovar sobria (HG8/10), *A. hydrophila* (HG1, HG2, and HG3), and *A. caviae* (HG4), whether from clinical or environmental sources, are potentially enterotoxigenic.

Hemolysins

Aeromonads produce at least two major classes of hemolysins, α and β. As discussed above, aerolysin (β-hemolysins) are enterotoxigenic cytolysins. β-Hemolysins form during log phase (except in strains of *A. caviae*). They cause complete lysis of erythrocytes (clear zones of hemolysis on blood agar). In spite of the controversy regarding the pathogenicity of *Aeromonas* spp., aerolysin is possibly the best characterized of the bacterial hemolysins in terms of its molecular properties (160). As mentioned above, the protein ultimately produces a transmembrane channel that destroys cells by breaking their permeability barriers. There are two inactive precursor forms of the toxin. The first, preproaerolysin, contains a typical 23-amino-acid signal sequence that directs transport across the inner bacterial membrane and is removed during transit. The resulting protoxin is exported and then activated by proteolysis of about 25 amino acids from the carboxy-terminal end. Aerolysin binds to the eukaryotic receptor glycophorin. It has a novel protein fold as a result of its relatively hydrophobic primary structure, which lacks extended regions of hydrophobic residues such as those normally found with transmembrane proteins. After binding, it oligomerizes, which is an essential step in channel formation (20, 34, 133, 162, 163). Oligomerization seems to precede membrane insertion. The *Aeromonas* β-hemolysins are heat-labile (56°C, 5 min) proteins with molecular masses of 49 to 53 kDa. Aerolysin genes from a number of mesophilic aeromonads have been cloned and sequenced (Table 14.4) (62, 71, 72, 75). The amino acid sequences of these proteins are very similar.

The second type of hemolysin, α-hemolysin (molecular mass, 65 kDa), is elaborated during the late stationary phase of growth and causes incomplete lysis of erythrocytes (double-zone lysis on blood agar). It is not expressed at temperatures above 30°C, and it has not been associated with enterotoxic properties.

Another extracellular hemolysin gene has been cloned from the type strain (ATCC 7966) of *A. hydrophila* (HG1) and sequenced (60). This hemolysin showed 46% homology at the DNA level with aerolysin (72). However, its amino acid composition was different from that reported for the Asao hemolysin (Table 14.4). It contains some regions homologous in sequence to the *Vibrio vulnificus* and *V. cholerae* cytolysin-hemolysin (60, 61). The relative molecular mass of the protein was similar to that of the α-hemolysin, yet it caused complete, not partial, hemolysis of erythrocytes. By colony hybridization analysis, the gene for this hemolysin was detected in 43 of 62 hemolysin-producing *Aeromonas* strains (60). Thus, it appears that *Aeromonas* strains possess a heterogeneous family of cytolytic hemolysins.

Cytotonic Enterotoxins

A variety of cytotonic enterotoxins have been described, and several have now been cloned (Table 14.4). They cause elongation of Chinese hamster ovary (CHO) cells and increased levels of cyclic AMP in CHO cells. They are thus all thought to act similarly to CT, despite their different molecular sizes and variable reactivity with cholera antitoxin. Several non-CT-reactive cytotonic enterotoxins have been described for *Aeromonas* spp., including a 15 to 20-kDa heat-stable protein identified by Ljungh et al. (108) and a 44-kDa, heat-labile protein identified by Chopra and Houston (38). Other cytotonic enterotoxins, such as that purified from a strain of *A. sobria* by Potomski et al. (141), have exhibited cross-reactivity to CT. This latter purified toxin exhibited bands of 43.5, 29.5, and 26 kDa on sodium dodecyl sulfate-PAGE. Schultz and McCardell similarly demonstrated that in cell lysates of *Aeromonas* strains, some strains had three protein bands (sizes of 89, 37, and 11 kDa) which reacted with CT antitoxin on an immunoblot (149). Chakraborty et al. cloned a cytotonic enterotoxin gene from an isolate [now known to be *A. trota* (HG14)] and showed conclusively that this strain produced a cytotonic enterotoxin that was distinct from cytotoxic/hemolytic enterotoxin (33). Chopra et al. have cloned genes encoding two cytotonic enterotoxins from *A. hydrophila* (40). One toxin was heat labile at 56°C, and the other was heat stable at this temperature. Expression of the heat-labile cytotonic enterotoxin gene yielded a major 35-kDa protein band. At the DNA level, there was no homology with the *A. trota* cytotonic toxin gene studied by Chakraborty and colleagues (33). However, a 4.8-kb *Sal*I-*Bam*HI DNA fragment encoding the heat-stable toxin did hybridize to a 3.5-kb *Bam*HI DNA fragment of a plasmid that contained this gene.

Aeromonas spp., therefore, may produce different types of cytotonic enterotoxins that are functionally similar, but these enterotoxins may be produced by only a minority of strains. Seidler et al. reported their production in <6% (20 of 330) of strains (150). Shimada et al. demonstrated the production of a CT-like enterotoxin by *Aeromonas* spp. in ~4.5% (8 of 179) of strains tested (151).

As mentioned earlier, detection of toxins is difficult, and there is still little known of their roles, if any, in extraintestinal and gastrointestinal infections. Isogenic mutant bacteria designed to knock out different toxins one at a time and appropriate animal models will be required for further investigations. Despite advances in taxonomy and molecular genetic studies, much remains to be done to clarify the picture with regard to *Aeromonas* toxins.

For extraintestinal infections (septicemia), it has been demonstrated in a mouse infection model that aerolysin is required for both the initiation and maintenance of septicemic infection (32). Jin et al. (86) have shown that intraperitoneal pretreatment of mice with cytolytic toxin/aerolysin predisposed them to nonlethal quantities of *A. hydrophila* that was translocated into the liver, spleen, and heart blood, whereas in mice not given toxin, the organism did not spread to these organs. The toxin impairs the phagocytic ability of mouse phagocytes. It is also a potent chemoattractant, inducing leukocyte infiltration in vivo and resulting in inflammation (86). The discovery of these pathogenic roles for cytolytic enterotoxin may help explain why *Aeromonas* species are capable of causing wound infections, diarrhea, and/or septicemia.

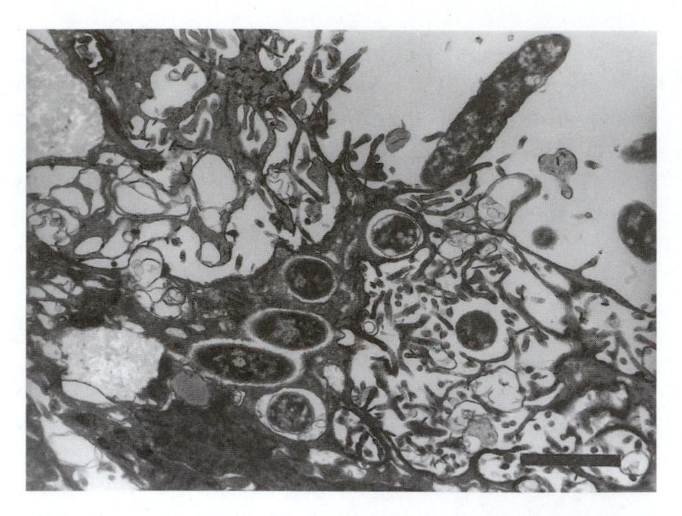

Figure 14.2 Invasion of Caco-2 cells by aeromonads. Bacteria are enclosed within membrane-bound vacuoles inside a Caco-2 cell. Bar = 1 μm. (Reprinted from reference 124.)

Invasins

The Sereny test for tissue invasiveness has been universally negative for aeromonads (43). Only a few studies have examined the ability of strains to penetrate epithelial cell lines, such as HEp-2 of epipharyngeal origin and the intestinal cell line Caco-2 (Fig. 14.2) (107, 124, 161). Invasive ability has been reported most commonly for strains of *A. sobria*, followed by *A. hydrophila* (HG1, HG2, and HG3), particularly strains isolated from patients with symptoms of dysentery (107, 161). (The possible link between tissue culture cell invasiveness and dysentery needs to be confirmed with larger numbers of patients, however.) The mechanisms of *Aeromonas* cell line adhesion and invasion thus remain to be elucidated. Invasin determinants are probably chromosomally located (107). Nishikawa et al. showed that the DNA of four invasive *Aeromonas* strains that they studied did not hybridize with probes to the genes associated with the attaching and effacing (*eae*) and invasion (*ipaB*) abilities of *E. coli* (124).

Plasmids

Plasmids carry important virulence determinants in a number of bacterial enteropathogens, but little is known of their role (if any) in *Aeromonas* virulence. Most strains of *A. salmonicida* carry at least one large plasmid (60 to 150 kb) and two small plasmids (5.2 and 5.4 kb); additional plasmids are also frequently encountered (123). Plasmids are detected less commonly (17 to 27% of strains) in human diarrhea-associated *Aeromonas* isolates but have been found in up to 95% of isolates from patients with bacteremia (3, 17, 159). Plasmids of clinical isolates are predominantly cryptic (function not known), but some have been shown to be involved in antibiotic resistance and possibly siderophore production (17). Mini pilin (see below), a putative adhesin, has been localized to a 7.6-kb plasmid in a strain of *A. hydrophila* (HG1, HG2, and HG3) (64). There has also been a report that adherence and hemolytic properties of a strain of *Aeromonas* spp. may be regulated by a plasmid (40 MDa, ~62 kb) which also codes for antibiotic resistance (55). Further study is needed to confirm this finding and to determine whether such regulation occurs in other *Aeromonas* strains.

Adhesins

Adhesion is likely to be an essential virulence factor for aeromonads which infect through mucosal surfaces or cause gastroenteric disease (91). Limited work with cell culture and animal models has been used to study this aspect of *Aeromonas* virulence. Most investigations of *Aeromonas* adhesion to cultured mammalian cell lines, such as mouse Y_1 adrenal, HEp-2, and human intestinal cell lines such as INT407 (embryonic human intestinal cell line) and Caco-2 (colon carcinoma cells which differentiate in culture to express characteristics of small

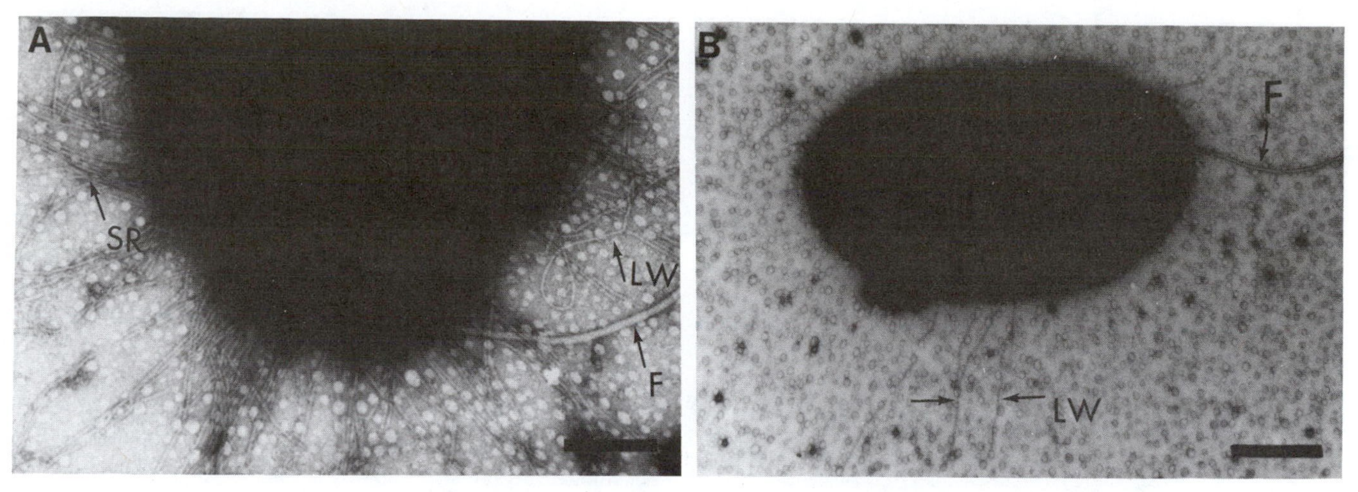

Figure 14.3 Transmission electron micrographs of negatively stained *A. veronii* biotype sobria strains showing pili. (A) A heavily piliated bacterium showing pili of short, rigid (SR) and long, flexible or wavy (LW) morphology and a flagellum (F). Environmental strains tend to be heavily piliated. Bar = 0.2 μm. (B) Fecal isolates of this species tend to be poorly piliated. Generally a few long, flexible (LW) pili (arrows) are seen. Bar = 0. 4 μm.

intestinal enterocytes), have shown that strains of *A. veronii* biovar sobria were the most adhesive (the highest proportion of adherent strains and the strains with the greatest number of adherent bacteria per cell) (30, 54, 91). Clinical strains of *A. caviae* (>30%) have also been reported to be adherent (54, 121, 122). *A. hydrophila* (HG1, HG2, and HG3) adhesion to cell lines seems to be less efficient than that of *A. veronii* biovar sobria and *A. caviae*. Thus, these models suggest that there may be species differences in adhesive mechanisms. Cell line models have proved useful in investigations of adhesins of other bacterial enteropathogens, but they need further validation for *Aeromonas* species (91, 100). Limited work with animal models also points to possible species-related differences in the mechanisms by which strains adhere to the mucosal surface (100). As yet, there are virtually no studies of *Aeromonas*-mucus interactions.

Receptors in the host are usually carbohydrate moieties which may also be expressed on erythrocytes. Hemagglutination has therefore been used as a screening model to detect adhesins. The hemagglutinins of *Aeromonas* species are numerous. Some have been associated with filamentous appendages (fimbriae or pili); others have been associated with outer membrane proteins. Some hemagglutination patterns (for example, that not inhibited by fucose, galactose, or mannose) have been reported to correlate with diarrhea, but hemagglutination screening has not proved useful for the identification of potentially significant strains (21).

Filamentous Adhesins

Aeromonas filamentous surface structures are diverse. Their number and type vary according to the source of the isolate and the bacterial growth conditions (63, 91, 99). Environmental strains of *A. veronii* biovar sobria tend to be heavily piliated on isolation (99). Short, rigid pili are the predominant type expressed on heavily piliated aeromonads (Fig. 14.3A). These pili cause autoaggregation of bacteria but are not hemagglutinating and do not bind to intestinal cells (69). The amino acid sequence of one, purified from a clinical *A. hydrophila* (HG1, HG2, and HG3) isolate, showed homology with *E. coli* type I and Pap pilin (63). The role of the short, rigid pili in colonization is unknown.

In contrast, many *Aeromonas* isolates (in particular strains of *A. veronii* biovar sobria) from diarrheal stools are poorly piliated (<10 pili per cell). Short, rigid pili can be induced on some clinical strains grown under environmental conditions (liquid medium, <22°C) (99), but long, thin (4- to 7-nm), flexible or wavy pili are the predominant type seen on strains isolated from feces (Fig. 14.3B) (30, 99). On some strains, bundles of flexible pili and thin (<2-nm) filamentous networks are also seen on small numbers of bacteria (<10%), particularly when strains are grown in liquid medium at temperatures of ≤22°C (Fig. 14.4) (99, 100).

There is indirect evidence that at least some of these surface structures are colonization factors. Adhesion to cell lines is optimal when bacterial strains are grown under the same conditions as those which induce maxi-

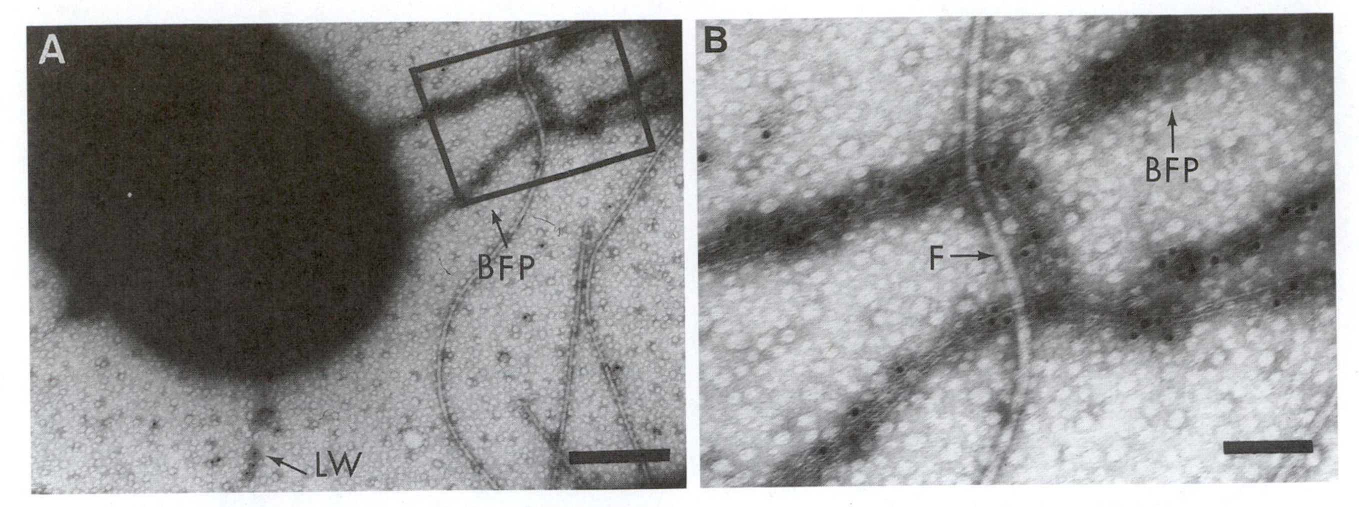

Figure 14.4 (A) Immunoelectron micrograph of a fecal strain of *A. veronii* biotype sobria. The gold particles bind specifically to pili and stain bundles of long, flexible pili (BFP) as well as isolated long, wavy (LW) pili (arrow). Bar = 0.3 μm. (B) Enlargement of BFP and flagellum (F) shown in panel A. Bar = 0.1 μm.

mal expression of the aforementioned structures. Removal of the surface structures by mechanical or enzymatic means decreases (by 60 to 80%) bacterial adhesive ability for some strains (96). In addition, purified long, flexible pili from three isolates of *A. sobria* and one of *A. hydrophila* (HG1, HG2, and HG3) were able to adhere to human intestinal tissue and could block (to various degrees) adhesion of the source bacteria to intestinal tissue (65–67, 79). Analysis of the N-terminal amino acid sequences of these flexible pili, and others subsequently purified by two other groups (100, 135a), has shown they are type IV pili. Type IV pili are known to be important in the binding of a number of gram-negative-pathogens to epithelial cells (100, 154). These structures are therefore likely to be important virulence determinants for *Aeromonas* spp. As yet, there has been very little genetic analysis of them or evaluation of their roles in virulence. Further genetic characterization of adhesins may shed further light on distinctions between pathogenic and nonpathogenic strains and may lead to rapid methods for detection of strains of public health significance. Pepe et al. have recently cloned a type IV pilus biogenesis gene cluster from an *Aeromonas* strain (135a).

Not all pili of flexible morphology are type IV pili. The plasmid-coded, flexible pilus mentioned earlier was characterized by Ho et al. and found to be composed of a novel 46-amino-acid polypeptide (mini pilin) (63). The mini pilin structural gene (*fxp*) has not, however, been found in any other strains (66 strains tested in the original study; >100 strains tested in our laboratory). Its significance as a colonization factor is unclear. Mini pilin is also environmentally regulated and is maximally expressed in liquid medium at 22°C in conditions of iron depletion (63, 64).

Nonfilamentous Adhesins

Outer membrane proteins and a lipopolysaccharide-protein complex have also been implicated as adhesins for *Aeromonas* species. A nonfimbrial hemagglutinin associated with a 43-kDa outer membrane protein of a strain of *A. hydrophila* has been purified. The receptor for this latter protein was human blood group H antigen, an oligosaccharide found lining the enteric tract. H-antigen terminal trisaccharides were used to prepare carbohydrate-reactive outer membrane proteins from solubilized outer membranes by affinity chromatography. Carbohydrate-reactive outer membrane proteins of this strain were indistinguishable from a porin subset isolated from the same strain. It has been proposed that porins may act as lectin-like adhesins for attachment of this strain to carbohydrate-rich surfaces, such as erythrocytes, and possibly the human gut (91, 142).

Virulence and Temperature

Bacterial growth temperature has been shown to affect the expression of several virulence-associated factors. Numbers of flexible pili (intestinal colonization factors) were greater when bacteria were grown at 22 and 7°C than when they were grown at 37°C, and adhesive ability to cell lines was also optimal at 22°C (or 7°C for some environmental strains) (96, 100). The titers of extracellular products (hemolysin, cytotoxin, and proteases) reportedly increased for some strains, serum sensitivity decreased, lipopolysaccharide levels increased, and virulence for animals (mice and fish) was greater in strains grown at 20°C than in those grown at 37°C (129, 157). Increased expression of virulence determinants at low temperature may have implications for the pathogenicity of aeromonads in stored foods and requires further investigation.

VIRULENCE FACTORS AND MECHANISMS OF PATHOGENICITY OF *PLESIOMONAS* SPECIES

There are no virulence factors that are widely accepted as being important for *Plesiomonas*-associated infections. None is tested for routinely (43, 82, 120). Potential virulence factors that have been described are summarized in Table 14.5. Very few have been investigated in any detail.

A glycocalyx has been detected on the outer surface of *P. shigelloides* by ruthenium red staining and transmission electron microscopy. It has been speculated that this material may be involved in the attachment of plesiomonads to epithelial cells. Endotoxin may play a role in *Plesiomonas* virulence as it does for other pathogens (19). Plesiomonads do produce extracellular enzymes, such as elastase, which may be involved in connective tissue degradation. However, the ability of strains to produce toxins has been controversial. It may be that toxin assays used for *Aeromonas* species are not suitable for detecting *Plesiomonas* toxins. Several reports concluded that plesiomonads did not produce toxins since rabbit ileal loop and suckling mouse assays, cell (Y_1, Vero, CHO) culture assays, and rabbit skin permeability toxin assays all yielded negative results (19, 43, 115). Later reports, however, described the production of both heat-stable and heat-labile cytotoxic enterotoxins from all strains, irrespective of origin or serotype (148). Manorama et al. attempted to purify and characterize these toxins (110). The key to detecting *Plesiomonas* enterotoxins may be serial passage of isolates in vivo (19). Bioassays and enzyme-linked immunosorbent assays used to detect *E. coli* enterotoxins have been negative for *P. shigelloides,* as have nucleic acid probes for such toxins (128). Iron has been shown to regulate production of a factor that causes elongation of CHO cells similar to that produced by *V. cholerae* enterotoxin (46). It has also been reported that more than 90% of 36 *Plesiomonas* strains tested produced a β-hemolysin, as judged by the results of agar overlay and contact-dependent hemolysis assays. The hemolysin was cell associated and was produced at both 25 and 35°C (82). It may play a role in iron acquisition in vivo or have another, as yet unknown role in gastrointestinal disease.

Table 14.5 Potential virulence factors for *P. shigelloides*[a]

Extracellular enzymes (e.g., elastase)
Enterotoxins/cytotoxins
SMA, RIL, lysis of Y_1 cells (after passage in RIL)
CHO elongating factor (when grown in iron-poor medium)
β-Hemolysin (cell associated)
Endotoxin
Invasins (invasion of HeLa cells)
Adhesins (glycocalyx?)

[a]See references in text. SMA, suckling mouse assay; RIL, rabbit ileal loop.

Little is known about intestinal colonization and mucus interactions by plesiomonads. Some strains reportedly adhere to HeLa and HEp-2 cells. Invasiveness in the Sereny test does not occur, but invasion of HeLa and HEp-2 cells has been seen. With freshly isolated strains from children with acute diarrhea, invasion of HeLa cells has been reported to be comparable with that of *Shigella sonnei* (19, 43, 128, 148). As mentioned, the patients from whom plesiomonads were isolated in the study by Holmberg et al. tended to have symptoms typical of an invasive pathogen (68). Twenty-three of 27 isolates from these patients contained plasmids. A single, very large (>150-MDa, ~230-kb) plasmid was found in 12 isolates, but it did not appear to be the same as the large virulence plasmid described for *Shigella* species or enteroinvasive *E. coli.* Large plasmids have also been detected in strains from patients with *Plesiomonas*-associated colitis in other studies. It is possible that unstable virulence plasmids are involved in *Plesiomonas* pathogenicity.

SUMMARY

Aeromonas species are commonly found in foods and water. Despite the essentially negative result obtained in a single human trial experiment, there is strong evidence that some strains can be enteropathogenic. Molecular genetic studies have led to considerable advances in the taxonomy of the genus *Aeromonas* in recent years. Good typing schemes for accurately identifying clinical isolates now exist, but there is a need for further work to develop schemes that will identify all environmental isolates. Disease-causing strains appear to be only a subset of the diversity of strains present in the environment. There is still much to be learned about *Aeromonas* virulence determinants and how they may combine to result in the virulent subsets within each *Aeromonas* species that can cause disease. At present it is not possible to identify the disease-causing strains because of our lack of knowledge about *Aeromonas* virulence mechanisms. Some of the confusion surrounding *Aeromonas* toxins has been resolved. It is likely that future molecular genetic investigations will shed further light on adhesive and invasive mechanisms. We may then be able to develop tests which will identify those strains which pose a risk to human health.

The case for *Plesiomonas* being an enteric pathogen is less convincing than that for *Aeromonas* spp. Although it has been isolated from diarrheal patients and has been incriminated in several large water- and foodborne outbreaks, no definite virulence mechanism has yet been identified in a majority of strains associated with gastrointestinal infections. The negative results obtained in human volunteer studies and the lack of an intestinal sIgA response in infected patients cast further doubt on the enteropathogenicity of the species. Establishment of the role of *P. shigelloides* in human disease, therefore, must await further studies.

References

1. **Abbott, S. L., W. K. W. Cheung, S. Kroske-Bystrom, T. Malekzadeh, and J. M. Janda.** 1992. Identification of *Aeromonas* strains to the genospecies level in the clinical laboratory. *J. Clin. Microbiol.* **30**:1262–1266.
2. **Abeyta, C., Jr., S. A. Palumbo, and G. N. Stelma.** 1994. *Aeromonas hydrophila* group, p. 1–27. *In* Y. H. Hui, J. R. Gorham, K. D. Murrel, and D. O. Cliver (ed.), *Foodborne Disease Handbook*, vol. 1. Marcel Dekker Inc., New York.
3. **Alabi, S. A., and T. Odugbemi.** 1990. Plasmid screening amongst *Aeromonas* species and *Plesiomonas shigelloides* isolated from subjects with diarrhoea in Lagos, Nigeria. *Afr. J. Med. Med. Sci.* **19**:303–306.
4. **Aldová, E., D. Danesová, J. Postupa, and T. Shimada.** 1994. New serovars of *Plesiomonas shigelloides* 1992. *Cent. Eur. J. Public Health* **1**:32–36.

4a. **Aldová, E., and R. H. W. Schubert.** 1996. Serotyping of *Plesiomonas shigelloides*—a tool for understanding ecological relationships. *Med. Microbiol. Lett.* **5**:33–39.

5. **Altwegg, M.** 1993. A polyphasic approach to the classification and identification of *Aeromonas* strains. *Med. Microbiol. Lett.* **2**:200–205.
6. **Altwegg, M., G. Martinetti-Lucchini, J. Lüthy-Hottenstein, and M. Rohrbach.** 1991. *Aeromonas*-associated gastroenteritis after consumption of contaminated shrimp. *Eur. J. Clin. Microbiol. Infect. Dis.* **10**:44–45.
7. **Altwegg, M., A. G. Steigerwalt, R. Altwegg-Bissig, J. Lüthy-Hottenstein, and D. J. Brenner.** 1990. Biochemical identification of *Aeromonas* genospecies isolated from humans. *J. Clin. Microbiol.* **28**:258–264.
8. **Anguita, J., L. B. Rodriguez Aparicio, and G. Naharro.** 1993. Purification, gene cloning, amino acid sequence analysis, and expression of an extracellular lipase from an *Aeromonas hydrophila* isolate. *Appl. Environ. Microbiol.* **59**:2411–2417.
9. **Arai, T., N. Ikejima, T. Itoh, S. Sakai, T. Shimada, and R. Sakazaki.** 1980. A survey of *Plesiomonas shigelloides* from aquatic environments, domestic animals, pets, and humans. *J. Hyg.* (Cambridge) **84**:203–211.
10. **Araujo, R. M., R. M. Arribas, F. Lucena, and R. Pares.** 1989. Relation between *Aeromonas* and fecal coliforms in fresh waters. *J. Appl. Bacteriol.* **67**:213–217.
11. **Ashbolt, N. J., A. Ball, M. Dorsch, C. Turner, P. Cox, A. Chapman, and S. M. Kirov.** 1995. The identification and human health significance of environmental aeromonads. *Water Sci. Technol.* **31**:263–269.
12. **Ashdown, L. R., and J. M. Koehler.** 1993. The spectrum of *Aeromonas*-associated diarrhea in tropical Queensland, Australia. *Southeast Asian J. Trop. Med. Public Health* **24**:347–353.
13. **Ballard, J., J. Crabtree, B. A. Roe, and R. K. Tweten.** 1995. The primary structure of *Clostridium septicum* alpha-toxin exhibits similarity with that of *Aeromonas hydrophila* aerolysin. *Infect. Immun.* **63**:340–344.
14. **Bargouthi, S., S. M. Payne, J. E. L. Arceneaux, and B. R. Byers.** 1991. Cloning, mutagenesis, and nucleotide sequence of a siderophore biosynthetic gene (*amoA*) from *Aeromonas hydrophila. J. Bacteriol.* **173**:5121–5128.
15. **Berrang, M. E., R. E. Brackett, and L. R. Beuchat.** 1989. Growth of *Aeromonas hydrophila* on fresh vegetables grown under a controlled atmosphere. *Appl. Environ. Microbiol.* **55**:2167–2171.
16. **Beuchat, L. R.** 1991. Behaviour of *Aeromonas* species at refrigeration temperatures. *Int. J. Food Microbiol.* **13**:217–224.
17. **Borrego, J. J., M. A. Morinigo, E. Martinez-Manzanares, M. Bosca, D. Castro, J. L. Barja, and A. E. Toranzo.** 1991. Plasmid associated virulence properties of environmental isolates of *Aeromonas hydrophila. J. Med. Microbiol.* **35**:264–269.
18. **Bottone, E. J.** 1993. Correlation between known exposure to contaminated food or surface water and development of *Aeromonas hydrophila* and *Plesiomonas shigelloides* diarrheas. *Med. Microbiol. Lett.* **2**:217–225.
19. **Brenden, R. A., M. A. Miller, and J. M. Janda.** 1988. Clinical disease spectrum and pathogenic factors associated with *Plesiomonas shigelloides* infections in humans. *Rev. Infect. Dis.* **10**:303–316.
20. **Buckley, J. T.** 1991. Secretion and mechanism of action of the hole-forming toxin aerolysin. *Experientia* **47**:418–419.
21. **Burke, V., M. Cooper, J. Robinson, M. Gracey, M. Lesmana, P. Echeverria, and J. M. Janda.** 1984. Hemagglutination patterns of *Aeromonas* spp. in relation to biotype and source. *J. Clin. Microbiol.* **19**:39–43.
22. **Burke, V., J. Robinson, M. Gracey, D. Peterson, and K. Partridge.** 1984. Isolation of *Aeromonas hydrophila* from a metropolitan water supply: seasonal correlation with clinical isolates. *Appl. Environ. Microbiol.* **48**:361–366.
23. **Byers, B. R., and J. E. L. Arceneaux.** 1993. Siderophore diversity in the genus *Aeromonas. Med. Microbiol. Lett.* **2**:281–285.
24. **Byers, B. R., and J. E. L. Arceneaux.** 1994. Iron acquisition and virulence of the bacterial genus *Aeromonas*, p. 29–40. *In* J. A. Mantley, D. E. Crowley, and D. G. Luster

(ed.), *Biochemistry of Metal Micronutrients in the Rhizosphere.* CRC Press, Inc., Boca Raton, Fla.

25. **Cahill, M. M., and I. C. MacRae.** 1992. Characteristics of O/129-sensitive motile *Aeromonas* strains isolated from freshwater on starch-ampicillin agar. *Microb. Ecol.* **24:** 215–226.
26. **Callister, S. M., and W. A. Agger.** 1987. Enumeration and characterization of *Aeromonas hydrophila* and *Aeromonas caviae* isolated from grocery store produce. *Appl. Environ. Microbiol.* **53:**249–253.
27. **Carnahan, A. M., S. Behram, and S. W. Joseph.** 1991. Aerokey II: a flexible key for identifying clinical *Aeromonas* species. *J. Clin. Microbiol.* **29:**2843–2849.
28. **Carnahan, A. M., T. Chakraborty, G. R. Fanning, D. Verma, A. Ali, J. M. Janda, and S. W. Joseph.** 1991. *Aeromonas trota* sp. nov., an ampicillin-susceptible species isolated from clinical specimens. *J. Clin. Microbiol.* **29:** 1206–1210.
29. **Carnahan, A. M., M. A. Marii, G. R. Fanning, M. A. Pass, and S. W. Joseph.** 1989. Characterization of *Aeromonas schubertii* strains recently isolated from traumatic wound infections. *J. Clin. Microbiol.* **27:**1826–1830.
30. **Carrello, A., K. A. Silburn, J. R. Budden, and B. J. Chang.** 1988. Adhesion of clinical and environmental *Aeromonas* isolates to HEp-2 cells. *J. Med. Microbiol.* **26:** 19–27.
31. **Chakraborty, T., B. Huhle, H. Bergbauer, and W. Goebel.** 1986. Cloning, expression, and mapping of the *Aeromonas hydrophila* aerolysin gene determinant in *Escherichia coli* K-12. *J. Bacteriol.* **167:**368–374.
32. **Chakraborty, T., B. Huhle, H. Hof, H. Bergbauer, and W. Goebel.** 1987. Marker exchange mutagenesis of the aerolysin determinant in *Aeromonas hydrophila* demonstrates the role of aerolysin in *A. hydrophila*-associated systemic infections. *Infect. Immun.* **55:**2274–2280.
33. **Chakraborty, T., M. A. Montenegro, S. C. Sanyal, R. Helmuth, E. Bulling, and K. N. Timmis.** 1984. Cloning of enterotoxin gene from *Aeromonas hydrophila* provides conclusive evidence of production of a cytotonic enterotoxin. *Infect. Immun.* **46:**435–441.
34. **Chakraborty, T., A. Schmid, S. Notermans, and R. Benz.** 1990. Aerolysin of *Aeromonas sobria:* evidence for formation of iron-permeable channels and comparison with alpha-toxin of *Staphylococcus aureus. Infect. Immun.* **58:**2127–2132.
35. **Champsaur, H., A. Andremont, D. Mathieu, E. Rottman, and P. Auzepy.** 1982. Cholera-like illness due to *Aeromonas sobria. J. Infect. Dis.* **145:**248–254.
36. **Chang, M. C., S. Y. Chang, S. L. Chen, and S. M. Chuang.** 1992. Cloning and expression in *Escherichia coli* of the gene encoding an extracellular deoxyribonuclease (DNase) from *Aeromonas hydrophila. Gene* **122:**175–180.
37. **Chopra, A. K., M. R. Ferguson, and C. W. Houston.** 1993. Molecular characterization of enterotoxins from *Aeromonas hydrophila. Med. Microbiol. Lett.* **2:**261–268.
38. **Chopra, A. K., and C. W. Houston.** 1989. Purification and partial characterization of a cytotonic enterotoxin produced by *Aeromonas hydrophila. Can. J. Microbiol.* **35:** 719–727.
39. **Chopra, A. K., C. W. Houston, J. W. Peterson, and G.-F. Jin.** 1993. Cloning, expression and sequence analysis of a cytolytic enterotoxin gene from *Aeromonas hydrophila. Can. J. Microbiol.* **39:**513–523.
40. **Chopra, A. K., R. Pham, and C. W. Houston.** 1994. Cloning and expression of putative cytotonic enterotoxin-encoding genes from *Aeromonas hydrophila. Gene* **139:**87–91.
41. **Dorsch, M., N. J. Ashbolt, P. T. Cox, and A. E. Goodman.** 1994. Rapid identification of *Aeromonas* using 16S rDNA targeted oligonucleotide primers: a molecular approach based on screening of environmental isolates. *J. Appl. Bacteriol.* **77:**722–726.
42. **Eley, A., I. Geary, and M. H. Wilcox.** 1993. Growth of *Aeromonas* spp. at 4°C and related toxin production. *Lett. Appl. Microbiol.* **16:**36–39.
43. **Farmer, J. J., III, M. J. Arduino, and F. W. Hickman-Brenner.** 1992. The genera *Aeromonas* and *Plesiomonas,* p. 3012–3028. *In* A. Balows, H. G. Trüper, M. Dworkin, W. Harder, and K.-H. Schleifer (ed.), *The Prokaryotes,* 2nd ed. Springer-Verlag, New York.
44. **Freij, B. J.** 1987. Extraintestinal *Aeromonas* and *Plesiomonas* infections of humans. *Experientia* **43:**359–360.
45. **Fricker, C. R., and S. Tompsett.** 1989. *Aeromonas* spp. in foods: a significant cause of food poisoning? *Int. J. Food Microbiol.* **9:**17–23.
46. **Gardner, S. E., S. E. Fowlston, and W. L. George.** 1990. Effect of iron on production of a possible virulence factor by *Plesiomonas shigelloides. J. Clin. Microbiol.* **28:**811–813.
47. **George, W. L., M. M. Nakata, J. Thompson, and M. L. White.** 1985. *Aeromonas*-related diarrhea in adults. *Arch. Intern. Med.* **145:**2207–2211.
48. **Gobat, P.-F., and T. Jemmi.** 1993. Distribution of mesophilic *Aeromonas* species in raw and ready-to-eat fish and meat products in Switzerland. *Int. J. Food Microbiol.* **20:**117–120.
49. **Gobius, K. S., and J. M. Pemberton.** 1988. Molecular cloning, characterization, and nucleotide sequence of an extracellular amylase gene from *Aeromonas hydrophila. J. Bacteriol.* **170:**1325–1332.
50. **Gold, W. L., and I. E. Salit.** 1993. *Aeromonas hydrophila* infections of skin and soft tissue: report of 11 cases and review. *Clin. Infect. Dis.* **16:**69–74.
51. **Gracey, M., V. Burke, and J. Robinson.** 1982. *Aeromonas*-associated gastroenteritis. *Lancet* **ii:**1304–1306.
52. **Gracey, M., V. Burke, J. Robinson, P. L. Masters, J. Stewart, and J. Pearman.** 1984. *Aeromonas* spp. in travellers' diarrhoea. *Br. Med. J.* **289:**658.
53. **Gray, S. J., and D. J. Stickler.** 1989. Some observations on the faecal carriage of mesophilic *Aeromonas* in cows and pigs. *Epidemiol. Infect.* **103:**523–537.
54. **Grey, P. A., and S. M. Kirov.** 1993. Adherence to HEp-2 cells and enteropathogenic potential of *Aeromonas* spp. *Epidemiol. Infect.* **110:**279–287.
55. **Hanes, D. E., and D. K. F. Chandler.** 1993. The role of a 40-megadalton plasmid in the adherence and hemolytic properties of *Aeromonas hydrophila. Microb. Pathog.* **15:**313–317.
56. **Hänninen, M.-L.** 1993. Occurrence of *Aeromonas* spp. in samples of ground meat and chicken. *Int. J. Food Microbiol.* **18:**339–342.

57. **Hänninen, M.-L., S. Salmi, L. Mattila, R. Taipalinen, and A. Siitonen.** 1995. Association of *Aeromonas* spp. with travellers' diarrhoea in Finland. *J. Med. Microbiol.* **42**:26–31.

58. **Havelaar, A. H., F. M. Schets, A. van Silhouf, W. H. Jansen, G. Wieten, and D. van der Kooij.** 1992. Typing of *Aeromonas* strains from patients with diarrhoea and from drinking water. *J. Appl. Bacteriol.* **72**:435–444.

59. **Havelaar, A. H., J. F. M. Versteegh, and M. During.** 1990. The presence of *Aeromonas* in drinking water supplies in the Netherlands. *Zentralbl. Hyg.* **190**:236–256.

60. **Hirono, I., and T. Aoki.** 1991. Nucleotide sequence and expression of an extracellular hemolysin gene of *Aeromonas hydrophila. Microb. Pathog.* **11**:189–197.

61. **Hirono, I., and T. Aoki.** 1993. Cloning and characterization of three hemolysin genes from *Aeromonas salmonicida. Microb. Pathog.* **15**:269–282.

62. **Hirono, I., T. Aoki, T. Asao, and S. Kozaki.** 1992. Nucleotide sequences and characterization of haemolysin genes from *Aeromonas hydrophila* and *Aeromonas sobria. Microb. Pathog.* **13**:433–446.

63. **Ho, A. S. Y., T. A. Mietzner, A. J. Smith, and G. K. Schoolnik.** 1990. The pili of *Aeromonas hydrophila:* identification of an environmentally regulated "mini pilin." *J. Exp. Med.* **172**:795–806.

64. **Ho, A. S. Y., I. Sohel, and G. K. Schoolnik.** 1992. Cloning and characterization of *fxp*, the flexible pilin gene of *Aeromonas hydrophila. Mol. Microbiol.* **6**:2725–2732.

65. **Hokama, A., Y. Honma, and N. Nakasone.** 1990. Pili of an *Aeromonas hydrophila* strain as a possible colonization factor. *Microbiol. Immunol.* **34**:901–915.

66. **Hokama, A., and M. Iwanaga.** 1991. Purification and characterization of *Aeromonas sobria* pili, a possible colonization factor. *Infect. Immun.* **9**:3478–3483.

67. **Hokama, A., and M. Iwanaga.** 1992. Purification and characterization of *Aeromonas sobria* Ae24 pili: a possible new colonization factor. *Microb. Pathog.* **13**:325–334.

68. **Holmberg, S. D., I. K. Wachsmuth, F. W. Hickman-Brenner, P. A. Blake, and J. J. Farmer III.** 1986. *Plesiomonas* enteric infections in the United States. *Ann. Intern. Med.* **105**:690–694.

69. **Honma, Y., and N. Nakasone.** 1990. Pili of *Aeromonas hydrophila:* purification, characterization and biological role. *Microbiol. Immunol.* **34**:83–98.

70. **Howard, S. P., and J. T. Buckley.** 1985. Activation of the hole-forming toxin aerolysin and extracellular processing. *J. Bacteriol.* **163**:336–340.

71. **Howard, S. P., and J. T. Buckley.** 1986. Molecular cloning and expression in *Escherichia coli* of the structural gene for the hemolytic toxin aerolysin from *Aeromonas hydrophila. Mol. Gen. Genet.* **204**:289–295.

72. **Howard, S. P., W. J. Garland, M. J. Green, and J. T. Buckley.** 1987. Nucleotide sequence of the gene for the hole-forming toxin aerolysin of *Aeromonas hydrophila. J. Bacteriol.* **169**:2869–2871.

73. **Huq, A., A. Akhtar, M. A. R. Chowdhury, and D. A. Sack.** 1991. Optimal growth temperature for the isolation of *Plesiomonas shigelloides,* using various selective and differential agars. *Can. J. Microbiol.* **37**:800–802.

74. **Husslein, V., T. Chakraborty, A. Carnahan, and S. W. Joseph.** 1992. Molecular studies on the aerolysin gene of *Aeromonas* species and discovery of a species-specific probe for *Aeromonas trota* species nova. *Clin. Infect. Dis.* **14**:1061–1068.

75. **Husslein, V., B. Huhle, T. Jarchau, R. Lurz, W. Goebel, and T. Chakraborty.** 1988. Nucleotide sequence and transcriptional analysis of the *aerCaerA* region of *Aeromonas sobria* encoding aerolysin and its regulatory region. *Mol. Microbiol.* **2**:507–517.

76. **Husslein, V., S. H. E. Notermans, and T. Chakraborty.** 1988. Gene probes for the detection of aerolysin in *Aeromonas* spp. *J. Diarrhoeal Dis. Res.* **6**:124–130.

77. **Huys, G., R. Coopman, P. Janssen, and K. Kersters.** 1996. High-resolution genotypic analysis of the genus *Aeromonas* by AFLP fingerprinting. *Int. J. Syst. Bacteriol.* **46**:572–580.

78. **Ingham, S. C.** 1990. Growth of *Aeromonas hydrophila* and *Plesiomonas shigelloides* on cooked crayfish tails during cold storage under air, vacuum, and a modified atmosphere. *J. Food Prot.* **53**:665–667.

79. **Iwanaga, M., and A. Hokama.** 1992. Characterization of *Aeromonas sobria* TAP 13 pili: a possible new colonization factor. *J. Gen. Microbiol.* **138**:1913–1919.

80. **Jahagirdar, R., and P. Howard.** 1994. Isolation and characterization of a second *exe* operon required for extracellular protein secretion in *Aeromonas hydrophila. J. Bacteriol.* **176**:6819–6826.

81. **Janda, J. M.** 1991. Recent advances in the study of the taxonomy, pathogenicity, and infectious syndromes associated with the genus *Aeromonas. Clin. Microbiol. Rev.* **4**:397–410.

82. **Janda, J. M., and S. L. Abbott.** 1993. Expression of hemolytic activity by *Plesiomonas shigelloides. J. Clin. Microbiol.* **31**:1206–1208.

83. **Janda, J. M., S. L. Abbott, and A. M. Carnahan.** 1995. *Aeromonas* and *Plesiomonas,* p. 477–482. *In* P. R. Murray, E. J. Baron, M. A. Pfaller, F. C. Tenover, and R. H. Yolken (ed.), *Manual of Clinical Microbiology,* 6th ed. American Society for Microbiology, Washington, D.C.

84. **Jeppesen, C.** 1995. Media for *Aeromonas* spp., *Plesiomonas shigelloides* and *Pseudomonas* spp. from food and environment. *Int. J. Food Microbiol.* **26**:25–41.

85. **Jiang, Z. D., A. C. Nelson, J. J. Mathewson, C. D. Ericsson, and H. L. DuPont.** 1991. Intestinal secretory immune response to infection with *Aeromonas* species and *Plesiomonas shigelloides* among students from the United States in Mexico. *J. Infect. Dis.* **164**:979–982.

86. **Jin, G.-F., A. K. Chopra, and C. W. Houston.** 1992. Stimulation of neutrophil leukocyte chemotaxis by a cloned cytolytic enterotoxin of *Aeromonas hydrophila. FEMS Microbiol. Lett.* **98**:285–290.

87. **Joseph, S. W., and A. Carnahan.** 1994. The isolation, identification and systematics of the motile *Aeromonas* species. *Annu. Rev. Fish Dis.* **4**:315–343.

88. **Kaper, J. B., H. Lockman, and R. R. Colwell.** 1981. *Aeromonas hydrophila:* ecology, and toxigenicity of isolates from an estuary. *J. Appl. Bacteriol.* **50**:359–377.

89. **Kelly, K. A., M. Koehler, and L. R. Ashdown.** 1993. Spectrum of extraintestinal disease due to *Aeromonas* spe-

cies in tropical Queensland, Australia. *Clin. Infect. Dis.* **16:**574–579.

90. **Kirov, S. M.** 1993. The public health significance of *Aeromonas* spp. in foods. *Int. J. Food Microbiol.* **20:**179–198.
91. **Kirov, S. M.** 1993. Adhesion and piliation of *Aeromonas* spp. *Med. Microbiol. Lett.* **2:**274–280.
92. **Kirov, S. M., M. J. Anderson, and T. A. McMeekin.** 1990. A note on *Aeromonas* spp. from chickens as possible foodborne pathogens. *J. Appl. Bacteriol.* **68:**327–334.
93. **Kirov, S. M., E. K. Ardestani, and L. J. Hayward.** 1993. The growth and expression of virulence factors at refrigeration temperature by *Aeromonas* strains isolated from foods. *Int. J. Food Microbiol.* **20:**159–168.
94. **Kirov, S. M., and F. Brodribb.** 1993. Exotoxin production by *Aeromonas* spp. in foods. *Lett. Appl. Microbiol.* **17:**208–211.
95. **Kirov, S. M., and L. J. Hayward.** 1993. Virulence traits of *Aeromonas* in relation to species and geographic region. *Aust. J. Med. Sci.* **14:**54–58.
96. **Kirov, S. M., L. J. Hayward, and M. A. Nerrie.** 1995. Adhesion of *Aeromonas* sp. to cell lines used as models for intestinal colonization. *Epidemiol. Infect.* **115:**465–473.
97. **Kirov, S. M., J. A. Hudson, L. J. Hayward, and S. J. Mott.** 1994. Distribution of *Aeromonas hydrophila* hybridization groups and their virulence properties in Australasian clinical and environmental strains. *Lett. Appl. Microbiol.* **18:**71–73.
98. **Kirov, S. M., D. S. Hui, and L. J. Hayward.** 1993. Milk as a potential source of *Aeromonas* gastrointestinal infection. *J. Food Prot.* **56:**306–312.
99. **Kirov, S. M., I. Jacobs, L. J. Hayward, and R. Hapin.** 1995. Electron microscopic examination of factors influencing the expression of filamentous surface structures on clinical and environmental isolates of *Aeromonas veronii* biotype sobria. *Microbiol. Immunol.* **39:**329–338.
100. **Kirov, S. M., and K. Sanderson.** 1995. *Aeromonas* cell line adhesion, surface structures and in vivo models of intestinal colonization. *Med. Microbiol. Lett.* **4:**305–315.
101. **Knøchel, S.** 1990. Growth characterisitics of motile *Aeromonas* spp. isolated from different environments. *Int. J. Food Microbiol.* **10:**235–244.
102. **Kohbayashi, K., and T. Ohnaka.** 1989. Food poisoning due to newly recognized pathogens. *Asian Med. J.* **32:**1–12.
103. **Kokka, R. P., N. A. Vedros, and J. M. Janda.** 1992. Immunochemical analysis and possible biological role of an *Aeromonas hydrophila* surface array protein in septicaemia. *J. Gen. Microbiol.* **138:**1229–1236.
104. **Krovacek, K., S. Dumontet, E. Eriksson, and S. B. Baloda.** 1995. Isolation, and virulence profiles, of *Aeromonas hydrophila* implicated in an outbreak of food poisoning in Sweden. *Microbiol. Immunol.* **39:**655–661.
105. **Kühn, I., T. Lindberg, K. Olsson, and T. Stentström.** 1992. Biochemical fingerprinting for typing of *Aeromonas* strains from food and water. *Lett. Appl. Microbiol.* **15:**261–265.
106. **Kuijper, E. J., P. Bol, M. F. Peeters, A. G. Steigerwalt, H. C. Zanen, and J. G. Brenner.** 1989. Clinical and epidemiologic aspects of members of *Aeromonas* DNA hybridization groups isolated from human feces. *J. Clin. Microbiol.* **27:**1531–1537.
107. **Lawson, M. A., V. Burke, and B. J. Chang.** 1985. Invasion of HEp-2 cells by fecal isolates of *Aeromonas hydrophila. Infect. Immun.* **47:**680–683.
108. **Ljungh, Å., P. Enroth, and T. Wadström.** 1982. Cytotonic enterotoxin from *Aeromonas hydrophila. Toxicon* **20:**787–794.
109. **Majeed, K. N., and I. C. MacRae.** 1991. Experimental evidence for toxin production by *Aeromonas hydrophila* and *Aeromonas sobria* in a meat extract at low temperatures. *Int. J. Food Microbiol.* **12:**181–188.
110. **Manorama, V. Taneja, R. K. Agarwal, and S. C. Sanyal.** 1983. Enterotoxins of *Plesiomonas shigelloides:* partial purification and characterization. *Toxicon* **3**(Suppl.):269–272.
111. **Martínez, M. J., D. Simon-Pujol, F. Congregado, S. Merino, X. Rubires, J. M. Tomás.** 1995. The presence of capsular polysaccharide in mesophilic *Aeromonas hydrophila* serotypes O:11 and O:34. *FEMS Microbiol. Lett.* **128:**69–74.
112. **Martinez-Murcia, A. J., S. Benlloch, and M. D. Collins.** 1992. Phylogenetic interrelationships of members of the genera *Aeromonas* and *Plesiomonas* as determined by 16S ribosomal RNA sequencing: lack of congruence with results of DNA-DNA hybridizations. *Int. J. Syst. Bacteriol.* **42:**412–421.
113. **Medema, G., and C. Schets.** 1993. Occurrence of *Plesiomonas shigelloides* in surface water: relationship with faecal pollution and trophic state. *Zentralbl. Hyg.* **194:** 398–404.
114. **Merino, S., S. Camprubí, and J. M. Tomás.** 1993. Detection of *Aeromonas hydrophila* in food with an enzyme-linked immunosorbent assay. *J. Appl. Bacteriol.* **74:** 149–154.
115. **Miller, M. L., and J. A. Kohburger.** 1985. *Plesiomonas shigelloides:* an opportunistic food and waterborne pathogen. *J. Food Prot.* **48:**449–457.
116. **Miller, M. L., and J. A. Kohburger.** 1986. Tolerance of *Plesiomonas shigelloides* to pH, sodium chloride and temperature. *J. Food Prot.* **49:**877–879.
117. **Millership, S. E., S. R. Curnow, and B. Chattopadhyay.** 1983. Faecal carriage rate of *Aeromonas hydrophila. J. Clin. Pathol.* **36:**920–923.
118. **Morgan, D. R., P. C. Johnson, H. L. DuPont, T. K. Satterwhite, and L. V. Wood.** 1985. Lack of correlation between known virulence properties of *Aeromonas hydrophila* and enteropathogenicity for humans. *Infect. Immun.* **50:**62–65.
119. **Moyer, N. P.** 1987. Clinical significance of *Aeromonas* species isolated from patients with diarrhea. *J. Clin. Microbiol.* **25:**2044–2048.
120. **Namdari, H., and E. J. Bottone.** 1988. Correlation of the suicide phenomenon in *Aeromonas* species with virulence and enteropathogenicity. *J. Clin. Microbiol.* **26:**2615–2619.
121. **Namdari, H., and E. J. Bottone.** 1991. *Aeromonas caviae:* ecologic adaptation in the intestinal tract of infants cou-

pled to adherence and enterotoxin production as factors in enteropathogenicity. *Experientia* **47**:434–436.

122. **Neves, M. S., M. P. Nunes, and A. M. Milhomen.** 1994. *Aeromonas* species exhibit aggregative adherence to HEp-2 cells. *J. Clin. Microbiol.* **32**:1130–1131.

123. **Nielsen, B., J. E. Olsen, and J. L. Larsen.** 1993. Plasmid profiling as an epidemiological marker within *Aeromonas salmonicida*. *Dis. Aquat. Org.* **15**:129–135.

124. **Nishikawa, Y., A. Hase, J. Ogawasara, S. M. Scotland, H. R. Smith, and T. Kimura.** 1994. Adhesion to and invasion of human colon carcinoma Caco-2 cells by *Aeromonas* strains. *J. Med. Microbiol.* **40**:55–61.

125. **Nishikawa, Y., and T. Kishi.** 1988. Isolation and characterization of motile *Aeromonas* from human, food and environmental specimens. *Epidemiol. Infect.* **101**:213–223.

126. **Nishikawa, Y., J. Ogasawara, and T. Kimura.** 1993. Heat and acid sensitivity of motile *Aeromonas:* a comparison with other food-poisoning bacteria. *Int. J. Food Microbiol.* **18**:271–278.

127. **O'Brien, M., and M. Tandy.** 1994. L-Histidine decarboxylase: a new diagnostic test for *Plesiomonas shigelloides*. *Aust. Microbiol.* **15**:A-69.

128. **Olsvik, Ø., K. Wachsmuth, B. Kay, K. A. Birkness, A. Yi, and B. Sack.** 1990. Laboratory observations on *Plesiomonas shigelloides* strains isolated from children with diarrhea in Peru. *J. Clin. Microbiol.* **28**:886–889.

129. **Palumbo, S. A.** 1993. The occurrence and significance of organisms of the *Aeromonas hydrophila* group in food and water. *Med. Microbiol. Lett.* **2**:339–346.

130. **Palumbo, S. A., J. E. Call, P. H. Cooke, and A. C. Williams.** 1995. Effect of polyphosphates and NaCl on *Aeromonas hydrophila* K144. *J. Food Safety* **15**:77–87.

131. **Palumbo, S. A., F. Maxino, A. C. Williams, R. L. Buchanan, and D. W. Thayer.** 1985. Starch-ampicillin agar for the quantitative detection of *Aeromonas hydrophila*. *Appl. Environ. Microbiol.* **50**:1027–1030.

132. **Palumbo, S. A., A. C. Williams, R. L. Buchanan, and J. G. Philips.** 1987. Thermal resistance of *Aeromonas hydrophila*. *J. Food Prot.* **50**:761–764.

133. **Parker, M. W., J. T. Buckley, J. P. M. Postma, A. D. Tucker, K. Leonard, F. Pattus, and D. Tsernoglou.** 1994. Structure of *Aeromonas* toxin proaerolysin in its water-soluble and membrane-channel states. *Nature* (London) **367**:292–295.

134. **Parras, F., M., D. Díaz, J. Reina, S. Morena, and C. Guerro.** 1993. Meningitis due to *Aeromonas* species: case report and review. *Clin. Infect. Dis.* **17**:1058–1060.

135. **Pazzaglia, G., R. B. Sack, A. L. Bourgeois, J. Froehlich, and J. Eckstein.** 1990. Diarrhea and intestinal invasiveness of *Aeromonas* strains in the removable intestinal tie rabbit model. *Infect. Immun.* **58**:1924–1931.

135a. **Pepe, C. M., M. W. Eklund, and M. S. Strom.** 1996. Cloning of an *Aeromonas hydrophila* type IV pilus biogenesis gene cluster: complementation of pilus assembly functions and characterization of a type IV leader peptidase/*N*-methyltransferase required for extracellular secretion. *Mol. Microbiol.* **19**:857–869.

136. **Picard, B., and P. Goullet.** 1985. Comparative electrophoretic profiles of esterases, and of glutamate, lactate and malate dehydrogenases, from *Aeromonas hydrophila, A. caviae* and *A. sobria*. *J. Gen. Microbiol.* **131**:3385–3391.

137. **Picard, B., and P. Goullet.** 1987. Seasonal prevalence of nosocomial *Aeromonas hydrophila* infection related to aeromonas in hospital water. *J. Hosp. Infect.* **10**:152–155.

138. **Picard, B., and P. Goullet.** 1987. Epidemiological complexity of hospital aeromonas infections revealed by electrophoretic typing of esterases. *Epidemiol. Infect.* **98**:5–14.

139. **Pitarangsi, C., P. Echeverria, R. Whitmire, C. Tiripat, C. Formal, G. J. Dammin, and M. Tingtalapong.** 1982. Enteropathogenicity of *Aeromonas hydrophila* and *Plesiomonas shigelloides:* prevalence among individuals with and without diarrhea in Thailand. *Infect. Immun.* **35**: 666–673.

140. **Popoff, M.** 1984. Genus III. *Aeromonas* Kluyver and Van Neil 1936, 398, p. 545–548. *In* N. R Krieg and J. G. Holt (ed.), *Bergey's Manual of Systematic Bacteriology,* vol. 1. The Williams and Wilkins Co., Baltimore.

141. **Potomski, J., V. Burke, J. Robinson, D. Fumarola, and G. Miragliotta.** 1987. *Aeromonas* cytotonic enterotoxin cross-reactive with cholera toxin. *J. Med. Microbiol.* **23**:179–186.

142. **Quinn, D. M., H. M. Atkinson, A. H. Bretag, M. Tester, T. J. Trust, C. Y. F. Wong, and R. L. P. Flower.** 1994. Carbohydrate-reactive, pore-forming outer membrane proteins of *Aeromonas hydrophila*. *Infect. Immun.* **62**:4054–4058.

143. **Radomyski, T., E. A. Murano, D. G. Olson, and P. S. Murano.** 1994. Elimination of pathogens of significance in food by low dose irradiation: a review. *J. Food Prot.* **57**:73–86.

144. **Ratkowsky, D. A., J. Olley, T. A. McMeekin, and A. Ball.** 1982. Relationship between temperature and growth rate of bacterial cultures. *J. Bacteriol.* **149**:1–5.

145. **Rivero, O., J. Anguta, C. Paniagua, and G. Naharro.** 1990. Molecular cloning and characterization of an extracellular protease gene from *Aeromonas hydrophila*. *J. Bacteriol.* **172**:3905–3908.

146. **Rutala, W. A., F. A. Sarubbi, Jr., C. S. Finch, J. N. MacCormack, and G. E. Steinkraus.** 1982. Oyster-associated outbreak of diarrhoeal disease possibly caused by *Plesiomonas shigelloides*. *Lancet* **i**:739.

147. **San Joaquin, V. H., and D. A. Pickett.** 1988. *Aeromonas*-associated gastroenteritis in children. *Pediatr. Infect. Dis. J.* **7**:53–57.

148. **Saraswathi, B., R. K. Agarwal, and S. C. Sanyal.** 1983. Further studies on enteropathogenicity of *Plesiomonas shigelloides*. *Indian J. Med. Res.* **78**:12–18.

149. **Schultz, A. J., and B. A. McCardell.** 1988. DNA homology and immunological cross-reactivity between *Aeromonas hydrophila* cytotonic enterotoxin and cholera toxin. *J. Clin. Microbiol.* **26**:57–61.

150. **Seidler, R. J., D. A. Allen, H. Lockman, R. R. Colwell, S. W. Joseph, and O. P. Daily.** 1980. Isolation, enumeration and characterization of *Aeromonas* from polluted waters encountered in diving operations. *Appl. Environ. Microbiol.* **39**:1010–1018.

151. **Shimada, T., R. Sakazaki, K. Horigome, Y. Ueska, and K. Niwano.** 1984. Production of cholera-like enterotoxin by *Aeromonas hydrophila*. *Jpn. J. Med. Sci. Biol.* **37**:141–144.

152. **Singh, D. V., and S. C. Sanyal.** 1992. Enterotoxicity of clinical and environmental isolates of *Aeromonas* spp. *J. Med. Microbiol.* **36:**269–272.
153. **Snower, D. P., C. Ruef, A. P. Kuritza, and S. C. Edberg.** 1989. *Aeromonas hydrophila* infection associated with the use of medicinal leeches. *J. Clin. Microbiol.* **27:**1421–1422.
154. **Tennent, J. M., and J. S. Mattick.** 1994. Type IV fimbriae. p. 127–146. *In* P. Klemm (ed.), *Fimbriae: Adhesion, Genetics, Biogenesis, and Vaccines.* CRC Press Inc., Boca Raton, Fla.
155. **Thomas, L. V., R. J. Gross, T. Cheasty, and B. Rowe.** 1990. Extended serogrouping scheme for motile, mesophilic *Aeromonas* species. *J. Clin. Microbiol.* **28:**980–984.
156. **Trust, T. J.** 1993. Molecular, structural and functional properties of *Aeromonas* S-layers, p. 159–171. *In* T. J. Beveridge and S. F. Koval (ed.), *Advances in Bacterial Paracrystalline Surface Layers.* Plenum Press, New York.
157. **Tso, M. D., and J. S. G. Dooley.** 1995. Temperature-dependent protein and lipopolysaccharide expression in clinical *Aeromonas* strains. *J. Med. Microbiol.* **42:**32–38.
158. **Tsukamoto, T., Y. Kinoshita, T. Shimada, and R. Sakazaki.** 1978. Two epidemics of diarrhoeal disease possibly caused by *Plesiomonas shigelloides. J. Hyg.* (Cambridge) **80:**275–280.
159. **Vadivelu, J., S. D. Puthucheary, M. Phipps, and Y. M. Chee.** 1995. Possible virulence factors involved in bacteraemia caused by *Aeromonas hydrophila. J. Med. Microbiol.* **42:**171–174.
160. **van der Goot, F. G., F. Pattus, M. Parker, and J. T. Buckley.** 1994. The cytolytic toxin aerolysin: from the soluble form to the transmembrane channel. *Toxicology* **87:**19–28.
161. **Watson, I. M., J. O. Robinson, V. Burke, and M. Gracey.** 1985. Invasiveness of *Aeromonas* spp. in relation to biotype, virulence factors and clinical features. *J. Clin. Microbiol.* **22:**48–51.
162. **Wilmsen, H. U., J. T. Buckley, and F. Pattus.** 1991. Site-directed mutagenesis at histidines of aerolysin from *Aeromonas hydrophila:* a lipid planar bilayer study. *Mol. Microbiol.* **5:**2745–2751.
163. **Wilmsen, H. U., K. R. Leonard, W. Tichelaar, J. T. Buckley, and F. Pattus.** 1992. The aerolysin membrane channel is formed by heptamerization of the monomer. *EMBO J.* **11:**2457–2463.

Karen L. Dodds
John W. Austin

15

Clostridium botulinum

INTRODUCTION

Since the first recognition of botulism as a foodborne disease in the late 1800s (36), it has been a major concern of food processors and consumers alike. Currently, four categories of human botulism are recognized. Foodborne botulism is caused by eating food contaminated with preformed botulinum neurotoxin (BoNT). Infant botulism is caused by ingestion of viable spores that germinate, colonize, and produce neurotoxin in the intestinal tracts of infants under 1 year of age. Infant botulism was first recognized in 1976 and is now the most common form of botulism in the United States (22). Wound botulism results from infection of a wound with spores of *Clostridium botulinum*, which grow and produce neurotoxin in the wound. While wound botulism is rare, it is being increasingly associated with intravenous drug use (36a). The fourth category, unclassified, includes cases of unknown origin and adult cases which resemble infant botulism. Animal botulism, which will not be discussed here, affects cattle and birds worldwide and occurs to a lesser extent in sheep, horses, zoo animals, and various other animals (20).

As a result of the severity of the disease, botulism is extensively reported. In the United States in 1994, there were 42 cases of foodborne botulism, 86 cases of infant botulism, and 11 cases of wound botulism, all with no deaths (36a). In Canada in 1995, there were 13 cases of foodborne botulism (representing seven outbreaks), 1 case of infant botulism, and no cases of wound botulism (11a).

CHARACTERISTICS OF *C. BOTULINUM*

Classification

C. botulinum is a gram-positive, anaerobic, rod-shaped, spore-forming bacterium (36). Originally, all organisms known to produce botulinum neurotoxin were included in this species (36). There are seven types of *C. botulinum*, A through G, based on the serological specificity of the neurotoxin produced. Human botulism, including foodborne, wound, and infant botulism, is associated with types A, B, E, and, very rarely, F. Types C and D cause botulism in animals. To date, there is no direct evidence linking type G to disease.

The species is also divided into four groups based on physiological differences (Table 15.1) (36), as follows: group I, all type A strains and proteolytic strains of types B and F; group II, all type E strains and nonproteolytic strains of types B and F; group III, type C and D strains; and group IV, *C. botulinum* type G, for which the new name *C. argentinense* has been proposed (94). This grouping agrees with results of DNA homology studies and of 16S and 23S rRNA gene sequence studies (44–46, 74) that show a high degree of relatedness among strains within each group but little relatedness among groups (36).

Group I strains are proteolytic and are typified by strains that produce neurotoxin type A (36). The optimal temperature for growth is 37°C, with growth occurring between 10 and 48°C. High levels of neurotoxin (10^6 mouse LD_{50}/ml [1 LD_{50} is the amount of neurotoxin required to kill 50% of injected mice within 4 days]) are

Table 15.1 Grouping and characteristics of strains of *C. botulinum*[a]

Characteristic	Result for group: I	II	III	IV
Neurotoxin type(s)	A, B, F	B, E, F	C, D	G
Growth temp (°C)				
Minimum	10	3.3	15	ND
Optimum	35–40	18–25	40	37
Minimum pH for growth	4.6	5.0	ND	ND
Inhibitory [NaCl] (%)	10	5	ND	ND
Minimum a_w for growth	0.94	0.97	ND	ND
$D_{100°C}$ of spores (min)	25	<0.1	0.1–0.9	0.8–1.12
$D_{121°C}$ of spores (min)	0.1–0.2	<0.001	ND	ND

[a]From reference 36. ND, not determined.

typically produced in cultures. Spores have a high heat resistance, with $D_{100°C}$ values of approximately 25 min (the *D* value is the time required to inactivate 90% of the population at a given temperature). To inhibit growth, the pH must be below 4.6, the salt concentration above 10%, or the water activity (a_w) below 0.94.

Group II strains are nonproteolytic, have a lower optimum growth temperature (30°C), and will grow at temperatures as low as 3.3°C. The spores have a much lower heat resistance, with $D_{100°C}$ values of less than 0.1 min. Group II strains are inhibited by a pH below 5.0, salt concentrations above 5%, or a_w below 0.97. Since these organisms lack proteolytic enzymes, the toxicity of cultures is usually increased by treating with trypsin, which activates the neurotoxin.

Group III includes types C and D strains, which are not involved in human botulism but cause animal botulism. Consequently, they have been studied in less detail. These strains are nonproteolytic and grow optimally at 40°C and at temperatures only as low as 15°C. Group IV strains, which produce type G neurotoxin, grow optimally at 37°C and have a minimal growth temperature of 10°C. Spores are rarely seen and have a fairly low resistance to heat, with $D_{104°C}$ values of 0.8 to 1.12 min.

These four groups, plus neurotoxin-producing strains of *C. butyricum* and *C. barati*, make a total of six distinct genomic groups which produce BoNT (26). The focus in this chapter will be on groups I and II, since they include the strains most commonly involved in human illness.

Tolerance to Preservation Methods

Temperature, pH, a_w, redox potential (E_h), added preservatives, and the presence of other microorganisms are the main factors controlling growth of *C. botulinum* in foods. Historically, studies have established maximum and/or minimum limits for these parameters that control growth of *C. botulinum* (Table 15.1). These factors seldom function independently; usually they act in concert, often having synergistic effects.

Low Temperature

Refrigerated storage is used to prevent or inhibit growth of *C. botulinum*. The minimum temperatures allowing growth have been determined, largely to assess the impact of refrigeration as a control. The established lower limits are 10°C for group I and 3.3°C for group II (36). However, these limits apply to few strains and depend on otherwise optimal growth conditions. Irrespective of the actual minimum growth temperature, production of neurotoxin generally requires weeks at the lower temperature limits for group I and group II organisms. The optimum growth temperatures are between 35 and 40°C for group I organisms and between 25 and 30°C for group II organisms.

Thermal Inactivation

Thermal processing is used to inactivate spores of *C. botulinum* and is the most common method of producing shelf stable foods. *C. botulinum* spores of group I, which are very heat resistant, are the target organisms for most thermal processes. *D* values vary considerably among *C. botulinum* strains (49). *D* values depend on how the spores are produced and treated, the heating environment, and the recovery system (49). Spores of types A and B are the most heat resistant, having $D_{121°C}$ values of between 0.1 and 0.2 min. These spores are of particular concern in the sterilization of canned low-acid foods. The canning industry has adopted a *D* value of 0.2 min at 121°C as a standard for calculating thermal processes. The *z* value (the temperature change necessary to cause a 10-fold change in the *D* value) for the most resistant strains is approximately 10°C, which has also been adopted as a standard. Despite variations in *D* and *z* values, the adoption of a 12-*D* process as the minimum thermoprocess applied to commercial canned low-acid foods by the canning industry has ensured the production of safe products (37).

Strains of group II are considerably less heat resistant ($D_{100°C}$ < 0.1 min) than those of group I, but survival of these spores in pasteurized, refrigerated products is of concern because of their ability to grow at refrigeration temperatures (60). $D_{82°C}$ values of type E in neutral phosphate buffer are generally in the range of 0.2 to 1.0 min. Values ranging from 0.15 to greater than 4.90 min have been reported for type E strains, depending on the heating menstruum, strain, and recovery medium. Various regulations and guidelines for the safe production, distribution, and sale of refrigerated foods of extended durability have been published (60). The Sous Vide Advisory Committee of the United Kingdom has produced guidelines for sous vide (vacuum-packed) foods specifying that the vacuum packaging and heat treatment must ensure destruction of vegetative organisms and significantly reduce the number of psychrotrophic *C. botulinum*. American recommendations from the National Advisory Committee on Microbiological Criteria for Foods include inoculated pack studies with *C. botulinum* to determine shelf life.

pH

The minimum pH allowing growth of *C. botulinum* group I is 4.6; for group II, it is about 5.0. Thus, many fruits and vegetables are sufficiently acidic to inhibit *C. botulinum* by their pH alone, while acidulants are used to preserve other products. Substrate, temperature, nature of the acidulent agent, presence of preservatives, a_w, and E_h are all factors that influence the acid tolerance of *C. botulinum*. Acid-tolerant microorganisms such as yeasts and molds may grow in acidic products and raise the pH in their immediate vicinity to a level that allows growth of *C. botulinum*. *C. botulinum* can also grow in some acidified foods if excessively slow pH equilibration occurs.

Salt and a_w

Salt (NaCl) is one of the most important factors controlling *C. botulinum* in foods. It acts primarily by decreasing the a_w. Consequently, its concentration in the aqueous phase, called the brine concentration (% brine = % NaCl × 100/% H_2O + % NaCl), is critical. The growth-limiting brine concentrations are about 10% for group I and 5% for group II under otherwise optimal conditions. These concentrations correspond well to the limiting a_w of 0.94 for group I and 0.97 for group II in foods in which NaCl is the main a_w depressant. The solute used to control a_w may influence these limits. Generally, NaCl, KCl, glucose, and sucrose show similar effects, while glycerol allows growth at lower a_w (37). The limiting a_w may be raised significantly by other factors, such as increased acidity or the use of preservatives.

Atmosphere and E_h

Modified atmosphere packaging (MAP) is being increasingly used to extend the shelf life and improve the quality of foods. MAP has been a concern because conditions might promote growth of *C. botulinum*. The high incidence of type E spores in fish (21) has made the safety of MAP fish, in particular, a concern. *C. botulinum* can grow and produce neurotoxin in MAP fish, and depending on conditions, neurotoxin may be present before the fish is considered spoiled (34, 71). While it is commonly assumed that *C. botulinum* cannot grow in foods exposed to O_2, the E_h of most such foods is usually low enough to allow its growth (49). Initial atmospheres containing 20% O_2 did not delay neurotoxin production by *C. botulinum* in inoculated pork compared with samples packaged with 100% N_2, and some toxic samples contained 15% residual O_2 (56). CO_2 is used in MAP to inhibit spoilage and pathogenic microorganisms, but CO_2 may stimulate *C. botulinum* (33). Initial levels of 15 to 30% CO_2 did not inhibit *C. botulinum* in inoculated pork; only 75% CO_2 showed significant inhibition (57). The safety of different atmospheres with respect to *C. botulinum* should be carefully investigated before use.

C. botulinum grows optimally at an E_h of –350 mV, but growth initiation may occur in the E_h range of +30 to +250 mV (49). The presence of other inhibitory factors lowers this upper limit. Once growth is initiated, the E_h declines rapidly.

Preservatives

Nitrite has several functions in cured meat products; an important role is the inhibition of *C. botulinum*, while effects on color and flavor are important organoleptic considerations. The exact mechanism of botulinal inhibition by nitrite is not known. Its effectiveness is dependent on complex interactions among pH, salt, heat treatment, time and temperature of storage, and the composition of food (49). Nitrite is depleted from cured foods, and the depletion rate is dependent on product formulation, pH, and time and temperature during processing and storage. However, a significant contribution of nitrite to the inhibition of *C. botulinum* continues even when nitrite is no longer detectable (37). Nitrite reacts with many cellular constituents and appears to inhibit *C. botulinum* by more than one mechanism, including reaction with essential iron-sulfur proteins to inhibit energy-yielding systems in the cell (49). The reaction of nitrite, or nitric oxide, with secondary amines in meats to produce nitrosamines, some of which are carcinogenic, has led to regulations limiting the amount of nitrite used.

Sorbates, parabens, nisin, phenolic antioxidants, polyphosphates, ascorbates, EDTA, metabisulfite, *n*-monoalkyl maleates and fumarates, and lactate salts are also active against *C. botulinum* (49). The use of natural or liquid smoke has a significant inhibitory effect against *C. botulinum* in fish but appears to have an insignificant effect in meats.

Other Microorganisms

The growth of other microorganisms in foods has a very significant effect on the growth of *C. botulinum* (37, 49). Acid-tolerant yeasts and molds may make the environment more favorable for growth of *C. botulinum* (43). Other microorganisms may inhibit *C. botulinum*, either by changing the environment or by producing specific inhibitory substances or both. Lactic acid bacteria, including *Lactobacillus*, *Pediococcus*, and *Streptococcus* species, can inhibit growth of *C. botulinum* in meat products, largely by reducing the pH but also by the production of bacteriocins (47, 64, 67). The use of lactic acid bacteria and a fermentable carbohydrate (the Wisconsin process) is permitted for producing bacon with a decreased level of nitrite in the United States (95). The growth of other microorganisms may also protect consumers by causing spoilage that would make a toxic product less likely to be consumed.

Inactivation by Irradiation

C. botulinum spores are probably the most radiation-resistant spores of public health concern. *D* values (irradiation doses required to inactivate 90% of the population) of group I strains at −50 to −10°C are between 2.0 and 4.5 kGy in neutral buffers and in foods (49). Spores of type E are only marginally more sensitive, having *D* values of between 1 and 2 kGy. Radappertization is designed to reduce the number of viable spores of the most radiation-resistant *C. botulinum* by 12 log cycles. Any pretreatment such as the presence of O_2, change in irradiation temperature, and irradiation and recovery environments affect the *D* values of spores. Generally, spores are more sensitive in the presence of O_2 or preservatives and at temperatures above 20°C.

RESERVOIRS

Occurrence of *C. botulinum* in the Environment

Because contamination of food largely depends on the incidence of *C. botulinum* in the environment, many surveys for spores of *C. botulinum* have been undertaken worldwide. Results show that spores of *C. botulinum* are commonly present in soils and sediments, but their numbers and types vary depending on the location (Table 15.2) (20).

In North America, *C. botulinum* spores are widely distributed, but the spore load varies considerably, as does the predominating type. Soils in the United States east of the rise of the Rocky Mountains usually contain spores of type B proteolytic *C. botulinum*. Type A spores predominate in the western United States. Overall, type E is found infrequently and only in damp or wet locations. However, type E predominates, and high numbers are found in the region around the Great Lakes, particularly around Green Bay of Lake Michigan, and in the coastal areas of Washington and Alaska. The distribution of types on the Pacific coast changes with latitude; south of 36°N, the prevalent types shift from E to A and B.

In Europe, survey results show that type B predominates in the terrestrial environments of Britain, Ireland, Iceland, Denmark, and Switzerland and in the aquatic environments of the United Kingdom. Disease outbreaks also confirm its prevalence in Spain, Portugal, Italy, France, Belgium, Germany, Poland, Hungary, and the former Czechoslovakia and Yugoslavia. Most European type B strains are nonproteolytic. Type E is the predominant serotype from other aquatic environments, with the highest numbers found in Scandinavian waters, particularly in the Sound and Kattegat, between Denmark and Sweden. In the former USSR, type E spores also predominate, except in the central region, where type B spores predominate.

Surveys of Asia report lower numbers, with the exceptions of a high incidence of type E spores around the Caspian Sea and a high incidence of all types in the Sinkiang district of China. In Japan, the incidence of type E spores is high in northern areas, while type C predominates in other areas. In the tropical regions of Asia, types C and D replace type E as the predominant type in aquatic environments.

Fewer surveys have been done in the southern hemisphere. Type A spores predominate in Brazilian and Argentine soils, but types B, C, F, and G are also present. In Paraguay, the prevalent spore type was F, followed by A and C. In Africa, few surveys have been done, but one following an outbreak of type A botulism found type A and C spores in Kenyan soil. In South Africa, type B

Table 15.2 Incidence of *C. botulinum* in soils and sediments[a]

Location	Sample size (g)	% Positive samples	MPN[b]/kg	Type (% of types identified) A	B	C/D	E	F
United States								
Eastern, soil	10	19	21	12	64	12	12	0
Western, soil	10	29	33	62	16	14	8	0
Green Bay, sediment	1	77	1,280	0	0	0	100	0
Alaska, soil	1	41	660	0	0	0	100	0
Britain, soil	50	6	2	0	100	0	0	0
Britain, coast, sediment	2	4	18	0	100	0	0	0
Scandinavian coast, sediment	6	100	>780	0	0	0	100	0
The Netherlands, soil	0.5	94	2,500	0	22	46	32	0
Switzerland, soil	12	44	48	28	83	6	0	27
Rome, Italy, soil	7.5	1	2	86	14	0	0	0
Iran, Caspian sea, sediment	2	17	93	0	8	0	92	0
Sinkiang, China, soil	10	70	25,000	47	32	19	2	0
Japan								
Hokkaido, soil	5–10	4	4	0	0	0	100	0
Ishikawa, soil	40–50	56	16	0	0	100	0	0
Brazil, soil	5	35	86	57	7	29	0	7
Paraguay, soil	5	24	10	14	0	14	0	71
South Africa, soil	30	3	1	0	100	0	0	0
Thailand, sediment	10	3	3	0	0	83	17	0
New Zealand, sediment	20	55	40	0	0	100	0	0

[a]From reference 20.
[b]MPN, most probable number.

spores were found in soil and type D spores were found in sediment. In Indonesia, types C and D predominate, but types A, B, and E have also been identified. In Australia and New Zealand, the environmental incidence is very low, with both type A and type B spores found.

In summary, type A spores predominate in soils in the western United States, China, Brazil, and Argentina. Type B spores predominate in soils in the eastern United States, the United Kingdom, and much of continental Europe. Most American type B strains are proteolytic, while most European strains are nonproteolytic. Type E predominates in northern regions and in most temperate aquatic regions and their surroundings. Types C and D are found more frequently in warmer environments.

Occurrence of *C. botulinum* in Foods

Various surveys have been carried out to learn the incidence of *C. botulinum* spores in foods (Table 15.3) (21). However, considering that the risk of foodborne botulism is directly related to the contamination of foods, surprisingly there have been fewer surveys of foods than of the environment. As well, food surveys have largely focused on fish, meats, and infant foods, primarily honey.

C. botulinum type E spores are commonly found in fish and aquatic animals. In North America, the incidence and level of contamination appear highest in samples from the Pacific coast, followed by samples from the Great Lakes and then from the Atlantic seaboard. In Europe, fish shows a lower level of contamination, except for fish from Scandinavia and from the Caspian Sea. In Indonesia, most positive fish samples contain either type C or D spores, but types A, B, and F are also present.

Meat and meat products have been examined less frequently than fish and fish products, and the level of contamination is generally low. These products are less likely than fish to be contaminated with spores before slaughter of the animals, since contamination of the farm environment is far lower than contamination of the aquatic environment. In North America, very low levels of contamination have been found in raw pork, beef, and chicken in processing plants. Studies have either failed to detect or detected a very low incidence of botulinum spores in cured meats and meat trimmings and in vacuum-packed sliced processed meat such as bologna, smoked beef, turkey and chicken, liver sausage, luncheon loaf, salami, and pastrami. In the United Kingdom the incidence in both raw and finished product

Table 15.3 Incidence of *C. botulinum* spores in food[a]

Product	Origin	Sample size (g)	% Positive samples	MPN[b]/kg	Type(s) identified
Eviscerated whitefish chubs	Great Lakes	10	12	14	E, C
Vacuum-packed frozen flounder	Atlantic Ocean	1.5	10	70	E
Dressed rockfish	California	10	100	2,400	A, E
Salmon	Alaska	24–36	100	190	A
Vacuum-packed fish	Viking Bank		42	63	E
Smoked salmon	Denmark	20	2	<1	B
Salted carp	Caspian Sea	2	63	490	E
Fish and seafood	Osaka, Japan	30	8	3	C, D
Raw meat	North America	3	<1	0.1	C
Cured meat	Canada	75	2	0.2	A
Raw pork	United Kingdom	30	0–14	<0.1–5	A, B, C
Cooked, vacuum-packed potatoes	The Netherlands		0	0.63	
Mushrooms	Canada			2,100	B
Random honey samples	United States	30	1	0.4	A, B
Honey samples associated with infant botulism	United States	30	100	8×10^4	A, B

[a]From reference 21.
[b]MPN, most probable numbers.

varies considerably between sampling occasions. In North America, the average most probable number is ~0.1 spore per kg, while in Europe, the average most probable number is ~2.5 spores per kg. The types most often associated with meats are A and B.

C. botulinum, usually type A or B, may be present on fruits and vegetables, particularly those in close contact with the soil. Different agricultural practices, such as the use of manure for fertilizer, may affect the level of contamination. Products in which contamination has often been detected include asparagus, beans, cabbage, carrots, celery, corn, onions, potatoes, turnips, olives, apricots, cherries, peaches, and tomatoes. A product of particular concern because of the high number of spores found is cultivated mushrooms, in which up to 2.1×10^3 type B spores per kg have been detected (39). The potential presence of spores in honey and other infant foods is problematic because in some infants, the spores can colonize the intestines, produce neurotoxin, and cause infant botulism. Honey is the only food that has ever been implicated in infant botulism. Surveys show that the botulinum spore level in random samples of honey is between 1 and 10 spores per kg (21). The level is higher in honey samples associated with infant botulism, approximately 10^4 spores per kg (21). While spores have been detected in other infant foods, namely, corn syrup and rice cereal, these foods do not seem to present the same risk as honey because the level of contamination is low and unlikely to increase during production and storage. Only a very low incidence of *C. botulinum* spores has been found in other foods, including dairy products, vacuum-packed products, and convenience foods (21).

FOODBORNE OUTBREAKS

Botulism is more likely to be detected and reported than other, milder forms of food poisoning. Therefore, the epidemiological data for botulism are probably more complete than for most other foodborne illnesses. Although botulism from commercial foods is rare, many countries report relatively frequent outbreaks (Table 15.4). Unrecognized and misdiagnosed cases of botulism do occur, as shown by a 1985 outbreak in Vancouver, Canada, in which the initial diagnoses for 28 patients included psychiatric illness, viral syndrome, and a variety of other maladies (93). As well, in large areas of the world, particularly where botulism occurs infrequently, epidemiological data are scarce.

In many northern areas, including the Canadian north, Alaska, Scandinavia, and northern Japan, most botulism outbreaks involve fish, especially traditional native dishes such as raw and parboiled meats from sea mammals, fermented meats such as muktuk (meat, blubber, and skin of the beluga whale), and fermented salmon eggs (38). *C. botulinum* type E has been implicated in the vast majority of outbreaks involving northern native foods (102). Many of these foods are fermented products, but the level of fermentable carbohydrates is too low to ensure a pH reduction sufficiently rapid to prevent growth of *C. botulinum.*

Table 15.4 Foodborne botulism outbreaks[a]

Country	Period	No. of outbreaks	No. of cases	No. of deaths	Usual type	Usual food
Poland	1984–1987	1,301	1,791	3	B	Meats
China	1958–1983	986	4,377	13	A	Vegetables
United States[b]	1971–1993	302	655	16	A	Vegetables[c]
Italy	1979–1987	Unknown	310	Unknown	B	Vegetables
France	1978–1989	175	304	2	B	Meats
Japan	1951–1987	97	479	23	E	Fish
Former USSR	1958–1964	95	328	29	B	Fish
Iran	1972–1974	Unknown	314	11	E	Fish
Canada	1971–1994	99	231	18	E	Meats[d]
Spain	1969–1988	63	198	6	B	Vegetables
Germany	1984–1989	96	206	12	B	Meats
Hungary	1985–1989	31	57	2	B	Meats
Portugal	1970–1989	24	80	0	B	Meats
Norway	1961–1990	19	42	7	B, E	Fish
Former Czechoslovakia	1979–1984	17	20	0	B	Meats
Argentina	1980–1989	16	36	13	B	Meats
Former Yugoslavia	1984–1989	12	51	Unknown	Unknown	Meats
Belgium	1982–1989	11	25	4	B	Mrats
Denmark and Greenland	1984–1989	11	16	12	E	Meats

[a]From reference 23 unless noted otherwise.
[b]From refrence 36a.
[c]In Alaska, most outbreaks are due to traditional Aleut dishes.
[d]Mostly traditional Inuit meat dishes.

Uneviscerated, salt-cured fish (ribyetz or kapchunka) has caused several outbreaks of botulism. In 1981, a California man became ill, and in 1985, two Russian immigrants died after eating uneviscerated, salt-cured fish (4). In 1987, an international outbreak involving eight individuals in Israel and the United States, with one fatality, was caused by consumption of kapchunka distributed in New York City (54, 96). The largest outbreak of type E botulism ever reported, and the first recorded outbreak of botulism in Egypt, was caused by uneviscerated salted mullet fish (103). Ninety-one patients were hospitalized in Cairo after consuming faseikh purchased from the same shop. The viscera of the ungutted, salted fish may provide a low-salt environment for *C. botulinum* to germinate and produce neurotoxin.

In many areas, home-preserved vegetables are the foods most often implicated (38). This is true for the continental United States and China, where most outbreaks are caused by type A, and for Italy and Spain, where type B is usually implicated.

In Poland and several other European countries, including France, Germany, Hungary, Portugal, the former Czechoslovakia, and Belgium, the foods most often implicated are home-preserved meats such as ham, fermented sausages, and canned products, and the predominant type involved is B (38). Temperature abuse of home-prepared foods continues to be an important cause of botulism. Recently, a 47-year-old Oklahoma resident was hospitalized for 49 days, including 42 days on mechanical ventilation, after eating stew that had been cooked and left at room temperature for 3 days before being eaten (53).

Commercial products generally have had a good safety record since the early days of canning in the 1920s. Bottled garlic in oil, which caused an outbreak in Canada and one in the United States, is a recently implicated commercial product (23). As a result of these two outbreaks, garlic in oil can be sold in North America only if a second barrier, such as acidification, is present in addition to refrigeration. Two outbreaks of type B botulism in Italy were associated with commercially prepared sliced roasted eggplant in oil (15). In the first outbreak, two waitresses working at a sandwich bar ate ham, cheese, and eggplant sandwiches. Two days later they were admitted to a hospital with dysphagia, diplopia, and constipation. In the second outbreak, four of nine members of an extended family who had consumed a meal of green olives, prosciutto, bean salad, green salad, mozzarella cheese, sausages, and commercially prepared roasted eggplant in oil were admitted to a hospital and diagnosed with botulism.

In the United Kingdom, hazelnut yogurt was the cause of a recent outbreak (23). Type B neurotoxin was produced in an underprocessed hazelnut purée that was then used to flavor a yogurt produced by a local dairy. Two outbreaks in Japan were associated with commercial food (23). In one, vacuum-packaged, stuffed lotus rhizome caused a type A outbreak and involved 36 cases with 11 deaths. Imported bottled caviar caused the second outbreak, with 21 cases and 3 deaths, due to type B neurotoxin. Taiwan reported an unusual botulism outbreak involving commercially canned peanuts (14).

Several outbreaks have also occurred at food service establishments (19). Temperature abuse of either food ingredients or the final product is often the problem. The use of leftover baked potatoes kept at room temperature has caused two reported outbreaks, one in which the potatoes were used for potato salad and one in which they were used in a Greek food known as skordalia.

CHARACTERISTICS OF DISEASE

Foodborne botulism varies from a mild illness, which may be disregarded or misdiagnosed, to a serious disease that may be fatal within 24 h (23). Symptoms typically appear 12 to 36 h after ingestion of neurotoxin but may appear within a few hours or not for up to 14 days. The earlier symptoms appear, the more serious the disease. The first symptoms are generally nausea and vomiting, followed by neurological signs and symptoms, including visual impairments (blurred or double vision, ptosis, fixed and dilated pupils), loss of normal mouth and throat functions (difficulty in speaking and swallowing; dry mouth, throat, and tongue; sore throat), general fatigue and lack of muscle coordination, and respiratory impairment. Other gastrointestinal symptoms may include abdominal pain, diarrhea, or constipation. Nausea and vomiting appear more often in cases associated with types B and E than in those associated with type A. Dysphagia and muscle weakness are more common in outbreaks of types A and B than in outbreaks of type E. Dry mouth, tongue, and throat are observed most frequently in type B cases. Respiratory failure and airway obstruction are the main causes of death. Fatality rates in the first half of the century were about 50% or higher, but with the availability today of antisera and modern respiratory support systems, they have decreased to about 10%.

Botulism may be confused with other illnesses, including other forms of foodborne poisoning, myasthenia gravis, and carbon monoxide poisoning, but most commonly with Guillain-Barré syndrome. The neurological signs in botulism appear first in the cranial nerve area (eyes, mouth, and throat) and then descend; Guillain-Barré syndrome progresses in an ascending fashion, beginning in the extremities.

The initial symptoms of infant botulism are less clear-cut. The most common, and usually the earliest, symptom is constipation (22). Medical attention is generally requested several days to a week later. The infants usually show a generalized weakness and a weak cry. Other symptoms may include feeding difficulty and poor sucking, lethargy, lack of facial expression, irritability, and progressive "floppiness." Respiratory arrests occur frequently but are seldom fatal.

Initially, treatment of foodborne botulism tries to remove or inactivate the neurotoxin by (i) neutralization of circulating neurotoxin with antiserum, (ii) use of enema to remove residual neurotoxin from the bowel, and (iii) gastric lavage or treatment with emetics (23). Antiserum is most effective in the early stages of the illness. The impact of antiserum is obvious from the Chinese data: before the availability of antisera in 1960, the death rate in China was approximately 50%, but it was only 8% in the nearly 4,000 patients who have received antitoxin since then. None of the 139 cases of botulism in the United States in 1994, or the 13 cases in Canada in 1995, resulted in death. Subsequent treatment is mainly to counteract the paralysis of the respiratory muscles by artificial ventilation. Optimal treatment for infant botulism consists primarily of high-quality supportive care (22). Approximately 25% of all affected infants require mechanical ventilation, and many require gavage feeding. The use of equine antitoxin and antibiotics is not recommended and does not appear to change the course or outcome of infant botulism.

INFECTIVE DOSE AND SUSCEPTIBLE POPULATIONS

Little is known concerning the minimum toxic (infective) dose of *C. botulinum* and its neurotoxins. From a food safety perspective, there is no tolerance for the presence of neurotoxin or for conditions permitting growth of *C. botulinum*. The mouse LD_{50} for BoNT is approximately 0.1 ng/kg. A minimum infective dose has not been established for infant or wound botulism. Two separate studies have found the spore load in honey samples implicated in infant botulism to be quite high, 8×10^3 and 8×10^4 spores per kg, approximately 1,000-fold higher than in other honey samples (22). However, this does not indicate a high infective dose since the babies affected consumed very little honey.

While infants are most susceptible to germination of spores in their intestines, with subsequent growth, toxigenesis, and illness, some adults are also susceptible (22). Susceptibility in infants is related to their diet and indigenous intestinal microflora. Susceptibility in adults appears to be associated with major perturbations in their intestinal microflora caused by such treatments as chemotherapy and antibiotic therapy.

The only effective means of preventing foodborne botulism is by preventing neurotoxin production in foods. Immunization of high-risk populations with botulinal toxoids has been considered, but it is not considered effective; now only laboratory workers at risk are normally immunized. Usually, the preservation of high-moisture foods is geared toward control of *C. botulinum.* In shelf-stable canned foods, the thermal process generally ensures destruction of spores. Control in minimally processed foods is achieved by inhibition, typically by a combination of factors. Such control generally also ensures control of other foodborne pathogens and of many spoilage microorganisms.

VIRULENCE FACTORS AND MECHANISMS OF PATHOGENICITY

Eight antigenically distinct toxins, designated types A, B, C_1, C_2, D, E, F, and G, are produced by *C. botulinum* (36). All of the toxins except C_2 are neurotoxins. C_2 toxin and exoenzyme C3 have been identified as ADP-ribosylating enzymes (3).

As mentioned previously, *C. botulinum* is divided into four broad groups (36). Group I strains produce type A, B, and F neurotoxins or combinations of A_B, A_F, B_A, and B_F neurotoxins, where the minor component is designated by the subscript letter. Group II strains produce type B, E, and F neurotoxins. Group III strains produce C_1 and D neurotoxins and C_2 toxin. These toxins can be produced separately or in combinations of C_1 and C_2 or D and C_2 toxins. Group IV strains produce type G neurotoxin. Other species of *Clostridium* can produce neurotoxins. *C. butyricum* can produce type E neurotoxin (61), *C. barati* can produce type F neurotoxin (35), and *C. novyi* can produce type C_1 or D neurotoxin (28).

C. botulinum C_2 toxin and exoenzyme C3 are produced by type C and D serotypes and belong to a class of bacterial ADP-ribosylating toxins which modify actin or the small GTP-binding proteins of the Rho family (3). Other ADP-ribosylating toxins include diphtheria toxin, *Pseudomonas aeruginosa* exotoxin A, exoenzyme S of *P. aeruginosa,* cholera toxin, pertussis toxin, and the heat-labile enterotoxin of *Escherichia coli* (70). These toxins are all capable of transferring the ADP-ribosyl moiety of NAD to specific protein substrates.

C. botulinum C_2 toxin is a binary toxin, composed of a binding component and an enzyme component with actin ADP-ribosylating activity (3). Monomeric G-actin is ADP ribosylated at arginine 177 by the enzymatic component of C_2 toxin (C_2-I) (101). The ribosylated actin is no longer capable of polymerization, resulting in destabilization of the cytoskeleton.

C. botulinum exoenzyme C3 has been purified as a protein with a molecular mass of 23 kDa. Exoenzyme C3 ADP ribosylates a specific asparagine residue of the low-molecular-mass GTP-binding proteins of the Rho family. The Rho proteins are involved in regulation of the actin cytoskeleton (3). No role in pathogenesis has been attributed to either C_2 toxin or exoenzyme C3.

Neurotoxins

All seven of the neurotoxins are similar in structure and mode of action. *C. botulinum* neurotoxins are high-molecular-mass (150-kDa) two-chain proteins which are among the most toxic substances known (mouse $LD_{50} < 0.1$ ng/kg). Botulinum neurotoxin types A and B possess toxicities of 2×10^8 LD_{50}/mg. The neurotoxins block neurotransmission at peripheral motor nerve terminals by selectively hydrolyzing proteins involved in the fusion of synaptic vesicles with the presynaptic plasma membrane, thereby preventing acetylcholine release.

Structure of Neurotoxins

BoNTs are water-soluble proteins produced as a single polypeptide with an approximate M_r of 150,000. They are cleaved by a protease approximately one-third of the distance from the N terminus to produce the active neurotoxin which is composed of one heavy (H; M_r = 100,000) and one light (L; M_r = 50,000) chain linked by a single disulfide bond (17).

Endogenous bacterial proteases or proteases such as trypsin can cause the proteolytic cleavage. The two chains individually are nontoxic. The L chain remains bound to the N-terminal half of the H chain by a disulfide bond between Cys-429 and Cys-453 (55) and noncovalent bonds (16).

Native gel electrophoresis and chemical cross-linking experiments have been used to determine the quaternary structure of the BoNTs (59). These studies indicate that BoNT type A exists in several forms, including a dimer, trimer, and a larger species, BoNT type E exists as a monomer and primarily as a dimer, and BoNT type B exists as a dimer in aqueous solution. The oligomerization of BoNT molecules may be involved in possible channel formation and translocation of the L chain into the cytoplasm.

While very little information exists concerning the secondary or tertiary structures of the BoNTs, the recent determination of the complete nucleotide sequences of the neurotoxins and the crystallization of type A neurotoxin (92) should facilitate elucidation of the secondary and tertiary structures. Predictions of the secondary structures of BoNTs have been made on the basis of DNA sequences (58). The amounts of helical, extended, and loop conformations from circular dichroism studies (90) generally agree with the computer-predicted secondary structures.

BoNTs form complexes with nontoxic proteins in naturally contaminated foods and culture supernatants to form what is referred to as progenitor toxin (75). The non-toxic proteins can be dissociated from BoNT by pH values greater than 7.2 and spontaneously reassociate when the pH is lowered. Three forms of progenitor neurotoxin have been distinguished and are referred to as M (medium-sized), L (large), and LL (extra-large) toxins (75). M toxin is produced by all strains producing neurotoxin except those producing type G neurotoxin, has a molecular mass of ca. 300 kDa, and is composed of the neurotoxin and a nontoxic nonhemagglutinin (NTNH) protein with a size similar to that of the neurotoxin. L toxin has a molecular mass of approximately 500 kDa and is produced by strains which produce type A, B, C, D, and G neurotoxins. L toxin contains hemagglutinins with molecular masses 33 kDa (HA-33) and 17 kDa (HA-II) in addition to BoNT and NTNH. The nontoxic proteins have been shown to consist of several polypeptides with different molecular masses (27, 35, 55, 115, and 120 kDa, or 33, 53, and 130 kDa) (99, 100). It has been speculated that the nontoxic proteins are important for protection of the neurotoxin from low pH and proteases during passage through the stomach.

Genetic Regulation of Neurotoxins

The neurotoxins are arranged as part of a transcriptional unit which includes the genes encoding BoNT as well as genes encoding NTNH components and hemagglutinins. This transcriptional unit may be referred to as the BoNT gene complex (26).

Complete gene sequences have been determined for the neurotoxins produced by *C. botulinum* types A, B (proteolytic and nonproteolytic), C, D, E, F, and G, *C. barati* type F, and *C. butyricum* type E (7, 8, 13, 25, 42, 50, 51, 73, 97, 98, 104, 105). The degree of relatedness of the various neurotoxins has been determined on the basis of sequence homologies (13). Different serotype neurotoxins display less sequence homology than the same serotypes even if the same serotypes are from different species. Type E BoNTs from *C. botulinum* and *C. butyricum* have 97% identical amino acids, suggesting gene transfer between *C. botulinum* and *C. butyricum* (43, 73). Type F BoNTs from nonproteolytic *C. botulinum* and *C. barati* have approximately 70% identity (98). Type E and type F BoNTs have approximately 63% identity (98), while type G and type B BoNTS are approximately 58% identical (13).

The locations of the genes coding for BoNTs and the associated nontoxic proteins vary depending on the serotype. The genes coding for BoNTs A, B, E, and F and the associated nontoxic proteins are located on the bacterial chromosome (8, 97, 104, 105). The genes coding for BoNTs C_1 and D and the associated nontoxic proteins are encoded by bacteriophages (27, 30, 31, 41, 42, 99), whereas the genes coding for BoNT G and the associated nontoxic proteins are located on a plasmid (29, 108).

The organization and sequences of the genes for BoNTs A, B, and C_1 and the genes encoding the associated nontoxic proteins have been determined (26, 41, 99, 100). Six genes encoding BoNT C_1 and its associated nontoxic proteins (Antp) are organized into three transcriptional units (41). One cluster encodes the structural gene for BoNT C_1 and an associated nontoxic protein (Antp 139/C1 or NTNH). A second cluster contains three genes, two of which encode proteins similar to HA-33 and HA-II. The third cluster consists of a single gene encoding a protein which displays homology with a regulatory protein of *C. perfringens* (*uviA* gene product) and could regulate the expression of the genes in the other two clusters (41).

C. botulinum type A and proteolytic and nonproteolytic type B have very similar gene clusters encoding the neurotoxins and the associated nontoxic proteins (26). Strains producing type A and B neurotoxins possess genes encoding HA-33 and HA-II upstream of the genes encoding the BoNT and NTNH (26, 99). In addition, type A and B strains have an open reading frame (ORF) similar to the ORF from type C strains which consists of a single gene displaying homology with the regulatory protein of *C. perfringens* (*uviA* gene product). In the case of type A and B strains, this third ORF is located between the genes encoding the hemagglutinins (HA-II and HA-33) and the NTNH protein (26), while in type C strains it is the furthest upstream ORF in the gene cluster (41).

Mode of Action of Neurotoxins

BoNTs block the exocytic release of the excitatory neurotransmitter acetylcholine from synaptic vesicles at peripheral motor nerve terminals. This results in the flaccid paralysis observed in botulism poisoning and has been utilized in the therapy of several neurologic disorders (48). BoNTs are also being used as tools to study the molecular mechanisms of vesicle docking and membrane fusion mechanisms involved in exocytosis (12, 82).

The mechanism of action of BoNTs can be divided into three steps: binding, internalization, and intracellular action (88). A recent review on the mode of action of BoNTs by Montecucco and Schiavo (63) includes membrane translocation in addition to these three steps. The H chains are responsible for selective binding of the neurotoxin to neurons, internalization of the entire neurotoxin, intraneuronal sorting, and translocation of the L chains into the cytosol. The L chains block exocytosis as soon as they are released into the cytoplasm (2, 63, 65).

Binding

The potency of the BoNTs results from the specificity of the toxins for neurons. Neurotoxin receptors are located at the motor neuron plasma membrane at the neuromuscular junction (24). Receptors may be different for the various neurotoxins (9, 107). Binding experiments with ^{125}I-labeled BoNTs and synaptosomes indicate that type A, B, C_1, and D neurotoxins all have a small number of high-affinity binding sites and a large number of sites with much lower affinity (1, 32, 106).

While the neurotoxins appear not to share receptors, the receptors may all possess at least one sialic acid residue. Gangliosides, which are sialic acid-containing glycosphingolipids, bind to the BoNTs (52, 76, 89). Sialic acid-specific lectins from the plants *Triticum vulgaris* and *Limax flavus* are able to block binding of BoNTs to brain membranes (5). Putative receptors for BoNT type A and tetanus neurotoxin have been identified from blots of synaptosomal proteins (77). Both neurotoxins adhered to 80- and 116-kDa glycoproteins. The H chain of BoNT type A specifically adhered to the same proteins, further evidence that attachment to target cell membranes is mediated through the C-terminal half of the H chain (69, 72, 77).

A double-receptor model for binding of botulinum and tetanus neurotoxins has been used to explain the high-affinity binding (62). In this model, the neurotoxins first bind to the negatively charged surface of the presynaptic membrane, which contains large amounts of acidic lipids. After binding to the negatively charged lipids, the neurotoxin may diffuse laterally in the membrane to bind to a protein receptor. The high binding affinity of the neurotoxins for the presynaptic plasma membrane is explained in this model because the association constant is the product of the association constants of the neurotoxin with the negatively charged lipids and the protein receptor (63).

Internalization

After the neurotoxin has bound to its receptors at the neuromuscular junctions, the entire neurotoxin is internalized by receptor-mediated endocytosis. Once the neurotoxin has been internalized, it can no longer be neutralized by antineurotoxin. Studies using chemical cross-linking and high-resolution electron microscopy suggest that the H chain aggregates and forms channels in the endosomal membrane to allow the L chain to pass into the cytoplasm. Cross-linking studies have shown that in aqueous solution, type A neurotoxin exists as a dimer, a trimer, and a larger species, type E neurotoxin exists as a monomer and a dimer, and type B neurotoxin exists as a dimer (59). Three-dimensional reconstructions of ordered arrays of type B neurotoxin on surfaces of ganglioside/phosphocholine lipid vesicles show that four neurotoxin molecules combine to form channels which cross the vesicle membrane (87). In addition, synthetic peptides of the predicted transmembrane sequence of BoNT type A form cation-selective channels in planar lipid bilayers (66). Thus, it appears that the L chain exits the endosome by passing through channels created by the H chain.

Intracellular Action

The L chain is capable of inhibiting neurotransmitter release independently of the H chain (65). This has been demonstrated by intracellular injection of isolated L chain (72), by intracellular delivery of the L chain by liposome fusion with motor nerve endings (18), and by exposure of permeabilized cells to the L chain (91).

Comparison of the nucleotide sequences of the genes encoding the BoNTs with that of the gene encoding tetanus neurotoxin demonstrated a conserved segment in the central region of the L chain, which contained the His-Glu-x-x-His zinc-binding motif of metallo-endoproteinases (80, 83, 86). L chains act as zinc-dependent endopeptidases, whose substrates are components of the synaptic vesicle docking and fusion complex (for a review, see reference 63). BoNT types B, F, D, and G and tetanus neurotoxin cause selective degradation of vesicle-associated membrane protein (VAMP) (also called syn-

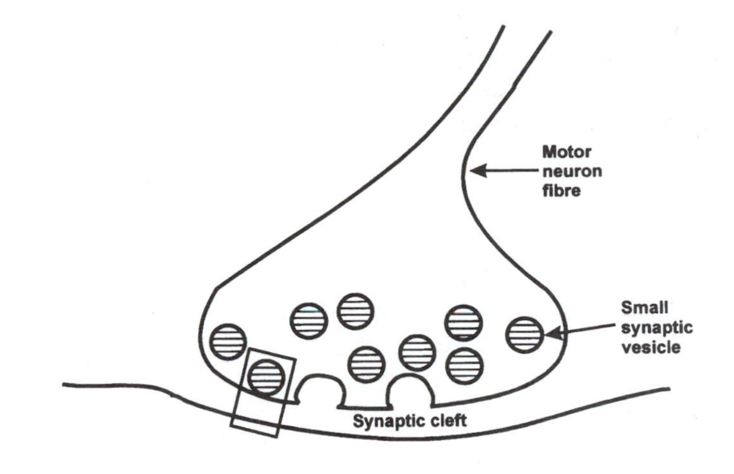

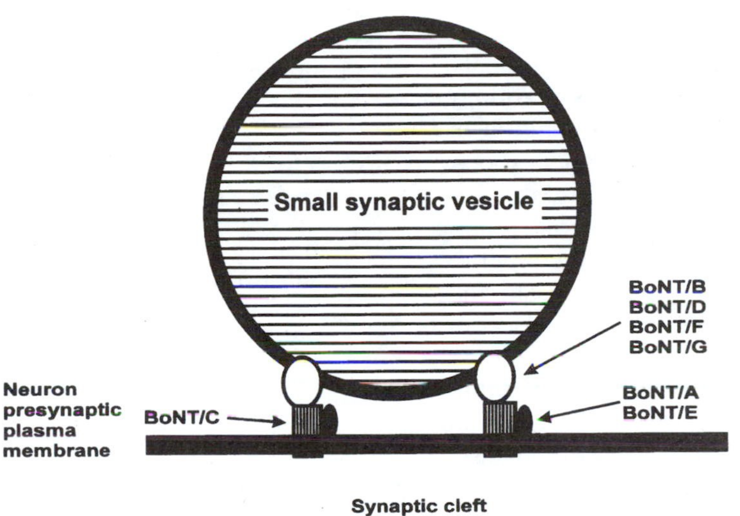

Figure 15.1 Schematic representation of the peripheral motor nerve terminal and synapse, depicting sites of action of BoNTs. The upper diagram shows the anatomy of the presynaptic terminal and synapse. The boxed area is enlarged beneath, showing details concerning the small synaptic vesicle. Synaptobrevin (also called VAMP) (open ovals), SNAP-25 (solid ovals), and syntaxin (also called HPC-1) (striped boxes) are proteins involved in the fusion of synaptic vesicles with the presynaptic plasma membrane. Synaptobrevin is embedded in the synaptic vesicle membrane, while SNAP-25 and syntaxin are located on the cytoplasmic surface of the neuron plasma membrane. The arrows indicate the protein which is the specific target of the different BoNTs.

aptobrevin) (78, 79, 81, 86). Type A and E neurotoxins degrade the synaptosome-associated protein SNAP-25 (6, 10, 84). Type C1 neurotoxin degrades syntaxin/HPC-1 (11, 85).

The zinc-dependent protease activity of the BoNTs appears to be very specific. BoNT type B and tetanus neurotoxin hydrolyze a Gln-Phe peptide bond in VAMP-2, BoNT type F cleaves VAMP at a Gln-Lys bond present in both VAMP-1 and VAMP-2, and BoNT type G hydrolyzes VAMP at a single Ala-Ala peptide bond (79, 85).

VAMP, syntaxin, and SNAP-25 form the core of a multicomponent complex which mediates fusion of carrier vesicles to target membranes in eukaryotic cells. Proteolysis of these proteins involved in docking and fusion of synaptic vesicles blocks neuroexocytosis and subsequent neurotransmitter release. The exact mechanism of exocytosis blockage remains unknown. A schematic representation of the peripheral motor nerve terminal and synapse, showing the sites of action of the BoNTs, is presented in Fig. 15.1. Type A BoNT does not prevent the formation or the disassembly of the synaptosomal fusion complex in rat brain synaptosomes (68) but still blocks exocytosis. Otto et al. (68) speculate that inhibition of exocytosis by type A BoNT occurs before the fusion of the vesicle with the plasma membrane but after the disassembly of the fusion complex.

CONCLUDING REMARKS

C. botulinum is a spore-forming bacterium that is widely distributed, occurring in soils, freshwater and marine sediments, and the intestinal tracts of many animals. Control of foodborne botulism is based almost entirely on thermal destruction of the spores or inhibition of spore germination and bacterial cell growth in foods. Outbreaks continue to occur in home-preserved foods (including vegetables and meats), commercial foods, and restaurant foods. Some traditional northern native foods, including whale, seal, and walrus meat and salmon eggs, are responsible for a high proportion of cases of botulism, especially in Canada.

The increased consumption of ready-to-eat, minimally processed foods, especially those packaged in modified atmospheres, is of concern. The combination of lack of a heat treatment sufficient to destroy *C. botulinum* spores, lack of heating food prior to consumption, and packaging in an atmosphere with reduced or no oxygen increases the risk of botulism from these foods. The reliance on refrigeration for storage of these foods does not preclude the growth of, and toxin production by, nonproteolytic *C. botulinum.* Novel preservatives or bacteriocin-producing bacteria to inhibit the germination and growth of *C. botulinum* in these foods are possible alternatives for controlling this pathogen.

Complete nucleotide sequences are now known for all of the BoNTs. In addition, several accessory genes, including those encoding the hemagglutinins, have been sequenced and localized into transcriptional units. The isolation of neurotoxin-producing strains of *C. butyricum* and *C. barati* and the high sequence similarity between neurotoxins produced in these strains and *C. botulinum* strains suggest lateral transfer of the genes encoding BoNTs and their associated polypeptides. Future research in this area may include elucidation of possible transcriptional or translational control of synthesis of the neurotoxins, organization of gene clusters in unusual strains, the role of the nontoxic proteins, and transfer of the genes encoding the neurotoxin and associated proteins between species of *Clostridium.*

The modes of action of the BoNTs have recently been elucidated. BoNT types B, D, F, and G cleave VAMP/synaptobrevin, an integral membrane protein of synaptic vesicles, whereas BoNT types A and E cleave SNAP-25 and botulinum neurotoxin type C cleaves syntaxin. Both SNAP-25 and syntaxin are synaptic proteins located on the plasma membrane of the neuron. Recent identification of zinc-dependent endopeptidase activity of BoNTs and the discovery of the substrates of the neurotoxins have facilitated the study of synaptic vesicle docking, exocytosis, and neurotransmitter release. Future research will address selective alteration of neurotoxins by site-directed mutagenesis to determine which amino acid residues are involved in binding, uptake, proteolysis, and antigenicity.

References

1. **Agui, T., B. Syuto, K. Oguma, H. Iida, and S. Kubo.** 1983. Binding of *Clostridium botulinum* type C neurotoxin to rat synaptosomes. *J. Biochem.* **94:**521–527.
2. **Ahnert-Hilger, G., and H. Bigalke.** 1995. Molecular aspects of tetanus and botulinum neurotoxin poisoning. *Prog. Neurobiol.* **46:**83–96.
3. **Aktories, K.** 1994. Clostridial ADP-ribosylating toxins: effects on ATP and GTP-binding proteins. *Mol. Cell. Biochem.* **138:**167–176.
4. **Badhey, H., D. J. Cleri, R. F. D'Amato, J. R. Vernaleo, V. Veinni, J. Tessler, A. A. Wallman, A. J. Mastellone, M. Guiliani, and L. Hochstein.** 1986. Two fatal cases of type E adult food-borne botulism with early symptoms and terminal neurologic signs. *J. Clin. Microbiol.* **23:**616–618.
5. **Bakry, N., Y. Kamata, and L. L. Simpson.** 1991. Lectins from *Triticum vulgaris* and *Limax flavus* are universal antagonists of botulinum neurotoxin and tetanus toxin. *J. Pharmacol. Exp. Ther.* **258:**830–836.
6. **Binz, T., J. Blasi, S. Yamasaki, A. Baumeister, E. Link, T. C. Sudhof, R. Jahn, and H. Niemann.** 1994. Proteolysis of SNAP-25 by types E and A botulinal neurotoxins. *J. Biol. Chem.* **269:**1617–1620.
7. **Binz, T., H. Kurazono, M. R. Popoff, M. W. Eklund, G. Sakaguchi, S. Kozaki, K. Krieglstein, A. Henschen, D. M. Gill, and H. Niemann.** 1990. Nucleotide sequence of the gene encoding *Clostridium botulinum* neurotoxin type D. *Nucleic Acids Res.* **18:**5556.
8. **Binz, T., H. Kurazono, M. Wille, J. Frevert, K. Wernars, and H. Niemann.** 1990. The complete sequence of botulinum neurotoxin type A and comparison with other clostridial neurotoxins. *J. Biol. Chem.* **265:**9153–9158.
9. **Black, J. D., and J. O. Dolly.** 1986. Interaction of ^{125}I-labelled botulinum neurotoxins with nerve terminals. I.

Ultrastructural autoradiographic localization and quantitation of distinct membrane acceptors for types A and B on motor nerves. *J. Cell Biol.* **103**:521–534.

10. **Blasi, J., E. R. Chapman, E. Link, T. Binz, S. Yamasaki, P. De Camilli, T. C. Sudhof, H. Niemann, and R. Jahn.** 1993. Botulinum neurotoxin A selectively cleaves the synaptic protein SNAP-25. *Nature* (London) **365**:160–163.
11. **Blasi, J., E. R. Chapman, S. Yamasaki, T. Binz, H. Niemann, and R. Jahn.** 1993. Botulinum neurotoxin C1 blocks neurotransmitter release by means of cleaving HPC-1/syntaxin. *EMBO J.* **12**:4821–4828.

11a. **Botulism Reference Service for Canada.** Unpublished data.

12. **Boyd, R. S., M. J. Duggan, C. C. Shone, and K. A. Foster.** 1995. The effect of botulinum neurotoxins on the release of insulin from the insulinoma cell lines HIT-15 and RINm5F. *J. Biol. Chem.* **270**:18216–18218.
13. **Campbell, K., M. D. Collins, and A. K. East.** 1993. Nucleotide sequence of the gene coding for *Clostridium botulinum* (*Clostridium argentinense*) type G neurotoxin: genealogical comparison with other clostridial neurotoxins. *Biochim. Biophys. Acta* **1216**:487–491.
14. **Chou, J. H., P. H. Hwang, and M. D. Malison.** 1988. An outbreak of type A foodborne botulism in Taiwan due to commercially preserved peanuts. *Int. J. Epidemiol.* **17**: 899–902.
15. **D'Argenio, P., F. Palumbo, R. Ortolani, R. Pizzuti, M. Russo, R. Carducci, M. Soscia, P. Aureli, L. Fenicia, G. Franciosa, A. Parella, and V. Scala.** 1995. Type B botulism associated with roasted eggplant in oil. *Morbid. Mortal. Weekly Rep.* **44**:33–36.
16. **DasGupta, B. R.** 1990. Structure and biological activity of botulinum neurotoxin. *J. Physiol.* (Paris) **84**:220–228.
17. **DasGupta, B. R., and H. Sugiyama.** 1972. Role of a protease in natural activation of *Clostridium botulinum* neurotoxin. *Infect. Immun.* **6**:587–590.
18. **dePaiva, A., and J. O. Dolly.** 1990. Light chain of botulinum neurotoxin is active in mammalian motor nerve terminals when delivered via liposomes. *FEBS Lett.* **277**:171–174.
19. **Dodds, K. L.** 1990. Restaurant-associated botulism outbreaks in North America. *Food Control* **1**:139–141.
20. **Dodds, K. L.** 1993. *Clostridium botulinum* in the environment, p. 21–51. *In* A. H. W. Hauschild and K. L. Dodds (ed.), *Clostridium botulinum: Ecology and Control in Foods.* Marcel Dekker, Inc., New York.
21. **Dodds, K. L.** 1993. *Clostridium botulinum* in foods, p. 51–68. *In* A. H. W. Hauschild and K. L. Dodds (ed.), *Clostridium botulinum: Ecology and Control in Foods.* Marcel Dekker, Inc., New York.
22. **Dodds, K. L.** 1993. Worldwide incidence and ecology of infant botulism, p. 105–117. *In* A. H. W. Hauschild and K. L. Dodds (ed.), *Clostridium botulinum: Ecology and Control in Foods.* Marcel Dekker, Inc., New York.
23. **Dodds, K. L.** 1993. *Clostridium botulinum,* p. 97–131. *In* Y. H. Hui, J. R. Gorham, K. D. Murrell, and D. O. Clover (ed.), *Foodborne Pathogens.* Marcel Dekker, Inc., New York.
24. **Dolly, J. O., J. Black, R. S. Williams, and J. Melling.** 1984. Acceptors for botulinum neurotoxin reside on motor nerve terminals and mediate its internalization. *Nature* (London) **307**:457–460.
25. **East, A. K., P. T. Richardson, D. Allaway, M. D. Collins, T. A. Roberts, and D. E. Thompson.** 1992. Sequence of the gene encoding type F neurotoxin of *Clostridium botulinum. FEMS Microbiol. Lett.* **75**:225–230.
26. **East, A. K., J. M. Stacey, and M. D. Collins.** 1994. Cloning and sequencing of a hemagglutinin component of the botulinum neurotoxin complex encoded by *Clostridium botulinum* types A and B. *Syst. Appl. Microbiol.* **17**: 306–312.
27. **Eklund, M. W., F. T. Poysky, and W. H. Habig.** 1989. Bacteriophages and plasmids in *Clostridium botulinum* and *Clostridium tetani* and their relationship to production of toxins, p. 25–51. *In* L. L. Simpson (ed.), *Botulinum Neurotoxin and Tetanus Toxin.* Academic Press, New York.
28. **Eklund, M. W., F. T. Poysky, J. A. Meyers, and G. A. Pelroy.** 1974. Interspecies conversion of *Clostridium botulinum* type C to *Clostridium novyi* type A by bacteriophage. *Science* **186**:456–458.
29. **Eklund, M. W., F. T. Poysky, L. M. Mseitif, and M. S. Strom.** 1988. Evidence for plasmid-mediated toxin and bacteriocin production in *Clostridium botulinum* type G. *Appl. Environ. Microbiol.* **54**:1405–1408.
30. **Eklund, M. W., F. T. Poysky, and S. M. Reed.** 1972. Bacteriophages and toxigenicity of *Clostridium botulinum* type D. *Nature* (London) *New Biol.* **235**:16–17.
31. **Eklund, M. W., F. T. Poysky, S. M. Reed, and C. A. Smith.** 1971. Bacteriophage and the toxigenicity of *Clostridium botulinum* type C. *Science* **172**:480–482.
32. **Evans, D., R. S. Williams, C. C. Shone, P. Hambleton, J. Melling, and J. O. Dolly.** 1986. Botulinum neurotoxin type B. Its purification, radioiodination and interaction with rat-brain synaptosomal membranes. *Eur. J. Biochem.* **154**:409–416.
33. **Foegeding, P. M., and F. F. Busta.** 1983. Effect of carbon dioxide, nitrogen and hydrogen gases on germination of *Clostridium botulinum* spores. *J. Food Prot.* **46**:987–989.
34. **Garcia, G., and C. Genigeorgis.** 1987. Quantitative evaluation of *Clostridium botulinum* nonproteolytic types B, E, and F growth risk in fresh salmon tissue homogenates stored under modified atmospheres. *J. Food Prot.* **50**:390–397.
35. **Hall, J. D., L. M. McCroskey, B. J. Pincomb, and C. L. Hatheway.** 1985. Isolation of an organism resembling *Clostridium barati* which produces type F botulinal toxin from an infant with botulism. *J. Clin. Microbiol.* **21**:654–655.
36. **Hatheway, C. L.** 1993. *Clostridium botulinum* and other clostridia that produce botulinum neurotoxin, p. 3–20. *In* A. H. W. Hauschild and K. L. Dodds (ed.), *Clostridium botulinum: Ecology and Control in Foods.* Marcel Dekker, Inc., New York.

36a. **Hatheway, C. L.** Personal communication.

37. **Hauschild, A. H. W.** 1989. *Clostridium botulinum,* p. 111–189. *In* M.P. Doyle (ed.), *Foodborne Bacterial Pathogens.* Marcel Dekker, Inc., New York.
38. **Hauschild, A. H. W.** 1993. Epidemiology of human foodborne botulism, p. 69–104. *In* A. H. W. Hauschild

and K. L. Dodds (ed.), *Clostridium botulinum: Ecology and Control in Foods.* Marcel Dekker, Inc., New York.

39. **Hauschild, A. H. W., B. J. Aris, and R. Hilsheimer.** 1975. *Clostridium botulinum* in marinated products. *Can. Inst. Food Sci. Technol. J.* **8**:84–87.

40. **Hauser, D., M. W. Eklund, P. Boquet, and M. R. Popoff.** 1994. Organization of the botulinum neurotoxin C1 gene and its associated non-toxic protein genes in *Clostridium botulinum* C 468. *Mol. Gen. Genet.* **243**:631–40.

41. **Hauser, D., M. W. Eklund, H. Kurazono, T. Binz, H. Niemann, D. M. Gill, P. Boquet, and M. R. Popoff.** 1990. Nucleotide sequence of *Clostridium botulinum* C1 neurotoxin. *Nucleic Acids Res.* **18**:4924.

42. **Hauser, D., M. Gibert, P. Boquet, and M. R. Popoff.** 1992. Plasmid localization of a type E botulinal neurotoxin gene homologue in toxigenic *Clostridium butyricum* strains, and absence of this gene in non-toxigenic *C. butyricum* strains. *FEMS Microbiol. Lett.* **78**:251–255.

43. **Huhtanen, C. N., J. Naghski, C. S. Custer, and R. W. Russell.** 1976. Growth and toxin production by *Clostridium botulinum* in moldy tomato juice. *Appl. Environ. Microbiol.* **32**:711–715.

44. **Hutson, R. A., M. D. Collins, A. K. East, and D. E. Thompson.** 1994. Nucleotide sequence of the gene coding for non-proteolytic *Clostridium botulinum* type B neurotoxin: comparison with other clostridial neurotoxins. *Curr. Microbiol.* **28**:101–110.

45. **Hutson, R. A., D. E. Thompson, and M. D. Collins.** 1993. Genetic interrelationships of saccharolytic *Clostridium botulinum* types B, E and F and related clostridia as revealed by small-subunit rRNA gene sequences. *FEMS Microbiol. Lett.* **108**:103–110.

46. **Hutson, R. A., D. E. Thompson, P. A. Lawson, R. P. Schocken-Itturino, E. C. Bottger, and M. D. Collins.** 1993. Genetic interrelationships of proteolytic *Clostridium botulinum* types A, B, and F and other members of the *Clostridium botulinum* complex as revealed by small-subunit rRNA gene sequences. *Antonie Leeuwenhoek* **64**:273–283.

47. **Hutton, M. T., P. A. Chehak, and J. H. Hanlin.** 1991. Inhibition of botulinum toxin production by *Pediococcus acidilactici* in temperature abused refrigerated foods. *J. Food Safety* **11**:255–267.

48. **Jankovic, J., and M. F. Brin.** 1991. Therapeutic uses of botulinum toxin. *N. Engl. J. Med.* **324**:1186–1194.

49. **Kim, J., and P. M. Foegeding.** 1993. Principles of control, p. 121–176. *In* A. H. W. Hauschild and K. L. Dodds (ed.), *Clostridium botulinum: Ecology and Control in Foods.* Marcel Dekker, Inc., New York.

50. **Kimura, K., N. Fujii, K. Tsuzuki, T. Murakami, T. Indoh, N. Yokosawa, and K. Oguma.** 1991. Cloning of the structural gene for *Clostridium botulinum* type C1 toxin and whole nucleotide sequence of its light chain component. *Appl. Environ. Microbiol.* **57**:1168–1172.

51. **Kimura, K., N. Fujii, K. Tsuzuki, T. Murakami, T. Indoh, N. Yokosawa, K. Takeshi, B. Syuto, and K. Oguma.** 1990. The complete nucleotide sequence of the gene coding for botulinum type C1 toxin in the C-ST phage genome. *Biochem. Biophys. Res. Commun.* **171**:1304–1311.

52. **Kitamura, M., M. Iwamori, and Y. Nagai.** 1980. Interaction between *Clostridium botulinum* neurotoxin and gangliosides. *Biochim. Biophys. Acta* **628**:328–335.

53. **Knubley, W., T. C. McChesney, and J. Mallonee.** 1995. Foodborne botulism—Oklahoma, 1994. *JAMA* **273**: 1167. [Reprinted from *Morbid Mortal. Weekly Rep.* **44**: 200–202, 1995.]

54. **Kotev, S., A. Leventhal, A. Bashary, H. Zahavi, A. Cohen, et al.** 1987. International outbreak of type E botulism associated with ungutted, salted whitefish. *Morbrd. Mortal. Weekly Rep.* **36**:812–813.

55. **Krieglstein, K. G., B. R. Dasgupta, and A. H. Henschen.** 1994. Covalent structure of botulinum neurotoxin type-A—location of sulfhydryl groups, and disulfide bridges and identification of C-termini of light and heavy chains. *J. Protein Chem.* **13**:49–57.

56. **Lambert, A. D., J. P. Smith, and K. L. Dodds.** 1991. Combined effect of modified atmosphere packaging and low-dose irradiation on toxin production by *Clostridium botulinum* in fresh pork. *J. Food Prot.* **54**:94–101.

57. **Lambert, A. D., J. P. Smith, and K. L. Dodds.** 1991. Effect of headspace CO_2 concentration on toxin production by *Clostridium botulinum* in MAP, irradiated fresh pork. *J. Food Prot.* **54**:588–592.

58. **Lebeda, F. J., and M. A. Olson.** 1994. Secondary structural predictions for the clostridial neurotoxins. *Proteins Struct. Funct. Genet.* **20**:293–300.

59. **Ledoux, D. N., X. H. Be, and B. R. Singh.** 1994. Quaternary structure of botulinum and tetanus neurotoxins as probed by chemical cross-linking and native gel electrophoresis. *Toxicon* **32**:1095–1104.

60. **Lund, B. M., and S. H. W. Notermans.** 1993. Potential hazards associated with REPFEDS, p. 279–303. *In* A. H. W. Hauschild and K. L. Dodds (ed.), *Clostridium botulinum: Ecology and Control in Foods.* Marcel Dekker, Inc., New York.

61. **McCroskey, L. M., C. L. Hatheway, L. Fenicia, B. Pasolini, and P. Aureli.** 1986. Characterization of an organism that produces type E botulinal toxin but which resembles *Clostridium butyricum* from the feces of an infant with type E botulism. *J. Clin. Microbiol.* **23**:201–202.

62. **Montecucco, C.** 1986. How do tetanus and botulinum toxins bind to neuronal membranes? *Trends Biochem.* **11**:314–317.

63. **Montecucco, C., and G. Schiavo.** 1994. Mechanism of action of tetanus and botulinum neurotoxins. *Mol. Microbiol.* **13**:1–8.

64. **Montville, T. J., A. M. Rogers, and A. Okereke.** 1992. Differential sensitivity of *Clostridium botulinum* strains to nisin is not biotype-associated. *J. Food Prot.* **55**:444.

65. **Niemann, H.** 1991. Molecular biology of clostridial neurotoxins, p. 303–348. *In* J. E. Alouf and J. H. Freer (ed.), *Sourcebook of Bacterial Protein Toxins.* Academic Press, London.

66. **Oblatt-Montal, M., M. Yamazaki, R. Nelson, and M. Montal.** 1995. Formation of ion channels in lipid bilayers by a peptide with the predicted transmembrane sequence of botulinum neurotoxin A. *Protein Sci.* **4**:1490–1497.

67. **Okereke, A., and T. J. Montville.** 1991. Bacteriocin-mediated inhibition of *Clostridium botulinum* spores by lactic acid bacteria at refrigeration and abuse temperatures. *Appl. Environ. Microbiol.* **57:**3423–3428.
68. **Otto, H., P. I. Hanson, E. R. Chapman, J. Blasi, and R. Jahn.** 1995. Poisoning by botulinum neurotoxin A does not inhibit formation or disassembly of the synaptosomal fusion complex. *Biochem. Biophys. Res. Commun.* **212:**945–952.
69. **Park, M. K., H. H. Jung, and K. H. Yang.** 1990. Binding of *Clostridium botulinum* type B toxin to rat brain synaptosome. *FEMS Microbiol. Lett.* **60:**243–247.
70. **Passador, L., and W. Iglewski.** 1994. ADP-ribosylating toxins. *Methods Enzymol.* **235:**617–631.
71. **Post, L. S., D. A. Lee, M. Solberg, D. Furgang, J. Specchio, and C. Graham.** 1985. Development of botulinal toxin and sensory deterioration during storage of vacuum and modified atmosphere packaged fish fillets. *J. Food Sci.* **50:**990–996.
72. **Poulain, B., S. Mochida, U. Weller, B. Hogy, E. Habermann, J. D. Wadsworth, C. C. Shone, J. O. Dolly, and L. Tauc.** 1991. Heterologous combinations of heavy and light chains from botulinum neurotoxin A and tetanus toxin inhibit neurotransmitter release in *Aplysia. J. Biol. Chem.* **266:**9580–9585.
73. **Poulet, S., D. Hauser, M. Quanz, H. Niemann, and M. R. Popoff.** 1992. Sequences of the botulinal neurotoxin E derived from *Clostridium botulinum* type E (strain Beluga) and *Clostridium butyricum* (strains ATCC 43181 and ATCC 43755). *Biochem. Biophys. Res. Commun.* **183:**107–113.
74. **Ronner, S. G. E., and E. Stackebrandt.** 1994. Further evidence for the genetic heterogeneity of *Clostridium botulinum* as determined by 23S rDNA oligonucleotide probing. *Syst. Appl. Microbiol.* **17:**180–188.
75. **Sakaguchi, G.** 1990. Molecular structure of *Clostridium botulinum* progenitor toxins, p. 173–180. *In* A. E. Pohland, V. R. Dowell, and J. L. Richard (ed.), *Microbial Toxins in Foods and Feeds. Cellular and Molecular Modes of Action.* Plenum Press, New York.
76. **Schengrund, C. L., B. R. DasGupta, and N. J. Ringler.** 1991. Binding of botulinum and tetanus neurotoxins to ganglioside GT1b and derivatives thereof. *J. Neurochem.* **57:**1024–1032.
77. **Schengrund, C. L., N. J. Ringler, and B. R. Dasgupta.** 1992. Adherence of botulinum and tetanus neurotoxins to synaptosomal proteins. *Brain Res. Bull.* **29:**917–924.
78. **Schiavo, G., F. Benfenati, B. Poulain, O. Rossetto, P. Polverino de Laureto, B. R. DasGupta, and C. Montecucco.** 1992. Tetanus and botulinum-B neurotoxins block neurotransmitter release by proteolytic cleavage of synaptobrevin. *Nature* (London) **359:**832–835.
79. **Schiavo, G., C. Malizio, W. S. Trimble, P. Polverino de Laureto, G. Milan, H. Sugiyama, E. A. Johnson, and C. Montecucco.** 1994. Botulinum G neurotoxin cleaves VAMP/synaptobrevin at a single Ala-Ala peptide bond. *J. Biol. Chem.* **269:**20213–20226.
80. **Schiavo, G., O. Rossetto, F. Benfenati, B. Poulain, and C. Montecucco.** 1994. Tetanus and botulinum neurotoxins are zinc proteases specific for components of the neuroexocytosis apparatus. *Ann. N. Y. Acad. Sci.* **710:**65–75.
81. **Schiavo, G., O. Rossetto, S. Catsicas, P. Polverino de Laureto, B. R. DasGupta, F. Benfenati, and C. Montecucco.** 1993. Identification of the nerve terminal targets of botulinum neurotoxin serotypes A, D, and E. *J. Biol. Chem.* **268:**23784–23787.
82. **Schiavo, G., O. Rossetto, and C. Montecucco.** 1994. Clostridial neurotoxins as tools to investigate the molecular events of neurotransmitter release. *Semin. Cell Biol.* **5:**221–229.
83. **Schiavo, G., O. Rossetto, A. Santucci, B. R. DasGupta, and C. Montecucco.** 1992. Botulinum neurotoxins are zinc proteins. *J. Biol. Chem.* **267:**23479–23483.
84. **Schiavo, G., A. Santucci, B. R. Dasgupta, P. P. Mehta, J. Jontes, F. Benfenati, M. C. Wilson, and C. Montecucco.** 1993. Botulinum neurotoxins serotypes A and E cleave SNAP-25 at distinct COOH-terminal peptide bonds. *FEBS Lett.* **335:**99–103.
85. **Schiavo, G., C. C. Shone, M. K. Bennett, R. H. Scheller, and C. M. Montecucco.** 1995. Botulinum neurotoxin type C cleaves a single Lys-Ala bond within the carboxyl-terminal region of syntaxins. *J. Biol. Chem.* **270:**10566–10570.
86. **Schiavo, G., C. C. Shone, O. Rossetto, F. C. Alexander, and C. Montecucco.** 1993. Botulinum neurotoxin serotype F is a zinc endopeptidase specific for VAMP/synaptobrevin. *J. Biol. Chem.* **268:**11516–11519.
87. **Schmid, M. F., J. P. Robinson, and B. R. DasGupta.** 1993. Direct visualization of botulinum neurotoxin-induced channels in phospholipid vesicles. *Nature* (London) **364:**827–830.
88. **Simpson, L. L.** 1986. Molecular pharmacology of botulinum toxin and tetanus toxin. *Annu. Rev. Pharmacol. Toxicol.* **26:**427–453.
89. **Simpson, L. L., and M. M. Rapport.** 1971. Ganglioside inactivation of botulinum toxin. *J. Neurochem.* **18:**1341–1343.
90. **Singh, B. R., and B. R. DasGupta.** 1989. Molecular topography and secondary structure comparisons of botulinum neurotoxin types A, B. and E. *Mol. Cell Biochem.* **86:**87–95.
91. **Stecher, B., U. Weller, E. Habermann, M. Gratzl, and G. Ahnert-Hilger.** 1989. The light chain but not the heavy chain of botulinum A toxin inhibits exocytosis from permeabilized adrenal chromaffin cells. *FEBS Lett.* **255:**391–394.
92. **Stevens, R. C., M. L. Evenson, W. Tepp, and B. R. DasGupta.** 1991. Crystallization and preliminary X-ray analysis of botulinum neurotoxin type A. *J. Mol. Biol.* **222:**877–880.
93. **St. Louis, M. E., S. H. S. Peck, G. B. Morgan, J. Blatherwick, S. Banerjee, G. D. M. Kettyls, W. A. Black, M. E. Milling, A. H. W. Hauschild, R. V. Tauxe, and P. A. Blake.** 1988. Botulism from chopped garlic: delayed recognition of a major outbreak. *Ann. Intern. Med.* **108:**363–368.
94. **Suen, J. C., C. L. Hatheway, A. G. Streigerwalt, and D. J. Brenner.** 1988. *Clostridium argentinense* sp. nov.: a genetically homogeneous group composed of all strains of *Clostridium botulinum* toxin type G and some nontoxige-

nic strains previously identified as *Clostridium subterminale* or *Clostridium hastiforme*. *Int. J. Syst. Bacteriol.* **38**:375–385.

95. **Tanaka, N., L. Meske, M. P. Doyle, E. Traisman, D. W. Thayer, and R. W. Johnston.** 1985. Plant trials of bacon made with lactic acid bacteria, sucrose and lowered sodium nitrite. *J. Food Prot.* **48**:679–686.
96. **Telzak, E. E., E. P. Bell, D. A. Kautter, L. Crowell, L. D. Budnick, D. L. Morse, and S. Schultz.** 1990. An international outbreak of type E botulism due to uneviscerated fish. *J. Infect. Dis.* **161**:340–342.
97. **Thompson, D. E., J. K. Brehm, J. D. Oultram, T. J. Swinfield, C. C. Shone, T. Atkinson, J. Melling, and N. P. Minton.** 1990. The complete amino acid sequence of the *Clostridium botulinum* type A neurotoxin, deduced by nucleotide sequence analysis of the encoding gene. *Eur. J. Biochem.* **189**:73–81.
98. **Thompson, D. E., R. A. Hutson, A. K. East, D. Allaway, M. D. Collins, and P. T. Richardson.** 1993. Nucleotide sequence of the gene coding for *Clostridium barati* type F neurotoxin: comparison with other clostridial neurotoxins. *FEMS Microbiol. Lett.* **108**:175–182.
99. **Tsuzuki, K., K. Kimura, N. Fujii, N. Yokosawa, T. Indoh, T. Murakami, and K. Oguma.** 1990. Cloning and complete nucleotide sequence of the gene for the main component of hemagglutinin produced by *Clostridium botulinum* type C. *Infect. Immun.* **58**:3173–3177.
100. **Tsuzuki, K., K. Kimura, N. Fujii, N. Yokosawa, and K. Oguma.** 1992. The complete nucleotide sequence of the gene coding for the nontoxic-nonhemagglutinin component of *Clostridium botulinum* type C progenitor toxin. *Biochem. Biophys. Res. Commun.* **183**:1273–1279.
101. **Vandekerckhove, J., B. Schering, M. Barmann, and K. Aktories.** 1988. Botulinum C2 toxin ADP-ribosylates cytoplasmic beta/gamma-actin in arginine 177. *J. Biol. Chem.* **263**:696–700.
102. **Wainwright, R. B.** 1992. Hazards from northern native foods, p. 305–322. *In* A. H. W. Hauschild and K. L. Dodds (ed.), *Clostridium botulinum. Ecology and Control in Foods.* Marcel Dekker, Inc., New York.
103. **Weber, J. T., R. G. Hibbs, A. Darwish, B. Mishu, et al.** 1993. A massive outbreak of type E botulism associated with traditional salted fish in Cairo. *J. Infect. Dis.* **167**: 451–454.
104. **Whelan, S. M., M. J. Elmore, N. J. Bodsworth, T. Atkinson, and N. P. Minton.** 1992. The complete amino acid sequence of the *Clostridium botulinum* type-E neurotoxin, derived by nucleotide-sequence analysis of the encoding gene. *Eur. J. Biochem.* **204**:657–667.
105. **Whelan, S. M., M. J. Elmore, N. J. Bodsworth, J. K. Brehm, T. Atkinson, and N. P. Minton.** 1992. Molecular cloning of the *Clostridium botulinum* structural gene encoding the type B neurotoxin and determination of its entire nucleotide sequence. *Appl. Environ. Microbiol.* **58**:2345–2354.
106. **Williams, R. S., C.-K. Tse, J. O. Dolly, P. Hambleton, and J. Melling.** 1983. Radioiodination of botulinum neurotoxin type A with retention of biological activity and its binding to brain synaptosomes. *Eur. J. Biochem.* **131**:437–445.
107. **Yokosawa, N., K. Tsuzuki, B. Syuto, N. Fujii, K. Kimura, and K. Oguma.** 1991. Binding of botulinum type Cl, D and E neurotoxins to neuronal cell lines and synaptosomes. *Toxicon* **29**:261–264.
108. **Zhou, Y. T., H. Sugiyama, H. Nakano, and E. A. Johnson.** 1995. The genes for the *Clostridium botulinum* type G toxin complex are on a plasmid. *Infect. Immun.* **63**:2087–2091.

Bruce A. McClane

16 Clostridium perfringens

Clostridium perfringens was first recognized as an important cause of foodborne disease in the 1940s and 1950s, following the pioneering work of Knox, McDonald, McClung, and Hobbs (see references 31 and 43 for reviews; because of page limitations, readers will be referred to reviews for well-established points). It gradually became appreciated that *C. perfringens* actually causes two quite different human diseases that can be transmitted by food, i.e., *C. perfringens* type A food poisoning and necrotic enteritis (also known as Darmbrand and Pig-Bel). Given the relative rarity of foodborne necrotic enteritis in industrialized societies, this chapter will focus on *C. perfringens* type A food poisoning; readers interested in necrotic enteritis are directed to a review on this topic (63).

CHARACTERISTICS OF THE ORGANISM

General

C. perfringens is a gram-positive, rod-shaped, encapsulated, nonmotile bacterium of variable size that is capable of causing a broad spectrum of human and veterinary diseases (31, 43). The bacterium's pathogenicity is largely derived from its prolific ability to express protein toxins, including at least two toxins, *C. perfringens* enterotoxin (CPE) and β-toxin, that are active on the human gastrointestinal tract.

Besides this ability to produce gastrointestinal tract-active toxins, *C. perfringens* possesses several other characteristics that significantly contribute to its ability to cause foodborne diseases. First, vegetative cells of *C. perfringens* can double in as little as 10 min (31), allowing the organism to multiply very rapidly in food. Second, under certain still incompletely understood conditions, *C. perfringens* will form spores (Fig. 16.1) that are highly resistant to environmental stresses such as radiation, desiccation, and heat (31). The heat resistance of its spores often allows *C. perfringens* to survive incomplete cooking of food, with the surviving bacteria then able to cause food poisoning.

C. perfringens is considered to be anaerobic since it does not produce colonies on agar plates continuously exposed to air (31). Being anaerobic does not represent a significant impediment to *C. perfringens's* ability to cause foodborne disease since the bacterium tolerates some exposure to air and, compared with many other anaerobes, requires only relatively modest reductions in oxidation-reduction potential (E_h) for growth (31). The implications of this E_h tolerance for *C. perfringens* type A food poisoning are discussed in later sections of this chapter.

Toxin Typing

While at least 13 different toxins are known to be expressed by *C. perfringens*, an individual *C. perfringens* cell will produce only a defined subset of these toxins (43). This observation forms the basis for a toxin typing system that is used to classify *C. perfringens* isolates into five

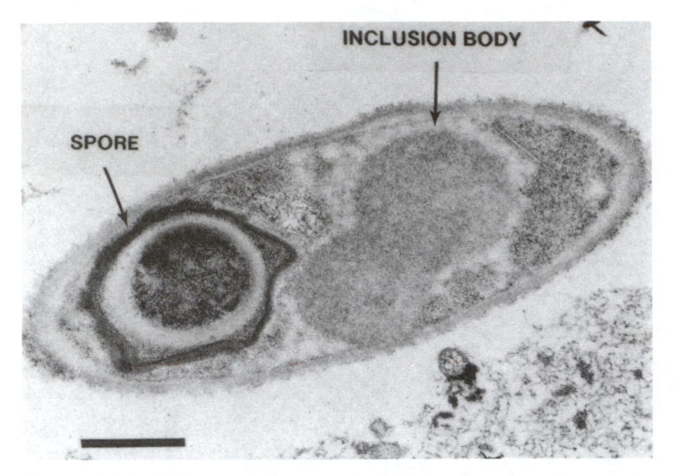

Figure 16.1 Electron micrograph of thin sections of a sporulating cell of *C. perfringens.* Magnification, ×40,000; bar, 0.5 μm. Arrows indicate the endospore and an inclusion body in the cytoplasm of the mother cell. (Reproduced with the author's and publisher's permission from reference 15.)

types (A through E), according to each isolate's ability to express 4 (alpha, beta, epsilon, and iota) of the 13 *C. perfringens* toxins (Table 16.1). Traditionally, toxin typing of *C. perfringens* has involved laborious toxin antiserum neutralization tests in mice (31, 43) and thus could be done in only a few laboratories. However, the recent development of PCR-based schemes (12) for toxin typing of *C. perfringens* isolates should now simplify this process considerably.

Each of the foodborne diseases caused by *C. perfringens* is associated with a distinct type of *C. perfringens.* Necrotic enteritis is caused by type C isolates, with the β-toxin produced by type C isolates considered to be the primary virulence factor involved in this disease (43, 63). As implied by its name, *C. perfringens* type A food poisoning is almost always associated with type A isolates of *C. perfringens* (31). There are also occasionally "*C. perfringens* type A food poisoning" cases caused by α-toxin-negative *C. perfringens* isolates (5), which perhaps may be considered "atypical" *C. perfringens* type A isolates that have lost their ability to express α-toxin.

Table 16.1 Toxin typing of *C. perfringens*

C. perfringens type	Toxins produced			
	Alpha	Beta	Epsilon	Iota
A	+	−	−	−
B	+	+	+	−
C	+	+	−	−
D	+	−	+	−
E	+	−	−	+

The explanation for the strong association between type A isolates and *C. perfringens* type A food poisoning is not particularly clear since at least some type C and type D isolates can also express an enterotoxin that is physically, serologically, and biologically similar, if not identical, to CPE (31, 43), the factor believed to be responsible for the characteristic symptoms of *C. perfringens* type A food poisoning. It is possible that *C. perfringens* type A food poisoning cases usually involve type A isolates because there simply are more of the enterotoxigenic type A isolates than the other types of enterotoxigenic *C. perfringens* isolates present in environments conducive to food poisoning, but this question requires further epidemiologic study for resolution (see Reservoirs for *C. perfringens* Type A Food Poisoning, below).

Tolerance or Susceptibility to Preservation Methods

Since it has become well established that the growth of pathogens in food is affected by sensitivity to such factors as temperature, E_h, pH, and water activity (a_w) levels, the effects of these factors on the growth of *C. perfringens* will be discussed briefly.

Temperature

As mentioned above, the heat resistance of *C. perfringens* spores can clearly contribute to the organism's ability to cause food poisoning by allowing it to survive incomplete cooking of foods. The heat resistance properties of *C. perfringens* spores depend on both environmental and genetic factors. With respect to environmental factors, the medium in which a *C. perfringens* spore is heated clearly influences its heat resistance (31). In a relatively protective medium such as cooked meat medium, many *C. perfringens* spores will survive boiling for an hour or longer (31). The involvement of genetic factors in spore heat resistance has been established by observations indicating that spores made by different *C. perfringens* strains vary considerably in their heat resistance properties (interested readers can consult reference 31 for a listing of decimal reduction values comparing the heat resistances of spores made by different *C. perfringens* strains). Perhaps because of selective pressure, spores made by food poisoning isolates are generally more heat resistant than spores made by *C. perfringens* isolates from other sources (31). It is important to appreciate that incomplete cooking of foods not only may fail to rid foods of *C. perfringens* spores but can actually facilitate the development of *C. perfringens* type A food poisoning, since heating (e.g., to 70 to 80°C for 20 min) is an excellent way to induce the germination of *C. perfringens* spores (31).

It should also be noted that even the vegetative cells of *C. perfringens* are somewhat heat tolerant. Although not truly thermophilic, *C. perfringens* vegetative cells have a relatively high optimal growth temperature (43 to 45°C) and will continue to grow at temperatures up to at least 50°C (31). However, vegetative *C. perfringens* cells are not particularly tolerant of either refrigeration or freezing (31). *C. perfringens* growth rapidly decreases at temperatures below ~15°C and usually stops by 6°C (31). In contrast to its vegetative cells, *C. perfringens* spores are considerably more cold resistant (31). Food poisoning may result if viable spores in refrigerated or frozen foods are induced to germinate as this food is warmed for serving (see below).

Other Factors

The growth of *C. perfringens* cells in food is affected by factors other than temperature, including a_w levels, E_h, pH, and possibly the presence of curing agents (25, 31). *C. perfringens* is less tolerant of low-a_w environments than another common gram-positive foodborne pathogen, *Staphylococcus aureus* (31). The lowest a_w supporting vegetative growth of *C. perfringens* is reported to be 0.93 to 0.97, depending on the solute used to control the a_w of the medium (25, 31). As mentioned above, relative to many other anaerobes, *C. perfringens* does not require an extremely reduced environment for its growth. Provided that the environmental E_h is low enough to initiate growth (the exact E_h value needed to initiate *C. perfringens* growth depends on several environmental factors, e.g., pH [31]), *C. perfringens* will then modify the E_h of its surrounding environment (by producing reducing molecules such as ferredoxin) to produce more optimal growth conditions (31). As a practical guide for food microbiologists, the E_h of many common foods (e.g., raw meats and gravies) is often low enough to permit the growth of *C. perfringens* (31). Growth of *C. perfringens* is also sensitive to pH extremes, with optimal growth occurring at pH 6 to 7 (i.e., at pHs that are commonly found in the meat and poultry products that usually serve as food vehicles for *C. perfringens* type A food poisoning [see below]) and severe inhibition of growth occurring at pHs of ≤5 and ≥8.3 (25, 31).

The effectiveness of curing agents on limiting *C. perfringens* growth in foods is a somewhat unsettled, if not controversial, subject that (with one exception mentioned below) has not received much recent research attention (25, 31). Older laboratory studies (25) indicate that the concentrations of curing salts needed to significantly inhibit survival of *C. perfringens* cells may exceed commercially acceptable levels; i.e., inhibition of *C. perfringens* growth may require at least 6 to 8% NaCl, 10,000 ppm of $NaNO_3$, or 400 ppm of $NaNO_2$. However, the following observations have led to some belief (31) that curing salts can be at least partially effective in preventing *C. perfringens* growth in food, even when used at commercially acceptable levels: (i) the simultaneous coapplication of other preservation factors such as heating and nonneutral pHs increases the sensitivity of *C. perfringens* to curing salts (31), (ii) the simultaneous use of several curing agents often produces a synergistic inhibition of *C. perfringens* growth (31), and (iii) foods may contain initial burdens of *C. perfringens* cells and spores lower than those used in laboratory studies evaluating the effectiveness of curing agents for inhibiting *C. perfringens* growth (31). A recent study (28) now offers some additional support for proponents of the view that the presence of curing salts helps to inhibit *C. perfringens* growth in foods, even when used at commercially acceptable levels, by demonstrating that the presence of 3% NaCl delays the growth of *C. perfringens* in vacuum-packed beef. As a final comment, the best argument that curing agents are at least partially effective in reducing *C. perfringens* levels in treated foods perhaps may be the relatively uncommon involvement of commercially cured meat products in *C. perfringens* type A food poisoning outbreaks (31).

One of the more important roles that food preservation factors such as pH, a_w, and perhaps curing agents play in controlling *C. perfringens* levels in foods is to inhibit the outgrowth of germinating *C. perfringens* spores in foods (31). However, ungerminated spores in foods may remain viable despite the application of these preservation factors and could later undergo germination and outgrowth should these growth-limiting factors be removed during food preparation.

RESERVOIRS FOR *C. PERFRINGENS* TYPE A FOOD POISONING

C. perfringens is ubiquitous throughout the natural environment (25, 31). For example, this organism is commonly encountered in soils (at levels of 10^3 to 10^4 CFU/g), foods (e.g., approximately 50% of raw or frozen meat contain some *C. perfringens*), dust, and the intestinal tracts of humans and domestic animals (e.g., human feces typically contain 10^3 to 10^6 *C. perfringens* cells per g). This widespread distribution of *C. perfringens* has long been considered the primary explanation for the prevalence of *C. perfringens* type A food poisoning.

However, is understanding the reservoir(s) for this illness really so straightforward? Two independent surveys (29, 62) of large numbers of global *C. perfringens* isolates from various animal and other environmental sources have recently suggested that <5% of all *C. perfringens* isolates actually carry the *cpe* gene, which is considered essential for producing *C. perfringens* type A food poisoning symptoms (see below). Since it now appears that only a tiny minority of all *C. perfringens* isolates can cause *C. perfringens* type A food poisoning, the pertinent question for understanding *C. perfringens* type A food poisoning reservoirs becomes, not where do *any C. perfringens* reside in the environment, but instead, where are those *enterotoxigenic C. perfringens* isolates capable of causing food poisoning found?

To help address this question, Saito recently used highly sensitive and specific modern CPE serologic assays to survey human and animal feces, as well as food, for the presence of CPE-positive *C. perfringens* isolates (52). This survey found that while *C. perfringens* cells are commonly present in poultry, pork, and shellfish, only some of the shellfish isolates were actually enterotoxigenic, i.e., CPE positive. While meat or poultry should not be excluded as potential reservoirs for CPE-positive *C. perfringens* isolates on the basis of this single limited survey, Saito's results do indicate that shellfish (and possibly other marine life) may be reservoirs for enterotoxigenic *C. perfringens.* More importantly, the results clearly illustrate how surveys will reach different conclusions about *C. perfringens* type A food poisoning reservoirs depending on whether the criterion is the detection of any *C. perfringens* isolates or only the detection of CPE-positive *C. perfringens* isolates.

Saito's study (52) also found that ~6% of feces from presumably healthy professional food handlers contain some CPE-positive *C. perfringens* isolates, supporting previous surveys (25) that had suggested that healthy humans may serve as reservoirs for enterotoxigenic *C. perfringens* isolates. Can animals, particularly food animals, also serve as reservoirs for CPE-producing *C. perfringens* capable of causing human food poisoning? The answer to this question appears to be increasingly complicated. Although no enterotoxigenic isolates were detected in chickens, swine or cattle in Saito's survey (52), *cpe*-positive *C. perfringens* isolates were recovered from a variety of animals (including some food animals) in another recent survey (29). However, even with these results, it remains premature to conclude that animals represent a significant reservoir for the CPE-positive isolates that actually cause human food poisoning, since new findings (9) now suggest that the *cpe* gene is plasmid borne in veterinary isolates but is chromosomal in human food poisoning isolates. These observations open the possibility that there are at least two distinct populations of *cpe*-positive *C. perfringens* isolates, with only one of these populations (i.e., those isolates containing a chromosomal *cpe* gene) being capable of causing *C. perfringens* type A food poisoning. These recent findings also raise several other questions about potential reservoirs for *C. perfringens* food poisoning isolates that should now be addressed; e.g., do CPE-positive isolates from healthy human carriers have a chromosomal or plasmid *cpe* gene?

From the foregoing discussion, it should be obvious that we now have only a limited understanding about the environmental reservoirs for those *C. perfringens* isolates capable of causing human food poisoning. If these reservoirs can be identified through further molecular epidemiology studies, this information should help us to control *C. perfringens* type A food poisoning by providing insights into how and when food becomes contaminated with food poisoning isolates, knowledge that could open the possibility of intervention to decrease food contamination by these isolates.

C. PERFRINGENS TYPE A FOOD POISONING OUTBREAKS

Incidence

C. perfringens type A food poisoning annually ranks among the most common foodborne diseases in the United States and Europe. The Centers for Disease Control and Prevention (CDC) reported (2) that from 1973 to 1987 there were 190 outbreaks (10.2% of total bacterial foodborne disease outbreaks) of *C. perfringens* type A food poisoning in the United States, involving 12,234 cases (11.2% of total cases of bacterial foodborne diseases) and 12 deaths (~5% of total deaths from bacterial foodborne diseases). However, since most cases of this disease go unreported (see below), these official statistics significantly understate the true prevalence and impact of *C. perfringens* type A food poisoning. Todd has estimated (60, 61) there are actually 652,000 cases of *C. perfringens* type A food poisoning in the United States each year, with an average of 7.6 deaths per year and annual costs of $123 million.

Identified cases of *C. perfringens* type A food poisoning usually involve large outbreaks (the median outbreak size is 25 cases [2]), often in institutionalized settings. This epidemiologic pattern results from at least two factors. First, large institutions often prepare food in advance and then hold this food for later serving (allowing the possible growth of *C. perfringens* in any temperature-abused foods). Second, given the relatively mild and nondistinguishing symptoms (discussed below) of most cases of *C. perfringens* type A food poisoning, it is only when a significant number of people become simultaneously sickened with diarrheal symptoms that public health officials are sufficiently motivated to investigate, identify, and report this illness. *C. perfringens* type A food poisoning can occur at any time of year but is slightly more common during summer months (56), perhaps because higher ambient temperatures facilitate temperature abuse of foods during cooling and holding.

Food Vehicles

In the 1973–1987 CDC statistics (2), meat and poultry continued their traditional roles as the most common food vehicles for *C. perfringens* type A food poisoning in the United States. Beef accounted for nearly 30% of all *C. perfringens* type A food poisoning outbreaks during this period, with turkey and chicken together accounting for another ~15%. The CDC statistics also indicate that Mexican foods containing meats are emerging as another important vehicle for *C. perfringens* type A food poisoning.

Contributing Factors

C. perfringens type A food poisoning almost always results from temperature abuse during the cooking, cooling, or holding of foods. The CDC reports (2) that improper storage or holding temperatures contributed to 97% of all recent *C. perfringens* type A food poisoning outbreaks, while improper cooking was a factor in 65% of these outbreaks. Other major contributing factors for *C. perfringens* type A food poisoning include contaminated equipment and poor personal hygiene, which were involved in 28 and 26%, respectively, of recent outbreaks of this food poisoning. As introduced earlier, the importance of temperature abuse in *C. perfringens* type A food poisoning is not surprising given the relative heat tolerance of *C. perfringens* vegetative cells and, more particularly, the high heat tolerance of *C. perfringens* spores. As also mentioned previously, incomplete cooking actually promotes this illness by increasing the germination rates of *C. perfringens* spores present in foods (31); after outgrowth of these spores into new vegetative cells, *C. perfringens* can multiply rapidly in temperature-abused foods that are cooled or stored improperly.

Prevention and Control

From the previous discussion, it should be obvious that the best way to prevent and control *C. perfringens* type A food poisoning is, first, to cook foods thoroughly. This is particularly important for large roasts and turkeys, since because of their size, it is difficult to generate internal temperatures high enough to kill *C. perfringens* spores. This difficulty is undoubtedly a major reason why large roasts and poultry products are such common food vehicles for *C. perfringens* type A food poisoning outbreaks. Another step for controlling *C. perfringens* type A food poisoning is to quickly cool cooked food and then store or serve the food at conditions nonpermissive for vegetative growth of *C. perfringens* (e.g., either at refrigeration temperatures or temperatures above 70°C).

Examples of Recent Outbreaks

Since consideration of recent outbreaks often illustrates important concepts underlying bacterial food poisonings, a few recent reports describing *C. perfringens* type A food poisoning outbreaks in the United States and the United Kingdom will be discussed briefly.

In 1994, the CDC published a report on an investigation of two outbreaks of *C. perfringens* type A food poisoning that were associated with St. Patrick's Day meals (8). The first of these outbreaks occurred in Cleveland, Ohio, and involved 156 persons, all of whom acquired *C. perfringens* type A food poisoning from ingesting corned beef that had been prepared at a local delicatessen. During its preparation, the corned beef had been boiled for 3 h and then allowed to cool slowly (at room temperature) before refrigeration. Four days later, portions of this corned beef were warmed to 48.8°C and served; some sandwiches prepared with this corned beef were held at room temperature from the time of their preparation (at 11:00 a.m.) until they were eaten throughout the afternoon.

The second St. Patrick's Day outbreak described in the CDC report occurred in Virginia and involved 86 persons who attended a traditional St. Patrick's Day dinner. This dinner also included corned beef, which was later found to contain large numbers of *C. perfringens* cells. The corned beef involved in the Virginia outbreak was a frozen, commercially prepared, brined product that had been thawed, cooked in large (10-lb [ca. 4.5-kg]) pieces, stored in a refrigerator, and held for 90 min under a heat lamp before being served.

These two outbreaks illustrate the typical association between *C. perfringens* type A food poisoning and meat (particularly beef) vehicles and the importance of temperature abuse as a contributing factor to this foodborne disease. Specifically, in the Ohio delicatessen outbreak, the corned beef was clearly cooled too slowly after cooking and then was not reheated at sufficiently high temperatures before being served. In the Virginia outbreak, the beef was first cooked in excessively large portions and then not adequately reheated before being served. Interestingly, the Virginia outbreak is somewhat unusual in that it involved a commercially prepared, brined meat product.

Another recent report described a *C. perfringens* type A food poisoning outbreak that occurred in a British hospital and sickened 17 patients (50). Epidemiologic investigation of this outbreak found large numbers of *C. perfringens* cells present both in the stools from ill patients and in vacuum-packed pork served to the patients, implicating this pork as the presumed food vehicle for the outbreak. The pork had apparently become contaminated with large numbers of *C. perfringens* as a result of very slow cooling after cooking at a commercial meat preparation facility. This British outbreak appears typical for a *C. perfringens* type A food poisoning outbreak with respect to the implicated food vehicle (meat), epidemiology (like the Virginia St. Patrick's Day outbreak mentioned above, the British hospital outbreak involved a large number of cases in an institutional setting), and cause (temperature abuse). However, this British outbreak, like the Virginia outbreak, is somewhat unusual in that it involved a commercially prepared food product. Perhaps the involvement of vacuum-packed meat in the British outbreak presages future problems with increasingly popular, vacuum-packed, precooked meat products serving as common vehicles for *C. perfringens* type A food poisoning. Vacuum-packed foods would appear to provide an ideal anaerobic environment for growth of clostridia, including *C. perfringens.*

Identification of Outbreaks

Public health agencies usually consider several criteria, including incubation time and symptoms (described in the next section), type and history of food vehicles (e.g., is temperature-abused meat or poultry involved?), and laboratory results for identifying *C. perfringens* type A food poisoning outbreaks. Because of their potential specificity, laboratory results are often critical for definitively identifying such outbreaks. However, merely isolating *C. perfringens* from suspect food or from feces from ill individuals does not suffice to establish this particular illness, since this bacterium is ubiquitous throughout the environment (including having a normal presence in food and feces). Traditionally, more vigorous bacteriologic criteria have been applied by public health agencies to establish an outbreak of *C. perfringens* type A food poisoning, including (36) (i) finding $>10^5$ *C. perfringens* cells per g of contaminated food, (ii) demonstrating $>10^6$ *C. perfringens* spores per g of feces from an ill individual, (iii) finding the same capsular serotype of *C. perfringens* in all ill individuals in an outbreak, or (iv) demonstrating that the same serotype of *C. perfringens* is present in both contaminated food and feces associated with a single outbreak. (*Note:* "Serotyping" refers to an identification system [56] based on *C. perfringens* capsular polysaccharides and should not be confused with the toxin typing system used for *C. perfringens* classification described earlier in this chapter.)

While these traditional bacteriologic criteria have been used successfully for identifying some *C. perfringens* type A food poisoning outbreaks, their limitations are becoming increasingly recognized. To cite only two of the problems now associated with these criteria (limitations of these criteria are discussed in more detail elsewhere [36]): (i) feces from many elderly people normally contain high numbers of *C. perfringens* spores, and (ii) in both the United States and Japan, most *C. perfringens* isolates do not react with available serotyping reagents (36, 52).

Given the increasingly recognized limitations of traditional bacteriologic methods and the emerging appreciation that only a select subset (at most, the <5% of *C. perfringens* isolates that apparently carry the *cpe* gene) of all *C. perfringens* isolates are actually capable of causing *C. perfringens* type A food poisoning, there is currently considerable interest in updating the diagnostic criteria for *C. perfringens* type A food poisoning outbreaks to include demonstrating the presence of CPE in feces of food poisoning victims or demonstrating the presence of specifically enterotoxigenic *C. perfringens* isolates in food or feces associated with a food poisoning outbreak.

Several serologic assays (e.g., enzyme-linked immunosorbent assays [ELISAs] and radioimmunoassays) have been developed for detection of fecal CPE. These assays are described in detail elsewhere (31, 36, 56), but it is worth noting that at least two assays, including a reverse-passive latex agglutination assay (Oxoid) and a rapid ELISA (Tech Lab), are now available commercially for fecal CPE detection. Demonstration of the presence of CPE in feces of ill individuals appears to be a very good indicator of *C. perfringens* type A food poisoning (3); further, healthy individuals rarely, if ever, contain detectable levels of CPE in their feces (1, 3). Unfortunately, a drawback to this diagnostic approach is that fecal samples should be collected soon after the onset of food poisoning symptoms to ensure meaningful results (1).

A potential alternative or supplemental approach to fecal CPE detection for diagnosing *C. perfringens* type A food poisoning outbreaks would be to look for enterotoxigenic *C. perfringens* isolates in outbreak-associated food or feces. Either CPE serologic detection or *cpe* gene detection assays could be useful for evaluating isolate enterotoxigenicity. Gene detection assays have an advantage over serologic assays in that they do not require isolates to sporulate in vitro. This advantage becomes relevant when one considers that sporulation is necessary for CPE expression (see below) but is often difficult to achieve for *C. perfringens* isolates grown in laboratory media (29, 31). However, gene detection assays also have a disadvantage compared with serologic assays since they can only identify isolates as potentially enterotoxigenic; i.e., perhaps some *C. perfringens* isolates carry silent, unexpressed *cpe* genes.

At this point, it seems premature to depend on identification of food or fecal enterotoxigenic (or potentially enterotoxigenic) *C. perfringens* isolates for definitive identification of *C. perfringens* type A food poisoning outbreaks. For this approach to be more useful, at least two issues should be resolved. First, since there is increasing evidence that feces from some healthy individuals may contain some CPE-positive *C. perfringens* isolates, it seems important to establish the levels of specifically enterotoxigenic *C. perfringens* isolates that may be present in feces from individuals ill with *C. perfringens* type A versus feces from healthy individuals or in normal versus contaminated foods. These threshold values could then be applied to developing quantitative assays capable of demonstrating the presence of sufficient numbers of enterotoxigenic *C. perfringens* isolates in food or feces to be consistent with a diagnosis of *C. perfringens* type A food poisoning. Second, it needs to be definitively determined whether only those *C. perfringens* isolates carrying chromosomal *cpe* genes are capable of causing food poisoning. If this relationship holds up to further analysis, it will then become necessary to develop genetic assays capable of distinguishing between *cpe*-positive isolates carrying chromosomal *cpe* genes and those carrying plasmid-borne *cpe* genes.

CHARACTERISTICS OF *C. PERFRINGENS* TYPE A FOOD POISONING

Symptoms of *C. perfringens* type A food poisoning develop 8 to 24 h after ingestion of contaminated food (31, 43) and usually resolve spontaneously within 12 to 24 h. Victims of *C. perfringens* type A food poisoning typically suffer only from diarrhea and severe abdominal cramps; vomiting and fever are not commonly associated with this food poisoning. While death rates from *C. perfringens* type A food poisoning are low, death is more prevalent in debilitated or elderly individuals afflicted with this illness.

The typical pathogenesis of *C. perfringens* type A food poisoning is illustrated in Fig. 16.2. Initially, as a result of temperature abuse, vegetative cells of enterotoxigenic *C. perfringens* multiply rapidly in the food and are consumed when the food vehicle is ingested. Many of the ingested *C. perfringens* vegetative cells probably die when exposed to stomach acidity (25), but if the food vehicle is sufficiently contaminated, some vegetative cells survive passage through the stomach and enter the small intestine, where they multiply and sporulate. It is during this sporulation in the small intestine that CPE is expressed. Once released into the intestinal lumen, CPE quickly binds to intestinal epithelial cells, where it exerts its unique action and produces morphologic damage to intestinal epithelial cells (Fig. 16.3). There is mounting evidence (53) that it is this CPE-induced intestinal tissue damage that causes the intestinal fluid loss (clinically manifested as diarrhea) typically noted in victims of *C. perfringens* type A food poisoning.

The relatively mild, self-limiting nature of most cases of *C. perfringens* type A food poisoning probably stems from two major factors (53): (i) the diarrhea associated with *C. perfringens* type A food poisoning helps to mitigate the severity of this illness by flushing unbound CPE and many *C. perfringens* cells (containing additional unreleased CPE) from the small intestine, and (ii) CPE preferentially affects villus tip cells, which are the oldest intestinal cells and can be rapidly replaced in young, healthy individuals by the normal turnover of intestinal cells.

INFECTIVE DOSE AND SUSCEPTIBLE POPULATIONS

Because many *C. perfringens* cells are killed by exposure to stomach acidity after ingestion (25), cases of *C. perfringens* type A food poisoning usually develop only when heavily contaminated food (i.e., food containing $>10^6$ to 10^7 *C. perfringens* vegetative cells per g of food) is consumed (36). As mentioned above, the enterotoxin responsible for the disease symptoms of *C. perfringens* type A food poisoning is produced in vivo during sporulation of *C. perfringens* in the intestine; i.e., this illness is typically an infection, not an intoxication. While there are a few literature reports (31) of early symptom onsets

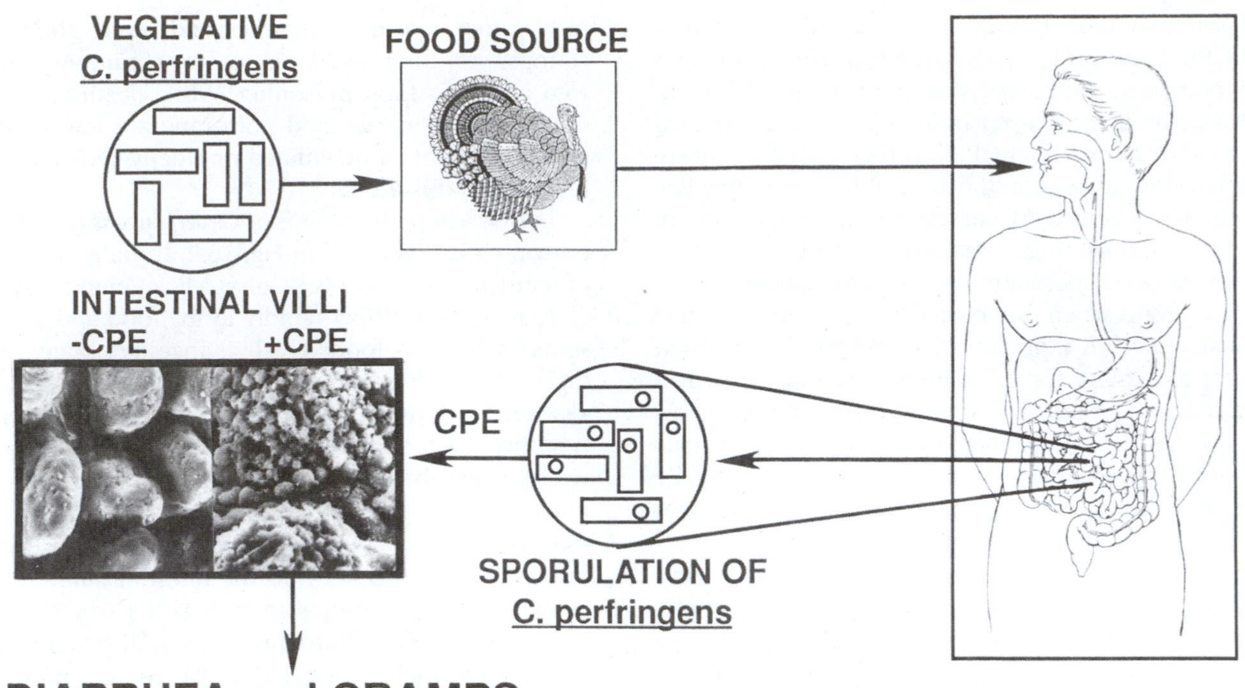

Figure 16.2 Pathogenesis of *C. perfringens* type A food poisoning. Vegetative *C. perfringens* cells multiply rapidly in contaminated food (usually a meat or poultry product) and, after ingestion, sporulate in the small intestine. Sporulating *C. perfringens* cells produce an enterotoxin (CPE) that causes morphologic damage to the small intestine, resulting in diarrhea and abdominal cramps. (Reproduced with the publisher's permission from reference 36.)

that appear consistent with the possibility that preformed CPE in foods may occasionally contribute to this food poisoning, the long incubation period typical of this food poisoning (despite the quick action of CPE) indicates that the involvement, if any, of preformed CPE in *C. perfringens* type A food poisoning symptoms must be very rare.

While everyone appears to be susceptible to *C. perfringens* type A food poisoning, the illness tends to be more serious in elderly or debilitated individuals (43). When individuals become sickened by this food poisoning, they often develop at least a transient serum antibody response to CPE (3). However, there is no evidence that previous exposure to *C. perfringens* type A food poisoning results in future protection against this illness (25).

CPE

Evidence for the Involvement of CPE in *C. perfringens* Type A Food Poisoning

Molecular Koch's postulates have not yet been applied to formally prove CPE involvement in *C. perfringens* type A food poisoning. However, a large number of epidemiologic studies have provided compelling evidence that CPE plays the major role in this disease: (i) a strong positive correlation exists between illness and the presence of CPE in a victim's feces (depending on the sensitivity of the assay used and how quickly the fecal sample was collected after the onset of symptoms, 80 to 100% of feces from individuals ill with *C. perfringens* type A food poisoning will test CPE positive, while virtually no feces from well individuals test CPE positive [1, 3]), (ii) CPE is often present in the feces of food poisoning victims at levels (1, 3) known to cause serious intestinal effects in experimental animals (45), (iii) human volunteers fed highly purified CPE will develop all of the symptoms characteristic of *C. perfringens* type A food poisoning (54), (iv) *C. perfringens* food poisoning isolates later shown to be CPE positive (29) were dramatically more effective than CPE-negative *C. perfringens* isolates at producing either fluid accumulation in rabbit ileal loops or diarrhea in human volunteers (57), and (v) rabbit ileal loop effects produced by CPE-positive isolates can be neutralized with CPE-specific antisera (23).

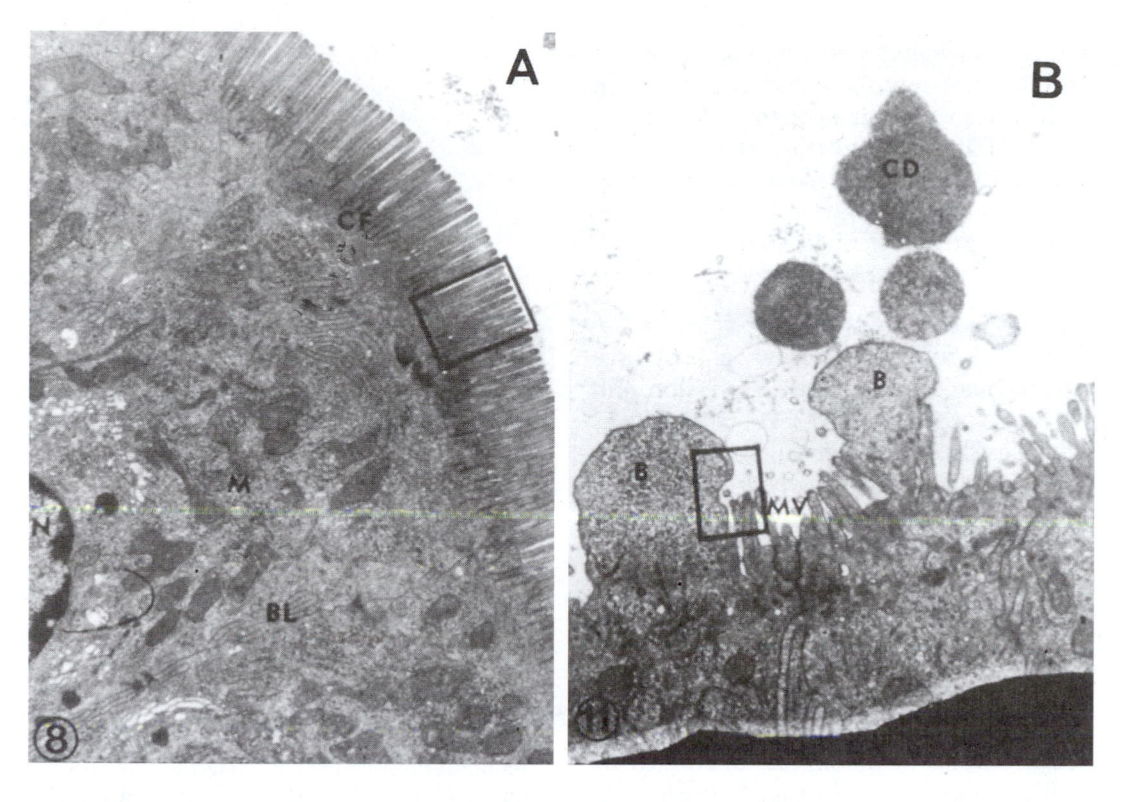

Figure 16.3 Morphologic damage to rabbit small intestinal cells in response to treatment with CPE. The control specimen (left) shows an intestinal epithelial cell with normal BBMs (box). CPE-treated intestinal cells (right) show visible damage, including bleb formation (B), to their BBMs. This BBM damage precedes the development of damage to internal organelles in CPE-treated cells (44). (Reproduced with the author's and publisher's permission from reference 44.)

Is *C. perfringens* type A food poisoning the only disease that CPE contributes to? There is increasing evidence that CPE is also involved in several nonfoodborne human gastrointestinal illnesses (e.g., antibiotic-associated diarrhea) and veterinary diarrheas (37). Further, CPE is also now being associated with some cases of sudden infant death syndrome. It has been suggested that CPE-associated sudden infant death syndrome may involve CPE absorbed from an infant's gastrointestinal tract into the circulation; this systemically distributed CPE may then induce lethal effects in infants similar to those observed in animals administered CPE by intravenous or intraperitoneal injections (33, 43, 49). The mechanism behind CPE-induced lethality in animals (and perhaps in infants) is not clear but could involve CPE effects on the immune system (33, 49), possibly triggered by the putative superantigenic effects of the enterotoxin (4).

Genetics of CPE

Cloning of the intact *cpe* gene proved elusive (with at least three groups obtaining only partial *cpe* clones in early attempts), but the complete *cpe* gene has now been cloned (9, 10) from at least two CPE-positive strains (NCTC 8239 and 8-6). The availability of *cpe* gene probes generated by these *cpe* cloning studies has recently produced significant new advances in our understanding of *cpe* genetics. As mentioned earlier, two independent surveys both suggest that only ~5% of *C. perfringens* isolates carry the *cpe* gene (29, 62). This result is consistent with the possibility that, like many bacterial virulence factor genes, the *cpe* gene is present on a mobile genetic element(s).

Some direct experimental support exists for the involvement of mobile genetic elements in the mobilization or transfer of the *cpe* gene. First, as mentioned earlier, it was recently demonstrated that the *cpe* gene localizes to a large plasmid in veterinary *C. perfringens* isolates (9). Second, while the *cpe* gene appears to be present as a single chromosomal copy in food poisoning isolates of *C. perfringens*, the *cpe* gene maps to a hypervariable region of these isolates' chromosomes, a result consistent with the *cpe* gene being present on either a lysogenized phage or a transposon that has integrated into a chromosomal hot spot for transposon insertion (7). Finally, it has been shown that an open reading frame, which may be part of an insertion sequence or transposable element, immediately precedes the *cpe* gene, whether the *cpe* gene is present on the chromosome of food poisoning isolates or on plasmids in veterinary strains (6).

Expression and Release of CPE

As discussed in more detail below, there are at least three interesting and unusual aspects to the expression and release of CPE by *C. perfringens:* (i) CPE expression is tightly regulated (i.e., this toxin is strongly expressed by sporulating but not by vegetative *C. perfringens* cells); (ii) during the sporulation of many CPE-positive isolates, CPE is often produced in extremely large amounts; and (iii) CPE is not actually secreted by sporulating *C. perfringens* cells but instead is released into the intestine when the mother cell lyses during sporulation.

Regulation of CPE Synthesis

A series of classic studies in the 1960s and 1970s by Duncan et al. first established a linkage between CPE expression and sporulation (for a review, see reference 43). For example, one of these early studies showed (13) that *C. perfringens* mutants blocked at stage 0 of sporulation completely lost the ability to produce CPE. This relationship between sporulation and CPE expression has recently been confirmed by highly sensitive and specific Western blot (immunoblot) analyses (Fig. 16.4) (11, 29). These studies demonstrated that while vegetative *C. perfringens* cells produce trace amounts of CPE (probably as a result of "leaky" gene regulation), sporulating cells of the same *C. perfringens* strain express at least 1,500-fold more CPE. In light of these findings, it is worth mentioning that occasional literature reports of significant levels of CPE expression by vegetative cultures have never held up to close scrutiny. For example, while one report claimed detection of significant CPE expression by vegetative cultures of *C. perfringens* ATCC 3624 (16), this result has to be considered erroneous since two independent groups have recently shown that this strain does not even carry the *cpe* gene (29, 47).

The sporulation-associated pattern of CPE expression raises the interesting question of whether CPE expression is repressed (negatively regulated) in vegetative *C. perfringens* cells or positively regulated (activated) during sporulation. One approach to answering this question has been to determine whether *cpe*-positive recombinant *Escherichia coli* (which is a nonsporulating organism) can express CPE. Although many non-

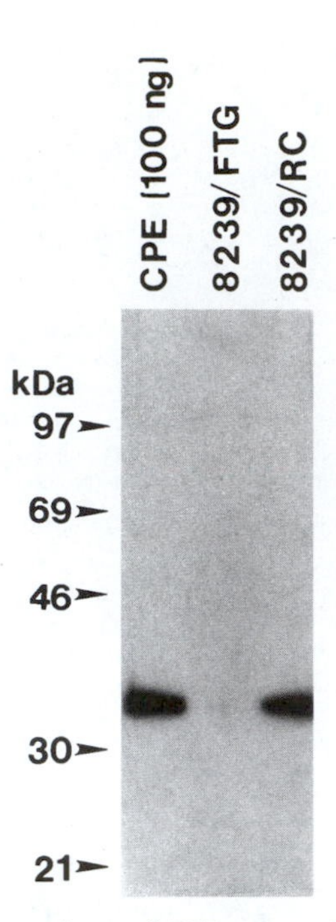

Figure 16.4 Comparison of CPE expression between vegetative and sporulating *C. perfringens* cultures. Western immunoblot results for CPE detection are shown for 100 ng of purified CPE (CPE [100 ng]), 80 μl of a lysate from an 8-h vegetative culture of enterotoxigenic *C. perfringens* NCTC 8239 (8239/FTG), and 2 μl of a lysate from an 8-h sporulating culture of NCTC 8239 (8239/RC). Western immunoblotting was performed as described in reference 29.) *Note:* In order to visualize trace CPE expression by vegetative cultures, special conditions (long autoradiography, lysate concentration, etc.) must be used (11).

sporulation-associated clostridial toxins are expressed well by *E. coli*, no CPE expression by recombinant *E. coli* containing 50 copies of a *cpe*-containing shuttle plasmid is observed (10). Presuming it unlikely that *E. coli* would produce a *cpe*-specific repressor, these results provide some evidence that the regulation of CPE expression does not only involve negative regulation during *C. perfringens* vegetative growth and, conversely, are consistent with CPE expression involving positive regulation during sporulation of *C. perfringens.*

Recent studies (10) also compared CPE expression between vegetative and sporulating cultures of naturally *cpe*-negative *C. perfringens* type A, B, and C isolates that had been transformed with the *cpe*-containing shuttle plasmid used in the *E. coli* studies described above. While no CPE expression was detected for any of these *C. perfringens* transformants during vegetative growth, all three transformants were able to produce CPE during sporulation. These results indicate that most, if not all, *C. perfringens* isolates (including type B isolates that are not known to naturally carry the *cpe* gene) produce the regulatory factor(s) leading to normal sporulation-associated CPE expression. The widespread distribution of this regulatory factor(s) among *C. perfringens* isolates suggests that regulatory factors involved in CPE expression may be global regulators (e.g., sporulation-associated sigma factors) that also help regulate transcription of other genes in *C. perfringens.*

Synthesis of CPE appears to start soon after the induction of sporulation and then progressively increases for at least the first 6 to 8 h of sporulation (43, 47, 55). After 6 to 8 h of sporulation, CPE represents up to 15% (11), or even 30% (30), of total cell protein in some sporulating cells. Why do sporulating cells of *C. perfringens* produce so much CPE? Only now are studies starting to examine this question. RNA slot blot studies have indicated (47) that CPE expression involves transcriptional regulation, with significant levels of *cpe* mRNA made during sporulation but little or no *cpe* message produced during vegetative growth of *C. perfringens.* Recent Northern (RNA) blot studies (10) confirm this conclusion and also suggest that *cpe* mRNA may be transcribed as a monocistronic message of ~1.2 kb.

Consistent with this apparent size for the CPE message, primer extension analysis studies have indicated that *cpe* mRNA transcription starts ~200 bp upstream of the CPE translation start site (47). Interestingly, this putative *cpe* promoter region does not appear to have significant sequence homology with other bacterial promoters, including those recognized by sporulation-specific sigma factors in *Bacillus subtilis* (47). Support for this ~200-bp putative *cpe* promoter region serving as a functional *cpe* promoter was obtained when this sequence was fused to a reporter gene on a shuttle plasmid. After this plasmid was transformed into an enterotoxigenic *C. perfringens* isolate, expression of this reporter construct became sporulation associated, as would be expected if expression were being driven from the true *cpe* promoter.

Posttranscriptional effects may also help to regulate CPE expression levels. An older study (32) suggests a functional half-life of 58 min for *cpe* mRNA in sporulating *C. perfringens* cells, indicating that *cpe* mRNA may be unusually long-lived for a bacterial message. This exceptional message stability could be a strong contributor to the abundant CPE expression noted in many sporulating *C. perfringens* cells. Given that stem-loop structures are believed to contribute to message stability (11), it is possible that *cpe* mRNA's unusual stability results from a putative stem-loop structure lying 36 bp downstream of the 3′ end of the *cpe* open reading frame [this same stem-loop structure is followed by an oligo(dT) tract, and so it may also function as a rho-independent transcriptional terminator (11)].

Two misconceptions about CPE synthesis should be briefly mentioned since they persist in the literature. First, studies conducted in the early 1980s suggested that CPE may be a posttranslationally processed product of a larger (52-kDa) protein precursor (55). However, results from DNA sequence analysis (11) (and Northern blot analysis [10]) are not consistent with this possibility. Second, Frieben and Duncan hypothesized in the 1970s that CPE may be made in large amounts by sporulating *C. perfringens* cells so it can be used as a structural spore coat protein (14). However, a major structural role for CPE in spores now seems less likely since Ryu and Labbe found that only relatively small amounts of CPE are associated with spores (51). This spore-associated CPE may simply represent CPE that has become trapped in developing spores during spore assembly and maturation (51).

Release of CPE from *C. perfringens*

What happens to the large amount of CPE produced during sporulation? Unlike most *C. perfringens* toxins, CPE is not secreted outside the *C. perfringens* cell (31, 43); i.e., CPE is not an exotoxin in the classic sense. Consistent with this, the *cpe* gene does not encode a 5′ signal peptide that usually contributes to toxin secretion (11).

Instead of being secreted immediately following its synthesis, CPE continues to accumulate in the cytoplasm of the mother cell. In some strains, cytoplasmic CPE levels reach concentrations high enough that CPE-containing paracrystalline inclusion bodies form in the cytoplasm of the mother cell (Fig. 16.1) (31). Eventually intracellular CPE is released into the intestine when the mother cell lyses to free the mature spore; i.e., CPE is released at the completion of the sporulation process. This dependence on mother cell lysis for CPE release undoubtedly explains, at least in part, why (despite CPE's quick intestinal action [see below]) *C. perfringens* type A food poisoning symptoms take 8 to 24 h to develop after ingestion of contaminated foods. Before CPE can be released into the intestine to produce its effects, the sporulating *C. perfringens* cells must complete sporulation, a process that takes at least 8 to 12 h (31).

The Biochemistry of CPE

CPE was initially purified and characterized by several laboratories during the early 1970s (43). Results from these early studies showed CPE to be a single polypeptide of ~35,000 Da with an isoelectric point of 4.3. Recent *cpe* sequencing results (9, 11, 62) from three

C. perfringens strains (NCTC 8239, 8–6, and F3686) have expanded on this basic understanding of the CPE molecule by indicating that (i) CPE is 319 amino acids in length and has an M_r of 35,317 and (ii) the deduced CPE sequence appears to be highly conserved in CPEs made by different isolates. In fact, the CPE sequence was found to be identical for CPEs made by strains NCTC 8239 and 8–6, although there may be some microsequence variations between the CPE expressed by NCTC 8329 and 8–6 and the CPE made by strain F3686. Computer searches have indicated that the CPE of NCTC 8239 and 8–6 appears to lack significant sequence homology with other proteins, except for some limited homology with one of the complexing proteins produced by *C. botulinum* (24). The significance, if any, of this limited homology remains unclear.

CPE is not a heat-stable enterotoxin and can be inactivated by heating for 5 min at 60°C (43). The toxin is also quite sensitive to pH extremes but resistant to some proteolytic treatments (43). In fact, Granum et al. have shown (17, 18) that limited trypsinization or chymotrypsinization actually causes a two- to threefold activation of CPE activity (see below), suggesting that intestinal proteases may produce an activated CPE during food poisoning. However, no direct in vivo evidence supporting this hypothesis has been presented.

CPE Action

Over the past 20 years, considerable research effort has been directed toward understanding CPE's action during *C. perfringens* type A food poisoning. From these studies, it has become possible to generate a time line depicting the overall sequence of events that are believed to lead to CPE's intestinal effects (Fig. 16.5). These in vivo and in vitro effects of CPE, discussed in detail below, represent a novel mechanism of action for a bacterial enterotoxin.

Effects on the Gastrointestinal Tract

CPE is classified as an enterotoxin since it causes fluid and electrolyte losses from the gastrointestinal tracts of a number of mammalian species (43). The target organ for CPE is believed to be the small intestine, with the ileum being particularly sensitive to this toxin (43). Interestingly, the rabbit colon is relatively insensitive to CPE even though this toxin binds well to rabbit colonic cells (43). Unlike cholera and *E. coli* heat-labile enterotoxins, CPE (43) (i) does not increase intestinal cyclic AMP levels, (ii) does inhibit glucose absorption, and (iii) does produce direct histopathologic damage to the small intestine (Fig. 16.3), with the villus tips being particularly CPE sensitive (43). While some other bacterial enterotoxins (e.g., Shiga toxin and *C. difficile* toxins) are also cytotoxic

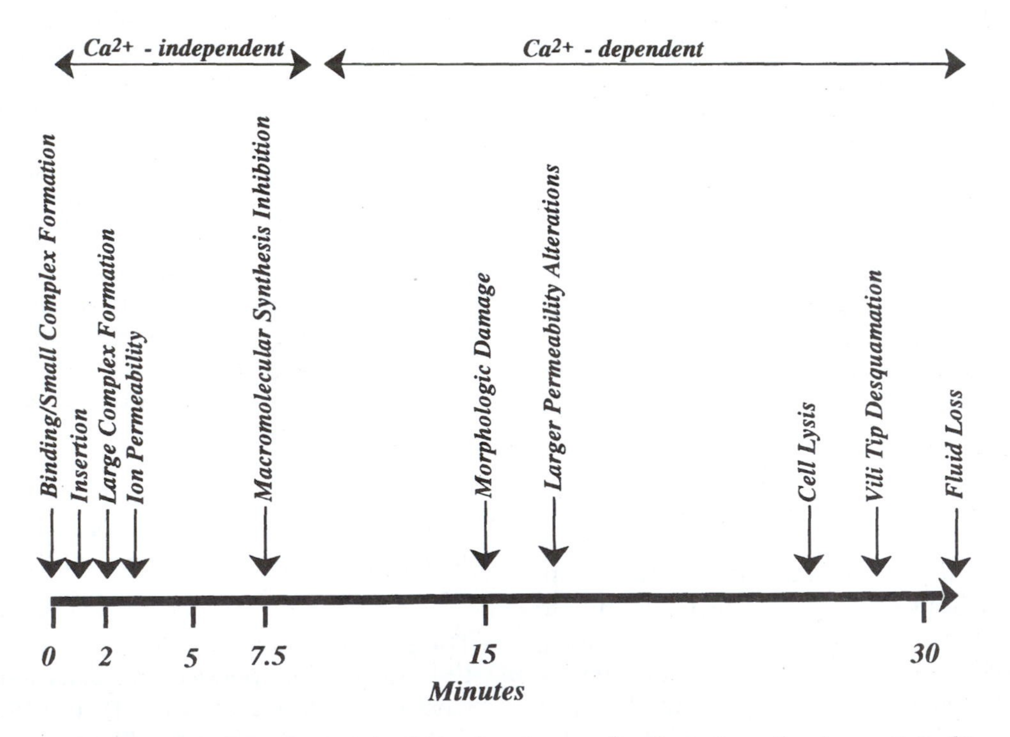

Figure 16.5 Time line of events in CPE action (see text for discussion of each event). In this scheme, it is assumed that CPE has been added at time zero. (Modified, with the publisher's permission, from reference 40.)

and produce intestinal damage, CPE-induced intestinal tissue damage is distinctive for its rapid development; i.e., CPE-induced intestinal damage can develop within 15 to 30 min of toxin treatment (53).

This tissue damage appears to help initiate CPE-induced fluid and electrolyte transport alterations, as inferred from two correlations (45, 53). First, the onset of fluid transport changes closely coincides with the development of tissue damage in the CPE-treated rabbit ileum. Second, only those CPE doses producing tissue damage will elicit intestinal fluid transport alterations in the rabbit ileum. Therefore, it seems likely that during *C. perfringens* type A food poisoning, CPE causes tissue damage that leads to a loss of villus integrity and a consequent breakdown of the normal intestinal secretion/absorption equilibrium (an effect that is clinically manifested as diarrhea).

Current Model for Early Steps in CPE Action

Recent CPE studies have focused primarily on understanding early events in CPE action, with the rationale that these should be the primary events leading to CPE's cytotoxic effects on mammalian cells and that this CPE-induced cytotoxicity causes the tissue damage associated with CPE treatment of the small intestine (37). The current overview model for early steps in CPE's action (64), presented diagrammatically in Fig. 16.6, predicts that CPE action is a multistep process involving at least four early events. The initial event in CPE action is the binding of CPE to its receptor (believed to be a 50-kDa membrane protein), resulting in formation of a 90-kDa small complex. This is almost immediately followed by some physical change to the CPE molecule sequestered in small complex; this step could correspond either to the insertion of CPE (or small complex) into the membrane bilayer or to a conformational change in small complex. The third early event in CPE action involves the formation of a large (160-kDa) complex, whose formation apparently results from an interaction between CPE-containing small complex and a 70-kDa protein. In the fourth and final early step in CPE action, large complex causes the plasma membrane to lose its normal permeability properties. This effect could result from the large complex serving directly as a pore, or it could result from a less direct mechanism.

This model emphasizes the uniqueness of CPE action at the molecular level by predicting that membrane proteins are intimately involved in every early step of CPE action. No other membrane-active toxin is known to involve eukaryotic proteins so heavily in its action.

Early Steps in the Molecular Action of CPE

Binding of CPE to Its Receptor (Step 1)

The initial interaction between CPE and mammalian plasma membranes exhibits characteristics typical of a receptor-mediated process; i.e., CPE binding is specific, in that binding of ^{125}I-CPE to membranes can be abolished by coincubation with excess native CPE (41) and CPE binding is clearly saturable, with ~10^6 CPE receptors per cell (37, 46, 66). CPE's binding to its receptor(s) occurs rapidly and is temperature sensitive (there is less CPE binding at 4°C than at 37°C [37]). Specific binding of CPE to its receptor is an essential step in initiating CPE cytotoxicity; for example, cell types (e.g., CHO cells) that do not specifically bind CPE also do not respond to CPE treatment (64).

The CPE receptor is clearly proteinaceous, since protease pretreatment of cells or isolated intestinal brush border membranes (BBMs) destroys their ability to bind CPE (41, 66). CPE receptors have been found in the small intestines of several mammalian species (37, 42, 58), which helps to explain why so many mammalian species are CPE sensitive (43). However, intestinal cells are not the only mammalian cells expressing CPE receptors (37, 42, 46). The broad distribution of the CPE receptor among many different cell types may indicate that the receptor plays an important normal physiologic role for some mammalian cells, but, as noted previously, the CPE receptor apparently is not an essential molecule for the viability of all mammalian cells since certain cell types will not bind CPE (64).

How many different types of CPE receptors exist? Kinetic analysis of CPE binding was unable to resolve the number of CPE receptor types. Some kinetic studies gave results consistent with a single CPE receptor type, while other studies claimed detection of multiple CPE receptor types (37, 43). Two possible candidates for the CPE receptor(s) in BBMs and Vero cells were identified by affinity chromatography of BBM or Vero cell extracts on CPE immobilized to Sepharose 4B. Depending on the detergent used for extraction of BBMs or Vero cells in these studies, mammalian proteins of 50 or 70 kDa could be shown to specifically bind to CPE affinity columns (58, 65, 67). Was either of these CPE-binding proteins a functional CPE receptor capable of conveying the cytotoxic response to CPE treatment? Recently Wieckowski et al. (64) have answered this question, at least in part, by detecting a 90-kDa CPE-containing complex (called CPE small complex) that forms quickly in CPE-treated mammalian membranes (Fig. 16.7). This 90-kDa species occurs only in cell types capable of binding and responding to CPE, a correlation that strongly

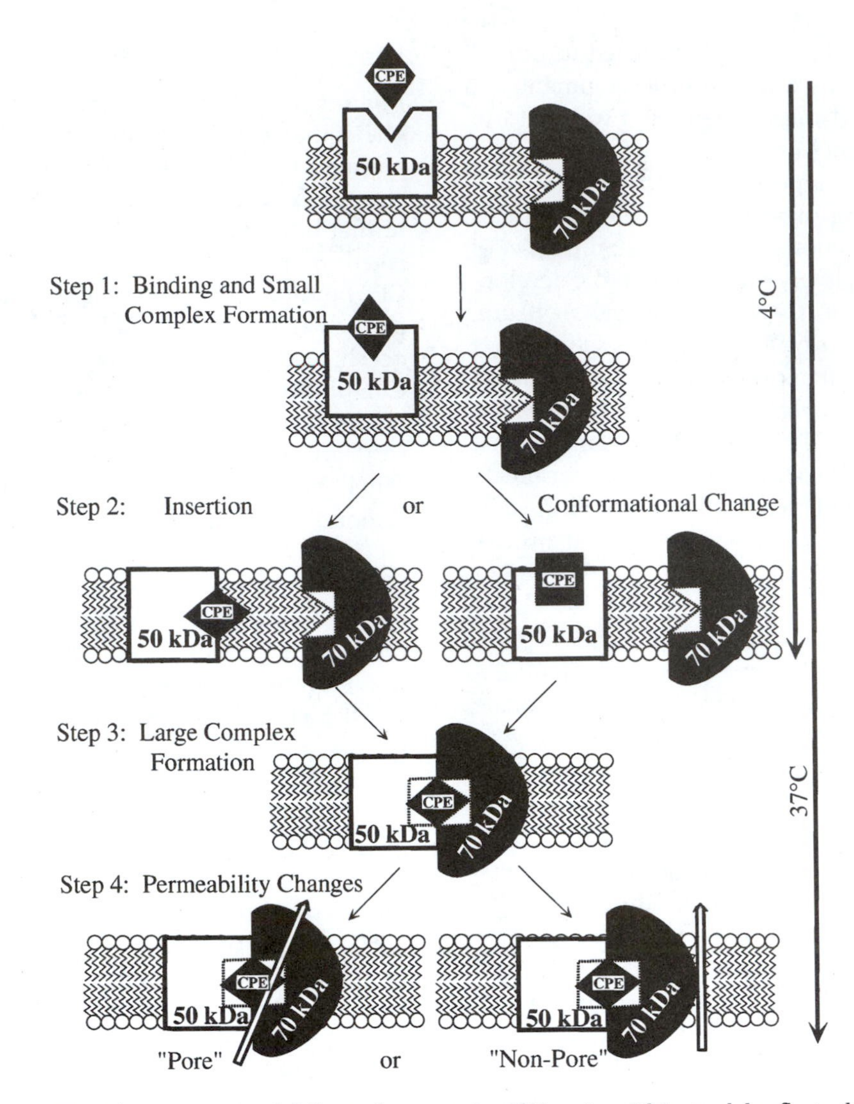

Figure 16.6 The current model for early events in CPE action. This model reflects the four known early events in CPE's action: binding of CPE to its receptor to form a small (90-kDa) CPE-containing complex (step 1), a physical change in this CPE sequestered in small complex (an effect which could correspond to either CPE insertion or a conformational change in small complex) (step 2), formation of a large CPE-containing complex resulting from an interaction between small complex and a 70-kDa membrane protein (step 3), and the development of permeability changes in large-complex-containing plasma membranes of mammalian cells (step 4). Note that steps 3 and 4 do not occur at low temperature. See text for full discussion of each event.

suggests that small complex represents the binding of CPE to its functional receptor. Immunoprecipitation analysis of two CPE-sensitive cell lines demonstrated that small complex is stoichiometrically composed of one CPE molecule and one 50-kDa eukaryotic membrane protein. Collectively these results strongly implicate this 50-kDa protein as a (or the) functional CPE receptor.

CPE Insertion (Step 2)

Many membrane-active toxins besides CPE can damage the normal permeability properties of mammalian plasma membranes (37). The actions of these other membrane-active toxins often involve their insertion into the lipid bilayer of the plasma membrane of mammalian cells (37). As early as 1980, McDonel proposed that CPE action may also involve an insertion step (42).

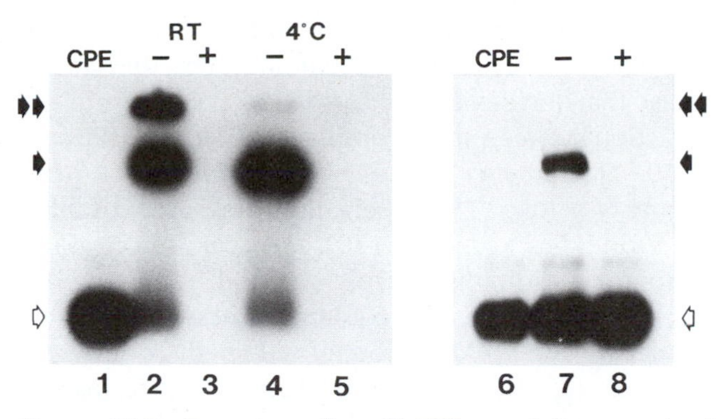

Figure 16.7 Detection of small CPE-containing complex in rabbit intestinal BBMs. Lanes 1 to 5, small-complex formation in CPE-treated intact BBMs. Samples: free ^{125}I-CPE (lane 1) and ^{125}I-CPE incubated with intact BBMs in the presence (+) or absence (–) of 50-fold-excess unlabeled CPE at either room temperature (RT; lanes 2 and 3) or 4°C (lanes 4 and 5) prior to Triton X-100 extraction, electrophoresis in a nondenaturing gel system, and autoradiography (60). Lanes 6 to 8, small-complex formation in CPE-treated BBM extracts. Samples: free ^{125}I-CPE (lane 6) and BBM Triton X-100 extracts incubated with ^{125}I-CPE in the presence (+; lane 8) or absence (–; lane 7) of 50-fold-excess unlabeled CPE prior to electrophoresis in a nondenaturing gel system and autoradiography (64). Note that small complex (single solid arrow) is readily distinguished from large complex (double solid arrow) and free CPE (single open arrow) and that unlike large-complex formation, small-complex formation occurs at both 4°C and room temperature and does not require the presence of intact BBMs. (Reproduced from reference 64 with the publisher's permission.)

Since McDonel's CPE insertion hypothesis was originally offered, several pieces of indirect evidence have accumulated in support of possible CPE insertion into membranes (37). First, although CPE remains plasma membrane associated after binding (i.e., it is not internalized into the cytoplasm), bound enterotoxin does not dissociate from either intact cells or isolated BBMs, an effect consistent with CPE being entrapped by its insertion into lipid bilayers (37). Bound CPE does not dissociate from either CPE-containing cells or BBMs even when these cells or BBMs are incubated for long periods in the presence of chaotropic salts or EDTA, agents that are known to induce the release of peripherally bound membrane proteins but not inserted toxins (37). Second, the kinetics of CPE's association with membranes indicate that a rapid two-step process occurs early in CPE action. These results are consistent with CPE binding being rapidly followed by a second process such as insertion (41). Third, once CPE becomes bound to membranes, it acquires the amphiphilic characteristics expected of an insertion-capable toxin (64). Finally, the longer CPE associates with membranes after binding, the more difficult it becomes to release this toxin from membranes by using external proteases (41). The progressive nature of this resistance of membrane-associated CPE to protease-induced release argues that this phenomenon (often referred to as CPE insertion) does not represent simply the static binding of CPE to its receptor to form a protease-resistant structure but instead represents some postbinding physical change that occurs in bound CPE, a change such as the insertion of CPE into membranes. CPE's development of resistance to protease-induced release (i.e., CPE insertion) is temperature independent (occurring at both 4°C and higher temperatures [40]) and is considered to be an essential step in CPE action, since a nontoxic C-terminal CPE fragment appears to be specifically blocked at this step in CPE action (19).

Is the evidence accumulated to date sufficient to conclude that CPE actually inserts into lipid bilayers in membranes? Not necessarily, since most of this evidence also remains compatible with an alternative possibility, i.e., that CPE-containing small complex undergoes a conformational change entrapping CPE on the cell surface. Further, these "insertion versus conformational change" hypotheses should not be considered mutually exclusive, since it is possible that a conformational change in small complex contributes to the insertion of CPE into membranes.

If CPE insertion does occur, it will be intriguing to determine whether only the CPE molecule itself is inserted into membranes or whether one or more eukaryotic complex proteins also become embedded in lipid bilayers.

Formation of Large Complex (Step 3)

At high temperatures (either 24 or 37°C), membrane-bound CPE becomes sequestered in a large (160-kDa) complex (Fig. 16.8) within the first few minutes of CPE treatment (40, 67). Formation of this large complex does not require intact, viable mammalian cells since both intact cells and isolated BBMs will form large complex (40, 67). However, large-complex formation does require intact membranes, as can be seen in Fig. 16.8 (in this assay, complex did not form when CPE was added to detergent extracts of BBMs). Both CPE small complex and CPE large complex may exist simultaneously in the same plasma membrane (64), as can be readily demonstrated by nondenaturing polyacrylamide gel electrophoresis (PAGE) analysis of CPE-treated BBMs (Fig. 16.7), but these two CPE-containing complexes are readily distinguishable (64) on the basis of (i) size (160 kDa for large complex and 90 kDa for small complex), (ii) stability in sodium dodecyl sulfate (SDS) (large complex is

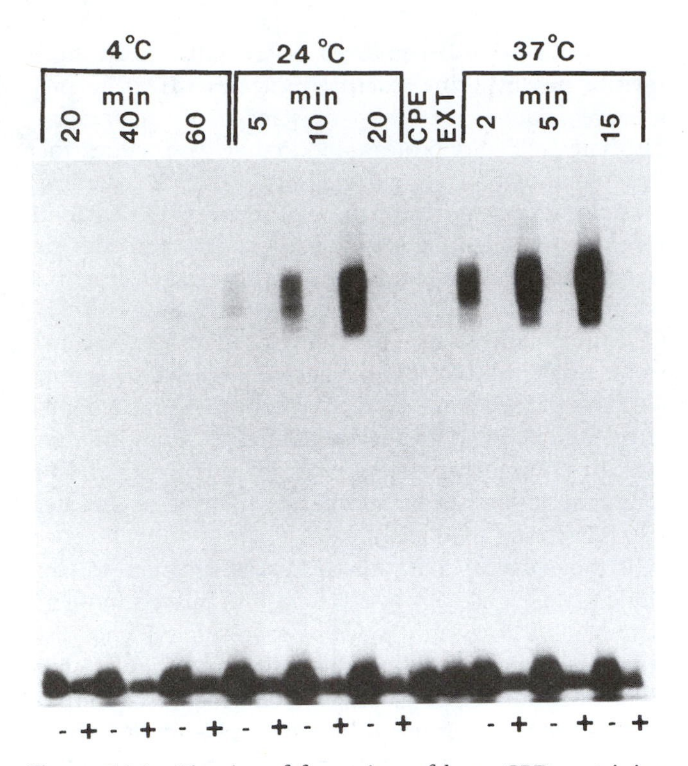

Figure 16.8 Kinetics of formation of large CPE-containing complex in BBMs. ^{125}I-CPE was added to BBMs in the presence (+) or absence (–) of a 50-fold excess of unlabeled CPE for the indicated times at 4, 24, or 37°C. After removal of unbound enterotoxin, large complex was extracted from the BBMs with SDS and analyzed by SDS-PAGE (no sample boiling) using 6% acrylamide gels. Radioactivity was detected by autoradiography. Other samples: ^{125}I-CPE alone (no BBMs) (CPE) and ^{125}I-CPE added to SDS-extracted BBMs (EXT). Note that (i) the absence of large complex in the EXT sample indicates that large-complex formation requires intact BBMs and (ii) there are greater amounts of radioactivity at the bottom of the gel in – versus + matched samples (this radioactivity corresponds to small complex, which falls apart in SDS [64]). (Reproduced from reference 40 with the publisher's permission.)

stable in SDS, while small complex falls apart in the presence of SDS; note that the lability of small complex to SDS explains the absence of a 90-kDa species in the SDS-PAGE result shown in Fig. 16.8), and (iii) temperature sensitivity (formation of large complex is blocked at 4°C, while small complex forms and remains stable in membranes treated with CPE at low temperatures).

CPE large complex is at least partially sensitive to boiling, suggesting that it is not held together by covalent bands (67). Large complex appears to be composed of one CPE molecule, one 50-kDa membrane protein, and one 70-kDa protein (67). The identities and normal functions of both the 50- and 70-kDa membrane proteins of large complex are currently unknown but are under investigation. However, if it is presumed that the 50-kDa protein present in large complex is the same 50-kDa protein detected in small complex, this would suggest that large-complex formation results from an interaction between small complex (containing CPE and the 50-kDa protein) and the 70-kDa protein (further comments below). The observation that large-complex formation is blocked at 4°C (40), a temperature at which membrane fluidity is sharply reduced, suggests that large-complex formation requires membrane diffusion in order for the 70-kDa protein and small complex to interact. Finally, both 70- and 50-kDa BBM proteins can bind directly to CPE affinity columns (67), which suggests that the CPE molecule directly interacts with both proteins when it is sequestered in large complex (see below).

The observation (40) that large-complex formation does not occur at 4°C provides two important insights into CPE action. First, since both CPE binding/small-complex formation and insertion can occur at 4°C (although CPE binding is somewhat reduced at low temperatures), the lack of large-complex formation at 4°C strongly suggests that large-complex formation must follow both binding/small-complex formation and insertion during CPE's early action (40). Second, both CPE cytotoxicity (i.e., the onset of small-molecule permeability alterations in CPE-treated cells [see below]) and large-complex formation are completely blocked at 4°C (40), a correlation which suggests that large-complex formation is a necessary step in CPE-induced cytotoxicity.

A role for large-complex formation in CPE cytotoxicity is also supported by experiments (40) using Vero cells treated with CPE at 4°C (a temperature at which the only CPE-containing membrane species formed will be small complex). After unbound CPE was removed and these cells were warmed to 37°C, there was a simultaneous onset of cytotoxicity and formation of large complex, as would be expected if large complex causes cytotoxicity. This important experiment also provides additional support for two previously mentioned hypotheses regarding early events in CPE action. First, by demonstrating that Vero cells treated with CPE at 4°C (i.e., cells containing CPE sequestered only in small complex) develop cytotoxic effects after warming to 37°C, this experiment provides additional evidence that small complex represents CPE bound to a receptor that can convey a cytotoxic response. Second, by demonstrating that large complex forms very rapidly after Vero cells containing only CPE in small complex are warmed to 37°C, this experiment also supports the hypothesis that large-complex formation directly involves small complex.

Finally, if it is assumed that the 70-kDa protein involved in large-complex formation is the same 70-kDa protein that binds to CPE affinity columns (67), why does this 70-kDa protein not serve directly as a CPE receptor? There are at least two possible answers to this question. First, it remains conceivable that the 70-kDa protein can function as a receptor but the 70-kDa protein–CPE complex is very labile and thus has not been detected with current techniques (however, even if true, this finding would not negate the accumulating evidence that the 50-kDa protein can also serve as a functional CPE receptor). An alternative, and perhaps more likely, explanation is that direct access of CPE to the 70-kDa protein in intact membranes may be sterically inhibited prior to the occurrence of the second step in CPE action. For example, perhaps either the insertion of CPE into bilayers or the onset of a conformational change in small complex results in the CPE molecule being brought into closer contact with the 70 kDa protein in intact membranes.

CPE-Induced Alterations in Plasma Membrane Permeability for Small Molecules (Step 4)

In the late 1970s and early 1980s, it was shown that CPE directly affects plasma membrane permeability (34–38, 43, 46). These CPE effects on membrane permeability can develop within as little as 5 min of CPE treatment (Fig. 16.9) and are initially restricted to small molecules of <200 Da, e.g., ions and amino acids (37). This finding suggests that CPE produces an initial membrane lesion of ~50 to 100 Å^2 (1 Å = 0.1 nm) that is neither directional nor discriminating (except for size); i.e., CPE-treated cells rapidly become permeable to the influx and efflux of a variety of small molecules (34–38, 43, 46).

As mentioned previously, there is an extremely close temporal correlation between large-complex formation and the onset of CPE-induced small-molecule permeability alterations (40). This correlation precludes the existence of extensive intermediate steps between these two early CPE events and is consistent with large complex directly causing CPE's effects on plasma membrane permeability. How might large complex disrupt the normal permeability properties of the plasma membrane? Although unproven, the simplest hypothesis (Fig. 16.6) is that large complex directly functions as a unique membrane pore. While many other membrane-active toxins can form pores, eukaryotic proteins are not known to be involved in the pore structures formed by any of these other toxins. In possible support of the hypothesis that CPE can form pores, it has been shown that CPE sonicated into lipid bilayers forms channels (59). However, this result needs to be interpreted cautiously since the artificial lipid bilayer model system

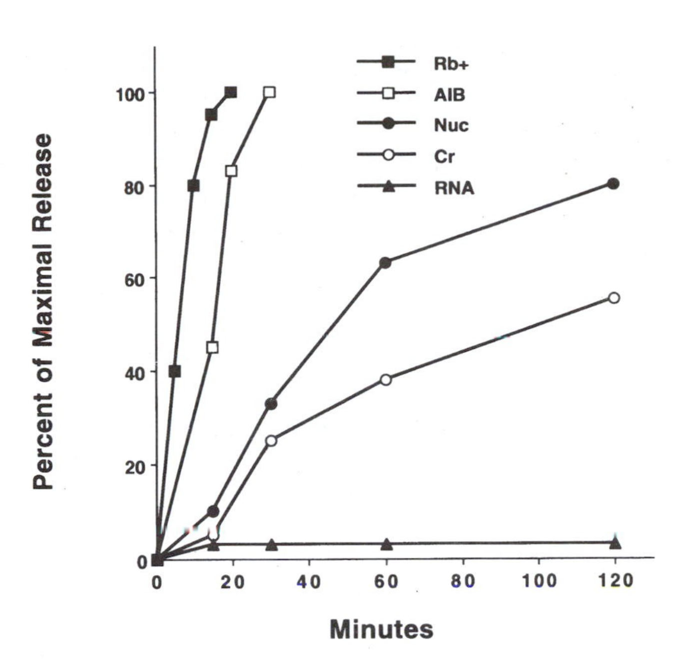

Figure 16.9 Effects of CPE on plasma membrane permeability in Vero cells. CPE effects on Vero cell membrane permeability were measured by specific CPE-induced release of radioactive cytoplasmic markers of defined sizes, including ^{86}Rb (Rb$^+$; M_r, ~100), [^{14}C]aminoisobutyric acid (AIB; M_r, ~100), ^{3}H-labeled nucleotides (Nuc; M_r, <1,000), ^{51}Cr (Cr; M_r, ~3,000), and [^{3}H]RNA (RNA; M_r, >25,000). Note the rapid release of small molecules such as ions (Rb$^+$) and amino acids (AIB) following CPE treatment and that very large molecules (e.g., RNA) are never released from intact CPE-treated cells. Compiled from studies discussed in references 37 and 43.

used in these studies does not accurately reflect in vivo conditions in which CPE is always associated with eukaryotic proteins following its binding to membranes.

As an alternative possibility to large complex representing a pore, large complex could function indirectly via a nonpore mechanism; e.g., perhaps large-complex formation interferes with regulation of existing membrane pumps in such a manner that these pumps become continuously activated (or inactivated). Considerable insights into how CPE disrupts membrane permeability may be gleaned when normal physiologic functions are elucidated for the 50-kDa and 70-kDa membrane protein components of large complex; e.g., perhaps these proteins are components or regulators of ion pumps.

Later Events in CPE Action

With time, CPE-treated mammalian cells develop a number of secondary effects (27, 38, 39, 43): they become inhibited for DNA, RNA, and protein synthesis; they develop permeability alterations for larger molecules of up to 3 to 5 kDa; they develop morphologic damage (manifested as blebs [Fig. 16.3); and with sufficient dosage and time, some CPE-treated cells will lyse.

It has become apparent that these effects are secondary consequences of the disruption of the normal colloid-osmotic equilibrium of the mammalian cell resulting from CPE's initial membrane effects on small-molecule permeability (35, 37, 38). Specifically, while mammalian cells must normally maintain a lower intracellular ion concentration than exists in the external medium in order to balance osmolarity contributed by their cytoplasmic macromolecules, CPE treatment somehow breaks down the normal osmotic barrier properties of the plasma membrane (see the previous section) and thereby allows a rapid net ion influx into the cytoplasm of CPE-treated cells. Water naturally follows this ion influx, causing the plasma membrane of CPE-treated cells to "stretch" (an effect visualized as membrane blebs) and become more permeable to larger molecules. CPE's effects on both cell morphology and plasma membrane permeability for larger (i.e., 3- to 5-kDa) molecules specifically require a net influx of Ca^{2+} into mammalian cells (37, 41). Elevated intracellular Ca^{2+} levels in CPE-treated cells probably leads to collapse of the cytoskeleton, an effect which could contribute to morphologic changes and stretching of the now unanchored plasma membrane. Similarly, the shutdown of macromolecular synthesis in CPE-treated cells also results from small-molecule permeability changes that cause CPE-treated cells to lose some precursors (e.g., amino acids) needed for macromolecular synthesis and also alter the intracellular milieu so it becomes incapable of supporting macromolecular synthesis (27). Considering this information collectively, it becomes apparent that CPE's cytotoxic effect is to induce small-molecule permeability changes that render the CPE-treated cell nonviable.

Although much of the initial work on CPE action used easily manipulated cultured mammalian cells, it increasingly appears that similar events also occur in the intestine during food poisoning. In support of this view, the first three early events in CPE action have now been confirmed to occur in intestinal BBMs, the apparent target site for CPE action in vivo (40, 41, 64, 67). Further, CPE-induced morphologic changes in cultured Vero cells and intestinal epithelial cells are qualitatively similar and develop within similar time frames (39).

In light of this detailed discussion of CPE's cellular and molecular action, it may be worthwhile to briefly revisit CPE's in vivo action and reconsider how these molecular and cellular effects may induce *C. perfringens* type A food poisoning symptoms. Current thinking is that, through the molecular and cellular process described above, CPE-induced small-molecule permeability changes cause the morphologic damage noted in CPE-treated intestinal epithelial cells. This damage leads to both intestinal cell lysis and desquamation of intestinal cells from the villus tips, effects resulting in a disruption of normal villus integrity and function. Consequently, net secretion of fluids and electrolytes develops in the CPE-damaged intestine, an effect corresponding to the diarrheal symptoms typical of *C. perfringens* type A food poisoning.

Structure-Function Relationships

As understanding of CPE's action increases, it becomes increasingly important to understand how the CPE molecule mediates this action. Toward this goal, studies have been undertaken to map important functional regions on the CPE molecule. Generally, these studies (17–22, 26) have involved producing defined CPE fragments (usually by recombinant DNA approaches) and then assaying each fragment for its retention of such CPE activities as receptor binding. An overview of CPE functional regions, as currently understood, is shown in Fig. 16.10.

To date, these studies indicate that most, if not all, receptor binding activity appears to localize to the extreme C terminus of the CPE molecule (20–22, 26). These studies have also demonstrated that C-terminal CPE fragments, which are fully capable of receptor binding, are not themselves cytotoxic (i.e., these fragments do not induce small-molecule permeability alterations in mammalian cells) and do not insert (i.e., these fragments remain susceptible to protease-induced release). These observations indicate that mere occupancy of the CPE receptor does not trigger CPE-like cytotoxicity (a result supporting the involvement of postbinding steps in CPE action). They also demonstrate that sequences in the N-terminal half of the CPE molecule must be essential for insertion and cytotoxicity, suggesting that, like many bacterial toxins, CPE segregates its binding and activity regions. Since it has also been shown that the 36 N-terminal amino acids of CPE are not required for cytotoxicity (17, 18), key sequences required for cytotoxicity must exist between residues 37 and 171 of the CPE molecule. In fact, as previously mentioned, removal of the 36 N-terminal amino acids with chymotrypsin actually causes a twofold activation in CPE cytotoxicity (17, 18); i.e., the extreme N-terminal region of the full-length CPE molecule apparently interferes with CPE's cytotoxic effects.

Studies are currently ongoing to extend these preliminary observations, e.g., to map CPE regions required for large-complex formation. Once construction of crude functional maps for CPE activities is completed, more

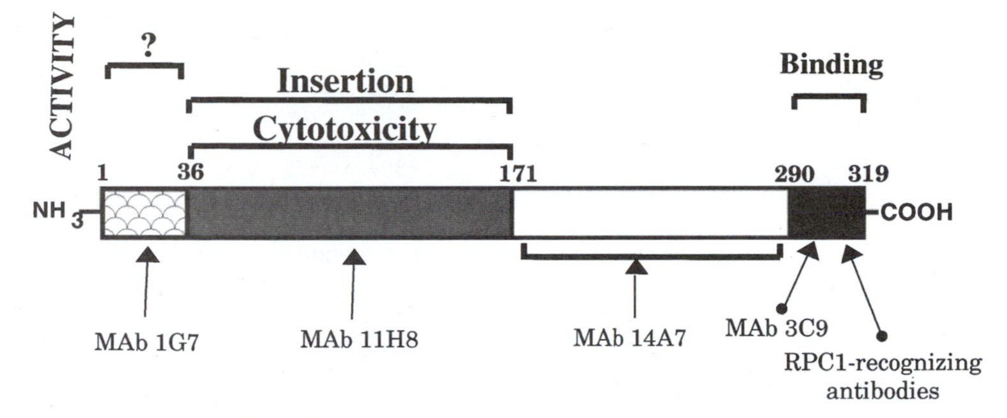

Figure 16.10 Map of CPE functional regions. CPE regions required for insertion, cytotoxicity, and binding are depicted (compiled from references 17 to 22 and 48), as are sequences required for presentation of MAbs 1G7, 11H8, 14A7, and 3C9 (48). The C-terminal region of CPE also appears to contain at least one other epitope, named RPC-1 (48). Note that, of the four MAbs shown, only MAb 3C9 neutralizes CPE cytotoxicity.

precise mapping will be undertaken to identify key individual amino acids involved in each CPE-associated activity. The eventual goal of this work is to relate the information generated by these structure-function mapping studies to physical data on CPE's three-dimensional structure (once CPE's structure is solved by crystallography) to learn precisely how this enterotoxin exerts its effects.

CPE Epitopes: Is a CPE Vaccine Possible?

The CPE fragments prepared for structure-function mapping studies were also reacted (21) with a series of CPE-specific monoclonal antibodies (MAbs) (68) to identify regions of the enterotoxin that are required for epitope presentation. In support of earlier studies indicating that native CPE contains at least four or five epitopes (68), these MAb mapping studies (21) found that disparate regions scattered throughout the enterotoxin appear to be involved in presentation of CPE epitopes. Of most significance, it was demonstrated that 3C9, a MAb which neutralizes the cytotoxicity of native CPE by blocking the binding of the enterotoxin to its receptor (68), recognizes sequences localized to the extreme C terminus of the CPE molecule.

Besides providing additional evidence confirming that the extreme C terminus of the CPE molecule contains receptor binding activity, this result also maps a neutralizing epitope to the C-terminal region of the CPE molecule. Since the C-terminal CPE fragments are not themselves cytotoxic (19–22) but contain a neutralizing epitope(s) for native CPE, these CPE fragments appeared to possess properties desirable for CPE vaccine candidates. To further evaluate the use of these C-terminal CPE fragments for developing a CPE vaccine, a 30-mer synthetic peptide corresponding to the extreme C-terminal CPE sequence was chemically conjugated to a thyroglobulin carrier. Immunization of mice with this conjugate was shown to induce the formation of serum antibodies capable of neutralizing the cytotoxicity of native CPE (48).

Of course, since effective immunity to *C. perfringens* type A food poisoning would likely require a secretory immunoglobulin A response localized to the intestinal lumen, even high titers of anti-CPE serum antibodies may afford little protection against this illness. Therefore, should further development of a CPE food poisoning vaccine be deemed desirable, future studies will have to pursue approaches to specifically stimulate the development of intestinal immunoglobulin A immunity against CPE. However, the ability of the conjugate vaccine to induce CPE-neutralizing serum antibodies still may have practical applications, particularly if the linkage between circulating CPE (which may be absorbed from the infant's intestine) and sudden infant death syndrome becomes better established (33).

CONCLUDING REMARKS

There has been considerable recent progress toward understanding the unique molecular pathogenesis of *C. perfringens* type A food poisoning. However, there is no shortage of future challenges facing CPE researchers; for example, what are the reservoirs for the *cpe*-positive *C. perfringens* isolates that cause food poisoning? How does *C. perfringens* regulate the timing and amount of CPE that it produces during sporulation? Is the second step in CPE action insertion or a conformational change in small complex? How does large-complex formation lead to membrane permeability alterations? What is the three-dimensional structure of CPE?

If these and other questions can be answered through continued basic research, the information obtained may identify better approaches to control CPE-associated diseases, including *C. perfringens* type A food poisoning. For example, once the reservoirs for food poisoning isolates are identified, specific hygenic measures can be used to reduce food contamination with these organisms. Additional molecular epidemiology studies may also lead to the development of improved detection assays that are capable of identifying and distinguishing between *C. perfringens* isolates causing specific CPE-associated illnesses. Finally, continued research on CPE structure-function relationships and the regulation of CPE expression may ultimately provide new insights toward developing CPE vaccines and/or specific inhibitors of CPE expression that can be used to prevent or treat illness in people who do become exposed to enterotoxigenic *C. perfringens* isolates.

Acknowledgments
Preparation of this chapter was supported by Public Health Service grant AI 19844-12 from the National Institute of Allergy and Infectious Diseases. I thank Marilyn Williams for secretarial assistance and Timothy Mietzner, Ewa Wieckowski, and John Kokai-Kun for help with the computer graphics.

References

1. **Bartholomew, B. A., M. F. Stringer, G. N. Watson, and R. J. Gilbert.** 1985. Development and application of an enzyme-linked immunosorbent assay for *Clostridium perfringens* type A enterotoxin. *J. Clin. Pathol.* **38**:222–228.
2. **Bean, N. H., and P. M. Griffin.** 1990. Foodborne disease outbreaks in the United States, 1973–1987; pathogens, vehicles and trends. *J. Food Prot.* **53**:804–817.
3. **Birkhead, G., R. L. Vogt, E. M. Heun, J. T. Snyder, and B. A. McClane.** 1988. Characterization of an outbreak of *Clostridium perfringens* food poisoning by quantitative fecal culture and fecal enterotoxin measurement. *J. Clin. Microbiol.* **26**:471–474.
4. **Bowness, P., P. A. Moss, H. Tranter, J. T. Bell, and A. J. McMichael.** 1992. *Clostridium perfringens* enterotoxin is a superantigen reactive with human T cell receptors Vβ 6.9 and Vβ 22d. *J. Exp. Med.* **176**:893–896.
5. **Brett, M. M.** 1994. Outbreaks of food poisoning associated with lecithinase-negative *Clostridium perfringens. J. Med. Microbiol.* **41**:405–407.
6. **Brynestad, S., L. A. Iwanejko, G. S. A. B. Stewart, and P. E. Granum.** 1994. A complex array of Hpr consensus DNA recognition sequences proximal to the enterotoxin gene in *Clostridium perfringens* type A. *Microbiology* **140**:97–104.
7. **Canard, B., B. Saint-Joanis, and S. T. Cole.** 1992. Genomic diversity and organization of virulence genes in the pathogenic anaerobe *Clostridium perfringens. Mol. Microbiol.* **6**:1421–1429.
8. **Centers for Disease Control and Prevention.** 1994. *Clostridium perfringens* gastroenteritis associated with corned beef served at St. Patrick's Day meals; Ohio and Virginia. 1993. *Morbid. Mortal. Weekly Rep.* **43**:137–144.
9. **Cornillot, E., B. Saint-Joanis, G. Daube, S. Katayama, P. E. Granum, B. Canard, and S. T. Cole.** 1995. The enterotoxin gene (*cpe*) of *Clostridium perfringens* can be chromosomal or plasmid-borne. *Mol. Microbiol.* **15**: 639–647.
10. **Czeczulin, J. R., R. C. Collie, and B. A. McClane.** 1996. Regulated expression of *Clostridium perfringens* enterotoxin in naturally *cpe*-negative type A, B, and C isolates of *C. perfringens. Infect. Immun.* **64**:3301–3309.
11. **Czeczulin, J. R., P. C. Hanna, and B. A. McClane.** 1993. Cloning, nucleotide sequencing and expression of the *Clostridium perfringens* enterotoxin gene in *Escherichia coli. Infect. Immun.* **61**:3429–3439.
12. **Daube, G., B. China, P. Simon, K. Hvala, and J. Mainil.** 1994. Typing of *Clostridium perfringens* by *in vitro* amplification of toxin genes. *J. Appl. Bacteriol.* **77**:650–655.
13. **Duncan, C. L., D. H. Strong, and M. Sebald.** 1972. Sporulation and enterotoxin production by mutants of *Clostridium perfringens. J. Bacteriol.* **110**:378–391.
14. **Frieben, W. R., and C. L. Duncan.** 1973. Homology between enterotoxin protein and spore structural protein in *Clostridium perfringens* type A. *Eur. J. Biochem.* **39**:393–401.
15. **Garcia-Alvarado, J. S., R. G. Labbe, and M. A. Rodriguez.** 1992. Sporulation and enterotoxin production by *Clostridium perfringens* type A at 37 and 43°C. *Appl. Environ. Microbiol.* **58**:1411–1414.
16. **Goldner, S. B., M. Solberg, S. Jones, and L. S. Post.** 1986. Enterotoxin synthesis by nonsporulating cultures of *Clostridium perfringens. Appl. Environ. Microbiol.* **52**:407–412.
17. **Granum, P., and M. Richardson.** 1991. Chymotrypsin treatment increases the activity of *Clostridium perfringens* enterotoxin. *Toxicon* **29**:898–900.
18. **Granum, P. E., J. R. Whitaker, and R. Skjelkvale.** 1981. Trypsin activation of enterotoxin from *Clostridium perfringens* type A. *Biochim. Biophys. Acta* **668**:325–332.
19. **Hanna, P. C., and B. A. McClane.** 1991. A recombinant C-terminal toxin fragment provides evidence that membrane insertion is important for *Clostridium perfringens* enterotoxin cytotoxicity. *Mol. Microbiol.* **5**:225–230.
20. **Hanna, P. C., T. A. Mietzner, G. K. Schoolnik, and B. A. McClane.** 1991. Localization of the receptor-binding region of *Clostridium perfringens* enterotoxin utilizing cloned toxin fragments and synthetic peptides. *J. Biol. Chem.* **266**:11037–11043.
21. **Hanna, P. C., E. U. Wiekowski, T. A. Mietzner, and B. A. McClane.** 1992. Mapping functional regions of *Clostridium perfringens* type A enterotoxin. *Infect. Immun.* **60**:2110–2114.
22. **Hanna, P. C., A. P. Wnek, and B. A. McClane.** 1989. Molecular cloning of the 3′ half of the *Clostridium perfringens* enterotoxin gene and demonstration that this region encodes receptor-binding activity. *J. Bacteriol.* **171**:6815–6820.

23. **Hauschild, A. H., L. Niilo, and W. J. Dorward.** 1971. The role of enterotoxin in *Clostridium perfringens* type A enteritis. *Can. J. Microbiol.* **17**:987–991.

24. **Hauser, D., M. W. Eklund, P. Boquet, and M. R. Popoff.** 1994. Organization of the botulinum neurotoxin C1 gene and its associated non-toxin protein genes in *Clostridium botulinum* C468. *Mol. Gen. Genet.* **243**:631–640.

25. **Hobbs, B. C.** 1979. *Clostridium perfringens* gastroenteritis, p. 131–167. *In* H. Riemann and F. L. Bryan (ed.), *Foodborne Infections and Intoxications,* 2nd ed. Academic Press, New York.

26. **Horiguchi, Y., T. Akai, and G. Sakaguchi.** 1987. Isolation and function of a *Clostridium perfringens* enterotoxin fragment. *Infect. Immun.* **55**:2912–2915.

27. **Hulkower, K. I., A. P. Wnek, and B. A. McClane.** 1989. Evidence that alterations in small molecule permeability are involved in the *Clostridium perfringens* type-A enterotoxin-induced inhibition of macromolecular synthesis in Vero cells. *J. Cell. Physiol.* **140**:498–504.

28. **Junja, V. K., and W. M. Majka.** 1995. Outgrowth of *Clostridium perfringens* spores in cook-in-bag beef products. *J. Food Safety* **15**:21–34.

29. **Kokai-Kun, J. F., J. G. Songer, J. R. Czeczulin, F. Chen, and B. A. McClane.** 1994. Comparison of Western immunoblots and gene detection assays for identification of potentially enterotoxigenic isolates of *Clostridium perfringens. J. Clin. Microbiol.* **32**:2533–2539.

30. **Labbe, R. G.** 1981. Enterotoxin formation by *Clostridium perfringens* type A in a defined medium. *Appl. Environ. Microbiol.* **41**:315–317.

31. **Labbe, R. G.** 1989. *Clostridium perfringens,* p. 192–234. *In* M. P. Doyle (ed.), *Foodborne Bacterial Pathogens.* Marcel Dekker, New York.

32. **Labbe, R. G., and C. L. Duncan.** 1977. Evidence for stable messenger ribonucleic acid during sporulation and enterotoxin synthesis by *Clostridium perfringens* type A. *J. Bacteriol.* **129**:843–849.

33. **Lindsay, J., A. Mach, M. Wilkinson, L. Martin, L. Wallace, F. Keller, and M. Wojciechowski.** 1993. *Clostridium perfringens* type A cytotoxic-enterotoxin(s) as triggers for death in sudden infant death syndrome: development of a toxico-infection hypothesis. *Curr. Microbiol.* **27**:51–59.

34. **Matsuda, M., K. Ozutsumi, H. Iwashi, and N. Sugimoto.** 1986. Primary action of *Clostridium perfringens* type A enterotoxin on HeLa and Vero cells in the absence of extracellular calcium: rapid and characteristic changes in membrane permeability. *Biochem. Biophys. Res. Commun.* **141**:704–710.

35. **McClane, B. A.** 1984. Osmotic stabilizers differentially inhibit permeability alterations induced in Vero cells by *Clostridium perfringens* enterotoxin. *Biochim. Biophys. Acta* **777**:99–106.

36. **McClane, B. A.** 1992. *Clostridium perfringens* enterotoxin: structure, action and detection. *J. Food Safety* **12**:237–252.

37. **McClane, B. A.** 1994. *Clostridium perfringens* enterotoxin acts by producing small molecule permeability alterations in plasma membranes. *Toxicology* **87**:43–67.

38. **McClane, B. A., P. C. Hanna, and A. Wnek.** 1988. *Clostridium perfringens* type A enterotoxin. *Microb. Pathog.* **4**:317–323.

39. **McClane, B. A., and J. L. McDonel.** 1979. The effects of *Clostridium perfringens* enterotoxin on morphology, viability and macromolecular synthesis in Vero cells. *J. Cell. Physiol.* **99**:191–200.

40. **McClane, B. A., and A. P. Wnek.** 1990. Studies of *Clostridium perfringens* enterotoxin action at different temperatures demonstrate a correlation between complex formation and cytotoxicity. *Infect. Immun.* **58**:3109–3115.

41. **McClane, B. A., A. P. Wnek, K. I. Hulkower, and P. C. Hanna.** 1988. Divalent cation involvement in the action of *Clostridium perfringens* type A enterotoxin. *J. Biol. Chem.* **263**:2423–2435.

42. **McDonel, J. L.** 1980. Binding of *Clostridium perfringens* ^{125}I-enterotoxin to rabbit intestinal cells. *Biochemistry* **21**: 4801–4807.

43. **McDonel, J. L.** 1986. Toxins of *Clostridium perfringens* types A, B, C, D and E, p. 477–517. *In* F. Dorner and H. Drews (ed.), *Pharmacology of Bacterial Toxins.* Pergamon Press, Oxford.

44. **McDonel, J. L., L. W. Chang, J. L. Pounds, and C. L. Duncan.** 1978. The effects of *Clostridium perfringens* enterotoxin on rat and rabbit ileum: an electron microscopy study. *Lab. Invest.* **39**:210–218.

45. **McDonel, J. L., and C. L. Duncan.** 1975. Histopathological effects of *Clostridium perfringens* enterotoxin in the rabbit ileum. *Infect. Immun.* **12**:1214–1218.

46. **McDonel, J. L., and B. A. McClane.** 1979. Binding vs. biological activity of *Clostridium perfringens* enterotoxin in Vero cells. *Biochem. Biophys. Res. Commun.* **87**:497–504.

47. **Melville, S. B., R. G. Labbe, and A. L. Sonenshein.** 1994. Expression from the *Clostridium perfringens cpe* promoter in *C. perfringens* and *Bacillus subtilis. Infect. Immun.* **62**: 5550–5558.

48. **Mietzner, T. A., J. F. Kokai-Kun, P. C. Hanna, and B. A. McClane.** 1992. A conjugated synthetic peptide corresponding to the C-terminal region of *Clostridium perfringens* type A enterotoxin elicits an enterotoxin-neutralizing antibody response in mice. *Infect. Immun.* **60**:3947–3951.

49. **Murrell, W. G., B. J. Stewart, C. O'Neill, S. Siarakas, and S. Kariks.** 1993. Enterotoxigenic bacteria in the sudden infant death syndrome. *J. Med. Microbiol.* **39**:114–127.

50. **Regan, C. M., Q. Syed, and P. J. Tunstall.** 1995. A hospital outbreak of *Clostridium perfringens* food poisoning—implications for food hygiene review in hospital. *J. Hosp. Infect.* **29**:69–73.

51. **Ryu, S., and R. G. Labbe.** 1989. Coat and enterotoxin-related proteins in *Clostridium perfringens* spores. *J. Gen. Microbiol.* **135**:3109–3118.

52. **Saito, M.** 1990. Production of enterotoxin by *Clostridium perfringens* derived from humans, animals, foods and the natural environment in Japan. *J. Food Prot.* **53**:115–118.

53. **Sherman, S., E. Klein, and B. A. McClane.** 1994. *Clostridium perfringens* type A enterotoxin induces tissue damage and fluid accumulation in rabbit ileum. *J. Diarrheal Dis. Res.* **12**:200–207.

54. **Skjelkvale, R., and T. Uemura.** 1977. Experimental diarrhea in human volunteers following oral administration of *Clostridium perfringens* enterotoxin. *J. Appl. Bacteriol.* **46**:281–286.

55. **Smith, W. P., and J. L. McDonel.** 1980. *Clostridium perfringens* type A: in vitro systems for sporulation and enterotoxin synthesis. *J. Bacteriol.* **144**:306–311.

56. **Stringer, M. F.** 1985. *Clostridium perfringens* type A food poisoning, p. 117–144. *In* S. P. Boriello (ed.), *Clostridia in Gastrointestinal Disease.* CRC Press, Boca Raton, Fla.

57. **Strong, D. H., C. L. Duncan, and G. Perna.** 1971. *Clostridium perfringens* type A food poisoning. II. Response of the rabbit ileum as an indication of enteropathogenicity of strains of *Clostridium perfringens* in human beings. *Infect. Immun.* **3**:171–178.

58. **Sugii, S., and Y. Horiguchi.** 1988. Identification and isolation of the binding substance for *Clostridium perfringens* enterotoxin on Vero cells. *FEMS Microbiol. Lett.* **52**:85–90.

59. **Sugimoto, W., M. Takagi, K. Ozutsumi, S. Harada, and M. Matsuda.** 1988. Enterotoxin of *Clostridium perfringens* type A forms ion-permeable channels in a lipid bilayer membrane. *Biochem. Biophys. Res. Commun.* **156**:551–556.

60. **Todd, E. C. D.** 1989. Cost of acute bacterial foodborne disease in Canada and the United States. *Int. J. Food Microbiol.* **9**:313–326.

61. **Todd, E. C. D.** 1989. Preliminary estimates of costs of foodborne disease in the United States. *J. Food Prot.* **52**:595–601.

62. **Van Damme-Jongsten, M., K. Werners, and S. Notermans.** 1989. Cloning and sequencing of the *Clostridium perfringens* enterotoxin gene. *Antonie Leeuwenhoek J. Microbiol.* **56**:181–190.

63. **Walker, P. D.** 1985. Pig-Bel, p. 93–116. *In* S. P. Borriello (ed.), *Clostridia in Gastrointestinal Disease.* CRC Press, Boca Raton, Fla.

64. **Wieckowski, E. U., A. P. Wnek, and B. A. McClane.** 1994. Evidence that an ~50-kDa mammalian plasma membrane protein with receptor-like properties mediates the amphiphilicity of specifically bound *Clostridium perfringens* enterotoxin. *J. Biol. Chem.* **269**: 10838–10848.

65. **Wnek, A. P., and B. A. McClane.** 1983. Identification of a 50,000 Mr protein from rabbit brush border membranes that bind *Clostridium perfringens* enterotoxin. *Biochem. Biophys. Res. Commun.* **112**:1099–1105.

66. **Wnek, A. P., and B. A. McClane.** 1986. Comparison of receptors for *Clostridium perfringens* type A and cholera enterotoxins in isolated rabbit intestinal brush border membranes. *Microb. Pathog.* **1**:89–100.

67. **Wnek, A. P., and B. A. McClane.** 1989. Preliminary evidence that *Clostridium perfringens* type A enterotoxin is present in a 160,000-M_r complex in mammalian membranes. *Infect. Immun.* **57**:574–581.

68. **Wnek, A. P., R. J. Strouse, and B. A. McClane.** 1985. Production and characterization of monoclonal antibodies against *Clostridium perfringens* type A enterotoxin. *Infect. Immun.* **50**:442–448.

Per Einar Granum

17 Bacillus cereus

Bacillus cereus causes two different types of food poisoning: the diarrheal type, first recognized after a hospital outbreak (caused by contaminated vanilla sauce) in Oslo, Norway, in 1948 (25, 26), and the emetic type, described about 20 years later after several outbreaks (rice) in London (33). The diarrheal type of food poisoning is caused by an enterotoxin(s) produced during vegetative growth of *B. cereus* in the small intestine (19), while emetic toxin is produced by cells growing in the food (32). For both types of foodborne illness, the food involved has usually been heat treated, and surviving spores are the source of the illness. *B. cereus* is not a competitive microorganism but grows well after cooking and cooling (<48°C). The heat treatment will cause spore germination, and in the absence of competing flora, *B. cereus* grows well. *B. cereus* is a common soil saprophyte and is easily spread to many types of foods, especially of plant origin, but is also frequently isolated from meat, eggs, and dairy products (32). The development of psychrotrophic strains in the dairy industry has led to increasing surveillance of *B. cereus* in recent years (17, 20, 24, 43, 44).

B. cereus foodborne illness is underreported, as both types of illness are relatively mild and usually last less than 24 h (32). However, a more severe form of the diarrheal type of *B. cereus* foodborne illness that lasts longer has occasionally been reported (18, 19).

CHARACTERISTICS OF THE ORGANISM

B. cereus is a gram-positive, spore-forming, motile, aerobic rod, but it grows well anaerobically. The genus *Bacillus* as it is characterized in *Bergey's Manual of Systematic Bacteriology* (12) is probably too diverse for a single genus. As more genetic information about the different species is made available, new divisions of the genus and species will probably result. For the latest review on the genus *Bacillus*, see reference 34; here, this genus is divided into six different subgroups, and *B. cereus* is classified in the *Bacillus subtilis* group. Four members of this group are closely related: *B. cereus*, *B. thuringiensis*, *B. anthracis*, and *B. mycoides*. On the basis of both phenotypic and genetic properties, the latter three species should be considered subspecies of *B. cereus* (7, 14). It seems that the variation among these four species is due mainly to genes on episomes rather than genes on the chromosome (6, 14, 15).

The four above-mentioned species are all large (cell width, >0.9 μm) and produce central to terminal ellipsoid or cylindrical spores that do not distend the sporangia (12). These *Bacillus* species sporulate easily after 2 to 3 days on most media, and *B. cereus* and *B. thuringiensis* lose their motility during the early stages of sporulation. The four species can be differentiated by using the criteria defined in Table 17.1.

Table 17.1 Criteria for differentiating between the four closely related *Bacillus* species

Species	Colony morphology	Hemolysis	Motility	Susceptibility to penicillin	Parasporal crystal inclusion
B. cereus	White	+	+	–	–
B. anthracis	White	–	–	+	–
B. thuringiensis	White/grey	+	+	–	+
B. mycoides	Rhizoid	(+)	–	–	–

RESERVOIRS

B. cereus is widespread in nature, being frequently isolated from soil and growing plants (32). From this natural environment it is easily spread to foods, especially those of plant origin. Through cross-contamination, it may then be spread to other foods, such as meat products (32). The problems in milk and milk products are caused by *B. cereus* which is spread from soil and grass to the udders of cows and into the raw milk. Through sporulation, *B. cereus* spores survive pasteurization, and after germination, the cells are free from competition from other vegetative cells (5). According to the classic literature (12), *B. cereus* is unable to grow at temperatures below 10°C and cannot grow in milk and milk products stored at temperatures between 4 and 8°C. However, the psychrotrophic strains that have developed can grow at temperatures as low as 4 to 6°C (17, 20, 44). In addition to rice and spices, dairy products are among the foods most frequently contaminated with *B. cereus.*

The closely related species *B. thuringiensis* is reported to produce enterotoxin (13, 35) and has been shown to cause foodborne illness in human volunteers (35). This characteristic may develop into a serious problem, as spraying of this organism to protect crops against insect attacks has become common in several countries. *B. thuringiensis* was recently reported to have caused an outbreak of foodborne illness (31). However, since the normally used procedures for confirmation of *B. cereus* would not differentiate between the two organisms (Table 17.1), undetected outbreaks may have occurred. To ensure safe spraying with *B. thuringiensis,* the organism in use should be unable to produce enterotoxin(s).

FOODBORNE OUTBREAKS

As already pointed out, the number of outbreaks of *B. cereus* foodborne illness is highly underestimated in the literature and official statistics. The main reason for this is the relatively short duration of both types of diseases (usually <24 h); also, foodborne illness transmitted through milk may affect only a few persons in many homes and thus would not be considered an outbreak. Another factor may be that many milk drinkers are partially protected against this type of foodborne illness through immunity acquired by continuous consumption of small amounts of *B. cereus.* Toward the end of its shelf life, milk frequently contains enough *B. cereus* cells to cause foodborne illness, given that the strains are high enterotoxin producers. However, *B. cereus* can produce a protease that results in off flavors that might limit the consumption of the product containing high numbers of the organism.

The dominant type of illness caused by *B. cereus* differs from county to country. In Japan, the emetic type is reported about 10 times more frequently than the diarrheal type (36), while in Europe and North America, the diarrheal type is most frequently reported (1, 32). This variance is probably due to differences in eating habits, although milk is reported to have caused at least one large outbreak of the emetic type in Japan (36). Several reports indicate that some patients have experienced both types of *B. cereus* foodborne illness simultaneously (32). It is also clear that many *B. cereus* strains have the ability to produce both types of toxins (20, 32).

Countries differ in their reporting procedures for foodborne illnesses, and therefore it is difficult to compare numbers of outbreaks among countries. The percentages of reported outbreaks and cases attributed to *B. cereus* in Japan, North America, and Europe vary from about 1 to 22% of outbreaks and from about 0.7 to 33% of cases (reports from different periods between 1960 and 1992) (1, 32). The greatest number of reported outbreaks and cases were from The Netherlands and Norway. *B. cereus* foodborne illness has been a focus of research and food control authorities in both of these relatively small countries. However, compared with many countries, Norway is relatively free from outbreaks of *Salmonella* spp. and *Campylobacter* spp. (1), the two most frequently reported causes of foodborne illness in the United Kingdom (40).

CHARACTERISTICS OF DISEASE

As noted, there are two types of *B. cereus* foodborne illness. The first type is caused by an emetic toxin that causes vomiting, while the second type, caused by en-

Table 17.2 Characteristics of the two types of disease caused by *B. cereus*[a]

Characteristic	Diarrheal	Emetic syndrome
Infective dose (cells)	10^5–10^7 (total)	10^5–10^8 (per g)
Toxin produced	In the small intestine of host	Preformed in foods
Type of toxin	Protein	Cyclic peptide
Incubation period (h)	8–16 (occasionally >24)	0.5–5
Duration of illness (h)	12–24 (occasionally several days)	6–24
Symptoms	Abdominal pain, watery diarrhea, occasionally nausea	Nausea, vomiting, malaise (sometimes followed by diarrhea, due to additional enterotoxin production?)
Foods most frequently implicated	Meat products, soups, vegetables, puddings and sauces, milk and milk products	Fried and cooked rice, pasta, pastry, noodles

[a]From references 18, 19, 32, 36, and 42.

terotoxin(s), causes diarrhea (32). In a small number of cases, patients have exhibited symptoms of both types of illness (32), probably as a result of production of both types of toxins. There has been some debate about whether the enterotoxin(s) can be preformed in foods and cause an intoxication. This is unlikely because the incubation time is too long (>6 h; average, 12 h) (32), and studies in model experiments have revealed that enterotoxin(s) is degraded in the gastrointestinal tract before reaching the ileum (19). Studies have revealed that the enterotoxin(s) can be preformed in foods; however, the number of *B. cereus* cells in food containing preformed enterotoxin is at least 2 orders of magnitude higher than that necessary for causing foodborne illness (16, 17, 19). Such products would be considered spoiled and no longer acceptable to the consumer. Characteristics of the two types of *B. cereus* foodborne illness are shown in Table 17.2.

Two recent outbreaks in Norway were associated with eating stew. The infective dose was estimated at 10^4 to 10^5 cells. Of 17 affected people, 3 were hospitalized, 1 for 3 weeks (18). The onset of symptoms for the three hospitalized patients was more than 24 h after consumption of the stew. The second outbreak, in February 1995, involved 152 cases among competitors of the Norwegian junior ski championship (22). The young atheletes (16 to 19 years old) had the most severe symptoms, whereas their coaches and the officials were not affected. The time to onset of symptoms for some patients was more than 24 h, and the duration of illness was from 2 to several days. Perhaps some strains of *B. cereus* colonize in the small intestine of some patients and cause more severe symptoms by producing enterotoxin at the site of colonization.

INFECTIVE DOSE AND SUSCEPTIBLE POPULATIONS

After the first recognized diarrheal outbreak of *B. cereus* foodborne illness in Oslo (caused by contaminated vanilla sauce), Hauge isolated the bacterium, grew it to 4×10^6 cells per ml, and drank 200 ml of the bacterial culture (25). After about 13 h, he developed abdominal pain and watery diarrhea that lasted for about 8 h. The infective dose was about 8×10^8 cells. Later studies of outbreak-related incriminated foods revealed *B. cereus* counts ranging from 200 to 10^9/g (or ml) (18, 19, 26, 32), with calculated infective doses ranging from 5×10^4 to 10^{11} cells per g. The infective doses may vary from about 10^5 to 10^8 viable cells or spores per g in part because of the large differences in the amounts of enterotoxin produced by different strains (19). Hence, food containing more than 10^4 *B. cereus* cells per g may not be safe for consumption.

Little is known about susceptible populations, but more severe symptoms have been associated with young athletes (<19 years old) and the elderly (>60 years old) (18, 22).

VIRULENCE FACTORS AND MECHANISMS OF PATHOGENICITY

The two types of *B. cereus* foodborne illness are caused by very different toxins. The emetic toxin, causing vomiting, has recently been isolated and characterized (2), whereas the diarrheal illness can be caused by more than one enterotoxin (10, 11, 19, 23, 32, 39, 41, 42).

The Emetic Toxin

The emetic toxin causes emesis (vomiting) only, and its structure has been elusive because until recently the only

Table 17.3 Properties of the emetic toxin cereulide[a]

Determination	Property or activity
Molecular mass	1.2 kDa
Structure	Ring-form peptide
Isoelectric point	Uncharged
Antigenic	No (?)
Biological activity on living primates	Vomiting
Receptor	5-HT_3 (stimulation of the vagus afferent)
Ileal loop tests (rabbit, mouse)	None
Cytotoxic	No
HEp-2 cells	Vacuolation activity
Stability to heat	90 min at 121°C
Stability to pH	Stable at pH 2–11
Effect of proteolysis (trysin, pepsin)	None
Toxin produced	In food: rice and milk at 25–32°C
Production	Not known (probably enzymatically)

[a]From references 2, 4, 32, 36, and 37.

detection method available involved primates (28, 32). The recent observation that the toxin could be detected in HEp-2 cells (vacuolation activity) (28) led to its isolation and characterization of its structure (2). Although there has been some uncertainty regarding the relationship between the emetic toxin and the vacuolating factor (36, 38), they are now confirmed to be the same toxin (4, 37). The emetic toxin, named cereulide, consists of a ring structure of three repeats of four amino and/or oxy acids: [D-*O*-Leu-D-Ala-L-*O*-Val-L-Val]$_3$. This ring structure (dodecadepsipeptide) has a molecular mass of 1.2 kDa and is closely related to the potassium ionophore valinomycin (2). The biosynthetic pathway and mechanism of action of the emetic toxin are presently undefined, although the toxin has been determined recently to stimulate the vagus afferent through binding to the 5-HT_3 receptor (4). It is not known if the toxin is a modified gene product or is enzymatically produced through modification of components in the growth medium. However, considering the structure of cereulide, it is more likely to be an enzymatically synthesized peptide and not a gene product.

It was previously hypothesized that the emetic toxin was a lipid (19, 32), likely because of cereulide's hydrophobic nature. Now that its structure is known, it is not surprising that it is not antigenic or resistant to heat, pH, or proteolysis (32) (Table 17.3).

Enterotoxins

The number of enterotoxins formed by *B. cereus* and their properties have been debated for many years (19, 32, 39, 41, 42). Gene cloning and sequencing studies have revealed that *B. cereus* produces at least two different enterotoxins (3, 27). However, there is no evidence that the recently reported enterotoxin T (3) causes illness.

Although several proteins may be involved in *B. cereus* foodborne illness, only one type of enterotoxin is likely responsible for the major symptom (8, 10, 19, 32, 39, 41). However, whether the enterotoxicity is due to a single protein or a multicomponent enterotoxin is unknown (8, 10, 19, 23, 39, 41). Although the structure of the enterotoxin is not certain, central to its activity is a protein whose gene has recently been cloned and sequenced. The protein has a molecular mass of 38 kDa as a mature protein (344 amino acids), after a signal peptide of 31 amino acids is split off (27). It has been shown in our laboratory by Western blotting (immunoblotting) with antienterotoxin (obtained from J. Kramer, Public Health Laboratory Service, London, England), partial protein sequencing, and use of the same purification procedures in different laboratories (Table 17.4) that this enterotoxin component (B component of the hemolysin BL) is probably the same protein purified by several other groups (23, 32, 41) (Table 17.4). It remains unknown whether this protein is functional by itself as an enterotoxin or jointly with other proteins. It is known that the enterotoxin does not possess multiple subunits because the molecular mass of the active fractions after column chromatography is not large enough for a multisubunit toxin (19a, 39).

At least four research groups have studied the same enterotoxin, with some differences reported in the number of constituents of the fully active enterotoxin. These differences could be due to variations in the methods used to purify the toxin, to differences in the toxicity assays used, or to strain differences. It is also possible that the single components purified by Shinagawa et al.

Table 17.4 Properties of enterotoxins purified in several laboratories

Laboratory, strain	Reference(s)	Enterotoxin	Relationship between different proteins	Biological test systems used	Hemolytic
Kramer, Turnbull, and coworkers, *B. cereus* F4433/73	32, 42	One main protein and two other proteins	The main protein, the same as the B component, and the two other proteins, probably sphingomyelinase and the nontoxic protein identified by Granum and Nissen (23), based on Western blotting with Kramer's antienterotoxin	Positive in rabbit loop, vascular permeability, and mouse lethality tests	Yes (weakly)
Thompson et al., *B. cereus* B-4ac	41	Three proteins	Not known, but the middle-sized protein (39.5 kDa) is probably the same as the B component (or enterotoxin T [same strain]). It is impossible to relate the other two proteins to the other work.	Positive in rabbit loop, vascular permeability, monkey feeding, and Vero cell tests	Yes
Shinagawa et al., *B. cereus* FM-1	39	One protein	The same as the B component, based on purification of this component in Granum's laboratory by Shinagawa's method and tested with Kramer's antiserum (21)	Positive in mouse loop, vascular permability, and mouse lethality tests	No
Granum and coworkers, *B. cereus* 1230-88	19, 20, 23	One protein	The same as the B component, based on the N-terminal amino acid sequence. The two other proteins that copurify with the enterotoxin until the final purification step are a nontoxic protein and sphingomyelinase (based on N-terminal amino acid sequencing)	Positive in Vero cell test and on Caco-2 cells (human intestinal cells)	No
Wong, Beecher, Heinrichs, and coworkers, *B. cereus* F837/76	10, 27	Three proteins	The B component cloned and sequenced; two other components (L_1 and L_2) necessary for biological activity, purified and N-terminal sequenced	Positive in rabbit loop and vascular permeability tests	Yes
Agata et al., *B. cereus* B- 4ac	3	One protein	Cloned and sequenced: unrelated to the other enterotoxin, based on the amino acid sequence (yet not known if this protein plays any role in food poisoning)	Positive in mouse loop, Vero cell, and mouse lethality tests	?

(39) and Granum and Nissen (23) contained sufficient amounts of other components needed for biological activity. However, from the N-terminal amino acid sequences in Table 17.5, one may conclude that the L_2 component described by Beecher and Wong (10) is the same protein as the 48-kDa protein described by Granum and Nissen (23). The L_2 component is the protein detected by the Oxoid (Basingstoke, England) *B. cereus* enterotoxin reversed passive latex agglutination assay kit (11).

The activity of hemolysin BL (L_1, L_2, and the B components) has been studied in the rabbit ileal loop assay (9). All three components were required for maximal fluid accumulation. When the B and L_1 components were tested in the vascular permeability assay, minor edema and vascular permeability were observed (9, 10). However, additional necrosis and a 30- to 60-fold increase in activity over the band L_1 component combination were observed when the L_2 component was added (10).

In spite of evidence that all three components of hemolysin BL must be present for full enterotoxin activity (at least for hemolysin BL from one strain), the 48-kDa protein (L_2 component, which is nontoxic in cell culture tests [20, 23]) may not be necessary for enterotoxin activity. Some strains of *B. cereus* associated with foodborne illness do not produce this protein (22). (See Addendum.) Sphingomyelinase also is not necessary for enterotoxin activity (B component); the en-

Table 17.5 Properties of the proteins in the enterotoxin fractions

Laboratory	Reference(s)	Protein(s) in purified fraction	N-terminal sequence of protein (aligned if possible)	Comment
Turnbull, Kramer, and coworkers	32, 42	50 kDa and pI 4.9; pI 5.1; pI 5.6	Not known	
Thompson et al.	41	43, 39.5, and 38 kDa	Not known	
Shinagawa et al.	39	1.45 kDa	Not known	Contained trace amounts of one contaminating protein
Granum and coworkers	23	48 kDa 40 kDa (the main protein) 34 kDa	xTQQEGMDISSSLxK IEQTNNGDTALSANE EASTNENDTLKVMT	Fraction before the last purification step, after the signal sequence is split off
Wong, Beecher, Heinrichs, and coworkers	10, 27	L_2 (43.2 kDa) L_1 (38.5 kDa) B component (37.8 kDa)	ETQXENMDIXS xETIAQEQKVGNYALGPE SEIEQTNNGDTALSANEA	After the signal sequence is split off
Agata et al.	3	Enterotoxin T (41 kDa with signal sequence)	MKELVSTARIGRSLGIHL	The start of the signal sequence

terotoxin has been purified free from this activity (23). Additionally, Ca^{2+} ions that inhibit sphingomyelinase do not alter the activity of enterotoxin (B component) in cell culture assays (23).

Presently, there is no consensus on the components necessary for enterotoxin activity. The B component alone may have enterotoxin activity at least for some strains, or it may require the L_1 and L_2 components. It is likely that differences in results among investigators are at least in part due to differences among strains. PCR assays have revealed that the structure of the B component varies among strains (19b).

The B component (enterotoxin) has been cloned and sequenced (27); the nucleotide and amino acid sequences are shown in Fig. 17.1. Northern (RNA) blot analysis revealed a 5.1-kb transcript that hybridizes with a 0.5-kb internal probe to the B-component coding sequence (27). It was suggested that this fragment contains the coding region for the two L components (27). Interestingly, the first 158 amino acids of the open reading frame following the B component (Fig. 17.1 and 17.2) are 73% homologous to the first 159 amino acids of the B component (83% homologous in nucleotide sequence), suggesting that this protein may be an essential part of the enterotoxin or a related but different enterotoxin.

A few studies have addressed the mode of action of the enterotoxin (19, 23, 32). Studies with ileal loop assays have revealed that receptor binding is weak, with antibodies able to reverse the effect of enterotoxin in ileal loops within 10 min after injection (32). *B. cereus* enterotoxin is at least 100-fold more potent than *Clostridium perfringens* enterotoxin (19, 21), and it acts independently of divalent cations (19). Although *B. cereus* and *C. perfringens* enterotoxins both disrupt the membrane of epithelial cells, their mechanisms of action are different (19).

Another toxin, enterotoxin T, which is composed of a single protein, has a molecular mass of 41 kDa and was recently identified by Agata et al. (3). Enterotoxin T is different from any of the known proteins previously associated with the multicomponent enterotoxin described above. Until more is known about the biological activity and strain variation in production of this enterotoxin, its significance in *B. cereus* foodborne illness is uncertain. Enterotoxin T has been cloned and sequenced, and the nucleotide and amino acid sequences are shown in Fig. 17.3. Enterotoxin T has limited homology to the B component of BL hemolysin (Fig. 17.4). Whether the areas of homology have any biological significance is yet to be determined.

Other Toxins

Other proteins with enterotoxin activity have been reported (32), but little is known about their properties. One enterotoxin-like protein reportedly had a molecular mass of about 57 kDa, and another was found to have a molecular mass of about 100 kDa (32). The smaller protein may be the same as one of those described above.

```
                                                                          M  I  K  K  I  P  Y  K  L  L
   1 GATATTGCTTTTAAACAGGAGTAGAACTGAAATTTAGAACCTAAATTGGAGGAAAATGAAATGATAAAAAAAATCCCTTACAAATTACTC
      A  V  S  T  L  L  T  I  T  T  A  N  V  V  S  P  V  A  T  F  A  S  E  I  E  Q  T  N  N  G
  91 GCTGTATCGACGTTATTAACTATTACAACCGCTAATGTAGTTTCACCTGTAGCAACTTTTGCAAGTGAAATTGAACAAACGAACAATGGA
      D  T  A  L  S  A  N  E  A  K  M  K  E  T  L  Q  K  A  G  L  F  A  K  S  M  N  A  Y  S  Y
 181 GATACGGCTCTTTCTGCAAATGAAGCGAAGATGAAAGAAACTTTGCAAAAGGCTGGATTATTTGCAAAATCTATGAATGCCTATTCTTAT
      M  L  I  K  N  P  D  V  N  F  E  G  I  T  I  N  G  Y  V  D  L  P  G  R  I  V  Q  D  Q  K
 271 ATGTTAATTAAAAATCCTGATGTGAATTTTGAGGGAATTACTATTAATGGATATGTAGATTTACCTGGTAGAATCGTACAAGATCAAAAG
      N  A  R  A  H  A  V  T  W  D  T  K  V  K  K  Q  L  L  D  T  L  T  G  I  V  E  Y  D  T  T
 361 AATGCAAGAGCACATGCTGTTACTTGGGATACGAAAGTGAAAAAACAGCTTTTAGATACATTGACTGGTATTGTTGAATATGATACGACG
      F  D  N  Y  Y  E  T  M  V  E  A  I  N  T  G  D  G  E  T  L  K  E  G  I  T  D  L  R  G  E
 451 TTTGACAATTATTATGAAACAATGGTAGAGGCAATTAATACAGGGGATGGAGAAACTTTAAAAGAAGGGATTACAGATTTGCGAGGTGAA
      I  Q  Q  N  Q  K  Y  A  Q  Q  L  I  E  E  L  T  K  L  R  D  S  I  G  H  D  V  R  A  F  G
 541 ATTCAACAAAATCAAAAGTATGCACAACAACTAATAGAAGAATTAACTAAATTAAGAGACTCTATTGGACACGATGTTAGAGCATTTGGA
      S  N  K  E  L  L  Q  S  I  L  K  N  Q  G  A  D  V  D  A  D  Q  K  R  L  E  E  V  L  G  S
 631 AGTAATAAAGAGCTCTTGCAGTCAATTTTAAAAAATCAAGGTGCAGATGTTGATGCCGATCAAAAGCGTCTAGAAGAAGTATTAGGATCA
      V  N  Y  Y  K  Q  L  E  S  D  G  F  N  V  M  K  G  A  I  L  G  L  P  I  I  G  G  I  I  V
 721 GTAAACTATTATAAACAATTAGAATCTGATGGGTTTAATGTAATGAAGGGTGCTATTTTGGGTCTACCAATAATTGGCGGTATTATAGTG
      G  V  A  R  D  N  L  G  K  L  E  P  L  L  A  E  L  R  Q  T  V  D  Y  K  V  T  L  N  R  V
 811 GGAGTAGCAAGGGATAATTTAGGTAAGTTAGAGCCTTTATTAGCAGAATTACGTCAGACCGTGGATTATAAAGTAACCTTAAATCGTGTA
      V  G  V  A  Y  S  N  I  N  E  I  D  K  A  L  D  D  A  I  N  A  L  T  Y  M  S  T  Q  W  H
 901 GTTGGAGTTGCTTACAGTAATATTAATGAAATCGACAAGGCGCTTGATGATGCTATTAACGCTCTTACTTATATGTCCACGCAGTGGCAT
      D  L  D  S  Q  Y  S  G  V  L  G  H  I  E  N  A  A  Q  K  A  D  Q  N  K  F  K  F  L  K  P
 991 GATTTAGATTCTCAATATTCGGGCGTTCTAGGGCATATTGAGAATGCAGCTCAAAAAGCCGATCAAAATAAATTTAAATTCTTAAAACCT
      N  L  N  A  A  K  D  S  W  K  T  L  R  T  D  A  V  T  L  K  E  G  I  K  E  L  K  V  E  T
1081 AATTTAAATGCAGCGAAAGATAGTTGGAAAACATTACGAACAGATGCTGTTACATTAAAAGAAGGAATAAAGGAGTTAAAAGTAGAAACT
      V  T  P  Q  K  *
1171 GTTACTCCACAAAAATAGGGAAATATTAATTCTGTTGTAAAGTGAACTAAAACATAGAAAGTCTATGATTACACTGTGAAACAGAAAAGT

1261 GATAATTAATAAATCCTATATAACAGAAAGGCGGAGTCTCATTAAAGACTTCGCCTTTCAATTATATATAAGTATGATTCGAATAAAAGA

1351 ATAACCTTGTAGTTTAATTGCTTTTTTTTAATTTGCAACAGAATCGGTTAATTTGAAGCTGATTATATACATTTTCTTTAATACTTAATG

1441 ATTTGAAGACTGCTAAAGAAAGTTCAAACGATACTAATAAAAGATGTAGAACTCATTTTGAAAGACATCGCTTTTTAATAAGATTAGAAC
                                              M  K  K  I  P  N  K  L  L  A  V  S  A  F  L  T  I  T
1531 TTGAACCCATACCCTATATAGGAGGAATACGAAAAGATGAAAAAAATCCCTAATAAACTACTCGCTGTATCAGCGTTTTTAACTATAACA
      T  T  Y  A  V  I  P  I  E  T  F  A  I  E  I  Q  Q  T  N  T  E  N  R  S  L  S  A  N  E  E
1621 ACTACTTATGCAGTCATACCAATAGAAACTTTTGCAATTGAAATTCAACAAACGAACACTGAAAATAGGTCTCTTTCAGCAAATGAAGAA
      Q  M  K  K  A  L  Q  D  A  G  L  F  V  K  A  M  N  E  Y  S  Y  L  L  I  H  N  P  D  V  S
1711 CAGATGAAAAAAGCTTTGCAAGATGCTGGTTTATTTGTAAAAGCTATGAATGAATATTCTTATTTGCTAATTCATAATCCAGATGTGAGT
      F  E  G  I  T  I  N  G  N  T  D  L  P  S  K  I  V  Q  D  Q  K  N  A  R  A  H  A  V  T  W
1801 TTTGAAGGAATAACTATTAATGGAAATACAGATTTACCTAGTAAAATTGTACAAGATCAAAAGAATGCAAGAGCACATGCTGTTACATGG
      N  T  H  V  K  K  Q  L  L  D  T  L  T  G  I  I  E  Y  D  T  K  F  E  N  H  Y  E  T  L  V
1891 AATACACACGTAAAAAAACAGCTTTTAGATACATTGACAGGCATTATAGAATACGATACAAAATTTGAAAATCATTATGAAACATTAGTA
      E  A  I  N  T  G  N  G  D  T  L  K  K  G  I  T  D  L  Q  G
1981 GAGGCGATCAATACTGGAAATGGAGATACTTTAAAAAAAGGGATTACAGATTTACAAGGAG
```

Figure 17.1 Nucleotide and deduced amino acid sequences of the B component and the start of an unidentified second gene, as determined by Heinrichs et al. (27). The amino acids in boldface indicate the signal sequence. The potential ribosomal binding sites are underlined.

```
B component MIKKIPYKLLAVSTLLTITTANVVSPVATFASEIEQTNNGDTALSANEAKMKETLQKAGL  60
              |||| ||||||  |||||   | |  ||| || |||     |||||  ||  || |||
3´gene       MKKIPNKLLAVSAFLTITTTYAVIPIETFAIEIQQTNTENRSLSANEEQMKKALQDAGL  59

B component FAKSMNAYSYMLIKNPDVNFEGITINGYVDLPGRIVQDQKNARAHAVTWDTKVKKQLLDT 120
            | | || ||| || |||| ||||||||  |||  ||||||||||||||| | ||||||||
3´gene      FVKAMNEYSYLLIHNPDVSFEGITINGNTDLPSKIVQDQKNARAHAVTWNTHVKKQLLDT 119

B component LTGIVEYDTTFDNYYETMVEAINTGDGETLKEGITDLRG 159
            |||| |||| | | ||| ||||||| | ||| ||||| |
3´gene      LTGIIEYDTKFENHYETLVEAINTGNGDTLKKGITDLQG 158
```

Figure 17.2 Amino acid sequence homology between the B component and the unidentified gene 3′ to the B component (from Fig. 17.1). Signal sequence is shown in boldface type.

```
       M  K  E  L  V  S  T  A  R  I  G  R  S  L  G  I  H  L  I  L  A  T  Q  K  P  S  G  V  V  D
   1  ATGAAAGAGTTAGTTTCAACAGCGCGTATCGGTCGTTCACTCGGGATCCATTTAATATTAGCTACGCAAAAACCGAGTGGTGTTGTAGAT
       D  Q  I  W  S  N  S  K  F  K  L  A  L  K  V  Q  N  T  S  D  S  N  E  I  L  K  T  P  D  A
  91  GATCAAATTTGGAGTAACTCGAAATTCAAACTAGCATTAAAAGTTCAAAATACGTCAGATAGTAATGAAATCTTAAAAACGCCAGATGCT
       A  E  I  T  L  P  G  R  A  Y  L  Q  V  G  N  N  E  I  Y  E  L  F  Q  S  A  W  S  G  A  D
 181  GCTGAAATTACATTACCAGGACGTGCTTACTTACAAGTTGGGAATAATGAAATTTATGAACTATTCCAATCAGCTTGGAGCGGAGCAGAC
       Y  V  E  N  K  E  D  K  E  H  L  D  A  T  I  Y  A  I  N  D  L  G  Q  Y  E  I  L  S  E  D
 271  TATGTAGAAAACAAAGAGGATAAAGAACATTTAGACGCAACAATCTATGCAATAAATGATCTAGGACAATATGAAATATTAAGTGAAGAT
       L  S  G  L  G  S  S  K  E  V  I  S  V  P  S  E  L  D  A  V  I  D  Y  I  H  D  Y  A  E  I
 361  TTAAGTGGCCTTGGTAGCAGTAAAGAAGTAATAAGCGTACCATCTGAACTGGATGCTGTTATTGACTACATTCACGATTACGCAGAAATA
       N  E  I  E  A  L  A  R  P  W  L  P  P  L  P  E  S  V  Y  L  Q  D  L  H  A  I  Q  F  K  E
 451  AATGAAATTGAAGCGTTAGCTAGACCGTGGTTACCACCACTTCCAGAAAGCGTATATTTACAAGACTTACATGCAATTCAGTTCAAAGAA
       A  W  A  K  E  K  K  P  L  Q  A  T  V  G  L  L  D  Q  P  E  L  Q  S  Q  T  P  L  T  L  D
 541  GCATGGGCGAAAGAAAAGAAACCATTACAAGCAACAGTTGGTCTACTAGATCAGCCTGAATTACAATCACAAACACCATTAACATTAGAT
       I  S  K  D  G  H  V  A  V  F  S  S  P  G  Y  G  K  S  T  F  L  Q  S  V  V  M  D  V  A  R
 631  ATTAGTAAAGACGGACACGTAGCGGTCTTCTCAAGCCCAGGCTACGGAAAATCAACATTCTTACAATCAGTCGTTATGGATGTAGCTCGT
       Q  H  S  P  E  H  L  H  V  Y  L  V  D  L  G  T  N  G  L  L  P  L  K  G  L  P  H  V  A  D
 721  CAGCATAGTCCGGAGCATTTGCATGTGTATTTAGTGGACCTTGGAACAAATGGTCTTCTACCCTTGAAAGGATTACCTCATGTAGCGGAT
       T  I  T  I  D  E  S  E  K  C  L  K  F  V  E  R  L  T  Q  E  M  K  N  R  K  R  L  L  S  E
 811  ACGATTACGATTGATGAATCTGAAAAATGTTTAAAGTTTGTTGAAAGATTAACTCAAGAAATGAAAAATCGTAAACGATTATTAAGTGAA
       Y  D  V  A  N  I  E  M  Y  E  K  A  S  G  K  E  I  P  H  I  I  I  A  I  D  N  Y  D  A  V
 901  TATGACGTTGCAAATATTGAAATGTATGAAAAGGCAAGCGGGAAAGAAATACCACATATTATTATTGCAATCGACAATTATGATGCAGTA
       K  E  A  K  F  Y  E  S  F  E  M  L  I  M  Q  I  V  R  D  G  A  S  L  G  I  L  F  F  Y  V
 991  AAAGAAGCGAAGTTCTATGAAAGCTTTGAAATGCTAATTATGCAAATTGTCCGAGATGGTGCAAGTTTAGGAATTCTATTCTTCTATGTC
       F  F  Y  F  G  G  *
1081  TTCTTTTACTTCGGCGGATAA
```

Figure 17.3 Nucleotide and deduced amino acid sequences of the coding region for enterotoxin T, as determined by Agata et al. (2).

The Spore

The spore of *B. cereus* is an important factor in foodborne illness, as it is more hydrophobic than any other *Bacillus* spp. spores, which enables it to adhere to several types of surfaces (29, 30). Hence, it is difficult to remove from equipment during cleaning. *B. cereus* spores also possess appendages and/or pili (5, 30) that are, at least in part, involved in adhesion (30). These adherence properties not only enable spores to resist normal sanitation procedures, and thus contaminate foods during processing, but also aid in binding to epithelial cells. Studies have shown that *B. cereus* spores can adhere to Caco-2 cells in culture and that these properties are associated with their hydrophobicity and possibly their appendages (5a). Spore adherence to epithelial cells followed by germination and production of enterotoxin may explain the long incubation periods observed in some food-associated outbreaks (18, 19).

Commercial Methods for Detection of *B. cereus* Toxins

No commercial kit for detection of the emetic toxin (cereulide) is available. However, two commercial kits for detection of *B. cereus* enterotoxin are available. One is a reversed passive latex agglutination kit (Oxoid) based on the nontoxic (L_2) component of the enterotoxin complex (11, 20). This component has been identified by Western blot analysis (16) and N-terminal amino acid sequencing of the protein used as the control in the kit (11, 23). Some outbreak-associated strains that produce large amounts of enterotoxin do not produce this protein, and hence it has limited value in enterotoxin detection (20, 22). Christiansson (16) has also shown that the amount of toxin detected by this kit is not correlated with cytotoxicity.

The second *B. cereus* enterotoxin kit, an enzyme-linked immunosorbent assay method developed by

Enterotoxin T

```
149  ATAGTAATGAAATCTTAAAAACGCCAGATGCTGCTGAAATTACATTACCAGGACGTGCTTACTTACAAGT  218
     ||| ||||||||||   ||  |||  |||| ||||        ||||     |||  |||||||| | ||
920  ATATTAATGAAATCGACAAGGCGCTTGATGATGCT........ATTA.....ACGCTCTTACTTATATGT  976
```

B component

Figure 17.4 Nucleotide sequence homology between the B component (from Fig. 17.1) and enterotoxin T (from Fig. 17.3).

Tecra (Roseville, Australia), is based on proteins that are not toxic (11). The dominant of the two proteins has been N-terminal sequenced, and no homology to any known *B. cereus* protein was found (11). However, Christiansson (16) determined that protein detected by the Tecra kit correlated with cytotoxicity. Neither of the two kits specifically detects enterotoxin of *B. cereus.* (See Addendum.)

CONCLUDING REMARKS

B. cereus is a normal inhibitant of the soil and is frequently isolated from a variety of foods, including vegetables, dairy products, and meat. It causes an emetic or diarrheal type of food-associated illness that is becoming increasingly important in the industrialized world. The diarrheal type of illness is most prevalent in the western hemisphere, whereas the emetic type is most prevalent in Japan. Desserts, meat dishes, and dairy products are the foods most frequently associated with diarrheal illness, whereas rice is the most common vehicle of emetic illness.

A *B. cereus* emetic toxin has been isolated but is yet to be characterized. At least two types of *B. cereus* enterotoxins have been identified, although the involvement of enterotoxin T in foodborne illness has yet to be determined. The best-characterized enterotoxin is hemolytic and is composed of three proteins.

Some *B. cereus* strains are psychrotrophic and capable of growing at refrigeration temperature. These variants raise concern about the safety of cooked, refrigerated foods with extended shelf lives. *B. cereus* foodborne illness is likely highly underreported because its relatively mild symptoms are of short duration. However, consumer interest in precooked chilled foods with long shelf lives may lead to products well suited for *B. cereus* survival, growth, and toxin formation. Such foods could increase the prominence of *B. cereus* as a foodborne pathogen.

Addendum

Two recent papers (Granum et al., *FEMS Microbiol. Lett.* **141:** 145–149, 1996; Lund and Granum, *FEMS Microbiol. Lett.* **141:** 151–156, 1996) have shown that *B. cereus* produces a further three-component nonhemolytic enterotoxin. One of the components of this complex is recognized by the Tecra kit.

References

1. **Aas, N., B. Gondrosen, and G. Langeland.** 1992. *Norwegian Food Control Authority's Report on Food Associated Diseases in 1990.* SNT report 3. Norwegian Food Control Authority, Oslo.
2. **Agata, N., M. Mori, M. Ohta, S. Suwan, I. Ohtani, and M. Isobe.** 1994. A novel dodecadepsipeptide, cereulide, isolated from *Bacillus cereus* causes vacuole formation in HEp-2 cells. *FEMS Microbiol. Lett.* **121:**31–34.
3. **Agata, N., M. Ohta, Y. Arakawa, and M. Mori.** 1995. The *bceT* gene of *Bacillus cereus* encodes an enterotoxic protein. *Microbiology* **141:**983–988.
4. **Agata, N., M. Ohta, M. Mori, and M. Isobe.** 1995. A novel dodecadepsipeptide, cereulide, is an emetic toxin of *Bacillus cereus. FEMS Microbiol. Lett.* **129:**17–20.
5. **Andersson, A., U. Rönner, and P. E. Granum.** 1995. What problems does the food industry have with the sporeforming pathogens *Bacillus cereus* and *Clostridium perfringens*? *Int. J. Food Microbiol.* **18:**145–157.

5a. **Anderson, A., U. Rönner, and P. E. Granum.** Unpublished results.

6. **Ash, C., and M. D. Collins.** 1992. Comparative analysis of 23S ribosomal RNA gene sequence of *Bacillus anthracis* and emetic *Bacillus cereus* determined by PCR-direct sequencing. *FEMS Microbiol. Lett.* **73:**75–80.
7. **Ash, C., J. A. Farrow, M. Dorsch, E. Steckebrandt, and M. D. Collins.** 1991. Comparative analysis of *Bacillus anthracis, Bacillus cereus,* and related species on the basis of reverse transcriptase sequencing of 16S rRNA. *Int. J. Syst. Bacteriol.* **41:**343–346.
8. **Beecher, D. J., and J. D. Macmillan.** 1991. Characterization of the components of hemolysin BL from *Bacillus cereus. Infect. Immun.* **59:**1778–1784.
9. **Beecher, D. J., J. L. Schoeni, and A. C. L. Wong.** 1995. Enterotoxin activity of hemolysin BL from *Bacillus cereus. Infect. Immun.* **63:**4423–4428.
10. **Beecher, D. J., and A. C. L. Wong.** 1994. Improved purification and characterization of hemolysin BL, a hemolytic dermonecrotic vascular permeability factor from *Bacillus cereus. Infect. Immun.* **62:**980–986.
11. **Beecher, D. J., and A. C. L. Wong.** 1994. Identification and analysis of the antigens detected by two commercial *Bacillus cereus* diarrheal enterotoxin immunoassay kits. *Appl. Environ. Microbiol.* **60:**4614–4616.
12. **Claus, D., and R. C. W. Berkeley.** 1986. Genus *Bacillus,* p. 1105–1139. *In* P. H. A. Sneath (ed.), *Bergey's Manual of Systematic Bacteriology,* vol. 2. The Williams & Wilkins Co., Baltimore.
13. **Drobniewski, F. A.** 1993. *Bacillus cereus* and related species. *Clin. Microbiol. Rev.* **6:**324–338.
14. **Carlson, C. R.** 1994. Physical mapping of the *Bacillus cereus* and *Bacillus thuringiensis* chromosomes by pulsed field gel electrophoresis. Ph.D. thesis. Institute of Pharmacy, University of Oslo, Oslo, Norway.
15. **Carlson, C. R., D. A. Caugant, and A.-B. Kolstø.** 1994. Genotypic diversity among *Bacillus cereus* and *Bacillus thuringiensis* strains. *Appl. Environ. Microbiol.* **60:**1719–1725.
16. **Christiansson, A.** 1993. Enterotoxin production in milk by *Bacillus cereus:* a comparison of methods for toxin detection. *Bull. Int. Dairy Fed.* **287:**54–59.
17. **Christiansson, A., A. S. Naidu, I. Nilsson, T. Wadström, and H.-E. Pettersson.** 1989. Toxin production by *Bacillus cereus* dairy isolates in milk at low temperatures. *Appl. Environ. Microbiol.* **55:**2595–2600.

18. **Granum, P. E.** 1994. *Bacillus cereus* and food hygiene. *Norsk Vet. Tidskr.* **106**:911–915. (In Norwegian.)

19. **Granum, P. E.** 1994. *Bacillus cereus* and its toxins. *J. Appl. Bacteriol. Symp. Suppl.* **76**:61S-66S.

19a. **Granum, P. E.** 1995. Unpublished results.

19b. **Granum, P. E., et al.** 1995. Unpublished results.

20. **Granum, P. E., S. Brynestad, and J. M. Kramer.** 1993. Analysis of enterotoxin production by *Bacillus cereus* from dairy products, food poisoning incidents and non-gastrointestinal infections. *Int. J. Food Microbiol.* **17**:269–279.

21. **Granum, P. E., S. Brynestad, K. O'Sullivan, and H. Nissen.** 1993. The enterotoxin from *Bacillus cereus:* production and biochemical characterization. *Neth. Milk Dairy J.* **47**:63–70.

22. **Granum, P. E., A. Næstvold, and K. N. Gundersby.** 1995. *Bacillus cereus* food poisoning during the Norwegian Ski Championship for juniors. *Norsk Vet. Tidskr.* **107**:945–948. (In Norwegian, English abstract.)

23. **Granum, P. E., and H. Nissen.** 1993. Sphingomyelinase is part of "enterotoxin complex" produced by *Bacillus cereus. FEMS Microbiol. Lett.* **110**:97–100.

24. **Griffiths, M. W.** 1990. Toxin production by psychrotropic *Bacillus* spp. present in milk. *J. Food Prot.* **53**:790–792.

25. **Hauge, S.** 1950. Matforgiftninger framkalt av *Bacillus cereus. Nord. Hyg. Tidskr.* **31**:189–206.

26. **Hauge, S.** 1955. Food poisoning caused by aerobic spore forming bacilli. *J. Appl. Bacteriol.* **18**:591–595.

27. **Heinrichs, J. H., D. J. Beecher, J. M. MacMillan, and B. A. Zilinskas.** 1993. Molecular cloning and characterization of the *hblA* gene encoding the B component of hemolysin BL from *Bacillus cereus. J. Bacteriol.* **175**:6760–6766.

28. **Hughes, S., B. Bartholomew, J. C. Hardy, and J. M. Kramer.** 1988. Potential application of a HEp-2 cell assay in the investigation of *Bacillus cereus* emetic-syndrome food poisoning. *FEMS Microbiol. Lett.* **52**:7–12.

29. **Husmark, U.** 1993. Adhesion mechanisms of bacterial spores to solid surfaces. Ph.D. thesis. Department of Food Science, Chalmers University of Technology and SIK, The Swedish Institute for Food Research, Göteborg, Sweden.

30. **Husmark, U., and U. Rönner.** 1992. The influence of hydrophobic, electrostatic and morphologic properties on the adhesion of Bacillus spores. *Biofouling* **5**:335–344.

31. **Jackson, S. G., R. B. Goodbrand, R. Ahmed, and S. Kasatiya.** 1995. *Bacillus cereus* and *Bacillus thuringiensis* isolated in a gastroenteritis outbreak investigation. *Lett. Appl. Microbiol.* **21**:103–105.

32. **Kramer, J. M., and R. J. Gilbert.** 1989. *Bacillus cereus* and other *Bacillus* species, p. 21–70. *In* M. P. Doyle (ed.), *Foodborne Bacterial Pathogens.* Marcel Dekker, New York.

33. **Mortimer, P. R., and G. McCann.** 1974. Food poisoning episodes associated with *Bacillus cereus* in fried rice. *Lancet* **i**:1043–1045.

34. **Preist, F. G.** 1993. Systematics and ecology of *Bacillus,* p. 3–16. *In* A. L. Sonenshein, J. A. Hoch, and R. Losick (ed.), *Bacillus subtilis and Other Gram-Positive Bacteria.* American Society for Microbiology, Washington, D.C.

35. **Ray, D. E.** 1990. Pesticides derived from plants and other organisms, p. 585–636. *In* W. J. Hayes, Jr., and E. R. Laws, Jr. (ed.), *Handbook of Pesticide Toxicology.* Academic Press, Inc., New York.

36. **Shinagawa, K.** 1993. Serology and characterization of *Bacillus cereus* in relation to toxin production. *Bull. Int. Dairy Fed.* **287**:42–49.

37. **Shinagawa, K., H. Konuma, H. Sekita, and S. Sugii.** 1995. Emesis of rhesus monkeys induced by intragastric administration with the HEp-2 vacuolation factor (cereulide) produced by *Bacillus cereus. FEMS Microbiol. Lett.* **130**:87–90.

38. **Shinagawa, K., S. Otake, N. Matsusaka, and S. Sugii.** 1992. Production of the vacuolation factor of *Bacillus cereus* isolated from vomiting-type food poisoning. *J. Vet. Med. Sci.* **54**:443–446.

39. **Shinagawa, K., S. Ueno, H. Konuma, N. Matsusaka, and S. Sugii.** 1991. Purification and characterization of the vascular permeability factor produced by *Bacillus cereus. J. Vet. Med. Sci.* **53**:281–286.

40. **Skirrow, M. B.** 1990. Foodborne illness: Campylobacter. *Lancet* **336**:921–923.

41. **Thompson, N. E., M. J. Ketterhagen, M. S. Bergdoll, and E. J. Schantz.** 1984. Isolation and some properties of an enterotoxin produced by *Bacillus cereus. Infect. Immun.* **43**:887–894.

42. **Turnbull, P. C. B.** 1986. *Bacillus cereus* toxins, p. 397–448. *In* F. Dorner and J. Drews (ed.), *Pharmacology of Bacterial Toxins. International Encyclopedia of Pharmacology and Therapeutics,* section 119. Pergamon Press, Oxford.

43. **Väisäinen, O. M., N. J. Mwaisumo, and M. S. Salkinoja-Salonen.** 1991. Differentiation of dairy strains of the *Bacillus cereus* group by phage typing, minimum growth temperature, and fatty acid analysis. *J. Appl. Bacteriol.* **70**:315–324.

44. **van Netten, P., A. van de Moosdijk, P. van Hoensel, D. A. A. Mossel, and L. Perales.** 1990. Psychrotropic strains of *Bacillus cereus* producing enterotoxin. *J. Appl. Bacteriol.* **69**:73–79.

J. Rocourt
P. Cossart

18

Listeria monocytogenes

Listeriosis has emerged as one of the major foodborne diseases during the last decade. However, it is not a new disease, as are AIDS or legionellosis; the first reported human case occurred in a soldier of the First World War who suffered from meningitis. From that time to 1950, a few human cases were reported; now, hundreds are reported every year (75). The emergence of listeriosis is the result of complex interactions between various factors reflecting changes in social patterns. These factors include (i) medical progress and consequent demographic changes, such as the increasing proportion of immunocompromised people and the elderly; (ii) change in primary food production (large-scale production of raw materials, modifications in food processing technology, expansion of the agrifood industry, and development of cold storage systems); and (iii) changes in food habits (increased consumer demand for convenience food that has a fresh-cooked taste, can be purchased ready to eat, refrigerated, or frozen, can be prepared rapidly, and requires essentially little cooking before consumption) and changes in handling and preparation practices.

Listeriosis is an atypical foodborne disease of major public health concern because of the severity and the nonenteric nature of the disease (meningitis, septicemia, and abortion), a high case-fatality rate (around 20 to 30% of cases), a frequently long incubation time, and a predilection for individuals who have underlying conditions which lead to impairment of T-cell-mediated immunity. *Listeria monocytogenes* differs in many respects from most other foodborne pathogens: it is ubiquitous, is resistant to diverse environmental conditions such as low pH and high NaCl concentrations, and is microaerobic and psychrotrophic. The various ways in which the bacterium can enter into food processing plants, its ability to survive for long periods of time in the environment (soil, plants, and water) and on foods and in food processing plants, and its ability to grow at very low temperatures (2 to 4°C) and to survive in or on food for prolonged periods under adverse conditions have made *L. monocytogenes* a major concern for the agrifood industry during the last decade. The significance of *L. monocytogenes* as a foodborne pathogen is complex. The severity and case-fatality rate of the disease require appropriate preventive measures, but the characteristics of the microorganism are such that it is unrealistic to expect all food to be *Listeria* free. This dilemma has generated animated debate and prompted research in various areas, such as conventional and rapid methods for detection of *L. monocytogenes* in food, its behavior in food, the genetics of virulence and molecular typing of the organism, and epidemiologic investigations of the disease.

CHARACTERISTICS OF THE ORGANISM

Classification

The Genus *Listeria*

The genus *Listeria* belongs to the *Clostridium* subbranch together with *Staphylococcus, Streptococcus, Lactobacillus,*

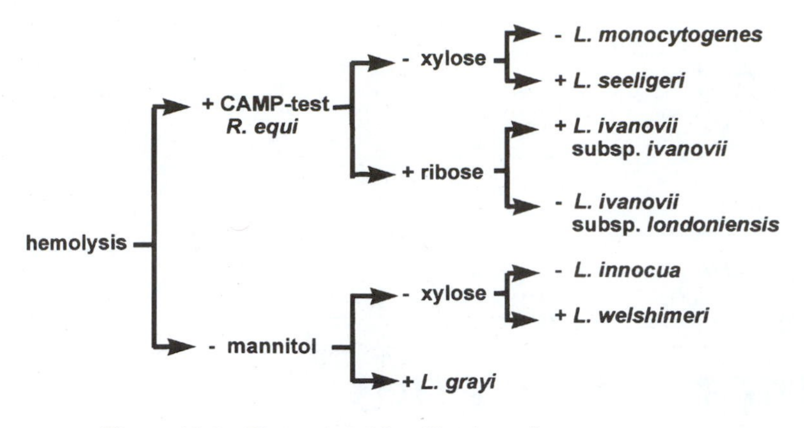

Figure 18.1 Phenotypic identification of *Listeria* species.

and *Brochothrix*. This phylogenetic position of *Listeria* is consistent with its low G+C DNA content (36 to 42%). On the basis of DNA-DNA hybridization, multilocus enzyme analysis, and 16S rRNA sequencing, the genus *Listeria* presently comprises six species divided into two lines of descent: (i) *L. monocytogenes* and the closely related species *L. innocua*, *L. ivanovii* (subspecies *ivanovii* and subspecies *londoniensis*), *L. welshimeri*, and *L. seeligeri* and (ii) *L. grayi* (*L. murrayi* was recently included in this species). Within the genus *Listeria*, only *L. monocytogenes* and *L. ivanovii* are considered virulent, with respect to both the 50% lethal dose in mice and the ability to grow in mouse spleen and liver. Only one species, *L. monocytogenes*, is a public health concern. The increase in testing of foods since 1986 has led to the study of numerous strains isolated throughout the world and discovery of time-saving methods based on biochemical tests which have been commercialized.

The identification of *Listeria* species is based on a limited number of biochemical markers, among which hemolysis is used to differentiate between *L. monocytogenes* and the most frequently encountered nonpathogenic *Listeria* species, *L. innocua* (82) (Fig. 18.1).

L. monocytogenes

Until recently, *L. monocytogenes* typing, which is used to characterize strains beyond the species level, relied mainly on serotyping and phage typing. Serotyping has proven its value over many years. There are 13 serovars which can cause disease, but 95% of human isolates belong to 3 serovars: 1/2a, 1/2b, and 4b. Because of the low discriminatory power of this method, phage-typing systems were developed and were the only means to distinguish between strains of the same serovar before the introduction of molecular typing methods. Since 1989, various molecular typing methods, including multilocus enzyme electrophoresis, ribotyping, DNA microrestriction (using high-frequency cutting enzymes and conventional electrophoresis) and macrorestriction (using rarely cutting enzymes and pulsed-field gel electrophoresis) profile analysis, and random amplification of polymorphic DNA, have been applied to *L. monocytogenes* (10, 37, 61). Because of their ability to type all strains and the discriminatory power of some of them, these methods have become invaluable tools for epidemiologic investigations. In addition, unlike serotyping and phage typing, which are practicable only in appropriately equipped reference laboratories, these methods do not require specialized reagents. However, they are not currently suitable for routine testing, and thus serotyping and phage typing are used for screening large numbers of strains.

Serovar 4b strains are responsible for 33 to 50% of sporadic human cases worldwide and for all major foodborne outbreaks since 1981. In contrast, isolates recovered from food in numerous countries mostly belong to serogroup 1/2. Classification of strains by using multilocus enzyme electrophoresis analysis and ribotyping is based on two groups, each containing two serovars. One group consists of most serovar 4b and 1/2b strains of human origin, and the other consists of serovar 1/2a and 1/2c strains, mainly from food and the environment (8, 37). However, the significance of this observation to public health and food contamination remains unclear.

Tolerance and Susceptibility to Preservation Methods

Conflicting results have been published on the heat resistance of *L. monocytogenes*. An informal working group of the World Health Organization concluded that "pasteurization is a safe process which reduces the number of

L. monocytogenes in raw milk to levels that do not pose an appreciable risk to human health" (96). The efficacy of the lactoperoxidase system as an antimicrobial agent against *L. monocytogenes* is dependent on the number of bacterial cells and the incubation temperature. However, this could be a feasible procedure for controlling *L. monocytogenes* in raw milk at refrigeration temperatures. *L. monocytogenes* is unable to grow at a pH below 4.5. Experimentally, the presence of up to 0.1% acetic, citric, and lactic acids in tryptose broth inhibits the growth of *L. monocytogenes*, with inhibition increasing as the incubation temperature decreases (1). The antilisterial activity of these acids is related to their degree of dissociation, citric and lactic acids being less detrimental for the pathogen at an equivalent pH than acetic acid. *L. monocytogenes* is able to survive in the presence of high levels of NaCl (30%) and at nitrite concentrations that are allowed in foods (21). Freezing and storage at −18°C, and even repeated freezing, have little effect, and these conditions are more likely to injure than to inactivate *L. monocytogenes*.

New trends in food preservation, including the use of biopreservatives, vacuum packaging, and modified atmospheres, have recently been developed (57). The antagonistic effect of nisin on *L. monocytogenes* has been demonstrated, and its activity in foods is strongly dependent on the chemical composition of the food to which it is added. Pediocins (from *Pediococcus pentosaceus* and *P. acidilactici*) inhibit *L. monocytogenes* growth. Bacteriocin from *Lactobacillus bavaricus* transiently affects *L. monocytogenes* in various beef systems, especially at low temperatures. A similar effect was observed with a bacteriocin from *Carnobacterium piscicola* in broth and skim milk (60). However, a drawback to the use of bacteriocins is the emergence of resistant mutants, as has been observed with bavaricin A and nisin (18, 52).

A survey of vacuum-packaged processed meat in retail stores revealed that 53% of the samples tested were contaminated with *L. monocytogenes* and that 4% contained more than 1,000 CFU/g (36). This observation corroborates experimental evidence that the growth of *L. monocytogenes* (a facultative anaerobe) is not significantly affected by vacuum packaging. There has been considerable interest in modified atmosphere packaging of meat products (low oxygen and high carbon dioxide concentrations) over recent years because of the increasing demand for refrigerated convenience foods with extended shelf lives. Experiments with meat juice, raw chicken, and precooked chicken nuggets indicated that such atmospheres do not significantly affect the growth of *L. monocytogenes*. However, *L. monocytogenes* is sensitive to low doses of irradiation.

A substantial portion of surviving cells are injured by heating to 54°C or above, freezing, or various other treatments. Heat-stressed *L. monocytogenes* may be considerably less pathogenic than nonstressed cells. On nonselective agar, injured cells can repair the damage induced by stress and can grow, but they are subject to additional stress in selective agar and variable recovery is observed. Thus, the presence of sublethally damaged cells in food samples may lead to differences in cell recovery rates.

RESERVOIRS

Two properties of *L. monocytogenes* in particular contribute to its widespread distribution: (i) despite not forming spores, it can survive for long periods of time in many different environments; and (ii) it is psychrotrophic. Food can therefore become contaminated at any step of the food chain, and cold storage does not inhibit the growth of listeriae.

Environment

L. monocytogenes has been isolated from various environments for decades: the role of silage in the transmission of animal disease was bacteriologically documented in 1960, and strains were isolated from natural decaying vegetation in 1968 and thereafter (41, 95). The bacterium is able to survive and grow in soil and water. *Listeria* species have been detected in various aqueous environments: surface water of canals and lakes, ditches of polders in The Netherlands, freshwater tributaries draining into a California bay, and sewage (13, 19, 38). Alfalfa plants and other crops grown on soil treated with sewage sludge are contaminated with listeriae (3). Half of the radish samples grown in soil inoculated with *L. monocytogenes* were confirmed positive 3 months later (91). Similarly, the presence of *L. monocytogenes* in pasture grasses and grass silages has often been documented (41).

Food Processing Plants

Entry of *L. monocytogenes* into food processing plants occurs through soil on workers' shoes and clothing and on transport equipment, animals which excrete the bacterium or have contaminated hides or surfaces, raw plant tissue, raw food of animal origin, and possibly healthy human carriers. Growth of listeriae is favored by high humidity and the presence of nutrients. *L. monocytogenes* is most often detected in moist areas such as floor drains, condensed and stagnant water, floors, residues, and processing equipment (17). *L. monocytogenes* can attach to various kinds of surfaces (including stainless steel, glass, and rubber), and biofilms have been found in meat and dairy processing environments (45). Listeriae survive on fingers after hand washing and in aerosols. The presence of *L. monocytogenes* in the food processing chain is evidenced by the widespread distribution of the organism in processed products. Contaminated effluents from food processing plants increase the spread of *L. monocytogenes* in the environment.

Sources of *L. monocytogenes* in dairy processing plants are the environment (floors and floor drains, especially in areas in and around coolers or places subject to outside contamination) and raw milk. Efforts to ensure that milk is safe from *L. monocytogenes* contamination should focus on promoting appropriate methods of pasteurization and on identifying and eliminating sources of postpasteurization contamination (94).

The presence of *L. monocytogenes* on carcasses is usually attributed to contamination by fecal matter during slaughter. A high percentage (11 to 52%) of animals are healthy fecal carriers. A recent study of a slaughter line indicated that 45% of pigs harbor *L. monocytogenes* in tonsils and 24% of cattle have contaminated internal retropharyngeal nodes (12, 85). *L. monocytogenes* has been recovered from both unclean and clean zones (especially on workers' hands) in slaughterhouses; the most heavily contaminated working areas appear to be the cow dehiding and the pig stunning and hoisting areas. Studies in turkey and poultry slaughterhouses failed to detect *L. monocytogenes* in feather samples, scalding tank water overflow, neck skin, liver, heart, cecum, or large intestine. In contrast, *L. monocytogenes* was recovered from feather plucker drip water, chill water overflow, recycling water for cleaning gutters, and mechanically deboned meat. These findings demonstrate the importance of the defeathering machine, chillers, and recycled water in product cross-contamination (33).

Food

L. monocytogenes is present in a wide variety of foods, both raw and processed. It is able to survive and grow in many foodstuffs during storage. A survey by the Public Health Laboratory Service, London, England, indicated that 6% of 18,000 food samples were contaminated with *L. monocytogenes* and 5% of the positive samples had more than 1,000 CFU/g (62). In another study, 11% of foods sampled from the refrigerators of patients suffering from listeriosis in the United States were *L. monocytogenes* positive, and 10% of the positive samples had more than 100 CFU/g (71).

Milk and dairy products were the first and are among the most extensively studied foods. *L. monocytogenes* was detected in about 2 to 5% of bulk tank raw milk samples, without evident seasonal trends (56). Possible sources of exogenous contamination include feeds (which can contain up to 10^5 CFU/g), fecal shedding by healthy carrier cows, and the farm environment (40). Cows with *L. monocytogenes* mastitis are uncommon, but in such cases up to 10^5 CFU/ml of milk may be excreted. *L. monocytogenes* populations in bulk tank raw milk are usually low (<1 to 10 CFU/ml). A wide variety of cheeses can be contaminated, but soft cheeses, with a frequency of about 2 to 10% and populations of *L. monocytogenes* ranging from 10 to 10^7 CFU/g, are of greatest concern. Soft cheeses have a pH that allows the growth of *L. monocytogenes* to large populations. There is a significant correlation of *L. monocytogenes* growth with pH values (>5.5) and absence of starter cultures during cheese manufacturing. Ice cream is infrequently contaminated with *L. monocytogenes* (0.3 to 2% of samples tested), and when present, populations are usually low (<1 CFU/g) (32). Similarly, yogurt (thermophilic lactic cultures inhibit *Listeria* growth) and butter are infrequently contaminated.

A wide variety of meats and meat products, including beef, pork, minced meat, ham, smoked and fermented sausages, salami, and paté, have been associated with *L. monocytogenes* contamination. Most is surface contamination. The prevalence of *L. monocytogenes* contamination of raw and processed meat products can be high (from <1 to 70%). *L. monocytogenes* is capable of growing on meat, depending on the pH, tissue type (lean or fat), type and amount of indigenous microflora, temperature, and curing ingredients (46). Poultry (broiler, ready-to-eat, precooked, chilled, or frozen chicken) is also frequently contaminated, with up to 60% of samples *L. monocytogenes* positive in some studies. The populations of *L. monocytogenes* present in raw or processed meat products are usually low, with 80 to 90% of samples containing <10 to 100 CFU/g. However, higher populations have been reported for some ready-to-eat products, including those implicated in outbreaks of listeriosis.

L. monocytogenes is often present on fresh vegetables (radishes, cucumbers, cabbage, and potatoes), but usually in low numbers. Unlike low-acid salad vegetables, tomatoes and carrots are not good substrates for growth of *L. monocytogenes* (5, 6). Sources of contamination include soil, water, animal manure, decaying vegetation, and effluents from sewage treatment plants.

The prevalence of *L. monocytogenes* on raw and ready-to-eat seafood and fish products (especially smoked fish) can be high (up to 25%) (25). Low populations have been observed in ready-to-eat lobster and shrimps (0.2 to 2 CFU/g) and frozen fish fingers (<100 CFU/g), but higher populations (up to 10^4 CFU/g) have been detected in smoked fish (25, 43). Studies of *L. monocytogenes* during the production and storage of smoked salmon revealed that the populations of *L. monocytogenes* remained unchanged during marination and smoking, but substantial growth occurred during storage at 4 to 10°C (39).

Human Carriage

Asymptomatic fecal carriage of *L. monocytogenes* has been studied in a variety of human populations, including pregnant women, healthy people, patients undergoing renal transplantation or hemodialysis, and patients with symptoms of gastroenteritis. *L. monocytogenes* was isolated from 2 to 6% of fecal samples from healthy people. Patients with listeriosis often excrete high populations of listeriae; specimens from 21% of patients had ≥10^4 CFU/g of feces, and 18% of household contacts of patients with listerosis fecally shed the same serovar and isoenzyme profile of *L. monocytogenes* as the corresponding index case (44, 77). In addition, results of investigation of an outbreak in California in 1985 revealed that community-acquired outbreaks might be amplified through secondary transmission by fecal carriers (59). *L. monocytogenes* has not been isolated from oropharyngeal samples of healthy people, and the presence of the organism in cervicovaginal specimens is always associated with pregnancy-related listeriosis. The role of healthy carriers in the epidemiology of listeriosis is unclear and warrants further study.

FOODBORNE OUTBREAKS

Foodborne transmission of listeriosis was suggested early in the medical literature but was first demonstrated in 1981 during an outbreak in Canada with the simultaneous use of a case-control study and strain typing. Since 1981, epidemiologic investigations have repeatedly indicated that the consumption of contaminated food is a primary vehicle of transmission of listeriosis. Food has been identified as the vehicle of several major (>30 cases) outbreaks of listeriosis investigated since 1981.

The first confirmed foodborne outbreak of listeriosis occurred in 1981 in Nova Scotia, Canada, and involved 41 patients (total number of cases [the number of epidemic cases is not known]). A case-control study implicated a locally prepared coleslaw as the vehicle, and the epidemic strain was subsequently isolated from an unopened package of this product. Cabbage fertilized with manure from sheep suspected to have had *Listeria* meningitis was the probable source (76). The next outbreak was in Boston, Mass., in 1983 and included 49 cases (total number of cases [the number of epidemic cases is not known]) over a 2-month period (27); a case-control study implicated pasteurized milk as the vehicle. In 1985 in California, an outbreak of listeriosis with 142 cases (total number of cases [the number of epidemic cases is not known]) was traced to a Mexican-style cheese (55). An *L. monocytogenes*-contaminated soft cheese was responsible for a 4-year (1983 to 1987) outbreak of 122 cases in Switzerland (7), and a contaminated paté caused a 300-case outbreak in the United Kingdom in 1989 to 1990 (63). In France, contaminated pork tongue in aspic was the principal vehicle of 279 cases of listeriosis in 10 months in 1992 (42), potted pork ("rillettes") was associated with 39 cases in 1993 (24), and soft cheese was the vehicle of 33 cases in 1995 (35). Recalling the implicated food, advising the general population through the mass media to avoid eating contaminated product, and taking appropriate action to prevent *L. monocytogenes* contamination at product processing and handling facilities terminated the outbreaks (California in 1985, Switzerland in 1987, the United Kingdom in 1989, France in 1993 and 1995).

Experience during investigations of recent outbreaks indicates that several different methods of typing (phenotypic and molecular) should be used to evaluate relatedness between isolates from patients and foods (42). Strains responsible for some major outbreaks since 1981 (Canada in 1981, California in 1985, Switzerland in 1983 to 1987, France in 1992) belong to a small number of closely related clones, as evidenced by ribotyping, isoenzyme analysis, and DNA macrorestriction pattern analysis (10).

All outbreaks were associated with an industrially processed food. However, cross-contamination at the distribution point was strongly suspected during the French outbreak in 1992 (42), and it is likely that similar cross-contamination occurs in the kitchen (17).

The data obtained during the past 10 years regarding the sources of outbreaks suggest that some foods are more hazardous than others. Highest-risk foods are (i) ready to eat and stored at refrigeration temperatures for a long period of time, thereby enabling listeriae to grow, and (ii) contaminated with a high population of *L. monocytogenes* (>100 CFU/g or ml). This observation was confirmed by an analysis of sporadic listeriosis cases (71, 78).

CHARACTERISTICS OF DISEASE

Listeriosis is characterized by a variety of severe syndromes. Pregnant women are most frequently affected in the third trimester. The infection of the mother may be asymptomatic or characterized by a flulike illness with fever, myalgia, or headache. Consequences for the the fetus or infant are more serious, including spontaneous abortion, fetal death, stillbirth, severe neonatal septicemia, and meningitis. In nonpregnant adults, *L. monocytogenes* has a particular tropism for the central nervous system. Meningitis and meningoencephalitis (rarely brain abscess and brain stem) frequently occur in cases of listeriosis (11). In nonperinatal patients with predisposing factors, bacteremia is more frequent, and infections of the central nervous system are the predominant clinical syndromes diagnosed in patients with no underlying condition. Focal infections, including endocarditis, pulmonary infection, septic arthritis, osteomyelitis, peritonitis, hepatitis, and arterial infections, are rare and may by preceded by bacteremia (67). Cases of mild gastrointestinal illness following the ingestion of *L. monocytogenes* also have been documented (74). The overall case-fatality rate for systemic or invasive listeriosis is usually about 20 to 30% for both epidemic and sporadic cases. The prognosis of listeriosis depends on both the type of infection and the underlying conditions. Mortality is usually higher (38 to 40%) among immunocompromised, elderly patients and patients suffering from central nervous system infections. Residual syndromes due to invasive listeriosis have been reported sporadically. Up to 11% of neonates and 30% of adult survivors of central nervous system infections suffer from residual symptoms (11, 28). The severity of illness, the incidence of residual symptoms, and the high case-fatality rate make listeriosis a costly foodborne disease.

Listeriosis is a rare disease; the incidence evaluated over recent years by passive surveillance in some European countries and Canada is between two and seven cases per million population. A more precise estimate of 7.4 cases per million population was obtained through an active surveillance study in the United States in 1989 and 1990 in a population of 19 million people. A decrease in the number of cases (44%) and deaths (48%) attributable to listerosis has been observed in the United States since 1990, suggesting that the measures taken in the food industry and the recommendations for people at greatest risk for the disease have been effective (89). Listeriosis is mainly reported from industrialized countries; the prevalences in Africa, Asia, and South America are unknown or low. Whether this difference reflects different consumption patterns and dietary habits, differences in host susceptibility, different food processing and storage technologies, or lack of awareness or laboratory facilities is not known (75).

Most cases of human listeriosis are sporadic, although a portion of these sporadic cases may be unrecognized common-source clusters. The source and route of infection of most of these cases remain unknown, although foodborne transmission is demonstrated in some cases. Many cases have not been linked to food because of the difficulties of prospectively investigating sporadic cases of the disease: long incubation times (up to 5 weeks) make accurate food histories difficult to obtain and examination of incriminated foodstuffs difficult or impossible because the foodstuffs are usually consumed or discarded. Understanding the epidemiology of sporadic cases is critical to the development of effective control strategies. In association with active surveillance of listeriosis, a case-control study of dietary risk factors undertaken in the United States demonstrated that patients with listeriosis were more likely than controls to have eaten soft cheeses or food purchased from store delicatessen counters. Data suggested that foodborne transmission could be responsible for about one-third of cases (71, 78).

Although exposure to *L. monocytogenes* is common, invasive listeriosis is rare. It is unclear whether this is because of early acquired protection or because most strains are only weakly virulent. The pathogenesis of human listeriosis is poorly understood. It is estimated that 2 to 6% of healthy individuals are asymptomatic fecal carriers of *L. monocytogenes:* the risk of clinical disease in these individuals is unknown. Endogenous infection by *L. monocytogenes* in the gut is plausible, especially in patients receiving immunosuppressive therapy, which not only impairs resistance to infection but also can result in alterations of intestinal defense mechanisms favoring listerial invasion. Nevertheless, asymptomatic fecal carriage has been observed in pregnant women who proceed to normal birth at term, and women who have given birth to infected infants do not necessarily suffer the same problem in later pregnancies. Similarly, recent transplant recipients may harbor *L. monocytogenes* in the gut without developing the disease.

Epidemiologic studies since 1981 have focused on the role of contaminated food in transmission of listeriosis. However, two unusual transmission routes have been described. Hospital-acquired listeriosis is found sporadically, mainly in nursery mates, with equipment serving as the vehicle (amniotic fluid during intrauterine infections could contain as many as 10^8 CFU/ml [64]). Mineral oil has been implicated in an outbreak of neonatal listeriosis (79). Primary cutaneous infections without systemic involvement have been observed as an occupational disease in veterinarians and farmers, and most cases are caused by manipulation of presumably infected bovine fetuses or cows.

INFECTIVE DOSE AND SUSCEPTIBLE POPULATIONS

Infective Dose

The infective dose of *L. monocytogenes* depends on many factors, including the immunological status of the host. In addition to host factors and exposure to particular foods, it is likely that microbial characteristics are important risk factors for disease. The occurrence and the course of infection may well depend on virulence factors and infective dose. The severity of the disease is such that tests with human volunteers are impossible. Studies in monkeys and in mice suggest that reducing levels of exposure will reduce clinical disease (26). However, these experiments do not help to determine the minimal infective dose for humans. Published data indicate that the numbers of *L. monocytogenes* in contaminated food responsible for epidemic and sporadic foodborne cases were more than 100 CFU/g. However, because enumeration procedures are not fully reliable and the time between consumption and analysis of the contaminated food can enable growth or death of listeriae, results may not always be indicative of the numbers consumed. Hence, these data do not preclude the possibility that lower doses are infective. More epidemiologic information is needed for an accurate assessment of the infective dose.

Susceptible Populations

Most human cases of listeriosis occur in individuals who have a predisposing disease which leads to impairment of their T-cell-mediated immunity. The percentage of patients suffering from a known underlying condition varies greatly among studies, ranging from 70 to 85% of the cases in some surveys to nearly all cases in others (11, 84). The most commonly affected populations include neonates and the elderly, pregnant women, and those who are immunosuppressed by medication (corticosteroids, cytotoxic drugs), especially after organ transplantation or illness (hematologic malignancies such as leukemia, lymphoma, and myeloma as well as solid malignancies). Listeriosis is 300 times more frequent in people with AIDS than in the general population (47, 80). In addition to T-cell immunity impairment, a small percentage of listeriosis patients suffer from chronic diseases not usually associated with immunosuppression, such as congestive heart failure, diabetes, cirrhosis, alcoholism, and systemic lupus erythematosus, alone or in association with known predisposing diseases.

A concurrent infection could influence susceptibility to listeriosis. This was exemplified by a cluster of cases in 1987 in Philadelphia characterized by a number of different strains isolated from patients. A single food vehicle could not be identified because of the diversity of strains. Clinical and epidemiologic investigations suggested that individuals who were previously asymptomatic for listerial infection but whose gastrointestinal tracts harbored *L. monocytogenes* became symptomatic possibly because of a coinfecting agent (81).

VIRULENCE FACTORS AND MECHANISMS OF PATHOGENICITY

L. monocytogenes can infect laboratory animals and mammalian cells in tissue culture, can grow well in culture media, and can be genetically manipulated and used to experimentally infect animals. Hence, during the last decade there have been major advances in studying and identifying the virulence factors of this highly invasive intracellular pathogen (14–16, 23, 73, 83, 90).

Pathogenicity of *L. monocytogenes*

Many tests for studying *L. monocytogenes* pathogenicity have been developed, including tissue culture assays and tests using laboratory animals, particularly mice, which are either immunocompetent or immunocompromised. In these studies, mice are infected intraperitoneally, intravenously, or intragastrically and virulence is evaluated either by comparing the 50% lethal dose or by enumerating bacteria in the spleen or liver. Considerable heterogeneity in levels of virulence of various *L. monocytogenes* strains has been observed: some show high virulence, whereas some show low or even, rarely, no virulence. No clear correlation between origin (human, animal, category of food, environment) or strain characteristics (serovars, phagovars, ribovars, DNA macrorestriction patterns) and virulence has been observed (9, 70). Similarly, epidemic strains demonstrate a wide range of virulence in the mouse model. Although rare nonpathogenic or weakly pathogenic *L. monocytogenes* isolates have been reported, all strains of the species are considered to be potentially capable of causing human disease.

Experimental Infection and Cell Biology of the Infectious Process

Experimental infection of rodents with *L. monocytogenes* has been widely used for the study of cell-mediated immunity. When mice are exposed to a sublethal inoculum of *L. monocytogenes,* the ensuing infection follows a well-defined course, lasting for approximately 1 week. Routinely, mice are injected intravenously and bacterial growth kinetics is monitored in the spleen and liver. Within 10 min after intravenous injection, 90% of the inoculum is taken up by the liver and 5 to 10% is taken up by the spleen. During the first 6 h, the number of viable listeriae in the liver decreases 10-fold, indicating rapid destruction of most of the bacteria. Surviving organisms then multiply within susceptible macrophages and grow exponentially in the spleen and liver for the next 48 h, peaking at day 2 or 3 postinfection (4). Rapid inactivation ensues during the next 3 to 4 days, indicating recovery of the host. Convalescent mice are resistant to challenge and have a delayed-type hypersensitivity with swelling of the footpads injected with crude cell preparations of *L. monocytogenes.*

L. monocytogenes crosses the intestinal barrier in animals infected by the oral route. However, the site of entry, i.e., the epithelial cells or the M cells in the Peyer's patches, has yet to be elucidated. Bacteria are then internalized by resident macrophages, in which they can survive and replicate. They are subsequently transported via the blood to regional lymph nodes. When they reach the liver and the spleen, most listeriae are rapidly killed. In the initial phase of infection, infected hepatocytes are the target for neutrophils and later for mononuclear phagocytes, which are responsible for control and resolution of infection. Depending on the level of T-cell response induced in the first days following initial infection, further dissemination via the blood to the brain or, in the pregnant animal, the placenta may subsequently occur. Hence, infection is not localized at the site of entry but involves entry and multiplication in a wide variety of cell types and tissues. The principal site of infection is the liver. In vitro, *L. monocytogenes* invades many cell lines of different types (macrophages, fibroblasts, hepatocytes, and epithelial cells) and is one of the most invasive bacteria known. It has presumably evolved specific strategies allowing entry into different cell types but retaining some tropism for particular organs.

L. monocytogenes enters both phagocytic cells and nonprofessional phagocytes by phagocytosis. This first step in infection is prevented by addition of cytochalasin D, a drug which inhibits actin polymerization and hence active participation of the mammalian cell. In the case of nonprofessional phagocytic cells, this process is triggered by the bacterium and is therefore called induced phagocytosis. Soon after entry, bacteria are internalized in membrane-bound vacuoles, which are lysed in less than 30 min. Intracellular bacteria are released to the cytosol and begin to multiply with a doubling time of about 1 h. These intracytoplasmic bacteria become progressively covered by a "cloud" of cell actin filaments which later rearranges into a polarized "comet tail" up to 40 μm in length (15, 14, 90). The actin comet tail is made of actin microfilaments that are continuously assembled in the vicinity of the bacterium and are then released and cross-linked. The actin comet tail is stationary in the cytosol and left behind by moving bacteria. The length of the tail is thus proportional to the speed of movement, with faster-moving bacteria having longer tails. Their speed ranges from 0.1 to 1 μm/s. When bacteria reach the plasma membrane, they put out long protrusions, each with a bacterium at the tip. These protrusions are then internalized by a neighboring cell, giving rise to a two-membrane-bound vacuole. After lysis of this vacuole, a new cycle of replication, movement, and spreading of the bacteria begins. The entire cycle is completed in about 5 h. If cytochalasin D is added after entry, bacteria do not spread within the cytosol; they replicate and form microcolonies in the vicinity of the nucleus. Hence, actin polymerization is essential to intracellular movement and cell-to-cell spread (Fig. 18.2).

The strategy of direct cell-to-cell spread allows bacteria to disseminate within host tissues and induce the formation of infectious foci while being sheltered from host defenses such as circulating antibodies. This may explain why antibodies play no role in recovery from infection or protection against secondary infection (72).

Genetic Approaches to the Study of Virulence Factors

Genetic manipulation of *L. monocytogenes* in vitro is now feasible; however, some useful techniques applicable to other bacterial genera, e.g., transduction, are not available. The first transposon mutagenesis experiment with *L. monocytogenes* was conducted in 1985 (31). The conjugative transposons Tn*1545* and Tn*916*, or the nonconjugative transposon Tn*917* and its derivative Tn*917-lac*, can all be used for transposon mutagenesis. Transformation, originally involving protoplast formation, is now possible by electroporation. Efficiency varies, and complementation of spontaneous mutants with plasmid libraries is not possible. However, complementation can be performed with cloned genes, and allelic exchange is available for several thermosensitive plasmids. Plasmids can also be introduced into *Listeria* spp. by conjugation. Vectors such as pAT18 have been very useful.

Entry into Mammalian Cells

Following invasion of the intestinal barrier, two types of cells, macrophages and hepatocytes, are critical to infection. In vivo observations have led to conflicting conclusions concerning the primary site of entry (epithelial

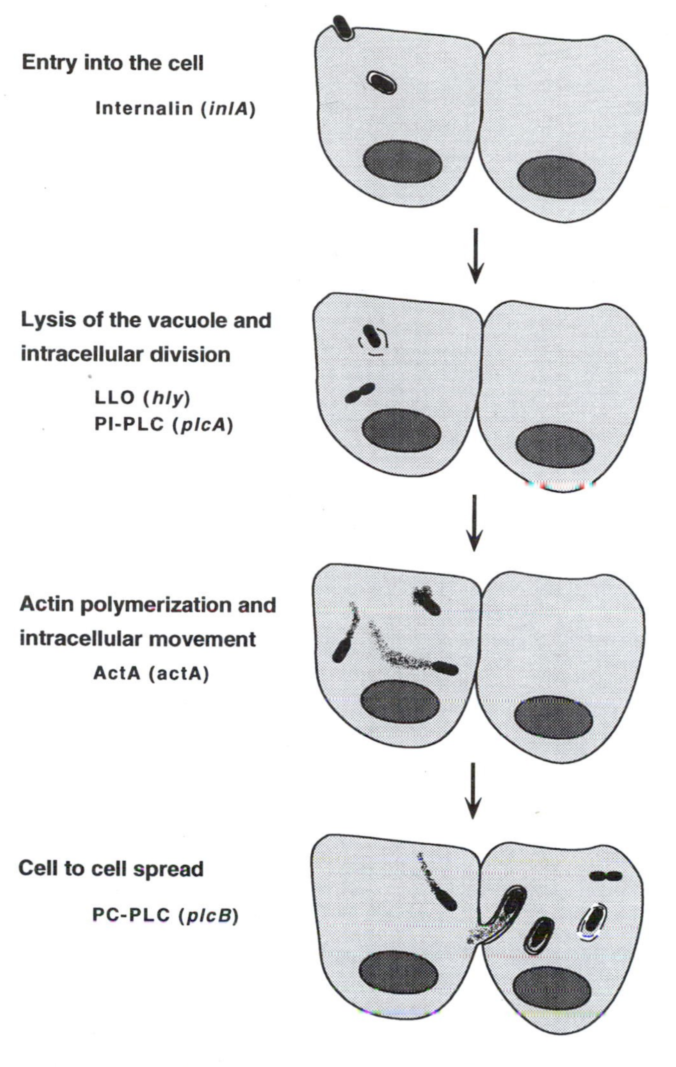

Figure 18.2 Schematic representation of the successive steps in the infectious process.

cells or M cells) of *L. monocytogenes* into the host. Entry into epithelial cells has been studied by genetic approaches (23), leading to the identification of internalin, a protein required for entry into epithelial cells and encoded by gene *inlA* (30). *inlA* is part of a multigene family. A second gene of this family, *inlB*, contributes to entry into hepatocytes in vitro (22) (Fig. 18.3). Analysis of spontaneous rough mutants has led to the identification of a second protein, p60, associated with invasion. Its gene was named *iap*, for invasion-associated protein (51).

Internalin and the inl Gene Family

A library of Tn*1545* mutants was developed and screened for mutants unable to invade the human enterocyte-like cell line Caco-2 (30). In three mutants, the transposon had inserted in nearly the same position upstream from two open reading frames, *inlA* and *inlB*. *inlA* encodes an 800-amino-acid protein with a signal sequence that has two regions of repeats. These include region A, which is characterized by leucine-rich repeats and a C-terminal hydrophobic region that may serve as a membrane anchor because it is preceded by the hexapeptide LPTTGD, which is the signature of gram-positive bacterial membrane-anchored proteins. *inlA* confers invasiveness to the noninvasive species *L. innocua;* hence, the *inlA* gene product was named internalin. Antibodies raised against *L. monocytogenes* were used to demonstrate that *L. innocua* transformed with *inlA* produces a novel surface protein, in agreement with the predicted structural features of internalin and its putative role in interacting with mammalian cells. Recent studies have revealed that internalin is not always present on the surface of *L. monocytogenes* and can be released into culture supernatants. Release of internalin begins during the exponential phase of growth, when the cell wall-associated form is most abundant. Since cultures in the exponential phase are most invasive, it was suggested that the cell wall-associated form of internalin was the most important in invasion. Recent experiments with mutant strains lacking the membrane anchor and expressing only a soluble form of internalin have revealed that anchoring of internalin in the cell wall is required for it to function.

inlB encodes a 630-amino-acid protein with a signal sequence and a region of leucine-rich repeats analogous to those of internalin (same length [22 amino acids] and same consensus sequence) but without a membrane anchor. *inlB* also encodes a surface protein that has a very minor role in the entry of *L. monocytogenes* into intestinal epithelial cells but is essential for entry into cultured hepatocytes (22). In contrast to internalin, the *inlB* gene product is not sufficient to enable *L. innocua* to enter cultured hepatocytes. Hence, entry into hepatocytes requires other *L. monocytogenes*-specific factors. Five other *inl* genes have been cloned and sequenced, and their properties are currently under investigation. One hypothesis is that the *inl* family of genes encodes surface proteins, each with a tropism for a specific cell type. Internalin expression is greater at 37°C than at 20°C and is under the control of the pleiotropic activator gene *prfA* (see below).

p60 and iap

Spontaneously rough mutants produce reduced levels of a 60-kDa extracellular protein termed p60. These mutants form long chains separated by double septa and

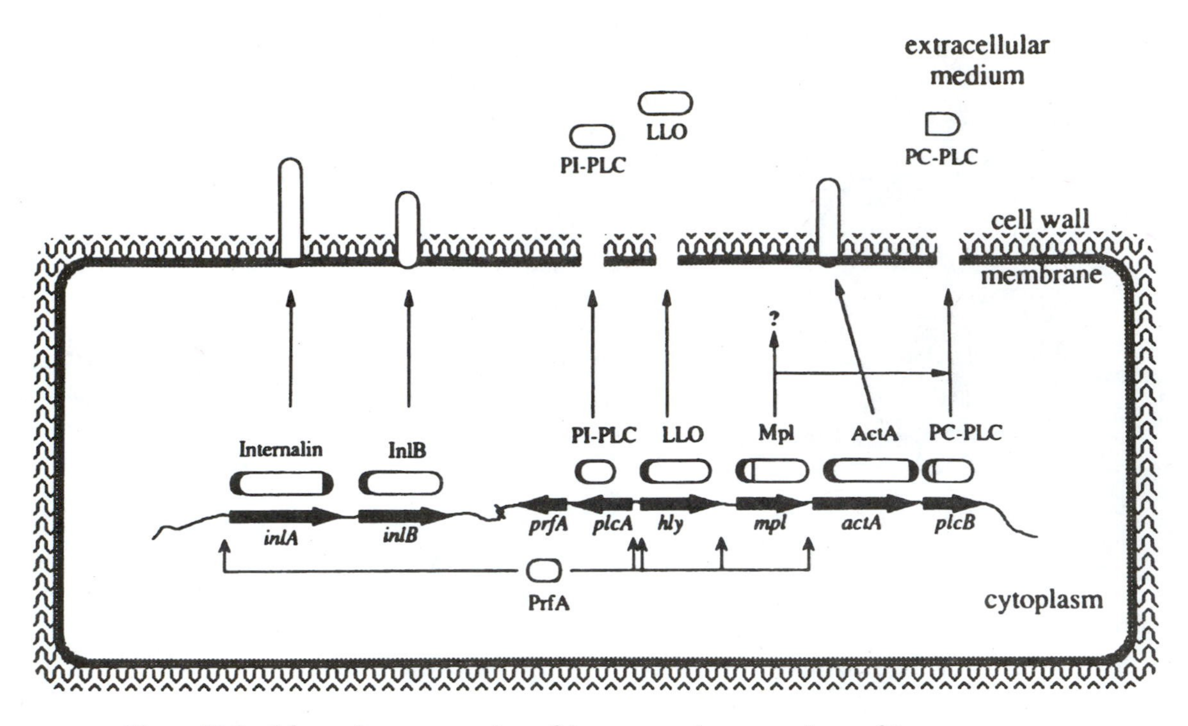

Figure 18.3 Schematic representation of the genes and gene products of *L. monocytogenes*.

have reduced virulence in mice. They also invade 3T6 fibroblasts poorly. Adding partially purified p60 to rough mutants disaggregates the chains, and the resulting bacteria become invasive. In contrast, if the chains are disrupted by ultrasonication, the single cells are not invasive. The gene coding for p60 has been cloned; and the deduced protein consists of 484 amino acids with a signal sequence but no membrane anchor (51). A peculiar sequence of 19 Thr-Asn pairs is near the C terminus of the protein. In cultures of wild-type *L. monocytogenes*, p60 is present both on the cell surface and in the culture supernatant. However, the protein is present only on the surface of the rough mutant. p60 is an essential protein with bacteriolytic activity, and it may be a murein hydrolase involved in septum formation (97). The exact role of p60 in invasion is unclear, but it primarily affects entry into fibroblasts. Rough mutants are thought to be noninvasive for fibroblasts not because of their large size but rather because of their inability to adhere to these cells. In contrast, when rough mutants are in contact with Caco-2 cells, the bacteria are adherent but not invasive. After addition of p60 or after disruption of the chains, these mutants invade Caco-2 cells. Hence, neither adherence nor uptake of *L. monocytogenes* by epithelial cells is dependent on p60.

L. monocytogenes invasion of mammalian cells is inhibited by tyrosine kinase inhibitors (88, 93). In agreement with these data, entry of mammalian cells by *L. monocytogenes* induces phosphorylation of a 37-kDa host cell protein (88).

Escape from the Phagosome and Intracellular Growth

Once internalized in a vacuole, *L. monocytogenes* rapidly accesses the cytosol, where it multiplies. The main factor involved in lysis of the vacuole is a protein that has a pore-forming activity, listeriolysin O (LLO). However, a second factor, a phosphatidylinositol-specific phospholipase C (PI-PLC), may also be involved (73, 83).

LLO is a toxin that belongs to the class of thiol-activated toxins whose prototype is streptolysin O, which is produced by *Streptococcus pyogenes*. These thiol-activated toxins are produced by several gram-positive bacteria, including *Streptococcus*, *Clostridium*, and *Bacillus* species. They have close similarities in sequence and probably have a common mode of action. They are active only on cholesterol-containing membranes, with cholesterol likely acting as the receptor in the membrane. It has been suggested that after LLO binds to the surface of the cell membrane, the protein forms oligomers which lead to pore formation. The activity of pore-forming toxins is usually determined by their lytic activity on erythrocytes; hence, they are termed hemolysins. Colonies of bacteria producing this type of toxin are hemolytic on blood agar plates, a phenotype that has been used to identify the gene encoding LLO.

Nonhemolytic *L. monocytogenes* strains, which are very rare, are avirulent in experimentally infected mice. This finding led to analyses of the role of hemolysin in virulence. By using transposon Tn*1545*, a nonhemolytic mutant that was avirulent in mice was obtained. Both the wild type and the nonhemolytic strains were able to enter mammalian (Caco-2) cells; however, the nonhemolytic strain was impaired in its ability to grow intracellularly. Electron microscopy revealed that the nonhemolytic mutant, although invasive, was unable to escape from the phagocytic vacuole.

The locus containing the transposon was cloned and determined to be a structural gene, named *hlyA*, later shortened to *hly*. The insertion had no polar effect since *hly* is a monocistronic unit. *hly* encodes a 58-kDa protein that, like all thiol-activated toxins, contains a single cysteine residue in the amino acid sequence ECTGLAWEWWR, located toward the C terminus of the protein. The cysteine residue is not essential for hemolysin activity as previously thought. Studies have determined that LLO is needed for the escape of *L. monocytogenes* from the vacuole and intracellular growth, hence contributing to virulence.

LLO has been purified from culture supernatants. Unlike all other thiol-activated toxins, it has maximal activity at pH 5 and is inactive at pH 7 (34). It has been proposed that this property is particularly important in the acidic environment that results from phagosome-lysosome fusion. The most interesting feature of LLO is not its high activity at low pH but its low activity at high or neutral pH, which prevents the deleterious effect of LLO on cellular membranes when the bacterium is free in the cytosol.

The gene *plcA* is upstream from *hly*. *plcA* encodes the PI-PLC that is present in culture supernatants of *L. monocytogenes*. This 37-kDa protein hydrolyzes glycosylphosphatidylinositol anchors. Analysis of in-frame deletions of *plcA* revealed that this gene is not required for virulence in mice; however, *plcA* is involved in escape from the vacuole in primary macrophages. These results indicate that *L. monocytogenes* has evolved the use of specific genes not only for particular steps of the infection but also for specific cell types.

A priori, it was believed that the cytoplasm of eukaryotic cells may not be permissive for the growth of bacteria and that intracellular pathogens would have to induce expression of several genes for survival. However, infection of macrophages with a *Bacillus subtilis* strain expressing *hly* has shown that the cytosol is permissive for growth of at least this species. Additional studies revealed that most auxotrophic mutants of *L. monocytogenes* in cell culture have growth rates unaffected by the intracellular environment and are only slightly affected in virulence (2, 49, 58).

Intracytoplasmic Movement and Cell-to-Cell Spread

Induced phagocytosis, escape from the phagosome, and intracellular multiplication are essential steps for infection of individual cells but are not sufficient to achieve infection in mice. *L. monocytogenes* must efficiently invade its preferred tissues by direct cell-to-cell spread to infect mice. Direct cell-to-cell spread is the result of intracellular movement, protrusion formation, phagocytosis of the protrusion by a neighboring cell, formation of a double-membrane-bound vacuole, and lysis of the vacuole. Some of the bacterial genes involved in these steps are clustered in an operon located immediately downstream from *hly*.

Intracytoplasmic movement is strictly coupled to continuous actin assembly that provides the force for bacterial propulsion (14) (Fig. 18.4). Actin polymerization requires expression of the *actA* gene (20, 50). *actA* mutants are invasive, escape from the phagosome, replicate, and form intracellular microcolonies but are not covered with actin, do not move intracellularly, and do not spread from cell to cell. *actA* encodes a 610-amino-

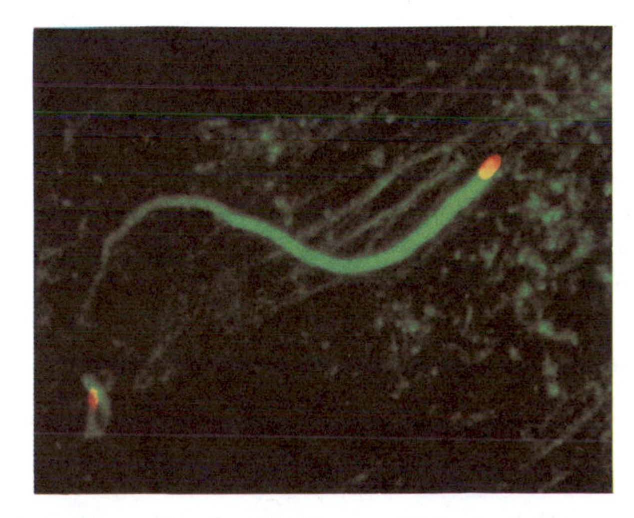

Figure 18.4 Directional actin assembly by *L. monocytogenes*, with filamentous actin assembled at one bacterial extremity. Shown is a double immunofluorescence confocal laser scanning micrograph of a Vero cell infected for 5 h and treated with anti-*L. monocytogenes* antiserum (bacterium; red) and fluorescein isothiocyanate-phalloidin (F-actin; green). (Reprinted with permission from *Molecular Microbiology*, vol. 13, no. 3, 1994, front cover.)

acid surface protein (ActA) anchored in the bacterial cytoplasmic membrane by its hydrophobic C-terminal region. Half of the molecule protrudes from the cell wall and can interact with cellular components. Immunolocalization of the ActA protein in infected cells has shown that ActA is asymmetrically distributed on the bacterial surface such that it is weakly detectable at one pole, with an increasing concentration towards the other pole, which is the site of comet tail formation. Even before comet tail formation, ActA colocalizes with F-actin. This finding strongly suggests that ActA is involved in actin filament nucleation from the bacterial surface. In addition, the asymmetrical distribution of ActA appears to be essential for intracellular *L. monocytogenes* movement in the direction of the non-ActA-expressing pole. ActA seems to be sufficient to trigger actin-based motility, because expression of *actA* in *L. innocua* is sufficient to produce movement in cell-free extracts.

Factors involved in the formation of the protrusion and its subsequent phagocytosis are unknown. Lysis of the two membrane vacuoles involves a lecithinase (phospholipase C) encoded by the *plcB*. *plcB* is a gene located immediately downstream from *actA* (92). The optimum activity of phospholipase C is between pH 5.5 and 7, the enzyme can cleave phosphatidyl choline (in which case it is termed PC-PLC), phosphatidylserine, and phosphatidylethanolamine. Phospholipase C is synthesized as a precursor that must be cleaved by the product of *mpl* to be functional. *mpl* is the first gene of the operon that encodes a metalloprotease. The protease has not been purified, but results obtained with an *mpl* mutant suggest that the *plcB* gene product and ActA can be its substrates. *plcB* mutants produce small plaques in cell monolayers, and electron microscopy studies have shown substantial accumulation of these mutants inside the two membrane vacuoles. It is not known whether LLO also contributes to the lysis of the two-membrane vacuole.

Coregulation of Virulence Factors

The genes encoding the bacterial factors involved in the different steps of the infectious process, i.e., internalin, LLO, the two phospholipases (PI-PLC and PC-PLC), and ActA, are clustered in two regions of the chromosome (66). Their expression is modulated by the same environmental conditions, being repressed at low temperatures and maximally expressed at 37°C (53). In addition, they are coregulated by *prfA*, a gene coding for a 27-kDa protein (54, 65).

Coregulation of virulence genes was originally predicted on the basis of detection of similar palindromic sequences present in the *hly* promoter and in the promoters of the two adjacent operons. These palindromes are probable targets of PrfA, because in *B. subtilis*, PrfA activates the *hly* promoter if the palindrome is intact, whereas activation is eliminated by point mutations in the sequence. In addition, recombinant PrfA shifts the electrophoretic mobility of DNA fragments containing specific PrfA target sites, such as the *hly* promoter.

Expression of *prfA* is a complex phenomenon. *prfA* is the second gene of the *plcA-prfA* operon whose primary promoter is regulated by *prfA* (65). Hence, *prfA* autoregulates (activates) its own synthesis. *prfA* can also be transcribed from its own promoter region. Since *prfA* transcripts are fewer in stationary growth, it has been proposed that *prfA* may also downregulate its own synthesis. It has been shown in a *prfA* mutant that transcription at the *prfA*-specific promoter is upregulated (29).

PrfA is similar to CAP, the cyclic AMP receptor in *E. coli* (83). It was suggested that this protein contains a helix-turn-helix motif that can interact with DNA. Mutagenesis in the region coding for this motif and gel shift assays strongly suggest that the product of the *prfA* gene is a DNA-binding protein of the CAP/FNR family that mediates transcription activation by binding to a conserved palindromic region located in the −35 region of promoters under its control.

The mechanism of thermoregulation of *prfA* expression is unclear. At low temperatures, the small *prfA*-specific transcript is present (54). However, no apparent transcription of virulence genes is observed. For activation, the *prfA* gene product may act in concert with another regulatory protein or be posttranscriptionally modified. It is also possible that oligomerization of *prfA* plays a role in activation and that the concentration of *prfA* may be critical for its activity. In addition, the various *prfA*-regulated promoters have different affinities for *prfA*. Clearly, the expression of virulence genes is highly regulated. There is evidence that heat shock (86, 87) or various nutrients in broth medium (69) can also affect virulence gene expression.

A group of Tn*917-lac* mutants was screened to identify genes specifically induced in the intracellular environment so as to test for strains in which *lacZ* expression was greater inside cells than in broth culture. Five genes were identified as preferentially expressed within cells (49). Three were genes involved in the biosynthesis of purines and pyrimidines; one encoded an arginine ABC transporter, and another was the *plcA* gene, coding for PI-PLC, which is cotranscribed with *prfA*, the pleiotropic regulator of all virulence genes. However, *prfA* expression does not appear to be regulated at the level of transcription in vitro. This observation raises the possibility that virulence gene activation does not result from an increase in *prfA* transcription and may result either from an upregulation of *prfA* translation, posttranslational modification, or the coexpression of *prfA* with an unidentified coregulator.

Conclusion

A multidisciplinary approach has enabled the identification of several virulence factors. Research under way is principally focusing on (i) identification of the internalin receptor of mammalian cells, (ii) elucidation of the role of the various internalin genes, (iii) understanding of *actA* function, and (iv) analysis of *prfA* in the global regulation of virulence. *L. monocytogenes* has become one of the best-understood intracellular pathogens at the cellular and molecular levels. The immune response to this pathogen continues to be a topic of intensive investigations. Such studies have defined the epitopes involved in the protective CD8 response (48, 68).

CONCLUDING REMARKS

Considerable research has been done during the last decade to characterize the disease listeriosis, define the magnitude of the public health problem and its impact on the food industry, identify the risk factors associated with disease, and identify appropriate control strategies. The food processing industry has made progress in reducing the prevalence of *L. monocytogenes* in processing plant environments and in high-risk foods, and preventive measures have been developed and implemented for persons at increased risk of infection. Recent data from the United States indicate that there has been a large reduction in the incidence of illness (44%) and death (48%) between 1989 and 1993 (89).

References

1. **Ahamad, N., and E. H. Marth.** 1989. Behavior of *Listeria monocytogenes* at 7, 13, 21, and 35°C in tryptose broth acidified with acetic, citric, or lactic acid. *J. Food Prot.* **52**:688–695.
2. **Alexander, J., P. Andrew, D. Jones, and I. Roberts.** 1993. Characterization of an aromatic amino acid-dependent *Listeria monocytogenes* mutant: attenuation, persistence, and ability to induce protective immunity in mice. *Infect. Immun.* **61**:2245–2248.
3. **Al-Ghazali, M. R., and S. K. Al-Azawi.** 1990. *Listeria monocytogenes* contamination of crops grown on soil treated with sewage sludge cake. *J. Appl. Bacteriol.* **69**:642–674.
4. **Audurier, A., P. Pardon, J. Marly, and F. Lantier.** 1980. Experimental infection of mice with *Listeria monocytogenes* and *L. innocua*. *Ann. Microbiol.* **131B**:47–57.
5. **Beuchat, L. R., and R. E. Brackett.** 1990. Inhibitory effects of raw carrots on *Listeria monocytogenes*. *Appl. Environ. Microbiol.* **56**:1734–1742.
6. **Beuchat, L. R., and R. E. Brackett.** 1991. Behavior of *Listeria monocytogenes* inoculated into raw tomatoes and processed tomato products. *Appl. Environ. Microbiol.* **57**:1367–1371.
7. **Bille, J.** 1989. Anatomy of a foodborne listeriosis outbreak, p. 29–36. *In Foodborne Listeriosis.* Proceedings of a symposium on September 7, 1988, in Wiesbaden, Germany. B. Behr's Gmbh & Co., Hamburg, Germany.
8. **Boerlin, P., and J. C. Piffaretti.** 1991. Typing of human, animal, food, and environmental isolates of *Listeria monocytogenes* by multilocus enzyme electrophoresis. *Appl. Environ. Microbiol.* **57**:1624–1629.
9. **Brosch, R., B. Catimel, G. Milon, C. Buchrieser, E. Vindel, and J. Rocourt.** 1993. Virulence heterogeneity of *Listeria monocytogenes* strains from various sources (food, human, animal) in immunocompetent mice and its association with typing characteristics. *J. Food Prot.* **56**:301–312.
10. **Buchrieser, C., R. Brosch, B. Catimel, and J. Rocourt.** 1993. Pulsed-field gel electrophoresis applied for comparing *Listeria monocytogenes* strains involved in outbreaks. *Can. J. Microbiol.* **39**:395–401.
11. **Bula, C. J., J. Bille, and M. P. Glauser.** 1995. An epidemic of food-borne listeriosis in Western Switzerland: description of 57 cases involving adults. *Clin. Infect. Dis.* **20**:66–72.
12. **Buncić, S.** 1991. The incidence of *Listeria monocytogenes* in slaughtered animals, in meat, and in meat products in Yugoslavia. *Int. J. Food Microbiol.* **12**:173–180.
13. **Colburn, K. G., C. A. Kaysner, C. Abeyta, Jr., and M. M. Wekell.** 1990. *Listeria* species in a California estuarine environment. *Appl. Environ. Microbiol.* **56**:2007–2011.
14. **Cossart, P.** 1995. Actin-based bacterial motility. *Curr. Top. Cell Biol.* **7**:94–101.
15. **Cossart, P., and C. Kocks.** 1994. The actin-based motility of the intracellular pathogen *Listeria monocytogenes*. *Mol. Microbiol.* **13**:395–402.
16. **Cossart, P., and J. Mengaud.** 1989. *Listeria monocytogenes:* a model system for the molecular study of intracellular parasitism. *Mol. Biol. Med.* **6**:463–474.
17. **Cox, L. J., T. Kleiss, J. L. Cordier, C. Cordellana, P. Konkel, C. Pedrazzini, R. Beumer, and A. Siebenga.** 1989. *Listeria* spp. in food processing, non-food and domestic environments. *Food Microbiol.* **6**:49–61.
18. **Davies, E. A., and M. R. Adams.** 1994. Resistance of *Listeria monocytogenes* to the bacteriocin nisin. *Int. J. Food Microbiol.* **21**:341–347.
19. **Dijkstra, R. G.** 1982. The occurrence of *Listeria monocytogenes* in surface water of canals and lakes, in ditches of one big polder and in the effluents and canals of a sewage treatment plant. *Zentralbl. Bakteriol. Hyg. Abt. 1 Orig. Reihe B* **176**:202–205.
20. **Domann, E., J. Wehland, M. Rohde, S. Pistor, M. Hartl, W. Goebel, M. Leimester-Wächter, M. Wuenscher, and T.**

Chakraborty. 1992. A novel bacterial gene in *Listeria monocytogenes* required for host cell microfilament interaction with homology to the proline-rich region of vinculin. *EMBO J.* **11:**1981–1990.

21. **Doyle, M. P.** 1988. Effect of environmental and processing conditions on *Listeria monocytogenes. Food Technol.* **42:** 169–171.
22. **Dramsi, S., I. Biswas, L. Braun, E. Maguin, P. Mastroenni, and P. Cossart.** 1995. Entry into hepatocytes requires expression of the *inlB* gene product. *Mol. Microbiol.* **16:**251–261.
23. **Dramsi, S., M. Lebrun, and P. Cossart.** 1996. Molecular and genetic determinants involved in invasion of mammalian cells by *Listeria monocytogenes. Curr. Top. Microbiol. Immunol.* **203:**61–77.
24. Epidémie de listériose à lysovar 2671-108-312 en France—résultats préliminaires de l'enquête épidémiologique coordonnée par le Réseau National de Santé Publique. 1993. *Bull. Epidemiol. Hebd.* **34:**157–158.
25. **Farber, J. M.** 1991. *Listeria monocytogenes* in fish products. *J. Food Prot.* **54:**922–934.
26. **Farber, J. M., E. D. F. Coates, N. Beausoleil, and J. Fournier.** 1991. Feeding trials of *Listeria monocytogenes* with a nonhuman primate model. *J. Clin. Microbiol.* **29:**2606–2608.
27. **Fleming, D. W., S. L. Cochi, K. L. MacDonald, J. Brondum, P. S. Hayes, B. D. Plikaytis, M. B. Holmes, A. Audurier, C. V. Broome, and A. L. Reingold.** 1985. Pasteurized milk as a vehicle of infection in an outbreak of listeriosis. *N. Engl. J. Med.* **312:**404–407.
28. **Frederiksen, B., and S. Samuelsson.** 1992. Foeto-maternal listeriosis in Denmark 1981–1988. *J. Infect.* **24:**277–287.
29. **Freitag, N. E., L. Rong, and D. A. Portnoy.** 1993. Regulation of the *prfA* transcriptional activator of *Listeria monocytogenes:* multiple promoter elements contribute to intracellular growth and cell-to-cell spread. *Infect. Immun.* **61:**2537–2544.
30. **Gaillard, J.-L., P. Berche, C. Frehel, E. Gouin, and P. Cossart.** 1991. Entry of *L. monocytogenes* into cells is mediated by internalin, a repeat protein reminiscent of surface antigens from Gram-positive cocci. *Cell* **65:**1127–1141.
31. **Gaillard, J. L., P. Berche, and P. Sansonetti.** 1986. Transposon mutagenesis as a tool to study the role of hemolysin in the virulence of *Listeria monocytogenes. Infect. Immun.* **52:**50–55.
32. **Genigeorgis, C., M. Carnicu, D. Dutulescu, and T. B. Farver.** 1991. Growth and survival of *Listeria monocytogenes* in market cheeses stored at 4 to 30°C. *J. Food Prot.* **54:**662–668.
33. **Genigeorgis, C. A., D. Dutulescu, and J. Fernandez Garayzabal.** 1989. Prevalence of *Listeria* spp. in poultry meat at the supermarket and slaughterhouse level. *J. Food Prot.* **52:**618–624.
34. **Geoffroy, C., J.-L. Gaillard, J. E. Alouf, and P. Berche.** 1987. Purification, characterization, and toxicity of the sulfhydryl-activated hemolysin listeriolysin O from *Listeria monocytogenes. Infect. Immun.* **55:**1641–1646.
35. **Goulet, V., C. Jacquet, V. Vaillant, I. Rebière, E. Mouret, E. Lorente, F. Steiner, and J. Rocourt.** 1995. Listeriosis from consumption of raw milk cheese. *Lancet* **345:**1581–1582.
36. **Grau, F. H., and P. B. Vanderlinde.** 1992. Occurrence, numbers and growth of *Listeria monocytogenes* on some vacuum-packaged processed meats. *J. Food Prot.* **55:**4–7.
37. **Graves, L. M., B. Swaninathan, M. W. Reeves, S. B. Hunter, R. E. Weaver, B. D. Plikaytis, and A. Schuchat.** 1994. Comparison of ribotyping and multilocus enzyme electrophoresis for subtyping of *Listeria monocytogenes* isolates. *J. Clin. Microbiol.* **32:**2936–2943.
38. **Guenich, H.-H., H. E. Muller, A. Schrettenbrunner, and H. P. R. Seeliger.** 1985. The occurrence of different *Listeria* species in municipal waste water. *Zentralbl. Bakteriol. Hyg. Abt. 1 Orig. Reihe B* **181:**563–565.
39. **Guyer, S., and T. Jemmi.** 1991. Behavior of *Listeria monocytogenes* during fabrication and storage of experimentally contaminated smoked salmon. *Appl. Environ. Microbiol.* **57:**1523–1527.
40. **Husu, J. R., J. T. Seppänen, S. K. Sivelä, and A. L. Rauramaa.** 1990. Contamination of raw milk by *Listeria monocytogenes* on dairy farms. *J. Vet. Med. B* **37:**268–275.
41. **Husu, J. R., S. K. Sivelä, and A. L. Rauramaa.** 1990. Prevalence of *Listeria* species as related to chemical quality of farm-ensiled grass. *Grass Forage Sci.* **45:**309–314.
42. **Jacquet, C., B. Catimel, R. Brosch, C. Buchrieser, P. Dehaumont, V. Goulet, V. Lepoutre, P. Veit, and J. Rocourt.** 1995. Investigations related to the epidemic strain involved in the French listeriosis outbreak in 1992. *Appl. Environ. Microbiol.* **61:**2242–2246.
43. **Jemmi, T.** 1990. Stand der Kenntisse über Listerien bei Fleisch und Fischprodukten. *Mitt. Geb. Lebensmittelunters. Hyg.* **81:**144–157.
44. **Jensen, A.** 1993. Excretion of *Listeria monocytogenes* in faeces after listeriosis: rate, quantity and duration. *Med. Microbiol. Lett.* **2:**176–182.
45. **Jeong, D. K., and J. F. Frank.** 1994. Growth of *Listeria monocytogenes* at 10°C in biofilms with microorganisms isolated from meat and dairy processing environments. *J. Food Prot.* **57:**576–586.
46. **Johnson, J. L., M. P. Doyle, and R. G. Cassens.** 1990. *Listeria monocytogenes* and other *Listeria* spp. in meat and meat products—a review. *J. Food Prot.* **53:**81–91.
47. **Jurado, R. L., M. M. Farley, E. Pereira, R. C. Harvey, A. Schuchat, J. D. Wenger, and D. S. Stephens.** 1993. Increased risk of meningitis and bacteriemia due to *Listeria monocytogenes* in patients with human immunodeficiency virus infection. *Clin. Infect. Dis.* **17:**224–227.
48. **Kaufmann, S. H. E.** 1993. Immunity to intracellular bacteria. *Annu. Rev. Immunol.* **11:**129–163.
49. **Klarsfeld, A. D., P. Goossens, and P. Cossart.** 1994. Five *Listeria monocytogenes* preferentially expressed in mammalian cells. *Mol. Microbiol.* **13:**585–597.
50. **Kocks, C., E. Gouin, M. Tabouret, P. Berche, H. Ohayon, and P. Cossart.** 1992. *Listeria monocytogenes* induced actin assembly requires the ActA gene product, a surface protein. *Cell* **68:**521–531.
51. **Köhler, S., M. Leimeister-Wächter, T. Chakraborty, F. Lottspeich, and W. Goebel.** 1990. The gene coding for protein p60 of *Listeria monocytogenes* and its use as a specific probe for *Listeria monocytogenes. Infect. Immun.* **58:**1943–1950.

52. **Larsen, A. G., and B. Norrung.** 1993. Inhibition of *Listeria monocytogenes* by bavaricin A, a bacteriocin produced by *Lactobacillus bavaricus* Ml401. *Lett. Appl. Microbiol.* **17:**132–134.

53. **Leimeister-Wächter, M., E. Domann, and T. Chakraborty.** 1992. The expression of virulence genes in *L. monocytogenes* is thermoregulated. *J. Bacteriol.* **174:**947–952.

54. **Leimeister-Wächter, C. Haffner, E. Domann, W. Goebel, and T. Chakraborty.** 1990. Identification of a gene that positively regulates expression of listeriolysin, the major virulence factor of *Listeria monocytogenes*. *Proc. Natl. Acad. Sci. USA* **87:**8336–8340.

55. **Linnan, M. J., L. Mascola, X. D. Lou, V. Goulet, S. May, C. Salminen, D. W. Hird, M. L. Yonekura, P. Hayes, R. Weaver, A. Audurier, B. D. Plikaytis, S. L. Fannin, A. Kleks, and C. V. Broome.** 1988. Epidemic listeriosis associated with Mexican-style cheese. *N. Engl. J. Med.* **319:**823–828.

56. **Lovett, J., D. W. Francis, and J. M. Hunt.** 1987. *Listeria monocytogenes* in raw milk: detection, incidence, and pathogenicity. *J. Food Prot.* **50:**188–192.

57. **Luchansky, J. B.** 1994. Use of biopreservatives to control pathogenic and spoilage microbes in food, p. 253–262. *In* A. Amgar (ed.), *Food Safety 94*. ASEPT, Laval, France.

58. **Marquis, H., H. A. Bouwer, D. Hinrichs, and D. Portnoy.** 1993. Intracytoplasmic growth and virulence of *Listeria monocytogenes* auxotrophic mutants. *Infect. Immun.* **61:**3756–3760.

59. **Mascola, L., F. Sorvillo, V. Goulet, B. Hall, R. Weaver, and M. Linnan.** 1992. Fecal carriage of *Listeria monocytogenes*—observations during a community wide, common-source outbreak. *Clin. Infect. Dis.* **15:**557–558.

60. **Matthieu, F., M. Michel, A. Lebrihi, and G. Lefebvre.** 1994. Effect of the bacteriocin carnocin CP5 and of the producing strain *Carnobacterium piscicola* CP5 on the viability of *Listeria monocytogenes* ATCC 15313 in salt solution, broth and skimmed milk, at various incubation temperatures. *Int. J. Food Microbiol.* **22:**155–172.

61. **Mazurier, S. I., and K. Wernars.** 1992. Typing of *Listeria* strains by random amplification of polymorphic DNA. *Res. Microbiol.* **143:**499–505.

62. **McLauchlin, J., and R. J. Gilbert.** 1990. *Listeria* in food. *PHLS Microbiol. Dig.* **7:**54–55.

63. **McLauchlin, J., S. M. Hall, S. K. Velani, and R. J. Gilbert.** 1991. Human listeriosis and pate—a possible association. *Br. Med. J.* **303:**773–775.

64. **McLauchlin, J., and P. N. Hoffman.** 1989. Neonatal cross-infection from *Listeria monocytogenes*. *Commun. Dis. Rep.* **6:**3–4.

65. **Mengaud, J., S. Dramsi, E. Gouin, J. A. Vasquez-Boland, G. Milon, and P. Cossart.** 1991. Pleiotropic control of *Listeria monocytogenes* virulence factors by a gene which is autoregulated. *Mol. Microbiol.* **5:**2273–2283.

66. **Michel, E., and P. Cossart.** 1992. Physical map of the *Listeria monocytogenes* chromosome. *J. Bacteriol.* **174:**7098–7103.

67. **Nieman, R. E., and B. Lorber.** 1980. Listeriosis in adult: a changing pattern—report of eight cases and review of the literature, 1968–1978. *Rev. Infect. Dis.* **2:**207–227.

68. **Pamer, E. G., J. T. Harty, and M. J. Bevan.** 1991. Precise prediction of a dominant class I IMC restricted epitope of *Listeria monocytogenes*. *Nature* (London) **353:**852–856.

69. **Park, S. F., and R. G. Kroll.** 1993. Expression of listeriolysin and phosphatidylinositol-specific phospholipase C is repressed by the plant-derived molecular cellobiose in *Listeria monocytogenes*. *Mol. Microbiol.* **8:**653–661.

70. **Pine, L., S. Kathariou, F. Quinn, V. George, J. D. Wenger, and R. E. Weaver.** 1991. Cytopathogenic effects in enterocytelike Caco-2 cells differentiate virulent from avirulent *Listeria* strains. *J. Clin. Microbiol.* **29:**990–996.

71. **Pinner, R. W., A. Schuchat, B. Swaminathan, P. S. Hayes, K. A. Deaver, R. E. Weaver, B. D. Plikaytis, M. Reeves, C. V. Broome, and J. D. Wenger.** 1992. Role of foods in sporadic listeriosis. 2. Microbiologic and epidemiologic investigation. *JAMA* **267:**2046–2050.

72. **Portnoy, D. A.** 1992. Innate immunity to a facultative intracellular bacterial pathogen. *Curr. Top. Immunol.* **4:**20–24.

73. **Portnoy, D. A., T. Chakraborty, W. Goebel, and P. Cossart.** 1992. Molecular determinants of *Listeria monocytogenes* pathogenesis. *Infect. Immun.* **60:**1263–1267.

74. **Riedo, F. X., R. W. Pinner, M. D. Tosca, M. L. Carter, L. M. Graves, M. W. Reeves, R. E. Weaver, B. D. Plikaytis, and C. V. Broome.** 1994. A point-source foodborne listeriosis outbreak: documented incubation period and possible mild illness. *J. Infect. Dis.* **170:**693–696.

75. **Rocourt, J., and R. Brosch.** 1992. *Human Listeriosis—1990.* WHO/HPP/FOS/92.3. World Health Organization, Geneva.

76. **Schlech, W. F., III, P. M. Lavigne, R. A. Bortolussi, A. C. Allen, E. V. Haldane, A. J. Wort, A. W. Hightower, S. E. Johnson, S. H. King, E. S. Nicholls, and C. V. Broome.** 1983. Epidemic listeriosis—evidence for transmission by food. *N. Engl. J. Med.* **308:**203–206.

77. **Schuchat, A., K. Deaver, P. S. Hayes, L. Graves, L. Mascola, and J. D. Wenger.** 1993. Gastrointestinal carriage of *Listeria monocytogenes* in household contacts of patients with listeriosis. *J. Infect. Dis.* **167:**1261–1262.

78. **Schuchat, A., K. A. Deaver, J. D. Wenger, B. D. Plikaytis, L. Mascola, R. W. Pinner, A. L. Reingold, and C. V. Broome.** 1992. Role of foods in sporadic listeriosis. 1. Case-control study of dietary risk factors. *JAMA* **267:**2041–2045.

79. **Schuchat, A., C. Lizano, C. V. Broome, B. Swaminathan, C. Kim, and K. Win.** 1991. Outbreak of neonatal listeriosis associated with mineral oil. *Pediatr. Infect. Dis. J.* **10:**183–189.

80. **Schuchat, A., R. W. Pinner, K. Deaver, B. Swaminathan, R. Weaver, P. S. Hayes, M. Reeves, P. Pierce, J. D. Wenger, C. V. Broome, and *Listeria* Study Group.** 1991. Epidemiology of listeriosis in the USA, p. 69–73. *In* A. Amgar (ed.), *Listeria and Food Safety*. ASEPT, Laval, France.

81. **Schwartz, B., D. Hexter, C. V. Broome, A. W. Hightower, R. B. Hischorn, J. D. Porter, P. S. Hayes, W. F. Bibb, B. Lorber, and D. G. Faris.** 1989. Investigation of an outbreak of listeriosis: new hypotheses for the etiology of epidemic *Listeria monocytogenes* infections. *J. Infect. Dis.* **159:**680–685.

82. **Seeliger, H. P. R., and D. Jones.** 1986. *Listeria*, p. 1235–1245. *In* P. H. A. Sneath, N. S. Mair, M. E. Sharpe, and J. G. Holt (ed.), *Bergey's Manual of Systematic Bacteriology*, vol. 2. Williams & Wilkins, Baltimore.

83. **Sheehan, B., C. Kocks, S. Dramsi, E. Gouin, A. Klarsfeld, J. Mengaud, and P. Cossart.** 1994. Molecular and genetic determinants of the *Listeria monocytogenes* infectious process. *Curr. Top. Microbiol. Immunol.* **192:**187–216.

84. **Skogberg, K., J. Syrjanen, M. Jahkola, O. V. Renkonen, J. Paavonen, J. Ahonen, S. Kontiainen, P. Ruutu, and V. Valtonen.** 1992. Clinical presentation and outcome of listeriosis in patients with and without immunosuppressive therapy. *Clin. Infect. Dis.* **14:**815–821.

85. **Skovgaard, N., and B. Norrung.** 1989. The incidence of *Listeria* spp. in faeces of Danish pigs and in minced pork meat. *Int. J. Food Microbiol.* **8:**59–63.

86. **Sokolovic, Z., A. Fuchs, and W. Goebel.** 1990. Synthesis of species-specific stress proteins by virulent strains of *Listeria monocytogenes. Infect. Immun.* **58:**3582–3587.

87. **Sokolovic, Z., and W. Goebel.** 1989. Synthesis of listeriolysin in *Listeria monocytogenes* under heat shock conditions. *Infect. Immun.* **57:**295–298.

88. **Tang, P., I. Rosenshine, and B. B. Finley.** 1994. *Listeria monocytogenes,* an invasive bacterium stimulates MAP kinase upon attachment to epithelial cells. *Mol. Biol. Cell* **5:**455–464.

89. **Tappero, J. W., A. Schuchat, K. A. Deaver, L. Mascola, and J. D. Wenger.** 1995. Reduction in the incidence of human listeriosis in the United States. Effectiveness of prevention efforts. *JAMA* **273:**1118–1122.

90. **Tilney, L. G., and M. S. Tilney.** 1993. The wily ways of a parasite: induction of actin assembly by *Listeria. Trends Microbiol.* **1:**25–31.

91. **Van Renterghem, B., F. Huysman, R. Rygole, and W. Verstraete.** 1991. Detection and prevalence of *Listeria monocytogenes* in the agricultural ecosystem. *J. Appl. Bacteriol.* **71:**211–217.

92. **Vazquez-Boland, J. A., C. Kocks, S. Dramsi, H. Ohayon, C. Geoffroy, J. Mengaud, and P. Cossart.** 1992. Nucleotide sequence of the lecithinase operon of *Listeria monocytogenes* and possible role of lecithinase in cell-to-cell spread. *Infect. Immun.* **60:**219–230.

93. **Velge, P., E. Bottreau, B. Kaeffer, N. Yurdusev, P. Pardon, and V. Van Langendonck.** 1994. Protein tyrosine kinase inhibitors block the entries of *Listeria monocytogenes* and *Listeria ivanovii* into epithelial cells. *Microbiol. Pathol.* **17:**37–50.

94. **Walker, R. L., L. H. Jensen, H. Kinde, A. V. Alexander, and L. S. Owen.** 1991. Environment survey for *Listeria* species in frozen milk plants in California. *J. Food Prot.* **54:**178–182.

95. **Weis, J., and H. P. R. Seeliger.** 1975. Incidence of *Listeria monocytogenes* in nature. *Appl. Microbiol.* **30:**29–32.

96. **World Health Organization.** 1988. *Food Listeriosis—Report of the WHO Informal Working Group.* WHO/EHE/FOS/88.5. World Health Organization, Geneva.

97. **Wuenscher, M., S. Kohler, A. Bubert, U. Gerike, and W. Goebel.** 1993. The *iap* gene of *Listeria monocytogenes* is essential for cell viability and its gene product, p60, has bacteriolytic activity. *J. Bacteriol.* **175:**3491–3501.

Lynn M. Jablonski
Gregory A. Bohach

19

Staphylococcus aureus

CHARACTERISTICS OF THE ORGANISM

Historical Aspects and General Considerations

Staphylococcal food poisoning (SFP) ranks as one of the most prevalent causes of gastroenteritis worldwide. It results from ingestion of one or more preformed staphylococcal enterotoxins (SEs) in staphylococcus-contaminated food. The etiological agents of SFP are members of the genus *Staphylococcus,* predominantly *Staphylococcus aureus.* This form of food poisoning is considered an intoxication since it does not require growth of the organism in the host.

The association of staphylococci with foodborne illness was made more than a century ago. Barber, in 1914, was the first to implicate a toxin in SFP (7). He reported that repeated ingestion of contaminated milk produced symptoms typical of the illness. Barber cultured the milk, demonstrated the presence of a putative causative staphylococcal agent, and provided the first evidence that a soluble toxin was responsible for the disease. The next major advance in understanding SFP etiology was reported in 1930 by Dack et al. (25), who voluntarily consumed supernatants from cultures of "a yellow hemolytic staphylococcus" grown from contaminated sponge cake. Upon ingestion of the filtrates, they became ill with vomiting, abdominal cramps, and diarrhea. At that time, the only other foodborne toxin that had been recognized was clostridial botulinum toxin. However, the staphylococcal toxin, which exerted an effect on the gastrointestinal tract, was the first true enterotoxin described. It seemed to be particularly unique in comparison with botulinum toxin since its activity was "not entirely destroyed by heating even for 30 minutes at 100°C."

S. aureus has been extensively characterized. This organism produces a variety of extracellular products. Many of these, including the SEs, are virulence factors which have been implicated in diseases of humans and animals. As a group, the SEs elaborate a set of biological properties that enable staphylococci to cause at least two common human diseases, toxic shock syndrome (TSS) and SFP. This chapter deals primarily with SFP. However, in regard to the SEs, there is significant overlap in the natural histories of the two diseases. Thus, TSS is also discussed when this overlap is most relevant.

Nomenclature, Characteristics, and Distribution of SE-Producing Staphylococci

The term "staphylococci" informally describes a group of small spherical, gram-positive bacteria. Depending on the species and culture conditions, their cells have a diameter ranging from approximately 0.5 to 1.5 μm. They are catalase-positive chemoorganotrophs with a DNA composition of 30 to 40 mol% G+C content. Staphylococci have a typical gram-positive cell wall containing peptidoglycan and teichoic acids. Except for clinical isolates and strains exposed to antimicrobial therapy, most staphylococci are sensitive to β-lactams, tetracyclines, macrolides, lincosamides, novobiocin, and

Table 19.1 General characteristics of selected species of *Staphylococcus*

Characteristic	*S. aureus*	*S. chromogenes*	*S. hyicus*	*S. intermedius*	*S. epidermidis*	*S. saprophyticus*
Coagulase	+	–	+	+	–	–
Thermostable nuclease	+	–	+	+	+/–	–
Clumping factor	+	–	–	+	–	–
Yellow pigment	+	+	–	–	–	+/–
Hemolytic activity	+	–	–	+	+/–	–
Phosphatase	+	+	+	+	+/–	–
Lysostaphin	Sensitive	Sensitive	Sensitive	Sensitive	Slightly sensitive	ND[a]
Hyaluronidase	+	–	+	–	+/–	ND
Mannitol fermentation	+	+/–	–	+/–	–	+/–
Novobiocin resistance	–	–	–	–	–	+

[a]ND, not determined.

chloramphenicol but are resistant to polymyxin and polyene. Some differential characteristics of *S. aureus* and several other selected species of staphylococci are summarized in Table 19.1.

There have been many useful schemes for classification of the staphylococci. According to *Bergey's Manual of Determinative Bacteriology* (53a), staphylococci have been placed in the family *Micrococcaceae*. This family includes the genera *Micrococcus, Staphylococcus,* and *Planococcus.* The genus *Staphylococcus* is further subdivided into more than 23 species and subspecies. Many of these are found in food as a result of human, animal, or environmental contamination. Several species of *Staphylococcus*, including both coagulase-negative and coagulase-positive isolates, have been reported to produce SEs. Although several species have the potential to produce enterotoxin that causes gastroenteritis, nearly all cases of SFP can be attributed to *S. aureus.* This is a reflection of the higher incidence of SE production by *S. aureus* isolates than by other staphylococcal species. Although the reason for this specificity has not been confirmed, the SEs are known to act as superantigens and therefore are potential immunomodulating agents (see below). Thus, SE production seems to provide a selective advantage to *S. aureus*, a species that is common to both humans and animals, the two most common sources of food contamination.

Enterotoxigenic strains of staphylococci have been well characterized on the basis of a number of genotypic and phenotypic characteristics. An extensive phage typing system is available for *S. aureus.* Most SE-producing isolates belong to phage group I or III or are nontypeable. Although SE production by other phage groups is less common, it has been documented. Hajek and Marsalek (36) developed a classification scheme based largely on the animal host of origin. They were able to differentiate *S. aureus* into at least six biotypes. By far, SE production was most prevalent among human isolates within biotype A. SE production by other biotypes is reported to be rare except for biotype C bovine and ovine mastitis isolates. SEs may also be produced by *S. intermedius* and *S. hyicus* (formerly *S. aureus* biotypes E and F, respectively), albeit less frequently.

Introduction and Nomenclature of the SEs

Structure-function and mechanisms of pathogenicity of the SEs are discussed later in this chapter. This section will introduce the SE nomenclature, evolution, and regulation by *S. aureus.*

Current Classification Scheme Based on Antigenicity

Major advances in characterization of SEs were made approximately two decades after Dack and colleagues (25) associated SFP with an exotoxin. Bergdoll and coworkers at the Food Research Institute were the first investigators to produce purified SE preparations and develop specific antisera (reviewed in reference 9). They and others, using purified or partially purified toxins, showed that protective antibodies could be induced in several species of animals. However, this immunological protection was strain specific. Some animals possessing immunity to symptoms induced by the toxin from one strain were not protected against the toxin produced by other strains (11). It soon became apparent that *S. aureus* can produce multiple toxins with similar molecular weights as well as similar biological and physicochemical properties.

Initially, differentiation between the multiple antigenic forms of SE was based on the observation that many food isolates produce one common antigenic type of toxin, tentatively designated the "F" toxin. Most other enterotoxigenic strains, such as those from enteritis patients, also produced the "F" toxin in addition to a second antigenic "E" toxin. The discovery of additional isolates that did not conform to this pattern prompted adoption of an improved nomenclature system. A committee was assembled in 1962 to establish the current alphabetical nomenclature (22). Accordingly, SEs are sequentially assigned a letter of the alphabet in the order of their discovery. The "F" and "E" toxins were designated SEA and SEB, respectively. Between 1962 and 1972, three additional SE serotypes (SEC, SED, and SEE) were reported (10). Protein sequencing and recombinant DNA methods have resulted in our current detailed knowledge of the primary sequences of all of the classical SEs (5, 24, 42, 50, 57, 75) (Fig. 19.1). More recent additions to the SE family include SEG and SEH. A partial characterization of the SEG biochemical and physical properties has been published (12). SEH was reported by two groups of investigators. Ren et al. (70) sequenced the gene for SEH in 1994. Although they did not report its ability to induce vomiting, an emetic SE with the same N terminus as SEH was purified and characterized by Su and Wong (82). SEF was initially used in reference to an exotoxin commonly produced by TSS isolates. This designation was later dropped when it was confirmed that SEF was not emetic. To avoid confusion, SEF has been retired from use in the SE nomenclature system and is now referred to as TSS toxin 1 (TSST-1) (14).

The incidence of SE involvement in SFP appears to change with time. Prior to 1971, SEA was the predominant toxin identified in cases of SFP, followed in frequency by SED and SEC. In some cases, SEA was also identified as the causative agent in combination with SEC or SED; SEB was only rarely associated with SFP (59). In a study examining more recent outbreaks of SFP from 1977 through 1981, SEA remained the most common toxin implicated (40). However, in contrast to previous surveys, SEB was the only other SE identified. Holmberg and Blake (40) suggested that the observed decrease in cases attributed to other SEs was due to improved conditions for the processing and storage of milk, which had been commonly contaminated by SEC- and SED-producing strains of *S. aureus* in the past. Presently, SEE is the least common associated with SFP (45).

SE Antigenic Subtypes and Molecular Variants

Designation of SEs based on serological typing has been very useful. However, sequence analysis and detailed immunological studies have produced some examples in which the antigenic characteristics of the proteins do not reflect their molecular or biological uniqueness. The best-documented examples are with SEC. It had been noted for some time that the SEC serological variant can be further divided into at least three subtypes (SEC1, SEC2, and SEC3) based on minor differences in immunological reactivity (Fig. 19.2). However, within each subtype, significant sequence variability may occur. For example, the SECs produced by strains FRI-909 and FRI-913 were both designated SEC3 according to their immunological reactivity. However, it was later shown that the sequences of the two toxins differ by nine residues (57). The SEC variants produced by bovine and ovine isolates of *S. aureus* have very similar sequences and are apparently indistinguishable from SEC1 in immunological assays. In contrast, they behave differently from SEC1 in biological assays. For example, although SEC-bovine differs from SEC1 by only three residues, the potencies of the two toxins differ by several orders of magnitude in lymphocyte proliferation assays (57).

Staphylococcal Genetics and Evolutionary Aspects of SE Production

SEs Are Superantigens and Belong to a Large PT Family

In discussing the genetics and evolution of the SEs, one must also consider additional staphylococcal toxins, plus some toxins produced by other organisms, especially group A streptococci. SEs are part of a large family of related toxins produced by *S. aureus* and *Streptococcus pyogenes* (15). This family of toxins has been termed the pyrogenic toxin (PT) family; members of this family are grouped together on the basis of shared biological and biochemical properties. The one feature in common to all PTs, including SEs, is their unique ability to act as superantigens (58).

Superantigens are molecules that have the ability to stimulate an exceptionally high percentage of T cells. The mechanism by which this occurs distinguishes them from mitogens and conventional antigens. In regard to T-cell stimulation, superantigens are bifunctional molecules that interact initially with major histocompatibility complex (MHC) class II molecules on antigen-presenting cells. Unlike the situation with conventional antigens, this interaction does not require processing and occurs outside the MHC peptide-binding groove (Fig. 19.3). Once formed, the MHC-superantigen complex interacts with the T-cell receptor (TCR). The interaction with the TCR is also nonconventional and relatively nonspecific; it occurs at a variable location on the TCR β chain (the Vβ region). Since superantigens bind outside

```
SEC-bovine    -ESQPDPTPDELHKASKFTGL-MENMKVLY-DDRYVSATKVKSVDKFLAH   47
SEC-ovine     -ESQPDPTPDELHKASKFTGL-MENMKVLY-DDRYVSATKVKSVDKFLAH   47
SEC1          -ESQPDPTPDELHKASKFTGL-MENMKVLY-DDHYVSATKVKSVDKFLAH   47
SEC2          -ESQPDPTPDELHKSSEFTGT-MGNMKYLY-DDHYVSATKVMSVDKFLAH   47
SEC3-FRI913   -ESQPDPMPDDLHKSSEFTGT-MGNMKYLY-DDHYVSATKVKSVDKFLAH   47
SEC3-FRI909   -ESQPDPMPDDLHKSSEFTGT-MGNMKYLY-DDHYVSATKVKSVDKFLAH   47
SEB           -ESQPDPKPDELHKSSKFTGL-MENMKVLY-DDNHVSAINVKSIDQFLYF   47
SEA           SEKSEEINEKDLRKKSELQGTALGNLKQIYYYNEKAKTENKESHDQFLQH   50
SEE           ---SEEINEKDLRKKSELQRNALSNLRQIYYYNEKAITENKESDDQFLEN   47
SED           -----SVKEKELHKKSELSSTALNNMKHSYADKNPIIGENKSTGDQFLEN   45
SEH           ---------EDLHDKSELTDLALANAYGQY--NHPFIKENIKSDEISGEK   39

SEC-bovine    DLIYNISDKKLKNYDKVKTELLNEDLAKKYKDEVVDVYGSNYYVNCYFSS   97
SEC-ovine     DLIYNISDKKLKNYDKVKTELLNEDLAKKYKDEVVDVYGSNYYVNCCFSS   97
SEC1          DLIYNISDKKLKNYDKVKTELLNEGLAKKYKDEVVDVYGSNYYVNCYFSS   97
SEC2          DLIYNISDKKLKNYDKVKTELLNEDLAKKYKDEVVDVYGSNYYVNCYFSS   97
SEC3-FRI913   DLIYNISDKKLKNYDKVKTELLNEDLAKKYKDEVVDVYGSNYYVNCYFSS   97
SEC3-FRI909   DLIYNINDKKLNNYDKVKTELLNEDLANKYKDEVVDVYGSNYYVNCYFSS   97
SEB           DLIYSIKDTKLGNYDNVRVEFKNKDLADKYKDKYVDVFGANYYYQCYFSK   97
SEA           TILFKGFFTDHSWYNDLLVDFDSKDIVDKYKGKKVDLYGNYYGYQCA---   97
SEE           TLLFKGFFTGHPWYNDLLVDLGSKDATNKYKGKKVDLYGAYYGYQCA---   94
SED           TLLYKKFFTDLINFEDLLINFNSKEMAQHFKSKNVDVYPIRYSINCY---   92
SEH           DLIFRN---QGDSGNDLRVKFATADLAQKFKNKNVDIYGASFYYKCE---   83

SEC-bovine    KDNVGKVTGG---KTCMYGGITKHEGNHFDNGKLQNVLIRVYENKRNTIS  144
SEC-ovine     KDNVGKVTGG---KTCMYGGITKHEGNHFDNGNLQNVLIRVYENKRNTIS  144
SEC1          KDNVGKVTGG---KTCMYGGITKHEGNHFDNGNLQNVLIRVYENKRNTIS  144
SEC2          KDNVGKVTGG---KTCMYGGITKHEGNHFDNGNLQNVLIRVYENKRNTIS  144
SEC3-FRI913   KDNVGKVTGG---KTCMYGGITKHEGNHFDNGNLQNVLVRVYENKRNTIS  144
SEC3-FRI909   KDNVGKVTSG---KTCMYGGITKHEGNHFDNGNLQNVLIRVYENKRNTIS  144
SEB           KTNDINSHQTDKRKTCMYGGVTEHNGNQLD--KYRSITVRVFEDGKNLLS  145
SEA           -------GGTPNKTACMYGGVTLHDNNRLTEEKKVPINLWL-DGKQNTVP  139
SEE           -------GGTPNKTACMYGGVTLHDNNRLTEEKKVPINLWI-DGKQTTVP  136
SED           -------GGEIDRTACTYGGVTPHEGNKLKERKKIPINLWI-NGVQKEVS  134
SEH           -------KISENISECLYGGTTL-NSEKLAQERVIGANVWV-DGIQKETE  124

SEC-bovine    FE-VQTDKKSVTAQELDIKARNFLINKKNLY--EFNSSPYETGYIKFIEN  191
SEC-ovine     FE-VQTDKKSVTAQELDIKARSFLINKKNLY--EFNSSPYETGYIKFIEN  191
SEC1          FE-VQTDKKSVTAQELDIKARNFLINKKNLY--EFNSSPYETGYIKFIEN  191
SEC2          FE-VQTDKKSVTAQELDIKARNFLINKKNLY--EFNSSPYETGYIKFIEN  191
SEC3-FRI913   FE-VQTDKKSVTAQELDIKARNFLINKKNLY--EFNSSPYETGYIKFIEN  191
SEC3-FRI909   FE-VQTDKKSVTAQELDIKARNFLINKKNLY--EFNSSPYETGYIKFIES  191
SEB           FD-VQTNKKKVTAQELDYLTRHYLVKNKKLY--EFNNSPYETGYIKFIE-  191
SEA           LETVKTNKKNVTVQELDLQARRYLQEKYNLYNSDVFDGKVQRGLIVFHTS  189
SEE           IDKVKTSKKEVTVQELDLQARHYLHGKFGLYNSDSFGGKVQRGLIVFHSS  186
SED           LDKVQTDKKNVTVQELDAQARRYLQKDLKLYNNDTLGGKIQRGKIEFDSS  184
SEH           L--IRTNKKNVTLQELDIKIRKILSDKYKIYYKD---SEISKGLIEFDMK  169

SEC-bovine    NGNTFWYDMMPAPGDKFDQSKYLMMYNDNKTVDSKSV-KIEVHLTTKNG  239
SEC-ovine     NGNTFWYDMMPAPGDKFDQSKYLMMYNDNKTVDSKSV-KIEVHLTTKNG  239
SEC1          NGNTFWYDMMPAPGDKFDQSKYLMMYNDNKTVDSKSV-KIEVHLTTKNG  239
SEC2          NGNTFWYDMMPAPGDKFDQSKYLMMYNDNKTVDSKSV-KIEVHLTTKNG  239
SEC3-FRI913   NGNTFWYDMMPAPGDKFDQSKYLMMYNDNKTVDSKSV-KIEVHLTTKNG  239
SEC3-FRI909   NGNTFWYDMMPAPGDKFDQSKYLMIYKDNKMVDSKSV-KIEVHLTTKNG  239
SEB           NENSFWYDMMPAPGDKFDQSKYLMMYNDNKMVDSKDV-KIEVYLTTKKK  239
SEA           TEPSVNYDLFGAQGQ--YSNTLLRIYRDNKTINSENM-HIDIYLYTS--  233
SEE           EGSTVSYDLFDAQGQ--YPDTLLRIYRDNKTINSENL-HIDLYLYTT--  230
SED           DGSKVSYDLFDVKGD--FPEKQLRIYSDNKTLSTEHL-HIDIYLYEK--  228
SEH           TPRDYSFDIYDLKGE--NDYEIDKIYEDNKTLKSDDISHIDVNLYTKKV  216
```

Figure 19.1 Alignment of primary sequences of mature SE proteins in the current literature. Also shown are residue sequence numbers (on the right) and dashes to indicate gaps in the sequences made by alignment. See the text for references. Sequence alignment and output were conducted by using the PileUp and PrettyPlot programs, respectively (31).

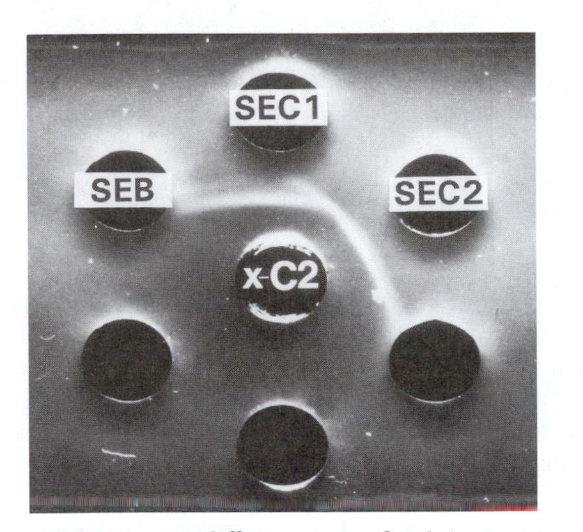

Figure 19.2 Immunodiffusion assays for demonstrating SE uniqueness. A precipitin line of partial identity forms when SEC1 and SEC2 diffuse from separate wells toward the center well containing hyperimmune SEC2 rabbit antiserum. While this demonstrates the highly related nature of the SEC subtypes, less related antigenic types such as SEB are more immunologicallly distinct and do not react.

the area on the TCR used for antigen recognition, they activate a much higher percentage of T cells than can be activated by conventional antigens. However, compared with mitogens which stimulate T cells in an indiscriminate manner, there is some degree of specificity in superantigen action since only certain Vβ sequences are recognized. Hence, not all T cells are stimulated.

The staphylococcal and streptococcal PTs are prototype microbial superantigens that exert a variety of immunomodulatory effects leading to shock, immunosuppression, and other systemic abnormalities associated with TSS. While the SEs are included with the PTs, they have the unique distinction of possessing an additional ability to induce an emetic response upon oral ingestion and are thus solely responsible for SFP. It is generally agreed that many of the toxins in this family, including the SEs, arose from a common ancestral gene which crossed the genus barrier and became stably introduced in both *Staphylococcus* and *Streptococcus* genera. Evidence for this idea is strongly supported by the observation that the structural genes for some of the SEs and related streptococcal PTs are carried on discrete genetic elements (see below).

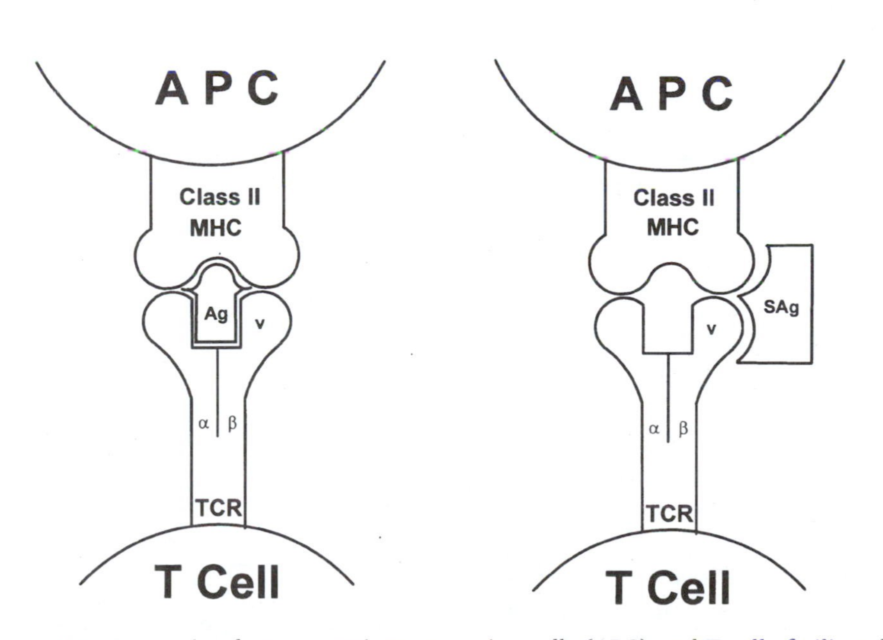

Figure 19.3 Interaction between antigen-presenting cells (APC) and T cells facilitated by conventional antigens (Ag) and superantigens (SAg). Following processing by the antigen-presenting cells, conventional antigens are presented to highly specific TCR molecules in association with the antigen-binding groove of the MHC class II molecule. Superantigens interact with MHC class II molecules (without processing) outside the antigen-binding groove. The superantigen-MHC bimolecular complex next interacts with the TCR through specificity determined only by the variable (V) region of the receptor β chain.

Role of Mobile Genetic Elements in Generation and Dissemination of the SEs

Some staphylococcal and streptococcal toxins are encoded by structural genes located on bacteriophage genomes. Although the streptococcal toxins have been best studied, the genetic mobility of SEA and SEE appears to be similar to the situation in group A streptococci. Betley and Mekalanos (13) confirmed that the SEA structural gene (*sea*) is carried by a lysogenic phage. *sea* was cloned directly from the induced bacteriophage genome (13). Both *sea* and *see* map near the *att* site on their respective phage genomes. In a high percentage of cases, the toxin-encoding phages cannot be induced to a lytic cycle and appear to be defective. The most likely explanation for these observations is that the toxin genes originally were located on the bacterial genome but subsequently were obtained by the phage upon abnormal excision from the chromosome. This phenomenon is documented for some bacterial toxins that are transferred by lysogenic conversion in other genera of bacteria such as *Streptococcus* and *Corynebacterium* species (49).

The role of plasmids has received considerable attention in relation to transmission of PT genes. However, of the PTs, only SED is plasmid encoded. Bayles and Iandolo reported that in more than 20 characterized Sed$^+$ isolates, the *sed* structural gene is localized to a stable 27.6-kb plasmid (pIB485) which also encodes penicillin and cadmium resistance (5). The literature also includes reports associating SEC and SEB genes with transmissible penicillin resistance plasmids. However, it is now generally agreed that both of these toxins are chromosomally encoded. This does not preclude the possibility that these genes are harbored on a variable genetic element as predicted by several investigations. SEB is the better characterized of the two toxins in this regard. Evidence suggests that *seb* is located on a DNA element that is at least 26.8 kb in length (47). It remains to be determined whether this element represents a phage or integrated plasmid. While SEB production has not been associated with lysogenic conversion, from what is known about SEA, it is possible that the *seb* gene resides on a defective phage. Circumstantial evidence for this possibility includes variable upstream regions in different SEB-producing strains of *S. aureus.*

Some staphylococcal toxin genes are insertion sites for genetic elements carrying other virulence determinants (12). The expression of several SEs is affected by this feature. For example, SEB and TSST-1 syntheses are mutually exclusive in *S. aureus.* However, the SEB and TSST-1 genes (*seb* and *tst,* respectively) often coexist in *S. aureus.* The lack of Seb$^+$ Tst$^+$ strains is due to the insertion of *tst* into the *seb* locus. Likewise, the phage which harbors *sea* utilizes the β-toxin locus, *hlb,* as its insertion site, and thus Sea$^+$ isolates do not produce β-toxin.

Mechanisms and Rationale for Generation of SE Diversity

On the basis of amino acid sequences, the currently known SEs can be divided into three groups (Fig. 19.4). Group 1 contains SEB and the SEC subtypes and molecular variants. Toxins in this group are highly related to each other (66 to 99% identity) and to several streptococcal PTs. Group 2 contains the highly related SEA and SEE (84% identity) and SED, which is more distantly related. SEH alone currently forms group 3. It shares only 38% identity with its closest SE relative, SEE.

Most staphylococcal and streptococcal PTs, even of those with no significant overall homology, contain four highly conserved stretches of primary sequence (39). This suggests that there is a selective advantage in host-parasite interactions for these organisms to maintain certain toxin characteristics. At the same time, modifica-

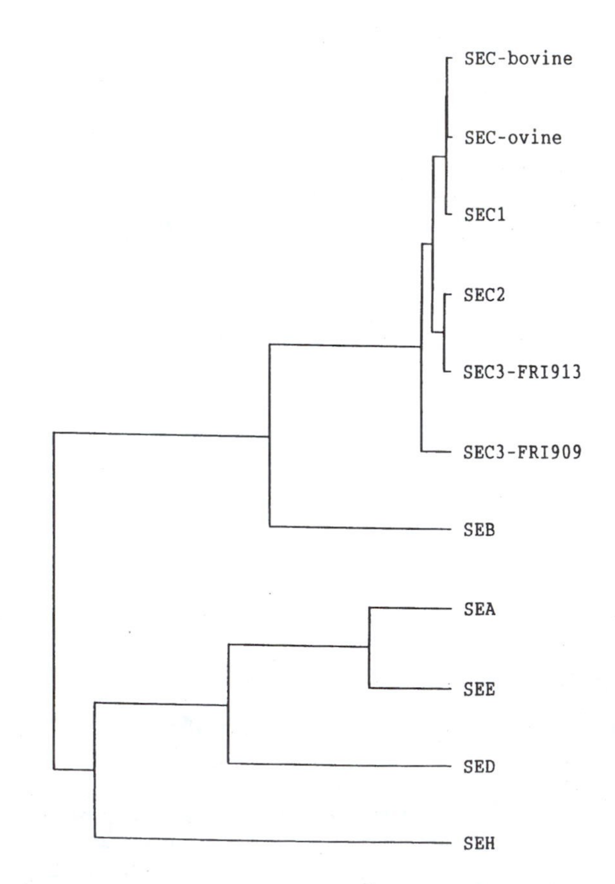

Figure 19.4 Tree representation demonstrating molecular relatedness of the currently known SE family. This tree was created with the clustering feature of the PileUp program (31).

tion of selected regions of the proteins could allow the microorganisms to broaden their host range. This molecular diversity may explain how a group of toxins with the same function but different host specificities could have arisen for the purpose of exploiting a broader repertoire of receptors.

Sequence comparison of members within the PT family has provided several examples in which diversity among the toxins appears to have arisen through gene duplication and/or homologous recombination. For example, SEC1 is most similar to SEC2 and SEC3. However, residues 14 through 26 of SEC1 (Fig. 19.1) are identical to the analogous region of SEB but significantly different from analogous regions of the other SEC subtypes (12). Genetic recombination between *seb* and *sec* in a strain producing both SEB and SEC2 (or SEC3) could explain the generation of SEC1.

Additional minor variabilities have resulted by point mutations; even closely related SEs display some sequence differences. This may reflect fine-tuning of the toxin sequences for interacting with cells from a variety of hosts, as demonstrated with the SEC subtype variants. The sequences of SEC toxins produced by strains of *S. aureus* isolated from humans differ slightly (>95% identity) from sequences of SEC variants produced by bovine and ovine isolates (57).

Staphylococcal Regulation of SE Expression

General Considerations

SEs are produced in extremely low quantities throughout most of the exponential growth phase (7, 60). There is generally a large increase in expression during the late exponential or early stationary phase of growth, with SEA and SED accumulating somewhat earlier than other SEs (12). Production of SE is dependent on de novo synthesis within the cell. The quantity of toxin produced is strain dependent. However, SEB and SEC are most consistently produced in high quantities, up to 350 µg/ml. Other SEs are usually produced in much lower quantities than SEB and SEC. In most instances, even SEA, SED, and SEE are easily detectable by gel diffusion assays, which detect as little as 100 ng of SE per ml of culture. Some strains produce very low levels of toxins and require more sensitive analysis methods for detection. SED is the toxin most likely to be undetected in cultures of enterotoxigenic *S. aureus*, followed in order of decreasing incidence by SEC, SEA, and SEB (81).

Molecular Regulation of SE Production

Four loci implicated in regulation of virulence factor expression in *S. aureus* include *agr* (accessory gene regulator) (61), *xpr* (exoprotein regulator) (77), *sar* (staphylococcal accessory regulator) (23), and *sae* (*S. aureus* exoprotein expression) (33). The best characterized of these is *agr*. Mutations in the *agr* locus result in decreased expression of several SE and other exoproteins. Regulation of gene expression by *agr* can be transcriptional or translational. It has been found to regulate α-hemolysin at both the transcriptional and translational levels, while SEB and SEC are regulated at the transcriptional level. Not all SEs are regulated by *agr*. SEA expression does not appear to be affected by *agr* mutations (90). Since at least 15 genes are under *agr* control, *agr* fits into the general class of global regulators.

The *agr* locus maps to approximately 4 o'clock on the staphylococcal genome, near *purA*, *bla*, and *sea*, on the standard map of *S. aureus*. It contains two divergent operons separated by approximately 120 bp (Fig. 19.5). Transcription can be initiated from three promoters (P1, P2, and P3). P1 is weakly constitutive and transcribes *agrA*. P2 and P3 are induced strongly during late exponential and early stationary phases but are only weakly expressed earlier. The P2 transcript, RNAII, encodes four proteins designated AgrA, AgrB, AgrC, and AgrD. Strains with mutations in any of these have an Agr$^-$ phenotype and do not initiate transcription from either P2 or P3.

The 514-nucleotide transcript from P3, designated RNAIII, encodes the 26-residue staphylococcal δ-hemolysin and contains a significant amount of untranslated sequence. Interestingly, Agr$^-$ phenotypes can be complemented with plasmids encoding RNAIII under control of an inducible promoter, even when the δ-hemolysin gene, *hld*, has been inactivated (55). Therefore, it appears that RNAIII is a diffusible element that plays a key role in the *agr* regulation of exoprotein structural genes, including those for the SEs (44, 55, 61). This RNA molecule can replace the regulatory function of the entire *agr* locus and regulates expression predominantly at the transcriptional level. The exact mechanism by which transcriptional regulation occurs has not been determined, although it has been proposed to require an association with peptide factors and formation of an RNAIII-peptide complex.

The dissection of the properties of *agr* has helped explain several characteristics of SE production. For example, *agr* expression coincides temporally with expression of SEB and SEC during the bacterial growth cycle. All are maximally expressed during late-exponential and postexponential growth. SEA, which is not regulated by *agr*, is produced earlier (12). Furthermore, the production of several of SEs is negatively regulated by growth in media containing glucose, which is now known to affect

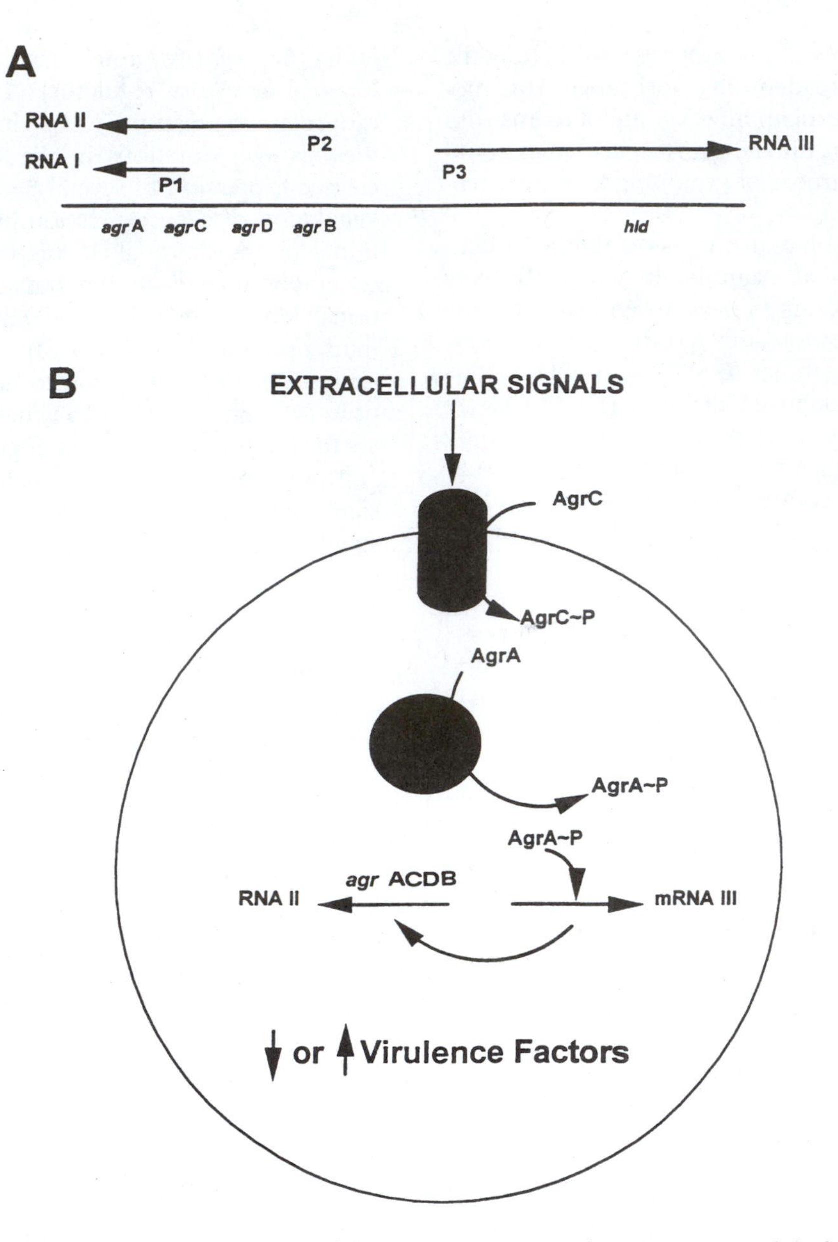

Figure 19.5 General characteristics of the accessory gene regulator in *S. aureus.* (A) Physical map of the *agr* locus showing the relative locations of genes within the locus (not drawn to scale). The nomenclature used for this diagram and in the text differs from that previously published for the four genes on the P2 transcript (55) as agreed mutually by R. Novick and S. Arvidson and as published in reference 4. (B) Potential role of AgrA and AgrC (formerly AgrB in reference 55) in environmental sensing and signal transduction as a classical two-component system.

agr. Most of the work in this area has been done with SEC. SEC expression is affected by glucose through at least two different mechanisms. First, metabolism of glucose indirectly influences SEC production through *agr* by reducing the pH (69). Since *agr* is maximally expressed at neutral pH, growth in a nonbuffered environment containing glucose lowers pH levels, which directly reduces *agr* expression. Consequently, expression of *sec* and other *agr* target genes is affected correspondingly (69). Glucose also reduces *sec* expression in *agr* mutant strains. This observation points to the existence of a second glucose-dependent mechanism for reduction of SE expression, independent of *agr* and apparently not involving pH (see below). The tight regulation of *agr* expression can also be demonstrated under alkaline conditions. Regardless of glucose levels, RNAIII is not expressed efficiently above pH 8.0.

Less is known about the roles of other regulatory loci in SE expression. Researchers have speculated that *agr* and *xpr* may interact (77), since RNAII and RNAIII levels are reduced in *xpr* mutants and strains with mutations in either locus are phenotypically similar. *sar* mutants display phenotypes distinct from those of mutants produced by affecting either the *agr* or the *xpr* locus. In contrast to *agr* mutants, *sar* mutants express an increased amount of α-hemolysin and decreased amounts of δ- and β-hemolysins. Analysis of the DNA sequence of the *sar* locus suggests that the protein encoded by *sar* has characteristics similar to those of other DNA-binding proteins (23). In addition, *sar* may actually regulate the expression of the *agr* locus, specifically RNAIII, at the transcriptional level. Whether *sar* interacts with *xpr* is not known. The recently described *sae* locus is a positive effector of cell-free β- and α-hemolysins, coagulase, DNase, and protein A but apparently does not affect SEA, protease, lipase, staphylokinase, or cell-bound protein A. These results suggest that the *sae* locus is distinct from other previously identified regulatory loci and may not act at the level of transcription.

Sigma factors may also affect temporal expression of SEs. RNA polymerase purified from exponential-phase cultures of *S. aureus* contains a σ^{70}-related factor similar to that of *Escherichia coli* (66). The holoenzyme containing this sigma factor transcribed *sea* more efficiently than either *sec* or *agr* P2. While *agr* and most exoprotein genes (including *sec*) are expressed as the cells enter stationary phase, *sea* is expressed earlier during the exponential phase, simultaneously with the σ^{70}-like factor. Thus, differential SE expression, at specific points of the bacterial growth phase, may coincide with availability of compatible sigma factors.

Environmental Signaling in Regulation of SE Gene Expression

Virulence factor expression is subject to environmental factors. The ability of *S. aureus* to respond to changes in the environment involves at least two components of *agr* (55). These components make up a signal transduction pathway conforming to the general pattern of bacterial two-component systems (80). The sensory component of these system is usually a histidine protein kinase. The N-terminal domain transmits the signal across the membrane to the C terminus. Sensors usually pass on the signal by phosphorylating an intracellular activator protein, thereby continuing the cascade.

As judged from its homology to conserved domains of histidine protein kinases, in particular a conserved histidine autophosphorylation site (80), the staphylococcal environment sensor is probably the 423-residue *agrC* gene product. Its N-terminal membrane-spanning segments suggest that it is a membrane-associated regulator protein. Likewise, the 241-residue *agrA* gene product appears to function as the activator and has homology to activator proteins in other systems such as OmpR. AgrA and other typical response regulators exhibit a highly conserved N terminus approximately 120 residues in length. Three amino acids, two aspartic acids and a lysine, are conserved among all response regulators. A potential model for signal transduction through the *agr* system, supported by existing evidence, is shown in Fig. 19.5. According to predictions, autophosphorylation of the *agrC* product at the end of the exponential phase allows AgrC-P to phosphorylate AgrA. An activated AgrA-P subsequently induces RNAIII expression.

Balaban and Novick (4) have described autocrine control of exotoxin synthesis in *S. aureus* through a mechanism involving the *agr* regulation and signal transduction functions. They showed that *agr* is autoinduced by RNAIII activator protein (RAP), which is produced and secreted by the organism and accumulates in the supernatant during staphylococcal growth. Interestingly, concentrated supernatants from postexponential *S. aureus* cultures induce transcription of RNAIII. Presumably, following secretion, RAP activates its putative receptor, AgrC, initiating this central signal transduction system and induction of *agr* transcription. Another key component of this system is the RNAIII-inhibitory peptide, which competes with RAP and blocks activation of *agr*. It was proposed that the RNAIII-inhibitory peptide may be a mutational derivative of RAP and that both regulatory molecules bind to, with opposite effects, the AgrC receptor.

Other Relevant Molecular Aspects of SE Expression

Although many chemical and physical factors selectively inhibit SE expression (30, 43, 51), their effects on regulation and signal transduction are only beginning to be defined. *agr* is not the only signal transduction mechanism for *S. aureus*, as suggested by investigations into the inhibitory effect of glucose. The negative regulatory effect of glucose described above cannot be attributed entirely to lower pH levels in cultures grown on glucose since cultures containing the carbohydrate produce less

SE, even when the pH is stably maintained. Although this effect has some attributes of catabolite repression, there are major differences between the inhibitory effect of glucose in *S. aureus* and catabolite repression in *E. coli*. For example, inhibition by glucose cannot be reversed by adding cyclic AMP (cAMP) to staphylococcal cultures. The significance of these differences is still unclear. Even the catabolite-repressible staphylococcal *lac* operon has features that are significantly different from those of the analogous operon in *E. coli*, and it is apparently unresponsive to cAMP.

S. aureus is an osmotolerant organism (see below). While it is able to survive and grow in environments with low water activity, production of some SEs, especially SEB and SEC, is reduced when the organism is grown under osmotic stress. In experiments performed with SEC-producing strains, levels of *sec* mRNA and SEC protein are both reduced in response to high NaCl concentrations. However, addition of osmoprotectants reverses the effect. This reduced expression is also seen in Agr^- strains, indicating that the signal transduction pathway used in this mechanism occurs by an alternative pathway.

Low concentrations of the commonly used emulsifier glycerol monolaurate (GML) inhibit transcription of many exoprotein genes, including *sea*, without inhibiting *S. aureus* growth. Inhibition of SE production is not associated with a simultaneous effect on *agr* transcription and occurs in *agr* mutants as well as wild-type strains. These results, plus the finding that constitutive expression of some genes is not affected, suggest that GML interferes with non-*agr*-mediated signal transduction. It has been proposed that GML and a variety of related food additives exert this effect by inserting into the staphylococcal membrane, altering the membrane protein conformation, and thereby interfering with signal transduction.

RESERVOIRS

Sources of Staphylococcal Food Contamination

Humans are the main reservoir for staphylococci involved in human disease, including *S. aureus*. Although most species are considered to be normal inhabitants of the external regions of the body, *S. aureus* is a leading human pathogen. Colonized individuals are carriers and provide the main source for dissemination of the organism to others and to food. In humans, the anterior nares are the predominant site of colonization, although *S. aureus* can be found on other sites such as the skin or perineum. Dissemination of *S. aureus* among humans and from humans to food can occur by direct contact, indirectly by skin fragments, or through respiratory tract droplet nuclei.

Today, most sources of SFP are traced to humans who inoculate food during preparation. In addition to contamination by food preparers who are carriers, *S. aureus* may be introduced into food from fomites used in food processing such as meat grinders, knives, storage utensils, cutting blocks, and saw blades. A survey of over 700 foodborne disease outbreaks revealed the following conditions most often associated with food poisoning: (i) inadequate refrigeration; (ii) preparation of foods far in advance; (iii) poor personal hygiene, e.g., not washing either hands or instruments properly; (iv) inadequate cooking or heating of food; or (v) prolonged use of warming plates when serving foods, a practice that promotes staphylococcal growth and SE production (20).

Animals, also an important source of *S. aureus*, are often heavily colonized with the organism. Predisposing factors which facilitate the survival of the organism are major concerns in the maintenance and processing of domestic animals and their products. For example, one very serious problem for the dairy industry is mastitis, an infectious disease often caused by *S. aureus*. The combined losses and expenses associated with bovine mastitis make it the single most costly disease of agriculture in the United States. Colonization of animals by this organism is also a public health concern since it may result in contamination of food and milk with *S. aureus* prior to or during processing.

It is not always been possible to trace the source of staphylococcal food contamination to human or animal origin. Regardless of its source, numerous studies have demonstrated the common presence of *S. aureus* in many types of food products (Table 19.2). Obviously the levels of staphylococci are often low initially. However, their widespread presence clearly provides a potential source of organisms capable of inducing SFP if conditions appropriate for SE expression are provided.

Resistance to Adverse Environmental Conditions

Some unique resistance properties of *S. aureus* facilitate its contamination and growth in food. Outside the body, *S. aureus* is one of the most resistant non-spore-forming human pathogens and can survive for extended periods in a dry state. Its survival is facilitated by organic material, which is likely to be associated with the organism from an inflammatory lesion. Isolation of the organism from air, dust, sewage, and water is relatively easy, and

Table 19.2 Prevalence of *S. aureus* in several common food products

Product	No. of samples tested	% Positive for *S. aureus*	No. of *S. aureus* organisms/g[a]	Reference
Ground beef	74	57	≥100	84
	1,830	8	≥1,000	21
	1,090	9	>100	63
Big game	112	46	≥10	78
Pork sausage	67	25	100	85
Ground turkey	50	6	>10	34
	75	80	>3.4	35
Salmon steaks	86	2	>3.6	27
Oysters	59	10	>3.6	27
Blue crabmeat	896	52	≥3	94
Peeled shrimp	1,468	27	≥3	87
Lobster tail	1,315	24	≥3	87
Assorted cream pies	465	1	≥25	88
Tuna pot pies	1,290	2	≥10	95
Delicatessen salads	517	12	≥3	62

[a]Determined by either direct plate count or most-probable-number techniques.

environmental sources of food contamination have been documented in several outbreaks of SFP.

S. aureus is known for acquiring genetic resistance to heavy metals and antimicrobial agents used in clinical medicine. However, generally the resistance of this organism to common food preservative methods is unremarkable. One noteworthy exception is its osmotolerance, which permits growth in medium containing the equivalent of 3.5 M NaCl and survival at water activities less than 0.86. This is especially problematic since other organisms, with which *S. aureus* does not compete efficiently, are likely to be inhibited under these conditions.

The molecular basis for staphylococcal osmotolerance has received much interest in recent years, although systems for responding to osmotic stress have been more intensively studied in less tolerant microorganisms. Considering the unique resistance of staphylococci, it would not be surprising to find they have developed a highly efficient osmoprotectant system. As in other organisms, several compounds accumulate in the cell or enhance staphylococcal growth under osmotic stress. Glycine betaine appears to be the most important osmoprotectant for *S. aureus*. To various degrees, other compounds, including L-proline, proline betaine, choline, and taurine, can also act as compatible solutes for this organism. In *S. aureus*, intracellular levels of proline and glycine betaine accumulate to very high levels in response to increased concentrations of NaCl in the environment. Although the signal transduction pathway is not known for staphylococci, in other organisms, it involves a loss in turgor pressure in the cell and activation of required transport systems. High-affinity and low-affinity transport systems operate in *S. aureus* for both proline and glycine betaine (89, 93). The low-affinity systems are primarily stimulated by osmotic stress, have broad substrate specificity, and may in fact be the same transporter shared by both osmoprotectants.

By itself, the demonstration of a stress response system in *S. aureus* does not explain the unusual staphylococcal osmotolerance. Other, less tolerant microorganisms possess mechanisms for counteracting osmotic stress. The efficiency of the staphylococcal system may reflect an unusually high endogenous level of intracellular K^+ and lack of need for de novo transporter synthesis. For example, in other well-studied systems such as that of *E. coli*, changes in osmotic stress activate K^+ transport systems. Elevated intracellular K^+ levels that result are required for induction of *proU* and eventually lead to synthesis of transporters for glycine betaine and other osmoprotectants. In *S. aureus*, K^+ levels are high in unstressed cells. Therefore, the transport system is constitutively present in this organism and is preformed when high-salt conditions are encountered. The net result is a very rapid and efficient response. *S. aureus* cells accumulate a 21-fold increase in proline after less than 3 min of exposure to high salt concentrations (89).

FOODBORNE OUTBREAKS

Incidence of Staphylococcal Food Poisoning

SFP occurs as either isolated cases or outbreaks affecting a large number of individuals. Since SFP is self-limiting, the incentive to report cases has not been as great as for other foodborne diseases. Although there is a national surveillance system for SFP, it is not an officially reportable disease. It has been estimated that only 1 to 5% of all SFP cases are reported in the United States, usually at the state health department level. The majority of these are highly publicized outbreaks. Isolated cases occurring in the home are not usually reported. Staphylococci account for an estimated 14% of the total of foodborne disease outbreaks within the United States (40). There has been an average of approximately 25 major outbreaks of SFP annually within the United States. Occurrence of SFP is cyclical in nature (40). The highest incidence is typically in the late summer, when temperatures are warm and food is more likely to be stored improperly. A second peak occurs in November and December. Approximately one-third of these outbreaks are associated with leftover holiday food.

SFP may be the leading cause of foodborne illness worldwide, although reporting in other countries is even less complete than in the United States. In one study, 40% of outbreaks of foodborne gastroenteritis in Hungary were due to SFP (7). The percentage is slightly lower in Japan (approximately 20 to 25%), where contamination of rice balls during preparation is a potential problem (7). Outbreaks due to improper manufacturing of canned corned beef have been reported in England, Brazil, Argentina, Malta, northern Europe, and Australia. Cases in Great Britain have been attributed to contaminated milk and cheese resulting from sheep mastitis (45). In some countries, ice cream has been a major cause of SFP.

Characteristics of a Recent Large Typical SFP Outbreak

The following is a summary of an outbreak of SFP reported by the Food and Drug Administration (3). Many of the aspects of this outbreak such as type of food involved, mechanism of contamination, inadequate safety measures in food handling, and clinical pictures are typical of the usual SFP outbreak. This particular outbreak originated from one meal that was fed to 5,824 elementary school children at 16 sites in Texas. Of the 5,824 children exposed, 1,364 developed typical signs of SFP. Investigation into the source of the illness showed that 95% of the children who became ill had eaten chicken salad. Large numbers of *S. aureus* were cultured from the chicken salad implicated in the outbreak.

The series of events leading up to the outbreak is as follows. Preparation of the meals was performed in a centralized kitchen facility and was begun on the preceding day. Frozen chickens used for the salad were boiled for 3 h. After cooking, the chickens were deboned, cooled to room temperature with a fan, ground into small pieces, placed into 12-in. (30.48-cm)-deep aluminum pans, and stored overnight in a walk-in refrigerator at 42 to 45°F (ca 5.6 to 7.2°C). The next morning, the remaining ingredients of the salad were added and the mixture was blended with an electric mixer. The food was placed in thermal containers and transported by truck to the various schools between 9:30 and 10:30 a.m. It was kept at room temperature until served between 11:30 a.m. and noon. It is believed that the chicken became contaminated after cooking when it was deboned. Most likely, the storage of the warm chicken in the deep aluminum pans did not permit rapid cooling and provided an environment favorable for staphylococcal growth and SE production. Further growth of the bacteria probably occurred during the period when the food was kept in the warm classrooms. Prevention of the incident would have entailed more rapid cooling of the chicken and refrigeration of the salad after preparation.

CHARACTERISTICS OF DISEASE

SFP is usually described as a self-limiting illness presenting with emesis following a short incubation period. However, vomiting is not the only symptom that is commonly observed. Likewise, a significant number of patients with SFP do not vomit. Other common symptoms include nausea, abdominal cramps, diarrhea, headaches, muscular cramping, and/or prostration. In a summary of clinical symptoms involving 2,992 patients diagnosed with SFP, 82% complained of vomiting, 74% felt nauseated, 68% had diarrhea, and 64% exhibited abdominal pain (40). In all cases of diarrhea, vomiting was present. Diarrhea is usually watery but may contain blood as well. The absence of high fever is consistent with the absence of infection in this type of food poisoning, although some patients present with low-grade fever. Other potential symptoms include headaches, general weakness, dizziness, chills, and perspiration.

Symptoms typically develop within 6 h after ingestion of contaminated food. In one report, 75% of the exposed individuals exhibited symptoms of SFP within 6 to 10 h postingestion (56). The mean incubation rate is 4.4 h, although incubation periods as short as 1 h have been reported. In another outbreak, symptoms lasted for 1 to 88 h, with a mean of 26.3 h. Death due to SFP is not common, but the fatality rate ranges from 0.03% for the general public to 4.4% for more susceptible populations such as children and the elderly (40). Approximately 10% patients with confirmed SFP seek medical attention. Treatment is minimal in most cases, although administration of fluids is indicated when diarrhea and vomiting are severe.

INFECTIVE DOSE AND SUSCEPTIBLE POPULATIONS

Numbers of Staphylococci Required

Since many variables affect the amount of SE produced, one cannot predict with certainty the number of *S. aureus* in food required to cause SFP. Factors contributing to the levels of toxin concentration have been extensively studied and include environmental conditions such as food composition, temperature, other physical and chemical parameters, and the presence of inhibitors. Bacterial factors to be considered include potential differences in the types, amounts, and numbers of different SEs that the strain in question has the physiological ability to produce. It is likely that these combined conditions are unique for each isolated case or outbreak of SFP. Despite this variability, there are several general guidelines that are useful for assessing general risk. According to the Food and Drug Administration, effective doses of SE may be achieved when populations of *S. aureus* are greater than 10^5 organisms per g of contaminated food (3). In other studies, 10^5 to 10^8 organisms were noted to be the typical range, despite the fact that lower levels were sometimes implicated (40).

Toxin Dose Required

A large number of investigations have been conducted to assess SE potency and the amount of toxin in food required to initiate SFP symptoms. Perhaps the most valid information in this regard has come from analysis of food recovered from outbreaks of the illness. Although the SEs are quite potent, the amount required to induce symptoms is large in comparison with many other exotoxins that are acquired through contaminated food. Regarding concentration, a basal level of approximately 1 ng of SE per g of contaminated food is sufficient to cause symptoms associated with SFP. Although levels of 1 to 5 μg of ingested toxin are typically associated with many outbreaks, the actual levels of detectable SE were even less (<0.01 μg) in 16 SFP outbreaks (32). One of the most useful studies for predicting the minimal oral dose of SE required to induce SFP in humans was a well-documented investigation of an outbreak caused by ingestion of contaminated chocolate milk (27). In that study, the minimal dose of SEA required to cause SFP in school children was 144 ± 50 ng.

Many factors contribute to the likelihood of developing symptoms and their severity. The most important include susceptibility of the individual to the toxin, the total amount of food ingested, and the overall health of the affected person. The toxin type may also influence the likelihood and severity of disease. Though SFP outbreaks attributed to ingestion of SEA are much more common, individuals exposed to SEB exhibit more severe symptoms. Forty-six percent of 2,291 individuals exposed to SEB exhibited SFP symptoms severe enough to warrant admittance to the hospital (40). Only 5% of 1,813 individuals exposed to SEA required such treatment. It is possible that these observations reflect differences in levels of toxin expression, since SEB is generally produced at higher levels than SEA.

Human volunteers and several species of macaque monkeys have been used to determine the minimal amount of purified SEs required to induce emesis when administered orally. Generally, monkeys are reported to be less susceptible to SE-induced enterotoxicity than humans. The study of purified SEs has provided useful comparative information and has important research applications, but its direct relevance to SFP is uncertain since potential stabilization of unpurified SEs by food is an important consideration. From a study in which human volunteers ingested partially purified toxin, Raj and Bergdoll estimated that 20 to 25 μg of SEB (0.4 μg/kg of body weight) is sufficient to cause vomiting in humans (65). In the rhesus monkey model, the 50% emetic dose is between 5 and 20 μg per animal (or approximately 1 μg/kg) when administered intragastrically. In our investigations, the minimal emetic dose of SEC1 for pigtail monkeys (*Macaca nemestrina*) is consistently between 0.1 and 1.0 μg/kg.

VIRULENCE FACTORS AND MECHANISMS OF PATHOGENICITY

SE Structure-Function Associations

Basic Structural and Biophysical Features

SEs are single polypeptides of approximately 26 to 28 kDa. Most are neutral or basic proteins with pIs ranging from 7 to 8.6, but they display a high degree of microheterogeneity when assessed by isoelectric focusing. For example, SEC2 focuses into at least eight bands ranging in pI from 5.50 to 7.35 (26). Although this has been attributed to enzymatic deamidation, the putative enzyme has not been identified. Based on what is known about their sequences, biochemistry, and functional aspects, all of the SE antigenic variants exist as monomeric proteins. Thus far, evidence does not suggest that SE molecules are arranged functionally into a function-dependent A-B subunit organization. Typical of most other exotoxins, the SEs are translated as larger precursors with a classical signal peptide that is cleaved during export.

SE sequence analysis, relatedness, and diversity have been discussed above (Fig. 19.1 and 19.4). Indicative of their partial sequence conservation, structural studies have long suggested that all SEs are similar in molecular topology (86). Spectral techniques and computer-assisted structural predictions revealed that SEA, SEB, SEC, and SEE contain a low content of α-helix (<10%) compared with β-pleated sheet/β-turn structures (approximately 60 to 85%) (19, 76). These predictions have been recently confirmed with the reported three-dimensional crystal structures of SEA, SEB, and SEC3 (see below) (17, 19, 39, 72, 86).

Although detailed studies have not been performed with every SE, as a group they are stable molecules in many respects. Their recognition as heat-stable toxins arose from early studies, in which enterotoxicity and antigenicity were not completely destroyed upon boiling crude preparations. Furthermore, less extreme elevations of temperature, such as those used for pasteurization of milk, had little or no effect on SE toxicity. The heat resistance of the SEs has been extensively studied. The general conclusion from this combined work is that SEs are difficult to inactivate by heating and are even more stable when present in high concentrations or in crude states such as in the environment of food. Since temperatures required to inactivate the SEs are much higher than those needed to kill *S. aureus* under the same environmental conditions, toxic food involved in many cases of SFP is devoid of viable organisms at the time of serving.

One additional property of SEs that has potential significance toward development of SFP is their resistance to inactivation by proteases found in the gastrointestinal tract. Resistance to pepsin, especially in a relatively low-pH environment, is a key requirement for SE activity in vivo. All of the SEs have some resistance to pepsin, a property not shared by at least one nonemetic staphylococcal PT, TSST-1. SEB is susceptible to degradation by pepsin at very low pH levels, but partial neutralization of the gastric acidity by food intake is presumed to temporarily provide a protective environment for the toxin (7).

The SEs may be cleaved by other common proteases. However, unless the fragments generated are separated in the presence of denaturing agents, proteolysis alone may not be sufficient to cause a loss of biological activity. This is apparently representative of inherent SE molecular stability, which can be demonstrated by renaturing studies. For example, denaturation occurs only under strong denaturing conditions using high concentrations of urea or guanidine hydrochloride. If the denaturing conditions are removed, the SEs may spontaneously renature and regain biological activity (79). Differences in stability do exist among the toxins. For example, SEB and SEC1 are approximately 50-fold more stable under denaturing conditions than SEA (91). There has been some speculation that the SE disulfide bond is responsible for an inherent molecular stability. Experiments by several investigators suggest that the closed disulfide bond does contribute at least some degree of conformational stabilization, but its disruption has only minimal effects on the overall stability and activity of the molecule (92) (see below).

Three-Dimensional Structure

The crystal structure of SEC3, determined at a resolution of 1.9 Å (1 Å = 0.1 nm) and depicted in Fig. 19.6, is very similar to that of SEA and SEB (17, 19, 39, 72, 86). The molecule has an overall ellipsoidal shape with maximal dimensions of 43 by 38 by 32 Å and is folded into two

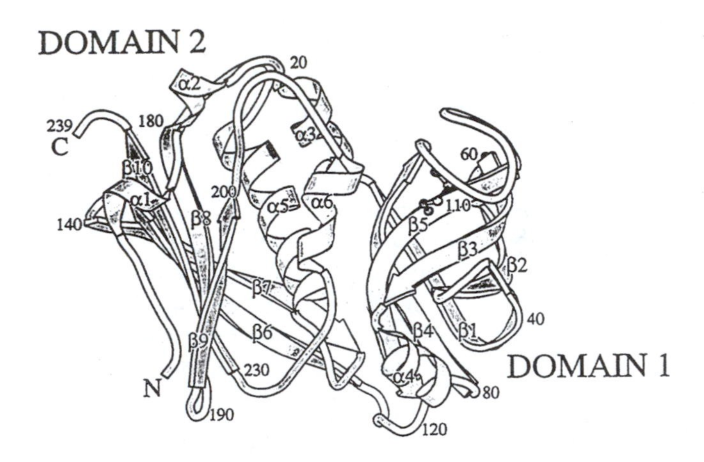

Figure 19.6 Schematic diagrams of the SEC3 crystal structure illustrating major structural features. Numerical designation of the locations of select residues and each α- and β-strand is shown within the two major domains. Also indicated are the N and C termini. The intramolecular disulfide linkage between Cys residues 93 and 110 (ball-and-stick figure) connects the disulfide loop to the β5-strand containing the conserved residues (see Fig. 19.7) potentially important for emesis. The zinc atom bound by SEC3 faces toward the back of the SEC3 molecule between domains 1 and 2 and is coordinated by D-83, H-118, and H-122. In contrast, the zinc atom in SEA has been proposed to be positioned on the opposite edge of domain 2. The conformational topology of domain 1 is the same as for the oligonucleotide/oligosaccharide-binding domains of other proteins described in the text.

domains containing a mixture of α and β structures. Domain 1, the smaller of the two domains, contains residues near the N terminus but not the N-terminal residues themselves. The residues in SEC3 that are located in domain 1 include those in positions 35 through 120 (Fig. 19.1). The folding conformation of this domain may have potential significance for the function of the toxin. Its topology, in which a β-barrel structure is capped at one end by an α-helix, is known as the oligonucleotide/oligosaccharide-binding fold. The internal portion of the β-barrel is rich in hydrophobic residues, and the potential oligomer-binding surface is covered with mainly hydrophilic residues. This same conformational folding pattern is found in several bacterial enzymes and exotoxins. Although members of this diverse group of proteins are not related according to their amino acid sequences, they share the common feature of exerting their activity by interacting with either oligosaccharides or oligonucleotides. Staphylococcal nuclease and *E. coli* Shiga toxin are examples of proteins that interact through their oligonucleotide/oligosaccharide-binding fold domains with oligonucleotides and oligosaccharides, respectively. The other prominent feature of domain 1 is that it contains two cysteine residues responsible for forming the disulfide linkage characteristic of all SEs. This bond and the cystine loop are located at the end of the domain opposite its α-helix cap. Crystallographic data for SEA, SEB, and SEC3 indicate that the loop regions of all three toxins are quite flexible (19, 72, 86).

The larger domain 2 contains the N and C termini and encompasses residues 1 through 33 and 123 through 239 of SEC3. It can be described as a five-strand antiparallel β-sheet wall, overlaid with a group of α-helices. The N-terminal 20 residues of SEC3 form a loosely attached structure which folds over the edge of this domain. Residues immediately downstream from the N terminus form α-helices which mark the interfaces between the two domains. Specifically, these α-helices form a long groove on one side of the molecule (α5 groove) and a shallow α3 cavity near the top of the molecule (39).

Binding of Zinc by SEs

The first evidence that a cation could affect SE structure or function was the demonstration that binding of SEA, SEE, and possibly SED to MHC class II requires zinc. Fraser et al. (29) showed that SEA and SEE bind zinc via a single site with a dissociation constant of 1 to 2 μM. The binding site was subsequently predicted by mutagenesis of SEA to be composed of a nonlinear stretch of residues (H-187, H-225, and D-227). SED also appears to bind zinc through the same site, but it contains aspartic acid at position 187 instead of histidine. The crystal structure for SEA reported by Schad et al. (72) has confirmed that these three residues bind zinc and also implicates the N-terminal serine in the metal coordination.

SEC3 also binds zinc, albeit through a different mechanism compared with SEA and SEE (17, 19). The α5 groove of SEC3 contains a zinc atom bound by the classical motif (H-E-X-X-H), typically found in the catalytic site of metalloenzymes such as thermolysin (38). The significance of a zinc-binding site with potential enzymatic activity in SEC3 is being actively investigated. Its presence raises several interesting questions. None of the SEs is known to possess protease activity. However, zinc-mediated metalloprotease activity that is dependent on a conserved H-E-X-X-H motif has been demonstrated for other exotoxins produced by gram-positive bacteria (53). In the tetanus and botulinum neurotoxin group, protease activity involving this zinc-binding motif is central to the toxins' mechanism of action. Future work will determine whether similarities between botulism and SFP are coincidental or whether zinc binding represents a related mechanism of action. Interestingly, both diseases are usually acquired as intoxications. The responsible toxins are ingested orally but subsequently exert their effects through an action on nerves (see below). In any regard, if this highly speculative possibility were true for SEC3, the heterogeneity in zinc binding among the staphylococcal toxins would make it unlikely that zinc-dependent protease activity is a uniform mechanism for all the SEs.

Molecular Regions of SEs Responsible for Enterotoxicity

The structural aspects of SEs that enable them to survive degradation by pepsin and other enzymes in the gastrointestinal tract are required for the toxins to induce SFP. However, stability alone is not sufficient. SEs must also be able to interact with the appropriate target, leading to emesis, diarrhea, and other gastrointestinal tract symptoms. Initial attempts to define molecular regions responsible for enterotoxicity involved testing biological activity of protease-generated fragments derived from SEA, SEB, or SEC1 (79). Three main conclusions were drawn from this work. First, only large toxin fragments containing central and C-terminal portions of the SEs retained enough of the native structure to cause emesis. Second, N-terminal residues of the SEs were not required for emesis. SEC1, modified by removal of the 59 N-terminal residues, retains the ability to cause emesis

		Cystine loop or analogous region	Conserved Downstream Sequences
Enterotoxigenic PTs:			
SEC1	93	CYFSSKDNVGKVTGG---KTC	111 M Y G G I T K H E G N H
SEC2	93	CYFSSKDNVGKVTGG---KTC	111 M Y G G I T K H E G N H
SEC3	93	CYFSSKDNVGKVTGG---KTC	111 M Y G G I T K H E G N H
SEB	93	CYFSKKTNDINSHQTDKRKTC	114 M Y G G V T E H N G N Q
SED	92	CYGGEIDRTAC	103 T Y G G V T P H E G N K
SEA	96	CAGGTPNKTAC	107 M Y G G V T L H D N N R
SEE	93	CAGGTPNKTAC	104 M Y G G V T L H D N N R
SEH	82	CEKISENISEC	93 L Y G G T T L - N S E K
Nonenterotoxigenic PTs:			
SPEA	87	CYLCENAERSAC	99 I Y G G V T N H E G N H
SPEC	74	GLFYILNSHTGE	86 Y I Y G G I T P A Q N N
TSST-1	69	KRTKKSQHTSEGTYYH	85 Q I S G V T N T E K L P

Figure 19.7 Comparison of cysteine loop and adjacent sequences for SEs and the analogous regions of nonenterotoxin PTs. Numbers designate positions within the primary sequences of the mature proteins. Evidence suggests that proper positioning of the critical downstream residues by a stable disulfide bond is required for emesis. The streptococcal superantigen sequence (68) is not included here since its enterotoxic ability has not been investigated. SPEA and SPEC refer to streptococcal pyrogenic exotoxins types A and C, respectively. SPEA is nonemetic (71) despite having the potential to form a disulfide bond in this region. Although it also contains multiple cysteines, the nature of the putative SPEA disulfide bond is unclear since it has an additional cysteine at position 90.

in monkeys. Smaller toxin fragments from SEC1 and other SEs were inactive. The third conclusion was that emesis seemed to require preservation of structure in the area of the conserved SE disulfide loop.

The disulfide bond is a structural feature that is characteristic of all SEs but is not found uniformly in nonemetic staphylococcal and streptococcal exotoxins. With one known exception, all SEs contain exactly two cysteine residues which could potentially form an intramolecular linkage and a spacer disulfide loop. Only the SEC molecular variant produced by staphylococcal isolates from cattle (SEC-bovine) deviates from this pattern and possesses an additional third cysteine (Fig. 19.1). The disulfide bond, located roughly in the middle of every SE regardless of its antigenic type, has led to speculation about its importance in emesis. Potential structural contributions from the Cys-Cys bond that could contribute to the SE conformation necessary for emetic activity could include one or more of the following features: (i) proper positioning of cysteine residues upon formation of the disulfide bond, (ii) exposure and/orientation of crucial residues in the loop formed between the two linked cysteines, (iii) exposure and/or orientation of residues immediately adjacent to the disulfide linkage but not contained within the cystine loop, and (iv) contributions to the overall SE conformation by the linkage of the two cysteine residues.

Each of these possibilities has been considered. The cysteine residues and the loop probably do not play a direct role in the emetic response. It has been possible to substitute the cysteine residues in several SEs by site-directed mutagenesis and show that neither of these two residues is absolutely critical. Although all of the SEs have a cystine loop, the lengths and composition of residues within the loops of different SE antigenic types vary greatly. This lack of consistency among SE loop properties suggests that they are unlikely to have a shared enterotoxigenic function (Fig. 19.7). Furthermore, proteolytic nicking of toxins in their loops has no effect on their ability to cause emesis (79). Warren et al. (92) showed that the disulfide bond contributes only minimally to overall protein conformation. Presumably then, if the disulfide linkage is important in emesis, the effect is probably to provide a particular orientation of residues near the disulfide linkage but not in the loop. The most convincing evidence in support of this possibility has been provided by mutagenesis of the SEC1 in which its cysteine residues were substituted by either serine or alanine (41). It was found that mutants with serine substitutions were emetic, whereas the analogous mutants with alanine substitutions were nonemetic. Although serine and alanine are both considered to be conservative substitutions for cysteine, one difference between these two amino acids is that serine has the ability to hydrogen bond. Thus, hydrogen bonding by serine may be able to replace the disulfide linkage stabilization of local structure.

Which critical local residues require proper orientation by the disulfide bond (or hydrogen bonding at the same positions) in order for the SEs to induce emesis? Possible candidates are those within a highly conserved stretch of residues directly adjacent to, and downstream from, the disulfide loop (Fig. 19.7). In SEC3, these residues are located on the β5-strand. An attractive hypothesis is that in addition to stability in the gut, two other structural requirements need to be met for enterotoxicity. First, the appropriate conserved residues must be present in the toxin. Many PTs, emetic and nonemetic, have similar sets of highly conserved residues in a location analogous to the β5-strand of SEC3. Second, they must be positioned properly for interacting with their target in the gut. The unique SE disulfide bond may serve this function. Of the entire PT family, only two non-SE toxins produced by *Streptococcus pyogenes* (streptococcal pyrogenic exotoxin and streptococcal superantigen) could potentially form a disulfide linkage (68). However, the presence of more than two cysteines in these toxins suggests that the structure in this area and the degree of local stabilization by their putative disulfide linkage may not be identical to those for the SEs.

Additional work with SEA showed that single-site substitutions of residues closer to the N terminus also influence emesis (37). This was especially the case for mutants constructed by substitution with glycine. For example, mutagenesis of residues 25, 47, and 48 causes significant reduction in the emetic potency of SEA. Although these residues are far from the disulfide bond in the SEA primary sequence, they are located near or within domain 1 of SEA and could potentially influence the area around the disulfide bond.

SE Antigenic Epitopes

The need for reagents that could detect SEs in food and clinical samples plus a desire to differentiate antigenically different toxins in the family has been the impetus for considerable effort directed toward epitope characterization and mapping. Individually, each SE type and subtype has a sufficient degree of antigenic distinctness to allow its differentiation from other PTs by using highly specific polyclonal antisera and monoclonal antibodies. However, some degree of cross-reactivity can often be demonstrated between several SEs. The level of cross-reactivity generally correlates with shared primary sequences. The type C SE subtypes and their molecular variants exhibit a substantial amount of cross-reactivity, as do SEA and SEE, the two major serological types with the greatest sequence relatedness. For these toxins, cross-reactivity can even be demonstrated in relatively insensitive assays such as immunodiffusion assays, in which these two toxins produce lines of partial identity with the heterologous antiserum. SEB and the SEC subtypes and molecular variants (see above) are also highly related at the amino acid level. Cross-reactivity between SEB and SEC may be demonstrated occasionally by immunodiffusion but more consistently by using sensitive methods such as radioimmunoassays or immunoblotting. Generally, it has not been possible to produce useful antibodies that cross-react among less related SEs. Although one investigator has produced a monoclonal antibody that cross-reacts with all five major SE antigenic types, A through E, this antibody has low affinity and cross-reacts with other staphylococcal proteins (8). The two most distantly related SEs that can be recognized by a common epitope are SEA and SED (10).

The mapping of conserved and specific antigenic epitopes on SEs, and their differentiation from potentially toxic regions, has potential applications for rational development of nontoxic vaccines. Considering the array of SE antigenic types, the most efficient toxoid would presumably contain one or more epitopes that are shared by multiple toxins. Several approaches have been used to partially localize antigenic epitopes on the SEs and differentiate them from toxic regions. One of the earlier methods used for this purpose was to identify protease-generated toxin fragments from several SEs that bind to cross-reactive antibodies (16). These studies found both N- and C-terminal toxin fragments that contain cross-reactive epitopes but were unable to define shorter stretches of residues.

There is some evidence that immunization with short, highly conserved peptides could have merit. For example, the use of synthetic peptides from highly conserved stretches of primary sequence has resulted in the production of neutralizing antibody for several of the SEs. Immunization with synthetic peptides corresponding to residues 130 to 160 of SEB or the same region of SEC1 (residues 148 to 162) induced antibodies that neutralized both native toxins (39, 46). The highly conserved SE sequence K-K-X-V-T-X-Q-E-L-D (Fig. 19.1), encompassed by both peptides, may represent part of an epitope that could be useful for protective immunity. It remains to be determined if major epitopes identified on other toxins such as SEA also show promise (64).

SE Mode of Action in Induction of Emesis and Other Symptoms Related in SFP

SE-Induced Emesis Requires Nerve Stimulation

Except for rare SFP cases in which massive doses of SEs are consumed, systemic dissemination of the toxins does not contribute significantly to the illness. When fed to rodents, the SEs do enter the circulation but are rapidly removed by the kidneys (6). Most studies using the simian model indicate that the SE site of action, following ingestion, is the abdominal viscera. Early studies into the mechanism of action of the SEs tested the emetic responsiveness of animals to the toxins after disruption of well-defined neural systems or after visceral deafferation. The characteristic emetic response was noted to result from a stimulation of local neural receptors in the abdomen (83) which transmit impulses through the vagus and sympathetic nerves, ultimately stimulating the medullary emetic center.

Cellular Histopathology in the Gastrointestinal Tract

Information on the histological effects of oral doses of SEs in humans is extremely limited. Most of what is known has been derived from information obtained from experiments in rhesus monkeys. Upon ingestion of the toxin, pathological changes compatible with a definition of gastroenteritis are observed in several parts of the gastrointestinal tract (52).

The primate stomach becomes hyperemic and is marked by lesions which begin with the influx of neutrophils into the lamina propria and epithelium. A mucopurulent exudate in the gastric lumen is also typically observed. Also characteristic are mucus-filled surface cells which eventually release their contents. Later in the illness, neutrophils become replaced by macrophages. Upon resolution of the symptoms, the cellular infiltrate clears.

Similar cellular infiltrate and lumen exudate occur in the small intestine, although they decrease in severity in sections taken from lower, compared with upper, portions of the intestine. Clearly evident in the jejunum are extension of crypts, disruption or loss of the brush border, and an extensive infiltrate of neutrophils and macrophages into the lamina propria. Changes in the colon are minimal in the monkey model. Only a mild cellular exudate and mucus depletion are evident. The only other significant effect in monkeys is acute lymphadenitis in the mesenteric lymph nodes.

Search for the Gastrointestinal Tract Target

The specific cells that SEs interact with in the abdomen have not been clearly identified, nor has their receptor. Evidence suggests that interaction of SEs with their target directly or indirectly causes production of inflammatory mediators that induce SFP symptoms. Jett et al. (46) showed that oral administration of SEB produced elevated levels of arachidonic acid cascade products. Specifically, they observed significant increases in prostaglandin E_2, leukotreine B4, and 5-hydroxyeicosatetraenoic acid. These three compounds are potent vasoactive inflammatory mediators that can also act as chemoattractants for neutrophils. Both of these activities are consistent with histopathology described above for the SFP monkey model.

Scheuber et al. (73) could not demonstrate an effect of prostanoid inhibitors on SEB-induced emesis but were able to correlate gastrointestinal symptoms with cysteinyl leukotreine generation. Intoxication of animals with SEB resulted in a 10-fold increase in levels of leukotreine E4 in bile, plus an unidentified leukotreine in the urine. This group of investigators suggested a role for mast cells in pathogenesis of SFP. Although induction of histamine production by SEB was responsible for some secondary nonenteric immediate hypersensitivity skin reactions, it did not appear to correlate with emesis.

The currently available data suggest that ingestion of SEs causes a stimulation of mast cells and possibly other inflammatory cells in the abdomen. Thus far, the abdominal receptor has not been identified. Experiments using anti-idiotype antibodies in binding assays and protection assays have provided circumstantial evidence for its existence on monkey mast cells (67). Komisar et al. showed that SEB can stimulate rodent peritoneal mast cells as well and provided evidence for a protein receptor (54). It is possible that SEs act directly on their receptor and circumvent the typical two-stage mast cell immunoglobulin E antibody-antigen interaction (73).

Despite these observations, Alber et al. (2) were unable to directly stimulate monkey or human skin and intestinal mast cells with SEB to release inflammatory mediators. It was suggested that stimulation of mast cells in vivo occurs through a nonimmunological mechanism requiring the generation of neuropeptides released from peripheral terminals of primary sensory nerves. At least one putative mast cell-stimulating peptide, substance P, was implicated in SEB-induced toxicity by use of antibodies and a variety of inhibitors. The attractiveness of this explanation is that it is consistent with earlier predictions of a neural involvement in the pathogenesis of SFP.

Is There a Relationship Between Superantigenicity, TSS, and SFP?

The discovery of the mechanism of superantigen action and the unique properties of this class of proteins provided an explanation for the multiple systemic effects seen in TSS patients. The massive cellular stimulation induced by superantigenic PTs explained the long-recognized fact that TSS patients had elevated serum cytokine levels, which presumably mediate many symptoms of the disease. The realization that at least some of the pathogenesis of TSS could be attributed to immune cell stimulation led to the prediction by some investigators that SFP could also be a reflection of superantigen function. Consistent with this prediction was the fact that TSS patients often have a gastrointestinal tract component characterized by vomiting and diarrhea. Also, patients with endotoxin shock have elevated cytokine levels and similarly display vomiting and diarrhea. If superantigenicity is responsible for SFP, the SEs presumably act directly on T cells and antigen-presenting cells in the gut. Although some SEs enter the circulation, they appear to be rapidly cleared by the kidneys so that significant systemic concentrations are unlikely to be achieved (6). This finding and the fact that TSS-associated symptoms (i.e., shock and fever) are not observed in SFP patients suggest that SEs do not mediate SFP through systemic cytokines.

Despite their similarities and the evidence cited above, several lines of evidence suggest that the partial overlap between SFP and TSS symptoms is probably coincidental and that superantigenicity is not directly responsible for SFP. First, as discussed above, nonimmunological mast cell stimulation has been linked to the release of inflammatory mediators resulting from nerve interactions. The second line of evidence has come through mutagenesis of several SEs. Studies have shown that T-cell stimulation and induction of emesis are separable functions and are determined by distinct portions of the SE molecules (1, 37, 41). It has been possible to construct SEA, SEB, or SEC1 mutants that are deficient in T-cell-stimulatory activity but retain the ability to induce emesis, and vice versa. Finally, although all PTs have been reported to have superantigen function, only the SEs are emetic when ingested. The lack of emesis-inducing ability of some nonenterotoxic PTs has been attributed to instability in the gastrointestinal tract. However, at least one nonemetic PT, streptococcal pyrogenic exotoxin A, is very stable in gastric fluid (71).

If superantigenicity does not explain SFP, how do PTs cause vomiting and diarrhea in TSS? Several possibilities could explain how SEs and other PTs act on the gastrointestinal tract if they are not consumed through the oral route. First, the toxins' ability to induce TSS symptoms may be limited to the systemic circulation. If this is so, cytokines would need to enter the abdomen from the circulation and mediate gastroenteritis pathogenesis. Alternatively, the PTs could enter the gut from the circulation and act directly at the local level as superantigens or through other mechanisms as proposed for SFP. The latter possibility is less likely since even nonenterotoxic PTs, including those that are susceptible to degradation in the gut, are known to cause gastrointestinal symptoms when they are associated with TSS.

These complex and unresolved issues are relevant to interpreting models for studying the enterotoxic activity of SEs. Oral administration of staphylococcal culture filtrates or suspected food to monkeys is regarded as the preferred method for detecting enterotoxic properties of the SEs. However, intravenous administration of SEs to primates and other animals, especially cats, has been used as an alternative to feeding. Considering that systemic exposure to SEs mimics the situation leading to TSS symptoms, intravenous administration may not accurately reflect the pathogenesis and toxin properties required to induce emesis in SFP.

CONCLUDING REMARKS

Remarkable progress has been made toward understanding the molecular aspects relevant to SFP. *S. aureus* clearly has some unique properties that promote its ability to induce foodborne illness. However, further understanding of the molecular aspects of these unique properties should facilitate the implementation of more efficient ways to selectively inhibit staphylococcal survival, growth, and SE production in food. For example, the use of the newly discovered RNAIII-inhibitory peptide (4) to block exotoxin production has recently been suggested and could have potential applications in the food industry.

One issue that continues to puzzle the scientific community is the questionable rationale for staphylococci to produce an emetic toxin. Unlike enteric pathogens that inhabit the gut, the ability to induce emesis and diarrhea as a mechanism to promote exit from the host and dissemination does not appear to be important for staphylococci. One may ask a similar question regarding the ability of SEs to induce lethal shock in TSS, since it is generally agreed that lethality is not advantageous to microorganisms. Instead, long-term host-pathogen coexistence usually relies on adaptations that allow the organism to survive for extended periods without harming the host or being cleared by the immune response.

From what is now known regarding the superantigenic properties and proposed host-specific molecular adaptation of the SEs, one may propose that SEs are produced for the purpose of immunomodulating the host. Exposure of animals and peripheral blood mononuclear cell cultures to SEs and other superantigens has repeatedly been shown to induce at least a transient immunosuppression. Similarly, TSS patients often fail to produce a significant immune response to causative superantigenic PTs and remain susceptible to subsequent toxigenic illnesses. Thus, one may speculate that the harmful effects on the host (SFP and TSS) are merely secondary effects of the staphylococcus's attempt to affect immune cell function, allowing the organism to survive and persistently colonize its many potential animal hosts.

Acknowledgments

Our efforts in preparation of this chapter were supported by grants from the Public Health Service (AI28401 and RR00166), the U.S. Department of Agriculture (94-02399), and the Idaho Agriculture Experiment Station. We also thank Patrick Schlievert, Richard Novick, Amy Wong, Cynthia Stauffacher, Sibyl Munson, Brian Wilkinson, and Scott Minnich for comments, suggestions, and critical opinions on certain topics covered in this chapter. Jana Joyce, Scott Callantine, and Claudia Deobald are acknowledged for computer analysis, artwork, and organizational aspects of this chapter.

References

1. **Alber, G., D. K. Hammer, and B. Fleischer.** 1990. Relationship between enterotoxic- and T lymphocyte-stimulating activity of staphylococcal enterotoxin B. *J. Immunol.* **144:**4501–4506.
2. **Alber, G., P. H. Scheuber, B. Reck, B. Sailar-Kramer, A. Hartmann, and D. K. Hammer.** 1989. Role of substance P in immediate type skin reactions induced by staphylococcal enterotoxin B in unsensitized monkeys. *J. Allergy Clin. Immunol.* **84:**880–885.
3. **Anonymous.** 1992. *Foodborne Pathogenic Microorganisms and Natural Toxins.* Center for Food Safety and Applied Nutrition, U.S. Food and Drug Administration, Rockville, Md.
4. **Balaban, N., and R. P. Novick.** 1995. Autocrine synthesis of toxin synthesis by *Staphylococcus aureus. Proc. Natl. Acad. Sci. USA* **92:**1619–1623.
5. **Bayles, K. W., and J. J. Iandolo.** 1989. Genetic and molecular analyses of the gene encoding staphylococcal enterotoxin D. *J. Bacteriol.* **171:**4799–4806.
6. **Beery, J. T., S. L. Taylor, L. R. Schlunz, R. C. Freed, and M. S. Bergdoll.** 1984. Effects of staphylococcal enterotoxin A on the rat gastrointestinal tract. *Infect. Immun.* **44:**234–240.
7. **Bergdoll, M. S.** 1979. Staphylococcal intoxications, p. 443–494. *In* H. Riemann and F. L. Bryan (ed.), *Food-Borne Infections and Intoxications.* Academic Press, Inc., New York.
8. **Bergdoll, M. S.** 1985. The staphylococcal enterotoxins—an update, p. 247–254. *In* J. Jeljaszewicz (ed.), *The Staphylococci. Zentralbl. Bacteriol.* Suppl. 14. Gustav Fischer Verlag, Stuttgart.
9. **Bergdoll, M. S.** 1989. *Staphylococcus aureus,* p. 463–523. *In* M. P. Doyle (ed.), *Foodborne Bacterial Pathogens.* Marcel Dekker, Inc., New York.
10. **Bergdoll, M. S., C. R. Borja, R. Robbins, and K. F. Weiss.** 1971. Identification of enterotoxin E. *Infect. Immun.* **4:** 593–595.
11. **Bergdoll, M. S., M. J. Surgalla, and G. M. Dack.** 1959. Staphylococcal enterotoxin. Identification of a specific neutralizing antibody with enterotoxin neutralizing property. *J. Immunol.* **83:**334–338.
12. **Betley, M. J., D. W. Borst, and L. B. Regassa.** 1992. Staphylococcal enterotoxins, toxic shock syndrome toxin and streptococcal pyrogenic exotoxins: a comparative study of their molecular biology. *Chem. Immunol.* **55:**1–35.
13. **Betley, M. J., and J. J. Mekalanos.** 1985. Staphylococcal enterotoxin A is encoded by phage. *Science* **229:**185–187.
14. **Betley, M. J., P. M. Schlievert, M. S. Bergdoll, G. A. Bohach, J. J. Iandolo, S. A. Khan, P. A. Pattee, and R. R. Reiser.** 1990. Staphylococcal gene nomenclature. *ASM News* **56:**182.
15. **Bohach, G. A., D. J. Fast, R. D. Nelson, and P. M. Schlievert.** 1990. Staphylococcal and streptococcal pyrogenic toxins involved in toxic shock syndrome and related illnesses. *Crit. Rev. Microbiol.* **17:**251–272.
16. **Bohach, G. A., C. J. Hovde, J. P. Handley, and P. M. Schlievert.** 1988. Cross-neutralization of staphylococcal and streptococcal pyrogenic toxins by monoclonal and polyclonal antibodies. *Infect. Immun.* **56:**400–404.
17. **Bohach, G. A., L. M. Jablonski, C. F. Deobald, Y.-I. Chi, and C. V. Stauffacher.** 1995. Functional domains of staphylococcal enterotoxins, p. 339–356. *In* M. Ecklund, J. L. Richard, and K. Mise (ed.), *Molecular Approaches to Food Safety; Issues Involving Toxic Microorganisms.* Alaken Inc., Fort Collins, Colo.
18. **Bohach, G. A., and P. M. Schlievert.** 1987. Nucleotide sequence of the staphylococcal enterotoxin C gene and relatedness to other pyrogenic toxins. *Mol. Gen. Genet.* **209:**15–20.
19. **Bohach, G. A., C. V. Stauffacher, D. H. Ohlendorf, Y.-I. Chi, G. M. Vath, and P. M. Schlievert.** 1996. The staphylococcal and streptococcal pyrogenic toxin family, p. 131–154. *In* B. R. Singh and A. T. Tu (ed.), *Natural Toxins II.* Plenum Publishing Corporation, New York.
20. **Bryan, F. L.** 1976. *Staphylococcus aureus,* p. 12–128. *In* M. P. deFigueiredo and D. F. Splittstoesser (ed.), *Food Microbiology: Public Health and Spoilage Aspects.* AVI, Westport, Conn.
21. **Carl, K. E.** 1975. Oregon's experience with microbiological standards for meat. *J. Milk Food Technol.* **38:**483–486.
22. **Casman, E. P., M. S. Bergdoll, and J. Robinson.** 1963. Designation of staphylococcal enterotoxins. *J. Bacteriol.* **85:**715–716.
23. **Cheung, A. L., and S. J. Projan.** 1994. Cloning and sequencing of *sarA* of *Staphylococcus aureus,* a gene required for the expression of *agr. J. Bacteriol.* **176:**4168–4172.

24. **Couch, J. L., M. T. Soltis, and M. J. Betley.** 1988. Cloning and nucleotide sequence of the type E staphylococcal enterotoxin gene. *J. Bacteriol.* **170:**2954–2960.
25. **Dack, G. M., W. E. Cary, O. Woolper, and H. Wiggers.** 1930. An outbreak of food poisoning proved to be due to a yellow hemolytic staphylococcus. *J. Prev. Med.* **4:**167–175.
26. **Dickie, N., Y. Yang, H. Robern, and S. Stavric.** 1972. On the heterogeneity of staphylococcal enterotoxin C2. *Can. J. Microbiol.* **18:**801–804.
27. **Everson, M. L., M. W. Hinds, R. S. Bernstein, and M. S. Bergdoll.** 1988. Estimation of human dose of staphylococcal enterotoxin A from a large outbreak of staphylococcal food poisoning involving chocolate milk. *Int. J. Food Microbiol.* **7:**311–316.
28. **Foster, J. F., J. L. Fowler, and J. Dacey.** 1977. A microbial survey of various fresh and frozen seafood products. *J. Food Prot.* **40:**300–303.
29. **Fraser, J. D., S. Lowe, M. J. Irwin, N. R. J. Gascoigne, and K. R. Hudson.** 1993. Structural model of staphylococcal enterotoxin A interaction with MHC class II antigens, p. 7–30. *In* B. T. Huber and E. Palmer (ed.), *Superantigens: a Pathogen's View of the Immune System.* Cold Spring Harbor Laboratory Press, Plainview, N.Y.
30. **Friedman, M. E.** 1966. Inhibition of staphylococcal enterotoxin B formation in broth cultures. *J. Bacteriol.* **92:**277–278.
31. **Genetics Computer Group.** 1994. *Program Manual for the Wisconsin Package Version 8.* Genetics Computer Group, Madison, Wis.
32. **Gilbert, R. J., and A. A. Wieneke.** 1973. Staphylococcal food poisoning with special reference to the detection of enterotoxin in food, p. 273–285. *In* B. C. Hobbs and J. H. B. Christian (ed.), *The Microbiological Safety of Food.* Academic Press, Inc., New York.
33. **Giraudo, A. T., C. G. Raspanti, A. Calzolari, and R. Nagel.** 1994. Characterization of Tn551-mutant of *Staphylococcus aureus* defective in the production of several exoproteins. *Can. J. Microbiol.* **40:**677–681.
34. **Guthertz, L. S., J. T. Fruin, R. L. Okoluk, and J. L. Fowler.** 1977. Microbial quality of frozen comminuted turkey meat. *J. Food Sci.* **42:**1344–1447.
35. **Guthertz, L. S., J. T. Fruin, D. Spicer, and J. L. Fowler.** 1976. Microbial quality of fresh comminuted turkey meat. *J. Milk Food Technol.* **39:**823–829.
36. **Hajek, V., and E. Marsalek.** 1973. The occurrence of enterotoxigenic *Staphyococcus aureus* strains in hosts of different animal species. *Zentralbl. Bakteriol. Hyg. Abt. 1 Orig. Reihe A* **217:**176–182.
37. **Harris, T. O., and M. J. Betley.** 1995. Biological activities of staphylococcal enterotoxin type A mutants with N-terminal substitutions. *Infect. Immun.* **63:**2133–2140.
38. **Hase, C. C., and R. A. Finkelstein.** 1993. Bacterial extracellular zinc-containing metalloproteases. *Microbiol. Rev.* **57:**823–837.
39. **Hoffmann, M. L., L. M. Jablonski, K. K. Crum, S. P. Hackett, Y.-I. Chi, C. V. Stauffacher, D. L. Stevens, and G. A. Bohach.** 1994. Predictions of T cell receptor and major histocompatibility complex binding sites on staphylococcal enterotoxin C1. *Infect. Immun.* **62:**3396–3407.
40. **Holmberg, S. D., and P. A. Blake.** 1984. Staphylococcal food poisoning in the United States. *JAMA* **251:**487–489.
41. **Hovde, C. J., J. C. Marr, M. L. Hoffmann, S. P. Hackett, Y.-I. Chi, K. K. Crum, D. L. Stevens, C. V. Stauffacher, and G. A. Bohach.** 1994. Investigation of the role of the disulfide bond in activity and structure of staphylococcal enterotoxin C1. *Mol. Microbiol.* **13:**897–909.
42. **Huang, I. Y., and M. S. Bergdoll.** 1970. The primary sequence of staphylococcal enterotoxin B. The cyanogen bromide peptides of reduced and aminoethylated enterotoxin B, and the complete amino acid sequence. *J. Biol. Chem.* **245:**3518–3525.
43. **Iandolo, J. J., and W. M. Shafer.** 1977. Regulation of staphylococcal enterotoxin B. *Infect. Immun.* **16:**610–616.
44. **Janzon, L., and S. Arvidson.** 1990. The role of the delta-lysin gene (*hld*) in the regulation of virulence genes by the accessory gene regulator (*agr*) in *Staphylococcus aureus. EMBO J.* **9:**1391–1399.
45. **Jay, J. M.** 1986. Staphylococcal gastroenteritis, p. 437–458. *In Modern Food Microbiology,* 3rd ed. Van Nostrand Reinhold Company, New York.
46. **Jett, M. R., C. Neill, T. Welch, E. Boyle, D. Bernton, G. Hoove, R. E. Lowell, S. Hunt, Chatterjee, and P. Gemski.** 1994. Identification of staphylococcal enterotoxin B sequences important for induction of lymphocyte proliferation by using synthetic peptide fragments of the toxin. *Infect. Immun.* **62:**3408–3415.
47. **Johns, M. B., Jr., and S. A. Khan.** 1988. Staphylococcal enterotoxin B gene is associated with a discrete genetic element. *J. Bacteriol.* **170:**4033–4039.
48. **Johnson, H. M., J. K. Russell, and C. H. Pontzer.** 1992. Superantigens in human disease. *Sci. Am.* **266:**92–101.
49. **Johnson, L. P., and P. M. Schlievert.** 1983. A physical map of the group A streptococcal pyrogenic exotoxin bacteriophage T12 genome. *Mol. Gen. Genet.* **189:**251–255.
50. **Jones, C. L., and S. A. Khan.** 1986. Nucleotide sequence of the enterotoxin B gene from *Staphylococcus aureus. J. Bacteriol.* **166:**29–33.
51. **Katsuno, S., and M. Kondo.** 1973. Regulation of staphylococcal enterotoxin B synthesis and its relation to other extracellular proteins. *Jpn. J. Med. Sci. Biol.* **26:**26–29.
52. **Kent, T. H.** 1966. Staphylococcal enterotoxin gastroenteritis in rhesus monkeys. *Am. J. Pathol.* **48:**387–405.
53. **Klimpel, K. R., N. Arora, and S. H. Leppla.** 1994. Anthrax toxin lethal factor contains a zinc metalloprotease consensus sequence which is required for lethal toxin activity. *Mol. Microbiol.* **13:**1093–1100.
53a. **Kloos, W. E., and K. H. Schleifer.** 1986. Genus IV. *Staphylococcus,* p. 1013–1035. *In* P. A. Sneath (ed.), *Bergey's Manual of Systematic Bacteriology,* 8th ed. The Williams & Wilkins Co., Baltimore.
54. **Komisar, J., J. Rivera, A. Vega, and J. Tseng.** 1992. Effects of staphylococcal enterotoxin B on rodent mast cells. *Infect. Immun.* **60:**2969–2975.
55. **Kornblum, J., B. Kreiswirth, S. J. Projan, H. Ross, and R. P. Novick.** 1990. *Agr:* a polycistronic locus regulating exoprotein synthesis *Staphylococcus aureus,* p. 373–402. *In* R. P. Novick and R. Skurray (ed.), *Molecular Biology of the Staphylococci.* VCH Publishers, New York.

56. **Lovejoy, H. F. Jr.** 1991. Contributed information on staphylococcal food poisoning to poison center data base. [Aug. 1981. Revised by Poisondex editorial staff, Denver, Colo., in October 1991.]

57. **Marr, J. C., J. D. Lyon, J. R. Roberson, M. Lupher, W. C. Davis, and G. A. Bohach.** 1993. Characterization of novel type C staphylococcal enterotoxins: biological and evolutionary implications. *Infect. Immun.* **61:**4254–4262.

58. **Marrack, P., and J. Kappler.** 1990. The staphylococcal enterotoxins and their relatives. *Science* **248:**705–711.

59. **Merson, M. H.** 1973. The epidemiology of staphylococcal foodborne disease, p. 20–37. *In* Proceedings staphylococci in foods (conference). Pennsylvania State University Press, University Park, Pa.

60. **Noleto, A. L., and M. S. Bergdoll.** 1982. Production of enterotoxin by a *Staphylococcus aureus* strain that produces three identifiable enterotoxins. *J. Food Prot.* **45:**1096–1097.

61. **Novick, R. P., H. F. Ross, S. J. Projan, J. Kornblum, B. Kreiswirth, and S. Moghazeh.** 1993. Synthesis of staphylococcal virulence factors is controlled by a regulatory RNA molecule. *EMBO J.* **12:**3967–3975.

62. **Pace, P. J.** 1975. Bacteriological quality of delicatessen foods. *J. Milk Food Technol.* **38:**347–353.

63. **Pivnick, H., I. E. Erdman, D. Collins-Thompson, G. Roberts, M. A. Johnston, D. R. Conley, G. Lachapelle, U. T. Purvis, R. Foster, and M. Milling.** 1976. Proposed microbiological standards for ground beef based on a Canadian survey. *J. Milk Food Technol.* **39:**408–412.

64. **Pontzer, C. H., J. K. Russell, and H. M. Johnson.** 1989. Localization of an immune functional site on staphylococcal enterotoxin A using the synthetic peptide approach. *J. Immunol.* **143:**280–284.

65. **Raj, H. D., and M. S. Bergdoll.** 1969. Effect of enterotoxin B on human volunteers. *J. Bacteriol.* **98:**833–834.

66. **Rao, L., R. K. Karls, and M. J. Betley.** 1995 In vitro transcription of pathogenesis-related genes by purified RNA polymerase from *Staphylococcus aureus. J. Bacteriol.* **177:**2609–2614.

67. **Reck, B., P. H. Scheuber, W. Londong, B. Sailer-Kramer, K. Bartsch, and D. K. Hammer.** 1988. Protection against the staphylococcal enterotoxin-induced intestinal disorder in the monkey anti-idiotypic antibodies. *Proc. Natl. Acad. Sci. USA* **85:**3170–3174.

68. **Reda, K. B., V. Kapur, V. Mollick, J. A. Lamphear, J. M. Musser, and R. R. Rich.** 1994. Molecular characterization and phylogenetic distribution of the streptococcal superantigen gene (*ssa*) from *Streptococcus pyogenes. Infect. Immun.* **62:**1867–1874.

69. **Regassa, L. B., J. L. Couch, and M. J. Betley.** 1991. Steady-state staphylococcal enterotoxin type C mRNA is affected by a product of the accessory gene regulator (*agr*) and by glucose. *Infect. Immun.* **59:**955–962.

70. **Ren, K., J. D. Bannan, V. Pancholi, A. L. Cheung, J. C. Robbins, V. A. Fischetti, and J. B. Zabriskie.** 1994. Characterization and biological properties of a new staphylococcal exotoxin. *J. Exp. Med.* **180:**1675–1683.

71. **Roggiani, M., L. Jablonski, G. A. Bohach, and P. M. Schlievert.** 1994. Localization of biological activities of streptococcal pyrogenic exotoxin A, abstr. B-93, p. 45. *In Abstracts of the 94th General Meeting of the American Society for Microbiology 1994.* American Society for Microbiology, Washington, D.C.

72. **Schad, E. M., I. Zaitseva, V. N. Zaitsev, M. Dohlsten, T. Kalland, P. M. Schlievert, D. H. Ohlendorf, and L. A. Svensson.** 1995. Crystal structure of the superantigen staphylococcal enterotoxin A. *EMBO J.* **14:**3292–3301.

73. **Scheuber, P. H., C. Denzlinger, D. Wilker, G. Beck, D. Keppler, and D. K. Hammer.** 1987. Staphylococcal enterotoxin B as a nonimmunological mast cell stimulus in primates: the role of endogenous cysteinyl leukotrienes. *Int. Arch. Allergy Appl. Immunol.* **82:**289–291.

74. **Schlievert, P. M.** 1993. Role of superantigens in human disease. *J. Infect. Dis.* **167:**997–1002.

75. **Schmidt, J. J., and L. Spero.** 1983. The complete amino acid sequence of staphylococcal enterotoxin C1. *J. Biol. Chem.* **258:**6300–6306.

76. **Singh, B. R., and M. J. Betley.** 1989. Comparative structural analysis of staphylococcal enterotoxins A and E. *J. Biol. Chem.* **264:**4404–4411.

77. **Smeltzer, M. S., M. E. Hart, and J. J. Iandolo.** 1993. Phenotypic characterization of *xpr*, a global regulator of extracellular virulence factors in *Staphylococcus aureus. Infect. Immun.* **61:**919–925.

78. **Smith, F. C., R. A. Field, and J. C. Adams.** 1974. Microbiology of Wyoming big game meat. *J. Milk Food Technol.* **37:**129–131.

79. **Spero, L., B. Y. Griffin, J. L. Middlebrook, and J. F. Metzger.** 1976. Effect of single and double peptide bond scission by trypsin on the structure and activity of staphylococcal enterotoxin C. *J. Biol. Chem.* **251:**5580–5588.

80. **Stock, J. B., A. J. Ninfa, and A. M. Stock.** 1989. Protein phosphorylation and regulation of adaptive responses in bacteria. *Microbiol. Rev.* **53:**450–490.

81. **Su, Y.-C., and A. C. L. Wong.** 1993. Optimal condition for the production of unidentified staphylococcal enterotoxins. *J. Food Prot.* **56:**313–316.

82. **Su, Y.-C., and A. C. L. Wong.** 1995. Identification and purification of a new staphylococcal enterotoxin H. *Appl. Environ. Microbiol.* **61:**1438–1443.

83. **Sugiyama, H., and T. Hayama.** 1965. Abdominal viscera as site of emetic action for staphylococcal enterotoxin in the monkey. *J. Infect. Dis.* **115:**330–336.

84. **Surkiewicz, B. F., M. E. Harris, R. P. Elliott, J. F. Macaluso, and M. M. Strand.** 1975. Bacteriological survey of raw beef patties produced at establishments under federal inspection. *Appl. Microbiol.* **29:**331–334.

85. **Surkiewicz, B. F., R. W. Johnston, R. P. Elliott, and E. R. Simmons.** 1972. Bacteriological survey of fresh pork sausage produced at establishments under federal inspection. *Appl. Microbiol.* **23:**515–520.

86. **Swaminathan, S., W. Furey, J. Pletcher, and M. Sax.** 1992. Crystal structure of staphylococcal enterotoxin B, a superantigen. *Nature* (London) **359:**801–806.

87. **Swartzentruber, A., A. H. Schwab, A. P. Duran, B. A. Wentz, and R. B. Read, Jr.** 1980. Microbiological quality of frozen shrimp and lobster tail in the retail market. *Appl. Environ. Microbiol.* **40:**765–769.

88. **Todd, E. C. D., G. A. Jarvis, K. F. Weiss, G. W. Riedell, and S. Charbonneau.** 1983. Microbiological quality of frozen cream-type pies sold in Canada. *J. Food Prot.* **46:**34–40.

89. **Townsend, D. E., and B. J. Wilkinson.** 1992. Proline transport in *Staphylococcus aureus:* a high-affinity system

and a low-affinity system involved in osmoregulation. *J. Bacteriol.* **174**:2702–2710.

90. **Tremaine, M. T., D. K. Brockman, and M. J. Betley.** 1993. Staphylococcal enterotoxin A gene (*sea*) expression is not affected by the accessory gene regulator (*agr*). *Infect. Immun.* **61**:356–359.
91. **Warren, J. R.** 1977. Comparative kinetic stabilities of staphylococcal enterotoxin types A, B, and C1. *J. Biol. Chem.* **252**:6831–6834.
92. **Warren, J. R., L. Spero, and J. F. Metzger.** 1974. Stabilization of the native structure by the closed loop of staphylococcal enterotoxin B. *Biochim. Biophys. Acta* **359**:351–363.
93. **Wengender, P. A., and K. J. Miller.** 1995. Identification of a PutP proline permease gene homolog from *Staphylococcus aureus* by expression cloning of the high-affinity proline transport system in *Escherichia coli*. *Appl. Environ. Microbiol.* **61**:252–259.
94. **Wentz, B. A., A. P. Duran, A. Swartzentruber, A. H. Schwab, and R. B. Read, Jr.** 1983. Microbiological quality of fresh blue crabmeat, clams and oysters. *J. Food Prot.* **46**:978–981.
95. **Wentz, B. A., A. P. Duran, A. Swartzentruber, A. H. Schwab, and R. B. Read, Jr.** 1984. Microbiological quality of frozen onion rings and tuna pot pies. *J. Food Prot.* **47**:58–60.

Morris E. Potter
Silvia Gonzalez Ayala
Narumol Silarug

20

Epidemiology of Foodborne Diseases

Infectious foodborne diseases are of major importance worldwide and will not be eliminated in the foreseeable future (25). Microbial contamination of food presents challenges that are different from those posed by toxins and physical hazards to the scientific, public health, and food authority communities in all countries. First, microorganisms are complex and ever-changing. They can evolve quickly by exchanging genetic material to acquire properties that help them colonize new niches or create new diseases. Second, many microorganisms are able to grow and multiply on food by using it as an energy source. Third, the potential for foods to transmit infectious hazards is high because food safety can be compromised at so many points between farm or catch and table. Our food supply brings us into intimate contact with a variety of microorganisms that are present in the production and processing environments from around the world.

Food safety is a complex matter that depends on a number of interrelated environmental, cultural, and socioeconomic factors. If one recognizes that ensuring food safety is inherently uncertain, foodborne illnesses become opportunities to learn about how these factors interrelate rather than failures to predict. Foodborne disease will occur, and we must be prepared to react quickly to reduce the risk of new foodborne hazards. Our prevention goal should be the rapid identification and constant reduction of conditions under which foodborne disease occurs. In developing countries where foodborne disease surveillance is particularly weak, the attention of policy makers and the food industry is frequently not directed toward the role of food in transmission of disease, and important opportunities for control are missed. For example, the risk of botulism associated with a traditionally consumed salted fish, faseikh, was unrecognized until a very large outbreak occurred in 1991 (61).

This chapter is intended to provide an explanation of epidemiology and epidemiologic techniques, basic information on foodborne disease, and a description of how epidemiologic data can be used in qualitative and quantitative risk assessments of foodborne hazards. Epidemiologic data are also very important in establishing food safety priorities, allocating food safety resources, stimulating public and industry interest in food safety issues, establishing risk reduction strategies (including technical and consumer educational materials), and evaluating the effectiveness of food safety programs (58). The examples in the chapter are drawn from, and therefore limited by, the authors' experiences in foodborne disease in three different geographic regions. Relevant epidemiologic data specific to most foodborne pathogens are presented in the chapters on those agents.

WHAT IS EPIDEMIOLOGY?

Epidemiology has been defined as the study of the occurrence and distribution of disease in populations and the factors that account for this distribution (30). Those who teach epidemiology in academic settings frequently prefer even longer definitions; however, those who use epidemiologic techniques to combat threats to the public's health generally view epidemiology as a tool used to study the health status of the world around us. Epidemiologists identify problems in the food supply that cause illness and then characterize these problems through surveillance and investigations. Epidemiology can identify new hazards, associate new foods with recognized foodborne pathogens, and characterize the risk factors that increase the likelihood of illness in consumers. New foodborne hazards might be newly evolved microorganisms, previously unrecognized pathogens that have become identifiable because of changes in laboratory technology, pathogens whose presence in a food has only recently become possible because of a change in food production, or pathogens formerly geographically restricted which have been introduced into new geographic areas through international trade or travel. Risk factors that increase the likelihood of disease include those related to human behavior at all stages of the food chain, from production to consumption, and those related to host susceptibility.

One hundred fifty years ago, nearly 30 years before the bacterial cause of cholera was identified, John Snow investigated an outbreak of cholera in London (48). He used the principles of descriptive epidemiology, including locating patients' homes on a map of London, to characterize the cases of cholera, and he observed that the cases were clustered around the Broad Street pump, the community's source of water. He removed the pump handle, forcing the local residents to find other sources of water, and the outbreak ended.

In descriptive epidemiology, illnesses are characterized by who became ill, what symptoms the patients suffered, when symptoms began, and where the illnesses occurred. This was nearly all that John Snow needed to know to end the Broad Street cholera epidemic in London. Descriptive epidemiology involves collecting, collating, analyzing, and interpreting information on the features of the occurrence of a disease in a community (30). The key characteristics of interest include time, place, person, incidence data, and clinical features. The element of time is generally displayed as a graph of date of onset of symptoms on one axis and number of cases on the other, called an epidemic curve. The period between the time of exposure and onset of symptoms of illness in each person is the incubation period and is measured in minutes, hours, or days, depending on the etiologic agent. When exposures to a foodborne agent are restricted to a single place and time, called a point source, the times of onset of symptoms for many of the cases occur close together and the epidemic curve has a distinct peak. Depicting place by creating a spot map can help determine where exposure occurred by demonstrating geographic clustering in a neighborhood, place of work, restaurant, or picnic ground or, as in the London cholera outbreak, around a source of water. The characteristics of the patient play a key role in behavioral and cultural patterns that affect exposure, in the frequency and severity of clinical illness following exposure, and in secondary transmission of disease. Therefore, in descriptive epidemiology, the element of person is generally categorized by age, gender, and the other demographic characteristics mentioned above that influence the distribution of disease.

In analytical epidemiologic investigations, persons who have been exposed to a hazard or have become ill are compared with those who have not. Determining other differences between the case-patients and their comparison group, such as what foods have been consumed by these two groups, enables analytical epidemiologic investigations to identify risk factors for foodborne disease. Case-control studies are the most common type of analytical epidemiologic study for infectious foodborne diseases, although prospective and retrospective cohort studies and cross-sectional studies are also performed (30). In another phase of the cholera epidemic in London, John Snow was able to show that the cholera death rate was higher among households supplied by two water companies than in neighboring households supplied by a third source. The latter company drew its water from the Thames River above London, while the former two drew Thames water below the main sewage discharge for London. This difference in rates was critical to demonstrating that epidemic cholera was waterborne and led to the sanitary reform movement to improve the safety of drinking water even before bacteriologic evidence was available.

Analytical epidemiology also has been used to investigate the current cholera epidemic in Latin America. In late January 1991, cholera appeared in Peru, and it spread rapidly to neighboring countries. In less than 3 years, 21 countries in the western hemisphere had reported more than 800,000 cholera cases and thousands of deaths. A case-control investigation of epidemic cholera in Ecuador identified several risk factors for infection: drinking unboiled contaminated water, drinking a beverage from a street vendor, eating raw seafood, and eating cooked crab (62). In this study, responses to questions on possible sources of exposure to *Vibrio cholerae* from patients were compared with those from age- and gender-matched neighborhood controls who had not had diarrhea in the preceding 3 months to identify the risk factors.

Answering some questions, however, requires epidemiologic techniques more highly refined than those used in the past. For example, many of the 132 cholera cases imported into the United States from the Latin American epidemic have resulted from consumption of contaminated food. However, in addition to the foodborne cholera cases that have occurred in the United States as a result of the epidemic in Latin America or importation from other parts of the world, an environmental reservoir of *V. cholerae* exists along the coast of the United States in the Gulf of Mexico. That source has been responsible for an additional 65 foodborne cholera cases in the United States in the past 20 years. Designing strategies to prevent cholera requires knowing its origins and determining its mechanisms of spread in the community, and distinguishing the Latin American and Asian cholera strains from our domestic strains requires separating isolates of *V. cholerae* into biologically related groups so that epidemiologic information from related cases can be analyzed. Molecular biology has provided epidemiologists the laboratory tools to make these separations; this combination of laboratory and epidemiologic techniques has been termed molecular epidemiology. Plasmid profile analysis, restriction endonuclease DNA analysis, and DNA hybridization are among the methods used singly or in combination to characterize foodborne pathogens genetically for molecular epidemiologic analysis (60). These methods have demonstrated that at least four distinct toxigenic *V. cholerae* clones exist: the seventh-pandemic clone in the eastern hemisphere, the United States Gulf Coast clone, a clone in Australia, and the isolates associated with the outbreak in Latin America (59).

EPIDEMIOLOGIC METHODS FOR STUDYING FOODBORNE DISEASE

The patterns of relative importance of reported food-pathogen combinations vary throughout the world (52, 58). For example, in the United States, salmonellae and beef is a commonly reported food-pathogen combination (3). Salmonellae are also important causes of outbreaks in most of Europe, but in the United Kingdom and The Netherlands, poultry-associated campylobacteriosis sometimes is more frequently reported, and in Finland and Sweden, the incidence of salmonellosis is decreasing while that of campylobacteriosis is increasing (9, 52). In Japan, *Vibrio parahaemolyticus* infections associated with seafood frequently top the list of food safety issues (63). However, with increasing international travel and trade, the patterns of foodborne diseases may become more homogeneous.

A number of epidemiologic techniques, including laboratory-based surveillance of specific foodborne microorganisms, regional or national outbreak surveillance, intensive epidemiologic and laboratory investigations, and studies of sporadic foodborne diseases, can be used to identify and characterize foodborne disease (57). Each of these epidemiologic techniques is incomplete in itself; data from each must be viewed within the greater context of the total body of microbiologic and epidemiologic knowledge.

Clinical Laboratory-Based Surveillance of Foodborne Microorganisms

For diagnostic purposes, clinical laboratories identify pathogens that may have been foodborne in clinical specimens from ill persons, such as stool samples or blood samples. Characterizing the patients and tracking the frequency of isolation of specific pathogens provide important epidemiologic information. Laboratory-based surveillance for foodborne pathogens requires an infrastructure of competent diagnostic laboratories that receive clinical specimens routinely and that are willing to provide information to a central data manager; when this infrastructure is present, laboratory-based surveillance is neither difficult nor resource-intensive. While these data have many uses, this method of surveillance, at least as it is constituted in the United States, frequently does not determine the role of specific foods as a source of exposure, and little other patient information generally is collected.

To improve public health detection of outbreaks and trends, some pathogens may be further characterized or subtyped at a reference laboratory. For example, *Salmonella* isolates can be serotyped in state or regional public health laboratories. While determining serotype does not affect the management of an individual patient, it is critical to the ability of public health to detect and track outbreaks. For example, data from the United States demonstrate that infections with *Salmonella* serotype Enteritidis (SE) increased markedly during the 1980s, from 8% of all human *Salmonella* isolates in 1979 to 25% in 1994; SE now constitutes more than a third of all *Salmonella* isolates in many parts of the United States. Foodborne infections with SE have also become very common in many European countries (52). For example, in the early 1990s, more than half of all *Salmonella*

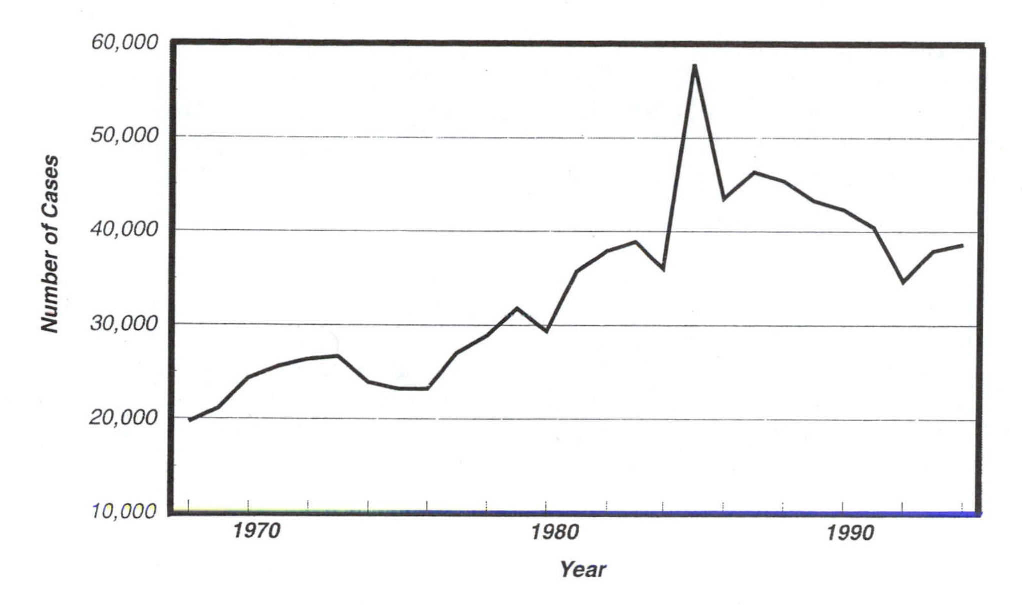

Figure 20.1 National *Salmonella* surveillance: human isolates by year, United States, 1968 to 1994.

infections were due to SE in France, Germany, Hungary, Italy, Norway, Romania, and the United Kingdom; in the Czech Republic, Greece, and Lithuania, more than three-quarters of all infections were due to SE.

Human salmonellosis has been a reportable disease in the United States since 1943, and a national *Salmonella* surveillance system was established at the Centers for Disease Control and Prevention (CDC) in 1963 to collate the data from public health laboratories conducting serotyping of isolates (55). This laboratory-based surveillance system depends on voluntary reporting by state health departments of *Salmonella* isolates serotyped by reference laboratories. Routine determination and reporting of serotypes has been helpful in detecting outbreaks, determining their source, and monitoring long-term trends. Between 1970 and 1990, reported infections with nontyphoidal *Salmonella* strains steadily increased, from 20,000 to 25,000 isolates per year to 40,000 to 45,000 isolates per year (Fig. 20.1). Since the early 1990s, reports of human salmonellosis in the United States have declined, with 38,619 isolates reported to CDC in 1994.

In addition to tracking secular incidence trends and identifying clusters of cases caused by specific serotypes, laboratory-based surveillance of human infections can provide specimens for additional laboratory analysis of agents of foodborne disease. For example, *Salmonella* isolates from humans in the United States have been examined periodically for resistance to antimicrobial drugs. CDC has examined resistance to a standard panel of agents at 5-year intervals since 1980 (32, 36). Overall, the proportion of strains resistant to any of the agents doubled in the last decade, from 16% to 31%. The proportion that is resistant to two or more of the agents has more than doubled, from 12% to 25%.

Laboratory-based surveillance of *Salmonella* isolates has shown salmonellosis to be one of the most frequently reported foodborne diseases in Latin America, and in that area, more laboratory isolations of salmonellae than of any other foodborne pathogen are reported. Mexico reported more than 40,000 laboratory isolations of salmonellae annually in 1987 to 1989. Pathogenic *Escherichia coli*, also a major foodborne agent in South and Central America, has been isolated from dairy and meat products, vegetables, and other products at the site of production or at retail. *E. coli*, *Shigella* species, and *Yersinia enterocolitica* are important causes of diarrheal disease in children in Latin America.

Another use of clinical isolates for surveillance is applying a battery of isolation and identification techniques to a set of clinical specimens to determine the relative importance of various enteric pathogens that may be foodborne. Karmali and Fleming (29) examined all stools and rectal swab specimens submitted to the laboratory at the Hospital for Sick Children in Toronto, Canada, for a number of bacterial pathogens, and they isolated nontyphoidal salmonellae from 74, *Campylobacter* species from 39, shigellae from 11, and *Y. enterocolitica* from 7. In a year-long survey in western Washington State in 1985 to 1986 that included culturing

6,485 stools, *Campylobacter* was isolated from 195 stools, salmonellae were isolated from 110, shigellae were isolated from 39, and *E. coli* serotype O157:H7 (Eco157) was isolated from 25 (37). In a 1987 study in Minnesota, 2,164 stool specimens were examined for a number of bacterial pathogens; *Campylobacter* was isolated from 25, salmonellae were isolated from 21, Eco157 was isolated from 10, *Y. enterocolitica* was isolated from 5, and shigellae were isolated from 1 (38).

The reported incidence of acute diarrhea in young children in Thailand was 1 to 2% in 1992. In 1985, Echeverria et al. (13) found that rotavirus was the most commonly detected enteric pathogen in Thai children (20%); however, in children less than 1 year old, *Campylobacter jejuni* was most common (13%), and *Salmonella* species, generally serotype Typhimurium, was isolated most commonly from children under 3 months of age (12%). *Shigella* isolation rates increased progressively with age during childhood and became a common cause of diarrhea among children over 4 years old. As is frequently the case for laboratory-based surveillance systems, the role of food in these illnesses in Thailand is poorly defined; combining these data with information from other sources is necessary to create an accurate perception of the relative importance of food in the transmission of various pathogens so that education and other control programs can be efficiently targeted.

The distribution of parasitic foodborne diseases is influenced by the natural distribution of the parasites and the sociocultural practices that bring susceptible consumers into contact with them (1). Inadequate containment of human feces, the culinary practice of eating raw foods of animal origin, and the use of human feces as fish food in aquaculture are important determinants in the foodborne transmission of various parasites in different parts of the world. The most commonly reported foodborne parasitic diseases in the United States are trichinellosis, toxoplasmosis, taeniasis/cysticercosis, diphyllobothriasis, and anisakiasis (50). Although these diseases are less common than many other foodborne diseases in the United States, toxoplasmic encephalitis has become the most commonly recognized cause of central nervous system opportunistic infection in AIDS patients, cases of neurocysticercosis are being diagnosed with increased frequency in immigrants from regions of endemicity in Mexico and Central America, and diphyllobothriasis and anisakiasis now occur in the United States among consumers of Asian-style foods containing raw fish.

Parasitic foodborne diseases are very common in South and Central America, and laboratory-based surveillance can be used to study their characteristics. In some areas, foodborne and waterborne parasitic infections are more frequent than illness due to bacterial or viral pathogens (46). More than 10% of patients with neurologic disease in Peru had serologic evidence of infection with *Taenia solium*, and Schenone et al. (51) reported rates of infection by *T. solium* of approximately 1/100 population in Mexico, Ecuador, and Guatemala. *Entamoeba histolytica* is a frequent cause of foodborne disease in tropical and subtropical Latin American countries; raw vegetables and untreated water are the major vehicles (44). *Giardia lamblia* is probably the most frequently identified cause of parasitic diarrhea in Latin America, and the disease is generally attributed to waterborne transmission (47). Cryptosporidiosis is an increasing problem in many countries of the Americas (16). A children's hospital in Santa Fe, Argentina, studied 2,431 fecal samples from children with diarrhea in 1 year and found 2% of the children to be infected with *Cryptosporidium* species (5). In Costa Rica, the parasite was found in the feces of 4% of urban and rural children with gastroenteritis (40). While cryptosporidiosis is more frequently waterborne than foodborne, its foodborne transmission has been documented (41), and contaminated processing water can contaminate ready-to-eat foods like fresh produce.

A number of parasitic diseases, including amebiasis, giardiasis, trichinellosis, gnathostomiasis, *Enterobius* infections, angiostrongyliasis, taeniasis, opisthorchiasis, and fasciolopsiasis are important in Thailand (35). Transmission of *Gnathostoma spinigerum* has been demonstrated in food prepared from fermented shredded fish and rice wrapped in banana leaves, as well as raw frogs, snakes, and eels. *Opisthorchis viverrini* infections are commonly associated with eating a spicy dish of mashed raw fish in Thailand (1).

Microbiologic Monitoring of Food Animals and Foods

Monitoring food-producing animals on the farm for microorganisms that may cause foodborne illness in humans is critical in providing producers and the food safety authority scientifically sound and statistically valid information on the occurrences, distribution, and trends of these agents. In 1983, the U.S. Department of Agriculture initiated the National Animal Health Monitoring System (NAHMS) as an animal production information-gathering program. Through nationwide studies, NAHMS collects on-farm data and generates descriptive statistics on animal health, productivity, management, and frequency of infection with selected foodborne pathogens. Collection of biologic specimens from animals and their environment is an integral component of NAHMS data collection. These samples include fecal specimens to determine animal and herd or flock prevalence and evaluate factors related to the presence of the pathogen. Prevalence of Eco157 in dairy calves was identified through the NAHMS National Dairy Heifer Evaluation Project in 1990 to 1991. Animal and herd or flock prevalence of *Salmonella* and *Yersinia* species also has been determined.

After epidemiologic investigations demonstrated the role of eggs in outbreaks of human infection with SE in the United States, the U.S. Department of Agriculture conducted surveys of retired laying hens (spent hens) and of bulk prepasteurized liquid egg. In the survey of spent hens in 1991, 23,431 pooled cecum samples were collected from 406 layer houses (11). Salmonellae were recovered from 24% of the pooled cecum samples; SE was recovered from 3%. Geographically, 45, 3, and 17% of the layer houses were SE positive in the northern, southeastern, and central/western regions of the United States. A survey of unpasteurized liquid egg in the United States was conducted in 1991 to 1992 and demonstrated that the geographic distribution of SE-positive liquid egg samples paralleled the distribution of infected laying hens found in the national spent hen survey (12).

The presence of microbial pathogens on fresh produce has been recognized for many years, and recent produce-associated outbreaks in the United States and Europe have brought into focus the potential importance of fresh produce as a vehicle of foodborne pathogens (7). Advances in agronomics, processing, preservation, distribution, and marketing have created a global marketplace for fresh produce (limited only by willingness and ability to pay) and have increased concern over the microbial safety of fresh produce in domestic and international trade. This has resulted in many surveys of all manner of fresh and minimally processed produce for all production areas in the world (7, 43). These studies provide information on the range of likely exposures to pathogens from consumption of fresh produce.

Microbiologic surveys of foods can help define the risk for exposure to potential pathogens and the relative importance of various commodities as vehicles of foodborne pathogens. In the absence of epidemiologic data on human illness, food microbiology data collected for international trade or other purchase specifications may be the only available indication of the risk for foodborne disease. Surveys in Latin America have isolated *Yersinia* species from fish, eggs, raw and pasteurized milk, vegetables, and meat products. *C. jejuni* and *Campylobacter coli* have been isolated frequently from uncooked poultry in Argentina, Brazil, Chile, and Trinidad (10, 23) and appear to cause as much as 5 to 20% of the gastrointestinal disease in Latin American children (17, 19, 39). A survey for *Salmonella* contamination of poultry products in Thailand demonstrated high rates of contamination at various points from slaughter to retail and identified a number of high-risk conditions amenable to control efforts (26). *V. parahaemolyticus* has been found to be a common contaminant of raw fish and other seafood in Thailand and is a frequent cause of foodborne disease. Salmonellae, *E. coli*, *V. parahaemolyticus*, various clostridia, and marine biotoxins have been identified in shellfish in various microbiologic surveys in Thailand.

Active follow-up of cases identified through laboratory diagnostics provides additional epidemiologic data, including information on possible sources of infection and whether the case is sporadic or is associated with other cases in an outbreak. Without this follow-up, the use of this kind of surveillance data is more limited but still useful for program planning, designing food safety educational materials, and evaluating control programs. Because microorganisms may be killed (or multiply) between the time of sample collection and the time of food consumption, these survey data are only an indirect measure of exposure potential.

Investigation and Surveillance of Foodborne Disease Outbreaks

One of the fundamentals of epidemiology is that the study of the departure of the observed patterns of occurrence of disease from the expected may lead to identification of the causes of the deviation (57). When the observed rate is higher than expected and cases are unusually close together in time or space or within the same demographic group, that group of cases is called a cluster, an outbreak, or an epidemic. Inasmuch as one would not expect illness to result from consumption of food, any illness associated with a meal is a departure from expected rates. For the purposes of CDC's foodborne disease outbreak surveillance system in the United States, an outbreak is defined as an incident in which two or more persons experience a similar illness after ingestion of a common food or meal. A sporadic case is represented by an ill person whose illness is not known to be related by common exposure to other ill persons.

One of the most helpful ways to observe trends in foodborne diseases is to examine data from investigations of foodborne disease outbreaks. Outbreaks are easier to investigate and characterize than sporadic illnesses, and large clusters of cases provide an opportunity to rapidly identify a common source of exposure by epidemiologic methods. However, because outbreak surveillance focuses on identifiable clusters of illnesses, it may miss pathogens, such as *Vibrio vulnificus,* which cause sporadic illnesses rather than large outbreaks, and small clusters of cases associated with foods consumed at home. In addition, low-level contamination of commonly consumed, widely distributed products that results in apparently unrelated illnesses across broad geographic areas tests the capabilities of this approach. For example, in 1989 a multistate outbreak of salmonellosis occurred in residents of the United States who had eaten contaminated mozzarella cheeses and shredded cheese products (21). The outbreak was initially identified because of increased incidence in Minnesota of clinical isolates of *Salmonella* serotype Javiana, a serotype rarely identified in that state. Although more than 100 persons had documented infections, the outbreak would have gone unnoticed if serotyping was not performed by the state public health laboratory and if the serotype had not been unusual in Minnesota. The low levels of contamination and the diffuse distribution of the millions of pounds of product shipped by the producer would have caused the cases of human illness to blend into the background of anonymous cases of sporadic salmonellosis if the causative organism had been the more common serotype Typhimurium (22). Likewise, the cheese-associated outbreak of listeriosis in California in 1985 would have been missed if most patients had not been seen in one hospital. In this outbreak, product distribution was also broad, extending to 18 or 20 states, but 80% of the cheese was marketed in the city of Los Angeles, Calif., largely to one ethnic group, and so a degree of clustering that was critical to recognition of the outbreak was achieved (53).

Data from surveillance of foodborne disease outbreaks do not accurately reflect the universe of foodborne disease, and they should not be used as the sole source of information for ranking the public health importance of foodborne pathogens, food commodities, and food preparation practices (45). In addition, because thorough outbreak investigations generally require more public resources than are available, this system usually lacks sensitivity. It may not detect an outbreak of diarrheal illness involving thousands of cases randomly distributed in a large urban area (6). Nonetheless, surveillance systems that have been in place for an extended period provide a valid impression of trends in the occurrence of foodborne disease outbreaks in the geographic area under surveillance. They also identify dominant human behaviors associated with illness that can be tracked over time to judge the effectiveness of public health efforts to modify those high-risk behaviors.

Table 20.1 Reported foodborne disease (outbreaks by vehicle, United States, 1973 to 1991)[a]

Vehicle	% of reported outbreaks	% of outbreak-associated cases
Beef	8.2	9.4
Pork	7.5	5.4
Chicken	4.4	5.6
Turkey	3.8	8.1
Shellfish	5.4	3.0
Finfish	15.0	3.3
Eggs	1.2	1.6
Dairy products	2.7	1.9
Fruit and vegetable	6.7	3.6
Other, multiple	45.1	58.1

[a]Includes reports of outbreaks involving known vehicles only.

In 1973 through 1991, 9,502 outbreaks of foodborne disease were reported in the United States. In almost 4,000 of the outbreaks, the causative food was identified (Table 20.1). These 4,000 outbreaks affected nearly 200,000 persons. Seafood accounted for 20% of the outbreaks, compared with 8% for beef, 7% for poultry, and 6% for pork. However, because most of the outbreaks attributed to seafood involved fewer persons than those due to other foods, seafood accounted for only 5% of all reported foodborne outbreak-associated illnesses, compared with 10% each for poultry and beef. In these 19 years, 400 to 700 outbreaks of foodborne diseases were reported in the United States annually. Reported outbreaks peaked in 1982, declined during the early to mid-1980s, and began increasing during the latter part of the 1980s.

From 1973 to 1991, salmonellae caused 34% of reported foodborne outbreaks and 42% of outbreak-associated illnesses in the United States. Other commonly reported causes of illness were *Staphylococcus aureus,* shigellae, and *Clostridium perfringens* (Table 20.2). It is not surprising that bacterial agents were the most commonly reported causes of outbreaks with known etiology, because the laboratory diagnosis of these bacterial infections is easier than for most other causes of foodborne disease. Salmonellae accounted for an ever-greater percentage of outbreaks during this period, whereas *S. aureus* and *C. perfringens* decreased in importance. For specific food vehicles, beef and pork became less commonly associated with outbreaks of disease. Among *Salmonella* outbreaks, beef and turkey became less important, whereas a greater proportion of out-

Table 20.2 Reported foodborne disease (outbreaks by etiology, United States, 1973 to 1991)

Etiologic agent(s)	% of outbreaks
Salmonellae	13.3
Staphylococcus aureus	4.3
Clostridium botulinum	3.0
C. perfringens	2.3
Shigellae	1.3
Other bacteria	2.8
Trichinella spiralis	1.5
Other parasites	0.2
Hepatitis A virus	1.5
Other viruses	0.3
Ciguatoxin	2.9
Scombrotoxin	2.8
Other biotoxins	1.0
Other toxins	1.8
Unknown	61.0

breaks were associated with chicken and eggs; many of the egg-associated outbreaks were due to infections by SE.

From 1985 through 1993, 504 outbreaks of infection by SE were reported in the United States. These outbreaks accounted for 17,925 illnesses, 1,978 hospitalizations, and 62 deaths. The number of outbreaks reported each year increased from 26 in 1985 to 77 in 1989. Sixty-three SE outbreaks were reported in 1993. Outbreaks have continued to occur predominantly in the northeastern United States, but an increasing proportion is occurring outside this area. Of the 233 SE outbreaks for which epidemiologic evidence was sufficient to implicate a food vehicle, 193 (83%) were associated with shell eggs; no reported SE outbreaks have been traced to the use of pasteurized eggs. The most common place of consumption, accounting for approximately 61% of the reported SE outbreaks, has been a commercial venue, i.e., a restaurant, delicatessen, cafeteria, or catered event. Health institutions (including hospitals and nursing homes) accounted for 14% of the outbreaks, 9% of the cases, and 87% of the deaths. The case-fatality rate in these institutions was 3.1%, compared with 0.05% in other settings, a 62-fold difference.

At least half of all outbreaks of foodborne illness reported in the United States have unknown causes, and a percentage of these outbreaks is almost certainly due to microorganisms that we do not yet recognize as foodborne pathogens. However, for many outbreaks, specimens are not cultured or processed properly, and good laboratory procedures are not available for all potential foodborne pathogens.

While not all countries in South and Central America have well-organized epidemiologic surveillance systems for foodborne disease, those that do provide important information on the characteristics of foodborne disease in this region. Cuba reported 186 outbreaks involving 8,813 patients in the first 6 months of 1990. The primary food vehicles were beef, pork, chicken, and cake. The most frequently identified etiologic agents were *S. aureus*, *C. perfringens*, salmonellae, *Entamoeba histolytica*, and *Bacillus cereus*. In 1989, 23 outbreaks involving 293 patients were investigated in Venezuela; implicated food vehicles were dairy products in 57% of outbreaks and seafood in 22%. The incidence rate of foodborne disease in 1988 in Columbia was 29.1/100,000 population, but available data did not specify etiology. Mexico also has an established program for investigating foodborne disease. In 1981 to 1990, 393 outbreaks were investigated, for an average of 39 outbreaks per year. The average outbreak size was 41 persons, and the foods most frequently involved were cheese, pasteurized milk, chicken, and beef. Most of the cases were due to bacterial contamination.

In recent years in Argentina, salmonellosis due to SE has been an important public health problem. This serotype was isolated mainly from poultry products and egg-based foods such as homemade mayonnaise, and these foods were incriminated in many outbreaks of human salmonellosis. From 1986 to 1988, Eiguer et al. (14) studied 39 outbreaks of salmonellosis and isolated 210 strains from feces and 59 from suspected foods; 2,500 persons were affected. The most commonly implicated food was potato salad containing homemade mayonnaise. One outbreak, caused by commercially prepared sandwiches containing homemade mayonnaise, involved hundreds of people. In other outbreaks, the pathogen was isolated from raw milk and water. In 1990 to 1991, 14 outbreaks of infections by various serotypes of *Salmonella* were reported in Buenos Aires, Argentina; as in other countries, most outbreaks occurred during warmer months. Hofer and dos Reis (24) investigated 25 foodborne outbreaks of salmonellosis in southeast and south Brazil from 1982 to 1992 and found that the most frequently detected serotype was Typhimurium (52%) and the food most frequently involved in *Salmonella* transmission was homemade mayonnaise. In contrast to rates of hemolytic-uremic syndrome of 0.2 to 2.8/100,000/year in North America and western Europe, 20 cases of hemolytic-uremic syndrome per 100,000 children 0 to 4 years of age occur in Buenos Aires; the role of Eco157 and other Shiga toxin-producing strains is being investigated.

Foodborne infections with *C. perfringens* and shigellae continue to be a problem in the Americas. For example, an outbreak of shigellosis occurred on a cruise ship in the Caribbean in 1989 (33). A case-control investigation of the outbreak implicated potato salad prepared by a food worker from a country where multidrug-resistant shigellae are common, and it is likely that this employee contaminated the potato salad during its preparation. Limited availability of toilet facilities for the galley crew may have facilitated the spread of infection. The sharp rise of the epidemic curve produced during the investigation suggested a point source epidemic and helped narrow the field of inquiry during interviews of the patients and controls.

Staphylococcal intoxications continue to be the most frequent foodborne disease in some Latin American countries, including Cuba and Venezuela. In Brazil, 30 staphylococcal outbreaks were investigated in 1988 to 1990, and most could be attributed to white cheese and cream-filled cakes; enterotoxigenic *S. aureus* producing toxins A, B, and D was isolated from various foods, including marine and dairy products. As in other countries, the real frequency of staphylococcal intoxication in Brazil is not known because most cases are never reported. In Mexico, 34 foodborne staphylococcal outbreaks were reported in 1981 to 1990.

Although botulism is rarely reported in Latin America and the Caribbean, outbreaks have occurred in Argentina, Brazil, Chile, Mexico, and Peru. Implicated vehicles generally are homemade preserved foods, but commercial products have been implicated in some outbreaks. In Argentina, botulism type A outbreaks have been associated with commercially prepared cheese with onions, spinach, peaches, tomatoes, palm hearts, and pickled vizcacha, but familial outbreaks due to pickled red or green pepper, cucumbers, and vizcacha were more frequent. The first three outbreaks of botulism type B reported in Peru occurred in 1988 and involved pork sausages; the epidemiologic investigation demonstrated that of at least 70 persons exposed, 15 became ill and 2 died.

As in North America, biotoxins transmitted by fish or shellfish are very important in some parts of Latin America and the Caribbean. Outbreaks of paralytic shellfish poisoning have occurred in Argentina, Chile, El Salvador, Guatemala, Honduras, Mexico, Nicaragua, and the United States. Ciguatera is most common in the Caribbean, and outbreaks have occurred in Cuba, the Dominican Republic, and Puerto Rico.

Trichinella spiralis remains an important problem in much of Latin America. Many outbreaks occur, most frequently during the winter in Argentina, and outbreaks also have been reported in Mexico, Chile, and Bolivia. In contrast to infection rates in swine in the United States of approximately 0.1%, in a 1991 survey in one area of Bolivia, *T. spiralis* larvae were observed in tissues of 11.2% of pigs (2, 8). While *Trichinella* infection of domestic pigs is well controlled in much of Europe, recently in Lithuania trichinellosis outbreaks in humans have become more frequent, increasing from 16 reports in 1988 to 76 reports in 1992 (52). Incidence of trichinellosis also increased sharply in Romania in the early 1990s.

Systems for foodborne disease outbreak investigations and surveillance are being developed in Southeast Asia. For example, in Thailand in 1991 to 1992, 19 foodborne disease outbreaks involving 1,033 persons were investigated. *S. aureus, V. parahaemolyticus,* and *E. coli* were the most commonly reported etiologic agents. The most commonly reported food vehicles included seafoods, various desserts, beef, and chicken. For these 19 outbreaks, the incriminated food was prepared most often in private homes, but kitchens in schools, hotels, community centers, and temples were also implicated more than once each.

T. spiralis is an important public health problem in Thailand, especially in swine of the hill tribe and in wild pigs. Of 35 outbreaks involving 1,714 cases and five deaths that were reported in 1985 to 1992, 23 (66%) were associated with pigs from the hill tribe region. In Thailand, trichinellosis is most common in men 25 to 44 years of age, especially those involved in agriculture. Between 1983 and 1993, the prevalence of trichinellosis declined from 1.13 to 0.30/100,000 population, and the case-fatality ratio dropped from 5.26 in 1983 to 0 in 1993. Determining the causes of these decreases will require more intensive epidemiologic investigations.

While this review touches on only a limited number of the causes of outbreaks of infectious foodborne disease in a few countries, it demonstrates some of the differences and similarities of foodborne disease outbreaks in several regions of the world that can be observed through surveillance. As producers all over the world begin to market their products to consumers all over the world, similarities in the distribution of foodborne diseases throughout the world will increase.

The routine collection and collation of information on foodborne disease outbreaks from medical care providers and diagnostic laboratories has several advantages. It does not require costly systems of primary diagnosis and data generation, it can provide nationwide or regional statistics, and it identifies many of the common causes of acute epidemic foodborne disease. However, foodborne disease outbreak surveillance depends on the existence of a reasonably well developed public health infrastructure capable of detecting and appropriately investigating outbreaks and a mechanism for reporting and summarizing the results of those investigations. As currently supported in most countries, foodborne disease outbreak surveillance tends to be insensitive, especially for conditions without good diagnostic tests and for diseases with incubation periods of longer than 1 week. The data obtained are frequently incomplete and of poor quality, and the surveillance fails to completely characterize the epidemiologic features of the outbreaks that it describes.

Investigating some foodborne disease outbreaks by using more extensive epidemiologic and laboratory resources than are required for routine public health efforts addresses some of these deficiencies and is especially useful for emerging foodborne diseases with unknown etiology. For example, Eco157 was first identified as a foodborne pathogen in 1982 during two intensively investigated outbreaks in the United States that were found to be associated with eating hamburgers (18). In the decade since those outbreaks, a number of additional outbreaks have been intensively investigated. Infection with Eco157 has occurred in all age groups, and the distribution of cases by gender generally has reflected the population exposed. Community outbreaks have been most common; other sites include child care centers, schools, and nursing homes. Outbreaks in communities are usually traced to food, generally of bovine origin, such as ground beef, roast beef, and unpasteurized milk. Outbreaks in child care centers have not been traced to food but are instead due to direct spread from person to person, usually after introduction of the etiologic agent by a person with foodborne disease. Contaminated fruits and vegetables, recreational and drinking water, and apple cider have also transmitted illness. Thus, intensive investigations of foodborne disease outbreaks have identified risk factors, food vehicles, and consumer characteristics that increase the likelihood of foodborne illness. For Eco157, person-to-person and waterborne transmission suggests a low infectious dose, and several meat-borne outbreaks appear to have been caused by very low levels of product contamination as well.

In January 1993, an unexpected number of patients with symptoms consistent with Eco157 infection saw physicians at a children's hospital in Washington State (4). Because the medical staff was familiar with Eco157 and the state health department had required reporting of infections, this unusual event was reported and investigated. The investigation identified additional cases and associated infection with eating ground beef at several restaurants of one quick-service restaurant chain. Approximately 600 patients with hemorrhagic colitis and/or hemolytic-uremic syndrome were reported in Washington, and an additional 200 cases were identified, mostly in retrospect, in other western states.

A number of clusters of Eco157 infections have been reported in the United States since this large outbreak stimulated interest in the microorganism and increased the number of stools being examined for Eco157. Some of these clusters may have been unrelated cases that occurred during a short period rather than outbreaks of related cases; making that distinction requires subtyping of isolates and more intensive molecular epidemiologic investigations (15). As clinical laboratories in more areas begin to culture stools for Eco157, more clusters and sporadic cases are being recognized.

Epidemiologic studies in the Philippines have documented the importance of food as a vehicle of infection for cholera. A large study that focused specifically on the role of street vendors found that consumers of mussel soup and a dish of rice noodles with shrimp, meat, and vegetables were at particularly high risk for illness (34). In this study, the difficulty of preparing safe food from heavily contaminated raw ingredients was stressed. In a similar study of cholera associated with street-vended foods in Guatemala, the importance of vendor and consumer knowledge and safe food preparation practices was the focus. Thus, epidemiologic studies can be directed at the food, at its origins, processing, and preparation, and at the behavior of consumers to identify risk factors for foodborne disease.

International transport of foods contaminated with recognized and new pathogens is a consequence of increased global commerce. An intensive epidemiologic investigation by Kapperud et al. (27) showed that contaminated iceberg lettuce imported from Spain caused a large outbreak of *Shigella sonnei* infections in Norway; concurrent investigations identified related cases in other European countries as well. In addition to shigellosis, many of the patients also were infected with enterotoxigenic *E. coli*, *G. lamblia*, salmonellae, or *Campylobacter* species; these infections suggest that the lettuce might have been irrigated with incompletely treated sewage effluent or fertilized with inadequately treated sewage sludge.

Epidemiologic Studies of Laboratory-Confirmed Sporadic Foodborne Disease

Sometimes the impressions created by studying sporadic illness are different from the picture painted by investi-

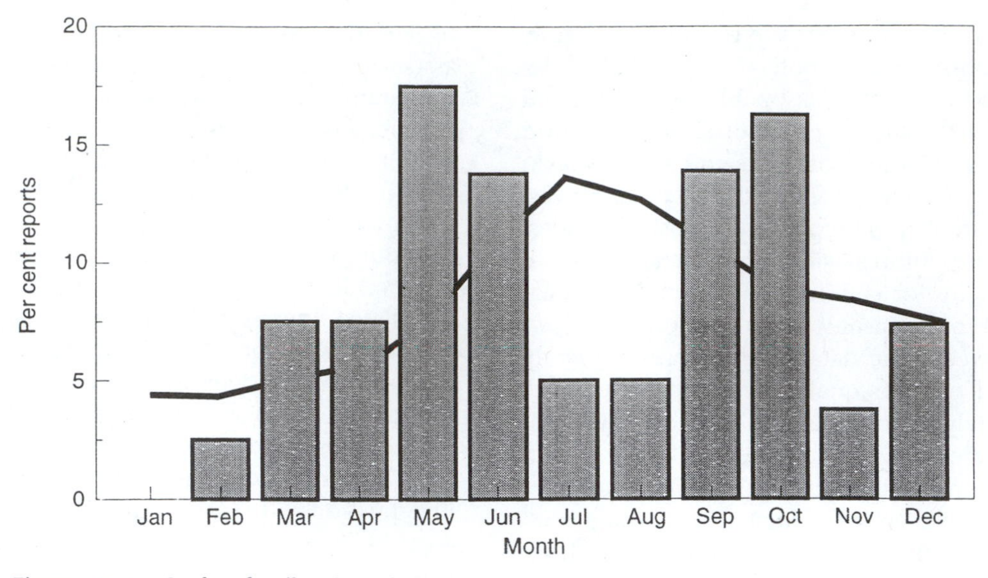

Figure 20.2 Outbreaks (bars) and sporadic cases (line): campylobacteriosis by season, United States.

gating outbreaks. For example, outbreak investigations of *C. jejuni* infections in the United States found that most are associated with the rather uncommon practice of drinking unpasteurized milk. Outbreaks have a marked bimodal distribution, with peaks in May and October and the low point during the summer, a temporal distribution that parallels the occurrence of visits to farms by urban school children. In contrast, most sporadic cases of *Campylobacter* infections are reported during the summer, suggesting that other epidemiologic characteristics, such as the food involved in transmission, may be different as well (Fig. 20.2). Several case-control studies in the United States and Europe have implicated poultry as an important source of sporadic *C. jejuni* infections (56).

In Norway, Kapperud et al. (28) found that sporadic campylobacteriosis was associated with eating sausage cooked at barbecues, contact with pets, and eating poultry brought raw into home kitchens. Handling raw poultry and consuming undercooked poultry meat have been implicated as risk factors for sporadic campylobacteriosis by many investigators in a number of countries; however, the relative importance of various foods as vehicles of sporadic infections by *C. jejuni* has not been consistent among the studies (42).

In the Czech Republic, salmonellosis accounts for more than three-quarters of all reported foodborne infections (52). However, with the exception of a few commercial food service outbreaks and small family clusters, foodborne salmonellosis in the Czech Republic is largely sporadic. Therefore, sporadic case surveillance is critically important to understanding foodborne salmonellosis there and has shown that most cases involve children less than 5 years of age, occur in August to October, and are associated with eggs, egg products, sweets, and meat products.

During the past decade, investigators have learned that *Listeria monocytogenes* can be found in a wide variety of processed foods (53). It survives well in the processing environment and thus is available to contaminate foods between lethal processing steps and protective packaging. If contaminated milk is used in cheese making, *L. monocytogenes* can survive the process and for months afterwards. Growth of the organism during prolonged refrigerated storage of food products presents a considerable risk to susceptible consumers, even though the disease caused by the microorganism is uncommon in the population as a whole. In the United States, early case-control studies of sporadic listeriosis identified undercooked chicken and nonreheated hot dogs as risk factors for infection. While these results stimulated considerable industry interest in determining critical control points during food processing, there was a reluctance to accept epidemiologic data that were unsubstantiated microbiologically. To extend these observations and provide microbiologic support for the link between contaminated foods and sporadic disease, further studies were conducted in the United States in 1988 to 1990. These studies implicated soft cheeses and foods purchased from store delicatessen counters, identified the patient's strain of *L. monocytogenes* in foods from his or her refrigerator for a number of cases, and, in general, validated the epidemiologic conclusions of the earlier investigations. These efforts resulted in (i) an educational campaign on specific dietary recommendations for consumers at high risk and (ii) industry and regulatory measures to prevent contamination of ready-to-eat foods. The most recent epidemiologic studies of sporadic listeriosis in the United States suggest that the combination of risk management strategies has resulted in a 44% reduction in illness and a 48% reduction in deaths associated with listeriosis (54). Thus, epidemiology is also valuable for program evaluation.

ROLE OF EPIDEMIOLOGY IN RISK ASSESSMENT FOR INFECTIOUS FOODBORNE DISEASE

Epidemiology is a field science. Its laboratory is not a controlled setting but the noisy, changing world of human actions and other nonrandom events. Nonetheless, correct inferences from epidemiologic data derived from clinical experiences produce useful estimates of risk. Epidemiologic data linking infections with food sources, information from microbiological surveys of foods, and data from controlled studies together provide the basis for qualitative and quantitative risk assessments. By clarifying the data needs, these studies also are a step in protocol and mathematical model development necessary for improved quantitative risk assessment for infectious foodborne disease.

Risk assessment begins with showing that an infection can, in fact, be foodborne. Investigations of foodborne disease outbreaks are the primary tool for identifying hazards in foods (45). For example, a large outbreak in Nova Scotia in 1981 provided the first evidence for transmission of *L. monocytogenes* by uncooked vegetables. The epidemiologically implicated coleslaw was shown to be contaminated by the epidemic strain of *L. monocytogenes*, providing microbiologic support for the epidemiologic conclusions. Outbreak investigations have identified other foodborne hazards. Epidemiologic investigations in 1982 identified the previously unknown pathogen Eco157 in ground beef as a cause of hemorrhagic colitis. In 1988, an outbreak investigation identified pork chitterlings as a new source of *Y. enterocolitica* infections (31). In addition, outbreak investigations can provide information on rates of hospitalization and death, economic costs of illness, and other indicators of severity of disease. Prospective studies of sporadic foodborne diseases provide additional data on infection rates in the populations studied, which can help rank foodborne pathogens; the proportion of illness attributable to specific foods, which can help rank food vehicles; and other information useful for characterizing risk.

Microbial characteristics and patterns of food handling complicate exposure assessment for infectious foodborne hazards. Although microbiologic data from surveys of foods can be modeled to predict the frequency and amount of pathogenic microorganisms going into the kitchen, prediction of exposure cannot automatically lead to an assumption of human health risk because the hazard may be eliminated by events between the point of its identification and time of consumption of the food. Because it is not always clear how the level of contamination during production, slaughter, processing, and other stages from farm to kitchen correlates with the potential for disease transmission, microbiologic survey data alone are insufficient to judge the risk of exposure to a foodborne hazard.

Data needs for dose response assessment include information on the variability in host susceptibility, attack rates, and pathogenicity and the effects of the characteristics of the food matrix and competing microflora. Epidemiologic studies of foodborne disease outbreaks and sporadic illnesses can help characterize who becomes ill and under which conditions and can therefore sometimes provide valuable clues to dose response. Epidemiologic investigations of outbreaks caused by common foodborne pathogens like shigellae, *Campylobacter* species, and Eco157 have also identified their waterborne and person-to-person transmission; this finding suggests that small numbers of these microorganisms typically can cause disease. Salmonellae appear generally to cause disease when present in higher numbers, but waterborne and person-to-person outbreaks also have occurred, and foods like cheese and chocolate that protect microorganisms during gastric transit have caused outbreaks with very low levels of product contamination. Therefore, although some foodborne pathogens characteristically require few organisms to cause infection and illness, others may generally require a larger inoculum but can, under some conditions, cause illness at lower levels of exposure.

Mathematical approaches have been developed to predict the likelihood of survival of a *Giardia* cyst in a glass of water as it passes through the body defenses to cause infection, and it should be possible to apply the same logic for foodborne pathogens, even though factoring in host susceptibility, vehicle, microorganism, and preparation variables will complicate the model (20). Rose and Sobsey (49) performed a quantitative risk assessment for pathogenic viruses in shellfish, demonstrating the difficulties that these variables present. The quality and quantity of infectious dose data are rarely of the precise form and format required for classic statistical methods, and so expert judgment must be used to create a range of estimates for the proportion of the population that will become ill at increasing levels of exposure. Well-studied outbreaks for which quantitation of the consumed dose is possible will help determine the shape of the distribution of dose response estimates.

To conclude, epidemiologic data are useful for hazard identification, dose response assessment, and severity assessment. Thus, these data help to characterize risk, including identifying the products, processes, and people most likely to be associated with foodborne disease. However, epidemiologic data must be integrated with information from clinical and food microbiologic studies and consumption data to produce credible risk assessments.

SUMMARY

Current epidemiologic data from industrialized and developing countries suggest that the frequency of both illness and death associated with foodborne disease are unacceptably high and that a broad array of potential pathogens contaminate our food supply. These data are generally more qualitative than quantitative, even in the richest nations, and they form an imperfect picture of the occurrence of foodborne disease.

However, provided that we have the analytical capacity to select from available information those data that best characterize the current foodborne disease problems, we can proceed toward a safer food supply while we improve the scientific underpinnings of our risk management decisions. Well-designed and well-implemented surveillance programs will identify the important agents of foodborne disease, determine which foods and behavior are causing disease, determine populations at highest risk for severe foodborne infections, and assess the success of specific intervention efforts. Data from these epidemiologic studies will permit accurate ranking of the public health impact of disease associated with various foods, practices, and behaviors so that we can improve the effectiveness and efficiency of risk-based food safety programs.

References

1. **Arambulo, P. V., and N. Moran.** 1980. Food-transmitted parasitic zoonoses—socio-cultural and technological determinants. *Int. J. Zoonoses* **7**:135–141.
2. **Bailey, T. M., and P. M. Schantz.** 1990. Trends in the incidence and transmission patterns of trichinosis in humans in the United States: comparisons of the periods 1975–1981 and 1982–1986. *Rev. Infect. Dis.* **12**:5–11.
3. **Bean, N. H., and P. M. Griffin.** 1990. Foodborne disease outbreaks in the United States, 1973–1987: pathogens, vehicles, and trends. *J. Food Prot.* **53**:804–817.
4. **Bell, B. P., M. Goldoft, P. M. Griffin, M. A. Davis, D. C. Gordon, P. I. Tarr, C. A. Bartleson, J. H. Lewis, T. J. Barrett, J. G. Wells, R. B. Baron, and J. Kobayashi.** 1994. A multistate outbreak of *Escherichia coli* 0157:H7-associated bloody diarrhea and hemolytic uremic syndrome from hamburgers. *JAMA* **272**:1349–1353.
5. **Beltramino, J. C., H. Sosa, A. Hernandez, L. Ghirardi, and M. Wagener.** 1991. Criptosporidiosis en ninos internados. *Arch. Argent. Pediatr.* **89**:148–154.
6. **Berkelman, R. L., R. T. Bryan, M. T. Osterholm, J. W. LeDuc, and J. M. Hughes.** 1994. Infectious disease surveillance: a crumbling foundation. *Science* **264**:368–370.
7. **Beuchat, L. R.** 1996. Pathogenic microorganisms associated with fresh produce. *J. Food Prot.* **59**:204–216.
8. **Bioland, J., D. Brown, H. R. Gamble, and J. B. McAuley.** 1993. *Trichinella spiralis* in pigs in the Bolivian altiplano. *Vet. Parasitol.* **47**:349–354.
9. **Cooke, E. M.** 1990. Epidemiology of foodborne illness: U.K. *Lancet* **336**:790–793.
10. **Dias, T. C., D. M. Queiroz, E. N. Mendes, and J. N. Peres.** 1990. Chicken carcasses as a source of *Campylobacter jejuni* in Belo Horizonte, Brazil. *Rev. Inst. Med. Trop. Sao Paulo* **32**:414–418.
11. **Ebel, E. D., M. J. David, and J. Mason.** 1992. Occurrence of *Salmonella* enteritidis in the U. S. commercial egg industry: report on a national spent hen survey. *Avian Dis.* **36**:646–654.
12. **Ebel, E. D., J. Mason, L. A. Thomas, K. E. Ferris, M. G. Beckman, D. R. Cummins, L. Scroeder-Tucker, W. D. Sutherlin, R. L. Glasshoff, and N. M. Smithhisler.** 1993. Occurrence of *Salmonella* enteritidis in unpasteurized liquid egg in the United States. *Avian Dis.* **37**:135–142.
13. **Echeverria, P., D. N. Taylor, U. Lexsomboon, M. Bhaibulaya, N. R. Blacklow, K. Tamura, and R. Sakazaki.** 1989. Case-control study of endemic diarrhea disease in Thai children. *J. Infect. Dis.* **159**:543–548.
14. **Eiguer, T., M. I. Caffer, and G. B. Fronchkowsky.** 1990. Importancia de la *Salmonella enteritidis* en brotes de enfernedades transmitidas por alimentos en Argentina, anos 1986–1988. *Rev. Argent. Microbiol.* **22**:41–46.
15. **Feng, P.** 1995. *Escherichia coli* serotype O157:H7: novel vehicles of infection and emergence of phenotypic variants. *Emerging Infect. Dis.* **1**:47–52.
16. **Garcia, V. E., L. M. Legaspi, P. Coello Ramirez, P. J. Gonzalez, and S. Aguilar Benavides.** 1991. *Cryptosporidium* sp. in 300 children with and without diarrhea. *Arch. Invest. Med.* (Mexico) **22**:329–332.
17. **Grados, O., N. Bravo, R. E. Black, and J. P. Butzler.** 1989. Diarrhea pediatrica por *Campylobacter* debida a la exposicion domestica a pollos vivos en Lima, Peru. *Bol. Of. Sanit. Panam.* **109**:205–213.
18. **Griffin, P. M., and R. V. Tauxe.** 1991. The epidemiology of infections caused by *Escherichia coli* O157:H7, other enterohemorrhagic *E. coli*, and the associated hemolytic uremic syndrome. *Epidemiol. Rev.* **13**:60–98.

19. **Guderian, R. H., R. G. Ordonez, and R. Bossano.** 1987. Diarrhea aguda asociada a *Campylobacter* y otros agentes patogenos en Quito, Ecuador. *Bol. Of. Sanit. Panam.* **102:** 333–339.

20. **Haas, C. N.** 1983. Estimation of risk due to low doses of microorganisms: a comparison of alternative methodologies. *Am. J. Epidemiol.* **118:**573–582.

21. **Hedberg, C. W., J. A. Korlath, J. Y. D'Aoust, K. E. White, W. L. Schell, M. R. Miller, D. N. Cameron, K. L. MacDonald, and M. T. Osterholm.** 1992. A multistate outbreak of *Salmonella javiana* and *Salmonella oranienburg* infections due to consumption of contaminated cheese. *JAMA* **268:**3203–3207.

22. **Hedberg, C. W., K. L. MacDonald, and M. T. Osterholm.** 1994. Changing epidemiology of food-borne disease: a Minnesota perspective. *Clin. Infect. Dis.* **18:**671–682.

23. **Hernandez, H., K. Kahler, R. Salazar, and M. A. Rios.** 1994. Prevalence of thermotolerant species of *Campylobacter* and their biotype in children and domestic birds and dogs in southern Chile. *Rev. Inst. Med. Trop. Sao Paulo* **36:**433–436.

24. **Hofer, E., and E. M. F. dos Reis.** 1994. *Salmonella* serovars in food poisoning episodes recorded in Brazil from 1982 to 1991. *Rev. Inst. Med. Trop. Sao Paulo* **36:**7–9.

25. **Institute of Medicine.** 1992. *Emerging Infections: Microbial Threats to Health in the United States.* National Academy Press, Washington, D.C.

26. **Jerngklinchan, J., C. Koowatunanukul, K. Daengprom, and K. Saitanu.** 1994. Occurrence of salmonellae in raw broilers and their products in Thailand. *J. Food Prot.* **57:**808–810.

27. **Kapperud, G., L. M. Rorvik, V. Hasseltvedt, E. A. Hoiby, B. G. Iversen, K. Staveland, G. Johnson, J. Leitao, H. Herikstad, Y. Andersson, G. Langeland, B. Gondrosen, and J. Lassen.** 1995. Outbreak of *Shigella sonnei* infection traced to imported iceberg lettuce. *J. Clin. Microbiol.* **33:**609–614.

28. **Kapperud, G., E. Skjerve, N. H. Bean, S. M. Ostroff, and L. Lassen.** 1992. Risk factors for sporadic *Campylobacter* infections: results of a case-control study in southeastern Norway. *J. Clin. Microbiol.* **30:**3117–3121.

29. **Karmali, M. A., and P. C. Fleming.** 1979. *Campylobacter* enteritis in children. *J. Pediatr.* **94:**527–533.

30. **Kelsey, J. L., W. D. Thompson, and A. S. Evans.** 1986. *Methods in Observational Epidemiology.* Oxford University Press, New York.

31. **Lee, L. A., A. R. Gerber, D. R. Lonsway, J. D. Smith, G. P. Carter, N. D. Puhr, C. M. Parrish, R. K. Sikes, R. J. Finton, and R. V. Tauxe.** 1990. *Yersinia enterocolitica* O:3 infections in infants and children associated with the household preparation of chitterlings. *N. Engl. J. Med.* **322:**984–987.

32. **Lee, L. A., N. D. Puhr, E. K. Maloney, N. H. Bean, and R. V. Tauxe.** 1994. Increase in antimicrobial-resistant *Salmonella* infections in the United States, 1989–1990. *J. Infect. Dis.* **170:**128–134.

33. **Lew, J. F., D. L. Swerdlow, M. E. Dance, P. M. Griffin, C. A. Bopp, M. J. Gillenwater, T. Mercatante, and R. I. Glass.** 1991. An outbreak of shigellosis aboard a cruise ship caused by a multiple-antibiotic-resistant strain of *Shigella flexneri. Am. J. Epidemiol.* **134:**413–420.

34. **Lim-Quizon, M. C., R. M. Benabaye, F. M. White, M. M. Dayrit, and M. E. White.** 1994. Cholera in metropolitan Manila: foodborne transmission via street vendors. *Bull. W.H.O.* **72:**745–749.

35. **Looareesuwan, S.** 1990. *Textbook of Tropical Medicine.* Raumtus Ltd., Bangkok, Thailand.

36. **MacDonald, K. L., M. L. Cohen, and N. T. Hargrett-Bean.** 1987. Changes in antimicrobial resistance of *Salmonella* isolated from humans in the United States. *JAMA* **258:**1496–1499.

37. **MacDonald, K. L., M. J. O'Leary, M. L. Cohen, P. Norris, J. G. Wells, E. Noll, J. M. Kobayashi, and P. A. Blake.** 1988. *Escherichia coli* O157:H7, an emerging gastrointestinal pathogen: results of a one-year, prospective, population-based study. *JAMA* **259:**3567–3570.

38. **Marshall, W. F., C. A. McLimans, P. K. W. Yu, F. J. Allerberger, R. E. Van Scoy, and J. P. Anhalt.** 1990. Results of a 6-month survey of stool cultures for *Escherichia coli* O157:H7. *Mayo Clin. Proc.* **65:**787–792.

39. **Mata, L., H. Bolanos, D. Pizarro, and M. Vives.** 1983. Diarrhea associated with rotaviruses, enterotoxigenic *Escherichia coli, Campylobacter,* and other agents in Costa Rica children. *Am. J. Med. Hyg.* **32:**146–153.

40. **Mata, L., H. Bolanos, D. Pizarro, et al.** 1984. Cryptosporidiosis in children from some highland Costa Rica rural and urban areas. *Am. J. Trop. Med. Hyg.* **33:**24–29.

41. **Millard, P. S., K. F. Gensheimer, D. G. Addiss, D. M. Sosin, G. A. Beckett, A. Houck-Jankoski, and A. Hudson.** 1994. An outbreak of cryptosporidiosis from fresh-pressed apple cider. *JAMA* **272:**1592–1596.

42. **National Advisory Committee on Microbiological Criteria for Foods.** 1995. *Campylobacter jejuni/coli. J. Food Prot.* **57:**1101–1121.

43. **Nguyen-the, C., and F. Carlin.** 1994. The microbiology of minimally processed fresh fruits and vegetables. *Crit. Rev. Food. Sci. Nutr.* **34:**371–401.

44. **PAHO/WHO Caribbean Epidemiology Center.** 1989. Giardiasis in the Caribbean. *CSR CAREC Surveillance Rep.* **15:**1.

45. **Potter, M. E.** 1994. The role of epidemiology in risk assessment: a CDC perspective. *Dairy Food Environ. Sanit.* **14:**738–741.

46. **Quevedo, F., and A. S. Thakur.** 1990. *Food-Borne Parasitic Diseases.* Scientific and Technical Monographs 12/Rev 1. PAHO/WHO, Martinez, Buenos Aires.

47. **Ramirez, R., H. Schenone, and M. Galdames.** 1972. Frecuencia en Chile de las infecciones humanas por protozoos y helmintos intestinales (1962–1972). *Bol. Chil. Parasitol.* **27:**116–118.

48. **Richardson, B. W., and W. H. Frost.** 1936. *Snow on Cholera.* The Commonwealth Fund, New York.

49. **Rose, J. B., and M. D. Sobsey.** 1993. Quantitative risk assessment for viral contamination of shellfish and coastal waters. *J. Food Prot.* **56:**1043–1050.

50. **Schantz, P. M., and J. McAuley.** 1991. Current status of food-borne parasitic zoonoses in the United States. *Southeast Asian J. Trop. Med. Public Health* **22**(Suppl.):65–71.

51. **Schenone, H., R. Villarroel, A. Rojas, and R. Ramirez.** 1982. Epidemiology of human cysticercosis in Latin America, p. 25–38. *In* A. Flisser, K. Williams, J. P. Laclette,

C. Larralde, C. Ridauta, and F. Beltran (ed.), *Cysticercosis: Present State of Knowledge and Perspectives.* Academic Press, New York.

52. **Schmidt, K.** 1995. *WHO Surveillance Programme for Control of Foodborne Infections and Intoxications in Europe. Sixth Report, 1990–1992.* FAO/WHO Collaborating Centre for Research and Training in Food Hygiene and Zoonoses, Berlin.
53. **Schuchat, A., B. Swaminathan, and C. V. Broome.** 1991. Epidemiology of human listeriosis. *Clin. Microbiol. Rev.* **4**:169–183.
54. **Tappero, J. W., A. Schuchat, K. A. Deaver, L. Mascola, and J. D. Wenger.** 1995. Reduction in the incidence of human listeriosis in the United States. *JAMA* **273**:1118–1122.
55. **Tauxe, R. V.** 1991. *Salmonella:* a postmodern pathogen. *J. Food Prot.* **54**:563–568.
56. **Tauxe, R. V.** 1992. Epidemiology of *Campylobacter jejuni* infections in the United States and other industrialized nations, p. 9–19. *In* I. Nachamkin, M. J. Blaser, and L. S. Tompkins (ed.), *Campylobacter jejuni: Current Status and Future Trends.* American Society for Microbiology, Washington, D.C.
57. **Teutsch, S. M., and R. E. Churchill.** 1994. *Principles and Practice of Public Health Surveillance.* Oxford University Press, New York.
58. **Todd, E. C. D.** 1996. Worldwide surveillance of foodborne disease: the need to improve. *J. Food Prot.* **59**:82–92.
59. **Wachsmuth, I. K., G. M. Evins, P. I. Fields, O. Olsvik, T. Popovic, C. A. Bopp, J. G. Wells, C. Carrillo, and P. A. Blake.** 1993. The molecular epidemiology of cholera in Latin America. *J. Infect. Dis.* **167**:621–626.
60. **Wachsmuth, K.** 1986. Molecular epidemiology of bacterial infections: examples of methodology and investigation of outbreaks. *Rev. Infect. Dis.* **8**:682–692.
61. **Weber, J. T., R. G. Hibbs, A. Darwish, B. Mishu, A. L. Corwin, M. Rakha, C. L. Hatheway, S. E. Sharkawy, S. A. El-Rahim, M. F. S. El-hamd, J. E. Sarn, P. A. Blake, and R. V. Tauxe.** 1993. A massive outbreak of type E botulism associated with traditional salted fish in Cairo. *J. Infect. Dis.* **167**:451–454.
62. **Weber, J. T., E. D. Mintz, R. Canizares, A. Semiglia, I. Gomez, R. Sempertegui, A. Davila, K. D. Greene, N. D. Puhr, D. N. Cameron, F. C. Tenover, T. J. Barrett, N. H. Bean, C. Ivey, R. V. Tauxe, and P. A. Blake.** 1994. Epidemic cholera in Ecuador: multidrug-resistance and transmission by water and seafood. *Epidemiol. Infect.* **112**:1–11.
63. **Yamazaki, S., T. Inada, S. Inouye, N. Kagie, K. Miyamura, A. Nakamura, T. Shimada, H. Watanabe, S. Yamadera, and K. Yamashita.** 1992. Annual report on findings of infectious agents in Japan. *Jpn. J. Med. Sci. Biol.* **45**(Suppl.):1–143.

Mycotoxigenic Molds

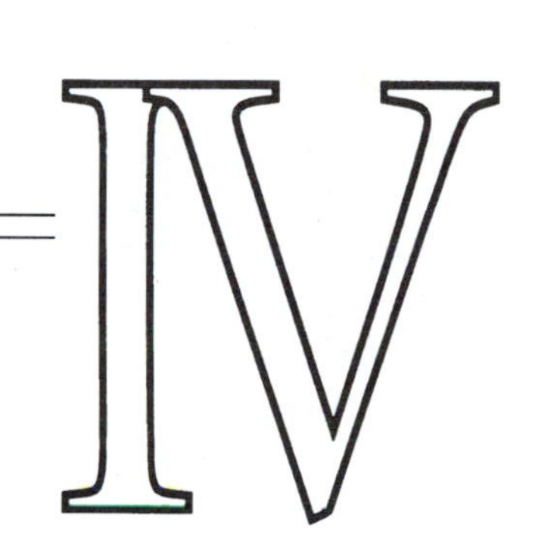

Ailsa D. Hocking

21 Toxigenic *Aspergillus* Species

For centuries it has been known that eating particular species of mushrooms can cause illness and death, but the recognition that some common food spoilage molds are responsible for animal and human diseases and deaths has come relatively recently. Although it has been known since 1940 that a *Penicillium* species was the source of toxicity in "yellow" rice in Japan (67; chapter 22, this volume), mycotoxins were brought to the attention of scientists in the Western world in the early 1960s with the outbreak of turkey X disease in England, which killed about 100,000 turkeys and other farm animals. The cause of this disease was traced to peanut meal in the feed, which was heavily contaminated with *Aspergillus flavus*. Analysis of the feed revealed that a group of fluorescent compounds, later named aflatoxins, were responsible for this outbreak, for the deaths of large numbers of ducklings in Kenya, and for widespread hepatoma in hatchery-reared trout in California which occurred more or less simultaneously (2, 5, 94).

Aspergillus was first described almost 300 years ago and is an important genus in foods. Although a few species have been harnessed in production of food (e.g., *A. oryzae* in soy sauce manufacture), most *Aspergillus* species occur in foods as spoilage or biodeterioration fungi. They are extremely common in stored commodities such as grains, nuts, and spices and occur more frequently in tropical and subtropical than in temperate climates (82).

The genus *Aspergillus* contains several species capable of producing mycotoxins, although the mycotoxin literature of the past 35 years has been dominated by papers on aflatoxins: chemistry, detection, toxicology, production, occurrence in foods, and regulatory aspects. The existence of other toxigenic *Aspergillus* species means that correct identification of isolates from foods and knowledge of the ecology of these molds are of paramount importance.

TAXONOMY

Aspergillus is a large genus containing more than 100 recognized species, most of which grow well in laboratory culture. There are a number of teleomorphic (ascosporic) genera which have *Aspergillus* conidial states (anamorphs), but the only two of real importance in foods are the xerophilic genus *Eurotium* (previously known as the *Aspergillus glaucus* group) and *Neosartorya* species, which produce heat-resistant ascospores and cause spoilage in heat-processed foods, mainly fruit products (82).

The most widely used taxonomy for *Aspergillus* is that of Raper and Fennell (88), although some of their concepts are now out of date (91, 92), and many new species have since been described (87). Gams et al. (41) proposed a nomenclaturally correct classification for species within the genus *Aspergillus*, grouping species into six subgenera which are subdivided into sections. A modern taxonomy of the most common *Aspergillus* species, including those important in foods, is provided by Klich and Pitt (50).

In addition to traditional morphological taxonomic techniques, chemical techniques such as isoenzyme patterns (23), secondary metabolites (36, 37), ubiquinone systems (58), and molecular techniques (30, 68, 69, 71) have been used to clarify relationships within the genus *Aspergillus*.

ISOLATION, ENUMERATION, AND IDENTIFICATION

Techniques for the isolation and enumeration of *Aspergillus* species from foods have been described in detail (82, 90). Antibacterial media containing compounds to inhibit or reduce mold colony spreading, such as dichloran rose bengal chloramphenicol (DRBC) agar (48) or dichloran–18% glycerol (DG18) agar (47), are recommended as media for enumerating fungi in foods (49, 89).

The most recent complete taxonomy is that of Raper and Fennell (88), but keys and descriptions of the most common foodborne *Aspergillus* species can be found elsewhere (50, 54, 82, 90). Identification of *Aspergillus* species requires growth on media developed for this purpose, including Czapek agar, a defined medium based on mineral salts, or a derivative such as Czapek yeast extract agar, and malt extract agar. Growth on Czapek yeast extract–20% sucrose agar can be a useful aid in identifying species of *Aspergillus* (50).

Unlike *Penicillium* species, *Aspergillus* species are conveniently "color coded," and the color of the conidia can be a very useful starting point in identification, at least to the section level. As well as conidial color, microscopic morphology is important in identification. Phialides (cells producing conidia) may be produced directly from the swollen apex (vesicle) of long stalks (stipes), or there may be an intermediate row of supporting cells (metulae). Correct identification of *Aspergillus* species is an essential prerequisite to assessing the potential for mycotoxin contamination in a commodity, food, or feedstuff.

SIGNIFICANT *ASPERGILLUS* MYCOTOXINS

Observations on significant *Penicillium* mycotoxins (chapter 22) can also be applied to *Aspergillus* mycotoxins, although identification of *Aspergillus* species is less frequently incorrect. The exception to this is the confusion in the literature over the identity of *Aspergillus* isolates reported to produce aflatoxins, with *A. flavus* and *A. parasiticus* frequently misidentified or misreported (51). The close taxonomic relationship between the genera *Penicillium* and *Aspergillus* is reflected in the common production of many mycotoxins.

Almost 50 species of *Aspergillus* have been listed as capable of producing toxic metabolites (19), but the *Aspergillus* mycotoxins of greatest significance in foods and feeds are aflatoxins (produced by *A. flavus, A. parasiticus,* and *A. nomius*), ochratoxin A from *A. ochraceus* and related species, sterigmatocystin (produced primarily by *A. versicolor* but also by *Emericella* species), and cyclopiazonic acid (*A. flavus* is the primary source, but it is also reported to be produced by *A. tamarii*). Citrinin, patulin, and penicillic acid may also be produced by certain *Aspergillus* species, and tremorgenic toxins are produced by *A. terreus* (territrems), *A. fumigatus* (fumitremorgens), and *A. clavatus* (tryptoquivaline) (35, 55, 96).

Like *Penicillium* species, *Aspergillus* species produce toxins that exhibit a wide range of toxicities, with the most significant effects being long term. Aflatoxin B_1 is perhaps the most potent liver carcinogen known for a wide range of animal species, including humans. Ochratoxin A and citrinin both affect kidney function. Cyclopiazonic acid has a wide range of effects (18), and tremorgenic toxins such as territrems affect the central nervous system. Table 21.1 lists the most significant toxins produced by *Aspergillus* species and their toxic effects.

A. FLAVUS AND *A. PARASITICUS*

Undoubtedly, the most important group of toxigenic aspergilli are the aflatoxigenic molds, *A. flavus, A. parasiticus,* and the recently described but much less common species *A. nomius,* all of which are classified in *Aspergillus* sect. *Flavi* (41). Although these three species are closely related and have many similarities, a number of characteristics may be used in their differentiation (Table 21.2). The suite of toxins produced by these three species is species specific (51, 56). *A. flavus* can produce aflatoxins B_1 and B_2 and cyclopiazonic acid, but only a proportion of isolates are toxigenic. *A. parasiticus* produces aflatoxins B_1, B_2, G_1, and G_2 but not cyclopiazonic acid, and almost all isolates are toxigenic. *A. nomius* is morphologically similar to *A. flavus* but, like *A. parasiticus,* produces B and G aflatoxins without cyclopiazonic acid. Because this species appears to be uncommon, it has been little studied; thus, the potential toxigenicity of isolates is not known, and the practical importance of this species is hard to assess. *A. flavus* and *A. parasiticus* are closely related to *A. oryzae* and *A. sojae,* species that

Table 21.1 Significant mycotoxins produced by *Aspergillus* species and their toxic effects

Mycotoxin(s)	Toxicity	Species producing
Aflatoxins B_1 and B_2	Acute liver damage, cirrhosis, carcinogenic (liver), teratogenic, immunosuppressive	*A. flavus, A. parasiticus, A. nomius*
Aflatoxins G_1 and G_2	Effects similar to those of B aflatoxins: G_1 toxicity less than that of B_1 but greater than that of B_2	*A. parasiticus, A. nomius*
Cyclopiazonic acid	Degeneration and necrosis of various organs, tremorgenic, low oral toxicity	*A. flavus*
Ochratoxin A	Kidney necrosis (especially pigs), teratogenic, immunosuppressive, ?carcinogenic	*A. ochraceus* and related species
Sterigmatocystin	Acute liver and kidney damage, carcinogenic (liver)	*A. versicolor, Emericella* spp.
Fumitremorgens	Tremorgenic (rats and mice)	*A. fumigatus*
Territrems	Tremorgenic (rats and mice)	*A. terreus*
Tryptoquivalines	Tremorgenic	*A. clavatus*
Cytochalasins	Cytotoxic	*A. clavatus*
Echinulins	Feed refusal (pigs)	*Eurotium chevalieri, E. amstelodami*

are used in the manufacture of fermented foods and do not produce toxins (30). Obviously, accurate differentiation of related species within section *Flavi* is important in order to determine the potential for toxin production and the types of toxins likely to be present.

Detection and Identification

The most effective medium for rapid detection of aflatoxigenic molds is Aspergillus flavus and parasiticus agar (85), a medium formulated specifically for this purpose. Under the incubation conditions specified for this medium (30°C for 42 to 48 h), *A. flavus, A. parasiticus,* and *A. nomius* produce a bright orange-yellow colony reverse which is diagnostic and readily recognized, even by untrained observers. *A. flavus* and related species also grow well on general purpose yeast and mold enumeration media such as DRBC agar or DG18 agar and are relatively easily recognized by experienced personnel. Moderately deep, yellow-green colonies with moplike fruiting structures (best observed under a stereomicroscope) can be counted as *A. flavus* or *A. parasiticus.*

A. flavus and *A. parasiticus* (Fig. 21.1) are easily distinguished from other *Aspergillus* species by using appropriate media and methods (50, 82, 90). Though differentiating between these two species and *A. nomius* is more difficult, it is important because of their different mycotoxin profiles and toxin-producing potentials. The texture of the conidial walls is a reliable differentiating feature: conidia of *A. flavus* (Fig. 21.1A and 21.1B) are usually smooth to finely roughened, while those of *A. parasiticus* (Fig. 21.1C and 21.1D) are clearly rough when observed under an oil immersion lens. Screening isolates for aflatoxin production can also be used to differentiate the species. Cultures can be grown on coconut-cream agar and observed under UV light (29), or a simple agar plug technique coupled with thin layer chromatography (TLC) (34) can be used to screen cultures for aflatoxin production as an aid to identification. The combinations of characteristics most useful in differentiation between the three aflatoxigenic species are summarized in Table 21.2.

Aflatoxins

Aflatoxins are difuranocoumarin derivatives (12). Aflatoxins B_1, B_2, G_1, and G_2 are produced in nature by the molds discussed above. The letters B and G refer to the fluorescent colors (blue and green, respectively) observed under long-wave UV light, and the subscripts 1 and 2 refer to their separation patterns on TLC plates. Aflatoxins M_1 and M_2 are produced from their respective B aflatoxins by hydroxylation in lactating animals and are excreted in milk at a rate of approximately 1.5% of the rate of ingested B aflatoxins (39).

Table 21.2 Distinguishing features of *A. flavus, A. parasiticus,* and *A. nomius*

Species	Conidia	Sclerotia	Toxins
A. flavus	Smooth to moderately roughened, variable in size	Large, globose	Aflatoxins B_1 and B_2, cyclopiazonic acid
A. parasiticus	Conspicuously roughened, little variation in size	Large, globose	Aflatoxins B and G
A. nomius	Similar to *A. flavus*	Small, elongated (bullet shaped)	Aflatoxins B and G

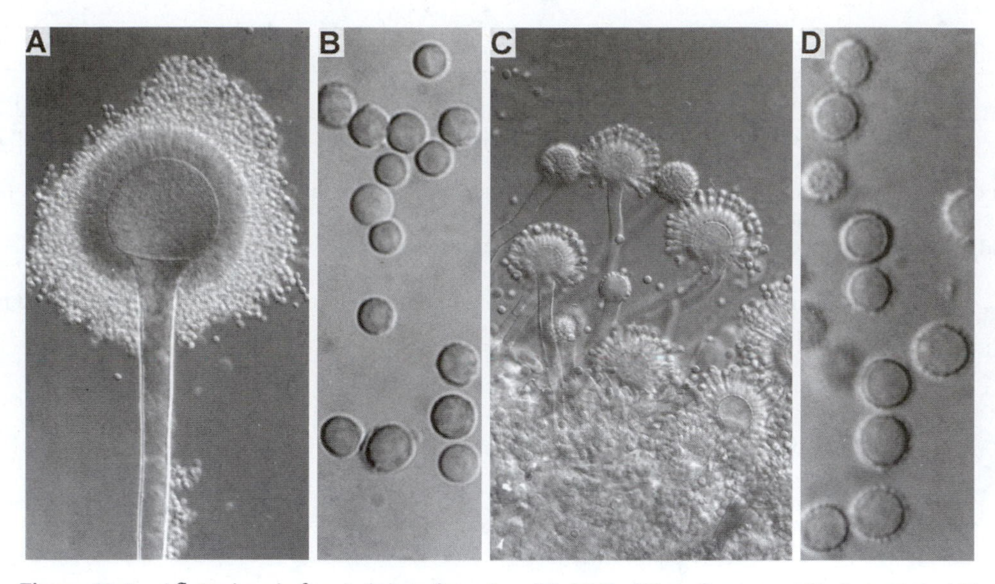

Figure 21.1 Aflatoxigenic fungi. (A) *A. flavus* head (×216); (B) *A. flavus* conidia (×1,350); (C) young *A. parasiticus* heads (×216); (D) *A. parasiticus* conidia (×1,350).

Aflatoxins are synthesized through the polyketide pathway, beginning with condensation of an acetyl unit with two malonyl units and the loss of carbon dioxide (8). The resultant hexanoate is enzyme bound and reacts successively with seven malonate units to yield a polyketide intermediate which undergoes cyclization and aromatization to give norsolorinic acid. Norsolorinic acid undergoes several metabolic conversions to form aflatoxin B_1 (31).

Toxicity

Aflatoxins are both acutely and chronically toxic in animals and humans, producing acute liver damage, liver cirrhosis, tumor induction, and teratogenesis (101). Perhaps of greater significance to human health are the immunosuppressive effects of aflatoxins, either alone or in combination with other mycotoxins (80). Immunosuppression can increase susceptibility to infectious diseases, particularly in populations in which aflatoxin ingestion is chronic (25), and can interfere with production of antibodies in response to immunization in animals (80) and perhaps also in children.

Acute aflatoxicosis in humans is rare (97); however, several outbreaks have been reported. In 1967, 26 people in two farming communities in Taiwan became ill with apparent food poisoning. Nineteen were children, three of whom died. Although postmortems were not performed, rice from affected households contained about 200 μg of aflatoxin B_1 per kg, which was probably responsible for the outbreak. An outbreak of hepatitis in India in 1974 which affected 400 people, 100 of whom died, almost certainly was caused by aflatoxins (57). The outbreak was traced to corn heavily contaminated with *A. flavus* and containing up to 15 mg of aflatoxins per kg. It was calculated that affected adults may have consumed 2 to 6 mg on a single day, implying that the acute lethal dose for adult humans is of the order of 10 mg. More recently, deaths of 13 Chinese children in the northwestern Malaysian state of Perak, apparently due to ingestion of contaminated noodles, were reported (66). Aflatoxins were confirmed in postmortem tissue samples from the patients.

Undoubtedly, the greatest direct impact of aflatoxins on human health is their potential to induce liver cancer (65). Human liver cancer has a high incidence in central Africa and parts of Southeast Asia, and studies in several African countries and Thailand have shown a correlation between aflatoxin intake and the occurrence of primary liver cancer (108). No such correlation could be demonstrated for populations in rural areas of the United States, despite the occurrence of considerable amounts of aflatoxins in corn (102, 103). This apparent anomaly may be explained by the relationship between the roles of hepatitis B virus and aflatoxins in induction of human liver cancer. Aflatoxins and hepatitis B virus are apparently cocarcinogens, and the occurrence of both predisposing factors greatly increases the probability of human liver cancer. However, evidence supports the hypothesis that high aflatoxin intakes are causally related to high incidences of cancer, even in the absence of hepatitis B (44, 45, 78).

In animals, aflatoxins have been shown to cause various syndromes, including cancer of the liver, colon, and kidneys in rats, mice, monkeys, ducklings, and trout (31). Regular low-level intake of aflatoxins can lead to poor feed conversion, low weight gain, and poor milk yields in cattle (10).

Mechanism of Action

Aflatoxin B_1 is metabolized by the microsomal mixed function oxidase system in the liver, leading to the formation of highly reactive intermediates, one of which is 2,3-epoxy-aflatoxin B_1 (70). Binding of these reactive intermediates to DNA results in disruption of transcription and abnormal cell proliferation, leading to mutagenesis or carcinogenesis (31). Aflatoxins also inhibit oxygen uptake in the tissues by acting on the electron transport chain and inhibiting various enzymes, resulting in decreased production of ATP (31).

Toxin Detection

Aflatoxins can be detected by chemical or biological methods. Chemical determination of aflatoxins has become fairly standardized (4, 72). Samples are extracted with organic solvents such as chloroform or methanol in combination with small amounts of water. The presence of fats, lipids, or pigments in extracts reduces the efficiency of the separation, and solvents such as hexane may be used to partition these components from the extract (31). Extracts are further cleaned up by passage through a silica gel column or proprietary separating cartridge. The extract is then concentrated, usually by evaporation under nitrogen, and separated by TLC or high-performance liquid chromatography (HPLC). TLC is most often used, and aflatoxins are visualized under UV light and quantified by visual comparison with known concentrations of standards or by fluorimetry. HPLC with detection by UV light or absorption spectrometry provides a more readily quantifiable (although not necessarily more accurate or sensitive) technique. Immunoassay techniques including enzyme-linked immunosorbent assays (ELISA) and dipstick tests for aflatoxin detection have been developed (14, 61, 79, 95, 105), and a number of kits are now commercially available.

Occurrence of Aflatoxigenic Molds and Aflatoxins

A. flavus is widely distributed in nature, but *A. parasiticus* is probably less widespread, the actual extent of its occurrence being complicated by the tendency for both species to be reported indiscriminately as *A. flavus.* In a wide-ranging survey of the mycoflora of commodities in Thailand (83, 84), *A. flavus* was one of the most commonly occurring molds in nuts and oilseeds, but *A. parasiticus* was rarely encountered. *A. flavus* was the most common species in peanuts and the second most common (after *Fusarium moniliforme*) in corn. *A. nomius* was reported from both commodities, the first published report of the occurrence of this species in food (83). Soybeans, mung beans, sorghum, and other commodities also contained considerable populations of *A. flavus,* but *A. parasiticus* was rare (84).

A. flavus and *A. parasiticus* have a strong affinity for nuts and oilseeds. Corn, peanuts, and cottonseed are the most important crops invaded by these molds, and in many instances, invasion takes place before harvest, not during storage as was once believed. Peanuts are invaded while still in the ground if the crop suffers drought stress or related factors (20, 81, 93). In corn, insect damage to developing kernels allows entry of aflatoxigenic molds, but invasion can also occur through the silks of developing ears (62). Cotton seeds are invaded through the nectaries (52).

Cereals and spices are common substrates for *A. flavus* (82), but aflatoxin production in these commodities is almost always a result of poor drying, handling, or storage, and aflatoxin levels are rarely significant. Significant amounts of aflatoxins can occur in peanuts, corn, and other nuts and oilseeds, particularly in some tropical countries where crops may be grown under marginal conditions and where drying and storage facilities are limited (3, 26, 65).

Factors Affecting Growth and Toxin Production

A. flavus and *A. parasiticus* have similar growth patterns. Both grow at temperatures ranging from 10 to 12°C to 42 to 43°C, with an optimum near 32 to 33°C (6), with aflatoxins being produced at 12 to 40°C (27, 53, 75). The optimum water activity (a_w) for growth is near 0.99, with minima reported as 0.80 (6) to 0.82 to 0.83 (75, 86). Aflatoxins are generally produced in greater quantity at higher a_w values (0.98 to 0.99), with toxin production apparently ceasing at or near a_w 0.85 (27, 53, 75). Although growth can take place over the pH range of just above 2.0 up to 10.5 (*A. parasiticus*) or 11.2 (*A. flavus*) (111), aflatoxin production has been reported for *A. parasiticus* only between pH 3.0 and 8.0, with an optimum near pH 6.0 (11). Reduction of available oxygen by modified atmosphere packaging of foods in barrier film or with oxygen scavengers can inhibit aflatoxin formation by *A. flavus* and *A. parasiticus* (32, 33).

Control and Inactivation

Control of aflatoxins in commodities generally relies on screening techniques which separate affected nuts, grains, or seeds. In corn, cottonseed, and figs, screening for aflatoxins can be done by examination under UV light: those particles which fluoresce may be contaminated. Because all peanuts fluoresce when exposed to UV light, this method is not useful for detecting kernels that are contaminated with aflatoxins. Peanuts containing aflatoxins are segregated by electronic color-sorting machines which detect discolored kernels.

Aflatoxins can be partially destroyed by various chemical treatments. Oxidizing agents such as ozone and hydrogen peroxide have been demonstrated to remove aflatoxins from contaminated peanut meals (110). Although ozone was effective in removal of aflatoxins B_1 and G_1 under the conditions applied (100°C for 2 h), there was no effect on aflatoxin B_2, and the treatment decreased the lysine content of the meal (28). Treatment with hydrogen peroxide resulted in 97% destruction of aflatoxin in defatted peanut meal (100). The most practical chemical method of aflatoxin destruction appears to be the use of anhydrous ammonia gas at elevated temperatures and pressures, with a 95 to 98% reduction in total aflatoxin in peanut meal reported (107). This technique is used commercially for detoxification of animal feeds in Senegal, France, and the United States (70, 107).

The ultimate method of control of aflatoxins in commodities, particularly peanuts, is to prevent the plants from becoming infected with aflatoxigenic strains of molds. Progress toward this goal is being made by a biological control strategy: early infection of plants with nontoxigenic strains of *A. flavus* to prevent the subsequent entry of toxigenic strains (13, 21, 22, 24).

Aflatoxins are one of the few mycotoxins covered by legislation. Statutory limits are imposed by some countries on the amount of aflatoxin that can be present in particular foods. The limit imposed by most Western countries is 5 to 20 μg of aflatoxin B_1 per kg in several human foods, including peanuts and peanut products, with the amount allowed in animal feeds varying, but up to 300 μg/kg allowed in feedstuffs for beef cattle and sheep in the United States (42, 65). There are currently no statutory limits for aflatoxins in foods or feeds in Southeast Asia (65, 107).

Cyclopiazonic Acid

Cyclopiazonic acid is produced by *A. flavus* and has also been reported to be produced by *A. tamarii* (98) and *A. versicolor* (19). The toxin is an indole tetramic acid which can occur in naturally contaminated agricultural commodities and compounded animal feeds. It is acutely toxic to rats and other test animals, causing severe gastrointestinal and neurological disorders (73). Degenerative changes and necrosis may occur in the digestive tract, liver, kidney, and heart (74). Cyclopiazonic acid may have been responsible for many of the symptoms observed in turkey X disease in the 1960s, originally attributed to aflatoxins (9). Cyclopiazonic acid can be detected by TLC with visualization by spraying with dimethylaminobenzaldehyde and 50% ethanolic H_2SO_4 followed by heating for 5 min at 100°C (19).

A. OCHRACEUS

A. ochraceus (Fig. 21.2A) is a widely distributed mold, particularly common on dried foods (82). It is the most commonly occurring species in what was known as the *Aspergillus ochraceus* group of Raper and Fennell (88), now correctly known as *Aspergillus* sect. *Circumdati* (41). Ochratoxin was first isolated from *A. ochraceus* in 1965 (106), not as a result of a toxicosis but in a laboratory study on toxigenic molds. Three toxins were found, the major toxin being ochratoxin A, with minor components of lower toxicity designated ochratoxins B and C. *A. ochraceus* also produces penicillic acid, a mycotoxin of lower toxicity and of uncertain importance in human health. Other reported toxic metabolites are xanthomegnin and viomellein (38).

Other *Aspergillus* species closely related to *A. ochraceus* can also produce ochratoxin A. *A. sclerotiorum*, *A. alliaceus*, *A. melleus* and *A. sulphureus* have all been reported to produce ochratoxins (16), but in *A. ochraceus* and these related species, only a proportion of isolates are toxigenic. Production of ochratoxin A by *A. niger* var. *niger* has been reported (1).

Isolation and Identification

A. ochraceus and other species in *Aspergillus* sect. *Circumdati* are xerophilic and are best isolated on reduced-a_w media such as DG18 agar (82, 89), although DRBC agar should also give satisfactory results. Colonies of *A. ochraceus* and related species are relatively deep ochre-brown to yellow-brown in color, with long stipes bearing radiate *Aspergillus* heads. The vesicles are spherical, bearing densely packed metulae and phialides with small, smooth, pale brown conidia. Differentiating these closely related species can be difficult (50, 88) but is rarely necessary, as *A. ochraceus* is by far the most commonly occurring.

Growth and Toxin Production

A. ochraceus is widely distributed in dried foods such as nuts (peanuts, pecans), beans, dried fruit, biltong, and

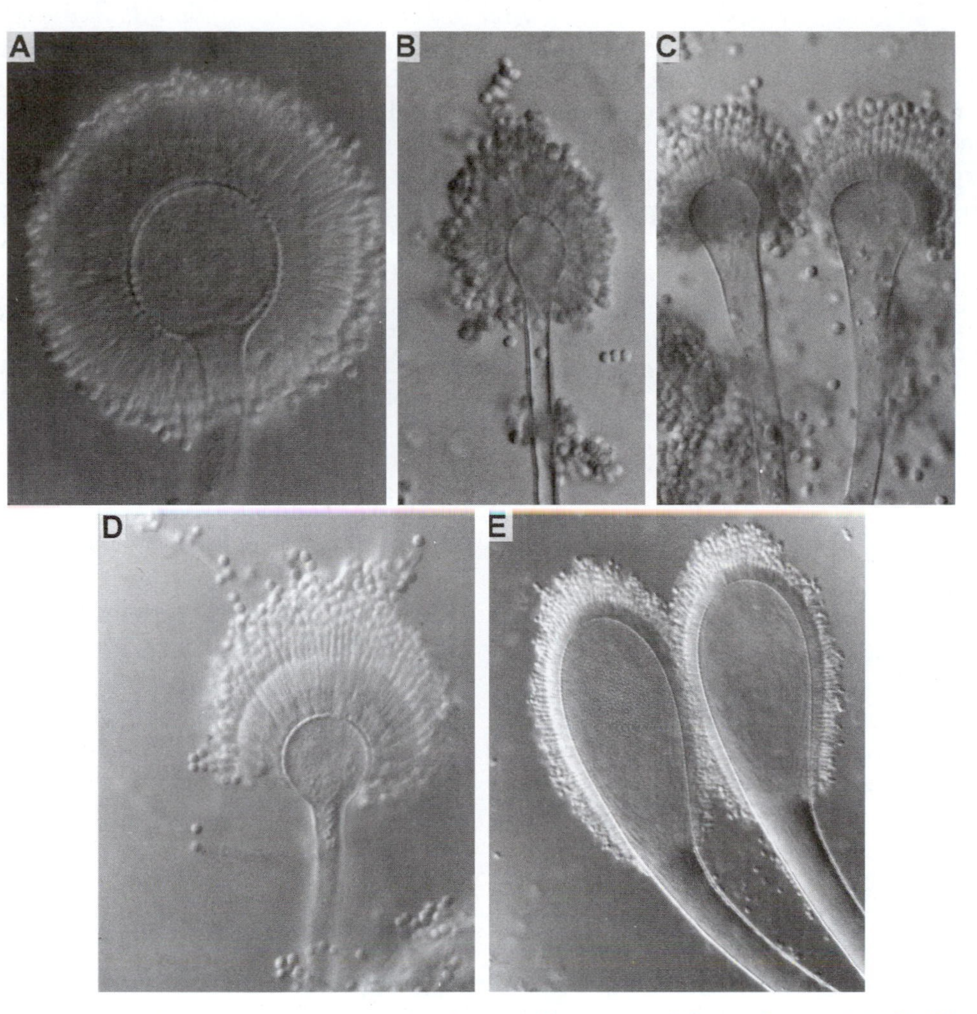

Figure 21.2 Some common mycotoxigenic *Aspergillus* species. (A) *A. ochraceus* (×540); (B) *A. versicolor* (×540); (C) *A. fumigatus* (×540); (D) *A. terreus* (×540); (E) *A. clavatus* (×216).

dried fish (82). It is a xerophile, capable of growth down to a_w 0.79 with an optimum near a_w 0.99 (82). *A. ochraceus* grows at temperatures of 8 to 37°C (76, 77) and within a wide pH range (2.2 to 10.3) (111). Ochratoxin A is produced at quite high temperatures, optimally between 25 and 30°C; by comparison, the optimum for penicillic acid production is 10 to 20°C (7, 16, 76, 77).

Ochratoxin A contamination of foods is of greatest concern in Scandinavia and possibly in the Baltic states, where the source is *Penicillium verrucosum*. Ochratoxin A from *A. ochraceus* or related species would occur only in warmer climates, but ochratoxin is rarely found in tropical commodities (112). The toxic effects of ochratoxin A are discussed in chapter 22.

Ochratoxins can be assayed by routine chromatographic methods (TLC, HPLC) with detection of green fluorescence under UV light at 333 nm (19). Immunoassay techniques, including an ELISA for detection of ochratoxin A in swine kidneys (17), have been developed for detection of ochratoxin.

A. VERSICOLOR

A. versicolor (Fig. 21.2B) is the most important food spoilage and toxigenic species in *Aspergillus* sect. *Versicolores* (41), previously known as the *Aspergillus versicolor* group (88). *A. versicolor* is the major producer of sterigmatocystin, a carcinogenic dihydrofuranoxanthone which is a precursor of the aflatoxins (19). Sterigmatocystin is also produced by members of *Aspergillus* sect. *Nidulantes,* including *A. nidulans* and a number of *Emericella* species (98).

A. versicolor is a xerophile, with a minimum a_w for growth of 0.74 to 0.78 (82). It is very widely distributed in foods, particularly stored cereals, cereal products, nuts, spices, and dried meat products (82). *A. versicolor* grows slowly on all media, but because it is most commonly found in reduced-a_w foods and feeds, a medium of reduced a_w, such as DG18 agar, is recommended for isolation. Small, grey-green colonies showing pinkish or reddish colors in the mycelium and/or reverse and with moplike heads are indicative of *A. versicolor.* The reported minimum temperature for growth is 9°C at a_w 0.97, and the maximum temperature is 39°C at a_w 0.87, with optimum growth at 27°C at a_w 0.97 (99). Little is known about factors affecting production of sterigmatocystin.

Occurrence, Toxicity, and Detection of Sterigmatocystin

Natural occurrence of sterigmatocystin has been found in rice in Japan, wheat and barley in Canada, and cereal-based products in the United Kingdom (112). Although the toxin has not been found in significant quantities, its acute and chronic toxicity are such that it should be regarded as a major potential hazard in foods (104). Sterigmatocystin has low acute oral toxicity because it is relatively insoluble in water and gastric juices. However, even low doses can cause tumors in mice (40) and pathological changes in rat livers (101). As a liver carcinogen, sterigmatocystin appears to be only about 1/150 as potent as aflatoxin B_1, but it is still much more potent than most other known liver carcinogens. Doses as low as 15 μg/day fed continuously or a single dose of 10 mg caused liver cancer in ≥30% of male Wistar rats (101).

Sterigmatocystin can be detected by TLC with chloroform-methanol (98:2) as a solvent system. The toxin is visualized as an orange-red spot under UV light. A light yellow fluorescence develops after spraying with acetic acid (19). An ELISA method for detection of sterigmatocystin has been described (15).

A. FUMIGATUS

A. (sect. *Fumigati*) *fumigatus* (Fig. 21.2C) is best recognized as a human pathogen, causing aspergillosis of the lung (55). It is thermophilic, with a temperature range for growth of between 10 and 55°C and an optimum between 40 and 42°C. It is one of the least xerophilic of the common aspergilli, with a minimum a_w for growth of 0.85 (59). Its prime habitat is decaying vegetation, in which it causes spontaneous heating. *A. fumigatus* is isolated frequently from foods, particularly stored commodities, but is not regarded as a serious spoilage mold (82).

A. fumigatus is capable of producing several toxins which affect the central nervous system, causing tremors. Fumitremorgens A, B, and C are toxic cyclic dipeptides which are produced by *A. fumigatus* and *A. caespitosus* (19). When fed to laboratory rats and mice, fumitremorgens caused tremors and death in 70% of animals tested (43). Verruculogen, also produced by these species (19, 60), has a structure similar to that of the fumitremorgens, but with a different side chain. Verruculogen is tremorgenic when fed to mice and day-old chicks (43) and appears to cause inhibition of the alpha motor cells of the anterior horn (19).

A. TERREUS

A. (sect. *Terrei*) *terreus* (Fig. 21.2D) produces rapidly growing pale brown colonies, with *Aspergillus* heads bearing densely packed metulae and phialides with minute conidia borne in long columns. *A. terreus* occurs commonly in soil and in foods, particularly stored cereals and cereal products, beans, pulses, and nuts, but is not regarded as an important spoilage mold (82). It is probably thermophilic, growing much more strongly at 37°C than at 25°C (82), but little information has been published on its physiology. The reported minimum a_w for growth is 0.78 at 37°C (6).

A. terreus can produce a group of tremorgenic toxins known as territrems (64). The most distinctive structural feature of these toxins is that they do not contain nitrogen. Territrems are acutely toxic, with an intraperitoneal dose of 1 mg of territrem B injected into 20-g Swiss mice causing whole body tremors within 5 min and other neurological symptoms within 20 to 30 min, all of which subside within 1 h (63). A 2-mg dose can cause death after convulsions and apnea. Territrem B appears to act by blocking acetylcholinesterase activity (63).

Territrems can be detected in chloroform extracts by TLC, exhibiting blue fluorescence under UV light. Natural contamination of food or feeds with territrems has not been reported (63).

A. CLAVATUS

A. clavatus (Fig. 21.2E) is found in soil and decomposing plant materials. It is the most common member of the section *Clavati*, subgenus *Clavati*, and is easily recognizable by its large, blue-green clavate (club-shaped) heads (50, 88). Although *A. clavatus* has been reported to be present in various stored grains (82), it is especially common in malting barley, an environment particularly suited to its growth and sporulation (35).

A. clavatus produces patulin, cytochalasins, and the tremorgenic mycotoxins tryptoquivaline, tryptoquivalone, and related compounds. Outbreaks of *A. clavatus*-associated mycotoxicoses have been reported in stock fed on culms from distillery maltings in Europe, the United Kingdom, and South Africa (35).

A. clavatus is a recognized health hazard to workers in the malting industry. Inhalation of large numbers of highly allergenic spores can cause respiratory diseases such as bronchitis, emphysema, or malt worker's lung, a serious occupational extrinisic allergic alveolitis (35).

EUROTIUM SPECIES

The genus *Eurotium* is an Ascomycete genus, characterized by the formation of bright yellow cleistothecia, often enmeshed in yellow, orange, or red hyphae, overlaid by the grey-green (glaucus) *Aspergillus* heads of the anamorphic state. Members of this genus were (and often still are) referred to as the *Aspergillus glaucus* group (88). All *Eurotium* species are xerophilic. They are important spoilage molds in all types of stored commodities, often being the primary invaders of stored grains, spices, nuts, and animal feeds. The four most common species are *Eurotium chevalieri, E. repens, E. rubrum,* and *E. amstelodami.*

Since *Eurotium* species grow poorly on all high-a_w media, DG18 agar is recommended for their isolation and enumeration from foods (46, 82). Identification to species is based on colony color and ascospore morphology, as observed on Czapek yeast extract–20% sucrose agar (50, 82).

E. chevalieri and *E. amstelodami* have been reported to produce toxic alkaloid metabolites called echinulin and neoechinulins (19). Echinulin was identified as the compound responsible for refusal by swine of moldy feed containing high populations of *E. chevalieri* and *E. amstelodami.* The toxin was extracted in acetone and identified by TLC using ethyl acetate-hexane (8:2) as the solvent system. Echinulin turned blue in the presence of *p*-anisaldehyde reagent at 110°C (109). There is little published information on the toxicity of echinulins.

CONCLUSION

Aspergillus is one of the most important genera in the spoilage of foods and animal feeds, particularly in warm-temperate climates and the tropics. The genus contains a number of highly mycotoxigenic molds, the aflatoxigenic species clearly being the most important from the point of view of human health. There is still much to learn about the significance of many of the other mycotoxins produced by aspergilli, particularly their roles in cancer induction and immunosuppression. The interactive effects of naturally occurring mixtures of mycotoxins, e.g., aflatoxins and *Fusarium* toxins (particularly fumonisins) in corn, are poorly understood and may be of much greater significance to human and animal health than we yet realize.

References

1. **Abarca, M. L., M. R. Bragulat, G. Castellá, and F. J. Cabañes.** 1994. Ochratoxin production by strains of *Aspergillus niger* var. *niger. Appl. Environ. Microbiol.* **60:**2650–2652.
2. **Allcroft, R., and R. B. A. Carnaghan.** 1963. Toxic products in groundnuts—biological effects. *Chemy Ind.* **1963:** 50–53.
3. **Arim, R. H.** 1995. Present status of the aflatoxin situation in the Philippines. *Food Addit. Contam.* **12:**291–296.
4. **Association of Official Analytical Chemists.** 1984. *Official Methods of Analysis,* 14th ed. Association of Official Analytical Chemists, Washington, D.C.
5. **Austwick, P. K. C., and G. Ayerst.** 1963. Groundnut mycoflora and toxicity. *Chemy Ind.* **1963:**55–61.
6. **Ayerst, G.** 1969. The effects of moisture and temperature on growth and spore germination in some fungi. *J. Stored Prod. Res.* **5:**127–141.
7. **Bacon, C. W., J. G. Sweeney, J. D. Robbins, and D. Burdick.** 1973. Production of penicillic acid and ochratoxin A on poultry feed by *Aspergillus ochraceus:* temperature and moisture requirements. *J. Food Prot.* **44:**450–454.
8. **Bennett, J. W., and L. S. Lee.** 1979. Mycotoxins: their biosynthesis in fungi—aflatoxins and other bisfuranoids. *J. Food Prot.* **42:**805–809.
9. **Bradburn, N., R. D. Coker, and G. Blunden.** 1994. The aetiology of turkey 'X' disease. *Phytochemistry* **35:**817.
10. **Bryden, W. L.** 1982. Aflatoxins and animal production: an Australian perspective. *Food Technol. Aust.* **34:**216–223.
11. **Buchanan, R. L., and J. C. Ayres.** 1976. Effect of sodium acetate on growth and aflatoxin production in *Aspergillus parasiticus* NRRL 2999. *J. Food Sci.* **41:**128–132.
12. **Büchi, G., and I. D. Rae.** 1969. The structure and chemistry of aflatoxins, p. 55–75. *In* L. A. Goldblatt (ed.), *Aflatoxins.* Academic Press, New York.
13. **Chourasia, H. K., and R. K. Sinha.** 1994. Potential of the biological control of aflatoxin contamination in developing peanut (*Arachis hypogaea* L.) by atoxigenic strains of *Aspergillus flavus. J. Food Sci. Technol. Mysore* **31:**362–366.
14. **Chu, F. S.** 1984. Immunoassays for analysis of mycotoxins. *J. Food Prot.* **47:**562–569.
15. **Chung, D. H., M. I. Abouzied, and J. J. Pestka.** 1989. Immunochemical assay applied to mycotoxin biosynthesis: ELISA comparison of sterigmatocystin production by *Aspergillus versicolor* and *Aspergillus nidulans. Mycopathologia* **107:**93–100.

16. **Ciegler, A.** 1972. Bioproduction of ochratoxin A and penicillic acid by members of the *Aspergillus ochraceus* group. *Can. J. Microbiol.* **18**:631–636.

17. **Clarke, J. R., R. R. Marquardt, A. A. Frohlich, and R. J. Pitura.** 1994. Quantification of ochratoxin A in swine kidneys by enzyme-linked immunosorbent assay using a simplified sample preparation procedure. *J. Food Prot.* **57**:991–995.

18. **Cole, R. J.** 1986. Etiology of turkey "X" disease in retrospect: a case for the involvement of cyclopiazonic acid. *Mycotoxin Res.* **2**:3–7.

19. **Cole, R. J., and R. H. Cox.** 1981. *Handbook of Toxic Fungal Metabolites.* Academic Press, New York.

20. **Cole, R. J., R. A. Hill, P. D. Blankenship, T. H. Sanders, and H. Garren.** 1982. Influence of irrigation and drought on invasion of *Aspergillus flavus* in corn kernels and peanut pods. *Dev. Ind. Microbiol.* **23**:299–326.

21. **Cotty, P. J.** 1994. Influence of field application of an atoxigenic strain of *Aspergillus flavus* on the populations of *A. flavus* infecting cotton bolls and on aflatoxin content of cottonseed. *Phytopathology* **84**:1270–1277.

22. **Cotty, P. J., P. Bayman, D. S. Engel, and K. S. Elias.** 1994. Agriculture, aflatoxins and *Aspergillus*, p. 1–27. *In* K. A. Powell, A. Renwick, and J. F. Peberdy (ed.), *The Genus Aspergillus. From Taxonomy and Genetics to Industrial Application.* Plenum Press, New York.

23. **Cruickshank, R., and J. I. Pitt.** 1990. Isoenzyme patterns in *Aspergillus flavus* and closely related taxa, p. 259–265. *In* R. A. Samson and J. I. Pitt (ed.), *Modern Concepts in Penicillium and Aspergillus Classification.* Plenum Press, New York.

24. **Daigle, D. J., and P. J. Cotty.** 1995. Formulating atoxigenic *Aspergillus flavus* for field release. *Biocontrol Sci. Technol.* **5**:175–184.

25. **Denning, D. W., S. C. Quiepo, D. G. Altman, K. Makarananda, G. E. Neal, E. L. Camellere, M. R. A. Morgan, and T. E. Tupasi.** 1995. Aflatoxin and outcome from acute lower respiratory infection in children in the Philippines. *Ann. Trop. Paediatr.* **15**:209–216.

26. **Dhavan, A. S., and M. R. Choudary.** 1995. Incidence of aflatoxins in animal feedstuff: a decade's scenario in India. *J. Assoc. Off. Anal. Chem. Int.* **78**:693–698.

27. **Diener, U. L., and N. D. Davis.** 1967. Limiting temperature and relative humidity for growth and production of aflatoxins and free fatty acids by *Aspergillus flavus* in sterile peanuts. *J. Am. Oil Chem. Soc.* **44**:259–263.

28. **Dollear, F. G., G. E. Mann, L. P. Codifer, H. K. Gardner, S. P. Koltun, and H. L. E. Vix.** 1968. Elimination of aflatoxin from peanut meal. *J. Am. Oil Chem. Soc.* **45**:862–865.

29. **Dyer, S. K., and S. McCammon.** 1994. Detection of toxigenic isolates of *Aspergillus flavus* and related species on coconut cream agar. *J. Appl. Bacteriol.* **76**:75–78.

30. **Egel, D. S., P. J. Cotty, and K. S. Elias.** 1994. Relationships among isolates of *Aspergillus* sect. *Flavi* that vary in aflatoxin production. *Phytopathology* **84**:906–912.

31. **Ellis, W. O., J. P. Smith, B. K. Simpson, and J. H. Oldham.** 1991. Aflatoxins in food: occurrence, biosynthesis, effects on organisms, detection, and methods of control. *Crit. Rev. Food Sci. Nutr.* **30**:403–439.

32. **Ellis, W. O., J. P. Smith, B. K. Simpson, H. Ramaswamy, and G. Doyon.** 1994. Effect of gas barrier characteristics of films on aflatoxin production by *Aspergillus flavus* in peanuts packaged under modified atmosphere packaging (MAP) conditions. *Food Res. Int.* **27**:505–512.

33. **Ellis, W. O., J. P. Smith, B. K. Simpson, H. Ramaswamy, and G. Doyon.** 1994. Novel techniques for controlling growth of and aflatoxin production by *Aspergillus parasiticus* in packaged peanuts. *Food Microbiol.* **11**:357–368.

34. **Filtenborg, O., J. C. Frisvad, and J. A. Svendsen.** 1983. Simple screening method for molds producing intracellular mycotoxins in pure culture. *Appl. Environ. Microbiol.* **45**:581–585.

35. **Flannigan, B., and A. R. Pearce.** 1994. *Aspergillus* spoilage: spoilage of cereals and cereal products by the hazardous species *A. clavatus*, p. 115–127. *In* K. A. Powell, A. Renwick, and J. F. Peberdy (ed.), *The Genus Aspergillus. From Taxonomy and Genetics to Industrial Application.* Plenum Press, New York.

36. **Frisvad, J. C.** 1985. Secondary metabolites as an aid to *Emericella* classification, p. 437–444. *In* R. A. Samson and J. I. Pitt (ed.), *Advances in Penicillium and Aspergillus Systematics.* Plenum Press, New York.

37. **Frisvad, J. C.** 1989. The connection between the penicillia and aspergilli and mycotoxins with species emphasis on misidentified isolates. *Arch. Environ. Contam. Toxicol.* **18**:452–467.

38. **Frisvad, J. C., and U. Thrane.** 1995. Mycotoxins production by food-borne fungi, p. 251–260. *In* R. A. Samson, E. S. Hoekstra, J. C. Frisvad, and O. Filtenborg (ed.), *Introduction to Food-Borne Fungi*, 4th ed. Centraalbureau voor Schimmelcultures, Baarn, The Netherlands.

39. **Frobish, R. A., B. D. Bradley, D. D. Wagner, P. E. Long-Bradley, and H. Hairston.** 1986. Aflatoxin residues in milk of dairy cows after ingestion of naturally contaminated grain. *J. Food Prot.* **49**:781–785.

40. **Fujii, K., H. Kurata, S. Odashirna, and Y. Hatsuda.** 1976. Tumour induction by a single subcutaneous injection of sterigmatocystin in newborn mice. *Cancer Res.* **36**:1615–1618.

41. **Gams, W., M. Christensen, A. H. S. Onions, J. I. Pitt, and R. A. Samson.** 1985. Intrageneric taxa of *Aspergillus*, p. 55–62. *In* R. A. Samson and J. I. Pitt (ed.), *Advances in Penicillium and Aspergillus Systematics.* Plenum Press, New York.

42. **Gilbert, J.** 1991. Regulatory aspects of mycotoxins in the European Community and USA, p. 194–197. *In* B. R. Champ, E. Highley, A. D. Hocking, and J. I. Pitt (ed.), *Fungi and Mycotoxins in Stored Products: Proceedings of an International Conference, Bangkok, Thailand 23–26 April, 1991.* ACIAR Proceedings no. 36. Australian Centre for International Agricultural Research, Canberra, Australia.

43. **Golinski, P.** 1991. Secondary metabolites (mycotoxins) produced by fungi colonizing cereal grain in store—structure and properties, p. 355–403. *In* J. Chelkowski (ed.), *Cereal Grain. Mycotoxins, Fungi and Quality in Drying and Storage.* Elsevier, Amsterdam.

44. **Groopman, J. D., L. G. Cain, and T. W. Kensler.** 1988. Aflatoxin exposure in human populations: measurements and relation to cancer. *Crit. Rev. Toxicol.* **19**:113–145.

45. **Hatch, M. C., C.-J. Chen, B. Levin, B.-T. Ji, G.-Y. Yang, S.-W. Hsu, L.-W. Wang, L.-L. Hsieh, and R. M. Santella.** 1993. Urinary aflatoxin levels, hepatitis B virus infection and hepatocellular carcinoma in Taiwan. *Int. J. Cancer* **54**:931–934.

46. **Hocking, A. D.** 1992. Collaborative study on media for enumeration of xerophilic fungi, p. 121–125. *In* R. A. Samson, A. D. Hocking, J. I. Pitt, and A. D. King (ed.), *Modern Methods in Food Mycology.* Elsevier, Amsterdam.

47. **Hocking, A. D., and J. I. Pitt.** 1980. Dichloran-glycerol medium for enumeration of xerophilic fungi from low moisture foods. *Appl. Environ. Microbiol.* **39**:488–492.

48. **King, A. D., A. D. Hocking, and J. I. Pitt.** 1979. Dichloran-rose bengal medium for enumeration and isolation of molds from foods. *Appl. Environ. Microbiol.* **37**:959–964.

49. **King, A. D., J. I. Pitt, L. R. Beuchat, and J. E. L. Corry (ed.).** 1986. *Methods for the Mycological Examination of Food.* Plenum Press, New York.

50. **Klich, M. A., and J. I. Pitt.** 1988. *A Laboratory Guide to Common Aspergillus Species and Their Teleomorphs.* CSIRO Division of Food Processing, North Ryde, New South Wales, Australia.

51. **Klich, M. A., and J. I. Pitt.** 1988. Differentiation of *Aspergillus flavus* from *A. parasiticus* and closely related species. *Trans. Br. Mycol. Soc.* **91**:99–108.

52. **Klich, M. A., S. H. Thomas, and J. E. Mellon.** 1984. Field studies on the mode of entry of *Aspergillus flavus* into cotton seeds. *Mycologia* **76**:665–669.

53. **Koehler, P. E., L. R. Beuchat, and M. S. Chinnan.** 1985. Influence of temperature and water activity on aflatoxin production by *Aspergillus flavus* in cowpea (*Vigna unguiculata*) seeds and meal. *J. Food Prot.* **48**:1040–1043.

54. **Kozakiewicz, Z.** 1989. *Aspergillus* species on stored products. *Mycol. Pap.* **161**:1–188.

55. **Kozakiewicz, Z.** 1994. *Aspergillus,* p. 575–616. *In* Y. H. Hui, J. R. Gorham, K. D. Murrell, and D. O. Cliver (ed.), *Foodborne Disease Handbook,* vol. 2. Marcel Dekker, New York.

56. **Kozakiewicz, Z.** 1994. *Aspergillus* toxins and taxonomy, p. 303–311. *In* K. A. Powell, A. Renwick, and J. F. Peberdy (ed.), *The Genus Aspergillus. From Genetics to Industrial Application.* Plenum Press, New York.

57. **Krishnamachari, K. A. V., R. V. Bhat, V. Nagarajan, and T. B. G. Tilak.** 1975. Investigations into an outbreak of hepatitis in parts of Western India. *Indian J. Med. Res.* **63**:1036–1048.

58. **Kuraishi, H., M. Ito, M. Tsuzaki, Y. Katayama, T. Yokohama, and J. Sugiyama.** 1990. Ubiquinone systems as a taxonomic tool in *Aspergillus* and its teleomorphs, p. 407–421. *In* R. A. Samson and J. I. Pitt (ed.), *Modern Concepts in Penicillium and Aspergillus Classification.* Plenum Press, New York.

59. **Lacey, J.** 1994. Aspergilli in feeds and seeds, p. 73–92. *In* K. A. Powell, A. Renwick, and J. F. Peberdy (ed.), *The Genus Aspergillus. From Taxonomy and Genetics to Industrial Application.* Plenum Press, New York.

60. **Land, C. J., H. Lundstrom, and S. Werner.** 1993. Production of tremorgenic mycotoxins by isolates of *Aspergillus fumigatus* from sawmills in Sweden. *Mycopathologia* **124**:87–93.

61. **Lawellin, D. W., D. W. Grant, and B. K. Joyce.** 1977. Enzyme-linked immunosorbent analysis of aflatoxin B_1. *Appl. Environ. Microbiol.* **34**:94–96.

62. **Lillehoj, E. B., W. F. Kwolek, E. S. Horner, N. W. Widstrom, L. M. Josephson, A. O. Franz, and E. A. Catalano.** 1980. Aflatoxin contamination of preharvest corn: role of *Aspergillus flavus* inoculum and insect damage. *Cereal Chem.* **57**:255–257.

63. **Ling, K. H.** 1994. Territrems, tremorgenic mycotoxins isolated from *Aspergillus terreus. J. Toxicol. Toxin Rev.* **13**:243–252.

64. **Ling, K. H., C. K. Yang, and F. T. Peng.** 1979. Territrem, tremorgenic mycotoxin of *Aspergillus terreus. Appl. Environ. Microbiol.* **37**:355–357.

65. **Lubulwa, A. S. G., and J. S. Davis.** 1994. Estimating the social costs of the impacts of fungi and aflatoxins, p. 1017–1042. *In* E. Highley, E. J. Wright, H. J. Banks, and B. R. Champ (ed.), *Stored Product Protection. Proceedings of the 6th International Working Conference on Stored-Product Protection.* CAB International, Wallingford, Oxford.

66. **Lye, M. S., A. A. Ghazali, J. Mohan, N. Alwin, and R. C. Nair.** 1995. An outbreak of acute hepatic encephalopathy due to severe aflatoxicosis in Malaysia. *Am. J. Trop. Med. Hyg.* **53**:68–72.

67. **Miyake, I., H. Naito, and H. Sumeda.** 1940. *Rep. Res. Inst. Rice Improvement* **1**:1. (In Japanese.)

68. **Moody, S. F., and B. M. Tyler.** 1990. Restriction enzyme and mitochondrial DNA of the *Aspergillus flavus* group, *Aspergillus flavus, Aspergillus parasiticus,* and *Aspergillus nomius. Appl. Environ. Microbiol.* **56**:2441–2452.

69. **Moody, S. F., and B. M. Tyler.** 1990. Use of DNA restriction length polymorphisms to analyze the diversity of the *Aspergillus flavus* group, *Aspergillus flavus, Aspergillus parasiticus,* and *Aspergillus nomius. Appl. Environ. Microbiol.* **56**:2453–2461.

70. **Moss, M. O., and J. E. Smith.** 1985. *Mycotoxins: Formation, Analysis and Significance.* John Wiley & Sons, Chichester, England.

71. **Mullaney, E. J., and M. A. Klich.** 1990. A review of molecular biological techniques for systematic studies of *Aspergillus* and *Penicillium,* p. 301–307. *In* R. A. Samson and J. I. Pitt (ed.), *Modern Concepts in Penicillium and Aspergillus Classification.* Plenum Press, New York.

72. **Nesheim, S., and M. Trucksess.** 1986. Thin-layer chromatography/high-performance thin-layer chromatography as a tool for mycotoxin determination. *In* R. J. Cole (ed.), *Modern Methods for Analysis and Structural Elucidation of Mycotoxins.* Academic Press, Orlando, Fla.

73. **Nishie, K., R. J. Cole, and F. W. Dorner.** 1985. Toxicity and neuropharmacology of cyclopiazonic acid. *Food Chem. Toxicol.* **23**:831–839.

74. **Norred, W. P., R. E. Morrisey, R. T. Riely, R. J. Cole, and L. W. Dorner.** 1985. Distribution, excretion and skeletal muscle effects of the mycotoxin (^{14}C) cyclopiazonic acid in rats. *Food Chem. Toxicol.* **23**:1069–1076.

75. **Northolt, M. D., H. P. van Egmond, and W. E. Paulsch.** 1977. Differences in *Aspergillus flavus* strains in growth

and aflatoxin B_1 production in relation to water activity and temperature. *J. Food Prot.* **40**:778–781.

76. **Northolt, M. D., H. P. van Egmond, and W. E. Paulsch.** 1979. Ochratoxin A production by some fungal species in relation to water activity and temperature. *J. Food Prot.* **42**:485–490.
77. **Northolt, M. D., H. P. van Egmond, and W. E. Paulsch.** 1979. Penicillic acid production by some fungal species in relation to water activity and temperature. *J. Food Prot.* **42**:476–484.
78. **Peers, F., X. Bosch, J. Kaldor, A. Linsell, and M. Pluumen.** 1987. Aflatoxin exposure, hepatitis B virus infection and liver cancer in Swaziland. *Int. J. Cancer* **39**:545–553.
79. **Pestka, J. J., M. N. Abouzied, and Sutikno.** 1995. Immunological assays for mycotoxin detection. *Food Technol.* **49**:120–128.
80. **Pier, A. C.** 1991. The influence of mycotoxins on the immune system, p. 489–497. *In* J. E. Smith, and R. S. Henderson (ed.), *Mycotoxins and Animal Foods.* CRC Press, Boca Raton, Fla.
81. **Pitt, J. I., S. K. Dyer, and S. McCammon.** 1991. Systemic invasion of developing peanut plants by *Aspergillus flavus. Lett. Appl. Microbiol.* **13**:16–20.
82. **Pitt, J. I., and A. D. Hocking.** 1985. *Fungi and Food Spoilage.* Academic Press, Sydney, Australia.
83. **Pitt, J. I., A. D. Hocking, K. Bhudhasamai, B. F. Miscamble, K. A. Wheeler, and P. Tanboon-Ek.** 1993. The normal mycoflora of commodities from Thailand. 1. Nuts and oilseeds. *Int. J. Food Microbiol.* **20**:211–226.
84. **Pitt, J. I., A. D. Hocking, K. Bhudhasamai, B. F. Miscamble, K. A. Wheeler, and P. Tanboon-Ek.** 1994. The normal mycoflora of commodities from Thailand. 2. Beans, rice, small grains and other commodities. *Int. J. Food Microbiol.* **23**:35–53.
85. **Pitt, J. I., A. D. Hocking, and D. R. Glenn.** 1983. An improved medium for the detection of *Aspergillus flavus* and *A. parasiticus. J. Appl. Bacteriol.* **54**:109–114.
86. **Pitt, J. I., and B. F. Miscamble.** 1995. Water relations of *Aspergillus flavus* and closely related species. *J. Food Prot.* **58**:86–90.
87. **Pitt, J. I., and R. A. Samson.** 1993. Species names in current use in the *Trichocomaceae* (Fungi, Eurotiales). *Regnum Veg.* **128**:13–57.
88. **Raper, K. B., and D. I. Fennell.** 1965. *The Genus Aspergillus.* The Williams & Wilkins Co., Baltimore.
89. **Samson, R. A., A. D. Hocking, J. I. Pitt, and A. D. King (ed.).** 1992. *Modern Methods in Food Mycology.* Elsevier, Amsterdam.
90. **Samson, R. A., E. S. Hoekstra, J. C. Frisvad, and O. Filtenborg (ed.).** 1995. *Introduction to Food-Borne Fungi,* 4th ed. Centraalbureau voor Schimmelcultures, Baarn, The Netherlands.
91. **Samson, R. A., and J. I. Pitt (ed.).** 1985. *Advances in Penicillium and Aspergillus Systematics.* Plenum Press, New York.
92. **Samson, R. A., and J. I. Pitt (ed.).** 1990. *Modern Concepts in Penicillium and Aspergillus Classification.* Plenum Press, New York.
93. **Sanders, T. H., R. A. Hill, R. J. Cole, and P. D. Blankenship.** 1981. Effect of drought on the occurrence of *Aspergillus flavus* in maturing peanuts. *J. Am. Oil Chem. Soc.* **58**:966A–970A.
94. **Sargeant, K., R. B. A. Carnaghan, and R. Allcroft.** 1963. Toxic products in groundnuts—chemistry and origin. *Chemy Ind.* **1963**:53–55.
95. **Schneider, E., E. Usleber, E. Martl-Bauer, R. Dietrich, and G. Terplan.** 1995. Multimycotoxin dipstick enzyme immunoassay applied to wheat. *Food Addit. Contam.* **12**:387–393.
96. **Scott, P. M.** 1994. *Penicillium* and *Aspergillus* toxins, p. 261–285. *In* J. D. Miller and H. L. Trenholm (ed.), *Mycotoxins in Grain. Compounds Other than Aflatoxin.* Eagan Press, St. Paul, Minn.
97. **Shank, R. C.** 1978. Mycotoxicoses of man: dietary and epidemiological considerations, p. 1–12. *In* T. D. Wyllie and L. G. Morehouse (ed.), *Mycotoxigenic Fungi, Mycotoxins, Mycotoxicoses: an Encyclopedic Handbook.* Marcel Dekker, New York.
98. **Smith, J. E., and K. Ross.** 1991. The toxigenic aspergilli, p. 101–118. *In* J. E. Smith and R. S. Henderson (ed.), *Mycotoxins and Animal Foods.* CRC Press, Boca Raton, Fla.
99. **Smith, S. L., and S. T. Hill.** 1982. Influence of temperature and water activity on germination and growth of *Aspergillus restrictus* and *A. versicolor. Trans. Br. Mycol. Soc.* **79**:558–460.
100. **Sreenivasamurthy, V., H. A. B. Parpia, S. Srikanta, and A. Shankarmurti.** 1967. Detoxification of aflatoxin in peanut meal by hydrogen peroxide. *J. Assoc. Off. Anal. Chem.* **50**:350–354.
101. **Stoloff, L.** 1977. Aflatoxins—an overview, p. 7–28. *In* J. V. Rodricks, C. W. Hesseltine, and M. A. Mehlman (ed.), *Mycotoxins in Human and Animal Health.* Pathatox Publishers, Park Forest South, Ill.
102. **Stoloff, L.** 1983. Aflatoxin as a cause of primary liver-cell cancer in the United States: a probability study. *Nutr. Cancer* **5**:165–168.
103. **Stoloff, L., and L. Friedman.** 1976. Information bearing on the evaluation of the hazard to man from aflatoxin ingestion. *PAG Bull.* **6**:21–32.
104. **Terao, K.** 1983. Sterigmatocystin—a masked potent carcinogenic mycotoxin. *J. Toxicol. Toxin Rev.* **2**:77–100.
105. **Trucksess, M. W., and M. E. Stack.** 1994. Enzyme-linked immunosorbent assay of total aflatoxins B_1, B_2 and G_1 in corn: follow-up collaborative study. *J. Assoc. Off. Anal. Chem.* **77**:655–658.
106. **van der Merwe, K. J., P. S. Steyn, L. Fourie, D. B. Scott, and J. J. Theron.** 1965. Ochratoxin A, a toxic metabolite produced by *Aspergillus ochraceus Wilh. Nature* (London) **205**:1112–1113.
107. **Van Egmond, H. P.** 1991. Regulatory aspects of mycotoxins in Asia and Africa, p. 198–204. *In* B. R. Champ, E. Highley, A. D. Hocking, and J. I. Pitt (ed.), *Fungi and Mycotoxins in Stored Products: Proceedings of an International Conference, Bangkok, Thailand 23–26 April, 1991.* ACIAR Proceedings no. 36. Australian Centre for International Agricultural Research, Canberra, Australia.
108. **van Rensburg, S. J.** 1977. Role of epidemiology in the elucidation of mycotoxin health risks, p. 699–711. *In* J. V.

Rodricks, C. W. Hesseltine, and M. A. Mehlman (ed.), *Mycotoxins in Human and Animal Health.* Pathatox Publishers, Park Forest South, Ill.

109. **Vesonder, R. F., R. Lambert, D. T. Wicklow, and M. L. Biehl.** 1988. *Eurotium* spp. and echinulin in feed refusal by swine. *Appl. Environ. Microbiol.* **54:**830–831.

110. **Weng, C. Y., A. J. Martinez, and D. L. Park.** 1994. Efficacy and permanency of ammonia treatment in reducing aflatoxin levels in corn. *Food Addit. Contam.* **11:**649–658.

111. **Wheeler, K. A., B. F. Hurdman, and J. I. Pitt.** 1991. Influence of pH on the growth of some toxigenic species of *Aspergillus, Penicillium* and *Fusarium. Int. J. Food Microbiol.* **12:**141–150.

112. **Yoshizawa, T.** 1991. Natural occurrence of mycotoxins in small grain cereals (wheat, barley, rye, oats, sorghum, millet, rice), p. 301–324. *In* J. E. Smith and R. S. Henderson (ed.), *Mycotoxins and Animal Foods.* CRC Press, Boca Raton, Fla.

John I. Pitt

22 Toxigenic *Penicillium* Species

Experimental evidence that common microfungi can produce toxins is popularly believed to date only from about 1960. However, the study of mycotoxicology began 100 years ago. In 1891, in Japan, Sakaki demonstrated that an ethanol extract from moldy, unpolished "yellow" rice was fatal to dogs, rabbits, and guinea pigs, with symptoms indicating paralysis of the central nervous system (94). In consequence, the sale of yellow rice was banned in Japan in 1910. In 1913, an extract from a *Penicillium puberulum* culture isolated from moldy corn in Nebraska was found to be toxic to animals when injected at 200 to 300 mg/kg of body weight (1). This was the first reliable account of toxin production by a mold in pure culture. This careful study aimed to resolve the question of whether common molds or mold products could have an injurious effect on animals, produced positive results, and was far ahead of its time. Such direct evidence that common molds could be toxic was largely ignored.

In 1940, *P. toxicarium* was isolated from yellow rice and described (60). It produced a highly toxic metabolite, subsequently named citreoviridin. Perhaps because it was published in wartime, this study also failed to alert the world to the potential or actual danger of the toxicity of common molds.

The discovery of penicillin in 1929 gave impetus to a search for other *Penicillium* metabolites with antibiotic properties and ultimately to the recognition of citrinin, patulin, and griseofulvin as "toxic antibiotics" or, later, mycotoxins.

The literature on toxigenic Penicillia is now quite vast. In a comprehensive review of literature on fungal metabolites, about 120 from common molds were demonstrably toxic to higher animals (15). Forty-two were reported to be produced by one or more *Penicillium* species. No fewer than 85 *Penicillium* species were listed as toxigenic (15). The literature on this subject has accumulated in a rather random fashion, with emphasis in the majority of papers on chemistry or toxicology rather than mycology. The impression gained from the literature is that toxin production by penicillia lacks species specificity, i.e., most toxins are produced by a variety of species. This viewpoint is commonly accepted. For example, citrinin has been reported to be produced by at least 22 species (15, 34, 52). One reason for this perception is the fact that the taxonomy of toxigenic *Penicillium* species was uncertain, and many reports on toxigenicity of particular species were based on misidentifications.

Recent developments in *Penicillium* taxonomy have profoundly changed this picture. The principal milestones have been the development of a new morphological classification scheme based on a broader range of standardized characters than used previously (66), the development of simple screening methods for mycotoxins and other secondary metabolites and the recognition that such compounds are of taxonomic value (29, 30, 33, 35), reexamination of a large collection of penicillia (1,500 cultures) with emphasis on accurate identification and mycotoxin production (25, 77), and the use of electrophoretic fingerprinting of certain isoenzymes as an aid to taxonomy (18, 19). For a review, see reference 71; for more details, see reference 82. A classification of the more common species has been published (69), a computer-based key to common species has been produced (70), and an accurate picture of the more important species-mycotoxin relationships has been developed (77). More recently, a list of currently accepted names in *Penicillium* and related genera has been published (78).

TAXONOMY

Penicillium is a large genus, with 150 species recognized in the most recent complete taxonomy (66). Subsequent studies, especially the developments in taxonomic concepts noted above, indicate that this number is too low. At least 50 species are of common occurrence (69). All common species grow and sporulate well on synthetic or semisynthetic media and are usually readily recognizable to the genus level.

Classification of the penicillia is based primarily on microscopic morphology (Fig. 22.1). The genus *Penicillium* is divided into subgenera based on the number and arrangement of phialides (elements producing conidia) and metulae and rami (elements supporting phialides) on the main stalk cells (stipes). The classification of Pitt (66) includes four subgenera: *Aspergilloides*, in which phialides are borne directly on the stipes without intervening supporting elements; *Furcatum* and *Biverticillium*, in which phialides are supported by metulae; and *Penicillium*, in which both metulae and rami are usually present (Fig. 22.1). The majority of important toxigenic and food spoilage species are found in subgenus *Penicillium*.

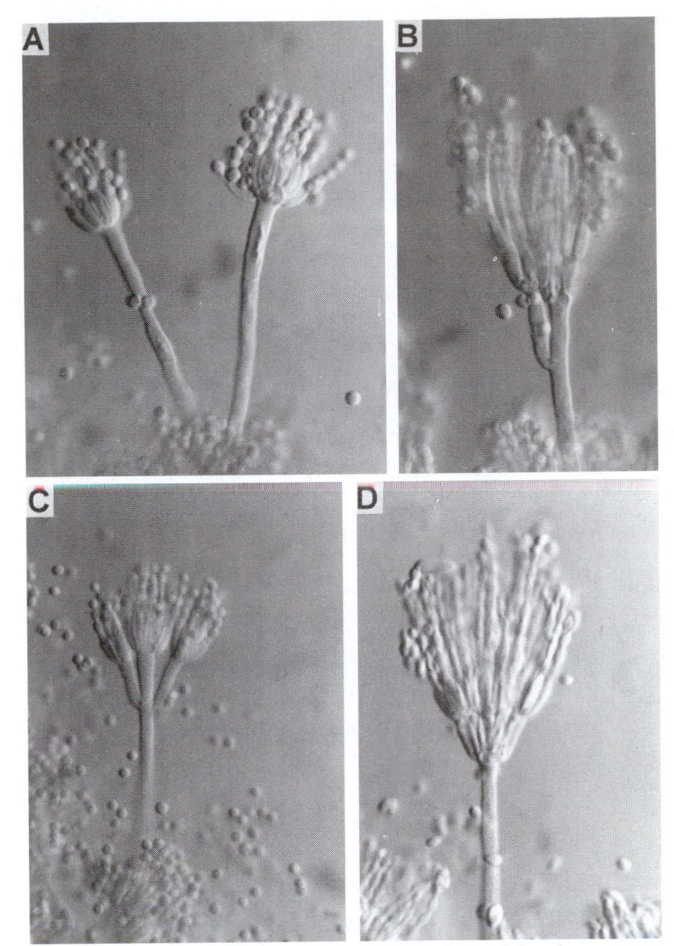

Figure 22.1 Penicilli (×555) representative of the four *Penicillium* subgenera. (A) *Penicillium* subg. *Aspergilloides* (*P. glabrum*); (B) *Penicillium* subg. *Penicillium* (*P. expansum*); (C) *Penicillium* subg. *Furcatum* (*P. citrinum*); (D) *Penicillium* subg. *Biverticillium* (*P. variabile*).

ENUMERATION

General enumeration procedures suitable for foodborne molds are effective for enumerating all common *Penicillium* species. Many antibacterial enumeration media can be expected to give satisfactory results. However, some *Penicillium* species grow rather weakly on very dilute media such as potato-dextrose agar. Dichloran-rose bengal-chloramphenicol (DRBC) agar (74) or dichloran–18% glycerol (DG18) agar (74) is recommended (42, 46).

IDENTIFICATION

For a comprehensive taxonomy of *Penicillium* see reference 66; for keys and descriptions to common species, see reference 69; for foodborne species, see references 74 and 81. A computer-assisted key to common *Penicillium* species is also available (70).

Identification of *Penicillium* isolates to the species level is not easy and is preferably carried out under carefully standardized conditions of media, incubation time, and temperature. In addition to microscopic morphology, gross physiological features, including colony diameters and colors of conidia and colony pigments, are used to distinguish species (66, 69).

SIGNIFICANT *PENICILLIUM* MYCOTOXINS

As already noted, nearly 100 *Penicillium* species have been reported as toxin producers. In assessing the relevance of these reports, several points need to be kept in

Table 22.1 Significant mycotoxins produced by *Penicillium* species

Mycotoxin	Toxicity (LD_{50})[a]	Species producing[b]
Citreoviridin	Mice, 7.5 mg/kg i.p.	*Penicillium citreonigrum* Dierckx
	Mice, 20 mg/kg oral	*Eupenicillium ochrosalmoneum* Scott & Stolk
Citrinin	Mice, 35 mg/kg i.p.	*P. citrinum* Thom
	Mice, 110 mg/kg oral	*P. expansum* Link
		P. verrucosum Dierckx
Cyclopiazonic acid	Rats, 2.3 mg/kg i.p.	*P. camemberti* Thom
	Male rats, 36 mg/kg oral	*P. commune* Thom
	Female rats, 63 mg/kg oral	*P. chrysogenum* Thom
		P. griseofulvum Dierckx
		P. hirsutum Dierckx
		P. viridicatum Westling
Ochratoxin A	Young rats, 22 mg/kg oral	*P. verrucosum*
Patulin	Mice, 5 mg/kg i.p.	*P. expansum*
	Mice, 35 mg/kg oral	*P. vulpinum* (Cooke & Massee) Seifert & Samson[c]
		P. griseofulvum
		P. roqueforti Thom
Penitrem A	Mice, 1 mg/kg i.p.	*P. crustosum* Thom
		P. glandicola (Oudem.) Seifert & Samson[d]
PR toxin	Mice, 6 mg/kg i.p.	*P. roqueforti*
	Rats, 115 mg/kg oral	
Roquefortine C	Mice, 340 mg/kg i.p.	*P. roqueforti*
		P. chrysogenum
		P. crustosum
Secalonic acid D	Mice, 42 mg/kg i.p.	*P. oxalicum* Currie & Thom

[a]References 15 and 77. LD_{50}, 50% lethal dose. Dosing: i.p., intraperitoneal injection; s.c., subcutaneous injection; i.v., intravenous injection.
[b]Reference 27 and personal observations.
[c]Now the accepted name for *P. claviforme* Bainier (85).
[d]Now the accepted name for *P. granulatum* Bainier (85).

mind. (i) Many of the identifications reported in the literature are incorrect, and this has resulted in confusion. (ii) A number of genuinely toxic compounds are produced by species not usually occurring in commodities, foods, or feeds. Such toxins are of little practical importance and are not discussed here. (iii) Some compounds often cited as mycotoxins in the literature are of low toxicity and again are of little practical importance (iv) Growth of mold does not always mean production of toxin. The conditions under which toxins are produced are often narrower than the conditions for growth. Appropriate information is provided where possible. Taking these factors into account, a recent review listed 27 mycotoxins, produced by 32 species, which possessed demonstrated toxicity to humans or domestic animals (77).

The range of mycotoxin classes produced by *Penicillium* species is broader than for any other fungal genus. Moreover, molecular composition is diverse in the extreme. Patulin is an unsaturated lactone with a molecular weight of 150, while penitrem A has nine adjacent rings with four to eight atoms in each and a molecular weight of more than 650 (15, 20).

Toxicity due to *Penicillium* species is also very diverse. However, most toxins can be placed in two broad groups: those that affect liver and kidney function, and those that are neurotoxins. In general terms, the *Penicillium* toxins which affect liver or kidney function are asymptomatic or cause generalized debility in humans or animals. In contrast, toxicity of the neurotoxins in animals is often characterized by sustained trembling. However, individual toxins show wide variations from these generalizations.

Table 22.1 lists nine mycotoxins, produced by 17 *Penicillium* species, which I consider to be significant or potentially significant in human health. Relative toxicities of the various compounds are shown for comparative purposes. For this reason, toxicities in a standard test animal (mouse) and by a standard route of administration (oral) have been provided where possible. The list of species shown as producing each toxin results from my own experience and is undoubtedly conservative. Some general notes follow on these toxins and the species which produce them.

Ochratoxin A

Undoubtedly the most important toxin produced by a *Penicillium* species, ochratoxin A has immunosuppressive, embryonic, and probably carcinogenic effects. Its source is barley and wheat crops infected by mold in the field or in storage, crops used both for bread making and for animal feeds. Ochratoxin A plays a major role in the etiology of nephritis (kidney disease) in pigs in Scandinavia (47–49) and indeed in much of northern Europe. This a serious animal health problem.

Because ochratoxin A is fat soluble and not readily excreted, it accumulates in the depot fat of affected animals and from there is ingested by humans eating pork. A second source is bread made from barley or wheat containing the toxin. Ochratoxin A is now considered to be a serious human health risk in northern Europe, especially in rural areas, where domestic consumption of pig meats not subject to government inspection systems may occur. Ochratoxin A has been found in human blood over wide areas of Europe (9), with levels up to 35 μg/kg reported (65), and in human milk at up to the same concentrations (4). Although clear evidence of human disease is still elusive, such levels indicate a widespread problem with ochratoxin A in Europe.

It has been suggested that ochratoxin A is a causal agent of Balkan endemic nephropathy, a kidney disease with a high mortality rate in certain areas of Bulgaria, Yugoslavia, and Romania (2, 50). However, it is now considered that other, less well characterized toxins may be responsible (56, 57).

Ochratoxin A was originally described as a metabolite of *Aspergillus ochraceus* (96). Later work showed that production of ochratoxin A by *A. ochraceus* is not very common; in one study, only 2 of 33 isolates were found to be producers (62). Later, ochratoxin A (and sometimes citrinin) was reported to be produced by *P. viridicatum* (99), and this view prevailed for more than a decade. A complicating factor was that some isolates identified as *P. viridicatum* did not produce ochratoxin A but rather produced a group of other, unrelated toxic metabolites (89). Eventually it became clear that isolates regarded as *P. viridicatum* but producing ochratoxin and citrinin were more correctly classified in a separate species, *P. verrucosum* (68). The majority of *P. verrucosum* isolates are ochratoxin A producers. In a comprehensive study, 48 of 84 isolates produced ochratoxin A in the laboratory, and 6 of these same isolates produced citrinin (68).

P. viridicatum has generally been regarded as the mold producing ochratoxin A (and citrinin) in barley in Scandinavia; however, this cannot be correct (68). One isolate producing ochratoxin A and citrinin and regarded as *P. viridicatum* (no. 67B) (48) has been identified by me as *P. verrucosum*, and other *P. viridicatum* isolates from Danish barley have also proved to be *P. verrucosum* (77). In an extensive survey of 70 Danish barley samples from farms where pigs were suffering from nephritis, 67 contained high populations of *P. verrucosum* and 66 contained ochratoxin A (33, 37). It seems certain that *P. verrucosum* is the major source of ochratoxin A in Scandinavia and other cool temperate zone areas. The implications of this finding for human health in Europe should not be underestimated.

Taxonomy

Described originally in 1901, *P. verrucosum* was ignored until revived by Samson et al. in 1976 (83). Their concept was very broad, encompassing several species previously considered distinct. *P. verrucosum* was restricted to a much narrower concept by Pitt (66). With the recognition that this concept is linked to ochratoxin production (68), it has now been widely accepted. *P. verrucosum* is classified in subgenus *Penicillium* sect. *Penicillium*, which includes many mycotoxigenic species of common occurrence in foods. *P. verrucosum* (Fig. 22.2A) is distinguished by slow growth on Czapek yeast extract agar (CYA) and malt extract agar (MEA) at 25°C (17 to 24 mm and 10 to 20 mm, respectively, after 7 days) (68), bright green conidia, clear to pale yellow exudate, and rough stipes (69). It is similar in general appearance to *P. viridicatum*, differing most obviously by slower growth, and to *P. solitum*, from which it differs by having green rather than blue conidia.

Enumeration and Identification

The media specified above for general enumeration of *Penicillium* species are effective for *P. verrucosum*. On dichloran-rose bengal-yeast extract-sucrose agar, a selective medium for the enumeration of *P. verrucosum* and *P. viridicatum*, *P. verrucosum* produces a violet-brown reverse coloration (32). Isolation and identification of *P. verrucosum* in pure culture are essential for confirmation.

Ecology

P. verrucosum has been found almost exclusively in grain from temperate zones. It is associated with Scandinavian barley and wheat and has also been isolated quite fre-

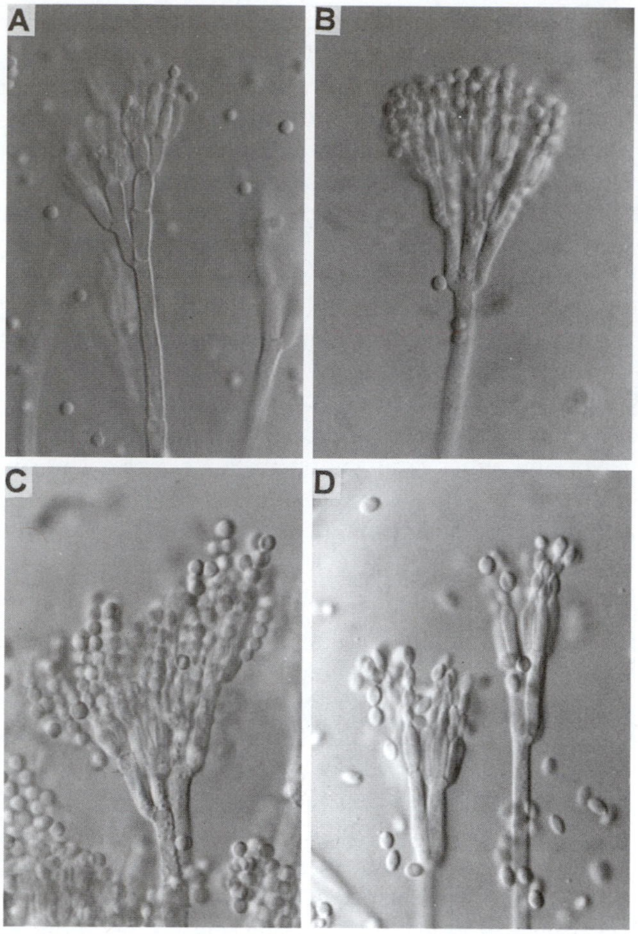

Figure 22.2 Some toxigenic *Penicillium* species (×555). (A) *P. verrucosum;* (B) *P. crustosum;* (C) *P. roqueforti;* (D) *P. oxalicum.*

quently from meat products in Germany and other European countries. It does not appear to be common elsewhere (74).

P. verrucosum grows most strongly at relatively low temperatures, down to 0°C, with a maximum near 31°C (63). It is xerophilic, growing down to water activity (a_w) 0.80. Maximum ochratoxin A production occurs at about 20°C, and is possible down to about a_w 0.85. Cereals in storage only marginally above levels considered safe from mold growth appear especially vulnerable.

Citrinin

Citrinin is a significant renal toxin affecting monogastric domestic animals such as pigs and dogs (8, 31). It is also an important toxin in domestic birds, in which it produces watery diarrhea, increased water consumption, and reduced weight gain due to kidney degeneration (87). The importance of citrinin in human health is currently difficult to assess. When it is ingested in the absence of other toxins, significant effects appear unlikely. However, citrinin is produced in nature along with ochratoxin A by some isolates of *P. verrucosum* (49) and with patulin by some isolates of *P. expansum*. The possibility of synergy between these toxins must not be discounted.

Primarily recognized as a metabolite of *P. citrinum*, citrinin has been reported to be produced by more than 20 other fungal species (15, 84). However, apart from *P. citrinum*, only *P. expansum* and *P. verrucosum* are certain producers (27). These three species are among the more commonly occurring penicillia, and so citrinin is probably the most widely occurring *Penicillium* toxin. It appears to be abundantly produced in nature (66).

Taxonomy

Classified in subgenus *Furcatum* sect. *Furcatum* (66), *P. citrinum* is a very well circumscribed species, accepted without controversy for many years. The most distinctive feature of *P. citrinum* is its penicillus (Fig. 22.1C), which consists of a cluster of three to five divergent metulae, usually apically swollen. Under the stereomicroscope, the phialides from each metula bear conidia as long columns, producing a distinctive pattern which can be of diagnostic value. Colonies of this species on CYA and MEA are of moderate size (25 to 30 mm and 14 to 18 mm, respectively), with the smaller size on MEA also a distinctive feature. Growth normally occurs at 37°C, but colonies seldom exceed 10 mm after 7 days (69).

Enumeration and Identification

P. citrinum can be enumerated effectively on any of the enumeration media mentioned in the introduction. In all cases, confirmation requires isolation and identification by the standard methods. No selective media have been developed for this species.

Ecology

P. citrinum is a ubiquitous mold and has been isolated from nearly every kind of food surveyed for fungi. The most common sources are cereals, especially rice, wheat, and corn, milled grains, and flour (74). Toxin production is also likely to be a common occurrence.

P. citrinum grows from 5 to 7 to 40°C, with an optimum near 30°C (74). It is xerophilic, with a minimum a_w for growth near 0.82 (41). Citrinin is produced over most of the temperature growth range, but the effect of a_w is unknown (72).

Patulin

Patulin was once considered to be a relatively important mycotoxin (91), but early reports of carcinogenicity (21) have not been substantiated. More recent evidence suggests adverse affects on the rodent fetus, as well as immunological, neurological, and gastrointestinal effects (87). Its major significance is that it occurs in a product, apple juice, which is consumed by children as well as adults. In consequence, some countries have set an upper limit of 50 μg/liter for patulin in apple juice and other apple products (87, 97). The most important *Penicillium* species producing patulin is *P. expansum*, best known as a fruit pathogen but also of widespread occurrence in other fresh and processed foods (74). *P. expansum* produces patulin as it rots apples and pears. The use of rotting fruit in juice or cider manufacture can result in high concentrations of patulin, 350 μg/liter (3) or even 630 μg/liter (100) having been found in the resultant juice. Levels in commercial practice are usually much lower than this, and given that patulin appears to lack chronic toxic effects in humans, low levels in juices are perhaps of little concern (61). It is certainly important, however, as an indicator of the use of poor-quality raw materials in juice manufacture.

Patulin is produced by several *Penicillium* species other than *P. expansum* (Table 22.1), but the potential for production of unacceptable levels in foods appears to be much lower.

Taxonomy

One of the oldest described *Penicillium* species, *P. expansum* has been established as the principal cause of spoilage of pome fruits throughout this century. It belongs to subgenus *Penicillium* sect. *Penicillium* (66). *P. expansum* colonies usually grow quite rapidly on CYA and MEA (30 to 40 mm in diameter in 7 days) and show orange-brown or cinnamon colors in exudate, soluble pigment, and reverse (69). Penicilli are terverticillate, with smooth stipes (Fig. 22.1B). Destructive rots of pomaceous fruits are almost always caused by this species.

Enumeration and Identification

P. expansum can be enumerated effectively on DRBC or DG18. No selective media have been developed; however, isolation from decaying apples and pears provides presumptive identification.

Ecology

Although *P. expansum* is a broad-spectrum pathogen on fresh fruits, its major source is rotting apples and pears. It is less common in cereals and cereal products than many related species.

Like other species in subgenus *Penicillium*, *P. expansum* is psychrotrophic, growing at −2 to −3°C, with an optimum near 25°C and a maximum about 35°C (74). The minimum a_w for germination of conidia is 0.82 to 0.83 (41). Nothing appears to be known about factors affecting toxin production.

Cyclopiazonic Acid

A highly toxic compound, cyclopiazonic acid causes fatty degeneration and hepatic cell necrosis in the liver and kidneys of domestic animals. Chickens are particularly susceptible (23). Mammals may be less affected (98). When the compound is injected, central nervous dysfunction occurs, and high doses may result in death of experimental animals (87).

Cyclopiazonic acid is produced by *Aspergillus flavus* and six *Penicillium* species (Table 22.1). The most common of these was until recently regarded as *P. cyclopium;* however, isolates producing this toxin and previously assigned to that species are correctly identified as *P. commune,* a common cause of cheese spoilage (73). Along with *A. flavus, P. commune* is probably the most common source of cyclopiazonic acid in foods. It has been shown (73) that *P. commune* is the wild ancestor of *P. camemberti,* used in the production of Camembert-type cheeses. Nearly all isolates of *P. camemberti* produce cyclopiazonic acid (90), but apparently not under commercial cheese-manufacturing conditions. Nevertheless, the hunt for nontoxigenic *P. camemberti* strains suitable for use as starter cultures continues (51).

Taxonomy

The *Penicillium* species producing cyclopiazonic acid are all classified in *Penicillium* subg. *Penicillium* (Table 22.1).

Enumeration and Identification

All of the species producing cyclopiazonic acid can be enumerated effectively on the media discussed in the introduction. No selective media are available for any of these species. Identification of any of them requires specialist methods and information (69, 81).

Ecology

The principal *Penicillium* species producing cyclopiazonic acid, *P. commune,* is a common cause of cheese spoilage around the world (40, 55). Other species producing this toxin have a wide range of habitats.

All of the *Penicillium* species producing cyclopiazonic acid have similar physiologic characteristics, including the ability to grow at or near 0°C, optima around 25°C, and maxima at or below 37°C. Most are able to grow at or below a_w 0.85 (66).

Citreoviridin

The role of citreoviridin in the human disease acute cardiac beriberi has been well documented (95). Acute cardiac beriberi was a common disease in Japan in the second half of the 19th century. Symptoms were heart distress, labored breathing, nausea, and vomiting, followed by anguish, pain, restlessness, and sometimes maniacal behavior. In extreme cases, progressive paralysis leading to respiratory failure occurred. It is notable that victims of this disease were often young healthy adults. In 1910 the incidence of acute cardiac beriberi suddenly decreased. This decrease coincided with implementation of a government inspection scheme which dramatically reduced the sale of moldy rice in Japan (95).

The major source of citreoviridin is *P. citreonigrum* (synonyms *P. citreoviride, P. toxicarium*), a species which usually occurs in rice, less commonly in other cereals, and rarely in other foods (66, 74). Recent studies (76) failed to find more than occasional infections of *P. citreonigrum* in Southeast Asian rice. The threat of citreoviridin toxicosis at least in advanced Southeast Asian countries appears now to be very low. However, the possibility cannot be discounted that *P. citreonigrum* still occurs in less developed African or Asian countries where adequate rice-drying systems and controls are not yet in place.

Citreoviridin is also produced by *Eupenicillium ochrosalmoneum* (anamorph *P. ochrosalmoneum*), which is a relatively uncommon though widespread species, usually associated with cereals (66). The reported formation of citreoviridin by this species in standing corn in the United States is potentially of great concern (102), but it has not been found in Southeast Asian corn (75). The ecology of this mold-commodity-mycotoxin relationship remains to be fully elucidated.

Taxonomy

Classified in subgenus *Aspergilloides, P. citreonigrum* was described in 1901 but subsequently ignored because the description was meager. However, in 1979 the name was revived (66) on the basis of the 1923 neotypification by Biourge, as it has priority over *P. citreoviride*, the more widely used name. *E. ochrosalmoneum* is a relatively recently described species, with no taxonomic complications.

When grown on standard identification media, *P. citreonigrum* is a distinctive species. Colonies grow quite slowly (after 7 days at 25°C, 20 to 28 mm in diameter on CYA and 22 to 26 mm in diameter on MEA; 0 to 10 mm at 37°C), produce low numbers of pale grey-green conidia, and exhibit yellow mycelial, soluble pigment, and reverse colors. Penicilli consist of small clusters of phialides only; stipes are slender and not apically enlarged, and conidia are spherical, smooth walled, and tiny (66, 69).

E. ochrosalmoneum forms slowly growing colonies (usually less than 30 mm in diameter) which are bright yellow on both CYA and MEA at 25°C. Growth on CYA at 37°C is similar to that at 25°C. Penicilli are biverticillate, with few, divergent metulae (69).

Enumeration and Identification

Little specific information exists on techniques for enumerating *P. citreonigrum* or *E. ochrosalmoneum*. Both would be expected to grow satisfactorily on media such as DRBC or DG18 agar. No selective media have been developed, however.

Ecology

The physiology of *P. citreonigrum* and of citreoviridin production has been little studied. Cardinal temperatures are below 5°C, 20 to 24°C, and 37 to 38°C (72). The minimum a_w for growth is not known, but this species is undoubtedly xerophilic (74). Citreoviridin is produced at temperatures from 10 to 37°C, with a maximum near 20°C (93).

Penitrem A

Chemicals capable of inducing a tremorgenic (trembling) response in vertebrate animals are regarded as rare except for mold metabolites, in which case at least 20 such compounds have been reported (14, 104). Tremorgens are neurotoxins; in low doses they appear to cause no adverse effects in animals, which are able to feed and function more or less normally while sustained trembling occurs, even over periods as long as 18 days (45). However, relatively small increases in dosage (5- to 20-fold) can be rapidly lethal (17, 43, 44). The presence of tremorgens is exceptionally difficult to diagnose postmortem because no discernible pathological effects are produced.

Several tremorgenic mycotoxins are produced by *Penicillium* species, the most important being the highly toxic penitrem A. Verruculogen, equally toxic, is not produced by species of common occurrence in foods. Less toxic compounds include fumitremorgen B, paxilline, verrucosidin, and janthitrems (77). Although the tremorgenic toxicosis produced by penitrem A in animals is well documented, the response in humans is unclear. The acute toxicity of penitrem A to animals precludes experimental dosing to humans. Circumstantial evidence suggests that penitrem A may produce an emetic effect, which would render serious toxic effects much less likely.

The classification of *Penicillium* species producing penitrem A has been particularly confusing. Early literature described penitrem production by *P. cyclopium, P. palitans, P. puberulum, P. martensii,* and *P. crustosum* (10, 12, 103, 104). It was later concluded that this confusion was due to misidentification and that the only common foodborne species producing penitrem A is *P. crustosum* (67). Nearly all isolates of *P. crustosum* produce penitrem A at high levels; therefore, the presence of this species in food or feed is a warning signal (12, 26, 27).

Taxonomy

The validity of *P. crustosum* as a species was in doubt until 1980 (66). However, it has been accepted without any confusion since. A member of *Penicillium* subg. *Penicillium, P. crustosum* produces large penicilli, with rami, metulae, and phialides characteristic of this subgenus (Fig. 22.2B). It is one of the faster-growing species in section *Penicillium,* producing dull green colonies with a granular texture on both CYA and MEA. Microscopically, *P. crustosum* is characterized by large rough-walled stipes and smooth-walled, usually spherical conidia. However, the most distinctive feature of typical isolates is the production of enormous numbers of conidia on MEA, which become detached from the colony when the petri dish is jarred (66, 69).

Enumeration and Identification

Since *P. crustosum* grows relatively rapidly, enumeration is best carried out with a modern medium such as DRBC, which inhibits spreading of colonies (74). Confirmation of this species requires isolation and growth on standard identification media. No selective media have been developed for this species.

Ecology

P. crustosum is a ubiquitous spoilage mold, having been found in the majority of cereal and animal feed samples examined over a period of more than a decade (74). *P. crustosum* causes spoilage of corn, processed meats, nuts, cheese, and fruit juices, as well as being a weak pathogen on pomaceous fruits and cucurbits (74). The occurrence of penitrem A in animal feeds is well documented (10, 44, 80, 104). Its occurrence in human foods appears equally certain.

As has been noted earlier, *P. crustosum* has been poorly recognized until recently, and so a lack of information about growth and toxin production is to be expected. Like nearly all species in *Penicillium* subg. *Penicillium, P. crustosum* does not grow at 37°C. A recent study has shown that penitrem A is produced only at high a_w (72).

PR Toxin and Roquefortine

Treated together here are toxins produced by *P. roqueforti,* a species that is used in cheese manufacture but also can be a spoilage mold. Roquefortine has a relatively high 50% lethal dose (Table 22.1) but has been responsible for the deaths of dogs in Canada, with symptoms similar to strychnine poisoning (54). PR toxin is apparently much more toxic than roquefortine (Table 22.1), but has not been implicated in animal or human disease.

In addition to being produced by *P. roqueforti,* roquefortine has been shown to be produced by *P. crustosum* (16) and *P. chrysogenum* (25, 27). PR toxin is produced only by *P. roqueforti* (34).

Taxonomy

P. roqueforti is classified in subgenus *Penicillium* sect. *Penicillium* (Fig. 22.2C). This species has been shown to comprise two distinct varieties; *P. roqueforti* var. *roqueforti* is widely used as a starter culture for mold-ripened cheeses, while *P. roqueforti* var. *carneum* occurs in meats and silage (36). *P. roqueforti* var. *roqueforti* produces PR toxin, *P. roqueforti* var. *carneum* produces patulin, and both varieties produce roquefortine (36).

P. roqueforti grows very rapidly on CYA and MEA and produces green reverse colors on one or both media. Stipes are often very rough; conidia are large, spherical, and smooth walled (69).

Enumeration and Identification

Like other similar species, *P. roqueforti* can be enumerated on DRBC or DG18. A valuable property in some circumstances is the fact that *P. roqueforti* is more tolerant of acetic acid than are other *Penicillium* species. In consequence, *P. roqueforti* can be selectively enumerated on a medium such as malt acetic agar (MEA plus 0.5% glacial acetic acid) (74).

Ecology

P. roqueforti and the other two *Penicillium* species producing roquefortine are common in foods. As well as growing rapidly at refrigeration temperatures, *P. roqueforti* is able to grow in oxygen concentrations below 0.5%, even in the presence of up to 20% carbon dioxide (92); hence it is a common cause of spoilage in chilled meats, cheese, and other products (6, 7, 64). The presence of roquefortine in cheese and other foods is to be expected. Disease from PR toxin or roquefortine in association with cheese has not been reported (74). However, roquefortine has been implicated in a case of human intoxication from moldy beer infected with *P. crustosum* (16).

Because of their potential occurrence in staple foods, PR toxin and roquefortine are of considerable significance from the public health viewpoint. PR toxin is unstable in cheese in storage, but roquefortine has been isolated from finished products (84). Extensive searches for nontoxic strains for use as cheese starter cultures have so far been largely unsuccessful (51, 58).

Like other species in subgenus *Penicillium*, *P. roqueforti* is psychrotrophic, growing vigorously at temperatures as low as 2°C. It is notably tolerant of weak acid preservatives such as sorbic acid and is able to grow in 0.5% acetic acid or more (28). This species also has the lowest oxygen requirement for growth of any *Penicillium* species, being capable of growth in less than 0.5% oxygen in the presence of 20% carbon dioxide (92).

Secalonic Acid D

Secalonic acids are dimeric xanthones produced by a range of taxonomically distant molds (15). Secalonic acid D, the only secalonic acid produced by *Penicillium* species, has significant animal toxicity (11, 88). Secalonic acid D is produced as a major metabolite of *P. oxalicum*. It has been found in grain dusts at levels of up to 4.5 mg/kg (24). The possibility that such concentrations can be toxic to grain handlers by inhalation should not be ignored.

Taxonomy

P. oxalicum is a distinctive species in subgenus *Furcatum* sect. *Furcatum* (Fig. 22.2D). It grows rapidly on CYA at 25 and 37°C, forming a continuous layer of conidia which, under a low-power microscope, can be seen to lie in closely packed, readily fractured sheets, with a uniquely shiny appearance. Penicilli are terminal and biverticillate, with long phialides producing large ellipsoidal conidia (69).

Enumeration and Identification

P. oxalicum can be enumerated without difficulty on standard media such as DRBC and DG18. Selective media have not been developed, however.

Ecology

A major habitat for *P. oxalicum* is corn at harvest, from which it is the most common *Penicillium* species isolated (39, 53, 59, 74). The occurrence of secalonic acid D in corn and hence in grain dusts is a potential hazard (24). *P. oxalicum* grows at temperatures of about 8 to about 40°C (74). The minimum a_w for conidial germination is 0.86 (74).

QUANTIFYING MYCOTOXINS

Quantifying most *Penicillium* mycotoxins is relatively straightforward, but procedures vary with commodity and toxin. As mycotoxins are usually present in minute amounts, best practice is essential to obtain good results. Scrupulous attention to safety in analyst training, in laboratory equipment and procedures, and in toxin containment and disposal is necessary.

Basic procedures for mycotoxin assays are sampling and subsampling, extraction and cleanup, and detection, quantification, and confirmation. Each of these areas is described in general terms below.

Sampling and Subsampling

The incidence of mycotoxins in a food commodity is usually heterogeneous, and so sampling plays an important part in assay accuracy. Adequate (representative) sample sizes depend on particle size. While a 500-g sample is adequate for oils or milk powder, 3 kg of flour or peanut butter is necessary. To be representative, corn samples should be 10 kg and peanut samples should be 20 kg per batch (87).

Samples of raw materials should be taken with online samplers during processing or be composites from several sites in raw material lots. Much attention has been paid to sample plans for aflatoxins in peanuts (5, 13, 79, 101), corn (22), and figs (86). Such detailed studies have not been carried out on sampling for other mycotoxins, but the aflatoxin systems should have broad applicability.

To obtain a representative subsample for analysis, samples of particulate foods should be ground or blended. After thorough mixing, subsamples (20 to 50 g) are taken. To check on the adequacy of sampling procedures, assays are often performed in duplicate.

Extraction and Cleanup

To release toxins from the ground food material, extraction in a solvent system is necessary. This is usually carried out by blending the subsample with a suitable solvent in an explosionproof blender for 1 to 3 min or shaking it on a wrist-action shaker (30 min). Various solvent systems are used, depending on circumstances and likely toxins, the most common being methanol plus acidified water or chloroform (38).

The extract usually requires some form of cleanup procedure, in particular to remove lipids and pigments which would interfere with analysis. Dialysis, precipitation, chromatographic columns, solvent partitioning, and specific antibody layers are all in use (87). Concentration of the purified extract in a steam bath or vacuum rotary evaporator is then carried out.

Detection and Quantification

One of the earliest and still the most common techniques for mycotoxin assays is thin-layer chromatography (TLC). Concentrated extracts are spotted (1 to 10 µl) on glass or aluminum plates coated with a thin layer of activated silica gel and then developed in tanks containing suitable solvents designed to separate toxins from interfering chemical of all types. For some substrates, two-dimensional TLC (developing the plate twice at right angles) is of value. Appropriate standards are also run at the same time.

After development, toxins are visualized by comparison of spots on the chromatogram with standard spots. Many mycotoxins, including aflatoxins, fluoresce under UV light and are readily visualized. Others require reaction with spray reagents, such as sulfuric acid or aluminum chloride. Comparison of the R_f and color of unknown spots against standards enables separation of a wide variety of toxins. The intensity of spots compared with that of suitable standards provides quantification. Confirmation is important and usually involves use of a spray reagent to produce a different color.

A newer technique, more precise but not necessarily more accurate than TLC, is high-performance liquid chromatography (HPLC). Passage of extracts through a long column packed with an inert layer and a suitable adsorbent causes separation of molecules in a manner similar to that for TLC. Visualization is by spectrophotometric detectors. Sometimes compounds must be derivatized before analysis, to improve sensitivity. HPLC is more sensitive than TLC and more suited to automation.

Rapid Methods of Analysis

Assay techniques such as TLC and HPLC are effective but slow. The search for more rapid methods continues. Observation of cracked corn kernels under UV light has long been used as a screening technique for aflatoxins. This method is also of value for aflatoxin in figs but ineffective for peanuts, which autofluoresce under UV. Peanuts are screened for aflatoxins by using minicolumns, in which aflatoxin is bound to an absorbent material in a small tube and then assayed by UV light.

The newest approach to rapid assays involves the use of antibodies developed to specific toxins. Spot or minicolumn tests using antibodies, called immunoassays, have been developed for several major toxins (87).

Specific assay methods exist for most *Penicillium* toxins, using the general principles outlined above (38, 87). However, the literature for the less common and less well known toxins is widely scattered.

CONCLUSIONS

Penicillium species produce a very wide range of toxic compounds. The role of one toxin, citreoviridin, in a historical human health problem is well established. Of the other toxins discussed here, only ochratoxin is currently regarded as a serious threat to human health. However, *Penicillium* species are so widespread and abundant in foods and feeds that they must be considered to be a potential hazard to both human and animal health. Many species make several compounds known to be toxic, and the possibility of synergy also cannot be ignored. Much more research is needed to improve detection methods and understand the ecology of toxigenic *Penicillium* species and to evaluate the significance of *Penicillium* toxins in human health.

References

1. **Alsberg, C. L., and O. F. Black.** 1913. Contributions to the study of maize deterioration: biochemical and toxicological investigations of *Penicillium puberulum* and *Penicillium stoloniferum*. *Bull. Bur. Anim. Ind. U.S. Dept. Agric.* **270:**1–47.
2. **Austwick, P. K. C.** 1981. Balkan nephropathy. *Practitioner* **1981**(1):1038.
3. **Brackett, R. E., and E. H. Marth.** 1979. Patulin in apple juice from roadside stands in Wisconsin. *J. Food Prot.* **42:**862–863.
4. **Bretholtz-Emanuelsson, A., M. Olsen, A. Oskarsson, I. Palminger, and K. Hult.** 1993. Ochratoxin A in cow's milk and human milk with corresponding human blood samples. *J. Assoc. Off. Anal. Chem. Int.* **76:**842–846.
5. **Brown, G. H.** 1982. Sampling for 'needles in haystacks'. *Food Technol. Aust.* **34:**224–227.
6. **Bullerman, L. B.** 1980. Incidence of mycotoxic molds in domestic and imported cheeses. *J. Food Safety* **2:**47–58.
7. **Bullerman, L. B.** 1981. Public health significance of molds and mycotoxins in fermented dairy products. *J. Dairy Sci.* **64:**2439–2452.

8. **Carlton, W. W., G. Sansing, G. M. Szczech, and J. Tuite.** 1974. Citrinin mycotoxicosis in beagle dogs. *Food Cosmet. Toxicol.* **12:**479–490.
9. **Castegnaro, M., R. Pleština, G. Dirheimer, I. N. Chernozemsky, and H. Bartsch.** 1991. *Mycotoxins, Endemic Nephropathy and Urinary Tract Tumours.* IARC Scientific Publications no. 115. World Health Organization/International Agency for Research on Cancer, Lyon, France.
10. **Ciegler, A.** 1969. Tremorgenic toxin from *Penicillium palitans. Appl. Microbiol.* **18:**128–129.
11. **Ciegler, A., A. W. Hayes, and R. F. Vesonder.** 1980. Production and biological activity of secalonic acid D. *Appl. Environ. Microbiol.* **39:**285–287.
12. **Ciegler, A., and J. I. Pitt.** 1970. Survey of the genus *Penicillium* for tremorgenic toxin production. *Mycopathol. Mycol. Appl.* **42:**119–124.
13. **Coker, R. D.** 1989. Control of aflatoxin in groundnut products with emphasis on sampling, analysis, and detoxification, p. 123–132. *In Aflatoxin Contamination of Groundnut: Proceedings of the International Workshop, 6–9 October, 1987, ICRISAT Centre, India.* ICRISAT, Patancheru, India.
14. **Cole, R. J.** 1981. Tremorgenic mycotoxins: an update, p. 17–33. *In* R. L. Ory (ed.), *Antinutrients and Natural Toxicants in Foods.* Food and Nutrition Press, Westport, Conn.
15. **Cole, R. J., and R. H. Cox.** 1981. *Handbook of Toxic Fungal Metabolites.* Academic Press, New York.
16. **Cole, R. J., J. W. Dorner, R. H. Cox, and L. W. Raymond.** 1983. Two classes of alkaloid mycotoxins produced by *Penicillium crustosum* Thom isolated from contaminated beer. *J. Agric. Food Chem.* **31:**655–657.
17. **Cole, R. J., J. W. Kirksey, J. H. Moore, B. R. Blankenship, U. L. Diener, and N. D. Davis.** 1972. Tremorgenic toxin from *Penicillium verruculosum. Appl. Microbiol.* **24:**248–250.
18. **Cruickshank, R. H., and J. I. Pitt.** 1987. The zymogram technique: isoenzyme patterns as an aid in *Penicillium* classification. *Microbiol. Sci.* **4:**14–17.
19. **Cruickshank, R. H., and J. I. Pitt.** 1987. Identification of species in *Penicillium* subgenus *Penicillium* by enzyme electrophoresis. *Mycologia* **79:**614–620.
20. **De Jesus, A. E., P. S. Steyn, F. R. van Heerden, R. Vleggaar, and P. L. Wessels.** 1981. Structure and biosynthesis of the penitrems A-F, six novel tremorgenic mycotoxins from *Penicillium crustosum. Chem. Commun. (J. Chem. Soc. Sect. D)* **1981:**289–291.
21. **Dickens, F., and H. E. H. Jones.** 1961. Carcinogenic activity of a series of reactive lactones and related substances. *Br. J. Cancer* **15:**85–100.
22. **Dickens, J. W., and T. B. Whitaker.** 1983. Sampling, BGYF, and aflatoxin analysis in corn, p. 35–37. *In* U. L. Diener, R. L. Asquith, and J. W. Dickens (ed.), *Aflatoxin and Aspergillus flavus in Corn.* Alabama Agricultural Experiment Station, Auburn University, Auburn, Ala.
23. **Dorner, J. W., R. J. Cole, L. G. Lomax, H. S. Gosser, and U. L. Diener.** 1983. Cyclopiazonic acid production by *Aspergillus flavus* and its effect on broiler chickens. *Appl. Environ. Microbiol.* **46:**698–703.
24. **Ehrlich, K. C., L. S. Lee, A. Ciegler, and M. S. Palmgren.** 1982. Secalonic acid D: natural contaminant of corn dust. *Appl. Environ. Microbiol.* **44:**1007–1008.
25. **El-Banna, A. A., J. Fink-Gremmels, and L. Leistner.** 1987. Investigation of *Penicillium chrysogenum* isolates for their suitability as starter cultures. *Mycotoxin Res.* **3:**77–83.
26. **El-Banna, A. A., and L. Leistner.** 1988. Production of penitrem A by *Penicillium crustosum* from foodstuffs. *Int. J. Food Microbiol.* **7:**9–17.
27. **El-Banna, A. A., J. I. Pitt, and L. Leistner.** 1987. Production of mycotoxins by *Penicillium* species. *Syst. Appl. Microbiol.* **10:**42–46.
28. **Engel, G., and M. Teuber.** 1978. Simple aid for the identification of *Penicillium roqueforti* Thom. *Eur. J. Appl. Microbiol. Biotechnol.* **6:**107–111.
29. **Filtenborg, O., and J. C. Frisvad.** 1980. A simple screening method for toxigenic moulds in pure culture. *Lebensm. Wiss. Technol.* **13:**128–130.
30. **Filtenborg, O., J. C. Frisvad, and J. A. Svendsen.** 1983. Simple screening method for molds producing intracellular mycotoxins in pure cultures. *Appl. Environ. Microbiol.* **45:**581–585.
31. **Friis, P., E. Hasselager, and P. Krogh.** 1969. Isolation of citrinin and oxalic acid from *Penicillium viridicatum* Westling and their nephrotoxicity in rats and pigs. *Acta Pathol. Microbiol. Scand.* **77:**559–560.
32. **Frisvad, J. C.** 1983. A selective and indicative medium for groups of *Penicillium viridicatum* producing different mycotoxins in cereals. *J. Appl. Bacteriol.* **54:**409–416.
33. **Frisvad, J. C.** 1986. Taxonomic approaches to mycotoxin identification, p. 415–457. *In* R. J. Cole (ed.), *Modern Methods in the Analysis and Structural Elucidation of Mycotoxins.* Academic Press, Orlando, Fla.
34. **Frisvad, J. C.** 1987. High performance liquid chromatographic determination of profiles of mycotoxins and other secondary metabolites. *J. Chromatogr.* **392:**333–347.
35. **Frisvad, J. C., and O. Filtenborg.** 1983. Classification of terverticillate penicillia based on profiles of mycotoxins and other secondary metabolites. *Appl. Environ. Microbiol.* **46:**1301–1310.
36. **Frisvad, J. C., and O. Filtenborg.** 1989. Terverticillate Penicillia: chemotaxonomy and mycotoxin production. *Mycologia* **81:**837–861.
37. **Frisvad, J. C., and B. T. Viuf.** 1986. Comparison of direct and dilution plating for detection of *Penicillium viridicatum* in barley containing ochratoxin, p. 45–47. *In* A. D. King, J. I. Pitt, L. R. Beuchat, and J. E. L. Corry (ed.), *Methods for the Mycological Examination of Food.* Plenum Press, New York.
38. **Helrich, K. (ed.).** 1990. *Official Methods of Analysis of the Association of Official Analytical Chemists,* 15th ed. Association of Official Analytical Chemists, Arlington, Va.
39. **Hesseltine, C. W., R. F. Rogers, and O. L. Shotwell.** 1981. Aflatoxin and mold flora in North Carolina in 1977 corn crop. *Mycologia* **73:**216–228.
40. **Hocking, A. D., and M. Faedo.** 1992. Fungi causing thread mould spoilage of vacuum packaged Cheddar

cheese during maturation. *Int. J. Food Microbiol.* **16**:123–130.

41. **Hocking, A. D., and J. I. Pitt.** 1979. Water relations of some *Penicillium* species at 25°C. *Trans. Br. Mycol. Soc.* **73**:141–145.
42. **Hocking, A. D., J. I. Pitt, R. A. Samson, and A. D. King.** 1992. Recommendations from the closing session of SMMEF II, p. 359–364. *In* R. A. Samson, A. D. Hocking, J. I. Pitt, and A. D. King (ed.), *Modern Methods in Food Mycology.* Elsevier, Amsterdam.
43. **Hou, C. T., A. Ciegler, and C. W. Hesseltine.** 1971. Tremorgenic toxins from Penicillia. II. A new tremorgenic toxin, tremortin B, from *Penicillium palitans. Can. J. Microbiol.* **17**:599–603.
44. **Hou, C. T., A. Ciegler, and C. W. Hesseltine.** 1971. Tremorgenic toxins from penicillia. III. Tremortin production by *Penicillium* species on various agricultural commodities. *Appl. Microbiol.* **21**:1101–1103.
45. **Jortner, B. S., M. Ehrich, A. E. Katherman, W. R. Huckle, and M. E. Carter.** 1986. Effects of prolonged tremor due to penitrem A in mice. *Drug Chem. Toxicol.* 9:101–116.
46. **King, A. D., J. I. Pitt, L. R. Beuchat, and J. E. L. Corry (ed.).** 1986. *Methods for the Mycological Examination of Food.* Plenum Press, New York.
47. **Krogh, P., N. H. Axelsen, F. Elling, N. Gyrd-Hansen, B. Hald, J. Hyldgaard-Jensen, A. E. Larsen, A. Madsen, H. P. Mortensen, T. Møller, O. K. Petersen, U. Ravnskov, M. Rostgaard, and O. Aalund.** 1974. Experimental porcine nephropathy. *Acta Pathol. Microbiol. Scand. Sect. A Suppl.* **246**:1–21.
48. **Krogh, P., B. Hald, P. Englund, L. Rutqvist, and O. Swahn.** 1974. Contamination of Swedish cereals with ochratoxin A. *Acta Pathol. Microbiol. Scand. Sect. B* **82**: 301–302.
49. **Krogh, P., B. Hald, and E. J. Pedersen.** 1973. Occurrence of ochratoxin A and citrinin in cereals associated with mycotoxic porcine nephropathy. *Acta Pathol. Microbiol. Scand. Sect. B* **81**:689–695.
50. **Krogh, P., B. Hald, R. Plestina, and S. Ceovic.** 1977. Balkan (endemic) nephropathy and foodborn ochratoxin A: preliminary results of a survey of foodstuffs. *Acta Pathol. Microbiol. Scand. Sect. B* **85**:238–240.
51. **Leistner, L., R. Geisen, and J. Fink-Gremmels.** 1989. Mould-fermented foods of Europe: hazards and developments, p. 145–154. *In* S. Natori, K. Hashimoto, and Y. Ueno (ed.), *Mycotoxins and Phytotoxins '88.* Elsevier Science, Amsterdam.
52. **Leistner, L., and J. I. Pitt.** 1977. Miscellaneous *Penicillium* toxins, p. 639–653. *In* J. V. Rodricks, C. W. Hesseltine, and M. E. Mehlman (ed.), *Mycotoxins in Human and Animal Health.* Pathotox Publishers, Park Forest South, Ill.
53. **Lichtwardt, R. W., and L. H. Tiffany.** 1958. Mold flora associated with shelled corn in Iowa. *Iowa State Coll. J. Sci.* **33**:1–11.
54. **Lowes, N. R., R. A. Smith, and B. E. Beck.** 1992. Roquefortine in the stomach contents of dogs suspected of strychnine poisoning. *Can. Vet. J.* **33**:535–538.
55. **Lund, F., O. Filtenborg, and J. C. Frisvad.** 1995. Associated mycoflora of cheese. *Food Microbiol.* **12**:173–180.
56. **Macgeorge, K. M., and P. G. Mantle.** 1990. Nephrotoxicity of *Penicillium aurantiogriseum* and *P. commune* from an endemic nephropathy area of Yugoslavia. *Mycopathologia* **112**:139–145.
57. **Mantle, P. G., and K. M. Macgeorge.** 1991. Nephrotoxic fungi in a Yugoslav community in which Balkan nephropathy is hyperendemic. *Mycol. Res.* **95**:660–664.
58. **Medina, M., P. Gaya, and M. Nunez.** 1985. Production of PR toxin and roquefortine by *Penicillium roqueforti* isolates from Cabrales blue cheese. *J. Food Prot.* **48**:118–121.
59. **Mislivec, P. B., and J. Tuite.** 1970. Species of *Penicillium* occurring in freshly-harvested and in stored dent corn kernels. *Mycologia* **62**:67–74.
60. **Miyake, I., H. Naito, and H. Sumeda.** 1940. *Rep. Res. Inst. Rice Improvement* **1**:1. (In Japanese.)
61. **Mortimer, D. N., I. Parker, M. J. Shepherd, and J. Gilbert.** 1985. A limited survey of retail apple and grape juices for the mycotoxin patulin. *Food Addit. Contam.* 2:165–170.
62. **Natori, S., S. Sakaki, H. Kurata, S. Udagawa, M. Ichinoe, M. Saito, and M. Umeda.** 1970. Chemical and cytotoxicity survey on the production of ochratoxins and penicillic acid by *Aspergillus ochraceus* Wilhelm. *Chem. Pharm. Bull.* **18**:2259–2268.
63. **Northolt, M. D., H. P. van Egmond, and W. E. Paulsch.** 1979. Ochratoxin A production by some fungal species in relation to water activity and temperature. *J. Food Prot.* **42**:485–490.
64. **Northolt, M. D., H. P. van Egmond, P. Soentoro, and E. Deijll.** 1980. Fungal growth and the presence of sterigmatocystin in hard cheese. *J. Assoc. Off. Anal. Chem.* **63**:115–119.
65. **Petkova-Bocharova, T., and M. Castegnaro.** 1991. Ochratoxin A in human blood in relation to Balkan endemic nephropathy and urinary tract tumours in Bulgaria, p. 135–137. *In* M. Castegnaro, R. Pleština, G. Dirheimer, I. N. Chernozemsky, and H. Bartsch (ed.), *Mycotoxins, Endemic Nephropathy and Urinary Tract Tumours.* IARC Scientific Publications no. 115. World Health Organization/ International Agency for Research on Cancer, Lyon, France.
66. **Pitt, J. I.** 1979. *The Genus Penicillium and Its Teleomorphic States Eupenicillium and Talaromyces.* Academic Press, London.
67. **Pitt, J. I.** 1979. *Penicillium crustosum* and *P. simplicissimum,* the correct names for two common species producing tremorgenic mycotoxins. *Mycologia* **71**:1166–1177.
68. **Pitt, J. I.** 1987. *Penicillium viridicatum, Penicillium verrucosum,* and production of ochratoxin A. *Appl. Environ. Microbiol.* **53**:266–269.
69. **Pitt, J. I.** 1988. *A Laboratory Guide to Common Penicillium Species,* 2nd ed. CSIRO Division of Food Processing, North Ryde, N.S.W., Australia.
70. **Pitt, J. I.** 1988. *PENNAME, a New Computer Key to Common Penicillium Species.* CSIRO Division of Food Processing, North Ryde, N.S.W., Australia.

71. **Pitt, J. I.** 1989. Recent developments in the study of *Penicillium* and *Aspergillus* systematics. *J. Appl. Bacteriol. Symp. Ser. Suppl.* **1989:**37S–45S.

72. **Pitt, J. I.** 1996. Toxigenic fungi: *Penicillium*, p. 397–413. *In* International Commission on Microbiological Specifications for Foods (ed.), *Microorganisms in Foods. 5. Microbiological Specifications of Food Pathogens.* Blackie, London.

73. **Pitt, J. I., R. H. Cruickshank, and L. Leistner.** 1986. *Penicillium commune, P. camembertii,* the origin of white cheese moulds, and the production of cyclopiazonic acid. *Food Microbiol.* **3:**363–371.

74. **Pitt, J. I., and A. D. Hocking.** 1985. *Fungi and Food Spoilage.* Academic Press, Sydney.

75. **Pitt, J. I., A. D. Hocking, K. Bhudhasamai, B. F. Miscamble, K. A. Wheeler, and P. Tanboon-Ek.** 1993. The normal mycoflora of commodities from Thailand. 1. Nuts and oilseeds. *Int. J. Food Microbiol.* **20:**211–226.

76. **Pitt, J. I., A. D. Hocking, K. Bhudhasamai, B. F. Miscamble, K. A. Wheeler, and P. Tanboon-Ek.** 1994. The normal mycoflora of commodities from Thailand. 2. Beans, rice and other commodities. *Int. J. Food Microbiol.* **23:**35–53.

77. **Pitt, J. I., and L. Leistner.** 1991. Toxigenic *Penicillium* species, p. 91–99. *In* J. E. Smith and R. S. Henderson (ed.), *Mycotoxins and Animal Foods.* CRC Press, Boca Raton, Fla.

78. **Pitt, J. I., and R. A. Samson.** 1993. Species names in current use in the *Trichocomaceae* (Fungi, Eurotiales). *Regnum Veg.* **128:**13–57.

79. **Read, M.** 1989. Removal of aflatoxin from the Australian groundnut crop, p. 133–140. *In Aflatoxin Contamination of Groundnut: Proceedings of the International Workshop, 6–9 October, 1987, ICRISAT Centre, India.* ICRISAT, Patancheru, India.

80. **Richard, J. L., and L. H. Arp.** 1979. Natural occurrence of the mycotoxin penitrem A in moldy cream cheese. *Mycopathologia* **67:**107–109.

81. **Samson, R. A., E. S. Hoekstra, J. C. Frisvad, and O. Filtenborg.** 1995. *Introduction to Food-Borne Fungi,* 4th ed. Centraalbureau voor Schimmelcultures, Baarn, The Netherlands.

82. **Samson, R. A., and J. I. Pitt (ed.).** 1990. *Modern Concepts in Penicillium and Aspergillus Classification.* Plenum Press, New York.

83. **Samson, R. A., A. C. Stolk, and R. Hadlok.** 1976. Revision of the Subsection Fasciculata of *Penicillium* and some allied species. *Stud. Mycol.* (Baarn) **11:**1–47.

84. **Scott, P. M.** 1977. *Penicillium* mycotoxins, p. 283–356. *In* T. D. Wyllie and L. G. Morehouse (ed.), *Mycotoxic Fungi, Mycotoxins, Mycotoxicoses, an Encyclopedic Handbook,* vol. 1. *Mycotoxigenic Fungi.* Marcel Dekker, New York.

85. **Seifert, K. A., and R. A. Samson.** 1985. The genus *Coremium* and the synnematous Penicillia, p. 143–154. *In* R. A. Samson and J. I. Pitt (ed.), *Advances in Penicillium and Aspergillus Systematics.* Plenum Press, New York.

86. **Sharman, M., S. Macdonald, A. J. Sharkey, and J. Gilbert.** 1994. Sampling bulk consignments of dried figs for aflatoxin analysis. *Food Addit. Contam.* **11:**17–23.

87. **Smith, J. E., C. W. Lewis, J. G. Anderson, and G. L. Solomons.** 1994. *Mycotoxins in Human Nutrition and Health.* Report EUR 16048 EN. European Commission Directorate-General XII, Brussels.

88. **Sorenson, W. G., F. H. Y. Green, V. Vallyathan, and A. Ciegler.** 1982. Secalonic acid D toxicity in rat lung. *J. Toxicol. Environ. Health* **9:**515–525.

89. **Stack, M. E., R. M. Eppley, P. A. Dreifuss, and A. E. Pohland.** 1977. Isolation and identification of xanthomegnin, viomellein, rubrosulphin, and viopurpurin as metabolites of *Penicillium viridicatum. Appl. Environ. Microbiol.* **33:**351–355.

90. **Still, P., C. Eckardt, and L. Leistner.** 1978. Bildung von Cyclopazonsaure durch *Penicillium camembertii*-isolate von Kase. *Fleischwirtschaft* **58:**876–878.

91. **Stott, W. T., and L. B. Bullerman.** 1975. Patulin: a mycotoxin of potential concern in foods. *J. Milk Food Technol.* **38:**695–705.

92. **Taniwaki, M.** 1995. Growth and mycotoxin production by fungi under modified atmospheres. Ph.D. thesis. University of New South Wales, Kensington, N.S.W., Australia.

93. **Ueno, Y.** 1972. Temperature-dependent production of citreoviridin, a neurotoxin of *Penicillium citreo-viride* Biourge. *Jpn. J. Exp. Med.* **42:**107–114.

94. **Ueno, Y., and I. Ueno.** 1972. Isolation and acute toxicity of citreoviridin, a neurotoxic mycotoxin of *Penicillium citreo-viride* Biourge. *Jpn. J. Exp. Med.* **42:**91–105.

95. **Uraguchi, K.** 1969. Mycotoxic origin of cardiac beriberi. *J. Stored Prod. Res.* **5:**227–236.

96. **Van der Merwe, K. J., P. S. Steyn, L. Fourie, D. B. Scott, and J. J. Theron.** 1965. Ochratoxin A, a toxic metabolite produced by *Aspergillus ochraceus Wilh. Nature* (London) **205:**1112–1113.

97. **Van Egmond, H. P.** 1989. Current situation on regulations for mycotoxins. Overview of tolerances and status of standard methods of sampling and analysis. *Food Addit. Contam.* **6:**139–188.

98. **Van Rensburg, S. J.** 1984. Subacute toxicity of the mycotoxin cyclopiazonic acid. *Food Chem. Toxicol.* **22:**993–998.

99. **Van Walbeek, W., P. M. Scott, J. Harwig, and J. W. Lawrence.** 1969. *Penicillium viridicatum* Westling: a new source of ochratoxin A. *Can. J. Microbiol.* **15:**1281–1285.

100. **Watkins, K. L., G. Fazekas, and M. V. Palmer.** 1990. Patulin in Australian apple juice. *Food Aust.* **42:**438–439.

101. **Whitaker, T. B., F. E. Dowell, W. M. Hagler, F. G. Giesbrecht, and J. Wu.** 1994. Variability associated with sampling, sample preparation, and chemical testing for aflatoxin in farmers' stock peanuts. *J. Assoc. Off. Anal. Chem. Int.* **77:**107–116.

102. **Wicklow, D. T., and R. J. Cole.** 1984. Citreoviridin in standing corn infested by *Eupenicillium ochrosalmoneum. Mycologia* **76:**959–961.

103. **Wilson, B. J., T. Hoekman, and W. D. Dettbarn.** 1972. Effects of a fungus tremorgenic toxin (penitrem A) on transmission in rat phrenic nerve-diaphragm preparations. *Brain Res.* **40:**540–544.

104. **Wilson, B. J., C. H. Wilson, and A. W. Hayes.** 1968. Tremorgenic toxin from *Penicillium cyclopium* grown on food materials. *Nature* (London) **220:**77–78.

Lloyd B. Bullerman

23 Fusaria and Toxigenic Molds Other than Aspergilli and Penicillia

Toxigenic molds in genera other than *Aspergillus* and *Penicillium* are most often found as contaminants of plant-derived foods, especially cereal grains. As such, these molds and their metabolites (mycotoxins) find their way into animal feeds and human foods. Animals, both food-producing animals and pets, are more often affected by the toxins of these molds than are humans because of the nature of animal feed and the way it is stored and handled. Human food generally is higher in quality and more carefully protected, particularly in the so-called developed countries in temperate climates. In other areas with more tropical and subtropical climates, this may not be the case. Nevertheless, there is evidence that grains and processed human foods are occasionally contaminated with mycotoxins produced by molds other than aspergilli and penicillia (30). Also, exposure of food-producing animals to mycotoxigenic molds may have an impact on the human food supply by causing death of animals, reducing their rate of growth, or depositing toxins in meats, milk, and eggs. Diseases in animals caused by mycotoxins may also suggest that similar conditions can occur in humans.

The most important group of mycotoxigenic molds other than *Aspergillus* and *Penicillium* species are species of the genus *Fusarium* (42). Many *Fusarium* species are plant pathogens, while others are saprophytic; most can be found in the soil (15, 16). In terms of human foods, *Fusarium* species are most often encountered as contaminants of cereal grains, oil seeds, and beans. Corn, wheat, and products made from these grains are most commonly contaminated. However, barley, rye, triticale, millet, and oats can also be contaminated.

Another mold genus other than *Aspergillus* and *Penicillium* that can produce mycotoxins is *Alternaria* (78). *Alternaria* species are widely distributed in the environment and can be found in soil, decaying plant materials, and dust. Both plant pathogenic and saprophytic species occur. Toxigenic molds also belong to several other genera. These include species of *Acremonium, Chaetomium,* and *Claviceps, Diplodia maydis, Myrothecium* species, *Phoma herbarum, Phomopsis* species, *Pithomyces chartarum,* species of *Rhizoctonia, Rhizopus,* and *Stachybotrys,* and *Trichothecium roseum* (36).

DISEASES OF HUMANS ASSOCIATED WITH *FUSARIUM* SPECIES

Alimentary Toxic Aleukia

During World War II, a very severe human disease occurred in the former Soviet Union, particularly in the Orenburg area of Russia. The disease, known as alimentary toxic aleukia, is believed to be caused by T-2 and HT-2 toxins produced by *Fusarium sporotrichioides* and *F. poae* of *Fusarium* sect. *Sporotrichiella* (32, 38). Because of the war, there was such a shortage of farm workers that much grain was not harvested in the fall and overwintered in the field. By spring, near-famine conditions

Table 23.1 Human diseases associated with *Fusarium* species and toxins

Disease	Food(s)	Organism(s)	Toxin(s)	Symptoms or effects
Alimentary toxic	Cereal grains, wheat, rye, bread	*F. sporotrichioides*, *F. poae*	T-2 toxin	Burning sensation in mouth and throat, vomiting, diarrhea, abdominal pain, bone marrow destruction, hemorrhaging, death
Urov or Kasin-Bek disease	Cereal grains	*F. poae*	Unknown	Osteoarthritis, shortened long bones, deformed joints, muscular weakness
Drunken bread	Cereal grains, wheat, bread	*F. graminearum*	Unknown	Headache, dizziness, tinnitus, trembling, unsteady gait, abdominal pain, nausea, diarrhea
Akakabi-byo; scabby grain intoxication	Cereal grains, wheat, barley, noodles	*F. graminearum*	Unknown, possibly deoxynivalenol	Anorexia, nausea, vomiting, headache, abdominal pain, diarrhea, chills, giddiness, convulsions
Foodborne illness outbreaks	Cereal grains, wheat, barley, corn, bread	*F. graminearum*	Deoxynivalenol, acetyldeoxynivalenol, nivalenol, T-2 toxin	Irritation of throat, nausea, headaches, vomiting, abdominal pain, diarrhea
Esophageal cancer	Corn	*F. moniliforme*	Unknown, possibly fumonisins and other toxins	Precancerous and cancerous lesions in the esophagus

existed and people were forced to eat the overwintered cereal grains that were milled into flour and made into bread. This resulted in a severe toxicosis manifested as alimentary toxic aleukia. The disease developed over several weeks, with increasingly severe symptoms upon continued consumption of the toxic grain.

There are actually four stages to the disease (32). The first stage is characterized by symptoms that appear a short time after ingestion of the toxic grain. These symptoms include a burning sensation in the mouth, tongue, esophagus, and stomach (Table 23.1). The stomach and intestinal mucosa become inflamed, resulting in vomiting, diarrhea, and abdominal pain. This first stage may appear and disappear rather quickly, within 3 to 9 days. During the second stage, the individual experiences no outward signs of the disease and feels well. However, during this stage the hematopoietic system is being damaged or destroyed by progressive leukopenia, granulopenia, and lymphocytosis. The blood-making capacity of the bone marrow is being destroyed, the platelet count decreases, and anemia develops. The leukocyte count decreases, and secondary bacterial infections occur. There are also disturbances in the central and autonomic nervous systems. The third stage develops suddenly and is marked by petechial hemorrhage on the skin and mucous membranes. The hemorrhaging becomes more severe, with bleeding from the nose and gums and hemorrhaging in the stomach and intestines. Necrotic lesions also develop in the mouth, gums, mucosa, larynx, and vocal cords. At this stage the disease is highly fatal, with up to 60% mortality. If death does not occur, the fourth stage of the disease is recovery or convalescence. It takes 3 to 4 weeks of treatment for the necrotic lesions, hemorrhaging, and bacterial infections to clear up and 2 months or more for the blood-making capacity of the bone marrow to return to normal. Alimentary toxic aleukia has been reproduced in cats and monkeys administered T-2 toxin isolated from the strain of *F. sporotrichioides* involved in the fatal human outbreaks of the disease in Russia. The related compound HT-2 toxin may also contribute to the disease.

Urov or Kashin-Bek Disease

Urov or Kashin-Bek disease has been observed among the Cossack people of eastern Russia for well over 100 years (29). The disease has been endemic in areas along the Urov river in Siberia and was studied extensively by two Russian scientists, Kashin and Bek, thus the name Urov or Kashin-Bek disease (32). The disease is a deforming bone-joint osteoarthrosis that is manifested by a shortening of long bones and a thickening deformation of joints, plus muscular weakness and atrophy (Table 23.1). The disease occurs most commonly in preschool and school-age children.

Several hypotheses have been advanced to describe possible causes of the disease, including nutritional deficiencies caused by low levels of calcium and other minerals in water and food crops of the region, and hereditary considerations. These ideas have been discounted by Russian scientists in favor of a mycotoxic origin of the disease (38). This conclusion is based on the fact that the disease has been reproduced in puppies and rats by feeding *F. poae* isolates obtained only from regions where the disease is endemic. No specific toxin(s) has yet been isolated and proven to be the definite cause, and so the actual etiology has not been completely resolved. The same disease has also been reported in North Korea and northern China.

Drunken Bread

Another human mycotoxicosis reported in the former Soviet Union is known as drunken bread (29). This syndrome is apparently caused by consumption of bread made from rye grain infected with *F. graminearum*. The illness is milder than alimentary toxic aleukia and is a nonfatal self-limiting disorder. Symptoms associated with this disease are headache, dizziness, tinnitus, trembling, and shaking of the extremities, with an unsteady or stumbling gait, hence the name (Table 23.1). There is also flushing of the face and gastrointestinal symptoms including abdominal pain, nausea, and diarrhea. Victims may appear to be euphoric and confused. The duration of the illness is 1 to 2 days after consumption of the toxic food is ceased. The infected grain may seem normal or appear shriveled and light in weight with a white to pinkish coloration suggestive of fusarium head blight, or scab.

Akakabi-byo

Akakabi-byo is also called scabby grain intoxication or red mold disease. It has been observed in Russia, Japan, and China (38). In China the disease is known as michum. In all cases the disease is associated with eating bread made from scab-infested wheat, barley, or other grains infected with *F. graminearum*. Symptoms of this illness are anorexia, nausea, vomiting, headache, abdominal pain, diarrhea, chills, giddiness, and convulsions (Table 23.1). Clinical signs are similar to those of the so called drunken bread syndrome discussed above. No specific mycotoxins have been shown to cause this illness, but deoxynivalenol and nivalenol naturally occur in scabby grain from the endemic regions.

Foodborne Illness Outbreaks

Outbreaks of foodborne illness associated with *Fusarium* species have involved foods made from wheat or barley infected mainly with *F. graminearum*. These outbreaks resemble scabby grain intoxication or red mold disease and may be essentially the same thing. Foodborne illnesses reported in Japan, Korea, and China have involved foods, particularly noodles, made from scabby wheat (75). The onset of illness is usually rapid, from 5 to 30 min, suggesting the presence of a preformed toxin, most likely deoxynivalenol. The most common symptoms included nausea, vomiting, abdominal pain, diarrhea with headache, fever, chills, and throat irritation in some victims (Table 23.1). An outbreak in India was characterized by an onset time of 15 to 60 min, abdominal pain, irritation of the throat, vomiting, and diarrhea, some with blood in the stools (7). In addition, some victims had a facial rash, nausea, and flatulence. The food involved was bread made from molded wheat. Apparently, flour millers had mixed infected wheat with sound wheat and milled it into flour. *Fusarium* species were isolated from the wheat and flour, and trichothecene mycotoxins including deoxynivalenol, acetyldeoxynivalenol, nivalenol, and T-2 toxin were detected in many samples.

Outbreaks of foodborne illness that have occurred in China were reviewed by Kuiper-Goodman (34). These have involved corn and wheat contaminated with *Fusarium* species, deoxynivalenol, and zearalenone. An outbreak of precocious pubertal changes in thousands of young children in Puerto Rico in which zearalenone or its derivatives or other exogenous estrogenic substances were the suspected cause has also been reported (57, 58). Affected persons experienced premature pubarche, prepubertal gynecomastia, and precocious pseudopuberty. Zearalenone or a derivative was found in the blood of some of the patients, and food was believed to be the source of the estrogenic substances.

Esophageal Cancer

F. moniliforme has been associated with high rates of esophageal cancer in certain parts of the world, particularly the Transkei region of South Africa, northeastern Italy, and northern China (26, 34, 45, 51). In these regions, corn is a dietary staple and is the main or only food consumed. Corn and corn-based foods from these regions may contain significant amounts of fumonisins and possibly other metabolites of this fungus.

Immunotoxic Effects of *Fusarium* Toxins

The foregoing discussions of human diseases that have been associated with various *Fusarium* species in grains and grain-based foods illustrate the range of illnesses caused by *Fusarium* toxins. Members of the genus *Fusarium* produce a diverse array of biologically active compounds that differ in chemical structure and activity. However, it is believed that many of these compounds, especially the trichothecenes, can affect the immune system by suppressing immune functions. T-2 toxin, for example, is known to be highly immunosuppressive

(27). There are reports of so-called sick houses, i.e., houses where individuals have contracted diseases such as leukemia, where *Fusarium* species have been detected in dust (38). In one report, a husband and wife both contracted leukemia while living in a house that contained dust-associated spores of *F. equiseti* (83). In another report, *Fusarium* spores were detected in a house in which four cases of leukemia occurred (82). *F. equiseti* is a known producer of diacetoxyscirpenol, a trichothecene that has known immunosuppressive properties (38). Other *Fusarium* isolates from the house were toxic to ducklings, hamsters, and mice. The conclusion was that mycotoxigenic *Fusarium* species, including *F. equiseti*, were possibly involved in the development of leukemia by their immunosuppressive effects (83).

Immunotoxicity can cause two general types of adverse effects on the immune system (48). In the first type, a toxin or chemical suppresses one or more functions of the immune system. This can result in increased susceptibility to infection or neoplastic disease. In the second type of immunotoxicity, the toxin or chemical may stimulate an immune function, resulting in autoimmune types of disorders. Trichothecenes, for example, are known to inhibit protein and DNA synthesis and to interact with cell membranes, causing weakening and damage. Exposure to trichothecenes can cause damage to bone marrow, spleen, thymus, lymph nodes, and intestinal mucosa (48, 64). T-2 toxin and deoxynivalenol have been shown to affect B-cell and T-cell mitogen responses in lymphocytes. Dietary exposures to deoxynivalenol at concentrations as low as 2 μg/g for 5 weeks or 5 μg/g for 1 week cause decreased mitogen responses (48). T-2 toxin and diacetoxyscirpenol cause increased susceptibility to *Candida* infections, as well as to *Listeria, Salmonella, Mycobacterium,* and *Cryptococcus* infection, in experimental animals (48).

Dietary deoxynivalenol has been shown to stimulate immunoglobin production, causing elevated immunoglobin A levels in mice. Among the harmful effects of this stimulation are kidney damage that is very similar to a common human kidney condition known as glomerulonephritis or immunoglobulin A nephropathy (48). While the cause of this condition is unknown, there is an association with grain-based diets. Other *Fusarium* mycotoxins, e.g., zearalenone and fumonisins, may also have immunotoxic effects, but less information is available about these toxins. Of all the harmful effects of mycotoxins, immunotoxicity or immunomodulation may have the most significant impact on human health. It appears that relatively low levels of the toxins can cause these responses.

TOXIGENIC *FUSARIUM* SPECIES AND THEIR TOXINS

According to the key of Nelson et al. (47), there are 12 sections, or groupings, within the genus *Fusarium*. Only four sections, containing the most common toxic species, will be discussed here: *Sporotrichiella (F. sporotrichioides* and *F. poae), Gibbosum (F. equiseti), Discolor (F. graminearum* and *F. culmorum),* and *Liseola (F. moniliforme, F. proliferatum,* and *F. subglutinans).* The key of Nelson et al. (47) should be consulted for identifying *Fusarium* species. In addition, the book by Samson et al. (59) can be used to identify most isolates encountered in food. The book by Pitt and Hocking (50) is also a good general reference.

The genus *Fusarium* is characterized by production of septate hyphae that generally range in color from white to pink, red, or brown as a result of pigment production. The most common characteristic of the genus is the production of large septate, crescent-shaped, fusiform, or sickle-shaped spores known as macroconidia. The macroconidia exhibit a foot-shaped basal cell and beak-shaped or snout-like apical cell (Fig. 23.1A). The macroconidia are produced from phialides in a stroma known as a sporodochium or in mucoid or slimy masses known as pionnotes. Macroconidia can also be produced in the hyphae, but these are less typical and more variable. Some species also produce smaller one- or two-celled conidia known as microconidia (Fig. 23.1B, panel a). Some species also produce swollen, thick-walled chlamydospores in the hyphae or in the macroconidia (Fig. 23.1B, panel d). *Fusarium* species are highly variable because of their genetic makeup and can undergo mutations and morphological changes in culture after isolation.

Section *Sporotrichiella*

F. poae

F. poae is widespread in soils of temperate climate regions and is found on grains such as wheat, corn, and barley (38, 50). It exists primarily as a saprophyte, but may be weakly parasitic, and is most commonly found in temperate regions of Russia, Europe, Canada, and the northern United States; *F. poae* has also been found in warmer regions, such as Australia, India, Iraq, and South Africa. Its optimum growth temperature is 22 to 27°C, but it can grow at temperatures as low as 2 to 3°C. *F. poae* can be isolated from overwintered grain. Diseases that have been associated with *F. poae* include alimentary toxic aleukia, a hemorrhagic syndrome, Urov or Kashin-Bek disease, and human esophageal cancer in northern China. *F. poae* produces rapid, profuse white to pink mycelial growth on potato-dextrose agar (PDA), with a

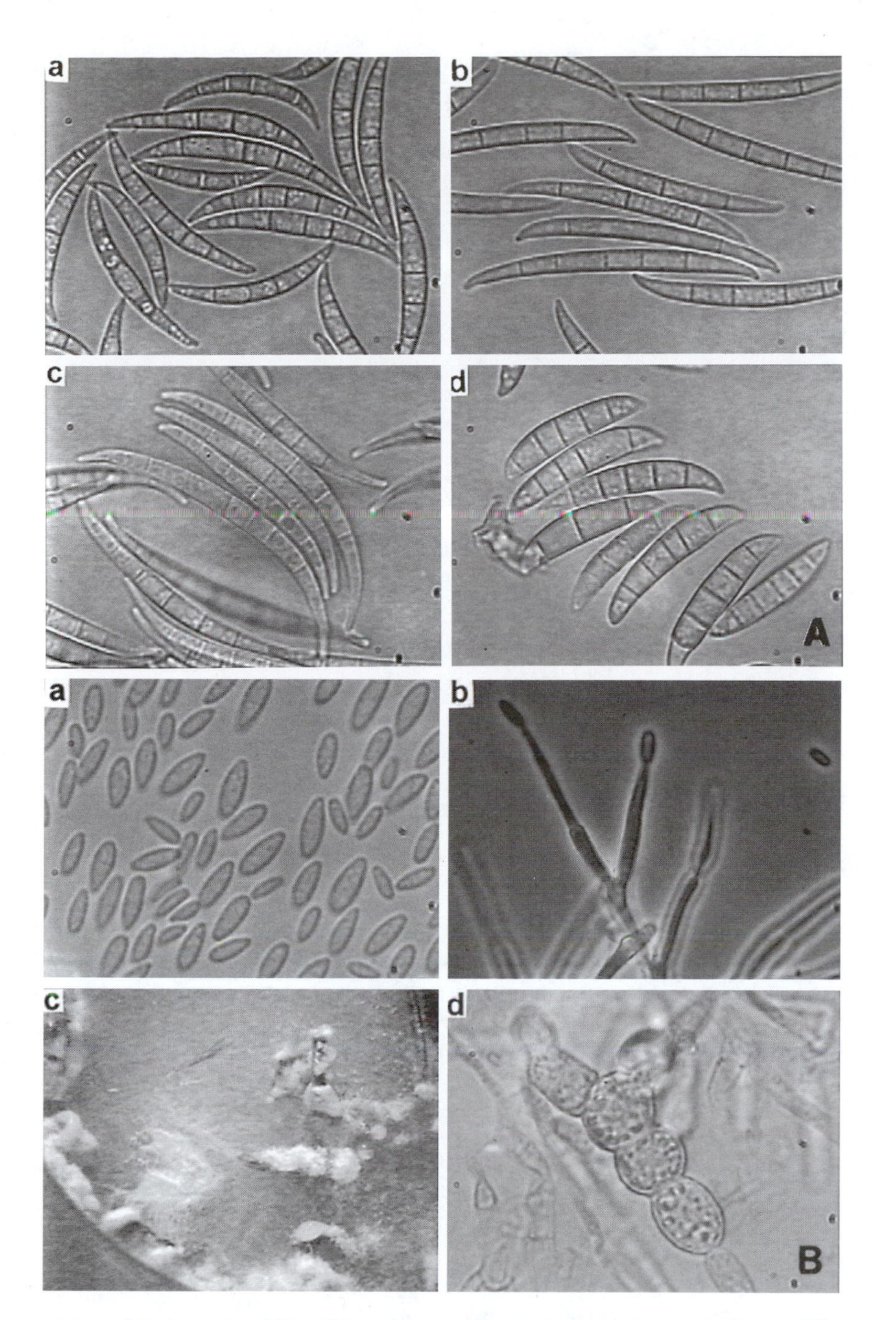

Figure 23.1 (A) Macroconidia of *Fusarium* species. a, *F. graminearum;* b, *F. moniliforme;* c, *F. equiseti;* d, *F. culmorum.* Magnification, all ×1,000. (B) Micro- and macroscopic structures of *Fusarium* species. a, Microconidia; b, monophialides; c, sporodochia; d, chlamydospores. Magnification: a, b, and d, ×1,000; c, ×10.

red to very deep carmine red reverse coloration (47). The most distinctive feature of *F. poae* is its production of abundant globose to oval, almost pyriform (pear-shaped) microconidia, with few macroconidia. Conidia are produced on branched or unbranched monophialides. Chlamydospores are produced infrequently (47). The teleomorphic state has not been observed in *F. poae.*

F. sporotrichioides

F. sporotrichioides is found in soil and a wide variety of plant materials. This mold is found in the temperate to

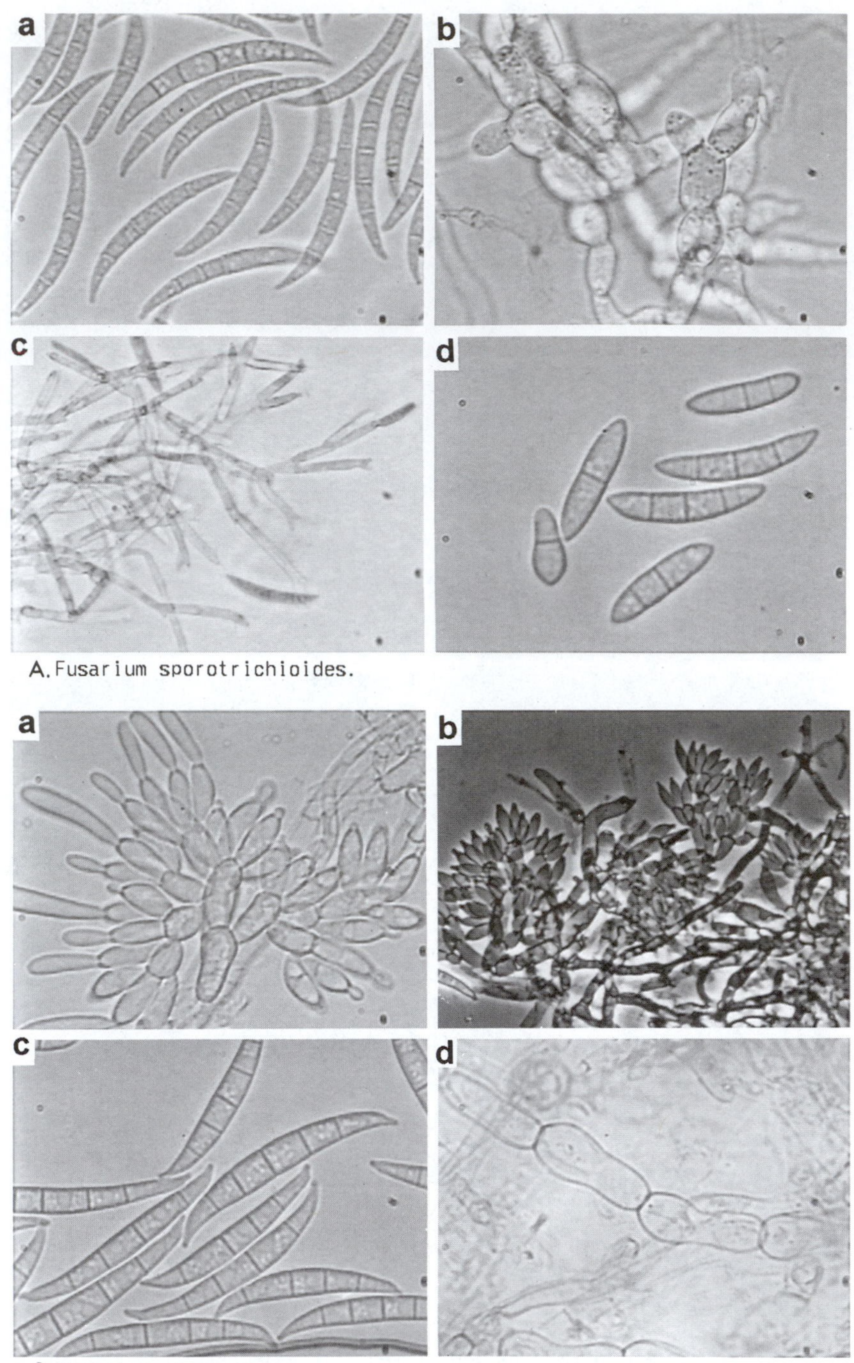

Figure 23.2 Microscopic structures. (A) *F. sporotrichioides.* a, Macroconidia; b, chlamydospores; c, phialides; d, microconidia. Magnification: a, b, and d, ×1,000; c, ×550. (B) *F. graminearum.* a, Conidiophores (monophialides); b, monophialides in sporodochia; c, macroconidia; d, chlamydospores. Magnification: a, c, and d, ×1,000; b, ×550. (C) *F. moniliforme.* a, Microconidia in chains; b, microconidia; c, monophialides producing microconidia; d, macroconidia. Magnification: a, ×165; b, ×550; c and d, ×1,000. (D) *F. proliferatum.* a, Microconidia in chains; b, microconidia; c, polyphialides; d, macroconidia. Magnification: a, ×165; b to d, ×1,000.

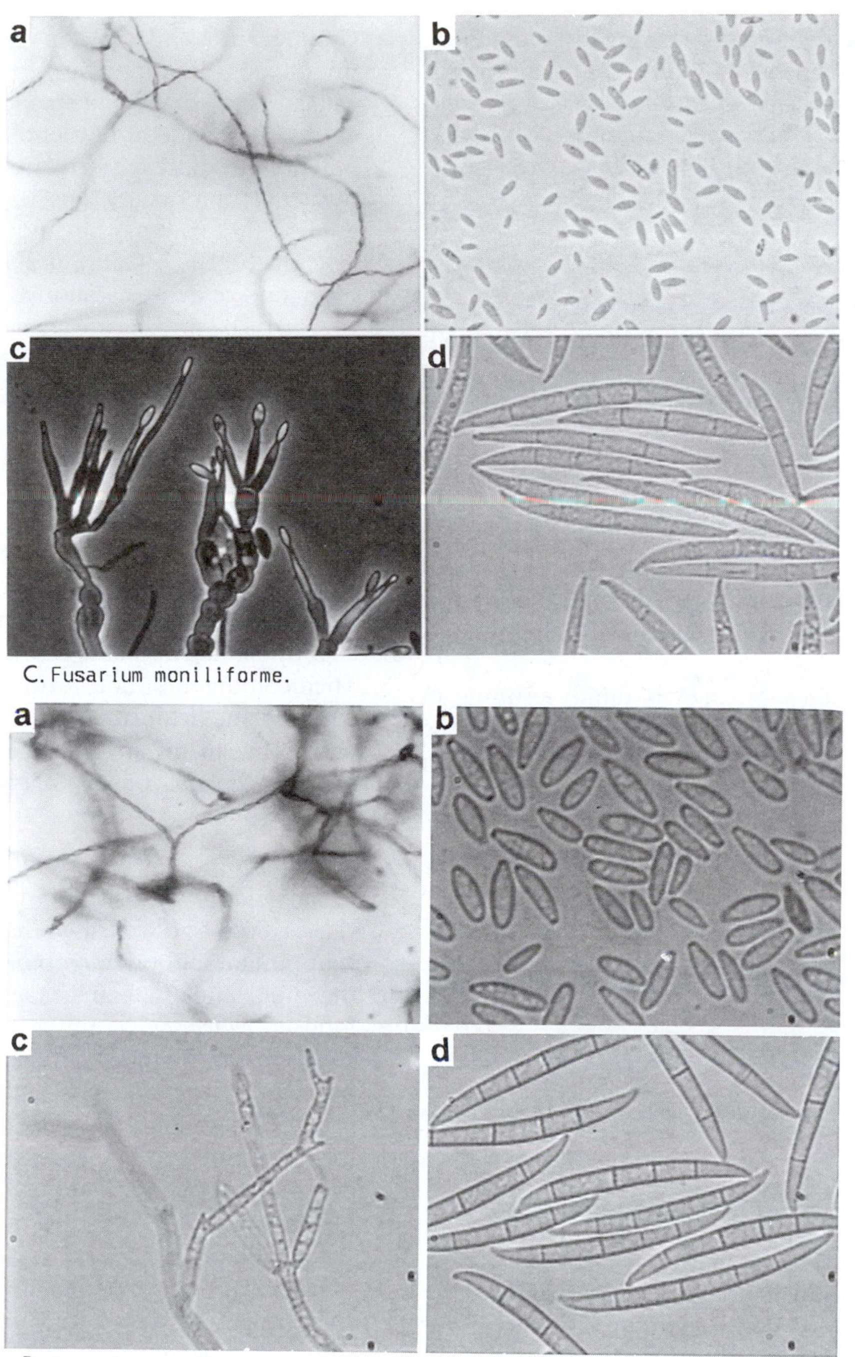

Figure 23.2 *Continued*

colder regions of the world, including Russia, northern Europe, Canada, northern United States, and Japan (38). It can grow at low to very low temperatures, e.g., at −2°C on grain overwintering in the field. Its optimum growth temperature is 22 to 27°C. Diseases that have been associated with *F. sporotrichioides* include alimentary toxic aleukia, a hemorrhagic syndrome, and akakabi-byo.

F. sporotrichioides produces dense white to pink or brown mycelia on PDA, with a deep red reverse. The most distinctive feature of *F. sporotrichioides* is the production of branched and unbranched polyphialides that produce two kinds of microconidia, oval to pear shaped and multiseptate spindle shaped that resemble macroconidia. Macroconidia are also produced, but on monophialides (Fig. 23.2A). *F. sporo-*

trichioides produces abundant chlamydospores singly and in chains and bunches (47, 50). The teleomorphic state has not been observed in *F. sporotrichioides. F. tricinctum* and *F. chlamydosporum* are also members of the section *Sporotrichiella*, and both have been reported to be toxigenic (47).

Section *Gibbosum*

F. equiseti

F. equiseti is a very cosmopolitan mold found in the soil. It is particularly common in tropical and subtropical areas but also occurs in temperate regions (38, 50). For the most part, *F. equiseti* is saphrophytic, but it may be pathogenic to plants such as bananas, avocados, and curcubits. *F. equiseti* has been found in soils from Alaska to tropical regions and has been isolated from cereal grains and overwintered cereals in Europe, Russia, and North America. It has been suggested that *F. equiseti* may contribute to leukemia in humans by affecting the immune system (38, 82, 83).

Growth of *F. equiseti* on PDA is rapid, resulting in dense, cottony aerial mycelia that fill the petri dish. The color of the colony is white with a pale salmon to almost brown reverse (47, 50). As cultures age, orange sporodochia may be produced. The most distinctive characteristic of *F. equiseti* is the shape of the macroconidia, which are long, slender, and curved, with five to seven septa (Fig. 23.1A, panel c). The apical cell is elongated, and the basal cell has a distinctive foot shape. Microconidia and chlamydospores are also produced (47, 50). The teleomorph of *F. equiseti* is *Gibberella intricans*, but its occurrence in nature is rare (47, 50). Other species in the section *Gibbosum* include *F. scirpi, F. acuminatum*, and *F. longipes*, with *F. acuminatum* also being reported to be toxigenic (47).

Section *Discolor*

F. graminearum

F. graminearum, a plant pathogen found worldwide in the soil, is the most widely distributed toxigenic *Fusarium* species (30). It causes various diseases of cereal grains, including gibberella ear rots in corn and fusarium head blight or scab in wheat and other small grains (38, 42). These two diseases are important to food microbiology and food safety because the mold and two of its main toxic metabolites, deoxynivalenol and zearalenone, may contaminate grain and subsequent food products made from the grain. Some confusion may surround the taxonomy of *F. graminearum*. In some of the older literature the name *Fusarium roseum* was used for *F. graminearum* and other species, e.g., *F. roseum* Graminearum and *F. roseum* var. *graminearum*. The major mycotoxins produced by *F. graminearum* are deoxynivalenol and zearalenone. In addition, 3-acetyldeoxynivalenol, 15-acetyldeoxynivalenol, diacetyldeoxynivalenol, butenolide, diacetoxyscirpenol, fusarenon-X (4-acetylnivalenol), monoacetoxy scirpenol, neosolaniol, nivalenol, or T-2 toxin may be produced by some strains (38).

Growth of *F. graminearum* on PDA is rapid, with the formation of dense aerial mycelia that fill the petri dish (47, 50). The mycelia are grayish with yellow and brown tinges and white margins. The reverse is usually a deep carmine red. The formation of sporodochia on PDA is sparse and may take 30 days to appear. Sporodochia appear red-brown to orange. The macroconidia of *F. graminearum* are distinctive of the species, in that they are almost cylindrical, with the central dorsal and ventral surfaces parallel (Fig. 23.2B). They most often have five septa. The ends of the macroconidia are slightly and unequally curved, and the apical cell is cone shaped and slightly bent. The basal cell has a distinct foot shape. Microconidia are not formed, and chlamydospores, when formed, most often occur in the macroconidia with some in the mycelia (47, 50). The teleomorphic state of *F. graminearum* is *Gibberella zeae*.

F. culmorum

F. culmorum is also widely distributed in the soil and causes diseases of cereal grains, of which ear rot in corn is important in food microbiology and food safety, since the mold and its toxins may contaminate corn-based foods. *F. culmorum* is in many ways similar to *F. graminearum* and in the past has been lumped with *F. graminearum* in the nonexistent species *F. roseum*, e.g., *F. roseum* Culmorum and *F. roseum* var. *culmorum* (47, 50). Mycotoxins reported to be produced by *F. culmorum* include deoxynivalenol, zearalenone, and acetyldeoxynivalenol (38, 42).

Growth of *F. culmorum* on PDA is rapid, with the formation of dense aerial mycelia that are white with some tinges of yellow and brown. The reverse of the colony is a deep carmine red color (47). Sporodochia that are orange to red-brown may be produced in older cultures. While *F. culmorum* may resemble *F. graminearum* in macroscopic appearance when growing on PDA, the macroconidia are quite different. Macroconidia of *F. culmorum* are very short and stout (Fig. 23.1A, panel d), compared with those of other members of the section, and have three to five septa (47, 50). The macroconidia have curved dorsal and ventral surfaces, and the basal cell varies from being foot shaped to having a notched appearance. Chlamydospores are formed readily in the macroconidia and mycelia. Microconidia are not formed. No teleomorphic state of *F. culmorum* has been observed. Other species in the section *Discolor* include *F. heterosporum, F. reticulatum, F. sambucinum,* and *F. crookswellense.* Of these, *F. heterosporum* and *F. sambucinum* have been reported to form toxins (47).

Section *Liseola*

F. moniliforme

F. moniliforme is a soil-borne plant pathogen that is found in corn growing in all regions of the world. It is the most prevalent mold associated with corn. It often produces symptomless infections of corn plants, but may infect the grain as well, and has been found worldwide on food- and feed-grade corn. The presence of *F. moniliforme* in corn grain is often not discernible (e.g., the grain does not appear infected), yet it is not uncommon to find lots of shelled corn with 100% internal kernel infection (38). *F. moniliforme* may also be associated with other cereals such as sorghum and rice. The presence of *F. moniliforme* in corn is a major concern in food microbiology and food safety because of the possible widespread contamination of corn and corn-based foods with its toxic metabolites, especially the fumonisins.

F. moniliforme has long been suspected of being involved in animal and human diseases. In the 1880s, a mold found on corn in Italy and called *Oospora verticillioides* was associated with pellagra (39). In the United States, *F. moniliforme* growing on corn was linked to diseases of farm animals in Nebraska and other parts of the Midwest (49, 65). In more recent years, animal diseases associated with *F. moniliforme* have included equine leukoencephalomalacia, a liquefactive necrosis of the brain of horses and other equidae (33), pulmonary edema and hydrothorax in swine (28), and experimental liver cancer in rats (26). In addition, *F. moniliforme* has been associated with abnormal bone development in chicks and pigs, manifested as leg deformities and rickets-like diseases (38). Experimental toxicity has been induced in animals by feeding culture material of *F. moniliforme.* These animals include baboons, in which acute congestive heart failure and cirrhosis of the liver were observed (38), monkeys, in which atherogenic and hypercholesterolemic responses occurred (22), and chickens, donkeys, ducklings, geese, horses, mice, pigeons, pigs, rabbits, rats, and sheep (38).

The main human disease associated with *F. moniliforme* is esophageal cancer. Several studies have linked the presence of *F. moniliforme* and fumonisins in corn to high incidences of esophageal cancer in humans in certain regions of the world, including the Transkei of South Africa, northeastern Italy, northern China, and an area around Charleston, S.C., in the United States (24, 25, 40, 51). Mycotoxins that have been associated with *F. moniliforme* include fumonisins, fusaric acid, fusarins, and fusariocins.

F. moniliforme grows rapidly and produces dense white mycelia that might be tinged with purple on PDA (47). The reverse of the colony can range from colorless to purple. Macroconidia are long, slender, and almost straight to slightly curved, especially near the ends (Fig. 23.2C, panel d). Macroconidia have three to five septa, a snout-shaped apical cell, and a foot-shaped basal cell. The most distinctive microscopic feature of *F. moniliforme* is the formation of long chains of oval, single-celled microconidia on monophialides (Fig. 23.2C, panel a). Microconidia can also be formed in false heads. Chlamydospores are not formed. *F. moniliforme* can grow over a wide temperature range from 2.5 to 37°C, with an optimum of 22 to 27°C, and at water activity above 0.87 (50).

Besides being isolated from corn, *F. moniliforme* has also been isolated from rice, sorghum, yams, hazelnuts, pecans, and cheeses (50). *F. moniliforme* has a teleomorphic state known as *Gibberella fujikuroi,* which has six different mating populations. Two of these mating populations, A and F, are found within the *F. moniliforme* anamorph, are very distinct, and differ in their ability to produce fumonisins (45). Mating type A is found in corn and is capable of producing high levels of fumonisins, while mating type F is found in sorghum and produces little or no fumonisin. On the basis of information concerning the mating populations, the taxonomy of the *F. moniliforme* biological species is undergoing considerable change (35). Older synonyms for *F. moniliforme* are *F. verticillioides* and *F. fujikuroi.*

F. proliferatum

F. proliferatum is closely related to *F. moniliforme,* yet less is known about this species, possibly because of its frequent misidentification as *F. moniliforme* (38). It is also frequently isolated from corn, in which it probably occurs in much the same way as *F. moniliforme. F. proliferatum* is capable of producing fumonisins but as yet has not been associated with animal or human diseases (38). In addition to producing fumonisins, *F. proliferatum* has been reported to produce moniliformin and fusaric acid.

On PDA, *F. proliferatum* grows rapidly to produce heavy white aerial mycelia that may become tinged with purple. Macroconidia are produced abundantly and are long and thin, have three to five septa, and are only slightly curved to almost straight (Fig. 23.2D). The basal cell is foot shaped. Microconidia are single celled with a flattened base and are produced in short to varying-length chains or in false heads from polyphialides (more than one opening), which distinguishes *F. proliferatum* from *F. moniliforme* (47). Chlamydospores are not produced. *F. proliferatum* is widely distributed in the soil and may contaminate several types of food grains. The teleomorphic state has not been observed, but *F. proliferatum* equates with mating population D of *G. fujikuroi* and appears to be a significant producer of fumonisins (35, 45).

F. subglutinans

F. subglutinans is very similar to *F. moniliforme* as well as *F. proliferatum*. It is widely distributed on corn and other grains. Little information is available about this mold, again probably because of its misidentification as *F. moniliforme* (43, 50). *F. subglutinans* has not been specifically associated with any reported animal or human diseases but has been found in corn from regions with high incidences of human esophageal cancer. Cultures of *F. subglutinans* have been shown to be acutely toxic to ducklings and rats and to be dermatoxic to rabbit skin (38). However, *F. subglutinans* does not produce fumonisins (46, 71). In these studies, isolates of *F. subglutinans* from the United States, Mexico, Nigeria, and South Africa were examined. Thus, toxicity attributed to *F. subglutinans* must be due to other toxic metabolites, such as moniliformin.

F. subglutinans grows rapidly on PDA, forming white aerial mycelia that may be tinged with purple. The reverse of the colony may range from colorless to a dark purple (47, 50). Macroconidia are long, slender, and almost straight to slightly curved, with a foot-shaped basal cell and three to five septa. Microconidia are oval and usually single celled (but may have one to three septa) and are produced only in false heads. The teleomorphic state is *Gibberella subglutinans* (synonym *G. fujikuroi* var. *subglutinans*); synonyms of the anamorphic state are *F. moniliforme* var. *subglutinans*, *F. moniliforme* Subglutinans, and *F. sacchari* (47). Mating populations B and E of *G. fujikaroi* appear to be equivalent to those of *F. subglutinans* (35). Another species in the *Liseola* section is *F. anthophilum*.

DETECTION, ISOLATION, AND IDENTIFICATION OF *FUSARIUM* SPECIES

Fusarium species are most often associated with cereal grains, seeds, milled cereal products such as flour and corn meal, barley malt, animal feeds, and necrotic plant tissue. These substrates may also contain or be colonized by many other microorganisms, and *Fusarium* species may be present in low numbers. To isolate *Fusarium* species from these products, it is necessary to use selective media. The basic techniques for detection and isolation of *Fusarium* species employ plating techniques, either as plate counts of serial dilutions of products or by the placement of seeds or kernels of grain directly on the surface of agar media in petri dishes, i.e., direct plating.

Several culture media have been used to detect and isolate *Fusarium* species. These include Nash-Snyder medium (44), modified Czapek-Dox agar (47), Czapek iprodione-dichloran (CZID) agar (2), potato-dextrose-iprodione-dichloran (PDID) agar (72), and dichloran-chloramphenicol-peptone agar (DCPA) (3, 50). The Nash-Snyder medium and modified Czapek-Dox agar contain pentachloronitrobenzene, a known carcinogen, and are not favored for routine use in food microbiology laboratories. However, these media can be useful for evaluating samples that are heavily contaminated with bacteria and other fungi. CZID agar is becoming a regularly used medium for isolating *Fusarium* species from foods, but rapid identification of *Fusarium* isolates to the species level is difficult if not impossible on this medium. Isolates must be subcultured on other media such as carnation leaf agar (CLA) for identification. However, CZID agar is a good selective medium for *Fusarium* species. While some other molds may not be completely inhibited on CZID agar most are, and *Fusarium* species can be readily distinguished. Thrane et al. (72) reported that PDID agar is as selective as CZID agar for *Fusarium* species, with the advantage that it supports *Fusarium* growth with morphological and cultural characteristics that are the same as on PDA, which facilitates more rapid identification, since various monographs and manuals for *Fusarium* identification describe characteristics of colonies grown on PDA. Thrane et al. (72) compared several media for their suitability to support colony development by *Fusarium* species and found that PDID and CZID agars were better than DCPA. Growth rates were much higher on DCPA, making colony counts more difficult. Conner (18), however, modified DCPA by adding 0.5 μg of crystal violet per ml and reported increased selectivity by inhibiting *Aspergillus* and *Penicillium* species but not *Fusarium* species.

Identification of *Fusarium* species is based largely on the production and morphology of macroconidia and microconidia. Identification keys described by Nelson et al. (47) rely heavily on the morphology of conidia and conidiophores, the structures on which conidia are produced. *Fusarium* species do not readily form conidia on all culture media, and conidia formed on high-carbohydrate media such as PDA are often more variable and less typical. A medium that supports abundant and consistent spore production is CLA. Carnation leaves from actively growing, disbudded, young carnation plants free of pesticide residues are cut into small pieces (5 mm^2), dried in an oven at 45 to 55°C for 2 h, and sterilized by irradiation (47). CLA is prepared by placing a few pieces of carnation leaf on the surface of 2.0% water agar (23, 47). *Fusarium* isolates are then inoculated on the agar and leaf interface, where they form abundant and typical conidia and conidiophores in sporodochia, rather than mycelia. CLA is low in carbohydrates and rich in other complex naturally occurring substances that apparently stimulate spore production.

Since many *Fusarium* species are plant pathogens and all are found in fields where crops are grown, these molds respond to light. Growth, pigmentation, and spore production are most typical when cultures are grown in alternating light and dark cycles of 12 h each. Fluorescent light or diffuse sunlight from a north window is best. Fluctuating temperatures such as 25°C (day) and 20°C (night) also enhance growth and sporulation. Identification keys are provided by Nelson et al. (47), Samson et al. (59), Marasas et al. (39), and Marasas (37).

DETECTION AND QUANTITATION OF *FUSARIUM* TOXINS

Fusarium species produce several toxic or biologically active metabolites. The trichothecenes are a group of closely related compounds that are esters of sesquiterpene alcohols that possess a basic trichothecene skeleton and an epoxide group (69). The trichothecenes are divided into three groups: the type A trichothecenes, which include diacetoxyscirpenol, T-2 toxin, HT-2 toxin, and neosolaniol; the type B trichothecenes, which include deoxynivalenol, 3-acetyldeoxynivalenol, 15-acetyldeoxynivalenol, nivalenol, and fusarenon-X; and the type C or so-called macrocyclic trichothecenes known as satratoxins. Of these, the toxin most commonly found in cereal grains or most often associated with human illness is deoxynivalenol (42). Other *Fusarium* toxins associated with diseases are zearalenone and the fumonisins. T-2 toxin occurs rarely in grain in the United States but has been associated with alimentary toxic aleukia in Russia in the 1940s and earlier. Moniliformin, fusarin C, and fusaric acid are also of interest and concern but have not been shown to commonly occur or be specifically associated with diseases.

If present, *Fusarium* toxins are usually found at low levels in cereal grains and processed grain-based foods. Their concentrations may range from less than nanogram to microgram quantities per gram (parts per billion to parts per million, respectively). *Fusarium* toxins vary in their chemical structures and properties, making it difficult to develop a single method for quantitating all toxins. The basic steps involved in detection of *Fusarium* mycotoxins are similar to those for other mycotoxins. These include sampling, size reduction and mixing, subsampling, extraction, filtration, cleanup, concentration, separation of components, detection, quantification, and confirmation (9, 69).

The first problem encountered in the analysis of grains for *Fusarium* toxins is the same as for other mycotoxins, i.e., sampling. Obtaining a representative sample from a large lot of cereal grain can be very difficult if the toxin is present in a relatively small percentage of the kernels, which may be the case with toxins such as deoxynivalenol and zearalenone. On the other hand, fumonisins appear to be more evenly distributed in grain such as corn. Processed grain-based foods may contain a more even distribution of toxins as a result of grinding and mixing. Samples are usually ground and mixed further, and a subsample of 50 to 100 g is taken for extraction. *Fusarium* toxins, like all mycotoxins, must be extracted from the matrix in which they are found.

Most mycotoxins are more soluble in slightly polar organic solvents than in water. The most commonly used extraction solvents consist of combinations of water with organics such as methanol, acetone, and acetonitrile. Following extraction, the extract is filtered to remove solids and subjected to a cleanup step to remove interfering substances. Cleanup can be done in several ways, but the most common method used for *Fusarium* toxins is to pass the extract through a column packed with sorbant packing materials. In recent years, the use of small prepacked commercially available disposable columns or cartridges such as Sep-Pak, Bond Elut, and MycoSep has become more common. After the extract has been cleaned, the sample may need to be concentrated before analysis in order to detect the toxin. This may be accomplished by mild heating such as in a water bath, heating block, or rotary evaporator under reduced pressure or a stream of nitrogen. Detection and quantification of the toxins are done after the toxins are separated from other components by chromatographic means. The most common chromatographic separation techniques used are thin-layer chromatography (TLC) and high-performance liquid chromatography (HPLC).

Gas chromatography (GC) also has some applications, particularly when coupled with mass spectrometry.

A commonly used method for quantitating deoxynivalenol is TLC (21, 52, 74). GC is more sensitive than TLC but is also more laborious. While HPLC methods employing UV absorbance at 219 nm for detection are fairly sensitive, they require purification of deoxynivalenol by passage through high-capacity activated charcoal columns (13). The method of choice for quantitation of zearalenone is HPLC with fluorescence detection (6, 79). A TLC method for zearalenone has been tested collaboratively and is useful as a screening method (67). The methods most commonly used for T-2 toxin are GC methods (14, 17, 53). Because type A trichothecenes lack a UV chromophore and are not fluorescent, TLC and HPLC methods are unsuitable, resulting in the reliance on GC methods. The most widely used analytical methods for fumonisins are HPLC methods involving the formation of fluorescent derivatives (52). Methods using derivatizing agents, such as *o*-phthalaldehyde (54, 55, 66), fluorescamine (80), and naphthalene dicarboxaldehyde (5), have been developed. A TLC method for fumonisins has also been developed but is used mainly for screening (52, 56). For discussions of methods most commonly used for analysis of *Fusarium* and other mycotoxins, see references 52 and 69.

Immunoassays have been developed for *Fusarium* toxins. Enzyme-linked immunosorbent assay kits for *Fusarium* toxins are commercially available (Neogen Corporation, Lansing, Mich.). Qualitative kits for screening as well as kits for quantitative analyses are available for deoxynivalenol, zearalenone, T-2 toxin, and fumonisins. A rapid screening TLC kit for deoxynivalenol is available from Romer Labs (Union, Mo.). This method uses a special cleanup column that requires only 10 s per sample. An antibody-based affinity column for fumonisins is also available (Vicam, Watertown, Mass.).

OCCURRENCE OF *FUSARIUM* TOXINS IN FOODS

Fusarium toxins, particularly deoxynivalenol and fumonisins, have been found in finished human food products. Various *Fusarium* toxins have been found naturally occurring in numerous cereal grains, but most of these grains have been destined for animal feed. Deoxynivalenol is the most common trichothecene found in commodity grains; therefore, the greatest potential exists for it to occur in finished foods. Food products such as bread, pasta, and beer may contain at least trace amounts of the toxin. Scott et al. (63) reported finding trace amounts (ca. 5 ng/ml) of deoxynivalenol in 29 of 50 samples of commercially available domestic and imported beer in Canada. In the United States, deoxynivalenol has been found in breakfast cereals in average amounts of 100 ng/g (73), while corn syrup and beer samples were negative. In another study (1), the toxin was found in corn breakfast cereals, wheat flour muffin mixes, wheat- and oat-based cookies, crackers, corn chips, popcorn, and mixed grain cereals. Deoxynivalenol concentrations ranged from 4 to 19 μg/g. Brumley et al. (8) found deoxynivalenol in wheat, flour, corn, corn meal, and snack foods in levels ranging from 0.08 to 0.3 μg/g. Thus, there is evidence that deoxynivalenol is a contaminant of processed human food products and that levels sometimes exceed the United States government guideline of 1.0 μg/g in finished food products. Deoxynivalenol is quite heat stable, probably tolerating most thermal processes to some degree (61, 62).

Fumonisins have also been found in processed or finished food products. Food products that have been examined include corn meal, corn grits, corn breakfast cereals, tortillas, tortilla chips, corn chips, popcorn, and hominy corn (10, 11, 70, 76). The products most consistently contaminated with the greatest amounts of fumonisins are those foods which receive only physical processing such as milled products, e.g., corn meal and corn muffin mixes. Sydenham et al. (70) determined fumonisin levels in corn meal obtained from Canada, Egypt, Peru, South Africa, and the United States. Corn meal from Egypt and the United States had the highest levels. Corn meal from United States contained fumonisins in concentrations of less than 1.0 to up to 2.8 μg/g of corn meal. In another survey, the highest levels of fumonisins in corn-based foods were found in corn meal and corn grits (68). Corn flakes and corn pops cereals, corn chips, and corn tortilla chips were negative for fumonisins, and very low levels were found in tortillas, popcorn, and hominy. Fumonisins have been detected in processed corn products in Germany, Italy, Japan, Spain, and Switzerland (20, 60, 76, 77, 84).

OTHER TOXIC MOLDS

Other potentially toxic molds that may contaminate foods include species of the genera *Acremonium, Alternaria, Chaetomium, Cladosporium, Claviceps, Myrothecium, Phomopsis, Rhizoctonia,* and *Rhizopus.* Molds such as *D. maydis, Phoma herbarum, Pithomyces chartarum, Stachybotrys chartarum,* and *T. roseum* are also potentially toxic

(36). However, most of these molds are more likely to be present in animal feeds, and their significance to food safety may be minimal. Some have been shown to produce toxic secondary metabolites in vitro which have yet to be found to occur naturally.

The ergot mold, *Claviceps purpurea,* is the cause of the earliest recognized human mycotoxicosis, ergotism (4). Ergotism has been reported in sporadic outbreaks in Europe since 857, with near-epidemic outbreaks occurring in the Middle Ages. The disease was known as St. Anthony's fire during the Middle Ages because it was believed that pilgrimages to a shrine of St. Anthony could lead to cure of the disease. It is probable that as pilgrims traveled to the shrine, they left areas where ergot was endemic and traveled through areas where it was not a problem. Consequently, when they stopped eating toxic bread from the area where ergot was endemic and ate nontoxic bread obtained along the way, by the time they reached the shrine, symptoms subsided. Ergot is a disease of rye in which the rye grains are replaced by ergot sclerotia that contain toxic alkaloids. If sclerotia are not removed when the rye is milled into flour, the flour, and subsequent bread made from the flour, become contaminated. Ergotism can be manifested as a convulsive condition or a necrotic gangrenous condition of the extremities. During the Middle Ages, necrotic gangrenous ergotism was characterized by swollen limbs and alternating cold and burning sensations in fingers, hands, and feet, hence the term "fire" in St. Anthony's fire. The main ergot alkaloid, ergotamine, has vasoconstrictive properties that cause these symptoms. In severe cases, the extremities, such as feet in humans and hooves in animals, were sloughed off. Convulsive ergotism may have been the reason for the Salem witchcraft trials of 1692 in Salem, Mass. (12, 41). In more recent times, outbreaks of ergotism have occurred in Russia in 1926, Ireland in 1929, France in 1953, India in 1958, and Ethiopia in 1973 (4).

Alternaria species may be especially significant as potential toxic contaminants of food. *Alternaria* infects plants in the field and may contaminate wheat, sorghum, and barley (19). *Alternaria* species also infect various fruits and vegetables, including apples, pears, citrus fruits, peppers, tomatoes, and potatoes. *Alternaria* species can cause spoilage of these foods in refrigerated storage. Several *Alternaria* toxins, including alternariol, alternariol monomethyl ether, altenuene, tenuazonic acid, and the altertoxins, have been described. Relatively little is known about the toxicity of these toxins; however, cultures of *Alternaria* that have been grown on corn or rice and fed to rats, chicks, turkey poults, and ducklings have been shown to be quite toxic. An *Alternaria* toxin was also implicated in the alimentary toxic aleukia toxicoses in Russia in the 1940s (33). Toxicity of various *Alternaria* toxins has not been studied extensively, but there is evidence that the toxins may have synergistic activity; i.e., mixtures of *Alternaria* toxins or culture extracts are more toxic than the individual toxins. Species of *Alternaria* that produce one or more of these toxins include *A. tenuis, A. alternata, A. citri,* and *A. solani. A. alternata* f. sp. *lycopersici,* a pathogen of tomatoes, produces a host-specific phytotoxin known as *Alternaria alternata lycopersici* (AAL) toxin which is nearly identical in structure to fumonisins. It has been shown that fumonisins can cause lesions in tomatoes identical to those caused by AAL toxin and that AAL toxin is toxic to animal cells in tissue culture (81).

References

1. **Abbouzied, M. M., J. I. Azcona, W. E. Braselton, and J. J. Pestka.** 1991. Immunochemical assessment of mycotoxins in 1989 grain foods: evidence for deoxynivalenol (vomitoxin) contamination. *Appl. Environ. Microbiol.* **57**:672–677.
2. **Abildgren, M. P., F. Lund, U. Thrane, and S. Elmholt.** 1987. Czapek-Dox agar containing iprodione and dichloran as a selective medium for the isolation of *Fusarium* species. *Lett. Appl. Microbiol.* **5**:83–86.
3. **Andrews, S., and J. I. Pitt.** 1986. Selective medium for isolation of *Fusarium* species and dematiaceous hyphomycetes from cereals. *Appl. Environ. Microbiol.* **51**: 1235–1238.
4. **Beardall, J. M., and J. D. Miller.** 1994. Diseases in humans with mycotoxins as possible causes, p. 487–539. *In* J. D. Miller and H. L. Trenholm (ed.), *Mycotoxins in Grain. Compounds Other than Aflatoxin.* Eagan Press, St. Paul, Minn.
5. **Bennett, G. A., and J. L. Richard.** 1992. High performance liquid chromatographic method for naphthalene dicarboxaldehyde derivative of fumonisins, p. 143. *In Proceedings of the AOAC International Meeting,* Cincinnati, OH. American Association of Official Analytical Chemists, Washington, D.C. (Abstract.)
6. **Bennett, G. A., O. L. Shotwell, and W. F. Kwolek.** 1985. Liquid chromatographic determination of α zearalenol and zearalenone in corn: collaborative study. *J. Assoc. Off. Anal. Chem.* **68**:958–962.
7. **Bhat, R. V., S. R. Beedu, Y. Ramakrisna, and K. L. Munshi.** 1989. Outbreak of trichothecene mycotoxicosis associated with consumption of mould-damaged wheat products in Kashmir Valley, India. *Lancet* **7**(Jan.):35–37.
8. **Brumley, W. C., M. W. Trucksess, S. H. Adler, C. K. Cohen, K. D. White, and J. A. Sphon.** 1985. Negative ion chemical ionization mass spectrometry of deoxynivalenol (DON): application to identification of DON in grains and snack foods after quantitation/isolation by thin-layer chromatography. *J. Agric. Food Chem.* **33**:326–330.

9. **Bullerman, L. B.** 1987. Methods for detecting mycotoxins in foods and beverages, p. 571–598. *In* L. R. Beuchat (ed.), *Food and Beverage Mycology,* 2nd ed. Van Nostrand Reinhold Company Inc., New York.
10. **Bullerman, L. B.** 1996. Occurrence of *Fusarium* and fumonisins on food grains and in foods, p. 27–38. *In* L. Jackson, J. DeVries, and L. Bullerman (ed.), *Fumonisins in Foods.* Plenum Publishing Corp., New York.
11. **Bullerman, L. B., and W. Y. J. Tsai.** 1994. Incidence and levels of *Fusarium moniliforme, Fusarium proliferatum* and fumonisins in corn and corn-based foods and feeds. *J. Food Prot.* **57:**541–546.
12. **Caporeal, L. R.** 1976. Ergotism: the Satan loosed in Salem? *Science* **192:**21–26.
13. **Chang, H. L., J. W. Devries, P. A. Larson, and H. H. Patel.** 1984. Rapid determination of deoxynivalenol (vomitoxin) by liquid chromatography using modified Romer column clean-ups. *J. Assoc. Off. Anal. Chem.* **67:**52–54.
14. **Chaytor, J. P., and M. J. Saxby.** 1982. Development of a method for the analysis of T-2 toxin in maize by gas-chromatography-mass spectrometry. *J. Chromatogr.* **237:**107–111.
15. **Chelkowski, J.** 1989. Mycotoxins associated with corn cob fusariosis, p. 53–62. *In* J. Chelkowski (ed.), *Fusarium. Mycotoxins, Taxonomy and Pathogenicity.* Elsevier Science Publishing Company, New York.
16. **Chelkowski, J.** 1989. Formation of mycotoxins produced by fusaria in heads of wheat triticale and rye, p. 63–84. *In* J. Chelkowski (ed.), *Fusarium. Mycotoxins, Taxonomy and Pathogenicity.* Elsevier Science Publishing Company, New York.
17. **Cohen, H., and M. Lapointe.** 1984. Capillary gas chromatographic determination of T-2 toxin, HT-2 toxin and diacetoxyscirpenol in cereal grains. *J. Assoc. Off. Anal. Chem.* **67:**1105–1109.
18. **Conner, D. E.** 1992. Evaluation of methods for selective enumeration of *Fusarium* species in feedstuffs, p. 299–302. *In* R. A. Samson, A. D. Hocking, J. I. Pitt, and A. D. King (ed.), *Modern Methods in Food Mycology.* Elsevier Scientific Publishers, Amsterdam.
19. **Coulombe, R. A.** 1991. *Alternaria* toxins, p. 425–433. *In* R. P. Sharma and D. K. Salunke (ed.), *Mycotoxins and Phytoalexins.* CRC Press, Inc., Boca Raton, Fla.
20. **Doko, M. B., and A. Visconti.** 1994. Occurrence of fumonisins B_1 and B_2 in corn and corn-based human foodstuffs in Italy. *Food Addit. Contam.* **11:**433–439.
21. **Eppley, R. M., M. W. Trucksess, S. Nesheim, C. W. Thorpe, and A. E. Pohland.** 1986. Thin layer chromatographic method for detection of deoxynivalenol in wheat: collaborative study. *J. Assoc. Off. Anal. Chem.* **69:**37–40.
22. **Fincham, J. E., W. F. O. Marasas, J. J. F. Taljaard, N. P. J. Kriek, C. J. Badenhorst, W. C. A. Gelderblom, J. V. Seier, C. M. Smuts, M. Faber, M. J. Weight, W. Slazus, C. W. Woodroof, M. J. van Wyk, M. Kruger, and P. G. Thiel.** 1992. Atherogenic effects in a non-human primate of *Fusarium moniliforme* cultures added to a carbohydrate diet. *Atherosclerosis* **94:**13–25.
23. **Fisher, N. L., L. W. Burgess, T. A. Toussoun, and P. E. Nelson.** 1982. Carnation leaves as a substrate and for preserving cultures of *Fusarium* species. *Phytopathology* **72:**151–153.
24. **Franceschi, S., E. Bidoli, A. E. Baron, and C. LaVecchia.** 1990. Maize and the risk of cancers of the oral cavity, pharynx and esophagus in Northeastern Italy. *J. Natl. Cancer Inst.* **82:**1407–1411.
25. **Gelderblom, W. C. A., W. F. O. Marasas, R. Vleggaar, P. G. Thiel, and M. E. Cawood.** 1992. Fumonisins: isolation, chemical characterization and biological effects. *Mycopathologia* **117:**11–16.
26. **Gelderblom, W. C. A., N. P. J. Kriek, W. F. O. Marasas, and P. G. Thiel.** 1991. Toxicity and carcinogenicity of the *Fusarium moniliforme* metabolite fumonisin B_1 in rats. *Carcinogenesis* **12:**1247–1251.
27. **Graveson, S., J. C. Frisvad, and R. A. Samson.** 1994. *Microfungi.* Munksgaard International Publishers Ltd., Copenhagen.
28. **Harrison, L. R., B. M. Colvin, J. T. Greens, L. E. Newman, and J. R. Cole.** 1990. Pulmonary edema and hydrothorax in swine produced by fumonisin B_1 a toxic metabolite of *Fusarium moniliforme. J. Vet. Diagn. Invest.* **2:**217–221.
29. **Hayes, A. W.** 1981. Involvement of mycotoxins in animal and human health, p. 11–40. *In Mycotoxin Teratogenicity and Mutagenicity.* CRC Press, Inc., Boca Raton, Fla.
30. **International Agency for Research on Cancer.** 1993. *Some Naturally Occurring Substances: Food Items and Constituents, Heterocyclic Aromatic Amines and Mycotoxins.* Monograph 56. International Agency for Research on Cancer, Lyon, France.
31. **Joffe, A. Z.** 1960. The mycoflora of overwintered cereals and its toxicity. *Bull. Res. Counc. Isr.* **90:**101–126.
32. **Joffe, A. Z.** 1986. Effects of fusariotoxins in humans, p. 225–298. *In Fusarium Species: Their Biology and Toxicology.* John Wiley & Sons, New York.
33. **Kellerman, T. S., W. F. O. Marasas, P. G. Thiel, W. C. A. Gelderblom, M. Cawood, and and J. A. W. Coetzer.** 1990. Leukoencephalomalacia in two horses induced by oral dosing of fumonisin B_1. *Onderstepoort J. Vet. Res.* **57:**269–275.
34. **Kuiper-Goodman, T.** 1994. Prevention of human mycotoxicoses through risk assessment and risk management, p. 439–469. *In* J. D. Miller and H. L. Trenholm (ed.), *Mycotoxins in Grain. Compounds Other than Aflatoxin.* Eagan Press, St. Paul, Minn.
35. **Leslie, J. F., R. D. Plattner, A. E. Desjardins, and C. J. R. Klittich.** 1992. Fumonisin B_1 production by strains from different mating populations of *Gibberella fujikuroi* (*Fusarium* section Liseola). *Phytopathology* **82:**341–345.
36. **Mantle, P. G.** 1991. Miscellaneous toxigenic fungi, p. 141–152. *In* J. E. Smith and R. S. Henderson (ed.), *Mycotoxins and Animal Feeds.* CRC Press, Inc., Boca Raton, Fla.
37. **Marasas, W. F. O.** 1991. Toxigenic fusaria, p. 119–139. *In* J. E. Smith and R. S. Henderson (ed.), *Mycotoxins and Animal Feeds.* CRC Press, Inc., Boca Raton, Fla.
38. **Marasas, W. F. O., P. E. Nelson, and T. A. Tousson.** 1984. *Toxigenic Fusarium Species: Identity and Mycotoxicology.* The Pennsylvania State University Press, University Park, Pa.
39. **Marasas, W. F. O., P. E. Nelson, and T. A. Tousson.** 1985. Taxonomy of toxigenic fusaria, p. 3–14. *In* J. Lacey (ed.), *Trichothecenes and Other Mycotoxins. In Proceedings of the International Mycotoxin Symposium Sidney, Australia, 1984.* John Wiley & Sons, New York.

40. **Marasas, W. F. O., F. C. Wehner, S. J. van Rensberg, and D. J. van Schalkwyk.** 1981. Mycoflora of corn produced in human esophageal cancer areas in Transkei, Southern Africa. *Phytopathology* **71**:792–796.

41. **Matossian, M. K.** 1982. Ergot and the Salem witchcraft affair. *Am. Sci.* **70**:355–357.

42. **Miller, J. D.** 1995. Fungi and mycotoxins in grain: implications for stored product research. *J. Stored Prod. Res.* **31**:1–16.

43. **Mills, J. T.** 1989. Ecology of mycotoxigenic *Fusarium* species on cereal seeds. *J. Food Prot.* **52**:737–742.

44. **Nash, S. M., and W. C. Snyder.** 1962. Quantitative estimations by plate counts of propagules of the bean root rot *Fusarium* in field soils. *Phytopathology* **52**:567–572.

45. **Nelson, P. E., A. E. Desjardins, and R. D. Plattner.** 1993. Fumonisins, mycotoxins produced by *Fusarium* species: biology, chemistry and significance. *Annu. Rev. Phytopathol.* **31**:233–252.

46. **Nelson, P. E., R. D. Plattner, D. D. Shackelford, and A. E. Desjardins.** 1992. Fumonisin B_1 production by *Fusarium* species other than *F. moniliforme* in section Liseola and by some related species. *Appl. Environ. Microbiol.* **58**:984–989.

47. **Nelson, P. E., T. A. Tousoun, and W. F. O. Marasas.** 1983. *Fusarium species: an Illustrated Manual for Identification.* The Pennsylvania State University Press, University Park, Pa.

48. **Pestka, J. J., and G. S. Bondy.** 1994. Immunotoxic effects of mycotoxins, p. 339–359. *In* J. D. Miller and H. L. Trenholm (ed.), *Mycotoxins in Grain. Compounds Other than Aflatoxin.* Eagan Press, St. Paul, Minn.

49. **Peters, A. T.** 1904. A fungus disease in corn. *Agric. Exp. Stn. Nebr. Annu. Rep.* **17**:13–22.

50. **Pitt, J. I., and A. D. Hocking.** 1985. *Fungi and Food Spoilage.* Academic Press, Inc., Sydney, Australia.

51. **Rheeder, J. P., W. F. O. Marasas, P. G. Thiel, E. W. Sydenham, G. S. Shepard, and D. J. van Schalkwyk.** 1992. *Fusarium moniliforme* and fumonisins in corn in relation to human esophageal cancer in Transkei. *Phytopathology* **82**:353–357.

52. **Richard, J. L., G. A. Bennett, P. F. Ross, and P. E. Nelson.** 1993. Analysis of naturally occurring mycotoxins in feedstuffs and foods. *J. Anim. Sci.* **71**:2563–2574.

53. **Romer, T. R., T. M. Boling, and J. L. McDonald.** 1978. Gas-liquid chromatographic determination of T-2 toxin and diacetoxyscirpenol in corn and mixed feeds. *J. Assoc. Off. Anal. Chem.* **61**:801–805.

54. **Ross, P. F., P. E. Nelson, J. L. Richard, G. D. Osweiler, L. G. Rice, R. D. Plattner, and T. M. Wilson.** 1990. Production of fumonisins by *Fusarium moniliforme* and *Fusarium proliferatum* isolates associated with equine leukoencephalomalacia and a pulmonary edema syndrome in swine. *Appl. Environ. Microbiol.* **56**:3225–3226.

55. **Ross, P. F., L. G. Rice, R. D. Plattner, G. O. Osweiler, T. M. Wilson, D. L. Owens, P. A. Nelson, and J. L. Richard.** 1991. Concentrations of fumonisin B_1 in feeds associated with animal health problems. *Mycopathologia* **114**:129–135.

56. **Rottinghaus, G. E., C. F. Coatney, and Harry C. Minoir.** 1992. A rapid, sensitive thin layer chromatography procedure for the detection of fumonisin B_1 and B_2. *J. Vet. Diagn. Invest.* **4**:326–329.

57. **Saenz de Rodriguez, C. A.** 1984. Environmental hormone contamination in Puerto Rico. *N. Engl. J. Med.* **310**:1741–1742.

58. **Saenz de Rodriguez, C. A., A. M. Bongiovanni, and L. Conde de Borrego.** 1985. An epidemic of precocious development in Puerto Rican Children. *J. Pediatr.* **107**:393–396.

59. **Samson, R. A., E. S. Hoekstra, J. C. Frisvad, and O. Filtenborg (ed.).** 1995. *Introduction to Food-Borne Fungi.* Centraalbureau voor Schimmelcultures, Baarn, The Netherlands.

60. **Sanchis, V., M. Abadias, L. Oncins, N. Sala, I. Vinas, and R. Canela.** 1994. Occurrence of fumonisins B_1 and B_2 in corn-based products from the Spanish Market. *Appl. Environ. Microbiol.* **60**:2147–2148.

61. **Scott, P. M.** 1984. Effects of food processing on mycotoxins. *J. Food Prot.* 47:489–499.

62. **Scott, P. M., S. R. Kanhere, P.-Y. Lau, J. E. Dexter, and R. Greenhalgh.** 1983. Effects of experimental flour milling and bread baking on retention of deoxynivalenol (vomitoxin) in hard red spring wheat. *Cereal Chem.* **60**:421–424.

63. **Scott, P. M., S. R. Kanhere, and D. Weber.** 1993. Analysis of Canadian and imported beers for *Fusarium* mycotoxins by gas chromatography-mass spectrometry. *Food Addit. Contam.* **10**:381–389.

64. **Sharma, R. P., and Y. W. Kim.** 1991. Trichothecenes, p. 339–359. *In* R. P. Sharma and D. K. Salunkhe (ed.), *Mycotoxins and Phytoalexins.* CRC Press, Inc., Boca Raton, Fla.

65. **Sheldon, J. L.** 1904. A corn mold (*Fusarium moniliforme* n. sp.). *Agric. Exp. Stn. Nebr. Annu. Rep.* **17**:23–43.

66. **Shepard, G. W., E. W. Sydenham, P. G. Thiel, and W. C. A. Gelderblom.** 1990. Quantitative determination of fumonisins B_1 and B_2 by high performance liquid chromatography with fluorescence detection. *J. Liquid Chromatogr.* **13**:2077–2087.

67. **Shotwell, O. L., M. L. Goulden, and G. A. Bennett.** 1976. Determination of zearalenone in corn: collaborative study. *J. Assoc. Off. Anal. Chem.* **59**:666–669.

68. **Stack, M. E., and R. M. Eppley.** 1992. Liquid chromatographic determination of fumonisins B_1 and B_2 in corn and corn products. *J. Assoc. Off. Anal. Chem. Int.* **75**:834–837.

69. **Steyn, P. S., P. G. Thiel, and D. W. Trinder.** 1991. Detection and quantification of mycotoxins by chemical analysis, p. 165–221. *In* J. E. Smith and R. S. Henderson (ed.), *Mycotoxins and Animal Foods.* CRC Press, Inc., Boca Raton, Fla.

70. **Sydenham, E. W., G. S. Shephard, P. G. Thiel, W. F. O. Marasas, and S. Stockenstrom.** 1991. Fumonisin contamination of commerical corn-based human foodstuffs. *J. Agric. Food Chem.* **25**:767–771.

71. **Thiel, P. G., W. F. O. Marasas, E. W. Sydenham, G. S. Shepard, W. C. A. Gelderblom, and J. J. Nieuwenhuis.** 1991. Survey of fumonisin production by *Fusarium* species. *Appl. Environ. Microbiol.* **57**:1089–1093.

72. **Thrane, U., O. Filtenborg, F. C. Frisvad, and F. Lund.** 1992. Improved methods for the detection and identifica-

tion of toxigenic *Fusarium* species, p. 285–291. *In* R. A. Samson, A. D. Hocking, J. I. Pitt, and A. D. King (ed.), *Modern Methods in Food Mycology.* Elsevier Science Publishers, New York.

73. **Trucksess, M. W., M. T. Flood, and S. W. Page.** 1986. Thin layer-chromatography determination of deoxynivalenol in processed grain products. *J. Assoc. Off. Anal. Chem.* **69:**35–36.
74. **Trucksess, M. W., S. Nesheim, and R. M. Eppley.** 1984. Thin layer chromatographic determination of deoxynivalenol in wheat and corn. *J. Assoc. Off. Anal. Chem.* **67:** 40–44.
75. **Ueno, Y. (ed.).** 1983. Toxicoses, natural occurrence and control, p. 195–307. *In Trichothecenes. Chemical, Biological and Toxicological Aspects. Developments in Food Science.* Elsevier Science Publishing Company, Inc., New York.
76. **Ueno, Y., S. Aoyama, Y. Sugiura, D. S. Wang, U.S. Lee, E. Y. Hirooka, S. Hara, T. Karki, G. Chen, and S. Z. Y.** 1993. A limited survey of fumonisins in corn and corn-based products in Asian countries. *Mycotoxin Res.* **9:**27–34.
77. **Usleber, E., M. Straka, and G. Terplan.** 1994. Enzyme immunoassay for fumonisin B_1 applied to corn-based food. *J. Agric. Food Chem.* **42:**1392–1396.
78. **Visconti, A., and A. Sibilia.** 1994. *Alternaria* toxins, p. 315–336. *In* J. D. Miller and H. L. Trenholm (ed.), *Mycotoxins in Grain. Compounds Other than Aflatoxin.* Eagan Press, St. Paul, Minn.
79. **Ware, G. M., and C. W. Thorp.** 1978. Determination of zearalenone in corn by high-pressure liquid chromatography and fluorescence detection. *J. Assoc. Off. Anal. Chem.* **61:**1058–1061.
80. **Wilson, T. M., P. F. Ross, L. G. Rice, G. D. Osweiler, H. A. Nelson, D. L. Owens, R. D. Plattner, C. Reggiardo, T. H. Noon, and J. W. Pickrell.** 1990. Fumonisin B_1 levels associated with an epizootic of equine leukoencephalomalacia. *J. Vet. Diagn. Invest.* **2:**213–216.
81. **Winter, C. K., D. G. Gilchrist, M. B. Dickman, and C. J. Jones.** 1996. Chemistry and biological activity of AAL toxins, p. 307–316. *In* L. Jackson, J. DeVries, and L. Bullerman (ed.), *Fumonisins in Foods.* Plenum Publishing Corp., New York.
82. **Wray, B. B., and K. G. O'Steen.** 1975. Mycotoxin-producing fungi from a house associated with leukemia. *Arch. Environ. Health* **30:**571–573.
83. **Wray, B. B., E. J. Rushings, R. C. Boyd, and A. M. Schindel.** 1979. Suppression of phytohemagglutinin response by fungi from a "leukemia" house. *Arch. Environ. Health* **34:**350–353.
84. **Zoller, O., F. Sager, and B. Zimmerli.** 1994. Occurrence of fumonisins in foods. *Mitt. Geb. Lebensmittelunters. Hyg.* **85:**81–99.

Viruses

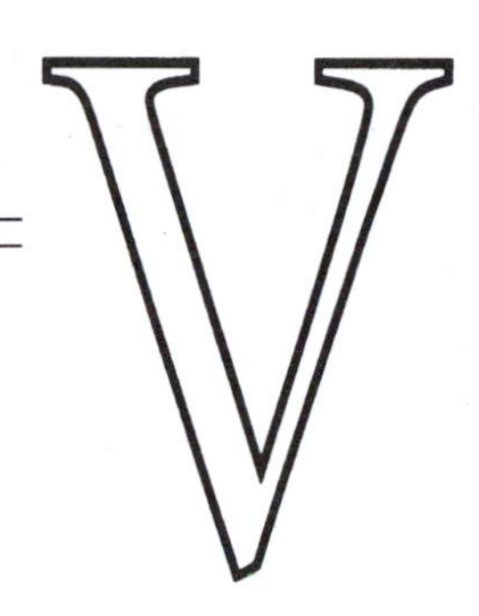

Dean O. Cliver

24 Foodborne Viruses

CHARACTERISTICS OF VIRUSES: GENERAL

Viruses occupied 3 of the top 10 places (based on number of persons ill) among causes of foodborne disease outbreaks reported in the United States during the period 1983 to 1987 (Table 24.1). Viruses are transmitted as particles of submicroscopic size. Particles of viruses known to be foodborne range generally from 25 to 75 nm in diameter; though they have icosahedral symmetry, they are approximately spherical, as seen in electron micrographs. Particles of foodborne viruses generally contain RNA (rather than DNA) genomes. With the exception of the rotaviruses, the genome is made up of a single strand with "plus sense"—which is to say, is capable of serving as messenger and being directly translated into protein. At least part of the virus-specific protein that results from this initial translation is an enzyme that catalyzes RNA synthesis on an RNA template; reverse transcription is unnecessary in that DNA has no role in the replicative cycles of these viruses. Viral particles are totally inert and thus cannot multiply in food or anywhere outside of susceptible living host cells.

Foodborne viruses are generally *enteric:* they infect perorally (by ingestion) and are shed with feces. This means that they may also be transmitted by person-to-person "contact" and via water, as are other enteric infectious agents. By no means do all foodborne viral infections occur as part of outbreaks (defined by the U.S. Centers for Disease Control and Prevention as a cluster of two or more cases), but there has been until now no means of attributing single illnesses to food, except in a few studies in which shellfish consumption was recorded. A distinctive property of enteric foodborne viruses is that they are generally adapted exclusively to humans as hosts; alternative animal hosts are unknown or of no practical significance. With the exception of tick-borne encephalitis virus, to be discussed briefly, neither nonhuman reservoirs nor biological vectors are significant in the transmission of these agents. Mechanical vectors, such as flies, that transmit other feces-associated agents have been mentioned occasionally but have not been directly implicated in recent outbreaks.

After enteric viruses are ingested, some have their principal site of action in the lining of the small intestine. Others infect the liver, and still others may affect other parts of the host's body. Disease results from killing of infected cells by the viral replicative process or from destruction of infected cells by the host's immune response. Although the genomic organizations of some of the foodborne viruses are known, virulence factors as such have not been described. Common elements of the genome include a region that codes for the structural (coat) protein of the virus and another that codes for the RNA-dependent RNA polymerase that is required for transcription of the viral nucleic acid. Other regions may not be translated, may code for additional virus-specific enzymes, may interfere with normal activities of the host cell, or may be of unknown function.

Table 24.1 Top 10 identified causes of foodborne disease outbreaks, United States, 1983 to 1987 (5, 6)

Rank	Causative agent	Illnesses No.	Illnesses %	Outbreaks No.	Outbreaks %
1	*Salmonella* spp.	31,245	57.3	342	37.2
2	*Shigella* spp.	9,971	18.3	44	4.8
3	*Staphylococcus aureus*	3,181	5.8	47	5.2
4	*Clostridium perfringens*	2,743	5.0	24	2.6
5	Norwalk virus	1,164	2.1	10	1.1
6	Hepatitis A virus	1,067	2.0	29	3.2
7	*Streptococcus*, group A	1,001	1.8	7	0.8
8	*Campylobacter* spp.	727	1.3	28	3.1
9	*Escherichia coli*	640	1.2	7	0.8
10	Other virus	558	1.0	2	0.2

The peroral infectious dose of virus is subject to debate. A certain amount of research on this subject has been done with human subjects, but doses were generally measured in cell culture infectious units rather than individual viral particles. There is little doubt that one mature viral particle contains all of the information needed to produce a human infection. However, there are some intrinsic inefficiencies in the process, so that ingestion of a single viral particle is unlikely to result in infection. The effects of vehicles (e.g., specific foods, drinking water) on the efficiency with which viruses produce infection have not been studied either. Given that peak rates of shedding of some viruses exceed 10^8 particles per g of feces, 10 μg of fecal contamination, which could well occur by handling food with feces-soiled fingers, would contain 1,000 particles and would carry a very serious threat of infection. If the vehicle were 100 g of food or water, the dilution factor would be 10^{-7}, which is why viruses are almost never detected in foods implicated in outbreaks.

CHARACTERISTICS OF SPECIFIC FOODBORNE VIRUSES

Hepatitis Viruses

The hepatitis viruses transmitted enterically are types A and E (16). These differ serologically, but in many other ways as well. The viruses of human hepatitis generally have in common only that they all infect the liver (Table 24.2); other taxonomic properties and even modes of transmission vary widely. One or more mutants of the hepatitis A virus that replicate and produce cytopathic effects in cell cultures have been identified; this has enabled research on many aspects of the virus and led to the development of promising vaccines.

Size and Shape of Particle

The hepatitis A virus looks like other picornaviruses: diameter of ~28 nm, apparently spherical, with an approximately smooth surface as seen in transmission electron micrographs. The hepatitis E virus looks like other caliciviruses: diameter of ~32 nm, apparently spherical, with visible cup-shaped concavities on its surface ("calicivirus" was coined from the Greek *kalyx* [κάλυξ], for cup). Thus, although a skilled electron microscopist could distinguish between hepatitis viruses A and E, neither could be distinguished morphologically from many other members of its taxonomic group that infect humans.

Characteristics of Disease, Shedding

Once either hepatitis A or E causes illness, the two diseases are not clinically distinguishable. Either of these viruses can produce illness that lasts several weeks and includes jaundice, anorexia, vomiting, and profound

Table 24.2 Viruses of human hepatitis

Type	Former name	Mode of transmission	Remarks
A	Infectious hepatitis	Fecal-oral	"Killed" vaccine available
B	Serum hepatitis	Parenteral	Recombinant vaccine available
C	Non-A, non-B hepatitis	Parenteral	Now screened for in U.S. blood banks
D	Delta agent	Parenteral	Satellite of hepatitis B virus
E	Non-A, non-B hepatitis	Fecal-oral	Not yet in North America

malaise. One clue that is sometimes observed is that hepatitis E is much more serious in pregnant women (17% to 33% mortality) than hepatitis A. Diagnosis of hepatitis A is usually based on demonstration of immunoglobulin M (IgM)-class antibody against the virus in the blood of the affected person, using one of several commercial kits. Hepatitis A is more likely to cause asymptomatic or mild infections in young children than in adolescents or adults, whereas serologic surveys indicate that hepatitis E less often infects young children. The incubation period of hepatitis A ranges from 15 to 50 days, with a median of 28 or 30 days. The virus is often shed in feces for 10 to 14 days before the onset of illness, during which time an infected food handler has ample opportunity to contaminate food unless personal hygiene (specifically, hand washing) has been scrupulous. Shedding may continue for 1 to 2 weeks after onset of illness (16). The incubation period of hepatitis E is 22 to 60 days, with an average of 40 days; shedding during the incubation period has not been known to lead to transmission via food.

Mechanisms of Pathogenesis; Genome

Virulence factors as such are poorly characterized. The hepatitis A virus ordinarily does not kill the liver cells that it infects (16). Rather, the body's immune response to the infection leads eventually to destruction of infected hepatocytes by cytotoxic T cells, hence the long incubation period. Death is rare, though permanent sequelae and occasional relapses are recorded. Immunity is durable, probably lifelong. Pathogenesis in hepatitis E is probably similar. Genomic organizations are not remarkable among single (plus-sense)-strand RNA viruses: consistent features are an extensive 5′ nontranslated region and specific sequences that code for structural (coat, or capsid) protein and for the required RNA-dependent RNA polymerase mentioned previously. The hepatitis A capsid comprises 60 copies of each of four different structural peptides; the genome also codes for one or more proteases that process the translated large polypeptide into four small units. Because many more molecules of coat protein than of the enzymes are needed during replication, translation of the capsid region of the genome is selectively enhanced.

Stability in Food

Food-associated outbreaks of hepatitis A have been recorded in great numbers. In most instances, these outbreaks have not appeared to depend on extreme stability of the virus in the food, in that the food was eaten uncooked or was thought to have been contaminated just before it was eaten. However, limited experiments to date have shown that hepatitis A virus is much more resistant to heat and drying than are other picornaviruses (16), whereas resistance to acid (e.g., pH 2 for short periods) and to gamma rays is common among picornaviruses. More research relating these properties to situations in food is needed. Hepatitis A virus is not remarkably resistant to chlorine in water. Comparable information regarding hepatitis E virus is not yet available. The virus is often transmitted via water, but under circumstances that chlorination could not have been expected to prevent; transmission of hepatitis E via foods has been proposed on various occasions, but conclusive epidemiologic evidence has not resulted. No cell culture-adapted mutant of hepatitis E virus has been identified. There are reports of infection of swine with this agent, but the epidemiologic significance of this is not known, and swine have not yet been used to study the stability of the virus as it might pertain to transmission via food or water.

Susceptible Populations

Susceptibility to hepatitis A is general. In countries where fecal-oral transmission of disease agents is common, most children have experienced infection and acquired active immunity by 5 years of age. Hepatitis A is more likely to cause symptoms in adults in developed countries, with more severe consequences. Other than age, host factors affecting the severity of hepatitis A are not well characterized; immune impairment may be less significant than with some other foodborne diseases, but liver abnormalities from other causes (e.g., cirrhosis) may be significant. Pooled human immune serum globulin has been used to afford passive immunity to susceptible persons who are going into "at risk" situations and to persons recently exposed to the virus (e.g., via food) if it can be administered within 2 weeks of probable exposure. One or more vaccines already in use in Europe have recently been licensed in the United States (10); food handlers have been proposed as an important target group for vaccination in countries where they are unlikely to have acquired immunity by earlier infection. Means of immunization against hepatitis E virus are not yet available. No vaccine is in prospect, and in North America especially, infection is evidently so rare that significant levels of antibody against hepatitis E virus are not found in pooled human immune serum globulin. Whereas hepatitis A is virtually worldwide, outbreaks of hepatitis E are seen principally in Asia, Africa, and Latin America.

Outbreaks

The reported annual incidence of hepatitis A in the United States declined from ~31,000 cases in 1990 to ~23,000 in 1992 (9); this rate of decline has not continued since. For the years 1990, 1991, and 1992, suspected food- or waterborne outbreaks were cited as possible sources of these infections at crude rates of 9.4, 6.0, and 8.0%, respectively, but at rates of 4.4, 3.0, and 4.7% when sources were assessed on a mutually exclusive (only one possible source per infection) basis. Another report includes estimates of 4,800 to 35,000 cases of food-associated hepatitis A in the United States annually (15).

Hundreds of outbreaks of food-associated hepatitis A have been reported since 1943 (11). Shellfish (bivalve mollusks) taken from waters contaminated with human feces have been the vehicle in a plurality of outbreaks, but any food handled by an infected person may become contaminated and, if not cooked subsequently before being eaten, transmit the infection (12).

Four outbreaks will be summarized here, to give some idea of the scope of the problem:

- In Sweden during the Christmas season of 1955, oysters that had been held in a harbor awaiting sale were contaminated by feces from a privy that overhung the water. These oysters were eaten raw as part of a traditional holiday celebration and gave rise, over the next few weeks, to 629 illnesses among people in many cities and villages and some who had by then left the country. This was the first recorded instance of hepatitis A transmission via shellfish (19).
- In St. Louis, Mo., in 1962, 28 cases of hepatitis A occurred among the staff of a hospital (18). The vehicle was found to be orange juice which had been thawed and reconstituted by a worker who was inapparently infected by the virus. Although she never became ill, her husband, who had not been at the hospital, became ill at the time of the staff outbreak. Much more conclusive diagnostic methods for such infections are now available.
- In 1978, outbreaks comprising 82 cases in Arkansas and 58 cases in Texas were traced to a single worker (39). The man's work consisted of preparing sandwiches and salads in vegetarian-health food restaurants. He continued to work at the second location while he was visibly ill.
- In 1988, clams taken from waters off Shanghai, China, and eaten raw were the source of nearly 300,000 cases of hepatitis A (25). The waters from which the shellfish were taken had apparently become contaminated with human sewage.

Another four recent outbreaks of foodborne hepatitis A illustrate the variety of vehicles that may transmit the disease (the list is far from exhaustive):

- In late 1988 and early 1989, a cluster of 38 cases of hepatitis A in Italy was traced to consumption of raw mussels and to drinking a particular brand of mineral water (38).
- In 1989, 50 people in several villages in the United Kingdom acquired hepatitis A by eating foods from a single bake shop (37).
- In 1990, 110 people in Missouri, got hepatitis A by eating lettuce apparently handled by an infected worker (8).
- A 1994 report tells of four people in Denmark who acquired hepatitis A by eating caviar that had been illegally imported from Latvia (20).

Norwalk-Like Viruses

Despite difficulties of diagnosis, the Norwalk-like viruses are being recognized as a major cause of foodborne disease in the United States and elsewhere. During the period 1983–1987, the Norwalk virus ranked fifth as a cause of foodborne illnesses among outbreaks reported in the United States (5, 6). The virus does not multiply in any known laboratory host. Diagnoses were based on detection of the viral particles in patients' stool samples by electron microscopy, immune electron microscopy, or enzyme immunoassay, or on demonstration of antibody against the virus in patients' serum samples by enzyme immunoassay. Viral antigen has been produced biosynthetically (27). Several serologic varieties, forming at least two major groups, occur among the "Norwalk-like" viruses (1). In general, diagnostic tests have been directed to the Norwalk virus proper, with detection of novel types of Norwalk-like viruses most easily done by immune electron microscopy using virus in acute-phase stool extract and antibody in convalescent-phase serum from the same patient (3). At least during 1983 to 1987, outbreaks attributed to Norwalk virus were probably due to the virus itself and very close antigenic relatives, rather than the entire group of Norwalk-like viruses.

Size and Shape of Particle

The Norwalk-like viruses have been assigned to the calicivirus group described above in conjunction with the hepatitis E virus (3). Although the Norwalk-like viruses are not uniformly reported as 32 nm in diameter, they do regularly reveal the pattern of surface depressions from which the name calicivirus derives. When detected by electron microscopy, they have been called "small, round, structured viruses" (3). Like most other foodborne viruses, they contain a single, plus-sense strand of RNA.

Characteristics of Disease; Shedding

The disease includes nausea, vomiting, diarrhea, and other symptoms common to gastroenteritis. If enough people are involved to make comparisons valid, vomiting is usually as common as diarrhea; before transmission via food and water had been described, "winter vomiting disease" was a commonly reported syndrome. Both the incubation period and the duration of symptoms usually range from 24 to 48 h. The ill person sheds the virus both in vomitus and feces, but emesis has not yet been identified as a source of food contamination. Now that more sensitive methods are available to detect the virus, it has been determined that fecal shedding may last a week after the gastroenteritis ends (22).

Mechanisms of Pathogenesis; Genome

Virulence factors as such have not been identified for the Norwalk and Norwalk-like agents. The viruses infect and kill cells of the small intestinal mucosa (3). Progeny virus passes down the digestive tract and is shed in the feces. This is an exceptionally fast-acting virus: volunteers have become ill as early as 10 h after ingesting the virus. Because the virus particle contains plus-sense RNA, it must code for an RNA-dependent RNA polymerase that can be translated early to permit replication. In the expectation that the genomic sequence coding for this polymerase would be highly conserved, sequences of many outbreak strains have been determined and compared. It was surprising to learn that each strain differed to some extent from the others, though at least two groups with substantial homology were identified (1, 31, 41).

Stability in Food

Because the infectivity of the virus can be determined only by feeding it to people, its stability in food (e.g., in processing) has not been determined. However, qualitative studies in human volunteers have shown that infectivity persists during 3 h at pH 2.7 at room temperature and for 60 min at neutral pH at 60°C (17). Vehicles in reported outbreaks have been shellfish eaten raw or slightly cooked, or other foods that were not heated after having undergone hands-on contamination, so there is no reason to suppose that the virus is exceptionally heat stable in food. Human volunteers have been used to study the depuration of Norwalk virus-contaminated oysters, which led to the determination that the virus persists longer than fecal coliform bacterial indicators in contaminated shellfish (24).

Susceptible Populations

Attack rates in reported outbreaks have often exceeded 50% of those at risk, which suggests that susceptibility to the viruses is widespread (32). Antibody produced in response to the infection is of diagnostic significance, but evidently does not afford protection to the person who has been infected. Indeed, in volunteer studies, those who had preexisting antibody often became ill, and although their antibody levels often rose as a result of the infection, they were again susceptible to the same agent when challenged over a year later. Given the lack of protection afforded by antibody produced as a result of infection, the hope of producing a useful vaccine against these agents seems dim.

Outbreaks

The many recorded outbreaks of food-associated gastroenteritis that have been attributed to Norwalk-like viruses show that the pattern of transmission resembles that of hepatitis A. Shellfish are frequently vehicles, but foods handled by infected persons often transmit these viruses, as well.

- A series of outbreaks of gastroenteritis associated with consuming cockles in the United Kingdom in late 1976 and early 1977 led to the detection of small, round viruses in the feces of several patients (4).
- The first fully documented foodborne outbreak of Norwalk virus gastroenteritis occurred in 1978 in New South Wales, Australia (32). Over 2,000 people who ate oysters, principally from the Georges River, were affected. Some groups of people who ate oysters showed attack rates of 85%.
- Norwalk virus was the leading cause of reported foodborne disease in the United States in 1982, due largely to an outbreak that involved some 3,000 people in the area of Minneapolis and St. Paul, Minn. (30). A baker's assistant who was ill apparently contaminated butter cream frosting that was later applied on a great number and variety of pastries.
- In 1987, ice made with contaminated water was used to cool drinks served at a college football game and a fundraising gathering; approximately 5,000 people became ill with gastroenteritis associated with a small, round virus in feces (7).
- The Snow Mountain agent, a small, round, structured virus that is antigenically distinct from the

Norwalk virus, caused 155 illnesses (2 deaths) among residents and 28 among employees of a retirement facility in the area of San Francisco Bay, Calif. (21). The vehicles were evidently shrimp dishes.

- Small, round, structured viruses resembling the Norwalk agent were detected in the stools of persons in a large gastroenteritis outbreak in Toyota City, Japan (29). School lunches from a preparation center that served nine schools were implicated as vehicles. Those affected included 3,236 students and 117 teachers.
- Various outbreaks of viral gastroenteritis have been reported aboard cruise ships. A Hawaiian cruise ship was the site of one in 1990, in which the vehicle was fresh cut fruit, that affected 238 people (26); another Hawaiian cruise ship served contaminated ice in 1992 that led to infections in ~200 people (28).

Astroviruses

Like the rotaviruses to be discussed below, the astroviruses most often affect the very young and the elderly, but there have been significant recorded instances of transmission via foods.

Size and Shape of Particle

The astroviruses are among the small, round, structured viruses seen with the electron microscope. Their size is in the range of of the Norwalk-like caliciviruses, but they are distinguished by an apparently raised, five- or six-pointed star pattern on the protein coat surface, whereas the caliciviruses have depressions (3). The nucleic acid is a single, plus-sense strand of RNA.

Characteristics of Disease; Shedding

Astrovirus illness differs from that of the Norwalk-like viruses in that the incubation period is 3 to 4 days, diarrhea is more typical than vomiting (which is rare), and the duration of illness is 2 to 3 days but may be 7 to 14 days (3). Shedding of the virus in feces lasts at least during the period of diarrhea.

Mechanisms of Pathogenesis; Genome

The astrovirus infects and kills mature enterocytes on the villi of the small intestine. These are replaced after a few days by new enterocytes generated in the crypts. An antibody response, which may be protective, results from infection. At least six serotypes have been identified, and several of these are able to infect laboratory cell cultures and cause cytopathic effects (3).

Stability in Food

The astroviruses resemble other enteric viruses in being resistant to pH 3 for periods of time. They are resistant to lipid solvents and show residual infectivity after 30 min at 50°C (3). Decimal reduction times at 60°C are in the range of 2 to 3 min (3).

Susceptible Populations

Because of the high rate of infection in children under 1 year old, 70% of children in the United Kingdom are said to have antibody against astrovirus by 3 to 4 years of age. However, resistance is likely to be serotype specific, so that susceptibility to some types of astroviruses continues. A foodborne outbreak of astrovirus gastroenteritis involving upwards of 4,700 people was reported in Katano City, Osaka, Japan (34).

Rotaviruses

Rotaviral illness, especially that caused by serogroup A, is common in infants throughout the world and is a significant cause of infant death in developing countries (36).

Size and Shape of Particle

The rotavirus particle is 70 to 75 nm in diameter. Although its shape is demonstrably icosahedral, it appears roughly spherical in most electron micrographs. The coat protein is double-layered, and the nucleic acid is double-stranded RNA in 11 segments.

Characteristics of Disease; Shedding

After an incubation period of 1 to 3 days, illness is characterized by fever, vomiting, and diarrhea that may last for 4 to 6 days. The virus is shed at least during illness, but usually not longer than 8 days. Infection begins in the enterocytes of the proximal small intestine but eventually involves the jejunum and ileum as well.

Stability in Food

Rotavirus has been found to lose 99% of its infectivity titer during 30 min at 50°C (36). It is unstable outside the pH range of 3 to 10 (36). It survives for many days at 4 or 20°C on vegetables and has been shown experimentally to withstand the process of making soft cheese (36).

Susceptible Populations

Although most rotaviral illness involves infants, outbreaks of foodborne and waterborne disease affecting a broad spectrum of ages have been recorded. For whatever reasons, the majority of foodborne outbreaks have been recorded in Japan and in New York State, whereas waterborne outbreaks have been in Germany, Israel, Russia, Sweden, and China. Serogroups other than A are sometimes involved, and animals may also be infected with some non-A serogroups but have not been shown to transmit their infections to humans. Shellfish growing in sewage-polluted waters undoubtedly become contaminated but have not yet been implicated as a vehicle in a recorded outbreak of human illness.

Other Viruses

A few other viruses have been reported to be transmitted via foods on occasion (13). Each will be discussed briefly here.

Tick-Borne Encephalitis Virus

In Slovakia and perhaps adjacent regions, dairy animals (particularly goats, but perhaps also cattle and sheep) bitten by infected ticks become infected with an encephalitis virus that is then shed in milk (23). This tick-borne encephalitis virus is perhaps the only known viral zoonosis that is transmissible via food. The virus is readily inactivated by pasteurizing the milk, but may be found in several milk products made with unpasteurized milk. Fortunately, infection is rarely contracted in this way because of the limited distribution of the virus; however, small outbreaks continue to occur even though the problem has been recognized for many years. For example, a group of seven people who drank goats' milk raw contracted the disease in Slovakia in 1993.

Enteroviruses

Poliomyelitis, caused by any of three serotypes of enteroviruses (a subset of the picornavirus group that includes the hepatitis A virus), was the first virus disease reported to be foodborne. However, no outbreak of foodborne poliomyelitis has been reported in developed countries since 1949, *before* the introduction of preventive vaccines (11). Members of the enterovirus group other than the polioviruses include the coxsackieviruses and the echoviruses. One outbreak of food-associated coxsackievirus B1, B3, and B5 infection has been reported from the then U.S.S.R., in association with widespread contamination in a kitchen that prepared food for children in a day-care facility (35). Echovirus outbreaks that have been reported include one of type 4 that occurred in 1976 in Pennsylvania (40) and one of type 5 that occurred in 1988 in New York State (33). Eighty people who ate contaminated cole slaw were affected in the former outbreak; the vehicle in the latter outbreak was not determined, but foods were the common element among the 161 persons affected.

Parvoviruses

The parvoviruses are perhaps the smallest of enteric viruses, with diameters of 20 to 26 nm (2). Their protein coat appears smooth, and they contain single-stranded DNA. Several (e.g., cockle, Wollan, Ditchling, and Parramatta agents) have been detected in the stools of persons involved in food-associated outbreaks of gastroenteritis. However, Appleton (3) says that the role of these agents in causation of human gastroenteritis has yet to be confirmed and that parvoviruses are often detected in diarrheal stools in association with other viruses that are known to cause gastroenteritis. Replication of the viral genome from a single-stranded DNA template presents some special challenges. Parvoviruses certainly cause illnesses other than gastroenteritis in humans and cause gastroenteritis in other animal species, but these viruses may be just satellites in human gastroenteritis, as they are known to be in some other contexts.

Other Gastroenteritis Viruses

Other human enteric viruses, particularly adenoviruses 40 and 41 and coronaviruses, are known to cause gastroenteritis in humans, occasionally in significant outbreaks. The adenoviruses are ~75 nm in diameter and contain double-stranded DNA. They have not been known to be transmitted via food or water. The coronaviruses are among the very rare enteric viruses that have lipid-containing envelopes; in this case, the envelope surrounds a "nucleocapsid" of single-stranded RNA wrapped in protein. One gastroenteritis outbreak, among persons who ate at a sports club in Scotland, has been attributed to coronavirus (2). The specific food by which the virus was transmitted could not be identified.

Viruses Probably Not Transmitted via Foods

Because they represent a threat to human health that is quite serious in some instances, some other viruses have been suspected of being transmissible via foods. Despite the inability of these viruses to infect humans perorally, the perceived risk of their introduction into food has at times led to detention or destruction of food that presented no danger to consumer health. Included are the human immunodeficiency virus (HIV), herpesvirus, hantavirus, rabies virus, rhinoviruses, and the agent of bovine spongiform encephalopathy.

HIV is not transmissible via food because the virus evidently does not infect perorally. Not only are HIV-positive persons not a risk as food handlers, but people with frank symptoms of AIDS are not a threat when handling food, as long as they do not have diarrhea (e.g., as caused by *Cryptosporidium*). Neither are the herpesviruses, the hantavirus, the rabies virus, and the rhinoviruses (which cause the "common cold") infectious perorally, with or without food. The agent (a prion) of bovine spongiform encephalopathy may infect cattle perorally, but there is no reason to believe that it can infect humans under any circumstances. What is said here does not suggest that it is reasonable to eat meat from sick animals nor to allow people to handle food (professionally or as amateurs) while ill. It simply means that food that is suspected of contamination with one of the viruses mentioned (e.g., milk from a cow that subsequently showed symptoms of rabies) is unlikely to injure human consumers.

OTHER CONSIDERATIONS

Once diagnostic methods for a foodborne viral illness are well established, it is reasonable to aspire to detect the virus in food. Whatever other considerations there are, one would rather be able to detect virus in food before the food was eaten than to detect virus in the feces of those who became ill by eating the food. When diagnostic methods for enteric viral infections were based on detection of the virus in patient's stools, one might have aspired to increase the sensitivity of the same test by at least two orders of magnitude (assuming that feces never make up more than 1% of a contaminated food) and apply it to food testing. However, diagnostic tests have now evolved in the direction of detecting antiviral antibody (usually of the IgM class, which denotes recent or current infection) in the patient's serum. This approach offers no help in detecting viruses in food.Viruses in food might be detected on the basis of their infectivity in living hosts (most likely laboratory cell cultures), their size and shape as seen by electron microscopy, the unique reactions of their coat antigens in some serologic test, or the specific reactions of their nucleic acids with selected probes or primers (14). The first two options have not worked well because the most important foodborne viruses do not cause demonstrable infections in cell cultures and are likely to be present at such low levels as to be undetectable by electron microscopy. Many methods for detecting viruses in food and water by specific reactions with the viral protein or nucleic acid have been described; but none has yet gained general acceptance and all are exacting, time-consuming, and expensive.

Some tests that have been described might be applied to test for viruses in foods implicated in an outbreak, but these methods are unlikely to be used in routine monitoring of foods destined for human consumption. Indicators of fecal contamination have been used to monitor the sanitary quality of some foods (e.g., shellfish) and might be pertinent to virus contamination because viruses in foods are shed with human feces. However, the established indicator systems (e.g., fecal coliforms) are bacterial, and their presence or absence is not a good predictor of viruses, for a number of reasons. Alternative tests, such as detection of bacteriophages that infect enteric bacteria (*Escherichia coli, Bacteroides fragilis,* etc.), have been under study for a few years, but none has yet emerged as a solution to the problem of routine monitoring of foods for possible virus contamination.

Avoiding fecal contamination of food is the surest means of preventing food-associated transmission of viruses. However, the epidemiologic record shows that fecal and viral contamination occurs at times, either from an infected food handler or with wastewater. Now that a vaccine against hepatitis A is available, it might be well to immunize professional food handlers who are not already immune. Of course, this would not protect against contamination of food by amateurs infected with hepatitis A virus, nor by anyone infected with a gastroenteritis virus. Although it is extremely difficult to document, it seems likely that a good deal of viral disease is transmitted via foods contaminated in the home. Proper hand washing and exclusion of ill persons from handling food are important preventive measures, as is avoidance of contaminating food with polluted water or wastewater. Proper use of chlorine or of UV light will inactivate viruses in water and on surfaces, but viruses within a food are best inactivated by cooking to temperatures that will kill vegetative bacterial pathogens. Enteric viruses are likely to persist for weeks in refrigerated foods and indefinitely in frozen foods.

SUMMARY

Viruses have been shown to be a frequent cause of foodborne disease in the United States and probably in many other countries. Almost all of the viruses transmissible to humans via foods are produced in the human body and shed with feces. Almost all of these enteric viruses contain RNA (usually single-stranded) coated only with protein. They are more acid stable, but not more heat stable, than most enteric bacteria. These viruses enter food as a result of fecal contamination. Because the particles are inert outside of living human cells, they cannot multiply in food. Viruses in food are inactivated by high temperatures and preserved by low temperatures.

References

1. **Ando, T., M. N. Mulders, D. C. Lewis, M. K. Estes, S. S. Monroe, and R. I. Glass.** 1994. Comparison of the polymerase region of small round structured virus strains previously classified in three antigenic types by solid-phase immune electron microscopy. *Arch. Virol.* **135:**217–226.
2. **Anonymous.** 1991. Outbreak of diarrhoeal illness associated with coronovirus [sic] infection. *Commun. Dis. Environ. Health Scotland Weekly Rep.* **25**(46):1.
3. **Appleton, H.** 1994. Norwalk virus and the small round viruses causing foodborne gastroenteritis, p. 57–79. *In* Y. H. Hui, J. R. Gorham, K. D. Murrell, and D. O. Cliver (ed.), *Foodborne Disease Handbook,* vol. 2. *Diseases Caused by Viruses, Parasites, and Fungi.* Marcel Dekker, New York.
4. **Appleton, H., and M. S. Pereira.** 1977. A possible virus aetiology in outbreaks of food-poisoning from cockles. *Lancet* **i:**780–781.
5. **Bean, N. H., and P. M. Griffin.** 1990. Foodborne disease outbreaks in the United States, 1973–1987: pathogens, vehicles, and trends. *J. Food Prot.* **53:**804–817.
6. **Bean, N. H., P. M. Griffin, J. S. Goulding, and C. B. Ivey.** 1990. Foodborne disease outbreaks, 5-year summary, 1983–1987. *J. Food Prot.* **53:**711–728.
7. **Centers for Disease Control.** 1987. Outbreak of viral gastroenteritis—Pennsylvania and Delaware. *Morbid. Mortal. Weekly Rep.* **36:**709–711.
8. **Centers for Disease Control and Prevention.** 1993. Multistate outbreak of viral gastroenteritis related to consumption of oysters—Louisiana, Maryland, Mississippi, and North Carolina, 1993. *Morbid. Mortal. Weekly Rep.* **42:**945–948.
9. **Centers for Disease Control and Prevention.** 1994. Viral hepatitis surveillance program, 1990–1992, p. 19–34. *Hepatitis Surveillance Report no. 55.* Centers for Disease Control and Prevention, Atlanta, Ga.
10. **Cimons, M.** 1995. Hepatitis A vaccine licensed, progress for rotavirus, parvovirus. *ASM News* **61:**324–326.
11. **Cliver, D. O.** 1983. *Manual on Food Virology.* VPH/83.46. World Health Organization, Geneva.
12. **Cliver, D. O.** 1985. Vehicular transmission of hepatitis A. *Public Health Rev.* **13:**235–292.
13. **Cliver, D. O.** 1994. Other foodborne viral diseases, p. 137–143. *In* Y. H. Hui, J. R. Gorham, K. D. Murrell, and D. O. Cliver (ed.), *Foodborne Disease Handbook,* vol. 2. *Diseases Caused by Viruses, Parasites, and Fungi.* Marcel Dekker, New York.
14. **Cliver, D. O., R. D. Ellender, G. S. Fout, P. A. Shields, and M. D. Sobsey.** 1992. Foodborne viruses, p. 763–787. *In* C. Vanderzant and D. F. Splittstoesser (ed.), *Compendium of Methods for the Microbiological Examination of Foods,* 3rd ed. American Public Health Association, Washington, D.C.
15. **Council for Agricultural Science and Technology.** 1994. *Foodborne Pathogens: Risks and Consequences.* Task Force Report No. 122, September 1994. Council for Agricultural Science and Technology, Ames, Iowa.
16. **Cromeans, T., O. V. Nainan, H. A. Fields, M. O. Favorov, and H. S. Margolis.** 1994. Hepatitis A and E viruses, p. 1–56. *In* Y. H. Hui, J. R. Gorham, K. D. Murrell, and D. O. Cliver (ed.), *Foodborne Disease Handbook,* vol. 2. *Diseases Caused by Viruses, Parasites, and Fungi.* Marcel Dekker, New York.
17. **Dolin, R., N. R. Blacklow, H. DuPont, R. F. Buscho, R. G. Wyatt, J. A. Kasel, R. Hornick, and R. M. Chanock.** 1972. Biological properties of Norwalk agent of acute infectious nonbacterial gastroenteritis. *Proc. Soc. Exp. Biol. Med.* **140:**578–583.
18. **Eisenstein, A. B., R. D. Aach, W. Jacobson, and A. Goldman.** 1963. An epidemic of infectious hepatitis in a general hospital. *J. Am. Med. Assoc.* **185:**171–174.
19. **Gard, S.** 1957. Discussion, p. 241–243. *In* F. W. Hartman (ed.), *Hepatitis Frontiers.* Little, Brown & Co., Boston.
20. **Glerup, H., H. T. Sorensen, A. Flyvbjerg, P. Stokvad, and H. Vilstrup.** 1994. A "mini epidemic" of hepatitis A after eating Russian caviar. *J. Hepatol.* **21:**479.
21. **Gordon, S. M., L. S. Oshiro, W. R. Jarvis, D. Donenfeld, M. S. Ho, F. Taylor, H. B. Greenberg, R. Glass, H. P. Madore, R. Dolin, and O. Tablan.** 1990. Foodborne Snow Mountain agent gastroenteritis with secondary person-to-person spread in a retirement community. *Am. J. Epidemiol.* **131:**702–710.
22. **Graham, D. Y., X. Jiang, T. Tanaka, A. R. Opekun, H. P. Madore, and M. K. Estes.** 1994. Norwalk virus infection of volunteers: new insights based on improved assays. *J. Infect. Dis.* **170:**34–43.
23. **Grešíková, M.** 1994. Tickborne encephalitis, p. 113–135. *In* Y. H. Hui, J. R. Gorham, K. D. Murrell, and D. O. Cliver (ed.), *Foodborne Disease Handbook,* vol. 2. *Diseases Caused by Viruses, Parasites, and Fungi.* Marcel Dekker, New York.
24. **Grohmann, G. S., A. M. Murphy, P. J. Christopher, E. Auty, and H. B. Greenberg.** 1981. Norwalk virus gastroenteritis in volunteers consuming depurated oysters. *Austr. J. Exp. Biol. Med.* **59**(Pt. 2):219–228.
25. **Halliday, M. L., L.-Y. Kang, T.-K. Zhou, M.-D. Hu, Q.-C. Pan, T.-Y. Fu, Y.-S. Huang, and S.-L. Hu.** 1991. An epidemic of hepatitis A attributable to the ingestion of raw clams in Shanghai, China. *J. Infect. Dis.* **164:**852–859.
26. **Herwaldt, B. L., J. F. Lew, C. L. Moe, D. C. Lewis, C. D. Humphrey, S. S. Monroe, E. W. Pon, and R. I. Glass.** 1994. Characterization of a variant strain of Norwalk virus from a food-borne outbreak of gastroenteritis on a cruise ship in Hawaii. *J. Clin. Microbiol.* **32:**861–866.
27. **Jiang, X., J. Wang, D. Y. Graham, and M. K. Estes.** 1992. Expression, self-assembly, and antigenicity of the Norwalk virus capsid protein. *J. Virol.* **66:**6517–6531.
28. **Khan, A. S., C. K. Moe, R. I. Glass, S. S. Monroe, M. K. Estes, L. E. Chapman, X. Jiang, C. Humphrey, E. Pon, J. K. Iskander, and L. B. Schonberger.** 1994. Norwalk virus-associated gastroenteritis traced to ice consumption aboard a cruise ship in Hawaii: comparison and application of molecular method-based assays. *J. Clin. Microbiol.* **32:**318–322.
29. **Kobayashi, S., T. Morishita, T. Yamashita, K. Sakae, O. Nishio, T. Miyake, Y. Ishihara, and S. Isomura.** 1991. A large outbreak of gastroenteritis associated with a small round structured virus among schoolchildren and teachers in Japan. *Epidemiol. Infect.* **107:**81–86.

30. **Kuritsky, J. N., M. T. Osterholm, H. B. Greenberg, J. A. Korlath, J. R. Godes, C. W. Hodberg, J. C. Forfang, A. Z. Kapikian, J. C. McCullough, and K. E. White.** 1984. Norwalk gastroenteritis: a community outbreak associated with bakery product consumption. *Ann. Intern. Med.* **100:**519–521.

31. **Lew, J. F., A. Z. Kapikian, J. Valdesuso, and K. Y. Green.** 1994. Molecular characterization of Hawaii virus and other Norwalk-like viruses: evidence for genetic polymorphism among human caliciviruses. *J. Infect. Dis.* **170:**535–542.

32. **Murphy, A. M., G. S. Grohmann, P. J. Christopher, W. A. Lopez, G. R. Davey, and R. H. Millsom.** 1979. An Australia-wide outbreak of gastroenteritis from oysters caused by Norwalk virus. *Med. J. Austr.* **2:**329–333.

33. **New York Department of Health.** 1989. *A Review of Foodborne Disease Outbreaks in New York State 1988.* Bureau of Community Sanitation and Food Protection, Albany, N.Y.

34. **Oishi, I., K. Yamazaki, T. Kimoto, Y. Minekawa, E. Utagawa, S. Yamazaki, S. Inouye, G. S. Grohmann, S. S. Monroe, S. E. Stine, C. Carcamo, T. Ando, and R. I. Glass.** 1994. A large outbreak of acute gastroenteritis associated with astrovirus among students and teachers in Osaka. *Jpn. J. Infect. Dis.* **170:**439–443.

35. **Osherovich, A. M., and G. S. Chasovnikova.** 1967. Study on the isolation of enteroviruses from environmental objects, p. 89–90. *In* M. P. Chumakov (ed.), *Materialy Problemnoi Komissii Akad. Med. Nauk SSSR "Poliomyelit i Virusnye Entsefality,"* Vypusk 1, *Enterovirusy.* Akad. Med. Nauk SSSR, Moscow. (In Russian.)

36. **Sattar, S. A., V. S. Springthorpe, and S. A. Ansari.** 1994. Rotavirus, p. 81–111. *In* Y. H. Hui, J. R. Gorham, K. D. Murrell, and D. O. Cliver (ed.), *Foodborne Disease Handbook,* vol. 2. *Diseases Caused by Viruses, Parasites, and Fungi.* Marcel Dekker, New York.

37. **Sockett, P. N., J. M. Cowden, S. Le Baigue, D. Ross, G. K. Adak, and H. Evans.** 1993. Foodborne disease surveillance in England and Wales: 1989–1991. *Commun. Dis. Rep.* **3:**159–172.

38. **Stroffolini, T., W. Biagini, L. Lorenzoni, G. P. Palazzesi, M. Divizia, and R. Frongillo.** 1990. An outbreak of hepatitis A in young adults in central Italy. *Eur. J. Epidemiol.* **6:**156–159.

39. **U.S. Department of Health, Education, and Welfare.** 1978. Unpublished data.

40. **U.S. Department of Health, Education, and Welfare.** 1979. Aseptic meningitis outbreak at a military installation in Pennsylvania, p. 11. *Aseptic Meningitis Surveillance, Annual Summary 1976,* HEW-CDC publication no. 79-8231. U.S. Department of Health, Education, and Welfare, Atlanta, Ga.

41. **Wang, J. X., X. Jiang, H. P. Madore, J. Gray, U. Desselberger, T. Ando, Y. Seto, I. Oishi, J. F. Lew, K. Y. Green, and M. Estes.** 1994. Sequence diversity of small, round-structured viruses in the Norwalk virus group. *J. Virol.* **68:**5982–5990.

Foodborne and Waterborne Parasites

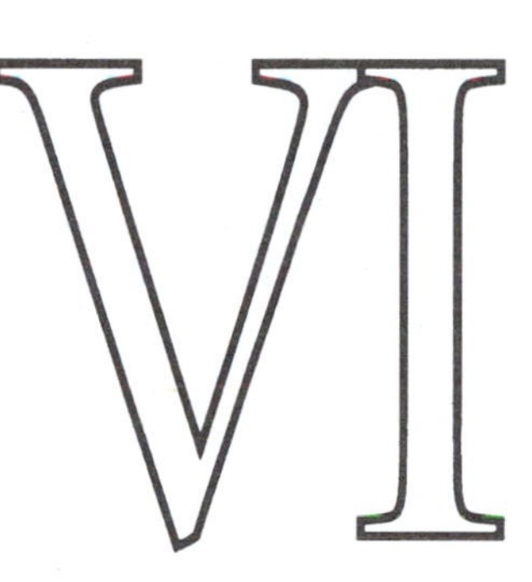

Charles W. Kim

25 Helminths in Meat

Foodborne parasites have had an impact on humans throughout history. They have been important from the standpoint of the health of livestock animals because of consequences for the meat supply and, more importantly, from their direct effect on the well-being of humans, who almost universally consume animal meat as a source of protein and other nutrients. Three helminths of medical significance are *Trichinella spiralis* and *Taenia solium,* which occur primarily in pork, and *Taenia saginata,* which is present in beef. There are other helminths that are primarily transmitted by other means; e.g., the application of meat as a poultice is a minor mode of transmission.

Despite the availability of sensitive, specific diagnostic tests and effective chemotherapeutic agents, these parasites continue to be a threat to public health throughout the world. This is partly due to demographic changes in human populations that directly affect cultural and agricultural practices by introducing new culinary procedures for preparing beef, pork, and other meats. Thus, current control and preventive procedures are often inadequate, and more effective control measures are needed to ensure safe meat for human consumption.

TRICHINOSIS

The history of trichinosis is fascinating, going back to when infection by *Trichinella* species was presumed more than proven. Whether the commandment in the Bible (Leviticus 11 and Deuteronomy 14) to not eat the flesh of cloven-footed animals (swine) was due in part to the potential danger of contracting trichinosis is only speculative. In the 7th century A.D., Mohammed prohibited the eating of pork. Hence, to this day, trichinosis is rare in Jews and Moslems. The earliest known case of trichinosis may be evidenced by the mummy of a person who probably lived near River Nile circa 1200 B.C., an observation made only in 1980 when supposedly *Trichinella* larvae were identified in the mummy's intercostal muscle (41).

T. spiralis was first observed in 1835 in the muscles of a man during a postmortem dissection (9). A landmark in the history of clinical trichinosis was Zenker's demonstration in 1860 that encapsulated larvae in the arm muscle caused the illness and death of a young woman. Even after 1860, many cases of trichinosis were undoubtedly not diagnosed because of the difficulty in recognizing the infection clinically.

Epidemiology

Trichinosis (or trichinellosis) is worldwide in its distribution because its etiologic agent, *Trichinella* species, is ubiquitous in the animal population, both domestic and wild. Until recently, it was assumed that there was only one species of *Trichinella, T. spiralis.* However, several isolates that differ in certain respects have been reported. A new isolate from arctic carnivores was reported in 1972 and designated *Trichinella nativa* (8).

Trichinella nelsoni, first obtained from a hyena in Kenya, was found to have low infectivity in the domestic pig, the primary hosts being the bush pig and warthog (46). Another new isolate, *Trichinella pseudospiralis*, believed to be a parasite of birds and hence not pathogenic for humans, was reported to be smaller than other isolates and more importantly lacked a capsule (23). In 1994, however, the first human case of trichinosis attributed to *T. pseudospiralis* was reported (1). The most recent isolate, *Trichinella britovi*, is reported to be similar to *T. nativa* but different in certain aspects (50). Attempts to characterize various isolates by biochemical and immunological means were inconclusive until isoenzyme patterns of isolates from different geographic areas and different hosts revealed different gene pools (51) and significantly different restriction enzyme fragment lengths of the small ribosomal DNA subunit (63). Thus, from biochemical and molecular analyses, it is apparent that there are several species or subspecies of *Trichinella* that are responsible for infection.

Trichinosis is considered a zoonosis because it occurs in domestic and wild animals, and humans become infected only when they ingest raw or poorly cooked animal meat containing the larvae. The domestic cycle consists of carnivorous animals consuming the flesh of animals, primarily domestic pigs. Pigs become infected by eating uncooked scraps of infected pork. The high incidence of trichinosis in the past was largely due to garbage-fed, rather than grain-fed or pasture-raised, pigs. Uncooked pork scraps were either household garbage or abattoir offal. Pork products such as fresh sausage, summer sausage, and dried or smoked sausage have been implicated as major vehicles of human trichinosis in the United States. Surprisingly, human infections resulting from ingestion of meat of an herbivorous animal, the horse, have been reported (3). It was postulated that infection of the horse resulted from ingestion of an infected, dead rodent along with grain. Three outbreaks in southern suburbs of Paris were also attributed to the consumption of raw horse meat in the form of steak tartare (6). An outbreak in China was reported to be due to the consumption of mutton (14). An unusual outbreak in northern Germany was attributed to the consumption of air-dried meat known as pastyrma that had been purchased as camel meat in Egypt (5). It was not certain whether this delicacy was indeed camel meat, since pastyrma is usually made from beef. If beef was the source, there may have been adulteration with pork. In Thailand, dog meat has been implicated as the source of infection (16). Dog meat is prepared as a special dish and is favored by the Vietnamese, Chinese, and hill tribe Thais. Dog meat has also been implicated in an outbreak of trichinosis in China (31).

The sylvatic cycle involves more than 100 species of mammals that can serve as sources for trichinosis. Wild carnivorous or omnivorous animals eat the carrion of dead carnivores, as exemplified by the polar bear, in which the infection rate is high; after hunters pelt the animal, infected flesh is left for scavengers. Apparently, *Trichinella* sp. remains infective for years in the flesh of the arctic black bear. Trichinous polar bear meat may conceivably have been responsible for the deaths of three Swedish explorers on an expedition to the North Pole in 1897. More than 50 years later, laboratory examination verified the presence of larvae in minute particles of bear meat still remaining on the explorers' equipment (53). What makes this theory so intriguing and plausible is the fact that arctic isolates of *T. nativa* are known for their ability to survive freezing in bear muscle. In Alaska, all human cases of trichinosis have been traced to the consumption of bear or walrus meat (55). Polar bear meat is considered a delicacy among the Eskimos. Among the Inuit Eskimos in northeastern Canada, there have been outbreaks resulting from the consumption of primarily raw walrus meat (60) and probably other mammals (39). In the former Soviet Union, up to 96% of human trichinosis cases have been traced to the consumption of wild animals, particularly boars (4). In the United States, bear meat is the most commonly incriminated wild meat, and wild pig is second. An isolated case in Oregon was attributed to consumption of cougar meat (13).

In the United States, trichinosis in humans has been declining. Trends in incidence and transmission patterns have been changing, especially with a large influx of people from Southeast Asia (34). In the past, trichinosis was more common in individuals of German, Italian, and Polish descent because of culinary preferences for raw or undercooked pork. Between 1975 and 1984, trichinosis among Southeast Asians, who prepare some dishes containing essentially raw pork, was 25 times more frequent than in the general population (59). Of the 1,260 cases of trichinosis reported to the Centers for Disease Control and Prevention during this period, 60 (4.8%) were among refugees of Southeast Asia. One of the largest outbreaks ever reported in the United States occurred in 1990 when 70% of Southeast Asian refugees in six states and Canada developed trichinosis after eating pork sausage at a wedding held in Iowa (40).

Asian countries are new foci of trichinosis, with epidemics reported from Thailand, Laos, Japan, China, and Hong Kong. A dish known as nahm is a common source among the inhabitants of northern Thailand (32). There are also other Thai dishes, such as lu (lahb [raw spiced meat] mixed with fresh blood) and satay (small pieces of spiced pork grilled rare to medium on a bamboo skewer), that are sources of infection. Sometimes satay is eaten raw. In both dishes, the larvae may remain alive. Outbreaks of trichinosis in Laos have been due to the consumption of pork dishes known as som-mou, lap mou, and lap leuat. Laotian immigrants in the United States still prepare these traditional dishes, resulting in outbreaks (12).

It had been suspected for many years that trichinosis can be contracted via contaminated utensils, such as butcher knives, meat grinders, butcher blocks, or even human hands after handling infected pork. Circumstantial evidence for this mode of transmission was strong for an outbreak that occurred in Liverpool, England (57). Trichinosis associated with beef that had been in contact with the meat grinder or butcher block used for pork, or from mixing of beef and pork, has been reported in the United States (2, 11) and Canada (21).

Life Cycle

Trichinella species completes its life cycle within one host; it does not require an intermediate host or any extrinsic development and thus is completely parasitic. When larvae encysted in raw or inadequately cooked meat are ingested, the muscle fibers and capsules that enclose the parasite are digested within a few hours. The liberated larvae burrow into the lamina propria of the villi in the jejunum and ileum. Four molts occur within 48 h, and by the third day, the worms are sexually mature. The small tapered head of the adult worm has a round, unarmed mouth that opens into a tubular esophagus. Immediately below the esophagus and the stichosome lies a thin-walled intestine, the hind portion of which terminates in the rectum, a muscular tube lined with chitin. The female worm measures about 3.5 mm in length and possesses a vulval opening about one-fifth the body length from the anterior end (Fig. 25.1). The male measures 1.3 mm to 1.6 mm in length and possesses a single testis that originates in the posterior portion of the body and extends anteriorly to near the posterior end of the esophagus, where it turns posteriorly to form the vas deferens, which becomes the enlarged vesicula seminalis. The vesicula seminalis becomes the ejaculatory duct to join the copulatory tube in the cloaca. The copulatory tube forms the copulatory bell that is extruded during copulation (Fig. 25.2). There are two ventrally located copulatory appendages on each side of the cloacal opening which possibly serve to clasp the female in copula. Between these appendages lie four tubercles or papillae. Variations in size and differences in

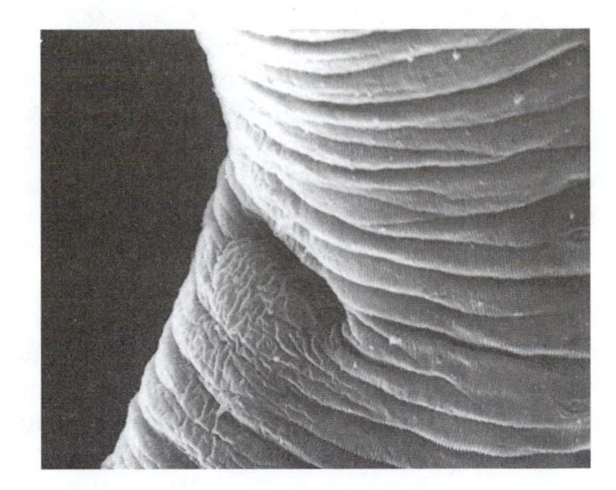

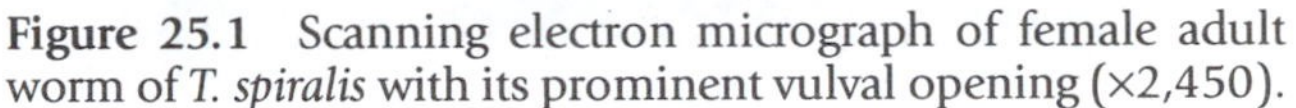

Figure 25.1 Scanning electron micrograph of female adult worm of *T. spiralis* with its prominent vulval opening (×2,450).

the cuticle have been noted for different isolates. For example, *T. pseudospiralis* is as much as one-fourth to one-third smaller than other isolates. Sexually mature adult worms reenter the lumen of the small intestine, where copulation takes place. The male adults die shortly after copulation, but the female worms reburrow into the mucosa and begin to larviposit, usually into the central lacteals of the villi, about 7 days after infection and may continue to do so for a period up to a few weeks. Each female worm bears approximately 1,500 newborn larvae.

The tiny newborn larvae (100 by 6 μm) are carried from the intestinal lymphatic vessels to the regional lymph nodes and into the thoracic duct and the venous blood, passing through the right side of the heart, through the pulmonary capillaries back to the left side of the heart, and into peripheral circulation. During migration, the larvae are known to enter many tissues,

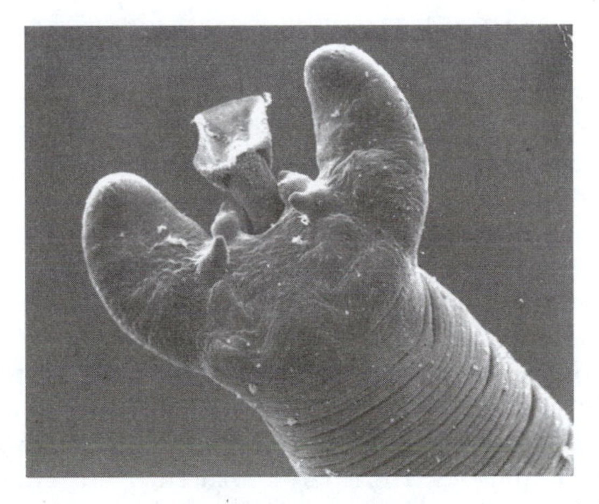

Figure 25.2 Scanning electron micrograph of male adult of *T. spiralis* with its copulatory bell (×1,400).

including those of the myocardium, brain, and other sites, but they either are destroyed or reenter the bloodstream. Generally, only larvae that reach striated muscles are able to continue the cycle. They penetrate the sarcolemma of the fibers, where they mature, reaching approximately 1,250 μm in length. They become coiled within the fibers and are encapsulated as a result of the host's cellular response, except for *T. pseudospiralis*, which lacks a capsule. The encapsulated cyst eventually becomes calcified, usually within 6 months. Since cannibalism among humans is not practiced, humans are a blind alley for the parasite. However, in animal hosts, the carcass serves as a source of infection.

Pathogenesis and Pathology

Damage caused by *Trichinella* infection varies with the intensity of the infection and the tissues invaded. The in-and-out movements of the adult worms, especially the females, in the intestinal tissue cause an acute inflammatory response and petechial hemorrhages. The cellular response consists of primarily neutrophils with eosinophils. This is followed by an infiltration of lymphocytes, plasma cells, and macrophages that peaks at about 12 days after infection, gradually declining thereafter. Thus, lesions in the intestine are due to the host's response to the adult worms or their protein products.

The newly released newborn larvae also cause an acute inflammatory response as they pass through or become lodged in various tissues and organs. The infiltration consists of lymphocytes, neutrophils, and especially eosinophils. Although there is myocarditis, viable larvae are more numerous in the pericardial fluid than in the myocardium. Pulmonary hemorrhage and bronchopneumonia may be observed during this stage of larval migration through the capillaries. Rarely, encephalitis may result if the larvae migrate through cerebral capillaries.

When larvae encyst in striated muscle fibers, there is an immediate tissue response consisting of inflammation of the sarcolemma of the involved muscle fibers. The disturbance of ultrastructure and metabolic processes in muscle fibers results in basophilic transformation. This is followed by destruction of the muscle fibers and the eventual formation of a capsule. The larvae gradually die, provoking an intense granulomatous reaction or foreign body cellular response that culminates in calcification. The most heavily parasitized striated muscles include the diaphragm and intercostal, ocular, and masseter muscles. Those larvae that enter tissues other than striated muscles disintegrate and are eventually absorbed.

Clinical Manifestations

The vast majority of individuals infected with *Trichinella* sp. are asymptomatic, probably because low numbers of larvae are ingested. Classical trichinosis is usually described as a febrile disease with gastrointestinal symptoms, periorbital edema, myalgia, petechial hemorrhage, and eosinophilia. During the intestinal stage of infection, gastrointestinal symptoms such as nausea, vomiting, "toxic" diarrhea or dysentery, fever (over 38°C), and sweating may be observed. Malaise can be severe and lasts longer than fever. Onset of symptoms occurs usually within 72 h after infection and may last for 2 weeks or longer.

Generally, from the second week, when the newborn larvae are migrating and can reach almost any tissue within the body, a characteristic edema is noted around the eyes and in some cases around the sides of the nose, at the temples, and even on the hands. Periorbital edema is present in about 85% of patients and is often complicated by subconjunctival and subungual splinter hemorrhages which disappear within 2 weeks. Eosinophilia greater than 50% is the most consistent manifestation. In severe cases, however, the number of eosinophils may be low. As the larvae enter striated muscle fibers, a striking myositis and muscular pain are evident. In addition to inflammation and pain, eosinophilia is also observed during this phase. Serum creatine phosphokinase and transaminase can be elevated as a result of leakage of these enzymes from muscle fibers into the serum. Higher levels of serum enzymes may not correlate well with severity of the clinical condition. Hypoalbuminemia, which is believed to be due to a demand for protein by the parasite, is usually accompanied by hypopotassemia. Muscle atrophy and contractures may also occur.

Respiratory symptoms, including dyspnea, cough, and hoarseness, may either be due to myositis of respiratory muscles or be secondary to pulmonary congestion. Neurologic manifestations are rare but may result from invasion of the brain by migrating larvae. In severe cases, death may result following cardiac decompensation or respiratory failure, cyanosis, and coma.

Intimate contact between the parasite and host tissues stimulates the production of antibodies that can be demonstrated in the serum. The exact role of antibodies in acquired immunity is not clear. On the basis of experimental animal data, it would be safe to assume that in most cases in humans, the severity of infection is reduced considerably by the development of immunity from previous subclinical infections. The local gut immunity in mice has been shown to be T-cell dependent (27).

Diagnosis

It is difficult to clinically diagnose trichinosis because it simulates so many other infections as a result of its wide distribution in the body. Diagnosis is based on changes in the blood and recovery of larvae from muscles. Although eosinophilia is not restricted to trichinosis, its presence in 30 to 85% of infected individuals is a constant and important diagnostic aid. Eosinophilia is initiated in the second week of infection, reaching its peak by about day 20. It may be absent not only in patients with very severe and fatal cases but also in individuals with secondary infection. The most definitive diagnostic method is muscle biopsy to detect unencysted or encysted larvae. Diagnostic success depends on chance distribution of larvae in the particular striated muscle that is sampled. Gastrocnemius, pectoralis major, deltoid, or biceps muscles are commonly used because of easy accessibility. The muscle strip can be compressed tightly between two microscope slides and examined under a microscope (Fig. 25.3). Part of the biopsied sample can be either digested or fixed and then sectioned, stained, and examined. The presence of active larvae following digestion with artificial gastric juice indicates a recent infection. Recovery of adult worms from diarrheic stool is less reliable and impractical, since adult worms are present early in the infection, when trichinosis is usually not suspected.

There are numerous immunological tests available for diagnosis of trichinosis (33, 35, 44). With the introduction of marker assays, such as immunofluorescence and enzyme immunoassays, the sensitivity of tests has greatly improved. Common methods in which fluorescence-labeled antibodies are used include direct, inhibition, and indirect staining techniques. The enzyme-linked immunosorbent assay (ELISA) has been shown to be sensitive. With the aid of class-specific conjugates, specific antibodies of various classes that are important for diagnosis of acute trichinosis can be detected; an increase in immunoglobulin M and A titers is indicative of a recent infection. Various types of ELISA can be used for measuring antigen. The sandwich ELISA has been used to detect circulating antigens in sera. Other serologic tests, including complement fixation, precipitin, latex agglutination, and various flocculation tests, have been replaced by newer techniques that are more sensitive and specific. It is advisable to take multiple serum samples at intervals of several weeks in order to demonstrate seroconversion in patients whose sera were initially negative.

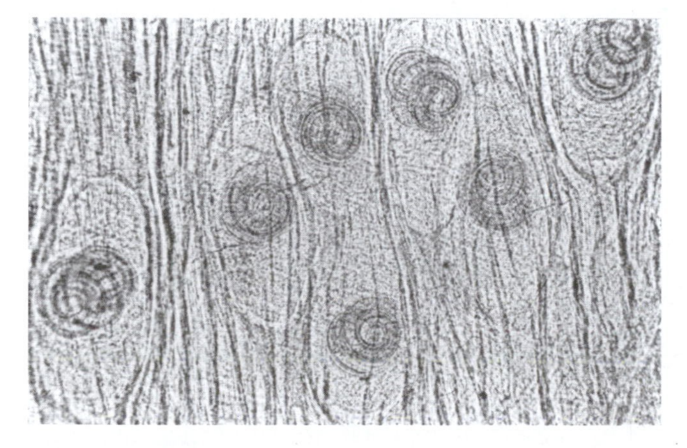

Figure 25.3 Pressed muscle containing *T. spiralis* larvae.

Treatment

The efficacy of treatment of trichinosis depends on the intensity of infection, the species of *Trichinella*, the stage of infection, and the character and intensity of the host response. The purpose of treatment during the intestinal phase is to destroy adult worms and to interfere with production of newborn larvae. The drug of choice is mebendazole (Vermox) at dosages of 200 to 400 mg three times a day for 3 days and then 400 to 500 mg three times a day for 10 days. Mebendazole is also believed to be active against developing and encysted larvae but at dosages higher than those for adult worms. Mebendazole is administered during the first week of infection before newborn larvae migrate, in moderate or severe infections treated with corticosteroids, and in infections with *T. nativa*, which responds poorly to treatment with nonbenzimidazole compounds. Mebendazole is not recommended for women in the first trimester of pregnancy. Thiabendazole (Mintezol) is no longer used because of adverse reactions. Albendazole (Valbazen, Zentel) is better absorbed and is probably as active as or even more active than mebendazole. Pyrantel (Antiminth) at 10 mg/kg of body weight per day for 4 days and levamisole at 2.5 mg/kg/day (maximum of 150 mg) are active only against adult worms.

To minimize the provocation of hypersensitivity, it is recommended that corticosteroids be given with the drug. Corticosteriods are recommended for acute severe trichinosis not only for antiallergic action but also for anti-inflammatory and antishock actions. Supportive therapy, such as bed rest, is very important.

Prevention and Control

There are several measures to prevent human infection that can be directed against the principal sources of human trichinosis (pigs) before they are slaughtered for market. In the United States, Public Law 96-468 of 1980 requires treatment of garbage if it is to be fed to pigs (38). The Swine Health Protection Act of 1983 stipulates that all garbage must be boiled for 30 min in a licensed facility before being fed to pigs (43). The other, more desirable alternative is to strictly prohibit feeding of garbage to pigs, a measure taken by 34 states. Unfortunately, these preventive measures do not apply to the feeding of raw scraps of wild animal meat to pigs on farms or of scraps of raw household garbage to pigs.

In the United States, as in many countries, each food animal is examined by a federal or state inspector at the time of slaughter to determine if the animal is healthy and suitable for use as human food. Unlike in some European countries where the pigs are inspected for *Trichinella* infection, in the United States pigs are not inspected individually for larvae because of unusually large numbers that are slaughtered under federal inspection (45). In many European countries, human trichinosis has been drastically reduced by compulsory trichinoscopic examination of each pork carcass. In the United States, pig slaughtering plants do not use the trichinoscope to examine carcasses for *Trichinella* infection. However, all commercially processed pork products in the United States are controlled by state and federal inspection programs. Pork products that are ready to eat, such as cold-smoked sausage, must be processed to kill all larvae by heating pork muscle to at least 58°C, freezing to -35°C in its center, or freeze-drying (38, 43). Curing procedures are specified in the Federal Meat Inspection Regulations for most classes of pork products (38). The effectiveness of curing, however, depends on the salt concentration, temperature, and time (36).

The U.S. Department of Agriculture and the Food and Drug Administration have approved irradiation of pork to destroy the larvae (44). However, commercially produced irradiated pork is not yet available. The success of application of irradiation will depend on cost-effectiveness for the meat industry and willingness on the part of consumers to eat irradiated pork.

The most effective control measure after pigs have been slaughtered and distributed is education of the public. The final responsibility lies with consumers to ensure that larvae are killed before the pork is eaten. It is very important that fresh pork be cooked or processed properly before consumption in the home. The U.S. Department of Agriculture recommends that consumers assume that all pork is infected with *Trichinella* larvae. Thus, it is recommended that all pork be cooked to an internal temperature of 160°F (71°C) throughout (44). This recommendation is based on observations that rapid cooking methods, such as the use of microwave ovens, may not heat the pork uniformly or for sufficient time to destroy the larvae (37). The U.S. Department of Agriculture recommends that pork be frozen for 20 days at 5°F (-15°C), 10 days at -10°F (-23°C), or 6 days at -20°F (-29°C) to kill the larvae, provided that the meat is less than about 6 in. (15 cm) thick (44). These freezing temperature and time requirements do not appear to be effective for killing the arctic isolate, *T. nativa*. In fact, Alaskan bear meat has been shown to be infective for laboratory animals even after 35 days of storage at -15°C (61). Thus, freezing meat may not be adequate for eliminating the potential source of infection by this species.

T. SAGINATA TAENIASIS

The history of taeniasis caused by *T. saginata* has been described in ancient literature with intriguing theories as to the parasite's origin and nature. *T. saginata* was recognized as a distinct species by Goeze in 1782. The relationship between the adult worm in humans and the larval bladder worm in cattle was not established until 1861, by Leuckart. There has been some controversy regarding the generic term for the unarmed tapeworm of humans acquired from the consumption of beef, but *T. saginata* is the currently accepted name.

Epidemiology

Taeniasis can result when raw or inadequately cooked beef is consumed. Raw or rare beef appears to be popular in many countries, particularly in the form of beef tartare (basterma) in the Middle East, shish kebab in India, lahb in Thailand, yuk hoe in Korea (42), and raw or slightly roasted meat in parts of Africa, especially Ethiopia (10), where the beef is diced into cubes, skewered, and then roasted or is served raw as kitfo, the equivalent of steak tartare. Although the meat of reindeer in northern Siberia, as well as meat of other herbivorous animals, has been reported to contain cysticerci of *T. saginata* and has probably been responsible for taeniasis in humans, there has been no evidence of wild intermediate hosts in the Americas, except for possibly the llama and pronghorn antelope reported at the turn of the century (49). Thus, it can be assumed that at present beef is the only important source of *T. saginata* taeniasis in humans.

With respect to the incidence of taeniasis in humans, geographic areas can be classified into three groups: countries or regions where infection is highly endemic, with an incidence in the human population exceeding 10%; countries with moderate infection rates; and countries with a very low prevalence, below 0.1%, or even no infections caused by the parasite. The areas of high endemicity are Central and East African countries, such as Ethiopia, Kenya, and Zaire, Caucasian South Central Asian republics of the former Soviet Union, Middle Eastern countries, such as Syria and Lebanon, and parts of the former Yugoslavia. In the Asian republics of the former Soviet Union, the prevalence rate may reach 45% (47). The incidence is low in the United States. There were reportedly only 443 patients treated for taeniasis due to *T. saginata* in 1981 (45). Critical factors responsible for the persistence and sometimes increase in human *T. saginata* infection depend on multiple factors that govern infection in cattle. Cattle have an increased risk of infection if they have access to pastures contaminated directly with human feces or sewage effluent, feed contaminated by laborers, and irrigation ditches and cattle pens directly contaminated by farm workers. Eggs of *T. saginata* are capable of surviving long periods in the environment and are known to be resistant to moderate desiccation, disinfectants, and low temperatures (4 to 5°C). A longevity of 71 days in liquid manure, 33 days in river water, and 154 days in pastures has been reported (58). The eggs can also survive many sewage treatment processes and remain viable in fluid effluent or dried sludge. The very high percentage of cattle harboring cysticerci in their musculature is reflected in the number of human infections. The sheer greater number of cattle than of pigs accounts for *T. saginata* infection being more prevalent than *T. solium* infection, which originates from pigs. Some of the other reasons for the prevalence of infection have included population displacements due to wars, increased consumption of rare beef, and inadequate treatment of sewage. An interesting observation is that *T. saginata* is a problem in developing countries, where relatively lower standards of hygiene and sanitation are practiced, and in economically developed countries, where people can afford to consume more beef and sewage treatment facilities are overtaxed. Thus, both the practice of defecating in the bush in poor countries and placing stress on sewage treatment facilities in rich countries ensure a continuity of the infection. Among affluent Americans, the high infection rate has been attributed to greater freedom of international travel in addition to a preference for imaginative diets, such as those including rare or raw beef. The penchant for raw beef and steak tartare is not limited to affluent Americans and Europeans but is also found among more affluent Ethiopians who can afford beef. In many countries of the Middle East and East Africa, a combination of poor sanitation and culinary preference for raw and rare beef has contributed to a higher risk of infection for everyone at some point in his or her life. It should be remembered that transmission can also result from tasting the meat during preparation and cooking.

Life Cycle

Following ingestion of raw or rare beef containing the viable cyst in the larval stage (cysticercus bovis) of *T. saginata*, the cysticercus is released from its surrounding muscle tissue by digestion in the small intestine. The scolex of the cysticercus evaginates from the vesicle or bladder and attaches to the mucosa of the jejunum, where the larva develops into a mature adult worm in 8 to 10 weeks. A mature adult worm is characterized by having a scolex with four hemispherical suckers of 0.7 to 0.8 mm in diameter situated at the four angles of the head. The entire strobila measures 4 to 12 m and possesses 1,000 to 2,000 proglottids or segments. Immediately posterior to the neck is a series of immature, mature, and gravid proglottids. The mature and gravid proglottids are hermaphroditic, containing both male and female reproductive organs. The adult worm attaches to the mucosal surface of the upper jejunum by means of the four suckers but, because of its length, might extend down to the terminal ileum. The developing proglottids extend down the small intestine. The most distal gravid proglottids, which measure about 20 mm long and 5 to 7 mm wide, detach singly from the rest of the strobila and independently migrate to the outside. Each gravid proglottid contains about 80,000 eggs, which are expressed from the proglottids and are deposited on the perianal skin.

When gravid proglottids come to rest on the ground, eggs are extruded. The eggs also may be present as a result of promiscuous defecation. Cattle ingest the eggs, which then hatch in the duodenum, liberating a six-hooked embryo that penetrates mesenteric venules or lymphatics and reaches skeletal muscles or the heart, where it develops into the cysticercus larva or cysticercus bovis. The cysticercus is essentially a miniature scolex and neck invaginated into a fluid-filled bladder that measures about 10 by 6 mm. The bladder larva becomes infective within 8 to 10 weeks following ingestion of eggs by cattle and remains infective for more than a year, and thus a person eating the raw beef from these cattle within this period is subject to infection.

Pathogenesis and Pathology

The scolex of the adult worm generally is lodged in the upper part of the jejunum. Usually, only a single worm is present, although multiple infections have been reported. The adult tapeworm does not cause pathologic changes. However, mucosal biopsies have shown minimal inflammatory reactions, suggesting that the worm can have irritative action that results in bowel distention or spasm.

Migration by the adult worm to unusual sites is rare, but complications such as appendicitis, invasion of pancreatic and bile ducts, intestinal obstruction, or perforation and vomiting of proglottids with aspiration have been reported.

Clinical Manifestations

Although most cases of taeniasis are asymptomatic, up to one-third of patients complain of nausea or abdominal "hunger" pain that is often relieved by eating. The epigastric pain may be accompanied by weakness, weight loss, increased appetite, headache, constipation, dizziness, and diarrhea. The patient usually becomes aware of the infection when a proglottid is passed in the stool or is found on the perianal area or even on underclothing.

The adult worm is weakly immunogenic, as manifested by a moderate eosinophilia and increased levels of serum immunoglobulin E. Allergic reactions, such as urticaria and pruritus, may be due to the worm and its metabolites. The adult worms induce the production of antibodies. However, the persistence for years of a large, actively growing worm does not seem to be consistent with the development of protective immunity.

Diagnosis

Definitive diagnosis is based on identifying the proglottid, since the eggs of *T. saginata* cannot be distinguished from those of other species of *Taenia* or those of *Multiceps* and *Echinococcus* species. The gravid proglottid of *T. saginata* has 15 to 20 lateral branches of the uterus on each side of the main uterine stem, a characteristic feature (Fig. 25.4). If the gravid proglottid is treated with 10% formaldehyde and injected with India ink, the uterine branches are very prominent. Uterine branches also can be seen by gently pressing the proglottid between two microscope slides and holding them in front of a bright light. If the scolex is present, the four characteristic hookless suckers can be used as a distinguishing feature for identification.

The egg is nearly spherical in shape, measuring 30 to 40 μm in diameter, has characteristic radial striations on its thick shell, and contains a hexacanth embryo with delicate lancet-shaped hooklets (Fig. 25.5). Rather than looking for the eggs in the stool, it is better to use the commercial Scotch tape method to obtain the eggs or proglottids from the perianal region. Since detection of eggs is nonspecific in terms of species of origin, DNA probes that allow differentiation of genomic DNA of tapeworm species by the hybridization technique have been developed (22, 28). PCR has been used to discriminate between various *Taenia* and *Echinococcus* species at the genomic level (26). The amplification of a 0.55-kb DNA fragment is a unique *T. saginata* genomic template DNA (25).

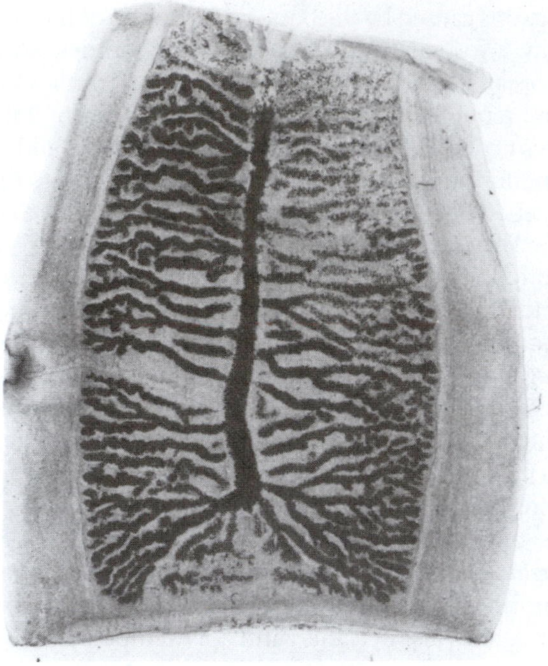

Figure 25.4 Gravid proglottid of *T. saginata*. Note at least 16 lateral uterine branches.

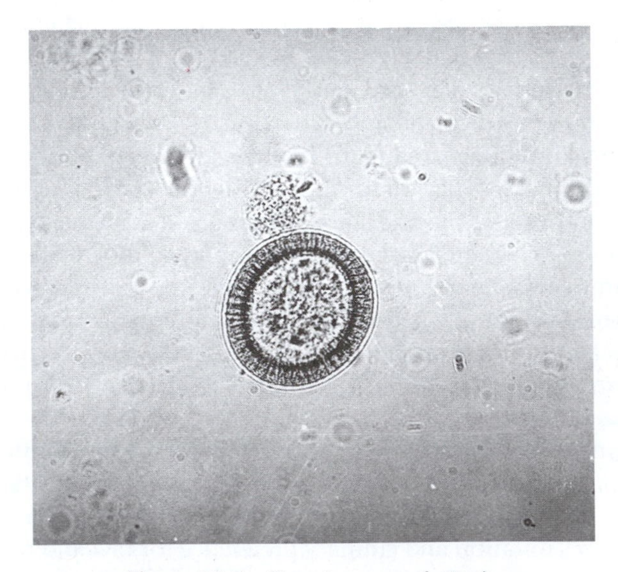

Figure 25.5 *T. saginata* egg (×590).

Although an antibody response is induced by the adult worm, specific serologic tests such as are used for diagnosis of trichinosis have not been routinely used for diagnosis of taeniasis. However, a highly specific immunoassay, using a <12-kDa antigen prepared from the cyst fluid of *Taenia hydatigena*, that is able to differentiate between the types of human taeniasis has been developed (29).

Treatment

The drug of choice in treating taeniasis is niclosamide (Niclocide, Yomesan), a taeniacide that is effective in damaging the worm to such an extent that a purge following therapy will often produce the scolex. Cure rates are very high. A single oral dose of 10 mg of praziquantel (Biltricide) per kg also has been reported to be highly effective.

Prevention and Control

In addition to effective drug therapy against the adult *Taenia* worms, there are measures that can be taken to prevent cysticercosis in cattle. The most important measure is improved sanitation for adequate sewage disposal so that the eggs are unavailable to cattle. Since humans are the only definitive hosts of *T. saginata* and thus the only disseminators of eggs, which can number 480,000 to 720,000 daily, the success of this measure depends on educating livestock producers and their employees about modes of transmission, examining all workers' stools for *Taenia* species, and providing adequate toilet facilities in cattle-feeding establishments (56). Where toilet facilities are available, overloading of systems should be avoided.

Meat inspection to detect cysticercosis in slaughtered cattle can reduce the transmission of taeniasis, but at present there are limitations to meat inspection. In the United States, the percentage of cattle inspected by state and municipal authorities varies from state to state. Infected cattle for intrastate and even for interstate commerce may pass undetected. Moreover, cattle from small farms where sanitary conditions may often be poor are frequently slaughtered in local abattoirs without inspection requirements. Even routine examination of the heart and masseter muscles may miss a significant percentage of infected cattle. All infected carcasses should be condemned, but approximately 70% of infected carcasses are trimmed of visible cysts and then passed unrestricted for consumption (56). However, regulations require infected carcasses to be frozen for a minimum of 10 days at 15°F (−9.4°C) or for 5 days at 0°F (−18°C) in the United States (47).

An indirect approach to assessing cysticercosis in cattle might be to use a sensitive serologic method. The ELISA technique has been suggested (54). A positive ELISA would warrant gross inspection of the carcass for the larval bladder worm (beef measles) at slaughter; a negative result from direct examination would indicate that the carcass is very likely free of infection. The development of an effective vaccine would not only prevent initial infection but also protect cattle against challenge infections.

On the part of the consumer, mass public health education concerning risks of eating raw or inadequately cooked beef is important. The cysticerci in meat are inactivated by freezing the meat at −10°C for 10 days or −18°C for 5 days, heating to an internal temperature of 56°C, or salt curing under appropriate conditions. Major obstacles to this approach have been a reluctance by consumers to modify preferred culinary habits and to break with long-established cultural traditions. Thus, attempts to eradicate *T. saginata* taeniasis have been unsuccessful.

T. SOLIUM TAENIASIS

Taeniasis caused by *T. solium* was known at the time of Hippocrates and possibly even earlier. The Greeks described the larval stage in the tongue of swine as resembling a hailstone. *T. solium*, the pork tapeworm, in contrast to *T. saginata*, possesses an armed rostellum and a smaller number of lateral uterine branches and lacks a vaginal sphincter (52).

Epidemiology

Humans are the only known natural definitive hosts for *T. solium* taeniasis, which has a worldwide distribution. It is very difficult to evaluate the prevalence of infection caused by this species because the coproscopical methods used for surveys cannot differentiate between infection caused by *T. saginata* and that caused by *T. solium*. *T. solium* is important in pork-eating countries, especially those with suboptimal practices of sanitation and pig husbandry. Human infection is practically nonexistent in Moslem countries (48). It is most common in Slavic countries, Mexico, Central and South America, Central and South Africa, Asia, non-Islamic Southeast Asia, and southern and eastern Europe. In developed societies, raw pork is eaten less frequently than raw beef, which may be responsible for the lower incidence of *T. solium* than of *T. saginata* taeniasis.

T. solium taeniasis has much more serious consequences than *T. saginata* taeniasis because the larval or cysticercus stage of *T. solium*, which usually infects pigs, can also infect humans. Cysticercosis in humans is an important public health problem in developing countries where hygiene and sanitary conditions are deficient or nonexistent and where swine-rearing practices are primitive. The relatively high prevalence of *T. solium* taeniasis in the Bantu population of South Africa has been attributed to a widespread practice of using tapeworm proglottids as a component of muti, a medication used by native herbalists for the treatment of intestinal worms (30). Infection of humans with larvae results from inadvertent ingestion of eggs in contaminated food, e.g., vegetables fertilized with nightsoil or contaminated water, from hands of individuals already infected with the adult worm (external autoinfection), and from direct contact with a tapeworm carrier. Another important mode of transmission involves patients infected with intestinal taeniasis, in whom the eggs are carried by reversed peristalsis back to the duodenum or stomach, where they are stimulated to hatch and subsequently invade extraintestinal tissues to become cysticerci (internal autoinfection).

Life Cycle

When meat, usually pork, containing the bladder larva (cysticercus cellulosae) is ingested, the head of the larva evaginates from the fluid-filled milky white bladder. The scolex bears four suckers and an apical crown of hooklets. The cysticerci are referred to as pork measles and are larger (5 to 20 mm in diameter) than cysticercus bovis. They attach to the wall of the small intestine and mature into adult worms in 5 to 12 weeks. The adult worm measures up to 8 m (usually 1.5 to 5 m) and has a scolex that is roughly quadrate, possessing a conspicuous rounded rostellum armed with a double row of large and small hooklets, numbering 22 to 32. A short cervical region is anterior to a series of proglottids or segments. Immature proglottids are broader than long, mature ones being nearly square, while gravid ones are longer than broad. The total number of distinct proglottids is less than 1,000. The terminal proglottids become separated from the rest of the strobila and migrate out of the anus or are passively expelled in the stool. A single gravid proglottid contains fewer than 50,000 eggs. Upon escape from the ruptured uterus of the gravid proglottid and after deposition on the soil, the eggs may remain viable for many weeks. The eggs of *T. solium* are more apt than those of *T. saginata* to appear in the stool.

In the normal cycle, the eggs are ingested by pigs, the usual intermediate host. The hexacanth embryo hatches in the duodenum, migrates through the intestinal wall to reach the blood and lymphatic channels, and is carried to the skeletal muscles and the myocardium. The embryo develops in 8 to 10 weeks into a cysticercus or cysticercus cellulosae. In the abnormal cycle, humans can serve as an intermediate hosts and harbor cysticerci by ingesting the eggs. The cysticercus develops most commonly in striated muscles and subcutaneous tissues but also in the brain, eye, heart, lung, and peritoneum.

Pathogenesis and Pathology

As with *T. saginata* infection, usually there is only a single worm, and pathological changes are similar. There is a mild local inflammation of the intestinal mucosa as a result of attachment by the suckers and especially the hooklets. However, because of its smaller size, *T. solium* is less likely to cause intestinal obstruction. Rare instances of intestinal perforation with secondary peritonitis and gallbladder infection have been reported.

Pathological changes due to cysticerci can be serious, depending on the tissue invaded and the number of cysticerci that become established. Damage results from pressure caused by encapsulated larvae on surrounding tissue, since the cysticerci produce occupying lesions. There may be no prodromal symptoms or only slight muscular pain and a mild fever during invasion of the muscles and subcutaneous tissues. In ocular cysticercosis, which accounts for about 20% of neurocysticercosis cases, there may be loss of vision. Invasion of the meninges, cortex, cerebral substance, and ventricles evokes tissue reactions leading to focal epileptic attacks or other motor or sensory involvement. The reasons for the predilection of cysticerci for the central nervous system are still obscure.

Clinical Manifestations

Since only a single adult worm is usually present, there are no symptoms of epigastric fullness. Patients are asymptomatic and become aware of the infection only when they find proglottids in their stools or on perianal skin. However, there may be vague abdominal discomfort, hunger pains, anorexia, and nervous disorders. Rare instances of intestinal perforation with secondary peritonitis and gallbladder infection have been reported. A moderate eosinophilia, as high as 28%, and leukopenia can occur. The persistence for years of large, actively growing tapeworms does not appear to be consistent with the development of protective immunity.

Diagnosis

Diagnosis is based on stool examination and perianal scrapings. The sensitivity of both methods is increased

Figure 25.6 Gravid proglottid of *T. solium*. Note fewer lateral uterine branches than in *T. saginata* in Fig. 25.4.

by three examinations daily. Since *T. solium* eggs cannot be distinguished from those of *T. saginata*, specific diagnosis is based on the identification of the gravid proglottid, which has fewer lateral branches (7 to 12) on each side of the main uterine stem (Fig. 25.6) than that of *T. saginata*. If the scolex is obtained, it possesses hooklets in addition to four suckers. The development of DNA probes has made it possible to distinguish *T. solium* from *T. saginata* (28). Sensitivity of serologic tests varies, depending on the particular method and the clinical form of infection. A highly specific immunoassay for diagnosis of cysticercosis, using the <12-kDa *T. hydatigena* antigen, is reported to be so specific as to distinguish between human clinical cases of cysticercosis and taeniasis. The assay is positive only with serum from cases of cysticercosis (29).

Treatment

Niclosamide is the drug of choice because of its effectiveness against the scolex and proliferative zone of strobila. A single dose of praziquantel is also highly effective. Mebendazole also has been reported to be effective. It is imperative to treat patients harboring the adult worm, since cysticercosis can occur from internal autoinfection. The major concerns in treating patients with the adult worm are to prevent vomiting and to ensure rapid expulsion of disintegrated proglottids from the intestine.

Prevention and Control

Prevention and eradication of *T. solium* adults in the intestine and of cysticerci in various tissues in humans and in pigs are a problem in areas of endemicity where economic, social, and sanitary conditions are substandard. Needed changes are monumental in scope. Without fundamental changes, the most scrupulous personal hygiene and eating habits will not prevent or eradicate infection in underdeveloped or developing countries. In developed countries, the incidence of infection can be maintained in an extinction status with modern animal husbandry practices (24). The best preventive measure in interrupting the transmission from humans to animals is to introduce and maintain proper sanitary facilities to dispose of contaminated feces. Even with proper toilet facilities, measures must be taken to make sure that sewage treatment is adequate to kill the eggs.

Since one infected individual can infect literally thousands of market hogs via contaminated feedlots, management personnel and employees must be educated regarding the parasite and means of avoiding transmission. Prospective employees for animal care should be examined for tapeworm infection before employment and semiannually during employment, and chemical toilets should be installed and properly maintained at convenient locations on slaughtering premises. The most important practice is to keep pigs in enclosures or indoors to prevent access to human fecal matter.

Meat inspection programs can be an effective tool by condemning infected carcasses. Meat inspection for measly pork is done in the United States only for interstate commerce, but light inspections can miss infected carcasses. Cysticerci can be detected in measly pork, but it is too expensive to institute routine examination of all pork. Thus, the consumer should make sure that the meat is thoroughly cooked to an internal temperature of 56°C or higher or is frozen at −10°C for at least 14 days to render it incapable of transmitting cysticerci (45). Other means of inactivating cysticerci in pork, such as irradiation, have not been commercially applied. The best preventive measure is to avoid eating raw, uninspected pork.

ASIAN TAENIASIS

Asian taeniasis was first described in the aboriginal population in the mountainous regions of Taiwan (17). Initially, the etiologic agent was considered to be *T. saginata* because of morphologic similarities of the adult worms, although notable differences, including its shorter length, fewer number of proglottids, wider diameter of the scolex, and fewer number of testes in the mature proglottid, were found (17). Cloned ribosomal DNA fragments and sequence amplification by PCR showed that the Asian *Taenia* species is similar but genetically distinct from *T. saginata* (62). Also, sequence variation in the 28S rRNA and mitochondrial cytochrome *c* oxidase I genes and the cytochrome *c* oxidase I and ribosomal DNA internal transcribed spacer I PCR restriction fragment length polymorphism pattern in the Asian *Taenia* species have been used to identify it as an entity genetically distinct from other taeniid cestodes (7). However, the Asian *Taenia* species is closely related to *T. saginata*, suggesting that it should be classifed as a subspecies or strain of *T. saginata*. The significant public health implication to this finding is that the Asian form is unlikely to be an important cause of human cysticercosis.

Epidemiology

Asian *Taenia* infection has been found in several Asian countries. In Taiwan, of 1,661 aboriginal cases of Asian taeniasis, the overall clinical infection rate was 76% among nine aboriginal tribes in 10 counties in mountainous areas (20). Pigs, cattle, goats, wild boars, and monkeys can serve as intermediate hosts. Of these, the wild boar appears to be the probable natural intermediate host in Taiwan. This may be true also in Indonesia, where people have become infected presumably from consuming pork (15). The infectivity for pigs, calves, goats, and monkeys has been confirmed experimentally, with the pig as the most favorable laboratory intermediate host (18).

The cysticercus is armed with tiny rostellar hooklets like that of *T. solium* and develops in a period shorter than that for either *T. saginata* or *T. solium*. Interestingly, the cysticercus of the Asian *Taenia* species is found mainly in the parenchyma of the liver (17), whereas cysticerci of *T. saginata* and *T. solium* are found primarily in the muscles of cattle and pigs, respectively. Thus, the custom of eating the viscera, especially the liver, and blood of fresh-killed animals appears to be a contributing factor in the infection.

Clinical Manifestations

Infected individuals pass proglottids in their feces, even for 30 years or more, suggesting that the life span of this form of *Taenia* is probably longer than 30 years (20). Common clinical manifestations include pruritis ani, nausea, abdominal pain, and dizziness. Abdominal pain is usually localized on the midline of the epigastrium or in the umbilical region and varies in intensity from a dull, aching, gnawing, or burning to intense colic-like, sharp pain. There may be either an increased appetite or a lack of appetite, headache, or diarrhea, as well as constipation.

Treatment

Clinical trials in Taiwan have shown that a single dose (150 mg) of praziquantel (Biltricide) is highly effective against the Asian *Taenia* infection (19).

Prevention and Control

The best preventive measure is to avoid eating raw, uninspected pork, or other meat that contains the cysticerci. Since most cases of Asian taeniasis have been reported to result from the consumption of infected pork, the meat should be cooked thoroughly until it turns gray or should be frozen for at least 4 days at −10°C, to kill the cysticerci, the same procedures as used to prevent *T. solium* taeniasis.

References

1. **Andrews, J. R. H., R. Ainsworth, and D. Abernethy.** 1994. *Trichinella pseudospiralis* in humans: description of a case and its treatment. *Trans. R. Soc. Trop. Med. Hyg.* **88:**200–203.
2. **Bailey, T. M., and P. M. Schantz.** 1990. Trends in the incidence and transmission patterns of trichinosis in humans in the United States: comparison of the periods 1975–1981 and 1982–1986. *Rev. Infect. Dis.* **12:**5–11.
3. **Bellani, L., A. Mantovani, S. Pampiglione, and I. Fillippini.** 1978. Observations on an outbreak of human trichinellosis in Northern Italy, p. 535–539. *In* C. W. Kim and Z. S. Pawlowski (ed.), *Trichinellosis.* University Press of New England, Hanover, N.H.
4. **Bessonov, A. S.** 1981. Changes in the epizootic and epidemic situation of trichinellosis in the USSR, p. 365–368. *In* C. W. Kim, E. J. Ruitenberg, and J. S. Teppema (ed.), *Trichinellosis.* Reedbooks, Chertsey, England.
5. **Bommer, W., H. Kaiser, W. Mannweiler, H. Mergerian, and G. Pottkämper.** 1985. An outbreak of trichinellosis in northern Germany caused by imported air-dried meat from Egypt, p. 314–317. *In* C. W. Kim (ed.), *Trichinellosis.* State University of New York Press, Albany, N.Y.
6. **Bouree, P., J. L. Leymarie, and C. Aube.** 1989. Epidemiological study of two outbreaks of trichinosis in France, due to horse meat, p. 382–386. *In* C. E. Tanner, A. R. Martinez-Fernandez, and F. Bolas-Fernandez (ed.), *Trichinellosis.* CSIC Press, Madrid.
7. **Bowles, J., and D. P. McManus.** 1994. Genetic characterization of the Asian *Taenia*, a newly described taeniid cestode of humans. *Am. J. Trop. Med. Hyg.* **50:**33–34.
8. **Britov, V. A., and S. N. Boev.** 1972. Taxonomic rank of various strains of *Trichinella* and their circulation in nature. *Vestn. Akad. Nauk SSSR* **28:**27–32.
9. **Campbell, W. C.** 1983. Historical introduction, p. 1–30. *In* W. C. Campbell (ed.), *Trichinella and Trichinosis.* Plenum, New York.
10. **Carmichael, J.** 1952. Animal-man relationships in tropical disease in Africa. *Trans. R. Soc. Trop. Med. Hyg.* **46:**385–394.
11. **Centers for Disease Control.** 1976. Trichinosis surveillance annual summary—1975.
12. **Centers for Disease Control.** 1982. Common-source outbreaks of trichinosis—New York City, Rhode Island. *Morbid. Mortal. Weekly Rep.* **31:**161–164.
13. **Centers for Disease Control.** 1988. Trichinosis surveillance, United States, 1986. CDC surveillance summaries, December 1988. *Morbid. Mortal. Weekly Rep.* **37:**1–8.

14. **Coordinating Group for Prevention and Treatment of Trichinosis, Harbin City.** 1981. *A Survey of Trichinosis Due to Eating Scalded Mutton.* WHO/HELM/82.5. (English abstr.) World Health Organization, Geneva.
15. **Depary, A. A., and M. L. Kosman.** 1990. Taeniasis in Indonesia with special reference to Samosir Island, North Sumatra. *Southeast Asian J. Trop. Med. Public Health* **22** (Suppl.):239–241.
16. **Dissamarn, R., and P. Indrakamhang.** 1985. Trichinosis in Thailand during 1962–1983. *Int. J. Zoon.* **12**:257–266.
17. **Fan, P. C.** 1988. Taiwan *Taenia* and taeniasis. *Parasitol. Today* **4**:86–88.
18. **Fan, P. C.** 1990. Asian *Taenia saginata:* species or strain? *Southeast Asian J. Trop. Med. Public Health* **22**(Suppl.):245–250.
19. **Fan, P. C., W. C. Chung, C. H. Chan, Y. A. Chen, F. Y. Cheng, and M. C. Hsu.** 1986. Studies on taeniasis in Taiwan. V. Field trial on evaluation of therapeutic efficacy of mebendazole and praziquantel against taeniasis. *Southeast Asian J. Trop. Med. Public Health* **17**:82–90.
20. **Fan, P. C. W. C. Chung, C. Y. Lin, and C. H. Chan.** 1992. Clinical manifestations of taeniasis in Taiwan aborigines. *J. Helminthol.* **66**:118–123.
21. **Faubert, G. M., J. C. Pechere, R. Delisle, H.-C. Smith, and Y. Brindle.** 1981. An outbreak of trichinellosis in Canada: the enzyme linked immunosorbent assay (ELISA) and clinical findings, p. 269–273. *In* C. W. Kim, E. J. Ruitenberg, and J. S. Teppema (ed.), *Trichinellosis.* Reedbooks, Chertsey, England.
22. **Flisser, A., A. Reid, E. Garcia-Zepeda, and D. P. McManus.** 1988. Specific detection of Taenia saginata eggs by DNA hybridisation. *Lancet* **2**:1429–1430.
23. **Garkavi, B. L.** 1972. Species of *Trichinella* from wild carnivores. *Veterinariya* (Moscow) **49**:90–101.
24. **Gemmell, M. A.** 1987. A critical approach to the concepts of control and eradication of echinococcosis/hydatidosis and taeniasis/cysticercosis. *Int. J. Parasitol.* **17**:465–472.
25. **Gottstein, B., P. Deplazes, I. Tanner, and J. S. Skaggs.** 1991. Diagnostic identification of *Taenia saginata* with the polymerase chain reaction. *Trans. R. Soc. Trop. Med. Hyg.* **85**:248–249.
26. **Gottstein, B., and M. R. Mowatt.** 1991. Sequencing and characterization of an *Echinococcus multilocularis* DNA probe and its use in the polymerase chain reaction. *Mol. Biochem. Parasitol.* **44**:183–194.
27. **Grencis, R. K., and D. Wakelin.** 1985. Analysis of lymphocyte subsets involved in mediation of intestinal immunity to *Trichinella spiralis* in the mouse, p. 26–30. *In* C. W. Kim (ed.), *Trichinellosis.* State University of New York Press, Albany, N.Y.
28. **Harrison, L. J. S., J. Delgado, and R. M. E. Parkhouse.** 1990. Differential diagnosis of *Taenia saginata* and *Taenia solium* with DNA probes. *Parasitology* **100**:459–461.
29. **Hayunga, E. G., M. P. Sumner, M. L. Rhoads, K. D. Murrell, and R. S. Isenstein.** 1991. Development of a serologic assay for cysticercosis, using an antigen isolated from *Taenia* spp. cyst fluid. *Am. J. Vet. Res.* **52**:462–470.
30. **Heinz, H., and G. Macnab.** 1965. Cysticercosis in the Bantu of Southern Africa. *S. Afr. J. Med. Sci.* **30**:19–31.
31. **Hou, H. W., et al.** 1983. *A Survey of an Outbreak of Trichinosis Caused by Eating Roasted Dog Meat.* WHO/HELM/84.15. (English abstr.) World Health Organization, Geneva.
32. **Khamboonruang, C.** 1990. The present status of trichinellosis in Thailand. *Southeast Asian J. Trop. Med. Public Health* **22**(Suppl.):312–315.
33. **Kim, C. W.** 1975. The diagnosis of parasitic diseases. *Prog. Clin. Pathol.* **6**:267–288.
34. **Kim, C. W.** 1991. The significance of changing trends in Trichinellosis. *Southeast Asian J. Trop. Med. Public Health* **22**(Suppl.):316–320.
35. **Kim, C. W.** 1994. A decade of progress in trichinellosis, p. 35–47. *In* W. C. Campbell (ed.), *Trichinellosis.* Istituto Superiore di Sanita Press, Rome.
36. **Kotula, A. W.** 1983. Postslaughter control of *Trichinella spiralis. Food Technol.* **37**:91–94.
37. **Kotula, A. W., K. D. Murrell, L. Acosta-Stein, L. Lamb, and L. Douglass.** 1983. Destruction of *Trichinella spiralis* during cooking. *J. Food Sci.* **48**:765–768.
38. **Leighty, J. C.** 1983. Regulatory action to control *Trichinella spiralis. Food Technol.* **37**:95–97.
39. **MacLean, J. D., J. Viallet, C. Law, and M. Staudt.** 1989. Trichinosis in the Canadian arctic: report of five outbreaks and a new clinical syndrome. *J. Infect. Dis.* **160:** 513–520.
40. **McAuley, J. B., M. K. Michelson, A. W. Hightower, S. Engeran, L. A. Wintermeyer, and P. M. Chantz.** 1992. A trichinosis outbreak among Southeast Asian refugees. *Am. J. Epidemiol.* **135**:1404–1410.
41. **Millet, N. B., G. D. Hart, T. A. Reyman, M. R. Zimmerman, and P. K. Lewin.** 1980. ROM I: mummification for the common people, p. 71–84. *In* A. Cockburn and E. Cockburn (ed.), *Mummies, Disease and Ancient Cultures.* Cambridge University Press, Cambridge.
42. **Moon, J. R.** 1976. Public health significance of zoonotic tapeworms in Korea. *Int. J. Zoon.* **3**:1–18.
43. **Murrell, K. D.** 1985. Strategies for the control of human trichinosis transmitted by pork. *Food Technol.* **39**:65–68, 110–111.
44. **Murrell, K. D., and F. Bruschi.** 1994. Clinical trichinellosis. *Prog. Clin. Parasitol.* **4**:117–150.
45. **Murrell, K. D., R. Fayer, and J. P. Dubey.** 1986. Parasitic organisms. *Adv. Meat Res.* **2**:311–377.
46. **Nelson, G. S., and J. Mikundi.** 1963. A strain of *Trichinella spiralis* from Kenya of low infectivity to rats and domestic pigs. *J. Helminthol.* **37**:329–338.
47. **Pawlowski, Z. S.** 1982. Taeniasis and cysticercosis, p. 313–348. *In* J. H. Steele (ed.), *Parasitic Zoonosis,* vol. I. CRC Handbook Series in Zoonoses. CRC Press, Boca Raton, Fla.
48. **Pawlowski, Z. S.** 1990. Cestodiasis, p. 490–504. *In* K. S. Warren and A. A. F. Mahmoud (ed.), *Tropical and Geographic Medicine,* 2nd ed. McGraw-Hill, New York.
49. **Pawlowski, Z., and M. G. Schultz.** 1972. Taeniasis and cysticercosis (*Taenia saginata*). *Adv. Parasitol.* **10**:269–343.
50. **Pozio, E., G. La Rosa, K. D. Murrell, and J. R. Lichtenfels.** 1992. Taxonomic revision of the genus *Trichinella. J. Parasitol.* **78**:654–659.
51. **Pozio, E., G. La Rosa, P. Rossi, and K. D. Murrell.** 1989. New taxonomic contribution to the genus *Trichinella*

(Owen, 1835). Biochemical identification of seven clusters by gene-enzyme systems, p. 76–82. *In* C. E. Tanner, A. R. Martinez-Fernandez, and F. Bolas-Fernandez (ed.), *Trichinellosis.* CSIC Press, Madrid.

52. **Proctor, E. M.** 1972. Identification of tapeworms. *S. Afr. Med. J.* **46:**234–238.
53. **Roberts, D.** 1986. The last trace. *Am. Photogr.* **16:**64–68.
54. **Ruitenberg, E. J., F. Van Knapen, and J. W. Weiss.** 1979. Foodborne parasitic infections—a review. *Vet. Parasitol.* **5:**1–10.
55. **Schantz, P. M.** 1983. Trichinosis in the United States, 1947–1981. *Food Technol.* **37:**83–86.
56. **Schultz, M. G., J. A. Hermos, and J. H. Steele.** 1970. Epidemiology of beef tapeworm infection in the United States. *Public Health Rep.* **85:**169–176.
57. **Semple, A. B., J. B. M. Davies, W. E. Kershaw, and C. A. St. Hill.** 1954. An outbreak of trichinosis in Liverpool in 1953. *Br. Med. J.* **1:**1002–1006.
58. **Snyder, G. R., and K. D. Murrell.** 1983. Bovine cysticercosis, p. 161–170. *In* G. Woods (ed.), *Practices in Veterinary Public Health and Preventive Medicine.* Iowa State University Press, Ames, Iowa.
59. **Stehr-Green, J. K., and P. M. Schantz.** 1986. Trichinosis in Southeast Asian refugees in the United States. *Am. J. Public Health* **76:**1238–1239.
60. **Viens, P., and P. Auger.** 1981. Clinical and epidemiological aspects of trichinosis in Montreal, p. 275–277. *In* C. W. Kim, E. J. Ruitenberg, and J. S. Teppema (ed.), *Trichinellosis.* Reedbooks, Chertsey, England.
61. **Woods, G. T.** 1986. Trichinosis surveillance in the United States and the swine industry's action plan, p. 190–198. *In* G. T. Woods (ed.), *Practices in Veterinary Public Health and Preventive Medicine in the United States.* Iowa State University Press, Ames, Iowa.
62. **Zarlenga, D. S., D. P. McManus, P. C. Fan, and J. H. Cross.** 1991. Characterization and detection of a newly described Asian Taeniid using cloned ribosomal DNA fragments and sequence amplification by the polymerase chain reaction. *Exp. Parasitol.* **72:**174–183.
63. **Zarlenga, D. S., and K. D. Murrell.** 1989. Molecular cloning of *Trichinella spiralis* ribosomal RNA genes: application as genetic markers for isolate classification, p. 35–40. *In* C. E. Tanner, A. R. Martinez-Fernandez, and F. Bolas-Fernandez (ed.), *Trichinellosis.* CSIC Press, Madrid.

Eugene G. Hayunga

26 Helminths Acquired from Finfish, Shellfish, and Other Food Sources

A variety of human helminthic infections can be acquired by eating food products from infected animals and plants, by the accidental ingestion of infected invertebrates in foodstuffs or drinking water, or by inadvertent fecal contamination (Table 26.1). Effective prevention involves exploiting parasite vulnerabilities (32) and requires a sound understanding of parasite life cycles and modes of transmission. Unlike bacteria, the infective stages of helminths generally do not propagate. As a result, the critical control point for foodborne helminthiases is initial food preparation, not subsequent storage, reheating, or processing. Although these helminthic infections can readily be prevented, the reality is that safe water supplies, adequate sanitation, and reliable food handling simply do not exist for much of the world's population, a fact generally not appreciated by tourists "who explore tropical countries with a zeal undamped by any knowledge of preventive medicine" (42).

Foodborne helminths, although taxonomically diverse, share the common characteristic of requiring more than one host to complete their life cycles. The transmission of these organisms, termed biohelminths by Kisielewska (40), requires close behavioral contact between hosts. Typically, the definitive host of a biohelminth occupies the highest trophic level of the food chain. Prevention of biohelminth infections can be accomplished by avoiding the intermediate hosts or by adequate cooking of foods. In contrast, helminths with eggs or free-living stages that can survive a certain length of time in the external environment, termed geohelminths (40), are typically transmitted via contaminated water or foods and are best controlled by improved sanitation. In addition, any parasite capable of penetrating the skin can also penetrate the buccal epithelium.

HELMINTHS ACQUIRED FROM FINFISH AND SHELLFISH

Anisakis Species and Related Roundworms

Several related nematodes of the genera *Anisakis, Pseudoterranova,* and *Contracaecum* may be acquired by eating raw fish or squid in seafood dishes such as sushi, sashimi, ceviche, and lomi-lomi (58). The noninvasive form of anisakiasis is generally asymptomatic, resulting in "tingling throat syndrome" when worms are released from seafood following digestion and migrate up the esophagus into the pharynx, where they subsequently may be expectorated (59). The invasive form typically penetrates the mucosa or submucosa of the stomach or small intestine (Fig. 26.1) resulting in epigastric pain, nausea, vomiting, and diarrhea, usually 12 h after the infected seafood is eaten. Chronic anisakiasis may mimic peptic ulcer, appendicitis, enteritis, or gastric carcinoma. Most human infections of the so-called sushi parasite have been reported from Japan and the Netherlands; the growing number of documented cases in the

Table 26.1 Sources of infection for some foodborne helminths of humans

Helminth	Beef or pork[a]	Finfish	Shellfish	Other invertebrates	Fruits or vegetables	Other food source	Fecal contamination
Alaria americana						X[b]	
Angiostrongylus sp.			X	X	X		
Anisakis sp.		X	X[c]				
Ascaris lumbricoides							X
Baylisascaris procyonis							X
Capillaria philippinensis		X					
Clonorchis sinensis		X	X				
Dicrocoelium dendriticum				X	X[d]		
Diphyllobothrium sp.		X					
Dipylidium caninum				X			
Dracunculus medinensis				X			
Echinococccus granulosus							X
Echinococcus multilocularis							X
Echinostomum ilocanum				X			
Enterobius vermicularis							X
Eustrongylides sp.		X					
Fasciola hepatica					X	X[e]	
Fasciolopsis buski					X		
Gnathostoma spinigerum	X	X		X		X[f]	
Heterophyes heterophyes		X					
Hymenolepis diminuta				X			
Hymenolepis nana							X
Ligula intestinalis		X					
Macracanthorhynchus hirudinaceus				X			
Metagonimus yokogawai		X					
Moniliformis moniliformis				X			
Multiceps multiceps							X
Nanophyetus salmincola		X					
Nybelinia surmenicola			X[c]				
Opisthorchis sp.		X					
Paragonimus westermani			X			X[g]	
Phaeneropsolos bonnei				X			
Philometra sp.		X					
Prosthodendrium molenkampi				X			
Spirometra sp.	X			X		X[h]	
Strongyloides sp.						X[i]	
Taenia saginata	X						
Taenia solium	X						X
Toxocara canis							X
Trichinella spiralis	X					X[j]	
Trichostrongylus sp.							X
Trichuris trichiura							X

[a]Described in chapter 25.
[b]Frog, raccoon, and opossum.
[c]Squid.
[d]Acquired by ingestion of ants on unwashed herbs and vegetables.
[e]The condition halzoun occurs when adult worms in uncooked sheep liver attach to the pharynx.
[f]Pork, chicken, duck, frog, eel, snake, and rat.
[g]Wild boar.
[h]Acquired by ingestion of procercoids in copepods and plerocercoids in frogs, birds, and mammals; infection has also been reported from eating undercooked pork.
[i]Transmammary infection has been reported.
[j]Found in a variety of animals including bear, walrus, and horse.

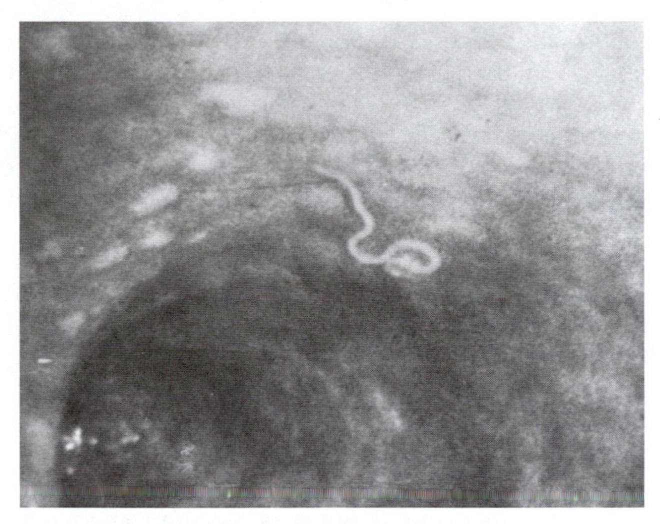

Figure 26.1 *Anisakis* roundworm embedded in the human gastric mucosa as visualized by gastroscopy. (Photograph contributed by Tomoo Oshima; illustration courtesy of the Armed Forces Institute of Pathology, AFIP 76-2118.)

United States, between 25 and 50, is still very small (61). In Japan, hypochlorhydria or achlorhydria has been found in more than half of anisakiasis patients and may predispose them to infection (22).

The life cycles of these helminths are not completely known. The adult worms are intestinal parasites of dolphins, whales, seals, and sea lions. Thus, the important reservoir hosts for this disease include several protected species, and there is evidence that the prevalence of such fishborne helminthiases is closely related to the local geographic distribution of the reservoir hosts (16). Eggs passed in the feces of marine mammals embryonate in seawater and develop into larvae that are eaten by krill. The infected crustaceans are next eaten by fish or squid, and the larvae develop further. The life cycle is completed in marine mammals, but when fish or squid are eaten by the unsuitable human host, the parasites do not develop further or reproduce.

Anisakidae larvae have been found in rockfish, herring, cod, halibut, mackerel, and salmon. It has been estimated that 90% of cod fillet sold in the Washington, D.C., area may be infected (22). The majority of human infections acquired in the United States have been associated with dishes prepared at home. The rare occurrence of restaurant-acquired cases has been attributed to the training and expertise of professional sushi chefs (61). The prominent reddish-brown larvae of *Pseudoterranova* are readily visible in contrast to the whitish fish tissue, but the smaller, lighter-colored *Anisakis* larvae are more difficult to detect (58). Comparison of salmon steaks with salmon fillets indicates predilection sites for the larvae and suggests that certain cuts may pose lesser risk for infection (18). In some fishes, most of the juvenile larvae are found in the viscera (15), suggesting that immediate evisceration would prevent their subsequent migration into the musculature. However, only thorough cooking or prolonged freezing will kill the parasite and completely eliminate the risk of infection. Cold smoking and most methods of brining fish are not reliable preventive measures (61). The only effective treatment for anisakiasis is surgical removal of the worms.

Capillaria philippinensis

C. philippinensis, a nematode found in freshwater fishes, causes a severe and potentially fatal infection in humans (9). In 1967 and 1968, the disease reached epidemic proportions in the Philippines, with over 1,000 confirmed cases and more than 100 deaths (11); cases have also been reported from Thailand. The infection appears to be acquired as a result of eating raw fish that contain infective larvae. The freshwater fishes *Ambassis miops, Eleotris melanosoma,* and *Hypseleotris bipartita* have been implicated in experimental infections, but the life cycle of *C. philippinensis* has not been fully elucidated, nor have natural reservoir hosts been identified. Massive infections are believed to result from internal autoinfection, but it is not known whether the disease can be acquired by the ingestion of *C. philippinensis* eggs.

Both larvae and adult worms may be found embedded in the intestine (Fig. 26.2). The long slender adult worms are dioecious. Males are approximately 2 to 3 mm long and 20 to 30 μm in diameter, while the larger females are approximately 3 to 5 mm long and 30 to

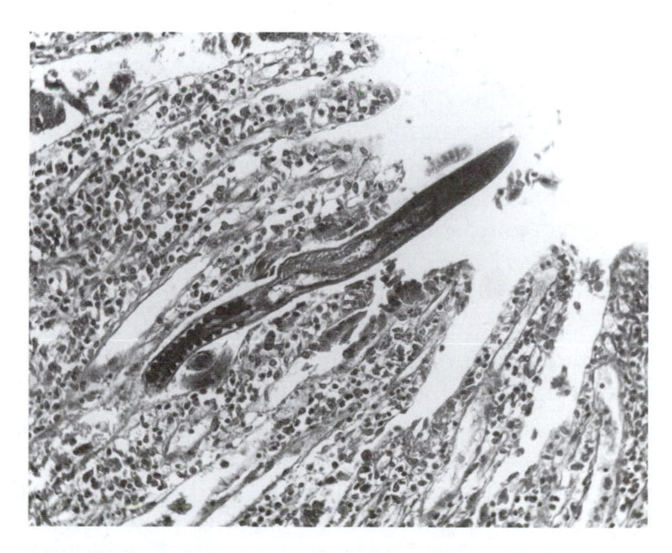

Figure 26.2 Transverse section of larval *C. philippinensis* embedded in human intestinal glands (×95). (Illustration courtesy of the Armed Forces Institute of Pathology, AFIP 69-1066.)

50 µm in diameter. *C. philippinensis* eggs are distinctive thick-shelled, capsule-shaped structures with bipolar plugs, and they measure approximately 45 to 21 µm, differing in appearance from *Trichuris* eggs of the same size that also have bipolar plugs. Diagnosis is made by identifying eggs in the feces. Symptoms of capillariasis include borborygmus, abdominal pain, nausea, vomiting, diarrhea, and anorexia during the acute phases. If the infection is untreated, intestinal malabsorption and untractable diarrhea lead to cachexia and possibly death. Treatment consists of a regimen of mebendazole or thiabendazole with electrolyte and protein supplementation.

Gnathostoma Species

Roundworms of the genus *Gnathostoma* reside in the stomach walls of a variety of carnivorous mammals. The life cycle of the parasite typically involves two intermediate hosts, a copepod and a freshwater fish; however, a variety of animals may serve as paratenic or transport hosts. Unable to mature in human hosts, the parasite wanders aimlessly through the tissues, causing a severe larva migrans that may persist for years. Cases have been reported throughout the world, with greatest prevalence in Thailand.

Gnathostoma spinigerum has been found in tigers, leopards, lions, domestic cats, minks, and dogs. The stout, reddish female worms range in length from 25 to 54 mm and are characterized by spines covering the anterior half of the body and by a prominent cephalic bulb covered by rows of sharp hooklets. Male worms are about half as long as females. Unembryonated eggs are passed in the feces and hatch in about a week, releasing motile first-stage larvae. When eaten by a copepod, the larva penetrates into the hemocoel and develops further. Infected copepods are next ingested by fish, frogs, or snakes, and the parasite continues its development into a third-stage larva, which is the infective stage for the final mammalian host.

Human infection typically results from eating raw, marinated, or poorly cooked freshwater fish, as once happened at a diplomatic dinner for visiting dignitaries (61). Recent cases in Mexico have been attributed to substituting marinated freshwater tilapia for the more expensive marine fish normally used in ceviche (60). In addition, infection can be acquired from pork, chicken, duck, frog, eel, snake, or rat or by the accidental ingestion of infected copepods in drinking water. It is also possible for larvae to penetrate the skin during food handling (13). Larvae can be killed by cooking or by immersion in strong vinegar for 5 h or longer; immersion in lime juice or chilling at 4°C for 1 month is not effective (4). Raw foods, particularly fish and chicken, should be avoided in areas where infection is endemic, and drinking water should be filtered before consumption.

Nausea, abdominal pain, and vomiting usually develop between 24 and 48 h following the ingestion of infected meat or fish. Symptoms of larva migrans or creeping eruption include pruritus, urticaria, tenderness, and painful subcutaneous swelling. Invasion of the central nervous system may result in meningitis and neuropathy. Diagnosis is difficult, and chemotherapy is of questionable value. As with anisakiasis, the only effective treatment is surgical removal of the larvae.

Diphyllobothrium latum

The broad tapeworm, *D. latum*, occurs in northern temperate regions of the world where raw or undercooked freshwater fishes are eaten; infection is closely related to dietary and cultural practices in food preparation (20). Infective larvae may be found in whitefish, trout, pike, and salmon. Cases have been reported throughout Europe, particularly in the Baltic countries, and from the Great Lakes region of the United States, Canada, Japan, South America, and Australia. *D. latum* was introduced into the United States by European immigrants in the middle of the 19th century (16). Although it is the largest of the human tapeworms, measuring up to 9 m in length, infections may be mild or asymptomatic. Nausea, abdominal pain, diarrhea, and weakness are common manifestations. *D. latum* may also cause pernicious anemia and vitamin B_{12} deficiency because the worm is highly efficient in competing with its host for available vitamin.

D. latum eggs shed in feces require approximately 12 days to embryonate, at which time the ciliated coracidium exits the egg through the operculum. When eaten by *Cyclops* or *Diaptomus* species, the coracidium penetrates the body cavity of the freshwater crustacean and develops into the procercoid stage of the parasite. When small fish such as minnows eat infected crustaceans, the procercoid penetrates into the viscera or muscles of the fish and develops into another unsegmented stage called the plerocercoid. The carnivorous fish that eat these fish are termed paratenic hosts because although the plerocercoid penetrates into their viscera or muscles, it does not develop beyond this stage. Further development from plerocercoid into adult worm requires ingestion by the human definitive host.

Diagnosis is made on the basis of eggs in the feces. The distinctive *D. latum* egg is ovoid, approximately 45 to 70 µm, with a prominent operculum at one end and a characteristic knob at the abopercular end; proglottids may occasionally be found in the feces. Infection is best prevented by adequate cooking or freezing of fish before consumption. Improved sanitation measures can also help reduce prevalence by interrupting the parasite's life cycle. Niclosamide and praziquantel are two of several effective anthelminthic drugs. *Diphyllobothrium dendriticum* and *Ligula intestinalis,* tapeworms of piscivorous birds, and *Diphyllobothrium pacificum,* a tapeworm of seals, have also been found in humans.

Clonorchis sinensis

Although the correct scientific name for the Chinese liver fluke is *Opisthorchis sinensis,* its original name, *Clonorchis sinensis,* remains more commonly accepted. Widespread throughout Asia, *C. sinensis* has been reported from Hong Kong, Japan, Korea, Taiwan, Vietnam, and large areas of the People's Republic of China. The parasite is typically acquired by eating infected freshwater fishes; the infective stage of *C. sinensis* has also been found in crayfish (23). Reservoir hosts include dogs, cats, foxes, pigs, rats, mink, badgers, and tigers.

Adult worms, measuring approximately 1.2 to 2.4 cm in length and 0.3 to 0.5 cm in width, reside in the bile duct. When eggs passed in the feces are eaten by *Parafossarulus manchouricus* or other hydrobiid snails such as *Bulimus, Semisulcospira, Alocinma,* or *Melanoides* species, the miracidium is released in the digestive tract and penetrates into the hemocoel, where it develops into first the sporocyst and then the redia stage. The redia gives rise to free-swimming cercariae that leave the snail and penetrate the second intermediate host, a cyprinid fish, where they encyst as metacercariae in the gills, fins, or muscles or under the skin. When the definitive host eats an infected fish, metacercariae excyst in the duodenum, migrate into the bile duct, and develop into adult worms.

C. sinensis may live in the human host for as long as 25 to 30 years, and massive infections with as many as 500 to 1,000 parasites have been reported. The severity of symptoms is related to the intensity and duration of infection. Diarrhea, epigastric pain, and anorexia are typical manifestations of acute clonorchiasis. The adult worm produces localized tissue damage that may result in hyperplasia or metaplasia of the bile duct epithelium, duct thickening, fibrosis, bilary stasis, and secondary bacterial infection. Pancreatitis may occur when worms enter the pancreatic duct. An association between cholangiocarcinoma and *C. sinensis* infection has also been reported (35). Diagnosis is made by identifying eggs in the feces. The operculate *C. sinensis* eggs measure approximately 27 to 16 μm and are characterized by distinctive opercular shoulders and a very small knob at the abopercular end. Infection is best prevented by avoiding raw, undercooked, or pickled finfish and shellfish. Snail control and improved sanitation can also help reduce transmission. Praziquantel is an effective anthelminthic drug.

Two closely related bile duct flukes of dogs and cats may also be acquired by eating uncooked cyprinid fishes. *Opisthorchis felineus* occurs throughout Eastern Europe and portions of the former Soviet Union, while *Opisthorchis viverrini* has been reported from Southeast Asia. The clinical picture for human infection by these species is very similar to that seen for clonorchiasis.

The Human Lung Fluke, *Paragonimus westermani*

P. westermani is the best-known and most widely distributed lung fluke, although other species of this genus may parasitize humans (69). *P. westermani* infection is common throughout the Far East and occurs to a lesser extent in parts of Africa and the Indian subcontinent. It has been found in Japan, Korea, the People's Republic of China, Manchuria, Taiwan, the Philippines, Indonesia, the Solomon Islands, both American and British Samoa, India, Sri Lanka, Cameroon, and Nigeria and in American troops in the South Pacific during World War II. It is a common parasite of mink in Canada and the eastern United States. Other reservoir hosts include a variety of animals that may eat crustaceans. Infection of dogs, cats, tigers, and cattle has been reported.

The reddish-brown, thick-bodied flatworms are found encapsulated in cystic structures adjacent to the bronchi or bronchioles. They measure approximately 0.8 to 1.6 cm in length by 0.4 to 0.8 cm in width and are 0.3 to 0.5 cm thick. Eggs are released by the parasite into the bronchioles, where they may be either expectorated or swallowed and passed in the feces. Eggs require several weeks in an aqueous environment for development. Upon hatching, the free-swimming miracidium penetrates a snail of the genus *Brotia, Semisulcospira, Tarebia,* or *Thiara,* where it undergoes further development into a sporocyst and then two generations of rediae. Approximately 11 weeks after the snail is infected, cercariae are shed and then penetrate any of a variety of freshwater crabs and crayfish, where they become encysted in the muscles, gills, and other organs as metacercariae. Important shellfish hosts include the freshwater and brackish water crabs of the genera *Eriocheir, Potamon,* and *Sundathelphusa* and the crayfish *Procambarus.*

Humans become infected by eating raw or improperly cooked freshwater crabs or crayfish. Dishes such as raw crayfish salad, jumping salad (which contains live shrimp), drunken crab (live crabs in wine), and crayfish curd are popular throughout the Orient. Pickling does not kill the parasite. The infective metacercariae can be killed by boiling the crabs for several minutes until the meat has congealed and turned opaque (38). Infection may also be acquired from shellfish juices used in food dishes or folk remedies, from food prepared by using contaminated utensils or chopping blocks, or by drinking water contaminated with metacercariae released from dead or injured crustaceans.

Migration of parasites through host tissues produces localized hemorrhage and infiltration of lymphocytes. Pulmonary symptoms include dyspnea, chronic cough, chest pain, night sweats, hemoptysis, and persistent rales. The severity of symptoms appears to be related to the number of parasites present. Pleural effusion and fibrosis may occur in long-standing infections, although there is also evidence that lesions may resolve without treatment. Neurological complications result from migration of the parasite into the spinal cord or brain. *P. westermani* has also been found in the intestinal wall, peritoneum, pleural cavity, and testes. *Paragonimus szechuanensis* and *Paragonimus hueitungensis,* two species which apparently are incapable of developing in the lungs, cause cutaneous larva migrans, characterized by eosinophilia, anemia, and low-grade fever. Diagnosis is made on the basis of eggs in expectorant or feces. The characteristic golden brown eggs are 80 to 120 μm long by 48 to 60 μm wide. Chest X rays may show nodular shadows or calcified spots. Effective anthelminthic drugs include praziquantel and nicloforan.

Other Helminths Associated with Seafood

Heterophyes heterophyes, an intestinal fluke acquired from mullet, has been found in Egypt, in Israel, and throughout Asia. The parasite may cause nausea, diarrhea, and abdominal pain, but light infections are often asymptomatic. There have been reports of fatal myocarditis and neurological complications when helminth eggs penetrate the intestine and enter the circulatory system. The closely related trematode *Metagonimus yokogawai,* acquired from salmonid fishes, has been reported from Asia, the Balkans, Israel, Spain, and portions of the former Soviet Union. Fish-eating mammals and birds, such as dogs, cats, and pelicans, serve as reservoir hosts. Diagnosis of both species is made by identifying eggs passed in the feces. Intestinal symptoms are mild and depend on the number of worms present. Praziquantel is an effective anthelminthic drug.

Salmon poisoning, a severe and frequently fatal disease of dogs in the Pacific Northwest, is associated with the intestinal trematode *Nanophyetus* (=*Troglotrema*) *salmincola.* This disease does not result directly from the fluke itself but occurs because *N. salmincola* serves as a vector for the rickettsial pathogen *Neorickettsia helmintheca.* Human *N. salmincola* infection, characterized by nausea, diarrhea, and intestinal discomfort, may be acquired by eating uncooked salmonid fishes and has also been attributed to handling freshly killed coho salmon (31).

The nematode *Eustrongylides* is a common parasite of piscivorous birds. In addition to fish, reptiles and amphibians may also serve as intermediate hosts and thus contain infective larvae. An infection in New York City was attributed to eating raw fish prepared at home (67). In Maryland and New Jersey, infections were found in fishermen who had eaten live bait (8, 25). As with *Anisakis* infections, surgical removal of worms is the only effective treatment.

Infection by *Philometra,* a nematode closely related to *Dracunculus,* was found in a fisherman in Hawaii as a result of filleting a carangid fish (17). Infection by the tapeworm *Nybelinia surmenicola* has been attributed to eating raw squid (39). Sparganosis may be transmitted by a variety of animals, including fish. Although typically associated with snails, *Angiostrongylus* infection may also be acquired from freshwater prawns and land crabs.

HELMINTHS ACQUIRED FROM VEGETATION

The Sheep Liver Fluke, *Fasciola hepatica*

Sheep liver rot, caused by the digenetic trematode *F. hepatica,* was recognized as early as the 14th century (19). It was the first trematode for which a complete life cycle was elucidated (41). Sheep, cattle, and other herbivores acquire the infection by eating metacercariae encysted on aquatic plants. *F. hepatica* has been found in goats, horses, deer, rabbits, camels, vicuna, swine, dogs, and squirrels. Human infection is prevalent in sheep-raising areas throughout the world. In the United States, human fascioliasis has been reported from California (49) and Puerto Rico (33).

F. hepatica is a large fluke, measuring approximately 3 cm in length by 1.5 cm in width, and is readily identified by its characteristic "cephalic zone," a distinct conical projection at the anterior end. Adult worms reside in the biliary passages and gallbladder. Eggs passed with feces into the water require 9 to 15 days to mature but may remain viable for several months in soil if they remain moist. Upon hatching, the free-swimming miracidium penetrates a lymnaeid snail of the genus *Lym-*

naea, Succinea, Fossaria, or *Practicolella,* where it develops into a sporocyst and then two generations of rediae. Cercariae are shed from the snail, swim freely, and then attach to aquatic vegetation, where they encyst as metacercariae. Encysted metacercariae are susceptible to drying but can survive over winter (52).

Humans become infected by eating infested freshwater plants or free metacercariae in drinking water. Human cases are frequently traced to watercress, *Nasturtium officinalis* (37). A pharyngeal form of disease, called halzoun, may occur following the ingestion of raw liver from infected animals, when worms present in liver attach to the pharynx. Transmission can be controlled by the use of molluscicides, by draining ponds, and by protecting crops and water supplies from contact with livestock.

Some inflammation is associated with migration of the parasite, but mild infections are often asymptomatic. Tissue destruction occurs when worms penetrate the liver. In sheep, liver rot causes massive damage. There may be mechanical obstruction of bile ducts, hyperplasia of biliary epithelium, and proliferation of the connective tissue. Worms may erode the walls of the bile ducts and invade the liver parenchyma. Secondary bacterial infection and portal cirrhosis have been reported, but liver calcification appears rare. Pain, bleeding, and edema of the face and neck are associated with halzoun.

Diagnosis is made on the basis of eggs in the feces. The relatively large, operculate *F. hepatica* eggs are approximately 130 to 150 µm long by 63 to 90 µm wide. Ingestion of liver from infected sheep or cattle may result in spurious infection when eggs present in the food are passed in feces. Such false-positive findings can be ruled out by subsequent stool examinations. Triclabendazole now replaces bithionol as the anthelminthic of choice.

The Giant Intestinal Fluke, *Fasciolopsis buski*

F. buski is the largest trematode parasite of humans. Endemic throughout Asia, it has been reported from China, Taiwan, Korea, Vietnam, Cambodia, Laos, Myanmar, Thailand, Borneo, Sumatra, Malaysia, Indonesia, India, and Bangladesh. The pig is an important reservoir host. Infection has also been found in dogs and rabbits.

Although only 0.8 to 3.0 mm thick, *F. buski* can grow to 7.5 cm in length by 2 cm in width. Adult worms attach to the bowel walls, primarily along the duodenum and jejunum. Eggs passed in the feces are unembryonated and require 3 to 7 weeks in fresh water to develop. Upon hatching, the free-swimming miracidium penetrates a snail of the genus *Segmentina* or *Hippeutis,* where it develops into a sporocyst and then two generations of rediae. Approximately 4 to 7 weeks after the snail is infected, cercariae are shed into the water and then attach to vegetation, where they encyst as metacercariae.

Humans become infected by eating infested water chestnuts, bamboo, caltrop, or lotus. Individuals may also acquire the parasite by peeling the hulls of plants with their teeth. The metacercariae excyst in the small intestine, attach to the mucosa, and develop into adult worms in about 3 months. Drying or cooking the plants before eating kills the metacercariae (5). Immersion of vegetables in boiling water for a few seconds, or even peeling and washing them in clear water, is sufficient to preclude infection (4). Poor sanitation and the use of human and swine feces as fertilizer are major factors in disease transmission (57).

Adult worms feed not only on intestinal contents but also on intestinal epithelium, leading to local ulceration and hemorrhage. Nausea, abdominal pain, diarrhea, and hunger pangs are common. In heavy infections, stools are profuse and light yellow in color, suggestive of malabsorption; intestinal obstruction and ascites have been reported. Diagnosis is made on the basis of eggs in the feces. The yellow-brown eggs have a small operculum and are approximately 130 to 140 µm long by 80 to 85 µm wide. Adult worms are seen in feces only following chemotherapy or purgation. Effective anthelminthic drugs include praziquantel and niclosamide.

Other Helminths Associated with Vegetation

Fresh vegetables grown in areas where nightsoil is used as fertilizer are frequently contaminated and thus may facilitate transmission of any of a number of geohelminths. Human infection by *Dicrocoelium* is explained by the accidental ingestion of ants on vegetation, and *Angiostrongylus costaricensis* infection is thought to be acquired by eating raw fruits and vegetables on which snails have left larvae in mucus deposits, or by accidentally ingesting infected snails on unwashed vegetation. *Trichostrongylus* and *Echinococcus granulosus* infections have also been attributed to contaminated vegetation (4).

HELMINTHS ACQUIRED FROM INVERTEBRATES IN DRINKING WATER

The Guinea Worm, *Dracunculus medinensis*

The long, threadlike roundworm *D. medinensis* has plagued humans since antiquity. It is believed to be the "fiery serpent" described in the Bible (50) and is symbolically depicted in the caduceus, the insignia of the medical profession. Prevalence has declined markedly as a result of an aggressive worldwide eradication campaign, but dracunculiasis still occurs sporadically among rural populations in parts of India, in Pakistan, and in several African countries (34). Although rarely fatal, this parasite causes considerable discomfort and disability, particularly in areas where poverty is severe. The female nematode is almost a meter in length but less than 2 mm thick; males are inconspicuous and only 2 cm long. Worms develop to maturity in the body cavity or deep connective tissues. Females then migrate to the subcutaneous tissues, become gravid, and stimulate the formation of a blister that eventually ruptures to expose part of the worm. The uterus accounts for most of the parasite's volume and is distended with as many as 1 million to 3 million larvae. Upon contact with fresh water, the worm bursts to release larvae. Larvae are ingested by copepods of the genera *Cyclops, Mesocyclops,* and *Thermocyclops* and mature into the infective form in approximately 3 weeks.

Humans become infected by swallowing infected copepods present in drinking water (Fig. 26.3). The larvae penetrate the digestive tract and take about a year to reach maturity. Transmission is clearly related to poverty, the quality of drinking water, and water contact by infected individuals. A measure as simple as sieving drinking water through a piece of cloth will remove copepods and prevent infection. The provision of safe drinking water supplies in a village is usually followed by disappearance of the disease (47, 54). *Dracunculus* infections have been found in dogs in China and the former Soviet Union, but it is not clear whether canines play any significant role as reservoir hosts.

Subcutaneous migration of the female nematode results in localized erythema, tenderness, and pruritus. Other symptoms include nausea, vomiting, diarrhea, generalized urticaria, and asthma. A papule forms under the skin, where the anterior end of the parasite lies, and becomes vesicular; it then ulcerates, exposing the worm. There may be a painful local reaction when larvae are discharged. Secondary bacterial infection may occur when the parasite is resorbed and is a frequent complication when the worm breaks following unsuccessful attempts at removal. On rare occasions, nervous system involvement resulting in paraplegia (45), quadriplegia (21), and death (55) has been reported.

The traditional treatment of winding the worm around a small stick and slowly extruding can still be effective provided that asepsis is maintained. Surgical removal of *D. medinensis* has been practiced in India and Pakistan. Metronidazole and thiabendazole, although not curative, decrease inflammation and facilitate removal of the worm, but the former is mutagenic and the latter is toxic in the dosage recommended; mebendazole has been reported to kill the parasite directly (1).

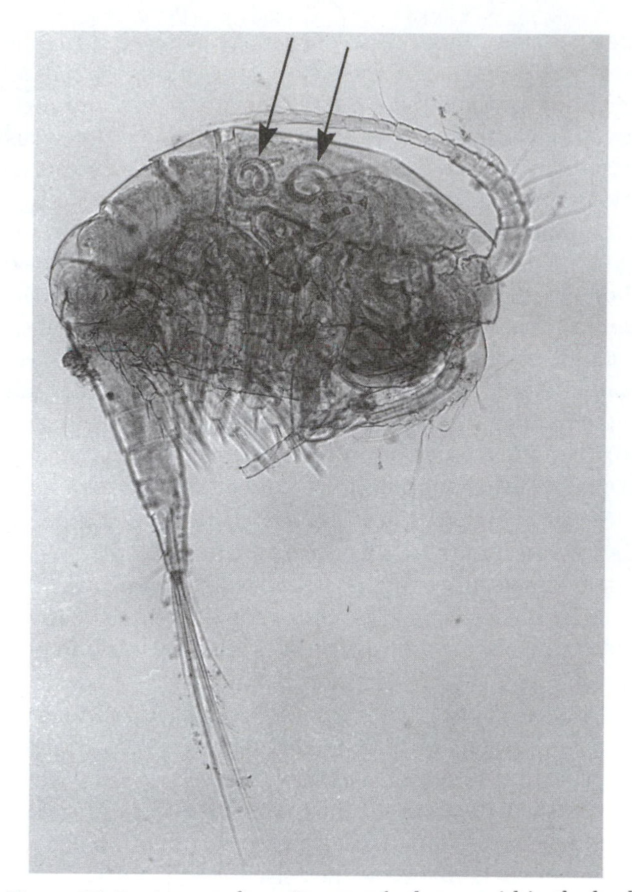

Figure 26.3 Arrows show *Dracunculus* larvae within the body cavity of the intermediate host, *Cyclops* species (×60). (Specimen contributed by E. L. Schiller, Johns Hopkins University School of Public Health, Baltimore, Md.; illustration courtesy of the Armed Forces Institute of Pathology, AFIP 68-4629.)

Other Helminths Acquired from Copepods

There is evidence that the tissue-dwelling nematode *G. spinigerum* may be acquired directly from infected copepods (12). Sparganosis, although typically acquired from a variety of vertebrate hosts, may also result from ingesting copepods infected with procercoids.

HELMINTHS ACQUIRED FROM OTHER INVERTEBRATES

Snails

The rodent lungworm, *Angiostrongylus cantonensis,* is a slender roundworm about 25 mm in length that typically resides in the pulmonary arteries of rodents.

Human cases have been reported from Taiwan, Thailand, India, the Philippines, Hawaii, several Pacific islands, Cuba, and the Ivory Coast. An autochthonous *A. cantonensis* infection was recently found in New Orleans in a child who had consumed a raw snail (48). Migration of the parasite through the brain causes meningitis, and fatal cases have been reported.

The adult nematodes lay eggs that hatch in the lungs. First-stage larvae migrate to the trachea, are swallowed, and pass in the feces, where they are ingested by any of a variety of slugs, land snails, or planarians. They then develop into infective third-stage larvae in about 2 weeks. When rodents eat infected molluscs, the larvae migrate to the brain, where they develop into adults in about 4 weeks, then to the pulmonary arteries, where they begin to lay eggs after an additional 2 weeks.

Humans become infected by eating raw or undercooked molluscs such as the freshwater snail *Pila* or the giant African land snail *Achatina fulica*. Freshwater prawns, which are frequently eaten raw in Thailand, Vietnam, and Tahiti, have been implicated in human infection (56). Natural infections have also been found in the land crab *Cardisoma hirtipes* and the coconut crab *Birgus latro* (2). Shrimp, crabs, and frogs may serve as paratenic hosts. Infection can also be acquired by the accidental ingestion of slugs on lettuce or of parasite larvae left by snails in mucus deposits on fruits and vegetables; contaminated drinking water may be another source of infection. Transmission can be prevented by thorough cooking and appropriate attention to food-handling practices, particularly the careful washing of fruits and vegetables and washing of hands after exposure to molluscs while gardening.

Angiostrongyliasis usually presents as a self-limiting meningitis during the migratory phase. Symptoms of meningitis or meningoencephalitis are characterized by abrupt onset and include headache, stiff neck, and sensorial changes. Cerebrospinal fluid may contain 100 to 2,000 leukocytes per mm^3, and there is marked eosinophilia. Nausea, vomiting, fever, abdominal pain, malaise, and constipation have also been reported (68). While larvae may be found in cerebrospinal fluid, diagnosis is often presumptive, based upon eosinophilia and symptoms of meningitis in an endemic area. Analgesics or corticosteroids may provide symptomatic relief. The effectiveness of mebendazole or thiabendazole has been documented only in animals (1). Since the disease is self-limiting, anthelminthic drugs may not be necessary. The killing of parasites while in the central nervous system could result in even greater complications from an increased inflammatory reaction (4).

A. costaricensis is a closely related roundworm found in the mesenteric arteries of the cotton rat *Sigmodon hispidus*. Human abdominal angiostrongyliasis has been reported from Costa Rica, Honduras, Panama, Mexico, Brazil, and Venezuela. The veronicellid slug, *Vaginulus plebius*, has been implicated in the transmission of *A. costaricensis*. Human infection is acquired from accidental ingestion of slugs or by contact with contaminated fruits, vegetables, or grass.

Echinostomum ilocanum is an intestinal trematode found in Thailand, Java, and the Philippines that may be acquired by eating metacercariae encysted in snails. The Norway rat is an important reservoir host. Diagnosis is based on identification of unembryonated, operculate eggs, measuring approximately 83 to 116 μm in length by 58 to 69 μm in width, in stool specimens. Praziquantel appears to be an effective anthelminthic drug.

Ants

Dicrocoelium dendriticum is a trematode found in the biliary passages of sheep, deer, and other herbivores. The intermediate hosts are land snails and ants. Cercariae are shed from snails in slime balls that are deposited on the grass; when eaten by ants, they encyst as metacercariae. The mammalian host becomes infected by ingestion of infected ants. Reports of human infection are frequently spurious. Ingestion of liver from infected sheep can result in a false-positive diagnosis when eggs present in the food are passed in feces. However, genuine human cases have been reported in Europe, Asia, and Africa. Transmission of *Dicrocoelium dendriticum* can best be prevented by careful washing of herbs and vegetables to remove ants.

Fleas

Dipylidium caninum is a common tapeworm of dogs and cats throughout the world. Tapeworm eggs are ingested by the flea *Ctenocephalides*, in which they develop into the cysticercoid stage. Human infection appears limited to young children and results from accidental ingestion of infected fleas. Infection can best be prevented by controlling fleas on pets and by periodic worming of animals when necessary.

Beetles, Cockroaches, and Other Insects

Hymenolepis diminuta is a tapeworm of rats, mice, and other rodents. Eggs passed in rodent feces are ingested by flour beetles (*Tribolium*, *Tenebrio*), cockroaches (*Blattella*, *Periplaneta*), or fleas, in which they develop into the cysticercoid stage. Human infection, acquired by the accidental ingestion of infected insects, is typically asymptomatic, although nausea, abdominal pain, and diarrhea may occur. Preventive measures include protecting grains and foodstuffs from insects and instituting rodent control.

The acanthocephalan *Moniliformis moniliformis* is an intestinal parasite of rats that also utilizes cockroaches and beetles as intermediate hosts. As with *H. diminuta*, human infection is usually found in young children. The giant leech-like acanthocephalan *Macracanthorhynchus hirudinaceus* is a cosmopolitan parasite of swine. The spiny proboscis embeds itself in the intestinal mucosa; female worms measure up to 65 cm in length. Human infections are rare but have been directly attributed to eating raw beetles. The trematodes *Prosthodendrium molenkampi* and *Phaeneropsolos bonnei* may be acquired from ingesting dragonfly and damselfly aquatic larvae.

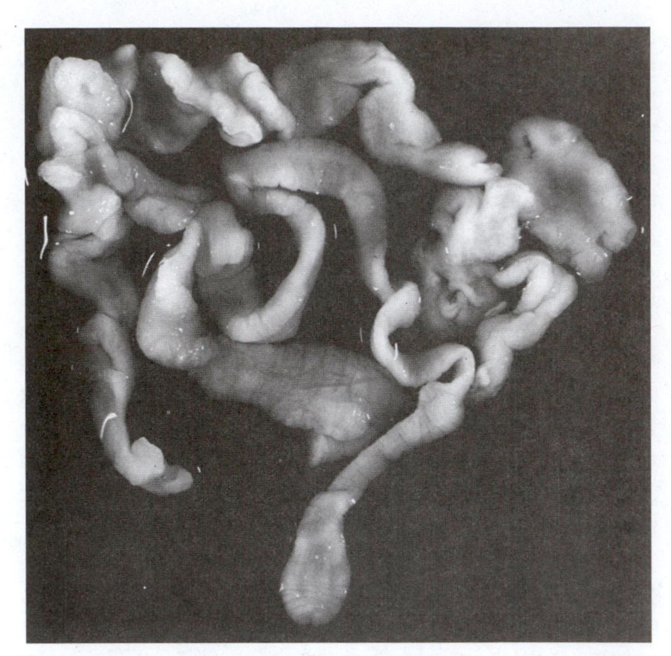

Figure 26.4 A sparganum removed from a subcutaneous nodule in the inguinal region (×1.7). (Illustration courtesy of the Armed Forces Institute of Pathology, AFIP 70-7392.)

HELMINTHS ACQUIRED FROM OTHER FOOD SOURCES

Tapeworms Causing Sparganosis

Sparganosis refers to infection by the plerocercoid stage of the diphyllobothrid tapeworm *Sparganum* or *Spirometra*. Human infection may be acquired by the accidental ingestion of copepods infected with the procercoid stage of the tapeworm. Alternatively, humans may serve as a paratenic hosts by eating any of a variety of animals infected with the plerocercoid stage, such as frogs, tadpoles, lizards, snakes, birds, and mammals. Foodborne transmission has been attributed to eating raw pork (10). Infection may also result from the folk medicine practice of applying poultices of frog or snake to the eyes, skin, or vagina, as spargana are capable of migrating out of the infected animal flesh and penetrating the lesion.

The sparganum is a wrinkled, ivory-white, ribbon-like flatworm that can grow up to 30 cm in length but is only about 3 mm wide (Fig. 26.4). Histological sections reveal a parenchymal tissue typical of undifferentiated cestode plerocercoids. Although usually found in subcutaneous tissue, spargana are highly motile and may migrate into muscle, viscera, brain, and eye. *Sparganum proliferum* is an especially hyperplastic species, exhibiting extensive branching and the capability of asexual reproduction by budding into separate organisms.

Migration of spargana is characterized by a painful, localized inflammatory reaction. Excessive lacrimation, periorbital edema, and swelling of the eyelids is associated with ocular sparganosis. Subcutaneous lesions may develop into abscesses; if the lesion ulcerates, the sparganum may be mistaken for a guinea worm. Diagnosis is presumptive and usually not confirmed until the worm is removed and identified in histological section. Surgical removal of spargana is the treatment of choice. Praziquantel has been proven to be an effective anthelminthic drug for laboratory animals.

Other Biohelminths

Alaria americana is a strigeid trematode typically found in the intestines of dogs and foxes. The normal life cycle involves sporocysts in snails and mesocercariae (a nonencysted stage) in tadpoles and frogs. Human infection has been attributed to eating inadequately cooked frog legs, raccoon, and opossum. Parasites have been recovered from the eye and from intradermal lesions, and at least one massive, fatal, infection with mesocercariae disseminated into the lungs and other internal organs has been reported (27).

Pentasomids, or "tongue worms" (Fig. 26.5), are larval arthropods that have been recovered from the liver, spleen, lungs, and eye. Human infection has been attributed to eating raw snake, lizard, goat, and sheep. Ocular involvement probably results from direct contact with pentastomid eggs in water. Most infections are asymptomatic, although respiratory discomfort and intestinal obstruction have been reported (43). Hoarseness and coughing due to young worms attached to the pharynx, a condition known as halzoun in the Near East, has been

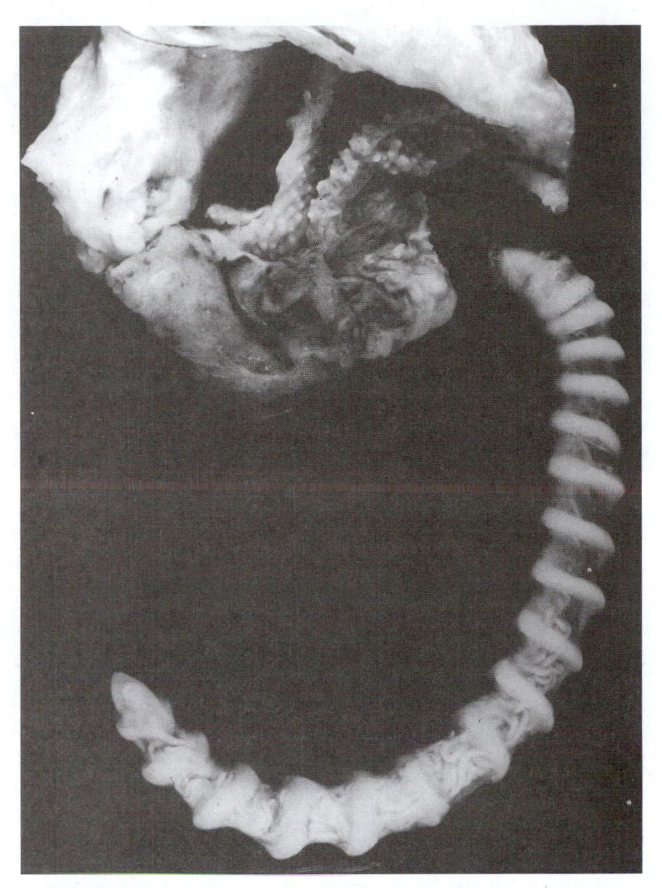

Figure 26.5 *Armillifer armillatus* adult female attached to the respiratory surface of the lung of a rock python from Zaire (life size). (Illustration courtesy of the Armed Forces Institute of Pathology, AFIP 72-881.)

attributed to pentastomid infection (5). Halzoun may also be caused by *F. hepatica* in sheep liver. Ingestion of liver from sheep infected with adult *F. hepatica* or *Dicrocoelium dendriticum* may result in false-positive diagnoses of these parasites when their eggs are released into the alimentary canal. Although typically associated with fish and copepods, *Gnathostoma* infection may also be acquired from pork, chicken, duck, frog, eel, snake, or rat.

Trichinella spiralis, described in detail in chapter 25, is typically associated with ingestion of raw or undercooked pork. However, human infection has also been acquired from sources as diverse as bear and walrus. An outbreak of trichinellosis in France was traced to horsemeat imported from the United States (3). Wild boar, a paratenic host for *P. westermani*, has been implicated as another potential source of human infection by that species (46). Prenatal and transmammary transmission of infective larval stages of helminths is more common than generally realized and in some instances may be the major route of infection. It has been observed for cestodes, trematodes, and, most frequently, nematodes (44). Larvae of *Strongyloides fuelleborni* have been recovered from human milk (6), and *Strongyloides stercoralis* may be similarly transmitted (4). Strongyloidiasis is a common cause of infant death in Papua New Guinea (66).

HELMINTHS ACQUIRED FROM FECAL CONTAMINATION

Inadequate washing of produce or poor hygiene among food handlers can result in a variety of helminthic infections. Personal hygiene is critical because helminth eggs are often adherent, and contamination may be found not only on hands but under fingernails, on clothing, and in washwater (51). Parasites can be acquired from both animal and human waste, and the use of nightsoil to fertilize crops represents a major source of infection. In addition to the direct transmission of zoonoses from animals to humans, experimental studies with seagulls and *Taenia* spp. suggest that birds may play an important ancillary role in disseminating helminth eggs (62).

A variety of geohelminth species utilize the fecal-oral route for person-to-person and animal-to-person transmission. *Ascaris lumbricoides*, the largest intestinal roundworm of humans (Fig. 26.6), has been known since antiquity. *Ascaris* eggs have been found in mummified remains in Egypt and in archeological artifacts from the sites of Roman legion encampments. In parts of Central

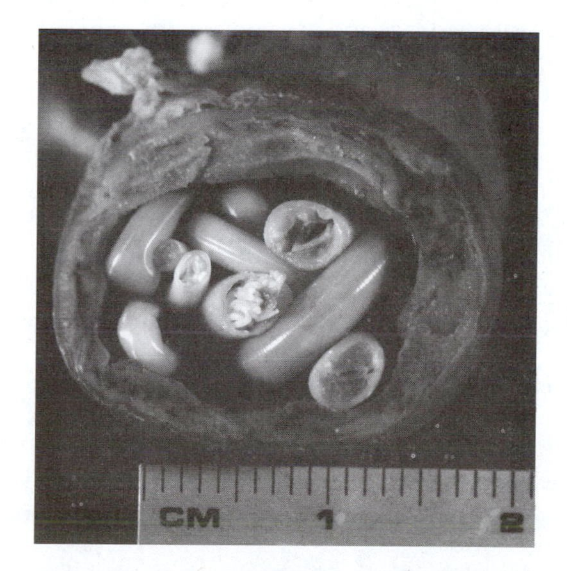

Figure 26.6 Numerous adult *A. lumbricoides* obstructing the jejunum of a 13-year-old Zairian (×2.5). (Illustration courtesy of the Armed Forces Institute of Pathology, AFIP 72-13204.)

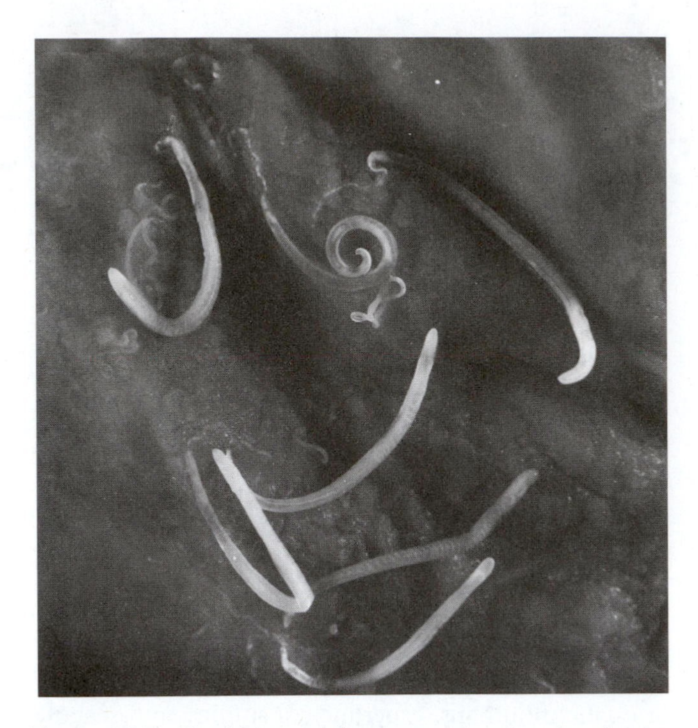

Figure 26.7 *T. trichiura* adult worms, showing the slender anterior ends threaded beneath the colonic epithelium (×3.7). (Illustration courtesy of the Armed Forces Institute of Pathology, AFIP 69-3583.)

and South America, infection rates may approach 45% (4). It has been estimated that in China alone, 18,000 tons of *Ascaris* eggs were produced each year (64). Unembryonated eggs shed in the feces are resistant to desiccation, moderate freezing temperatures, and chemical treatment of sewage and may remain dormant in the soil for years (7). The swine ascarid, *Ascaris suum,* can also infect humans but does not develop to maturity (14).

The whipworm, *Trichuris trichiura* (Fig. 26.7), occurs throughout the world but is most common in the tropics and in regions where sanitation is poor. Infections have been reported from the southeastern United States. *Trichuris* eggs can survive in the soil for several years. There is no reservoir host for *T. trichiura,* but *Trichuris suis* and *Trichuris vulpis,* parasites of swine and dogs, respectively, have been reported in humans.

Trichostrongylus is a common intestinal nematode of herbivores that resembles the hookworm. Human trichostrongyliasis, caused by the accidental ingestion of larvae on contaminated vegetation or in drinking water, occurs primarily in Asia but has been reported throughout the world.

Oesophagostomum is a common intestinal nematode of primates, swine, and domestic animals throughout Asia, Africa, and South America. In human infections, the accidentally ingested larvae produce nodular lesions and abscesses, approximately 1 to 2 cm in diameter, in the intestinal wall and sometimes other organs (30). As with anisakiasis, there is no effective anthelminthic drug, and surgical removal of the parasite is the only effective treatment. Definitive diagnosis is based on identifying worms in surgical or biopsy specimens. In livestock, severe diarrhea may kill the animal, and parasitic nodules render the intestine unsuitable for making sausage casings, thus representing a serious economic loss to farmers (24).

The dwarf tapeworm, *Hymenolepis nana,* is the only human tapeworm that does not require an intermediate host; its life cycle is maintained by person-to-person contact and by autoinfection, although rodents may serve as reservoir hosts. Distributed throughout the world, *H. nana* is common in southern Europe, the former Soviet Union, India, and Latin America. Young children are most frequently infected. Eggs shed in feces are fully embryonated and immediately infective but have poor resistance outside the host. Internal autoinfection may occur when eggs hatch in the small intestine and penetrate the villi to repeat the developmental cycle without leaving the host. Eggs ingested by fleas and flour beetles develop into cysticercoids, but insects do not appear to play a major role in transmission. Mild infections are typically asymptomatic, and even large numbers of *H. nana* are well tolerated.

Toxocara canis, an intestinal roundworm of dogs and foxes, appears to be the principal cause of visceral larva migrans in the United States (53). *Toxocara cati* is a related species in domestic felines. Following the accidental ingestion of soil contaminated with embryonated *Toxocara* eggs, the larvae migrate throughout the body and may cause serious complications when they invade the central nervous system or the eye. Eggs may survive for years in the environment, and the typically high rates of contamination in municipal parks, playgrounds, and schoolyards pose a serious potential risk for young children (65). Appropriate preventive measures include animal control, particularly leash laws and waste disposal regulations, the periodic worming of pets, and scrupulous encouragement of hand washing before meals. Ocular involvement is a serious complication and, tragically, a number of eyes with benign *Toxocara* inflammatory lesions have been unnecessarily enucleated because of suspected retinoblastoma (28). Ocular lesions may be more likely to occur in mild infections (29).

Baylisascaris procyonis, an ascarid of raccoons, can cause severe neurological complications in humans because this species is aggressive and grows considerably during migration. Fatal human infections have been found in children believed to have been exposed to *Baylisascaris* eggs present on pieces of firewood or in raccoon feces deposited by animals nesting in unused hearths and chimneys (26, 36).

Echinococcus granulosus is a tapeworm of dogs and other canids. Human infection following the accidental ingestion of eggs in canine feces is common in sheep-raising areas throughout the world and among indigenous populations such as Eskimos, where there is a close ecological relationship between humans and dogs. *Echinococcus multilocularis* is a closely related species producing an alveolar cyst that grossly resembles an invading neoplasm. The course of the disease resembles that of a growing carcinoma, and it is among the most lethal of all helminthic infections (4). Risk factors for *Echinococcus* infection include exposure to infected dogs, consumption of contaminated water, ice, or snow, contact with infected foxes or their skins, and ingestion or handling of contaminated strawberries, huckleberries, cranberries, or other vegetation (63).

Multiceps multiceps develops into a similar proliferative stage called a coenurus. In sheep, coenuriasis of the brain or spinal cord is relatively common and is known as gid or staggers. Human infections may occur in any organ but frequently involve subcutaneous tissue, brain, and eye. Cysticercosis, described in chapter 25, results from the accidental ingestion of *Taenia solium* eggs. Racemose cysticercosis has been described as an aberrant cysticercus form of *T. solium* but may represent coenurus infection.

Acknowledgments

I thank Judy A. Sakanari, University of California, San Francisco, for sharing recent research findings about *Anisakis* species and other fishborne parasites and John H. Cross, Uniformed Services University of the Health Sciences, Bethesda, Md., for providing background information about capillariasis and other helminthic diseases in the Far East. I am especially grateful to John S. Mackiewicz, State University of New York at Albany, for encouraging attention to the importance of ecological factors when considering parasitic diseases. Illustrations for this chapter were obtained from the extensive slide collection of the Armed Forces Institute of Pathology, Washington, D.C.

References

1. **Abramowicz, M. (ed.).** 1993. Drugs for parasitic diseases. *Med. Lett. Drugs Ther.* **35**:111–122.
2. **Alicata, J. E.** 1965. Notes and observations on murine angiostrongylosis and eosinophilic meningoencephalitis in Micronesia. *Can. J. Zool.* **43**:667–672
3. **Ancelle, T., J. Dupouy-Camet, M. E. Bougnoux, V. Fourestie, H. Petit, G. Mougeot, J. P. Nozais, and J. LaPierre.** 1988. Two outbreaks of trichinosis caused by horsemeat in France in 1985. *Am. J. Epidemiol.* **127**:1302–1311.
4. **Beaver, P. C., R. C. Jung, and E. W. Cupp.** 1984. *Clinical Parasitology,* 9th ed. Lea & Febiger, Philadelphia.
5. **Bogitsh, B. J., and T. C. Cheng.** 1990. *Human Parasitology.* Saunders College Publications; Holt, Rinehart & Winston, New York.
6. **Brown, R. C., and M. H. F. Girardeau.** 1977. Transmammary passage of *Strongyloides* sp. larvae in the human host. *Am. J. Trop. Med. Hyg.* **26**:215–219.
7. **Bryan, F. D.** 1977. Diseases transmitted by foods contaminated by waste water. *J. Food Prot.* **40**:45–56.
8. **Centers for Disease Control.** 1982. Intestinal perforation caused by larval *Eustrongylides*—Maryland. *Morbid. Mortal. Weekly Rep.* **31**:383–389.
9. **Chitwood, M. B., C. Valesquez, and N. G. Salazar.** 1968. *Capillaria philippinensis* sp.n. (Nematoda:Trichinellida) from the intestine of man in the Philippines. *J. Parasitol.* **54**:368–371.
10. **Corkum, K. C.** 1966. Sparganosis in some vertebrates of Louisiana and observations on a human infection. *J. Parasitol.* **52**:444–448.
11. **Cross, J. H.** 1992. Intestinal capillariasis. *Clin. Microbiol. Rev.* **5**:120–129.
12. **Daengsvang, S.** 1971. Infectivity of *Gnathostoma spinigerum* larvae in primates. *J. Parasitol.* **57**:476–578.
13. **Daengsvang, S., B. Sermswatsri, P. Youngyi, and D. Guname.** 1970. Development of adult *Gnathostoma spinigerum* in the definitive host (cat and dog) by skin penetration of the advanced third-stage larvae. *Southeast Asia J. Trop. Med. Public Health* **1**:187–192.
14. **Davies, N. J., and J. M. Goldsmid.** 1978. Intestinal obstruction due to *Ascaris suum* infection. *Trans. R. Soc. Trop. Med. Hyg.* **72**:107.
15. **Deardorff, T. L., and R. M. Overstreet.** 1981. Larval *Hysterothylacium* (=*Thynnascaris*) (Nematoda:Anisakidae) from fishes and invertebrates in the Gulf of Mexico. *Proc. Helminthol. Soc. Wash.* **48**:113–126.
16. **Deardorff, T. L., and R. M. Overstreet.** 1990. Seafood-transmitted zoonoses in the United States: the fishes, the dishes, and the worms, p. 211–265. *In* D. R. Ward and D. R. Hackney (ed.), *Microbiology of Marine Food Products.* Van Nostrand Reinhold, New York.
17. **Deardorff, T. L., R. M. Overstreet, M. Okihiro, and R. Tam.** 1986. Piscine adult nematode invading an open lesion in a human hand. *Am. J. Trop. Med. Hyg.* **35**:827–830.
18. **Deardorff, T. L., and R. Throm.** 1988. Commercial blast freezing of third-stage *Anisakis simplex* larvae encapsulated in salmon and rockfish. *J. Parasitol.* **74**:600–603.
19. **de Brie, J.** 1379. *Le Bon Berger ou le Vray Regime et Gouvenement de Bergers et Bergeres: Compose par le Rustique Jehan de Brie le Bon Berger.* Isidor Liseux, Paris [Reprint, 1879.]
20. **Desowitz, R. S.** 1981. *New Guinea Tapeworms and Jewish Grandmothers.* Norton and Company, New York.
21. **Donaldson, J. R, and T. A. Angelo.** 1961. Quadriplegia due to guinea-worm abscess. *J. Bone Joint Surg.* **43A**:197–198.
22. **Dooley, J. R., and R. C. Neafie.** 1976. Anisakiasis, p. 475–481. *In* C. H. Binford and D. H. Connor (ed.), *Pathol-*

ogy of Tropical and Extraordinary Diseases, vol. 2. Armed Forces Institute of Pathology, Washington, D.C.

23. **Dooley, J. R., and R. C. Neafie.** 1976. Clonorchiasis and opisthorchiasis, p. 509–516. *In* C. H. Binford and D. H. Connor (ed.), *Pathology of Tropical and Extraordinary Diseases,* vol. 2. Armed Forces Institute of Pathology, Washington, D.C.
24. **Dooley, J. R., and R. C. Neafie.** 1976. Oesophagostomiasis, p. 440–445. *In* C. H. Binford and D. H. Connor (ed.), *Pathology of Tropical and Extraordinary Diseases,* vol. 2. Armed Forces Institute of Pathology, Washington, D.C.
25. **Eberhard, M. L., H. Hurwitz, A. M. Sun, and D. Coletta.** 1989. Intestinal perforation caused by larval *Eustrongylides* (Nematode:Dioctophymatoidae) in New Jersey. *Am. J. Trop. Med. Hyg.* **40:**648–650.
26. **Fox, A. S., K. R. Kazacos, N. S. Gould, P. T. Heydemann, C. Thomas, and K. M. Boyer.** 1985. Fatal eosinophilic meningoencephalitis and visceral larva migrans caused by the raccoon ascarid *Baylisascaris procyonis. N. Engl. J. Med.* **312:**1619–1623.
27. **Freeman, R. S., P. F. Stuart, J. B. Cullen, A. C. Ritchie, A. Mildon, B. J. Fernandes, and R. Bonin.** 1976. Fatal human infection with mesocercariae of the trematode *Alaria americana. Am. J. Trop. Med. Hyg.* **25:**803–807.
28. **Glickman, L. T.** 1984. Toxocariasis, p. 431–437. *In* K. S. Warren and A. A. F. Mahmoud (ed.), *Tropical and Geographical Medicine.* McGraw-Hill, New York.
29. **Glickman, L. T., P. M. Schantz, and R. H. Cypress.** 1979. Canine and human toxocariasis: review of transmission, pathogenesis and clinical disease. *J. Am. Vet. Med. Assoc.* **175:**1265–1269.
30. **Gordon, J. A., C. M. D. Ross, and H. Affleck.** 1969. Abdominal emergency due to an oesophagostome. *Ann. Trop. Med. Parasitol.* **63:**161–164.
31. **Harrell, L. W., and T. L. Deardorff.** 1990. Human nanophyetiasis: transmission by handling naturally infected coho salmon (*Oncorhynchus kisutch*). *J. Infect. Dis.* **161:**146–148.
32. **Hayunga, E. G.** 1989. Parasites and immunity: tactical considerations in the war against disease—or, how did the worms learn about Clausewitz? *Perspect. Biol. Med.* **32:**349–370.
33. **Hillyer, G. V.** 1981. Fascioliasis in Puerto Rico: a review. *Bol. Asoc. Med. P. R.* **73:**94–101.
34. **Hopkins, D. R., E. Ruiz-Tiben, T. Ruebush III, A. N. Agle, and P. C. Withers, Jr.** 1995. Dracunculiasis eradication: March 1994 update. *Am. J. Trop. Med. Hyg.* **52:**14–20.
35. **Hou, P. C.** 1965. Hepatic clonorchiasis and carcinoma of the bile duct in a dog. *J. Pathol. Bacteriol.* **89:**365.
36. **Huff, D. S., R. C. Neafie, M. J. Binder, G. A. DeLeon, L. W. Brown, and K. R. Kazacos.** 1984. Case 4: the first fatal *Baylisascaris* infection in humans: an infant with eosinophilic meningoencephalitis. *Pediatr. Pathol.* **2:**345–352.
37. **Jones, E. A., J. M. Key, H. P. Milligan, and D. Owens.** 1977. Massive infection with *Fasciola hepatica* in man. *JAMA* **63:**836–842.
38. **Katz, M., D. D. Despommier, and R. W. Gwadz.** 1982. *Parasitic Diseases.* Springer-Verlag, New York.
39. **Kikuchi, Y., T. Takenouchi, M. Kamiya, and H. Ozake.** 1981. Trypanorhynchiid cestode larva found on the human palatine tonsil. *Jpn. J. Parasitol.* **30:**3497–3499.
40. **Kisielewska, K.** 1970. Ecological organization of intestinal helminth groupings in *Clethrionomys glareolus* (Schreb.) (Rodentia). I. Structure and seasonal dyamics of helminth groupings in a host population in the Bialowieza National Park. *Acta Parasitol. Pol.* **18:**121–147.
41. **Leuckart, K. G.** 1882. Zur Entwickelungsgeschichte des Leberegels. *Zweite Mitt. Zool. Anz.* **5:**524–528.
42. **MacArthur, W. P.** 1933. Cysticercosis as seen in the British Army, with special reference to the production of epilepsy. *Trans. R. Soc. Trop. Med. Hyg.* 27:343–363. (Quotation at p. 357.)
43. **Meyers, W. M., R. C. Neafie, and D. H. Connor.** 1976. Pentastomiasis, p. 546–550. *In* C. H. Binford and D. H. Connor (ed.), *Pathology of Tropical and Extraordinary Diseases,* vol. 2. Armed Forces Institute of Pathology, Washington, D.C.
44. **Miller, G. C.** 1981. Helminths and the transmammary route of infection. *Trends Perspect. Parasitol.* **82:**335–342.
45. **Mitra, A. K., and D. R. W. Haddock.** 1970. Paraplegia due to guinea worm infection. *Trans. R. Soc. Trop. Med. Hyg.* **64:**102–106.
46. **Miyazaki, I., and S. Habe.** 1976. A newly recognized mode of human infection with the lung fluke, *Paragonimus westermani* (Kerbert 1978). *J. Parasitol.* **62:**646–648.
47. **Muller, R.** 1971. *Dracunculus* and dracunculiasis. *Adv. Parasitol.* **9:**73–151.
48. **New, D., M. D. Little, and J. Cross.** 1995. *Angiostrongylus cantonensis* infection from eating raw snails. *N. Engl. J. Med.* **332:**1105–1106.
49. **Norton, R. A., and L. Monroe.** 1961. Infection by *Fasciola hepatica* acquired in California. *Gastroenterology* **41:**46–48.
50. **Numbers 21:**4–9.
51. **Ockert, G., and J. Obst.** 1973. Ausstreuung umhüllter Onkosphären durch Bandwurmträger. *Monatsschr. Veterinaermed.* **28:**97–98.
52. **Ollerenshaw, C. B.** 1980. Forecasting liver fluke disease, p. 33–52. *In* A. E. R. Taylor and R. Muller (ed.), *12th Annual Symposium of the British Society for Parasitology.* Blackwell, London.
53. **Paul, A. J., K. S. Todd, Jr., and J. A. Dipietro.** 1988. Environmental contamination by eggs of *Toxocara* species. *Vet. Parasitol.* **26:**339–342.
54. **Rao, C. K., R. C. Paul, and M. I. D. Sharma.** 1981. Guinea worm disease in India—control status and strategy of its eradication. *J. Commun. Dis.* **13:**1–7.
55. **Reddy, C. R. R. M., and V. V. Valli.** 1967. Extradural guinea-worm abscess. *Am. J. Trop. Med. Hyg.* **16:**23–25.
56. **Rosen, L., G. Loison, J. Laigret, and G. D. Wallace.** 1967. Studies on eosinophilic meningitis. 3. Epidemiologic and clinical observations on Pacific islands and the possible etiologic role of *Angiostrongylus cantonensis. Am. J. Epidemiol.* **85:**17–44.
57. **Sadun, E. H., and C. Maiphoom.** 1953. Studies in the epidemiology of the human intestinal fluke, *Fasciolopsis buski* (Lankester) in Central Thailand. *Am. J. Trop. Med. Hyg.* **2:**1070–1084.
58. **Sakanari, J. A., H. M. Loinaz, T. L. Deardorff, R. B. Raybourne, H. H. McKerrow, and J. G. Frierson.** 1988. Intestinal anisakiasis: a case diagnosed by morphologic

and immunologic methods. *Am. J. Clin. Pathol.* **90**:107–113.

59. **Sakanari, J. A., and J. H. McKerrow.** 1989. Anisakiasis. *Clin. Microbiol. Rev.* **2**:278–284.
60. **Sakanari, J. A., M. Moser, and T. L. Deardorff.** 1995. *Fish Parasites and Human Health, Epidemiology of Human Helminthic Infections.* Report no. T-CSGCP-034. California Sea Grant College, University of California, La Jolla, Calif.
61. **Schantz, P. M.** 1989. The dangers of eating raw fish. *N. Engl. J. Med.* **320**:1143–1145.
62. **Silverman, P. H., and R. B. Griffiths.** 1955. A review of methods of sewage disposal in Great Britain with special reference to the epizootiology of *Cysticercus bovis*. *Ann. Trop. Med. Parasitol.* **49**:436–450.
63. **Stehr-Green, J. K., P. A. Stehr-Green, P. M. Schantz, J. F. Wilson, and A. Lanier.** 1988. Risk factors for infection with *Echinococcus multilocularis* in Alaska. *Am. J. Trop. Med. Hyg.* **38**:380–385.
64. **Stoll, N. R.** 1947. This wormy world. *J. Parasitol.* **33**:1–18.
65. **Uga, S., and N. Kataoka.** 1995. Measures to control *Toxocara* egg contamination in sandpits of public parks. *Am. J. Trop. Med. Hyg.* **52**:21–24.
66. **Vince, J. D., R. W. Ashford, M. J. Gratten, and J. Bana-Koiri.** 1979. *Strongyloides* species infestation in young infants of Papua New Guinea: association with generalized oedema. *Papua New Guinea Med. J.* **22**:120–127.
67. **Wittner, M., J. W. Turner, G. Jacquotte, L. R. Ash, M. P. Salgo, and H. B. Tanowitz.** 1989. Eustrongylidiasis—a parasitic infection acquired by eating sushi. *N. Engl. J. Med.* **320**:1124–1126.
68. **Yii, C-Y.** 1976. Clinical observations of eosinophilic meningitis and meningoencephalitis caused by *Angiostrongylus cantonensis* in Taiwan. *Am. J. Trop. Med. Hyg.* **25**:233–249.
69. **Yokogawa, M.** 1969. *Paragonimus* and paragonimiasis. *Adv. Parasitol.* **7**:375–387.

C. A. Speer

27

Protozoan Parasites Acquired from Food and Water

Protozoan parasites that contaminate food and water occur principally in the phyla Apicomplexa, Ciliophora, Sarcomastigophora, and Microspora (Table 27.1). The most medically important foodborne and waterborne apicomplexans include the coccidians *Cryptosporidium parvum, Cyclospora cayetanensis, Toxoplasma gondii,* and *Isospora belli.* The medically important food and waterborne sarcomastigophorans include the flagellate *Giardia intestinalis* and the sarcodine *Entamoeba histolytica.* The only ciliophoran of medical importance is *Balantidium coli.* Medically important microsporidians include *Enterocytozoon bieneusi* and *Septata intestinalis.*

Foodborne and waterborne members of the phyla Ciliophora and Sarcomastigophora have relatively simple life cycles in that they are homoxenous, i.e., complete their cycle in a single host, with endogenous stages consisting of trophozoites and cysts, which are environmentally resistant forms that serve to transmit infections among hosts. Microsporidia that contaminate food and water are also homoxenous and have slightly more complex life cycles consisting of trophozoites (uninucleate forms), plasmodia (also called schizonts or meronts), and spores. The spores are highly complex structurally and serve to transmit infections. All apicomplexan contaminates of food and water are in the order Eucoccidiorida, commonly referred to as coccidia. They have relatively complex life cycles consisting of alternating phases of asexual and sexual reproduction. Apicomplexans have an apical complex at the anterior end at some stage in their life cycle that consists of polar rings, a conoid (may be absent), rhoptries, micronemes, and subpellicular microtubules. The apical complex is used to penetrate cells of the host. Some coccidia are homoxenous; others are facultatively or obligatorily heteroxenous, with certain stages occurring in an intermediate host and other stages occurring in a definitive host.

The key to the transmission of most coccidian diseases is the oocyst, an environmentally resistant stage, which is shed in the feces by the definitive host and is infectious to other hosts via oocyst-contaminated food and water. Oocysts contain sporozoites which, after ingestion, excyst from oocysts, penetrate cells of the host, and undergo one or more asexual reproductive cycles via schizogony or endodyogeny (an internal budding mechanism in which the mother tachyzoite is completely consumed by the formation of two daughter zoites) to form merozoites or tachyzoites. In homoxenous coccidians, these zoites eventually form gametocytes, fertilization occurs, and oocysts are shed in the feces of the host. Generally, in heteroxenous coccidia, merozoites or tachyzoites eventually form tissue cysts (containing bradyzoites) in the intermediate host that are infectious to appropriate definitive hosts in which gametocytes form oocysts that are shed in the feces.

Table 27.1 Parasitic protozoa acquired from food and water

Phylum	Protozoan	Infective stage
Apicomplexa	*Cryptosporidium parvum*[a]	Oocyst
	Cyclospora cayetanensis[a]	Oocyst
	Isospora belli[a]	Oocyst
	Toxoplasma gondii[a]	Oocyst/tissue cyst
	Sarcocystis hominis[a]	Oocyst
	Sarcocystis suihominis[a]	Oocyst
Ciliophora	*Balantidium coli*[a]	Cyst
Microspora	*Enterocytozoon bieneusi*[a]	Spore
	Septata intestinalis[a]	Spore
Sarcomastigophora	*Acanthamoeba* spp.[b]	Trophozoite
	Chilomastix mesnili[c]	Cyst
	Dientamoeba fragilis[a]	Trophozoite
	Endolimax nana[c]	Cyst
	Entamoeba coli[c]	Cyst
	Entamoeba dispar[c]	Cyst
	Entamoeba histolytica[a]	Cyst
	Enteromonas hominis[c]	Cyst
	Giardia intestinalis[a]	Cyst
	Iodamoeba buetschlii[c]	Cyst
	Naegleria fowleri[b]	Trophozoite
	Retortamonas intestinalis[c]	Cyst
	Retortamonas sinensis[c]	Cyst
	Trichomonas hominis[c]	Cyst

[a]Pathogen, capable of inducing disease.

[b]Normally free-living protozoa that rarely infect the central nervous system of humans. Infections are acquired via water used in bathing or swimming, with trophozoites entering the central nervous system through the cribiform plate at the base of the cranium. Although a few individuals have survived, amoebic meningitis is nearly always fatal.

[c]Commensal, not normally capable of inducing disease.

CRYPTOSPORIDIOSIS

Although the first cases of human cryptosporidiosis were reported in 1976 (45, 48), there was little interest and few cases were reported until 1982, when severe diarrheal disease caused by *Cryptosporidium* sp. was reported in immunologically healthy persons and in human immunodeficiency virus-infected patients. The Centers for Disease Control and Prevention (Atlanta, Ga.) reported that 14 of 21 patients with AIDS and infected with *Cryptosporidium* sp. had died. By 1986, the Centers for Disease Control and Prevention reported that 697 of the first 19,187 AIDS patients were infected with *Cryptosporidium* sp. and had higher case mortality rates than AIDS patients without crytosporidiosis. In immunocompetent patients, *Cryptosporidium* infection usually causes a short-term diarrheal illness that resolves spontaneously, whereas in immunocompromised patients, it may cause a life-threatening, prolonged, cholera-like illness. To date, there is no effective therapy for cryptosporidiosis and the prognosis is poor for immunocompromised patients, especially those with AIDS.

One of the major reasons that *Cryptosporidium* infection went unnoticed for so long was the lack of reliable techniques for microscopic diagnosis. In the early 1980s, diagnosis became much easier with the development of various techniques to concentrate and to chemically stain *Cryptosporidium* oocysts in stool samples (reviewed in reference 13). The public is becoming increasingly aware of the medical importance of cryptosporidiosis as a result of municipal waterborne outbreaks of cryptosporidiosis in Texas (1984), Georgia (1987), Oregon (1992), and most recently Wisconsin (1993), in which more than 400,000 people became infected and several individuals died (9, 11, 41). In the 1987 outbreak in Georgia, 36% of a population of nearly 36,000 people were infected with *C. parvum.* A survey of 107 surface water samples collected in six western states revealed that 72% were positive for *C. parvum* oocysts (28). More recent surveys have revealed that *Cryptosporidium* oocysts are present in 65 to 97% of rivers, lakes, and streams in the United States (38, 58).

Twenty-one species of *Cryptosporidium* have been named, primarily on the basis of the host that the para-

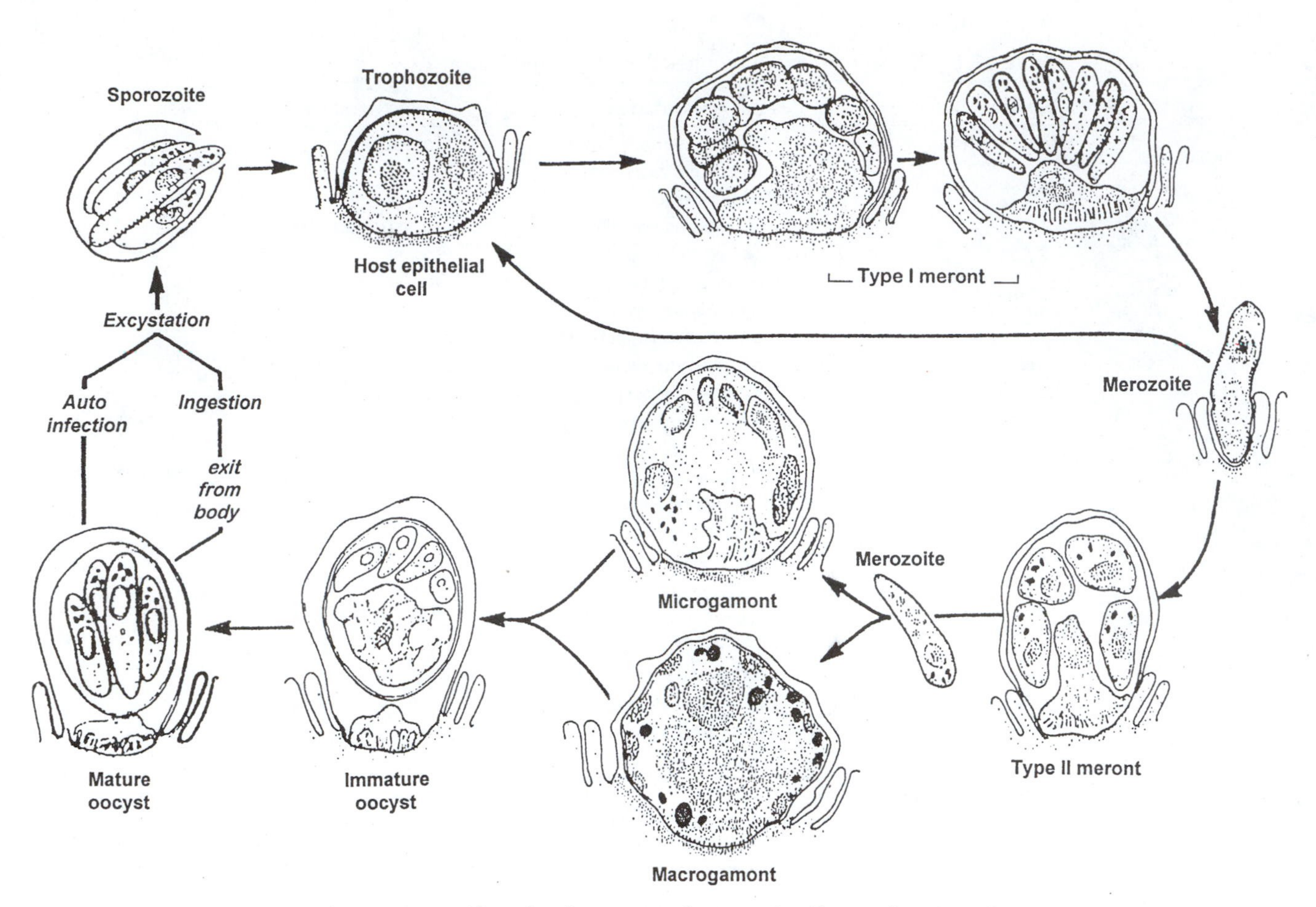

Figure 27.1 Life cycle of *Cryptosporidium* species (from reference 24).

site infects, but cross-transmission studies have placed the validity of many of these classifications in doubt. Generally, two species, *C. parvum* and *C. muris,* occur in mammals, with the former species being infectious to humans (24).

Cryptosporidium is transmitted via oocysts (Fig. 27.1 to 27.3), which are resistant to environmental stresses, even to chlorine routinely used in municipal water supplies. Each oocyst contains four sporozoites, which, after ingestion or inhalation, excyst through a gap in the oocyst wall created by dissolution of a suture located at one pole of the oocyst (Fig. 27.2) (55). Sporozoites parasitize epithelial cells of the gastrointestinal or respiratory tract and appear by light microscopy to be attached superficially to their host (Fig. 27.3B), but they are actually situated inside a parasitophorous vacuole at the apex of the epithelial cell (Fig. 27.4A, B, and G). A feeder organelle is formed at the interface between all stages of the parasite and the host cell and probably serves to increase the surface area between the parasite and the host cell, facilitating exchange of nutrients and wastes (Fig. 27.4D and G). The slender sporozoite transforms into a spheroidal trophozoite and then a schizont in which two or three nuclear divisions occur, and merozoites bud at the schizont surface. Merozoites escape from the host cell (Fig. 27.4C), penetrate other epithelial cells, and undergo several cycles of schizogony. Eventually, some merozoites enter the sexual phase of the life cycle (formation of male and female gametocytes). Microgametocytes undergo three nuclear divisions and give rise to eight microgametes (Fig. 27.4D and E), which are small (1.4 by 0.5 μm) and bullet shaped and fertilize large single-nucleate macrogametocytes (Fig. 27.4F) to form zygotes that develop into oocysts with thin or thick oocyst walls (Fig. 27.4G). Thin-walled oocysts give rise to autoinfection, with sporozoites undergoing excystation and penetrating other host cells without leaving the host. Thick-walled oocysts are shed in the feces, sputum, or respiratory droplets and transmit the infection to other hosts.

Infection usually occurs during contact with infected individuals or animals (especially calves) and via fecal

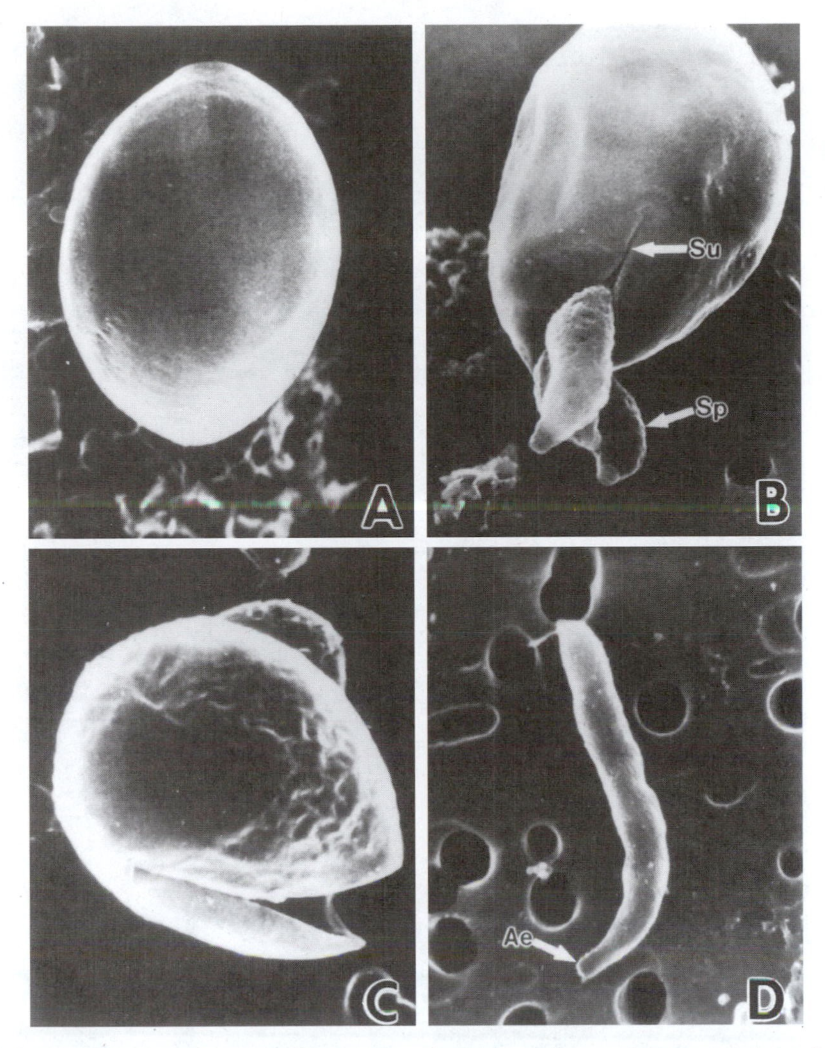

Figure 27.2 Scanning electron micrographs of oocysts and excysting sporozoites of *C. parvum* (from reference 55). (A) Intact oocyst prior to excystation (×11,200). (B) Three sporozoites (Sp) excysting from oocyst simultaneously via the cleaved suture (Su) (×11,200). (C) Empty oocyst (×11,200). (D) Excysted sporozoite (×9,800). Ae, apical end.

contamination of water used for drinking or swimming. Respiratory infections probably occur via airborne transmission of the parasite. Other sources of infection include contaminated food, fomites, and arthropods such as flies.

The overall effects of cryptosporidiosis are malabsorption and impaired respiration, but just how cryptosporidia cause disease is not known. Young children are usually more susceptible to infection and have more severe clinical signs, but previously unexposed adults are routinely susceptible to *Cryptosporidium* infection. The most prominent signs are voluminous watery diarrhea (as much as 12 to 17 liters passed per day, especially by immunocompromised patients), anorexia, weight loss, dehydration, and abdominal discomfort. Microscopically, infected epithelial cells change from columnar to cuboidal or squamous and occasionally slough from the epithelium (Fig. 27.3C).

The key to control and prevention of cryptosporidiosis is elimination or reduction of the oocysts in the environment or avoidance of contact with known sources of oocysts. Since cryptosporidia are highly resistant to chemical disinfectants used to treat municipal water supplies, physical removal of the parasites from water by filtration appears to be the only viable treatment process. However, many water treatment facilities in United States cities do not use filtration, and filtration is no guarantee for complete removal of oocysts. Nevertheless, the risk of transmission of *Cryptosporidium* infection can be reduced by water filtration if it is done

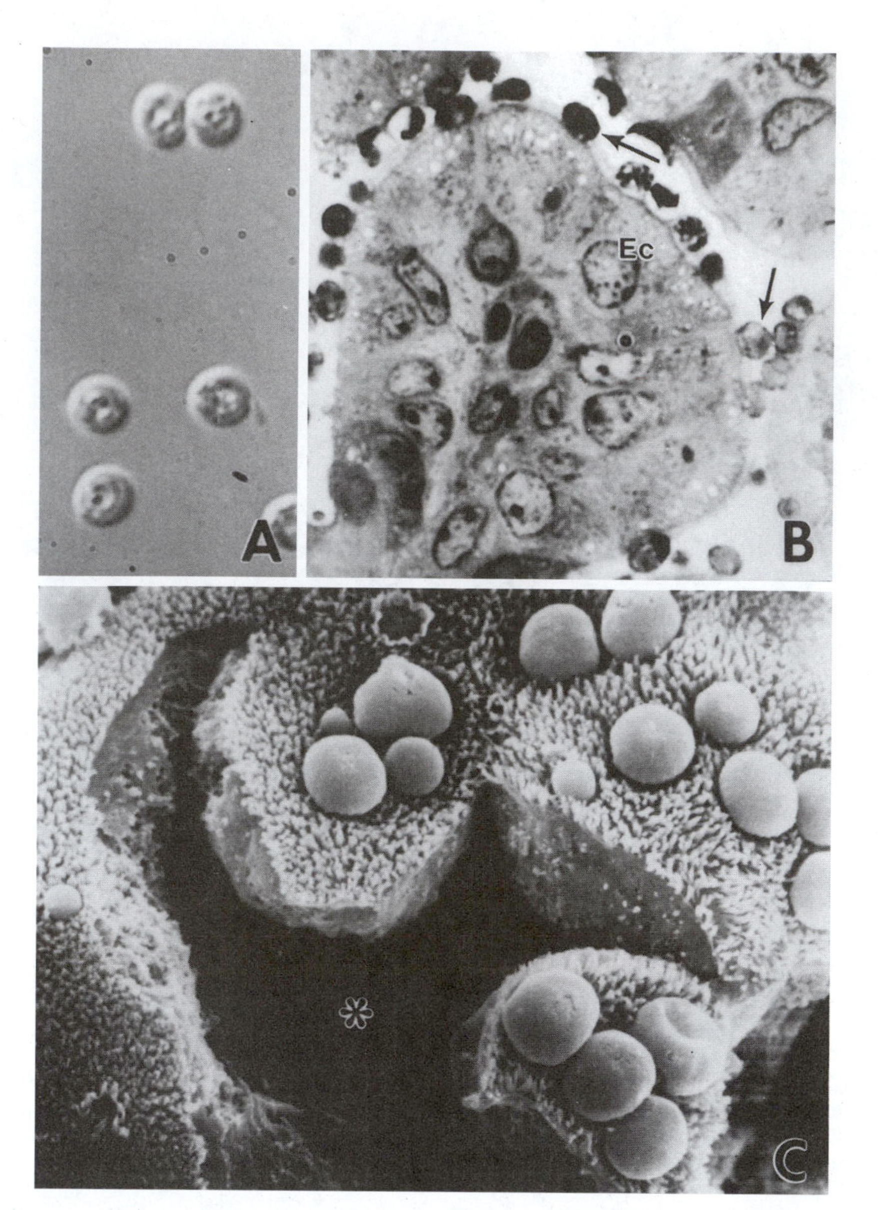

Figure 27.3 Photomicrographs (A and B) and scanning electron micrograph (C) of *C. parvum*. (A) Interference contrast microscopy of live oocysts (×1,135). (B) Numerous parasites (arrows) at the luminal surface of the epithelium of mouse small intestine (×970). Ec, epithelial cell. (C) Numerous cryptosporidia on the surface of sheep intestinal epithelium (×3,160). Note that some of the epithelial cells have sloughed (*). Panel C is from reference 24.

properly. Oocysts are susceptible to freezing, desiccation, high temperatures, and the chemical agents hydrogen peroxide, ozone, ammonia, and chlorine dioxide. Although more than 100 therapeutic and preventive modalities have been tested for efficacy against cryptosporidiosis, only a few, including amprolium, arpinocid, dinitolmide, salinomycin, sulfaquinoxalinen, and dehydroepiandrosterone (24), have been found to have any appreciable effect.

CYCLOSPORIDIOSIS

A newly discovered protozoan parasite with a *Cryptosporidium*-like oocyst (63) is becoming more frequently recognized as a cause of diarrheal illness in humans. This parasite is classified as a coccidian in the phylum Apicomplexa and named *Cyclospora cayetanensis* (9). Its oocysts are spherical, are 8 to 10 μm in diameter, stain variably with acid-fast stains, and autofluoresce under UV light (40, 49). Most outbreaks have occurred

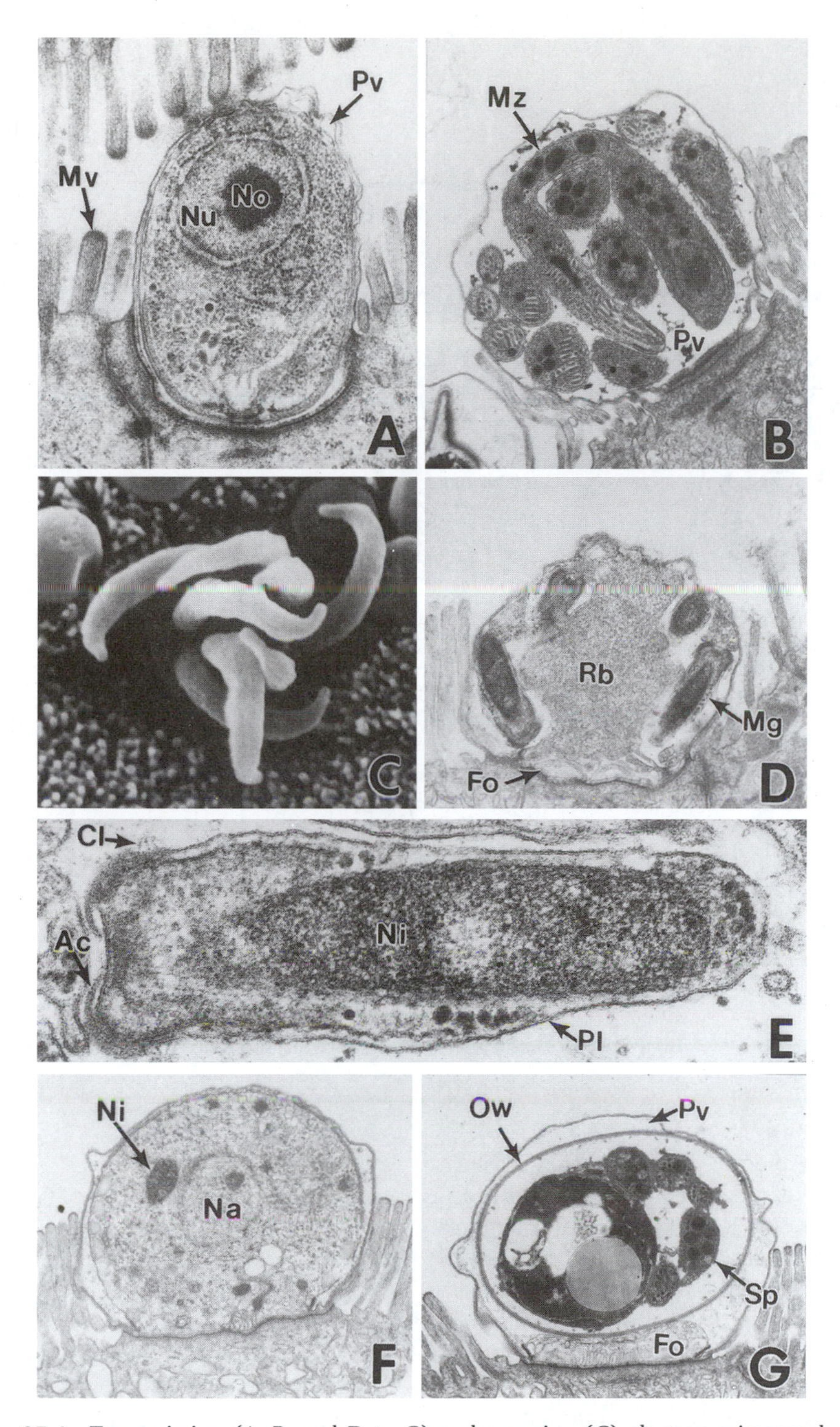

Figure 27.4 Transmission (A, B, and D to G) and scanning (C) electron micrographs of *C. parvum* in small intestinal epithelium. (A) Merozoite in early stage of attachment to an intestinal epithelial cell (×16,320). Mv, microvilli of intestinal epithelial cell; No, nucleolus of parasite; Nu, nucleus of parasite; Pv, parasitophorous vacuole. (B) Mature schizont with merozoites (Mz) (×8,910). Pv, parasitophorous vacuole. (C) Eight merozoites escaping from sheep intestinal epithelial cell (×7,370). (D) Mature microgametocyte showing portions of four microgametes (Mg) (×10,530). Fo, feeder organelle; Rb, residual body. (E) Longitudinal section of microgamete (×53,460). Pl, plasmalemma of microgamete; Ni, nucleus; Ac, apical cap; Cl, concentric lamellae. (F) Macrogametocyte containing nucleus (Ni) of microgamete (×9,315). Na, nucleus of macrogametocyte. (G) Thick-wall oocyst (×7,060). Ow, thick oocyst wall; Pv, parasitophorous vacuole; Sp, sporozoite; Fo, feeder organelle. Panels A, C, and G are from reference 24.

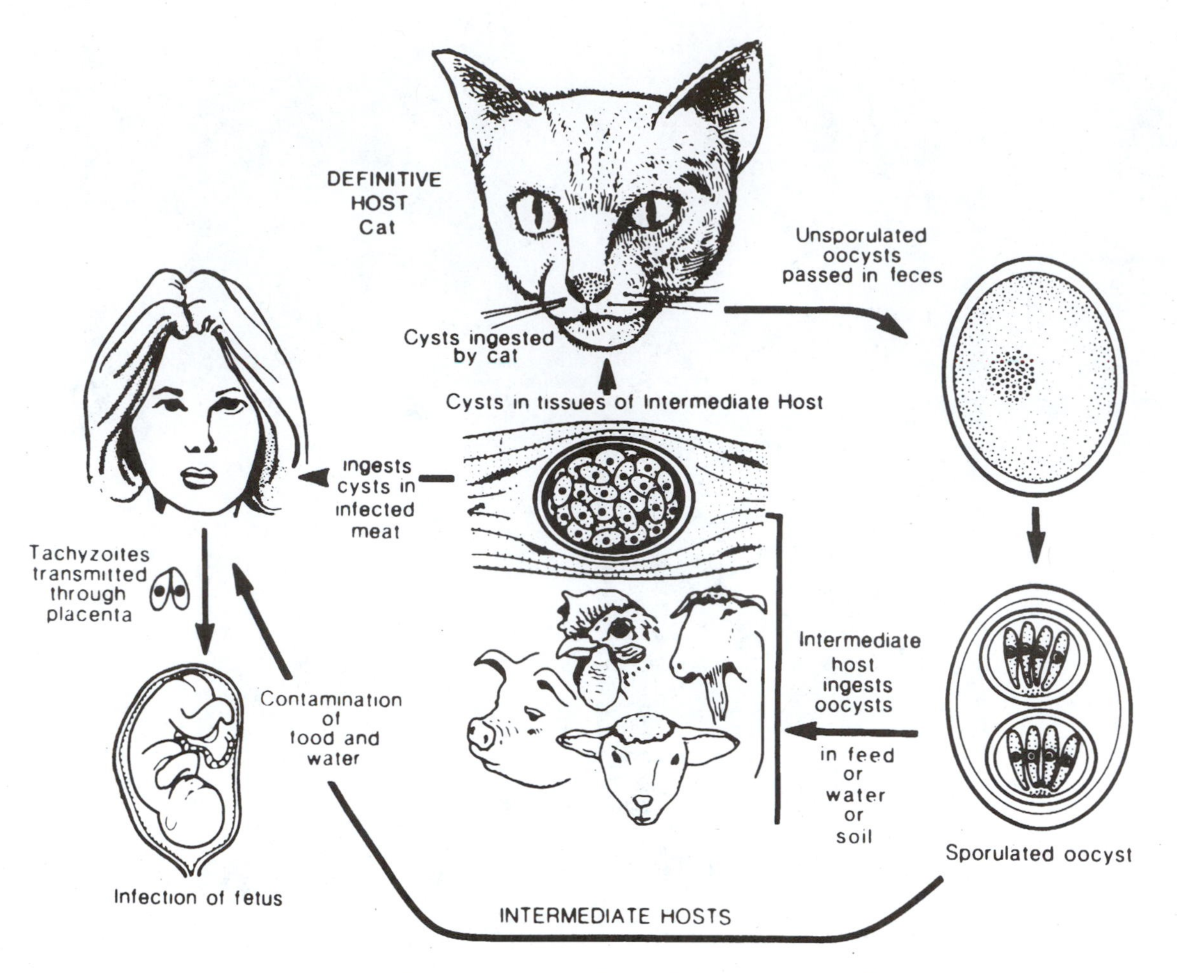

Figure 27.5 Life cycle of *T. gondii* (from reference 19).

in Central and South America, Southeast Asia, Eastern Europe, and the Caribbean islands. Most infections appear to be acquired via oocyst-contaminated water supplies. Cyclosporidiosis causes a self-limiting diarrhea in immunocompetent patients, most frequently in children. Patients may present an explosive and watery diarrhea accompanied by fatigue, nausea, vomiting, abdominal cramping, anorexia, myalgia, and weight loss. Symptoms may last a few days or as long as 6 weeks or more. Microscopically, the parasites are intracytoplasmic, residing within a parasitophorous vacuole near the luminal end of the cell (3). Except for a single report of successful treatment with co-trimoxazole (42), no antimicrobial therapy has been found.

Currently, there is no information concerning the effects of heat, freezing, or various chemical agents on *Cyclospora cayetanensis* oocysts, but susceptibility or resistance to such treatments might be similar to those for other coccidian parasites.

TOXOPLASMOSIS

T. gondii is a ubiquitous parasite infecting humans and many warm-blooded animals (16). Although *T. gondii* was first discovered as a parasite of rodents in Africa nearly 100 years ago (47), the complete life cycle including the oocysts and the definitive host was established in 1970 (22, 23, 31, 34, 61). *T. gondii* is an obligate intracellular parasite that can be transmitted transplacentally or by ingestion of infected meat or water (Fig. 27.5). *T. gondii* is a facultative two-host parasite that uses domestic cats and wild members of the Felidae as definitive hosts and warm-blooded animals as intermediate hosts. Cats excrete in their feces oocysts of *T. gondii,* which undergo sporulation in 1 to 5 days to become infective to felines or intermediate hosts. The oocysts are resistant to most disinfectants and can survive for several years in moist, shaded conditions (19). Infections are acquired principally by ingesting food or water contaminated with oocysts or by ingesting animal tissues containing cysts.

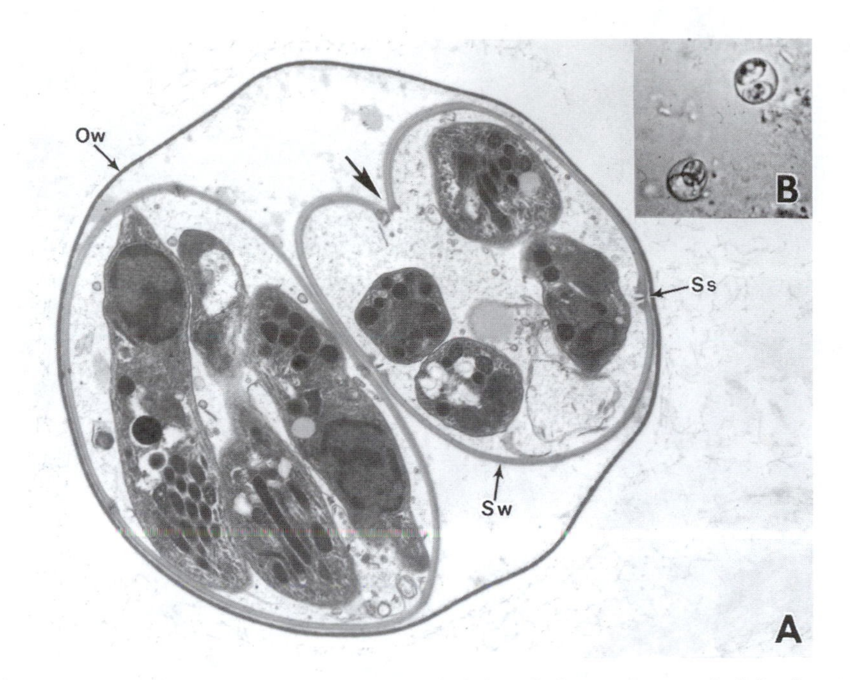

Figure 27.6 Transmission electron micrograph (A) and photomicrograph (B) of sporulated oocysts of *T. gondii.* (A) Oocyst with two sporocysts exposed to excysting fluid consisting of bile salts and trypsin (×10,980). Note that one of the sporocysts has partially collapsed by infolding (arrow) at the sites of the sutures (Ss) in the sporocyst wall (Sw). Ow, oocyst wall. (Courtesy of D. Lindsay, Auburn University.) (B) Oocysts (×510).

Sporulated oocysts consist of an oocyst wall and two sporocysts, each of which contains four sporozoites (Fig. 27.6). When oocysts are ingested, the oocyst wall ruptures, and trypsin and bile salts in the intestinal tract cause the sporocyst wall to release the sporozoites (Fig. 27.6A). Sporozoites penetrate cells (Fig. 27.7A) in the intestine and associated lymph nodes and transform into rapidly multiplying tachyzoites (Fig. 27.7B) that are dispersed to other areas of the body via the blood and lymph, where they continue to multiply. Tachyzoites multiply by endodyogeny (Fig. 27.7B). Eventually, tachyzoites encyst in the brain, liver, and skeletal and cardiac muscles to form microscopic cysts containing bradyzoites (cystozoites), which are slowly multiplying forms. Cysts persist for the duration of the life of the host. In general, tachyzoites occur during the acute phase of toxoplasmosis, whereas bradyzoites occur during the chronic phase.

When tissue cysts are ingested, proteolytic enzymes dissolve the cyst wall, releasing the bradyzoites, which then infect intestinal epithelial cells of the host and transform into tachyzoites. Tachyzoites multiply rapidly and become dispersed via the blood and lymph to other tissues of the body, where they encyst. When tissue cysts are ingested by felines, bradyzoites begin the enteroepithelial phase of the life cycle, in which asexual reproduction by schizogony and sexual reproduction by gametogony occur. After fertilization of the female gamete by a male gamete, the zygote forms the oocyst, which is passed in the feces.

Although most infections are acquired by ingestion of tissues containing bradyzoites or food and water contaminated with oocysts, *T. gondii* can be transmitted during organ transplantation or transfusion of blood or its components. Fatal disseminated toxoplasmosis may occur in individuals who acquire infections during organ transplantation and are given routine immunosuppressive therapy. In 1995, an outbreak of toxoplasmosis involving more than 100 adults who exhibited various symptoms, especially severe chorioretinitis, occurred in Vancouver, British Columbia. Although the source has not been identified, it appears that most of the infections may have been acquired from oocysts contaminating the municipal water supply (18).

Transplacental transmission can occur when previously noninfected women become infected during pregnancy. After multiplying in the placenta, tachyzoites spread to all fetal tissues and continue to multiply. Infections can occur during any stage of pregnancy, but the fetus is most severely affected when infection occurs during the first 4 to 5 months of pregnancy. Approximately 30% of the women of childbearing age in the

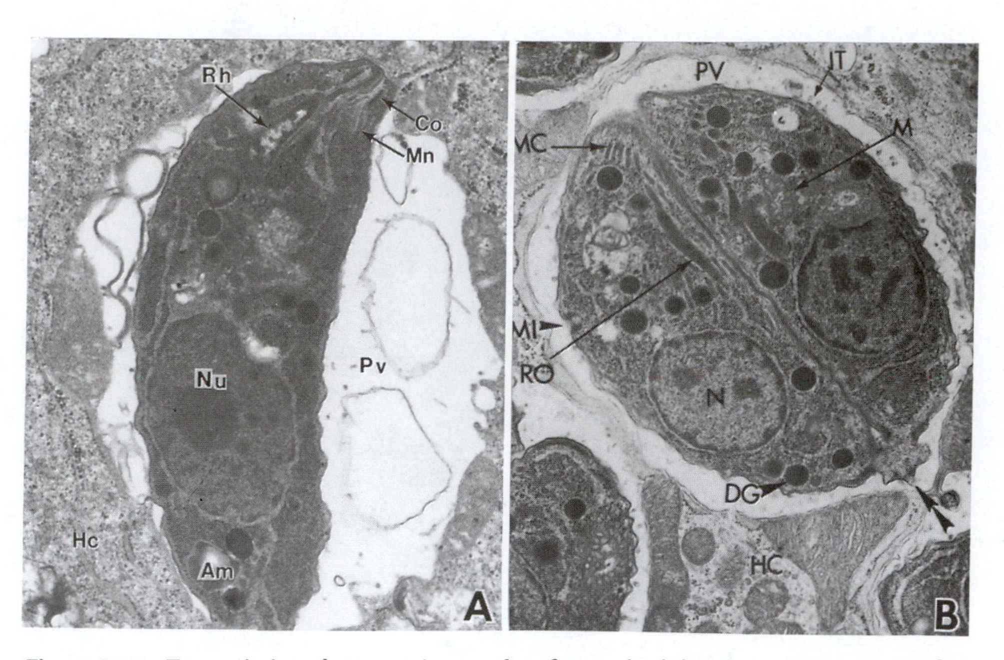

Figure 27.7 Transmission electron micrographs of *T. gondii*. (A) Sporozoite in parasitophorous vacuole (Pv) of host cell (Hc) at 24 h after inoculation (×13,930). Am, amylopectin granule; Co, conoid; Mn, microneme; Nu, nucleus of sporozoite; Rh, rhoptry. (B) Final stage of endodyogeny to form two daughter tachyzoites that are still attached at their posterior ends (arrowheads) (×12,815). (From reference 19, with permission.) DG, dense granule; HC, host cell cytoplasm; IT, intravacuolar tubules; M, mitochondrion; MC, microneme; MI, micropore; N, nucleus of tachyzoite; RO, rhoptry.

United States have antibodies against *T. gondii* and are immune to toxoplasmosis (19), whereas the remaining 70% are at risk of acquiring an infection during pregnancy. If a noninfected woman acquires a *T. gondii* infection during pregnancy, then the fetus has a 20 to 50% chance of becoming infected (56). The number of *T. gondii*-infected children born each year in the United States is estimated to be 1 per 1,000 to 10,000 births (56, 57). Most infected children do not exhibit clinical signs at birth but are likely to have manifestations of the disease such as chorioretinitis and mental retardation later in life (33).

The prevalence of *T. gondii* infection varies among countries, depending on the culinary customs and eating habits. Infection can be acquired by ingestion of tissue cysts in meat or by ingestion of oocysts. Approximately 30 to 50% of adults in the United States have antibodies to *T. gondii*, with most infections acquired by ingesting cysts in undercooked or uncooked meat (35). Fresh pork appears to be the main source of *T. gondii* infection (21, 70). Surveys of *T. gondii*-infected swine in Iowa and Illinois showed that seroprevalence rates among sows housed in total confinement were substantially less than in sows with outdoor access (62). Confined housing can decrease or eliminate exposure of swine to cat feces or rodents that may be infected with *T. gondii*.

Infections may also be acquired by ingesting meat from lamb, poultry, horses, and wild game animals, e.g., deer, elk, pronghorn, and moose. Although the role of beef in the transmission of *T. gondii* has yet to be determined, it appears that it is not an important source of infection because cattle have a high resistance to the parasite and *T. gondii* has not been isolated from the edible tissues of cattle (17, 25). Thorough cooking and freezing will kill *Toxoplasma* oocysts and cysts in meat. Temperatures of 61°C or higher for 3.6 min will inactivate cysts in meat (20), and freezing to −13°C will usually render tissue cysts nonviable (36).

In immunosuppressed patients, toxoplasmosis may result from reactivation from latent infection, i.e., bradyzoites escaping from tissue cysts and giving rise to tachyzoites, from ingesting oocysts in water, or from ingesting tissue cysts in raw or undercooked meat. In a recent retrospective neuropathological survey, *T. gondii* was found in 34% of the brains from 200 patients who died of AIDS in Berlin, Germany, making it the most prevalent pathogen or abnormality observed in the survey (43).

The most effective treatment for toxoplasmosis has been found to be a combination of pyrimethamine and sulfonamides that acts synergistically against tachyzoites (reviewed in reference 19). Spirimycin and clindamycin are also effective, but all these drugs have serious side effects.

ISOSPORIASIS

Human isosporiasis is an important cause of protracted diarrhea, especially in immunocompromised hosts, including patients with AIDS. The disease is caused by *I. belli,* which is similar to cryptosporidia and other coccidia in that it has alternating asexual and sexual cycles in the enterocytes of the small intestine. Oocysts are shed unsporulated (Fig. 27.8) in the feces and after 1 to 2 days become sporulated and infectious. Infection occurs by ingestion of sporulated oocysts containing two sporocysts, each with four sporozoites. Sporozoites excyst from oocysts in the small intestine, infect enterocytes, and undergo schizogony followed by gametogony and oocyst formation.

Isosporiasis occurs worldwide but is more prevalent in the tropics and subtropics (60) and is occasionally incriminated as a cause of traveler's diarrhea (30). Isosporiasis is an opportunistic infection in only 0.2% of human immunodeficiency virus-infected individuals in the United States. Much higher prevalence rates occur in the tropics, where surveys in Africa and Haiti have revealed that more than 15% of AIDS patients are infected with *I. belli* (14, 64).

Humans are recognized as the only host for *I. belli* infection. The disease is transmitted by ingestion of food or water contaminated with mature sporulated oocysts. It is possible for infections to be transmitted directly from person to person, but this means of transmission appears to be uncommon. In immunocompetent individuals, isosporiasis is characterized by watery diarrhea, abdominal pain, and low-grade fever which may last for less than 1 month. In immunodeficient persons, particularly patients with AIDS, isosporiasis is associated with protracted diarrhea, nausea, abdominal pain, malabsorption, and weight loss and may persist for years if left untreated.

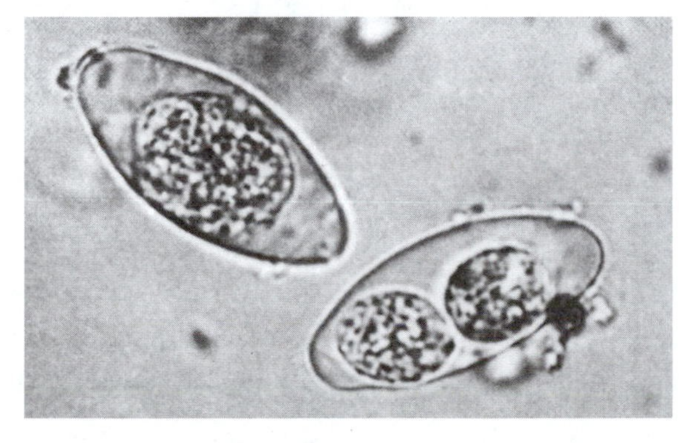

Figure 27.8 Bright-field photomicrograph of two partially sporulated oocysts of *I. belli* (×1,200). Note that one of the sporocysts has two sporoblasts.

Many therapeutic drugs and antibiotics have been found to be useless for treatment of isosporiasis. In several cases, furadantin (30), fansidar (pyrimethamine-sulfadoxine) (64), and trimethoprim-sulfamethoxazole (32) have been found to induce disappearance of organisms in the stool and to reduce diarrhea and malabsorption. Favorable treatment has been accomplished with trimethoprim-sulfamethoxazole, considered an investigational drug for this condition by the U.S. Food and Drug Administration (51), but symptomatic disease recurs in approximately 50% of patients (14).

GIARDIASIS

Worldwide, giardiasis is one of the most common protozoal infections of the human intestine, with prevalence rates of 2 to 5% in industrialized nations and up to 20 and 30% in developing nations. Giardiasis is caused by *Giardia* species, which are flagellated protozoa that inhabit the intestinal tracts of humans and various animals. More than 40 species of *Giardia* have been described, based primarily on the basis of hosts that they infect. It is still not known how many of these species are infectious to humans, but it is likely that several animal hosts can serve as reservoir hosts for human infections. Infections in humans are usually caused by *G. intestinalis* (synonym *G. lamblia*). On the basis of rRNA analysis, it has been suggested that the genus *Giardia* represents the missing link between prokaryotes and eukaryotes, because the gene sequence encoding the small subunit rRNA of *Giardia* appears to be more closely related to those of the archaebacteria than to those of other eukaryotes (65). *Giardia* species also differ from eukaryotes by being aerotolerant anaerobes that lack mitochondria and mitochondrial enzymes, and they respire in the presence of oxygen by a flavin, iron-sulfur protein-mediated electron transport system (39).

The life cycle of *Giardia* species consists of a binucleate, flagellated, motile trophozoite and a thick-walled cyst stage (Fig. 27.9). Cysts are environmentally resistant and serve to transmit infections principally via the fecal-oral route. Transmission occurs via soiled hands, contaminated drinking water, and food contaminated with feces. Because *Giardia* species are resistant to chlorina-

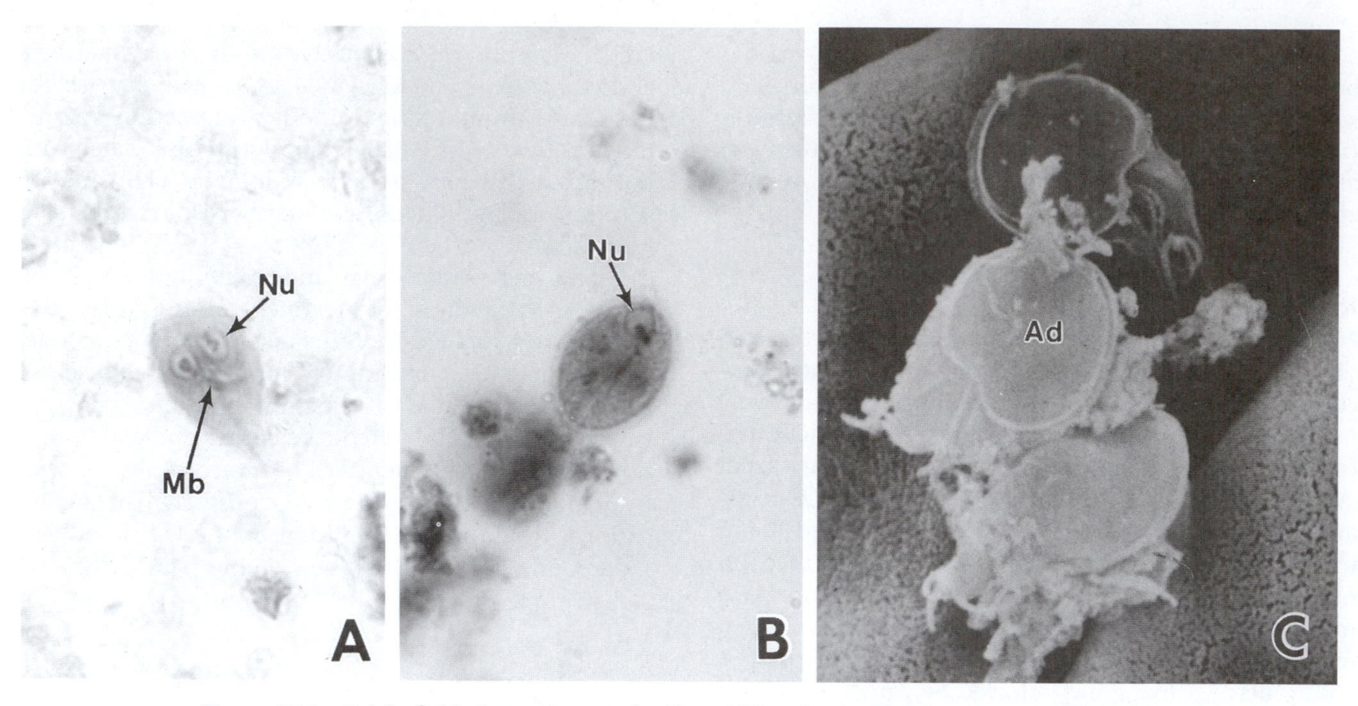

Figure 27.9 Bright-field photomicrographs (A and B) and scanning electron micrograph (C) of *G. intestinalis.* (A) Trophozoite showing two nuclei (Nu) and median body (Mb) (×1,200). (B) Cyst with two nuclei (Nu) visible at one pole (×1,200). (C) Three trophozoites on intestinal epithelium (×3,000). Ad, adhesive disc.

tion, sand filtration is required to eliminate the majority of cysts from municipal water supplies.

The trophozoite is approximately 12 to 15 µm in length and 5 to 9 µm in width, and it contains two nuclei, four pairs of symmetrically arranged flagella, a prominent dark-staining median body in the posterior portion of the trophozoite, and a ventrally located adhesive disc (Fig. 27.9). The adhesive disc is presumed to be involved in attachment of the trophozoite to the intestinal epithelium (29). Cysts are ovoid or elliptical in shape, measuring 7 to 10 µm in length, and contain four nuclei, a median body, and various cytoskeletal components.

Infection is initiated by ingestion of *Giardia* cysts. Trophozoites, which excyst in the small intestine and multiply exclusively by binary fission, are luminal parasites that adhere closely to the intestinal epithelium (Fig. 27.9C). As trophozoites pass along the intestinal tract, they undergo encystment in the distal small intestine to form cysts that are shed in the feces (Fig. 27.9B). Humans may shed as many as 20,000 cysts per g of feces per day (53).

High-risk groups include infants and young children and immunodeficient persons. Infection during the first 6 months is rare when breast feeding is common (27). Malnutrition in children probably represents an additional risk factor that may contribute to the development of chronic disease. Numerous outbreaks of giardiasis have occurred in day care centers, residential institutions, and schools, where prevalence may be as high as 35% (59, 66). Although there have been several outbreaks attributed to foodborne *Giardia* species (50), the relative importance of foodborne giardiasis has not yet been well characterized. There is also little information on the survival of *Giardia* cysts in food. In most cases of foodborne outbreaks, infected food handlers appeared to transmit cysts to freshly prepared food (50).

Giardiasis has been long recognized as a cause of traveler's diarrhea and was for many years commonly referred to as Leningrad's curse, at a time when approximately 30% of the travelers to the Soviet Union acquired *Giardia* infections (5). Today, the risk of visitors acquiring giardiasis is relatively low, representing less than 5% of the cases of traveler's diarrhea (26). Chronic giardiasis occurs more frequently in male homosexuals than in the general population (37).

An enzyme-linked immunosorbent assay is available to detect *Giardia* antigen in feces (Alexon, Santa Clara, Calif.), and two fluorescent antibody tests are available for parasite detection in feces (Meridian Diagnostics, Cincinnati, Ohio) and in water samples (Biovir Laboratories, Benicia, Calif.). A cDNA probe has also been used to detect *Giardia* species in water samples (46).

Metronidazole, tinidazole, mepacrine, and furazolidone are widely used in the treatment of giardiasis, but none is ideal because of adverse side effects, and none is safe for use during pregnancy (12).

AMOEBIASIS

Although several amoebae are infectious to humans, only *E. histolytica* is considered to be a consistent pathogen. Amoebiasis due to *E. histolytica* is the third leading parasitic cause of death worldwide (68). Genetic differences exist between pathogenic and nonpathogenic strains of *E. histolytica* (10), leading to a recent suggestion that the name *Entamoeba dispar* be used for the nonpathogenic strain (15).

The life cycle of *E. histolytica* is relatively straightforward, consisting of trophozoites and cysts (Fig. 27.10). The cyst is the infective form which can survive several weeks in an external environment because of its thick chitinous wall. After ingestion, stomach acid stimulates the parasite to excyst in the small intestine, where it multiplies by binary fission before passing on to colonize the large intestine, feeding on bacteria, multiplying, or encysting, depending on local conditions. Trophozoites are not infective because they disintegrate rapidly outside the host and are susceptible to the low pH of the stomach. Transmission of infections usually occurs via ingestion of cysts in contaminated food and water or direct passage from person to person. Cysts can also be transmitted by contaminated hands of infected food handlers with poor personal hygiene. There are no known reservoir hosts for *E. histolytica*.

Identification of *E. histolytica* is based on microscopic examination of fecal samples, water samples, or sediment obtained either from backflushing filters through which water has passed or from the pellet of particulate matter from centrifuged water. Mature cysts are spheroidal, 10 to 20 μm in diameter, and contain four nuclei and rod-shaped bodies with rounded ends (Fig. 27.10B). Cysts can be differentiated from those of *Entamoeba coli*, a nonpathogen with mature cysts usually having eight nuclei and rod-shaped bodies with pointed ends. However, cysts and trophozoites of the nonpathogen *E. dispar* cannot be distinguished microscopically from those of pathogenic *E. histolytica*.

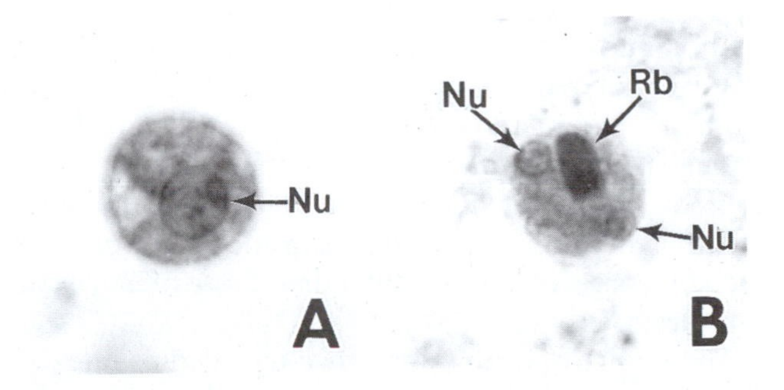

Figure 27.10 Bright-field photomicrographs of *E. histolytica*. (A) Trophozoite (×1,000). Nu, nucleus. (B) Cyst showing two of four nuclei (Nu) and rod-shaped body (Rb) with rounded ends (×1,000).

There are two general forms of amoebiasis, referred to as intestinal amoebiasis and amoebic liver abscess, the latter being a sequela of the intestinal form. Clinically, patients with intestinal amoebiasis can be separated into four groups: asymptomatic cyst passers, and those with acute colitis, fulminant colitis, and amoeboma. Ninety percent of patients infected with *E. histolytica* are asymptomatic cyst passers of pathogenic and nonpathogenic strains. In developed countries, infections are most frequently seen in homosexual men, in whom as high as 30% of routine stool examinations have been found to be cyst positive (54). A variable percentage of asymptomatic individuals will be carriers of pathogenic *E. histolytica*, and it is important to identify and treat these individuals, as they are potential sources of disease transmission. Patients with acute amoebic colitis experience a gradual onset of abdominal pain and frequent, loose, watery stools containing blood and mucus; back pain; tenesmus; occasional flatulence; dehydration; and abdominal tenderness often localized to the lower abdomen. Fulminating colitis is seen more frequently in children than in adults. This form is an unusual complication of amoebic dysentery and carries a grave prognosis, with survival rates of approximately 40% (2). These patients have severe bloody diarrhea, fever, and abdominal tenderness which are associated with the pathological progression from superficial ulceration to transmural necrosis of the bowel. Amoeboma occurs in less than 1% of patients with invasive intestinal disease (1). Most of these patients exhibit an abdominal mass which is sometimes tender but otherwise asymptomatic. Radiographically, the amoeboma resembles a carcinoma with an apple core-like lesion.

Amoebic liver abscess is the most common complication associated with invasive amoebiasis and is difficult to diagnose because of the nonspecific nature of the symptoms and the potential for presentation months after the patient leaves an area where the disease is (6). An enlarged, painful liver is most frequently seen but is not diagnostic. Generally, patients are most likely to present a wasting disease with hepatomegaly, weight loss, and anemia; the death rate due to amoebic abscess is approximately 1%, but should the abscess rupture, the death rate increases to 10%.

As trophozoites multiply within the epithelium of the large intestine, an ulceration develops that progresses into the lamina propria to the muscularis mucosa and then laterally under normal-appearing mucosa, forming a flask-shaped ulcer (4). In amoebic liver abscess, the liver parenchyma is gradually replaced with necrotic debris, inflammatory cells, and amoebic trophozoites.

Basically, there are two classes of drugs to treat amoebiasis: (i) luminal agents that are poorly absorbed by the intestinal tract and (ii) absorptive agents with good tissue penetration. Luminal agents include iodoquinol, diloxamide furoate, and paromomycin, which have efficacy rates of 85 to 95% for eradication of *E. histolytica* cyst passage (44). The most effective absorptive agent is metronidazole, which is the drug most commonly used for treatment of intestinal and extraintestinal amoebiasis, but it is not effective against cysts. The most difficult aspect of controlling amoebiasis is the lack of an effective means of treating asymptomatic cyst passers.

BALANTIDIOSIS

B. coli is the only ciliated protozoan of medical importance, and although it is distributed worldwide, it is not a prevalent parasite of humans. However, *B. coli* is a common commensal of pigs. Its life cycle consists of a trophozoite and cyst (Fig. 27.11). Trophozoites reside in the large intestine, where they multiply by binary fission and cause ulcerative colitis and diarrhea. The cyst is considered to be the infective form, which is capable of surviving in moist environments for several weeks.

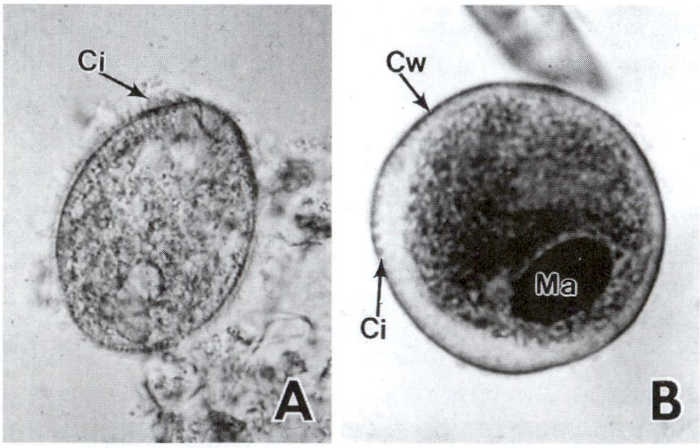

Figure 27.11 Bright-field photomicrographs of *B. coli.* (A) Trophozoite showing cilia (Ci) (×590). (B) Cyst showing large macronucleus (Ma) and cilia (Ci) beneath cyst wall (Cw) (×720).

Water obtained from drainage areas contaminated with human or pig feces is considered to be the primary source of human infection.

Identification of the parasite in water or fecal samples is based on microscopy. Trophozoites and cysts of *B. coli* are relatively large and ovoid to oblong, and they measure 45 to 65 μm in diameter. Cysts and trophozoites contain a large nucleus which is usually bean shaped. Cysts are surrounded by a distinct wall, and cilia are usually visible immediately beneath the wall. Positive stools usually contain trophozoites as well as cysts, but the former will disintegrate within a few hours after passage. Trophozoites move by cilia and rotate on their longitudinal axis. In formalin-fixed and stained preparations, trophozoites can be confused microscopically with debris, artifacts, or helminth eggs (69). Although tetracycline is the treatment of choice, diiodohydroxyquin or paromomycin is also effective (69).

MICROSPORIDIOSIS

Protozoa belonging to the phylum Microspora are primitive eukaryotes that lack mitochondria, peroxisomes, and Golgi complexes. Sequence analyses of rRNA have demonstrated that microsporidia are similar to prokaryotes and have little homology to other eukaryotes (67), leading to the speculation that microsporidia branched off early from prototype eukaryotes. Most human infections have been reported in patients with AIDS and are attributed to two species of microsporidia, *Enterocytozoon bieneusi* and *S. intestinalis.* Infections due to *Enterocytozoon bieneusi* are 10 times more prevalent than those due to *S. intestinalis.* Microsporidia have a unique mode of transmission in which their spores contain an extrusion apparatus consisting of a coiled polar filament and anchoring disc. Infection of host cells occurs through extrusion of the polar filament and then ejection of the sporoplasm through a hollow tube and penetration into the host cell cytoplasm (7). After entering a host cell, the parasite appears to multiply as a plasmodium or schizont and eventually forms spores that erupt through the lysing enterocyte membrane and are shed in the feces. Infections due to these two microsporidia can be differentiated because the parasitic stages of *S. intestinalis* are several times larger than those of *Enterocytozoon bieneusi.*

These microsporidia were once regarded as rare, but they are now considered to be common enteric pathogens in patients with AIDS. Prevalence among AIDS patients varies between 2 and 50%, with the majority of cases occurring in homosexual males. Although the method of transmission has not been defined, infections are believed to be acquired by ingestion of spores contaminating food and water.

These microsporidia are most numerous in enterocytes of the small intestine, where they cause an enteritis of varying intensity. Some bowel movements are watery and of large volume, but small, formed stools may occasionally be passed. Bowel movements tend to be clustered during one part of the day, usually during late evening or early morning, and nocturnal diarrhea is uncommon. Microsporidiosis is occasionally associated with biliary tract disease (52) and may be a potential cause of cholangiopathy that often accompanies AIDS (8).

Until recently, diagnosis of microsporidiosis was done almost exclusively by transmission electron microscopy of infected tissues. Microsporidia can now be identified in tissue sections and in fecal specimens by special stains and light microscopy.

Acknowledgments

I thank Joan Haynes for assisting with the literature searches and proofreading the manuscript, and Andy Blixt and Kathy Prokop for technical assistance. Published as Journal Series no. 4057, Montana Agricultural Experiment Station.

References

1. **Adams, E. B., and I. N. MacLeod.** 1977. Invasive amebiasis. I. Amebic dysentery and its complications. *Medicine* (Baltimore) **56**:315–323.
2. **Aristizabal, J., J. Acevedo, and M. Botero.** 1991. Fulminant amebic colitis. *World J. Surg.* **15**:216–221.
3. **Bendall, R. P., S. Lucas, A. Moody, G. Tovey, and P. L. Chiodini.** 1993. Diarrhoea associated with cyanobacterium-like bodies: a new coccidian enteritis of man. *Lancet* **341**:590–592.
4. **Brandt, H., and R. Perez Tamayo.** 1970. Pathology of human amebiasis. *Hum. Pathol.* **1**:351–385.
5. **Brodsky, R. E., H. C. Spencer, and M. G. Schultz.** 1974. Giardiasis in American travelers to the Soviet Union. *J. Infect. Dis.* **130**:319–323.
6. **Bruckner, D. A.** 1992. Amebiasis. *Clin. Microbiol. Rev.* **5**:356–369.
7. **Canning, E. U., and W. S. Hollister.** 1987. Microsporidia of mammals—widespread pathogens or opportunistic curiosities? *Parasitol. Today* **3**:267–273.
8. **Cello, J.** 1989. Acquired immunodeficiency syndrome cholangiopathy: spectrum of disease. *Am. J. Med.* **86**:539–546.
9. **Centers for Disease Control.** 1995. Assessing the public health threat associated waterborne cryptosporidiosis: report of a workshop. *Morbid. Mortal. Weekly Rep.* **44**:1–16.
10. **Clark, C. G., and L. S. Diamond.** 1991. Ribosomal RNA genes of "pathogenic" *Entamoeba histolytica* are distinct. *Mol. Biochem. Parasitol.* **49**:297–302.
11. **Colley, D. G.** 1995. Waterborne cryptosporidiosis threat addressed. *Emerging Infect. Dis.* **1**:67–68.
12. **Craft, J. C., T. Murphy, and J. D. Nelson.** 1981. Furazolidone and quinacrine: comparative study of therapy for giardiasis in children. *Am. J. Dis. Child.* **135**:164–166.
13. **Current, W. L.** 1990. Techniques and laboratory maintenance of *Cryptosporidium*, p. 31–49. *In* J. P. Dubey, C. A. Speer, and R. Fayer (ed.), *Cryptosporidiosis of Man and Animals.* CRC Press, Inc., Boca Raton, Fla.
14. **DeHovitz, J. A., J. W. Pape, M. Boncy, and W. D. Johnson.** 1986. Clinical manifestations and therapy of *Isospora belli* infection in patients with the acquired immunodeficiency syndrome. *N. Engl. J. Med.* **315**:87–90.
15. **Diamond, L. S., and C. G. Clark.** 1993. A redescription of *Entamoeba histolytica* Schaudinn, 1903 (Emended Walker 1911) separating it from *Entamoeba dispar* (Brumpt, 1925). *J. Eukaryotic Microbiol.* **40**:340–344.
16. **Dubey, J. P.** 1977. *Toxoplasma, Hammondia, Besnoitia, Sarcocystis,* and other tissue cyst-forming coccidia of man and animals, p. 101–237. *In* J. P. Kreier (ed.), *Parasitic Protozoa,* vol. III. *Gregarines, Haemogregarines, Coccidia, Plasmodia and Haemoproteids.* Academic Press, New York.
17. **Dubey, J. P.** 1992. Isolation of *Toxoplasma gondii* from a naturally infected beef cow. *J. Parasitol.* **78**:151–153.
18. **Dubey, J. P.** 1995. Personal communication.
19. **Dubey, J. P., and C. P. Beattie.** 1988. *Toxoplasmosis of Animals and Man.* CRC Press, Inc., Boca Raton, Fla.
20. **Dubey, J. P., A. W. Kotula, A. Sharar, C. D. Andrews, and D. S. Lindsay.** 1990. Effect of high temperature on infectivity of *Toxoplasma gondii* cysts in pork. *J. Parasitol.* **76**:201–204.
21. **Dubey, J. P., J. C. Leighty, V. C. Beal, W. R. Anderson, C. D. Andrews, and P. Thulliez.** 1991. National seroprevalence of *Toxoplasma gondii* in pigs. *J. Parasitol.* **77**:517–521.
22. **Dubey, J. P., N. L. Miller, and J. K. Frenkel.** 1970. The *Toxoplasma gondii* oocysts from cat feces. *J. Exp. Med.* **132**:636–662.
23. **Dubey, J. P., N. L. Miller, and J. K. Frenkel.** 1970. Characterization of new fecal form of *Toxoplasma gondii. J. Parasitol.* **56**:447–456.
24. **Dubey, J. P., C. A. Speer, and R. Fayer (ed.).** 1990. *Cryptosporidiosis of Man and Animals.* CRC Press, Inc., Boca Raton, Fla.
25. **Dubey, J. P., and P. Thulliez.** 1993. Persistence of tissue cysts in edible tissues of cattle fed *Toxoplasma gondii* oocysts. *Am. J. Vet. Res.* **54**:270–273.
26. **Farthing, M. J. G.** 1995. *Giardia lamblia,* p. 1081–1105. *In* M. J. Blaser, P. D. Smith, J. I. Ravdin, H. B. Greenberg, and R. L. Guerrant (ed.), *Infections of the Gastrointestinal Tract.* Raven Press, Ltd., New York.
27. **Farthing, M. J. G., L. Mata, J. J. Urrutia, and R. A. Kronmal.** 1986. Natural history of *Giardia* infection of infants and children in rural Guatemala and its impact on physical growth. *Am. J. Clin. Nutr.* **43**:393–403.
28. **Fayer, R., H. R. Gamble, J. R. Lichtenfels, and J. W. Bier.** 1992. Waterborne and foodborne parasites, p. 789–809. *In* C. Vanderzant and D. F. Splittstoesser (ed.), *Compendium of Methods for the Microbiological Examination of Foods,* 3rd ed. American Public Health Association, Washington, D.C.

29. **Feely, D. E., J. V. Schollmeyer, and S. L. Erlandsen.** 1982. *Giardia* spp.: distribution of contractile proteins in the attachment organelle. *Exp. Parasitol.* **53:**145–154.

30. **French, J. M., J. L. Whitby, and A. G. W. Whitfield.** 1964. Steatorrhea in man infected with coccidiosis (*Isospora belli*). *Gastroenterology* **47:**642–648.

31. **Frenkel, J. K., J. P. Dubey, and N. L. Miller.** 1970. *Toxoplasma gondii* in cats: fecal stages identified as coccidian oocysts. *Science* **167:**893–895.

32. **Greenberg, S. J., M. P. Davey, W. S. Zierat, and T. A. Waldman.** 1988. *Isospora belli* enteric infection in patients with human T-cell leukemia virus type 1-associated adult T-cell leukemia. *Am. J. Med.* **85:**435–438.

33. **Guerina, N. G., H. W. Hsu, H. C. Meissner, J. H. Maguire, R. Lynfield, B. Stechenberg, I. Abroms, M. S. Pasternack, R. Hoff, R. B. Eaton, G. F. Grady, and the New England Regional *Toxoplasma* Working Group.** 1994. Neonatal serologic screening and early treatment for congenital *Toxoplasma gondii* infection. *N. Engl. J. Med.* **330:**1858–1863.

34. **Hutchinson, W. M., J. F. Dunachie, J. C. Siims, and K. Work,** 1970. Coccidian-like nature of *Toxoplasma gondii. Br. Med. J.* **1:**142–144.

35. **Kimball, A. C., B. H. Kean, and F. Fuchs.** 1974. Toxoplasmosis: risk variations in New York City obstetric patients. *Am. J. Obstet. Gynecol.* **119:**203–214.

36. **Kotula, A. W., J. P. Dubey, A. K. Skarar, C. D. Andrews, S. K. Shen, and D. S. Lindsay.** 1991. Effect of freezing on infectivity of *Toxoplasma gondii* tissue cysts in pork. *J. Food Prot.* **54:**687–690.

37. **Laughon, B. E., D. A. Druckman, A. Vernon, T. C. Quinn, B. F. Polk, J. F. Modlin, R. H. Yolken, and J. G. Barlett.** 1988. Prevalence of enteric pathogens in homosexual men with and without acquired immunodeficiency syndrome. *Gastroenterology* **94:**984–993.

38. **LeChevallier, M. W., W. D. Norton, and R. G. Lee.** 1991. Occurrence of *Giardia* and *Cryptosporidium* spp. in surface water supplies. *Appl. Environ. Microbiol.* **57:**2610–2616.

39. **Lindmark, D. G.** 1980. Energy metabolism of the anaerobic protozoan *Giardia lamblia. Mol. Biochem. Parasitol.* **1:**1–12.

40. **Long, E. G., E. H. White, W. W. Carmichael, P. M. Quinlisk, R. Raja, B. L. Swisher, H. Daugharty, and M. T. Cohen.** 1991. Morphologic and staining characteristics of a cyanobacterium-like organism associated with diarrhea. *J. Infect. Dis.* **164:**199–202.

41. **MacKenzie, W. R., M. S. Hoxie, and M. E. Proctor.** 1994. A massive outbreak in Milwaukee of *Cryptosporidium* infection transmitted through the public water supply. *N. Engl. J. Med.* **331:**161–167.

42. **Madico, G., R. H. Gilman, E. Miranda, L. Cabrera, and C. R. Sterling.** 1993. Treatment of *Cyclospora* infections with co-trimoxazole. *Lancet* **342:**122–123.

43. **Martinez, A. J., M. Sell, T. Mitrovics, G. Stoltenberg-Didinger, J. R. Inglesias-Rozas, M. A. Giraldo-Velasquez, G. Gosztonyi, V. Schneider, and J. Cervos-Navarro.** 1995. The neuropathology and epidemiology of AIDS. A Berlin experience. A review of 200 cases. *Pathol. Res. Pract.* **191:**427–443.

44. **McAuley, J. B., and D. D. Juranek.** 1992. Luminal agents in the treatment of amebiasis. *Clin. Infect. Dis.* **14:**1161–1162.

45. **Meisel, J. L., D. R. Perera, C. Meligro, and C. E. Rubin.** 1976. Overwhelming watery diarrhea associated with *Cryptosporidium* in an immunosuppressed patient. *Gastroenterology* **70:**1156–1160.

46. **Nakhforoosh, M., and J. B. Rose.** 1989. Detection of *Giardia* with a gene probe, abstr. Q-236, p. 369. *In Abstracts of the 89th Annual Meeting of the American Society for Microbiology, 1989.* American Society for Microbiology, Washington, D.C.

47. **Nicolle, C., and L. Manceaux.** 1908. Sur une infection à corps de Leishman (ou organismes voisins) du gondii. *C. R. Acad. Sci.* **147:**763–768.

48. **Nime, F. A., J. C. Burek, D. L. Page, M. A. Hoscher, and J. H. Yardley.** 1976. Acute enterocolitis in human being infected with the protozoan *Cryptosporidium. Gastroenterology* **70:**592–598.

49. **Ortega, Y. R., C. R. Sterling, R. H. Gilman, V. A. Cama, and F. Diaz.** 1993. *Cyclospora* species—a new protozoan pathogen of humans. *N. Engl. J. Med.* **328:**1308–1312.

50. **Osterholm, M. T., J. C. Forgang, T. L. Ristinen, A. G. Dean, J. W. Washburn, J. R. Godes, R. A. Rude, and J. G. McCullough.** 1981. An outbreak of foodborne giardiasis. *N. Engl. J. Med.* **304:**24–28.

51. **Pape, J. W., R. I. Verdier, and W. D. Johnson, Jr.** 1989. Treatment and prophylaxis of *Isospora belli* infection in patients with the acquired immunodeficiency syndrome. *N. Engl. J. Med.* **320:**1044–1047.

52. **Pol, S., C. A. Romana, S. Richard, P. Amouyal, I. Deportes-Lirage, F. Carnot, J. F. Pays, and P. Bertholot.** 1993. Microsporidia infection in patients with the human immunodeficiency virus and unexplained cholangitis. *N. Engl. J. Med.* **328:**95–99.

53. **Porter, A.** 1961. An enumerative study of the cysts of *Giardia (lamblia) intestinalis* in human dysenteric faeces. *Lancet* **i:**1116–1169.

54. **Quinn, T. C., W. E. Stamm, S. E., Goodell, E. Mkrtichian, J. Benedetti, L. Corey, M. D. Schuffler, and K. K. Holmes.** 1983. The polymicrobial origin of intestinal infections in homosexual men. *N. Engl. J. Med.* **309:**576–582.

55. **Reduker, D. W., C. A. Speer, and J. A. Blixt.** 1985. Ultrastructure of *Cryptosporidium parvum* oocysts and excysting sporozoites as revealed by high resolution scanning electron microscopy. *J. Protozool.* **32:**708–711.

56. **Remington, J. S., and G. Desmonts.** 1990. Toxoplasmosis, p. 89–195. *In* J. S. Remington and J. O. Klein (ed.), *Infectious Diseases of the Fetus and Newborn Infant.* The W. B. Sauders Co., Philadelphia.

57. **Roberts, T., and J. K. Frenkel.** 1990. Estimating income losses and other preventable costs caused by congenital toxoplasmosis in people in the United States. *J. Am. Vet. Med. Assoc.* **196:**249–256.

58. **Rose, J. B., C. P. Gerba, and W. Jakubowski.** 1991. Survey of potable water supplies for *Cryptosporidium* and *Giardia. Environ. Sci. Technol.* **25:**1393–1400.

59. **Sealy, D. P., and S. H. Schuman.** 1983. Endemic giardiasis and daycare. *Pediatrics* **72:**154–158.

60. **Shaffer, N., and L. Moore.** 1989. Chronic traveler's diarrhea in a normal host due to *Isospora belli. J. Infect. Dis.* **159:**596–597.

61. **Sheffield, H. G., and M. L. Melton.** 1970. *Toxoplasma gondii:* the oocyst, sporozoites and infection of cultured cells. *Science* **167:**892–893.

62. **Smith, K. E., J. J. Zimmerman, S. Patton, G. W. Beran, and H. T. Hill.** 1992. The epidemiology of toxoplasmosis on Iowa swine farms with an emphasis on the roles of free-living mammals. *Vet. Parasitol.* **42:**199–211.

63. **Soave, R., J. P. Dubey, L. J. Ramos, and M. Tummings.** 1986. A new intestinal pathogen? *Clin. Res.* **34:**533A.

64. **Soave, R., and W. D. Johnson, Jr.** 1988. *Cryptosporidium* and *Isospora belli* infections. *J. Infect. Dis.* **157:**225–229.

65. **Sogin, M. L., J. H. Gunderson, H. J. Elwood, and R. A. Alonso.** 1981. Phylogenetic meaning of the kingdom concept: an unusual ribosomal RNA from *Giardia lamblia. Science* **243:**75–77.

66. **Thacker, S. B.** 1981. Parasitic disease control in a residential facility for the mentally retarded: failure of selected isolation procedures. *Am. J. Public Health* **71:**303–305.

67. **Vossbrinck, C. R., J. V. Maddox, S. Friedman, B. A. Debrunner-Vossbrinck, and C. R. Woese.** 1987. Ribosomal RNA sequence suggests microsporidia are extremely ancient eukaryotes. *Nature* (London) **326:**411–414.

68. **Walsh, J. A.** 1988. Prevalence of *Entamoeba histolytica* infection, p. 93–105. *In* J. I. Ravdin (ed.), *Amebiasis: Human Infection by Entamoeba histolytica.* Churchill Livingstone, New York.

69. **Walzer, P. D., and G. R. Healy.** 1982. Balantidiasis, p. 15–24. *In* L. Jacobs and P. Arambulo (ed.), *Handbook Series on Zoonoses,* section C. *Parasitic zoonoses,* vol. II. CRC Press, Inc., Boca Raton, Fla.

70. **Weigel, R. M., J. P. Dubey, A. M. Siegel, D. Hoefling, D. Reynolds, L. Herr, U. D. Kitron, S. K. Shen, P. Thulliez, R. Fayer, and K. S. Todd.** 1995. Prevalence of antibodies to *Toxoplasma gondii* in swine in Illinois in 1992. *J. Am. Vet. Med. Assoc.* **206:**1747–1751.

Preservatives and Preservation Methods

József Farkas

28 Physical Methods of Food Preservation

Whenever values of environmental stress factors discussed in chapter 2 are outside the vital range for growth and survival of a specific microorganism, they cause cellular damage. Depending on the severity of this effect, growth can be inhibited (microbistatic effect or reversible inhibition) or cells can be killed (microbicidal effect). Food spoilage can be prevented by applying one or more of the following strategies: inhibition of microbial growth, destruction (irreversible inactivation) of microbial cells, and mechanical removal of microorganisms from the food.

Physical methods of food preservation are those that utilize physical treatments to inhibit, destroy, or remove undesirable microorganisms without involving antimicrobial additives or products of microbial metabolism as preservative factors. Microbial growth can be inhibited by physical dehydration processes (drying, freeze-drying, and freeze concentration), cold storage, or freezing and frozen storage. Microorganisms can be destroyed (irreversibly inactivated) by established physical microbicidal treatments such as heating (including microwave heat treatment); UV or ionizing radiation; or emerging methods of new nonthermal treatment, such as application of high hydrostatic pressure, pulsed electric fields, oscillating magnetic fields, photodynamic effects, or a combination of physical processes such as heat-irradiation, dehydro-irradiation, or mano-thermo-sonication. Mechanical removal of microorganisms from food can be accomplished by filtration of liquid foods. This chapter discusses the microbiological fundamentals of the physical preservation methods outlined above, with the exception of removal. Wherever possible, the mechanisms and underlying principles are explained.

PHYSICAL DEHYDRATION PROCESSES

Water is one of the most important factors controlling the rate of deterioration of food by either microbial or nonmicrobial effects. A knowledge of the moisture content alone is not sufficient to predict the stability of foods, because it is not the total moisture content but rather the availability of water that determines the shelf life of a food. A proportion of the total water in food is strongly bound to specific sites.

The availability of water is measured by the water activity (a_w) of a food, defined as the ratio of the vapor pressure of water in a food, P, to the vapor pressure of pure water, P_0, at the same temperature:

$$a_w = \frac{P}{P_0}$$

The relation of water activity to solute concentration is expressed by Raoult's law for "ideal" solutions:

$$a_w = \frac{P}{P_0} = \frac{n_2}{n_1 + n_2}$$

where n_1 and n_2 are the number of moles of solute and solvent, respectively. An ideal solution may be defined as one in which the molecules of a solute are affected by their environment in exactly the same manner as they are in a pure state at the same temperature and pressure as the solution. In reality, there are no such solutions, because interactions between molecules may reduce the effective number of particles in solution; on the other hand, dissociation may greatly increase the number. Departures from the ideal are large for nonelectrolytes at concentrations above 1 molal (molality is the number of gram moles per kilogram of water) and for electrolytes at all concentrations.

The movement of water vapor from a food to the surrounding air depends on the moisture content, the composition of the food, the temperature, and the humidity of the air. At constant temperature, the moisture content of food changes until it comes into equilibrium with water vapor in the surrounding air. This is called the equilibrium moisture content of the food. At the equilibrium moisture content, the food neither gains nor loses water on storage under those conditions. The relative humidity of the surrounding air is then the equilibrium relative humidity (ERH%). When different values of relative humidity versus equilibrium moisture content are plotted, a curve known as a water sorption isotherm is obtained. ERH is related to a_w by the expression ERH% = $a_w \times 100$.

Fresh, raw, high-moisture food materials such as fruits, vegetables, meats, and fish have a_w levels of 0.98 or higher. Food preservation by dehydration is based on the principle that microbial growth is inhibited if the water available for growth is removed, i.e., if the a_w is reduced (79). Technologies for drying solid foods remove water by hot air; freeze-drying does so by sublimation after freezing. Liquid foods can be dried by freeze concentration, whereby freezing is followed by mechanical removal of frozen water.

Table 28.1 Water activities of various foods[a]

Food	a_w
Fresh, raw fruits, vegetables, meat, fish	≥0.98
Cooked meat, bread	0.95–0.98
Cured meat products, cheeses	0.91–0.95
Sausages, syrups	0.87–0.91
Rice, beans, peas	0.80–0.87
Jams, marmalades	0.75–0.80
Candies	0.65–0.75
Dried fruits	0.60–0.65
Dehydrated vermicelli, spices, milk powder	0.20–0.60

[a]Adapted from references 26 and 141.

Table 28.2 Moisture contents of various dry or dehydrated food products when their a_w is 0.7 at 20°C[a]

Food	% Moisture content
Various seed grains	4–9
Milk powder	7–10
Cocoa powder	7–10
Whole egg powder	10–11
Skim milk powder	10–15
Dried, fat-free meat	10–15
Rice and legume seeds	12–15
Dehydrated vegetables	12–22
Dried soups	13–21
Dried fruits	18–25

[a]Adapted from references 23, 26, 98, 99, and 141.

These processes reduce the a_w of both the food and the microbial cells to levels insufficient for microbial growth. Water activities of various raw materials and food products are compared in Table 28.1. The microbiological stability of many intermediate-moisture foods, i.e., those with a_w levels of 0.7 to 0.85 (jams, sausages, etc.), depends also on other preservative factors (reduced pH, preservatives, pasteurization, etc.); however, their reduced a_w is of major importance. Because of their chemical composition and the water-binding capacity of their components, various foods may have very different moisture contents at the same a_w level, as shown in Table 28.2.

The a_w requirements of various microorganisms vary significantly (66). In the vital range of growth, decreasing the a_w increases the lag phase of growth and decreases the growth rate. Foodborne microorganisms are grouped according to their minimal a_w requirements in Table 28.3.

In general, among bacteria, gram-negative species have the highest a_w requirement. Important gram-negative bacteria such as *Pseudomonas* spp. and most of the members of the family *Enterobacteriaceae* can usually grow only above a_w 0.96 and 0.93, respectively. Less sensitive to

Table 28.3 Minimal a_w levels required for growth of foodborne microorganisms at 25°C[a]

Group of microorganisms	Minimal a_w required
Most bacteria	0.91–0.88
Most yeasts	0.88
"Regular" molds	0.80
Halophilic bacteria	0.75
Xerotolerant molds	0.71
"Xerophilic" molds and "osmophilic" yeasts	0.62–0.60

[a]Adapted from references 26, 66, 98, and 141.

Table 28.4 Minimal a_w requirements for growth and mycotoxin production of some toxigenic molds[a]

Mycotoxin(s)	Mold	Minimal a_w requirements for: Growth	Toxin production
Aflatoxins	*Aspergillus flavus*	0.82	0.83–0.87
	A. parasiticus	0.82	0.87
Ochratoxin	*A. ochraceus*	0.77	0.85
	Penicillium cyclopium	0.82–0.85	0.87–0.90
Patulin	*P. expansum*	0.81	0.99
	P. patulum	0.81	0.95
Stachybotryn	*Stachybotrys atra*	0.94	0.94

[a]Adapted from references 23, 26, and 115.

reduced a_w are gram-positive non-spore-forming bacteria. While many members of the family *Lactobacilliaceae* have a minimum a_w near 0.94, some representatives of the family *Micrococcaceae* are capable of growth below a_w 0.90. Staphylococci are unique among nonhalophilic bacteria in being able to grow at high levels of NaCl. The generally recognized minimum a_w for growth of *Staphylococcus aureus* is about 0.86 (66); however, under otherwise ideal conditions, its growth has been demonstrated at as low as a_w 0.83. Production of staphylococcal enterotoxin in food slurries has not been observed below a_w 0.93 (142). Most spore-forming bacteria do not grow below a_w 0.93. Germination and outgrowth of spores from food-poisoning strains of *Bacillus cereus* are prevented at a_w 0.97 to 0.93, depending on the nature of the bulk solute (67). Various serological types of *Clostridium botulinum* differ in their a_w requirements. The lower a_w limits in salt-adjusted media for growth from spore inocula are 0.95 for type A, 0.94 for type B, and 0.97 for type E (104). Toxin production occurs at a_w levels closely approaching these growth minima. The minimum a_w for the growth and spore germination of *Clostridium perfringens* is between 0.97 and 0.95 in complex media when sucrose or NaCl is used to adjust the a_w (72). For both *C. botulinum* and *C. perfringens,* growth proceeds at lower a_w levels when it is controlled by glycerol instead of by salt (7).

Several species of yeasts have a_w requirements much lower than those of bacteria. Salt-tolerant species such as *Debaryomyces hansenii, Hansenula anomala,* and *Candida pseudotropicalis* may grow well on cured meats and pickles at NaCl concentrations of up to 11% (a_w = 0.93). Some "osmophilic" species (such as *Zygosaccharomyces rouxii, Z. baillii,* and *Z. bisporus*) are able to grow in food of high sugar content (jams, honey, syrups). Terms such as "xerotolerant" or "saccharotolerant" would be more appropriate for these "osmophilic" yeasts.

In general, molds have a_w requirements much lower than those for other groups of foodborne microorganisms (6). However, molds with sporangiospores have greater moisture requirements, and germination of their spores may be inhibited at a_w = 0.93. The most common xerotolerant molds belong to the genus *Eurotium* (their asexual forms are the members of the *Aspergillus glaucus* group). The minimal a_w for their growth is in the range of 0.71 to 0.77, while the optimal a_w is 0.96. True xerophilic molds (113) such as *Monascus* (*Xeromyces*) *bisporus* do not grow at a_w levels greater than 0.97 to 0.99. The relationship of a_w to mold growth and toxin formation is complex (124). It is of great public health importance that the minimal a_w requirement for mycotoxin production is generally higher than that for growth of toxigenic molds (Table 28.4). That is, under many conditions in which molds grow, they do not produce toxins.

Differences in a_w limits for growth reflect differences in osmoregulatory capacities (48). Mechanisms of tolerance to low a_w are different in bacteria and fungi; the key point, however, is that the cell osmoregulation mechanism operates to maintain homeostasis with respect to water content (48). In general, the strategy employed by microorganisms as protection against osmotic stress is the intracellular accumulation of compatible solutes, i.e., compounds that have the general property of interfering minimally with the metabolic activities of the cell. While some bacteria accumulate K^+ ions and amino acids, such as proline, as a response to low a_w (12), halotolerant and xerotolerant fungi concentrate polyols such as glycerol, erythritol, and arabitol (34). These polyols not only act as osmoregulators but also prevent inhibition or inactivation of enzymes and probably serve as nutrient reserves (13). In the case of halophilic bacteria, KCl is a requirement, while osmophilic yeasts have a high tolerance for high solute concentrations (11).

Drying

During hot-air drying of vegetables, vermicelli, and similar solid foods, the airstream has two functions. It transfers heat to evaporate the water from the raw materials, and it carries away the vapor produced. In the course of drying, microorganisms on the raw materials are affected by both the drying temperature and changes of a_w. Microbiological consequences of the drying technology are influenced by a number of other factors (size and composition of food pieces, time-temperature combinations, etc.).

During the warming-up period of drying, the temperature is still low and the relative humidity is high. The length of this phase depends mainly on the size of food particles. If this phase is long, microorganisms may even grow during the slow increase of temperature in the range of 20 to 40°C under the existing high a_w. During later phases of drying, the temperature exceeds 50 to 70°C. Here there is no opportunity for growth, but heat destruction is not significant either, because the higher temperature develops parallel to decreased moisture content. Wet heat is more lethal than dry heat. If the drying lasts long enough (at least 30 min), time-temperature combinations which damage microbial cells irreversibly occur. The temperature and moisture distribution within the food are important with respect to the thermal inactivation of microorganisms. With certain drying technologies, the surface temperature may reach 100°C, while the internal temperature remains lower. When microorganisms are mainly on the product surface, their inactivation is increased. On the other hand, the moisture content of the surface decreases rapidly, thereby decreasing the heat sensitivity of microbial cells. Therefore, the deadly combination of high temperature and high humidity is rare in drying technologies. The major microbicidal effect is due to high-temperature–low-humidity conditions which last for a long time. Loss of viability continues during storage because reversibly damaged cells, being unable to regenerate at low a_w, gradually die. Further viability losses may occur during rehydration of dried products, especially when the rehydration is rapid, causing large differences in osmotic pressure.

The a_w of dried foods is usually considerably lower than the critical value for microbial growth. Therefore, dried products are stable microbiologically, provided that the relative humidity of storage atmosphere is less than 70%. If the ERH increases or the temperature changes, conditions on the product surface may change and permit the growth of xerotolerant molds (141). Of course, the microbiological quality of dried foods depends also on the microbiological quality of raw materials and their handling prior to drying (92).

Freeze-Drying

Freeze-drying (lyophilization) is a combined method of food preservation. Its principle is dehydration of the food in the frozen state through the vacuum sublimation of its ice content (45). This method of water removal is very gentle. In other methods for dehydrating solid foods, the distribution of solutes in the food changes as a result of a continuous transport of solutions toward the surface and increased solute loss when they reach the surface. During freeze-drying, the sublimation front moves in the direction of the core, and the ice sublimates in the location where it is formed. Thus, solutes remain inside the food at their original location, their loss is insignificant, and the food retains its original form and structure very well. As a result, rehydratability of freeze-dried foods is very fast and almost complete; such foods are able to regain 90 to 95% of their original moisture content. On the other hand, the very large specific surface of freeze-dried products makes them vulnerable to other damage such as lipid oxidation. To prevent spontaneous rehydration, freeze-dried foods should be kept in vapor-impermeable packaging. They should be packaged in an inert atmosphere to prevent oxidation processes.

The effect of freeze-drying on the microbial contaminants of foods is a combination of decreased a_w by freezing and further reduction by sublimation of the ice. The extents of cell damage in these two phases may be different, depending on the conditions (temperatures and rates) of freezing and sublimation. Microbial survival also depends on the composition of freeze-dried food. Carbohydrates, proteins, and collodial substances, in general, have a protective effect on microorganisms. During storage of freeze-dried products having a 2 to 5% residual moisture content, a gradual, slow decrease of cell viability occurs, similar to that of conventionally dried foods, especially in the presence of atmospheric oxygen. Again, there may be additional microbial destruction during the rehydration of freeze-dried products because large differences in solution concentrations develop suddenly.

Because of the mild manner of moisture removal, lyophilization conditions can be intentionally optimized for maximal survival of microbial cells (i.e., preservation of stock cultures). Low dehydration temperature, protective additives (such as glycerol) in the microbe suspension, and storage of lyophilized cultures in a nitrogen atmosphere or in vacuo increase culture viability. Gram-positive bacteria survive freeze-drying better than gram-negative bacteria.

COOL STORAGE

Cool (or chill) storage generally refers to storage at temperatures above freezing, from about 16°C down to –2°C (118). While pure water will freeze at 0°C, most foods do not begin to freeze until a temperature of about –2°C or lower is reached. Depending on the inherent storability of raw foods, the duration of cool storage may vary from a few days to several weeks (Table 28.5).

The preservation effect of cool storage (refrigeration) is based on the fact that reducing the temperature decreases the rate of chemical reactions and growth of microorganisms. The temperature of proper cool storage is generally less than the minimal growth temperature of most foodborne microorganisms (see chapter 2), and the generation time of psychrotrophic microorganisms is also considerably increased. The usual refrigeration temperatures permit growth of psychrophilic and psychrotrophic microorganisms (123). If their initial population is high, refrigerated foods can spoil in a short time. Since the temperature requirements of various microorganisms differ, refrigeration may considerably change the qualitative composition of the microbiota.

Because growth rates of microorganisms increase significantly with an increase in temperature of only a few degrees, variation of storage temperatures should be avoided. Psychrotrophic pathogens such as *Yersinia enterocolitica, Vibrio parahaemolyticus, Listeria monocytogenes,* and *Aeromonas hydrophila* (see section III of this volume) are important, especially in "minimally processed, extended shelf-life" products.

There is some evidence that the minimum growth temperature of microorganisms is determined by the inhibition of solute transport. In this regard, it has been proposed that the growth temperature range of an organism depends on how well the organism can regulate its lipid fluidity within a given range (41). Psychrotrophs contain increased amounts of unsaturated fatty acid residues in their lipids when grown at low temperatures.

Table 28.5 Extension of shelf life of raw foods by cool storage[a]

Food	Avg useful storage life (days) at: 0°C (32°F)	22°C (72°F)
Meat	6–10	1
Fish	2–7	1
Poultry	5–18	1
Fruits	2–180	1–20
Leafy vegetables	3–20	1–7
Root crops	90–300	7–50

[a]Adapted from reference 118.

This increase in the degree of unsaturation of fatty acids leads to a decrease in lipid melting point, suggesting that increased synthesis of unsaturated fatty acids at low temperatures acts to maintain the lipid in a fluid and mobile state, thereby allowing membrane proteins to continue to function (68). In addition, the transport permeases of psychrotrophs are apparently more operative at low temperatures than those of mesophiles (9), and a cold-resistant transport system characterizes psychrotrophic bacteria (146).

Controlled Atmosphere Storage

The minimal temperature of microbial growth is influenced by many factors and is lowest when other growth conditions are optimal. Under unfavorable values of other environmental factors, the minimal growth temperature increases considerably. Very important environmental factors include pH, a_w, and oxygen concentration. Therefore, when suboptimal values of these factors are combined with low temperature, the refrigerated stability of many commodities increases greatly. Controlled atmosphere storage of certain fruits and vegetables is widely used. A reduced (2 to 5%) oxygen content and generally increased (8 to 10%) carbon dioxide content are established and maintained in airtight chilled storage rooms. A modified storage atmosphere is advantageous because it depresses the respiration and adverse changes in sensory and textural qualities of stored fruits and vegetables while inhibiting the growth of certain spoilage organisms.

Growth inhibition is evident both in the extension of the lag phase and the reduction of biomass formed. The inhibitory effect of carbon dioxide is due to several factors. It reduces after-ripening, thereby maintaining greater phytoimmunity of plant tissues. When dissolved in the aqueous phase, carbon dioxide reduces the pH. But the primary mechanism is direct inhibition of microbial respiration. The simultaneous reduction of oxygen concentration contributes significantly to carbon dioxide's inhibitory effect. Psychrotrophic spoilage bacteria of flesh foods such as members of the aerobic genera *Pseudomonas* and *Acinetobacter* are particularly sensitive to carbon dioxide, while lactic acid bacteria and yeasts are not sensitive.

MAP

Modified atmosphere packaging (MAP) operates by a principle similar to that for controlled atmosphere storage. In the case of vacuum-packaged (VP) meat products, not the composition but the pressure of air is changed (the residual air pressure is only 0.3 to 0.4 bar

[1 bar = 10^5 Pa] instead of the 1 bar for normal air), thereby reducing the amount of available oxygen. If packaging films of very low gas permeability are used with fresh meats, fruits, and vegetables, the composition of the package atmosphere will change as a result of the enzymatic activity of the tissue and its associated microorganisms. This causes oxygen levels to decrease as carbon dioxide levels increase. This modified atmosphere is very inhibitory to certain microorganisms and greatly increases the keeping quality of foods. However, with regard to vacuum packaging of chilled foods, one must remember that *C. botulinum* (and some other pathogens) grow well in the absence of O_2. Thus, a maximum temperature of 3.3°C is advisable for some VP foods, especially pasteurized foods.

Almost any combination of carbon dioxide, nitrogen, and oxygen may be used in MAP to sustain visual appearance and/or to extend the shelf life of meat and meat products (44, 103). In meats, aerobic spoilage bacteria such as the *Pseudomonas/Acinetobacter/Moraxella* group are inhibited (35). The lactic acid bacteria then become the dominant components of the microbiota, but they grow more slowly than the former group and produce less offensive sensory changes (44).

MAP for respiring ("live") product, i.e., fresh fruits and vegetables, is rather complicated since proper gas permeability of the packaging film is required to achieve an equilibrium atmosphere in the package while both respiration rate and the gas permeability change with temperature (73). Despite an increasing interest in the use of MAP to extend the shelf life of many perishable products, the concern about the potential growth of pathogenic bacteria at refrigeration temperatures (107) remains the limiting factor to further expansion of the method. Therefore, recent studies have examined the effect of VP/MAP processing on the growth and survival of psychrotrophic pathogens and the effect on physicochemical changes occurring in VP/MAP foods (48, 102).

The effects of epiphytic flora on the growth of *L. monocytogenes* Scott A were assessed on different types of ready-to-eat leafy vegetables under a modified atmosphere at 10°C (46). The highest growth rate was obtained on butterhead lettuce, and no growth occurred on lamb lettuce. Strain Scott A grew faster on young, yellow leaves than on older, green leaves of broad-leaf chicory. There is a positive correlation between the extent of spoilage of the leaves and the number of listeriae as well as of the number of epiphytes after storage. *L. monocytogenes* inoculated on leaves disinfected with H_2O_2 (10%), which reduced the number of epiphytic bacteria by 1 to 2 log units, grew faster and to higher number than on leaves rinsed with water only.

Survival and growth of *B. cereus* and *L. monocytogenes* were determined on mung bean sprouts and chicory endive (46). *B. cereus* proliferated during storage at 10°C under air at atmospheric pressure conditions but diminished in a moderate vacuum (reduced oxygen tension, 400 millibars of pressure) system. In the latter system *L. monocytogenes* expired on mung bean sprouts but grew to significant levels on chicory endive.

Historically, fish and fish products have been of the greatest concern with respect to *C. botulinum* growth in MAP products. The U.S. National Academy of Sciences recommends that fish not be packed under modified atmospheres until the safety of the system has been established. In European Community projects, the microbiological safety of MAP fishes (cod and rainbow trout) was studied at 9, 5, and 12°C. In no instance was the growth or survival of *Aeromonas* spp., *Y. enterocolitica*, and *Salmonella typhimurium* greater under MAP than in the aerobically stored control, and frequently growth was reduced under MAP (20). However, these investigations did not include studies of *C. botulinum*.

FREEZING AND FROZEN STORAGE

Freezing normally lowers the temperature of a food to −18°C and then stores it at that temperature or below. Many convenience food products are made possible by this technology, which preserves foods without causing major changes in their consumer quality. Freezing and frozen transportation and storage are, however, highly energy-intensive. There are three basic freezing methods in commercial use: (i) in cold air, (ii) by placing the food or the food package in contact with a surface that is cooled by a refrigerant (indirect contact freezing), and (iii) by submerging the food or spraying the cold refrigerant liquid onto the food or package surface (118).

The freezing of foods occurs not at one defined temperature but over a broad temperature range. Depending on the composition of the foods, the water starts freezing at −1 to −3°C. This increases the solute concentration in the water not yet frozen, which decreases further the freezing point of the solution. At the so-called eutectic temperature, the concentration of solutes reaches the solubility limit (the point at which they precipitate), and the residual water freezes too. A totally frozen state, which is a complex system of ice crystals and crystallized soluble substances, results at −15 to −20°C for fruits and vegetables and at less than −40°C for meats.

As an effect of freezing, both the temperature and the water activity decrease. Thus, in frozen foods, only those microorganisms which are cold tolerant and xerotolerant can grow. While the spoilage of refrigerated nonfrozen meat is due mainly to bacteria, on frozen meats those molds which are able to grow at lower a_w levels may be problematic.

During freezing, microorganisms suffer multiple damage which may cause their inactivation immediately or later (82). This is the outcome of several effects, the magnitude of which is influenced by, for example, the rate and temperature of freezing and the composition of the food. The rapid cooling of mesophilic bacteria from the normal growth temperature brings about immediate death to a proportion of the culture. Freezing also involves a temperature shock and may cause metabolic injury, presumably by virtue of damage to the plasma membrane. Gram-negative bacteria, particularly mesophiles, appear to be more susceptible to this cold shock than gram-positive bacteria (63, 129). Rapid chilling results in membrane phase transitions from liquid crystal to gel status without allowing lateral phase separation of phospholipid and protein domains. This results in loss of permeability control by the cytoplasmic and outer membranes (87). Cold shock sensitizes cells to various forms of oxidative stress (43, 83).

During freezing of water, an osmotic shock occurs, and intercellular ice crystal formation causes mechanical injury. The concentration of cellular liquids changes the pH and ionic strength, thereby inactivating enzymes, denaturing proteins, and hampering the functioning of DNA, RNA, and cellular organelles. The main targets of injury are cell membranes. Not only do injured cells die gradually during frozen storage, but surviving cells are reexposed to osmotic effects during thawing. The injury of microbial cells may be reversible or irreversible. In the former case, cells are able to repair if nutrients, energy sources, and specific ions (mainly Mg^{2+} and phosphate) are available and metabolism can commence. The extent of injury and repair and the death and survival rates vary according to the conditions of freezing, frozen storage, and thawing.

Freezing rate is an important determinant of microbial viability. During slow freezing, crystallization occurs extracellularly, as the cell cytoplasmic fluid becomes supercooled at −5 to −10°C. Because their vapor pressure is greater than that of ice crystals, cells lose water, which freezes extracellularly. As a result of crystallization, the concentration of the external solution increases, also removing water from the cells. On the whole, microorganisms are exposed to osmotic effects for a relatively long time, which causes increased injury. With increased freezing rates, the period of osmotic effects decreases, thereby increasing survival to a maximum. When freezing rates are too high, crystal formation also occurs intracellularly, injuring the cells drastically and causing a decreasing survival rate.

The most important factor influencing the effect of freezing on microbial cells is the composition of the suspending medium. Certain compounds enhance, and others diminish, the lethal effects of freezing. Sodium chloride is very important in this regard because it reduces the freezing point of solutions, thereby extending the time period during which cells are exposed to high concentrations of solutions before freezing occurs. Other compounds, e.g., glycerin, saccharose, gelatin, and proteins in general, have a cryoprotective effect.

At temperatures below around −8°C there is, in practice, no microbial growth (64). The fate of microorganisms surviving freezing may vary during frozen storage. Often the death of survivors is fast initially and then slows gradually, and finally the survival level stabilizes. During frozen storage, the death rate of microbial populations is usually lower than during the freezing process. Death during frozen storage is probably due to the not yet frozen, very concentrated residual solution formed by freezing. The concentration and composition of this residual solution may change during the course of storage, and the size of the ice crystals may increase, especially at fluctuating storage temperatures. These processes also influence the survival of microorganisms. In general, there is less loss of viability during frozen storage when the storage temperature is stable rather than fluctuating. The lower the temperature of frozen storage, the slower is the death rate of survivors (64). Gram-positive microorganisms survive frozen storage better than gram-negative bacteria.

Although freezing and frozen storage reduce considerably the viable cell counts in food, freezing preservation cannot be considered a sterilization process; the extent of microbial destruction may be significantly reduced by the components of food. Under some conditions, survivors can grow during the thawing of frozen food. Their populations may then equal or exceed the level prior to destruction during freezing and frozen storage. The cells and the tissue structure of foods are also damaged by freezing, frozen storage, and thawing. During thawing, a solution rich in nutrients is released from the food cells, the microorganisms can penetrate the damaged tissue more easily, and a liquid film condenses on the surface of the product. These conditions favor microbial growth. Therefore, thawed products may be vulnerable to quick microbial spoilage. Refreezing of thawed products may be a dangerous practice and should be avoided.

PRESERVATION BY HEAT TREATMENTS

The most effective and most widely used method for destroying microorganisms and inactivating enzymes is heat treatment. The heat resistance of various microorganisms is therefore one of their most important characteristics from the point of view of the food industry.

The heat processing method called pasteurization (named after Louis Pasteur) is a relatively mild heat treatment aimed at inactivating enzymes and destroying a large population (99 to 99.9%) of vegetative bacterial cells. The main objective of this treatment is to eliminate non-spore-forming pathogenic bacteria. Because a significant proportion of the spoilage microorganisms is also destroyed by pasteurization, pasteurized products have an extended shelf life. For safety and keeping quality, the pasteurization must be complemented with packaging which prevents recontamination. Pasteurized food must be stored at a low temperature to prevent the growth of sporeformers whose spores survive the heat treatment. The shelf life of pasteurized products depends on the type of food and conditions of pasteurization and storage.

Sterilization refers to the complete destruction of microorganisms. Industrially, the emphasis is not on absolute sterility but rather on freedom from pathogenic microorganisms and shelf stability of hermetically packaged products. Shelf-stable and microbiologically safe products may contain low number of viable, but dormant, bacterial spores. These products are "commercially sterile."

Technological Fundamentals

Heat preservation practices can be divided into two broad categories: those that heat foods in their final containers, and those that use heat prior to packaging. The latter technology requires separate sterilization of the food and of the packaging material before the food is placed aseptically (in a sterile environment) into the package (aseptic technology) (122). Commercial sterilization of food in hermetically closed containers (cans or bottles) is called canning, or appertization (named after Nicholas Appert, who invented the basic process in the early 1800s).

Major technological developments in heat processing are the use of ultrahigh-temperature, short-time treatments (rapid heating to temperatures about 140°C, holding for several seconds, and then rapid cooling) for the production of foods shelf-stable at ambient temperature while providing possibilities for increasing the flexibility of packaging and product types as well as improving product quality and processing efficiency (81). The use of higher temperatures for shorter times results in less damage to important nutrients and functional ingredients. Thermobacteriology of ultrahigh-temperature-processed foods has been reviewed (14).

Aseptic technology is widely used for fruit juices, dairy products, creams, and sauces; continuous heat treatment with heat exchangers enables an efficient process with improved product quality. Significant progress has been made also for the adoption of this technology to low-acid soups, including particulate products.

Ohmic heating (128) and microwave energy treat foods in unique fashions because these processes generate heat throughout the food mass simultaneously, which causes very rapid heating.

Recently, a combined process of microbial inactivation including heat (112°C) and ultrasonication (20 kHz) under pressure (300 kPa) (mano-thermo-sonication) was proposed (120) for some liquid foods. The lethality of this combined process is much greater than that of heat treatments at the same temperature.

Fundamentals of Thermobacteriology

Heat in the presence of water (so-called wet-heat treatment) kills microorganisms by the denaturation of nucleic acids, structural proteins, and enzymes. Identification of primary sites for lethal heat-induced damage is difficult because of the many changes which heat brings about in cells. Nevertheless, there is substantial evidence that DNA damage, directly or indirectly, may be the key lethal event in heated cells (49). However, heat damages many other, less critical sites. In general, the thermal stability of ribosomes corresponds to the maximal growth temperature of microorganisms (68), and cytoplasmic membranes seem to be major sites of injury of mild heating (42), with a consequent increase in sensitivity to environmental stress factors. However, degradation of rRNA in membrane-damaged, heated cells often precedes loss of viability and therefore is not thought to be a prime cause of cell death (61). There is also evidence that some part of the bacterial spore germination system (a specific enzyme[s]) may be the least heat-resistant component and thus a key site of heat inactivation (49). Dry heat is less lethal and kills the microorganisms by dehydration and oxidation. Dry heat necessitates higher temperatures and longer heating times to generate the same lethality as wet heat.

The calculation of the heat treatment required to kill a given number of bacteria is based on the assumption that the destruction of a pure bacterial culture by wet heat at constant temperature exhibits a negative exponential kinetics (the order of death is logarithmic), $N = N_0 \cdot e^{-k \cdot t}$, where N is the actual number of viable cells, N_0 is the initial number of viable cells, k is the rate constant (time^{-1}) of heat destruction at constant temperature T, and t is the heat processing time. That is, equal periods at lethal temperatures result in equal percentage of destruction regardless of the microbial population size. A theoretical treatment of this relationship is given in chapter 2.

The destruction rate is inversely related to the decimal reduction time (D value), $D_T = \ln 10/k_T$. The D value is defined as the time in minutes required to destroy 90% of the population as shown in Fig. 28.1 (survival curve). Survival curves provide data on the rate of destruction of a specific organism in a specific medium or food at a specific temperature. Under given conditions, the rate of death is constant at any given temperature and independent of the initial viable cell number. The logarithms of D values plotted against exposure temperatures give the thermal death time curve (Fig. 28.2). Thermal death time is the time necessary to kill a given number of organisms at a specified temperature.

While the D value represents the measure of heat resistance of a given microorganism at a given temperature, the thermal death time curve indicates the relative resistances at different temperatures. This enables one to calculate the effects of different temperatures. The slope of the thermal death time curve is expressed by the z value, which is the temperature increase required to reduce the thermal death time by a factor of 10.

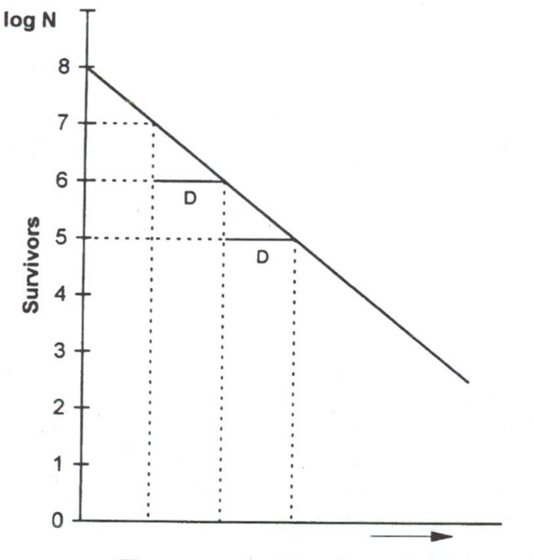

Figure 28.1 Bacterial survival curve showing logarithmic order of death and the concept of the D value.

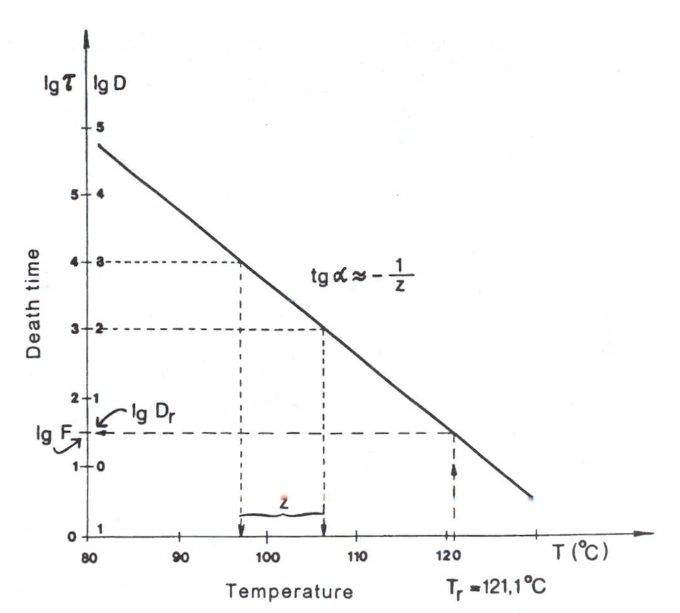

Figure 28.2 Thermal death time curve showing definition of z value and F value. T_r, reference temperature; τ, death time.

Heat resistances (D values) of various microorganisms and effects of temperature changes on the death rate are different and are influenced by many factors (64). There are inherent differences among species, among strains within the same species, and between spores and vegetative cells. Bacterial spores are more heat resistant than vegetative cells of the same strain, some being capable of surviving for minutes at 120°C or for several hours at 100°C. Less heat-resistant spores may have $D_{100°C}$ values of less than 1 min (95). Vegetative cells of bacteria, yeasts, and molds are killed after a few minutes at 70 to 80°C.

The age of the cells and their stage of growth, the temperature at which the microorganisms were grown, and the composition of the growth medium also affect heat resistance. Physicochemical factors such as pH, a_w, salt content, and organic composition of the suspension medium in which microorganisms are heated also profoundly influence heat resistance (22, 95, 133).

Tables 28.6 and 28.7 provide selected heat resistance data for vegetative bacteria and bacterial endospores, respectively. Vegetative cells of spore-forming bacteria are as heat sensitive as other vegetative cells; the high heat resistance of spores relates to their specific structure and composition, and it is basically the consequence of

Table 28.6 Relative heat resistances of some vegetative bacteria[a]

Organism	Heating medium	*D* value (min) at: 70°C	65°C	60°C	55°C	*z* value (°C)
Escherichia coli	Ringer solution, pH 7.0				4	
Lactobacillus plantarum	Trypticase soy broth + sucrose, (a_w = 0.95)		4.7–8.1			
	Tomato juice, pH 4.5	11.0				12.5
Pseudomonas aeruginosa	Nutrient agar				1.9	
P. fluorescens	Nutrient agar				1–2	
Salmonella senftenberg 775W	Skim milk			10.8		6.0
	Phosphate buffer (0.1 M), pH 6.5		0.29			
	+Sucrose, 30:70 (% wt/vol)		1.4:43			
	+Glucose, 30:70 (% wt/vol)		2.0:17.0			
	Milk chocolate	440				18.0
Salmonella senftenberg	Heart infusion broth (a_w = 0.99), pH 7.4			6.1		6.8
	+NaCl (a_w = 0.90, pH 7.4)			2.7		13
	+Sucrose (a_w = 0.90. pH 7.4)			75.2		8.9
Salmonella typhimurium	Phosphate buffer (0.1 M), pH 6.5		0.056			
	Milk chocolate	816				19.0
Staphylococcus aureus	Custard or pea soup			7.8		4.5

[a]Adapted from reference 140.

Table 28.7 Approximate heat resistances (*D* values) of some bacterial spores[a]

Type of food and typical organism(s)	*D* value (min) at: 121°C	100°C	*z* value (°C)
Low-acid (pH >4.6) foods			
Thermophilic aerobe			
Bacillus stearothermophilus	4.0–4.5	3,000	7
Thermophilic anaerobes			
Clostridium thermosaccharolyticum	3.0–4.0		12–18
Desulfotomaculum nigrificans	2.0–3.0		
Mesophilic anaerobes			
C. sporogenes	0.1–1.5		9–13
C. botulinum types A and B	0.1–0.2	50	10
C. perfringens		0.3–20	10–30
Mesophilic aerobes			
B. licheniformis		13	6
B. subtilis		11	7
B. cereus		5	10
B. megaterium		1	9
Acid (pH ≤4.6) foods			
Thermotolerant aerobe			
B. coagulans	0.01–0.1		
Mesophilic aerobes			
B. polymyxa		0.1–0.5	
B. macerans		0.1–0.5	
C. butyricum (or *C. pasteurianum*)		0.1–0.5	

[a]Adapated from references 64 and 137.

the dehydrated state of the spore core (49). The molecular basis of heat resistance in spores is covered more fully in chapter 3.

Bacterial spores suspended in oils are more heat resistant than those suspended in aqueous systems. This increased heat resistance was ascribed to the low a_w of oils, and the difference in *z* values suggests that the mechanism of inactivation differs for spores suspended in lipids and in aqueous systems (1).

Vegetative forms of yeasts can usually be killed by a 10- to 20-min heat treatment at 55 to 60°C (119). Given the same period of heat treatment, yeast ascospores require a temperature only 5 to 10°C higher than that required by the mother cells to effect similar lethality (25, 69). Vegetative propagules of most molds and conidiospores can be inactivated by 5 to 10 min of wet-heat treatment at 60°C (32). Several mold species forming sclerotia or ascospores are much more heat resistant (3, 33, 134). Mold spores survive dry-heat treatment for 30 min at 120°C.

Calculating Heat Processes for Foods

Containers of food do not heat instantaneously, and since all temperatures (above a minimum value) contribute to microbial destruction, a method to determine the relative effect of changing temperature while a food is being heated and cooled during thermal processing is necessary.

For determination of heat processes for canned foods, the *F* value was introduced. The *F* value is defined as the number of minutes at a specific temperature required to destroy a specific number of viable cells having a specific *z* value (Fig. 28.2).

Since *F* values represent the number of minutes to diminish a homogeneous population with a specific *z* value at a specific temperature, and because *z* values as well as temperatures vary, it is necessary to designate a reference *F* value. This reference is the F_0 value, which is the number of minutes at 121°C (250°F) required to destroy a specific number of cells whose *z* value is 10°C (18°F). The F_0 value of a heat treatment, i.e., its sterilization value, is a measure of a given heat treatment's lethality and is related to τ, the time required for identical lethality at temperature *T*, shown by the equation

$$\log F - \log \tau = \frac{T - 121.1}{z}$$

F_0 requirements of various foods differ. Since heat sensitivities of microorganisms and the characteristics of the thermal death curve are affected by many factors, thermal death curves should be established in the specific food for which a heat process is designed. Because the spores of *C. botulinum* types A and B are the most heat-resistant spores of a foodborne pathogen (Table 28.7), commercial sterilization of low-acid (pH >4.6) foods must be sufficient to destroy these spores. (The targets for the heat processing of acid foods [pH <4.6] are heat-resistant fungi [114].)

A "botulinum cook" is a heat process which reduces the population of the most resistant *C. botulinum* spores by an arbitrarily established factor of 12 decimal values (12 *D* concept). This provides a substantial margin of safety in low-acid canned foods. In practice, 2.45 min is the 12 *D* value of botulinal spores in phosphate buffer (64). The botulinum cook will produce a product that is safe but not necessarily commercially sterile. Other heat-resistant spores, which may cause spoilage but are not pathogenic, may be present and viable (see Table 28.7). Thus, their higher heat resistance often is the determining factor in establishing commercial heat processes.

Detailed guidance for calculating heat processes on the basis of semi-log-linear kinetics is provided by several monographs (111, 137, 138). Recently, kinetic models taking into account the concurrent activation of dormant spores and distinct inactivation of dormant and activated spores during heat treatments have been developed for advanced process design (80, 127). In the case of vegetative cells, the death rate coefficient depends significantly on environmental factors. Reichart and Mohácsi-Farkas (121) developed several kinetic models for describing the combined effect of a_w, pH, and redox potential on the thermal destruction rate of seven foodborne microorganisms (*Lactobacillus plantarum, Lactobacillus brevis, Saccharomyces cerevisiae, Z. bailii, Yarrowia lipolytica, Paecilomyces varioti,* and *Neosartoria fisheri*).

Recent Developments in Thermobacteriology of Food

Previous concepts about the heat preservation of foods assumed implicitly that heat resistance values of microorganisms measured under isothermal conditions are constant and unaffected by the heating rate in the sublethal temperature range. Work during the last two decades has demonstrated that the rate of temperature elevation affects the subsequent isothermal death of various vegetative bacteria (84, 143). Exposure of cells to sublethal temperatures above their growth maximum (to a sublethal heat shock) induces resistance to heating at higher temperatures (85). These effects were observed in nutritionally rich media. In recent years, the induction or enhanced synthesis of a family of proteins (heat shock proteins [hsps]) in response to heat shock and also to other environmental stresses and chemical or mechanical treatments has been reported as a phenomenon that occurs in most if not all organisms (5, 10, 130). Many of the hsp inducers have in common the capacity to produce protein damage (54). It has been proposed that the signal for hsp induction is protein denaturation and increase in the concentration of unfolded protein in the cell (108). However, the mechanism responsible for the development of thermotolerance has not been completely elucidated, and the connection between hsp induction and the acquisition of thermotolerance is unclear (5, 10). The lesions produced by various chemicals triggering thermotolerance are different from those produced by heat (10). Some investigators propose two states of tolerance to heat, one not requiring hsps and another (induced by more severe damage) requiring them (77). These observations lead to the conclusion that environmental stresses by industrial processes can induce protective responses in microorganisms. Thus, the possibility that heat resistance of vegetative cells can increase during heating must be taken into account when one is establishing safe heat treatments for foods that may be heated relatively slowly.

The traditional model of thermal inactivation kinetics, which assumes that the reduction in log numbers of survivors decreases in a linear manner with time, has served the food industry and regulatory agencies well for many years. Indeed, in situations in which bacterial spores are the critical targets or death is rapid, a good fit to this hypothesis is generally observed. However, as a result of the introduction of milder thermal processes that use low processing temperatures under a wide range of environmental conditions, the food industry is increasingly requiring accurate predictions for death kinetics for vegetative bacteria. Under these conditions, significant deviations from log-linear kinetics are encountered, "shoulder" and/or "tailing" of survival curves occurs frequently, and the method incorporating *D* and *z* values cannot be relied upon. The theory underlying log-linear death kinetics assumes that inactivation results from random single hits of key targets within microbial cells. Explanations offered as to why the variations in the straight-line survivor curve occur include (i) experimental errors and artifacts such as mixed populations of test organisms, flocculation and deflocculation of cells during the heating period (137), heat activation for spore germination, aggregation and absorption of cells to walls of the vessel, or the presence of cells in aerosol droplets above the surface of the suspension (53, 110); (ii) a multiple-hit destruction mechanism (49); (iii) variability of heat sensitivities (resistance distribution) in the heated population (17, 52); and (iv) heat adaptation during treatment (52).

Recently, mathematical models were developed to describe the observed inactivation kinetics (21) which allow for variability of heat sensitivity throughout a population of cells, e.g., normal distribution of heat sensitivity. A further, improved model describes the heat sensitivity of cells as being distributed by a logistic function, which allows for accurate predictions of thermal inactivation to be made at a relatively low heating rate (136) and even when parameters other than temperature, e.g., pH and NaCl concentration, are varied (21).

Microwave Heat Treatment

Electromagnetic heating of food by using radio frequency (1- to 500-MHz) or microwave (500-MHz to 10-GHz) energy has the advantage of internal heat generation. Water molecules, because of their negatively charged oxygen atoms and positively charged hydrogen atoms, are electric dipoles. When a rapidly oscillating radio frequency or microwave field is applied to a biological tissue or food rich in water, the water molecules reorient with each change in the field direction. This creates intermolecular friction, which produces heat (100).

Microwave heating modes may reduce process times and energy and water usage in some food processing areas, and microwave energy is suited for heat processing of food or pest control in the food industry (27, 125). However, the geometry of the food product, its thermal, physical, and dielectric properties, its mass, the power input, and the radiation frequency all influence the interaction of electromagnetic waves with a food (116). These factors can cause differential microwave heating of different food constituents or of different parts of a meal. This thermal nonuniformity has limited the commercial application of microwave heating for microbial inactivation, even though microwave heating has gained widespread application in home food preparation. The formation of heterogeneous temperatures within a food might create problems if it results in insufficient heating and thereby allows survival of potentially pathogenic microorganisms (57). Therefore, metal shielding and other packaging designs which increase the heating uniformity of food are desirable (28, 55).

Regarding specific electromagnetic effects, electromagnetic fields cause ion shifts on cellular membranes, leading to changes in permeability, functional disturbances, and cell rupture (116). Theoretically, microbial cells in food may be differentially heated compared with the food matrix, which could result in specific inactivation. However, these theoretical findings need to be validated in food. So far, experimental data on the specific effects of radio frequency or microwave frequency electromagnetic energies as reviewed by Ponne and Bartels (116) are contradictory, but in the majority of studies, no nonthermal effects were detected.

A more fundamental approach in food-related research concerning microwave effects is needed, and more research has to be conducted to determine these effects and utilize the advantages of various frequencies. The choice of the most widely used frequency, 2.45 GHz, was a compromise of various factors and is not necessarily the most suitable frequency for a certain application (116). Sophisticated temperature registration methods and dielectric measurements can be important tools for better understanding interactions in studies of specific effects of electromagnetic fields.

Ohmic Heating

The use of direct electrical heating known as ohmic heating is another development in heating technology. An electric current is passed through the material as it is pumped through a tube. The heating rate is a function of the electrical conductance of the solid and liquid phases. If these are about the same, the two phases heat at the same rate. The design of equipment, products, and processes for ohmic heating are still in the developmental stages (128). In the limited thermobacteriological stud-

Table 28.8 Preservative effects of ionizing radiation[a]

Effect(s)	Result(s)
Inhibition of sprouting	Increased shelf life of root and bulb crops; reduction of malting losses
Decrease of after-ripening and delaying senescence of some fruits and vegetables	Increased shelf life of fruits and vegetables
Killing and sterilization of insects	Insect disinfestation of food
Prevention of growth and reproduction of parasites transmitted by food	Prevention of parasitic diseases
Reduction of microbial populations	Decreased contamination of food; increased shelf life of foods; prevention of food poisoning

[a]From reference 36.

ies conducted to date, microbial death appears to be caused solely by thermal effect (106).

PRESERVATION BY IRRADIATION

UV Radiation

The microbiologically most destructive wavelength range of UV radiation is between 240 and 280 nm. The great microbicide efficiency of the so-called germicidal radiation of 253.7 nm is based on the fact that this radiation can damage most efficiently nucleic acids which have an absorption maximum at 265 nm. Gram-negative bacteria are most easily killed by UV, while bacterial endospores and molds are much more resistant.

Viruses are more UV resistant than bacterial cells (139). UV resistance of various microorganisms also depends on their pigment formation. Cocci that form colored colonies are less susceptible to UV radiation than those with colorless colonies. Dark conidia of certain molds are highly UV resistant (96).

The very low penetration of UV radiation (78) and the difficulty in attaining an even exposure level over all points of the food surface are major technological problems which limit the feasibility of UV radiation in food preservation. Therefore, UV sources are used mainly for disinfection of air, e.g., in aseptic filling of liquid food such as milk, beer, and fruit juices, in packaging of sliced bread, and in the ripening rooms of cheese and dry sausages, or for disinfection of water when the use of chlorine is undesirable (18). Possible uses of UV irradiation of packaging materials and preformed containers as a single treatment and in combination with hydrogen peroxide for aseptic packaging have been described by Maunder (88), Narasimhan et al. (101), and Stannard et al. (135), respectively.

Ionizing Radiation

The potential application of ionizing radiation in food products is based mainly on the fact that ionizing radiation very effectively inhibits DNA synthesis so that cell division is impaired. Given the right doses, this impairment is obtained without serious effects on the food itself. Therefore, microorganisms, insect gametes, and plant meristems can be prevented from reproducing, thus rendering the food stable (Table 28.8). Excellent general reviews on various aspects, prospects, and problems of food irradiation are numerous (e.g., 29, 30, 62, 76, 144, 147).

In those parts of the world where the transport of food is difficult and where refrigerated storage of food is scarce and extremely expensive, the use of ionizing radiation may facilitate wider distribution of some food than would otherwise be feasible. In this way, a more varied and possibly nutritionally superior diet may become available to the residents.

The ionizing radiation used in food irradiation is limited to high-energy electromagnetic radiation (gamma rays of ^{60}Co or X rays) with energies of up to 5 MeV or electrons from electron accelerators with energies of up to 10 MeV. These types of radiation are chosen because (i) they produce the desired effects with respect to the food, (ii) they do not induce radioactivity in foods or packaging materials, and (iii) they are available in quantities and at costs that allow commercial use of the process. Other kinds of ionizing radiation, in some respects, do not suit the needs of food irradiation.

The penetration of electron beams and X or gamma rays into matter occurs in different ways. The practical usable depth limit for 10 MeV electrons in water-equivalent material is 3.9 cm, while the so-called half-thickness value for 5-MeV X rays is 23.0 cm (144). Except for differences in penetration, electromagnetic radiation and electrons are equivalent in food irradiation and can be used interchangeably.

Microbiological Fundamentals

One of the main reasons for using ionizing radiation is to kill organisms that cause spoilage or are a health hazard to the consumer. The biological effects of ionizing radiation on cells can be due both to direct interactions with critical cell components and to indirect actions on these critical targets caused by radiolytic products of other molecules in the cell, particularly free radicals formed from water. The DNA in the chromosomes is the most critical target of ionizing radiation. Effects on cytoplasmic membrane appear to play an additional important role in radiation-induced damage of cells (51). Although the changes by radiation are at the cellular level, the consequences of these changes vary with the organism. The correlation of radiation sensitivity is roughly inversely proportional to the size and complexity of an organism (Table 28.9). This correlation is related to genome size: the DNA in the nuclei of the insect cells represents a target much larger than the genomes of bacteria, while the cells of mammalian organisms containing DNA molecules, which must provide much more genetic information than those of insects, are correspondingly larger and even more sensitive to radiation (29). Differences in radiation sensitivities within groups of similar organisms are related to differences in their chemical and physical structure and the ability to recover from radiation injury. The amount of radiation energy required to control microorganisms in foods, therefore, varies according to the resistance of the particular species and according to the number of organisms present.

The relative radiation resistances of various foodborne microorganisms can be found in comparable forms of presentation in several reviews (e.g., 29, 62, 65, 94). Summarizing data from many of references, Table 28.10 presents typical radiation resistances of some foodborne microorganisms in fresh and frozen foods of animal origin. In general, gram-negative bacteria, including the common spoilage organisms of many foods (e.g., pseudomonads) and particularly enteric species including pathogens (salmonellae, shigellae, etc.), are generally more sensitive than vegetative gram-positive bacteria. The spores of the genera *Bacillus* and *Clostridium* are more resistant still. Rarely, even more highly resistant vegetative forms may be encountered, typified by *Deinococcus* (formerly *Micrococcus*) *radiodurans* (2) and the gram-negative, rod-shaped *Deinobacter* sp. (50). The radiation sensitivity of many molds is of the same order as that of vegetative bacteria. However, fungi with melanized hyphae have a radiation resistance comparable to that of bacterial spores (126). Yeasts are as resistant as the more resistant bacteria. Viruses are highly radiation resistant (144).

The death of microorganisms by ionizing radiation is a result of damage to DNA. Ionizing radiation can affect DNA directly by the ionizing ray or indirectly by the primary water radicals $^{\cdot}H$, $^{\cdot}OH$, and e_{aq}^{-}. The most im-

Table 28.9 Approximate doses of radiation needed to kill various organisms[a]

Organisms	Dose (kGy)[b]
Higher animals	0.005–0.1
Insects	0.01–1
Non-spore-forming bacteria	0.5–10
Bacterial spores	10–50
Viruses	10–200

[a]From reference 144.

[b]Gray is the SI (Système International de Unités) unit of absorbed radiation dose (1 J/kg); 1 Gy = 100 rad.

Table 28.10 Typical radiation resistances of some foodborne microorganisms in fresh and frozen foods of animal origin

Microorganism(s)	D_{10} (kGy) Fresh food	D_{10} (kGy) Frozen food
Vibrio spp.	0.03–0.12	0.11–0.75
Yersinia enterocolitica	0.04–0.21	0.4
Pseudomonas putida	0.06–0.11	
Campbylobacter jejuni	0.08–0.20	0.21–0.32
Shigella spp.		0.2–0.4
Aeromonas hydrophila	0.14–0.19	
Bacillus cereus (vegetative cells)	0.17	
Proteus vulgaris	0.2	
Escherichia coli	0.23–0.35	0.3–0.6
E. coli O157:H7	0.24–0.27	0.31–0.44
Staphylococcus aureus	0.26–0.6	0.3–0.45
Brucella abortus	0.34	
Salmonella spp.	0.3–0.8	0.4–1.3
Listeria monocytogenes	0.27–1.0	0.52–1.3
Lactobacillus spp.	0.3–0.9	
Streptococcus faecalis	0.65–1.0	
Clostridium perfringens (vegetative cells)	0.59–0.83	
Moraxella phenylpyruvica	0.63–0.88	
Clostridium botulinum type E (spores)	1.25–1.40	
Bacillus cereus (spores)	1.6	
Clostridium sporogenes (spores)	1.5–2.2	
Clostridium botulinum types A and B (spores)	1.0–3.6	
Deinococcus radiodurans	2.5–3.1	
Deinobacter sp.	5.05	
Coxsackievirus		6.8–8.1

portant of these three radicals in DNA damage is the ˙OH radical. The ˙OH radicals formed in the hydration layer around the DNA molecule are responsible for 90% of DNA damage. Thus, in living cells, the indirect radiation damage is predominant. The principal lesions induced by ionizing radiation in intracellular DNA are chemical damage to the purine and pyrimidine bases and to the deoxyribose sugar and physicochemical damage resulting in a break in the phosphodiester backbone in one strand of the molecule (single-strand break) or in both strands at the same place (double-strand break) (97). Double-strand breaks are produced by ionizing radiation at about 5 to 10% of the rate for single-strand breaks.

Most microorganisms can repair single-strand breaks. The more sensitive organisms, such as *Escherichia coli*, cannot repair double-strand breaks, while highly resistant species (e.g., *Deinococcus* sp.) are able to repair them. The low water content of the spore protoplast is a major factor in the radiation resistance of bacterial spores. During germination, the water content of the spore protoplast increases, and therefore the radiation resistance disappears.

Radiation survival is conveniently represented with the logarithm of the number of surviving organisms plotted against radiation dose. Typical radiation survival curves are shown in Fig. 28.3. Type 1 is an exponential survival curve, quite common for the more radiation sensitive organisms. Type 2 is characterized by an initial shoulder, indicating that equal increments of radiation are more effective at high doses than at low doses. The shoulder in the curve may be explained by a certain repair at low doses. Type 3 is characterized as concave with a resistant tail.

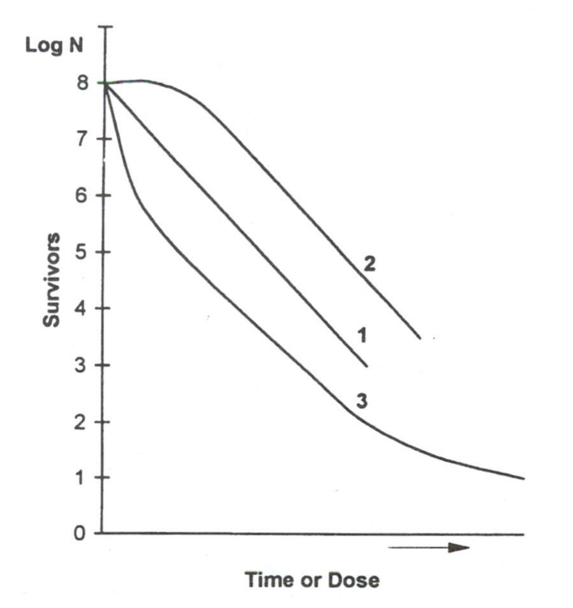

Figure 28.3 Typical radiation survival curves (modified from reference 24). N, survivors of radiation dose; N_0, original number of viable cells; 1, 2, and 3, see text.

Similar to heat resistance, the radiation response in microbial populations can be expressed by the decimal reduction dose (D_{10} value). When the dose-survival curve is a straight line, D_{10} is the reciprocal of its slope,

$$D_{10} = \frac{\text{radiation dose}}{\log N_0 - \log N}$$

where N_0 is the initial number of organisms and N is the number of organisms surviving the radiation dose. Survival curves of type 2 and type 3 are often best represented by quoting an inactivation dose, for example, the dosage to inactivate 90 or 99% of the initial viable cell count.

Several environmental factors influence the radiation resistance of microorganisms (24, 51). (i) The composition of the medium surrounding the microorganisms plays an important role in the dose requirement for a given microbiological effect. Cells irradiated in phosphate buffer are much more sensitive than those irradiated in foodstuffs. In general, the more complex the medium, the greater the competition of the medium components for the free radicals formed from water and activated molecules produced by the radiation, thus protecting the microorganisms. It is impossible to predict in which particular foods bacterial cells will be more radiation sensitive or radiation resistant.

(ii) A reduction in the moisture content (reduction of a_w) of the food protects microorganisms against the lethal effect of ionizing radiation, similar to what is known for heat treatment. In dry conditions, the yield of free radicals produced from water by radiation is lower, and thus the indirect DNA damage that they may cause is decreased.

(iii) The temperature at which a product is treated may influence the radiation resistance of microorganisms. For vegetative cells, elevated temperature treatments, generally in the sublethal range above 45°C, synergistically enhance the lethal effect of radiation. This is thought to occur because the repair systems, which normally operate at ambient and slightly above ambient temperatures, are damaged at higher temperatures.

(iv) Freezing causes a striking increase in the radiation resistance of vegetative cells. Microbial radiation resistance in frozen foods is about two- or threefold higher than at ambient temperature. This increase is due

Table 28.11 Dose requirements of various applications of food irradiation[a]

Application	Dose requirement (kGy)
Inhibition of sprouting of potatoes and onions	0.03–0.12
Insect disinfestation of seed products, flours, fresh and dried fruits, etc.	0.2–0.8
Parasite disinfestation of meat and other foods	0.1–3.0
Radurization of perishable food items (fruits, vegetables, meat, poultry, fish)	0.5–10
Radicidation of frozen meat, poultry, eggs, and other foods and feeds	3.0–10
Reduction or elimination of microbial population in dry food ingredients (spices, starch, enzyme preparations, etc.)	3.0–20
Radappertization of meat, poultry, and fish products	25–60

[a]From reference 36.

to the immobilization of the free radicals and prevention of their diffusion when the medium is frozen. Thus, in the frozen state, the indirect DNA damage by ˙OH radicals is nearly prevented. The change in resistance with temperature demonstrates the importance of the indirect action in high-moisture foods.

(v) The lethal effect of ionizing radiation on microbial cells increases in the presence of oxygen. In the total absence of oxygen, and in wet conditions, radiation resistance usually increases by a factor of 2 to 4. In dry conditions, radiation resistance might increase by a factor of 8 to 17. Data reported for cell suspensions irradiated in sealed tubes can frequently be plotted as type 3 survival curves (Fig. 28.3). These tailing effects probably represent a shift to anaerobic conditions; that is, irradiation of aerated water will consume all of the oxygen after a dose of 500 Gy.

The foregoing discussion emphasizes the necessity for specifying the environmental conditions in measuring and citing the radiation sensitivity of a microorganism.

Technological Fundamentals

Generally, there is a minimum dose requirement. Whether every mass element of a food requires irradiation will depend on the purpose of the treatment. In some cases, irradiation of the surface will suffice. In others, the entire food must receive the minimum dose. Guidelines for dose requirements are given in Table 28.11. Detailed requirements must again be considered as specific for a given food.

Radurization is the term for substantial reduction in the number of spoilage microorganisms, thereby extending the refrigerated shelf life of a food. With relatively high doses, approaching 5 kGy, the normal spoilage organisms are often eliminated, and the more resistant and metabolically less active species, e.g., resistant *Moraxella* spp., lactic acid bacteria, and yeasts, remain. During chilled storage, this surviving biota grows less rapidly than the normal spoilage biota, and the storage life is correspondingly further extended to three or four times normal. As with the unirradiated foods, the nature of the spoilage biota is strongly influenced by the nature of the packaging. In anaerobic packages, lactic acid bacteria and yeasts predominate.

Radicidation is the term for reduction in number of specific viable non-spore-forming pathogenic microorganisms (other than viruses) and parasites, thereby improving the hygienic quality and reducing the risk to the public from certain pathogens.

Radappertization is the term for application of ionizing radiation to prepackaged, enzyme-inactivated foods in doses sufficient to reduce the number and/or activity of viable microorganisms to such an extent that in the absence of postprocessing contamination, no microbial spoilage or toxicity should occur, no matter how long or under what conditions the food is stored. The dose requirement for radappertization is determined by the most radiation resistant microorganism associated with the enzyme-inactivated food. For non-acid, low-salt foods, this is the *C. botulinum* type A spore. By analogy with thermal processing, the safety of the radappertization process is based on a 12 *D* reduction in the viability of botulinal spores. For low-salt, non-acid foods, the required dose is about 45 to 50 kGy (29, 114).

The usefulness of ionizing radiation in overcoming microbiological problems has been clearly demonstrated with a variety of foodstuffs and bacteria of public health significance, especially pathogenic bacteria such as salmonellae, *L. monocytogenes*, campylobacters, and *Y. enterocolitica*, (37). Radiation doses of 3 to 10 kGy are sufficient to reduce viable cell counts in spices, herbs, and enzyme preparations to satisfactory levels (38).

Radiation treatment causes practically no temperature rise in the product. The largest dose likely to be used is about 50 kGy. This amount of energy is equivalent to about 12 cal (50 J). Hence, if all of the ionizing radiation degrades to heat, the temperature of a food will rise about 12°C. For this reason, irradiation has been termed a cold process.

Ionizing irradiation can be applied through any type of packaging materials, including those which cannot withstand heat. Radiation can be applied after packaging, thus avoiding recontamination and reinfestation. Most standard packaging materials are satisfactory for use with irradiated foods.

One should keep in mind that the quality of the irradiated foods, as for any other preserved food, is a function of the quality of the original material and that to obtain good results, good manufacturing practice is needed. The longest shelf life may be obtained if the quality of the raw material is good and proper hygienic conditions are maintained. In certain cases, the use of more thorough inspection and sorting to remove substandard-quality raw materials before treatment (or treatments such as washing, blanching, or chemical rinse) can be used to reduce the radiation dose required to produce a high-quality product. The benefits of irradiation should never be considered as a substitute for product quality or as compensation for poor handling and storage conditions.

The basic irradiation process applies a prescribed amount of ionizing radiation of either electrons or electromagnetic rays and uses certain other procedures which may be needed. Such other procedures will vary according to the food being treated and the purpose of the treatment. For radappertized foods intended to have shelf stability without refrigeration, packaging suitable to protect the food from microbial contamination after irradiation is essential, and inactivation of the radiation-resistant food enzymes by mild heat processing may be necessary. For radurized foods, enzyme inactivation is not necessary, and protective packaging may or may not be needed. Protection to prevent reinfestation may be needed for products irradiated to control insects. Specific storage conditions could be required, depending on the nature of the products treated.

Some products may require irradiation under special conditions, such as at low temperature or in an oxygen-free atmosphere, or with combination treatments such as heat and irradiation (39). Sensitization of microorganisms with the aim of lowering the dose can be obtained by physical (heat) and chemical means. The microbiota surviving irradiation is more sensitive to subsequent food processing treatments than the microbiota of unirradiated products.

The actual radiation dose used is a balance between what is needed and what can be tolerated by the product without unwanted changes. High doses cause organoleptic changes (off flavors and/or texture changes) in high-moisture food. These undesirable effects can be limiting factors. In the case of protein foods of animal origin, the maximum dose permitted usually is set by the development of the "irradiation flavor."

Radiation-induced changes in the texture of some foods, such as fresh fruits and vegetables, can be a limiting factor. Polysaccharides such as pectin can be degraded by irradiation. This is often accompanied by a release of calcium, which causes softening of the product. These observations, and also the observed increase in permeability, can explain the fact that the natural resistance of the irradiated plant tissues is often reduced, resulting in an accelerated spoilage when microorganisms are given a chance to attack the products after irradiation. However, this radiation-induced initial softening in several commodities is often compensated for by the fact that during extended postirradiation storage, this softening (after-ripening) takes place considerably more slowly than in the untreated plant tissues.

In some cases, the chemical and physical effects of irradiation can be regarded as improvements. For example, the radiation-induced softening and increase in permeability are advantageous when shortening of the cooking time (e.g., for dried vegetables) is required; an increased yield of juice can be obtained from irradiated fruits.

NEW NONTHERMAL PHYSICAL PRESERVATION TREATMENTS

During recent years, several other nonthermal physical treatments such as high hydrostatic pressure and use of electric and magnetic fields have attracted much interest as promising tools for food processing and preservation and for creating new types of food products (90).

High Hydrostatic Pressure for Food Preservation

One advantage of the application of high or ultrahigh (300 to 1000 MPa) hydrostatic pressure is that it is transferred throughout food instantly and uniformly (Pascal principle) and thus is independent of sample size and geometry. It can be applied at ambient or moderate temperatures. Because only noncovalent bonds are affected by the treatment, there is good retention of color, nutrients and flavor (19). High-pressure-treated fruit juices and other fruit products were first marketed in

Japan, but research and development work is ongoing in several countries (132, 145). Several reviews (4, 40, 89) summarize high-pressure treatment of microorganisms, foods, and food components or developments in high-pressure food processing technologies. The chemical principle behind the effect of ultrahigh pressure is that noncovalent bonds, such as hydrogen, ion, and hydrophobic bonds in proteins and carbohydrates, undergo changes which alter their molecular structures.

Microbial inactivation is probably due to a number of factors. One likely cause is altered permeabilities of cell membranes, leading to cell leakage. Microbial resistances to heat and pressure are often associated. For example, the relatively heat resistant *L. monocytogenes* was also more resistant to pressure than *Salmonella thompson,* and their pressure-treated cells showed marked differences in ultrastucture (86). However, the heat-resistant strain of *Salmonella senftenberg* is less resistant to pressure than other strains of *Salmonella* (91). Nevertheless, vegetative cells are relatively pressure sensitive, being readily killed in the range of 500 to 700 MPa, while bacterial spores are resistant to pressures far exceeding 300 MPa at ambient temperatures (47, 58). However, the combination of elevated temperatures and pressures of, e.g., 400 MPa is effective against spores (131). Spores can be inactivated by alternating pressure cycles which initiate spore germination under moderate pressure (e.g., 30 min at 80 MPa) and then kill the germinated spores by high pressure (e.g., 30 min at 350 MPa) (15, 47). Pressure germination of spores is affected by pH and ionic environment (47).

High pressure on foods and food components (19) can cause protein denaturation, altered gellation properties of biopolymers, increased susceptibility to enzymatic degradation, and increased drying rate of high-pressure-pretreated plant tissues as a result of permeabilization of cell walls (31, 75). Activation and inactivation of enzymes can be controlled by using selected pressure and temperature conditions (74). However, various enzymes have considerable differences in barosensitivity, and further work is required to study possible regeneration of enzyme activities in foods during subsequent storage (75). The efficiency of high-pressure preservation, like that of other treatments, depends on the food composition and the pH value.

Electric Field Effects

High-voltage discharges (electroporation treatment) can kill microorganisms (105). An external pulsed electric field (PEF) charges cells in such a way that they behave like small dipoles and an electric potential develops over the cell membranes. When the induced electric potential exceeds some critical value, a reversible increase in membrane permeability results (8). For vegetative bacteria, this critical electric field strength is about 15 kV/cm. Much higher field strength is required for inactivation of asco- and endospores (90). When the critical electric field is greatly exceeded, the ion channels and pores of membranes are irreversibly extended to such a degree that cell contents leak out and cells die (dielectric rupture theory). The critical electric field is affected by pulse duration and resistivity of the suspension material (149). Because electric resistivity decreases as ionic strength increases, the bactericidal effect of PEF is generally inversely proportional to the ionic strength of the suspension material (59). Microbial inactivation is proportional to the number and duration of pulses (148). Gram-positive bacteria are less sensitive to electric pulse treatment than gram-negative bacteria (60). PEF inactivation is also a function of the treatment temperature and the microbial growth stage (148).

Over the last decade, high-voltage PEF has been investigated as a potential nonthermal treatment for food preservation (16, 93, 148). Microbiological test results show that bipolar square-wave pulses are the most efficient in terms of microbial inactivation for commercial PEF pasteurization of fluid foods. Both electrical stimulation as used in the meat industry and pulsed electricity significantly reduce the viability of microorganisms on beef surfaces (8). Recent studies have proposed kinetic models for PEF inactivation of microorganisms, calculating the relationship between survival rate, field strength, and the number of pulses (109).

Magnetic Field Effects

Magnetic fields cause a change in the orientation of biomolecules and biomembranes to a direction parallel or perpendicular to the applied magnetic field and induce a change in the ionic drift across the plasma membrane (117). Thus, magnetic fields may affect growth and reproduction of microorganisms. It was found that oscillating magnetic fields with an intensity (flux density) of 5 to 50 T with short pulses of 25 μs to 10 ms and frequencies of 5 to 500 kHz can inactivate vegetative microorganisms and pasteurize food material of high electrical resistivity, packaged in plastic packaging, without a significant temperature increase (56, 117). (Microwave cooking also involves application of magnetic fields but differs in that it produces a thermal effect.) The explanation of the microbicidal effect of oscillating magnetic field pulses suggests that the oscillating magnetic field couples enough energy into the magneto-active parts (magnetic dipoles) of large biological molecules to break bonds in DNA or proteins critical for the survival and reproduction of microorganisms. As with high-voltage pulse treatment, the total exposure time is very short and produces no significant temperature increase (90). Much further research is required, however, to overcome technical difficulties, understand the mechanisms of microbial inactivation, and establish the commercial feasibility of this treatment.

Future Work Needed

Like the established physical preservation methods discussed earlier in this chapter, the nonthermal physical treatments discussed above induce sublethal injury to microbial cells surviving the treatments. The injury makes them sensitive to different physical and chemical environments to which normal cells are resistant. Thus, both conventional and newly developed physical treatments can be used in combination with other preservative factors to increase antimicrobial efficiency and enhance the safety and shelf life of foods (71). The various newer nonthermal physical treatments appear to be promising tools for food preservation. However, conflicting data about their effects on bacterial spores and enzymes and conflicting kinetic data need to be reconciled. The mechanistic understanding of the effects of these processes on foods, food components, and microbial cells, as well as of storage-dependent changes after nonthermal treatments, is limited. Thus, further research should concentrate on these questions.

References

1. **Ababouch, L., and F. F. Busta.** 1987. Effect of thermal treatments in oils on bacterial spore survival. *J. Appl. Bacteriol.* **62**:491–502.
2. **Anderson, A. W., H. C. Nordan, R. F. Cain, G. Parnish, and D. Duggan.** 1956. Studies on a radio-resistant micrococcus. I. The isolation, morphology, cultural characteristics and resistance to gamma radiation. *Food Technol.* **10**:575–577.
3. **Anderson, J. G., and J. E. Smith.** 1976. Effects of temperature on filamentous fungi. p. 191–218. *In* F. A. Skinner and W. B. Hugo (ed.), *Inhibition and Inactivation of Vegetative Microbes.* Academic Press, London.
4. **Anonymous.** 1993. Use of hydrostatic pressure in food processing. *Food Technol.* **47**(6):149.
5. **Auffray, Y., E. Lecesne, A. Hartky, and P. Boulibounes.** 1995. Basic features of the *Streptococcus thermophilus* heat shock response. *Curr. Microbiol.* **30**:87–91.
6. **Ayerst, G.** 1969. The effects of moisture and temperature on growth and spore germination in some fungi. *J. Stored Prod. Res.* **5**:127–141.
7. **Baird-Parker, A. C., and B. Freame.** 1967. Combined effect of water activity, pH and temperature on the growth of *Clostridium botulinum* from spore and vegetative cell inocula. *J. Appl. Bacteriol.* **30**:420–429.
8. **Bawcom, D. W., L. D. Thompson, M. F. Miller, and C. B. Ramsey.** 1995. Reduction of microorganisms on beef surfaces utilizing electricity. *J. Food Prot.* **58**:35–38.
9. **Baxter, R. M., and N. E. Gibbons.** 1962. Observations on the physiology of psychrophilism in a yeast. *Can. J. Microbiol.* **8**:511–517.
10. **Boutibonnes, P., V. Bisson, B. Thammavongs, A. Hartke, J. M. Panoff, A. Benackour, and Y. Auffray.** 1995. Induction of thermotolerance by chemical agents in *Lactococcus lactis* subsp. *lactis* IL 1403. *Int. J. Food Microbiol.* **25**:83–94.
11. **Brown, A. D.** 1974. Microbial water relations: features of the intracellular composition of sugar-tolerant yeasts. *J. Bacteriol.* **118**:769–777.
12. **Brown, A. D.** 1976. Microbial water stress. *Bacteriol. Rev.* **40**:803–846.
13. **Brown, A. D., and J. R. Simpson.** 1972. Water relations of sugar-tolerant yeasts: the role of intracellular polyols. *J. Gen. Microbiol.* **72**:589–591.
14. **Brown, K. L., and C. A. Ayres.** 1982. Thermobacteriology of UHT processed foods. *Dev. Food Microbiol.* **1**:119–152.
15. **Butz, P., J. Ries, U. Traugott, H. Weber, and H. Ludwig.** 1990. Hochdruckinaktivierung von Bakterien und Bakteriensporen. *Pharm. Ind.* **52**:487–491.
16. **Castro, A. J., G. V. Barbosa-Canovas, and B. G. Swanson.** 1993. Microbial inactivation of foods by pulsed electric fields. *J. Food Process. Preserv.* **17**:47–73.
17. **Cerf, O.** 1977. Tailing of survival curves of bacterial spores. *J. Appl. Bacteriol.* **42**:1–19.
18. **Chang, J. C. H., S. F. Ossoff, D. C. Lobe, M. H. Dorfman, C. M. Dumais, R. G. Qualls, and J. D. Johnson.** 1985. UV inactivation of pathogenic and indicator microorganisms. *Appl. Environ. Microbiol.* **49**:1361–1365.
19. **Cheftel, J. C.** 1992. Effects of high hydrostatic pressure on food constituents: an overview, p. 195–209. *In* C. Balny, R. Hayashi, K. Heremans, and P. Masson (ed.), *High Pressure and Biotechnology.* John Libbey & Co. Ltd., London.
20. **Church, N.** 1994. Developments in modified-atmosphere packaging and related technologies. *Trends Food Sci. Technol.* **5**:345–352.
21. **Cole, M. B., K. W. Davies, G. Munro, C. D. Holyoak, and D. C. Kilsby.** 1993. A vitalistic model to describe the thermal inactivation of *Listeria monocytogenes. J. Ind. Microbiol.* **12**:232–239.
22. **Condon, S., and F. J. Sala.** 1992. Heat resistance of *Bacillus subtilis* in buffer and foods of different pH. *J. Food Prot.* **55**:605–608.
23. **Corry, J. E. L.** 1987. Relationship of water activity to fungal growth, p. 51–99. *In* L. R. Beuchat (ed.), *Food and Beverage Mycology.* AVI Van Nostrand Reinhold Co., New York.
24. **Davies, R.** 1976. The inactivation of vegetative bacterial cells by ionizing radiation, p. 239–255. *In* F. A. Skinner and W. B. Hugo (ed.), *Inhibition and Inactivation of Vegetative Microbes.* Academic Press, London.
25. **Dawes, I. W.** 1976. Inactivation of yeast, p. 279–304. *In* F. A. Skinner and W. B. Hugo (ed.), *Inhibition and Inactivation of Vegetative Microbes.* Academic Press, London.

26. **Deák, T.** 1991. Preservation by dehydration, p. 70–82. *In* E. Szenes and M. Oláh (ed.), *Handbook of Canning Industry.* Integra Projekt, Budapest. (In Hungarian.)
27. **Decereau, R. V.** 1985. *Microwaves in the Food Processing Industry.* Academic Press, Orlando.
28. **Decereau, R. V.** 1992. *Microwave Foods: New Product Development.* Food & Nutrition Press, Trumbull, Conn.
29. **Diehl, J. F.** 1990. *Safety of Irradiated Foods.* Marcel Dekker, Inc., New York.
30. **Diehl, J. F.** 1992. Food irradiation: is it an alternative to chemical preservatives? *Food Addit. Contam.* **9**:409–416.
31. **Dörnenburg, H., and D. Knorr.** 1992. Cellular permeabilization of cultured plant tissues by high electric field pulses or ultra high pressure for the recovery of secondary metabolites. *Food Biotechnol.* **7**:35.
32. **Doyle, M. P., and E. H. Martihn.** 1975. Thermal inactivation of conidia from *Aspergillus flavus* and *Aspergillus parasiticus. J. Milk Food Technol.* **38**:678.
33. **Eckardt, C., and E. Ahrens.** 1977. Untersuchungen über *Byssochlamys fulva* Olliver & Smith als potentiellen Verderbniserreger in Erdbeerkonserven. II. Hitzeresistenz der Ascosporen von *Byssochlamys fulva. Chem. Mikrobiol. Technol. Lebensm.* **5**:76–80.
34. **Edgey, M., and A. D. Brown.** 1978. Response of xerotolerant and nontolerant yeasts to water stress. *J. Gen. Microbiol.* **104**:343–345.
35. **Erichsen, I., and G. Molin.** 1981. Microbial flora of normal and high pH beef stored at 4°C in different gas environments. *J. Food Prot.* **44**:866.
36. **Farkas, J.** 1980. Principles of food irradiation, p. 1–7. *In* Handouts of IFFIT International Training Course on Food Irradiation. International Facility for Food Irradiation Technology, Wageningen, The Netherlands.
37. **Farkas, J.** 1987. Decontamination, including parasite control, of dried, chilled and frozen food by irradiation. *Acta Aliment.* **16**:351–384.
38. **Farkas, J.** 1988. *Irradiation of Dry Food Ingredients.* CRC Press, Inc., Boca Raton, Fla.
39. **Farkas, J.** 1990. Combination of irradiation with mild heat treatment. *Food Control* **1**:223–229.
40. **Farr, D.** 1990. High pressure technology in the food industry. *Trends Food Sci. Technol.* **1**:14–16.
41. **Finne, G., and J. R. Matches.** 1976. Spin-labeling studies on the lipids of psychrophilic, psychrotrophic, and mesophilic clostridia. *J. Bacteriol.* **125**:211–219.
42. **Flowers, R. S., and S. E. Martin.** 1980. Ribosome assembly during recovery of heat-injured *Staphylococcus aureus. J. Bacteriol.* **141**:645–651.
43. **George, A. M., W. A. Cramp, and M. B. Yatwin.** 1980. The influence of membrane fluidity on the radiation induced changes in the DNA of *E. coli* K 1060. *Int. J. Radiat. Biol.* **38**:427–438.
44. **Gill, C. O., and G. Molin.** 1991. Modified atmospheres and vacuum packaging, p. 172–199. *In* N. J. Russell and G. W. Gould (ed.), *Food Preservatives.* Blackie, Glasgow, Scotland.
45. **Goldblith, S. A., L. Rey, and W. W. Rothmayr.** 1975. *Freeze Drying and Advanced Food Technology.* Academic Press, London.
46. **Gorris, L. G. M.** 1994. Improvement of the safety and quality of refrigerated ready-to-eat foods using novel mild preservation techniques, p. 57–72. *In* R. P. Singh and F. A. R. Oliveira (ed.), *Minimal Processing of Foods and Process Optimization. An Interface.* CRC Press, Inc., Boca Raton, Fla.
47. **Gould, G. W.** 1973. Inactivation of spores in food by combined heat and hydrostatic pressure. *Acta Aliment.* **2**:377–383.
48. **Gould, G. W.** 1989. Drying, raised osmotic pressure and low water activity, p. 97–117. *In* G. W. Gould (ed.), *Mechanisms of Action of Food Preservation Procedures.* Elsevier Applied Science, London.
49. **Gould, G. W.** 1989. Heat induced injury and inactivation, p. 11–42. *In* G. W. Gould (ed.), *Mechanisms of Action of Food Preservation Procedures.* Elsevier Applied Science, London.
50. **Grant, I. R., and M. F. Patterson.** 1989. A novel radiation-resistant *Deinobacter* sp. isolated from irradiated pork. *Lett. Appl. Microbiol.* **8**:21–24.
51. **Grecz, N., D. B. Rowley, and A. Matsuyama.** 1983. The action of radiation on bacteria and viruses. *In* E. S. Josephson and M. S. Peterson (ed.), *Preservation of Food by Ionizing Radiation,* vol. 2. CRC Press, Inc., Boca Raton, Fla.
52. **Han, Y. W.** 1975. Death rates of bacterial spores: nonlinear survivor curves. *Can. J. Microbiol.* **21**:1464–1467.
53. **Hiatt, C. W.** 1964. Kinetics of the inactivation of viruses. *Bacteriol. Rev.* **28**:150–163.
54. **Hightower, L. E.** 1991. Heat shock, stress proteins, chaperones, and proteotoxicity. *Cell* **66**:191–197.
55. **Ho, Y. C., and K. L. Yam.** 1993. Effect of metal shielding on microwave heating uniformity of a cylindrical food model. *J. Food Process. Preserv.* **16**:337–359.
56. **Hofman, G. A.** 1985. Deactivation of microorganisms by an oscillating magnetic field. U.S. patent 4,524,079.
57. **Hollywood, N. W., Y. Varabioff, and G. E. Mitchell.** 1991. The effect of microwave and conventional cooking on the temperature profiles and microbial flora of minced beef. *Int. J. Food Microbiol.* **14**:67–76.
58. **Hoover, D. G., C. Metrick, A. M. Papineau, D. F. Farkas, and D. Knorr.** 1989. Application of high hydrostatic pressure on foods to inactivate pathogenic and spoilage organisms for extension of shelf life. *Food Technol.* **43(3)**:99.
59. **Hülsheger, H., J. Potel, and E. G. Niemann.** 1981. Killing of bacteria with electric pulses of high field strength. *Rad. Environ. Biophys.* **20**:53–65.
60. **Hülsheger, H., J. Potel, and E. G. Neumann.** 1983. Electric field effects on bacteria and yeast cells. *Rad. Environ. Biophys.* **22**:149–162.
61. **Hurst, A.** 1984. Reversible heat damage, p. 303–318. *In* A. Hurst and A. Nasim (ed.), *Repairable Lesions in Microorganisms.* Academic Press, London.
62. **Ingram, M., and J. Farkas.** 1977. Microbiology of foods pasteurized by ionising radiation. *Acta Aliment.* **6**:123–185.
63. **Ingram, M., and B. M. Mackey.** 1976. Inactivation by cold. *Soc. Appl. Bacteriol. Symp. Ser.* **5**:111–151.
64. **International Commission on Microbiological Specifications for Foods.** 1980. Temperature, p. 1–37. *In* J. H.

Silliker et al. (ed.), *Microbial Ecology of Foods,* vol. 1. *Factors Affecting Life and Death of Microorganisms.* Academic Press, New York.

65. **International Commission on Microbiological Specifications for Foods.** 1980. Ionizing radiation, p. 46–69. *In* J. H. Silliker et al. (ed.), *Microbial Ecology of Foods,* vol. 1. *Factors Affecting Life and Death of Microorganisms.* Academic Press, New York.
66. **International Commission on Microbiological Specifications for Foods.** 1980. Reduced water activity, p. 70–91. *In* J. H. Silliker et al. (ed.), *Microbial Ecology of Foods,* vol. 1. *Factors Affecting Life and Death of Microorganisms.* Academic Press, New York.
67. **Jakobsen, M., O. Filtenborg, and F. Bramsnaes.** 1972. Germination and outgrowth of the bacterial spore in the presence of different solutes. *Lebensm. Wiss. Technol.* **5:**159–162.
68. **Jay, J. M.** 1992. *Modern Food Microbiology,* 4th ed. Chapman & Hall, New York.
69. **Jermini, M. F. G., and W. Schmidt-Lorenz.** 1987. Heat resistance of vegetative cells and asci of two *Zygosaccharomyces* yeasts in broth at different water activity values. *J. Food Prot.* **50:**835–841.
70. **Josephson, E. S., and M. S. Peterson (ed.).** 1983. *Preservation of Foods by Ionizing Radiation.* CRC Press, Inc., Boca Raton, Fla.
71. **Kalchayanand, N., T. Sikes, C. P. Dunne, and B. Ray.** 1994. Hydrostatic pressure and electroporation have increased bactericidal efficiency in combination with bacteriocins. *Appl. Environ. Microbiol.* **60:**4174–4177.
72. **Kang, C. K., M. Woodburn, A. Pagenkopf, and R. Cheney.** 1969. Growth, sporulation and germination of *Clostridium perfringens* in media of controlled water activity. *Appl. Microbiol.* **18:**798–805.
73. **Keteleer, A., and P. P. Tobback.** 1994. Modified atmosphere storage of respiring product, p. 59–64. *In* L. Leistner and L. G. M. Gorris (ed.), *Food Preservation by Combined Processes.* Final Report, FLAIR Concerted Action no. 7, Subgroup B. EUR 15776 EN. European Commission, Brussels.
74. **Knorr, D.** 1993. Effects of high hydrostatic pressure processes on food safety and quality. *Food Technol.* **47**(6): 156–161.
75. **Knorr, D.** 1994. Non-thermal processes for food preservation, p. 3–15. *In* R. P. Singh and F. A. R. Oliveira (ed.), *Minimal Processing of Foods and Process Optimization. An Interface.* CRC Press, Inc., Boca Raton, Fla.
76. **Lagunas-Solar, M. C.** 1995. Radiation processing of foods. An overview of scientific principles and current status. *J. Food Prot.* **58:**186–192.
77. **Laszlo, A.** 1988. Evidence for two states of thermotolerance in mammalian cells. *Int. J. Hyperthermia* **4:**513–526.
78. **Lee, B. H., S. Kermasha, and B. E. Baker.** 1989. Thermal, ultrasonic and ultraviolet inactivation of *Salmonella* in thin films of aqueous media and chocolate. *Food Microbiol.* **6**(3):143–152.
79. **Leistner, L., and N. J. Russell.** 1991. Solutes and low water activity, p. 111–134. *In* N. J. Russell and G. W. Gould (ed.), *Food Preservatives.* Blackie, Glasgow, Scotland.
80. **Le Jean, G., G. Abraham, E. Debray, Y. Candau, and G. Piar.** 1994. Kinetics of thermal destruction of *Bacillus stearothermophilus* spores using a two reaction model. *Food Microbiol.* **11:**229–241.
81. **Lewis, M.** 1993. UHT processing: safety and quality aspects. *Food Technol. Int. Eur.* **1993:**47–51.
82. **Löndahl, G., and T. E. Nilsson.** 1978. Microbiological aspects of the freezing of meat and prepared foods. *Int. J. Refrig.* **1**(1):53–56.
83. **Mackey, B. M., and C. M. Derrick.** 1986. Peroxide sensitivity of cold-shocked *Salmonella typhimurium* and *Escherichia coli* and its relationship to minimal medium recovery. *J. Appl. Bacteriol.* **60:**501–511.
84. **Mackey, B. M., and C. M. Derrick.** 1986. Elevation of heat resistance of *Salmonella typhimurium* by sublethal heat shock. *J. Appl. Bacteriol.* **61:**389–394.
85. **Mackey, B. M., and C. M. Derrick.** 1987. Changes in the heat resistance of *Salmonella typhimurium* during heating at rising temperatures. *Lett. Appl. Microbiol.* **4:**13–16.
86. **Mackey, B. M., K. Forestiere, N. S. Isaacs, R. Stenning, and B. Brooker.** 1994. The effect of high hydrostatic pressure on *Salmonella thompson* and *Listeria monocytogenes* examined by electron microscopy. *Lett. Appl. Microbiol.* **19:**429–432.
87. **Macleod, R. A., and P. H. Calcott.** 1976. Cold shock and freezing damage to microbes. *Symp. Soc. Gen. Microbiol.* **26:**81–109.
88. **Maunder, D. T.** 1977. Possible use of ultraviolet sterilization of containers for aseptic packaging. *Food Technol.* **31**(4):36–37.
89. **Mertens, B.** 1993. Developments in high pressure food processing. *ZFL Int. J. Food Technol. Mark. Packag. Anal.* **44:**100, 182.
90. **Mertens, B., and D. Knorr.** 1992. Development of nonthermal processes for food preservation. *Food Technol.* **46**(5):124–133.
91. **Metrick, C., D. Hoover, and D. Farkas.** 1989. Effects of high hydrostatic pressure on heat-resistant and heat-sensitive strains of *Salmonella. J. Food Sci.* **54:**1547–1549.
92. **Michels, M. J. M.** 1978. Die mikrobiologische Qualitat von Trockengemuse. *Lebensm. Technol. Verfahrenstech.* **29**(1):14–18.
93. **Mittal, G. S., S. Ho, and J. Cross.** 1994. Food pasteurization using high voltage electric pulses. *J. Food Phys.* (Budapest) *Suppl.* **1:**99–101.
94. **Monk, D. J., L. R. Beuchat, and M. P. Doyle.** 1995. Irradiation inactivation of food-borne microorganisms. *J. Food Prot.* **58:**197–208.
95. **Montville, T. J., and G. M. Sapers.** 1981. Thermal resistance of spores from pH elevating strains of *Bacillus licheniformis. J. Food Sci.* **46:**1710–1712.
96. **Moreno, M. A., M. del Carmen Ramos, A. Gonzalez, and G. Suarez.** 1987. Effect of ultraviolet light irradiation on viability and aaflatoxin production by *Aspergillus parasiticus. Can. J. Microbiol.* **33:**927–929.
97. **Moseley, B. E. B.** 1989. Ionizing radiation: action and repair, p. 43–70. *In* G. W. Gould (ed.), *Mechanisms of Action of Food Preservation Procedures.* Elsevier Applied Science, London.

98. **Mossel, D. A. A.** 1975. Occurrence, prevention and monitoring of microbial quality loss of foods and dairy products. *Crit. Rev. Environ. Control* **5**:1–139.
99. **Mossel, D. A. A., and M. Ingram.** 1955. The physiology of the microbial spoilage of foods. *J. Appl. Bacteriol.* **18**:232–268.
100. **Mudgett, R. E.** 1985. Dielectrical properties of foods, p. 15–37. *In* R. V. Decareau (ed.), *Microwaves in the Food Processing Industry.* Academic Press, Orlando, Fla.
101. **Narasimhan, R., M. M. Habibullah-Khan, J. Ernest, and K. Thangavel.** 1989. Effect of ultraviolet radiation on the bacterial flora of the packaging materials of milk and milk products. *Cherion* **18**(2):89–92.
102. **Nychas, G. J. E.** 1994. Modified atmosphere packaging of meats, p. 417–436. *In* R. P. Singh and F. A. R. Oliveira (ed.), *Minimal Processing of Foods and Process Optimization. An Interface.* CRC Press, Inc., Boca Raton, Fla.
103. **Nychas, G. J. E., and J. S. Arkoudelos.** 1990. Microbiological and physico-chemical changes in minced meat under carbon dioxide, nitrogen or air at 3°C. *Int. J. Food Sci. Technol.* **25**:389.
104. **Ohye, D. F., and J. H. B. Christian.** 1967. Combined effects of temperature, pH and water activity on growth and toxin production by *C. botulinum* types A, B and E, p. 217–223. *In* M. Ingram and T. A. Roberts (ed.) *Botulism 1966.* Chapman & Hall, London.
105. **Palaniappan, S., S. K. Sastry, and E. R. Richter.** 1990. Effects of electricity on microorganisms: a review. *J. Food Process. Preserv.* **14**:393–414.
106. **Palaniappan, S., S. K. Sastry, and E. R. Richter.** 1992. Effects of electroconductive heat treatment and electrical pretreatment on thermal death kinetics of selected microorganisms. *Biotechnol. Bioeng.* **39**:225–232.
107. **Palumbo, S. A.** 1987. Is refrigeration enough to restrain foodborne pathogens? *J. Food Prot.* **49**:1003.
108. **Parsell, D. A., and R. T. Sauer.** 1989. Induction of heat shock-like response by unfolded protein in *Escherichia coli:* dependence on protein level not protein degradation. *Genes Dev.* **3**:1226–1232.
109. **Peleg, M.** 1995. A model of microbial survival after exposure to pulsed electric fields. *J. Sci. Food Agric.* **67**:93–99.
110. **Perkin, A. G., F. L. Davies, P. Neaves, B. Jarvis, C. A. Ayres, K. L. Brown, W. C. Falloon, H. Dallyn, and P. G. Bean.** 1980. Determination of bacterial spore inactivation at high temperatures, p. 173–188. *In* G. W. Gould and J. E. L. Corry (ed.), *Microbial Growth and Survival in Extreme Environments.* Academic Press, London.
111. **Pflug, I. J.** 1990. *Microbiology and Engineering of Sterilization Process,* 7th ed. Environmental Sterilization Laboratory, Minneapolis.
112. **Pichhardt, K.** 1993. *Lebensmittelmikrobiologie.* 3. *Auflage.* Springer-Verlag, Berlin.
113. **Pitt, J. I.** 1975. Xerophilic fungi and the spoilage of foods of plant origin, p. 273–307. *In* R. B. Duckworth (ed.), *Water Relations of Foods.* Academic Press, London.
114. **Pitt, J. I., and A. D. Hocking.** 1985. *Fungi and Food Spoilage.* Academic Press, Sidney, New South Wales, Australia.
115. **Pitt, J. I., and B. F. Miscamble.** 1995. Water relations of *Aspergillus flavus* and closely related species. *J. Food Prot.* **58**:86–90.
116. **Ponne, C. T., and P. V. Bartels.** 1995. Interaction of electromagnetic energy with biological material—relation to food processing. *Radiat. Phys. Chem.* **45**:591–607.
117. **Pothakamury, U. R., G. V. Barbosa-Canovas, and B. G. Swanson.** 1993. Magnetic field inactivation of microorganisms and generation of biological changes. *Food Technol.* **47**(12):85–93.
118. **Potter, N. N.** 1986. *Food Science,* 4th ed. AVI-Van Nostrand Reinhold, New York.
119. **Put, H. M. C., J. de Jong, F. E. M. Sand, and A. M. van Grinsven.** 1976. Heat resistance studies on yeast spp. causing spoilage in soft drinks. *J. Appl. Bacteriol.* **40**:135–152.
120. **Raso, J., S. Condon, and F. J. Sala Trepat.** 1994. Mano-thermo-sonication: a new method of food preservation, p. 37–41. *In* L. Leistner and L. G. M. Gorris (ed.), *Food Preservation by Combined Process.* Final report, FLAIR Concerted Action no. 7, Subgroup B. EUR 15776. European Commission, Brussels.
121. **Reichart, O., and C. Mohácsi-Farkas.** 1994. Mathematical modelling of the combined effect of water activity, pH and redox potential on the heat destruction. *Int. J. Food Microbiol.* **24**:103–112.
122. **Reuter, H.** 1989. *Aseptic Packaging of Food.* Technomic Publishing, Lancaster, Pa.
123. **Roberts, T. A., G. Hobbs, J. H. B. Christian, and N. Skovgaard.** 1981. *Psychrotrophic Microorganisms in Spoilage and Pathogenicity.* Academic Press, London.
124. **Rockland, L. B., and L. R. Beuchat (ed.).** 1987. *Water Activity: Theory and Applications to Food.* Marcel Dekker, Inc., New York.
125. **Rosenberg, U., and W. Bogl.** 1987. Microwave pasteurization, sterilization, blanching and pest control in the food industry. *Food Technol.* **41**(6):92–99.
126. **Saleh, Y. G., M. S. Mayo, and D. G. Ahearn.** 1988. Resistance of some common fungi to gamma irradiation. *Appl. Environ. Microbiol.* **54**:2134–2135.
127. **Sapru, V., A. A. Teixeira, G. H. Smerage, and J. A. Lindsay.** 1992. Predicting thermophilic spore population dynamics for U. H. T. sterilization processes. *J. Food Sci.* **57**:1248–1257.
128. **Sastry, S. K.** 1994. Ohmic heating, p. 17–33. *In* R. P. Singh and F. A. R. Oliveira (ed.), *Minimal Processing of Foods and Process Optimization. An Interface.* CRC Press, Inc., Boca Raton, Fla.
129. **Sato, M., and H. Takahashi.** 1968. Cold shock of bacteria. I. General features of cold shock in *Escherichia coli. J. Gen. Appl. Microbiol.* **14**:417–428.
130. **Schlesinger, M., M. Ashburner, and A. Tissieres.** 1982. *Heat Shock from Bacteria to Man.* Cold Spring Harbor Laboratory Press, Cold Spring Harbor, N.Y.
131. **Seyderhelm, I., and D. Knorr.** 1992. Reduction of *Bacillus stearothermophilus* spores by combined high pressure and temperature treatments. *ZFL Int. J. Food Technol. Mark. Packag. Anal.* **43**(4):17.
132. **Shigehisa, T., T. Ohmori, A. Saito, S. Taji, and R. Hayashi.** 1991. Effects of high hydrostatic pressure on characteristics of pork slurries and inactivation of microorganisms associated with meat and meat products. *Int. J. Food Microbiol.* **12**:207–216.

133. **Splittstoesser, D. F., S. B. Leasor, and K. M. J. Swanson.** 1986. Effect of food composition on the heat resistance of yeast ascospores. *J. Food Sci.* **51:**1265–1267.

134. **Splittstoesser, D. F., and C. M. Splittstoesser.** 1977. Ascospores of *Byssochlamys fulva* compared with those of a heat resistant *Aspergillus*. *J. Food Sci.* **42:**685–688.

135. **Stannard, C. J., J. S. Abbiss, and J. M. Wood.** 1983. Combined treatment with hydrogen peroxide and ultra-violet irradiation to reduce microbial contamination levels in preformed packaging cartons. *J. Food Prot.* **46:**1060–1064.

136. **Stephens, P. J., M. B. Cole, and M. V. Jones.** 1994. Effect of heating rate on the thermal inactivation of *Listeria monocytogenes*. *J. Appl. Bacteriol.* **77:**702–708.

137. **Stumbo, C. R.** 1973. *Thermobacteriology in Food Processing,* 2nd ed. Academic Press, New York.

138. **Stumbo, C. R., K. S. Purohit, T. V. Ramakrishnan, D. A. Evans, and F. J. Francis.** 1983. *Handbook of Lethality Guides for Low-Acid Canned Foods,* vol. 1 and 2. CRC Press, Inc., Boca Raton, Fla.

139. **Tartera, C., A. Bosch, and J. Jofre.** 1988. The inactivation of bacteriophages infecting *Bacteroides fragilis* by chlorine treatment and UV-irradiation. *FEMS Microbiol. Lett.* **56:**313–316.

140. **Tomlins, R. I., and Z. J. Ordal.** 1976. Thermal injury and inactivation in vegetative bacteria. *Soc. Appl. Bacteriol. Symp. Ser.* **5:**153–190.

141. **Troller, J. A.** 1983. Effect of low moisture environments on the microbial stability of foods, p. 173–198. *In* A. H. Rose (ed.), *Economic Microbiology,* vol. 8. Academic Press, London.

142. **Troller, J. A., and J. V. Stinson.** 1975. Influence of water activity on growth and enterotoxin formation by *Staphylococcus aureus* in foods. *J. Food Sci.* **40:**802–804.

143. **Tsuchido, T., M. Takano, and I. Shibasaki.** 1974. Effect of temperature-elevating process on the subsequent isothermal death of Escherichia coli K-12. *J. Ferment. Technol.* **52:**788–792.

144. **Urbain, W. M.** 1986. *Food Irradiation.* Academic Press, Orlando, Fla.

145. **Vardag, T., and P. Korner.** 1995. High pressure: a real alternative in food processing. *Int. Food Mark. Technol.* 9(1):42–47.

146. **Wilkins, P. O.** 1973. Psychrotrophic gram-positive bacteria: temperature effects on growth and solute uptake. *Can. J. Microbiol.* **19:**909–915.

147. **World Health Organization.** 1988. *Food Irradiation. A Technique for Preserving and Improving the Safety of Food.* World Health Organization, Geneva.

148. **Zhang, Q., F.-J. Chang, G. V. Barbosa-Canovas, and B. G. Swanson.** 1994. Inactivation of microorganisms in a semisolid model food using high voltage pulsed electric fields. *Lebensm. Wiss. Technol.* **27:**538–543.

149. **Zhang, Q., B.-L. Quin, G. V. Barbosa-Cánovas, and B. G. Swanson.** 1994. Growth stage and temperature affect the inactivation of *E. coli* by pulsed electric fields, p. 104. *In Proceedings of the Canadian Institute of Food Science Technology 37th Annual Conference.* Vancouver, British Columbia, Canada.

P. Michael Davidson

29 Chemical Preservatives and Natural Antimicrobial Compounds

The quality of a food product decreases from the time of harvest or slaughter until it is consumed. Quality loss may be due to microbiological, enzymatic, chemical, or physical changes. The consequences of quality loss caused by microorganisms are consumer hazards due to the presence of microbial toxins or pathogenic microorganisms or economic losses due to spoilage. Many food preservation technologies, some in use since ancient times, protect foods from the effects of microbial and inherent deterioration. Microorganisms may be inhibited by chilling, freezing, water activity reduction, nutrient restriction, acidification, modification of packaging atmosphere, or fermentation or through addition of antimicrobial compounds.

Food antimicrobials are chemical compounds added to or present in foods that retard microbial growth or kill microorganisms, thereby resisting deterioration in safety or quality. The major targets for antimicrobials are food-poisoning microorganisms (infective agents and toxin producers) and food spoilage microorganisms whose metabolic end products or enzymes cause off odors, off flavors, texture problems, discoloration, slime, or haze. Food antimicrobials are sometimes referred to as food preservatives. However, food preservatives include not only antimicrobials but also antibrowning agents (e.g., citric acid) and antioxidants (e.g., butylated hydroxyanisole [BHA]) (80). The use of food antimicrobials is changing because of consumer trends favoring consumption of "natural" foods, which contain fewer "chemical" additives. However, approximately 37.5 million kg of preservatives were consumed in the United States in 1991, and utilization is expected to increase to 47.3 million kg by 2000 (376). Worldwide use of traditional food preservatives is expected to increase by 4.1%/year through 2002 (376). The most widely used traditional preservatives are propionates, sorbates, and benzoates (183, 376).

Most food antimicrobials are only bacteriostatic or fungistatic and not bactericidal or fungicidal. Because they are generally bacteriostatic or fungistatic, they will not preserve a food indefinitely. Depending on storage conditions, the food product eventually spoils or becomes hazardous. Food antimicrobials are often used in combination with other food preservation procedures.

Cellular targets of food antimicrobials include the cell wall, cell membrane, metabolic enzymes, protein synthesis systems, and genetic systems. Since all are essential, inactivation of one can inactivate cells. The exact mechanisms or targets for food antimicrobials are often not known or well defined. There are several possible reasons for this. Researchers often focus on a single target such as an enzyme or the cell membrane without determining the effect on other cellular functions (99). It is difficult to pinpoint a target when many interacting reactions take place simultaneously. For example, membrane-disrupting compounds could cause leakage, interfere with active transport or metabolic enzymes, or dissipate cellular energy in the form of ATP. Food antimicrobials generally have multiple targets with concentration-dependent thresholds for inactivation or inhibition. A given target is important only if its sensitivity is within the inhibitory concentration range of the antimicrobial (99).

FACTORS AFFECTING ACTIVITY

The effectiveness of food antimicrobials depends on many factors associated with the food product, its storage environment, its handling, and the target microorganisms. Food preservation is best achieved when the antimicrobial type and concentration, storage time and temperature, food pH and buffering capacity, and presence of other agents which may influence shelf life are known and taken into account (183). Gould (136) classified the factors that affect the activity of antimicrobials into microbial, intrinsic, extrinsic, and process related.

Microbial factors that affect antimicrobial activity include inherent resistance (e.g., vegetative cells versus spores; strain differences), initial number and growth rate, interaction with other microorganisms (e.g., antagonism), cellular composition (Gram stain reaction), and cellular status (injury). Intrinsic factors affecting activity are those associated with a food product and include composition, pH, buffering capacity, oxidation reduction potential, and water activity. Extrinsic (environmental) factors affecting antimicrobial activity include temperature and time of storage, atmosphere, and relative humidity. Processing factors include changes in food composition, shifts in microflora, changes in microbial numbers, and change in microstructure. Most factors influence microorganisms in an interactive manner.

pH is the most important factor influencing the effectiveness of most food antimicrobials. Many food antimicrobials are weak acids and are most effective in their undissociated form. This is because weak acids are able to penetrate the cytoplasmic membrane of a microorganism more effectively in the protonated form. Therefore, the pK_a value of these compounds is important in selecting a particular compound for an application. The lower the pH of a food product, the greater the proportion of acid in the undissociated form and the greater the antimicrobial activity. Some researchers have suggested that only the undissociated form of a weak acid has activity. However, Eklund (96) demonstrated that while the undissociated species has significantly greater activity, the anion does contribute to antimicrobial activity.

Another important factor affecting activity is polarity. This relates both to the ionization of the molecule and to the contribution of any alkyl side groups or hydrophobic parent molecules. According to Freese (124), antimicrobials must be lipophilic to attach and pass through the cell membrane and also be soluble in the aqueous phase. Because most food antimicrobials are at least partially hydrophobic, the presence of lipid in a food may decrease the activity as a result of solubilization or binding of the compound. Proteins in foods may also decrease activity of antimicrobials through hydrophobic interactions (294).

TRADITIONAL ANTIMICROBIALS

It is a difficult, and somewhat arbitrary, task to classify antimicrobial compounds. This chapter divides preservatives into two classes, traditional and naturally occurring. Antimicrobials are classified as traditional when they fall into one or more of the following categories: they have been used for many years, they are approved by many countries for inclusion as antimicrobials in foods, or they are produced by synthetic means or are inorganic (as opposed to natural extracts or organic compounds, respectively). Ironically, many synthetic traditional antimicrobials are found in nature. These include benzoic acid from cranberries and sorbic acid from mountain ash berries (rowanberries). Classification within the broad categories, especially traditional antimicrobials, is even more difficult. For example, while acetic, sorbic, and benzoic acids are all organic acids, they have markedly different structures and physical properties. Other factors that make useful classification difficult include solubility, importance or rate of use, and lack of knowledge concerning spectrum of action or applications. This chapter discusses each antimicrobial's physical characteristics, spectra of action, applications, and mechanisms of action. Because many compounds have similar mechanisms of action, these will be explained for the first antimicrobial and cross-referenced for subsequent compounds with similar mechanisms.

Organic Acids and Esters

Many organic acids are used as food additives, but not all have antimicrobial activity. The most active antimicrobials are acetic, lactic, propionic, sorbic, and benzoic acids. Citric, caprylic, malic, fumaric, and other organic acids have limited activity but are used for flavorings. Esters of fatty acids are discussed here because they are derivatives of organic acids and have similar mechanisms.

The activity of organic acids is highly pH dependent. Early research demonstrated that the activity of organic acids is related to pH and that the undissociated form of the acid is primarily responsible for antimicrobial activity (76, 90, 112, 163, 210). Therefore, in selecting an organic acid for use as an antimicrobial food additive, both the product pH and the pK_a of the acid must be taken into account. The use of organic acids is generally limited to foods with pH less than 5.5, since most organic acids have pK_as of pH 3.0 to 5.0 (90).

The mechanisms of action of organic acids and their esters have some common elements. There is little evidence that the organic acids and related esters influence cell wall synthesis in prokaryotes or that they significantly interfere with protein synthesis or genetic mechanisms. As stated previously, in the undissociated form, organic acids can penetrate the cell membrane lipid bilayer more easily. Once inside the cell, the acid dissociates because the cell interior has a higher pH than the exterior (161). Bacteria maintain internal pH near neutrality to prevent conformational changes to the cell structural proteins, enzymes, nucleic acids, and phospholipids. Protons generated from intracellular dissociation of the organic acid acidify the cytoplasm and must be extruded to the exterior. According to the chemiosmotic theory (242) discussed in chapter 2, the cytoplasmic membrane is impermeable to protons, and the protons must be transported to the exterior. This proton extrusion creates an electrochemical potential across the membrane called the proton motive force (PMF). The PMF is a function of the differences in membrane potential ($\Delta\Psi$) and pH (ΔpH) and is defined as $\Delta p = \Delta\Psi - Z\Delta\text{pH}$, where $Z = 2.3RT/F$ (R = gas constant; T = absolute temperature; F = Faraday constant).

Since protons generated by the organic acid inside the cell must be extruded by using energy in the form of ATP, the constant influx of these protons will eventually deplete cellular energy (Fig. 29.1). It must be noted that this same phenomenon could be caused by interference with membrane permeability as well. Other related mechanisms involving the cytoplasmic membrane were studied by Freese, Sheu, and coworkers in the 1970s. Sheu and Freese (334) suggested that short-chain organic acids interfere with energy metabolism by altering the structure of the cytoplasmic membrane through interaction with membrane proteins. They further hypothesized that the interference with membrane proteins reduces ATP regeneration by uncoupling the electron transport system or by inhibiting active transport of nutrients into the cell. Sheu et al. (335) and Freese et al. (125) determined that short-chain fatty acids act as uncouplers of amino acid carrier proteins from the electron transport system. As proof, they showed that transport of

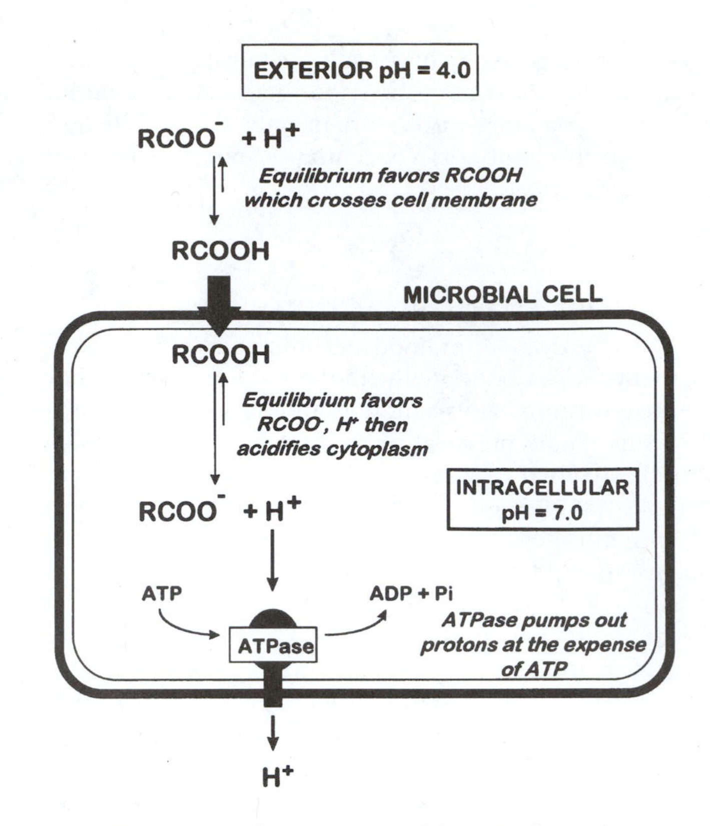

Figure 29.1 Fate of an organic acid (RCOOH) in a low-pH environment in the presence of a microbial cell.

L-serine, L-leucine, and malate is inhibited in membrane vesicles of *Bacillus subtilis* upon exposure to acetate and other fatty acids. Later, Sheu et al. (336) found that active transport inhibition influenced energy metabolism only indirectly; cells were not necessarily ATP depleted. They suggested that inhibition of active transport was due to destruction of the PMF, which in turn caused active transport to cease. Somewhat contradictory results were obtained by Eklund (95) for both organic acids and parabens. He studied the growth inhibition and uptake of amino acids and glucose for *Escherichia coli*, *B. subtilis*, and *Pseudomonas aeruginosa* and concluded that uptake inhibition could not account for growth inhibition by propionate, benzoate, or sorbate. In a later study, Eklund (98) evaluated the effect of sorbic acid and *p*-hydroxybenzoic acid esters (parabens) on the ΔpH and $\Delta\Psi$ components of the PMF in *E. coli* membrane vesicles. Both compounds eliminated the ΔpH but did not significantly affect the $\Delta\Psi$ component of the PMF, and so active transport of amino acids continued. Eklund (99) concluded that neutralization of the PMF and subsequent transport inhibition was not the sole mechanism of action of the parabens. To summarize, the organic acids and their esters have significant effects on bacterial cytoplasmic membranes, interfering with metabolite transport and maintenance of membrane potential. There is also considerable evidence that many organic acids and esters affect activities of microbial enzymes. However, because many of these studies are done with whole cells, it is not clear whether these are direct or indirect effects (99).

Acetic Acid and Acetates

Acetic acid (Fig. 29.2; $pK_a = 4.75$), the primary component of vinegar, and its sodium, potassium, and calcium salts, sodium and calcium diacetate, and dehydroacetic acid (methylacetopyranone) are some of the oldest food antimicrobials. Only *Acetobacter* species (microorganisms involved in vinegar production), lactic acid bacteria, and butyric acid bacteria are tolerant to acetic acid (13, 90). Bacteria inhibited by acetic acid include *Bacillus* species, *Clostridium* species, *Listeria monocytogenes*, salmonellae, *Staphylococcus aureus*, *E. coli*, *Campylobacter jejuni*, and pseudomonads (90, 99, 151, 210, 241, 426). Molds and yeasts are generally more resistant to acetic acid than bacteria (99). Yeasts and molds sensitive to acetic acid include *Aspergillus*, *Penicillium*, and *Rhizopus* species and some strains of *Saccharomyces* (76, 90, 193, 210).

Acetic acid and its salts have shown variable success as antimicrobials in food applications. Acetic acid can increase poultry shelf life when added to cut-up chicken parts in cold water at pH 2.5 (244). Addition of acetic acid at 0.1% to scald tank water used in poultry processing decreases the heat resistance of *Salmonella newport*, *Salmonella typhimurium*, and *Campylobacter jejuni* (254). In contrast, Lillard et al. (214) found that 0.5% acetic acid in the scald water has no significant effect on salmonellae, total aerobic bacteria, or members of the family *Enterobacteriaceae* on unpicked poultry carcasses. At 1 to 3% as a dip for beef or lamb, acetic acid reduces counts of both pathogenic and spoilage microorganisms (7, 24). Acetic acid has shown variable effectiveness as an antimicrobial for use as a spray sanitizer on meat carcasses.

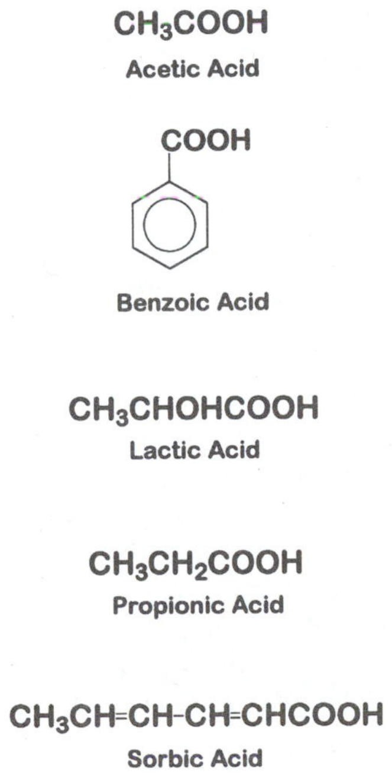

Sodium acetate at 1.0% increases the shelf life of catfish fillets by 6 days when stored at 4°C compared with the control (191). Sodium diacetate ($pK_a = 4.75$) is effective at 0.1 to 2.0% in inhibiting mold growth in cheese spread (90). Sodium acetate is also an effective inhibitor of rope-forming bacteria (*B. subtilis*) in baked goods and of the molds *Aspergillus flavus*, *A. fumigatus*, *A. niger*, *A. glaucus*, *Penicillium expansum*, and *Mucor pusillus* at pH 3.5 to 4.5 (133). It is useful in the baking industry because it has little effect on the yeast used in baking. Sodium diacetate is inhibitory to *L. monocytogenes*, *E. coli*, *Pseudomonas fluorescens*, *Salmonella enteritidis*, and *Shewanella putrefaciens* but not *S. aureus*, *Yersinia enterocolitica*, *Pseudomonas fragi*, *Enterococcus faecalis*, or *Lactobacillus fermentans* (328). Dehydroacetic acid has a high pK_a of 5.27 and is therefore active at higher pH values. It is inhibitory to bacteria at 0.1 to 0.4% and fungi at 0.005 to 0.1% (90).

Acetic acid is used commercially in baked goods, cheeses, condiments and relishes, dairy product analogs, fats and oils, gravies and sauces, and meats. The sodium and calcium salts are used in breakfast cereals, cheeses, fats and oils, gelatin, hard candy, jams and jellies, meats, soft candy, snack foods, soup mixes, and sweet sauces. Sodium diacetate is used in baked goods, candy, cheese spreads, gravies, meats, sauces, and soup mixes.

The general mechanism by which acetic acid inhibits microorganisms is related to that of other organic acids discussed previously. Sheu and Freese (334) and Freese et al. (125) observed that acetic acid inhibits oxygen uptake and resultant ATP production by 76 to 77% in whole cells of *B. subtilis*. The compound does not, however, inhibit NADH oxidation by isolated membranes. Further, they found that α-glycerol phosphate- or NADH-energized uptake of serine transport in membrane vesicles of *B. subtilis* is inhibited by acetic acid. *E. coli* whole cells and vesicles give similar results. They concluded that acetate inhibits growth by uncoupling substrate

CH_3COOH

Acetic Acid

COOH (on benzene ring)

Benzoic Acid

$CH_3CHOHCOOH$

Lactic Acid

CH_3CH_2COOH

Propionic Acid

$CH_3CH{=}CH{-}CH{=}CHCOOH$

Sorbic Acid

Figure 29.2 Structures of the organic acids.

transport and oxidative phosphorylation from the electron transport system. This inhibits uptake of metabolites into the cell. Later, Sheu et al. (336) determined that the short-chain fatty acids, such as acetic acid, interfere with the PMF. In addition to a demonstrated effect on the cell membrane, acetic acid may also act on cellular enzymes by reducing the intracellular pH (157).

Benzoic Acid and Benzoates

Benzoic acid (Fig. 29.2) and sodium benzoate were the first antimicrobial compounds permitted in foods by the U.S. Food and Drug Administration (170). Benzoic acid occurs naturally in cranberries, plums, prunes, cinnamon, cloves, and most berries (50). Sodium benzoate is highly soluble in water (66.0 g/100 ml at 20°C), while benzoic acid is much less so (0.27% at 18°C). As the undissociated form of benzoic acid ($pK_a = 4.19$) is the most effective antimicrobial agent, their most effective pH range is 2.5 to 4.5 (61, 215). Eklund (97) demonstrated that both the dissociated and undissociated forms of benzoic acid inhibit various bacteria and a yeast but that the MIC of the undissociated acid is 15 to 290 times lower.

Benzoic acid and sodium benzoate are used primarily as antifungal agents. The inhibitory concentration of benzoic acid at pH less than 5.0 against most yeasts ranges from 20 to 700 μg/ml, while for molds it is 20 to 2,000 μg/ml (13, 63). *Zygosaccharomyces bailii* is particularly resistant to benzoic acid. While some bacteria associated with food poisoning are inhibited by 1,000 to 2,000 μg of undissociated acid per ml, the control of many spoilage bacteria requires much higher concentrations (13). *L. monocytogenes* is inhibited by 1,000 to 3,000 μg of benzoic acid per ml at pH 5.6, depending on incubation temperature (105, 432). Benzoic acid at 0.1% is effective in reducing viable *E. coli* O157:H7 in apple cider (pH 3.6 to 4.0) by 3 to 5 log units in 7 days at 8°C (439). Sodium benzoate is used as an antimicrobial at up to 0.1% in carbonated and still beverages, syrups, cider, margarine, olives, pickles, relishes, soy sauce, jams, jellies, preserves, pie and pastry fillings, fruit salads, and salad dressings and in the storage of vegetables (63).

There are multiple cellular targets for benzoic acid. Most research has focused on the cytoplasmic membrane and cellular enzymes. Macris (223) studied the effect of benzoic acid on *Saccharomyces cerevisiae* and observed that rapid uptake of the compound reaches saturation in about 2 min. Only the undissociated form is taken up by the cells because of its lipophilic character and ability to cross the cytoplasmic membrane. When the yeast cells are exposed to >60°C, the uptake rate decreases. Heat inactivation of this process suggests enzymatic inactivation. Freese et al. (125) hypothesized that benzoic acid uncouples both substrate transport and oxidative phosphorylation from the electron transport system. Freese (124) suggested that benzoic acid destroys the PMF by continuous transport of protons into the cell, causing disruption of the transport system.

Benzoates inhibit various microbial enzymes and enzyme systems. Acetic acid metabolism and oxidative phosphorylation (39), α-ketoglutarate and succinate dehydrogenases (39), and trimethylamine-*N*-oxide reductase activity of *E. coli* (205) are inhibited by benzoic acid. Aflatoxin production by *A. flavus* (64, 390, 391) and 6-phosphofructokinase activity (123) are inhibited in fungi.

Lactic Acid and Lactates

Lactic acid (Fig. 29.2; $pK_a = 3.79$) is produced naturally during fermentation of foods by lactic acid bacteria. While the acid and salts act as preservatives in food products, their primary uses are as pH control agents and flavorings. Lactic acid inhibits spore-forming bacteria, *S. aureus*, and *Y. enterocolitica* (40, 241, 426). The antimicrobial activity of lactic acid depends on the food application and the target microorganism. Lactic acid is more effective than malic, citric, propionic, or acetic acid in inhibiting growth of *Bacillus coagulans* in tomato juice (292). Smulders et al. (346) and Snijders et al. (347) found that lactic acid at 1 to 2% reduces *Enterobacteriaceae* and aerobic mesophilic microorganisms on beef, veal, pork, and poultry and delays growth of spoilage microflora during long-term storage of products. Sodium lactate (2.5 to 5.0%) inhibits *Clostridium botulinum*, *C. sporogenes*, *L. monocytogenes*, and spoilage bacteria in various meat products (60, 222, 268, 332, 389, 411).

Very little research has been done specifically on the mechanism of action of lactic acid against foodborne microorganisms. Presumably, it functions similarly to other organic acids and has a primary mechanism involving disruption of the cytoplasmic membrane PMF (99). There has been some speculation concerning the mechanism of lactate salts used at concentrations of 2.5% and higher. Lactate salts have minimal effects on product pH; most of the lactate remains in the less effective anionic form. There is some evidence that high concentrations of the salts reduce water activity sufficiently to inhibit microorganisms (84). However, Chen and Shelef (60) and Weaver and Shelef (411) measured the water activity of cooked meat model systems and liver sausage, respectively, containing lactate salts up to

4% and concluded that water activity reduction is not sufficient to inhibit *L. monocytogenes.* It is most likely that at the high concentrations of lactate used, sufficient undissociated lactic acid is present, possibly in combination with a slightly reduced pH and water activity, to inhibit some microorganisms.

Propionic Acid and Propionates

Up to 1% propionic acid (Fig. 29.2; pK_a = 4.87) is produced naturally in Swiss cheese by *Propionibacterium freudenreichii* subsp. *shermanii.* The activities of propionates depend on the pH of the substance to be preserved, with the undissociated acid the most active form. Eklund (97) demonstrated that undissociated propionic acid is 11 to 45 times more effective than the dissociated form.

Propionic acid and sodium, potassium, and calcium propionates are used primarily against molds; however some yeasts and bacteria are also inhibited. The microorganism in bread dough responsible for rope formation, *B. subtilis,* is inhibited by propionic acid at pH 5.6 to 6.0 (255, 426). Propionates (0.1 to 5.0%) retard the growth of the bacteria *E. coli, S. aureus, Sarcina lutea, Salmonella* species, *Proteus vulgaris, Lactobacillus plantarum,* and *L. monocytogenes* and of the yeasts *Candida* species and *Saccharomyces cerevisiae* (65, 106). Propionic acid and propionates are used as antimicrobials in baked goods and cheeses. Propionates may be added directly to bread dough because they have no effect on the activity of baker's yeast. There is no limit to the concentration of propionates allowed in foods, but amounts used are generally less than 0.4% (299).

Early work on the mechanism of propionic acid inhibition showed that sodium propionate inhibition of *E. coli* is overcome by the addition of β-alanine (427). However, this interference with β-alanine synthesis is probably not a universal mechanism, as no effect was observed with *B. subtilis,* pseudomonads, or *Aspergillus clavatus.* The primary mode of action of propionic acid action is probably similar to that of other organic acids, i.e., interference with cytoplasmic membrane or cellular enzymes. Propionic acid inhibits amino acid uptake and inhibits growth of various bacteria (125, 334–336). Hunter and Segel (161) also showed that amino acid transport in the mold *Penicillium chrysogenum* is inhibited by propionic acid. These researchers theorized that propionic acid neutralized the PMF of the microbial membrane by passing through the membrane as the undissociated molecule and dissociating intracellularly. This leads cells to utilize energy to pump out the excess protons and eventually depletes cellular ATP. Eklund (95) suggested that inhibition is more complex since inhibition of alanine and glucose uptake by whole cells does not correlate well with growth inhibition. He concluded that uptake inhibition is only partially responsible for growth inhibition.

Sorbic Acid and Sorbates

Sorbic acid (Fig. 29.2) was first identified in 1859 by A. W. van Hoffman, a German chemist, from the berries of the mountain ash tree (rowanberry) (351). Sorbic acid is a *trans-trans,* unsaturated monocarboxylic fatty acid which is slightly soluble in water (0.16 g/100 ml) at 20°C. The potassium salt of sorbic acid is readily soluble in water (58.2 g/100 ml at 20°C). As with other organic acids, the antimicrobial activity of sorbic acid is greatest when the compound is in the undissociated state. With a pK_a of 4.75, activity is greatest at pH less than 6.0 to 6.5. The undissociated acid is 10 to 600 times more effective than the dissociated form (96).

Sorbates are the best characterized of all food antimicrobials as to their spectrum of action. They inhibit fungi and certain bacteria (25, 41, 299, 350, 351). Food-related yeasts inhibited by sorbates include species of *Brettanomyces, Byssochlamys, Candida, Cryptococcus, Debaryomyces, Hansenula, Pichia, Rhodotorula, Saccharomyces, Torulaspora,* and *Zygosaccharomyces* (351). Food-related mold species inhibited by sorbates belong to genera including *Alternaria, Aspergillus, Botrytis, Cephalosporium, Fusarium, Geotrichum, Helminthosporium, Mucor, Penicillium, Pullularia (Aureobasidium), Sporotrichum,* and *Trichoderma* (351). Sorbates inhibit the growth of yeasts and molds in microbiological media, cheeses, fruits, vegetables and vegetable fermentations, sauces, and meats (14, 121, 147, 250, 267, 344). Sorbates inhibit growth and mycotoxin production by the mycotoxigenic molds *A. flavus, A. parasiticus, Byssochlamys nivea, Penicillium expansum,* and *Penicillium patulum* (47, 48, 209, 314, 319). High initial mold populations can degrade sorbic acid in cheese (239). A number of *Penicillium, Saccharomyces,* and *Zygosaccharomyces* species can grow in the presence of and degrade potassium sorbate (35, 116, 212, 409). Sorbates may be degraded through a decarboxylation reaction resulting in the formation of 1,3-pentadiene, a compound having a kerosene-like or hydrocarbon-like odor (212, 229).

Sorbate inhibits the growth of many foodborne pathogenic and spoilage bacteria, including salmonellae and *S. aureus* in cooked, uncured sausage (382), *S. aureus* in bacon (273), *Pseudomonas putrefaciens* and *P. fluorescens* in Trypticase soy broth (297, 298), *Vibrio parahaemolyticus* in crabmeat and flounder homogenates

(300), salmonellae, *S. aureus*, and *E. coli* in poultry (139, 299, 303), *Y. enterocolitica* in pork (248), *Salmonella typhimurium* in laboratory media and in milk and cheese (261, 262), histamine production by *Proteus morgani* and *Klebsiella pneumoniae* strains in a Trypticase soy broth (371), and listeriolysin O production by *L. monocytogenes* (235).

Sorbic acid inhibits primarily catalase-positive bacteria (272, 429, 430). Despite exceptions (74, 145), catalase-negative lactic acid bacteria are generally resistant to sorbates. This allows use of sorbates in products fermented by lactic acid bacteria.

Sorbates are effective anticlostridial agents in cured meats and other meat and seafood products (38, 120, 219, 303, 309, 352). Sorbate prevents spores of *C. botulinum* from germinating and forming toxin in beef, pork, poultry, and soy protein frankfurters and emulsions and in bacon (160, 169, 352–354). Potassium sorbate is a strong inhibitor of both *Bacillus* and *Clostridium* spore germination at pH 5.7 but much less so at pH 6.7 (345).

Sorbate is applied to foods by direct addition, dipping, spraying, dusting, or incorporation into packaging (351). Cakes, pie crust, tortillas, pastries, rolls, doughnuts, icing, fruit fillings, and cream fillings can be protected from yeast and molds through the use of 0.05 to 0.10% potassium sorbate applied either as a spray after baking or by direct addition. Sorbates may be used in or on beverages, jams, jellies, preserves, margarine, chocolate syrup, salads, dried fruits, dry sausages, salted and smoked fish, cheeses, and various lactic acid fermentations.

One of sorbic acid's primary targets in vegetative cells appears to be the cytoplasmic membrane. Sorbic acid was included in studies by Sheu and Freese (334), Sheu et al. (335, 336), and Freese et al. (125) on the effect of organic acids on amino acid uptake. Sorbic acid inhibits amino acid uptake, which in turn was theorized to be responsible for eliminating the membrane PMF through nutrient depletion. Ronning and Frank (316) also showed that sorbic acid reduces the cytoplasmic membrane electrochemical gradient and consequently the PMF. They concluded that the sorbic acid-induced loss of PMF inhibits amino acid transport, which could eventually result in the inhibition of many cellular enzyme systems. In contrast, Eklund (98) showed that while low concentrations of sorbic acid reduce the ΔpH of the PMF of *E. coli* vesicles, concentrations much greater than those required for inhibition reduce, but do not eliminate, the $\Delta\Psi$ component. Since the $\Delta\Psi$ component alone could energize active uptake of amino acids, the amino acid uptake inhibition theory does not entirely explain the mechanism of inhibition by sorbic acid (99).

The mechanism by which sorbic acid inhibits microbial growth may also be partially due to its effect on enzymes. Melnick et al. (239) theorized that sorbic acid inhibits dehydrogenases involved in fatty acid oxidation. Addition of sorbic acid results in the accumulation of β-unsaturated fatty acids that are intermediate products in the oxidation of fatty acids by fungi. This prevents the function of dehydrogenases and inhibits metabolism and growth. Sorbic acid also inhibits sulfhydryl enzymes, including fumarase, aspartase, succinic dehydrogenase, ficin, and alcohol dehydrogenase (230, 418, 431). Whitaker (418) first suggested that sorbate activity is due to the formation of stable thiohexenoic acid complexes with sulfhydryl-containing enzymes. York and Vaughn (431) showed that sorbate reacts with the thiol group of cysteine and suggested that this is a mechanism of inactivation of sulfhydryl enzymes. Wedzicha and Brook (414) also demonstrated a 1:1 reaction between sorbate and cysteine. Martoadiprawito and Whitaker (230) proposed that sorbate inhibits the enzymes by formation of a covalent bond between the sulfhydryl or zinc hydroxide groups of the enzyme and the α and/or β carbons of sorbate. Other suggested mechanisms for sorbate have involved interference with enolase, proteinase, and catalase or the inhibition of respiration by competitive action with acetate in acetyl coenzyme A formation (10, 77, 387). Sorbate inhibits activation of the hemolytic activity of listeriolysin O of *L. monocytogenes* by reacting with cysteine (204).

The mechanism of sorbate action against bacterial spores has been studied extensively. In some of the only reports involving effects on the cell wall, Gould (135) and Seward et al. (325) demonstrated that sorbic acid inhibits cell division of germinated spores of bacilli and *C. botulinum* type E. Sorbic acid at pH 5.7 competitively inhibits L-alanine and L-α-NH_2-*n*-butyric acid-induced germination of *Bacillus cereus* T spores and L-alanine- and L-cysteine-induced germination of *C. botulinum* 62A (345).

Miscellaneous Organic Acids

Many organic acids and their esters have been examined as potential antimicrobials, but most have little or no activity. They are used in foods as acidulants or flavoring agents rather than as preservatives. Fumaric acid is used to prevent the malolactic fermentation in wines (260) and as an antimicrobial agent in wines (275). Esters of fumaric acid (monomethyl, dimethyl, and ethyl) at 0.15 to 0.2% have been tested as substitutes or adjuncts for nitrite in bacon. Citric acid retards growth and toxin production by *A. parasiticus* and *A. versicolor* but not *Penicillium expansum* (287). It is inhibitory to salmonellae in media and on poultry carcasses, growth and toxin production by *C. botulinum* in shrimp and tomato products, and *S. aureus* in microbiological medium (241, 278, 364, 376). Branen and Keenan (42) were the first to suggest that inhibition by citrate may be due to chelation, in studies with *Lactobacillus casei*. In contrast, Buchanan and Golden (43) found that while undissociated citric acid is inhibitory against *L. monocytogenes*, the dissociated molecule protects the microorganism. They theorized that this protection is due to chelation by the anion.

Fatty Acid Esters

Certain fatty acid esters have antimicrobial activity in foods. One of the most effective of the fatty acid esters is glyceryl monolaurate (monolaurin). Razavi-Rohani and Griffiths (283) found that monolaurin was effective against all seven gram-positive bacteria tested at 8 to 96 μg/ml but was ineffective against any of nine gram-negative strains at >3,170 μg/ml. Presence of EDTA caused the monolaurin to be inhibitory to all gram-negative strains (90 to 1,500 μg/ml) and decreased the MICs against the gram-positive strains. Monolaurin at 3 to 10 μg/ml was inhibitory to *L. monocytogenes* in tryptic soy broth with yeast extract at pH 5.5 to 7.0 (251, 252).

Nitrites

Sodium ($NaNO_2$) and potassium (KNO_2) nitrite have a specialized use in cured meat products. Meat curing utilizes salt, sugar, spices, and ascorbate or erythorbate in addition to nitrite. In addition to serving as an antimicrobial, nitrite has many functions in cured meats. As nitric oxide, it reacts with the meat pigment, myoglobin, to form the characteristic cured meat color, nitrosomyoglobin. It also contributes to the flavor and texture of cured meats and serves as an antioxidant (162). Meat curing is often combined with drying, heating, smoking, or fermentation as preservation adjuncts. At one time, sodium nitrate ($NaNO_3$) and potassium nitrate (KNO_3), also known as saltpeter, were used extensively in cured meat production. Their use was diminished when it was discovered that nitrate is converted to nitrite and that nitrite is the effective antimicrobial agent. The specific contribution of nitrite to the antimicrobial effects of curing salt was not recognized until the late 1920s (67a), and evidence that nitrite is an effective antimicrobial agent came even later (360).

The primary use for sodium nitrite as an antimicrobial is to inhibit *C. botulinum* growth and toxin production in cured meats. In association with other components in the curing mix, such as salt, and reduced pH, nitrite exerts a concentration-dependent antimicrobial effect on the outgrowth of spores from *C. botulinum* and other clostridia. The use of nitrites in cured meat products to control *C. botulinum* has been studied extensively (156, 307, 379). Nitrite inhibits bacterial sporeformers by inhibiting outgrowth of the germinated spore (72, 92). Only very high nitrite concentrations significantly inhibit spore germination.

Nitrite's effectiveness depends on several environmental factors. It was first suggested in the 1920s that nitrites were more effective at a lower pH (425). The interaction of nitrite and reduced pH against bacteria is well established (55, 340, 368–370). Nitrite is more inhibitory under anaerobic conditions (45, 55, 207). Ascorbate and isoascorbate enhance the antibotulinal action of nitrite, probably by acting as reducing agents (313, 380). Storage and processing temperatures, salt concentration, and initial inoculum size also significantly influence the antimicrobial effectiveness of nitrite (296, 306, 310, 312). Roberts and Ingram (311) and Duncan and Foster (92) demonstrated that nitrite addition prior to heating does not increase inactivation of spores but inhibits outgrowth following heating.

Nitrite has variable effects on microorganisms other than *C. botulinum*. *Clostridium perfringens* growth at 20°C is inhibited by 200 μg of nitrite per ml and 3% salt or 50 μg of nitrite per ml and 4% salt at pH 6.2 in a laboratory medium (131). Nitrite is inhibitory to bacteria such as *E. coli* and *Achromobacter*, *Enterobacter*, *Flavobacterium*, *Micrococcus*, and *Pseudomonas* species at 200 μg/g and pH 6.0 (368–370). *L. monocytogenes* growth is inhibited for 40 days at 5°C by 200 ppm of sodium nitrite with 5% NaCl in vacuum-packaged and film-wrapped smoked salmon (269). Gibson and Roberts (130, 131) found limited inhibition by nitrite (400 μg/ml) and salt (up to 6%) against fecal streptococci, salmonellae, and enteropathogenic *E. coli*. *C. perfringens* (141, 270) and species of *Salmonella* (55), *Lactobacillus* (55, 355), and *Bacillus* (141) are resistant to nitrite.

Meat products that may contain nitrites include bacon, bologna, corned beef, frankfurters, luncheon meats, ham, fermented sausages, shelf-stable canned cured meats, and perishable canned cured meat (e.g., ham). Nitrite is also used in a variety of fish and poultry products. The concentrations used in these products are specified by governmental regulations but are generally limited to 156 ppm (mg/kg) for most products and 100 to 120 ppm (mg/kg) in bacon. Sodium erythorbate or isoascorbate is required in products containing nitrites as a cure accelerator and as an inhibitor to the formation of nitrosamines, carcinogenic compounds formed by reactions of nitrite with secondary or tertiary amines. Sodium nitrate is used in certain European cheeses to prevent spoilage by *Clostridium tyrobutyricum* or *C. butyricum* (379).

The mechanism of nitrite inhibition of microorganisms has been studied for 50 years (379, 425). Since Ingram (162) first hypothesized that nitrite inactivated enzymes associated with respiration, the compound has been shown to inactivate a variety of enzymes or enzyme systems. However, the likely targets of clostridial inhibition by nitrite have been elucidated only in the past 20 years. Woods et al. (424) showed that nitrite caused a reduction in intracellular ATP and excretion of pyruvate in cells of *C. sporogenes.* Since these cells oxidize pyruvate to acetate to produce ATP by using the phosphoroclastic system, it was theorized that this enzyme system was being inhibited by nitrite. Two enzymes in the system, pyruvate-ferredoxin oxidoreductase (PFR) and ferredoxin, were suspected to be susceptible to nitrite. Inhibition was due to reaction of nitrite in the form of nitric oxide with the nonheme iron of the proteins. PFR was most susceptible. Later, Woods and Wood (423) showed that the phosphoroclastic system of *C. botulinum* is also inhibited by nitrite. Carpenter et al. (54) confirmed that nitrite inhibits the phosphoroclastic system of *C. botulinum* and *C. pasteurianum* but that the iron-sulfur enzyme, ferredoxin, is more susceptible than PFR. McMindes and Siedler (237) reported that nitric oxide is the active antimicrobial principle of nitrite and that pyruvate decarboxylase may be an additional target for growth inhibition by nitrite. Roberts et al. (313) also confirmed inhibition of the phosphoroclastic system and found that ascorbate enhanced inhibition. In addition, they showed that other iron-containing enzymes of *C. botulinum,* including other oxidoreductases and the iron-sulfur protein, hydrogenase, are inhibited. It has been suggested that inhibition of clostridial ferredoxin and/or PFR is the ultimate mechanism of growth inhibition for clostridia (54, 379). These observations are supported by the fact that addition of iron to meats containing nitrite reduces the inhibitory effect of the compound (381).

The mechanism of inhibition against non-sporeforming microorganisms may be different from that for sporeformers. Nitrite inhibits active transport, oxygen uptake, and oxidative phosphorylation of *P. aeruginosa* by oxidizing ferrous iron of an electron carrier, such as cytochrome oxidase, to the ferric form (318). Muhoberac and Wharton (245) and Yang (428) also found inhibition of cytochrome oxidase of *P. aeruginosa. Enterococcus faecalis* and *Lactococcus lactis,* which do not depend on active transport or cytochromes, are not inhibited by nitrite (318). Woods et al. (425) theorized that nitrites inhibit aerobic bacteria by binding the heme iron of cytochrome oxidase.

Parabens

Alkyl esters of parabens (Fig. 29.3) were first reported to possess antimicrobial activity in the 1920s (280). Esterification of the carboxyl group of benzoic acid allows the molecule to remain undissociated up to pH 8.5, giving the parabens an effective range of pH 3.0 to 8.0 (1, 61). In most countries, the methyl, propyl, and heptyl parabens are allowed for direct addition to foods as antimicrobials, while the ethyl and butyl esters are approved in some countries.

The antimicrobial activity of parabens is, in general, directly proportional to the chain length of the alkyl component (1, 337). As the alkyl chain length of the parabens increases, inhibitory activity generally increases. Increasing activity with decreasing polarity is more evident against gram-positive than against gram-negative bacteria. Parabens are generally more active against molds and yeasts than against bacteria (Table 29.1). Against bacteria, the parabens are more active against gram-positive genera (Table 29.2). Little research on the activity of the *n*-heptyl ester in foods seems to be available. Chan et al. (58) did show that this

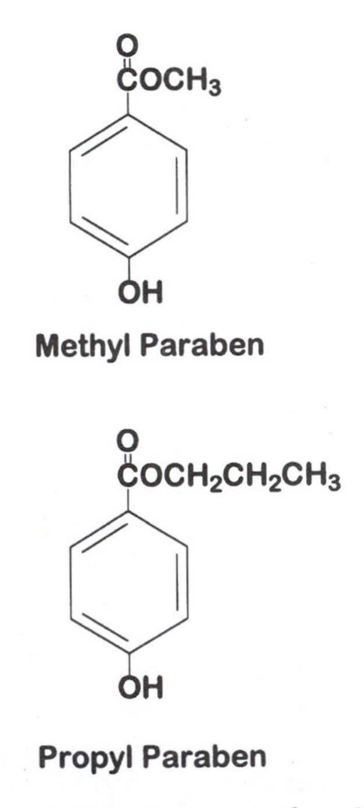

Figure 29.3 Structures of parabens.

Table 29.1 Concentration ranges of esters of *p*-hydroxybenzoic acid necessary for total inhibition of various fungi (pH, incubation temperature, and time vary)[a]

	Concn (µg/ml)				
Fungus	Methyl	Ethyl	Propyl	Butyl	Heptyl
Alternaria sp.			100		50–100
Aspergillus flavus	>608	330–500	180–360	388	
Aspergillus niger	1,000	400–500	200–250	125–200	
Aspergillus parasiticus	530–>608	415–500	270–360	388	
Byssochlamys fulva			200		
Candida albicans	1,000	500–1,000	125–250	125	
Debaryomyces hansenii		400			
Fusarium graminearum	530	330	270	194	
Fusarium moniliforme	>608	330	180	290	
Fusarium oxysporum	608	330	180	290	
Penicillium citrinum	608	250	180	195	
Penicillium digitatum	500	250	63	<32	
Penicillium chrysogenum	500–>608	250–330	125–270	63–388	
Rhizopus nigricans	500	250	125	63	
Saccharomyces bayanus	930		220		
Saccharomyces cerevisiae	1,000	500	125–200	32–200	25–100
Torula utilis			200		25
Torulaspora delbrueckii		700			
Zygosaccharomyces bailii		900			
Zygosaccharomyces bisporus		400			
Zygosaccharomyces rouxii		700			

[a]Sources: references 1, 173, 178, 187, 211, 233, 374, and 386.

Table 29.2 Concentration ranges of esters of *p*-hydroxybenzoic acid necessary for total inhibition of growth of various bacteria (pH, incubation temperature, and time vary)[a]

	Concn (µg/ml)				
Microorganism	Methyl	Ethyl	Propyl	Butyl	Heptyl
Gram positive					
Bacillus cereus	1,000–2,000	830–1,000	125–400	63–400	12
Bacillus megaterium	1,000		320	100	
Bacillus subtilis	1,980–2,130	1,000–1,330	250–450	63–115	
Clostridium botulinum	1,000–1,200	800–1,000	200–400	200	
Lactococcus lactis			400		12
Listeria monocytogenes	1,430–1,600		512		
Micrococcus sp.		60–110	10–100		
Sarcina lutea	4,000	1,000	400–500	125	12
Staphylococcus aureus	1,670–4,000	1,000–2,500	350–540	120–200	12
Streptococcus faecalis		130	40		
Gram negative					
Aeromonas hydrophila	550		100		
Enterobacter aerogenes	2,000	1,000	1,000	4,000	
Escherichia coli	1,200–2,000	1,000–2,000	400–1,000	1,000	
Klebsiella pneumoniae	1,000	500	250	125	
Pseudomonas aeruginosa	4,000	4,000	8,000	8,000	
Pseudomonas fluorescens	1,310		670		
Pseudomonas fragi			4,000		
Pseudomonas putida	450				
Salmonella typhosa	2,000	1,000	1,000	1,000	
Salmonella typhimurium			180–>300		
Vibrio parahaemolyticus			50–100		
Yersinia enterocolitica	350				

[a]Sources: references 1, 20, 79, 93, 97, 100, 178, 179, 187, 208, 211, 218, 243, 266, 284, 285, and 301.

compound was very effective in inhibiting bacteria involved in the malolactic fermentation of wines.

To take advantage of their respective solubilities and increased activities, methyl and propyl parabens are normally used in a combination of 2:1 to 3:1 (methyl:propyl). The compounds may be incorporated into foods by being dissolved in water, ethanol, propylene glycol, or the food product itself. The *n*-heptyl ester is used in fermented malt beverages (beers) and noncarbonated soft drinks and fruit-based beverages. Parabens are used in a variety of foods, including baked goods, beverages, fruit products, jams and jellies, fermented foods, syrups, salad dressings, wine, and fillings.

As parabens are phenolic derivatives, a discussion of their mechanism includes research done on phenol and related phenolic compounds, such as the phenolic antioxidants. Much of the research on mechanisms of phenolic compounds has centered on their effects on cellular membranes. Vas (396) proposed that phenol attacked the cytoplasmic membrane of microorganisms, causing the release of cytoplasmic constituents. Judis (177) demonstrated that *E. coli* lost radioactively labeled intracellular constituents (^{14}C, ^{32}P, ^{35}S) in the presence of phenol and some of its halogenated derivatives. Judis (177) also proposed that these compounds cause physical damage to the membrane or permeability barrier. Furr and Russell (128) detected similar leakage of intracellular RNA by *Serratia marcescens* in the presence of the parabens. The amount of leakage is proportional to the alkyl chain length of the paraben. The phenolic antioxidant butylated hydroxytoluene (BHT) causes leakage in *Tetrahymena pyriformis* (365) and disruption of a viral envelope (membrane) (407). Davidson and Branen (81) and Degré and Sylvestre (86) found that butylated hydroxyanisole (BHA) causes leakage of ^{14}C-labeled intracellular compounds from *P. fragi* and *P. fluorescens* and leakage of nucleotides from *S. aureus* Wood 46.

Other studies on the effect of phenolic compounds on the cytoplasmic membrane have evaluated nutrient uptake. Freese et al. (125) found that the parabens inhibit serine uptake as well as the oxidation of α-glycerol phosphate and NADH in membrane vesicles of *B. subtilis*. They concluded that the parabens inhibit both membrane transport and the electron transport system. Eklund (95) did a similar study using *E. coli*, *B. subtilis*, and *P. aeruginosa*. He determined the uptake of alanine by whole cells and uptake of alanine, serine, phenylalanine, and glucose by vesicles. Parabens generally caused a decrease in amino acid uptake but not in glucose uptake. Eklund (95) postulated that since the parabens are known to cause leakage of cellular contents, they are capable of neutralizing chemical and electrical forces which establish a normal membrane gradient. In continued work with *E. coli*, Eklund (98) found that parabens eliminated the ΔpH of the cytoplasmic membrane. In contrast, the compounds did not significantly affect the $\Delta\Psi$ component of the PMF. He concluded that neutralization of the PMF and subsequent transport inhibition could not be the only mechanism of inhibition for the parabens. Both Eklund (95) and Freese et al. (125) stated that gram-negative bacteria are probably resistant to the parabens because of a screening effect by the cell wall lipopolysaccharide layer.

Degré et al. (85) measured the effect of BHA on the cytochrome system and ability to oxidize various substrates by *S. aureus*. They found that BHA causes a reduction of the cytochromes and theorized that it reduces the Fe^{3+} in the cytochromes to Fe^{2+}. To measure oxidation, they assayed cell extracts and whole cells for oxygen uptake, with and without inhibitor. BHA inhibited respiration compared with other electron donors such as NADH and succinate. They concluded that since BHA interferes with lipid synthesis, and lipids are involved in the respiratory process, inhibition of electron transfer reactions is due to lipid interference. Oka (253) suggested that parabens act on yeast cells by absorbing on the solid phase of cells rather than in the cell fluid or lipid layers. This conclusion was reached even though there was a direct relationship between the antimicrobial dissolved in the lipid phase and the minimum concentration necessary to inhibit the yeast.

If the phenolic compounds act on the cell membrane of the microorganism, the question arises as to what portion of the membrane they attack. Kaye and Proudfoot (188) investigated the effect of phenolics on phosphatidylethanolamine monolayers. They found that various phenolics cause disruption of the integrity of the monolayers in direct correlation with their antimicrobial activities. Studies with the phenolic antioxidant BHT have shown that it causes disruption of the symmetry and increases fluidity of lipid alkyl chains in phospholipids (102, 341). BHA causes alterations of the incorporation of [^{14}C]acetate into phospholipids, as well as changes in the ratios of tetrahymenol to polar lipids in *T. pyriformis* (366). Al-Issa et al. (4) increased the lipid content of *L. monocytogenes* by subculturing in glycerol medium. Cells grown in this manner have increased resistance to BHA. They suggested that glycerol was involved in surface lipid synthesis, which acts as a screening mechanism against BHA. Al-Issa et al. (4) also found that strains of *L. monocytogenes* which are more resistant to BHA have lower levels of unsaturated and ante-iso fatty acids. Post and Davidson (279) also studied the effect of BHA on the fatty acid composition of *B. cereus, C. perfringens, S. aureus, P. fluorescens, P. fragi, E. coli,* and salmonellae. They found a relationship between susceptibility to BHA and saturated/unsaturated fatty acid ratio of the gram-positive strains for the total lipid fraction. No relationships were found for the polar lipid fraction of gram-positive bacteria or either fraction of the gram-negative bacteria. Bargiota et al. (20) examined the relationship between lipid composition of *S. aureus* and resistance to parabens. A paraben-resistant strain had a greater percentage of total lipid, greater percentage of phosphatidylglycerol, and fewer cyclopropane fatty acids than sensitive strains. These changes could influence membrane fluidity. Juneja and Davidson (178) altered the lipid composition of *L. monocytogenes* by growth in the presence of added fatty acids. In the presence of added $C_{14:0}$ or $C_{18:0}$ fatty acids, the microorganism showed increased resistance to tertiary butyl hydroquinone (TBHQ) and parabens. Growth in the presence of $C_{18:1}$ increased sensitivity to the antimicrobial agents. Results indicated that for *L. monocytogenes,* a correlation existed between lipid composition of the cell membrane and susceptibility to antimicrobials.

Early studies on phenol reported that high concentrations precipitated all proteins but that lower concentrations selectively inhibited essential enzymes (280). A study by Chipley (62) may be closely linked to the findings on membrane disruption by phenolics discussed previously. Chipley (62) found that 0.1 mM 2,4-dinitrophenol noncompetitively inhibits enzymes associated with the cell envelopes of both *E. coli* and *Salmonella enteritidis.* In contrast, with the same enzymes in a purified isolated state, no inhibition was demonstrated. These results suggested that a conformational change in the membrane of the vesicle causes inhibition of the enzymes and that there is no direct effect of 2,4-dinitrophenol. Exterkate (110) obtained similar results with membrane-perturbing solvents and *Lactococcus cremoris.* Rico-Muñoz et al. (293) investigated the effects of BHA, BHT, TBHQ, propyl gallate, *p*-coumaric acid, ferulic acid, caffeic acid, and methyl and propyl parabens on the membrane-bound ATPase of two strains of *S. aureus.* Only BHA stimulates the activity of the enzyme. BHT or the parabens have no effect on the enzyme. It was suggested that BHA in the membrane interferes with membrane potential, causing the ATPase to increase the hydrolysis of ATP in an attempt to regenerate membrane potential. The authors concluded that phenolic compounds probably do not have the same mechanism of action and there may be several targets which lead to inhibition of microorganisms by these compounds (293).

Sodium Chloride

Sodium chloride (NaCl), or common salt, is probably the oldest known food preservative. While a very few foods, such as raw meats and fish products, are preserved directly by high concentrations of sodium chloride, salt is used primarily as an adjunct to other processing methods such as canning and pasteurization (348). In general, foodborne pathogenic bacteria are inhibited by a water activity of 0.92 or less (equivalent to an NaCl concentration of 13% [wt/vol]). The exception to that generality is *S. aureus,* which has a minimum water activity for growth of 0.83 to 0.86. Enterotoxin production by the organism is more restricted (236). Another relatively salt-tolerant foodborne pathogen is *L. monocytogenes,* which can survive in saturated salt solutions at low temperatures. Fungi are more tolerant to low water activity than bacteria. The minimum for growth of xerotolerant fungi is 0.61 to 0.62, but most are inhibited by 0.85 or lower (73).

Sodium chloride inhibits microorganisms primarily by its plasmolytic effect. The antimicrobial activity of sodium chloride is related to its ability to reduce water activity and create unfavorable conditions for microbial growth. As the water activity of the external medium is reduced, cells are subjected to osmotic shock and rapidly lose water through plasmolysis. During plasmolysis, a cell ceases to grow and either dies or remains dormant. In order to resume growth, the cell must reduce its intracellular water activity (356). Aside from the osmotic influence on growth, other possible mechanisms include limiting oxygen solubility, alteration of pH, toxicity of sodium and chloride ions, and loss of magnesium ions (17). Certain isolated enzymes are more susceptible to inhibition by sodium chloride than is the growth of the host species (238).

Sulfites

Sulfur dioxide (SO_2) and its salts have been used as disinfectants since the time of the ancient Greeks and Romans (165, 259). The use of sulfur dioxide as a food preservative was reported in the 17th century by Evelyn (109), who suggested that casks should be filled with cider that contained sulfur dioxide (produced by burning sulfur). Salts of sulfur dioxide include potassium sulfite (K_2SO_3), sodium sulfite (Na_2SO_3), potassium bisulfite ($KHSO_3$), sodium bisulfite ($NaHSO_3$), potassium metabisulfite ($K_2S_2O_5$), and sodium metabisulfite (NaS_2O_5). While sulfites now have multiple uses as food additives, they were originally used as antimicrobials (317). As antimicrobials, sulfites are used primarily in fruit and vegetable products to control three groups of microorganisms: spoilage and fermentative yeasts and molds on fruits and fruit products (e.g., wine), acetic acid bacteria, and malolactic bacteria (259). In addition to use as antimicrobials, sulfites act as antioxidants and inhibit enzymatic and nonenzymatic browning in a variety of foods (413). Their primary application is in fruits and vegetable products, but they are also used to a limited extent in meats.

The most important factor affecting the antimicrobial activity of sulfites is pH. Sulfur dioxide and its salts exist as a pH-dependent equilibrium mixture when dissolved in water:

$$SO_2 \cdot H_2O \rightleftharpoons HSO_3^- + H^+ \rightleftharpoons SO_3^{2-} + H^+$$

Aqueous solutions of sulfur dioxide theoretically yield sulfurous acid (H_2SO_3); however, evidence indicates that the actual form is more likely $SO_2 \cdot H_2O$ (137). As the pH decreases, the proportion of $SO_2 \cdot H_2O$ increases and the bisulfite (HSO_3^-) ion concentration decreases. The pK_a values for sulfur dioxide, depending upon temperature, are 1.76 to 1.90 and 7.18 to 7.20 (137, 259, 317). The inhibitory effect of sulfites is most pronounced when the acid or $SO_2 \cdot H_2O$ is in the undissociated form (144). Therefore, their most effective pH range is less than 4.0. King et al. (192) demonstrated this when they found that undissociated H_2SO_3 ($SO_2 \cdot H_2O$) was the only form active against yeasts and that neither HSO_3^- or SO_3^{2-} had antimicrobial activity. Similarly, $SO_2 \cdot H_2O$ is 1,000, 500, and 100 times more active than HSO_3^- or SO_3^{2-} against *E. coli*, yeasts, and *A. niger*, respectively (286). Increased effectiveness at low pH is likely due to the ability of un-ionized sulfur dioxide to pass across the cell membrane (163, 281, 317).

Sulfites, especially as the bisulfite ion, are very reactive, forming addition compounds (α-hydroxysulfonates) with aldehydes and ketones. It is generally agreed that these bound forms have much less, or no, antimicrobial activity compared with the free forms. For example, Ough (259) reported that addition compounds with sugars completely neutralize the antimicrobial activity of sulfites against yeasts. However, Stratford and Rose (362) demonstrated that pyruvate-sulfite complexes retained some activity against *Saccharomyces cerevisiae.*

Sulfur dioxide is fungicidal even in low concentrations against yeasts and molds. The inhibitory concentration range of sulfur dioxide is 0.1 to 20.2 μg/ml for *Saccharomyces, Zygosaccharomyces, Pichia, Hansenula,* and *Candida* species (286). Roland et al. (315) and Roland and Beuchat (314) found that 25 to 100 μg of sulfur dioxide per ml inhibits *Byssochlamys nivea* growth and patulin production in grape and apple juices.

Sulfites may be used to inhibit acetic acid-producing and lactic acid bacteria in wines and fruit products and spoilage bacteria in meat products. At 100 to 200 mg/liter, sulfites inhibit *Acetobacter* species that cause wine spoilage (259, 286). The concentration of sulfur dioxide required to inhibit lactic acid bacteria varies significantly, depending on conditions, but can be 1 to 10 μg/ml in fruit products at pH 3.5 or less (419). Sulfur dioxide is more inhibitory to gram-negative rods than to gram-positive rods (305). Banks and Board (15) tested several genera of *Enterobacteriaceae* isolated from sausage for their metabisulfite sensitivity. The microorganisms tested and the concentrations of free sulfite (micrograms per milliliter) necessary to inhibit their growth at pH 7.0 were as follows: salmonellae, 15 to 109; *E. coli,* 50 to 195; *Citrobacter freundii,* 65 to 136; *Y. enterocolitica,* 67 to 98; *Enterobacter agglomerans,* 83 to 142; *Serratia marcescens,* 190 to 241; and *Hafnia alvei,* 200 to 241.

Sulfur dioxide is used to control the growth of undesirable microorganisms in fruits, fruit juices, wines, sausages, fresh shrimp, and acid pickles and during extraction of starches. It is added at 50 to 100 mg/liter to expressed grape juices used for making wines to inhibit molds, bacteria, and undesirable yeasts (6). At appropriate concentrations, sulfur dioxide does not interfere with wine yeasts or with the flavor of wine. During fermentation, sulfur dioxide also serves as an antioxidant, clarifier, and dissolving agent. The optimum level of sulfur dioxide (50 to 75 mg/liter) is maintained to prevent postfermentation changes by microorganisms. In some countries, sulfites may be used to inhibit the growth of microorganisms on fresh meat and meat products (190). Sulfite or metabisulfite added in sausages is effective in delaying the growth of molds, yeasts, and salmonellae during storage at refrigerated or room temperature (15, 163). Sulfur dioxide restores a bright color but may give a false impression of freshness.

Because of their extreme reactivity, it is difficult to pinpoint the exact antimicrobial mechanism for sulfites. This reactivity is due to the ability of sulfites to act as reducing agents or take part in nucleophilic attack (137). Sulfites react with disulfide bonds of proteins and with glutathione-forming thiosulfonates:

$$R\text{-}S\text{-}S\text{-}R + SO_3^{2-} \rightarrow R\text{-}S\text{-}SO_3^- + RS^-$$

This reaction can inactivate enzymes which have disulfide links and effect conformation of proteins. Inactivation of the cellular redox agent, glutathione (γ-glutamyl-cysteinyl-glycine), can result in oxidation of protein thiol groups and oxidation of membrane lipids (137). Sulfites react with coenzymes and enzyme prosthetic groups. The coenzymes NAD and the related NADP are inactivated by sulfite addition. Prosthetic groups including thiamin, folic acid, pyridoxal, flavins, and hemes are all susceptible to sulfite inactivation. The pyramidines cytosine and uracil, components of DNA and RNA, are susceptible to addition reactions. In vitro, bisulfite converts cytosine and 5-methylcytosine to uracil and thymine, respectively (150). In addition, the purine adenine, which is a component of nucleic acids ATP and NAD, can form an addition compound with sulfite. These reactions may terminate DNA polymerization, transcription, and translation of RNA. Compounds containing carbonyls (aldehydes and ketones) are also subject to attack by sulfites. These compounds form α-hydroxysulfonates. Sugars that react with sulfites include arabinose, mannose, galactose, lactose, glucose, maltose, and raffinose (164). Sucrose does not react. Sulfites can also form addition products with quinones.

The most likely targets for inhibition by sulfites include the cytoplasmic membrane, DNA replication, protein synthesis, membrane-bound or cytoplasmic enzymes or individual components in metabolic pathways. In the cases of *Saccharomyces cerevisiae, Saccharomycodes ludwigii,* and *Z. bailii,* sulfites gain access to the cell by free diffusion (274, 361, 363). Other fungi have active transport systems for the compound. Sulfites dissipate the PMF and inhibit solute active transport (317). Once inside the cell, sulfur dioxide acts on enzymes and individual components of metabolism. For example, dehydrogenases and transaminases are particularly susceptible to sulfites as a result of their cofactors of NAD and pyridoxal phosphate, respectively (271). In *E. coli,* NAD-dependent formation of oxaloacetate from malate is inhibited. Hinze and Holzer (154) did elaborate experiments to demonstrate that the most sensitive enzyme in the Embden-Meyerhof-Parnas pathway of *Saccharomyces cerevisiae* was glyceraldehyde-3-phosphate dehydrogenase. Inactivation of this enzyme led to a decrease in cellular ATP. It is still uncertain whether the target is the enzyme, its cofactor NAD, or the substrate glyceraldehyde-3-phosphate. One or more of these factors may result in microbial death or inhibition (146).

Indirect and Miscellaneous

DMDC

Dimethyl dicarbonate (DMDC; Fig. 29.4) is a colorless liquid which is slightly soluble in water (3.6%). The

$$H_3C-O-\overset{O}{\overset{\|}{C}}-O-\overset{O}{\overset{\|}{C}}-O-CH_3$$

Figure 29.4 Structure of DMDC.

compound is very reactive with many substances, including water, ethanol, alkyl and aromatic amines, and sulfhydryl groups (258). The primary target microorganisms for DMDC are yeasts, including *Saccharomyces, Zygosaccharomyces, Rhodotorula, Candida, Pichia, Torulopsis, Torula, Endomyces, Kloeckera,* and *Hansenula* species. The compound is also bactericidal at 30 to 400 mg/liter to a number of species, including *Acetobacter pasteurianus, E. coli, P. aeruginosa, S. aureus,* several *Lactobacillus* species, and *Pediococcus cerevisiae* (258). Molds are generally more resistant to DMDC than yeasts or bacteria. The mechanism by which DMDC acts is most likely related to inactivation of enzymes. A related compound, diethyl dicarbonate, reacts with imidazole groups, amines, or thiols of proteins (94). In addition, diethyl dicarbonate readily reacts with histidyl groups of proteins. This can cause inactivation of the enzyme lactate dehydrogenase or alcohol dehydrogenase by reacting with the histidine in the active site (258).

Phenolic Antioxidants

Phenolic antioxidants, including BHA (Fig. 29.5), BHT, propyl gallate, and TBHQ (Fig. 29.5), are used in foods primarily to delay autoxidation of unsaturated lipids.

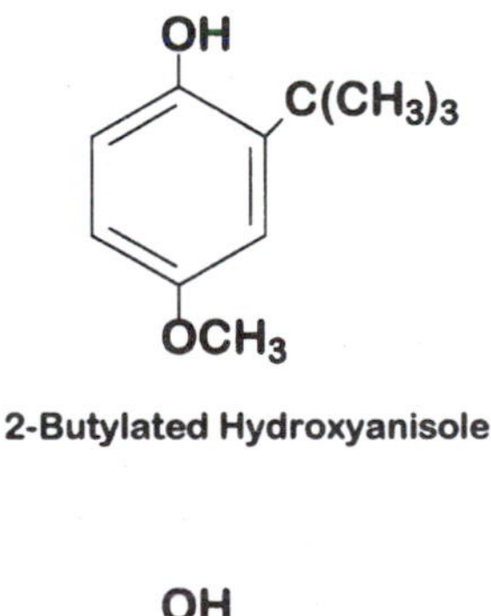

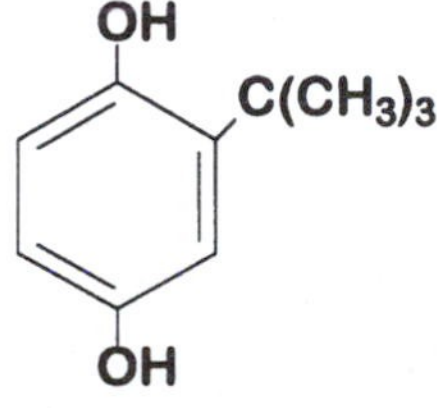

Tertiary Butylhydroquinone

Figure 29.5 Structures of phenolic antioxidants.

Table 29.3 Concentrations of the phenolic antioxidants BHA and TBHQ necessary for complete inhibition or delayed growth of bacteria in laboratory media (pH, incubation temperature, and time vary)[a]

Microorganism	Concn (μg/ml)	
	BHA	TBHQ
Gram positive		
Arthrobacter sp.	100	100
Bacillus cereus	100–200	
Bacillus megaterium	100–200	
Bacillus pumilis		50–100
Bacillus subtilis	100–2,500	50–100
Clostridium botulinum	25–50	
Clostridium perfringens	150	
Lactobacillus brevis	100	300
Listeria monocytogenes	128–300	64–105
Pediococcus pentosaceus		15–20
Pediococcus sp.	100	100
Sarcina lutea	150	100
Staphylococcus aureus	50–400	25–100
Streptococcus faecalis	400	>500
Gram negative		
Aeromonas hydrophila	100–150	
Arizona sp.	200	300
Citrobacter freundii	350	>500
Edwardsiella tarda	150	150
Enterobacter aerogenes	350	500
Escherichia coli	200–400	450
Klebsiella pneumoniae	400	500
Proteus vulgaris	250	250
Pseudomonas aeruginosa	400	500
Pseudomonas fluorescens	100–300	500
Pseudomonas fragi	100–>400	100
Salmonella senftenberg	>10,000	
Salmonella typhimurium	150–400	
Serratia marcescens	>500	400
Vibrio angularum	250	100
Vibrio parahaemolyticus	50	
Yersinia enterocolitica	250	250

[a]Source: reference 79.

This is accomplished by interrupting the free radical chain mechanism of hydroperoxide formation during the autoxidation process (333). Because these compounds are phenolic in nature, it was theorized that they may have antimicrobial activity. The first report on the antibacterial effectiveness of BHA was that of Chang and Branen (59), who performed a study in which *E. coli*, *Salmonella typhimurium*, and *S. aureus* were inhibited by 400, 400, and 150 μg/ml, respectively, in nutrient broth. This and further studies generally demonstrate that gram-positive bacteria are more susceptible to BHA than gram-negative bacteria (Table 29.3). BHT is generally less effective than other phenolic antioxidants (79). TBHQ is an extremely effective inhibitor of gram-positive bacteria, including *S. aureus* and *L. monocytogenes* (82, 107, 266, 304, 433). Research on TBHQ has shown it to be generally less effective against gram-negative bacteria than gram-positive bacteria. Generally, BHA is a more effective antifungal agent than TBHQ, BHT, or propyl gallate. The compound inhibits *A. flavus*, *A. parasiticus*, *Penicillium*, *Geotrichum*, and *Byssochlamys* species, and *Saccharomyces cerevisiae* (2, 33, 59, 83, 108, 127, 196, 294, 373).

Many environmental factors influence the antimicrobial activity of the phenolic antioxidants. The presence of lipid or protein dramatically decreases the activity of phenolic antioxidants (100, 194, 294, 302). Calcium, but not magnesium, increases the antimicrobial activity of the phenolic antioxidants (295, 407).

A number of studies have been carried out to determine the antimicrobial effectiveness of phenolic antioxidants in foods (79). In nearly all studies, the concentration of phenolic antioxidants required for inhibition in a food is significantly higher than that needed for in vitro inhibition, especially in meat products (79). Application studies of lower-fat and protein products or studies using surface applications have shown slightly more promise (2, 196, 266).

The mechanism of inhibition by phenolic antioxidants is most likely related to other phenolics and was discussed in the section on the parabens (see above).

Phosphates

Some phosphate compounds, including sodium acid pyrophosphate (SAPP), tetrasodium pyrophosphate (TSPP), sodium tripolyphosphate (STPP), sodium tetrapolyphosphate, sodium hexametaphosphate (SHMP), and trisodium phosphate (TSP), have variable levels of antimicrobial activity in foods (331). Phosphates have important uses in food processing, including buffering or pH stabilization, acidification, alkalization, sequestration or precipitation of metals, formation of complexes with organic polyelectrolytes (e.g., proteins, pectin, and starch), deflocculation, dispersion, peptization, emulsification, nutrient supplementation, anticaking, antimicrobial preservation, and leavening (103).

Gram-positive bacteria are generally more susceptible to phosphates than gram-negative bacteria (277). Kelch and Bühlmann (189) tested commercial mixtures of TSPP, SAPP, STPP, and SHMP with and without heat against gram-positive bacteria. *S. aureus* and *Enterococcus faecalis* were completely inhibited in nutrient medium plus heat (50°C). Without heat, susceptibility was variable. A mixture of TSPP, STPP, and SHMP inhibited *B. subtilis, C. sporogenes,* and *C. bifermentans* at 0.5%. In a similar study, Jen and Shelef (172) tested seven phosphate derivatives against the growth of *S. aureus* 196E. Only 0.3% SHMP (with a phosphate chain length *n* of 21) and 0.5% STPP or SHMP (n = 13 or 15) were effective growth inhibitors. Magnesium reversed the growth-inhibiting effect. Wagner and Busta (401) found that SAPP has no effect on the growth of *C. botulinum* but delayed or prevented toxicity to mice. It was theorized that this was due to binding of the toxin molecule or inactivation of the protease responsible for protoxin activation. Gould (135) showed that 0.2 to 1.0% SHMP permitted germination of *Bacillus* spores but prevented outgrowth. Zaika and Kim (434) found that 1% sodium polyphosphates inhibited lag and generation times of *L. monocytogenes* in brain heart infusion broth, especially in the presence of NaCl.

Phosphate derivatives also have antimicrobial activity in food products. The effect of phosphates in curing mixtures for meat was reviewed by Wagner (399). Several studies demonstrated that SAPP, SHMP, or polyphosphates enhance the effect of nitrite, pH, and salt against *C. botulinum* (168, 249, 308, 400). Post et al. (276) preserved cherries against fungal growth by *Penicillium, Rhizopus,* and *Botrytis* species with a 10% sodium tetrapolyphosphate dip. Tanaka (367) showed that phosphate along with sodium chloride, water activity, water content, pH, and lactic acid interact to prevent the outgrowth of *C. botulinum* in pasteurized process cheese. Various phosphate salts also have various antimicrobial activities against rope-forming *B. subtilis* in bread and salmonellae in pasteurized egg whites (378).

TSP at levels of 8 to 12% reduces the number of pathogens, especially salmonellae, on poultry (132). This was suggested to be a physical removal process rather than a chemical inactivation. Lillard (213) determined that while 10% TSP appeared to reduce viable salmonellae by 2 log units, this effect was a function of the high pH (11 to 12) of the of the system. Waldroup (403) and Waldroup et al. (404) concluded that TSP had no effect on pathogens or indicator microorganisms on poultry.

Several mechanisms have been suggested for bacterial inhibition by polyphosphates (349). The ability of polyphosphates to chelate metal ions appears to play an important role in their antimicrobial activity. Post et al. (277) found that the presence of magnesium reverses inhibition of gram-positive bacteria by polyphosphates. Jen and Shelef (172) also found that magnesium reverses inhibition of *S. aureus* by SHMP or STPP and that calcium and iron are partially effective in reversing inhibition. Knabel et al. (195) stated that the chelating ability of polyphosphates is responsible for growth inhibition of *B. cereus, L. monocytogenes, S. aureus,* lactobacilli, and *A. flavus.* Orthophosphates had no inhibitory activity against any of the microorganisms and have no chelating ability. Further, Knabel et al. (195) reported that inhibition is reduced at lower pH as a result of protonation of the chelating sites on the polyphosphates. They concluded that polyphosphates inhibited gram-positive bacteria and fungi by removal of essential cations from binding sites on the cell walls of these microorganisms. Polyphosphates may also interfere with RNA function or metabolic activities of cells (103, 349).

NATURALLY OCCURRING COMPOUNDS AND SYSTEMS

Many food products contain naturally occurring compounds which have antimicrobial activity. In the natural state, these compounds may play a role in extending the shelf life of a food product. In addition, many of these naturally occurring compounds have been studied for their potential as direct food antimicrobials. There are many problems associated with use of natural compounds as direct food additives. According to Beuchat and Golden (32), the challenge is to isolate, purify, stabilize, and incorporate natural antimicrobials into foods without adversely affecting sensory, nutritional, or safety characteristics. This must be done without significant increased costs for formulation, processing, or marketing (32).

Animal

Lactoperoxidase System

Lactoperoxidase is an enzyme that occurs in raw milk, colostrum, saliva, and other biological secretions. The enzyme is a glycoprotein with one heme group that contains Fe^{3+}. It has 610 amino acids and a molecular weight of 78,000 (101). The enzyme is similar to other biologically significant peroxidases such as human myeloperoxidase. Bovine milk naturally contains 10 to 60 mg of lactoperoxidase per liter (101, 420). This enzyme reacts with thiocyanate (SCN^-) in the presence of hydrogen peroxide and forms antimicrobial compound(s); this is termed the lactoperoxidase system (LPS). Thiocyanate is found in many biological secretions, including milk. It is formed by the detoxification of thiosulfates, by metabolism of sulfur containing amino acids, and in the diet through metabolism of various glucosides (420). Fresh milk contains 1 to 10 mg of thiocyanate per liter, which is not always sufficient to activate the LPS. Wilkins and Board (420) recommended addition of 10 to 12 mg of thiocyanate per liter. Hydrogen peroxide, the third component of the LPS, is not present in fresh milk because of the action of natural catalase, peroxidase, or superoxide dismutase. Approximately 8 to 10 mg of hydrogen peroxide per liter is required for LPS. This can be added directly, through the action of lactic acid bacteria, or through the enzymatic action of xanthine oxidase, glucose oxidase, or sulfhydryl oxidase. The amount of hydrogen peroxide required is much lower than that used in pasteurization of raw milk (ca. 300 to 800 mg/liter). In the LPS reaction, thiocyanate is oxidized to the antimicrobial hypothiocyanate ($OSCN^-$), which exists in equilibrium with hypothiocyanous acid (pK_a = 5.3) (16, 129, 290).

The LPS is more effective against gram-negative bacteria, including pseudomonads, than against gram-positive bacteria (36). However, it does inhibit both gram-positive and gram-negative foodborne pathogens, including salmonellae, *S. aureus*, *L. monocytogenes*, and *Campylobacter jejuni* (34, 184, 343). There is variable activity against catalase-negative microorganisms, including the lactic acid bacteria. In resistant strains, the enzyme NADH: $OSCN^-$ oxidoreductase oxidizes NADH with reduction of $OSCN^-$ (52). The LPS is theorized to be a protective mechanism against mastitis or a method for selection of beneficial microorganisms in the gut of the newborn calf. The LPS can increase the shelf life of raw milk in countries that have a poorly developed refrigerated storage system (36, 37, 149, 438). Zajac et al. (438) found that the keeping quality of raw milk is best at temperatures below 15°C. In addition, the LPS has been used as a preservation process in infant formula, ice cream, cream, and cheeses (101).

There are several potential mechanisms for inhibition by the LPS. The thiocyanate ion can oxidize sulfhydryl groups to disulfides, sulfenylthiocyanates (-S-SCN), or sulfenic acid (-S-OH) (101). Therefore, the compound may react with cysteine side chains on proteins and inactivate enzymes with functionally important sulfhydryl groups. Enzymes that are inhibited in vitro by thiocyanate include aldolase, glyceraldehyde phosphate dehydrogenase, hexokinase, lactate dehydrogenase, and 6-phosphogluconate dehydrogenase (101). Thiocyanate may also oxidize NADH or NADPH to the corresponding NAD or NADP (101). Through reaction with NADH or interference with membrane proteins, thiocyanate may cause leakage of the membrane or loss of electrochemical potential. This will eventually lead to inhibition of transport of amino acids and sugars.

Lactoferrin and Other Iron-Binding Proteins

Iron-binding proteins with potential for use as food antimicrobials occur in milk and eggs. In milk and colostrum, the primary iron-binding protein is lactoferrin. This protein is also found in other physiological fluids and polymorphonuclear leukocytes (234). Milk contains low concentrations of another iron-binding protein, transferrin, that is found in larger amounts in blood. Lactoferrin is a glycoprotein with a molecular weight of around 76,500. It exists in milk primarily as a tetramer with Ca^{2+} (101). Lactoferrin has two iron-binding sites per molecule. For each Fe^{3+} bound by lactoferrin, one bicarbonate (HCO_3^-) is required. Citrate, another chelator in milk, inhibits the antimicrobial activity of lactoferrin (288, 289) while the presence of bicarbonate reverses inhibition. Lactoferrin must be low in iron saturation and bicarbonate must be present for the compound to be an effective antimicrobial. Lactoferrin is potentially active in milk because of the low iron concentration and presence of bicarbonate (420). The exact biological role of lactoferrin is unknown; however, it may act as a barrier to infection of the nonlactating mammary gland and may help protect the gastrointestinal tract of the newborn against infection (372).

Lactoferrin is inhibitory to a number of microorganisms, including *B. subtilis*, *B. stearothermophilus*, *L. monocytogenes*, *Micrococcus* species, *E. coli*, and *Klebsiella* species (202, 225, 257, 265, 288). Payne et al. (265) found that bovine lactoferrin had to be reduced to 18% iron saturation (apo-lactoferrin) by using dialysis to have bacteriostatic activity against four strains of *L. monocytogenes* and an *E. coli* strain at concentrations of 15 to 30 mg/ml in ultrahigh-temperature-treated milk. At 2.5 mg/ml, the compound has no activity against *Salmonella typhimurium* or *P. fluorescens* and little activity against *E. coli* O157:H7 or *L. monocytogenes* VPHI (264). Some gram-negative bacteria may be resistant because they adapt to low-iron environments by producing siderophores such as phenolates and hydroxamates (75). Microorganisms with a low-iron requirement, such as lactic acid bacteria, would not be inhibited by lactoferrin (291).

Production of an iron-deficient environment by lactoferrin may be part of its mechanism of inhibition. Iron stimulates the growth of bacteria in many genera, including *Clostridium*, *Escherichia*, *Listeria*, *Pseudomonas*, *Salmonella*, *Staphylococcus*, *Vibrio*, and *Yersinia* (415). However, there is evidence that lactoferrin may have effects on gram-negative bacteria in addition to iron deprivation. In a study by Arnold et al. (8), lactoferrin inhibited many microorganisms that were not inhibited by a low iron environment. Ellison et al. (104) studied the potential mechanisms of lactoferrin against gram-negative bacteria. They found that both lactoferrin and EDTA, another chelator, cause release of anionic lipopolysaccharides from the outer membrane of *E. coli.* Lactoferrin causes this loss by chelation of cations, including magnesium, calcium, and iron, that stabilize lipopolysaccharides in the membrane. The latter was demonstrated by a reversal of lipopolysaccharide loss by addition of iron. Since it is cationic, lactoferrin may increase the outer membrane permeability to hydrophobic compounds, including other antimicrobials.

Lactoferricin B is a small peptide produced by acid-pepsin hydrolysis of bovine lactoferrin (27). It contains 25 amino acid residues and has a molecular weight of 3,000. Lactoferricin may be the antimicrobial fraction of the lactoferrin molecule, or lactoferrin may be the in vitro precursor of lactoferricin. Jones et al. (176) determined that the compound is inhibitory and cidal to *Shigella, Salmonella,* and *Candida* species, *Y. enterocolitica, E. coli* O157:H7, *S. aureus,* and *L. monocytogenes,* at concentrations ranging from 1.9 to 125 μg/ml. *Proteus* and *Serratia* species and *Pseudomonas cepacia* are resistant to 500 μg of lactoferricin B per ml. These results confirm and expand earlier in vitro inhibition studies with lactoferricin B (26, 377, 402). The mode of action of this compound has not been fully elucidated. Since it is a cationic peptide, it may act on the cell membrane by altering permeability and causing leakage (68, 176).

Another iron-binding molecule, ovotransferrin or conalbumin, occurs in egg albumen. This 77,000- to 80,000-Da glycoprotein makes up 10 to 13% of the total egg white protein (263, 420). Ovotransferrin has 49% sequence homology with lactoferrin (383). Each ovotransferrin molecule has two iron-binding sites and, like lactoferrin, it binds anions, such as bicarbonate or carbonate, with each ferric iron bound. The compound is 80% denatured by heating at 60°C for 5 min. The compound was first shown to be antimicrobial by Schade and Caroline (322), who demonstrated that raw egg white in nutrient broth inhibits *E. coli, Shigella* species, *S. aureus,* and *Saccharomyces cerevisiae.* Inhibition was reversed by iron. To be inhibitory, ovotransferrin must be in stoichiometric excess of iron and the pH must be in the alkaline range (above 7.5) (385, 420). Ovotransferrin is inhibitory against both gram-positive and gram-negative bacteria, but gram-positive bacteria are generally more sensitive, with *Bacillus* and *Micrococcus* species being particularly sensitive (383). Some yeasts are also sensitive. The compound is bacteriostatic rather than bactericidal. As with lactoferrin, one of the primary mechanisms suggested for ovotransferrin is deprivation of iron from microorganisms that require the element. However, Tranter and Board (384) first suggested that inhibition may be partially or completely related to the cationic nature of the protein. Certain gram-positive bacteria and yeasts are inhibited in egg albumen with or without iron (384). Later, Valenti et al. (393, 394) demonstrated that iron-saturated ovotransferrin inhibits *Candida albicans* and that fluorescence-labeled ovotransferrin (and lactoferrin) binds the yeast cell wall.

Avidin

Avidin is a glycoprotein present in egg albumen. The concentration varies with the hen's age, but the mean is 0.05% of the total egg albumen protein (383). The protein has a molecular mass of 66,000 to 69,000 Da and has four identical subunits of 128 amino acids each (87). It is stable to heat and a wide pH range (140). Avidin strongly binds the cofactor biotin at a ratio of four molecules of biotin per molecule of avidin. Biotin is a cofactor for enzymes in the tricarboxylic acid cycle and fatty acid biosynthesis. While the biological role of the protein is unknown, it may be secreted by macrophages and play a role in the immune system (383). Avidin inhibits growth of some bacteria and yeasts that have a requirement for biotin, with the primary mechanism being nutrient deprivation. However, Korpela (203) investigated another potential mechanism. He found that avidin can bind porin proteins in the outer membrane of *E. coli.* This finding suggested that avidin may inhibit microorganisms in vitro by interfering with transport through porins.

Lysozyme

Lysozyme (1,4-β-*N*-acetylmuramidase; EC 3.2.1.17) is a 14,600-Da enzyme present in avian eggs, mammalian milk, tears and other secretions, insects, and fish. While tears contain the greatest concentration of lysozyme, dried egg white (3.5%) is the commercial source (383). Lysozyme c, the enzyme in hen eggs, has 129 amino acids. It is stable to heat (100°C) at low pH (<5.3) but is inactivated at lower temperatures when the pH is increased. The enzyme catalyzes hydrolysis of the β-1,4 glycosidic bonds between *N*-acetylmuramic acid and *N*-acetylglucosamine of the peptidoglycan of bacterial cell walls. This causes cell wall degradation and lysis in hypotonic solutions. Depending on the enzyme source, chitin and certain esters are also susceptible to lysozyme.

Lysozyme is most active against gram-positive bacteria, most likely because the peptidoglycan of the cell wall is more exposed. Inhibition by the enzyme was first observed by Laschtschenko (206) with egg white and *Bacillus* species and later by Fleming (117) against *S. aureus*. The enzyme inhibits *C. botulinum, C. thermosaccharolyticum, C. tyrobutyricum, Bacillus stearothermophilus, B. cereus, Micrococcus lysodeikticus,* and *L. monocytogenes* (53, 91, 158, 159, 392). Lysozyme is the primary antimicrobial compound in egg albumen, but its activity is enhanced by ovotransferrin, ovomucoid, and alkaline pH (408). Johansen et al. (174) suggested that low pH (5.5) causes increased inhibition of *L. monocytogenes* by lysozyme because the organism has a slower growth rate which allows enzymatic hydrolysis of the cell wall to exceed the cell proliferation rate. Variation in susceptibility of gram-positive bacteria is likely due to the presence of teichoic acids and other materials that bind the enzyme and the fact that certain species have greater proportions of 1,6 or 1,3 glycosidic linkages in the peptidoglycan which are more resistant than the 1,4 linkage (383). For example, four strains of *L. monocytogenes* were not inhibited by lysozyme alone but were inhibited when EDTA was added (158, 264). Hughey and Johnson (158) hypothesized that the peptidoglycan of the microorganism may be partially masked by other cell wall components and that EDTA enhances penetration of the lysozyme to the peptidoglycan. Lysozyme is less effective against gram-negative bacteria because of their reduced peptidoglycan content (5 to 10%) and presence of the outer membrane of lipopolysaccharide and lipoprotein (420). Gram-negative cell susceptibility can be increased by pretreatment with chelators (e.g., EDTA) that bind Ca^{2+} or Mg^{2+}, which are essential for maintaining integrity of the lipopolysaccharide layer, or by the antibiotics polymyxin B or aminoglycosides, which disrupt the lipopolysaccharide (321, 392, 420). In addition, gram-negative cells may be sensitized to lysozyme if the cells are subjected to pH shock, heat shock, osmotic shock, drying, and freeze-thaw cycling (282, 383, 420). Samuelson et al. (321) found that EDTA plus lysozyme inhibits *Salmonella typhimurium* on poultry. In contrast, no inhibition was demonstrated with up to 2.5 mg of EDTA per ml and 200 μg of lysozyme per ml in milk against either *Salmonella typhimurium* or *P. fluorescens* in a study by Payne et al. (264). It was theorized that there may be a significant influence of the food product on activities of lysozyme and EDTA. There is some evidence that lysozyme obtained from milk is more inhibitory to both gram-positive and gram-negative bacteria than the enzyme from hen egg albumen (289).

Lysozyme is one of the few naturally occurring antimicrobials approved by regulatory agencies for use in foods. In Europe, lysozyme is used to prevent gas formation ("blowing") in cheeses such as Edam and Gouda by *C. tyrobutyricum* (51, 410). Lysozyme is used to a great extent in Japan to preserve seafood, vegetables, pasta, and salads. The enzyme has shown potential for use as an antimicrobial with EDTA to control the growth of *L. monocytogenes* in vegetables but is less effective in refrigerated meat and soft cheese products (159).

Plant

Spices and Their Essential Oils

Spices are roots, bark, seeds, buds, leaves, or fruit of aromatic plants added to foods as flavoring agents. However, it has been known since ancient times that spices and their essential oils have varying degrees of antimicrobial activity. The earliest report on use of spices as preservatives was around 1550 B.C., when the ancient Egyptians used spices for food preservation and embalming the dead. Cloves, cinnamon, oregano, and thyme and to a lesser extent sage and rosemary have the strongest antimicrobial activity among spices.

The major antimicrobial components of clove and cinnamon are eugenol [2-methoxy-4-(2-propenyl)-phenol; Fig. 29.6] and cinnamic aldehyde (3-phenyl-2-propenal; Fig. 29.6), respectively. Cinnamon contains 0.5 to 1.0% volatile oil, of which 75% is cinnamic aldehyde and 8% is eugenol, while cloves contain 14 to 21% volatile oil, 95% of which is eugenol (49). One of the first studies on the antimicrobial activity of spices was by Chamberland (57), who showed that cinnamon inhibits

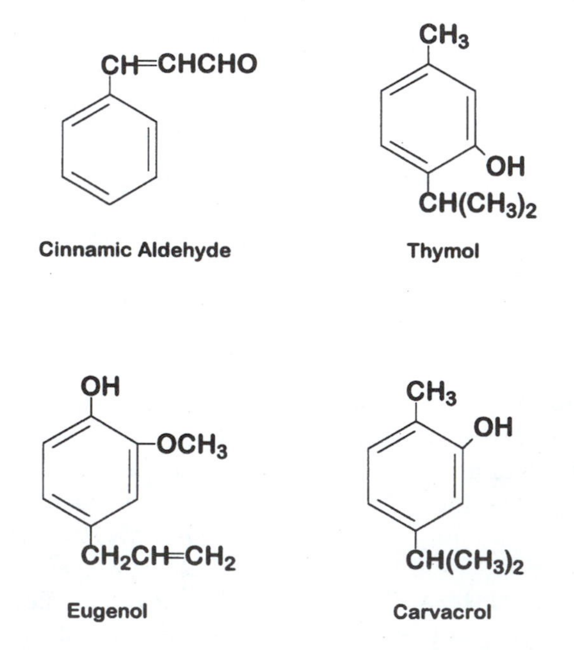

Figure 29.6 Structures of antimicrobial compounds found in spices

spores of *Bacillus anthracis.* Fabian et al. (111) tested 10% extracts of cinnamon and clove against *B. subtilis* and *S. aureus* and found cinnamon only slightly inhibitory but clove extract very inhibitory at 1:100 and 1:800, respectively. Zaika and Kissinger (435) showed that clove at 0.4% in a Lebanon bologna formulation inhibits growth and acid production by a lactic acid bacterial starter culture, while cinnamon at 0.8% inhibits growth slightly. Both spices stimulated acid production at low concentrations. Aqueous clove infusions (0.1 to 1.0% [wt/vol]) and eugenol (0.06%) inhibit outgrowth of germinated spores of *B. subtilis* in nutrient agar (5). Stecchini et al. (359) showed that essential oil of clove at 500 µg/ml inhibits *Aeromonas hydrophila* growth in microbiological media and in vacuum-packaged and air-packed cooked pork. Similarly, 0.5% clove oil in combination with 2% NaCl completely inhibits growth and biogenic amine formation by *Enterobacter aerogenes* in mackerel broth (416). Bullerman (46) determined that 1.0% cinnamon in raisin bread inhibits growth and aflatoxin production by *A. parasiticus.* In a related study (49), eugenol and cinnamic aldehyde were more effective inhibitors of *A. parasiticus* growth and toxin production than the parent spices. Azzouz and Bullerman (11) evaluated 16 ground herbs and spices at 2% (wt/vol) against nine mycotoxin-producing *Aspergillus* and *Penicillium* species. The most effective antimicrobial spice evaluated was clove, which inhibited growth initiation at 25°C by all species for over 21 days. Cinnamon was the next most effective spice, inhibiting three *Penicillium* species for over 21 days.

The antimicrobial activities of oregano and thyme have been attributed to their essential oils which contain the terpenes carvacrol [2-methyl-5-(1-methylethyl)phenol; Fig. 29.6] and thymol [5-methyl-2-(1-methylethyl)-phenol; Fig. 29.6], respectively. These compounds have inhibitory activities against a number of bacterial species, molds, and yeasts including *B. subtilis, E. coli, Lactobacillus plantarum, Pediococcus cerevisiae, P. aeruginosa, Proteus* species, *Salmonella enteritidis, S. aureus, V. parahaemolyticus,* and *A. parasiticus* (28, 44, 70, 186, 231, 232, 436, 437).

Sage and rosemary also have antimicrobial activity. The active fractions of the sage and rosemary were suggested to be the terpene fractions of the essential oils. Rosemary contains primarily borneol (endo-1,7,7-trimethylbicyclo[2.2.1] heptan-2-ol) along with pinene, camphene, and camphor, while sage contains thujone {4-methyl-1-(1-methylethyl)bicyclo[3.1.0]-hexan-3-one}. At 2% in growth medium, sage and rosemary are more active against gram-positive than gram-negative bacterial strains (330). The inhibitory effect of these two spices at 0.3% is bacteriostatic, while at 0.5% they are bactericidal to the gram-positive strains. The sensitivity of *B. cereus, S. aureus,* and pseudomonads to sage is greatest in microbiological medium and significantly reduced in rice and in chicken and noodles (329). It was theorized that loss of activity was due to solubilization of the antimicrobial fraction in the lipid of the foods.

Vanillin (4-hydroxy-3-methoxybenzaldehyde) is a major constituent of vanilla beans, the fruit of an orchid (*Vanilla planifola, Vanilla pompona,* or *Vanilla tahitensis*). Vanillin is most active against molds and non-lactic acid gram-positive bacteria (171). López-Malo et al. (216) prepared fruit-based agars containing mango, papaya, pineapple, apple, and banana with up to 2,000 µg of vanillin per ml and inoculated each with *A. flavus, A. niger, A. ochraceus,* or *A. parasiticus.* Vanillin at 1,500 µg/ml significantly inhibited all strains of *Aspergillus* in all media. The compound was least effective in banana and mango agars. The authors attributed this to binding of the phenolic vanillin by protein or lipid in these fruits, a phenomenon demonstrated for other antimicrobial phenolic compounds (294).

Many other spices have been tested and shown to have limited or no activity. They include allspice, anise, bay leaf, black pepper, cardamom, celery seed, chili powder, coriander, cumin, curry powder, dill, fenugreek, ginger, juniper oil, mace, marjoram, mustard, nutmeg, orris root, paprika, parsley, red pepper, sesame, spearmint, tarragon, and white pepper (111, 228, 412, 435).

Few studies have focused on the mechanism by which spices or their essential oils inhibit microorganisms. Since it has been concluded that the terpenes in essential oils of spices are the primary antimicrobials, the mechanism most likely involves these compounds. Many of the most active terpenes, e.g., eugenol, thymol, and carvacrol, are phenolic in nature. Therefore, it would seem reasonable that their modes of action might be related to those of other phenolic compounds. As was discussed for parabens, the mode of action of phenolic compounds is generally thought to involve interference with functions of the cytoplasmic membrane, including PMF and active transport (79, 98). In addition, terpenes may have other antimicrobial mechanisms. Conner and Beuchat (70) suggested that spice essential oils may inhibit enzyme systems in yeasts, including those involved in energy production and synthesis of structural components.

Onions and Garlic

Probably the best-characterized antimicrobial system in plants is that found in the juice and vapors of onions

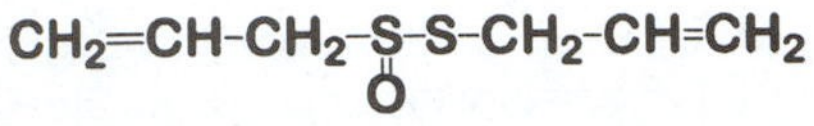

$CH_2{=}CH{-}CH_2{-}\underset{\underset{O}{\|}}{S}{-}S{-}CH_2{-}CH{=}CH_2$

Figure 29.7 Structure of allicin.

(*Allium cepa*) and garlic (*Allium sativum*). Walton et al. (406) and Lovell (217) reported that volatile agents from both onion and garlic inhibit *B. subtilis, Serratia marcescens*, and mycobacteria in microbiological media. Since those early reports, the growth and toxin production of many microorganisms, including *B. cereus, C. botulinum* type A, *E. coli, Lactobacillus plantarum, Leuconostoc mesenteroides*, salmonellae, shigellae, *S. aureus*, and the fungi *A. flavus, A. parasiticus, Candida albicans*, and *Cryptococcus, Penicillium, Rhodotorula, Saccharomyces, Torulopsis*, and *Trichosporon* species, have been shown to be inhibited by onion and garlic (3, 21, 69, 70, 78, 89, 134, 138, 175, 185, 320, 326, 327, 397, 422).

Cavallito and Bailey (56) isolated the major antimicrobial compounds from garlic by using steam distillation of ethanolic extracts. They identified the antimicrobial component as allicin (diallyl thiosulfinate; thio-2-propene-1-sulfinic acid-S-allyl ester; Fig. 29.7). Allicin is formed by the action of the enzyme allinase on the substrate alliin [S-(2-propenyl)-L-cysteine sulfoxide]. The reaction occurs only when cells of the garlic are disrupted, releasing the enzyme to act on the substrate. A similar reaction occurs in onion except the substrate is S-(1-propenyl)-L-cysteine sulfoxide and one of the major products is thiopropanal-*S*-oxide. The products apparently responsible for antimicrobial activity are also responsible for the flavor of onions and garlic. In addition to antimicrobial sulfur compounds, onions contain the phenolic compounds protocatechuic acid and catechol, which could contribute to their antimicrobial activities (405).

The mechanism of action of allicin is most likely inhibition of sulfhydryl-containing enzymes (29). Wills (421) found that 0.5 mM allicin inhibits many sulfhydryl enzymes, including alcohol dehydrogenase, choline esterase, choline oxidase, glyoxylase, hexokinase, papain, succinic dehydrogenase, urease, and xanthine oxidase. Carboxylase, ATPase, and β-amylase are not affected by allicin. Some nonsulfhydryl enzymes, including lactic dehydrogenase, tyrosinase, and alkaline phosphatase, were also inhibited. Barone and Tansey (21) theorized that allicin inactivates proteins by oxidizing thiols to disulfides and inhibiting the intracellular reducing activity of glutathione and cysteine. Conner et al. (71) showed that garlic oil inhibits ethanol production by *Saccharomyces cerevisiae* and delays sporulation of *Hansenula anomala* and *Lodderomyces elongisporus*.

Other Plant Extracts

Isothiocyanate (R-N=C=S) derivatives from glucosinolates in cells of plants of the Cruciferae, or mustard family (cabbage, kohlrabi, Brussels sprouts, cauliflower, broccoli, kale, horseradish, mustard, turnips, rutabaga), are potent antifungal and antimicrobial agents (88). These compounds are formed from the action of the enzyme myrosinase (thioglucoside glucohydrolase; EC 3.2.3.1) on the glucosinolates when plant tissue is injured or disrupted. Common isothiocyanate side groups include allyl, ethyl, methyl, benzyl, and phenyl. Most of the antimicrobial studies with these compounds have been with plant pathogens. The compounds are inhibitory to fungi, yeasts, and bacteria in the range of 0.016 to 0.062 μg/ml in the vapor phase (167) or 10 to 600 μg/ml in liquid media (227, 398). Inhibition depends on the compound type against bacteria, but generally gram-positive bacteria are less sensitive to the allyl side group than are gram-negative bacteria. The mechanism by which isothiocyanates inhibit cells may involve enzymes by direct reaction with disulfide bonds or through thiocyanate anion reaction to inactivate sulfhydryl enzymes (88). The isothiocyanates may act as uncouplers of oxidative phosphorylation (197). While these compounds have very low sensory thresholds, they may be useful as food antimicrobials because of their low inhibitory concentrations.

Beuchat and Brackett (30) and Beuchat et al. (31) found that an aqueous extract of carrots inhibits *L. monocytogenes* and that inhibition is dependent on juice concentration, pH, and presence of NaCl. Marchetti et al. (226) found slight growth inhibition of *P. fluorescens* and *Aeromonas hydrophila* grown in carrot extract compared with controls. Babic et al. (12) prepared a purified methanolic extract of carrots and found it to be bactericidal against *Leuconostoc mesenteroides, S. aureus, L. monocytogenes, E. coli, P. fluorescens*, and the yeast *Candida lambica* at concentrations ranging from 55 to 220 mg/ml. Inhibition by the carrot extract was not due to the phenolic compounds (chlorogenic acid; *p*-hydroxybenzoic acid and its esters) or to the phytoalexin 6-methoxymellein. Rather, the compounds dodecanoic (lauric) acid and methyl esters of dodecanoic (monolaurin) and pentadecanoic acids were identified by gas chromatography-mass spectrometry in the extract and suggested to be responsible for inhibition. Batt et al. (22) showed that an extract of carrots that inhibited sporulation and aflatoxin production by *A. parasiticus*. The inhibitor was not identified but was determined not to be 6-methoxymellein, *p*-hydroxybenzoic acid, or falcarindiol but rather to be part of the volatile carrot seed oil, a mixture of terpenoid compounds (23).

Phenolic Compounds

A phenolic compound can be defined as any substance that possesses an aromatic ring with one or more hydroxyl groups and can include functional derivatives. Phenolic compounds occurring in foods are classified as simple phenols and phenolic acids, hydroxycinnamic acid derivatives, and the flavonoids (155). Some of the important phenolic compounds, including alkyl esters of parabens, phenolic antioxidants (e.g., BHA and TBHQ), and certain of the terpene fraction of the essential oils of spices (e.g., thymol, carvacrol, eugenol, vanillin), have already been discussed. Only the latter compounds are naturally occurring.

Much of the work on the mechanisms of phenolic compounds centers on their effects on cellular membranes (79). Simple phenols disrupt the cytoplasmic membrane and cause leakage of cells (81, 86, 128, 177, 396, 407). A phenolic glycoside isolated from green olives, oleuropein, causes losses of labeled glutamate, potassium, and phosphate from *Lactobacillus plantarum* cells (181). The compound also decreases cellular ATP but has no effect on the glycolysis rate. Juven et al. (181) concluded that the leakage of these compounds is caused by loss in membrane permeability. By interfering with the permeability of the cytoplasmic membrane, the phenolics inhibit cells by interfering with components of the PMF and thereby inhibiting active transport of nutrients (98, 125). Phenolics may also inhibit cellular proteins directly (280). Rico-Muñoz et al. (293) investigated the effect of the naturally occurring propyl gallate, *p*-coumaric acid, ferulic acid, and caffeic acids on the activity of membrane-bound ATPase of *S. aureus*. All of the compounds inhibit ATPase activity. It was suggested that compounds that inhibited ATPase interact directly with the enzyme, possibly through binding. The authors concluded that all phenolic compounds probably do not have the same mechanism of action and that there may be several targets which lead to inhibition of microorganisms.

Simple Phenols and Phenolic Acids

Simple phenolic compounds include monophenols (e.g., *p*-cresol), diphenols (e.g., hydroquinone), and triphenols (e.g., gallic acid). Gallic acid occurs in plants as quinic acid esters or hydrolyzable tannins (tannic acid) (155).

The only practical use of simple phenols for preservation is found in the application of wood smoke. Smoking of foods such as meats, cheeses, fish, and poultry not only imparts a desirable flavor but also has a preservative effect through drying and the deposition of chemicals (17). While many chemicals are deposited on smoked foods, the major contributors of flavor and antioxidant and antimicrobial effects are phenol and cresol (126). Fretheim et al. (126) determined that beechwood smoke distillate is completely inhibitory to *S. aureus*, *E. coli*, and *Saccharomyces cerevisiae* at 667, 1,250, and 5,000 μg/ml, respectively. Their results are consistent with other studies using phenolic compounds, in that the gram-positive *S. aureus* is more susceptible than *E. coli*. Several commercial smoke preparations at 0.25 and 0.5% cause a reduction in viable cells of *L. monocytogenes* in saline-phosphate buffer (240). Dipping frankfurters inoculated with *L. monocytogenes* in full-strength liquid smoke eliminates viable cells within 72 h at 4°C. Liquid smoke on the surface of Cheddar cheese inhibits growth of *A. oryzae*, *Penicillium camemberti*, and *P. roqueforti* (417). Of the eight major phenolic compounds in liquid smoke, isoeugenol is the most effective antifungal compound, followed by *m*-cresol and *p*-cresol. Faith et al. (113) also identified isoeugenol as the most effective component of liquid smoke against the growth of *L. monocytogenes*.

The phenolic acids, including derivatives of *p*-hydroxybenzoic acid (protocatechuic, vanillic, gallic, syringic, ellagic) and *o*-hydroxybenzoic acid (salicylic) may be found in plants and foods (148, 153, 224). Schanderl (323) found that 100 μg of gallic, caffeic (a hydroxycinnamic acid), and quinic acids per ml halt wine yeast fermentations in the presence of 6% ethanol. Reddy et al. (285) tested gallic acid and its esters for their effectiveness against *C. botulinum* in tryptone-yeast extract-glucose medium at 37°C and found inhibition only with the esters butyl gallate, isobutyl gallate, isoamyl gallate, *p*-octyl gallate, and *p*-dodecyl gallate. Chung et al. (67) also found that gallic and ellagic acid had no effect on *Alcaligenes faecalis*, *Enterobacter aerogenes*, *E. coli*, *Klebsiella pneumoniae*, *L. monocytogenes*, *Proteus vulgaris*, *Salmonella enteritidis*, *Salmonella paratyphi*, *Shigella flexneri*, *S. aureus*, or *Enterococcus faecalis*. In contrast, propyl gallate inhibited all of the microorganisms except *S. aureus* and *Enterococcus faecalis*. Addition of 250 μg of propyl gallate per ml and 250 μg of tannic acid per ml to cabbage juice inhibited growth of *L. monocytogenes*. Payne et al. (266) showed that tannic acid at 256 μg/ml was inhibitory to *L. monocytogenes* in tryptose phosphate agar incubated 18 h at 35°C. Chung and Murdock (66) found tannic acid to have antimicrobial activity against *Aeromonas hydrophila, E. coli, L. monocytogenes, Salmonella enteritidis, S. aureus,* and *Enterococcus faecalis.* Stead (358) demonstrated stimulation of growth of the lactic acid spoilage bacteria *Lactobacillus collinoides* and *Lactobacillus brevis* by gallic, quinic, and chlorogenic acids. He suggested that the compounds were being metabolized by the microorganisms and concluded that the presence of these compounds in fruit-based beverages may result in antagonistic reactions with other food-grade antimicrobials.

The phenolic glycoside oleuropein, or its aglycone, inhibits *Lactobacillus plantarum, Leuconostoc mesenteroides, P. fluorescens, B. subtilis, Rhizopus* sp., and *Geotrichum candidum* (118, 119, 180, 182). *Enterobacter aerogenes, E. coli,* or *Candida, Pichia,* or *Saccharomyces* species are not inhibited at concentrations up to 2,000 μg/ml (180, 182).

Resin from the flowers of the hop vine (*Humulus lupulus* L.) is used in the brewing industry for imparting a desirable bitter flavor to beer. About 3 to 12% of the resin is composed of α-bitter acids, including humulone (humulon), cohumulone, and adhumulone, and β-bitter acids, including lupulone (lupulon), colupulone, xanthohumol, and adlupulone (256). Both types of bitter acids possess antimicrobial activity primarily at low water activity against gram-positive bacteria and fungi (29). Lactic acid bacteria that spoil beer, including species of *Lactobacillus* and *Pediococcus,* were resistant to the antimicrobial effects of humulone, colupulone, and *trans*-isohumulone, while the same species that did not spoil beer were sensitive (114, 115). Haas and Barsoumian (143) found *E. coli* and *B. subtilis* naturally resistant to both the iso-α-acid and β-acids, while strains of *Enterococcus salivarius, S. aureus,* and *Bacillus megaterium* were susceptible. Resistance to *trans*-isohumulone does not confer resistance to sorbic acid or benzoic acid (115), indicating a potential difference in targets for mechanism of antimicrobial action.

Hydroxycinnamic Acids

Hydroxycinnamic acids include caffeic, *p*-coumaric, ferulic, and sinapic acids. They occur frequently as esters and less often as glucosides (155). They are also found in plants and foods (148, 153, 224). In tests with chlorogenic (3-caffeoyl quinic acid), isochlorogenic (dicaffeoyl quinic acid), caffeic, gallic, and ellagic acids, Sikovec (338, 339) determined that a mixture of the compounds at 800 μg/ml inhibited respiration (CO_2 production) by wine yeasts approximately 50%. Grodzinska-Zachwieja and Kahl (142) discovered that an extract isolated from chicory root (*Chicorium intybus* L) and containing chlorogenic and isochlorogenic acids was relatively effective against both gram-negative and gram-positive bacteria. They concluded that the active portion of the molecule was caffeic acid. Marchetti et al. (226) found that sterile juices of green and red chicory were bacteriostatic and bactericidal, respectively, to strains of *P. fluorescens* and *Aeromonas hydrophila.* Herald and Davidson (152) demonstrated that ferulic acid at 1,000 μg/ml and *p*-coumaric acid at 500 and 1,000 μg/ml inhibit the growth of *B. cereus* and *S. aureus.* The compounds were much less effective against *P. fluorescens* and *E. coli.* In contrast, alkyl esters of hydroxycinnamic acids, including methyl caffeoate, ethyl caffeoate, propyl caffeoate, methyl *p*-coumarate, and methyl cinnamate, were effective inhibitors of the growth of *P. fluorescens* (19). Lyon and McGill (220) tested the antimicrobial activities of caffeic, cinnamic, ferulic, salicylic, sinapic, and vanillic acids on *Erwinia carotovora* and demonstrated total inhibition at pH 6.0 in nutrient broth with 0.5 mg of cinnamic, ferulic, salicylic, or vanillic acid per ml. While effective as growth inhibitors of *Erwinia carotovora,* none of these phenolic compounds inhibit the pectolytic enzymes, polygalacturonic acid lyase and polygalacturonase, produced by the organism except caffeic acid, which inhibits only the latter (221). Stead (357) determined the effects of caffeic, coumaric, and ferulic acids against the wine spoilage lactic acid bacteria *Lactobacillus collinoides* and *Lactobacillus brevis.* At pH 4.8 in the presence of 5% ethanol, *p*-coumaric and ferulic acids were the most inhibitory compounds at 500 and 1,000 μg/ml. At 100 μg/ml, all three hydroxycinnamic acids stimulated growth of the microorganisms, suggesting that these compounds may play a role in initiating the malolactic fermentation of wines.

Many of the studies with hydroxycinnamic acids have involved their antifungal properties. Valle (395) reported that 500 μg of caffeic acid per ml and 1,000 μg of chlorogenic acid per ml inhibit a species of *Fusarium.* Chipley and Uraih (64) showed that ferulic acid at 5.0 mg per flask (ca. 26 ml) inhibits aflatoxin B_1 and G_1 production of *A. flavus* by approximately 50% and that of *A. parasiticus* by 75%. Salicylic and *trans*-cinnamic acids totally inhibit aflatoxin production at the same 5.0-mg-per-flask level. Baranowski et al. (18) studied the effects of caffeic, chlorogenic, *p*-coumaric, and ferulic acids at pH 3.5 on the growth of *Saccharomyces cerevisiae.* Caffeic and chlorogenic acid had little effect on the organism at 1,000 μg/ml. In the presence of *p*-coumaric acid, however, the organism was completely inhibited by 1,000 μg/ml. Ferulic acid was the most effective growth inhibitor tested. At 50 μg/ml, this compound extended the lag phase of *Saccharomyces cerevisiae,* and at 250 μg/ml, growth of the organism was completely inhibited. The degree of inhibition was inversely related to the polarity of the compounds.

The furocoumarins are related to the hydroxycinnamates. These compounds, including psoralen and its derivatives, are phytoalexins in citrus fruits, parsley, carrots, celery, and parsnips at levels of 2 to 6 μg/ml. They inhibit many gram-positive bacteria, including *B. subtilis* and *S. aureus,* following irradiation with long-wave (365 nm) UV light (9, 122). These compounds inhibit growth by interfering with DNA replication. This occurs when the furocoumarin and DNA are exposed to UV light at 365 nm, which causes monoadducts of DNA and cross-linking of the furocoumarin and DNA (246, 247).

Flavonoids

The flavonoids consist of catechins and flavons, flavonols, and their glycosides (155). Proanthocyanidins or condensed tannins are polymers of favan-3-ol and are found in apples, grapes, strawberries, plums, sorghum, and barley (155). Wine yeasts, wild yeasts, and vinegar bacteria were unaffected or affected only slightly by tannins at concentrations as high as 20,000 μg/ml (388). Schanderl (323), however, reported that 2,000 μg of tannin per ml halted wine yeast fermentations earlier than in a control. Singleton and Esau (342) stated that the increased antimicrobial activity of wines at 8 to 10 years was due to the formation of tannins. They predicted that the antibacterial effects of red and white wines would be proportional to the amount of flavonoid tannin present. Marwan and Nagel (233) did a very interesting study to identify the antimicrobial compounds associated with cranberries. Benzoic acid, proanthocyanidins, and flavonols account for 66% of cranberry microbial inhibition against *Saccharomyces bayanus,* with the latter two being the most important. While inhibition by benzoic acid decreased with increasing pH, there was no effect on the activity of the flavonols and increased activity of the proanthocyanidins at higher pH levels. Konowalchuk and Speirs (198–201) published a series of reports on the antiviral activities of extracts of blueberries, crabapples, strawberries, red wines, grape juice, apple juice, and tea. Some of these extracts were able to inactivate poliovirus, coxsackievirus, echovirus, reovirus, and herpes simplex virus. They concluded that the primary viral inhibitors in these products are the condensed or hydrolyzable tannins.

The flavones, flavonols, and their glycosides occur naturally in fruits and vegetables. One of the most common flavonols in the plant kingdom is quercetin. Schraufstatter (324) investigated the antimicrobial effects of chalcone, flavonone, flavone, and flavonol. All compounds except flavonol have a bacteriostatic effect on *S. aureus.* Natural sources of these compounds also have slight bacteriostatic activity against the microorganism. Payne et al. (266) found no inhibition of any of six strains of *L. monocytogenes* by quercetin at 205 mg/ml.

CONCLUSIONS

Traditional food antimicrobials are an important tool in preserving foods from the hazardous and detrimental effects of microorganisms. In the future, they will continue to play an important role in food preservation. However, it is doubtful that any newly discovered synthetic antimicrobials will be approved for use in foods by worldwide regulatory agencies. Therefore, the future of traditional antimicrobials lies in novel uses (e.g., new foods or in combination with other preservation procedures) and in "designer" combinations, i.e., combinations of antimicrobials designed on the basis of their mechanisms so as to obtain a wide spectrum of activity against microorganisms.

Extensive research on sources and activities of natural antimicrobials is currently taking place. The vast number of potential antimicrobial compounds from nature and food sources precludes their discussion in the short space of this chapter. In the future, more information needs to be gathered on these natural compounds, including their activity spectra, mechanisms, targets, and interactions with environmental conditions and food components. Prior to their regulatory approval as food antimicrobials, they will have to be proven toxicologically safe. The latter point could be the greatest hurdle to their future use.

References

1. **Aalto, T. R., M. C. Firman, and N. E. Rigler.** 1953. *p*-Hydroxybenzoic acid esters as preservatives. I. Uses, antibacterial and antifungal studies, properties and determination. *J. Am. Pharm. Assoc. Sci. Ed.* **42:**449–458.
2. **Ahmad, S., and A. L. Branen.** 1981. Inhibition of mold growth by butylated hydroxyanisole. *J. Food Sci.* **46:**1059–1063.
3. **Al-Delaimy, K. S., and S. H. Ali.** 1970. Antibacterial action of vegetable extracts on the growth of pathogenic bacteria. *J. Sci. Food Agric.* **21:**110–112.
4. **Al-Issa, M., D. R. Fowler, A. Seaman, and M. Woodbine.** 1984. Role of lipid in butylated hydroxyanisole resistance of *Listeria monocytogenes. Zentralbl. Bakteriol. Hyg. A* **258:**42–50.

5. **Al-Khayat, M. A., and G. Blank.** 1985. Phenolic spice components sporostatic to *Bacillus subtilis*. *J. Food Sci.* **50:**971–974.
6. **Amerine, M. A., and M. A. Joslyn.** 1970. *Table Wines: the Technology of Their Production,* 2nd ed. University of California Press, Berkeley, Calif.
7. **Anderson, M. E., H. E. Huff, H. D. Naumann, and R. T. Marshall.** 1988. Counts of six types of bacteria on lamb carcasses dipped or sprayed with acetic acid at 25°C or 55°C and stored vacuum packaged at 0°C. *J. Food Prot.* **51:**874–877.
8. **Arnold, R. R., J. E. Russell, W. J. Champion, M. Brewer, and J. J. Gauthier.** 1982. Bactericidal activity of human lactoferrin: differentiation from the stasis of iron deprivation. *Infect. Immun.* **35:**792–797.
9. **Ashwood-Smith, M. J., O. Ceska, S. K. Chaudhary, P. J. Warrington, and P. Woodcock.** 1986. Detection of furocoumarins in plants and plant products with an ultrasensitive biological photoassay employing a DNA-repair-deficient bacterium. *J. Chem. Ecol.* **12:**915–932.
10. **Azukas, J. J., R. N. Costilow, and H. L. Sadoff.** 1961. Inhibition of alcoholic fermentation by sorbic acid. *J. Bacteriol.* **81:**189–194.
11. **Azzouz, M. A., and L. B. Bullerman.** 1982. Comparative antimycotic effects of selected herbs, spices, plant components and commercial fungal agents. *J. Food Sci.* **45:**1298–1301.
12. **Babic, I., C. Nguyen-the, M. J. Amiot, and S. Aubert.** 1994. Antimicrobial activity of shredded carrot extracts on food-borne bacteria and yeast. *J. Appl. Bacteriol.* **76:**135–141.
13. **Baird-Parker, A. C.** 1980. Organic acids, p. 126–135. *In* International Commission on Microbiological Specifications for Foods (ed.), *Microbial Ecology of Foods,* vol. I. *Factors Affecting Life and Death of Microorganisms.* Academic Press, New York.
14. **Baldock, J. D., P. P. Frank, P. P. Graham, and F. J. Ivey.** 1979. Potassium sorbate as a fungistatic agent in country ham processing. *J. Food Prot.* **42:**780–783.
15. **Banks, J. G., and R. G. Board.** 1982. Sulfite-inhibition of *Enterobacteriaceae* including *Salmonella* in British fresh sausage and in culture systems. *J. Food Prot.* **45:**1292–1297.
16. **Banks, J. G., R. G. Board, and N. H. C. Sparks.** 1986. Natural antimicrobial systems and their potential in food preservation of the future. *Biotechnol. Appl. Biochem.* **8:**103–147.
17. **Banwart, G. J.** 1979. *Basic Food Microbiology.* AVI Publishers, Westport, Conn.
18. **Baranowski, J. D., P. M. Davidson, C. W. Nagel, and A. L. Branen.** 1980. Inhibition of *Saccharomyces cerevisiae* by naturally occurring hydroxycinnamates. *J. Food Sci.* **45:** 592–594.
19. **Baranowski, J. D., and C. W. Nagel.** 1983. Properties of alkyl hydroxycinnamates and effects on *Pseudomonas fluorescens. Appl. Environ. Microbiol.* **45:**218–222.
20. **Bargiota, E. E., E. Rico-Muñoz, and P. M. Davidson.** 1987. Lethal effect of methyl and propyl parabens as related to *Staphylococcus aureus* lipid composition. *Int. J. Food Microbiol.* **4:**257–266.
21. **Barone, F. E., and M. R. Tansey.** 1977. Isolation, purification, identification, synthesis and kinetics of the activity of the anticandidal component of *Allium sativum,* and a hypothesis for its mode of action. *Mycologia* **69:**793–825.
22. **Batt, C., M. Solberg, and M. Ceponis.** 1980. Inhibition of aflatoxin production by carrot root extract. *J. Food Sci.* **45:**1210–1213.
23. **Batt, C., M. Solberg, and M. Ceponis.** 1983. Effect of volatile components of carrot seed oil on growth and aflatoxin production by *Aspergillus parasiticus. J. Food Sci.* **48:**762–768.
24. **Bell, M. F., R. T Marshall, and M. E. Anderson.** 1986. Microbiological and sensory tests of beef treated with acetic and formic acids. *J. Food Prot.* **49:**207–210.
25. **Bell, T. A., J. L. Etchells, and A. F. Borg.** 1959. Influence of sorbic acid on the growth of certain species of bacteria, yeasts, and filamentous fungi. *J. Bacteriol.* **77:**573–580.
26. **Bellamy, W., M. Takase, H. Wakabayashi, K. Kawase, and M. Tomita.** 1992. Antibacterial spectrum of lactoferricin B, a potent bactericidal peptide derived from the *N*-terminal region of bovine lactoferrin. *J. Appl. Bacteriol.* **73:**472–479.
27. **Bellamy, W., M. Takase, K. Yamauchi, H. Wakabayashi, K. Kawase, and M. Tomita.** 1992. Identification of the bactericidal domain of lactoferrin. *Biochim. Biophys. Acta* **1121:**130–136.
28. **Beuchat, L. R.** 1976. Sensitivity of *Vibrio parahaemolyticus* to spices and organic acids. *J. Food Sci.* **41:**899–902.
29. **Beuchat, L. R.** 1994. Antimicrobial properties of spices and their essential oils, p. 167–179. *In* V. M. Dillon and R. G. Board (ed.), *Natural Antimicrobial Systems and Food Preservation.* CAB Intl., Wallingford, England.
30. **Beuchat, L. R., and R. E. Brackett.** 1990. Inhibitory effect of raw carrots on *Listeria monocytogenes. Appl. Environ. Microbiol.* **56:**1734–1742.
31. **Beuchat, L. R., R. E. Brackett, and M. P. Doyle.** 1994. Lethality of carrot juice to *Listeria monocytogenes* as affected by pH, sodium chloride and temperature. *J. Food Prot.* **57:**470–474.
32. **Beuchat, L. R., and D. A. Golden.** 1989. Antimicrobials occurring naturally in foods. *Food Technol.* **43:**134–142.
33. **Beuchat, L. R., and W. K. Jones.** 1978. Effects of food preservatives and antioxidants on colony formation by heated conidia of *Aspergillus flavus. Acta Aliment. Acad. Sci. Hung.* **7:**373.
34. **Beumer, R. R., A. Noomen, J. A. Marijs, and E. H. Kampelmacher.** 1985. Antibacterial action of the lactoperoxidase system on *Campylobacter jejuni* in cow's milk. *Neth. Milk Dairy J.* **39:**107–114.
35. **Bills, S., L. Restaino, and L. M. Lenovich.** 1982. Growth response of an osmotolerant sorbate-resistant yeast, *Saccharomyces rouxii* at different sucrose and sorbate levels. *J. Food Prot.* **45:**1120–1124.
36. **Björck, L.** 1978. Antibacterial effect of the lactoperoxidase system on psychrotrophic bacteria in milk. *J. Dairy Res.* **45:**109–118.
37. **Björck, L., O. Claesson, and W. Schulthess.** 1979. The lactoperoxidase/thiocyanate/hydrogen peroxide system as a temporary preservative for raw milk in developing countries. *Milchwissenschaft* **34:**726–729.

38. **Blocher, J. C., and F. F. Busta.** 1983. Influence of potassium sorbate and reduced pH on the growth of vegetative cells of four strains of type A and B *Clostridium botulinum*. *J. Food Sci.* **48:**574–580.

39. **Bosund, L.** 1962. The action of benzoic and salicylic acids on the metabolism of microorganisms. *Adv. Food Res.* **11:**331–353.

40. **Brackett, R. E.** 1987. Effect of various acids on growth and survival of *Yersinia enterocolitica*. *J. Food Prot.* **50:**598–601.

41. **Bradley, R. L., L. G. Harmon, and C. M. Stine.** 1962. Effect of potassium sorbate on some organisms associated with cottage cheese spoilage. *J. Milk Food Technol.* **25:**318.

42. **Branen, A. L., and T. W. Keenan.** 1970. Growth stimulation of *Lactobacillus casei* by sodium citrate. *J. Dairy Sci.* **53:**593.

43. **Buchanan, R. L., and M. H. Golden.** 1994. Interaction of citric acid concentration and pH on the kinetics of *Listeria monocytogenes* inactivation. *J. Food Prot.* **57:**567–570.

44. **Buchanan, R. L., and A. J. Sheperd.** 1981. Inhibition of *Aspergillus parasiticus* by thymol. *J. Food Sci.* **46:**976–977.

45. **Buchanan, R. L., and M. Solberg.** 1972. Interaction of sodium nitrite, oxygen and pH on growth of *Staphylococcus aureus*. *J. Food Sci.* **37:**81–85.

46. **Bullerman, L. B.** 1974. Inhibition of aflatoxin production by cinnamon. *J. Food Sci.* **39:**1163–1165.

47. **Bullerman, L. B.** 1983. Effects of potassium sorbate on growth and aflatoxin production by *Aspergillus parasiticus* and *Aspergillus flavus*. *J. Food Prot.* **46:**940–942.

48. **Bullerman, L. B.** 1984. Effects of potassium sorbate on growth and patulin production by *Penicillium patulum* and *Penicillium roqueforti*. *J. Food Prot.* **47:**312–315.

49. **Bullerman, L. B., F. Y. Lieu, and S. A. Seier.** 1977. Inhibition of growth and aflatoxin production by cinnamon and clove oils, cinnamic aldehyde and eugenol. *J. Food Sci.* **42:**1107–1109.

50. **Busta, F. F., and P. M. Foegeding.** 1983. Chemical food preservatives, p. 656–694. *In* S. S. Block (ed.), *Disinfection, Sterilization, and Preservation,* 3rd ed. Lea & Febiger, Philadelphia.

51. **Carini, S., and R. Lodi.** 1982. Inhibition of germination of clostridial spores by lysozyme. *Ind. Latte* **18:**35–48.

52. **Carlsson, J., Y. Iwami, and T. Yamada.** 1983. Hydrogen peroxide excretion by oral streptococci and effect of lactoperoxidase-thiocyanate-hydrogen peroxide. *Infect. Immun.* **40:**70–80.

53. **Carminiti, D., E. Nevianti, and G. Muchetti.** 1985. Activity of lysozyme on vegetative cells of *Clostridium tyrobutyricum*. *Latte* **10:**194–198.

54. **Carpenter, C. E., D. S. A. Reddy, and D. P. Cornforth.** 1987. Inactivation of clostridial ferredoxin and pyruvate-ferredoxin oxidoreductase by sodium nitrite. *Appl. Environ. Microbiol.* **53:**549–552.

55. **Castellani, A. G., and C. F. Niven.** 1955. Factors affecting the bacteriostatic action of sodium nitrite. *Appl. Microbiol.* **3:**154–159.

56. **Cavallito, C. J., and J. H. Bailey.** 1944. Allicin, the antibacterial principle of *Allium sativum*. I. Isolation, physical properties and antibacterial action. *J. Am. Chem. Soc.* **16:**1950–1951.

57. **Chamberland, R.** 1887. Les essences au point de vue de leurs proprietes antiseptiques. *Ann. Inst. Pasteur* **1:**152.

58. **Chan, L., R. Weaver, and C. S. Ough.** 1975. Microbial inhibition caused by p-hydroxybenzoate esters in wines. *Am. J. Enol. Vitic.* **26:**201–207.

59. **Chang, H. C., and A. L. Branen.** 1975. Antimicrobial effects of butylated hydroxyanisole (BHA). *J. Food Sci.* **40:**349–351.

60. **Chen, N., and L. A. Shelef.** 1992. Relationship between water activity, salts of lactic acid, and growth of *Listeria monocytogenes* in a meat model system. *J. Food Prot.* **55:**574–578.

61. **Chichester, D. F., and F. W. Tanner.** 1972. Antimicrobial food additives, p. 115–184. *In* T. W. Furia (ed.), *Handbook of Food Additives,* 2nd ed. CRC Press, Cleveland.

62. **Chipley, J. R.** 1974. Effects of 2, 4-dinitrophenol and N,N′-cyclohexylcarbodiimide on cell-envelope associated enzymes of *Escherichia coli* and *Salmonella enteritidis*. *Microbios* **10:**115.

63. **Chipley, J. R.** 1993. Sodium benzoate and benzoic acid, p. 11–48. *In* P. M. Davidson and A. L. Branen (ed.), *Antimicrobials in Foods,* 2nd ed. Marcel Dekker, Inc., New York.

64. **Chipley, J. R., and N. Uraih.** 1980. Inhibition of *Aspergillus* growth and aflatoxin release by derivatives of benzoic acid. *Appl. Environ. Microbiol.* **40:**352–357.

65. **Chung, K. C., and J. M. Goepfert.** 1970. Growth of *Salmonella* at low pH. *J. Food Sci.* **35:**326–328.

66. **Chung, K. T., and C. A. Murdock.** 1991. Natural systems for preventing contamination and growth of microorganisms in foods. *Food Microstruct.* **10:**361–374.

67. **Chung, K. T., S. E. Stevens, W. F. Lin, and C. I. Wei.** 1992. Antimicrobial properties of tannic acid, propyl gallate and related compounds, abstr. P-63, p. 334. *In Abstracts of the 92nd General Meeting of the American Society for Microbiology 1992*. American Society for Microbiology, Washington, D.C.

67a. **Committee on Nitrite and Alternative Curing Agents, National Research Council.** 1981. *The Health Effects of Nitrate, Nitrite and n-Nitroso compounds.* National Academy Press, Washington, D.C.

68. **Conner, D. E.** 1993. Naturally occurring compounds, p. 441–468. *In* P. M. Davidson and A. L. Branen (ed.), *Antimicrobials in Foods,* 2nd ed. Marcel Dekker, Inc., New York.

69. **Conner, D. E., and L. R. Beuchat.** 1984. Effects of essential oils from plants on growth of food spoilage yeasts. *J. Food Sci.* **49:**429–434.

70. **Conner, D. E., and L. R. Beuchat.** 1984. Sensitivity of heat-stressed yeasts to essential oils of plants. *Appl. Environ. Microbiol.* **47:**229–233.

71. **Conner, D. E., L. R. Beuchat, R. E. Worthington, and H. L. Hitchcock.** 1984. Effects of essential oils and oleoresins of plants on ethanol production, respiration and sporulation of yeasts. *Int. J. Food Microbiol.* **1:**63–74.

72. **Cook, F. K., and M. D. Pierson.** 1983. Inhibition of bacterial spores by antimicrobials. *Food Technol.* **37:**115–126.

73. **Corry, J. E. L.** 1987. Relationships of water activity to fungal growth, p. 51–99. *In* L. R. Beuchat (ed.), *Food and Beverage Mycology,* 2nd ed. AVI/Van Nostrand Reinhold, New York.
74. **Costilow, R. N., W. E. Ferguson,and S. Ray.** 1955. Sorbic acid as a selective agent in cucumber fermentations. I. Effect of sorbic acid on microorganisms associated with cucumber fermentations. *Appl. Microbiol.* **3**:341–345.
75. **Crichton, R. R., and M. Charloteux-Wauters.** 1987. Review: iron transport and storage. *Eur. J. Biochem.* **164:** 485–506.
76. **Cruess, W. V., and J. H. Irish.** 1932. Further observations on the relation of pH value to toxicity of preservatives to microorganisms. *J. Bacteriol.* **23**:163–166.
77. **Dahl, H. K., and J. Nordal.** 1972. Effect of benzoic acid and sorbic acid on the production and activities of some bacterial proteinases. *Acta Agric. Scand.* **22**:29.
78. **Dankert, J., F. J. T. Tromp, H. DeVries, and H. J. Klasen.** 1979. Antimicrobial activity of crude juices of *Allium ascalonicum, Allium cepa* and *Allium sativum. Zentralbl. Bakteriol. Parasitenkd. Infektionskr. Hyg.* **245**:229.
79. **Davidson, P. M.** 1993. Parabens and phenolic compounds, p. 263–306. *In* P. M. Davidson and A. L. Branen (ed.), *Antimicrobials in Foods,* 2nd ed. Marcel Dekker, Inc., New York.
80. **Davidson, P. M.** 1994. Food preservatives, p. 341–354. *In* C. J. Arntzen (ed.), *Encyclopedia of Agricultural Science,* vol. II. Academic Press, Inc., San Diego, Calif.
81. **Davidson, P. M., and A. L. Branen.** 1980. Antimicrobial mechanisms of butylated hydroxyanisole against two *Pseudomonas* species. *J. Food Sci.* **45**:1607–1613.
82. **Davidson, P. M., C. J. Brekke, and A. L. Branen.** 1981. Antimicrobial activity of butylated hydroxyanisole, tertiary butylhydroquinone and potassium sorbate in combination. *J. Food Sci.* **46**:314–316.
83. **Davis, J. C., F. A. Draughon, and P. M. Davidson.** 1982. Inhibition of toxin production in patulin-producing fungi by phenolic antioxidants and insecticides. Presented at the Annual Meeting of the Institute of Food Technologists, Las Vegas, Nev.
84. **Debevere, J. M.** 1989. The effect of sodium lactate on the shelflife of vacuum-packed coarse liver pate. *Fleischwirtschaft* **69**:223–224.
85. **Degré, R., M. Ishaque, and M. Sylvestre.** 1983. Effect of butylated hydroxyanisole on the electron transport system of *Staphylococcus aureus* Wood 46. *Microbios* **37**:7–13.
86. **Degré, R., and M. Sylvestre.** 1983. Effect of butylated hydroxyanisole on the cytoplasmic membrane of *Staphylococcus aureus* Wood 46. *J. Food Prot.* **46**:206–209.
87. **DeLange, R. J., and T. S. Huang.** 1971. Egg white avidin. III. Sequence of the 78-residue middle cyanogen bromide peptide: complete amino acid sequence of the protein subunit. *J. Biol. Chem.* **246**:698–709.
88. **Delaquis, P. J., and G. Mazza.** 1995. Antimicrobial properties of isothiocyanates in food preservation. *Food Technol.* **49**:73–84.
89. **De Wit, J. C., S. Notermans, N. Gorin, and E. H. Kampelmacher.** 1979. Effect of garlic oil or onion oil on toxin production by *Clostridium botulinum* in meat slurry. *J. Food Prot.* **42**:222–224.
90. **Doores, S.** 1993. Organic acids, p. 95–136, *In* P. M. Davidson and A. L. Branen (ed.), *Antimicrobials in Foods,* 2nd ed. Marcel Dekker, Inc., New York.
91. **Duhaiman, A. S.** 1988. Purification of camel milk lysozyme and its lytic effect on *Escherichia coli* and *Micrococcus lysodeikticus. Comp. Biochem. Physiol.* **91B**:793.
92. **Duncan, C. L., and E. M. Foster.** 1968. Effect of sodium nitrite, sodium chloride, and sodium nitrate on germination and outgrowth of anaerobic spores. *Appl. Microbiol.* **16**:406–411.
93. **Dymicky, M., and C. N. Huhtanen.** 1979. Inhibition of *Clostridium botulinum* by *p*-hydroxybenzoic acid n-alkyl esters. *Antimicrob. Agents Chemother.* **15**:798–801.
94. **Ehrenberg, L., I. Fedorscsak, and F. Solymosy.** 1976. Diethyl pyrocarbonate in nucleic acid research. *Prog. Nucleic Acid Res. Mol. Biol.* **16**:189.
95. **Eklund, T.** 1980. Inhibition of growth and uptake processes in bacteria by some chemical food preservatives. *J. Appl. Bacteriol.* **48**:423–432.
96. **Eklund, T.** 1983. The antimicrobial effect of dissociated and undissociated sorbic acid at different pH levels. *J. Appl. Bacteriol.* **54**:383–389.
97. **Eklund, T.** 1985. Inhibition of microbial growth at different pH levels by benzoic and propionic acids and esters of p-hydroxybenzoic acid. *Int. J. Food Microbiol.* **2**:159–167.
98. **Eklund, T.** 1985. The effect of sorbic acid and esters of p-hydroxybenzoic acid on the protonmotive force in *Escherichia coli* membrane vesicles. *J. Gen. Microbiol.* **131:** 73–76.
99. **Eklund, T.** 1989. Organic acids and esters, p. 161–200. *In* G. W. Gould (ed.), *Mechanisms of Action of Food Preservation Procedures.* Elsevier Applied Science, London.
100. **Eklund, T., I. F. Nes, and R. Skjelkvåle.** 1981. Control of *Salmonella* at different temperatures by propyl paraben and butylated hydroxyanisole, p. 377–384. *In* T. A. Roberts, G. Hobbs, J. H. B. Christian, and N. Skovgaard (ed.), *Psychrotrophic Microorganisms in Spoilage and Pathogenicity.* Academic Press, London.
101. **Ekstrand, B.** 1994. Lactoperoxidase and lactoferrin, p. 15–63. *In* V. M. Dillon and R. G. Board (ed.), *Natural Antimicrobial Systems and Food Preservation.* CAB Intl., Wallingford, England.
102. **Eletr, S., M. A. Williams, T. Watkins, and A. D. Keith.** 1974. Perturbations of the dynamics of lipid alkyl chains in membrane systems: effect on the activity of membrane-bound enzymes. *Biochim. Biophys. Acta* **339**:190–201.
103. **Ellinger, R. H.** 1972. Phosphates in food processing, p. 617–807. *In* T. E. Furia (ed.), *Handbook of Food Additives,* 2nd ed. CRC Press, Cleveland.
104. **Ellison, R. T., T. G. Giehl, and F. M. LaForce.** 1988. Damage of the outer membrane of enteric gram-negative bacteria by lactoferrin and transferrin. *Infect. Immun.* **56**:2774–2781.
105. **El-Shenawy, M. A., and E. H. Marth.** 1988. Sodium benzoate inhibits growth of or inactivates *Listeria monocytogenes. J. Food Prot.* **51**:525–530.
106. **El-Shenawy, M. A., and E. H. Marth.** 1989. Behavior of *Listeria monocytogenes* in the presence of sodium propionate. *Int. J. Food Microbiol.* **8**:85–94.

107. **Erickson, D. R., and R. B. Tompkin.** 1977. Antimicrobial and antirancidity agent. U.S. patent 4,044,160, August 23.
108. **Eubanks, V. L., and L. R. Beuchat.** 1982. Effects of antioxidants on growth, sporulation and pseudomycelium production by *Saccharomyces cerevisiae. J. Food Sci.* **47:** 1717–1722.
109. **Evelyn, J.** 1664. Pomona, or an appendix concerning fruit trees, p. 24. *In* P. Beale (ed.), *Relation to Cider and Several Ways of Ordering It,* supplement. *Aphorisms Concerning Cider.* J. Martin and J. A. Allestry, London.
110. **Exterkate, F. A.** 1979. Effect of membrane perturbing treatments on the membrane-bound peptidases of *Streptococcus cremoris* HP. *J. Dairy Res.* **46:**473–484.
111. **Fabian, F. W., C. F. Krehl, and N. W. Little.** 1939. The role of spices in pickled food spoilage. *Food Res.* **4:**269–286.
112. **Fabian, F. W., and C. K. Wadsworth.** 1939. Experimental work on lactic acid in preserving pickles and pickle products. I. Rate of penetration of acetic and lactic acid in pickles. *Food Res.* **4:**499–510.
113. **Faith, N. G., A. E. Yousef, and J. B. Luchansky.** 1992. Inhibition of *Listeria monocytogenes* by liquid smoke and isoeugenol, a phenolic component found in smoke. *J. Food Safety* **12:**303–314.
114. **Fernandez, J. L., and W. J. Simpson.** 1992. Factors affecting antibacterial activity of hop compounds and their derivatives. *J. Appl. Bacteriol.* **72:**327–334.
115. **Fernandez, J. L., and W. J. Simpson.** 1993. Aspects of the resistance of lactic acid bacteria to hop bitter acids. *J. Appl. Bacteriol.* **75:**315–319.
116. **Finol, M. L., E. H. Marth, and R. C. Lindsay.** 1982. Depletion of sorbate from different media during growth of *Penicillium* species. *J. Food Prot.* **45:**398–404.
117. **Fleming, A.** 1922. On a remarkable bacteriolytic element found in tissues and secretions. *Proc. R. Soc. Lond.* **93:**306–317.
118. **Fleming, H. P., and J. L. Etchells.** 1967. Occurrence of an inhibitor of lactic acid bacteria in green olives. *Appl. Microbiol.* **15:**1178.
119. **Fleming, H. P., W. M. Walter, and J. L. Etchells.** 1969. Isolation of a bacterial inhibitor from green olives. *Appl. Microbiol.* **18:**856.
120. **Fletcher, G. C., W. G. Murrell, J. A. Statham, B. J. Stewart, and H. A. Bremner.** 1988. Packaging of scallops with sorbate: an assessment of the hazards from *Clostridium botulinum. J. Food Sci.* **53:**349–352.
121. **Flores, L. M., L. S. Palomar, P. A. Roh, and L. B. Bullerman.** 1988. Effect of potassium sorbate and other treatments on the microbial content and keeping quality of a restaurant-type Mexican hot sauce. *J. Food Prot.* **51:**4–7.
122. **Fowlks, W. L., D. G. Griffith, and E. L. Oginsky.** 1958. Photosensitization of bacteria by furocoumarins and related compounds. *Nature* (London) **181:**571–572.
123. **Francois, J., E. Vam Scjaftingen, and H. G. Hers.** 1986. Effect of benzoate on the metabolism of fructose 2,6-biphosphate in yeast. *Eur. J. Biochem.* **154:**141–145.
124. **Freese, E.** 1978. Mechanism of growth inhibition by lipophilic acids, p. 123–131. *In* J. J. Kabara (ed.), *The Pharmacological Effect of Lipids.* American Oil Chemists Society, Champaign, Ill.
125. **Freese, E., C. W. Sheu, and E. Galliers.** 1973. Function of lipophilic acids as antimicrobial food additives. *Nature* (London) **241:**321–327.
126. **Fretheim, K., P. E. Granum, and E. Vold.** 1980. Influence of generation temperature on the chemical composition, antioxidative, and antimicrobial effects of wood smoke. *J. Food Sci.* **45:**999–1002.
127. **Fung, D. Y. C., S. Taylor, and J. Kahan.** 1977. Effect of butylated hydroxyanisole (BHA) and butylated hydroxytoluene (BHT) on growth and aflatoxin production of *Aspergillus flavus. J. Food Safety* **1:**39–51.
128. **Furr, J. R., and A. D. Russell.** 1972. Some factors influencing the activity of esters of p-hydroxybenzoic acid against *Serratia marcescens. Microbios* **5:**189–195.
129. **Gaya, P., M. Medina, and M. Nunez.** 1991. Effect of the lactoperoxidase system on *Listeria monocytogenes* behavior in raw milk at refrigeration temperatures. *Appl. Environ. Microbiol.* **57:**3355–3360.
130. **Gibson, A. M., and T. A. Roberts.** 1986. The effect of pH, water activity, sodium nitrite and storage temperature on the growth of *Escherichia coli* and salmonellae in a laboratory medium. *Int. J. Food Microbiol.* **3:**183–194.
131. **Gibson, A. M., and T. A. Roberts.** 1986. The effect of pH, sodium chloride, sodium nitrite and storage temperature on the growth of *Clostridium perfringens* and faecal streptococci in laboratory media. *Int. J. Food Microbiol.* **3:**195–210.
132. **Giese, J.** 1992. Experimental process reduces *Salmonella* on poultry. *Food Technol.* **46**(4)**:**112.
133. **Glabe, E. F., and J. K. Maryanski.** 1981. Sodium diacetate: an effective mold inhibitor. *Cereal Foods World* **26:** 285–289.
134. **González-Fandos, E., M. L. García-López, M. L. Sierra, and A. Otero.** 1994. Staphylococcal growth and enterotoxins (A-D) and thermonuclease synthesis in the presence of dehydrated garlic. *J. Appl. Bacteriol.* **77:**549–552.
135. **Gould, G. W.** 1964. Effect of food preservatives on the growth of bacteria from spores, p. 17–24. *In* N. Molin (ed.), *Microbial Inhibitors in Food.* Almqvist and Miksell, Stockholm.
136. **Gould, G. W. (ed.).** 1989. *Mechanisms of Action of Food Preservation Procedures.* Elsevier Applied Science, London.
137. **Gould, G. W., and N. J. Russell.** 1991. Sulphite, p. 72–88 *In* N. J. Russell and G. W. Gould (ed.), *Food Preservatives.* Blackie and Son Ltd., Glasgow, Scotland.
138. **Graham, H. D., and E. J. F. Graham.** 1987. Inhibition of *Aspergillus parasiticus* growth and toxin production by garlic. *J. Food Safety* **8:**101–108.
139. **Gray, R. J. H., P. H. Elliott, and R. I. Tomlins.** 1984. Control of two major pathogens on fresh poultry using a combination potassium sorbate/carbon dioxide packaging treatment. *J. Food Sci.* **49:**142–145.
140. **Green, N. M.** 1963. Avidin. IV. Stability at extremes of pH and dissociatoin into subunits by GuHCl. *Biochem. J.* **89:**609–620.
141. **Grever, A. B. G.** 1974. Minimum nitrite concentrations for inhibition of clostridia in cooked meat products, p. 103. *In* B. Krol and B. J. Tinbergen (ed.), *Proceedings of*

the International Symposium on Nitrite in Meat Products. Pudoc, Wageningen, The Netherlands.

142. **Grodzinska-Zachwieja, A., and W. Kahl.** 1966. Bacteriostatic action of chicory (*Chicorium intybus L.*). II. The action of chlorogenic and isochlorogenic acid. *Acta Biol. Cracov. Ser. Bot.* **9**:87–97.
143. **Haas, G. J., and R. Barsoumian.** 1994. Antimicrobial activity of hop resins. *J. Food Prot.* **57**:59–61.
144. **Hailer, E.** 1911. Experiments on the properties of free sulfurous acid of sulfites and a few complex compounds of sulfurous acid in killing germs and retarding their development. *Arb. Kais. Gesundh.* **36**:297. [*Chem. Abstr.* **5**:1805, 1911.]
145. **Hamden, I. Y., D. D. Deane, and J. E. Kunsman.** 1971. Effect of potassium sorbate on yogurt cultures. *J. Milk Food Technol.* **34**:307.
146. **Hammond, S. M., and J. C. Carr.** 1976. The antimicrobial activity of SO_2—with particular reference to fermented and non-fermented fruit juices. *Soc. Appl. Bacteriol. Symp. Ser.* **5**:86–110.
147. **Harada, K., R. Hizuchin, and I. Utsumi.** 1968. Studies on sorbic acid. IV. Inhibition of the respiration in yeasts. *Agric. Biol. Chem.* **32**:940–946.
148. **Harborne, J. B., and N. W. Simmonds.** 1964. The natural distribution of the phenolic aglycones, p. 77–127. *In* J. B. Harborne (ed.), *Biochemistry of Phenolic Compounds.* Academic Press, London.
149. **Härnulv, B. G., and C. Kandasamy.** 1982. Increasing the keeping quality of milk by activation of its lactoperoxidase system. Results from Sri Lanka. *Milchwissenschaft* **37**:454–457.
150. **Hayatsu, H., Y. Wataya, K. Kai, and S. Iida.** 1970. Reaction of sodium bisulfite with uracil, cytosine and their derivatives. *Biochemistry* **9**:2858.
151. **Hedberg, M., and J. K. Miller.** 1969. Effectiveness of acetic acid, betadine, amphyll, polymyxin B, colistin and gentamicin against *Pseudomonas aeruginosa. Appl. Microbiol.* **18**:854.
152. **Herald, P. J., and P. M. Davidson.** 1983. The antibacterial activity of selected hydroxycinnamic acids. *J. Food Sci.* **48**:1378–1379.
153. **Herrmann, K.** 1989. Occurrence and content of hydroxycinnamic and hydroxybenzoic compounds in foods. *Crit. Rev. Food Sci. Nutr.* **28**:315.
154. **Hinze, H., and H. Holzer.** 1985. Effect of sulfite or nitrite on the ATP content and the carbohydrate metabolism in yeast. *Z. Lebensm. Unters. Forsch.* **181**:87–91.
155. **Ho, C. T.** 1992. Phenolic compounds in food. An overview. *Am. Chem. Soc. Symp. Ser.* **506**:2–7.
156. **Holley, R. A.** 1981. Review of the potential hazard from botulism in cured meats. *Can. Inst. Food Sci. Technol. J.* **14**:183–195.
157. **Huang, L., C. W. Forsberg, and L. N. Gibbins.** 1986. Influence of external pH and fermentation products on *Clostridium acetobutylicum* intracellular pH and cellular distribution of fermentation products. *Appl. Environ. Microbiol.* **51**:1230–1234.
158. **Hughey, V. L., and E. A. Johnson.** 1987. Antimicrobial activity of lysozyme against bacteria involved in food spoilage and food-borne disease. *Appl. Environ. Microbiol.* **53**:2165–2170.
159. **Hughey, V. L., R. A. Wilger, and E. A. Johnson.** 1989. Antibacterial activity of hen egg white lysozyme against *Listeria monocytogenes* Scott A in foods. *Appl. Environ. Microbiol.* **55**:631–638.
160. **Huhtanen, C. M., and J. Feinberg.** 1980. Sorbic acid inhibition of *Clostridium botulinum* in nitrite-free poultry frankfurters. *J. Food Sci.* **45**:453–457.
161. **Hunter, D. R., and I. H. Segel.** 1973. Effect of weak acids on amino acid transport by *Penicillium chrysogenum.* Evidence for a proton or charge gradient as the driving force. *J. Bacteriol.* **113**:1184–1192.
162. **Ingram, M.** 1939. The endogenous respiration of *Bacillus cereus.* II. The effect of salts on the rate of absorption of oxygen. *J. Bacteriol.* **24**:489.
163. **Ingram, M., F. J. H. Ottoway, and J. B. M. Coppock.** 1956. The preservative action of acid substances in food. *Chem. Ind.* (London) **42**:1154–1163.
164. **Ingram, M., and K. Vas.** 1950. Combination of sulfur dioxide with concentrated orange juice. I. Equilibrium states. *J. Sci. Food Agric.* **1**:21–27.
165. **Institute of Food Technologists.** 1975. Sulfites as food additives: a scientific status summary. *Food Technol.* **29**:117–120.
166. **Institute of Food Technologists.** 1987. Nitrate, nitrite and nitroso compounds in foods. *Food Technol.* **41**:127–136.
167. **Isshiki, K., K. Tokuora, R. Mori, and S. Chiba.** 1992. Preliminary examination of allyl isothiocyanate vapor for food preservation. *Biosci. Biotechnol. Biochem.* **56**:1476–1477.
168. **Ivey, F. J., and M. C. Robach.** 1978. Effect of sorbic acid and sodium nitrite on *Clostridium botulinum* outgrowth and toxin production in canned comminuted pork. *J. Food Sci.* **43**:1782–1785.
169. **Ivey, F. J., K. J. Shaver, L. N. Christiansen, and R. B. Tompkin.** 1978. Effect of potassium sorbate on toxigenesis by *Clostridium botulinum* in bacon. *J. Food Prot.* **41**:621–625.
170. **Jay, J. M.** 1992. *Modern Food Microbiology,* 4th ed. Van Nostrand Reinhold, New York.
171. **Jay, J. M., and G. M. Rivers.** 1984. Antimicrobial activity of some food flavoring compounds. *J. Food Safety* **6**:129–139.
172. **Jen, C. M. C., and L. A. Shelef.** 1986. Factors affecting sensitivity of *Staphylococcus aureus* 196E to polyphosphate. *Appl. Environ. Microbiol.* **52**:842–846.
173. **Jermini, M. F. G., and W. Schmidt-Lorenz.** 1987. Activity of Na-benzoate and ethyl-paraben against osmotolerant yeasts at different water activity values. *J. Food Prot.* **50**:920–927.
174. **Johansen, C., L. Gram, and A. S. Meyer.** 1994. The combined inhibitory effect of lysozyme and low pH on growth of *Listeria monocytogenes. J. Food Prot.* **57**:561–566.
175. **Johnson, M. G., and R. H. Vaughn.** 1969. Death of *Salmonella typhimurium* and *Escherichia coli* in the presence of freshly reconstituted dehydrated garlic and onion. *Appl. Microbiol.* **17**:903–905.

176. **Jones, E. M., A. Smart, G. Bloomberg, L. Burgess, and M. R. Millar.** 1994. Lactoferricin, a new antimicrobial peptide. *J. Appl. Bacteriol.* **77:**208–214.

177. **Judis, J.** 1963. Studies on the mechanism of action of phenolic disinfectants. II. Patterns of release of radioactivity from *Escherichia coli* labeled by growth on various compounds. *J. Pharm. Sci.* **52:**261–264.

178. **Juneja, V. K., and P. M. Davidson.** 1992. Influence of altered fatty acid composition on resistance of *Listeria monocytogenes* to antimicrobials. *J. Food Prot.* **56:**302–305.

179. **Jurd, L., A. D. King, K. Mihara, and W. L. Stanely.** 1971. Antimicrobial properties of natural phenols and related compounds. I. Obtusastyrene. *Appl. Microbiol.* **21:**507–510.

180. **Juven, B., and Y. Henis.** 1970. Studies on the antimicrobial activity of olive phenolic compounds. *J. Appl. Bacteriol.* **33:**721–732.

181. **Juven, B., Y. Henis, and B. Jacoby.** 1972. Studies on the mechanism of the antimicrobial action of oleuropein. *J. Appl. Bacteriol.* **35:**559–567.

182. **Juven, B., Z. Samish, and Y. Henis.** 1968. Identification of oleuropein as a natural inhibitor of lactic fermentation of green olives. *Isr. J. Agric. Res.* **18:**137–138.

183. **Kabara, J. J., and T. Eklund.** 1991. Organic acids and esters, p. 44–71. *In* N. J. Russell and G. W. Gould (ed.), *Food Preservatives.* Blackie and Son Ltd., Glasgow, Scotland.

184. **Kamau, D. N., S. Doores, and K. M. Pruitt.** 1990. Antibacterial activity of the lactoperoxidase system against *Listeria monocytogenes* and *Staphylococcus aureus* in milk. *J. Food Prot.* **53:**1010–1014.

185. **Karaioannoglou, P. G., A. J. Mantis, and A. G. Panetsos.** 1977. The effect of garlic extract on lactic acid bacteria (*Lactobacillus plantarum*) in culture media. *Lebensm. Wiss. Technol.* **10:**148–150.

186. **Katayama, T., and I. Nagai.** 1960. Chemical significance of the volatile components of spices from the food preservation standpoint. IV. Structure and antibacterial activity of some terpenes. *Nippon Suisan Gakkaishi* **26:**29–32.

187. **Kato, A., and I. Shibasaki.** 1975. Combined effect of different drugs on the antibacterial activity of fatty acids and their esters. *J. Antibacterial Antifungal Agents* (Japan) **8:**355–361.

188. **Kaye, R. C., and S. G. Proudfoot.** 1971. Interactions between phosphatidyl-ethanolamine monolayers and phenols in relation to antibacterial activity. *J. Pharm. Pharmacol. Suppl.* **23:**223S.

189. **Kelch, F., and X. Bühlmann.** 1958. Effect of commercial phosphates on the growth of microorganisms. *Fleischwirtschaft* **10:**325–328.

190. **Kidney, A. J.** 1974. The use of sulfite in meat processing. *Chem. Ind.* (London) **1974:**717–718.

191. **Kim, C. R., J. O. Hearnsberger, A. P. Vickery, C. H. White, and D. L. Marshall.** 1995. Extending shelf life of refrigerated catfish fillets using sodium acetate and monopotassium phosphate. *J. Food Prot.* **58:**644–647.

192. **King, A. D., J. D. Ponting, D. W. Sanshuck, R. Jackson, and K. Mihara.** 1981. Factors affecting death of yeast by sulfur dioxide. *J. Food Prot.* **44:**92–97.

193. **Kirby, G. W., L. Atkin, and C. N. Frey.** 1973. Further studies on the growth of bread molds as influenced by acidity. *Cereal Chem.* **14:**865.

194. **Klindworth, K. J., P. M. Davidson, C. J. Brekke, and A. L. Branen.** 1979. Inhibition of *Clostridium perfringens* by butylated hydroxyanisole. *J. Food Sci.* **44:**564–567.

195. **Knabel, S. J., H. W. Walker, and P. A. Hartman.** 1991. Inhibition of *Aspergillus flavus* and selected gram-positive bacteria by chelation of essential metal cations by polyphosphates. *J. Food Prot.* **54:**360–365.

196. **Knox, T. L., P. M. Davidson, and J. R. Mount.** 1984. Evaluation of selected antimicrobials in fruit juices as sodium metabisulfite replacements or adjuncts. Presented at the 44th Annual Meeting of the Institute of Food Technologists, Anaheim, Calif.

197. **Kojima, M., and K. Ogawa.** 1971. Studies on the effect of isothiocyanates and their analogues on microorganisms I. Effects of isothiocyanates on the oxygen uptake of yeasts. *J. Ferment. Technol.* **49:**740–746.

198. **Konowalchuk, J., and J. I. Speirs.** 1976. Antiviral activity of fruit extracts. *J. Food Sci.* **41:**1013–1017.

199. **Konowalchuk, J., and J. I. Speirs.** 1976. Virus inactivation by grapes and wines. *Appl. Environ. Microbiol.* **32:**757–763.

200. **Konowalchuk, J., and J. I. Speirs.** 1978. Antiviral effect of commercial juices and beverages. *Appl. Environ. Microbiol.* **35:**1219–1220.

201. **Konowalchuk, J., and J. I. Speirs.** 1978. Antiviral effect of apple beverages. *Appl. Environ. Microbiol.* **36:**798–801.

202. **Korhonen, H.** 1978. Effect of lactoferrin and lysozyme in milk on the growth inhibition of *Bacillus stearothermophilus* in the Thermocult method. *Suomen Eläinlááhárilehti* **84:**255–267.

203. **Korpela, J.** 1984. Avidin, a high affinity biotin-binding protein, as a tool and subject of biological research. *Med. Biol.* **65:**5–26.

204. **Kouassi, Y., and L. A. Shelef.** 1995. Listeriolysin O secretion by *Listeria monocytogenes* in the presence of cysteine and sorbate. *Lett. Appl. Microbiol.* **20:**295–299.

205. **Kruk, M., and J. S. Lee.** 1982. Inhibition of *Escherichia coli* trimethylamine-*N*-oxide reductase by food preservatives. *J. Food Prot.* **45:**241–243.

206. **Laschtschenko, P.** 1909. Über die Keimtötende und Entwicklungshemmende Wirkung von Hühnereiweiss. *Z. Hyg. Infektionskr.* **64:**419–427.

207. **Lechowich, R. V., J. B. Evans, and C. F. Niven.** 1956. Effect of curing ingredients and procedures on the survival and growth of staphylococci in and on cured meats. *Appl. Microbiol.* **4:**360.

208. **Lee, J. S.** 1973. What seafood processors should know about *Vibrio parahaemolyticus*. *J. Milk Food Technol.* **36:**405–408.

209. **Lennox, J. E., and L. J. McElroy.** 1984. Inhibition of growth and patulin synthesis in *Penicillium expansum* by potassium sorbate and sodium propionate in culture. *Appl. Environ. Microbiol.* **48:**1031–1033.

210. **Levine, A. S., and C. R. Fellers.** 1940. Action of acetic acid on food spoilage microorganisms. *J. Bacteriol.* **39:**499–514.

211. **Lewis, J. C., and L. Jurd.** 1972. Sporostatic action of cinnamylphenols and related compounds on *Bacillus megaterium*, p. 384–389. *In* H. O. Halvorson, R. Hewson, and L. L. Campbell (ed.), *Spores V.* American Society for Microbiology, Washington, D.C.

212. **Liewen, M. B., and E. H. Marth.** 1985. Growth of sorbate-resistant and -sensitive strains of *Penicillium roquefortii* in the presence of sorbate. *J. Food Prot.* **48:**525–529.

213. **Lillard, H. S.** 1994. Effect of trisodium phosphate on salmonellae attached to chicken skin. *J. Food Prot.* **57:** 465–469.

214. **Lillard, H. S., L. C. Blankenship, J. A. Dickens, S. E. Craven, and A. D. Shackelford.** 1987. Effect of acetic acid on the microbiological quality of scalded picked and unpicked broiler carcasses. *J. Food Prot.* **50:**112–114.

215. **Lloyd, A. G., and J. J. P. Drake.** 1975. Problems posed by essential food preservatives. *Br. Med. Bull.* **31:**214.

216. **López-Malo, A., S. M. Alzamora, and A. Argaiz.** 1995. Effect of natural vanillin on germination time and radial growth of moulds in fruit-based agar systems. *Food Microbiol.* **12:**213–219.

217. **Lovell, T. H.** 1937. Bactericidal effects of onion vapors. *Food Res.* **2:**435.

218. **Lueck, E.** 1980. *Antimicrobial Food Additives.* Springer-Verlag, Berlin.

219. **Lund, B. M., S. M. George, and J. G. Franklin.** 1987. Inhibition of type A and type B (proteolytic) *Clostridium botulinum* by sorbic acid. *Appl. Environ. Microbiol.* **53:**935–941.

220. **Lyon, G. D., and F. M. McGill.** 1988. Inhibition of growth of *Erwinia carotovora* in vitro by phenolics. *Potato Res.* **31:**461–467.

221. **Lyon, G. D., and F. M. McGill.** 1989. Inhibition of polygalacturonase and polygalacturonic acid lyase from *Erwinia carotovora* subsp. *carotovora* by phenolics in vitro. *Potato Res.* **32:**267–274.

222. **Maas, M. R., K. A. Glass, and M. P. Doyle.** 1989. Sodium lactate delays toxin production by *Clostridium botulinum* in cook-in-bag turkey products. *Appl. Environ. Microbiol.* **55:**2226–2229.

223. **Macris, B. J.** 1975. Mechanism of benzoic acid uptake by *Saccharomyces cerevisiae. Appl. Microbiol.* **30:**503–506.

224. **Maga, J. A.** 1978. Simple phenol and phenolic compounds in food flavor. *Crit. Rev. Food Sci. Nutr.* **10:**323–372.

225. **Mandel, I. D., and S. A. Ellison.** 1985. The biological significance of the nonimmunoglobulin defense factors, p. 1–14. *In* K. M. Pruitt and J. O. Tenovuo (ed.), *The Lactoperoxidase System: Its Chemistry and Biological Significance.* Marcel Dekker, Inc., New York.

226. **Marchetti, R., M. A. Casadei, and M. E. Guerzoni.** 1992. Microbial population dynamics in ready-to-use vegetable salads. *Ital. J. Food Sci.* **2:**97–108.

227. **Mari, M., R. Iori, O. Leoni, and A. Marchi.** 1993. *In vitro* activity of glucosinolate derived isothiocyanates against postharvest fruit pathogens. *Ann. Appl. Biol.* **123:**155–164.

228. **Marth, E. H.** 1966. Antibiotics in foods—naturally occurring, developed and added. *Residue Rev.* **12:**65–161.

229. **Marth, E. H., C. M. Capp, L. Hasenzah, H. W. Jackson, and R. V. Hussong.** 1966. Degradation of potassium sorbate by pencillium species. *J. Dairy Sci.* **49:**1197.

230. **Martoadiprawito, W., and J. R. Whitaker.** 1963. Potassium sorbate inhibition of yeast alcohol dehydrogenase. *Biochim. Biophys. Acta* **77:**536–544.

231. **Maruzella, J. C., and P. A. Henry.** 1958. The in vitro antibacterial activity of essential oils and oil combinations. *J. Am. Pharm. Assoc.* **47:**294.

232. **Maruzella, J. C., and L. Liguori.** 1958. The *in vitro* antifungal activity of essential oils. *J. Am. Pharm. Assoc.* **47:**331–336.

233. **Marwan, A. G., and C. W. Nagel.** 1986. Microbial inhibitors of cranberries. *J. Food Sci.* **51:**1009–1013.

234. **Masson, P. L., J. F. Heremans, and E. Schonne.** 1969. Lactoferrin, an iron-binding protein in neutrophilic leukocytes. *J. Exp. Med.* **130:**643–656.

235. **McKellar, R. C.** 1993. Effect of preservatives and growth factors on secretion of listeriolysin O by *Listeria monocytogenes. J. Food Prot.* **56:**380–384.

236. **McLean, R. A., H. D. Lilly, and J. A. Alford.** 1968. Effect of meat curing salts and temperature on production of staphylococcus enterotoxin B. *J. Bacteriol.* **95:**1207–1211.

237. **McMindes, M. K., and A. J. Siedler.** 1988. Nitrite mode of action: inhibition of yeast pyruvate decarboxylase (E.C. 4.1.1.1) and clostridial pyruvate:oxidoreductase (E.C. 1.2.7.1) by nitric oxide. *J. Food Sci.* **53:**917–919.

238. **Measures, J. C.** 1975. Role of amino acids in osmoregulation of nonhalophilic bacteria. *Nature* (London) **257:**398–400.

239. **Melnick, D., F. H. Luckmann, and C. M. Gooding.** 1954. Sorbic acid as a fungistatic agent for foods. VI. Metabolic degradation of sorbic acid in cheese by molds and the mechanism of mold inhibition. *Food Res.* **19:**44–58.

240. **Messina, M. C., H. A. Ahmad, J. A. Marchello, C. P. Gerba, and M. W. Paquette.** 1988. The effect of liquid smoke on *Listeria monocytogenes. J. Food Prot.* **51:**629–631.

241. **Minor, T. E., and E. H. Marth.** 1970. Growth of *Staphylococcus aureus* in acidified pasteurized milk. *J. Milk Food Technol.* **33:**516–520.

242. **Mitchell, P., and J. Moyle.** 1969. Estimation of membrane potential and pH difference across the cristae membrane of rat liver mitochondria. *Eur. J. Biochem.* **7:** 471–484.

243. **Moir, C. J., and M. J. Eyles.** 1992. Inhibition, injury and inactivation of four psychrotrophic foodborne bacteria by the preservatives methyl p-hydroxybenzoate and potassium sorbate. *J. Food Prot.* **55:**360–366.

244. **Mountney, G. J., and J. O'Malley.** 1965. Acids as poultry meat preservatives. *Poultry Sci.* **44:**582.

245. **Muhoberac, B. B., and D. C. Wharton.** 1980. EPR study of heme-NO complexes of ascorbic acid-reduced *Pseudomonas* cytochrome oxidase and corresponding model complexes. *J. Biol. Chem.* **255:**8437–8442.

246. **Musajo, L., F. Bordin, and R. Bevilacqua.** 1967. Photoreactions at 3655Å linking the 3–4 double bond of furocoumarins with pyrimidine bases. *Photochem. Photobiol.* **6:**927–931.

247. **Musajo, L., F. Bordin, G. Caporale, S. Marciani, and G. Rigatti.** 1967. Photoreactions at 3655Å between pyrimidine bases and skin-photosensitizing furocoumarins. *Photochem. Photobiol.* **6**:711–719.

248. **Myers, B. R., J. E. Edmondson, M. E. Anderson, and R. T. Marshall.** 1983. Potassium sorbate and recovery of pectinolytic psychrotrophs from vacuum-packaged pork. *J. Food Prot.* **46**:499–502.

249. **Nelson, K. A., F. F. Busta, J. N. Sofos, and C. E. Allen.** 1983. Effect of polyphosphates in combination with nitrite-sorbate or sorbate on *Clostridium botulinum* growth and toxin production in chicken frankfurter emulsions. *J. Food Prot.* **46**:846–850.

250. **Nury, F. S., M. W. Miller, and J. E. Brekke.** 1960. Preservative effect of some antimicrobial agents on high moisture dried fruits. *Food Technol.* **14**:113–155.

251. **Oh, D.-H., and D. L. Marshall.** 1992. Effect of pH on the minimum inhibitory concentration of monolaurin against *Listeria monocytogenes. J. Food Prot.* **55**:449–450.

252. **Oh, D.-H., and D. L. Marshall.** 1993. Influence of temperature, pH, and glycerol monolaurate on growth and survival of *Listeria monocytogenes. J. Food Prot.* **56**:744–749.

253. **Oka, S.** 1960. Studies on the transfer of antiseptics to microbes and their toxic effect. Part III. Adsorption of esters of p-hydroxybenzoic acid on yeast cell and their toxic effect. *Bull. Agric. Chem. Soc. Jpn.* **24**:412–417.

254. **Okrend, A. J., R. W. Johnston, and A. B. Moran.** 1986. Effect of acetic acid on the death rates at 52°C of *Salmonella newport, Salmonella typhimurium,* and *Campylobacter jejuni* in poultry scald water. *J. Food Prot.* **49**:500–503.

255. **O'Leary, D. K., and R. D. Kralovec.** 1941. Development of *Bacillus mesentericus* in bread and control with calcium acid phosphate or calcium propionate. *Cereal Chem.* **18**:730–741.

256. **Omar, M. M.** 1992. Phenolic compounds in botanical extracts used in foods, flavors, cosmetics, and pharmaceuticals. *Am. Chem. Soc. Symp. Ser.* **506**:154–168.

257. **Oram, J. D., and B. Reiter.** 1968. Inhibition of bacteria by lactoferrin and other iron chelating agents. *Biochim. Biophys. Acta* **170**:351–365.

258. **Ough, C. S.** 1993. Dimethyl dicarbonate and diethyl dicarbonate, p. 343–368. *In* P. M. Davidson and A. L. Branen (ed.), *Antimicrobials in Foods,* 2nd ed. Marcel Dekker, Inc., New York.

259. **Ough, C. S.** 1993. Sulfur dioxide and sulfites, p. 137–190. *In* P. M. Davidson and A. L. Branen (ed.), *Antimicrobials in Foods,* 2nd ed. Marcel Dekker, Inc., New York.

260. **Ough, C. S., and R. E. Kunkee.** 1974. The effect of fumaric acid on malolactic fermentation in wines from warm areas. *Am. J. Enol. Vitic.* **25**:188–190.

261. **Park, H. S., and E. H. Marth.** 1972. Inactivation of *Salmonella typhimurium* by sorbic acid. *J. Milk Food Technol.* **35**:532–539.

262. **Park, H. S., E. H. Marth, and N. F. Olson.** 1970. Survival of *Salmonella typhimurium* in cold-pack cheese food during refrigerated storage. *J. Milk Food Technol.* **33**:383–388.

263. **Parkinson, T. L.** 1966. The chemical composition of eggs. *J. Sci. Food Agric.* **17**:101.

264. **Payne, K. D., P. M. Davidson, and S. P. Oliver.** 1994. Comparison of EDTA and apo-lactoferrin with lysozyme on the growth of foodborne pathogenic and spoilage bacteria. *J. Food Prot.* **57**:62–65.

265. **Payne, K. D., P. M. Davidson, S. P. Oliver, and G. L. Christen.** 1990. Influence of bovine lactoferrin on the growth of *Listeria monocytogenes. J. Food Prot.* **53**:468–472.

266. **Payne, K. D., E. Rico-Muñoz, and P. M. Davidson.** 1989. The antimicrobial activity of phenolic compounds against *Listeria monocytogenes* and their effectiveness in a model milk system. *J. Food Prot.* **52**:151–153.

267. **Pederson, M., N. Albury, and M. D. Christensen.** 1961. The growth of yeasts in grape juice stored at low temperature. IV. Fungistatic effect of organic acids. *Appl. Microbiol.* **9**:162–167.

268. **Pelroy, G. A., M. E. Peterson, P. J. Holland, and M. W. Eklund.** 1994. Inhibition of *Listeria monocytogenes* in cold-process (smoked) salmon by sodium lactate. *J. Food Prot.* **57**:108–113.

269. **Pelroy, G. A., M. E. Peterson, R. Paranjpye, J. Almond, and M. W. Eklund.** 1994. Inhibition of *Listeria monocytogenes* in cold-process (smoked) salmon by sodium nitrite and packaging method. *J. Food Prot.* **57**:114–119.

270. **Perigo, J. A., and T. A. Roberts.** 1968. Inhibition of clostridia by nitrite. *J. Food Technol.* **3**:91–94.

271. **Pfleiderer, G., D. Jeckel, and T. Wieland.** 1956. Uber die Eingirkung von sulfit auf einige DPN hydrierende enzyme. *Biochem. Z.* **328**:187–194.

272. **Phillips, G. F., and J. O. Mundt.** 1950. Sorbic acid as an inhibitor of scum yeast in cucumber fermentations. *Food Technol.* **4**:291.

273. **Pierson, M. D., L. A. Smoot, and N. J. Stern.** 1979. Effect of potassium sorbate on growth of *Staphylococcus aureus* in bacon. *J. Food Prot.* **42**:302–304.

274. **Pilkington, B. J., and A. H. Rose.** 1988. Reactions of *Saccharomyces cerevisiae* and *Zygosaccharomyces bailii* to sulphite. *J. Gen. Microbiol.* **134**:2823–2830.

275. **Pilone, G. J.** 1975. Control of malo-lactic fermentation in table wines by addition of fumaric acid, p. 121–138. *In* J. G. Carr, C. V. Cutting, and G. C. Whiting (ed.), *Lactic Acid Bacteria in Beverages and Foods.* Academic Press, London.

276. **Post, F. J., W. S. Coblentz, T. W. Chou, and D. K. Salunhke.** 1968. Influence of phosphate compounds on certain fungi and their preservative effect on fresh cherry fruit (*Prunus cerasus* L.). *Appl. Microbiol.* **16**:138–142.

277. **Post, F. J., G. B. Krishnamurty, and M. D. Flanagan.** 1963. Influence of sodium hexametaphosphate on selected bacteria. *Appl. Microbiol.* **11**:430–435.

278. **Post, L. S., T. L. Amoroso, and M. Solberg.** 1985. Inhibition of *Clostridium botulinum* type E in model acidified food systems. *J. Food Sci.* **50**:966–968.

279. **Post, L. S., and P. M. Davidson.** 1986. Lethal effect of butylated hydroxyanisole as related to bacterial fatty acid composition. *Appl. Environ. Microbiol.* **52**:214–216.

280. **Prindle, R. F.** 1983. Phenolic compounds, p. 197–224. *In* S. S. Block (ed.), *Disinfection, Sterilization, and Preservation,* 3rd ed. Lea & Febiger, Philadelphia.

281. **Rahn, O., and J. E. Conn.** 1944. Effect of increase in acidity on antiseptic efficiency. *Ind. Eng. Chem.* **36:**185.

282. **Ray, B., C. Johnson, and B. Wanismail.** 1984. Factors influencing lysis of frozen *Escherichia coli* cells by lysozyme. *Cryo Lett.* **5:**183–190.

283. **Razavi-Rohani, S. M., and M. W. Griffiths.** 1994. The effect of mono and polyglycerol laurate on spoilage and pathogenic bacteria associated with foods. *J. Food Safety* **14:**131–151.

284. **Reddy, N. R., and M. D. Pierson.** 1982. Influence of pH and phosphate buffer on inhibition of *Clostridium botulinum* by antioxidants and related phenolic compounds. *J. Food Prot.* **45:**925–927.

285. **Reddy, N. R., M. D. Pierson, and R. V. Lechowich.** 1982. Inhibition of *Clostridium botulinum* by antioxidants, phenols and related compounds. *Appl. Environ. Microbiol.* **43:**835–839.

286. **Rehm, H. J., and H. Wittman.** 1962. Beitrag zur Kenntnis der antimikrobiellen Wirkung der schwefligen Saure. I. Ubersicht uber einflussnehmende Factoren auf die antimikrobielle Wirkung der schwefligen Saure. *Z. Lebens. Unters. Forsch.* **118:**413–429.

287. **Reiss, J.** 1976. Prevention of the formation of mycotoxins in whole wheat bread by citric acid and lactic acid. *Experientia* **32:**168.

288. **Reiter, B.** 1978. Review of the progress of dairy science: antimicrobial systems in milk. *J. Dairy Res.* **45:**131–147.

289. **Reiter, B.** 1984. The biological significance and exploitation of some of the immune systems in milk: a review. *Microbiol. Aliments Nutr.* **2:**1–20.

290. **Reiter, B., and B. G. Harnulv.** 1984. Lactoperoxidase antibacterial system: natural occurrence, biological functions and practical applications. *J. Food Prot.* **47:**724–732.

291. **Reiter, B., and J. D. Oram.** 1986. Iron and vanadium requirements of lactic acid streptococci. *J. Dairy Res.* **35:**67–69.

292. **Rice, A. C., and C. S. Pederson.** 1954. Factors influencing growth of *Bacillus coagulans* in canned tomato juice. II. Acidic constituents of tomato juice and specific organic acids. *Food Res.* **19:**124.

293. **Rico-Muñoz, E., E. E. Bargiota, and P. M. Davidson.** 1987. Effect of selected phenolic compounds on the membrane-bound adenosine triphosphatase of *Staphylococcus aureus. Food Microbiol.* **4:**239–249.

294. **Rico-Muñoz, E., and P. M. Davidson.** 1983. The effect of corn oil and casein on the antimicrobial activity of phenolic antioxidants. *J. Food Sci.* **48:**1284–1288.

295. **Rico-Muñoz, E., and P. M. Davidson.** 1984. The effect of calcium and magnesium on the antibacterial activity of phenolic antioxidants against *Staphylococcus aureus* A100. *J. Food Sci.* **49:**282–283.

296. **Riemann, H., W. H. Lee, and C. Genigeorgis.** 1972. Control of *Clostridium botulinum* and *Staphylococcus aureus* in semipreserved meat products. *J. Milk Food Technol.* **35:**514–523.

297. **Robach, M. C.** 1978. Effect of potassium sorbate on the growth of *Pseudomonas fluorescens. J. Food Sci.* **43:**1886–1887.

298. **Robach, M. C.** 1979. Influence of potassium sorbate on growth of *Pseudomonas putrefaciens. J. Food Prot.* **42:**312–313.

299. **Robach, M. C.** 1980. Use of preservatives to control microorganisms in food. *Food Technol.* **34:**81–84.

300. **Robach, M. C., and C. S. Hickey.** 1978. Inhibition of *Vibrio parahaemolyticus* by sorbic acid in crab meat and flounder homogenates. *J. Food Prot.* **41:**699–702.

301. **Robach, M. C., and M. D. Pierson.** 1978. Influence of *para*-hydroxybenzoic acid esters on the growth and toxin production of *Clostridium botulinum* 10755A. *J. Food Sci.* **43:**787–789.

302. **Robach, M. C., L. A. Smoot, and M. D. Pierson.** 1977. Inhibition of *Vibrio parahaemolyticus* O4:K11 by butylated hydroxyanisole. *J. Food Prot.* **40:**549–551.

303. **Robach, M. C., and J. N. Sofos.** 1982. Use of sorbate in meat products, fresh poultry and poultry products: a review. *J. Food Prot.* **45:**374–383.

304. **Robach, M. C., and C. L. Stateler.** 1980. Inhibition of *Staphylococcus aureus* by potassium sorbate in combination with sodium chloride, tertiary butylhydroquinone, butylated hydroxyanisole or ethylenediamine tetracetic acid. *J. Food Prot.* **43:**208–211.

305. **Roberts, A. C., and D. J. McWeeny.** 1972. The uses of sulfur dioxide in the food industry. A review. *J. Food Technol.* **7:**221–238.

306. **Roberts, T. A.** 1975. The microbiological role of nitrite and nitrate. *J. Sci. Food Agric.* **26:**1775–1760.

307. **Roberts, T. A., and A. M. Gibson.** 1986. Chemical methods for controlling *Clostridium botulinum* in processed meats. *Food Technol.* **40:**163–171.

308. **Roberts, T. A., A. M. Gibson, and A. Robinson.** 1981. Factors controlling the growth of *Clostridium botulinum* types A and B in pasteurized cured meats. II. Growth in pork slurries prepared from high pH meat (pH ranges 6.3–6.8). *J. Food Technol.* **16:**267–281.

309. **Roberts, T. A., A. M. Gibson, and A. Robinson.** 1982. Factors controlling the growth of *Clostridium botulinum* types A and B in pasteurized cured meats. III. The effect of potassium sorbate. *J. Food Technol.* **17:**307–326.

310. **Roberts, T. A., R. L. Gilbert, and M. Ingram.** 1966. The effect of sodium chloride on heat resistance and recovery of heated spores of *C. sporogenes* (PA 3679/S2). *J. Appl. Bacteriol.* **29:**549–555.

311. **Roberts, T. A., and M. Ingram.** 1966. The effect of sodium chloride, potassium nitrate and sodium nitrite on the recovery of heated bacterial spores. *J. Food Technol.* **1:**147–163.

312. **Roberts, T. A., B. Jarvis, and A. C. Rhodes.** 1976. Inhibition of *Clostridium botulinum* by curing salts in pasteurized pork slurry. *J. Food Technol.* **11:**25–40.

313. **Roberts, T. A., L. F. J. Woods, M. J. Payne, and R. Cammack.** 1991. Nitrite, p. 89–111 *In* N. J. Russell and G. W. Gould (ed.), *Food Preservatives.* Blackie and Son Ltd., Glasgow, Scotland.

314. **Roland, J. O., and L. R. Beuchat.** 1984. Biomass and patulin production by *Byssochlamys nivea* in apple juice as affected by sorbate, benzoate, SO_2 and temperature. *J. Food Sci.* **49:**402–406.

315. **Roland, J. O., L. R. Beuchat, R. E. Worthington, and H. L. Hitchcock.** 1984. Effects of sorbate, benzoate, sulfur

dioxide and temperature on growth and patulin production by *Byssochlamys nivea* in grape juice. *J. Food Prot.* **47**:237–241.

316. **Ronning, I. E., and H. A. Frank.** 1987. Growth inhibition of putrefactive anaerobe 3679 caused by stringent-type response induced by protonophoric activity of sorbic acid. *Appl. Environ. Microbiol.* **53**:1020–1027.

317. **Rose, A. H., and B. J. Pilkington.** 1989. Sulphite, p. 201–224. *In* G. W. Gould (ed.), *Mechanisms of Action of Food Preservation Procedures.* Elsevier Applied Science, London.

318. **Rowe, J. J., J. M. Yabrough, J. B. Rake, and R. G. Eagon.** 1979. Nitrite inhibition of aerobic bacteria. *Curr. Microbiol.* **2**:51.

319. **Rusul, G., and E. H. Marth.** 1987. Growth and aflatoxin production by *Aspergillus parasiticus* NRRL 2999 in the presence of potassium benzoate or potassium sorbate at different initial pH values. *J. Food Prot.* **50**:820–825.

320. **Saleem, Z. M., and K. S. Al-Delaimy.** 1982. Inhibition of *Bacillus cereus* by garlic extracts. *J. Food Prot.* **45**:1007–1009.

321. **Samuelson, K. J., J. H. Rupnow, and G. W. Froning.** 1985. The effect of lysozyme and ethylenediaminetetraacetic acid on *Salmonella* on broiler parts. *Poultry Sci.* **64**:1488–1490.

322. **Schade, A. L., and L. Caroline.** 1944. Raw egg white and the role of iron in growth inhibition of *Shigella dysenteriae, Staphylococcus aureus, Escherichia coli* and *Saccharomyces cerevisiae. Science* **100**:14–15.

323. **Schanderl, H.** 1962. Der Einfluss von Polyphenolen und Gerbstoffen auf die Physiologie der Weinhefe und der Wert des pH-7 Testes fur die Auswahl von Sektgrundweinen. *Mitt. Rebe Wein Ostbau Freuchtverwert Klosterneuburg* **12A**:265–274.

324. **Schraufstatter, E.** 1948. Die bakteriostatische Wirkung von Chalkon, Flavanon, Flavon und Flavonol. *Experientia* **4**:484–486.

325. **Seward, R. A., R. H. Dielbel, and R. C. Lindsay.** 1982. Effects of potassium sorbate and other antibotulinal agents on germination and outgrowth of *Clostridium botulinum* type E spores in microculture. *Appl. Environ. Microbiol.* **44**:1212–1221.

326. **Sharma, A., G. M. Tewari, A. J. Shrikhande, S. R. Padwal-Desai, and C. Bandyopadhyay.** 1979. Inhibition of aflatoxin-producing fungi by onion extracts. *J. Food Sci.* **46**:741–744.

327. **Shashikanth, K. N., S. C. Basappa, and V. S. Murthy.** 1981. Studies on the antimicrobial and stimulatory factors of garlic (*Allium sativum* Linn.). *J. Food Sci. Technol.* **18**:44–47.

328. **Shelef, L. A., and L. Addala.** 1994. Inhibition of *Listeria monocytogenes* and other bacteria by sodium diacetate. *J. Food Safety* **14**:103–115.

329. **Shelef, L. A., E. K. Jyothi, and M. A. Bulgarelli.** 1984. Growth of enteropathogenic and spoilage bacteria in sage-containing broth and foods. *J. Food Sci.* **49**:737–740.

330. **Shelef, L. A., O. A. Naglik, and D. W. Bogen.** 1980. Sensitivity of some common foodborne bacteria to the spices sage, rosemary, and allspice. *J. Food Sci.* **45**:1042–1044.

331. **Shelef, L. A., and J. A. Seiter.** 1993. Indirect antimicrobials, p. 539–570. *In* P. M. Davidson and A. L. Branen (ed.), *Antimicrobials in Foods,* 2nd ed. Marcel Dekker, Inc. New York.

332. **Shelef, L. A., and Q. Yang.** 1991. Growth suppression of *Listeria monocytogenes* by lactates in broth, chicken, and beef. *J. Food Prot.* **54**:283–387.

333. **Sherwin, E. R.** 1990. Antioxidants, p. 139–193. *In* A. L. Branen, P. M. Davidson, and S. Salminen (ed.), *Food Additives.* Marcel Dekker, Inc., New York.

334. **Sheu, C. W., and E. Freese.** 1972. Effects of fatty acids on growth and envelope proteins of *Bacillus subtilis. J. Bacteriol.* **111**:516–524.

335. **Sheu, C. W., W. N. Konings, and E. Freese.** 1972. Effects of acetate and other short-chain fatty acids on sugars and amino acid uptake of *Bacillus subtilis. J. Bacteriol.* **111**:525–530.

336. **Sheu, C. W., D. Salomon, J. L. Simmons, T. Sreevalsan, and E. Freese.** 1975. Inhibitory effects of lipophilic acids and related compounds on bacteria and mammalian cells. *Antimicrob. Agents Chemother.* **7**:349–363.

337. **Shibasaki, I.** 1969. Antimicrobial activity of alkyl esters of *p*-hydroxybenzoic acid. *J. Ferment Technol.* **47**:167–177.

338. **Sikovec, S.** 1966. Der Einfluss einiger Polyphenole aud die Physiologie von Weinhefen. I. Der Einfluss von Polyphenolen auf den Verlauf der alkoholischen Gärung insbesondere von Emgärungen. *Mitt. Rebe Wein Ostbau Freuchtverwert Klosterneuburg* **16**:127–138.

339. **Sikovec, S.** 1966. Der Einfluss von Polyphenolen auf die Physiologie von Weinhefen. II. Der Einfluss von Polyphenolen auf die Vermehrung and Atmung von Hefen. *Mitt. Rebe Wein Ostbau Freuchtverwert Klosterneuburg* **16**:272–281.

340. **Silliker, J. H., R. A. Greenberg, and W. R. Schack.** 1958. Effect of individual curing ingredients on the shelf stability of canned comminuted meats. *Food Technol.* **12**:551–554.

341. **Singer, M., and J. Wan.** 1977. Interaction of butylated hydroxytoluene (BHT) with phospholipid bilayer membranes: effect on ^{22}Na permeability and membrane fluidity. *Biochem. Pharmacol.* **26**:2259–2268.

342. **Singleton, V. L., and P. Esau.** 1969. Phenolic substances in grapes and wine, and their significance. Academic Press, New York.

343. **Siragusa, G. R., and M. G. Johnson.** 1989. Inhibition of *Listeria monocytogenes* growth by the lactoperoxidase-thiocyanate-H_2O_2 antimicrobial system. *Appl. Environ. Microbiol.* **55**:2802–2805.

344. **Smith, D. P., and N. J. Rollin.** 1954. Sorbic acid as a fungistatic agent for foods. VII. Effectiveness of sorbic acid in protecting cheese. *Food Res.* **19**:59–65.

345. **Smoot, L. A., and M. D. Pierson.** 1981. Mechanisms of sorbate inhibition of *Bacillus cereus* T and *Clostridium botulinum* 62A spore germination. *Appl. Environ. Microbiol.* **42**:477–483.

346. **Smulders, F. J. M., P. Barendsen, J. G. van Logtestjin, D. A. A. Mossel, and G. M. Van der Marel.** 1986. Review: lactic acid: considerations in favour of its acceptance as a meat decontaminant. *J. Food Technol.* **21**:419.

347. **Snijders, J. M. A., J. G. van Logtestjin, D. A. A. Mossel, and F. J. M. Smulders.** 1985. Lactic acid as a decontaminant in slaughter and processing procedures. *Vet. Q.* **7:** 277–282.

348. **Sofos, J. N.** 1983. Antimicrobial effects of sodium and other ions in foods. *J. Food Safety* **6:**45–78.

349. **Sofos, J. N.** 1986. Use of phosphates in low-sodium meat products. *Food Technol.* **40:**52–68.

350. **Sofos, J. N., and F. F. Busta.** 1981. Antimicrobial activity of sorbates. *J. Food Prot.* **44:**614–622.

351. **Sofos, J. N., and F. F. Busta.** 1993. Sorbic acid and sorbates, p. 49–94. *In* P. M. Davidson and A. L. Branen (ed.), *Antimicrobials in Foods,* 2nd ed. Marcel Dekker, Inc., New York.

352. **Sofos, J. N., F. F. Busta, and C. E. Allen.** 1979. Botulism control by nitrite and sorbate in cured meats. A review. *J. Food Prot.* **42:**739–770.

353. **Sofos, J. N., F. F. Busta, and C. E. Allen.** 1980. Influence of pH on *Clostridium botulinum* control by sodium nitrite and sorbic acid in chicken emulsions. *J. Food Sci.* **45:**7–12.

354. **Sofos, J. N., F. F. Busta, K. Bhothipaksa, C. E. Allen, M. C. Robach, and M. W. Paquette.** 1980. Effects of various concentrations of sodium nitrite and potassium sorbate on *Clostridium botulinum* toxin production in commercially prepared bacon. *J. Food Sci.* **45:**1285–1292.

355. **Spencer, R.** 1971. Nitrite in curing: microbiological implications, p. 161. *In Proceedings of the 17th European Meeting of Meat Research Workers Conference,* Bristol, U.K.

356. **Sperber, W. H.** 1983. Influence of water activity on foodborne bacteria—a review. *J. Food Prot.* **46:**142–150.

357. **Stead, D.** 1993. The effect of hydroxycinnamic acids on the growth of wine-spoilage lactic acid bacteria. *J. Appl. Bacteriol.* **75:**135–141.

358. **Stead, D.** 1994. The effect of chlorogenic, gallic and quinic acids on the growth of spoilage strains of *Lactobacillus collinoides* and *Lactobacillus brevis. Lett. Appl. Microbiol.* **18:**112–114.

359. **Stecchini, M. L., I. Sarais, and P. Giavedoni.** 1993. Effect of essential oils on *Aeromonas hydrophila* in a culture medium and in cooked pork. *J. Food Prot.* **56:**406–409.

360. **Steinke, P. D. W., and E. M. Foster.** 1951. Botulism toxin formation in liver sausage. *Food Res.* **16:**477.

361. **Stratford, M., P. Morgan, and A. H. Rose.** 1987. Sulphur dioxide resistance in *Saccharomyces cerevisiae* and *Saccharomycodes ludwigii. J. Gen. Microbiol.* **133:**2173–2179.

362. **Stratford, M., and A. H. Rose.** 1985. Hydrogen sulphide production from sulphite by *Saccharomyces cerevisiae. J. Gen. Microbiol.* **131:**1417–1424.

363. **Stratford, M., and A. H. Rose.** 1986. Transport of sulphur dioxide by *Saccharomyces cerevisiae. J. Gen. Microbiol.* **132:**1–6.

364. **Subramanian, C. S., and E. H. Marth.** 1968. Multiplication of *Salmonella typhimurium* in skim milk with and without added hydrochloric, lactic and citric acids. *J. Milk Food Technol.* **31:**323.

365. **Surak, J. G., R. L. Bradley, A. L. Branen, E. Shrago, and W. E. Ribelin.** 1976. Effect of butylated hydroxytoluene on *Tetrahymena pyriformis. Food Cosmet. Toxicol.* **14:**541–549.

366. **Surak, J. G., and R. G. Singh.** 1980. Butylated hydroxyanisole (BHA) induced changes in the synthesis of polar lipids and in the molar ratio of tetrahymenol to polar lipids in *Tetrahymena pyriformis. J. Food Sci.* **45:**1251–1255.

367. **Tanaka, N.** 1982. Challenge of pasteurized process cheese spreads with *Clostridium botulinum* using in-process and post-process inoculation. *J. Food Prot.* **45:**1044–1050.

368. **Tarr, H. L. A.** 1941. Bacteriostatic action of nitrites. *Nature* (London) **147:**417–418.

369. **Tarr, H. L. A.** 1941. The action of nitrites on bacteria. *J. Fish Res. Bd. Can.* **5:**265–275.

370. **Tarr, H. L. A.** 1942. The action of nitrites on bacteria, further experiments. *J. Fish Res. Bd. Can.* **6:**74–82.

371. **Taylor, S. L., and M. W. Speckhard.** 1984. Inhibition of bacterial histamine production by sorbate and other antimicrobial agents. *J. Food Prot.* **47:**508–511.

372. **Teraguchi, S., K. Shin, T. Ogata, M. Kingaku, A. Kaino, H. Miyauchi, Y. Fukuwatari, and S. Shimamura.** 1995. Orally administered bovine lactoferrin inhibits bacterial translocation in mice fed bovine milk. *Appl. Environ. Microbiol.* **61:**4131–4134.

373. **Thompson, D. P.** 1991. Effect of butylated hydroxyanisole on conidial germination of toxigenic species of *Aspergillus flavus* and *Aspergillus parasiticus. J. Food Prot.* **54:**375–377.

374. **Thompson, D. P.** 1994. Minimum inhibitory concentrations of esters of p-hydroxybenzoic acid (paraben) combinations against toxigenic fungi. *J. Food Prot.* **57:** 133–135.

375. **Thomson, J. E., G. J. Banwart, D. H. Sanders, and A. J. Mercuri.** 1967. Effect of chlorine, antibiotics, β-propiolactone, acids and washing on *Salmonella typhimurium* on eviscerated fryer chickens. *Poult. Sci.* **46:**146.

376. **Tollefson, C.** 1995. Stability preserved. *Chem. Mark. Rep.* **May 29:**SR28–SR31.

377. **Tomita, M., W. Bellamy, M. Takase, K. Yamauchi, H. Wakabayashi, and K. Kawase.** 1992. Potent antibacterial peptides generated by pepsin digestion of bovine lactoferrin. *J. Dairy Sci.* **74:**4137–4142.

378. **Tompkin, R. B.** 1983. Indirect antimicrobial effects in foods: phosphates. *J. Food Safety* **6:**13–27.

379. **Tompkin, R. B.** 1993. Nitrite, p. 191–262. *In* P. M. Davidson and A. L. Branen (ed.), *Antimicrobials in Foods,* 2nd ed. Marcel Dekker, Inc., New York.

380. **Tompkin, R. B., L. N. Christiansen, and A. B. Shaparis.** 1978. Antibotulinal role of isoascorbate in cured meat. *J. Food. Sci.* **43:**1368–1370.

381. **Tompkin, R. B., L. N. Christiansen, and A. B. Shaparis.** 1978. The effect of iron on botulinal inhibition in perishable canned cured meat. *J. Food Technol.* **13:**521–527.

382. **Tompkin, R. B., L. N. Christiansen, A. B. Shaparis, and H. Bolin.** 1974. Effect of potassium sorbate on salmonellae, *Staphylococcus aureus, Clostridium perfringens,* and *Clostridium botulinum* in cooked uncured sausage. *Appl. Microbiol.* **28:**262–264.

383. **Tranter, H. S.** 1994. Lysozyme, ovotransferrin and avidin, p. 65–97. *In* V. M. Dillon and R. G. Board (ed.), *Natural Antimicrobial Systems and Food Preservation.* CAB Intl., Wallingford, England.

384. **Tranter, H. S., and R. G. Board.** 1982. Review: the antimicrobial defense of avian eggs: biological perspective and chemical basis. *J. Appl. Biochem.* **4:**295–338.

385. **Tranter, H. S., and R. G. Board.** 1984. Influence of incubation temperature and pH on the antimicrobial properties of hen egg albumen. *J. Appl. Bacteriol.* **56:**53–61.

386. **Tri-K Industries, Inc.** No date. Technical data, esters of parahydroxybenzoic acid (parabens). Tri-K Industries, Inc., Westwood, N.J.

387. **Troller, J. A.** 1965. Catalase inhibition as a possible mechanism of the fungistatic action of sorbic acid. *Can. J. Microbiol.* **11:**611–617.

388. **Turbovsky, M. W., F. Filipello, W. V. Cruess, and P. Esau.** 1934. Observations on the use of tannin in wine making. *Fruit Prod. J.* **14:**106.

389. **Unda, J. R., R. A. Mollins, and H. W. Walker.** 1991. *Clostridium sporogenes* and *Listeria monocytogenes:* survival and inhibition in microwave-ready beef roasts containing selected antimicrobials. *J. Food Sci.* **56:**198–205.

390. **Uriah, N., T. R. Cassity, and J. R. Chipley.** 1977. Partial characterization of the action of benzoic acid on aflatoxin biosynthesis. *Can. J. Microbiol.* **23:**1580–1584.

391. **Uriah, N., and J. R. Chipley.** 1976. Effects of various acids and salts on growth and aflatoxin production by *Aspergillus flavus. Microbios* **17:**51–59.

392. **Vakil, J. R., R. C. Chandan, R. M. Parry, and K. M. Shahani.** 1969. Susceptibility of several microorganisms to milk lysozymes. *J. Dairy Sci.* **52:**1192–1197.

393. **Valenti, P., P. Visca, G. Antonini, and N. Orsi.** 1985. Antifungal activity of ovotransferrin towards genus *Candida. Mycopathology* **89:**165–175.

394. **Valenti, P., P. Visca, G. Antonini, and N. Orsi.** 1986. Interaction between lactoferrin and *Candida* cells. *FEMS Microbiol. Lett.* **33:**271–275.

395. **Valle, E.** 1957. On anti-fungal factors in potato leaves. *Acta Chem. Scand.* **11:**395.

396. **Vas, K.** 1953. Mechanism of antimicrobial action. Interference with the cytoplasmic membrane. *Agrokem. Talajtan* **2:**1–16. [*Chem. Abstr.* **48:**794d, 1954.]

397. **Vaughn, R. H.** 1951. The microbiology of dehydrated vegetables. *Food Res.* **16:**429.

398. **Virtanen, A. I.** 1965. Studies on organic sulfur containing compounds and other labile substances in plants. *Angew. Chem.* **1:**207–228.

399. **Wagner, M. K.** 1986. Phosphates as antibotulinal agents in cured meats. A review. *J. Food Prot.* **49:**482–487.

400. **Wagner, M. K., and F. F. Busta.** 1983. Effect of sodium acid pyrophosphate in combination with sodium nitrite or sodium nitrite/potassium sorbate on *Clostridium botulinum* growth and toxin production in beef/pork frankfurter emulsions. *J. Food Sci.* **48:**990–991.

401. **Wagner, M. K., and F. F. Busta.** 1985. Inhibition of *Clostridium botulinum* 52A toxicity and protease activity by sodium acid pyrophosphate in media systems. *Appl. Environ. Microbiol.* **50:**16–20.

402. **Wakabayashi, H., W. Bellamy, M. Takase, and M. Tomita.** 1992. Inactivation of *Listeria monocytogenes* by lactoferricin, a potent antimicrobial peptide derived from cow's milk. *J. Food Prot.* **55:**238–240.

403. **Waldroup, A.** 1995. Evaluating reduction technologies. *Meat Poult.* **41**(8):10.

404. **Waldroup, A., J. Marcy, M. Doyle, and M. Scantling.** 1995. TSP: a market survey. *Meat Poult.* **41**(12):18–20.

405. **Walker, J. C., and M. A. Stahmann.** 1955. Chemical nature of disease resistance in plants. *Annu. Rev. Plant Physiol.* **6:**351–366.

406. **Walton, L., M. Herbold, and C. C. Lindegren.** 1936. Bactericidal effects of vapors from crushed garlic. *Food Res.* **1:**163.

407. **Wanda, P., J. Cupp, W. Snipes, A. Keith, T. Rucinsky, L. Polish, and J. Sands.** 1976. Inactivation of the enveloped bacteriophage $\phi6$ by butylated hydroxytoluene and butylated hydroxyanisole. *Antimicrob. Agents Chemother.* **10:**96–104.

408. **Wang, C., and L. Shelef.** 1991. Factors contributing to antilisterial effects of raw egg albumen. *J. Food Sci.* **56:**1251–1254.

409. **Warth, A. D.** 1977. Mechanism of resistance of *Saccharomyces bailii* to benzoic, sorbic, and other weak acids used as food preservatives. *J. Appl. Bacteriol.* **43:**215–230.

410. **Wasserfall, F., E. Voss, and D. Prokopek.** 1976. Experiments on cheese ripening: the use of lysozyme instead of nitrite to inhibit late blowing of cheese. *Kiel. Milchwirtsch. Forschungsber.* **238:**3–16.

411. **Weaver, R. A., and L. A. Shelef.** 1993. Antilisterial activity of sodium, potassium or calcium lactate in pork liver sausage. *J. Food Safety* **13:**133–146.

412. **Webb, A. H., and F. W. Tanner.** 1944. Effect of spices and flavoring materials on growth of yeasts. *Food Res.* **10:**273–282.

413. **Wedzicha, B. L.** 1981. Sulphur dioxide: the reaction of sulphite species with food components. *Nutr. Food Sci.* **72:**12–13, 19.

414. **Wedzicha, B. L., and M. A. Brook.** 1989. Reaction of sorbic acid with nucleophiles: preliminary studies. *Food Chem.* **31:**29–40.

415. **Weinberg, E.** 1978. Iron and infection. *Bacteriol. Rev.* **42:**45–65.

416. **Wendakoon, C. N., and M. Sakaguchi.** 1993. Combined effect of sodium chloride and clove on growth and biogenic amine formation of *Enterobacter aerogenes* in mackerel muscle extract. *J. Food Prot.* **56:**410–413.

417. **Wendorff, W. L., W. E. Riha, and E. Muehlenkamp.** 1993. Growth of molds on cheese treated with heat or liquid smoke. *J. Food Prot.* **56:**963–966.

418. **Whitaker, J. R.** 1959. Inhibition of sulfhydryl enzymes with sorbic acid. *Food Res.* **24:**37–43.

419. **Wibowo, D., R. Eschenbruch, C. R. Davis, G. H. Fleet, and T. H. Lee.** 1985. Occurrence and growth of lactic acid bacteria in wine. A review. *Am. J. Enol. Vitic.* **36:**302–313.

420. **Wilkins, K. M., and R. G. Board.** 1989. Natural antimicrobial systems, p. 285–362. *In* G. W. Gould (ed.), *Mechanisms of Action of Food Preservation Procedures.* Elsevier Applied Science, London.

421. **Wills, E. D.** 1956. Enzyme inhibition by allicin, the active principle of garlic. *Biochem. J.* **63**:514–520.

422. **Wilson, D. C., and H. D. Brown.** 1953. Heat-induced inhibitory agents obtained from processed fruits and vegetables. *Food Technol.* **7**:250.

423. **Woods, L. F. J., and J. M. Wood.** 1982. The effect of nitrite inhibition on the metabolism of *Clostridium botulinum*. *J. Appl. Bacteriol.* **52**:109–110.

424. **Woods, L. F. J., J. M. Wood, and P. A. Gibbs.** 1981. The involvement of nitric oxide in the inhibition of the phosphoroclastic system in *Clostridium sporogenes* by sodium nitrite. *J. Gen. Microbiol.* **125**:399–406.

425. **Woods, L. F. J., J. M. Wood, and P. A. Gibbs.** 1989. Nitrite, p. 225–246. *In* G. W. Gould (ed.), *Mechanisms of Action of Food Preservation Procedures.* Elsevier Applied Science, London.

426. **Woolford, M. K.** 1975. Microbiological screening of the straight chain fatty acids (C_1-C_{12}) as potential silage additives. *J. Sci. Food Agric.* **26**:219–228.

427. **Wright, L. D., and H. R. S. Skeggs.** 1946. Reversal of sodium propionate inhibition of *Escherichia coli* with β-alanine. *Arch. Biochem.* **10**:383.

428. **Yang, T.** 1985. Mechanism of nitrite inhibition of cellular respiration in *Pseudomonas aeruginosa*. *Curr. Microbiol.* **12**:35–40.

429. **York, G. K., and R. H. Vaughn.** 1954. Use of sorbic acid enrichment media for species of *Clostridium*. *J. Bacteriol.* **68**:739–744.

430. **York, G. K., and R. H. Vaughn.** 1955. Resistance of *Clostridium parabotulinum* to sorbic acid. *Food Res.* **20**:60–65.

431. **York, G. K., and R. H. Vaughn.** 1964. Mechanisms in the inhibition of microorganisms by sorbic acid. *J. Bacteriol.* **88**:411–417.

432. **Yousef, A. E., M. A. El-Shenawy, and E. H. Marth.** 1989. Inactivation and injury of *Listeria monocytogenes* in a minimal medium as affected by benzoic acid and incubation temperature. *J. Food Sci.* **54**:650–652.

433. **Yousef, A. E., R. J. Gajewski, and E. H. Marth.** 1991. Kinetics of growth and inhibition of *Listeria monocytogenes* in the presence of antioxidant food additives. *J. Food Sci.* **56**:10–13.

434. **Zaika, L. L., and A. H. Kim.** 1993. Effect of sodium polyphosphates on growth of *Listeria monocytogenes*. *J. Food Prot.* **56**:577–580.

435. **Zaika, L. L., and J. C. Kissinger.** 1979. Effect of some spices on acid production by starter cultures. *J. Food Prot.* **42**:572–576.

436. **Zaika, L. L., and J. C. Kissinger.** 1981. Inhibitory and stimulatory effects of oregano on *Lactobacillus plantarum* and *Pediococcus cerevisiae*. *J. Food Sci.* **46**:1205–1210.

437. **Zaika, L. L., J. C. Kissinger, and A. E. Wasserman.** 1983. Inhibition of lactic acid bacteria by herbs. *J. Food Sci.* **48**:1455–1459.

438. **Zajac, M., J. Gladys, M. Skarzynska, B. G. Härnulv, and L. Björck.** 1983. Changes in bacteriological quality of raw milk stabilized by activation of its lactoperoxidase system and stored at different temperatures. *J. Food Prot.* **46**:1065–1068.

439. **Zhao, T., M. P. Doyle, and R. E. Besser.** 1993. Fate of enterohemorrhagic *Escherichia coli* O157:H7 in apple cider with and without preservatives. *Appl. Environ. Microbiol.* **59**:2526–2530.

Thomas J. Montville
Karen Winkowski

30 Biologically Based Preservation Systems and Probiotic Bacteria

The chemical and physical methods of food preservation covered in this volume are well established and easily identified by consumers. Physical changes in the food or label declarations help the consumer determine how a food is processed. A little knowledge can be a dangerous thing; consumers are increasingly wary of chemical preservatives and "processed" foods, even though these things provide the unparalleled safety and diversity of our food supply. One result of these consumer trends is the increased reliance on refrigeration to assure the safety of foods that are minimally processed and free of chemical preservatives. However, two factors make it unwise to rely too heavily on refrigeration. The first is that 20% of commercial and residential refrigerators maintain temperatures of >10°C (50°F) (73, 163). The second factor is the ability of *Listeria monocytogenes* and other pathogens to grow at near-freezing temperatures (118). Therefore, the National Food Processors Association recommends that additional barriers to microbial growth be incorporated into refrigerated foods. Such a barrier may be provided through the use of lactic acid bacteria (LAB). Their use for food preservation is accepted by consumers as "natural" and "health-promoting." The fermentation of food may be the oldest form (after drying) of processing for preservation. These factors have generated tremendous interest in biologically based preservation methods.

This chapter provides an overview of the biologically based preservation technologies which can be classified as "biopreservation." Biopreservation is defined here as the use of LAB, their metabolic products, or both to improve or assure the safety and quality of foods that are not generally considered fermented. The preservative, nutritional, and functional properties of fermented foods are covered in section VIII of this book. Acid production by LAB in temperature-abused foods ("controlled acidification") is covered in the first part of this chapter.

Some strains of LAB associated with fermented foods also produce antimicrobial proteins called bacteriocins. Bacteriocins inhibit spoilage and pathogenic bacteria without changing (i.e., through acidification, protein denaturation, etc.) the physical-chemical nature of the food being preserved. Because the use of bacteriocins is a newer and emerging area of food microbiology, the largest section of this chapter deals with this topic.

The appropriateness of discussing probiotic bacteria in a chapter on biological preservation may not be intuitively obvious. Probiotic bacteria may biologically preserve *us*, but this is a facetious explanation. In the market-driven food industry, marketing professionals try to turn marketing negatives (i.e., the need to preserve a food) into marketable positive attributes (i.e., the method of preservation may improve consumer health). Thus there is increasing interest in foods that are not merely nutritious, but are distinctly health-promoting. Indeed, the terms "nutraceuticals" and "pharmafoods" have entered the food science lexicon (56). Since LAB are not only "natural" but widely perceived as health-promoting, foods preserved by biological techniques may be marketed as health-promoting. For these reasons, this chapter closes by examining the claims for the therapeutic action of LAB.

BIOPRESERVATION BY CONTROLLED ACIDIFICATION

It is widely recognized, and discussed more fully in chapter 29, that organic acids inhibit microbial growth. While this is usually done by adding organic acids to foods, LAB can produce lactic acid in situ. The controlled production of acid in situ is an important form of biopreservation. Many factors determine the effectiveness of in situ acidification. These include the product's initial pH; its buffering capacity; the type and level of challenge organism; the nature and concentration of the fermentable carbohydrate; ingredients that might influence the viability and growth rate of the LAB; and the growth rates of the LAB and target pathogen at refrigerated and abuse temperatures (73). Clearly, such applications require customization and a high level of technical support. The production of bacteriocins, diacetyl, and hydrogen peroxide may also contribute to the overall inhibition. For example, Microgard is a cultured milk product added to much of the cottage cheese in the United States as a Generally Recognized as Safe (GRAS) food preservative. It is made by fermenting milk using *Propionibacterium shermanii* to produce acetate, propionic acid, and low-molecular-weight proteins. Although Microgard does contain a bacteriocin, the propionic acid certainly plays a major role in its activity (3, 97).

The idea of using LAB to prevent botulinal toxigenesis through in situ acid production dates back to the 1950s. This technology uses the inability of *Clostridium botulinum* to grow at pH <4.8 as a defense against temperature abuse. LAB and a fermentable carbohydrate are added to the food. The LAB grow and produce acid in situ only under conditions of temperature abuse. Under proper refrigeration, the LAB cannot grow and no acid is formed. Saleh and Ordal (133) used a "normal cheese culture," *Lactobacillus bulgaricus*, or *Lactococcus lactis* in experiments designed to prevent toxigenesis from spore inocula in chicken à la king containing a fermentable carbohydrate. When incubated at 30°C in the absence of the LAB, 16 out of 16 samples rapidly became toxic. In the presence of the "normal cheese culture," *Lb. bulgaricus*, or *Lc. lactis*, only three or fewer samples became toxic after 5 days at 30°C. In all samples to which the LAB had been added, including those positive for botulinal toxin, the pH was reduced to <4.5. The differences between the LAB cultures were not statistically significant, and the inhibition was attributed to acid production.

The discovery that carcinogenic nitrosamines are formed from nitrites used as curing agents in meats initiated a search for nitrite substitutes. Tanaka et al. (151) sought to reduce nitrite concentrations in bacon by using controlled acidification. When bacon was inoculated with 10^3 botulinal spores per g and incubated at 28°C, toxin was produced in 58% of the conventional bacon samples prepared with 120 ppm nitrite but no sucrose or starter culture. When similarly challenged, only ≤2% of bacon prepared with 80 or 40 ppm nitrite, 0.7% sucrose, and starter cultures became toxic. The United States Department of Agriculture (USDA) approved what is now known as the "Wisconsin process" for bacon manufacture in 1986.

BACTERIOCINS

The interest in the bacteriocins produced by LAB has grown dramatically in the last decade and is documented by many major reviews and books (70, 74, 84, 112, 124, 147, 150). One reason for this interest is that many bacteriocins inhibit food spoilage and pathogenic bacteria, such as *L. monocytogenes*, which are recalcitrant to traditional food preservation methods. In addition, bacteriocinogenic LAB are associated with, and are used as, starter cultures. The use of bacteriocins, the organisms which produce them, or both is attractive to the food industry because it is facing both increasing consumer demand for natural products and increasing concern about foodborne disease. However, this interest is tempered by regulatory uncertainty and concerns that the development of bacteriocin-resistant pathogens might render the technology ineffective.

General Characteristics

The bacteriocins produced by LAB are a relatively heterogeneous group of ribosomally synthesized small proteins. They normally act against "closely related" bacteria, but not against the producing organism. In some cases, "closely related" covers a wide range of gram-positive bacteria. Chelating agents, hydrostatic pressure, or injury can render gram-negative bacteria bacteriocin sensitive (82, 83, 145, 146). Enzymes, such as lysozyme, which act exclusively through degradative enzymatic activities are not considered bacteriocins. However, in addition to their action at the target cell's membrane, some bacterio-

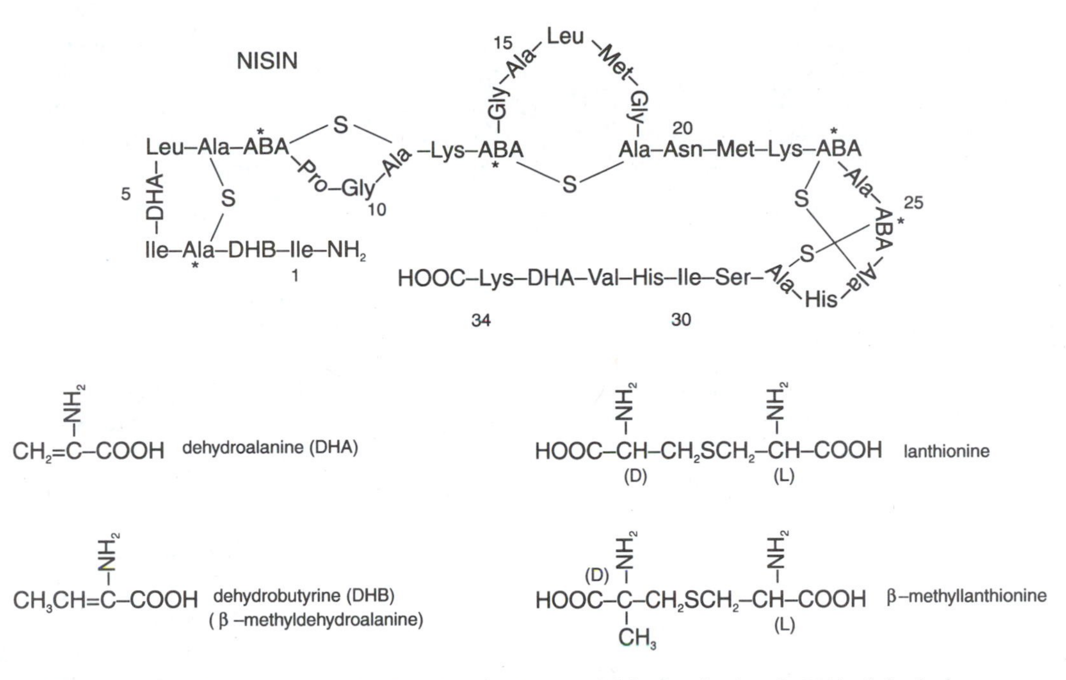

Figure 30.1 Structure of nisin showing positions of dehydroalanine (DHA), dehydrobutyrine (DHB), lanthionine (Ala-S-Ala), and methyl lanthionine (ABA-S-Ala). ABA, aminobutyric acid; *, D-stereo configuration for the α-carbon. (From reference 94, with permission.)

cins, such as the colicins produced by *Escherichia coli*, inhibit protein synthesis, degrade RNA, or have other biological functions (87).

Early bacteriocin researchers often applied to bacteriocins the seven characteristics of colicins cited by Tagg's influential review (150). They would restrict the designation "bacteriocins" to those plasmid-mediated proteins produced by lethal biosynthesis which are bactericidal to a narrow range of closely related bacteria having specific binding sites for that bacteriocin. It is now clear that few LAB bacteriocins meet all of these criteria. A proposal to designate as bacteriocins only those proteins having all seven characteristics and use "bacteriocin-like inhibitory substance (BLIS) proteins" to describe other antimicrobial proteins (149) has not been widely adopted. However, the BLIS designation is useful for another purpose. There are at least five separate examples where bacteriocins have been independently isolated and named by several different investigators but were later found to be identical (based on their amino acid sequences). The resulting confusion of having many different names for the same molecule impedes research progress. In order to prevent this confusion, Jack et al. (74) proposed using "BLIS protein" followed by a strain designation as a provisional designation when proteinaceous inhibitors are first identified. A new bacteriocin name would be given only when the amino acid sequence indicates that the bacteriocin is, in fact, unique.

Bacteriocins differ in their spectra of activity, biochemical characteristics, and genetic determinants (84, 85, 112). Most bacteriocins are small (3 to 10 kDa), have a high isoelectric point, and contain both hydrophobic and hydrophilic domains. Klaenhammer (85) further classified bacteriocins into four major groups. These groupings provide a useful conceptual framework for bacteriocin researchers.

Group I bacteriocins contain the unusual amino acids dehydroalanine (Dha), dehydrobutyrine (Dhb), lanthionine, and β-methyllanthione (Fig. 30.1). These amino acids are produced by posttranslational modification of serine and threonine to their dehydro forms. The dehydro amino acids react with cysteine to form thioether (single sulfhydryl) lanthionine rings. Bacteriocins containing these lanthionine rings are commonly referred to as lantibiotics. There are many structurally similar lantibiotics. Nisin, the first and best-characterized LAB bacteriocin, is produced in two related forms. Nisin A contains a histidine at position 27, whereas nisin Z has an asparagine. While subtilin, produced by *Bacillus subtilis*, also contains five lanthionine rings and a conformation similar to nisin, it has other amino acid substitutions including a carboxy terminus 2 amino acids shorter than that of nisin. The 12 amino acids at the amino terminus

Table 30.1 Proposed numerical system for bacteriocin nomenclature[a]

BACTERIOCIN NAME [Bac. II.1.1.2.35]

Klaenhammer class no. ← (II)
Inhibitory spectrum: ← (1)
1. Broad
2. Narrow
Mode of action: ← (1)
1. Pore forming
2, 3, etc., Other
Number of peptides required for activity: ← (2)
1. One component
2. Two component
Number to describe how many bacteriocins of the same type have been described: e.g., 35th bacteriocin of this type described. ← (35)

Examples: Nisin (Bac. I.1.1.1.1)
Lactacin F (Bac. II.2.1.2.1)

[a]Table originally presented at Workshop on the Bacteriocins of Lactic Acid Bacteria—Applications and Fundamentals, Alberta, Canada, 17–22 April 1995.

of nisin and epidermin are similar, but epidermin lacks the central lanthionine ring common to nisin and subtilin and has a cyclized carboxy terminus. Lacticin 481, lactococcin, lactocin S, and carnocin are other lantibiotics produced by LAB.

Many bacteriocins belong in group II. Group II bacteriocins are a large group of small heat-stable proteins with a consensus leader sequence containing a Gly-Gly^{-1} -Xaa^{+1} cleavage site important for processing the prebacteriocin during export. Group II is subdivided into three groups. Bacteriocins active against *L. monocytogenes* and having a -Tyr-Gly-Asn-Gly-Val-Xaa-Cys amino-terminal consensus sequence are classified in the subgroup IIa. Pediocin PA-1 (whose amino acid sequence is identical to that of pediocin AcH), sakacins A and P, leucocin A, bavaricin MN, and curvacin A are members of this subgroup. Subgroup IIb contains bacteriocins, such as lactococcin G, lactococcin M, and lactacin F, which require two different peptides for activity (2). Bacteriocins in subgroup IIc, such as lactacin B, require reduced cysteines for activity.

Group III and IV bacteriocins differ markedly from other bacteriocins. The larger (>30-kDa) heat-labile antimicrobial proteins such as helveticins J and V and lactacins A and B are classified as group III bacteriocins. Group IV bacteriocins such as leuconocin S, lactocin 27, and pediocin SJ-1 have lipid or carbohydrate moieties. The composition and function of the nonprotein portions are largely unknown.

Classification systems and nomenclature for bacteriocins are still evolving. Klaenhammer's groupings are incorporated into a proposed numerical system analogous to the numbering system used in enzyme nomenclature (Table 30.1). An alternate nomenclature system (74) is based on sulfhydryl chemistry. Just as bacteriocins containing lanthionine rings are called lantibiotics, those containing disulfide bonds would be called cystibiotics and those requiring reduced sulfhydryls would be called thiobiotics. Bacteriocin research is very dynamic; some time may be required before a system of definitive nomenclature is in place.

Methodological Considerations

Bacteriocin-producing bacteria are easy to isolate from foods. The methods used for their initial isolation and characterization are relatively simple (101). The most common method for demonstrating bacteriocin production (the Bac$^+$ phenotype) is to overlay a colony of the putative bacteriocin producer with agar medium containing the target organism to be inhibited. An inhibition zone in the confluent growth of the target organism is presumptive evidence for bacteriocin production (Fig. 30.2). Such zones can also be produced by acid, bacteriophage, hydrogen peroxide, or other nonspecific inhibitors. Negative control experiments are required to exclude these. The positive control, confirming the inhibitor's protease sensitivity, is equally important and can often be done on the same petri dish.

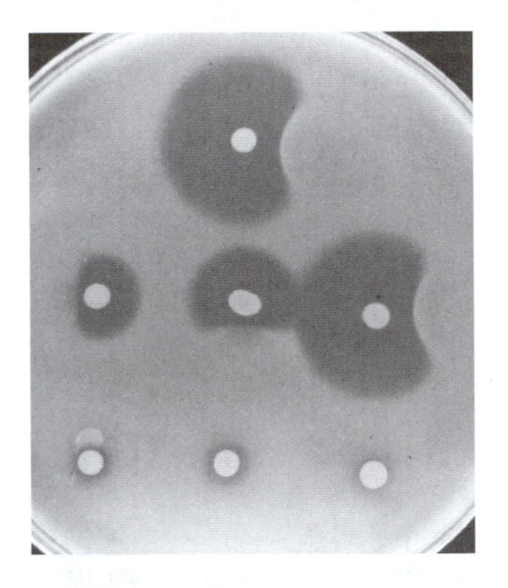

Figure 30.2 Evidence of bacteriocin production using a differed spot-on-the-lawn method. A bacteriocin-producing colony is cultured on agar and then overlaid with an agar seeded with a bacteriocin-sensitive organism. After further incubation, diffusion of bacteriocin away from the colony causes clearing zones in the lawn of sensitive bacteria. When a protease is spotted near the inhibition zone, a "dimple" of growth in the clearing zone indicates that the inhibitor is a protein. The small diffuse zones around the bottom three colonies of bacteria that do not produce bacteriocins are due to acid production (photo courtesy of M. Stiles).

Research beyond the initial isolation of the Bac+ bacteria and characterization of the inhibitor as a bacteriocin is more difficult. Some isolates produce large inhibition zones on agar media, but have no detectable activity in broth. High bacteriocin activities (10,000 to >50,000 arbitrary units [AU]/ml; see below for definitions) in the culture supernatants facilitate bacteriocin purification. Considerable effort in optimizing the liquid medium, incubation conditions, pH, and other factors may be needed before starting purification studies. The purification usually involves salting-out the protein followed by some combination of gel filtration, ion exchange, and affinity and hydrophobic interaction chromatography. Amino acid sequences are then determined from the electrophoretically pure protein. As more bacteriocins are being purified to the sequence level, it is becoming common to discover that independently purified bacteriocins are identical.

The lack of recognized standards for bacteriocin activity is a major impediment to progress in this field. Only for nisin has an international unit of activity been adopted. One gram of commercially available nisin preparation (Nisaplin) contains 25 mg of pure nisin and is defined as having 10^6 International Units (IU; earlier known as Reading Units) of activity. Nisin activity is measured by the well diffusion assay of Tramer and Fowler (154) using *Micrococcus luteus* ATCC 10420 as the sensitive organism. *M. luteus* is also used to measure the activity of other peptide antimicrobial agents such as scorpion defensins (24). Investigators frequently use other indicator strains in their assays for other bacteriocins if they are more sensitive (generate larger zone sizes) than *M. luteus*.

Bacteriocin activity can be estimated from the size of the inhibition zones produced by the diffusion of the bacteriocin in the confluent lawns of bacteriocin-sensitive bacteria. The bacteriocin can be placed in a well (such as the one made by a number 3 cork borer) in the assay medium which contains the sensitive organism. Diffusion assays can also be conducted by "spotting" a small volume (10 to 50 µl) of test material directly on a lawn of sensitive bacteria. In either case, the activity is reported as the radius or diameter of the inhibition zone, either before or after subtracting the dimension of the well or spot. The zone sizes obtained in diffusion assays are proportional to the log of the bacteriocin activity. Results are easily misinterpreted when zone sizes are reported without considering the log-linear nature of the assay.

AU are more useful quantitative measurements of bacteriocin activity. These are usually determined by assaying twofold serial dilutions of the sample. The reciprocal of the highest dilution producing inhibition becomes the number of AU. This can be divided by the sample volume to yield AU per milliliter. The arbitrary nature of this system for measuring activity cannot be overemphasized. By virtue of the twofold dilutions, the assay is insensitive to activity differences that are less than twofold. The assumption of linearity with volume used to calculate AU per milliliter is rarely verified. The choice of indicator organism, assay technique, length of incubation time, assay medium, etc., is so idiosyncratic as to make it virtually impossible to compare the arbitrary units used by different investigators. There is no easy solution to this problem. However, by assaying a known concentration of nisin using the same experimental conditions, an approximation of the relationship between AU and equivalent nisin IU can be generated.

Some investigators use conceptually different assays to generate an "arbitrary unit" unrelated to those described above. For example, the ability to reduce growth rate by 50% or decrease viability by 50% can be a measure of bacteriocin activity. These assays do not require the 12- to 24-h incubation period of the diffusion assays, but nonetheless are not widely used.

Bacteriocin Applications in Foods

There are several distinct applications for the use of bacteriocins in foods. Bacteriocins can be added directly to the food to inhibit spoilage or pathogenic organisms. Only nisin is commercially available for addition in pure form, but the efficacy of pediocin addition has also been demonstrated. Spoilage or pathogenic organisms can also be inhibited by bacteriocinogenic cultures added to the food. Bacteriocinogenic cultures can be added to nonfermented foods or used as starter cultures in fermented foods to improve safety and quality. The final application for bacteriocins is not related to spoilage or safety, but rather to improving the quality of fermented foods. The use of defined starter cultures offers many benefits, such as improved quality and consistency, when compared to the use of indigenous bacteria as inocula. However, unless the indigenous microbiota can be inactivated, it usually predominates over the defined inocula. Because of this, the benefits of defined starter cultures are most pronounced in the dairy industry, where the indigenous microbiota is inactivated during pasteurization. The use of bacteriocin-producing starter cultures in fermented meats and vegetables might inactivate the indigenous microbiota and allow the use of defined starter cultures in these products.

Addition of Nisin

Nisin is added to milk, cheese, and dairy products, a variety of canned foods, mayonnaise, and baby foods throughout the world (72, 93). It is designated GRAS (43) as an antibotulinal agent in certain cheese spreads and is used commercially as an antimastitis teat dip (138). Nisin has potential as a treatment for ulcers, in personal hygiene, and as a general sanitizing agent. When nisin is absorbed onto surfaces, it inhibits listerial growth (28) and prevents biofilm formation (13). Nisin also sensitizes spores to heat, so that the thermal treatment can be reduced (72). This application is not approved in the United States.

Many nisin applications target *C. botulinum*. The spores of this organism are much less nisin sensitive than are *L. monocytogenes* vegetative cells. While as little as 200 IU of nisin per ml can reduce *L. monocytogenes* viability by 6 logs, concentrations of up to 10,000 IU/ml are required to obtain similar results with botulinal spores (109).

Temperature is the major determinant of nisin's inhibitory action. Nisin is much less effective at elevated temperatures than at refrigeration temperatures against *C. botulinum* 56A spores in a model food system (128). Eventually, growth in this system occurs when nisin activity falls below some threshold level. The threshold nisin levels are lower at decreasing temperatures. For example, botulinal growth occurred when residual nisin concentrations fell below 154 IU/ml at 35°C, but not until nisin levels were <12 IU/ml at 15°C. Nisin also inhibits listeria and staphylococcal vegetative cells better at refrigerated temperatures than at elevated temperatures (23).

Many other factors influence nisin's sporostatic efficacy (137). In the studies used to support the GRAS affirmation of nisin in pasteurized process cheese, between 500 and 2,000 IU of nisin per ml inhibited botulinal spore outgrowth by 50% in broth, but levels of up to 10,000 IU/ml were ineffective in cooked-meat medium (136). Nisin at 100 to 250 ppm allows some salt reduction or increased moisture levels in pasteurized process cheese spreads without elevating the risk of botulism (141). Specific phospholipids also decrease nisin's antibotulinal efficacy (128), and butterfat decreases nisin's ability to inhibit *Staphylococcus aureus* (79).

In most applications, nisin serves as one part of a multiple-barrier inhibitory system. Nisin may be a useful adjunct to modified atmosphere storage. Nisin increases the shelf life and delays toxin production by type E botulinal strains in fresh fish packaged in a carbon dioxide atmosphere. However, toxin can sometimes be detected before samples are obviously spoiled (152). The combination of nisin and modified atmosphere to prevent *L. monocytogenes* growth in pork is more effective than either used alone (42). Nisin added at 5 mg/liter to liquid whole egg prior to pasteurization extends its refrigerated shelf life from 6 to 11 days to 17 to 20 days (32).

While most of the research has been directed against botulinal spores, nisin's inhibition of spores from other bacterial species also depends on many factors. Salt antagonizes the action of nisin against *Bacillus licheniformis* spores (9). The ability of nisin to inhibit thermally stressed bacillus spores is influenced by the time-temperature combination used to affect the thermal stress, the subsequent incubation temperature, the pH, and even the type of acidulant used (116). In general, nisin is more effective at lower temperatures, against lower spore loads, and under acidic conditions.

Addition of Pediocin

While inactive against spores, pediocins inhibit *L. monocytogenes* vegetative cells. European patents cover the use of Pediocin PA-1 in the form of a dried powder or culture liquid to extend the shelf life of salads and salad dressing (58) and as an antilisterial agent in foods such as cream, cottage cheese, meats, and salads (119, 160).

Pediocins are more effective than nisin in meat and are even more effective in dairy products. Dipping meat in 5,000 AU of crude pediocin PA-1 per ml decreases the viability of attached *L. monocytogenes* 100- to 1,000-fold. Pretreating meat with pediocin reduces subsequent *L. monocytogenes* attachment (114). Pediocin AcH at 1,350 AU/ml reduces listeriae in ground beef, sausage, and other products by between 1 and 7 log cycles. Pediocin AcH is more effective at 4°C than at 25°C against *L. monocytogenes* in hot dog exudate. Emulsifiers such as Tween 80, or the entrapment of the pediocin in multilamellar vesicles, increases pediocin effectiveness in fatty foods (29–31). In most cases, the bacteriocin rapidly reduces listeria viability and delays growth of the survivors.

Addition of Bacteriocin-Producing Bacteria to Nonfermented Foods

Many successful antilisterial applications of bacteriocins add the bacteriocinogenic culture rather than the pure bacteriocin. In wieners held at 4°C, *L. monocytogenes* grows after a 20- to 30-day lag period and increases from 10^4 to 10^6 by the end of 60 days (10). At 10^7 CFU/g, both Bac^+ and derivative strains of *Pediococcus acidilactici* inhibit *L. monocytogenes* growth in hot dogs for 60 days. The inhibition by the Bac^- strain was not due to acidification or production of hydrogen peroxide. The control wieners had pH values similar to those treated with Bac^+ and Bac^- pediococci, and results under anaerobic conditions were similar to those under aerobic conditions.

However, at low inoculum levels, the Bac$^+$ strain, but not its Bac$^-$ derivative, extended the lag period of listeria cells. The degree of inhibition increases with decreasing temperatures and, in this case, is greater under anaerobic conditions than aerobic conditions.

Lactobacillus bavaricus MN (which produces bavaricin MN) inhibits listerial growth in model gravy at 4°C, even in the absence of a fermentable carbohydrate (168). The addition of a fermentable carbohydrate, reduction of incubation temperatures, and increased lactobacillus/listeria inoculation ratios all increase the degree of inhibition. These variables also influence the success of this preservation technology in *sous vide* beef cubes (167). Under the most favorable conditions (4°C, high inoculation ratio, and beef packed in gravy which contains a fermentable carbohydrate), *Lb. bavaricus* causes *L. monocytogenes* viability to decline 10-fold over 6 weeks at 4°C. In the least favorable conditions (10°C, low inoculation ratio, beef cubes without gravy), listerial growth is inhibited by the bavaricin for 1 week, but increases 100-fold in the beef without the *Lb. bavaricus*.

Carnobacterium piscicola LK5 is also more effective against *L. monocytogenes* at 5°C than at 19°C in UHT milk, dog food made from beef, pasteurized crabmeat, creamed corn, and wieners (16). *C. piscicola* LK5 inhibits *L. monocytogenes* Scott A, even when the listeriae are present at levels 100-fold higher than the *C. piscicola*.

Use of Bacteriocin-Producing Starter Cultures to Improve Safety of Fermented Foods

If a food is going to be fermented by LAB anyway, the use of a bacteriocin-producing starter culture can provide added value to the product. For example, the presence of a nisin producer among the strains used to make cheddar cheese provides enough nisin to increase the shelf life of pasteurized processed cheese made from it from 14 to 87 days at 22°C (126).

Bacteriocinogenic pediococci appear especially effective in fermented meats. In situ pediocin production by *P. acidilactici* PAC 1.0 during the manufacture of fermented dry sausage reduces *L. monocytogenes* viability >10-fold relative to the acid-induced decrease caused by nonbacteriocinogenic control (46). When wild-type Bac$^+$ *P. acidilactici* H is used to ferment summer sausage, 5,000 AU of pediocin is produced per g, reducing *L. monocytogenes* viability by 3.4 log units (96).

Use of Bacteriocinogenic Starter Cultures to Direct the Fermentation of Fermented Foods

The use of undefined indigenous bacteria to ferment foods compromises product quality, makes true process control difficult, and introduces an uncontrolled variable in the manufacturing of fermented foods. These problems can be overcome by the use of bacteriocinogenic starter cultures. Their ability to outgrow the indigenous microbiota (67, 121, 122, 165) and therefore facilitate the use of defined starter cultures in unpasteurized foods is an extremely important and promising application.

There are several novel applications of nisin or nisin-producing strains in a variety of fermented foods. The use of a nisin-producing *Lc. lactis* paired with nisin-resistant *Leuconostoc mesenteroides* in sauerkraut fermentations retards the growth of *Lactobacillus plantarum* (67). This allows the *Ln. mesenteroides* to establish itself and results in a higher-quality product. The LAB found in wine are very sensitive to nisin and can be inhibited without inhibiting the yeast or influencing the taste (121, 122). At 100 IU/ml, nisin inhibits the bacteria which, by their malolactic fermentation, spoil wine. If the malolactic fermentation is desired, nisin-resistant *Leuconostoc oenos* can be added with nisin. This promotes the malolactic fermentation and suppresses the indigenous LAB (27).

Genetics of LAB Bacteriocins

Location of Bacteriocin Genes

The genetic information encoding bacteriocin production and immunity is located on plasmids, on the chromosome, or both. To show that bacteriocin production is plasmid mediated, researchers must provide (i) phenotypic evidence (e.g., the spontaneous loss of the Bac$^+$ phenotype), (ii) physical evidence (e.g., the isolation of a plasmid from the Bac$^+$ strain which is missing in Bac$^-$ strains) and (iii) genetic confirmation (e.g., the introduction of the plasmid isolated from the Bac$^+$ strain into a Bac$^-$ strain reestablishes the Bac$^+$ phenotype).

Research on lactococcins (86) provides an example of how a group of bacteriocins was determined to be plasmid mediated. Geis et al. (53) screened 280 lactococcal strains for bacteriocin production against four indicator strains. On the basis of secretion into liquid medium, sensitivity to proteolytic enzymes, and precipitation by ammonium sulfate, 16 of these strains were classified as producing bacteriocin-like compounds. Some of these Bac$^+$ strains easily lost their ability to produce bacteriocin and became sensitive to their own bacteriocin, thus establishing the instability of the Bac$^+$ phenotype. Thirteen of these Bac$^+$ strains were tested for their ability to transfer the Bac$^+$ trait to plasmid-free (Bac$^-$) recipient cells; four Bac$^+$ transconjugants were detected (113). Plasmid analysis of these transconjugants revealed the presence of a 60-kb or a 113-kb plasmid in three of the four Bac⁺ transconjugants. Additional small plasmids were observed in some isolates. Curing experiments of these transconjugants resulted in Bac⁻ variants which had lost their 60- or 113-kb plasmid. Restriction endonuclease analysis showed that two 60-kb plasmids isolated from *Lc. lactis* subsp. *cremoris* strains 9B4 and 4G6 were almost identical. Genetic analysis of the 60-kb plasmid from the 9B4 strain revealed that it carried the genes for three different bacteriocins, lactococcins A, B, and M (155, 157). In addition, pediocin A, pediocin PA-1, sakacin A, lactocin S, and the carnobacteriocins A and B are plasmid encoded.

Nisin production was initially thought to be plasmid mediated. Frequent spontaneous loss of the Nis^+ phenotype was observed, and conjugal transfer of the Nis^+ phenotype to Nis^- was reported (38, 47, 59, 143). However, no physical evidence linking the Nis^+ phenotype to a distinctive plasmid was obtained. In addition, attempts to transform Nis^- strains with plasmid DNA isolated from Nis^+ strains did not result in Nis^+ transformants, although the plasmid pool was successfully transferred (143). By means of nucleic acid hybridization techniques, the nisin structural gene was finally located on the chromosome of *Lc. lactis* (17, 37, 123). The nisin gene resides within a 70-kb conjugative transposon and is genetically linked to the genes encoding sucrose fermentation. The integration of this transposable element into the chromosome of a Nis^- strain was observed after conjugation by probing the digested total DNA of transconjugants for different regions of the nisin-coding transposon (71). Other chromosomally encoded LAB bacteriocins are helveticin J (77) and lactacin B (7).

Organization of Bacteriocin Operons: a Generic Operon

The DNA sequence or amino acid composition of most bacteriocins is unknown. However, a general picture of the genetic organization of bacteriocin genes is emerging. The structural genes for many bacteriocins appear to be located in an operon-like structure (85). The organization of a generic operon is shown in Fig. 30.3. The operon is organized so that it clusters the genes containing the structural information together with an array of genes involved in immunity, maturation, processing, and export of the bacteriocin molecules as well as genes encoding products involved in the regulation of bacteriocin biosynthesis. Not all operons contain all of the genes depicted, nor is the organization of the genes in each operon identical, but they share many similarities.

The *structural gene* usually codes for a prepeptide which comprises the precursor of the mature bacteriocin preceded by an N-terminal extension or "leader sequence." The secondary structure of this N-terminal extension is predicted to be an α-helix that is cleaved

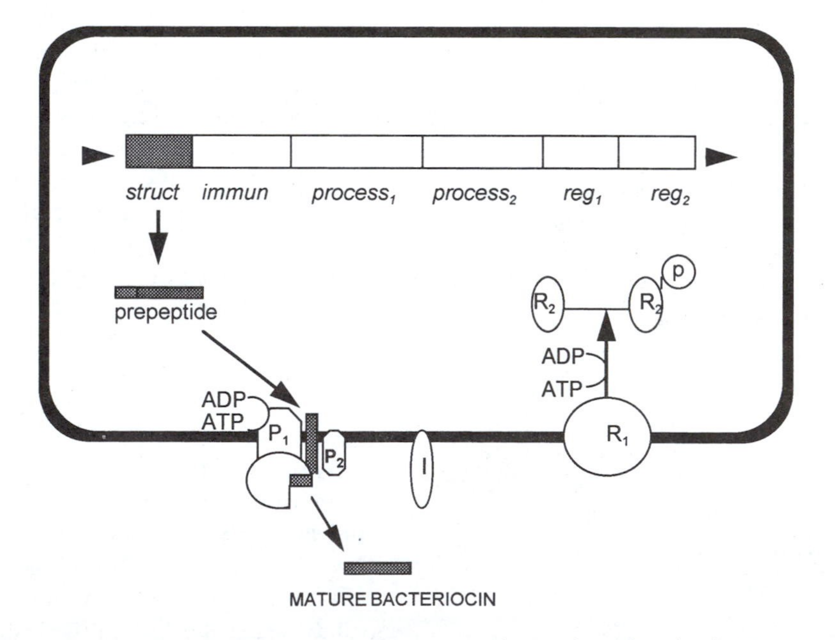

Figure 30.3 A generic bacteriocin operon. The structural gene (*struct*) codes for a prepropeptide which is modified and excreted by the processing gene products (P_1 and P_2) and may be regulated by a signal transduction pathway containing proteins R_1 and R_2 coded for by reg_1 and reg_2. The immunity gene (*immun*) codes for a protein (I) that confers immunity against the specific bacteriocin being produced.

during the maturation or export process. Class II bacteriocins encode a prepeptide containing a consensus sequence Gly^{-2}-Gly^{-1} at the cleavage site and present strong homology in their hydrophobicity profile. The role of the N-terminal extension is still unknown. It may be required for recognition by the maturation or export machinery. In this regard, most bacteriocin leader sequences, except for divergicin A (169), do not exhibit characteristics of *sec*-dependent export proteins. Rather, they may be exported by a *sec*-independent pathway involving a transport protein encoded within the operon (see below). Alternatively, the N-terminal extension may play a role in neutralizing bacteriocin activity within the cell to protect the producer strain.

The structural genes of several LAB bacteriocins have been described and reviewed (74, 85, 132). These include the structural genes coding for nisin (*nisA*), pediocin PA-1 (*pedA*), lactococcins A and B (*lcnA, lcnB*), sakacin A (*sakA*), leucocin A (*leuA*), lactacin F (*lafA, lafX*), and lactococcin M (*lcnM, lcnN*).

The *immunity gene* confers on Bac^+ cells immunity to their own bacteriocin. Immunity is specific and seems to be coordinated with bacteriocin production. Several immunity genes have been sequenced, and their ability to confer immunity to the producer strains has been proven (69, 89, 100, 156, 161). In the case of nisin, the *nisI* gene product is predicted to be an extracellular lipoprotein which, by anchoring to the membrane through its lipid moiety, confers immunity to the producer cells. Additional genes (*nisE, nisF,* and *nisG*) thought to be involved in immunity have been described in the nisin gene cluster (139).

The *processing and export genes* encode at least two proteins that insure that the mature bacteriocin is formed and exported from the cytoplasm. One of these proteins is an ATP-binding cassette translocator of the hemolysin B (HlyB) subfamily. These translocators utilize ATP hydrolysis as an energy source for protein translocation and are characterized by the presence of two domains: a membrane-spanning domain and a cytoplasmic ATP-binding domain (44). Examples of genes encoding ATP-binding transporter proteins in LAB bacteriocins are *nisT* (nisin), *lcnC* (lactococcin A), and *pedD* (pediocin PA-1) (40, 100, 148). The second protein of the secretion apparatus is a structural homolog of hemolysin D (HlyD), an accessory protein which may facilitate transport. The HlyD homologous protein does not seem to be present in the nisin gene cluster.

Some of the ABC transporters found in bacteria that produce class II bacteriocins have similar N-terminal amino acid sequences. This domain may contain the proteolytic cleavage site for the N-terminal extension of these bacteriocins. In contrast, the nisin gene cluster contains the *nisP* gene. Its gene product, a serine protease, is thought to be involved in the cleavage of nisin's N-terminal extension. In addition, the nisin operon contains the genes *nisB* and *nisC,* which encode two membrane-associated proteins involved in catalyzing the posttranslational modifications of the precursor molecules (40, 161).

The *regulatory genes* encode proteins homologous to proteins of the two-component regulatory systems. In this signal transduction pathway, a histidine kinase located in the membrane (R_1) senses an external signal and transduces it to the cell's interior by phosphorylating a second, cytoplasmic protein (R_2) referred to as a "response regulator." The phosphorylation of the response regulator activates the transcription of the bacteriocins' biosynthetic genes. Such regulatory genes occur in the nisin operon (*nisKR*) (41, 161) and have been postulated for the sakacin A operon (*sakKR*) (6), the carnobacteriocins BM1 and B2 (120), and the plantaricin operon (*plnBCD*) (35). The bacteriocin molecule itself may be the signal molecule, playing an autoinductive role in a cell-to-cell communication pathway (33).

Genetic "Modification" of Bacteriocins

Natural variants of bacteriocins and advances in protein engineering provide insight into the role of specific amino acids in the biological activity of bacteriocins. This information can be used for rational development of bacteriocins with superior activity. The two natural nisin variants, nisin A and nisin Z, differ by one amino acid (His or Asn at position 27, respectively) but otherwise share the same structure. Nisin Z is more soluble at neutral pH values than nisin A (34). The conserved presence of modified amino residues in lantibiotics may play an important role in biological activity. However, the evidence is not conclusive: (i) while the substitution of Dha-5 (dehydroalanine 5) → Dhb (dehydrobutyrine) in nisin Z decreases activity 2- to 10-fold (90) and the exchange of Dha-5 → Ala in subtilin leads to the loss of inhibitory activity against outgrowth of spores, the latter mutation (both in nisin and subtilin) does not decrease activity against vegetative cells (36, 95); (ii) the exchange of Dhb-2 → Dha in nisin increases biological activity (52), but deletion of the Dha-33 yields a molecule with activity similar to that of the wild-type nisin molecule (19). Rather, it seems that the role of dehydro residue depends on whether the action of the bacteriocin is sporicidal or bactericidal. The unusual amino acids play an important role in maintaining structural elements of the lantibiotics (132). Mutations which disrupt these structures are prone to affect the antimicrobial activity of bacteriocins. Nisin is a flexible molecule in solution except for those regions constricted by the five thioether-closed rings. The nisin molecule may be regarded as a two-fragment peptide connected by a hinge region at amino acid positions 20 to 22. The N-terminal fragment is mainly hydrophobic. The C-terminal fragment seems to adopt an α-helix configuration in membrane-like environments (162). Mutation of Asn-21/Met-22 → Pro reduces the flexibility of the molecule and decreases biological activity (52). Mutations which lead to the opening of the first (destruction of Dha-5) or third (Met-17 → Lys) rings also decrease antimicrobial activity. On the other hand, Dha-5 in subtilin seems to be essential for sporicidal activity (95). This dehydro group may act as a Michael acceptor towards sulfhydryl groups in the envelope of germinated spores.

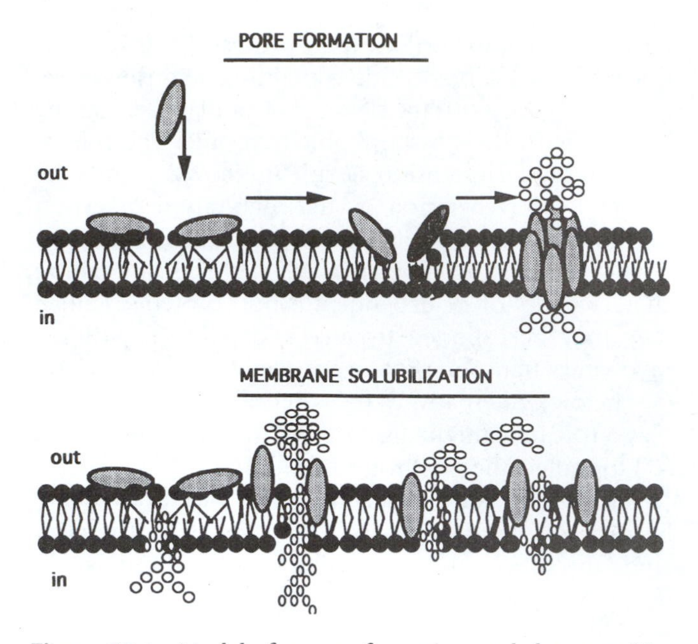

Figure 30.4 Models for pore formation and detergent-like mechanisms of bacteriocin action.

Mechanism of Action against Vegetative Cells

LAB bacteriocins act at the cell membrane. They disrupt the integrity of the cytoplasmic membrane, thus increasing its permeability to small compounds. The addition of bacteriocins to vegetative cells results in a rapid and nonspecific efflux of preaccumulated ions, amino acids, and, in some cases (nisin, lactostrepcin 5) but not others (pediocin PA-1), ATP molecules (20, 21, 98, 129, 158, 164, 166, 170). This increased flux of compounds across the membrane rapidly dissipates chemical and electrical gradients across the membrane. The proton motive force (PMF), an electrochemical gradient which serves as the major driving force of many vital energy-dependent processes, is dissipated within minutes of bacteriocin addition (14, 15). The chemical gradient (ΔpH) component of the PMF is dissipated faster than the electrical gradient (ΔΨ) component, probably as a result of rapid influx of protons into the cytoplasm. Bacteriocin-treated cells have decreased intracellular ATP levels. The loss of ATP can be caused by ATP efflux or ATP hydrolysis. While the efflux mechanism may be a direct result of membrane disruption, hydrolysis could result from a shift in the ATP equilibrium due to P_i efflux or as a futile attempt of the cell to regenerate the PMF (1, 166). Ultimately, these changes in permeability render the cell unable to protect its cytoplasm from the environment. This leads to cell inhibition and possibly death.

Given the diversity in the biochemical attributes of the LAB bacteriocin molecules and genetic determinants, it is surprising that they all act by the common mechanism of membrane permeabilization. A structural motif similar to other antimicrobial peptides occurring in nature (see below) may explain their interaction with membranes (115). In this regard, most bacteriocins with known sequences are amphiphilic cationic peptides showing α-helix, β-sheet, or, in the case of lantibiotics, screwlike secondary structures. Based on the amphiphilic characteristics of bacteriocins, there are at least two different mechanisms which may explain their membrane-permeabilization action (Fig. 30.4). Bacteriocins may act by a multiple-step poration complex in which bacteriocin monomers bind, insert, and oligomerize in the cytoplasmic membrane to form a pore with the hydrophilic residues facing inward and the hydrophobic ones facing outward. (The small size of bacteriocins makes it is unlikely that one molecule could form a pore.) Alternatively, bacteriocins may disrupt the membrane integrity by a detergent-like membrane solubilization action. The action of nisin, pediocin PA-1, lactococcins A and B, and lactacin F in vivo is concentration and time dependent, supporting a poration complex mechanism (1, 21, 156, 164, 166). A generalized solubilization mechanism would result in an all-or-none lysis of the cells, with a sudden collapse of the bioenergetic parameters and no saturation kinetics. Further evidence supporting the mechanism of pore formation is obtained from in vitro observations. Studies in black-lipid membranes suggest that nisin, Pep 5, and other lantibiotics form transient, potential-dependent pores with a predicted diameter of 0.2 to 2 nm and lifetimes of from milliseconds to seconds (131). Moreover, the addition of bacteriocins to membrane vesicles or liposomes loaded with different-size probes does not always result in leakage of the probe. Rather, leakage depends on both the size of the probe and the amount of bacteriocin added (21, 50).

Bacteriocins act similarly in that they breach the permeability barrier of the cytoplasmic membrane. However, they may require different conditions in the membrane target to establish a successful interaction. The lantibiotic bacteriocins act on energized membranes and do not require receptor proteins to exert their action (49, 88). Alternatively, most non-lantibiotic bacteriocins act on nonenergized membranes but may require a membrane receptor protein (21, 156, 164). The activity of some bacteriocins (including lactacin F and lactococcins M and G) occurs through the action of two peptides. These two-component bacteriocins are postulated to form a pore in the cytoplasmic membrane (85, 86).

Mechanism of Action against Spores

Information about the mechanisms of bacteriocin action against spores deals primarily with nisin. Spore germination and outgrowth is a multistep process detailed elsewhere in this volume. Nisin allows spores to germinate but inhibits outgrowth of the preemergent spore. Heat resistance and refractility under phase microscopy are lost at this point. The growth of vegetative botulinal cells is inhibited at much lower nisin concentrations than those required to inhibit outgrowth of spores. *Bacillus* species whose spore coats are opened by mechanical pressure are much more nisin sensitive than species whose spore coats are opened by lysis (93). At a molecular level, nisin modifies the sulfhydryl groups in the envelopes of germinated spores (111), presumably because the dehydro residues of nisin act as electron acceptors (61, 62).

Similarity to Other Antimicrobial Proteins

The antimicrobial proteins of LAB are not unique in structure or function. Many very different organisms produce similar "cytolytic pore-forming proteins" (115) including the colicins, cecropins, mellittins, magainins, and defensins. In many cases, these have been better studied than the bacteriocins produced by LAB. Their widespread occurrence attests to the "naturalness" of bacteriocins; nature is full of similar antimicrobial proteins.

A brief review of the properties of these other antimicrobial proteins provides an important opportunity to compare and contrast these with the bacteriocins produced by LAB. Colicins made by *Escherichia coli* have molecular masses of 35 to 70 kDa and form ion-permeable channels in the cytoplasmic membrane of sensitive bacteria (87). Cecropins are small (~25 amino acids) peptides of the moth *Hyalophora cecropia* which are active against gram-negative and gram-positive bacteria (4). Cecropins form voltage-dependent ion channels in planar lipid membranes (22) and contain amphiphilic helices consistent with the structural requirements for membrane insertion (144). Mellittin is the major toxic component of bee venom. It contains 26 amino acids, with the N-terminal region rich in hydrophobic residues and the C-terminal end predominantly hydrophilic (153). Mellittin associates with lipid bilayers, causing membrane lysis, permeabilization, and inhibition of membrane-bound enzyme systems (63). Magainins are antimicrobial peptides of the frog *Xenopus laevis* and contain approximately 23 amino acids which exert their microbicidal action by permeabilizing sensitive membranes by generalized membrane destabilization, the formation of channels, or both mechanisms simultaneously (60). Defensins are small antimicrobial peptides found in cytoplasmic granules of phagocytes isolated from humans, rabbits, guinea pigs, and rats (39, 91). These small (29 to 34 amino acids) peptides have similar amino acid sequences, three-dimensional structures, and a highly conserved cysteine motif consisting of six cysteine residues forming three disulfide bonds (91). Defensins also exert their antimicrobial activity by permeabilization or disruption of cell membranes and play a major role in the oxygen-independent killing of bacteria, fungi, and certain viruses. Clearly, antimicrobial proteins are common in the biosphere and are used by many organisms to protect themselves.

Unresolved Issues

Bacteriocin Resistance

While nisin-resistant starter cultures can be used to great advantage, the appearance of nisin-resistant bacteria can undermine the use of nisin as an inhibitor of spoilage and pathogenic bacteria. Genetically stable nisin-resistant *L. monocytogenes* can be isolated at a frequency of 10^{-6} (66, 106). Nisin-resistant isolates are generated from vegetative cells of *S. aureus, B. licheniformis, B. subtilis, Bacillus cereus* (106), and *C. botulinum* (102) at similar frequencies. The use of multiple bacteriocins to overcome this problem has been suggested (43, 65), but will be effective only if resistance to each bacteriocin is conferred by different mechanisms. *L. monocytogenes* resistant to mesenterocin 52, curvaticin 13, or plantaricin C19 can be isolated at frequencies of 10^{-3} to 10^{-8}. Strains resistant to any of these bacteriocins are cross-resistant

Table 30.2 Parallel mechanisms of antibiotic and bacteriocin resistance[a]

Mechanism of resistance	Specificity	Example	
		Antibiotic	Bacteriocins
Destruction	Specific or general	β-Lactamases	Protease, specific "bacteriocinase," nisinase (93)
Modification	Specific	Methylation of aminoglycosides	Dehydroreductase (76)
Altered receptors	Specific	Penicillin-binding proteins	Probable, but not reported to date
Membrane composition	General	Altered membranes in resistant *E. coli* and bacilli	Demonstrated for nisin resistance (106)

[a]Adapted from reference 110.

to the other two. When all three bacteriocins are used together, resistant strains are isolated at frequencies similar to those obtained when the bacteriocins are used alone (125). Isolates resistant to all three of these bacteriocins, however, are not resistant to nisin.

Some mechanisms through which bacteria can become bacteriocin resistant parallel mechanisms of antibiotic resistance (110) (Table 30.2). This comparison is made to provide a conceptual context for bacteriocin resistance and does not suggest that bacteriocin-resistant foodborne pathogens will be cross-resistant to antibiotics. Mechanisms characterized as "specific" have a biochemical specificity to a particular antimicrobial and cannot generate cross-resistance. General mechanisms of resistance, such as changes in membrane permeability, can be viewed as affecting the intrinsic resistance of the bacteria and might cause cross-resistance to other food preservatives. Membranes from nisin-resistant *L. monocytogenes* have more straight-chain fatty acids, resulting in a higher phase transition temperature (T_c) than that of the wild-type strain (106). Presumably, the lack of fluidity hinders nisin insertion into the membrane. Membrane fluidity plays an important role in listerial resistance to other antimicrobial agents (80). *L. monocytogenes* cells grown in the presence of $C_{14:0}$ or $C_{18:0}$ fatty acids have higher T_c and increased resistance to four common antimicrobial agents relative to cells grown in the presence of $C_{18:1}$, which have lower T_c and are more sensitive to other antimicrobial agents.

The study of bacteriocin resistance is in its infancy. Under conditions closer to those of actual applications, the frequency of bacteriocin resistance may be much lower than that obtained in laboratory media optimal for growth (66, 102). Nonetheless, it is prudent to conduct additional research on bacteriocin resistance so that it might be advantageously manipulated. "Resistance management" is already a component in other antimicrobial applications ranging from the use of antibiotics in hospitals to the use of *Bacillus thuringiensis* toxin as an agricultural insecticide.

Regulatory Status

The regulatory status of bacteriocins and bacteriocinogenic bacteria is unclear. LAB are GRAS (Generally Recognized as Safe) for the production of fermented foods. GRAS status, which is conferred by the U.S. Food and Drug Administration (FDA), is especially desirable because it allows a compound to be used in a specific application without additional regulatory approval. The linkage of GRAS status to a specific application is often overlooked. Thus, the GRAS status of LAB for the production of fermented foods does not automatically make LAB, or their metabolic products, GRAS for uses such as the preservation of foods which are not fermented. Nisin is the only bacteriocin that has GRAS affirmation. The 1988 GRAS affirmation for the use of nisin in pasteurized process cheese (45) was supported by toxicological data. This affirmation is the foundation for additional GRAS affirmations. The FDA has accepted for filing a request that nisin be affirmed as GRAS to restore the shelf life of reduced-fat pasteurized egg to a level comparable to that of conventional eggs. A subsequent request (*Federal Register* of August 17, 1994 [*Fed. Regist.* 59:42277–42278]) to extend the affirmation to conventional processed eggs was also accepted for filing. Both of these requests rely on the toxicological and safety data in the 1988 affirmation and are subject to its 10,000-IU/g (250-ppm) usage limit.

Bacteriocins produced by GRAS organisms are not automatically GRAS themselves. Bacteriocins that are not GRAS are regulated as food additives (105) and require premarket approval by the FDA. A food fermented by bacteriocin-producing starters can be used as an ingredient in a second food product. Its use *as an ingredient* might coincidentally extend shelf life of the product without necessitating preservative declarations. However, if the ingredient is added *for the purpose of extending shelf life,* the FDA would probably consider it an additive and require both premarket clearance and label declaration. There is no doubt that purified bacteriocins used as preservatives require premarket approval by the FDA.

GRAS status for bacteriocins can be based on documented use prior to 1958, a consensus of scientific opinion, or a formal GRAS affirmation from the FDA. The presence of bacteriocins in foods prior to 1958 might be inferred from the ease with which bacteriocinogenic bacteria are isolated from a variety of foods. This suggests that they are long-standing members of the natural microbiota of food. Furthermore, strains that produce nisin (75, 96, 127) and strains that produce pediocin PA-1 (11, 26, 51, 67) have been independently isolated from foods in different parts of the world. This demonstrates widespread occurrence of bacteriocinogenic bacteria in nature and suggests that bacteriocins have been consumed for decades. While these are reasonable arguments, the FDA might nonetheless require isolation of both the bacteriocinogenic organism and its bacteriocin from food produced prior to 1958 before accepting them as GRAS under the prior use clause.

The international regulation of bacteriocins is complex and beyond the scope of this book. Nisin is the only bacteriocin approved internationally for use in foods. The Joint Food and Agriculture/World Health Organization accepted nisin as a food additive in 1969 and set maximum intake levels as 33,000 IU/kg of body weight. Based on this, many countries allow nisin in a variety of products, sometimes with no restrictions as to maximum level. In addition to milk, cheese, and dairy products, these uses include canned tomatoes, canned soups, other canned vegetables, mayonnaise, and baby food (159).

PROBIOTIC BACTERIA

The widely held belief that consumption of LAB is beneficial to health first appeared in print in 1907. In his book *Prolongation of Life,* the Nobel Laureate Elie Metchnikoff attributed the long life spans of people in the Balkans to the bacteria in the yogurt they consumed. He hypothesized that consumption of the bacteria in the yogurt alters the balance of the intestinal microbiota and suppresses putrefactive bacteria. Current usage of the term "probiotic" dates to 1974, when it was used to describe the beneficial effects of microbial feed supplements in animal husbandry (117), an area that will not be discussed here. In 1989, Fuller proposed a narrower definition emphasizing the use of viable bacteria. This chapter adopts his view and defines therapeutic or probiotic bacteria as those viable organisms which, when consumed, act in the gastrointestinal tract to the benefit of the host organism. While it is generally accepted that fermentations can improve digestibility, generate free amino acids, and produce vitamins and cofactors in the food substrate, there is a dearth of hard data for claims that probiotic bacteria promote "intestinal well-being" or are otherwise "health-promoting" in humans (54). There are many reports of the probiotic effect of LAB. However, because there are so few well-controlled studies in humans, a major review of this area (99) was forced to conclude that "There is . . . no proven medical indication of lactic acid bacteria (LAB) for therapy or immunomodulation in man."

The Human Gastrointestinal Tract Is a Microbial Ecosystem

The complex nature of the human gastrointestinal tract makes research on probiotic bacteria difficult. Over 400 species of bacteria colonize the average human gastrointestinal tract. Thirty or 40 of these species make up 99% of the bacteria in healthy humans. It is existentially interesting to note that we contain as many bacterial cells as humans cells ($\approx 10^{13}$ to 10^{14}) (135). The growth rate of these organisms in the gastrointestinal tract is 1 to 4 doublings per day (140). The variety of organisms and environmental conditions present in the gastrointestinal tract requires that it be studied as an ecosystem. Humans, like other ecosystems, experience a progression of inhabitants. We are born with no indigenous microbiota, but are colonized by bifidobacteria, when breast fed, or lactobacilli when fed with cow milk. Strict anaerobes are introduced with the consumption of solid food (135). Once established, the composition and distribution of indigenous organisms are remarkably resistant to change. In ecological terms, the microbiota of the adult human gastrointestinal tract is a climax community (i.e., an ecosystem where the progression of species has stopped and a characteristic biota is stably maintained) where all available niches are filled. The gross metabolic activity of the resident microbiota is, however, quite variable and strongly influenced by the host physiology and by everything that affects host physiology. Boddy and Wimpenny (12) provide a detailed review of how food microbiology can be viewed in ecological terms sensu strictu.

The human gastrointestinal tract is about 350 cm long from the mouth to the anal orifice and is divided into three major sections, each having its own distinct microbiota (135, 140). The stomach is highly acidic (pH 1 to 3) and is populated primarily by $<10^3$ CFU of aerobic gram-positive organisms per g. The intestinal pH (6.8 to 8.6) is highly favorable to microbial growth. The small intestine is a transitional zone inhabited by 10^3 to 10^4 CFU of the genera *Lactobacillus, Bifidobacterium, Bac-*

teroides, and *Streptococcus* per g. Microbial growth in the large intestine is luxuriant. There are 10^{11} to 10^{12} CFU/g, with anaerobes outnumbering aerobes by 100- to 10,000-fold. The large intestine is populated by a varied and diverse microbial population. Bacteria from the genera *Bacteroides, Fusobacterium, Lactobacillus, Bifidobacterium,* and *Eubacterium* are present in large numbers. The *Lactobacillus acidophilus* complex of organisms is especially important. This group contains *Lb. acidophilus, Lb. crispatus, Lb. amylovorus, Lb. gasseri,* and *Lb. johnsonii. Streptococcus, Peptococcus, Peptostreptococcus,* and *Ruminococcus* species are frequently isolated from the large intestine, and the *Enterobacteriaceae* are present at relatively lower levels (18).

Cultures used in fermented products are usually chosen on the basis of some technological characteristics rather than their suitability for the gastrointestinal ecosystem and health-promoting attributes. For example, the *Lb. bulgaricus* and *Streptococcus thermophilus* traditionally used in yogurt manufacture have poor resistance to acid and bile salts (99). Conscious of this fact, many yogurt manufacturers now use *Lb. acidophilus* and *Bifidobacterium* spp. resistant to acid and bile salts.

The "normal" human biota contains autochthonous (indigenous) bacteria, which have colonized the host by adhering to the intestinal mucosa, and allochthonous (nonindigenous) bacteria from sources such as foods, that are just passing through (135). Dietary factors such as carnivorous or vegetarian diet, or even starvation, make surprisingly little difference in the gastrointestinal tract's microbiota but cause marked changes in enzyme activities (see below). Dietary supplements containing high levels of specific bacteria can establish these bacteria as members of the allochthonous segment of the normal biota. The theory which explains the steady-state maintenance of nonadherent bacteria in continuous culture vessels (107) can be applied to this situation. Like bacteria in other continuous bioreactors, nonadherent bacteria will be retained only if their specific growth rate (number of doublings per day) exceeds the dilution rate (flow rate ÷ volume) of the system. If they cannot grow fast enough, they wash out of the system. Assuming that contents of the gastrointestinal tract have a residence time of 48 h, and that the gastrointestinal tract has a nominal working volume of 8 liters, any organism that doubles less than twice per day will wash out. Organisms which double more frequently, or which are consumed with enough regularity to compensate for the cells washed out of the system, can be maintained in the gut without becoming part of the indigenous microbiota.

The complexity of the gastrointestinal ecosystem makes colonization difficult to study. Colonization is influenced by gastric acidity, bile salt concentration, peristalsis, digestive enzymes, and the immune response (81). Information about colonization cannot be inferred from studies which examine only fecal microbiota. Feces constitute an ecosystem distinct from the bacterial population that actually colonizes the mucosa. The fecal ecosystem is subject to dehydration and a myriad of enzyme activities; feces are 33 to 50% bacterial biomass on a dry weight basis (134). Colonization per se can only be determined by biopsy of the mucosa followed by culturing of the sample under stringent anaerobic conditions. In one well-controlled study (78), 19 different bacterial strains were fermented in oatmeal soup and fed to humans at doses of 10^8 per day for 10 days. Biopsies were done at 1 day prefeeding, 1 day after feeding started, and 1 day after the termination of the feeding. There were marked differences in the microbiota of some individual subjects, but when results were averaged across all subjects, the only statistically significant changes were a significant increase in the number of lactobacilli in the jejunum and a decreased number of anaerobes and gram-negative bacteria in the rectum. Two of the 19 strains ingested were isolated most frequently. Both were *Lb. plantarum,* one being originally isolated from sourdough bread.

Applications of LAB for Health-Related Purposes

Use in Treatment and Prevention of Lactose Intolerance and Diarrhea

The intestinal mucosa of people of Asian and African descent is deficient in β-galactosidase activity. This results in the inability to metabolize the lactose in dairy products and causes flatulence, pain, and diarrhea when lactose-containing products are consumed. Thus these populations are deprived of an important source of calcium and protein. The use of LAB as a dietary adjunct to ameliorate lactose intolerance is well documented (68, 99, 117). Active cultures are effective in this action whereas cultures killed by pasteurization are not. The mechanism by which LAB alleviate lactose intolerance is uncertain. The active bacteria may metabolize the unfermented lactose in the gut, or may stimulate lactase production by the intestinal mucosa. The ability to utilize the lactose in the fermented product is limited and does not confer the ability to consume other (nonfermented) lactose-containing foods.

While there is evidence that consuming LAB prevents and reduces the severity of diarrheal illnesses, the mechanism is again unknown. The symptomatic relief may be just an osmotic effect. This would explain why the consumption of LAB alleviates the stomach distress that sometimes accompanies antibiotic treatments. One well-controlled trial fed hospitalized children formula or formula supplemented with *Bifidobacterium bifida* at 10^8/g and *S. thermophilus* at 10^7/g (130). During a total of 4,447 patient days, 88% of the infants fed the formula augmented with these bacteria remained free of diarrhea, compared to 33% of the control group fed unsupplemented formula. Both groups were similar in all other clinical aspects (i.e., weight gain, length of hospitalization, etc.).

Reduction of Serum Cholesterol

Many LAB reduce the cholesterol content of laboratory media through assimilation or oxidation (55, 142). These laboratory results are difficult to relate to in vivo applications. The consumption of large quantities (680 to 5,000 ml/day) of fermented dairy products decreases serum cholesterol, but the physiologic impact of "normal" consumption is unknown (99). Indeed, the hypocholesterolemic activity of fermented milk is similar to that of unfermented milk (68). The lack of well-controlled clinical trials is problematic. In a study on the efficacy of commercially available lactobacillus tablets containing 2×10^6 CFU, their consumption four times a day had no effect on serum lipoprotein concentrations (92).

Prevention of Colointestinal Cancer

Most human cancers are triggered by chemicals or events in the environment (5). This makes them at least theoretically preventable. The development of breast, colon, and prostate cancers is associated with diets that are high in fat and low in fiber. Most dietary carcinogens are consumed as procarcinogens which require chemical modification to induce carcinogenesis (18). The chemical modifications are made by the enzyme activities of the gastrointestinal microbiota and, to a much smaller degree, by the enzymes of the intestinal mucosa. The gastrointestinal microbiota is a metabolic interface between humans and their diet, playing an important role in the carcinogenic process (64).

The enzymatic reactions of the gastrointestinal tract are classified as either procarcinogenic (phase I) reactions or anticarcinogenic (phase II) reactions (18). Both phase I and phase II enzymes help the body excrete compounds by rendering them less lipid soluble. Phase I reactions do this by hydrolysis, oxidation, or reduction of the parent compound. These reactions activate procarcinogens and are heavily influenced by diet. Cycasin, rutin, digoxin, and numerous plant glycosides are typical natural substrates for these reactions (57).

The gross activity of the colonic enzymes varies widely and is affected by many factors. For example, high-fat diets induce increased secretion of bile acids. The bile acids induce the microbial production of 7-α-dehydrolase. Western diets increase β-glucuronidase activity. The activities of phase I enzymes are markedly higher in conventional rats compared to germ-free rats, suggesting that the rat microbiota contributes to this procarcinogenic activity (18). However, some bacteria produce enzymes having phase II activity. Phase II reactions generally detoxify chemicals through acylation, methylation, glutathione conjugation, sulfate conjugation, or glucuronidation. Phase II reactions provide the major defense against electrophilic and free radical intermediates. While the enzymes of the intestinal mucosa contribute to these activities, the intestinal microbiota can have a major influence. Ingestion of bacteria which produce enzymes having increased phase II activities should decrease the risk of certain cancers. As previously noted, yogurt consumption decreases β-glucuronidase activity in the intestine, but the activity returns to normal levels when the yogurt is withdrawn from the diet (48).

CONCLUSIONS AND OUTLOOK FOR THE FUTURE

The industrial processing and preservation of foods did not start until the late 1800s. From this time until the 1940s, the food industry was driven by physical methods for food preservation. From the 1950s until the present, chemical methods of food preservation became increasingly important, gaining a status equal to engineering. The current advances in nonnutritive sweeteners, fat mimetics, and functional polysaccharides are a testament to the ascendance of food chemistry. In contrast, biological methods of food preservation are still in their infancy. The first reports of genetic recombination (25), transduction in LAB, and the importance of their plasmids occurred only in the 1970s (103, 104). The "shock wave" from the explosion of knowledge in molecular biology is just now hitting food microbiology. In the early 1970s, one could imagine a resurrected Pasteur entering a food microbiology laboratory and resuming his work after the briefest of orientations. This is not true of the 1990s. Our increasing reliance on the concepts and tools of molecular biology would leave Pasteur completely befuddled.

Table 30.3 Analogy between the use of insecticides in production agriculture and the use of antimicrobials for food safety[a]

Control of insects in crops	Control of bacteria in foods
Chemical pesticides	Chemical preservatives
Integrated pest management	Microbial competition
Bioinsecticides	Bacteriocins
Insecticidal plants	Antimicrobial foods

[a]Modified from reference 108.

Having made its mark in analytical food microbiology, it is only a matter of time before molecular biology provides new methods for ensuring food quality and safety. The biological methods of food preservation covered in this chapter mark only the crude beginning of this new era in the life of the food industry, the era of food biology. Controlled acidification is conceptually straightforward, but its successful application depends on a variety of product-specific factors. This has limited both its commercial use and academic interest in controlled acidification. The use of antimicrobial proteins, in one form or another, is sure to increase. In production agriculture, *B. thuringiensis* insecticidal proteins have been applied to plants for the last 20 years. The genes for this protein are now being cloned into the plant itself (8). By analogy (Table 30.3), the day may come when we can genetically engineer pathogen resistance into microbially sensitive foods.

The future use of probiotic bacteria in the arena of human health is more difficult to predict. The literature clearly suggests that *something* happens in the gastrointestinal tract when LAB are consumed in large numbers. Exactly *what* happens and *why* is difficult to determine. The answers to these questions may lie outside the scope of pure-culture microbiology. A new perspective of the gastrointestinal tract and its associated microbiota may be required if the probiotic effects of LAB are ever to be fully understood. This may require collaboration with ecologists, gastroenterologists, statisticians, and epidemiologists. Even then, the difficulty of obtaining the large number of human subjects required to obtain statistical significance may be insurmountable.

Acknowledgments

This is manuscript F10974-2-96 of the New Jersey Agricultural Experiment Station. Research in the authors' laboratory and preparation of this manuscript were supported by state appropriations, U.S. Hatch Act Funds, the U.S. Israel Bilateral Research and Development Fund (BARD, Project no. US-2113-92), and grants from the United States Department of Agriculture CSRS NRI Food Safety Program (no. 91-37201-6796 and 94-37201-0994).

References

1. **Abee, T., F. M. Rombouts, J. Hugenholtz, G. Guihard, and L. Letellier.** 1994. Mode of action of nisin Z against *Listeria monocytogenes* Scott A grown at high and low temperatures. *Appl. Environ. Microbiol.* **60:**1962–1968.
2. **Allison, G., C. Fremaux, C. Ahn, and T. R. Klaenhammer.** 1994. Expansion of bacteriocin activity and host range upon complementation of two peptides encoded within the lactacin F operon. *J. Bacteriol.* **176:**2235–2241.
3. **Al-Zoreky, N., J. W. Ayres, and W. Sandine.** 1991. Antimicrobial activity of Microgard(R) against food spoilage and pathogenic microorganisms. *J. Dairy Sci.* **74:**758–763.
4. **Andreu, D., and R. B. Merrifield.** 1985. N-terminal analogues of cecropin A: synthesis, antibacterial activity, and conformation properties. *Biochemistry* **24:**1683–1688.
5. **Armstrong, D., and R. Doll.** 1975. Environmental factors and cancer incidence and mortality in different countries, with special reference to dietary practices. *Int. J. Cancer* **15:**617–631.
6. **Axelsson, L., and A. Holck.** 1995. Molecular analysis of sakacin A gene cluster. Presented at *Workshop on the Bacteriocins of Lactic Acid Bacteria—Applications and Fundamentals.* Alberta, Canada, April 17–22, 1995.
7. **Barefoot, S. F., and T. R. Klaenhammer.** 1984. Purification and characterization of the *Lactobacillus acidophilus* bacteriocin lactacin B. *Antimicrob. Agents Chemother.* **26:**328–334.
8. **Barton, B., M. Miller, M. Maffitt, J. Kofron, S. Cannon, and P. Umbeck.** 1989. Development of insect resistant plants. *Dev. Ind. Microbiol.* **30:**195–202.
9. **Bell, R. G., and K. M. De Lacy.** 1985. The effect of nisin-sodium chloride interactions on the outgrowth of *Bacillus licheniformis* spores. *J. Appl. Bacteriol.* **59:**127–132.
10. **Berry, E. D., R. W. Hutkins, and R. Mandigo.** 1991. The use of bacteriocin producing *Pediococcus acidilactici* to control post processing *Listeria monocytogenes* contamination of frankfurters. *J. Food Prot.* **54:**681–686.
11. **Bhunia, A. K., and M. C. Johnson.** 1992. Monoclonal antibody-colony immunoblot method specific for isolation of *Pediococcus acidilactici* from foods and correlation with pediocin (bacteriocin) production. *Appl. Environ. Microbiol.* **58:**2315–2320.
12. **Boddy, L., and J. W. T. Wimpenny.** 1992. Ecological concepts in food microbiology. *J. Appl. Bacteriol.* **73:**23S-38S.
13. **Bower, C. K., J. McGuire, and M. A. Daeschel.** 1995. Suppression of *Listeria monocytogenes* colonization following adsorption of nisin onto silica surfaces. *Appl. Environ. Microbiol.* **61:**992–997.
14. **Bruno, M. E. C., A. Kaiser, and T. J. Montville.** 1992. Depletion of proton motive force by nisin in *Listeria monocytogenes* cells. *Appl. Environ. Microbiol.* **58:**2255–2259.
15. **Bruno, M. E. C., and T. J. Montville.** 1993. Common mechanistic action of bacteriocins from lactic acid bacteria. *Appl. Environ. Microbiol.* **59:**3003–3010.
16. **Buchanan, R. L., and L. A. Klawitter.** 1992. Effectiveness of *Carnobacterium piscicola* LK5 for controlling the growth

of *Listeria monocytogenes* Scott A in refrigerated foods. *J. Food Safety* **12**:217–224.

17. **Buchman,G. W., S. Banergee, and J. N. Hansen.** 1988. Structure, expression and evolution of a gene encoding the precursor of nisin, a small protein antibiotic. *J. Biol. Chem.* **263**:16260–16266.
18. **Chadwick, R. W., S. E. George, and L. D. Claxton.** 1992. Role of gastrointestinal mucosa and microflora in the bioactivation of dietary and environmental carcinogens. *Drug Metab. Rev.* **24**(4):425–492.
19. **Chan, W. C., B. W. Bycroft, L. Y. Lian, and G. C. Roberts.** 1989. Isolation and characterization of two degradation products derived from the peptide antibiotic nisin. *FEBS Lett.* **252**:29–36.
20. **Chen, Y., and T. J. Montville.** 1995. Efflux of ions and ATP depletion induced by pediocin PA-1 are concomitant with cell death in *Listeria monocytogenes* Scott A. *J. Appl. Bacteriol.* **79**:684–690.
21. **Chikindas, M. L., M. J. Garcia-Garcera, A. J. M. Driessen, A. M. Ledeboer, J. Nissen-Meyer, I. F. Nes, T. Abee, W. N. Konings, and G. Venema.** 1993. Pediocin PA-1, a bacteriocin from *Pediococcus acidilactici* PAC1.0, forms hydrophilic pores in the cytoplasmic membrane of target cells. *Appl. Environ. Microbiol.* **59**:3577–3584.
22. **Christensen, B., J. Fink, R. B. Merrifield, and D. Mauzerall.** 1988. Channel-forming properties of cecropins and related model compounds incorporated into planar lipid membranes. *Proc. Natl. Acad. Sci. USA* **85**:5072–5076.
23. **Chung, K. T., J. S. Dickson, and J. D. Crouse.** 1989. Effects of nisin on growth of bacteria attached to meat. *Appl. Environ. Microbiol.* **55**:1329–1333.
24. **Cociancich, S., M. Goyffon, F. Bontems, P. Bulet, F. Bouet, A. Menez, and J. Hoffman.** 1993. Purification and characterization of a scorpion defensin, a 4kDa antibacterial peptide presenting structural similarity with insect defensins and scorpion toxins. *Biochem. Biophys. Res. Commun.* **194**:17–22.
25. **Cohen, S. N., A. C. Y. Chang, H. W. Boyers, and R. B. Helling.** 1973. Construction of biologically functional bacterial plasmids *in vitro*. *Proc. Natl. Acad. Sci. USA* **70**:32–40.
26. **Daba, H., C. Lacroix, J. Huang, R. E. Simard, and L. Lemieux.** 1994. Simple method of purification and sequencing of a bacteriocin produced by *Pediococcus acidilactici* UL5. *J. Appl. Bacteriol.* **77**:682–698.
27. **Daeschel, M. A.** 1990. Controlling wine malolactic fermentation with nisin and nisin-resistant strains of *Leuconostoc oenos*. *Appl. Environ. Microbiol.* **51**:601–603.
28. **Daeschel, M. A., J. McGuire, and H. Al-Makhlafi.** 1992. Antimicrobial activity of nisin adsorbed to hydrophilic and hydrophobic silicon surfaces. *J. Food Prot.* **55**:731–735.
29. **Degnan, A. J., N. Buyong, and J. B. Luchansky.** 1993. Antilisterial activity of pediocin AcH in model food systems in the presence of an emulsifier or encapsulated within liposomes. *Int. J. Food Microbiol.* **18**:127–138.
30. **Degnan, A. J., and J. B. Luchansky.** 1992. Influence of beef tallow and muscle on the antilisterial activity of pediocin AcH and liposome-encapsulated pediocin AcH. *J. Food Prot.* **55**:552–554.
31. **Degnan, A. J., A. E. Yousef, and J. B. Luchansky.** 1992. Use of *Pediococcus acidilactici* to control *Listeria monocytogenes* in temperature-abused vacuum-packaged wieners. *J. Food Prot.* **55**:98–103.
32. **Delves-Broughton, J., G. C. Williams, and S. Williamson.** 1992. The use of the bacteriocin, nisin, as a preservative in pasteurized white egg. *Lett. Appl. Bacteriol.* **15**:133–136.
33. **de Vos, W. M.** 1995. Expression of bacteriocin genes: genetic and physiological control. Presented at *Workshop on the Bacteriocins of Lactic Acid Bacteria—Applications and Fundamentals*. Alberta, Canada, April 17–22, 1995.
34. **de Vos, W. M., J. W. M. Mulders, R. J. Siezen, J. Hugenhoetz, and O. P. Kuipers.** 1993. Properties of nisin Z and distribution of its gene, *nisZ*, in *Lactococcus lactis*. *Appl. Environ. Microbiol.* **59**:3683–3693.
35. **Diep, D. B., L. S. Havarstein, J. Nissen-Meyer, and I. F. Nes.** 1994. The gene encoding plantaricin A, a bacteriocin from *Lactobacillus plantarum* C11, is located on the same transcription unit as an *agr*-like regulatory system. *Appl. Environ. Microbiol.* **60**:160–166.
36. **Dodd, H. M., and M. J. Gasson.** 1994. Bacteriocins of lactic acid bacteria, p. 211–251. *In* M. J. Gasson and W. M. DeVos (ed.), *Genetics and Biotechnology of Lactic Acid Bacteria*. Blackie Academic & Professional, Glasgow.
37. **Dodd, H. M., N. Horn, and M. J. Gasson.** 1990. Analysis of the genetic determinant for the production of the peptide antibiotic nisin. *J. Gen. Microbiol.* **136**:555–556.
38. **Donkersloot, J. A., and J. Thompson.** 1990. Simultaneous loss of N5-carboxyethylornithine synthase, nisin production, and sucrose-fermenting ability by *Lactococcus lactis* K1. *J. Bacteriol.* **172**:4122–4126.
39. **Elsbach, P.** 1990. Antibiotics from within: antibacterials from human and animal sources. *Trends Biotechnol.* **8**:26–30.
40. **Engelke, G., Z. Gutowski-Eckel, M. Hammelmann, and K. D. Entian.** 1992. Biosynthesis of the lantibiotic nisin: genomic organization and membrane localization of the NisB protein. *Appl. Environ. Microbiol.* **58**:3730–3743.
41. **Engelke, G., Z. Gutowski-Eckel, P. Kiesau, K. Siegers, M. Hammelmann, and K. D. Entian.** 1994. Regulation of nisin biosynthesis and immunity in *Lactococcus lactis* 6F3. *Appl. Environ. Microbiol.* **60**:814–825.
42. **Fang, T. J., and L. W. Lin.** 1994. Inactivation of *Listeria monocytogenes* on raw pork treated with modified atmosphere packaging and nisin. *J. Food Drug Anal.* **2**(3):189–200.
43. **Farber, J. M.** 1993. Current research on *Listeria monocytogenes* in foods: an overview. *J. Food Prot.* **56**:640–643.
44. **Fath, M. J., and R. Kolter.** 1993. ABC transporters: bacterial exporters. *Microbiol. Rev.* **57**:995–1017.
45. **Federal Register.** 1988. Nisin preparation: affirmation of GRAS status as a direct human food ingredient. 21CFR Part 184. *Fed. Regist.* **53**:11247–11251.
46. **Foegeding, P. M., A. B. Thomas, D. H. Pinkerton, and T. R. Klaenhammer.** 1992. Enhanced control of *Listeria monocytogenes* by in situ-produced pediocin during dry fermented sausage production. *Appl. Environ. Microbiol.* **58**:884–890.

47. **Fuchs, P. G., J. Zajdel, and W. T. Dobrzanski.** 1975. Possible plasmid nature of the determinant for production of the antibiotic nisin in some strains of *Streptococcus lactis*. *J. Gen. Microbiol.* **88:**1899–1902.
48. **Fuller, R.** 1991. Probiotics in human medicine. *Gut* **32:**433–439.
49. **Gao, F. H., T. Abee, and W. N. Konings.** 1991. The mechanism of action of the peptide antibiotic nisin in liposomes and cytochrome C oxidase proteoliposomes. *Appl. Environ. Microbiol.* **57:**2164–2170.
50. **Garcia-Garcera, M. J. G., G. L. Elferink, J. M. Driessen, and W. N. Konings.** 1993. In vitro pore-forming activity of the lantibiotic nisin. Role of PMF force and lipid composition. *Eur. J. Biochem.* **212:**417–422.
51. **Garver, K. I., and P. M. Muriana.** 1993. Detection, identification and characterization of bacteriocin-producing lactic acid bacteria from retail food products. *Int. J. Food Microbiol.* **19:**241–258.
52. **Gasson, M. J., H. M. Dodd, and N. Horn.** 1995. Protein engineering of bacteriocins. Presented at *Workshop on the Bacteriocins of Lactic Acid Bacteria—Applications and Fundamentals.* Alberta, Canada, April 17–22, 1995.
53. **Geis, A., J. Singh, and M. Teuber.** 1983. Potential of lactic streptococci to produce bacteriocins. *Appl. Environ. Microbiol.* **45:**205–211.
54. **Gilliland, S. E.** 1990. Health and nutritional benefits from lactic acid bacteria. *FEMS Microbiol Rev.* **87:**175–188.
55. **Gilliland, S. E., C. R. Nelson, and C. Maxwell.** 1985. Assimilation of cholesterol by *Lactobacillus acidophilus. Appl. Environ. Microbiol.* **49:**377–380.
56. **Goldberg, I. G.** 1994. *Functional Foods: Designer Foods, Pharmafoods, Nutraceuticals.* Chapman & Hall, New York.
57. **Goldin, B. R.** 1990. Intestinal microflora: metabolism of drugs and carcinogens. *Ann. Med.* **22:**43–48.
58. **Gonzalez, C. F.** February 1988. Method for inhibiting bacterial spoilage and composition for this purpose. European Patent Application 88101624.
59. **Gonzalez, C. F., and B. S. Kunka.** 1985. Transfer of sucrose-fermenting ability and nisin production phenotype among lactic streptococci. *Appl. Environ. Microbiol.* **49:**627–633.
60. **Grant, E., Jr., T. J. Beeler, K. M. P. Taylor, K. Gable, and M. A. Roseman.** 1992. Mechanism of magainin 2a induced permeabilization of phospholipid vesicles. *Biochemistry* **31:**9912–9918.
61. **Gross, E., and J. L. Morell.** 1967. The presence of dehydroalanine in the antibiotic nisin and its relationship to activity. *J. Am. Chem. Soc.* **89:**2791–2792.
62. **Gross, E., and J. L. Morell.** 1971. The structure of nisin. *J. Am. Chem. Soc.* **93:**4634–4635.
63. **Haberman, E.** 1972. Bee and wasp venoms. *Science* **177:**314–322.
64. **Hall, M.** 1987. The role of bacteria in human carcinogenesis. *Anticancer Res.* **7:**1079–1084.
65. **Hanlin, M. B., N. Kalchayan, P. Ray, and B. Ray.** 1993. Bacteriocins of lactic acid bacteria in combination have greater antibacterial activity. *J. Food Prot.* **56:**252–255.
66. **Harris, L. J., H. P. Fleming, and T. R. Klaenhammer.** 1991. Sensitivity and resistance of *Listeria monocytogenes* ATCC 19115 Scott A and VAL 500 to nisin. *J. Food Prot.* **54:**836–840.
67. **Harris, L. J., H. P. Fleming, and T. R. Klaenhammer.** 1992. Novel paired starter culture system for sauerkraut, consisting of a nisin-resistant *Leuconostoc mesenteroides* strain and a nisin-producing *Lactococcus lactis* strain. *Appl. Environ. Microbiol.* **58:**1484–1489.
68. **Hitchins, A. D., and F. E. McDonough.** 1989. Prophylactic and therapeutic aspects of fermented milk. *Am. J. Clin. Nutrit.* **49:**675–684.
69. **Holo, H., O. Nissen, and I. F. Nes.** 1991. Lactococcin A, a new bacteriocin from *Lactococcus lactis* subsp. *cremoris:* isolation and characterization of the protein and its gene. *J. Bacteriol.* **173:**3879–3887.
70. **Hoover, G., and L. R. Steenson.** 1993. *Bacteriocins of Lactic Acid Bacteria.* Academic Press, Inc., New York.
71. **Horn, N., S. Swindell, H. Dodd, and M. Gasson.** 1991. Nisin biosyntheis genes are encoded by a novel conjugative transposon. *Mol. Gen. Genet.* **228:**129–135.
72. **Hurst, A.** 1981. Nisin. *Adv. Appl. Microbiol.* **27:**85–123.
73. **Hutton, M. T., P. A. Chehak, and J. H. Hanlin.** 1991. Inhibition of botulism toxin production by *Pediococcus acidilactici* in temperature abused refrigerated foods. *J. Food Safety* **11:**255–267.
74. **Jack, R. N., J. R. Tagg, and B. Ray.** 1995. Bacteriocins of gram-positive bacteria. *Microbiol. Rev.* **59:**171–200.
75. **Jager, K., and S. Harlander.** 1992. Characterization of a bacteriocin from *Pediococcus acidilactici* PC and comparison of bacteriocin-producing stains using molecular typing procedures. *Appl. Microbiol. Biotechnol.* **37:**631–637.
76. **Jarvis, B., and J. Farr.** 1971. Partial purification, specificity and mechanism of the nisin-inactivating enzyme from *Bacillus cereus. Biochim. Biophys. Acta* **227:**232–240.
77. **Joerger, M. C., and T. R. Klaenhammer.** 1986. Characterization and purification of helveticin J and evidence for a chromosomally determined bacteriocin produced by *Lactobacillus helveticus* 481. *J. Bacteriol.* **167:**439–446.
78. **Johansson, M. L., G. Molin, B. Jeppsson, S. Noback, A. Ahrne, and B. Bergmark.** 1993. Administration of different *Lactobacillus* strains in fermented oatmeal soup: in vivo colonization of human intestinal mucosa and effect on indigenous flora. *Appl. Environ. Microbiol.* **59:**15–20.
79. **Jones, L. W.** 1974. Effect of butterfat on inhibition of *Staphylococcus aureus* by nisin. *Can. J. Microbiol.* **20:**1257–1260.
80. **Juneja, V. K., and P. M. Davidson.** 1993. Influence of altered fatty acid composition on resistance of *Listeria monocytogenes* to antimicrobials. *J. Food Prot.* **56:**302–305.
81. **Juven, B. J., R. J. Meinersman, and N. J. Stern.** 1991. Antagonistic effects of lactobacilli and pediococci to control intestinal colonization by human enteropathogens in live poultry. *J. Appl. Bacteriol.* **70:**95–103.
82. **Kalchayanand, N., M. B. Hanlin, and B. Ray.** 1992. Sublethal injury makes Gram-negative and resistant Gram-positive bacteria sensitive to the bacteriocins, pediocin AcH and nisin. *Lett. Appl. Microbiol.* **15:**239–243.
83. **Kalchayanand, N., T. Sikes, C. P. Dunne, and B. Ray.** 1994. Hydrostatic pressure and electroporation have increased bactericidal efficiency in combination with bacteriocins. *Appl. Environ. Microbiol.* **60:**4174–4177.

84. **Klaenhammer, T. R.** 1988. Bacteriocins of lactic acid bacteria. *Biochimie* **70**:337–349.

85. **Klaenhammer, T. R.** 1993. Genetics of bacteriocins produced by lactic acid bacteria. *FEMS Microbiol. Rev.* **12**:39–86.

86. **Kok, J., H. Holo, M. J. van Belkum, A. J. Haandrikman, and I. F. Nes.** 1993. Nonnisin bacteriocins in lactococci: biochemistry, genetics and mode of action, p. 121–150. *In* D. G. Hoover and L. R. Steenson (ed.), *Bacteriocins of Lactic Acid Bacteria.* Academic Press, Inc., New York.

87. **Konisky, J.** 1982. Colicins and other bacteriocins with established modes of action. *Annu. Rev. Microbiol.* **36**: 125–144.

88. **Kordel, M., and H. G. Sahl.** 1986. Susceptibility of bacterial, eukaryotic and artificial membranes to the disruptive action of the cationic peptide Pep 5 and nisin. *FEMS Microbiol. Lett.* **34**:139–144.

89. **Kuipers, O. P., M. M. Beerthuyzen, R. J. Siezen, and W. M. de Vos.** 1993. Characterization of the nisin gene cluster *nisABTCIPR* of *Lactococcus lactis:* requirement of expression of the *nisA* and *nisI* gene for producer immunity. *Eur. J. Biochem.* **216**:281–292.

90. **Kuipers, O. P., H. S. Rollema, W. M. Yap, H. J. Boot, R. J. Siezen, and W. M. de Vos.** 1992. Engineering dehydrated amino acid residues in the antimicrobial peptide nisin. *J. Biol. Chem.* **267**:2430–2436.

91. **Lehrer, R. I., A. Barton, K. A. Daher, S. S. L. Harwig, T. Ganz, and M. E. Selsted.** 1989. Interaction of human defensins with *Escherichia coli,* mechanism of bactericidal activity. *J. Clin. Invest.* **84**:553–561.

92. **Lin, S. Y., J. W. Ayers, W. Winkler, Jr., and W. E. Sandine.** 1989. *Lactobacillus* effects on cholesterol: *in vitro* and *in vivo* results. *J. Dairy Sci.* **72**:2885–2899.

93. **Lipinska, E.** 1977. Nisin and its applications, p. 103–130. *In* M. Woodbine (ed.), *Antibiotics and Antibiosis in Agriculture.* Butterworths, London.

94. **Liu, W., and N. Hansen.** 1990. Some chemical and physical properties of nisin, a small protein antibiotic produced by Lactococcus lactis. *Appl. Environ. Microbiol.* **56**:2551–2558.

95. **Liu, W., and N. Hansen.** 1992. Enhancement of the chemical and antimicrobial properties of subtilin by site-directed mutagenesis. *J. Biol. Chem.* **267**:25078–25085.

96. **Luchansky, J. B., K. A. Glass, K. D. Harsono, A. J. Degnan, N. G. Faith, B. Cauvin, G. Baccus-Taylor, K. Arihara, B. Bater, A. J. Maurer, and R. G. Cassens.** 1992. Genomic analysis of *Pediococcus* starter cultures used to control *Listeria monocytogenes* in turkey summer sausage. *Appl. Environ. Microbiol.* **58**:3053–3059.

97. **Lyon, W. J., J. E. Sethi, and B. A. Glatz.** 1993. Inhibition of psychrotrophic organisms by propionicin PLG-1, a bacteriocin produced by *Propionibacterium thoenii. J. Dairy Sci.* **76**:1506–1513.

98. **Maftah, A., D. Renault, C. Vignoles, Y. Hechard, P. Bressollier, M. H. Ratinaud, Y. Cenatiempo, and R. Julien.** 1993. Membrane permeabilization of *Listeria monocytogenes* and mitochondria by the bacteriocin mesentericin Y105. *J. Bacteriol.* **175**:3232–3235.

99. **Marteau, P., and J. C. Rambaud.** 1993. Potential of using lactic acid bacteria for therapy and immunomodulation in man. *FEMS Microbiol. Rev.* **12**:207–220.

100. **Marugg, J. D., C. F. Gonzalez, B. S. Kunka, A. M. Ledeboer, M. J. Pucci, M. Y. Toonen, S. A. Walker, L. C. M. Zoetmulder, and P. A. Vandenbergh.** 1992. Cloning, expression, and nucleotide sequence of genes involved in production of pediocin PA-1, a bacteriocin from *Pediococcus acidilactici* PAC 1.0. *Appl. Environ. Microbiol.* **58**: 2360–2367.

101. **Mayr-Harting, A., A. J. Hedges, and R. C. W. Beerkley.** 1972. Methods for studying bacteriocins. *Methods Microbiol.* **7A**:313–342.

102. **Mazzotta, A. S., A. D. Crandall, and T. J. Montville.** 1995. Resistance of *Clostridium botulinum* spores and vegetative cells to nisin, abstr. P32, p. 387. *Abstracts of the 95th General Meeting of the American Society for Microbiology 1995.* American Society for Microbiology, Washington, D.C.

103. **McKay, L. L.** 1983. Functional properties of plasmids in lactic streptococci. *Antonie van Leeuwenhoek* **49**:259–274.

104. **McKay, L. L., B. R. Cords, and K. A. Baldwin.** 1973. Transduction of lactose metabolism in *Streptococcus lactis* C2. *J. Bacteriol.* **115**:810–815.

105. **McNamara, S.** 1986. Regulatory issues in the food biotechnology area, p. 15–28. *In* S. Harlander and T. P. Labuza (ed.), *Biotechnology in Food Processing.* Noyes Publications, Park Ridge, N.J.

106. **Ming, X., and M. A. Daeschel.** 1993. Nisin resistance of foodborne bacteria and the specific resistance responses of *Listeria monocytogenes* Scott A. *J. Food Prot.* **11**:944–948.

107. **Montville, T. J.** 1987. Continuous culture: theory and applications, p. 165–186. *In* T. J. Montville (ed.), *Food Microbiology,* vol. 2. *New and Emerging Technologies.* CRC Press, Inc., Boca Raton, Fla.

108. **Montville, T. J.** 1989. The evolving impact of biotechnology on food microbiology. *J. Food Safety* **10**:87–97.

109. **Montville, T. J., A. M. Rogers, and A. Okereke.** 1992. Differential sensitivity of *Clostridium botulinum* strains to nisin. *J. Food Prot.* **56**:444–448.

110. **Montville, T. J., K. Winkowski, and R. D. Ludescher.** 1995. Models and mechanisms for bacteriocin action and application. *Int. Dairy J.* **5**:797–815.

111. **Morris, S. L., R. C. Walsh, and J. N. Hansen.** 1984. Identification and characterization of some bacterial membrane sulfhydryl groups which are targets of bacteriostatic and antibiotic action. *J. Biol. Chem.* **259**:13590–13594.

112. **Nettles, C. G., and S. F. Barefoot.** 1993. Biochemical and genetic characteristics of bacteriocins of food-associated lactic acid bacteria. *J. Food Prot.* **56**:338–356.

113. **Neve, H., A. Geis, and M. Teuber.** 1984. Conjugal transfer and characterization of bacteriocin plasmids in group N (lactic acid) streptococci. *J. Bacteriol.* **157**:833–838.

114. **Nielsen, J. W., J. S. Dickson, and J. D. Crouse.** 1990. Use of a bacteriocin produced by *Pediococcus acidilactici* to inhibit *Listeria monocytogenes* associated with fresh meat. *Appl. Environ. Microbiol.* **56**:2142–2145.

115. **Ojcius, D. M., and J. D. E. Young.** 1991. Cytolytic pore-forming proteins and peptides: is there a common structural motif? *Trends Biochem. Sci.* **16**:225–229.

116. **Oscroft, C. A., J. G. Banks, and S. McPhee.** 1990. Inhibition of thermally-stressed *Bacillus* spores by combinations of nisin, pH and organic acids. *Lebensm.-Wiss. Technol.* **23**:538–544.
117. **O'Sullivan, M. G., G. Thornton, G. C. O'Sullivan, and J. K. Collins.** 1993. Probiotic bacteria: myth or reality? *Trends Food Sci. Technol.* **21**:309–313.
118. **Palumbo, S. A.** 1987. Can refrigeration keep our food safe? *Dairy Food Sanitarian* **7**:56–60.
119. **Pucci, M. J., E. R. Vedamuthu, B. S. Kunka, and D. A. Vandenbergh.** 1988. Inhibition of *Listeria monocytogenes* by using bacteriocin PA-1 produced by *Pediococcus acidilatici* PAC 1.0. *Appl. Environ. Microbiol.* **54**:2349–2353.
120. **Quadri, L. E. N., K. L. Roy, J. C. Vederos, and M. E. Stiles.** 1995. Immunity and gene organization of carnobacteriocius BM1 and B2 produced by *Carnobacterium piscicole* LV17B. Presented at *Workshop on the Bacteriocins of Lactic Acid Bacteria—Applications and Fundamentals.* Alberta, Canada, April 17–22, 1995.
121. **Radler, F.** 1990. Possible use of nisin in winemaking. I. Action of nisin against lactic acid bacteria and wine yeasts in solid and liquid media. *Am. J. Enol. Vitic.* **41**:1–6.
122. **Radler, F.** 1990. Possible use of nisin in winemaking. II. Experiments to control lactic acid bacteria in the production of wine. *Am. J. Enol. Vitic.* **41**:7–11.
123. **Rauch, P. J. G., and W. M. deVos.** 1992. Characterization of the novel nisin-sucrose conjugative transposon Tn*5276* and its insertion in *Lactococcus lactis*. *J. Bacteriol.* **174**:1280–1287.
124. **Ray, B., and M. A. Daeschel.** 1992. *Food Biopreservation of Microbial origin.* CRC Press, Boca Raton, Fla.
125. **Rekhif, N., A. Atrih, and G. Lefebvre.** 1994. Selection and properties of spontaneous mutants of *Listeria monocytogenes* ATCC 15313 resistant to different bacteriocins produced by lactic acid bacteria strains. *Curr. Microbiol.* **28**(4):237–242.
126. **Roberts, R. E., and E. A. Zottola.** 1993. Shelf-life of pasteurized process cheese spreads made from cheddar cheese manufactured with a nisin producing starter culture. *J. Dairy Sci.* **76**:1830–1836.
127. **Rodriguez, J. M., L. M. Cintas, P. Casaus, N. Horn, H. M. Dodd, P. E. Hernandez, and M. J. Gasson.** 1995. Isolation of nisin-producing *Lactococcus lactis* strains from dry fermented sausages. *J. Appl. Bacteriol.* **78**:109–115.
128. **Rogers, A. M., and T. J. Montville.** 1994. Quantification of factors influencing nisin's inhibition of *Clostridium botulinum* 56A in a model food system. *J. Food Sci.* **59**: 663–668, 686.
129. **Ruhr, E., and H.-G. Sahl.** 1985. Mode of action of the peptide antibiotic nisin and influence on the membrane potential of whole cells and on cytoplasmic and artificial membrane vesicles. *Antimicrob. Agents Chemother.* **27**: 841–845.
130. **Saaverda, J. M., N. A. Bauman, I. Oung, J. A. Perzman, and R. Yolken.** 1994. Feeding of *Bifidobacterium bifidum* and *Streptococcus thermophilus* to infants in hospitals for prevention of diarrhoea and shedding of rotavirus. *Lancet* **344**:1046–1049.
131. **Sahl, H. G., M. Grossgarten, W. R. Widger, W. A. Cramer, and H. Brandis.** 1985. Structural similarities of the staphylococcin-like peptide Pep 5 to the antibiotic nisin. *Antimicrob. Agents Chemother.* **27**:836–840.
132. **Sahl, H. G., R. W. Jack, and G. Bierbaum.** 1995. Biosynthesis and biological activities of lantibiotics with unique post-translational modifications. *Eur. J. Biochem.* **230**: 827–853.
133. **Saleh, M. A., and Z. J. Ordal.** 1955. Studies on growth and toxin production of *Clostridium botulinum* in precooked frozen food. II. Inhibition by lactic acid bacteria. *Food Res.* **20**:340–346.
134. **Salminen, S.** 1990. The role of intestinal microflora in preserving intestinal integrity and health with specific reference to lactic acid bacteria. *Ann. Med.* **22**:35.
135. **Savage, D.** 1977. Microbiology of the gastrointestinal tract. *Annu. Rev. Microbiol.* **31**:107–133.
136. **Scott, V. N., and S. L. Taylor.** 1981. Effect of nisin on outgrowth of *Clostridium botulinum* spores. *J. Food Sci.* **46**:117–120.
137. **Scott, V. N., and S. L. Taylor.** 1981. Temperature, pH, and spore load on the ability of nisin to prevent the outgrowth of *Clostridium botulinum* spores. *J. Food Sci.* **46**:121–126.
138. **Sears, P. M., B. S. Smith, W. K. Stewart, R. Gonzalez, S. O. Rubino, S. A. Gusik, E. S. Kulisek, S. J. Projan, and P. Blackburn.** 1992. Evaluation of a nisin-based germicidal formulation on teat skin of live cows. *J. Dairy Sci.* **75**:3185–3190.
139. **Siegers, K., and K. D. Entian.** 1995. Genes involved in immunity to the lantibiotic nisin produced by *Lactococcus lactis* 6F3. *Appl. Environ. Microbiol.* **61**:1082–1089.
140. **Simon, G. L., and S. L. Gorbach.** 1984. Intestinal flora in health and disease. *Gastroenterology* **86**:174–193.
141. **Somers, E. B., and S. L. Taylor.** 1987. Antibotulinal effectiveness of nisin in pasteurized process cheese spreads. *J. Food Prot.* **50**:842–848.
142. **Somkuti, G. A., D. K. Y. Solaiman, T. L. Johnson, and D. H. Steinberg.** 1991. Transfer and expression of a *Streptomyces* cholesterol oxidase gene in *Streptococcus thermophilus*. *Biotechnol. Appl. Biochem.* **12**:238–245.
143. **Steele, J. L., and L. L. McKay.** 1986. Partial characterization of the genetic basis for sucrose metabolism and nisin production in *Streptococcus lactis*. *Appl. Environ. Microbiol.* **51**:57–64.
144. **Steiner, H.** 1982. Secondary structure of the cecropins: antibacterial peptides from the moth *Hyalophora cecropia*. *FEBS Lett.* **137**:283–287.
145. **Stevens, K. A., B. W. Sheldon, N. A. Klapes, and T. R. Klaenhammer.** 1991. Nisin treatment for inactivation of *Salmonella* species and other gram-negative bacteria. *Appl. Environ. Microbiol.* **57**:3613–3615.
146. **Stevens, K. A., B. W. Sheldon, N. A. Klapes, and T. R. Klaenhammer.** 1992. Effect of treatment conditions on nisin inactivation of Gram-negative bacteria. *J. Food Prot.* **55**:763–767.
147. **Stiles, M. E., and J. W. Hastings.** 1991. Bacteriocin production by lactic acid bacteria: potential for use in meat preservation. *Trends Food Sci. Technol.* **2**:247–251.

148. **Stoddard, G. W., J. P. Petzel, M. J. van Belkum, J. Kok, and L. L. McKay.** 1992. Molecular analyses of the lactococcin A gene cluster from *Lactococcus lactis* subsp. *lactis* biovar *diacetylactis* WM4. *Appl. Environ. Microbiol.* **58:**1952–1961.
149. **Tagg, J. R.** 1991. Bacterial BLIS. *ASM News* **57:**611.
150. **Tagg, J. R., A. S. Dajani, and L. W. Wannamaker.** 1976. Bacteriocins of gram-positive bacteria. *Bacteriol. Rev.* **40:**722–756.
151. **Tanaka, N., E. E. Traisman, M. H. Lee, and R. Casses.** 1980. Inhibition of botulism toxin formation in bacon by acid development. *J. Food Prot.* **43:**450–452.
152. **Taylor, L. Y., O. O. Cann, and B. J. Welch.** 1990. Antibotulinal properties of nisin in fresh fish packaged in an atmosphere of carbon dioxide. *J. Food Prot.* **53:**953–957.
153. **Terwilliger, T. C., and D. Eisenberg.** 1982. The structure of melittin. *J. Biol. Chem.* **257:**6016–6022.
154. **Tramer, J., and G. G. Fowler.** 1964. Estimation of nisin in foods. *J. Sci. Food Agric.* **15:**522–528.
155. **van Belkum, M. J., B. J. Hayema, A. Geis, J. Kok, and G. Venema.** 1989. Cloning of two bacteriocin genes from a lactococcal bacteriocin plasmid. *Appl. Environ. Microbiol.* **55:**1187–1191.
156. **van Belkum, M. J., B. J. Hayema, R. E. Jeeninga, J. Kok, and G. Venema.** 1991. Organization and nucleotide sequence of two lactococcal bacteriocin operons. *Appl. Environ. Microbiol.* **57:**492–498.
157. **van Belkum, M. J., J. Kok, and G. Venema.** 1992. Cloning, sequencing, and expression in *Escherichia coli* of *lenB*, a third bacteriocin determinant from the lactococcal bacteriocin plasmid p984-6. *Appl. Environ. Microbiol.* **58:**572–577.
158. **van Belkum, M. J., J. Kok, G. Venema, H. Holo, I. F. Nes, W. N. Konings, and T. Abee.** 1991. The bacteriocin lactococcin A specifically increases the permeability of lactococcal cytoplasmic membranes in a voltage-independent, protein-mediated manner. *J. Bacteriol.* **173:** 7934–7941.
159. **Vandenbergh, P. A.** 1993. Lactic acid bacteria, their metabolic products and interference with microbial growth. *FEMS Microbiol. Rev.* **12:**221–238.
160. **Vandenbergh, P. A., M. J. Pucci, B. S. Kunka, and E. R. Vedamuthu.** January 1989. Method for inhibiting *Listeria monocytogenes* using a bacteriocin. European Patent Application 89101126.6.
161. **van der Meer, J. R., J. Polman, M. M. Beerthuyzen, R. J. Siezen, O. P. Kuipers, and W. M. de Vos.** 1993. Characterization of the *Lactococcus lactis* nisin A operon genes *nisP*, encoding a subtilisin-like serine protease involved in precursor processing, and *nisR*, encoding a regulatory protein involved in nisin biosynthesis. *J. Bacteriol.* **175:**2578–2588.
162. **van de Ven, F. J. M., H. W. van den Hooven, R. N. H. Konings, and C. W. Hilbers.** 1991. The spatial structure of nisin in aqueous solution, p. 35–42. *In* G. Jung and H. G. Sahl (ed.), *Nisin and Novel Lantibiotics.* Escom Publishers, Leiden, The Netherlands.
163. **Van Garde, S. J., and M. J. Woodburn.** 1987. Food discard practices of householders. *J. Am. Diet. Assoc.* **87:** 322–329.
164. **Venema, K., T. Abee, A. J. Haandrikman, K. J. Leenhouts, J. Kok, W. N. Konings, and G. Venema.** 1993. Mode of action of lactococcin B, a thiol-activated bacteriocin from *Lactococcus lactis. Appl. Environ. Microbiol.* **59:**1041–1048.
165. **Vogel, R. F., B. S. Pohle, P. S. Tichaczek, and W. Hammes.** 1993. The competitive advantage of *Lactobacillus curvatus* LTH 1174 in sausage fermentations is caused by formation of curvacin. *Syst. Appl. Microbiol.* **16:**457–462.
166. **Winkowski, K., M. E. C. Bruno, and T. J. Montville.** 1994. Correlation of bioenergetic parameters with cell death in *Listeria monocytogenes* cells exposed to nisin. *Appl. Environ. Microbiol.* **60:**4186–4187.
167. **Winkowski, K., A. D. Crandall, and T. J. Montville.** 1993. Inhibition of *Listeria monocytogenes* by *Lactobacillus bavaricus* MN in meat systems at refrigeration temperatures. *Appl. Environ. Microbiol.* **59:**2552–2557.
168. **Winkowski, K., and T. J. Montville.** 1992. Use of a meat isolate, *Lactobacillus bavaricus* MN, to inhibit *Listeria monocytogenes* growth in a model meat gravy system. *J. Food Safety* **13:**19–31.
169. **Worobo, R. W., M. J. van Belkum, H. Sailer, K. L. Roy, J. C. Vederas, and M. E. Stiles.** 1995. A signal peptide secretion-dependent bacteriocin from *Carnobacterium divergens. J. Bacteriol.* **177:**3143–3149.
170. **Zajdel, J. K., P. Ceglowski, and W. T. Dobrzanski.** 1985. Mechanism of action of lactostrepcin 5, a bacteriocin produced by *Streptococcus cremoris* 202. *Appl. Environ. Microbiol.* **49:**969–974.

Food Fermentations VIII

Mark E. Johnson
James L. Steele

31 Fermented Dairy Products

Several groups of microorganisms participate in the manufacture and ripening of fermented milk products (Table 31.1). The primary microflora are the homofermentative lactic acid bacteria, which are intentionally added to milk and are commonly referred to as starter cultures. Their function is to ferment lactose to lactic acid. These bacteria include both mesophilic (optimal growth at 25 to 30°C) and thermophilic (optimal growth at 37 to 42°C) species. Mesophilic lactic acid bacteria include *Lactococcus lactis* subsp. *lactis* and *Lc. lactis* subsp. *cremoris.* Thermophilic lactic acid bacteria include *Streptococcus thermophilus, Lactobacillus delbrueckii* subsp. *bulgaricus,* and *Lactobacillus helveticus.* A protocooperative relationship exists between *S. thermophilus* and starter lactobacilli which results in enhanced growth and increased rate and extent of acid production.

In some fermented dairy products, additional bacteria, referred to as secondary microflora, are added to influence flavor and alter texture of the final product by producing carbon dioxide. Two lactic acid bacteria, *Leuconostoc* species and strains of *Lc. lactis* subsp. *lactis* capable of metabolizing citric acid (Cit^+ strains), are added to produce aroma compounds and carbon dioxide in cultured buttermilk and certain cheeses (Gouda, Edam, blue, and Havarti). When used together, these bacteria make up 10 to 20% of the total starter culture, with *Leuconostoc* species being present at about three times the population of Cit^+ *Lc. lactis* subsp. *lactis.* Heterofermentative lactobacilli (*Lactobacillus brevis, Lactobacillus fermentum,* and *Lactobacillus kefir*) are part of the varied microflora (including several yeast species) found in the more exotic cultured milks such as kefir and koumiss, in which they produce ethanol, carbon dioxide, and lactic acid. They are not used in other fermented dairy products because of the copious quantities of carbon dioxide produced. *Propionibacterium freudenreichii* subsp. *shermanii* is added to Swiss-type cheeses primarily to metabolize L-lactic acid to propionic acid, acetic acid, and carbon dioxide. The carbon dioxide forms the "eyes" in Swiss-type cheeses.

Other types of secondary microflora include undefined mixtures of yeasts, molds, and bacteria. These microorganisms are added directly to the milk or are smeared, sprayed, or rubbed onto the cheese surface. This group of microorganisms has extremely varied and complex metabolic activities, their main function being to produce unique flavors. The use of these secondary microflora is limited to a few types of surface-ripened and mold-ripened cheeses. Yeasts (*Debaryomyces* and *Trichosporon* species) and bacteria (*Brevibacterium linens* and *Micrococcus* species) are employed in the aging of surface-ripened cheeses such as Limburger and Gruyère. Molds (*Penicillum camemberti* and *Penicillium roqueforti*) are used in Camembert and blue-veined cheeses, respectively.

Fermented dairy products are not commercially produced in an environment free of microorganisms other than those added in starter cultures. However, the metabolism of certain contaminants, e.g., lactobacilli, may

Table 31.1 Microorganisms involved in the manufacture of cheeses and fermented milks

Products	Principal acid producers	Intentionally introduced secondary microflora
Cheeses		
Colby, Cheddar, cottage, cream	*Lactococcus lactis* subsp. *cremoris, Lc. lactis* subsp. *lactis*	None
Gouda, Edam, Havarti	*Lactococcus lactis* subsp. *cremoris, Lc. lactis* subsp. *lactis*	*Leuconostoc* sp., Cit^+ *Lactococcus lactis* subsp. *lactis*
Brick, Limburger	*Lactococcus lactis* subsp. *cremoris, Lc. lactis* subsp. *lactis*	*Geotrichum candidum, Brevibacterium linens, Micrococcus* sp.
Camembert	*Lactococcus lactis* subsp. *cremoris, Lc. lactis* subsp. *lactis*	*Penicillium camemberti*, sometimes *Brevibacterium linens*
Blue	*Lactococcus lactis* subsp. *cremoris, Lc. lactis* subsp. *lactis*	Cit^+ *Lactococcus lactis* subsp. *lactis, Penicillium roqueforti*
Mozzarella, provolone, Romano, Parmesan	*Streptococcus thermophilus, Lactobacillus delbrueckii* subsp. *bulgaricus, Lb. helveticus*	None; animal lipases added to Romano for picante or rancid flavor
Swiss	*Streptococcus thermophilus, Lactobacillus helveticus, Lb. delbrueckii* subsp. *bulgaricus*	*Propionibacterium freudenreichii* subsp. *shermanii*
Fermented milks		
Yogurt	*Streptococcus thermophilus, Lactobacillus delbrueckii* subsp. *bulgaricus*	None
Buttermilk	*Lactococcus lactis* subsp. *cremoris, Lc. lactis* subsp. *lactis*	*Leuconostoc* sp., Cit^+ *Lactococcus lactis* subsp. *lactis*
Sour cream	*Lactococcus lactis* subsp. *cremoris, Lc. lactis* subsp. *lactis*	None

be essential for the development of flavor in certain cheese varieties such as Cheddar. Rapid acid development by the starter within 4 to 8 h lowers the pH to less than 5.3 in cheese and to less than 4.6 in fermented milk products. After fermentation is complete, only acid-tolerant bacteria can grow. However, if acid development is slow or if the pH does not decrease sufficiently, contaminants which otherwise would have been inhibited may be able to grow. In some cheeses, the pH can increase during ripening and permit the growth of previously inhibited bacteria. Dairy products may contain yeasts, molds, and many different genera of bacteria whose metabolic activities destroy quality. Spoilage of dairy products is described in chapter 6.

IMPORTANCE OF THE STARTER CULTURE

The key to commercial development of fermented milk products is the consistent and predictable rate of acid development by lactic acid bacteria. The rate and extent of pH decrease during manufacture and in the finished product are critical. pH has profound effects on moisture control during cheese manufacture, retention of coagulants, loss of minerals, and electrochemical interactions between protein molecules. These, in turn, have consequences for the development of flavor and physical properties (body and texture) of cheeses and fermented milks. The reader is referred to other texts (17, 29) for discussion of these complex and interrelated phenomena.

In the past, antibiotic residues and overmature starters have been causes of inconsistent acid production. However, these problems have been overcome through monitoring of the milk supply and the use of improved starter media (66). Today bacteriophage infection is the most common cause of inconsistent acid development, causing significant loss of revenue to the cheese industry.

In cultured milks, desired flavors are derived directly from the metabolism of starter cultures and deliberately added aroma-producing secondary microflora. Thus, the desired flavor of the product dictates the choice of microorganisms. With cultured milks and some cheeses such as Mozzarella, cream, and cottage cheese, the short time from processing to consumption (1 day to 4 weeks) and refrigerated storage precludes the development of flavor other than that produced by starter and secondary microflora. In other cheeses, the choice of microorganism(s) depends primarily on parameters of the manufacturing protocol such as the temperature to which the product will be subjected, the desired rate and extent of acid development, and the desired physical properties of the finished product.

There is considerable debate as to the exact contribution of the starter culture to flavor development in cheese, especially in cheeses to which no secondary microflora is added. In these cheeses, nonstarter lactic acid bacteria, particularly lactobacilli, are the dominant adventitious microflora during ripening (50), and it is generally believed that nonstarter lactic acid bacteria

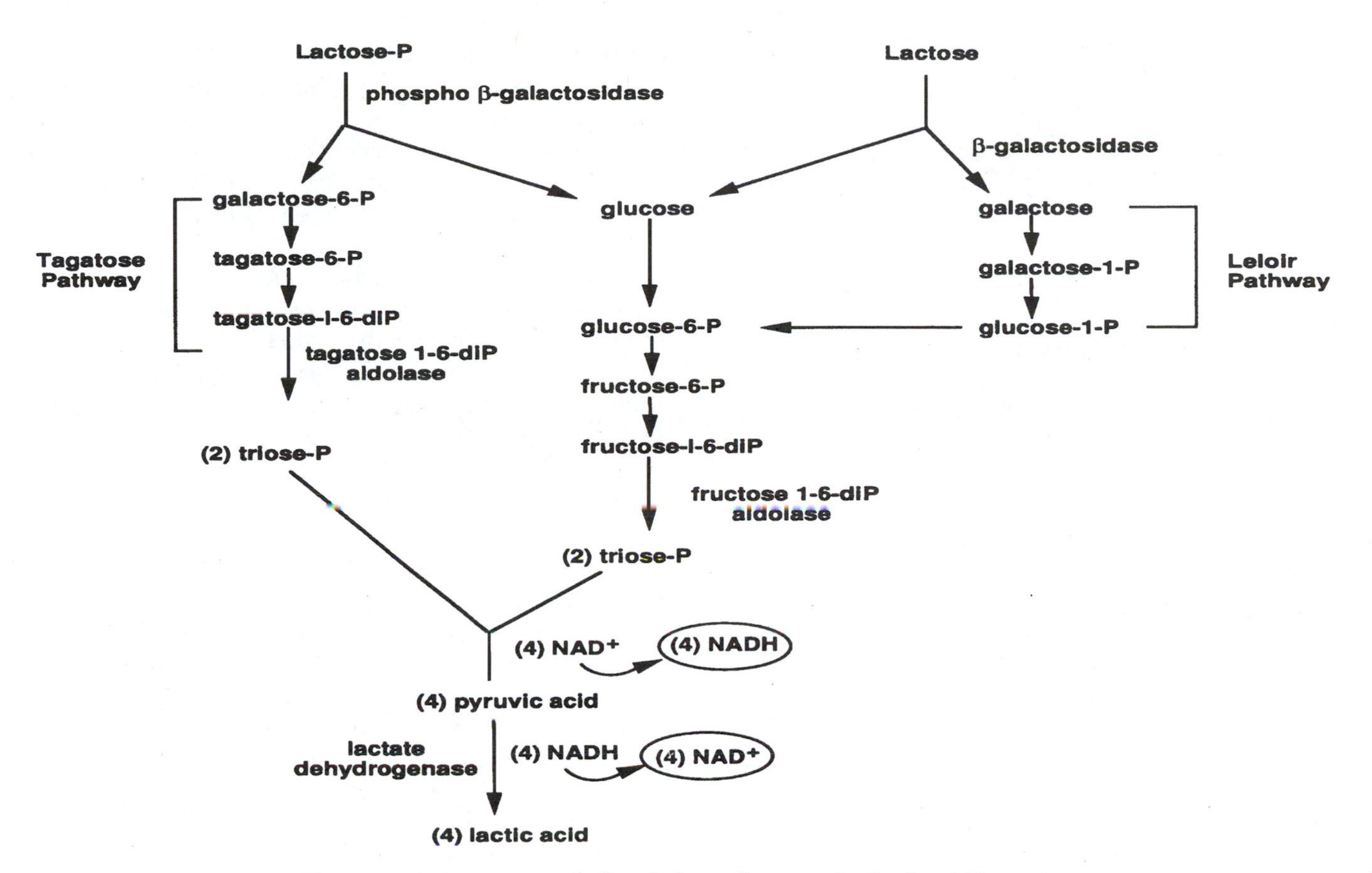

Figure 31.1 Lactose metabolism in homofermentative lactic acid bacteria.

play a significant role in the ripening of these cheeses. The use of nonstarter lactic acid bacteria, especially lactobacilli, as adjunct flavor cultures is a burgeoning research area and has many proponents. Unfortunately, selection of appropriate cultures is a trial-and-error process since agreement on specific compounds which contribute to desired cheese flavor is often lacking. Differences in descriptions in desired flavor arise from the sheer complexity of cheese flavor as well as individual flavor perceptions and taste preferences. It should be understood that flavor development in cheese is a dynamic process and occurs in an environment that is constantly changing: it is not just that cheese develops stronger flavor with age, but the very nature of the flavor evolves. The development of flavor in different cheese varieties has been described elsewhere (18).

LACTOSE METABOLISM

Energy transduction and fermentation pathways for carbohydrate metabolism in lactic acid bacteria have been described in detail elsewhere (19, 51). Lactose, a disaccharide composed of glucose and galactose, is the only free-form sugar present in milk (45 to 50 g/liter). Lactococci translocate lactose into the cell by a phosphoenolpyruvate phosphotransferase system. The lactose is phosphorylated during translocation and then cleaved by phospho-β-galactosidase into glucose and galactose 6-phosphate (Fig. 31.1). The glucose moiety enters the glycolytic pathway, and galactose 6-phosphate is converted to tagatose 6-phosphate via the tagatose pathway. Both sugars are cleaved by specific aldolases into triose phosphates, which are converted to pyruvic acid at the expense of NAD^+. For continued energy production, NAD^+ must be regenerated. This is usually accomplished by reduction of pyruvic acid to lactic acid.

S. thermophilus and some thermophilic lactobacilli transport lactose via a lactose-galactose antiport system driven by an electrochemical proton gradient (51). Lactose is not phosphorylated but is cleaved by β-galactosidase to yield glucose and galactose. The glucose moiety enters the glycolytic pathway, but galactose is excreted from the cells and accumulates in milk or cheese. Thermophilic lactobacilli that do not excrete galactose and *Lb. helveticus* strains able to transport excreted galactose utilize the Leloir pathway to metabolize galactose.

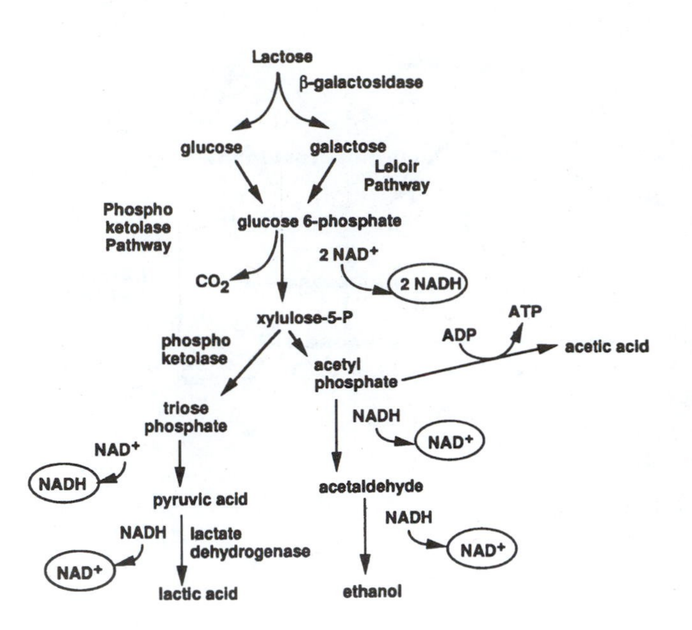

Figure 31.2 Lactose metabolism in heterofermentative lactic acid bacteria.

Lb. delbrueckii subsp. *bulgaricus* and most strains of *S. thermophilus* cannot metabolize galactose. This presents a problem in cheese manufacture, since residual sugar can be metabolized heterofermentatively by other bacteria. Rapid production of carbon dioxide by heterofermentative bacteria causes cheese to crack and packages to swell. Residual sugar can also react with amino groups and form pink or brown pigments, i.e., Maillard-browning reaction products. Thermophiles may not metabolize the residual sugar during cold storage (4 to 7°C) of the cheese. Therefore, lactococci are sometimes included in the starter to ensure that all residual sugar is fermented. In pasta filata cheese manufacture, the curd is heated (52 to 66°C) and molded. The heat treatment may inactivate starter bacteria and prevent further sugar metabolism.

It is not known how lactose is transported into cells by *Leuconostoc* species or heterofermentative lactobacilli; however, lactose is known to be hydrolyzed by β-galactosidase (26). The galactose moiety is transformed into glucose 6-phosphate (Leloir pathway) and, together with glucose, is metabolized through the phosphoketolase pathway (Fig. 31.2). Heterofermentative lactic acid bacteria lack aldolase, but, through a dehydrogenation-decarboxylation system, a pentose sugar (xylulose 5-phosphate) and carbon dioxide are formed. Xylulose 5-phosphate is then cleaved by phosphoketolase to yield glyceraldehyde and acetyl-phosphate. Lactic acid and ethanol, respectively, are formed from these intermediates, facilitating the regeneration of NAD^+. However, during cometabolism of lactose and citric acid, *Leuconostoc* species convert acetyl-phosphate into acetic acid and generate ATP.

Two enzymes, L-lactic acid dehydrogenase and D-lactic acid dehydrogenase, are responsible for the conversion of pyruvate to L-lactic acid and D-lactic acid, respectively. Lactococci produce only L-lactic acid, while *Lb. delbrueckii* subsp. *bulgaricus* and *Leuconostoc* species form only D-lactic acid. Other lactobacilli possess both enzymes and produce both D- and L-lactic acid.

The key to end product formation in lactose metabolism is the regeneration of reducing equivalents. Lactococci have the enzymatic potential to produce compounds other than lactic acid to regenerate NAD^+, but these activities are not usually expressed under aerobic conditions (16). Oxygen in milk is used as an electron acceptor by lactic acid bacteria through the activity of oxidases and peroxidases (10, 58). As a consequence, hydrogen from NADH is transferred to oxygen to produce hydrogen peroxide, and NAD^+ is regenerated. However, under anaerobic conditions and low sugar levels, lactococci (4, 16, 60) produce formic acid and ethanol (Fig. 31.3) to regenerate NAD^+. Heterofermentative lactic acid bacteria convert acetyl-phosphate into acetic acid rather than ethanol under aerobic conditions and regenerate $NAD(P)^+$ through NAD(P)H oxidases (43, 65).

Ethanol has a very high flavor threshold value and would not be expected to contribute directly to flavor in the amounts produced by homofermentative lactic acid bacteria. However, subsequent esterification of ethanol with short-chain fatty acids yields esters with very low flavor thresholds. These compounds are responsible for the fruity flavor defects of Cheddar cheese (5). Short-chain fatty acids are probably generated by nonstarter lactic acid bacteria or exogenous sources, since starter bacteria have limited lipase activity (32).

The potential for production of diacetyl and carbon dioxide from lactose metabolism in lactococci with reduced lactic acid dehydrogenase activity has been described (45). Formation of ethanol, acetic acid, and formic acid would regenerate NAD^+ (Fig. 31.3).

As a result of lactose metabolism (and oxidase activity), the environment becomes anaerobic and the oxidation-reduction potential is reduced. In cheese, further metabolism by nonstarter lactic acid bacteria may be needed to maintain this low potential. It has been postulated that a low oxidation-reduction potential is necessary for the production and stability of reduced sulfur-containing compounds thought to be vital for the development of certain cheese flavors (22, 45). Sugar and citric acid metabolism may result in the formation of

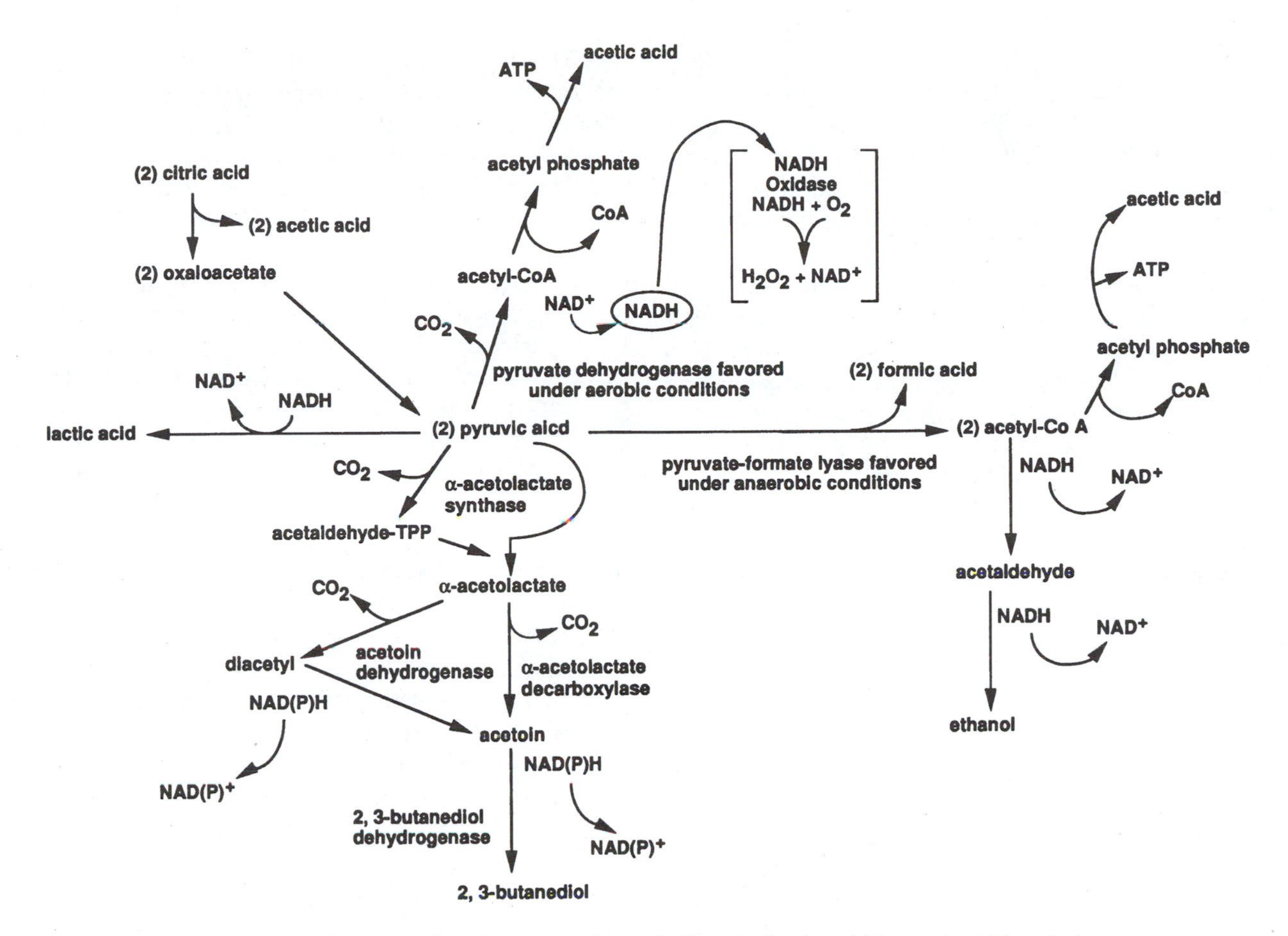

Figure 31.3 Pyruvic acid and citric acid metabolism in lactic acid bacteria. Abbreviations: CoA, coenzyme A; TPP, thiamin pyrophosphate.

α-dicarbonyls such as glyoxal, methylglyoxal, and diacetyl. These compounds readily react with amino acids and produce numerous compounds that contribute to cheese flavor (23).

PRODUCTION OF AROMA COMPOUNDS

Although lactic acid is the main metabolic end product of lactose metabolism in cultured diary products and is responsible for the acid taste, it is nonvolatile and odorless and does not contribute to the aroma. The main aroma of volatile flavor components of fermented milks is produced by acetic acid, acetaldehyde, and diacetyl. In yogurt, these volatile compounds are formed by the lactic acid starter bacteria, *S. thermophilus* and *Lb. delbrueckii* subsp. *bulgaricus*. *Leuconostoc* species and Cit$^+$ *Lc. lactis* subsp. *lactis* are added to produce aroma compounds in buttermilk and some cheese varieties. Literature describing the production of aroma compounds in fermented dairy products has been compiled by Imhof and Bosset (28).

Diacetyl Production

The production of diacetyl, acetic acid, and carbon dioxide from citric acid by lactic acid bacteria has been reviewed by Hugenholtz (27) and is shown in Fig. 31.3. The carbon dioxide produced is responsible for the holes (eyes) in Gouda and Edam cheeses and the effervescent quality of buttermilk. Milk contains 0.15 to 0.2% citric acid, but not all lactic acid bacteria can metabolize it. However, *Leuconostoc* species, Cit$^+$ *Lc. lactis* subsp. *lactis*, and facultative heterofermentative lactobacilli (41) metabolize citric acid. *Leuconostoc* species and Cit$^+$ *Lc. lactis* subsp. *lactis* strains utilize citric acid and lactose simultaneously but do not derive energy from citric acid. Citric acid is transported into the cell by a citric acid permease, which is plasmid encoded in lactococci and *Leuconostoc* species (34, 62) and metabolized to pyruvic acid without generation of NADH. The result is an excess of pyruvic acid which does not have to be reduced to lactic acid to regenerate NAD$^+$; therefore, it is available for other reactions. Ramos et al. (53) appear to have resolved conflicting reports on the pathway leading to the formation of diacetyl. Using ^{13}C nuclear magnetic resonance, they verified that diacetyl formation involves nonenzymatic decarboxylative oxidation of α-acetolactate (an unstable intermediate derived from two molecules of pyruvic acid) and that the alternative suggested pathway, via a diacetyl synthase, is highly unlikely. α-Acetolactate can be produced only when pyruvic acid accumulates within the cell. *Leuconostoc* species metabolize citric acid during growth but do not form diacetyl until the pH is below 5.4. α-Acetolactate synthase is inhibited at pH 5.4 or higher by many intermediates of lactose metabolism; however, the inhibition is relieved at lower pH values (9, 54). When diacetyl is not formed, *Leuconostoc* species form lactic acid from pyruvic acid derived from citric acid and regenerate NAD^+. Since NAD^+ is regenerated, there is less demand to form ethanol from the acetyl-phosphate that is generated via the phosphoketolase pathway. Consequently, acetic acid is formed with the generation of ATP (Fig. 31.3) and growth is enhanced (8, 56).

Diacetyl can be reduced by 2,3-butanediol dehydrogenases (11) to acetoin and 2,3-butanediol, both flavorless compounds. The presence of citric acid inhibits these reactions, but reduction begins when citric acid is exhausted. To ensure residual levels of citric acid in cultured milks, the ratio of starter culture to Cit^+ bacteria must be controlled (14). The cultured milk is stirred when the pH is reduced to 4.6. This introduces oxygen and helps to increase and maintain the desired diacetyl content. The introduction of oxygen is required for nonenzymatic oxidative decarboxylation of α-acetolactate to diacetyl. In addition, lactic acid bacteria produce NADH oxidase, which transfers hydrogen to oxygen and regenerates NAD^+. The NADH oxidase activity replaces the role of the 2,3-butanediol dehydrogenases in regenerating NAD^+ and allows the accumulation of diacetyl, rather than its reduction to acetoin and 2,3-butanediol (2). NADH oxidase is more active at lower temperatures, while dehydrogenases are less active (3). For rapid acidification, non-citrate-metabolizing lactococci must be the dominant acid producers; otherwise, cometabolism of citric acid and lactose by Cit^+ *Lc. lactis* subsp. *lactis* would quickly consume the citric acid and result in the reduction of diacetyl to acetoin and 2,3-butanediol before pH 4.6 is reached. Hugenholtz (27) described the use of genetic engineering to construct strains of lactococci which elevated levels of diacetyl.

Acetaldehyde Production

There are several metabolic pathways in lactic acid bacteria which can lead to the formation of acetaldehyde (38, 39). This finding has resulted in some controversy over the primary pathway utilized by lactic acid bacteria. Cleavage of threonine by threonine aldolase to glycine and acetaldehyde has been suggested to be the most important mechanism for acetaldehyde production in yogurt and buttermilk (39, 69). However, using radiolabeled threonine, Wilkins et al. (67) demonstrated that only 2% of the acetaldehyde produced by mixed cultures of *Lb. delbrueckii* subsp. *bulgaricus* and *S. thermophilus* originated from threonine, even though both bacteria possess threonine aldolase. Acetaldehyde is also formed by Cit^+ *Lc. lactis* subsp. *lactis* strains (33). When the ratio of diacetyl to acetaldehyde in fermented milks is lower than 3:1, a yogurt or green apple flavor defect is observed (42). The defect is due to excess metabolic activity by Cit^+ *Lc. lactis* subsp. *lactis*. Excessive acetaldehyde in yogurt is the result of overripening and is always associated with high acid content. Prevention of an excessive amount of acetaldehyde may be accomplished by the use of a *Leuconostoc* species which metabolizes acetaldehyde to ethanol. Obviously, to prevent overripening, the product must be cooled rapidly and stored at lower temperatures. The trend for faster acid development and larger fermentation vessels may limit the ability of the manufacturer to cool the product fast enough to prevent overripening.

PROTEOLYTIC SYSTEMS IN LACTIC ACID BACTERIA

Proteolytic systems in lactic acid bacteria contribute to their ability to grow in milk and are necessary for the development of flavor in ripened cheeses. Lactic acid bacteria are amino acid auxotrophs typically requiring several amino acids for growth. Well-characterized examples include strains of lactococci and lactobacilli which require 6 to 15 amino acids, respectively. The quantities of free amino acids present in milk are not sufficient to support the growth of these bacteria to high cell density; therefore, they require a proteolytic system capable of utilizing the peptides present in milk and hydrolyzing milk proteins (α_{S1}-, α_{S2}-, κ-, and β-caseins) to obtain essential amino acids. Peptides and amino acids formed by proteolysis may impart flavor directly or serve as flavor precursors in fermented dairy products. Additionally, the resulting flavors may have either positive or negative impact. The production of high-quality fermented dairy products is dependent on proteolytic systems of lactic acid bacteria.

Proteolytic Systems and Their Physiological Role

Since the mid-1980s, significant progress has been made in defining the biochemistry and genetics of proteolytic

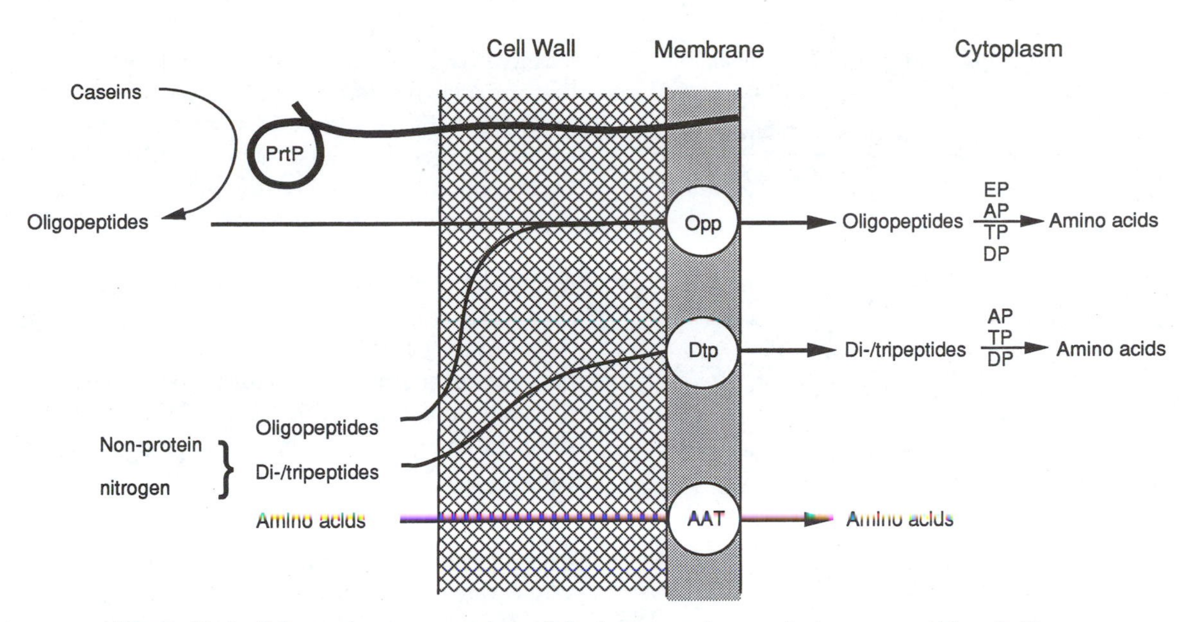

Figure 31.4 Schematic representation of the lactococcal proteolytic system. Abbreviations: PrtP, cell envelope-associated proteinase; Opp, oligopeptide transport system; Dtp, di/tripeptide transport systems; AAT, amino acid transport systems; EP, endopeptidases; AP, aminopeptidases; TP, tripeptidases; DP, dipeptidases.

systems from lactic acid bacteria. This is especially true for lactococcal systems, in which many of the enzymes have been purified and characterized, and numerous genes encoding components of proteolytic systems have been sequenced. For extensive reviews of the proteolytic enzyme systems of lactic acid bacteria, see references 52 and 36. A model of the lactococcal proteolytic enzyme system is presented in Fig. 31.4. While growing in milk, lactic acid bacteria obtain essential amino acids in a variety of ways. They first utilize nonprotein nitrogen sources such as free amino acids and small peptides. Casein, which composes 80% of all proteins present in milk, becomes the primary nitrogen source after nonprotein nitrogen is depleted. Proteolytic systems in lactic acid bacteria can be divided into three components: enzymes outside the cytoplasmic membrane, transport systems, and intracellular enzymes.

Extensive investigations have revealed that a cell envelope-associated proteinase, designated PrtP, is the only extracellular proteolytic enzyme present in lactococci. The enzyme is a serine-protease which is expressed as a pre-pro-proteinase. A signal peptidase removes the signal peptide upon transport across the cytoplasmic membrane. Subsequently, a lipoprotein maturase (PrtM) is thought to cause a conformational change in the pro-proteinase, resulting in release of the pro-region, via autoproteolysis, and an active PrtP. The activated enzyme remains associated with the cell as a result of the presence of a C-terminal membrane anchor sequence. Genes encoding lactococcal proteinases with different substrate specificities have been sequenced, and the amino acid residues involved in substrate binding and catalysis have been determined. A critical feature of the enzyme is its broad cleavage specificity, which results in the release of more than 100 oligopeptides from soluble β-casein, 20% of which are small enough to be transported by the oligopeptide transport system (31). Loss of PrtP, which is typically plasmid encoded in lactococcal strains, results in derivatives capable of reaching only about 10% of the final cell density of the parental strain, indicating that PrtP is essential for growth of lactococci to high cell density in milk.

Transport of nitrogenous compounds across the lactococcal cytoplasmic membrane takes place via group-specific amino acid transport systems, di/tripeptide transport systems, and an oligopeptide transport system (Opp). Of these systems, Opp is of greatest importance during growth of lactococci in milk. Growth studies with Opp derivatives have shown that Opp is essential for the uptake of PrtP-generated peptides from β-casein and oligopeptides present in the nonprotein nitrogen component of milk (31). This system, which is organized in an operon, consists of two ATP-binding proteins, two integral membrane proteins, and a substrate-binding protein. The system has broad specificity, being capable of transporting at least 18 distinct pep-

Table 31.2 Peptidases purified and characterized from lactococci

Peptidase(s)	Abbreviation	Specificity[a]
X-prolyl dipeptidyl aminopeptidase	PepX	X-Pro↓Y-...
Aminopeptidase N	PepN	X↓Y-Z...
Aminopeptidase C	PepC	X↓Y-Z...
Aminopeptidase A	GAP	Asp(Glu)↓Y-Z...
Pyrrolidone carboxyl peptidase	PCP	pGlu↓Y-Z...
Prolyl iminopeptidase	PIP	Pro↓Y-Z...
Dipeptidase	DIP	X↓Y
Prolidase	PRD	X↓YPro
Tripeptidase	PepT	X↓Y-Z
Endopeptidases	PepO	...W-X↓Y-Z...
	PepF	
	LEPI	
	MEP	

[a]↓ indicates which peptide bond is hydrolyzed.

tides generated by PrtP from β-casein (37). The system is capable of transporting peptides containing between four and eight amino acid residues, although longer peptides may also be transported. Two distinct di/tripeptide transport systems, designated DtpT and DtpP, have been characterized. The characterization of DtpT derivatives has revealed that DtpT is not essential for lactococcal growth on the nonprotein nitrogen component of milk- or casein-derived peptides. The physiological role of DtpP has not been defined.

Once inside the cell, peptides are hydrolyzed by peptidases. Peptidase classes which have been identified in lactococci include exopeptidases and endopeptidases. The greatest variety of enzymes are from the exopeptidase class, which includes aminopeptidases, tripeptidases, and dipeptidases. No carboxypeptidases have been detected in lactococci. Peptidases identified in lactococci and their cleavage specificities are summarized in Table 31.2. This combination of endopeptidases, aminopeptidases, tripeptidases, and dipeptidases converts the transported peptides into free amino acids which lactococci require for growth. Examination of the growth of strains lacking one of the peptidases (PepX, PepN, PepO, or PepT) in milk has revealed that only the PepN single mutant grows significantly more slowly. Characterization of lactococcal derivatives lacking more than one of these peptidases indicated that, in general, the more peptidases that were inactivated, the more slowly the strain grew in milk. The only possible exception to this general rule is PepX, which has not been observed to have a significant effect on growth rate in milk. Inactivation of all five of these peptidases results in a derivative that grows more than 10-fold more slowly that the wild-type strain (47). These strains confirm the hypothesis that a variety of peptidases are required to hydrolyze casein-derived oligopeptides transported by Opp to obtain essential amino acids.

Proteolytic systems of other lactic acid bacteria have not been as extensively studied. However, numerous components of proteolytic enzyme systems of *Lb. helveticus* and *Lb. delbrueckii* have been characterized. In general, these components have homologs in the lactococcal system. However, unlike in the lactococcal system, inactivation of PepX in *Lb. helveticus* has been shown to significantly reduce this bacterium's growth rate in milk (68).

Proteolysis and Cheese Flavor Development

While flavor development in various types of cheese remains a poorly defined process, it is generally agreed that proteolysis is essential for flavor development in bacterially ripened cheeses (18, 63). Proteolytic enzymes present in this group of products include chymosin, plasmin, and proteolytic enzymes from starter cultures, adjunct cultures, and nonstarter lactic acid bacteria. The specificities and relative activities of the proteolytic enzymes present in the cheese matrix determine which peptides and amino acids accumulate and, hence, how flavor develops. The use of isogenic strains, differing only in the specificity of activity of a single proteolytic enzyme, will have great utility in elucidating the role of individual enzymes in cheese flavor development.

Free amino acids and peptides in the cheese matrix can contribute to cheese flavor either directly or indirectly and with positive or negative effects. Cheese flavor development has been the subject of numerous comprehensive reviews (18, 59, 61). A major negative effect of proteolytic products is bitterness, which is believed to be caused by hydrophobic peptides ranging in length from 3 to 27 amino acid residues (40). They are believed to be generated from casein principally by the joint action of chymosin and the lactic acid bacteria proteinase, and they can be hydrolyzed to nonbitter peptides and amino acids by lactic acid bacteria peptidases. Therefore, the accumulation of bitter peptides is dependent on the relative rates of their formation and hydrolysis. A variety of volatile compounds can be derived from catabolism of amino acids (24). Numerous sulfur-containing compounds, particularly methanethiol, are thought to be important in cheese flavor. The formation of methanethiol is believed to result from the action of cystathionine β-lyase on methionine (1). Alternatively, amino acid catabolism can give rise to compounds which have a negative impact on cheese flavor. For example, catabolism of aromatic amino acids can give rise to compounds such as indole and skatole, which contribute to "unclean" flavors in cheese. Overall, proteolysis is believed to be essential for development of characteristic flavor compounds in bacterial ripened cheeses; however, the mechanism(s) by which products of proteolysis give rise to beneficial flavor compounds remains unknown. Additionally, other than bitterness, the mechanisms by which proteolysis affects the development of undesirable flavor compounds are unknown.

BACTERIOPHAGES AND BACTERIOPHAGE RESISTANCE

Bacteriophage infection may lead to a decrease or complete inhibition of lactic acid production by the starter culture. This has a major impact on the manufacture of fermented dairy products, as lactic acid synthesis is required to produce these products. Additionally, slow acid production disrupts manufacturing schedules and typically results in products which are of lower economic value. The severe consequences of bacteriophage infection have led to extensive investigations into bacteriophages and the mechanisms by which lactic acid bacteria resist infection. Bacteriophage infection of lactic acid bacteria was first identified in lactococcal starter cultures in the 1930s. Since then, the dairy industry has instituted improved sanitation regimens, used sophisticated starter culture propagation vessels, developed starter culture systems to minimize the impact of phage infection, and isolated or constructed starter strains with enhanced bacteriophage resistance. However, bacteriophage infection of the starter culture has remained a significant problem in the dairy industry. Klaenhammer and Fitzgerald (35) listed four reasons why dairy product fermentations are particularly susceptible to bacteriophage infection: (i) phage contamination can occur when fermentations are not protected from environmental contaminants and take place in a nonsterile fluid medium, pasteurized milk; (ii) processing efficiency is easily disrupted because batch culture fermentations occur under increasingly stringent manufacturing schedules; (iii) increasing reliance on specialized strains limits the number and diversity of available dairy starter cultures; and (iv) continuous use of defined cultures provides an ever-present host for bacteriophage attack.

Historically, most bacteriophage-related problems have occurred with lactococcal starter cultures; however, problems with starter systems which use *S. thermophilus* and *Lb. delbrueckii* subsp. *bulgaricus* have been increasing. The principal problem has been bacteriophage infection of *S. thermophilus* strains (49). For more extensive reviews of bacteriophage and bacteriophage resistance in lactic acid bacteria, readers are referred to other reports (13, 25, 35).

Bacteriophages of Lactic Acid Bacteria

Significant progress has been made on the characterization of bacteriophages from lactic acid bacteria. All of the bacteriophages examined contain double-stranded linear DNA genomes with either cohesive or circularly permuted terminally redundant ends. Both lytic and temperate bacteriophages have been characterized. A major outcome of this characterization has been a clear classification, based on DNA homology and morphological studies, of lactococcal bacteriophages (30). Twelve species have been described, and type phages have been proposed for each species. Less comprehensive DNA homology data are also available for bacteriophages of lactobacilli, *S. thermophilus*, and *Leuconostoc* species. Additionally, complete nucleotide sequence information is now available for several lactococcal bacteriophages. This has made it possible to determine how bacteriophage-host interactions evolve over time and to construct novel bacteriophage defense mechanisms.

Lactococcal Bacteriophage Resistance Mechanisms

Selective environmental pressure placed on lactococci by bacteriophages over thousands of years has resulted in strains which contain numerous bacteriophage defense mechanisms. The best-characterized bacteriophage-resistant strain is *Lc. lactis* ME2. This strain has been shown to contain at least five distinct phage defense loci, including one which interferes with bacteriophage adsorption, two restriction-modification systems, and two abortive infection (Abi) mechanisms (35). These defense loci are encoded by plasmids capable of conjugal transfer, which suggests that genetic exchange between starter cultures has had an important role in the development of bacteriophage-resistant starter cultures. Recombinant DNA techniques have also been used to construct lactococcal strains with enhanced bacteriophage resistance. These include the use of antisense RNA derived from conserved bacteriophage genes and the cloning of bacteriophage origins of replication on multicopy plasmids. The latter approach is thought to titrate bacteriophage replication factors and result in a bacteriophage defense mechanism similar to abortive bacteriophage infection. The remainder of this section will cover the previously mentioned three natural gene bacteriophage defense mechanisms.

Interference with Bacteriophage Adsorption

The adsorption of a bacteriophage to a host cell is determined by bacteriophage specificity, physicochemical properties of the cell envelope, accessibility and density of bacteriophage receptor material, and electrical potential across the cytoplasmic membrane (57). The complexity of this interaction has facilitated the isolation of bacteriophage-resistant starter cultures by exposing them to bacteriophages and isolating resistant variants. Mutants isolated in such a fashion typically have a reduced capacity to adsorb the bacteriophage(s) used in the challenge and are referred to as bacteriophage-insensitive mutants. This is a common practice for deriving starter cultures with reduced bacteriophage sensitivity.

Researchers have begun to characterize host components required for bacteriophage adsorption. It is now well established that bacteriophages initially interact reversibly with cell envelope-associated polysaccharide and then interact irreversibly with a cell membrane protein(s) (48). The best characterized example of a lactococcal mechanism which interferes with adsorption of bacteriophages to the cell envelope-associated polysaccharide is *Lc. lactis* subsp. *cremoris* SK110. Reduction in the ability of a bacteriophage to adsorb to cells of this strain is due to masking of the phage receptors by a galactose-containing lipoteichoic acid, rather than the absence of receptors.

Inhibition of adsorption of lactococcal bacteriophage as a result of the lack of a membrane-bound protein has also been clearly demonstrated. Characterization of phage-resistant mutants of *Lc. lactis* subsp. *lactis* C2 has revealed that mutants bind normally with bacteriophage, but no plaques are formed. Subsequently, it was demonstrated that a 32-kDa membrane protein essential for bacteriophage infection was lacking in these mutants. Similarly, it has been suggested that a membrane protein is involved in bacteriophage adsorption to *Lc. lactis* subsp. *lactis* ML3.

Restriction-Modification Systems

Restriction-modification systems are widely distributed in lactococci and are often plasmid encoded. In fact, it is not uncommon for strains to contain two or more restriction-modification systems, and at least 23 plasmids which encode these systems have been identified. The two components of a restriction-modification system are a site-specific modifying enzyme and a corresponding site-specific restriction endonuclease. This system enables the cell to differentiate bacteriophage DNA from its own DNA and inactivate the foreign DNA by hydrolysis. The typical end result of bacteriophage infection of a culture containing a restriction-modification system is a reduction in the number of progeny bacteriophages produced. The extent of reduction is dependent on both the activity of the restriction-modification system and the number of unmodified restriction endonuclease sites on the bacteriophage genome. It is important to note that bacteriophages which escape restriction will give rise to modified progeny phage which are immune to the corresponding restriction endonuclease. Therefore, while this is an important and widely distributed mechanism of bacteriophage defense, it is also very fragile.

Two mechanisms by which bacteriophages have evolved resistance to restriction-modification systems have been identified. Characterization of lactococcal bacteriophages has revealed that they contain far fewer restriction endonuclease sites than expected for genomes of their size, suggesting that evolutionary pressure has selected for bacteriophages with few restriction endonuclease sites. This view is supported by the observation that lactococcal bacteriophages which have recently emerged in the dairy industry are more sensitive to restriction-modification systems. The characterization of bacteriophages which have evolved to overcome a specific restriction-modification system has revealed that they have acquired a functional copy of the modification enzyme from that system. These examples illustrate that bacteriophage-host interactions are continually changing, with bacterial strains acquiring new defense mechanisms and bacteriophages evolving mechanisms to overcome them.

Abi Systems

Like restriction-modification systems, Abi systems are widely distributed in lactococci and are frequently plasmid encoded, although some are also encoded by episomes. By definition, these mechanisms inhibit bacteriophage infection following adsorption, DNA penetration, and the early stages of the bacteriophage lytic cycle. Few infections successfully release viable progeny, and those which are successful result in less progeny bacteriophage being released. The end result for the host, even those which do not release viable bacteriophages, is death.

Five genes encoding Abi mechanisms have been isolated from lactococci and sequenced; comparison of deduced amino acid sequences indicates that two of the genes are related. The gene product of *abiA* inhibits or severely reduces bacteriophage DNA replication. The other characterized *abi* genes result in significant reduction in burst size, i.e., the number of progeny bacteriophages released per infection. Frequently, *abi* genes are associated with restriction-modification systems. These systems work well together; the restriction-modification system reduces the number of viable bacteriophage genomes, and the Abi system reduces the number of progeny bacteriophages released from infected cells which evade the restriction-modification system.

Some bacteriophages have the ability to overcome Abi mechanisms. These phages either have a relatively small change in their genomes or have acquired a piece of chromosomal DNA from their hosts. Durmaz and Klaenhammer (15) proposed use of a culture system in which derivatives of a single strain containing different Abi and restriction-modification systems are rotated. This strategy would reduce the selection for bacteriophages which have evolved resistance to a specific defense mechanism by rotating these mechanisms.

GENETICS OF LACTIC ACID BACTERIA

Research on the genetics of lactic acid bacteria began in the early 1970s. Initially, research focused on plasmids and natural gene transfer systems in lactococci. Interest in the genetics of lactic acid bacteria has expanded rapidly, and there are now hundreds of researchers active in this area. This interest has resulted in the development of a relatively detailed understanding of the basic genetics of these bacteria, natural gene transfer systems, and the development of tools required for application of recombinant DNA techniques. For a more complete description of the genetics of lactic acid bacteria, the reader is referred to other publications (7, 12, 21, 46, 64).

General Genetics of Lactic Acid Bacteria

Genetic elements of lactic acid bacteria which have been characterized include chromosomes, transposable elements, and plasmids. Chromosomes of lactic acid bacteria are smaller than those of other eubacteria, ranging from 1.1 to 2.6 Mbp, depending on the species (6). Several chromosomal genetic and physical maps have been constructed for this group of bacteria. These maps have revealed the genetic organization of lactic acid bacteria and will be useful in future genetic characterization. Transposable elements, genetic elements capable of moving as discrete units from one site to another in the genome, have been found in lactic acid bacteria (21). Insertion sequences, the simplest of transposable elements, are widely distributed in bacteria and have also been identified in numerous lactic acid bacteria. Their ability to mediate molecular rearrangements and affect gene regulation has had both positive and negative implications for dairy product fermentations. In one case, the incorporation of an insertion sequence into a prophage of *Lactobacillus casei* resulted in the conversion of a temperate bacteriophage into a virulent bacteriophage. Alternatively, insertion sequence-mediated cointegration has played a pivotal role in the dissemination of many beneficial characteristics via conjugation. More complex transposable elements, such as self-transmissible conjugal transposons which encode the production of the bacteriocin nisin, the ability to metabolize sucrose, and a bacteriophage defense mechanism, have also been described in detail (21). Plasmids, i.e., autonomous replicating extrachromosomal circular DNA molecules, have been identified in several lactic acid bacteria. These are of particular importance in lactococci, in which they encode numerous characteristics essential for dairy product fermentations, including lactose metabolism, proteinase activity, oligopeptide transport, bacteriophage resistance mechanisms, bacteriocin production and immunity, bacteriocin resistance, exopolysaccharide production, and citrate utilization (64).

An indication of the extent of research activity in the area of molecular biology of lactic acid bacteria is the fact that 535 nucleotide sequences from this group, the majority of which are from lactobacilli and lactococci, have been submitted to the GenBank genetic sequence data bank. Research has resulted in the identification of sequences responsible for the initiation and termination of transcription and initiation of translation in lactobacilli and lactococci. Additionally, regulation or transcription of the lactococcal lactose operon has been partially characterized. The cumulative knowledge gained by the characterization of these genes, their regulatory signals, and excretion mechanisms has set the stage for the use of molecular biology techniques to construct strains of lactic acid bacteria which express proteins of interest in a controlled fashion.

Natural Gene Transfer Mechanisms

Transduction

Transduction, the transfer of bacterial genetic material by a bacteriophage, has been demonstrated in lactococci, lactobacilli, and *S. thermophilus*. Transduction plays an important role in determining that lactose metabolism and proteinase activity are plasmid encoded in lactococci. However, its use in the construction of strains to be used in industry is limited by the relatively narrow host range of transducing bacteriophages.

Conjugation

Conjugation, the transfer of genetic material from one bacterial cell to another via cell-to-cell contact, has been characterized in detail in lactococci. In fact, most plasmid-encoded characteristics important in the manufacture of fermented dairy products can be transferred by conjugation. By utilizing an approach which does not require antibiotic-resistance markers, conjugation has been used to transfer plasmids which encode bacteriophage resistance genes into commercial lactococcal strains. These strains have enhanced resistance to bacteriophages and have been used successfully in the dairy industry for several years. Conjugation is likely to be used in the future to construct other lactococcal strains with enhanced industrial utility.

Genetic Modification of Lactic Acid Bacteria by Using Recombinant DNA Techniques

Experiments using recombinant DNA techniques have led to most of the recent advances in the understanding of the physiology and genetics of lactic acid bacteria. The power of recombinant DNA approaches is that strains which differ in a single defined genetic alteration, e.g., inactivation of a specific gene, can be constructed. By comparing the wild-type culture with its isogenic derivative, the role of that gene in the phenotype being examined can be unequivocally determined. This general approach has resulted in a detailed understanding of how these bacteria utilize lactose, obtain essential amino acids, produce diacetyl, and resist bacteriophage infection. Additionally, recombinant DNA approaches have been used to construct novel bacteriophage resistance mechanisms and to overproduce enzymes of interest in dairy fermentations. However, because of regulatory and consumer concerns, recombinant DNA techniques have not yet been used for the construction of commercial strains. In the future, commercial strains will likely be constructed by using recombinant DNA techniques. Readers interested in more comprehensive reviews on the physiology and genetics of lactic acid bacteria should consult texts (20, 55).

References

1. **Alting, A. C., W. J. M. Engels, S. van Schalkwijk, and F. A. Exterkate.** 1995. Purification and characterization of cystathionine β-lyase from *Lactococcus lactis* subsp. *cremoris* B78 and its possible role in flavor development in cheese. *Appl. Environ. Microbiol.* **61**:4037–4042.
2. **Bassit, N., C. Y. Boquien, D. Picque, and G. Corrieu.** 1993. Effect of initial oxygen concentration on diacetyl and acetoin production by *Lactococcus lactis* subsp. *lactis* biovar *diacetylactis*. *Appl. Environ. Microbiol.* **59**:1893–1897.
3. **Bassit, N., C. Y. Boquien, D. Picque, and G. Corrieu.** 1995. Effect of temperature on diacetyl and acetoin production by *Lactococcus lactis* subsp. *lactis* biovar *diacetylactis* CNRZ 483. *J. Dairy Res.* **62**:123–129.
4. **Bills, D. D., and E. A. Day.** 1966. Dehydrogenase activity of lactic streptococci. *J. Dairy Sci.* **49**:1473–1477.
5. **Bills, D. D., M. E. Morgan, L. M. Libby, and E. A. Day.** 1965. Identification of compounds responsible for fruity flavor defect of experimental cheeses. *J. Dairy Sci.* **48**:1168–1173.
6. **Bourgeois, P. L., M. Lautier, and P. Ritzenthaler.** 1993. Chromosome mapping in lactic acid bacteria. *FEMS Microbiol. Rev.* **12**:109–124.
7. **Chassy, B. M., and C. M. Murphy.** 1993. *Lactococcus* and *Lactobacillus*, p. 65–82. *In* A. L. Sonenshein, J. A. Hoch, and R. Losick (ed.), *Bacillus subtilis and Other Gram-Positive Bacteria.* American Society for Microbiology, Washington, D.C.
8. **Cogan, T. M.** 1987. Co-metabolism of citrate and glucose by *Leuconostoc* spp.: effects on growth, substrates and products. *J. Appl. Bacteriol.* **63**:551–558.
9. **Cogan, T. M., R. J. Fitzgerald, and S. Doonan.** 1984. Acetolactate synthase of *Leuconostoc lactis* and its regulation of acetoin production. *J. Dairy Res.* **51**:597–604.
10. **Condon, S.** 1987. Responses of lactic acid bacteria to oxygen. *FEMS Microbiol. Rev.* **46**:269–280.
11. **Crow, V. L.** 1990. Properties of 2,3-butanediol dehydrogenases from *Lactococcus lactis* subsp. *lactis* in relation to citrate fermentation. *Appl. Environ. Microbiol.* **56**:1656–1665.
12. **de Vos, W. M., and G. F. M. Simons.** 1994. Gene cloning and expressions systems in lactococci, p. 52–105. *In* M. J. Gasson and W. M. de Vos (ed.), *Genetics and Biotechnology of Lactic Acid Bacteria.* Chapman and Hall, Ltd., London.
13. **Dinsmore, P. K. and T. R. Klaenhammer.** 1995. Bacteriophage resistance in *Lactococcus. Mol. Biotechnol.* **4**:297–314.
14. **Driesson, F. M., and Z. Puhan.** 1988. Technology of mesophilic fermented milks. *Int. Dairy Fed. Bull.* **227**:75–81.
15. **Durmaz, E., and T. R. Klaenhammer.** 1995. A starter culture rotation strategy incorporating paired restriction/modification and abortive infection bacteriophage defenses in a single *Lactococcus lactis* strain. *Appl. Environ. Microbiol.* **61**:1266–1273.
16. **Fordyce, A. M., V. L. Crow, and T. D. Thomas.** 1984. Regulation of product formation during glucose or lactose limitation in nongrowing cells of *Streptococcus lactis. Appl. Environ. Microbiol.* **48**:332–337.
17. **Fox, P. F. (ed.).** 1993. *Cheese: Chemistry, Physics and Microbiology,* vol. 1 and 2. Chapman and Hall, Ltd., London.
18. **Fox, P. F., J. Law, P. L. H. McSweeney, and J. Wallace.** 1993. Biochemistry of cheese ripening, p. 389–438. *In* P. F. Fox (ed.), *Cheese: Chemistry, Physics and Microbiology,* vol. 1. Chapman and Hall, Ltd., London.
19. **Fox, P. F., J. A. Lucey, and T. M. Cogan. 1990.** Glycolysis and related reactions during cheese manufacture and ripening. *Food Sci. Nutr.* **29**:237–253.
20. **Gasson, M. J., and W. M. de Vos (ed.).** 1994. *Genetics and Biotechnology of Lactic Acid Bacteria.* Chapman and Hall, Ltd., London.

21. **Gasson, M. J., and G. F. Fitzgerald.** 1994. Gene transfer systems and transposition, p. 1–51. *In* M. J. Gasson and W. M. de Vos (ed.), *Genetics and Biotechnology of Lactic Acid Bacteria.* Chapman and Hall, Ltd., London.
22. **Green, M. L., and D. J. Manning.** 1982. Development of texture and flavor in cheese and other fermented products. *J. Dairy Res.* **49:**737–748.
23. **Griffith, R., and E. G. Hammond.** 1989. Generation of Swiss cheese flavor components by the reaction of amino acids with carbonyl compounds. *J. Dairy Sci.* **72:**604–613.
24. **Hemme, D., C. Boulianne, F. Métro, and M.-J. Desmazeaud.** 1982. Microbial catabolism of amino acids during cheese ripening. *Sci. Aliments* **2:**113–123.
25. **Hill, C.** 1993. Bacteriophage and bacteriophage resistance in lactic acid bacteria. *FEMS Microbiol. Rev.* **12:**87–108.
26. **Huang, D. Q., H. Prévost, and C. Divies.** 1995. Principal characteristics of β-galactosidase from *Leuconostoc* spp. *Int. Dairy J.* **5:**29–43.
27. **Hugenholtz, J.** 1993. Citrate metabolism in lactic acid bacteria. *FEMS Microbiol. Rev.* **12:**165–178.
28. **Imhof, R., and J. O. Bosset.** 1994. Review: relationships between micro-organisms and formation of aroma compounds in fermented dairy products. *Lebensm. Unters. Forsch.* **198:**267–276.
29. **International Dairy Federation.** 1988. Fermented milks: science and technology. *Int. Dairy Fed. Bull.* 227.
30. **Jarvis, A. W., G. F. Fitzgerald, M. Mata, A. Mercenier, H. Neve, I. B. Powell, C. Ronda, M. Saxelin, and M. Teuber.** 1991. Species and type phages of lactococcal bacteriophages. *Intervirology* **32:**2–9.
31. **Juillard, V., D. Le Bars, E. R. S. Kunji, W. N. Konings, J-.C. Gripon, and J. Richard.** 1995. Oligopeptides are the main source of nitrogen for *Lactococcus lactis* during growth in milk. *Appl. Environ. Microbiol.* **61:**3024–3030.
32. **Kamaly, M. K., and E. H. Marth.** 1989. Enzyme activities of lactic streptococci and their role in maturation of cheese: a review. *J. Dairy Sci.* **72:**1945–1966.
33. **Keenan, T. W., R. C. Lindsay, M. E. Morgan, and E. A. Day.** 1966. Acetaldehyde production by single strain lactic streptococci. *J. Dairy Sci.* **49:**10–14.
34. **Kempler, G. M., and L. L. McKay.** 1981. Biochemistry and genetics of citrate utilization in *Streptococcus lactis* spp. *diacetylactis. J. Dairy Sci.* **64:**1527–1539.
35. **Klaenhammer, T. R., and G. F. Fitzgerald.** 1994. Bacteriophages and bacteriophage resistance, p. 106–168. *In* M. J. Gasson and W. M. de Vos (ed.), *Genetics and Biotechnology of Lactic Acid Bacteria.* Chapman and Hall, Ltd., London.
36. **Kok, J., and W. M. de Vos.** 1994. The proteolytic system of lactic acid bacteria, p. 169–210. *In* M. J. Gasson and W. M. de Vos (ed.), *Genetics and Biotechnology of Lactic Acid Bacteria.* Chapman and Hall, Ltd., London.
37. **Kunji, E. R. S., A. Hagting, C. J. De Vries, V. Juillard, A. J. Haandrikman, B. Poolman, and W. N. Konings.** 1995. Transport of β-casein-derived peptides by the oligopeptide transport system is a crucial step in the proteolytic pathway of *Lactococcus lactis. J. Biol. Chem.* 270:1569–1574.
38. **Lees, G. J., and G. R. Jago.** 1978. Role of acetaldehyde in metabolism: a review. 1. Enzymes catalyzing reactions involving acetaldehyde. *J. Dairy Sci.* **61:**1205–1215.
39. **Lees, G. J., and G. R. Jago.** 1978. Role of acetaldehyde in metabolism: a review. 2. The metabolism of acetaldehyde in cultured dairy products. *J. Dairy Sci.* **61:**1216–1224.
40. **Lemieux, L., and R. E. Simard.** 1992. Bitter flavour in dairy products. II. A review of bitter peptides from caseins: their formation, isolation and identification, structure masking and inhibition. *Lait* **72:**335–382.
41. **Lindgren, S. E., and L. T. Axelsson.** 1990. Anaerobic L-lactate degradation by *Lactobacillus plantarum. FEMS Microbiol. Lett.* **66:**209–214.
42. **Lindsay, R. C., E. A. Day, and W. E. Sandine.** 1965. Green flavor defect in lactic starter cultures. *J. Dairy Sci.* **48:**863–869.
43. **Lucey, C. A., and S. Condon.** 1986. Active role of oxygen and NADH oxidase in growth and energy metabolism of *Leuconostoc. J. Gen. Microbiol.* **132:**1789–1796.
44. **Manning, D. J.** 1979. Chemical production of essential Cheddar flavor compounds. *J. Dairy Res.* **46:**531–537.
45. **McKay, L. L., and K. A. Baldwin.** 1974. Altered metabolism in a *Streptococcus lactis* C2 mutant deficient in lactate dehydrogenase. *J. Dairy Sci.* **57:**181–186.
46. **Mercenier, A., P. H. Pouwels, and B. M. Chassy.** !994. Genetic engineering of lactobacilli, leuconostocs and *Streptococcus thermophilus,* p. 252–294. *In* M. J. Gasson and W. M. de Vos (ed.), *Genetics and Biotechnology of Lactic Acid Bacteria.* Chapman and Hall, Ltd., London.
47. **Mierau, I.** 1996. Peptide degradation in *Lactococcus lactis in vivo:* a first exploration. Ph.D. thesis. University of Groningen, Groningen, The Netherlands.
48. **Montville, M. R., B. Ardestani, and B. L. Geller.** 1994. Lactococcal bacteriophage require a host cell wall carbohydrate and a plasma membrane protein for adsorption and ejection of DNA. *Appl. Environ. Microbiol.* **60:**3204–3211.
49. **Oberg, C. J., and J. R. Broadbent.** 1993. Thermophilic starter cultures: another set of problems. *J. Dairy Sci.* **76:**2392–2406.
50. **Peterson, S. D., and R. T. Marshall.** 1990. Nonstarter lactobacilli in Cheddar cheese: a review. *J. Dairy Sci.* **73:**1395–1410.
51. **Poolman, B.** 1993. Energy transduction in lactic acid bacteria. *FEMS Microbiol. Rev.* **12:**125–148.
52. **Pritchard, G. G., and T. Coolbear.** 1993. The physiology and biochemistry of the proteolytic system in lactic and bacteria. *FEMS Microbiol. Rev.* **12:**179–206.
53. **Ramos, A., K. N. Jordan, T. M. Cogan, and H. Santos.** 1994. ^{13}C nuclear magnetic resonance studies of citrate and glucose cometabolism by *Lactococcus lactis. Appl. Environ. Microbiol.* **60:**1739–1748.
54. **Ramos, A., J. S. Lolkema, W. N. Konings, and H. Santos.** 1995. Enzyme basis for pH regulation of citrate and pyruvate metabolism by *Leuconostoc oenos. Appl. Environ. Microbiol.* **61:**1303–1310.
55. **Salminen, S., and A. von Wright (ed.).** 1993. *Lactic Acid Bacteria.* Marcel Dekker, Inc., New York.
56. **Schmitt, P., and C. Divies.** 1991. Co-metabolism of citrate and lactose by *Leuconostoc mesenteroides* subsp. *cremoris. J. Ferment. Bioeng.* **71:**72–74.
57. **Sijtsma, L., N. Jansen, W. C. Hazeleger, J. T. M. Wouters, and K. J. Hellingwerf.** 1990. Cell surface characteristics of bacteriophage-resistant *Lactococcus lactis* subsp. *cremoris*

SK110 and its bacteriophage-sensitive variant SK112. *Appl. Environ. Microbiol.* **56**:3230–3233.

58. **Smart, J. B., and T. D. Thomas.** 1987. Effect of oxygen of lactose metabolism in lactic streptococci. *Appl. Environ.* **53**:533–541.

59. **Steele, J. L.** 1995. Contribution of lactic acid bacteria to cheese ripening, p. 209–220. *In* E. L. Malin and M. H. Tunick (ed.), *Chemistry of Structure-Function Relationships in Cheese.* Plenum Publishing Corp., New York.

60. **Thomas, T. D., D. C. Ellwoos, and V. M. C. Longyear.** 1979. Change from homo- to heterolactic fermentation by *Streptococcus lactis* resulting from glucose limitation in anaerobic chemostat cultures. *J. Bacteriol.* **138**:109–117.

61. **Urbach, G.** 1995. Contribution of lactic acid bacteria to flavor compound formation in dairy products. *Int. Dairy J.* **5**:877–903.

62. **Vaughan, E. E., S. David, A. Harrington, C. Daly, G. F. Fitzgerald, and W. M. De Vos.** 1995. Characterization of plasmid-encoded citrate permease (*citP*) genes from *Leuconostoc* species reveals high sequence conversation with the *Lactococcus lactis citP* gene. *Appl. Environ. Microbiol.* **61**:3172–3176.

63. **Visser, S.** 1993. Proteolytic enzymes and their relation to cheese ripening and flavor: an overview. *J. Dairy Sci.* **76**: 329–350.

64. **von Wright, A., and M. Sibakov.** 1993. Genetic modification of lactic acid bacteria, p. 161–198. *In* S. Salminen and A. von Wright (ed.), *Lactic Acid Bacteria.* Marcel Dekker, Inc., New York.

65. **Warriner, K. S. R., and J. G. Morris.** 1995. The effects of aeration on the bioreductive abilities of some heterofermentative lactic acid bacteria. *Lett. Appl. Microbiol.* **20**: 322–327.

66. **Whitehead, W. E., J. W. Ayres, and W. E. Sandine.** 1993. A reviewer of starter media for cheese making. *J. Dairy Sci.* **76**:2344–2353.

67. **Wilkins, D. W., R. H. Schmidt, R. B. Shireman, K. L. Smith, and J. J. Jezeski.** 1986. Evaluating acetaldehyde synthesis from L-[^{14}C(U)] threonine by *Streptococcus thermophilus* and *Lactobacillus bulgaricus. J. Dairy Sci.* **69**:1219–1224.

68. **Yüksel, G. Ü., and J. L. Steele.** 1996. DNA sequence analysis, expression, distribution, and physiological role of the X-prolyl dipeptidyl aminopeptidase (PepX) gene from *Lactobacillus helveticus* CNRZ32. *Appl. Microbiol. Biotechnol.* **44**:766–773.

69. **Zourari, A., J. P. Accolas, and M. J. Desmazeaud.** 1992. Metabolism and biochemical characteristics of yogurt bacteria. A review. *Lait* **72**:1–34.

Herbert J. Buckenhüskes

32 Fermented Vegetables

In vegetable processing, two preservation methods, brining (or salting) and fermentation, are closely related to each other. Depending on the amount of salt (sodium chloride) added, the raw material will be preserved due to a decrease of water activity and high ionic activity (equilibrium salt content of >10%). Otherwise, vegetables will undergo fermentation. Brining is rarely practiced today due to problems of salt removal and decreased consumer acceptability of high-salt foods.

Fermentation is a very ancient process of preserving plant materials while retaining their nutritive value. The fermentation of vegetables may be affected by various groups of microorganisms (Table 32.1). Lactic acid bacteria and yeasts are preferentially used in the western hemisphere (Europe and America), whereas in the Orient a great number of victuals are fermented by molds (see chapter 34). However, the most extensively used procedure for biopreservation of vegetables involves lactic acid fermentation.

Lactic acid fermentation of vegetables is believed to have first been practiced by the Chinese in prehistoric times. However, the oldest written evidence dates from the first century A.D. when Pliny described the preservation of white cabbage in earthen vessels. Today we are aware that this procedure results in lactic acid fermentation of cabbage into sauerkraut. Due to the development of efficient heat sterilizing and refrigeration systems, lactic acid fermentation has, to some extent, lost its importance as a preservation method in industrialized countries, but it is still extensively used in developing countries, where in recent years it has gained in importance. This is due to the fact that fermentations today are more than just preservation methods. They are used as well

- to preserve vegetables and fruits
- to develop characteristic sensory properties, i.e., flavor, aroma, and texture
- to destroy naturally occurring toxins and undesirable components in raw materials (for examples, see Table 32.2)
- to improve digestibility, especially of some legumes (Table 32.2)
- to enrich products with desired microbial metabolites, e.g., L-(+)-lactic acid or amino acids (see Table 32.3)
- to create new products for new markets (for examples, see Table 32.3)
- to enhance dietary value (26)

An up-to-date review concerning the status of vegetable fermentation in Europe was prepared by the Cooperation in Science and Technology (COST) 91 Program of the European Communities (10). This publication lists 21 different vegetables that are commercially fermented (Table 32.4). In addition, an unspecified number of variably formulated vegetable blends as well as fermented vegetable juices from cabbage, carrots, celery, tomatoes, red beets, and turnips are available in the market. However, at present only olives, cabbage for

Table 32.1 Types of vegetable fermentation

Type of fermentation	Product	Raw material	Major microorganisms involved
Lactic acid	Fermented vegetables	See Table 4	Lactic acid bacteria, e.g., *Ln. mesenteroides*, *Lb. plantarum*
	Fermented vegetable juice	Sauerkraut, carrots, celery, tomatoes, red beets, turnips	*Lb. plantarum*, *Lb. casei*, *Lb. xylosus*, *Lb. bavaricus*
Acetic acid	Vinegar	Grapes, potatoes, various fruits after alcoholic fermentation	*Acetobacter aceti*, *Acetobacter pasteurianus*, *Acetobacter hansenii*, *Gluconobacter oxydans*
Alcoholic	Spirits	Potatoes, horseradish (for rimanto), peas (fen-djin)	*Saccharomyces cerevisiae*, *Kluyveromyces marxianus*
	Soy sauce	Soybeans	*Aspergillus oryzae*, *Tetragenococcus halophilus* yeasts

Table 32.2 Application of microorganisms in vegetable processing to achieve technological or nutritional advantages

Application	Species	Effect	Reference(s)
Sauerkraut	Paired starter cultures	Aroma formation	23, 24
	Ln. mesenteroides	Nisin resistant	
	Lc. lactis	Nisin producing	
Fermented vegetables	Various species	Increased bioavailability of iron by reducing the phytate content	2
Lupins, peas, lentils, and other legumes	Various species, e.g., *Lb. acidophilus*, *Lb. buchneri*, *Lb. fermentum*	Improved digestibility, e.g., by reduction in flatulence-causing oligosaccharides	12, 13
Cassava	*Lb. plantarum*	Reduction in linamarine	22
Delicacy salads, potato salads, ready-to-eat salads	Various	Extension of shelf life	25, 41
Salads, baby food, vegetable juices	Various	Reduction of nitrate content	44

Table 32.3 Development of new fermented vegetable products

Application	Species	Effect	Reference
Fermented fruit juices	*Lb. casei*	Supplementation with L-(+)-lactic acid	43
Tempeh-like products from beans (*Vicia faba*)	*Rhizopus oligosporus*	Shelf life, aroma formation	3
Yogurt-like product from soy proteins	Combined starter: *Streptococcus salivarius* subsp. *thermophilus*	Acidification	4
	Lb. delbrueckii subsp. *bulgaricus*	Aroma formation	
Camembert cheese-like product from advanced soy proteins	Combined starter: *Lc. lactis* subsp. *cremoris*, *Lc. lactis* subsp. *lactis* biovar. *diacetylactis*, *Ln. mesenteroides* subsp. *cremoris*	Acidification and aroma formation	4

Table 32.4 Lactic acid-fermented vegetables and fruits produced commercially (10)

Artichokes	Melons
Capers	Olives
Carrots	Red beets
Cauliflower	Red cabbage
Celery	Sauerkraut (white cabbage)
Cucumbers	Silver-skinned onions (levant garlic)
Eggplants	Swedes (*Brassica napus*)
Fungi (not specified)	Tomato-shaped peppers (green and red) (paprika)
Green beans	Turnips (*Brassica rapa*)
Green tomatoes	Waxy peppers (paprika)
Green peppers (paprika)	Whole cabbage (white cabbage)
Lupinus beans	

sauerkraut, and cucumbers for pickles are of real economic importance.

TECHNOLOGY OF FERMENTATION

The fermentation of vegetables represents a very complex network of independent and interactive microbiological, biochemical, enzymatic, chemical, and physical processes and reactions. Factors influencing major fermentation reactions and interactions and quality attributes affected are shown in Fig. 32.1. To complicate matters, fermentation is influenced by a multitude of exogenous factors which can be classified into four groups, namely, technological factors, nature and amount of admixed ingredients and additives, quality of the raw material, which in turn depends on numerous agricultural factors, and the nature of microflora on raw materials. An overview of the relevant influencing factors and their important interactions is given in Fig. 32.2.

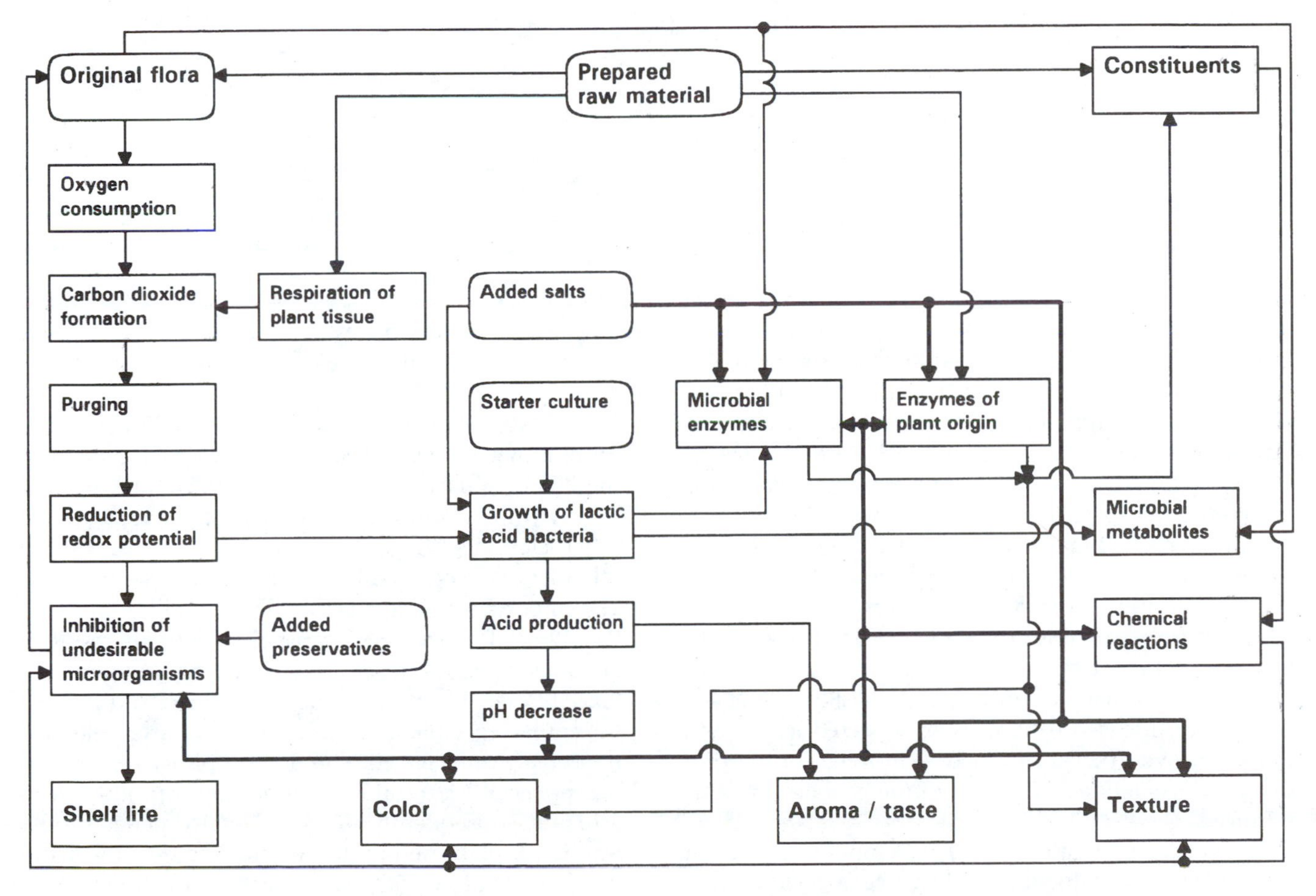

Figure 32.1 Dynamics of fermented vegetables.

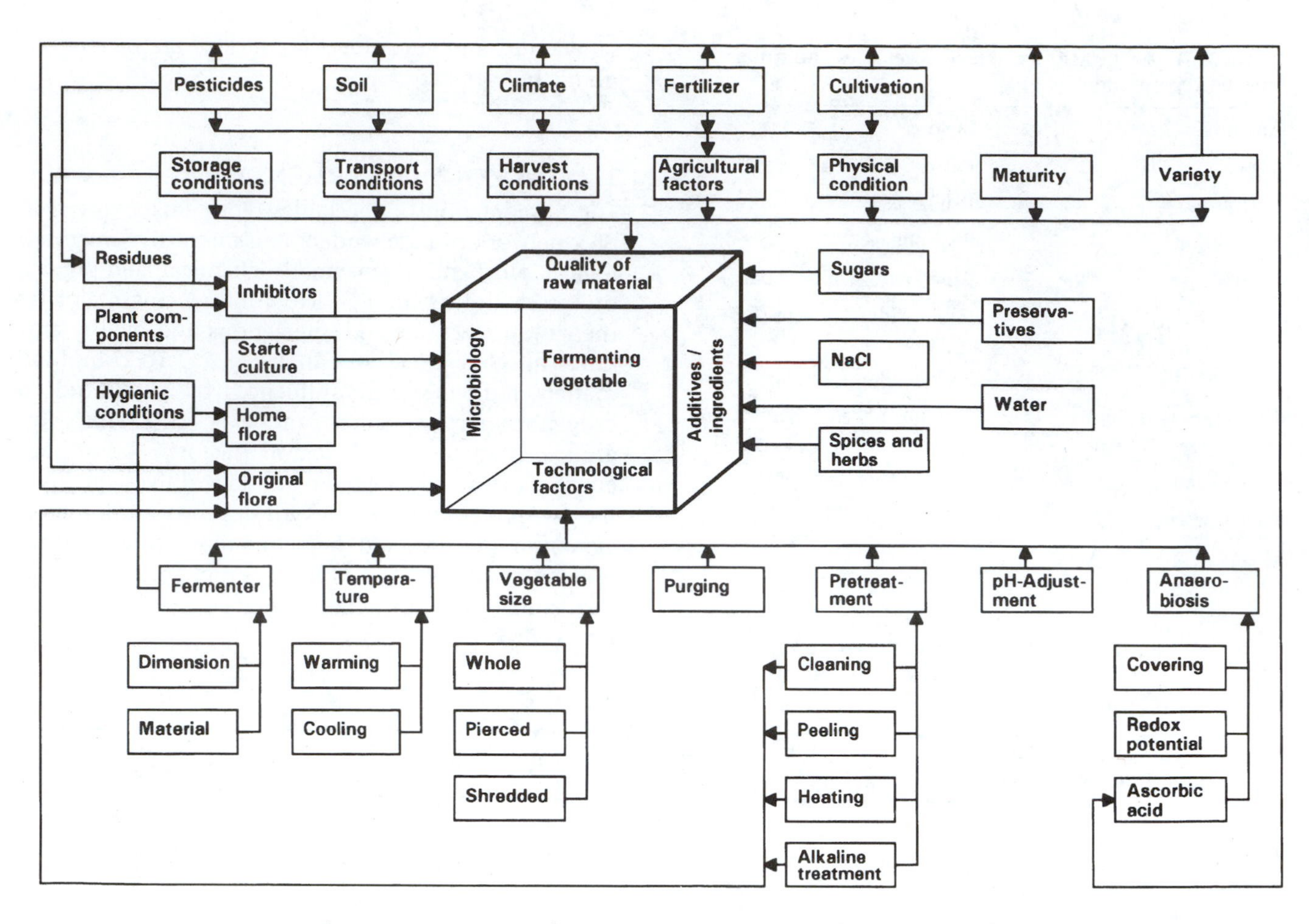

Figure 32.2 Factors influencing the quality of fermented vegetables.

Technological Factors

Botanical, physical, chemical, and textural properties of various vegetables differ widely, resulting in differences in technologies applied to fermentation. A simplified flow sheet for manufacturing lactic acid-fermented vegetables is shown in Fig. 32.3. The essential processing requirements can be described as follows.

1. To produce high-quality fermented products, vegetables must be sound, undamaged, and at the proper stage of maturity. Raw materials that are to be fermented whole must be suitably size-graded to achieve a homogeneous fermentation.

2. In the first instance, mechanically damaged, bruised, or diseased fruit, dirty or withered leaves, and unripe or overripe fruit must be removed. For the production of sauerkraut, the core of the cabbage must be removed.

3. Further pretreatment depends on the particular vegetable and may include peeling (e.g., beets, carrots), blanching (green beans), cooking (beets, to modify the texture as well as to destroy the characteristic earthy flavor), or an alkaline treatment (green Spanish-style olives).

4. If the raw material is to be fermented in an altered form, it may be pierced, shredded, or sliced. Piercing of the skin or surface of plant materials facilitates access of brine to inner tissues as well as movement of sugars and other substances from the tissues into the brine, which otherwise can occur only by relatively slow diffusion. The importance of this step is illustrated by onions in Fig. 32.4. More extensive wounding of tissue results in an accelerated decrease of pH during fermentation. Since fermentation mainly occurs in the liquid phase of the entire vegetable and brine system, a faster release of nutrients from pierced, shredded, or sliced onions or other vegetables results in a more rapid production of acid and, consequently, in an accelerated penetration of acid from the brine into the vegetable tissue. In cucumber fermentation, lactic acid bacteria can enter tissue from the brine via stomata and grow within. Yeasts pres-

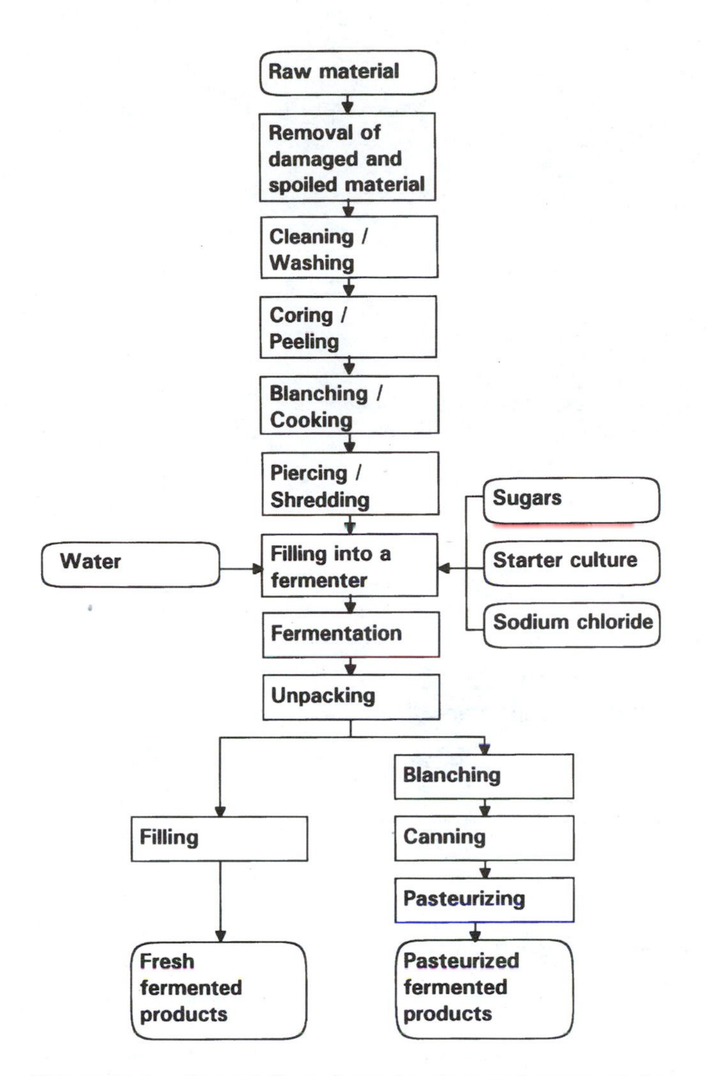

Figure 32.3 General flow sheet for the production of fermented vegetables (9).

ent in the brine are unable to enter cucumbers through stomata, presumably because of their larger size (14).

5. Whole, pierced, shredded, or sliced vegetables are placed in suitable fermentation vessels, which may range from 100-liter drums to tanks or silos with approximately 100 metric tons capacity. Filling must be done very carefully. To achieve favorable fermentation conditions, vegetables such as shredded cabbage must be salted homogeneously on a conveyor belt between shredding machines and tanks or during filling of silos. Whole fruits such as cucumbers or tomatoes must be placed into vessels already containing some brine to prevent mechanical damage. The filled vessels must be sealed in such a way that the plant material is totally covered by brine. To achieve anaerobic conditions, open tanks or silos are normally covered with wooden plates weighted with stones or by water-filled balloons. This process results in submersion of materials and, thus, exclusion of oxygen and spoilage microorganisms from the air.

6. The vast majority of vegetables are spontaneously fermented by lactic acid bacteria naturally present on their surfaces or from contact with equipment during preparation in processing facilities. Fermentative processes leading to good-quality products can occur at 10°C, but the optimum temperature is between 15 and 20°C. The fermentation time depends primarily on the temperature, the type of vegetable, the degree of disintegration, and the desired kind and quality of the final product. During fermentation, tanks must be carefully monitored for acid development and microbial defects. Sometimes it is necessary to remove yeasts and molds that have grown on the surface of brines. *Debaryomyces hansenii* and *Pichia membranaefaciens* are among the most common spoilage yeasts in brine.

7. The final fermented product may either be distributed fresh, packaged or unpackaged, or pasteurized in pouches, cans, or jars. Products such as sauerkraut can be stored without any further treatment in fermentation tanks until needed for sale in the retail market. For products that easily soften, the original fermentation liquid should be replaced by a brine containing an appropriate amount of salt and lactic acid.

If fermented vegetables are to be distributed unpasteurized, it is important that all fermentable carbohydrates have been metabolized. Otherwise, secondary fermentation by yeasts may result in gaseous spoilage, brine turbidity, and probably alcoholic fermentation. In the case of cucumbers, secondary fermentation may cause bloater formation (15). Butyric acid spoilage of fermented cucumbers can also occur as a result of secondary fermentation of lactic acid by bacteria (18).

In the case of pasteurized products, desired fermentation may be achieved as soon as the pH value is reduced to 4.1. Products are then blanched, placed in cans or jars, topped with fermentation liquid or a liquid containing salt and sometimes herbs, spices, and/or wine, and pasteurized. Until recently, pasteurization has been carried out by traditional methods of trial and error or by following the rule that "much helps much." To obtain less heat-treated and therefore higher-quality products, as well as to save energy, pasteurization processes should be evaluated and defined using *P*-value calculations. However, successful process applications are difficult because of a lack of information about heat inactivation kinetics of several enzymes. Pectinolytic enzymes or lipoxygenases can cause marked changes in sensory qualities (8).

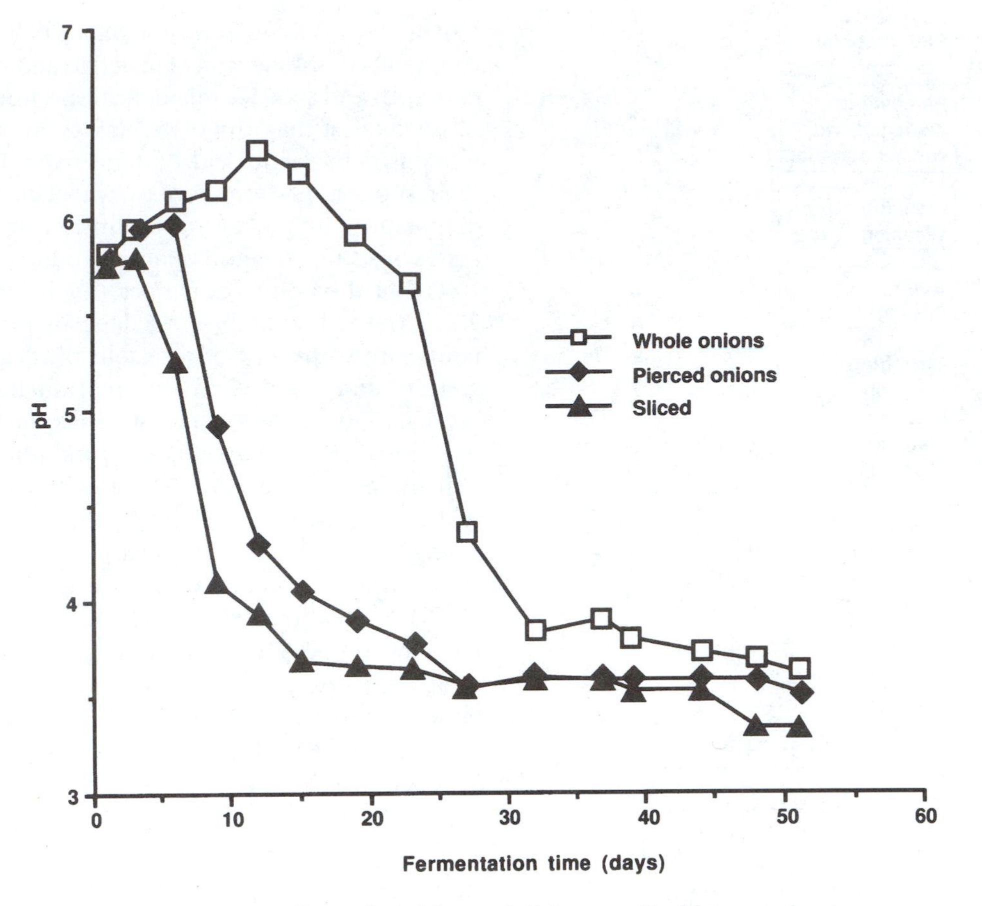

Figure 32.4 Effects of stitching and slicing on pH of fermented onions.

Ingredients and Additives

Sodium Chloride

In principle, vegetables can be fermented without the addition of sodium chloride (19). However, salt is often a very important ingredient for many reasons, a major one being its contribution to flavor. The amount of salt used for flavoring depends on the particular vegetable and on consumer demands. For example, the average salt content of canned sauerkraut is 11.3 g/kg in Germany and 16.7 g/kg in the United States (5).

Salt supports the development of anaerobic conditions in fermentation vessels. In sauerkraut production, salt enhances the release of tissue fluids from shredded cabbage during filling into fermentation vessels. To increase the fermentation capacity, a part of this brine is removed while the balance takes up space between shredded cabbage pieces, thus excluding oxygen and promoting growth of lactic acid bacteria.

Salt exerts a selective effect on microorganisms naturally present on vegetables. Increasing amounts of salt inhibit the growth of undesirable bacteria and fungi and select for lactic acid bacteria. In sauerkraut fermentation, heterofermentative lactic acid bacteria are favored by low salt concentration (ca. 1%) and greatly inhibited at 3%. Higher amounts of salt favor homofermentative species, resulting in accelerated fermentation. From a microbiological safety point of view, this may be of interest; however, an unbalanced fermentation by homofermentative lactic acid bacteria results in sauerkraut with poor flavor because of the lack of acetic acid and other desirable fermentative by-products. It is generally agreed that salt content below 0.8% often results in undesirable fermentation as well as in soft sauerkraut. The latter quality defect may be due to the fact that salt has an effect on the texture of final products as well as on several plant and microbial enzyme systems (Fig. 32.1). For example, the effect of increasing the amount of salt at pH 4 and pH 6 on the activity of cucumber pectinesterase has been demonstrated (Fig. 32.5) (35). Pectinesterase and other enzymes are involved in the softening of cabbage, cucumbers, olives, and other vegetables. Normally, cell wall degradation starts with enzyme-

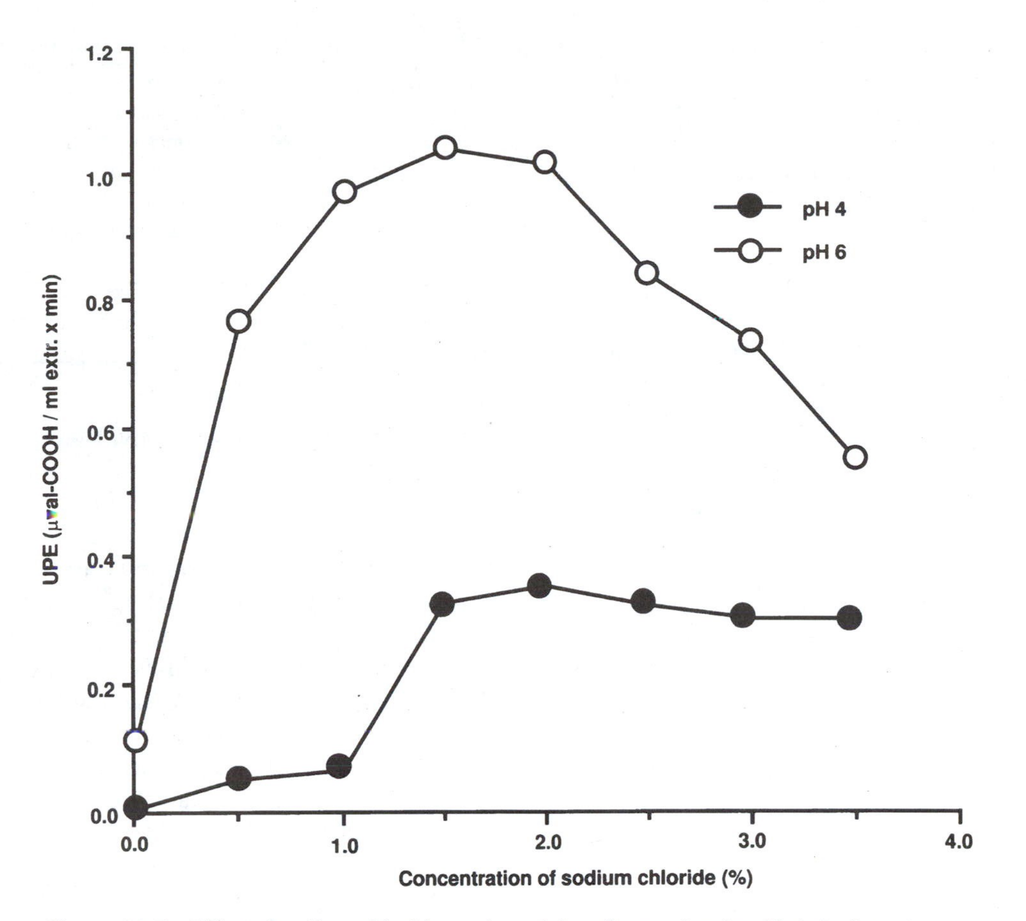

Figure 32.5 Effect of sodium chloride on the activity of cucumber (cv. Christine) pectinesterase.

catalyzed deesterification of carboxyl groups of pectin molecules. The polysaccharide chain is then hydrolyzed by polygalacturonases which are specifically active at the deesterified positions within the chain.

While these processes may normally cause a more or less rapid disintegration of the cell wall constituents with consequent softening of the plant material, investigations have shown that deesterification may also enhance the texture of fermented vegetables. The biochemical mechanism involves linking deesterified carboxyl groups of adjoining pectin molecules with bivalent cations, e.g., calcium. In cucumber fermentation, this can be achieved by stimulating pectinesterase by using raw materials of optimal maturity or by process control (e.g., adjustment of sodium chloride and calcium chloride concentration), by inhibiting polygalacturonases through the addition of sufficient amounts of sodium chloride or by heat treatment before fermentation, and finally by adding calcium chloride to fermented products (34).

It is not clear whether lactic acid bacteria are extensively involved in the softening of fermented vegetables. Pectinesterase and endopolygalacturonase activity in *Lactobacillus plantarum* has been demonstrated (38, 39), but this phenomenon has not been confirmed by other researchers.

Carbohydrates

Considering the variety of fermented vegetables produced in homes and available in the marketplace, as well as the number of crops ensilaged for animal feed, it is assumed that most if not all plant material will undergo lactic acid fermentation if properly prepared. The major requirement is the availability of sufficient amounts of fermentable carbohydrates to be converted to acetic and lactic acid. The extent of pH decrease in fermented plant material as caused by production of acids depends on the buffering capacity of the product, which is positively correlated with protein content (6). If the buffering capacity is known, it is possible to calculate the minimum amount of fermentable carbohydrate needed to produce sufficient acid to reduce the pH to 4.1. Numerous tests with sauerkraut have shown that in practice approximately twice the calculated minimum amount of carbohydrate is needed to meet the desired pH reduction. Using this as a basis for sufficient fermentation, the addition of fermentable carbohydrates such as sucrose or glucose is required to achieve desired fermentation of certain vegetables such as cauliflower and celery.

Other Additives

The use of other food additives in fermented vegetables depends on the particular product and is restricted by law in nearly all countries. The addition of ascorbic acid to sauerkraut is commonly done to prevent gray or brown discoloration of fresh or canned products. For the same purpose, citric acid or sulfur dioxide is permitted in several countries.

To prevent microbial spoilage and especially the growth of yeasts and molds on the surface of brines or in fermenting vegetables, sorbic acid is permitted in several countries. In some East European countries, the use of restricted amounts of tartaric acid, benzoic acid, or acetic acid is allowed.

Microbiology

Each particular type of vegetable provides a unique environment in terms of type, availability, and concentration of substrate, buffering capacity, competing microorganisms, and perhaps natural plant antagonists (14). As shown by characteristics compiled in Table 32.5, wounded

Table 32.5 Ecological factors affecting fermentation of vegetables

1. The fermentation substrate consists of solids in a liquid environment; the brine can be circulated by a pump, distributing the microorganisms as well as the sodium chloride and the released nutrients; within the vegetable pieces, mass transfer can only occur by diffusion.
2. Microorganisms originating from the raw material are distributed throughout the fermentation stock.
3. *Salmonella, Clostridium, Listeria,* and other undesirable microorganisms are probably present.
4. Raw materials are rich in nutrients, growth factors, and minerals; however, for microbial growth, these substances must move from the plant tissue into the surrounding liquid.
5. Water activity (a_w): 0.95–0.99
6. pH: 5.9–6.5 for cabbage; 8.5 is maximum for olives
7. Temperature: 5–20°C for sauerkraut, 25–25°C for vegetable juice
8. Sugar content: 25–100 g/kg of vegetable
9. Buffer capacity: 0.15–0.90 g of lactic acid/100 g of vegetable
10. Sodium chloride content: 0.6–2% for sauerkraut, 5–10% for cucumber brine

Table 32.6 Lactic acid bacteria associated with plants (14)

Lactobacillus spp.	*Pediococcus* spp.
Lb. arabinosus	*P. acidilactici*
Lb. brevis	*P. pentosaceus* (formerly *P. cerevisiae*)
Lb. buchneri	*Enterococcus* spp.
Lb. casei	*E. faecalis*
Lb. curvatus	*E. faecalis* var. *liquefaciens*
Lb. fermentum	*E. faecium*
Lb. plantarum	*Lactococcus lactis*
Lb. sake	
Leuconostoc mesenteroides	

or disintegrated plant tissues provide excellent substrates for microbial growth.

Fresh plant material harbors numerous and varied types of microorganisms. Although an extremely small population of lactic acid bacteria are present, it is assumed that plants are a natural habitat for some species. An analysis of 30 different samples of white cabbage from four growing seasons has shown that the microflora normally is dominated by aerobic bacteria (e.g., pseudomonads, enterobacteria, and coryneforms) and yeasts, while lactic acid bacteria represent 0.15 to 1.5% of the total bacterial population (40). Lactic acid bacteria mainly consist of *Leuconostoc mesenteroides* subsp. *mesenteroides;* streptococci have been isolated only in a few cases. Lactic acid bacteria reported to be associated with plants are summarized in Table 32.6. Although the composition of natural microflora is affected by methods of preparing vegetables for fermentation, this does not seem to have a significant effect on the fermentation process.

The microbial population in general, as well as the population of lactic acid bacteria, undergoes considerable change during the course of vegetable fermentation. The spontaneous fermentation of cabbage, for example, has been categorized into four distinct stages, as follows.

1. Fermentation starts as soon as the cabbage is placed into vessels. When cabbage is tightly packed, the number of strictly aerobic bacteria decreases immediately, while facultatively anaerobic enterobacteria grow for the first 2 or 3 days (Table 32.7). During this period, oxygen dissolved in the substrate is consumed by these microorganisms and by plant respiratory processes. The change in pH is influenced by the formation of lactic, acetic, formic, and succinic acids. The production of carbon dioxide at the beginning of fermentation can result in the formation of foam.

2. Due to a lack of oxygen, non-lactic acid bacteria are suppressed and overgrown by the facultatively anaerobic lactic acid bacteria. Lactic acid fermentation is initiated

Table 32.7 Development of bacteria during sauerkraut fermentation carried out at 17°C in laminated plastic pouches (40)

Time (days)	pH	Redox potential	Aerobic bacteria: Total (CFU/g)	Aerobic bacteria: Strictly aerobic (%)	Aerobic bacteria: Entero-bacteria (%)	Total no. (CFU/g): Lactic acid bacteria	Total no. (CFU/g): Yeasts
0	6.48	ND[a]	1.9×10^5	95.0	1.5	6.8×10^2	3.9×10^4
1	5.72	23.0	3.0×10^5	50.0	50.0	2.8×10^4	1.1×10^4
2	5.63	22.6	1.7×10^6	7.0	93.0	6.6×10^6	1.2×10^3
3	4.38	17.5	6.9×10^6		100.0	4.4×10^8	9.5×10^2
4	4.23	15.5	1.6×10^5		100.0	9.7×10^8	1.0×10^2
5	4.04	14.1	5.0×10^3		100.0	8.3×10^8	$<10^2$
7	4.02	14.5	1.5×10^3		100.0	3.3×10^8	$<10^2$
9	4.00	15.3	$< 10^2$	ND	ND	2.6×10^7	$<10^2$
11	3.93	15.4	$< 10^2$	ND	ND	4.2×10^7	$<10^2$
14	3.96	15.7	$< 10^2$	ND	ND	8.0×10^6	$<10^2$

[a]ND, not determined.

by heterofermentative lactic acid bacteria, namely, *Ln. mesenteroides*, followed by *Lactobacillus brevis*. Depending on the temperature, this succession of microorganisms is complete after 3 to 6 days, during which the concentration of lactic acid will increase to approximately 1%.

3. The third stage of fermentation is dominated by homofermentative lactic acid bacteria which are selectively favored by the complete lack of oxygen, lowered pH, and an elevated salt content. Approximately 90% of the homofermentative bacteria consist of streptobacteria, whereas streptococci and pediococci represent usually less than 10% of the total number of lactic acid bacteria. In the older literature streptobacteria are described as a single species, namely, *Lactobacillus plantarum* (formerly *Lactobacillus cucumeris*). Recent investigations have revealed that only 30 to 80% of the streptobacteria in sauerkraut are truly *Lb. plantarum*. During the early part of the third stage of fermentation, only 1 or 2% are true members. In addition to *Lb. plantarum*, two other closely related species are of importance, namely, *Lactobacillus sake* and *Lactobacillus curvatus*. Of special interest is a species designated *Lactobacillus bavaricus* which is characterized by the exclusive formation of L-(+)-lactic acid (27). Growth of homofermentative lactic acid bacteria will increase the total acid content of the substrate to 1.5 to 2.0%. Most sauerkraut is pasteurized upon reaching pH of 3.8 to 4.1.

4. Only sauerkraut which is stored in the fermentation vessel and distributed as fresh sauerkraut undergoes the final stage of fermentation, which is dominated by *Lb. brevis* and some heterofermentative species that are able to metabolize pentoses such as arabinose and xylose. Living plant material usually does not contain free pentoses. These sugars are liberated after harvest as a result of hydrolysis of cell wall material. The acid content may increase to 2.5%.

The initial population and growth rate of microorganisms, as well as salt and acid tolerance, are important factors that influence the sequential development of various lactic acid bacteria in most vegetable fermentations. In cucumber fermentation, excessive growth of heterofermentative *Ln. mesenteroides* is particularly undesirable, since the production of carbon dioxide may contribute to gaseous spoilage. The same situation occurs in the production of Spanish-style olives, which in general represent a very special type of fermentation. To destroy oleuropein, an extremely bitter-tasting glucoside in olives, freshly harvested fruits are treated with sodium hydroxide (lye) before fermentation. When the debittering process is finished, the lye must be removed by washing and neutralization, causing extensive losses in nutrients. For detailed information, see Garrido Fernández (21).

BIOGENIC AMINES

Most biogenic amines present in foods are formed by decarboxylation of corresponding amino acids through substrate-specific microbial enzymes. Biogenic amines and their precursors in fermented vegetables include ethanolamine (serine), putrescine (ornithine), cadaverine (lysine), spermidine (reaction of putrescine with a propylamine residue which is derived from methionine), phenylethylamine (phenylalanine), tyramine (tyrosine), and histamine (histidine). Concentrations of

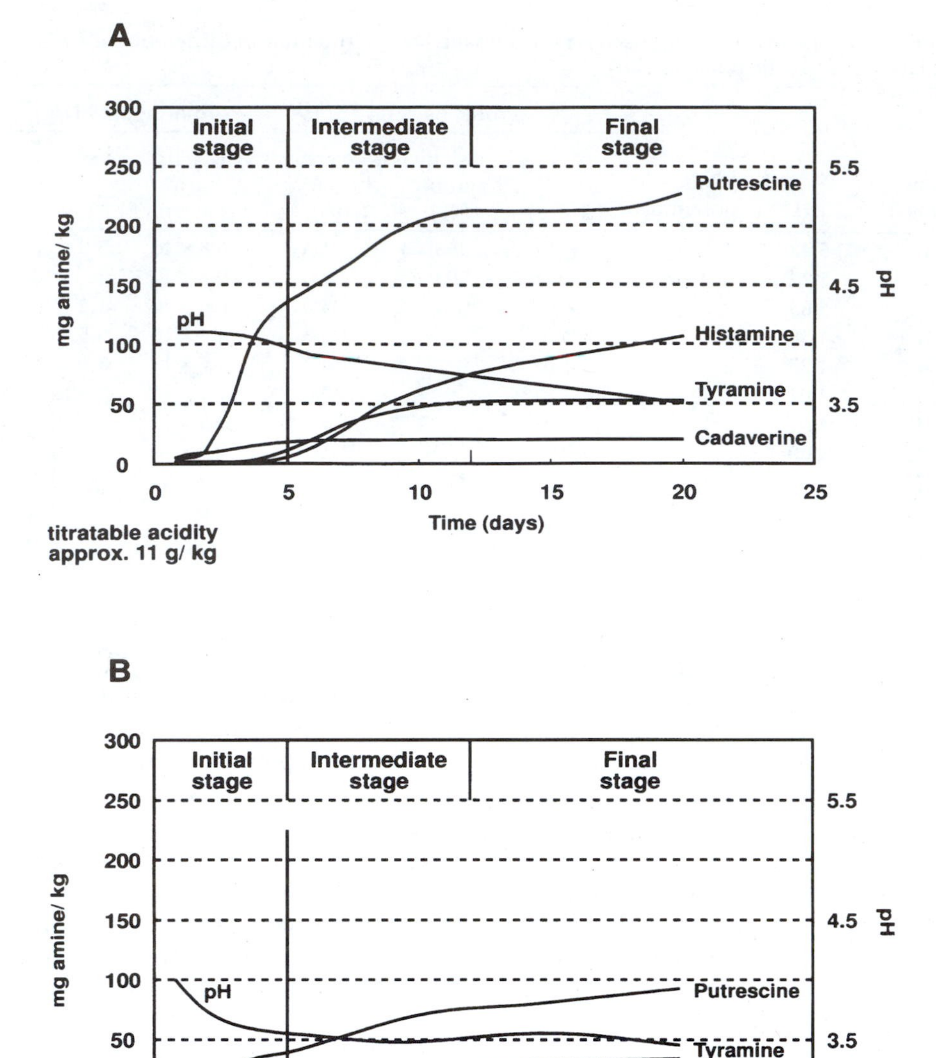

Figure 32.6 Formation of biogenic amines during fermentation of sauerkraut. (A) Spontaneous fermentation; (B) fermentation after inoculation with *Lb. plantarum.*

biogenic amines detected in sauerkraut are as high as 104 μg of histamine, 192 μg of tyramine, 311 μg of cadaverine, and 550 μg of putrescine per g (42). Since vegetables normally are fermented by an undefined spontaneous microflora, decarboxylase-positive microorganisms are involved in fermentation. Genera of the families *Enterobacteriaceae* and *Bacillaceae* present at beginning of fermentation, as well as species of *Lactobacillus, Pediococcus,* and *Streptococcus,* are known to be capable of decarboxylating one or more amino acids.

Biogenic amine formation during a spontaneous sauerkraut fermentation of 25-kg test batches of sauerkraut is shown in Fig. 32.6A. Although this investigation (29) was not attended by a detailed microbiological analysis, it is probable that the initial fermentation phase was dominated by *Ln. mesenteroides.* On the other hand, the formation of histamine was accompanied by a vigorous growth of *Pediococcus* species. The effect of a commercial *Lb. plantarum* starter culture (2×10^6 CFU/g) on the development of biogenic amines is illustrated in Fig. 32.6B. While

Table 32.8 Problems which may occur during production and storage of lactic acid-fermented vegetables (10)

Vegetable	Discoloration	Softening[a]	Other problems
Cucumbers	White or gray internal spots; surface discoloration	+++	Bloater formation, off flavor, butyric acid fermentation (18)
Sauerkraut	Gray, pink, or brown	+++	Off flavor
Olives	Gray, brown, or yellow spots	+++	Off flavor, butyric acid fermentation, propionic acid fermentation
Cauliflower	Yellow or pink		
Green tomatoes	Gray	++	
Peppers (paprika)	Gray	+++	Bitter taste
Carrots	Bleaching		Oxidized taste
Celery	Brown		
Green beans		++	Off flavor
Swedes	Gray or pink		

[a]++, relevant; +++, serious problem.

the amount of histamine, cadaverine, and tyramine remained on the same level as shown in Fig. 32.6A, the formation of putrescine was significantly suppressed. Despite inoculation with a starter culture, *Ln. mesenteroides* dominated the initial stage of fermentation; however, the growth of pediococci was not observed (29).

NITRATE REDUCTION

Depending on the plant species, the part of the plant utilized, several horticultural factors, and various processing and fermentation conditions, the nitrate content of fermented vegetables can vary considerably. Microbiological activity during the early fermentation stage converts nitrate into nitrite (1). In an acidic environment, nitrite may react with secondary or tertiary amines or with amides to form undesirable nitrosamines recognized as having potent and organ-specific carcinogenicity. To improve the safety of products, attempts have been made to reduce the initial nitrate content by biotechnological means. Various microorganisms, including lactic acid bacteria, are capable of reducing nitrate. *Paracoccus denitrificans* has been reported to reduce nitrate in commercial carrot juice (28). Nitrite accumulation was not observed.

A decrease in nitrate and nitrite content in plant products due to metabolic activity of lactic acid bacteria has been described by several researchers. Using MRS agar containing nitrate and molybdenum, it is possible to induce nitrate reductase formation by *Lactobacillus pentosus* (44). Investigations with cabbage juice have revealed that only 10% of the original nitrate is reduced and that nitrite accumulates. Several strains of *Lb. plantarum, Lb. pentosus,* and *Ln. mesenteroides* are capable of producing nitrite reductases. The final product of nitrite reduction by these bacteria is ammonia. In an acidic environment, nitrate or nitrite reduction is not complete, so the process is not applicable in the fermented-vegetable industry (17).

BACTERIOPHAGES

Bacteriophages are not a problem in spontaneous fermentation of vegetables or after application of starter cultures. This is because pure-culture fermentations seldom occur. If a starter or a naturally fermented batch is infected with phages, other strains of naturally occurring lactic acid bacteria will take over and carry out the fermentation. Generally, the problem is much more important in natural fermentation of cucumbers and olives than in sauerkraut fermentation. Cucumbers are fermented in brine which sometimes is circulated or purged with nitrogen. In the event of an infection, both treatments cause a homogeneous dissemination of phages in the brine. On the other hand, the absence of circulating brine in sauerkraut fermentation precludes the dissemination of phages.

MICROBIAL SPOILAGE AND OTHER PROBLEMS

A summary of economically important problems which may occur during the fermentation and storage of fermented vegetables is presented in Table 32.8. Although some problems can be prevented by good manufacturing practice, their molecular or microbial background is not always clear.

Bloater formation is a serious problem in cucumber fermentation. Damage is accentuated with larger cucum-

Table 32.9 Traits considered relevant in starter cultures for vegetable fermentation[a] (9, 15)

Criteria	Sauerkraut	Cucumbers	Olives	Vegetable juices[b]
Technologically relevant criteria				
Rapid and predominant growth	++	++	++	+
Homofermentative metabolism	–	++	++	++
Salt tolerance	+	++	++	0
Acid production and tolerance	++	++	++	+
Inability to metabolize organic acids	++	++	++	+
Growth at low temperatures	++	++	++	0
Few growth factors required	0	0	+	0
Tolerance of phenolic glycosides	0	0	++	0
Formation of dextrans	–	–	–	–
Pectinolytic activities	–	–	–	–
Formation of bacteriocins	+	+	+	0
Bacteriophage resistance	0	+	+	++
Sensorially relevant criteria				
Heterofermentative metabolism	++	–	–	–
Formation of flavor precursors	++	++	+	0
Nutritionally relevant criteria				
Reduction of nitrate and nitrite	+	0	0	++
Formation of L-(+)-lactate	+	+	0	++
Formation of biogenic amines	–	–	–	–

[a]++, important; +, advantageous; 0, not relevant; –, detrimental.
[b]Except sauerkraut juice.

bers and at higher fermentation temperatures, and the problem usually increases with an increase of dissolved carbon dioxide in the brine. Measures should be taken to prevent or reduce the carbon dioxide content in brine and fermenting products, e.g., through the control of the initial population of bacteria and yeasts that are capable of producing carbon dioxide (30), through control of excessive growth of heterofermentative lactic acid bacteria, by removal of carbon dioxide by purging the brine with nitrogen gas (19), or by brine circulation.

A further source of carbon dioxide sufficient to contribute to bloater damage is malolactic fermentation by certain homofermentative lactic acid bacteria, whereby malate is converted to lactate and carbon dioxide (33). Malic acid is the predominant organic acid naturally present in pickling cucumbers. The concentration ranges from 0.2 to 0.3% among various cultivars (32). The use of mutants of *Lb. plantarum*, obtained by *N*-methyl-*N′*-nitro-*N*-nitrosoguanidine mutagenesis, which lack the ability to produce carbon dioxide from malic acid has been described but not exploited commercially (16, 31).

STARTER CULTURES

Fermentation of the vast majority of vegetables is done by naturally occurring lactic acid microflora rather than by defined starter cultures. Economically interesting exceptions are the so-called L-(+)-sauerkraut and most fermented vegetable juices, except sauerkraut juice. L-(+)-Sauerkraut is directly fermented in small containers (300 to 400 ml) using homofermentative *Lb. bavaricus*, which mainly produces L-(+)-lactic acid, as a starter culture. The product is made in Germany and distributed directly in the fermentation container through health stores. The *Lb. bavaricus* strain used is highly competitive so that the initial *Leuconostoc* population is suppressed and even the homofermentative lactic acid bacteria are not able to compete. In freshly made sauerkraut, the L-isomer represents more than 90% of the total lactic acid content. The suppression of the heterofermenters leads to a less developed flavor which, however, is accepted by some consumers (27).

Most fermented vegetable juices are manufactured according to the "lactoferment process" (7). In contrast to other vegetable products, mash or raw juice is pasteurized before fermentation so that pure culture fermentation can be achieved through the addition of a starter. Lactic acid bacteria used for this purpose include *Lactobacillus* species (*acidophilus, delbrueckii, helveticus, plantarum, salivarius, xylosus, bifidus, brevis,* and *casei*), *Lactococcus lactis,* and *Ln. mesenteroides* (9). The species most often used are *Lb. plantarum* and sometimes *Lb. casei.*

Because of requirements in some countries for certification according to ISO 9000–9004 (International Organization for Standardization, Geneva), modern consumer demands, and economic reasons, it is anticipated that the development of controlled fermentation processes will become necessary in order to consistently provide safe products with high quality. However, the fermentation of vegetables is difficult to control, primarily because of variation in shape, the large number and types of naturally occurring microorganisms, and variability in nutrient content. Since it is not possible to eliminate the natural microflora by appropriate methods, consistent products made by pure culture fermentation realistically cannot be achieved. One promising approach might be the application of defined starter cultures which are capable of growing rapidly and are highly competitive under environmental conditions used to ferment products. Depending on the particular material to be fermented as well as the desired quality of the final product, starter cultures must possess numerous characteristics. Selection criteria for starter cultures used in the fermentation of sauerkraut, cucumbers, olives, and vegetable juices are compiled in Table 32.9. However, due to modest profit margins in the vegetable fermentation industry, it is questionable whether the use of starter cultures is economically acceptable.

Extensive investigations of lactic acid bacteria used in sauerkraut fermentation have shown that strains of *Lb. plantarum* may cause a rapid decrease in pH but lead to poorly flavored products. Sauerkraut with perfect sensory characteristics has been obtained using selected strains of *Ln. mesenteroides* which simultaneously improve the uniformity of sauerkraut from different batches (11).

Investigations to find bacteriocin-producing strains of *Ln. mesenteroides* with selective advantages over competing but sensitive strains have not been successful. A new approach involves using a paired starter culture system consisting of a nisin-resistant *Ln. mesenteroides* strain and a nisin-producing *Lactococcus lactis* strain, both of which were isolated from commercial sauerkraut fermentations (23). Whether the presence of nisin can actually promote and extend the dominance of heterofermentative *Ln. mesenteroides* in a spontaneous fermentation has yet to be demonstrated (24).

PROSPECTS

Systematic scientific research on fermented vegetables dates back to the early 1900s. Several hundred research papers and reports have been published on this subject, the majority of which have dealt with microbiological, technological, and analytical aspects of sauerkraut, cucumber, and olive fermentation. Still, we are far from understanding the complete array of biochemical reactions and interactions that lead to reproducible quality of fermented vegetables. Attempts are being made to develop models for cucumber fermentation. Such models can be helpful for the prediction of microbiological safety and shelf life of pickles, detection of critical control points in production and distribution processes, optimization of production and distribution chains, and control of fermentation processes (36, 37).

A summary of the major future areas of research needs published by the COST program is still relevant (10):

- prevention of softening during fermentation and storage
- prevention of discoloration
- reducing the nitrate content, especially of green leafy and root vegetables
- prevention of formation of biogenic amines
- outlining controlled fermentation processes
- development and use of starter cultures
- identification and controlled formation of typical aroma components

To solve these problems, a better knowledge of ecological factors prevailing in fermenting substrates and an improved understanding of the role of various microorganisms in the development of sensory properties will be necessary. In addition, a better understanding of the genetics of lactic acid bacteria and perhaps of other microorganisms involved in the fermentation of vegetables will enhance the ability of the vegetable fermentation industry to consistently produce high-quality products.

Lactic acid fermentation is a valuable tool for the production of a wide range of vegetable products. Due to its long tradition, lactic acid fermentation is more readily accepted by consumers than many other food preservation techniques. Given this situation and considering public opinion with regard to gene manipulation, it is questionable whether genetic engineering should be pursued to solve the problems summarized in Table 32.8. Most of the physiological properties hitherto recognized to be essential for vegetable fermentation are inherent in lactic acid bacteria, and several of the desired properties depend on plasmid-encoded traits which facilitate genetic manipulation. However, respecting consumer opinion, at present it appears expedient to take advantage of the pool of naturally available strains of lactic acid bacteria rather than to apply genetically modified microorganisms.

References

1. **Andersson, R.** 1984. Characteristics of the bacterial flora isolated during spontaneous lactic acid fermentation of carrots and red beets. *Lebensm.-Wiss. Technol.* **17**:282–286.
2. **Andersson, R., A.-S. Svanberg, and U. Svanberg.** 1990. Effect of lactic acid fermentation of vegetables on the availability of iron. *FEMS Microbiol. Rev.* **9**:100.
3. **Berghofer, E., and A. Werzer.** 1986. Production of tempeh from domestic beans. *Chem. Mikrobiol. Technol. Lebensm.* **10**:54–62. (In German.)
4. **Brückner, H., A. Salmen, A. Amar, and H. J. Buckenhüskes.** 1993. Fermented dairy-like products from advanced soy proteins, p. 215–222. *In Proceedings: Euro Food Chem VII,* 20–22 September 1993, Valencia, Spain.
5. **Buckenhüskes, H., A. Gessler, and K. Gierschner.** 1988. Analytical characterization of canned and pasteurized sauerkraut. *Ind. Obst-Gemueseverwert.* **73**:454–463. (In German.)
6. **Buckenhüskes, H., and K. Gierschner.** 1985. Determination of the buffering capacity of some vegetables. *Alimenta* **24**:83–88. (In German.)
7. **Buckenhüskes, H., and K. Gierschner.** 1989. Vegetable juice production—state of the art. *Flüssiges Obst* **56**:751–764. (In German.)
8. **Buckenhüskes, H., K. Gierschner, and W. P. Hammes.** 1988. Theory and practice of pasteurization—optimization of pasteurizing of pickled vegetables. *Ind. Obst-Gemueseverwert.* **73**:315–322. (In German.)
9. **Buckenhüskes, H. J.** 1993. Selection criteria for lactic acid bacteria to be used as starter cultures for various food commodities. *FEMS Microbiol. Rev.* **12**:253–272.
10. **Buckenhüskes, H. J., H. Aabye Jensen, R. Andersson, A. Garrido Fernandez, and M. Rodrigo.** 1990. Fermented vegetables, p. 167–187. *In* P. Zeuthen, J. C. Cheftel, C. Eriksson, T. R. Gormley, P. Linko, and K. Paulus (ed.), *Food Biotechnology: Avenues to Healthy and Nutritious Products,* vol. 2. *Processing and Quality of Foods.* Elsevier Applied Science, London.
11. **Buckenhüskes, H. J., and W. P. Hammes.** 1990. Starter cultures in fruit and vegetable processing. *BioEngineering* **6**:34–42. (In German.)
12. **Camacho, L., C. Sierra, D. Marcus, E. Guzman, M. Andrade, R. Campos, N. Diaz, M. Parraguez, and L. Trugo.** 1991. Nutritional improvement of commonly consumed vegetables fermented by cultures of *Lactobacillus* spp. *Alimentos* **16**:5–11.
13. **Camacho, L., C. Sierra, D. Marcus, E. Guzman, R. Campos, D. von Bäer, and L. Trugo.** 1991. Nutritional quality of lupine (*Lupinus albus* cv. multolupa) as affected by lactic acid fermentation. *Int. J. Food Microbiol.* **14**:277–286.
14. **Daeschel, M. A., R. E. Andersson, and H. P. Fleming.** 1987. Microbial ecology of fermenting plant materials. *FEMS Microbiol. Rev.* **46**:357–367.
15. **Daeschel, M. A., and H. P. Fleming.** 1984. Selection of lactic acid bacteria for use in vegetable fermentations. *Food Microbiol.* **1**:303–313.
16. **Daeschel, M. A., R. F. McFeeters, and H. P. Fleming.** 1985. Modification of lactic acid bacteria for cucumber fermentations: elimination of carbon dioxide production from malate. *Dev. Ind. Microbiol.* **26**:339–346.
17. **Emig, J.** 1989. Microbial nitrate reduction and lactic acid fermentation of vegetable products. Doctoral dissertation. Hohenheim University, Stuttgart, Germany. (In German.)
18. **Fleming, H. P., M. A. Daeschel, R. F. McFeeters, and M. D. Pierson.** 1989. Butyric acid spoilage of fermented cucumbers. *J. Food Sci.* **54**:636–639.
19. **Fleming, H. P., J. L. Etchells, R. L. Thompson, and T. A. Bell.** 1975. Purging of CO_2 from cucumber brines to reduce bloater damage. *J. Food Sci.* **40**:1304–1310.
20. **Fleming, H. P., L. C. McDonald, R. F. McFeeters, R. L. Thompson, and E. G. Humphries.** 1995. Fermentation of cucumbers without sodium chloride. *J. Food Sci.* **60**:312–315, 319.
21. **Garrido Fernández, A.** 1990. Controlled fermentation of green table olives, p. 211–218. *In* P. Zeuthen, J. C. Cheftel, C. Eriksson, T. R. Gormley, P. Linko, and K. Paulus (ed.), *Food Biotechnology: Avenues to Healthy and Nutritious Products,* vol. 2. *Processing and Quality of Foods.* Elsevier Applied Science, London.
22. **Giraud, E., L. Gosselin, and M. Raimbault.** 1992. Degradation of cassava linamarin by lactic acid bacteria. *Biotechnol. Lett.* **14**:593–598.
23. **Harris, L. J., H. P. Fleming, and T. R. Klaenhammer.** 1992. Characterization of two nisin-producing *Lactococcus lactis* strains isolated from a commercial sauerkraut fermentation. *Appl. Environ. Microbiol.* **58**:1477–1483.
24. **Harris, L. J., H. P. Fleming, and T. R. Klaenhammer.** 1992. Novel paired starter culture system for sauerkraut, consisting of a nisin-resistant *Leuconostoc mesenteroides* strain and a nisin-producing *Lactococcus lactis* strain. *Appl. Environ. Microbiol.* **58**:1484–1489.
25. **Holzapfel, W. H., R. Geisen, and U. Schillinger.** 1995. Protective cultures—potentiality and limitations, p. 143–161. *In* H. J. Buckenhüskes (ed.), *Proceedings, 13th Filderstädter Colloquium, Fertiggerichte—Fast Food—Catering,* 12–13 March 1992, Stuttgart Hohenheim, Germany. Gesellschaft Deutscher Lebensmitteltechnologen e.V., Bonn. (In German.)
26. **Huis in't Veld, J. H. J., H. Hose, G. J. Schaafsma, H. Silla, and J. E. Smith.** 1990. Health aspects of food biotechnology, p. 273–297. *In* P. Zeuthen, J. C. Cheftel, C. Eriksson, T. R. Gormley, P. Linko, and K. Paulus (ed.), *Food Biotechnology: Avenues to Healthy and Nutritious Products,* vol. 2. *Processing and Quality of Foods.* Elsevier Applied Science, London.
27. **Kandler, O., W. P. Hammes, M. Schneider, and K. O. Stetter.** 1986. Microbial interaction in sauerkraut fermentation, p. 302–308. Proceedings of the 4th International Symposium on Microbial Ecology, Ljubljana, Yugoslavia.
28. **Kerner, M., E. Mayer-Miebach, A. Rathjen, and H. Schubert.** 1990. Reduction of nitrate content in vegetable food using denitrifying microorganisms, p. 236–239. *In* P. Zeuthen, J. C. Cheftel, C. Eriksson, T. R. Gormley, P. Linko, and K. Paulus (ed.), *Food Biotechnology: Avenues to Healthy and Nutritious Products,* vol. 2. *Processing and Quality of Foods.* Elsevier Applied Science, London.
29. **Künsch, U., H. Schärer, and A. Temperli.** 1989. Biogenic amines—a quality indicator of sauerkraut. *In Proceedings,* XXIV Vortragstagung Qualitätsaspekte von Obst und Gemüse, 13–14 March 1989, Ahrensburg, Germany. Deutsche Gesellschaft für Qualitätsforschung, Kiel. (In German.)

30. **McDonald, L. C., H. P. Fleming, and M. A. Daeschel.** 1991. Acidification effects on microbial populations during initiation of cucumber fermentation. *J. Food Sci.* **56:**1353–1356, 1359.

31. **McDonald, L. C., D. H. Shieh, H. P. Fleming, R. F. McFeeters, and R. L.Thompson.** 1994. Evaluation of malolactic-deficient strains of *Lactobacillus plantarum* for use in cucumber fermentation. *Food Microbiol.* **10:**489–499.

32. **McFeeters, R. F., H. P. Fleming, and R. L. Thompson.** 1982. Malic and citric acids in pickling cucumbers. *J. Food Sci.* **47:**1859–1861, 1865.

33. **McFeeters, R. F., H. P. Fleming, and R. L. Thompson.** 1982. Malic acid as a source of carbon dioxide in cucumber juice fermentation. *J. Food Sci.* **47:**1862–1865.

34. **Meurer, P.** 1991. Effect of plant origin enzymes and other factors on the texture of fermented cucumbers. Doctoral dissertation. Hohenheim University, Stuttgart, Germany. (In German.)

35. **Omran, H., H. Buckenhüskes, E. Jäckle, and K. Gierschner.** 1991. Some properties of pectinesterase, exo polygalacturonase and endo-β-1,4-gluconase of pickling cucumbers. *Dtsch. Lebensm.-Rundsch.* **87:**151–156. (In German.)

36. **Passos, F. V., H. P. Fleming, D. F. Ollis, H. M. Hassan, and R. M. Felder.** 1993. Modeling the cucumber fermentation: growth of *Lactobacillus plantarum*. *J. Ind. Microbiol.* **12:**341–345.

37. **Passos, F. V., D. F. Ollis, H. P. Fleming, H. M. Hassan, and R. M. Felder.** 1993. Modeling the specific growth rate of *Lactobacillus plantarum* in cucumber extract. *Appl. Microbiol. Biotechnol.* **40:**143–150.

38. **Sakellaris, G., and A. E. Evangelopoulos.** 1989. Production, purification and characterization of extracellular pectinesterase from *Lactobacillus plantarum* (str. BA 11). *Biotechnol. Appl. Biochem.* **11:**503–507.

39. **Sakellaris, G., S. Nikolaropoulos, and A. E. Evangelopoulos.** 1989. Purification and characterization of an extracellular polygalacturonase from *Lactobacillus plantarum* strain BA 11. *J. Appl. Bacteriol.* **7:**77–85.

40. **Schneider, M.** 1988. Microbiology of sauerkraut fermented in small ready-to-sell containers. Doctoral dissertation. Hohenheim University, Stuttgart, Germany. (In German.)

41. **ten Brink, B.** 1993. Bacteriocins from lactic acid bacteria: natural preservatives? *Voedingsmiddelentechnologie* **26:**37–38.

42. **ten Brink, B., C. Damink, H. M. L. J. Joosten, and J. H. J. Huis in't Veld.** 1990. Occurrence and formation of biologically active amines in foods. *Int. J. Food Microbiol.* **11:**73–84.

43. **Wiesenberger, A., E. Kolb, J. A. Schildmann, and H. M. Dechent.** 1986. Lactic acid fermentation of natural substrates with low pH. *Chem. Mikrobiol. Technol. Lebensm.* **10:**32–36. (In German.)

44. **Wolf, G., and W. P. Hammes.** 1987. Lactic acid bacteria as agents for reduction of nitrate and nitrite in food. *Dechema Monogr.* **105:**271–272.

Steven C. Ricke
Jimmy T. Keeton

33

Fermented Meat, Poultry, and Fish Products

Fermented meat products are defined as meats that are deliberately inoculated during processing to ensure sufficient controlled microbial activity to alter the product characteristics (5). If fresh meat is not preserved or cured in some manner, it spoils rapidly due to the growth of indigenous gram-negative bacteria and subsequent putrefaction resulting from their metabolic activities (26). Although some manufacturers still depend upon naturally occurring microflora to ferment meat, most use starter cultures consisting of a single species or multiple-species combinations of lactic acid bacteria and/or micrococci which have been selected for metabolic activities especially suited for fermentation in meat ecosystems. Understanding the technological, microbiological, and biochemical processes that occur during meat, poultry, and fish fermentation is essential to ensure safe, palatable products.

MANUFACTURE OF FERMENTED MEAT AND POULTRY PRODUCTS

Sausage Categories and Meat Fermentation

Dry and semidry sausages represent the largest category of fermented meat products, with many of the present-day processing practices having their origin in the Mediterranean region. Traditionally, dry sausages acquired their particular sensory characteristics from exposure to salt and the rapid drying conditions existing in the warm, dry Mediterranean climate. These products were heavily seasoned, typically not smoked, and derived their name, "sausage," from the Latin term *salsus*, meaning salted. Sausage-processing practices later spread to Northern Europe, and by the Middle Ages, hundreds of varieties of dry and semidry sausages were manufactured across the continent. In contrast to the Mediterranean variety, Northern European sausages were prepared during the cold winter months and stored until summer and were thus called summer sausages. These sausages contained more water than their Mediterranean counterparts and were lightly spiced, heavily smoked at cool temperatures, and less susceptible to spoilage due to colder ambient temperatures. Summer sausages are similar to present day semidry sausages. Examples of the most common dry and semidry sausage varieties are listed in Table 33.1 and are categorized as salamis or cervelats.

Product Categories and Compositional Endpoint Characteristics

Compositional characteristics of dry and semidry sausage categories produced in the United States, and some processing criteria, have been defined by the United States Department of Agriculture Food Safety and Inspection Service (USDA-FSIS). Examples of the compositional characteristics of dry and semidry fermented meat products are given in Table 33.2.

Table 33.1 Categories and origins of selected dry and semidry sausages[a]

Category	Origin	Description and unique characteristics
Dry sausages		
Salamis (Genoa, Milano, Siciliano)	Italy	Lean pork (coarse), some with wine or cured beef (fine), garlic, large-diameter casing (beef or hog bungs) with flax twine, some with coating of white mold
Lombardi salami	Italy	Coarse cut, high fat content, brandy, twine wrap over casing
Cappicola	Italy	Boneless pork shoulder butt combined with red hot or sweet peppers, mildly cured
Mortadella	Italy	Finely chopped cured beef and pork with added cubes of backfat, mildly spiced, smoked, encased in beef bladders
Pepperoni	Italy	Cured pork, some beef, cubed fat, red peppers, small-diameter casing
Chorizos	Spain or Portugal	Pork (coarse), highly spiced, hot, small-diameter casing
D'Arles	France	Similar to Italian salamis, coarse, large-diameter casings (hog bungs)
Lyons	France	All pork (fine), diced fat (fine), spices and garlic, cured, large-diameter casing
Allesandri and Alpino	United States	Similar to Italian salami
Katenrauchwurst, Dauerwurst, Plockwurst, Zerevelat	Germany	Dry sausages, smoked, air dried
Semidry sausages		
Summer sausages (cervelat, farmer cervelat)	Generic	Mildly seasoned soft cervelat, beef and pork (coarse), no garlic, cured, small-diameter casing
Holsteiner cervelat	Germany	Similar to farmer cervelat, packed in ring-shaped casing
Thüringer cervelat	Germany	Medium dry to soft, tangy flavor, mildly spiced, smoked, some beef and veal added
Gothaer cervelat	Germany	Very lean pork, fine chopped, cured, soft texture, mild seasoning
Goteborg cervelat, medwurst	Sweden	Coarse, salty, soaked in brine before heavy smoke
Landjaeger cervelat	Switzerland	Small diameter (frankfurter size), pressed flat, smoked, flavored with garlic and caraway seeds
Teewurst, frische Mettwurst	Germany	Undried, spreadable, smoked
Lebanon bologna	United States	Smoked, coarse, large diameter, very acid

[a]From references 5, 71, 78, and 80.

Manufacturing Procedures and Processing Conditions

Dry and semidry sausage manufacture using starter cultures involves the following basic steps: (i) reducing the particle size of high-quality raw meat trimmings; (ii) incorporating salt, nitrate (Europe mostly) or nitrite (United States and Europe), glucose, spices, seasonings, and a specific inoculum selected on the basis of incubation temperature optimum and level of lactic acid desired; (iii) uniformly blending all ingredients and further reducing the particle size; (iv) vacuum stuffing into a semipermeable casing to minimize the presence of oxygen; (v) incubating (ripening) at or near the temperature optimum of the starter culture until a specific pH endpoint is achieved or carbohydrate utilization is complete; (vi) heating (usually) the product to inactivate the inoculum and ensure pathogen destruction; and (vii) drying (aging) the product to the required moisture or moisture-protein endpoint. A generic scheme for manufacturing dry and semidry sausages using starter cultures is given in Table 33.3.

Table 33.2 Composition of two types of fermented sausages[a]

Parameter	Composition (%, wt/wt): Summer sausage, Thüringer cervelat	Composition (%, wt/wt): Pepperoni, hard salami
Moisture	50	30
Fat	24	39
Protein	21	21
Salt	3.4	4.2
pH	4.9	4.7
Total acidity	1.0	1.3
Yield (from raw state)	90	64

[a]Adapted from Terrell et al. (94).

Factors Affecting Color, Texture, Flavor, and Appearance of Fermented Meats

Raw Meat Tissues

Fresh or frozen raw meats to be used for fermented sausages should be chilled to <4.5°C, have low microbial

Table 33.3 Generic manufacturing scheme for dry and semidry fermented sausages with starter cultures[a]

Processing sequence	Semidry sausage	Dry sausage (North America)	Dry sausage (Europe)
Meat tissue selection	Fresh or frozen meats with low microbial populations, no discoloration, no off odors, limited age, no dark, firm and dry (DFD) tissue, trimmed free of blood clots, glands, sinews, gristle, bruises, refrigerated to <4.5°C (<40°F)	Same as semidry	Same as semidry
Comminution, grinding, blending of ingredients; inoculum level at ca. 10^7 CFU/g	Fresh or tempered −3°C (26°F) meats: coarse grind, chop, or mince lean (0.25–0.5 in. [6.35–12.7 mm]) and fat (0.5–1 in. [12.7–25.4 mm]) meats separately, combine to a specified fat endpoint; blend with seasonings and cure ingredients to uniformly distribute ingredients; avoid overmixing and excessive protein extraction or fat smearing	Same as semidry	Same as semidry
Ingredients used Salt (2.5–3%) Glucose (0.4–0.8%) Nitrite (<150 mg/kg) Sodium erythorbate (550 mg/kg) Antioxidants (natural or synthetic) Spices (sterilized)	Rehydrate frozen or lyophilized culture with nonchlorinated, distilled water at ambient temperature <1 h before use; add inoculum (high-temperature optimum for smokehouse incubation) to meat batch; blend, but avoid excessive mixing (causes fat smearing and coating of lean particles); fine grind or chop (0.125–0.187 in. [3.2–4.8 mm]) to specified particle size	Same as semidry. Exceptions: Use high-temperature inoculum (>32.5°C, 90°F) for smokehouse incubation, low-temperature inoculum (21.3°C, 70°F) for "green" or "ripening" room incubation	Same as semidry. Exceptions: Use nitrate (200–600 mg/kg) alone or in combination with nitrite (nitrate used mostly in Europe; only for Lebanon bologna and country-style hams in United States); low-temperature fermentation (21.3°C, 70°F) used most often

Vacuum encasing (stuffing)	Keep at 2°C (ca. 34°F); vacuumize to remove oxygen and encase in fibrous or natural casing; oxygen exclusion accelerates anaerobic fermentation, favors lactic acid bacteria growth, color and flavor development	Same as semidry	Same as semidry
Incubation and fermentation (ripening)	High-temperature smokehouse incubation at 32.5–38.1°C (90–100°F) and 90% relative humidity for ≥18 h (dependent upon sausage diameter) to an endpoint pH of <4.7. Air movement of >1 m/s; smoke at the end of fermentation	Low-temperature smokehouse incubation or "ripening" room at 15–26°C (60–78°F) and 90% relative humidity for ca. 72 h to an endpoint pH of <4.7. Air movement of >1 m/s; smoke application at the end of fermentation if desired	Low-temperature "ripening" at 26°C (78°F) and 88% relative humidity for 3 days to an endpoint pH of 4.7–4.8. Chamber relative humidity held at 5–10% lower than relative humidity within the sausage, or use the following schedule: a. 22.2–23.9°C (72–75°F), 94–95% relative humidity for 24 h b. 20.0–22.2°C (68–72°F), 90–92% relative humidity for 24 h c. 18.3–20.0°C (65–68°F), 85–88% relative humidity for 24 h
Drying (aging)	Drying chamber at 12.9–15.7°C (55–60°F) and 65–70% relative humidity for ≥12 days (dependent upon sausage diameter) to specified moisture:protein ratio	Drying chamber at 10–11.2°C (50–52°F) and 68–72% relative humidity for ≥21 days (dependent upon sausage diameter) to specified moisture:protein ratio	Remains in the ripening room or is moved to drying chamber at 20°C (68°F) and 88% relative humidity for 10 days, then 15.7°C (60°F) and 82% relative humidity for 14 days to a specified endpoint or hold at 11.7–15.0°C (53–59°F) at 75–80% relative humidity to specified endpoint

[a] Adapted from references 5 and 56.

populations, be free of physical and chemical defects, and meet the compositional specifications for the product being manufactured. The predominant bacteria that develop in meats not held under vacuum are typically gram-negative, oxidase-positive, aerophilic rods composed of psychrotrophic pseudomonads along with psychrotrophic *Enterobacteriaceae* (56). Small numbers of lactic acid bacteria and other gram-positive microorganisms are initially present in meat and become the dominant microflora if oxygen is excluded, as in the case of vacuum-packaged or vacuum-encased meat products.

Post-rigor pH and the residual glycogen concentration of muscle tissues influence the quality of fermented sausages. ATP levels in muscle tissues average 1 μmol/g 24 h post slaughter, while pH values range from 5.5 to 5.7 for beef, 5.7 to 5.9 for pork, and 5.8 to 6.0 for poultry. Pork meat sometimes exhibits pale, soft, and exudative (PSE) characteristics, which include tissue pH values of 5.3 to 5.5, a wet or watery meat surface, and a pale pink color. However, these tissues may be incorporated into dry sausages at levels up to 50% of the meat block without impairing sensory qualities (95). Use of 100% PSE meats in sausages, however, will likely result in products with a pale, yellowish cured color, poor water-holding capacity, lower a_w, increased susceptibility to oxidative rancidity, and poor (soft, grainy, uncohesive) textural characteristics. Meats exhibiting a dark, firm, and dry (DFD) condition are characterized by a dry surface, dark red color, and high pH (>6.0 to 6.2). This condition occurs more frequently in beef than in pork. DFD trimmings are not suitable for dry sausage because of their excessive water-binding capacity and potential for accelerated microbial spoilage. Dark red meat from more mature animals, however, may be desirable due to its contribution to product appearance.

Use of lamb and mutton meats in fermented sausages is limited (2), but studies indicate that acceptable sausage products can be produced when combined with appropriate seasonings and limited amounts of mutton fat (10, 101). Fat tissues from beef, lamb, or pork have a high proportion of saturated fatty acids and yield products that are firmer and have a more desirable texture than products containing poultry and turkey fats, which have a predominance of polyunsaturated fatty acids. Polyunsaturated fatty acids are more susceptible to autooxidation and rancidity, which can lead to the development of off flavors. Thus, poultry meats may be a less desirable source of fat for fermented sausage formulations.

Lean poultry meat, if used, is often supplemented with pork or beef meat to avoid sausages that appear too light in color and to ensure an appropriate level of textural attributes. Poultry sausages are usually lower in fat (15%) compared to red meat sausages (23 to 45%), initially have a higher pH, and contain more moisture, which can affect product uniformity (diameter) during drying. Incorporation of turkey thigh meat into sausages can result in darker red products due to a higher myoglobin content in the lean tissue, but turkey, as with poultry, has a higher proportion of polyunsaturated fatty acids and moisture compared to red meats. Slightly higher levels of fermentable carbohydrate should be used in formulas containing higher-pH meats such as poultry. Heating to >68.9°C (155°F), applying heating schedules as outlined in 9CFR Ch. III 318.10 (95a), or heating for the same time/temperature combinations (318.17 [95b]) as specified for roast beef may be required (USDA) to ensure destruction of *Salmonella* spp. potentially present on poultry tissues. Mechanically deboned poultry meat is an acceptable meat source in fermented dry sausages when limited to 10% of the meat block (35), as sausages tend to become soft at higher levels.

Ingredients

Incorporation of sodium chloride, sodium or potassium nitrite and/or nitrate, glucose, and homofermentative lactic acid starter cultures (Table 33.3) in sausage formulas dramatically alters the ecology of the culture environment and chemical characteristics of finished products (Table 33.4). Incubation of sausages at the optimum inoculum growth temperature under a reduced oxygen environment causes a reduction in pH by the rapid conversion of glucose to lactic acid as a consequence of the exponential growth of lactic acid bacteria and subsequent suppression of indigenous microflora such as psychrotrophic pseudomonads, *Enterobacteriaceae,* and most pathogens. *Pseudomonas* species are sensitive to salt, nitrite, elevated incubation temperatures (>37°C), and reduced oxygen tension, while competitiveness of the *Enterobacteriaceae* is restricted at reduced oxygen levels, low pH, and the presence of salt.

The glucose content of post-rigor meats (4.5 to 7 μmol/g) is not sufficient to reduce pH significantly; therefore, 0.4 to 0.8% fermentable carbohydrate in the form of glucose, sucrose, or maltose is added to sausage formulas to enable reduction of the pH to 4.6 to 5.0. About 1 oz of glucose per 100 lb of meat (0.62 g of glucose/kg of meat) is required to reduce the pH by 0.1 unit. Carbohydrates such as lactose, raffinose, trehalose, dextrins, and maltodextrins, when used in place of glucose, yield less lactic acid due to incomplete utilization (47). In the United States, fermented sausages having final pH values of 4.8 to 5.0 contain approximately

Table 33.4 Chemical characteristics of selected fermented sausage products

Category	Final pH	Lactic acid (%)	Moisture: protein ratio	Moisture loss (%)	Moisture[a] (%)	Comments
Dry sausages[b]	5.0–5.3 (<5.3)	0.5–1.0	<2.3:1	25–50	<35	Heat processed (optional[c]); dried or aged after fermentation for moisture loss; smoked
Cervelat			1.9:1		32–38	Shelf stable
Cappicola			1.3:1		23–29	Shelf stable
German Dauerwurst	4.7–4.8		1.1:1		25–27	Shelf stable
German salami	4.7–4.8		1.6:1	1	34–35	Shelf stable
Pepperoni	4.5–4.8	0.8–1.2	1.6:1	35	25–32	Shelf stable
Italian salami, hard or dry			1.9:1	30	32–38	Shelf stable
Genoa salami	4.9	0.79	2.3:1	28	33–39	Shelf stable
Thüringer, dry	4.9	1.0	2.3:1	28	46–50	Shelf stable
Semidry sausages[b]	4.7–5.1 (<5.3)	0.5–1.3	>2.3<3.7:1	8–15	45–50	Heat processed[c]; typically smoked; packaged after processing and chilling
Lebanon bologna	4.7	1.0–1.3	2.5:1	10–15	56–62	Refrigerate
Cervelat, soft			2.6:1	10–15		Refrigerate
Salami, soft			2.3–3.7:1	10–15	41–51	Refrigerate
Summer sausage	<5.0	1.0	3.1:1	10–15	41–52	Refrigerate
Thüringer, soft			3.7:1		46–50	Refrigerate
For comparison						
Drief beef			2.04:1	29		
Beef jerky			0.75:1	>50	28–30	
Air-dried sausage			2.1:1			

[a]Water activity ranges for dry and semidry sausages are <0.85 to 0.91 and 0.90 to 0.94, respectively. The European Economic Directive 77/99 requests a_w of <0.91 or pH < 4.5 for dry sausages to be shelf stable, or a combined a_w and pH of <0.95 and <5.2, respectively.

[b]From references 5, 49, 69, 80, and 96.

[c]USDA-FSIS Title 9 CFR may be amended to require specified time and temperature heating combinations after fermentation, or verification that processing conditions destroy all pathogenic microorganisms.

25 g of lactic acid per kg (dry weight), while some Italian and Hungarian salamis that are fermented with limited amounts of carbohydrate may show a decrease in pH of only 0.5 unit.

Nitrates and nitrites (40 to 50 mg/kg, minimum) in fermented sausages react with the heme moiety of myoglobin to facilitate the development of cured color, retard lipid oxidation, inhibit the growth of *Clostridium botulinum* (through a synergistic relationship with salt), and enhance cured flavor. In the United States, sodium or potassium chloride (2.5 to 3.0%) and sodium or potassium nitrite (<156 mg/kg added, not to exceed 200 mg/kg in the final product), in combination, serve as the primary curing agents in fermented sausages. Sodium or potassium nitrate are also allowed in products except bacon and are typically added to dry sausages such as Lebanon bologna and to dry-cured hams at maximum levels of 172 and 218 mg/kg, respectively. In Europe, sodium or potassium nitrate (200 to 600 mg/kg), in combination with nitrate-reducing microorganisms such as *Micrococcus varians* and *Staphylococcus carnosus*, is utilized during low-temperature fermentation (ripening) to enable metabolism by these acid-sensitive, nitrate-reducing bacteria. Sodium ascorbate is often used in combination with nitrates and nitrites in dry sausages at concentrations up to 550 and 600 mg/kg in the United States and Europe, respectively. Ascorbates, isoascorbates, or erythorbates accelerate the reduction of nitrous acid to nitric oxide, thus enhancing color development, reducing residual nitrite, and retarding the formation of *N*-nitrosamines, a class of potent carcinogens.

Ground pepper, paprika, garlic, mace, pimento, cardamom, red pepper, and mustard are commonly used in fermented sausages but, as with all spices, should be sterilized to avoid wild fermentations. Red pepper and mustard are known to stimulate lactic acid formation, possibly due to the available manganese in these spices which enhances the glycolytic enzyme fructose-1,6-diphosphate aldolase (56). Garlic, rosemary, and sage contain antioxidant and antimicrobial compounds and may assist in preserving flavor, color, and microbial shelf life of fermented sausages.

Glucono-delta-lactone (GDL), a chemical acidulant, is hydrolyzed to gluconic acid which is then converted to lactic acid and acetic acid by indigenous lactobacilli. The acidulant is sometimes used at concentrations ranging from 0.25 to 0.5%, although up to 1% is allowed in the United States. Improved color and consistency in conjunction with a rapid decrease in the pH of dry sausage caused by natural starter culture fermentation have been observed at GDL levels of 0.25% (35). At higher concentrations, GDL can inhibit growth of lactobacilli, produce undesirable aromas, and impart a sweet flavor from the unfermented sugar.

Smoke contains phenols, carbonyls, and organic acids which may act to preserve sausage products. Phenols are effective antioxidants and microbial inhibitors, while carbonyls contribute a desirable amber color by combining with free amino groups to form brown furfural compounds. Organic acids in smoke, such as formic, acetic, propionic, butyric, and isobutyric acids, inhibit the growth of microorganisms on the surface of sausages and promote coagulation of surface proteins.

MANUFACTURE OF FERMENTED FISH PRODUCTS

Fermented fish products include a variety of fish sauces, fish pastes, and fish-vegetable blends that have been salted, packed whole in layers, or ground into small particles, and then fermented in their own "pickle." These products are eaten as a proteinaceous staple or condiment in Southeast Asia, but are consumed as a condiment in Northern Europe (11). Fish fermentation involves minimal bacterial conversion of carbohydrates to lactic acid, but entails extensive tissue degradation by proteolytic and lipolytic enzymes derived from viscera and muscle tissues. Low-molecular-weight compounds from fish tissue degradation are the primary contributors to aroma and flavor characteristics of sauces. Indigenous microorganisms, however, do contribute to aroma and flavor, but are limited to those species tolerant of high salt concentrations (10 to 20%) in the curing brine. Partial tissue hydrolysis is responsible for the unique textural attributes of pastes and fish-vegetable blends.

Fish Sauces

Fish sauces such as nuoc-mam (Vietnam), patis (Philippines), nam-pla (Thailand), budu (Malaysia), nuoc-mam-nuoc (Thailand), and shottsuru (Japan) are liquids consumed as a condiment with rice and vary in color from clear brown to yellow-brown. These sauces have a predominantly salty taste and are derived from decanting or pressing fermented fish or shrimp after a 9-month to 1-year fermentation (64). Products fermented over a 1- to 2-year period have a distinctive sharp, meaty aroma and may range in protein content from 9.6 to 15.2%.

Commercial fish sauce production begins with layering seine-netted fish, shrimp, or shellfish with salt in concrete vats in an approximate ratio of 3:1 (fish:salt), sealing the vat, allowing supernatant liquor to develop, and carefully decanting this liquid. Enzymatic digestion or fermentation may range from 6 months for small fish to 18 months for larger species. The first liquid removed from the fermenting fish contains an abundance of peptides, amino acids, ammonia, and volatile fatty acids and is considered the highest quality. Extracts may be supplemented with caramel, caramelized sugar, molasses, roasted corn, or roasted barley to enhance the color and keeping qualities. For some products, the sauce is ripened in the sun for 1 to 3 months and blended with bacterial by-products from the manufacture of monosodium glutamate.

The nitrogen content of the supernatant liquor increases as a result of proteolysis during fermentation. Initially, salt penetrates the tissues by osmosis (0 to 25 days), enabling a protein-rich liquid to develop through autolysis (80 to 120 days), and, ultimately, the fish tissue is transformed into a nitrogen-containing liquid (140 to 200 days). Proteases such as cathepsin B and trypsin-like enzymes have been shown to increase soluble protein content of the liquor during the first 2 months, but their activity gradually decreases through an inhibition feedback mechanism with the buildup of amino acids and polypeptides. New polypeptide formation occurs during the last fermentation stage as the level of free amino acids increases.

Aseptically produced fish sauces do not have a typical aroma (12), suggesting that some microbial involvement is required for aroma development. Bacterial populations in raw fish, predominantly facultative anaerobes, have been shown to be high (2.7×10^4 CFU/g) initially, but decline (2×10^3 CFU/g) over a 6- to 9-month period. *Bacillus*, *Lactococcus*, *Micrococcus*, and *Staphylococcus* spp. have been confirmed as indigenous microflora, but as fermentation progresses, populations and number of species decline with the changing brine environment. Crisan and Sands (19) identified *Bacillus cereus* and *Bacillus licheniformis* as the dominant bacteria in nam-pla, but after 7 months of fermentation, another strain of *B. licheniformis*, *Bacillus megaterium*, and *Bacillus subtilis* were the dominant species. *Micrococcus copoyenes*, *M. vari-*

ans, Bacillus pumilus, and *Candida clausenii* are other halotolerant microorganisms (those capable of growth in 10% brine, but not 20%) that have been isolated from sauces.

The characteristic aroma and flavor of fish sauce are complex and cannot be attributed to specific volatile fatty acids, peptides, or amino acids derived from bacterial fermentation alone. One theory suggests a complex interaction of enzymatic activity and oxidation during the fermentation; however, some evidence exists for bacterial production of volatile fatty acids in fresh fish before salting and in the brief period following salting (13). Dougan and Howard (25) characterized nam-pla aroma as being ammoniacal-trimethylamine, cheesy-ethanoic, and *n*-butanoic, but having other notes attributable to low-molecular-weight volatile fatty acids, meaty-ketones, keto-acids, and amino acids, especially glutamic acid. The flavor of fish sauce has a strong salt component with contributions from a combination of monoamino acids and possibly aspartic and glutamic acids. Other factors such as pH of the brine, fermentation temperature, and salt concentration also affect flavor.

Fish Paste

Fish pastes, more widely produced than sauces, are consumed raw or cooked as a condiment with rice and vegetables. In some countries, fish pastes may be the primary source of dietary protein for low-income families. A wide variety of fermented fish pastes are produced and require shorter processing periods than sauces. These include bagoong, tinabal, and balbakwa (Philippines), par-hoc (Cambodia), padec (Laos), blachan or bleachon and trassi-udang (Malaysia), sidal (Pakistan), and shiokara (Japan) (11). Paste production consists of mixing cleaned, eviscerated, whole or ground fish, shrimp, plankton, or squid with salt in a ratio of 3:1 (fish:salt) and then placing the mixture in vats to ferment. Proteolytic enzymes from viscera tissue, and to some extent bacteria, break down the tissue until it attains a pasty consistency. Aging in hermetically sealed containers may follow, or the paste may be hand kneaded and then aged. Pickle (liquid exudate), which forms due to the osmotic differential of the brine solution and the fish tissue, is decanted, aged, and consumed as fish sauce. When pickle no longer forms, the fish paste is ready for use or aged further. Typical paste has a salty, cheese-like aroma, but other characteristics vary depending upon the method and region of production.

Bacteria do not appear to play a major role in the proteolysis of fish paste, but may contribute to aroma and flavor or to spoilage. Populations of 6.5×10^3 CFU/g, representing 40 bacterial species, have been reported to occur in bagoong (20% salt), of which *Bacillus, Micrococcus,* and *Moraxella* species were dominant (29). Eighteen strains of bacteria, some of which were halophilic, were found in shiokara (12, 102), and *Micrococcus* species appeared to be responsible for ripening. Halophilic *Vibrio* and *Achromobacter* species were also isolated and were likely responsible for spoilage.

Proteolytic enzymes from fish viscera, stomach, pancreas, and intestine are more active than muscle enzymes and are most often responsible for release of free amino acids and polypeptides which are believed to undergo further microbial synthesis to yield specific flavor compounds. Some of the compounds derived from proteolytic degradation include volatile fatty acids (formic, ethanoic, propanoic, isobutanoic, and *n*-pentanoic acids), ammonia, trimethylamine, and mono- and dimethylamines. Aminobutane and 2-methyl propylamine are thought to result from microbial action. If carbohydrates are added to the raw materials and fermented to alcohol, this may suppress proteolytic and microbial activity.

Bogoong, the residue of partially hydrolyzed fish or shrimp (64), has a pH of 5.2 and contains 65 to 68% moisture, 13 to 15% protein, 2% fat, 20 to 30% salt, 1.45% nitrogen, 0.18% volatile nitrogen, 0.015% trimethylamine, and 0.011% hydrogen sulfide in a 35% solids base. Pra-hoc, used in dishes such as soups, may contain as much as 24% protein and 17% salt. Shiokara, a Japanese fish or squid product with or without malted rice, is texturally between a sauce and paste and characterized as a dark brown liquid containing lumps of solid tissue. In its final form, the squid product contains 74.2% water, 7.8% salt, 11.6% protein, and 8.7% ash.

Fermented Rice and Shrimp or Rice and Fish

Balao balao, a traditional food of the Philippines, is a cooked rice and shrimp product which is fermented at room temperature for 7 to 10 days in a 20% salt brine (64). A similar Filipino fermented product in which fish is substituted for shrimp is known as burong isda, while the Japanese version is called naresushi or funasushi. Bacterial isolates involved in the fermentation of burong have been demonstrated to be capable of starch hydrolysis and have been identified (64) as having characteristics similar to *Lactobacillus plantarum* and *Lactobacillus coryneformis* subsp. *coryneformis,* but with the capacity to convert oligosaccharides and reducing sugars to lactic acid. An enzyme with a pH optimum of 4.0 has been isolated from these bacteria and found to hydrolyze amylose to oligosaccharides. However, during the initial stages of fermentation, other bacteria may hydrolyze starch for use by lactic acid bacteria.

STARTER CULTURES IN MEATS

The use of starter cultures in fermented meat products is a relatively recent practice compared to their use in fermented dairy foods and alcoholic beverages (52). The rationale for the use of a starter culture in meat fermentation is similar in concept to the use of starter cultures in dairy products. In the United States, lactobacilli or pediococci are the predominant culture microorganisms, while in Europe, these genera are most often used in combination with micrococci and staphylococci (7, 52). Proper inoculation and incubation procedures are among the most critical steps for the uniform production of safe, flavorful, and wholesome fermented sausages. Inoculation is accomplished by one of three methods: natural fermentation, which relies on indigenous microflora in the meat to serve as the inoculum; back-inoculation, which involves transfer of a portion of raw meat from a previous batch of sausage to the present batch, i.e., transfer from a mother batch of raw sausage; or the use of starter cultures, i.e., inoculation of unfermented meat with a pure strain or strains of lactic acid bacteria. Use of commercial starter cultures is the predominant method of inoculation in the United States.

Development of Commercial Starter Cultures

In fresh meat, lactic acid bacteria are a minor component of the microflora, but when meat is packaged and stored under vacuum, the resulting microenviroment facilitates the growth of lactic acid bacteria (46). Commercial application of starter cultures in the cheese industry during the 1930s led investigators to identify potential starter cultures for fermented meats in the 1940s. Use of pure cultures in sausages began after several studies in the 1950s demonstrated that lactic acid bacteria are responsible for lactic acid production (21, 66–68). An essential requirement for starter cultures is that they be produced and preserved in a viable and metabolically active form suitable for commercial distribution. In the United States, *Lactobacillus* species were first used as starter cultures for meat fermentations in the temperature range of 20 to 25°C. Original starter cultures were logically derived from the predominant microflora of fermented meat products, but when attempts were made to preserve these cultures via lyophilization as was routinely done with dairy starter cultures, these strains invariably died. Deibel et al. (22) were able to surmount this problem by using *Pediococcus cerevisiae* (now *Pediococcus acidilactici*), the first commercially available meat starter culture. Although *P. cerevisiae* was not a predominant bacterium in naturally fermented meats, it did, in addition to surviving lyophilization, possess characteristic lactic acid fermentation ability, had a higher optimal growth temperature, and tolerated salt concentrations up to at least 6.5% (5, 7, 22). When Deibel et al. (22) developed a lyophilization procedure for maintaining starter cultures consisting of pediococci, it was considered the best method for distributing a viable culture in a reliable and economical manner (9). Eventually, lyophilization proved commercially problematic because rehydration procedures were time-consuming and introduced unacceptable variation (5). Also, lyophilized cultures exhibited inordinately long lag phases, thus lengthening fermentation times.

Implementation of frozen-culture technology in the late 1960s, along with improvements in conditions for storing, handling, and shipping, led to commercial acceptance of frozen culture concentrates (9, 27). This approach eliminated the need for a rehydration step and provided higher numbers of viable cells than did lyophilization (75). Widespread use of pure starter culture strains in the United States occurred during the late 1970s and was a consequence of the development of efficient, high-volume dry sausage production technologies that required short ripening times and consistently uniform products with low defect levels.

Foodborne illness outbreaks associated with coagulase-positive staphylococci also focused attention on the need to control the fermentation process and ensure the production of safe products. In addition to solving these problems, the introduction of frozen culture concentrates also renewed interest in lactobacilli as starter cultures (5, 8). In 1974, *Lb. plantarum* was patented as a starter culture to be used alone or in combination with *P. acidilactici* for dry and semidry sausages (5, 28).

Currently, the predominant genera of bacteria used either singly or as mixed starter cultures in the United States and Europe are *Pediococcus*, *Lactobacillus*, *Micrococcus*, and *Staphylococcus* (5, 36). These cultures are available either fresh, frozen, freeze-dried, or in a low-temperature stabilized liquid form, with the frozen culture being the most often used (6). *Lb. plantarum*, *P. acidilactici*, and *Pediococcus pentosaceus* (6, 20) are favored in the United States for their rapid and nearly complete (>90%) conversion of glucose to lactic acid (pH 4.6 to 5.1) at high temperatures (32°C). In Europe, less fermentative (pH 5.2 to 5.6) *Staphylococcus xylosus*, *S. carnosus*, *Staphylococcus simulans*, *Staphylococcus saprophyticus*, and, to a lesser extent, *Micrococcus* spp. (59), which have a lower temperature optimum (<24°C), are preferred for flavor development and red color enhancement. Chemical reduction of nitrate and nitrite to nitric oxide via nitrate and nitrite reductase, followed by reaction with the singular heme moiety of myoglobin, forms dinitrosylhemochrome, which develops into a characteristic pink cure color when heated (70). *P. pentosaceus* and *P. acidilactici* have temperature optima of 32°C and 35°C, respectively, while lactobacilli have lower optima, in the range of 21 to 24°C. Rapidly fermented products such as beef sticks and pepperoni utilize *P. acidilactici* and *P. pentosaceus* alone or in combination with *Lactobacillus* spp. or *Micrococcus* spp., which permits the use of elevated fermentation temperatures ranging from 32 to 46°C. *Lactobacillus curvatus, Lactobacillus sake, Lb. plantarum, P. pentosaceus,* and *P. acidilactici* produce bacteriocins which may find broader application in starter cultures for meat fermentations in the future.

Characteristics of Commercial Meat Starter Cultures

Bacterial starter cultures for meat and poultry products in the United States have been selected on the basis of being homofermentative, capable of rapidly converting glucose or sucrose to DL-lactic acid anaerobically with sustained growth to pH 4.5, tolerant of salt brines (NaCl, KCl) up to 6%, and capable of growth in the presence of sodium or potassium nitrate or nitrite in the range of 600 mg/kg (nitrate) and 150 mg/kg (nitrite). These cultures are aciduric, capable of growth in the range of 21 to 43°C, nonproteolytic, nonlipolytic, inactivated at 60.5 to 63.2°C, resistant to phage infection and mutation, and able to outgrow and/or suppress pathogens. Specific advantages favoring the use of commercial starter cultures over natural fermentations, i.e., back-inoculation, are consistency of the inoculum, reduced risk of bacterial cross-contamination, uniformity of lactic acid development, production of desirable flavor components, predictability of pH endpoint (as regulated by carbohydrate level and incubation temperature), and reduced risk of proteolytic and pathogenic bacterial outgrowth. Other benefits include the acceleration of fermentation time, to increase commercial plant throughput, and the reduction of product defects such as off flavors, lack of tangy flavor, excessive softness, crumbly texture, gas pockets, and pinholes, all of which can be attributed to heterofermentative bacteria.

An inoculum population of 10^7 CFU/g of raw product is sufficient for rapid lactic acid development within 6 to 18 h under controlled temperature, airflow, and relative humidity conditions (6), but incubation time is also dependent upon carbohydrate type and concentration, spice composition, exclusion of oxygen, and product diameter or thickness. In the United States, commercial processors inoculate meats with homofermentative, gram-positive bacteria that have been isolated from and adapted to specific meat products. These inocula (starter cultures) may consist of a single species of *Lb. plantarum, Lactobacillus pentosus, Lb. sake, Lb. curvatus, M. varians, P. acidilactici,* or *P. pentosaceus,* or more commonly a combination of these bacteria. Use of starter cultures assures dominance of desirable microorganisms, production of acids consisting of >90% lactic acid, and modulation of fermentation based on combinations of inoculum level, glucose concentration, and incubation time and temperature intervals (5, 6). Thus the combined effects of low pH, increased acidity, concomitant loss of moisture during drying, reduction of a_w, concentration of curing salts such as sodium chloride and sodium nitrite, bacterial inhibition of spoilage or pathogenic microorganisms, and heat processing (if applied) preserve fermented meat and poultry products against spoilage by inactivating indigenous tissue and bacterial enzymes.

Mold and yeast starter cultures are not widely used in the United States, with the exception of dry sausages produced in the San Francisco area. White or gray-colored molds are typical on casing surfaces of sausages produced in Hungary, Italy, Spain, Greece, Yugoslavia, Romania, Slovakia, and the Czech Republic. Their effect is primarily cosmetic, but it has been reported that the mycelial coat can reduce moisture loss and facilitate uniform drying (5). Catalase produced by molds may serve as an antioxidant by reacting with surface oxygen to prevent it from entering the product, while nitrate reductase promotes the development of red surface color (59). Green mold on the surface of sausage is typically the result of excessive humidity and sporulation, but *Penicillium chrysogenum* on French sausage is noted for its ability to produce a preferred ebony color. Molds are capable of decomposing lactic acid, which increases pH and results in a milder flavor (33). *Penicillium nalgiovense* is most commonly used for this purpose, but application of *Penicillium chrysogenum* and *Penicillium camemberti* may also be used. In Germany, only *Penicillium candidum, Penicillium nalgiovense,* and *Penicillium roqueforti* are approved for application to sausages. The use of nontoxigenic molds may reduce the risk of mycotoxin production by other molds (33, 51).

Yeasts such as *Candida famata* and *Debaryomyces hansenii* are used alone or in combination with bacterial cultures to produce a powdery surface or a fruity or alcoholic aroma which may be construed as spoiled if uncontrolled (5). These cultures are added at a population of 10^6 CFU/g and are characterized by a high salt tolerance (a_w 0.87) (37). *Streptomyces griseus* subsp. *hutter* produces a cellar-ripened sausage aroma and enhances color because of its nitrate reductase and catalase activities (59).

Classification of Bacterial Starter Cultures for Meat

The genera most commonly used as meat starter cultures are *Lactobacillus, Pediococcus, Micrococcus,* and *Staphylococcus* (5, 36); *Pediococcus, Lactobacillus,* and other lactic acid bacteria are preferred when acid production is of primary importance. Specific strains of *Micrococcus* and coagulase-negative staphylococci are used in meat curing when lower incubation temperatures and less acid production are required for flavor, as found in many European sausages (5). Eight genera (*Lactobacillus, Leuconostoc, Pediococcus, Streptococcus, Carnobacterium* [formerly *Lactobacillus*], *Enterococcus, Lactococcus,* and *Vagococcus* [the latter three genera were formerly *Streptococcus*]) are most commonly used as starter cultures (41). *Lactobacillus hordniae* and *Lactobacillus xylosus* are now in the genus *Lactobacillus,* while *Streptococcus diacetilactis* has been classified as a citrate-utilizing strain of *Lactococcus lactis* subsp. *lactis.*

Lactobacilli and pediococci are gram-positive, nonsporing rods or cocci which produce lactic acid as a major end product during the fermentation of carbohydrates (4). Originally classified on the basis of morphology, glucose fermentation pathway, optimal growth temperature, and stereoisomer of lactic acid produced, this phenotypic grouping has remained largely intact but may change considerably as molecular characterization is completed (4). Ongoing research on oligonucleotide cataloging and 16S rRNA sequencing indicates that all gram-positive bacteria cluster in 1 of the 11 major eubacterial phyla and can be further divided into two main groups or clusters (4, 100). One cluster, designated as the Actinomycetes subdivision, comprises bacteria possessing DNA of above 55 mol% G+C and includes meat fermentation genera such as *Micrococcus,* while the cluster with low moles percent G+C, designated as the *Clostridium* subdivision, includes meat fermentation genera such as the staphylococci (4, 92, 100). Lactic acid bacteria are thought to form a large related cluster which phylogenetically lies between strictly anaerobic species such as the clostridia and the facultative or strictly aerobic staphylococci and bacilli (4, 44, 45).

At the species level, lactobacilli appear to fall into three clusters which do not appear to correlate with current classification schemes (4). Most of the species involved or associated with meat fermentations fall within the *Lactobacillus casei/Pediococcus* subgroup, comprising obligately homofermentative, some heterofermentative, and all of the facultative heterofermentative lactobacilli (4). The pediococci are gram-positive cocci and are the only lactic acid bacteria capable of dividing in two planes. Consequently, they can appear as pairs, tetrads, or other formations (30, 75). The latest taxonomic classification places pediococci as members of the phylogenetic cluster of gram-positive facultative anaerobic cocci that contains 15 genera, including the staphylococci, streptococci, micrococci, and peptococci (84). This designation will probably change because comparison studies using 16S rRNA and nucleotide sequencing have indicated that pediococci are more closely aligned with the phylogenetic cluster that includes the lactobacilli and leuconostocs (18, 91, 98).

The lactobacilli associated with meat fermentations are members of the genus *Lactobacillus.* They are characterized as gram-positive, non-spore-forming rods which are catalase negative on media not containing blood, are usually nonmotile, usually reduce nitrate, and ferment glucose (38, 45). The genus *Lactobacillus* currently consists of more than 50 species. DNA hybridization and sequencing comparisons indicate that despite the numerous species, as a whole they comprise a well-defined group of bacteria (38). Part of the reason for the large number of species is that the genus *Lactobacillus* has been exhaustively studied not only from the perspective of taxonomy and identification, but also from the standpoint of nutritional requirements for application in studies on biochemistry and metabolism (85). Most research on meat fermentations has emphasized *Lb. plantarum* as the predominant species (5). However, attempts to identify lactobacilli isolated from meat are usually less than successful because most of the documented descriptions and schemes of identification are based on isolates from other food sources (46, 77, 82, 86). A series of investigations on atypical lactic acid streptococci isolated from fermented meats resulted in their identification as *Lb. sake* and *Lb. curvatus* (reference 42, as cited in reference 36). These species outnumbered the typical lactobacilli identified as *Lb. plantarum, Lactobacillus brevis, Lactobacillus alimentarius, Lb. casei, Lactobacillus farciminis, Lactobacillus viridescens,* unspecified leuconostocs, and pediococci by 1,000-fold and were typically psychrotrophic and less acid tolerant (36, 76). Efforts to accurately classify and identify these strains are becoming more important as various isolates of lactic acid bacteria become more commonly used as starter cultures.

Starter Culture Metabolism

The term "fermentation" was first defined by Pasteur as life in environments devoid of oxygen. However, meat fermentations are more accurately defined as bacterial or microbial metabolic processes in which carbohydrates and related compounds are oxidized with the release of energy in the absence of oxygen as a final electron acceptor. Thus, by definition, fermentative microorganisms cannot use oxygen as a terminal electron acceptor to generate ATP and must either conduct anaerobic respiration using more electronegative electron acceptors (carbon dioxide, sulfate, nitrate, or fumarate) or ferment intermediate metabolites formed from the substrate as an electron acceptor (32). The bacteria responsible for fermentation are either facultative or obligate anaerobes.

As a collective group, the gram-positive acidogenic (predominantly lactic acid) bacteria include the genera *Lactobacillus, Streptococcus, Pediococcus, Leuconostoc, Lactococcus,* and *Enterococcus,* which can metabolize a large number of mono- and oligosaccharides, polyalcohols, aliphatic compounds, mono-, di-, and tricarboxylic acids, and some amino acids, although individual species characteristically have a limited range of carbon and energy sources (54). During fermentation of sausages, members of this group of bacteria are responsible for two basic microbiological processes that occur simultaneously and are interdependent, namely, the production of nitric oxide by nitrate- and nitrite-reducing bacteria and a decrease in pH as a result of anaerobic glycolysis. These two activities are synergistic due to the pH dependency of nitrite and nitrate reduction. The following discussion details reactions that are important in meat, poultry, and fish fermentations and is based on several excellent reviews (4, 43, 52).

Carbohydrate Fermentation Pathways

Acidogenic bacteria ferment indigenous and added carbohydrates primarily to form DL-lactic acid. The formation of lactic acid, which is an anaerobic process, helps to create a reduced pH environment in the meat, and this in turn contributes to the development of cured meat color. The formation of lactic acid also causes coagulation of meat proteins and, in combination with drying, gives sausages their characteristic firm texture. There are three potential pathways that the lactic acid bacteria may use to form lactic acid from carbohydrates. The homofermentative pathway yields 2 mol of lactic acid per mol of glucose; the heterofermentative pathway yields 1 mol of lactic acid, 1 mol of ethanol, 1 mol of acetic acid, and 1 mol of carbon dioxide per mol of glucose; and the bifidum pathway (which will not be discussed at length here, since it is generally not found in the bacteria important in meat starter cultures) yields 3 mol of acetic acid and 2 mol of lactate per 2 mol of glucose (32). The homofermentative pathway or glycolysis is the primary means of generating lactic acid in meat fermentations. Both *P. acidilactici* and *P. pentosaceus* are microaerophilic and, under anaerobic growth conditions, are homofermentative, yielding DL-lactic acid (30, 75). Glucose and most other monosaccharides are fermented, and, unlike other pediococci, both species can ferment pentoses (30). It is probable that *P. acidilactici* and *P. pentosaceus* transport glucose via the phosphoenolpyruvate phosphotransferase system and derive ATP using the Embden-Meyerhof pathway (glycolysis). The Embden-Meyerhof pathway features the formation of fructose 1,6-diphosphate (FDP), which is cleaved by FDP aldolase to form dihydroxyacetonephosphate and glyceraldehyde 3-phosphate. These three-carbon intermediates are converted to pyruvate, which also energetically favors ATP formation via substrate-level phosphorylation at two separate metabolic transformation steps (two ATP per glucose). Because the resulting NADH formed during glycolysis requires oxidizing to regenerate NAD^+ and maintain oxidation-reduction balance, pyruvate is reduced to lactate by a NAD^+-dependent lactate dehydrogenase. Since lactic acid is virtually the only end product, the fermentation is referred to as a homolactic fermentation (4).

Heterofermentation is also characterized by dehydrogenation steps yielding 6-phosphogluconate, followed by decarboxylation and formation of a pentose 5-phosphate. The pentose is cleaved by phosphoketolase to form glyceraldehyde 3-phosphate, which eventually becomes lactic acid through glycolysis, and acetyl phosphate (which is required to maintain electron balance) is reduced to ethanol (4, 32). In theory, homofermentative and heterofermentative bacteria can be distinguished by the pattern of fermentation end products and the presence or absence of FDP aldolase and phosphoketolase (43). Thus, obligately homofermentative species possess a constitutive FDP aldolase and lack phosphoketolase, while the reverse is true of obligately heterofermentative species (4, 43, 45). Species of lactic acid bacteria used to ferment meat can be regarded essentially as facultative heterofermenters because they not only possess a constitutive FDP aldolase and consequently use glycolysis for hexose fermentation, but also possess a pentose-inducible phosphoketolase. The type of fermentation is dependent on growth conditions. In an ecosystem such as meats with a wide variety of complex substrates serving as sources of pentoses, organic acids, and other fermentable compounds in addition to hexoses, it is possible for lactic acid bacteria to be homofermentative when using hexose, but heterofermentative when using pentoses and other substrates. Lactic acid is the primary metabolite derived from both homo- and heterofermentation, but during heterofermentation mixtures of lactic acid, acetic acid, ethanol, and carbon dioxide are also produced.

ATP Generation, Pyruvate Metabolism, and Energy Recycling

The most important energy requirements of the bacterial cell are macromolecular synthesis and the transport of essential solutes against a concentration gradient (4). The fermentation characteristic of lactic acid bacteria is essentially the oxidation of a substrate to generate energy-rich intermediates that in turn are used for ATP production by substrate-level phosphorylation. Although lactic acid bacteria can tolerate a more minimal and broad range of internal pHs, there still must be a substantial amount of ATP generated to keep the cytoplasmic pH above a threshold level to offset the massive external acid production (4).

Bacteria have considerable flexibility in the pathways they can use to generate ATP and regulate its distribution. In aerobic bacteria which rely on an electron transport chain, large quantities of ATP can be generated via a proton motive force across the electron transport chain coupled to a H^+-translocating ATP synthase (4). Even though lactic acid bacteria do not possess ATP-generating capabilities via electron transport, lactobacilli do have some metabolic flexibility by altering pyruvate pathways and conserving energy by end product efflux. Some lactic acid bacteria have alternative pathways of pyruvate metabolism other than direct reduction to lactic acid. Oxidation involves the formation of NADH from NAD^+, which requires regeneration back to NADH, and pyruvate serves as the key intermediate for this by acting as the electron acceptor (4). Pyruvate-formate lyase generates formate and acetyl coenzyme A (CoA) from pyruvate and CoA, and the acetyl CoA can be used as either a precursor for substrate-level phosphorylation or direct reduction to ethanol or both. This alteration in pyruvate metabolism has been demonstrated in studies where lactobacilli have been shown to shift fermentation patterns, and concomitantly the amount of potential ATP formed, as growth rate changes (24). In these studies, mixed acid-forming strains of lactobacilli produced less lactic acid and more acetic acid with decreasing growth rate, and for every mole of acetic acid, an extra mole of ATP (2 versus 1 mol of ATP produced per mol of lactic acid) was potentially formed (24). During fermentation, lactic acid bacteria form sufficient end products in the cytoplasm to result in a very high internal-to-external cell gradient (72). It has been suggested that when a fermentation product such as lactic acid exceeds the electrochemical proton gradient, additional metabolic energy is gained by product efflux through carrier-mediated transport in symport with protons (58, 72, 93). Metabolic energy is conserved because a product gradient is essentially a proton gradient which can be used to directly produce ATP by proton-driven ATPase, and it helps to maintain a proton motive force without consumption of ATP (4, 72).

Nitrate and Nitrite Reduction

Cured meat derives its characteristic pink color from the reaction of myoglobin (Fe^{2+}) or metmyoglobin (Fe^{3+}) with nitric oxide to ultimately form nitric oxide myoglobin, an unstable, bright pink compound (70). Nitric oxide myoglobin is formed directly from the reaction of purple-red myoglobin with nitric oxide. Another proposed route of formation may be through the oxygenation of myoglobin to bright red oxymyoglobin (Fe^{2+}) and subsequent oxidation to brown metmyoglobin (Fe^{3+}). Nitric oxide reduces metmyoglobin to gray-brown nitric oxide metmyoglobin (Fe^{2+}), which transmutates to nitric oxide myoglobin with the loss of oxygen. Nitric oxide myoglobin, when heated, forms nitrosohemochrome (Fe^{2+}), the stable pink pigment present in commercially processed cured meats. Nitrate- and nitrite-reducing bacteria transform added nitrates into nitrites and eventually into more reduced forms for reaction with the dominant myoglobin or metmyoglobin pigment forms. Micrococcaceae possess nitrate reductase to convert nitrate to nitrite, which in turn is converted to nitric oxide via nitrous acid under acidic conditions. Nitric oxide is a strong electron donor and rapidly reacts with the heme moiety of myoglobin to form the characteristic pink meat pigment dinitrosylhemochrome. Some strains of lactic acid bacteria may produce metabolites such as carbon dioxide, acetate, formate, succinate, and acetoin during fermentation that cause off flavors; however, nitrate and nitrite prevent the formation of certain off flavors from compounds such as formate by their inhibition of pyruvate:formate lyase activity in lactobacilli (37). The formation of nitric oxide from the reduction of nitrates and nitrites is crucial because it reduces nitrosometmyoglobin to nitrosomyoglobin and induces red color formation (52). The advantage of microbial activity is the reduction of nitrates, thus removing excess nitrate and nitrite from the meat.

Metabolic Activities Important in Commercial Starter Cultures

Strains of *P. acidilactici* were initially selected for use as starter cultures because they ferment carbohydrates rapidly to lactic acid at higher temperatures and effectively lower the pH of the meat from a range of 5.6 to 6.2 down to 4.7 to 5.2 (9). Traditionally, semidry sausages (summer sausage, Thüringer, beef sticks) and some types of pepperoni are allowed to undergo greening, i.e., fermentation at 16 to 38°C, which favors the development of indigenous microflora and extends fermentation times (9). Although indigenous microflora have the advantage of imparting unique flavor(s) and other desirable sensory properties to sausages, there is also sufficient variation in lactic acid production to be at a considerable disadvantage for large-scale production of sausage. The use of starter cultures consisting of pediococci allows for the inoculation of large numbers of one microorganism into the raw meat followed by a uniform fermentation. More importantly, meat inoculated with *P. acidilactici* can be incubated at higher temperatures (43 to 50°C), which precludes the growth of most indigenous microorganisms so that flavor and pH characteristics are predictable (9). Consequently, commercial production is enhanced since flavor development can be controlled and the fermentation process can be hastened.

Strains of *P. acidilactici* that have been selected and developed for use as commercial starter cultures are well suited for the production of semidry sausages which are fermented and/or smoked at higher temperatures (26 to 50°C). However, when used for production of dry sausages which require lower fermentation temperatures (15 to 27°C), these strains generally produce lactic acid at a much slower rate (9). Since *P. pentosaceus* has a lower optimum temperature than *P. acidilactici* for growth (28 to 32°C versus 40°C) (30), it has been promoted as an effective meat starter culture. *P. pentosaceus* is also more attractive because it has a 25% lower Arrhenius energy of activation for fermentation than that of *P. acidilactici* (74). Thus, *P. pentosaceus* has a lower optimum growth temperature and a higher capacity for rapid production of lactic acid compared to *P. acidilactici.*

Metabolic Contributions of Starter Cultures to Sausage Sensory Qualities

Fermentation of sausages most often involves inoculation of ground meats with a starter culture, followed by an incubation period to allow enzymatic conversion of available carbohydrate to approximately equimolar concentations of DL-lactic acid. Fermentable carbohydrates in the form of residual muscle glucose (0.1 mg/g) and added glucose or sucrose are the primary sources of energy available for metabolism. The decline in pH to 4.5 to 4.7 with the buildup of lactic acid results in a dramatic loss of water-holding capacity as the pH nears the isoelectric point (pI = ~5.1) of myofibrillar proteins. Product texture becomes firmer, density increases with dehydration during aging, and partial denaturation of the myofibrillar proteins occurs due to the presence of lactic acid. Thus, the tangy flavor and chewy texture of fermented sausages are consequences of the dominant fermentation metabolite, lactic acid, and dehydration.

Lactic acid is the dominant flavor component in fermented sausage, but spices, salt, sugar, sodium nitrite reaction products, smoke, and meat components such as fatty acids, amino acids, and peptides are also contributors to the flavor profile (71). Secondary flavor contributions from metabolites of lactobacilli and pediococci are minor in fermented sausages produced in the United States; however, natural heterofermentative lactobacilli (*Lb. brevis, Lactobacillus buchneri*) from back-inoculations, when combined with staphylococci and micrococci, give many European sausages their characteristic flavor as a result of volatile acids, alcohols, and carbon dioxide. Members of *Micrococcaceae,* for example, possess lipolytic and proteolytic enzymes which generate aldehydes, ketones, and short-chain fatty acids that give characteristic aromas and flavors to sausages (56). They also produce catalase, which decomposes hydrogen peroxide. *Leuconostoc mesenteroides* and *Lb. brevis* produce ethanol, acetic acid, lactic acid, pyruvic acid, acetoin (which imparts a nutty flavor and aroma), and carbon dioxide to give an effervescent sensation, but fermentation conditions must be controlled to avoid excessive pinholes, gas pockets, and off flavors.

In natural fermentations as well as in sausages inoculated with lactobacilli, *Lb. sake* and *Lb. curvatus* have been observed to be the dominant bacteria at temperatures of 25°C and below, while *Lb. plantarum* tends to dominate at higher fermentation temperatures (42). *Lb. sake, Lb. curvatus,* and *Lb. plantarum,* to a lesser extent, form DL-lactic acid and have flavin-dependent oxidases capable of forming hydrogen peroxide. Tolerance to peroxide may be one of the factors contributing to their dominance during fermentation. Hydrogen peroxide, if not decomposed, can induce oxidation of unsaturated fatty acids, leading to rancid flavor development, and/or oxidize the heme component of myoglobin, leading to fading or the formation of green, yellow, gray, or other off-color pigments (59). *Lb. curvatus* does not possess pseudocatalase or true catalase and may permit the accumulation of hydrogen peroxide. *Lb. sake, Lb. plantarum,* and pediococci, on the other hand, exhibit catalase activity and can therefore decompose hydrogen peroxide formed during fermentation.

Sensory analysis of unspiced fermented dry sausages produced using various starter culture combinations (14) has revealed that sausage flavor is influenced by culture composition. Berdague et al. (14) reported that the butter odor of dry sausages was largely dependent upon degradation of sugars by way of pyruvic acid and that curing and rancid odors were correlated with compounds resulting from lipid oxidation. Their work indicated that *S. saprophyticus* and *Staphylococcus warneri* were most associated with butter odor, which in turn was correlated to the presence of acetoin, diacetyl, 1,3-butanediol, and 2,3-butanediol. Distinctive curing odors were associated with combinations of *S. carnosus* and *P. acidilactici, S. carnosus* and *Lb. sake,* and *S. carnosus* and *P. pentosaceus* and were correlated with 2-pentanone, 2-hexanone, 2-heptanone, and an unknown compound. A combination of *S. saprophyticus* and *Lb. sake,* which produced the most acetic acid, was associated with fruity odors derived from esters and a less intense rancid odor. Meat proteins are hydrolyzed by endogenous and microbial proteases to yield peptides and amino acids which are in turn degraded to ammonia and amines, causing a slight rise in the pH. Demeyer and Samejima (23), however, suggested that the major protease activity in fermented meat is derived from the meat enzymes. Amino acid degradation products and nucleotide IMP intensify meat flavors and contribute to the overall sausage flavor. Thus, it appears that compounds produced via endogenous proteolysis in combination with starter culture fermentation play a significant role in the nonacidic flavor and aroma of fermented sausages.

Genetics and Biotechnology of Meat Starter Cultures

To utilize biotechnological approaches with meat starter cultures, genetic transfer systems and mobile genetic elements for carrying a potentially important gene(s) must be identified. The standard approach to gene cloning in lactic acid bacteria is to develop plasmid vectors based on either indigenous cryptic plasmids or heterologous plasmids resistant to a broad range of antibiotics (31). Since Chassy et al. (17) first demonstrated the presence of plasmids in lactobacilli, significant progress has been made, particularly in elucidating genetic systems in dairy lactobacilli and *Lc. lactis* (31, 38).

Plasmids have been detected in many strains of *P. pentosaceus, P. cerevisiae,* and *P. acidilactici,* and some may be transferable between strains. Some properties that are characteristic of a particular species of *Pediococcus* have been shown to be plasmid linked. The production of bacteriocin and bacteriocin-like substances and the ability to ferment some sugars have been shown to be linked to or encoded by plasmids in strains of *P. cerevisiae, P. pentosaceus,* and *P. acidilactici.* Numerous plasmids have been isolated from meat lactobacilli, and some have been functionally identified. Nes (65) observed that 8 of the 10 strains of *Lb. plantarum* examined for the presence of extrachromosomal DNA contained from one to six plasmids. Six strains containing plasmids were commonly used as starter cultures in dry sausage. Metabolic functions in *Lb. plantarum* and four atypical *Lactobacillus* species isolated from fresh meat (53) that have been shown to be plasmid linked include maltose utilization and cysteine metabolism (87). When Schillinger and Lücke (83) screened 221 strains of lactobacilli from meat and meat products for antibacterial compounds, they linked immunity and production of a bacteriocin in a strain of *Lb. sake* with an 18-MDa plasmid by showing that cured variants no longer possessed either trait. *Lb. sake* also produces a bacteriocin, lactocin S, which inhibits species of *Lactobacillus, Pediococcus,* and *Leuconostoc.* Both bacteriocin production by *Lb. sake* and its immunity factor have been shown to be associated with an unstable 50-kb plasmid (61–63).

Genetic transfer has been demonstrated in several species of lactobacilli and lactococci, but limited studies have been performed with species associated directly with meat fermentations. In vivo gene transfer systems have involved either conjugation, which requires cell-to-cell contact (bacterial mating) for transfer of genetic material, or transduction (transfection or phage-mediated genetic exchange), while in vitro physiological transformation requires uptake of naked DNA by the recipient cell (39, 97). Intergenic conjugation of an antibiotic-resistant streptococcal plasmid was demonstrated in *Lb. plantarum* (99), and successful inter- and intragenic conjugation of this same plasmid has also been reported for *Lb. plantarum* (88) as well as a strain of *Lactobacillus* isolated from fermented sausage (79). Conjugative transfer of an *Escherichia-Streptococcus* nonconjugative shuttle plasmid has also been achieved in *Lb. plantarum* by cointegration formation (89). Further genetic analysis and manipulation of meat lactobacilli have been handicapped by a lack of natural competence and transformation procedures (38, 60). This will change with the emergence of electroporation methods that have been successfully used to transform plasmids in *Lb. plantarum* (3, 15, 55, 73).

Although shuttle vectors have been used to construct and introduce heterologous genes into lactobacilli, they generally contain antibiotic resistance markers which greatly facilitate the process, but may not receive regulatory approval for use in foods (39). Furthermore, considerable instability and loss of function occurs during replication because these plasmids for the most part belong to a group which replicates via single-stranded DNA intermediates or rolling-circle replication (31, 34). This form of replication is a potential source of plasmid segregational instability. With meat lactobacilli, vectors containing replicons from *Lactobacillus, Staphylococcus,* or *Escherichia coli* sequences have all been shown to contribute to this instability (50, 73). Therefore, plasmids containing a *Lactobacillus* replicon can be stabilized under selective conditions and used to express genes from multiple copies of the plasmid, but under nonselective conditions are frequently segregationally unstable and become lost (50, 73). Consequently, efforts have focused on gene cloning strategies that enable heterologous genes to integrate in a stable fashion into the bacterial chromosome (31, 40).

Recombination in chromosomes can occur as general recombination when chromosomal DNA is transferred from one bacterium to the next via conjugation or transduction, which necessitates extensive sequence homology between externally introduced DNA and chromosomal DNA as well as specific bacterial recombination factors (97). Recombination can also be mediated by transposable genetic elements which involve short target sequences of 10 nucleotide base pairs or less and occur independent of bacterial host function (97). Transposons have been introduced into *Lb. curvatus* (48) and *Lb. plantarum* (3), and integration of plasmids by single crossover events within regions of homology can be accomplished by the constructing plasmid suicide vectors which lack the ability to replicate in lactic acid bacteria (31). Scheirlinck et al. (81) used a suicide plasmid to integrate *Bacillus stearothermophilus* α-amylase genes and *Clostridium thermocellum* endoglucanase genes into *Lb. plantarum.* This homologous insertion into the chromosome was made possible by generating plasmid constructs containing an unknown portion of a *Lactobacillus* DNA sequence as part of the suicide vector to facilitate site-specific integration. Complete replacement of a gene(s) can be achieved with homologous double cross-over recombination events. An alternative to using homologous sequences is to locate insertion sequences (IS elements) already present on the chromosome that can transpose chromosomal DNA to plasmids and use these in the construction of integrative vectors (39). Only a few such elements have been identified in meat lactobacilli, and only one, IS*1163* from *Lb. sake,* has been isolated, sequenced, and described in any detail (90).

Development of genetically engineered strains of lactobacilli that have large-scale use in fermented meats or other foods has remained elusive. Part of the problem is that transfer of structural genes to a new host does not guarantee that the genes will be expressed (39). Optimal expression of cloned genes requires not just cloning of the structural gene but also efficient promoters, ribosome binding sites, and termination sites, all of which must be identified, isolated, cloned, and sequenced (39). Integration systems that have been used do not allow high enough levels of heterologous gene expression for practical application to fermented meats and give an unsatisfactory stabilization of the foreign gene (40). One approach to solving this problem essentially involves using the structural gene *amyL,* which encodes the production and secretion of α-amylase from *B. licheniformis,* as a reporter gene for the cloning of

Table 33.5 Research needs for meat starter cultures

Improving current starter culture technology
Substrate specificity of starter culture microorganisms
Immobilized cells for meat fermentation
Increased rate of lactic acid production (reduction of lag phase)
Control of pathogens without heating
Use of starter cultures in combination with acidulants, e.g., glucono-delta-lactone (GDL)
Identification and testing of nitrite substitutes (development of cured color with protection against *C. botulinum* outgrowth)
New cryoprotectant (antifreeze) solutions for frozen starter cultures
Development of new pediocins that are not bound by fat, subject to proteolysis, or inactivated by other meat components
Use of cultures to enhance meat product nutrition and quality
Enhancement of nutritional quality by in situ production of critical dietary nutrients for selected populations
Greater utilization of nontraditional meat sources via fermentation
Accelerated curing of traditional meat products (non-nitrite hams, bacon, comminuted meats, meat snacks, natural acidification against *C. botulinum*)
Generation of natural antioxidants in situ (α-tocopherol production by starter cultures)
Production of antimycotic agents for mold inhibition
Development of starter cultures unique to fish pastes, sauces

expression-secretion sequences from *Lb. plantarum* (40). The key requirement for this approach was that the *amyL* gene did not possess a specific and/or regulated promoter that was operational in any gram-positive bacterium outside its genus of origin, *Bacillus*. Thus, the high expression and secretion of over 90% of the *Bacillus* extracellular amylase in *Lb. plantarum* served as an expression reporter to indicate that replacement of the *Bacillus* promoter by a *Lb. plantarum* promoter had taken place. The plasmid containing the silent *amyL* coding frame could subsequently be used as a probe to locate expression (transcription and translation) or expression-secretion regions on the chromosome of *Lb. plantarum* strains expressing and secreting α-amylase.

Numerous opportunities are available for advances in meat starter culture research. Potential research areas are summarized in Table 33.5. The number of groups conducting research on genetic systems, particularly in lactobacilli, has multiplied severalfold in recent years. Earlier predictions (16, 57) for biotechnological approaches to optimize fermentations are nearing reality.

References

1. **Acton, J. C.** 1977. The chemistry of dry sausages. *Proc. Recip. Meat Conf.* **30**:49–62.
2. **Al-Sheddy, I. A., D. Y. C. Fung, and C. L. Kastner.** 1995. Microbiology of fresh and restructured lamb meat: a review. *Crit. Rev. Microbiol.* **21**:31–52.
3. **Aukrist, T., and I. F. Nes.** 1988. Transformation of *Lactobacillus plantarum* with the plasmid pTV1 by electroporation. *FEMS Microbiol. Lett.* **52**:127–132.
4. **Axelsson, L. T.** 1993. Lactic acid bacteria: classification and physiology, p. 127–159. *In* S. Salminen and A. von Wright (ed.), *Lactic Acid Bacteria*. Marcel Dekker, Inc., New York.
5. **Bacus, J. N.** 1986. Fermented meat and poultry products. *Adv. Meat Res.* **2**:123–164.
6. **Bacus, J. N.** 1995. Personal communication.
7. **Bacus, J. N., and W. L. Brown.** 1981. Use of microbial cultures: meat products. *Food Technol.* **35(1)**:74–78, 83.
8. **Bacus, J. N., and W. L. Brown.** 1985. The lactobacilli: meat products, p. 58–71. *In* S. E. Gilliland (ed.), *Bacterial Starter Cultures for Foods*. CRC Press, Inc., Boca Raton, Fla.
9. **Bacus, J. N., and W. L. Brown.** 1985. The pediococci: meat products, p. 86–95. *In* S. E. Gilliland (ed.), *Bacterial Starter Cultures for Foods*. CRC Press, Inc., Boca Raton, Fla.
10. **Bartholomew, D. R., and C. I. Osuala.** 1986. Acceptability of flavor, texture, and appearance of mutton processed meat products made by smoking, curing, spicing, adding starter cultures and modifying fat source. *J. Food Sci.* **51**:1560–1562.
11. **Beddows, C. G.** 1985. Fermented fish and fish products, p. 1–39. *In* B. J. B. Wood (ed.), *Microbiology of Fermented Foods*, vol. II. Elsevier Applied Science Publishers, London.
12. **Beddows, C. G., A. G. Ardeshir, and W. Johari bin Daud.** 1979. Biochemical changes occurring during the manufacture of Budu. *J. Sci. Food Agric.* **30**:1097–1103.
13. **Beddows, C. G., A. G. Ardeshir, and W. Johari bin Daud.** 1980. Development and origin of the volatile fatty acids in Budu. *J. Sci. Food Agric.* **31**:86–92.
14. **Berdague, J. L., P. Monteil, M. C. Montel, and R. Talon.** 1993. Effects of starter cultures on the formation of flavor compounds in dry sausage. *Meat Sci.* **35**:275–287.
15. **Bringle, F., and J.-C. Hubert.** 1990. Optimized transformation by electroporation of *Lactobacillus plantarum* strains by plasmid vectors. *Microbiol. Biotechnol.* **33**:664–670.
16. **Chassy, B. M.** 1987. Prospect for the genetic manipulation of lactobacilli. *FEMS Microbiol. Lett.* **46**:297–312.
17. **Chassy, B. M., E. Gibson, and A. Giuffrida.** 1976. Evidence for extrachromosomal elements in *Lactobacillus*. *J. Bacteriol.* **127**:1576–1578.
18. **Collins, M. D., A. M. Williams, and S. Wallbanks.** 1990. The phylogeny of *Aerococcus* and *Pediococcus* as determined by 16 rRNA sequence analysis: description of *Tetragenococcus* gen. nov. *FEMS Microbiol. Lett.* **70**:255–262.
19. **Crisan, E. V., and A. Sands.** 1975. The microbiology of four fermented fish sauces. *Appl. Microbiol.* **29**:106.
20. **Curtis, S. I.** 1995. Personal communication.
21. **Deibel, R. H., and C. F. Niven, Jr.** 1957. *Pediococcus cerevisiae:* a starter culture for summer sausage. *Bacteriol. Proc.* 1957, p. 14–15.
22. **Deibel, R. H., G. D. Wilson, and C. F. Niven, Jr.** 1961. Microbiology of meat curing. IV. A lyophilized *Pediococcus cerevisiae* starter culture for fermented sausages. *Appl. Microbiol.* **9**:239–243.
23. **Demeyer, D., and K. Samejima.** 1991. Animal biotechnology and meat processing, p. 127–143. *In* L. O. Fiems, B. G. Cottyn, and D. I. Demeyer (ed.), *Animal Biotechnology and the Quality of Meat Production*. Elsevier, Amsterdam.
24. **de Vries, W., W. M. C. Kapteijn, E. G. van der Beek, and A. H. Stouthamer.** 1970. Molar growth yields and fermentation balance of *Lactobacillus casei* L3 in batch cultures and continuous cultures. *J. Gen. Microbiol.* **63**:333–345.
25. **Dougan, J., and G. Howard.** 1975. Some favouring constituents of fermented fish sauces. *J. Sci. Food Agric.* **26**:887–894.
26. **Egan, A. F.** 1983. Lactic acid bacteria of meat and meat products. *Antonie van Leeuwenhoek* **49**:327–336.
27. **Everson, C. W., W. E. Danner, and P. A. Hammes.** 1970. Bacterial starter cultures in sausage products. *J. Agric. Food Chem.* **18**:570–571.
28. **Everson, C. W., W. E. Danner, and P. A. Hammes.** 1974. Process for curing dry and semidry sausages. U.S. Patent 3,814,817.
29. **Fujii, T., S. D. Basuki, and H. Tozawa.** 1980. Microbiological studies on the ageing of fish sauce; chemical composition and microflora of fish sauce produced in the Philippines. *Nippon Suissan Gakkaishi* **46**:1235–1240.
30. **Garvie, E. I.** 1986. Genus *Pediococcus*, p. 1075–1079. *In* P. H. A. Sneath, N. S. Mair, M. E. Sharpe, and J. G. Holt (ed.), *Bergey's Manual of Systematic Bacteriology*, vol. 2. The Williams and Wilkins Co., Baltimore.
31. **Gasson, M. J.** 1993. Progress and potential in the biotechnology of the lactic acid bacteria. *FEMS Microbiol. Rev.* **12**:3–20.
32. **Gottschalk, G.** 1986. *Bacterial Metabolism*, 2nd ed. Springer-Verlag, New York.

33. **Grazia, L., P. Romano, A. Bagni, D. Roggiani, and G. Guglielmi.** 1986. The role of moulds in the ripening process of salami. *Food Microbiol.* **3**:19–25.
34. **Gruss, A., and S. D. Ehrlich.** 1989. The family of highly interrelated single-stranded deoxyribonucleic acid plasmids. *Microbiol. Rev.* **53**:231–241.
35. **Hammer, G. F.** 1987. Meat processing: ripened products. *Fleischwirtschaft* **67**:71–74.
36. **Hammes, W. P., A. Bantleon, and S. Min.** 1990. Lactic acid bacteria in meat fermentation. *FEMS Microbiol. Lett.* **87**:165–174.
37. **Hammes, W. P., and H. J. Knauf.** 1994. Starters in the processing of meat products. *Meat Sci.* **36**:155–168.
38. **Hammes, W. P., N. Weiss, and W. Holzapfel.** 1992. The genera *Lactobacillus* and *Carnobacterium*, p. 1535–1594. *In* A. Balows, H. G. Trüper, M. Dworkin, W. Harder, and K.-H. Schleifer (ed.), *The Prokaryotes—a Handbook on the Biology of Bacteria: Ecophysiology, Isolation, Identification, Applications*, 2nd ed. Springer-Verlag, New York.
39. **Harlander, S. K.** 1992. Genetic improvement of microbial starter cultures, p. 20–26. *In Applications of Biotechnology to Traditional Fermented Foods.* National Academy Press, Washington, D.C.
40. **Hols, P., T. Ferain, D. Garmyn, N. Bernard, and J. Delcour.** 1994. Use of homologous expression-secretion signals and vector-free stable chromosomal integration in engineering of *Lactobacillus plantarum* for α-amylase and levanase expression. *Appl. Environ. Microbiol.* **60**:1401–1413.
41. **Jay, J. M.** 1992. Fermented foods and related products, p. 371–409. *In Modern Food Microbiology,* 4th ed. Chapman and Hall, New York.
42. **Kagermeier, A.** 1981. Taxonomie und Vorkommen von Milchsaurebakterien in Fleischprodukten. Dissertation, Fakultat fur Biologie. Ludwig-Maximilian-Universität München, Munich.
43. **Kandler, O.** 1983. Carbohydrate metabolism in lactic acid bacteria. *Antonie van Leeuwenhoek* **49**:209–224.
44. **Kandler, O.** 1984. Current taxonomy of lactobacilli. *Dev. Ind. Microbiol.* **25**:109–123.
45. **Kandler, O., and N. Weiss.** 1986. Regular, non-sporing Gram-positive rods, p. 1208–1234. *In* P. H. A. Sneath, N. S. Mair, M. E. Sharpe, and J. G. Holt (ed.), *Bergey's Manual of Systematic Bacteriology,* vol. 2. The Williams and Wilkins Co., Baltimore.
46. **Kitchell, A. G., and B. G. Shaw.** 1975. Lactic acid bacteria in fresh and cured meat, p. 209–220. *In* J. G. Carr, C. V. Cutting, and G. C. Whiting (ed.), *Lactic Acid Bacteria in Beverages and Food.* Academic Press, Inc., New York.
47. **Klettner, P.-G., and D. List.** 1980. Beitrag zum Einfluss der Kohlenhydratart auf den Verlauf der Rohwurstreifung. *Fleischwirtschaft* **60**:1589–1593.
48. **Knauf, H. J., R. F. Vogel, and W. P. Hammes.** 1989. Introduction of the transposon Tn919 into *Lactobacillus curvatus. FEMS Microbiol. Lett.* **65**:101–104.
49. **Languer, H. J.** 1972. Aromastoffe in der Rohwurst. *Fleischwirtschaft* **52**:1299–1306.
50. **Leer, R. J., N. van Luijk, M. Posno, and P. H. Pouwels.** 1992. Structural and functional analysis of two cryptic plasmids from *Lactobacillus pentosus* MD353 and *Lactobacillus plantarum* ATCC 8014. *Mol. Gen. Genet.* **234**:265–274.
51. **Leistner, L.** 1986. Mould-ripened foods. *Fleischwirtschaft* **66**:1385–1388.
52. **Liepe, H. U.** 1983. Starter cultures in meat production, p. 400–424. *In* H.-J. Rehm and G. Reed (ed.), *Biotechnology, Food and Feed Production with Microorganisms,* vol. 5. Verlag Chemie, Weinheim, Germany.
53. **Liu, M.-L., J. K. Kondo, M. B. Barnes, and D. T. Bartholomew.** 1988. Plasmid-linked maltose utilization in *Lactobacillus* spp. *Biochimie* **70**:351–355.
54. **London, J.** 1990. Uncommon pathways of metabolism among lactic acid bacteria. *FEMS Microbiol. Lett.* **87**:103–112.
55. **Luchansky, J. B., P. M. Muriana, and T. R. Klaenhammer.** 1988. Application of electroporation for transfer of plasmid DNA to *Lactobacillus, Lactococcus, Leuconostoc, Listeria, Pediococcus, Bacillus, Staphylococcus, Enterococcus* and *Propionibacterium. Mol. Microbiol.* **2**:637–646.
56. **Lücke, F.-K.** 1985. Fermented sausages, p. 41–83. *In* B. J. B. Wood (ed.), *Microbiology of Fermented Foods,* vol. 2. Elsevier Applied Science Publishing Co., Inc., London.
57. **McKay, L. L., and K. A. Baldwin.** 1990. Applications for biotechnology: present and future improvements in lactic acid bacteria. *FEMS Microbiol. Lett.* **87**:3–14.
58. **Michels, P. A. M., J. P. J. Michels, J. Boonstra, and W. N. Konings.** 1979. Generation of electrochemical proton gradient in bacteria by the extrusion of metabolic end products. *FEMS Microbiol. Lett.* **5**:357–364.
59. **Mogensen, G.** 1993. Starter cultures, p. 1–22. *In* J. Smith (ed.), *Technology of Reduced-Additive Foods.* Blackie Academic and Professional, Chapman and Hall, New York.
60. **Morelli, L., P. S. Cocconcelli, V. Bottazzi, G. Damiani, L. Ferretti, and V. Sgaramella.** 1987. *Lactobacillus* protoplast transformation. *Plasmid* **17**:73–75.
61. **Mortvedt, C. I., and I. F. Nes.** 1989. Bacteriocin production by a *Lactobacillus* strain isolated from fermented meat. *Eur. Food Chem. Proc.* **1**:336–341.
62. **Mortvedt, C. L., and I. F. Nes.** 1990. Plasmid-associated bacteriocin production by a *Lactobacillus sake* strain. *J. Gen. Microbiol.* **136**:1601–1607.
63. **Mortvedt, C. I., J. Nissen-Meyer, K. Sletten, and I. F. Nes.** 1991. Purification and amino acid sequence of lactocin S, a bacteriocin produced by *Lactobacillus sake* L45. *Appl. Environ. Microbiol.* **57**:1829–1834.
64. **National Research Council.** 1992. *Applications of Biotechnology to Traditional Fermented Foods,* p. 121–149. National Academy Press, Washington, D.C.
65. **Nes, I. F.** 1984. Plasmid profiles of ten strains of *Lactobacillus plantarum. FEMS Microbiol. Lett.* **21**:359–361.
66. **Niinivaara, F. P.** 1955. The influence of pure cultures of bacteria on the maturing and reddening of raw sausage. *Acta Agr. Fenn.* **85**:95–101.
67. **Niven, C. F., Jr.** 1951. Sausage discolorations of bacterial origin. *Am. Meat Inst. Fndn. Bull.* no. 13.
68. **Niven, C. F., Jr., R. H. Deibel, and G. D. Wilson.** 1958. *The AMIF Sausage Starter Culture.* Circular no. 41. American Meat Institute Foundation, Chicago.
69. **Palumbo, S. A., and J. L. Smith.** 1977. Lebanon bologna processing. *Proc. Recip. Meat Conf.* **30**:63–68.
70. **Pearson, A. M., and W. F. Tauber.** 1984. *Processed Meats,* 2nd ed. AVI Publishing Co., Inc. Westport, Conn.

71. **Pederson, C. S.** 1979. Fermented sausage, p. 210–234. *In Microbiology of Food Fermentations.* AVI Publishing Co., Inc., Westport, Conn.

72. **Poolman, B.** 1993. Energy transduction in lactic acid bacteria. *FEMS Microbiol.* **12**:125–148.

73. **Posno, M., R. J. Leer, N. Van Luijk, M. J. F. van Giezen, P. T. H. M. Heuvelmans, B. C. Lokman, and P. H. Pouwels.** 1991. Incompatibility of *Lactobacillus* vectors with replicons derived from small cryptic *Lactobacillus* plasmids and segregational instability of the introduced vectors. *Appl. Environ. Microbiol.* **57**:1822–1828.

74. **Raccach, M.** 1984. Method for selection of lactic acid bacteria and determination of minimum temperature for meat fermentations. *J. Food Prot.* **47**:670–671.

75. **Raccach, M.** 1987. Pediococci and biotechnology. *Crit. Rev. Microbiol.* **14**:291–309.

76. **Reuter, G.** 1975. Classification problems, ecology and some biochemical activities of lactobacilli in meat products, p. 221–229. *In* J. G. Carr, C. V. Cutting, and G. C. Whiting (ed.), *Lactic Acid Bacteria in Beverages and Food.* Academic Press, Inc., New York.

77. **Rogosa, M., and M. E. Sharpe.** 1959. An approach to the classification of the lactobacilli. *J. Appl. Bacteriol.* **22**:329–340.

78. **Romans, J. R., W. J. Costello, C. W. Carlson, M. L. Greaser, and K. W. Jones.** 1994. Sausages, p. 773–886. *In The Meat We Eat.* Interstate Publishers, Inc., Danville, Ill.

79. **Romero, D. A., and L. L. McKay.** 1986. Isolation and plasmid characterization of a Lactobacillus species involved in the manufacture of fermented sausage. *J. Food Prot.* **48**:1028–1035.

80. **Rust, R. E.** 1976. *Sausage and Processed Meats Manufacturing.* American Meat Institute, Washington, D.C.

81. **Scheirlinck, T., J. Mahillon, H. Joos, P. Dhaese, and F. Michiels.** 1989. Integration and expression of α-amylase and endoglucanase genes in the *Lactobacillus plantarum* chromosome. *Appl. Environ. Microbiol.* **55**:2130–2137.

82. **Schillinger, U., and F.-K. Lücke.** 1987. Identification of lactobacilli from meat and meat products. *Food Microbiol.* 4:199–208.

83. **Schillinger, U., and F.-K. Lücke.** 1989. Antibacterial activity of *Lactobacillus sake* isolated from meat. *Appl. Environ. Microbiol.* **55**:1901–1906.

84. **Schleifer, K. H.** 1986. Gram-positive cocci, p. 999–1002. *In* P. H. A. Sneath, N. S. Mair, M. E. Sharpe, and J. G. Holt (ed.), *Bergey's Manual of Systematic Bacteriology,* vol. 2. The Williams and Wilkins Co., Baltimore.

85. **Sharpe, M. E.** 1981. The genus *Lactobacillus,* p. 1653–1679. *In* M. P. Starr, H. G. Trüper, A. Balows, and H. G. Schlegel (ed.), *The Prokaryotes—a Handbook on Habitats, Isolation, and Identification of Bacteria,* vol. 2. Springer-Verlag, New York.

86. **Sharpe, M. E., T. F. Fryer, and D. G. Smith.** 1966. Identification of the lactic acid bacteria. *In* B. M. Gibbs and F. A. Skinner (ed.), *Identification Methods for Microbiologists,* part A. Academic Press, Ltd., London.

87. **Shay, B. J., A. F. Egan, M. Wright, and P. J. Rogers.** 1988. Cysteine metabolism in an isolate of *Lactobacillus sake:* plasmid composition and cysteine transport. *FEMS Microbiol. Lett.* **56**:183–188.

88. **Shrago, A. W., B. M. Chassy, and W. J. Dobrogosz.** 1986. Conjugal plasmid transfer (pAMb1) in *Lactobacillus plantarum. Appl. Environ. Microbiol.* **52**:574–576.

89. **Shrago, A. W., and W. J. Dobrogosz.** 1988. Conjugal transfer of group B streptococcal plasmids and comobilization of *Escherichia coli-Streptococcus* shuttle plasmids to *Lactobacillus plantarum. Appl. Environ. Microbiol.* **54**:824–826.

90. **Skaugen, M., and I. F. Nes.** 1994. Transposition in *Lactobacilli sake* and its abolition of lactocin S production by insertion of IS*1163*, a new member of the IS3 family. *Appl. Environ. Microbiol.* **60**:2818–2825.

91. **Stackebrandt, E., V. J. Fowler, and C. R. Woese.** 1983. A phylogenetic analysis of lactobacilli, *Pediococcus pentosaceus* and *Leuconostoc mesenteroides. Syst. Appl. Microbiol.* **4**:326–337.

92. **Stackebrandt, E., and M. Teuber.** 1988. Molecular taxonomy and phylogenetic position of lactic acid bacteria. *Biochimie* **70**:317–324.

93. **ten Brink, R. Otto, U. P. Hansen, and W. N. Konings.** 1985. Energy recycling by lactate efflux in growing and nongrowing cells of *Streptococcus cremoris. J. Bacteriol.* **162**:383–390.

94. **Terrell, R. N., G. C. Smith, and Z. L. Carpenter.** 1977. Practical manufacturing technology for dry and semi-dry sausage. *Proc. Recip. Meat Conf.* **30**:39–44.

95. **Townsend, W. E., C. E. Davis, and C. E. Lyon.** 1978. Some properties of fermented dry sausage prepared from PSE and normal pork, 24(2):G9:1–G9:6. *In* Kongressdokumentation, 24th Europäischer Fleischforscher-Kongress, Kulmbach, Germany.

95a. **USDA-FSIS.** 1995. Prescribed treatment for pork and products containing pork to destroy trichinae, Part 318.10. *Code of Federal Regulations, Title 9.* Office of the Federal Register, Washington, D.C.

95b. **USDA-FSIS.** 1996. Requirements for the production of cooked beef, roast beef, and cooked corn beef, Part 318.17. *Code of Federal Regulations, Title 9.* Office of the Federal Register, Washington, D.C.

96. **Vandekerckhove, P., and D. Demeyer.** 1975. Die Zusammernstzung belgischer Rohwurst (Salami). *Fleischwirtschaft* **55**:680–682.

97. **von Wright, A., and M. Sibakov.** 1993. Genetic modification of lactic acid bacteria, p. 161–198. *In* S. Salminen and A. von Wright (ed.), *Lactic Acid Bacteria.* Marcel Dekker, Inc., New York.

98. **Weiss, N.** 1992. The genera *Pediococcus* and *Aerococcus,* p. 1502–1507. *In* A. Balows, H. G. Trüper, M. Dworkin, W. Harder, and K.-H. Schleifer (ed.), *The Prokaryotes—a Handbook on the Biology of Bacteria: Ecophysiogy, Isolation, Identification, Applications,* 2nd ed. Springer-Verlag, New York.

99. **West, C. A., and P. J. Warner.** 1985. Plasmid profiles and transfer of plasmid-encoded antibiotic resistance in *Lactobacillus plantarum. Appl. Environ. Microbiol.* **50**:1319–1321.

100. **Woese, C. R.** 1987. Bacterial evolution. *Microbiol. Rev.* **51**: 221–271.

101. **Wu, W. J., D. C. Rule, J. R. Busboom, R. A. Field, and B. Ray.** 1991. Starter culture and time/temperature of storage influences on quality of fermented mutton sausage. *J. Food Sci.* **56**:919–925.

102. **Zenitani, B.** 1955. Studies on fermented fish products. I. On the aerobic bacteria in "Shiokara." *Bull. Jpn. Soc. Sci. Fish.* **21**:280–283.

Larry R. Beuchat

34 Traditional Fermented Foods

Fermented foods of plant and animal origin are an essential part of the diet of people in many countries. While the preparation of many traditional or "indigenous" fermented foods and beverages is a household art, the preparation of others, e.g., soy sauce, has evolved to a biotechnological process on a commercial scale. Several books (23, 76, 85, 105) and reviews (9, 11, 12, 18, 19, 31, 33, 35, 55, 56, 63, 64, 67, 69, 79, 83, 84, 86, 91, 106, 109, 112) have been published on the subject of traditional fermented foods. A book describing potential applications of biotechnology to traditional fermented foods was published by the U.S. National Research Council (77). A dictionary and guide to fermented foods of the world (17) and a glossary of indigenous fermented foods (102) provide excellent descriptions of known biochemical and microbiological processes associated with fermented foods.

Table 34.1 lists some of the more common traditional fermented foods prepared and consumed in various countries. The information summarized in this table was compiled from several reviews of the subject but is not necessarily complete or even accurate in all aspects. However, the diversity of types of fermented foods consumed by various cultures is evident. Only a few traditional fermented foods will be covered here, mainly to illustrate the complexity of biochemical, sensory, and nutritional changes that can result from more or less controlled microbial activity in a range of raw materials.

BAKERY PRODUCTS

One of the oldest traditional fermented foods is bread, which, in various forms, has been a staple in the diets of many population groups for several centuries. The history of bread can be traced back about 6 millennia. The production of loaf bread supposes the development of the oven and the discovery of dough fermentation. The Egyptians developed a baking oven approaching the design of a modern oven about 2700 B.C. (83). Since 1750 B.C. there were professional bakers in Egypt. The development of cereal foods has proceeded through several stages, from roasted grain to gruels to flat breads and finally to leavened bread loaves. All stages are practiced today. About 60% of the world population, mainly in Central America, parts of South America, Africa, the Middle East, and the Far East, eat flat breads and gruels made from grains. The production of bakery products consists of five major steps, i.e., the preparation of raw materials, dough formation, dough processing and fermentation, baking, and preservation. Attention will be given here to breads owing their sensorial properties, at least in part, to fermentative activities of microorganisms.

Preparation of Raw Materials

Wheat or rye flours are the major ingredients in basic, leavened bread recipes. Wheat breads are leavened with yeasts while rye breads require, in addition to yeasts, acidification either by the use of sourdough starter or by

Table 34.1 Traditional fermented foods[a]

Product	Geography	Substrate	Microorganism(s)	Nature of product	Product use
Ang-kak (anka, red rice)	China, Southeast Asia	Rice	*Monascus purpureus*	Dry red powder	Colorant
Bagoong	Philippines	Fish	Unknown	Paste	Seasoning agent
Bagni	Caucasus	Millet	Unknown	Liquid	Drink
Banku	Ghana	Maize, cassava	Lactic acid bacteria, yeasts	Dough	Staple
Bonkrek	Central Java (Indonesia)	Coconut press cake	*Rhizopus oligosporus*	Solid	Roasted or fried in oil, used as a meat substitute
Bouza	Egypt	Wheat	Yeasts	Liquid	Thick acidic beverage
Braga	Romania	Millet	Unknown	Liquid	Drink
Bread	International	Wheat, rye, other grains	*Saccharomyces cerevisiae*, other yeasts, lactic acid bacteria	Solid	Staple
Burukutu	Savannah regions of Nigeria	Sorghum and cassava	Lactic acid bacteria, *Candida* spp., *Saccharomyces cerevisiae*	Liquid	Creamy drink with suspended solids
Busa	Tartars of Krim, Turkestan, Egypt	Rice or millet, sugar	*Lactobacillus* and *Saccharomyces* spp.	Liquid	Drink
Chee-fan	China	Soybean-wheat curd	*Mucor* sp., *Aspergillus glaucus*	Solid	Eaten fresh, cheeselike
Chicha	Peru	Maize	*Aspergillus* and *Penicillium* spp., yeasts, bacteria	Liquid	Alcoholic beverage
Chickwangue	Congo	Cassava roots	Bacteria	Paste	Staple
Chinese yeast	China	Soybeans	Mucoraceous molds and yeasts	Solid	Eaten fresh or canned, used as a side dish with rice
Darassum	Mongolia	Millet	Unknown	Liquid	Drink
Dawadawa (daddowa, iru, kpalugu,kinda)	West Africa, Nigeria	African locust bean	*Bacillus subtilis*, *B. licheniformis*	Solid, sun-dried	Eaten fresh or canned, used as a side dish with rice
Dhokla	India	Bengal gram and wheat	Lactic acid bacteria	Spongy	Staple
Dosai (doza)	India	Black gram and rice	*Leuconostoc mesenteroides*, yeasts	Spongy, pancake-like	Breakfast food
Fish sauce (nuoc-mam, patis, mampla, ngam-pya-ye)	Southeast Asia	Fish	Bacteria	Liquid	Seasoning agent
Gari	West Africa	Cassava root	*Corynebacterium manihot*, *Geotrichum candidum*	Flour	Eaten boiled as staple with stews, vegetables
Hama-natto	Japan	Whole soybeans, wheat flour	*Aspergillus oryzae*, *Streptococcus*, *Pediococcus*	Beans retain individual form, raisin-like, soft	Flavoring agent for meat and fish, eaten as snack
Idli	Southern India	Rice and black gram	Lactic bacteria (*Leuconostoc mesenteroides*), *Torulopsis*, *Candida*, and *Trichosporon pullulans*	Spongy, moist, steamed bread	Bread substitute
Injera	Ethiopia	Teff or maize, wheat, barley, sorghum	*Candida guilliermondii*	Moist, breadlike pancake	Bread substitute
Inyu	Taiwan, China, Hong Kong	Black soybeans	*Aspergillus oryzae*	Liquid	Flavor enhancer
Jamin-bang	Brazil	Maize	Yeasts and bacteria	Bread- or cakelike	Bread substitute
Kaanga-kopuwai	New Zealand	Maize	Bacteria and yeasts	Soft, slimy	Eaten as vegetable

Kanji	India	Rice and carrots	*Hansenula anomala*	Liquid	Sour, added to vegetables
Katsuobushi	Japan	Whole fish	*Aspergillus glaucus*	Solid, dry	Seasoning agent
Kecap	Indonesia and vicinity	Soybeans, wheat	*Aspergillus oryzae, Lactobacillus, Hansenula, Saccharomyces*	Liquid	Condiment, seasoning agent
Kenkey	Ghana	Maize	Unknown	Mush	Steamed, eaten with vegetables
Ketjap	Indonesia	Black soybeans	*Aspergillus oryzae*	Sirup	Seasoning agent
Khaman	India	Bengal gram	Lactic acid bacteria	Solid, cakelike	Breakfast food
Kimchi (kim-chee)	Korea	Cabbage,vegetables, sometimes seafoods, nuts	Lactic acid bacteria	Solid and liquid	Condiment
Kinema	Nepal, Sikkim, Darjeeling district of India	Soybeans	*Bacillus subtilis, Enterococcus faecium,* yeasts	Solid	Snack
Kishk (kushuk, kushik)	Egypt, Syria, Arab world	Wheat, milk	Lactic acid bacteria, *Bacillus* spp.	Solid	Dried balls dispersed rapidly in soups
Lafun	West Africa, Nigeria	Cassava root	Lactic acid bacteria	Paste	Staple food
Lao-chao	China, Indonesia	Rice	*Rhizopus oryzae, R. chinensis, Chlamydomucor oryzae, Saccharomycopsis* sp.	Soft, juicy, glutinous	Eaten as such as dessert or combined with eggs or seafood
Mahewu (magou)	South Africa	Maize	Lactic acid bacteria (*Lactobacillus delbrueckii*)	Liquid	Drink, sour and nonalcoholic
Meitauza	China, Taiwan	Soybean cake	*Actinomucor elegans, Mucor meitauza*	Solid	Fried in oil or cooked with vegetables
Meju	Korea	Black soybeans	*Aspergillus oryzae, Rhizopus* spp.	Paste	Seasoning agent
Merissa	Sudan	Sorghum	*Saccharomyces* sp.	Liquid	Drink
Minchin	China	Wheat gluten	*Paecilomyces, Aspergillus, Cladosporium, Fusarium, Syncephalastum, Penicillium, Trichothecium* spp.	Solid	Condiment
Miso (chiang, jang, doenjang, tauco, tao chieo)	Japan, China	Rice and soybeans or rice and other cereals such as barley	*Aspergillus oryzae, Torulopsis etchellsii, Lactobacillus*	Paste	Soup base, seasoning
Munkoyo	Africa	Millet, maize or kaffir corn plus roots of munkoyo	Unknown	Liquid	Drink
Natto	Northern Japan	Soybeans	*Bacillus natto*	Moist, mucilaginous	A meat substitute
Ogi	Nigeria, West Africa	Maize	Lactic bacteria *Cephalosporium, Fusarium, Aspergillus, Penicillium* spp., *Saccharomyces cerevisiae, Candida mycoderma, C. valida,* or *C. vini*	Paste, porridge	Staple, eaten for breakfast; for weaning babies
Oncom (ontjom, lontjom)	Indonesia	Peanut press cake	*Neurospora intermedia,* less often *Rhizopus oligosporus*	Solid	Roasted or fried in oil, used as meat substitute

(continued)

Table 34.1 Traditional fermented foods[a] *(continued)*

Product	Geography	Substrate	Microorganism(s)	Nature of product	Product use
Papadam	India	Black gram	*Saccharomyces* spp.	Solid, crisp	Condiment
Peujeum	Java	Banana, plantain	Unknown	Solid	Eaten fresh or fried
Pito	Nigeria	Guinea corn or maize or both	Yeasts, lactic acid bacteria	Liquid	Drink
Poi	Hawaii	Taro corms	*Lactobacillus* bacteria, *Candida vini (Mycoderma vini), Geotrichum candidum*	Semisolid	Side dish with fish, meat
Pozol	Southeastern Mexico	Maize	Molds, yeasts, bacteria	Dough, spongy	Diluted with water, drunk as basic food
Prahoc	Cambodia	Fish	Unknown	Paste	Seasoning agent
Puto	Philippines	Rice	Lactic acid bacteria, *Saccharomyces cerevisiae*	Solid	Snack
Rabadi	India	Maize and buttermilk	Lactic acid bacteria	Semisolid	Mush, eaten with vegetables
Sierra rice	Ecuador	Unhusked rice	*Aspergillus flavus, A. candidus, Bacillus subtilis*	Solid	Brownish-yellow, seasoning
Sorghum beer (Ibantu beer, kaffir beer, leting, joala, utshivala, mqom-boti, igwelel)	South Africa	Sorghum, maize	Lactic acid bacteria, yeasts	Liquid	Drink, acidic and weakly alcoholic
Soybean milk yogurt	China, Japan	Soybeans	Lactic acid bacteria	Liquid	Drink
Soy sauce (chaing-yu, shoyu, toyo, kanjang, kecap, seeieu)	Japan, China, Philippines, other parts of Orient	Soybeans and wheat	*Aspergillus oryzae* or *A. soyae, Lactobacillus* bacteria, *Zygosaccharo-myces rouxii*	Liquid	Seasoning for meat, fish, cereals, vegetables
Sufu (tahur, taokaoan, tao-hu-yi)	China, Taiwan	Soybean whey curd	*Actinomucor elegans, Mucor hiemalis, M. silvaticus, M. subtilissimus*	Paste	Soybean cheese, condiment

Tao-si	Philippines	Soybeans plus wheat flour	*Aspergillus oryzae*	Semisolid	Seasoning agent, side dish
Taotjo	East Indies	Soybeans plus roasted wheat meal or glutinous rice	*Aspergillus oryzae*	Semisolid	Condiment
Tapé	Indonesia and vicinity	Cassava or rice	*Saccharomyces cerevisiae, Hansenula anomala, Rhizopus oryzae, Chlamydomucor oryzae, Mucor* sp., *Endomycopsis fibuliger* (*Saccharomycopsis* sp.)	Soft solid	Eaten fresh as staple
Tarhana	Turkey	Parboiled wheat meal and yogurt (2:1)	Lactic acid bacteria	Biscuits or powder	Dried seasoning for soups
Tauco	West Java (Indonesia)	Soybeans, cereals	*Rhizopus oligosporus, Aspergillus oryzae*	Paste	Condiment
Tempeh (tempe kedeke)	Indonesia and vicinity, Surinam	Soybeans	*Rhizopus* spp., principally *R. oligosporus*	Solid	Fried in oil, roasted, or used as meat substitute in soup
Thumba (bojah)	West Bengal	Millet	*Endomycopsis fibuliger*	Liquid	Drink, mildly alcoholic
Torani	India	Rice	*Hansenula anomala, Candida guilliermondii, C. tropicalis, Geotrichum candidum*	Liquid	Seasoning for vegetables
Ugba	Nigeria, West Central Africa	African oil bean	*Bacillus subtilis, Micrococcus luteus, M. roseus, Bacillus* sp.	Solid	Condiment, flavoring agent
Waries	India, Pakistan	Black gram flour, Bengal gram flour	*Candida* spp., *Saccharomyces* spp.	Spongy	Spicy condiment eaten with vegetables, legumes, rice

[a]Compiled from references 17, 29, 30, 34, 35, 69, 76, 81, 83–85, and 92 and used with permission (12).

the addition of acid. The function of flours is to absorb water and to form a cohesive, viscoelastic mass, i.e., dough. In wheat breads, crackers, and other bakery products consisting largely of wheat flour, functionality is complemented by starch. However, the most important functional component is gluten, upon which the texture of bakery products is most dependent. Other ingredients include water, fat, and salt, which greatly influence the texture of bakery products, sugar to promote fermentation, color development, and flavor, and optional ingredients such as eggs, milk, skim milk powder, spices, cocoa, fruits, oil seeds, and dough conditioners (69, 83). The latter ingredients impart desired flavor, aroma, color, and texture characteristics to specific types of products.

Top-fermenting strains of *Saccharomyces cerevisiae*, known as baker's yeast, are widely used for commercial production of baked goods. Leavening of doughs requires the addition of 1 to 6% yeast based on the weight of flour; the optimum temperatures for growth and fermentation are 28 and 32°C, respectively (83). The optimum pH is between 4 and 5. Baker's yeast is available in the form of yeast cakes, bulk or crumbled yeast, yeast cream, active dry baker's yeast, and instant active dry baker's yeast. Yeast cakes are the most popular form and are similar to bulk or crumbled yeast (30 to 32% yeast solids), the main difference being that bulk yeast is not extruded to form uniform blocks but rather broken into irregular pieces after removal from a filter. Yeast cream or liquid yeast is essentially a water suspension of baker's yeast with about 18% yeast solids. Active dry baker's yeast (92 to 96% yeast solids) is available in granular or powder forms in hermetically sealed containers under vacuum or an inert gas atmosphere. It may be stored without substantial loss of activity for up to 1 year. Its use is mainly in bakeries in tropical and subtropical countries and for home use. Rehydration at 35 to 42°C is required before use in bakery products. Instant active dry yeast (95% yeast solids) may be added directly to flour or dough during mixing.

Sourdough starters contain 10^7 to 10^9 lactic acid bacteria per g and 10^5 to 10^7 yeasts per g. Both homofermentative and heterofermentative *Lactobacillus* species, in various combinations, as well as *S. cerevisiae*, *Pichia saitoi*, *Issatchenkia orientalis*, and *Torulopsis holmii*, may be present. Homofermentative lactobacilli are less likely to produce the desired sensory qualities of sourdough breads than are heterofermentative species such as *Lactobacillus brevis*, *Lactobacillus fermentum*, and *Lactobacillus sanfrancisco*. The predominant lactic acid bacterium in San Francisco sourdough French bread is *Lb. sanfrancisco*, which ferments maltose but not sucrose, glucose, xylose, arabinose, galactose, rhamninose, or raffinose (39). Yeasts include *T. holmii*, which ferments glucose, sucrose, galactose, and raffinose, but not maltose, *Saccharomyces inusitus*, which ferments maltose, glucose, sucrose, and galactose, and *Saccharomyces exiguus*, which grows in the presence of lactic acid bacteria (89).

Dough Formation

Raw materials are mixed to form the base for development of the dough to eventually produce bakery products. The baking quality of the flour depends on the variety of grain and milling processes (83). Conditions under which flour is stored and sieved before being used in doughs can also affect the quality of bakery products.

Doughs containing largely flour and water are prepared in mixers. Incorporation of air in small bubbles is essential for the leavening of dough and the grain of the bread. The number of bubbles does not increase during fermentation, but the size does. The choice of mixer and the conditions during mixing largely determine the final volume of bakery products and the structure of the crumb (83). Conditions such as pressure, speed, and intensity of mixing, temperature, and time of mixing all affect the dough characteristics and subsequent baked product quality (69).

Dough Processing

Principal methods of processing and fermenting doughs to produce bread include those described as sponge dough, straight dough, liquid ferment, continuous mix, and short-time dough. Various methods to produce sourdoughs also exist.

Sponge Dough Process

The sponge dough process involves mixing part of the flour, water, and yeast to produce a sponge or pre-dough. After the sponge has fully fermented at 25°C, it is mixed with the remaining flour, water, and other ingredients to form the final dough formulation and is held at 26 to 28°C. The time required for preparation of the sponge and final dough varies depending on the formulation and desired sensorial qualities of the final baked product. However, sponge fermentation time is generally in the range of 3.5 to 5 h.

Straight Dough Process

In the straight dough process, flour, water, yeast, salt, and all other ingredients are combined to form a dough. After mixing, the dough is fermented for 2 to 4 h at 26 to 32°C. If an overnight fermentation (8 to 12 h) is desired, the amount of yeast added to the dough is reduced and the dough is held at 18 to 20°C. Straight doughs require less labor, time, and equipment to prepare compared to sponge doughs. However, breads made from straight doughs have a characteristically blander flavor than do breads made from sponge doughs, which may be less acceptable to some consumers.

Liquid Ferment Process

The liquid ferment or brew is prepared using yeast, sugar, salt, and up to 70% of the flour of the ingredient formula. The temperature of fermentation is 23 to 31°C, with the temperature rising 2 to 5°C during a 2- to 3-h fermentation period. The pH of the fermented liquid is 4.6 to 5.2, depending upon the type of dough and bread desired. Lactic acid bacteria are responsible for the reduction in pH. Liquid ferments have the same function as sponge doughs. They are mixed with the remaining formulation ingredients to produce the final dough.

Continuous Mix Process

Continuous mix processes for preparing dough use a liquid ferment, with (Am Flow process) or without (Do-Maker process) flour, a premixer in which all of the ingredients of the formula are combined, and a developer in which the dough is formed under pressure and with intensive mixing. The dough is softer than that produced by the sponge dough process and contains too much water to be worked by hand. Continuous mix processes are time-saving compared to conventional processes.

Short-Time Dough Process

Short-time or no-time doughs are obtained by mechanical or chemical means rather than by fermentation. In the Chorleywood bread process, dough is made by combining all ingredients and mixing intensively in a batch, high-speed mixer (69, 83). Mechanical dough development is enhanced by the addition of oxidants such as potassium iodate, potassium bromate, or ascorbic acid. The process may be carried out with continuous extrusion of dough directly into baking pans, producing bread with very fine grain. In the Brimec process, dough mixing is carried out with high- or low-speed mixers. The addition of 1% shortening and high levels of oxidants is required to produce good bread.

Sourdough Processes

Bread doughs containing more than 20% rye flour require acidification. Otherwise, the development of desirable flavor, aroma, and texture cannot be achieved. Commercial sourdough cultures, spontaneous (natural) souring, or a portion of sourdough used in a preceding batch can be used to achieve acidification. Modifications or processes to make sourdoughs are numerous (83), but biological modifications are brought about largely by the growth of lactic acid bacteria. In addition to acid production, protease and peptidase activity results in the production of amino acids and peptides which are used by yeasts during sourdough fermentation.

Regardless of the method used to process doughs, a fermentation period is essential. Enzymatic activity of *S. cerevisiae* increases and the production of carbon dioxide causes the dough to rise. Fermentation is complete when the formation of a colloidal dough structure and the intensity of yeast activity have reached their optima. If fermentation is extended, the gluten absorbs too much water and the dough becomes too bulky. Fully fermented dough is divided into pieces, rounded, and permitted to rest for 5 to 30 min, the intermediate proof. Doughs containing rye flour may require only a 1- to 7-min intermediate proof. The dough is the placed in a molder for final proofing, which is done in a proof box or cabinet for periods ranging from 30 to 60 min, depending on the size of the dough piece and the activity of the yeast. The temperature of the proof box is maintained at 30 to 40°C. A relative humidity of 65 to 85% is required to promote the development of a desired surface appearance of baked products.

Baking

Baking stabilizes dough structure and results in the formation of the characteristic aroma, flavor, and color of baked products. The production of gas causes a 40% increase in volume of dough, although this can be controlled within limits by temperature of baking, type of baking oven, relative humidity, and time of baking (83).

Preservation

Aside from preserving sensorial qualities of baked products, steps must be taken to prevent microbial spoilage. A spoilage defect called "rope" results from the growth of *Bacillus* species, principally in wheat breads that have not been acidified or in breads with high concentrations of sugar, fat, or fruits. A slightly bitter taste and fruity aroma are followed by yellow, brown, and finally red discoloration and a repulsive aroma. Molds associated with spoilage include species of *Aspergillus, Penicillium, Mucor,* and *Rhizopus.* A defect called "red bread" results from the growth of *Neurospora intermedia.* Yeasts responsible for white, chalky spots include *Pichia burtonii* and *Geotrichum candidum.*

Prevention or retardation of microbial spoilage can be achieved by the use of appropriate packaging, by application of chemical preservatives, or by freezing bakery products (69). Propionic acid and its calcium and sodium salts are effective against molds as well as against *Bacillus* species that cause rope. The development of rope can be prevented by addition of acetic acid or calcium acetate to dough. Sorbic acid and its potassium or sodium salts are highly effective in controlling the growth of yeasts and molds.

SOY SAUCE

The principles of soy sauce fermentation have been reviewed (31, 108, 112). There are five main types of soy sauce recognized in Japan (27). Koikuchi soy sauce is an all-purpose seasoning characterized by a strong aroma and dark reddish brown color. Usukuchi type soy sauce is lighter in color and milder in flavor and is used mainly for cooking when preservation of the original color and flavor of foods is desired. Tamari style soy sauce has a strong flavor and is dark brown in color. Saishikomi soy sauce contains a trace of alcohol, and Shiro soy sauce is characterized by a high level of reducing sugars and a yellowish tan color. All of these types of soy sauce contain relatively high levels of salt, in the range of 17 to 19%, and all are used as seasoning agents to enhance the flavor of meats, seafoods, and vegetables. Typical ranges in other characteristics are: pH 4.6 to 4.8, 2.2 to 30 Bé, 0.5 to 2.5 g of total nitrogen per 100 ml, 0.2 to 1.1 g of formal nitrogen per 100 ml, 3.8 to 2.0 g of reducing sugar per 100 ml, and from a trace to 2.2 ml of ethanol per 100 ml.

Preparation of Soybeans and Wheat

Soaking and cooking of soybeans and roasting and crushing (cracking) of wheat are separate processes in the early stages of soy sauce production. Whole soybeans or defatted soybean meal or flakes can be used. If whole beans are used, oil must eventually be removed from the fermented mash; otherwise an inferior product will result. Utilization of nitrogen is higher and fermentation time is shorter with defatted beans than with whole soybeans. This may be due to a lower surface:volume ratio in whole beans versus meal, hence a more pronounced physical restraint in whole beans with regard to access of microbial enzymes to soybean components during fermentation. Whole beans or meal are soaked for 12 to 15 h at ambient temperature or, preferably, at about 30°C until the weight is doubled. Soaking is done either by changing still water every 2 or 3 h or by constantly running water over the beans. If water is not changed, sporeforming *Bacillus* species may proliferate and eventually be deleterious to end product quality. Depending upon the depth of soybeans in the water, the temperature of those in the bottom layer of tanks may increase if water is not changed or circulated during soaking. Hydrated soybeans or meal are then drained, covered with water again, and steamed to achieve further softening and pasteurization. If pressure is used during steaming, the soybeans can be sterilized. The conditions for cooking soybeans influence enzyme activity during subsequent fermentation and may affect the turbidity of the final product (65). Turbidity is caused by insoluble undenatured protein which is dispersed in concentrated salt solution. With increased moisture or pressure during steaming, the soy protein tends to denature more readily; excessively denatured soy protein, on the other hand, is less accessible for enzyme reaction. Consequently, the yield of soluble nitrogen and other soluble compounds will be reduced. Rapid cooling on an industrial scale is done by spreading the soybeans in about a 30-cm layer on perforated tray-like platforms and forcing air through them (112). It is important to reduce the temperature to less than 40°C within a few hours. Otherwise, proliferation of naturally occurring microorganisms may spoil the soybeans before controlled fermentation can be initiated.

Concurrent with the preparation of soybeans is the roasting and crushing (cracking) of wheat. The roasting of wheat contributes to the aroma and flavor of soy sauce. Wheat flour or wheat bran may be used in place of whole wheat kernels. Breakdown and conversion products resulting from the roasting process include guaiacyl series compounds such as vanillin, vanillic acid, ferulic acid, and 4-ethylguiacol (5). The free phenolic compound content increases with heating due to degradation of lignin and glycosides. Roasting causes the formation of several brown-colored reaction products that contribute to the desired sensory properties of soy sauce.

Koji Process

Koji is an enzyme preparation produced from cereals or sometimes pulses that is used as a starter for larger batches of traditional fermented foods. In the case of soy sauce, seed (tane) koji is produced by culturing single or mixed strains of *Aspergillus oryzae* or *Aspergillus sojae* on either steamed polished rice or a mixture of wheat bran and soybean flour (112). Seed koji is added to a soybean-wheat mixture at a level of 0.1 to 0.2% to produce what is then simply called koji.

Strains of *A. oryzae* or *A. sojae* used to prepare koji must have high proteolytic and amylolytic activities and should contribute to the characteristic aroma and flavor of soy sauce. Lipase (114), cellulase (28), and several acid, neutral, and alkaline proteases as well as peptidases (49–53) may also be produced by *A. oryzae* and *A. sojae.* On a commercial scale, a mixture of equal weights of cooked soybeans and roasted wheat is spread in 5-cm layers in trays made from bamboo strips or in stainless steel trays, inoculated with the seed koji, and stacked in such a way as to allow good air circulation. Temperature (25 to 35°C) and moisture (27 to 37%) control is important to the development of good koji; a temperature of 30°C for 2 to 3 days is desirable. A high moisture content is required at the beginning of the incubation period when mycelial growth occurs, followed by a lower moisture content in later stages when spores are being formed (112). The incubation period should be long enough to achieve adequate production and accumulation of enzymes but not so long as to encourage excess sporulation, which may impart undesirable flavors to the finished product. Good-quality koji is clear yellow to yellowish green in color.

Enzymes produced by *A. oryzae* that contribute to desirable soybean fermentation have been described (28, 49–53, 113). The effects of diisopropyl-phosphorofluoridate, sulfhydryl reagents such as *p*-chloromercuribenzoate and monoiodoacetate, and metal-chelating agents such as ethylenediaminetetraacetate, α,α′-dipyridyl, and *p*-phenanthroline, as well as carboxypeptidases, alkaline and neutral proteinases, leucine aminopeptidases, and lipases, on soy sauce fermentation have been studied. Koji infected with molds such as *Rhizopus* or *Mucor* species should be discarded because of the undesirable flavor and aroma it will impart to soy sauce.

Mash (Moromi) Stage

Mature koji is mixed with an equal amount or more (up to 120% by volume) of brine to form the mash (moromi). The sodium chloride content of the mash should be 17 to 19%. Concentrations less than 16% salt may enable growth of undesirable putrefactive bacteria during subsequent fermentation and aging. Higher salt concentrations retard the growth of desirable osmophilic yeasts, namely *Zygosaccharomyces rouxii,* and halophilic bacteria. The mycelium of the koji mold is not tolerant of the high concentration of salt and, consequently, dies during the very early stage of mash preparation.

If mash fermentation is allowed to proceed naturally without controlling temperature, as would be the situation in the family home, 12 to 14 months is required for the fermentation and aging process. If the mash is kept in large wooden, concrete, or metal containers such as those used by commercial manufacturers, the temperature is usually maintained at 35 to 40°C, thus reducing the fermentation and aging period to 2 to 4 months. Regardless of temperature controls, it is important to stir the mash intermittently with a wooden stick on a small scale or with metal paddles or compressed air in modern commercial facilities. Stirring must be correlated to a certain extent with the rate of carbon dioxide production. Elevated levels of carbon dioxide will enhance the growth of certain anaerobic microorganisms which may impart undesirable flavor and aroma to the finished product. Excessive aeration, on the other hand, will hinder proper fermentation. Koji enzymes hydrolyze proteins to yield peptides and free amino acids during the early stages of fermentation. Starch is converted to simple sugars which in turn are fermented by microorganisms to yield lactic, glutamic, and other acids as well as alcohols and carbon dioxide. As a consequence, the pH of the mash drops from near neutrality to 4.5 to 4.8.

Various groups of bacteria and yeasts predominate in sequence during mash fermentation and aging (113). *Pediococcus halophilus,* a salt-tolerant bacterium, grows readily in the first stage of fermentation, converting simple sugars to lactic acid and causing a decrease in pH. Later, *Z. rouxii, Torulopsis* species, and other yeasts dominate. Molds that may grow on the surface of the mash are believed to have no relation to proper fermentation or aging (109). Soy sauce owes its pleasant aroma or flavor largely to the enzymatic activities of microorganisms. A partial list of flavor components identified in soy sauce is given in Table 34.2. *Pediococcus halophilus* and, perhaps, *Lactobacillus* species produce lactic and other organic acids which contribute to aroma and flavor. However, yeasts probably make the greatest contribution to characteristic sensory qualities of soy sauce. By-products of fermentation such as 4-ethylguaiacol, 4-ethylphenol and 2-phenylethanol (110), furfuryl alcohol (44, 49), pyrazines, furanones (59–62), and ethyl acetate (111) are among the main flavor-contributing compounds.

The extent of aging can be determined by measuring the glutamic acid content. Liquid (sauce) is removed from the mash with a press or by siphoning off the top of the mash. Fresh brine is sometimes added to the residue and a second, lower-quality fermentation is allowed to proceed for 1 or 2 months before a second drawing is made. Oil is removed from the filtrate by decantation.

Table 34.2 Some flavor components in soy sauce[a]

Acetaldehyde
Acetic acid
Acetone
2-Acetyl furan
2-Acetylpyrrole
Benzaldehyde
Benzoic acid
Benzyl alcohol
Borneol
Bornyl acetate
Butanoic acid
1-Butanol
2,6-Dimethoxyphenol
2,3-Dimethylpyrazine
2,6-Dimethylpyrazine
Diethyl succinate
Ethanol
Ethyl acetate
Ethyl benzoate
3-Ethyl-2,5-dimethylpyrazine
Ethyl-2-hydroxypropanoate (ethyl lactate)
2-Ethyl-6-methylpyrazine
Ethyl myristate
4-Ethylphenol
Ethyl phenylacetate
Furfural
Furfuryl acetate
Furfuryl alcohol
2,3-Hexanedione
2-Hexanone
3-Hydroxyl-2-butanone (acetoin)
4-Hydroxy-2-ethyl-5-methyl-3(2H)-furanone
4-Hydroxy-5-ethyl-2-methyl-3(2H)-furanone
4-Hydroxy-5-methyl-3(2H)-furanone
3-Hydroxyl-2-methyl-4-pyrone (maltol)
2-Methoxy-4-ethylphenol (4-ethylguaiacol)
2-Methoxyphenol (guaiacol)
3-Methylbutanal
3-Methylbutanoic acid
3-Methyl-1-butanol
3-Methylbutylacetate
2-Methylpropanal
2-Methyl propanoic acid
2-Methyl-1-propanol
2-Methylpyrazine
3-Methyl-3-tetrahydrofuranone
3-Methylthio-1-propanol (methional)
4-Pentanolide
Phenyl acetaldehyde
2-Phenylethanol
2-Phenylethyl acetate
Propanal
2-Propanol

[a]Adapted from Nunomura et al. (59–62).

Pasteurization

Raw soy sauce is pasteurized at 70 to 80°C, thus killing the vegetative cells of microorganisms and denaturing most enzymes and other proteins. Alum or kaolin may be added to enhance clarification, after which the sauce is filtered and bottled. Preservatives may be added to prevent growth of yeasts during storage (112). Butyl-*p*-hydroxybenzoate and sodium benzoate are most widely used.

MISO

Fermented soybean pastes are known as miso in Japan, chiang in China, jang or doenjang in Korea, tauco in Indonesia, and tao chieo in Thailand. In addition to soybeans and salt, most of these products also contain cereals such as rice or barley. In Japan, miso is mainly used as a base for soups.

Methods for preparing fermented soybean pastes differ somewhat, but the basic process is the same (12). Rice miso is made from rice, soybeans, and salt, barley miso is made from barley, soybeans, and salt, and soybean miso is made from soybeans and salt. These major types of miso are further classified on the basis of degree of sweetness and saltiness. The procedure for making miso consists of first preparing the koji and the soybeans (simultaneous processes), then brining or fermentation, and finally aging. The preparation of rice miso is considered here to illustrate the general procedure for making miso.

Koji Preparation

Polished rice is washed and soaked in water overnight at about 15°C to bring the moisture content to about 35%. Excess water is removed and the rice is steamed at atmospheric pressure for 40 min to 1 h. The cooked rice is then deposited in trays or on platforms and cooled to about 35°C before seed koji, prepared as described for use in soy sauce manufacturing, is added at a ratio of 1 g/kg of rice based on a viable spore count of 10^9/g of seed koji (24). The mixture is placed in a rotating fermentor drum in which the temperature, air circulation, and atmospheric relative humidity are controlled. The temperature is maintained at 30 to 35°C to promote growth of *A. oryzae* and maximum production of proteases and carbohydrases. Overheating is usually caused by rapid growth of undesirable bacteria which may be present on the uncooked rice and survive the steaming process. Ventilation should be adequate to supply sufficient oxygen and to remove carbon dioxide, and humidity should be such that neither drying nor sticking of the rice occurs. Fermentation is complete after 40 to 50 h or when koji is characterized by a sweet aroma and flavor; musty odors indicate an inferior quality koji. The addition of salt to koji as it is removed from fermentors or trays retards further growth of *A. oryzae.*

Preparation of Soybeans

Whole soybeans are prepared for fermentation concurrently with the preparation of koji. The soybeans should also be large and uniform in size and have an ability to absorb water and cook very rapidly. After extraneous materials are removed by mechanical equipment or by hand, soybeans are washed and soaked in water for 18 to 22 h. Water should be changed during the soaking period, especially in the summer months when temperatures are elevated, to control the proliferation of bacteria. At the end of the soaking period, soybeans have increased in volume by about 240% and weight by 220 to 260%. Drained soybeans are cooked in water or steamed at 115°C for about 20 min or until they are sufficiently soft to be easily pressed flat between the thumb and finger. Flavor and color development can be achieved by varying the heating temperature and time.

Fermentation and Aging

Cooked, cooled soybeans are then mixed with salted koji and an inoculum consisting of a portion of miso from a previous batch or pure cultures of osmophilic yeasts and bacteria. Strains of *Z. rouxii, Torulopsis* sp., and *Pediococcus halophilus* are the most important microorganisms in miso fermentation (101). The mixture, known as green miso, is packed into vats or tanks to undergo anaerobic fermentation and aging at 25 to 30°C. White miso takes about 1 week, salty miso 1 to 3 months, and soybean miso over 1 year. White miso contains 4 to 8% salt which permits rapid fermentation, and yellow or brown misos contain 11 to 13% salt. Moisture content ranges from 44 to 52%, protein from 8 to 19%, carbohydrate from 6 to 30%, and fat from 2 to 10%, depending on the ratio of soybeans, rice, and barley used as ingredients.

During fermentation and aging, soybean protein is digested by proteases produced by *A. oryzae* in the koji. Amino acids and their salts, particularly sodium glutamate, contribute to flavor. The addition of commercial enzyme preparations to enhance fermentation has met with some success. The relative amount of carbohydrates in miso is a reflection of the amount of rice in the product. Starch is extensively saccharified by koji amylases to yield glucose and maltose, some of which is utilized as an source of energy by the microorganisms responsible for fermentation. Miso contains 0.6 to 1.5% acids, mainly lactic, succinic, and acetic. Esters that are formed with ethyl and higher alcohols, together with fatty acid esters derived from fatty acids of soybean lipid, are important in giving miso its characteristic aroma (82). Total tocopherol content is decreased in the cooking process but is unaffected during aging (115). Substantial hydrolysis of triglycerides may occur during the early stages of fermentation. Antioxidative activity in miso is attributed in part to the existence of isoflavones, tocopherols, lecithin, compounds of amino-carbonyl reactions, and living microbial cells which tend to have a reducing activity (24). Cooking and steaming reduce thiamin and riboflavin content of soybeans.

NATTO

Natto is a Japanese name given to fermented whole soybeans, but related products are known as tou-shih by the Chinese, tao-tjo by the East Indians, and tao-si by the Filipinos (76, 90). Color, aroma, and flavor of these products vary, depending upon the microorganisms used to ferment the soybeans; however, products are generally dark in color and have a pungent but pleasant aroma and often a harsh flavor due to their relatively high free-fatty-acid and low-molecular-weight protein content. Fermented whole soybeans are eaten with boiled rice or as a seasoning agent with cooked meats, seafoods, and vegetables.

Three major types of natto are prepared in Japan (38, 76). Itohiki-natto, produced in large quantities in eastern Japan, is referred to simply as natto. Washed soybeans are soaked overnight or until they are approximately doubled in weight, then steamed for about 15 min and inoculated with *Bacillus natto,* a variant strain of *Bacillus subtilis.* The beans are packaged in approximately 150-g quantities and allowed to ferment at 40 to 45°C for 18 to 20 h (29). Production of polymers of glutamic acid by *B. natto* causes the surface of the final product to have a viscous appearance and texture. A second type of fermented whole soybean is known as yuki-wari-natto. This product is made by mixing itohiki-natto with salt and rice koji and then aging at 25 to 30°C for about 2 weeks. Hama-natto is a third major type of whole fermented soybean. Soybeans are soaked in water for about 4 h, steamed without pressure for 1 h, cooled, inoculated with a koji prepared from roasted wheat and barley, and fermented for about 20 h or until covered with the green mycelium of *A. oryzae* (76). After drying to a moisture content of about 12%, soybeans are submerged in a salt brine along with strips of ginger and allowed to age under pressure for 6 to 12 months. Breakdown of proteins, carbohydrates, and lipids during aging contributes to desirable sensory qualities. In addition to hydrolytic enzymes originating from *A. oryzae* in the koji, enzymes produced by bacteria such as *Micrococcus, Lactococcus,* and *Pediococcus* may also contribute to hydrolysis of soybean components.

The proximate composition of natto varies greatly, but the ranges (as percentage of wet weight) (76) are: water, 55.0 to 60.8; protein, 16.7 to 22.7; fat, 0.7 to 8.5; carbohydrate, 5.4 to 6.6; and ash, 2.1 to 3.0.

SUFU

Sufu is a mold-fermented soybean curd produced and consumed largely in the Orient. Preparation consists of three major steps, namely, making a soybean milk curd, fermenting the curd with an appropriate mold(s), and finally brining the fermented curd. Soybeans are washed, soaked overnight in water, and ground in a fashion similar to that followed to make tofu. After boiling or steaming for 20 to 30 min, the ground mass is strained through a fine sieve to separate the soybean milk from the insoluble residue. Alternatively, soybeans may be soaked, ground, and strained without heating. The milk is then heated to boiling to inactivate trypsin inhibitors and reduce some of the undesirable beany flavor. Coagulation of the milk is achieved by adding calcium sulfate or magnesium sulfate, and occasionally acid. The curd is then transferred to a cloth-lined wooden box and pressed to remove the whey. The resulting curd (tofu) contains 80 to 85% water, 10% protein, and 4% lipid. Tofu is prepared for fermentation by cutting into 3-cm cubes and soaking for 1 h in a brine containing about 6% sodium chloride and 2.5% citric acid. This treatment retards or prevents the growth of bacterial contaminants but has little effect on growth of desired molds during subsequent stages of sufu preparation. The cubes are boiled in brine for about 15 min, cooled, placed in perforated trays, surface inoculated with a selected mold, and incubated at 12 to 25°C for 2 to 7 days, depending upon the type and rate of growth of the mold and the desired flavor characteristics of the final product. White to yellowish white mycelium covers each cube, known as pehtze, at the end of the incubation period. Pehtze contains about 74% water, 12% protein, and 4.3% lipid (98).

Molds isolated from sufu include *Mucor corticolus, Mucor hiemalis* (*M. dispersus*), *Mucor praini, Mucor racemosus, Mucor silvaticus, Mucor subtilissimus, Actinomucor elegans,* and *Rhizopus chinensis* (27, 29, 31), all of which secrete proteases to hydrolyze soybean protein and yield peptides and amino acids, thus contributing to flavor development.

The last step of making sufu involves brining and aging. Depending upon the desired flavor and color, pehtzes may be submerged in salted, fermented rice or soybean mash, fermented soybean paste, or a solution containing 5 to 12% sodium chloride, red rice, and 10% ethanol. Red rice and soybean mash impart a red color to sufu. Use of brine containing high levels of ethanol results in sufu with a marked alcoholic bouquet. In addition to imparting taste, sodium chloride also enhances the release of mycelial enzymes which penetrate the molded cubes and hydrolyze soybean components. The aging period ranges from 1 to 12 months, after which time the sufu is consumed as a condiment or used to season vegetables or meat.

MEITAUZA

Meitauza is a fermented product made from the waste from ground, steeped, strained soybeans resulting from preparation of tofu and sufu. Soybean cakes approximately 10 to 14 cm in diameter and 2 to 3 cm thick are fermented for 10 to 15 days with moderate aeration (29). During fermentation, cakes become covered with white mycelium of *Mucor meitauza* (*A. elegans*), a principal mold involved in sufu fermentation. At the end of fermentation, cakes are partially sun-dried. Meitauza is cooked in vegetable oil or with vegetables as a flavoring agent.

LAO-CHAO

Lao-chao, also known as chiu-niang or tien-chiu-niang by the Chinese, is a fermented rice product. Glutinous rice is first steamed and cooked, then mixed with a small amount of commercial starter known as chiu-yueh or peh-yueh (100). The mass is incubated at ambient temperature for 2 to 3 days, during which time yeasts hydrolyze the starch, rendering the product soft, juicy, sweet, fruity, and slightly alcoholic (1 to 2%). Mucoraceous molds, including *Rhizopus oryzae, R. chinensis,* and *Chlamydomucor oryzae,* can be consistently isolated from lao-chao. *Endomycopsis,* one of the few yeasts capable of producing amylases and utilizing starch, is also an integral part of the necessary microflora. Lao-chao is consumed as such or it may be cooked with eggs and served as a dessert.

ANG-KAK

Red rice (ang-kak, ankak, anka, ang-quac, beni-koji, aga-koji) has been used in the fermentation industry for preparing red rice wine and foods such as sufu, fish sauce, fish paste, and red soybean curd. Pigments produced by *Monascus purpureus* and *Monascus anka* on a rice substrate are used as household and industrial food colorants in many Oriental countries. The major pigments produced by *Monascus* species are monascorubrin, rubropunctatin, and monascorubramine (104). Monascorubrin (red) and monascin (yellow) pigments produced by *M. purpureus* have been studied most extensively. The optimum cultural conditions for the production of pigments by a *Monascus* species isolated from the solid koji of Kaoliang liquor are reported to be pH 6.0 for a 3-day incubation at 32°C (42). Among the carbon sources tested, starch, maltose, and galactose were suitable for pigment production; a starch content of 3.5% (5% rice powder) and sodium nitrate or potassium nitrate content of 0.5% gave maximum yield of pigment in laboratory media. Zinc may act as a growth inhibitor of *M. purpureus* and concomitantly as a stimulant for glucose uptake and the synthesis of secondary metabolites such as pigments (7).

PUTO

Puto is a fermented rice cake prepared in the Philippines from 1-year-old rice that is ground with sufficient water to allow fermentation before steaming (78). The product is similar to idli, prepared from rice and black gram mungo (*Phaseolus mungo*) in India. The quality of puto depends on the microflora present in the milled rice as well as the variety of rice used as a starting material. There is a high degree of correlation between amylose content (within limits of 20 to 27%) and general acceptability.

RAGI

The Indonesian word "ragi" connotes the starter or inoculum used to initiate various kinds of fermentations (23). Thus, for example, Indonesian bread (roti) is made with a baker's yeast preparation (ragiroti), and fermented glutinous rice or cassava products called tapé ketan and tapé ketella, respectively, are prepared using ragi-tapé. Each type of ragi has a distinct mycological profile and is a source of enzymes necessary for the breakdown of carbohydrates and proteins in grains, legumes, and roots used as main fermentation substrates. Of particular importance is amylase produced by *Endomycopsis fibuliger* in ragi-tapé. Yeasts isolated from ragi-tapé are known to produce α-D-1,4-gluconoglucohydrolase, which releases the β form of glucose by hydrolysis and has high specific activities toward maltodextrins with four degrees of polymerization, amylose, amylopectin, and glycogen but little or no activity toward α-methyl- or *p*-nitrophenyl-α-glucoside (36).

TAPÉ

Indonesian tapé ketan is a fermented, partially liquefied, sweet-sour, mildly alcoholic rice paste (4, 21). In the traditional process, fermentation is initiated by the addition of powdered ragi made from rice flour containing the desired fungi. Tapé ketan (tepej) is prepared by fermenting glutinous rice whereas tapé ketella (Indonesian), tapé télo (Javanese), or peujeum (Sudanese) are prepared by fermenting cassava roots. Rice or peeled, chopped cassava is steamed or boiled until soft, spread in thin layers in bamboo trays, inoculated with powdered ragi, covered with a banana or other suitable leaf, and allowed to ferment for 1 to 2 days. The product will take on a white appearance, soft texture, and pleasant, sweet, alcoholic aroma and flavor. Amylolytic molds such as *Amylomyces rouxii* and alcohol-producing yeasts, particularly *Endomycopsis burtonii*, appear to be necessary for preparing good tapé.

TEMPEH

Tempeh is a fermented soybean product consumed largely in Indonesia, New Guinea, and Surinam. Kedelee or kedele, meaning soybean, is used to differentiate tempeh made using soybeans from tempeh bongkrek, a product prepared from coconut press cake (copra). Other beans, peas, and cereals are also used to make tempeh (58). Tempeh, as described in the following text, will be synonymous with tempeh kedelee.

Preparation of Soybeans and Fermentation

Soybeans are soaked in water at ambient temperature overnight or until hulls (testae) can be easily removed by hand. Lactic acid bacteria and yeasts are predominant in water in which soybeans have been soaked, suggesting their involvement in the fermentation (57). A comprehensive, quantitative study of the ecology of soybean soaking for tempeh fermentation has been made (46). *Lactobacillus casei* and *Lactococcus* species dominate the fermentation but significant contributions are made by other bacteria and yeasts, e.g., *Pichia burtonii, Candida diddensiae,* and *Rhodotorula mucilaginosa*.

After the hulls are removed from the soaked soybeans, cotyledons may be pressed slightly to remove more water and then mixed with small amount of tempeh from a previous batch or a commercial starter. The inoculated beans are then spread onto bamboo frames, wrapped in banana leaves, and allowed to ferment at ambient temperature for 1 to 2 days. At this point, the soybeans are covered with white *Rhizopus oligosporus* mycelium and bound together as a cake (32).

The temperature and moisture content of the fermenting substrate are critical if a good-quality tempeh is to be obtained. The most desirable temperature is between 30 and 38°C. The fermenting beans should be kept covered during fermentation to retard the rate of loss of moisture. However, slow diffusion of air to and release of gas from the product are essential to promote proper growth and metabolic activities of *R. oligosporus.* Production of black sporangia and spores is undesirable and usually indicates inadequate environmental conditions during fermentation or a product which has been kept beyond its normal shelf life of about 1 day. Fermented, yeasty off odors are often produced during storage (103). The production of ammonia as a result of enzymatic breakdown of mycelia and soybean protein causes the tempeh to be inedible within a very short period of time. Tempeh is either sliced, dipped in salt solution, and deep-fat fried in coconut oil, or cut into pieces and used in soups.

Biochemical Changes

While other genera of molds are occasionally found in tempeh, only *Rhizopus* can produce acceptable tempeh (29). *Rhizopus* species are known to produce carbohydrases, lipases, proteases, phytases, and other enzymes (58), *R. oligosporus, R. stolonifer, R. arrhizus, R. oryzae, R. formosaensis,* and *R. achlamydosporus* are among the species. The principal species used in Indonesia is *R. microsporus.* Among the carbohydrases produced by *R. oligosporus* are endocellulase, xylanase, and arabinase.

The hemicellulose content (as glucose) is reduced from 2.8% in raw soybeans to 2.0% as the beans are cleaned and cooked, and to 1.1% after fermentation (95). The fiber content, however, may increase upon fermentation due to the production of mold mycelium (71). Protease production by *R. oligosporus* is substantial and may play an important role in developing good-quality tempeh. Two proteolytic enzyme systems, one with an optimum pH at 3.0 and the other at 5.5, have been described (99). The pH 5.5 system dominates in tempeh fermentation. Amino acid profiles are not changed significantly upon fermentation; however, free amino acids may increase as much as 85-fold (48). Prolonged fermentation may result in losses of lysine (103).

R. oligosporus possesses strong lipase activity (14) and has been shown to hydrolyze over one-third of the neutral fat in soybeans during a 3-day fermentation period (97). Except for the depletion of as much as 40% of the linolenic acid in the late stages of fermentation, there apparently is no preferential utilization of any fatty acid.

ONCOM

Oncom is a fermented peanut press cake product prepared and consumed largely in Indonesia (8, 10, 12). According to Hesseltine (29), the flavor of fermented peanut press cake is fruitlike and somewhat alcoholic but takes on a mincemeat or almond character if the product is deep-fried.

Preparation of Peanuts and Fermentation

After extraction of oil from peanuts, the press cake is broken up and soaked in water for about 24 h (8, 29, 31). Oil that rises to the surface of the water during the soaking period is removed, and the press cake is then steamed for 1 to 2 h and pressed into a layer about 3 cm deep in a bamboo frame. The mass is inoculated with a portion of a previous batch of oncom containing either *Neurospora intermedia* (*N. sitophila)* or, less often, *R. oligosporus,* covered with banana leaves, and allowed to ferment for 1 to 2 days, during which time the internal portion of the mass becomes invaded by mycelia. Aeration is important in the production of oncom, as are temperature, moisture content, and degree of press cake granulation. A temperature range of 25 to 30°C is suitable for producing the best oncom. Sporulation may occur, resulting in an orange- to apricot-colored product if *N. intermedia* is used or a gray to black product if *R. oligosporus* is used. Only the surface of the fermented press cake is covered with colored conidia or spores.

Cassava (tapioca), potato peels, or other high-carbohydrate materials may be added to the press cake before inoculation to enhance fermentation. The addition of cassava to peanuts has been shown to promote the growth of *Rhizopus* (29), and the addition of citric acid (1.25% by weight), cassava (1%), and sodium chloride (0.63%) to defatted peanut flour has been demonstrated to enhance the growth of *N. intermedia, R. oligosporus,* and *Rhizopus delemar* (14).

Biochemical Changes

N. intermedia and *R. oligosporus* produce hydrolytic enzymes that act upon peanut constituents during fermentation. Crude protein is elevated slightly while true protein is decreased from 94% of the crude protein in peanut press cake to 74% of the crude protein in oncom (94). Large-molecular-weight globulins are hydrolyzed to smaller components, and the percentages of specific amino acids and proportions of specific amino acids within the free amino acid fraction change during fermentation (15). The nitrogen solubility of the peanut substrate is increased from 5% to 24% and 19%, respectively, when fermented with *N. intermedia* and *R. oligosporus* (70). Maximal protease activity of *N. intermedia* is at pH 6.5 (13). The protein efficiency ratio of fermented peanuts apparently is not increased over that of heated unfermented peanuts (71, 94, 96).

N. intermedia has strong α-galactosidase activity (107). The increased digestibility of oncom compared with unfermented peanuts may be attributable in part to the decrease in levels of these sugars. Lipid is hydrolyzed as a consequence of fermentation (14). The free fatty acid fraction of fermented peanuts contains a significantly higher level of saturated fatty acids, particularly palmitic and stearic acids, and lower amounts of linoleic acid than does the total lipid of oncom. The phytic acid content of peanut press cake is reduced by fermentation (26).

IDLI

Idli is a steamed fermented dough made in India from various proportions (1:4 to 4:1) of rice and black gram flours (76). Other ingredients such as cashew nuts, ghee, chili peppers, ginger, fried cumin seeds, or curry may be added to the dough in small quantities to impart additional flavor. To prepare idli, dehulled black gram and rice are washed and soaked in water separately for 5 to 10 h at ambient temperature (30). After soaking, the black gram is ground with water to give a coarse paste, whereas the rice is ground to give a smooth gelatinous paste. Salt (about 0.8%) is added to a mixture of pastes and fermentation is allowed to proceed for 15 to 24 h. The fermented mass is then steamed to yield a soft, spongy product. The open texture of idli is attributed to the protein (globulin) and polysaccharide (arabinogalactan) in black gram (90).

Idli batter volume increases 1.6- to 3.1-fold and the pH decreases from an initial 6.0 to 4.3 during fermentation (87). Bacteria identified as part of the microflora responsible for production of good idli include *Leuconostoc mesenteroides, Lactobacillus delbrueckii, Lb. fermentum, Lactobacillus lactis, Streptococcus faecalis,* and *Pediococcus cerevisiae* (6, 41, 45, 72, 73). Yeasts that may be involved in idli fermentation include *Oidium lactis* (*Geotrichum candidum*), *Torulopsis holmii, Torulopsis candida,* and *Trichosporon pullulans* (6, 22, 41). Lactic acid bacteria are responsible for pH reduction and may also contribute to improvement of the nutritional value of unfermented black gram and rice. An increase in thiamin and riboflavin contents occurs as a result of fermentation (40). Bacteria may also play a role in the breakdown of phytate present in black gram. *Leuconostoc mesenteroides* isolated from soybean idli secretes β-*N*-acetylglucosaminidase and α-D-mannosidase, which are involved in the hydrolysis of hemagglutinin (74).

WARIES AND PAPADAM

Waries is a spicy condiment shaped in the form of a ball 3 to 8 cm in diameter which is used in cooking with vegetables, beans, or rice in India (6). Dehulled beans are soaked in water, ground into a coarse paste, and mixed with spices such as asafoetida, caraway, cardamom, cloves, fenugreek, ginger, red pepper, and salt and a small amount of paste from a previous fermented batch. The paste is fermented for 4 to 10 days at ambient temperature, then formed into balls and air dried in the sun. The surface of the balls becomes sealed with a mucilaginous coating during the drying process, thus entrapping gases produced by *Candida* species and *S. cerevisiae* (76).

Papadam is similar to waries but does not contain fenugreek or ginger. Circular wafers are prepared from a mixture of black gram paste and spices which have been fermented for 4 to 6 h (6). Yeasts responsible for fermentation are the same as those found in waries. Papadams are served roasted or deep-fat fried and consumed as a condiment.

DAWADAWA

Locust beans (*Parkia filicoidea*) are fermented to produce a seasoning agent in West Africa. The product is also known as daddawa in Nigeria, kpalugu in Northern Ghana, kinda in Sierra Leone, and neteton in Gambia (64, 76). To prepare dawadawa, the yellow powdery pulp is removed from the dark brown to black locust bean seeds, and the seeds are boiled in water with the possible addition of potash until slightly soft. The beans are then stored overnight in earthenware, metallic pots, or baskets to further soften the seed coat. The black seed coats are then removed and the swollen cotyledons are washed in water, boiled for about 30 min, and deposited in a tray, pot, or basket. The preparation is covered with leaves or sheets of polyethylene and left to ferment for 2 to 3 days at ambient temperature. Microorganisms responsible for fermentation undoubtedly include spore-forming bacilli, lactic acid bacteria, and yeasts (63). Metabolic activities of the microflora result in a mucilaginous, strongly proteolytic, ammoniacal-smelling substance which covers and binds the individual beans (63). Moisture is partially removed from the fermented, dark brown mass by sun drying before it is pounded into flattened cakes and further dried to prolong shelf life. Darkening during sun drying is due to polyphenol oxidation (16).

GARI

Fermented root of the cassava plant (also known as manioc, mandioca, apiun, yuca, cassada, or tapioca) is known as gari in the rain forest belt of West Africa. Preparation of gari consists of the following stages (61). First, the corky outer peel and the thick cortex of the root are removed and the inner portion is grated. The pulp is then packed into jute bags and weights are applied to express some of the juice. After 3 to 4 days of fermentation, cassava is sieved and heated while constantly turning over a hot steel pan or in an oven. This process has been termed "garifying" (66). The final product contains 10 to 15% moisture, 80 to 85% starch, 0.1% fat, 1 to 1.5% crude protein, and 1.5 to 2.5% crude fiber. Palm oil may be added as a colorant just before or after drying.

Fresh cassava roots contain cyanogenic glucosides, viz., linamarin and lotaustralin, that decompose during the fermentation of gari with the liberation of gaseous hydrocyanic acid (20). *Lactobacillus plantarum* and other lactic acid bacteria also contribute significantly to decreasing the pH (68), which causes hydrolysis of cyanogenic glucosides. The acid condition favors the growth of *G. candidum*, which produces aldehydes and esters that give gari its characteristic aroma and flavor. Other yeasts and molds (25, 80) and lactic acid bacteria (54) undoubtedly contribute to flavor development. The fermentation of gari is reported to be self-sterilizing, exothermic, and anaerobic and to proceed in two stages at an optimum temperature of about 35°C (1). Lactic and formic acids are produced with a trace of gallic acid.

OGI

Corn is eaten in West Africa principally as a porridge known as ogi (Nigeria) or kenkey (Ghana). The Bantu equivalent of ogi is called mahewu in Southern Africa. To prepare ogi, kernels of corn are soaked in warm water for 1 to 3 days, after which they are wet-milled and sieved with water through a screen to remove fiber, hulls, and much of the germ (2). The filtrate is fermented to yield a sour, white, starchy sediment. Ogi may be diluted in water to 8 to 10% solids and boiled into a pap or cooked and turned into a stiff gel (eke) before eating. Ogi is a major breakfast cereal for adults and a traditional food for weaning babies. Lactic acid bacteria are largely responsible for fermentation of ogi, although other bacteria, yeasts, and molds may be involved.

INJERA

Injera is an Ethiopian bread made from teff, sorghum, wheat, barley, corn, or a mixture of these grains (88). Teff flour and water are combined with irsho, a fermented yellow fluid saved from a previous batch. The resultant thin, watery paste is generally incubated for 1 to 3 days. A portion of the fermented paste is then mixed with three parts water and boiled to give a product called absit, which is in turn mixed with a portion of the original fermented flour to yield a thin injera. Thick injera (aflegna) is teff paste which has undergone only minimal fermentation (12 to 24 h) and is characterized by a sweet flavor and a reddish color. A third type of injera (komtata) is made from overfermented paste and consequently has a sour taste, probably due to extensive growth of lactic acid bacteria. While the microflora responsible for fermentation of the sweeter types of injera have not been fully determined, *Candida guilliermondii* apparently is a primary yeast in this process. Regardless of the method used to prepare injera, the paste is baked or grilled to result in a breadlike product similar in appearance to pancakes.

POI

Corms of the taro plant are the principal material used to prepare poi in Hawaii and islands in the South Pacific (3). Cooked corms are peeled and ground or pounded to a fine consistency. The addition of water at this point results in fresh poi. The second phase of poi preparation involves fermentation at ambient temperature for 1 to 3 days or longer. As fermentation progresses, the texture changes from a sticky mass to one having a more watery and fluffy consistency. Lactic acid bacteria are the predominant microflora during early stages of fermentation. *Lb. delbrueckii*, *Lactobacillus pastorianus*, *Lactobacillus pentosus*, *Lactococcus lactis*, and *Saccharomyces kefir* produce large amounts of lactic acid and moderate amounts of acetic, propionic, succinic, and formic acids. *Candida vini* and *G. candidum* are prevalent in the latter stages of fermentation and are thought to be responsible for imparting the pleasant fruity aroma and flavor to older poi.

FERMENTED FISH PRODUCTS

Fermented fish sauce and paste are popular condiments in the Orient. Whole small fish, with or without entrails, or shrimp are heavily salted (up to 30% sodium chloride), packed into containers, and fermented from periods ranging from a few days to over a year (101). Roasted cereals, glutinous or red rice flour, or bran may be added in varying amounts to prepare fish pastes. The fermentation is essentially anaerobic, involving bacterial and autolytic breakdown of proteins and lipids in fish tissue to result in highly flavored products. Because of their high salt content, consumption of large quantities of fish sauce and paste is limited. However, these products represent an important dietary source of calcium.

Fish sauce and paste are known as nuoc-mam and mams, respectively, in Cambodia and Vietnam, ngampya-ye and ngapi in Myanmar, patis and bagoong in the Philippines, and mampla and kapi in Thailand (93, 101). Traditional methods for preparing fish sauce or paste are similar, but variations exist to result in differences in appearance, aroma, texture, and flavor characteristics in the final products. Preparation of various fermented fish sauces and pastes is described in chapter 33.

SAFETY AND NUTRITIONAL ASPECTS

Of interest to food microbiologists and sanitarians is the possibility of microorganisms producing toxic substances during fermentation or storage of indigenous fermented foods. An investigation of the aflatoxin-forming ability of 238 strains of *Aspergillus* used in the Japanese food industry has revealed that 52 strains produced fluorescent compounds, but none produced aflatoxin (43). None of the 46 domestic rice, 11 imported rice, 108 miso, and 28 rice-koji samples examined contained aflatoxin. Over 200 strains of industrial koji molds have been reported to be nonaflatoxigenic (47). Molds isolated from 24 samples of miso, katuobushi, and tane koji are capable of producing koji acid and β-nitropropionic acid, but none produced aflatoxin (37). While opportunistic mycotogenic molds and pathogenic bacteria may occasionally grow in improperly fermented or stored products, traditional fermented foods must be considered as microbiologically safe. Otherwise, diseases and illnesses implicating the presence of these microorganisms in traditional fermented foods would be more evident.

Many traditional fermented foods are staples in the diets of vast populations of people who would otherwise have less than minimum intakes of protein and/or calories. While the quality or quantity of proteins in vegetable-based fermented foods generally is not dramatically increased over raw substrates, the digestibility may be improved. Antinutritional and toxic components in plant materials may actually be reduced by fermentation (75). The positive contribution of traditional fermented foods to the nutritional well-being of those who consume them on a regular basis is recognized.

References

1. **Akinrele, I. A.** 1964. Fermentation of cassava. *J. Sci. Food Agric.* **15**:589–594.
2. **Akinrele, I. A.** 1970. Fermentation studies on maize during the preparation of a traditional African starch-cake food. *J. Sci. Food Agric.* **21**:619–625.
3. **Allen, O. N., and E. K. Allen.** 1933. The manufacture of poi from taro in Hawaii: with special emphasis upon its fermentation. *Hawaii Agric. Exp. Stn. Bull.* no. 70.
4. **Ardhana, M. M., and G. H. Fleet.** 1989. The microbial ecology of tape ketan fermentation. *Int. J. Food Microbiol.* **9**:157–165.
5. **Asao, Y., and T. Yokotsuka.** 1958. Studies on flavorous substances in soy sauce. Part XVI. Flavorous substances in raw soy sauce. *J. Agric. Chem. Soc. Jpn.* **32**:617–623.
6. **Batra, L. R., and P. D. Millner.** 1974. Some Asian fermented foods and beverages and associated fungi. *Mycologia* **66**:942–950.
7. **Bau, Y.-S., and H.-C. Wong.** 1979. Zinc effects on growth, pigmentation and antibacterial activity of *Monascus purpureus* fungi. *Physiol. Plant.* **46**:63–67.
8. **Beuchat, L. R.** 1976. Fungal fermentation of peanut press cake. *Econ. Bot.* **30**:227–234.
9. **Beuchat, L. R.** 1978. Microbial alterations of grains, legumes and oilseeds. *Food Technol.* **32(5)**:193–198.
10. **Beuchat, L. R.** 1982. Flavor chemistry of fermented peanuts. *Ind. Eng. Chem. Res. Dev.* **21**:533–536.
11. **Beuchat, L. R.** 1987. Traditional fermented food products, p. 209–306. *In* L. R. Beuchat (ed.), *Food and Beverage Mycology*, 2nd ed. Van Nostrand Reinhold, New York.
12. **Beuchat, L. R.** 1995. Indigenous fermented foods, p. 505–559. *In* H.-J. Rehm, G. Reed, A. Puhler, and P. Stadler (ed.), *Biotechnology, a Multi-Volume Comprehensive Treatise*, 2nd ed., vol. 9. VCH, Weinheim, Germany.
13. **Beuchat, L. R., and S. M. M. Basha.** 1976. Protease production by the ontjom fungus, *Neurospora sitophila*. *Eur. J. Appl. Microbiol.* **2**:195–203.
14. **Beuchat, L. R., and R. E. Worthington.** 1974. Changes in lipid content of fermented peanuts. *J. Agric. Food Chem.* **22**:509–512.
15. **Beuchat, L. R., C. T. Young, and J. P. Cherry.** 1975. Electrophoretic patterns and free amino acid composition of peanut meal fermented with fungi. *Can. Inst. Food Sci. Technol. J.* **8**:40–45.
16. **Campbell-Platt, G.** 1980. African locus bean (*Parkia* species) and its West African fermented food product, dawadawa. *Ecol. Food Nutr.* **9**:123–132.

17. **Campbell-Platt, G.** 1987. *Fermented Foods of the World: a Dictionary and Guide.* Buttersworth, London.
18. **Campbell-Platt, G., and P. E. Cook.** 1989. Fungi in the production of foods and ingredients. *J. Bacteriol. Symp. Ser. 18* **67**(Suppl.):117S–131S.
19. **Chavan, J. K., and S. S. Kadam.** 1989. Nutritional improvement of cereals by fermentation. *Crit. Rev. Food Sci. Nutr.* **28**:349–400.
20. **Collard, P., and S. Levi.** 1959. A two-stage fermentation of cassava. *Nature* (London) **183**:620–621.
21. **Cronk, T. C., K. H. Steinkraus, L. R. Hackler, and L. R. Mattick.** 1977. Indonesian tapé ketan fermentation. *Appl. Environ. Microbiol.* **33**:1067–1073.
22. **Desikachar, H. S. R., R. Radhakrishnamurthy, G. Ramarao, S. B. Kadkol, M. Srinivasan, and V. Subrahmanyan.** 1960. Studies on idli fermentation. I. Some accompanying changes in the batter. *J. Sci. Ind. Res.* **19**:168–172.
23. **Dwidjoseputro, D., and F. T. Wolf.** 1970. Microbiological studies of Indonesian fermented foodstuffs. *Mycopathol. Mycol. Appl.* **41**:211–222.
24. **Ebine, H.** 1971. Miso, p. 127. *In* T. Kawabata, M. Fujimaki, and H. Mitsuda (ed.), *Conversion and Manufacture of Foodstuffs by Microorganisms.* Saikon Publishing Co., Tokyo.
25. **Ekunsanmi, T. J., and S. A. Odunfa.** 1990. Ethanol tolerance, sugar tolerance and invertase activities of some yeast strains isolated from steep water of fermenting cassava tubers. *J. Appl. Bacteriol.* **69**:672–675.
26. **Fardiaz, D., and P. Markakis.** 1981. Degradation of phytic acid in oncom (fermented peanut press cake). *J. Food Sci.* **46**:523–525.
27. **Fukushima, D.** 1979. Fermented vegetable (soybean) protein and related foods of Japan and China. *J. Am. Oil Chem. Soc.* **56**:357–362.
28. **Goel, S. K., and B. J. B. Wood.** 1978. Cellulose and exo-amylase in experimental soy sauce fermentations. *J. Food Technol.* **13**:243–248.
29. **Hesseltine, C. W.** 1965. A millennium of fungi, food, and fermentation. *Mycologia* **57**:149–197.
30. **Hesseltine, C. W.** 1979. Some important fermented foods of Mid-Asia, the Middle East, and Africa. *J. Am. Oil Chem. Soc.* **56**:367–374.
31. **Hesseltine, C. W.** 1983. Microbiology of oriental fermented foods. *Annu. Rev. Microbiol.* **37**:575–601.
32. **Hesseltine, C. W., M. Smith, B. Bradle, and K. S. Djien.** 1963. Investigation of tempeh, an Indonesian food. *Dev. Ind. Microbiol.* **4**:275–287.
33. **Hesseltine, C. W., M. Smith, and H. L. Wang.** 1967. New fermented cereal products. *Dev. Ind. Microbiol.* **8**:179–186.
34. **Hesseltine, C. W., and H. L. Wang.** 1980. The importance of traditional fermented foods. *BioScience* **30**:402–404.
35. **Hesseltine, C. W., and H. L. Wang.** 1986. Indigenous fermented food of non-western origin. *Mycologia Memoir* no. 11. Cramer, Berlin.
36. **Kato, K., K. Kuswanto, L. Banno, and T. Harada.** 1976. Identification of *Endomycopsis fibuligera* isolated from ragi in Indonesia and properties of its crystalline glucoamylase. *J. Ferment. Technol.* **54**:831–837.
37. **Kinosita, R., T. Ishiko, S. Sugiyama, T. Seto, S, Igarasi, and I. E. Goetz.** 1968. Mycotoxins in fermented food. *Cancer Res.* **28**:2296–2311.
38. **Kiuchi, K., T. Ohta, H. Itoh, T. Takabayashi, and H. Ebine.** 1976. Studies on lipids of natto. *J. Agric. Food Chem.* **24**:404–407.
39. **Kline, L., and T. F. Sugahara.** 1971. Microorganisms of the San Francisco sour dough bread process. II. Isolation and characterization of undescribed bacterial species responsible for the souring activity. *Appl. Microbiol.* **21**: 459–465.
40. **Lakshmi, I.** 1978. Studies on fermented foods. M.S. project. University of Baroda, Baroda, India.
41. **Lewis, Y. S., and D. S. Johar.** 1953. Microorganisms in fermenting grain mashes used for food preparations. *Cent. Food Technol. Res. Inst.* **2**:228.
42. **Lin, C.-F.** 1973. Isolation and cultural conditions of *Monascus* sp. for the production of pigment in a submerged culture. *J. Ferment. Technol.* **51**:407–414.
43. **Matsuura, S., M. Manabe, and T. Sato.** 1970. Surveillance for aflatoxins of rice and fermented-rice products in Japan, p. 48–55. *In* M. Herzberg (ed.), *Proceedings of the First U.S.-Japan Conference on Toxic Microorganisms, Mycotoxins, Botulism.* U.S. Government Printing Office, Washington, D.C.
44. **Morimoto, S., and N. Matsutani.** 1969. Studies on the flavor components of soy sauce: isolation of furfuryl alcohol and the formation of furfuryl alcohol by yeasts and molds. *J. Ferment. Technol.* **47**:518–525.
45. **Mukherjee, S. K., M. N. Albury, C. S. Pederson, A. G. Van Veen, and K. H. Steinkraus.** 1965. Role of *Leuconostoc mesenteroides* in leavening the batter of idli, a fermented food of India. *Appl. Microbiol.* **13**:227–231.
46. **Mulyowidarso, R. K., G. H. Fleet, and K. A. Buckle.** 1989. The microbial ecology of soybean soaking for tempe production. *Int. J. Food Microbiol.* **8**:35–46.
47. **Murakami, H. S., S. Takase, and T. Ishii.** 1967. Production of fluorescent substances in rice koji, and their identification by absorption spectrum. *J. Gen. Appl. Microbiol.* **14**:97–110.
48. **Murata, K., H. Ikehata, and T. Miyamoto.** 1967. Studies on the nutritional value of tempeh. *J. Food Sci.* **32**:580–584.
49. **Nakadai, T., S. Nasuno, and N. Iguchi.** 1972. The action of peptidases from *Aspergillus sojae* on soybean proteins. *Agric. Biol. Chem.* **36**:1239–1243.
50. **Nakadai, T., S. Nasuno, and N. Iguchi.** 1972. Purification and properties of acid carboxypeptidase I from Aspergillus oryzae. *Agric. Biol. Chem.* **36**:1343–1352.
51. **Nakadai, T., S. Nasuno, and N. Iguchi.** 1973. Purification and properties of leucine aminopeptidase I from *Aspergillus oryzae. Agric. Biol. Chem.* **37**:757–765.
52. **Nakadai, T., S. Nasuno, and N. Iguchi.** 1973. Purification and properties of alkaline proteinase from *Aspergillus oryzae. Agric. Biol. Chem.* **37**:2685–2694.
53. **Nakadai, T., S. Nasuno, and N. Iguchi.** 1973. Purification and properties of neutral proteinase from *Aspergillus oryzae. Agric. Biol. Chem.* **37**:2695–2701.

54. **Ngaba, P. R., and J. S. Lee.** 1979. Fermentation of cassava (*Manihot escuelenta* Crantz). *J. Food Sci.* **44**:1570–1571.

55. **Nout, M. J. R.** 1994. Fermented foods and food safety. *Food Res. Int.* **27**:291–298.

56. **Nout, M. J. R.** 1995. Useful role of fungi in food processing, p. 295–303. *In* R. S. Samson, E. S. Hoekstra, J. C. Frisvad, and O. Filtenberg (ed.), *Introduction to Food-Borne Fungi*, 4th ed. Centraalbureau voor Schimmelcultuur Baarn, Netherlands.

57. **Nout, M. J. R., M. A. De Dreu, A. M. Zuurbier, and T. M. G. Bonants-van Laarhoven.** 1987. Ecology of controlled soyabean acidification for tempeh manufacture. *Food Microbiol.* **4**:165–172.

58. **Nout, M. J. R., and F. M. Rombouts.** 1990. Recent developments in tempe research. *J. Appl. Bacteriol.* **69**:609–633.

59. **Nunomura, N., M., Sasaki, Y. Asao, and T. Yokotsuka.** 1976. Identification of volatile components in shoyu (soy sauce) by gas chromatography. *Agric. Biol. Chem.* **40**:485–491.

60. **Nunomura, N., M. Sasaki, Y. Asao, and T. Yokotsuka.** 1976. Isolation and identification of 4-hydroxy-2(or 5)ethyl-5(or 2)methyl-3(2H)furanone as a flavor component in shoyu (soy sauce). *Agric. Biol. Chem.* **40**:491–496.

61. **Nunomura, N., M. Sasaki, and T. Yokotsuka.** 1979. Isolation of 4-hydroxy-5-methyl-3(2H)furanone, a flavor component in shoyu (soy sauce). *Agric. Biol. Chem.* **43**:1361–1367.

62. **Nunomura, N., M. Sasaki, and T. Yokotsuka.** 1980. Shoyu (soy sauce) flavor components: acidic fractions and the characteristic flavor component. *Agric. Biol. Chem.* **44**:339–345.

63. **Odunfa, S. A.** 1985. African fermented foods, p. 155–191. *In* B. J. B. Wood (ed.), *Microbiology of Fermented Foods.* Elsevier, London.

64. **Odunfa, S. A.** 1988. Review: African fermented foods. From art to science. *Mircen J.* **4**:259–273.

65. **Ogawa, G., and A. Fujita.** 1980. Recent progress in soy sauce production in Japan, p. 381–394. *In* G. E. Inglett and G. E. Munck (ed.), *Recent Progress in Cereal Chemistry.* Academic Press, Inc., New York.

66. **Ogunsua, A. O.** 1980. Changes in some chemical constituents during the fermentation of cassava tubers (*Manihot esculenta*, Crantz). *Food Chem.* **5**:249–255.

67. **Onishi, H.** 1990. Yeasts in fermented foods, p. 167–198. *In* J. F. T. Spencer and D. M. Spencer (ed.), *Yeast Technology.* Springer-Verlag, Berlin.

68. **Oyewole, O. B., and S. A. Odunfa.** 1990. Characterization and distribution of lactic acid bacteria in cassava fermentation during fufu production. *J. Appl. Bacteriol.* **68**:145–152.

69. **Ponte, J. G., and C. C. Tsen.** 1987. Bakery products, p. 233–267. *In* L. R. Beuchat (ed.), *Food and Beverage Mycology*, 2nd ed. Van Nostrand Reinhold, New York.

70. **Quinn, M. R., and L. R. Beuchat.** 1975. Functional property changes resulting from fungal fermentation of peanut flour. *J. Food Sci.* **40**:475–478.

71. **Quinn, M. R., L. R. Beuchat, J. Miller, C. T. Young, and R. E. Worthington.** 1975. Fungal fermentation of peanut flour: effects on chemical composition and nutritive value. *J. Food Sci.* **40**:470–474.

72. **Rajalakshmi, R., and K. Vanaja.** 1967. Chemical and biological evaluation of the effects of fermentation on the nutritive value of foods prepared from rice and grams. *Br. J. Nutr.* **21**:467–473.

73. **Ramakrishnan, C. V.** 1976. Preschool child malnutrition. Pattern, prevalence and prevention. *Baroda J. Nutr.* **3**:1–39.

74. **Rao, G. S.** 1978. Studies on fermented foods with reference to hemagglutin in hydrolyzing bacteria isolated from rice-soy idli batter. Ph.D. thesis. University of Baroda, Baroda, India.

75. **Reddy, N. R., and M. D. Pierson.** 1994. Reduction in antinutritional and toxic components in plant foods by fermentation. *Food Res. Int.* **27**:281–290.

76. **Reddy, N. R., M. D. Pierson, and D. K. Salunkhe.** 1986. *Legume-Based Fermented Foods.* CRC Press, Inc., Boca Raton, Fla.

77. **Ruskin, F. R.** 1992. *Applications of Biotechnology to Traditional Fermented Foods.* National Academy Press, Washington, D.C.

78. **Sanchez, P. C.** 1975. Varietal influence on the quality of Philippine rice cake (puto). *Philippine Agric. J.* **58**:376–382.

79. **Sanni, A. I.** 1993. The need for process optimization of African fermented foods and beverages. *Int. J. Food Microbiol.* **18**:85–95.

80. **Sanni, M. O.** 1989. The mycoflora of gari. *J. Appl. Bacteriol.* **67**:239–242.

81. **Sarkar, P. K., J. P. Tamang, P. E. Cook, and J. D. Owens.** 1994. Kinema—a traditional soybean fermented food: proximate composition and microflora. *Food Microbiol.* **11**:47–55.

82. **Shibasaki, K., and C. W. Hesseltine.** 1962. Miso fermentation. *Econ. Bot.* **16**:180–195.

83. **Spicher, G., and J.-M. Brummer.** 1995. Baked goods, p. 241–319. *In* H.-J. Rehm, G. Reed, A. Puhler, and P. Stadler (ed.), *Biochemistry, a Multi-Volume Comprehensive Treatise*, 2nd ed., vol. 9. VCH, Weinheim, Germany.

84. **Stanton, W. R., and A. Wallbridge.** 1969. Fermented food processes. *Process Biochem.* **4(4)**:45–51.

85. **Steinkraus, K. H.** 1983. *Handbook of Indigenous Fermented Foods.* Marcel Dekker, New York.

86. **Steinkraus, K. H.** 1994. Nutritional significance of fermented foods. *Food Res. Int.* **27**:259–267.

87. **Steinkraus, K. H., A. G. Van Veen, and D. B. Thiebeau.** 1967. Studies on idli—an Indian fermented black gram-rice food. *Food Technol.* **21**:110–113.

88. **Stewart, R. B., and A. Getachew.** 1962. Investigations of the nature of injera. *Econ. Bot.* **16**:127–130.

89. **Sugihara, T. F., L. Kline, and M. W. Miller.** 1971. Microorganisms of the San Francisco sour dough bread process. I. Yeasts responsible for the leavening action. *Appl. Microbiol.* **21**:456–458.

90. **Susheelamma, N. S., and M. V. L. Rao.** 1979. Functional role of the arabinogalactan of black gram (*Phaseolus*

mungo) in the texture of leavened foods (steamed puddings). *J. Food Sci.* **44**:1309–1312, 1316.

91. **Tamang, J. P., P. K. Sarkar, and C. W. Hesseltine.** 1988. Traditional fermented foods and beverages of Darjeeling and Sikkim—a review. *J. Sci. Food Agric.* **44**:375–385
92. **Ulloa-Sosa, M.** 1974. Mycofloral succession in pozol from Tabasco, Mexico. *Biol. Soc. Mex. Microbiol.* **8**:17–48.
93. **van Veen, A. G.** 1953. Fish preservation in Southeast Asia. *Adv. Food Res.* **4**:209–231.
94. **van Veen, A. G., D. C. W Graham, and K. H. Steinkraus.** 1968. Fermented peanut press cake. *Cereal Sci. Today* **13**:96–98.
95. **van Veen, A. G., and G. Schaefer.** 1950. The influence of the tempeh fungus on the soya bean. *Doc. Neerl. Indones. Morbis Trop.* **2**:270–281.
96. **van Veen, A. G., and K. H. Steinkraus.** 1970. Nutritive value and wholesomeness of fermented foods. *J. Agric. Food Chem.* **18**:576–578.
97. **Wagenknecht, A. C., L. R. Mattick, L. M. Lewin, D. B. Hand, and K. H. Steinkraus.** 1961. Changes in soybean lipids during tempeh fermentation. *J. Food Res.* **26**:373–376.
98. **Wai, N. S.** 1968. Investigation of the various processes used in preparing Chinese cheese by the fermentation of soybean curd with *Mucor* and other fungi, p. 89. Tech. Rept., Public Law 480 Proj. UR-A6-(40)-1. U.S. Department of Agriculture, Washington, D.C.
99. **Wang, H. L., and C. W. Hesseltine.** 1965. Studies on the extracellular proteolytic enzymes of *Rhizopus oligosporus*. *Can. J. Microbiol.* **11**:727–732.
100. **Wang, H. L., and C. W. Hesseltine.** 1970. Sufu and lao-chao. *J. Agric. Food Chem.* **18**:572–575.
101. **Wang, H. L., and C. W. Hesseltine.** 1982. Oriental fermented foods, p. 492. *In* G. Reed (ed.), *Prescott and Dunn's Industrial Microbiology*, 4th ed. AVI Publishing Co., Westport, Conn.
102. **Wang, H. L., and C. W. Hesseltine.** 1986. Glossary of indigenous fermented foods, p. 317–344. *In* C. W. Hesseltine and H. L. Wang (ed.), *Indigenous Fermented Food of Non-Western Origin.* Cramer, Berlin.
103. **Winarno, F. G., and N. R. Reddy.** 1986. Tempe, p. 95–117. *In* N. R. Reddy, M. D. Pierson, and D. K. Salunke (ed.), *Legume-Based Fermented Foods.* CRC Press, Boca Raton, Fla.
104. **Wong, H.-C.** 1982. Antibiotic and pigment production by *Monascus purpureus*. Ph.D. dissertation. University of Georgia, Athens, Ga.
105. **Wood, B. J. B.** 1965. *Microbiology of Fermented Foods*, vol 2. Elsevier, London.
106. **Wood, B. J. B.** 1994. Technology transfer and indigenous fermented foods. *Food Res. Int.* **27**:269–280.
107. **Worthington, R. E., and L. R. Beuchat.** 1974. α-Galactosidase activity of fungi on intestinal gas-forming peanut oligosaccharides. *J. Agric. Food Chem.* **22**:1063–1066.
108. **Yokotsuka, T.** 1960. Aroma and flavor of Japanese soy sauce. *Adv. Food Res.* **10**:75.
109. **Yokotsuka, T.** 1971. Shoyu, p. 117–125. *In* M. Fujimaki and H. Mitsuda (ed.), *Conversion and Manufacture of Foodstuffs by Microorganisms.* Saikon Publishing Co., Tokyo.
110. **Yokotsuka, T., T. Sakasai, and Y. Asao.** 1967. Studies on flavorous substances in shoyu. Part 35. Flavorous compounds produced by yeast fermentation. *J. Agric. Chem. Soc. Jpn.* **41**:428–433.
111. **Yong, F. M., K. H. Lee, and H. A. Wong.** 1981. The production of ethyl acetate by soy yeast (*Saccharomyces rouxii* Y-1096). *J. Food Technol.* **16**:177–185.
112. **Yong, F. M., and B. J. B. Wood.** 1974. Microbiology and biochemistry of the soy sauce fermentation. *Adv. Appl. Microbiol.* **17**:157–194.
113. **Yong, F. M., and B. J. B. Wood.** 1976. Microbial succession in experimental soy sauce fermentation. *J. Food Technol.* **11**:525–536.
114. **Yong, F. M., and B. J. B. Wood.** 1977. Biochemical changes in experimental soy sauce koji. *J. Food Technol.* **12**:163–175.
115. **Yoshida, H., and G. Kajimoto.** 1972. Changes in lipid components during miso making. Studies on the lipids of fermented foodstuffs (Part I). *J. Jpn. Soc. Food Nutr.* **25**:415–421.

Sterling S. Thompson
Kenneth B. Miller
Alex S. Lopez

35

Cocoa and Coffee

Cocoa and coffee are two of the many foods that rely on a microbial curing process or fermentation for flavor development. The popularity and worldwide appeal of these products are due primarily to their unique flavor and aroma. Although a primary curing process is conducted in the preparation of each product before marketing, fermentation of cocoa is absolutely essential for flavor development, whereas with coffee, the curing process is less crucial to flavor and more important for the removal of pulp. Consequently, this chapter will focus mainly on the more comprehensive role of fermentation in cocoa curing and to a lesser extent in the production of coffee.

COCOA PROCESSING

Commercial cocoa is derived from the seeds (beans) of the ripe fruit (pods) of the plant *Theobroma cocoa,* which is native to the Amazon region of South America. It has been used by the South American Indians to produce a beverage since time immemorial and was introduced to Europe in the 15th century by Cortez during the period of discovery and colonization of the Americas. Its popularity and demand led to the establishment and spread of rootstock to virtually all of the European colonies located between 15 degrees north and south of the equator with climates that could support cocoa production.

Of the *Theobroma* species, only *T. cocoa* produces beans suitable for chocolate manufacturing. Immediately following harvesting of the ripe fruit, or following a brief storage period, the seeds are removed and subjected sequentially to a fermentation and drying process often referred to as "curing," which is carried out on farms, estates, or cooperatives in the producing countries. The origin of this process has been lost in antiquity, but it was believed at one time that fermentation was conducted simply to aid in removing the mucilaginous pulp surrounding the seed so as to facilitate drying and storage, as in the case of coffee. This in fact is true, but the main reason for fermentation of cocoa is to induce biochemical transformations within the beans that lead to formation of the aroma, flavor, and color precursors of chocolate. Without this treatment, cocoa beans are excessively bitter and astringent and when processed do not develop the flavor that is characteristic of chocolate. The character and strength of chocolate flavor are governed primarily by the genetic constitution of the cocoa variety, while the fermentation process releases and develops this flavor potential (28). The inherited characteristics of the bean therefore set a limit to what can be achieved by fermentation. It is impossible to improve genetically inferior material by superior processing techniques, yet, on the other hand, it is quite easy to ruin good-quality cocoa by careless or inadequate curing.

The cocoa fruit varies among varieties in size, shape, external color, and appearance. These characters have often been used in classifying cocoa, but as far as the flavor quality is concerned, the only really important morphological differences are those that distinguish between the white-seeded Criollo variety of South and Central America and the purple-seeded Forastero variety of the Amazon. The former type is the source of the original "fine" cocoa, which has almost disappeared from the market because of its susceptibility to disease, its lower productivity, and its replacement by the hardier, more prolific Forastero variety and their varietal crosses, which now account for over 95% of the world production. Hence the following discussion refers primarily to the processing of Forastero cocoa.

Flowers are produced seasonally from cushions that emerge on the bark of the trunk and stems. Fertilized flowers bear fruit 170 days from pollination, a period during which the fruit grows to maturity and changes color from green or dark red-purple to yellow, orange, or red, depending on the variety. The mature fruits are thick-walled and contain 30 to 40 beans, each enveloped in a sweet, white, mucilaginous pulp and loosely attached to an axial placenta. Only the beans are used in chocolate manufacturing. For the purpose of describing the curing process, the bean may be envisaged as comprising two main parts, namely the testa (seed coat), together with the attached sugary, mucilaginous pulp that surrounds it, and the embryo or the cotyledons contained within. The mucilage, containing sugars and citric acid, serves as a substrate for microorganisms that are involved in the fermentation process; the cotyledons, referred to as the "nib" in the cured bean, are used in chocolate manufacturing.

Processing begins with the harvesting of healthy ripe fruits, an operation carried out over a period of 3 to 4 days at a frequency which varies according to the size of the farm and yield. Fermentation is a batch-type process, and harvesting is conducted to allow for the accumulation of sufficient material for each batch while taking precautions that, in the process, pods do not over-ripen and the seeds within do not germinate. The pods are usually collected in piles in the field and broken open on site or at the processing plant (fermentary) at the end of the harvesting operation. The beans are removed manually or mechanically on some large estates in West African countries, Mexico, and Brazil, where pod breaking and bean extraction are mechanized (33). Once removed from the pods, the seeds are aggregated in heaps or in receptacles of one kind or another and left to ferment for a period of 2 to 8 days. During this interval, microorganisms which are transferred to the seeds from laborers' hands, fruit surfaces, and containers used in transporting and fermentation degrade the cells of the mucilage that surrounds the bean (37). The collective microbial activity resulting from the accidental inoculation by a multitude of microorganisms is referred to as "fermentation" or "sweating." This process results in the liberation of pulp juices from which alcohol and acetic acid are produced with the evolution of heat. Together, these factors provoke changes and affect the curing of the bean. The two principal objectives of fermentation are to remove mucilage, thus provoking aeration during fermentation of the beans and facilitating drying later on, and to provide heat and acetic acid necessary for inhibiting germination, which assures proper curing of the beans (29).

Methods of Fermentation

The manner of fermenting cocoa varies considerably from country to country, and in many instances even adjacent farms may adopt different curing methods. There has been a great deal of effort to standardize fermentation practices. Today, many of the primitive methods such as fermentation in banana leaf-lined holes in the ground, in derelict canoes, and in makeshift banana and bamboo frames are the exception rather than the rule. In general, large farms with an adequate production of cocoa fruits will opt for permanent facilities specifically constructed for this purpose. In such instances, fermentation is carried out in batteries of wooden or fiberglass boxes. However, most of the world's cocoa is produced on small holdings under very rural conditions, and the relatively small volumes produced do not always merit permanent processing facilities. In this case, cocoa is fermented in any convenient receptacle such as fruit boxes, baskets, plastic buckets, or fertilizer bags, or, when these are not readily available, the beans are simply piled on a sheet and covered with any handy material. On the whole, however, the majority of the world's cocoa is fermented on drying platforms, in heaps covered with banana leaves, in baskets, or in an assortment of wooden boxes (12, 24, 48). Approximately one-half of the world crop is fermented in some type of box, and the remaining half is fermented in heaps or by other primitive methods.

Fermentation on Drying Platforms

Fermentation on drying platforms is practiced in parts of Central America where Criollo cocoa was once grown. Wet cocoa beans are spread directly on drying platforms where they ferment and dry during the day and are heaped into piles each night to conserve heat and retard the growth of surface molds. Criollo cocoa requires only a short fermentation for flavor development, and this slow drying process is sufficient for that purpose. Forastero varieties require a fermentation time of 5 to 8 days for the development of flavor, whereas Criollo varieties require 2 to 3 days (13). Although Criollo cocoa has been largely replaced by Forastero hybrids in these countries, in many instances the old method of fermentation still persists (48). This practice preserves the fine flavor characteristics of Criollo beans; however, it is inappropriate for Forastero varieties which require longer fermentation times for optimal flavor development. Fermenting cocoa on the drying floor is convenient, but unless properly managed, the process tends to produce under-fermented cocoa with the added danger of undesirable mold growth and its consequences of off-flavor development.

Fermentation in Heaps

Fermentation in heaps is a popular method among small holding farmers in Ghana and many other African cocoa-producing countries. It is also observed sporadically in the Amazon region of Brazil. This method does not require a permanent structure and is well suited to family holdings with a small production. Judging from Ghanaian cocoa, fermentation in heaps can produce good-quality products. Varying quantities of cocoa beans from 25 to 1,000 kg are heaped in the field on plantain leaves and covered with the same material. The beans are mixed (turned) periodically to ensure even fermentation and to decrease the potential for mold growth. This is often done daily or every other day by forming another heap. Mixing is laborious and small heaps may not be turned at all. The duration of fermentation is from 4 to 7 days.

Fermentation in Baskets

Fermentation in baskets is practiced principally by small-scale producers in Nigeria, the Amazon region, the Philippines, and some parts of Ghana. Small lots of cocoa are placed in woven baskets lined with plantain leaves. The surface is covered with plantain leaves and weighted down. The turning procedure and the fermentation are similar to those used for small heaps. Basket fermentation is often used when fermentation in heaps is susceptible to predial larceny.

Fermentation in Boxes

Fermentation in boxes is considered to be an improvement over other methods. This batch process requires a fixed volume of cocoa and is the method of choice on large estates. The size of the containers varies from region to region, but the design and function are standard. The container or "sweat box" may be a single unit or one of a number of compartments within a large box created by subdividing the space into units measuring approximately 1 by 1 by 1 m, with either fixed or movable internal partitions. These boxes hold between 600 and 700 kg of freshly harvested (wet) cocoa beans. The box is always raised above ground level, over a drain which carries away the pulp juices (sweatings) liberated by the degradation of the mucilage during fermentation. The wooden floor of the box generally has holes or spaces between the boards or slats to facilitate drainage and aeration. Sweat boxes vary considerably in size from that of a small fruit box (0.4 by 0.4 by 0.5 m) to some measuring 7 by 5 by 1 m, used on some Malaysian estates (22). Large estates and cooperatives often have batteries of 20 to 30 sweat boxes arranged in tiers in three to seven rows, one below the other to facilitate mixing or turning. Mixing is achieved by simply removing a dividing wall and shoveling the beans into the next box or, in the case of the tier design, into the box below. On some Malaysian estates, boxes are built on pallets and a fork lift is used to transfer the contents into an empty box. Variations occur not only in the size of sweat boxes but also in the type of wood used in their construction and in methods of drainage, aeration, and duration of fermentation. The recommendation is to ferment a 1-m^3 volume for 6 to 7 days with two to three mixings during this period. In the majority of cases, boxes are filled to within 10 cm of the top and the surface is covered with a padding of banana leaves or jute sacking to help retain the heat and prevent the surface beans from drying.

In some countries, fermentation norms have been modified in an attempt to overcome problems such as acidity by varying the prefermentation treatment and the depth of beans in the sweat boxes (28). In Malaysia, for instance, harvested cocoa beans are stored in the pod up to 15 days before breaking, or the beans may be pressed or predried to reduce the pulp volume before fermentation (4, 10).

The progress of fermentation is assessed by the odor and the external and internal color changes in the beans. When the process is judged complete, the beans are dried in the sun or in mechanical dryers.

Microbiology of Cocoa Fermentation

Fermentation begins immediately after beans are removed from the pods, as they become inoculated with a variety of microorganisms from the pod surface, knives, laborers' hands, containers used to transport the beans to the fermentary, dried mucilage on surfaces of the fermentation box (tray, platform, or basket) from the previous fermentation, insects, and banana or plantain leaves (17, 18, 24, 34, 37, 46, 53). It is the pulp surrounding the beans, not the cocoa bean, that undergoes microbial fermentation. Chemical changes take place within the bean as a result of the fermentation of the pulp. The testa of the bean acts as a natural barrier between microbial fermentation activities outside the bean and chemical reactions within the bean. However, there is a migration of ethanol, acetic acid, and water of microbial origin from the outside to the inside of the bean. After the bean dies, soluble bean components are leached through the skin and are lost in the drainings. The pulp consists of about 85% water, 2.7% pentosans, 0.7% sucrose, 10% glucose and fructose, 0.6% protein, 0.7% acids, and 0.8% inorganic salts (19), making it a rich substrate for microbial growth. The concentration of sucrose, glucose, and fructose is influenced by the age of the pod (54).

The initial microbial population is variable in number and type; however, the key groups active during fermentation of beans are yeasts, lactic acid bacteria, acetic acid bacteria, and the genus *Bacillus*. Climatic conditions may influence the sequence of microorganisms involved in the fermentation (57). It is theorized that *Bacillus* species play an important role during the latter stages of the fermentation and become the dominant group during drying (57). More research is needed to confirm this theory. Over 100 aerobic spore-forming bacteria were isolated from cocoa bean fermentations in Bahia. Bacteria were identified as *Bacillus subtilis, B. licheniformis, B. firmus, B. coagulans, B. pumilus, B. macerans, B. polymyxa, B. laterosporus, B. stearothermophilus, B. circulans, B. pasteurii, B. megaterium, B. brevis,* and *B. cereus. B. subtilis, B. circulans,* and *B. licheniformis* were encountered more frequently than the other *Bacillus* species during fermentation (57).

During fermentation, yeasts, lactic acid bacteria, and acetic acid bacteria develop in succession. Species of microorganisms that have been detected in cocoa during fermentation in Ghana, Malaysia, and Belize are listed in Table 35.1. At the onset of fermentation, a pH of 3.4 to 4.0, a sugar content of 10 to 12%, and a low oxygen tension favor the growth of yeasts (54, 56). Yeasts utilize the carbohydrates in the pulp under aerobic and anaerobic conditions and may make up 40 to 65% of the microflora when the fermentation begins (9, 38). The yeast phase lasts 24 to 48 h, during which populations may increase to 90% of the total microflora. Yeast populations have been determined in several investigations (26).

Some yeasts produce various pectinolytic enzymes that degrade the cocoa pulp, thereby aiding in the drainage of juices (15, 46). In addition to metabolizing sugar to produce ethanol, yeasts utilize citric acid, causing the pH to increase (46). Not all yeast species that contribute to fermentation are present simultaneously, but follow a succession which is influenced by the turning step (aeration) and the fact that fermenting bean masses are not homogeneous (26). Several genera of yeasts are involved in fermentation (Table 35.1). In one study (9), of the 142 yeast genera detected in fermenting bean masses, 105 were asporogenous and 37 were ascosporogenous. Between 48 and 72 h of fermentation, the yeast population begins to decrease so that, by the third day, it is reduced to 10% of the total microbial population (4). Three factors are responsible for the rapid decline in the dominance of yeasts. First, yeasts rapidly metabolize sucrose, glucose, and fructose in the pulp to form carbon dioxide and ethanol, causing a reduction in energy source. Second, the production of ethanol produces a toxic environment that suppresses yeast growth. For example, Schwan et al. (57) reported that a decline in the population of *Kloeckera apiculata* was associated with an ethanol concentration greater than 4%. A small amount of heat is developed simultaneously with ethanol production (2). Third, acetic acid, which is produced from ethanol by the acetic acid bacteria, is also toxic to yeasts. The acetic acid concentration may reach 1 to 2% (6).

The anaerobic conditions created by yeasts make the environment suitable for lactic acid bacteria (38). Lactic acid bacteria prefer a low oxygen concentration or, if oxygen is present, a high concentration of carbon dioxide (26). Such an environment develops as the pulp collapses and the yeast population decreases. The population of lactic acid bacteria increases rapidly, but large numbers may be present for only a brief period (28, 46, 48). The lactic acid bacteria population has been observed to reach 10^6 to 10^7 CFU/g in a typical fermentation in Belize and a high of 20% of the total microflora after 1.5 days in fermentations in Trinidad (53). In Brazil, the lactic acid bacteria population is about 65% of the total microflora after 14 h of fermentation. The lactic acid bacteria population remained high up to 3 days, at which time it decreased to less than 10% of the total microflora (38).

Both homofermentative and heterofermentative lactic acid bacteria occur in cocoa fermentations; however, the majority are homofermentative (57). Lactic acid bacteria detected in traditional box fermentation of cocoa beans in Brazil were isolated and characterized as homofermentative and heterofermentative. The homofermentative species included *Lactobacillus plantarum, Lactobacillus casei, Lactobacillus delbrueckii, Lactobacillus*

Table 35.1 Microorganisms isolated from fermenting cocoa beans

Group	Country		
	Ghana[a]	Malaysia[a]	Belize[b]
Lactic acid bacteria	*Lactobacillus plantarum*	*Lb. plantarum*	*Lb. plantarum*
	Lb. mali	*Lb. collinoides*	*Lb. fermentum*
	Lb. collinoides		*Lb. brevis*
	Lb. fermentum		*Lb. buchneri*
			Lb. cellobiosus
			Lb. casei pseudoplantarum
			Lb. delbrueckii
			Lb. fructivorans
			Lb. kandleri
			Lb. gasseri
			Leuconostoc mesenteroides
			Ln. paramesenteroides
			Ln. oenos
Acetic acid bacteria	*Acetobacter rancens*	*A. rancens*	*Acetobacter* spp.
	A. ascendens	*A. lovaniensis*	*G. oxydans*
	A. xylinum	*A. xylinum*	
	Gluconobacter oxydans	*G. oxydans*	
Yeasts	*Candida* spp.	*Candida* spp.	*Brettanomyces claussenii*
	Hansenula spp.	*Debaryomyces* spp.	*Candida* spp.
	Kloeckera spp.	*Hanseniaspora* spp.	*C. boidinii*
	Pichia spp.	*Hansenula* spp.	*C. cocoai*
	Saccharomyces spp.	*Kloeckera* spp.	*C. intermedia*
	Saccharomycopsis spp.	*Rhodotorula* spp.	*C. guilliermondii*
	Schizosaccharomyces spp.	*Saccharomyces* spp.	*C. krusei*
	Torulopsis spp.	*Torulopsis* spp.	*C. reukaufii*
			Kloeckera apis
			K. javanica
			Pichia membranaefaciens
			Saccharomyces cerevisiae
			Saccharomyces chevalieri
			Schizosaccharomyces spp.
			Schizosaccharomyces malidevorans

[a]From Carr et al. (7).
[b]S. S. Thompson, unpublished data.

acidophilus, Pediococcus cerevisiae, Pediococcus acidilactici, and *Streptococcus (Lactococcus) lactis.* The heterofermentative species included *Leuconostoc mesenteroides* and *Lactobacillus brevis* (39). Citric acid is metabolized either to acetic acid, carbon dioxide, and lactic acid by heterofermentative species or to acetylmethylcarbinol and carbon dioxide by homofermentative lactics. Lactic acid bacteria may be more important in cocoa fermentation in Brazil where, during the first 48 h of the fermentation, their population is consistently larger than the yeast population (38). This differs from most other fermentations, where yeasts are the dominant microorganism during the first 48 h. If lactic acid bacteria remain as a high percentage of the total microbial population during the fermentation, high concentrations of lactic acid will be produced. Since lactic acid is not volatile, it will remain in the chocolate after manufacturing, producing an undesirable chocolate (50, 59, 67).

As the beans are turned to aerate the mass, more of the pulp is metabolized, and conditions become aerobic. The population of lactic acid bacteria decreases and the population of acetic acid bacteria increases with increased aeration. The population of acetic acid bacteria generally reaches 10^5 to 10^6 CFU/g in a typical fermentation in Belize. Two genera of acetic acid bacteria, *Acetobacter* and *Gluconobacter,* have been isolated from fermenting cocoa beans. *Acetobacter* species occur more frequently than *Gluconobacter* species (6, 7, 37). The acetic acid bacteria population has been observed to form 80 to 90% of the total microflora after 2 days in fermentations in Trinidad (53). Acetic acid bacteria oxidize ethanol to acetic acid exothermally (2), causing the temperature of the bean mass to rise to 45 to 50°C. Turning the beans periodically facilitates oxidation of the ethanol to acetic acid and conserves the high temperature of the bean mass. When all of the ethanol is oxidized to acetic acid and then carbon dioxide and water, fermentation subsides and the temperature of the bean mass decreases quickly.

During the later stages of fermentation and while drying, aerobic spore-forming *Bacillus* species develop and may become dominant (6, 37). *Bacillus* species are present during the first 72 h of the fermentation, but during this early stage, their population remains constant. They become dominant later in the fermentation, making up over 80% of the microbial population (56, 58). Development of *Bacillus* species in the bean mass is favored by increased aeration, an increased pH (3.5 to 5.0) of the pulp, and an increase in temperature to 45 to 50°C (6, 37, 58).

Bacillus species can produce several compounds that may contribute to the acidity and off flavors of fermented cocoa. The C_3, C_4, and C_5 free fatty acids that are present in the bean mass during the aerobic phase of fermentation may contribute to development of some of the off flavors of chocolate (31, 42). The importance of *Bacillus* species in cocoa bean fermentation is not well established, but they are reputed to produce acetic and lactic acids, 2,3-butanediol, and tetramethylpyrazine, which can affect the flavor of chocolate (31, 68).

A key factor that must be considered when deciding on a fermentation scheme is when to remove the beans from their fermentation environment and begin drying. Extending the fermentation can result in undesirable microbial activity, leading to putrefaction and the production of compounds such as butyric and valeric acids that contribute to off flavors (32). Forsyth and Quesnel (12) suggested that the following factors may collectively indicate when fermentation is optimum:

- external color of the beans
- time schedule
- decrease in temperature
- bean cut test, and the internal color used as a criterion
- aroma of the fermenting mass
- plumping or swelling of the beans

A more desirable indication of optimum fermentation would be a chemical method that is relatively rapid, inexpensive, and easy to perform and interpret. We have observed that the end point of fermentation can be determined by the pH of the beans, provided a normal temperature curve is established. The minimum pH that gives acceptable cocoa liquor is 5.2; however, the actual fermentation pH may be slightly lower.

Biochemistry of Cocoa Fermentation

The actual production of chocolate flavor precursors occurs within the cocoa bean and is primarily the result of biochemical changes that take place during fermentation and drying. The mode of fermentation and the microbial environment during these stages of cocoa production provide the necessary conditions for complex biochemical reactions to occur. Although flavor compounds such as lactic and acetic acids are produced external to the bean by microbial activity, chocolate flavor development is largely dependent on the enzymatic formation of flavor precursors within the cotyledon that are unique to cocoa. Such classes of compounds include free amino acids, peptides, reducing sugars, and polyphenols. When fermented dried cocoa beans containing these flavor precursors are subjected to roasting during chocolate manufacture, a necessary step in flavor development, a series of complex nonenzymatic browning reactions occurs to produce flavor and colored compounds characteristic of chocolate (21, 27, 49, 51, 52). However, if unfermented cocoa beans lacking these precursor compounds are roasted, very little chocolate flavor is produced. It is, therefore, important that these flavor precursor compounds are formed inside the cocoa bean during fermentation.

The initiation of fermentation also corresponds to an incipient germination phase which is necessary for mobilization of the enzymes and hydration of bean components in preparation for growth. However, the germination phase is undesirable in cocoa beans used to make chocolate.

Bean Death

Bean death is a critical event during cocoa fermentation which allows the biochemical reactions responsible for flavor development to occur within the cocoa bean. Although rising temperatures and increasing acetic acid concentrations during fermentation have been implicated in causing seed death (40), more recent data (25) indicate that the production of ethanol during the anaerobic yeast growth phase correlates very closely with death of the seed. Total inability of the seed to germinate occurs about 24 h after maximum concentrations of ethanol are attained within the cotyledon (Fig. 35.1). As a result, events associated with germination and certain quality defects, e.g., the utilization of valuable seed components such as cocoa butter and the opening of the

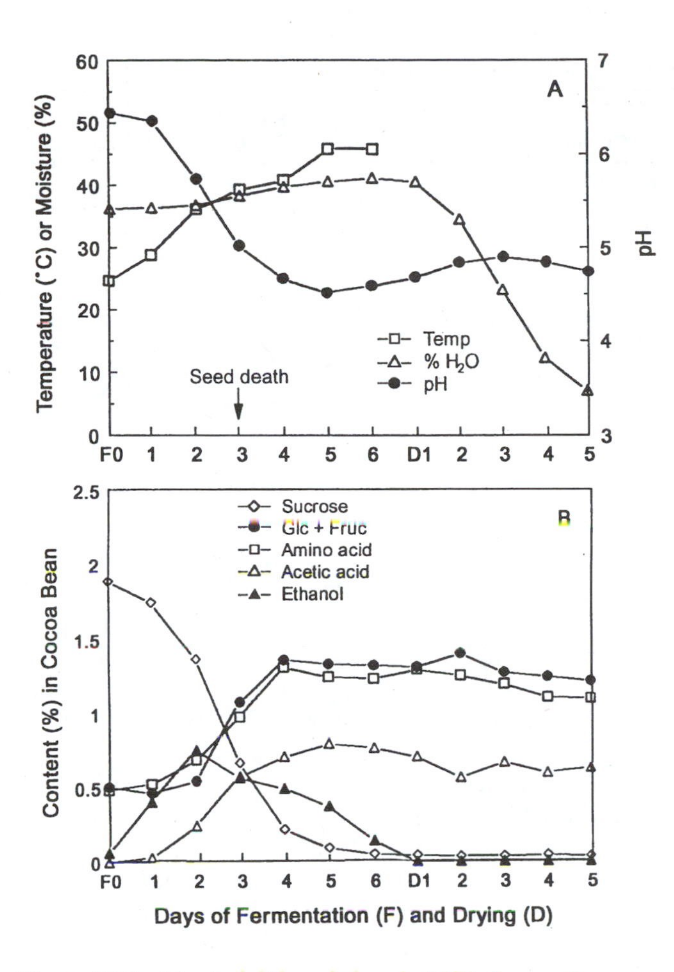

Figure 35.1 Physical (A) and chemical (B) changes in cocoa beans during fermentation and drying in Belize. Fermentation was conducted with 2,000 pounds of wet cocoa beans from ripe pods in wooden boxes that were turned daily. Drying was conducted in flat-bed dryers indirectly heated with hot air. Data represent an average of 11 fermentation trials using composite samples collected daily. (A) Temperature was measured in the whole bean mass. Moisture (%) and pH analyses are based on shell-free cotyledons. (B) Sucrose, glucose (Glc), fructose (Fruc), total amino acid, acetic acid, and ethanol contents (%) were determined by analysis of water extracts of shell-free cotyledon samples. Data from Lehrian (25).

testa by hypocotyl extension, will not occur. This produces a more stable, desirable end product.

From a flavor perspective, events associated with death of the seed also cause cellular membranes to leak and permit enzymes and substrates to react to form flavor precursor compounds important to chocolate flavor development. As shown in Fig. 35.1, activity of these enzymes in the cotyledon results in significant increases in free amino acid and reducing sugar contents (glucose and fructose). While the temperature of the beans increases, the concentration of organic acids increases, causing a decrease in pH. All of these factors influence the biochemistry within the bean and have an impact on cocoa flavor and quality.

Environmental Factors

There are several environmental factors, viz., pH, temperature, and moisture, in the fermenting mass that influence cocoa bean enzyme reactions. Each enzyme has an optimum pH at which it is most active, and within a defined range, an enzyme reaction accelerates as temperature increases. In addition, a certain amount of moisture is necessary to allow enzymes and their substrates to react to form products. Significant changes in pH, temperature, and moisture occur during cocoa fermentation and drying processes (Fig. 35.1) that influence the type and quantity of flavor precursor compounds produced by enzymatic action.

Moisture content within the cotyledon during fermentation is usually more than 35% and will permit adequate migration of enzymes and substrates for enzymatic activity. However, once the drying process begins, moisture content gradually decreases, making it increasingly difficult for enzymes and substrates to react. When a moisture content of 6 to 8% is achieved, virtually all enzyme activity ceases.

The pH of the unfermented cotyledon is about 6.5 and may decrease to as low as 4.5 by the end of the fermentation. This lowering of pH occurs after seed death and is primarily due to the diffusion into the bean of organic acids produced by lactic acid and acetic acid bacteria. It is the growth of these and other microorganisms that also contributes to the increasing temperature of the mass of fermenting beans. Typically, the bean mass temperature will rise from 25°C to about 50°C, followed by a slight decrease as bacterial growth subsides. An increase in temperature of more than 20°C during fermentation can have a profound impact on enzyme activity. If very little change in temperature occurs, enzyme activity is reduced, resulting in fewer flavor precursors and poor chocolate flavor. Likewise, if appropriate amounts of organic acids are not produced during fermentation, the pH of the cotyledon will not be suitable for optimal enzyme activity and the flavor profile of the resulting cocoa will be affected. However, too much acid will produce excessive sourness that can mask the chocolate flavor.

Consequently, there is a delicate balance among the length of fermentation, environmental factors, and microbial activity that influences enzyme activity within the cotyledon. Hydrolytic and oxidative enzymes play a major role in reactions that produce flavor precursors. A

Table 35.2 Characteristics of the principal enzymes active during the curing of the cocoa bean[a]

Enzyme	Location	Substrate	Product	pH	Temp (°C)	Reference(s)
Invertase	Testa	Sucrose	Glucose and fructose	4.0	52	30
				5.25	37	
Glycosidases (β-galactosidase)	Bean	Glycosides (3-β-D-galactosidyl cyanidin and 3-α-L-arabinosidyl cyanidin)	Cyanidin and sugars	3.8–4.5	45	11
Proteases	Bean	Proteins	Peptides and amino acids	4.7	55	5, 41
Polyphenol-oxidases	Bean	Polyphenols (epicatechin)	σ-quinones and σ-diquinones	6.0	31.5, 34.5	43

[a]From Lopez (28).

summary of cocoa bean enzymes, their substrates, and pH optimum is given in Table 35.2 (28).

Hydrolytic Enzyme Reactions

Hydrolytic enzymes such as invertase, glycosidases, and proteases have highest activity during the anaerobic phase of cocoa fermentation. The products of these enzyme activities during cocoa fermentation fall into three basic categories: sugars, amino acids/peptides, and cyanidins. Sugars and amino acids/peptides participate in nonenzymatic browning reactions during roasting to form important chocolate flavor precursors, whereas the cyanidins have more of an impact on color development and some minor flavor components.

Sucrose is the major sugar in unfermented cocoa beans. It is not a reducing sugar and therefore does not participate in nonenzymatic browning reactions that occur during roasting to contribute to chocolate flavor. However, sucrose is converted to glucose and fructose by invertase during the fermentation process. These reducing sugars represent more than 95% of the total reducing monosaccharides in cocoa beans, and their concentrations increase almost threefold during fermentation, while sucrose is depleted (Fig. 35.1).

Another class of hydrolytic enzymes within the cocoa bean that contributes to both flavor and color during fermentation is the glycosidases. The substrates for these enzymes are the purple-colored anthocyanins located in specialized vacuoles within the cotyledon and are responsible for the characteristic deep purple color of the unfermented bean. The actions of specific glycosidase enzymes begin at seed death and are responsible for cleaving the sugar moieties, galactose, and arabinose attached to the anthocyanins. This results in a bleaching of the purple color of the beans as well as the release of reducing sugars that can participate in flavor precursor reactions during roasting (11). Pigments themselves do not carry any flavor potential (11, 31, 47). Cocoa beans that still contain significant purple color are considered to have been poorly fermented and are less desirable.

Although there are small amounts of free amino acids present in unfermented cocoa beans, the total free amino acid pool increases significantly during fermentation due to the action of both endo- and exoproteases on cocoa bean proteins. After seed death occurs, these proteolytic enzymes are free to act on protein substrates within the bean, and their activity becomes dependent on pH and temperature. A vicilin-like globular storage protein within the cotyledon is the primary target of these proteolytic enzymes, and ratios of free amino acids and peptides that are unique to cocoa are produced (65). These flavor precursor compounds contribute to the development of cocoa flavor when roasted in the presence of reducing sugars.

Oxidative Enzyme Reactions

Significant oxidative enzyme activity also occurs, being most prevalent late in the aerobic phase of fermentation but continuing well into the drying of cocoa. Polyphenol oxidase is the major oxidase in cocoa and is responsible for much of the brown color that occurs during fermentation as well as some flavor modifications. This enzyme becomes active during the aerobic phase of the fermentation as a result of oxygen permeating the cotyledon. Events that contribute to its activity include seed death, subsequent breakdown of cellular membranes, reduction in the amount of seed pulp, and aeration of the bean mass by agitation. Oxygen continues to penetrate the beans during the drying process, enabling polyphenol oxidase activity to continue until rising temperatures and insufficient moisture become inhibiting factors.

Catachins and leucocyanidins are the major classes of polyphenols that are subject to oxidation in cocoa beans. Epicatechin comprises more than 90% of the total catechin fraction and is the major substrate of polyphenol oxidase (16). Oxidation of epicatechin during the aerobic phase of fermentation and drying is largely responsible for the characteristic brown color of fermented cocoa beans. Polyphenols in the dihydroxy configuration are oxidized to form quinones which in turn can polymerize with other polyphenols or complex with amino acids and proteins to yield characteristic colored compounds and high-molecular-weight insoluble material. This complexation also has an impact on flavor. The formation of these less soluble polyphenolic complexes reduces the astringency and bitterness associated with native polyphenols present in unfermented cocoa (13, 45). In addition, the ability of polyphenols to complex with proteins results in the reduction of off flavors associated with the roasting of peptide and protein material (20, 66).

Flavor and Quality Implications

The ultimate goal of biochemical changes during fermentation is to produce cocoa beans with desirable flavor and color characteristics. Good chocolate flavor potential is achieved by the production of specific amino acids, peptides, and sugars through the action of proteases and invertases on cocoa bean substrates. Proper control of fermentation conditions and microflora assures that concentrations of organic acids are maintained at reasonable levels to minimize sour and putrid off flavors while still developing the pH and temperature environment for enzyme-substrate reactions that produce chocolate flavor precursors. Enzyme-mediated conversion of polyphenol materials during fermentation and drying processes reduces astringency and bitterness and produces the desirable brown color typical of properly fermented cocoa beans. The drying process will then preserve the flavor and color characteristics of the beans until they are made into chocolate.

Although cocoa has been successfully produced for centuries, the flavor characteristics of specific varieties of cocoa are becoming diluted due to the prevalence of genetic hybrids. While these hybrids are being selected for high crop yield and disease resistance, certain flavor attributes are being lost. The actual identity of the flavor precursors responsible for chocolate flavor has not yet been confirmed. However, recent work has focused on characterizing cocoa bean enzymes and proteins that yield breakdown products unique to cocoa during the fermentation process (64). This work needs to continue in order to understand flavor development and maintain the high quality of chocolate flavor the consumer expects.

Drying

After the beans are fermented, they have a moisture content of about 40 to 50% which must be reduced to 6 to 8% for safe storage. A higher final moisture content will result in mold growth during storage. The drying process relies on air movement to remove water. This environment favors aerobic microorganisms which proliferate at rates that decrease with moisture loss. Sun drying is the preferred method, but in regions where harvesting coincides with frequent rainfall, some form of artificial drying is necessary and desirable. In general, sun drying is employed on small farms, whereas large estates may resort to both natural and artificial drying. Sun drying allows a slow migration of moisture throughout the bean, which transports flavor precursors that had been formed during fermentation.

During sun drying, beans are placed on wooden platforms, mats, polypropylene sheets, or concrete floors in layers ranging from 5 to 7 cm thick. The beans are constantly mixed to promote uniform drying, to break agglomerates that may form, and to discourage mold growth. Under sunny conditions, the beans dry in about a week, but under cloudy or rainy conditions drying times may be prolonged to 3 or 4 weeks, increasing the risk of mold development and spoilage.

Various types of artificial dryers employed to overcome dependence on weather conditions have been described by McDonald et al. (35). Hot air dryers of one form or another, fueled by wood or oil as a source of cheap, readily available energy, are generally employed. The beans may be heated by direct contact with the flue gases; however, the preferred method relies on indirect heating via heat exchangers. Improperly used or poorly maintained heating systems present the danger of contamination with smoke, which results in smoky or hammy off flavors characteristic of beans from some countries. Platforms, trays, and rotary dryers of various designs, coupled to furnaces, are used, but in every instance the initial drying must be slow and with frequent mixing to obtain uniform removal of water. This results in volatilization of acids and sufficient time for oxidative, biochemical reactions to occur. For this reason, temperatures should not exceed 60°C and drying times should take at least 48 h. Elevated temperatures also tend to produce cocoa with brittle shells and cotyledons which crumble during handling. In short, the drying rate should be controlled so as to remove moisture at a rate that will avoid case hardening (rapid drying on the bean surface with moisture retention inside the bean) or excessive mold growth, while still allowing sufficient time for biochemical oxidative reactions and loss of acid to occur.

Storage

Due to marketing practices and manufacturing procedures, fermented, cured, dried beans are stored for periods of 3 to 12 months in warehouses on farms, at wharfs in exporting and receiving countries, and at factories before being processed into chocolate. The efficiency of the drying process will determine the shelf life of the product. Uniformly dried beans with a moisture content of 7 to 8%, which are stored at a relative humidity of 65 to 70%, will generally maintain that moisture, resist mold growth and insect infestation, and not require repeated fumigation. The cocoa quality can change during storage depending on temperature, relative humidity, and ventilation conditions. Slow oxidation and acid loss continue to enhance product quality somewhat, but prolonged storage results in a noticeable staling (29).

COFFEE PROCESSING

Introduction

Coffee beans may also undergo a fermentation step to prepare the fruit for commercial use (3, 8, 60). Like cocoa, the fermentation of coffee has the important goal of breaking down the pulp layer surrounding the beans in order to aid in the processing of the fruit into a desirable finished product. Unlike cocoa, however, the role of fermentation in coffee is less critical in the development of coffee flavor, although improperly fermented fruit can result in undesirable off flavors. Coffee beans are produced by the genus *Coffea*. Over 40 species are known, but only a few are used to produce coffee. *Coffea arabica, C. canephora, C. robusta, C. liberica,* and *C. excelsa* are the most important species. The coffee fruit is a fleshy berry approximately the size of a small cherry (8).

It takes approximately 1 year from the flowering stage for a fruit, or coffee cherry as it is commonly called, to reach maturity. As the fruit ripens it changes color from green to cherry-red. The ripe, red coffee cherry consists of two green beans surrounded by a pulp layer and enclosed in an outer skin. It is necessary to first remove the outer skin and pulp layer to obtain the green coffee beans, which are then dried, roasted, milled, and used for making coffee beverages. The removal of the pulp layer can be accomplished by either a natural drying step yielding "natural coffee" or a wet fermentation step yielding "washed coffee." The percent yield of dried coffee from ripe cherries varies between species: arabica produces 12 to 18%, robusta produces 17 to 22%, and liberica produces about 10%.

Natural Coffee

The "natural" drying process relies on partial dehydration of the pulp layer while the coffee fruit is ripening on the tree. The ripe fruit is then harvested and subjected to additional drying, and the dried pulp layer and outer skin are removed by hulling. Some mucilage will penetrate the coffee bean during drying, and for this reason the natural process produces a light-brown-colored bean instead of the blue-green color of wet-processed coffee. Drying must be carefully controlled to avoid excessive fermentation during the initial stages which can result in reduced product quality. Molds that may develop on slowly drying coffee can produce off flavors typical of a butyric fermentation. Coffee is dried to a moisture content below 13%.

Washed Coffee

The wet process is accomplished by harvesting ripe coffee fruit and immediately removing the majority of pulp by mechanical squeezing or depulping. This is followed by a fermentation step which is used to convert the remaining mucilage to water-soluble products that can be removed by washing. In general, wooden or concrete bins are used for this fermentation step. The white, sticky, partially depulped beans are held under water for 12 to 60 h, depending on environmental conditions, ripeness of the cherries, and the variety being processed. Microbial by-products are periodically washed away during the fermentation step to avoid the development of excessive off flavors. Once the sticky pulp layer is converted to water-soluble products, the beans are washed with water and dried to a moisture content of about 12%.

Microbiology and Biochemistry of Coffee Fermentation

As stated above, the purpose of fermentation is removal of the mucilage from around the seed. A natural fermentation, which involves several different microorganisms, is the primary method used to remove the mucilage. Three nonmicrobial methods, namely the addition of enzymes, chemical treatment (sodium hydroxide), and mechanical force, may also be used to remove mucilage (23). However, only the natural fermentation used in the wet process to produce washed coffee will be discussed.

Coffee is generally fermented in wooden or concrete tanks. Mucilage is easily metabolized by most of the microorganisms that have been identified in coffee fermentation studies. Mucilage is composed of pectin, pectic acids, reducing sugars, sucrose, caffeine, chlorogenic acid, amino acids, and hydrolytic and oxidative enzymes (23).

Many coffee fermentation studies have been conducted over the years, but researchers do not agree on which microorganism(s) is responsible for digestion of the mucilage. However, there is agreement that several microorganisms are present during the fermentation. Molds, yeasts, several species of lactic acid bacteria, coliforms, and other gram-negative bacteria have been isolated. These microorganisms originate from the surface of the fruit and the soil (1, 14, 62).

Since the mucilage composition consists largely of pectic substances (23), the microorganisms responsible for colonization and utilization of this material must be capable of producing pectinases. Coffee fermentation studies have demonstrated that the highest microbial activity occurs during the first 12 to 24 h after the beans are harvested (63). Most of the microorganisms detected during this early stage belong to the genera *Aerobacter (Enterobacter)* and *Escherichia*. Populations of these bacteria increase from 10^2 to 10^9 CFU/g during that first 24-h period. Pectinolytic species of *Bacillus, Fusarium, Penicillium,* and *Aspergillus* have also been detected. Only a few yeast species have been identified. Fermentation studies conducted on Kona coffee beans demonstrated that *Erwinia dissolvens* was the main cause of mucilage decomposition (14).

The pH of unfermented coffee beans can range from 5.4 to 6.4. During fermentation the pH may decrease to 3.7 (1, 36). Low levels of ethanol have been detected under both aerobic and anaerobic conditions (36). However, ethanol may reach slightly higher levels under anaerobic fermentation conditions (23).

Duration of a natural fermentation will depend on climate conditions, regional factors, coffee variety, degree of anaerobiosis of the ferment, and the microorganisms present (23). An optimal fermentation time is considered to be 16 to 24 h (61). Extending the fermentation will result in development of flavor defects and bean discoloration (36, 44). Once the fermentation is complete, any remaining mucilage is removed by washing. Washing the beans is required to avoid excessive fermentation and enhance the drying process.

CONCLUSION

Although serving somewhat different purposes, microbial fermentation plays a critical role in the production of both cocoa and coffee. A common goal of fermentation of both of these products is the breakdown and removal of fruit pulp. This process aids in the proper drying of coffee and provides a suitable environment for flavor and color development in cocoa. As in most natural curing processes, there is a delicate balance between environmental factors and conditions enabling microbial activities, all of which influence the biochemical changes that take place in processed foods. Coffee and cocoa are no exceptions, and it is the proper control of the fermentation process that largely determines the color and flavor quality of final products. Consequently, understanding the microbiology and biochemistry of cocoa and coffee production, as well as the factors that influence them, is critical to quality control.

References

1. **Agate, A. D., and J. V. Bhat.** 1966. Role of pectinolytic yeasts in the degradation of the mucilage layer of *Coffee* and *Robusta* cherries. *Appl. Microbiol.* **14:**256–260.
2. **Anonymous.** 1979. *The Fermentation of Cocoa.* National Union of Cocoa Producers, Mexico.
3. **Arunga, R.** 1982. Coffee, p. 259–274. *In* A. H. Rose (ed.), *Economic Microbiology,* vol. 7. *Fermented Foods.* Academic Press, Inc., New York.
4. **Biehl, B., B. Meyer, G. Crone, L. Pollman, and M. B. Said.** 1984. Chemical and physical changes in the pulp during ripening and post-harvest storage of cocoa pods. *J. Sci. Food Agric.* **48:**189–208.
5. **Biehl, B., U. Passern, and D. Passern.** 1977. Subcellular structures in fermenting cocoa beans—effect of aeration and temperature during seed and fragment incubation. *J. Sci. Food Agric.* **28:**41–52.
6. **Carr, J. G., P. A. Davies, and J. Dougan.** 1979. *Cocoa Fermentation in Ghana and Malaysia* (Part I), p. 573–576. 7th International Cocoa Research Conference, Douala, Cameroon.
7. **Carr, J. G., P. A. Davies, and J. Dougan.** 1980. *Cocoa Fermentation in Ghana and Malaysia. II. Further Microbiological Methods and Results.* Cocoa Research Institute, Taso, Ghana.
8. **Castelein, J., and H. Verachtert.** 1983. Coffee fermentation, p. 587–615. *In* H. J. Rehm and G. Reed (ed.), *Biotechnology,* vol. 5. Verlag Chemie, Weinheim, Germany.
9. **Camargo, R. J., J. Leme, and A. M. Filho.** 1963. General observations on the microflora of fermenting cocoa beans (*Theobroma cocoa*) in Bahia (Brazil). *Food Technol.* **17:**116–118.
10. **Duncan, R. J. E., G. Godfrey, T. N. Yap, G. L. Pettipher, and T. Tharumarajah.** 1989. Improvement of Malaysian cocoa bean flavor by modification of harvesting, fermentation and drying methods—the Sime-Cadbury process. *Planter* **65:**157–173.
11. **Forsyth, W. G. C., and V. C. Quesnel.** 1957. Cocoa glycosidase and colour changes during fermentation. *J. Sci. Food Agric.* **8:**505–509.

12. **Forsyth, W. G. C., and V. C. Quesnel.** 1957. Variations in cocoa preparation, p. 157–168. Sixth Meeting of the Inter-american Cocoa Committee, Bahia, Brazil.
13. **Forsyth, W. G. C., and V. C. Quesnel.** 1963. Mechanisms of cocoa curing. *Adv. Enzymol.* **25**:457–492.
14. **Frank, H. A., N. A. Jum, and A. S. Dela Cruz.** 1965. Bacteria responsible for mucilage-layer deposition in Kona coffee cherries. *Appl. Microbiol.* **13**:201–207.
15. **Gauthier, B., J. Guiraud, J. C. Vincent, J. P. Porvais, and P. Galzy.** 1977. Comments on yeast flora from the traditional fermentation of cocoa in the Ivory Coast. *Rev. Ferment. Ind. Aliment.* **32**:160–163.
16. **Griffiths, L. A.** 1957. Detection of the substrate of enzymatic browning in cocoa by a post-chromatographic enzymatic technique. *Nature* (London) **180**:1373–1374.
17. **Grimaldi, J.** 1954. *The Cocoa Fermentation Process in the Cameroon.* V. Reunion del Comite Técnico Interamericano del Cocoa, 5RTIC/DOC 24. Bioq./2-6.
18. **Grimaldi, J.** 1978. The possibilities of improving techniques of pod breaking and fermentation in the traditional process of the preparation of cocoa. *Café Cocoa Thé* **22**:303–316.
19. **Hardy, F.** 1960. *Cocoa Manual,* p. 350. Inter American Institute of Agricultural Science, Turrialba, Costa Rica.
20. **Hardy, F., and C. Rodrigues.** 1953. Quantitative variations in nitrogenous components of the cocoa bean: effects of genetic type and soil type, p. 89–91. *A Report on Cocoa Research 1945–1951.* The Imperial College of Tropical Agriculture, St. Augustine, Trinidad, B.W.I.
21. **Hodge, J.** 1953. Chemistry of browning reactions in model systems. *J. Agric. Food Chem.* **1**:928–943.
22. **Hoi, O. K.** 1977. Cocoa bean processing—a review. *Planter* (Kuala Lumpur) **53**:507–530.
23. **Jones, K. L., and S. E. Jones.** 1984. Fermentations involved in the production of cocoa, coffee and tea, p. 433–446. *In* M. E. Bushell (ed.), *Modern Applications of Traditional Biotechnologies.* Elsevier, New York.
24. **Knapp, A. W.** 1937. *Cocoa Fermentation. A Critical Survey of Its Scientific Aspects.* John Bale, Sons and Curnow Ltd., London.
25. **Lehrian, D. W.** 1989. Recent developments in the chemistry and technology of cocoa processing, p. 22–33. *In* Y. B. Che Man, M. N. B. Abdul Karim, and B. A. Amabi (ed.), *Food Processing Issues and Prospects.* Faculty of Food Science and Biotechnology, Universiti Pertanian, Malaysia.
26. **Lehrian, D. W., and G. R. Patterson.** 1983. Cocoa fermentation, p. 529–575. *In* H. J. Rehm and G. Reed (ed.), *Biotechnology,* vol. 5. Chemie-Verlag, Weinheim, Germany.
27. **Lopez, A. S.** 1972. The development of chocolate aroma from non-volatile precursors, p. 640–646. Fourth International Cocoa Research Conference. Government of Trinidad and Tobago.
28. **Lopez, A. S.** 1986. Chemical changes occurring during the processing of cocoa, p. 19–53. *In* P. S. Dimick (ed.), *Proceedings of the Cocoa Biotechnology Symposium.* Department of Food Science, The Pennsylvania State University, University Park, Pa.
29. **Lopez, A. S., and P. S. Dimick.** 1995. Cocoa fermentation, p. 562–577. *In* H. J. Rehm and G. Reed (ed.), *Biotechnology,* vol. 9. Chemie Verlag, Weinheim, Germany.
30. **Lopez, A. S., D. W. Lehrian, and L. V. Lehrian.** 1978. Optimum temperature and pH of invertase of the seeds of *Theobroma cocoa* L. *Rev. Theobromo* (Brazil) **8**:105–112.
31. **Lopez, A. S., and V. C. Quesnel.** 1971. An assessment of some claims relating to the production and composition of chocolate aroma. *Int. Choc. Rev.* **26**:19–24.
32. **Lopez, A. S., and V. C. Quesnel.** 1973. Volatile fatty acid production in cocoa fermentation and the effect on chocolate flavor. *J. Sci. Food Agric.* **24**:319–326.
33. **Lozano, A. R.** 1958. Mechanical pod breakers. *Cocoa* (Turrialba) **3(14)**:35.
34. **Martelli, H. L., and H. F. K. Dittmar.** 1961. Cocoa fermentation. V. Yeasts isolated from cocoa beans during the curing process. *Appl. Microbiol.* **9**:370–371.
35. **McDonald, C. R., R. A. Lass, and A. S. Lopez.** 1981. Cocoa drying—a review. *Cocoa Growers' Bull.* **31**:5–39.
36. **Menchu, J. F., and C. Rolz.** 1973. Coffee fermentation technology. *Thé Cafe Cocoa* **17**:53–61.
37. **Ostovar, K., and P. G. Keeney.** 1973. Isolation and characterization of microorganisms involved in the fermentation of Trinidad's cocoa beans. *J. Food Sci.* **38**:611–617.
38. **Passos, F. M. L., A. S. Lopez, and D. O. Silva.** 1984. Aeration and its influence on the microbial sequence in cocoa fermentations in Bahia with emphasis on lactic acid bacteria. *J. Food Sci.* **49**:1470–1474.
39. **Passos, F. M. L., D. O. Silva, A. S. Lopez, C. L. L. F. Ferreira, and W. V. Guimaraes.** 1984. Characterization and distribution of lactic acid bacteria from traditional cocoa bean fermentations in Bahia. *J. Food Sci.* **49**:205–208.
40. **Quesnel, V. C.** 1965. Agents inducing the death of the cocoa seeds during fermentation. *J. Sci. Food Agric.* **16**:441–447.
41. **Quesnel, V. C.** 1972. Biochemistry and processing, p. 48. *In Annual Report of Cocoa Research.* University of the West Indies, St. Augustine, Trinidad.
42. **Quesnel, V. C.** 1972. Cacao curing in retrospect and prospect, p. 602–606. Fourth International Cocoa Research Conference. Government of Trinidad and Tobago.
43. **Quesnel, V. C., and K. Jugmohunsingh.** 1970. Browning reaction in drying cocoa. *J. Sci. Food Agric.* **21**:537–541.
44. **Rodriquez, D. B., and H. A. Frank.** 1969. Acetaldehyde as a possible indicator of spoilage in coffee Kona (Hawaiian coffee). *J. Sci. Food Agric.* **20**:15–19.
45. **Roelofsen, P. A.** 1953. Polygalacturonase activity of yeast, *Neurospora* and tomato extract. *Biochim. Biophys. Acta* **10**:410–413.
46. **Roelofsen, P. A.** 1958. Fermentation, drying, and storage of cocoa beans. *Adv. Food Res.* **8**:225–296.
47. **Rohan, T. A.** 1957. Cocoa preparation and quality. *West Afr. Cocoa Res. Inst.* **1956/57**:76–79.
48. **Rohan, T. A.** 1963. *Processing of Raw Cocoa from the Market.* FAO Agric. Studies no. 60, FAO, Rome.
49. **Rohan, T. A.** 1969. The flavor of chocolate; its precursors and a study of their reactions. *Gordian* **69**:443–447, 500–501, 542–544, 587–589.
50. **Rohan, T. A., and T. Stewart.** 1965. The precursors of chocolate aroma: the distribution of free amino acids in different commercial varieties of cocoa beans. *J. Food Sci.* **30**:416–419.

51. **Rohan, T. A., and T. S. Stewart.** 1967. The precursors of chocolate aroma: production of free amino acids during fermentation of cocoa beans. *J. Food Sci.* **32:**396–398.
52. **Rohan, T. A., and T. S. Stewart.** 1967. The precursors of chocolate aroma: production of reducing sugars during fermentation of cocoa beans. *J. Food Sci.* **32:**399–402.
53. **Rombouts, J. E.** 1952. Observations on the microflora of fermenting cocoa beans in Trinidad. *Trinidad Proc. Soc. Appl. Bacteriol.* **15:**103–111.
54. **Rombouts, J. E.** 1953. Critical review of the yeast species previously described from cocoa. *Trop. Agric.* (Trinidad) **30:**34–41.
55. **Saposhnikova, K.** 1952. Changes in the acidity and carbohydrates during ripening of the cocoa fruit: variations of acidity and weight of seeds during fermentation of cocoa in Venezuela. *Agron. Trop. Maracay* **2:**185–195.
56. **Schwan, R. F., A. S. Lopez, D. O. Silva, and M. C. D. Vanetti.** 1990. Influência da frequência e intervalos de revolvimentos sobre a fermentacao de cocoa e qualidade do chocolate. *Rev. Agratrop.* **2:**22–31.
57. **Schwan, R. F., A. H. Rose, and R. G. Board.** 1995. Microbial fermentation of cocoa beans, with emphasis on enzymatic degradation of the pulp. *J. Appl. Bacteriol.* **79:**96–107.
58. **Schwan, R. F., M. C. D. Vanetti, D. O. Silva, A. S. Lopez, and C. A. deMoraes.** 1986. Characterization and distribution of aerobic spore-forming bacteria from cocoa fermentations in Bahia. *J. Food Sci.* **51:**1583–1584.
59. **Sieki, K.** 1973. Chemical changes during cocoa fermentation using the tray method in Nigeria. *Rev. Int. Choc.* **28:**38–42.
60. **Sivetz, M., and N. Desrosier.** 1979. Harvesting and handling green coffee beans, p. 74–116. *In Coffee Technology.* AVI Publishing Company Inc., Westport, Conn.
61. **Suryakantha Raju, K., S. Vishveshwara, and C. S. Srinivasan.** 1978. Association of some characters with cup quality in *Coffea canephora* × *Coffea arabica* hybrids. *Indian Coffee* **42:**195–197.
62. **Van Pee, W., and J. M. Castelein.** 1972. Study of the pectinolytic microflora, particularly the Enterobacteriaceae, from fermenting coffee in the Congo. *J. Food Sci.* **37:**171–174.
63. **Vaughn, R. H., R. DeCamargo, H. Falanghe, G. Mello-Ayres, and A. Serzedello.** 1958. Observations on the microbiology of the coffee fermentation in Brazil. *Food Technol.* **12**(Suppl. 4):57.
64. **Voigt, J., and B. Beihl.** 1995. Precursors of the cocoa specific aroma components are derived from the vicilin class (7S) globulin of the cocoa seeds by proteolytic processing. *Bot. Acta* **108:**283–289.
65. **Voigt, J., D. Wrann, H. Heinrichs, and B. Biehl.** 1994. The proteolytic formation of essential cocoa-specific aroma precursors depends on particular chemical structures of the vicilin-class globulin of the cocoa seeds lacking in the globular storage proteins of coconuts, hazelnuts and sunflower seeds. *Food Chem.* **51:**197–205.
66. **Wadsworth, R. V.** 1955. The preparation of cocoa, p. 131–142. 1955 Cocoa Conference, London, U.K.
67. **Weissberger, W., T. E., Kavanagh, and P. G. Keeney.** 1971. Identification and quantification of several non-volatile organic acids of cocoa beans. *J. Food Sci.* **36:**877–879.
68. **Zak, D. K., K. Ostovar, and P. G. Keeney.** 1972. Implication of *Bacillus subtilis* in the synthesis of tetramethylpyrazine during fermentation of cocoa beans. *J. Food Sci.* **37:**967–968.

Iain Campbell

36

Beer

This chapter is a general overview of the scientific principles of the brewing industry. More detailed information on the science and technology of beer production is available in the textbooks by Lewis and Young (9) and Hough et al. (2, 7).

The legal definition of beer varies among countries. The strictest definition, as in Germany, limits the ingredients to hops, yeast, water, and malt, not necessarily of barley, although that cereal is understood if no other is specified. In many other countries either sugars or unmalted cereals are permitted as adjuncts, providing up to 30% of the fermentable sugar. Figure 36.1 shows a general outline of the brewing process.

Archeological evidence has shown that beer has been produced since at least 3000 BC, perhaps as a fortuitous discovery from the baking of bread. *Saccharomyces cerevisiae* is unable to ferment the starch of cereal grains, so a preliminary germination is required in which starch and protein are hydrolyzed enzymically to simple sugars and amino acids which provide the main nutrients for fermentation. This could have happened accidentally with moist grain, and naturally occurring fermentative yeasts would have produced a primitive beer. In subsequent developments over the millennia, barley became the principal cereal for beer production because of its husk, which provided an excellent natural filter for clarification of the extracted wort. However, oats, rye, sorghum, and wheat have been successfully malted and, occasionally alone but usually in combination with barley malt, are used in local beers in various parts of the world.

MALTING

The first stage of the brewing process is the production of malt, now almost always carried out by specialist maltsters. Not all barley is suitable for malting: one important property is the nitrogen content. If the level of nitrogen is too low, growth of *S. cerevisiae* is restricted, but normally the problem is too high a level of nitrogen in the barley, more than is necessary for yeast growth, with the resulting excess nitrogen encouraging microbial, particularly bacterial, spoilage of the final beer. Also, grain intended for malting must be carefully dried and stored to avoid risk of either dormancy or death of the barley (5, 11).

The malting process occurs in three stages: steeping, germination, and kilning. Steeping barley grain in water over 24 to 48 h, usually in two stages with an "air rest" between, stimulates the growth of the embryo plant, as would occur if it had been planted in soil. Germination is controlled by temperature, aeration, and humidity to the point where the stem is just about to emerge from the grain, by which time rootlets have already formed. In the traditional malting process, grain transferred from the steeping vessel is spread on the floor as a layer about 0.8 m deep and manually turned daily to prevent the developing rootlets from binding together over the

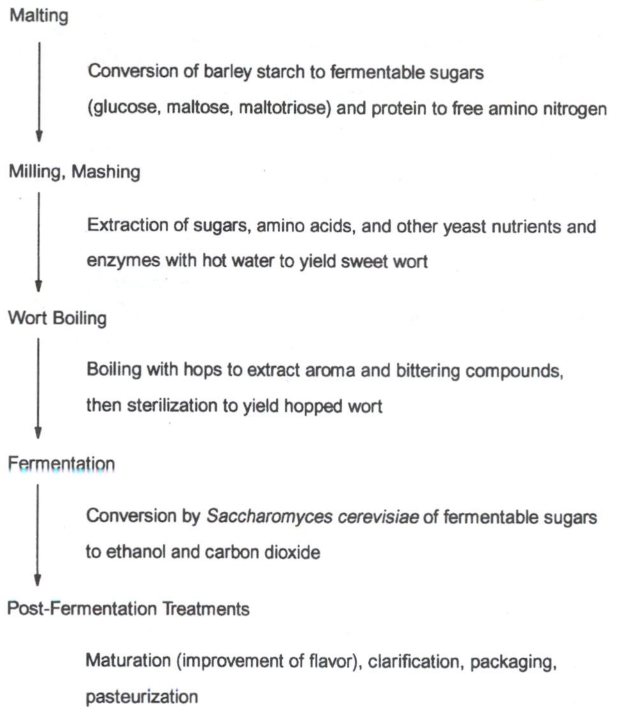

Malting

↓ Conversion of barley starch to fermentable sugars (glucose, maltose, maltotriose) and protein to free amino nitrogen

Milling, Mashing

↓ Extraction of sugars, amino acids, and other yeast nutrients and enzymes with hot water to yield sweet wort

Wort Boiling

↓ Boiling with hops to extract aroma and bittering compounds, then sterilization to yield hopped wort

Fermentation

↓ Conversion by *Saccharomyces cerevisiae* of fermentable sugars to ethanol and carbon dioxide

Post-Fermentation Treatments

Maturation (improvement of flavor), clarification, packaging, pasteurization

Figure 36.1 The brewing process.

2 weeks of germination. Since late in the 19th century, numerous designs of malting plants have been developed to speed up the process and reduce labor requirements (2, 5, 9).

At the end of germination the grain has a moisture content of about 50%, so drying to less than 6% moisture is required for storage. However, kilning is more complex than simply drying the grain. Malt is not only a source of fermentable sugars and other yeast nutrients. It also provides amylolytic and proteolytic enzymes for hydrolysis of any additional unmalted cereal in the formula. These enzymes are moderately heat resistant, but are increasingly inactivated by drying temperatures above 70°C. Also, higher temperatures give darker-colored malt as a result of "browning" reactions, first explained by Maillard (2, 9), between sugars and nitrogen components of the grain. For some beers, darker malts are desirable, but with the penalty of reduced enzymatic activity.

The production of malt is more the concern of the botanist and biochemist, but there are important microbiological aspects (4). First, as malt is a food product, high standards of hygiene are required. Although some bacterial contamination is inevitable, potentially the most troublesome contaminants are molds, some of which have public health implications or specific spoilage effects. Examples include the production of mycotoxins or, in the case of some species of *Fusarium*, of a polypeptide "gushing factor" which, by creating nuclei for development of carbon dioxide bubbles, causes violent frothing of the beer when the bottle or can is opened. However, since growth of any mold on barley gives an obvious weathered appearance and moldy aroma which persists in the final beer, simple sensory assessment is sufficient for the maltster to decide on acceptance or rejection of a barley sample. This is fortunate, since it is impracticable to carry out any form of microbiological analysis while the truck is waiting to unload the barley at the malting plant.

MASHING AND WORT PRODUCTION

Wort is the sugary solution prepared from malt, and other unmalted cereal if appropriate, by milling or crushing the grain and extracting the resulting grist with warm water. Details of the process vary among breweries, and it is impracticable to deal with all possibilities here (2, 9).

In milling, although the contents of the grains must be ground finely to maximize the yield of fermentable extract, it is important that the outermost layer, the husk, is only cracked rather than ground to powder. In a later stage of the process the husk functions as a natural filter. Barley has become the dominant cereal of the brewing industry largely because its husk structure is ideal for this purpose.

Infusion mashing is associated particularly with British ales and similar beers of Belgium and northern Germany. At its simplest, the mash tun has a floating bed of grist, suspended on entrapped air, with wort drawn off at the base and continuously replaced by a top spray of warm sparge water until further extraction of sugar is impracticable. The traditional decoction mashing process for Bavarian and Czech beers has origins predating the invention of the thermometer. Consistent brewing requires reproducible conditions, and consistent temperatures were achieved by mixing the grist with a measured volume of well water, heating a measured proportion to boiling, and returning the boiled slurry to the mash tun. Well water is of constant temperature throughout the year and, at a given location, water always boils at the same temperature, so with the same volumes each time, mashing temperatures would always be the same. Experience showed that several steps of increasing temperature were required, which we now know are the temperature optima for activity of α- and β-amylases and protease of the malt. This was achieved by a succession of accurately measured decoctions. Although double- or triple-decoction mashing undoubtedly produces a good-quality wort, the process is expensive and now is used only for prestigious products, for which the traditional process has sales appeal. Finally, since all cereal solids must be removed before the next stage of the process, the wort is clarified by running the grist slurry into a lauter tun (German: lauter = clear, pure) with a slotted base on which the bed of husk material functions as a filter.

The majority of modern breweries use a mashing vessel where the grist is mixed with warm water, heated in steps through the optimum temperatures or various mashing enzymes, and finally filtered in the lauter tun. When unmalted cereal forms part of the recipe, the ground cereal is heated in a cereal cooker, at least to the gelatinization temperature of the starch of the cereal, but often to boiling, and then transferred to the mashing vessel to allow hydrolysis of its starch and protein by the malt enzymes.

HOPS AND WORT BOILING

Many different herbs have been used in beer throughout brewing history, but hops have been the preferred flavoring for the past five centuries. The hop plant, *Humulus lupulus,* requires a temperate climate, between 35° and 55° north or south of the equator, but in practice its cultivation is chiefly restricted to certain areas, e.g., the county of Kent in Britain and Washington and Oregon in the United States. Hop varieties in common use vary in their content of bitter acids, resins, and oils, which contribute to beer flavor and aroma, and whether they contain seeds. Many brewers believe that hop seeds impart a harsh flavor to the beer.

In appearance, hop cones resemble those of pine but are smaller and have a softer texture. Cone hops are still used by some traditional or specialty breweries, but the majority of breweries now use processed hops, either pellets prepared from ground cone material or hop extract. Modern hop processing technology uses liquid carbon dioxide as solvent to avoid potential problems of residual organic phase. In whatever form, the hops are added to the wort and boiled for 50 to 90 min, according to the practice of a particular brewery. Boiling has the incidental advantages of sterilizing and concentrating the wort and purging the wort of harsh grainy flavors.

To the brewer, the most important hop components are the flavor compounds, made up of resins, tannins, and essential oils. Typically, these constitute approximately 15, 4, and 0.5%, respectively, of the weight of mature hop cones. Bitterness of beer is due to the α-acid (humulone) and β-acid (lupulone) components of the resins, but not directly; these complex acids are isomerized during hop boiling to the bitter iso-α- and iso-β-acids, isohumulone and hulupone, respectively (7, 9, 18). Within these general groups of isoacids are numerous analogs according to the acyl side chains on the resin molecule. Essential oils contribute aroma to the beer, but most of the oil is lost during boiling. Therefore, aroma hops, as distinct from high-α-acid bitterness hops, are added late in the boil or even during fermentation or conditioning to ensure maximum aroma effect. Also, if glucose, sucrose, or maltose crystals or syrup are included in the formula, late in the boil is the appropriate stage for addition. Hop tannins have little direct influence on flavor, but react with malt protein during the boil to form hot break, a precipitate of protein and tannin complexes and insoluble calcium salts and phosphates. This must be removed before fermentation, since the particulate material would adversely affect fermentation and flavor development by *S. cerevisiae.*

After traditional boiling with cone hops, the wort is clarified by filtration through a settled bed of spent hops. Hop pellets or extract do not provide a suitable filter medium; with their use, wort must be clarified by centrifugation. Alternatively, in modern breweries the hopped wort is run tangentially into a circular tank called the whirlpool hop separator. Hop debris and hot break collect at the center of the resulting vortex, while clear wort is drawn from the side of the vessel.

Obviously, the wort, still at approximately 100°C, has to be cooled before "pitching" with yeast inoculum. Less obvious, but nevertheless essential, is the requirement to aerate the wort to 6 to 8 ppm dissolved oxygen. After cooling to about 20°C and aeration, the wort is ready for the fermentation stage of the process.

FERMENTATION

Brewing Yeast (*S. cerevisiae*)

Formerly, the actively fermenting yeasts used in the beverage fermentation industry—both culture yeasts and common contaminant "wild yeasts"—were classified as different *Saccharomyces* species, i.e., *S. cerevisiae* (ale yeast) and *S. carlsbergensis* (lager yeast) (8, 19). The wine yeasts *Saccharomyces bayanus, Saccharomyces ellipsoideus,* and *Saccharomyces uvarum* are also possible brewery contaminants, as is *Saccharomyces diastaticus,* an amylolytic yeast. All were included in a single species, *S. cerevisiae,* in the most recent classification of yeasts (18), but still it is convenient to distinguish the different types by their former species names.

Contrary to popular belief, *S. cerevisiae* is not a facultative anaerobe (15, 20). Although brewing yeast does change between oxidative and fermentative metabolism according to aerobic or anaerobic conditions, respectively, it cannot grow anaerobically indefinitely. As in all eukaryotic cells, cell membranes of *S. cerevisiae* contain unsaturated fatty acids and sterols which can be synthesized only under aerobic conditions. The amounts of unsaturated fatty acids and sterols from malt naturally present in wort are too low to support yeast growth. Therefore initial aeration of the wort is required to allow brewing yeast to synthesize these compounds.

"Pitching yeast" in satisfactory condition must be added in the correct amount to the wort in the fermentation vessel: usually to 1×10^7 to 2×10^7 cells per ml, although, on a production scale, measurement by weight is more convenient. In the course of a typical fermentation, the population of *S. cerevisiae* increases by a factor of about 8, i.e., only three cell divisions. Subsequent multiplication is inhibited by the lack of oxygen, unsaturated fatty acids, and sterols, and since aeration late in the fermentation would cause flavor problems, no further cell growth is possible.

Progress of Fermentation

Traditional fermentation vessels are open rectangular tanks of 2 to 3 m in depth (7). Such vessels are particularly useful for "top yeasts," i.e., those brewing strains which rise to the surface of the fermenting beer, from where they are skimmed off as the inoculum for the next fermentation. Since about 1970 the cylindroconical vessel or unitank has become the preferred type (10). This vessel is enclosed to reduce risk of contamination and facilitate recovery of carbon dioxide and has a conical lower section to facilitate recovery of "bottom yeast," which settles out late in the fermentation. The shape of the vessel also encourages a vigorous mixing of yeast cells in the fermenting wort, giving a fast fermentation.

Brewery fermentations cover the lag, logarithmic, and early stationary phases of the yeast growth curve (Fig. 36.2) (7, 9). Brewing is unusual in modern biotechnology in reusing the culture from the previous fermentation, although a new culture is prepared at regular intervals, e.g., after 15 successive fermentations. During the lag phase of 6 to 12 h, *S. cerevisiae* utilizes the dissolved oxygen in the wort to restore its unsaturated fatty acid and sterol supply and to adjust from the anaerobic, acidic, alcoholic conditions at the end of the previous

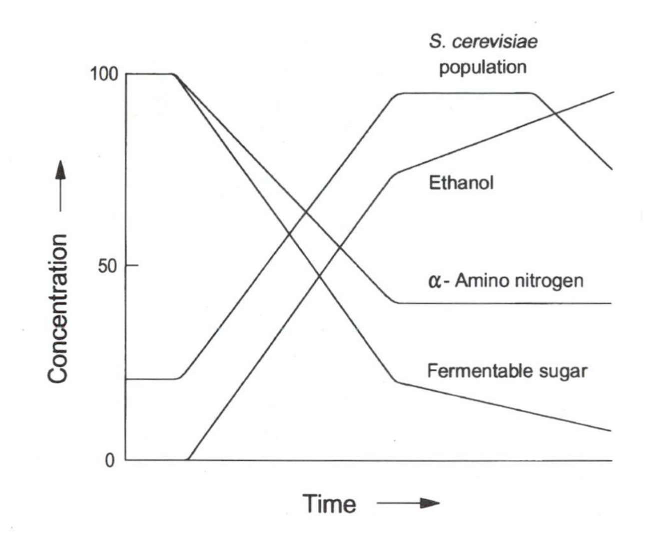

Figure 36.2 Theoretical progress of a typical brewery fermentation, showing changes in population of *S. cerevisiae* and concentration of sugar, amino nitrogen, and ethanol. The graph shows yeast cells in suspension and the start of settling of cells from the beer at the end of fermentation. The time axis is not calibrated, since fermentation rates differ widely between breweries. Other variables are shown as percentages of the initial or final value, expressed as 100%.

fermentation to the very different environment of fresh wort.

Yeast nutrients, including sugars, are taken up from the wort to generate new yeast cells; fermentable sugars also provide energy for this process. A wort of 10° Plato, i.e., equivalent to 10% sugar, has a specific gravity of 1.040 which is mainly due to dissolved sugars. Amino acids, vitamins, and inorganic salts account for less than 1%. Typically, the approximate sugar composition of such a wort would be 4% maltose, 2% glucose, and 2% maltotriose, with 2% higher dextrins which are not utilized by brewing yeast. However, the fermentable sugars are not utilized simultaneously. This is related to the mechanism of transport of the sugar into the cells. Most strains of *S. cerevisiae* transport and metabolize the glucose first, with maltose transport beginning only after most of the glucose has been metabolized. For a similar reason, utilization of maltotriose begins late in the fermentation, when most of the maltose has been consumed. Ethanol and carbon dioxide are the main products of yeast metabolism, but small amounts of other compounds make an important contribution to the flavor of the beer (15, 20).

By-products of the Embden-Meyerhof pathway, the major metabolic route in the anaerobic conditions of fermentation, contribute to flavor, but by-products of nitrogen metabolism of yeasts have a more important effect on flavor. *S. cerevisiae* has a limited number of

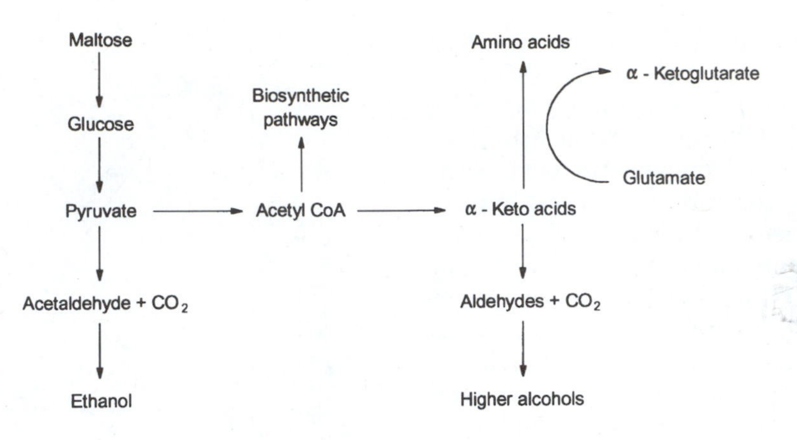

Figure 36.3 Formation of higher alcohols (fusel alcohols) as by-products of amino acid biosynthesis. Note the similarity of the reactions pyruvate → acetaldehyde → ethanol and α-keto acid → aldehyde → alcohol; both are important for redox balance under anaerobic conditions. Acetyl CoA, acetyl coenzyme A.

amino acid permeases; most amino acids have to be synthesized, since they are either not absorbed from the wort or not transported in adequate amount. Keto acids which are intermediates in the biosynthetic pathway may be converted to higher alcohols (Fig. 36.3). Also related to transport is the decrease in pH during fermentation, resulting from excretion of H^+ during uptake of nutrients. The pH of wort is normally 5.1 to 5.2 and decreases during fermentation to 3.8 to 4.0, mostly during the period of yeast growth. Note also in Fig. 36.2 that fermentation of sugars continues after yeast growth ceases, due to the continuing action of the relevant enzymes, but more slowly than before. However, without yeast growth and protein synthesis, there is no further requirement for amino nitrogen.

An important property of brewing yeasts is the ability to flocculate, i.e., spontaneously aggregate into clumps late in the fermentation. Too early flocculation stops the fermentation, as the yeast cells settle out of partly fermented beer, and nonflocculent yeasts have to be removed by filtration or centrifugation. The current explanation of flocculation is a lectin-like activity of the cell wall which develops late in the fermentation, but other factors are also involved, e.g., divalent ions, particularly Ca^{2+}, decreasing amounts of flocculation-inhibiting fermentable sugar, and increasing amounts of stimulatory ethanol (15, 16).

POST-FERMENTATION TREATMENTS

Conditioning

"Green beer" directly from the fermentation vessel contains acetaldehyde, diacetyl, and other unwanted by-products of yeast metabolism which must be removed. Although present in only small amounts, these compounds have a pronounced effect on the flavor, or more correctly, aroma of the beer. In traditional "cask-conditioning," the beer, still containing about 1% fermentable sugar, undergoes a secondary fermentation in casks and generates sufficient carbon dioxide for draft dispense. Also, the fermenting yeast absorbs unwanted flavor compounds. Similar changes are associated with low-temperature (0 to 2°C) secondary fermentation of pilsner and other lager beers (German: lager = store) over several months. It is now known that a low temperature is not essential for development of flavor. Beneficial changes in flavor of both ales and lagers require simply a few days' storage of the beer in contact with 10^5 to 10^6 yeast cells per ml at 15°C, followed by cooling to 0°C only long enough to precipitate "chill haze" material and allow yeast cells to sediment. It is important to avoid accidental exposure to air at this stage, since diacetyl is formed from acetolactate by spontaneous chemical reactions in the presence of oxygen, and yeast is no longer present to remove it.

Filtration

With the exception of cask-conditioned beers, which are clarified by addition of fining agents, it is usual practice now to filter beer to complete clarity. Various designs of filters are in common use, most using either cellulose fibers or powders, diatomite, or pumice as filter media. Such materials, which absorb microbial and inert haze-forming material in the depths of the filter, do not sterilize the beer, since sufficiently fine filters would cause unacceptably slow flow rates, but typically the yeast population is reduced to less than 10^2 cells per liter. At least 10^5 yeast cells per ml are required to form a visible haze.

Membrane filtration, which does sterilize beer, is becoming increasingly popular. Pasteurization is unnecessary for sterile-filtered beer, so membrane filtration avoids substantial energy cost and the potential flavor defects that result from heating the beer. Two filters are used in series. A rough prefilter is first used to remove as much particulate material as possible, and a second membrane filter is used to achieve sterilization. With this system, an acceptable flow rate and long filter life are possible.

Pasteurization and Packaging

Draft beer is pasteurized by passage through a heat exchanger before filling into clean sterilized kegs. Beer for sale in bottle or can is pasteurized after filling. In both systems the heat treatment is equal in terms of pasteurization units (1 PU = 60°C for 1 min), but the "tunnel" pasteurizer for bottles and cans uses a lower temperature (typically 60 to 62°C) for a longer time. Individual breweries have their own preferred pasteurization treatment. In theory, 5 PU is sufficient to kill the small number of *S. cerevisiae* cells likely to pass through a cellulose filter, but to eliminate the slightly more heat-resistant bacterial or wild yeast contaminants which could be present, up to 30 PU may be applied. Since increasing heat treatment could adversely affect flavor, choice of PU value is a compromise between potential risks of oxidized flavors and microbial spoilage (7, 9).

MICROBIAL CONTAMINANTS OF THE BREWING PROCESS

Beer production carries a risk of microbial contamination at all stages of the process. On barley, the most important contaminants are molds. The "field fungi" which develop during growth of barley are seldom a problem, but "storage fungi," principally *Aspergillus*, *Penicillium*, and *Fusarium* species, can have serious effects, e.g., forming mycotoxins or gushing factor polypeptide (4).

In the brewery itself, yeasts and bacteria are potentially the troublesome contaminants. In beer, its acid (usually about pH 3.9), alcoholic, anaerobic characteristics, with the additional antibacterial effects of carbon dioxide and hop acids and oils, restrict the range of spoilage microorganisms. Yeasts on grain are mainly aerobes which are unable to grow in fermenting wort or beer. The potential spoilage yeasts are species of the genera *Brettanomyces*, *Candida*, *Debaryomyces*, *Pichia*, *Saccharomyces*, *Torulaspora*, and *Zygosaccharomyces* (3). Presumably their natural habitat and original source are plants, but once established in a brewery they are difficult to eradicate and are often transferred from one fermentation to the next in the pitching yeast, in increasing numbers each time. The fermentative genera *Saccharomyces*, *Torulaspora*, and *Zygosaccharomyces* and their anamorphic form (*Candida* species) may contaminate the fermentation or persist through filtration and pasteurization to produce turbidity and off flavors in the beer. *S. cerevisiae* (5 to 7 μm diameter) is larger than most microbial contaminants, so filters designed to retain these yeasts are less efficient with contaminants. *Dekkera* and its anamorph *Brettanomyces* cause turbidity and acetic acid spoilage. *Debaryomyces*, *Pichia*, and *Candida* species are oxidative yeasts and so are limited to the early stages of fermentation, or to beer to which air has gained access post-fermentation. These yeasts cause turbidity and yeasty or estery off flavor and, in bottled or canned beer, often form a surface film fragmenting to flaky particles or deposit.

Lactic acid bacteria (*Lactobacillus* and *Pediococcus*) may cause beer spoilage, but only a few species are sufficiently resistant to the antibacterial properties of beer to grow at any stage of the process (12), causing turbidity and an off flavor often caused by diacetyl. Diacetyl is a strongly flavored minor by-product of their metabolism which is highly valued in the dairy industry but usually considered to be a spoilage defect of beer. Superattenuation by starch-fermenting lactobacilli and slime formation by pediococci are other possible defects. *Zymomonas* also grows at all stages, but its most usual effect is in beer, causing further fermentation with turbidity and off flavors and often "rotten apple" or dimethyl sulfide aroma.

Other spoilage bacteria are limited by their properties to specific stages of the brewing process (17). *Acetobacter* and *Gluconobacter* are strict aerobes and grow, with characteristic acetic acid production, only on beer accidentally exposed to air. Enterobacteria (including *Obesumbacterium*, the most important of that group in the brewery environment) cause turbidity and off flavors, often indole, phenols, diacetyl, hydrogen sulfide, and dimethyl sulfide, but grow well in the early stages of fermentation until inhibited by the decreasing pH and increasing ethanol content. Even so, these bacteria survive to be transferred in increased number with the yeast recovered for the next fermentation. Finally, *Megasphera* (cocci) and *Pectinatus* (rods) are recently discovered strictly anaerobic gram-negative bacteria which form acetic, butyric, and propionic acids, hydrogen sulfide, dimethyl sulfide, and turbidity and have become troublesome only because of modern advances in maintaining low dissolved oxygen levels in beer.

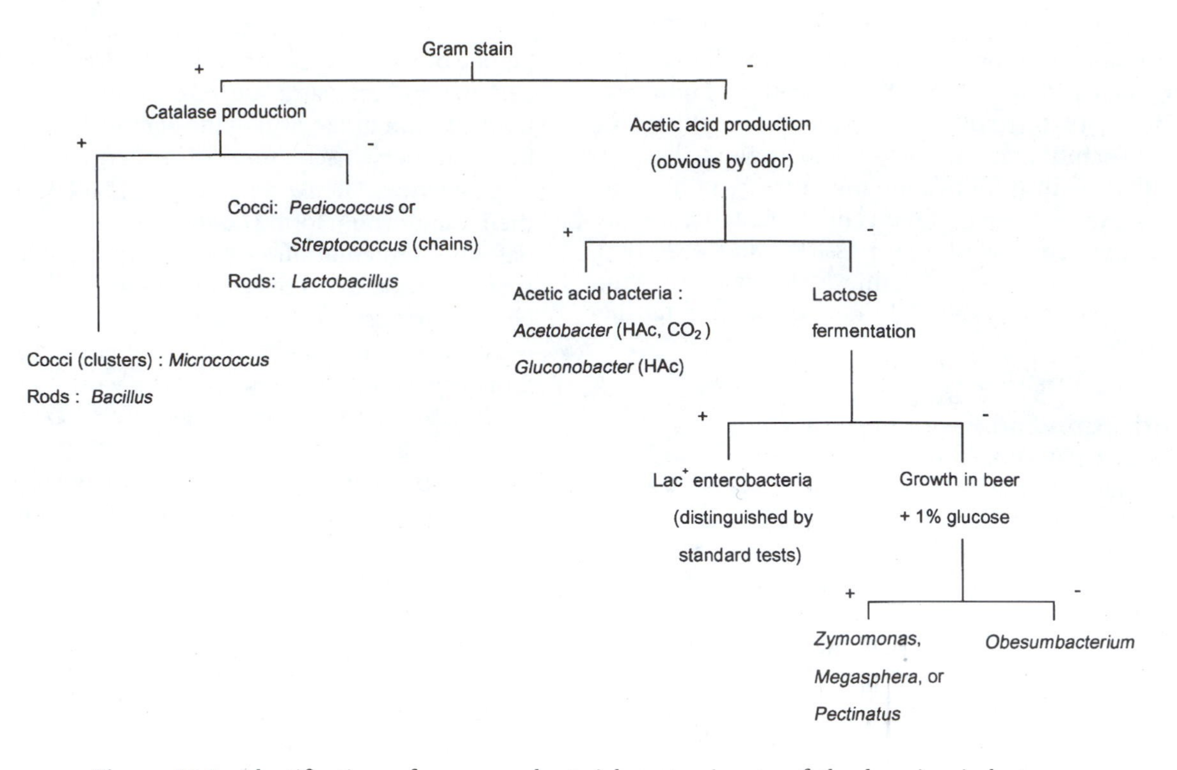

Figure 36.4 Identification of common bacterial contaminants of the brewing industry. *Micrococcus* and *Bacillus* may be present but are unlikely to grow in beer, and *Megasphera* and *Pectinatus* grow only under strictly anaerobic conditions.

Other bacteria are occasionally isolated, e.g., *Bacillus* and *Micrococcus* species, but are unable to grow in wort or beer. Their main nuisance is the further laboratory testing to confirm that they are not the more troublesome *Lactobacillus* or *Pediococcus,* which are similar in appearance (12). A simple distinguishing characteristic is that the lactic bacteria do not produce catalase (Fig. 36.4).

Isolation of Microbial Contaminants

Culture media for microbiological quality control can be either nonselective or selective. All brewing yeasts and bacteria should grow on nonselective media, usually either malt extract agar or Wallerstein Laboratories nutrient (WLN) agar, a semisynthetic medium of more consistent composition. These media are useful for examination, by plating or membrane filtration, of samples where no microorganisms should be found or should be present only in low numbers, e.g., in pasteurized beer or swabs or rinsings of recently cleaned equipment. Any microorganism, whether bacterium, culture yeast, or wild yeast, can be regarded as a contaminant in these situations.

Selective media are intended to suppress the growth of culture yeasts but allow the growth of any contaminants which may be present. Such media should normally be used to examine samples of pitching yeast or samples taken during or immediately after fermentation, where culture yeast is present. A commonly used selective medium is lysine agar, a synthetic medium composed of glucose, inorganic salts, vitamins, and L-lysine as the sole nitrogen source.

The most useful selective agent for detection or enumeration of bacterial contaminants is the antifungal antibiotic cycloheximide (actidione), which can be added to any suitable culture medium, but usually to WLN agar as "actidione agar." Some wild yeasts are sufficiently resistant to grow, but in general a more effective way to recover wild *Saccharomyces* contaminants relies on their sporulation. Sporulation, which is stimulated by starvation conditions, is presumably advantageous in the wild, but most brewing yeasts, after innumerable generations in a rich medium, have lost the ability to form viable spores. Yeast spores, unlike those of bacteria, are only marginally more heat resistant than their vegetative cells, but the difference is sufficient to recover contaminants on nonselective media after heat treatment. Practi-

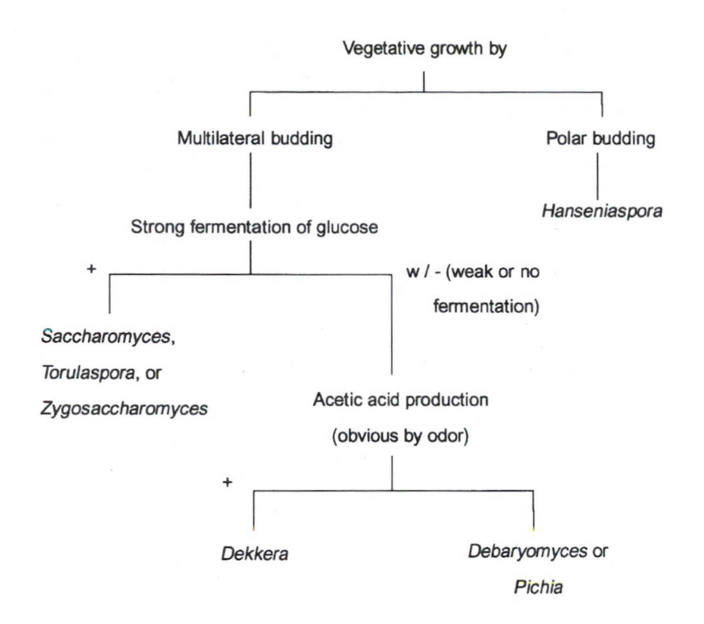

Figure 36.5 Identification of common yeast contaminants of the brewing industry. All genera listed are teleomorphic; i.e., they form spores. Anamorphic (nonsporing) forms of *Dekkera* and *Hanseniaspora* are *Brettanomyces* and *Kloeckera*, respectively. The anamorph for all other genera listed is *Candida*.

cal details of microbiological analyses are fully explained in analytical manuals (1).

In general, identification of recovered contaminants is unnecessary. Bacteria and non-*Saccharomyces* wild yeasts need be identified to genus only, as shown in Fig. 36.4 and 36.5. Only fermentative yeasts need further identification to determine if they are *Saccharomyces* species and, if so, *S. cerevisiae*, which ferments glucose, sucrose, maltose, and raffinose but not lactose. In this case even identification to species level is insufficient; it is important to determine whether the isolate is the culture yeast itself, possibly persisting on the surface of an improperly cleaned fermentor or other equipment or surviving through filtration and pasteurization. Remedial action to that situation would obviously be different from the discovery of a wild yeast strain as a contaminant. A quick and simple test to distinguish industrial yeast strains is their reaction to a range of antifungal compounds on a "multidisk" as used in medical mycology. Alternatively, a small-scale fermentation can distinguish the culture yeast from others, but the delay in obtaining useful results may be unacceptably long. With recent advances in genetic methods, DNA fingerprinting by analysis of DNA fragments is a possible rapid method to distinguish different strains, but at present is not sufficiently developed for routine use (6, 13).

Sterilization

In large-scale fermentation processes, steam is the usual method of sterilization. In the brewery, however, chemical sterilants are preferred, mainly because they can be used at ambient temperature and avoid attemperation problems in adjacent working fermentors (14). Modern breweries use automatic in-place cleaning and sterilization equipment to remove deposits of yeast and organic soil derived from beer foam by a powerful jet of water containing detergent, followed by a spray of chemical sterilant, and finally by a spray of sterile rinse water. Until recently, caustic sterilants containing 2% NaOH and additives were widely used, but acid sterilants based on phosphoric or peracetic acids have now become popular, not least because they are unaffected by carbon dioxide and can be used in closed vessels without the long delay and expense of draining off valuable gas. Chlorine-based sterilants are also used, but not widely because of the possibility of flavor problems. Residual chlorine reacts with unknown compounds in beer to produce a strong medicinal flavor. Quaternary ammonium compounds and biguanides are also regarded with suspicion. These compounds are favorably regarded in other food industries because of their protective persistence on treated surfaces, even after rinsing, but this is unacceptable in the brewing industry because of the possible adverse effects of residual sterilant on head retention in beer.

MODERN DEVELOPMENTS

For most of its history, brewing has been either a domestic enterprise or associated with monasteries. Developments over the past 300 years have coincided with increasingly large-scale commercial production and with general improvements in scientific knowledge and technology. Research in brewing industry has improved our understanding of the processes involved, and better-quality malts and beers can be produced more rapidly and efficiently. For example, at the start of this century it took at least 14 days to produce a batch of malt, mashing required up to 8 h, ale and lager fermentations could last 7 and 14 days, respectively, and subsequent lager maturation many weeks longer. By 1980 these times had been halved, with improvement in consistency and quality, although some traditionalist consumers might disagree.

New products offer possibilities for increasing market share. One recent example is "diet beer" of lower carbohydrate content and, by implication, lower calorie content. Up to 20% of the carbohydrate of wort can be dextrins, which natural *S. cerevisiae* does not utilize and thus remain in the beer. Addition of hydrolytic enzymes is unlikely to be permitted, but dextrin-fermenting yeasts can legally be used to ferment that carbohydrate. One possible source of such strains is hybridization between *S. diastaticus* and brewing yeasts, and successful results have indeed been achieved in that way. Recent genetic engineering research has also produced potentially useful diastatic yeasts, but at present commercial brewers are reluctant to risk public disapproval of their use. Another obvious use of diastatic yeasts is to produce beers of higher alcohol content, by fermentation of the dextrins (6, 15). In "ice beer," the alcohol content is increased by selectively freezing out part of the water content of the beer, with the incidental benefit of improved flavor and stability due to the chilling process.

There is no doubt that yet more novel fermentation methods and products will be developed in the brewing industry in the future. We can also be sure that there will be a continuing demand for traditional beers. Small breweries specializing in such products are certainly enjoying much success at the present time.

References

1. **Baker, C. D. (ed.).** 1991. *Recommended Methods of Analysis.* Institute of Brewing, London.
2. **Briggs, D., J. S. Hough, R. Stevens, and T. W. Young.** 1981. *Malting and Brewing Science,* 2nd ed., vol. 1, *Malt and Sweet Wort.* Chapman and Hall, London.
3. **Campbell, I.** 1996. Wild yeasts in brewing and distilling, p. 193–208. *In* F. G. Priest and I. Campbell (ed.), *Brewing Microbiology,* 2nd ed. Chapman and Hall, London.
4. **Flannigan, B.** 1996. The microflora of barley and malt, p. 83–125. *In* F. G. Priest and I. Campbell (ed.), *Brewing Microbiology,* 2nd ed. Chapman and Hall, London.
5. **Gibson, G.** 1989. Malting plant technology, p. 279–325. *In* G. H. Palmer (ed.), *Cereal Science and Technology.* Aberdeen University Press, Aberdeen, U.K.
6. **Hammond, J. R. M.** 1996. Yeast genetics, p. 43–82. *In* F. G. Priest and I. Campbell (ed.), *Brewing Microbiology,* 2nd ed. Chapman and Hall, London.
7. **Hough, J. S., D. Briggs, R. Stevens, and T. W. Young.** 1982. *Malting and Brewing Science,* 2nd ed., vol. 2, *Hopped Wort and Beer.* Chapman and Hall, London.
8. **Kreger-van Rij, N. J. W.** 1987. Classification of yeasts, p. 5–61. *In* A. H. Rose and J. S. Harrison (ed.), *The Yeasts,* 2nd ed., vol. 1. Academic Press, London.
9. **Lewis, M. J., and T. W. Young.** 1995. *Brewing.* Chapman and Hall, London.
10. **Maule, D. R.** 1986. A century of fermenter design. J. Inst. Brewing **92:**137–145.
11. **Palmer, G. H.** 1989. Cereals in malting and brewing, p. 61–242. *In* G. H. Palmer (ed.), *Cereal Science and Technology.* Aberdeen University Press, Aberdeen, U.K.
12. **Priest, F. G.** 1996. Gram-positive brewery bacteria, p. 127–161. *In* F. G. Priest and I. Campbell (ed.), *Brewing Microbiology,* 2nd ed. Chapman and Hall, London.
13. **Schofield, M. A., S. M. Rowe, J. R. M. Hammond, S. W. Molzahn, and D. R. Quain.** 1995. Differentiation of yeast strains by DNA fingerprinting. *J. Inst. Brewing* **101:**75–78.
14. **Singh, M., and J. Fisher.** 1996. Cleaning and disinfection in the brewing industry, p. 271–300. *In* F. G. Priest and I. Campbell (ed.), *Brewing Microbiology,* 2nd ed. Chapman and Hall, London.
15. **Stewart, G. G., and I. Russell.** 1986. One hundred years of yeast research and development in the brewing industry. *J. Inst. Brewing* **92:**537–558.
16. **Stratford, M.** 1992. Yeast floccoulation: a new perspective. *Adv. Microb. Physiol.* **33:**1–71.
17. **Van Vuuren, H. J. J.** 1996. Gram-negative spoilage bacteria, p. 163–191. *In* F. G. Priest and I. Campbell (ed.), *Brewing Microbiology,* 2nd ed. Chapman and Hall, London.
18. **Verzele, M.** 1986. 100 years of hop chemistry and its relevance to brewing. *J. Inst. Brewing* **92:**32–48.
19. **Yarrow, D.** 1984. Genus 22. *Saccharomyces* Meyen ex Reess, p. 379–395. *In* N. J. W. Kreger-van Rij (ed.), *The Yeasts, a Taxonomic Study,* 3rd ed. Elsevier, Amsterdam.
20. **Young, T. W.** 1996. The biochemistry and physiology of yeast growth, p. 13–42. *In* F. G. Priest and I. Campbell (ed.), *Brewing Microbiology,* 2nd ed. Chapman and Hall, London.

Graham H. Fleet

37 Wine

Winemaking is a bioprocess that has its origins in antiquity. Scientific understanding of the process commenced with the studies of Louis Pasteur, who demonstrated that wines were the product of an alcoholic fermentation of grape juice by yeasts (1). Since the time of Pasteur, winemaking has developed into a modern, multinational industry with a strong research and development base in the disciplines of viticulture and enology. Viticulture concerns the study of grapes and grape cultivation, while enology covers postharvest processing of the grapes from crushing through fermentation to packaging and retailing of the wine.

Microorganisms are fundamental to the winemaking process. To understand their contribution, it is necessary to know the taxonomic identity of each species associated with the process, the kinetics of growth and survival of each species throughout the process, the biochemical activities of these species and how such activities determine the physicochemical properties of the wine, the influence of winemaking practices upon microbial growth and activity, and the combined impact of microbial action and process influences on sensory quality and consumer acceptability of the wine. This chapter will focus on the occurrence, growth, and significance of microorganisms in winemaking. It covers wines produced only from grapes and includes table wines, sparkling wines, and fortified wines.

THE PROCESS OF WINEMAKING

Details of the process of winemaking are beyond the scope of this chapter and are described elsewhere (11, 98). Figure 37.1 outlines the main steps in the production of white and red table wines.

Grapes

Numerous grape varieties are used in winemaking, with the particular variety determining the fruity or floral characteristics of the final product. Some main varieties used in white winemaking are Riesling, Traminer, Muller-Thurgau, Chardonnay, Semillon, and Sauvignon Blanc, while those used in red winemaking include Cabernet Sauvignon, Merlot, Cabernet Franc, Pinot Noir, Shiraz, Gamay, Grenache, and Barbera. The grapes are harvested at an appropriate stage of maturity which determines the chemical composition of the juice extracted from them. Particularly important are the concentrations of sugars and acids which are the major constituents of the juice (Table 37.1) and have an important impact on its fermentation properties. Other preharvest conditions that affect the chemical composition of the grape and its juice include climate, sunlight exposure, soil, use of fertilizers, availability of water, vine age, and use of fungicides and insecticides. Traditionally, grapes were harvested manually but now there is increasing use of mechanical harvesters which are often oper-

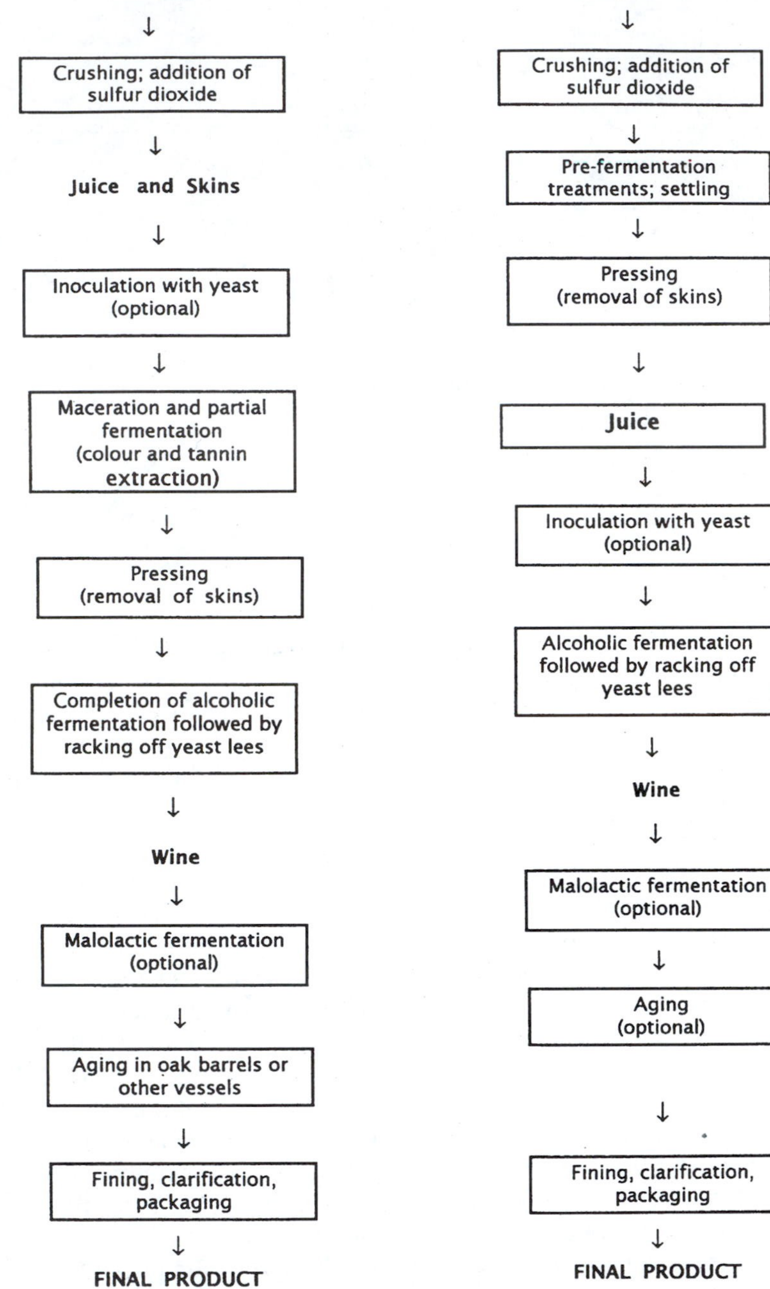

Figure 37.1 Outline of processes for making red and white wines.

Table 37.1 Main components of grape juice[a]

Substance	Concn
Glucose	75–150 mg/ml
Fructose	75–150 mg/ml
Pentose sugars	0.8–2 mg/ml
Pectin	0.1–1 mg/ml
Tartaric acid	2–10 mg/ml
Malic acid	1–8 mg/ml
Citric acid	0.1–0.5 mg/ml
Ammonia	5–150 μg/ml
Amino acids (total)	150–2,500 μg/ml
Protein	10–100 μg/ml
Vitamins (varies with vitamin)	μg–mg/ml
Anthocyanins	0.5 mg/ml
Flavonoids, non-flavonoids	0.1–1.0 mg/ml

[a]Data obtained from references 4, 11, 62, 64, and 98.

ated at night to minimize the temperature of the berries at the time of crushing.

Crushing and Prefermentation Treatments

For white wines, the grapes are mechanically destemmed and crushed, and the juice is drained away from the skins. If required, clarification of the juice is done by cold settling, filtration, centrifugation, or combinations of these methods. Cold settling is generally done at 5 to 10°C for 24 to 48 h with the addition of pectolytic enzymes to help break down grape material. The juice is then transferred to the fermentation tank, where the fermentation may commence naturally or may be initiated by inoculation with selected yeasts (34).

Red grapes are mechanically destemmed and crushed, and the juice plus skins (must) are directly transferred to the fermentation tank. Fermentation begins either naturally or after inoculation, and during the first few days the skins rise to the top of the juice to form a cap. Throughout this early stage, often described as maceration, juice is regularly pumped over the cap. The purpose of this step is to extract purple and red anthocyanin pigments, as well as other phenolic substances, from the grape skins to give color, tannic, and astringent character to the wine. The extraction process is assisted by the production of ethanol during this preliminary fermentation. When sufficient extraction has been achieved, the partially fermented wine is drained and pressed from the skins to another tank for completion of the fermentation. Some variations to this process include thermovinification and carbonic maceration. In thermovinification, the juice plus skins are heated to 45 to 55°C with pumping over to accelerate color and tannin extraction, after which the juice is separated from the skins and transferred to the fermentation tank. In carbonic maceration, uncrushed grapes are placed in a tank which is gassed with carbon dioxide to remove oxygen. The temperature is maintained at 25 to 35°C for several days, during which the grapes undergo endogenous metabolism that extracts color and phenolics from the skin. After about 8 to 10 days, the grapes are pressed to yield a partially fermented juice (1 to 1.5% ethanol) that is transferred to a tank for subsequent fermentation (10, 11).

Other pretreatments to the juice or must include the adjustment of pH and sugar concentration (where permitted), addition of diammonium phosphate or other nutrients to assist yeast growth, addition of sulfur dioxide (50 to 75 μg/ml) as an antioxidant, antimicrobial agent, and inhibitor of phenol oxidase activity, and addition of ascorbic acid or erythorbic acid as an antioxidant.

Alcoholic Fermentation

Traditionally, fermentation of the juice was conducted in large wooden barrels or concrete tanks, but most modern wineries now use relatively sophisticated stainless steel tanks with facilities for temperature control (principally cooling), cleaning in place, and other features for process management (11, 28). White wines are generally fermented at 10 to 18°C for 7 to 14 days or more, where the lower temperature and slower fermentation rate favor the retention of desirable volatile flavor compounds. Red wines are fermented for about 7 days at 20 to 30°C, where the higher temperature is necessary to extract color from the grape skins.

The alcoholic fermentation can be conducted either as a "natural" fermentation or as an induced, so-called "pure culture" fermentation. With natural fermentation, yeasts resident in the grape juice initiate and complete the fermentation. With pure culture fermentation, selected strains of yeasts, generally species of *Saccharomyces cerevisiae*, are inoculated into the juice at initial populations of 10^6 to 10^7 cells per ml. Such yeasts have been commercially available as active dry preparations for the last 30 to 40 years and are now used extensively throughout the world, especially in the newer wine-producing countries of the United States, Australia, and South Africa (26). The advantages and disadvantages of natural and pure culture fermentations have been well discussed (1, 4, 60, 101, 102). Essentially, the pure culture approach gives a more rapid and more predictable fermentation, while the natural approach has a more varied outcome, with the potential of failures but with the prospect of wines with more interesting character due to contributions from a greater range of yeast species.

Alcoholic fermentation is considered complete when the fermentable sugars, glucose and fructose, of the juice are completely utilized. The wine is then drained or pumped (racked) from the sediment of yeast and grape material (lees) and transferred to stainless steel tanks or wooden barrels for malolactic fermentation, if desired, and aging. Clarification by filtration or centrifugation may be done at this stage. Leaving the wine in contact with the lees for long periods is not encouraged because the yeast cells autolyze, with the potential of adversely affecting wine flavor and providing nutrients for the subsequent growth of spoilage bacteria.

Malolactic Fermentation

It has been known since the early 1900s that, after alcoholic fermentation, wines frequently undergo another fermentation which has been termed the malolactic fermentation (11, 23, 49, 131). The malolactic fermentation generally commences naturally about 2 to 3 weeks after completion of the alcoholic fermentation and lasts about 2 to 4 weeks. Lactic acid bacteria resident in the wine are responsible for the malolactic fermentation, but many winemakers now choose to encourage this reaction by inoculation of commercial cultures of *Leuconostoc oenos*. The main reaction is decarboxylation of L-malic acid to L-lactic acid, giving a decrease in acidity of the wine and an increase in its pH by about 0.3 to 0.5 unit. Wines produced from grapes cultivated in cool climates tend to have higher concentrations of malic acid, which can mask the varietal character of the wine. A decrease in acidity by malolactic fermentation gives a wine with a softer and more mellow taste. Moreover, there is an increasing view that growth of malolactic bacteria in wine contributes additional metabolites that confer complex and interesting flavor characteristics. Apart from flavor considerations, there are practical reasons for having wines complete malolactic fermentation. Wines that have not undergone malolactic fermentation before bottling risk this reaction occurring at some later stage in the bottle. If this happens, the wine becomes gassy and cloudy and is considered to be spoiled. There is also a view that wines with completed malolactic fermentation have greater microbiological stability and are less prone to spoilage by other species of lactic acid bacteria. After malolactic fermentation by *Ln. oenos*, it is believed that fewer nutrients are available for microbial growth and that bacteriocin production may be a further inhibitory factor.

Malolactic fermentation is not necessarily beneficial to all wines. Wines produced from grapes grown in warmer climates tend to be less acid (pH >3.5) and further reduction in acidity by malolactic fermentation is deleterious to overall sensory balance; also, it increases the pH to values where spoilage bacteria are more likely to grow. However, preventing the natural occurrence of malolactic fermentation in these wines (as might occur after bottling) is an extra technical burden. Consequently, many winemakers prefer to encourage the malolactic fermentation and later adjust wine acidity, if necessary. Nevertheless, there are winemakers who prefer not to have the malolactic fermentation occur in their wines.

Postfermentation Processes

Most white wines are not stored for lengthy periods after completion of the alcoholic fermentation or malolactic fermentation. If storage is necessary, it is generally done in stainless steel tanks. Some white wines (e.g., Chardonnay) may be aged in wooden barrels. Most red wines are aged for periods of 1 to 2 years by storage in wooden (generally oak) barrels. During this time, chemical reactions that contribute to flavor development occur between wine constituents and components extracted from the wood of the barrels. Critical points for control during storage and aging are exclusion of oxygen and addition of sulfur dioxide to free levels of 20 to 25 μg/ml. These controls are necessary to prevent the growth of spoilage bacteria and yeasts and to prevent unwanted oxidation reactions.

Just before bottling, the wines may be cold stored at 5 to 10°C to precipitate excess tartrate and then clarified by application of one or more processes which include addition of fining agents (bentonite, albumen, isinglass, gelatin), centrifugation, pad filtration, and membrane filtration. For some white wines with residual sugar, potassium sorbate up to 100 to 200 μg/ml may be added to control yeast growth (34, 98).

Wine Flavor

The distinctive flavors of wine originate from the grape, as raw material, and the processing operations which include alcoholic fermentation, malolactic fermentation, and aging (19, 86, 113). The grapes contribute trace amounts of many volatile components (e.g., terpenes) that give the wine its distinctive varietal, fruity character. In addition, they contribute nonvolatile acids (tartaric and malic), which affect flavor, and tannins (flavonoid phenols) that give bitterness and astringency. The fermentation steps, especially alcoholic fermentation, increase the chemical and flavor complexity by assisting extraction of compounds from the grapes, modifying some grape-derived substances, and producing a vast array of volatile and nonvolatile metabolic end products. Further chemical alterations occur during aging; also, enzymes derived from the grapes and excreted by yeasts and malolactic bacteria, as well as those added at prefermentation, might be expected to participate in chemical-flavor transformations. Thus, the final flavor represents contributions from many compounds and cannot be attributed to any one "impact substance."

YEASTS

Yeasts are significant in winemaking because they carry out the alcoholic fermentation, they may cause spoilage of the wine, and their autolytic products may affect sensory quality and influence the growth of malolactic and spoilage bacteria. Procedures for their isolation, enumeration, and identification have been reviewed (38).

Origin

Wine yeasts originate from any of three sources, namely, the surfaces of grapes, the surfaces of winery equipment (crushers, presses, fermenters, tanks, pipes, pumps, barrels, filtration units), and inoculum cultures.

The main yeasts found on mature, sound grapes are species of *Kloeckera* and *Hanseniaspora*, with lesser representations from species of *Candida*, *Metschnikowia*, *Cryptococcus*, *Pichia*, *Kluyveromyces*, and *Hansenula*. *S. cerevisiae*, the principal wine yeast, occurs at very low populations (<50 CFU/ml) on sound grapes and is rarely isolated from this source by direct plating methods. Freshly crushed grapes generally yield a must or juice with a yeast population of 10^3 to 10^5 CFU/ml, of which *Kloeckera* and *Hanseniaspora* species make up 50 to 70%. However, numerous factors affect the total yeast population and relative proportions of individual species on grapes. These include temperature, rainfall, and other climatic conditions, degree of maturity at harvest, physical damage due to mold or insect or bird attack, and grape variety (9, 39, 62, 77).

The surfaces of winery equipment that come into contact with the grape juice and wine are locations for development of the so-called residential or winery yeast flora. The extent of this development depends upon the nature of the surface and the effectiveness of cleaning and sanitizing operations. *S. cerevisiae* is prevalent on these surfaces, which are considered to be the main source of this species in wine fermentations. Such sources are likely to harbor multiple strains of *S. cerevisiae* that have accumulated from the grapes and starter cultures used in previous vintages (39, 77).

Starter cultures, if used to inoculate the juice, will be a principal source of yeasts. Presently, various strains of *S. cerevisiae* are used, but the future could see the development and use of other species. Commercial yeast preparations used for inoculation are not necessarily pure and could contain a proportion of species other than *S. cerevisiae* (26, 97, 101).

Growth during Fermentation

Over the last 100 years, many studies have described the yeast populations that grow during alcoholic fermentation. Early studies were essentially qualitative descriptions of the main species isolated from different stages (early, mid, final) of fermentation. Subsequent studies followed the growth of individual species throughout the entire course of fermentation. Recently, molecular techniques have made it possible to follow the development of particular strains throughout fermentation. Virtually all ecological studies to date have shown that *S. cerevisiae* is the principal wine yeast. Invariably, this species predominates during the middle to final stages of fermentation, but there are important contributions of other yeast species that must be considered in conjunction with the influences of winemaking practices upon yeast growth (4, 9, 39, 63, 64).

Figure 37.2 gives a general representation of the growth of yeasts during the fermentation of grape juice, whether it be conducted by natural or inoculated processes. As mentioned already, freshly extracted grape

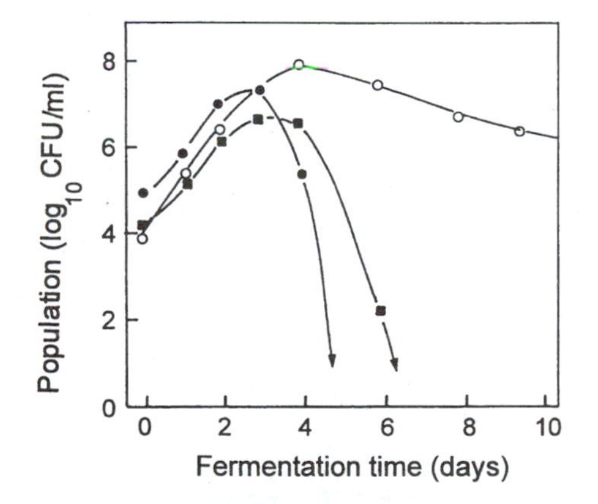

Figure 37.2 Generalized growth of yeast species during alcoholic fermentation of wine. ○, *S. cerevisiae*; ●, *Kloeckera* and *Hanseniaspora* species; ■, *Candida* species. Variations will occur in the initial and maximum populations for each species; for fermentations inoculated with *S. cerevisiae*, the initial population is approximately 10^6 CFU/ml (39).

juice harbors a yeast population of 10^3 to 10^5 CFU/ml, made up mostly of *Kloeckera* and *Hanseniaspora* species, but species of *Candida, Metschnikowia, Pichia, Hansenula, Kluyveromyces,* and *Rhodotorula* also occur. These species are often referred to as the non-*Saccharomyces* yeasts or wild flora. The juice will also contain low populations of *S. cerevisiae,* depending on the extent of contamination during crushing and juice pretreatment. Fermentation is initiated by the growth of various species of *Kloeckera, Hanseniaspora, Candida,* and *Metschnikowia* (e.g., *Kloeckera apiculata, Hanseniaspora valbyensis, Candida stellata, Candida colliculosa, Metschnikowia pulcherrima*) as well as *S. cerevisiae.* The growth of the non-*Saccharomyces* species is generally limited to the first 2 to 4 days of fermentation, after which they die off. Nevertheless, they achieve maximum populations of 10^6 to 10^7 CFU/ml before death, and such growth would be metabolically significant in terms of substrates utilized (hence not available to *S. cerevisiae*) and end products released into the wine. Also, the dead cells become part of the yeast pool for subsequent autolysis. Their death is attributed to an inability to tolerate the increasing concentrations of ethanol, which is largely produced by *S. cerevisiae.* After 4 days, the fermentation is continued and completed by *S. cerevisiae,* which reaches final populations of about 10^8 CFU/ml. Wine strains of *S. cerevisiae* are particularly noted for their strong production and tolerance of ethanol (up to 15% [vol/vol] or more). By comparison, strains of *K. apiculata* or *C. stellata* rarely tolerate ethanol concentrations greater than 5% or 8%, respectively, at 20 to 25°C (40).

Recent applications of molecular techniques, such as PCR, karyotyping by pulsed-field gel electrophoresis, and restriction endonuclease analysis of mitochondrial DNA, to the ecological study of wine yeasts have demonstrated that different strains of the one species may contribute to the fermentation (68, 85, 95a, 130). Different phases (early, mid, final) of the fermentation may be dominated by different strains of *S. cerevisiae,* and strain variation can also occur within the non-*Saccharomyces* species (94, 95, 114, 115). Overall, these finer technologies are revealing substantial strain diversity within wine isolates of *S. cerevisiae* and suggest the potential for chromosomal reorganization during growth (73, 81).

Factors Affecting Yeast Growth during Fermentation

The process of winemaking involves many intrinsic and extrinsic variables that determine the duration and completeness of fermentation (Table 37.2) (9, 12, 39, 63, 103). Within the context of the overall reaction, these factors will influence the rate and extent of growth of individual yeast species or strains, but detailed information on these effects remains limited.

Table 37.2 Factors affecting the growth of yeasts during alcoholic fermentation

Composition of grape juice	Settling and clarification of juice
Sulfur dioxide	Temperature of fermentation
Inoculation of juice	Killer yeasts
Interactions with fungi and bacteria	Fungicide residues
Stuck fermentations	

Juice Composition

In most circumstances, grape juices provide all of the nutrients and conditions necessary for a vigorous and complete fermentation. However, chemical and physical properties of the juice vary according to grape variety, climatic influences, viticulture practices, and maturity at harvest. Relevant properties include sugar concentration, amount of nitrogenous substances, concentrations of vitamins, dissolved oxygen content, amount of soluble solids, fungicide or pesticide residues, pH, and presence of any yeast-inhibitory or stimulatory substance produced by growth of fungi and bacteria on the grapes.

The concentration of fermentable hexoses in grape juice varies between 150 and 300 mg/ml (Table 37.1) and may be as high as 400 mg/ml in juice prepared from grapes infected with *Botrytis cinerea* ("pourriture noble"; 29, 64). This variation in sugar concentration might be expected to affect the relative rates of growth of the different species and strains of wine yeasts and, consequently, the extent to which they contribute to the overall fermentation. There are suggestions that the presence and growth of *C. stellata* may be favored in juices with higher sugar concentrations (4, 64).

Free amino acids and ammonium ions are the principal nitrogen sources used by yeasts during fermentation. For a long time, it was thought that most juices contained sufficient nitrogen substrates to allow rapid and complete fermentation. However, it is now known that some juices, especially those that have been heavily processed or clarified, may not have sufficient nitrogen nutrients to allow maximum yeast growth and complete fermentation (7, 52). Moreover, the nitrogen demand by yeasts increases significantly with increasing sugar concentration in the juice and also varies with the strain of *S. cerevisiae.* Consequently, supplementation of juices with various yeast foods or diammonium phosphate is now common practice to ensure that nitrogen availability is not a factor that limits yeast growth. Most studies on the nitrogen requirements of wine yeasts have been conducted with *S. cerevisiae,* and little to nothing is known about the nitrogen demands of non-*Saccharomyces* species or their ability to remove specific nitrogen substrates from the juice before the growth of *S. cerevisiae.* The ability of wine yeasts to utilize grape juice proteins as a source of nitrogen requires further consideration. Strains of *S. cerevisiae* generally do not produce extracellular proteolytic enzymes (37). However, some strains of the non-*Saccharomyces* wine yeasts, e.g., *K. apiculata* and *M. pulcherrima* (37; C. Charoenchai and G. H. Fleet, unpublished data), are proteolytic.

Grape juices generally contain sufficient concentrations of vitamins (inositol, thiamine, biotin, pantothenic acid, and nicotinic acid) to permit maximum growth of *S. cerevisiae* (89). Vitamin losses may occur in heavily processed juices, in which case supplementation can improve the growth of *S. cerevisiae.* Species of non-*Saccharomyces* are more demanding of vitamins than *S. cerevisiae,* and vitamin availability could be a factor which limits their contribution to fermentation. The pH of grape juice varies between 3.0 and 4.0, depending on the concentration of tartaric and malic acids. Although growth and fermentation rates by *S. cerevisiae* are decreased as the pH is decreased from 3.5 to 3.0 (48, 87), it is not known how juice pH affects the relative growth rates of the non-*Saccharomyces* yeasts and their potential to influence alcoholic fermentation.

Treatment of grapes with fungicides and pesticides before harvest can give juices that contain residues of these substances. Depending on their concentration and chemical nature, these residues may decrease yeast growth, leading to slow or incomplete fermentations, and even change the ecology of the fermentation (39, 103).

Conditions which stimulate yeast growth and fermentation include aeration of the juice before or during the early stages of fermentation (56) and the presence of grape solids and particulate materials (27, 90). Different yeast species may be selectively adsorbed to such particles to form a biofilm of immobilized biomass. The ecological and metabolic properties of such biofilms could be different from those of yeasts growing as free cells in the juice, but this concept has not been researched.

Clarification of Grape Juice

The procedures used to clarify juices, especially for white wine fermentations, will influence the populations of indigenous yeasts in the juice and their potential contribution to the fermentation. Centrifugation and filtration remove yeast cells, thereby decreasing or eliminating the contribution of indigenous species to the fermentation. In contrast, clarification by cold-settling presents opportunities for the growth of indigenous yeasts, especially those species or strains that grow well at low temperatures (80).

Sulfur Dioxide

The addition of sulfur dioxide to grapes or juice for controlling oxidation reactions and restricting the growth of indigenous microflora is a well-established practice (88, 108). However, a general view that addition of sulfur dioxide (50 to 100 μg/ml, total) to juice will suppress the growth of non-*Saccharomyces* yeasts relative to *S. cerevisiae* is not supported by recent studies. Strong growth of *K. apiculata* and various *Candida* species during the early stages of fermentation is frequently noted in wines produced under commercial conditions where the usual amounts of sulfur dioxide have been added. In one study, growth of *K. apiculata* was not inhibited by total sulfur dioxide concentrations of 100 to 250 μg/ml (38). These findings question the efficacy of sulfur dioxide in controlling indigenous yeasts and challenge one of the basic reasons for using sulfur dioxide in winemaking. Despite the importance of sulfur dioxide in winemaking, good comparative data on the responses of wine yeasts to this agent are lacking.

Temperature of Fermentation

Temperature control has become an important practice in modern winemaking. Temperature affects the rate of growth and metabolic activities of yeasts. It also affects the ability of individual yeast species to tolerate ethanol (126), thereby determining their survival and contribution to the fermentation. Fastest yeast growth and fermentation occur at 25 to 30°C, and the ecology of the fermentation follows the pattern outlined in Fig. 37.2. However, when the temperature is decreased below 20°C, there is an increased contribution of the non-*Saccharomyces* species to the fermentation. Species such as *K. apiculata* and *C. stellata* exhibit increased tolerance to ethanol and do not die off as shown in Fig. 37.2. They can produce maximum populations of 10^7 to 10^8 CFU/ml which remain viable until the end of fermentation (48). Moreover, they may have faster growth rates than *S. cerevisiae* at low temperatures. The impact of such ecological shifts on the chemical composition and sensory quality of wines has yet to be determined.

Inoculation of the Juice

Perhaps the most significant technological innovation in winemaking during the last 20 to 30 years has been the

Table 37.3 Some desirable and undesirable features of wine yeasts (18, 26)

Desirable	Undesirable
High tolerance of alcohol	Production of sulfur dioxide
Complete and rapid fermentation of sugars	Production of hydrogen sulfide
Resistance to sulfur dioxide	Production of volatile acidity
Fermentation at low temperatures	High formation of acetaldehyde, pyruvate, and esters
Malic acid degradation	Foaming properties
Ferment under pressure	Formation of ethyl carbamate precursors
Production of glycerol	Production of polyphenol oxidase
Production of β-glycosidases	Inhibition of malolactic fermentation
Killer phenomenon	
Good sedimentation properties	
Production of good flavor and aroma	
Suitability for mass culture, freeze-drying distribution, and rehydration	

seeding (inoculation) of the juice with selected strains of *S. cerevisiae.* These strains have been selected according to criteria that will facilitate the process of winemaking and will yield a product of desired quality (Table 37.3) (18, 26). Seeding of the fermentation is undertaken with the assumption and expectation that the inoculated strain will out-compete and dominate over indigenous strains of *S. cerevisiae* and the non-*Saccharomyces* yeasts. While much evidence shows that inoculated strains dominate at the end of fermentation, the view that early growth of the indigenous species is suppressed or insignificant cannot be supported. Growth of the indigenous non-*Saccharomyces* species, according to Fig. 37.2, still occurs (47), and, moreover, indigenous strains of *S. cerevisiae* may grow despite massive competition from the seeded strain. Indeed, if conditions in the juice do not favor growth of the seeded strain, indigenous *S. cerevisiae* may dominate the fermentation, and this can be verified by molecular techniques that allow differentiation of yeast strains. Although there is high probability that inoculated *S. cerevisiae* will dominate the fermentation, seeding will not necessarily guarantee the dominance of any particular strain or its exclusive contribution to the fermentation (94, 95, 114, 115). Significant factors that affect this outcome will be the population of indigenous yeasts already in the juice and the extent to which they have adapted to grow in that juice.

Interactions with Other Microorganisms

Various species of molds, acetic acid bacteria, and lactic acid bacteria naturally occur on grapes and on winery equipment. Conditions which allow their proliferation on the grape or in the juice before yeast growth have the potential to affect the ecology and success of the alcoholic fermentation and are discussed in later sections.

Killer yeasts are certain strains which produce extracellular proteins or glycoproteins, termed killer toxins, that can destroy other yeasts (117, 128). Usually strains of one species only kill strains within that species, but as more yeasts are examined, it is becomirg evident that killer interactions between different species may occur. Killer toxin-producing strains of *S. cerevisiae* and killer-sensitive strains of *S. cerevisiae* occur as part of the natural flora of wine fermentations. In some wineries, killer strains may predominate at the end of fermentation, suggesting that they have asserted their killer property and taken over the fermentation. Killer strains have also been found within wine species of *Candida, Pichia, Hanseniaspora,* and *Hansenula,* and indeed some of these strains can assert their killer action against wine strains of *S. cerevisiae.* Expression of the killer phenomenon during wine fermentations is affected by many factors which include ethanol concentration, pH, temperature, presence of fining agents, and the relative populations of killer and killer-sensitive strains. There are several implications of killer yeasts in winemaking. First, inoculated strains of *S. cerevisiae* could be destroyed by indigenous killer strains of *S. cerevisiae* or non-*Saccharomyces* species, leading to premature cessation of the fermentation, slower fermentation, or completion of the fermentation by a less desirable species. Second, there may be advantage in conducting the fermentation with selected or genetically engineered killer strains of *S. cerevisiae* for the purposes of controlling the growth of less desired indigenous species. Moreover, strains could be selected or constructed to produce stable, broad-spectrum killer toxins that would protect the wine from infection by spoilage yeasts. Finally, strains might be selected to have immunity against the killing action of indigenous yeasts,

Table 37.4 Concentrations of volatile compounds produced by different species of wine yeasts[a]

Yeast species	Concn (μg/ml)						
	Propanol	Isobutanol	Isoamyl alcohol	2-Phenyl ethanol	Ethyl acetate	Isoamyl acetate	Acetoin
Saccharomyces cerevisiae	0.4–170	5–666	17–769	5–83	10–205	0.1–16	0–29
Kloeckera apiculata	4–25	3–60	10–117	10–35	40–870	0.04–1.1	56–187
Candida stellata	4–8	13–21		6–11	7–25	0.1–0.4	35–254
Hansenula anomala	3–15	18–29	11–25	27	137–2150	1–11	
Metschnikowia pulcherrima	1–43	37–123	21–243	22	150–382	0.1–0.8	
Zygosaccharomyces bailii	18–25	20–30	48–85	13–22	23–53	0.1–0.5	17–24
Pichia membranaefaciens	<1	1–9	0.5–9.5		16–21	1–6	
Brettanomyces bruxellensis	2–3	19	46	26	36–860	<1	

[a]Data compiled from references 4, 19, 20, 31, 37, 40, 55, 64, 75, 78, 86, 107–109, 109a, 110–116, 118, 119, and 129.

thereby giving them a greater chance of dominating the fermentation.

Stuck Fermentations

A sporadic but serious problem is the premature cessation of yeast growth and alcoholic fermentation, giving wine with residual unfermented sugar and a lower concentration of ethanol than expected. Such fermentations are commonly referred to as being "stuck" or "sluggish" (52, 56, 122). Factors considered to cause this problem include excessive clarification and processing of the juice, fermentation temperature too high, deficiency in nutrients or growth factors in the juice, presence of fungicide residues, influences from other microorganisms such as molds, acetic acid bacteria, and killer yeasts, and accumulation of medium-chain-length fatty acids such as octanoic and decanoic acids that can become toxic to yeast growth. Initiatives to overcome stuck fermentations include the addition of nitrogen-containing yeast foods, controlled aeration of the juice or wine, and the addition of yeast cell wall hulls or other bioadsorbents to remove toxic substances (14, 32, 52, 56, 83).

Biochemistry

Yeasts utilize grape juice constituents as substrates for their growth, generating metabolic end products that are excreted into the wine (11). The main products are carbon dioxide and ethanol and, to a lesser extent, glycerol and succinic acid. In addition, many hundreds of volatile and nonvolatile secondary metabolites are produced in small amounts that, collectively, contribute to the sensory quality of the wine. These substances include a vast range of organic acids, higher alcohols, esters, aldehydes, ketones, sulfur compounds, and amines. The chemical identities of individual substances, their flavor or aroma sensation, their flavor threshold, and their concentrations in wines are well documented (19, 86, 113), and the biochemistry of their formation in *S. cerevisiae*, at least, is reasonably well known (5, 64, 75, 109a). The production of these metabolite varies considerably depending on the yeast strain, yeast species, and conditions of fermentation. Tables 37.4 and 37.5 show some of the main metabolites produced by yeasts in wine and the concentrations produced by different yeast species.

Table 37.5 Concentrations of ethanol, glycerol, acetaldehyde, acetic acid, and succinic acid produced by different species of wine yeasts[a]

Yeast	Ethanol (%)	Glycerol (mg/ml)	Acetaldehyde (μg/ml)	Acetic acid (mg/ml)	Succinic acid (mg/ml)
Saccharomyces cerevisiae	6–23	3.7–6.8	15–30	0.1–2.0	0.6–1.7
Kloeckera apiculata	2–7	5.5–8.2	8–54	0.1–1.2	0.3
Candida stellata	6–7			1.08–1.3	
Hansenula anomala	0.5–5	0.2–2.2	3.2–8.1	1.6	0.2
Metschnikowia pulcherrima	2–4	2.7–4.2	23–40	0.1–0.2	
Zygosaccharomyces bailii	5–13			0.1–0.3	1.6
Pichia membranaefaciens	0.1–0.5	4.1–5.4	2.9	0.3	
Brettanomyces bruxellensis	9–12			1–7	

[a]Data compiled from references 4, 19, 20, 31, 37, 40, 55, 64, 75, 78, 86, 107–116, 118, and 119.

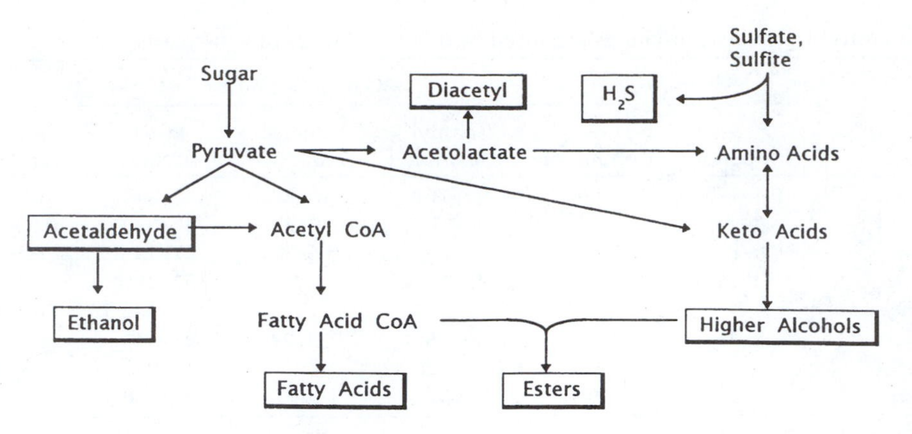

Figure 37.3 Derivation of flavor compounds from sugar, amino acids, and sulfur metabolism by yeasts (5, 52).

Carbohydrates

Glucose and fructose in juice are metabolized by the glycolytic pathway to pyruvate, which is decarboxylated by pyruvate decarboxylase to acetaldehyde. Acetaldehyde is reduced to ethanol by alcohol dehydrogenase (8). Although most of the pyruvate is converted to ethanol and carbon dioxide, small proportions are converted to secondary metabolites (Fig. 37.3). Glycerol, which imparts desirable smoothness and viscosity to the wine, is derived from the glycolytic intermediate dihydroxy acetone. Its production is increased by the presence of sulfur dioxide, higher incubation temperature, and increased sugar concentration (44). Transport of sugars into the cell and factors which affect this could be important rate-limiting steps in the fermentation process (8).

The potential for wine yeasts to degrade grape juice pectins needs more consideration given that some strains of *S. cerevisiae* may produce pectin-degrading enzymes (37). The ability of the non-*Saccharomyces* species to degrade pectin requires study.

Nitrogen Compounds

Wine yeasts take up and metabolize ammonium ions and amino acids in the juice as sources of nitrogen (52). The mechanisms of transport of these compounds into the cells and biochemical pathways for their metabolism, involving decarboxylation, transamination, reduction, and deamination reactions, have been reviewed (66). There is no strong evidence that *S. cerevisiae* produces extracellular proteolytic enzymes to degrade grape juice proteins. However, some of the non-*Saccharomyces* species such as *K. apiculata* and *M. pulcherrima* do produce extracellular proteases that could be involved in the breakdown of juice proteins (37).

The availability of nitrogen in the juice and its metabolism by yeasts have important implications in winemaking (7, 52). Juices that are limited in nitrogen content can give stuck or sluggish fermentations and wines with unacceptably high contents of hydrogen sulfide. Metabolism of arginine (the most predominant amino acid in grape juices) by *S. cerevisiae* can lead to the production and release of urea, which is able to react with ethanol to form ethyl carbamate, a suspected carcinogen. The amount of urea produced depends on many factors including the concentration of arginine relative to other nitrogen substances in the juice and the strain of *S. cerevisiae* (53). Juice nitrogen can have a significant impact on the amounts of higher alcohols, fatty acids, esters, and carbonyl compounds, such as aldehydes and diacetyl, produced by yeasts (52, 75).

In addition to utilization, yeasts can release amino acids into the wine. This occurs during the final stages of alcoholic fermentation, by mechanisms not fully understood, and later when the cells have died and there is autolytic degradation of yeast proteins (15). These amino acids can serve as nutrients for the growth of malolactic bacteria or spoilage bacteria.

Sulfur Compounds

Yeasts produce a range of volatile sulfur compounds which, above certain threshold concentrations, have a negative impact on wine flavor (91, 99). The most predominant compounds are sulfur dioxide (sulfide), hydrogen sulfide, and dimethyl sulfide, with lesser amounts of other organic sulfites, mercaptans, and thioesters. The production of sulfite and hydrogen sulfide is linked to the biosynthesis of cysteine and methionine by the sulfate reduction pathway, the details of which are well known for *S. cerevisiae* (45, 57). These sulfur-containing amino acids exert feedback control on the operation of this pathway and the production of sulfite and hydrogen sulfide. The formation of sulfite by *S. cerevisiae* depends on the strain, with most strains producing less than 10 μg/ml. However, a small percentage of strains produce sulfite at levels up to 100 μg/ml (108). The use of such strains is avoided in winemaking because of the negative effect of these high concentrations on wine quality and the potential of such concentrations to cause allergic reactions in some consumers.

At concentrations exceeding 50 μg/liter, hydrogen sulfide gives an unpleasant "rotten egg" aroma to wine. Many chemical and biological factors affect the production of hydrogen sulfide during wine fermentation, the most significant of which is the strain of *S. cerevisiae.* Some strains produce hydrogen sulfide at concentrations exceeding 1 μg/ml. This hydrogen sulfide production is genetically based, but it is also influenced by factors such as the composition of the grape juice and fermentation conditions. Elemental sulfur used as a fungicide on grapes prior to harvest, metabisulfite added to the grapes and juice at crushing, and sulfate which occurs naturally in the juice are all significant precursors of hydrogen sulfide in the wine. Deficiencies in readily available nitrogen in the juice are an important cause of hydrogen sulfide production by *S. cerevisiae.* Under these conditions, the intracellular pool of cysteine and methionine is low, allowing the sulfate reduction pathway to operate with consequent production of hydrogen sulfide. If the juice contains an adequate supply of assimilable nitrogen (e.g., ammonium ions, amino acids), cysteine and methionine are produced at concentrations which, through feedback inhibition, decrease the activity of the sulfate reduction pathway and production of hydrogen sulfide. There is also a view that, under nitrogen-limited conditions, *S. cerevisiae* degrades intracellular proteins to provide essential amino acids including cysteine and methionine from which hydrogen sulfide is formed (52, 57). The potential for the non-*Saccharomyces* yeasts to produce sulfur compounds during wine fermentations has not been investigated.

Organic Acids

Of the numerous organic acids produced in wine by yeasts, succinic and acetic acids are the most significant (96). Succinic acid has a bitter salty taste and is produced by *S. cerevisiae* at concentrations up to 2.0 mg/ml, depending on the strain. Lower concentrations are produced by non-*Saccharomyces* species (116). The production of succinic acid is not associated with any major defects in wine quality. In contrast, acetic acid becomes detrimental to wine quality at concentrations exceeding 1.5 mg/ml. Most strains of *S. cerevisiae* produce only small amounts (<0.75 mg/ml) of this acid, but some can produce greater than 1.0 mg/ml and are unsuitable for winemaking. Factors which limit yeast growth such as low temperature, high sugar concentrations, low pH, deficiency in available nitrogen, and excessive clarification cause increased acetic acid production by *S. cerevisiae* (27, 82, 116). *Candida, Kloeckera,* and *Hanseniaspora* species may produce higher amounts of acetic acid than *S. cerevisiae* (Table 37.5), but there is substantial strain variation in this property (110).

The production of lactic acid by wine yeasts is considered insignificant (<0.1 mg/ml), but there are species of *Saccharomyces* (96) and some strains of *Kluyveromyces thermotolerans* (79) and *C. colliculosa* (Charoenchai and Fleet, unpublished data) that can produce this acid in amounts of 5 to 10 mg/ml. Such strains could be used to increase the acidity of some wines. Although tartaric acid is prevalent in grape juice and wine, it is not metabolized by wine yeasts (96). However, malic acid is partially (5 to 50%) metabolized by *S. cerevisiae* and other wine yeasts. It is completely degraded by some species of *Schizosaccharomyces* and some strains of *Zygosaccharomyces bailii* (43, 96), where it is oxidatively decarboxylated to pyruvate which is then converted to ethanol. The possibility of using species of *Schizosaccharomyces* to deacidify wines in place of the malolactic fermentation has attracted considerable interest, but such use must be carefully controlled as these yeasts can produce off flavors (43, 133).

Yeasts produce small amounts (1 to 15 μg/ml) of free fatty acids in wines (32, 96). Of special note are hexanoic, octanoic, and decanoic acids which, on accumulation, may become toxic to *S. cerevisiae* and contribute to stuck fermentations.

Autolysis

The autolytic degradation of yeast cells at the end of alcoholic fermentation and during cellar storage of the wines is often underestimated as a significant biochemical event. During autolysis, yeast proteins, nucleic acids, and lipids are extensively degraded, releasing peptides, amino acids, nucleotides, bases, and free fatty acids into the wine. These products affect wine flavor and serve as nutrients for the growth of bacteria. In addition to *S. cerevisiae,* the non-*Saccharomyces* species would be involved in autolytic reactions (15, 54).

Flavor Compounds

Yeasts produce many organic acids, higher alcohols, esters, aldehydes, ketones, sulfur compounds, and amines that, in conjunction with grape constituents, determine wine flavor (19, 86, 113). It has been often stated that wines produced by natural fermentations are perceived as having more complex and interesting flavors compared with those produced by inoculation with selected strains of *S. cerevisiae* (4, 62, 101, 102). Such views imply distinctive contributions by indigenous, non-*Saccharomyces* yeasts to the flavor profile, but definitive studies which connect their growth to a chemical profile and a particular sensory outcome remain wanting. Nevertheless, there are numerous studies which show that non-*Saccharomyces* species produce some compounds in greater or lesser amounts than *S. cerevisiae* (Tables 37.4 and 37.5) and that mixed-culture fermentations involving these species as well as *S. cerevisiae* have different chemical profiles from fermentations conducted solely by *S. cerevisiae* (55, 78, 114, 134). However, these conclusions are often complicated by the fact that metabolite production by particular yeast species can vary considerably depending on the strain and conditions of fermentation (107, 110–112).

Spoilage Yeasts

Growth of inappropriate species or strains of yeasts during alcoholic fermentation can give an inferior wine (e.g., high in content of esters, acetic acid, hydrogen sulfide) and the product is considered spoiled. Spoilage can also occur during storage of wine in the cellar and after bottling (11, 120, 124). Wine that is exposed to air, as in incompletely filled barrels or tanks, quickly develops a film or surface flora of weakly fermentative or oxidative yeasts of the genera *Candida, Pichia,* and *Hansenula.* Particularly significant is *Pichia membranaefaciens.* These species oxidize ethanol, glycerol, and acids, giving wines with unacceptably high levels of acetaldehyde, esters, and acetic acid. Fermentative species that grow in bottled wines and in wines during cellar storage include *Z. bailii, Brettanomyces* and *Dekkera* species (*Brettanomyces intermedia*), and *Saccharomycodes ludwigii.* In addition to causing excessive carbonation, sediments, and haze, these species produce acid and estery off flavors. The growth of *Brettanomyces* species is also associated with the production of unpleasant mousy or medicinal taints due to the formation of tetrahydropyridines and volatile phenolic substances such as 4-ethyl guaiacol and 4-ethyl phenol (120, 124). Wines that contain residual sugars are prone to refermentation unless properly managed. In such cases, the growth of *S. cerevisiae* constitutes a form of spoilage.

Genetic Improvement of Wine Yeasts

There is a vast pool of naturally occurring wine yeasts, and selection from this reservoir has generally met the needs of winemaking. Nevertheless, the process of strain selection and development can be accelerated and more specifically targeted through the use of classical and modern genetic improvement technologies (2, 93, 95a). Classical mutagenesis has yielded strains of *S. cerevisiae* that give decreased levels of higher alcohols and better fermentation performance at low temperatures. Traditional mating and hybridization methods have given strains with decreased production of hydrogen sulfide and improved flocculation characteristics. Protoplast fusion has been used in attempts to introduce the killer property into desirable strains of *S. cerevisiae* and to introduce the malic acid-degrading properties of *Schizosaccharomyces* species into *S. cerevisiae.* The more precise and efficient recombinant DNA technologies are being explored to produce strains of *S. cerevisiae* with a range of desirable properties such as malic acid degradation, broad-spectrum killer activity, improved flocculation and foaming characteristics, decreased production of unwanted metabolites, and secretion of enologically advantageous enzymes such as pectinases and proteases. In addition to giving a pleasing product and an efficient fermentation, any new strain must be acceptable to legislative authorities and consumers on the basis of safety criteria.

LACTIC ACID BACTERIA

Lactic acid bacteria are significant in winemaking because they cause spoilage and are responsible for malolactic fermentation. In addition, they will release autolytic products into the wine. Wine lactic acid bacteria have the unique ability to tolerate the stresses of the wine environment, namely, low pH, presence of ethanol and sulfur dioxide, low temperature, and dilute concentrations of nutrients. Their occurrence, growth, and significance in wines have been comprehensively reviewed (23, 49, 50, 61, 131). The main species of concern occur in the genera *Leuconostoc, Pediococcus,* and *Lactobacillus* and include *Ln. oenos, Pediococcus parvulus, Pediococcus pentosaceus, Pediococcus damnosus* (formerly *P. cerevisiae*), and various species of *Lactobacillus* (e.g., *Lb. brevis, Lb. fermentum, Lb. buchneri, Lb. hilgardii,* and *Lb. trichodes*). *Ln. oenos* is uniquely found in wines and is intimately associated with the occurrence of malolactic fermentation in wines produced in many countries. Consequently, the taxonomy, biochemistry, and physiology of this species have been specifically studied (127).

Ecology

The lactic acid bacteria of wines originate from the grapes and winery equipment, but inoculation of se-

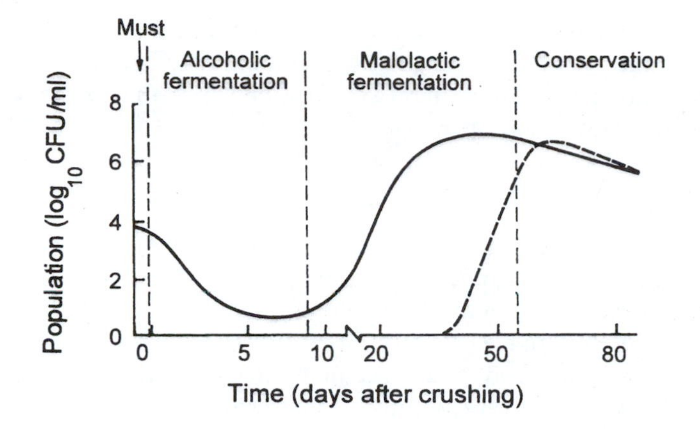

Figure 37.4 Growth of lactic acid bacteria during vinification of red wines, pH 3.0 to 3.5. The solid line shows the growth of *Ln. oenos*, often the only species present. Occasionally species of *Lactobacillus* and *Pediococcus* develop towards the end of malolactic fermentation or at later stages during conservation (broken line). For wines of pH 3.5 to 4.0, a similar growth curve is obtained but there may be slight growth and death of lactic acid bacteria during the early stages of alcoholic fermentation. Also, there is a greater chance that species of *Lactobacillus* and *Pediococcus* will grow and conduct malolactic fermentation.

lected species to conduct the malolactic fermentation is widely practiced. Freshly extracted grape juice contains lactic acid bacteria at populations of 10^3 to 10^4 CFU/ml, but they undergo little or no growth during the alcoholic fermentation and tend to die off because of competition from yeasts (Fig. 37.4). Nevertheless, these bacteria are capable of abundant growth in the juice, and if yeast growth is delayed, they could grow and spoil the juice. About 1 to 3 weeks after completion of the alcoholic fermentation, the surviving lactic acid bacteria commence vigorous growth to conduct the malolactic fermentation. Final populations of 10^6 to 10^8 CFU/ml are produced. The onset, duration, and ecology of this growth are determined by many factors which include the properties of the wine, vinification variables, and influences of other microorganisms. Consequently, the natural occurrence of malolactic fermentation and its completion by the preferred species, *Ln. oenos*, can be unpredictable.

Of the wine properties, pH, concentration of ethanol, and concentration of sulfur dioxide have strong influences on the growth of lactic acid bacteria. Different species, and even strains within species, show different responses to these properties (24, 127). Wines of low pH (e.g., pH 3.0), high ethanol content (>12% [vol/vol]), and high total sulfur dioxide (>50 µg/ml) are less likely to support the growth of lactic acid bacteria and may not undergo successful malolactic fermentation. Strains of *Ln. oenos* are more tolerant of low pH than those of *Pediococcus* and *Lactobacillus* species and generally predominate in wines of pH 3.0 to 3.5. Wines with pH values exceeding 3.5 tend to have a mixed microflora consisting of *Ln. oenos* and various species of *Pediococcus* and *Lactobacillus*. Species of *Pediococcus* and *Lactobacillus* are more tolerant of higher concentrations of sulfur dioxide than *Ln. oenos* and more likely occur in wines with higher amounts of this substance (24, 25). Thus, winemaker management of pH and sulfur dioxide content is important if it is desired to have malolactic fermentation conducted solely by *Ln. oenos*.

Microbiological factors that affect the growth of *Ln. oenos* in wine and successful completion of malolactic fermentation include excessive growth of molds and acetic acid bacteria on grapes, yeast species and strains responsible for the alcoholic fermentation, and bacteriophages. Substances produced by the growth of fungi or acetic acid bacteria on damaged grapes could either stimulate or inhibit malolactic fermentation (59, 131). During alcoholic fermentation and subsequent autolysis, yeasts release nutrients that are believed to encourage the growth of lactic acid bacteria. However, the growth of some strains of *S. cerevisiae* during alcoholic fermentation can be inhibitory to the subsequent growth of *Ln. oenos* (50, 132). Production of high concentrations of sulfur dioxide, inhibitory proteins, or toxic fatty acids (hexanoic, octanoic, decanoic, and dodecanoic) by these strains can be inhibitory to the bacteria (13, 32, 51, 74). It is not known how the growth of non-*Saccharomyces* species affects development of the malolactic fermentation. Bacteriophages active against *Ln. oenos* have been isolated from wines and can interrupt and delay the malolactic fermentation (22). However, bacteriophage-resistant strains of *Ln. oenos* have been described (24).

The fate of lactic acid bacteria after malolactic fermentation depends on the wine and winemaking practices. These bacteria may survive in wine for long periods, or they may die. Because wine pH is increased by malolactic fermentation, it becomes a more favorable environment for bacterial growth, and spoilage microorganisms such as *Pediococcus* and *Lactobacillus* species could develop, especially in wines of pH 3.5 to 4.0. As noted already, wines after malolactic fermentation may have better microbiological stability. Nevertheless, there are reports that *Ln. oenos* and various species of *Lactobacillus* and *Pediococcus* can reestablish growth in such wines (33, 132). The significance of bacteriocin production by the growth of lactic acid bacteria in wines requires more study. The death of *Ln. oenos* in wines subsequent to the growth of *Pediococcus* and *Lactobacillus* species can occur (25, 106). A strain of *P. pentosaceus* isolated from wines produces pediocin which is active against *Leuconostoc, Lactobacillus,* and *Pediococcus* isolated from wine (121). However, Edwards et al. (33) noted the production of substances by *Ln. oenos* that are inhibitory to the growth of *P. parvulus.* Strains of *Ln. oenos* that produce bacteriocins with broad-spectrum action against pediococci and lactobacilli would have obvious value in controlling spoilage by these bacteria.

Biochemistry

Lactic acid bacteria change the chemical composition of wine by utilizing some constituents for growth and generating metabolic end products. Depending upon the species that develop, such changes will have a positive or negative impact on wine quality. Detailed studies which correlate the growth of lactic acid bacteria with chemical changes in the composition of wine are few. Consequently, the biochemical mechanisms by which lactic acid bacteria grow in wines and affect flavor are poorly understood.

Carbohydrates

Wines contain residual amounts of glucose and fructose (0.5 to 1 mg/ml) and smaller amounts (<0.5 mg/ml) of other hexose and pentose sugars. Concentrations of many of these sugars decrease with the growth of lactic acid bacteria, but consistent trends have not emerged (25). This finding agrees with the ability of these bacteria to ferment a wide range of hexose and pentose sugars and the substantial variation between species and strains in conducting these reactions (24, 71, 92, 105, 127). The pathways utilized by lactic acid bacteria for sugar metabolism are well known (17). *Pediococcus* species as well as some species of *Lactobacillus* ferment hexoses by the Embden-Meyerhof-Parnas pathway (homofermenters), while *Ln. oenos* and some species of *Lactobacillus* use the hexose monophosphate or phosphoketolase pathway (heterofermenters). This latter pathway is used by all species to metabolize pentose sugars. It is not known how the wine conditions of low pH and high ethanol concentration affect the kinetics of transport of different hexose and pentose sugars into the cells and their subsequent metabolism. We have observed that the sugar fermentation profiles of *Ln. oenos, P. parvulus, Lactobacillus plantarum,* and *Lb. brevis* were substantially different at pH 6.0 and 4.0 (24).

Some strains of *Ln. oenos* as well as other lactic acid bacteria can produce extracellular viscous polysaccharide materials that have the potential to affect the filterability of wines (127, 131). It is not known if species of wine lactic acid bacteria produce enzymes that break down grape juice pectins.

Nitrogen Compounds

The concentrations of some wine amino acids (e.g., arginine and histidine) decrease with the growth of lactic acid bacteria and are consistent with their utilization as a nitrogen source (25, 49). The ability of wine lactic acid bacteria to produce extracellular proteases and peptidases seems very limited. Concentrations of some amino acids are increased after malolactic fermentation, but it is not clear if this is due to enzymatic hydrolysis of wine proteins and peptides or to bacterial autolysis (127). Decarboxylation of amino acids to produce amines has been reported for wine lactic acid bacteria, especially for species of *Pediococcus,* but this property is not shared by all species or strains (36, 61, 64, 131). Some wine lactic acid bacteria can metabolize arginine to citrulline, which is a precursor of ethyl carbamate (72).

Organic Acids

The decarboxylation of L-malic acid to L-lactic acid is one of the most significant metabolic reactions conducted by lactic acid bacteria in wines. These bacteria possess mechanisms for the transport of malic acid into the cell, efflux of lactic acid, and the malolactic enzyme for decarboxylation. Purification and properties of the malolactic enzyme have been studied for several species (49, 50, 61). The ability to degrade tartaric acid is limited to a few species of lactobacilli. The metabolism of tartaric acid is not usually encountered during or after malolactic fermentation, but when it occurs the wine is spoiled (120). The concentration of citric acid can be completely or partially metabolized during malolactic fermentation, depending on the wine pH and species of lactic acid bacteria. Its degradation is frequently correlated with small increases in the concentrations of acetic acid and diacetyl, and this observation is consistent with the action of citrate lyase which is produced by some but not all species of wine lactic acid bacteria (25, 76). The concentrations of fumaric, gluconic, and pyruvic acids can decrease during malolactic fermentation. Gluconic acid, which is significantly increased in wines made from grapes infected with *Botrytis cinerea,* is metabolized by most lactic acid bacteria except the pediococci (131). Sorbic acid, which may be added to wines in some countries to control yeast growth, can be metabolized by

Ln. oenos to form 3,5-hexadien-2-ol and 2-ethoxyhexa-3-diene, which cause geranium-like off flavors (131).

Flavor Compounds

Much has been written about the impact of malolactic fermentation on wine flavor (23, 49, 50). Malolactic fermentation not only affects the taste of wine through deacidification, but also contributes other flavor characteristics (often described as buttery, nutty, fruity, or vegetative) that may enhance or detract from overall acceptability. Obviously, such changes will be related to the wine constituents metabolized by the malolactic bacteria and the nature, concentration, and flavor threshold of the products generated. Attempts to connect sensory impression to flavor substances produced by particular species or strains of malolactic bacteria have proved elusive because of confounding influences of grape variety, yeasts, and vinification variables. However, diacetyl production during malolactic fermentation has been linked to the evolution of desirable buttery aromas (67, 76). Substances responsible for these aromas are produced by the metabolism of citrate, but can also be a by-product of sugar metabolism. Other flavor constituents that increase in concentration during malolactic fermentation include acetic acid, diethyl succinate, and ethyl lactate as well as other volatile acids, esters, and higher alcohols (23).

Spoilage Reactions

Uncontrolled growth of lactic acid bacteria during or after malolactic fermentation results in wine spoilage, the type of spoilage depending upon the particular wine and species of lactic acid bacteria which grow (11, 120, 131). Wines containing high concentrations of residual glucose and fructose will undergo acidification with the production of unacceptable amounts of acetic acid and D-lactic acid. Mannitol taint is caused by some strains of heterofermentative lactobacilli (e.g., *Lb. brevis*) due to enzymatic reduction of fructose to mannitol. Excessive production of diacetyl (>1 to 5 μg/ml) can give overpowering buttery flavors. *Lb. hilgardii, Lb. brevis*, and *Lactobacillus cellobiosus* have been implicated in the formation of mousy taints due to their production of acetyltetrahydropyridines. Degradation of glycerol by lactic acid bacteria, especially by species of *Pediococcus* and *Lactobacillus*, can lead to the production of acrolien and associated development of bitterness. Spoilage arising from the degradation of tartaric and sorbic acids has already been mentioned. However, the ability to metabolize glycerol and tartaric acid is not widespread among the lactic acid bacteria.

Control of Malolactic Fermentation

Control of malolactic fermentation has emerged as one of the challenges of modern winemaking (23, 50). Because natural occurrence of malolactic fermentation can be unpredictable, commercial cultures of malolactic bacteria have become available for inoculation into the wine to induce this reaction. Generally, these are strains of *Ln. oenos* that have been selected for a range of desirable criteria (Table 37.6) (26, 49, 60). Many factors affect the successful induction of malolactic fermentation by inocula, the most important of which include proper reactivation and preculture of the freeze-dried concentrate, level of inoculum, and timing of inoculation. Generally, wines are inoculated to give 10^6 to 10^7 CFU of bacteria per ml just after alcoholic fermentation is completed. Arguments have been advanced for inoculating malolactic bacteria into the juice either before or simultaneously with the yeast culture. Under these conditions, bacterial cells are not exposed to the stresses of high ethanol concentration, giving a higher probability of successful growth and malic acid degradation. Risks associated with early inoculation of malolactic bacteria include their potential metabolism of grape juice sugars to yield unacceptable concentrations of acetic acid, and potential interference with yeast growth and alcoholic fermentation (51, 61).

Even under "optimum" conditions, inoculation with malolactic bacteria does not assure successful completion of malolactic fermentation. In some instances, the particular properties of the wine may not be suitable for growth of the bacterial strain or mixture of strains inoculated. Several biotechnological innovations have been considered to overcome this problem. Bioreactor systems consisting of high concentrations of malolactic bacteria immobilized in beads of alginate or carrageenan have been developed (18, 23, 28). An alternative system uses high densities (ca. 10^9 to 10^{10} CFU/ml) of *Ln. oenos* retained within a cell-recycle membrane bioreactor

Table 37.6 Desirable properties of malolactic bacteria

Strong malolactic activity under wine conditions
Strong ability to grow in wines including those of low pH (3.0–3.2) or high ethanol (14%), containing sulfur dioxide (50 μg/ml total), and at low temperatures (15–20°C)
Resistance to destruction by bacteriophages; nonlysogenic
Production of desirable malolactic flavors and no off flavors
Nonproduction of biogenic amines and precursors of ethyl carbamate
Nonproduction of yeast inhibitory factors if used before alcoholic fermentation
Suitability for mass culture, freeze-drying, distribution, and rehydration

(41, 42). Under these conditions, cells of malolactic bacteria act as biocatalysts and, in the absence of growth, rapidly convert malic acid to lactic acid in wine that is passed through the reactor on a continuous basis. Such technologies give rapid, continuous deacidification of wines, but so far have not proved commercially successful mainly because of instability of the malolactic activity of cells in the reactor.

Yeast species within the genus *Schizosaccharomyces* (*Schizosaccharomyces pombe* and *Schizosaccharomyces malidevorans*) rapidly metabolize malic acid under anaerobic conditions. Malic acid is decarboxylated to pyruvate by the malic enzyme, and then pyruvate is decarboxylated to acetaldehyde which is reduced to ethanol (96). Many attempts have been made to exploit this property to deacidify wines in place of the bacterial malolactic fermentation. In one approach, these yeasts are cultured in grape juice either with or before inoculation with *S. cerevisiae*. In another strategy, juices or wines are passed through reactors containing immobilized cells of *S. pombe*. While successful malic acid degradation has been achieved with this yeast, it may produce undesirable off flavors (28, 43, 133). *Z. bailii* is another yeast that gives good degradation of malic acid, and there may be potential in using this species for wine deacidification (96, 109).

The availability of strains of *S. cerevisiae* that could carry out malolactic fermentation simultaneously with alcoholic fermentation would be most attractive to winemakers. Using recombinant DNA technology, attempts have been made to transfer the malolactic gene from lactic acid bacteria into a wine strain of *S. cerevisiae*. While expression of such activity within the yeast was obtained, activity was low and unstable and revealed the need for increased understanding of the molecular biology of the malolactic reaction and its regulation, including transport of malic acid into the cell (2, 61).

Complete prevention of malolactic fermentation is an option preferred by some winemarkers. Wines of low pH (<3.2), high ethanol content (>14%), and high sulfur dioxide content (>50 mg/ml) are less prone to malolactic fermentation. The commercially available bacteriocin, nisin, at 100 U/ml inhibits the growth of malolactic and spoilage lactic acid bacteria in wines (21, 35).

ACETIC ACID BACTERIA

Acetic acid bacteria are well known for their ability to cause the vinegary spoilage of wines through the oxidation of ethanol to yield acetaldehyde and acetic acid. In addition, their growth on grapes before fermentation can produce substances that not only affect wine flavor but can interfere with the growth of yeasts, leading to stuck fermentations (30, 65, 120). The main species of concern are *Gluconobacter oxydans*, *Acetobacter pasteurianus*, and *Acetobacter aceti*, where strains have evolved that can tolerate the wine environment.

Ecology

Sound, unspoiled grapes harbor low populations of acetic acid bacteria, generally less than 10^2 CFU/g, with *G. oxydans* being the predominant species. Damaged, spoiled grapes and those infected with the mold *B. cinerea* harbor much higher populations (>10^6 CFU/g) that are characterized by a dominance of *A. pasteurianus* and *A. aceti*. The populations of acetic acid bacteria in freshly extracted grape juice reflect those present on the grapes.

In the absence of yeasts, acetic acid bacteria quickly grow in grape juice, reaching populations of 10^6 to 10^8 CFU/ml. The extent of their growth during alcoholic fermentation depends on initial populations, relative to yeasts, in the juice, but this question has not been conclusively studied. In juice prepared from sound grapes where the natural yeast population is 10 to 100 times greater than that of acetic acid bacteria, there appears to be little growth or influence of these bacteria on the alcoholic fermentation (31, 59). However, when initial populations of these bacteria exceed about 10^4 CFU/ml, they may grow in conjunction with yeasts during the early stages of fermentation. Populations as high as 10^8 CFU/ml can develop but die off due to the combined influences of ethanol and anaerobiosis caused by yeast growth. Under these conditions, acetic acid and other substances produced by acetic acid bacteria become inhibitory to the yeast, causing premature cessation of the alcoholic fermentation (30).

At the end of alcoholic fermentation, the population of acetic acid bacteria is generally less than 10^2 CFU/ml. *G. oxydans* is rarely isolated from wines at this stage, and *A. pasteurianus* and *A. aceti* are the main species found. Subsequent transfer of the wine from fermentation tanks to other storage vessels may produce sufficient agitation and aeration to encourage the growth of surviving acetic acid bacteria to populations of 10^5 CFU/ml or higher. It is not uncommon to isolate *A. pasteurianus* or *A. aceti* from wines during bulk storage in barrels or tanks presumably being kept under anaerobic conditions (30, 38, 65). The mechanisms by which these aerobic bacteria survive for long periods under apparently anaerobic or semianaerobic conditions require explanation. Their growth is activated by exposure of the wine to air and the wine is quickly spoiled.

Biochemistry

Oxidation of ethanol to acetic acid is a key reaction of acetic acid bacteria. Ethanol is first oxidized by ethanol dehydrogenase to acetaldehyde, which is then oxidized by acetaldehyde dehydrogenase to acetic acid. There is substantial variation between strains of *G. oxydans*, *A. aceti*, and *A. pasteurianus* in the strengths of these reactions. Species of *Acetobacter* can further oxidize acetic acid to carbon dioxide and water by the tricarboxylic acid (TCA) cycle, but this reaction is inhibited in the presence of ethanol. Species of *Gluconobacter* do not have a fully functional TCA cycle and are unable to completely oxidize acetic acid. Sulfur dioxide, which is generally present in wine, can chemically trap acetaldehyde, causing an accumulation of this intermediate at the expense of its further oxidation to acetic acid. Ethanol concentrations above 10% become increasingly inhibitory to the growth of acetic acid bacteria and their ability to oxidize ethanol. Aldehyde dehydrogenase is less stable than ethanol dehydrogenase at high concentrations of ethanol, and such conditions give an increased accumulation of acetaldehyde. Lower oxygen concentrations also favor the accumulation of acetaldehyde. In addition to acetic acid and acetaldehyde, ethyl acetate is another end product that is significant in the vinegary spoilage of wines (30, 31, 65).

Acetobacter and *Gluconobacter* lack a functional Embden-Meyerhof-Parnas pathway and metabolize hexose and pentose sugars by the hexose monophosphate pathway to acetic and lactic acids. However, *Acetobacter* species give only weak metabolism of sugars, which appears to be associated with a weak ability to phosphorylate these substrates. The carbohydrate metabolism of *G. oxydans* has been the subject of considerable research due to the ability of this species to oxidize various sugars and sugar alcohols to products of industrial significance such as sorbose, dihydroxyacetone, gluconic acid, and ketogluconic acids. At pH values below 3.5 or glucose concentrations above 5 to 15 mM, as occurs in grapes or grape juice, metabolism of glucose by the hexose monophosphate pathway is inhibited and glucose is directly oxidized to gluconic and ketogluconic acids. Consequently, grapes heavily infected with *G. oxydans* may yield juices with high concentrations (50 to 70 mg/ml) of gluconic acid. Strains of *A. aceti* and *A. pasteurianus* also produce gluconic acid in grape juice but to a lesser extent than *G. oxydans* (31).

Strains of *G. oxydans* and *A. aceti* oxidize glycerol to dihydroxyacetone. Glycerol is not normally present in grape juice, but its concentration may be significant (up to 20 mg/ml) in juices prepared from grapes infected with *B. cinerea* or other molds. Metabolism of glycerol by acetic acid bacteria, either on grapes during the early stages of alcoholic fermentation or in the final wine, leads to significant production of dihydroxyacetone that could affect wine quality (30, 120).

Species of *Acetobacter* are able to use the TCA cycle for metabolism of organic acids, causing decreases in concentrations of citric, succinic, malic, tartaric, and lactic acids in wines. Moreover, lactate may be oxidized to acetoin. The metabolism of amino acids and protein by wine acetic acid bacteria has not been examined. One other property of these bacteria that could be relevant in wine production is their ability to produce extracellular fibrils of cellulose and other polysaccharides (30).

OTHER BACTERIA

There are scattered reports that *Bacillus* and other bacterial species can survive and grow in wines and contribute to spoilage (11). Species of *Bacillus* implicated include *B. coagulans*, *B. circulans*, and *B. subtilis* isolated from spoiled sweet wines, *B. megaterium* isolated from spoiled brandy, and *B. coagulans* and *B. badius* which were isolated from wine corks and had limited ability to grow in wines (38, 69, 120). Juices and wines of high pH (e.g., 4.0) have, on rare occasions, been spoiled by the growth of *Clostridium butyricum* and have elevated concentrations of butyric acid, isobutyric acid, propionic acid, and acetic acid (120). Species of *Actinomyces* and *Streptomyces* have been isolated from wines, corks, and wooden barrels and could contribute to the musty, earthy, or corky taints sometimes found in wines (38).

MOLDS

Although molds do not grow during wine fermentations, they can affect wine quality under some circumstances (29, 38). Grapes harbor a range of mold species, the diversity and population of which depend on grape variety, degree of berry maturity and physical damage, viticultural practices, and climatic conditions. These molds include species of *Botrytis*, *Penicillium*, *Aspergillus*, *Mucor*, *Rhizopus*, *Cladosporium*, *Alternaria*, *Uncinula*, and *Plasmospora*. Although mold mycelium and spores are significant contaminants of grape juice, the conditions of anaerobiosis, increasing ethanol concentration, and presence of sulfur dioxide are strong deterrents to their growth during wine fermentation and conservation.

Mold spoilage or rotting of grapes before harvest is a major threat to the production of quality wines and is controlled through the careful application of fungicides (84). A novel approach could be the use of wine yeasts or other microorganisms for the biocontrol of this spoilage problem (123). Grapes heavily infected with molds have altered chemical composition and mold enzymes that may adversely affect wine flavor and color and the growth of yeasts during alcoholic fermentation (3). *B. cinerea* is particularly significant as a grape contaminant. Although it commonly causes spoilage known as bunch rot, its controlled development on grapes is the basis for producing the distinctive and highly prized botrytized sweet wines: Sauternes, Trockenbeerenauslese, and Tokay (29, 104). Under certain climatic and viticultural conditions, this species can parasitize healthy grape berries without disrupting the general integrity of their skin, causing the so-called "pourriture noble" or noble rot. Mold growth on and in the berry leads to its dehydration, with consequent concentration of its chemical constituents. This concentration effect, along with fungal metabolism of grape sugars and acids (especially tartaric), gives a juice that, typically, has increased sugar concentration (300 to 400 mg/ml), high concentration of glycerol (3 mg/ml) and other polyols, high concentration of gluconic acid, and less tartaric acid. The fermentation of such juices requires particular attention as it is prone to become stuck, possibly due to nitrogen deficiency combined with the higher sugar content; also, there is evidence that the mold secretes anti-yeast substances. *B. cinerea* also produces various phenolic oxidases and glycosidases that can affect wine color and flavor, and it produces extracellular soluble glucans that block membranes during filtration processes.

Mold contamination of wine corks and wooden barrels can be a cause of earthy, moldy, or corky taints in wines (16, 69). Molds use the cork or wood as growth substrate, generating potent aroma compounds such as trichloroanisoles, 1-octen-3-one, geosmin, and guaiacol that are subsequently leached into the wine. Molds isolated from these sources include species of *Penicillium, Aspergillus, Trichoderma, Cladosporium, Paecilomyces,* and *Monilia.*

SPARKLING WINES OR CHAMPAGNE

Sparkling wine production is characterized by secondary fermentation of a base wine in a closed system, whether it be in a bottle or a tank (58, 98). The secondary fermentation produces carbon dioxide which dissolves in the wine and is perceived as bubbles when the pressure is released, hence the term "sparkling." Three types of secondary fermentation processes are used. The traditional *méthode champenoise* requires fermentation in the bottle in which the wine is sold. A base wine (usually white wine) is selected to which sugar or sugar syrup is added to give a final concentration of about 2.4 mg/ml. Yeast nutrients (diammonium phosphate and vitamins) may also be added. A suitable strain of *S. cerevisiae* is inoculated into the wine, which is thoroughly mixed and transferred to the bottles (levurage, tirage). The bottles are sealed with crown caps or corks and then stored at 12 to 15°C for fermentation. The bottles are specially made to withstand the high pressure (about 600 kPa) of carbon dioxide produced during fermentation. Fermentation is completed in 3 to 6 months, after which the wine is aged in the bottles in contact with the yeast lees for at least 6 to 12 months, or longer for premium quality wines. After storage, the bottles are restacked with their necks downwards and periodically twisted or shaken to facilitate settling of the yeast sediment into the neck. This process is called "riddling" or *remuage.* Finally, the sedimented yeast plug is removed from the bottle neck by a process termed *disgorgement.* The yeast plug is frozen by dipping the neck of the bottle into calcium chloride solution at −24°C and then the cork or cap is removed, allowing the internal pressure to force out the yeast plug with little loss of wine. The lost wine is replaced by the addition of base wine (*dosage*) which may contain a small amount of sugar, to give a desired amount of sweetness, and sulfur dioxide. Remuage, disgorgement, and dosage are automated processes in the modern winery. Following these operations the bottles are corked and an *agraffe* (wire clamp) is applied.

Variations in the *méthode champenoise* include the transfer method and bulk fermentation or Charmat process. In the transfer process, fermentation is conducted in the bottle as described already, followed by a short maturation on lees. The bottle contents are chilled and emptied into a pressurized tank. Dosage is added to the wine, which is filtered to remove the yeasts and bottled. In the Charmat process, the base wine, with added sugar and yeast, is fermented in a pressurized tank. After fermentation, the wine is chilled and clarified by centrifugation before addition to a second tank (pressurized) containing the dosage. The wine is finally filtered and bottled.

Special strains of *S. cerevisiae* are required for the secondary fermentation. Criteria for these strains include: give complete sugar fermentation under conditions of low temperature (10 to 15°C), relatively high ethanol concentration (8 to 12%), low pH (as low as 3.0), presence of up to 20 μg/ml of free sulfur dioxide, low nutrient availability, and increasing pressure of carbon dioxide (up to 600 kPa); flocculate and sediment to facilitate the riddling process; give good flavor; and undergo autolysis during aging on the lees. The yeast may be added directly to the wine base as a rehydrated commercial culture, or it may be propagated for several days in a grape juice–wine mixture as a method for preadaptation to growth in low temperature, low pH, and the presence of ethanol. Usually, the yeast is inoculated into the wine base at a population of 1×10^6 to 5×10^6 CFU/ml. The secondary fermentation is completed over the next 30 to 40 days, during which time the yeast grows to a population of about 10^7 CFU/ml. After this time, the population of viable yeast cells slowly decreases, and by about 100 days all the cells have died. Autolysis of the yeast cells then commences (6, 15).

Chemical changes occur in two phases during production of sparkling wine. Changes occur during the short, relatively distinct phase of secondary fermentation and also occur during the period of aging and yeast autolysis. During secondary fermentation, the added sugar is utilized with the production of carbon dioxide and about 1% ethanol. In the early stages, amino nitrogen is consumed, but after exhaustion of sugar, yeast cells give a small efflux of amino nitrogen into the wine. Minor changes in the concentrations of glycerol, some organic acids, and some esters have been noted, but the data are not consistent (6). It must be appreciated that, during secondary fermentation, the yeast cells are operating under extreme environmental conditions, especially increasing concentration of carbon dioxide, and these circumstances are likely to affect their metabolic behavior.

Yeast autolysis is considered to have a distinctive (mostly positive) impact upon sparkling wine quality in terms of aroma, flavor, bubble size, and bubble persistence. As mentioned already, the autolysis of yeast cells is characterized by the degradation of cellular macromolecules and release of their degradation products as well as other cell constituents into the external medium. Consistent with these reactions, the aging stage of sparkling wine production has been correlated with gradual increases in concentrations of proteins, amino acids, nucleic acids, lipids, free fatty acids, and mannoproteins (originating from the yeast cell wall) in wine. In addition, changes occur in the concentrations of various esters, higher alcohols, carbonyl compounds, terpenes, and lactones which may or may not be related to yeast autolysis. By mechanisms not fully understood, these changes influence wine aroma, flavor, and bubble properties (15, 125).

An innovation in sparkling wine production is the use of yeasts immobilized in beads of alginate, principally for the purposes of accelerating removal of yeasts by the riddling process (18, 70). The entrapped yeast cells perform satisfactorily during the secondary fermentation and subsequent aging, giving products with sensory quality comparable to that produced by the traditional process. Strategies to accelerate yeast autolysis and decrease the time and cost of the aging stage would be worthy of development.

FORTIFIED WINES

Fortified wines such as sherry, port, and Madeira have ethanol concentrations of 15 to 22%. This higher ethanol concentration is achieved by the addition of ethanol (usually derived from the distillation of wine products) at certain stages during the process. Details of the processes for producing fortified wines are given elsewhere (11, 46, 100).

With sherry-style wines, an essentially dry white wine base is produced from particular grape varieties. The yeast ecology and biochemistry of this fermentation are similar to those described already. At the completion of alcoholic fermentation, the wine is fortified to about 15 to 16% ethanol and transferred to large oak casks for aging and maturation by the so-called *solera* system, where portions of the wine in the cask are systematically removed and replaced with an amount of younger wine of the same style so that the wine is continuously blended and emerges with a consistent character. The frequency of transfers and the amount of wine removed from each cask at any one time vary according to the producer. With the *finos* style of sherry, a surface microflora of yeast (*flor*) naturally develops at the air-wine interface. Essentially, the flor is a wrinkled layer or film of yeast biomass about 5 to 10 mm thick. The main species involved are *Torulaspora delbrueckii* or its imperfect form *C. colliculosa* and, to a lesser extent, *Zygosaccharomyces rouxii*, but the ecology of the flor requires more thorough study. While conditions in the flor are strongly oxidative, the interface between the wine and the flor is effectively anaerobic and is likely to influence its ecology. Through its oxidative metabolism, the flor gives a gradual decrease in volatile acidity, glycerol, and alcohol content and a substantial increase in the concentration of acetaldehyde. These changes, as well as many other less quantitative chemical changes caused by the yeast, give the final character of the sherry. Not all sherries develop a flor. The *oloroso* sherries are fortified to 18 to 19% ethanol, which is sufficient concentration of ethanol to prevent flor formation.

Port-style wines are prepared from a red wine base, while the Madeira-style wines are reproduced from a blend of red and white wines. In these processes, fortification with ethanol is done at an appropriate stage during the alcoholic fermentation so that the fermentation is arrested to give a wine with residual unfermented sugar and a desired amount of sweetness. These wines are then subjected to particular processes for aging and maturation which do not involve any secondary fermentations because of the high concentration of ethanol.

Despite their high ethanol concentrations, fortified wines may undergo spoilage from ethanol-tolerant strains of yeasts (*Zygosaccharomyces bisporus, S. cerevisiae,* and *Rhodotorula, Candida,* and *Brettanomyces* species) and lactic acid bacteria (*Lb. hilgardii* and other species) (120, 124).

References

1. **Amerine, M. A.** 1985. Winemaking, p. 67–81. *In* H. Koprowski and S. A. Plotin (ed.), *World's Debt to Pasteur.* Alan R. Liss Inc., New York.
2. **Barre, P., F. Vezinhet, S. Dequin, and B. Blondin.** 1993. Genetic improvement of wine yeasts, p. 265–287. *In* G. H. Fleet (ed.), *Wine Microbiology and Biotechnology.* Harwood Academic Publishers, Chur, Switzerland.
3. **Bayonove, C.** 1989. Incidences des attaques parasitaires fongiques sur la composition qualitative du raisin et des vins. *Rev. Fr. Oenol.* **116:**29–39.
4. **Benda, I.** 1982. Wine and brandy, p. 293–402. *In* G. Reed (ed.), *Prescott and Dunns Industrial Microbiology,* 4th ed. AVI Publishing Co., Westport, Conn.
5. **Berry, D. R., and D. C. Watson.** 1987. Production of organoleptic compounds, p. 345–366. *In* D. R. Berry, I. Russell, and G. Stewart (ed.), *Yeast Biotechnology.* Allen and Unwin, London.
6. **Bidan, P., M. Feuillat, and J. Moulin.** 1986. Rapport de la France. Les vins Mousseux. *Bull. OIV (Off. Int. Vigne Vin)* **59:**563–626.
7. **Bisson, L. F.** 1991. Influence of nitrogen on yeast and fermentation of grapes, p. 172–184. *In* J. Rantz (ed.), *Proceedings of the International Symposium on Nitrogen in Grapes and Wines.* American Society for Enology and Viticulture, Davis, Calif.
8. **Bisson, L. F.** 1993. Yeasts—metabolism of sugars, p. 55–75. *In* G. H. Fleet (ed.), *Wine Microbiology and Biotechnology.* Harwood Academic Publishers, Chur, Switzerland.
9. **Bisson, L. F., and R. E. Kunkee.** 1991. Microbial interactions during wine production, p. 37–68. *In* J. G. Zeikus and E. A. Johnson (ed.), *Mixed Cultures in Biotechnology.* McGraw-Hill, New York.
10. **Boulton, R.** 1995. Red wines, p. 121–158. *In* A. G. H. Lea and J. R. Piggott (ed.), *Fermented Beverage Production.* Blackie Academic & Professional, Glasgow.
11. **Boulton, R. B., V. L. Singleton, L. F. Bisson, and R. E. Kunkee.** 1995. *Principles and Practices of Winemaking.* Chapman and Hall, New York.
12. **Cantarelli, C.** 1989. Factors affecting the behaviour of yeast in wine fermentation, p. 127–151. *In* C. Cantarelli and G. Lanzarini (ed.), *Biotechnology Applications in Beverage Production.* Elsevier Applied Science, London.
13. **Capucho, I., and M. V. San Ramão.** 1994. Effect of ethanol and fatty acids on malolactic activity of *Leuconostoc oenos. Appl. Microbiol. Biotechnol.* **42:**391–395.
14. **Charpentier, C.** 1993. Stuck fermentations: the role of ethanol and resistance of the yeast. *Rev. Fr. Oenol.* **140:**49–52.
15. **Charpentier, C., and M. Feuillat.** 1993. Yeast autolysis, p. 225–242. *In* G. H. Fleet (ed.), *Wine Microbiology and Biotechnology.* Harwood Academic Publishers, Chur, Switzerland.
16. **Chatonnet, P., N. J. Boidron, and D. Dubourdieu.** 1994. The microflora of cork wood and their development during the seasoning and ageing in the open air. *J. Int. Sci. Vigne Vin* **28:**185–201.
17. **Cogan, T. M.** 1995. Flavor production by dairy starter cultures. *J. Appl. Bacteriol.* (Symp. Suppl.) **79:**49S–64S.
18. **Colagrande, O., A. Silva, and M. D. Fumi.** 1994. Recent applications of biotechnology in wine production. *Biotechnol. Prog.* **10:**2–18.
19. **Cole, V. C., and A. C. Noble.** 1995. Flavor chemistry and assessment, p. 361–385. *In* A. G. H. Lea and J. R. Piggott (ed.), *Fermented Beverage Production.* Blackie Academic and Professional, Glasgow.
20. **Cottrell, T. H. E., and M. R. McLellan.** 1986. The effect of fermentation temperature on chemical and sensory characteristics of wine from seven white grape cultivars grown in New York State. *Am. J. Enol. Viticult.* **37:**190–194.
21. **Daeschel, M. A., D.-S. Jung, and B. T. Watson.** 1991. Controlling wine malolactic fermentation with nisin and nisin-resistant strains of *Leuconostoc oenos. Appl. Environ. Microbiol.* **57:**601–603.
22. **Davis, C. R., N. F. A. Silveira, and G. H. Fleet.** 1985. Occurrence and properties of bacteriophages of *Leuconstoc oenos* in Australian wines. *Appl. Environ. Microbiol.* **50:**872–876.
23. **Davis, C. R., D. Wibowo, R. Eschenbruch, T. H. Lee, and G. H. Fleet.** 1985. Practical implications of malolactic fermentation—a review. *Am. J. Enol. Viticult.* **36:**209–301.
24. **Davis, C. R., D. Wibowo, G. H. Fleet, and T. H. Lee.** 1988. Properties of wine lactic acid bacteria: their potential enological significance. *Am. J. Enol. Viticult.* **39:**137–142.
25. **Davis, C. R., D. Wibowo, T. H. Lee, and G. H. Fleet.** 1986. Growth and metabolism of lactic acid bacteria during and after malolactic fermentation of wines at different pH. *Appl. Environ. Microbiol.* **51:**539–545.
26. **Degré, R.** 1993. Selection and cultivation of wine yeast and bacteria, p. 421–447. *In* G. H. Fleet (ed.), *Wine Mi-*

crobiology and Biotechnology. Harwood Academic Publishers, Chur, Switzerland.

27. **Delfini, G., and A. Costa.** 1993. Effects of grape must lees and insoluble materials on the alcoholic fermentation rate and the production of acetic acid, pyruvic acid and acetaldehyde. *Am. J. Enol. Viticult.* **44:**86–92.
28. **Divies, C.** 1993. Bioreactor technology and wine fermentation, p. 449–475. *In* G. H. Fleet (ed.), *Wine Microbiology and Biotechnology.* Harwood Academic Publishers, Chur, Switzerland.
29. **Doneche, B.** 1993. Botrytized wines, p. 327–351. *In* G. H. Fleet (ed.), *Wine Microbiology and Biotechnology.* Harwood Academic Publishers, Chur, Switzerland.
30. **Drysdale, G. S., and G. H. Fleet.** 1988. Acetic acid bacteria in winemaking—a review. *Am. J. Enol. Viticult.* **39:** 143–154.
31. **Drysdale, G. S., and G. H. Fleet.** 1989. The effect of acetic acid bacteria upon the growth and metabolism of yeasts during the fermentation of grape juice. *J. Appl. Bacteriol.* **67:**471–481.
32. **Edwards, C. G., R. B. Beelman, C. E. Bartley, and A. L. McConnell.** 1990. Production of decanoic acid and other volatile compounds and the growth of yeast and malolactic bacteria during vinification. *Am. J. Enol. Viticult.* **41:**48–56.
33. **Edwards, C. G., J. C. Peterson, T. D. Boylston, and J. D. Vasile.** 1994. Interactions between *Leuconostoc oenos* and *Pediococcus* spp. during vinification of red wines. *Am. J. Enol. Viticult.* **45:**49–55.
34. **Ewart, A.** 1995. White wines, p. 95–120. *In* A. G. H. Lea and J. R. Piggot (ed.), *Fermented Beverage Production.* Blackie Academic and Professional, Glasgow.
35. **Faia, A. M., and F. Radler.** 1990. Investigation of the bactericidal effect of nisin on lactic acid bacteria of wine. *Vitis* **29:**233–238.
36. **Faith, K. P., and F. Radler.** 1994. Investigation of the formation of amines by lactic acid bacteria. *Wein-Wiss.* **49:**11–16.
37. **Fleet, G. H.** 1992. Spoilage yeasts. *Crit. Rev. Biotechnol.* **12:**1–44.
38. **Fleet, G. H.** 1993. The microorganisms of winemaking—isolation, enumeration and identification, p. 1–26. *In* G. H. Fleet (ed.), *Wine Microbiology and Biotechnology.* Harwood Academic Publishers, Chur, Switzerland.
39. **Fleet, G. H., and G. M. Heard.** 1993. Yeasts—growth during fermentation, p. 27–54. *In* G. H. Fleet (ed.), *Wine Microbiology and Biotechnology.* Harwood Academic Publishers, Chur, Switzerland.
40. **Gao, C., and G. H. Fleet.** 1988. The effects of temperature and pH on the ethanol tolerance of the wine yeasts, *Saccharomyces cerevisiae, Candida stellata* and *Kloeckera apiculata. J. Appl. Bacteriol.* **65:**405–410.
41. **Gao, C., and G. H. Fleet.** 1994. Degradation of malic acid by high density cell suspensions of *Leuconostoc oenos. J. Appl. Bacteriol.* **76:**632–637.
42. **Gao, C., and G. H. Fleet.** 1995. Cell-recycle membrane bioreactor for conducting continuous malolactic fermentation. *Aust. J. Grape Wine Res.* **1:**32–38.
43. **Gao, C., and G. H. Fleet.** 1995. Degradation of malic and tartaric acids by high density cell suspensions of wine yeasts. *Food Microbiol.* **12:**65–71.
44. **Gardner, N., N. Rodrigue, and C. P. Champagne.** 1993. Combined effects of sulfites, temperature and agitation time on production of glycerol in grape juice by *Saccharomyces cerevisiae. Appl. Environ. Microbiol.* **59:** 2022–2028.
45. **Giudici, P., and R. E. Kunkee.** 1994. The effect of nitrogen deficiency and sulfur-containing amino acids on the reduction of sulfate to hydrogen sulfide by wine yeasts. *Am. J. Enol. Viticult.* **45:**107–112.
46. **Goswell, R. W., and R. E. Kunkee.** 1977. Fortified wines, p. 478–533. *In* A. H. Rose (ed.), *Economic Microbiology,* vol. 1. Academic Press, London.
47. **Heard, G. M., and G. H. Fleet.** 1985. Growth of natural yeast flora during the fermentation of inoculated wines. *Appl. Environ. Microbiol.* **50:**727–728.
48. **Heard, G. M., and G. H. Fleet.** 1988. The effects of temperature and pH on the growth of yeast species during the fermentation of grape juice. *J. Appl. Bacteriol.* **65:**23–28.
49. **Henick-Kling, T.** 1993. Malolactic fermentation, p. 289–326. *In* G. H. Fleet (ed.), *Wine Microbiology and Biotechnology.* Harwood Academic Publishers, Chur, Switzerland.
50. **Henick-Kling, T.** 1995. Control of malolactic fermentation in wine: energetics, flavor modification and methods of starter culture preparation. *J. Appl. Bacteriol.* (Symp. Suppl.) **79:**29S–37S.
51. **Henick-Kling, T., and Y. H. Park.** 1994. Considerations for the use of yeast and bacterial starter cultures: sulfur dioxide and timing of inoculation. *Am. J. Enol. Viticult.* **45:**464–469.
52. **Henschke, P., and V. Jiranek.** 1993. Yeasts—metabolism of nitrogen compounds, p. 77–164. *In* G. H. Fleet (ed.), *Wine Microbiology and Biotechnology.* Harwood Academic Publishers, Chur, Switzerland.
53. **Henschke, P. A., and C. S. Ough.** 1991. Urea accumulation in fermenting grape juice. *Am. J. Enol. Viticult.* **42:**317–321.
54. **Hernawan, T., and G. H. Fleet.** 1995. Chemical and cytological changes during the autolysis of yeasts. *J. Ind. Microbiol.* **14:**440–450.
55. **Herraiz, T., G. Reglero, M. Herraiz, P. J. Martin-Alvarez, and M. D. Cabezudo.** 1990. The influence of the yeast and type of culture on the volatile composition of wines fermented without sulfur dioxide. *Am. J. Enol. Viticult.* **41:**313–318.
56. **Ingledew, M., and R. E. Kunkee.** 1985. Factors influencing sluggish fermentation of grape juice. *Am. J. Enol. Viticult.* **36:**65–76.
57. **Jiranek, V., P. Langridge, and P. A. Henschke.** 1995. Regulation of hydrogen sulfide liberation in wine-producing *Saccharomyces cerevisiae* strains by assimilable nitrogen. *Appl. Environ. Microbiol.* **61:**461–467.
58. **Jordan, A. D.** 1994. Style and quality of Australian sparkling wines over the past 20 years. *Food Aust.* **46:**184–188.
59. **Joyeux, A., S. Lafon-Lafourcade, and P. Ribéreau-Gayon.** 1984. Metabolism of acetic acid bacteria in grape must. Consequences on alcoholic and malolactic fermentation. *Sci. Aliments* **4:**247–255.

60. **Kunkee, R. E.** 1984. Selection and modification of yeasts and lactic acid bacteria for wine fermentation. *Food Microbiol.* **1**:315–332.
61. **Kunkee, R. E.** 1991. Some roles of malic acid in the malolactic fermentation in winemaking. *FEMS Microbiol. Rev.* **88**:55–72.
62. **Kunkee, R. E., and M. A. Amerine.** 1970. Yeasts in winemaking, p. 50–71. *In* A. H. Rose and J. S. Harrison (ed.), *The Yeasts,* vol. 3. *Yeast Technology.* Academic Press, London.
63. **Kunkee, R. E., and L. Bisson.** 1993. Wine-making yeasts, p. 69–128. *In* A. H. Rose and J. S. Harrison (ed.), *The Yeasts,* 2nd ed., vol. 5. *Yeast Technology.* Academic Press, London.
64. **Lafon-Lafourcade, S.** 1983. Wine and brandy, p. 81–163. *In* H. J. Rehm and G. Reed (ed.), *Biotechnology,* vol. 5, *Food and Feed Production with Microorganisms.* Verlag Chemie, Weinheim.
65. **Lafon-Lafourcade, S., and A. Joyeux.** 1981. Les bactéries acétique du vin. *Bull. OIV (Off. Int. Vigne Vin)* **608**:803–829.
66. **Large, P. J.** 1986. Degradation of organic nitrogen compounds by yeasts. *Yeast* **2**:1–34.
67. **Laurent, M. H., T. Henick-Kling, and T. Acree.** 1994. Changes in the aroma and odour of Chardonnay wine due to malolactic fermentation. *Wein-Wiss.* **49**:3–10.
68. **Lavalée, F., Y. Salvas, S. Lamy, D. Y. Thomas, R. Degré, and L. Dulav.** 1994. PCR and DNA fingerprinting used as quality control in the production of wine yeast strains. *Am. J. Enol. Viticult.* **45**:86–91.
69. **Lee, T. H., and R. F. Simpson.** 1993. Microbiology and chemistry of cork taints in wine, p. 353–372. *In* G. H. Fleet (ed.), *Wine Microbiology and Biotechnology.* Harwood Academic Publishers, Chur, Switzerland.
70. **Lemonnier, J., and B. Duteurtre.** 1989. Un progrès important pour le champagne et les vins de methode traditionnelle. *Rev. Franc. Oenol.* **121**:16–26.
71. **Liu, S. Q., C. R. Davis, and J. D. Brooks.** 1995. Growth and metabolism of selected lactic acid bacteria in synthetic wine. *Am. J. Enol. Viticult.* **46**:166–174.
72. **Liu, S. Q., G. G. Pritchard, M. Hardman, and G. J. Pilone.** 1994. Citrulline production and ethyl carbamate (urethane) precursor formation from arginine degradation by wine lactic acid bacteria. *Leuconostoc oenos* and *Lactobacillus buchneri. Am. J. Enol. Viticult.* **45**:235–242.
73. **Longo, E., and F. Vezinhet.** 1993. Chromosomal rearrangements during vegetative growth of a wild strain of *Saccharomyces cerevisiae. Appl. Environ. Microbiol.* **59**:322–326.
74. **Lonvaud-Funel, A., A. Joyeux, and C. Desens.** 1988. Inhibition of malolactic fermentation by products of yeast metabolism. *J. Sci. Food Agric.* **44**:183–191.
75. **Margalith, P. Z.** 1981. *Flavor Microbiology.* Charles C Thomas Publisher, Springfield, Ill.
76. **Martineau, B., and T. Henick-Kling.** 1995. Performance and diacetyl production of commercial strains of malolactic bacteria in wine. *J. Appl. Bacteriol.* **78**:526–536.
77. **Martini, A.** 1993. Origin and domestication of the wine yeast *Saccharomyces cerevisiae. J. Wine Res.* **4**:165–176.
78. **Mateo, J. J., M. Jimenez, T. Huerta, and A. Pastor.** 1991. Contribution of different yeasts isolated from musts of monastrell grapes to the aroma of wine. *Int. J. Food Microbiol.* **14**:153–160.
79. **Mora, J., J. I. Barbas, and A. Mulet.** 1990. Growth of yeast species during the fermentation of musts inoculated with *Kluyveromyces thermotolerans* and *Saccharomyces cerevisiae. Am. J. Enol. Viticult.* **41**:156–159.
80. **Mora, J., and A. Mulet.** 1991. Effects of some treatments of grape juice on the population and growth of yeast species during fermentation. *Am. J. Enol. Viticult.* **42**: 133–136.
81. **Mortimer, R. K., P. Romano, G. Suzzi, and M. Polsnelli.** 1994. Genome renewal: a phenomenon revealed from genetic study of 43 strains of *Saccharomyces cerevisiae* derived from natural fermentation of grape musts. *Yeast* **10**:1543–1552.
82. **Moruno, E. G., C. Delfini, E. Pessione, and C. Giunta.** 1993. Factors affecting acetic acid production by yeasts in strongly clarified grape musts. *Microbios* **74**:249–256.
83. **Munoz, E., and M. Ingeldew.** 1990. Yeast hulls in wine fermentations—a review. *J. Wine Res.* **1**:197–210.
84. **Nair, N. G.** 1990. Strategies for fungicidal control of bunch rot of grapes caused by *Botrytis cinerea* in the Hunter Valley. *Aust. New Zeal. Wine Ind. J.* **1990**(May): 218–220.
85. **Ness, F., F. Lavellée, D. Dubourdieu, M. Aigle, and L. Dulau.** 1993. Identification of yeast strains using the polymerase chain reaction. *J. Sci. Food Agric.* **62**:89–94.
86. **Nykanen, L.** 1986. Formation and occurrence of flavor compounds in wine and distilled alcoholic beverages. *Am. J. Enol. Viticult.* **37**:84–96.
87. **Ough, C. S.** 1966. Fermentation rates of grape juice. II. Effects of initial Brix, pH and fermentation temperature. *Am. J. Enol. Viticult.* **17**:74–81.
88. **Ough, C. S., and E. A. Crowell.** 1987. Use of sulfur dioxide in winemaking. *J. Food Sci.* **52**:386–388, 393.
89. **Ough, C. S., M. Davenport, and K. Joseph.** 1989. Effects of certain vitamins on growth and fermentation rate of several commercial active dry wine yeasts. *Am. J. Enol. Viticult.* **40**:208–213.
90. **Ough, C. S., and M. L. Groat.** 1978. Particle nature, yeast strain, and temperature interactions on the fermentation rates of grape juice. *Appl. Environ. Microbiol.* **35**:881–885.
91. **Park, S. K., R. B. Boulton, E. Bartra, and A. C. Noble.** 1994. Incidence of volatile sulfur compounds in California wines. A preliminary survey. *Am. J. Enol. Viticult.* **45**:341–344.
92. **Pimentel, M. S., M. H. Silva, I. Cortes, and A. M. Faia.** 1994. Growth and metabolism of sugar and acids of *Leuconostoc oenos* under different conditions of temperature and pH. *J. Appl. Bacteriol.* **76**:42–48.
93. **Pretorius, I., and T. J. van der Westiuzen.** 1991. The impact of yeast genetics and recombinant DNA technology on the wine industry—a review. *S. Afr. J. Enol. Viticult.* **12**:3–31.
94. **Querol, A., E. Barrio, T. Huerta, and D. Ramon.** 1992. Molecular monitoring of wine fermentations conducted by active dry yeast strains. *Appl. Environ. Microbiol.* **58**: 2948–2953.

95. **Querol, A., E. Barrio, and D. Ramon.** 1994. Population dynamics of natural *Saccharomyces* strains during wine fermentation. *Int. J. Food Microbiol.* **21**:315–323.

95a. **Querol, A., and D. Ramón.** 1996. The application of molecular techniques in wine microbiology. *Trends Food Sci. Technol.* **7**:73–78.

96. **Radler, F.** 1993. Yeasts—metabolism of organic acids, p. 165–182. *In* G. H. Fleet (ed.), *Wine Microbiology and Biotechnology.* Harwood Academic Publishers, Chur, Switzerland.

97. **Radler, F., and B. Lotz.** 1990. The microflora of active dry yeast and the quantitative changes during fermentation. *Wein Wiss.* **45**:114–122.

98. **Rankine, B. L.** 1989. *Making Good Wine. A Manual of Winemaking Practices for Australia and New Zealand.* Sun Books, Melbourne.

99. **Rauhut, D.** 1993. Yeasts—production of sulfur compounds, p. 183–223. *In* G. H. Fleet (ed.), *Wine Microbiology and Biotechnology.* Harwood Academic Publishers, Chur, Switzerland.

100. **Reader, H. P., and M. Dominguez.** 1995. Fortified wines: sherry, port and madeira, p. 159–207. *In* A. G. H. Lea and J. R. Piggott (ed.), *Fermented Beverage Production.* Blackie Academic and Professional, Glasgow.

101. **Reed, G., and T. W. Nagodawithana.** 1988. Technology of yeast usage in wine making. *Am. J. Enol. Viticult.* **39**:83–90.

102. **Reed, G., and T. W. Nagodawithana (ed.).** 1992. *Yeast Technology,* 2nd ed. AVI Publishing Co., Westport, Conn.

103. **Regueiro, L. A., C. L. Costas, and J. E. L. Rubio.** 1993. Influence of viticultural and enological practices on the development of yeast populations during winemaking. *Am. J. Enol. Viticult.* **44**:405–408.

104. **Ribéreau-Gayon, J., P. Ribéreau-Gayon, and G. Seguin.** 1980. *Botrytis cinerea* in enology, p. 251–274. *In* J. R. Coley-Smith, K. Verhoeff, and W. R. Jarvis (ed.), *The Biology of Botrytis.* Academic Press, London.

105. **Rodriguez, A. V., and M. C. Marca de Nadra.** 1994. Sugar and organic acid metabolism in mixed cultures of *Pediococcus pentosaceus* and *Leuconostoc oenos* isolated from wine. *J. Appl. Bacteriol.* **77**:61–66.

106. **Rodriguez, A. V., and M. C. Manca de Nadra.** 1995. Mixed culture of *Lactobacillus hilgardii* and *Leuconostoc oenos* isolated from Argentine wine. *J. Appl. Bacteriol.* **78**:521–525.

107. **Romano, P., and G. Suzzi.** 1993. Acetoin production in *Saccharomyces cerevisiae* wine yeasts. *FEMS Microbiol. Lett.* **108**:23–26.

108. **Romano, P., and G. Suzzi.** 1993. Sulfur dioxide and wine microorganisms, p. 373–394. *In* G. H. Fleet (ed.), *Wine Microbiology and Biotechnology.* Harwood Academic Publishers, Chur, Switzerland.

109. **Romano, P., and G. Suzzi.** 1993. Potential use for *Zygosaccharomyces* species in winemaking. *J. Wine Res.* **4**:87–94.

110. **Romano, P., G. Suzzi, G. Comi, and R. Zironi.** 1992. Higher alcohol and acetic acid production by apiculate wine yeasts. *J. Appl. Bacteriol.* **73**:126–130.

111. **Romano, P., G. Suzzi, L. Turbanti, and M. Polsinelli.** 1994. Acetaldehyde production in *Saccharomyces cerevisiae* wine yeasts. *FEMS Microbiol. Lett.* **118**:213–218.

112. **Romano, P., G. Suzzi, R. Zironi, and G. Comi.** 1993. Biometric study of acetoin production in *Hanseniaspora guilliermondii* and *Kloeckera apiculata. Appl. Environ. Microbiol.* **59**:1838–1841.

113. **Schreier, P.** 1979. Flavor composition of wines: a review. *Crit. Rev. Food Sci. Nutr.* **12**:59–111.

114. **Schutz, M., and J. Gafner.** 1993. Analysis of yeast diversity during spontaneous and induced alcoholic fermentations. *J. Appl. Bacteriol.* **75**:551–558.

115. **Schutz, M., and J. Gafner.** 1994. Dynamics of the yeast strain population during spontaneous alcoholic fermentation determined by CHEF gel electrophoresis. *Lett. Appl. Microbiol.* **19**:253–259.

116. **Shimazu, Y., and M. Watanabe.** 1981. Effects of yeast strains and environmental conditions on formation of organic acids in must during fermentation. *J. Ferment. Technol.* **59**:27–32.

117. **Shimizu, K.** 1993. Killer yeasts, p. 243–264. *In* G. H. Fleet (ed.), *Wine Microbiology and Biotechnology.* Harwood Academic Publishers, Chur, Switzerland.

118. **Shinohara, T.** 1984. L'importance des substances volatiles dur vin. Formation et effets sur la qualité. *Bull. Off. Int. Vin* **57**:606–618.

119. **Soufleros, E., and A. Bertrand.** 1979. Rôle de la souche de levure dans la production des substances volatiles au cours de la fermentation du jus de raisin. *Conn. Vigne Vin* **13**:181–198.

120. **Sponholz, W. R.** 1993. Wine spoilage by microorganisms, p. 395–420. *In* G. H. Fleet (ed.), *Wine Microbiology and Biotechnology.* Harwood Academic Publishers, Chur, Switzerland.

121. **Strasser de Saad, A. M., S. E. Pasteris, and M. C. Manca de Nadra.** 1995. Production and stability of pediocin N5p in grape juice medium. *J. Appl. Bacteriol.* **78**:473–476.

122. **Strehaiano, P.** 1993. Stuck fermentations: causes and physiological factors. *Rev. Fr. Oenol.* **140**:41–48.

123. **Suzzi, G., P. Romano, I. Ponti, and C. Montuschi.** 1995. Natural wine yeasts as biocontrol agents. *J. Appl. Bacteriol.* **78**:304–309.

124. **Thomas, S. D.** 1993. Yeasts as spoilage organisms in beverages, p. 517–562. *In* A. H. Rose and J. S. Harrison (ed.), *The Yeasts,* 2nd ed., vol. 5. *Yeast Technology.* Academic Press, London.

125. **Tvetanov, O. S., and G. K. Bambalov.** 1994. Effect of yeast strain on the content of nitrogenous compounds and the quality of red sparkling wines stored with yeasts. *J. Wine Res.* **5**:41–52.

126. **van Uden, N.** 1989. Effect of alcohols on the temperature relations of growth and death in yeasts, p. 77–88. *In* V. van Uden (ed.), *Alcohol Toxicity in Yeasts and Bacteria.* CRC Press Inc., Boca Raton, Fla.

127. **van Vuuren, H. J. J., and L. M. T. Dicks.** 1993. *Leuconostoc oenos*—a review. *Am. J. Enol. Viticult.* **44**:99–112.

128. **van Vuuren, H. J., and C. J. Jacobs.** 1992. Killer yeasts in the wine industry. A review. *Am. J. Enol. Viticult.* **43**:119–128.

129. **van Zyl, J. A., M. J. De Vries, and A. S. Zeeman.** 1963. The microbiology of South African winemaking. III. The

effect of different yeasts on the composition of fermented musts. *S. Afr. J. Agric. Sci.* **6**:165–180.

130. **Versavaud, A., P. Courcoux, C. Roulland, L. Dulau, and J. N. Hallet.** 1995. Genetic diversity and geographical distribution of wild *Saccharomyces cerevisiae* strains from the wine-producing area of Charentes, France. *Appl. Environ. Microbiol.* **61**:3521–3529.
131. **Wibowo, D., R. Eschenbruch, C. Davis, G. H. Fleet, and T. H. Lee.** 1985. Occurrence and growth of lactic acid bacteria in wine—a review. *Am. J. Enol. Viticult.* **36**: 302–313.
132. **Wibowo, D., G. H. Fleet, T. H. Lee, and R. E. Eschenbruch.** 1988. Factors affecting the induction of malolactic fermentation in red wines with *Leuconostoc oenos*. *J. Appl. Bacteriol.* **64**:421–428.
133. **Yokotsuka, K., A. Otaki, A. Naitoh, and H. Tanaka.** 1993. Controlled deacidification and alcohol fermentation of a high-acid grape must using two immobilized yeasts, *Schizosaccharomyces pombe* and *Saccharomyces cerevisiae*. *Am. J. Enol. Viticult.* **44**:371–377.
134. **Zironi, R., P. Romano, G. Suzzi, F. Battistutta, and G. Comi.** 1993. Volatile metabolites produced in wine by mixed and sequential cultures of *Hanseniaspora guilliermondii* or *Kloeckera apiculata* and *Saccharomyces cerevisiae*. *Biotechnol. Lett.* **15**:235–238.

Advanced Techniques in Food Microbiology

Servé Notermans
Rijkelt Beumer
Frank Rombouts

38

Detecting Foodborne Pathogens and Their Toxins
Conventional versus Rapid and Automated Methods

The genesis of food microbiology is chronicled in the first chapter of this volume. By the end of the 19th century, it was recognized that microorganisms were responsible for a variety of foodborne diseases. Scientists then demonstrated that treatments such as heating, drying, and salting of foods could destroy or inhibit many of the microorganisms present. This led to the introduction of production processes aimed at reducing foodborne disease. At the same time, techniques were developed for the microbiological analysis of foods. Initially, these were relatively simple and included plating techniques, using blood agar, and enrichment methods for detecting organisms such as *Clostridium botulinum* (49). It also became apparent that foodborne disease resulted from unhygienic practices in food production and preparation. In particular, fecal contamination, as occurs in abattoirs, came to be understood as a means of transmitting infectious diseases.

The cumulative experience of this period laid the foundations for modern food microbiology and the development of more extensive and diverse analytical techniques. The purpose of these techniques was to evaluate the new food processing systems and to help improve process hygiene. It soon became clear that in both cases, the direct search for pathogens could be replaced by testing for so-called indicator organisms. In the early days of food microbiology, it was difficult to detect pathogenic organisms. Initially, only simple techniques such as nonselective plating and enrichment cultures were used for detecting pathogens. When they were isolated, most foodborne pathogens were not identified. Gradually, these techniques were improved by developing selective and differential media. The introduction of specific biochemical tests, immunoassays, and DNA techniques now allows the detection and identification of nearly all known foodborne pathogens.

This chapter considers the need of current detection methods for foodborne microorganisms and their toxins, with special reference to rapid and automated methods. Table 38.1 presents an overview of how microbiological methods are used in safe food production and their relative importance. Additionally, the suitability of conventional versus rapid and automated methods for achieving these ends is indicated. The remainder of this chapter discusses factors involved in the detection of important foodborne pathogens and their toxins. Finally, the often overlooked issues of validation and quality assurance of microbial analytical techniques are considered.

NEED FOR MICROBIOLOGICAL METHODS

Production of Safe Food

Over the last 70 to 80 years, many different methods have been developed for detecting pathogenic microor-

Table 38.1 Use of microbiological methods in safe food production, their relative importance, and use of conventional versus rapid and automated methods

Use of microbiological methods	Relative importance[a]	Most suited methods: Conventional[b]	Rapid and automated[c]
Safe food production			
Monitoring and surveillance			
Detection of pathogens	–	–	++
Detection of indicator organisms	+	+	–
Detection of bacterial toxins	–	–	++
Storage tests	++	+	+
MCT	++	+	+
Predictive models			
Performance testing	+	+	+
Mathematical models	–	–	–
Management of safe food production			
GMP	+	++	++
HACCP			
Hazard analysis	+	+	–
Identification of critical control points	±	+	–
Monitoring	–	–	–
Verification	–	–	–
Failure analysis	+	++	–
Foodborne disorders			
Testing of reported outbreaks	+	++	–
Sentinel studies	++	–	++
Risk assessment studies	++	++	+

[a]Necessity and convenience of a microbiological technique for obtaining reliable and/or applicable results.

[b]Methods based on enumeration of organisms, such as determination of colony-forming particles, and methods allowing organisms to be obtained in a pure state for a further characterization. Conventional methods for detecting bacterial toxins are those using animal models.

[c]Methods that detect organisms on the basis of production of metabolic products or compounds. Such methods for bacterial toxins use a direct test system for the toxin itself.

ganisms or their toxins in food. Since the beginning of that period, statutory requirements relating to food safety have been established in most countries. These requirements are based on the testing of prepared foods for the presence of pathogenic organisms and toxins. Thus, surveillance testing and monitoring were born. In addition to product monitoring, other kinds of tests (e.g., storage tests and microbiological challenge testing [MCT]) may provide relevant information. When organisms of concern are present with sufficient frequency, storage tests can be carried out to determine the potential for multiplication in the final product. If, however, the organism in question is present infrequently or at very low levels, MCT may be more appropriate.

Monitoring and Surveillance

The traditional approach to controlling food safety has been based on education and training of personnel, inspection of production facilities and operations, and microbiological testing of the finished product. The testing of food products has attracted much attention and is usually carried out routinely as part of the overall monitoring system. This monitoring requires the establishment of criteria by which to judge the acceptability of foods. Microbial criteria and the use of indicator organisms are covered extensively in chapter 4. The International Commission on Microbiological Specification of Foods (18) has established microbiological criteria and numerical limits for different organisms in various kinds of foods. It is important to realize that the analytical method(s) to be used for detection and/or enumeration of the target pathogen or toxin is an integral component of the criterion (18).

It is increasingly recognized, however, that although the sampling plans are statistically sound and based on the principles of hazard identification and risk assessment (19), they cannot guarantee that a given batch of food is safe. This is because in practice, the relatively small number of samples that can be taken from an individual batch of food and examined in the laboratory is unlikely to give a reliable indication of the true level of microbial contamination (2). The distribution of microorganisms in foods is rarely homogeneous, and therefore results obtained from samples may not be representative of the batch as a whole (22). Although increasing the number of test samples from a lot will improve the probability of detecting unsafe batches, safety can never be guaranteed by end product testing. Therefore, the relative importance of monitoring in safe food production is low. When monitoring and surveillance are required, the desirability of rapid and automated methods should be obvious since large numbers of samples must be tested to obtain relevant information.

Storage Tests

Storage testing simulates the changes that can occur in a food product after processing. It can be used to determine the microbiological consequences of processing, distribution, and subsequent handling by the consumer. Usually a storage test is carried out only on finished products in which microorganisms of concern are present in sufficient numbers. The test can provide information on the effects of both intrinsic and extrinsic factors on the microbiological safety of a particular product. For this purpose, the product is processed as usual and then held under a range of controlled conditions. If the conditions relate to normal handling of the product during distribution, storage, and preparation by the consumer, then the results indicate probable exposure of the consumer to the target pathogens. Storage testing is not routinely done on a continuous basis and therefore does not require rapid and automated methods.

MCT

MCT requires that the food product be inoculated (i.e., challenged) with one or more relevant organisms. The processing, distribution, and subsequent handling of the product are simulated to determine the fate of the microbes. When the decision has been taken to carry out MCT and the relevant challenge organisms have been chosen, a trial protocol is needed. A user's guide to MCT for ensuring the safety of food products has been developed by Notermans et al. (34). MCT tests yield information on the types of microorganisms that can grow in the product, so that the risk of food poisoning can be assessed. Like storage tests, MCT is not routine and can use a traditional approach based on cultural methods.

Management of Safe Food Production

As indicated above, monitoring of the final product is no guarantee of safety. Even in the best postprocess monitoring plan, unsafe batches may escape detection. Also, even storage tests and MCT may give misleading results if, for example, the food is recontaminated after processing. Food production is subject to variables such as seasonal variation in the composition and microbial quality of raw materials. Also, the formulation of the food product may change gradually in response to consumer demand. Furthermore, there will be variations between factories in processing procedures. Thus, the production of safe food must depend on the use of appropriate quality assurance systems.

GMP and Microbiological Methods

One of the first quality assurance systems developed by the food industry is the application of good manufacturing practices (GMPs), which supplement end product testing. Such an approach is essentially proactive and preventive. GMPs provide general rules based on practical experience over a long period of time and include attention to environmental conditions, e.g., requirements for plant design and control of working procedures. Nevertheless, in a food production environment, there are various other factors that contribute to food safety, such as hand washing, cleaning and disinfection of the plant, avoiding condensation, and excluding sick workers from the factory. Because the effects of such factors on product safety cannot be quantified, they are outside the requirements of a quantitative hazard analysis and critical control point (HACCP) plan (see below and chapter 41) and are best included in the category of GMP. Although most GMP activities can be checked by visual inspection, there is one important application of a rapid method for verifying GMPs: the use of techniques which quantify ATP levels to monitor hygiene and cleanliness of food processing environments. For example, food contact surfaces are swabbed, the contents of the swab are mixed with special reagents, and the amount of ATP is estimated through the use of enzyme-mediated bioluminescence. Modern hand-held instruments quantify the amount of ATP as relative light units in a matter of minutes. In these tests, it is unimportant whether the ATP is microbial in origin or due to food debris; neither should be present on properly sanitized surfaces.

HACCP and Microbiological Methods

The use of GMPs constitutes a general means of controlling food safety, which has led to the introduction of the

HACCP concept (6). The HACCP system is a structured approach to the identification, assessment, and control of microbial (and other) hazards in any particular food operation. It aims to identify problems before they occur and establishes suitable control measures. Increasingly, the system is finding favor with both food producers and legislative authorities. In the future, it will play an increasingly important part in controlling the safety of food products. There are now several practical working documents to assist the introduction of HACCP into a food operation (5–7, 19, 21, 51).

In the hazard identification step of HACCP, it is necessary to identify any hazardous organism that may be present in food. In general, however, identification of such organisms is a matter of experience, and if testing is required, it is usually only for a short period. As a consequence, there is not much need for rapid and automated methods in this step.

Verification is step 6 of the HACCP system and involves appropriate supplementary tests, together with a review of the operation to confirm that HACCP is working effectively. The verification step should review the entire HACCP system and all relevant records (6). The study team should specify the methods to be used and the frequency of verification procedures. The methods may include internal auditing systems, microbiological examination of the product at different stages in the process, and more stringent tests at selected critical control points. However, here again it is the functioning HACCP system, not the microbiological testing of end products, that guarantees that the food produced is safe.

Failure analysis is important if, despite all that has been done to produce safe food, a health hazard still occurs. By definition, these failures are unexpected or not predictable. Because of the incidental and unpredictable nature of the failures, it is appropriate to conduct the necessary microbiological testing by conventional methodology.

Methods for Investigating Foodborne Diseases

The production of safe food requires adequate information on foodborne diseases. Establishing the patterns of disease facilitates policy decisions and provides a basis for legislative action, the development of intervention strategies, and the setting of research priorities. In addition, new or emerging foodborne diseases need to be recognized rapidly. Last but not least, the incidence of foodborne diseases can be used to judge the success of legislative measures and control programs. There are several ways in which the necessary information can be obtained: (i) analysis of reported food-associated incidents of disease, (ii) sentinel and population studies, and (iii) calculation of human exposure to disease agents via dose-response relationships. This last item, termed risk assessment, is, however, only an approximation based on past experience and the use of healthy volunteers. The three options for gaining information on foodborne diseases are discussed below.

Reporting Systems for Foodborne Diseases

In most countries, outbreaks and single cases of foodborne disease are reportable to the relevant authority. Mostly, however, reporting is done on a voluntary basis. In The Netherlands, for example, foodborne illness can be reported to the regional food inspection services. All such reports are taken seriously, and laboratory investigations are carried out whenever possible. The causative agents in reported cases of known etiology are mainly bacteria. These organisms account for 80 to 90% of single cases and cases in outbreaks for which the cause is established. Even after laboratory investigation, however, the source is identified in only about 20% of the cases investigated. Thus, statistics for foodborne disease based on systems of reporting and laboratory investigations grossly underestimate the true incidence because most of the affected individuals do not consult a general practitioner and probably do not need medical treatment. Because reported outbreaks are underestimated, incidental in nature, and unpredictable, their investigation is clearly not a routine activity amenable to rapid methods. The requirement that these outbreaks be investigated by using conventional methods should be obvious.

Sentinel Studies

Because of the lack of essential information on the incidence and causes of foodborne illness, sentinel and population studies may be considered. The sentinel approach is based on the assumption that a combination of clinical data and laboratory findings would provide more adequate information on foodborne illness. If, for example, the fecal sample yielded salmonellae, the most likely source of the illness would appear to be contaminated food or water.

In sentinel studies, large numbers of samples (almost always fecal) have to be tested. For example, the sentinel study carried out in The Netherlands demonstrated that only 4 to 5% of fecal samples obtained from persons with acute enteritis contain *Salmonella* organisms (33). Rapid and automated methods are highly suited for such type of researches. However, in the case of *Salmonella* and *Campylobacter* infections, large numbers of organisms will be present in fecal samples. As a result, spreading of diluted feces on a selective agar plate and incubation for 18 to 24 h will be sufficient to detect their presence.

Risk Assessment Based on Dose-Response Relationships

A new approach to assessing the microbiological safety of food is based on the utilization of dose-response relationships for microorganisms and their toxins. Organisms such as host-nonspecific *Salmonella* spp., *Salmonella typhi, Campylobacter* spp., *Vibrio* spp., *C. botulinum,* and *Escherichia coli* O157:H7 have caused well-documented foodborne infections. From the data available, dose-response curves can be plotted, provided that the number of organisms present in the food at the time of consumption can be determined. It must be recognized, however, that several factors may influence both the probability of infection and the mortality rate. Factors such as the host individual, strain of organism, and the type of food in which the organism occurs can have a considerable effect on susceptibility to infection. Thus, the probability of infection varies from pathogen to pathogen, from one kind of food to another, and among different groups of people in the community. With *Staphylococcus aureus* type A enterotoxin, as little as 500 ng, ingested with food, can be hazardous (40). For botulinum toxin, lethal doses ingested orally by humans are thought to be about 5,000 to 10,000 mouse intraperitoneal 50% lethal dose units (15).

In general, only low levels of bacterial toxins are needed to produce human illness, and it is widely accepted that ready-to-eat food products should be free of toxins. However, as further practical and scientific data become available, a more realistic use of risk assessment will be possible for all types of foodborne pathogens. At present, relatively few experimental data concerning dose-response relationships are available (23). The use of rapid and automated methods is not essential to obtain these experimental data but may be useful.

PRINCIPLES OF CURRENT METHODS FOR DETECTING MICROORGANISMS

General Aspects of Classical Methods

The production of safe food requires analytical techniques which can detect pathogenic organisms. However, it is clear that safe food can be produced only by using hygienic practices, including prevention of recontamination. In establishing and validating suitable procedures and processes, indicator organisms, which provide indirect information about the safety of the food product in question, can be used. Indicator organisms are discussed more fully in chapter 4.

Pathogenic Organisms

Because even small numbers of infectious organisms may cause disease, they should be absent from ready-to-eat foods. This pathogen-free concept implies that analytical tests for these agents need only be qualitative (presence/absence tests). For toxin-forming organisms, however, much higher numbers (>10^5 to 10^6 organisms per ml) are usually required to produce foodborne illness. Thus, low numbers of these pathogens are usually tolerated in food, and the relevant analytical tests must be quantitative (i.e., enumerate the pathogens).

For both qualitative and quantitative tests, the physiological state of the target organisms must be taken into account. For example, organisms present in food may be sublethally injured as a result of the treatments applied during production, the composition of the food, and its pH and water activity. Therefore, it will often be necessary to allow the organisms to repair the injury before transferring them to selective media. In the case of presence/absence testing, food samples are usually incubated first in a nonselective enrichment medium. However, organisms present in dry products may be susceptible to osmotic damage as a result of this procedure. To avoid the lethal effect of osmotic shock, samples must be hydrated gradually. With quantitative tests, plating should be carried out on a nonselective agar medium, which is then overlaid with the selective agar.

Detection of pathogenic organisms in foods furthermore requires the use of selective techniques to separate the target organisms from others that may be present in much greater numbers. Selectivity can be achieved by addition of certain agents to the medium and/or by using specific culture conditions (aerobic or anaerobic, a particular incubation temperature). Most selective agents show a degree of toxicity toward the target organisms, and therefore substrates may be added to stimulate specifically the growth of the target organisms. Modern isolation media also contain substances that allow differentiation of target organisms from others present in the food. The separation is based mainly on one or more specific biochemical characteristics of the organisms. Such formulations are called differential media. Because of these reasons, the detection of pathogens requires some fixed period of time before reliable results can be obtained.

Indicator Organisms

The detection of pathogenic organisms may involve relatively complex and sometimes imperfect analytical methods. In short, testing of foods for pathogens alone is not cost-effective, nor does it offer a guarantee of safety. Therefore, testing for indicator organisms has been introduced as a simpler means of controlling the hygienic status of foods and helps to ensure production of safe food.

Initially, *E. coli* and coliform organisms were used as indicators only in relation to water, to determine whether water had been polluted with fecal material. This concept was important because it recognized feces as a significant source of human pathogens. The use of *E. coli* and other fecal coliforms (able to grow and produce gas at 44 to 45°C) as indicators was based on the assumption of early "microbe hunters," such as Escherich (11), Schardinger (42), and Smith (46), that feces of warm-blooded animals were the sole source of these organisms. Thereafter, the use of indicator organisms came into vogue for evaluating hygiene in food processing. For example, members of the family *Enterobacteriaceae* are used as indicators of effective food pasteurization, and enterococci are used to evaluate cleaning and disinfection of processing equipment. Another example is *S. aureus*, which is a normal inhabitant of the skin of humans and animals. This organism may persist on processing equipment (27). Thus, the presence of *S. aureus* in food ingredients generally indicates contamination from human handling and/or inadequate cleaning and disinfection of processing equipment. However, the absence of *S. aureus* from the end product does not necessarily mean that the food has been produced hygienically. Thus, the indicator organisms used should be selected carefully, and the results should be interpreted with caution. Only when the organisms are used correctly can they provide information on fecal contamination of ingredients, the effect of heat treatment, recontamination, and various other microbiological aspects of food production.

The estimation of indicator organisms can generally be carried out by simple, relatively inexpensive plating techniques and almost always within a period of 20 to 24 h. Furthermore, modern tools such as automatic dilution and automatic plating and counting facilitate this estimation and save labor.

Rapid Methods: Principles and Applications

Rapid methods for detecting microorganisms are usually based on the early recognition of characteristic metabolites or biomass. Metabolites and/or biomass are recognized by sensitive immunological methods, such as enzyme-linked immunosorbent assay (ELISA) and latex agglutination, or by monitoring electrical properties of the culture medium. Another approach detects DNA sequences which are characteristic for specific microorganisms. Besides immunoassays and DNA techniques, several other useful rapid methods have been developed. These include microscopic techniques (DEFT, BACTOSCAN, and CHEMSCAN), ATP assays for estimating biomass, and techniques for measuring bacterial activity (impedance, CO_2 production, etc.). The use of rapid methods is not limited to the detection of specific pathogenic organisms.

Immunoassays

The most appropriate metabolites for the detection of some pathogenic microorganisms (i.e., *C. botulinum*, *S. aureus*, *C. perfringens*, and *E. coli*) are the toxins that they produce. For organisms with no clear pathogenic mechanism, appropriate metabolites are usually selected by trial and error. In this case, culture fluid and/or bacterial cells are injected into mice. By using hybridoma cell technology, monoclonal antibodies are produced and selected according to their suitability. The antigens reacting with these selected monoclonal antibodies are mostly extracellular substances released from the cell membrane and are often heat stable.

Various immunoassays have been developed for detecting almost all of the important foodborne pathogens. To determine the presence of any of these pathogens in food, there must be sufficient metabolite or biomass formation to give a positive test result. In practice, this usually requires that the organisms be cultured before testing, which is a disadvantage of rapid methods based on immunological detection of specific metabolites.

A particular problem in using immunological methods to detect microorganisms in foods is the setting of a baseline value. From the work of Beckers et al. (3), it appears that the baseline extinction values for ELISA systems depend, for example, on the type of food being tested, and this may lead to misinterpretation of the results. In studies using reference samples (milk powder containing small but known numbers of salmonellae), it was observed that both immunological and traditional culture methods could give false-negative results. A further disadvantage of most rapid methods based on immunological detection of metabolites is that they do not lead to isolation of the organisms present, and so further identification, e.g., by serotyping, is impossible.

DNA Assays

Because of rapid progress in molecular biology, DNA techniques may now be used as analytical tools in food microbiology. Nucleic acid hybridization, in particular, is an exciting new approach for the detection of microorganisms in foods. The presence of the organisms is demonstrated by hybridization of their genetic material with specific gene probes (e.g., probes for detecting genes involved in virulence or toxin production, or genes that are characteristic of a particular organism). So far, the techniques have been used mainly for identification of strains isolated by traditional culture methods. For this purpose, a suspension of the food product is plated on a defined agar medium and incubated under standard conditions. The colonies developing on the surface of the agar are hybridized with a specific gene probe. This can be done simultaneously for all of the colonies present on the petri dish. The process is often referred to as a colony hybridization test.

A serious initial drawback in the practical application of gene probe technology to food analysis was the need to use radioactive labels such as ^{32}P. Now, however, nonradioactive labels are available (25). Another early disadvantage of the colony hybridization technique was the inability to detect very low numbers of pathogens among large numbers of contaminating organisms that are generally present in raw foods. The recently developed technique PCR (50) has greatly improved the use of DNA hybridization methods in food microbiology. The PCR method is based on in vitro amplification of target DNA sequences and involves the application of primers (carefully selected synthetic oligonucleotides) and heat-stable polymerase. With this technique, it is theoretically possible to detect pathogens in the food sample directly, but so far this has been done in very few investigations. Major problems must be solved before PCR can be used routinely. These include the inability to distinguish between live and dead cells, the presence of polymerase inhibitors in food samples, and the accessibility of the target organisms. Preenrichment of test samples overcomes most of these problems and is presently needed for detection of specific pathogens in food by PCR. Again, DNA hybridization assays do not lead to the isolation of the organism, and so no further characterization of the contaminant can be done.

PRINCIPLES OF CURRENT METHODS FOR DETECTING MICROBIAL TOXINS

Apart from microbial infections, foodborne illness may be caused by organisms that produce a toxico-infection or from toxin that is present in the food prior to ingestion. Thus, an intoxication results. Bacteria that cause toxico-infections, such as *Bacillus cereus* (diarrheal type) and *C. perfringens*, usually need to be ingested in large numbers (ca. 10^5 vegetative cells or more) to produce illness. Once in the intestinal tract, the organisms release their toxin, which then causes diarrhea. Organisms such as *C. botulinum*, *S. aureus*, *B. cereus* (emetic type), *Pseudomonas cocovenenans*, and molds produce toxins in food under certain conditions; however, only the preformed toxin causes human illness in most cases.

For both groups of organisms, toxin detection is important. For diagnostic purposes, it is necessary to test the organisms for the ability to produce toxin and to screen serum or fecal material for the presence of toxin. Where relevant, food samples can be examined quantitatively for the toxin in question.

The principles of methods used currently for detecting some of the important microbial toxins are discussed below.

Botulinum Toxin

Foodborne botulism (see also chapter 15) is caused by ingesting food that contains the preformed toxin. Because only low levels of toxin are needed to produce illness and the symptoms are severe, no amount of toxin can be tolerated in foods. Safe food production is therefore based on either destroying *C. botulinum* (e.g., by heat) or inhibiting growth of the organism in food (e.g., by drying or acidification of the food), in combination with low storage temperatures and limited storage times.

The customary procedure for detecting *C. botulinum* in food is based on an enrichment procedure. Samples are enriched in suitable media, and after appropriate incubation, culture supernatants are tested for the presence of toxin. The amount of toxin produced depends on several factors, such as type of sample, the presence of competing microorganisms, and incubation temperature (8). Generally only small amounts of toxin are produced, and production of various toxic substances by microorganisms other than *C. botulinum* may affect the results of the test. Detection of botulinum toxin in food or in human serum requires highly sensitive methods. While the sensitivity of the mouse bioassay made this assay the method of choice, immunoassays that are equally sensitive to the toxin are now available and becoming widely used.

S. aureus Enterotoxin

Some strains of *S. aureus* (see also chapter 19) produce one or more enterotoxins, which are proteins with molecular weights of 28,000 to 30,000. There are five immunologically distinct types, A to E. The toxins are released by *S. aureus* during growth in food. For many years, oral dosing of monkeys or kittens was the only means available to study these toxins. Later, sensitive immunoassays were developed (35, 36). Small amounts of toxin (<1 μg) may have an emetic effect in humans (4). Therefore, it is necessary to be able to detect at least 0.2 ng of toxin per 100 g of food. Such small amounts are easily detected by immunoassays, and because the tests are highly sensitive, only a simple extraction procedure is usually needed. Because of the simple extraction procedure, the toxin can be detected quantitatively by ELISA, unless there is interference from the food product itself. In this case, the level of recovery must be checked by taking uncontaminated food of the same type and adding known amounts of purified toxin (43). Although staphylococcal enterotoxins are generally considered to be heat stable, there is evidence that heat resistance depends on several factors, including the type of food, pH, and heating temperature (44). Reactivation of the toxin after heat treatment has been noted by Fung et al. (13), Tatini (47), and Schwabe et al. (44). It was shown by Tatini (47) that treatment of heated toxin with urea increased the toxin recovery almost fourfold. Schwabe et al. (44) found that loss of immunological activity by heat treatment in pea extract resulted in a concomitant loss of biological activity by the toxin (monkey feeding test). Adjusting the pH of heat-treated food products containing toxin to 11.0 restored both the immunological and biological activities in most samples tested.

An additional problem in toxin detection is that other substances present in food extracts (e.g., staphylococcal protein A) may yield false-positive immunological reactions (24). Such false-positive reactions can be recognized easily by examining the reaction kinetics of the ELISA. It was demonstrated by Notermans et al. (38) that regression coefficients obtained from plotting ELISA extinction values against $\log_{10}$ reciprocal sample dilutions were identical for type A toxin produced by several different strains of *S. aureus*. Regression coefficients obtained by cross-reacting other substances differed from that obtained with type A toxin, as did the maximum extinction values obtained for cross-reacting substances.

C. perfringens Enterotoxin

Food poisoning caused by *C. perfringens* (see chapter 16) results from the ingestion of large numbers of vegetative cells. Subsequent sporulation in the intestine results in the release of *C. perfringens* enterotoxin (CPE). So far, only one immunological type of CPE has been recognized, and only a small proportion of randomly tested strains of *C. perfringens* possess the gene encoding CPE (39, 48). The toxin is produced only sporadically in food, and the potential health hazard from preformed CPE in food is thought to be minimal. There is considerable loss of activity during passage through the stomach. However, CPE released in the intestinal tract maintains its immunological stability in stools over a long period of time (31). Demonstration of CPE in the stools of patients is therefore a suitable means of confirming foodborne gastroenteritis caused by *C. perfringens*. The amount of CPE detected depends on various factors, including the timing of sample collection. Four days after the onset of symptoms, CPE can no longer be demonstrated. It should be noted that feces may contain substances capable of destroying the immunoglobulin G used in coating ELISA trays (31).

B. cereus Toxins

B. cereus (see chapter 17) causes either diarrheal- or vomiting-type food poisoning (14). The diarrheal type is attributed to the production of a heat-labile enterotoxin (14); the vomiting type is related to a heat-tolerant emetic toxin (29).

The diarrheal-type toxin has a molecular weight of about 50,000. The toxin is strongly labile and is susceptible to proteolytic enzymes. Commercial immunoassays are available to detect preformed toxin in foods and in culture fluids of *B. cereus*. Initially, results obtained with some of these assays were inconsistent. However, research carried out by Day et al. (10) indicates that almost all strains of *B. cereus* have the capacity to produce enterotoxin, although the amounts produced may differ significantly (37). Production of toxin in foods has been observed under nearly all conditions that allow growth of the organism to $>10^7$/g (37), but, because the toxin is unstable and readily inactivated by gastric juices, it is questionable whether preformed toxin is of any significance in diarrhea. It is most likely that the illness caused by *B. cereus* results from production of toxin in the intestine after large numbers of cells have been ingested (toxico-infection).

Although the diarrheal enterotoxin produced by *B. cereus* has been studied extensively, relatively little is known about the emetic toxin (14). One reason for this is that few methods are available for detecting this toxin. Currently, the most reliable method is oral dosing of primates, as described by Melling and Capel (28). More recently, Hughes et al. (17) have shown that the vacuole response in HEp-2 cells caused by culture supernatants of *B. cereus* correlates with emetic activity. Recently, Agata et al. (1) purified a dodecadepsipeptide from the culture fluid of a *B. cereus* strain that caused vacuole formation in HEp-2 cells. The structure is (D-O-Leu-D-

Ala-L-O-Val-L-Val)$_3$, and this cyclic depsipeptide is identified as cereulide. In the future, detection of cereulide may be simplified and may be carried out routinely. However, routine detection of cereulide is not a component of the basic method for the production of safe food. A detection methodology will be needed only for testing the conditions under which the toxin is produced in food.

Mycotoxins

Mycotoxins (covered in section IV of this volume) are secondary metabolites produced by specific filamentous fungi that cause a toxic response when introduced by a natural route in low concentrations to higher vertebrates and other animals (45). While some mycotoxins are produced by only a limited number of fungal species, others may be produced by a relatively large range of species from several genera (12). It is now increasingly apparent that most toxigenic fungi have the potential to produce more than one mycotoxin (9). Several rapid techniques, such as thin-layer chromatography and direct ELISA systems, have been developed for detection of mycotoxins in food. The drawback of these methods is that information is obtained about the presence of only one mycotoxin. Therefore, a more general detection of molds may be a more applicable approach for screening purposes. Rapid detection of fungi in foods is also important because of food spoilage. Within the past decade, immunological detection of fungi in foods has resulted in several research projects that have shed new light on the use of immunoassays to detect fungi in foods and have indicated some of the important immunodominant sites. The early research assessed simply the ability of immunoassays to detect fungi in foods (30). Attention then focused on the characterization of the fungal antigens in an effort to determine immunodominant sites. Sugar and protein compositions were identified for species of *Aspergillus, Botrytis, Cladosporium, Fusarium, Geotrichum, Monascus, Mucor, Penicillium,* and *Rhizopus.* There has been success in partially characterizing the immunodominant sites in some of these fungal antigens (30). Experimental work carried out by Notermans et al. (32), among others, showed that in samples containing aflatoxin, the immunological presence of aspergilli could always be demonstrated.

VALIDATION AND QUALITY ASSURANCE OF MICROBIOLOGICAL METHODS

The validity of microbiological data depends on the methods used and the laboratory performing the analysis. Errors in the detection of microorganisms may cause false-positive results due to the detection of nontarget organisms and false-negative results if the target organisms are present but not detected. In quantitative tests, errors can result in an over- or underestimation of the true incidence of the target organisms. Both the accuracy of a test method and the performance of the laboratory carrying out the test determine the final test result.

According to ISO/DIS 6107–8 (21), the accuracy of a single test result is defined as the degree of similarity between the measurements obtained and the true value of the measured quantity. The accuracy of test results also depends on the precision of the method. For optimal performance, a quality assurance system is necessary. The accuracy of microbiological methods and the required means of quality assurance are discussed below.

Accuracy of Methods for the Microbiological Examination of Foods

As mentioned above, microbiological methods need to be both accurate and precise. In relation to accuracy, the concept described by Havelaar et al. (16) is widely used. Generally, the true extent to which foods are contaminated with microorganisms is unknown, and the numbers of organisms detectable by a particular method depend on the characteristics of the test method itself. Most analytical methods for determining microorganisms in foods are based on the use of selective culture conditions. Selectivity is obtained by means of chemical agents that are added to the isolation medium and the use of appropriate culture conditions (incubation temperature, atmosphere). These factors should permit growth of the target organisms while suppressing the development of others. Thus, the accuracy of a test method is related to the success of the above-mentioned approach. Normally, some inhibition of the target organisms occurs, while not all nontarget organisms will be entirely inhibited, or nontarget organisms may give reactions typical of the target organisms. Therefore, confirmatory tests are often necessary.

The accuracy of a particular method depends on (i) recovery of the required organisms; (ii) the inhibitory capability of the isolation medium; (iii) the sensitivity of the medium; and (iv) the medium's specificity (involving differential characteristics).

Recovery

The physiological state of bacteria in foods may vary from fully viable to dead. However, various states between the two extremes have been distinguished, includ-

Table 38.2 Comparison of rapid methods for *Listeria* species and *L. monocytogenes* detection in samples of naturally contaminated meat products

Method (enrichment)[a]	No. of samples	Samples positive (total/method)[b] for: *Listeria* sp.	*L. monocytogenes*	%[c]
Transia (UVM/m-FB, 25°C)	34	31/10		32.3
Tecra (UVM/m-FB, 30°C)	34	31/30		96.8
USDA	34	31/21		67.7
Oxoid (1/2 FB/BLEB, 30°C)	46	27/19		70.4
bioMérieux (m-FB/m-FB, 30°C)	46		14/13	92.9
USDA	46	27/24		88.8
USDA (*L. monocytogenes*)	46		14/12	85.7
Organon Teknika (m-FB/m-FB, 30°C)	76	38/13		34.2
bioMérieux (m-FB/m-FB, 30°C)	76	38/26		68.4
Gene-Trak (UVM/plate, 37°C)	76		13/13	100
USDA	76	38/26		68.4
USDA (*L. monocytogenes*)	76		13/8	61.5

[a]Abbreviations: UVM, University of Vermont broth; m-FB, modified Fraser broth; FB, Fraser broth; BLEB, buffered *Listeria* enrichment broth.

[b]Number of samples in which either *Listeria* sp. or *L. monocytogenes* was confirmed by using any of the methods listed for the given subset of samples.

[c]Percentage of samples in which either *Listeria* sp. or *L. monocytogenes* was detected with the indicated method.

ing cells exhibiting sublethal injury (i.e., cells incapable of growth on a selective medium) and viable but nonculturable forms (i.e., morphologically intact cells with demonstrable metabolic activity but not culturable on conventional, nonselective media) (see chapter 2). Different methods of isolation may recover different proportions of the total population, thus introducing a method-specific bias. The proportion of organisms recovered on a particular medium may be very different for an injured population than for a population comprising mainly viable cells. The most direct way of assessing recovery is to analyze one or more pure suspensions of the target organisms. Some relevant type of injury should be induced to mimic the physiological state in which microorganisms may be present in food. The type of injury depends on the kind of food and may include damage caused by heat, dryness, acidification, etc. It is evident that treatment of cultures to cause injury should be carried out in a standardized manner. Assessment of recovery should be an integral part of the validation process for a particular method. For this purpose, use of carefully standardized and preferably certified reference materials, containing known numbers of target organisms, is an advantage.

Table 38.2 presents recovery results of seven commercially available *Listeria* test kits. Various series of samples of meat products were used, none of which were spiked with the target organisms. In addition, the conventional U.S. Department of Agriculture (USDA) method (26) was used for all samples. It is obvious that the USDA method for meat products does not detect all samples which could be positively confirmed with any of the methods applied to a given subset of samples. Furthermore, considerable differences (32 to 96%) were found in the yields of positive samples of the various ELISAs.

Detection Limit

Table 38.3 provides the detection limits for the various rapid methods used in the above-mentioned study. These values were determined with pure cultures. It is obvious that the combination of a suitable enrichment

Table 38.3 Detection limits for the various rapid methods for detecting listeriae

Test	*Listeria* concn in test fluid (mean $\log_{10}$ CFP[a]/ml ± SD)
Gene-Trak[b]	6.8 ± 0.3
Transia	6.5 ± 0.3
Vidas	5.8 ± 0.3
Tecra	5.7 ± 0.3
Organon Teknika	5.5 ± 0.2
Oxoid	5.2 ± 0.2
Vidas-LMO[b]	5.2 ± 0.2

[a]CFP, colony-forming particles.

[b]Detects specifically *L. monocytogenes*.

Table 38.4 Comparison of sensitivity and specificity of rapid methods for the detection of salmonellae

Type	Manufacturer	Test kit, format	% Sensitivity	% Specificity
Immunological method	Lumac	Path-Stik, dipstick	91.4	98.5
	Transia	*Salmonella* immunoassay	95.5	85.5
	Organon-Teknika	Salmonella-TEK ELISA	88.9	82.6
DNA assay	Gene-Trak	*Salmonella* assay, DNA probe	77.5	100

technique with a test with a low detection level will lead to the highest number of positive results.

Inhibitory Capability

Specific detection of certain target organisms usually depends on selective conditions, involving incorporation of inhibitory substances in the isolation medium. The inhibitory capability is directed toward nontarget organisms; however, this should not have any adverse effect on detection of the organisms being sought. Inhibitory properties of the medium can be tested with pure cultures by comparing counts on the basal medium without inhibitors with those on the complete selective medium. The inhibitory effect of the medium on interfering organisms can easily be assessed by means of natural samples that can be spiked with standardized suspensions of target organisms.

Sensitivity and Specificity

The sensitivity of a method may be defined (16) as the proportion of target organisms that can develop under the test conditions and give characteristic reactions. Specificity is reflected in the proportion of nontarget organisms showing such reactions. For presence/absence testing, percent sensitivity is defined as number of true-positive samples × 100/number of true-positive samples + number of false-negative samples, and percent specificity is defined as number of true-negative samples × 100/number of true-negative samples + number of false-positive samples.

For practical purposes, the aim is to select methods of high specificity and sensitivity. If specificity is low, extensive confirmation of colonies will be necessary. In a study of Rombouts and Beumer (41), the sensitivities and specificities of several rapid methods for detection of salmonellae in 135 samples (e.g., poultry and egg products, smoked sausages, and dried foods) were estimated. Again, none of the samples were spiked. Sixty-four samples were found to be positive. The results of the various rapid methods were related to the results of the ISO 6579 standard method and are presented in Table 38.4. The highest sensitivity was obtained with the Transia immunoassay (95.5%). The highest specificity was obtained with the Gene-Trak assay. However, the best choice is the Lumac Path-Stik, because of its high scores for sensitivity and specificity. In addition, this test was the most convenient to use.

Precision of Methods for Microbiological Analysis of Foods

The general principles involved in determining the precision of microbiological test methods are given in ISO 5725 (20). In this standard, two measures of precision are described: (i) repeatability, which is defined as the closeness of agreement between independent test results obtained by the same method and using identical test material in the same laboratory, with the same operator using the same equipment; and (ii) reproducibility, which is defined as the closeness of agreement between test results obtained with the same method on identical test material in different laboratories, with different operators using their own equipment. Taken together, these definitions make it clear that precision depends on several factors. First, samples should be representative of the food under investigation and should be held under conditions that maintain all relevant properties of the food until it is analyzed. Second, because of differences in the degree of homogenization and errors in sample dilution, there may be variation between subsamples. Also, there may be between-operator effects involving human factors, such as differences in learning capability, and laboratory effects due to technical variation. All of these factors can become apparent in international collaborative studies involving the use of standardized reference materials.

Quality Assurance

As apparent from the foregoing discussion, quality assurance is needed for all microbiological methods concerned with food analysis. Quality assurance requires (i) clear and well-described test methods, (ii) the use of reference materials, and (iii) testing in collaborative studies (involving both the development of selective isolation media and their use in routine testing). Quality assurance is an integral part of good laboratory practices and is necessary for accreditation of analytical laboratories. In consequence, certified reference materials have become available, and it is possible to participate in international collaborative studies that seek to evaluate the test methods.

SUMMARY AND FUTURE TRENDS

Production of safe food is a matter of knowledge and the use of preventive measures as described in chapter 41. Production of safe foods will increasingly be based on identification of possible problems before they occur and the establishment of control measures at the optimal stages in processing. Thus, the use of microbiological techniques will be focused on identification of the possible problems and on developing active intervention strategies. For this last use, microbiological methods will need to characterize the organisms in detail, allowing prediction of growth, sensitivity to heat, etc. As a result, routine testing, and thus the application of microbiological methods, will decrease, leaving only a need for reliable and sophisticated methodologies. There is no special requirement that these methods be rapid and automated.

At present, however, assurance of food safety is still largely based on meeting established microbiological criteria. Therefore, single-sample testing is needed, particularly for testing the microbiological status of raw materials. But microbial testing must also fulfill the requirements set by customers, consumers, and other interested parties. In addition to inspection carried out under GMP and HACCP programs, single-sample testing will be always be a part of these activities.

References

1. **Agata, N., M. Mori, M. Ohta, S. Suwan, I. Ohtani, and M. Isobe.** 1994. A novel dodecadepsipeptide, cereulide, isolated from *Bacillus cereus* causes vacuole formation in HEp-2 cells. *FEMS Microbiol. Lett.* **121**:31–34.
2. **Baird-Parker, A. C.** 1994. Foods and microbiological risks. *Microbiology* **140**:695–701.
3. **Beckers, H. J., P. D. Tips, P. S. S. Soentoro, E. H. M. Delfgau-van Asch, and R. Peters.** 1988. The efficacy of enzyme immuno-assays for the detection of samonellas. *Food Microbiol.* **5**:147–156.
4. **Bergdoll, M. S.** 1979. Staphylococcal intoxications, p. 443–494. *In* M. S. Bergdoll (ed.), *Foodborne Infections and Intoxications.* Academic Press, Inc., New York.
5. **Campden Food and Drink Research Association.** 1992. *HACCP: A Practical Guide.* Technical manual no. 38. Campden Food and Drink Research Association, Chipping Campden, England.
6. **Codex Alimentarius Commission, Committee on Food Hygiene.** 1991. *Draft Principles and Application of the Hazard Analysis Critical Control Point (HACCP) System, Alinorm 93/13, Appendix VI.* Food and Agriculture Organization/World Health Organization, Rome.
7. **Codex Committee on Food Hygiene.** 1993. *Report of the 26th Session of the Codex Committee on Food Hygiene,* paragraphs 81–86 (Alinorm 93/13A). Food and Agriculture Organization/World Health Organization, Rome.
8. **Committee on the Microbiological Safety of Food.** 1990 and 1992. *Report of the Committee on the Microbiological Safety of Food,* parts I and II. *Microbiological Safety of Food.* Her Majesty's Stationery Office, London.
9. **Coker, R.** 1995. Controlling mycotoxins in oilseeds and oilseed cakes. *Chem. Ind.* **7**:260–264.
10. **Day, T. L., S. R. Tatini, S. Notermans, and R. W. Bennet.** 1994. A comparison of ELISA and RPLA for detection of *Bacillus cereus* diarrhoeal enterotoxin. *J. Appl. Bacteriol.* **77**:9–13.
11. **Escherich, T.** 1885. Die Darmbakteriën des Neugeborenen und Säuglings. *Fortschr. Med.* **3**:515.
12. **Frisvad, J. C., and O. Filtenborg.** 1983. Classification of terverticillate penicillia based on profiles of mycotoxins and other metabolites. *Appl. Environ. Microbiol.* **46**:1301–1310.
13. **Fung, D. Y. C., D. H. Steinberg, R. D. Miller, M. J. Kurantnick, and T. F. Murphy.** 1973. Thermal inactivation of staphylococcal enterotoxins B and C. *Appl. Microbiol.* **26**:938–942.
14. **Gilbert, R. J., and J. M. Kramer.** 1984. *Bacillus cereus* enterotoxins. *Biochem. Soc. Trans.* **12**:198–200.
15. **Hauschild, A. H. W.** 1989. *Clostridium botulinum,* p. 111–189. *In* M. P. Doyle (ed.), *Foodborne Bacterial Pathogens.* Marcel Dekker, New York.
16. **Havelaar, A. H., S. M. Heisterkamp, J. A. Hoekstra, and K. A. Mooijman.** 1993. Performance characteristics of methods for the bacteriological examination of water. *Water Sci. Technol.* **27**:1–13.
17. **Hughes, S., B. Bartholomew, J. C. Hardy, and J. M. Kramer.** 1988. Potential application of a HEp-2 cell assay in the investigation of *Bacillus cereus* emetic syndrome food poisoning. *FEMS Microbiol. Lett.* **52**:7–12.
18. **International Commission on Microbiological Specification of Foods.** 1986. *Microorganisms in Food,* vol. 2. *Sampling for Microbiological Analysis: Principles and Specific Application.* Blackwell Scientific Publications, London.
19. **International Life Science Institute, Europe.** 1993. *A Simple Guide to Understanding and Applying the Hazard Analysis and Critical Control Point Concept.* Concise monograph series. ILSI Press, Brussels.
20. **International Organization for Standardization.** 1986. *Precision of Test Methods—Determination of Repeatability and Reproducibility for a Standard Test Method by Inter-Laboratory Tests.* ISO 5725, 2nd ed. International Organization for Standardization, Geneva.
21. **International Organization for Standardization.** 1991. *Water Quality—Vocabulary,* part 8. IOS/DIS 6107-8. International Organization for Standardization, Geneva.
22. **Kilsby, D.** 1982. Sampling schemes and limits, p. 387–421. *In* M. H. Brown (ed.), *Meat Microbiology.* Applied Science Publishing, London.
23. **Klipstein, F. A., R. F. Engert, H. Short, and E. A. Schenk.** 1985. Pathogenic properties of *Campylobacter jejuni:* assay

and correlation with clinical manifestations. *Infect. Immun.* **50**:43–49.

24. **Koper, J. W., A. M. Hagenaars, and S. Notermans.** 1980. Prevention of cross-reactions in the ELISA for the detection of *Staphylococcus aureus* enterotoxin type B in culture filtrates and foods. *J. Food Safety* **2**:35–45.
25. **Mathews, J. A., and L. J. Kricka.** 1988. Analytical strategies for the use of DNA probes. *Anal. Biochem.* **169**:1–25.
26. **McClain, D., and W. H. Lee.** 1988. Development of USDA-FSIS method for isolation of *Listeria monocytogenes* from raw meat. *J. Assoc. Off. Anal. Chem.* **71**:660–664.
27. **Mead, G. C., and C. R. Dodd.** 1990. Incidence, origin and significance of staphylococci on processed poultry. *Soc. Appl. Bacteriol. Symp. Ser.* **19**:81S–91S.
28. **Melling, J., and B. J. Capel.** 1978. Characteristics of *Bacillus cereus* emetic toxin. *FEMS Microbiol. Lett.* **4**:133.
29. **Melling, J., B. J. Capel, P. C. B. Turnbull, and R. J. Gilbert.** 1976. Identification of a novel enterotoxigenic activity associated with *Bacillus cereus. J. Clin. Pathol.* **29**: 938.
30. **Notermans, S. H. W., M. A. Cousin, G. A. de Ruiter, and F. M. Rombouts.** *Fungal Immunotaxonomy,* in press.
31. **Notermans, S., C. J. Heuvelman, H. Beckers, and T. Uemura.** 1984. Evaluation of the ELISA as tool in diagnosing *Clostridium perfringens* enterotoxins. *Zentralbl. Bakteriol. Hyg. Abt. 1 Orig. Reihl B* **179**:225–234.
32. **Notermans, S., C. J. Heuvelman, H. P. van Egmond, W. E. Paulsch, and J. R. Besling.** 1986. Detection of molds in food by ELISA. *J. Food Prot.* **49**:786–791.
33. **Notermans, S., and A. M. M. Hoogenboom-Verdegaal.** 1992. Existing and emerging foodborne diseases. *Int. J. Food Microbiol.* **15**:197–205.
34. **Notermans, S., P. in 't Veld, T. Wijtzes, and G. C. Mead.** 1993. A user's guide to microbiological challenge testing for ensuring the safety and stability of food products. *Food Microbiol.* **10**:145–157.
35. **Notermans, S., H. L. Verjans, J. Bol, and M. van Schothorst.** 1978. Enzyme linked immunosorbent assay (ELISA) for determination of *Staphylococcus aureus* enterotoxin type B. *Health Lab. Sci.* **15**:28–31.
36. **Notermans, S., and K. Wernars.** 1991. Immunological methods for detection of foodborne pathogens and their toxins. *Int. J. Food Microbiol.* **12**:91–102.
37. **Notermans, S., and S. Tatini.** 1993. Characterization of *Bacillus cereus* in relation to toxin production. *Neth. Milk Dairy J.* **47**:71–77.
38. **Notermans, S., K. Wernars, and J. Dufrenne.** 1991. Recognition of false positive results obtained with immunoassays by analysing the reaction kinetics. *Food Agric. Immunol.* **3**:85–92.
39. **Olsen, J. E., S. Aabo, W. Hill, S. Notermans, K. Wernars, P. E. Granum, T. Popovic, H. N. Rasmussen, and O. Olsvik.** 1995. Probes and polymerase chain reaction for the detection of foodborne bacterial pathogens. *Int. J. Food Microbiol.* **28**:1–78.
40. **Reiser, R., D. Conaway, and M. S. Bergdoll.** 1974. Detection of staphylococcal enterotoxin in food. *Appl. Microbiol.* **27**:83–85.
41. **Rombouts, F. M., and R. R. Beumer.** 1995. Rapid detection methods of foodborne pathogens. Lecture, University of Gent.
42. **Schardinger, F.** 1892. Ueber das Vorkommen Gährungerregender Spaltpilzer im Trinkwasser und ihre Bedeutung für die hygienische Beurteilung desselben. *Wien. Klin. Wochenschr.* **5**:403–405.
43. **Schönwälder, H., J. J. Haaijman, R. Holbrook, J. Huis in 't Veld, S. Notermans, I. M. Schäffers, and R. Zschaler.** 1988. A collaborative study comparing three ELISA systems for detecting *Staphylococcus aureus* enterotoxin A in sausage extracts. *J. Food Prot.* **51**:680–684.
44. **Schwabe, M., S. Notermans, R. Boot, S. R. Tatini, and J. Krämer.** 1990. Inactivation of staphylococcal enterotoxins by heat and reactivation by high pH treatment. *Int. J. Food Microbiol.* **10**:33–42.
45. **Smith, J. E., and M. O. Moss.** 1985. *Mycotoxins: Formation, Analysis, and Significance.* John Wiley & Sons, Chichester, England.
46. **Smith, T.** 1895. Notes on *Bacillus coli*-communis and related forms, together with the suggestions concerning the bacteriological examination of drinking water. *Am. J. Med. Sci.* **110**:283–302.
47. **Tatini, S. R.** 1976. Thermal stability of enterotoxins in food. *J. Milk Food Technol.* **39**:432–438.
48. **van Damme-Jongsten, M., J. Rodhouse, R. J. Gilbert, and S. Notermans.** 1990. DNA probes for detection of enterotoxigenic *Clostridium perfringens* strains isolated from outbreaks of food poisoning. *J. Clin. Microbiol.* **28**:131–133.
49. **van Ermengem, E.** 1898. Ueber einen neuen anäroben Bacillus und seine Beziehungen zum Botulismus. *Z. Hyg. Infektionskr.* **26**:1–86.
50. **Vries, M., D. E. Verlaan, M. E. Bogaard, M. van den Elst, J. H. van Boom, A. J. van der Eb, and J. L. Bos.** 1986. A dot blot screening procedure for mutated *ras* oncogenes using synthetic oligonucleotides. *Gene* **50**:313–320.
51. **World Health Organization.** 1993. *Training Considerations for the Application of the Hazard Analysis Critical Control Point System to Food Processing and Manufacture.* WHO/FNU/FOS/93.3. World Health Organization, Geneva.

W. Mark Barbour
George Tice

39

Genetic and Immunologic Techniques for Detecting Foodborne Pathogens and Toxins

COMPARISON OF TECHNOLOGIES

A wide range of foodborne pathogens and toxins threaten consumer health. Up to 33 million people per year in the United States suffer from the effects of microbial foodborne disease (9a). Despite the magnitude of this problem, methods for detecting these hazards have historically been of limited diversity and relied on cultural techniques. During the last decade, there has been a dramatic change in this situation as a result of the rapid development of immunologic and genetic technologies. These technologies are highly specific and allow detection of pathogens and toxins at low concentrations. The resulting rapid, inexpensive, and simple methodologies allow for extensive testing, increased surveillance, and a reduced health risk to consumers.

This chapter examines modern molecular detection technologies, highlighting the breadth and scientific basis of the methods that are used to detect foodborne targets. Promising methods not yet applied to food pathogens and toxins are also discussed.

Mechanistic Comparison

Technologies can be compared on several different levels. The most appropriate place to begin is with an understanding of the underlying mechanisms of the methodologies. For example, as will be discussed in detail in this chapter, the fundamental distinction between immunologic and genetic methods for detecting pathogens and toxins is the use of two fundamentally different mechanisms for target discrimination. Immunologic methods rely on the binding specificity of an antibody to an antigen. In comparison, genetic methods rely on the specificity of nucleotide base pair formation between a probe and a target DNA or RNA. All of the other differences between these methods result from the myriad schemes for detecting the specific antibody or nucleic acid interaction.

Comparisons are complicated when methods relying on different mechanisms are combined. Immunologic tools are often used in combination with genetic methods. For example, antibodies are used for capture of target cells that are then detected by a genetic method. In other cases, antibodies may be used to detect the label on a hybridized DNA probe. The key to understanding and comparing these methods is a determination of the type of interaction (nucleic acid or antibody-antigen) that confers specificity. In the former example, the specificity relies on both interactions (and consequently is only as good as the least specific interaction), while in the latter case, the specificity is derived only from the DNA probe interaction with the target nucleic acid.

Despite the difference in mechanism, the two groups of methods share a critical characteristic. Given sufficient target concentrations, there is no requirement for cellular activity by the target pathogen. Conventional methods for detecting pathogens rely on the organisms' physiological responses to substrates or selective agents. This response is in the form of growth on the substrate,

Table 39.1 Applicability of genetic and immunologic tests

Sample	Relevant characteristics
Food	Very low target concentration
	High potential for food interference
	Generally not amenable to direct assay for low target concentration
	Sample must be concentrated and purified; possible only with beverages
	Dilution followed by immunocapture and plating an option
	Accuracy mainly dependent on sensitivity
Food, preenrichment culture	Potentially low target concentration (10^4 CFU/ml)
	High potential for food interference
	PCR testing applicable; some foods require purification to remove inhibitory components
	Immunoassays generally not directly applicable; immunocapture used to enrich sample prior to testing
	Accuracy dependent on sensitivity and specificity
Food, selective enrichment culture	High target concentration ($>10^5$ CFU/ml)
	Low to moderate potential for food interference
	Immunoassays, hybridization, and PCR all applicable
	Accuracy most dependent on specificity of test
Environmental	Target level and interference variable with sample
	Applicability of method dependent on sample type and required sensitivity
Colony	Highest target level; no interference from sample
	All methods applicable; simplest and least expensive formats based on agglutination
	Specificity of assay critical if the assay is used as final confirmation

growth or survival in the presence of selective agents, conversion of the substrate to a detectable product, etc. Finding easily assayed physiological responses on which to base a test can be a serious challenge. Furthermore, such responses are often dependent on many factors (temperature, pH, the concentrations and types of competing organisms, etc.) The basis of more rapid testing is the direct detection of existing antigens or nucleic acids without need for time-consuming analysis of metabolic response.

While they do not require measurement of a physiological response, current immunologic and genetic food pathogen detection schemes do not fully eliminate the need for growth of the target organism. Most often, pathogen screening tests are "zero tolerance," meaning that the required sensitivity is one pathogen cell in a test sample (generally 25 g). Consequently, for even the most sensitive tests, several hours of growth are required to increase the target cell concentration (typically greater than 10^4-fold) to the test's threshold detection level.

Applicability of Methodologies

The applicability of a given detection method depends on the nature of the target and the nature of the sample. For example, in the case of foods contaminated with preformed toxins, the presence of the organism that produces the toxin is not the best indicator of the existence of a hazard. Therefore, a test for the presence of a particular DNA sequence contained in the organism is not appropriate, while an antibody-based test for the toxin itself may work well. Conversely, when detection of a low concentration of a bacterium or virus is required, a DNA-based test may be the better choice.

The availability of the nucleic acid sequence data necessary to construct probes has increased dramatically in recent years, but such data, especially for the less common pathogens, may still be insufficient. A researcher may choose to avoid a search for a unique sequence on which to base a probe by developing an immunoassay.

Other, less obvious reasons for developing a particular method are based on sample characteristics such as food matrix inhibition, the relative concentrations of target and nontarget organisms, and parameters related to the user's requirements such as sensitivity, specificity, speed, and cost. Table 39.1 describes the characteristics of the common sample types which determine the applicability of the various methods. These parameters are discussed in detail below.

Sensitivity and Specificity

The parameters most commonly used to evaluate a method are sensitivity and specificity. Sensitivity characterizes the number or concentration of a target organism or toxin required for a positive result. The theoretical sensitivity of a given method is seldom obtained in practice, especially when complex samples such as foods are tested. The actual sensitivity of a method will vary depending on many factors, such as the choice of reporter or detection method, the presence of other organisms, and the presence of inhibitory substances.

Increasing assay sensitivity will improve the accuracy of a test that is not sensitive enough to detect the lowest levels of target encountered. In other cases, improved sensitivity can be used to influence other parameters such as the speed and simplicity of a test. A potential pitfall of increased sensitivity, particularly with methods which detect very low target levels (e.g., <10 targets per assay), is the detection of dead target cells. In practice, this pitfall is often not a problem. Even the most sensitive assays for food pathogens still incorporate a growth step in which, of course, only live cells multiply. Consequently, the actual level of dead cells required for detection must be very high (on the order of 10^5 cells per g of food) for the detection of dead cells to become a real problem.

Specificity is a measure of an assay's ability to discriminate between the target organism or toxin and other organisms or substances. A positive result in the absence of the target is a false-positive result, while a failure to detect the target when present is a false-negative result. The relative importance of the two types of incorrect results depends on the application of the test. In food microbiology, only a very low frequency of false-negative results can be tolerated for obvious safety reasons. Alternatively, a rapid screening method that is routinely followed by confirmation of all positive results may still have value even though it has a significant false-positive frequency.

Specificity is generally measured by testing a group of bacterial strains within the target genus or species and a group of related nontarget strains. The resulting frequency of target organism detection and the frequency of absence of nontarget organism detection are referred to as inclusivity and exclusivity, respectively. Less than 100% inclusivity is a source of false-negative results, while less than 100% exclusivity is a source of false-positive results. Maintaining appropriate specificity for detection of pathogens in foods represents a serious challenge for genetic and immunologic test methods. Depending on the sample matrix, there can be 10^8 to 10^9 bacteria per ml in a preenrichment or enrichment broth. These numbers can represent a huge excess of bacteria closely related to the target, for example, 10^4 *Salmonella* cells per ml in a background of 10^8 *Citrobacter* cells per ml. This challenge mandates a need for high specificity in order to achieve the desired sensitivity of target detection without sacrificing specificity.

Sensitivity and specificity requirements determine the form of sample treatment required prior to analysis. Methods for the detection of bacterial pathogens typically require 1 or more days of enrichment culturing prior to testing. Therefore, when various methods are being assessed, it is critical to determine the total time to result, not just the few hours required for the final genetic and immunologic procedure.

Preparation of Samples

Genetic and immunologic assays require intricate molecular interaction and enzymatic function. Problems with assay inhibition often arise in moving away from the "clean" system used to develop an assay. For example, one might use pure cultures of a target organism to develop an assay and then apply the assay to a food homogenate containing thousands of food chemicals and nontarget organisms. Some of the most powerful, rapid, and simple assays require lengthy and cumbersome procedures for preparing clean cell or DNA samples. This results in assays that are not so simple or rapid. The degree of difficulty often depends on the stage of sampling and whether it be directly from a food sample or after one or more dilution and enrichment steps. The enrichment steps increase the concentration of target organisms relative to nontarget organisms and in addition remove 90% or more of the food matrix through dilution. Consequently, the development of more rapid methods that allow the detection of targets from samples prior to enrichment imposes a much greater burden on sample preparation (28).

Inhibitors present in foods are a common problem, especially in assays using enzymes. The inhibiting substance can be a proteinase which degrades the required assay enzyme, as has been found in an assay for detection of *Listeria monocytogenes* in milk (60). Other, less obvious inhibitors include polyphenolic compounds in plant-derived foods (27) and hemoglobin degradation products in blood or animal tissue (31).

IMMUNOASSAYS

Introduction

Immunoassays are analytical tests that use antibodies. These tests can be quantitative or qualitative and take advantage of the unique binding characteristics of the immunoglobulin molecule. Early immunoassays used immunodiffusion techniques to determine the presence of a specific antigen (71). This technique involves the antibody and antigen diffusing through a semisolid medium and the formation of a precipitin line where optimal concentration of the two reactants occurs. The immunoassay was further advanced by attaching antibodies to latex particles (61). When antibody-coated latex particles are mixed with specific antigen, they agglutinate in a reaction that can be seen as clumping. More recently, a family of immunoassays that use a capture antibody, reporter antibody, and reporter substrate has been developed (15). This type of assay is generally referred to as a sandwich assay. The remainder of this section will discuss in more detail the foundation of the immunoassays, the antibody, and the various formats used.

The Antibody

Antibodies are glycoproteins that are involved in the recognition function of host defense mechanisms. There are five major isotypes of immunoglobulins, immunoglobulin G (IgG), IgM, IgA, IgD, and IgE; most of those used in immunoassays are of the IgG class. The IgG molecule is made of two heavy chains and two light chains that are connected through a disulfide linkage, each of which contains a constant region and a variable region. The variable regions of the heavy and light chains determine the binding specificity of the antibody (5). Each IgG molecule has two identical antigen binding sites. This characteristic allows the molecule to bind to antigens from either two distinct sources or a single multiantigenic source. Sources of antibodies used in immunoassays can be polyclonal or monoclonal (82).

Polyclonal Antibodies

Polyclonal antisera contain an assortment of antibodies having different cellular origins and, therefore, somewhat different specificities. Most polyclonal antibodies used in immunoassays are derived from either rabbit or goat serum. An animal is immunized with a specific immunogen, typically the analyte of interest or a conjugated hapten of the analyte. Approximately 14 days after a second booster immunization, a small volume of blood is drawn and the immune response is measured. If the specific titer of the serum is adequate, a large volume of blood is collected. Blood cells are removed by coagulation, and the resulting antisera are then tested for titer and specificity before being pooled. The antisera are typically purified by ammonium sulfate precipitation, protein A chromatography, or ion-exchange chromatography. The purified antibody is then used as a reagent in immunoassays.

Monoclonal Antibodies

One of the disadvantages of using polyclonal antisera in immunoassays is the variability found in the animal's immune response. This disadvantage was overcome when a technique to immortalize lymphocytes was described by Kohler and Milstein (37). Briefly, this procedure is performed by immunizing a mouse with the desired antigen. Once an immune response has been elicited, B-lymphocytes from the spleen are isolated and fused with a myeloma cell. The resulting cells, called hybridomas, continue to produce a specific antibody. By using limiting-dilution techniques, individual hybridomas that secrete a single antibody derived from one lymphocyte are isolated. The hybridoma is injected into a mouse abdomen, causing the formation of a solid tumor. As a result of this tumor, ascites fluid that contains high levels of the monoclonal antibody is produced. Antibodies produced in this manner have a constant affinity for the antigen. The development of monoclonal antibodies greatly enhanced the field of immunoassays by providing a consistent and reliable source of characterized antibodies.

Recombinant Antibodies

Although monoclonal antibodies overcame the obstacle of antibody supply, they did little to control the specificity and affinity of the produced antibody. Methods now being developed employ molecular biology techniques so that theoretically all physical attributes of an antibody can be manipulated (48). The techniques involve isolation, amplification, and expression of the genes responsible for the variable region of the antibody. The heavy- and light-chain variable regions are cloned in a bacteriophage M13. *Escherichia coli* is then infected with the phage and subsequently expresses the protein. The expressed protein has full antibody function and can be used in immunoassays. The variable region is short and can be easily sequenced and manipulated to derive mutants having altered affinity and specificity geared for a specific immunoassay.

Immunoassay Formats

Immunoassays can be classified as homogeneous or heterogeneous. The major difference between the two formats is that a homogeneous assay eliminates the need to separate the bound antibody from the unbound antibody and requires shorter incubation times. Two examples of homogeneous assays are agglutination reactions and the particle-enhanced turbidimetric inhibition immunoassay (described in more detail below) (44). In a heterogeneous assay, the unbound antibody must be separated from the bound antibody. An example of a heterogeneous immunoassay is an enzyme-linked im-

Table 39.2 Applications of immunoassays in food safety[a]

Format	Sensitivity	Reference(s)
Latex agglutination		
E. coli heat-labile enterotoxin	32 ng/ml	22
Salmonellae	NA	34
Immunodiffusion		
Staphylococcus enterotoxins	0.3 μg/ml	50
Salmonellae	NA	55
Direct microscopic detection (FA)		
Salmonellae	NA	19, 68
E. coli O157:H7	16 CFU/g	72
Enzyme-linked immunosorbent assay		
Toxins		
Aflatoxin	10–250 pg/ml	42, 57, 58
Botulinum	50–100 mouse LD_{50}	54, 69
Staphylococcus enterotoxins	0.1 ng/ml	25, 67
Organisms		
Salmonellae	10^5–10^6/ml	11, 59
Listeriae	1 CFU/g	12, 47

[a]NA, not available; FA, fluorescent antibody; LD_{50}, 50% lethal dose.

munoassay. Table 39.2 lists published sensitivities for various assay formats for the detection of foodborne pathogens and toxins.

Homogeneous Assays

As stated previously, homogeneous assays do not need to separate bound from unbound antibody, involve relatively short reaction times, and are therefore easier to automate than heterogeneous assays. In agglutination reactions (53), usually for larger analytes such as proteins or whole cells, the antibody is mixed with the sample and allowed to incubate. After the incubation period, the mixture is checked for clumping, which indicates the presence of the analyte. Conversely, in the absence of the analyte, there is no clumping. The clumping may be monitored visually for qualitative assays.

Agglutination assays can be quantitative when large analytes, such as proteins or cells, form antigen-antibody complexes that are measured spectrophotometrically. The divalent (two binding sites) nature of the antibody allows it to bind two different molecules, thus forming a cross-linked complex (Fig. 39.1). The formation of this complex is measured by shining a beam of light through the solution and measuring scattered light. The larger the antibody-antigen complex, the more light is scattered.

Agglutination assays can also be quantified turbidimetrically. Rather than measuring the amount of light scattered, this method measures the amount of light lost as a result of absorption, scatter, and reflection. This technique can be used for small analytes, as shown with the particle-enhanced turbidimetric inhibition assay

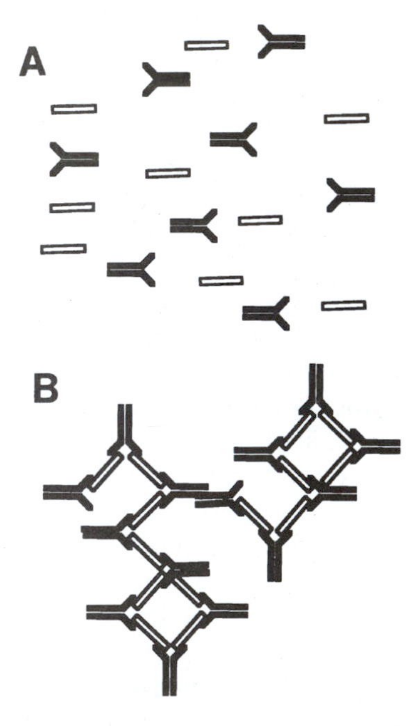

Figure 39.1 Schematic of an agglutination reaction. (A) Antibody (solid bars) is mixed with sample containing antigen (open bars) and allowed to incubate. The antibody recognizes and binds to the specific target (B). This causes clumping of the solution or the formation of an immunoprecipitin line in the gel. Lack of change in the physical characteristics of the solution indicates that the target antigen is not present in the sample.

(53). In this assay, a latex particle is coated with a target analyte. Sample containing the analyte is mixed with coated latex particles and with the antibody specific for the analyte. The free analyte competes with the analyte bound on the latex particle for binding sites on the antibody. If there is no analyte in the sample, the antibody can react only with the latex particle. This results in agglutination of the particle and a rapid increase in the absorbance of the solution. If there is analyte in the sample, the agglutination reaction slows and the absorbance does not change as rapidly. Therefore, the rate of change in the absorbance is inversely related to the concentration of the analyte in the sample.

Nonagglutination homogeneous assays may be mediated by reporter molecules. One technique, as described by Rubinstein et al. (63), uses an enzyme as a reporter. In this assay, the test analyte is conjugated to an enzyme so that when an antibody is bound to the analyte, there is a decrease in enzyme activity. The sample, conjugated enzyme, and enzyme substrate are mixed. The analyte in the sample competes with the conjugated analyte for the binding sites on the antibody. Antibody binding to enzyme-conjugated analyte decreases enzyme activity. Enzyme activity is measured by the turnover of enzyme substrate, which is usually accompanied by a colorimetric change. The amount of substrate turnover is inversely related to the amount of analyte in the sample.

Fluorescence polarization is another method used in homogeneous assays. An analyte is labeled with a fluorescent compound. Sample, antibody, and the labeled analyte are mixed. Polarized light at the correct excitation wavelength is passed through the solution. The labeled analyte and free analyte compete for the binding sites on the antibody. The unbound labeled analyte is rotating very fast in solution, and since the emitted light is in a different plane than the incident light, negligible light is detected through a polarized filter. Conversely, rotation of the antibody bound to labeled analyte is restricted such that the emitted light is almost in the same plane as the incident light and will be detected. The amount of signal detected is inversely related to the amount of analyte in the sample (13).

Heterogeneous Assays

Heterogeneous assays usually include the following components: analyte capture, reporter addition, washes, and reporter detection. A typical sandwich assay (Fig. 39.2) entails immobilization of a capture antibody on a solid support, the addition of test sample, and an incubation period. Following the incubation, the unbound sample is removed by washing and a reporter antibody conjugate is added. After a period of time, the unbound reporter conjugate is washed away and a reporting substance is added. The amount of reporter signal is thus related to the amount of analyte in the sample. This immunoassay format is the predominant format used for the detection of foodborne pathogens. Through evolution of the sandwich assay, a number of different solid supports and reporting systems have been developed.

Immunoassay Components and Configurations

Solid Supports

The solid support chosen for an immunoassay sets the format for assay delivery. The selection of a material and the manner in which it is formed dictate the washing and detection needs of the assay. The solid support should ideally be able to irreversibly bind capture antibodies and then, once treated, not bind reporter antibodies.

One of the most common solid supports used for immunoassays is polystyrene. Polystyrene has the property of irreversibly adsorbing antibodies and proteins (6). Polystyrene tubes, wells, balls, and microtiter plates are used as solid supports for immunoassays. Once the antibody is adsorbed to the polystyrene surface, all nonspecific binding sites are blocked by treating the surface with a blocking agent such as bovine serum albumin, casein, or human serum albumin. Many immunoassays used for foodborne pathogen detection use polystyrene as the solid support.

Another common solid support is magnetic particles (16). The attachment of antibodies to magnetic particles can be passive, as with polystyrene, or covalent. The magnetic particle offers advantages in terms of available surface area for antigen capture as well as facilitating the wash steps. Separation is achieved by using an external magnet to draw the particles to the edge of a tube and then aspirating the supernatant.

Membranes made of nylon or nitrocellulose are also used as solid supports. The assay is performed by immobilizing the capture antibody, covalently or through adsorption, onto the membrane (39). As with polystyrene, the membrane is then treated with a blocking agent to bind any remaining nonspecific binding sites. The wash steps usually consist of passing liquid through the membrane with the help of absorbent paper or vacuum.

Reporter Antibodies

Once the analyte has been captured by the capture antibody, a labeled reporter antibody is added to the solid support. The reporter antibody conjugate binds to the previously captured analyte, forming a sandwich. The

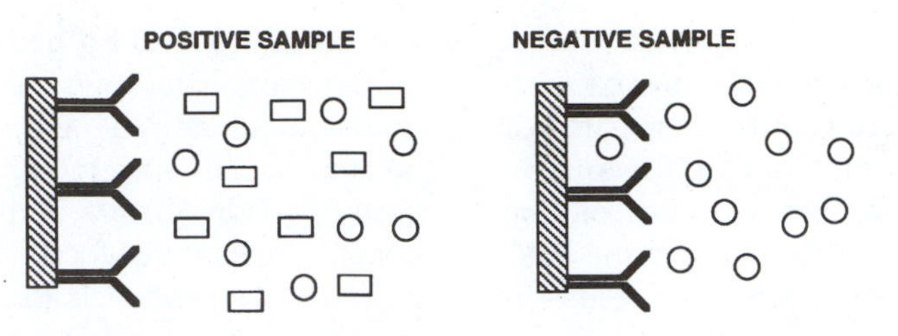

The sample is added to an immobilized capture antibody on a solid support.

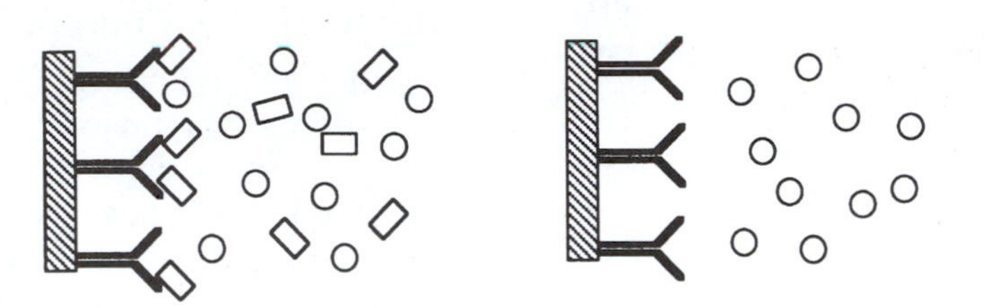

Target is bound by the capture antibody. Washing removes sample matrix

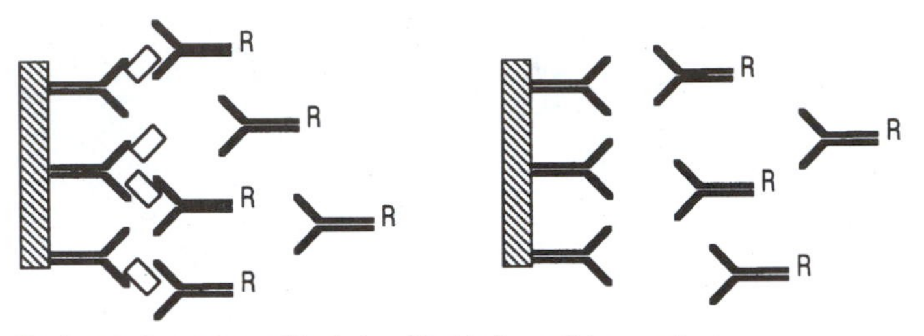

Conjugated reporter antibody is added to the solid support.

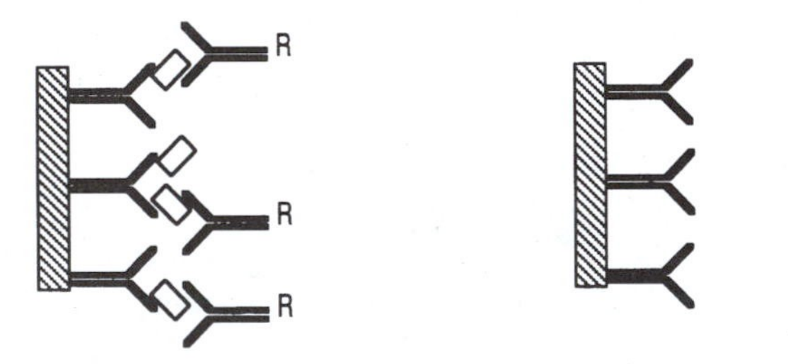

Washing removes unbound reporter antibody.
Detection of reporter indicates a positive test result.

Figure 39.2 Generic sandwich immunoassay format. The solid support shown could be made of polystyrene, magnetic particles, or membrane. The reporter antibody could be conjugated with an enzyme, radioisotope, or fluorescent compound. The detection of the specific reporter compound would indicate that the target antigen is present in the sample. The rectangles represent target antigen, and the circles represent nontarget antigen.

sandwich thus contains captured analyte with a bound reporter system. When there is no analyte in the sample, no reporter antibody can bind and remain on the solid support. The presence or absence of reporter antibody is then determined, either directly or indirectly, allowing a determination of the amount of analyte in the sample.

Reporter antibody can be measured directly with the use of radioisotopes and fluorescence. The isotope used most commonly with reporter antibodies is ^{125}I; also used are ^{3}H, ^{57}Co, and ^{14}C. Disadvantages of using an isotope as the reporter include instability due to isotopic decay, the inherent dangers of using radioisotopes, and the problems associated with radioactive waste disposal. For these reasons, and because of the development of equally good or better alternative reporter systems, the use of radioisotopes has declined. The reporter antibody can be labeled with a fluorescent molecule such as fluorescein and measured directly by the amount of fluorescence.

Advances in antibody-enzyme conjugation procedures have led to the development of indirect measurements of the reporter antibody via an enzymatic reaction (75). Reporter antibodies successfully conjugated with several enzymes include horseradish peroxidase, alkaline phosphatase, and β-galactosidase. With antibody-enzyme conjugates, the reporter antibody is detected by adding the substrate specific for the reporter enzyme into the reaction well. If the enzyme is present, the substrate will be modified, resulting in a product that can be detected by colorimetric, fluorometric, or chemiluminescent techniques.

A technique that uses DNA amplification as the reporter system for a sandwich assay has been described (30). In this system, an oligonucleotide is attached to the reporter antibody. The reporter antibody-oligonucleotide is then detected by using PCR. If the analyte was in the sample, the oligonucleotide would be carried to the PCR process as an analyte-reporter antibody-oligonucleotide. The presence or absence of amplified DNA following PCR determines whether the analyte was present. By using oligonucleotides of different lengths, a multianalyte immunoassay higher in sensitivity than enzyme-linked assays has been developed. This hybrid technology may lead to the development of the next generation of high-sensitivity immunoassays.

Assay Configurations

An immunoassay comprises a variety of individual components. Therefore, it is not surprising that immunoassays can be configured in many ways. Commercial assay configurations use a variety of supports and reagents, as described earlier, along with many specialized devices which facilitate use by nonexperts. The formats of two commercial immunoassays for the detection of food pathogens will be described.

TECRA Diagnostics (Roseville, Australia) has developed a rapid immunoassay for the detection of salmonellae. The assay uses a plastic dipstick-type device as the solid support, an anti-*Salmonella* antibody-enzyme conjugate, and a colorimetric substrate. The assay is performed by placing the dipstick carrying a *Salmonella* capture antibody in the overnight preenrichment culture. Following a short incubation to allow capture of *Salmonella* cells, the dipstick is removed from the preenrichment culture, washed, placed in a tube containing growth medium, and incubated for 4 h. During the incubation, captured *Salmonella* cells replicate, increasing the target concentration. The dipstick is then placed in a tube containing an anti-*Salmonella* antibody-enzyme conjugate. Target cells bound to the dipstick are bound by the conjugate, forming the classic antibody-antigen-antibody "sandwich." The dipstick is then washed and placed in a tube containing a substrate for the enzyme. The enzyme metabolizes the substrate, causing the dipstick color to change from white to purple, indicating a positive test for *Salmonella* species.

A second example of a commercial immunoassay format is the single-step immunoassay for the detection of *E. coli* O157:H7 available from Neogen Corporation (Lansing, Mich.). This assay is an example of the lateral-flow format that is popular in home pregnancy tests. The lateral-flow format uses immunoassay components immobilized along the length of an absorbent pad. The assay is performed by adding three drops of sample enrichment broth to a sample port near one end of the device. The sample wicks through the pad, by capillary action, into a zone containing dried, colloidal gold-labeled anti-*E. coli* O157 antibody. Target cell antigen in the sample forms a complex with the labeled antibody, and then the complex continues to wick through the pad into a zone containing a second capture antibody. The complex is bound in this capture zone, causing a concentration of the gold particles along a line perpendicular to the liquid flow. Thus, the presence of a dark band in the capture zone indicates a positive test.

NUCLEIC ACID HYBRIDIZATION-BASED DETECTION

Introduction

All nucleic acid-based detection systems rely on the discrimination of target organisms from closely related organisms on the basis of characteristic DNA or RNA sequences. An advantage of this genotypic approach is the essentially invariant nature of the DNA sequence. Phenotypic approaches (those which rely on expressed characteristics) require gene expression by the target organism. An example of this approach is immunologic detection of a cell surface component. Depending on the phenotypic character assayed, there can be quantitative differences in expression resulting from differences in growth conditions.

All genetic methods used to detect pathogens are based on the formation of a nucleic acid hybrid. This

Table 39.3 Examples of nucleic acid hybridization-based assays for foodborne pathogens

Target organism[a]	Gene or nucleic acid sequence targeted by probe	Detection	Food(s)	Reference(s)
Clostridium perfringens	Enterotoxin	Colony blot[b]	Raw beef	3
Hepatitis A virus	cDNA clones	^{32}P dot blot	None	35
Escherichia coli				
ETEC	Heat-labile enterotoxin	Colony blot	Many	62
SLTEC	Shiga-like toxins	Colony blot	Many	66
Listeria monocytogenes	Hemolysin	Colony blot	None	14
	16S rRNA	Colorimetric dipstick	Many	36
Salmonellae	Library clones	Filter hybridization	Many	23
	23S rRNA	Colorimetric dipstick	Many	7, 10
Vibrio vulnificus	Cytotoxin-hemolysin	^{32}P filter hybridization	Oysters	51
Yersinia enterocolitica	44-MDa plasmid	^{32}P colony blot	Scallops	33

[a]ETEC, enterotoxigenic *E. coli;* SLTEC, Shiga-like toxin-producing *E. coli.*
[b]Unless indicated otherwise, colony blotting used nonradioactive chromagenic detection.

nucleic acid hybridization is typically between a DNA or RNA molecule present in the target organism and a diagnostic, or probe, DNA which has a sequence complementary to the target sequence. Most differences among hybridization methodologies lie in the strategies used to detect the nucleic acid hybrids. The initial step(s) in all of the genetic methods necessarily includes some means of cell lysis and, often, purification to free the nucleic acid so that it can participate in hybridization. These critical steps are often more tedious than those typical for immunoassays.

Standard Probe Hybridization

In the simplest form, standard probe hybridization uses a labeled DNA probe to hybridize to nucleic acid within a sample. Hybridization is a highly specific method and has been applied extensively to detection of foodborne pathogens (Table 39.3). Commercial tests exist for only a fraction of the targets which have been detected by hybridization in research laboratories. The approaches to hybridization differ in whether one or both of the probe and target are in solution and in the mechanism of detection of the hybridized probe (hybrid).

Probes

Probe DNAs can contain hundreds of nucleotides but usually are limited to 15 to 30 nucleotides (the so-called oligonucleotide probes). The specificity of a hybridization assay is ultimately controlled by the nucleotide sequence of the probe, but since hybridization can occur in the presence of mismatched base pairs, specificity is also affected by other assay conditions. The probe's length and nucleotide base composition, as well as the hybridization temperature and salt concentration (ionic strength), combine to determine the stringency of the hybridization. Stringency is a measure of the degree of base pair mismatch that can be tolerated in the formation of a probe-target hybrid. Requirements for the degree of base pair matching of probe and target sequences differ between assays. In some cases, not all target sequences are identical and a certain tolerance of mismatch is necessary for detection of all targets. Longer probes (up to approximately 30 bases), lower temperature, or higher salt concentration may all be used to decrease the stringency. Once the probe sequence is determined, the most favorable stringency of hybridization is usually empirically determined by altering the temperature or salt concentration. Generally, the minimum stringency required for the desired specificity is chosen since the kinetics of hybrid formation, and therefore the sensitivity of the assay, are most favorable at the lowest acceptable stringency. Typically, the sensitivity of standard probe hybridizations is approximately 10^5 to 10^7 targets. For this reason, hybridization is normally performed on food samples only after selective enrichment during which the target concentration reaches the required level.

Hybridization

Differences among hybridization formats are based on considerations of the kinetics of probe-target hybrid formation and on the mechanism used to separate unbound probe from hybrids. Immobilization of either the target DNA or a capture probe prior to hybridization allows for simple removal of unbound reporter probe. Solid supports are membranes, typically nylon or nitrocellulose, and polymer particles. Historically, the most common solid-phase formats are the Southern

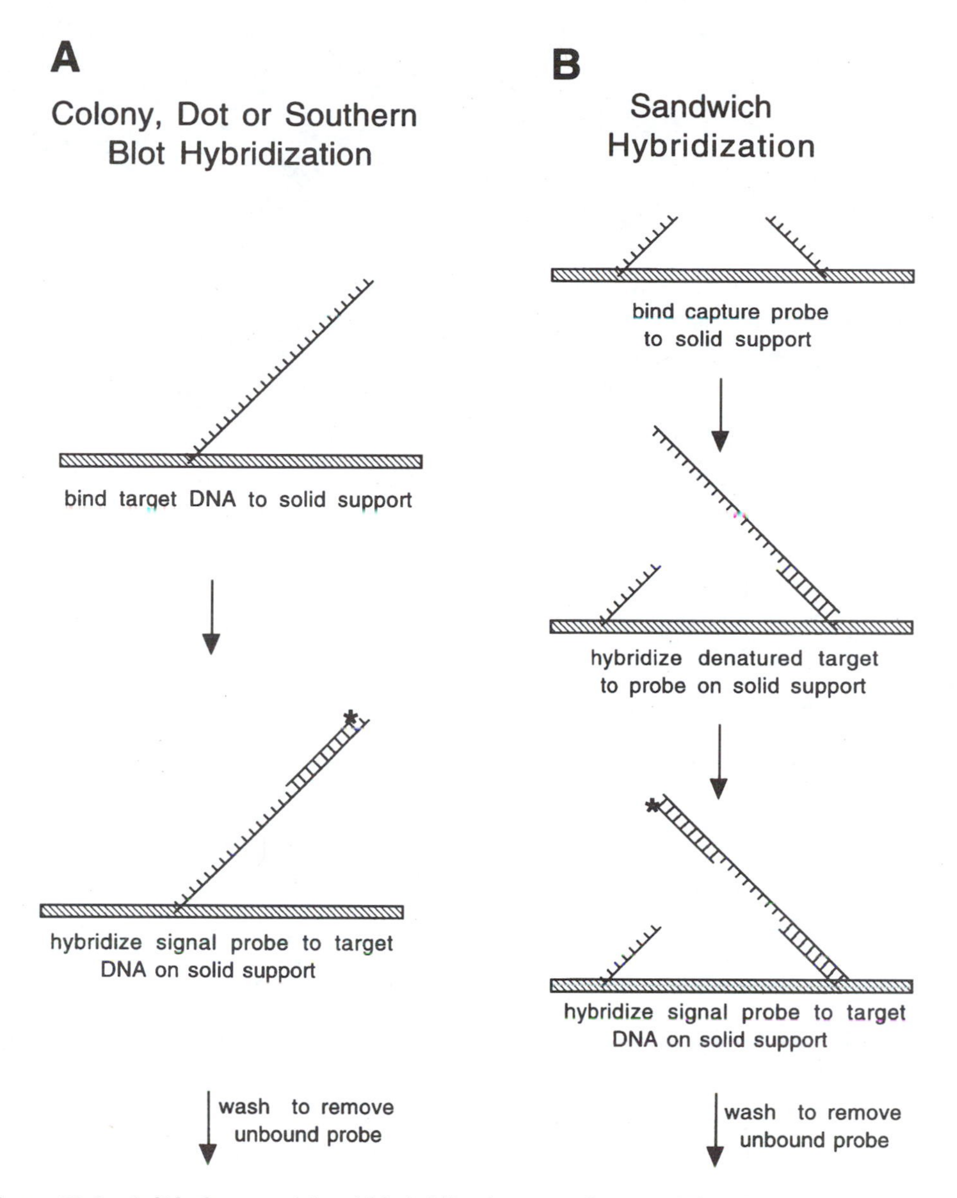

Figure 39.3 Solid-phase nucleic acid hybridization assay formats. (A) Colony, dot, or Southern blot hybridization, in which the denatured target nucleic acid is immobilized to a support which is then soaked in a solution containing a labeled probe. Detection is by any of several direct or indirect methods. (B) Sandwich hybridization, in which a capture probe is first immobilized and then used to capture the denatured target nucleic acid from solution. Detection is accomplished following hybridization of a labeled probe.

blot and the various dot blot formats, in which the target nucleic acids are immobilized on a membrane, either after separation in an electrophoresis gel (Southern blot) or from solution (dot blot) (Fig. 39.3A). Sandwich hybridizations (Fig. 39.3B) offer more sensitivity by using an immobilized capture probe to bind target nucleic acid to the solid phase followed by hybridization of a reporter probe to an adjacent sequence on the same target DNA molecule. Detection of the reporter follows a wash step to remove unbound probe.

Solution hybridization, in which the probe and target are free in solution, offers more favorable kinetics than solid support hybridization because both probe and target are free to diffuse (Fig. 39.4). More rapid hybridization results in shorter incubation times. However, these methods are not widely used because of difficulty in

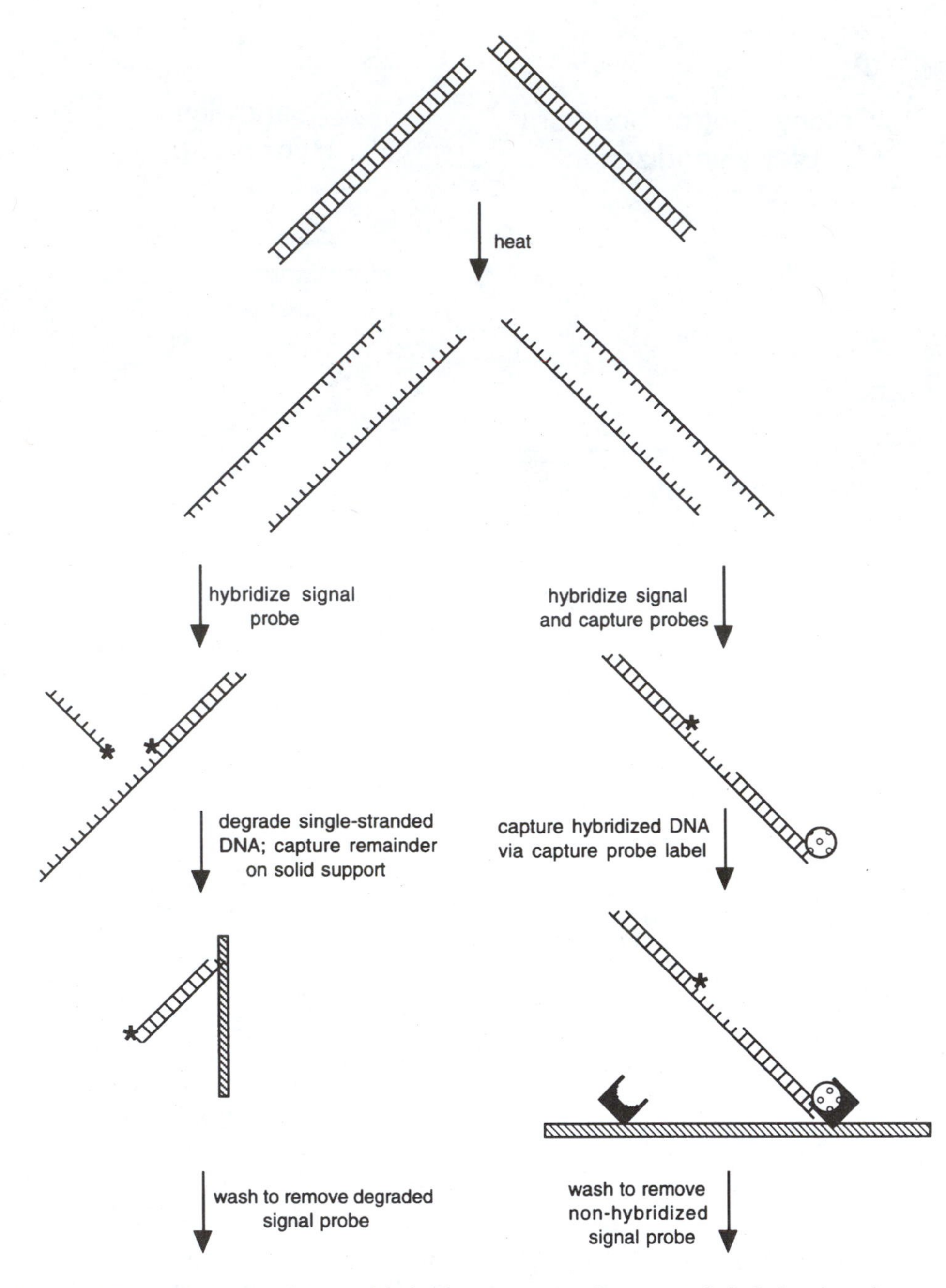

Figure 39.4 Solution nucleic acid hybridization assay formats. Labeled signal probes are hybridized in solution to the denatured target DNA. Detection follows one of two paths. In the first path, degradation of all single-stranded DNA allows specific capture of only those probes which hybridized to target DNA. In the second path, simultaneous hybridization of a second labeled probe allows capture of the target DNA to a solid support and detection of only those targets which are hybridized with signal probe.

removing signal produced by nonhybridized probe. The hybrids in solution are most easily detected in assays in which unhybridized probe molecules are degraded. In nuclease protection assays, single-stranded nucleases are used for this purpose, while hybridization protection assays use a chemiluminescent ester label which is protected from degradation only when present in double-stranded (hybrid) DNA.

In a variation of sandwich hybridization, the separation of probe hybrids occurs by specific capture of the

Table 39.4 Examples of amplification-based assays for food pathogen detection

Amplification method[a]	Target	Detection method	Food(s)	Reference
PCR	*Clostridium botulinum* type A	Gel, dot blot	Meat, canned foods	17
	C. botulinum type A	Gel, Southern blot	Canned vegetables	20
	C. perfringens	Gel, Southern blot	Raw ground beef	2
	Enterotoxigenic *Escherichia coli*	Gel, Southern blot	Minced meat	78
	Enterotoxigenic *E. coli*	Gel, dot blot	Soft cheese	49
	Listeria monocytogenes	Gel, Southern, dot blot	Sausage, milk	26
	L. monocytogenes	Gel	Milk	70
	L. monocytogenes	Gel, dot blot	Poultry	77
	L. monocytogenes	Gel	Soft cheese	79
	Shigella flexneri	Gel, Southern blot	Lettuce	41
	Vibrio parahaemolyticus	Gel	Shellfish	43
	V. vulnificus	Gel	Oysters	32
Reverse transcription-PCR	Enteric viruses	Gel	Oysters	1
Immuno-PCR	Salmonellae	Gel	Raw beef	24
PCR-LCR	*L. monocytogenes*	Microplate capture and reporter	None	80
LCR	*L. monocytogenes*	^{32}P label, gel detection	None	81
NASBA[b]	*Campylobacter jejuni*	ELGA[c]	Meats, milk, etc.	74

[a]If available, examples in which foods were tested are presented.
[b]NASBA, nucleic acid sequence-based amplification.
[c]ELGA (enzyme-linked gel assay) is a nonradioactive solution hybridization with gel-based analysis.

probe hybrids after solution hybridization (Fig. 39.4). Again, two probes which bind to adjacent sequences on the target nucleic acid can be used. One probe carries a reporter label, while the other probe carries a capture label. Following hybridization, binding of the capture probes to a solid support allows collection of only those reporter probes which are hybridized to the target.

Detection

Once the hybrid is formed, many subsequent detection techniques can be used. Most of these techniques are similar to those used for antibody-antigen detection in immunoassays. The various probe labels allow either direct or indirect detection of the probe hybrids. Radioactive and fluorescent probes allow direct detection of hybrids. However, the flexibility, safety, and simplicity of handling have made the indirect methods more popular. Central to the indirect methods are binding pairs such as biotin-avidin, biotin-streptavidin, digoxigenin-antidigoxigenin, and fluorescein-antifluorescein. Biotinylated probes can be prepared by chemical or enzymatic means and take advantage of the high affinity of biotin for avidin or streptavidin. Following hybridization, avidin- or streptavidin-reporter conjugates are allowed to bind to the biotin on the hybrids. If an enzyme reporter is used, then after washing to remove unbound conjugate, an enzyme substrate is added. Enzymes such as alkaline phosphatase and horseradish peroxidase have a range of substrates which allow detection of the activity and, therefore, the presence of the enzyme. Cleavage or modification of the substrates generates measurable changes such as color (chromagenic substrate) or production of light (chemiluminescent substrate). The use of different labels or reporters on different probes allows for simultaneous screening for multiple targets.

Amplification-Based Methods

Because of a need for greater sensitivity, DNA-based methods which include an amplification step have been the focus of intense research and development in recent years. The amplified component of the assay can be either the target nucleic acid, the probe, or the signal. Depending on the amplification type (some are logarithmic, others are linear), the level of target, probe, or signal may be increased several million-fold. Table 39.4 lists examples of amplification-based methods. With few exceptions, to date only methods involving PCR, which use a DNA polymerase enzyme to amplify a DNA target sequence, have been employed to detect foodborne pathogens. These assays are largely restricted to research laboratories. Recently, however, a commercial PCR-based assay, the Bax system for *Salmonella* screening of preenrichment samples, was introduced by Qualicon, a DuPont subsidiary (Wilmington, Del.). Because several different amplification methods have been or are being developed for clinical pathogen screening (often a precursor to use of a method for foodborne pathogens), each approach is described below.

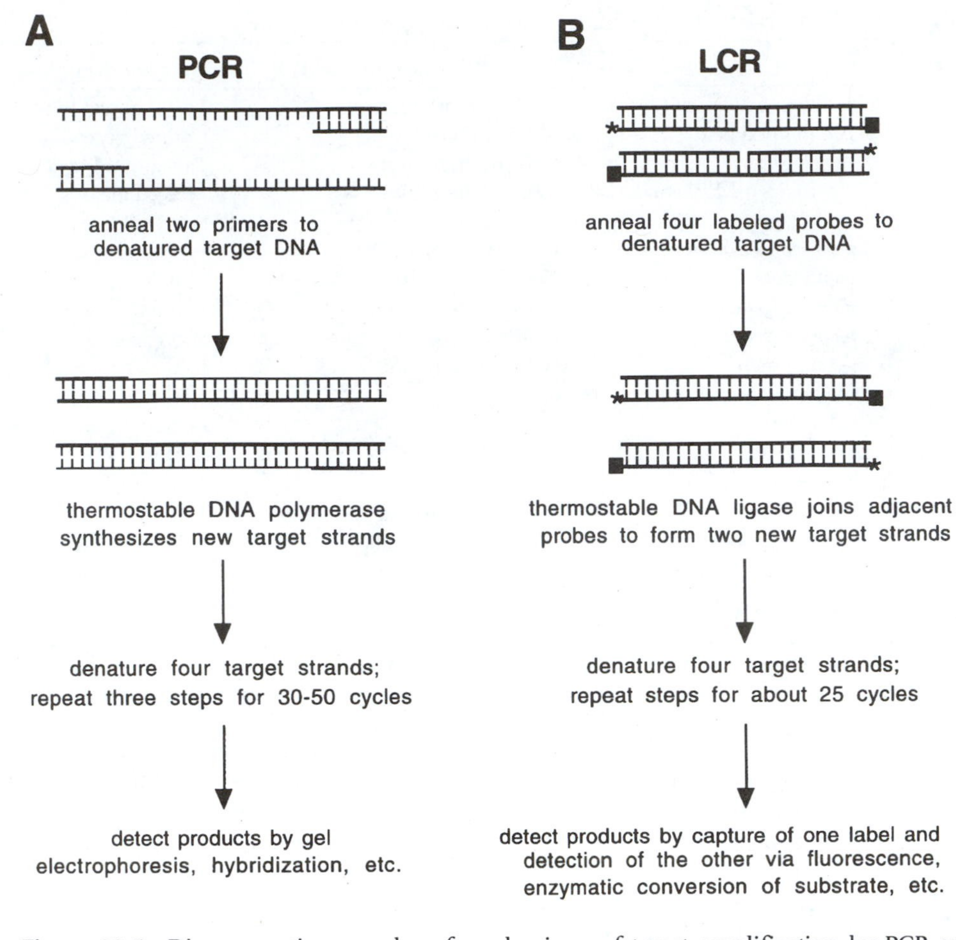

Figure 39.5 Diagrammatic examples of mechanisms of target amplification by PCR and probe amplification by LCR. In PCR, DNA polymerase catalyzes the synthesis of new target DNA; in LCR, new target DNA molecules are formed by the joining of probe molecules.

Target Amplification

The most popular and currently most powerful method of amplification is that of target amplification. There are a number of different methods of target amplification. Some are fundamentally different, while others are variations of a fundamental method.

All target amplification methods require the identification of two conserved sequences which can be used to design amplification probes, or "primers." The ideal distance between the primer sequences varies depending on the particular method and the detection system used. The primer sequences must be unique to the target organism (or group of organisms) and highly conserved (i.e., nearly invariant within the target group). The sequences between the primers are amplified and used for detection; however, it is not necessary that they be unique. The ability to locate useful primer sequences is prerequisite to developing a method and can influence the choice of method as described below.

That products of target amplification methods represent new targets is a critical facet, but also a disadvantage, of amplification methods. The potential for product from one reaction contaminating subsequent reactions and causing false-positive results is a major concern. The huge numbers of products made compared with the small number required for detection necessitate special means to prevent carryover or to inactivate contaminant molecules. There are two ways to prevent carryover. The sample preparation area can be physically separated from the amplification and detection area, or the amplified product can be physically contained (21). Though not 100% effective, inactivation of products following PCR can been accomplished through chemical or enzymatic treatments which render the nucleic acids incapable of being amplified. UV irradiation (56) and photochemical inactivation with psoralen or isopsoralen (8) are used extensively. The enzyme uracil-*N*-glycosylase can be used to pretreat DNA amplification mixtures in order to inactivate contaminating product from previous amplifications. The method involves the incorporation of uracil in place of thymine in all reactions and then starting each reaction with a uracil glycosylase treatment, which degrades any uracil-containing (amplified) DNA but not sample DNA (46).

PCR

Starting from a single target DNA or RNA sequence, more than 1 billion product DNA sequences can routinely be synthesized by PCR. This quantity of DNA can be visualized as a band on an ethidium bromide-stained electrophoresis gel, the most common means of detecting PCR products. The method (Fig. 39.5) (65) involves the use of two primers, each one complementary to sequences on the opposite DNA strands generally separated by 100 or more nucleotides. In cyclical fashion, the two strands of the target DNA are separated by heat (94°C) denaturation, and the primers are allowed to anneal to their complementary sequences at a reduced temperature (usually 50 to 70°C). Then the enzyme DNA polymerase catalyzes the synthesis of second strands and in so doing produces two new target molecules from the initial double-stranded sequence. The use of a thermostable DNA polymerase such as *Taq* polymerase allows the entire process to proceed uninterrupted for as many as 50 cycles (64).

For RNA viruses to be detected by PCR, the RNA targets must first be converted to their cDNAs. This conversion can be accomplished by the enzyme reverse transcriptase; however, the use of *Tth* polymerase, a thermostable DNA polymerase which also has reverse transcriptase activity, makes the process more specific and efficient (52).

A deterrent to routine use of PCR in food testing laboratories is the relatively tedious and exacting nature of the reaction setup, principally the requirement for multiple, microliter-level reagent additions. Innovative reagent delivery systems, such as the PCR reagent and positive control tablets used in the Bax system, have greatly alleviated this problem. With the PCR tablets, PCR setup involves only a single 50-μl addition of sample.

SDA

Strand displacement amplification (SDA) (76) exploits the ability of DNA polymerases to initiate synthesis at a single-stranded nick of a duplex molecule and synthesize a new strand while displacing the existing strand. In so doing, two new templates, one single stranded and one double stranded, are produced from the starting template. The new strands can act as substrates for additional synthesis and displacement, thereby causing a geometric amplification. Special primers which carry a homologous priming region as well as a site for single-strand enzymatic nick formation are required along with DNA polymerase and the restriction enzyme to produce the nicks. The initial step in SDA involves the denaturation of the target DNA, annealing of the special primers, and second-strand synthesis as in PCR. However, in contrast to PCR, products of the first step are amplified without additional denaturation. Thus, while the technique is more complex than PCR, it does allow isothermal (37°C) amplification, eliminating the need for a thermal cycler instrument. The use of a thermostable polymerase and restriction enzyme in SDA has been described (83). The use of elevated temperatures allows higher specificity in the annealing of primers.

Transcription-Based Amplification

Nucleic acid sequence-based amplification, also referred to as self-sustaining sequence replication and TAS, represents transcription-based amplification systems (9, 29, 40) and differs from PCR and SDA in that RNA is the amplification product. Again, these methods are more complex than PCR, involving multiple enzymes and specialized primers, but they are isothermal. All methods couple the reverse transcriptase-mediated synthesis of a cDNA template from the target RNA sequence with in vitro transcription by RNA polymerase from the cDNA template. The initial target sequence can therefore be either RNA or DNA.

Probe Amplification

Methods which amplify only the probe or a portion of the probe used for hybridization are classified as probe amplification techniques. The idea is similar to target amplification in that probes complementary to a unique region of the DNA of a target are required. The difference is that no other target DNA is amplified, but rather various means are used to amplify only the probe sequences into a detectable DNA or RNA product. These methods are most appropriate for the detection of specific sequences and do not have the general research utility of PCR. Thus, although the methods have application for detection of pathogens, the amount of general interest in the methods has not been as great.

LCR

The method commonly referred to as ligase chain reaction (LCR) (4) produces diagnostic DNA product molecules without the DNA replication described for other amplification methods. Instead, new DNA molecules are produced by joining two DNA probes which are complementary to specific sequences on the same strand of target DNA (Fig. 39.5). The ability to join the probes is attained by the selection of probes which will anneal to the target such that the 3′ end of one probe is immediately adjacent to the 5′ end of the other probe. At that point, the enzyme DNA ligase can covalently join the probes, effectively forming a new target sequence. After thermal denaturation, two targets are produced, thus allowing for a geometric amplification. As with the use of a thermostable polymerase in PCR, the use of a thermostable DNA ligase is key to the ease and fidelity of LCR, alleviating the original need to add fresh enzyme after each cycle. LCR is particularly adept at distinguishing single-base-pair differences in target sequences, as even a one-base mismatch at the adjacent ends of the probes will prevent efficient amplification. This allows precise discrimination of closely related sequences.

Products can be detected by gel electrophoresis or by using specific labels to create a capture and a reporter probe. A convenient scheme is the incorporation of an affinity label on one probe which can be used for capturing the ligated product on a solid support. Detection of the captured product is accomplished by detecting the label on the reporter probe (Fig. 39.5). Biotin and digoxigenin labels on the two probes were successfully used in this manner to detect *L. monocytogenes* (80).

Qβ Replicase

Qβ replicase is an RNA polymerase which replicates the RNA genome of the bacteriophage Qβ. The enzyme recognizes a specially folded structure of the genome. The rate of replication is such that as many as 10^9 copies of the genome can be produced from a single copy, in vitro, in 30 min at 37°C. It is possible to replace a small section of the genome with a diagnostic probe RNA sequence without altering the correct folded structure of the molecule. Probe molecules produced in this way will bind specifically to their target and be replicated. The basic assay (38, 45) is carried out by hybridizing a probe with the target DNA, capturing target-probe complexes, and removing unbound probes. Finally, after the addition of the replicase and an incubation period, the replicated RNA probe molecules are detected.

Signal Amplification

The third class of amplification technique, signal amplification, increases the sensitivity of hybridization by facilitating the expression of multiple signal molecules from hybridization to a single target. This expression is accomplished by the use of specially designed sets of probes with multiple functionalities.

Compound probes are designed with a binding sequence specific for the target nucleic acid plus adjacent DNA containing multiple binding sequences for the reporter probe (18). The binding of numerous reporter probes adds to the strength of the assay signal. Unfortunately, there is typically a concurrent increase in background signal.

Branched DNA (bDNA) hybridization is a related form of signal amplification which exploits the ability to synthesize bDNA molecules (73). The bDNA assay utilizes four probe types: capture, extender, bDNA, and reporter. The capture probe hybridizes and immobilizes target DNA to a solid support. Extender probes bind to a separate DNA sequence on the target DNA. Extender probes carry additional sequence complementary to a sequence on the bDNA probe. The bDNA probe is also multifunctional, with sequence for binding to the extender probe and numerous sequences arranged in the shape of a comb, which allows hybridization of up to 3,000 of the reporter probes. In this way, a dramatic increase in sensitivity can be achieved. Because two separate hybridization probes (the capture probe and the extender probe) must bind to the same target DNA in order to form the final complex, higher specificity is maintained and the background signal is lower. Compound probes for food pathogens have not yet been reported.

SUMMARY AND FUTURE TRENDS

An extensive array of targets and samples having various sensitivity and specificity requirements are used for the detection of foodborne pathogens and toxins. Diverse methodologies have been, and will continue to be, developed and used in food diagnostics to meet the needs described. Among these methodologies, the immunologic and genetic techniques have brought completely new potential to food testing. Exploitation of the sensitivity and specificity of the antibody-antigen interaction and the nucleic acid hybridization now allows conclusive results to be obtained more quickly (by 1 day to several days) than with classical methods. Additionally, these methods are more amenable to automation.

Development of food testing technology will continue to benefit from advances in clinical nucleic acid- and antibody-based diagnostics, although key improvements are needed for these technologies to reach full potential for routine application in food diagnostics. The most critical improvements will come in sample preparation and assay simplification, miniaturization, and automation. The continued development of multianalyte immunoassays and multiplex amplification assays will eventually allow routine, simultaneous screening of more than one pathogen or toxin.

References

1. **Altmar, R. L. T. G. Metcalf, F. H. Neill, and M. K. Estes.** 1993. Detection of enteric viruses in oysters by using the polymerase chain reaction. *Appl. Environ. Microbiol.* **59:**631–635.
2. **Baez, L. A., and V. K. Juneja.** 1995. Detection of enterotoxigenic *Clostridium perfringens* in raw beef by polymerase chain reaction. *J. Food Prot.* **58:**154–159.
3. **Baez, L. A., and V. K. Juneja.** 1995. Nonradioactive colony hybridization assay for detection and enumeration of enterotoxigenic *Clostridium perfringens* in raw beef. *Appl. Environ. Microbiol.* **61:**807–810.
4. **Barany, F.** 1991. Genetic disease detection and DNA amplification using cloned thermostable ligase. *Proc. Natl. Acad. Sci. USA* **88:**189–193.
5. **Barstad, P., J. Hubert, M. Hunicapiller, A. Goetze, J. Schilling, B. Black, B. Eaton, J. Richards, M. Weigart, and L. Hood.** 1978. Immunoglobulins with hapten binding activity: structure-function correlations and genetic implication. *Eur. J. Immunol.* **8:**497–503.
6. **Cantareno, L. A., J. E. Butler, and J. W. Osborne.** 1980. The adsorptive characteristics of proteins for polystyrene and their significance in solid phase immunoassays. *Anal. Biochem.* **105:**375–382.
7. **Chan, S. W., S. G. Wilson, M. Vera-Garcia, K. Whippie, M. Ottavian, A. Whilby, A. Shah, A. Johnson, M. A. Mozola, and D. N. Halbert.** 1990. Comparative study of colorimetric DNA hybridization method and conventional culture procedure for detection of *Salmonella* in foods. *J. Assoc. Off. Anal. Chem.* **73:**419–424.
8. **Cimino, G. D., K. C. Metchette, J. W. Tessman, J. E. Hearst, and S. T. Isaacs.** 1991. Post-PCR sterilization: a method to control carryover contamination for the polymerase chain reaction. *Nucleic Acids Res.* **19:**99–107.
9. **Compton, J.** 1991. Nucleic acid sequence-based amplification. *Nature* (London) **350:**91–92.

9a. **Council for Agriculture Science and Technology.** 1994. *Interpretive Summary of Foodborne Pathogens: Risks and Consequences,* p. 1. Council for Agriculture Science and Technology, Ames, Iowa.

10. **Curiale, M. S., and M. J. Klatt.** 1990. Colorimetric deoxyribonucleic acid hybridization assay for rapid screening of *Salmonella* in foods: collaborative study. *J. Assoc. Off. Anal. Chem.* **73:**248–256.
11. **Curiale, M. S., M. J. Klatt, B. J. Robison, and L. T. Beck.** 1990. Comparison of colorimetric monoclonal enzyme immunoassay screening methods for detection of *Salmonella* in foods. *J. Assoc. Off. Anal. Chem.* **73:**43–50.
12. **Curiale, M. S., W. Lepper, and B. J. Robison.** 1994. Enzyme-linked immunoassays for detection of *Listeria monocytogenes* in dairy products, seafoods, and meats: collaborative study. *J. Assoc. Off. Anal. Chem.* **77:**1472–1489.
13. **Dandliker, W. B., R. J. Kelly, J. Farquhar, and J. Levin.** 1973. Fluorescence polarization immunoassay theory and experimental method. *Immunochemistry* **10:**219.
14. **Datta, A. R., B. A. Wentz, and W. E. Hill.** 1987. Detection of hemolytic *Listeria monocytogenes* by using DNA colony hybridization. *Appl. Environ. Microbiol.* **53:**2256–2259.
15. **Davies, C.** 1994. Principles, p. 1–6. *In* D. Wild (ed.), *Handbook of Immunoassays.* Stockton Press, New York.
16. **Eystein, S., and O. Olsvik.** 1991. Immunomagnetic separation of *Salmonella* from foods. *Int. J. Food Microbiol.* **14:**11–18.
17. **Fach, P., D. Hauser, J. P. Guillou, and M. R. Popoff.** 1993. Polymerase chain reaction for the rapid identification of *Clostridium botulinum* type A strains and detection in food sample. *J. Appl. Bacteriol.* **75:**234–239.
18. **Fahrlander, P. D., and A. Klausner.** 1988. Amplifying DNA probe signals: a "Christmas tree" approach. *Bio/Technology* **6:**1165–1168.
19. **Fantasia, L. D., J. P. Schrade, J. F. Yager, and D. Debler.** 1975. Fluorescent antibody method for the detection of *Salmonella:* development, evaluation, and collaborative study. *J. Assoc. Off. Anal. Chem.* **58:**828–844.
20. **Ferreira, J. L., B. R. Baumstark, M. K. Hamdy, and S. G. McCay.** 1993. Polymerase chain reaction for detection of type A *Clostridium botulinum* in foods. *J. Food Prot.* **56:**18–20.
21. **Findlay, J. B., S. M. Atwood, L. Bergmeyer, J. Chemelli, K. Christy, T. Cummins, W. Donish, T. Ekeze, J. Falvo, D. Patterson, J. Puskas, J. Quenin, J. Shah, D. Sharkey, J. W. H. Sutherland, R. Sutton, H. Warren, and J. Wellman.** 1993. Automated closed-vessel system for in vitro diagnostics based on polymerase chain reaction. *Clin. Chem.* **39:**1927–1933.
22. **Finkelstein, R. A., and Z. Yang.** 1983. Rapid test for identification of heat-labile enterotoxin-producing *Escherichia coli* colonies. *J. Clin. Microbiol.* **18:**23–28.
23. **Fitts, R., M. Diamond, C. Hamilton, and M. Neri.** 1983. DNA-DNA hybridization assay for detection of *Salmonella* spp. in foods. *Appl. Environ. Microbiol.* **46:**1146–1151.
24. **Fluit, A. C., M. N. Widjojoatmodjo, A. T. A. Box, R. Torensma, and J. Verhoef.** 1993. Rapid detection of salmonellae in poultry with the magnetic immuno-polymerase chain reaction assay. *Appl. Environ. Microbiol.* **59:**1342–1346.
25. **Freed, R. C., M. L. Evenson, R. F. Reiser, and M. S. Bergdoll.** 1982. Enzyme-linked immunosorbent assay for detection of staphylococcal enterotoxins in foods. *Appl. Environ. Microbiol.* **44:**1349–1355.
26. **Furrer, B., U. Candrian, C. Hoefelein, and J. Luethy.** 1991. Detection and identification of *Listeria monocytogenes* in cooked sausage products and in milk by in vitro amplification of haemolysin gene fragments. *J. Appl. Bacteriol.* **70:**372–379.
27. **Gibb, K., and A. Padovan.** 1994. A DNA extraction method that allows reliable PCR amplification of MLO DNA from "difficult" plant host species. *PCR Methods Appl.* **4:**56–58.
28. **Greenfield, L., and T. J. White.** 1993. Sample preparation methods, p. 122–137. *In* D. H. Persing, T. F. Smith, F. C. Tenover, and T. J. White (ed.), *Diagnostic Molecular Microbi-*

ology: Principles and Applications. ASM Press, Washington, D.C.

29. **Guatelli, J. C., K. M. Whitfield, D. Y. Kwoh, K. J. Barringer, D. D. Richman, and T. R. Gingeras.** 1990. Isothermal, *in vitro* amplification of nucleic acids by multienzyme reaction modeled after retroviral replication. *Proc. Natl. Acad. Sci. USA* **87:**1874–1878.
30. **Hendrickson, E. R., T. M. Truby, R. D. Joerger, W. R. Majarian, and R. C. Ebersole.** 1995. High sensitivity multianalyte immunoassay using covalent DNA-labeled antibodies and polymerase chain reaction. *Nucleic Acids Res.* **23:**522–529.
31. **Higuchi, R.** 1989. Simple and rapid preparation of sample for PCR, p. 31–38. *In* H. A. Erlich (ed.), *PCR Technology.* Stockton Press, New York.
32. **Hill, W. E., S. P. Keasler, M. W. Trucksess, P. Feng, C. A. Kaysner, and K. A. Lampel.** 1991. Polymerase chain reaction identification of *Vibrio vulnificus* in artificially contaminated oysters. *Appl. Environ. Microbiol.* **57:**707–711.
33. **Hill, W. E., W. L. Payne, and C. C. G. Aulisio.** 1983. Detection and enumeration of virulent *Yersinia enterocolitica* in food by DNA colony hybridization. *Appl. Environ. Microbiol.* **46:**636–641.
34. **Holbrook, R., J. M. Anderson, A. C. Baird-Parker, and S. H. Stuchbury.** 1989. Comparative evaluation of the Oxoid *Salmonella* rapid test with three other rapid *Salmonella* methods. *Lett. Appl. Microbiol.* **9:**161–164.
35. **Jiang, X., M. K. Estes, T. G. Metcalf, and J. L. Melnick.** 1986. Detection of hepatitis A virus in seeded estuarine samples by hybridization with cDNA probes. *Appl. Environ. Microbiol.* **52:**711–717.
36. **King, W., S. Raposa, J. Warshaw, A. Johnson, D. Halbert, and J. D. Klinger.** 1989. A new colorimetric nucleic acid hybridization assay for *Listeria* in foods. *Int. J. Food Microbiol.* **8:**225–232.
37. **Kohler, G., and C. Milstein.** 1975. Continuous cultures of fused cells secreting antibody of predefined specificity. *Nature* (London) **236:**495–497.
38. **Kramer, F. R., and P. M. Lizardi.** 1989. Replicatable RNA reporters. *Nature* (London) **339:**401–402.
39. **Kricka, L. J.** 1985. Separation procedures, p. 53–110. *In Ligand Binder-Assay.* Marcel Dekker, Inc., New York.
40. **Kwoh, D. Y., G. R. Davis, K. M. Whitfield, H. L. Chappelle, L. J. DiMichele, and T. R. Gingeras.** 1989. Transcription-based amplification system and detection of amplified human immunodeficiency virus type 1 with a bead-based sandwich hybridization format. *Proc. Natl. Acad. Sci. USA* **86:**1173–1177.
41. **Lampel, K. A., J. A. Jagow, M. Trucksess, and W. E. Hill.** 1990. Polymerase chain reaction for detection of invasive *Shigella flexneri* in food. *Appl. Environ. Microbiol.* **56:**1536–1540.
42. **Lawellin, D. W., D. W. Grany, and B. K. Joyce.** 1977. Enzyme-linked immunosorbent analysis for aflatoxin B1. *Appl. Environ. Microbiol.* **34:**94–96.
43. **Lee, C.-Y., S.-F. Pan, and C.-H. Chen. 1995.** Sequence of a cloned pR72H fragment and its use for detection of *Vibrio parahaemolyticus* in shellfish with PCR. *Appl. Environ. Microbiol.* **61:**1311–1317.
44. **Litchfield, W. J., A. R. Craig, W. A. Frey, C. C. Leflar, C. E. Looney, and M. A. Luddy.** 1984. Novel shell/core particles for automated turbidimetric immunoassays. *Clin. Chem.* **30:**1498.
45. **Lizardi, P. M., C. E. Guerra, H. Lomeli, I. Tussie-Luna, and F. R. Kramer.** 1988. Exponential amplification of recombinant-RNA hybridization probes. *Bio/Technology* **6:**1197–1202.
46. **Longo, M. C., M. S. Berninger, and J. L. Hartley.** 1990. Use of uracil DNA glycosylase to control carry-over contamination in polymerase chain reactions. *Gene* **93:**125–128.
47. **Mattingly, J. A., B. T. Butman, M. C. Plank, and R. J. Durham.** 1988. Rapid monoclonal antibody-based enzyme-linked immunosorbent assay for the detection of *Listeria* in food products. *J. Assoc. Off. Anal. Chem.* **71:**679–681.
48. **McCafferty, J., A. D. Griffiths, G. Winter, and D. J. Chiswell.** 1990. Phage antibodies: filamentous phage displaying antibody variable domains. *Nature* (London) **348:**552–554.
49. **Meyer, R., J. Luthy, and U. Candrian.** 1991. Direct detection by polymerase chain reaction (PCR) of *Escherichia coli* in water and soft cheese and identification of enterotoxigenic strains. *Lett. Appl. Microbiol.* **13:**268–271.
50. **Meyer, R. F., and M. J. Palmieri.** 1980. Single radial immunodiffusion method for screening staphylococcal isolates for enterotoxin. *Appl. Environ. Microbiol.* **40:**1080–1085.
51. **Morris, J. G., A. C. Wright, D. M. Roberts, P. K. Wood, L. M. Simpson, and J. D. Oliver.** 1987. Identification of environmental *Vibrio vulnificus* isolates with a DNA probe for the cytotoxin-hemolysin gene. *Appl. Environ. Microbiol.* **53:**193–195.
52. **Myers, T. W., and D. H. Gelfand.** 1991. Reverse transcription and DNA amplification by a *Thermus thermophilus* DNA polymerase. *Biochemistry* **30:**7661–7666.
53. **Nilsson, L.-A.** 1981. Precipitation and related immunoassay techniques, p. 43–68. *In* A. Voller, A. Bartlett, and D. Bidwell (ed.), *Immunoassays for the 80's.* University Park Press, Baltimore.
54. **Notersmans, S., J. Dufrenne, and S. Kozaki.** 1979. Enzyme-linked immunosorbent assay for detection of *Clostridium botulinum* type E toxin. *Appl. Environ. Microbiol.* **37:**1173–1175.
55. **Oggel, J. J., D. C. Nundy, and C. J. Randall.** 1990. Modified 1–2 test(TM) sytem as a rapid screening method for detection of *Salmonella* in foods and feeds. *J. Food Prot.* **53:**656–658.
56. **Ou, C. Y., J. J. Moore, and G. Schochetman.** 1991. Use of UV irradiation to reduce false positivity in the polymerase chain reaction. *BioTechniques* **10:**442–446.
57. **Pestka, J. J., P. K. Gaur, and F. S. Chi.** 1980. Quantitation of aflatoxin B_1 and aflatoxin B_1 antibody by an enzyme-linked immunosorbent microassay. *Appl. Environ. Microbiol.* **40:**1027–1031.
58. **Pestka, J. J., Y. Li, W. O. Harder, and F. S. Chu.** 1981. Comparison of radioimmunoassay and enzyme-linked immunosorbent assay for determining aflatoxin M_1 in milk. *J. Assoc. Off. Anal. Chem.* **64:**294–301.

59. **Poucke, L. G.** 1990. Salmonella-TEK, a rapid screening method for *Salmonella* species in food. *Appl. Environ. Microbiol.* **56**:924–927.
60. **Powell, H. A., C. M. Gooding, S. D. Garrett, B. M. Lund, and R. A. McKee.** 1994. Proteinase inhibition of the detection of *Listeria monocytogenes* in milk using the polymerase chain reaction. *Lett. Appl. Microbiol.* **18**:59–61.
61. **Robbins, J. L., G. A. Hill, and B. N. Carle.** 1962. Latex agglutination reactions between human chorionic gonadotropin and rabbit antibody. *Proc. Soc. Exp. Biol. Med.* **109**:321–325.
62. **Romick, T. L., J. A. Lindsay, and F. F. Busta.** 1989. Evaluation of a visual DNA probe for enterotoxigenic *E. coli* detection in foods and wastewater by colony hybridization. *J. Food Prot.* **52**:466–470.
63. **Rubinstein, K. E., R. S. Schneider, and E. F. Ullman.** 1972. "Homogeneous" enzyme immunoassay. A new immunochemical technique. *Biochem. Biophys. Commun.* **47**:846–851.
64. **Saiki, R. K., D. H. Gelfand, S. Stoffel, S. Scharf, R. Higuchi, G. T. Horn, K. B. Mullis, and H. A. Erlich.** 1988. Primer-directed enzymatic amplification of DNA with a thermostable DNA polymerase. *Science* **239**:487–491.
65. **Saiki, R. K., S. Scharf, F. Faloon, K. B. Mullis, G. T. Horn, H. A. Erlich, and N. Arnheim.** 1985. Enzymatic amplification of beta-globin genomic sequences and restriction site analysis for the diagnosis of sickle-cell anemia. *Science* **230**:1350–1354.
66. **Samadpour, M., J. Liston, J. E. Ongerth, and P. I. Tarr.** 1990. Evaluation of DNA probes for detection Shiga-like-toxin-producing *Escherichia coli* in food and calf fecal samples. *Appl. Environ. Microbiol.* **56**:1212–1215.
67. **Saunders, G. C., and M. L. Bartlett.** 1977. Double-antibody solid-phase enzyme immunoassay for the detection of staphylococcal enterotoxin A. *Appl. Environ. Microbiol.* **34**:518–522.
68. **Schultz, S. J., J. S. Witzeman, and W. M. Hall.** 1968. Immunofluorescent screening for *Salmonella* in foods: comparison with cultural methods. *J. Assoc. Off. Anal. Chem.* **51**:1334–1338.
69. **Shone, C., P. Wilton-Smith, N. Appleton, P. Hambleton, N. Modi, S. Gatley, and J. Melling.** 1985. Monoclonal antibody-based immunoassay for type A *Clostridium botulinum* toxin is comparable to the mouse bioassay. *Appl. Environ. Microbiol.* **50**:63–67.
70. **Starbuck, M. A. B., P. J. Hill, and G. S. A. B. Stewart.** 1992. Ultrasensitive detection of *Listeria monocytogenes* in milk by the polymerase chain reaction (PCR). *Lett. Appl. Microbiol.* **15**:248–252.
71. **Sternberger, L. A.** 1986. The precipitin reaction, p. 19–20. *In Immuno-Chemistry.* John Wiley & Sons, New York.
72. **Tortorello, M. L., and D. S. Stewart.** 1994. Antibody-direct epifluorescent filter technique for rapid, direct enumeration of *Escherichia coli* O157:H7 in beef. *Appl. Environ. Microbiol.* **60**:3553–3559.
73. **Urdea, M. S., T. Fultz, T. J. Anderson, M. Running, J. A. Hamren, S. Ahle, and C. A. Chang.** 1991. Branched amplification multimers for the sensitive, direct detection of human hepatitis viruses. *Nucleic Acids Symp. Ser.* **24**:197–200.
74. **Uyttendaile, M., R. Schukkink, B. van Gemen, and J. Debevere.** 1995. Detection of *Campylobacter jejuni* added to foods by using a combined selective enrichment and nucleic acid sequence-based amplification (NASBA). *Appl. Environ. Microbiol.* **61**:1341–1347.
75. **Van Weemen, B. K., and A. H. W. M. Schuurs.** 1971. Immunoassay using antigen-enzyme conjugates. *FEBS Lett.* **15**:232–235.
76. **Walker, G. T., M. C. Little, J. G. Nadeau, and D. D. Shank.** 1992. Isothermal in vitro amplification of DNA by a restriction enzyme/DNA polymerase system. *Proc. Natl. Acad. Sci. USA* **89**:392–396.
77. **Wang, R. F., W. W. Cao, and M. G. Johnson.** 1992. 16S rRNA-based probes and polymerase chain reaction method to detect *Listeria monocytogenes* cells added to foods. *Appl. Environ. Microbiol.* **58**:2827–2831.
78. **Werners, K., E. Delfgou, P. S. Soentoro, and S. H. W. Notermans.** 1991. Successful approach for detection of low numbers of enterotoxigenic *Escherichia coli* in minced meat by using the polymerase chain reaction. *Appl. Environ. Microbiol.* **57**:1914–1919.
79. **Werners, K., C. J. Heuvelman, T. Chakraborty, and S. H. W. Notermans.** 1991. Use of the polymerase chain reaction for direct detection of *Listeria monocytogenes* in soft cheese. *J. Appl. Bacteriol.* **70**:121–126.
80. **Wiedmann, M., F. Barany, and C. A. Batt.** 1993. Detection of *Listeria monocytogenes* with a nonisotopic polymerase chain reaction-coupled ligase chain reaction assay. *Appl. Environ. Microbiol.* **59**:2743–2745.
81. **Wiedmann, M., J. Czajka, F. Barany, and C. A. Batt.** 1992. Discrimination of *Listeria monocytogenes* from other *Listeria* species by ligase chain reaction. *Appl. Environ. Microbiol.* **58**:3443–3447.
82. **Wild, D., and C. Davies.** 1994. Components, p. 49–80. *In* D. Wild (ed.), *Handbook of Immunoassays.* Stockton Press, New York.
83. **Wright, D. J., M. S. Fraiser, C. M. Nycz, C. A. Spargo, P. A. Spears, M. D. Van Cleve, and G. T. Walker.** 1995. Thermophilic strand displacement amplification: a rapid method for amplification of DNA sequences to extremely high copy number, abstr. U-96, p. 133. *In Abstracts of the 95th General Meeting of the American Society for Microbiology 1995.* American Society for Microbiology, Washington, D.C.

Richard C. Whiting
Robert L. Buchanan

40

Predictive Modeling

MODELING

Modeling in food microbiology began about 1920 with the development of methods for calculating thermal death times. These models revolutionized the canning industry (16). The recent resurgence of predictive modeling is driven by a proliferation of refrigerated and limited-shelf-life foods, the development of multiple-hurdle preservation systems, and the advent of the personal computer. The microbiological, mathematical, and statistical tools existed prior to this expansion in modeling efforts. However, without the ability to distribute cumbersome equations and to solve them rapidly and repeatedly at one's desk, the extensive research effort necessary to create microbial models would not have been worthwhile (6).

In classic microbial research, data handling merely presents graphs showing growth or inactivation for several levels of one or perhaps two factors. It is left to the readers to subjectively interpolate these results to their situation. This became increasingly difficult when three or more factors were involved and all were changing. Standard deviations or other estimates of error ranges which could inform the user about the precision of the estimates were seldom included.

Modeling assumes that the growth or inactivation of microorganisms in a food can be satisfactorily estimated by measuring microorganism behavior in broth cultures or other model systems having the same levels of the environmental factors (30). Typically, models include factors such as temperature, pH, salt or water activity, sodium nitrite, organic acid, and aerobic-anaerobic atmosphere (22). These can explain most microbial behavior in foods (30). Exceptions do occur, however, if a food has another significant factor not in the model. For example, high phosphate concentrations in processed cheese or reduced water activity from carbohydrates in jellies would make inoperable a model which did not include these factors.

Mathematical models based upon Monod or substrate-based kinetics are used in studying fermentation and related biological processes. Fermentation models are usually concerned with high concentrations of cells growing at optimal conditions. However, in foods, the identities and concentrations of the nutrients are not known and are typically not limiting. Instead, foodborne microbes, especially pathogens, are initially present in low numbers and often under suboptimal conditions. Therefore, mathematical modeling in food microbiology has taken a more empirical approach than fermentation processes, focusing on batch instead of continuous-culture kinetics.

Empirical though these models may be, they are based upon linear and nonlinear regression techniques (27). Growth data or model parameters are fitted to equations using interactive least-squares computer algorithms. Assumptions about randomness, normal distributions, interpolations within tested ranges rather than extrapolations outside the ranges, parsimony, and stochastic specifications have to be made in microbial modeling as they do for any statistical application of regression (27).

All models are simplifications that represent the complex biochemical processes controlling microbial growth. They must be simplified to a reasonable number of input parameters. These parameters need to be easily measured (e.g., temperature and pH) or known for a food (e.g., added salt level). An input parameter of glucose or peptide concentration would necessitate an analysis of the food before the model could be used. Therefore, a model must simultaneously be sufficiently complex to provide a useful prediction, but simple enough to be usable. This balance between simplicity and complexity means that no model will be "best" for all situations. Broad-range, multiparameter models based upon broth cultures have wider applications than models for a single variable such as temperature or those based on a specific food, but the multiparameter models are probably less accurate than the more specific models.

LEVELS OF MODELS

Models can be conceptualized as having three levels (36). The first or primary level is mathematical expressions that describe the changes in microbial numbers with time. An example is a growth model that calculates the change in log colony-forming units (CFU) per milliliter with time. Another primary-level model is one that describes the decreasing counts with time during thermal processing such as the widely used decimal reduction time or *D* value. Alternatively, the measure of microbial numbers may be made by turbidity or conductance measurements. The formation of a microbial toxin or other metabolic product with time could constitute another type of primary-level model.

The secondary level is equations describing how the parameters of primary models change with changes in environmental factors. These equations might be based on Arrhenius or square root equations (24, 31), particularly if temperature is the primary factor of concern, as is often the case if a specific group of foods is being modeled. These equations were expanded to include pH and water activity. When other factors such as organic acid concentration or nitrite are included in the model, polynomial regression equations can be used. These are very flexible, with squared, cubic, and cross-product terms, but have less mechanistic interpretation than the other two secondary models.

To obtain a prediction or estimation, this modeling process is reversed. Environmental values are entered into the secondary models to obtain specific values for the primary model. The primary model is then solved for increasing time periods to obtain the growth or inactivation curve expected from that combination of environmental values. Because of the unwieldiness of these equations, both the secondary- and primary-level models are typically placed into spreadsheet software. The use of spreadsheets avoids reentering the equations, takes advantage of graphics capabilities, and allows performance of other calculations. Such applications software is designated as the tertiary level. Different tertiary systems vary in complexity from a single equation on a spreadsheet to expert systems or risk assessment simulation programs. Two commercially available tertiary systems that contain a variety of growth and other models are the Food MicroModel (Leatherhead Food Research Association, Surrey, U.K.; tel: +44 [0] 372 376761) and the Pathogen Modeling Program (U.S. Department of Agriculture, Wyndmoor, Pa.; tel: [215] 233-6437; see appendix below).

METHODOLOGY

A mixture of four to six strains is frequently used to inoculate the growth medium which will be modeled. This, in effect, creates models for the fastest-growing or longest-surviving strains at a particular combination of conditions. Using a mixture of strains increases the probability of creating a representative, yet conservative, model while still being experimentally practical. The risk is that if a frequently occurring strain having properties exceeding those of the other strains is not included, the model could be "fail-dangerous." Conversely, inclusion of a rare strain with exceptional properties, such as the unusually high thermal resistance of *Salmonella seftenberg* 775W, which has not been implicated in an outbreak, would result in an excessively conservative model. Data on a large number of strains isolated during outbreaks, and obtained at conditions representing rapid and slow change, are needed to determine the natural distribution of *D* values, growth rates, or other parameter values.

Most growth modeling uses 18- to 24-h cultures grown at optimal conditions as the inoculum. The lag phase thus represents the time for adjustment of the inoculum from late growth-early stationary phases to the new, and probably less favorable, environmental conditions. The length of the lag phase depends upon both the prior temperature of incubation and the new growth condition (Table 40.1; 18). Cells adapted to high temperatures and then transferred to low temperatures have the longest lag periods. Minimum lag times at each

Table 40.1 Effect of preincubation temperatures on the subsequent lag time and generation time of *Aeromonas hydrophila*[a]

Test culture temp (°C)	Lag time (h) at temp:				Generation time (h) at temp:			
	5°C	15°C	25°C	35°C	5°C	15°C	25°C	35°C
5	12.27	34.60	316.4	335.7	13.33	16.76	15.64	15.07
15	2.89	1.00	2.42	3.50	1.61	1.55	1.50	1.48
25	0.78	0.22	0.16	0.30	0.59	0.55	0.59	0.55
35	1.71	0.33	0.36	0.12	0.39	0.34	0.33	0.35

[a]From Hudson (18), with permission.

growth temperature occur in inocula preadapted to the same temperature. Once the cells are adapted, the prior temperatures have no effect on the growth rates. Similar effects will probably be found for pH and other environmental changes. Heat shock, cold adaptation, and other stress proteins, as well as membrane changes and synthesis of new RNA and enzymes, are involved in the adaptations occurring during the lag phase (see chapter 2). The duration of lag phases from cells that are starved, desiccated, in biofilms, or injured has not been modeled yet. Such conditions are often more typical of the physiological state of cells introduced into a food. The υ and q_0 parameters in the recent model by Baranyi and Roberts (2) are intended to represent the physiological state of the cell and the change from the prior to the modeled environment, respectively (see below for further description of this model).

Plate counts are the preferred method for enumerating cells despite the potential for undercounting due to chains or clusters. The use of shaking flasks minimizes this, and simple microscopic examination can determine whether more vigorous techniques are needed. One of the advantages of plating is that the lower limit of detection is as low as 20 CFU/ml. Repeated sampling of the same flask produces more precise curves than having a separate vessel for each sampling time, but increases the potential for contamination.

Turbidity techniques are generally insensitive until population densities of 10^5 to 10^6 cells per ml are reached. These levels are well beyond those that limit most food safety and many food quality attributes (12, 23). If the optical density has been correlated with plate counts, the initial increase in the optical density curve can be used to estimate the growth rate. Inoculating several tubes with known numbers of cells and determining the time to the initial turbidity change can be used, along with the growth rate, to estimate the duration of the lag period by extrapolation.

Using conductivity or other electrical measures has similar considerations as the turbidity measurements (26). This approach measures changes in the medium's electrical properties resulting from cell growth or metabolism. High cell numbers (ca. 10^7 to 10^8/ml) are necessary for a detectable signal, and the magnitude of the signal per amount of growth varies with medium pH and salts. Conductivity or turbidity techniques may have certain applications for modeling because of their capability for simultaneously following many samples and automated operation.

GROWTH MODELS

Primary Level

The principal primary model for growth curves is the Gompertz equation (equation 1; Fig. 40.1) (15):

$$Y_t = Y_0 + C\exp\{-\exp[-B(t - M]\} \quad (1)$$

where Y_t is the log CFU per milliliter at time t, Y_0 is the log inoculum, C is the change in cell numbers between inoculum and stationary phase, B is the relative growth rate, and M is the time when the maximum growth rate

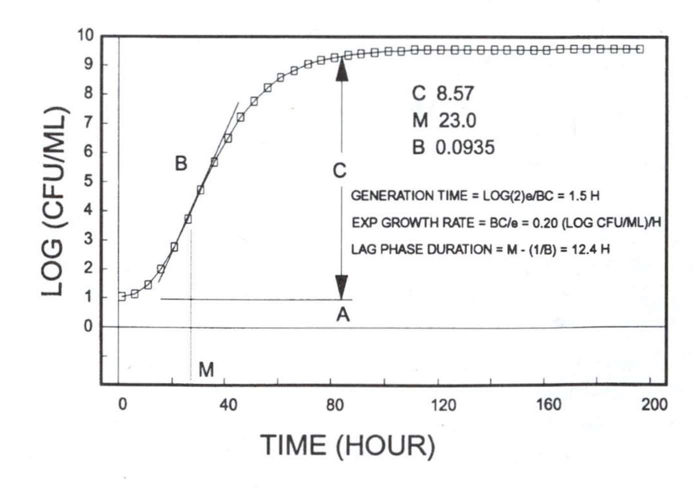

Figure 40.1 Gompertz equation growth model: $Y(t) = A + C\exp\{-\exp[-B(t - M)]\}$. Parameter values and calculated parameters for this curve are indicated.

is achieved. The lag time and exponential growth rate are calculated from the following equations:

$$\text{Lag} = M - 1/B \quad (2)$$

$$\text{Exponential growth rate} = BC/e \quad (3)$$

This equation (equation 1) was recently reparameterized to include parameters for growth rate and lag time directly in the equation (39).

Although there are a variety of sigmoidal curves, the Gompertz equation is used because it gives better overall fits to microbiological data (40). This equation is used in the United Kingdom effort behind Food MicroModel and the USDA Pathogen Modeling Program. Despite its wide use, the Gompertz equation has several flaws. The lag period is not strictly horizontal, and the mathematical asymptote (inoculum number) is off-scale in negative time. The equation is continually curving and does not contain a period of linear increase (log CFU/ml) during the exponential growth phase as observed with most growth curves. The exponential growth rate is determined by the curve's inflection point. The fitting process tends to yield values that are slightly larger than the corresponding growth rates determined from the linear growth period.

To overcome these objections and provide a more mechanistic or biological basis, a model proposed by Baranyi and colleagues (2) includes a linear exponential growth phase, $\mu(x)$, and a lag phase which is determined by an adjustment function. The adjustment function, $\alpha(t)$, is multiplied by the maximum specific growth rate to give the growth rate at a specified time (equation 4). The exponential growth rate with the adjustment function is:

$$dx/dt = \alpha(t)\mu(x)x \quad (4)$$

The cell populations, x, are in arithmetic values, not their logarithm.

Initially the adjustment function is 0 because the cells are not growing. With time, the cells adjust to their new environment, the adjustment factor increases to 1, and the cells are at the maximum growth rate for that combination of environmental conditions.

The maximum cell population is included by:

$$\mu(x) = \mu_{max}\,[1 - (x/x_{max})] \quad (5)$$

Combining these equations, rearranging, and converting to the $\log_{10}$ CFU/ml gives:

$$Y_T = Y_0 + \mu_{max}A(t) - \ln(1 + \{[e^{\mu_{max}\alpha(t)} - 1]/e^{(Y_{max} - Y_0)}) \quad (6)$$

where μ_{max} is the maximum exponential growth rate, $A(t)$ is the integral of adjustment function $\alpha(t)$, and Y_{max} is the log maximum cell density (stationary phase).

An important concept presented by this model is that the observed lag phase is a combination of the physiological state of the cell (q_0) and the adjustment to the new environment (υ). One form for the adjustment function is

$$\alpha(t) = q_0/[q_0 + \exp(-\upsilon t)] \quad (7)$$

The familiar lag phase duration is calculated by

$$\lambda = \ln(1 + 1/q_0)/\upsilon \quad (8)$$

The lag phase is not a single process or event. It is the result of a series of processes as the cell adapts to the new conditions. If the cells are not ready to grow (small q_0) or the adjustment is slow (small υ), the log phase will be lengthy.

Secondary Level

The relationship of growth rate to temperature can be modeled based on chemical reaction rates using the Arrhenius relationship, where the log of the growth rate is inversely proportional to the reciprocal of the absolute temperature. This assumes that the growth rate is controlled by a single rate-limiting enzymatic reaction. Typical Arrhenius plots have been observed at suboptimal temperatures for some bacteria. Unfortunately, plots with two distinct slopes are more frequently found (28).

Expanded versions of the Arrhenius relationship have added parameters to account for the break or curvilinearity of the line (39). Other versions include terms for pH and a_w to create a more useful, if less theoretical, model. This approach can be used for lag time by using its reciprocal as a rate term (13):

$$\ln(1/\text{lag time}) = b_0 + b_1/T + b_2/T^2 + b_3 a_w \quad (9)$$

where T is the temperature, a_w is water activity, and b_x are parameter values.

The Bélerádek or square root model is based on the linear relationship between the square root of the growth rate below the optimum growth rate and the temperature (24):

$$\mu^{0.5} = b(T - T_0) \quad (10)$$

where μ is the growth rate, b is the parameter value for a specific condition, and T_0 is the temperature where the extrapolated growth rate would be zero. The T_0 is characteristic of a microorganism, and a group of lines are obtained with common T_0 in broths having various pH or a_w levels. This allows the expansion of the model to

$$\mu^{0.5} = b(T - T_0)(pH - pH_0)^2(a_w - a_{w_0})^2 \qquad (11)$$

The pH_0 and a_{w_0} are the extrapolated values for when the growth rate is zero.

The temperature range above the optimum can be modeled by:

$$\mu^{0.5} = b_1(T - T_{min})\{1 - \exp[b_2(T - T_{max})]\} \qquad (12)$$

where T_{min} is the extrapolated zero rate for the low-temperature range and T_{max} is the extrapolated zero rate for the high-temperature range. The square root model assumes that each environmental factor is independent and there are no interactions between environmental factors. Whether this assumption is broadly applicable remains to be determined.

Polynomial regression equations make no assumptions about the relationship between independent and modeled variables. The equation is the best fit to the particular data set. The more complex the equation, with interactions and quadratic or cubic terms, the more "flex" in the multidimensional surface and the closer to the data it would be. Whether this complexity is warranted, or whether the model is overparameterized, can be determined by testing the fit of the equation with a new data set. These equations can be simplified by removing terms that are not statistically significant. Because the variation usually increases with increasing growth rate or increasing time, the logarithm, square root, or other transformation of the values is frequently used to normalize the variance.

INACTIVATION/SURVIVAL MODELS

Thermal Death Time Models

The log-linear thermal death time model assumes first-order kinetics; a constant proportion of organisms is inactivated in each successive time period. This model assumes that all cells have the same heat resistance and that the death of an individual cell results from a random inactivation of a critical molecule. The *D* value is the time for a 1-log decrease in viability at a defined temperature in a specific food matrix. The *z* value is the change in the log of the *D* value with changing temperature, a second-level model. This model has a long history of successfully modeling the heat inactivation of spores in retort processing. Additional information on *D* and *z* values is contained in chapter 28.

However, deviations from this linear thermal death time model occur in two forms, a shoulder or lag period before any death begins and a tailing of an apparently more resistant portion of the population (9, 25). Tailing can also be an artifact of the experimental method used, as explained in chapter 2. Smerage and Teixeira (32) proposed a population dynamics theory of thermal inactivation of spores which links first-order activation and inactivation processes. The shoulder has been ascribed to the need to activate spores before they become susceptible to thermal destruction. A proposed mechanism for the tailing was that different cells (spores) in a population have different degrees of heat resistance. Alternatively, at different times in a cell's life cycle, or in response to the environment, a small portion of a population may be in a resistant physiological state.

Inactivation of vegetative cells is also modeled with the linear *D* value model. However, Cole et al. (10) found that the inactivation data did not fit the traditional log-linear relationship, particularly at low temperatures. A logistic function with log survivor versus log time was employed to describe the results in broth:

$$Y_t = \alpha_1 + [(\omega_2 - \alpha_1)/\{1 + \exp[4\sigma(\tau - t)/(\omega_2 - \alpha_1)]\}] \qquad (13)$$

where Y_t is the log number of survivors, α_1 is the upper asymptote ($\approx Y_0$), ω_2 is the lower asymptote, τ is the time of the maximum slope, σ is the maximum slope, and t is the $\log_{10}$ time. For *Listeria monocytogenes* data, α_1, ω_2, and σ were not significantly different for the various factors and were modeled with fixed parameter values (10). A polynomial regression equation for τ, with temperature, salt, and hydrogen ion concentration (not pH) as parameters, successfully modeled this system.

The thermal destruction of *L. monocytogenes* in a liver sausage slurry had nonlinear kinetics (4). A shoulder period and tailing were observed. The length of the shoulder period was inversely related to the temperature over the range of 57.2 to 62.8°C. Because of the increasing number of thermally pasteurized foods, additional research to understand the biochemistry of this process and develop appropriate models is needed.

Nonthermal Inactivation Models

For modeling *L. monocytogenes* survival in unfavorable environments, where cells die slowly over a long period, Buchanan et al. (7) employed a simple, discontinuous model:

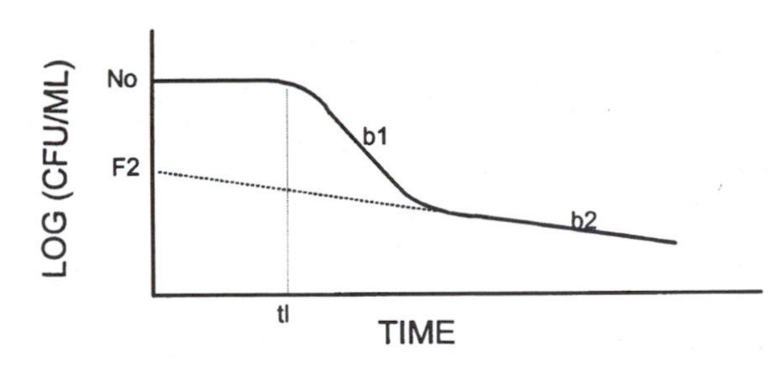

Figure 40.2 Survival/inactivation model with terms for shoulder period and long-lived subpopulation.

$$Y = Y_0 \qquad \text{when } t < t_l \qquad (14)$$

and

$$Y = -b(t - t_l) \qquad \text{when } t > t_l \qquad (15)$$

The shoulder (t_l) parameter is not related to the environmental parameters. This suggests that other factors in the cell's history affected its short-term survival. However, when the time for 4 logs decline (shoulder plus 99.99% death) is modeled, the regression equations are good.

A logistic model for lethality curves exhibiting a shoulder and a tailing population was used by Whiting (34) for *Salmonella* and *Staphylococcus aureus* (Fig. 40.2):

$$Y = Y_0 + \log[F_1([1 + \exp(-b_1 t_1)]/\{1 + \exp[b_1(t - t_1)]\}) + (1 - F_1)([1 + \exp(-b_2 t_l)]/\{1 + \exp[b_2(t - t_1)]\})] \quad (16)$$

where F_1 is the fraction of the cells in the major population and $(1 - F_1)$ is the fraction in the subpopulation; b_1 and b_2 are the rate of declines for the major and subpopulation, respectively, and t_l is the shoulder or lag period. The D values are estimated by $D = 2.3/b$. Again the appearance of a shoulder and a tailing subpopulation do not correlate well with the environmental parameters, but the times for 4 logs lethality do.

C. botulinum Models

Various time-to-toxin or time-to-turbidity models for *Clostridium botulinum* have been proposed, because the time for spore germination and initial toxin formation is a more critical parameter for this pathogen than is growth rate or total growth. A most-probable-number (MPN) technique can be utilized for serial dilutions of spores inoculated into defined media (20). From the patterns of turbidity, the probability of a single spore germinating and growing to cause turbidity is

$$P(\%) = (\text{MPN} \times 100)/s \qquad (17)$$

where P is the probability in percent, MPN is the most probable number of spores which germinate and grow, and s is the inoculum size. The P with time of a single spore germinating, growing (as vegetative cells), and producing toxin is modeled for temperature with a cubic regression equation.

The interactions of a series of factors in pork on the time-to-toxin were modeled with a logistic probability function and a polynomial equation (29):

$$P = 1/(1 + e^{-Y}) \qquad (18)$$

$$Y = a_0 + a_1x_1 + a_2x_2 + a_{12}x_1x_2 + \ldots \qquad (19)$$

where a_n and x_n represent the parameter values and respective factor levels for storage temperature, NaCl, and nitrite with presence or absence of isoascorbate and polyphosphate and high or low heat treatments.

Another description of the spore germination process has the time-to-turbidity within a set of identical samples characterized by (i) an initial period without any turbid samples, (ii) a period of increasing numbers of turbid samples, and then (iii) no additional turbidities (38). The probability can be modeled by a logistic function (Fig. 40.3):

$$P = P_{max}/[1 + e^{k(\tau - t)}] \qquad (20)$$

where P is the probability of a turbid sample (0.0 to 1.0), P_{max} is the maximum probability of being turbid at the end of the storage period, k is the slope term for the rate of increasing turbid samples, τ is the time of the inflection point where P equals half of P_{max}, and t is the time. Polynomial regression equations were then calcu-

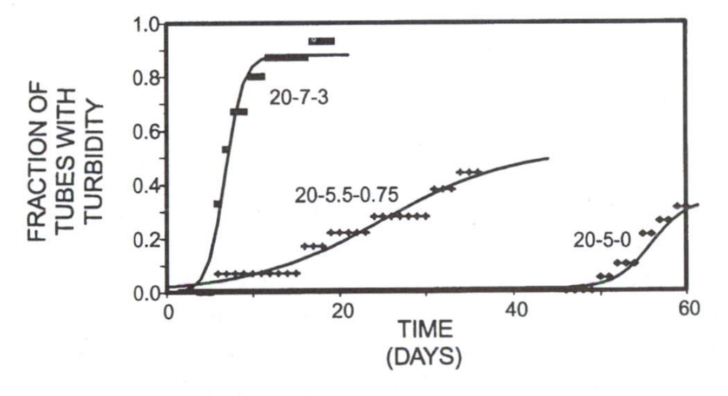

Figure 40.3 Probability-of-growth model showing the increasing probability of turbid samples with increasing storage times. The three lines represent data points and fitted models for three conditions at the indicated temperature, pH value, and percent NaCl. All samples had 10^4 spores.

lated for each of the three parameters (P_{max}, k, and τ) to describe their changes with temperature, pH, and NaCl.

USES OF MODELS

These microbial models are useful tools for microbiologists to obtain an initial estimate of a microorganism's behavior. However, they are only one of several information sources the microbiologist can use. *It is critical that these models be validated before they are relied upon to make safety or policy decisions.* Validation is done by comparing the prediction with inoculated-pack data, preferably for the specific foods of interest. Comparisons of model predictions with data from published inoculated-pack studies reveal that their predictive ability is generally good and usually somewhat conservative (22). However, it remains critical that the user understand and respect the limitations of a specific model. Some foods will have environmental values outside those used to build the model. Foods may contain additional factors not included in the model that have a significant influence on the pathogen's behavior. Use of the model in these situations is inappropriate and will yield predictions that are probably inaccurate. Once validated, however, models can quickly provide information for making decisions in many situations.

Prediction of Safety

Models can estimate the risk of growth or survival of a pathogen after a period of normal or abuse storage. Growth models can assist in establishing "pull dates" by estimating the growth of likely pathogens and comparing this with the growth of spoilage flora. A food should be designed so that spoilage occurs before the growth of potential pathogens. Estimating the microbial behavior for a range of potential environmental and compositional factors can quickly highlight problem areas and guide product development.

Quality Control

Models can aid the development of Hazard Analysis Critical Control Points (HACCP; see chapter 41) programs by demonstrating environmental conditions which allow microbial growth and survival. Quantitative estimates at different levels of the environmental factors indicate the acceptable ranges and suggest critical limit values. This is particularly useful in situations where several factors interact to control microbial growth and subjective evaluations are difficult to make. Ultimately the criteria for critical control points must be based upon microbial growth or death and the risk associated with the probable pathogen populations. Models are also useful when an out-of-process event, such as an unexpected temperature rise, must be evaluated for microbial consequences. Decisions on whether to rework, utilize rapidly, or discard a food or ingredient can be made based on models rather than waiting for microbial testing.

Product Development

Modeling aids in product development because the microbial consequences from changes in the composition or processing can quickly be evaluated. New formulations can be compared to the old.

Education

Models can help explain microbiological behavior to nontechnical people. Graphs of times to reach critical populations can dramatically demonstrate the importance of critical control points or the importance of obtaining raw materials with low microbial counts. The models are especially useful in teaching food microbiology because they quickly illustrate the effects of environmental conditions, particularly in relation to the use of multiple-barrier technologies.

Data Analysis and Laboratory Planning

Modeling techniques are becoming a routine tool for the description and analysis of microbial data. Curve-fitting routines provide an unbiased determination of growth rates or other parameters. Laboratory efficiency is increased when models guide the design of experimental protocols. The models can indicate the importance of a factor and whether extensive testing with multiple levels and replicates is needed. The models can suggest appropriate sampling times and thereby save resources, time, and money.

RISK ASSESSMENTS

The ultimate goal for predictive modeling of foodborne pathogens is to assess the probability that a food will cause illness (11). To accomplish this, modelers must have quantitative information on the initial numbers of the pathogen at the beginning of the process to be modeled. This requires a mean and variance, or a histogram of the frequency at which various populations are present. The next step is to determine the changes in microbial population through the process. This calls for growth, survival, or thermal death models. A series of steps with changing food composition, various temperatures, and alternating growth and death periods results in a frequency distribution of the number of samples having various pathogen concentrations. At the conclusion of this process, a specified quantity of the food will be consumed to give a distribution of total number of pathogens ingested. Last, these numbers are entered into an infectious-dose model to yield the frequencies with which various probabilities of infection and illness occur.

The infectious dose depends upon several factors including the susceptibility of particular subpopulations, the particular pathogen strain, its virulence when consumed, and the food matrix. For some pathogens, such as *Salmonella*, the first step is an infection, i.e., the establishment of the pathogen in the gastrointestinal tract in an asymptotic or carrier state. This is a more conservative model than one based upon numbers-to-cause-illness. For pathogens which produce toxins during growth in the food, the models would involve just morbidity and mortality.

The infectious-dose relationship for *Shigella flexneri* was described by:

$$P = 1 - (1 + N/\beta)^{-\alpha} \qquad (21)$$

where P is the probability, N is the number of shigellae, and the parameter values for β and α are 2,000 and 0.2, respectively (17). These models do not have threshold pathogen populations; one cell has a quantifiable probability of causing an infection. Because infection probabilities of 10^{-4} to 10^{-7} are of concern for semipreserved foods, pathogen populations may be predicted to have values such as 10^{-3} CFU. Depending on the context, this value may be interpreted as one cell in 10^3 g or as a viable cell consumed by one person when the food is eaten by 1,000 people.

The distribution of the probabilities is more important than the mean probability of infection. With all of these steps, it is the cells at the edges of the distribution, not the average, that are most important for causing illness and assessing the safety of a food process. It is the few food samples with highest initial numbers, the first spores to germinate, the most heat-resistant cells, or the samples held longest in the refrigerator that are critical. To determine the probabilities of an infectious dose through a multiple-stage process, particularly when some distributions are defined by a histogram, simulation or Monte Carlo modeling is used. The modeled process is repeatedly calculated, perhaps 1,000 or more times; each time through the process, the simulation picks a value from the respective distribution or defined variation (e.g., normal or log normal distribution) of each step. The resulting distribution of infectious-dose probabilities clusters around the mean value of each step. However, a process which on average has an acceptably low risk may still have a certain percentage of simulations with an unacceptably high risk. There is a quantifiable chance that a high initial population will grow rapidly and receive a mild thermal treatment to produce a high probability of being infectious. This situation would make the process unacceptable.

A prototype of a risk assessment model for *Salmonella* in a refrigerated, cooked chicken patty is illustrated in Fig. 40.4. The four areas are the initial population of *Salmonella* in the meat, changes in numbers through a period of storage followed by cooking, the total number of salmonellae consumed, and the likelihood that the number consumed is an infectious dose. The initial distribution (inset figure) was adapted from data on the surface counts of *Salmonella* on poultry (33), which had an average count of 0.002 CFU/g, but in which 3.5% of the samples had high initial numbers (>0.44 CFU/g). The exponential growth rate was calculated using the *Salmonella* model of Gibson et al. (15), and the increase in numbers after 5 h at 21°C was determined to be 2.9 logs. The thermal death model was determined from published D values for *Salmonella* in eggs (1). Input was for cooking temperature (60°C) and time (6 min) to give a 4.6-log decrease. The amount of food consumed (100 g) represents a typical serving, although a distribution of serving sizes and frequency could be incorporated into the model. The infectious-dose model has a probability equal to 0.007 that one *Salmonella* cell is an infectious dose. For demonstration purposes, the standard deviation of the growth period was assumed to be 0.2 and that for the cooking was assumed to be $0.1D$. When this process was simulated 1,000 times, the mean probability of the process resulting in an infectious dose was $10^{-4.6}$. This appears to be a moderate-risk process. However, the distribution of the probabilities of an infectious dose shows a tailing towards the high risk probabilities (Fig. 40.5). About 2.7% of the simulations exceeded a risk of 10^{-3}, clearly an unacceptable situation. Examination of the individual simulations reveals that the high-risk samples originated with the meat having high initial numbers. The model can easily be rerun with the high initial samples eliminated or with increased cooking time or temperature, and the change in the distribution can be interpreted.

This model can be expanded by incorporating distributions for storage time, cooking temperature, or amount of meat consumed. Additional processing steps and storage periods could be added. This type of model

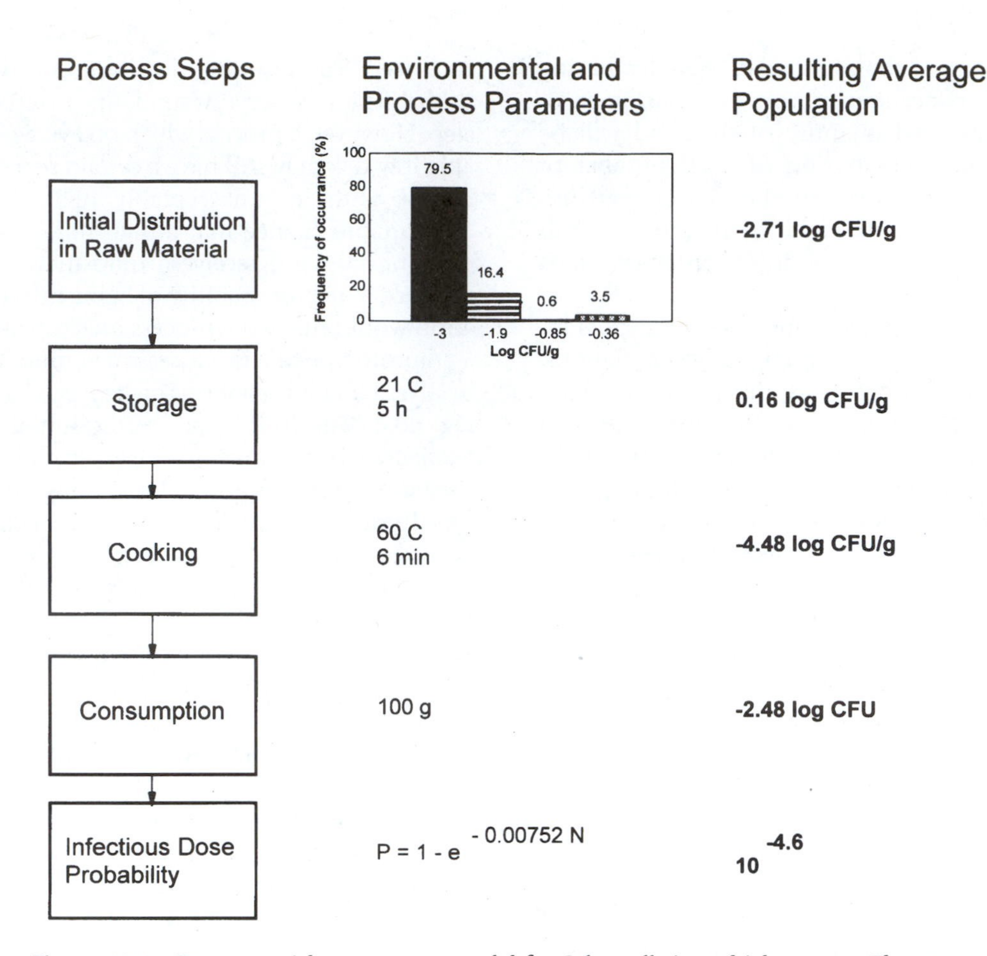

Figure 40.4 Prototype risk assessment model for *Salmonella* in a chicken patty. The process flow is down the left side, various parameters are in the center, and the respective *Salmonella* populations are on the right.

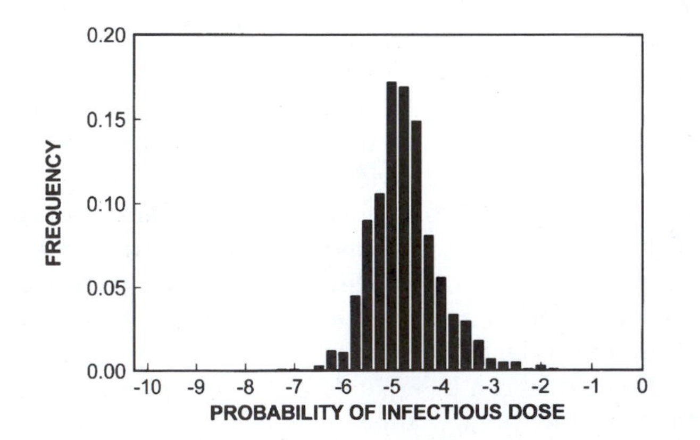

Figure 40.5 Frequency chart for the probable risks of *Salmonella* in a chicken patty.

can quickly show which factors have a major influence on the final probability of infection and show where process control must be maintained. A key point is the importance of the variance or other distribution about each step. It is readily apparent that the outliers of a distribution are much more important than the average in establishing the risk of a process.

What constitutes a tolerable risk of foodborne illness is not just a scientific question but one that involves societal value judgments. Risk factors of 10^{-6} are usually used for carcinogens, and estimates of overall foodborne illness in the United States would give the current risk from all sources to be about 10^{-1} per year or 10^{-4} to 10^{-5} per meal (11, 21). These latter figures could be used as a baseline with the goal of reducing the risk of foodborne illness.

FUTURE DEVELOPMENTS IN MODELING

The accuracy of any model depends upon the quality of the data used for fitting the primary- and secondary-level equations. Additional data covering expanded ranges will improve the present models. Both the Food MicroModel and the Pathogen Modeling Program model constant environments. More information is needed to link a series of growth, thermal death, and survival stages together and have confidence in the final population estimate. Modeling changing environments, such as fluctuating temperatures or the declining pH during a fermentation, will require a more complex approach. Heterogeneous foods, microenvironments, surface growth, and biofilm growth are other situations in which the appropriate parameter values to be entered can only be estimated. Current models do not attempt to factor in microbial competition or bacteriocin production.

The risk assessment model demonstrates the importance of the variation or distribution for each of the steps. Current models usually give a predicted value but no indication of the variation. These variations can be relatively large at the extremes of a parameter's range, particularly at growth-limiting extremes such as low temperature or pH. Without an estimate of variation, a false sense of precision may be conveyed and the average value may be given misleading importance. The source of variation is partially in the collecting of model data and statistical processes. However, strain and other differences exist which need to be incorporated into the models. For example, the lag phase duration for 39 strains of *L. monocytogenes* at 4°C was 151 ± 30 h and the generation times were 43 ± 11 h (3). The values for the strains may not always be normally distributed. Providing a confidence interval allows users to understand the precision of the prediction and interpret it better.

Confidence intervals are necessary for growth, thermal death, and survival parameters for the most prevalent strains of each pathogen. A risk assessment model that links the frequency of occurrence with the parameter value for each strain will come closer to modeling the real world than current models.

ADDITIONAL READING

General reviews of microbial modeling have been written by Farber (14), Buchanan (5), Labuza et al. (19), Skinner et al. (31), Whiting (35), and Whiting and Buchanan (37). The development of the square root model and its uses are in a book by McMeekin et al. (24). The proceedings of the First International Workshop on the Application of Predictive Modeling for the Food Industry (8) are available from Stockton Press, Basingstoke, U.K.

Appendix
USDA Pathogen Modeling Program
Thomas J. Montville

The growth of most foodborne pathogens is influenced by an interaction of many different factors, including time, temperature, atmosphere, pH, etc. Changes in these factors not only influence the growth of a particular pathogen, but may actually determine which pathogen is of primary concern. This simple concept (that the interaction of factors governs microbial growth) is often difficult to comprehend or to apply to a specific situation. Recent advances in microbial modeling (as described in this chapter) have made these effects much easier to visualize. Scientists at the United States Department of Agriculture (USDA) Eastern Regional Research Center (ERRC) have now made the microbial models generated from their research available to the larger scientific community in a series of programs that can be run on IBM-compatible personal computers. The USDA's Pathogen Modeling Program provides powerful demonstrations of how the interactions of environmental factors influence microbial growth and thus reinforces many concepts covered throughout this book. The effort required to download and to learn how to use this program will be repaid many times over by an improved understanding of food microbiology. We thank Robert Buchanan and Richard Whiting for their generosity in making this program available to the readers of this volume.

Instructions for obtaining this program are as follows.

To Download the Pathogen Modeling Program

The current version of the UDSA Pathogen Modeling Program can be obtained by using the anonymous file transfer protocol (FTP) resident on your Internet server. If you have never used FTP before, you should obtain documentation about your FTP program from your Internet provider, and you may need some instruction on the use of file transfer programs. Using FTP, you can connect with the ERRC FTP server over the Internet and download the program. If you do not have Internet access, you may obtain the program on disk by writing to Dr. Buchanan or Dr. Whiting at the address given in the Contributors list in this volume.

The following are general instructions for downloading the program using a command line FTP transfer. The exact commands and protocols will vary somewhat depending on your FTP utility; many utilities automate the steps indicated below. Commands to be typed are indicated in boldface print. Comments that the user will see are indicated in italics.

1. At the *ftp*> prompt, log onto the ERRC computer by typing **Open ceres.arserrc.gov** and then RETURN or ENTER. The server will respond, *"Connect to server"*.

2. At the prompt for your *"User name"* respond by typing **Anonymous** and then RETURN or ENTER.
3. At the prompt for a *"Password"* respond by typing in your e-mail address, followed by RETURN or ENTER.
4. Type **cd mfsmodel** to move to the directory where the program is located.
5. Type **binary** to transfer as a binary file. The server will respond with *"Download as a binary file"*.
6. At the prompt to *"Choose local directory"* type **lcd c:\temp** (or another directory on your computer to which you want the files downloaded).
7. Type **mget.** The computer will respond, *"Get all files in the current directory"*. At this command, seven files will be transferred to the directory specified (i.e., to c:\temp). The files include install.bat, mfsmodel.bat, ovr.zip, pkunzjr.com, run.zip, setdisp.exe, and wkb.zip. Downloading takes approximately 10 to 15 min with a 28.8 baud modem (approximately 1.2 megabytes) or up to half an hour using a 14.4 baud modem.
8. Type **quit** to terminate your connection to ceres.arserrc.gov.
9. You can now install the program on your computer by transferring to the directory to which the program was downloaded and typing **install.** This can also be done through the **Run** command in the program or file manager of Windows 3.1, or the **Start** icon in Windows 95. The install program will make a new directory (c:\mfs.model), copy the programs to the directory, and then expand them. After a successful installation, the seven files can be removed from the temporary directory where they were initially downloaded.

References

1. **Annellis, A., J. Lubas, and M. M. Rayman.** 1954. Heat resistance in liquid eggs of some strains of the genus *Salmonella. Food Res.* **19:**377–395.
2. **Baranyi, J., and T. A. Roberts.** 1994. A dynamic approach to predicting bacterial growth in food. *Int. J. Food Microbiol.* **23:**277–294.
3. **Barbosa, W. B., L. Cabedo, H. J. Wederquist, J. N. Sofos, and G. R. Schmidt.** 1994. Growth variation among species and strains of Listeria in culture broth. *J. Food Prot.* **57:**765–769, 775.
4. **Bhaduri, S., P. W. Smith, S. A. Palumbo, C. O. Turner-Jones, J. L. Smith, B. S. Marmer, R. L. Buchanan, L. L. Zaika, and A. C. Williams.** 1991. Thermal destruction of *Listeria monocytogenes* in liver sausage slurry. *Food Microbiol.* **8:**75–78.
5. **Buchanan, R. L.** 1992. Predictive microbiology: mathematical modeling of microbial growth in foods, p. 250–260. *In* J. W. Finley, S. F. Robinson, and D. J. Armstrong (ed.), *Food Safety Assessment.* ACS Symposium Series 484. American Chemical Society, Washington, D.C.
6. **Buchanan, R. L.** 1993. Developing and distributing user-friendly application software. *J. Ind. Microbiol.* **12:**251–255.
7. **Buchanan, R. L., M. H. Golden, and R. C. Whiting.** 1993. Differentiation of the effects of pH and lactic or acetic acid concentration on the kinetics of *Listeria monocytogenes* inactivation. *J. Food Prot.* **56:**474–478, 484.
8. **Buchanan, R. L., R. C. Whiting, and S. A. Palumbo.** 1993. Proceedings of workshop on the application of predictive microbiology and computer modeling techniques to the food industry. *J. Ind. Microbiol.* **12:**137–359.
9. **Cerf, O.** 1977. Tailing of survival curves of bacterial spores—a review. *J. Appl. Bacteriol.* **42:**1–19.
10. **Cole, M. B., K. W. Davies, G. Munro, C. D. Holyoak, and D. C. Kilsby.** 1993. A vitalistic model to describe the thermal inactivation of *Listeria monocytogenes. J. Ind. Microbiol.* **12:**232–239.
11. **Council for Agricultural Science and Technology.** 1994. *Foodborne Pathogens: Risks and Consequences.* Task Force Report no. 122. Council for Agricultural Science and Technology, Ames, Iowa.
12. **Dalgaard, P., T. Ross, L. Kamperman, K. Neumeyer, and T. A. McMeekin.** 1994. Estimation of bacterial growth rates from turbidimetric and viable count data. *Int. J. Food Microbiol.* **23:**391–404.
13. **Davey, K. R.** 1991. Applicability of the Davey (linear Arrhenius) predictive model to the lag phase of microbial growth. *J. Appl. Bacteriol.* **70:**253–257.
14. **Farber, J. M.** 1986. Predictive modeling of food deterioration and safety, p. 57–90. *In* M. D. Pierson and N. J. Stern (ed.), *Foodborne Microorganisms and Their Toxins: Developing Methodology.* Marcel Dekker, New York.
15. **Gibson, A. M., N. Bratchell, and T. A. Roberts.** 1988. Predicting microbial growth: growth responses of salmonellae in a laboratory medium as affected by pH, sodium chloride, and storage temperature. *Int. J. Food Microbiol.* **6:**155–178.
16. **Goldblith, S. A., M. A. Joslyn, and J. T. R. Nickerson.** 1961. *An Introduction to the Thermal Processing of Foods,* vol. 1, p. 1128. AVI Publishing Co., Westport, Conn.
17. **Haas, C. N.** 1983. Estimation of risk due to low doses of microorganisms: a comparison of alternative methodologies. *Am. J. Epidemiol.* **118:**573–582.
18. **Hudson, J. A.** 1993. Effect of pre-incubation temperature on the lag time of *Aeromonas hydrophila. Lett. Appl. Microbiol.* **16:**274–276.
19. **Labuza, T. P., B. Fu, and P. Taoukis.** 1992. Prediction for shelf life and safety of minimally processed CAP/MAP chilled foods: a review. *J. Food Prot.* **55:**741–750.
20. **Lindroth, S. E., and C. A. Genigeorgis.** 1986. Probability of growth and toxin production by nonproteolytic *Clostridium botulinum* in rockfish stored under modified atmospheres. *Int. J. Food Microbiol.* **3:**167–181.
21. **Macler, B. A., and S. Regli.** 1993. Use of microbial risk assessment in setting US drinking water standards. *Int. J. Food Microbiol.* **18:**245–256.
22. **McClure, P. J., C. W. Blackburn, M. B. Cole, P. S. Curtis, J. E. Jones, J. D. Legan, I. D. Ogden, M. W. Peck, T. A. Roberts, J. P. Sutherland, and S. J. Walker.** 1994. Modelling the growth, survival and death of microorganisms in foods: the UK Food Micromodel approach. *Int. J. Food Microbiol.* **23:**265–275.

23. **McClure, P. J., M. B. Cole, K. W. Davies, and W. A. Anderson.** 1993. The use of automated turbimetric data for the construction of kinetic models. *J. Ind. Microbiol.* **12**:277–285.
24. **McMeekin, T. A., J. Olley, T. Ross, and D. A. Ratkowsky.** 1993. *Predictive Microbiology: Theory and Application.* John Wiley & Sons, New York.
25. **Moats, W. A., R. Dabbah, and V. M. Edwards.** 1971. Interpretation of nonlogarithmic survivor curves of heated bacteria. *J. Food Sci.* **36**:523–526.
26. **Owens, J. D., D. S. Thomas, P. S. Thompson, and J. W. Timmerman.** 1989. Indirect conductimetry: a novel approach to the conductimetric enumeration of microbial populations. *Lett. Appl. Microbiol.* **9**:245–249.
27. **Ratkowsky, D. A.** 1993. Principles of nonlinear regression modelling. *J. Ind. Microbiol.* **12**:195–199.
28. **Reichardt, W., and R. Y. Morita.** 1982. Temperature characteristics of psychrotrophic and psychrophilic bacteria. *J. Gen. Microbiol.* **128**:565–568.
29. **Roberts, T. A., A. M. Gibson, and A. Robinson.** 1982. Factors controlling the growth of *Clostridium botulinum* types A and B in pasteurized, cured meats. III. The effect of potassium sorbate. *J. Food Technol.* **17**:307–326.
30. **Ross, T., and T. A. McMeekin.** 1994. Predictive microbiology. *Int. J. Food Microbiol.* **23**:241–264.
31. **Skinner, G. E., J. W. Larkin, and E. J. Rhodehamel.** 1994. Mathematical modeling of microbial growth: a review. *J. Food Safety* **14**:175–217.
32. **Smerage, G. H., and A. A. Teixeira.** 1993. A new view on the dynamics of heat destruction of microbial spores. *J. Ind. Microbiol.* **12**:211–220.
33. **Surkiewicz, B. F., R. W. Johnston, A. B. Moran, and G. W. Krumm.** 1969. A bacteriological survey of chicken eviscerating plants. *Food Technol.* **23**:1066–1069.
34. **Whiting, R. C.** 1993. Modeling bacterial survival in unfavorable environments. *J. Ind. Microbiol.* **12**:240–246.
35. **Whiting, R. C.** 1995. Microbial modeling in foods. *Crit. Rev. Food Sci. Nutr.* **35**:467–494.
36. **Whiting, R. C., and R. L. Buchanan.** 1993. A classification of models for predictive microbiology. *Food Microbiol.* **10**:175–177.
37. **Whiting, R. C., and R. L. Buchanan.** 1994. Microbial modeling. Inst. Food Technologists, Scientific Status Summary. *Food Technol.* **48**:113–119.
38. **Whiting, R. C., and J. E. Call.** 1993. Time of growth model for proteolytic *Clostridium botulinum. Food Microbiol.* **10**:295–301.
39. **Zwietering, M. H., J. T. de Koos, B. E. Hasenack, J. C. de Wit, and K. van't Riet.** 1991. Modeling of bacterial growth as a function of temperature. *Appl. Environ. Microbiol.* **57**:1094–1101.
40. **Zwietering, M. H., I. Jongenburger, F. M. Rombouts, and K. van't Riet.** 1990. Modeling of the bacterial growth curve. *Appl. Environ. Microbiol.* **56**:1875–1881.

Dane T. Bernard

41

Hazard Analysis and Critical Control Point System

Use in Controlling Microbiological Hazards

INTRODUCTION

The preceding chapters present state-of-the-art information regarding microorganisms that are associated with food or water. Under certain conditions, many of these organisms present significant food safety hazards. The food industry bears the responsibility of raising, transporting, processing, and preparing foods that present the minimum level of risk from foodborne hazards, including pathogenic microorganisms, that is practical and achievable within the limits of current technology. To assist in accomplishing this task, the Hazard Analysis and Critical Control Point (HACCP) system has been developed. HACCP is currently regarded as the best system available for designing programs to assist food firms to produce safe foods. Recognizing this, the U.S. Food and Drug Administration (FDA) (5) and the U.S. Department of Agriculture (USDA) (6) have published final regulations mandating the development and implementation of HACCP plans to cover a variety of domestic and imported foods.

Paralleling these U.S. Government initiatives is an international movement toward a food safety assurance and regulatory scheme based on the principles of HACCP. For example, the European Union (E.U.) has adopted several product-specific ("vertical") directives that will require specific foods introduced into commerce within the E.U. (including imports) to be produced in accordance with HACCP principles. Examples include the directive on fishery products (91/493/EEC), the directive on milk, heat-treated milk, and milk-based products (92/46/EEC), and the directive on meat products (92/5/EEC). In addition to these, the E.U. has adopted a "horizontal" directive (93/43/EEC) which requires consistency with HACCP principles for a wide range of food items. Countries outside the E.U. such as Australia, New Zealand, Canada, Japan, Egypt, South Africa, and many others have also adopted or are considering HACCP-based food safety control systems.

The growing importance of HACCP as a tool to judge the acceptability of foods traded internationally is illustrated by the agreements and treaties that support the General Agreement on Tariffs and Trade (GATT). These agreements give guidance to GATT nations about how to establish sanitary and phytosanitary requirements without creating technical trade barriers (an illegal practice under the terms of these agreements). For the establishment of food safety standards, these agreements reference the standards, guidelines, and recommendations developed and adopted by the Codex Alimentarius Commission as an acceptable scientific baseline for consumer protection.

At the 20th Session of the Codex Alimentarius Commission (July 1993), the document "Guidelines for the Application of the Hazard Analysis Critical Control

Point (HACCP) System" was adopted. The Commission noted that the text was "urgently needed" in order to incorporate it into the "Draft Revised Recommended International Code of Practice—General Principles of Food Hygiene." Even as this chapter was being prepared, however, the Codex Committee on Food Hygiene was considering a redraft of the Codex guidance on HACCP. This activity is reflective of the rapid development of the understanding of HACCP and its worldwide application. Thus, the actions of official international bodies have raised the importance of HACCP to a new level.

Origin of HACCP

The acronym "HACCP" was first coined by the Pillsbury Company to describe the systematic approach to food safety it developed in response to requirements imposed by NASA for foods produced for manned space flights. In order to approach the goal of 100% assurance that these foods would be safe to consume, a system had to be developed which went well beyond the limited effectiveness of sampling and analyzing finished goods for the presence of hazardous microorganisms, foreign materials, or chemicals. Methods used by others, including "Haz-Ops" studies and "Modes of Failure Analysis," were considered in the search for new ways to assure that foods were of adequate hygienic quality. Although the HACCP concept reflects much of this earlier thinking, these programs did not take into account the special needs of the food industry. From these approaches, Pillsbury adopted the concept that if one understands how a product (in this case a food) becomes unsafe, then control measures can be developed to prevent or detect such failures and keep them from reaching consumers. This concept formed the foundation for development of the HACCP system of food safety control.

The HACCP concept was presented to the public by Dr. Howard Bauman in 1971 at the first National Conference on Food Protection (11). While the concept created some interest in the early 1970s, initial attempts at incorporating HACCP into food operations were not very successful. It is likely that a major contributor to these early unsuccessful attempts was a lack of description and understanding of the HACCP concept. Early HACCP plans typically had far more critical control points than were needed to assure production of a safe food, were not "plant friendly," and were too cumbersome to be sustained. However, in 1985, a Subcommittee of the Food Protection Committee of the National Academy of Sciences (NAS) issued a report on microbiological criteria that included a strong endorsement of HACCP (9). Because of NAS recommendations, the National Advisory Committee on Microbiological Criteria for Foods (NACMCF) was formed in 1988 as an advisory group to the Secretaries of Agriculture, Commerce, Defense, and Health and Human Services. Part of the mission of the NACMCF was to develop material to promote an understanding of HACCP and to encourage adoption of the HACCP approach to food safety. The NACMCF has been a leader in elaboration of HACCP principles and in providing guidance for their application to food processing operations. In recognition that the ideas underlying the successful application of HACCP continue to evolve, the NACMCF is updating its 1992 guidance document (10), with expected publication in 1996. Both the Codex and the NACMCF HACCP documents were used as reference for preparing this chapter

HACCP Concept

The approach taken in this chapter is that HACCP plans should only address significant food safety hazards. It is these safety hazards which warrant the extensive monitoring, record-keeping, and verification activities required by the HACCP food safety management system. The FDA's final rule on seafood safety and the USDA Food Safety Inspection Service's regulation on pathogen reduction and HACCP, referenced above, also limit HACCP to issues of food *safety*. This may be viewed as an unnecessarily narrow interpretation by some; the application of a similar systematic approach can be used to control other aspects of food quality. However, the safety-only approach is increasingly viewed as a necessity because HACCP is likely to be used to judge acceptability of product in domestic and international trade. While quality is a negotiable item, safety should be nonnegotiable. Mixing safety and nonsafety aspects in the same plan will result in difficulties for food regulatory officials in determining acceptability of food products. If the seven principles are applied to develop a control strategy for *quality*, this should not be labeled as a HACCP plan.

The HACCP principles describe a format for identifying and assuring control of those factors (hazards) which are *reasonably likely* to cause a food to be unsafe for consumption. It is based on a common-sense application of technical and scientific principles to the food production process from field to table. The process of developing a HACCP plan can be viewed as the process of evaluating the risks (hazards) associated with a food, then using the results of the evaluation to develop a management plan to address the hazards "which are of such a nature that their elimination or reduction to acceptable levels is essential to the production of a safe food" (2). In the hazard analysis step (principle 1), hazards of potential significance are identified and evaluated. The evaluation results in a list of hazards that are significant, e.g., reasonably likely to present consumer safety problems if not properly controlled. The remaining six HACCP principles describe the steps necessary to develop a management strategy to control significant hazards.

Most contemporary references on HACCP point to its applicability to all segments of the food industry from basic agriculture through food preparation and handling, food processing, food packaging, food service, distribution systems, and consumer handling and use. While in a broad sense this is true, the way in which the HACCP principles are interpreted and applied will differ depending on which segment in the food chain is being addressed. This subject is discussed further in a paper published by the Food and Agriculture Organization of the United Nations (7). While the major focus of this chapter is the application of HACCP to the food manufacturing segment, other segments will be noted where appropriate.

The most basic notion underlying HACCP is to prevent hazards during production rather than rely on inspection of finished goods in hopes of detecting and correcting problems. While testing finished products usually plays a role in a HACCP system, the main focus is on prevention through process control. As most food microbiologists will attest, results derived from sampling and testing only support definite conclusions concerning the units tested. When interpreting test results, the limits of the methodology and the sampling plan must be considered, because statements about the nature of units not tested can only be made by inference within the context of these limitations. By contrast, if the hazards of concern are properly identified and effective control measures are followed, HACCP provides greater confidence in the safety of each unit produced than can be obtained through any practical sampling and testing program. Thus, firms that produce foods in accord with a properly designed HACCP plan will typically have a reduced need for testing and analysis of finished product.

A Cautionary Note

In the opinion of many HACCP authorities, the most important (and least understood) activity in the development of a HACCP plan is the hazard analysis step. Although this process will be discussed in more depth below, it must be emphasized that a HACCP plan will not be effective in minimizing risks associated with a food unless the hazard analysis is done correctly and effective control measures are adopted. A note of caution, however: HACCP is not a "zero" defects program, even though this should always be the ultimate goal in food safety. Any hazard analysis will be limited by the information available to the experts conducting the analysis. Some risk factors, such as a pathogen that adapts to a new environment, cannot be anticipated and thus cannot be addressed in a HACCP plan. HACCP is not a magic bullet that will cure all food safety problems. However, adoption of HACCP will put a system in place that has the potential to be more responsive to newly identified problems. Because the natural world is always presenting us with new hazards, it is vital to the success of a HACCP system that plans are reviewed in light of new scientific information. To be effective over time, the hazard analysis and resulting HACCP plans must be reviewed frequently.

Further, HACCP is not a system which will be effective unless supported by general Good Hygienic Practice. These general programs include effective systems to assure clean food contact surfaces, adequate facilities for the production and handling of foods, programs which encourage prevention of cross-contamination, good personal hygiene, employee training, etc. A food operation contemplating the development of HACCP plans for its products should begin with an assessment of these necessary prerequisites.

USING HACCP TO CONTROL MICROBIOLOGICAL HAZARDS

The systematic approach to food safety embodied by HACCP is based on seven principles. The principles adopted by the NACMCF are used in this chapter. Although variations of these principles have been adopted and published by other groups and official bodies, there is surprising agreement on the basic concepts. However, in reviewing most of the publications describing HACCP, e.g., those by the International Life Sciences Institute (8), the Food Processors Institute (12), Campden and Chorleywood Food Research Association (3), etc., some variation is noted in the interpretation of the principles and in the resulting guidance about how they are to be applied in developing and implementing a HACCP plan. In dealing with these apparent differences we should not lose sight of the intended common-sense approach to food safety management and follow through with this thought.

As noted earlier, HACCP is an evolving concept. As additional real-world experience is gained with the system, more techniques will be developed which will make it easier to understand and apply. The guidance documents about HACCP will also continue to be modified. Thus, each firm, besides following the legal requirements that pertain to its operation, should do what it feels is appropriate to develop a workable, plant-friendly, and effective HACCP plan. Remember that many of the substantial benefits of HACCP accrue because plans will be plant and line specific, therefore fitting the unique needs of each operation. Each operation therefore should view the HACCP guidance available today as just that: "guidance."

Preliminary Steps

After review of the adequacy of prerequisite programs is conducted, the NACMCF recommends that certain preliminary steps be undertaken before beginning the hazard analysis. The steps that precede the hazard analysis include: (i) gain management support, (ii) assemble the HACCP team, (iii) describe the food and the method of its distribution, (iv) identify the intended use and consumers of the food, (v) develop a flow diagram, and (vi) verify the flow diagram.

Regarding point (i) from the above list, most who have attempted the development and implementation of HACCP will agree that it is difficult to put such systems in place without firm support and understanding from company management. HACCP often requires basic changes in management styles and requires the commitment of resources to the process. Without management support, these changes will be difficult to achieve.

Assembling a HACCP team (point [ii]) can be especially challenging for small firms which do not have the luxury of a large technical staff. Most now agree that individuals with the necessary expertise must be involved in the process, even though it may be a challenge for some firms. However, these individuals need not necessarily be employed at the firm. All firms developing HACCP plans must find a way to have the requisite expertise reflected in their plans. Points (iii) through (vi) from the list are necessary to allow for development of HACCP plans which are tailored to specific products and food operations. In order to address these preliminary steps, the food firm must have sufficient information about the ingredients, the processing methods, the distribution system, and the expected consumers so that it can determine where and how significant food safety hazards may be introduced or enhanced. If the "where" and "how" are known, control is simplified.

Establishing Food Safety Objectives

The food safety objectives of the HACCP plan should be identified by the HACCP team and clearly stated in the plan. These objectives will typically be clarified as the team conducts its hazard analysis. An example would be the target of producing a canned food that is commercially sterile and poses a negligible risk from *Clostridium botulinum*. A similar objective would be to produce a pasteurized product that poses a negligible risk from vegetative pathogens such as *Salmonella*. Such public health-related food safety objectives can typically be met by process steps designed to inactivate pathogenic organisms which may be present in the preprocess product. For raw products, one objective may be to produce a product that poses minimal risk from vegetative pathogens. This objective is usually accomplished by application of treatments designed to reduce microorganisms present, to minimize potential for contamination of ordinarily sterile tissue, and, by control of conditions (e.g., refrigeration), specifically to reduce or prevent the proliferation of pathogens. Similar public health objectives should be identified for chemical and physical hazards as well. Prevention of the formation of significant amounts of histamine during the storage of scombrotoxin-producing species of fish is an example of a food safety objective for a chemical hazard, and minimization of presence of metal fragments in a ground meat product may be an appropriate objective for control of physical hazards.

Principle no. 1: Conduct a hazard analysis. Prepare a list of steps in the process where significant hazards occur, and describe the preventive measures.

After addressing the preliminary steps discussed above, the HACCP team conducts a hazard analysis or assures that an analysis is conducted. The NACMCF defines a hazard (for HACCP) as a "biological, chemical or physical property which may cause a food to be unsafe for consumption." Based on the results of the hazard analysis, the points or steps in the process where significant hazards can be controlled are identified. For a hazard to be listed among the significant hazards which the HACCP system must address, the NACMCF recommended that the hazard ". . . be of such a nature that its prevention, elimination or reduction to acceptable levels is *essential* to the production of a safe food. Hazards which are of a *low risk* and *not likely* to occur would not require further consideration." (Emphasis added.)

The emphasis is clearly on those hazards whose control is essential for safety. However, note that *HACCP does not allow food operators to ignore low-risk hazards nor to ignore existing legal requirements,* including those legal constraints which define adulteration. Operations must be in compliance with all legal requirements. Hazards which pose a low risk should not be dismissed as insignificant. The food operation should assure that they will be addressed within other management systems, like Good Manufacturing Practice (GMP) programs, which are considered prerequisites for the HACCP system. Most authorities agree that if a food operation does not have sound prerequisite programs which comply with GMPs and other regulatory requirements, the HACCP system will have little chance for success. It is clear, however, that the NACMCF recognized that not all biological, chemical, or physical properties that *may* cause a food to be unsafe for consumption are *significant* enough to warrant being addressed within a HACCP plan. Making the decision as to whether or not a hazard is reasonably likely to present a significant consumer safety problem is at the heart of the hazard assessment.

Conducting a Hazard Analysis

The process of conducting a hazard analysis involves at least two steps, hazard identification and hazard evaluation. Hazard identification can be regarded as a brainstorming session. During this first step, the HACCP team reviews the ingredients used in the product, the activities conducted at each step described in the process flow, the method of storage and distribution of the product, and its intended use and consumers. Based on this review, the team develops a list of *potential* biological, chemical, or physical hazards which may be introduced or enhanced at each step of the production process. While the review focuses on developing a list of hazards associated with the process steps under the direct control of the food operation, this is done within the context of hazards present in the food as consumed. The preceding chapters of this book are an excellent source of the information needed to assemble a list of potential biological hazards. For the food under consideration, a list of pathogenic bacteria, viruses, and parasites that may be of concern should be assembled along with a list of potential chemical and physical hazards. While the list should be complete, it need not be so detailed as to be impractical. Experts familiar with the epidemiological history of the product being considered will be of great value in this exercise.

After the list of potential hazards is assembled, each hazard is evaluated based on *risk* (likelihood of occurrence) and *severity.* The NACMCF notes that "severity is the seriousness of the hazard." When conducting the hazard analysis, it may also be helpful to consider the concept of the likelihood of exposure and severity of the potential consequences *if the hazard is not properly controlled.* During the hazard evaluation, each potential hazard, the food, its methods of preparation, transportation, and storage, and the nature of the consumers likely to purchase the product should be considered to determine how each of these factors may enhance or diminish the public health impact of the hazard being considered. The team must determine the effect on food safety of the manner in which the food is likely to be stored and prepared and whether the food is specifically intended for consumption by a group which may be more susceptible to a particular agent. For biological hazards, each individual organism on the list of potentially significant hazards will need to be evaluated to determine if a particular organism needs to be addressed in the HACCP plan. Practically speaking, however, an in-depth evaluation may not be required for each individual organism provided the control measure is the same for an entire group of pathogens. For example, if the product under evaluation is sold as fully cooked, all vegetative pathogens would be expected to be eliminated by the same control measure. Thus, in such cases, each organism does not need to be evaluated individually. Alternatively, if the control measure being applied is an intrinsic property of the food such as water activity, its effect on individual pathogens may need to be evaluated.

Information in this text will be of great value when evaluating potential significance of hazards. For example, a firm that makes a product containing liquid eggs as an ingredient will find in chapter 8 that eggs have been associated with outbreaks of salmonellosis. Chapter 8 also gives the information necessary to evaluate the risk that this potential hazard may pose to consumers of the product. In this case, the firm can find that the frequency of *Salmonella* contamination in eggs is sufficient to determine that some units of the product are likely to contain the organism if control measures are not applied. In addition, we find in this text that consuming only a few cells of certain strains of *Salmonella* can result in illness to consumers and that the disease syndrome affects the general population. Considering this information, the HACCP team or their food safety expert may determine that this hazard is *reasonably likely* to pose an unacceptable health risk to consumers if not effectively controlled. Therefore, the firm decides that this is a *significant* hazard that will be addressed in their HACCP plan.

The hazard analysis process, consisting of hazard identification and evaluation, is the key to preparing an effective HACCP plan. If this is not done correctly and the significant hazards warranting control within the

Table 41.1 Comparison of steps employed in a qualitative risk assessment with the steps of a hazard analysis for a product containing eggs[a]

Risk assessment	Explanation	Hazard analysis
1. Hazard identification	The potential hazard to be evaluated is the possibility of *Salmonella* in the finished product.	Identify hazard
2. Hazard characterization	Salmonellosis is a foodborne infection causing a moderate to severe illness that can be caused by ingestion of only a few cells of *Salmonella*.	Evaluate hazard based on severity of consequences if not controlled
3. Exposure assessment	Product is formulated with liquid eggs, which have been associated with past outbreaks of salmonellosis. Recent problems with *Salmonella* sp. serotype Enteritidis in eggs cause increased concern. Probability of *Salmonella* in raw eggs cannot be ruled out. If the hazard is not effectively controlled, some consumers are likely to be exposed to *Salmonella* organisms from this food.	Evaluate hazard based on risk (likelihood of exposure)
4. Risk characterization	HACCP team determines that if hazard is not properly controlled, consumption of product is likely to result in unacceptable health risk. This hazard is significant.	Determine if hazard needs to be addressed in HACCP plan (significance).

[a]This table is presented for illustrative purposes only. The terms used in the risk assessment column are defined by the FAO/WHO Expert Consultation referenced earlier (13) and are as follows.

Risk assessment: The scientific evaluation of known or potential adverse health effects resulting from human exposure to foodborne hazards. The process consists of the following steps: (i) hazard identification, (ii) hazard characterization, (iii) exposure assessment, and (iv) risk characterization. The definition includes quantitative risk assessment, which emphasizes reliance on numerical expressions of risk, and also qualitative expressions of risk, as well as an indication of attendant uncertainties.

Hazard identification: The identification of known or potential health effects associated with a particular agent.

Hazard characterization: The qualitative and/or quantitative evaluation of the nature of the adverse effects associated with biological, chemical, and physical agents which may be present in food. For chemical agents, a dose-response assessment should be performed. For biological or physical agents, a dose-response assessment should be performed if the data are obtainable.

Exposure assessment: The qualitative and/or quantitative evaluation of the degree of intake likely to occur.

Risk characterization: Integration of hazard identification, hazard characterization, and exposure assessment into an estimation of the adverse effects likely to occur in a given population, including attendant uncertainties.

HACCP system are not identified, the plan will not be effective regardless of how well it is followed.

Another way of understanding the hazard analysis process is by comparison with a similar system such as that for conducting a *qualitative* risk assessment as outlined by a joint FAO/WHO Expert Consultation (13). While the two processes differ in their purpose, the logic sequence involved is virtually identical, and it may be helpful to draw some parallels between these two processes. Table 41.1 summarizes the rationale from the above example and compares a hazard analysis with the steps in a qualitative risk assessment.

The purpose of the hazard analysis process is to identify hazards that are reasonably likely to make consumers ill if not effectively controlled. These are the significant hazards that must be addressed within the HACCP plan. Hazards eliminated from further consideration within the HACCP system may be assigned to other programs (i.e., GMP). For example, when preparing a HACCP plan for fresh pork sausage, the results of our hazard analysis may indicate that we do not need to be concerned about control of *Vibrio vulnificus* because this organism is not likely to be associated with the raw materials. A potential hazard from *V. vulnificus* would not warrant further consideration within the HACCP plan being prepared. If there were a lingering concern over potential for cross-contamination, this should be addressed within the GMP program.

As another example, consider a formulated product that contains precooked, boned chicken. Such products might occasionally contain a few *Staphylococcus aureus* cells as a result of human handling during the boning process. The hazard associated with this contaminant is the formation of an enterotoxin which will only be produced in amounts capable of producing illness if the microbial population reaches at least 10^5 to 10^6 organisms per g (see chapter 19). If the handling practices used for a particular item are satisfactory, the potential for enterotoxin formation is very low. However, it is still desirable to keep the initial number of *S. aureus* organisms low. Therefore, the GMP program should include instructions and employee hygiene procedures which assure minimum contamination from this organism.

Preventive (Control) Measures

Once the hazard analysis has yielded a list of significant hazards, the HACCP team must consider what measures can be applied to control each hazard. Note that the

NACMCF uses the term "preventive measure" where the term "control measure" is used here. In future guidance documents, the term "control measure" is likely to be adopted, because not all hazards can be prevented. Thus, control seems a more appropriate term. Control measures are physical, chemical, or other factors that can be used to prevent or minimize a hazard. More than one control measure may be required to control a specific hazard. More than one hazard may be addressed by a specified control measure.

To determine an appropriate control strategy for *Salmonella* in the production of a product formulated with egg, the intrinsic properties of the food and its method of preparation can be compared with those physical parameters which permit growth and survival of *Salmonella* and with those conditions needed to inactivate it. Again, information such as that provided in this book should be valuable in identifying control options. Based on the HACCP team's review of control options for *Salmonella*, the team may determine that sufficient heat must be applied to kill the organism and that, to serve this purpose, only pasteurized liquid eggs will be used in formulation. This step will reduce the likelihood of *Salmonella* occurrence in finished goods to an acceptable level. Thus the control measure selected is egg pasteurization.

A summary of the HACCP team deliberations and the rationale developed during the hazard evaluation and identification of the control measures should be kept for future reference.

Principle no. 2: Identify the CCPs in the process.

Once the significant hazards and control measures are identified, Critical Control Points (CCPs) within the process scheme are identified. A control point is any point, step, or procedure at which biological, physical, or chemical factors can be controlled. A CCP is defined as a point, step, or procedure at which control can be applied and a food safety hazard (significant hazard) can be prevented, eliminated, or reduced to acceptable levels. Each significant hazard identified during the hazard analysis must be addressed by at least one CCP. Examples of CCPs include cooking, chilling, product formulation control, application of a bactericidal rinse, a decontamination step, etc.

While adequate control of a hazard should be reflected in the HACCP plan, in order to keep HACCP programs plant-friendly and sustainable, CCPs should not be redundant. Redundant CCPs typically add little to the margin of safety but will add to the record-keeping and administrative burden of a firm's management structure. They will also add significant cost without concomitant benefit. Experience has shown that HACCP plans which are unnecessarily cumbersome have less chance for support over extended periods.

As an example of this concept, a firm may have multiple magnets and metal detectors in a line to protect production equipment from damage and consumers from harm. If the product is passed through a metal detection/reject device after it is in its final package, the last detector will typically be regarded as the CCP for this hazard while up-line metal detectors or magnets will be considered control points.

For the product referred to above which was formulated with pasteurized liquid eggs, the pasteurization step occurs at a supplier's facility. In this case, the CCP for the firm purchasing the eggs occurs when they are received, with monitoring being conducted through review of documentation confirming that the eggs have been pasteurized and had been protected from recontamination until received by the customer. Alternately, if the firm was producing an egg-containing item that is cooked in its facility, the CCP for *Salmonella* control could be located at the cooking step rather than receiving. While this firm may choose to continue using pasteurized eggs as an ingredient, confirmation on receipt that the liquid eggs had been pasteurized would be considered as a control point rather than a CCP.

Considerations for Raw Products

The objective of a HACCP plan will be different for raw products than for products that are processed by cooking or applying some other bactericidal treatment. For products that receive a defined bactericidal treatment, the risk from microbial pathogens can be reduced to negligible through proper design and application of the treatment. For raw products, however, the risk associated with the product can, at best, be only minimized.

The degree to which risk can be minimized for raw products will often be limited by the technology available to the firm. The HACCP team should consider this and determine whether the technology available is adequate or whether some redesign of the line or process might be in order to reduce risk appropriately. This type of evaluation is typically one of the hidden values of conducting the hazard analysis. Many firms will find that simple adjustments in a line configuration or conduct of a particular operation may pay large dividends in terms of product safety.

Processors of raw products should establish CCPs that are within their control. For foods intended to be cooked after leaving the manufacturer's control, the firm should evaluate the potential for inadequate cooking, either intentionally or unintentionally. If a significant possibility exists that the product may be inadequately cooked, producers should minimize their reliance on the consumer's cooking step as a CCP.

Principle no. 3: Establish critical limits for preventive measures associated with each identified CCP.

A "critical limit" is defined as a criterion that must be met for each control (preventive) measure associated with a CCP. Each CCP will have one or more control measures that must be properly applied to assure prevention, elimination, or reduction of hazards to acceptable levels. Each control measure will have an associated critical limit(s) that serves as a safety boundary for the control measure(s) applied at each CCP and should be viewed as the dividing line between product which is unacceptable versus that which is acceptable. Critical limits may be set for such factors as temperature, time, physical dimension, humidity, moisture level, water activity (a_w), pH, titratable acidity, salt concentration, available chlorine, viscosity, and presence or concentration of preservatives.

Sensory information such as texture, aroma, and visual appearance is rarely a critical limit. Although product condition is often important, it is typically related more to product quality than to its potential for producing illness or injury. Remember a point made earlier in this chapter that HACCP does not change existing laws regarding adulteration. It is illegal to process a food which "consists in whole or in part of any filthy, putrid or decomposed substance ." Processors are bound by existing statute (4) to meet this mandate. However, food quality should not be addressed in a HACCP plan unless it is directly linked to the safety of a product.

Setting Critical Limits

Each firm must ensure that the critical limits identified in its HACCP plan are adequate for the intended purpose. Thus the firm or its process authority should base critical limits on a benchmark or performance standard which the treatment should achieve. There are many possible sources for these benchmarks. They may be derived from sources such as regulatory standards and guidelines (when they relate directly to a health hazard), literature surveys, experimental studies, and experts such as process authorities. Referring again to our earlier example of pasteurized eggs, United States regulatory authorities require a treatment targeted to achieve a 9-log cycle reduction of *Salmonella typhimurium* (1). Thus, a cooking process could be judged adequate if it achieves this performance standard. The firm should have access to adequate scientific studies to validate the critical limits it has selected and the proper delivery of the control measure. The normal variation in applying control measures must be considered when establishing critical limits. Target or operational values should be set which help compensate for normal variations in application of the control measures so that critical limits are not routinely violated. Validation, which includes assurance that the critical limits are well founded, will be discussed in more detail in the verification section of this chapter.

Microbiological Criteria as Critical Limits

Just as microbiological testing is not a good tool for monitoring, microbiological criteria are typically not useful as critical limits in a HACCP program. Because HACCP focuses on process control, factors that lend themselves to "real-time" monitoring and quick feedback are preferred as control measures. Thus, the critical limits used in a HACCP program do not typically relate to a criterion that is directly associated with microbiological testing of products or ingredients. However, there will be instances where the only option open to a processor is to hold lots of an ingredient and perform microbiological testing before release. This may arise when a particular ingredient can only be obtained from sources where little control is exercised over factors that affect the contaminants or pathogens associated with the ingredient. In this case, the CCP is located at receiving, where the incoming ingredient would be sampled for analysis and placed in controlled storage until the results of the analysis are available. Microbiological testing is the monitoring activity, and a microbiological criterion would be the critical limit. Chapter 4, which addresses microbiological criteria and their use, should be consulted for additional information.

Principle no. 4: Establish CCP monitoring requirements. Establish procedures for using the results of monitoring to adjust the process and maintain control.

Monitoring is the act of conducting a planned sequence of observations or measurements to assess whether a CCP is under control and to use the information developed to produce an accurate record for future use in verification. Typically, monitoring involves measurement of the physical factor identified as the control (preventive) measure for a significant hazard. Examples of monitoring activities include:

- measuring temperature
- tracking elapsed time, e.g., time at a specific temperature
- determining pH
- determining moisture level or water activity

Monitoring serves three main purposes. First, monitoring tracks the systems operation in a manner essential to food safety management. If monitoring detects a trend towards loss of control, that is, approaching a target level, then action can be taken to bring the process back into control before a deviation occurs. Second, monitoring is used to determine when a deviation occurs at a CCP, i.e., violation of the critical limit which will initiate an appropriate corrective action. Third, it provides the basis for written documentation of process control for use in verifying that the HACCP plan has been followed.

When specifying monitoring activities, the firm should specify:

- what control measure(s) is being monitored
- how often the monitoring needs to be conducted
- what procedures will be followed to collect data and what methods and equipment will be used
- who will be responsible for performing each monitoring activity

To assure adequate control, the frequency at which activities are monitored must be consistent with the needs of the operation in relation to the variation inherent in the control step. Frequency of monitoring should be integrated with a product coding system which is designed to prevent excessive amounts of product from being involved in a corrective action if problems arise and a critical limit is violated. For example, assume that a food operation is making a product which must reach a certain temperature during a cooking step applied on a continuous basis, and the temperature is only monitored every 2 h. If the temperature in the cooker drops below the critical limit, the processors may have up to 2 h of production which is in violation of the parameters of the HACCP plan. All product produced since the last acceptable check may need to be reworked or destroyed. If the temperature is monitored more frequently, then the amount of product involved in the deviation will be reduced.

In most instances, control measures that can be monitored on a continuous basis are preferred over those which cannot be monitored continuously. Those which can be monitored continuously are typically amenable to continual recording and electronic monitoring with automated trend analysis. Such monitoring can provide advanced warning to allow for correction of a problem before violation of a critical limit. In the example above, constant monitoring of temperature with an alarm to warn of a trend toward violation of a critical limit may have avoided a deviation or at least would minimize the amount of product which needed corrective action.

Certain characteristics, however, cannot be measured on a continuous basis, or they may not need continuous monitoring. Products that are mixed in a batch, such as some acidified products, do not need continuous monitoring of pH. A control measure like water activity in a food cannot easily be monitored continually. Such items will be monitored through a planned sequence of sampling and testing performed at a frequency determined to be adequate to assure control.

One of the underlying concepts of HACCP is to promote to the line level in an operation an awareness of food safety and individual responsibility. One way to promote this is to involve line operators in monitoring activities. In many operations, monitoring activities are assigned to quality control when they could be accomplished just as effectively by line operators, with the extra benefit of line worker involvement. The role of quality control personnel may be more appropriate in verification, or "checking the checker."

Those responsible for monitoring a control measure at a CCP must:

- be trained in the appropriate monitoring techniques
- understand the importance of monitoring
- accurately report and record results of the monitoring activity
- immediately report deviations from critical limits so that corrective actions can be taken, or initiate the appropriate corrective action

Returning again to our example of a baked item formulated with liquid eggs, it was determined that the control measure for *Salmonella* would be heat applied during cooking and that the CCP would be a cooking step. The parameters being monitored by the firm are temperature of the oven and the duration of the cooking cycle. Critical limits for these physical factors must be set such that the performance criterion identified earlier (minimum of 9 log cycle inactivation of *Salmonella*) is met. If this firm chooses to use equipment that provides continuous monitoring (recording) of time and temperature, then the operators' duty is to ensure that the monitoring equipment is working and to make a manual reading periodically as a verification activity. If there is a deviation from either the time or temperature components of the cooking step, the operator should initiate the proper corrective action.

Principle no. 5: Establish corrective action to be taken when monitoring indicates that there is a deviation from an established critical limit.

The HACCP system is designed to identify situations where health hazards are reasonably likely to occur and to establish strategies to prevent or minimize their occurrence. However, ideal circumstances do not always prevail. When there is a deviation from established critical limits, corrective actions must be taken. Corrective actions, in the HACCP context, are procedures to be followed when a deviation or failure to meet a critical limit occurs. Corrective action plans should be in place which address how the firm will (i) fix or correct the cause of noncompliance to assure that the process is brought under control and the critical limits are being met, and (ii) determine the disposition of noncompliant product.

Because of the variations in CCPs for different foods and the diversity of possible deviations, corrective action plans must be developed that are specific to the particular operation. A corrective action must be developed for deviations that may occur at each CCP. The actions outlined in the corrective action must demonstrate that the process has been brought under control. Individuals who have a thorough understanding of the process, product, and HACCP plan are to be assigned responsibility for assuring that the appropriate corrective actions have been implemented. Records of the corrective actions that have been taken must be maintained. Long-term solutions should be sought when a particular critical limit appears to be violated frequently. Such repeated events should trigger a review of the plan, the CCP, the identified critical limit, and the method of process control to determine what improvements are needed to reduce the deviations.

Corrective actions for inadequately cooking the egg-containing product considered earlier would include correcting the equipment problem that caused the temperature or time error. In addition, any product involved in the deviation must not be released until it is evaluated to determine whether it is safe for consumption. In this case, microbiological testing is seldom appropriate; the number of samples that would need analysis to detect a low-level contaminant may be excessive. Instead, the preferred method would be an expert interpretation of the microbicidal effectiveness of the time and temperature the product was exposed to during the deviation period. If this is determined to be adequate, the product may be safely consumed. Alternately, the product may be reprocessed, diverted for another use in which it will be rendered safe, or destroyed.

Principle no. 6: Establish effective record-keeping procedures that document the HACCP system.

Record keeping is integral to maintaining a HACCP system. Without effective record keeping and review, the HACCP system will not be sustained as an ongoing practice. Records provide the basis for management assessment of the safety of products produced and for documenting safety of products to customers. Records provide a means to trace the production history of foods produced, to document that critical limits were met, and to prove that any needed corrective actions were carried out.

The HACCP plan and associated support documents should be on file at the food establishment. Generally, the records utilized in the HACCP system will include the following:

1. The HACCP plan

- listing of the HACCP team and assigned responsibilities
- description of the product and its intended use
- block flow diagram for the entire manufacturing process indicating CCPs
- food safety objectives to be accomplished through implementation of the plan
- hazards associated with each CCP and preventive measures
- critical limits for each CCP
- monitoring procedures (who, what, when, where, how)
- corrective action plans for deviations from critical limits
- record-keeping procedures
- procedures for verification of the HACCP system

2. Records obtained during the operation of the plan

- records of data collected by monitoring of control measures at CCPs
- corrective action records
- records of certain verification activities

Examples of operational records for a cooked product might include a computerized log of the operational details of the oven used for cooking. Such records would document the time the unit was turned on, when it reached temperature, when product began to move through, the record of temperature and time during the production run, etc. A record of operator checks of a separate temperature monitoring device as well as confirmation of through-put or residence time in the oven would also be expected. If any deviations occurred, a record of the time and extent of the deviation should be available along with a description of actions needed to bring the cooker back into control and the disposition of any product produced while the deviation occurred. In this case, the processor may chose to automatically stop the line if a time or temperature parameter is violated. This eliminates the need to determine product disposition. Other operational records would include the those documenting calibration of the temperature monitoring equipment, time measurements, and any other practices needed to assure the accuracy of the primary monitoring activities.

In addition to the above, there should be a HACCP master file which will contain a record of the deliberations of the HACCP team and documentation for various aspects of the plan. This should include justification for the critical limits identified, details on sampling plans and methods of analysis, standard operating procedures for monitoring activities and corrective actions, and other details that may need further review and modification to support the plan. The forms used, format, practices to be followed, and review procedures should be explained in the HACCP master file.

Principle no. 7: Establish procedures for verification that the HACCP system is working correctly.

There are several HACCP-related activities that will be conducted under the label of verification. Verification is the use of methods, procedures, or tests, in addition to those used in monitoring, to determine if the HACCP plan is being followed and whether the records of monitoring activities are accurate. Verification represents a second level of review, beyond the primary review done by line personnel who are actually conducting the monitoring and making the records. This secondary level of review is typically the responsibility of quality control personnel and will be aimed at verifying, through auditing and by direct observation, that records are being kept accurately. Verification may also include a program of sampling and testing where appropriate. Product may be tested for chemical or microbiological contaminants or analyzed for physical hazards to verify that planned control measures are being effectively carried out.

Verification often involves a third level of review utilizing outside audit teams to review all aspects of the HACCP plan, the monitoring procedures being used, record-keeping and review practices, etc.

The concept of validation is evolving as a distinct activity to be conducted as a subset of verification. Validation is described as the process of assuring that the plan reflects identification of all significant hazards, that the control measures identified are appropriate, that the designated critical limits are adequate, that monitoring frequencies and procedures are adequate to assure proper control, and that the record-keeping practices and forms used are appropriate. Most now feel that validation should be conducted as a separate step and that it should be completed before the plan is put into use. The final step in the initial validation of a plan will typically involve acknowledgment via signature by a management representative that the plan is accepted by the firm. Plans will need to be revalidated whenever significant changes occur in product formulations or equipment. Plans should be revalidated on a periodic basis even if no significant changes have been made. In this way, the plan will retain its support base.

Validation of the control procedures for *Salmonella* identified in the earlier example of a cooked product would include a determination of whether the time and temperature of the cook process were adequate to inactivate an appropriate number of *Salmonella* cells. This determination involves reviewing appropriate literature or consulting with experts in the field to determine approximate thermal inactivation kinetics of the organism in the product. It will also be necessary to determine the heating rate of the product in the oven to be used for cooking so that the cooking time and temperature needed to achieve an adequate kill can be established. Uniform delivery of cooking in the oven must also be validated so that each unit of product passing through the oven is assured of receiving at least the minimum treatment. Alternatively, the cook step can be validated microbiologically by using inoculated packs with a nonpathogenic simulator organism if these are statistically designed to assure that "worst-case" conditions are accounted for and enough sample units are run.

Verification activities for the cook step would include calibrations necessary to assure that the time and temperature measurements are accurate. The firm may also wish to review all HACCP records, especially the cook records, before final release of any lot of product. Verification activities should take place on a regular and predetermined schedule. Some activities will be conducted daily, others will be weekly or monthly. Some activities will be conducted even less frequently. The need for accuracy and precision and the effects of equipment wear and normal variance should be considered when establishing schedules.

SUMMARY

The HACCP concept provides a systematic, structured approach to assuring the safety of food products. However, there is no blueprint or universal formula for putting together the specific details of a HACCP plan. The plan must be dynamic. It must allow for modifications to production, substitution of new materials and ingredients, and development of new products. In order for HACCP to work well, it should be a participatory endeavor at all levels of an organization, both in formulating and implementing as well as in managing the plan. The strength a successful HACCP program provides is a system that a company can use effectively to organize and efficiently manage the safety of the products it prepares.

References

1. **Anonymous.** 1968. *Egg Pasteurization Manual.* U.S. Department of Agriculture, Agricultural Research Service, Washington, D.C.
2. **Anonymous.** 1996. *Hazard Analysis and Critical Control Point (HACCP) System and Guidelines for Its Application.* Proposed Draft Annex to Revised Recommended International Code of Practice—General Principles of Food Hygiene. Codex Alimentarius Commission, Alinorm97/13. Food and Agriculture Organization, Rome.
3. **Campden and Chorleywood Food Research Association.** 1992. *HACCP: a Practical Guide.* Technical Bulletin no. 38. Campden and Chorleywood Food Research Association, Chipping Campden, Gloucestershire, U.K.
4. **Federal Food Drug and Cosmetic Act.** 1958. Title 21, U.S. Code. Section 402 (a) (4).
5. **Federal Register.** 1995. Procedures for the safe and sanitary processing and importing of fish and fishery products; final rule. *Fed. Regist.* **60**:65096–65202 (December 18, 1995).
6. **Federal Register.** 1996. Pathogen reduction; hazard analysis and critical control point (HACCP) systems: final rule. *Fed. Regist.* **61**:38806–38989 (July 25, 1996).
7. **Food and Agriculture Organization of the United Nations.** 1995. The use of hazard analysis critical control point (HACCP) principles in food control. Report of an FAO Expert Technical Meeting, Vancouver, Canada, 12–16 December 1994. Food and Nutrition Paper 58. Food and Agriculture Organization of the United Nations, Rome.
8. **International Life Sciences Institute.** 1993. *A Simple Guide to Understanding and Applying the Hazard Analysis Critical Control Point Concept.* International Life Sciences Inst., Washington, D.C.
9. **National Academy of Sciences/National Research Council.** 1985. *An Evaluation of the Role of Microbiological Criteria for Foods and Food Ingredients.* National Academy Press, Washington, D.C.
10. **National Advisory Committee on Microbiological Criteria for Foods.** 1992. Hazard analysis and critical control point system. *Int. J. Food Microbiol.* **16**:1–23.
11. **National Conference on Food Protection.** 1971. *Proceedings of the National Conference on Food Protection.* Department of Health Education and Welfare, Public Health Service, Washington, D.C.
12. **Stevenson, K., and D. T. Bernard (ed.).** 1995. *HACCP, Establishing Hazard Analysis Critical Control Point Programs: Workshop Manual,* 2nd ed. The Food Processors Institute, Washington, D.C.
13. **World Health Organization.** 1995. *Application of Risk Analysis to Food Issues.* Report of the joint FAO/WHO Expert Consultation, 13–17 March 1995, Geneva, Switzerland. World Health Organization, Geneva.

Index

D

I

J

K

Q

R

W

X

Y

Z